Help Us Improve This Dictionary.

In a world of ever-changing technology, we know you need reference materials that reflect the state-of-the-art in electrical and electronics engineering.

That's why at the Standards Office of the IEEE, we're trying to determine how quickly engineers, researchers, librarians, and other users need revisions or updates to the **IEEE Standard Dictionary.**

Should we change the revision cycle? Should we sell annual supplements instead?

We'd like to hear your thoughts on this. Please fill in the postage-paid research card on the reverse side, and return it to us at the address below. If you give us your address, we'll also be able to keep you informed about developments in glossary standards.

Thank you in advance for sharing your point of view.

If your primary field of interest is:	Please fill in this code on the card below:
Acoustics, Speech, & Signal Processing	(01)
Aerospace & Electronic Systems	(10)
Antennas & Propagation	(03)
Broadcast Technology	(02)
Circuits & Systems	(04)
Communications	(19)
Components, Hybrids, & Manufacturing Technology	(21)
Computer	(16)
Consumer Electronics	(08)
Control Systems	(23)
Dielectrics & Electrical Insulation	(32)
Education	(25)
Electromagnetic Compatibility	(27)
Electron Devices	(15)
Engineering Management	(14)
Engineering in Medicine & Biology	(18)
Geoscience & Remote Sensing	(29)
Industrial Electronics	(13)

If your primary field of interest is:	Please fill in this code on the card below:
Industry Applications	(34)
Information Theory	(12)
Instrumentation & Measurement	(09)
Lasers & Electro-Optics	(36)
Library	(NNL)
Magnetics	(33)
Microwave Theory & Techniques	(17)
Nuclear & Plasma Sciences	(05)
Oceanic Engineering	(22)
Power Electronics	(35)
Power Engineering	(31)
Professional Communication	(26)
Reliability	(07)
Social Implications of Technology	(30)
Systems, Man, & Cybernetics	(28)
Ultrasonics, Ferroelectrics & Frequency Control	(20)
Vehicular Technology	(06)
Other	(NNK)

▶ Please complete reverse side, detach here and mail ▶

NO POSTAGE
NECESSARY
IF MAILED
IN THE
UNITED STATES

BUSINESS REPLY MAIL

FIRST CLASS   PERMIT NO. 103   PISCATAWAY, N.J.

POSTAGE WILL BE PAID BY ADDRESSEE

IEEE SERVICE CENTER
Att: **The Standards Office**
445 Hoes Lane
P.O. Box 1331
Piscataway NJ 08855-9916

Define Your Own Terms.

Please give us your opinion on how often you think the IEEE Standard Dictionary should be published, and how it should be sold.

1. *How often do you think the **IEEE Standard Dictionary** should be published? Every —*

2. *Would you purchase a yearly supplement to the **IEEE Standard Dictionary**? —*

3. *If we offered you an incentive for buying the next edition, would you rather receive a free gift, or a coupon for buying the dictionary at a discount? —*

4. *Did you purchase this dictionary directly from the IEEE Service Center in Piscataway, New Jersey? —*

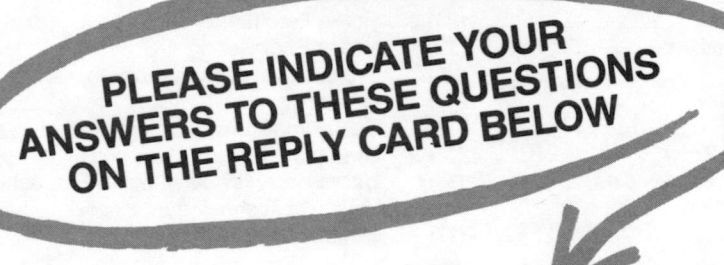

PLEASE INDICATE YOUR ANSWERS TO THESE QUESTIONS ON THE REPLY CARD BELOW

NNM

Stay Current.
Please fill in your name and address, so we can contact you about updates to the IEEE Standard Dictionary.

Name _____

Address _____

_____

City _____ State _____ Zip / Postal Code _____

Country _____

Please select a code from the list on the reverse side, that best describes your primary field of interest.
My primary field of interest is _____ *(Please refer to list of codes on the reverse side.)*

My answers to the questions above are:

1. ☐ 1 yr.   ☐ 2 yrs.   ☐ 3 yrs.   ☐ 4 yrs.
   ☐ 5 yrs.   ☐ 6 or more yrs.

3. ☐ I'd rather receive a free gift.
   ☐ I'd rather get a coupon for buying at a discount.
   ☐ Neither of these incentives would make
      me buy the next edition.

2. ☐ Yes   ☐ No

4. ☐ Yes   ☐ No

An American National Standard

Acknowledged as An American National Standard
July 8, 1988

IEEE
Standard Dictionary
of
Electrical and
Electronics
Terms

621. 303—

I 8

3-26-92

Fourth Edition

ANSI/IEEE Std 100-1988
Fourth Edition

IEEE Standard Dictionary of Electrical and Electronics Terms

Frank Jay
Editor in Chief

J. A. Goetz
Chairman
Standards Coordinating Committee
on Definitions (SCC 10)

Membership

Ashcroft, D. L.	Gelperin, D.	Radatz, J.
Azbill, D. C.	Guifridda, T. S.	Reymers, H. E.
Ball, R. D.	Goldberg, A. A.	Roberts, D. E.
Balaska, T. A.	Graube, M.	Rosenthal, S. W.
Bauer, J. T., Jr.	Griffin, C. H.	Rothenbukler, W. N.
Blasewitz, R. M.	Heirman, D. N.	Sabath, J.
Boberg, R. M.	Horch, J. W.	Shea, R. F.
Boulter, E. A.	James, R. E.	Showers, R. M.
Frewin, L. F.	Karady, G. G.	Skomal, E. N.
Bucholz, W.	Key, T. S.	Smith, T. R.
Buckley, F. J.	Kieburtz, R. B.	Smith, E. P.
Cannon, J. B.	Kincaid, M. R.	Smolin, M.
Cantrell, R. W.	Klein, R. J.	Snyder, J. H.
Chartier, V. L.	Klopfenstein, A.	Spurgin, A. J.
Cherney, E. A.	Koepfinger, J. L.	Stephenson, D.
Compton, O. R.	Lensner, W.	Stepniak, F.
Costrell, L.	Masiello, R. D.	Stewart, R. G.
Davis, A. M.	Meitzler, A. H.	Swinth, K. L.
Denbrock, F.	Michael, D. T.	Tice, G. D.
DiBlasio, R.	Michaels, E. J.	Turgel, R. S.
Donnan, R. A.	Migliaro, H. W.	Thomas, L. W., Sr.
Duvall, L. M.	Mikulecky, H. W.	Vance, E. E.
Elliott, C. J.	Moore, H. R.	Wagner, C. L.
Erickson, C. J.	Mukhedir, D.	Walter, F. J.
Flick, C.	Muller, C. R.	Weinschel, B. O.
Freeman, M.	O'Donnell, R. M.	Zitovsky, S. A.
	Petersons, O.	

Published by
The Institute of Electrical and Electronics Engineers, Inc
New York, NY

Library of Congress Catalog Number 88-082198

ISBN: 1-55937-000-9

© Copyright 1988

The Institute of Electrical and Electronics Engineers, Inc

November 3, 1988

SH12070

Introduction

From their earliest years, both the American Institute of Electrical Engineering (AIEE) (1884) and the Institute of Radio Engineers (IRE) (1912) published standards defining technical terms. They have maintained this practice since they were combined in 1963 to become the IEEE (Institute of Electrical and Electronics Engineers).

In 1928 the AIEE organized Sectional Committee C42 on Definitions of Electrical Terms under the procedures of the American Standards Association, now the American National Standards Institute. In 1941 AIEE published its first edition of *American Standard Definitions of Electrical Terms* in a single volume. However, by the time a second edition was ready, the highly accelerated development of new terms made it impracticable to publish in a single volume, and 17 separate documents, each limited to a specific field, were published from 1956 to 1959.

Over the years, IRE published a large number of standards that either included definitions or were devoted entirely to definitions. In 1961 it published all of its then-approved definitions in an alphabetically arranged single volume.

The 1972, 1977, and 1984 editions of the *IEEE Standard Dictionary of Electrical and Electronics Terms* included all terms and definitions that had been standardized previously by IEEE, as well as many from American National Standards and the International Electrotechnical Vocabulary.

Most of the definitions that have been continued from the earlier editions have been reaffirmed because of their inherent usefulness. The committees originally responsible for these definitions still consider them to be equally appropriate at the present time.

The current 1988 edition includes terms from standards generated since 1984.

Acknowledgments

The 1988 IEEE Dictionary was edited at the Secretariat of the Standards Coordinating Committee 10 (Definitions), completely by means of the Microsoft WORD software system.

Appreciation is extended to Jack Goetz, Chairman, and to all other members of SCC 10 for their guidance and willing assistance.

We also appreciate the patient efforts and support of our data processor, Sedgwick Printout Systems, of Princeton, New Jersey, and the services of its representatives, John Liwok, Sr., W. Thomas Brush, Jackie Doan, Jim Glenn, Stan Weiss, and their staff of page makeup artists.

How to Use This Dictionary

The terms defined in this dictionary are listed in alphabetical order. Terms made up of more than two words appear in the order most familiar to the people who use them. In some cases cross-references are given.

Some terms take on different meanings in different fields. When this happens the different definitions are numbered, identified as to area of origin, coded, and listed under the main entry.

If a reader wants to know the source of a definition he need only look up the code number following the definition in the SOURCES section that appears at the back of the book between pages 1112 and 1129.

List of Contributing Editors

The following lists are made up of the names of engineers who, over the years, chaired the many committees that have generated standards containing definitions. To a large extent the definitions in this dictionary come from those standards.

The first list contains the names of the chairpeople of committees responsible for the development of standards published since the last edition of the IEEE Dictionary in 1984. The second list was published in the 1984 edition and contains the names of the chairpeople of standards committees that contributed definitions to the 1984 Dictionary and all previous editions.

IEEE has profound appreciation for the work that these people so willingly undertook and so carefully fulfilled.

IEEE also remembers that behind each name printed here were committees and subcommittees of volunteers (the sum total numbering in the thousands) without whose painstaking efforts the IEEE standards and this dictionary would never have been possible.

Addiss, R. R., Jr.
Adolphson, E. J.
Allen, R. E.
Atkinson, S. C.

Bailey, B. G.
Baker, D. S.
Baldwin, G.
Barker, J.
Barnard, H. J.
Bates, T. M., Jr.
Bauer, J. T.
Behrens, G. D.
Belanger, E. D.
Bellack, J. H.
Bellian, J.
Bhavaraju, M. P.
Bizub, E. J.
Blackburn, J. L.
Blasewitz, R. M.
Blasiak, L. J.
Boberg, R. W.
Bobo, P. O.
Boettger, J. T.
Bogner, P. W.
Booth, L. E.
Boulter, E. A.
Brench, B. L.
Bridges, J. E.
Bronaugh, E. L.
Brown, E. M.
Bruns, J. H.
Buckley, F. J.
Burke, J. J.
Burstein, N. M.
Burt, W.
Butt, A. E., Jr.

Cady, C. A.
Cannon, J. B.
Cantrell, R. W.
Capra, R. L.
Castenschiold, R.
Chamberlain, A. B.
Chartier, V. L.
Cherney, E. A.

Cody, W. J.
Cohen, E. J.
Cook, D. C.
Cooper, J.
Cooper, K.
Costrell, L.
Cox, V. L.
Crowdes, G. J.

Dalio, H. J.
Dandeno, P. L.
Davis, A. M.
Davis, B.
DeBlasio, R. R.
Denbrock, F. A.
Denkowski, W. J.
Dietrich, F. M.
Dougherty, J. J.
Drew, T. C.
Durham, M. O.

Easley, J. H.
Erickson, C. J.
Estcourt, V. F.
Ewer, M. H.

Fagan, E. J.
Fantozzi, G. V.
Farrell, F. C.
Few, R. A.
Flack, M.
Fogarty, E. P.
Foster, S. L.
Frank, J.
Frederick, R.
Fricke, C. A.
Fujii, R.

Gaibrois, G. L.
Gangloff, W. C.
Garetz, M.

Gelperin, D.
Godber, A. S.
Goetz, J. A.
Goldberg, A. A.

Grooms, F. H.
Grose, C. W.
Guy, A. W.

Haack, W. R.
Healy, J. W.
Heimer, G.
Hendrix, K. D.
Henry, G. K.
Herrera, J. J.
Hetrick, J. A.
Horch, J. W.

Jackson, D. L.
Jackson, R. M.
James, R. E., III
James, C.
Jensen, A. G.
Jimenez, H. M.
Johnson, B. C.
Johnson, G. S.
Jordan, R. E.

Karady, G. G.
Kelly, L. J.
Kerchner, W. J.
Klopfenstein, A.
Koenig, D. F.
Koepfinger, J.
Koponen, I. H.

Laubach, W. E.
Lee, R. H.
Leib, M. J.
Leidich, G. R.
Lester, G. N.
Levitsky, F. J.
Long, L. W.
Loughry, D. C.
Love, D. J.
Luehring, E. L.
Lyons, P. C.

Madison, L. C.
Mallay, J. F.
Mallinson, G. H.

Mambuca, J. A.
Marrow, E. H., Jr.
Marshall, G.
Marta, F. A.
Martin, F.
Martzloff, F. D.
Maruvada, P. S.
Masse, F. X.
Mathur, B. K.
Matthey, A.
McDermott, T. J.
McMurray, W.
McNutt, W. J.
Meisel, H. E.
Meitzler, A. H.
Michaels, E. J.
Michalec, J. R.
Moore, H. R.
Mukhedkar, D.

Nathanson, F. E.
Nault, J. G.
Nevins, P. A.
Nichols, N.
Nicholson, C. H.
Nikolakakos, S.
Nippes, P. I.
Niskode, P. M.

O'Donnell, P.
Ohlson, R. O.
Olivier, I. E.
Olken, M. I.

Pai, M.
Parris, J. L.
Peach, N.
Pearce, H. A.
Peterson, W. K.
Pettit, D. L.
Popeck, C. A.
Porter, N. S.
Potts, C. D.

Radatz, J. W.
Ramsey, M. H.

Ranga, H.
Raquet, D. A.
Rautio, W. S.
Reifschneider, P. J.
Reymers, H. E.
Riley, E. W.
Roberts, D. E.
Roby, A. R.
Rosenthal, S.
Rowell, R. M.

Sacks, N.
Sandiforth, D.
Schultz, B. H.
Scott, J. C.
Scott, J. V.

Shea, R. F.
Shipway, G. D.
Simonsen, K.
Skomal, E. N.
Smith, B. E.
Smith, J. K.
Smith, T. R.
Spurgin, A. J.
Stepniak, F.
Stevenson, D.
Stewart, J. R.
Stitzer, S. N.
Stoner, J. E., Jr.
Stratford, R. P.
Swallows, R. L.
Swinyard, W. O.

Szabados, P.

Taylor, E. R., Jr.
Taylor, R. E.
Telender, S. H.
Thomas, J. E.
Thomas, R. G.
Tice, G. B.
Toppeto, A. A.
True, W. F.
Turgel, R. S.

Vaillancourt, G. H.
Vallario, E. J.

Wagner, C. L.

Weinschel, B. O.
White, H. F.
Whitman, B.
Wilkens, W. B.
Wilson, J. C.
Wiot, E. R.
Wrenn, M. E.

York, J. W.
Younkins, T. D.

Zipse, D. W.
Zitovsky, S. A.
Zweigler, R.

Abrahams, I. C.
Allen, G. Y. R.
Amato, C. J.
Andrews, F. T.
Angelo, S. J.
Angus, A. C.
Armstrong, J. H.
Arndt, R. H.
Arthur, M. G.
Avins, J.
Axelby, G. S.

Baker, J. M.
Baldwin, M. S.
Baljet, A.
Ball, J. E. D.
Bangert, J. T.
Baracket, A. J.
Bargellini, P. L.
Barrow, B. B.
Barstow, J. M.
Bartheld, R. G.
Bates, T. M., Jr.
Bauer, J. T.
Baum, J. F.
Bellack, J. H.
Blachman, N. M.
Bloomquist, W. C.
Bixby, W. E.
Blackburn, J. L.
Blodgett, E. D.
Bloom, L. R.
Bobo, P.O.
Bochnak, P.M.
Boice, C. W., Jr.
Booth, L. E.
Borowski, R. R.
Borst, D. W.
Bowers, G. H.
Bowers, W. W.
Brainerd, J. G.
Brereton, D. S.
Brociner, V.
Brockwell, K. C.
Broome, W. M.
Brown, R. D.
Buhl, H. A.

Cadwell, C. L.
Callahan, D. P.
Cameron, F. L.
Cameron, A. W. W.
Campbell, A. T.
Caslake, S. G.
Chappell, J. F.
Chase, A. A.
Chiappetta, C. M.
Christensen, P.
Clark, D. M.
Clark, R. A.
Clarridge, C. H.
Clevenger, C. M.
Cohen, E. J.
Cohn, S. I.
Coley, R. F.
Conroy, J. F.
Cook, W. H.
Copel, M.
Costrell, L.
Cotta, R. E.
Cottony, H. V.
Cox, V. L.
Curdts, E. B.

Dallas, J. P.
Davidoff, F.
Delaplace, L. R.
Denkowski, W. J.
Desch, R. F.
Deschamps, G. A.
Dietrich, R. E.
Dietsch, C. G.
Doba, S., Jr.
Doble, F. C.
Donahoe, F. J.
Dowling, E. F.
Drown, J. L.
Dumper, W. C.
Duncan, R. O.
Dutton, J. C.

Early, J. M.
Easley, G. J.
Ecker, H. A.
Edwards, R. F.

Edwards, A. T.
Eiselein, J. E.
Elias, P.
Eliasson, I. E.
Ellerbruch, D. A.
Elwood, J. R.
Espersen, G. A.
Estcourt, V. F.
Evans, C. T.
Evendorff, S.
Ewing, J. S.

Farr, N. C.
Faust, L. G.
Feng, Tse-yun
Ferencsik, J. J.
Ferris, R.
Fields, C. V.
Fink, L.
Fischer, R. C.
Fitzpatrick, A. R.
Fogarty, E. P.
Fornwalt, M. B.
Forster, J.
Fox, A. G.
Fredendall, G. L.
Fricke, C. A.
Frihart, H. N.
Fulks, R. G.

Gaibrois, G. L.
Garrity, T. F.
Garschick, A.
Gelperin, D.
Gerber, E. A.
Giles, W. F.
Gillespie, E. S.
Gloss-Soler, S. A.
Graham, J. D.
Greenspan, A. M.
Gressitt, T. J.
Griffith, M. S.
Grube, C. W.
Gubbins, H. L.

Hackley, R. A.
Haddad, E. E.

Hall, J. R.
Hannan, P. W.
Hansell, C. D.
Hansen, R. C.
Hanver, G. N.
Harper, W. E.
Harvey, F. K.
Haymes, T. W.
Hedrick, D. L.
Hefele, J. R.
Hendrix, K. D.
Hilibrand, J.
Hillen, R. J.
Hirtler, R. W.
Hissey, T. W.
Holland, M. G.
House, D. L.
Hovey, L. M.
Hubbs, J. C.
Huber, R. F.
Hvizd, A.

Jacobs, I. M.
Jasik, H.
Jesch, R. L.
Jocz, A.
Johnson, L. M.
Johnson, I. B.
Johnson, W. J.
Johnston, F. C.
Jones, J. L.

Kaenel, R. A.
Kaufmann, R. H.
Keay, F. W.
Keezer, D. C.
Kelley, R. J.
Kerny, I.
Kerwein, A. E.
Killen, T. S.
Kirkwood, L. W.
Kitchens, J.
Klein, P. H.
Knowles, C. H.
Koch, F. W.
Koen, H. R., Jr.
Koepfinger, J.

Kolb, F. J., Jr.
Kolcio, N.
Komb, K. W.
Kotter, F. R.
Kreer, J. G., Jr.
Kroll, J. W.
Kummer, W. H.
Kurth, C. F.
Kurtz, S. K.
Kurtz, L. W.

Laidig, J. R.
Lambert, C. D.
Lang, W. W.
LeBrun, A. A.
Lee, R. H.
Lee, R.
Lester, G. N.
Levin, R. E.
LeVine, D. J.
Liguori, F.
Liu, C. H.
Lloyd, R.
Lokken, G.
Louden, V. J.
Lougher, E. H.
Luehring, E. L.
Lynch, R. D.
Lynch, W. A.
Lynn, G. E.

Madison, L. C.
Marieni, G. J.
Marta, F. A.
Martin, G.
Martin, T. J.
Masse, F. X.
Mattingly, R. L.
Mayer, R. P.
McClain, R. D.
McConnell, K. R.
McFarlin, V. F.
McGee, A. A.
McGrath, I. N.
McGrath, T. J.
McHugh, Jr., G. M.
McKean, A. L.
McKiernan, J.
McKnight, J. G.
McMaster, R. A.
McWilliams, D. W.
Meindl, J. D.

Meitzler, A. H.
Mertz, P.
Michael, D. T.
Michaels, E. J.
Mikulecky, H. W.
Miles, H. C.
Miller, G. L.
Mitsanas, H.
Mobley, M. D.
Moran, R. J.
Morris, C. R.
Morris, R. M.
Morrison, G. E.
Morrison, S. M.
Mortenson, K.
Morton, G. A.
Moses, G. L.
Mulhem, W. J.
Muller, C. R.

Nalley, L. J.
Nathanson, H. C.
Neiswender, W. J.
Neubauer, J. R.

O'Neal, W. E.
Oliner, A. A.

Page, C. H.
Palmer, W.
Paniri, Z. S.
Parker, J. C.
Pearce, J. G.
Pelc, T.
Penn, W. B.
Persyk, D. E.
Phillips, V. E.
Piccione, N.
Poston, R. M.
Potter, D. H.
Powers, K. H.
Priebe, E. P.
Pritchard, R. L.

Ray, K. A.
Redhead, P. A.
Rediker, R. H.
Reilly, J. W.
Reynolds, J. N.
Rice, B. M.
Rietz, E. B.
Roberts, D. E.

Roberts, W. K.
Roe, J. H.
Rohlfs, A. F.
Rook, L. E.
Rose, R. H., 2nd
Rothauser, E. H.
Rubin, S.
Ruete, R. C.
Russ, J. C.

Samuel, C. H.
Schaufelberger, F.
Schmidt, P. L.
Schuermeyer, F. L.
Schwalbe, C. A.
Schwartz, J. D.
Schwertz, F. A.
Scott, L. H.
Scoville, M. E.
Selby, J. M.
Seyer, C. F., III
Shackman, N.
Sharrow, R. F.
Shea, R. F.
Sheckler, A. C.
Shields, F. J.
Shipman, W. A.
Shores, R. B.
Showers, R. M.
Sidway, C. L.
Silbiger, H. R.
Simmons, A. J.
Simonson, K.
Simpson, H. A.
Singer, G. A.
Sinnott, N. F.
Skolnik, M. I.
Sledge, C.
Smith, J. H.
Smith, P. H.
Smith, T. R.
Sommers, R. A.
Sorensen, D. K.
Sorger, G. U.
Spitzer, C. F.
Spurgin, A. J.
Stadtfeld, N., Jr.
Stewart, J. A.
Stewart, R. G.
Stuckert, P. E.
Sullivan, J. B.

Sullivan, R. J.
Swanson, C. O.

Talaat, M. E.
Taub, J. J.
Tebo, J. D.
Test, L. D.
Thomas, R. F., Jr.
Thomas, R. C.
Thurell, J. R.
Tillinger, H. I.
Tilston, W. V.
Tjepkema, S.
Toman, K.
Toppeto, A. A.
Truitt, J. R.
Tuller, W. G.

Unnewehr, L. E.
Ure, R. W.

Vadersen, C. W.
Vallario, E. J.
Vlahos, P.
von Recklinghausen, D. R.
von Roeschlaub, F.

Wagner, C. L.
Wagner, S.
Wahlgren, W. W.
Weber, E.
Weddendorf, W. A.
Weinberg, L.
Weinschel, B. O.
Weitzel, H. B.
Whistler, J. P.
White, H. F.
White, J. C.
Wickham, W. H.
Williamson, R. A.
Wintringham, W. T.
Wolfe, P. N.
Woods, D. E.
Wroblewski, J. J.

Yasuda, E. J.
Yates, E. S.
York, J. W.
Younkin, G. W.

Zucker, M.

A

aa auxiliary switch. *See:* **auxiliary switch; aa contact.**

aa contact (power switchgear). A contact that is open when the operating mechanism of the main device is in the standard reference position and that is closed when the operating mechanism is in the opposite position. 103

A and R display (radar). An *A* display, any portion of which may be expanded. *See:* **navigation.** 13

a auxiliary switch. *See:* **auxiliary switch; a contact.**

abampere. The unit of current in the centimeter-gram-second (cgs) electromagnetic system. The abampere is 10 amperes. 172

abandoned call (telephone switching systems). A call during which the calling station goes on-hook prior to its being answered. 55

A battery. A battery designed or employed to furnish current to heat the filaments of the tubes in a vacuum-tube circuit. *See:* **battery (primary or secondary).** 328

abbreviated dialing (telephone switching systems). A feature permitting the establishment of a call with an input of fewer digits than required under the numbering plan. 55

abbreviation. A shortened form of a word or expression. *See:* **functional designation; graphic symbol; letter combination; mathematical symbol; reference designation; symbol for a quantity; symbol for a unit.** 173

abnormal decay (charge-storage tubes). The dynamic decay of multiply-written, superimposed (integrated) signals whose total output amplitude changes at a rate distinctly different from that of an equivalent singly-written signal. *Note:* Abnormal decay is usually very much slower than normal decay and is observed in bombardment-induced conductivity type of tubes. *See:* **charge-storage tube.** 174

abnormal glow discharge (gas tube). The glow discharge characterized by the fact that the working voltage increases as the current increases. *See:* **discharge.** 175

abort (software). To terminate a process prior to completion. *See:* **process.** 434

abort sequence (token ring access method). A sequence that terminates the transmission of a frame prematurely. 472

above threshold firing time (microwave switching tubes)(nonlinear, active, and nonreciprocal waveguide components). The time to establish an above-threshold discharge in the gas tube after the application of radio frequency power. This time delay is responsible for the spike in the leading edge of the output leakage waveform. *See:* **receiver protector; gas tube; duplexer.** 530

abrupt junction (semiconductor)(nonlinear, active,

and nonreciprocal waveguide components). A semiconductor crystal having an n-region containing a near constant net concentration of donor impurities adjoining a p-region with a near constant net concentration of acceptors; used primarily in microwave frequency multipliers , dividers, and parametric circuits. 530

ABS (cable systems in power generating stations). Conduit fabricated from Acrylonitrile-Butadiene-Styrene. 35

absolute accuracy. Accuracy as measured from a reference that must be specified. 224, 207

absolute address (computing machines). (1) An address that is assigned by the machine designer to a physical storage location. (2) A pattern of characters that identifies a unique storage location without further modification. *See:* **machine address.** 255, 77

absolute altimeter (navigation aid terms). A device that measures altitude above local terrain. In its usual form it does this by measuring the time interval between transmission of a signal and the return of its echo, or by measuring the phase difference between the transmitting signal and the echo. 526

absolute block (automatic train control). A block governed by the principle that no train shall be permitted to enter the block while it is occupied by another train. 328

absolute code (microprocessor object modules). Data or executable machine code in memory or an image thereof. Distinguish from relocatable code. 466

absolute delay (loran)(navigation aid terms). The interval of time between the transmission of a signal from the master station and transmission of the next signal from the slave station. 526

absolute dimension. A dimension expressed with respect to the initial zero point of a coordinate axis. *See:* **coordinate dimension word.** 224, 207

absolute error. (1) The amount of error expressed in the same units as the quantity containing the error. (2) Loosely, the absolute value of the error, that is, the magnitude of the error without regard to algebraic sign. 255, 77, 54

absolute gain. *See:* **gain.**

absolute loader (microprocessor object modules). A process which can load one or more sections of absolute code only at the locations specified by the sections. *See:* **relocator.** 466

absolute luminance threshold (illuminating engineering). Luminance threshold for a bright object like a disk on a totally dark background. 167

absolute machine code (software). Machine language code that must be loaded into fixed storage locations at each use and may not be relocated. *See:* **machine language code; relocatable machine code.** 434

absolute permissive block (automatic train control). A term used for an automatic block signal system on a track signaled in both directions. For opposing movements the block is from siding to siding and the signals governing entrance to this block indicate STOP. For following movements the section between sidings is divided into two or more blocks and train movements into these blocks, except the first one, are governed by intermediate signals usually displaying STOP: then proceed at restricted speed, as their most restrictive indication.   328

absolute photocathode spectral response (diode-type camera tube). The ratio of the photocathode current, measured in amperes, to the radiant power incident on the photocathode face, measured in watts, as a function of the photon energy, frequency, or wavelength. Units: amperes watt^{-1} (A W^{-1}).   380

absolute refractory state (medical electronics). The portion of the electrical recovery cycle during which a biological system will not respond to an electric stimulus.   192

absolute steady-state deviation (control). The numerical difference between the ideal value and the final value of the directly controlled variable (or another variable if specified). *See:* **deviation (control); percent steady-state deviation.**   206

absolute system deviation (control). At any given point on the time response, the numerical difference between the ideal value and the instantaneous value of the directly controlled variable (or another variable if specified). *See:* **deviation.**   206

absolute threshold. The luminance threshold or minimum perceptible luminance (photometric brightness) when the eye is completely dark adapted. *See:* **visual field.**   167

absolute transient deviation (control). The numerical difference between the instantaneous value and the final value of the directly controlled variable (or another variable if specified). *See:* **deviation; percent transient deviation.**   206

absolute-value circuit (analog computers). A transducer or circuit that produces an output signal equal in magnitude to the input signal but always of one polarity.   9

absolute-value device. A transducer that produces an output signal equal in magnitude to the input signal but always of one polarity. *See:* **electronic analog computer.**   9

absorptance, $a = \Phi_a / \Phi_i$ **(illuminating engineering).** The ratio of the absorbed flux to the incident flux. *Note:* The sum of the hemispherical reflectance, the hemispherical transmittance, and the absorptance is one.   167

absorption (1) (fiber optics). In an optical waveguide, that portion of attenuation resulting from conversion of optical power into heat. *Note:* Intrinsic components consist of tails of the ultraviolet and infrared absorption bands. Extrinsic components include (a) impurities, for example, the OH$^-$ion and transition metal ions and, (b) defects, for example , results of thermal history and exposure to nuclear radiation. *See:* **attenuation.**   433

(2) (illuminating engineering). A general term for the process by which incident flux is converted to another form of energy, usually and ultimately to heat. *Note:* All of the incident flux is accounted for by the processes of reflection, transmission, and absorption.   167

(3) (laser-maser). The transfer of energy from a radiation field to matter.   363

(4) (radio wave propagation). The irreversible conversion of the energy of an electromagnetic wave into another form of energy as a result of its interaction with matter.   146

absorption coefficient (power station noise control). The ratio of the energy absorbed by the surface to the energy incident upon it.   500

absorption current (or component) (1) (rotating machinery). A reversible component of the measured current, which changes with time of voltage application, resulting from the phenomenon of "dielectric absorption" within the insulation when stressed by direct voltage. *See:* **IEEE Std 62-1958, Guide for Making Dielectric Measurements in the Field, Section 6.**   6

(2) (power cable systems). Current resulting from charge absorbed in the dielectric as a result of polarization.   437

(3)(electric submersible pump cable). Current resulting from charge absorbed in the dielectric as a result of polarization.   484

absorption frequency meter (reaction frequency meter) (waveguide). A one-port cavity frequency meter that, when tuned, absorbs electromagnetic energy from a waveguide. *See:* **waveguide.**   179

absorption loss (data transmission). The loss of signal energy in a communication circuit that results from coupling to a neighboring circuit or conductor.   59

absorption modulation. A method for producing amplitude modulation of the output of a radio transmitter by means of a variable-impedance (principally resistive) device inserted in or coupled to the output circuit.   211

absorptive attenuator (waveguide). *See:* **resistive attenuator.**

absorptive loss. *See:* **arc loss.**

abstraction (software). (1) A view of a problem which extracts the essential information relevant to a particular purpose and ignores the remainder of the information. (2) The process of forming an abstraction.   434

abstract machine (software). (1) A representation of the characteristics of a process or machine. (2) A module which processes inputs as though it were a machine. *See:* **module; process.**   434

abstract quantity. *See:* **mathematico-physical quantity.**

ac. *See:* **alternating current.**

ACA. *See:* **adjacent channel attenuation.**

ac analog computer (analog computers). An analog computer in which electrical signals are of the form of amplitude-modulated suppressed carrier signals where the absolute value of a computer variable is represented by the amplitude of the carrier and the sign of a computer variable is represented by the phase (0 to 180 degrees) of the carrier relative to the reference ac signal. 9

ac breakdown voltage (gas tube surge-protective device). The minimum root-mean-square value of sinusodial voltage at frequencies between 15 hertz (Hz) and 62 Hz that results in arrester sparkover. 370

AC cable (armored cable) (National Electrical Code). A fabricated assembly of insulated conductors in a flexible metallic enclosure. 256

accelerated life test (test, measurement and diagnostic equipment). A test in which certain factors, such as voltage, temperature, and so forth, are increased or decreased beyond normal operating values to obtain observable deterioration in a reasonable period of time, and thereby afford some measure of the probable life under normal operating conditions or some measure of the durability of the equipment when exposed to the factors being aggravated. 54

accelerated test (evaluation of thermal capability) (thermal classification of electric equipment and electrical insulation). A functional test in which one or more factors of influence are increased in magnitude or frequency of application so as to decrease the time needed for the test. 506

accelerating (rotating machinery). The process of running a motor up to speed after breakaway. *See:* **asynchronous machine.** 63

accelerating electrode. An electrode to which a potential is applied to increase the velocity of the electrons or ions in the beam. 117, 125

accelerating grid (electron tubes). *See:* **accelerating electrode.**

accelerating or decelerating device (18) (power system device function numbers). A device that is used to close or to cause the closing of circuits which are used to increase or decrease the speed of a machine. 402

accelerating relay (power switchgear). A programming relay whose function is to control the acceleration of rotating electrical equipment. 103

accelerating time (control) (industrial control). The time in seconds for a change of speed from one specified speed to a higher specified speed while accelerating under specified conditions. *See:* **electric drive.** 219, 206

accelerating torque (rotating machinery). Difference between the input torque to the rotor (electromagnetic for a motor or mechanical for a generator) and the sum of the load and loss torques: the net torque available for accelerating the rotating parts. *See:* **rotor.** 63

accelerating voltage (oscilloscope). The cathode-to-viewing-area voltage applied to a cathode-ray tube for the purpose of accelerating the electron beam. *See:* **oscillograph.** 185

acceleration (electric drive). Operation of raising the motor speed from zero or a low level to a higher level. *See:* **electric drive.** 206

acceleration factor (reliability). The ratio between the times necessary to obtain the same stated proportion of failures in two equal saples under two different sets of stress conditions involving the same failure modes and mechanisms. 164

acceleration-insensitive drift rate (gyro). That component of systematic drift rate which has no correlation with acceleration. 46

acceleration, programmed. A controlled velocity increase to the programmed rate. 224, 207

acceleration-sensitive drift rate (gyro). Those components of systematic drift rates that are correlated with the first power of linear acceleration applied to the gyro case. The relationship of these components of drift rate to acceleration can be stated by means of coefficients having dimensions of angular displacement per unit time per unit acceleration for accelerations along each of the principal axes of the gyro, for example, drift rate caused by mass unbalance. 46

acceleration space (velocity-modulated tube). The part of the tube following the electron run in which the emitted electrons are accelerated to reach a determined velocity. *See:* **velocity-modulated tube.** 244, 190

acceleration-squared-sensitive drift rate (gyro). Those components of systematic drift rates that are correlated with the second power or product of linear acceleration applied to the gyro case. The relationship of these components of drift rate to acceleration squared can be stated by means of coefficients having dimensions of angular displacement per unit time per unit acceleration squared for accelerations along each of the principal axes of the gyro and angular displacement per unit time per the product of accelerations along combinations of two principal axes of the gyro for example, drift rate caused by anisoelasticity. 46

acceleration, timed (industrial control). A control function that accelerates the drive by automatically controlling the speed change as a function of time. *See:* **electric drive.** 219, 206

accelerometer (1) (gyro). A device that senses the inertial reaction of a proof mass for the purpose of measuring linear or angular acceleration. 46 **(2)(navigation aid terms).** A device that senses inertial reaction to measure linear or angular acceleration. *Note:* In its simplest form, an accelerometer consists of a case-mounted spring and mass arrangement where displacement of the mass from its rest position relative to the case is proportional to the total nongravitational acceleration experienced along the instrument's sensitive axes. 526

accent lighting (illuminating engineering). Directional lighting to emphasize a particular object or draw attention to a part of the field of view. 167

accept (logical link control). The condition assumed by an LLC (logical link control) upon accepting a correctly received PDU (protocol data unit) for processing. 585

acceptable (1)(**class 1E equipment and circuits**). Demonstrated to be adequate by the safety analysis(-ses) of the station. 1, 102, 428, 99, 131,541
(2)(**diesel-generator unit**). Demonstrated to be adequate by the safety analysis of the plant. 99

acceptable energized background noise level (partial discharge measurement in liquid-filled power transformers and shunt reactors). Energized background noise level present during test which is considered acceptable. It should not exceed 50% of the acceptable terminal partial discharge level and in any case should be below 100 %. 580

acceptable terminal partial discharge level (partial discharge measurementin liquid-filled power transformers and shunt reactors). That specified maximum terminal partial discharge value for which measured terminal partial discharge values exceeding said value are considered unacceptable. The method of measurement and the test voltage for a given test object must be specified with the acceptable terminal partial discharge level. 580

acceptance angle (fiber optics). Half the vertex angle of that cone within which optical power may be coupled into bound modes of an optical waveguide. *Notes:* (1) Acceptance angle is a function of position on the entrance face of the core when the refractive index is a function of radius in the core. In that case, the local acceptance angle is

$$\arcsin\sqrt{n^2(r)-n^2_2}$$

where n(r) is the local refractive index and$n_2$is the minimum refractive index of the cladding. The sine of the local acceptance angle is sometimes referred to as the local numerical aperture. (2) Power may be coupled into leaky modes at angles exceeding the acceptance angle. *See:* **launch numerical aperture; power-law index profile.** 433

acceptance criteria (1) (nuclear power quality assurance). Specified limits placed on characteristics of an item, process, or service defined in codes, standards, or other requirement documents. 417
(2) (**software**). The criteria a software product must meet to successfully complete a test phase or meet delivery requirements. *See:* **delivery requirements; software product; test phase.** 434

acceptance proof test (rotating machinery). A test applied to new insulated winding before commercial use. It may be performed at the factory or after installation, or both. *See:* **American National Standard General Requirements for Synchronous Machines, ANSI C50.10-1977.** 6

acceptance test (1) (general) (A). A test to demonstrate the degree of compliance of a device with purchaser's requirements. (B) A **conformance test** demonstrates the quality of the units of a consignment, without implication of contractual relations between buyer and seller. *See:* **routine test; test (instrument or meter); conformance test.** 91, 103
(2) (**lead storage batteries**). A constant current capacity test made on a new battery to determine that it meets specifications or manufacturer's ratings. 38

(3) (**power cable systems**). A test made after installation but before the cable is placed in normal service. This test is intended to detect shipping or installation damage and to show any gross defects or errors in workmanship on splicing and terminating. 437
(4) (**nuclear power generating stations**) (**lead storage batteries**). A capacity test made on a new battery to determine that it meets specifications or manufacturer's ratings, or both. 31, 38
(5)(**electric submersible pump cable**). Test intended to detect damage prior to the initial installation of new cable. 484

acceptance testing (1)(software). Formal testing conducted to determine whether a system satisfies its acceptance criteria and to enable the customer to determine whether to accept the system. *See:* **acceptance criteria; formal testing; qualification testing; system; system testing.** 434
(2)(**software verification and validation plans**). Formal testing conducted to determine whether or not a system satisfies its acceptance criteria and to enable the customer to determine whether or not to accept the system. 511

acceptance tests. *See:* **conformance tests.** *Note:* American National Standards should use the term *conformance test,* as directed by the Standards Council of ANSI, rather than the term **acceptance test.** Use of the term *conformance test* avoids the implication of contractual relations between buyer and seller. 103

acceptor (semiconductor). *See:* **impurity, acceptor; semiconductor.**

access. *See:* **random access; serial access.**

access code (telephone switching systems). One or more digits required in certain situations in lieu of or preceding an area or office code. 55

access-control mechanism (software). Hardware or software features, operating procedures, or management procedures designed to permit authorized access to a computer system. *See:* **computer system; hardware; procedures; software.** 434

access coupler (fiber optics). A device placed between two waveguide ends to allow signals to be withdrawn from or entered into one of the waveguides. *See:* **optical waveguide coupler.** 433

access fitting. A fitting permitting access to the conductors in a raceway at locations other than at a box. *See:* **raceway.** 328

accessibility (1) (software). The extent to which software facilitates selective use or maintenance of its components. *See:* **components; maintenance; software.** 434
(2) (**telephone switching systems**). The ability of a given inlet to reach the available outlets. 55
(3) (**National Electric Code**) (A) (As applied to equipment)Admitting close approach: not guarded by locked doors, elevation, or other effective means. *See:* **readily accessible.** (B) (As applied to wiring methods.) Capable of being removed or exposed without damaging the building structure or finish or not permanently closed in by the structure or finish of the building. *See:* **concealed; exposed.** 256

(4) (power and distribution transformer). Admitting close approach because not guarded by locked doors, elevation, or other effective means.   53

(5) (wiring methods). Not permanently closed in by the structure or finish of the building: capable of being removed without disturbing the building structure or finish.   328

(6) (transformer). Admitting close approach to contact by persons due to lack of locked doors, elevation or other effective safeguards.   53

accessible, readily. Capable of being reached quickly without requiring those to whom ready access is requisite to climb over or remove obstacles or to resort to portable ladders, chairs, etc.   53

accessories (1) (general). Devices that perform a secondary or minor duty as an adjunct or refinement to the primary or major duty of a unit of equipment.   103, 53

(2) (power and distribution transformer). Devices that perform a secondary or minor duty as an adjunct or refinement to the primary or major duty of a unit of equipment.   103, 53

(3)(raceway)(raceway systems for Class 1E circuits for nuclear power generating stations). Devices that are used to supplement the functions of raceway systems. These include such items as dropouts, covers, conduit adapters, fastening devices (items such as conduit clamps, support connections, and cable tray cover clamps), adjustable connectors, and dividers.   513

accessory (1) (test, measurement and diagnostic equipment). An assembly of a group of parts or a unit which is not always required for the operation of a test set or unit as originally designed but serves to extend the functions or capabilities of the test set: similarly as headphones for a radio set supplied with a loudspeaker: a vibrator power unit for use with a set having a built-in power supply, or a remote control unit for use with a set having integral controls.   54

(2) (electric and electronics parts and equipment). A basic part, subassembly, or assembly designed for use in conjunction with or to supplement another assembly, unit, or set, contributing to the effectiveness thereof without extending or varying the basic function of the assembly or set. An accessory may be used for testing, adjusting, or calibrating purposes. Typical examples: test instrument, recording camera for radar set, headphones, emergency power supply.   17

(3)(power line maintenance). A removable device attached to a major or primary operating tool allowing diversified operations. Example: universal tool.   458

accessory equipment (Class 1E motor) (nuclear power generating station). Devices other than the principal motor components that are furnished with or built as a part of the motor structure and are necessary for the operation of the motor.   104

access time (1). A time interval that is characteristic of a storage device, and is essentially a measure of the time required to communicate with that device. *Note:* Many definitions of the beginning and ending of this interval are in common use.   235

(2) (A) The time interval between the instant at which data are called for from a storage device and the instant delivery is completed, that is, the read time. **(B)** The time interval between the instant at which data are requested to be stored and the instant at which storage is completed, that is, the write time.   255

(3) (acousto-optic deflector). The minimum time to randomly deflect the light beam from one spot position to another. It is given by the time it takes the acoustic beam to cross the optical beam, viz, $\tau = S/V$, with τ the access time, S the optical beam dimension, and V the acoustic velocity.   72

(4) (acoustically tunable optical filter). The minimum time to randomly tune the filter from one wavelength to another. It is given by the time it takes the acoustic beam to cross the optical beam, namely: τ, S.V, with S the length of the optical beam along the acoustic beam direction and V the acoustic velocity.   72

access tools (tamper-resistant switchgear assembly) (relaying). Keys or other special accessories with unique characteristics that make them suitable for gaining access to the tamper-resistant switchgear assembly.   79

accommodation (1) (laser-maser). The ability of the eye to change its power and thus focus for different object distances.   363

(2) (illuminating engineering). The process by which the eye changes focus from one distance to another.   167

accommodation, electrical (biology) (electrobiology). A rise in the stimulation threshold of excitable tissue due to its electrical environment, often observed following a previous stimulation cycle. *See:* **excitability.**   192

ac converter (self-commutated converters). A converter for changing alternating current (ac) power of a given voltage, frequency, and phase number to ac power in which one or more of these paramenters are different.   584

accumulating stimulus (electrotherapy). A current that increases so gradually in intensity as to be less effective than it would have been if the final intensity had been abruptly attained. *See:* **electrotherapy.**   192

accumulator. (1) A device that retains a number (the augend) adds to it another number (the addend) and replaces the augend with the sum. (2) Sometimes only the part of (1) that retains the sum. *Note:* The term is also applied to devices that function as described but that also have other properties.   235

accumulator function. *See:* **supervisory control functions**

accuracy (1) (instrumentation and measurement). The quality of freedom from mistake or error, that is, of conformity to truth or to a rule. *Notes:* (A) Accuracy is distinguished from precision as in the following example: A six-place table is more precise than a four-place table. However, if there are errors in the six-place table, it may be more or less accurate than the four-place table. (B) The accuracy of an indicated or recorded value is expressed by the ratio of the error of the indicated value to the true value. It is usually expressed in percent. Since the true value cannot be determined exactly, the measured or calculated value of highest available accuracy is taken to be the true value or reference value. Hence, when a meter is calibrated in a given echelon, the measurement made on a meter of a higher-accuracy echelon usually will be used as the reference value. Comparison of results obtained by different measurement procedures is often useful in establishing the true value. *See:* **dynamic accuracy; electronic analog computer; measurement system; static accuracy.**   183

(2) (analog computers). (A) Conformity of a measured value to an accepted standard value. (B) A measure of the degree by which the actual output of a device approximates the output of an ideal device nominally performing the same function. *See:* **electronic analog computer.**   9

(3) (power supply). Used as a specification for the output voltage of power supplies, accuracy refers to the absolute voltage tolerance with respect to the stated nominal output. *See:* **power supply.**   186

(4) (numerically controlled machine). Conformity of an indicated value to the true value, that is, an actual or an accepted standard value. *Note:* Quantitatively, it should be expressed as an error or an uncertainty. The property is the joint effect of method, observer, apparatus, and environment. Accuracy is impaired by mistakes, by systematic bias such as abnormal ambient temperature, or by random errors (imprecision). The accuracy of a control system is expressed as the system deviation (the difference between the ultimately controlled variable and its ideal value), usually in the steady state or at sampled instants. *See:* **precision and reproducibility.**   207

(5) (electronic navigation). Generally, the quality of freedom from mistake or error: that is, of conformity to truth or a rule. Specifically, the difference between the mean value of a number of observations and the true value. *Note:* Often refers to a composite character including both accuracy and precision. *See:* **navigation; precision.**   13, 187

(6) (signal-transmission system). Conformity of an indicated value to an accepted standard value, or true value. *Note:* Quantitatively, it should be expressed as an error or uncertainty. The accuracy of a determination is affected by the method, observer, environment, and apparatus, including the working standard used for the determination. *See:* **signal.**   188

(7) (indicated or recorded value). The accuracy of an indicated or recorded value is expressed by the ratio of the error of the indicated value to the true value. It is usually expressed in percent. *See:* **accuracy rating of an instrument.**   328

(8) (instrument transformer). The extent to which the current or voltage in the secondary circuit reproduces the current or voltage of the primary circuit in the proportion stated by the marked ratio, and represents the phase relationship of the primary current or voltage.   53

(9) (test, measurement and diagnostic equipment). The degree of correctness with which a measured value agrees with the true value. *See:* **precision.**   54

(10) (electrothermic power meters). The degree of correctness with which a measurement device yields the true value of a measured quantity: quantitatively expressed by uncertainty. *See:* **uncertainty.**   47

(11) (nuclear power generating stations). The quality of freedom from mistake or error.   41

(12) (pulse measurement). The degree of agreement between the result of the application of a pulse measurement process and the true magnitude of the pulse characteristic, property, or attribute being measured.   15

(13) (excitation control system). The degree of correspondence between the controlled variable and the desired value under specified conditions such as load changes, ambient temperature, humidity, frequency, and supply voltage variations. Quantitatively, it is expressed as the ratio of difference between the controlled variable and the desired value to the desired value.   105

(14) (ionization chambers). The accuracy, usually described in terms of overall uncertainty, is the estimate of the overall possible deviation from the "true" value. As used in ANSI N42.13-1978 the overall uncertainty is a total of the estimated errors itemized in Section 5 plus the random error of the measurement.   396

(15) (metric practice). The degree of conformity of a measured or calculated value to some recognized standard or specified value. This concept involves the systematic error of an operation which is seldom negligible. *See:* **precision.**   21

(16) (nuclear power generating stations) (measuring and test equipment). A measure of the degree by which the actual output of a device approximates the output of an ideal device nominally performing the same function.   41

(17) (plutonium monitoring). The degree of agreement with the true value of the quantity being measured.   413

(18) (radiation protection). The degree of agreement of the observed value with the true or correct value of the quantity being measured.   399

(19) (software). (A) A quality of that which is free of error. (B) A qualitative assessment of freedom from error, a high assessment corresponding to a small error. (C) A quantitative measure of the magnitude of error, preferably expressed as a function of the relative error, a high value of this measure corresponding to a small error. (D) A quantitative assessment of freedom from error. *See:* **error; precision.**   434

(20)('dose calibrator' ionization chambers). Usually described in terms of overall uncertainty, accuracy is the estimate of the overall possible deviation from the stated value. As used in Standard N42.13-1986, the overall uncertainty is a total of the estimated error itemized in Section 5 plus the random uncertainty of the measurement.   499

(21)(excitation systems for synchronous machines). The degree of correspondence between the controlled variable and the desired value under specified conditions such as load changes, ambient temperature, humidity, frequency, and supply voltage variations. Quantitatively, it is expressed as the ratio of difference between the controlled variable and the desired value to the desired value.   507

(22)(mathematics of computing) A qualitative assessment of correctness, or freedom from error. Contrast with precision.   564

(23)(monitoring radioactivity in effluents). The degree of agreement with the true value of the quantity being measured. *Note:* Accuracy is subject to the influence of unknown systematic errors.   559

accuracy classes for metering (instrument transformer). Limits of a transformer correction factor, in terms of percent error, that have been established to cover specific performance ranges for line power-factor conditions between 1.0 and 0.6 lag.   303, 203

accuracy classes for relaying (instrument transformer). Limits, in terms of percent ratio error, that have been established.   303

accuracy rating (class) (1) (general) (electric instrument). The accuracy classification of the instrument. It is given as the limit, usually expressed as a percentage of full-scale value, that errors will not exceed when the instrument is used under reference conditions. *Notes:* (A) The accuracy rating is intended to represent the tolerance applicable to an instrument in an "as-received condition." Additional tolerances for the various influences are permitted when applicable. It is required that the accuracy, as received, be directly in terms of the indications on the scale and without the application of corrections from a curve, chart, or tabulation. Over that portion of the scale where the accuracy tolerance applies, all marked division points shall conform to the stated accuracy class. (B) Generally the accuracy of electrical indicating instruments is stated in terms of the electrical quantities to which the instrument responds. In instruments with the zero at a point other than one end of the scale, the arithmetic sum of the end-scale readings to the right and to the left of the zero point shall be used as the full-scale value. Exceptions: (1) The accuracy of frequency meters shall be expressed on the basis of the percentage of actual scale range. Thus, an instrument having a scale range of 55 to 65 hertz would have its error expressed as a percentage of 10 hertz. (2) The accuracy of a power-factor meter shall be expressed as a percentage of scale length. (3) The accuracy of instruments that indicate derived quantities, such as series type ohmmeters, shall be expressed as a percentage of scale length. (C) In the case of instruments having nonlinear scales, the stated accuracy only applies to those portions of the scale where the divisions are equal to or greater than two-thirds the width they would be if the scale were even divided. The limit of the range at which this accuracy applies may be marked with a small isosceles triangle whose base marks the limit and whose point is directed toward the portion of the scale having the specified accuracy. (D) Instruments having an accuracy rating of 0.1 percent

are frequently referred to as laboratory standards. Portable instruments having an accuracy rating of 0.25 percent are frequently referred to as portable standards.   280

(2) (automatic control system). The limit which the system deviation will not exceed under specified operating conditions.   56

accuracy ratings of instrument transformers. Means of classifying transformers in terms of percent error limits under specified conditions of operation.   305

accuracy, synchronous-machine regulating system. The degree of correspondence (or ratio) between the actual and the ideal values of a controlled variable of the synchronous-machine regulating system under specified conditions, such as load changes, drift, ambient temperature, humidity, frequency, and supply voltage.   63

accuracy test (instrument transformer). A test to determine the degree to which the value of the quantity obtained from the secondary reflects the value of the quantity applied to the primary.   203

ac-dc general-use snap switch (National Electric Code). A form of general-use snap switch suitable for use on either AC or DC circuits for controlling the following: (1) Resistive loads not exceeding the ampere rating of the switch at the voltage applied. (2) Inductive loads not exceeding 50 percent of the ampere rating of the switch at the applied voltage. Switches rated in horsepower are suitable for controlling motor loads within their rating at voltage applied. (3) Tungsten-filament lamp loads not exceeding the ampere rating of the switch at the applied voltage if 'T' rated.   256

acetate disks. Mechanical recording disks, either solid or laminated, that are made of various acetate compounds.   176

ac general-use snap switch (National Electric Code). A form of general-use snap switch suitable only for use on alternating-current circuits for controlling the following: (1) Resistive and inductive loads, including electric-discharge lamps, not exceeding the ampere rating of the switch at the voltage involved. (2) Tungsten-filament lamp loads not exceeding the ampere rating of the switch at 120 volts. (3) Motor loads not exceeding 80 percent of the ampere rating of the switch at its rated voltage. (4) Snap switches rated 20 amperes or less directly connected to aluminum conductors shall be approved for the purpose and marked CO/ALR.   256

achromatic locus (achromatic region) (television). A region including those points in a chromaticity diagram that represent, by common acceptance, arbitrarily chosen white points (white references). *Note:* The boundaries of the achromatic locus indefinite, depending on the tolerances in any specific application. Acceptable reference standards of illumination (commonly referred to as white light) are usually represented by points close to the locus of Planckian radiators having temperatures higher than about 2000 Kelvins. While any point in the achromatic locus may be chosen as the reference point for the determination of dominant wavelength, complementary wavelength, and purity for specification of object colors, it is usually advisable to adopt the point representing the chromaticity of the illuminator. Mixed qualities of illumination and luminators with chromaticities represented very far from the Planckian locus require special consideration. After a suitable reference point is selected, dominant wavelength may be determined by noting the wavelength corresponding to the intersection of the spectrum locus with the straight line drawn from the reference point through the point representing the sample. When the reference point lies between the sample point and the intersection, the intersection indicates the complementary wavelength. Any point within the achromatic locus chosen as a reference point may be called an achromatic point. Such points have also been called white points.   18

acid-resistant (industrial control). So constructed that it will not be injured readily by exposure to acid fumes.   225, 206

acknowledger (forestaller). A manually operated electric switch or pneumatic valve by means of which, on a locomotive equipped with an automatic train stop or train control device, an automatic brake application can be forestalled, or by means of which, on a locomotive equipped with an automatic cab signal device, the sounding of the cab indicator can be silenced.   328

acknowledging (forestalling). The operating by the engineman of the acknowledger associated with the vehicle-carried equipment of an automatic speed control or cab signal system to recognize the change of the aspect of the vehicle-carried signal to a more restrictive indication. The operation stops the sounding of the warning whistle, and in a locomotive equipped with speed control it also forestalls a brake application.   328

acknowledging device. *See:* **acknowledger (forestaller).**

acknowledging switch. *See:* **acknowledger (forestaller).**

acknowledging whistle. An air-operated whistle that is sounded when the acknowledging switch is operated. Its purpose is to inform the fireman that the engineman has recognized a more restrictive signal indication.   328

a contact (power switchgear). A contact that is open when the main device is in the standard reference position, and that is closed when the device is in the opposite position. *Notes:* (1) a contact has general appliction. However, this meaning for front contact is restricted to relay parlance. (2) For indication of the specific point of travel at which the contact changes position, an additional letter or percentage figure may be added to a as detailed in 9.4.4.1 and 9.4.4.2 of ANSI/IEEE C37.2-1979. (3) *See:* **standard reference position.**   103

acoustical. *See:* **acoustic.**

acoustical depth finder (navigation aid terms). *See:* **echo sounder.**

acoustically tunable optical filter. An optical filter which is driven by an acoustic wave and which is tunable by varying the acoustic frequency.   72

acoustic coupler (data communication). A type of data communication equipment which has sound transducers that permit the use of a telephone handset as a connection to a voice communication system for the purpose of data transmission.   12

acoustic delay line. A delay line whose operation is based on the time of propagation of sound waves.   255, 77

acoustic-gravity wave (radio wave propagation). In the atmosphere, a low frequency wave whose restoring forces are both compressional and gravitational.   146

acoustic input (transmission performance of telephone sets). The free field sound pressure level developed by an artificial mouth at a specified location, called the calibration position, measured in decibels referred to a root-mean-square (rms) sound pressure of 1 pascal (Pa).   491

acoustic interferometer. An instrument for the measurement of wavelength and attenuation of sound. Its operation depends upon the interference between reflected and direct sound at the transducer in a standing-wave column. *See:* **instrument.**   328

acoustic output (transmission performance of telephone sets). The sound pressure level developed in an artificial ear, measured in decibels referred to a root-mean-square (rms) sound pressure of 1 pascal (Pa).   491

acoustic radiator. A means for radiating acoustic waves. *See:* **loudspeaker.**   328

acoustic wave filter. A filter designed to separate acoustic waves of different frequencies. *Note:* Through electroacoustic transducers such a filter may be associated with electric circuits. *See:* **filter.**   328

acousto-optic device. A device which is used to modulate light in amplitude, frequency, phase, polarization, or spatial position by virtue of optical diffraction from an acoustically generated diffraction grating.   82

acousto-optic effect (fiber optics). A periodic variation of refractive index caused by an acoustic wave. *Note:* The acousto-optic effect is used in devices that modulate and deflect light. *See:* **modulation.**   433

acquisition (radar). The process of establishing a stable track on a target which is designated in one or more coordinates. A search of a given volume of coordinate space is usually required because of errors or incompleteness of the designation.   13

acquisition probability (radar). The probability of establishing a stable track on a designated target.   13

across-the-line starter (electric installations on shipboard). A device which connects the motor to the supply without the use of a resistance or autotransformer to reduce the voltage.It may consist of a manually operated switchor a master switch,which energizes an electromagnetically operated contactor.   3

across-the-line starting (rotating machinery). The

process of starting a motor by connecting it directly to the supply at rated voltage. 63

ACSR. *See:* **aluminum cable steel reinforced.**

acting stress (seismic design of substations). Maximum applied or expected stress in the material during normal operation of the apparatus of which it is a part, including the stresses caused by wind, seismic or short-circuit loading, acting either independently or simultaneously as determined by the user. 465

action potential (action current) (medical electronics). The instantaneous value of the potential observed between excited and resting portions of a cell or excitable living structure. *Note:* It may be measured direct or through a volume conductor. 192

action spike (medical electronics). The greatest in magnitude and briefest in duration of the characteristic negative waves seen during the observation of action potentials. 192

activate. *See:* **assert.**

activation (cathode) (thermionics). The treatment applied to a cathode in order to create or increase its emission. *See:* **electron emission.** 244, 190

activation polarization. The difference between the total polarization and the concentration polarization. *See:* **electrochemistry.** 328

activation time (gyro, accelerometer). *See:* **turn-on time.** 46

active (1) (electric generating unit reliability, availability, and productivity). The state in which a unit is in the population of units being reported on. *Note:* A unit generally enters the active state on its service date. 567

(2) (signals and paths)(696 interface devices). A signal in its logically true state. 538

(3) (power system measurement). The state in which a unit is in the population of units being reported on. 432

active antenna array. An array in which all or parts of the elements are equipped with their own transmitter or receiver or both. *Notes:* (1) Ideally, for the transmitting case, amplitudes and phases of the output signals of the various transmitters are controllable and can be coordinated in order to provide the desired aperture distribution. (2) Often it is only a stage of amplification or frequency conversion that is actually located at the array elements, with the other stages of the receiver or transmitter remotely located. 111

active area (solar cell). The illuminated area normal to light incidence, usually the face area less the contact area. *Note:* For the purpose of determining efficiency, the area covered by collector grids is considered a part of the active area. *See:* **semiconductor.** 113

active current (rotating machinery). The component of the alternating current that is in phase with the voltage. *See:* **asynchronous machine.** 63

active-current compensator (rotating machinery). A compensator that acts to modify the functioning of a voltage regulator in accordance with active current. 63

active dimension (of a position-sensitive detector)(-charged-particle detectors). A dimension (that is, length, width) of that region of a position-sensitive detector which is depleted. 119

active electric network. An electric network containing one or more sources of power. *See:* **network analysis.** 328

active electrode (electrobiology). (1) A pickup electrode that, because of its relation to the flow pattern of bioelectric currents, shows a potential difference with respect to ground or to a defined zero, or to another (reference) electrode on related tissue. (2) Any electrode, in a system of stimulating electrodes, at which excitation is produced. (3) A stimulating electrode (different electrode) applied to tissue for stimulation and distinguished from another (inactive, dispersive, diffuse, or indifferent) electrode by having a smaller area of contact thus affording a higher current density. *See:* **electrobiology.** 192

active homing guidance (navigation aid terms). A system of homing guidance wherein both the source of illuminating the target and the receiver for detecting the energy reflected from the target, as a result of illuminating the target, are carried within the vehicle. 526

active impedance (of an array element) (antennas). The ratio of the voltage across the terminals of an array element to the current flowing at those terminals when all array elements are in place and excited. 111

active laser medium (fiber optics). The material within a laser, such as crystal, gas, glass, liquid, or semiconductor, that emits coherent radiation (or exhibits gain) as the result of stimulated electronic or molecular transitions to lower energy states. *Syn:* **laser medium.** *See:* **laser; optical cavity.** 433

active maintenance time. The time during which maintenance actions are performed on an item either manually or automatically. *Notes:* (1) Delays inherent in the maintenance operation, for example those due to design or to prescribed maintenance procedures, shall be included. (2) Active maintenance may be carried out while the item is performing its intended function. 164

active materials (storage battery). The materials of the plates that react chemically to produce electric energy when the cell discharges and that are restored to their original composition, in the charged condition, by oxidation and reduction processes produced by the charging current. *See:* **battery.** 328

active power (1) (rotating machinery). A term used for power when it is necessary to distinguish among apparent power, complex power and its components, active and reactive power. *See:* **asynchronous machine.** 63

(2) (metering). The time average of the instantaneous power over one period of the wave. *Note:* For sinusoidal quantities in a two-wire circuit, it is the product of the voltage , the current, and the cosine of the phase angle between them. For nonsinusoidal quantities, it is the sum of all the harmonic components, each determined as above. In a poly-phase circuit it is the sum of the active powers of the individual phases. *See:* **power, active.** 212

active-power relay (1) (general). A power relay that responds to active power. *See:* **relay; power relay; reactive power relay.** 127

(2) (power switchgear). A power relay that responds to active power. *Note:* See **power relay** and **reactive power relay.** 103

active preventive maintenance time. That part of the active maintenance time in which preventive maintenance is carried out. *Notes:* (1) Delays inherent in the preventive maintenance operation, for example those due to design or prescribed maintenance procedures, shall be included. (2) Active preventive maintenance time does not include any time taken to maintain an item which has been replaced. 164

active redundancy. *See:* **redundancy, active.**

active reflection coefficient (of an array element) (antennas). The reflection coefficient at the terminals of an array element when all array elements are in place and excited. 111

active repair time. The time during which corrective maintenance actions are performed on an item either manually or automatically. *Notes:* (1) Delays inherent in the repair operation, for example those due to design or to prescribed maintenance procedures, shall be included. (2) Active repair time does not include any time taken to repair an item which has been replaced as part of the corrective maintenance action under consideration. 164

active segment interconnect (FASTBUS acquisition and control). A segment interconnect is said to be active if it is assering AS = 1 (address sync = 1) on the far-side segment. 480

active testing (test, measurement and diagnostic equipment). The process of determining equipment static and dynamic characteristics by performing a series of measurements during a series of known operating conditions. Active testing may require an interruption of normal equipment operations and it involves measurements made over the range of equipment operation. *See:* **interference testing.** 54

active transducer. *See:* **transducer, active.**

activity response (sodium iodide detector). The net number of counts registered by the detector system per unit of time divided by the activity of the radionuclide that is being measured during the same unit of time. 423

actual generation (AAG)(electric generating unit reliability, availability, and productivity). The energy that was generated by a unit in a given period. Actual generation can be expressed as gross actual generation (GAAG) or net actual generation (NAAG). 567

actual parameter (software). An argument or expression used within a call to a subprogram to specify data or program elements to be transmitted to the subprogram. *See:* **data; formal parameter; program; subprogram.** 434

actual transient recovery voltage (power switchgear). The TRV (transient recovery voltage) that actually occurs across the terminals of a pole of a switching device following current interruption. *Note:* The actual TRV may differ from the inherent TRV due to the modifying effects of device impedance and arc-circuit interaction. 487, 488, 577

actuated equipment (nuclear power generating station). (1) The assembly of prime movers and driven equipment used to accomplish a protective action. *Note:* Examples of prime movers are: turbines, motors, and solenoids. Examples of driven equipment are: control rods, pumps, and valves. 102, 387

(2) (A) A component or assembly of components that performs a protective function such as reactor trip, containment isolation, or emergency coolant injection. *Note:* The following are examples of actuated equipment: an entire control rod and its release mechanism, a containment isolation valve and its operator, and a safety injection pump and its prime mover. 31, 109

(B) A component or assembly of components that performs, or directly contributes to the performance of, a protective function such as reactor trip, containment isolation, or emergency coolant injection. The following are examples of actuated equipment: an entire control rod with its release or drive mechanism, a containment isolation valve with its operator, and a safety injection pump with its prime mover. 355

actuating current (of an automatic line sectionalizer) (power switchgear). The root-mean-square current which actuates a counting operation or an automatic operation. 103

actuating device (protective signaling). A manually or automatically operated mechanical or electric device that operates electric contacts to effect signal transmission. *See:* **protective signaling.** 328

actuating signal (industrial control). The reference input signal minus the feedback signal. *See:* **control system, feedback.** 206, 219

actuation device (actuator) (nuclear power generating stations). (1) A component or assembly of components that directly controls the motive power (electricity, compressed air, etcetera) for actuated equipment. *Note:* The following are examples of an actuation device: a circuit breaker, a relay, and a pilot valve used to control compressed air to the operator of a containment isolation valve. 109

(2) A component or assembly of components (or module) that directly controls the motive power (electricity, compressed air, etcetera) for actuated equipment. The following are examples of an actuation device: a circuit breaker, a relay, a valve (with its operator) used to control compressed air to the operator of a containment isolation valve, (and a module containing such equipment). 355

(3) A component or assembly of components that directly controls the motive power (electricity, compressed air, hydraulic fluid, etcetera) for actuated equipment. *Note:* Examples of actuation devices are: circuit breakers, relays, and pilot valves. 102, 387

actuation time, relay. *See:* **relay actuation time.**

actuator (automatic train control). A mechanical or electric device used for automatic operation of a brake valve. 328

actuator, centrifugal (rotating machinery). Rotor-mounted element of a centrifugal starting switch. *See:* **centrifugal starting switch.** 63

actuator, relay. *See:* **relay actuator.**

actuator valve. An electropneumatic valve used to control the operation of a brake valve actuator. 328

acyclic machine (homopolar machine*) (unipolar machine*) (rotating machinery). A direct-current machine in which the voltage generated in the active conductors maintains the same direction with respect to those conductors.

*Deprecated. 63

adaptability (software). The ease with which software allows differing system constraints and user needs to be satisfied. *See:* **software; system.** 434

adaptation (illuminating engineering). The process by which the retina becomes accustomed to more or less light than it was exposed to during an immediately preceding period. It results in a change in the sensitivity to light. *Note:* Adaptation is also used to refer to the final state of the process, as reaching a condition of adaptation to this or that level of luminance. *See:* **scotopic vision; photopic vision; chromatic adaptation.** 167

adapter (1) (general). A device for connecting parts that will not mate. An accessory to convert a device to a new or modified use. 185

(2) (test, measurement and diagnostic equipment). A device or series of devices designed to provide a compatible connection between the unit under test and the test equipment. May include proper stimuli or loads not contained in the test equipment. *See:* **interface.** 54

adapter kit (test, measurement and diagnostic equipment). A kit containing an assortment of cables and adapters for use with test or support equipment. 54

adapter, standard. A two-port device having standard connectors for joining together two waveguides or transmission lines with nonmating standard connectors. 110

adapter, waveguide (waveguide components). A structure used to interconnect two waveguides which differ in size or type. If the modes of propagation also differ, the adapter functions as a mode transducer. 166

adapting. *See:* **self-adapting.**

adaptive antenna system. An antenna system having circuit elements associated with its radiating elements such that some of the antenna properties are controlled by the received signal. *See:* **antenna.** 111

adaptive color shift (illuminating engineering). The change in the perceived object color caused solely by change of the state of chromatic adaptation. *See:* **state of chromatic adaptation.** 167

adaptive control system. *See:* **control system, adaptive.**

adaptive equalization (data transmission). A system that has a means of monitoring its own frequency response characteristics and a means of varying its own parameters by closed-loop action to obtain the desired overall frequency response. 59

adaptive maintenance (software). Maintenance performed to make a software product usable in a changed environment. *See:* **maintenance; software product.** 434

adaptive system. A system that has a means of monitoring its own performance and a means of varying its own parameters by closed-loop action to improve its performance. *See:* **system science.** 209

Adcock antenna. A pair of vertical antennas separated by a distance of one-half wavelength or less, and connected in phase opposition to produce a radiation pattern having the shape of the figure eight in all planes containing the centers of the two antennas. *See:* **antenna.** 111

adder. A device whose output is a representation of the sum of the two or more quantities represented by the inputs. *See:* **half-adder; electronic analog computer.** 235, 54

addition agent (electroplating). A substance that, when added to an electrolyte, produces a desired change in the structure or properties of an electrodeposit, without producing any appreciable change in the conductivity of the electrolytes, or in the activity of the metal ions or hydrogen ions. *See:* **electroplating.** 328

additive (insulating oil). A chemical compound or compounds added to an insulating fluid for the purpose of imparting new properties or altering those properties which the fluid already has. 461

address (A) (1) (semiconductor memory). Those inputs whose states select a particular cell or group of cells. 441

(2) (electronic computations and data processing). (A) An identification, as represented by a name, label, or number, for a register, location in storage, or any other data source or destination such as the location of a station in a communication network. (B) Loosely, any part of an instruction that specifies the location of an operand for the instruction. (C) (electronic machine-control system). A means of identifying information or a location in a control system. *Example:* The x in the command x 12345 is an address identifying the numbers 12345 as referring to a position on the x axis. 255

(3) (software). (A) A character or group of characters that identifies a register, a particular part of storage, or some other data source or destination. (B) To refer to a device or an item of data. *See:* **data.** 434

(4) (test pattern language). The identification of a specific memory word, usually expressed in X-, Y-, and Z-coordinates, and in binary code. *See:* **logic address; physical address.** 463

address cycle (FASTBUS acquisition and control). *Syn:* **primary address cycle.** 480

address, effective (computing systems). The address that is derived by applying any specified rules (such as rules relating to an index register or indirect address) to the specified address and that is actually used to identify the current operand. 77

address fields (DSAP [destination service access point] and SSAP [source service access point])(logical link control). The ordered pair of service access point addresses at the beginning of an LLC PDU (logical link control protocol data unit) which identifies the LLC(s) designated to receive the PDU and the LLC sending the PDU. Each address field is one octet in length. 585

address format (computing machines). The arrangement of the address parts of an instruction. *Note:* The expression plus-one is frequently used to indicate that one of the addresses specifies the location of the next instruction to be executed, such as one-plus-one, two-plus-one, three-plus-one, four-plus-one.
255, 77

addressing mode (microprocessor assembly language). The manner in which an operand is to be accessed during execution of an instruction. 466

address locked operation (FASTBUS acquisition and control). An operation directed to a single primary address containing a mixture of read and write cycles, possibly including block transfers as well. 480

address part. A part of an instruction that usually is an address, but that may be used in some instructions for another purpose. *See:* **instruction code.** 235

address register (computing machines). A register in which an address is stored. 255, 77

address space (software). The range of addresses available to a computer program. *See:* **addresses; computer program.** 434

address, tag. *See:* **symbolic address.**

ADF. *See:* **automatic direction finder.**

A-display (radar). A display in which targets appear as vertical deflections from a horizontal line representing a time base. Target distance is indicated by the horizontal position of the deflection from one end of the time base. The amplitude of the vertical deflection is a function of the signal intensity. 13

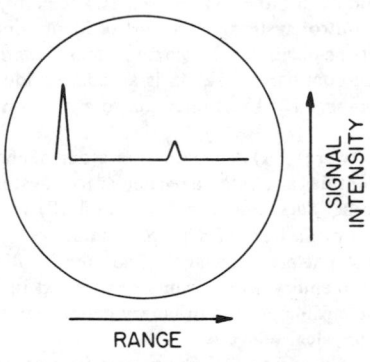

RANGE

A-Display

adjacent channel (data transmission). The channel whose frequency is adjacent to that of the reference channel. 59

adjacent-channel attenuation (receivers). *See:* **selectance.**

adjacent channel interference (data transmission). Interference, in a reference channel, caused by the operation of an adjacent channel. 59

adjacent-channel selectivity and desensitization (receiver performance) (receiver). A measure of the ability to discriminate against a signal at the frequency of the adjacent channel. Desensitization occurs when the level of any off-frequency signal is great enough to alter the useable sensitivity. *See:* **receiver performance.** 181

adjoint system. (1) A method of computation based on the reciprocal relation between a system of ordinary linear differential equation and its adjoint. *Note:* By solution of the adjoint system it is possible to obtain the weighting function (response to a unit impulse) $W(T, t)$ of the original system for fixed T (the time of observation) as a function of t (the time of application of the impulse). Thus, this method has particular application to the study of systems with time-varying coefficients. The weighting function then may be used in convolution to give the response of the original system to an arbitrary input. *See:* **electronic analog computer.** 9
(2) For a system whose state equations are $dx(t)/dt = f(x(t),u(t),t)$, the adjoint system is defined as that system whose state equations are $dy(t)/dt = -y(t)$, where A^* is the conjugate transpose of the matrix whose i,j element is $\partial f_i / \partial x_j$. *See:* **control system.** 198

adjust (instrument). Change the value of some element of the mechanism, or the circuit of the instrument or of an auxiliary device, to bring the indication to a desired value, within a specified tolerance for a particular value of the quantity measured. *See:* **instrument.** 328

adjustable (National Electric Code). (As applied to circuit breakers.) A qualifying term indicating that the circuit breaker can be set to trip at various values of current and/or time within a pre-determined range. 256

adjustable constant-speed motor. A motor, the speed of which can be adjusted to any value in the specified range, but when once adjusted the variation of speed with load is a small percentage of that speed. For example, a direct-current shunt motor with field-resistance control designed for a specified range of speed adjustment. *See:* **asynchronous machine.** 63

adjustable impedance-type ballast (illuminating engineering). A reference ballast consisting of an adjustable inductive reactor and a suitable adjustable resistor in series. These two components are usually designed so that the resulting combination has sufficient current-carrying capacity and range of impedance to be used with a number of different sizes of lamps. The impedance and power factor of the reactor-resistor combination are adjusted and checked each time the unit is used. 271

adjustable-speed drive (industrial control). An electric drive designed to provide easily operable means for speed adjustment of the motor, within a specified speed range. *See:* **electric drive.** 206

adjustable-speed motor. A motor the speed of which

can be varied gradually over a considerable range, but when once adjusted remains practically unaffected by the load: such as a direct-current shunt-wound motor with field resistance control designed for a considerable range of speed adjustment. *See:* **asynchronous machine.** 3

adjustable-varying-speed motor (electric installations onshipboard. A motor whose speed can be adjusted gradually,but when once adjusted for a given load will vary in considerable degree with change in load;such as a dc compound-wound motor adjusted by field control or a wound-rotor induction motor with rheostatic speed control. 3

adjustable varying-voltage control (industrial control). A form of armature-voltage control obtained by impressing on the armature of the motor a voltage that may be changed by small increments, but when once adjusted for a given load will vary considerably with change in load with a consequent change in speed, such as maybe obtained from a differentially compound-wound generator with adjustable field current or by means of an adjustable resistance in the armature circuit. *See:* **control.** 206

adjustable voltage control. A form of armature-voltage control obtained by impressing on the armature of the motor a voltage that maybe changed in small increments; but when once adjusted, it, and consequently the speed of the motor, are practically unaffected by a change in load. *Note:* Such a voltage may be obtained from an individual shunt-wound generator with adjustable field current, for each motor. *See:* **control.** 206

adjusted speed (industrial control). The speed obtained intentionally through the operation of a control element in the apparatus or system governing the performance of the motor.*Note:* The adjusted speed is customarily expressed in percent (or per unit) of basespeed (for direct-current shunt motors). *See:* **electric drive.** 206

adjustment (test, measurement and diagnostic equipment). The act of manipulating the equipment's controls to achieve a specified condition. 54

adjustment accuracy (direct current instrument shunts). The limit of error, expressed as a percentage of the rated output voltage, in the initial adjustment of the shunt made when employing a low-current measurement method. 527

adjustment accuracy of instrument shunts (electric power systems). The limit of error, expressed as a percentage of the rated voltage drop, of the initial adjustment of the shunt by resistance or low-current methods. 201

Adler tube*. *See:* **beam parametric-amplifier.**
 *Deprecated

administrative authority (NESC). An organization exercising jurisdiction over application of this code. 494

administrative controls (nuclear power generating station). Written rules, orders, instructions, procedures, policies, practices, and designations of authority and responsibility for the operation and mainte-nance of a nuclear generating station. 387, 102, 428

administrative data processing (ADP)(computer applications). The use of computers for administrative applications such as personnel, payroll, and accounting functions. 571

admissible control input set (control system). A set of control inputs that satisfy the control constraints. *See:* **control system.** 198

admittance (1)(data transmission). The reciprocal of impedance. 59

(2)(linear constant-parameter system). (A)The corresponding admittance function with *p* replaced by j in which is real. (B) The ratio of the phasor equivalent of a steady-state sine-wave current or current-like quantity (response) to the phasor equivalent of the corresponding voltage or voltage-like quantity (driving force). The real part is the conductance and the imaginary part is the susceptance. *Note:* Definitions (A) and (B) are equivalent.*Editor's Note:* The ratio *Y* is commonly expressed in terms of its orthogonal components, thus:

$$Y = G + jB$$

where *Y, G,* and *B* are respectively termed the admittance,conductance, and susceptance, all being measured in mhos (reciprocal ohms). In a simple circuit consisting of *R, L,* and *C* all in parallel, *Y* becomes

$$Y = G + j\left(\omega C - \frac{1}{\omega L}\right),$$

where $\omega = 2$ and $2\pi f$ and ω is the frequency. In this special case,$G = 1/R$. Historically, some authors have preferred an opposite sign convention for susceptance, thus$Y = G\text{-}jB$ (now deprecated). Thus, in the simple parallel circuit

$$Y = G - j\left(\frac{1}{\omega L} - \omega C\right)$$

the reader will note that according to either convention, the end result is that any predominantly capacitive susceptance will advancethe phase of the response, relative to the driving force. However, the *Y* $= G\text{-}jB$ convention* requires that a predominantly capacitive susceptance *B,* like a predominantly capacitive reactance *X,*shall be denoted as negative. The reader is therefore advised to become aware of his author's preference. *See:* **conductance; susceptance.** 150

admittance, effective input (electron tube or valve). The quotient of the sinusoidal component of the control-grid current by the corresponding component of the control voltage, taking into account the action of the anode voltage on the grid current; it is a function of the admittance of the output circuit and the inter-electrode capacitance. *Note:* It is the reciprocal of the effective input impedance. *See:* electron-tube admittances.   244, 189

admittance, effective output (electron tube or valve). The quotient of the sinusoidal component of the anode current by the corresponding component of the anode voltage, taking into account the output admittance and the interelectrode capacitance. *Note:* It is the reciprocal of the effective output impedance. *See:* electron-tube admittances.   244, 189

admittance, electrode (*j*th electrode of the *n*-electrode electron tube). *See:* electrode admittance.

admittance matrix, short-circuit (multiport network) (circuits and systems). A matrix whose elements have the dimension of admittance and, when multiplied into the vector of port voltages, gives the vector of port currents.   67

admittance, short-circuit (circuits and systems). An admittance of a network that has a specified pair or group of terminals short-circuited. (2) (four-terminal network or line) The input, output or transfer admittance parameters y_{11}, y_{22}, and y_{12} of a four-terminal network when the far-end is short cir-cuited.   67

admittance, short-circuit driving-point (*j*th terminal ofan *n*-terminal network). The driving-point admittance between that terminal and the reference terminal when all other terminal shave zero alternating components of voltage with respect to the reference point. *See:* electron-tube admittances.   125

admittance, short-circuit feedback (electron-device transducer). The short-circuit transfer admittance from the physically available output terminals to the physically available input terminals of a specified socket, associated filters, and electron device. *See:* electron-tube admittances.   125

admittance, short-circuit forward (electron-device transducer). The short-circuit transfer admittance from the physically available output terminals of a specified socket, associated filters, and electron device. *See:* electron-tube admittances.   125

admittance, short-circuit input (electron-device transducer). The driving-point admittance at the physically available input terminals of a specified socket, associated filters, and tube. All other physically available terminals are short-circuited. *See:* electron-tube admittances.   125

admittance, short-circuit output (electron-device transducer). The driving-point admittance at the physically available output terminals of a specified socket, associated filters, and tube. All other physically available terminals are short-circuited. *See:* electron-tube admittances.   125

admittance, short-circuit transfer (from the *j*th terminal to the *l*th terminal of an *n*-terminal network). The transfer admittance from terminal *j* to terminal *l* when all terminals except *j* have zero complex alternating components of voltage with respect to the reference point. *See:* electron-tube admittances.   125

advance ball (mechanical recording). A rounded support (often sapphire) attached to a cutter that rides on the surface of the recording medium so as to maintain a uniform depth of cut and correct for small irregularities of the disk surface.   176

adverse water conditions (power operations). Water conditions which limit hydroelectric energy production.   516

adverse weather (electric power systems). Weather conditions that cause an abnormally high rate of forced outages for exposed components during the periods such conditions persist. *Note:* Adverse weather conditions can be defined for a particular systemby selecting the proper values and combinations of conditions re-ported by the Weather Bureau: thunderstorms, tornadoes, wind velocities, precipitation, temperature, etcetera. *See:* outage.   200

adverse-weather lamps. *See:* fog lamps.

aeolight (optical sound recording). A glow lamp employing a cold cathode and a mixture of permanent gases in which the intensity of illumination varies with the applied signal voltage.   176

aeration cell. *See:* differential aeration cell.

aerial cable. An assembly of insulated conductors installed on a pole or similar overhead structures; it may be self-supporting or installed on a supporting messenger cable. *See:* cable.   64

aerial lug. *See:* external connector (pothead).

aerial platform (conductor stringing equipment). A device designed to be attached to the boom tip of a crane or aerial lift and support a workman in an elevated working position. Platforms may be constructed with surrounding railings, fabricated from aluminum, steel or fiber reinforced plastic. Occasionally, a platform is suspended from the load line of a large crane. *Syn:* cage; platform.   431

aerial work (power line maintenance). Work performed on equipment used for the transmission and distribution of electricity, which is performed in an elevated position on various structures, conductors, or associated equipment.   458

aerodrome beacon (illuminating engineering). An aeronautical beacon used to indicate the location of an aerodrome. *Note:* An aerodrome is any defined area on land or water--including any buildings, installations, and equipment--intended to be used either wholly or in part for the arrival, departure and movement of aircraft.   167

aerometeorgraph (navigation aid terms). A self-recording instrument for the simultaneous recording of atmospheric pressure, temperature, and humidity.   526

aeronautical beacon (illuminating engineering). An aeronautical ground light visible at all azimuths, either continuously or intermittently, to designate a particular location on the surface of the earth.   167

aeronautical ground light (illuminating engineering). Any light specially provided as an aid to air navigation, other than a light displayed on an aircraft.   167

aeronautical light (illuminating engineering). Any luminous sign or signal which is specially provided as an aid to air navigation. 167

aerophase (air operations)(navigation aid terms). A name for a radio beacon. 526

aerosol (laser-maser). A suspension of small solid or liquid particles in a gaseous medium. Typically, the particle sizes may range from 100 μm to 0.01 μm or less. 363

aerosol development (electrostatography). Development in which the image-forming material is carried to the field of the electrostatic image by means of a suspending gas. *See:* electrostatography. 191

aerospace support equipment (test, measurement and diagnosticequipment). All equipment (implements, tools, test equipment,devices (mobile or fixed), and so forth), both airborne and ground,required to make an aerospace system (aircraft, missile, and soforth) operational in its intended environment. Aerospace supportequipment includes ground support equipment. 54

AEW (navigation aid terms). Abbreviation for airborne early warning, describing an early-warning radar carried by an airborne or spaceborne vehicle. 526

AF. *See:* analog-to-frequency converter.

AFC. *See:* automatic frequency control.

AFCS. *See:* automatic flight-control system.

after image (illuminating engineering). A visual response that occurs after the stimulus causing it has ceased. 167

afterpulse (photo multipliers). A spurious pulse induced in a photomultiplier by a previous pulse. *See:* phototube. 117

AGC. *See:* automatic gain control.

aggressive carbon dioxide (corrosion). Free carbon dioxide in excess of the amount necessary to prevent precipitation of calcium as calcium carbonate. 205

aging (natural) (1) (class 1E static battery chargers and inverters). The change with passage of time in physical, chemical, or electrical properties of components or equipment under design range operating conditions that may result in degradation of significant performance characteristics. 408
(2)(nuclear power generating station). The effect of operational, environmental, and system conditions on equipment during a period of time up to, but not including design basis events, or the process of simulating these events. 120, 535, 492
(3)(thermal classification of electric equipment and electrical insulation). The irreversible change (usually degradation) that takes place with time. 506
(4) (valve actuators). The cumulative effect of operating cycles and environmental and system conditions imposed on the actuator during a period of time. 142

aging factor (thermal classification of electric equipment and electrical insulation). A factor of influence that causes aging. 506

aging mechanism (nuclear power generating stations). Any process attributable to service conditions which results in degradation of an equipment's ability to perform its Class 1E functions. 440

agitator (hydrometallurgy) (electrowinning). A receptacle in which ore is kept in suspension in a leaching solution. *See:* eletrowinning. 328

aided tracking (radar). A tracking technique in which the manual correction of the tracking error automatically corrects the rate of motion of the tracking mechanism. 13

aid to navigation. *See:* navigational aid.

air (industrial control) (prefix). Applied to a device that interrupts an electric circuit, indicates that the interruption occurs in air. 225, 206

AI radar (airborne intercept radar). A fire control radar for use in interceptor aircraft. 13

air baffle (rotating machinery). *See:* air guide.

air-blast circuit breaker. *See:* Note under circuit breaker.

air cell. A gas cell in which depolarization is accomplished by atmospheric oxygen. *See:* electrochemistry. 328

air circuit breaker. *See:* Note under circuit breaker. 103

air-conditioning or comfort-cooling equipment (National Electrical Code). All of that equipment intended or installed for the purpose of processing the treatment of air so as to control simultaneously its temperature, humidity, cleanliness and distribution to meet the requirements of the conditioned space. 256

air conduction (hearing). The process by which sound is conducted to the inner ear through the air in the outer ear canal as part of the pathway. 176

air-core inductance (winding inductance). The effective self-inductance of a winding when no ferromagnetic materials are present.*Note:* The winding inductance is not changed when ferromagnetic materials are present. 197

aircraft aeronautical light (illuminating engineering). Any aeronautical light specially provided on an aircraft. 167

aircraft bonding. The process of electrically interconnecting all parts of the metal structure of the aircraft as a safety precaution against the buildup of isolated static charges and as a means of reducing radio interference. 328

aircraft hangar (National Electrical Code). A location used for storage or servicing of aircraft in which gasoline, jet fuels, or other volatile flammable liquids or flammable gases are used. 256

air data system (navigation aid terms). A set of aerodynamic and thermodynamic sensors, and a computer which provide flight parameters such as airspeed, static pressure, air temperature, and Mach number. 526

air-derived navigation data (navigation aid terms). Data obtained from measurements made at an airborne vehicle. 526

air equivalent radiation dose (valve actuators). The energy that is absorbed per unit mass of air at the geometric center of the volume occupied by the specimen if it were replaced with air and a uniform flux were incident at the boundary of the volume, directed toward the center. 492

air gap (ferroresonant voltage regulators). A space in the magnetic core, void of magnetic material, used to establish the required reluctance of the flux path. 456

air gap field voltage (excitation systems for synchronous machines). The synchronous machine field voltage required to produce rated voltage on the air-gap line of the synchronous machine with its field winding at (1) 75 °C for field windings designed to operate at rating with a temperature rise of 60 °C or less; or (2) 100 °C for field windings designed to operate at rating with a temperature rise greater than 60 °C. *Note:* This defines one per unit excitation system voltage for use in computer representation of excitation systems. 507

air-gap line (excitation systems for synchronous machines). The extended straight line part of the no-load saturation curve of the synchronous machine. 507

air gap, relay. *See:* relay air gap.

air gap surge arrester (low-voltage air gap surge-protective devices). A gap or gaps, in air at ambient atmospheric pressure, designed to protect apparatus and personnel, or both, from high transient voltages. 556

air gap surge protector (low-voltage air gap surge-protective devices). A protective device, consisting of one or more air gap surge arresters; optional fuses, short-circuiting devices, etcetera; and a mounting assembly, for limiting surge voltages on low voltage (600 volts [V] root-mean-square [rms] or less) electrical and electronic equipment or circuits. 556

air horn. A horn having a diaphragm that is vibrated by the passage of compressed air. *See:* protective signaling. 328

air mass. The mass of air between a surface and the sun that affects the spectral distribution and intensity of sunlight. *See:* air mass one; air mass zero. 113

air mass one. A term that specifies the spectral distribution and intensity of sunlight on earth at sea level with the sun directly overhead and passing through a standard atmosphere. *See:* air mass; air mass zero. 113

air mass zero. A term that specifies the spectral distribution and intensity of sunlight in near-earth space without atmospheric attenuation. *Note:* The air mass must be specified when reporting the efficiency of solar cells; for example, 10 percent efficient at air mass zero, 60 degrees Celsius. *See:* air mass; air mass one. 113

air navigation (navigation aid terms). The navigation of aircraft. 526

airport beacon. *See:* aerodrome beacon.

airport surface-detection equipment (ASDE)(navigation aid terms). A ground-based radar for observation of the positions of aircraft and other vehicles on the surface of an airport. 13, 526

airport surveillance radar (ASR)(navigation aid terms). A medium range (for example, 60 nautical miles [NMI]) surveillance radar used to control aircraft in the vicinity of an airport. 13, 526

airposition indicator (API)(navigation aid terms). An airborne computing system which presents a continuous indication of the aircraft's position on the basis of aircraft heading, airspeed, and elapsed time. 13, 526

air-route surveillance radar (ARSR)(navigation aid terms). A long range (for example, 200 nautical miles [nmi]) surveillance radar used to control aircraft on airways beyond the coverage of airport surveillance radar (ASR). 13, 526

airspeed (navigation aid terms). The rate of motion of a vehicle relative to the air mass. 13, 526

airspeed indicator (navigation aid terms). An instrument for measuring airspeed. 526

air switch (1) (power switchgear). A switching device designed to close and open one or more electric circuits by means of guided separable contacts that separate in air. 103

(2) (high-voltage switchgear). A switch with contacts that separate in air. 443

air terminal (lightning protection). The combination of elevation rod and brace, or footing placed on upper portions of structures,together with tip or point if used. *See:* lightning protection and equipment. 328

airway beacon (illuminating engineering). An aeronautical beacon used to indicate a point on the airway. 167

alarm (electric pipe heating systems). A signal for attracting attention to some abnormal condition. Alarms associated with electric pipe heating systems can signal high temperature, loss of heater circuit voltage, etcetera. 448

alarm checking (telephone switching system). The identification of an alarm from a remote location by communicating with its point of origin. 55

alarm condition (supervisory control, data acquisition, and automatic control). A predefined change in the condition of equipment or the failure of equipment to respond correctly. Indication may be audible or visual, or both. 570

alarm function. *See:* supervisory control functions

alarm point (power-system communication). A supervisory control status point considered to be an alarm. *See:* supervisory control system. 59

alarm relay.(1) A monitoring relay whose function is to operate an audible or visual signal to announce the occurrence of an operation or a condition needing personal attention, and usually provided with a signaling cancellation device. *See:* relay. 127

(2)(74) (power system device function numbers). A relay other than an annunciator, as covered under device function 30, [annunciator relay] , that is used to operate, or to operate in connection with, a visual or audible alarm. 402

(3)(signal) (power switchgear). A monitoring relay whose function is to operate an audible or visual signal to announce the occurrence of an operation or a condition needing personal attention, and which is usually provided with a signaling cancellation device. 103

alarm sending (telephone switching system). The extension of alarms from an office to another location.
55

alarm signal. A signal for attracting attention to some abnormal condition.
193

alarm summary printout (sequential events recording systems). The recording of all inputs currently in the alarm state.
48, 58

alarm switch (of a switching device) (power switchgear). An auxiliary switch that actuates a signaling device upon the automatic opening of the switching device with which it is associated.
103

alarm system (protective signaling). An assembly of equipment and devices arranged to signal the presence of a hazard requiring urgent attention. *See:* **protective signaling.**
328

albedo (1) (photovoltaic power system). The reflecting power expressed as the ratio of light reflected from an object to the total amount falling on it. *See:* **solar cells (photovoltaic power system).**
186

(2) (radio wave propagation). The ratio of the reflected power to the incident power of radiation falling upon a body.
146

ALC. *See:* **automatic load (level) control.**

Alford loop antenna. A multielement antenna having approximately equal amplitude currents which are in phase and uniformly distributed along each of its peripheral elements and producing a substantially circular radiation pattern in its principal E plane (originally developed as a four-element horizontally polarized UHF loop antenna).
111

Alford loop (navigation aid terms). A multielement antenna, having approximately equal amplitude currents which are in phase and uniformly distributed along each of its peripheral elements, producing a substantially circular radiation pattern in the plan of polarization (originally developed as a four-element horizontally polarized very high frequency (VHF) loop antenna.
526

Alfven velocity (radio wave propagation). The characteristic propagation velocity of an Alfven wave, given by $H_0\sqrt{\mu/\rho}$ where μ is the permeability, H_0 is the static magnetic field strength, and ρ is the mass density of the conducting fluid.
146

Alfven wave (radio wave propagation). In a magnetoionic medium, the magnetohydrodynamic wave that propagates in the direction of the static magnetic field, with associated electric and magnetic fields and fluid particle velocities oriented perpendicular to the direction of propagation and with no associated density fluctuations in the medium.
146

algorithm (1) (general). A prescribed set of well-defined rules or processes for the solution of a problem in a finite number of steps, for example,a full statement of an arithmetic procedure for evaluating sin $\underline{x}$ to a stated precision. *See:* **heuristic.**
255, 77, 54

(2) (software). (A) A finite set of well-defined rules for the solution of a problem in a finite number of steps; for example, a complete specification of a sequence of arithmetic operations for evaluating sin x to a given precision. (B) A finite set of well-defined rules which gives a sequence of operations for performing a specific task.
434

(3)(mathematics of computing). A finite set of well-defined rules for the solution of a problem in a finite number of steps; for example, a complete specification of a sequence of arithmetic operations for evaluating sine x to a given precision.
564

algorithm analysis (software). The examination of an algorithm to determine its correctness with respect to its intended use, to determine its operational characteristics, or to understand it more fully in order to modify, simplify, or improve it. *See:* **algorithm.**
434

algorithmic language (test, measurement and diagnostic equipment). A language designed for expressing algorithms.
54

alias (software). (1) An additional name for an item. (2) An alternate label. For example, a label and one or more aliases may be used to refer to the same data element or point in a computer program. *See:* **computer program; data; label.**
434

align (test, measurement and diagnostic equipment). To adjust a circuit, equipment or system so that their functions are properly synchronized or their relative positions properly oriented. For example, trimmers, padders, or variable inductances in tuned circuits are adjusted to give a desired response for fixed tuned equipment or to provide tracking for tuneable equipment.
54

aligned bundle (fiber optics). A bundle of optical fibers in which the relative spatial coordinates of each fiber are the same at the two ends of the bundle. *Note:* The term "coherent bundle" is often employed as a synonym, and should not be confused with phase coherence or spatial coherence. *Syn:* **coherent bundle.** *See:* **fiber bundle.**
433

aligned-grid tube (or valve). A vacuum multigrid tube or valve in which at least two of the grids are aligned the one behind the other so as to obtain a particular effect (canalizing an electron beam; suppressing noise, etcetera). *See:* **electron tube.**
244, 190

alignment (1)(data transmission). In communication practice, alignment is the process of adjusting a plurality of components of a system for proper interrelationship. The term is applied especially to (A) the adjustment of the tuned circuits of an amplifier for desired frequency response, and (B) the synchronization of the components of a system.
59

(2)(inertial navigation equipment)(navigation aid terms). The orientation of the measuring axes of the inertial components with respect to the coordinate system in which the equipemt is used. *Note:* Inertial alignment refers to the result of either the process of bringing the measuring axis into a desired orientation or the computation of the angles between the measuring axis and the desired orientation with respect to the coordinate system in which the equipment is used. The initial alignment can be accomplished by the use of noninertial sensors. *See:* **gyrocompass alignment; transfer alignment.**
526

(3) (communication practice). The process of adjusting a plurality of components of a system for proper interrelationship. *Note:* The term is applied especially to (A) the adjustment of the tuned circuits of an amplifier for desired frequency response, and (B) the synchronization of components of a system. *See:* **radio transmission.**   59

alignment kit (test, measurement and diagnostic equipment). A kit containing all the instruments or tools necessary for the alignment of electrical or mechanical components.   54

alignment tool (test, measurement and diagnostic equipment). A small screwdriver, socket wrench, or special tool used for adjusting electronic, mechanical, or optical units, usually constructed of nonmagnetic materials.   54

alive (1) (electric system). Electrically connected to a source of potential difference, or electrically charged so as to have a potential different from that of the ground. *Note:* The term **alive** is sometimes used in place of the term **current-carrying,** where the intent is clear to avoid repetitions of the longer term. *Syn:* **live.** *See:* **insulated; energized**   262

(2)(transmission and distribution) (live). *See:* **energized.**

alkaline cleaning (electroplating). Cleaning by means of alkaline solutions. *See:* **electroplating.**   328

alkaline storage battery. A storage battery in which the electrolyte consists of an alkaline solution, usually potassium hydroxide. *See:* **battery (primary or secondary).**   328

allocation (computing machine). *See:* **storage allocation.**

allotting (telephone switching system). The preselecting by a common control of an idle circuit.   55

allowable continuous current (of a fuse link, fuse unit or refill unit) (high-voltage switchgear). The maximum root-mean-square (rms) current in amperes at rated frequency and at a specific ambient temperature which a device will carry continuously without exceeding the allowable total temperature as listed in the table below.   443

allowable stress (seismic design of substations). The maximum stress permitted by applicable standards or codes, or both.   465

alloy junction (semiconductor). A junction formed by recrystallization on a base crystal from a liquid phase of one or more components and the semiconductor. *See:* **semiconductor.**   237, 66

alloy plate. An electrodeposit that contains two or more metals codeposited in combined form or in intimate mixtures. *See:* **electroplating.**   328

all-pass function (linear passive network). A transmittance that provides only phase shift, its magnitude characteristic being constant. *Notes:* (1) For lumped-parameter networks, this is equivalent to specifying that the zeros of the function are the negatives of the poles. (2) A realizable all-pass function exhibits nondecreasing phase lag with increasing frequency. (3) A trivial all-pass function has zero phase at all frequencies.   238

Summary of Temperature Limitations

	Allowable Temperature Rise (and Total Temperature Shown in Parenthesis) Temperatures Shown in °C					
	All Conducting Parts, Except Conducting Element of Fuse Link				All Parts Made Up of or in Contact with Insulating Materials Except Fuse Link	
	Type of Contact					
Device	AG-AG	AG-CU	CU-CU	Sn-Sn	Class of Insulation	Temperature Limits
Distribution cutouts (except oil cutouts and open link cutouts) — With fuse link	40 (80)		30 (70)		Bone fiber 90 / 105 / 130	30 (70) / 50 (90) / 65 (105) / 90 (130)
With switch blade	40 (80)	35 (75)	35 (75)			
Distribution oil cutouts — With link or blade	45 (85)	30 (70)	30 (70)		90 / 105 / 130	50 (90) / 65 (105) / 90 (130)
Distribution air switches	40 (80)	35 (75)	35 (75)	40 (80)		
Distribution current limiting fuses	65 (105)		30 (70)	55 (95)	90 / 105	50 (90) / 65 (105)
Power Fuses — Current limiting	65 (105)		30 (70)	55 (95)	130 / 155 / 180 / 220	90 (130) / 115 (155) / 140 (180) / 180 (220)
Power Fuses — All others	45 (85)		30 (70)	45 (85)		

all-pass network (all-pass transducer). A network designed to introduce phase shift or delay without introducing appreciable attenuation at any frequency. *See also:* **network analysis.**   328

all-pass transducer. *See:* **all-pass network.**

all-relay system. An automatic telephone switching system in which all switching functions are accomplished by relays.   328

almanac (navigation aid terms). A periodic publication of astronomical data useful to a navigator.   526

alphabet. A character set arranged in certain order. *Note:* Character sets are finite quantities of letters of the normal alphabet, digits, punctuation marks, control signals, such as carriage return and other ideographs. Characters are usually represented by letters (graphics) or technically realized in the form of combinations of punched holes, sequences of electric pulses, etcetera.   194

alphameric. *See:* **alphanumeric.**

alphanumeric. Pertaining to a character set that contains both letters and digits, but usually some other characters such as punctuation symbols. *Syn:* **alphameric.**   77

alpha profile. *See:* **power-law index profile.**

$\alpha_{\min}$. *See:* **limiting angular subtense.**

alteration (elevator, dumbwaiter, or escalator.) Any change or addition to the equipment other than ordinary repairs or replacements. *See:* **elevator.**   328

alternate-channel interference (second-channel interference). Interference caused in one communication channel by a transmitter operating in a channel next beyond an adjacent channel. *See:* **radio transmission.**   328

alternate display (oscillography). A means of displaying out-put signals of two or more channels by switching the channels in sequence. *See:* oscillograph.   185

alternate power source (National Electric Code). One or more generator sets intended to provide power during the interruption of the normal electrical service or the public utility electrical service intended to provide power during interruption of service normally provided by the generating facilities on the premises.   256

alternate route (data transmission). A secondary communications path used to reach a destination if the primary path is unavailable.   59

alternate-route trunk group (telephone switching system). A trunk group which accepts alternate-routed traffic.   55

alternate routing (telephone switching system). A means of selectively distributing traffic over a number of routes ultimately leading to the same destination.   55

alternating charge characteristic (nonlinear capacitor). The function relating the instantaneous values of the alternating component of transferred charge, in a steady state, to the corresponding instantaneous values of a specified applied periodic capacitor-voltage. *Note:* The nature of this characteristic may depend upon the nature of the applied voltage. *See:* **nonlinear capacitor.**   191

alternating current (ac) (electric installations on shipboard). A periodic current the average value of which over a period is zero.(Unless distinctly specified otherwise,the term refers to a current which reverses at regularly recurring intervals of time and which has alternately positive and negative values.)   3

alternating-current (ac) circuit breaker (52) (power system device function numbers). A device that is used to close and interrupt an ac power circuit under normal conditions or to interrupt this circuit under fault or emergency conditions.   402

(ac) directional overcurrent relay (67) (power system device function numbers). A relay that functions on a desired value of ac overcurrent flowing in a predetermined direction.   402

alternating current (ac)-linked ac converter(self-commutated converters). A converter comprising two cascaded frequency changers in which the intermediate link is usually a high-frequency tank circuit.   584

alternating current (ac)-linked dc (direct current) converter (self-commutated converters). A converter comprising an inverter and a rectifier, with an intermediate ac link.   584

alternating-current (ac) reclosing relay (79) (power system device function numbers). A relay that controls the automatic reclosing and locking out of an ac circuit interrupter.   402

alternating-current (ac) standby power (low voltage varistor surge arresters). Varistor ac power dissipation measured at rated root-mean-square (rms) voltage.   62

alternating-current (ac) time overcurrent relay (51) (power system device function numbers). A relay that operates when its ac input current exceeds a predetermined value, and in which the input current and operating time are inversely related through a substantial portion of the performance range.   402

alternating-current and direct-current (ac-dc) ringing (telephone switching systems). Ringing in which alternating current activates the ringer and direct current controls the removal of ringing upon answer.   55

alternating-current breakdown voltage (gas-tube surge-protective devices). The minimum root-mean-square (rms) value of a sinusoidal voltage at frequencies between 15 hertz (Hz) and 62 Hz that results in arrester sparkover.   490

alternating-current circuit. A circuit that includes two or more interrelated conductors intended to be energized by alternating current.   210

alternating-current commutator motor. An alternating-current motor having an armature connected to a commutator and included in an alternating-current circuit. *See:* **asynchronous machine.**   63

alternating-current component (of a total current) (power switchgear).

alternating-current component. The current remaining when the average value has been subtracted from an alternating current. *See:* **symmetrical component (total current).**   103

alternating-current-direct-current general-use snap-switch. A form of general-use snap-switch suitable for use on either direct-or alternating-current circuits for controlling the following: (1) Resistive loads not exceeding the ampere rating at the voltage involved.(2) Inductive loads not exceeding one-half the ampere rating at the voltage involved, except that switches having a marked horsepower rating are suitable for controlling motors not exceeding the horse-power rating of the switch at the voltage involved. (3) Tungsten filament lamp loads not exceeding the ampere rating at 125 volts, when marked with the letter "T." Alternating-current-direct-current general use snap-switches are not generally marked alternating-current-direct-current, but are always marked with their electrical rating. *See:* **switch.**   256

alternating-current-direct-current ringing. Ringing in which a combination of alternating and direct currents is utilized, the direct current being provided to facilitate the functioning of the relay that stops the ringing.   328

alternating-current distribution. The supply to points of utilization of electric energy by alternating current from its source to one or more main receiving stations. *Note:* Generally a voltage is employed that is not higher than that which could be delivered or utilized by rotating electric machinery. Step-down transformers of a capacity much smaller than that of the line are usually employed as links between the moderate voltage of distribution and the lower volt-age of the consumer's apparatus.   64

alternating-current electric locomotive. An electric locomotive that collects propulsion power from an alternating-current distribution system. *See:* **electric locomotive.**   328

alternating-current erasing head (magnetic recording). A head that uses alternating current to produce the magnetic field necessary for erasing. *Note:* Alternating-current erasing is achieved by subjecting the medium to a number of cycles of a magnetic field of a decreasing magnitude. The medium is, therefore, essentially magnetically neutralized.   176

alternating-current floating storage-battery system. A combination of alternating-current power supply, storage battery, and rectifying devices connected so as to charge the storage battery continuously and at the same time to furnish power for the operation of signal devices.   328

alternating-current general-use snap-switch. A form of general-use snap-switch suitable only for use on alternating-current circuits for controlling the following: (1) Resistive and inductive loads(including electric discharge lamps) not exceeding the ampere rating at the voltage involved. (2) Tungsten filament lamp loads not exceeding the ampere rating at 120 volts. (3) Motor loads not exceeding 80 percent of the ampere rating of the switches at the rated voltage.*Note:* All alternating-current general-use snap-switches are marked ac in addition to their electrical rating. *See:* **switch.**   256

alternating-current generator. A generator for the production of alternating-current power.   63

alternating-current magnetic biasing (magnetic recording). Magnetic biasing accomplished by the use of an alternating current,usually well above the signal-frequency range. *Note:* The high-frequency linearizing (biasing) field usually has a magnitude approximately equal to the coercive force of the medium.   176

alternating-current motor. An electric motor for operation by alternating current.   63

alternating-current pulse. An alternating-current wave of brief duration. *See:* **pulse.**   328

alternating current root-mean-square voltage rating (semiconductor rectifiers). The maximum root-mean-square value of applied sinusoidal voltage permitted by the manufacturer under stated conditions. *See:* **semiconductor rectifier stack.**   208

alternating-current saturable reactor (power and distribution transformer). A reactor whose impedance varies cyclically with the alternating current (or voltage).   53

alternating-current transmission (1)(transmission and distribution). The transfer of electric energy by alternating current from its source to one or more main receiving stations for subsequent distribution. *Note:* Generally a voltage is employed that is higher than that which would be delivered or utilized by electric machinery. Transformers of a capacity comparable to that of the line are usually employed as links between the high voltage of transmission and the lower voltage used for distribution or utilization. *See:* **alternating-current distribution.**   64
(2)(television). That form of transmission in which a fixed setting of the controls makes any instantaneous value of signal correspond to the same value of brightness only for a short time. *Note:* Usually this time is not longer than one field period and may be as short as one line period.*See:* **television.**   328

alternating-current winding of a rectifier transformer (power and distribution transformer). The primary winding that is connected to the alternating-current circuit and usually has no conductive connection with the main electrodes of the rectifier.   53

alternating function. A periodic function whose average valueover a period is zero. For instance, $f(t) = B$ sin wt is an alternating function (w, B assumed constants).   210

alternating voltage. *See:* **alternating current.**

alternative (electric power system) (generating stations electric power system). A qualifying word identifying a power circuit equipment, device, or component available to be connected (or switched) into the circuit to perform a function when the preferred component has failed or is inoperative. *See:* **reserve.**   381

alternator (rotating electric machinery). An alternating-current generator.   424

alternator-rectifier exciter (excitation systems for synchronous machines). An exciter whose energy is derived from an alternator and converted to direct current by rectifiers. The exciter includes an alternator and power rectifiers which may be either noncon-

trolled or controlled, including gate circuitry. It is exclusive of input control elements. The alternator may be driven by a motor, prime mover, or by the shaft of the synchronous machine. The rectifiers may be stationary or rotating with the alternator shaft.
507

alternator transmitter. A radio transmitter that utilizes power generated by a radio-frequency alternator. *See:* **radio transmitter.** 111, 240

altimeter (navigation aid terms). An instrument which determines the height of an object with respect to a fixed level, such as sea level. There are two common types: an aneroid, or barometric altimeter, and the radio, or radar altimeter. 526

altitude (1)(illuminating engineering). The angular distance of a heavenly body measured on that great circle which passes, perpendicular to the plane of the horizon, through the body and through the zenith. It is measured positively from the horizon to the zenith, from 0 to 90 degrees. 167

(2)(navigation aid terms). (1) Angular distance above the horizon--the arc of a vertical circle between the horizon and a point on the celestial sphere. (2) Vertical distance above a given datum. 526

(3) (series capacitor). The elevation of the series capacitor above mean sea level. 86

altitude-treated current-carrying brush. A brush specially fabricated or treated to improve its wearing characteristics at high altitudes (over 6000 meters). 328

aluminum cable steel reinforced (ACSR). A composite conductor made up of a combination of aluminum wires surround the steel. *See:* **conductor.** 64

aluminum conductor. A conductor made wholly of aluminum. *See:* **conductor.** 64

aluminum-covered steel wire (power distribution underground cables). A wire having a steel core to which is bonded a continuous outer layer of aluminum. 57

aluminum sheaths (aluminum sheaths for power cables). An impervious aluminum or aluminum alloy tube, either smooth or corrugated, which is applied over a cable core to provide mechanical protection. 406

AMA. *See:* **automatic message accounting system.**

amalgam (electrolytic cells). The product formed by mercury and another metal in an electrolytic cell. 328

amateur band (overhead-power-line corona and radio noise). Any one of several frequency groups assigned for the transmission of signals by amateur radio operators. 411

ambient air temperature (1) (metal enclosed bus) (relaying). The temperature of the surrounding air which comes in contact with equipment. *Note:* Ambient air temperature, as applied to enclosed bus or switchgear assemblies, is the average temperature of the surrounding air that comes in contact with the enclosure. 78, 79,103

(2)(metal-enclosed low-voltage power circuit-breaker switchgear)(metal-clad and station-type cubicle switchgear). The temperature of the surrounding air that comes in contact with the equipment. *Note:* Ambient air temperature, as applied to enclosed switchgear assemblies, is the average temperature of the surrounding air that comes in contact with the enclosure. 579, 572, 573

ambient background (sodium iodide detector). Those counts that can be observed, and thereby allowed for, by measuring a source that is identical to the unknown source in all respects except for the absence of radioactivity. These counts are attributable to environmental radioactivity in the detector itself, the detector shielding material, and the sample container; cosmic rays; electronic noise pulses; etcetera. 423

ambient conditions. Characteristics of the environment, for example, temperature, humidity, pressure. *See:* **measurement system.** 185, 54

ambient level (1)(electromagnetic compatibility). The values of radiated and conducted signal and noise existing at a specified test location and time when the test sample is not activated. *See:* **electromagnetic compatibility.** 197

(2)(radio-noise emission). The levels of radiated and conducted signal and noise existing at a specific test location and time when the test sample is not activated. 418

ambient noise (mobile communication). The average radio noise power in a given location that is the integrated sum of atmospheric, galactic, and man-made noise. *See:* **telephone station.** 181

ambient operating-temperature range (power supply). The range of environmental temperatures in which a power supply can be safely operated. For units with forced-air cooling, the temperature is measured at the air intake. 228, 186

ambient radio noise. *See:* **ambient level.**

ambient temperature (1)(electrical heating systems). The environmental temperature surrounding the object under consideration. For objects enclosed in a thermal insulation, the ambient temperature is the temperature external to the thermal insulation. 476

(2)(electrical heat tracing for industrial applications). The temperature surrounding the object under consideration. Where electrical heating cable is enclosed in thermal insulation, the ambient temperature is the temperature exterior to the thermal insulation. 523

(3)(electric equipment)(thermal classification of electric equipment and electrical insulation). The temperature of the ambient medium.

(4)(series capacitors). The temperature of the medium such as air, water, or earth into which the heat of the equipment is dissipated. *Notes:* (A) For the self-ventilated equipment, the ambient temperature is the average temperature of the air in the immediate vicinity of the equipment. (B) For air or gas-cooled equipment with forced ventilation or secondary water cooling, the ambient temperature is taken as that of the in-going air or cooling gas. (C) For self-ventilated enclosed (including oil-immersed) equipment consid-

ered as a complete unit, the ambient temperature is the average temperature of the air outside of the enclosure in the immediate vicinity of the equipment.   474
(5) (neutral grounding devices)(power and distribution transformer)(shunt power capacitor). The temperature of the medium such as air, water, or earth into which the heat of the equipment is dissipated. *Notes:* (A) For self-ventilated equipment,the ambient temperature is the average temperature of the air in the immediate neighborhood of the equipment. (B) For air- or gas-cooled equipment with forced ventilation or secondary water cooling, the ambient temperature is taken as that of the ingoing air or cooling gas. (C) For self-ventilated enclosed (including oil-immersed) equipment considered as a complete unit, the ambient temperature is the average temperature of the air outside of the enclosure in the immediate neighborhood of the equipment.   91,53,138
(6)(outdoor apparatus bushing). The temperature of the surrounding air that comes in contact with the device or equipment in which the bushing is mounted.   168
(7) (light emitting diode) (free air temperature) (Ta). The air temperature measured below a device, in an environment of substantially uniform temperature, cooled only by natural air convention and not materially affected by reflective and radiant surfaces.   162
(8)(nuclear power generating stations). The average of air temperature readings at several locations in the immediate neighborhood of the equipment.   440
(9)(packaging machinery). The temperature of the surrounding cooling medium, such as gas or liquid, that comes into contact with the heated parts of the apparatus.   429
(10) (power switchgear). The temperature of the surrounding medium that comes in contact with the device or equipment.   103, 443
ambient temperature time constant. At a constant operating resistance, the time required for the change in (bolometer unit)bias power to reach 63 percent of the total change in bias power after an abrupt change in ambient temperature.   115
ambiguity (navigation)(navigation aid terms). The condition obtained when navigation coordinates define more than one point, direction, line of position or surface of position.   13,526
ambiguity function (radar). The squared magnitude

$$|\chi(\tau, f_d)|^2$$

of the function which describes the response of a radar receiver to targets displaced in range delay τ and Doppler frequency f_d from a reference position, where $|\chi(0,0)|$ is normalized to unity. Mathematically

$$\chi(\tau, f_d) = \int u(t)u^*(t+\tau)\exp(2\pi jf_dt)\,dt$$

where u(t) is the transmitted envelope waveform, suitably normalized; u*(t+τ) is the complex conju-

gate of the same waveform with argument (t+τ; positive τ indicates a target beyond the reference delay, and positive f_d indicates an incoming target. *Note:* Used to examine the suitability of radar waveforms for achieving accuracy, resolution, freedom from ambiguities, and reduction of unwanted clutter.   13
American Morse Code. *See:* **Morse Code.**
ammeter (1) (general). An instrument for measuring the magnitude of an electric current. *Note:* It is provided with a scale, usually graduated in either amperes, milliamperes, microamperes, or kiloamperes. If the scale is graduated in milliamperes,microamperes, or kiloamperes, the instrument is usually designated as a milliammeter, a microammeter, or a kiloammeter. *See:* **instrument.**   328
(2) (circuits and systems). An instrument for measuring electric current in amperes.   67
amortisseur (electric installations on shipboard). A permanently short-circuited winding consisting of conductors embedded in the poleshoes of a synchronous machine and connected together at the ends of the poles, but not necessarily connected between poles. This winding when used in salient-pole machines sometimes includes bars which do not pass through the pole shoes,but are supported in the interpolar spaces between the pole tips.   3
amortisseur bar (damper bar) (rotating machinery). A single conductor that is a part of an amortisseur winding or starting winding. *See:* **rotor (rotating machinery); stator.**   63
amortisseur winding. *See:* **damper winding; damping winding.**
ampacity (1) (power and distribution transformer). Current-carrying capacity, expressed in amperes, of a wire or cable under stated thermal conditions.   53
(2) (National Electric Code). Current-carrying capacity of electric conductors expressed in amperes.   256
(3) (packaging machinery). Current-carrying capacity expressed in amperes.   429
ampere (metric practice). That constant current which, if maintained in two straight parallel conductors of infinite length, of negligible circular cross section, and placed one meter apart in vacuum, would produce between these conductors a force equal to 2×10^{-7} newton per meter of length. (adopted by 9th General Conference on Weights and Measures 1948).   21
ampere-hour capacity (storage battery). The number of ampere-hours that can be delivered under specified conditions as to temperature, rate of discharge, and final voltage. *See:* **battery (primary or secondary).**   328
ampere-hour efficiency (storage cell) (storage battery). The electrochemical efficiency expressed as the ratio of the ampere-hours output to the ampere-hours input required for the recharge. *See:* **charge.**   328
ampere-hour meter. An electricity meter that measures and registers the integral, with respect to time, of the current of the circuit in which it is connected. *Note:*

The unit in which this integral is measured is usually the ampere-hour. *See:* **electricity meter (meter).**
                                        328
ampere rating (protection and coordination of industrial and commercial power systems). That current that the fuse will carry continuously without deterioration and without exceeding temperature rise limits specified for that fuse.   504
Ampere's law. *See:* **magnetic field strength produced by an electric current.**
ampere-turn per meter. The unit of magnetic field strength in SI units (International System of Units). The ampere-turn per meter is the magnetic field strength in the interior of an elongated uniformly wound solenoid that is excited with a linear current density in its winding of one ampere per meter of axial distance.   210
ampere turns (electrical heating systems). The product of the number of turns and the alternating-current (ac) amperes flowing in an induction heating coil.
                                        476
amplification, current. *See:* **current amplification.**
amplification factor. The factor for a specified electrode and the control grid of an electron tube under the condition that the anode current is held constant. *Notes:* (1) In a triode this becomes the μ factor for the anode and control-grid electrodes. (2)In multielectrode tubes connected as triodes the term anode applies to the combination of electrodes used as the anode. (3) The voltage gain of an amplifier with the output unloaded. *See:* **electron-tube admittances.**
                                        125
amplification factor, gas (gas phototube). *See:* **gas amplification factor (gas phototube).**
amplification, voltage. *See:* **voltage amplification.**
amplified spontaneous emission (laser-maser). The radiation resulting from amplification of **spontaneous emission.**   363
amplifier (1)(analog computers). A device that enables an input signal to control a source of power and thus is capable of delivering at its output a reproduction or analytic modification of the essential characteristics of the signal.   9
(2)(data transmission). A unidirectional device which is capable of delivering an enlargement of the wave form of the electric current, voltage, or power supplied to it.   59
(3)(antennas). A device that enables an input signal to control power from a source independent of the signal and thus be capable of delivering an output that bears some relationship to, and is generally greater than, the input signal.   111
(4) (photomultipliers for scintillation counting). A device whose output is an enlarged reproduction of the essential features of an input signal and which draws power from a source other than the input signal.
                                        117
amplifier, balanced (push-pull amplifier). An amplifier in which there are two identical signal branches connected so as to operate in phase opposition and with input and output connections each balanced to ground.   111, 240, 59

amplifier, bridging. An amplifier with an input impedance sufficiently high so that its input may be bridged across a circuit without substantially affecting the signal level of the circuit across which it is bridged. *See:* **amplifier.**   239
amplifier, buffer (signal-transmission system). *See:* **amplifier, isolating.**
amplifier, carrier (signal-transmission system). An alternating current amplifier capable of amplifying a prescribed carrier frequency and information sidebands relatively close to the carrier frequency. *See:* **signal.**   188
amplifier, chopper (signal-transmission system). A modulated amplifier in which the modulation is achieved by an electronic or electromechanical chopper, the resultant wave being substantially square. *See:* **signal.**   188
amplifier class ratings (electron tube). (1) **class-A amplifier.** An amplifier in which the grid bias and alternating grid voltages are such that anode current in a specific tube flows at all times. *Note:* The suffix 1 is added to the letter or letters of the class identification to denote that grid current does not flow during any part of the input cycle. The suffix 2 is used to denote that current flows during some part of the cycle. *See:* **amplifier.**
(2) class-AB amplifier. An amplifier in which the grid bias and alternating grid voltages are such that anode current in a specific tube flows for appreciably more than half but less than the entire electrical cycle. *Note:* The suffix 1 is added to the letter or letters of the class identification to denote that grid current does not flow during any part of the input cycle. The suffix 2 is used to denote that current flows during some part of the cycle. *See:* **amplifier.**   111, 125
(3) class-B amplifier. An amplifier in which the grid bias is approximately equal to the cutoff value so that the anode current is approximately zero when no exciting grid voltage is applied, and so that anode current in a specific tube flows for approximately one half of each cycle when an alternating grid voltage is applied. *Note:* The suffix 1 is added to the letter or letters of the class identification to denote that grid current does not flow during any part of the input cycle. The suffix 2 is used to denote that current flows during some part of the cycle. *See:* **amplifier.**
(4) class-C amplifier. An amplifier in which the grid bias is appreciably greater than the cutoff value so that the anode current in each tube is zero when no alternating grid voltage is applied, and so that anode current in a specific tube flows for appreciably less than one half of each cycle when an alternating grid voltage is applied. *Note:* The suffix 1 is added to the letter or letters of the class identification to denote that grid current does not flow during any part of the input cycle. The suffix 2 is used to denote that current flows during some part of the cycle. *See:* **amplifier.**
                                        111, 125
amplifier, difference. *See:* **differential amplifier.**
amplifier, differential. *See:* **differential amplifier.**
amplifier ground (signal-transmission system). *See:* **receiver ground.**

amplifier, horizontal. *See:* horizontal amplifier.

amplifier, integrating. *See:* integrating amplifier.

amplifier, intensity. *See:* intensity amplifier.

amplifier, inverting. *See:* inverting amplifier.

amplifier, isolating (signal-transmission system). An amplifier employed to minimize the effects of a following circuit on the preceding circuit. *Example:* An amplifier having effective direct-current resistance and/or alternating-current impedance between any part of its input circuit and any other of its circuits that is high compared to some critical resistance or impedance value in the input circuit. *See:* signal.   188

amplifier, isolation (buffer). An amplifier employed to minimize the effects of a following circuit on the preceding circuit. *See:* amplifier.   178

amplifier, line. An amplifier that supplies a transmission line or system with a signal at a stipulated level. *See:* amplifier.   178

amplifier, modulated (signal-transmission signal). *See:* modulated amplifier.

amplifier, monitoring (electroacoustics). An amplifier used primarily for evaluation and supervision of a program. *See:* amplifier.   178

amplifier, peak limiting. *See:* peak limiter.

amplifier, program. *See:* line amplifier.

amplifier, relay. *See:* relay amplifier.

amplifier, servo. *See:* servo amplifier.

amplifier, summing. *See:* summing amplifier.

amplifier, vertical. *See:* vertical amplifier.

amplifier, X-axis. *See:* horizontal amplifier.

amplifier, Y-axis. *See:* vertical amplifier.

amplifier, Z-axis. *See:* Z-axis amplifier; intensity amplifier.

amplitude (radio wave propagation). Of a sinusoidal wave, the maximum value of a field quantity in space or time.   146

amplitude characteristic. *See:* amplitude-frequency characteristic.

amplitude-comparison monopulse (radar). A form of monopulse employing receiving beams with different amplitude-versus-angle patterns. If the beams have a common phase center, the monopulse is pure amplitude-comparison; otherwise, it is a combination of amplitude-comparison and phase-comparison. *See:* monopulse; phase-comparison monopulse.   13

amplitude discriminator (radar). A circuit whose output is a function of the relative magnitudes of two signals. *See:* navigation.   13, 187

amplitude distortion (data transmission). Distortion caused by a deviation from a desired linear relationship between specified measures of the output and input of a system. *Note:* The related measures need not to be output and input values of the same quantity; for example, in a linear detector the desired relation is between the output signal voltage and the input modulation envelope; or the modulation of the input carrier and the resultant detected signal.   59

amplitude factor (power switchgear). The ratio of the highest peak of the transient recovery voltage to the peak value of the normal-frequency voltage. *Note:* In tests made under one condition to simulate duty under another, as in single-phase tests made to simulate duty on three-phase ungrounded faults, the amplitude factor is expressed in terms of the duty being simulated.   103

amplitude-frequency response (1)(data transmission). The variation of gain, loss,amplification, or attenuation as a function of frequency. *Note:* This response is usually measured in the region of operation in which the transfer characteristic of the system or transducer is essentially linear.   59

(2) (high voltage testing). The amplitude frequency response G(f) of a measuring system is the ratio as a function of the frequency f of the output amplitude to the input amplitude of the system when the input is a sinusoid. A convenient form is 'the normalized frequency response g(f)" in which the constant value of the output amplitude is denoted as unity when that amplitude, multiplied by the scale factor of the system, equals the input amplitude.   150

amplitude gate. *See:* slicer.

amplitude locus (control system, feedback) (for a non-linear system or element whose gain is amplitude dependent). A plot of the describing function, in any convenient coordinate system. *See:* control system, feedback.   56

amplitude-modulated transmitter. A transmitter that transmits an amplitude-modulated wave. *Note.* In most amplitude-modulated transmitters, the frequency is stabilized. *See:* radio transmitter.   111,240

amplitude modulation (1) (data transmission). The process by which a continuous high-frequency wave (carrier) is caused to vary in amplitude by the action of another wave containing information. The usual procedure is to key the carrier wave on and off in accordance with the data to be transmitted. For example, a 1170 Hz tone (the carrier) could be off for 'space' and on for 'mark'. This method has several disadvantages. It does not use bandwidth efficiently, since two sidebands of the carrier are produced and unlike single sideband voice communication methods, the carrier and one sideband cannot be completely eliminated and still do a satisfactory job. The information carrying characteristic of an AM signal is its amplitude.   59

(2) (signal-transmission system). The process, or the result of the process, whereby the amplitude of one electrical quantity is varied in accordance with some selected characteristic of a second quantity, which need not be electrical in nature. *See:* signal.   188

(3)(overhead-power-line corona and radio noise). Modulation in which the amplitude of a carrier is caused to depart from its reference value by an amount proportional to the instantaneous value of the modulating wave.   411

amplitude-modulation noise. The noise produced by undesired amplitude variations of a radio-frequency signal. *See:* radio transmission.   240

amplitude-modulation noise level. The noise level produced by undesired amplitude variations of a radio frequency signal in the absence of any intended modulation.   111

amplitude noise (radar). Used variously to describe **target fluctuation** and **scintillation error.** Use of one of these specific terms is recommended to avoid ambiguity. 13

amplitude permeability (magnetic core testing). The value of permeability at a stated value of field strength (or induction), the field strength varying periodically with time and with no static magnetic field being present.

$$\mu_a = \frac{1}{\mu_0} \frac{B}{H}$$

μ_a = relative amplitude permeability. Maximum permeability is the maximum value of the amplitude permeability as a function of the field strength (or of the induction). 165

amplitude probability distribution (APD)(control of system electromagnetic compatibility)(electromagnet site survey). A distribution showing the probability (commonly percentage of time) that an amplitude is exceeded as a function of the amplitude. 495

amplitude, pulse. A general term indicating the magnitude of a pulse. *Note:* Pulse amplitude is measured with respect to the nominally constant baseline, unless otherwise stated. For specific designation, adjectives such as average, instantaneous, peak, root-mean-square, etcetera, should be used to indicate the particular meaning intended. *See:* **pulse.** 185

amplitude range (electroacoustics). The ratio, usually expressed in decibels, of the upper and lower limits of program amplitudes that contain all significant energy contributions. 239

amplitude ratio. *See:* **gain; subsidence ratio.**

amplitude reference level (pulse techniques). The arbitrary reference level from which all amplitude measurements are made. *Note:* The arbitrary reference level normally is considered to be at an absolute amplitude of zero but may, in fact, have any magnitude of either polarity. If this arbitrary reference level is other than zero, its value and polarity must be stated. *See:* **pulse.** 185

amplitude resonance. Resonance in which amplitude is stationary with respect to frequency. 210

amplitude response (camera tubes). The ratio of (1) the peak-to-peak output from the tube resulting from a spatially periodic test pattern, to (2) the difference in output corresponding to large-area blacks and large-area whites, having the same illuminations as the test pattern minima and maxima, respectively. *Note:* The amplitude response is referred to as modulation transfer (sine-wave response) when a sinusoidal test pattern is used and as square-wave response when the pattern consists of alternate black and white bars of equal width. *See:* **camera tube.** 190

amplitude response characteristic (camera tubes). The relation between (1) amplitude response and (2) television line number (camera tubes) or (image tubes) test-pattern spatial frequency, usually in line pairs per millimeter. *See:* **camera tube.** 190

amplitude selection. A summation of one or more variables and a constant resulting in a sudden change in rate or level at the output of a computing element as the sum changes sign. *See:* **electronic analog computer.** 9

amplitude suppression ratio (frequency modulation). The ratio of the undesired output to the desired output of a frequency-modulation receiver when the applied signal has simultaneous amplitude modulation and frequency modulation. *Note:* This ratio is generally measured with an applied signal that is amplitude modulated 30 percent at a 400-hertz rate and is frequency modulated 30 percent of maximum system deviation at a 1000-hertz rate. *See:* **frequency modulation.** 328

AM radio broadcast band (overhead-power-line corona and radio noise). A band of frequencies assigned for amplitude-modulated transmission of communication intended to entertain or enlighten the general public. *Note:* In the United State and Canada the frequency band is 535 to 1605 kHz. This is also one of the International Telecommunications Union (ITU) frequency allocations, on a world-wide basis, for broadcasting. 411

AMTI (radar). Abbreviation for airborne moving target indication. *See:* **moving target indication (MTI).** 13

AM to FS converter. *See:* **transmitting converter, facsimile; facsimile (electrical communication).**

anaerobic. Free of uncombined oxygen. 205

analog (ADJ) (1)(analog computers). Pertaining to representation by means of continuously variable physical quantities, for example, to describe a physical quantity, such as voltage or shaft position, that normally varies in a continuous manner, or devices such as potentiometers and synchros that operate with such quantities. 9

(2)(data transmission). Used to describe a physical quantity such as voltage or shaft position, that normally varies in a continuous manner. 59

(3) (industrial control). Pertains to information content that is expressed by signals dependent upon magnitude. *See:* **control system, feedback.** 206

analog and digital data. Analog data implies continuity as contrasted to digital data that is concerned with discrete states. *Note:* Many signals can be used in either the analog or digital sense, the means of carrying the information being the distinguishing feature. The information content of an analog signal is conveyed by the value or magnitude of some characteristics of the signal such as the amplitude, phase, or frequency of a voltage, the amplitude or duration of a pulse, the angular position of a shaft, or the pressure of the fluid. To extract the information, it is necessary to compare the value or magnitude of the signal to a standard. The information content of the digital signal is concerned with discrete states of the signal, such as the presence or absence of a voltage, a contact in the open or closed position, or a hole or no hole in certain positions on a card. The signal is given meaning by assigning numerical values or other information to the various possible combinations of the discrete states of the signal. *See:* **analog data; digital data.** 224

analog channel (data transmission). A channel on which the information transmitted can take any value between the limits defined by the channel. Voice channels are analog channels. 59

analog computer (1) (general). An automatic computing device that operates in terms of continuous variation of some physical quantities, such as electrical voltages and currents, mechanical shaft rotations, or displacements, and which is used primarily to solve differential equations. The equations governing the variation of the physical quantities have the same or very nearly the same form as the mathematical equations under investigation and therefore yield a solution analogous to the desired solution of the problem. Results are measured on meters, dials, oscillograph recorders, or oscilloscopes. *See:* **simulator.** 9

(2) (direct-current). An analog computer in which computer variables are represented by the instantaneous values of voltages. 9

(3) (alternating-current). An analog computer in which electrical signals are of the form of amplitude modulated suppressed carrier signals where the absolute value of a computer variable is represented by the amplitude of the carrier and the sign of a computer variable is represented by the phase (0 or 180 degrees) of the carrier relative to the reference alternating-current signal. 9

analog device (station control and data acquisition)- (supervisory control, data acquisition, and automatic control). A device that operates with variables represented by continuously measured quantities such as voltages, resistances, rotations and pressures. 570

analog function. *See:* **supervisory control functions**

analog output. One type of continuously variable quantity used to represent another: for example, in temperature measurement, an electric voltage or current output represents temperature input. *See:* **signal.** 188

analog quantity (1)(station control and data acquisition). A variable represented by a scalar value. 403

(2)(supervisory control, data acquisition, and automatic control). A continuous variable that is typically digitized and represented as a scalar value. 570

analog signal (control) (industrial control). A signal that is solely dependent upon magnitude to express information content. *See:* **control system, feedback.** 219, 206

analog switch (telephone loop performance). A switch capable of switching analog and digital signals without converting them into a set digital format. Most analog end office switches are two-wire systems which have simple interfaces with the loop. 473

analog switching (telephone switching systems). Switching of continuously-varying-level information signals. 55

analog telemetering (power switchgear)(station control and data acquisition). Telemetering in which

some characteristic of the transmitter signal is proportional to the quantity being measured. 103, 403

analog-to-digital (a/d) conversion (supervisory control, data acquisition, and automatic control). Production of a digital output corresponding to the value of an analog input quantity. 570

analog-to-digital converter (1) (data processing). A device that converts a signal that is a function of a continuous variable into a representative number sequence. 54

(2) (analog-to-digital). (A–D). A circuit whose input is information in analog form and whose output is the same information in digital form. *See:* **analog; digital.** 59

(3) (digitizer) (power switchgear). A device or a group of devices that converts an analog quantity or analog position input signal into some type of numerical output signal or code. The input signal is either the measurand or a signal derived from it. 103

(4) (hybrid computer linkage components) (analog-to-digital converter). (ADC). Provides the means of obtaining a digital number representation of a specific analog voltage value. 10

(5)(X-ray energy spectrometers). A device whose input is information in analog form and whose output is the same information in digital form. 471

analog-to-frequency (A F) converter. A circuit whose input is information in an analog form other than frequency and whose output is the same information as a frequency proportional to the magnitude of the information. *See:* **analog.** 59

analog voice frequency circuits abbreviations. (1) dBm. Decibels relative to one milliwatt. This is the customary unit worldwide for measurement of communications signal power. (2) dBm0. Decibels relative to one milliwatt, referred to a zero transmission level point (0 TLP). (3) dBrn. Decibels to one picowatt reference noise level. This is the customary North American unit for measurement of noise power in communications signal circuits. (4) dBrnC. Decibels relative to one picowatt reference noise level, measured with C-message or C-notch frequency weighting. (5) TLP. Transmission level point. The symbol TLP is preceded by a number that indicates, for a particular point in a transmission system, the design signal level in decibels (dB) relative to the level at a reference point (0 TLP). 468

analysis (1)(electric penetration assemblies). A process of mathematical or other logical reasoning that leads from stated premises to the conclusion concerning the qualification of an assembly or components. 493

(2)(valve actuators)(safety systems equipment in nuclear power generating stations). A course of reasoning showing that a certain result is a consequence of assumed premises. 492, 535

(3) (nuclear power generating systems)(class 1E static battery chargers and inverters). A process of mathematical or other logical reasoning that leads from stated premises to the conclusion concerning specific capabilities of equipment and its adequacy for a particular application. *See:* **numerical analysis.**

31, 120, 142, 408

analysis phase (software). *See:* **requirements phase.**

434

analytical model (software). A representation of a process or phenomenon by a set of solvable equations. *See:* **process; simulation.** 434

analyzer. *See:* **differential analyzer; digital differential analyzer; network analyzer.**

anchor (conductor stringing equipment). A device that serves as a reliable support to hold an object firmly in place. The term anchor is normally associated with cone, plate, screw or concrete anchors, but the terms *snub, deadman* and *anchor log* are usually associated with pole stubs or logs set or buried in the ground to serve as temporary anchors. The latter are often used at pull and tension sites. *Syn:* **anchor log; deadman; snub.** 431

anchorage (raceway)(raceway systems for Class 1E circuits for nuclear power generating stations). The connection between the building structure and the raceway support. 513

anchor guy guard. A protective cover over the guy, often a length of sheet metal or plastic shaped to a semicircular or tubular section and equipped with means of attachment to the guy. It may also be of wood. *See:* **tower.** 64

anchor light (illuminating engineering). An aircraft light designed for use on a seaplane or amphibian to indicate its position when at anchored or moored. 167

anchor log (dead man). A piece of rigid material such as timber, metal, or concrete, usually several feet in length, buried in earth in a horizontal position and at right angles to anchor rod attachment. *See:* **tower.** 64

anchor rod. A steel or other metal rod designed for convenient attachment to a buried anchor and also to provide for one or more guy attachments above ground. *See:* **tower.** 64

ancillary equipment (test, measurement and diagnostic equipment). Equipment which is auxiliary or supplementary to an automatic test equipment installation. Ancillary equipment usually consists of standard off-the-shelf items such as an oscilloscope and distortion analyzer. 54

ancillary logic (ANC)(FASTBUS acquisition and control). Logic required for each segment but not part of any device. Operations associated with arbitration, geographical addressing, system handshake and run/halt control are carried out in the ancillary logic which may also contain segment terminators. 480

AND (mathematics of computing). A Boolean operator having the property that if P is a statement, Q is a statement, R is a statement, ..., then the AND of P, Q, R, ...is true if and only if all statements are true. *Note:* P AND Q is ofter represented by P • Q, P&Q, P∧Q, or PQ. 564

AND-circuit. *See:* *AND-gate.*

Anderson bridge. A 6-branch network in which an outer loop of 4 arms is formed by three nonreactive resis-

tors and the unknown inductor, and an inner loop of 3 arms is formed by a capacitor and a fourth resistor in series with each other and in parallel with the arm that is opposite the unknown inductor, the detector being connected between the junction of the capacitor and the fourth resistor and that end of the unknown inductor that is separated from a terminal of the capacitor by only one resistor, while the source is connected to the other end of the unknown inductor and to the junction of the capacitor with two resistors of the outer loop. *Note:* Normally used for the comparison of self-inductance with capacitance. The balance is independent of frequency. *See:* **bridge.** 328

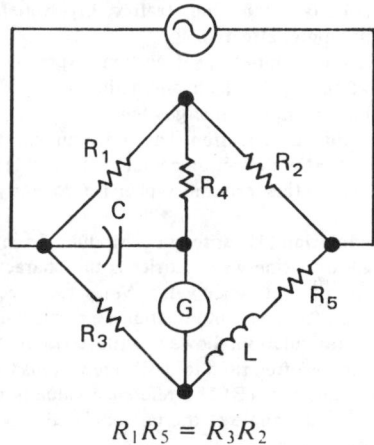

$$R_1 R_5 = R_3 R_2$$

$$L = CR_3 \left[R_4 \left(1 + \frac{R_2}{R_1} \right) + R_2 \right]$$

AND gate (1) (general). A combination logic element such that the output channel is in its ONE state if and only if each input channel is in its ONE state. 235

(2) A gate that implements the logic AND operator. 255

anechoic chamber. An enclosure especially designed with boundaries that absorb sufficiently well the sound incident thereon to create an essentially free-field condition in the frequency range of interest. 176

anechoic enclosure (radio frequency). An enclosure whose internal walls have low reflection characteristics. *See:* **electromagnetic compatibility.** 199

anelectrotonus (electrobiology). Electrotonus produced in the region of the anode. *See:* **excitability.** 192

anemometer (navigation aid terms). An instrument for measuring the speed of wind. 526

aneroid altimeter. *See:* **barometric altimeter.**

anesthetizing location (National Electric Code) (health care facilities). Any area in which it is intended to administer any flammable or nonflammable inhalation anesthetic agents in the course of examination or treatment and includes operating rooms, delivery rooms, emergency rooms, anesthetizing rooms, corridors, utility rooms and other areas which are intended for induction of anesthesia with flammable or nonflammable anesthetizing agents. 256

anesthetizing-location receptacle (National Electric Code) (health care facilities). A receptacle designed to accept the attachment plugs listed for use in such locations.   256

angel (radar). A false radar target echo caused by an atmospheric inhomogeneity, atmospheric refraction, insects, birds, or unknown phenomena. *Syn:* **angel echo.**   13

angle (of a waveform) (information theory). The phase of a periodic or approximately periodic waveform. *See:* **phase; phase angle (1).**   61

angle, bunching (electron stream). *See:* **bunching angle.**

angle, effective bunching (reflex klystrons). *See:* **bunching angle, effective.**

angle, flow (gas tube). That portion, expressed as an angle, of the cycle of an alternating voltage during which current flows. *See:* **gas tube.**   190

angle, maximum-deflection. The maximum plane angle subtended at the deflection center by the usable screen area. *Note:* In this term the hyphen is frequently omitted.   125

angle modulation (1) (antennas). Modulation in which the angle of a sine-wave carrier is the characteristic varied from its reference value. *Notes:* (A) Frequency modulation and phase modulation are particular forms of angle modulation: however, the term frequency modulation is often used to designate various forms of angle modulation. (B) The reference value is usually taken to be the angle of the unmodulated wave. *See:* **modulation index.**   111

(2) (data transmission). The process of causing the angle of the carrier wave to vary in accordance with the signal wave. Phase and frequency modulation are two particular types of angle modulation.   59

(3) (radar). The noise-like variation in the apparent angle of arrival of a signal received from a target, caused by changes in phase and amplitude of target scattering sources and including angular components of both glint and scintillation error.   13

angle of advance (1) (power inverter). The time interval in electrical degrees by which the beginning of anode conduction leads the moment at which the anode voltage would attain a negative value equal to that of the succeeding anode in the commutating group. *See:* **rectification.**   291

(2) (semiconductor rectifiers) (semiconductor power converter). The angle by which forward conduction is advanced by the control means only, in the incoming circuit element, ahead of the instant in the cycle at which the incoming commutating voltage passes through zero in the direction to produce forward conduction in the outgoing circuit element. *See:* **rectification; semiconductor rectifier stack.**   208

angle-of-approach lights (illuminating engineering). Aeronautical ground lights arranged so as to indicate a desired angle of descent during an approach to an aerodrome runway. *Syn:* **optical glide path lights.**   167

angle of attack (navigation aid terms). The angle between the mean chord or the wing and the line of flow of the air past the aircraft.   526

angle of climb (navigation aid terms). The angle between a climbing aircraft's flight path and the horizontal.   526

angle of collimation (illuminating engineering). The angle subtended by a luminaire on an irradiated surface.   167

angle of cut (navigation)(navigation aid terms). The angle at which two lines of position intersect. *Syn:* **crossing angle.**   526

angle of descent (navigation aid terms). The angle between a descending aircraft's flight path and the horizontal.   526

angle of deviation (fiber optics). In optics, the net angular deflection experienced by a light ray after one or more refractions or reflections. *Note:* The term is generally used in reference to prisms, assuming air interfaces. The angle of deviation is then the angle between the incident ray and the emergent ray. *See:* **reflection; refraction.**   433

angle of extinction (industrial control). The phase angle of the stopping (extinction) instant of anode-current flow in a glass tube with respect to the starting instant of the corresponding positive half cycle of the anode voltage of the tube. *See:* **electronic controller.**   206

angle of ignition (industrial control). The phase angle of the starting instant of anode-current flow in a gas tube with respect to the starting instant of the corresponding positive half cycle of the anode voltage of the tube. *See:* **electronic controller.**   206

angle of incidence (1)(acoustic-optic device). The angle in air between the acoustic wavefront and the normal to the optical wavefront. For operation in the Bragg region, maximum diffraction into the first order occurs when the angle of incidence is equal to the Bragg angle, θ_B, which is given by the equation $\sin \theta_B = \lambda_0/2\Lambda$.   82

(2)(fiber optics). The angle between an incident ray and the normal to a reflecting or refracting surface. *See:* **critical angle; total internal reflection.**   433

angle of protection (lightning). The angle between the vertical plane and a plane through the ground wire, within which the line conductors must lie in order to ensure a predetermined degree of protection against direct lightning strokes. *See:* **surge arrester (surge diverter).**   244, 62

angle of retard (1)(thyristor). The interval in electrical angular measure by which the trigger pulse is delayed in relation to operation that would occur with continuous gated control elements and a resistive load. See α in diagram below (Waveform of AC Power Controller Current).   445

(2) (semiconductor rectifiers) (semiconductor rectifier operating with phase control). The angle by which forward conduction is delayed by the control means only, beyond the instant in the cycle at which the incoming commutating voltage passes through zero in the direction to produce forward conduction in the incoming circuit element. *See:* **semiconductor rectifier stack.**   208

angle of retard unbalance (tracking unbalance) (thy-

ristor). The load voltage/current unbalance due to unequal angles of retard either between positive and negative half cycles of a single ac wave or between two or more phases in a three-phase system.   445

angle optimum bunching. *See:* **optimum bunching.**

angle or phase (sine wave). The measure of the progression of the wave in time or space from a chosen instant or position or both. *Notes:* (1) In the expression for a sine wave, the angle or phase is the value of the entire argument of the sine function. (2) In the representation of a sine wave by a phasor or rotating vector, the angle or phase is the angle through which the vector has progressed. *See:* **modulating systems; wavefront.**   111

angle, overlap. *See:* **overlap angle.**

angle, roll over (conductor stringing equipment). For tangent stringing, the sum of the vertical angles between the conductor and the horizontal on both sides of the traveler. Resultants of these angles must be considered when stringing through line angles. Under some stringing conditions, such as stringing large diameter conductors, excessive roll over angles can cause premature failure of a conductor splice if it is allowed to pass over the travelers.   45

angle tower. A tower located where the line changes horizontal direction sufficiently to require special design of the tower to withstand the resultant pull of the wires and to provide adequate clearance. *See:* **tower.**   64

angle tracking. *See:* **tracking.**

angle, transit. *See:* **transit angle.**

angstrom (A) (fiber optics). A unit of optical wavelength (obsolete).

$$1A = 10^{-10} \ meters$$

Note: The angstrom has been used historically in the field of optics, but it is not an SI (International System) unit.   433

angular-acceleration sensitivity (accelerometer) (inertial sensor). The output (divided by the scale-factor) of a linear accelerometer that is produced per unit of angular acceleration causing it. *Note:* In single-degree-of-freedom gyros, it is nominally equal to the effective moment of inertia of the gimbal assembly divided by the angular momentum.   46

angular accelerometer. A device that senses angular acceleration about an input axis. An output signal is produced by the reaction of the moment of inertia of a proof mass to an angular acceleration input. The output is usually an electrical signal proportional to applied angular acceleration.   46

angular accuracy (radar). The degree to which the

Waveform of AC Power Controller Current

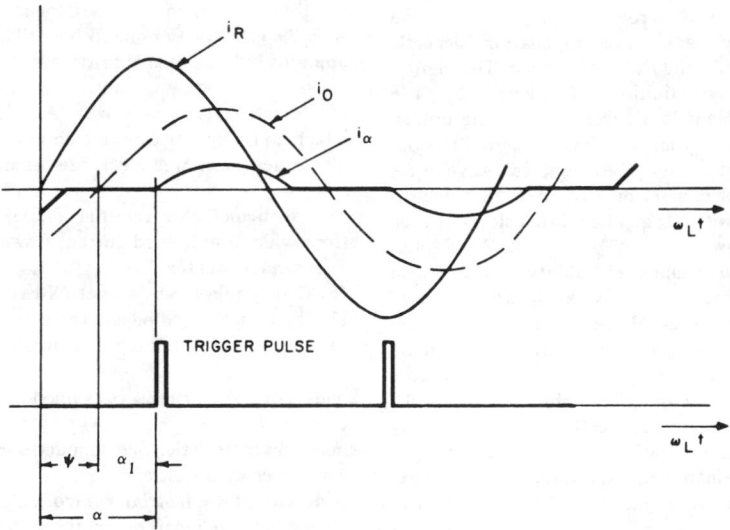

i_R = current in a control element with all control elements continuously gated and a resistive load

    NOTE: In the case of a single phase controller i_R is in phase with the line voltage. The latter may be used as a convenient reference voltage to measure α

i_O = current in a control element with all control elements continuously gated and at the specified load

i_α = current in a control element with a trigger delay angle of α and at the specified load

α = angle of retard

α_I = current delay angle

Ψ = angle between i_R and i_O

    NOTE: In the case of a single phase controller, Ψ is identical with the load power factor angle.

measurement of the angular location of a target with respect to a given reference represents the true angular location of the target with respect to this reference.   13

angular-case-motion sensitivity (dynamically tuned gyro) (inertial sensor) (tuned rotor gyro). The drift rate resulting from an oscillatory angular input about an axis normal to the spin axis at twice the rotor spin frequency. This effect is due to the single degree of freedom of the gimbal relative to the support shaft and is proportional to the input amplitude and phase relative to the flexure axes.   46

angular deviation loss (acoustic transducer). The ratio of the response in a specific direction to the response on the principal axis, usually expressed in decibels. *See:* **loudspeaker.**   176

angular deviation sensitivity (navigation aid terms). The ratio of change of course indication to the change of angular displacement from the course line.   526, 13

angular displacement (polyphase transformer). The phase angle expressed in degrees between the line-to-neutral voltage of the reference identified high-voltage terminal and the line-to-neutral voltage of the corresponding identified low-voltage of terminal. *Note:* The preferred connection and arrangement of terminal markings for polyphase transformers are those which have the smallest possible phase-angle displacements and are measured in a clockwise direction from the line-to-neutral voltage of the reference identified high-voltage terminal. Thus, standard three-phase transformers have angular displacements of either zero or 30 degrees. *See:* **routine test.**   53

angular frequency (radio wave propagation)(periodic function). 2π times the frequency. *Note:* This definition applies to any definition of frequency.   146

angular misalignment loss (fiber optics). The optical power loss caused by angular deviation from the optimum alignment of source to opti cal waveguide, waveguide to waveguide, or waveguide to detector. *See:* **extrinsic joint loss; gap loss; intrinsic joint loss; lateral offset loss.**   433

angular resolution (radar). The ability to distinguish between two targets solely by the measurement of angles; generally expressed in terms of the minimum angle by which two targets of equal strength within the same range resolution cell must be spaced to be separately distinguishable and measurable.   13

angular swing (acousto-optic deflector). The center-to-center angular separation between the deflected light beams obtained upon application of the maximum and minimum acoustic frequency of the frequency range.   72

angular-velocity-sensitivity (accelerometer) (inertial sensor). The output (divided by the scale factor) of a linear accelerometer that is produced per unit of angular velocity input about a specified axis.   46

angular vibration sensitivity (gyro, accelerometer) (inertial sensor). The ratio of the change in output due to angular vibration about a sensor axis divided by the amplitude of the angular vibration causing it.   46

angular width (electronic navigation). *See:* **course width.**

anisotropic (fiber optics). Pertaining to a material whose electrical or optical properties are different for different directions of propagation or different polarizations of a traveling wave. *See:* **isotropic.**   433

annealing (metal-nitride-oxide field-effect transistor). In the context of metal oxide semiconductor (MOS) device properties under irradiation, annealing refers to the sometimes observed reduction of the radiation induced threshold voltage change over a period of seconds to hours after exposure to radiation has ceased.   386

annunciator (thyristor). A visual signal device consisting of a number of pilot lights or drops, each one indicating the condition that exists or has existed in an associated circuit, accordingly labeled.   445

annunciator relay (30) (power system device function numbers). A nonautomatically reset device that gives a number of separate visual indications upon the functioning of protective devices, and which may also be arranged to perform a lockout function.   402

anode (1). An electrode through which current enters any conductor of the nonmetallic class. Specifically, an electrolytic anode is an electrode at which negative ions are discharged, or positive ions are formed, or at which other oxidizing reactions occur.   328

(2). An electrode or portion of an electrode at which a net oxidation-reaction occurs. *See:* **electrochemical cell.**   223, 186, 205

(3) (thyristor). The electrode by which current enters the thyristor, when the thyristor is in the ON state with the gate open-circuited. *Note:* This term does not apply to bidirectional thyristors.   243, 66, 208, 191

(4) (electron tube or valve). An electrode through which a principal stream of electrons leaves the interelectrode space. *See:* **electrode (electron tube).**   190, 117, 125

(5) (semiconductor rectifier diode). The electrode from which the foward current flows within the cell. *See:* **semiconductor.**   237, 66

(6) (X-ray tube). *See:* **target (X-ray tube).**

(7) (light emitting diodes). The electrode from which the foward current is directed within the device.   162

anode butt. A partially consumed anode. *See:* **fused electrolyte.**   328

anode characteristic. *See:* **anode-to-cathode voltage-current characteristic.**

anode circuit (industrial control). A circuit that includes the anode-cathode path of an electron tube in series connection with other elements. *See:* **electronic controller.**   206

anode circuit breaker (1) (power switchgear). A low-voltage power circuit breaker: (A) designed for connection in an anode of a mercury-arc power rectifier unit, (B) trips automatically only on reverse current and starts reduction of a current in a specified time when the arc-back occurs at the end of the forward current conduction, and (C) substantially interrupts

the arc-back current within one cycle of the fundamental frequency after the beginning of the arc-back. *Note:* The specified time in present practice is 0.008 second or less (at an ac frequency of 60 Hz).
                                                    103

(2) (power system device function numbers) (7). A device used in the anode circuits of a power rectifier circuit if an arc back should occur.   402

anode cleaning (reverse-current cleaning) (electroplating). Electrolytic cleaning in which the metal to be cleaned is made the anode. *See:* **battery (primary or secondary).**   328

anode corrosion efficiency. The ratio of the actual corrosion of an anode to the theoretical corrosion calculated from the quantity of electricity that has passed.
                                                    205

anode current (electron tubes). *See:* **electrode current; electronic controller.**

anode dark space (gas tubes) (gas). A narrow dark zone next to the surface of the anode. *See:* **discharge (gas).**   244

anode differential resistance. *See:* **anode resistance.**

anode effect. A phenomenon occurring at the anode, characterized by failure of the electrolyte to wet the anode and resulting in the formation of a more or less continuous gas film separating the electrolyte and anode and increasing the potential difference between them. *See:* **fused electrolyte.**   328

anode efficiency. The current efficiency of a specified anodic process. *See:* **electrochemistry.**   328

anode, excitation (pool-cathode tube). *See:* **excitation anode.**

anode (potential) fall (gas). The fall of potential due to the space charge near the anode. *See:* **discharge (gas).**
                                                  244, 190

anode firing (industrial control). The method of initiating conduction of an ignitron by connecting the ignitor through a rectifying element to the anode of the ignitron to obtain power for the firing current pulse. *See:* **electronic controller.**   206

anode glow (gas tubes) (gas). A very bright narrow zone situated at the near end of the positive column with respect to the anode. *See:* **discharge (gas).**
                                                  244, 190

anode layer. A molten metal or alloy, serving as the anode in an electrolytic cell, that floats on the fused electrolyte or upon which the fused electrolyte floats. *See:* **fused electolyte.**   328

anode, main (pool-cathode tube). *See:* **main anode.**

anode mud. *See:* **slime, anode.**

anode paralleling reactor (power and distribution transformer). A reactor with a set of mutually coupled windings connected to anodes operating in parallel from the same transformer terminal.   53

anode power supply (electron tube) (plate power supply). The means for supplying power to the plate at a voltage that is usually positive with respect to the cathode. *See:* **power pack.**   328

anode-reflected-pulse rise time. The rise time of a pulse reflected from the anode. *Note:* This time can be measured with a time-domain reflectometer.   117

anode region (gas tubes) (gas). The group of regions comprising the positive column, anode glow, and anode dark space. *See:* **discharge (gas).**   244, 190

anode relieving (pool-cathode tube) (gas tube). An anode that provides an alternative conducting path to reduce the current to another electrode. *See:* **electrode (electron tube).**   244, 190

anode resistance (anode differential resistance) (electron tube). The quotient of a small change in anode voltage by a corresponding small change of the anode current, all the other electrode voltages being maintained constant. It is equal to the reciprocal of the anode conductance. *See:* ON period (electron tubes).
                                                  244, 190

anode scrap. That portion of the anode remaining after the schedule period for the electrolytic refining of the bulk of its metal content has been completed. *See:* **electrorefining.**   328

anode slime. *See:* **slime, anode.**

anode strap (magnetron). A metallic connector between selected anode segments of a multicavity magnetron, principally for the purpose of mode separation. *See:* **magnetron.**   190, 125

anode supply voltage (industrial control). The voltage at the terminals of a source of electric power connected in series in the anode circuit. *See:* **electronic controller.**   206

anode terminal (1) (semiconductor device). The terminal by which current enters the device. *See:* **semiconductor; semiconductor device.**   245

(2) (thyristor). The terminal that is connected to the anode. *Note:* This term does not apply to bidirectional thyristors. *See:* **anode.**   191

anode-to-cathode voltage (anode voltage) (thyristor). The voltage between the anode terminal and the cathode terminal. *Note:* It is called positive when the anode potential is higher than the cathode potential and called negative when the anode potential is lower than the cathode potential. *See:* **electronic controller; principal voltage-current characteristic (principal characteristic).**   191

anode-to-cathode voltage-current characteristic (thyristor). A function, usually represented graphically, relating the anode-to-cathode voltage to the principal current with gate current, where applicable, as a parameter. *Note:* This term does not apply to bidirectional thyristors. *Syn:* **anode characteristic.** *See:* **principal voltage-current characteristic (principal characteristic).**   191

anode voltage (electron tubes). *See:* **electrode voltage; electronic controller.**

anode voltage drop (glow-discharge cold-cathode tube). The main gap voltage drop after conduction is established in the main gap.   190

anode voltage, forward, peak. *See:* **peak forward anode voltage.**

anode voltage, inverse, peak. *See:* **peak inverse anode voltage.**

anodic polarization. Polarization of an anode. *See also:* **electrochemistry.**   328

anolyte. The portion of an electrolyte in an electrolytic cell adjacent to an anode. If a diaphragm is present, it is the portion of electrolyte on the anode side of the diaphragm. *See:* **electrolytic cell.** 205

anomaly (software verification and validation plans). Anything observed in the documentation or operation of software that deviates from expectations based on previously verified software products or reference documents. A critical anomaly is one that must be resolved before the verification and validation (V&V) effort proceeds to the next lifecycle phase. 511

A-N radio range (navigation aid terms). A radio range providing four radial lines of position identified aurally as a continuous tone resulting from the interleaving of equal amplitude A and N International Morse Code. The sense of deviation from these lines is indicated by deterioration of the steady tone into audible A or N code signals. 526

answered call (telephone switching systems). A call on which an answer signal occurred. 55

answering plug and cord. A plug and cord used to answer a calling line. 328

answer signal (telephone switching systems). A signal that indicates that the call has been answered. 55

antenna (1) (general). That part of a transmitting or receiving system which is designed to radiate or to receive electromagnetic waves. 111
(2) (data transmission). A means for radiating or receiving radio waves. *See:* **dipole antenna; effective area antenna; effective height antenna; helical antenna; horn antenna; loop antenna; slot antenna.** 59
(3) (overhead-power-line corona and radio noise). A means for radiating or receiving radio waves. 411

antenna correction factor (land-mobile communication transmitters). A factor (usually supplied with the antenna) which, when properly applied to the meter reading of the measuring instrument, yields the electric field in volts/meters (V/m) or the magnetic field strength in amperes/meters (A/m). *Notes:* (1) This factor includes the effects of antenna effective length and impedance mismatch plus transmission line losses. (2) The factor for electric field strength is not necessarily the same as the factor for the magnetic field strength. 444

antenna effect (1)(radio direction finding)(navigation aid terms). The presence of output signals having no directional information and caused by the directioanl array acting as simple nondirectional antenna; the effect is manifested by angular displacement of the nulls, or a broadening of the nulls. 526
(2) (loop antenna) (old usage). Any spurious effect resulting from the capacitance of the loop to ground. *See:* **antenna.** 111

antenna, effective area (in a given direction). The ratio of the power available at the terminals of an antenna to the incident power density of a plane wave from that direction polarized coincident with the polarization that the antenna would radiate. *See:* **antenna.** 179

antenna, effective height (1) (data transmission). (A) The height of its center of radiation above the effective ground level. (B) In low-frequency applications the term effective height is applied to loaded or nonloaded vertical antennas and is equal to the moment of the current distribution in the vertical section, divided by the input current. *Note:* For an antenna with symmetrical current distribution the center of radiation is the center of distribution. For an antenna with asymmetrical current distribution the center of radiation is the center of current moments when viewed from directions near the direction of maximum radiation. *See:* **antenna.** 59
(2) (mobile communication). The height of the center of a vertical antenna (of at least 1.4 wavelength) above the effective ground plane of the vehicle on which the antenna is mounted. *See:* **antenna; mobile communication system.** 181

antenna, effective height base station (mobile communication). The height of the physical center of the antenna above the effective ground plane. *See:* **mobile communication system.** 181

antenna, effective length (effective height, low-frequency usage). (1) general. (A) For an antenna radiating linearly polarized waves, the length of a thin straight conductor oriented perpendicular to the direction of maximum radiation, having a uniform current equal to that at the antenna terminals and producing the same far field strength as the antenna. (B) Alternatively, for the same antenna receiving linearly polarized waves from the same direction, the ratio of the open-circuit voltage developed at the terminals of the antenna to the component of the electric field strength in the direction of antenna polarization. *Notes:* (1) The two definitions yield equal effective lengths. (2) In low-frequency usage the effective length of a ground-based antenna is taken in the vertical direction and is frequently referred to as effective height. Such usage should not be confused with effective height (of an antenna, high-frequency usage). *See:* **antenna.** 179
(2) (electromagnetic compatibility). The ratio of the antenna open-circuit voltage to the strength of the field component being measured. *See:* **electromagnetic compatibility.** 199

antenna efficiency (aperture-type antenna). For an antenna with a specified planar aperture, the ratio of the maximum effective area of the antenna to the aperture area. 111

antenna factor (field-strength meter). That factor that, when properly applied to the meter reading of the measuring instrument, yields the electric field strength in volts.meter or the magnetic field strength in amperes.meter. *Notes:* (1) This factor includes the effects of antenna effective length and mismatch and transmission line losses. (2) The factor for electric field strength is not necessarily the same as the factor for the magnetic field strength. *See:* **electromagnetic compatibility.** 199

antenna figure of merit (communication satellite). An antenna performance parameter equalling the antenna gain G divided by the antenna noise temperature T, measured at the antenna terminals. It can be expressed as a ratio, $M = G/T$ or logarithmically

$$M(dB) = 10 \log_{10} G - 10 \log_{10} T$$

See: **noise temperature, average operating.**   85

antenna (aperture) illumination efficiency. The ratio, usually expressed in percent, of the maximum directivity of an antenna (aperture) and its standard directivity. *Note:* For planar apertures the standard directivity is calculated by using the projected area of the actual antenna in a plane transverse to the direction of its maximum radiation intensity. *Syn:* **normalized directivity.** *See:* **standard directivity.**   111

antenna pattern. *See:* **radiation pattern.**

antenna resistance (1) (general). The real part of the input impedance of an antenna.   111
(2) (test procedures for antennas). The ratio of the power accepted by the entire antenna circuit to the mean-square antenna current referred to a specified point. *Note:* Antenna resistance is made up of such components as radiation resistance, ground resistance, radio-frequency resistance of conductors in the antenna circuit, and equivalent resistance due to corona, eddy currents, insulator leakage, and dielectric power loss. *See:* **antenna.**   246

antenna sensitivity-test input (amplitude-modulation broadcast receivers). The sensitivity input is the least signal-input voltage of a specified carrier frequency, modulated 30 percent at 400 cycles and applied to the receiver through a standard dummy antenna, which results in normal test output when all controls are adjusted for greatest sensitivity. It is expressed in decibels below 1 volt, or in microvolts.   524

antenna temperature (radio wave propagation). The temperature of a black body which, when placed around a matched antenna that is similar to the actual antenna but loss-free, produces from this antenna the same available power, in a specified frequency range, as the actual antenna in its normal electromagnetic environment.   146

antenna terminal conducted interference (electromagnetic compatibility). Any undesired voltage or current generated within a receiver, transmitter, or their associated equipment appearing at the antenna terminals. *See:* **electromagnetic compatibility.**   199

anticathode (X-ray tube). *See:* **anode.**

anticlutter circuits (radar). Circuits which attenuate undesired reflections to permit detection of targets otherwise obscured by such reflections.   13

anticlutter gain control (radar)(nonlinear, active, and nonreciprocal waveguide components). A device that automatically and smoothly increases the gain of a radar receiver from a low level to the maximum, within a specified period after each transmitter pulse, so that short-range echoes producing clutter are amplified less than long-range echoes.   530

anticoincidence (radiation counter). The occurrence of a count in a specified detector unaccompanied simultaneously or within an assignable time interval by a count in one or more other specified detectors.   190

anticoincidence circuit (pulse techniques). A circuit that produces a specified output pulse when one (frequently predesignated) of two inputs receives a pulse and the other receives no pulse within an assigned time interval.   117

anti-collision light (illuminating engineering). A flashing aircraft aeronautical light or system of lights designed to provide a red signal throughout 360 degrees of azimuth for the purpose of giving long-range indication of an aircraft's location to pilots of other aircraft.   167

antiferroelectric material. A material that exhibits structural phase changes and anomalies in the dielectric permittivity as do ferroelectrics, but has zero net spontaneous polarization, and hence exhibits no hysteresis phenomena. *Note:* In some cases it is possible to apply electric fields sufficiently high to produce a structural transition to a ferroelectric phase as evidenced by appearance of a double hysteresis loop. *See:* **ferroelectric material; ferroelectric domain; paraelectric.**   80

antifouling. Pertaining to the prevention of marine organism attachment and growth on a submerged metal surface (through the effects of chemical action).   205

antifreeze pin, relay. *See:* **relay antifreeze pin.**

antinode (standing wave). *See:* **loop (standing wave).**

antinoise microphone. A microphone with characteristics that discriminate against acoustic noise. *See:* **microphone.**   328

anti-overshoot (industrial control). The effect of a control function or a device that causes a reduction in the transient overshoot. *Note:* Anti-overshoot may apply to armature current, armature voltage, field current, etcetera. *See:* **control system, feedback.**   206

antioxidant (insulating oil). *See:* **oxidation inhibitor.**

antiplugging protection (industrial control). The effect of a control function or a device that operates to prevent application of counter torque by the motor until the motor speed has been reduced to an acceptable value. *See:* **control system, feedback.**   206

antipump (pump-free) device (power switchgear). A device which prevents reclosing after an opening operation as long as the device initiating closing is maintained in the position for closing.   103

antipump device (pump-free device). A device that prevents reclosing after an opening operation as long as the device initiating closing is maintained in the position for closing.   103

antireflection coating (fiber optics). A thin, dielectric or metallic film (or several such films) applied to an optical surface to reduce the reflectance and thereby increase the transmittance. *Note:* The ideal value of the refractive index of a single layered film is the square root of the product of the refractive indices on either side of the film, the ideal optical thickness being one quarter of a wavelength. *See:* **dichroic filter; Fresnel reflection; reflectance; transmittance.**   433

antiresonant frequency (circuits and systems). Usually in reference to a crystal unit or the parallel combination of a capacitor and inductor. The frequency at which, neglecting dissipation, the impedance of the object under consideration is infinite. 67

antisidetone induction coil. An induction coil designed for use in an antisidetone telephone set. *See:* **telephone station.** 328

antisidetone telephone set. A telephone set that includes a balancing network for the purpose of reducing sidetone. *See:* **telephone station; sidetone.** 328

anti-single-phase tripping device (power switchgear). A device that operates to open all phases of a circuit by means of a polyphase switching device, in response to the interruption of the current in one phase. *Notes:* (1) This device prevents single phasing of connected equipment resulting from the interruption of any one phase of the circuit. (2) This device may sense operation of a specific single-phase interrupting device or may sense loss of single phase potential. 103

antistatic (health care facilities). Adjective describing that class of materials which includes conductive materials and, also, those materials which, throughout their stated life, meet the requirements of 4-6.6.3 and 4-6.6.4 of NFPA-56A, 1978. 192

anti-transmit-receive box (ATR box) (electronic navigation). *See:* **anti-transmit-receive switch.**

anti-transmit-receive switch (ATR switch) (radar). A radio-frequency switch that automatically decouples the transmitter from the antenna during the receiving period; it is employed when a common transmitting and receiving antenna is used. 13

anti-transmit-receive tube (ATR tube)(1)(electron tubes). A gas-filled radio-frequency switching tube used to isolate the transmitter during the interval for pulse reception. *See:* **gas tube.** 125

(2)(nonlinear, active, and nonreciprocal waveguide components). A gas-filled radio-frequency switching tube used to isolate the transmitter during the interval for pulse reception. *See:* **gas tube.** 530

A **operator.** An operator assigned to an *A* switchboard. 328

APC. *See:* **automatic phase control.**

APD. *See:* **avalanche photodiode.** 433

aperiodically sampled equivalent time format (pulse measurement). A format which is identical to the aperiodically sampled real time format except that the time coordinate is equivalent to and convertible to real time. Typically, each datum point is derived from a different measurement on a different wave in a sequence of waves. *See:* **sampled format.** 15

aperiodically sampled real time format (pulse measurement). A format which is identical to the periodically sampled real time format except that the sampling in real time is not periodic and wherein the data exists as coordinate point pairs t_1, m_1; t_2, m_2; . . .; t_n, m_n. *See:* **sampled format.** 15

aperiodic antenna. An antenna which, over an extended frequency range, does not exhibit a cyclic behavior with frequency of either its input impedance or its pattern. *Note:* This term is often applied to a small monopole or small loop, each containing an active element as an integral component, with impedance and pattern characteristics varying but slowly over the extended frequency range. 111

aperiodic circuit. A circuit in which it is not possible to produce free oscillations. *See:* **oscillatory circuit.** 244, 59

aperiodic component of short-circuit current (rotating machinery). The component of current in the primary winding immediately after it has been suddenly short-circuited when all components of fundamental and higher frequencies have been subtracted. *See:* **asynchronous machine.** 63

aperiodic damping. *See:* **overdamping.**

aperiodic time constant (rotating machinery). The time constant of the aperiodic component when it is essentially exponential, or of the exponential which can most nearly be fitted. *See:* **asynchronous machine.** 63

aperture (1) (antenna). A surface, near or on an antenna, on which it is convenient to make assumptions regarding the field values for the purpose of computing fields at external points. *Notes:* (1) In some cases the aperture may be considered as a line. (2) In the case of a unidirectional antenna the aperture is often taken as that portion of a plane surface near the antenna, perpendicular to the direction of maximum radiation, through which the major part of the radiation passes. *See:* **antenna.** 246, 111

(2) (data transmission). For a unidirectional antenna, that portion of a plane surface near the antenna perpendicular to the direction of maximum radiation through which the major part of the radiation passes. 59

aperture blockage (antennas). A condition resulting from objects lying in the path of rays arriving at or departing from the aperture of an antenna. *Note:* For example, the feed, subreflector, or support structure produce aperture blockage for a symmetric reflector antenna. 111

aperture compensation (television). *See:* **aperture equalization.**

aperture correction (television). *See:* **aperture equalization.**

aperture efficiency (antenna) (for an antenna aperture). The ratio of its directivity to the directivity obtained when the aperture illumination is uniform. *See:* **antenna.** 179, 111

aperture equalization (television). Electrical compensation for the distortion introduced by the size of a scanning aperture. *See:* **television.** 178

aperture illumination (excitation). The field over the aperture as described by amplitude, phase, and polarization distributions. 111

aperture illumination efficiency. *See:* **antenna illumination efficiency.** 111

API. *See:* **air-position indicator.**

APL. *See:* **average picture level.**

apoapsis (communication satellite). The most distant point from the center of a primary body (or planet) to an orbit around it. 74

apogee (1)(navigation aid terms). That orbital point farthest from the earth, when the earth is the center of attraction. 526

(2)(communication satellite). The most distant point from the center of the earth to an orbit around it. 74

apparatus (power and distribution transformer). A general designation for large electrical equipment such as generators, motors, transformers, circuit breakers, etcetera. 53

apparatus insulator (cap and pin, post). An assembly of one or more apparatus-insulator units, having means for rigidly supporting electric equipment. *See:* **insulator.** 261

apparatus insulator unit. The assembly of one or more elements with attached metal parts, the function of which is to support rigidly a conductor, bus, or other conducting elements on a structure or base member. *See:* **tower.** 64

apparatus termination (high voltage AC cable terminations). A termination intended for use in apparatus where the ambient temperature of the medium immediately surrounding the termination may reach 55 °C. 4

apparatus thermal device (26) (power system device function numbers). A device that functions when the temperature of the shunt field or the amortisseur winding or a machine, or that of a load limiting or load shifting resistor, or of a liquid or other medium exceeds a predetermined value; or if the temperature of the protected apparatus, such as a power rectifier, or of any medium decreases below a predetermined value. 402

apparent altitude (navigation aid terms). That sextant altitude corrected for reading and reference level inaccuracies. 526

apparent bearing (direction finding)(navigation aid terms). A bearing from a direction-finder site to a target transmitter determined by averaging the readings made on a calibrated direction-finder test standard; the apparent bearing is then used in the calibration and adjustment of other direction finders at the same site. 526

apparent candlepower (extended source). At a specified distance, the candlepower of a point source that would produce the same illumination at that distance. 167

apparent charge (terminal charge)(q)(partial discharge measurementin liquid-filled power transformers and shunt reactors). That charge which, if it could be injected instantaneously between the terminals of the test object, would momentarily change the voltage between its terminals by the same amount as the partial discharge itself. The apparent charge should not be confused with the charge transferred across the discharging cavity in the dielectric medium. Apparent charge, within the terms of IEEE Draft Std 545-1987, is expressed in Coulombs which is abbreviated C. One pC is equal to 10^{-12} Coulombs. 580

apparent dead time. *See:* **dead time.**

apparent discharge magnitude (corona measurement). The charge transfer measured at the terminals of a sample caused by a corona pulse in a sample. 375

apparent horizon (navigation aid terms). Visible horizon. 526

apparent inductance. The reactance between two terminals of a device or circuit divided by the angular frequency at which the reactance was determined. This quantity is defined only for frequencies at which the reactance is positive. *Note:* Apparent inductance includes the effects of the real and parasitic elements that comprise the device or circuit and is therefore a function of frequency and other operating conditions. 197

apparent output power (converters having ac output)- (self-commutated converters). The product of fundamental current and fundamental phase voltage summed for all phases of the circuit. 584

apparent power (1) (rotating machinery). The product of the root-mean-square current and the root-mean-square voltage. *Note:* It is a scalar quantity equal to the magnitude of the phasor power. *See:* **asynchronous machine.** 63

(2)(metering). For sinusoidal quantities in either single-phase or polyphase circuits, apparent power is the square root of the sum of the squares of the active and reactive powers. *Note:* This is, in general, not true for nonsinusoidal quantities. 212

(3) (general). *See:* **power, apparent.**

apparent-power loss (volt-ampere loss) (electric instrument). Of the circuit for voltage-measuring instruments, the product of end-scale voltage and the resulting current: and for current-measuring instruments, the product of the end-scale current and the resulting voltage. *Notes:* (A) For other than current-measuring or voltage-measuring instruments, for example, wattmeters, the apparent power loss of any circuit is expressed at a stated value of current or of voltage. (B) Computation of loss: for the purpose of computing the loss of alternating-current instruments having current circuits at some selected value other than that for which it is rated, the actual loss at the rated current is multiplied by the square of the ratio of the selected current to the rated current. *Example:* A current transformer with a ratio of 500:5 amperes is used with an instrument having a scale of 0–300 amperes and, therefore, a 3-ampere field coil, and the allowable loss at end scale is as stated on the Detailed Requirement Sheet. The allowable loss of the instrument referred to a 5-amperes basis is as follows: Allowable loss in volt-amperes equals (allowable loss end-scale volt-amperes) $(5.3)^2$. *See:* **accuracy rating (instrument).** 280

apparent sag (wire in a span). (1) The maximum departure in the vertical plane of the wire in a given span from the straight line between the two points of support of the span, at 60 degrees Fahrenheit, with no wind loading. *Note:* Where the two supports are at the same level this will be the sag. *See:* **tower.** 64, 262

(2) The departure in the vertical plane of the wire at

the particular point in the span from the straight line between the two points of support. 64

apparent sag at any point in the span (transmission and distribution). The departure of the wire at the particular point in the span from the straight line between the two points of support of the span. 262

apparent sag of a span (transmission and distribution). The maximum departure of the wire in a given span from the straight line between the two points of support of the span. 262

apparent time constant (thermal converter) (63-percent response time) (characteristic time). The time required for 63 percent of the change in output electromotive force to occur after an abrupt change in the input quantity to a new constant value. *See:* Note 1 of **response time of a thermal converter.** *See:* **thermal converter.** 280

apparent vertical (navigation aid terms). The direction of the vector sum of the gravitational and all other accelerations. 526

apparent visual angle (laser-maser). The angular subtense of the source as calculated from the source size and distance from the eye. It is not the beam divergence of two sources. 363

appliance (1) (electric). A utilization item of electric equipment, usually complete in itself, generally other than industrial, normally built in standardized sizes or types that transforms electric energy into another form, usually heat or mechanical motion, at the point of utilization. For example, a toaster, flatiron, washing machine, dryer, hand drill, food mixer, air conditioner. 260

(2) (transmission and distribution). Current-consuming equipment, fixed or portable: for example, heating, cooking, and small motor-operated equipment. 262

(3) (National Electrical Code). Utilization equipment, generally other than industrial, normally built in standardized sizes or types, which is installed or connected as a unit to perform one or more functions such as clothes washing, air conditioning, food mixing, deep frying, etcetera. 256

(4) (National Electrical Safety Code). Current- conducting, energy-consuming equipment, fixed or portable; for example, heating, cooling, and small motor-operated equipment. 391

appliance branch circuit (1) (electric installations on shipboard). A circuit supplying energy to one or more outlets to which appliances are to be connected; such circuits to have no permanent connected lighting fixtures not a part of an appliance. 3

(2) (National Electrical Code). A branch circuit supplying energy to one or more outlets to which appliances are to be connected; such circuits to have no permanently connected lighting fixtures not a part of an appliance. 256

appliance, fixed (electric system). An appliance that is fastened or otherwise secured at a specific location. *See:* **appliances.** 256

appliance outlet (household electric ranges). An outlet mounted on the range and to which a portable appli-

ance may be connected by means of an attachment plug cap. 263

appliance, portable (electric system). An appliance that is actually moved or can easily be moved from one place to another in normal use. *See:* **appliances.** 256

appliance, stationary (electric system). An appliance that is not easily moved from one place to another in normal use. *See:* **appliances.** 256

application (computer applications). The use to which a computer system is put; for example, a payroll application, an airline application, or a network application. 571

application-oriented language (software). (1) A computer-oriented language with facilities or notations applicable primarily to a single application area, for example, a language for statistical analysis or machine design. (2) A problem-oriented language whose statements contain or resemble the terminology of the occupation or profession of the user. 434

application software. Software specifically produced for the functional use of a computer system, for example, software for navigation, gun fire control, payroll, general ledger. *See:* **computer system; software; system software.** 434

application valve (brake application valve). An air valve through the medium of which brakes are automatically applied. 328

applicator (electrodes) (dielectric heating). Appropriately shaped conducting surfaces between which is established an alternating electric field for the purpose of producing dielectric heating. *See:* **dielectric heating.** 14

applied fault protection. A protective method in which, as a result of relay action, a fault is intentionally applied at one point in an electric system in order to cause fuse blowing or further relay action at another point in the system. 103, 127

applied-fault protection (power switchgear). A protective method in which, as a result of relay action, a fault is intentionally applied at one point in an electrical system in order to cause fuse blowing or further relay action at another point in the system. 103

applied-potential tests (electric power). Dielectric tests in which the test voltages are low-frequency alternating voltages from an external source applied between conducting parts, and between conducting parts and ground. 91

applied voltage (corona measurement). Voltage which is applied across insulation. Applied voltage may be between windings or from winding(s) to ground. 375

applied voltage tests (power and distribution transformer). Dielectric tests in which the test voltages are low-frequency alternating voltages from an external source applied between conducting parts and ground without exciting the core of the transformer being tested. 53

approach circuit. A circuit used to announce the approach of trains at block or interlocking stations. 328

approach indicator. A device used to indicate the approach of a train.   328

approach-light beacon (illuminating engineering). An aeronautical ground light placed on the extended centerline of the runway at a fixed distance from the runway threshold to provide an early indication of position during an approach to a runway. *Note:* The runway threshold is the beginning of the runway usable for landing.   167

approach lighting. An arrangement of circuits so that the signal lights are automatically energized by the approach of a train.   328

approach-lighting relay. A relay used to close the lighting circuit for signals upon the approach of a train.   328

approach lights (illuminating engineering). A configuration of aeronautical ground lights located in extension of a runway or channel before the threshold to provide visual approach and landing guidance to pilots.   167

approach locking (electric approach locking). Electric locking effective while a train is approaching, within a specified distance, a signal displaying an aspect to proceed, and that prevents, until after the expiration of a predetermined time interval after such signal has been caused to display its most restrictive aspect, the movement of any interlocked or electrically locked switch, movable-point frog, or derail in the route governed by the signal, and that prevents an aspect to proceed from being displayed for any conflicting route. *See:* **interlocking.**   328

approach navigation (navigation aid terms). Navigation during the time that the approach to a dock, runway, or other terminal facility is of immediate importance.   526

approach path (navigation aid terms). That portion of the flight path between the point at which the descent for landing is normally started and the point at which the aircraft touches down on the runway.   526

approach signal. A fixed signal used to govern the approach to one or more other signals.   328

approval plate (mining). A label that the United States Bureau of Mines requires manufacturers to attach to every completely assembled machine or device sold as permissible mine equipment. *Note:* By this means, the manufacturer certifies to the permissible nature of the machine or device.   328

approval test (acceptance test). The testing of one or more meters or other items under various controlled conditions to ascertain the performance characteristics of the type of which they are a sample. *See:* **service test (field test).**   212

approval test (metering). A test of one or more meters or other items under various controlled conditions to ascertain the performance characteristics of the type of which they are a sample.   212

approved (1) (general). Approved by the enforcing authority.   328
(2) (National Electrical Code). Acceptable to the authority having jurisdiction.   256

approved supplier (replacement parts for Class 1E equipment in nuclear power generating stations). A supplier whose quality assurance (QA) system has been evaluated and found to meet the owner's QA requirements for the item or service to be purchased.   582

approximate value (metric practice). A value that is nearly but not exactly correct or accurate.   21

APT (numerically controlled machines). *See:* **automatic programmed tools.**

arbitration cycle (FASTBUS acquisition and control). The process by which the next master to be granted bus mastership is determined. It is initiated by the arbitration timing controller and is complete when the winning master assumes bus mastership.   480

arbitration locked sequence (FASTBUS acquisition and control). A sequence of operations by one master, directed to a number of different primary addresses, which is not interruptible by any other master because the originating master does not allow bus arbitration to take place.   480

arbitration timing control (ATC)(FASTBUS acquisition and control). Logic associated with each segment for the purpose of supervising and generating the arbitration control systems, run/halt control and broadcast system handshake. (This is part of the ancillary logic).   480

arc. A discharge of electricity through a gas, normally characterized by a voltage drop in the immediate vicinity of the cathode approximately equal to the ionization potential of the gas. *See:* **gas tube.**   125

arc (mode) current (gas tube surge arrester). The current that flows after breakdown when the circuit impedance allows a current that exceeds the glow-to-arc transition current.   62

arc (mode) voltage (gas tube surge arrester). The voltage drop across the arrester during the arc current flow.   62

arc-back (gas tube). A failure of the rectifying action that results in the flow of a principal electron stream in the reverse direction, due to the formation of a cathode spot on an anode. *See:* **gas tube; rectification.**   125

arc cathode (gas tube). A cathode the electron emission of which is self-sustaining with a small voltage drop approximately equal to the ionization potential of the gas.   244

arc chute (of a switching device) (power switchgear). A structure affording a confined space or passageway, usually lined with arc-resisting material, into or through which an arc is directed to extinction.   103

arc, clockwise (numerically controlled machines). An arc generated by the coordinated motion of two axes in which curvature of the path of the tool with respect to the workpiece is clockwise, when viewing the plane of motion in the negative direction of the perpendicular axis.   207

arc converter. A form of negative-resistance oscillator utilizing an electric arc as the negative resistance. *See:* **radio transmission.**   240

arc, counterclockwise (numerically controlled machines). An arc generated by the coordinated motion of two axes in which curvature of the path of the tool with respect to the workpiece is counter-clockwise, when viewing the plane of motion in the negative direction of the perpendicular axis.   207

arc current (gas-tube surge-protective devices). The current that flows after breakdown when the circuit impedance allows a current that exceeds the glow-to-arc transition current. *Syn:* **arc mode current.**
     370, 490

arc discharge (1) (illuminating engineering). An electric discharge characterized by high cathode current densities and a low voltage drop at the cathode.
     167

(2) (nonlinear, active, and nonreciprocal waveguide components). Commonly refers to weakly ionized plasma created by a radio-frequency (rf) discharge in gas tubes, receiver protectors, or duplexers.   530

arc-discharge tube (valve). A gas-filled tube or valve in which the required current is that of an arc discharge.
     90

arc-drop loss (gas tube). The product of the instantaneous values of the arc-drop voltage and current averaged over a complete cycle of operation. *See:* **gas tube.**
     125

arc-drop voltage (gas tube). The voltage drop between the anode and cathode of a rectifying device during conduction. *See:* **electrode voltage (electron tube); tube voltage drop.**
     190

arc-extinguishing medium (fuse filler) (of a fuse) (power switchgear). Material included in the fuse to facilitate current interruption.   103, 443

arc furnace. An electrothermic apparatus the heat energy for which is generated by the flow of electric current through one or more arcs internal to the furnace. *See:* **electrothermics.**   328

arc gap (microwave receiver protector). *See:* **resonant gap.**

architectural design (software). (1) The process of defining a collection of hardware and software components and their interfaces to establish a framework for the development of a computer system. (2) The result of the architectural design process. *See:* **computer system; hardware; interfaces; process; software components.**
     434

architecture. *See:* **program architecture; system architecture.**

arcing chamber (expulsion-type arrester). The part of an expulsion-type arrester that permits the flow of discharge current to the ground and interrupts the follow current. *See:* **surge arrester (surge diverter).**
     302

arcing contacts (power switchgear). The contacts of a switching device on which the arc is drawn after the main (and intermediate, where used) contacts have parted.   103

arcing horn (power switchgear). One of a pair of diverging electrodes on which an arc is extended to the point of extinction after the main contacts of the switching device have parted. *Syn:* **arcing runners.**
     103

arcing runners. *See:* **arcing horn.**   103

arcing time (1)(protection and coordination of industrial and commercial power systems). The arcing time of a fuse is the time elapsing from the melting of the current-responsive element (such as the link) to the final interruption of the circuit. This time will be dependent upon such factors as voltage and reactance of the circuit. See figure below.   504

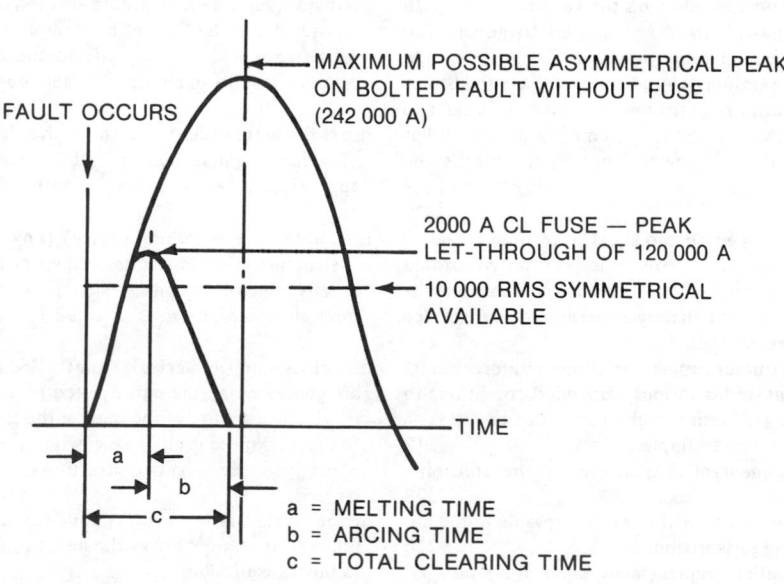

MAXIMUM POSSIBLE ASYMMETRICAL PEAK ON BOLTED FAULT WITHOUT FUSE (242 000 A)

FAULT OCCURS

2000 A CL FUSE — PEAK LET-THROUGH OF 120 000 A

10 000 RMS SYMMETRICAL AVAILABLE

TIME

a = MELTING TIME
b = ARCING TIME
c = TOTAL CLEARING TIME

(2)(power switchgear). (mechanical switching device). The interval of time between the instant of the first initiation of the arc and the instant of final arc extinction in all poles. *Note:* For switching devices that embody switching resistors, a distinction should be made between the arcing time up to the instant of the extinction of the main arc, and the arcing time up to the instant of the breaking of the resistance current. **(3) (fuse).** The time elapsing from the severance of the current-responsive element to the final interruption of the circuit. 443

arc loss (absorptive loss)(1)(nonlinear, active, and nonreciprocal waveguide components). Power absorbed in an active nonlinear device during above-threshold switching or limiting in gas tubes, duplexers, ferrite limiters, or diode limiters. 530
(2)(switching tube). The decrease in radio-frequency power measured in a matched termination when a fired tube, mounted in a series or shunt junction with a waveguide, is inserted between a matched generator and the termination. *Note:* In the case of a pretransmit-receive tube, a matched output termination is also required for the tube. *See:* **gas tube.** 125

arc mode current. *See:* **arc current.**

arc mode voltage. *See:* **arc voltage.**

arc-shunting-resistor-current arcing time (power switchgear). The interval between the parting of the secondary arcing contacts and the extinction of the arc-shunting-resistor current. 103

arc suppression (rectifier). The prevention of the recurrence of conduction, by means of grid or ignitor action, or both, during the idle period, following a current pulse. *See:* **rectification.** 291

arc-through (gas tube). A loss of control resulting in the flow of a principal electron stream through the rectifying element in the normal direction during a scheduled non-conducting period. *See:* **rectification.** 125

arc-tube relaxation oscillator. *See:* **gas-tube relaxation oscillator.**

arc voltage (gas-tube surge-protective devices). The voltage drop across the arrester duing arc current flow. *Syn:* **arc mode voltage.** 490

arc-welding engine generator. A device consisting of an engine mechanically connected to and mounted with one or more arc-welding generators. 264

arc-welding motor-generator. A device consisting of a motor mechanically connected to and mounted with one or more arc-welding generators. 264

area (volt time area) (1)(power fault effects). The area under a curve plotted with voltage versus time (or current versus time) between changes in polarity. This value is a measure of the cumulated flux in a transformer core only up to the first zero crossing and will not necessarily be indicative of the total degree of transformer core steel saturation. The volt time area that is generally of concern consists of the net accumulated volt time area that occurs during a certain number of power frequency cycles of the ground potential rise (GPR). The area is a function of the magni-

tude and decay rate of the off direct-current (dc) set. 404
(2)area *See:* **effective area of an antenna; equivalent flatplate area of a scattering object.**

area assist action (electric power system). The component of area supplementary control that involves the temporary assignment of generation changes to minimize the area control error prior to the assignment of generation changes on an economic dispatch basis. *See:* **speed-governing system.** 94

area code (telephone switching system). A one-, two-, or three-digit number that, for the purpose of distance dialing, designates a geographical subdivision of the territory covered by a separate national or integrated numbering plan. 55

area control error (isolated-power system consisting of a single control area). The frequency deviation (of a control area on an interconnected system) is the net interchange minus the biased scheduled-net interchange. *Note:* The above polarity is that which has been generally accepted by electric power systems and is in wide use. It is recognized that this is the reverse of the sign of control error generally used in servomechanism and control literature, which defines control error as the reference quantity minus the controlled quantity. 94,200

area frequency-response characteristic (control area). The sum of the change in total area generation caused by governor action and the change in total area load, both of which result from a sudden change in system frequency, in the absence of automatic control action. 94

areal beamwidth (antennas). For pencil-beam antennas the product of the two principal half-power beamwidths. *See:* **principal half-power beamwidth.** 111

area load-frequency characteristic (control area). The change in total area load that results from a change in system frequency. 94

area MTI (radar). A moving target indicator (MTI) system in which the moving target is selected on the basis of its change in time delay or angle between looks, rather than its Doppler frequency. 13

area supplementary control (electric power system). The control action applied, manually or automatically, to area generator speed governors in response to changes in system frequency, tie-line loading, or the relation of these to each other, so as to maintain the scheduled system frequency and/or the established net interchange with other control areas within predetermined limits. 94

area tie line (electric power system). A transmission line connecting two control areas. *Note:* Similar to interconnection tie. *See:* **transmission line.** 94

argand plane (ATLAS). A graphical representation of a vector used in complex notation. 400

arithmetic operations (test, measurement and diagnostic equip-ment). Operations in which nuumerical quantities form the elements of the calculation. 54

arithmetic reactive factor. The ratio of the reactive power to the arithmetic apparent power. 210

arithmetic shift (1)(mathematics of computing). A shift that affects all digit positions in a register, word, or numeral but does not affect the sign position. For example, $+231.702$ shifted two places to the left becomes $+170.200$. *Note:* The result is equivalent to multiplication or division by an integral power of the radix, except for the truncation effects. *See:* **logical shift.**   564

(2)(general) (1) A shift that does not affect the sign position. (2) A shift that is equivalent to the multiplication of a number by a positive or negative integral power of the radix.   255,54

arithmetic unit. The unit of a computing system that contains the circuits that perform arithmetic operations. *See:* **arithmetic element.**   77,255,54

arm. *See:* **branch; network analysis.**

armature (of a relay) (power switchgear). The moving element of an electromechanical relay that contributes to the designed response of the relay and which usually has associated with it a part of the relay contact assembly.   103

armature band (rotating machinery). A thin circumferential structural member applied to the winding of a rotating armature to restrain and hold the coils so as to counteract the effect of centrifugal force during rotation. *Note:* Armature bands may serve the further purpose of archbinding the coils. They may be on the end windings only or may be over the coils within the core.   63

armature band insulation (rotating machinery). An insulation member placed between a rotating armature winding and an armature band. *See:* **armature.**   63

armature bar (half coil) (rotating machinery). Either of two similar parts of an armature coil, comprising an embedded coil side and two end sections, that when connected together form a complete coil. *See:* **armature.**   63

armature coil (rotating machinery). A unit of the armature winding composed of one or more insulated conductors. *See:* **armature; asynchronous machine**   328

armature core (rotating machinery). A core on or around which armature windings are placed. *See:* **armature.**   63

armature I2R loss (synchronous machine). The sum of the I2R losses in all of the armature current paths. *Note:* The I2R loss in each current path shall be the product of its resistance in ohms, as measured with direct current and corrected to a specified temperature, and the square of its current in amperes.   63,298

armature quill. *See:* **armature spider.**

armature reaction (rotating machinery). The magnetomotive force due to armature-winding current.   63

armature-reaction excited machine. A machine having a rotatable armature, provided with windings and a commutator, whose load-circuit voltage is generated by flux that is produced primarily by the magnetomotive force of currents in the armature winding. *Notes:*

(1)By providing the stationary member of the machine with various types of windings different characteristics may be obtained, such as a constant-current characteristic or a constant-voltage characteristic.(2) The machine is normally provided with two sets of brushes, displaced around the commutator from one another, so as to provide primary and secondary circuits through the armature. (3) The primary circuit carrying the excitation armature current may be completed externally by a short-circuit connection, or through some other external circuit, such as a field winding or a source of power supply; and the secondary circuit is adapted for connection to an external load.   328

armature sleeve. *See:* **armature spider.**

armature spider (armature sleeve) (armature quill). A support upon which the armature laminations are mounted and which in turn is mounted on the shaft. *See:* **armature (rotating machinery).**   328

armature terminal (rotating machinery). A terminal connected to the armature winding. *See:* **armature.**   63

armature to field transfer function (G(s))(synchronous machine parameters by standstill frequency testing)(standstill frequency response testing). The ratio of the Laplace transform of the direct-axis armature flux linkages to the Laplace transform of the field voltage, with the armature open-circuited.   521

armature-voltage control. A method of controlling the speed of a motor by means of a change in the magnitude of the voltage impressed on its armature winding. *See:* **control.**   206

armature winding (rotating machinery). The winding in which alternating voltage is generated by virtue of relative motion with respect to a magnetic flux field. *See:* **asynchronous machine.**   63

armature winding cross connection. *See:* **armature winding equalizer.**

armature winding equalizer (armature winding cross connection)(rotating machinery.). An electric connection to normally equal-potential points in an armature circuit having more than two parallel circuits. *See:* **armature (rotating machinery).**   63

armed sweep. *See:* **single sweep mode.**

armor clamp (wiring methods). A fitting for gripping the armor of a cable at the point where the armor terminates or where the cable enters a junction box or other piece of apparatus.   64

armored cable (interior wiring). A fabricated assembly of insulated conductors and a flexible metallic covering. *Note:* Armored cable for interior wiring has its flexible outer sheath or armor formed of metal strip, helically wound and with interlocking edges. Armored cable is usually circular in cross section but may be oval or flat and may have a thin lead sheath between the armor and the conductors to exclude moisture, oil, etcetera, where such protection is needed. *See:* **nonmetallic sheathed cable.**   328

arm, thermoelectric. The part of a thermoelectric device in which the electric-current density and temperature gradient are approximately parallel or antiparallel and that is electrically connected only at its extremities to a part having the opposite relation between the direction of the temperature gradient and the electric-current density. *Note:* The term thermoelement is ambiguously used to refer to either a thermoelectric arm or to a thermoelectric couple,and its use is therefore not recommended. *See:* **thermoelectric device.** 248,191

arm, thermoelectric, graded. A thermoelectric arm whose composition changes continuously along the direction of the current density. *See:* **thermoelectric device.** 248,191

arm, thermoelectric, segmented. A thermoelectric arm composed of two or more materials having different compositions. *See:* **thermoelectric device.**
248,191

array (1) (photovoltaic converter). A combination of panels coordinated in structure and function. *See:* **semiconductor.** 186
(2) (solar cell). A combination of solar-cell panels or paddles coordinated in structure and function.
113

array antenna. An antenna comprised of a number of identical radiating elements in a regular arrangement and excited to obtain a prescribed radiation pattern. *Notes:* (1) The regular arrangements possible include ones in which the elements can be made congruent by simple translation or rotation. (2) This term is sometimes applied to cases where the elements are not identical or arranged in a regular fashion. For those cases qualifiers are added to distinguish from the usage implied in this definition. For example, if the elements are randomly located one may use the term "random array antennas". *Syn:* **antenna array.** 111

array control (terrestrial photovoltaic power systems). All electrical and mechanical controls that ensure proper electric and thermal performance of the array field. See figures 1 and 2. 496
array element (antennas). In an array antenna, a single radiating element or a convenient grouping of radiating elements that have fixed relative excitations.
111
array factor (antennas). The radiation pattern of an array antenna when each array element is considered to radiate isotropically. *Note:* When the radiation patterns of individual array elements are congruent under translation, then the product of the array factor and the element radiation pattern gives the radiation pattern of the entire array. 111
array field (terrestrial photovoltaic power systems). The aggregate of all array subfields. *See:* **array control**
496
array subsystem (terrestrial photovoltaic power systems). The array field and the controls that together produce dc electric and thermal energy. Associated thermal energy may be utilized or dissipated. *See:* **array control** 496
arrester alternating sparkover voltage. The root-

mean-square value of the minimum 60-hertz sine-wave voltage that will cause sparkover when applied between its line and ground terminals. *See:* **current rating, 60-hertz (arrester); surge arrester(surge diverter).** 62
arrester discharge capacity. The crest value of the maximum current of specified wave shape that the arrester can withstand without damage to any of its parts. *See:* **surge arrester (surge diverter).** 62
arrester discharge voltage-current characteristic. The variation of the crest values of discharge voltage with respect to discharge current. *Note:* This characteristic is normally shown as a graph based on three or more current surge measurements of the same wave shape but of different crest values. *See:* **lightning; surge arrester (surge diverter); current rating, 60-hertz (arrester).** 2,62
arrester discharge voltage-time curve. A graph of the discharge voltage as a function of time while discharging a current surge of given wave shape and magnitude. *See:* **surge arrester (surge diverter).** 62
arrester disconnector (1)(metal-oxide surge arresters for ac power circuits). A means for disconnecting an arrester in anticipation of, or after, a failure in order to prevent a permanent fault on the circuit and to give indication of a failed arrester. *Note:* Clearing of the fault current through the arrester during disconnection is generally done by the nearest source-side overcurrent-protective device. 583
(2)(surge arrester)(Same definition except for the following note.). (*Note:* Clearing of the power circuit through the arrester during disconnection generally is a function of the nearest source-side overcurrent-protective device. 430
arrester, expulsion-type. An arrester having an arcing chamber in which the follow-current arc is confined and brought into contact with gas-evolving or other arc-extinguishing material in a manner that results in the limitation of the voltage at the line terminal and the interruption of the follow current. *Note:* The term **expulsion arrester** includes any external series-gap or current-limiting resistor if either or both are used as a part of the complete device as installed for service.
308
arrester ground. An intentional electric connection of the arrester ground terminal to the ground. *See:* **surge arrester (surge diverter).** 62
arresters, classification of. Arrester classification is determined by prescribed test requirements. These classifications are: station valve arrester; intermediate valve arrester; secondary valve arrester; protector tube. 62
arrester, valve-type. An arrester having a characteristic element consisting of a resistor with a nonlinear volt-ampere characteristic that limits the follow current to a value that the series gap can interrupt. *Note:* If the arrester has no series gap the characteristic element limits the follow current to a magnitude that does not interfere with the operation of the system. *See:* **nonlinear-resistor type arrester (valve type); surge arrester (surge diverter).** 62

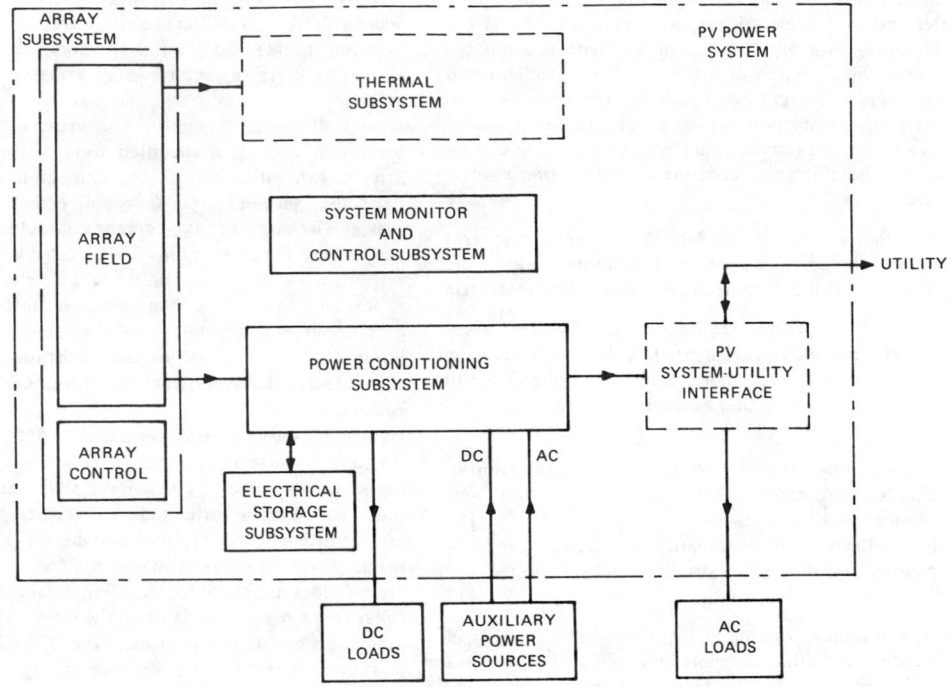

Figure 1. Block and Interface Diagram of Photovoltaic/Thermal Power System

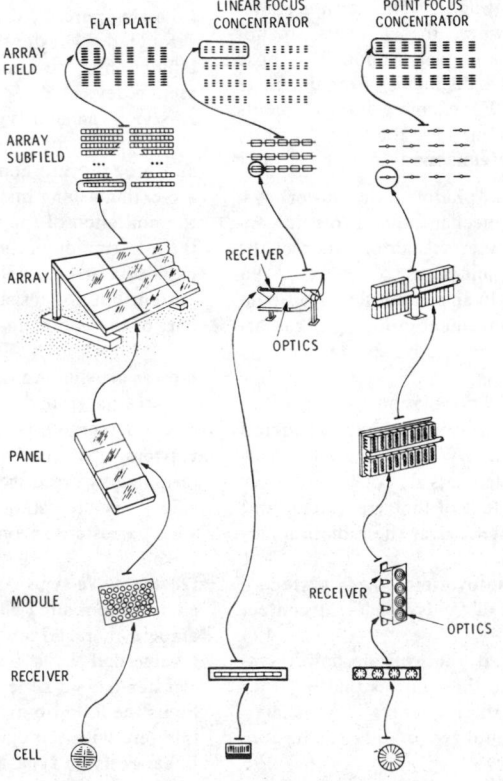

Figure 2. Illustration of Defined Array Terms

ARSR. *See:* **air-route surveillance radar.**

articulated unit substation (power switchgear) (1)power and distribution transformer.A unit substation in which the incoming, transforming, and outgoing sections are manufactured as one or more subassemblies intended for connection in the field. (2) radial type. One which has a single stepdown transformer and which has an outgoing section for the connection of one or more outgoing radial (stub end) feeders. (3) distributed-network type. One which has a single stepdown transformer having its outgoing side connected to a bus through a circuit breaker equipped with relays which are arranged to trip the circuit breaker on reverse power flow to the transformer and to reclose the circuit breaker upon the restoration of the correct voltage, phase angle and phase sequence at the transformer secondary. The bus has one or more outgoing radial (stub end) feeders and one or more tie connections to a similar unit substation.(4) spot-network type. One which has two stepdown transformers, each connected to an incoming high-voltage circuit. The outgoing side of each transformer is connected to a common bus through circuit breakers equipped with relays which are arranged to trip the circuit breaker on reverse power flow to the transformer and to reclose the circuit breaker upon the restoration of the correct voltage, phase angle and phase sequence at the transformer secondary. The bus has one or more outgoing radial (stub end) feeders. (5) secondary-selective type (low-voltage-selective type). One which has two stepdown transformers,each connected to an incoming high-voltage circuit. The outgoing side of each transformer is connected to a separate bus through a suitable switching and protective device. The two sections of bus are connected by a normally open switching and protective device. Each bus has one or more outgoing radial (stub end) feeders. (6) duplex type (breaker-and-a-half arrangement). One which has two stepdown transformers, each connected to an incoming high-voltage circuit. The outgoing side of each transformer is connected to a radial (stub end) feeder. These feeders are joined on the feeder side of the power circuit breakers by a normally-open-tie circuit breaker.   53

articulation (percent articulation) and intelligibility (percent intelligibility). The percentage of the speech units spoken by a talker or talkers that is correctly repeated, written down, or checked by a listener or listeners. *Notes:* (1) The word articulation is used when the units of speech material are meaningless syllables or fragments; the word intelligibility is used when the units of speech material are complete, meaningful words, phrases, or sentences. (2)It is important to specify the type of speech material and the units into which it is analyzed for the purpose of computing the percentage. The units may be fundamental speech sounds, syllables, words, sentences, etcetera. (3) The percent articulation or percent intelligibility is a property of the entire communication system; talker, transmission equipment or medium, and listener. Even when attention is focused upon one component

of the system (for example, a talker, a radio receiver), the other components of the system should be specified. (4) The kind of speech material used is identified by an appropriate adjective in phrases such as syllable articulation, individual sound articulation, vowel (or consonant) articulation, monosyllabic word intelligibility, discrete word intelligibility, discrete sentence intelligibility. *See:* **volume equivalent.**   176

articulation equivalent (complete telephone connection). A measure of the articulation of speech reproduced over it. The articulation equivalent of a complete telephone connection is expressed numerically in terms of the trunk loss of a working reference system when the latter is adjusted to give equal articulation. *Note:* For engineering purposes, the articulation equivalent is divided into articulation losses assignable to (1) the station set, subscriber line, and battery supply circuit that are on the transmitting end.(2) the station set, subscriber line, and battery supply circuit that are on the receiving end, (3) the trunk, and (4) interaction effects arising at the trunk terminals. *See:* **volume equivalent.**   328

artificial antenna (dummy antenna). A device that has the necessary impedance characteristics of an antenna and the necessary power-handling capabilities, but which does not radiate or receive radio waves. *See:* **antenna.**   111

artificial dielectric (antennas). A medium containing a distribution of scatterers, usually metallic, which react as a dielectric to radio waves. *Notes:* (1) The scatterers are usually small compared to a wavelength and embedded in a dielectric material whose effective permittivity and density are intrinsically low. (2) The scatterers may be in either a regular arrangement or a random distribution.   111

artificial ear (1)(transmission performance of telephone sets). A device for the measurement of the acoustic output of telephone-set receivers. It presents to the receiver an acoustic impedance approximating the impedance presented by the average human ear.
                                                        491

(2) (general). A device for the measurement of the acoustic output of earphones in which the artificial ear presents to the earphone an acoustic impedance approximating the impedance presented by the average human ear and is equipped with a microphone for measurement of the sound pressures developed by the earphone.   196,176

artificial hand (electromagnetic compatibility). A device simulating the impedance between an electric appliance and the local earth when the appliance is grasped by the hand. *See:* **electromagnetic compatibility.**   220, 199

artificial horizon (navigation aid terms). (1) A device for indicating the horizontal, as bubble, gyroscope, pendulum, or the flat surface of a liquid. (2) A gyroscopic flight instrument that shows the pitching and banking attitudes of a vehicle with respect to a reference line horizon. *Syn:* **gyro horizon.**   526

artificial language. *See:* **formal language.**

artificial line (1) (antennas) (data transmission). An

electric network that simulates the electrical characteristics of a line over a desired frequency range. *Note:* Although the term basically is applied to the case of simulation of an actual line, by extension it is used to refer to all periodic lines that may be used for laboratory purposes in place of actual lines, but that may represent no physically realizable line. For example, an artificial line may be composed of pure resistances. *See:* **network analysis.** 111, 59

(2) **(waveguide).** A network that simulates the electrical characteristics of a transmission line over a given frequency range. 267

artificial load. A dissipative but essentially nonradiating device having the impedance characteristics of an antenna, transmission line, or other practical utilization circuit. 111,59

artificial mains network (electromagnetic compatibility). A network inserted in the supply mains lead of the apparatus to be tested that provides a specified measuring impedance for interference voltage measurements and isolates the apparatus from the supply mains at radio frequencies. *See:* **electromagnetic compatibility.** 220, 199

artificial mouth (transmission performance of telephone sets). An electroacoustic transducer that produces a sound field simulating that of a typical human talker. The reference point for the handset and the headset is the center of the circular plane of contact of the handset ear-cap and the ear. If the handset ear-cap is not circular or has no external plane of contact, an effective center and an effective plane of contact is determined. 491

artificial pupil (illuminating engineering). A device or arrangement for confining the light passing through the pupil of the eye to an area smaller than the natural pupil. 167

artificial voice (close-talking pressure-type microphone). A sound source for microphone measurements consisting of a small loudspeaker mounted in a shaped baffle proportioned to simulate the acoustic constants of the human head. *See:* **close-talking pressure-type microphone; loudspeaker.** 249

as built curve (rotating electric machinery). A curve that is found on an individual machine during testing. *Syn:* **manufactured curve.** 424

A **scan (electronic navigation).** *See: A* **and *R* display.**

ascending node (communication satellite). The point on the line of nodes that the satellite passes through as the satellite travels from below to above the equatorial plane. 74

A-scope (radar). A cathode-ray oscilloscope arranged to present an A-display. 13

ASDE. *See:* **airport surface detection equipment.** 13

ash layer porosity(cm^3/cm^3)**(fly ash resistivity).** The ratio of the ash layer void volume to the test cell volume in a test cell used for the laboratory measurement of fly ash resistivity. 427

askarel (1)(handling and disposal of transformer grade insulating liquids containing PCBs). A generic term for a group of synthetic, fire-resistant, chlorinated

aromatic hydrocarbons used as electrical insulating liquid. They have a property under arcing conditions such that any gases produced will consist predominantly of noncombustible hydrogen chloride with lesser amounts of combustible gases. Askarel does not necessarily contain polychlorinated biphenyl (PCBs). 586

(2)**(insulating oil).** A generic term for a group of synthetic, fire-resistant, chlorinated aromatic hydrocarbons used as electrical insulating liquids. They have a property under arcing conditions so that any gases produced will consist predominantly of noncombustible hydrogen chloride with lesser amounts of combustible gases. The following trade names are some of the titles which have been used to identify askarels containing polychlorinated biphenyls (PCBs): Aroclor, Elemex, Chlorextol, Hyvol, Chlorphen, Inerteen, Diaclor, Noflamol, Dykanol, Pyranol. 461

(3) **(National Electrical Code).** A generic term for a group of nonflammable synthetic chlorinated hydrocarbons used as electrical insulating media. Askarels of various compositional types are used. Under arcing conditions the gases produced, while consisting predominantly of noncombustible hydrogen chloride, can include varying amounts of combustible gases depending upon the askarel type. 256

(4) **(power and distribution transformer).** A generic term for a group of synthetic, fire-resistant, chlorinated, aromatic hydrocarbons used as electrical insulating liquids. They have a property under arcing conditions such that any gases produced will consist predominantly of noncombustible hydrogen chloride with lesser amounts of combustible gases. 53

aspect ratio (television). (1) The ratio of the frame width to the frame height. The ratio of the frame width to the frame height as defined by the active picture. 18,372

asphalt (rotating machinery). A dark brown to black cementitious material, solid or semisolid in consistency, in which the predominating constituents are bitumens that occur in nature as such or are obtained as residue in refining of petroleum. 63

ASR. *See:* **airport surveillance radar.**

assay (sodium iodide detector). The determination of the activity of a radionuclide in a sample. 423

assemble (software). To translate a program expressed in an assembly language into a machine language and perhaps to link subroutines. Assembling is usually accomplished by substituting the machine language operation code for the assembly language operation code and by substituting absolute addresses, immediate addresses, relocatable addresses, or virtual addresses for symbolic addresses. *See:* **assembly language; compile; interpret; machine language; program; subroutines.** 434

assembler (1)(microprocessor assembly language). A utility program which translates symbolic assembly language instructions into machine instructions or data on a one-to-one basis. 466

(2) **(software).** A computer program used to assemble. *Syn:* **assembly program.** *See:* **assemble; compiler; computer program; interpreter.** 434

(3) (test, measurement and diagnostic equipment). A computer program that is one step more automatic than a translator: it translates not only operations but also data and input-output quantities from symbolic to machine language form in a one to one ratio. An assembler program may have the capability to assign locations within a storage device.   54

assembly (1) (electric and electronics parts and equipments). A number of basic parts or subassemblies, or any combination thereof, joined together to perform a specific function. The application, size, and construction of an item may be factors in determining whether an item is regarded as a unit, an assembly, a subassembly, or a basic part. A small electric motor might be considered as a part if it is not normally subject to disassembly. The distinction between an assembly and a subassembly is not always exact: an assembly in one instance may be a subassembly in another where it forms a portion of an assembly. Typical examples are: electric generator, audio-frequency amplifier, power supply.   17
(2) (nuclear power generating stations) (seismic qualification of class 1E equipment). Two or more devices sharing a common mounting or supporting structure. *Note:* Examples are control panels and diesel generators.   31,28

assembly language (software). (1) A computer-oriented language whose instructions are usually in one-to-one correspondence with computer instructions and that may provide facilities such as the use of macroinstructions. (2) A machine-specific language whose instructions are usually in one-to-one correspondence with computer instructions. *See:* **assemble; assembler; computer; higher order language; instructions; machine language; macroinstructions.**   434

assembly, microelectronic device (electric and electronics parts and equipments). An assembly of inseparable parts, circuits, or combination thereof. Typical examples are: microcircuit, integrated-circuit package, micromodule.   17

assert (signals and paths)(696 interface devices). To cause a signal line to transition from its logically false (inactive) state to its logically true (active) state. The true or active state is either a high or low state, as specified for each signal.   538

assertion (software). A logical expression specifying a program state that must exist or a set of conditions that program variables must satisfy at a particular point during program execution, for example, A is positive and A is greater than B. *See:* **execution; input assertion; output assertion; program; variables.**   434

assessed mean active maintenance time. The active maintenance time determined as the limit or the limits of the confidence interval associated with a stated confidence level, and based on the same data as the

observed mean active maintenance time or nominally identical items. *Notes:* (1) The source of the data shall be stated. (2) Results can be accumulated (combined) only when all conditions are similar. (3) It should be stated whether a one-sided or two-sided interval is being used. (4) The assumed underlying distribution of mean active maintenance times shall be stated with the reason for the assumption. (5) When one value is given this is usually the upper limit.   164

assessed reliability. The reliability of an item determined by a limiting value or values of the confidence interval associated with a stated confidence level, based on the same data as the observed reliability of nominally identical items. *Notes:* (1) The source of the data shall be stated. (2) Results can be accumulated (combined) only when all conditions are similar. (3) The assumed underlying distribution of failures against time shall be stated. (4) It should be stated whether a one-sided or a two-sided interval is being used. (5) Where one limiting value is given this is usually the lower limit.   164

assigned value. The best estimate of the value of a quantity. The assigned value may be from an instrument reading, a calibration result, a calculation, or other.   115, 47
assignment statement (software). An instruction used to express a sequence of operations, or used to assign operands to specified variables, or symbols, or both.   434
assistance call (telephone switching system). A call to an operator for help in making a call.   55

associated circuits (1)(Class 1E equipment and circuits). Non-Class 1E circuits that are not physically separated or are not electrically isolated from Class 1E circuits by acceptable separation distance, safety Class structures, barriers, or isolation devices. *Note:* Circuits include the interconnecting cabling and the connected loads as defined in ANSI/IEEE Std 100-1977.   131
(2) (design and installation of cable systems for Class 1E circuits in nuclear power generating stations). Non Class 1E circuits that share power supplies, signal sources, enclosures, or raceways with Class 1E circuits or are not physically separated or electrically isolated from Class 1E circuits by acceptable separation distance, barriers, or isolation devices.   536

associated equipment (packaging machinery). Any attachment or component part that is not necessarily located on the packaging machine but is directly associated with the performance of the machine. Limit switches and photoelectric devices are examples.   429
associated structural parts (insulation systems of synchronous machines). The associated structural parts of the installation system include the field collars, the slot wedges, the filler strips under the support ring insulation, the nonmetallic support for the winding, the space blocks used to separate the coil ends and

connections, the lead cleats, and the terminal boards.
298

associative storage. A storage device in which storage locations may be identified by specifying part or all of their contents. *Note:* Also called parallel-search storage or content-addressed storage. 235

assumed position (navigation aid terms). A point at which a craft is assumed to be located. 526

assured access protocol (FASTBUS acquisition and control). A potential master is operating in the assured access protocol if, on detecting an arbitration request inhibit (AI) assertion, it will not assert arbitration request (AR) and thus will not participate in subsequent arbitration cycles until all devices currently asserting AR have obtained bus mastership and completed their operations. 480

assured disruptive discharge voltage (high voltage testing). The assured disruptive discharge voltage is a specified voltage value which characterizes the insulation of an object with regard to a disruptive discharge test. 150

astern (navigation aid terms). Bearing approximately 180° relative. 526

aster rectifier circuit. A circuit that employs twelve or more rectifying elements with a conducting period of 30 electrical degrees plus the commutating angle. *See:* **rectification.** 328

astigmatism (electron optical). In an electron-beam tube, a focus defect in which electrons in different axial planes come to focus at different points. *See:* **oscillograph.** 125

Aston dark space (gas). The dark space in the immediate neighborhood of the cathode, in which the emitted electrons have a velocity insufficient to excite the gas. *See:* **discharge (gas).** 244, 190

astrocompass (navigation aid terms). An instrument which, when oriented to the horizontal and the celestial sphere, indicates horizontal reference direction relative to the earth. It is used to obtain true heading by reference to celestial bodies. 526

astrodynamics (communication satellite). Engineering application of celestial mechanics. 74

astro-inertial navigation equipment. *See:* **celestial-inertial navigation equipment.**

astronomical position (navigation aid terms). (1) A point on the earth where coordinates have been determined as a result of the observation of celestial bodies. (2) A point on the earth defined in terms of astronomical latitude and longitude. 526

astronomical unit (communication satellite). Abbreviated AU: the mean distance between the centers of the sun and the earth, 149.6×10^6 kilometers, 92.98×10^6 miles or 80.78×10^6 nautical miles. 74

astronomical unit of distance. The length of the radius of the unperturbed circular orbit of a body of negligible mass moving around the sun with a sidereal angular velocity of 0.017 202 098 950 radian per day of 86 400 ephemeris seconds. In the system of astronomical constants of the International Astronomical Union the value adopted for it is 1 AU = $149\,600 \times 10^6$ m. 21

astrotracker (navigation aid terms). An automatic sextant which has the ability to sight on and track selected stars throughout the day and night, providing heading and position data. The tracker may be optical or radiometric. *Syn:* **star tracker.** 526

A switchboard (telephone switching systems). A telecommunications switchboard in a local central office, used primarily to extend calls received from local stations. 55

asymmetrical cell. A cell in which the impedance to the flow of current in one direction is greater than in the other direction. *See:* **electrolytic capacitor.** 328

asymmetrical current. *See:* **total (asymmetrical) current.** 103

asymmetric terminal voltage (electromagnetic compatibility). Terminal voltage measured with a delta network between the midpoint of the resistors across the mains lead and ground. *See:* **electromagnetic compatibility.** 220, 199

asynchronous computer. A computer in which each event or the performance of each operation starts as a result of a signal generated by the completion of the previous event or operation, or by the availability of the parts of the computer required for the next event or operation. 77, 255

asynchronous impedance (rotating machinery). The quotient of the voltage, assumed to be sinusoidal and balanced, supplied to a rotating machine out of synchronism, and the same frequency component of the current. *Note:* The value of this impedance depends on the slip. *See:* **asynchronous machine.** 63

asynchronous machine (1) (rotating machinery). An alternating-current machine in which the rotor does not turn at synchronous speed. 63
(2) (electric installations on shipboard). A machine in which the speed of operation is not proportional to the frequency of the system to which it is connected. 3

asynchronous operation (rotating machinery). Operation of a machine where the speed of the rotor is other than synchronous speed. *See:* **asynchronous machine.** 63

asynchronous reactance (rotating machinery). The quotient of the reactive component of the average voltage at rated frequency, assumed to be sinusoidal and balanced, applied to the primary winding of a machine rotating out of synchronism, and the average current component at the same frequency. 63

asynchronous resistance (rotating machinery). The quotient of (a) the active component of the average voltage at rated frequency assumed to be sinusoidal and balanced, applied to the primary winding of a machine rotating out of synchronism, and (b) the average current component at the same frequency. 63

asynchronous transmission. *See:* **nonsynchronous transmission.**

ATC (radar). Abbreviation for air traffic control. 13

ATE (automatic test equipment) control software (test, measurement and diagnostic equipment). Software used during execution of a test program which controls the nontesting operations of the ATE. This software is used to execute a test procedure but does not contain any of the stimuli or measurement parameters used in testing the Unit Under Test (UUT). Where test software and control software are combined in one inseparable program, that program will be treated as test software not control software.   54

ATE (automatic test equipment) oriented language (test, measurement and diagnostic equipment). A computer language used to program an automatic test equipment to test units under test (UUT's), whose characteristics imply the use of a specific ATE system or family of ATE systems.   54

ATE (automatic test equipment) support software (test, measurement and diagnostic equipment). Computer programs which aid in preparing, analyzing, and maintaining test software. Examples are: ATE compilers, translation.analysis programs, and punch/print programs.   54

ATLAS. A standard abbreviated English language used in the preparation and documentation of test procedures or test programs which can be implemented either manually or with automatic or semiautomatic test equipment. The ATLAS language is defined in ANSI/IEEE Std 416-1978, which includes the material previously published in two separate volumes, namely, a general or functional definition and a formal definition written in a metanotation. *See:* **nonpreferred ATLAS.**   400

ATLAS compiler. A program that converts high-order ATLAS statements into executable machine code.   400

ATLAS vocabulary. The range of words and symbols used in standard ATLAS.   400

atmospheric absorption (1) (general). The loss of energy in transmission of radio waves, due to dissipation in the atmosphere. *See:* **radiation.**   328
(2) (communication satellite). Absorption, by the atmosphere, of electromagnetic energy traversing it.   85

atmospheric condition monitor (45) (power system device function numbers). A device that functions upon the occurrence of an abnormal atmospheric condition, such as damaging fumes, explosive mixtures, smoke, or fire.   402

atmospheric duct (radio wave propagation). A layer in the troposphere within which radio waves of sufficiently high frequency propagate with low rate of decrease of amplitude with distance. The duct extends from the level of a local minimum of the modified index of refraction as a function of height down to the level where the minimum value is again encountered, or down to the surface bounding the atmosphere if the minimum value is not again encountered.   146

atmospheric noise (communication satellite). Noise radiated by the atmosphere into a space communications receiver antenna.   85

atmospheric paths (atmospheric correction factors to dielectric tests). Paths entirely through atmospheric air, such as along the porcelain surface of an outdoor bushing.   50

atmospheric radio noise (electromagnetic compatibility)(control of system electromagnetic compatibility). Noise having its source in natural atmospheric phenomena. *See:* **electromagnetic compatibility.**   199,495

atmospheric radio wave. A radio wave that is propagated by reflection in the atmosphere. *Note:* It may include either the ionospheric wave or the tropospheric wave, or both. *See:* **radiation.**   328

atmospherics (radio wave propagation). Transient bursts of electromagnetic radiation arising from natural electrical disturbances in the lower atmosphere. (The term "static" to include atmospherics and other radio noise was in early usage.)   146
See: **static (atmospherics).**

atmospheric transmissivity. The ratio of the directly transmitted flux incident on a surface after passing through unit thickness of the atmosphere to the flux which would be incident on the same surface if the flux had passed through a vacuum.   167

A-trace (loran)(navigation aid terms). The first (upper) trace on the scope display.   526

ATR box. *See:* **anti-transmit-receive box.**

ATR switch. *See:* **anti-transmit-receive switch.**   13

ATR tube. *See:* **anti-transmit-receive tube.**

attached slave (FASTBUS acquisition and control). One that in the previous primary address cycle recognized its address and address type and as a result will participate in the ensuing data cycles.   480

attachment (electric and electronics parts and equipments). A basic part, subassembly, or assembly designed for use in conjunction with another assembly, unit, or set, contributing to the effectiveness thereof by extending or varying the basic function of the assembly, unit, or set. A typical example is: ultra-high-frequency (UHF) converter for very-high-frequency (VHF) receiver.   17

attachment plug (plug cap) (cap) (electrical installations on shipboard)(National Electrical Code). A device which, by insertion in a receptacle, establishes connection between the conductors of the attached (flexible) cord and the conductors connected permanently to the receptacle.   3, 256

attachments (power switchgear). Accessories to be attached to switchgear apparatus, as distinguished from auxiliaries.   103

attachment unit interface (AUI)(medium attachment units and repeater units). The cable, connectors, and transmission circuitry used to connect the physical signaling (PLS) and medium attachment unit (MAU).   543

attack time (electroacoustics). The interval required, after a sudden increase in input signal amplitude to a system or transducer, to attain a stated percentage (usually 63 percent) of the ultimate change in amplification or attenuation due to this increase.   239

attendant (telephone switching systems). A private branch exchange or centrex operator.   55

attenuating pad. *See:* **pad.**

attenuation (1) (data transmission). A general term

used to denote a decrease in signal magnitude in transmission from one point to another. Attenuation may be expressed as a scalar ratio of the input magnitude to the output magnitude or in decibels. 59
(2) (fiber optics). In an optical waveguide, the diminution of average optical power. *Note:* In optical waveguides, attenuation results from absorption, scattering, and other radiation. Attenuation is generally expressed in decibels (dB). However, attenuation is often used as a synonym for attenuation coefficient, expressed as dB/km. This assumes the attenuation coefficient is invariant with length. *See:* **attenuation coefficient; coupling loss; differential mode attenuation; equilibrium mode distribution; extrinsic joint loss; leaky modes; macrobend loss; material scattering; microbend loss; Rayleigh scattering; spectral window; transmission loss; waveguide scattering.**
433
(3) (laser-maser). The decrease in the radiant flux as it passes through an absorbing or scattering medium.
363
(4) (radio wave propagation). Of a quantity associated with a traveling wave in a slowly varying medium in which the refractive index changes are small over distances comparable to a vacuum wave-length, the decrease of its amplitude with increasing distance from the source, excluding the decrease due to spreading.
146
(5) (waveguide) (quantity associated with a traveling waveguide or transmission-line wave). The decrease with distance in the direction of propagation. *Note:* Attenuation of power is usually measured in terms of decibels or decibels per unit length. *See:* **loss.**
166, 267
(6) (control systems). (A) A decrease in signal magnitude between two points, or between two frequencies. (B) The reciprocal of gain. *Note:* It may be expressed · as a scalar ratio or in decibels as 20 times the log of that ratio. A decrease with time is usually called damping or "subsidence." *See* **subsidence ratio.**
56
(7) (waveguide. *See:* **loss).** 267
attenuation band (uniconductor waveguide). Rejection band. *See:* **waveguide.** 166, 267
attenuation coefficient (fiber optics). The rate of diminution of average optical power with respect to distance along the waveguide. Defined by the equation

$$P(z) = P(0)10^{(az/10)}$$

where P(z) is the power at distance z along the guide and P(0) is the power at $z = 0$; a is the attenuation coefficient in dB/km if z is in km. From this equation,

$$az = -10 \ \log_{10}[P(z)/P(0)].$$

This assumes that a is independent of z; if otherwise, the definition must be given in terms of incremental attenuation as:

$$P(z) = P(0)10^{-[\int_0^z a(z)dz} \ \text{over} \ 10]$$

or, equivalently,

$$a(z) = -10d/dz \ \log_{10}[P(z)/P(0)]$$

See: **attenuation; attenuation constant; axial propagation constant.** 433
attenuation constant (1) (general). The real part of the propagation constant. *Note:* Unit: Neper per unit length. (1 neper equals 8.686 decibels). *See:* **radio transmission.** 146
(2) (fiber optics). For a particular mode, the real part of the axial propagation constant. The attenuation coefficient for the mode power is twice the attenuation constant. *See:* **attenuation coefficient; axial propagation constant; propagation constant.** 433
(3) (waveguide). The rate of decrease in amplitude of a field component (or of voltage or current) of a traveling wave in a uniform transmission line at a given frequency in the direction of propagation as a function of distance; the real part of the propagation constant.
267
attenuation, current. Either (1) a decrease in signal current magnitude, in transmission from one point to another, or the process thereof, or (2) of a transducer, the scalar ratio of the signal input current to the signal output current. *Note:* By incorrect extension of the term decibel, this ratio is sometimes expressed in decibels by multiplying its common logarithm by 20. It may be correctly expressed in decilogs. *See:* **decibel; attenuation.** 239
attenuation distortion (frequency distortion) (data transmission). Either (1) departure in a circuit or system from uniform amplification or attenuation over the frequency range required for transmission, or (2) the effect of such departure on a transmitted signal.
59
attenuation equalizer (data transmission). A corrective network which is designed to make the absolute value of the transfer impedance, with respect to two chosen pairs of terminals, substantially constant for all frequencies within a desired range. 59
attenuation-limited operation (fiber optics). The condition prevailing when the received signal amplitude (rather than distortion) limits performance. *See:* **bandwidth-limited operation; distortion-limited operation.** 433
attenuation ratio (radio wave propagation). The magnitude of the propagation ratio. 146
attenuation vector (field quantity) (radio wave propagation). Of a field quantity, the vector pointing in the direction of maximum decrease of amplitude. The magnitude of this vector is the attenuation constant.
146
attenuation vector in physical media (antennas). The real part of the propagation vector. 111
attenuation, voltage. (1) (data transmission). An adjustable device for reducing the amplitude of a wave without introducing distortion. An adjustable passive network that reduces the power level of a signal without introducing appreciable distortion. 59
(2) (analog computer). A device for reducing the amplitude of a signal without introducing appreciable distortion. 9

attenuator tube (electron tubes). A gas-filled radio-frequency switching tube in which a gas discharge, initiated and regulated independently of radio-frequency power, is used to control this power by reflection or absorption. *See:* **gas tube.** 190, 125

attenuator, waveguide (waveguide components). A waveguide component that reduces the output power relative to the input, by any means, including absorption and reflection. *See:* **waveguide.** 166

attitude (1)(navigation aid terms). The position of a body as determined by the inclination of the axes to some frame of reference. 526

(2)(communication satellite). Orientation of a satellite vehicle with respect to a reference coordinate system. Deviations of the satellite axes from the reference system are called roll, pitch and yaw. The reference system is generally an orbital reference system with the x-axis (roll axis) in the orbital plane in direction of the satellite motion, the y-axis (pitch axis) normal to the orbital plane and the z-axis (yaw axis) in the orbital plane in direction of the center of the earth. 74

attitude and heading-reference system (AHRS)(navigation aid terms). *See:* **heading and attitude reference.** 526

attitude control (navigation aid terms). Devices or system that automatically regulates and corrects attitude. 526

attitude-effect error (navigation)(navigation aid terms). A manifestation of polarization error; an error in indicated bearing that is dependent upon the attitude of the vehicle with respect to the direction of signal propagation. *See:* **heading effect error.** 526

attitude gyro-electric indicator. An electrically driven device that provides a visual indication of an aircraft's roll and pitch attitude with respect to the earth. *Note:* It is used in highly maneuverable aircraft and differs from the gyro-horizon electric indicator in that the gyro is not limited by stops and has complete freedom about the roll and pitch axes. 328

attitude stabilized satellite (communication satellite). A satellite with at least one axis maintained in a specified direction, namely toward the center of the earth, the sun or a specified point in space. 74

attitude storage (gyro)(inertial sensor). The transient deviation of the output of a rate integrating gyro from that of an ideal integrator when the gyro is subjected to an input rate. It is a function of the gyro characteristic time. *See:* **float storage; torque command storage.** 46

audible busy signal (busy tone). An audible signal connected to the calling line to indicate that the called line is in use. 328

audible cab indicator. A device (usually an air whistle, bell, or buzzer) located in the cab of a vehicle equipped with cab signals or continuous train control designed to sound when the cab signal changes to a more restrictive indication. 328

audible signal device (protective signaling). A general term for bells, buzzers, horns, whistles, sirens, or other devices that produce audible signals. *See:* **protective signaling.** 328

audio (data transmission). Pertaining to frequencies corresponding to a normally audible sound wave. *Note:* These frequencies range roughly from 15 hertz to 20 000 hertz. 59

audio frequency (1) (general). Any frequency corresponding to a normally audible sound wave. *Notes:* (A) Audio frequencies range roughly from 15 hertz to 20 000 hertz. (B) This term is frequently shortened to audio and used as a modifier to indicate a device or system intended to operate at audio frequencies, for example, audio amplifier. 111, 176, 197

(2) (interference terminology). Components of noise having frequencies in the audio range. *See:* **signal.** 188

audio-frequency distortion. The form of wave distortion in which the relative magnitudes of the different frequency components of the wave are changed on either a phase or amplitude basis. 181

audio-frequency harmonic distortion. The generation in a system of integral multiples of a single audio-frequency input signal. *See:* **modulation.** 111

audio-frequency noise. *See:* **noise, audio frequency.**

audio-frequency oscillator (audio oscillator). A nonrotating device for producing an audio-frequency sinusoidal electric wave, whose frequency is determined by the characteristics of the device. *See:* **oscillatory circuit.** 239

audio-frequency peak limiter. A circuit used in an audio-frequency system to cut off peaks that exceed a predetermined value. 111

audio-frequency response (receiver performance). The measure of the relative departure of all audio-frequency signal levels within a specified bandwidth, from a specified reference frequency signal power level. 123

audio-frequency spectrum (audio spectrum). The continuous range of frequencies extending from the lowest to the highest audio frequency. 239

audio-frequency transformer. A transformer for use with audio-frequency currents. 197

audiogram (threshold audiogram). A graph showing hearing level as a function of frequency. 176

audio input power (transmitter performance). The input power level to the modulator, expressed in decibels referred to a 1 milliwatt power level. *See:* **audio-frequency distortion.** 181

audio input signal (transmitter performance). That composite input to the transmitter modulator that consists of frequency components normally audible to the human ear. *See:* **audio-frequency distortion.** 181

audiometer. An instrument for measuring hearing level. *See:* **instrument.** 176

audio oscillator. *See:* **audio-frequency oscillator.**

audio output power (receiver performance) (receiver). The audio-frequency power dissipated in a load across the output terminals. *See:* **receiver performance.** 181

audio power output (receiver performance). The measure of the audio-frequency energy dissipated in a specified output load. 123

audio-tone channel. *See:* **voice-frequency carrier-telegraph.**

audit (1) (nuclear power quality assurance). A planned and documented activity performed to determine by investigation, examination, or evaluation of objective evidence the adequacy of and compliance with established procedures, instructions, drawings, and other applicable documents, and the effectiveness of implementation. An audit should not be confused with surveillance or inspection activities performed for the sole purpose of process control or product acceptance.   417

(2) (software). (A) An independent review for the purpose of assessing compliance with software requirements, specifications, baselines, standards, procedures, instructions, codes, and contractual and licensing requirements. (B) An activity to determine through investigation the adequacy of, and adherence to, established procedures, instructions, specifications, codes, and standards or other applicable contractual and licensing requirements, and the effectiveness of implementation. *See:* **baselines; code audit; implementation; instructions; procedures; software requirements; specifications.**   434

auditable data (1)(safety systems equipment in nuclear power generating stations). Information which is documented and organized in a readily understandable and traceable manner that permits independent auditing of inferences or conclusions based on the information. *Note:* Examples of information include product catalog information, dimensional drawings, bills of material, engineering specifications, performance specifications, installation and calibration instructions and manuals, maintenance manuals, test reports and analyses.   535

(2)(valve actuators). Information which is documented and organized in a readily understandable and traceable manner that permits independent verification of inferences or conclusions based on the information. *Note:* Examples of information include product catalog information, dimensional drawings, bills of material, engineering specifications, installation and calibration instructions and manuals, maintenance manuals, test reports, and analyses.   492

(3) (nuclear power generating stations). Technical information which is documented and organized in a readily understandable and traceable manner that permits independent auditing of the inferences or conclusions based on the information.   120

auditory sensation area. (1) The region enclosed by the curves defining the threshold of feeling and the threshold of audibility each expressed as a function of frequency. (2) The part of the brain (temporal lobe of the cortex) that is responsive to auditory stimuli.   176

augment (information processing). An independent variable, for example, in looking up a quantity in a table, the number or any of the numbers, that identifies the location of the desired value.   77

A unit. A motive power unit so designed that it may be used as the leading unit of a locomotive, with adequate visibility in a forward direction, and which includes a cab and equipment for full control and observation of the propulsion power and brake applications for the locomotive and train. *See:* **electric locomotive.**   328

aural harmonic. A harmonic generated in the auditory mechanism.   176

aural radio range. *See:* **A-N radio range.**

aural transmitter. The radio equipment used for the transmission of the aural (sound) signals from a television broadcast station. *See:* **television.**   211, 111

aurora (radio wave propagation). Optical and electrical phenomena resulting from direct excitation or ionization of upper atmospheric constituents by energetic particles, generally at high latitudes.   146

auroral (power fault effects). Electrical voltages and currents on or around the earth due to emission of particle energy from the sun. *See:* **susceptibility.**   404

auroral attenuation (radio wave propagation). The attenuation of radio waves propagating through that portion of the atmosphere where additional ionization is associated with aurora.   146

auroral hiss (radio wave propagation). Audio-frequency electromagnetic noise associated with aurora.   146

austenitic. The face-centered cubic crystal structure of ferrous metals.   205

authorities (monitoring radioactivity in effluents). Any governmental agencies or recognized scientific bodies which by their charter define regulations or standards dealing with radiation protection.   559

authorized bandwidth (mobile communication). The frequency band containing those frequencies upon which a total of 99 percent of the radiated power appears. *See:* **mobile communication system.**   181

auto alarm. A radio receiver that automatically produces an audible alarm when a prescribed radio signal is received.   328

autocondensation (electrotherapy). A method of applying alternating currents of frequencies exceeding 100 kilohertz to limited areas near the surface of the human body through the use of one very large and one small electrode, the patient becoming part of the capacitor. *See:* **electrotherapy.**   192

autoconduction (electrotherapy). A method of applying alternating currents, of frequencies exceeding 100 kilohertz for therapeutic purposes, by electromagnetic induction, the patient being placed inside a large solenoid. *See:* **electro-therapy.**   192

autodyne reception. A system of heterodyne reception through the use of a device that is both an oscillator and a detector.   328

autoerection (gyro). The process by which gimbal axis friction causes the spin axis of a free gyro to tend to align with the axis about which the case is rotated. The resulting drift rate is a function of the angular displacement between the spin axis and the rotation axis.   46

automated design tool (software). A software tool which aids in the synthesis, analysis, modeling, or documentation of a software design. Examples include simulators, analytic aids design representation processors, and documentation generators. *See:* **design; documentation; simulators; software tool.** 434

automated test case generator. *See:* **automated test generator.**

automated test data generator. *See:* **automated test generator.** 434

automated test generator (software). A software tool that accepts as input a computer program and test criteria, generates test input data that meet these criteria, and, sometimes, determines the expected results. *See:* **computer program; data; software tool.** 434

automated verification system (software). A software tool that accepts as input a computer program and a representation of its specification, and produces, possibly with human help, a correctness proof or disproof of the program. *See:* **automated verification tools; computer program; correctness proof; program; software tool; specification.** 434

automated verification tools (software). A class of software tools used to evaluate products of the software development process. These tools aid in the verification of such characteristics as correctness, completeness, consistency, traceability, testability, and adherence to standards. Examples are design analyzers, automated verification systems, static analyzers, dynamic analyzers, and standards enforcers. *See:* **automated verification systems; correctness; design analyzers; dynamic analyzers; software development process; software tools; standards enforcers; static analyzers; testability; tools; verification.** 434

automatic (1)(computer applications). Pertaining to a function, operation, process, or device that, under specified conditions, functions without intervention by a human operator. 571

(2)(NESC). Self-acting, operating by its own mechanism when actuated by some impersonal influence--as, for example, a change in current stength; not manual; without personal intervention. Remote control that requires personal intervention is not automatic, but manual. 494

(3)(supervisory control, data acquisition, and automatic control)(station control and data acquisition)-(industrial control). Pertaining to a process or device that, under specified conditions, functions without intervention by a human operator. 570,403

(4) (National Electrical Code)(industrial control). Self-acting, operating by its own mechanism when actuated by some impersonal influence, as for example, a change in current strength, pressure, temperature, or mechanical configuration. *See:* **non-automatic.** 256,206

automatic acceleration (1) (automatic train control). Acceleration under the control of devices that function automatically to maintain, within relatively close predetermined values or schedules, current passing to the traction motors, the tractive force developed by them, the rate of vehicle acceleration, or similar factors affecting acceleration. *See:* **electric drive; multiple-unit control.** 328

(2) (industrial control). Acceleration under the control of devices that function automatically to raise the motor speed. *See:* **electric drive; multiple unit control.** 206

automatically regulated (rotating machinery). Applied to a machine that can regulate its own characteristics when associated with other apparatus in a suitable closed-loop circuit. 63

automatically reset relay. *See:* **self-reset relay.**

automatic approach control. A system that integrates signals, received by localizer and glide path receivers, into the automatic pilot system, and guides the airplane down the localizer and glide path beam intersection. 328

automatic bias nulling (washout) (gyro, accelerometer) (inertial sensor). A circuit or system technique for setting the mean value of sensor output, averaged over a defined time period, to zero, or to some defined value. 46

automatic block signal system. A series of consecutive blocks governed by block signals, cab signals, or both, operated by electric, pneumatic, or other agency actuated by a train or by certain conditions affecting the use of a block. *See:* **block-signal system.** 328

automatic cab signal system. A system that provides for the automatic operation of cab signals. *See:* **automatic train control.** 328

automatic call distributor (ACD) (telephone switching systems). The facility for allotting incoming traffic to idle operators or attendants. 55

automatic capacitor control equipment (power switchgear). An equipment that provides automatic control for functions related to capacitors, such as their connection to and disconnection from a circuit in response to predetermined conditions such as voltage, load, or time. 103

automatic carriage. A control mechanism for a typewriter or other listing device that can automatically control the feeding, spacing, skipping, and ejecting of paper or preprinted forms. 77

automatic chart-line follower (navigation aid terms). A device which automatically derives error signals proportional to the deviation of the position of a vehicle from a predetermined course line drawn on a chart. 526

automatic check. *See:* **check, automatic.**

automatic circuit closer (supervisory control, data acquisition, and automatic control). A self-controlled device for automatically interrupting and reclosing an alternating-current circuit, with a predetermined sequence of opening and reclosing followed by resetting, hold-closed, or lockout operation. 570

automatic circuit recloser (1) (power switchgear). A self-controlled device for automatically interrupting and reclosing an alternating-current circuit, with a predetermined sequence of opening and reclosing followed by resetting, hold-closed, or lockout operation. *Note:* When applicable it includes an assembly of control elements required to detect overcurrents and control the recloser operation. 103

(2) (station control and data acquisition). A self-controlled device for automatically interrupting and re-closing an alternating-current circuit, with a predetermined sequence of opening and reclosing, followed by resetting, hold-closed, or lockout operation.   403

automatic combustion control. A method of combustion control that is effected automatically by mechanical or electric devices.   64

automatic component interconnection matrix (analog computers). A hardware system for connecting inputs and outputs of parallel computing components according to a predetermined program. This system, which may consist of a matrix of mechanical or electronic switches, or both, replaces the manual program patch panel and patch cords on analog computers. *See:* **problem board.**   9

automatic computer. A computer that can perform a sequence of operations without intervention by a human operator.   77

automatic control (1) (excitation systems for synchronous machines). In excitation control system usage, automatic control refers to maintaining synchronous machine terminal voltage without operator action, over the operating range of the synchronous machine within its capabilities. *Note:* Voltage regulation under automatic control may be modified by the action of reactive or active load compensators or by var control elements; or may be constrained by the action of various limiters included in the excitation system.   507

(2) automatic control. *See:* **control (1) automatic.**   570

(3) (analog computers). In an analog computer, a method of computer operation using auxiliary automatic equipment to perform computer-control state selections, switching operations, or component adjustments in accordance with previously selected criteria. Such auxiliary automatic equipment usually consists of programmable digital logic which is part of the analog, a separate digital computer, or both. The case of the digital computer controlling the analog computer is an example of a hybrid computer.   9

(4) (power switchgear). An arrangement of controlling or both in an automatic sequence and under predetermined conditions the necessary devices comprising an equipment. These devices thereupon maintain the required character of service and provide adequate protection against all unusual operating emergencies.   103

(5) (station control and data acquisition). An arrangement of electrical controls that provides for switching or controlling, or both, of equipment in an automatic sequence and under predetermined conditions.   403

(6) (electrical controls). An arrangement that provides for switching or otherwise controlling, or both, in an automatic sequence and under predetermined conditions, the necessary devices comprising an equipment. *Note:* These devices thereupon maintain the required character of service and provide adequate protection against all usual operating emergencies.   202

automatic control equipment (1) (power switchgear). Equipment that provides automatic control for a specified type of power circuit or apparatus.   103
(2) (station control and data acquisition). An equipment that provides automatic control of power apparatus in response to predetermined conditions.   403

automatic controller (electrical heating applications to melting furnaces and forehearths in the glass industry)(process control)(emergency and standby power). A device that operates automatically to regulate a controlled variable in response to a command and a feedback signal. *See:* **controller, automatic.**   520, 512

automatic data processing (ADP)(computer applications). Data processing performed by a computer system.   571

automatic direct-control telecommunications system (telephone switching systems). A system in which the connections are set directly in response to pulsing from the originating calling device.   55

automatic direction finder (navigation aid terms). A direction finder which automatically and continuously provides a measure of the direction of arrival of the received signal. Data are usually displayed visually.   526

automatic dispatching system (electric power systems). A controlling means for maintaining the area control error or station control error at zero by automatically loading generating sources, and it also may include facilities to load the sources in accordance with a predetermined loading criterion.   94

automatic equipment (for a specified type of power circuit or apparatus). Equipment that provides automatic control.   103

automatic extraction or induction turbine, or both--condensing or noncondensing (control systems for steam turbine-generator units). Steam is extracted from or inducted into one or more stages with means for controlling the pressure(s) of the extraction or induction steam, or both.   522

automatic extraction turbine (condensing or noncondensing)(control systems for steam turbine-generator units). Steam is extracted from one or more stages with means for controlling the pressure(s) of the extracted steam.   522

automatic fire-alarm system. A fire-alarm system for automatically detecting the presence of fire and initiating signal transmission without human intervention. *See:* **protective signaling.**   328

automatic fire detector (fire protection devices). A device designed to detect the presence of fire and initiate action.   71

automatic flight control system (AFCS). An autopilot or automatic pilot. A system that controls the attitude, direction, and speed of a vehicle and directs it to travel along a selected course in response to manual or electronic commands. Stabilizes the dynamic response of the vehicle.   526

automatic frequency control (AFC) (data transmission). An arrangement whereby the frequency of an oscillator or the tuning of a circuit is automatically maintained within specified limits with respect to a reference frequency.   59

automatic-frequency-control synchronization. A process for locking the frequency (phase) of a local oscillator to that of an incoming synchronizing signal by the use of a comparison device whose output continuously corrects the local-oscillator frequency (phase).   328

automatic gain control (AGC) (1) (general). A process or means by which gain is automatically adjusted in a specified manner as a function of input or other specified parameters.   239, 178

(2) (data transmission). A method of automatically obtaining a substantially constant output of some amplitude characteristic of the signal over a range of variation of that characteristic at the input. The term is also applied to a device for accomplishing this result.   59

automatic generation control. The regulation of the power output of electric generators within a prescribed area in response to change in system frequency, tie-line loading, or the relation of these to each other, so as to maintain the scheduled system frequency or the established interchange with other areas within predetermined limits or both.   200

automatic grid bias. Grid-bias voltage provided by the difference of potential across resistance(s) in the grid or cathode circuit due to grid or cathode current or both. *See:* **radio receiver.**   111

automatic hold (analog computer). Attainment of the hold condition automatically through amplitude comparison of a problem variable or through an overload condition. *See:* **electronic analog computer.**   9

automatic holdup alarm system. An alarm system in which the signal transmission is initiated by the action of the robber. *See:* **protective signaling.**   328

automatic-identified outward dialing (AIOD) (telephone switching system). A method of automatically obtaining the identity of a calling station from a private branch exchange over a separate data link for use in automatic message accounting.   55

automatic indirect-control telecommunications system (telephone switching system). A system in which the pulsing from the originating calling device are stored in a register temporarily associated with the call, for the subsequent establishing of connections.   55

automatic interlocking. An arrangement of signals, with or without other signal appliances, that functions through the exercise of inherent powers as distinguished from those whose functions are controlled manually, and that are so interconnected by means of electric circuits that their movements must succeed one another in proper sequence. *See:* **interlocking (interlocking plant).**   328

automatic keying device. A device that, after manual initiation, controls automatically the sending of a radio signal that actuates the auto alarm. *Note:* The prescribed signal is a series of twelve dashes, each of four seconds duration, with one-second interval be-

tween dashes, transmitted on the radiotelegraph distress frequency in the medium-frequency band. This signal is used only to proceed distress calls or urgent warnings.   328

automatic line sectionalizer (supervisory control, data acquisition, and automatic control)(power switchgear). A self-contained circuit-opening device that automatically opens the main electrical circuit through it after sensing and responding to a predetermined number of successive main current impulses equal to or greater than a predetermined magnitude. It opens while the main electrical circuit is deenergized. It may also have provision to be manually operated to interrupt loads.   570, 403, 103

automatic load (armature current division) (industrial control). The effect of a control function or a device to automatically divide armature currents in a prescribed manner between two or more motors or two or more generators connected to the same load. *See:* **control system, feedback.**   225, 206

automatic load (level) control (ALC) (power-system communication). A method of automatically maintaining the peak power of a single-sideband suppressed-carrier transmitter at a constant level. *See:* **radio transmitter.**   59

automatic load throwover equipment (transfer or switchover)(supervisory control, data acquisition, and automatic control). An equipment that automatically transfers a load to another source of power when the original source to which it has been connected fails, and that automatically restores the load to the original source under desired conditions. *Note:* The restoration of the load to the preferred source from the emergency source upon reenergization of the preferred source after an outage may be of the continuous circuit restoration type or interrupted circuit restoration type. (1) Equipment of the nonpreferential type. Equipment that automatically restores the load to the original source only when the other source, to which it has been connected, fails. (2) Fixed preferential type. Equipment in which the original source always serves as the preferred source and other source as the emergency source. The automatic transfer equipment will restore the load to the preferred source upon its reenergization. (3) Selective preferential source. Equipment in which either source may serve as the preferred or the emergency source of preselection as desired, and which will restore the load to the preferred source upon its reenergization. (4) Semiautomatic load throwover equipment. An equipment that automatically transfers a load to another (emergency) source of power when the original (preferred) source to which it has been connected fails, but requires manual restoration of the load to the original source.   570

automatic machine control equipment (power switchgear). An equipment that provides automatic control for functions related to rotating machines or power rectifiers.   103

automatic message accounting (AMA) (telephone switching system). An arrangement for automatically collecting, recording, and processing information relating to calls for billing purposes.   55

automatic number identification (ANI) (telephone switching system). The automatic obtaining of a calling station directory or equipment number for use in automatic message accounting.   55

automatic opening (tripping)(1)(supervisory control, data acquisition, and automatic control)(station control and data acquisition). The opening of a switching device under predetermined conditions without operator intervention.   570,403

(2) (power switchgear). The opening of a switching device under predetermined conditions without the intervention of an attendant. *Syn:* **tripping.**   103

automatic operation (1) (elevator). Operation wherein the starting of the elevator car is effected in response to the momentary actuation of operating devices at the landing, and.or of operating devices in the car identified with the landings, and.or in response to an automatic starting mechanism, and wherein the car is stopped automatically at the landings. *See:* **control.**   328

(2) (of a switching device) (power switchgear). The ability to complete an assigned sequence of operations by automatic control without the assistance of an attendant.   103

automatic phase control (television). A process or means by which the phase of an oscillator signal is automatically maintained within specified limits by comparing its phase to the phase of an external reference signal and thereby supplying correcting information to the controlled source or a device for accomplishing this result. *Note:* Automatic phase control is sometimes used for accurate frequency control and under these conditions is often called automatic frequency control. *See:* **television.**   328

automatic pilot (electronic navigation). Equipment that automatically controls the attitude of a vehicle about one or more of its rotational axes (pitch, roll, and yaw), and may be used to respond to manual or electronic commands. *See:* **navigation.**   187

automatic-pilot servo motor. A device that converts electric signals to mechanical rotation so as to move the control surfaces of an aircraft.   328

automatic programmed tools (APT) (numerically controlled machines). A computer-based numerical control programming system that uses English-like symbolic descriptions of part and tool geometry and tool motion.   224, 203

automatic programming (analog computer). A method of computer operation using auxiliary automatic equipment to perform computer control state selections, switching operations, or component adjustments in accordance with previously selected criteria. *See:* **electronic analog computer.**   9

automatic reclosing equipment (1)(supervisory control, data acquisition, and automatic control)(station control and data acquisition). Equipment which initiates automatic closing of a switching device under predetermined conditions without operator intervention.   570,403

(2) (power switchgear). An automatic equipment that provides for reclosing a switching device as desired after it has opened automatically under abnormal conditions. *Note:* Automatic reclosing equipment may be actuated by conditions sensed on either or both sides of the switching device as designed.   103

automatic-reset manual release of brakes (control) (industrial control). A manual release that, when operated, will maintain the braking surfaces in disengagement but will automatically restore the braking surfaces to their normal relation as soon as electric power is again applied. *See:* **control system, feedback.**   206

automatic-reset relay. *See:* **relay, automatic-reset.**

automatic-reset thermal protector (rotating machinery). A thermal protector designed to perform its function by opening the circuit to or within the protected machine and then automatically closing the circuit after the machine cools to a satisfactory operating temperature. *See:* **starting switch assembly.**   63

automatic reversing (industrial control). Reversing of an electric drive, initiated by automatic means. *See:* **electric drive.**   206

automatic selective control or transfer relay (83) (power system device function numbers). A relay that operates to select automatically between certain sources or conditions in an equipment, or performs a transfer operation automatically.   402

automatic signal. A signal controlled automatically by the occupancy or certain other conditions of the track area that it protects.   328

automatic smoke alarm system. An alarm system designed to detect the presence of smoke and to transmit an alarm automatically. *See:* **protective signaling.**   328

automatic speed adjustment (industrial control). Speed adjustment accomplished automatically. *See:* **automatic; electric drive.**   206

automatic starter (electric installations on shipboard). A starter in which the influence directing its performance is automatic.   3

automatic station (1)(supervisory control, data acquisition, and automatic control)station control and data acquisition). A station that operates in automatic control mode. *Note:* An automatic station may go in and out of operation in response to predetermined voltage, load, time, or other conditions, or in response to a remote or locally manually operated control device.   570,403

(2) (power switchgear). A station (usually unattended) that under predetermined conditions goes into operation by an automatic sequence; that thereupon by automatic means maintains the required character of service within its capability; that goes out of operation by automatic sequence under other predetermined conditions, and includes protection against the usual operating emergencies. *Note:* An automatic station may go in and out of operation in response to predetermined voltage, load, time, or other conditions, or in response to supervisory control or to a remote or local manually operated control device.   103

automatic switchboard. A switchboard in which the connections are made by apparatus controlled from remote calling devices.   193

automatic switching system (telephone switching systems). The switching entity for an automatic telecommunication system.   55

automatic system. A system in which the operations are performed by electrically controlled devices without the intervention of operators.   194

automatic telecommunications exchange (telephone switching system). A telecommunications exchange in which connections between stations are automatically established as a result of signals produced by calling devices.   55

automatic telecommunications system (telephone switching system). A system in which connections between stations are automatically established as a results of signals produced by calling devices.   55

automatic telegraphy. That form of telegraphy in which transmission or reception of signals, or both, are accomplished automatically. *See:* **telegraphy.**   194

automatic test equipment (ATE) (test, measurement and diagnostic equipment). Equipment that is designed to conduct analysis of functional or static parameters to evaluate the degree of performance degradation and may be designed to perform fault isolation of unit malfunctions. The decision making, control, or evaluative functions are conducted with minimum reliance on human intervention.   54

automatic throw-over equipment. *See:* **automatic transfer equipment.**

automatic throw-over equipment of the fixed preferential type. *See:* **automatic transfer equipment of the fixed preferential type.**

automatic throw-over equipment of the nonpreferential type. *See:* **automatic transfer equipment of the nonpreferential type.**

automatic throw-over equipment of the selective-preferential type. *See:* **automatic transfer equipment of the selective-preferential type.**

automatic ticketing (telephone switching system). An arrangement for automatically recording information relating to calls, for billing purposes.   55

automatic TMDE (test, measurement and diagnostic equipment). *See:* **automatic test equipment.**

automatic track follower. *See:* **automatic chart-line follower.**

automatic tracking (navigation aid terms). Tracking in which a system employs some feedback mechanisms, such as a servo or computer, to follow automatically some characteristic of a signal or target such as range, angle, Doppler frequency, or phase. *See:* **tracking; tracking radar.**   526

automatic train control (automatic speed control) (train control). A system or an installation so arranged that its operation on failure to forestall or acknowledge will automatically result in either one or the other or both of the following conditions: (1) Automatic train stop: The application of the brakes until the train has been brought to a stop: (2) Automatic speed control: The application of the brakes when the speed of the train exceeds a prescribed rate and continued until the speed has been reduced to a predetermined and prescribed rate.   328

automatic train control application. An application of the brake by the automatic train control device.   328

automatic transfer equipment (power switchgear). (1) (General). An equipment that automatically transfers a load to another source of power when the original source to which it has been connected fails, and that will automatically retransfer the load to the original source under desired conditions. *Notes:* (A) It may be of the nonpreferential, fixed-preferential, or selective preferential type. (B) Compare with transfer switch where transfer is accomplished without current interruption. (2) (fixed preferential type). Automatic transfer equipment in which the original source always serves as the preferred source and the other source as the emergency source. The automatic transfer equipment will retransfer the load to the preferred source upon its reenergization. *Note:* The restoration of the load to the preferred source from the emergency source upon the reenergization of the preferred source after an outage may be of the continuous- circuit restoration type or the interrupted-circuit restoration type. (3) (nonpreferrential type). Automatic transfer equipment that automatically retransfers the load to the original source only when the other source, to which it has been connected, fails. (4) (selective preferential type). Automatic transfer equipment in which either source may serve as the preferred or the emergency source of preselection as desired, and which will retransfer the load to the preferred source upon its reenergization. *Note:* The restoration of the load to the preferred source from the emergency source upon the reenergization of the preferred source after an outage may be of the continuous-circuit restoration type or the interrupted-circuit restoration type. *Syn:* **throw-over equipment.**   103

automatic transfer equipment of the fixed-preferential type (automatic throw-over equipment of the fixed preferential type). Equipment in which the original source always serves as the preferred source and the other source as the emergency source. The automatic transfer equipment will retransfer the load to the preferred source upon its reenergization. *Note:* The restoration of the load to the preferred source from the emergency source upon the reenergization of the preferred source after an outage may be of the continuous-circuit restoration type or the interrupted-circuit restoration type.   103

automatic transfer equipment of the nonpreferential type (automatic throw-over equipment of the nonpreferential type). Equipment that automatically retransfers the load to the original source only when the other source, to which it has been connected, fails.   103

automatic transfer equipment of the selective-preferential type (automatic throw-over equipment of the selective-preferential type). Equipment in which either source may serve as the preferred or the emergency source of preselection as desired, and which will retransfer the load to the preferred source upon its reenergization. *Note:* The restoration of the load to the preferred source from the emergency source upon the reenergization of the preferred source after an outage may be of the continuous-circuit restoration type or the interrupted-circuit restoration type. 103

automatic transfer switch (emergency and standby power). Self-acting equipment for transferring one or more load conductor connections from one power source to another. 512

automatic transformer control equipment (power switchgear). An equipment that provides automatic control for functions relating to transformers, such as their connection, disconnection or regulation in response to predetermined conditions such as system load, voltage or phase angle. 103

automatic triggering (oscilloscope). A mode of triggering in which one or more of the triggering-circuit controls are preset to conditions suitable for automatically displaying repetitive waveforms. *Note:* The automatic mode may also provide a recurrent trigger of recurrent sweep in the absence of triggering signals. *See:* **oscillograph.** 185

automatic tripping. *See:* **automatic opening.**

automatic volume control (AVC) (data transmission). A method of automatically obtaining a substantially constant audio output volume over a range of input volume. The term is also applied to a device for accomplishing this result. 59

automation (computer applications). (1) The implementation of a process by automatic means. (2) The theory, art, or technique of making a process more automatic. 571

autonavigator (navigation aid terms). Navigation equipment which includes means for coupling the output navigational data derived from the navigation sensors to the control system of the vehicle. 526

autopatch. *See:* **automatic component interconnection matrix.**

autopilot. *See:* **automatic flight control system; automatic pilot.**

autopilot coupler (navigation)(navigation aid terms)(electronic navigation). The means used to link the navigation system-receiver output to the automatic pilot. 526,13

autoradar plot (electronic navigation). A particular chart comparison unit using a radar presentation of position. *See:* **navigation.** 187

autoregulation induction heater. An induction heater which a desired control is effected by the change in characteristics of a magnetic charge as it is heated at or near its Curie point. *See:* **dielectric heating; induction heating; coupling.** 14

autotrack (communication satellite). The capability of a space communications receiver antenna to automat-

ically track an orbiting satellite vehicle, for example, by using a monopulse system. 84

autotransformer (power and distribution transformer). A transformer in which at least two windings have a common section. 53

autotransformer, individual-lamp. *See:* **speciality transformer.**

autotransformer starter (electric installations on shipboard). A starter which includes an autotransformer to furnish a reducedvoltage for starting a motor.It includes the necessary switching mechanism and is frequently called a compensator or autostarter. 3

autotransformer starting (rotating machinery). The process of starting a motor at reduced voltage by connecting the primary winding to the supply initially through an autotransformer and reconnecting the winding directly to the supply at rated voltage for the running conditions. *See:* **asynchronous machine.** 63

auxilary supporting features (nuclear power generating station).(1) Systems or components which provide services (such as cooling, lubrication and energy supply) which are required for the safety system to accomplish its protective functions. 109
(2) Installed systems or components which provide services such as cooling, illumination and energy supply and which are required by the Post Accident Monitoring Instrumentation to perform its functions. 421

auxiliaries (1) (collective) (generating stations electric power system). For more than one auxiliary, that is, auxiliaries bus, auxiliaries power transformer, etcetera. 381
(2) **(power switchgear).** Accessories to be used with switching apparatus. 103

auxiliary (1) (controller) (thyristor). Apparatus peripheral to the main power flow but necessary for the operation of the controller. 445
(2) **(generating stations electric power system).** Any item not directly a part of a specified component or system but required for its functional operation. 381

auxiliary anode (industrial control). An anode located adjacent to the pool cathode in an ignitron to facilitate the maintenance of a cathode spot under conditions adverse to its maintenance by the main anode circuit. *See:* **electronic controller.** 206

auxiliary branch (converter circuit elements)(self-commutated converters). A branch other than a principal branch. *Note:* Examples of auxiliary branches are regenerative branches and turn-off branches. 584

auxiliary building(s) (radiological monitoring instrumentation). Building(s), near or adjacent to the reactor containment building in which primary system support equipment is housed. 398

auxiliary burden (capacitance potential device). A variable burden furnished, when required, for adjustment purposes. *See:* **outdoor coupling capacitor.** 341

auxiliary circuit breaker (ac high-voltage circuit

breaker). The circuit breaker used to disconnect the current circuit from direct connection with the test circuit breaker. 426

auxiliary compartment (metal-enclosed low-voltage power circuit-breaker switchgear)(metal-clad and station-type cubicle switchgear). That portion of the switchgear assembly that is assigned to the housing of auxiliary equipment, such as potential transformers, control power transformers, or other miscellaneous devices. 579, 572

auxiliary device (1) (general). Any electrical device other than motors and motor starters necessary to fully operate the machine or equipment. 429
(2) (packaging machinery). Any electrical device other than motors and motor starters necessary to fully operate the machine or equipment. 429

auxiliary devices (auxiliary devices for motors). Components installed either integrally within the motor, located adjacent to or mounted on the motor, or attached to its terminals for the purpose of monitoring the operating conditions or protecting the motor. 446

auxiliary device to an instrument. A separate piece of equipment used with an instrument to extend its range, increase its accuracy, or otherwise assist in making a measurement or to perform a function additional to the primary function of measurement. 328

auxiliary equipment (1) (nuclear power generating stations) (class 1E motor). Equipment that is not part of the motor but is necessary for the operation of the motor and will be installed within the containment. 31
(2) (test, measurement and diagnostic equipment). *See:* **ancillary equipment.**

auxiliary function (numerically controlled machine). A function of a machine other than the control of the coordinates of a workpiece or tool. Includes functions such as miscellaneous, feed, speed, tool selection, etcetera. *Note:* Not a preparatory function. 207

auxiliary generator. A generator, commonly used on electric motive power units, for serving the auxiliary electric power requirements of the unit. *See:* **traction motor.** 328

auxiliary generator set. A device usually consisting of a commonly mounted electric generator and a gasoline engine or gas turbine prime mover designed to convert liquid fuel into electric power. *Note:* It provides the aircraft with an electric power supply independent of the aircraft propulsion engines. 328

auxiliary lead (rotating machinery). A conductor joining an auxiliary terminal to the auxiliary device. 63

auxiliary means. A system element or group of elements that changes the magnitude but not the nature of the quantity being measured to make it more suitable for the primary detector. In a sequence of measurement operations it is usually placed ahead of the primary detector. *See:* **measurement system.** 328

auxiliary motor or motor generator (88) (power system device function numbers). One used for operating

auxiliary equipment, such as pumps, blowers, exciters, rotating magnetic amplifiers, etcetera. 402

auxiliary operation. An operation performed by equipment not under continuous control of the central processing unit. 77, 255

auxiliary or shunt capacitance (capacitance potential device). The capacitance between the network connection and ground, if present. *See:* **outdoor coupling capacitor.** 341

auxiliary power (thyristor). The power used by the controller to perform its various auxiliary functions as opposed to the principal power. 445

auxiliary power supply (industrial control). A power source supplying power other than load power as required for the proper functioning of a device. *See:* **electronic controller.** 206

auxiliary power transformer. A transformer having a fixed phase position used for supplying excitation to the rectifier station and essential power for the operation of rectifier equipment auxiliaries. *See:* **transformer.** 258

auxiliary relay (power switchgear). A relay whose function is to assist another relay or control device in performing a general function by supplying supplementary actions. *Notes:* (A) Some of the specific functions of an auxiliary relay are: (1) Reinforcing contact current-carrying capacity of another relay or device.
(2) Providing circuit seal-in functions.
(3) Increasing available number of independent contacts.
(4) Providing circuit-opening instead of circuit-closing contacts or vice-versa.
(5) Providing time delay in the completion of a function.
(6) Providing simple functions for interlocking or programming.
(B) The operating coil of the contacts of an auxiliary relay may be used in the control circuit of another relay or other control device. *Example:* An auxiliary relay may be applied to the auxiliary contact circuits of a circuit breaker in order to coordinate closing and tripping control sequences. (C) A relay which is auxiliary in its functions even though it may derive its driving energy from the power system current or voltage is a form of auxiliary relay. *Example:* A timing relay operating from current or potential transformers. (D) Relays which, by direct response to power system input quantities, assist other relays to respond to such quantities with greater discrimination are NOT auxiliary relays. *Example:* Fault detector relay. (E) Relays which are limited in function by a control circuit, but are actuated primarily by system input quantities, are NOT auxiliary relays. *Example:* Torque-controlled relays. 103

auxiliary relay driver (power switchgear). A circuit which supplies an input to an auxiliary relay. 103

auxiliary rope-fastening device (elevator). A device attached to the car or counterweight or to the overhead dead-end rope-hitch support that will function automatically to support the car or counterweight in case the regular wire-rope fastening fails at the point of connection to the car or counterweight or at the overhead dead-end hitch. *See:* **elevator.** 328

auxiliary secondary terminals. The auxiliary secondary terminals provide the connections to the auxiliary secondary winding, when furnished. *See:* **auxiliary secondary winding.** 341

auxiliary secondary winding (capacitance potential device). The auxiliary secondary winding is an additional winding that may be provided in the capacitance potential device when practical considerations permit. *Note:* It is a separate winding that provides a potential that is substantially in phase with the potential of the main winding. The primary purpose of this winding is to provide zero sequence voltage by means of a broken delta connection of three single-phase devices. *See:* **auxiliary secondary terminals; outdoor coupling capacitor.** 341

auxiliary storage (computing machine). A storage that supplements another storage. 77, 255

auxiliary supporting features (nuclear power generating station). (1) Systems or components which provide services (such as cooling, lubrication, and energy supply) which are required for the safety system(s) to accomplish their safety functions. 102
(2) Systems or components which provide services (such as cooling, lubrication, and energy supply) which are required for the safety system to accomplish its protective functions. *Note:* Examples of auxiliary supporting features are ventilation systems for Class 1E switchgear, cooling water systems for Class 1E motors, and fuel oil supply systems for emergency diesel generators. 131, 387, 428

auxiliary switch (power switchgear). A switch mechanically operated by the main device for signaling, interlocking, or other purposes. *Note:* Auxiliary switch contacts are classed as a, b, aa, bb, LC, etcetera, for the purpose of specifying definite contact positions with respect to the main device. 103

auxiliary terminal (rotating machinery). A termination for parts other than the armature or field windings. 63

auxiliary winding (single-phase induction motor). A winding that produces poles of a magnetic flux field that are displaced from those of the main winding, that serves as a means for developing torque during starting operation, and that, in some types of design, also serves as a means for improvement of performance during running operation. An auxiliary winding may have a resistor or capacitor in series with it and may be connected to the supply line or across a portion of the main winding. *See:* **asynchronous machine.** 63

availability (1)(emergency and standby power). The fraction of time within which a system is actually capable of performing its mission. 512
(2)(individual or system components)(reliable industrial and commercial power systems planning and design). The long-term average fraction of time that a component or system is in service satisfactorily performing its intended function. An alternative and equivalent definition for availability is the steady-state probability that a component or system is in service. *See:* **unavailability.** 561
(3)(supervisory control, data acquisition, and automatic control). The ratio of uptime and uptime plus downtime. 570
(4) (general). The fraction of time that the system is actually capable of performing its mission. *See:* **system.** 209, 89
(5) (software). (A) The probability that software will be able to perform its designated system function when required for use. (B) The ratio of system up-time to the total operating time. (C) The ability of an item to perform its designated function when required for use. *See:* **function; software; system; system function.** 434
(6) (station control and data acquisition). The ratio of uptime and uptime plus downtime. $A =$ uptime/ (uptime + downtime). Downtime in the above equation normally includes corrective maintenance, preventive maintenance, and system expansion downtimes if such times compromise the user's ability to operate apparatus normally controlled by the equipment being expanded. 403
(7) (nuclear power generating station). (A) The characteristic of an item expressed by the probability that it will be operational at a randomly selected future instant in time. 29, 31, 357
(B) Relates to the accessibility of information to the operator on a "continuous", "sequence" or "as called for" basis. 358
(8) (telephone switching system). The number of outlets of a group that can be reached from a given inlet in a switching stage or network. 55
(9) (reliability). The ability of an item–under combined aspects of its reliability, maintainability and maintenance support–to perform its required function at a stated instant of time or over a stated period of time. *Note:* The term availability is also used as an availability characteristic denoting either the probability of performing at a stated instant of time or the probability related to an interval of time. 164

availability factor. The ratio of the time a generating unit or piece of equipment is ready for or in service to the total time interval under consideration. *See:* **generating station.** 64

availability model (software). A model used for predicting, estimating, or assessing availability. *See:* **availability.** 434

available (electric generating unit reliability, availability, and productivity)(power system measurement). The state in which a unit is capable of providing service, whether or not it is actually in service and regardless of the capacity level that can be provided. 567,432

available (prospective) short-circuit test current (at the point of test) (power switchgear). The maximum short-circuit current for any given setting of a test circuit that the test power source can deliver at the point of test, with the test circuit short-circuited by a link of negligible impedance at the line terminals of the device to be tested. *Note:* This value can be in terms of either symmetrical or asymmetrical; peak or root-mean-square current, as specified. 103, 443

available (prospective) current (of a circuit with respect to a switching device situated therein) (power switchgear). The current that would flow in that circuit if each pole of the switching device were to be replaced by a link of negligible impedance without any other change in the circuit or the supply. 103

available (prospective) short-circuit current (at a given point in a circuit) (power switchgear). The maximum current that the power system can deliver through a given circuit point to any negligible impedance short circuit applied at the given point, or at any other point that will cause the highest current to flow through the given point. *Notes:* (1) This value can be in terms of either symmetrical or asymmetrical; peak or root-mean-square current, as specified. (2) In some resonant circuits the maximum available short-circuit current may occur when the short circuit is placed at some other point than the given one where the available current is measured. 103, 443

available accuracy (noise temperature of noise generators). An accuracy that is readily available to the public at large, such as may be announced in calibration service bulletins or instrument catalogs. This term shall not include accuracies that may be obtainable at any echelon by employing special efforts and expenditures over and above those invested in producing the advertised or announced accuracies, nor shall it include accuracies of calibration or measurement services that are not readily available to any and all customers and clients. 155

available capacity (electric generating unit reliability, availability, and productivity). The dependable capacity, modified for equipment at any time. 567

available conversion power gain (conversion transducer). The ratio of the available output-frequency power from the output terminals of the transducer to the available input-frequency power from the driving generator with terminating conditions specified for all frequencies that may affect the result. *Notes:* (1) This applies to outputs of such magnitude that the conversion transducer is operating in a substantially linear condition. (2) The maximum available conversion power gain of a conversion transducer is obtained when the input termination admittance, at input frequency, is the conjugate of the input-frequency driving-point admittance of the conversion transducer. *See:* **transducer.** 252, 125

available current (prospective current) (circuit) (1) (industrial control). The current that would flow if each pole of the breaking device under consideration were replaced by a link of negligible impedance without any change of the circuit or the supply. *See:* **contactor; prospective current.** 244, 206

(2) (of a circuit with respect to a switching device situated therein). The current that would flow in that circuit if each pole of the switching device were to be replaced by a link of negligible impedance without any other change in the circuit or the supply. 103

available generation (AG)(electric generating unit reliability, availability, and productivity). The energy that could have been generated by a unit in a given period if operated continuously at its available capacity. 567

available hours (AH)(1)(electric generating unit reliability, availability, and productivity). The number of hours a unit was in the available state. *Note:* Available hours is the sum of service hours and reserve shutdown hours, or may be computed from period hours minus unavailable hours. *See:* **unavailable hours.** 567

(2)(power system measurement). The number of hours a unit was in the available state. *Note:* Available hours is the sum of service hours (SH) and reserve shutdown hours (RH), or may be computed from period hours minus outage hours (planned and unplanned). 432

available line. The portion of the scanning line which can be used specifically for picture signals. 12

available power (at a port) (hydraulic turbines). The maximum power which can be transferred from the port to a load. *Note:* At a specified frequency, maximum power transfer will take place when the impedance of the load is the conjugate of that of the source. The source impedance must have a positive real part. 125

available power efficiency (electroacoustics). Of an electroacoustic transducer used for sound reception, the ratio of the electric power available at the electric terminals of the transducer to the acoustic power available to the transducer. *Notes:* (1) For an electroacoustic transducer that obeys the reciprocity principle, the available power efficiency in sound reception is equal to the transmitting efficiency. (2) In a given narrow frequency band the available power efficiency is numerically equal to the fraction of the open-circuit mean-square thermal noise voltage present at the electric terminals that contributed by thermal noise in the acoustic medium. *See:* **microphone.** 176

available power gain (1)(two-port linear transducer). At a specified frequency, the ratio of (A) the available signal power from the output port of the transducer, to (B) the available signal power from the input source. *Note:* The available signal power at the output port is a function of the match between the source impedance and the impedance of the input port. *See:* **network analysis.** 125

(2) (circuits and systems). The maximum power gain that can be obtained from a signal source. For a source of internal impedance Z_s, R_s jX_s the maximum power gain is obtained when the source is connected to a conjugate matched load: i.e. if Z_2, R_s mins jX_s. It is sometimes called completely matched power gain or available gain. *See:* **network analysis; transducer.** 67

available power response (electroacoustics) (electroacoustic transducer used for sound emission). The ratio of the mean-square sound pressure apparent at a distance of 1 meter in a specified direction from the effective acoustic center of the transducer to the available electric power from the source. *Notes:* (1) The sound pressure apparent at a distance of 1 meter can be found by multiplying the sound pressure observed at a remote point where the sound field is spherically divergent by the number of meters from the effective acoustic center to that point. (2) The available power response is a function not only of the transducer but also of some source impedances, either actual or nominal, the value of which must be specified. *See:* **loudspeaker.**   176

available short-circuit current (at a given point in a circuit) (prospective short-circuit current). The maximum current that the power system can deliver through a given circuit point to any negligible-impedance short-circuit applied at the given point, or at any other point that will cause the highest current to flow through the given point. *Notes:* (1) This value can be in terms of either symmetrical or asymmetrical: peak or root-mean-square current, as specified. (2) In some resonant circuits the maximum available short-circuit current may occur when the short circuit is placed at some other point than the given one where the available current is measured.   103

available short-circuit test current (at the point of test) (prospective short-circuit test current). The maximum short-circuit current for any given setting of a testing circuit that the test power source can deliver at the point of test, with the test circuit short-circuited by a link of negligible impedance at the line terminals of the device to be tested. *Note:* This value can be in terms of either symmetrical or asymmetrical, peak or root-mean-square current, as specified.   103

available signal-to-noise ratio (at a point in a circuit). The ratio of the available signal power at that point to the available random noise power. *See:* **signal-to-noise ratio.**   328

available time (electric drive) (industrial control). The period during which a system has the power turned on, is not under maintenance, and is known or believed to be operating correctly or capable of operating correctly. *See:* **electric drive.**   206

avalanche (gas-filled radiation counter tube). The cumulative process in which charged particles accelerated by an electric field produce additional charged particles through collision with neutral gas molecules or atoms. It is therefore a cascade multiplication of ions. *See:* **amplifier.**   96, 125

avalanche breakdown (of a semiconductor device)(-charged-particle detectors)(germanium gamma-ray detectors). A breakdown that is caused by the cumulative multiplication of charge carriers through field-induced impact ionization.   119,118, 245,528

avalanche impedance (semiconductor). *See:* **breakdown impedance; semiconductor.**

avalanche photodiode (APD)(fiber optics). A photodiode designed to take advantage of avalanche multiplication of photocurrent. *Note:* As the reverse-bias voltage approaches the breakdown voltage, hole-electron pairs created by absorbed photons acquire sufficient energy to create additional hole-electron pairs when they collide with ions; thus a multiplication (signal gain) is achieved. *See:* **photodiode; PIN photodiode.**   433

average absolute burst magnitude (audio and electroacoustics). The average of the instantaneous burst magnitude taken over the burst duration. *See:* figure under **burst duration.** *See:* **burst (audio and electroacoustics).**   176

average absolute pulse amplitude. The average of the absolute value of the instantaneous amplitude taken over the pulse duration.   254

average bundle gradient (overhead-power-line corona and radio noise). For a bundle of two or more subconductors, the arithmetic mean of the average gradients of the individual subconductors.   411

average crossing rate (ACR)(1)(electromagnetic site survey). The average number of crossings in the positive direction of a given level v_1 per unit time. (See Figure "Typical Noise Envelope of a Man-Made Radio-Noise Process".)   457
(2)(control of system electromagnetic compatibility). The average number of pulses crossing a specified level (zero, if not specified) in the positive-going direction per unit time.   495

average current (periodic current). The value of the current averaged over a full cycle unless otherwise specified. *See:* **rectification.**   237, 66

average detector (overhead-power-line corona and radio noise). A detector, the output voltage of which approximates the average value of the envelope of an applied signal or noise. *Notes:* (1) This detector function is often identified on radio noise meters as field intensity (FI). (Field intensity is deprecated; field strength should be used.) (2) Field intensity (FI) (field strength) setting on some radio noise meters produces on the meter scale the average value of the logarithmic detector.   411

average electrode current (electron tube). The value obtained by integrating the instantaneous electrode current over an averaging time and dividing by the averaging time. *See:* **electrode current (electron tube).**   125

average forward-current rating (rectifier circuit element). The maximum average value of forward current averaged over a full cycle, permitted by the manufacturer under stated conditions.   208

average information content (per symbol) (information rate from a source, per symbol). The average of the information content per symbol emitted from a source. *Note:* The term entropy rate is also used to designate average information content. *See:* **information theory.**   61

average inside air temperature (of enclosed switchgear) (power switchgear). The average temperature of the surrounding cooling air which comes in contact with the heated parts of the apparatus within the enclosure.   103

average luminance (illuminating engineering). Luminance is the property of a geometric ray. Luminance as measured by conventional meters is averaged with respect to two independent variables, area and solid angle; both must be defined for a complete description of a luminance measurement.   167

average magner. *See:* **average reactive power (single-phase, two-wire, or polyphase circuit).**

average maximum bundle gradient (overhead-power-line corona and radio noise). For a bundle of two or more subconductors, the arithmetic mean of the maximum gradients of the individual subconductors. For example, for a three-conductor bundle with individual maximum subconductor gradients of 16.5, 16.9, and 17.0 kV/cm, the average maximum bundle gradient would be$(1/3) (16.5 + 16.9 + 17.0) = 16.8$ kV/cm.
                                                             411

average mutual information (output symbols and input symbols). Mutual information averaged over the ensemble of pairs of transmitted and received symbols. *See:* **information theory.**   61

average noise factor. *See:* **average noise figure.**

average noise figure (transducer) (average noise factor). The ratio of total output noise power to the portion thereof attributable to thermal noise in the input termination, the total noise being summed over frequencies from zero to infinity, and the noise temperature of the input termination being standard (290 kelvins). *See:* **noise figure; signal-to-noise ratio.**
                                                             328

average noise temperature (antenna). The noise temperature of an antenna averaged over a specified frequency band.   111

average picture level (APL) (television). The average signal level, with respect to the blanking level, during the active picture scanning time (averaged over a frame period, excluding blanking intervals) expressed as a percentage of the difference between the blanking and reference white levels. *See:* **television.**   178

average power (in a waveguide). For a periodic wave, the time-average of the power passing through a given transverse section of the waveguide in a time interval equal to the fundamental period.   267
See: **average active power (single-phase, two-wire or polyphase circuit).**

average power output (amplitude-modulated transmitter). The radio-frequency power delivered to the transmitter output terminals averaged over a modulation cycle. *See:* **radio transmitter.**   111

average single-conductor (or subconductor) gradient. The value E_{av} obtained from

$$E_{av} = \frac{1}{2\pi} \int_{0}^{2\pi} E(\theta)\, d(\theta)$$

Approximately the average conductor gradient is given by

$$E_{av} = \frac{q}{2\pi\epsilon_0 r}$$

Note: For practical cases the average conductor gradient is approximately equal to the arithmetic mean of the maximum and minimum conductor gradients.
                                                             411

average voltage (rotating electric machinery). The value declared by the user to be the average of the system described, where externally powered.   424

average water conditions (power operations). Precipitation and runoff conditions which provide water for hydroelectric energy production approximating the average amount and distribution available over a long time period, usually the period of record.   516

averaging time, electrode current. *See:* **electrode current averaging time.**

A-weighted sound level (speech quality measurements). A weighted sound pressure level obtained by the use of a metering characteristic and the weighting *A* specified in USAS S1.4-1961 (General Purpose Sound Level Meters).   126

axially extended interaction tube (microwave tubes). A klystron tube utilizing an output circuit having more than one gap. *See:* **microwave tube or valve.**   190

axial magnetic centering force (rotating machinery). The axial force acting between rotor and stator resulting from the axial displacement of the rotor from magnetic center. *Note:* Unless other conditions are specified, the value of axial magnetic centering force will be for no load and rated voltage, and for rated no load field current and rated frequency as applicable.
                                                             63

axial mode (laser-maser). The mode in a beamguide or beam resonator which has one or more maxima for the transverse field intensities over the cross-section of the beam.   363

axial propagation constant (fiber optics). The propagation constant evaluated along the axis of a waveguide (in the direction of transmission). *Note:* The real part of the axial propagation constant is the attenuation constant while the imaginary part is the phase constant. *Syn:* **axial propagation wave number.** *See:* **attenuation; attenuation coefficient; attenuation constant; propagation constant.**   433

axial propagation wave number. *See:* **axial propagation constant.**   433

axial ratio (1) (antennas). The ratio of the major to the minor axes of a polarization ellipse. *Note:* The axial ratio sometimes carries a sign that is taken as plus if the sense of polarization is right-handed and minus if it is left-handed. *See:* **sense of polarization.**   111
(2) (waveguide). The ratio of the axes of the polarization ellipse. *Note:* The shape of the ellipse is defined by the axial ratio which is the major axis/minor axis. Sometimes the ratio is defined as the reciprocal of the above, that is, minor axis.   267

axial ratio pattern (antennas). A graphical representation of the axial ratio of a wave radiated by an antenna over a radiation pattern cut.   111

axial ray (fiber optics). A light ray that travels along the optical axis. *See:* **geometric optics; fiber axis; meridional ray; paraxial axis; skew ray.**   433

axial slab interferometry. *See:* **slab interferometry.**
                                                             433

axis-of-freedom (gyro). The axis about which a gimbal provides a degree-of-freedom.   46

axis *See:* **magnetic axis; direct axis; quadrature axis.**

axle bearing. A bearing that supports a portion of the weight of a motor or a generator on the axle of a vehicle. *See:* **bearing; traction motor.** 328

axle-bearing cap. The member bolted to the motor frame supporting the bottom half of the axle bearing. *See:* **bearing.** 328

axle-bearing-cap cover. A hinged or otherwise applied cover for the waste and oil chamber of the axle bearing. *See:* **bearing.** 328

axle circuit. The circuit through which current flows along one of the track rails to the train, through the wheels and axles of the train, and returns to the source along the other track rail. 328

axle current. The electric current in an axle circuit. 328

axle generator. An electric generator designed to be driven mechanically from an axle of a vehicle. *See:* **axle-generator system.** 328

axle-generator pole changer. A mechanically or electrically actuated changeover switch for maintaining constant polarity at the terminals of an axle generator when the direction of the rotation of the armature is reversed due to a change in direction of movement of a vehicle on which the generator is mounted. *See:* **axle-generator system.** 328

axle-generator regulator. A control device for automatically controlling the voltage and current of a variable-speed axle generator. *See:* **axle-generator system.** 328

axle-generator system. A system in which electric power for the requirements of a vehicle is supplied from an axle generator carried on the vehicle, supplemented by a storage battery. 328

axle-hung motor (or generator). A traction motor (or generator), a portion of the weight of which is carried directly on the axle of a vehicle by means of axle bearings. *See:* **traction motor.** 328

azimuth (1)(navigation aid terms). (A) The direction of a celestial point from a terrestrial point, expressed as the angle in the horizontal plane between a reference line and the horizontal projection of the line joining the two points. *Note:* True North is usually but not always implied where no reference direction is stated. (B) The angle between horizontal reference direction and the horizontal of the direction of boresight of the antenna. (3) Bearing. 526

(2) (illuminating engineering). The angular distance between the vertical plane containing a given line or celestial body and the plane of the meridian. 167

(3) (radar). The angle between a horizontal reference direction (usually north) and the horizontal projection of the direction of interest, usually measured clockwise. 13

azimuth discrimination (electronic navigation). *See:* **angular resolution.**

azimuth marks (markers)(navigation aid terms)(radar). Calibration marks for azimuth. 526,13

azimuth-stabilized plan position indicator (radar). A plan-position indicator (PPI) on which the reference direction remains fixed with respect to the indicator, regardless of the vehicle orientation. 13

B

babble. The aggregate crosstalk from a large number of interfering channels. *See:* **signal-to-noise ratio.** 239

back (motor or generator) (turbine or drive end). The end that carries the largest coupling or driving pulley. *See:* **armature.** 63

back-connected fuse (high-voltage switchgear). A fuse in which the current carrying conductors are fastened to the studs in the rear of the mounting base. 443

back connected switch (power switchgear). A switch in which the current carrying conductors are connected to studs in back of the mounting base. 103

back contact (1) (electric power apparatus relaying). A contact which is closed when the relay is reset. *Syn:* **"b" contact.** *See:* **b contact.** 103, 127

(2) (utility-consumer interconnections relaying). A contact that is closed when the relay is deenergized. 128

back course (ILS [instrument landing system])(navigation aid terms). The course which is located on the opposite side of the localizer from the runway. 526

backed stamper (phonograph techniques) (mechanical recording). A thin metal stamper that is attached to a backing material, generally a metal disk of desired thickness. *See:* **phonograph pickup.** 176, 256

backfill (noun)(NESC). Materials such as sand, crushed stone, or soil, which are placed to fill an excavation. 494

back filter (surge testing for equipment connected to low-voltage ac power circuits). A filter inserted in the power line feeding an equipment to be surge tested; this filter has a dual purpose: (1) To prevent the applied surge from being fed back to the power source where it may cause damage. (2) To eliminate loading effects of the power source on the surge generator. 578

backfire antenna. An antenna consisting of a radiating feed, a reflector element, and a reflecting surface such that the antenna functions as an open resonator, with radiation from the open end of the resonator. 111

back flashover (lightning). A flashover of insulation resulting from a lightning stroke to part of a network or electric installation that is normally at ground potential. *See:* **direct-stroke protection (lightning).** 64

back focal length (laser-maser). The distance from the last optical surface of a lens to the focal point. 363

background (associated with a spectral peak from a semiconductor detector)(1)(X-ray energy spectrometers). Non-ideal spectral response which results from radiation which is not part of the monoenergetic line of interest. 471

(2)(test, measurement and diagnostic equipment). Those effects present in physical apparatus or surrounding environment which limit the measurement or observation of low level signals or phenomenon: commonly referred to as noise (background acoustical noise, background electromagnetic radiation, background ionizing radiation). 54

background check source (liquid-scintillation counting)(liquid-scintillation counters). A sealed vial of liquid-scintillation solution containing no added radioactive material. 422,498

background count rate (in radioactive counters)(liquid-scintillation counters). Count rate recorded by the instrument when measuring a background check source. 498

background counts (1) (in radioactivity counters)(liquid-scintillation counting). Counts recorded in the instrument when measuring a background check source. 422

(2) (radiation counters). Counts caused by ionizing radiation coming from sources other than that to be measured: and by any electric disturbance in the circuitry that is used to record the counts. See: ionizing radiation. 190, 117, 96, 125

background ionization voltage (surge arresters). A high-frequency voltage appearing at the terminals of the apparatus to be tested that is generated by ionization extraneous to the apparatus. Note: While this voltage does not add arithmetically to the radio influence or internal ionization voltage, it affects the sensitivity of the test. See: surge arrester (surge diverter). 229, 62

background level (sound measurement). Any sound at the points of measurement other than that of the machine being tested. It also includes the sound of any test support equipment. 129

background noise (1)(radio noise from overhead power lines and substations). The total system noise independent of the presence or absence of radio noise from the power line or substation. Note: Background noise is presumed to be reduced to a level of insignificance. 509

(2) (data transmission). Noise due to audible disturbances or periodic random occurrence, or both. 59

(3) (overhead-power-line corona and radio noise). The total system noise independent of the presence or absence of radio noise from the power line. Note: Background noise is not to be included as part of the radio noise measured from the power line. See: (Revised 1976, ANSI.) 411

(4) (antennas). Noise due to audible disturbances of periodic and or random occurrence. See: modulation. 111

(5) (electroacoustics) (recording and reproducing). The total system noise in the absence of a signal. See: phonograph pickup. 176

(6) (receivers). The noise in the absence of signal modulation on the carrier. 339

(7) (telephone practice). The total system noise independent of the presence or absence of a signal. 24

(8) (communication satellite). That part of the receiving system noise power produced by noise sources in the celestial background of the radiation pattern of the receiving antenna. Typical sources are the galaxy (galactic noise) the sun and radio stars. 85

background response (radiation detectors). Response caused by ionizing radiation coming from sources other than that to be measured. See: ionizing radiation. 117

background returns (radar). See: clutter.

backing (planar structure) (rotating machinery). A fabric, mat, film, or other material used in intimate conjunction with a prime material and forming a part of the composite for mechanical support or to sustain or improve its properties. 63

backing lamp (backing light) (backup light). See: backup lamp. 328

backlash (1) (general). A relative movement between interacting mechanical parts, resulting from looseness. See: control system, feedback; industrial control. 266, 207, 219, 206

(2) (signal generator). The difference in actual value of a parameter when the parameter is set to an indicated value by a clockwise rotation of the indicator, and when it is set by a counterclockwise rotation. See: signal generator. 185

(3) (tunable microwave tube). The amount of motion of the tuner control mechanism (in a mechanically tuned oscillator) that produces no frequency change upon reversal of the motion. See: tunable microwave oscillators. 174, 190

back light (illuminating engineering). Illumination from behind the subject in a direction substantially parallel to a vertical plane through the optical axis of the camera. 167

back lobe (antennas). A radiation lobe whose axis makes an angle of approximately 180° with respect to the axis of the major lobe of an antenna. Note: By extension, a radiation lobe in the half-space opposed to the direction of peak directivity. 111

back pitch (rotating machinery). The coil pitch at the nonconnection end of a winding (usually in reference to a wave winding). 63

backplane (FASTBUS acquisition and control). Circuit board (typically printed) at the rear of a crate which, by means of its attached connectors, mates with the modules and constitutes the crate segment. 480

back plate (signal plate) (camera tubes). The electrode in an iconoscope or orthicon camera tube to which the stored charge image is capacitively coupled. See: television. 178

back porch (1) (monochrome composite picture sig-

nal). The portion that lies between the trailing edge of a horizontal synchronizing pulse and the trailing edge of the corresponding blanking pulse.
(2) (National Television System Committee composite color-picture signal). The portion that lies between the color burst and the trailing edge of the corresponding blanking pulse. *See:* **television.**   178
backscatter (radar). Energy reflected in a direction opposite to that of the incident wave.   13
backscatter coefficient (radar). A normalized measure of radar reflection from a distributed scatterer; usually applied to clutter. For area clutter, such as ground or sea clutter, it is defined as the monostatic radar cross section per unit surface area, which is dimensionless but is sometimes written in units of m^2/m^2 for clarity. For volume clutter, such as rain or chaff, it is defined as the monostatic radar cross section per unit volume; the units are m^2/m^3, or m^{-1}.   13
backscattering (fiber optics). The scattering of light into a direction generally reverse to the original one. *See:* **Rayleigh scattering; reflectance; reflection.**
    433
back scattering coefficient (B) (echoing area) (data transmission). Of an object for an incident plane wave is 4π the ratio of the reflected power per unit solid angle ϕ_r in the direction of the source to the power per unit area (W_1) in the incident wave:

$$B = 4\pi \frac{\phi_r}{W_1} = 4\pi r^2 \frac{W_r}{W_1}$$

where W_r is the power per unit area at distance r. *Note:* For large objects, the back scattering coefficient of an object is approximately the product of its interception area and its scattering gain in the direction of the source, where the interception area is the projected geometrical area and the scattering gain is the reradiated power gain relative to an isotropic radiator.
    59
backscattering cross section (1) (antenna). The scattering cross-section in the direction towards the source. *See:* **backscattering coefficient.**
(2) (radar). *See:* **radar cross section.**   13, 111
back-shunt keying. A method of keying a transmitter in which the radio-frequency energy is fed to the antenna when the telegraph key is closed and to an artificial load when the key is open. *See:* **radio transmission.**   111, 211
backstop, relay. *See:* **relay backstop.**
backswing (1) (last transition overshoot), A_{BS} (pulse transformers). The maximum amount by which the instantaneous pulse value is below the zero axis in the region following the fall time. It is expressed in amplitude units or as a percentage of A_M.   589
(2) (low power pulse transformers) (high power pulse transformers) (last transition overshoot). (A_{BS}) The maximum amount by which the instantaneous pulse value is below the zero axis in the region following the fall time. It is expressed in amplitude units or in percent of A_M. *See:* **input pulse shape.**   32, 33
backup (1)(supervisory control, data acquisition, and automatic control). Provision for an alternate means

of operation if the primary system is not available.
    570
back-up (2)(software). Provisions made for the recovery of data files or software and for restart of processing or use of alternative computer equipment after a system failure or disaster. *See:* **data files; software; system failure.**   434
back-up air gap devices (low-voltage air gap surge-protective devices)(gas-tube surge protective devices). An air gap device connected in parallel with a sealed gas tube device, having a higher breakdown voltage than the gas tube, which provides a secondary means of protection in the event of a venting to atmosphere by the primary gas tube device.   556, 490
backup current limiting fuse (1) (high-voltage switchgear). A fuse capable of interrupting all currents from the maximum rated interrupting current down to the rated minimum interrupting current. *See:* **function class "a" (backup) current-limiting fuse.**   443
backup gap (series capacitors). A supplementary gap which may be set to sparkover at a voltage level higher than the protective level of the primary protective device, and which is normally placed in parallel with the primary protective device.   474
back-up lamp (illuminating engineering). A lighting device mounted on the rear of a vehicle for illuminating the region near the back of the vehicle while moving in reverse. It normally can be used only while backing up.   167
backup overcurrent protective device or apparatus (nuclear power generating stations). A device or apparatus which performs the circuit interrupting function in the event the primary protective device or apparatus fails or is out of service.   26
backup programmer (software). The assistant leader of a chief programmer team; a senior-level programmer whose responsibilities include contributing significant portions of the software being developed by the team, aiding the chief programmer in reviewing the work of the other team members, substituting for the chief programmer when necessary, and having an overall technical understanding of the software being developed. *See:* **chief programmer; chief programmer team; software.**   434
backup protection (as applied to a relay system) (power switchgear). A form of protection that operates independently of specified components in the primary protective system. It may duplicate the primary protection or may be intended to operate only if the primary protection fails or is temporarily out of service.   103, 127
Backus-Naur form (BNF) (ATLAS). A particular metalanguage developed by Backus and Naur.
    400
Backus-Naur notation (ATLAS). A general term relating to metalanguages that use the concepts developed by Backus and Naur.   400
backward-acting regulator. A transmission regulator in which the adjustment made by the regulator affects the quantity that caused the adjustment. *See:* **transmission regulator.**   328

backward diode (nonlinear, active, and nonreciprocal waveguide components). A semiconductor device used primarily as a detector or mixer. Quantum-mechanical tunneling in this diode results in a current-voltage characteristic in which the reverse current is greater than the forward current for equal applied voltages of opposite polarity.　　530

backward-wave (BW) structure (microwave tubes). A slow-wave structure whose propagation is characterized on an ω/β diagram (sometimes called a Brillouin diagram) by a negative slope in the region $0 < \beta < \pi$ (in which the phase velocity is therefore of opposite sign to the group velocity). *See:* **microwave tube (or valve).**　　190

backward wave (traveling-wave tubes). A wave whose group velocity is opposite to the direction of electron-stream motion. *See:* **amplifier.**　　125

backward-wave oscillator. *See:* **carcinotron.**

back wave. A signal emitted from a radio telegraph transmitter during spacing portions of the code characters and between the code characters. *See:* **radio transmission.**　　211

bactericidal effectiveness (illuminating engineering). The capacity of various portions of the ultraviolet spectrum to destroy bacteria, fungi, and viruses. *Syn:* **germicidal effectiveness.**　　167

bactericidal efficency of radiant flux (illuminating engineering). The ratio of the bactericidal effectiveness of that wavelength to that of wavelength 265.0 nm (nanometers), which is rated as unity. *Note:* Tentative bactericidal efficiency of various wavelengths of radiant flux is given in the table under erythemal flux density. *Syn:* **germicidal efficency of radiant flux.**　　167

bactericidal exposure (illuminating engineering). The product of bactericidal flux density on a surface and time. It usually is measured in bactericidal microwatt-minutes per square centimeter or bactericidal watt-minutes per square foot. *Syn:* **germicidal exposure.**　　167

bactericidal flux (illuminating engineering). Radiant flux evaluated according to its capacity to produce bactericidal effects. It usually is measured in microwatts of ultraviolet radiation weighted in accordance with its bactericidal efficiency. Such quantities of bactericidal flux would be in bactericidal microwatts. *Note:* Ultraviolet radiation of wavelength 253.7 nm (nanometers) usually is referre d to as 'ultraviolet microwatts' or 'UV watts'. These terms should not be confused with 'bactericidal microwatts' because the radiation has not been weighted in accordance with the values given in the table under **erythemal flux density.** *Syn:* **germicidal flux.**　　167

bactericidal flux density (illuminating engineering). The bactericidal flux per unit area of the surface being irradiated. It is equal to the quotient of the incident bactericidal flux divided by the area of the surface when the flux is uniformly distributed. It usually is measured in microwatts per square centimeter or watts per square foot of bactericidally weighted ultraviolet radiation (bactericidal microwatts per square

centimeter or bactericidal watts per square foot). *Syn:* **germicidal flux density.**　　167

baffle (1) (audio and electroacoustics). A shielding structure or partition used to increase the effective length of the transmission path between two points in an acoustic system as for example, between the front and back of an electroacoustic transducer. *Note:* In the case of a loudspeaker, a baffle is often used to increase the acoustic loading of the diaphragm. *See:* **loudspeaker.**　　176

(2) (illuminating engineering). A single opaque or translucent element to shield a source from direct view at certain angles, or to absorb unwanted light.　　167

(3) (gas tube). An auxiliary member, placed in the arc path and having no separate external connection. *Note:* A baffle may be used for: (A) controlling the flow of mercury vapor or mercury particles: (B) controlling the flow of radiant energy: (C) forcing a distribution of current in the arc path: or (D) deionizing the

Erythemal and bactercidal efficiency of ultraviolet radiation

Wavelength (nanometers)	Erythemal Efficiency	Tentative Bactericidal Efficiency
*235.3	——	0.35
240.0	0.56	——
*244.6	0.57	0.58
*248.2	0.57	0.70
250.0	0.57	——
*253.7	0.55	0.85
*257.6	0.49	0.94
260.0	0.42	——
265.0	——	1.00
*265.4	0.25	0.99
*267.5	0.20	0.98
*270.0	0.14	0.95
*275.3	0.07	0.81
*280.4	0.06	0.68
285.0	0.09	0.68
*285.7	0.10	0.55
*289.4	0.25	0.46
290.0	0.31	——
*292.5	0.70	0.38
295.0	0.98	——
*296.7	1.00	0.27
300.0	0.83	——
*302.2	0.55	0.13
305.0	0.33	——
310.0	0.11	——
*313.0	0.03	0.01
315.0	0.01	——
320.0	0.005	——
325.0	0.003	——
330.0	0.000	——

*Emission lines in the mercury spectrum; other values interpolated.

mercury vapor following conduction. It may be of either conducting or insulating material. *See:* **electrode (electron tube).** 190

bag-type construction (dry cell) (primary cell). A type of construction in which a layer of paste forms the principal medium between the depolarizing mix, contained within a cloth wrapper, and the negative electrode. *See:* **electrolytic cell.** 328

balance beam (of a relay) (power switchgear). A lever form of relay armature, one end of which is acted upon by one input and the other end restrained by a second input. 127, 103

balance check (analog computers). The computer-control state in which all amplifier summing junctions are connected to the computer zero reference level (usually signal ground) to permit zero balance of the operational amplifiers. Integrator capacitors may be shunted by a resistor to permit the zero balance of an integrator. This control state may not be found in some analog computers. 9

balanced (1) (general). Used to signify proper relationship between two or more things, such as stereophonic channels.

(2) (data transmission). In communication practice, signifies (A) electrically alike and symmetrical with respect to ground, or (B) arranged to provide conjugate conductors between certain sets of terminals. 59

balanced amplifier (push-pull amplifier). *See:* **amplifier, balanced.**

balanced capacitance (between two conductors) (mutual capacitance between two conductors). The capacitance between two conductors when the changes in the charges on the two are equal in magnitude but opposite in sign and the potentials of the other $n - 2$ conductors are held constant. *See:* **direct capacitances (system of conductors).** 185

balanced circuit (1)(measuring longitudinal balance of telephone equipment operating in the voice band). A circuit in which two branches are electrically alike and symmetrical with respect to a common reference point, usually ground. 529

(2)(signal-transmission system). A circuit, in which two branches are electrically alike and symmetrical with respect to a common reference point, usually ground. *Note:* For an applied signal difference at the input, the signal relative to the reference at equivalent points in the two branches must be opposite in polarity and equal in amplitude. 185

(3) (electric power system). A circuit in which there are substantially equal currents, either alternating or direct, in all main wires and substantially equal voltages between main wires and between each main wire and neutral (if one exists). *See:* **center of distribution.** 64

balanced conditions (1) (time domain) (rotating machinery). A set of polyphase quantities (phase currents, phase voltages, etcetera) that are sinusoidal in time, that have identical amplitudes, and that are shifted in time with respect to each other by identical phase angles.

(2) (space domain). In space, a set of coils (for example, of a rotating machine) each having the same number of effective turns, with their magnetic axes shifted by identical angular displacements with respect to each other. *Notes:* (A) The impedance (matrix) of a balanced machine is balanced. A balanced set of currents will produce a balanced set of voltage drops across a balanced set of impedances. (B) If all sets of windings of a machine are balanced and if the magnetic structure is balanced, the machine is balanced. *See:* **asynchronous machine.** 63

balanced currents (on a balanced line) (waveguide). Currents flowing in the two conductors of a balanced line which, at every point along the line, are equal in magnitude and opposite in direction. 267

balanced duplexer (radar)(nonlinear, active, and nonreciprocal waveguide components). A dualized network using two quadrature hybrids on each side of a pair of self-switching elements used to interconnect the transmitter, receiver, and antenna in a radar. *See:* **duplexer.** 530

balanced line (two conductor) (waveguide). A transmission line consisting of two conductors in the presence of ground capable of being operated in such a way that the voltages on the two conductors at all transverse planes are equal in magnitude and opposite in direction. The ground may be a conducting sheath, forming a shielded transmission line. 267

balanced line system (waveguide). A system consisting of a generator, balanced line, and load adjusted so that the voltages of the two conductors at all transverse planes are equal in magnitude and opposite in polarity with respect to ground. 267

balanced mixer (1) (circuits and systems) (single, double). A type of mixer that forms from two signals A & B a third signal C having the form C = (a + A) (b + B). Single balanced implies a = 0, b ≠ 0: double balanced implies a = b = 0. Note: Such mixers can suppress a RF carrier and/or a local oscillator in their output spectrum. *Syn:* **balanced modulator.** 67

(2) A hybrid junction with crystal receivers in one pair of uncoupled arms, the arms of the remaining pair being fed from a signal source and a local oscillator. *Note:* The resulting intermediate-frequency signals from the crystals are added in such a manner that the effect of local-oscillator noise is minimized. *See:* **converter; hybrid junction; radio receiver; waveguide.** 244, 179

balanced modulator (signal-transmission system). A modulator, specifically a push-pull circuit, in which the carrier and modulating signal are so introduced that after modulation takes place the output contains the two sidebands without the carrier. *See:* **modulation; modulator, symmetrical.** 111

balanced oscillator. An oscillator in which at the oscillator frequency the impedance centers of the tank circuit are at ground potential and the voltages between either end and their centers are equal in magnitude and opposite in phase. *See:* **oscillatory circuit.** 111, 211

balanced polyphase load. A load to which symmetrical

currents are supplied when it is connected to a system having symmetrical voltages. *Note:* The term balanced polyphase load is applied also to a load to which are supplied two currents having the same wave form and root-mean-square value and differing in phase by 90 electrical degrees when it is connected to a quarter-phase (or two-phase) system having voltages of the same wave form and root-mean-square value. *See:* **generating station.** 64

balanced polyphase system. A polyphase system in which both the currents and voltages are symmetrical. *See:* **alternating-current distribution.** 64

balanced telephone-influence factor (three-phase synchronous machine). The ratio of the square root of the sum of the squares of the weighted root-mean-square values of the fundamental and the nontriple series of harmonics to the root-mean-square value of the normal no-load voltage wave. 63

balanced termination (system or network having two output terminals). A load presenting the same impedance to ground for each of the output terminals. *See:* **network analysis.** 267

balanced three-wire system. A three-wire system in which no current flows in the conductor connected to the neutral point of the supply. *See:* **three-wire system; alternating-current distribution.** 64

balanced voltages (1) (on a balanced line) (waveguide). Voltages relative to ground on the two conductors of a balanced line which, at every point along the line, are equal in magnitude and opposite in polarity. 267

(2) (signal-transmission system). The voltages between corresponding points of a balanced circuit (voltages at a transverse plane) and the reference plane relative to which the circuit is balanced. *See:* **signal.** 188

balanced wire circuit (data transmission). One whose two sides are electrically alike and symmetrical with respect to ground and other conductors. The term is commonly used to indicate a circuit whose two sides differ only by chance. 59

balancer. That portion of a direction-finder that is used for the purpose of improving the sharpness of the direction indication. *See:* **radio receiver.** 328

balance relay (power switchgear). A relay which operates by comparing the magnitudes of two similar input quantities. *Note:* The balance may be effected by counteracting electromagnetic forces on a common armature, or by counteracting magnetomotive forces in a common magnetic circuit, or by similar means, such as springs, levers, etcetera. 103

balance test (rotating machinery). A test taken to enable a rotor to be balanced within specified limits. *See:* **rotor (rotating machinery).** 63

balancing network. An electric network designed for use in a circuit in such a way that two branches of the circuit are made substantially conjugate, that is, such that an electromotive force inserted in one branch produces no current in the other branch. *See:* **network analysis.** 59

balancing of an operational amplifier (analog computers). The act of adjusting the output level of an opera-

tional amplifier to coincide with its input reference level, usually ground or zero voltage, in the "balance check" computer-control state. This operation may not be required in some amplifiers, and there may be no provision for performing it. 9

ballast (1) (fluorescent lamps or mercury lamps). Devices that by means of inductance, capacitance, or resistance, singly or in combination, limit the lamp current of fluorescent or mercury lamps, to the required value for proper operation, and also, where necessary, provide the required starting voltage and current and, in the case of ballasts for rapid-start lamps, provide for low-voltage cathode heating. *Note:* Capacitors for power-factor correction and capacitor-discharge resistors may form part of such a ballast. 268, 269

(2) (fixed-impedance type) (reference ballast). Designed for use with one specific type of lamp that, after adjustment during the original calibration, is expected to hold its established impedance through normal use. 270, 271

(3) (variable-impedance type). An adjustable inductive reactor and a suitable adjustable resistor in series. *Note:* These two components are usually designed so that the resulting combination has sufficient current-carrying capacity and range of impedance to be used with a number of different sizes of lamps. The impedance and power factor of the reactor-resistor combination are adjusted, or rechecked, each time the unit is used. 270, 271

(4) (illuminating engineering). A device used with an electric-discharge lamp to obtain the necessary circuit conditions (voltage, current, and wave form) for starting and operating. 167

ballast factor (illuminating engineering). The fractional loss of task illuminance due to use of a ballast other than the standard one. 167

ballast leakage. The leakage of current from one rail of a track circuit to another through the ballast, ties, earth, etcetera. 328

ballast resistance. The resistance offered by the ballast, ties, earth, etcetera, to the flow of leakage current from one rail of a track circuit to another. 328

ballast section (railroads)(NESC). The section of material, generally trap rock, which provides support under railroad tracks. 494

ballast tube (ballast lamp). A current-controlling resistance device designed to maintain substantially constant current over a specified range of variation in the applied voltage or the resistance of a series circuit. 328

ball bearing (rotating machinery). A bearing incorporating a peripheral assembly of balls. *See:* **bearing.** 63

ball burnishing. Burnishing by means of metal balls. *See:* **electroplating.** 328

ballistic deficit (germanium gamma-ray detectors). The loss in signal amplitude that occurs when the charge collection time in a detector is a significant fraction of the amplifier's differentiating time constant. 528

ballistic focusing (microwave tubes). A focusing system in which static electric fields cause an initial convergence of the beam and the electron trajectories are thereafter determined by momentum and space charge forces only. *See:* **microwave tube or valve.**
189

ball lightning. A type of lightning discharge reported from visual observations to consist of luminous, ball-shaped regions of ionized gases. *Note:* In reality ball lightning may or may not exist. *See:* **direct-stroke protection (lightning).**
64

balun. (1) A passive device having distributed electrical constants used to couple a balanced system or device to an unbalanced system or device. *Note:* The term is derived from **balance to unbalance transformer.**
197
(2) A network for the transformation from an unbalanced transmission line or system to a balanced line or system, or vice versa.
185

banana plug. A single-conductor plug with a spring metal tip that somewhat resembles a banana in shape.
329

band (1) (electronic computers). A group of circular recording tracks, on a moving storage device such as a drum or disc. *See:* **channel.**
54
(2) (data transmission). Range of frequency between two defined limits.
59

band, effective (facsimile). The frequency band of a facsimile signal wave equal in width to that between zero frequency and maximum keying frequency. *Note:* The frequency band occupied in the transmission medium will in general be greater than the effective band. *See:* **facsimile signal (picture signal).**
12

band-elimination filter. *See:* **filter, band-elimination.**

band gap (charged-particle detectors)(in a semiconductor)(germanium gamma-ray detectors). The difference in energy between the energy level of the bottom of the conduction band and the energy level of the top of the valence band.
119, 528

banding insulation (rotating machinery). Insulation between the winding overhang and the binding bands.
63

band of regulated voltage (excitation systems for synchronous machines). The band or zone, expressed in percent of the rated value of the regulated voltage, within which the excitation system will hold the regulated voltage of an electric machine during steady or gradually changing conditions over a specified range of load.
507

band-pass filter (data transmission). A wave filter which has a single transmission band, neither of the cutoff frequencies being zero or infinite.*See:* **optical filter**
59

bandpass tube (microwave gas tubes). *See:* **broad-band tube.**

band spreading. (1) The spreading of tuning indicators over a wide scale range to facilitate tuning in a crowded band of frequencies. (2) The method of double-sideband transmission in which the frequency band of the modulating wave is shifted upward in frequency so

that the sidebands produced by modulation are separated in frequency from the carrier by an amount at least equal to the bandwidth of the original modulating wave, and second-order distortion products may be filtered from the demodulator output. *See:* **radio receiver.**
328

band switch. A switch used to select any one of the frequency bands in which an electric transmission apparatus may operate.
328

bandwidth (1)(amplitude-modulation broadcast receivers). As applied to the selectivity of a radio receiver, the bandwidth is the width of a selectivity graph at a specified level on the scale of ordinates.
524
(2)(fiber optics). *See:* **fiber bandwidth.**
433
(3) (device). The range of frequencies within which performance, with respect to some characteristic, falls within specific limits. *See:* **radio receiver.**
(4) (wave). The least frequency interval outside of which the power spectrum of a time-varying quantity is everywhere less than some specified fraction of its value at a reference frequency. *Warning:* This definition permits the spectrum to be less than the specified fraction within the interval. *Note:* Unless otherwise stated, the reference frequency is that at which the spectrum has its maximum value.
339
(5) (burst) (burst measurements). The smallest frequency interval outside of which the integral of the energy spectrum is less than some designated fraction of the total energy of the burst. *See:* **burst.**
272
(6) (antenna). The range of frequencies within which its performance, in respect to some characteristics, conforms to a specified standard. *See:* **antenna.**
111, 179
(7) (facsimile). The difference in hertz between the highest and the lowest frequency components required for adequate transmission of the facsimile signals. *See:* **facsimile (electrical communication).**
12
(8) (industrial control) (excitation control systems). The interval separating two frequencies between which both the gain and the phase difference (of sinusoidal output referred to sinusoidal input) remain within specified limits. *Note:* For control systems and many of their components, the lower frequency often approaches zero. *See:* **control system, feedback.**
266, 219, 206, 329, 353
(9) (pulse terms). The two portions of a pulse waveform which represents the first nominal state from which a pulse departs and to which it ultimately returns. Typical closed-loop frequency response of an excitation control system with the synchronous machine open circuited.
(10) (signal-transmission system). The range of frequencies within which performance, with respect to some characteristic, falls within specific limits. *Note:* For systems capable of transmitting at zero frequency the frequency at which the system response is less than that at zero frequency by a specified ratio. For carrier-frequency systems: the difference in the frequencies at which the system response is less than that at the frequency of reference response by a specified

ratio. For both types of systems, bandwidth is commonly defined at the points where the response is 3 decibels less than the reference value (0.707 root-mean-square voltage ratio). *See:* **equivalent noise bandwidth.**   188

(11) (oscilloscope). The difference between the upper and lower frequency at which the response is 0.707 (mins3 decibels) of the response at the reference frequency. Usually both upper and lower limit frequencies are specified rather than the difference between them. When only one number appears, it is taken as the upper limit. *Notes:* (A) The reference frequency shall be at least 20 times greater for the lower bandwidth limit and at least 20 times less for the upper bandwidth limit than the limit frequency. The upper and lower reference frequencies are not required to be the same. In cases where exceptions must be made, they shall be noted. (B) This definition assumes the amplitude response to be essentially free of departures from a smooth roll-off characteristic. (C) If the lower bandwidth limit extends to zero frequency, the response at zero frequency shall be equal to the response at the reference frequency, not mins3 decibels from it.   185

(12) (dispersive and nondispersive delay lines). A specified frequency range over which the amplitude response does not vary more than a defined amount. *Note:* Typically amplitude range is 1 dB bandwidth, 3 dB bandwidth.   81

(13) (analog computer). (A) Of a signal, the difference between the limiting frequencies encountered in the signal. (B) Of a device, the range of frequencies within which performance in respect to some characteristic falls within specific limits.   9

(14) (data transmission). The range of frequencies within which performance, with respect to some characteristic falls within specific limits. Bandwidth is commonly defined at the points where the response is three decibels less than the reference value.   59

(15) (overhead-power-line corona and radio noise). The range of frequencies within which performance, with respect to some characteristic, falls within specific limits.   411

bandwidth, effective (bandpass filter in a signal transmission system). The width of an assumed rectangular bandpass filter having the same transfer ratio at a reference frequency and passing the same mean square of a hypothetical current and voltage having even distribution of energy over all frequencies. *Note:* For a nonlinear system, the bandwidth at a specified input level. *See:* **network analysis; signal.**   111, 188

bandwidth-limited operation (fiber optics). The condition prevailing when the system bandwidth, rather than the amplitude (or power) of the signal, limits performance. The condition is reached when the system distorts the shape of the waveform beyond specified limits. For linear systems, bandwidth-limited operation is equivalent to distortion-limited operation. *See:* **attenuation-limited operation; distortion-limited operation; linear optical element.**   433

bandwidth, root-mean-square. The root mean squared (rms) deviation of the power spectrum of the received signal relative to zero frequency or the spectral center, in units of radians per second. This bandwidth, β, may be defined as

$$\beta^2 = \frac{\int_{-\infty}^{\infty} [2\pi(f - f_0)]^2 \, |S(f)|^2 \, df}{\int_{-\infty}^{\infty} |S(f)|^2 \, df}$$

where $S(f)$ is the Fourier transform of the signal $s(t - \tau_0)$ with true time delay τ_0 and f_0 is the center frequency of the spectrum. *Note:* β^2 is the normalized second moment of the spectrum $|S(f)|^2$ about the mean.   13

bang snuffer (nonlinear, active, and nonreciprocal waveguide components). A switch used in radar receivers to suppress carrier leakage during the transmit period. *See:* **gate (microwave).**   530

bank. An aggregation of similar devices (for example, transformers, lamps, etcetera) connected together and used in cooperation. *Note:* In automatic switching, a bank is an assemblage of fixed contacts over which one or more wipers or brushes move in order to establish electric connections. *See:* **relay level.**   328

bank (navigation)(navigation aid terms). Lateral inclination of an aircraft in flight. *See:* **list.**   526

bank-and-wiper switch (telephone switching system). A switch in which an electromagnetic ratchet or other mechanisms are used, first, to move the wipers to a desired group of terminals, and second, to move the wipers over the terminals of this group to the desired bank contacts.   328

banked winding. *See:* **bank winding.**

bank winding (banked winding). A compact multilayer form of coil winding, for the purpose of reducing distributed capacitance, in which single turns are wound successively in each of two or more layers, the entire winding proceeding from one end of the coil to the other, without return.   329

bar (of lights) (illuminating engineering). A group of three or more aeronautical ground lights placed in a line transverse to the axis, or extended axis, of the runway.   167

bare conductor (National Electrical Code). A conductor having no covering or electrical insulation whatsoever. *See:* **covered conductor.**   256

bare lamp (illuminating engineering). A light source with no shielding. *Syn:* **exposed lamp.**   167

barette (illuminating engineering). A short bar in which the lights are closely spaced so that from a distance they appear to be a linear light. *Note:* Barettes are usually less than 4.6 m (15 ft) in length.   167

bar generator (television). A generator of pulses that are uniformly spaced in time and are synchronized to produce a stationary bar pattern on a television screen. *See:* **television.**   339

Barkhausen-Kurz oscillator. An oscillator of the retarding-field type in which the frequency of oscillation depends solely upon the electron transit-time within the tube. *See:* **oscillatory circuit.**   111

Barkhausen tube. *See:* positive-grid oscillator tube.

barometric altimeter (navigation aid terms). Essentially an aneroid barometer, an instrument which determines atmospheric pressure and is graduated in feet above sea level.   526

barothermograph (navigation aid terms). An instrument which automatically records pressure and temperature.   526

bar pattern (television). A pattern of repeating lines or bars on a television screen. When such a pattern is produced by pulses that are equally separated in time, the spacing between the bars on the television screen can be used to measure the linearity of the horizontal or vertical scanning systems. *See:* television.   328

barrel plating. Mechanical plating in which the cathodes are kept loosely in a container that rotates. *See:* electroplating.   328

barretter (waveguide components). A form of bolometer element having a positive temperature coefficient of resistivity which typically employs a power-absorbing wire or thin metal film.   166

barrier (1) (power switchgear). A partition for the insulation or isolation of electric circuits or electric arcs.   103, 443

(2) (in a semiconductor) (obsolete). *See:* depletion layer.

(3) (Class 1E equipment and circuits). A device or structure interposed between redundant Class 1E equipment or circuits, or between Class 1E equipment or circuits and a potential source of damage to limit damage to Class 1E systems to an acceptable level.   131

(4) (nuclear power generating stations). A device or structure interposed between Class 1E equipment or circuits and a potential source of damage to limit damage to Class 1E systems to an acceptable level.   131

barrier grid (charge-storage tubes). A grid, close to or in contact with a storage surface, which grid establishes an equilibrium voltage for secondary-emission charging and serves to minimize redistribution. *See:* charge-storage tube.   174, 190

barrier layer (fiber optics). In the fabrication of an optical fiber, a layer that can be used to create a boundary against OH⁻ ion diffusion into the core. *See:* core.   433

barrier wiring techniques (coupling in control systems). Those wiring techniques which obstruct electric or magnetic fields, excluding or partially excluding the fields from a given circuit. Barrier techniques are often effective against electromagnetic radiation also. In general, these techniques change the coupling coefficients between wires connected to a noise source and the signal circuit. Example: placement of signal lines within steel conduit to isolate them from an existing magnetic field. *See:* compensatory wiring techniques; suppressive wiring techniques.   43

barring hole (rotating machinery). A hole in the rotor to permit insertion of a pry bar for the purpose of turning the rotor slowly or through a limited angle. *See:* rotor (rotating machine).   63

bar, rotor (rotating machinery). *See:* rotor bar.

bar-type current transformer. One that has a fixed and straight single primary turn passing through the magnetic circuit. The primary and secondary(s) are insulated from each other and from the core(s) and are assembled as an integrated structure.   203

base (1) (number system). An integer whose successive powers are multiplied by coefficients in a positional notation system. *See:* positional notation; radix.   235

(2) (rotating machinery). A structure, normally mounted on the foundation, that supports a machine or a set of machines. In single-phase machines rated up through several horsepower, the base is normally a part of the machine and supports it through a resilient or rigid mounting to the end shields.   63

(3) (electron tube or valve). The part attached to the envelope, carrying the pins or contacts used to connect the electrodes to the external circuit and that plugs in to the holder. *See:* electron tube.   244, 190

(4) (basis or base metals) (electroplating). The object upon which the metal is electroplated. *See:* electroplating.   328

(5) (transistor). A region that lies between an emitter and a collector of a transistor and into which minority carriers are injected. *See:* transistor.   245, 66

(6) (high-voltage fuse). The supporting member to which the insulator unit or units are attached.   79, 103, 443

(7) (pulse terms). The two portions of a pulse waveform which represents the first nominal state from which a pulse departs and to which it ultimately returns.   254

base active power (synchronous generators and motors). The total (generator) output or (motor) input power at base voltage and base current with a power factor of unity.

base address. A given address from which an absolute address is derived by combination with a relative address.   255, 77

base ambient temperature (power distribution underground cables) (cable or duct). The no-load temperature in a group with no load on any cable or duct in the group.   57

base apparent power (1)(alternating-current (ac) rotating machinery)(basic per-unit quantities for ac rotating machines). A reference value expressing an electrical power rating of the machine. *Notes:* (A) Base apparent power may be either input or output power, and the numerical value may be either real power--watts (W)--or total apparent electrical power--voltamperes (VA), depending upon machine type. Base apparent power is usually expressed in voltamperes, but any consistent set of units may be used. For synchronous generators, induction generators, and synchronous motors, base apparent power is the total apparent electrical at rated voltage and rated current. In induction motors (preferred method), base apparent power is numerically equal to the rated power output. For induction motors (alternate method), base apparent power is the total apparent electrical power at rated voltage and rated current. (B) When the alternate method is used it should be identified as input voltampere base .   517

(2)(synchronous generators and motors). The total rated apparent power at rated voltage and rated current. *Note:* Base apparent power is usually expressed in volt-amperes, but any consistent set of units may be used.   5

baseband (1) (carrier or subcarrier wire or radio transmission system). The band of frequencies occupied by the signal before it modulates the carrier (or subcarrier) frequency to form the transmitted line or radio signal. *Note:* The signal in the baseband is usually distinguished from the line or radio signal by ranging over distinctly lower frequencies, which at the lower end relatively approach or may include direct current (zero frequency). In the case of a facsimile signal before modulation on a subcarrier, the baseband includes direct current. *See:* **facsimile transmission.**
    12, 178

(2) (data transmission). The band of frequencies occupied by the signal before it modulates the carrier (or subcarrier) frequency to form the transmitted line or radio signal.   59

baseband coaxial system (medium attachment units and repeater units). A system whereby information is directly encoded and impressed on the coaxial trasmission medium. At any point on the medium, only one information signal at a time can be present without disruption. *See:* **collision.**   543

baseband-multiplexed (data transmission). The frequency band occupied by the aggregate of the transmitted signals applied to the facility interconnecting the multiplexing and line equipment. The multiplex baseband is also defined as the frequency band occupied by the aggregate of the received signals obtained from the facility interconnecting the line and the multiplex equipment.   59

baseband response function. *See:* **transfer function (of a device).**   433

base current (ac rotating machinery)(basic per-unit quantities for ac rotating machines). The value of phase current corresponding to the value of base apparent power, base voltage, and the number of phases. *Note:* Base current is usually expressed in amperes (A), but any consistent set of units may be used. Base current equals the base apparent power divided by the product of base voltage and the number of phases.   517

base electrode (transistor). An ohmic or majority-carrier contact to the base region. *See:* **transistor base region; base.**   66

base group address (FASTBUS acquisition and control). The group address (GP) value which is used for geographical addressing on a segment. Normally the lowest GP assigned to the segment.   480

base impedance (ac rotating machinery)(basic per-unit quantities for ac rotating machines). The value of impedance corresponding to the value of the base

voltage divided by the value of the base current. *Note:* Base impedance is usually expressed in ohms (Ω), but any consistent set of units may be used.   517

base light (illuminating engineering). A uniform, diffuse illumination approaching a shadowless condition, which is sufficient for a television picture of technical acceptability, and which may be supplemented by other lighting.   167

baseline (1)(at pulse peak)(charged-particle detectors)(germanium gamma-ray detectors)(X-ray energy spectrometers). The instantaneous value that the voltage would have had at the time of the pulse peak in the absence of that pulse.   119, 528, 471

(2)(navigation)(navigation aid terms). The line joining the two points between which electrical phase or time is compared in determining navigation coordinates. For two ground stations, this is normally the great circle joining the two stations, and, in the case of a rotation collector system, it is the line joining the two sides of the collector.   526

(3) (pulse techniques). That amplitude level from which the pulse waveform appears to originate. *See:* **pulse.**   185

(4) (software). (A) A specification or product that has been formally reviewed and agreed upon, that thereafter serves as the basis for further development, and that can be changed only through formal change control procedures. (B) A configuration identification document or a set of such documents formally designated and fixed at a specific time during a configuration item's life cycle. Baselines, plus approved changes from those baselines, constitute the current configuration identification. For configuration management there are three baselines, as follows: (1) Functional baseline. The initial approved functional configuration. (2) Allocated baseline. The initial approved allocated configuration. (3) Product baseline. The initial approved or conditionally approved product configuration identification. *See:* **change control procedures; configuration identification document; configuration item; configuration management.**   434

baseline clipper–intensifier (spectrum analyzer). A means of changing the relative brightness between the signal and baseline portion of the display.   390

base line data (1) (electric pipe heating systems). Information retained for the purpose of evaluation against repeated information in order to establish trends in parameters.   405

(2) (nuclear power generating stations). Reference data that may be used to show acceptable functioning of the equipment during qualification testing.   440

baseline delay (navigation aid terms). The time interval needed for a signal from a loran master station to travel to the slave station.   526

baseline offset (pulse techniques). The algebraic difference between the amplitude of the baseline and the amplitude reference level. *See:* **pulse.**   185

baseline overshoot (pulse techniques). *See:* **distortion, pulse.**

baseline restoration (X-ray energy spectrometers).

Appropriate linear or nonlinear technique(s), or their combination, used to accelerate the return of a voltage to its baseline. 471

base load (power operations). The minimum load over a given period of time. 516

base load control (electric generating unit or station). A mode of operation in which the unit or station generation is held constant. *See:* speed-governing system. 94

base magnitude (pulse terms). The magnitude of the base as obtained by a specified procedure or algorithm. Unless otherwise specified, both portions of the base are included in the procedure or algorithm. *See:* The single pulse diagram below the **waveform epoch** entry. *See:* IEEE Std 181-1975, Pulse Measurement and Analysis by Objective Techniques, Section 4.3, for suitable algorithms. 254

basement. The rock region underlying the overburden largely comprising aged rock types, often crystalline and of low conductivity. 132

base-minus-ones complement. A number representation that can be derived from another by subtracting each digit from one less than the base. Nines complements and ones complements are base-minus-ones complements. 235

base-mounted electric hoist. A hoist similar to an overhead electric hoist except that it has a base or feet and may be mounted overhead, on a vertical plane, or in any position for which it is designed. *See:* hoist. 328

base page address (microprocessor assembly language). An address of reduced size which references a pre-specified portion of memory (which might be an on-board RAM). 466

base rate (telephone switching systems). A fixed amount charged each month for any one of the classes-of-service that is provided to a customer. 55

base-rate area (telephone switching systems). The territory in which the tariff applies. 55

base region (transistor). The interelectrode region of a transistor into which minority carriers are injected. *See:* transistor. 328

base repetition rate. *See:* basic repetition frequency.

base resistivity. The electrical resistivity of the material composing the base of a semiconductor device. 113

base speed (ac rotating machinery)(basic per-unit quantities for ac rotating machines). The rated synchronous speed. *Note:* Synchronous speed equals 120 times the value of line frequency, divided by the number of poles. Base speed is usually expressed in revolutions per minute (r/min), but any consistent set of units may be used. 517

base speed of an adjustable-speed motor (electric installationson shipboard). The lowest speed obtained at rated load and rated voltage at the temperature rise specified in the rating. 3

base station (mobile communication). A land station in the land-mobile service carrying on a radio communication service with mobile and fixed radio stations. *See:* mobile communication system. 181

base torque (ac rotating machinery)(basic per-unit quantities for ac rotating machines). The value of torque corresponding to the value of base apparent power and base synchronous speed. The value of base torque in pound-force feet (lbf·ft) is 7.043 times the value of the base apparent power in voltamperes (VA), divided by the value of base speed in revolutions per minute (r/min). The value of base torque in newton meters per radian (N·m/rad) is 9.549 times the value of the base apparent power (in voltamperes), divided by the value of the base speed in revolutions per minute. *Note:* Base torque has conventionally been expressed in pound-force feet or in newton meters (N·m). To avoid confusion with the unit of energy, which is also the newton meter, the designation newton meter per radian is recommended. 517

base value (rotating machinery). A normal or nominal or reference value in terms of which a quantity is expressed in per unit or percent. *See:* asynchronous machine. 63

base voltage (ac rotating machinery)(basic per-unit quantities for ac rotating machines). The rated phase voltage. *Note:* The value of the base voltage is the value of the rated line voltage for a delta-connected machine, and is the value of the rated line voltage divided by $\sqrt{3}$ for a wye-connected machine. Base voltage is usually expressed in volts (V), but any consistent set of units may be used. 517

basic alternating voltage (power rectifier). The sustained sinusoidal voltage that must be impressed on the terminal of the alternating-current winding of the rectifier transformer, when set on the rated voltage tap, to give rated output voltage at rated load with no phase control. *See:* rectification. 328

basic control element (thyristor). The basic thyristor or thyristor/diode circuit configuration, or both, employed as the principal means of power control. 445

basic current range (watthour meter). The current range of a multirange standard watthour meter designated by the manufacturer for the adjustment of the meter (normally the 5-ampere range). 212

BASIC definitions (real-time BASIC for CAMAC). The syntax definitions make use of a metalanguage that is the usual extension to BNF notation [4]. The meta-language contains symbols such as { or ::= or] or < which do not occur in BASIC. In addition, the meta-language contains words in angle brackets; for example, <numeric-expression>, where the meta-language symbols < and > indicate that the word between (in this case *numeric-expression*) is in the meta-language, not in BASIC. The symbols of the meta-language are listed below, together with their meanings. Any other symbols not enclosed by angle brackets stand for themselves and are part of BASIC.

::= means *is defined by*. It separates the left part from the right part of a definition

< opens a character string which constitutes a meta-language symbol

> terminates a character string which constitutes a meta-language symbol

/   separates alternatives in the right part of a definition

[opens an option, that is, the syntactic units enclosed by square brackets are optionally present

]   terminates an option

{   opens a group of elements which are to be considered a single syntactic unit for the purposes of the definition

}   terminates a group of elements to be considered as a single syntactic unit

...   means that the preceding syntactic unit may be repeated zero or more times

.is.   used in place of ::= in the formal semantic definition of a terminal symbol

Note that concatenation takes precedence over alternation, for example: F<integer>/<null> is equivalent to { F<integer> }/<null>.

The statement number is omitted in the formal definitions. It is mandatory for statements that form part of a program, but it may be omitted to indicate *immediate mode* execution of single statements in the usual way.

Tabulation, blanks, and new lines are used in the syntax definitions to make them easier to read, but they have no other significance. A program must follow the rules for the implementation concerning blanks.   389

basic device. *See:* **common device.**

basic element (measurement system). A measurement component or group of components that performs one necessary and distinct function in a sequence of measurement operations. *Note:* Basic elements are single-purpose units and provide the smallest steps into which the measurement sequence can be classified conveniently. Typical examples of basic elements are: a permanent magnet, a control spring, a coil, and a pointer and scale. *See:* **measurement system.**

     328

basic frequency. Of an oscillatory quantity having sinusoidal components with different frequencies, the frequency of the component considered to be the most important. *Note:* In a driven system, the basic frequency would, in general, be the driving frequency, and in a periodic oscillatory system, it would be the fundamental frequency.   176

basic functions (industrial control) (controller). The functions of those of its elements that govern the application of electric power to the connected apparatus. *See:* **electric controller.**   206

basic impulse insulation levels (BIL)(1)(metal-oxide surge arresters for ac power circuits). A reference impulse insulation strength expressed in terms of the crest value of withstand voltage of a standard full impulse voltage wave. *Note:* See ANSI C92.1-1982.

     583

(2) (electric power). Basic impulse insulation levels are reference levels expressed in impulse crest voltage with a standard wave not longer than 1.5×40 microsecond wave. *See:* **insulation.**

     91, 2, 103, 62, 274, 275, 276

(3) (surge arrester). A reference impulse insulation strength expressed in terms of the crest value of withstand voltage of a standard full impulse voltage wave. *Note:* See ANSI C92.1-1971 (R1978).   443, 430

(4) (outdoor apparatus bushings). A reference insulation level expressed as the impulse crest voltage of the 1.2×50 microsecond wave which the bushing will withstand when tested in accordance with specified conditions.   168

(5) (power cable systems). Impulse voltage which electrical equipment is required to withstand without failure or disruptive discharge when tested under specified conditions of temperature and humidity. Basic impulse levels are designated in terms of the crest voltage of $1.2 \cdot 50 \mu s$ full wave impulse voltage test.

     437

basic lightning impulse insulation level (BIL) (power and distribution transformer). A specific insulation level expressed in kilovolts of the crest value of a standard lightning impulse.   53

basic metallic rectifier. One in which each rectifying element consists of a single metallic rectifying cell. *See:* **rectification.**   328

basic numbering plan USA (telephony). The plan whereby every telephone station is identified for nationwide dialing by a code for routing and a number of digits.   193

basic part (electric and electronics parts and equipments). One piece, or two or more pieces joined together, which are not normally subject to disassembly without destruction of designed use. The application, size, and construction of an item may be factors in determining whether an item is regarded as a unit, an assembly, a subassembly, or a basic part. A small electric motor might be considered as a part if it is not normally subject to disassembly. Typical examples: electron tube, resistor, relay, power transformer, microelectronic device.   17

basic planned derating (electric generating unit reliability, availability, and productivity). The planned derating that is originally scheduled and of predetermined duration. *See:* **planned derating.**   567

basic planned outage (electric generating unit reliability, availability, and productivity). The planned outage state that is originally scheduled and of a predetermined duration.   567

basic reference designation (electric and electronics parts and equipments). The simplest form of a reference designation, consisting only of a class letter portion and a number (namely, without mention of the item within which the reference-designated item is located). The reference designation for a unit consists of only a number.   17

basic reference standards (metering). Those standards with which the values of the electrical units are maintained in the laboratory, and which serve as the starting point of the chain of sequential measurements carried out in the laboratory.   212

basic repetition frequency (loran)(navigation aid terms). The lowest pulse repetition frequency of each of the several sets of closely-spaced repetition frequencies employed.   526 basic repetition rate. *See:* basic repetition frequency.

basic series ferroresonant voltage regulator (fer-roresonant voltage regulators). This regulator consists of a series connection of a saturating inductor and a capacitor connected across the source. The load is inductively or conductively coupled to the saturating inductor. *Note:* Applications of this circuit are limited by the requisite large ratio of reactive to real powers. See figure 'Basic Series Ferroresonant Voltage Regulator' below.   456

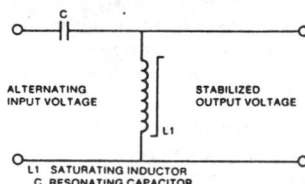

ALTERNATING
INPUT VOLTAGE

STABILIZED
OUTPUT VOLTAGE

L1 SATURATING INDUCTOR
C RESONATING CAPACITOR

basic series parallel ferroresonant voltage regulator (ferroresonant voltage regulators). This regulator consists of an essentially linear inductor connected in series with a parallel combination of a nonlinear inductor and a capacitor. This combination is connected across the source as shown in Figure 'Basic Series Parallel Ferroresonant Voltage Regulator' below. Load voltage is derived by inductive or conductive coupling to the nonlinear inductor.   456

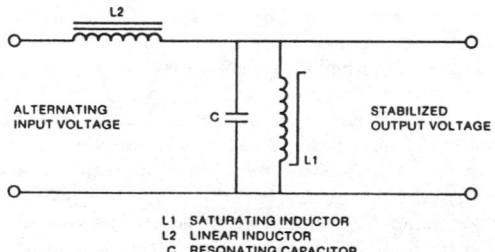

ALTERNATING
INPUT VOLTAGE

STABILIZED
OUTPUT VOLTAGE

L1 SATURATING INDUCTOR
L2 LINEAR INDUCTOR
C RESONATING CAPACITOR

basic status (logical link control). A logical link control's (LLC) capability to send or receive a PDU containing an information field.   585

basic switching impulse insulation level (BSL) (power and distribution transformer). A specific insulation level expressed in kilovolts of the crest value of a standard switching impulse.   53

basic voltage range (watthour meter). The voltage range of a multirange standard watthour meter designated by the manufacturer for the adjustment of the meter (normally the 120-volt range).   212

bass boost. An adjustment of the amplitude-frequency response of a system or transducer to accentuate the lower audio frequencies.   328

bath voltage. The total voltage between the anode and cathode of an electrolytic cell during electrolysis. It is equal to the sum of (1) equilibrium reaction potential, (2) *IR* drop, (3) anode polarization, and (4) cathode polarization. *See:* tank voltage; electrolytic cell.   328

bathythermograph (navigation aid terms). A recording

thermometer for determining the temperature of the sea at various depths.   526

battery (primary or secondary). Two or more cells electrically connected for producing electric energy. (Common usage permits this designation to be applied also to a single cell used independently. In this Dictionary, unless otherwise specified, the term battery will be used in this dual sense.)   328

battery-and-ground pulsing (telephone switching systems). Dial pulsing using battery-and-ground signaling.   55

battery-and-ground signaling (telephone switching systems). A method of loop signaling, used to increase the range, in which battery and ground at both ends of the loop are poled oppositely.   55

battery carry-over (magnetic tape pulse recorders for electricity meters). A device that maintains actual time of the interval recording from a standby power source for a specified period when the principal power source is inoperative.   551

battery chute. A small cylindrical receptacle for housing track batteries and so set in the ground that the batteries will be below the frost line.   328

battery-current regulation (generator). That type of automatic regulation in which the generator regulator controls only the current used for battery charging purposes. *See:* axle generator system.   328

battery duty cycle (large lead storage batteries). The load currents a battery is expected to supply for specified time periods.   377

battery, electric. A device that transforms chemical energy into electric energy. *See:* battery.   59

battery eliminator. A device that provides direct-current energy from an alternating-current source in place of a battery. *See:* battery.   59

battery feed (telephone loop performance). The direct current (dc) supply and coupling circuit powering the loop.   473

battery, power station (1) (communications). A battery that is a separate source of energy for communication equipment in power stations.
(2) (control). A battery that is a separate source of energy for the control of power apparatus in a power station. *See:* battery.   59

battery rack (1) (lead storage batteries). A structure used to support a group of cells.   38
(2) (nuclear power generating stations) (lead storage batteries). A rigid structure used to accommodate a group of cells.   38

baud (1)(supervisory control, data acquisition, and automatic control). Defines the signaling speed, that is, the keying rate, of the modem. The signaling speed in baud is equal to the reciprocal of the shortest element duration in seconds to be transmitted. For example, in the following table, the signaling speed is calculated from the signaling element duration. In addition, the distinction between bit rate and baud for two different types of modems is illustrated.   570
(2) (telegraphy). The unit of telegraph signaling speed, derived from the duration of the shortest signaling pulse. A telegraphic speed of one baud is one

Signaling Technique

	Modem One	Modem Two
Signaling element duration	0.833 ms	
Signaling speed	1200 baud	
Information transmitted per element duration	1 bit	2 bits
Bit rate	1200 bits per second	2400 bits per second

pulse per second. *Note:* The term unit pulse is often used for the same meaning as the baud. A related term, the dot cycle, refers to an ON-OFF or MARK-SPACE cycle in which both mark and space intervals have the same length as the unit pulse.   111

(3) (data transmission). A unit of signaling speed equal to the number of discrete conditions or signal events per second, or the reciprocal of the time of the shortest signal element in a character.   59

b auxiliary switch. *See:* **auxiliary switch; b contact.**   103

bay (computing system). *See:* **patch bay; electronic analog computer.**

Bayliss distribution (antennas) (1) (circular). A continuous distribution of a circular planar aperture which yields a difference pattern with a sidelobe structure similar to that of a sum pattern produced by a Taylor circular distribution. **(2) (linear).** A continuous distribution of a line source which yields a difference pattern with a sidelobe structure similar to that of a sum pattern produced by a Taylor linear distribution.   111

B battery. A battery designed or employed to furnish the plate current in a vacuum-tube circuit. *See:* **battery (primary or secondary).**   328

bb auxiliary switch. *See:* **auxiliary switch; bb contact.**   103

bb contact (power switchgear). A contact that is closed when the operating mechanism of the main device is

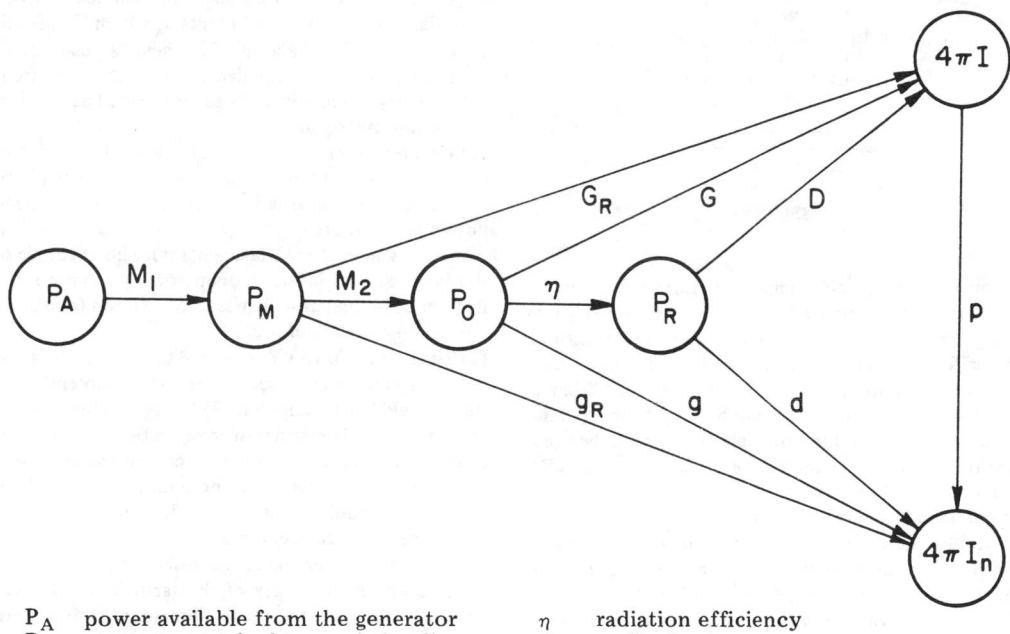

P_A	power available from the generator
P_M	power to matched transmission line
P_O	power accepted by the antenna
P_R	power radiated by the antenna
I	radiation intensity
I_n	partial radiation intensity†
M_1	Impedance mismatch factor 1
M_2	Impedance mismatch factor 2
η	radiation efficiency
G_R	realized gain
G	gain
D	directivity
g_R	partial realized gain
g	partial gain
d	partial directivity
p	polarization efficiency

†All partial quantities correspond to a specified polarization, n.

Gain and Directivity Flow Chart

in the standard reference position and that is open when the operating mechanism is in the opposite position. *See:* **standard reference position.** 103

b contact (power switchgear). A contact that is closed when the main device is in the standard reference position and that is open when the device is in the opposite position. *Notes:* (1) b contact has general application. However, this meaning for back contact is restricted to relay parlance. (2) For indication of the specific point of travel at which the contact changes position, an additional letter or percentage figure may be added to b and detailed in 9.4.4.1 and 9.4.4.2 of ANSI/IEEE C37.2-1979. *See:* **standard reference position.** *Syn:* **back contact.** 103

B-display (radar). A rectangular display in which each target appears as an intensity-modulated blip, with azimuth indicated by the horizontal coordinate and range by the vertical coordinate.

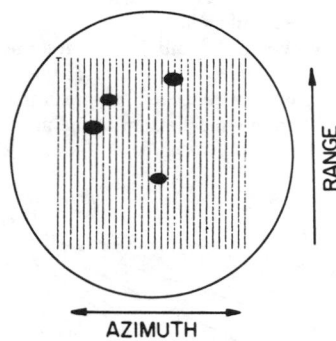

B-Display

13

beacon (navigation aid terms). (1) A fixed aid to navigation. (2) An unlighted aid to navigation. (3) Anything serving as a signal or conspicuous indication, either for guidance or warning. (4) In radar, a transponder used for replying to interrogations from a radar. *See:* **fan-marker beacon; homing beacon; identification beacon; landing beacon; lighted beacon; marker or marker beacon; racon; radar beacon; radio beacon; z-marker beacon.** 526

beacon equation (radar). An equation which gives maximum detection range or signal-to-noise ratio (SNR) of a transponder or secondary radar as a function of system parameters for a given set of conditions. It is the one-way counterpart of the two-way radar equation. 13

beacon receiver. A radio receiver for converting waves, emanating from a radio beacon, into perceptible signals. *See:* **radio receiver; radio beacon.** 328

beam (1) (antenna). The major lobe of the radiation pattern. *See:* **radiation.** 111
(2) **(laser-maser).** A collection of rays which may be parallel, divergent, or convergent. 363

beam alignment (camera tubes). An adjustment of the electron beam, performed on tubes employing low-velocity scanning, to cause the beam to be perpendicular to the target at the target surface. 125

beam angle. *See:* **scan angle.**

beam axis (of a pencil-beam antenna). The direction, within the major lobe of a pencil-beam antenna, for which the radiation intensity is a maximum. 111

beam axis of a projector (illuminating engineering). A line midway between two lines that intersect the candlepower distribution curve at points equal to a stated percent of its maximum (usually 50 percent). 167

beam bending (camera tubes). Deflection of the scanning beam by the electrostatic field of the charges stored on the target. 125

beam coverage solid angle (of an antenna over a specified surface). A solid angle, measured in steradians subtended by the antenna by the footprint of the antenna beam on a specified surface. *See:* **footprint (of an antenna beam on a specified surface); beam solid angle.** 111

beam current (storage tubes). The current emerging from the final aperture of the electron gun. *See:* **storage tube.** 174

beam-deflection tube. An electron-beam tube in which current to an output electrode is controlled by the transverse movement of an electron beam. 125

beam diameter (1) (fiber optics). The distance between two diametrically opposed points at which the irradiance is a specified fraction of the beam's peak irradiance; most commonly applied to beams that are circular or nearly circular in cross section. *Syn:* **beamwidth.** *See:* **beam divergence.** 433
(2) **(laser-maser).** The distance between diametrically opposed points in that cross section of a beam where the power per unit area is $1/e$ times that of the peak power per unit area. 363

beam divergence (ϕ) (1)(laser-maser). The full angle of the beam spread between diametrically opposed $1/e$ irradiance points; usually measured in mrad (one mrad $\approx \Delta$ 3.4 minutes of arc). 363
(2)**(fiber optics).** (A) For beams that are circular or nearly circular in cross section, the angle subtended by the far-field beam diameter. (B) For beams that are not circular or nearly circular in cross section, the far-field angle subtended by two diametrically opposed points in a plane perpendicular to the optical axis, at which points the irradiance is a specified fraction of the beam's peak irradiance. Generally, only the maximum and minimum divergences (corresponding to the major and minor diameters of the far-field irradiance) need be specified. *See:* **beam diameter; collimation; far-field region.** 433

beam error (1)(navigational systems using directionally propagated signals)(navigation aids). The lateral or angular distance between the mean direction of the actual course and the desired course direction. *Syn:* **course error.** 526

beam expander (laser-maser). A combination of optical elements which will increase the diameter of a laser beam. 363

beam finder (oscilloscope). A provision for locating the spot when it is not visible. 106

beamguide (laser-maser). A set of beam-forming elements spaced in such a way as to conduct a well-defined beam of radiation. Analogs are waveguides and fiber optic filaments.   363

beam-indexing color tube. A color-picture tube in which a signal, generated by an electron beam after deflection, is fed back to a control device or element in such a way as to provide an image in color.   125

beam landing error (camera tube). A signal non-uniformity resulting from beam electrons arriving at the target with a spatially varying component of velocity parallel to the target. *See:* **camera tube.**   190

beam locator. *See:* **beam finder.**

beam modulation, percentage (image orthicons). One hundred times the ratio of (1) the signal output current for highlight illumination on the tube to (2) the dark current.   125

beam noise (navigational systems using directionally propagated signals)(navigation aid terms). Extraneous disturbances tending to interfere with ideal system performance. *Note:* Beam noise is the aggregate effect of bends, scalloping, roughness, etcetera.   526

beam parametric amplifier. A parametric amplifier that uses a modulated electron beam to provide a variable reactance. *See:* **parametric device.**   191

beam pattern. *See:* **directional response pattern.**

beam pointing (communication satellite). The ability to orient the beam of a high gain antenna into a specific direction in a coordinate system.   84

beam power tube. An electron-beam tube in which use is made of directed electron beams to contribute substantially to its power-handling capability, and in which the control grid and the screen grid are essentially aligned.   125

beam resonator (laser-maser). A resonator which serves to confine a beam of raiation to a given region of space without continuous guidance along the beam.   363

beam rider guidance. That form of missile guidance wherein a missile, through a self-contained mechanism, automatically guides itself along a beam. *See:* **guided missile.**   328

beamshape loss (radar). A loss factor included in the radar equation to account for the use of the peak antenna gain in the radar equation instead of the effective gain that results when the received train of pulses is modulated by the two-way pattern of the scanning antenna. *Syn:* **antenna-pattern loss.**   13

beam shaping (communication satellite). Controlling the shape of an antenna beam, by design of the surfaces of the antenna or by controlling the phasing of the signals radiated from the antenna.   85

beam solid angle (antennas). The solid angle through which all the radiated power would stream if the power per unit solid angle were constant throughout this solid angle and at the maximum value of the radiation intensity.   111

beamsplitter (fiber optics). A device for dividing an optical beam into two or more separate beams; often a partially reflecting mirror.   433

beam splitter (laser-maser). An optical device which

uses controlled reflection to produce two beams from a single incident beam.   363

beam spread (in any plane) (1) (illuminating engineering). The angle between the two directions in the plane in which the intensity is equal to a stated percentage of the maximum beam intensity. The percentage typically is 10 percent for floodlights and 50 percent for photographic lights.   167

(2) (light emitting diodes). (source of light, ϑy, where y is the stated percent). *See:* **(1) (illuminating engineering)** above.   162

beam steering (antenna). Changing the direction of the major lobe of a radiation pattern. *See:* **radiation.**   179

beam waveguide. A quasioptical structure consisting of a sequence of lenses or mirrors used to guide an electromagnetic wave.   267

beamwidth. *See* **half-power beamwidth; beam diameter.**

bearing (1)(navigation aid terms). (A) The horizontal direction of one terrestrial point from another, expressed as the angle in the horizontal plane between a reference line and the horizontal projection of the line joining two points. (B) Azimuth. A bearing is often designated as true, magnetic, compass, grid, or relative and is dependent upon the reference direction.   526

(2) (rotating machinery). (A) A stationary member or assembly of stationary members in which a shaft is supported and may rotate. (B) In a ball or roller bearing, a combination (frequently preassembled) of stationary and rotating members containing a peripheral assembly of balls or rollers, in which a shaft is supported and may rotate.   63

bearing accuracy, instrumental (direction-finding). (1) The difference between the indicated and the apparent bearings in a measurement of the same signal source. (2) As a statement of overall system performance, a difference between indicated and correct bearings whose probability of being exceeded in any measurement made on the system is less than some stated value. *See:* **navigation.**   187

bearing bracket (rotating machinery). A bracket which supports a bearing, but including no part thereof. A bearing bracket is not specifically constructed to provide protection for the windings or rotating parts. *See:* **bracket-end shield.**   63

bearing cap (bearing bracket cap) (rotating machinery). A cover for the bearing enclosure of a bearing bracket type machine or the removable upper half of the enclosure for a bearing. *See:* **bearing.**   63

bearing cartridge (rotating machinery). A complete enclosure for a ball or roller bearing, separate from the bearing bracket or end shield. *See:* **bearing.**   63

bearing clearance (rotating machinery). (1) The difference between the bearing inner diameter and the journal diameter. (2) The total distance for axial movement permitted by a double-acting thrust bearing. *See:* **bearing.**   63

bearing distance heading indicator (BDHI)(navigation aid terms). A display device which presents continuous references as to course and distance to destination. 526

bearing dust-cap (rotating machinery). A removable cover to prevent the entry of foreign material into the bearing. *See:* **bearing.** 63

bearing error curve (navigation aid terms). (1) DF (direction finder) equipment. A plot of the instrumental bearing errors versus either indicated or correct bearing. (2) (In DF installations). A plot of the combined instrumental bearing error (of the equipment) and site error versus indicated bearings. 526

bearing housing (rotating machinery). A structure supporting the actual bearing liner or ball or roller bearing in a bearing assembly. *See:* **bearing.** 63

bearing insulation (rotating machinery). Insulation that prevents the circulation of stray currents by electrically insulating the bearing from its support. *See:* **bearing.** 63

bearing liner (rotating machinery). The assembly of a bearing shell together with its lining. *See:* **bearing.** 63

bearing lining (rotating machinery). The element of the journal bearing assembly in which the journal rotates. *See:* **bearing.** 63

bearing locknut (rotating machinery). A nut that holds a ball or roller bearing in place on the shaft. *See:* **bearing.** 63

bearing lock washer (rotating machinery). A washer between the bearing locknut and the bearing that prevents the locknut from turning. *See:* **bearing.** 63

bearing offset, indicated (direction-finding) (electronic navigation). The mean difference between the indicated and apparent bearings of a number of signal sources, the sources being substantially uniformly distributed in azimuth. *See:* **navigation.** 278, 187

bearing oil seal (bearing seal) (rotating machinery). *See:* **oil seal.**

bearing oil system (oil-circulating system) (rotating machinery). All parts that are provided for the flow, treatment, and storage of the bearing oil. *See:* **oil cup (rotating machinery).** 63

bearing pedestal (rotating machinery). A structure mounted from the bedplate or foundation of the machine to support a bearing, but not including the bearing. *See:* **bearing.** 63

bearing-pedestal cap (rotating machinery). The top part of a bearing pedestal. *See:* **bearing.** 63

bearing protective device (38) (power system device function numbers). A device that functions on excessive bearing temperature, or on other abnormal mechanical conditions associated with the bearing, such as undue wear, which may eventually result in excessive bearing temperature or failure. 402

bearing reciprocal. *See:* **reciprocal bearing.**

bearing reservoir (oil tank) (oil well) (rotating machinery). A container for the oil supply for the bearing. It may be a sump within the bearing housing. *See:* **oil cup (rotating machinery).** 63

bearing seal (bearing oil seal) (rotating machinery). *See:* **oil seal.**

bearing seat (rotating machinery). The surface of the supporting structure for the bearing shell. *See:* **bearing.** 63

bearing sensitivity (electronic navigation). The minimum field strength input to a direction-finder system to obtain repeatable bearings within the bearing accuracy of the system. *See:* **navigation.** 187, 172

bearing shell (rotating machinery). The element of the journal bearing assembly that supports the bearing lining. *See:* **bearing.** 63

bearing shoe (rotating machinery). *See:* **segment shoe.**

bearing-temperature detector (rotating machinery). A temperature detector whose sensing element is mounted at or near the bearing surface. *See:* **bearing.** 63

bearing-temperature relay (bearing thermostat) (rotating machinery). A relay whose temperature sensing element is mounted at or near the bearing surface. *See:* **bearing.** 63

bearing thermometer (rotating machinery). A thermometer whose temperature sensing element is mounted at or near the bearing surface. *See:* **bearing.** 63

bearing thermostat (rotating machinery). *See:* **bearing-temperature relay.**

beating (data transmission). A phenomenon in which two or more periodic quantities of different frequencies produce a resultant having pulsations of amplitude. 59

beat note. The wave of difference frequency created when two sinusoidal waves of different frequencies are supplied to a nonlinear device. *See:* **radio receiver.** 339

beat reception. *See:* **heterodyne reception.**

beats (1) (general). Periodic variations that result from the superposition of waves having different frequencies. *Note:* The term is applied both to the linear addition of two waves, resulting in a periodic variation of amplitude, and to the nonlinear addition of two waves, resulting in new frequencies, of which the most important usually are the sum and difference of the original frequencies. *See:* **signal wave.** 55

(2) (automatic control). Periodic variations that result from the superposition of periodic signals having different frequencies. 56

(3) (data transmission). Periodic variations that result from the superposition of waves having difference frequencies. *Note:* The term is applied both to the linear addition of two waves, resulting in a periodic variation of amplitude, and to the nonlinear addition of two waves, resulting in new frequencies, of which the most important usually are the sum and difference of the original frequencies. 59

becquerel (metric practice). The activity of a radionuclide decaying at the rate of one spontaneous nuclear transition per second. 21

begin-end block (software). A sequence of design or programming statements bracketed by "begin" and "end" delimiters and characterized by a single entrance and a single exit. *See:* **design.** 434

bel. The fundamental division of a logarithmic scale for expressing the ratio of two amounts of power, the number of bels denoting such a ratio being the logarithm to the base 10 of this ratio. *Note:* With P_1 and P_2 designating two amounts of power and N the number of bels denoting their ratio, $N = \log_{10}(P_1/P_2)$ bels. 111

bell box (ringer box). An assemblage of apparatus, associated with a desk stand or hand telephone set, comprising a housing (usually arranged for wall mounting) within which are those components of the telephone set not contained in the desk stand or hand telephone set. These components are usually one or more of the following: induction coil, capacitor assembly, signaling equipment, and necessary terminal blocks. In a magneto set a magneto and local battery may also be included. *See:* **telephone station.**
  328

bell crank (power switchgear). A lever with two arms placed at an angle diverging from a given point, thus changing the direction of motion of a mechanism.
  103

bell crank hanger (power switchgear). A support for a bell crank. 103

belt (rotating machinery). A continuous flexible band of material used to transmit power between pulleys by motion. 63

belt-drive machine (elevators). An indirect-drive machine having a single belt or multiple belts as the connecting means. *See:* **driving machine (elevators).**
  328

belted-type cable. A multiple-conductor cable having a layer of insulation over the assembled insulated conductors. 64

belt insulation (rotating machinery). A form of overhang packing inserted circumferentially between adjacent layers in the winding overhang. *See:* **rotor (rotating machinery); stator.** 63

belt leakage flux (rotating machinery). The low-order harmonic airgap flux attributable to the phase belts of a winding. The magnitude of this leakage flux varies with winding pitch. *See:* **rotor (rotating machinery); stator.** 63

benchboard (power switchgear). A combination of a control desk and a vertical switchboard in a common assembly. 103

benchmark problem (computers). A problem used to evaluate the performance of computers relative to each other. 255

bend (navigation aid terms). A departure of the course line from the desired direction at such a rate that it can be followed by the vehicle. 526

bend amplitude (navigation)(navigation aid terms). The measured maximum amount of course deviation due to bend; measurement is made from the nominal or bend-free position of the course. 526

bend frequency (navigation)(navigation aid terms). The frequency at which the course indicator oscillates when the vehicle track is straight and the course contains bends; bend frequency is a function of the vehicle velocity. 526

bend ratio (cable plowing). The radius of a bend (segment of a circle) divided by the outside diameter of a cable, pipe, etcetera. 52

bend-reduction factor (navigation)(navigation aid terms). The ratio of bend amplitude existing before the introduction of bend-reducing features to that existing afterward. 526

bend, waveguide (waveguide components). A section of waveguide or transmission line in which the direction of the longitudinal axis is changed. In common usage the waveguide corner formed by an abrupt change in direction is considered to be a bend. 166

beta (circuits and systems). The ratio of the collector current to the base current of a bipolar transistor, commonly referred to as either the common-emitter current gain or the current amplification factor.
  67

beta (B) circuit (feedback amplifier). That circuit that transmits a portion of the amplifier output back to the input. *See:* **feedback.** 328

beta (β) figure of merit (nonlinear, active, and nonreciprocal waveguide components). A figure of merit for parametric amplifier varactors that relates to capacitive nonlinearity. Historically, for silicon varactors,

$$\beta = \frac{C_J\,(+1\,\mu A)}{C_{J-3}} = \frac{C_{J\,V_s}}{C_{J-3}}$$

and for GaAs varactors

$$\beta = \frac{C_{J+0.5}}{C_{J-3}}$$

where
  $C_{J\,V}$ = junction capacitance at voltage V
  V_s = voltage at which the forward current is $1\,\mu A$.

  530

betatron. An electric device in which electrons revolve in a vacuum enclosure in a circular or a spiral orbit normal to a magnetic field and have their energies continuously increased by the electric force resulting from the variation with time of the magnetic flux enclosed by their orbits. 190

bevatron. A synchrotron designed to produce ions of a billion (10^9) electron-volts energy or more. 190

beveled brush corners (electric machines). Where material has been removed from a corner, leaving a triangular surface. *See:* **brush.** 279

beveled brush edges (electric machines). The removal of an edge to provide a slanting surface from which a shunt connection can be made or for clearance of pressure fingers or for any other purpose. *See:* **brush.**
  279

beveled brush ends and toes (electric machines). The angle included between the beveled surface and a plane at right angles to the length. The toe is the uncut or flat portion on the beveled end. When a brush has one or both ends beveled, the front of the brush is the short side of the side exposing the face level. *See:* **brush.**   279

Beverage antenna. A directional antenna composed of a system of parallel, horizontal conductors from one-half to several wavelengths long, and terminated to ground at the far end in its characteristic impedance. *Syn:* **wave antenna.** *See:* **antenna.**   111

bezel (cathode-ray oscilloscope). The flange or cover used for holding an external graticule or cathode-ray tube cover in front of the cathode-ray tube. It may also be used for mounting a trace recording camera or other accessory item.   106

BF. *See:* **ballistic focusing.**

bias (1)(of a semiconductor radiation detector)(germanium gamma-ray detectors)(X-ray energy spectrometers)(charged-particle detectors). The voltage applied to a detector to produce the electric field to sweep out the signal charge.   528, 471, 119

(2) (telegraph transmission). A uniform displacement of like signal transitions resulting in a uniform lengthening or shortening of all marking signal intervals. *See:* **telegraphy.**   194

(3) (accelerometer). An accelerometer output when no acceleration is applied.   46

(4) (gyro) (inertial sensor). *See:* **acceleration-insensitive drift rate.**   46

bias current or power. The direct and.or alternating current or power required to operate a bolometer at a specified resistance under specified ambient conditions.   115

bias distortion (data transmission). A measure of the difference in the pulse width of the positive and negative pulses of a dotting signal. Usually expressed in percent of a full signal.   59

biased amplifier (1)(charged-particle detectors)(germanium gamma-ray detctors)(X-ray energy spectrometers). An amplifier giving essentially zero output for all inputs below a threshold and having constant incremental gain for all inputs above the threshold up to a specified maximum amplitude.   528

(2)(semiconductor radiation detectors). An amplifier giving essentially zero output for all inputs below a threshold and having constant incremental gain for all inputs above the threshold up to a specified maximum amplitude.   23

biased exponent (binary floating-point arithmetic). The sum of the exponent and a constant (bias) chosen to make the biased exponent's range nonnegative.   469

biased scheduled net interchange of a control area (electric power systems). The scheduled net interchange plus the frequency and.or other bias.   94

bias error (radar). (1) A systematic error, whether due to equipment or propagation conditions. A nonzero mean value of a random error. (2) (radar). A systematic error, whether due to equipment or propagation conditions. Contrast with random or noise error.   13

bias, grid, direct. *See:* **direct grid bias.**

biasing (laser gyro). The action of intentionally impos-

ing a real or artificial rate into the laser gyro to avoid the region in which lock-in occurs.   46

bias magnet (magnetic tape pulse recorders for electricity meters). A device that provides a magnetic field used to orient the direction of magnetization on the magnetic tape to a predetermined polarity.   551

bias resistor (of a semiconductor radiation detector)(charged-particle detectors)(germanium gamma-ray detectors)(X-ray energy spectrometers). The resistor through which bias voltage is applied to a detector.   23, 528, 471

bias spectrum. At reference ambient a specification of the fractions of total bias power in the dc and ac components and the frequency of the ac component. *Note:* The polarity of the dc component should also be given.   115

bias telegraph distortion. Distortion in which all mark pulses are lengthened (positive bias) or shortened (negative bias). It may be measured with a steady stream of unbiased reversals, square waves having equal-length mark and space pulses. The average lengthening or shortening gives true bias distortion only if other types of distortion are negligible. *See:* **modulation.**   111

bias winding (relay). *See:* **relay bias winding.**

biaxial test (seismic testing of relays). The relay under test is subjected to acceleration in one principal horizontal axis and the vertical axis simultaneously.   392

biconical antenna (overhead-power-line corona and radio noise). An antenna consisting of two conical conductors having a common axis and vertex and excited or connected to the receiver at the vertex. When the vertex angle of one of the cones is 180 degrees, the antenna is called a discone.   411

biconical reflectance, $\rho(\omega_i;\omega_r)$. (illuminating engineering). Ratio of reflected flux collected through a conical solid angle to the incident flux limited to a conical solid angle. (See Figure below.) *Note:* The directions and extent of each cone must be specified.   167

Conical   Conical
Incident   Collected

biconical transmittance, $\tau(\omega_i;\omega_t)$. Ratio of transmitted flux, collected over an element of solid angle surrounding the direction, to the incident flux limited to a conical solid angle. (See Figure below.) *Note:* The directions and extent of each cone must be specified.   167

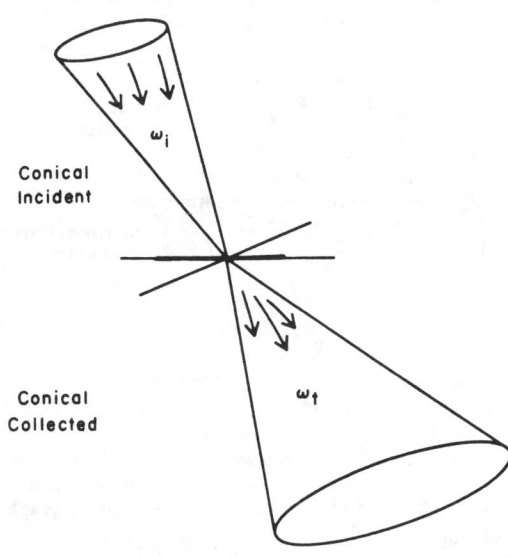

Conical
Incident

Conical
Collected

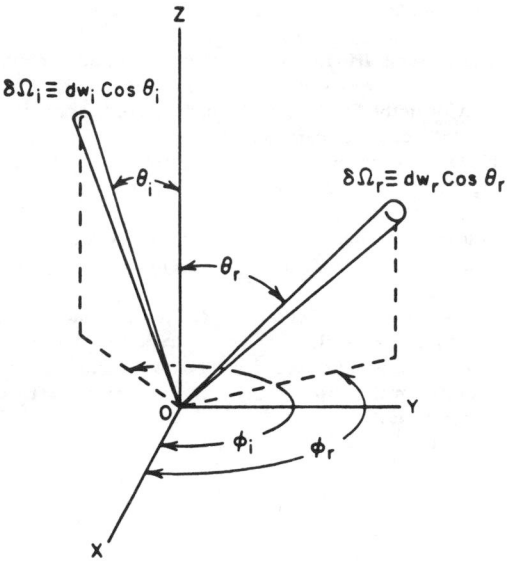

zation aspects must be defined for complete specification, since the BRDF as given above only defines the geometric aspects.   167

bidirectional antenna. An antenna having two directions of maximum response. *See:* **antenna.**   179

bidirectional bus (1)(programmable instrumentation). A bus used by any individual device for two-way transmission of messages, that is, both input and output.   378

(2)(signals and paths)(696 interface devices). A bus used by any individual device, or set of devices, for the two-way transmission of data, that is both input and output.   538

bidirectional diode-thyristor (thyristor ac power controllers)(diac). A two-terminal thyristor having substantially the same switching behavior in the first and third quadrants of the principal voltage-current characteristic.   445

bidirectional pulses. Pulses, some of which rise in one direction and the remainder in the other direction. *See:* **pulse.**   111

bidirectional reflectance-distribution function (BRDF), f_r.[17] The ratio of the differential luminance of a ray $dL_r(\theta_r,\phi_r)$ reflected in a given direction (θ_r,ϕ_r) to the differential luminous flux density $dE_i(\theta_i,\phi_i)$ incident from a given direction of incidence, (θ_i,ϕ_i) which produces it. (See Figure below.)

$$f_r(\theta_i,\phi_i;\theta_r,\phi_r) \equiv dL_r(\theta_r,\phi_r)/dE_i(\theta_i,\phi_i)\ (\text{sr})^{-1}$$
$$= dL_r(\theta_r,\phi_r)/L_i(\theta_i,\phi_i)d\Omega_i$$

where $d\Omega \equiv d\omega{\cdot}\cos\theta$

Notes: (1) This distribution function is the basic parameter for describing (geometrically) the reflecting properties of an opaque surface element (negligible internal scattering). (2) It may have any positive value and will approach infinity in the specular direction for ideally specular reflectors. (3) The spectral and polari-

bidirectional reflectance, $\rho(\theta_r,\phi_r;\theta_r,\phi_r)$. **(illuminating engineering).** Ratio of reflected flux collected over an element of solid angle surrounding the given direction to essentially collimated incident flux. (See Figure below.) *Notes:* (1) The directions of incidence and collections and the size of the solid angle "element" of collection must be specified. (2) In each case of conical incidence or collection, the solid angle is not restricted to a right circular cone, but may be of any cross section, including rectangular, a ring, or a combination of two or more solid angles.   167

bidirectional relay or add-and-subtract relay. A stepping relay in which the rotating wiper contacts may move in either direction.   329

bidirectional transducer (bilateral transducer). A transducer that is not a unidirectional transducer. *See:* **transducer.**   210

bidirectional transmission (fiber optics). Signal transmission in both directions along an optical waveguide or other component.   433

bidirectional transmittance, $\tau(\theta_i,\phi_i;\theta_r,\phi_t)$ (illuminating engineering). Ratio of incident flux collected over an element of solid angle surrounding the given direction to essentially collimated incident flux. *Note:* The directions of incidence and collection, and the size of the solid angle "elements" must be specified.   167

bidirectional triode-thyristor (triac). A three-terminal thyristor having substantially the same switching behavior in the first and third quadrants of the principal voltage-current characteristic.   445

bifilar suspension. A suspension employing two parallel ligaments, usually of conducting material, at each end of the moving element.   280

bifurcated feeder (power switchgear). A stub feeder that connects two loads in parallel to their only power source.   103

bihemispherical reflectance, $\rho(2\pi;2\pi)$. (illuminating engineering). Ratio of reflected flux collected over the entire hemisphere (See Figure below.) to the flux incident from the entire hemisphere. *See:* **hemispherical reflectance.**   167

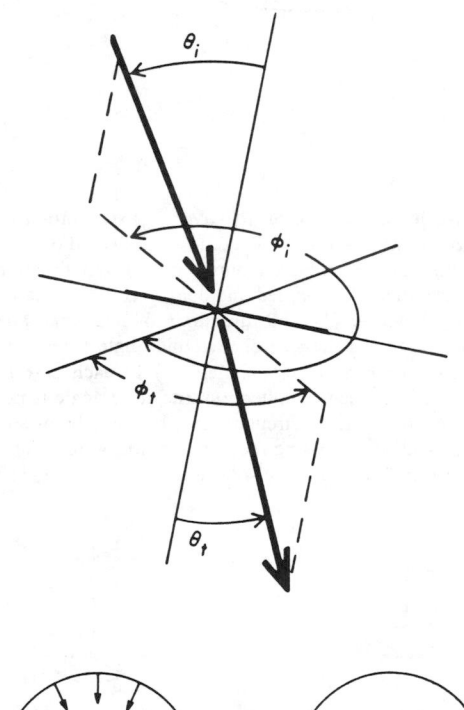

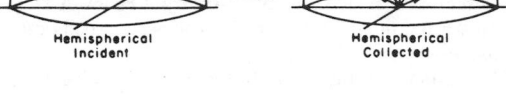

bihemispherical transmittance, $\tau(2\pi;2\pi)$. (illuminating engineering). Ratio of transmitted flux collected over the entire hemisphere to the incident flux from the entire hemisphere. (See Figure below.)   167

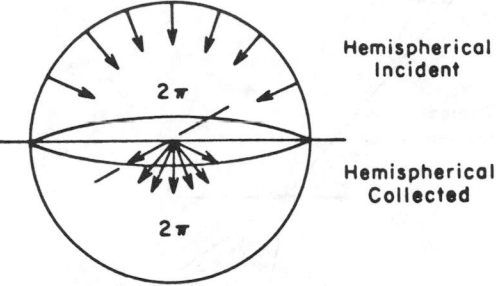

BIL (1). *See:* **basic impulse insulation level.**

(2) (insulation strength). *See:* **preferred basic impulse insulation level.**

bilateral-area track (electroacoustics). A photographic sound track having the two edges of the central area modulated according to the signal. *See:* **phonograph pickup.**   176

bilateral network (network analyzers). Network capable of transmission in both directions, not necessarily equal or symmetrical.   505

bilateral transducer. A transducer capable of transmission simultaneously in both directions between at least two terminations.   252

bilevel operation. Operation of a storage tube in such a way that the output is restricted to one or the other of two permissible levels. *See:* **storage tube.**   174

billing demand (power operations). The demand which is used to determine the demand charges in accordance with the provisions of a rate schedule or contract.   516

bimetallic element. An actuating element consisting of two strips of metal with different coefficients of thermal expansion bound together in such a way that the internal strains caused by temperature changes bend the compound strip. *See:* **relay.**   259

bimetallic thermometer. A temperature-measuring instrument comprising an indicating pointer and appropriate scale in a protective case and a bulb having a temperature-sensitive bimetallic element. The bimetallic element is composed of two or more metals mechanically associated in such a way that relative expansion of the metals due to temperature change produces motion.   7

binary (1)(data transmission). (A) Pertaining to a characteristic or property, involving a selection, choice or condition in which there are two possibilities. (B) Pertaining to the numeration system with a radix of two.   59

(2)(mathematics of computing). (A) Pertaining to a selection in which there are two possible outcomes. (B) Pertaining to the numeration system with a radix of two.   564

binary cell (computers). An elementary unit of storage that can be placed in either of two stable states. *Note:* It is therefore a storage cell of one binary digit capacity, for example, a single-bit register.   255, 77

binary code. (1) A code in which each code element may be either of two distinct kinds or values, for example, the presence or absence of a pulse.   59
(2) A code that makes use of members of an alphabet containing exactly two characters, usually 0 and 1. The binary number system is one of many binary codes. *See:* **information theory; pulse; reflected binary code.**   77, 207

binary coded decimal (BCD)(mathematics of computing). Pertaining to a number representation system in which each decimal digit is represented by a unique arrangement of binary digits (usually four); for example, the number 23 is represented as 0010 0011, whereas in binary notation, 23 is represented as 10111.   564

binary-coded-decimal number (BCD). The representation of the cardinal numbers 0 through 9 by 10 binary codes of any length. Note that the minimum length is four and that there are over 29×10^9 possible four-bit binary-coded-decimal codes. *Note:* an example of 8-4-2-1 binary-coded decimal code follows for numbers 0 thru 9.

Number	r4	r3	r2	r1
0	0	0	0	0
1	0	0	0	1
2	0	0	1	0
3	0	0	1	1
4	0	1	0	0
5	0	1	0	1
6	0	1	1	0
7	0	1	1	1
8	1	0	0	0
9	1	0	0	1

Where $r1$ is termed the **least significant binary digit (bit).** *See:* **digital.**   59

binary digit (data transmission). A character used to represent one of the two digits in the numeration system with a radix of two. Abbreviated "bit".   59

binary floating point number (binary floating-point arithmetic). A bit-string characterized by three components: a sign, a signed exponent, and a significand. Its numerical value, if any, is the signed product of its significand and two raised to the power of its exponent. In ANSI/IEEE Std 754-1985 a bit-string is not always distinguished from a number it may represent.
                                                        469

binary information (microprocessor object modules). Bit patterns to be loaded into memory.   466

binary number. Loosely, a binary numeral.
                                                    255, 77

binary number system. *See:* **positional notation.**

binary numeral. The binary representation of a number, for example, 101 is the binary numeral and V is the Roman numeral of the number of fingers on one hand.
                                                    255, 77

binary point. *See:* **point.**

binary pulse width modulation torquing (digital accelerometer). A torquing technique in which the time between (positive, negative) torquing transitions is constant.   383

binary search. A search in which a set of items is divided into two parts, one part is rejected, and the process is repeated on the accepted part until those items with the desired property are found. *See:* **dichotomizing search.**   255, 77

binary torquing (1) (digital accelerometer). System with two stable torquing states (for example, positive and negative).   383

(2) (gyro, accelerometer) (inertial sensor). A torquing mechanism which uses only two torquer current levels, usually positive and negative of the same magnitude; no sustained zero current or off condition exists. The positive and negative current periods can be either discrete pulses or duration-modulated pulses. In the case of zero input (acceleration or angular rate), a discrete pulse system will produce an equal number of positive and negative pulses. A pulse-duration-modulated system will produce positive and negative current periods of equal duration for zero input. Binary torquing delivers constant power to a sensor torquer (as compared to variable power ternary torquing) and results in stable thermal gradients for all inputs.
                                                         46

binder (bond) (1) (rotating machinery). A solid, liquid, or semiliquid composition that exhibits marked ability to act as an adhesive, and that, when applied to wires, insulation components, or other parts, will solidify, hold them in position, and strengthen the structure.
                                                         63

(2) (electroacoustics). A resinous material that causes the various materials of a record compound to adhere to one another. *See:* **phonograph pickup.**   176

binder, load. *See:* **load binder.**

binding (software). The assigning of a value or referent to an identifier; for example, the assigning of a value to a parameter, the assigning of an absolute address, virtual address, or device identifier to a symbolic address or label in a computer program. *See:* **computer program; dynamic binding; identifier; label; parameter; static binding.**   434

binding band (rotating machinery). A band of material, encircling stator or rotor windings to restrain them against radial movement. *See:* **rotor (rotating machinery).**   63

binding post. *See:* **binding screw.**

binding screw (binding post) (terminal screw) (clamping screw). A screw for holding a conductor to the terminal of a device or equipment.   328

binnacle (navigation aid terms). The stand in which a compass is mounted. (Marine navigation).   526

binocular (navigation aid terms). An optical instrument for use with both eyes simultaneously.   526

binocular portion of the visual field (illuminating engineering). That portion of space where the fields of the two eyes overlap.   167

binomial array. A linear array in which the currents in successive elements are made proportional to the binomial coefficients of $(x + y)^{n-1}$ for the purpose of reducing minor lobes. *See:* **antenna.**   179

bioelectric null (zero lead) (medical electronics). A region of tissue or other area in the system, which has such electric symmetry that its potential referred to infinity does not significantly change. *Note:* This may or may not be ground potential.   192

bionics (computer applications). A branch of technology relating the functions, characteristics, and phenomena of living systems to the development of mechanical systems.   571

Biot-Savart law. *See:* **magnetic field strength produced by an electric current.**

biparting door (elevator). A vertically sliding or a horizontally-sliding door, consisting of two or more sections so arranged that the sections or groups of sections open away from each other and so interconnected that all sections operate simultaneously. *See:* **hoistway (elevator or dumbwaiter).**   328

bipolar (power supplies). Having two poles, polarities, or directions. *Note:* Applied to amplifiers or power supplies, it means that the output may vary in either polarity from zero: as a symmetrical program it need not contain a direct-current component. *See:* **unipolar.**   228

bipolar device (circuits and systems). An electronic device whose operation depends on the transport of both holes and electrons.   67

bipolar electrode. An electrode, without metallic connection with the current supply, one face of which acts as an anode surface and the opposite face as a cathode surface when an electric current is passed through the cell. *See:* **electrolytic cell.**   328

bipolar electrode system (electrobiology). Either a pickup or stimulating system consisting of two electrodes whose relation to the tissue currents is roughly symmetrical. *See:* **electrobiology.**   192

bipolar pulse (pulse terms). Two pulse waveforms of opposite polarity which are adjacent in time and which are considered or treated as a single feature.   254

bipolar video (radar). A radar video signal whose amplitude can have both positive and negative values; derived from a synchronous phase detection process. Coherent detection produces one type of bipolar video.   13

biquinary. Pertaining to the number representation system in which each decimal digit N is represented by the digit pair AB, where $N = 5A + B$, and where $A = 0$ or 1 and $B = 0,1,2,3$, or 4; for example, decimal 7 is represented by biquinary 12. This system is sometimes called a mixed-radix system having the radices 2 and 5.   255

birefringence. *See:* **birefringent medium.**

birefringent medium (fiber optics). A material that exhibits different indices of refraction for orthogonal linear polarizations of the light. The phase velocity of a wave in a birefringent medium thus depends on the polarization of the wave. Fibers may exhibit birefringence. *See:* **refractive index (of a medium).**   433

bistable (1) (general). The ability of a device to assume either of two stable states.   206
(2) (industrial control). Pertaining to a device capable of assuming either one or two stable states. *See:* **control system, feedback.**   255

bistable amplifier (industrial control). An amplifier with an output that can exist in either of two stable states without a sustained input signal and can be switched abruptly from one state to the other by specified inputs. *See:* **control system, feedback; rating and testing magnetic amplifiers.**   206

bistable logic function (graphic symbols for logic functions). A sequential logic function that has two and only two internal output states. *Syn:* **flip-flop.**   88, 451

bistable operation. Operation of a charge-storage tube in such a way that each storage element is inherently held at either of two discrete equilibrium potentials. *Note:* Ordinarily this is accomplished by electron bombardment. *See:* **charge-storage tube.**   174

bistatic cross section (antennas). The scattering cross section in any specified direction other than back toward the source. *See:* **monostatic cross section; radar cross section**   111

bistatic radar. A radar using antennas at different locations for transmission and reception.   13

bit (1)(microprocessor operating systems). A contraction of the term 'binary digit'; a unit of information represented by either a zero or a one.   434, 478
(2)(supervisory control, data acquisition, and automatic control). (A) Least significant. In an n bit binary word its contribution is (0 or 1) toward the maximum word value of (2n-1). (B) Most significant. In an n bit binary word its contribution is (0 or 1 times 2(n-1) toward the maximum word value of (2n-1).   570
(3) (data transmission). (A) An abbreviation of 'binary digit'. (B) A single occurrence of a character in a language employing exactly two kinds of characters. (C) A unit of storage capacity. The capacity, in bits, of a storage device with logarithm to the base two of the number of possible states of the device.   59
(4) (information theory). A unit of information content equal to the information content of a message the *a priori* probability of which is one-half. *Note:* If, in the definition of information content, the logarithm is taken to the base two, the result will be expressed in bits. One bit equals $\log_{10} 2$ hartley. *See:* **information theory; check bit; parity bit.**   61
(5) (electronic computers). (A) An abbreviation of binary digit. (B) A single occurrence of a character in a language employing exactly two distinct kinds of characters. (C) A unit of storage capacity. The capacity, in bits, of a storage device is the logarithm to the base two of the number of possible states of the device. *See:* **storage capacity.**   54, 235

bit error. *See:* **error rate.**

bit-parallel (programmable instrumentation)(signals and paths)(696 interface devices). A set of concurrent data bits present on a like number of signal lines used to carry information. Bit-parallel data bits may be acted upon concurrently as a group or independently as individual data bits.   40, 378,538

bit rate (1)(supervisory control, data acquisition, and automatic control)(station control and data acquisition). The number of bits transferred in a given time interval. Bits per second is a measure of the rate at which bits are transmitted.   403, 570

(2) (data transmission). The speed at which bits are transmitted; usually expressed in bits per second.
59

bits per unit time (test, measurement and diagnostic equipment). Operating number of bits, handled by a device in a given unit of time, under specified conditions.
54

black and white. *See:* **monochrome.**

blackbody (1) (fiber optics). A totally absorbing body (which reflects no radiation). *Note:* In thermal equilibrium, a blackbody absorbs and radiates at the same rate; the radiation will just equal absorption when thermal equilibrium is maintained. *See:* **emissivity.**
433

(2) (illuminating engineering). A temperature radiator of uniform temperature whose radiant exitance in all parts of the spectrum is the maximum obtainable from any temperature radiator at the same temperature. Such a radiator is called a blackbody because it will absorb all the radiant energy that falls upon it. All other temperature radiators may be classed as non-blackbodies. They radiate less in some or all wavelength intervals than a blackbody of the same size and the same temperature. *Note:* the blackbody is practically realized over limited solid angles in the form of a cavity with opaque walls at a uniform temperature and with a small opening for observation purposes. *Syn:* **standard radiator; ideal radiator; complete radiator.**
167

blackbody (Planckian) locus (illuminating engineering). The locus of points on a chromaticity diagram representing the chromaticities of blackbodies having various (color) temperatures.
167

black compression (black saturation) (television). The reduction in gain applied to a picture signal at those levels corresponding to dark areas in a picture with respect to the gain at that level corresponding to the mid-range light value in the picture. *Notes:* (1) The gain referred to in the definition is for a signal amplitude small in comparison with the total peak-to-peak picture signal involved. A quantitative evaluation of this effect can be obtained by a measurement of differential gain. (2) The over-all effect of black compression is to reduce contrast in the low lights of the picture as seen on a monitor. *See:* **television.**
178

black level (television). The level of the picture signal corresponding to the maximum limit of black peaks. *See:* **television.**
178

black light (illuminating engineering). The popular term for ultraviolet energy near the visible spectrum. *Note:* For engineering purposes the wavelength range 320-400 nm (nanometers) has been found useful for rating lamps and their effectiveness upon fluorescent materials (excluding phosphors used in fluorescent lamps). By confining "black light" applications to this region, germicidal, and erythemal effects are, for practical purposes eliminated.
167

black light flux (illuminating engineering). Radiant flux within the wavelength range 320-400 nm (nanometers). It is usually measured in milliwatts. *Note:*

The fluoren is used as a unit of "black light" flux and is equal to one milliwatt of radiant flux in the wavelength range 320-400 nm. Because of the variability of the spectral sensitivity of materials irradiated by "black light" in practice, no attempt is made to evaluate "black light" flux according to its capacity to produce effects.
167

black light flux density (illuminating engineering). "Black light" flux per unit area of the surface being irradiated. It is equal to the incident "black light" flux divided by the area of the surface when the flux is uniformly distributed. It usually is measured in milliwatts per unit area of "black light" flux.
167

black peak (television). A peak excursion of the picture signal in the black direction.
178

black recording (1) (amplitude-modulation facsimile system). The form of recording in which the maximum received power corresponds to the maximum density of the record medium.

(2) (frequency-shift, facsimile system). The form of recording in which the lowest received frequency corresponds to the maximum density of the record medium. *See:* **recording (facsimile).**
12

black saturation (television). *See:* **black compression.**

black signal (at any point in a facsimile system). The signal produced by the scanning of a maximum-density area of the subject copy. *See:* **facsimile signal (picture signal).**
12

black transmission (1) (amplitude-modulation facsimile system). The form of transmission in which the maximum transmitted power corresponds to the maximum density of the subject copy.

(2) (frequency-modulation facsimile system). The form of transmission in which the lowest transmitted frequency corresponds to the maximum density of the subject copy. *See:* **facsimile transmission.**
12

blade (disconnecting blade) (1) (of a switching device). The moving contact member that enters or embraces the contact clips. *Note:* In cutouts the blade may be a fuse carrier or fuseholder on which a nonfusible member has been mounted in place of a fuse link. When so used the nonfusible member alone is also called a blade in fuse parlance.

(2) (high-voltage switchgear). *Note:* In distribution cutouts the blade may be a nonfusible member for mounting on a fuse carrier in place of a fuse link, or in a fuse support, in place of a fuse holder.
27, 443

blade antenna. A form of monopole antenna which is blade shaped for strength and low aerodynamic drag.
111

blade control deadband (hydraulic turbines). The magnitude of the change in the blade control cam follower position required to reverse the travel of the blade control servomotor. The deadband is expressed in percent of the change in cam follower position required to move the blades from extreme "flat" to extreme "steep."
8

blade guide (switch) (power switchgear). An attachment to insure proper alignment of the blade and contact clip when closing the switch.
103

blade latch (power switchgear). A latch used on a stick-operated switch to hold the switch blade in the closed position. 103, 443

blank (test, measurement and diagnostic equipment). (1) A place of storage where data may be stored (synonymous with space): (2) A character, used to indicate an output space on a printer in which nothing is printed: and (3) A condition of no information at all in a given column of a punched card or in a given location on perforated tape. 54

blank character. A character used to produce a character space on an output medium. 255, 77

blanked picture signal (television). The signal resulting from blanking a picture signal. *Notes:* (1) Adding synchronizing signal to the blanked picture signal forms the composite picture signal. (2) This signal may or may not contain setup. A blanked picture signal with setup is commonly called a noncomposite signal. *See:* **television.** 178, 87

blanketing. The action of a powerful radio signal or interference in rendering a receiving set unable to receive desired signals. *See:* **radiation.** 328

blanking (1) (general). The process of making a channel or device noneffective for a desired interval.

(2) (television). The substitution for the picture signal, during prescribed intervals, of a signal whose instantaneous amplitude is such as to make the return trace invisible. *See:* **television.** 328

(3) (oscilloscopes). Extinguishing of the spot. Retrace blanking is the extinction of the spot during the retrace portion of the sweep waveform. The term does not necessarily imply blanking during the holdoff interval or while waiting for a trigger in a triggered sweep system. 185

blanking, chopped. *See:* **blanking; chopping transient.**

blanking, chopping transient. The process of blanking the indicating spot during the switching periods in chopped display operation. 106

blanking, deflection (oscilloscope). Blanking by means of a deflection structure, in the cathode-ray tube electron gun which traps the electron beam inside the gun, to extinguish the spot, permitting blanking during retrace and between sweeps regardless of intensity setting. 106

blanking level (television). That level of a composite picture signal which separates the range containing picture information from the range containing synchronizing information. *Note:* This term should be used for controls performing this function. 87

blanking signal (television). A wave constituted of recurrent pulses, related in time to the scanning process, used to effect blanking. *Note:* In television, this signal is composed of pulses at line and field frequencies, which usually originate in a central synchronizing generator and are combined with the picture signal at the pickup equipment in order to form the blanked picture signal. The addition of synchronizing signal completes the composite picture signal. The blanking portion of the composite picture signal is intended primarily to make the return trace on a picture tube invisible. The same blanking pulses or others of some-

what shorter duration are usually used to blank the pickup device also. *See:* **television.** 337

blanking, transient. *See:* **blanking, chopping transient.**

blaster. *See:* **blasting unit.**

blasting circuit. A shot-firing cord together with connecting wires and electric blasting caps used in preparation for the firing of a blast in mines, quarries, and tunnels. *See:* **blasting unit.** 328

blasting switch. A switch used to connect a power source to a blasting circuit. *Note:* A blasting switch is sometimes used to short-circuit the leading wires as a safeguard against premature blasts. *See:* **blasting unit.** 328

blasting unit (blaster) (exploder) (shot-firing unit). A portable device including a battery or a hand-operated generator designed to supply electric energy for firing explosive charges in mines, quarries, and tunnels. 328

bleaching (laser-maser). The decrease of optical **absorption** produced in a medium by radiation or by external forces. 363

bleeder. A resistor connected across a power source to improve voltage regulation, to drain off the charge remaining in capacitors when the power is turned off, or to protect equipment from excessive voltages if the load is removed or substantially reduced. 328

blemish (television). A small area brightness gradient in the reproduced picture, not present in the original scene. 87

blemish charge (storage tubes). A localized imperfection of the storage assembly that produces a spurious output. *See:* **storage tube.** 174, 190, 125

blinder (power switchgear). A relay having a characteristic on an R-X diagram of one or more essentially straight lines, usually positioned at 75° to 90° from the R-axis and displaced from the origin. 103

blinding glare (illuminating engineering). Glare which is so intense that for an appreciable length of time after it has been removed, no object can be seen. 167

blind speed (radar systems using (MTI) moving target indicators). Radial velocity of a target with respect to the radar for which the MTI response is approximately zero; in a coherent MTI system using a uniform repetition rate, a blind speed is a radial velocity at which the target changes its distance by one-half wavelength, or a multiple thereof, during each pulse repetition interval. 13

blind spot (surge testing for equipment connected to low-voltage ac power circuits). A limited range within the total domain of application of a device, generally at values inferior to the maximum rating. Operation of the equipment or the protective device itself might fail in that limited range despite the device's demonstration of satisfactory performance at maximum ratings. 578

blinker signal. *See:* **Morse signal light.**

blinking (pulse systems)(navigation aid terms). A method of providing information by modifying the signal at its source so that the signal presentation on the display at the receiver alternately appears and disappears, for example, in loran, blinking is used to indicate that the signals of a pair of stations are out of synchronization. 526

blip (radar)(navigation aid terms). A deflection or a spot of contrasting luminescence on a radar display caused by the presence of a target. 13, 526

blip-scan ratio (radar). The fraction of scans on which a blip is observed at a given range. Corresponds to probability of detection when observer's integration time is less than scan period. 13

block (1)(microprocessor operating systems). A group of data that is contiguous in nature. 478

(2) (data transmission). (A) A set of things, such as words, characters, or digits handled as a unit. (B) A collection of contiguous records recorded as a unit. (C) In data communications, a group of contiguous characters formed for transmission purposes. 59

(3) (railway practice). A length of track of defined limits on which the movement of trains is governed by block signals, cab signals, or both. *See:* **absolute block.** 328

(4) (conductor stringing equipment). A device designed with one or more single sheaves, a wood or metal shell, and an attachment hook or shackle. When rope is reeved through two of these devices, the assembly is commonly referred to as a block and tackle. A set of 4s refers to a block and tackle arrangement utilizing two 4 inch double sheave blocks to obtain four load bearing lines. Similarly, a set of 5s or a set of 6s refers to the same number of load bearing lines obtained using two 5 inch or two 6 inch double sheave blocks, respectively. *Syn:* **set of 4s; set of 5s; set of 6s.** 431

(5) (relaying). An output signal of constant amplitude and specified polarity derived from an alternating input and with the duration controlled by the polarity of the input quantity. 79

(6) (software). (1) A string of records, a string of words, or a character string, formed for technical or logic reasons to be treated as an entity. (2) A collection of contiguous records recorded as a unit. Blocks are separated by interblock gaps and each block may contain one or more records. (3) A group of bits or N-ary digits for error-control purposes. (4) A set of things, such as words, characters, or digits, handled as a unit. *See:* **bit; error; N-ary; program block; records; string; words.** 434

(7) (as applied to static relay design) (power switchgear). An output signal of constant amplitude and specified polarity derived from an alternating input and with the duration controlled by the polarity of the input quantity. 103

(8) (city, town, or village) (National Electrical Code). A square or portion of a city, town, or village enclosed by streets and including the alleys so enclosed but not any street. 256

block-block element (power switchgear). A signal element in which two blocks are compared as to coincidence or sequence. 103

block cable (communication practice). A distribution cable installed on poles or outside building walls, in the interior of a block, including cable run within buildings from the point of entrance to a cross-connecting box, terminal frame, or point of connection to house cable. *See:* **cable.** 328

block count readout. Display of the number of blocks that have been read from the tape derived by counting each block as it is read. *See:* **sequence number readout.** 207

block diagram (software). A diagram of a system, a computer, or a device in which the principal parts are represented by suitably annotated geometrical figures to show both the basic functions of the parts and their functional relationships. *See:* **computer; flowchart; functions; system.** 434

blocked impedance (transducer). The input impedance of the transducer when its output is connected to a load of infinite impedance. *Note:* For example, in the case of an electromechanical transducer, the blocked electric impedance is the impedance measured at the electric terminals when the mechanical system is blocked or clamped: the blocked mechanical impedance is measured at the mechanical side when the electric circuit is open-circuited. *See:* **self-impedance.** 176

block error (data transmission). A discrepancy of information in a block as detected by a checking code or technique. 59

block indicator. A device used to indicate the presence of a train in a block. 328

blocking (1) (tube rectifier). The prevention of conduction by means of grid or ignitor action, or both, when forward voltage is applied across a tube. 204

(2) (semiconductor rectifier). The action of a semiconductor rectifier cell that essentially prevents the flow of current. *See:* **rectification.** 237, 66

(3) (power switchgear). A relay function which prevents action that would otherwise be initiated by the relay system. 103

(4) (rotating machinery). A structure or combination of parts, usually of insulating material, formed by hold coils in relative position for mechanical support. *Note:* Usually inserted in the end turns to resist forces during running and abnormal conditions. *See:* **stator.** 63

(5) (telephone switching system). The inability of a telecommunication system to establish a connection due to the unavailability of paths. 55

blocking capacitor (1) (blocking condenser). A capacitor that introduces a comparatively high series impedance for limiting the current flow of low-frequency alternating current or direct current without materially affecting the flow of high-frequency alternating current. *See:* **electrolytic capacitor.** 329

(2) (check valve) An asymmetrical cell used to prevent flow of current in a specified direction. *See:* **electrolytic capacitor.** 328

blocking condenser. *See:* **blocking capacitor.**

blocking contact (of a semiconductor radiation detector)(charged-particle detectors). That contact from which depletion proceeds into the semiconductor material under conditions of reverse bias. 119

blocking interval (circuit properties)(self-commutated converters). An interval during which voltage is impressed across a switching element in its off-state.
584

blocking oscillator (1). A relaxation oscillator consisting of an amplifier (usually single-stage) with its output coupled back to its input by means that include capacitance, resistance, and mutual inductance. *See:* **oscillatory circuit.**
328

(2) (squegging oscillator). An electron-tube oscillator operating intermittently with grid bias increasing during oscillation to a point where oscillations stop, then decreasing until oscillation is resumed. *Note:* Squegge rhymes with wedge. *See:* **oscillatory circuit.**
111

blocking period (1) (rectifier-circuit element). The part of an alternating-voltage cycle during which reverse voltage appears across the rectifier-circuit element. *Note:* The blocking period is not necessarily the same as the reverse period because of the effect of circuit parameters and semiconductor rectifier cell characteristics. *See:* **rectifier circuit element.**
237

(2) (gas tube). The part of the idle period corresponding to the commutation delay due to the action of the control grid.
244, 190

blocking relay (1) (power switchgear). A relay whose function is to render another relay or device ineffective under specified conditions.
103

(2) (68) (power system device function numbers). A relay that initiates a pilot signal for blocking of tripping on external faults in a transmission line or in other apparatus under predetermined conditions, or cooperates with other devices to block tripping or to block reclosing on an out-of-step condition or on power swings.
402

blocking switching network (telephone switching system). A switching network in which a given outlet cannot be reached from any given inlet under certain traffic conditions.
55

block, input (test, measurement and diagnostic equipment). (1) A section of internal storage of a computer, reserved for the receiving and processing of input information (synonymous with input area): (2) A block used as an input buffer: and (3) A block of machine words, considered as a unit and intended to be transferred from an external source or storage or storage medium to the internal storage of the computer.
54

block-interval demand meter. *See:* **integrated-demand meter; demand meter.**

block-interval demand register (mechanical demand registers). A demand register that indicates or registers the maximum demand obtained by arithmetically averaging the meter registration over a regularly repeated time interval.
548

block signal. A fixed signal installed at the entrance of a block to govern trains entering and using that block.
328

block-signal system. A method of governing the movement of trains into or within one or more blocks by block signals or cab signals.
328

block-spike element (power switchgear). A signal element in which a block and a spike are compared as to coincidence.
103

block station. A place at which manual block signals are displayed.
328

block-structured language (software). A design or programming language in which sequences of statements are demarcated, usually with "begin" and "end" delimiters. *See:* **design; program block; programming language.**
434

block transfer (FASTBUS acquisition and control). The portion of a FASTBUS operation in which a master either sends data to or receives data from an attached slave on every transition of data sync. The slave acknowledges receipt of or sends data with every transition of data acknowledge.
480

Blondel diagram (rotating machinery). A phasor diagram intended to illustrate the currents and flux linkages of the primary and secondary windings of a transformer, and the components of flux due to primary and secondary winding currents acting alone. *Note:* This diagram is also useful as an aid in visualizing the fluxes in an induction motor. *See:* **asynchronous machine.**
63

blooming (1) (radar). An increase in the blip size on the display as a result of an increase in signal intensity or duration.
13

(2) (television picture tube).
244, 190

(3) (diode-type camera tube). (A) The increase in the size of the displayed image of a bright source when its irradiance is sufficient to cause overload of the mosaic target. It is measured in the display of the video output as the ratio of the enlarged spot size to the dimension of the active raster diagonal.
(B) The ratio of the image device generated spot size at overload to the size of the active raster diagonal. The actual spot size imaged upon the device photosensitive surface is chosen as 1 percent of the active raster diagonal.
380

blower blade (rotating machinery). An active element of a fan or blower. *See:* **fan (rotating machinery).**
63

blower housing. *See:* **fan housing.**

blowoff valve (gas turbines). A device by means of which a part of the air flow bypasses the turbine(s) and/or the regenerator to reduce the rate of energy input to the turbine(s). *Note:* It may be used in the speed governing system to control the speed of the turbine(s) at rated speed when fuel flow permitted by the minimum fuel limiter would otherwise cause the turbine to operate at a higher speed. *See:* **asynchronous machine.**
98, 58

blowout coil. An electromagnetic device that establishes a magnetic field in the space where an electric circuit is broken and helps to extinguish the arc by displacing it, for example, into an arc chute. *See:* **contactor; relay.**
259, 244, 206

blowout magnet. A permanent-magnet device that establishes a magnetic field in the space where an electric circuit is broken and helps to extinguish the arc by displacing it. *See:* **relay.**
259

blue dip (electroplating). A solution containing a mercury compound, and used to deposit mercury upon an immersed metal, usually prior to silver plating. *See:* **electroplating.** 328

blur (null type DF [direction finder] systems)(navigation aid terms). The output (including noise) at the bearing of minimum response expressed as a percentage of the output at the bearing of maximum response. 526, 278, 13

B-message (analog voice frequency circuits). A frequency-weighting characteristic, used for measurement of noise in voice-frequency communications circuits and designed to weight noise frequencies in proportion to their perceived annoyance effect in telephone service. 468

board (computing system). *See:* **problem board.**

boatswain's chair (conductor stringing equipment). A seat designed to be suspended on a line reeved through a block and attached to a pulling device to hoist a workman to an elevated position. *Syn:* **bosun's chair.** 431

bobbin (1) (primary cell). A body in a dry cell consisting of a depolarizing mix molded around a central rod of carbon and constituting the positive electrode in the assembled cell. *See:* **electrolytic cell.** 328
(2) (rotating machinery). Spool-shaped ground insulation fitting tightly on a pole piece, into which field coil is wound or placed. *See:* **rotor (rotating machinery); stator.** 63

bobbin core. A tape-wound core in which the ferromagnetic tape has been wrapped on a form or bobbin that supplies mechanical support to the tape. *Note:* The dimensions of a bobbin are illustrated in the accompanying figure. Bobbin I.D. is the center-hole diameter (D) of the bobbin. Bobbin O.D. is the over-all diameter (E) of the bobbin. The bobbin height is the over-all axial dimension (F) of the bobbin. Groove diameter is the diameter (G) of the center portion of the bobbin on which the first tape wrap is placed. The groove width is the axial dimension (H) of the bobbin measured inside the groove at the groove diameter.

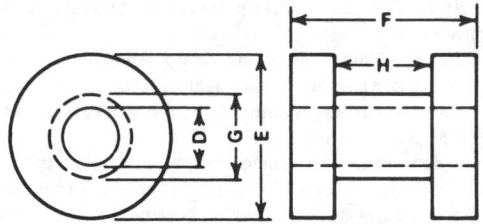

Dimensions of a bobbin.

331

bobbin height. *See:* **bobbin core; tape-wound core.**
bobbin I.D. *See:* **bobbin core; tape-wound core.**
bobbin O.D. *See:* **bobbin core; tape-wound core.**
Bode diagram (automatic control). A plot of log-gain and phase-angle values on a log-frequency base, for an element transfer function $G(j\omega)$, a loop transfer function $GH(j\omega)$, or an output transfer function $G(j\omega)/[1 + GH(j\omega)]$. The generalized Bode diagram comprises similar plots of functions of the complex variable $s = \sigma + j\omega$. *Note:* Except for functions containing lightly damped quadratic factors, the gain characteristic may be approximated by asymptotic straight-line segments that terminate at corner frequencies. The ordinate may be expressed as a gain, a log-gain, or in decibels as 20 times log-gain: the abscissa as cycles per unit time, radians per unit time, or as the ratio of frequency to an arbitrary reference frequency. *See:* **control system, feedback.** 56, 329

body (of an oil cutout). *See:* **housing (of an oil cutout).** 103

body capacitance. Capacitance introduced into an electric circuit by the proximity of the human body. 328

body-capacitance alarm system. A burglar alarm system for detecting the presence of an intruder through his body capacitance. *See:* **protective signaling.** 328

body effect (metal-nitride-oxide field-effect transistor). This effect occurs when the potential in the substrate of a (p-channel) insulated-gate field-effect transistor (IGFET) is more positive than the source potential. It can be expressed as an increment that increases the threshold voltage of an IGFET. The effect occurs routinely in integrated circuits. 386

body generator suspension. A design of support for an axle generator in which the generator is supported by the vehicle body. *See:* **axle generator system.** 328

bolometer (1) (fiber optics). A device for measuring radiant energy by measuring the changes in resistance of a temperature-sensitive device exposed to radiation. *See:* **radiant energy; radiometry.** 433
(2) (waveguide components). A term commonly used to denote the combination of a bolometer element and a bolometer mount; sometimes used imprecisely to refer to a bolometer element. 166
(3) (laser-maser). A radiation detector of the thermal type in which absorbed radiation produces a measurable change in the physical property of the sensing element. The change in state is usually that of electrical resistance. 363

bolometer bridge. A bridge circuit with provisions for connecting a bolometer in one arm and for converting bolometer-resistance changes to indications of power. *See:* **bolometric power meter.** 185

bolometer bridge, balanced. A bridge in which the bolometer is maintained at a prescribed value of resistance before and after radio-frequency power is applied, or after a change in radio-frequency power, by keeping the bridge in a state of balance. *Note:* The state of balance can be achieved automatically or manually by decreasing the bias power when the radio-frequency power is applied or increased and by increasing the bias power when the radio-frequency power is turned off, or decreased. The change in the bias power is a measure of the applied radio-frequency power. *See:* **bolometric power meter.** 115

bolometer bridge, unbalanced. A bridge in which the resistance of the bolometer changes after the radio-frequency power is applied and unbalances the bridge. The degree of bridge unbalance is a measure of the radio-frequency power dissipated in the bolometer. *See:* **bolometric power meter.**   185

bolometer-coupler unit. A directional coupler with a bolometer unit attached to either the side arm or the main arm, normally used as a feed-through power-measuring system. *Note:* Typically, a bolometer unit is attached to the side arm of the coupler so that the radio-frequency power at the output port can be determined from a measurement of the substitution power in the side arm. This system can be used as a terminating power meter by terminating the output port of the directional coupler. *See:* **bolometric power meter.**   115

bolometer element (bolometric detector) (waveguide components) A power-absorbing element which uses the resistance change related to the temperature coefficient of resistivity (either positive or negative) as a means of measuring or detecting the power absorbed by the element. *See:* **barretter, thermistor.**   166

bolometer mount (1) (general). A waveguide or transmission-line termination that houses a bolometer element(s). *Note:* It normally contains internal matching devices or other reactive elements to obtain specified impedance conditions when a bolometer element is inserted and appropriate bias power is applied. Bolometer mounts may be subdivided into tunable, fixed-tuned, and broad-band untuned types. *See:* **bolometric power meter.**   115
(2) (waveguide components). A waveguide or transmission line termination that can house a bolometer element.   166

bolometer unit. An assembly consisting of a bolometer element or elements and bolometer mount in which they are supported. *See:* **bolometric power meter.**   115

bolometer unit, dual element. An assembly consisting of two bolometer elements and a bolometer mount in which they are supported. *Note:* The bolometer elements are effectively in series to the bias power and in parallel to the radio frequency power.   185

bolometric detector (bolometer). The primary detector in a bolometric instrument for measuring power or current and consisting of a small resistor, the resistance of which is strongly dependent on its temperature. *Notes:* (1) Two forms of bolometric detector are commonly used for power or current measurement: (A) The barretter that consists of a fine wire or metal film: and (B) the thermistor that consists of a very small bead of semiconducting material having a negative temperature-coefficient of resistance: either is usually mounted in a waveguide or coaxial structure and connected so that its temperature can be adjusted and its resistance measured. (2) Bolometers for measuring radiant energy usually consist of blackened metal-strip temperature-sensitive elements arranged in a bridge circuit including a compensating arm for ambient temperature compensation. *See:* **instrument; bolometric instrument.**   185

bolometric instrument (bolometer). An electrothermic instrument in which the primary detector is a resistor, the resistance of which is temperature sensitive, and that depends for its operation on the temperature difference maintained between the primary detector and its surroundings. Bolometric instruments may be used to measure nonelectrical quantities, such as gas pressure or concentration, as well as current and radiant power. *See:* **instrument.**   328

bolometric power meter. A device consisting of a bolometer unit and associated bolometer-bridge circuit(s).   115

bolometric technique (power measurement). A technique wherein the heating effect of an unknown amount of radio-frequency power is compared with that of a measured amount of direct-current or audio-frequency power dissipated within a temperature sensitive resistance element (bolometer). *Note:* The bolometer is generally incorporated into a bridge network, so that a small change in its resistance can be sensed. This technique is applicable to the measurement of low levels of radio-frequency power, that is, below 100 milliwatts.   185

bolted fault (generating station grounding). A short circuit or electrical contact between two conductors at different potentials, in which the impedance or resistance between the conductors is essentially zero.   569

Boltzmann's constant (fiber optics). The number k that relates the average energy of a molecule to the absolute temperature of the environment. k is approximately 1.38×10^{-23} joules/kelvin.   433

bombardment-induced conductivity (storage tubes). An increase in the number of charge carriers in semiconductors or insulators caused by bombardment with ionizing particles. *See:* **storage tube.**   174

bomb-control switch. A switch that closes an electric circuit, thereby tripping the bomb-release mechanism of an aircraft, usually by means of a solenoid.   328

bond (electrolytic cell line working zone). A reliable connection to assure the required electrical conductivity between conductive parts required to be electrically connected.   133

bonded (conductor stringing equipment)(power line maintenance). The mechanical interconnection of conductive parts to maintain a common electrical potential. *Syn:* connected.   431, 458

bonded motor (rotating machinery). A complete motor in which the stator and end shields are held together by a cement, or by welding or brazing.   63

bonding (1)(generating station grounding). The permanent joining of metallic parts to form an electrically conductive path that will ensure electrical continuity and the capacity to conduct safely any current likely to be imposed.   569
(2)(transmission and distribution)(NESC). The electrical interconnecting of conductive parts, designed to maintain a common electrical potential.   262, 494
(3) (electric cables). The electric interconnecting of cable sheaths or armor to sheaths or armor of adjacent

conductors. *See:* **cable bond cross-cable bond; continuity-cable bond.** 64

bonding (4) (National Electrical Code). The permanent joining of metallic parts to form an electrically conductive path which will assure electrical continuity and the capacity to conduct safely any current likely to be imposed. 256

bonding jumper (generating station grounding). A reliable conductor to ensure the required electrical conductivity between metal parts required to be electrically connected. 569

bone conduction (hearing). The process by which sound is conducted to the inner ear through the cranial bones. 176

Boolean (1)(mathematics of computing). Pertaining to the rules of logic formulated by the Irish mathematician George Boole in 1847. 564

(2) (general). (A) Pertaining to the processes used in the algebra formulated by George Boole. (B) Pertaining to the operations of formal logic. 255

(3) (ATLAS). A labeled variable which can assume only two states, true or false. 400

boost. The act of increasing the power output capability of an operational amplifier by circuit modification in the output stage. *See:* **electronic analog computer.** 9

boost charge (quick charge) (storage battery). A partial charge, usually at a high rate for a short period. *See:* **charge.** 328

booster. An electric generator inserted in series in a circuit so that it either adds to or subtracts from the voltage furnished by another source. 328

booster coil. An induction coil utilizing the aircraft direct-current supply to provide energy to the spark plugs of an aircraft engine during its starting period. 328

booster dynamotor. A dynamotor having a generator mounted on the same shaft and connected in series for the purpose of adjusting the output voltage. *See:* **converter.** 328

bootleg (railway techniques). A protection for track wires when the wires leave the conduit or ground near the rail. 328

bootstrap (software). (1) A short computer program which is permanently resident or easily loaded into a computer, whose execution brings another, larger program, such as an operating system or its loader, into memory. (2) A set of instructions that cause additional instructions to be loaded until the complete computer program is in storage. (3) A technique or device designed to bring itself into a desired state by means of its own action, for example, a machine routine whose first few instructions are sufficient to bring the rest of itself into the computer from an input device. (4) That part of a computer program used to establish another version of the computer program. (5) To use a bootstrap. *See:* **computer; computer program; execution; instructions; loader; operating system; program; routine.** 434

bootstrap circuit (1) (general). A single-stage electron-tube amplifier circuit in which the output load is connected between cathode and ground or other common return, the signal voltage being applied between the grid and the cathode. *Note:* The name bootstrap arises from the fact that a change in grid voltage changes the potential of the input source with respect to ground by an amount equal to the output signal. 211

(2) (circuits and systems). A circuit in which an increment of the applied input signal is partially fed back across the input impedance resulting in a higher effective input impedance. 67

bootstrap loader (software). An input routine in which preset computer operations are used to load a bootstrap. *See:* **bootstrap; computer; routine.** 434

B **operator.** An operator assigned to a *B* switchboard. *See:* **telephone system.** 328

borderline between comfort and discomfort (BCD) (illuminating engineering). The average luminance of a source in a field of view which produces a sensation between comfort and discomfort. 167

bore (rotating machinery). The surface of a cylindrical hole (for example, stator bore). *See:* **stator.** 63

borehole cable (mining). A cable designed for vertical suspension in a borehole or shaft and used for power circuits in mines. *See:* **mine feeder circuit.** 328

bore-hole lead insulation (rotating machinery). Special insulation surrounding connections that pass through a hollow shaft. *See:* **rotor (rotating machinery).** 63

boresight. *See:* **electrical boresight; reference boresight.** 111

boresight error (antenna). The angular deviation of the electrical boresight of an antenna from its reference: boresight. *See:* **antenna.** 111, 246

boresighting (navigation aid terms). The process of aligning or determining the angle of the electrical or mechanical axes of a navigation system to a set of vehicle reference axes. Usually accomplished by an optical procedure. 526

borrow. In direct subtraction, a carry that arises when the result of the subtraction in a given digit place is less than zero. 235

bottom-car clearance (elevator). The clear vertical distance from the pit floor to the lowest structural or mechanical part, equipment, or device installed beneath the car platform, except guide shoes or rollers, safety jaw assemblies, and platform aprons or guards, when the car rests on its fully compressed buffers. *See:* **hoistway (elevator or dumbwaiter).** 328

bottom-coil slot (radially outer-coil side) (rotating machinery). The coil side of a stator slot farthest from the bore of the stator or from the slot wedge. *See:* **stator.** 63

bottom-half bearing (rotating machinery). The bottom half of a split-sleeve bearing. *See:* **bearing.** 63

bottom-terminal landing (elevators). The lowest landing served by the elevator that is equipped with a hoistway door and hoistway-door locking device that permits egress from the hoistway side. *See:* **elevator landing.** 328

bottom-up (software). Pertaining to an approach which

starts with the lowest level software components of a hierarchy and proceeds through progressively higher levels to the top level component, for example, bottom-up design, bottom-up programming, bottom-up testing. *See:* **bottom-up design; hierarchy; software components; testing; top-down.**   434

bottom-up design (software). The design of a system starting with the most basic or primitive components and proceeding to higher level components that use the lower level ones. *See:* **components; design; system; top-down design.**   434

bounce (television). A transient disturbance affecting one or more parameters of the display and having duration much greater than the periòd of one frame. *Note:* The term is usually applied to changes in vertical position or in brightness. *See:* **television.**   178

boundary, *p-n* (semiconductor). A surface in the transition region between *p-type* and *n-type* material at which the donor and acceptor concentrations are equal. *See:* **semiconductor; transistor.**   245

boundary lights (illuminating engineering). Aeronautical ground lights delimiting the boundary of a land aerodrome without runways.   167

boundary marker (ILS [instrument landing system]) (navigation aid terms). A radio-transmitting station near the approach end of the landing runway, which provides a fix on the localizer course.   526

boundary potential. The potential difference, of whatever origin, across any chemical or physical discontinuity or gradient. *See:* **electrobiology.**   192

bound mode (fiber optics). In an optical waveguide, a mode whose field decays monotonically in the transverse direction everywhere external to the core and which does not lose power to radiation. Specifically a mode for which

$$n(a)k = \beta = n(0)k$$

where β is the imaginary part (phase constant) of the axial propagation constant, n(a) is the refractive index at r = a, the core radius, n(0) is the refractive index at r = 0, k is the free-space wavenumber, $2\pi/\lambda$, and λ is the wavelength. Bound modes correspond to guided rays in the terminology of geometric optics. *Note:* Except in a monomode fiber, the power in bound modes is predominantly contained in the core of the fiber. *Syn:* **guided mode; trapped mode.** *See:* **cladding mode; guided ray; leaky mode; mode; normalized frequency; unbound mode.**   433

bound ray. *See:* **guided ray.**

Bourdon. A closed and flattened tube formed in a spiral, helix, or arc, which changes in shape when internal pressure changes are applied. *Note:* Bourdon tube, or simply Bourdon, has at times been used more restrictively to mean only the C-shaped member invented by Bourdon.   7

bowl (illuminating engineering). An open top diffusing glass or plastic enclosure used to shield a light source from direct view and to redirect or scatter the light.   167

box frame (rotating machinery). A stator frame in the form of a box with ends and sides and that encloses the stator core. *See:* **rotor (rotating machinery).**   63

BR (medium attachment units and repeater units). The rate of data throughput (bit-rate) on the trunk coaxial medium expressed in hertz.   543

bracket (1) (illuminating engineering). An attachment to a lamp post or pole from which a luminaire is suspended. *Syn:* **mast arm.**   167

(2) (rotating machinery). A solid or skeletal structure usually consisting of a central hub and a plurality of arms extending (often radially) outward from the hub to a supporting structure. The supporting structure usually is the stator frame when the axis of the shaft is horizontal. When the axis of the shaft is vertical, the stator usually supports the upper bracket and the foundation supports the lower bracket. *See:* **bearing bracket.**   63

bracket arm (rotating machinery). One of several structural members (beams) extending from the hub portion of a bracket to the supporting structure. The arms may be individual or parallel pairs extending radially or near-radially from the hub.   63

bracket-type handset telephone (suspended-type handset telephone). *See:* **hang-up hand telephone set.**

Bragg region (acousto-optic device). The region that occurs when the length of the acoustic column in the direction of light propagation, L, satisfies the inequality $L > n \Lambda^2/\lambda_0$, with n the index of refraction at wavelength λ_0 and Λ the acoustic wavelength.   82

brake assembly (rotating machinery). All parts that are provided to apply braking to the rotor. *See:* **rotor (rotating machinery).**   63

brake control (industrial control). The provision for controlling the operation of an electrically actuated brake. *Note:* Electrical energizing of the brake may either release or set the brake, depending upon its design. *See:* **electric controller.**   206

brake drum. *See:* **brake ring.**

brake ring (brake drum) (rotating machinery). A rotating ring mounted on the rotor that provides a bearing surface for the brake shoes. *See:* **rotor (rotating machinery).**   63

braking (industrial control). The control function of retardation by dissipating the kinetic energy of the drive motor and the driven machinery. *See:* **electric drive.**   206

braking magnet. *See:* **retarding magnet.**

braking resistor. A resistor commonly used in some types of dynamic braking systems, the prime purpose of which is to convert the electric energy developed during dynamic braking into heat and to dissipate this energy to the atmosphere. *See:* **dynamic braking.**   328

braking test (rotating machinery). (1) A test in which the mechanical power output of a machine acting as a motor is determined by the measurement of the shaft torque, by means of a brake, dynamometer, or similar device, together with the rotational speed. (2) A test performed on a machine acting as a generator, by means of a dynamometer or similar device, to determine the mechanical power input. *See:* **asynchronous machine.**   63

braking torque (synchronous motor). Any torque exerted by the motor in the same direction as the load torque so as to reduce its speed. 63

branch (1)(converter circuit elements)(self-commutated converters). *See:* **principal branch; auxiliary branch; turn-off branch; regenerative branch.** 584

(2) (network analysis). A line segment joining two nodes, or joining one node to itself. *See:* **directed branch; network analysis.** 282

(3) (electronic computer). (A) A set of instructions that are executed between two successive decision instructions. (B) To select a branch as in (A). (C) Loosely, a conditional jump. *See:* **conditional jump.** 255

(4) (circuits and systems). A portion of a network consisting of one or more two-terminal elements, comprising a section between two adjacent branch-points. 67

branch cable (medium attachment units and repeater units). The attachment unit interface (AUI) interconnecting the data terminal equipment (DTE) and medium attachment unit (MAU) system components. 543

branch circuit (1)(electrical heating applications to melting furnaces and forehearths in the glass industry). One, two, or more circuits whose main power is connected through the same main switch. 520

(2) (electric installations on shipboard). That portion of a wiring system extending beyond the final overcurrent device protecting the circuit. 3

(3) (National Electrical Code). The circuit conductors between the final overcurrent device protecting the circuit and the outlet(s). *See:* **thermal cutouts; thermal relays.** 256

(4) (packaging machinery). That portion of a wiring system extending beyond the final overcurrent device protecting the circuit. (A device not approved for branch circuit protection, such as a thermal cutout or motor overload protective device, is not considered as the overcurrent device protecting the circuit). 429

branch-circuit distribution center. A distribution center at which branch circuits are supplied. *See:* **distribution center.** 328

branch circuit, general purpose. A branch circuit that supplies a number of outlets for lighting and appliances. 256

branch circuit, individual. A branch circuit that supplies only one utilization equipment. 256

branch circuit, multiwire. A circuit consisting of two or more ungrounded conductors having a potential difference between them, and an identified grounded conductor having equal potential difference between it and each ungrounded conductor of the circuit and that is connected to the neutral conductor of the system. 256

branch-circuit selection current (National Electrical Code). The value in amperes to be used instead of the rated-load current in determining the ratings of motor branch-circuit conductors, disconnecting means, controllers and branch-circuit short-circuit and ground-fault protective devices wherever the running overload protective device permits a sustained current greater than the specified percentage of the rated-load current. The value of branch-circuit selection current will always be greater than the marked rated-load current. 256

branch circuits incorporating Type FCC cable (National Electrical Code).

(a) type FCC cable. Type FCC cable consists of 3 or more flat copper conductors placed edge to edge and separated and enclosed within an insulating assembly. *Note:* The wiring system is designed for installation under carpet squares.

(b) FCC system. A complete wiring system for branch circuits that is designed for installation under carpet squares. The FCC system includes Type FCC cable and associated shielding, connectors, terminators, adapters, boxes, and receptacles.

(c) cable connector. A connector designed to join Type FCC cables without using a junction box.

(d) insulating end. An insulator designed to electrically insulate the end of a Type FCC cable.

(e) top shield. A grounded metal shield covering under carpet components of the FCC system for the purposes of providing electrical safety and protection against physical damage.

(f) bottom shield. A shield mounted on the floor under the FCC system to provide protection against physical damage.

(g) transition assembly. An assembly to facilitate connection of the FCC system to other approved wiring systems, incorporating: (1) a means of electrical interconnection; and (2) a suitable box or covering for providing electrical safety and protection against physical damage.

(h) metal shield connections. Means of connection designed to electrically and mechanically connect a metal shield to another metal shield, to a receptacle housing or self contained device or to a transition assembly. 256

branch conductor (lightning protection). A conductor that branches off at an angle from a continuous run of conductor. 328

branch input signal (network analysis). The signal xj at the input end of branch jk. 282

branch instruction (test, measurement and diagnostic equipment). An instruction in the program that provides a choice between alternative subprograms in accordance with the test logic. 54

branch joint (1)(power cable joint). A cable joint used for connecting one or more cables to a main cable. A branch joint may be further designated by naming the cables between which it is made, eg. single conductor cable, three-conductor cable, three-conductor main cable to single-conductor branch, etc. It is customary to designate the various kinds of Y joint, T joint, H joint, cross joint, etcetera. 34

(2) (general). A joint used for connecting a branch conductor or cable to a main conductor or cable, where the latter continues beyond the branch. *Note:* A branch joint may be further designated by naming the cables between which it is made, for example, single-conductor cables: three-conductor main cable to single-conductor branch cable, etcetera. With the term **multiple joint** it is customary to designate the various kinds as 1-way, 2-way, 3-way, 4-way, etcetera, multiple joint. *See:* **cable joint; straight joint; reducing joint.** 64

(3) **(power cable joints).** A cable joint used for connecting one or more cables to a main cable. *Note:* A branch joint may be further designated by naming the cables between which it is made, for example, single conductor cable, three conductor cable, three conductor main cable to single conductor branch, etcetera. It is customary to designate the various kinds as Y joint, T joint, H joint, cross joint, etcetera. 34

branch number (b) (subroutines for CAMAC). The symbol *b* represents an integer which is the branch number component of a CAMAC address. It may represent a physical highway number in multiple highway systems, or it may represent sets of crates grouped together for functional or other reasons. In some systems it may be ignored, although it must be included in the parameter list for the sake of compatibility. 410

branch output signal (branch *jk*) (network analysis). The component of signal x_k contributed to node k via branch *jk*. 282

branch point (1) (electric network). A junction where more than two conductors meet. *See:* **network analysis; node.** 328

(2) **(computers).** A place in a routine where a branch is selected. *See:* **network analysis.** 255

branch, thermoelectric. Alternative term for thermoelectric arm. *See:* **thermoelectric device.** 248, 191

branch transmittance (network analysis). The ratio of branch output signal to branch input signal. 282

breadboard construction (communication practice). An arrangement in which components are fastened temporarily to a board for experimental work. 328

break (1) (circuit-opening device). The minimum distance between the stationary and movable contacts when these contacts are in the open position. (A) The length of a single break is as defined above. (B) The length of a multiple break (breaks in series) is the sum to two or more breaks. *See:* **contactor.** 206

(2) **(communication circuit).** For the receiving operator or listening subscriber to interrupt the sending operator or talking subscriber and take control of the circuit. *Note:* The term is used especially in connection with half-duplex telegraph circuits and two-way telephone circuits equipped with voice-operated devices. *See:* **telegraphy.** 328

breakaway. The condition of a motor at the instant of change from rest to rotation. 63

breakaway starting current (alternating-current motor) (rotating machinery). The highest root mean square current absorbed by the motor when at rest, and when it is supplied at the rated voltage and frequency. *Note:* This is a design value and transient phenomena are ignored. 63

breakaway torque (rotating machinery). The torque that a motor is required to develop to break away its load from rest to rotation. *See:* **asynchronous machine.** 63

break, % break (telephony)(dial-pulse address signaling systems). In dial-pulse signaling, that portion of the signal in which the dialing contacts are open (broken). % break is the ratio of break time to the total pulse period. (make + break) time. 540

break distance (of a switching device) (power switchgear). The minimum open-gap distance between the main-circuit contacts, or live parts connected thereto, when the contacts are in the open position. *Note:* In a multiple-break device, it is the sum of the breaks in series. 103, 443

breakdown (1)(low-voltage air gap surge-protective devices)(gas-tube surge-protective devices). The abrupt transition of the gap resistance from a practically infinite value to a relatively low value. In the case of a gap, this is sometimes referred to as sparkover or ignition. *See:* **sparkover.** 490, 556

(2)**(of a semiconductor diode)(charged-particle detectors)(germanium gamma-ray detectors)(X-ray energy spectrometers).** A phenomenon occurring in a reverse biased semiconductor diode, the initiation of which is observed as a transition from a region of high dynamic resistance to a region of substantially lower dynamic resistance for increasing magnitude of reverse current. 119, 528, 471

(3) **(rotating machinery).** The condition of operation when a motor is developing breakdown torque. *See:* **asynchronous machine.** 63

(4) **(sparkover) (gas tube surge-protective device).** The abrupt transition of the gap resistance from a practically infinite value to a relatively low value. In the case of a gap, this is sometimes referred to as sparkover. *See:* **sparkover** 370

(5) **(thyristor converter).** A failure that permanently deprives a rectifier diode or a thyristor of its property to block voltage in the reverse direction (reverse breakdown) or a thyristor in the forward direction (forward breakdown). 121

breakdown current (semiconductor). The current at which the breakdown voltage is measured. 66

breakdown impedance (semiconductor diode). The small-signal impedance at a specified direct current in the breakdown region. *See:* **semiconductor.** 245

breakdown region (of a semiconductor diode characteristic) (charged-particle detectors) (germanium gamma-ray detectors) (X-ray energy spectrometers). That entire region of the voltage-current characteristic beyond the initiation of breakdown for increasing magnitude of reverse current. 119, 528, 471

breakdown strength. *See:* **dielectric strength.**

breakdown torque (1) (rotating machinery). The maximum shaft-output torque that an induction motor (or a synchronous motor operating as an induction motor) develops when the primary winding is connected for running operation, at normal operating temperature, with rated voltage applied at rated frequency. *Note:* A motor with a continually increasing torque as the speed decreases to standstill, is not considered to have a breakdown torque. 63

(2) (electric installations on shipboard). The maximum torque of a motor which it will develop with rated voltage applied at rated frequency, without an abrupt drop in speed. 3

breakdown-torque speed (rotating machinery). The speed of rotation at which a motor develops breakdown torque. *See:* **asynchronous machine.** 63

breakdown transfer characteristic (gas tube). A relation between the breakdown voltage of an electrode and the current to another electrode. *See:* **gas tube.** 125

breakdown voltage (1)(diode)(nonlinear, active, and nonreciprocal waveguide components). The reverse voltage at which there is a conduction of current due to the Zener effect or the avalanche multiplication process. This voltage is usually specified at 10 μA of reverse current. 530

(2)(of a semiconductor diode)(charged-particle detectors)(germanium gamma-ray detectors)(X-ray energy spectrometers). The voltage measured at a specified current in the breakdown region.
119, 528, 471

(3) (rotating machinery). The voltage at which a disruptive discharge takes place through or over the surface of the insulation. 6

(4) (gas) (electron device). The voltage necessary to produce a breakdown. *See:* **gas tube.** 190

(5) (electrode of a gas tube). The voltage at which breakdown occurs to that electrode. *Notes:* (A) The breakdown voltage is a function of the other electrode voltages or currents and of the environment. (B) In special cases where the breakdown voltage of an electrode is referred to an electrode other than the cathode, this reference electrode shall be indicated. (C) This term should be used in preference to pickup voltage, firing voltage, starting voltage, etcetera, which are frequently used for specific types of gas tubes under specific conditions. *See:* **critical grid voltage (multi-electrode gas tubes).** 125

breakdown voltage, ac. *See:* **alternating-current breakdown voltage.**

breakdown voltage alternating current (ac) (gas tube surge arrester). The minimum root-mean-square (rms) value of sinusoidal voltage at frequencies between 15 hertz (Hz) and 62 Hz that results in arrester sparkover. 62

breakdown voltage, dc. *See:* **direct current (dc) breakdown voltage.**

breaking capacity (interrupting capacity) (industrial control). The current that the device is capable of breaking at a stated recovery voltage under prescribed conditions of use and behavior. *See:* **control.**
244, 206

breaking current (pole of a breaking device) (industrial control). The current in that pole at the instant of contact separation, expressed as a root-mean-square

value. *See:* **contactor; interrupting current.**
103, 244, 206

breaking point (transmission system or element thereof). A level at which there occurs an abrupt change in distortion or noise that renders operation unsatisfactory. *See:* **level.** 328

break-in keying. A method of operating a continuous-wave radio telegraph communication system in which the receiver is capable of receiving signals during transmission of spacing intervals. *See:* **modulation; radio transmission.** 328

break-in period (nuclear power generating stations) (class 1E static battery chargers and inverters). That early period, beginning at some stated time and during which the failure rate of some items is decreasing rapidly. *Syn:* **early failure period.** 29,408

breakover current (thyristor). The principal current at the breakover point. *See:* **principal current.**
191,208

breakover point (thyristor). Any point on the principal voltage-current characteristic for which the differential resistance is zero and where the principal voltage reaches a maximum value. *See:* **principal voltage-current characteristic (principal characteristic).**
191

breakover voltage (thyristor). The principal voltage at the breakover point. *See:* **principal voltage-current characteristic (principal characteristic).** 191,208

breakpoint (1) (data transmission) *See:* **change in slope.** 59

(2) (computer routine). (A) Pertaining to a type of instruction, instruction digit, or other condition used to interrupt or stop a computer at a particular place in a routine when manually requested. (B) A place in a routine where such an interruption occurs or can be made to occur. 235

breakthrough (thyristor converter). The failure of the forward-blocking action of an arm of a thyristor connection during a normal off-state period with the result that it allows on-state current to pass during a part of this period. *Note:* Breakthrough can occur in rectifier operation as well as inverter operation and for various reasons, for example, excessive virtual junction temperature, voltage surges in excess of rated peak off-state voltage, excessive rate of rise of off-state voltage, advance gating, or forward breakdown.
121

breather. A device fitted in the wall of an explosion-proof compartment, or connected by piping thereto, that permits relatively free passage of air through it, but that will not permit the passage of incendiary sparks or flames in the event of gas ignition inside the compartment. 328

breathing (carbon microphones). The phenomenon manifested by a slow cyclic fluctuation of the electric output due to changes in resistance resulting from thermal expansion and contraction of the carbon chamber. *See:* **close-talking pressure-type microphone.** 249

breezeway (television synchronizing waveform for color transmission). The time interval between the trailing edge of the horizontal synchronizing pulse and the start of the color burst.   178

Brewster's angle (fiber optics). For light incident on a plane boundary between two regions having different refractive indices, that angle of incidence at which the reflectance is zero for light that has its electric field vector in the plane defined by the direction of propagation and the normal to the surface. For propagation from medium 1 to medium 2, Brewster's angle is

$$\arctan(n_2/n_1).$$

See: **angle of incidence; reflectance; refractive index (of a medium).**   433

bridge (data transmission). (1) A network with minimum of two ports or terminal pairs capable of being operated in such a manner that when power is fed into one port, by suitable adjustment of the elements in the network or the element connected to one or more other ports, zero output can be obtained at another port. Under these conditions the bridge is balanced. (2) An instrument or intermediate means in a measurement system that embodies all or part of a bridge circuit, and by means of which one or more of the electrical constants of a bridge may be measured.   59

bridge (protection and coordination of industrial and commercial power systems). That narrowed portion of a fuse link that is expected to melt first. One link may have two or more bridges in parallel and in series as well. The shape and size of the bridge is a factor in determining the fuse characteristics under overload and fault current conditions.   504

bridge control. Apparatus and arrangement providing for direct control from the bridge or wheelhouse of the speed and direction of a vessel.   328

bridge current (power supply). The circulating control current in the comparison bridge. *Note:* Bridge current equals the reference voltage divided by the reference resistor. Typical values are 1 milliampere and 10 milliamperes, corresponding to control ratios of 1000 ohms per volt and 100 ohms per volt, respectively.   186

bridged tap (telephone loop performance). Any portion of a metallic circuit which is not in the path between the end office and the customer. The bridged tap may be connected at an intermediate location or be an extension of the circuit beyond the customer location. The pair associated with the bridged tap introduces a frequency-dependent bridging loss in the loop.   473

bridged-T network. A T network with a fourth branch connected across the two series arms of the T, between an input terminal and an output terminal. *See:* **network analysis.**   170

bridge duplex system. A duplex system based on the Wheatstone bridge principle in which a substantial neutrality of the receiving apparatus to the sent currents is obtained by an impedance balance. *Note:* Received currents pass through the receiving relay that is bridged between the points that are equipotential for the sent currents. *See:* **telegraphy.**   328

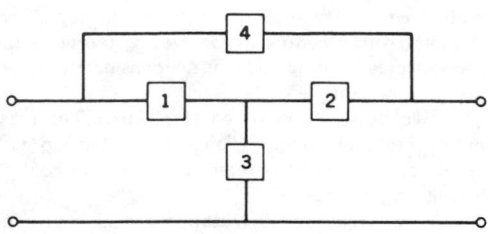

Bridged-T network. Branches 1, 2, and 3 comprise the T network and branch 4 is the fourth branch.

bridge limiter. *See:* **limiter circuit.**

bridge rectifier (power semiconductor). A rectifier unit which makes use of a bridge-rectifier circuit.   66

bridge rectifier circuit. A full-wave rectifier with four rectifying elements connected as the arms of a bridge circuit. *See:* **rectifier; single-way rectifier circuit; double-way rectifier circuit.**   111

bridge transition. A method of changing the connection of motors from series to parallel in which all of the motors carry like currents throughout the transfer due to the Wheatstone bridge connection of motors and resistors. *See:* **multiple-unit control.**   328

bridging (1) (signal circuits). The shunting of one signal circuit by one or more circuits usually for the purpose of deriving one or more circuit branches. *Note:* A bridging circuit often has an input impedance of such a high value that it does not substantially affect the circuit bridged.   239

(2) **(soldered connections).** Solder that forms an unwanted conductive path.   284

(3) **(relays).** *See:* **relay bridging.**

bridging amplifier. *See:* **amplifier, bridging.**

bridging connection (data transmission). A parallel connection by means of which some of the signal energy in a circuit may be withdrawn, frequently with imperceptible effect on the normal operation of the circuit.   59

bridging gain (data transmission). The ratio of the signal power a transducer delivers to its load(Z_B) to the signal power dissipated in the main circuit load(Z_M) across which the input transducer is bridged.   59

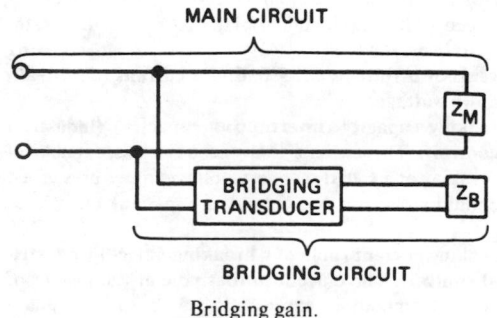

Bridging gain.

bridging loss (data transmission). (1) The ratio of the signal power delivered to that part of the system following the bridging point, before the insertion of the bridging element to this signal power delivered to the same part after the bridging. (2) The ratio of the power dissipated in a load B across which the input of a transducer delivers to its load A. *Note:* It is the inverse of bridging gain. Bridging loss is usually expressed in decibels.   59

bridle wire. Insulated wire for connecting conductors of an open wire line to associated pole-mounted apparatus.   328

bright dip (electroplating). A dip used to produce a bright surface on a metal. *See:* **electroplating.**   328

brightener (electroplating). An addition agent used for the purpose of producing bright deposits. *See:* **electroplating.**   328

brightness (1) (fiber optics). An attribute of visual perception, in accordance with which a source appears to emit more or less light; obsolete. *Notes:* (1) Usage should be restricted to nonquantitative reference to physiological sensations and perceptions of light. (2) "Brightness" was formerly used as a synonym for the photometric term "luminance" and (incorrectly) for the radiometric term "radiance". *See:* **radiance; radiometry.**   433
(2) (of a perceived aperture color) (illuminating engineering). The attribute by which an area of color of finite size is perceived to emit, transmit, or reflect a greater or lesser amount of light. No judgement is made as to whether the light comes from a reflecting, transmitting, or self-luminous object.   167
(3) (television). The attribute of visual perception in accordance with which an area appears to emit more or less light. *Note:* Luminance is recommended for the photometric quantity that has been called brightness. Luminance is a purely photometric quantity. Use of this name permits brightness to be used entirely with reference to the sensory response. The photometric quantity has often been confused with the sensation merely because of the use of one name for two distinct ideas. Brightness will continue to be used properly in nonquantitative statements, especially with reference to sensations and perceptions of light. Thus, it is correct to refer to a brightness match, even in the field of a photometer, because the sensations are matched, and only by inference are the photometric quantities (luminances) equal. Likewise, a photometer in which such matches are made will continue to be called an equality-of-brightness photometer. A photoelectric instrument, calibrated in footlamberts, should not be called a brightness meter. If correctly calibrated, it is a luminance meter. A troublesome paradox is eliminated by the proposed distinction of nomenclature. The luminance of a surface may be doubled, yet it will be permissible to say that the brightness is not doubled, since the sensation that is called brightness is generally judged not to be doubled.   18

brightness channel (television). *See:* **television.**
brightness contrast threshold (illuminating engineer-ing). When two patches of color are separated by a brightness contrast border as in the case of a bipartite photometric field or in the case of a disk shaped object surrounded by its background, the border between the two patches is a brightness contrast border. The contrast which is just detectable is known as the brightness contrast threshold.   167

brightness control (television). A control, associated with a picture display device, for adjusting the average luminance of the reproduced picture. *Note:* In a cathode-ray tube the adjustment is accomplished by shifting bias. This affects both the average luminance and the contrast ratio of the picture. In a color-television system, saturation and hue are also affected.   328

brightness of a surface (radio wave propagation). The power radiated per unit area, per hertz of bandwidth, per unit of solid angle subtended at a distance from the surface.   146

brightness signal. *See:* **luminance signal.**

brightness temperature (radio wave propagation). Of a region on an extended source at a given wavelength, the temperature of a black-body radiator which has the same brightness.   146

bright signal.* *See:* **luminance signal.**
*Deprecated

brine. A salt solution.   328

broadband interference (measurement) (electromagnetic compatibility). A disturbance that has a spectral energy distribution sufficiently broad, so that the response of the measuring receiver in use does not vary significantly when tuned over a specified number of receiver bandwidths. *See:* **electromagnetic compatibility.**   199

broadband radio noise. Radio noise having a spectrum broad in width as compared to the nominal bandwidth of the measuring instrument, and whose spectral components are sufficiently close together and uniform so that the measuring instrument cannot resolve them.   418

broadband response spectrum (seismic qualification of Class 1E equipment for nuclear power generating stations). A response spectrum that describes motion in which amplified response occurs over a wide (broad) range of frequencies.   581

broadband spurious emission (land-mobile communication transmitters). The term as used in this recommended practice is applicable to modulation products near the carrier frequency generated as a result of the normal modulation process of the transmitter and appearing in the spectrum outside the authorized bandwidth (FCC). The products may result from overdeviation or internal distortion and noise and may have a Gaussian distribution.   444

broadband tube (microwave gas tube). A gas-filled fixed-tuned tube incorporating a bandpass filter of geometry suitable for radio-frequency switching. *See:* **gas tube; pretransmit-receive tube; transmit-receive tube.**   125,190

broadcast (broadcast operation)(FASTBUS acquisition and control). An operation directed to one or more slaves on one or more segments.   480

broadcast address (FASTBUS acquisition and control). A primary address asserted by a master during a broadcast. 480

broadcast transmission (token ring access method). A transmission addressed to all stations. 472

broadside array. A linear or planar array antenna whose direction of maximum radiation is perpendicular to the line or plane,respectively, of the array. *See:* **antenna.** 111

bronze conductor. A conductor made wholly of an alloy of copper with other than pure zinc. *Note:* The copper may be alloyed with tin, silicon, cadmium, manganese, or phosphorus, for instance, or several of these in combination. *See:* **conductor.** 64

bronze leaf brush (rotating machinery). A brush made up of thin bronze laminations. *See:* **brush.** 63

brush (1) (electric machines). A conductor, usually composed in part of some form of the element carbon, serving to maintain an electric connection between stationary and moving parts of a machine or apparatus. *Note:* Brushes are classified according to the types of material used, as follows: carbon, carbon-graphite, electrographitic, graphite, and metal-graphite. 63

(2) (relay). *See:* **relay wiper.**

brush box (rotating machinery). the part of a brush holder that contains a brush. *See:* **brush.** 279

brush chamfer (electric machines). The slight removal of a sharp edge. *See:* **brush.** 279

brush contact loss (rotating machinery). The I2R loss in brushes and contacts of the field collector ring or the direct-current armature commutator. *See:* **brush.** 63

brush convex and concave ends (electric machines). Partially cylindrical surfaces of a given radius. *Note:* When concave bottoms are applied to bevels, both bevel angle and radius shall be given. *See:* **brush.** 279

brush corners (electric machines). The point of intersection of any three surfaces. *Note:* They are designated as top or face corners. *See:* **brush.** 279

brush diameter (electric machines). The dimension of the round portion that is at right angles to the length. *See:* **brush.** 279

brush edges (electric machines). The intersection of any two brush surfaces. *See:* **brush.** 279

brush ends (electric machines). The surface defined by the width and thickness of the brush. *Note:* They are designated as top or holder end and bottom or commutator end. The end that is in contact with the commutator or ring is also known as the brush face. *See:* **brush.** 279

brush friction loss (rotating machinery). The mechanical loss due to friction of the brushes normally included as part of the friction and windage loss. *See:* **brush.** 63

brush hammer, lifting, or guide clips (electric machines). Metal parts attached to the brush that serve to accommodate the spring finger or hammer or to act as guides. *Note:* Where these serve to prevent the wear of the carbon due to the pressure finger, they are called

hammer or finger clips. Rotary converter brushes may have clips that serve the dual purpose of lifting the brushes and of preventing wear from the spring finger. These are generally called lifting clips. *See:* **brush.** 279

brush holder (rotating machinery). A structure that supports a brush and that enables it to be maintained in contact with the sliding surface. *See:* **brush.** 63

brush-holder bolt insulation (rotating machinery). A combination of members of insulating materials that insulate the brush yoke mounting bolts, brush yoke, and brush holders. *See:* **brush.** 63

brush-holder insulating barriers (rotating machinery). Pieces of sheet insulation installed in the brush yoke assembly to provide longer leakage paths between live parts and ground, or between live parts of different polarities. *See:* **brush.** 63

brush-holder spindle insulation (rotating machinery). Insulation members that (when required by design) insulate the spindle on which the brush spring is mounted from the brush yoke and the brush holder. *See:* **brush.** 63

brush-holder spring. That part of the brush holder that provides pressure to hold the brush against the collector ring or commutator. *See:* **brush.** 63

brush holder stud (rotating machinery). An intermediate member between the brush holder and the supporting structure. *See:* **brush.** 63

brush-holder-stud insulation (rotating machinery). An assembly of insulating material that insulates the brush holder or stud from the supporting structure. *See:* **brush.** 63

brush-holder support (rotating machinery). The intermediate member between the brush holder or holders and the supporting structure. *Note:* This may be in the form of plates, spindles, studs, or arms. *See:* **brush.** 63

brush-holder yoke. A rocker arm, ring, quadrant, or other support for maintaining the brush holders or brush-holder studs in their relative positions. *See:* **brush.** 63

brush length (electric machines.) The maximum overall dimension of the carbon only, measured in the direction in which the brush feeds to the commutator or collector ring. *See:* **brush.** 279

brushless (rotating machinery). Applied to machines with primary and secondary or field windings that are constructed such that all windings are stationary, or in which the conventional brush gear is eliminated by the use of transformers having both moving and stationary windings, or by the use of rotating rectifiers. *See:* **brush.** 63

brushless exciter (1) (electric installations on shipboard). An ac (rotating armature type) exciter whose output is rectified by a semiconductor device to provide excitation to an electric machine. The semiconductor device would be mounted on and rotate with the ac exciter armature. 3

(2) (excitation systems for synchronous machines). An alternator-rectifier exciter employing rotating rectifiers with a direct connection to the synchronous machine field thus eliminating the need for field brushes. See figure below.   507

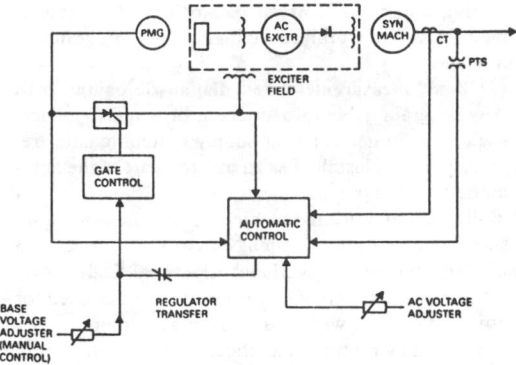

brushless synchronous machine (electric installations on shipboard) A synchronous machine having a brushless exciter with its rotating armature and semiconductor devices on a common shaft with the field of the main machine.This type of machine with its exciter has no collector, commutator, or brushes.   3

brush-operating or slip-ring short-circuiting device (35) (power system device function numbers). A device for raising, lowering, or shifting the brushes of a machine, or for short-circuiting its slip rings, or for engaging or disengaging the contacts of a mechanical rectifier.   402

brush or sponge plating (electroplating). A method of plating in which the anode is surrounded by a brush or sponge or other absorbent to hold electrolyte while it is moved over the surface of the cathode during the plating operation. *See:* **electroplating.**   328

brush rigging (rotating machinery). The complete assembly of parts whose main function is to position and support all of the brushes for a commutator or collector. *See:* **brush.**   63

brush rocker (brush yoke) (rotating machinery). The structure from which the brush holders are supported and fixed relative to each other and so arranged that the whole assembly may be moved circumferentially. *See:* **brush.**   63

brush-rocker gear (brush-yoke gear) (rotating machinery). The worm wheel or other gear by means of which the position of the brush rocker may be adjusted. *See:* **brush.**   63

brush shoulders (electric machines). When the top of the brush has a portion cut away by two planes at right angles to each other, this is designated as a shoulder. *See:* **brush.**   279

brush shunt (rotating machinery). The stranded cable or other flexible conductor attached to a brush to connect it electrically to the machine or apparatus. *Note:* Its purpose is to conduct the current that would

otherwise flow from the brush to the brush holder or brusholder finger. *See:* **brush.**   63

brush shunt length (electric machines). The distance from the extreme top of the brush to the center of the hole or slot in the terminal, or the center of the inserted portion of a plug terminal or, if there is no terminal, to the end of the shunt. *See:* **brush.**   63

brush sides (electric machines). (1) Front and back (bounded by width and length). *Note:* (1) If the brush has one or both ends beveled, the short side of the brush is the front. (2) If there are no top or bottom bevels and width is greater than thickness and there is a top clip, the side to which the clip is attached is the back, except in the case of angular clips where the front or short side is determined by the slope of the clip and not by the side to which it is attached. (3) Left side and right side (bounded by thickness and length). *See:* **brush.**   279

brush slots, grooves, and notches (electric machines). Hollows in the brush. *See:* **brush.**   279

brush spring (rotating machinery). The portion of a brush holder that exerts pressure on the brush to hold it in contact with the sliding surface.   63

brush thickness (electric machines). The dimension at right angles to the length in the direction of rotation. *See:* **brush.**   279

brush width (electric machines). The dimension at right angles to the length and to the direction of rotation. *See:* **brush.**   279

brush yoke. *See:* **brush rocker.**

BR/2 (medium attachment units and repeater units). One half of the BR in hertz. *See:* **BR.**   543

B scan (electronic navigation). *See:* **B display.**

B-scope (radar). A cathode ray oscilloscope arranged to present a B-display.   13

B stage. An intermediate stage in the reaction of certain thermosetting resin in which the material swells when in contact with certain liquids and softens when heated, but may not entirely dissolve or fuse. *Note:* The resin in an uncured thermosetting moulding compound is usually in this stage.   63

B switchboard (telephone switching systems.) A telecommunications switchboard in a local central office, used primarily to complete calls received from other central offices.   55

B-trace (loran)(navigation aid terms). The second (lower) trace on the scope display.   526

Buchmann-Meyer pattern (mechanical recording). *See:* **light pattern.**

buck arm. A crossarm placed approximately at right angles to the line crossarm and used for supporting branch or lateral conductors or turning large angles in line conductors. *See:* **tower.**   64

bucket (power line maintenance). A device designed to be attached to the boom tip of a line truck, crane, or aerial lift and support workers in an elevated working position. It is normally constructed of fiberglass to reduce its physical weight, maintain strength, and obtain good dielectric characteristics. *Syn:* **basket.**   458

buffer (1)(buffer storage)(supervisory control, data

acquisition, and automatic control). (A) A device in which data are stored tenporarily, in the course of transmission from one point to another; used to compensate for a difference in the flow of data, or time of occurrence of events, when transmitting data from one device to another. (B) An isolating circuit used to prevent a driven circuit from influencing a driving circuit.   570
(2) (data processing and computation). A storage device used to compensate for a difference in rate of flow of information or time of occurrence of events when transmitting information from one device to another.
                                                                 235
(3) (elevator design). A device designed to stop a descending car or counterweight beyond its normal limit of travel by storing or by absorbing and dissipating the kinetic energy of the car or counterweight. *See:* **elevator.**   328
(4) (relay). *See:* **relay spring stud.**
buffer amplifier (1) (general). An amplifier in which the reaction of output-load-impedance variation on the input circuit is reduced to a minimum for isolation purposes. *See:* **amplifier.**   111
(2) (analog computers). An amplifier in which the reaction of the output-load-impedance variation on the input circuit is reduced to a constant for isolation purposes on the input circuit. *See:* **unloading amplifier.**   9
buffered interconnect (BI)(FASTBUS acquisition and control). A device which implements an intersegment connection such that the FASTBUS protocol (FBP) on one segment is not synchronized with that on the other.   480
buffer memory (sequential events recording systems). The memory used to compensate for the difference in rate of flow of information or time of occurrence of events when transmitting information from one device to another. *See:* **buffer; event; storage.**   48
buffers (buffer salts). Salts or other compounds that reduce the changes in the pH of a solution upon the addition of an acid or alkali. *See:* **ion.**   328
buffer salts. *See:* **buffers.**
buffer storage. An intermediate storage medium between data input and active storage.   207
buffing (electroplating). The smoothing of a metal surface by means of flexible wheels, to the surface of which fine abrasive particles are applied, usually in the form of a plastic composition or paste. *See:* **electroplating.**   328
bug (1) (telegraphy). A semiautomatic telegraph key in which movement of a lever to one side produces a series of correctly spaced dots and movement to the other side produces a single dash.   328
(2) (software). *See:* **fault.**
bugduster. An attachment used on shortwall mining machines to remove cuttings (bugdust) from back of the cutter and to pile them at a point that will not interfere with operation.   328
bug seeding. *See:* **fault seeding.**
build (software). An operational version of a software product incorporating a specified subset of the capa-

bilities that the final product will include. *See:* **software product.**   434
building (National Electrical Code). A structure which stands alone or which is cut off from adjoining structures by fire walls with all openings therein protected by approved fire walls.   256
building block (1) (software). An individual unit or module which is utilized by higher-level programs or modules.   434
(2) (test, measurement and diagnostic equipment). Any programmable measurement or stimulus device, such as multimeter, power supply switching unit, frequency meter, installed as an integral part of the automatic test equipment.   54
building bolt (rotating machinery). A bolt used to insure alignment and clamping of parts.   63
building component (National Electrical Code). Any subsystem, subassembly, or other system designed for use in or integral with or as part of a structure, which can include structural, electrical, mechanical, plumbing and fire protection systems and other systems affecting health and safety.   256
building out (communication practice). The addition to an electric structure of an element or elements electrically similar to an element or elements of the structure, in order to bring a certain property of characteristics to a desired value. *Note:* Examples are building-out capacitors, building-out sections of line, etcetera.
                                                                 328
building-out capacitor (building-out condenser*). A capacitor employed to increase the capacitance of an electric structure to a desired value.   329
*Deprecated.
building-out network. An electric network designed to be connected to a basic network so that the combinations will simulate the sending-end impedance, neglecting dissipation, of a line having a termination other than that for which the basic network was designed. *See:* **network analysis.**   328
building pin (rotating machinery). A dowel used to insure alignment of parts.   63
building system (National Electrical Code). Plans, specifications and documentation for a system of manufactured building or for a type or a system of building components, which can include structural, electrical, mechanical, plumbing, and fire protection systems, and other systems affecting health and safety, and including such variations thereof as are specifically permitted by regulation, and which variations are submitted as part of the building system or amendment thereto.   256
building up (electroplating). Electroplating for the purpose of increasing the dimensions of an article. *See:* **electroplating.**   328
buildup or decay (diode-type camera tube). The response to the camera tube to a positive or negative step in irradiance.   380
built-in ballast (mercury lamp). A ballast specifically designed to be built into a lighting fixture.   269
built-in check (computers). *See:* **automatic check.**
built-in-test (BIT) (test, measurement and diagnostic

equipment). A test approach using **BITE** or self test hardware or software to test all or part of the unit under test. 54

built-in-test equipment (BITE) (test, measurement and diagnostic equipment). Any device which is part of an equipment or system and is used for the express purpose of testing the equipment or system. BITE is an identifiable unit of the equipment or system. *See:* **self-test.** 54

built-in transformer. A transformer specifically designed to be built into a luminaire. 274

built-up connection. A toll call that has been relayed through one or more switching points between the originating operator and the receiving exchange. *See:* **telephone system.** 328

bulb (electron tubes and electric lamp). (1) The glass envelope used in the assembly of an electron tube or an electric lamp. (2) The glass component part used in a bulb assembly. 286

bulk power system (power operations). An interconnected system for the movement or transfer of electric energy in bulk on transmission levels. 516

bulk storage (test, measurement and diagnostic equipment). A supplementary large volume memory or storage device. 54

bulk-storage plant (National Electrical Code). A location where gasoline or other volatile flammable liquids are stored in tanks having an aggregate capacity of one carload or more, and from which such products are distributed (usually by tank truck). 256

bull line (conductor stringing equipment). A high strength line, normally synthetic fiber rope, used for pulling and hoisting large loads. 431

bull ring. A metal ring used in overhead construction at the junction point of three or more guy wires. *See:* **tower.** 64

bullwheel (conductor stringing equipment). A wheel incorporated as an integral part of a bullwheel puller or tensioner to generate pulling or braking tension on conductors or pulling lines, or both, through friction. A puller or tensioner normally has one or more pairs arranged in tandem incorporated in its design. The physical size of the wheels will vary for different designs, but 17 in (43 cm) face widths and diameters of 5 ft (150 cm) are common. The wheels are power driven or retarded and lined with single or multiple groove neoprene or urethane linings. Friction is accomplished by reeving the pulling line or conductor around the groove of each pair. 431

bullwheel puller (conductor stringing equipment). A device designed to pull pulling lines and conductors during stringing operations. It normally incorporates one or more pairs of urethane- or neoprene-lined, power driven, single or multiple groove bullwheels where each pair is arranged in tandem. Pulling is accomplished by friction generated against the pulling line which is reeved around the grooves of a pair of the bullwheels. The puller is usually equipped with its own engine which drives the bullwheels mechanically, hydraulically or through a combination of both. Some of these devices function as either a puller or tensioner. *Syn:* **puller.** 431

bullwheel tensioner (conductor stringing equipment). A device designed to hold tension against a pulling line or conductor during the stringing phase. Normally, it consists of one or more pairs of urethane- or neoprene-lined, power braked, single or multiple groove bullwheels where each pair is arranged in tandem. Tension is accomplished by friction generated against the conductor which is reeved around the grooves of a pair of the bullwheels. Some tensioners are equipped with their own engines which retard the bullwheels mechanically, hydraulically or through a combination of both. Some of these devices function as either a puller or tensioner. Other tensioners are only equipped with friction type retardation. *Syn:* **retarder; tensioner.** 431

bump. *See:* **distortion pulse.**

bumper (elevators). A device other than an oil or spring buffer designed to stop a descending car or counterweight beyond its normal limit of travel by absorbing the impact. *See:* **elevator.** 328

buncher space (velocity-modulated tube). The part of the tube following the acceleration space where there is a high-frequency field, due to the input signal, in which the velocity modulation of the electron beam occurs. *Note:* It is the space between the input resonator grids. *See:* **velocity-modulated tube.** 244,190

bunching. The action in a velocity-modulated electron stream that produces an alternating convection-current component as a direct result of differences of electron transit time produced by the velocity modulation. *See:* **optimum bunching; overbunching; reflex bunching; space-charge debunching; underbunching; electron device.** 125

bunching angle (electron steam) (given drift space). The average transit angle between the processes of velocity modulation and energy extraction at the same or different gaps. *See:* **bunching angle, effective (reflex klystrons); electron device.** 125

bunching, optimum. *See:* **optimum bunching.**

bunching, parameter. One-half the product of: (1) the bunching angle in the absence of velocity modulation and (2) the depth of velocity modulation. *Note:* In a reflex klystron the effective bunching angle must be used. *See:* **electron device.** 125

bunching time, relay. *See:* **relay bunching time.**

bundle (1) (conductor stringing equipment). A circuit phase consisting of more than one conductor. Each conductor of the phase is referred to as a subconductor. A two conductor bundle has two subconductors per phase. These may be arranged in a vertical or horizontal configuration. Similarly, a three conductor bundle has three subconductors per phase. These are usually arranged in a triangular configuration with the vertex of the triangle up or down. A four conductor bundle has four subconductors per phase. These are normally arranged in a square configuration. Although other configurations are possible, those listed are the most common. *Syn:* **twin-bundle; tri-bundle; quad-bundle.** 431

(2) (fiber optics). *See:* **fiber bundle.** 433

bundled conductor (NESC)(transmission and distri-

bution). An assembly of two or more conductors used as a single conductor and employing spacers to maintain a predetermined configuration. The individual conductors of this assembly are called subconductors.
494, 262

B unit. A motive power unit designed primarily for use in multiple with an *A* unit for the purpose of increasing locomotive power, but not equipped for use as the leading unit of a locomotive or for full observation of the propulsion power and brake applications for a train. *Note:* *B* units are normally equipped with a single control station for the purpose of independent movement of the unit only, but are not usually provided with adequate instruments for full observation of power and brake applications. *See:* **electric locomotive.**
328

buoy (navigation aid terms). A floating object, other than a lightship, moored or anchored to the bottom of the sea, which is an aid to navigation. *See:* **buoys classified to location; combination buoy; danger or hazard buoy; lighted buoy; radio-beacon buoy; sonobuoy; sound buoy.**
526

buoys classified to location (navigation aid terms). Channel, mid-channel, turning, fairway, bifurcation, junction, sea. *See:* **buoy.**
526

burden (1) (instrument transformer) (metering). The impedance of the circuit connected to the secondary winding. *Note:* For voltage transformers it is convenient to express the burden in terms of the equivalent volt-ampere s and power factor at a specified voltage and frequency.
212

(2) (of a relay) (power switchgear). Load impedance imposed by a relay on an input circuit expressed in ohms and phase angle at specified conditions. *Note:* If burden is expressed in other terms such as volt-amperes, additional parameters such as voltage, current, and phase angle must be specified.
103

(3) (power system relaying). Load imposed by a relay on an input circuit, expressed in ohms or volt-amperes. *See:* **relay.**
60

burden of an instrument transformer. That property of the circuit connected to the secondary winding that determines the active and reactive power at the secondary terminals. The burden is expressed either as total ohms impedance with the effective resistance and reactance components, or as the total volt-amperes and power factor at the specified value of current or voltage, and frequency.
394

burden regulation (capacitance potential device). Refers to the variation in voltage ratio and phase angle of the secondary voltage of the capacitance potential device as a function of burden variation over a specified range, when energized with constant, applied primary line-to-ground voltage. *See:* **outdoor coupling capacitor.**
341

burglar-alarm system. An alarm system signaling an entry or attempted entry into the area protected by the system. *See:* **protective signaling.**
328

burial depth (cable plowing). The depth of soil cover over buried cable, pipe etcetera measured on level ground.
52

buried cable. A cable installed under the surface of the ground in such a manner that it cannot be removed without disturbing the soil.
328

burn in (1)(supervisory control, data acquisition, and automatic control)(station control and data acquisition). A period, usually prior to on-line operation, during which equipment is continuously energized for the purpose of forcing infant mortality failures.
570, 403

burn-in (2) (class 1E static battery chargers and inverters). The operation of components or equipment, prior to type test or ultimate application, intended to stabilize their characteristics and to identify early failures.
408

(3) (reliability). The operation of items prior to their ultimate application intended to stabilize their characteristics and to identify early failures. *See:* **reliability.**
182

burnishing. The smoothing of metal surfaces by means of a hard tool or other article. *See:* **electroplating.**
328

burnishing surface (mechanical recording). The portion of the cutting stylus directly behind the cutting edge, that smooths the the the groove. *See:* **phonograph pickup.**
176

burnout (nonlinear, active, and nonreciprocal waveguide components). The point at which a sensitive receiving device suffers a specified permanent degradation of noise figure or equivalent increase in noise temperature.
530

burnt deposit. A rough or noncoherent electrodeposit produced by the application of an excessive current density. *See:* **electroplating.**
328

burnup, nuclear (electric power supply). (1) A measure of nuclear reactor fuel consumption, usually expressed as energy produced per unit weight of fuel exposed (megawatt-days per metric ton of fuel.) (2) Percentage of fueled atoms that have undergone fission (atom percent burnup).
112

burst (1) (pulse techniques). A wave or waveform composed of a pulse train or repetitive waveform that starts as a prescribed time and/or amplitude, continues for a relatively short duration and/or number of cycles, and upon completion returns to the starting amplitude. *See:* **pulse.**
185

(2) (audio and electroacoustics). An excursion of a quantity (voltage, current, or power) in an electric system that exceeds a selected multiple of the long-time average magnitude of this quantity taken over a period of time sufficiently long that increasing the length of this period will not change the result appreciably. This multiple is called the upper burst reference. *Notes:* (A) If measurements are made at different points in a system, or at different times, the same quantity must be measured consistently. (B) The excursion may be an electrical representation of a change of some other physical variable such as pressure, velocity, displacement, etcetera.
176

(3) (radio wave propagation). A transient increase in intensity of radiation over a short period, such as is observed from the sun.
146

burst build-up interval. The time interval between the burst leading-edge time and the instant at which the upper burst reference is first equaled. *See:* The figure attached to the definition of **burst duration.** *See:* **burst.**
253,176

burst corona (overhead-power-line corona and radio noise). Burst corona may be considered as the initial stage of positive glow. It occurs at a positive electrode with field strengths at or slightly above the corona-inception gradient. Burst corona appears as a bluish film of velvet-like glow adhering closely to the electrode surface. The current pulses of burst corona are of low amplitude and may last for periods of milliseconds. *Note:* Occurrence of burst corona and positive onset streamers requires the same range of field strength.
411

burst decay interval (audio and electroacoustics). The time interval between the instant at which the peak burst magnitude occurs and the burst trailing-edge time. *See:* The figure attached to the definition of **burst duration.** *See:* **burst.**
253,176

burst duration (audio and electroacoustics). The time interval during which the instantaneous magnitude of the quantity exceeds the lower burst reference, disregarding brief excursions below the reference, provided the duration of any individual excursion is less than a burst safeguard interval of selected length. *Note:* (1) If the duration of an excursion is equal to or greater than the burst safeguard interval, the burst has ended. (2) These terms, as well as those defined below, are illustrated in the accompanying figure. (A) A burst is found with the aid of a "window" that is slid horizontally to the right with its base resting on the lower burst reference. The width of the window equals the burst safeguard interval and the height of the window equals the difference between the upper and lower burst references. The window is slid to the right until the trace crosses the top of the window. The upper burst reference has then been reached and a burst has occurred. (B) The burst leading-edge time is found by sliding the window to the left until the trace disappears from the window. The right-hand side of the window marks the burst leading-edge time. (C) The burst trailing edge time is found by a similar procedure. The window is slid to the right past its position in (A) until the trace disappears from the window. The left-hand side of the window marks the burst trailing-edge time. (D) Terms used in defining a burst: **burst leading-edge time,** t_1; **burst build-up interval,** $t_2 - t_1$; **burst rise interval,** $t_3 - t_1$; **burst trailing-edge time,** t_5; **burst decay interval,** $t_5 - t_3$; **burst fall-off interval,** $t_5 - t_4$; **burst duration,** $t_5 - t_1$; **upper burst reference, U; lower burst reference, L; long-time average power, P.** *See:* **burst.**
253,176

burst duty factor (audio and electroacoustics). The ratio of the average burst duration to the average spacing. *Note:* This is equivalent to the product of the average burst duration and the burst repetition rate. *See:* The figure attached to the definition of **burst duration.** *See:* **burst.**
253,176

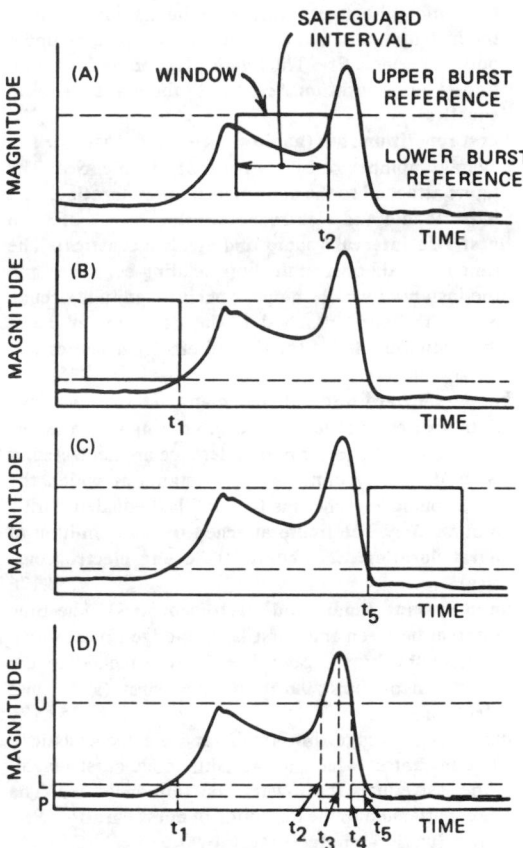

Plot of instantaneous magnitude versus time to illustrate terms used in defining a burst.

burst fall-off interval (audio and electroacoustics). The time interval between the instant at which the upper burst reference is last equaled and the burst trailing edge time. *See:* The figure attached to the definition of **burst duration.** *See:* **burst.**
253,176

burst flag (television). A keying or gating signal used in forming the color burst from a chrominance subcarrier source. *See:* **television.**
178

burst gate (television). A keying or gating device or signal used to extract the color burst from a color picture signal. *See:* **television.**
178

burst keying signal (television). *See:* **burst flag; television.**

burst leading-edge time (audio and electroacoustics). The instant at which the instantaneous burst magnitude first equals the lower burst reference. *See:* The figure attached to the definition of **burst duration.** *See:* **burst (audio and electroacoustics).**
253,176

burst magnitude, instantaneous. *See:* **instantaneous-burst magnitude.**

burst measurements. *See:* **energy density spectrum.**

burst quiet interval (audio and electroacoustics). The

time interval between successive bursts during which the instantaneous magnitude does not equal the upper burst reference. *See:* The figure attached to the definition of **burst duration**. *See:* **burst (audio and electroacoustics)**.   253,176

burst repetition rate (audio and electroacoustics). The average number of bursts per unit of time. *See:* The figure attached to the definition of **burst duration**. *See:* **burst (audio and electroacoustics)**.   253,176

burst rise interval (audio and electroacoustics). The time interval between the burst leading-edge time and the instant at which the peak burst magnitude occurs. *See:* The figure attached to the definition of **burst duration**. *See:* **burst (audio and electroacoustics)**.   253,176

burst safeguard interval (audio and electroacoustics). A time interval of selected length during which excursions below the lower burst reference are neglected; it is used in determining those instants at which the lower burst references is first and last equaled during a burst. *See:* The figure attached to the definition of **burst duration**. *See:* **burst (audio and electroacoustics)**.   253,176

burst spacing (audio and electroacoustics). The time interval between the burst leading-edge times of two consecutive bursts. *See:* The figure attached to the definition of **burst duration**. *See:* **burst (audio and electroacoustics)**.   253,176

burst trailing-edge time (audio and electroacoustics). The instant at which the instantaneous burst magnitude last equals the lower burst reference. *See:* The figure attached to the definition of **burst duration**. *See:* **burst (audio and electroacoustics)**.   253,176

burst train (audio and electroacoustics). A succession of similar bursts having comparable adjacent burst quiet intervals. *See:* The figure attached to the definition of **burst duration**. *See:* **burst (audio and electroacoustics)**.   253,176

bus (1)(power switchgear). As used in ANSI/IEEE C37.71-1984, A three-phase junction common to two or more ways.   489

(2)(signals and paths)(microcomputer system bus). A signal line or a set of lines used by an interface system to connect a number of devices and to transfer information.   542

(3)(signals and paths)(696 interface devices). A set of signal lines used by an interface system, to which a number of devices are connected, and over which information is transferred between the devices.   538

(4) (power switchgear). A conductor, or group of conductors, that serve as a common connection for two or more circuits.   103

(5) (programmable instrumentation). A signal line or set of signal lines used by an interface system to which a number of devices are connected and over which messages are carried.   378

(6) (electronic computers). One or more conductors used for transmitting signals or power from one or more sources to one or more destinations.   235

bus cycle (1)(general system terms)(microcomputer system bus). The process whereby digital signals effect the transfer of data bytes or words across the interface by means of an interlocked sequence of control signals. Interlocked denotes a fixed sequence of events in which one event must occur before the next event can occur.   542

(2)(signals and paths)(696 interface devices). The basic sequence of electrical events required to complete a transfer of data on the bus. A bus cycle contains at least three bus states.   538

bushing (1) (power switchgear). An insulating structure including a through conductor, or providing a passageway for such a conductor, with provision for mounting on a barrier, conducting or otherwise, for the purpose of insulating the conductor from the barrier and conducting current from one side of the barrier to the other.   103

(2) (power and distribution transformer). An insulating structure including a central conductor, or providing a central passage for a conductor, with provision for mounting on a barrier, conducting or otherwise, for the purpose of insulating the conductor from the barrier and conducting current from one side of the barrier to the other.   53

(3) (rotating machinery) (electrical). Insulator to permit passage of a lead through a frame or housing.   63

(4) (relay). *See:* **relay spring stud**.

bushing insert (separable insulated connectors). A connector component intended for insertion into a bushing well. *See:* **Fig 1 below**.   454

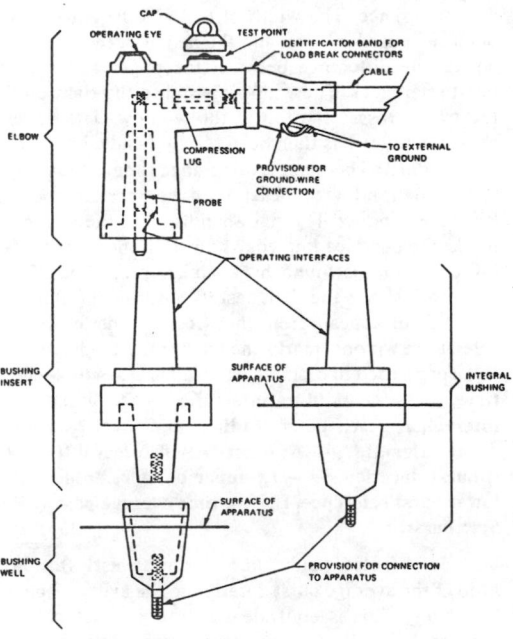

Fig 1
Typical Components of 200 A Separable
Insulated Connector System

bushing potential tap (outdoor apparatus bushings). An insulated connection to one of the conducting layers of a bushing providing a capacitance voltage divider to indicate the voltage applied to the bushing. 168

bushing, rotor. *See:* rotor bushing.

bushing tap (partial discharge measurementin liquid-filled power transformers and shunt reactors). Connection to a capacitor foil in a capacitively graded bushing designed for voltage or power factor measurement that also provides a convenient connecting point for partial discharge measurement. The tap-to-phase capacitance is generally designated as C1 and the tap-to-ground capacitance is designated as C2. *See:* bushing potential tap; bushing test tap; capacitance (of bushing). 580

bushing test tap (outdoor apparatus bushings). An insulated connection to one of the conduction layers of a bushing for the purpose of making insulation power factor tests. 168

bushing-type current transformer (power and distribution transformer). One which has an annular core and a secondary winding inslated from and permanently assembled on the core but has no primary winding nor insulation for a primary winding. This type of current transformer is for use with a fully insulated conductor as the primary winding. A bushing-type current transformer usually is used in equipment where the primary conductor is a component part of other apparatus. 53

bushing well (separable insulated conductors). An apparatus bushing having a cavity for insertion of a connector component, such as a bushing insert. See diagram below bushing insert. 454

business data processing (BDP)(computer applications). The use of computers for processing information to support the operational, logistical, and functional activities performed by an organization. 571

bus line (railway terminology). A continuous electric circuit other than the electric train line, extending through two or more vehicles of a train, for the distribution of electric energy. *See:* multiple-unit control. 328

bus reactor (power and distribution transformer). A current-limiting reactor for connection between two different buses or two sections of the same bus for the purpose of limiting and localizing the disturbance due to a fault in either bus. 53

bus state (signals and paths)(696 interface devices). A bus state is one clock cycle long and begins and ends just before the rising edge of φ. There are at least three bus states in every bus cycle. 538

bus structure (power switchgear)(substation rigid-bus structures). An assembly of bus conductors, with associated connection joints and insulating supports. 103,566

bus support (power switchgear)(substation rigid-bus structures). An insulating support for a bus. *Note:* It includes one or more insulator units with fittings for fastening to the mounting structure and for receiving the bus. 103, 566

bus-type shunt (direct current instrument shunts). An instrument shunt for switchboard use so that it can be installed in the bus or connection bar structure of the circuit whose current is to be measured. 527

busway (National Electrical Code). A grounded metal enclosure containing factory mounted, bare or insulated, conductors which are usually, copper or aluminum, bars, rods, or tubes. *See:* cablebus. 256

busy hour (1) (telephone switching systems). That uninterrupted period of sixty minutes during the day when the traffic offered is a maximum. 55
(2) (data transmission). The peak 60-minute period during a 24-hour period when the largest volume of communication traffic is handled. 59

busy test (telephone switching systems). A test made to determine if certain facilities, such as a line, link, junctor, trunk, or other servers are available for use. 55,193

busy tone. *See:* audible busy signal.

busy verification (telephone switching systems). A procedure for checking whether or not a called station is in use or out-of-order. 55

butt contacts (industrial control). An arrangement in which relative movement of the cooperating members is substantially in a direction perpendicular to the surface of contact. *See:* contactor. 244,206

butt joint (waveguides). A connection between two waveguides or transmission lines that provides physical contact between the ends of the waveguides in order to maintain electric continuity. *See:* waveguides. 166

buzz (electromagnetic compatibility). A disturbance of relatively short duration, but longer than a specified value as measured under specified conditions. *Note:* For the specified values and conditions, guidance should be found in documents of the International Special Committee on Radio Interference. *See:* electromagnetic compatibility. 226,199

buzzer. A signaling device for producing a buzzing sound by the vibration of an armature. 328

buzz stick. A device for testing suspension insulator units for fault when the units are in position on an energized line. *Note:* It consists of an insulating stick, on one end of which are metal prongs of the proper dimensions for spanning and short-circuiting the porcelain of one insulator unit at a time, and thereby checking conformity to normal voltage gradient. *See:* tower. 64

BW. *See:* backward wave structure.

by-link (telephone switching systems). A temporary connection between trunks and registers set up before the normal connection between them can be established. 55

bypass. *See:* jumper.

bypass capacitor (bypass condenser*). A capacitor for providing an alternating-current path of comparatively low impedance around some circuit element. 328

*Deprecated

bypass current (series capacitors). The current flowing through the bypass device or devices in parallel with the series capacitor.   474

bypass device (series capacitors). A device such as a switch or circuit breaker used in parallel with a series capacitor and its protective device to shunt line current for some specified time, or continuously. This device may also have the capability of inserting the capacitor into a circuit and carrying a specified level of current.   474

bypass gap (series capacitors). Gap, or system of gaps, capable of carrying load or fault current around the series capacitor for some specified time.   474

bypass interlocking device (series capacitors). Devices that require all three phases of the bypass devices to be in the same open or closed position.   474

bypass/isolation switch (emergency and standby power). A manually operated device used in conjunction with an automatic transfer switch to provide a means of directly connecting load conductors to a power source and of disconnecting the automatic transfer switch.   512

byproduct energy (power operations). Electric energy produced as a byproduct incidental to some other operation.   516

byte (1)(microprocessor operating systems). A binary bit string operated on as a unit and usually eight bits long and capable of holding one character in the local character set.   478

(2)(programmable instrumentation). A group of adjacent binary digits operated on as a unit and usually shorter than a computer word (frequently connotes a group of eight bits).   378

(3)(signals and paths)(microcomputer system bus). A group of eight adjacent bits operated on as a unit.   542

(4)(signals and paths)(696 interface devices). A set of bit-parallel signals corresponding to binary digits operated on as a unit. Connotes a group of eight bits where the most significant bit carries the subscript 7 and the least significant bit carries the subscript 0.   538

(5)(software). (A) A binary character string operated upon as a unit and usually shorter than a computer word. (B) A group of adjacent binary digits operated on as a unit and usually shorter than a computer word (frequently connotes a group of eight bits. *See:* **bits; computer word; string.**   434

byte-serial (programmable instrumentation)(signals and paths)(696 interface devices). A sequence of bit-parallel data bytes used to carry information over a common bus.   40, 538

C

cabinet (1)(National Electrical Code). An enclosure designed either for surface or flush mounting and provided with a frame, mat, or trim in which a swinging door or doors are or may be hung.   256

(2)(power system communication equipment). An enclosure provided with an internal equipment mounting rack and hinged doors.   453

cabinet for safe (burglar-alarm system). Usually a wood enclosure, having protective linings on all inside surfaces and traps on the doors, built to surround a safe and designed to produce an alarm condition in a protection circuit if an attempt is made to attack the safe. *See:* **protective signaling.**   328

cable (1)(NESC). A conductor with insulation, or a stranded conductor with or without insulation and other coverings (single-conductor cable) or a combination of conductors insulated from one another (multiple-conductor cable).   494

(2) (fiber optics). *See:* **optical cable.**   433

(3) (signal-transmission system). A transmission line or group of transmission lines mechanically assembled into a complex flexible form. *Note:* The conductors are insulated and are closely spaced and usually have a common outer cover which may be an electric portion of the cable. This definition also includes a twisted pair.   267

(4) (transmission and distribution). A conductor with insulation, or a stranded conductor with or without insulation and other coverings (single-conductor cable) or a combination of conductors insulated from one another (multiple-conductor cable). *See:* **spacer cable.**   262

cable accessories (power cable system). Those components of a cable system which cannot be readily disconnected from the cable and which will be subjected to the full test voltage applied to the cable system.   437

cable assembly. *See:* **multifiber cable; optical cable assembly.**   433

cable bedding (power distribution, underground cables.) A relatively thick layer of material, such as a jute serving, between two elements of a cable to provide a cushion effect, or gripping action, as between the lead sheath and wire armor of a submarine cable.   57

cable bond. An electric connection across a joint in the armor or lead sheath of a cable, or between the armor or lead sheath and the earth, or between the armor or sheath of adjacent cables. *See:* **continuity cable bond;**

cross cable bond.

328

cablebus (National Electrical Code). An approved assembly of insulated conductors with fittings and conductor terminations in a completely enclosed, ventilated protective metal housing. The assembly is designed to carry fault current and to withstand the magnetic forces of such current. Cablebus shall be permitted at any voltage or current for which the spaced conductors are rated. Cablebus is ordinarily assembled at the point of installation from components furnished or specified by the manufacturer in accordance with instructions for the specific job.

256

cable car (conductor stringing equipment). A seat or basket shaped device designed to be suspended by a framework and two or more sheaves arranged in tandem to enable a workman to ride a single conductor, wire or cable. *Syn:* **cable trolley.** 431

cable charging current (power switchgear). Current supplied to an unloaded cable. *Note:* Current is expressed in root-mean-square amperes. 103

cable clamp (conductor stringing equipment). A device designed to clamp cables together. It consists of a *U*-bolt threaded on both ends, two nuts and a base and is commonly used to make temporary bend back eyes on wire rope. *Syn:* **clip; Crosby; Crosby clip.**

431

cable complement (communication practice). A group of pairs in a cable having some common distinguishing characteristic. *See:* **cable.** 328

cable core (cable). The portion lying under other elements of a cable. *See:* **power distribution, underground construction.** 64

cable core binder. A wrapping of tapes or cords around the several conductors of a multiple-conductor cable used to hold them together. *Note:* Cable core binder is usually supplemented by an outer covering of braid, jacket, or sheath.

64

cable coupler (rotating machinery). A form of termination in which the ends of the machine winding are connected to the supply leads by means of a plug-and-socket device. 63

cable entrance fitting (pothead). A fitting used to seal or attach the cable sheath or armor to the pothead. *Note:* A cable entrance fitting is also used to attach and support the cable sheath or armor where a cable passes into a transformer removable cable terminating box without the use of potheads. *See:* **transformer.**

288,289

cable fill. The ratio of the number of pairs in use to the total number of pairs in a cable. *Note:* The maximum cable fill is the percentage of pairs in a cable that may be used safely and economically without serious interference with the availability and continuity of service. *See:* **cable.** 328

cable filler. The material used in multiple-conductor cables to occupy the interstices formed by the assembly of the insulated conductors, thus forming a cable core of the desired shape (usually circular). 64

cable-fire break (cable-penetration fire stops, fire breaks, and system enclosures)(design and installation of cable systems for Class 1E circuits in nuclear power generating stations). Material, devices, or an assembly of parts installed in a cable system, other than at a cable penetration of a fire-resistive barrier, to prevent the spread of fire along the cable system.

536

cableheads. *See:* **submersible entrance terminals (distribution oil cutouts).**

cable in the zone of influence (wire-line communication facilities). A high dielectric cable which provides high-voltage insulation between conductors and between conductors and shield. 414

cable jacket (NESC). A protective covering over the insulation, core, or sheath of a cable. 494

cable joint (cable splice) (1) (transmission and distribution). A connection between two or more separate lengths of cable with the conductors in one length connected individually to conductors in other lengths and with the protecting sheaths so connected as to extend protection over the joint. *Note:* Cable joints are designated by naming the conductors between which the joint is made, for example, 1 single-conductor to 2 single-conductor cables; 1 single-conductor to 3 single-conductor cables; 1 concentric to 2 concentric cables; 1 concentric to 1 single-conductor cable; 1 concentric to 2 single-conductor cables; 1 concentric to 4 single-conductor cables; 1 three-conductor to 3 single-conductor cables. *See:* **branch joint; regular straight joint; reducing joint.** 64

(2) (power cable joints). A complete insulated splice or group of insulated splices contained within a single protective covering or housing. In some designs, the insulating material may also serve as the protective covering. Insulated end caps are considered joints in this context. *See:* **straight joint; branch joint; insulating (isolating) joint; transition joint.** 34

cable Morse code. A three-element code, used mainly in submarine cable telegraphy, in which dots and dashes are represented by positive and negative current impulses of equal length, and a space by absence of current. *See:* **telegraphy.** 328

cable penetration (1)(cable penetration fire stop qualification test). An assembly or group of assemblies for electrical conductors to enter and continue through a fire-rated structural wall, floor, or floor-ceiling assembly. 368

(2)(cable-penetration fire stops, fire breaks, and system enclosures)(design and installation of cable systems for Class 1E circuits in nuclear power generating stations). An assembly or group of assemblies for electrical conductors to enter and continue through a fire-rated structural wall, floor, or floor-ceiling assembly. 536

cable-penetration fire stop (cable-penetration fire stops, fire breaks, and system enclosures)(design and installation of cable systems for Class 1E circuits in nuclear power generating stations). Material, devices, or an assembly of parts providing cable penetrations through fire-rated walls, floors, and floor-ceiling as-

semblies, and maintaining their required fire rating.
536

cable rack (hanger) (shelf*). A device usually secured to the wall of a manhole, cable raceway, or building to provide support for cables.   64
*Deprecated

cable reel (mining). A drum on which conductor cable is wound, including one or more collector rings and associated brushes, by means of which the electric circuit is made between the stationary winding on the locomotive or other mining device and the trailing cable that is wound on the drum. *Note:* The drum may be driven by an electric motor, a hydraulic motor, or mechanically from an axle on the machine. *See:* **mine feeder circuit.**   328

cable segment (FASTBUS acquisition and control). A FASTBUS segment consisting of a cable together with appropriate connectors for mating with devices.
480

cable separator (power distribution, underground cables). A serving of threads, tapes, or films to separate two elements of the cable, usually to prevent contamination or adhesion.   57

cable sheath (1)(NESC)(transmission and distribution). A conductive protective covering applied to cables. *Note:* A cable sheath may consist of multiple layers of which one or more is conductive.
494, 262

(2) (insulated conductor). A tubular impervious metallic protective covering applied directly over the cable core.   64,57

cable sheath insulator (pothead). An insulator used to insulate an electrically conductive cable sheath or armor from the metallic parts of the pothead or transformer removable cable terminating box in contact with the supporting structure for the purpose of controlling cable sheath currents. *See:* **transformer; transformer removable cable terminating box.**
288,289

cable shield. *See:* **duct edge fair-lead**

cable shielding (cable systems in power generating stations)(shielding and shield grounding)(design and installation of cable systems for Class 1E circuits in nuclear power generating stations). A nonmagnetic metallic material applied over the insulation of the conductor or conductors to confine the electric field of the cable to the insulation of the conductor or conductors.   35, 536

cable splicer (mining). A short piece of tubing or a specially formed band of metal generally used without solder in joining ends of portable cables for mining equipment. *See:* **mine feeder circuit.**   328

cable spreading room (cable systems). The cable spreading room is normally the area adjacent to the control room where cables leaving the panels are dispersed into various cable trays for routing to all parts of the plant.   35

cable-system enclosure (cocoon)(cable-penetration fire stops, fire breaks, and system enclosures)(design and installation of cable systems for Class 1E circuits in nuclear power generating stations). An assembly

installed around a cable system to maintain circuit integrity, for a specified time, of all circuits within the enclosure when it is exposed to the most severe fire that may be expected to occur in the area.   536

cable terminal (1)(NESC). A device which provides insulated egress for the conductors. *Syn:* **termination.**
494

(2) (pothead) (end bell) (power work). A device that seals the end of a cable and provides insulated egress for the conductors.   64

cable tray (raceway systems for Class 1E circuits for nuclear power generating stations). A prefabricated metal raceway with or without covers consisting of siderails and bottom support sections. Bottom support sections may be ladder, trough, or solid.   513

cable tray system (raceway systems for Class 1E circuits for nuclear power generating stations). An assembly of metallic cable tray sections, fittings, supports, anchorages, and accessories that form a structural system to support wire and cables.   513

cable type (nuclear power generating stations). A cable type for purposes of qualification testing shall be representative of those cables having the same materials, similar construction, and service rating, as manufactured by a given manufacturer.   31

cable value. *See:* **manhole.**

cab signal. A signal located in the engineman's compartment or cab indicating a condition affecting the movement of a train or engine and used in conjunction with interlocking signals and in conjunction with or in lieu of block signals. *See:* **automatic train control.**
328

CADF. *See:* **commutated-antenna direction finder.**

cage (Faraday cage). A system of conductors forming an essentially continuous conducting mesh or network over the object protected and including any conductors necessary for interconnection to the object protected and an adequate ground.   328

cage antenna. A multiwire element whose wires are so disposed as to resemble a cylinder, in general of circular cross section; for example, an elongated cage.
111

cage synchronous motor (rotating machinery). A salient pole synchronous motor having an amortisseur (damper) winding embedded in the pole shoes, the primary purpose of this winding being to start the motor.   63

cage winding. *See:* **squirrel-cage winding.**

caging (gyro). The process of orienting and mechanically locking one or more gyro axes or gimbal axes to a reference position.   46

calculations (International System of Units) (SI). Errors in calculations can be minimized if the base and the coherent derived SI units are used and the resulting numerical values are expressed in power-of-ten notation instead of using prefixes. *See:* **units and letter symbols; prefixes and symbols.**   21

calculator. (1) A device capable of performing arithmetic. (2) A calculator as in (1) that requires frequent manual intervention. (3) Generally and historically, a device for carrying out logic and arithmetic digital operations of any kind.   255

calibrate (1)(monitoring radioactivity in effluents). Adjustment of the system and the determination of system accuracy using one or more sources traceable to the National Bureau of Standards (NBS).   559
(2) (radiological monitoring instrumentation)(plutonium monitoring). To determine the response or reading of an instrument relative to a series of known radiation values over the range of the instrument.
398,413
(3) (radiation protection). To determine (A) the response or reading of an instrument relative to a series of known radiation values over the range of the instrument or (B) the strength of a radiation source relative to a standard.   399
calibrated (industrial control). Checked for proper operation at selected points on the operating characteristic.   206
calibrated-driving-machine test (rotating machinery). A test in which the mechanical input or output of an electric machine is calculated from the electric input or output of a calibrated machine mechanically coupled to the machine on test. See: **asynchronous machine.**   63
calibrated Marinelli beaker standard source (MBSS) (germanium semiconductor detector). A calibrated MBSS is an MBSS that has been calibrated by comparing its photon emission rate to that of a certified MBSS. Note: The photon emission rate as used in this standard is the number of photons per second resulting from the decay of radionuclides in the source and is thus higher than the detected rate at the surface.
388
calibrated-solution Marinelli beaker standard source (MBSS) (germanium semiconductor detector). A calibrated-solution MBSS is a standard beaker that contains as its radioactive filling material a solution that has been calibrated by comparing its photon emission rate at specified energies to that of a certified solution. Note: The photon emission rate as used in this standard is the number of photons per second resulting from the decay of radionuclides in the source and is thus higher than the detected rate at the surface.
388
calibration (1)(ionization chambers)('dose calibrator' ionization chambers). The process of determining the numerical relationship, within an overall stated uncertainty, between the observed output of a measurement system and the value, based on standard sources, of the physical quantity being measured.   396, 499
(2)(nuclear power generating stations). Comparison of items of measuring and test equipment with reference standards or with items of measuring and test equipment of equal or closer tolerance to detect and quantify inaccuracies and to report or eliminate those inaccuracies.   41
(3)(supervisory control, data acquisition, and automatic control). Adjustment of a device so that the output is within a specific range for particular values of the input.   570
(4) (device). The adjustment to have the designed operating characteristics and the subsequent marking

of the positions of the adjusting means, or the making of adjustments necessary to bring operating characteristics into substantial agreement with standardized scales or marking.   103,202,60,127
(5) (metering). Comparison of the indication of the instrument under test, or registration of the meter under test, with an appropriate standard.   212
(6) (power switchgear). The adjustment of a device to have the designed operating characteristics, and the subsequent marking of the positions of the adjusting means, or the making of adjustments necessary to bring operating characteristics into substantial agreement with standardized scales or marking.   103
(7) (station control and data acquisition). Adjustment of a device such that the output is within a specified range for particular values of the input.   403
calibration error (1)(electric pipe heating systems). In operation, the departure under specified conditions, of actual performance from performance indicated by scales, dials, or other markings on the device. See: **alarm signal.**   448
(2)(power switchgear). In the operation of a device the departure, under specified conditions, of actual performance from performance indicated by scales, dials, or other markings on the device. Note: The indicated performance may be by calibration markings in terms of input or performance quantities (amperes, ohms, seconds, etcetera) or by reference to a specific performance data recorded elsewhere. See: **setting error.**   103
calibration factor (1) (bolometer-coupler unit). The ratio of the substitution power in the bolometer attached to the side arm of the directional coupler to the microwave power incident on a nonreflecting load connected to the output port of the main arm of the directional coupler. Note: If the bolometer unit is attached to the main arm of the directional coupler, the calibration factor is the ratio of the substitution power in the bolometer unit attached to the main arm of the directional coupler to the microwave power incident upon a nonreflecting load connected to the output port of the side arm of the directional coupler.   185
(2) (bolometer unit). The ratio of the substitution power to the radio-frequency power incident upon the bolometer unit. The ratio of the bolometer-unit calibration factor to the effective efficiency is determined by the reflection coefficient of the bolometer unit. The two terms are related as follows:

$$K_b / \eta_e = 1 - |\Gamma|^2$$

where K_b, η_e, and Γ are the calibration factors effective efficiency, and reflection coefficient of the bolometer unit, respectively.   185
(3) (electrothermic unit). The ratio of the substituted reference power (dc, audio, or rf) in the electrothermic unit to the power incident upon the electrothermic unit for the same dc output voltage from the electrothermic unit at a prescribed temperature. Note: (A)

Calibration factor and effective efficiency are related as in the equation above, where K_b, η_c and Γ are the calibration factor, effective efficiency and reflection coefficient of the electrothermic unit, respectively. (B) The reference frequency is to be supplied with the calibration factor. 47

(4) (electrothermic-coupler unit). The ratio of the substituted reference power (dc, audio, or rf) in the electrothermic unit attached to the side arm of the directional coupler to the power incident upon a non-reflecting load connected to the output port of the main arm of the directional coupler for the same dc output voltage from the electrothermic unit is attached to the main arm of the directional coupler, the calibration factor is the ratio of the substituted reference (dc, audio, or rf) power in the electrothermic unit attached to the main arm of the directional coupler to the power incident upon a nonreflecting load connected to the output port of the side arm of the directional coupler for the same dc output voltage from the electrothermic unit at a prescribed temperature. *Note:* The reference frequency is to be supplied with the calibration factor. 47

calibration interval, period or cycle (test, measurement, and diagnostic equipment). The maximum length of time between calibration services during which each standard and test and measuring equipment is expected to remain within specific performance levels under normal conditions of handling and use. 42

calibration level (signal generator). The level at which the signal generator output is calibrated against a standard. *See:* **signal generator.** 185

calibration marks (navigation aid terms)(radar). Indications superimposed on a display to provide a numerical scale of the parameters displayed.

526, 13

calibration or conversion factor (calibration) (loosely called antenna factor). The factor or set of factors that, at given frequency, expresses the relationship between the field strength of an electromagnetic wave impinging upon the antenna of a field-strength meter and the indication of the field-strength meter. *Note:* The composite of antenna characteristics, balun and transmission line effects, receiver sensitivity and linearity, etcetera. *See:* **measurement system.** 213

calibration procedure (test, measurement and diagnostic equipment). A document which outlines the steps and operations to be followed by standards and calibration laboratory and field calibration activity personnel in the performance of an instrument calibration. 54

calibration programming (power supplies). Calibration with reference to power-supply programming describes the adjustment of the control-bridges current to calibrate the programming ratio in ohms per volt. *Note:* Many programmable supplies incorporate a calibrate control as part of the reference resistor that performs this adjustment.

228,186

calibration scale (power switchgear). A set of gradua-

tions marked to indicate values of quantities , such as current, voltage, or time, at which an automatic device can be set to operate. 103

calibration voltage. The voltage applied during the adjustment of a meter. *See:* **test (instrument or meter).**

212

calibrator (oscilloscopes). The signal generator whose output is used for purposes of calibration, normally either amplitude or time or both. 106

caliche (cable plowing). Common sedimentary rock normally formed from ancient marine life. 52

call (1) (communications) (computers). The action performed by the calling party, or the operations necessary in making a call, or the effective use made of a connection between two stations. 255

(2) (telephone switching systems). A demand to set up a connection. 55

call announcer (automatic telephone office). A device for receiving pulses and audibly reproducing the corresponding number in words so that it may be heard by a manual operator. 328

call circuit (manual switching). A communication circuit between switching points used by the traffic forces for the transmission of switching instructions.

328

called-line release (telephone switching systems). Release under the control of the line to which the call was directed. 55

call forwarding (telephone switching systems). A feature that permits a customer to instruct the switching equipment to transfer calls intended for his station to another station. 55

call indicator. A device for receiving pulses from an automatic switching system and displaying the corresponding called number before an operator at a manual switchboard. 328

calling device (telephone switching systems). An apparatus that generates the signals required for establishing connections in an automatic switching system.

55

calling line identification (telephone switching systems). Means for automatically identifying the source of calls. 55

calling-line release (telephone switching systems). Release under the control of the line from which the call originated. 55

calling-line timed release (telephone switching systems). Timed release initiated by the calling line.

55

calling plug and cord. A plug and cord that are used to connect to a called line. 328

calling sequence (computers). A specified arrangement of instructions and data necessary to set up and call a given subroutine. 255,77

call packing (telephone switching systems). A method of selecting paths in a switching network according to a fixed hunting sequence. 55

call rate (telephone switching systems). The number of calls per unit of time. 55

call splitting (telephone switching systems). Opening the transmission path between the parties of a call.

55

call tracing (telephone switching systems). A means for manually identifying the source of calls.   55

call waiting (telephone switching systems). A feature providing a signal to a busy called line to indicate that another call is waiting.   55

call-waiting tone (telephone switching systems). A tone used in the call-waiting feature.   55

calomel electrode. *See:* **calomel half-cell.**

calomel half-cell (calomel electrode). A half-cell containing a mercury electrode in contact with a solution of potassium chloride of specified concentration that is saturated with mercurous chloride of which an excess is present. *See:* **electrochemistry.**   328

calorimeter (laser-maser). A device for measuring the total amount of energy absorbed from a source of electromagnetic radiation.   363

calorimetric test (rotating machinery). A test in which the losses in a machine are deduced from the heat produced by them. The losses are calculated from the temperature rises produced by this heat in the coolant or in the surrounding media. *See:* **asynchronous machine.**   63

CAMAC (1)(Computer Automated Measurement and Control). CAMAC is a standard modular instrumentation and digital interface system.   51

(2)(FASTBUS acquisition and control). An internationally standardized modular instrumentation and digital interface system as defined in ANSI/IEEE Std 583-1982, IEEE Standard Modular Instrumentation and Digital Interface System (CAMAC), and the corresponding documents EUR 4100-1972, CAMAC: A Modular Instrumentation System for Data Handling, and IEC Pub 516-1975, A Modular Instrumentation System for Data Handling; CAMAC System (CAMAC is often treated as an acronym for 'Compiler Automated Measurement and Control').   480

CAMAC branch driver. *See:* **CAMAC parallel highway driver.**   51

CAMAC branch highway. *See:* **CAMAC parallel highway.**   51

CAMAC compatible crate. A mounting unit for CAMAC plug-in units that does not conform to the full requirements for a CAMAC crate but in which CAMAC modules can be mounted and operated in accordance with the dataway requirements of IEEE Standard 583.   51

CAMAC crate. A mounting unit for CAMAC plug-in units that includes a CAMAC dataway and conforms to the mandatory requirements for a CAMAC crate as specified in IEEE Standard 583.   51

CAMAC crate assembly. An assembly of a CAMAC crate controller and one or more CAMAC modules mounted in a CAMAC crate (or CAMAC compatible crate), and operable in conformity with the dataway requirements of IEEE Standard 583.   51

CAMAC crate controller. A functional unit that when mounted in the control station and one or more normal stations of a CAMAC crate (or CAMAC compatible crate) communicates with the dataway in accordance with IEEE Standard 583.   51

CAMAC data array (intc) (subroutines for CAMAC).

The symbol *intc* represents an array of CAMAC data words. Each element of *intc* has the same form as the CAMAC data word variable *int*. The length of *intc* is given by the value of the first element of *cb* at the time the subroutine is executed. *See:* **CAMAC data word; control block.**   410

CAMAC dataway. An interconnection between CAMAC plug-in units which conforms to the mandatory requirements for a CAMAC dataway as specified in IEEE Standard 583.   51

CAMAC data word (int) (subroutines for CAMAC). The symbol *int* represents a CAMAC data word stored in computer memory. The form is not specified, but the word must be stored in addressable storage entity capable of containing twenty-four bits. In a computer or programming system which does not have an addressable unit of storage which can contain twenty-four bits, multiple units must be used.   410

CAMAC external addresses (exta) (subroutines for CAMAC). The symbol *exta* represents an array of integers each of which is a CAMAC register address. The form and information content of each element of *exta* must be identical to the form and information content of the quantity *ext*. The length of *exta* is given by the value of the first element of *cb* at the time the subroutine is executed. *See:* **external address; control block.**   410

CAMAC module. A CAMAC plug-in unit that when mounted in one or more normal stations of a CAMAC crate is compatible with IEEE Standard 583.   51

CAMAC parallel highway. A standard highway (for a CAMAC system) in which the data is transferred in parallel and which conforms to the requirements of IEEE Standard 596. *Syn:* **CAMAC branch highway.**   51

CAMAC parallel highway driver. A unit that communicates via the CAMAC parallel highway with up to seven CAMAC crates and conforms to the requirements as specified in IEEE Standard 596. *Syn:* **CAMAC branch driver.**   51

CAMAC plug-in unit. A functional unit that conforms to the mandatory requirements for a plug-in unit as specified in IEEE Standard 583.   51

CAMAC serial highway. A standard highway (for a CAMAC system) in which the data is transferred in bit or byte serial and which conforms to the requirements of IEEE Standard 595.   51

CAMAC system. A system including at least one CAMAC crate assembly.   51

cam contactor (cam switch). A contactor or switch actuated by a cam. *See:* **control switch.**   1

camera storage tube. A storage tube into which the information is introduced by means of electromagnetic radiation, usually light, and read at a later time as an electric signal. *See:* **storage tube.**   174,190

camera tube (television). A tube for conversion of an optical image into an electrical signal.   18

cam-operated switch (industrial control). A switch consisting of fixed contact elements and movable contact elements operated in sequence by a camshaft. *See:* **switch.** 244,206

camping trailer (National Electrical Code). A vehicular portable unit mounted on wheels and constructed with collapsible partial side walls which fold for towing by another vehicle and unfold at the campsite to provide temporary living quarters for recreational, camping, or travel use. *See:* **recreational vehicle.** 256

camp-on busy (telephone switching systems). A feature whereby a call encountering a busy condition can be held and subsequently connected automatically when the busy condition is required. 55

cam-programmed (test, measurement and diagnostic equipment). (1) A programming technique that uses a rotating shaft, having specifically oriented, eccentric projections which control a series of switches that set up the proper circuits for a test; and (2) A cam-follower system used to set positions or values of a shafted instrument for programming instructions to the test system. 54

cam-shaft position (electric power system). The angular position of the main shaft directly operating the governor-controlled valves. *See:* **speed-governing system.** 94

can (dry cell). A metal container, usually zince, in which the cell is asserted and that serves as its negative electrode. *See:* **electrolytic cell.** 328

cancel (numerically controlled machines). A command that will discontinue any fixed cycles or sequence commands. 224,207

canceled (cancelled) video (radar). In MTI (moving target indication), the video output remaining after the cancellation process. *Note:* The preferred spelling is "canceled" in the United States, "cancelled" in Great Britain. 13

canceler (canceller)(radar). That portion of the system in which unwanted signals such as clutter, fixed targets, and other interference are suppressed by a process of linear subtraction. *Note:* The preferred spelling is "canceler" in the United States, "canceller" in Great Britain. 13

cancellation ratio (radar). In MTI (moving target indication), the ratio of canceler voltage amplification for fixed-target echoes received with a fixed antenna, to the gain for a single pulse passing through the unprocessed channel of the canceler. 13

cancelled video (radar moving-target indicator). The video output remaining after the cancellation process. *See:* **navigation.** 13, 187

candela (cd) (1) (illuminating engineering). The SI unit of luminous intensity. One candela is one lumen per steradian (lm/sr). Formerly, candle. *Note:* The fundamental luminous intensity definition in the SI is the candela in terms of a complete (blackbody) radiator. From this relation K_m and $K_{m'nd}$ consequently the lumen, are determined. One candela is defined as the luminous intensity of 1/600 000 of one square meter of projected area of a blackbody radiator operating at the temperature of solidification of platinum, at a pressure of 101 325 newtons per square meter

$(N/m^2 = \text{PA})$. From 1909 until the introduction of the present photometric system on January 1, 1948, the unit of luminous intensity in the United States, as well as in France and Great Britain, was the "international candle" which was maintained by a group of carbon-filament vacuum lamps. For the present unit as defined above, the internationally accepted term is candela. The difference between the candela and the old international candle is so small that only measurements of high precision are affected. The following resolution was adopted at the *Seizieme Conference Generale des Poids et Mesures* on October 11, 1979– The Sixteenth General Conference on Weights and Measures had decided: (A) The candela is the luminous intensity, in a given direction, of a source emitting monochromatic radiation of frequency 540 x10^{12} hertz and whose radiant intensity in this direction is 1/683 watt per steradian. (B) The candela so defined is the base unit applicable to photopic quantities, scotopic quantities, and quantities to be defined in the mesopic domain. *See:* **luminous flux.** 167

(2) (metric practice). The luminous intensity, in the perpendiculaur direction, of a surface of 1/600000 square meter of blackbody at thetemperature of freezing plantinum under a pressure of 101325 newtons per squaremeter (adopted by 13th General Conference on Weights and Measures). 21

(3) (television). The luminous intensity, in the perpendicular direction, of a 1/600 000 square meter surface of a blackbody at the freezing temperature of plantinum under a pressure of 101 325 pascals. *Note:* (A) Values for standards having other spectral distributions are derived by the use of accepted spectral luminous efficiency data for photopic vision. (B) From 1909 until the introduction of the present photometric system on January 1, 1948, the unit of luminous intensity in the United States, as well as in France and Great Britain, was the international candle which was maintained by a group of carbon-filament vacuum lamps. For the present unit as defined above, the internationally accepted term is candela. The difference between the candela and the old international candle is so small that only measurements of high precision are affected. 18

C and I (C I). *See:* **supervisory control point, control and indication.**

candle. *See:* **candela.** 18

candlepower (cp) (illuminating engineering) (television). Luminous intensity expressed in candelas. 167, 18

can loss (rotating machinery). Electric losses in a can used to protect electric components from the environment. *See:* **asynchronous machine.** 63

canned (rotating machinery). Completely enclosed and sealed by a metal sheath. 63

canned cycle (numerically controlled machines). *See:* **fixed cycle.**

capability (1)(microprocessor operating systems). A set of functions that are logically related by the common set of resources on which they operate. 478

(2)(power operations). The maximum load-carrying capability expressed in kilovolt-amperes (kVA) or kilowatts (kW) of generating equipment, other electrical apparatus, or system under specified conditions for a given time interval. *See:* **dependable capability; extended capability; first contingency incremental transfer capability; hydro capability; installed incremental transfer capability; maximum capability; normal transfer capability; pumped-storage hydro capability; second contingency incremental transfer capability; steam capability; system assured capability; system margin capability; total for load capability.**

capability margin (electric power supply). The difference between (A) total capability for load and (B) system load responsibility. It is the margin of capability available to provide for scheduled maintenance, emergency outages, adverse system operating requirements, and unforeseen loads. *See:* **generating station.**   200

capability module (microprocessor operating systems). A set of functions within a capability that provide a class of support. Each module is intended to be provided in its entirety.   478

capacitance (capacity)(1)(electric installations on shipboard). That property of a system of conductors and dielectrics which permits the storage of electricity when potential differences exist between the conductors. Its value is expressed as the ratio of a quantity of electricity to a potential difference. A capacitance value is always positive.   3

(2)(of a semiconductor radiation detector)(charged-particle detector)(germanium gamma-ray detectors)(X-ray energy spectrometers). The small-signal capacitance measured between terminals of the detector under specified conditions of bias and frequency.   528, 147

(3) (semiconductor diode) (semiconductor radiation conductor). The small-signal capacitance measured between the terminals of the diode or detector under specified conditions of bias and frequency. *See:* **rectification; semiconductor; semiconductor device.**   245, 23

(4) (control systems). A property expressible by the ratio of the time integral of the flow rate of a quantity, such as heat, or electric charge to or from a storage, divided by the related potential change. *Note:* Typical units are microfarads, Btu.deg F, lb.psi, gal.ft.   56

(5) (outdoor apparatus bushings). (A) The main capacitance, Cl, of a condenser bushing is the value in picofarads between the high-voltage conductor and the potential tap or the test tap. (B) The tap capacitance, C2, of a condenser bushing is the value in picofarads between the potential tap and mounting flange (ground). (C) The capacitance, C, of a bushing without a potential or test tap is the value in picofarads between the high-voltage conductor and the mounting flange (ground).   168

(6) (VLF insulation testing). Capacitance, as used here, and distinguished from power-frequency capacitance, is that value which would result from a mea-

surement at VLF, that is, 0.1 Hz $\pm$ 25 percent. In magnitude, it would tend to be greater than the power-frequency capacitance, to the extent of increased contributions made by dipole and interfacial polarizations.   135

capacitance between two conductors, balanced. *See:* **balanced capacitance.**

capacitance coupling (signal-transmission system). *See:* **coupling capacitance.**

capacitance current (1)(electric submersible pump cable). Current required to charge the capacitor formed by the dielectric of the cable under test.   484

(2)(rotating machinery). A reversible component of the measured current on charge or discharge of the winding and is due to the geometrical capacitance, that is, the capacitance as measured with alternating current of power or higher frequencies. With high direct voltage this current has a very short time constant and so does not affect the usual measurements. *See:* **insulation testing (large alternating-current rotating machinery).**   6

(3)(power cable systems). Current which charges the capacitor formed by the capacitance of the cable under test.   437

capacitance, discontinuity (waveguide or transmission line). The shunt capacitance that, when inserted in a uniform waveguide or transmission line, would cause reflected waves of the dominant mode equal to those resulting from the given discontinuity. *See:* **waveguide.**   194

capacitance, effective. The imaginary part of a capacitive admittance divided by the angular frequency.   194

capacitance, input (n-terminal electron tubes). The short-circuit transfer capacitance between the input terminal and all other terminals, except the output terminal, connected together. *Note:* This quantity is equivalent to the sum of the interelectrode capacitances between the input electrode and all other electrodes except the output electrode. *See:* **electron-tube admittances.**   125

capacitance meter. An instrument for measuring capacitance. *Note:* If the scale is graduated in microfarads the instrument is usually designated as a microfaradmeter. *See:* **instrument.**   328

capacitance, nonlinear element (varactor measurements). The capacitance of the high frequency equivalent series RC circuit.   136

capacitance, output (n-terminal electron tube). The short-circuit transfer capacitance between the output terminal and all other terminals, except the input terminal, connected together. *See:* **electron-tube admittances.**   190

capacitance potential device. A voltage-transforming equipment or network connected to one conductor of a circuit through a capacitance, such as a coupling capacitor or suitable high-voltage bushing, to provide a low voltage such as required for the operation of instruments and relays. *Notes:* (1) The term **potential device** applies only to the network and is exclusive of the coupling capacitor or high-voltage bushing. (2)

The term **coupling-capacitor potential device** indicates use with coupling capacitators. (3) The term **bushing potential device** indicates use with bushings. (4) The term **capacitance potential device** indicates use with any type of capacitance coupling. (5) Capacitance potential devices and their associated coupling capacitors or bushings are designed for line-to-ground connection, and not line-to-line connection. The potential device is a single-phase device, and, in combination with its coupling capacitor or bushing, is connected line-to-ground. The low voltage thus provided is a function of the line-to-ground voltage and the constants of the capacitance potential device. Two or more capacitance potential devices, in combination with their coupling capacitors or bushings, may be connected line-to-ground on different high-voltage phases to provide low voltages of other desired phase relationships. (6) Zero-sequence voltage may be obtained from the broken-delta connection of the auxiliary windings or by the use of one device with three coupling capacitors or bushings. In the latter case, the three operating-tap connection-points are joined together and one device connected between this common point and ground. Although used in combination with three coupling capacitors or bushings, the device output and accuracy rating standards are based on the single-phase conditions. *See:* **outdoor coupling capacitor.**									341

capacitance ratio (nonlinear capacitor). The ratio of maximum to minimum capacitance over a specified voltage range, as determined from a capacitance characteristic, such as a differential capacitance characteristic, or a reversible capacitance characteristic. *See:* **nonlinear capacitor.**									191

capacitance, short-circuit input (*n-terminal* electron device). The effective capacitance determined from the short-circuit input admittance. *See:* **electron-tube admittances.**									125

capacitance, short-circuit output (*n-terminal* electron device). The effective capacitance determined from the short-circuit output admittance. *See:* **electron-tube admittances.**									125

capacitance, short-circuit transfer (electron tubes). The effective capacitance determined from the short-circuit transfer admittance. *See:* **electron-tube admittances.**									125

capacitance, signal electrode. *See:* **electrode capacitance.**

capacitance, stray (electric circuits). Capacitance arising from proximity of component parts, wires, and ground. *Note:* It is undesirable in most circuits, although in some high-frequency applications it is used as the tuning capacitance. In bridges and other measuring equipment, its effect must be eliminated by preliminary balancing out, or known and included in the results of any measurement performed. *See:* **measurement system.**									185

capacitance-switching transient overvoltage ratio (power switchgear)(high voltage air switches, insulators, and bus supports). The ratio of the peak value of voltage above ground, during the transient conditions

resulting from the operation of the switch, to the peak value of the steady-state line-to-neutral voltage. *Note:* It is measured at either terminal of the switch whichever is higher, and is expressed in multiples of the peak values of the operating line-to-ground voltages at the switch with the capacitance connected.									103, 27

capacitance, target (camera tubes). *See:* **target capacitance (camera tubes).**

capacitance unbalance detection device (series capacitors). A device to detect objectionable unbalance in capacitance between capacitor groups within a phase, such as that caused by blown capacitor fuses or faulted capacitors, and to initiate an alarm or the closing of the capacitor bypass device, or both.									474

capacitator braking (rotating machinery). A form of dynamic braking for induction motors in which a capacitor is used to magnetize the motor. *See:* **asynchronous machine.**									63

capacitive current (or component) (rotating machinery). A reversible component of the measured current on charge or discharge of the winding which is due to the geometrical capacitance, that is, the capacitance as measured with alternating current of power or higher frequencies. With high direct voltage this current has a very short time constant and so does not affect the usual measurements. *See:* **IEEE Std 62-1958, Section 6.**									6

capacitive gap (microwave receiver protectors)(nonlinear, active, and nonreciprocal waveguide components). The distance between cone apexes (apices) in a waveguide resonant structure. *See:* **resonant gap.**									530

capacitor (1)(electric installations on shipboard). A device, the primary purpose of which is to introduce capacitance into an electric circuit. Capacitors are usually classified, according to their dielectrics, as air capacitors, mica capacitors, paper capacitors, etcetera.									3

(2)(series capacitors). An assembly of one or more capacitor elements in a single container, with one or more insulated terminals brought out.									474

capacitor antenna (condenser antenna). An antenna consisting of two conductors or systems of conductors, the essential characteristic of which is its capacitance. *See:* **antenna.**									197

capacitor bank (shunt power capacitors). An assembly at one location of capacitors and all necessary accessories, such as switching equipment, protective equipment, controls, etcetera, required for a complete operating installation. It may be a collection of components assembled at the operating site or may include one or more factory-assembled equipments.									138

capacitor bus (series capacitors). The main conductors which serve to connect the capacitor assemblies in series with the line.									474

capacitor bushing (condenser bushing) (outdoor apparatus bushings). A bushing in which cylindrical conducting layers are arranged coaxially with the conductor within the insulating material for the purpose of controlling the electric field of the bushing.									168

capacitor bypass switch (series capacitor). A switch

device with moving and stationary contacts that functions as a means of bypassing the capacitor. This switch may also have the capability of inserting the capacitor against a specified level of current.   86

capacitor-bypass-switch interlocking devices (series capacitor). Devices that perform the function of having all three integral bypass switches of a capacitor step take the same open or close position.   86

capacitor element (series capacitor). An individual part of a capacitor unit consisting of coiled conductors separated by dielectric material.   86

capacitor enclosure. The case in which the capacitor is mounted.   328

capacitor equipment (shunt power capacitors). An assembly of capacitors with associated accessories, such as fuses, switches, etcetera, all mounted on a common frame for handling, transportation, and operation as a single unit.   138

capacitor group (series capacitors). An assembly of more than one capacitor connected in series, in parallel, or both, between two buses or terminals.   86

capacitor, ideal. *See:* **ideal capacitor.**

capacitor indicating fuse (series capacitors). A capacitor fuse that provides an externally visible indication of fuse operation.   474

capacitor loudspeaker. *See:* **electrostatic loudspeaker.**

capacitor microphone. *See:* **electrostatic microphone.**

capacitor motor. A single-phase induction motor with a main winding arranged for direct connection to a source of power and an auxiliary winding connected in series with a capacitor. The capacitor may be directly in the auxiliary circuit or connected into it through a transformer. *See:* **asynchronous machine; capacitor-start motor; permanent-split capacitor motor; two-value capacitor motors.**   63

capacitor mounting strap. A device by means of which the capacitor is affixed to the motor.   328

capacitor pickup. A phonograph pickup that depends for its operation upon the variation of its electric capacitance. *See:* **phonograph pickup.**   290

capacitor platform (series capacitors). A structure that supports the capacitor- rack assemblies and all associated equipment and protective devices, and is supported on insulators compatible with line-to-ground insulation requirements.   474

capacitor rack (series capacitors). A frame that supports one or more capacitors.   474

capacitor segment (series capacitors). An assembly of groups of capacitors which has its own voltage-limiting device and relays to protect the capacitors from overvoltages and overloads. *See:* **Figure, capacitor group.**   474

capacitor start-and-run motor. A capacitor motor in which the auxiliary primary winding and series-connected capacitors remain in circuit for both starting and running. *See:* **permanent-split capacitor motor; asynchronous machine.**   63

capacitor-start motor. A capacitor motor in which the auxiliary winding is energized only during the starting operation. *Note:* The auxiliary-winding circuit is open-circuited during running operation. *See:* **asynchronous machine.**   63

Capacitor Group

1. Capacitor Segment
2. Capacitor Switching Step
3. Capacitor Group
4. Discharge Current Limiting Device
5. Protective Device
6. Bypass Device/Switch (See 6.3.6)
7. Additional Switching Steps When Required
8. External Bypass Disconnect Switch
9. External Isolating Disconnect Switch
10. External Grounding Disconnect Switch
11. Imbalance Detection Device

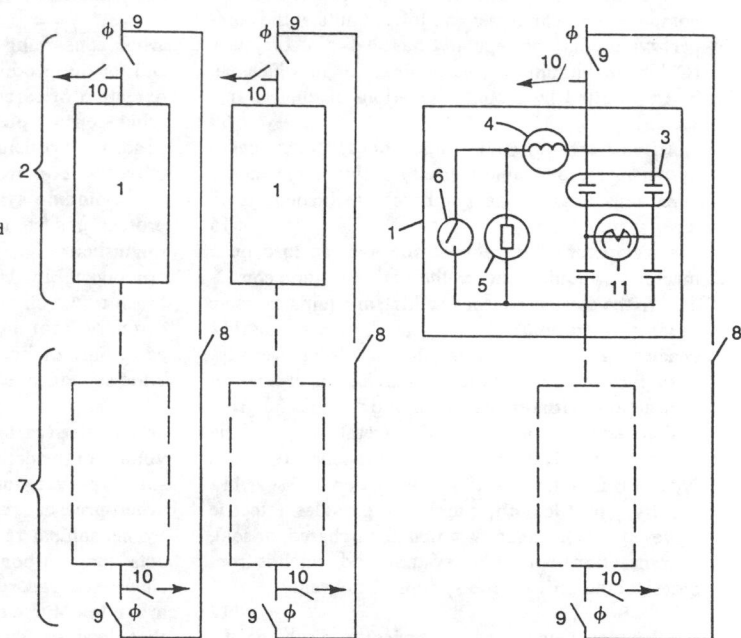

NOTE: All three phases, and other segments are as shown for one phase.

capacitor switch (power switchgear). A switch capable of making or breaking capacitive currents of capacitor banks.   103

capacitor switching step (series capacitors). A three-phase function that consists of one or more capacitor segments per phase with capacitor bypas devices, and provision for interlocked operation of the single-phase or three-phase switches when bypassing or inserting the capacitor segments. *See:* **Figure, capacitor group.**
474

capacitor unit (1) (general). A single assembly of dielectric and electrodes in a container with terminals brought out. *See:* **alternating-current distribution; indoor capacitor unit; outdoor capacitor unit.**   138
(2) (series capacitor). An assembly of one or more capacitor elements in a single container, with one or more insulated terminals brought out.   86

capacitor voltage. The voltage across two terminals of a capacitor.   191

capacity (C) (1)(lead-acid batteries for photovoltaic (PV) systems). Generally, the total number of ampere-hours that can be withdrawn from a fully charged battery at a specific discharge rate and electrolyte temperature, and to a specific cutoff voltage.   515
(2)(power operations). The rated load-carrying ability expressed in kilovolt-amperes (kVA) or kilowatts (kW) of generating equipment or other electrical apparatus. *See:* **capacity emergency; firm capacity; specific unit capacity.**   516
(3) (data transmission). (A)The number of digits or characters in a machine word regularly handled in a computer. (B) The upper and lower limits of the numbers which may be regularly handled in a computer. (C) The maximum number of binary digits which can be transmitted by a communications channel in one second.   59

capacity emergency (power operations). An emergency where the available capacity within a system (or pool) plus interchange purchases are inadequate to meet system load.   516

capacity factor. The ratio of the average load on a machine or equipment for the period of time considered to the capacity of the machine or equipment. *See:* **generating station.**   64,112

capacity, firm, purchases or sales (electric power supply). Firm capacity which is purchased, or sold, in transactions with other systems and which is not from designated units, but is from the overall system of the seller. *Note:* It is understood that the seller treats this type of transaction as a load obligation.   112

capacity, specific unit, purchases or sales (electric power supply). Capacity which is purchased, or sold, in transactions with other systems and which is from a designated unit on the system of the seller.
112

capacity test (lead storage batteries). A discharge of a battery to a designated terminal voltage.   38

cap-and-pin insulator. An assembly of one or more shells with metallic cap and pin, having means for direct and rigid mounting. *See:* **insulator.**   261

capture effect (1) (modulation systems). The effect occurring in a transducer (usually a demodulator) whereby the input wave having the largest magnitude controls the output.   242
(2) (radar). The tendency of a receiver to suppress the weaker of two signals within its passband.   13

capturing (gyro, accelerometer). The use of a torquer (forcer) in a servo loop to restrain a gyro gimbal, rotor, or accelerometer proof mass to a specified reference position.   46

car annunciator. An electric device in the car that indicates visually the landings at which an elevator-landing signal-registering device has been actuated. *See:* **elevator.**   328

carbon-arc lamp (illuminating engineering). An electric-discharge lamp employing an arc discharge between carbon electrodes. One or more of these electrodes may have cores of special chemicals which contribute importantly to the radiation.   167

carbon block protectors (wire-line communication facilities). An assembly of two carbon blocks and an air gap designed to a specific breakdown voltage. These devices are normally connected to communication circuits to provide overvoltage protection and a current path to ground during such overvoltage.
414

carbon brush (motors and generators). (1) A specific type of brush composed principally of amorphous carbon. *Note:* This type of brush is usually hard and is adapted to low speeds and moderate currents. (2) A broader classification of brush, containing carbon in appreciable amount. *See:* **brush (rotating machinery).**
63

carbon-consuming cell (carbon-combustion cell). A cell for the production of electric energy by galvanic oxidation of carbon. *See:* **electrochemistry.**   328

carbon-contact pickup. A phonograph pickup that depends for its operation upon the variation in resistance of carbon contacts. *See:* **phonograph pickup.**   328

carbon-dioxide system (rotating machinery). A fire-protection system using carbon-dioxide gas as the extinguisher.   63

carbon-graphite brush (electric machines). A carbon brush to which graphite is added. This type of brush can vary from medium hardness to very hard. It can carry only moderate currents and is adapted to moderate speeds. *See:* **brush (rotating machinery).**
279

carbon noise (carbon microphones). The inherent noise voltage of the carbon element. *See:* **close-talking pressure-type microphone.**   249

carbon-pressure recording facsimile. That type of electromechanical recording in which a pressure device acts upon carbon paper to register upon the record sheet. *See:* **recording (facsimile).**   12

carbon telephone transmitter. A telephone transmitter that depends for its operation upon the variation in resistance of carbon contacts. *See:* **telephone station.**
328

car, cable. *See:* **cable car.**

carcinotron (M-type backward-wave oscillator) (microwave tubes). A crossed-field oscillator tube in which an electron stream interacts with a backward wave on the nonreentrant circuit. The oscillation frequency is a function of anode-to-solve voltage. *See:* **microwave tube.**   190

car, conductor. *See:* **conductor car.**

card (computers). *See:* **magnetic card; punched card; tape to card.**

card extender (1) (power switchgear). A device which provides access to components on a circuit card for testing purposes while maintaining all the electrical connections to the card.   103

(2) (relaying). A device for testing static relay (circuit) cards which provides access to components on the card while maintaining all the electrical connections to the card.   79

card feed (test, measurement and diagnostic equipment). The mechanism which moves cards serially into a machine.   54

card field (test, measurement and diagnostic equipment). An area (one or more columns) of a card which is regularly assigned for the same information item.   54

card hopper (computers). A device that holds cards and makes them available to a card-feed mechanism. *See:* **input magazine; card stacker.**   255

card image (computers). A one-to-one representation of the contents of a punched card, for example, a matrix in which a 1 represents a punch and a 0 represents the absence of a punch.   255

cardinal plane (antennas). For an infinite planar array whose elements are arranged in a regular lattice, any plane of symmetry normal to the planar array and parallel to an edge of a lattice cell. *Notes:* (1) This term can be applied to a finite array, usually one containing a large number of elements, by the assumption that it is a subset of an infinite array with the same lattice arrangement. (2) This term is used to relate the regular geometrical arrangement of the array elements to the radiation pattern of the antenna.   111

cardiogram. *See:* **electrocardiogram.**

car-door or gate electric contact. An electric device, the function of which is to prevent operation of the driving machine by the normal operating device unless the car door or gate is in the closed position. *See:* **elevator.**   328

car-door or gate power closer. A device or assembly of devices that closes a manually opened car door or gate by power other than by hand, gravity, springs, or the movement of the car. *See:* **elevator.**   328

card-programmed (test, measurement and diagnostic equipment). The capability of performing a sequence of tests according to instructions contained in one or a deck of punched cards.   54

card reader (test, measurement and diagnostic equipment). A mechanism that senses an obtains information from punched cards.   54

card, relay. *See:* **relay armature card.**

card set function generator (analog computers). A diode function generator whose values are stored and set by means of a punched card and a mechanical card-reading device.   9

card stacker (computers). An output device that accumulates punched cards in a deck. *See:* **card hopper.**   255,77,54

card tester (test, measurement and diagnostic equipment). An instrument for testing and diagnosing printed circuit cards.   54

car enclosure (elevator). Consists of the top and the walls resting on and attached to the car platform. *See:* **elevator.**   328

car-frame sling. The supporting frame to which the car platform, upper and lower sets of guide shoes, car safety, and the hoisting ropes or hoisting-rope sheaves, or the plunger of a direct-plunger elevator are attached. *See:* **elevator.**   328

cargo vessel (electric installations on shipboard). A vessel that carries solids or liquids as freight, and no more than twelve passengers.   3

car or counterweight safety. A mechanical device attached to the car frame or to an auxiliary frame, or to the counterweight frame, to stop and hold the car or counterweight in case of predetermined overspeed of free fall, or if the hoisting ropes slacken. *See:* **elevator.**   328

car or hoistway door or gate (elevators). The sliding portion of the car or the hinged or sliding portion in the hoistway enclosure that closes the opening giving access to the car or to the landing. *See:* **hoistway (elevator or dumbwaiter).**   328

car platform (elevators). The structure that forms the floor of the car and that directly supports the load. *See:* **hoistway (elevator or dumbwaiter).**   328

car retarder. A braking device, usually power operated, built into a railway track and used to reduce the speed of cars by means of brake shoes that when set in braking position press against the sides of the lower portions of the wheels. *See:* **control machine; master controller (pressure regulator); switch machine; trimmer signal.**   328

carriage (typewriter). *See:* **automatic carriage.**

carriage return (typewriter). The operation that causes the next character to be printed at the left margin.   255

carried traffic (telephone switching systems). A measure of the calls served during a given period of time.   55

carrier (1) (data transmsission). (A) A wave having at least one characteristic that may be varied from a known reference value by modulation. (B) That part of the modulated wave that corresponds in a specified manner to the unmodulated wave, having, for example, the carrier-frequency spectral components. *Note:* Examples of carriers are a sine wave and a recurring series of pulses.   59

(2) (overhead-power-line corona and radio noise). A wave having at least one characteristic that may be varied from a known reference value by modulation. *Note:* Examples of carriers are a sine wave and a recurring series of pulses.   411

(3) (signal transmission system). (A) A wave having at least one characteristic that may be varied from a known reference value by modulation. (B) That part of the modulated wave that corresponds in a specified manner to the unmodulated wave, having, for example, the carrier-frequency spectral components. *Note:* Examples of carriers are a sine wave and a recurring series of pulses.   111

(4) (semiconductor). A mobile conduction electron or hole. *See:* **semiconductor device.**   54

(5) (electrostatography). The substance in a developer that conveys a toner, but does not itself become a part of the viewable record. *See:* **electrostatography.**   236,191

carrier-amplitude regulation. The change in amplitude of the carrier wave in an amplitude-modulated transmitter when modulation is applied under conditions of symmetrical modulation. *Note: The term* **carrier shift,** often applied to this effect, is deprecated.   111

carrier beat (facsimile). The undesirable heterodyne of signals each synchronous with a different stable reference oscillator causing a pattern in received copy. *Note:* Where one or more of the oscillators is fork controlled, this is called fork beat. *See:* **facsimile transmission.**   12

carrier chrominance signal. *See:* **chrominance signal.**

carrier-controlled approach system (CCA)(navigation aid terms)(radar). An aircraft-carrier radar system providing information by which aircraft approaches may be directed by way of radio communication.   526, 13

carrier current. The current associated with a carrier wave. *See:* **carrier.**   59

carrier-current choke coil (capacitance potential device). A reactor or choke coil connected in series between the potential tap of the coupling capacitor and the potential device transformer unit, to present a low impedance to the flow of power current and a high impedance to the flow of carrier-frequency current. Its purpose is to limit the loss of carrier-frequency current through the potential-device circuit. *See:* **outdoor-coupling capacitor.**   341

carrier-current grounding-switch and gap. Consists of a protective gap for limiting the voltage impressed on the carrier-current equipment and the line turning unit (if used); and a switch that, when closed, solidly grounds the carrier equipment for maintenance or adjustment without interrupting either high-voltage line or potential-device operation. *See:* **outdoor coupling capacitor.**   341

carrier frequency (1) (data transmission). (A) (in a periodic carrier). The reciprocal of its period. *Note:* The frequency of a periodic pulse carrier is often called the pulse repetition frequency in a signal transmission system. (B) (Modulated amplifier). The frequency that is used to modulate the input signal for amplification.   59

(2) (in a periodic carrier). The reciprocal of its period. *Note:* The frequency of a periodic pulse carrier often is called the pulse-repetition frequency in a signal-transmission system.

(3) (modulated amplifier). The frequency that is used to modulate the input signal for amplification. *See:* **carrier.**   111

carrier-frequency pulse. A carrier that is amplitude modulated by a pulse. *Note:* (1) The amplitude of the modulated carrier is zero before and after the pulse. (2) Coherence of the carrier (with itself) is not implied.   254

carrier frequency range (transmitter). The continuous range of frequencies within which the transmitter may be adjusted for normal operation. A transmitter may have more than one carrier-frequency range. *See:* **radio transmitter.**   111,240

carrier frequency stability (radio transmitter) (transmitter performance). The measure of the ability to remain on its assigned channel as determined on both a short term (1-second) and a long term (24-hour) basis.   181

carrier, fuse. *See:* **fuse carrier.**

carrier group (base group) (data transmission). The frequency band, 60 kHz to 108 kHz, containing twelve voice channels which serves as the basic building block of a larger system.   59

carrier isolating choke coil. An inductor inserted, in series with a line on which carrier energy is applied, to impede the flow of carrier energy beyond that point.   329

carrier modulation (data transmission). A process whereby a high frequency carrier wave is altered by a signal containing the information to be transmitted.   59

carrier noise level (residual modulation). The noise produced by undesired variations of radio-frequency signal in the absence of any intended modulation. *See:* **radio transmission modulation.**   111

carrier or pilot-wire receiver relay (85) (power system device function numbers). A relay that is operated or restrained by a signal used in connection with carrier-current or direct-current (dc) pilot-wire fault relaying.   402

carrier-pilot protection (power switchgear). A form of pilot protection in which the communication means between relays is a carrier current channel.   103

carrier power output (transmitter performance). The radio-frequency power available at the antenna terminal when no modulating signal is present. *See:* **audio-frequency distortion.**   181

carrier relaying protection. A form of pilot protection in which high-frequency current is used over a metallic circuit (usually the line protected) for the communicating means between the relays at the circuit terminals.   103

carrier sense (medium attachment units and repeater units). The signal provided by the physical layer to the access sublayer to indicate that one or more stations are currently transmitting on the trunk coaxial cable.   543

carrier shift (frequency shift) (data transmission). The difference between the steady state, mark, and space frequencies in a system utilizing frequency shift modulation.   59

carrier suppression (radio communication). The method of operation in which the carrier wave is not transmitted. *See:* **modulation.**   111 carrier system (data transmission). A communication system using frequency multiplexing to a number of channels over a single path by modulating each channel upon a different carrier frequency and demodulating at the receiving point to restore the signals to their original form. 59

carrier tap choke coil. A carrier-isolating choke coil inserted in series with a line tap. 329

carrier telegraphy. The form of telegraphy in which, in order to form the transmitted signals, alternating-current is supplied to the line after being modulated under the control of the transmitting apparatus. See: telegraphy. 194

carrier telephone channel. A telephone channel employing carrier transmission. See: channel. 328

carrier-to-noise ratio (1) (information theory). (A) The ratio of the powers of the carrier and the noise after specified band limiting and before any nonlinear process such as amplitude limiting and detection. Note: This ratio is expressed in many different ways, for example, in terms of peak values in the case of impulse noise and in terms of mean-squared values for other types of noise. See: amplitude modulation.
(B) A combination of transmission media and equipment capable of accepting signals at one point and delivering related signals at another point. See: information theory. 61
(2) (data transmission). The ratio of the magnitude of the carrier to that of the noise after selection and before any nonlinear process, such as amplitude limiting and detection. The bandwidth used for measurement of the noise should be specified when using this ratio. 59

carrier transmission. That form of electric transmission in which the transmitted electric wave is a wave resulting from the modulation of a single-frequency wave by a modulating wave. See: carrier. 328

carrier velocity (semiconductor)(nonlinear, active, and nonreciprocal waveguide components). The average velocity of the random thermal motion of electrons in n-type semiconductors and of holes in p-type semiconductors. 530

carrier wave. See: carrier

carry. (1) A character or characters, produced in connection with an arithmetic operation on one digit place of two or more number representations in positional notation, and forwarded to another digit place for processing there. (2) The number represented by the character or characters in (1). (3) Usually, a signal or expression as defined in (1) which arises in adding, when the sum of two digits in the same digit place equals or exceeds the base of the number system in use. Note: If a carry into a digit place will result in a carry out of the same digit place, and if the normal adding circuit is bypassed when generating this new carry, it is called a high-speed carry, or standing-on-nines carry. If the normal adding circuit is used in such a case, the carry is called a cascaded carry. If a carry resulting from the addition of carries is not allowed to propagate (for example, when forming the partial product in one step of a multiplication process), the

process is called a partial carry. If it is allowed to propagate, the process is called a complete carry. If a carry generated in the most-significant-digit place is sent directly to the least-significant place (for example, when adding two negative numbers using nines complements) that carry is called and end-around carry. (4) A carry, in direct subtraction, is a signal or expression as defined in (1) that aries when the difference between the digits is less than zero. Such a carry is frequently called a borrow. (5) To carry is the action for forwarding a carry. (6) A carry is the command directing a carry to be forwarded. See: cascaded carry; complete carry; end-around carry; high-speed carry; partial carry; standing-on-nines carry. 235

car-switch automatic floor-stop operation (elevators). Operation in which the stop is initiated by the operator from within the car with a definite reference to the landing at which it is desired to stop, after which the slowing down and stopping of the elevator is effected automatically. See: control. 328

car-switch operation (elevators). Operation wherein the movement and direction of travel of the car are directly and solely under the control of the operator by means of a manually operated car switch of continuous-pressure buttons in the car. See: control. 328

cartridge fuse (power switchgear). A low-voltage fuse consisting of a current-responsive element inside a fuse tube with terminals on both ends. 103

cartridge size (of a cartridge fuse) (power switchgear). The range of voltage and ampere ratings assigned to a fuse cartridge with specific dimensions and shape. 103

cartridge-type bearing (rotating machinery). A complete ball or roller bearing assembly consisting of a ball or roller bearing and bearing housing that is intended to be inserted into a machine endshield. See: bearing. 63

cart, splicing. See: splicing cart.

car-wiring apparatus. See: electric train-line; train-line coupler; wire; multiple-unit control.

cascade (electrolyte cells). A series of two or more electrolytic cells or tanks so placed that electrolyte from one flows into the next lower in the series, the flow being favored by differences in elevation of the cells, producing a cascade at each point where electrolyte drops from one cell to the next. See: electrowinning. 328
See: tandem.

cascade connection (cascade). A tandem arrangement of two or more similar component devices in which the output of one is connected to the input of the next. See: tandem. 328

cascade control (1) (street lighting system). A method of turning street lights on and off in sections, each section being controlled by the energizing and de-energizing of the preceding section. See: alternating-current distribution; direct-current distribution. 64
(2) (automatic control). See: control system, cascade. 56

cascaded carry (parallel addition). A carry process in which the addition of two numerals of results in a partial-sum numeral and a carry numeral that are in turn added together, this process being repeated until no new carries are generated. *See:* **carry; high-speed carry.** 77

cascade development (electrostatography). Development in which the image-forming material is carried to the field of the electrostatic images by means of gravitational forces, usually in combination with a granular carrier. *See:* **electrostatography.** 236,191

cascaded thermoelectric device. A thermoelectric device having two or more stages arranged thermally in series. *See:* **thermoelectric device.** 248,191

cascade node (branch) (network analysis). A node (branch) not contained in a loop. 282

cascade rectifier (cascade rectifier circuit). A rectifier in which two or more similar rectifiers are connected in such a way that their direct voltages add, but their commutations do not coincide. *Note:* When two or more rectifiers operate so that their commutations coincide, they are said to be in parallel if the direct currents add, and in series if the direct voltages add. *See:* **power rectifier; rectification; rectifier circuit element.** 237,291

cascade-type voltage transformer (1) (transformer). An insulated-neutral terminal type voltage transformer with the primary distributed on two or more cores electromagnetically coupled by coupling windings. The secondary is on the core at the neutral end of the primary. Each core is insulated from the other cores and is maintained at a fixed percentage of the voltage between the primary terminal and the neutral terminal. 203

(2) (instrument transformer) (power and distribution transformer). An insulated-neutral terminal type voltage transformer with the primary winding distributed on several cores with the cores electromagnetically coupled by coupling windings. The secondary winding is on the core at the neutral end of the high-voltage winding. Each core of this type of transformer is insulated from the other cores and is maintained at a fixed voltage with respect to ground and the line-to-ground voltage. 394, 53

cascading (of switching devices) (power switchgear). The application of switching devices in which the devices nearest the source of power have interrupting ratings equal to, or in excess of, the available short-circuit current, while devices in succeeding steps further from the source, have successively lower interrupting ratings. 103

case (1) (storage battery) (storage cell). A multiple compartment container for the elements and electrolyte of two or more storage cells. Specifically wood cases are containers for cells in individual jars. *See:* **battery (primary or secondary).** 328

(2) (electrotyping). A metal plate to which is attached a layer of wax to serve as a matrix. *See:* **electroforming.** 328

(3) (semiconductor device). The housing of a semiconductor device. 66

(4) (gyro; accelerometer). The structure which provides the mounting surfaces and establishes the reference axes. 46

(5) (software). A multi-branch conditional statement that allows for selective execution of bounded groups of program statements depending upon the value of a control expression. *See:* **control structure; execution; program.** 434

case capacitance (semiconductor)(nonlinear, active, and nonreciprocal waveguide components). The fixed capacitance of an empty enclosure (neither semiconductor chip nor connecting wires or straps are present). 530

case (frame) ground protection (power switchgear). Overcurrent relay protection used to detect current flow in the ground or earth connection of the equipment or machine. 103

case shift (telegraphy). The change-over of the translating mechanism of a telegraph receiving machine from letters-case to figures-case or vice versa. *See:* **telegraphy.** 194

Cassegrainian feed (communication satellite). A feed system used for parabolic reflector antennas, where a small hyperbolic subreflector is placed near the focus of the paraboloid. The cassegrainian feed system prevents spillover to the back of the reflector, thus a better noise performance is achieved. 83

Cassegrain reflector antenna. A paraboloidal-reflector antenna with a convex subreflector, usually hyperboloidal in shape, located between the vertex and the prime focus of the main reflector. *Notes:* (1) To improve the aperture efficiency of the antenna, the shapes of the main reflector and the subreflector are sometimes modified. (2) There are other alternate forms that are referred to as Cassegrainian. Examples include the following: one in which the subreflector is surrounded by a reflecting skirt and one which utilizes a concave hyperboloidal reflector. When referring to these alternate forms the term is modified in order to differentiate them from the antenna defined in the definition. 111

casting (electrotyping). The pouring of molten electrotype metal upon tinned shells. *See:* **electroforming.** 328

catastrophic failure (reliability). *See:* **failure, catastrophic.**

catcher (electron tubes). *See:* **output resonator.**

catcher space (velocity-modulated tube). The part of the tube following the drift space, and where the density modulated-electron beam excites oscillations in the output resonator. It is the space between the output-resonator grids. *See:* **velocity-modulated tube.** 244,190

catelectrotonus (electrobiology). Electrotonus produced in the region of the cathode. *See:* **excitability, electrotonus (electrobiology).** 192

cathode (or anode) sputtering (gas). The emission of fine particles from the cathode (or anode) produced by positive ion (or electron) bombardment. *See:* **discharge (gas).** 244,190

cathode (potential) fall (gas). The difference of potential due to the space charge near the cathode. *See:* **discharge (gas).** 244,190

cathode (1) (electron tube or valve). An electrode through which a primary stream of electrons enters the interelectrode space. *See:* **electrode (electron tube) anode.** 125,190
(2) (semiconductor rectifier diode). The electrode to which the forward current flows within the cell. *See:* **semiconductor.** 237,66
(3) (electrolytic). An electrode through which current leaves any conductor of the nonmetallic class. Specifically, an electrolytic cathode is an electrode at which positive ions are discharged, or negative ions are formed, or at which other reducing reactions occur. *See:* **electrolytic cell.** 328
(4) (thyristor). The electrode by which currents leaves the thyristor when the thyristor is in the ON state with the gate open-circuited. *Note:* This term does not apply to bidirectional thyristors. 191,208

cathode border (gas) (gas tube). The distinct surface of separation between the cathode dark space and the negative glow. *See:* **discharge (gas).** 244
cathode cleaning (electroplating). Electrolytic cleaning in which the metal to be cleaned is the cathode. *See:* **battery (primary or secondary).** 328
cathode coating impedance (electron tube). The impedance excluding the cathode interface (layer) impedance, between the the base metal and emitting surface of a coated cathode. 125,190
cathode, cold. *See:* **cold cathode.**
cathode current. *See:* **electrode current (electron tube); electronic controller.**
cathode current, peak (1) (fault). The highest instantaneous value of a nonrecurrent pulse of cathode current occurring under fault conditions. *See:* **electrode current (electron tube).** 190
(2) (steady state). The maximum instantaneous value of a periodically recurring cathode current. *See:* **electrode current (electron tube).** 190
(3) (surge). The highest instantaneous value of a randomly recurring pulse of cathode current. *See:* **electrode current (electron tube).** 190
cathode dark space (Crookes dark space) (gas tube). The relatively nonluminous region in a glow-discharge cold-cathode tube between the cathode glow and the negative glow. *See:* **gas tube.** 190
cathode efficiency. The current efficiency of a specified cathodic process. *See:* **electrochemistry.**
cathode follower. A circuit in which the output load is connected in the cathode circuit of an electron tube and the input is applied between the control grid and the remote end of the cathode load, which may be at round potential. *Note:* The circuit is characterized by low output impedance, high input impedance, gain less than unity, and negative feedback. 111,211
cathode glow (gas tube). The luminous glow that covers all, or part, of the surface of the cathode in a glow-discharge cold-cathode tube, between the cathode and the cathode dark space. *See:* **gas tube.** 125,190
cathode heating time (vacuum tube). The time required

for the cathode to attain a specified condition, for example: (1) a specified value of emission or (2) a specified rate of change of emission. *Note:* All electrode voltages are to remain constant during measurement. The tube elements must all be at room temperature at the start of the test. *See:* **operation time.** 125,190
cathode interface (layer) capacitance (electron tube). A capacitance that, in parallel with a suitable resistance, forms an impedance approximating the cathode interface impedance. *Note:* Because the cathode interface impedance cannot be represented accurately by the two-element resistance-capacitance circuit, this value of capacitance is not unique. 125,190
cathode interface (layer) impedance (electron tube). An impedance between the cathode base and coating. *Note:* This impedance may be the result of a layer of high resistivity or a poor mechanical bond between the cathode base and coating. 125,190
cathode interface (layer) resistance (electron tube). The low-frequency limit of cathode interface impedance. 125,190
cathode, ionic-heated. A hot cathode that is heated primarily by ionic bombardment of the emitting surface. 190
cathode layer. A molten metal or alloy forming the cathode of an electrolytic cell and which floats on the fused electrolyte, or upon which fused electrolyte floats. *See:* **fused electrolyte.** 328
cathode luminous sensitivity (multiplier phototube). *See:* **sensitivity, cathode luminous.**
cathode modulation. Modulation produced by application of the modulating voltage to the cathode of any electron tube in which the carrier is present. *Note:* Modulation in which the cathode voltage contains externally generated pulses is called cathode pulse modulation. 328
cathode, pool. A cathode at which the principal source of electron emission is a cathode spot on a metallic pool electrode. 190
cathode, preheating time (electron tube). The minimum period of time during which the heater voltage should be applied before the application of other electrod voltages. *See:* **heater current (electron device).** 190
cathode pulse modulation. Modulation produced in an amplifier or oscillator by application of externally generated pulses to the cathode circuit. *See:* **modulating function.** 111
cathode-ray charge-storage tube. A charge-storage tube in which the information is written by means of cathode-ray beam. *Note:* Dark-trace tubes and cathode-ray tubes with a long persistence are examples of cathode-ray storage tubes that are not charge-storage tubes. Most television camera tubes are examples of charge-storage tubes that are not cathode-ray storage tubes. *See:* **charge-storage tube.** 174
cathode-ray instrument. *See:* **electron beam instrument.**
cathode-ray oscillograph. An oscillograph in which a photographic or other record is produced by means of

the electron beam of a cathode-ray tube. *Note:* The term cathode-ray oscillograph has frequently been applied to a cathode-ray oscillograph but this usage is deprecated. *See:* **oscillograph.** 328

cathode-ray oscilloscope. An oscilloscope that employs a cathode-ray tube as the indicating device. *See:* **oscillograph.** 328

cathode-ray storage tube. A storage tube in which the information is written by means of a cathode-ray beam. *See:* **storage tube.** 174

cathode-ray tube (crt)(supervisory control, data acquisition, and automatic control)(station control and data acquisition). A display device in which controlled electron beams are used to present alphanumeric or graphical data on an electroluminescent screen. 403, 570

cathode-ray-tube display area. *See:* **graticule area.**

cathode region (gas) (gas tube). The group of regions that extends from the cathode to the Faraday dark space inclusively. *See:* **discharge (gas).** 244,290

cathode spot (arc). An area on the cathode of an arc from which electron emission takes place at a current density of thousands of amperes per square centimeter and where the temperature of the electrode is too low to account for such currents by thermionic emission. *See:* **gas tube.** 290

cathode terminal (1) (semiconductor device). The terminal from which forward current flows to the external circuit. *Note:* In the semiconductor rectifier components field, the cathode terminal is normally marked positive. *See:* **semiconductor; semiconductor rectifier cell.** 237,66,208

(2) (thyristor). The terminal that is connected to the cathode. *Note:* The term does not apply to birdirectional thyristors. *See:* **anode.** 243,66,191

cathodic corrosion. An increase in corrosion of a metal by making it cathodic. *See:* **stray current corrosion.** 221,205

cathodic polarization. Polarization of a cathode. *See:* **electrochemistry.** 221,205

cathodic protection. Reduction or prevention of corrosion by making a metal the cathode in a conducting medium by means of a direct electric current (which is either impressed or galvanic). 221,205

catholyte. The portion of an electrolyte in an electrolytic cell adjacent to a cathode. If a diaphragm is present, it is the portion of electrolyte on the cathode side of the diaphragm. *See:* **electrolytic cell.** 328

cation. A positively charged ion or radical that migrates toward the cathode under the influence of a potential gradient. *See:* **ion.** 221,205

catwhisker. A small, sharp-pointed wire used to make contact with a sensitive point on the surface of a semiconductor. 328

caustic embrittlement. Stress-corrosion cracking in alkaline solutions. 221,205

caustic soda cell. A cell in which the electrolyte consists primarily of a solution of sodium hydroxide. *See:* **electrochemistry.** 328

cavitation (liquid). Formation, growth, and collapse of gaseous and vapor bubbles due to the reduction of pressure of the cavitation point below the vapor pressure of the fluid at the working temperature. 176

cavitation damage. Deterioration caused by formation and collapse of cavities in a liquid. 205

cavity. *See:* **tuned grid; unloaded applicator impedance; optical cavity.**

cavity dumpers (acousto-optic device). Generally, a fast rise time pulse modulator used intracavity. 72

cavity magnetron*. *See:* **magnetron, cavity resonator.** *Deprecated

cavity ratio (CR) (illuminating engineering). A number indicating cavity proportions.

$$CR = \frac{5 \times (\text{Height of C.}) \times (\text{C.} + \text{C. Width})}{(\text{C. Length}) \times (\text{C. Width})}$$

"For cavities of irregular shape"

$$CR = \frac{2.5 \times (\text{C. Height}) \times (\text{C. Perimeter})}{(\text{Area of C. Base})}$$

Note: The relationship between **cavity ratio** and **room coefficient** should be noted. If the entire room is considered as a cavity, the room height becomes the cavity height and $CR = 10K_r$. 167

cavity resonator (1) (antenna). A space normally bounded by an electrically conducting surface in which oscillating electromagnetic energy is stored, and whose resonant frequency is determined by the geometry of the enclosure. *See:* **waveguide.** 111

(2) (waveguide components). A resonator formed by a volume of propagating medium bounded by reflecting surfaces. *See:* **resonator, waveguide.** 166

cavity resonator frequency meter (waveguide components). A cavity resonator used to determine frequency. *See:* **cavity resonator.** 166

CAX. *See:* **unattended automatic exchange.**

C-band (radar). A radar frequency band between 4 gigahertz (GHz) and 8 GHz, usually in the International Telecommunications Union (ITU) assigned band 5.2 GHz to 5.9 GHz. 13

C battery. A battery designed or employed to furnish voltage used as a grid bias in a vacuum-tube circuit. *See:* **battery (primary or secondary).** 328

CCA (electronic navigation). *See:* **carrier-controlled approach system.**

CCITT (data transmission). (International Consultative Committee for Telephone and Telegraph) An advisory committee established under the United Nations in accordance with International Tele-Communications Convention (Geneva 1959) Article 13, to study and recommend solutions for questions on technical operation and tariffs. The organization is attempting to establish standards for intercountry operation on a worldwide basis. 59

C-display (radar). A rectangular display iin which each target appears as an intensity-modulated blip with azimuth indicated by the horizontal coordinate and angle of elevation by the vertical coordinate. 13

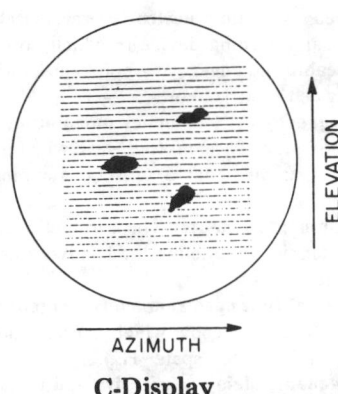

ELEVATION

AZIMUTH

C-Display

CDO. *See:* **unattended automatic exchange.**
CEI (television). The initials of the official French name, *Commission Electrotechnique Internationale,* of the International Electrotechnical Commission (IEC). 18
ceiling area lighting (illuminating engineering). A general lighting system in which the entire ceiling is, in effect, one large luminaire. *Note:* Ceiling area lighting includes luminous ceilings and louvered ceilings. *See:* **luminous ceiling; louvered ceiling.** 167
ceiling cavity ratio (CCR) (illuminating engineering). For a cavity formed by the ceiling, the plane of the luminaire, and the wall surfaces between these two planes, the CCR is computed by using the distance from the plane of the luminaire to the ceiling (h_c as the cavity height in the equations given in the *cavity ratio* definition. 167
ceiling current (excitation systems for synchronous machines). The maximum direct current which the excitation system is able to supply from its terminals for a specified time. 507
ceiling direct voltage (direct potential rectifier unit). The average direct voltage at rated direct current with rated sinusoidal voltage applied to the alternating-current line terminals, with the rectifier transformer set on rated voltage tap and with voltage regulating means set for maximum output. *See:* **power rectifier; rectification.** 291,208
ceiling projector (illuminating engineering). A device designed to produce a well-defined illuminated spot on the lower portion of a cloud for the purpose of providing a reference mark for the determination of the height of that part of the cloud. 167
ceiling ratio (illuminating engineering). The ratio of

the luminous flux which reaches the ceiling directly to the upward component of the luminaire. 167
ceiling voltage (excitation systems for synchronous machines). The maximum direct voltage which the excitation system is able to supply from its terminals under defining conditions. *Notes:* (1) The no-load ceiling voltage is determined with the excitation system supplying no current. (2) The ceiling voltage under load is determined with the excitation system supplying ceiling current. (3) For excitation systems whose supply depends on the synchronous machine voltage and (if applicable) current, the nature of power system disturbance and specific design parameters of the excitation system and the synchronous machine influence the excitation system output. For such systems, the ceiling voltage is determined considering an appropriate voltage drop and (if applicable) current increase. (4) For excitation systems employing a rotating exciter, the ceiling voltage is determined at rated speed. 507
ceiling voltage, exciter nominal. The ceiling voltage of an exciter loaded with a resistor having an ohmic value equal to the resistance of the field winding to be excited and with this field winding at a temperature of (1) 75 °C for field windings designed to operate at rating with a temperature rise of 60 °C or less; or (2) 100 ° C for field windings designed to operate at rating with a temperature rise greater than 60 °C. 105
ceilometer (navigation aid terms). An instrument for measuring the height of clouds. 526
celestial fix (navigation aid terms). A position fix established by observation of celestial bodies. 526
celestial-inertial navigation equipment (navigation aid terms). An equipment employing both celestial and inertial sensors. *Syn:* **astro-inertial navigation equipment; stellar-inertial navigation equipment.** 526
celestial mechanics (communication satellite). The mechanics of motion of celestial bodies, including satellites. 74
celestial navigation (navigation aid terms). Navigation with the aid of celestial bodies. Applied principally to the measurement of the altitudes of a celestial body. 526
cell (1)(lead-acid batteries for photovoltaic (PV) systems). The basic electrochemical unit, characterized by an anode and a cathode used to receive, store, and deliver electrical energy. For a lead-acid system, the cell is characterized by a nominal 2 volt (V) potential. 515
(2)(test pattern language). The element of a memory in which one bit is stored. 463
(3) (information storage). An elementary unit of storage, for example, binary cell, decimal cell. 235,210
(4) (National Electrical Code). A single, enclosed tubular space in a cellular metal floor member, the axis of the cell being parallel to the axis of the metal floor member. 256
cell cavity (electrolysis). The container formed by the cell lining for holding the fused electrolyte. *See:* **fused electrolyte.** 328

cell connector (storage cell). An electric conductor used for carrying current between adjacent storage cells. *See:* **battery (primary or secondary).**   328

cell constant (electrolytic cell). The resistance in ohms of that cell when filled with a liquid of unit resistivity.   328

cell cover. The transparent medium (glass, quartz, etc.) that protects the solar cells from space particulate radiation.   113

cell line (National Electrical Code). An assembly of electrically interconnected electrolytic cells supplied by a source of direct-current power.   133, 256

cell line attachments and auxiliary equipment (National Electrical Code). Include, but are not limited to: auxiliary tanks; process piping; duct work, structural supports, exposed cell line conductors; conduits and other raceways; pumps, positioning equipment and cell cutout or by-pass electrical devices. Auxiliary equipment includes tools, welding machines, crucibles, and other portable equipment used for operation and maintenance within the electrolytic cell line working zone. In the cell line working zone, auxiliary equipment includes the exposed conductive surfaces of the ungrounded cranes and crane- mounted cell-servicing equipment.   256

cell line potential (electrolytic cell line). The dc voltage applied to the positive and negative buses supplying power to a cell line.   133

cell line voltage (electrolytic cell line working zone). The dc voltage applied to the positive and negative buses supplying power to a cell line.   133

cell line working zone (electrolytic cell line working zone). The space envelope where operations or maintenance is normally performed on or in the vicinity of exposed energized surfaces of electrolytic cell lines or their attachments.   133

cell potential (electrolytic cell). The dc voltage between the positive and negative terminals of one electrolytic cell.   133

cell size (large lead storage batteries). The rated capacity of a lead storage cell or the number of plates in the cell.   377

cell type (class IE lead storage batteries). Cells of identical design, for example, plate size, alloy, construction details, but that may have differences in the number of plates and spacers, quantity of electrolyte, or length of container.   384

cell-type tube (microwave gas). A gas-filled radio-frequency switching tube that operates in an external resonant circuit. *Note:* A tuning mechanism may be incorporated in either the external resonant circuit or the tube. *See:* **gas tube.**   125

cellular metal floor raceway (National Electrical Code). The hollow spaces of cellular of cellular metal floors, together with suitable fittings, which may be approved as enclosures for electric conductors.   256

cell voltage (electrolytic cell line working zone). The dc voltage between the positive and negative terminals of one electrolytic cell.   133

cent (acoustics). The interval between two sounds whose basic frequency ratio is the twelve-hundredth root of 2. *Note:* The interval, in cents, between any two frequencies is 1200 times the logarithm to the base of 2 of the frequency ratio. Thus 1200 cents equal 12 equally tempered semitones equal 1 octave.   176,334

center-break switching device (power switchgear). A mechanical switching device in which both contacts are moveable and engage at a point substantially midway between their supports.   103

center frequency (1) (frequency modulation). The average frequency of the emitted wave when modulated by a symmetrical signal. *See:* **frequency modulation.**   111

(2) (burst measurements). The arithmetic mean of the two frequencies that define the bandwidth of a filter. *See:* **burst.**   292

(3) (non-real time spectrum analyzer) (spectrum analyzer). That frequency which corresponds to the center of a frequency span, (Hz).   390

center frequency delay (dispersive and nondispersive delay line). The frequency delay of the device at the center frequency, F_0, generally expressed in microseconds.   81

centerline (navigation aid terms). The lows of the points equidistant from two reference points or lines, as the perpendicular bisector of the baseline of a hyperbolic system of navigation, such as loran.   526

center of distribution (primary distribution). The point from which the electric energy must be supplied if the minimum weight of conducting material is to be used. *Note:* The center of distribution is commonly considered to be that fixed point that, in practice, most nearly meets the ideal conditions stated above.   64

center pivot irrigation machines (National Electrical Code). A center pivot irrigation machine is a multimotored irrigation machine which revolves around a central pivot and employs alignment switches or similar devices to control individual motors.   256

centimeter-gram-second (system of units). *See:* **cgs.**

central (foveal) vision (illuminating engineering). The seeing of objects in the central or foveal part of the visual field, approximately 2 degrees in diameter. It permits seeing much finer detail than does peripheral vision.   167

central alarm station(CAS)(nuclear security systems). A continuously manned alarm station that provides the primary security system functions.   464

central control room (nuclear power generating station). A continuously manned, protected enclosure from which actions are normally taken to operate the nuclear generating station under normal and abnormal conditions.   439

centralized accounting, automatic message (CAMA) (telephone switching systems). An arrangement at an intermediate office for collecting automatic message accounting information.   55

centralized computer network (data communication). A computer network configuration in which a central node provides computing power, control, or other services. *See:* **decentralized computer network.**   12

centralized control/alarm (electric pipe heating systems). A common (central) point where multiple control, alarm, or both signals or functions are brought together. With respect to electric pipe heating systems, centralized control/alarm stations usually consist of cabinets or panels where remote control, alarm, or both signals are brought together for a common output signal to the generating unit control room.   448

centralized test system (test, measurement and diagnostic equipment). A test system, which processes, records or displays at a central location, information gathered by test point data sensors at more than one remotely located equipment or system under test.   54

centralized traffic-control machine (railway practice). A control machine for operation of a specific type of traffic control system of signals and switches. *See:* **centralized traffic-control system.**   328

centralized traffic-control system (railway practice). A specific type of traffic control system in which the signals and switches for a designated section of track are controlled from a remotely located centralized traffic control machine. *See:* **block signal system; centralized traffic-control machine; control machine; electropneumatic interlocking machine.**   328

central office (CO)(1)(telephone loop performance). The building, one or more switching systems, and related equipment contained therein that provide telephone service.   473

(2) (data transmission). The place where communications common carriers terminate customer lines and locate the equipment which interconnects those lines. Usually the junction point between metallic pair and carrier system.   59

(3) (telephone switching systems). A switching entity that has one or more office codes and a system control serving a telecommunication exchange.   55

central office diagram (telephone switching systems). A simplified switching network plan for a given installation, specifying types and quantities of equipment and trunk groups and other parameters.   55

central office exchange (data transmission). The place where a communication common carrier locates the equipment which interconnects incoming subscribers and circuits.   59

central processing unit (computing system). The unit of a computing system that includes the circuits controlling the interpretation of instructions and their execution.   255,77

central station (protective signaling). An office to which remote alarm and supervisory signaling devices are connected, where operators supervise the circuits, and where guards are maintained continuously to investigate signals. *Note:* Facilities may be provided for transmission of alarms to police and fire departments or other outside agencies. *See:* **protective signaling.**   328

central station equipment (protective signaling). The signal receiving, recording, or retransmission equip-

ment installed in the central station. *See:* **protective signaling.**   328

central station switchboard (protective signaling). That portion of the central station equipment on or in which are mounted the essential control elements of the system. *See:* **protective signaling.**   328

central station system (protective signaling) (central office system). A system in which the operations of electric protection circuits and devices are signaled automatically to, recorded in, maintained, and supervised from a central station having trained operators and guards in attendance at all times. *See:* **protective signaling.**   328

central visual field (illuminating engineering). That region of the visual field which corresponds to the foveal portion of the retina.   167

centrex CO (company) (telephone switching systems). The provision of centrex service by switching, station equipment, and attendant facilities located on the premises of the customer.   55

centrex CU (customer) (telephone switching systems). The provision of centrex service by switching, station equipment, and attendant facilities located on the premises of the customer.   55

centrex service (telephone switching systems). A service that provides direct inward dialing and identified outward dialing in accordance with the national numbering plan for stations served as they would be by a private branch exchange.   55

centrifugal actuator. *See:* **actuator, centrifugal.**

centrifugal-mechanism pin (governor pin). A component of the linkage between the centrifugal mechanism weights and the short-circuiting device. *See:* **centrifugal starting switch.**   328

centrifugal-mechanism spring (governor spring). A spring that opposes the centrifugal action of the centrifugal-mechanism weights in determining the motor speed at which the switch or short-circuiting device is actuated. *See:* **centrifugal starting switch.**   328

centrifugal-mechanism weights (governor weights). Moving parts of the centrifugal-mechanism assembly that are acted upon by centrifugal force. *See:* **centrifugal staring switch.**   328

centrifugal relay. An alternating-current frequency-selective relay in which the contacts are operated by a fly-ball governor or centrifuge driven by an induction motor.   328

centrifugal starting-switch (rotating machinery). A centrifugally operated automatic mechanism used to perform a circuit-changing function in the primary winding of a single-phase induction motor after the rotor has attained a predetermined speed, and to perform the reverse circuit-changing operation prior to the time the rotor comes to rest. *Notes:* (1) One of the circuit changes that is usually performed is to open or disconnect the auxiliary winding circuit. (2) In the usual form of this device, the part that is mounted to the stator frame or end shield is the starting switch, and the part that is mounted on the rotor is the centrifugal actuator.   63

CEP. *See:* **circular probable error.**

certificate of conformance (replacement parts for Class 1E equipment in nuclear power generating stations)(nuclear power quality assurance). A document signed by an authorized individual certifying the degree to which items or services meet specified requirements. 582,417

certification (1) (nuclear power quality assurance). The act of determining, verifying, or attesting in writing to the qualifications of personnel, processes, procedures, or items in accordance with specified requirements. 417

(2) (software). (A) A written guarantee that a system or computer program complies with its specified requirements. (B) A written authorization which states that a computer system is secure and is permitted to operate in a defined environment with or producing sensitive information. (C) The formal demonstration of system acceptability to obtain authorization for its operational use. (D) The process of confirming that a system, software subsystem, or computer program is capable of satisfying its specified requirements in an operational environment. Certification usually takes place in the field under actual conditions, and is utilized to evaluate not only the software itself, but also the specifications to which the software was constructed. Certification extends the process of verification and validation to an actual or simulated operational environment. (E) The procedure and action by a duly authorized body of determining, verifying, and attesting in writing to the qualifications of personnel, processes, procedures, or items in accordance with applicable requirements. See: computer program; computer system; operational; procedures; processes; requirements; software subsystem; specifications; system; validation; verification. 434

(3) (test, measurement and diagnostic equipment). Attestation that a support test system is capable, at the time of certification demonstration, of correctly assessing the quality of the items to be tested. This attestation is based on an evaluation of all support test system elements and establishment of acceptable correlation among similar test systems. 54

certification tests (1)(metal-oxide surge arresters for ac power circuits). Tests run on a regular periodic basis to verify that selected, key performance characteristics of a product or representative samples thereof have remained within performance specifications. 583

(2)(surge arrester). Tests made, when required, to verify selected performance characteristics of a product or representative samples thereof. 430

certified design (station control and data acquisition). A test performed on a production model specimen of a generic type of equipment to establish a specific performance parameter of that genre of equipment. The condition and results of the test are described in a document that is signed and attested to by the testing engineer and other appropriate, responsible individuals. 403

certified Marinelli beaker standard source (MBSS) (germanium semiconductor detector). A certified MBSS is an MBSS that has been calibrated as to photon emission rate at specified energies by a laboratory recognized as a country's National Standardizing Laboratory for radioactivity measurements and has been so certified by the calibrating laboratory. Notes: (1) The photon emission rate as used in this standard is the number of photons per second resulting from the decay of radionuclides in the source and is thus higher than the detected rate at the surface. (2) For the United States, the US National Bureau of Standards is the National Standardizing Laboratory. 388

certified radioactivity standard source (germanium detectors). A calibrated radioactive source, with stated accuracy, whose calibration is certified by the source supplier as traceable to the National Radioactivity Measurements System. 397

certified solution (germanium semiconductor detector). A certified solution is a liquid radioactive filling material that has been calibrated by a laboratory (For the United States, the US National Bureau of Standards.) recognized as a country's National Standardizing Laboratory for radioactivity measurements and has been so certified by the calibrating laboratory. 388

certified-solution Marinelli beaker standard source (MBSS) (germanium semiconductor detector). A certified-solution MBSS is a standard beaker that contains a certified solution as its radioactive filling material. See: Marinelli beaker; certified solution. 388

certified unit (test, measurement and diagnostic equipment). A unit whose demonstrated ability to perform in accordance with preestablished criteria has been attested. 54

CFAR. See: constant false alarm rate. 13

cgs (centimeter-gram-second) electromagnetic system of units. A system in which the basic units are the centimeter, gram, second, and abampere. Notes: (1) The abampere is a derived unit defined by assigning the magnitude 1 to the unrationalized magnetic constant (sometimes called the permeability of space). (2) Most electrical units of this system are designated by prefixing the syllable "ab-" to the name of the corresponding unit in the mksa system. Exceptions are the maxwell, gauss, oersted, and gilbert. 210

cgs electrostatic system of units. The system in which the basic units are the centimeter, gram, second, and statcoulomb. Notes: (1) The statcoulomb is a derived unit defined by assigning the magnitude 1 to the unrationalized electric constant (sometimes called the permittivity of space). (2) Each electrical unit of this system is commonly designated by prefixing the syllable "stat-" to the name of the corresponding unit in the International System of Units. 210

cgs system of units. A system in which the basic units are the centimeter, gram, and second. 210

chad. The piece of material removed when forming a hole or notch in a storage medium such as punched tape or punched cards. 255,77,54

chadded. Pertaining to the punching of tape in which chad results. 255,77

chadless. Pertaining to the punching of tape in which chad does not result.   255,77,54

chadless tape. A punched tape wherein only partial perforation is completed and the chad remains attached to the tape.   207

chaff (radar). An airborne cloud of lightweight reflecting objects typically consisting of strips of aluminum foil or metal-coated fibers which produce clutter echoes in a region of space.   13

chafing strip. *See:* **drive strip.**

chain (navigation aid terms). A network of similar stations operating as a group for determination of position or for furnishing navigational information.   526

chain code (computing system). An arrangement in a cyclic sequence of some or all of the possible different *n*-bit words, in which adjacent words are related such that each word is derivable from its neighbor by displacing the bits one digit position to the left, or right, dropping the leading bit, and inserting a bit at the end. The value of the inserted bit needs only to meet the requirement that a word must not recur before the cycle is complete, for example, 000 001 010 101 011 111 110 100 000 ...   255,77

chain-drive machine (elevators). An indirect-drive machine having a chain as the connecting means. *See:* **driving machine (elevators).**   328

chained list (software). A list in which the items may be dispersed but in which each item contains an identifier for locating the next item. *Syn:* **linked list.** *See:* **indentifier; list.**   434

chain matrix (circuits and systems). The 2 X 2 matrix relating voltage and current at one port of a two port network to voltage and current at the other port, an indicated below:

Syn: **transmission or cascade matrix.**   67

chair, boatswain. *See:* **boatswain's chair.**

chalking (1)(composite insulators). The powdered surface on weathersheds consisting of particles of filler resulting from ultraviolet exposure.   483

(2)(corrosion). The development of loose removable powder at or just beneath a coating surface.   205

challenge (navigation aid terms). To cause an interrogator to transmit a signal which puts a transponder into operation.*See:* **interrogation.**   526

change control (software). The process by which a change is proposed, evaluated, approved or rejected, scheduled, and tracked.   434

changeover switch. A switching device for changing electric circuits from one combination to another. *Note:* It is usual to qualify the term changeover switch by stating the purpose for which it is used, such as a series-parallel changeover switch, trolley-shoe changeover switch, etcetera. *See:* **multiple-unit control.**   328

CHANHI. Abbreviation for upper channel corresponding to the half-amplitude point of a distribution.   117

CHANLO. Abbreviation for lower channel corresponding to the half-amplitude point of a distribution.   117

channel (1) (electric communication). (A) A single path for transmitting electric signals, usually in distinction from other parallel paths. (B) A band of frequencies. *Note:* The word path is to be interpreted in a broad sense to include separation by frequency division or time division. The term channel may signify either a one-way path, providing transmission in one direction only, or a two-way path, providing transmission in two directions.   328

(2) (electronic computers). (A) A path along which signals can be sent, for example, data channel, output channel. (B) The portion of a storage medium that is accessible to a given reading station. *See:* **track.**   235

(3) (information theory). A combination of transmission media and equipment capable of receiving signals at one point and delivering related signals at another point. *See:* **information theory.**   160

(4) (illuminating engineering). An enclosure containing the ballast, starter, lamp holders and wiring for a fluorescent lamp, or a similar enclosure on which filament lamps (usually tubular) are mounted.   167

(5) (nuclear power generating station). An arrangement of components and modules as required to generate a single protective action signal when required by a generating station condition. A channel loses its identity where single protective action signals are combined.   428, 387

(6) (metal-nitride-oxide field-effect transistor). A surface layer of carriers connecting source and drain in an insulated-gate field-effect transistor (IGFET). This channel was formed by inversion with the help of a gate voltage, or by the presence of charges in the gate insulator, or by deliberate doping of the region.   386

(7) (data transmission). In electric communication, (A) a single path for transmitting electric signals, usually distinct from other parallel paths, or (B) a band of frequencies. *Note:* The word "path" is to be interpreted in a broad sense to include separation by frequency division or time division. The term "channel" may signify either a one-way path, providing transmission in direction only, or a two-way path, providing transmission in two directions.   59

channel-busy tone (telephone switching system). A tone that indicates that a server other than a destination outlet is either busy or not accessible.   55

channel capacity (data transmission). The maximum possible information rate through a channel subject to the constraints of that channel. *Note:* Channel capacity may be either per second or per symbol.   59

channel failure alarm (power-system communication). A circuit to give an alarm if a communication channel should fail. *See:* **power-line carrier.**   59

channel group (group) (data transmission). A number

of channels regarded as a unit. *Note:* The term is especially used to designate part of a larger number of channels.   59

channeling, lattice (semiconductor radiation detector). A phenomenon that results in a crystallographic directional dependence of the rate of energy loss of ionizing particles.   118,119

channel lights (illuminating engineering). Aeronautical ground lights arranged along the sides of a channel of a water aerodrome.   167

channel, melting. *See:* **melting channel.**

channel multiplier. A tubular electron-multiplier with a continuous interior surface of secondary-electron emissive material. *See:* **amplifier; camera tube.**
                                                     190

channel, radio. *See:* **radio channel.**

channel spacing (radio communication). The frequency increment between the assigned frequency of two adjacent radio-frequency channels. *See:* **dispatch operation; radio transmission; single-frequency simplex operation; two-frequency simplex operation.**   181

channel supergroup (supergroup) (data transmission). A number of channel groups regarded as a unit. *Note:* The term is especially used to designate part of a larger number of channels.   59

channel, surface (semiconductor radiation detector). A thin region at a semiconductor surface of p- or n-type conductivity created by the action of an electric field; for example, that due to trapped surface charge.   118,119

channel utilization index (information theory). The ratio of the information rate (per second) through a channel to the channel capacity (per second). *See:* **information theory.**   160

character (1) (electronic computers). (A) An elementary mark or event that may be combined with others, usually in the form of a linear string, to form data or represent information. If necessary to distinguish from (B) below, such a mark may be called a character event. (B) A class of equivalent elementary marks or events as in (A) having properties in common, such as shape or amplitude. If necessary to distinguish from (A) above, such a class may be called a character design. There are usually only a finite set of character designs in a given language. *Notes:* (1) In "bookkeeper" there are six character designs and ten character events, while in "1010010" there are two character designs and seven character events. (2) A group of characters, in one context, may be considered as a single character in another, as in the binary-coeded-decimal system. *See:* **blank character; check character; control character; escape character; numerical control; special character.**   235,54

(2) (data transmission). One of a set of elementary symbols which normally include both alpha and numeric codes plus punctuation marks and any other symbol which may be read, stored, or written and is used for organization, control, or representation of data.   59

character density (test, measurement and diagnostic equipment). The number of characters that can be stored per unit area or length.   54

character distortion (data transmission). The normal and predictable distortion of data bit produced by characteristics of a given circuit at a particular transmission speed.   59

character form (microprocessor object modules). The (printable character representation of binary information as opposed to bit pattern information.   466

character-indicator tube (electron device). A blow-discharge tube in which the cathode glow displays the shape of a character, for example, letter, number, or symbol.   190

character interval (data transmission). In start-stop operation the duration of a character expressed as the total number of unit intervals (including information, error checking and control bits, and the start and stop elements) required to transmit any given character in any given communication system.   59

characteristic (1)(mathematics of computing). (A) The integer part of a logarithm. Contrast with 'mantissa (1)'. (B) For floating point arithmetic, see 'exponent (2)'.   564

See: **data characteristic; software characteristic.**

(2) (nuclear power quality assurance). Any property or attribute of an item, process, or service that is distinct, describable, or measurable.   417

(3) (semiconductor device). An inherent and measurable property of a device. Such a property may be electrical, mechanical, thermal, hydraulic, electromagnetic or nuclear and can be expressed as a value for stated or recognized conditions. A characteristic may also be a set of related values, usually shown in graphical form.   66

characteristic curve (1) (illuminating engineering). A curve which expresses the relationship between two variable properties of a light source, such as candle-power and voltage, flux and voltage, etcetera.
                                                      167

(2) (Hall generator). A plot of Hall output voltage versus control current, magnetic flux density, or the product of magnetic flux density and control current.
                                                      107

characteristic curves (rotating machinery). The graphical representation of the relationships between certain quantities used in the study of electric machines. *See:* **asynchronous machine.**   63

characteristic distortion (telegraphy). A displacement of signal transitions resulting from the persistence of transients caused by preceding transitions. *See:* **telegraphy.**   194

characteristic element (surge arresters). The element that in a valve-type arrester determines the discharge voltage and the follow current. *See:* **surge arrester (surge diverter).**   62

characteristic equation (feedback control system). The relation formed by equating to zero the denominator of a transfer function of a closed loop. *See:* **control system feedback.**   329

characteristic harmonic (1)(converter characteristics)(self-commutated converters). Those harmonics produced by semiconductor converter equipment in the course of normal operation. In a six-pulse converter, the characteristic harmonics are the nontriplen odd harmonics, for example, the 5th, 7th, 11th, 13th, etcetera.   584

(2)(harmonic control and reactive compensation of static power converters). Those harmonics produced by semiconductor converter equipment in the course of normal operation. In a six pulse converter, the characteristic harmonics are the nontriple odd harmonics, for example, the 5th, 7th, 11th, 13th, etc.

$h = kq \pm 1$

k = any integer

q = pulse number of converter   533

characteristic impedance (1) (data transmission). (A) (Two-conductor transmission line for a traveling transverse electromagnetic wave). The ratio of the complex voltage between the conductors to the complex current on the conductors in the same transverse plane with the sign so chosen that the real part is positive. (B) (Coaxial transmission line). The driving impedance of the forward traveling transverse electromagnetic wave.   59

(2) (circular waveguide). For a traveling wave in the dominant ($TE_{1,1}$) mode of a lossless circular waveguide at a specified frequency above the cutoff frequency, (A) the ratio of the square of the root-mean-square voltage along the diameter where the electric vector is a maximum to the total power flowing when the guide is match terminated, (B) the ratio of the total power flowing to the square of the total root-mean-square longitudinal current flowing in one direction when the guide is match terminated, (C) the ratio of the root-mean-square voltage along the diameter where the electric vector is a maximum to the total root-mean-square longitudinal current flowing along the half surface bisected by the diameter when the guide is match terminated. *Note:* Under definition (A) the power $W = V^2/Z_{(W,V)}$ where V is the voltage and $Z_{(W,V)}$ is the characteristic impedance defined in (A). Under definition (B) the power $W = I^2 Z_{(W,I)}$ where I is the current and $Z_{(W,I)}$ is the characteristic impedance defined in (B). The characteristic impedance $Z_{(V,I)}$ as defined in (C) is the geometric mean of the values given by (A) and (B). Definition (C) can be used also below the cutoff frequency. *See:* **self-impedance; waveguide.**

(3) (rectangular waveguide). For a traveling wave in the dominant ($TE_{1,0}$) mode of a lossless rectangular waveguide at a specified frequency above the cutoff frequency. (A) the ratio of the square of the root-mean-square voltage between midpoints of the two conductor's faces normal to the electric vector, to the total power flowing when the guide is match terminated, (B) the ratio of the total power flowing to the square of the root-mean-square longitudinal current, flowing on one face normal to the electric vector when the guide is match terminated, (C) the ratio of the root-mean-square voltage, between midpoints of the two conductor faces normal to the electric vector, to the total root-mean-square longitudinal current, flowing on one face when the guide is match terminated. *Note:* Under definition (A) the power $W = V^2/Z_{(W,V)}$

where V is the voltage, and $Z_{(W,V)}$ the characteristic impedance defined in (A). Under definition (B) the power $W = I^2 Z_{(W,I)}$ where I is the current and $Z_{(W,I)}$ the characteristic impedance defined in (B). The characteristic impedance $Z_{(V,I)}$ as defined in (C) is the geometric mean of the values given by (A) and (B). Definition (C) can be used also below the cutoff frequency. *See:* **waveguide; self-impedance.**

(4) (two-conductor transmission line) (for a traveling transverse electromagnetic wave). The ratio of the complex voltage between the conductors to the complex current on the conductors in the same transverse plane with the sign so chosen that the real part is positive. *See:* **self-impedance; transmission line; waveguide.**

(5) (coaxial transmission line). The driving impedance of the forward-traveling transverse electromagnetic wave. *See:* **self-impedance; transmission line.**   267

(6) (surge impedance) (surge arrester). The driving-point impedance that the line would have if it were of infinite length. *Note:* It is recommended that this term be applied only to lines having approximate electric uniformity. For other lines or structures the corresponding term is iterative impedance. *See:* **self-impedance.**   62

(7)(overhead-power-line corona and radio noise). The ratio of the complex voltage of a propagation mode (see 4.5) to the complex current of the same propagation mode in the same transverse plane with the sign so chosen that the real part is positive. *Note:* The characteristic impedance of a line with losses neglected is known as the surge impedance. *See:* **propagation mode.**   411

characteristic insertion loss (1) (waveguide and transmission line). The insertion loss in a transmission system that is reflectionless looking toward both the source and the load from the inserted transducer. *Notes:* (A) This loss is a unique property of the inserted transducer. (B) The frequency, internal impedance, and available power of the source and the impedance of the load have the same value before and after the transducer is inserted. *See:* **waveguide.**   185

(2) (fixed and variable attenuators).   96

P_{INPUT} = Incident power from Z_0 source
P_{OUTPUT} = Net power into Z_0 load

Characteristic insertion loss = $10 \log_{10} \dfrac{P_{INPUT}}{P_{OUTPUT}}$ (dB)

    110

characteristic insertion loss, incremental. The change in the characteristic insertion loss of an adjustable device between two settings. *See:* **waveguide.**

185

characteristic insertion loss, residual. The characteristic insertion loss of an adjustable device at an indicated minimum position. *See:* **waveguide.** 185

characteristic insertion phase shift (waveguide and transmission line)(network analyzers). The phase shift occurring upon insertion of a device in a transmission system that is reflectionless looking toward both the source and the load from the insertion plane. *Notes:* (1) The frequency, incident power from the source port, and impedance of the load port are the same before and after the device is inserted. (2) The connectors of source and load ports mate directly. (3) The device can be inserted and its connectors can mate directly with the connectors of the source and load ports. 505

characteristic phase shift. For a 2-port device inserted into a stable, nonreflecting system between the generator and its load, the magnitude of the phase change of the voltage wave incident upon the load before and after insertion of the device, or change of the device from initial to final condition. *Note:* The following conditions apply: (1) The frequency, the load impedance, and the generator characteristics, internal impedance and available power, initially have the same values as after the device is inserted; (2) the joining devices, connectors or adapters belonging to the system conform to some set of standard specifications, the same specifications to be used by different laboratories, if measurements are to agree precisely; (3) the nonreflecting conditions are to be obtained in uniform, standard sections of waveguide on the system sides of the connectors at the place of insertion. *See:* **measurement system.** 183

characteristics related to the voltage collapse during chopping (high voltage testing) (chopped impulses). The characteristics of the voltage collapse during chopping are defined in terms of two points C and D at 70 and 10 percent of the voltage at the instant of chopping The virtual duration of the voltage collapse is 1.67 times the time interval between points C and D. The virtual steepness of the voltage collapse is the ratio of the voltage at the instant of chopping to the virtual duration of voltage collapse. *Note:* The use of points C and D is for definition purposes only; it is not implied that the duration and steepness of chopping can be measured with any degree of accuracy using conventional measuring circuits. 150

characteristic telegraph distortion. Distortion that does not affect all signal pulses alike, the effect on each transistion depending upon the signal previously sent, due to remnants of previous transitions or transients that persist for one or more pulse lengths. *Note:* Lengthening of the mark pulse is positive, and shortening, negative. Characteristic distortion is measured by transmitting biased reversals, square waves having unequal mark and space pulses. The average lengthening or shortening of mark pulses, expressed in percent

of unit pulse length, gives a true measure of characteristic distortion only if other types of distortion are negligible. *See:* **modulation.** 111

chacteristic time (gyro, accelerometer) (inertial sensor). The time required for the output to reach 63 percent of its final value for a step input. *Note:* For a single degree of freedom rate integrating gyro, it is numerically equal to the ratio of the float moment of inertia to the damping coefficient about the output axis. For certain fluid-filled sensors, the float moment of inertia may include other effects, such as that of transported fluid. 46

characteristic wave impedance. The wave impedance of a traveling wave, with the sign so chosen that the real part is positive. *Note:* In a given mode, in a homogeneously-filled waveguide, this is constant for all points and all cross-sections. 267

character recognition (computer applications). The use of pattern recognition techniques to identify characters by automatic means. *See:* **magnetic ink character recognition; omni-font character recognition; optical character recognition; single-font character recognition.** 571

charge (1)(power operations). The amount paid for a service rendered or facilities used or made available for use. *See:* **connection charge; customer charge; demand charge; energy charge; facilities charge; terminations charge; wheeling charge.** 516

(2) (storage battery) (storage cell). The conversion of electric energy into chemical energy within the cell or battery. *Note:* This restoration of the active materials is accomplished by maintaining a unidirectional current in the cell or battery in the opposite direction to that during discharge; a cell or battery that is said to be charged is understood to be fully charged.

328

(3) (induction and dielectric-heating usage). *See:* **load.**

(4) (electric power supply). The amount paid for a service rendered or facilities used or made available for use. 112

charge, apparent (dielectric test). That charge of a partial discharge which, if injected instantaneously between the terminals of the test object, would momentarily change the voltage between its terminals by the same amount as the partial discharge itself. The apparent charge should not be confused with the charge transferred across the discharging cavity in the the dielectric. Apparent charge is expressed in coulombs. *Syn:* **terminal charge.** 139

charge carrier (of a semiconductor)(charged-particle detectors)(X-ray energy spectrometers)(germanium gamma-ray detectors). A mobile conduction electron or mobile hole. 23, 119, 147, 245, 528

charge collection time (of a semiconductor)(charged-particle detectors)(germanium gamma-ray detectors)(X-ray energy spectrometers). The time interval, after the passage of an ionizing particle, for the integrated current flowing between the terminals of the detector to increase from 10 percent to 90 percent of its final value. *Syn:* charge sweep-out time.

23, 471, 119, 528, 147

charge, connection (electric power supply). The amount paid by a customer for connecting the customer's facilities to the suppliers facilities.   112

charge, customer (electric power utilization). The amount paid periodically by a customer without regard to demand or energy consumption.   112

charge-delay interval (telephone switching systems). The recognition time for a valid answer signal in message charging.   55

charge, demand (electric power utilization). That portion of the charge for electric service based upon a customer's demand.   112

charge, energy (electric power utilization). That portion of the charge for electric service based upon the electric energy consumed or billed.   112

charge, facilities (electric power utilization). The amount paid by the customer as lump sum, or periodically, as reimbursement for facilities furnished. The charge may include operation and maintenance as well as fixed cost.   112

charge-resistance furnace. A resistance furnace in which the heat is developed within the charge acting as the resistor. *See:* **electrothermics.**

charge-sensitive preamplifier (germanium gamma-ray detectors). An amplifier, preceding the main amplifier, in which the output signal amplitude is proportional to the charge injected at the input. *See:* **voltage-sensitive preamplifier.**   528

charge,space. *See:* **space charge.**

charge storage (semiconductor)(nonlinear, active, and nonreciprocal waveguide components). An electrical property of step recovery, dual mode, and p-i-n diodes. As the diode is driven into forward conduction by the first half-cycle of the incident signal, it stores a charge and appears as a low impedance. As the polarity of the incident signal reverses, the charge is extracted, and the diode remains in its low-impedance state until virtually all of the charge is removed, whereupon the diode rapidly switches to a high-impedance state.   530

charge-storage tube (electrostatic memory tube). A storage tube in which the information is retained on the storage surface in the form of a pattern of electric charges.   174

charge sweep-out time (of a semiconductor radiation detector). *See:* **charge collection time.**   119

charge, termination (electric power utilization). The amount paid by a customer when service is terminated at the customer's request.   112

charge-to-third-number call (telephone switching systems). A call for which the charges are billed to a number other than that of the calling or called number.   55

charge transit time. *See:* **transit time.**

charge, wheeling (electric power supply). The amount paid to an intervening system for the use of its transmission facilities.   112

charging (electrostatography). *See:* **sensitizing; electrostatography.**

charging circuit (surge generator) (surge arresters). The portion of the surge generator connections through which electric energy is stored up prior to the production of a surge. *See:* **surge arrester (surge diverter).**   62, 64

charging current (transmission line). The current that flows in the capacitance of a transmission line when voltage is applied at its terminals. *See:* **transmission line.**   64

charging inductor. An inductive component used in the charging circuit of a pulse-forming network.   137

charging rack (mining). A device used for holding batteries for mining lamps and for connecting them to a power supply while the batteries are being recharged. *See:* **mine feeder circuit.**   328

charging rate (storage battery) (storage cell). The current expressed in amperes at which a battery is charged. *See:* **charge.**   328

charles or kino gun. *See:* **end injection.**

chart (1)(navigation aid terms). A map intended primarily for navigation use.   526

(2)(recording instrument). The paper or other material upon which the graphic record is made. *See:* **moving element (instrument).**   328

chart-comparison unit (navigation aid terms). A device for the simultaneous viewing of a navigational chart in such a manner that one appears superimposed upon the other.   526

chart mechanism (recording instrument). The parts necessary to carry the chart. *See:* **moving element (instrument).**   294

chart scale (recording instrument). The scale of the quantity being recorded, as marked on the chart. *Note:* Independent of and generally in quadrature with the chart scale is the time scale which is graduated and marked to correspond to the principal rate at which the chart is advanced in making the recording. This quadrature scale may also be used for quantities other than time. *See:* **moving element (instrument).**   328

chart scale length (recording instrument). The shortest distance between the two ends of the chart scale. *See:* **instrument.**   328

chassis (frame connection: equivalent chassis connection) (printed-wiring boards). A conducting connection to a chassis or frame, or equivalent chassis connection of a printed-wiring board. The chassis or frame (or equivalent chassis connection of a printed-wiring board) may be at substantial potential with respect to the earth or structure in which this chassis or frame (or printed-wiring board) is mounted.   25

chatter, relay. *See:* **relay chatter time; relay contact chatter.**

check (1)(monitoring radioactivity in effluents). The use of a source to determine if the detector and all electronic components of the system are operating correctly.   559

(2)(radiological monitoring instrumentation). To determine if the detector and all electronic components of a system are operating satisfactorily by determining consistent response to the same source.   398

(3) (standardize) (instrument or meter). Ascertain

the error of its indication, recorded value, or registration. *Note:* The use of the word **standardize** in place of **adjust** to designate the operation of adjusting the current in the potentiometer circuit to balance the standard cell is deprecated. *See:* **test (instrument or meter).** 328

(4) (computer-controlled machines). A process of partial or complete testing of (A) the correctness of machine operations, or (B) the existence of certain prescribed conditions within the computer. A check of any of these conditions may be made automatically by the equipment or may be programmed. 235, 210

(5) (nuclear power generating stations). The use of a source to determine if the detector and all electronic components of the system are operating correctly. 31

check back (power switchgear). The retransmission from the receiving end to the initiating end of a coded signal or message to verify, at the initiating end, the initial message before proceeding with the transmitting of data or a command. 103

checkback message (1)(station control and data acquisition). The response from the receiving end to the initiating end of a coded signal or message. (A) Partial. Message from the initiating end is mirrored by the receiving end back to the initiating end to verify error-free transmission of the message. (B) Complete. Message from the initiating end is interpreted by the receiving end. A new message is sent to the initiating end to verify both error-free transmission and proper interpretation of the message. *See:* **system security.** 403

(2)(supervisory control, data acquisition, and automatic control). The response from the receiving end to the initiating end of a coded signal or message. (A) Partial. Message from the initiating end is mirrored by the receiving end back to the initiating end to verify error-free transmission of the message. (B) Complete. Message from the initiating end is interpreted by the receiving end. 570

check before operate (checkback) (data transmission). A message and control technique providing for confirmation of control request before operation. 59

check bit. A binary check digit, for example, a parity bit. 77

check bits (data transmission). Associated with a code character or block for the purpose of checking the absence of error within the code character or block. *See:* **data processing.** 194

check character. A character used for the purpose of performing a check, but often otherwise redundant. 77

check digit. A digit used for the purpose of performing a check, but often otherwise redundant. *See:* **check, forbidden-combination.** 77

checking or interlocking relay (3) (power system device function numbers). A relay that operates in response to the position of a number of other devices (or to a number of predetermined conditions) in an equipment, to allow an operating sequence to proceed, or to stop, or to provide a check of the position of these devices or of these conditions for any purpose. 402

checkout (test, measurement and diagnostic equipment). A sequence of tests for determining whether or not a device or system is capable of, or is actually performing, a required operation or function. 54

checkout equipment (test, measurement and diagnostic equipment). Electric, electronic, optical, mechanical, hydraulic, or pneumatic equipment, either automatic, semiautomatic, or any combination thereof, which is required to perform the checkout function. 54

checkout time (test, measurement and diagnostic equipment). Time required to determine whether designated characteristics of a system are within specified values. 54

checkpoint (1) (electronic computation). A place in a routine where a check, or a recording of data for restart purposes, is performed.

(2) (electronic navigation). *See:* **way point.** 255, 77, 13

check problem (electronic computation). A routine or problem that is designed primarily to indicate whether a fault exists in the computer, without giving detailed information on the location of the fault. Also called check routine. *See:* **diagnostic; test; programmed check.** 235

check, programmed (electronic computation). *See:* **programmed check.**

check, redundant (checking code) (data transmission). A check which uses extra digits (check bits) short of complete duplication, to help detect the absence of error within the character or block. 59

check routine. Same as check problem. 54

check, selection (electronic computation). *See:* **selection check.**

check solution (analog computers). A solution to a problem obtained by independent means to verify a computer solution. 9

check source (germanium detectors) (liquid-scintillation counting) (radiation protection) (radiological monitoring instrumentation) (sodium iodide detector). A radioactive source, not necessarily calibrated, which is used to confirm the continuing satisfactory operation of an instrument.

Four types of check sources which are of the vial type may be used:

(1) Flame-sealed glass (activity known)
(2) Flame-sealed glass (activity unknown)
(3) Screw-capped glass or plastic (activity known)
(4) Screw-capped glass or plastic (activity unknown)

Check sources of the type (1) can be used for *all* measurements described in this standard. Such sources are available from instrument manufacturers of radiochemicals. They are often designated as *unquenched standards.* 397, 422, 399, 398, 423

check sum (mathematics of computing). A sum obtained by adding the digits in a numeral, or group of numerals, usually without regard to meaning, position, or significance. This sum may be compared with a previously computed value to verify that no errors have occurred.   564

checksum (microprocessor object modules). A deterministic function of a file's contents. If a file is copied and the checksum of the copy is different from the original, there has been an error in copying.   466

check summation (checksum) (data transmission). A redundant check in which groups of digits are summed usually without regard for overflow, and that sum checked against a previously computed sum to verify accuracy.   59

check, transfer (electronic computation). *See:* **transfer check.**

check valve. *See:* **blocking capacitor.**

cheek, field-coil flange (washer). *See:* **collar.**

cheese antenna. A reflector antenna having a cylindrical reflector enclosed by two parallel conducting plates perpendicular to the cylinder, spaced more than one wavelength apart. *Syn:* **pillbox antenna.**   111

chemical conversion coating. A protective or decorative coating produced in situ by chemical reaction of a metal with a chosen environment.   205

chemical vapor deposition (CVD) technique (fiber optics). A process in which deposits are produced by heterogeneous gas-solid and gas-liquid chemical reactions at the surface of a substrate. *Note:* The CVD method is often used in fabricating optical waveguide preforms by causing gaseous materials to react and deposit glass oxides. Typical starting chemicals include volatile compounds of silicon, germanium, phosphorus, and boron, which form corresponding oxides after heating with oxygen or other gases. Depending upon its type, the preform may be processed further in preparation for pulling into an optical fiber. *See:* **preform.**   433

chief programmer (software). The leader of a chief programmer team; a senior-level programmer whose responsibilities include producing key portions of the software assigned to the team, coordinating the activities of the team, reviewing the work of the other team members, and having an overall technical understanding of the software being developed. *See:* **chief programmer team; software.**   434

chief programmer team (software). A software development group that consists of a chief programmer, a backup programmer, a secretary/librarian, and additional programmers and specialists as needed and employs support procedures designed to enhance group communication and make optimum use of each member's skills. *See:* **backup programmer; chief programmer; procedures; secretary/librarian.**   434

Child-Langmuir equation (thermionics). An equation representing the cathode current of a thermionic diode in a space-charge-limited-current state.

$$I = GV^{3/2}$$

where I is the cathode current, V is the anode voltage of a diode or the equivalent diode of a triode or of a

multi-electrode value or tube, and G is a constant (perveance) depending on the geometry of the diode or equivalent diode. *See:* **electron emission.**
    244, 190

chip (1)(mechanical recording). The material removed from the recording medium by the recording stylus while cutting the groove. *See:* **phonograph pickup.**
    176

(2)(semiconductor)(nonlinear, active, and nonreciprocal waveguide components). A small unpackaged functional element made by subdividing a wafer of semiconductor material. Sometimes referred to as a die .   530

chip enable (E) (semiconductor memory). The inputs that when true permit input, internal transfer, manipulation, refreshing, and output of data, and when false cause the memory to be in a reduced power standby mode. *Note:* Chip enable is a clock or strobe that significantly affects the power dissipation of the memory. Chip select is a logical function that gates the inputs and outputs. For example, chip enable may be the cycle control of a dynamic memory or a power reduction input on a static memory.   441

chip select (S) (semiconductor memory). The inputs that when false prohibit writing into the memory and disable the output of the memory. *Note:* Chip enable is a clock or strobe that significantly affects the power dissipation of the memory. Chip select is a logical function that gates the inputs and outputs. For example, chip enable may be the cycle control of a dynamic memory or a power reduction input on a static memory.   441

chirp (radar). A technique for pulse compression which uses frequency modulation (usually linear) during the pulse.   13

chirping (fiber optics). A rapid change (as opposed to long-term drift) of the emission wavelength of an optical source. Chirping is often observed in pulsed operation of a source.   433

choice (telephone switching systems). The position of an outlet in a group with respect to the order of selection.   55

choke (waveguide). A device for preventing energy within a waveguide in a given frequency range from taking an undesired path. *See:* **waveguide.**   179

choke coil. An inductor used in a special application to impede the current in a circuit over a specified frequency range while allowing relatively free passage of the current at lower frequencies.   197

choke flange (1)(microwave technique). A flange in whose surface is cut a groove so dimensioned that the flange may form part of a choke joint. *See:* **waveguide.**
    179

(2)(waveguide components). A flange designed with auxiliary transmission-line elements to form a choke joint when used with a cover flange.   166

choke joint (waveguide components). A connection designed for essentially complete transfer of power between two waveguides without metallic contact between the inner walls of the waveguides. It typically consists of one cover flange and one choke flange. *Syn:* **choke coupling.**   166

choke piston (choke plunger) (noncontact plunger) (waveguide). A piston in which there is no metallic contact with the walls of the waveguide at the edges of the reflecting surface: the short-circuit to high-frequency currents is achieved by a choke system. *Note:* This definition covers a number of configurations: dumbbell; Z-slot; inverted bucket; etcetera. *See:* **waveguide.**   179

choke plunger. *See:* **choke piston.**

chopped display (oscilloscopes). a time-sharing method of displaying output signals of two or more channels with a single cathode-ray-tube gun, at a rate that is higher than and not referenced to the sweep rate. *See:* **oscillograph.**   185

chopped impulses (high voltage testing). Generally (but see Note), chopping of an impulse is characterized by an initial discontinuity, decreasing the voltage, which then falls to zero with or without oscillations (see Figures below). *Note:* With some test objects or test arrangements, there may be a flattening of the crest or a rounding off of the voltage before the final voltage collapse. Similar effects may also be observed, due to imperfections of the measuring system. Exact determination of the parameters related to the chopping then requires special consideration, but is not dealt with in this recommendation.   150

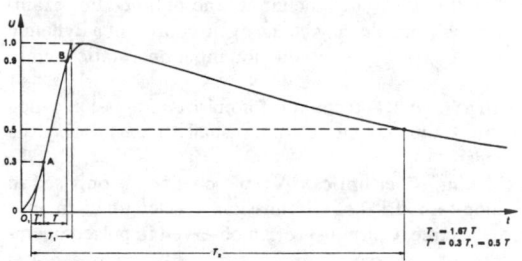

Full Lightning Impulse

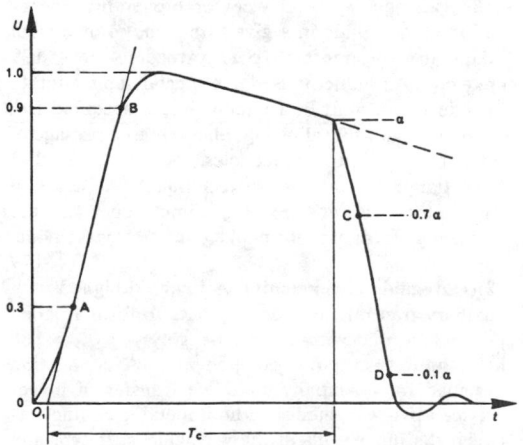

Lightning Impulse Chopped on the Tail

chopped impulse voltage. A transient voltage derived from a full impulse voltage that is interrupted by the disruptive discharge of an external gap or the external portion of the test specimen causing a sudden collapse in the voltage, practically to zero value. *Note:* The collapse can occur on the front, at the peak, or on the tail. *See:* **test voltage and current.**

chopped impulse wave (surge arresters). An impulse wave that has been caused to collapse suddenly by a flash-over.   244, 62

chopped lightning impulse (high voltage testing) (lightning impulse tests, general applicability). A chopped lightning impulse is a lightning impulse that is suddenly interrupted by a disruptive discharge, causing a rapid collapse of the voltage, practically to zero value. The collapse can occur on the front, at the crest, or on the tail. *Note:* The chopping can be accomplished with an external chopping gap or may occur on the internal or external insulation of a test object.   150

chopped wave. A voltage impulse that is terminated intentionally by sparkover of a gap.   64, 91

chopped-wave lightning impulse test (power and distribution transformer). A voltage impulse that is terminated intentionally by sparkover of a gap, which occurs subsequent to the maximum crest of the impulse wave voltage, with a specified minimum crest voltage, and a specified minimum time to flashover.
                                                                53

chopper (1)(analog computers). A mechanical, electrical, or electromechanical device that converts dc into a square wave. As applied to a direct-coupled operational amplifier, it is a modulator used to convert the dc at the summing junction to ac for amplifier and reinsertion as a correcting voltage to reduce offset.
                                                                 9

(2) (communications). A device for interrupting a current or a light beam at regular intervals. Choppers are frequently used to facilitate amplification.
                                                               328

(3) (capacitance devices). A special form of pulsing relay having contacts arranged to rapidly interrupt, or alternately reverse, the direct-current polarity input to an associated circuit.   351

(4) (analog computer). A mechanical, electrical, or electromechanical device that converts dc into a square wave. *Note:* As applied to a direct coupled operational amplifier, it is a modulator used to convert the direct current at the summing junction to alternating current for amplification and reinsertion as a correcting voltage to reduce offset.   9

chopping frequency. *See:* **chopping rate.**

chopping rate (oscilloscopes). The rate at which channel switching occurs in chopped-mode operation. *See:* **oscillograph.**   185

chroma. *See:* **Munsell chroma.**   18

chromatic adaptation (illuminating engineering). The process by which the chromatic properties of the visual system are modified by the observation of stimuli of various chromaticities and luminances.   167

chromatic color (illuminating engineering). Perceived color possessing a hue. In everyday speech, the word color is often used in this sense in contradistinction to white, grey, or black.   167

chromatic dispersion. *See:* dispersion.   433

chromaticity (1) (general). The color quality of light definable by its chromaticity coordinates, or by its dominant (or complementary) wavelength and its purity, taken together. *See:* color.   244, 178
(2) (television). That color attribute of light definable by its chromaticity coordinates. *Note:* When a specific white, the value of dominant (or complementary) wavelength, and saturation are given, there will be a corresponding set of unique chromaticity coordinates.   18

chromaticity coordinate (light) (television). The ratio of any one of the tristimulus values of a sample to the sum of the three tristimulus values. *Note:* In the standard calorimetric system (CEI, 1931), the symbols x, y, z are recommended for the chromaticity coordinates.   18

chromaticity coordinates of a color, x,y,z (illuminating engineering). The ratio of each of the tristimulus values of the color to the sum of the three tristimulus values. *See:* **tristimulus values of a light, X,Y,Z; color matching functions.**   167

chromaticity diagram (1) (illuminating engineering). A plane diagram formed by plotting one of the three chromaticity coordinates against another.   167
(2) (television). A plane diagram formed by plotting one chromaticity coordinate against another. *Notes:* (1) A commonly used chromaticity diagram is the 1931 CIE (x,y) diagram (2) Another chromaticity diagram coming into use is defined in the 1960 CIE (u,v) uniform chromaticity system (UCS). In contrast with the CIE (x,y) diagram, chromaticities that have just noticeable differences (j, n, d) are spaced by essentially equal distances over the entire diagram. Coordinate values in the two systems are related by the transformations: 18

chromaticity difference thresholds (illuminating engineering). The smallest difference in chromaticity, between two colors of the same luminance, that makes them perceptibly different. The difference may be a difference in hue or saturation, or a combination of the two.   167

chromaticity flicker (television). The flicker that results from fluctuation of chromaticity only.   18

chromaticity of a color (illuminating engineering). The dominant or complementary wavelength and purity aspects of the color taken together, or of the aspects specified by the chromaticity coordinates of the color taken together.   167

chrominance (television). The colorimetric difference between any color and a reference color of an equal luminance, the reference color having a specified chromaticity. *Notes:* (1) In three-dimensional color space, chrominance is a vector that lies in a plane of constant luminance. In that plane, it may be resolved into components called chrominance components. (2) In color television transmission, for example, the

$$u = \frac{4x}{-2x + 12y + 3}$$

$$v = \frac{6y}{-2x + 12y + 3}$$

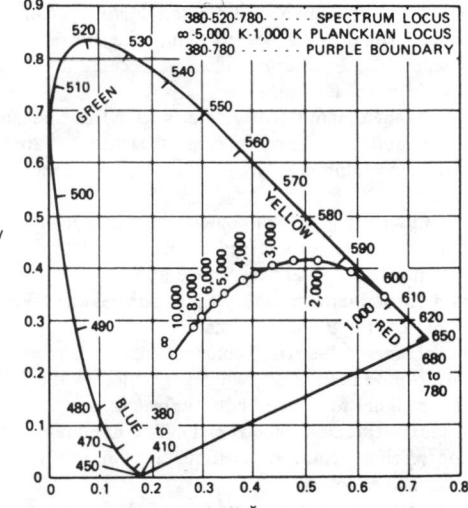

1931 CIE (x,y) Chromaticity Diagram

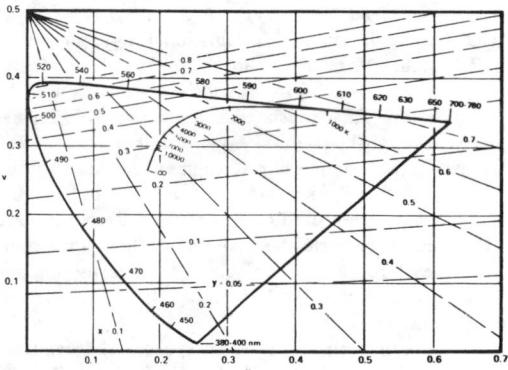

1960 CIE-UCS (u,v) Chromaticity Diagram

chromaticity of the reference color may be that of a specified white.   18

chrominance channel (color television). Any path that is intended to carry the chrominance signal.   18

chrominance channel bandwidth (color television). The bandwidth of the path intended to carry the chrominance signal.   18

chrominance components (television). *See:* **chrominance.**   18

chrominance demodulator (color television reception). A demodulator used for deriving video-frequency chrominance components from the chrominance signal and a sine wave of chrominance subcarrier frequency.   18

chrominance modulator (color television transmission). A modulator used for generating the chrominance signal from the video-frequency chrominance components and the chrominance subcarrier. 18

chrominance primary (color television). A transmission primary that is one of two whose amounts determine the chrominance of a color. *Note:* (1) Chrominance primaries have zero luminance and are nonphysical. (2) This term is obsolete because it is useful only in a linear system. 18

chrominance signal (color television). The sidebands of the modulated chrominance subcarrier that are added to the luminance signal to convey color information. 18

chrominance signal component (television). A signal resulting from suppressed-carrier modulation of a chrominance subcarrier voltage at a specified phase, by a chrominance primary signal such as the I Video Signal or the Q Video Signal. 163

chrominance subcarrier (color television). The carrier whose modulation sidebands are added to the luminance signal to convey color information. 18

chronaxie (medical electronics). The minimum duration of time required to stimulate with a current of twice the rheobase. 192

chronometer (navigation aid terms). A time piece with a nearly constant rate. Set approximately to Greenwich Mean Time. 526

chute. *See:* **feed tube.**

CIE. Abbreviation for Commission Internationale de l'Eclairage. *Note:* These are the initials of the official French name of the International Commission on Illumination. This translated name is approved for usage in English-speaking countries, but at its 1951 meeting the Commission recommended that only the initials of the French name be used. The initials ICI, which have been used commonly in this country, are deprecated because they conflict with an important trademark registered in England and because the initials of the name translated into other languages are different. 18

CIE (L*a*b*) uniform color space (Abbreviation: CIELAB) (illuminating engineering). A transformation of CIE tristimulus values X,Y,Z into three coordinates that define a space in which equal distances are more nearly representative of equal magnitudes of perceived color difference. This space is specially useful in cases of colorant mixtures, for example, dyestuffs, paints. 167

CIE (L*u*v*) uniform color space (abbreviation: CIELUV) (illuminating engineering). A transformation of CIE tristimulus values X,Y,Z into three coordinates that define a space in which equal distances are more nearly representative of equal magnitudes of perceived color difference. This space is specially useful in cases where colored lights are mixed additively for example, color television. 167

CIE (television). Abbreviation for *Commission Internationale de l'Eclairage. Note:* These are the initials of the official French name of the International Commission on Illumination. This translated name is approved

for usage in English-speaking countries, but at its 1951 meeting the Commission recommended that only the initials of the French name be used. The initials ICI, which have been used commonly in this country, are deprecated because they conflict with an important trademark registered in England and because the initials of the name translated into other languages are different. 18

CIE (1931) standard colorimetric observer. Receptor of radiation whose colorimetric characteristics correspond to the distribution coefficients $\bar{x}_\lambda, \bar{y}_\lambda, \bar{z}_\lambda$ adopted by the International Commission on Illumination in 1931. *See:* **color.** 244, 178

CIE standard chromaticity diagram (illuminating engineering). One in which the x and y chromaticity coordinates are plotted in rectangular coordinates. *Note:* The diagram may be based on the CIE 1931 Standard Observer or on the CIE 1964 Supplementary Standard Observer. *See:* **color matching functions.** 167

CIGRE *(Conference Internationale Des Grands Reseaux Electriques).* An international organization concerned with large high voltage electric power systems. 59

circle diagram (rotating machinery). (1) Circular locus describing performance characteristics (current, impedance, etcetera) of a machine or system. In case of rotating machinery, the term **circle diagram** has, in addition, some specific usages: The locus of the armature current phasor of an induction machine, or of some other type of asynchronous machine, displayed on the complex plane, with the shaft speed as the variable (parameter), when the machine operates at a constant voltage and at a constant frequency. (2) The locus of the current vector(s) of a nonsalient-pole synchronous machine, displayed in a synchronously rotating reference frame (Park transform, *d-q* coordinates), with the active component of the load, hence with the rotor displacement angle, as the variable (parameter), when the machine operates at a constant field current. (3) The locus of the current phasor(s) of (2) *See:* **asynchronous machine.** 63

circling guidance lights (illuminating engineering). Aeronautical ground lights provided to supply additional guidance during a circling approach when the circling guidance furnished by the approach and runway lights is inadequate. 167

circuit (1)(measuring longitudinal balance of telephone equipment operating in the voice band). A network providing one or more closed paths.
529

(2)(NESC). A conductor or system of conductors through which an electric current is intended to flow.
494

(3) (machine winding). The element of a winding that comprises a group of series-connected coils. A single-phase winding or one phase of a polyphase winding may comprise one circuit or several circuits connected in parallel. 63

(4) (circuits and systems). An interconnection of electrical elements. *See:* **network.** 67

(5) (data transmission). (A) A conductor or system of conductors through which an electric current is intended to flow. **(B)** A network providing one or more closed paths. 59

circuit analyzer (multimeter). The combination in a single enclosure of a plurality of instruments or instrument circuits for use in measuring two or more electrical quantities in a circuit. *See:* **instrument.** 328

circuitation. *See:* **circulation.**

circuit, balanced. *See:* **balanced circuit.**

circuit bonding jumper (National Electrical Code). The connection between portions of a conductor in a circuit to maintain required ampacity of the circuit. 256

circuit breaker (1) (general) (A). A device designed to open and close a circuit by nonautomatic means, and to open the circuit automatically on a predetermined overload of current, without injury to itself when properly applied within its rating. 256, 429 **(B)** A mechanical switching device capable of making, carrying, and breaking currents under normal circuit conditions and also, making, carrying for a specified time, and breaking currents under specified abnormal circuit conditions such as those of short-circuit. *Notes:* (1) A circuit breaker is usually intended to operate infrequently, although some types are suitable for frequent operation. (2) The medium in which circuit interruption is performed may be designated by suitable prefix, such as, air-blast circuit breaker, air circuit breaker, compressed-air circuit breaker, gas circuit breaker, oil circuit breaker, vacuum circuit breaker, etcetera. (3) Circuit breakers are classified according to their application or characteristics and these classifications are designated by the following modifying words or clauses delineating the several fields of application, or pertinent characteristics:
High-voltage power–Rated 1000 volts alternating current or above.
Molded-case–See separate definition.
Low-voltage power–Rated below 1000 volts alternating current or 3000 volts direct current and below, but not including molded-case circuit breakers.
Direct-current low-voltage power circuit breakers are subdivided according to their specified ability to limit fault-current magnitude by being called general purpose, high-speed, semi-high-speed, or anode. For specifications of these restrictions see the latest revision of the applicable American National Standard. *See:* **alternating-current distribution; switch.** 103

(2)(NESC). A switching device capable of making, carrying and breaking currents under normal circuit conditions and also making, carrying for a specified time, and breaking currents under specified abnormal conditions such as those of short circuit. 494

(3) (packaging machinery). An automatic device designed to open under abnormal conditions a current-carrying circuit without damage to itself. 429

(4) (power switchgear). A mechanical switching device, capable of making, carrying, and breaking currents under normal circuit conditions and also, mak-

ing, carrying for a specified time and breaking currents under specified abnormal circuit conditions such as those of short circuit. *Notes:* (A) A circuit breaker is usually intended to operate infrequently, although some types are suitable for frequent operation. (B) The medium in which circuit interruption is performed may be designated by suitable prefix, that is, air-blast circuit breaker, air circuit breaker, compressed-air circuit breaker, gas circuit breaker, oil circuit breaker, vacuum circuit breaker, oilless circuit breaker, etcetera. (C) Circuit breakers are classified according to their application or characteristics and these classifications are designated by the following modifying words or clauses delineating the several fields of application, or pertinent characteristics:
High-Voltage power–Rated above 1000 volts ac
Molded-Case–See separate definition
Low-Voltage power–Rated 1000 volts ac or below, or 3000 volts dc and below, but not including molded case circuit breakers.
Direct-current low-voltage power circuit breakers are subdivided according to their specified ability to limit fault current magnitude by being called General Purpose, High-Speed, Semi-High Speed, or Anode. For specificationsof these restrictions see latest revision of applicable American National Standard. 103

(5) (thyristor). A device designed to open and close a circuit by nonautomatic means, and to open the circuit automatically on a predetermined overload of current, without injury to itself when properly applied within its rating. *Note:* A circuit breaker is usually intended to operate infrequently, although some types are suitable for frequent operation. 445

(6) (transmission and distribution). A switching device capable of making, carrying and breaking currents under normal circuit conditions and also making, carrying for a specified time, and breaking currents under specified abnormal conditions such as those of short circuit. 262

circuit-breaker compartment (1)(metal-clad and station-type cubicle switchgear). That portion of the switchgear assembly that contains one circuit breaker or other removable primary interrupting device and the associated primary conductors. 572

(2)(metal-enclosed low-voltage power circuit-breaker switchgear). That portion of a switchgear assembly that contains one circuit breaker and the associated primary conductors and secondary control connection devices including current transformers. 579

circuit breaker, field discharge (enclosed field discharge circuit breakers for rotating electric machinery). A circuit breaker having main contacts for energizing and de-energizing the field of a generator, motor, synchronous condenser, or rotating exciter and having discharge contacts for short-circuiting the field through the discharge resistor at the instant preceding the opening of the circuit breaker main contacts. The discharge contacts also disconnect the field from the discharge resistor at the instant following the closing of the main contacts. For direct-current generator operation, the discharge contacts may open before the main contacts close. *Note:* When used in the main field circuit of an alternating or direct-current generator, motor, or synchronous condenser, the circuit breaker is designated as a main field discharge circuit breaker. When used in the field circuit of the rotating exciter of the main machine, the circuit breaker is designated as an exciter field discharge circuit breaker.   359

See: **field discharge circuit breaker.**

circuit breaker general purpose low-voltage dc power (low voltage dc power circuit breaker). A circuit breaker, which during interruption does not usually prevent the fault current from rising to its sustained value.   360

circuit breaker, high-speed low-voltage dc power (low voltage dc power circuit breaker). A circuit breaker, which, when applied in a circuit with the parameter values specified in ANSI Standard C37.16, Table 12, tests "b" (5 Amperes per microsecond initial rate-of-rise current), forces a current crest during interruption within 0.01 seconds after the current reaches the pick-up setting of the instantaneous trip device. *Note:* For total performance characteristics at other than test circuit parameter values, consult the manufacturer.   360

circuit breaker interrupting rating (rated interrupting current) (electric installations on shipboard). The highest rms current at a specified operating voltage which a circuit breaker is required to interrupt under the operating duty specified and with a normal frequency recovery voltage equal to the specified operating voltage. The current is the rms value, including the dc component, at the instant of contact separation as determined from the envelope of the current wave. Where limited by testing equipment, the maximum tolerance for normal frequency recovery voltage is 15 percent of the specified operating voltage. (For dc breakers the rated interrupting current is the maximum value of direct current.)   3

circuit breaker, semi-high-speed low voltage dc power (low voltage dc power circuit breaker). A circuit breaker which, when applied in a circuit with the parameter values specified in ANSI Standard C37.16, Table 11, tests "b" (1.7 amperes per microsecond initial rate-of-rise of current), forces a current crest during interruption within 0.030 seconds after the current reaches the pick-up setting of the instantaneous trip device. *Note:* For total performance at other than test circuit parameter values, consult the manufacturer.   360

circuit-commutated turn-off time (thyristor). The time interval between the instant when the principal current has decreased to zero after external switching of the principal voltage circuit and the instant when the thyristor is capable of supporting a specified principal voltage without turning on. *See:* **principal voltage-current characteristic (principal characteristic).**   243, 66, 208, 191

circuit components (thyristor). Those electrical controller devices that may conduct current during some part of the cycle. Instrumentation is excluded. *Note:* This definition may include devices within the con-

troller that are used for the suppression of voltage and current transients.   445

circuit controller. A device for closing and opening electric circuits.   328

circuit efficiency (output circuit of electron tubes). The ratio of (1) the power at the desired frequency delivered to a load at the output terminals of the output circuit of an oscillator or amplifier to (2) the power at the desired frequency delivered by the electron stream to the output circuit. *See:* **network analysis.**   125, 190

circuit element. A basic constituent part of a circuit, exclusive of interconnections.   328

circuit interrupter (packaging machinery). A manually operated device designed to open under abnormal conditions a current-carrying circuit without damage to itself.   429

circuit malfunction analysis (test, measurement and diagnostic equipment). The logical, systematic examination of circuits and their diagrams to (1) identify and analyze the probability and consequence of potential malfunctions and (2) for determining related maintenance and support requirements to investigate effects of failures.   54

circuit, multipoint. *See:* **multipoint circuit.**

circuit noise meter (noise measuring set). An instrument for measuring circuit noise level. Through the use of a suitable frequency-weighting network and other characteristics, the instrument gives equal readings for noises that are approximately equally interfering. The readings are expressed as circuit noise levels in decibels above reference noise. *See:* **circuit noise level; instrument.**   328

circuit properties (thyristor). Those conditions which exist, or actions which take place, inside the controller during its operating cycle.   445

circuit switch (data transmission). A communications switching system which completes a circuit from sender to receiver at the time of transmission (as opposed to a message switch).   59

circuit switching (data communication). A method of communications where an electrical connection between calling and called stations is established on demand for exclusive use of the circuit until the connection is released. *See:* **packet switching; store-and-forward switching; message switching.**   12

circuit switching element (inverters). A group of one or more simultaneously conducting thyristors, connected in series or parallel or any combination of both, bounded by no more than two main terminals and conducting principal current between these main terminals. *See:* **self-commutated inverters.**   208

circuit switching system (telephone switching systems). A switching system providing through connections for the exchange of messages.   55

circuit transient recovery voltage. The transient recovery voltage characterizing the circuit and obtained with 100-percent normal-frequency recovery voltage, symmetrical current, and no modifying effect of the interrupting device. *Note:* This voltage indicates the inherent severity of the circuit with respect to recovery voltage phenomena.   103

circuit voltage class (electric power system). A phase-to-phase reference voltage that is used in the selection of insultation class designations for neutral grounding devices.   91

circular array (antennas). An array of elements whose corresponding points lie on a circle. *Note:* Practical circular arrays may include arrangements of elements that are congruent under translation or rotation. *Syn:* **ring array.**   111

circular electric wave (waveguide). A transverse electric wave for which the lines of electric force form concentric circles.   267

circular grid array (antennas). An array of elements whose corresponding points lie on coplanar concentric circles.   111

circular interpolation (numerically controlled machines). A mode of contouring control that uses the information contained in a single block to produce an arc of a circle. *Note:* The velocities of the axes used to generate this arc are varied by the control.   224, 207

circularly polarized field vector (antennas). At a point in space, a field vector whose extremity describes a circle as a function of time. *Note:* Circular polarization may be viewed as a special case of elliptical polarization where the axial ratio has become equal to one.   111

circularly polarized plane wave (antennas). A plane wave whose electric field vector is circularly polarized.   111

circularly polarized wave (1) (general). An elliptically polarized wave in which the ellipse is a circle in a plane perpendicular to the direction of propagation. *See:* **radiation.**   328

(2) (radio wave propagation). An electromagnetic wave for which either the electric or the magnetic field vector at a fixed point describes a circle at the rate of the wave frequency. *Note:* This term is usually applied to transverse waves. *See* **left-handed polarized wave** and **right handed polarized wave.**   146

circular magnetic wave (waveguide). A transverse magnetic wave for which the lines of magnetic force form concentric circles.   267

circular mil. A unit of area equal to $\pi/4$ of a square mil (0.7854 square mil). The cross-sectional area of a circle in circular mils is therefore equal to the square of its diameter in mils. A circular inch is equal to one million circular mils. *Note:* A mil is one-thousandth part of an inch. There are 1974 circular mils in a square millimeter.   341

circular orbit (communication satellite). An orbit of a satellite in which the distance between the centers of mass of the satellite and of the primary body is constant.   74

circular probable error (CPE or CEP)(navigation aid terms). In two-dimensional error distribution, the radius of a circle encompassing half of all errors.   526

circular scanning (1) (antennas). Scanning when a beam axis of the antenna generates a conical surface. *Note:* This can include the special case where the cone degenerates to a plane.   111

(2) (radio). Scanning in which the direction of maximum response generates a plane or a right circular cone whose vertex angle is close to 180 degrees.   179

circular shift (mathematics of computing). A variation of a logical shift in which the digits moved out of one end of a register, word, or numeral are returned at the other end. For example, $+231.702$ shifted two places to the left becomes $3170.2+2$. *Note:* A circular shift may be applied to the multiple precision representation of a number.   564

circulating memory. *See:* **circulating register.**

circulating register (1) (data processing and computation). A register that retains data by inserting it into a delaying means, and regenerating and reinserting the data into the register.   235

(2) Shift register in which data moved out of one end of the register are reentered into the other end as in a closed loop. *See:* **cyclic shift.**   255, 77

circulation of electrolyte. A constant flow of electrolyte through a cell to facilitate the maintenance of uniform conditions of electrolysis. *See:* **electrorefining.**   328

circulator (waveguide system). A passive waveguide junction of three or more arms in which the arms can be listed in such an order that when power is fed into any arm it is transferred to the next arm on the list, the first arm being counted as following the last in order. *See:* **transducer; waveguide.**   244, 179

circulator coupled (isolated) port (nonlinear, active, and nonreciprocal waveguide components). With reference to a particular port of the circulator, a port to which waves pass from the reference port with low (high) insertion loss.   530

CIRGE. International Conference on Large High Voltage Electric Systems.   59

CISPR. International Special Committee on Radio Interference.

citizens bands (personal radio services bands) (overhead-power-line corona and radio noise). Frequency bands allocated for short-distance personal or business radio communication, radio signaling, and control of remote devices by radio. *Note:* The frequency bands may differ from country to country. Present United States bands are 26.965 to 17.405 MHz, 72 to 76 MHz, and 462.550 to 467.425 MHz.   411

civil twilight (morning and evening). *See:* **night.**   167

cladding (fiber optics). The dielectric material surrounding the core of an optical waveguide. *See:* **core; normalized frequency; optical waveguide; tolerance field.**   433

cladding center (fiber optics). The center of the circle that circumscribes the outer surface of the homogeneous cladding, as defined under tolerance field. *See:* **cladding; tolerance field.**   433

cladding diameter (fiber optics). The length of the longest chord that passes through the fiber axis and connects two points on the periphery of the homogeneous cladding. *See:* **cladding; core diameter; tolerance field.**   433

cladding mode (fiber optics). A mode that is confined by virtue of a lower index medium surrounding the cladding. Cladding modes correspond to cladding rays in the terminology of geometric optics. *See:* **bound mode; cladding ray; leaky mode; mode; unbound mode.**   433

cladding mode stripper (fiber optics). A device that encourages the conversion of cladding modes to radiation modes; as a result, the cladding modes are stripped from the fiber. Often a material having a refractive index equal to or greater than that of the waveguide cladding. *See:* **cladding; cladding mode.**   433

cladding ray (fiber optics). In an optical waveguide, a ray that is confined to the core and cladding by virtue of reflection from the outer surface of the cladding. Cladding rays correspond to cladding modes in the terminology of mode descriptors. *See:* **cladding mode; guided ray; leaky ray.**   433

clamp (converter circuit elements)(self-commutated converters). An auxiliary circuit element or combination of elements employed to limit the peak voltage or current of a semiconductor device.   584

clamp. *See:* **clamping circuit.**

clamper (data transmission). When used in broadband transmissions, it reinserts low frequency signal components that were not faithfully transmitted.   59

clamping (1) (control) (industrial control). A function by which the extreme amplitude of a waveform is maintained at a given level. *See:* **control system, feedback.**   206
(2) (pulse terms) (operations on a pulse). A process in which a specified instantaneous magnitude of a pulse is fixed at a specified magnitude. Typically, after clamping, all instantaneous magnitudes of the pulse are offset, the pulse shape remaining unaltered.

clamping circuit (clamper) (clamp) (1) (electronic circuits). A circuit that adds a fixed bias to a wave at each occurrence of some predetermined feature of the wave so that the voltage or current of the feature is held at or "clamped" to some specified level. The level may be fixed or adjustable.   328
(2) (analog computers). A circuit used to provide automatic hold and reset action electronically for the purposes of switching or supplying repetitive operation in an analog computer. *See:* **electronic analog computer.**   9

clamping screw. *See:* **binding screw.**

clamping voltage (low voltage varistor surge arresters). Peak voltage across the varistor measured under conditions of a specified peak pulse current and specified waveform. *Note:* Peak voltage and peak current are not necessarily coincident in time.   62

clapper. An armature that is hinged or pivoted.
                                                        259

class (electric instrument). *See:* **accuracy rating (electric instrument).**

class-A amplifier (electron tubes). *See:* **amplifier ratings.**

class-AB amplifier (electron tubes). *See:* **amplifier ratings.**

class-AB operation. *See:* **amplifier ratings.**

Class A component (seismic design of substations). Any component or system whose failure, malfunction, or need for repair prevents the proper operation of the substation during or after the design earthquake.
                                                        465

class A insulation. (nonpreferred term).   53

class-A modulator. A class-A amplifier that is used specifically for the purpose of supplying the necessary signal power to modulate a carrier. *See:* **modulation.**
                                                  111, 240

class-A operation. *See:* **amplifier class ratings.**

class-A push-pull sound track. A class-A push-pull photographic sound track consists of two single tracks side by side, the transmission of one being 180 degrees out of phase with the transmission of the other. Both positive and negative halves of the sound wave are linearly recorded on each of the two tracks. *See:* **phonograph pickup.**   176

class-B amplifier (electron tubes). *See:* **amplifier class ratings.**

Class B component (seismic design of substations). Any component or system whose failure, malfunction, or need for repair does not prevent the operation of the substation during or after the design earthquake.
                                                        465

class B insulation. (nonpreferred term).   53

class-B modulator. A class-B amplifier that is used specifically for the purpose of supplying the necessary signal power to modulate a carrier. *Note:* In such a modulator the class-B amplifier is normally connected in push-pull. *See:* **modulation.**   111, 240

class-B operation. *See:* **amplifier class ratings.**

class-B push-pull sound track. A class-B push-pull photographic sound track consists of two tracks side by side, one of which carries the positive half of the signal only, and the other the negative half. *Note:* During the inoperative half-cycle, each track transmits little or no light. *See:* **phonograph pickup.**
                                                        176

class-C amplifier (electron tubes). *See:* **amplifier class ratings.**

class C insulation. (non-preferred term).   53

class-C operation. *See:* **amplifier class ratings.**

class designation (watthour meter). The maximum of the load range in amperes. *See:* **load range (watthour meter).**   212

classes of grounding (neutral grounding in electrical utility systems). A specific range of degree of grounding; for example, effectively and noneffectively.
                                                        591

classes of insulation systems (insulation systems of synchronous machines). The insulation systems usually employed in synchronous machines covered by this standard are defined below. These definitions, in general, correspond with the principles set forth in IEEE Std 1-1969, General Principles for Temperature Limits in the Rating of Electric Equipment, which is also the accepted basis for interpretation.   298

class F insulation. (nonpreferred term).   53

class H insulation. (nonpreferred term).   53

classification current (metal-oxide surge arresters for ac power circuits). The designated current used to perform the classification tests. See tables below. 583

Table 3
Lightning Impulse Classifying Current

Classification of Arrester	Impulse Value Crest Amperes
Station (800 kV*)	20 000
Station (550 kV*)	15 000
Station (below 550 kV*)	10 000
Intermediate	5000
Distribution	
Heavy Duty	10 000
Normal Duty	5000
Secondary	1500

* Maximum system voltage

classification lamp (classification light). A signal lamp placed at the side of the front end of a train or vehicle, displaying light of a particular color to identify the class of service in which the train or vehicle is operating. 28

classification light. *See:* **classification lamp.**

classification of arresters (1)(metal-oxide surge arresters for ac power circuits). Arrester classification is determined by prescribed test requirements. These classifications are: (1) Station (2) Intermediate (3) Distribution (A) Heavy Duty (B) Normal duty (4) Secondary 583

(2) (surge protective devices). Arrester classification is determined by prescribed test requirements. These classifications are: station valve arrester, intermediate valve arrester, distribution valve arrester, secondary valve arrester, protector tube. 62

(3) (surge arrester). Arrester classification is determined by prescribed test requirements. These classifications are: station valve arrester, intermediate valve arrester, distribution valve arrester, distribution expulsion arrester, secondary valve arrester. *Note:* See section 5.2 of ANSI/IEEE C62.1-1981 for test requirements. 430

classification of insulation (high voltage testing). *See:* **external insulation; internal insulation; self-restoring insulation; non-self-restoring insulation.** 150

class-of-service indication (telephone switching systems). An indication of the features assigned to a switching network termination. 55

class-of-service tone (telephone switching systems). A tone that indicates to an operator that a certain class-of-service is appropriate to a call. 55

class-O insulation. *See:* **insulation class ratings.**

class-over-220 insulation. *See:* **insulation, class ratings.**

class over-220 insulation system. Materials consisting entirely of mica, porcelain, glass quartz, and similar inorganic materials. *Note:* Other materials or combinations of materials may be included in this class if by experience or accepted tests the insulation system can be shown to have the required thermal life at temperatures over 220 °C. 53

"class two" transformer. A step-down transformer of the low-secondary-voltage type, suitable for use in class 2 remote-control low-energy circuits. It shall be of the energy-limiting type, or of a non-energy-limiting type equipped with an overcurrent device. *Note:* Low-secondary-voltage, as used here, has a value of approximately 24 volts. 53

class 0 insulation. (nonpreferred term). 53

Class 0 unplanned outage (starting failure)(electric generating unit reliability, availability, and productivity). An outage that results from the unsuccessful attempt to place the unit in service. *See:* **starting failure.** 567

Class 1, Division 2 locations (auxiliary devices for motors). Basically those in which there may be flammable gas present due to a failure of a process system. Under normal conditions a flammable mixture of gas is not present. 446

Class 1E (nuclear power generating station). The safety classification of the electric equipment and systems that are essential to emergency reactor shutdown, containment isolation, reactor core cooling, and containment and reactor heat removal, or are otherwise essential in preventing significant release of radioactive material to the environment.
541, 492, 438, 387, 131, 143, 428, 440, 102, 142

Class 1E circuits (design and installation of cable systems for Class 1E circuits in nuclear power generating stations). The safety classification of circuits that are essential to emergency reactor shutdown, containment isolation, reactor core cooling, and containment and reactor heat removal, or are otherwise essential in preventing a significant release of radioactive material to the environment. 536

Class 1E control board, panel, or rack. A control board, panel, or rack fitted with Class 1E equipment. 140

Class 1E electric systems. *See:* **nuclear power generating stations, class ratings.**

class 1 electric equipment. *See:* **nuclear power generating stations, class ratings.**

class 1 structures and equipment. *See:* **nuclear power generating stations, class ratings.**

Class 1 unplanned outage (immediate)(electric generating unit reliability, availability, and productivity). An outage that requires immediate removal from the existing state. *Note:* A Class 1 unplanned outage can be initiated from either the in-service or shutdown states. A Class 1 unplanned outage can also be initiated from the planned outage state. *See:* **extended planned outage.** 567

class 2 structures and equipment. *See:* **nuclear power generating stations, class ratings.**

class 2 transformer (power and distribution transformer). A step-down transformer of the low-secondary-voltage type, suitable for use in class 2 remote-control low-energy circuits. It shall be of the energy-limiting type, or of a non-energy-limiting type equipped with an overcurrent device. *Note:* 'Low-secondary-voltage," as used here, has a value of approximately 24 V. 53

Class 2 unplanned outage (delayed)(electric generating unit reliability, availability, and productivity). An outage that does not require immediate removal from the in-service state but requires removal within 6 hours (h). 567

class 3 structures and equipment. *See:* **nuclear power generating stations, class ratings.**

Class 3 unplanned outage (postponed)(electric generating unit reliability, availability, and productivity). An outage that can be postponed beyond 6 hours (h) but requires that a unit be removed from the in-service state before the end of the next weekend. *Note:* Classes 2 and 3 can only be initiated from the in-service state. 567

Class 4 unplanned outage (deferred)(electric generating unit reliability, availability, and productivity). An outage that will allow a unit outage to be deferred beyond the end of the next weekend but requires that a unit be removed from the available state before the next planned outage. 567

class-90 insulation. *See:* **insulation, class ratings.**

class-105 insulation. *See:* **insulation, class ratings.**

class 105 insulation system. Materials or combinations of materials such as cotton, silk, and paper when suitably impregnated or coated or when immersed in a dielectric liquid. *Note:* Other materials or combinations may be included in this class if by experience or accepted tests the insulation system can be shown to have comparable themal life at 105 °C. 53

class 120 insulation system. Materials or combinations of materials such as cotton, silk, and paper when suitably impregnated or coated or when immersed in a dielectric liquid; and which possess a degree of thermal stability which allows them to be operated at a temperature 15 °C higher than temperature index 105 materials. *Note:* Other materials or combinations may be included in this class if by experience or accepted tests the insulation system can be shown to have comparable thermal life at 120 °C. 53

class-130 insulation. *See:* **insulation, class ratings.**

class 150 insulation system. Materials or combinations of sealed dry-type transformer, self-cooled (class GA) (power and distribution transformer) materials such as mica, glass fiber, asbestos, etcetera, with suitable bonding substances. *Note:* Other materials or combinations of materials may be included in this class if by experience or accepted tests the insulation system can be shown to have comparable thermal life at 150 °C. 53

class-155 insulation. *See:* **insulation, class ratings.**

class-180 insulation. *See:* **insulation, class ratings.**

class 185 insulation system. Materials or combinations of materials such as silicone elastometer, mica, glass fiber, asbestos, etc, with suitable bonding substances such as appropriate silicone resins. *Note:* Other materials or combinations of materials may be included in this class if by experience or accepted tests the insulation system can be shown to have comparable thermal life at 220 °C. 53

class-220 insulation. *See:* **insulation, class ratings.**

class 220 insulation system. Materials or combinations of materials such as silicone elastomer, mica, glass fiber, asbestos, etc, with suitable bonding substances such as appropriate silicone resins. *Note:* Other materials or combinations of materials may be included in this class if by experience or accepted tests, the insulation system can be shown to have comparable thermal life at 220 °C. 53

cleaner (electroplating). A compound or mixture used in degreasing, which is usually alkaline. 328

cleaning (electroplating). The removal of grease or other foreign material from a metal surface, chiefly by physical means. *See:* **electroplating.** 328

clear (electronic computers). (1) To preset a storage or memory device to a prescribed state, usually that denoting zero. (2) To place a binary cell in the zero state. *See:* **reset; always single-valued; nonlinear capacitor.** 191

clearance *See:* **minimum clearance.**

clearance *See:* **work permit.**

clearance (1)(navigation aid terms). (A) (ILS [instrument landing system]). The DDm (difference in depth of modulation) in excess of that required to produce full-scale deflection of the course-deviation indicator in flight areas outside the on-course sector; when the DDm is too low the indicator falls below full-scale deflection and the condition of low clearance exists. (B) (air traffic control). Permission by a control facility to the pilot to proceed in a mutually understood manner. 526

(2) (cable stringing equipment). (A) The condition where a circuit has been deenergized to enable work to be performed more safely. A clearance is normally obtained on a circuit presenting a source of hazard prior to starting work. *Syn:* **outage; permit; restriction.** (B) The minimum separation between two conductors, between conductors and supports or other objects, or between conductors and ground, or the clear space between any objects. 431

(3) (transmission and distribution). The minimum separation between two conductors, between conductors and supports or other objects, or between conductors and ground. *See:* **tower.** 64

(4) (conductor stringing equipment). (A) The de-energizing of a circuit to enable work to be performed safely. In the event of an accidental electrical interruption of a circuit where repairs will be required or where a potential hazard exists from contact with an energized circuit, a "clearance" is normally requested prior to starting work. *Syn:* **outage; permit; restriction.** (B) The minimum separation between two conductors, between conductors and supports or other

objects, or between conductors and ground, or the clear space between any objects. 45

(5) (power switchgear). *See:* **minimum clearance.** 103

clearance antenna array (directional localizer)(navigation aid terms). The antenna array that radiates a localizer signal on a separate frequency within the pass band of the receiver and provides the required signals in the clearance sectors as well as a back course. 526

clearance lamps (illuminating engineering). Lighting devices for the purpose of indicating the width and height of a vehicle. 167

clearance point. The location on a turnout at which the carrier's specified clearance is provided between tracks. 328

clearance sector (ILS [instrument landing system]) (navigation aid terms). The sector extending around either side of the localizer from the course sector to the back course sector, and within which the deviation indicator provides the required offcourse indication. 526

clearing (low-voltage air gap surge-protective devices) (low voltage surge protective devices). The characteristic of some types of air gap surge arresters to exhibit a low resistance and then to revert to a high resistance state as a result of an external influence. 556, 62

clearing circuit. A circuit used for the operation of a signal in advance of an approaching train. 328

clearing-out drop (cord circuit or trunk circuit). A drop signal that is operated by ringing current to attract the attention of the operator. 328

clearing source (low-voltage air gap surge-protective devices)low voltage surge protective devices). A defined electrical source which is intentionally applied as a clearing stimulus to an air gap surge protective device under laboratory test conditions. This stimulus is intended to simulate conditions encountered during normal usage. 556, 62

clearing time (1) (fuse) (total clearing time). The time elasping from the beginning of an overcurrent to the final circuit interruption. *Note:* The clearing time is equal to the sum of melting time and arcing time. 103,202

(2) (power switchgear). (A) (mechanical switching device). The interval between the time the actuating quantity in the main circuit reaches the value causing actuation of the release and the instant of final arc extinction on all poles of the primary arcing contacts. *Note:* Clearing time is numerically equal to the sum of contact parting time and arching time. (B) (total clearing time of a fuse). The time elapsing from the beginning of a specified overcurrent to the final circuit interruption, at rated maximum voltage. *Note:* The clearing time is equal to the sum of melting time and the arcing time. 102, 443

clear sky (illuminating engineering). A sky which has less than 30 percent cloud cover. 167

cleat. An assembly of two pieces of insulating material provided with grooves for holding one or more conductors at a definite spacing from the surface wired

over and from each other, and with screw holes for fastening in position. *See:* **raceway.** 328

clerestory (illuminating engineering). That part of a building which rises clear of the roofs or other parts and whose walls contain windows for lighting the interior. 167

click. A disturbance of a duration less than a specified value as measured under specified conditions. *Note:* For the specified values and conditions, guidance should be found in International Special Committee on Radio Interference (CISPR) publications. *See:* **electromagnetic compatibility.** 226,199

climbing space (1) (wiring system). The vertical space reserved along the side of a pole or tower to permit ready access for linemen to equipment and conductors located thereon. 64,252

(2) (transmission and distribution). The vertical space reserved along the side of a pole or structure to permit ready access for linemen to equipment and conductors located on the pole structure. 262

clinometer (navigation aid terms). An instrument for indicating the degree of slope of the angle of roll or pitch of a vehicle, according to the plane in which it is mounted. 526

clip (jargon)(X-ray energy spectrometers). A limiting operation such as (1) use of a high-pass filter, or (2) a nonlinear operation such as diode limiting of pulse amplitude. *Syn:* **clipping.** *See:* **differentiated.** 471, 23, 118, 119

clip. *See:* **contact clip; fuse clip.**

clipper (data transmission). A device that automatically limits the instantaneous value of the output to a predetermined maximum value. *Note:* The term is usually applied to devices which transmit only portions of an input wave lying on one side of an amplitude boundary. 59

clipper amplifier. *See:* **amplifier, clipper.**

clipper limiter. A transducer that gives output only when the input lies above a critical value and a constant output for all inputs above a second higher critical value. *Note:* This is sometimes called an amplitude gate, or slice. *See:* **transducer.** 111

clipping (voice-operated telephone circuit). The loss of initial or final parts of words or syllables due to nonideal operation of the voice-operated devices. 328

clipping-in (conductor stringing equipment). The transferring of sagged conductors from the travelers to their permanent suspension positions and the installing of the permanent suspension clamps. *Syn:* **clamping-in; clipping.** 431

clipping offset (conductor stringing equipment). A calculated distance, measured along the conductor from the plumb mark to a point on the conductor at which the center of the suspension clamp is to be placed. When stringing in rough terrain, clipping offsets may be required to balance the horizontal forces on each suspension structure. 431

clips. *See:* **contact clips; fuse clips.** 103

clock. (1) A device that generates periodic signals used for synchronization. (2) A device that measures and

indicates time. (3) A register whose content changes at regular intervals in such a way as to measure time.
77

clocked logic (power-system communication). The technique whereby all the memory cells (flip-flops) of a logic network are caused to change in accordance with logic input levels but at a discrete time. *See:* **digital.**
59

clocking (data transmission). The generation of periodic signals used for synchronization. *See:* **data processing.**
194

clock reference (digital accelerometer). Basic system timing reference.
383

close coupling (tight coupling). Any degree of coupling greater than the critical coupling. *See:* **coupling; critical coupling.**
328

closed air circuit (rotating machinery). A term referring to duct-ventilated apparatus used in conjunction with external components so constructed that while it is not necessarily airtight, the enclosed air has no deliberate connection with the external air. *Note:* The term must be qualified to describe the means used to circulate the cooling air and to remove the heat produced in the apparatus.
63

closed amortisseur. An amortisseur that has the end connections connected together between poles by bolted or otherwise separable connections.
328

closed circuit cooling (rotating machinery). A method of cooling in which a primary coolant is circulated in a closed circuit through the machine and if necessary a heat exchanger. Heat is transferred from the primary coolant to the secondary coolant through the structural parts or in the heat exchanger.
63

closed-circuit principle. The principle of circuit design in which a normally energized electric circuit, on being interrupted or de-energized, will cause the controlled function to assume its most restrictive condition.
328

closed circuit signaling (data transmission). That type of signaling in which current flows in the idle condition, and a signal is initiated by increasing or decreasing the current.
59

closed-circuit transition (industrial control). As applied to reduced-voltage controllers, including star-delta controllers, a method of starting in which the power to the motor is not interrupted during the starting sequence. *See:* **electric controller.**
206

closed circuit transition auto-transformer starting (rotating machinery). The process of auto-transformer starting whereby the motor remains connected to the supply during the transition from reduced to rated voltage.
63

closed-circuit voltage (working voltage) (battery). The voltage as its terminals when a specified current is flowing. *See:* **battery (primary or secondary); working voltage.**
328

closed construction (National Electrical Code). Any building, building component, assembly or system manufactured in such a manner that all concealed parts of processes of manufacture cannot be inspected before installation at the building site without disassembly, damage, or destruction.
256

closed loop (feedback loop) (automatic control) (industrial control). A signal path that includes a forward path, a feedback path, and a summing point and that forms a closed circuit. *See:* **control system feedback.**
219,206

closed loop control (station control and data acquisition). A type of automatic control in which control actions are based on signals fed back from the controlled equipment or system.
403

closed-loop control system (1)(control system feedback). A control system in which the controlled quantity is measured and compared with a standard representing the desired performance. *Note:* Any deviation from the standard is fed back into the control system in such a sense that it will reduce the deviation of the controlled quantity from the standard. *See:* **control; network analysis.**
151,94

(2)(high-power wide-band transformers). A system in which the controlled quantity is measured and compared with a standard representing the desired performance. Any deviation from the standard is fed back into the control system in such sense that it will reduce the deviation of the controlled quantity from the standard.
321

closed-loop gain (operational gain) (power supplies). The gain, measured with feedback, is the ratio of voltage appearing across the output terminal pair to the causative voltage required at the input resistor. If the open-loop gain is sufficiently large, the closed-loop gain can be satisfactorily approximated by the ratio of the feedback resistor to the input resistor. *See:* **open-loop gain.**
228,176

closed-loop series street lighting system. Street lighting system that employs two-wire series circuits in which the return wire is always adjacent. *See:* **alternating-current distribution.**
328

closed loop testing (test, measurement and diagnostic equipment). Testing in which the input stimulus is controlled by the equipment output monitor.
54

closed-numbering plan (telephone switching systems). A numbering plan in which a fixed number of digits is always dialed.
55

closed subroutine (computing system). A subroutine that can be stored at one place and can be connected to a routine by linkages at one or more location. *See:* **open subroutine; subroutine, closed.**
255,77

close-open operation (of a switching device) (power switchgear). A close operation followed immediately by an open operation without purposely delayed action. *Note:* The letters CO signify this operation: Close-Open.
103

close operation (of a switching device) (power switchgear). The movement of the contacts from the normally open to the normally closed position. *Note:* The letter C signifies this operation: Close.
103

close-talking microphone. A microphone designed particularly for use close to the mouth of the speaker. *See:* **microphone.**
328

close-talking pressure-type microphones. An acoustic transducer that is intended for use in close proximity

to the lips of the talker and is either hand-held or boom-mounted. *Notes:* (1) Various types of microphones are currently used for close-talking applications. These include carbon, dynamic, magnetic, piezoelectric, electrostrictive, and capacitor types. Each of these microphones has only one side of its diaphragm exposed to sound waves, and its electric output substantially corresponds to the instantaneous sound pressure of the impressed sound wave. (2) Since a close-talking microphone is used in the near sound field produced by a person's mouth, it is necessary when measuring the performance of such microphones to utilize a sound source that approximates the characteristics of the human sound generator.
249

close-time delay-open operation (of a switching device) (power switchgear). A close operation followed by an open operation after a purposely delayed action. *Note:* The letters CTO signify this operation: Close-Time Delay-Open.
103

closing coil (of a switching device) (power switchgear). A coil used in the electromagnet that supplies power for closing the device. *Note:* In an air-operated, or other stored-energy-operated device, the closing coil may be the coil used to release the air or other stored energy which in turn closes the device.
103

closing operating time (of a switch) (power switchgear). The interval during which the switch is being operated to move from the fully open position to the fully closed position.
103

closing relay (1) (electrically operated device). A form of auxiliary relay used to control the closing and opening of the closing circuit of the device so that the main closing current does not pass through the control switch or other initiating device.   103,202,60,127
(2) (power switchgear). A form of auxiliary relay used with an electrically operated device to control the closing and opening of the closing circuit of the device so that the main closing current does not pass through the control switch or other initiating device.   103

closing time (of a mechanical switching device). The interval of time between the initiation of the closing operation and the instant when metallic continuity is established in all poles. *Notes:* (1) It includes the operating time of any auxiliary equipment necessary to close the switching device, and which form an integral part of the switching device. (2) For switching devices that embody switching resistors, a distinction should be made between the closing time up to the instant of establishing a circuit at the secondary arcing contacts, and the closing time up to the establishment of a circuit at the main or primary arcing contacts, or both.
103

cloud chamber smoke detector (fire protection devices). A device which is a form of sampling detector. The air pump draws a sample of air into a high humidity chamber within the detector. After the air is in the humidity chamber, the pressure is lowered slightly. If smoke particles are present, the moisture in the air condenses on them forming a cloud in the chamber. The density of this cloud is then measured by the photoelectric principle. When the density is greater than a predetermined level, the detector responds to the smoke.
71

cloud pulse (charge-storage tubes). The output resulting from space-charge effects produced by the turning on or off of the electron beam. *See:* **charge-storage tube.**
174,125

cloudy sky (illuminating engineering). A sky which has more than 70 percent cloud cover.
167

CLR. *See:* **recording-completing trunk (combined line and recording trunk).**

clustered word processing (computer applications). Word processing performed on a system composed of multiple work stations, each with its own memory but operating under the control of a master work station. *See:* **dedicated word processing; shared-logic word processing; shared-resource word processing; stand-alone word processing.**
571

clutter (navigation aid terms). Atmospheric noise, extraneous signals, etcetera, which tend to obscure the reception of a desired signal.
526

clutter attenuation (radar). In moving target indication, the ratio of clutter power at the canceller input to clutter residue at the output, normalized to the attenuation for a single pulse passing through the unprocessed channel of the canceler. *Syn:* **clutter cancellation or suppression.** *See:* **moving-target indicator (MTI) improvement factor.**
13

clutter filter (radar). A filter or group of filters included in a radar for the purpose of rejecting clutter returns and passing target returns at other Doppler frequencies. Moving target indication (MTI) and pulsed Doppler processors are examples.
13

clutter improvement factor. *See:* **moving-target indicator (MTI) improvement factor.**
13

clutter reflectivity (radar). The backscatter coefficient of clutter. *See:* **backscatter coefficient.**
13

clutter residue (radar moving-target indicator). The clutter power remaining at the output of a moving-target indicator (MTI) system; it is the sum of several (generally uncorrelated) components resulting from radar instabilities, antenna scanning, relative motion of the radar with respect to the sources of clutter and fluctuations of the clutter reflectivity. *See:* **canceled video.**
13

clutter visibility factor (radar). The predetection signal-to-clutter ratio that provides stated probabilities of detection and false alarm on a display; in moving-target indicator systems, it is the ratio after cancellation or Doppler filtering.
13

C-message weighting (1) (data transmission). A noise weighting used in a noise measuring set to measure noise on a line that is terminated by a subset with a number 500 receiver or a similar subset.. The meter scale readings are in dBrn (C-message). *See:* **noise definitions.**
59
(2) (voice-frequency electrical-noise test). A weighting derived from listening tests, to indicate the relative annoyance or speech impairment by an interfering signal of frequency f as heard through a "500-type" telephone set. The result, called "c-message weighting," is shown below.
376

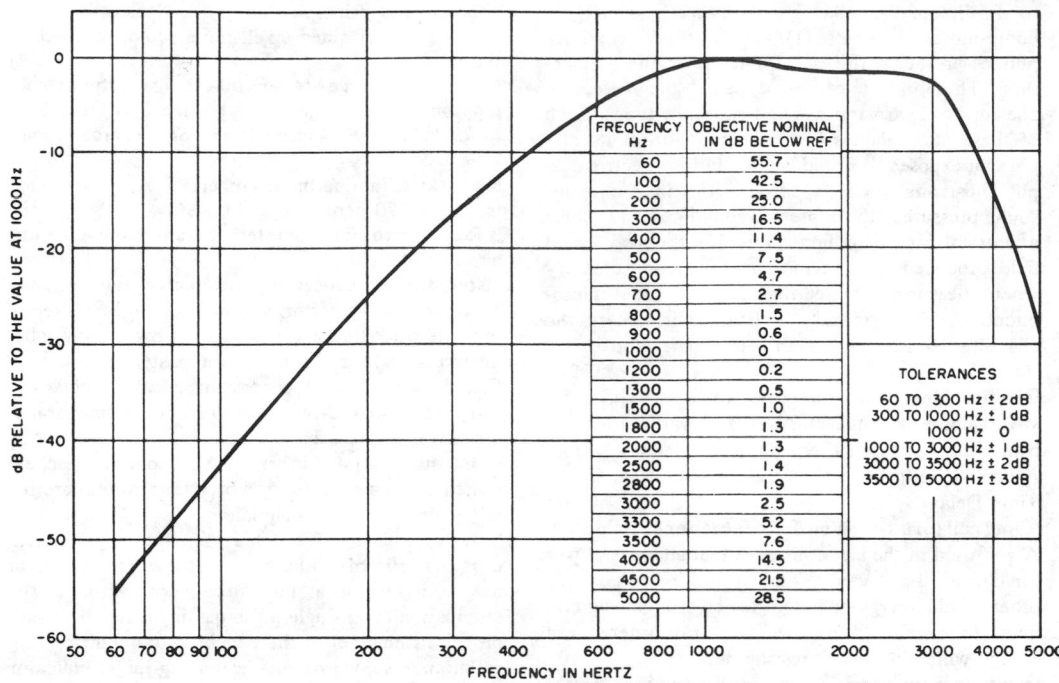

FREQUENCY Hz	OBJECTIVE NOMINAL IN dB BELOW REF
60	55.7
100	42.5
200	25.0
300	16.5
400	11.4
500	7.5
600	4.7
700	2.7
800	1.5
900	0.6
1000	0
1200	0.2
1300	0.5
1500	1.0
1800	1.3
2000	1.3
2500	1.4
2800	1.9
3000	2.5
3300	5.2
3500	7.6
4000	14.5
4500	21.5
5000	28.5

TOLERANCES

60 TO 300 Hz ± 2 dB
300 TO 1000 Hz ± 1 dB
1000 Hz 0
1000 TO 3000 Hz ± 1 dB
3000 TO 3500 Hz ± 2 dB
3500 TO 5000 Hz ± 3 dB

Response in Decibels Indicating Relative Interfering Effect,
c-Message Weighting

CMRR. *See:* **common-mode rejection ratio.**

C network. A network composed of three impedance branches in series, the free ends being connected to one pair of terminals, and the junction points being connected to another pair of terminals. *See:* **network analysis.**   328

C-notch (analog voice frequency circuits). The measure of noise on a channel when a signal is present. A very narrow band-elimination filter (notch filter) is used with a C-message filter to eliminate the holding tone at the measuring end of the circuit. *See:* **C-message; holding tone.**   468

C-notched filter (telephone loop performance). A filter used in front of the noise detector in conjunction with the measurement of noise in certain systems. A tone is transmitted in these systems to activate signal-dependent noise sources, but the tone power should not be included in the measurement. The C-notched filter has a C-message weighting transfer function with a sharp notch which removes this tone from the received signal before its power is measured.   473

CO. *See:* **close-open operation (switching device).**

coagulating current. *See:* **Tesla current (electrotherapy).**

coal cleaning equipment. Equipment generally electrically driven, to remove impurities from the coal as mined, such as slate, sulphur, pyrite, shale, fire clay, gravel, and bone.   328

coarse chrominance primary (National Television System Committee (NTSC) color television). An obsolete term. Use the preferred term, Q chrominance signal.   18

coast time (gyro). *See:* **run-down time.**   46

coated fabric (coated mat) (rotating machinery). A fabric or mat in which the elements and interstices may or may not in themselves be coated or filled but that has a relatively uniform compound or varnish finish on either one or both surfaces. *See:* **rotor (rotating machinery); stator.**   63

coated magnetic tape (magnetic powder-coated tape). A tape consisting of a coating of uniformly dispersed, powdered ferromagnetic material (usually ferromagnetic oxides) on a nonmagnetic base. *See:* **magnetic tape; phonograph pickup.**   176

coated mat. *See:* **coated fabric.**

coating (electroplating). The layer deposits by electroplating. *See:* **electroplating.**   328

coaxial antenna. An antenna comprised of a quarter-wavelength extension to the inner conductor of a coaxial line and a radiating sleeve which in effect is formed by folding back the outer conductor of the coaxial line for approximately one-quarter wavelength. *See:* **antenna.**   111

coaxial cable (medium attachment units and repeater units). A two-conductor (center conductor, shield system), concentric, constant impedance transmission line used as the trunk medium in the baseband system. 543

coaxial cable interface (medium attachment units and repeater units). The electrical and mechanical interface to the shared coaxial cable medium either contained within or connected to the medium attachment unit (MAU). *Syn:* **medium dependent interface (MDI).** 543

coaxial cable segment (medium attachment units and repeater units). A length of coaxial cable made up from one or more coaxial cable sections and coaxial connectors, and terminated at each end in its characteristic impedance. 543

coaxial conductor. An electric conductor comprising outgoing and return current paths having a common axis, one of the paths completely surrounding the other throughout its length. 14

coaxial detector (germanium gamma-ray detectors). A semiconductor radiation detector in which all or part of the two electrical contacts are substantially coaxial. Typically one end of each contact configuration is closed (closed-end coaxial detector), but both ends may be open (open-end coaxial detector). 528

coaxial line. *See:* **coaxial.**

coaxial pair. *See:* **coaxial.**

coaxial relay. A relay that opens and closes an electric contact switching high-frequency current as required to maintain minimum losses. *See:* **relay.** 341

coaxial stop filter (electromagnetic compatibility). A tuned movable filter set round a conductor in order to limit the radiating length of the conductor for a given frequency. *See:* **electromagnetic compatibility.** 220,199

coaxial stub. A short length of coaxial that is joined as a branch to another coaxial. *Note:* Frequently a coaxial stub is short-circuited at the outer end and its length is so chosen that a high or low impedance is presented to the main coaxial in a certain frequency range. *See:* **waveguide.** 328

coaxial switch. A switch used with and designed to simulate the critical electric properties of coaxial conductors. 346

coaxial transmission line (waveguide). A transmission line consisting of two essentially concentric cylindrical conductors. 267

co-channel interference. Interference caused in one communication channel by a transmitter operating in the same channel. *See:* **radio transmission.** 178

code (1)(microprocessor object modules). Data or executable machine code. *See:* **absolute code; relocatable code.** 466

(2) (electronic computers). (A) The characters or expressions of an orginating or source language, each correlated with its equivalent expression in an intermediate or target language, for example, alphanumeric characters correlated with their equivalent 6-bit expressions in a binary machine language. *Note:* For punched or magnetic tape; a predetermined arrangement of possible locations of holes or magnetized areas and rules for interpreting the various possible patterns. (B) Frequently, the set of expressions in the target language that represent the set of characters of the source language. (D) To encode is to express given information by means of a code. (E) To translate the program for the solution of a problem on a given computer into a sequence of machine-language or pseudo instructions acceptable to that computer. 235

(3) (software). (A) A set of unambiguous rules specifying the manner in which data may be represented in a discrete form. (B) To represent data or a computer program in a symbolic form that can be accepted by a processor. (C) To write a routine. (D) Loosely, one or more computer programs, or part of computer program. (E) A encryption of data for security purposes. *See:* **computer program; data; routine; security.** 434

code audit (software). An independent review of source code by a person, team, or tool to verify compliance with software design documentation and programming standards. Correctness and efficiency may also be evaluated. *See:* **audit; code; correctness; efficiency; inspection; software design documentation; static analysis; tool; walk-through.** 434

code character. A particular arrangement of code elements representing a specific symbol or value. 194,59

code classes (safety systems equipment in nuclear power generating stations). Levels of structural integrity and quality commensurate with the relative importance of the individual mechanical components of the nuclear power generating station. *Note:* For the recognized code classes, refer to the following documents:ANSI N18.2-1973, Nuclear Safety Criteria for the Design of Stationary Pressurized Water Reactor Plants.ANSI/ANS 51.8, Nuclear Safety Criteria for the Design of Stationary Pressurized Water Reactor Plants.ANSI/ASME BPV-III, Boiler and Pressure Vessel Code and its latest addenda, Section III.ANSI/ANS 52.1-1980, Nuclear Safety Criteria for Design of Stationary BWR Plants. 535

code conversion (telephone switching systems). The substitution of a routing code for a destination code. 55

coded-decimal code. The decimal number system with each decimal digit expressed by a code. 224,207

coded fire-alarm system. A local fire-alarm system in which the alarm signal is sounded in a predetermined coded sequence. *See:* **protective signaling.** 328

coded pulse (radar). A pulse with internal (intra-pulse) amplitude, frequency, or phase modulation, used for identification or for pulse compression. 13

coded track circuit. A track circuit in which the energy is varied or interrupted periodically. 328

code element. One of the discrete conditions or events in a code, for example, the presence or absence of a pulse. *See:* **data processing; information theory.** 194

code generator (software). A program or program

function, often part of a compiler, which transforms a computer program from some intermediate level of representation (often the output of a parser) into a lower level representation such as assembly code or machine code. *See:* **code; compiler; computer program; function; program.** 434

code inspection. *See:* **inspection.** 434

code letter (locked-rotor kilovolt-amperes). A letter designation under the caption "code" on the nameplate of alternating-current motors (except wound-rotor motors) rated 1/20 horsepower and larger to designate the locked-rotor kilovolt-amperes per horsepower as measured at rated voltage and frequency. 63

coder (1) (general). A device that sets up a series of signals in code form. 328

(2) (code transmitter). A device used to interrupt or modulate the track or line current periodically in various ways in order to establish corresponding controls in the other apparatus. 328

code ringing (telephone switching systems). Ringing wherein the number of rings or the duration, or both, indicate which system on a party line is being called. 55

code system. A system of control of wayside signals, cab signals, train stop or continuous train control in which electric currents of suitable character are supplied to control apparatus, each function being controlled by its own distinctive code. *See:* **block signal system.** 328

code translator. *See:* **digital converter.**

code walk-through. *See:* **walk-through.** 434

coding (1) (communication). The process of transforming messages or signals in accordance with a definite set of rules. 242

(2) (computing systems). Loosely, a routine. *See:* **relative coding; straight-line coding; symbolic coding.** 77

(3) (test, measurement and diagnostic equipment). A part of the programming process in which a completely defined, detailed sequence of operation is translated into computer-entry language. 54

coding delay* (loran)(navigation aid terms). An arbitrary time delay in the transmission of pulse signals from the slave station to permit the resolution of ambiguities; the term suppressed time delay more accurately represents what is being accomplished and should be used instead of coding delay. 526
*Deprecated

coding fan. *See:* **electrode radiator.**

coding siren. A siren having an auxiliary mechanism to interrupt the flow of air through the device, thereby enabling it to produce a series of sharp blasts as required in code signaling. *See:* **protective signaling.** 328

coefficient of attenuation, (illuminating engineering). The decrement in flux per unit distance in a given direction within a medium. It is defined by the relation: $\Phi_0 e - $ ux where Φ_x is the flux at any distance x from a reference point having flux Φ_0. 167

coefficient of beam utilization, CBU (illuminating en-gineering). The ratio of the luminous flux (lumens) reaching a specified area directly from a floodlight or projector to the total beam luminous flux (lumens). 167

coefficient of coupling. *See:* **coupling coefficient.**

coefficient of grounding (surge arrester) (power and distribution transformer). The ratio (E_{LG}/E_{LL}) expressed as a percentage, of the highest root-mean-square line-to-ground power-frequency voltage (E_{LG}) on a sound phase, at a selected location, during a fault to earth affecting one or more phases to the line-to-line power-frequency voltage (E_{LL}) that would be obtained, at the selected location, with the fault removed. *Notes:* (1) Coefficients of grounding for three-phase systems are calculated from the phase-sequence impedance components as viewed from the selected location. For machines use the subtransient reactance. (2) The coefficient of grounding is useful in the determination of an arrester rating for a selected location. (3) A value not exceeding 80 percent is obtained approximately when for all system conditions the ratio of zero-sequence reactance to positive-sequence reactance is positive and less than three and the ratio of zero-sequence resistance to positive-sequence reactance is positive and less than one. 62, 53

coefficient of performance (1) (thermoelectric cooling couple. The quotient of (A) the net rate of heat removal from the cold junction by the thermoelectric couple by (B) the electric power input to the thermoelectric couple. *Note:* This is an idealized coefficient of performance assuming perfect thermal insulation of the thermoelectric arms. *See:* **thermoelectric device.** 248,191

(2) (thermoelectric cooling device). The quotient of (A) the rate of heat removal from the cooled body by (B) the electric power input to the device. *See:* **thermoelectric device.** 248,191

(3) (thermoelectric heating device). The quotient of (A) the rate of heat addition to the heated body by (B) the electric power input to the device. *See:* **thermoelectric device.** 248,191

(4) (thermoelectric heating couple). The quotient of (A) the rate of heat addition to the hot junction by the thermoelectric couple by (B) the electric power input to the thermoelectric couple. *Note:* This is an idealized coefficient of performance assuming perfect thermal insulation of the thermoelectric arms. *See:* **thermoelectric device.** 248,191

coefficient of performance, reduced (thermoelectric device). The ratio of (1) a specified coefficient of performance to (2) the corresponding coefficient of performance of a Carnot cycle. *See:* **thermoelectric device.** 248,191

coefficient of trip point repeatability. *See:* **trip-point repeatability coefficient.**

coefficient of utilization, CU (illuminating engineering). The ratio of luminous flux (lumens) calculated as received on the work-plane to the rated luminous flux (lumens) emitted by the lamps alone. (It is equal to the product of *room utilization factor* and *luminaire efficiency*. 167

coefficient of zero error. *See:* **environmental coefficient.**

coefficient potentiometer. *See:* **parameter potentiometer.**   9

coefficient sensitivity. *See:* **sensitivity coefficient.**

coercive field (E_c)(primary ferroelectric terms). The electric field required to switch the polarization from $P = \pm P_R$ to $P = 0$. The coercive field of a ferroelectric crystal depends on its thermal and electrical history, temperature, pressure, type of electrodes, magnitude, and waveshape of the applied switching voltage (That is, E_c increases as a function of the rate of polarization reversal).   497

coercive force (magnetic core testing) (H_c). The magnetic field strength at which the magnetic induction is zero, when the core material is in a symmetrically cyclically magnetized condition, with a specified maximum value of field strength, (that is, loci of points on the hysteresis curve when $B = 0$).   165

cofactor (or path cofactor) (network analysis). *See:* **path (loop) factor.**

coffer (illuminating engineering). A recessed panel or dome in the ceiling.   167

cogging (rotating machinery). Variations in motor torque at very low speeds caused by variations in magnetic flux due to the alignment of the rotor and stator teeth at various positions of the rotor. *See:* **rotor (rotating machinery); stator.**   63

cohered video (in radar moving-target indicator). Video-frequency signal output employed in a coherent system. *See:* **navigation.**   13

coherence (1) (laser-maser). The correlation between electromagnetic fields at points which are separated in space or in time, or both.   363

(2) (metric practice). A characteristic of a coherent system as described in X1.9 of Appendix X1. In such a system the product or quotient of any two unit quantities is the unit of the resulting quantity. The SI base units, supplementary units, and derived units form a coherent set.   21

coherence area (1) (fiber optics). The area in a plane perpendicular to the direction of propagation over which light may be considered highly coherent. Commonly the coherence area is the area over which the degree of coherence exceeds 0.88. *See:* **coherent; degree of coherence.**   433

(2) (laser-maser). A quantitative measure of spatial coherence. The largest cross-sectional area of a light beam, such that light from this area (passing through any two pin holes placed in this area) will produce interference fringes.   63, 363

coherence function (seismic qualification of Class 1E equipment for nuclear power generating stations). Defines a comparative relationship between two time histories. It provides a statistical estimate of how much two motions are related, as a function of frequency. The numerical range is from zero for unrelated, to 1.0 for related motions.   581

coherence length (fiber optics). The propagation distance over which a light beam may be considered coherent. If the spectral linewidth of the source is $\Delta\lambda$ and the central wavelength is λ_0, the coherence length in a medium of refractive index n is approximately $\lambda^2_0 / n\Delta\lambda$. *See:* **degree of coherence; spectral width.**   433

coherence time (1) (fiber optics). The time over which a propagating light beam may be considered coherent. It is equal to coherence length divided by the phase velocity of light in a medium; approximately given by $\lambda^2_0 / c\Delta\lambda$, where λ_0 is the central wavelength, $\Delta\lambda$ is the spectral linewidth and c is the velocity of light in vacuum. *See:* **coherence length; phase velocity.**   433

(2) (laser-maser). A quantitative measure of temporal coherence. The maximum delay time which can be introduced between the two beams in a Michelson interferometer before the interference fringes disappear.   363

coherent (1) (fiber optics). Characterized by a fixed phase relationship between points on an electromagnetic wave. *Note:* A truly monochromatic wave would be perfectly coherent at all points in space. In practice, however, the region of high coherence may extend only a finite distance. The area on the surface of a wavefront over which the wave may be considered coherent is called the coherence area or coherence patch; if the wave has an appreciable coherence area, it is said to be spatially coherent over that area. The distance parallel to the wave vector along which the wave may be considered coherent is called the coherence length; if the wave has an appreciable coherence length, it is said to be phase or length coherent. The coherence length divided by the velocity of light in the medium is known as the coherence time; hence a phase coherent beam may also be called time (or temporally) coherent. *See:* **coherence area; coherence length; coherence time; degree of coherence; monochromatic.**   433

(2) (laser-maser). A light beam is said to be coherent when the electric vector at any point in it is related to that at any other point by a definite, continuous sinusoidal function.   363

coherent bundle. *See:* **aligned bundle.**

coherent interrupted waves. Interrupted continuous waves occurring in wave trains in which the phase of the waves is maintained through successive wave trains. *See:* **wave front.**   328

coherent MTI (moving-target indication). A form of moving-target indication in which a moving target is detected as a result of a pulse-to-pulse change in echo phase relative to the phase of a coherent reference oscillator.   13

coherent pulse operation. The method of pulse operation in which a fixed phase relationship is maintained from one pulse to the next. *See:* **pulse.**   328

coherent radiation. *See:* **coherent.**

coherent signal processing (radar). Echo integration, filtering or detection using amplitude and phase of the signal referred to a coherent oscillator.   13

coherent video (radar). Bipolar video obtained from a synchronous (coherent) detector.   13

cohesion (software). The degree to which the tasks

performed by a single program module are functionally related. *See:* **coupling; program module.**   434

coho (COHO) (radar). A coined word, derived from coherent oscillator, designating an oscillator used in a coherent radar to provide a reference phase by which changes in the radio-frequency phase of successively received pulses may be recognized. In practice, a coho usually operates at the receiver intermediate frequency.   13

coil (1) (general). An assemblage of successive convolutions of a conductor.

(2) (rotating machinery). A unit of a winding consisting of one or more insulated conductors connected in series and surrounded by common insulation, and arranged to link or produce magnetic flux. *See:* **rotor (rotating machinery); stator.**   63

coil brace (1) (coil support). A structure for the support or restraint for one or more coils.

(2) (V wedge, salient-pole construction). A trapezoidal insulated insert clamped between field poles, to provide radial restraint for the field coil turns against centrifugal force and to brace the coils tangentially. *See:* **stator.**   63

coil end-bracing (rotating machinery). *See:* **end-winding support; rotor (rotating machinery); stator.**

coil insulation (rotating machinery). The main insulation to ground or between phases surrounding a coil, additional to any conductor or turn insulation. *See:* **rotor (rotating machinery); stator.**   63

coil insulation with its accessories (insulation systems of synchronous machines). The coil insulation comprises all of the insulating materials that envelope the current-carrying conductors and their component turns and strands and form the insulation between them and the machine structure, and includes the armor tape, the tying cord, slot fillers, slot tube insulation, pole body insulation, and rotor-retaining ring insulation.   298

coil lashing (rotating machinery). The binding used to attach a coil end to the supporting structure. *See:* **rotor (rotating machinery); stator.**   63

coil loading. Loading in which inductors, commonly called loading coils, are inserted in a line at intervals. *Note:* The loading coils may be inserted either in series or in shunt. As commonly understood, coil loading is a series loading in which the loading coils are inserted at uniformly spaced recurring intervals. *See:* **loading.**   328

coil pitch (rotating machinery). The distance between the two active conductors (coil sides) of a coil, usually expressed as a percentage of the pole pitch. *See:* **armature.**   63

coil Q (dielectric heating usage). Ratio of reactance to resistance measured at the operating frequency. *Note:* The loaded-coil Q is that of a heater coil with the charge in position to be heated. Correspondingly, the unloaded-coil Q is that of a heater coil with the charge removed from the coil. *See:* **dielectric heating.**   14

coil section (rotating machinery). The basic electrical element of a winding comprising an assembly of one or more turns insulated from one another.   63

coil shape factor (dielectric heating usage). A correction factor for the calculation of the inductance of a coil based on its diameter and length. *See:* **dielectric heating.**   14

coil side (rotating machinery). Either of the two normally straight parts of a coil that lie in the direction of the axial length of the machine. *See:* **rotor (rotating machinery); stator.**   63

coil-side separator (rotating machinery). Additional insulation used to separate embedded coil sides. *See:* **rotor (rotating machinery); stator.**   63

coil space factor. The ratio of the cross-sectional area of the conductor metal in a coil to the total cross-sectional area of the coil. *Note:* If the overall insulation, such as spool bodies or stop linings, is omitted from consideration when the space factor is calculated, the omission should be specifically stated. *See:* **asynchronous machine.**   328

coil span (rotating machinery). *See:* **coil pitch; rotor (rotating machinery); stator.**

coil support bracket (rotating machinery). A bracket used to mount a coil support ring or binding band. *See:* **rotor (rotating machinery); stator.**   63

coin box. A telephone set equipped with a device for collecting coins in payment for telephone messages. *See:* **telephone station.**   328

coin call (telephone switching systems). A call in which a coil collection device is used.   55

coincidence (radiation counters). The practically simultaneous production of signals from two or more counter tubes. *Note:* A genuine or true coincidence is due to signals from related events (passage of one particle or of two or more related particles through the counter tubes); an accidental, spurious, or chance coincidence is due to unrelated signals that coincide accidentally. *See:* **anticoincidence (radiation counters).**   190

coincidence circuit. A circuit that produces a specified output pulse when and only when a specified number (two or more) or a specified combination of input terminals receives pulses within an assigned time interval. *See:* **anticoincidence (radiation counters); pulse.**   117

coincidence factor (electric power utilization). The ratio of the maximum coincident total demand of a group of consumers to the sum of the maximum power demands of individual consumers comprising the group both taken at the same point of supply for the same time. *See:* **generating station.**   64

coincident-current selection. The selection of a magnetic cell for reading or writing, by the simultaneous application of two or more currents.   77

coincident demand (power operations). Any demand that occurs simultaneously with any other demand; also the sum of any set of coincident demands.   516

coin-control signal (telephone switcing systems). On a coin call, one of the signals used for collecting or returning coins.   55

coin-denomination tone (telephone switching systems). The tone that indicates the value of coins when they are deposited in a coin telephone.   55

coin tone (telephone switching systems). A class-of-service tone that indicates to an operator that the call has originated from a coin telephone. 55

cold cathode. A cathode that functions without the application of heat. *See:* **electrode (electron tube).** 328

cold-cathode glow-discharge tube (glow tube). A gas tube that depends for its operation on the properties of a glow discharge. 190

cold-cathode lamp (illuminating engineering). An electric-discharge lamp whose mode of operation is that of a glow discharge, and which has electrodes so spaced that most of the light comes from the positive column between them. 167

cold-cathode stepping tube (electron device). A glow discharge tube having several main gaps with or without associated auxiliary gaps, and in which the main discharge has two or more stable positions and can be made to step in sequence, when a suitable shaped signal is applied to an input electrode, or a group of input electrodes. 190

cold-cathode tube. An electron tube containing a cold cathode. 125

cold-end termination (CET)(electrical heat tracing for industrial applications). The termination applied to the end of a heating cable where the power is supplied. 523

cold lead (electrical heat tracing for industrial applications). An electrically insulated conductor used to connect a heating conductor to the branch-circuit conductors and designed so as not to produce any appreciable heat. 523

cold reserve. Thermal generating capacity available for service but not maintained at operating temperature. 64

cold side. *See:* **unexposed side.** 368

cold test (test, measurement and diagnostic equipment). *See:* **passive test.**

collapsing loss (radar). Loss of information, measured by an increase in required input signal-to-noise ratio, occurring when envelope-detected noise from resolution elements not containing the signal is added to the signal during processing; for example; it occurs when radar returns containing range, azimuth, and elevation information are constrained to a two dimensional display. 13

collapsing ratio (radar). The total number of envelope detected noise samples added to the signal divided by the number which originated in the resolution cell containing the signal. 13

collar (cheek, field-coil flange) (washer) (rotating machinery). Insulation between the field coil and the pole shoe (top collar) and between the field coil and the member carrying the pole body (bottom collar). *See:* **rotor (rotating machinery).** 63

collate. To compare and merge two or more similarly ordered sets of items into one ordered set. 255,77

collating sequence. An ordering assigned to a set of items, such that any two sets in that assigned order can be collated. 255,77

collator. A device to collate sets of punched cards or other documents into a sequence. 255,77

collect call (telephone switching systems). A call for which the called customer agrees to pay. 55

collection efficiency (quantum yield). The number of carriers crossing the p-n junction per incident photon. 113

collector (1) (rotating machinery). An assembly of collector rings, individually insulated, on a supporting structure. *See:* **asynchronous machine.** 63
(2) (electron tube). An electrode that collects electrons or ions that have completed their functions within the tube. *See:* **electrode (electron tube).** 125
(3) (transistor). A region through which primary flow of charge carriers leaves the base. 245

collector grid (solar cells). A pattern of conducting material making ohmic contact to the active surface of a solar cell to reduce the series resistance of the device by reducing the mean path of the current carriers within the semiconductor. 113

collector junction. *See:* **junction, collector.**

collector plates. Metal inserts embedded in the cell lining to minimize the electric resistance between the cell lining and the current leads. *See:* **fused electrolyte.** 328

collector ring (slip ring). A metal ring suitably mounted on an electric machine that (through stationary brushes bearing thereon conducts curent into or out of the rotating member. *See:* **asynchronous machine.** 328

collector-ring (slip-ring) lead insulation (rotating machinery). Additional insulation, applied to the leads that connect the collector rings to the windings of the rotating member, to prevent grounding to the metallic parts of the rotating members, and to provide electrical separation between leads. *See:* **rotor (rotating machinery).** 63

collector-ring (slip-ring) shaft insulation (rotating machinery). The combination of insulating members that insulate the collector rings from the parts of the structure that are mounted on the shaft. *See:* **rotor (rotating machinery).** 63

collector rings (National Electrical Code). A collector ring is an assembly of slip rings for transferring electrical energy from a stationary to a rotating member. 256

collimate (storage tubes). To modify the paths of electrons in a flooding beam or of various rays of a scanning beam in order to cause them to become more nearly parallel as they approach the storage assembly. *See:* **storage tube.** 174

collimated beam (laser-maser). Effectively, a parallel beam of light with very low divergence or convergence. 363

collimating lens (storage tubes). An electron lens that collimates an electron beam. *See:* **storage tube.** 174

collimation (fiber optics). The process by which a divergent or convergent beam of radiation is converted into a beam with the minimum divergence possible for that system (ideally, a parallel bundle of rays). *See:* **beam divergence.**   433

collimator (laser-maser). An optical device for converting a diverging or converging beam of light into a collimted or parallel one.   363

collinear array (antenna). A linear array of radiating elements, usually dipoles, with their axes lying in a straight line. *See:* **antenna.**   111

collision (medium attachment units and repeater units). Multiple concurrent transmissions on the coaxial cable resulting in garbled data.   543

collision-avoidance system (navigation aid terms). A system providing the means of detection and prevention of impending collision between vehicles. The system performs one or more of the following functions: detection of intruders in surrounding vicinity, evaluation of miss distance of a collision hazard, determination of precise maneuver needed to avoid the hazard and specification of when an avoidance maneuver should be initiated.   526

collision frequency (radio wave propagation). In a plasma, the average number of collisions per second of a charged particle of a given species with particles of another or the same species.   146

collision presence (medium attachment units and repeater units). A signal provided by the physical layer to the media access sublayer (within the data link layer) to indicate that multiple stations are contending for access to the transmission medium.   543

colloidal ions. Ions suspended in a medium, that are larger than atomic or molecular dimensions but sufficiently small to exhibit Brownian movement.   328

color (television). That characteristic of visual sensation in the photopic range that depends on the spectral composition of light entering the eye.   18

color breakup (color television). Any transient or dynamic distortion of the color in a television picture. *Note:* This effect may originate in videotape equipment, in a television camera, or in a receiver. In videotape recording or playback, it occurs as intermittent misphasing or loss of the chrominance signal. In a field-sequential system, it may be caused at the camera by rapid motion of the image on the camera sensor through motion of either the camera or the subject. It may be caused at the receiver by rapid changes in viewing conditions such as blinking or motion of the eyes.   18

color burst (color television). The portion of the composite or noncomposite color-picture signal, comprising a few cycles of a sine wave of chrominance subcarrier frequency, that is used to establish a reference for demodulating the chrominance signal.   18

color-burst flag keying signal) (television). A keying signal used to form the color burst from a color-subcarrier signal source. *See:* **burst flag.**   328

color-burst gate (television). A keying or gating signal used to extract the color burst from a color-television signal. *See:* **burst gate; television.**   328

color-burst keying signal. *See:* **color-burst flag.**

color carrier (color television). *See:* **chrominance subcarrier.**   18

color cell (repeating pattern of phosphors on the screen of a color-picture tube). The smallest area containing a complete set of all the primary colors contained in the pattern. *Note:* If the cells are described by only one dimension as in the line type of screen, the other dimension is determined by the resolution capabilities of the tube.   125

color center (color-picture tubes). A point or region (defined by a particular color-selecting electrode and screen configuration) through which an electron beam must pass in order to strike the phosphor array of one primary color. *Note:* This term is not to be used to define the color-triad center of a color-picture tube screen.   125

color code (electrical). A system of standard colors adopted for identification of conductors for polarity, etcetera, and for identification of external terminals of motors and starters to facilitate making power connections between them. *See:* **mine feeder circuit.**   328

color coder (color television, transmission). *See:* **color encoder,** the preferred term in the United States. In Europe the term **color coder** is commonly used.   18

color comparison (illuminating engineering). The judgement of equality, or of the amount and character of difference, of the color of two objects viewed under identical illumination. *Syn:* **color grading.**   167

color contamination (color television). An error of color rendition caused by incomplete separation of paths carrying different color components of the picture. *Note:* Such errors can arise in the optical, electronic, or mechanical portions of a color television system as well as in the electrical portions.   18

color coordinate transformation (color television). Computation of the tristimulus values of colors in terms of one set of primaries from the tristimulus values of the same colors in another set of primaries. *Note:* This computation may be performing electrically in a color television system.   18

color correction (of a photograph or printed picture) (illuminating engineering). The adjustment of a color reproduction process to improve the perceived-color conformity of the reproduction to the original.   167

color decoder (color television). An apparatus for deriving the signals for the color display device from the color picture signal and the color burst.   18

color-difference signal (color television). An electrical signal that, when added to the luminance signal, produces a signal representative to one of the tristimulus values (with respect to a stated set of primaries) of the transmitted color.   18

color difference thresholds (illuminating engineering). The difference in chromaticity or luminance or both, between two colors, that makes them just perceptibly different. The difference may be a difference in hue, saturation, or brightness (lightness for surface colors) or a combination of the three.   167

color discrimination (illuminating engineering). The perception of differences between two or more colors.   167

color encoder (National Television System Committee (NTSC) color television). An apparatus for generating either the noncomposite or the composite color picture signal and the color burst from camera signals (or equivalents) and the chrominance subcarrier. 18

color-field corrector (electron tubes). A device located external to the tube producing an electric or magnetic field that affects the beam after deflection as an aid in the production of uniform color fields. 125

color flicker (television). The flicker that results from fluctuation of both chromaticity and luminance. 18

color fringing (color television). Spurious chromaticity at boundaries of objects in the picture. *Note:* Color fringing can be caused by a change in relative position of the televised object from field to field (in a field-sequential system), or by misregistration in either camera or receiver; in the case of small objects, it may cause them to appear separated into different color. 18

colorfulness of a perceived color (illuminating engineering). The attribute according to which it appears to exhibit more or less chromatic color. For a stimulus of a given chromaticity, colorfulness normally increases as the absolute luminance is increased. 167

color grading. *See:* color comparison.

colorimetric purity of a light, (illuminating engineering). The ratio L_1/L_2 where L_1 is the luminance of the single frequency component which must be mixed with a reference standard to match the color of the light and L_2 is the luminance of the light. 167

colorimetric shift (illuminating engineering). The change of chromaticity and luminance factor of an object color due to change of the light source. *See:* **adaptive color shift; resultant color shift.** 167

colorimetry (1) (illuminating engineering). The measurement of color. 167
(2) (television). The techniques for the measurement of color and for the interpretation of the results of such measurements. *Note:* The measurement of color is made possible by the properties of the eye, and is based on a set of conventions. 18

coloring (electroplating) (1) (chemical). The production of desired colors on metal surfaces by appropriate chemical action. *See:* **electroplating.** 328
(2) (buffing). Light buffing of metal surfaces, for the purpose of producing a high luster. *See:* **electroplating.** 328

color light signal. A fixed signal in which the indications are given by the color of a light only. 328

color match (colorimetry) (television). The condition in which the two halves of a structureless photometric field are judged by the observer to have exactly the same appearance. *Note:* A color match for the standard observer may be calculated. 18

color matching (illuminating engineering). Action of making a color appear the same as a given color. 167

color matching functions (spectral tristimulus values)

$x(\lambda) = X_\lambda/\Phi_{e\lambda};$
$y(\lambda) = Y_\lambda/\Phi_{e\lambda};$
$z(\lambda) = Z_\lambda/\Phi_{e\lambda}$ **(illuminating engineering).** The tristimulus value per unit wavelength interval and unit spectral radiant flux. *Note:* Color-matching functions have been adopted by the International Commission on Illumination. They are tabulated as functions of wavelength throughout the spectrum and are the basis for the evaluation of radiant energy as light and color. The standard values adopted by the CIE in 1931 are given in section 13.5 of ANSI/IES RP-16-1980. The $\overline{y}$ values are identical with the values of spectral luminous efficiency for photopic vision. *See:* **values of spectral luminous efficiency for photopic vision; section 13.4 of ANSI/IES RP-16-1980.** The $\overline{x}$, $\overline{y}$, and $\overline{z}$ values for the 1931 Standard Observer are based on a 2 degree bipartite field, and are recommended for predicting matches for stimuli subtending between 1 degree and 4 degree. Supplementary data based on a 10 degree bipartite field were adopted in 1964 for use for angular subtenses greater than 4 degrees (see section 13.5). Tristimulus computational data for CIE standard sources A and C are given in section 13.6 of ANSI/IES RP-16-1980. 167

color mixture (television). Color produced by the combination of lights of different colors. *Notes:* (1) The combination may be accomplished by successive presentation of the components, provided the rate of alternation is sufficiently high; or the combination may be accomplished by simultaneous presentation, either in the same area or on adjacent areas, provided they are small enough and close enough together to eliminate pattern effects. (2) A color mixture as here defined is sometimes denoted as an additive color mixture to distinguish it from combinations of dyes, pigments, and other absorbing substances. Such mixtures of substances are sometimes called subtractive color mixtures, but might more appropriately be called colorant mixtures. 18

color mixture data (television). *See:* **tristimulus values.** 18

color of a physical stimulus (illuminating engineering). One of the ways in which the word color may be used is to designate the property of light falling on the retina which causes it to generate an impression perceived as having or lacking a quality such as whiteness, redness, greenness, and the like. This property of light is determined by its spectral power distribution and may be specified in terms of its chromaticity and luminance. This same property of light may be imputed to a to a beam of light being propagated through space or originating at a distal stimulus. The distal stimulus itself may be described as colored because it gives off colored light. 167

color picture signal*. *See:* **noncomposite color picture signal; composite color picture signal.**
*Deprecated

color-picture tube. An electron tube used to provide an image in color by the scanning of a raster and by varying the intensity of excitation of phosphors to produce light of the chosen primary colors. 125

color plane (multibeam color-picture tubes). A surface approximating a plane containing the color centers.   125

color-position light signal. A fixed signal in which the indications are given by the color and the position of two or more lights.   328

color preference index (of a light source)(R_p)(illuminating engineering). Measure appraising a light source for enhancing the appearance of an object or objects by making their colors tend toward people's preferences. Judd's flattery index is an example.   167

color-purity magnet. A magnet in the neck region of a color-picture tube to alter the electron beam path for the purpose of improving color purity.   125

color rendering index (of a light source) (CRI) (illuminating engineering). Measure of the degree of color shift objects undergo when illuminated by the light source as compared with the color of those same objects when illuminated by a reference source of comparable color temperature.   167

color-selecting-electrode system. A structure containing a plurality of openings mounted in the vicinity of the screen of a color-picture tube (electron tubes), the function of this structure being to cause electron impingement on the proper screen area by using either masking, focusing, deflection, reflection, or a combination of these effects. *Note:* For examples see **shadow mask.**   125

color-selecting-electrode system transmission (electron tubes). The fraction of incident primary electron current that passes through the color-selecting-electrode system.   125

color signal (color television system). Any signal at any point for wholly or partially controlling the chromaticity values of a color television picture. *Note:* This is a general term that encompasses many specific connotations such as those conveyed by the words color picture signal (either composite or noncomposite), chrominance signal color carrier signal, luminance signal (in color television).   18

color sync signal (color television). A signal used to establish and maintain the same color relationships that are transmitted. *Note:* In Rules Governing Radio Broadcast Services, Part 3, of the Federal Communications Commission, the color sync signal consists of a sequence of color bursts that recur every line except for a specified time interval during the vertical interval, each burst occurring on the back porch.   18

color temperature (television). The absolute temperature of the full (black-body) radiator for which the ordinates of the spectral distribution curve of emission are proportional (or approximately so) in the visible regions, to those of the distribution curve of the radiation considered, so that both radiations have the same chromaticity. *Note:* In certain countries, by extension, the term color temperature is used in the case of a selective radiator when, for the colorimetric standard observer, this radiator has the same color (or at least approximately the same color) as a full radiator at a certain temperature; this temperature is then called the color temperature of the selective radiator.   18

color temperature of a light source (illuminating engineering). The absolute temperature of a blackbody radiator having a chromaticity equal to that of the light source.   167

color tracking (television). (1) The degree to which color balance is maintained over the complete range of the achromatic (neutral gray) scale. (2) A qualitative term indicating the degree to which constant chromaticity within the achromatic region in the chromaticity diagram is achieved on a color-display device over the range of luminances produced from a monochrome signal. *See:* **television.**   178

color transmission (color television). The transmission of a signal wave for controlling both the luminance values and the chromaticity values in a picture.   18

color triad (phosphor-dot screen). A color cell of a three-color phosphor-dot screen.   125

color triangle (television). A triangle drawn on a chromaticity diagram, representing the entire range of chromaticities obtainable as additive mixtures of three prescribed primaries represented by the corners of the triangle.   18

Colpitts oscillator. An electron tube or solid state circuit in which the parallel-tune tank circuit is connected between grid and plate, the capacitive portion of the tank circuit being comprised of two series elements, the connection between the two being at cathode potential with the feedback voltage obtained across the grid-cathode portion of the capacitor. *See:* **radio-frequency generator (electron tube type).**   211

column (1)(positional notation). (A) A vertical arrangement of characters or other expressions. (B) Loosely, a digital place. *See:* **place.**   235
(2)(test pattern language). A group of words or bits in a memory, identified by a common Y-address.   463

column binary. Pertaining to the binary representation of data on punched cards in which adjacent positions in a column correspond to adjacent bits of data, for example, each column in a 12-row card may be used to represent 12 consecutive bits of a 36-bit word.   255,77

column enable (CAS) (CE) (semiconductor memory). The input used to strobe in the column address in multiplexed address random access memories (RAM).   441

column, positive. *See:* **positive column.**

combinational logic element. (1) A device having zero or more input channels and one output channel, each of which is always in one of exactly two possible physical states, except during switching transients. *Note:* On each of the input channels and the output channel, a single state is designated arbitrarily as the "one" state, for that input channel or output channel, as the case may be. For each input channel and output channel, the other state may be referred to as the "zero" state. The device has the property that the output channel state is determined completely by the comtemporaneous input-channel-state combination, to within switching transients. (2) By extension, a device similar to (1) except that one or more of the input channels or the output channel, or both, have a finite number, but more than two, possible physical states each of which is designated as a distinct logic state. The output channel state is determined completely by the contemporaneous input-channel-state combination, to within switching transients. (3) A device similar to (1) or (2) except that it has more than one output channel. *See:* **AND gate; OR gate.**   235

combinational logic function (graphic symbols for logic functions). A logic function in which there exists one and only one resulting combination of states of the outputs for each possible combination of input states. *Note:* The terms combinative and combinatorial have also been used to mean combinational.   451

combination buoy (navigation aid terms). A buoy that has more than one means of conveying intelligence. *See:* **buoy.**   526

combination controller (industrial control). A full magnetic or semimagnetic controller with additional externally operable disconnecting means contained in a common enclosure. The disconnecting means may be a circuit breaker or a disconnect switch. *See:* **electric controller.**   206

combination current and voltage regulation. That type of automatic regulation in which the generator regulator controls both the voltage and current output of the generator. *Note:* This type of control is designed primarily for the purpose of insuring proper charging of storage batteries on cars or locomotives. *See:* **axle generator system.**   328

combination detector (fire protection devices). A device that either responds to more than one of fire phenomena (heat, smoke, or flame) or employs more than one operating principle to sense one of these phenomena.   71

combination electric locomotive. An electric locomotive, the propulsion power for which may be drawn from two or more sources, either located on the locomotive of elsewhere. *Note:* The prefix combination may be applied to cars, buses, etcetera, of this type. *See:* **electric locomotive.**   328

combination lighting and appliance branch circuit. A circuit supplying energy to one or more lighting outlets and to one or more appliance outlets. *See:* **branch circuit.**   328

combination microphone. A microphone consisting of a combination of two or more similar or dissimilar microphones. *Examples:* Two oppositely phased pressure microphones acting as a gradient microphone; a pressure microphone and velocity microphone acting as a unidirectional microphone. *See:* **microphone.**   328

combination rubber tape. The assembly of both rubber and friction tape into one tape that provides both insulation and mechanical protection for joints.   328

combinations of pulses and waveforms (pulse terms). *See:* **double pulse; bipolar pulse; staircase.**

combination starter (packaging machinery). A starter having manually-operated disconnecting means built into the same enclosure with the magnetic contactor.   429

combination support (raceway systems for Class 1E circuits for nuclear power generating stations). A support that serves either raceways or different types of raceway(s) and other mechanical or electric systems such as heating, ventilating, and air-conditioning (HVAC) ducts, piping, and lighting fixtures.   513

combination thermoplastic tape. An adhesive tape composed of a thermoplastic compound that provides both insulation and mechanical protection for joints.   328

combination watch-report and fire-alarm system. A coded manual fire-alarm system, the stations of which are equipped to transmit a single watch-report signal or repeated fire-alarm signals. *See:* **protective signaling.**   328

combined-line-recording trunk (CLR) (telephone switching). A one-way trunk for operator recording and extending of toll calls.   55

combined mechanical and electrical strength (insulator). The loading in pounds at which the insulator fails to perform its function either electrically or mechanically, voltage and mechanical stress being applied simultaneously. *Note:* The value will depend upon the conditions under which the test is made. *See:* **insulator; tower.**   64

combined telephone set. A telephone set including in a single housing all the components required for a complete telephone set except the handset which it is arranged to support. *Note:* Wall hand telephone sets are of this type, but the term is usually reserved for a self-contained desk telephone set to distinguish it from desk telephone sets requiring an associated bell box. A desk local-battery telephone set may be referred to as a combined set if it includes in its mounting all components except its associated local batteries. *See:* **telephone station.**   328

combined voltage and current influence (wattmeter). The percentage change (of full-scale value) in the indication of an instrument that is caused solely by a voltage and current departure from specified references while constant power at the selected scale point is maintained. *See:* **accuracy rating (instrument).**   280

combustible materials (power and distribution transformer). Materials which are external to the apparatus and made of or surfaced with wood, compressed paper, plant fibers, or other materials that will ignite and support flame.   53

combustion control. The regulation of the rate of combination of fuel with air in a furnace.   64

command (1)(logical link control). In data communications, an instruction represented in the control field of a protocol data unit (PDU) and transmitted by a logical link control (LLC). It causes the addressed LLC(s) to execute a specific data link control function.   585

(2) (electronic computation). (A) One of a set of several signals (or groups of signals) that occurs as a result of interpreting an instruction; the commands initiate the individual steps that form the process of executing the instruction's operation. (B) Loosely: an instruction in machine language. (C) Loosely: a mathematical or logic operator. (D) Loosely: an operation. 235,255,77,54

(3) (industrial control). An input variable established by means external to, and independent of, the feedback (automatic) control system. It sets, is equivalent to, and is expressed in the same units as the ideal value of the ultimately controlled variable. *See:* **control system, feedback; set point.** 206

command control (electric power system). A control mode in which each generating unit is controlled to reduce unit control error. *See:* **speed-governing system.** 94

command guidance (navigation aid terms). Guidance in which information transmitted to a craft from an outside source causes it to follow a prescribed path. 526

command language (software). A set of procedural operators with a related syntax, used to indicate the functions to be performed by an operating system. *Syn:* **control language.** *See:* **functions; operating system; operators; syntax.** 434

command link (communication satellite). A data transmission link (generally earth to spacecraft or satellite) used to command a satellite or spacecraft in space. 83

command PDU (protocol data unit)(logical link control). All PDU's transmitted by a logical link control (LLC) in which the C/R (command/response) bit is equal to 'O'. 585

command rate (gyro). The input rate equivalent of a torquer command signal. 46

command reference (servo or control system) (power supplies). The voltage or current to which the feedback signal is compared. As an independent variable, the command reference exercises complete control over the system output. *See:* **operational programming.** 186

comment (software). (1) Information embedded within a computer program, command language, or set of data which is intended to provide clarification to human readers, and that does not effect machine interpretation. (2) A description, reference, or explanation added to or interspersed among the statements of the source language, that has no effect in the target language. *See:* **command language; computer program; data; set; source language; target language.** 434

commercial data processing (computer applications). Data processing performed to support a commercial organization or function. 571

commercial grade part (replacement parts for Class 1E equipment in nuclear power generating stations). A part that is: (1) Not subject to design or specification requirements that are unique to nuclear power plants. (2) Used in applications other than nuclear power plants. (3) Ordered from the manufacturer/supplier on the basis of specifications set forth in the manufacturer's published product description (for example, a catalog). 582

commercial power (emergency and standby power). Power furnished by an electric power utility company; when available, it is usually the prime power source. However, when economically feasible, it sometimes serves as an alternative or standby source. *Syn:* **utility power** 512

commercial tank (electrorefining). An electrolytic cell in which the cathode deposit is the ultimate electrolytically refined product. *See:* **electrorefining.** 328

commissioning tests (rotating machinery). Tests applied to a machine at site under normal service conditions to show that the machine has been erected and connected in a correct manner and is able to work satisfactorily. *See:* **asynchronous machine.** 63

common-battery central office. *See:* **common-battery office.**

common-battery office (telephone switching systems). A central office that supplies transmitter and signaling currents for its associated stations and current for the central office equipment from a power source located in the central office. 55

common battery signaling (data transmission). A method of actuating a line or supervisory signal at the distant end of a telephone line by the closure of a direct-current (dc) circuit with the exchange providing the feeding current. 59

common-battery switchboard. A telephone switchboard for serving common-battery telephone sets. 193

common carrier (data communication). In telecommunications, a public utility company that is recognized by an appropriate regulatory agency as having a vested interest and responsibility in furnishing communication services to the general public. *See:* **specialized common carrier, value added service.** 12

common cause failure (reliability data for pumps and drivers, valve actuators, and valves). Two or more redundant component failures due to a single cause. The common cause events that cause multiple failures are usually secondary events or events which exceed the design envelope of the component. 502

common-channel interoffice (CCIS) signaling (telephone switching systems). The use of separate paths between switching entities to carry the signaling associated with a group of communication paths. 55

common control (telephone switching systems). An automatic switching arrangement in which the control equipment necessary for the establishment of connections is shared, being associated with a given call only during the period required to accomplish the control function. 55

common device (of a supervisory system) (power switchgear). A device in either the master or remote station that is required for the basic operation of the supervisory system and is not part of the equipment for the individual points. *Syn:* **basic device.** 103

common equipment (supervisory control, data acquisition, and automatic control)(station control and data acquisition). That complement of either the master or remote station supervisory equipment that interfaces with the interconnecting channel and is otherwise basic to the operation of the supervisory system, but is exclusive of those elements that are peculiar to and are required for the particular applications and uses of the equipment. 570, 403

common-mode (1) (general). The instantaneous algebraic average of two signals applied to a balanced circuit, both signals referred to a common reference. *See:* **oscillograph.** 185

(2) (in-phase signal) (medical electronics). A signal applied equally and in phase to the inputs of a balanced amplifier or other differential device. 192

common-mode conversion (interference terminology). The process by which differential-mode interference is produced in a signal circuit by a common-mode interference applied to the circuit. *Note:* See the accompanying figure.

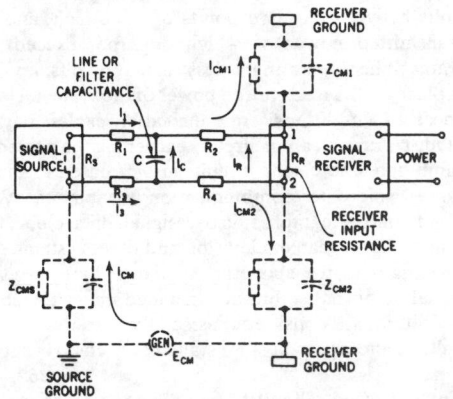

Common-mode conversion.

Common-mode currents are converted to differential-mode voltages by impedances R_1, R_2, R_3, R_4, R_S, R_R, and c. The differential-mode voltage at the receiver resulting from the conversion is the algebraic summation of the voltage drops produced by the various currents in these impedances. Various of the impedances may be neglected at particular frequencies. At direct current,

$$V_{CM} = I_r R_r \approx I_{CM1}(R_s + R_1 + R_2) - I_{CM2}(R_3 + R_4)$$

At

$$f > \frac{I}{c(R_1 + R_3 + R_s)},$$

$$V_{CM} \approx I_c X_c \frac{R_R}{R_2 + R_4 + R_R}$$

See: **interference.** 188

common mode failure (nuclear power generating stations)(safety systems equipment in nuclear power generating stations). Multiple failures attributable to a common cause. 102, 408, 440, 535

common-mode interference (automatic null-balanced electrical instruments). (1) Interference that appears between both signal leads and a common reference plane (ground) and causes the potential of both sides of the transmission path to be changed simultaneously and by the same amount relative to the common reference plane (ground). *See:* **interference.** 188
(2) A form of interference that appears between any measuring circuit terminal and ground. *See:* **accuracy rating (instrument).** 295

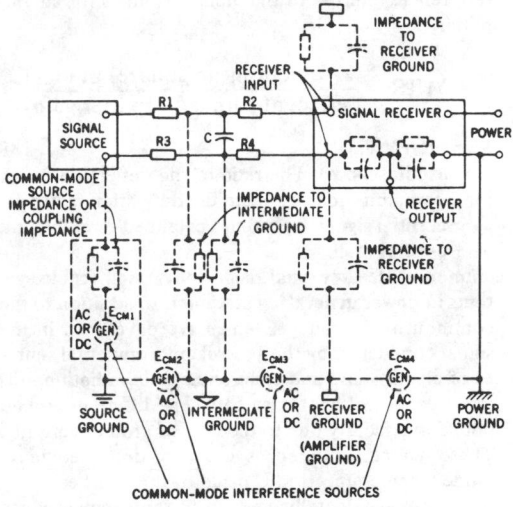

Common-mode interference–sources and current paths. The common-mode voltage V_{CM} in any path is equal to the sum of the common-mode generator voltages in that path; for example, in the source-receiver path,

$$V_{CM} = E_{CM1} + E_{CM2} + E_{CM3}.$$

common mode noise (longitudinal) (cable systems in power generating stations). The noise voltage which appears equally and in phase from each signal conductor to ground. Common mode noise may be caused by one or more of the following: (1) **Electrostatic induction.** With equal capacitance between the signal wires and the surroundings, the noise voltage developed will be the same on both signal wires. (2) **Electromagnetic induction.** With the magnetic field linking the signal wires equally, the noise voltage developed will be the same on both signal wires. 35

common-mode radio noise. Conducted radio noise that appears between a common reference plane (ground) and all wires of a transmission line causing their potentials to be changed simultaneously and by the same amount relative to the common reference plane (ground). 418

common-mode rejection (in-phase rejection). The ability of certain amplifiers to cancel a common-mode signal while responding to an out-of-phase signal. *See:* **degeneration negative feedback.** 192

common-mode rejection quotient (in-phase rejection quotient). The quotient obtained by dividing the response to a signal applied differentially by the response to the same signal applied in common mode, or the relative magnitude of a common-mode signal that produces the same differential response as a standard differential input signal. 192

common-mode rejection ratio (CMRR). (1) (signal transmission signal). The ratio of the common-mode interference voltage at the input terminals of the system to the effect produced by the common-mode interference, referred to the input terminals for an amplifier. For example,

$$CMRR = \frac{V_{CM}(\text{root-mean-square}) \text{ at input}}{\text{effect at output}/\text{amplifier gain}}$$

See: **interference.** 188

(2) (oscilloscopes). The ratio of the deflection factor for a common-mode signal to the deflection factor for a differential signal applied to a balanced-circuit input. *See:* **oscillograph.**

common mode to normal mode conversion (cable systems in power generating stations). In addition to the common mode voltages which are developed in the signal conductors by the general environmental sources of electrostatic and electromagnetic radiation, differences in voltage exist between different ground points in a facility due to the flow of ground currents. These voltage differences are considered common mode when connection is made to them either intentionally or accidentally, and the currents they produce are common mode. These common mode currents can develop normal mode noise voltage across circuit impedances. 35

common return. A return conductor common to several circuits. *See:* **center of distribution.** 64

common spectrum multiple access (CSMA) (communication satellite). A method of providing multiple access to a communication satellite in which all of the participating earth stations use a common time-frequency domain. Signal processing is employed to detect a wanted signal in the presence of others. Three typical approaches utilizing these techniques are spread spectrum, frequency-time matrix, and frequency-hopping. 84

common trunk (telephone switching systems). A trunk, link, or junctor accessible from all input groups of a grading. 55

common use (NESC). Simultaneous use by two or more utilities of the same kind. 494

common winding (autotransformer) (power and distribution transformer). That part of the autotransformer winding which is common to both the primary and the secondary circuits. 53

communication (telecommunication) (electric system) (data transmission). The transmission of information from one point to another by means of electromagnetic waves. 59

communication band. *See:* **frequency band of emission.**

communication circuits (electric installations on shipboard). Circuits used for audible and visual signals and communication of information from one place to another, within or on the vessel. 3

communication conductor (measuring longitudinal balance of telephone equipment operating in the voice band). A conductor used in a communication network. 529

communication control character (data communication). A functional character intended to control or facilitate transmission over data networks. Control characters form the basis for character-oriented communications control procedures. 12

communication facility (data transmission). Anything used or available for use in the furnishing of communication service. 59

communication lines (NESC)(transmission and distribution. . The conductors and their supporting or containing structures which are used for public or private signal or communication service, and which operate at potentials not exceeding 400 volts to ground or 750 volts between any two points of the circuit, and the transmitted power of which does not exceed 150 watts. When operating at less than 150 volts, no limit is place on the transmitted power of the system. Under specified conditions, communication cables may include communication circuits exceeding the preceding limitation where such circuits are also used to supply power solely to communication equipment. *Note:* Telephone, telegraph, railroad-signal, data, clock, fire, police-alarms, cable television and other systems conforming with the above are included. Lines used for signaling purposes, but not included under the above definition, are considered as supply lines of the same voltage and are to be so installed.*See:* **electric supply lines.** 262, 494

communication reliability (mobile communication). A specific criterion of system performance related to the percentage of times a specified signal can be received in a defined area during a given interval of time. *See:* **mobile communication system.** 181

communication satellite. A satellite used for communication between two or more ground points by transmitting the messages to the satellite and retransmitting them to the participating ground station. 83

communications common carrier (data transmission). A company recognized by an appropriate regulatory agency as having a vested interest in furnishing communications services to the public at large. 59

communications computer (data communications). A computer that acts as the interface between another computer or terminal and a network, or a computer controlling data flow in a network. *See:* **front end computer, concentrator.** 12

communications interface equipment. A portion of a relay system which transmits information from the relay logic to a communications link, or conversely to logic, for example, audio tone equipment, a carrier transmitter-receiver when an integral part of the relay system. 90

communications link. Any of the communications media, for example, microwave, power line carrier, wire line. 90

communication theory (data transmission) The mathematical theory underlying the communication of messages from one point to another.   59

community dial office (telephone switching systems). A small automatic central office that serves a separate exchange area which ordinarily has no permanently assigned central office operating or maintenance forces.   55

community-of-interest (telephone switching systems). A characteristic of traffic resulting from the calling habits of the customers.   55

commutated antenna direction finder (CADF)(navigation aid terms). A system using a multiplicity of antennas in a circular array and a receiver which is connected to the antennas in sequence through a commutating device for finding the direction of arrival of radio waves; the directional sensing is related to phase shift which occurs as a result of the communication.   526

commutating angle (1) (rectifier circuits). The time, expressed in degrees, during which the current is commutated between two rectifying elements. See: rectification; rectifier circuit element.   208

(2) (thyristor converter circuit). (μ) The time, expressed in degrees (1 cycle of the ac wave form - 360°, during which the current is commutated between two thyristor converter circuit elements.   121

commutating capacitor (converter circuit elements)-(self-commutated converters). A capacitor that provides commutating voltage for circuit-commutated thyristors in a self-commutated converter.   584

commutating-field winding. An assembly of field coils located on the commutating poles, that produces a field strength approximately proportional to the load current. The commutating field is connected in direction and adjusted in strength to assist the reversal of current in the armature coils for successful commutation. This field winding is used along, or supplemented by a compensating winding. See: asynchronous machine.   328

commutating group (rectifier circuit). A group of rectifier-circuit elements and the alternating-voltage supply elements conductively connected to them in which the direct current of the group is commutated between individual elements that conduct in succession. See: circuit element; rectification; rectifier.   208

commutating impedance (rectifier transformer). The impedance that opposes the transfer of current between two direct-current winding terminals of a commutating group, or a set of commutating groups. See: rectifier transformer.   258

commutating period (inverters). The time during which the current is commutated. See: self-commutated inverters.   208

commutating pole (interpole). An auxiliary pole placed between the main poles of a commutating machine. Its exciting winding carries a current proportional to the load current and produces a flux in such a direction and phase as to assist the reversal of the current in the short-circuited coil.   328

commutating reactance (thyristor converter). The reactance that effectively opposes the transfer of current

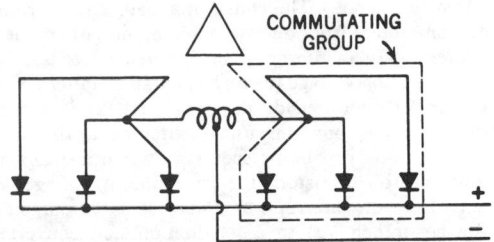

Commutating group in a single-way rectifier circuit.

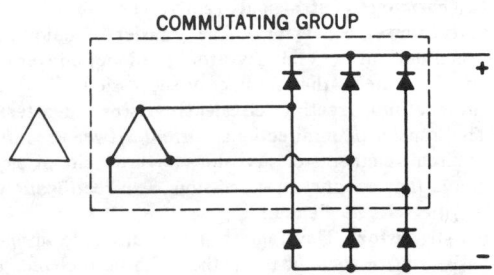

Commutating group in a double-way rectifier circuit.

between thyristor converter circuit elements of a commutating group or set of commutating groups. Note: For convenience, the reactance from phase to neutral, or one half the total reactance in the commutating circuit, is the value usually employed in computations, and is designated as the commutating reactance.   121

commutating reactance factor (rectifier circuit). The line-to-neutral commutating reactance in ohms multiplied by the direct current commutated and divided by the effective (root-mean-square) value of the line-to-neutral voltage of the transformer direct-current winding. See: circuit element; rectification; rectifier circuit.   208

commutating reactance transformation constant. A constant used in transforming line-to-neutral commutating reactance in ohms on the direct-current winding to equivalent line-to-neutral reactance in ohms referred to the alternating-current winding. See: rectification.   208

commutating reactor (1)(converter circuit elements)-(commutating inductor)(self-commutated converters). An inductor having one or more windings that modifies or couples the transient current produced by the commutating voltage.   584

(2)(power and distribution transformer). A reactor used primarily to modify the rate of current transfer between rectifying elements.   53

commutating resistance (rectifier transformer). The resistance component of the commutating impedance. See: rectifier transformer.   258

commutating voltage (circuit properties)(self-commutated converters). The voltage that causes the current to commutate from one switching branch to another. *Notes:* (1) In an internally commutated converter, the commutating voltage is supplied by an ac (alternating current) source outside the converter. (2) In a self-commutated converter using switching device that have turn-off capability, such as power transistors or gate turn-off thyristors, the commutating voltage results from the interruption of current in the outgoing device branch. (3) In a self-commutated converter using circuit-commutated thyristors, the commutating voltage is usually supplied by capacitors.   584

commutation (1)(circuit properties)(self-commutated converters). The transfer of current from one converter switching branch to another.   584

(2)(harmonic control and reactive compensation of static power converters). The transfer of unidirectional current between thyristor (or diode) converter circuit elements that conduct in succession.   533

(3) (rectifier) (rectifier circuit) (thyristor converter). The transfer of unidirectional current between rectifier circuit elements or thyristor converter circuit elements that conduct in succession. *See:* **rectification; rectifier circuit element.**   121

(4) (thyristor). The transfer of the current from one basic control element to another with both elements conducting current simultaneously.   445

commutation elements (semiconductor rectifiers). The circuit elements used to provide circuit-commutated turnoff time. *See:* **semiconductor rectifier stack.**   208

commutation factor (1) (rectifier circuits). The product of the rate of current decay at the end of conduction, in amperes per microsecond, and the initial reverse voltage, in kilovolts. *See:* **element; rectification; rectifier circuit.**   208

(2) (gas tubes). The product of the rate of current decay and the rate of the inverse voltage rise immediately following such current decay. *Note:* The rates are commonly states in amperes per microsecond and volts per microsecond. *See:* **heterodyne conversion transducer (converter); gas tube; rectification.**   125

commutation interval (circuit properties)(self-commutated converters). The time interval between the application of commutating voltage to a pair of commutating branches and the cessation of the resulting transient currents. *Note:* The commutation interval is the same as the overlap interval in an externally commutated converter in which the commutating voltage is supplied by the ac (alternating current) line.   584

commutation shrink ring. A member that holds the commutator-segment assembly together and in place by being shrunk on an outer diameter of and insulated from the commutator-segment assembly. *See:* **commutator.**   328

commutator (rotating machinery). An assembly of conducting members insulated from one another, in the radial-axial plane, against which brushes bear, used to enable current to flow from one part of the circuit to another by sliding contact.   63

commutator bars. *See:* **commutator segments.**

commutator bore. Diameter of the finished hole in the core that accommodates the armature shaft. *See:* **commutator.**   328

commutator brush track diameter. That diameter of the commutator segment assembly that after finishing on the armature is in contact with the brushes. *See:* **commutator.**   328

commutator core. The complete assembly of all of the retaining members of a commutator. *See:* **commutator.**   328

commutator-core extension. That portion of the core that extends beyond the commutator segment assembly. *See:* **commutator.**   328

commutator inspection cover. A hinged or otherwise attached part that can be moved to provide access to commutator and brush rigging for inspection and adjustment. *See:* **commutator.**   328

commutator insulating segments (rotating machinery). The insulation between commutator segments.   63

commutator insulating tube (rotating machinery). The insulation between the underside of the commutator segment assembly and the core. *See:* **commutator.**   63

commutator motor meter. A motor type of meter in which the rotor moves as a result of the magnetic reaction between two windings, one of which is stationary and the other assembled on the rotor and energized through a commutator and brushed. *See:* **electricity meter (meter).**   28

commutator nut. The retaining member that is used in combination with a vee ring and threaded shell to clamp the segment assembly. *See:* **commutator.**   328

commutator riser (rotating machinery). A conducting element for connecting a commutator segment to a coil. *See:* **commutator.**   63

commutator-segment assembly. A cylindrical ring or disc assembly of commutator segments and insulating segments that are bound and ready for installation. *Note:* The binding used may consist of wire, temporary assembly rings, shrink rings, or other means. *See:* **commutator.**   328

commutator segments (commutator bars). Metal current-carrying members that are insulated from one another by insulting segments and that make contact with the brushes. *See:* **commutator.**   328

commutator shell. The support on which the component parts of the commutator are mounted. *Note:* The commutator may be mounted on the shaft, on a commutator spider, or it may be integral with a commutator spider. *See:* **commutator.**   328

commutator-shell insulation (rotating machinery). The insulation between the under (or in the case of a disc commutator, the back) side of the commutator assembled segments and the commutator shell. *See:* **commutator.**   63

commutator vee ring. The retaining member that, in combination with a commutator shell, clamps or binds the commutator segments together. *See:* **commutator.**   328

commutator vee ring insulation (rotating machinery). The insulation between the V-ring and the commutator segments.   63

commutator vee-ring insulation extension (rotating machinery). The portion of the vee-ring insulation that extends beyond the commutator segment assembly. *See:* **commutator.**   63

companding (data transmission). A process in which compression is followed by expansion. *Note:* Companding is often used for noise reduction, in which case the compression is applied before the noise exposure and the expansion after the exposure.   59, 242

companding, instantaneous. *See:* **instantaneous companding.**

compandor (data transmission). A combination of a compressor at one point in a communication path for reducing the amplitude range of signals followed by an expander at another point for a complementary increase in the amplitude range. *Note:* The purpose of a compandor is to improve the ratio of the signal to the interference entering in the path between the compressor and expander.   59

comparative tests (test, measurement and diagnostic equipment). Comparative tests compare end item signal or characteristic values with a specified tolerance band and present the operator with a go/no-go readout; a go for signals within tolerances, and a no-go for signals out of tolerance.   54

comparator (1). A circuit for performing amplitude selection between either two variables or between a variable and a constant.   77

(2) (test, measurement and diagnostic equipment). A device capable of comparing a measured value with predetermined limits to determine if the value is within these limits.   54

(3) (analog computers). A circuit, having only two logic output states, for comparing the relative amplitudes of two analog variables, or of a variable and a constant, such that the logic signal output of the comparator uniquely determines which variable is the larger at all times.   9

(4) (software). A software tool used to compare two computer programs, files, or sets of data to identify commonalities or differences. Typical objects of comparison are similar versions of source code, object code, data base files, or test results. *See:* **code; computer program; data; data base; files; software tool.**   434

comparer (1) (power switchgear). A signal element which performs an AND logic function.   103

(2) (relaying). A signal element which performs an "AND" logic function.   79

comparison amplifier (power supplies). A high-gain non-inverting direct-current amplifier that, in a bridge-regulated power supply, has as its input the voltage between the null junction and the common terminal. The output of the comparison amplifier drives the series pass elements.   186

comparison bridge (power supplies). A type of voltage-comparison-circuit whose configuration and principle of operation resemble a four-arm electric bridge. See the accompanying figure.

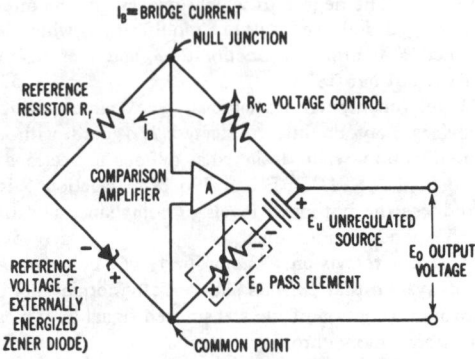

Comparison bridge connected as a voltage regulator. U.S. Patent No. 3,028,538–Foxbro Design Corp. *Note:* The elements are so arranged that, assuming a balance exists in the circuit, a virtual zero error signal is derived. Any tendency for the output voltage to change, in relation to the reference voltage, creates a corresponding error signal, that by means of negative feedback, is used to correct the output in the direction toward restoring bridge balance. *See:* **error signal.**   186

comparison lamp (luminous standards) (illuminating engineering). A light source having a constant, but not necessarily known, luminous intensity with which standard and test lamps are successively compared.   167

compartment (packaging machinery). A space within the base, frame, or column of the industrial equipment.   429

compass (navigation aid terms). An instrument for indicating a horizontal reference direction relative to the earth.   526

compass bearing (navigation aid terms). Bearing relative to compass north.   526

compass-controlled directional gyro. A device that uses the earth's magnetic field as a reference to correct a directional gyro. *Note:* The direction of the earth's field is sensed by a remotely located compass that is connected electrically to the gyro.   328

compass course (navigation aid terms). Course relative to compass north.   526

compass declinometer. *See:* **declinometer.**

compass deviation. *See:* **magnetic deviation.**

compass heading (navigation aid terms). Heading relative to compass north.   526

compass locator. *See:* **nondirectional beacon.**

compass north (navigation aid terms). The direction north as indicated by a magnetic compass.   526

compass repeater (navigation aid terms). That part of a remote-indicating compass system which repeats, at a distance, the indications of the master compass.   526

compass rose (navigation aid terms). A compass used to assist in aircraft magnetic compass compensation.   526

compatibility (general-system term)(1)(696 interface devices). The degree to which devices may be interconnected and used without modification, when designed to conform to Section 2, 3, and 4 of ANSI/ IEEE Std 696-1983.   538
(2)(microcomputer system bus). The degree to which devices may be interconnected and used without modification, when designed as defined in Sections 2 and 3 of ANSI/IEEE Std 796-1983. Section 5 introduces the notion of levels of compliance and the corresponding notation.   542
(3) (color television). The property of a color television system that permits substantially normal monochrome reception of the transmitted signal by typical unaltered monochrome receivers.   18
(4) (programmable instrumentation). The degree to which devices may be interconnected and used, without modification, when designed as defined throughout IEEE Std 488-1978 (for example, mechanical, electrical, or functional.   378
(5) (software). The ability of two or more systems to exchange information. *See:* **interoperability; systems.**   434

compatibility interfaces (medium attachment units and repeater units). The medium dependent interface (MDI) coaxial cable interface and the attachment unit interface (AUI) branch cable interface, the two points at which hardware compatibility is defined to allow connection of independently designed and manufactured components to the baseband transmission system.   543

compensated control system (control systems for steam turbine-generator units). An interconnected system that controls two or more variables (speed, load, pressure, etcetera) with compensation designed to minimize the interaction between the controlled variables.   522

compensated-loop direction-finder. A direction-finder employing a loop antenna and a second antenna system to compensate polarization error. *See:* **radio receiver.**   328

compensated repulsion motor. A repulsion motor in which the primary winding on the stator is connected in series with the rotor winding via a second set of brushes on the commutator in order to improve the power factor and commutation.   63

compensated semiconductor (charged-particle detectors)germanium gamma-ray detectors)(X-ray energy spectrometers). A semiconductor in which one type of impurity or imperfection (for example, donor) partially cancels the electric effects of the other type of impurity or imperfection (for example, acceptor).   119, 528, 147

compensated series-wound motor. A series-wound motor with a compensating-field winding. The compensating-field winding and the series-field winding may be combined into one field winding. *See:* **asynchronous machine.**   236,63

compensating-field winding (rotating machinery). Conductors embedded in the pole shoes and their end connections. It is connected in series with the commutating-field winding and the armature circuit. *Note:* A compensating-field winding supplements the commutating-field winding, and together they function to assist the reversal of current in the armature coils for successful commutation. *See:* **asynchronous machine.**   328

compensating-rope sheave switch. A device that automatically causes the electric power to be removed from the elevator driving-machine motor and brake when the compensating sheave approaches its upper or lower limit of travel. *See:* **hoistway (elevator or dumbwaiter).**   328

compensation (control system, feedback). A modifying of supplementary action (also, the effect of such action) intended to improve performance with respect to some specified characteristic. *Note:* In control usage, this characteristic is usually the system deviation. Compensation is frequently qualified as series, parallel, feedback, etcetera, to indicate the relative position of the compensating element. *See:* **control system, feedback; equalization.**   56,105

compensation theorem. States that if an impedance is inserted in a branch of a network, the resulting current increment produced in any branch in the network is equal to the current that would be produced at that point by a compensating voltage, acting in series with the modified branch, whose values is , where I is the original current that flowed where the impedance was inserted before the insertion was made.   328

compensator (1) (rotating machinery). An element or group of elements that acts to modify the functioning of a device in accordance with one or more variables. *See:* **asynchronous machine.**   63
(2) (radio direction-finders). That portion of a direction-finder that automatically applies to the direction indication all or a part of the correction for the deviation. *See:* **radio receiver.**   328
(3) (excitation systems). A feedback element of the regulator that acts to compensate for the effect of a variable by modifying the function of the primary detecting element. *Notes:* (A) Examples are reactive current compensator and active current compensator. A reactive current compensator is a compensator that acts to modify the function of a voltage regulator in accordance with reactive current. An active current compensator is a compensator that acts to modify the function of a voltage regulator in accordance with active current. (B) Historically, terms such as equalizing reactor and cross-current compensator have been used to describe the function of a reactive compensator. These terms are deprecated. (C) Reactive compensators are generally applied with generator voltage regulators to obtain reactive current sharing among generators operating in parallel. They function in the following two ways. (1) Reactive droop compensation is the more common method. It creates a droop in generator voltage proportional to reactive current and equivalent to that which would be produced by the

insertion of a reactor between the generator terminals and the paralleling point. (2) Reactive differential compensation is used where droop in generator voltage is not wanted. It is obtained by a series differential connection of the various generator current transformer secondaries and reactive compensators. The difference current for any generator from the common series current creates a compensating voltage in the input to the particular generator voltage regulator which acts to modify the generator excitation to reduce to minimum (zero) its differential reactive current. (D) Line drop compensators modify generator voltage by regulator action to compensate for the impedance drop from the machine terminals to a fixed point. Action is accomplished by insertion within the regulator input circuit of a voltage equivalent to the impedance drop. The voltage drops of the resistance and reactance portions of the impedance are obtained, respectively, in per unit quantities by an active compensator and a reactive compensator.   105

(4) (as applied to relaying) (power switchgear). A transducer with an air gapped core that produces an output voltage proportional to input current. The voltage modifies (or *compensates*) the voltage applied to the relay.   103

compensatory leads. Connections between an instrument and the point of observation so contrived that variations in the properties of leads, such as variations of resistance with temperature, are so compensated that they do not affect the accuracy of the instrument readings. *See:* **auxiliary device to an instrument.**
     328

compensatory wiring techniques (coupling in control systems). Those writing techniques which result in a substantial cancellation or counteracting of the effects of rates of change of electric or magnetic fields, without actually obstructing or altering the intensity of the fields. If the signal wires are considered to be part of the control circuit, these techniques change the susceptibility of the circuit. Example: twisting of signal and return wires associated with a susceptable instrument so as to cancel the voltage difference between wires caused by an existing varying magnetic field. *See:* **barrier wiring techniques; suppressive wiring techniques.**   43

compile (software). To translate a higher order language program into its relocatable or absolute machine code equivalent. *See:* **absolute machine code; assemble; higher order language program; interpret; relocatable.**   434

compiler (software). A computer program used to compile. A translator or an interpreter that is used to construct compilers. *Syn:* **metacompiler.** *See:* **assembler; compile; compiler generator; computer program; interpreter; metacompiler; translator.**   434

compiler compiler. *See:* **compiler generator.**   434

complement (1)(mathematics of computing). A numeral derived from a given numeral by a specified subtraction rule. Often used to represent the negative of the number represented by the given numeral. *See:* **radix complement; diminished-radix complement.**
     564

(2)(test pattern language). Another number in which each zero bit has been replaced by a one and each one bit has been replaced by a zero. Ones complement is formed by interchanging all ones and zeros. This is equivalent to logical inversion.   463

complementary commutation (circuit properties)-(self-commutated converters). Commutation occurs from one to the other of a complementary pair of principal switching branches arranged as a two-pulse group that conduct in alternate but not necessarily equal time intervals. The commutation may be direct or indirect. *Note:* An example of a converter employing complementary commutation is given in the figure below.   584

NOTE: The principal switching branches 1–6 are numbered in the order in which they begin conduction.

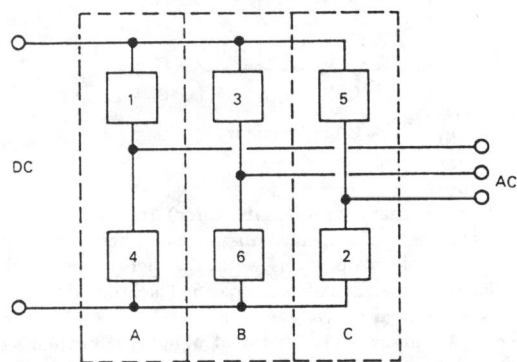

(a) Three 2-Pulse Commutating Groups: A, B, C

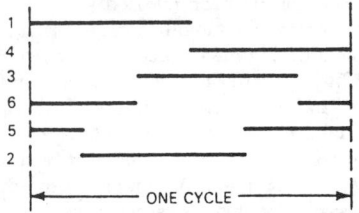

(b) Conducting Intervals of Principal Switching Branches 1–6

complementary function (automatic control). The solution of a homogeneous differential equation, representing a system or element, which describes a free motion.   56

complementary functions. Two driving-point functions whose sum is a positive constant. *See:* **linear passive networks.**   238

complementary tracking (power supplies). A system of interconnection of two regulated supplies in which one (the master) is operated to control the other (the slave). The slave supply voltage is made equal (or proportional) to the master supply voltage and of opposite polarity with respect to a common point. See the accompanying figure.   186

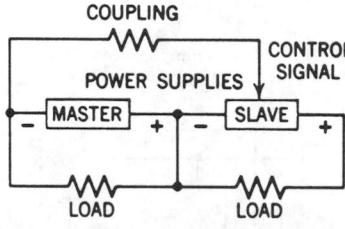

Complementary tracking.

complementary wavelength (color) (television). The wavelength of a spectrum light that, when combined in suitable proportions with the light considered, yields a match with the specified achromatic light. *See:* **dominant wavelength.**   18

complementary wavelength of a light, (illuminating engineering). The wavelength of radiant energy of a single frequency that, when combined in suitable proportion with the light, matches the color of a reference standard.   167

complete carry. A carry process in which a carry resulting from addition of carries is allowed to propagate. Contrasted with partial carry. *See:* **carry.**   235

completed call (telephone switching systems). An answered call that has been released.   55

complete diffusion (illuminating engineering). That in which the diffusing medium completely redirects the incident flux by scattering, that is, no incident flux can remain in an image-forming state.   167

complete failure. *See:* **failure, complete.**

complete operating test equipment (test, measurement and diagnostic equipment). Equipment together with the necessary detail parts, accessories, and components, or any combination thereof, required for the testing of a specified operational function.   54

complete reference designation (electric and electronics parts and equipments). A reference designation that consists of a basic reference designation and, as prefixes, all the reference designations that apply to the subassemblies or assemblies within which the item is located, including those of the highest level needed to designate the item uniquely. The reference designation for a unit consists of only a number.   17

complex capacitivity. *See:* **relative complex dielectric constant.**

complex conductivity (1) (antennas). For isotropic media, at a particular point, and for a particular frequency, the ratio of the complex amplitude of the total electric current density to the complex amplitude of the electric field strength. *Note:* The electric field strength and total current density are both expressed as phasors, with the latter composed of the conduction current density plus the displacement current density.   111

(2) (physical media). For isotropic media the ratio of the current density to the electric field, in which that current density comprises all currents at the point under consideration. *Note:* This term is used to describe both the conductive and dielectric properties of a medium   111

complex dielectric constant (antennas). The complex permittivity of a physical medium in ratio to the permittivity of free space. *See:* **relative permeability in physical media; relative complex dielectric constant.**   111

complexity (software). The degree of complication of a system or system component, determined by such factors as the number and intricacy of interfaces, the number and intricacy of conditional branches, the degree of nesting, the types of data structures, and other system characteristics. *See:* **component; data structures; interfaces; nesting; system.**   434

complex permeability (2) (magnetic core testing). Under stated conditions, the complex quotient of vectors representing induction and field strength inside the core material. One of the vectors is made to vary sinusoidally and the other referenced to it.

Series representation:

$$\bar{\mu} = \mu'_s - j\mu''_s = \frac{1}{\mu_0}\frac{\bar{B}}{H}$$

Parallel representation:

$$\frac{1}{\bar{\mu}} = \frac{1}{\mu'_p} - \frac{1}{j\mu''_p} = \mu_0\frac{\overline{H}}{\overline{B}}$$

165

complex permittivity (antennas). For isotropic media, the ratio of the complex amplitude of the electric displacement density to the complex amplitude of the electric field strength. *See:* **relative complex dielectric constant.**   111

complex permittivity in physical media (antenna). For isotropic media the ratio of the electric flux density to the electric field, in which the displacement current density represents the total current density. *Note:* This term is used to describe both the conductive and dielectric properties of a medium.   111

complex plane (automatic control). A plane defined by two perpendicular reference axes, used for plotting a complex variable or functions of this variable, such as a transfer function. 56

complex polarization ratio (antennas). For a given field vector at a point in space, the ratio of the complex amplitudes of two specified orthogonally polarized field vectors into which the given field vector has been resolved. *Note:* For these amplitudes to define definite phase angles particular unitary vectors (basis vectors) must be chosen for each of the orthogonal polarizations. *See:* **plane wave, Note 2; polarization vector, Note 1.** 111

complex power (rotating machinery). *See:* **power, phasor; phasor power.**

complex target (radar). A target composed of more than one scatterer within a single radar resolution cell. Note the distinction between this term and "distributed target". A target may be both complex and distributed. *See:* **distributed target.** 13

complex tone. (1) A sound containing simple sinusoidal components of different frequencies. (2) A sound sensation characterized by more than one pitch. 176

complex variable (automatic control). A convenient mathematical concept having a complex value, that is having a real part and an imaginary part. *Note:* In control systems, the pertinent independent variable is a generalized frequency $s = \sigma + j\omega$ used in the Laplace transform. 56

complex waveforms (pulse terms). *See:* **combinations of pulses and transitions; waveforms produced by magnitude superposition; waveforms produced by continuous time superposition of simpler waveforms; waveforms produced by noncontinuous time superposition of simpler waveforms; waveforms produced by operations on waveforms.**

compliance (industrial control). A property reciprocal to stiffness. *See:* **control system, feedback.** 206

compliance extension (power supply). A form of master/slave interconnection of two or more current-regulated power supplies to increase their compliance voltage range through series connection. *See:* **compliance voltage.** 186

compliance voltage (power supplies). The output voltage of a direct-current power supply operating in constant-current mode. *Note:* The compliance range is the range of voltages needed to sustain a given value of constant current throughout a range of load resistances. 186

component (1)(reliability data for pumps and drivers, valve actuators, and valves). The largest entity of hardware for which data are most generally collected and expected to be reliable (for example, pump with motor, valve with operator, amplifier, pressure transmitter). It is generally an off-the-shelf item procured by the system designer as a basic building block for his system. It should be distinguished from seals, materials, nuts, bolts, and other piece parts from which the component is made. 502

(2)(reliable industrial and commercial power systems planning and design). A piece of equipment, a line or circuit, or a section of a line or circuit, or a group of items which is viewed as an entity for purposes of reliability evaluation. 561

(3)(seismic design of substations). The devices and equipment which are assembled at the erection site, or readily removed or accessed for maintenance, and which perform a function (for example, power circuit breakers, disconnect switches, relays, sensors). 465

(4)(unique identification in power plants). A part or assembly of parts that is viewed as an entity for purposes of design, operation, and reporting. 544

(5)(unique identification in power plants and related facilities). A part or assembly of parts considered an entity for purpose of design, operation, and reporting. 545

(6) (software). A basic part of a system or program. *See:* **program; system.** 434

component assembly (1)(unique identification in power plants). An assembly of components, physically contiguous, which is viewed as a single entity for purposes of procurement, for example, boric-acid control panel. 544

(2)(unique identification in power plants and related facilities). An assembly of contiguous components, considered as a single entity for purpose of procurement, that is, boric acid control panel. 545

component function (1)(unique identification in power plants). The action performed by a component within a system. 544

(2)(unique identification in power plants and related facilities). The primary function performed by a component (element) within a system. 545

component function identifier (unique identification in power plants and related facilities). A one to four character alpha-numeric code that identifies the function the component performs within the system. 545

component hazard (reliability data). The instantaneous failure rate of a component or its conditional probability of failure versus time. 379

components (1)(electric pipe heating systems). Items from which a system is assembled; for example, resistors, capacitors, wires, connectors, tubes, switches, etcetera. 448

(2)(safety systems equipment in nuclear power generating stations). Items from which equipment is assembled (for example, attachments, bearings, bolts, capacitors, connectors, governors, inspection access ports, instrument sensors, locking devices, position indicators, resistors, seals, sight glasses, springs, switches, transistors, tubes, wires, etcetera. *Note:* Certain items, for example, instrument sensors, may satisfy the definition of the term component or the term equipment as used in IEEE Std 627-1980. Where such items are included within defined boundaries of equipment items, they are correctly referred to as components. Where such items are installed outside of defined boundaries for equipment items and perform independent functions, they are correctly referred to as equipment. 535

(3)(switchgear assemblies for Class 1E applications

in nuclear power generating stations). Items from which the switchgear assemblies are made (for example, power circuit breakers, instrument transformers, protective relays, control switches, primary insulation, etcetera).   576

(4) (accident monitoring instrumentation). Discrete items from which a system is assembled.   421

(5) (nuclear power generating station). Discrete items from which a system is assembled. *Note:* Examples of components are: wires, transistors, switches, motors, relays, solenoids, pipes, fittings, pumps, tanks, or valves.   120

(6) (nuclear power generating station). Items from which the system is assembled (for example, resistors, capacitors, wires, connectors, transistors, tubes, switches, springs, etcetera).   408

(7) (nuclear power generating stations). Items from which equipment is assembled. (For example, as used in this document, a component is a resistor, capacitor, wire, connector, spring, terminal block, bus support, etcetera.)   440

component testing (software verification and validation plans). Testing conducted to verify the implementation of the design for one software element (for example, unit, module) or a collection of software elements.   511

composite bushing (outdoor apparatus bushing). A bushing in which the major insulation consists of several coaxial layers of different insulation materials.
                                                                  168

composite cable (communication practice). A cable in which conductors of different gauges or types are combined under one sheath. *Note:* Differences in length of twist are not considered here as constituting different types. *See:* **cable.**   328

composite color picture signal (National Television System Committee (NTSC) color television). The electric signal that represents complete color picture information and all sync signals.   18

composite color signal (color television). The color-picture signal plus blanking and all synchronizing signals.   18

composite color sync (National Television System Committee (NTSC) color television). The signal comprising all the sync signals necessary for proper operations of a color receiver. *Note:* This includes the deflection sync signals to which the color sync signal is added in the proper time relationship.   18

composite conductor. A composite conductor consists of two or more strands consisting of two or more materials. *See:* **conductor.**   64

composite controlling voltage (electron tube). The voltage of the anode of an equivalent diode combining the effects of all individual electrode voltages in establishing the space-charge-limited currents. *See:* **excitation (drive).**   125,190

composited circuit. A circuit that can be used simultaneously for telephony and direct-current telegraphy or signaling, separation between the two being accomplished by frequency discrimination. *See:* **transmission line.**   328

composite error (gyro; accelerometer). The maximum deviation of the output data from a specified output function. Composite error is due to the composite effects of hysteresis, resolution, non-linearity, non-repeatability, and other uncertainties in the output data. It is generally expressed as a percentage of half the output span. *See:* **input-output characteristics.**
                                                                  46

composite level (measuring the performance of tone address signaling systems). In two-tone signaling systems, the total power of the two tones comprising a specific signal present condition.   508

composite picture signal (television). The signal that results from combining a blanked picture signal with the asynchronizing signal. *See:* **television.**   178

composite plate (electroplating). An electrodeposit consisting of two or more layers of metals deposited separately. *See:* **electroplating.**   328

composite pulse (pulse navigational systems)(navigation aid terms). A pulse composed of a series of overlapping pulses received from the same signal source but by way of different paths.   526

composite set. An assembly of apparatus designed to provide one end of a composited circuit.   328

composite signaling (CX) (telephone switching systems). A form of polar-duplex signaling capable of simultaneously serving a number of circuits using low-pass filters to separate the signaling currents from the voice currents.   55

composite supervision. The use of a composite signaling channel for transmitting supervisory signals between two points in a connection.   328

composite waveform (pulse terms). A waveform which is, or which for analytical or descriptive purposes is treated as, the algebraic summation of two or more waveforms. *See:* composite waveform diagram below.

composite waveform

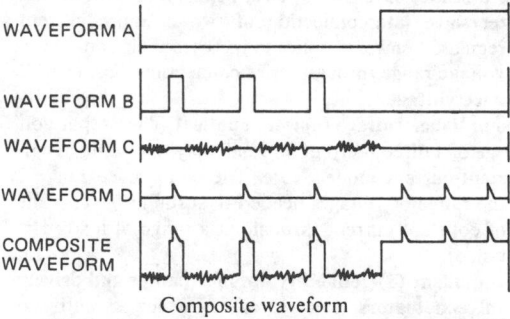

Composite waveform
                                                                  254

compound (rotating machinery). (1) A definite substance resulting from the combination of specific elements or radicals in fixed proportions: distinguished from mixture. (2) The intimate admixture of resin with ingredients such as fillers, softeners, plasticizers, catalysts, pigments, or dyes. *See:* **rotor (rotating machinery).**   63

compound circular horn antenna. A horn antenna of circular cross section with two or more abrupt changes of flare angle or diameter.   111

compound-filled (reactor, transformer) (grounding device). Having the coils/windings encased in an insulating fluid that becomes solid or remains slightly plastic at normal operating temperatures. *See: instrument transformer; reactor.* 91

compound-filled bushing (outdoor electric apparatus). A bushing in which the space between the inside surface of the porcelain/weather casing and the major insulation (or conductor where no major insulating is used) is filled with compound. 168

compound filled joints (power cable joints). Joints in which the joint housing is filled with an insulating compound that is non-fluid at normal operating temperatures. 34

compound-filled transformer (power and distribution transformer). A transformer in which the windings are enclosed with an insulating fluid which becomes solid, or remains slightly plastic, at normal operating temperatures. *Note:* The shape of the compound-filled transformer is determined in large measure by the shape of the contain or mold used to contain the fluid before solidification. 53

compound horn antenna * 111
*Deprecated

compounding curve (direct-current generator). A regulation curve of a compound-wound direct-current generator. *Note:* The shunt field may be either self or separately excited. 328

compound interferometer system. An antenna system consisting of two or more interferometer antennas whose outputs are combined using nonlinear circuit elements such that grating lobe effects are reduced. 111

compound rectangular horn antenna. A horn antenna of rectangular cross section in which at least one pair of opposing sides has two or more abrupt changes of flare angle or spacing. 111

compound source-rectifier exciter (excitation systems for synchronous machines). An exciter whose energy is derived from the currents and potentials of the ac terminals of the synchronous machine and converted to direct current by rectifiers. The exciter included the power transformers (current and potential), reactors, and rectifiers which may be either noncontrolled or controlled, including gate circuitry. It is exclusive of input control elements. 507

compound target* (radar). This term has been used to mean either "complex target" or "distributed target". Because of its ambiguity, it is deprecated.
*Deprecated 13

compound-wound. A qualifying term applied to a direct-current machine to denote that the excitation is supplied by two types of windings, shunt and series. *Note:* When the electromagnetic effects of the two windings are in the same direction, it is termed cumulative compound wound; when opposed, differential compound wound. 328

compound-wound generator (electric installations on shipboard). A dc generator which has two separate field windings: one, supplying the predominating excitation, is connected in parallel with the armature circuit and the other, supplying only partial excitation, is connected in series with the armature circuit and of such proportion as to require an equalizer connection for satisfactory parallel operation. 3

compound-wound motor (electric installations on shipboard). A dc motor which has two separate field windings: one, usually the predominating field, connected in parallel with the armature circuit, and the other connected in series with the armature circuit. (The characteristics as regards speed and torque are intermediate between those of shunt and series motors.) 3

compressed-air circuit breaker. *See: circuit breaker.* 103

compression (1) (data transmission). A process in which the effective gain applied to a signal is varied as a function of the signal magnitude, the effective gain being greater for small rather than for large signals. 59

(2) (television). The reduction in gain at one level of a picture signal with respect to the gain at another level of the same signal. *Note:* The gain referred to in the definition is for a signal amplitude small in comparison with the total peak-to-peak picture signal involved. A quantitative evaluation of this effect can be obtained by a measurement of differential gain. *See: black compression; white compression; television.* 178

(3) (oscillography). An increase in the deflection factor usually as the limits of the quality area are exceeded. *See: oscillograph.* 185

compressional wave. A wave in an elastic medium that is propagated by fluctuations in elemental volume, accompanied by velocity components along the direction of propagation only. *Note:* A compressional plane wave is a longitudinal wave. 176

compression joint (conductor stringing equipment). A tubular compression fitting designed and fabricated from aluminum, copper or steel to join conductors or overhead ground wires. It is usually applied through the use of hydraulic or mechanical presses. However, in some cases, automatic, wedge, and explosive type joints are utilized. *Syn: conductor splice; sleeve; splice.* 431

compression point (nonlinear, active, and nonreciprocal waveguide components). The level of the output signal at which the gain of a device is reduced by a specified amount, usually expressed in decibels, as in the 1 dB compression point. 530

compression ratio (gain or amplification). The ratio of (1) the magnitude of the gain (or amplification) at a reference signal level to (2) its magnitude at a higher stated signal level. *See: amplifier.* 125

compressor (data transmission). A transducer, which for a given amplitude range of input voltages, produces a smaller range of output voltages. One important type of compressor employs the envelope of speech signals to reduce their volume range by amplifying weak signals and attenuating strong signals. 59

compressor-stator-blade-control system (gas turbines). A means by which the turbine compressor stator blades are adjusted by vary the operating characteristics of the compressor. *See:* **speed-governing system (gas turbines).**   58

computation. *See:* **implicit computation.**

computer (1)(emergency and standby power analog computers). (A) A machine for carrying out calculations. (B) By extension, a machine for carrying out specified transformations on information.   512, 9
(2) (software). (A) A functional unit that can perform substantial computation, including numerous arithmetic operations, or logic operations without intervention by a human operator during a run. (B) A functional programmable unit that consists of one or more associated processing units and peripheral equipment, that is controlled by internally stored programs, and that can perform substantial computation, including numerous arithmetic operations or logic operations, without human intervention. *See:* **programs.**   434

computer-aided design (CAD)(computer applications). The use of computers to aid in design layout and analysis. May included modeling, analysis, simulation, or optimization of designs for production. Often used in combinations such as CAD/CAM. *See:* **computer-aided engineering; computer-aided manufacturing.**   571

computer-aided engineering (CAE)(computer applications). The use of computers to aid in engineering analysis and design. May included solution of mathematical problems, process control, numerical control, and execution of programs performing complex or repetitive calculations. *See:* **computer-aided design; computer-aided manufacturing.**   571

computer-aided inspection (CAI)(computer applications). The use of computers to inspect manufactured parts.   571

computer-aided instruction (CAI)(computer applications). The use of computers to present instructional material and to accept and evaluate student responses. *See:* **computer-based instruction.**   571

computer-aided management (CAM)(computer applications). The application of computers to business management activities. For example, database management, control reporting, and information retrieval. *See:* **decision support system; management information system.**   571

computer-aided manufacturing (CAM)(computer applications). The use of computers and numerical control equipment to aid in manufacturing processes. May include robotics, automation of testing, management functions, control, and product assembly. Often used in combinations such as CAD/CAM. *See:* **computer-aided design; computer-aided engineering.**   571

computer-aided typesetting (computer applications). The use of computers at any stage of the document composition process. This may involve text formatting, input from a word processing system, or computer-aided page makeup.   571

computer-assisted tester (test, measurement and diag-nostic equipment). A test not directly programmed by a computer but which operates in association with a computer by using some arithmetic functions of the computer.   54

computer-based instruction (CBI)(computer applications). The use of computers to support any process involving human learning.   571

computer code. A machine code for a specific computer.   255,77

computer component (analog computers). Any part, assembly, or subdivision of a computer, such as resistor, amplifier, power supply, or rack.   9

computer conferencing (computer applications). A form of teleconferencing that allows one or more users to exchange messages on a computer network.   571

computer control (physical process) (electric power systems). A mode of control wherein a computer, using as input the process variables, produces outputs that control the process. *See:* **power system.**   200

computer-control state (analog computers). One of several distinct and selectable conditions of the computer-control circuits. *See:* **balance check; hold; operate; potentiometer set; reset; static test.**   9

computer data (software). Data available for communication between or within computer equipment. Such data can be external (in computer-readable form) or resident within the computer equipment and can be in the form of analog or digital signals. *See:* **computer.**   434

computer diagram (analog computers). A functional drawing showing interconnections between computing elements, such interconnections being specified for the solution of a particular set of equations. *See:* **computer program; problem board.**   9

computer equation (machine equation) (analog computers). An equation derived from a mathematical model for use on a computer which is equivalent or proportional to the original equation. *See:* **scale factor.**   9

computer instruction. A machine instruction for a specific computer.   255, 77

computer-integrated manufacturing (CIM)(computer applications). Use of an integrated system of computer-controlled manufacturing centers. The centers may use robotics, design automation or CAD/CAM (computer-aided design/computer-aided manufacturing technologies.   571

computer interface equipment (surge withstand capability). A device which interconnects a protective relay system to an independent computer, for example, an analog to digital converter, a scanner, a buffer amplifier.   90

computer-managed instruction (CMI)(computer applications). The use of computers for management of student progress. Activities may include record keeping, progress evaluation, and lesson assignment. *See:* **computer-based instruction.**   571

computer network (1) (general). A complex consisting of two or more interconnected computing units.   255, 77

(2) (data communication). An interconnection of assemblies of computer systems, terminals and communications facilities. 12

(3) (software). A complex consisting of two or more interconnected computers. *See:* **computer.** 434

computer output microfilm (COM)(computer applications). The end result of a process that converts and records data from a computer directly to a microform. 571

computer program (1) (general). A plan or routine for solving a problem on a computer, as contrasted with such terms as fiscal program, military program, and development program. 255, 77, 54

(2) (analog computer). That combination of computer diagram, potentiometer list, amplifier list, trunk list, switch list, scaled equations, and any other documentation that defines the analog configuration for the particular problem to be solved. This term sometimes is used to include the problem patch board as well, and, in some loose usage, the computer program may be (incorrectly) used to refer solely to the program patch panel. 9

(3) (software). A sequence of instructions suitable for processing by a computer. Processing may include the use of an assembler, a compiler, an interpreter, or a translator to prepare the program for execution as well as to execute it. *See:* **assembler; compiler; computer; execution; instructions; interpreter; program.** 434

(4)(programmable digital computer systems in safety systems of nuclear power generating stations). A schedule or plan that specifies actions that may or may not be taken, expressed in a form suitable for execution by a programmable digital computer. 554

computer program abstract (software). A brief description of a computer program, providing sufficient information for potential users to determine the appropriateness of the computer program to their needs and resources. *See:* **computer program.** 434

computer program annotation. *See:* **comment.**

computer program certification. *See:* **certification.**

computer program configuration identification. *See:* **configuration identification.**

computer program development plan. *See:* **software development plan.**

computer program validation. *See:* **validation.**

computer program verification. *See:* **verification.**

computer system (software). A functional unit, consisting of one or more computers and associated software, that uses common storage for all or part of a program and also for all or part of the data necessary for the execution of the program; executes user-written or user-designated programs; performs user-designated data manipulation, including arithmetic operations and logic operations, and that can execute programs that modify themselves during their execution. A computer system may be a standalone unit or may consist of several interconnected units. *Syn:* **ADP system; computing system.** *See:* **computers; data; execution; functional unit; software.** 434

computer time. *See:* **time.** 9

computer variable (1). A dependent variable as represented on the computer. *See:* **time.** 9

(2) (machine variable). *See:* **scale factor.** 9

computer word. A sequence of bits or characters treated as a unit and capable of being stored in one computer location. *See:* **machine word.** 255, 77

computing elements (analog computers). A computer component that performs a mathematical operation required for problem solution. It is shown explicitly in computer diagrams, or computer programs. 9

concatenation (of optical waveguides) (fiber optics). The linking of optical waveguides, end to end. 433

concealed (National Electrical Code). Rendered inaccessible by the structure or finish of the building. Wires in concealed raceways are considered concealed, even though they may become accessible by withdrawing them. *See:* **accessible - (as applied to wiring methods).** 256

concealed knob-and-tube wiring (National Electrical Code). A wiring method using knobs, tubes, and flexible nonmetallic tubing for the protection and support of single insulated conductors concealed in hollow spaces of walls and ceilings of buildings. 256

concentrate (metallurgy). The product obtained by concentrating disseminated or lean ores by mechanical or other processes thereby eliminating undesired minerals or constituents. *See:* **electrowinning.** 328

concentrated winding (rotating machinery). A winding, the coils of which occupy one slot pole: or a field winding mounted on salient poles. *See:* **asynchronous machine.** 63

concentration cell. (1) An electrolyte cell, the electromotive force of which is due to differences in composition of the electrolyte at anode and cathode areas. 205

(2) A cell of the two-fluid type in which the same dissolved substance is present in differing concentrations at the two electrodes. *See:* **electrochemistry.** 328

concentration polarization. (1) That part of the total polarization that is caused by changes in the activity of the potential-determining components of the electrolyte. *See:* **electrochemistry.** 328

(2) That portion of the polarization of an electrode produced by concentration changes at the metal-environment interface. 205

concentrator (1)(telephone switching systems). A switching entity for connecting a number of inlets to a smaller number of outlets. 55

(2) (data communications). A device that provides communications capability between many low-speed, usually asynchronous channels and fewer high-speed, usually synchronous channels the sum of whose data rates is (usually) less than the sum of the data rates of the low-speed channels. 12

concentric electrode system (electrobiology) (coaxial electrode system). An electrode system that is geometrically coaxial but electrically unsymmetrical. *Ex-*

ample: One electrode may have the form of a cylindrical shell about the other so as to afford electrical shielding. *See:* **electrobiology.**   192

concentric groove (disk recording). *See:* **locked groove.**

concentricity error (fiber optics). When used in conjunction with a tolerance field to specify core/cladding geometry, the distance between the center of the two concentric circles specifying the cladding diameter and the center of the two concentric circles specifying the core diameter. *See:* **cladding; cladding diameter; core; core diameter; tolerance field.**   433

concentricity of coaxial connectors (fixed and variable attenuator). Total indicator runout between the diameter of outer conductor and that diameter of that portion of inner conductor which engages with the corresponding diameters of mating connector. *Note:* This does not apply to precision connectors with only butt contacts.   110

concentric-lay cable. (1) A concentric-lay conductor as defined below or (2) a multiple-conductor cable composed of a central core surrounded by one or more layers of helically laid insulated conductors.   64

concentric-lay cable.

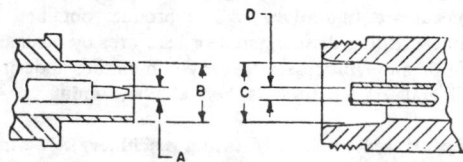

A ≡ Minimum diameter of circle capable to enclose maximum diameter of shank

B ≡ Minimum diameter of circle capable of enclosing maximum diameter of outer conductor

C ≡ Maximum diameter of circle fitting within minimum diameter of outer contact

D ≡ If contact is fully opened by insertion of nominal-size male shank, diameter of smallest circle enclosing inserted shank at end of female inner contact

Definitions of concentricity:

Male connectors: Total indicator runout between circles A and B

Female connectors: Total indicator runout between circles C and D

concentric-lay conductor. A conductor composed of a central core surrounded by one or more layers of helically laid wires. *Note:* In the most common type of concentric-lay conductor, all wires are of the same size and the central core is a single wire. *See:* **conductor.**   64

concentric line. *See:* **coaxial.**

concentric resonator (laser-maser). A beam resonator comprising a pair of spherical mirrors having the same axis of rotational symmetry and positioned so that their centers of curvature coincide on this axis.   363

concentric winding (rotating machinery). A winding in which the two coil sides of each coil of a phase belt, or of a pole of a field winding, are symmetrically located so as to be equidistant from a common axis. *See:* **asynchronous machine.**   63

concentric windings (of a transformer) (power and distribution transformer). An arrangement of transformer windings where the primary and secondary windings, and the tertiary winding, if any, are located in radial progression about a common core.   53

concept phase (software verification and validation plans). The initial phase of a software development project , in which user needs are described and evaluated through documentation (for example, statement of needs, advanced planning report, project initiation memo, feasibility studies, system definition documentation, regulations, procedures, or policies relevant to the project).   511

concrete quantity. *See:* **physical quantity.**

concrete-tight fitting (for conduit). A fitting so constructed that embedment in freshly mixed concrete will not result in the entrance of cement into the fitting.   296

concurrent processes (software). Processes that may execute in parallel on multiple processors or asynchronously on a single processor. Concurrent processes may interact with each other, and one process may suspend execution pending receipt of information from another process or the occurence of an external event. *See:* **execution; processes; sequential processes.**   434

condensed-mercury temperature (mercury-vapor tube). The temperature measured on the outside of the tube envelope in the region where the mercury is condensing in a glass tube or at a designated point on a metal tube. *See:* **gas tube.**   190

condenser (of a fuse). *See:* **fuse condenser.**

condenser (1). *See:* **capacitor.**

(2) (fuse). *See:* **fuse condenser.**

condenser antenna. *See:* **capacitor antenna.**

condenser box. *See:* **subdivided capacitor.**

condenser bushing. *See:* **capacitor bushing.**

condenser loudspeaker (capacitor loudspeaker). *See:* **electrostatic loudspeaker.**

condenser microphone (capacitor microphone). *See:* **electrostatic microphone.**

condition (computing system). *See:* **initial condition.**

condition adverse to quality (nuclear power quality assurance). An all inclusive term used in reference to any of the following: failures, malfunctions, deficiencies, defective items, and nonconformances. A significant condition adverse to quality is one which, if uncorrected, could have a serious effect on safety or operability.   417

conditional control structure (software). A programming control structure that allows alternative flow of control in a program depending upon the fulfillment of specified conditions, for example, case, if...then ... else,..., *See:* **case; control structure; flow of control; program.**   434

conditional jump. To cause, or an instruction that causes, the proper one of two (or more) addresses to be used in obtaining the next instruction, depending upon some property of one or more numerical expressions or other conditions. Sometimes called a branch. *See:* **jump.**   235

conditioning (data communications). The addition of equipment to or selection of communication facilities to provide the performance characteristics required for certain types of data transmission.   12

conditioning (replacement parts for Class 1E equipment in nuclear power generating stations). Any additional work or process imposed upon a part that makes it different from nominally similar parts. *Note:* Conditioning may include calibration, adjustment, tuning, selection testing, 'burn-in', heat treatment, machining, and similar processes. For example, if several parts are selected to test one that displays a special characteristic, the selected part is conditioned because it then displays a characteristic that makes it unique from parts with the same nominal description.
                                                     582

conditioning stimulus (medical electronics). A stimulus of given configuration applied to a tissue before a test stimulus.   192

conductance. (1) That physical property of an element, device, branch, network or system, that is the factor by which the mean square voltage must be multiplied to give the corresponding power lost by dissipation as heat or as other permanent radiation or loss of electromagnetic energy from the circuit. (2) The real part of admittance. *Note:* Definitions (1) and (2) are not equivalent but are supplementary. In any case where confusion may arise, specify the definition being used.
                                                     185

conductance, electrode. *See:* **electrode conductance.**

conductance for rectification (electron tube). The quotient of (1) the electrode alternating current of low frequency by (2) the in-phase component of the electrode alternating voltage of low frequency, a high frequency sinusoidal voltage being applied to the same or another electrode and all other electrode voltages being maintained constant. *See:* **rectification factor.**
                                                     190

conductance relay (power switchgear). A mho relay for which the center of the operating characteristic on the R-X diagram is on the R-axis. *Note:* The equation which describes such a characteristic is $Z-K \cos \theta$ where K is a constant and θ is the phase angle by which the input voltage leads the input current.
                                                     103

conducted heat. The thermal energy transported by thermal conduction. *See:* **thermoelectric device.**
                                                     191

conducted interference (electromagnetic compatibility)overhead power line corona and radio noise. Interference resulting from conducted radio noise or unwanted signals entering a transducer (receiver) by direct coupling. *See:* **electromagnetic compatibility.**   199, 411

conducted radio noise (1). Produced by equipment operation, which exists on the powerline of the equipment and measurable under specified conditions as a voltage or current.   418

(2) (electromagnetic compatibility). Radio noise propagated along circuit conductors. *Note:* It may enter a transducer (receiver) by direct coupling or by an antenna as by subsequent radiation from some circuit element. *See:* **electromagnetic compatibility.**
                                                     199

conducted spurious emission power (land-mobile communication transmitters). Any part of the spurious emission power output conducted over a tangible transmission path. Radiation is not considered a tangible path.   444

conducted spurious transmitter output (land-mobile communication transmitters). Any spurious output of a radio transmitter conducted over a tangible transmission path. *Note:* Power lines, control leads, radio frequency transmission lines and waveguides are all considered as tangible paths in the foregoing definition. Radiation is not considered a tangible path in this definition.   444

conducting (conduction) period (1) (rectifier circuit element) (semiconductor). That part of an alternating voltage cycle during which the current flows in the forward direction. *Note:* The forward period is not necessarily the same as the conducting period because of circuit parameters and semiconductor rectifier diode characteristics.   66

(2) (gas tube). That part of an alternating-voltage cycle during which a certain arc path is carrying current.
                                                     190

conducting element (of a fuse) (power switchgear). The conducting means, including the current-responsive element, for completing the electric circuit between the terminals of a fuseholder or fuse unit. *Syn:* **fuse link.**   103, 443

conducting interval (circuit properties)(self-commutated converters). An interval during which the principal current flows through a blocking element.
                                                     584

conducting mechanical joint (power switchgear). The juncture of two or more conducting surfaces held together by mechanical means. *Note:* Parts jointed by fusion processes, such as welding, brazing, or soldering, are excluded from this definition.   103

conducting paint (rotating machinery). A paint in which the pigment or a portion of pigment is a conductor of electricity and the composition is such that when it is converted to a solid film, the electric conductivity of the film approaches that of metallic substances.   63

conducting parts (industrial control). The parts that are designed to carry current or that are conductively connected therewith.   206

conducting salts. Salts that, when added to a plating solution, materially increase its conductivity. *See:* **electroplating.**   328

conduction (leakage) current (electric submersible pump cable). Current resulting from conduction through the cable insulating medium or over surfaces. Corona discharge from external energized elements will be indicated as conduction current.   484

conduction (leakage) current of the cable insulation

(power cable systems). Current resulting from conduction through the insulating medium or over surfaces (such as terminations). Corona discharge from external energized elements will be indicated as conduction current.   437

conduction band (semiconductor). A range of states in the energy spectrum of a solid in which electrons can move freely. See: semiconductor.   245

conduction electron. See: electrons, conduction.

conduction-through (thyristor converter). The failure to achieve forward blocking, during inverter operation, of an arm of a thyristor connection at the end of the normally conducting period, thus enabling the direct current to continue to pass during the period when the thyristor is normally in the off state. Note: A conduction-through occurs, for example, when the margin angle is too small or because of a misgating in the succeeding arm. Syn: shoot through.   121

conductive (health care facilities). Adjective describing not only those materials, such as metals, which are commonly considered as electrically conductive, but also that class of materials which, when tested in accordance with NFPA standard 56A, 1978, have a resistance not exceeding 1 000 000 ohms. Such materials are required where electrostatic interconnection is necessary.   192

conductive coating (rotating machinery). Conducting paint applied to the slot portion of a coil-side, to carry capacitive and leakage currents harmlessly between insulation and grounded iron.   63

conductive coupling (interference terminology). See: coupling, conductance.

conductivity (material). A factor such that the conduction-current density is equal to the electric-field intensity in the material multiplied by the conductivity. Note: In the general case it is a complex tensor quantity. See: transmission line.   185

conductivity, n-type (semiconductor). The conductivity associated with conduction electrons in a semiconductor. See: semiconductor.   245

conductivity, p-type (semiconductor). The conductivity associated with holes in a semiconductor. See: semiconductor.   245, 66, 210, 186

conductivity in physical media (antenna). The real part of the complex conductivity. See: complex permittivity.   111

conductivity modulation (semiconductor). The variation of the conductivity of a semiconductor by variation of the charge-carrier density. See: semiconductor; semiconductor device.   245

conductor (NESC). A material, usually in the form of a wire, cable, or bus bar, suitable for carrying an electric current. See: bundled conductor; covered conductor; grounded conductor; grounding conductor; insulated conductor; lateral conductor; line conductor; open conductor.   494

conductor (power line maintenance). A wire or combination of wires not insulated from one another, suitable for carrying an electric current. However, it may be bare or insulated. Syn: cable; wire.   458

conductor (substation grounding). A metallic substance that allows a current of electricity to pass continuously along it. As used in IEEE Std 837-1984, a conductor includes cable (wire), rods (electrodes), and metallic structures.   475

conductor (1) (general). (A) A substance or body that allows a current of electricity to pass continuously along it. See: conducting material.   63
(B) A wire or combination of wires not insulated from one another, suitable for carrying an electric current. It may be, however, bare or insulated.   345
(C) The portion of a lightning-protection system designed to carry the lightning discharge between air terminal and ground.   297

(2) (conductor stringing equipment). A wire or combination of wires not insulated from one another, suitable for carrying an electric current. It may be, however, bare or insulated. Syn: cable; wire.   431

(3) (transmission and distribution). A material, usually in the form of a wire, cable, or bus bar, suitable for carrying an electric current. See: bundled conductor; covered conductor; grounded conductor; grounding conductor; insulated conductor; lateral conductor; line conductor; open conductor (open wire).   262

(4) (National Electrical Safety Code). (A) A material, usually in the form of wire, cable, or bar, suitable for carrying an electric current. (B) covered conductor. A conductor covered with a dielectric having no rated insulating strength or having a rated insulating strength less than the voltage of the circuit in which the conductor is used. (C) insulated conductor. A conductor covered with a dielectric (other than air) having a rated insulating strength equal to or greater than the voltage of the circuit in which it is used. See: grounded conductor; grounding conductor; lateral conductor; line conductor; open conductor; open conductor, armless construction.   391

conductor, bare. One having no covering or insulation whatsoever.   256

conductor car (conductor stringing equipment). A device designed to carry workmen and ride on sagged bundle conductors, thus enabling them to inspect the conductors for damage and install spacers and dampers where required. These devices may be manual or powered. Syn: cable buggy; cable car; spacer buggy; spacer cart.   431

conductor clearance. See: ANSI C2, current edition.

conductor, coaxial. See: coaxial conductor.   54

conductor combination (substation grounding). The various conductors that may be joined by a connector.   475

conductor-cooled (rotating machinery). A term referring to windings in which coolant flows in close contact with the conductors so that the heat generated within the principal portion of the windings reaches the cooling medium without flowing through the major ground insulation.   244

conductor cover (power line maintenance). Electrical protection equipment designed specifically to cover conductors. Syn: cover-up; eel; hard cover; hose; snake; blanket. See: cover-up equipment.   458

conductor, covered (1) (general). A conductor having

one or more layers of nonconducting materials that are not recognized as insulation under the electric code.   256

(2) (transmission and distribution). A conductor covered with a dielectric having no rated insulating strength or having a rated insulating strength less than the voltage of the circuit in which the conductor is used.   64

conductor grip (power line maintenance). A device designed to permit the pulling of a conductor without splicing on fittings, eyes, etcetera. It permits the pulling of a continuous conductor where threading is not possible. The designs of these grips vary considerably. Grips such as the Klein (Chicago) and Crescent utilize an open-sided, rigid body with opposing jaws and swing latch. In addition to pulling conductors this type is commonly used to tension guys and, in some cases, pull wire rope. The design of the come-along (pocketbook, suitcase, four bolt, etcetera) incorporates a bail attached to the body of a clamp which folds to completely surround and envelop the conductor. Bolts are then used to close the clamp and obtain a grip. *Syn:* **buffalo; Chicago grip; come-along; Crescent; four bolt; grip; Klein; pocket book; seven bolt; six bolt; slip-grip; suitcase; Kellem grip.**   458

conductor insulation (rotating machinery). The insulation on a conductor or between adjacent conductors.   63

conductor lifting hook (power line maintenance). A device resembling an open boxing glove designed to permit the lifting of conductors from a position above the conductors. Normally used during clipping-in operations. Suspension clamps are sometimes used for this purpose. *Syn:* **boxing glove; conductor hook; lifting lip.**   458

conductor loading (mechanical). The combined load per unit length of a conductor due to the weight of the wire plus the wind and ice loads. *See:* **tower.**   64

conductor-loop resistance (telephone switching systems). The series resistance of the conductors of a line or trunk loop, excluding terminal equipment or apparatus.   55

conductor safety (conductor stringing equipment). A sling arranged in a vertical basket configuration, with both ends attached to the support-supporting structure and passed under the clipped-in conductor(s). These devices, when used, are normally utilized with bundled conductors, to act as a safety device in case of insulator failure while workmen in conductor cars are installing spacers between the subconductors, or as an added safety measure when crossing above energized circuits. These devices may be fabricated from synthetic fiber rope or wire rope.   431

conductor shielding (1) (power distribution underground cables). A conducting or semiconducting element in direct contact with the conductor and in intimate contact with the inner surface of the insulation so that the potential of this element is the same as the conductor. Its function is to eliminate ionizable voids at the conductor and provide uniform voltage stress at the inner surface of the insulating wall.   57

(2) (cable systems). A conducting material applied in manufacture directly over the surface of the conductor and firmly bonded to the inner surface of the insulation.   35

(3)(NESC). An envelope which encloses the conductor of a cable and provides an equipotential surface in contact with the cable insulation.   494

conductor support box. A box that is inserted in a vertical run of raceway to give access to the conductors for the purpose of providing supports for them. *See:* **cabinet.**   328

conductor temperature (electrical heat tracing for industrial applications). The temperature of the heat-producing element.   523

conduit (1) (electric power). A structure containing one or more ducts. *Note:* Conduit may be designated as iron pipe conduit, tile conduit, etcetera. If it contains one duct only it is called "single-duct conduit," if it contains more than one duct it is called "multiple-duct conduit," usually with the number of ducts as a prefix, namely, two-duct multiple conduit. *See:* **cable.**   64, 262

(2) (aircraft). An enclosure used for the radio shielding or the mechanical protection of electric wiring in an aircraft. *Note:* It may consist of either rigid or flexible, metallic or nonmetallic tubing. Conduit differs from pipe and metallic tubing in that it is not normally used to conduct liquids or gases. *See:* **flexible metal conduit; rigid metal conduit.**   328

(3) (packaging machinery). A tubular raceway for holding wires or cables, which is designed expressly for, and used solely for, this purpose.   429

(4)(NESC). A structure containing one or more ducts. *Note:* Conduit may be designated as iron pipe conduit, tile conduit, etcetera. If it contains one duct only it is called single-duct conduit; if it contains more than one duct it is called multiple-duct conduit, usually with the number of ducts as a prefix, for example, two-duct multiple conduit.   494

conduit body (National Electrical Code). A separate portion of a conduit or tubing system that provides access through a removable cover(s) to the interior of the system at a junction of two or more sections of the system or at a terminal point of the system.   256

conduit fitting. An accessory that serves to complete a conduit system, such as bushings and access fittings. *See:* **raceways.**   296

conduit knockout. *See:* **knockout.**   53

conduit run. *See:* **duct bank.**

conduit system (1)(NESC). Any combination of duct, conduit, conduits, manholes, handholes and vaults joined to form an integrated whole.   494

(2)(raceway systems for Class 1E circuits for nuclear power generating stations). Any assembly of conduit sections, fittings, supports, anchorages, and accessories that form a structural system to support wire and cable.   513

cone (1) (cathode-ray tube). The divergent part of the envelope of the tube. *See:* **cathode-ray tubes.**   190

(2) (vision). Retinal elements that are primarily con-

cerned with the perception of detail and color by the light-adapted eye. *See:* **retina.**   167

cone, leader. *See:* **leader cone.**

cone of ambiguity (navigation aid terms). A generally conical volume of airspace above a navigation aid within which navigational information from that facility is unreliable.   526

cone of nulls. A conical surface formed by directions of negligible radiation. *See:* **antenna.**   179

cone of protection (lightning). The space enclosed by a cone formed with its apex at the highest point of a lightning rod or protecting tower, the diameter of the base of the cone having a definite relation to the height of the rod or tower. *Note:* This relation depends on the height of the rod and the height of the cloud above the earth. The higher the cloud, the larger the radius of the base of the protecting cone. The ratio of radius of base to height varies approximately from one to two. When overhead ground wires are used, the space protected is called a zone of protection or protected zone.   64

cone of silence (navigation aid terms). A conically shaped region above an antenna where the field strength is relatively weak because of the configuration of the antenna system.   526

cones (illuminating engineering). Retinal receptors which dominate the retinal response when the luminance level is high and provide the basis for the perception of color.   167

conference call (telephone switching systems). A call in which communication is provided among more than two main stations.   55

conference connection. A special connection for a telephone conversation among more than two stations.   328

confidence test (test, measurement and diagnostic equipment). A test primarily performed to provide a high degree of certainty that the unit under test is operating acceptably.   54

confidence tester (test, measurement and diagnostic equipment). Any test equipment, either automatic, semiautomatic, or manual, which is used expressly for performing a test or series of tests to increase the degree of certainty that the unit under test is operating acceptably.   54

configuration (1)(microprocessor operating systems). A collection of capability modules.   478 **(2) (software).** (A) The arrangement of a computer system or network as defined by the nature, number, and the chief characteristics of its functional units. More specifically, the term configuration may refer to a hardware configuration or a software configuration. (B) The requirements, design, and implementation that define a particular version of a system or system component. (C) The functional or physical characteristics or both of hardware/software as set forth in technical documentation and achieved in a product. *See:* **component; computer system; design; documentation; functional units; hardware; implementation; network; requirements; software.**   434

configuration audit (software). The process of verifying that all required configuration items have been produced, that the current version agrees with specified requirements, that the technical documentation completely and accurately describes the configuration items, and that all change requests have been resolved.   434

configuration control (software). (1) The process of evaluating, approving or disapproving, and coordinating changes to configuration items after formal establishment of their configuration identification. (2) The systematic evaluation, coordination, approval or disapproval, and implementation of all approved changes in the configuration of a configuration item after formal establishment of its configuration identification. *See:* **configuration; configuration identification; configuration item; implementation.**   434

configuration control board (software). The authority responsible for evaluating and approving or disapproving proposed engineering changes, and ensuring implementation of the approved changes. *See:* **implementation.**   434

configuration factor, (illuminating engineering). The ratio of illuminance on a surface at point 2 (due to flux directly received from lambertian surface 1) to the exitance of surface 1. It is used in flux transfer theory.

$$C_{1-2} = (E_2)/(M_1).$$

167

configuration identification (software). (1) The process of designating the configuration items in a system and recording their characteristics. (2) The approved documentation that defines a configuration item. (3) The current approved or conditionally approved technical documentation for a configuration item as set forth in specifications, drawings and associated lists, and documents referenced therein. *See:* **configuration item; documentation; documents; specifications.**   434

configuration item (software). (1) A collection of hardware or software elements treated as a unit for the purpose of configuration management. (2) An aggregation of hardware/software, or any of its discrete portions, which satisfies an end use function and is designated for configuration management. Configuration items may vary widely in complexity, size and type, from an aircraft, electronic or ship system to a test meter or round of ammunition. During development and initial production, configuration items are only those specification items that are referenced directly in a contract (or an equivalent in-house agreement). During the operation and maintenance period, any reparable item designated for separate procurement is a configuration item. *See:* **complexity; configuration management; function; hardware; software; system.**   434

configuration management (software). (1) The process of identifying and defining the configuration items in a system, controlling the release and change of these items throughout the system life cycle, recording and reporting the status of configuration items and change requests, and verifying the completeness and correctness of configuration items. (2) A discipline applying technical and administrative direction and surveillance to (a) identify and document the functional and physical characteristics of a configuration item, (b) control changes to those characteristics, and (c) record and report change processing and implementation status. *See:* **change control; configuration audit; configuration control; configuration identification; configuration item; configuration status accounting; implementation; system.**　　　　434

configuration status accounting (software). The recording and reporting of the information that is needed to manage configuration effectively, including a listing of the approved configuration identification, the status of proposed changes to configuration, and the implementation status of approved changes. *See:* **configuration; configuration identification; implementation.**　　　　434

confinement (software). Prevention of unauthorized alteration, use, destruction, or release of data during authorized access. (2) Restriction on programs and processes so that they do not access or have influence on data, programs, or processes other than that allowed by specific authorization. *See:* **data; integrity; processes; programs.**　　　　434

conflict, antenna. *See:* **antenna conflict.**

confocal resonator (laser-maser). A beam resonator comprising a pair of spherical mirrors having the same axis of rotational symmetry and positioned so that their focal points coincide on this axis.　　　　363

conformal antenna. An antenna (an array) which conforms to a surface whose shape is determined by considerations other than electromagnetic, for example, aerodynamic or hydrodynamic. *Syn:* **conformal array.**　　　　111

conformance tests (acceptance tests) (1) (surge arrester)(metal-oxide surge arresters for ac power circuits)(surge arrester). Tests made, when required, to demonstrate selected performance characteristics of a product or representative samples thereof.
　　　　2, 62, 583, 430

(2)(metal-enclosed low-voltage power circuit-breaker switchgear)(metal-clad and station-type cubicle switchgear). Tests that demonstrate compliance with the applicable standards. The test specimen is normally subjected to all planned production tests prior to the initiation of the conformance test program. *Note:* The conformance test may, or may not, be similar to certain design tests. Demonstration of margin (capabilities) beyond the standards is not required.
　　　　579, 572, 573

(3) (general) (power and distribution transformer). Tests that are specifically made to demonstrate conformity with applicable standards.　　　　202

(4) (mechanical switching device)(X-radiation limits for ac high-voltage power vacuum interrupters used in power switchgear)(power switchgear). Those tests that are specifically made to demonstrate the conformity of switchgear or its component parts with applicable standards.　　　　92, 300, 103, 443, 553

(5) (transformer). Tests which are made by agreement

between the manufacturer and the purchaser at the time the order is placed. In some cases, by mutual agreement, certain Design Tests may be made as Conformance Tests.　　　　53

conformity (1) (potentiometer). The accuracy of its output: used especially in reference to a function potentiometer.　　　　9,10

(2) (curve) (automatic control). The closeness with which it approximates the specified functional curve (for example logarithmic, parabolic, cubic, etcetera). *Note:* It is usually expressed in terms of a nonconformity, for example the maximum deviation. For "independent conformity," any shift or rotation is permissible to reduce this deviation. For "terminal conformity," the specified functional curve must be drawn to give zero output at zero input and maximum output at maximum input, but the actual deviation at these points is not necessarily zero.　　　　56

conical array (antennas). A two-dimensional array of elements whose corresponding points lie on a conical surface.　　　　111

conical-directional reflectance, $\rho(\omega_i; \theta_r, \phi_r)$ (illuminating engineering). Ratio of reflected flux collected over an element of solid angle surrounding the given direction to the incident flux limited to a conical solid angle. (See Figure below.) *Note:* The direction and extent of the cone must be specified and the direction of collection and size of the solid angle "element" must be specified.　　　　167

conical-directional reflectance,

Conical　　　　　　　　Directional
Incident　　　　　　　　Collected

conical-directional transmittance, $\tau(\omega_i; \theta_r, \phi_t)$ (illuminating engineering). Ratio of transmitted flux, collected over an element of solid angle surrounding the direction, to the incident flux to a conical solid angle. (See Figure below.) *Note:* The direction and extent of the cone must be specified and the direction of collection and size of the solid angle "element" must be specified.　　　　167

conical-hemispherical reflectance, $\rho(\omega_i; 2\pi)$ (illuminating engineering). Ratio of reflected flux collected over the entire hemisphere to the incident flux limited to a conical solid angle. (See Figure below.) *Note:* The direction and extent of the cone must be specified.　　　　167

conical-hemispherical transmittance, $\tau(\omega_i; 2\pi)$ (illuminating engineering). Ratio of transmitted flux collected over the entire hemisphere to the incident flux limited to a conical solid angle. (See Figure below.) *Note:* The direction and extent of the cone must be specified.　　　　167

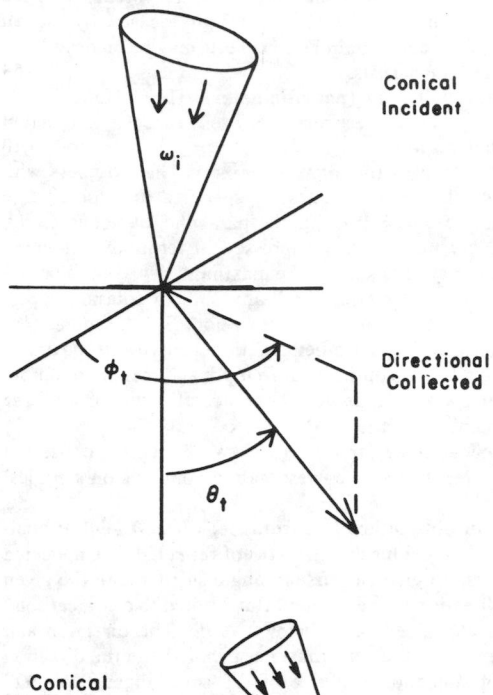

Conical
Incident

Directional
Collected

Conical
Incident

Hemispherical
Collected

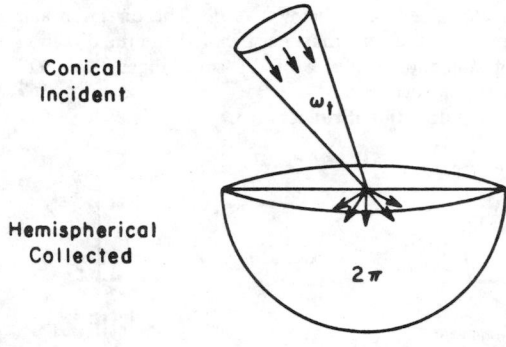

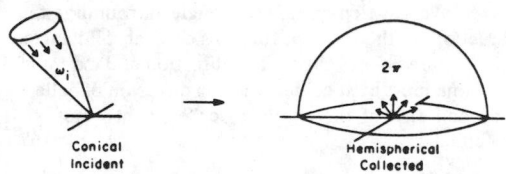

Conical
Incident

Hemispherical
Collected

conical horn. A horn whose cross-sectional area increases as the square of the axial length. *See:* **loudspeaker.** 176

conical scanning (1) (antenna). A form of sequential lobing in which the direction of maximum radiation generates a cone where the vertex angle is of the order of the antenna half-power beamwidth. *Note:* Such scanning may be either rotating or nutating according to whether the direction of polarization rotates or remains unchanged. *See:* **antenna.** 246

(2) (radar). A form of angular tracking in which the direction of maximum radiation is offset from the axis of the antenna. Rotation of the beam about the axis generates a cone whose vertex angle is of the order of the beamwidth: such scanning may be either rotating or nutating, according to whether the direction of polarization rotates or remains unchanged. 13

conical wave (radio wave propagation). A wave whose equiphase surfaces asymptotically form a family of coaxial circular cones. 146

coning effect (gyro)(inertial sensor). The apparent drift rate caused by motion of an input axis in a manner which generally describes a cone. This usually results from a combination of oscillatory motions about the principal axes. The apparent drift rate is a function of the amplitudes and frequencies of oscillations present and the phase angles between them. 46

conjugate bridge. The detector circuit and the supply circuit are interchanged as compared with a normal bridge of the given type. *See:* **bridge.** 328

connected (1) (network). A network is connected if there exists at least one path, composed of branches of the network, between every pair of nodes of the network. *See:* **network analysis.** 332

(2) (circuits and systems) (graph). A graph is connected if there exists at least one path between every pair of its vertices. 67

connected load. The sum of the continuous ratings of the load-consuming apparatus connected to the system or any part thereof. *See:* **generating station.** 64

connected position (of a switchgear-assembly removable element) (power switchgear). That position of the removable element in which both primary and secondary disconnecting devices are in full contact. 103

connected system (FASTBUS acquisition and control). All segments of a connected system are capable of communicating directly with one another through segment interconnects (SI). *Note:* Because of the route map table implementation of message paths segments of a system that are connected electrically by SIs are not necessarily also logically connected in the sense used here. 480

connecting rod or shaft (high voltage air switches, insulators, and bus supports) (power switchgear). A component of a switch operating mechanism designed to transmit motion from an offset bearing or bell crank to a switch pole unit. 27, 103

connecting wire (mining). A wire generally of smaller gauge than the shot-firing cord and used for connecting the electric blasting-cap wires from one drill hole to those of an adjoining one in mines, quarries, and tunnels. *See:* **mine feeder circuit.** 328

connection (1) (rotating machinery). Any low-impedance tie between electrically conducting components. 63

(2) (nuclear power generating stations)(cable, field splice, and connection qualification)(design and installation of cable systems for Class 1E circuits in nuclear power generating stations). A cable terminal, splice, or hostile environment boundary seal at the interface of cable and equipment.   141, 536

(3) (software). (A) A reference in one part of a program to the identifier of another part (that is, something found elsewhere). (B) An association established between functional units for conveying information. *See:* **functional units; identifier; interface; program.**   434

connection and winding support insulation (insulation systems of synchronous machines). The connection and winding support insulation includes all of the insulation materials that envelope the connections, which carry current from coil to coil or from bar to bar, and from field and armature coil terminals to the points of external circuit and attachment; and also the insulation of metallic supports for the winding.   298

connection assembly (Class 1E connection assemblies). Any connector or termination combined with related cables or wires as an assembly. This assembly may include environmental seals, but excludes fire stops, in-line splices, and containment electric penetration assemblies.   459

connection charge (power operations). The amount paid by a customer for connecting the customer's facilities to the supplier's facilities.   516

connection diagram (1). A diagram that shows the connection of an installation or its component devices, controllers, and equipment. *Notes:* (A) It may cover internal or external connections, or both, and shall contain such detail as is needed to make or trace connections that are involved. It usually shows the general physical arrangement of devices and device elements and also accessory items such as terminal blocks, resistors, etcetera. (B) A connection diagram excludes mechanical drawings, commonly referred to as wiring templates, wiring assemblies, cable assemblies, etcetera.   206, 301

(2) (packaging machinery). A diagram showing the electrical connections between the parts comprising the control and indicating the external connections.   429

(3) (power switchgear). A diagram showing the relation and connections of devices and apparatus of a circuit or a group of circuits.   103

connection insulation (joint insulation) (rotating machinery). The insulation at an electric connection such as between turns or coils or at a bushing connection. *See:* **stator.**   63

connections of polyphase circuits (rotating machinery). *See:* **mesh-connected circuit; star-connected circuit; zig-zag connection of polyphase circuits.**

connector (1). A coupling device employed to connect conductors of one circuit or transmission element with those of another circuit or transmission element. *See:* **auxiliary device to an instrument.**   185

(2) (wires). A device attached to two or more wires or cables for the purpose of connecting electric circuits without the use of permanent splices.   1

(3) (splicing sleeve). A metal sleeve, that is slipped over and secured to the butted ends of the conductors in making up a joint.   64

(4) (waveguides). A mechanical device, excluding an adapter, for electrically joining separable parts of a waveguide or transmission-line system.   179

(5) (power cable joints). A metallic device of suitable electric conductance and mechanical strength, used to splice the ends of two or more cable conductors, or as a terminal connector on a single conductor. Connectors usually fall into one of the following types: solder, welded, mechanical, and compression or indent. Conductors are sometimes spliced without connectors, by soldering, brazing or welding.   34

(6) (watthour meter). A coupling device employed to connect conductors of one circuit or transmission element with those of another circuit or transmission element.   419

(7) (fiber optics). *See:* **optical waveguide connector.**   433

(8)(substation grounding). A metallic device of suitable electric conductance and mechanical strength used to connect conductors.   475

connector base (motor plug) (motor attachment plug cap). A device, intended for flush or surface mounting on an appliance, that serves to connect the appliance to a cord connector.   328

connector insertion loss. *See:* **insertion loss.**

connector link (conductor stringing equipment). A rigid link designed to connect pulling lines and conductors together in series. It will not spin and relieve torsional forces. *Syn:* **bullet; connector; link; slug.**   431

connector, precision (waveguide or transmission line). A connector that has the property of making connections with a high degree of repeatability without introducing significant reflections, loss or leakage. *See:* **auxiliary device to an instrument.**   185

connector switch (connector). A remotely controlled switch for connecting a trunk to the called line.   328

connector thermal capacity (substation grounding). The ability of a connector to withstand the amount of current required to produce a specified temperature on the control conductor without increasing the resistance of the connector beyond that specified in IEEE Std 837-1984.   475

connector, waveguide (fixed and variable attenuator). A mechanical device, excluding an adapter, for electrically joining separable parts of a waveguide or transmission-line system.   110

conservation of radiance (fiber optics). A basic principle stating that no passive optical system can increase the quantity Ln^{-2} where L is the radiance of a beam and n is the local refractive index. Formerly called "conservation of brightness" or the "brightness theorem". *See:* **brightness; radiance.**   433

conservator (expansion tank system) (power and distribution transformer). A system in which the oil in the main tank is sealed from the atmosphere, over the temperature range specified, by means of an auxiliary tank partly filled with oil and connected to the completely filled main tank.   53

conservator/diaphragm system (power and distribu-

tion transformer). A system in which the oil in the main tank is completely sealed from the outside atmosphere, and is connected to an elastic diaphragm tank contained inside a tank mounted at the top of the transformer. As oil expands and contracts within a specified temperature range the system remains completely sealed with an approximately constant pressure.   53

consistency. See: precision.

consol (navigation aid terms). A keyed C-W (continuous wave) short-baseline-radio navigation system operating in the L/MF (low- and medium frequency) band, generally useful to about 1500 nmi (nautical miles)(2800 kilometers [km]), and using three radiators to provide a multiplicity of overlapping lobes of dot-and-dash patterns which form equisignal hyperbolic lines of position. These lines of position are moved slowly in azimuth by changing rf (radio frequency) phase, thus allowing a simple listening and counting of timing operation to be used to determine a line of position within the sector bounded by any pair of equisignal lines.   526

consolan (navigation aid terms). A form of consol using two radiators instead of three.   526

console (telephony) (1) (switchgear). A control cabinet located apart from the associated switching equipment arranged to control those functions for which an attendant or an operator is required.   193
(2) (computing system). The part of a computer used for communication between the operator or maintenance engineer and the computer.   255
(3) (telephone switching systems). A desk or desk-top cordless switchboard which may include display elements in addition to those required for supervisory purposes is required.   55
(4) (station control and data acquisition)(supervisory control, data acquisition, data control). That component of the system which provides facilities for control and observation of the system. Examples include operator's console, maintenance console. See: control panel; panel; control.   403, 570

consonant articulation (percent consonant articulation). The percent articulation obtained when the speech units considered are consonants (usually combined with vowels into meaningless syllables). See: volume equivalent.   328

conspicuity (illuminating engineering). The capacity of a signal to stand out in relation to its background so as to be readily discovered by the eye.   167

constancy (probe coupling). See: residual probe pickup.

constant. See: time constant of integrator.

constant-amplitude recording (mechanical recording). A characteristic wherein, for a fixed amplitude of a sinusoidal signal, the resulting recorded amplitude is independent of frequency. See: phonograph pickup.
                                                        176

constant available power source (transmission performance of telephone sets). A signal source with a purely resistive internal impedance and a constant open-circuit terminal voltage, independent of frequency.

Note: Receiving is tested under conditions of constant available power. A generator having an open-circuit voltage, E, and an internal resistance, R_θ, both constant with frequency, provides a constant available power of $E^2/4R_\theta$. When the load impedance is a resistance equal to R_θ, an impedance match exists and the maximum available power of $E^2/4R_\theta$, is dissipated in the load. When a generator with a constant open-circuit voltage and a constant internal resistance of the required value is not available, suitable test conditions for this type of measurement can be provided by a generator whose voltage is maintained at a constant value, E_θ, across its terminals in series with a resistance, R_θ. The generator and the resistance, R_θ, in series are then equivalent to a source of constant available power.   491

constant-current (Heising) modulation. A system of amplitude modulation wherein the output circuits of the signal amplifier and the carrier-wave generator or amplifier are directly and conductively coupled by means of a common inductor that has ideally infinite impedance to the signal frequencies and that therefore maintains the common plate-supply current of the two devices constant. Note: The signal-frequency voltage thus appearing across the common inductor appears also as modulation of the plate supply to the carrier generator or amplifier with corresponding modulation of the carrier output.   111

constant-current (series) incandescent filament lamp transformer. See: incandescent filament lamp transformer (series type).

constant-current (series) mercury-lamp transformer. A transformer that receives power from a current-regulated series circuit and transforms the power to another circuit at the same or different current from that in the primary circuit. Note: It also provides the required starting and operating voltage and current for the specified lamp. Further, it provides protection to the secondary circuit, casing, lamp, and associated luminaire from the high voltage of the primary circuit.
                                                        274

constant-current arc-welding power supply. A power supply that has characteristically drooping volt-ampere curves producing relatively constant current with a limited change in load voltage. Note: This type of supply is conventionally used in connection with manual-stick-electrode   or   tungsten-inert-gas   arc welding.   264

constant-current characteristic (electron tubes). The relation, usually represented by a graph, between the voltages of two electrodes, with the current to one of them as well as all other voltages maintained constant.
                                                        125

constant-current charge (storage battery) (storage cell). A charge in which the current is maintained at a constant value. Note: For some types of lead-acid batteries this may involve two rates called the starting and finishing rates. See: charge.   328

constant-current power supply. A power supply that is capable of maintaining a preset current through a variable load resistance. Note: This is achieved by automatically varying the load voltage in order to maintain the ratio V_{load}/R_{load} constant.   186

constant-current regulation (generator). That type of automatic regulation in which the regulator maintains a constant-current output from the generator. *See:* **axle generator system.**   328

constant current retention (metal-nitride-oxide field-effect transistor). Retention inherent in the metal-nitride-oxide-semiconductor (MNOS) transistor when gate and drain are biased to result in a constant drain current during information storage. The time period is defined by the intersection of the high conduction (HC) threshold voltage curve obtained under constant current condition, with the low conduction (LC) threshold voltage curve obtained under zero bias condition, when both are plotted against the logarithm of t_{rd}, the time elapsed between writing and the threshold voltage measurement.   386

constant-current street-lighting system (series street-lighting system). A street-lighting system employing a series circuit in which the current is maintained substantially constant. *Note:* Special generators or rectifiers are used for direct current while suitable regulators or transformers are used for alternating current. *See:* **alternating-current distribution; direct-current distribution.**   64

constant-current transformer (power and distribution transformer). A transformer that automatically maintains an approximately constant current in its secondary circuit under varying conditions of load impedance when supplied from an approximately constant-voltage source. *See:* **rated secondary current of a constant-current transformer; impedance voltage of a constant-current transformer; current regulation of a constant-current transformer; rated kilowatts of a constant-current transformer; rated primary voltage of a constant-current transformer.**   53

constant cutting speed (numerically controlled machine). The condition achieved by varying the speed of rotation of the workpiece relative to the tool inversely proportional to the distance of the tool from the center of rotation.   224

constant-delay discriminator. *See:* **pulse decoder.**

constant failure rate period (reliability). That possible period during which the failures occur at an approxi-

mately uniform rate. *Note:* The curve below shows the failure rate pattern when the terms of minor failure, early failure period, and constant failure rate period all apply to the item.   181

constant false alarm rate (CFAR) (radar). A property of threshold or gain control devices specially designed to suppress false alarms caused by noise, clutter, or ECM (electronic countermeasures) of varying levels.   13

constant fraction discriminator (germanium gamma-ray detectors). A pulse discriminator that gives an output when the amplitude of the input signal, delayed, matches a selected fraction of the undelayed signal. The instant of matching is invariant with input-signal amplitude, and if the delay is less than the fastest charge collection time, the timing spread in the instant of matching will be less than that in the charge collection time. The timing jitter is minimized by appropriate choices of fraction and delay. See figure below.   528

constant-frequency control (power system). A mode of operation under load-frequency control in which the area control error is the frequency deviation. *See:* **speed-governing system.**   94

constant-horsepower motor. *See:* **constant-power motor.**

constant-horsepower range (electric drive). The portion of its speed range within which the drive is capable of maintaining essentially constant horsepower. *See:* **electric drive.**   206

constant-luminance transmission (color television). A type of transmission in which the sole control of luminance is provided by the luminance signal, and no control of luminance is provided by the chrominance signal. *Notes:* (1) In such a system, noise signals falling within the bandwidth of the chrominance channel produce only chromaticity variations at the outputs of the chromiance demodulators. Coarse-structured chromaticity variations thus produced are subjectively less objectionable than correspondingly coarse-structured luminance variations. (2) Because of the use of gamma correction in the camera, these ideal conditions are not completely realized, especially for colors of high saturation.   18,178

constant multiplier (analog computers). A computing element that multiplies a variable by a constant factor.   9

constant-net-interchange control (power system). A mode of operation under load-frequency control in which the area control error is determined by the net interchange deviation. *See:* **speed-governing system.**   94

constant-power motor (constant-horsepower motor). A multispeed motor that develops the same related power output at all operating speeds. The torque then is inversely proportional to the speed. *See:* **asynchronous machine.**   63

constant-resistance (conductance) network. A network having at least one driving-point impedance (ad-

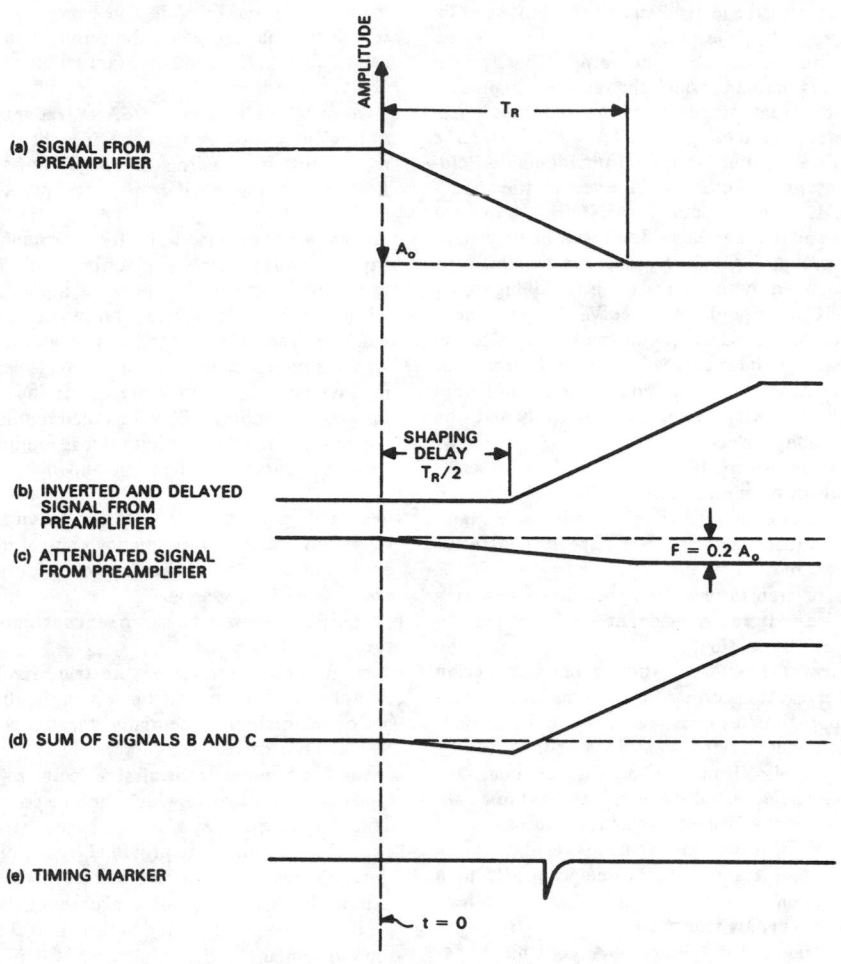

Timing Measurement Signals (Unshaped)

mittance) that is a positive constant. *See:* **linear passive networks.**   328

constant-speed motor (electric installations on shipboard). A motor, the speed of normal operation of which is constant or practically constant. For example, a synchronous motor, an induction motor with small slip, or an ordinary dc shunt-wound motor.   3

constant-torque motor. Multispeed motor that is capable of developing the same torque for all design speeds. The rated power output varies directly with the speed. *See:* **asynchronous machine.**   63

constant-torque range (electric drive). The portion of its speed range within which the drive is capable of maintaining essentially constant torque. *See:* **electric drive.**   206

constant torque resistor (electric installations on ship-
board). A resistor for use in the armature or rotor circuit of a motor in which the current remains practically constant throughout the entire speed range.   3

constant-torque speed range (industrial control). The portion of the speed range of a drive within which the drive is capable of maintaining essentially constant torque.   206

constant-velocity recording (mechanical recording). A characteristic wherein for a fixed amplitude of a sinusoidal signal, the resulting recorded amplitude is inversely proportional to the frequency. *See:* **phonograph pickup.**   176

constant-voltage arc-welding power supply. Power supply (arc welder) that has characteristically flat volt-amperer curves producing relatively constant voltage with a change in load current. This type of power supply is conventionally used in connection with welding processes involving consumable electrodes fed at a constant rate. 264

constant-voltage charge (storage battery) (storage cell). A charge in which the voltage at the terminals of the battery is held at a constant value. *See:* **charge.** 328

constant-voltage regulation (generator). That type of automatic regulation in which the regulator maintains constant voltage of the generator. *See:* **axle generator system.** 328

constant voltage retention (metal-nitride-oxide field-effect transistor). Retention inherent in the metal-nitride-oxide-semiconductor (MNOS) transistor when source, drain, and substrate are grounded, and a fixed read bias V_{GR} is maintained at the gate. The time period is defined by the intersection of the two high conduction (HC) and low conduction (LC) threshold voltage curves obtained under this condition when plotted versus the logarithm of t_{rd}, **the time elapsed between writing and threshold voltage measurement.** 386

constant-voltage transformer (power and distribution transformer). A transformer that maintains an approximately constant voltage ratio over the range from zero to rated output. 53

constitutive relations (radio wave propagation). Constraints imposed by the medium on the relationships between electric and magnetic field vectors and their respective flux density vectors. 146

constraints (1). Limits on the ranges of variables or system parameters because of physical or system requirements. *See:* **system.** 209
(2) (control system). A restriction placed on the control signal, control law, or state variables. *See:* **control system.** 329

construction diagram (industrial control). A diagram that shows the physical arrangement of parts, such as wiring, buses, resistor units, etcetera. *Example:* A diagram showing the arrangement of grids and terminals in a grid-type resistor. 206

construction test (Class 1E power systems). A test to verify proper installationand operation of individual components in a system prior to operation of the system as an entity. It is assumed that the construction test does not verify the interconnected-system equipment external to that component. For example, the protective relays are bench tested by simulating fault conditions to verify conformance with approved characteristics. During bench tests the alarm, trip, and permissive-interlock functions that the protective relay circuits are to perform are not verified. 455

contact (power switchgear). (1)(general) A conducting part that co-acts with another conducting part to make or break a circuit. (2)(of a relay) A conducting part that acts with another conducting part to make or break a circuit. 103, 127, 27

contact area (1) (photoelectric converter). The area of ohmic contact provided on either the p or n faces of a photoelectric converter for electric circuit connections. *See:* **semiconductor.** 186

(2) (solar cells). That area of ohmic contact provided on either the p or n surface of a solar cell for electric circuit connections. 113

contact bounce (telephony)(dial-pulse address signaling systems). The intermittent and undesired opening of contacts during the closure of open contacts or opening of closed contacts. An irregular wavefront during transition from one state to the other is implied. 540

contact chatter, relay. *See:* **relay contact chatter.**

contact clip (of a mechanical switching device) (power switchgear). The clip that the blade enters or embraces. 103

contact clips. *See:* **fuse clips.** 103

contact conductor (electric traction). The part of the distribution system other than the track rails, that is in immediate electric contact with current collectors of the cars or locomotives. *See:* **contact wire (trollery wire); trolley; underground collector or plow; multiple-unit control.** 1

contact converter (as applied to relaying) (power switchgear). A buffer element used to produce a prescribed output as the result of the opening or closing of a contact. 103

contact corrosion. *See:* **crevice corrosion.**

contact current-carrying rating of a relay (power switchgear). The current that can be carried continuously or for stated periodic intervals withoput impairment of the contact structure or interrupting capability. 127, 103

contact current-closing rating of a relay (power switchgear). The current that the device can close successfully with prescribed operating duty and circuit conditions without significant impairment of the contact structure. 103

contact flange (waveguide components). A flat flange used in conjunction with another flat flange to provide a contact joint. 166

contact follow-up (relays, switchgear, and industrial control). The distance between the position one contact face would assume, were it not blocked by the second (mating) contact, and the position the second contact removed, when the actuating member is fixed in its final contact-closed position. *See:* **electric controller; initial contact pressure.** 302,206

contact gap (break) (industrial control). The final length of the isolating distance of a contact in the open position. *See:* **contactor.** 244,206

contact high recombination rate (semiconductor). A semiconductor-semiconductor or metal-semiconductor contact at which thermal equilibrium charge-carrier concentrations are maintained substantially independent of current density. *See:* **semiconductor; semiconductor device.** 245

contact interrupting rating of a relay (power switchgear). The current that the device can interrupt successfully with prescribed operating duty and circuit conditions without significant impairment of the contact structure. 127, 103

contact joint (contact coupling) (waveguide components). A connection designed for essentially complete transfer of power between two waveguides by means of metallic contact between the inner walls of the waveguides. It typically consists of two contact flanges. 166

contactless vibrating bell. A vibrating bell whose continuous operation depends upon application of alternating-current power without circuit-interrupting contacts. *See:* **protective signaling.** 328

contact making clock. *See:* **demand meter (contact making clock).**

contact-making clock demand meter (metering). A device designed to close momentarily an electric circuit to a demand meter at periodic intervals. 212

contact mechanism (demand meter). A device for attachment to an electricity meter or to a demand-totalizing relay for the the purpose of providing electric impulses for transmission to a demand meter relay. *See:* **demand meter.** 328

contact nomenclature. *See:* **relay terms.**

contact opening time (of a relay) (power switchgear). The time a contact remains closed while in process of opening following a specified change of input. 103

contactor (1) (transformer)(thyristor). A device for repeatedly establishing and interrupting an electric power circuit. 53, 445
(2) (land transportation). A device which upon receipt of an electrical signal establishes or opens repeatedly an electrical circuit with a nominal current rating of 5 amperes minimum for its main contacts. 1

contactor, load. *See:* **load switch (load contactor).**

contactor or unit switch. A device operated other than by hand for repeatedly establishing and interrupting an electric power circuit under normal conditions. *See:* **control switch.** 1

contact parting time (of a mechanical switching device) (power switchgear). The interval between the time when the actuating quantity in the release circuit reaches the value causing actuation of the release and the instant when the primary arcing contacts have parted in all poles. *Note:* Contact parting time is the numerical sum of Release Delay and Opening Time. 103

contact piston (contact plunger) (waveguide). A piston with sliding metallic contact with the walls of a waveguide. *See:* **waveguide.** 179

contact plating. The deposition, without the application of an external electromotive force, of a metal coating upon a base metal, by immersing the latter in contact with another metal in a solution containing a compound of the metal to be deposited. *See:* **electroplating.** 328

contact plunger. *See:* **contact piston.**

contact potential. The difference in potential existing at the contact of two media or phases. *See:* **biological contact potential; electrolytic cell; depolarization (biological); depolarization front; injury potential electrobiology); negative after-potential (electrobiology); positive after-potential (electrobiology).** 328

contact-potential difference. The difference between the work functions of two materials divided by the electronic charge. 125

contact pressure, final (industrial control). The force exerted by one contact against the mating contact when the actuating member is in the final contact-closed position. *Note:* Final contact pressure is usually measured and expressed in terms of the force that must be exerted on the yielding contact while the actuating member is held in the final contact-closed position, and with the mating contact fixed in position, in order to separate the mating contact surfaces. *See:* **contactor.** 206

contact pressure, initial (industrial control). The force exerted by one contact against the mating contact when the actuating member is in the initial contact-touch position. *Note:* The initial contact pressure is usually measured and expressed in terms of the force that must be exerted on the yielding contact while the actuating member is held in the initial contact-touch position in order to separate the mating contact surface against the action of the spring or other contact pressure device. *See:* **electric controller.** 206

contact race (power system relaying). A circuit design condition wherein two or more independently operated contacts compete for the control of a circuit which they will open and close. 60

contact rectifier. A rectifier consisting of two different solids in contact, in which rectification is due to greater conductivity across the contact in one direction than in the other. *See:* **rectifier.** 111

contacts (1). Conducting parts which co-act to complete or to interrupt a circuit. 328
(2) (nonoverlapping) (industrial control). Combinations of two sets of contacts, actuated by a common means, each set closing in one of two positions, and so arranged that the contacts of one set open before the contacts of the other set close. *See:* **electric controller.** 206
(3) (auxiliary) (industrial control) (switching device). Contacts in addition to the main circuit contacts that function with the movement of the latter. *See:* **contactor.** 206
(4) (overlapping, industrial control). Combinations of two sets of contacts, actuated by a common means, each set closing in one of two positions, and so arranged that the contacts of one set open after the contacts of the other set have been closed. *See:* **electric controller.** 206

contact surface (power switchgear). That surface of a contact through which current is transferred to the co-acting contact. 103

contact voltage (human safety). A voltage accidentally appearing between two points with which a person can simultaneously make contact. 244,62

contact-wear allowance (industrial control)(power switchgear). The total thickness of material that may be worn away before the co-acting contacts cease to perform adequately. *See:* **contactor.** 103, 27

contact wire (trolley wire). A flexible contact conductor, customarily supported above or to one side of the vehicle. *See:* **contact conductor.** 1

containment (1)(safety systems equipment in nuclear power generating stations)(valve actuators). That (The) portion of the engineered safety features designed to act as the principal barrier, after the reactor system pressure boundary, to prevent the release, even under conditions of a reactor accident, of unacceptable quantities of radioactive material beyond a controlled zone.   492, 535, 104, 120 141, 408
(2) (radiological monitoring instrumentation). A structure or vessel which encloses the components of the reactor coolant pressure boundary or which serves as a leakage limiting barrier to radioactive material that could be released from the reactor coolant pressure boundary, or both.   398
contamination (rotating machinery). This deteriorates electrical insulation by actually conducting current over insulated surfaces, or by attacking the material reducing its electrical insulating quality or its physical strength, or by thermally insulating the material forcing it to operate at higher than normal temperatures. *Note:* Included here are: wetness or extreme humidity, oil or grease, conducting dusts and particles, non-conducting dusts and particles, and chemicals of industry.
37
content addresses storage (computing system). *See:* **associative storage.**
content, average information. *See:* **average information content; information theory.**
content, conditional information. *See:* **conditional information content; information theory.**
contention (1)(supervisory control, data acquisition, and automatic control). An operational condition on a data communication channel in which no station is designated a master station. In contention, each station on a channel monitors the signals on the channel and waits for a quiescent condition before initiating a bid for circuit control.   570
(2) (data transmission). A condition on a multipoint communication channel when two or more locations try to transmit at the same time.   59
(3) (station control and data acquisition). An operational condition on a data communication channel in which no station is designated a master station. In contention, each station on the channel must monitor the signals on the channel and wait for a quiescent condition before initiating a bid for circuit control.
403
(4) (data communication). A condition on a communications channel when two or more stations may try to seize the channel at the same time.   12
continuing current (lightning). The low-magnitude current that may continue to flow between components of a multiple stroke. *See:* **direct-stroke protection (lightning).**   64
continuity cable bond. A cable bond used for bonding of cable sheaths and armor across joints between continuous lengths of cable. *See:* **cable bond; cross cable bond.**   64
continuity test (test, measurement and diagnostic equipment). A test for the purpose of detecting broken or open connections and ground circuits in a network or device.   54

continuity tester (test, measurement and diagnostic equipment). An electrical tester used to determine the presence and location of broken or open connections and grounded circuits.   54
continuous corona (1) (corona measurement). Corona discharges which recur at regular intervals; for example, on approximately every cycle of an applied alternating voltage or at least once per minute for an applied direct voltage.   375
(2) (overhead-power-line corona and radio noise). Corona discharge that is either steady or recurring at regular intervals (approximately every cycle of an applied alternating voltage or at least several times per minute for an applied direct voltage).   411
continuous current rating (separable insulated connectors). The designated root-mean-square alternating or direct current which the connector can carry simultaneously under specified conditions.   454
continuous current tests (power switchgear). Tests made at rated current, until temperature rise ceases, to determine that the device or equipment can carry its rated continuous current without exceeding its allowable temperature rise.   103
continuous data. Data of which the information content can be ascertained continuously in time.   207
continuous duty (1) (electric installations on shipboard). A requirement of service that demands operation at a substantially constant load for an indefinitely long time.   3
(2) (National Electrical Code). Operation at a substantially constant load for an indefinitely long time.
256
(3) (rating of electric equipment) (power and distribution transformer). A duty that demands operation at a substantially constant load for an indefinitely long time. *See:* **asynchronous machine; transformer.**
3,257,53
continuous-duty current rating (watthour meter sockets). The rating in amperes which a meter socket will carry continuously under stated conditions, without exceeding the allowable temperature rise. A multiposition trough socket has an additional current rating which denotes the maximum ampere capacity of the line buses.   549
continuous-duty rating. The rating applying to operation for an indefinitely long time.   111
continuous electrode. A furnace electrode that receives successive additions in lengths at the end remote from the active zone of the furnace to compensate for the length consumed therein. *See:* **electrothermics.**
328
continuous enclosure (generating station grounding). Refers to a type of isolated-phase bus in which the enclosure is electrically continuous over the full length of the bus. All enclosures are electrically tied together at each end of the bus.   569
continuous inductive train control. *See:* **continuous train control.**
continuous lighting (railway practice). An arrangement of circuits so that the signal lights are continuously energized.   328

continuous load (watthour meter sockets). A load where the current continues for three hours or more.   549

continuous load rating (power inverter unit). Defines the maximum load that can be carried continuously without exceeding established limitations under prescribed conditions of test, and within the limitations of established standards. *See:* **self-commutated inverters.**   208

continuously acting regulator (excitation systems for synchronous machines). A regulator that initiates a corrective action for a sustained infinitesimal change in the controlled variable.   507

continuously adjustable inductor (continuously variable inductor) (variable inductor). An adjustable inductor in which the inductance can have every possible value within its range.   210

continuous noise (electromagnetic compatibility). Noise, the effect of which is not resolvable into a succession of discrete impulses. *See:* **electromagnetic compatibility.**   199

continuous periodic rating (industrial control). The load that can be carried for the alternate periods of load and rest specified in the rating and repeated continuously without exceeding the specified limitation.   206

continuous-pressure operation (elevators). Operation by means of buttons or switches in the car and at the landings, any one of which may be used to control the movement of the car as long as the button or switch is manually maintained in the actuating position. *See:* **control.**   328

continuous pulse (thyristor). A gate signal applied during the desired conducting interval, or parts thereof, as a dc signal.   445

continuous rating (1) (electric equipment). The maximum constant load that can be carried continuously without exceeding established temperature-rise limitations under prescribed conditions of test and within the limitations of established standards. *See:* **duty; rectification.**   257

(2) (packaging machinery). The rating that defines the load that can be carried continuously without exceeding the temperature rating.   429

(3) (power and distribution transformer). The maximum constant load that can be carried continuously without exceeding established temperature-rise limitations under prescribed conditions.   53

(4) (rotating electric machinery). The output that the machine can sustain for an unlimited period under the conditions of Section 4 of IEEE Std 11-1980 without exceeding the limits of temperature rise of Section 5.   424

continuous rating of diesel-generator unit (nuclear power generating stations). The electric power output capability that the diesel-generator unit can maintain in the service environment for 8760 hours of operation per (common) year with only scheduled outages for maintenance.   99

continuous-scan system, supervisory control. *See:* **supervisory control system, continuous-scan.**

continuous-speed adjustment (industrial control). Refers to an adjustable-speed drive capable of being adjusted with small increments, or continuously, between minimum and maximum speed. *See:* **electric drive.**   206

continuous test (battery). A service test in which the battery is subjected to an uninterrupted discharge until the cutoff voltage is reached. *See:* **battery (primary or secondary); cutoff voltage.**   328

continuous thermal burden (voltage transformer) (metering). The volt-ampere burden that the voltage transformer will carry continuously at rated voltage and frequency without causing the specified temperature limitations to be exceeded.   212

continuous thermal current rating factor (1) (RF) (instrument transformer). The factor by which the rated primary current of a current transformer is multiplied to obtain the maximum primary current that can be carried continuously without exceeding the limiting temperature rise from 30 °C average ambient air temperature. The RF of tapped-secondary or multi-ratio current transformers applies to the highest ratio, unless otherwise stated. (When current transformers are incorporated internally as parts of larger transformers or power circuit breakers, they shall meet allowable average winding and hot spot temperature limits under the specific conditions and requirements of the larger apparatus).   203

(2) (RF) (power and distribution transformer). The specified factor by which the rated primary current of a current transformer can be multiplied to obtain the maximum primary current that can be carried continuously without exceeding the limiting temperature rise from 30 °C ambient air temperature. (When current transformers are incorporated internally as parts of larger transformers or power circuit breakers, they shall meet allowable average winding and hot-spot temperatures under the specific conditions and requirements of the larger apparatus.)   53

continuous train control (continuous inductive train control). A type of train control in which the locomotive apparatus is constantly in operative relation with the track circuit and is immediately responsive to a change in the character of the current flowing in the track circuit of the track on which the locomotive is traveling. *See:* **automatic train control.**   328

continuous-type control (electric power system). A control mode that provides a continuous relation between the deviation of the controlled variable and the position of the final controlling element. *See:* **speed-governing system.**   94

continuous update. *See:* **supervisory control.**

continuous update supervisory system (station control and data acquisition). A system in which the remote station continuously updates indication and telemetering to the master station regardless of action taken by the master station. The remote station may interrupt the continuous data updating to perform a control operation.   403

continuous-voltage-rise test (rotating machinery). A controlled overvoltage test in which voltage is increased in continuous function of time, linear or otherwise. *See:* **asynchronous machine.**   63

continuous wave (1) (CW) (data transmission). Waves, the successive oscillations of which are identical under steady-state conditions.   59

(2) (cw) (laser-maser). The output of a laser which is operated in a continuous rather than pulsed mode. In this standard, a laser operating with a continuous output for a period greater than 0.25 s is regarded as a cw laser.   363

contoured beam antenna. A shaped-beam antenna designed in such a way that when its beam intersects a given surface, the lines of equal power flux density incident upon the surface form specified contours.   111

contouring control system (numerically controlled machines). A system in which the controlled path can result from the coordinated, simultaneous motion of two or more axes.   224

contract (C) (diode-type camera tube). The ratio of the difference between the peak and minimum values of irradiance to the peak irradiance of an image or specified portion of an image.   380

$$c = \frac{E_p - E_m}{E_p} \times 100 \ (\text{percent})$$

contract (software requirements specifications). A legally binding document agreed upon by the customer and supplier. This includes the technical, organizational, cost and schedule requirements of a product.   449

contract curve (rotating electric machinery). A specified machine characteristic curve that becomes part of the contract.   424

contract demand (power operations). The demand that the supplier of electric service agrees to have available for delivery.   516

contractor (power and distribution transformer). A device for repeatedly establishing and interrupting an electric power circuit.   53

contrast (display presentation). The subjective assessment of the difference in appearance of two parts of a field of view seen simultaneously or successively. (Hence: luminosity contrast, lightness contrast, color contrast, simultaneous contrast, successive contrast). *See:* **photometry; television.**   244,178

contrast control. A control, associated with a picture-display device, for adjusting the contrast ratio of the reproduced picture. *Note:* The contrast control is normally an amplitude control for the picture signal. In a monochrome-television system, both average luminance and the contrast ratio are affected. In a color-television system, saturation and hue also may be affected. *See:* **television.**   328

contrast ratio (1) (television). The ratio of the maximum to the minimum luminance values in a television picture or a portion thereof. *Note:* Generally the entire area of the picture is implied, but smaller areas may be specified as in detail contrast.   18

(2) (amplitude, frequency, and pulse modulation). For any diffraction order, the ratio of the maximum light intensity to the minimum light intensity in the order, so that $C = I_{max}/I_{min}$, where C is the contrast ratio. *Note:* In the limiting case when the depth of modulation is equal to 1, the minimum light intensity is due to background light, so that $C = I_{max}/I_b$. In the other extreme when m = 0, the contrast ratio is equal to 1.   72

(3) (acoustically tunable optical filter). The ratio of the dynamic transmission at a given acoustic frequency and power level to the dynamic transmission with no applied acoustic power. *Note:* The contrast ratio is a measure of light leakage through the device. It should be specified for either a monochromatic or white light source input, and the angular spread of the input light.   72

contrast rendition factor (CRF) (illuminating engineering). The ratio of visual task contrast with a given lighting environment to the contrast with sphere illumination.   167

contrast sensitivity (illuminating engineering). The ability to detect the presence of luminance differences. Quantitatively, it is equal to the reciprocal of the brightness contrast threshold. *See:* **brightness contrast threshold.**   167

contrast transfer function - square - wave response (diode-type camera tube). The contrast transfer function or CTF represents the response of the imaging system in the spatial frequency domain to a square-wave input. A bar pattern represents a one-dimensional input to a two-dimensional imaging sensor. CTF is synonymous with the square-wave amplitude response, $R_{sq}(N)$.   380

control (1)(supervisory control, data acquisition, and automatic control). The execution of a system change by manual means, remote means, automatic means, or partially automatic means. (A) Automatic. An arrangement of electrical controls that provides for switching or controlling, or both, of equipment in an automatic sequence and under predetermined conditions. (B) Closed loop. A type of automatic control in which control actions are based on signals fed back from the controlled equipment or system. For example, remote stations can manage local voltage conditions by control of load tap changers and volt amperes reactive (VAR) control compensation equipment. (C) Open loop. A form of control without feedback. (D) Manual. Control in which the system or main device, whether direct or power aided in operation, is directly controlled by an attendant. (E) Partial automatic. Control which is a combination of manual and automatic control. For example, to cause a voltage reduction the local automatic load tap changing closed-loop control may be biased by way of a supervisory control command. (F) Remote. Control of a device from a distant point.   570

(2) (electronic computation). (A) Usually, those parts of a digital computer that effect the carrying out of instructions in proper sequence, the interpretation of each instruction, and the application of the proper signals to the arithmetic unit and other parts in accordance with this interpretation. (B) In some business applications of mathematics, a mathematical check. *See:* **computer control state.** 235

(3) (cryotron). An input element of a cryotron. *See:* **super conductivity.** 191

(4) (packaging machinery). A device or group of devices that serves to govern in some predetermined manner the electric power delivered to the apparatus to which it is connected. 429

(5) (power switchgear). A designation of how the equipment is governed, that is, by an attendant, by automatic means, or partially by automatic means and partially by an attendant. *Note:* The word "control" is often used in a broad sense to include 'indication' also. 103

(6) (station control and data acquisition). The execution of a system change by manual means, remote means, automatic means, or partially automatic means. 403

control accuracy (industrial control). The degree of correspondence between the final value and the ideal value of the directly controlled variable. *See:* **control system, feedback.** 206

control action (automatic control). Of a control element or a controlling system, the nature of change of the output effected by the input. *Note:* The output may be a signal or the value of a manipulated variable. The input may be the control loop feedback signal when the command is constant, an actuating signal, or the output of another control element. One use of control action is to effect compensation. *See:* **compensation.** 56

control action, derivative (1) (industrial control). The component of control action for which the output is proportional to the rate of change to the input. *See:* **control system, feedback.** 206,94

(2) (automatic control). Action in which the output of the controller is proportional to the first time derivative of the input. 94

control action, integral (1) (industrial control). Control action in which the output is proportional to the time integral of the input. *See:* **control system, feedback.** 206,94

(2) (control system). Control action in which the output is proportional to the time integral of the input, that is the rate of change of output is proportional to the input. *Note:* In the practical embodiment of integral control action the relation between output and input, neglecting high frequency terms, is given by

$$\frac{Y}{X} = \pm \frac{I/s}{\frac{bI}{s} + 1} \quad 0 \leqq b \ll 1$$

where b = reciprocal of static gain
$I/2\pi$ = gain cross-over frequency in cycles per unit time
s = complex variable
X = input transform
Y = output transform

control action, lead. *See:* **lead, first order.**

control action, proportional (1) (industrial control). Control action in which there is a continuous linear relation between the output and the input. *Note:* This condition applies when both the output and input are within their normal operating ranges. 206

(2) (automatic control). Action in which there is a linear relation between the output and the input of the controller. *Note:* The ratio of the change in output produced by the proportional control action to the change in input is defined as the proportional gain. 94

control action, proportional plus derivative (control systems). Control action in which the output is proportional to a linear combination of the input and the time rate-of-change of input. *Syn:* **P.D.** *Note:* In the practical embodiment of proportional plus derivative control action the relationship between output and input, neglecting high frequency terms, is

$$\frac{Y}{X} = \pm P \frac{\frac{I}{s} + 1 + Ds}{\frac{bI}{s} + 1 + \frac{Ds}{a}} \quad \begin{array}{c} a > I \\ 0 \leqq b < 1 \end{array}$$

where a = derivative action gain
D = derivative action time constant
P = proportional gain
s = complex variable
X = input transform
Y = output transform 56

control action, proportional plus integral (control systems). Control action in which the output is proportional to a linear combination of the input and the time integral of the input. *Syn:* **P.I.** *Note:* In the practical embodiment of proportional plus integral action the relation between output and input, neglecting high frequency terms, is

$$\frac{Y}{X} = \pm P \frac{\frac{I}{s} + 1}{\frac{bI}{s} + 1} \quad 0 \leqq b \ll 1$$

where b = proportional gain/static gain
I = integral action rate
P = proportional gain
s = complex variable
X = input transform
Y = output transform 56

control action, proportional plus integral plus derivative (control systems). Control action in which the output is proportional to a linear combination of the input, the time integral of input and the time rate-of-change of input. *Syn:* **P.I.D.** *Note:* In the practical embodiment of proportional plus integral plus derivative control action the relationship of output and input, neglecting high frequency terms, is

$$\frac{Y}{X} = \pm P \frac{1 + sD}{1 + sD/a} \quad a > I$$

where a = derivative action gain
b = proportional gain/static gain
D = derivative action time constant
I = integral action rate
P = proportional gain
s = complex variable
X = input transform
Y = output transform   56

control and indication point, supervisory control. *See:* **supervisory control point, control and indication.**

control and instrumentation cables (cable systems in sub- stations). Insulated electrical conductors utilized to convey information or to intermittently operate devices controlling power switching or conversion equipment. The cross-sectional areas of the conductors are generally No. 6 American Wire Gage (AWG) or smaller, and the duty cycle is such that conductor heating is insignificant.   382

control and status register (CSR)(FASTBUS acquisition and control). A register used to control the operation of a device and record the status of an operation or both. It is accessible through a separate address space in a FASTBUS device. CSR#0, mandatory for all devices, contains the manufacturer's ID for the device and a number of device status bits as well as some user defined bits.   480

control apparatus. A set of control devices used to accomplish the intended control functions. *See:* **control.**   206

control area (electric power). A part of a power system or a combination of systems to which a common generation control scheme is applied.   94

control battery (industrial control). A battery used as a source of energy for the control of an electrically operated device.   206

control block (cb) (subroutines for CAMAC). The symbol *cb* represents an integer array having four elements. The contents of these elements are:

element 1	Repeat Count
element 2	Tally
element 3	LAM identification
element 4	Channel identification

The repeat count specifies the number of individual CAMAC actions or the maximum number of data words to be transferred. Some multiple action and block transfer subroutines permit termination of the sequence upon a signal from the addressed module. In such cases the repeat count represents an upper limit. The tally is the number of actions usually performed or the number of CAMAC data words actually transferred. If the block transfer or multiple action is terminated by the controller due to exhaustion of the repeat count, the tally will be equal to the repeat count; otherwise it may be less. The LAM identification is an integer value having the same form and information content as the variable *lam* (see 3.10). The channel

identification is an integer value which identifies system-dependent facilities which may be necessary to perform the block transfer or multiple action. This number, if it is required, has the same form and content as the parameter *chan* and can be created by the subroutine CDCHN.   410

control board (control boards, panels, and racks). An assembly of panels on which are installed components and modules for monitoring, measuring, and controlling remotely operated systems and equipment. It provides a visual and physical interface between the operator and the systems. *Syn:* **control switchboard; control panel; benchboard; console.**   140

control bus (power switchgear). A bus used to distribute power for operating electrically controlled devices.   103

control cable (cable systems in power generating stations). Cable applied at relatively low current levels or used for intermittent operation to change the operating status of a utilization device of the plant auxiliary system.   35

control center (generating stations electric power system). An assembly of devices for the purpose of switching and protecting a number of load circuits. The control center may contain transformers, contactors, circuit breakers, protective and other devices intended primarily for energizing and de-energizing load circuits.   381

control character. A character whose occurrence in a particular context initiates, modifies, or stops a control operation, for example, a character to control carriage return.   255,77

control characteristic (gas tube). A relation, usually shown by a graph, between critical grid voltage and anode voltage. *See:* **gas tube.**   125

control circuit (1) (electric installations on shipboard). The circuit which carries the electric signals of a control apparatus or system directing the performance of the controller but which does not carry the main power circuit.   3
(2) (packaging machinery). The circuit that carries the electric signals directing the performance of the controller but does not carry the main power circuit.   429

control-circuit limit switch. A limit switch the contacts of which are connected only into the control circuit. *See:* **control; switch.**   206

control circuit transformer (packaging machinery). A voltage transformer utilized to supply a voltage suitable for the operation of control devices.   429

control circuit voltage (packaging machinery). The voltage provided for the operation of shunt coil magnetic devices.   429

control compartment (packaging machinery). A space within the base, frame, or column of the machine, used for mounting the control panel.   429

control conductor (substation grounding). The conductor that is utilized to measure equivalent changes in temperature, size, etcetera, that are occurring in at least one of the conductors joined by the connector under test.   475

control current (Hall-effect devices). The current through the Hall plate that by its interaction with a magnetic flux density generates the Hall voltage. 107

control current sensitivity (Hall-effect devices). The ratio of the voltage across the Hall terminals to the control current for a given magnitude of magnetic flux density. 107

control current terminals (Hall-effect devices). The terminals through which the control current flows. 107

control cut-out switch (land transportation vehicles). An isolating switch that isolates the control circuits of a motor controller from the source of energy. 1

control data (software). Data that selects an operating mode or submode in a program, directs the sequential flow, or otherwise directly influences the operation of software. *See:* **data; program; software.** 434

control designation symbol. A symbol that identifies the particular manner, permissible or required, in which an input variable (possibly in combination with other variables) causes the logic element to perform according to its defined function. 88

control desk (power switchgear). A control switchboard consisting of one or more relatively short horizontal or inclined panels mounted on an assembly of such a height that the panel-mounted devices are within convenient reach of an attendant. 103

control device. An individual device used to execute a control function. *See:* **control.** 206

control electrode (electron tubes). An electrode used to initiate or vary the current between two or more electrodes. *See:* **electrode (electron tube).** 125

control-electrode discharge recovery time (attenuator tubes). The time required for the control-electrode discharge to deionize to a level such that a specified fraction of the critical high-power level is required to ionize the tube. *See:* **gas tube.** 125

control enclosure (packaging machinery). The metal housing for the control panel, whether mounted on the industrial equipment or separately mounted. 429

control exciter (rotating machinery). An exciter that acts as a rotary amplifier in a closed-loop circuit. *See:* **asynchronous machines.** 63

control field (C)(logical link control). The field immediately following the destination service access point (DSAP) and source service access point (SSAP) address fields of a protocol data unit (PDU). The content of the control field is interpreted by the receiving destination logical link control(s) (LLC) designated by the DSAP address field: (A) As a command, from the source LLC designated by the SSAP address field, instructing the performance of some specific function. (B) As a response, from the source LLC designated by the SSAP address field. 585

control function. *See:* **supervisory control functions**

control generator. A generator, commonly used on electric motive power units for the generation of electric energy in proportion to vehicle speed, prime mover speed, or some similar function, thereby serving as a guide for initiating appropriate control functions. *See:* **traction motor.** 328

control grid (electron tube). A grid, ordinarily placed between the cathode and an anode, for use as a control electrode. *See:* **electrode (electron tube); grid.** 125

control initiation. The function introduced into a measurement sequence for the purpose of regulating any subsequent control operations in relation to the quantity measured. *Note:* The system element comprising the control initiator is usually included in the end device but may be associated with the primary detector or the intermediate means. *See:* **measurement system.** 328

control interaction factors (control systems). In a proportional plus integral plus derivative control action unit, the ratio of the effective values to the values that would be measured when the product (integral action rate) (derivative action time constant) is zero. Example: Assume a control unit composed of elements whose ratios of output to input are $1 + D's$ and $P'(I'/s + 1)$ connected so that the output of one is the input of the other. The ratio of output to input of the combination is

$$\frac{Y}{X} = P'(1 + l'D') \left[\frac{l'/s}{1 + l'D'} + 1 + \frac{D's}{1 + l'D'} \right]$$

By comparison with the equation

$$\frac{Y}{X} = P \left[\frac{l}{s} + 1 + Ds \right]$$

it is seen that the effective values are
$P = P'(1 + l'D') = $ proportional gain
$I = l'/(1 + l'D') = $ integral action rate
$D = D'/(1 + l'D') = $ derivative action time constant
When either l' or D' is set equal to zero the factor $1 + l'D'$ equals unity and the measured values are P', l' and D'. Consequently, $1 + l'D'$ is the "proportional interaction factor" and $1/(1 + l'D')$ is both the "integral action rate interaction factor" and "derivative action time interaction factor." 56

controllability (control systems). In comparison of processes, a qualitative term indicating the relative ease with which they can be controlled. *Note:* The type of disturbance for which the comparison is made should be specified. *See:* **inherent regulation.** 56

controllable. A property of a component of a state whereby, given an initial value of the component at a given time, there exists a control input that can change this value to any other value at a later time. *See:* **control system.** 329

controllable, completely. The property of a plant whereby all components of the state are controllable within a given time interval. *See:* **control system.** 329

control law. A function of the state of a plant and

possibly of time, generated by a controller to be applied as the control input to a plant. *See:* **control system.**   329

control law, closed-loop. A control law specified in terms of some function of the observed state. *See:* **control system.**   329

control law, open-loop. A control law specified in terms of the initial state only and possibly of time. *See:* **control system.**   329

controlled access (communication satellite). A mode of operation of a communication satellite in which an earth station desiring access to the system must request and obtain access to the system via a network management facility.   84

controlled area (laser-maser). An area where the occupancy and activity of those within is subject to control and supervision for the purpose of protection from radiation hazards.   363

controlled-avalanche rectifier diode (semiconductor). A rectifier diode that has specified maximum and minimum breakdown-voltage parameters and is specified to operate under steady-state conditions in the breakdown region of its reverse characteristic. *See:* **breakdown.**   66

controlled carrier (floating carrier) (variable carrier). A system of compound modulation wherein the carrier is amplitude modulated by the signal frequencies in any conventional manner, and, in addition, the carrier is simultaneously amplitude modulated in accordance with the envelope of the signal so that the percentage of modulation, or modulation factor, remains approximately constant regardless of the amplitude of the signal.   111,211

controlled ferroresonant regulators (ferroresonant voltage regulators). A regulator consisting basically of an inductor connected in series with a parallel combination of a capacitor and controllable simulated inductor. This combination is connected across the source as shown in Figure Output Versus Input Voltage with Jump Resonance , below. Stabilized output voltage is derived by inductive or conductive coupling to the parallel combination of C and the controllable simulated inductor. In a controlled ferroresonant regulator the controllable simulated inductor can be a combination of switching devices (such as thyristors or transistors) and linear or saturating inductors. This circuit, in combination with a control input to the simulated inductor, controls the flux swing (or simulated flux swing) in the saturated (or simulated saturating) inductor, thereby controlling the stabilized output voltage.   456

controlled manual block signal system. A series of consecutive blocks governed by block signals, controlled by continuous track circuits, operated manually upon information by telegraph, telephone, or other means of communication, and so constructed as to require the cooperation of the signalmen at both ends of the block to display a clear or permissive block signal. *See:* **block signal system.**   328

controlled overvoltage test (dc leakage, measured current, or step voltage test) (rotating machinery). A

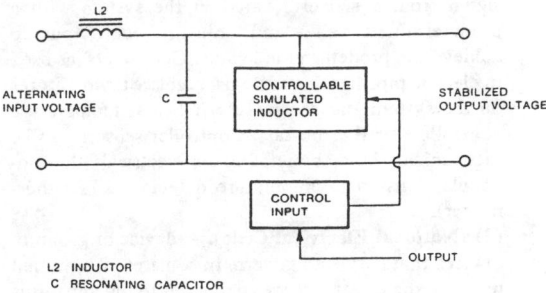

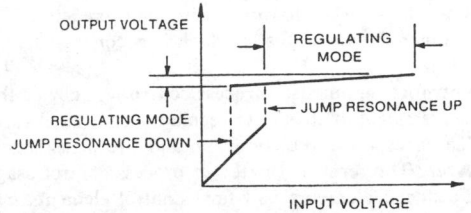

test in which the increase of applied direct voltage is controlled and measured currents are continuously observed for abnormalities with the intention of stopping the test before breakdown occurs.   6

controlled plasma switch (nonlinear, active, and nonreciprocal waveguide components). A triggered gas switch that uses an electron-beam-excited gaseous plasma in a waveguide to limit or switch radio frequency (rf) power.   530

controlled rectifier. A rectifier in which means for controlling the current flow through the rectifying devices is provided. *See:* **electronic controller; rectification.**   328

controlled-speed axle generator. An axle generator in which the speed of the generator is maintained approximately constant at all vehicle speeds above a predetermined minimum. *See:* **axle generator system.**   328

controlled system (automatic control). The apparatus, equipment, or machine used to effect changes in the value of the ultimately controlled variable. *See:* **control system.**   56

controlled vented power fuse (National Electrical Code) (installations and equipment operating at over 600 volts, nominal). A fuse with provision for controlling discharge circuit interruption such that no solid material may be exhausted into the surrounding atmosphere. The discharge gases shall not unite or damage insulation in the path of the discharge nor shall these gases propagate a flashover to or between grounded members or conduction members in the path of the discharge when the distance between the vent and such insulation or conduction members conforms to manufacturer's recommendations.   256

controller (CAMAC). *See:* **CAMAC crate.**

controller (1)(electric pipe heating systems). A device that regulates the state of a system by comparing a signal from a sensor located in the system with a predetermined value and adjusting its output to achieve the predetermined value. Controllers, as used in electric pipe heating systems, regulate temperatures on the system and can be referred to as temperature controllers or thermostats. Controller sensors can be mechanical (bulb, bimetallic) or electrical (thermocouple, resistance-temperature detector (RTD) thermistor).   448

(2) (National Electrical Code). A device or group of devices that serves to govern, in some predetermined manner, the electric power delivered to the apparatus to which it is connected.   256

(3) (packaging machinery). A device or group of devices that serves to control in some predetermined manner the apparatus to which it is connected.

     429

controller, automatic (process control). A device that operates automatically to regulate a controlled variable in response to a command and a feedback signal. *Note:* The term originated in process control usage. Feedback elements and final control elements may also be part of the device. *See:* **control system, feedback.**   56

controller characteristics (thyristor). The electrical characteristics of an ac power controller measured or observed at its input or output terminal.   445

controller current (thyristor). The current flowing through the terminals of the controller.   445

controller diagram (electric-power devices). A diagram that shows the electric connections between the parts comprising the controller and that shows the external connections.   210,206

controller equipment (thyristor). An operative unit for ac power control comprising one or more thyristor assemblies together with any input or output transformers, filters, other switching devices and auxiliaries required by the thyristor ac power controller to function.   445

controller faults (thyristor). A fault condition exists if the conduction cycles of some semiconductors are abnormal.   445

controller ON-state interval (thyristor). The time interval in which the controller conducts. *Note:* It is assumed that the starting instant of the controller ON-state interval is coincident with the starting instant of the trigger pulse.   445

controller power transformer (thyristor). A transformer within the controller employed to provide isolation or the transformation of voltage or current, or both.   445

controller section (thyristor). That part of a controller circuit containing the basic control elements necessary for controlling the load voltage.   445

controller, self-operated (automatic control). A control device in which all the energy to operate the final controlling element is derived from the controlled system through the primary detecting element.

     56

controllers for steel-mill accessory machines. Controllers for machines that are not used directly in the

processing of steel, such as pumps, machine tools, etcetera. *See:* **electric controller.**   206

controllers for steel-mill auxiliaries. Controllers for machines that are used directly in the processing of steel, such as screwdowns and manipulators but not cranes and main rolling drives. *See:* **electric controller.**   206

controller, time schedule (process control). A controller in which the command (or reference input signal) automatically adheres to a pre-determined time schedule. *Note:* The time schedule mechanism may be programmed to switch motors or other devices.

     56

controlling element, final (control system). That forward controlling element which directly changes the value of the manipulated variable.   56

controlling elements (control system, feedback). The functional components of a controlling system. *See:* **control system, feedback.**   56,329

controlling elements, forward (control system, feedback). The elements in the controlling system that change a variable in response to the actuating signal. *See:* **control system feedback.**   56,329

controlling means (of an automatic control system). Consists of those elements that are involved in producing a corrective action.   94

controlling section. A length of track consisting of one or more track circuit sections, by means of which the roadway elements or the device that governs approach to or movement within a block are controlled.

     328

controlling system (1) (automatic control system without feedback). That portion of the control system that manipulates the controlled system.   56,329

(2) (control system, feedback). The portion that compares functions of a directly controlled variable and a command and adjusts a manipulated variable as a function of the difference. *Note:* It includes the reference input elements; summing point; forward and final controlling elements; and feedback elements. *See:* **control system, feedback.**   56,329

controlling voltage, composite. *See:* **composite controlling voltage.**

control machine (railroad practice). (1) An assemblage of manually operated levers or other devices for the control of signals, switches, or other units, without mechanical interlocking, usually including a track diagram with indication lights. (2) A group of levers or equivalent devices used to operate the various mechanisms and signals that constitute the car retarder installation. *See:* **car retarder; centralized traffic control system.**   328

control, manual. Those elements in the excitation control system which provide for manual adjustment of the synchronous machine terminal voltage by open-loop control.   105

control mechanism (control systems for steam turbine-generator units). Includes all systems, devices, and mechanisms between a controller and the controlled valves.   522

control metering point (tie line) (electric power sys-

tems). The location of the metering equipment that it used to measure power on the tie line for the purpose of control. *See:* **center of distribution; power system.**
94, 200

control mode (thyristor). The starting instant of the controller ON-state interval is periodic. The control mode is defined only for steady state operation. *Note:* It is possible to combine several control modes, for example, ON-OFF control and phase control. *See:* **operation modes (Typical Examples of Operation Modes) G and H.**
445

control module (rotating machinery). Control-circuit subassembly. *See:* **starting switch assembly.**
63

control panel (station control and data acquisition)-(supervisory control, data acquisition, and automatic control). An assembly of man/machine interface devices.
403,570

control point selector (test, measurement and diagnostic equipment). A device capable of selecting and controlling the proper stimuli, power of loads, and applying it to the unit under test, in accordance with instructions from the programming device.
54

control position electric indicator. A deviced that provides an indication of the movement and position of the various control surfaces or structural parts of an aircraft. It may be used for wing flaps, cowl flaps, trim tabs, oil-cooler shutters, landing gears, etcetera.
328

control positioning accuracy, precision, or reproducibility (numerically controlled machines). Accuracy, precision, or reproducibility of position sensor or transducer and interpreting system and including the machine positioning servo. *Note:* May be the same as machine positioning accuracy, precision, or reproducibility in some systems.
207

control power disconnecting device (8) (power system device function numbers). A disconnecting device, such as a knife switch, circuit breaker, or pull-out fuse block, used for the purpose of respectively connecting and disconnecting the source of control power to and from the control bus or equipment. *Note:* Control power is considered to include auxiliary power which supplies such apparatus as small motors and heaters.
402

control-power winding (or transformer) (power and distribution transformer). The winding (or transformer) which supplies power to motors, relays, and other devices used for control purposes.
53

control precision (control system). Precision evidenced by either the directly or the indirectly controlled variable, as specified.
56

control procedure (data communications). The means used to control the orderly communication of information between stations on a data link.
12

control ratio (1) (gas tube). The ratio of the change in anode voltage to the corresponding change in grid voltage, with all other operating conditions maintained constant. *See:* **gas tube.**
125

(2) (power supplies). The required charge in control resistance to produce a one-volt change in the output voltage. The control ratio is expressed in ohms per volt

and is reciprocal of the bridge current. *See:* **power supply.**
186

control relay (power switchgear). An auxiliary relay whose function is to initiate or permit the next desired operation in a control sequence.
103

control ring. *See:* **grading ring.**

control room complex (nuclear power generating station). The complex which houses and protects plant operating personnel and control and instrumentation equipment. It includes the central control room, adjacent rooms which house supporting control equipment and instrumentation (sometimes known as the auxiliary equipment room), ventilation and life support equipment, and the cable spreading areas serving the equipment therein.
439

control sequence table (electric-power devices). A tabulation of the connections that are made for each successive position of the controller.
206

control statement (software). A programming language statement that affects the order in which operations are performed. *See:* **programming language.**
434

control station (mobile communication). A base station, the transmission of which is used to control automatically the emission or operation of another radio station. *See:* **mobile communication system.**
181

control structure (software). A construct that determines the flow of control through a computer program. *See:* **computer program; conditional control structure; flow of control.**
434

control switch (power switchgear). A manually operated switching device for controlling power-operated devices. *Note:* It may include signaling, interlocking, etcetera, as dependent functions.
103

control switchboard (power switchgear). A type of switchboard including control, instrumentation, metering, protective (relays) or regulating equipment for remotely controlling other equipment. Control switchboards do not include the primary power circuit-switching devices or their connections.
103

control-switching point (CSP) (telephone switching systems). A switching entity arranged for routing and control in the distance dialing network, at which intertoll trunks are interconnected.
55

control system (1) (broadly). An assemblage of control apparatus coordinated to execute a planned set of controls. *See:* **control.**
206

(2). A system in which a desired effect is achieved by operating on the various inputs to the system until the output, which is a measure of the desired effect, falls within an acceptable range of values. *See:* **closed-loop control system (control system, feedback); control; network analysis; open-loop control system; transfer function.**
151,304

(3) (automatic control). A system in which deliberate guidance or manipulation is used to achieve a prescribed value of a variable. *Note:* It is subdivided into a controlling system and a controlled system.
56

(4) (high-power wide-band transformers). A system in which a desired effect is achieved by operating on the various inputs to the system until the output, which is a measure of the desired effect, falls within an acceptable range of values.   321

control system, adaptive (industrial control). A control system within which automatic means are used to change the system parameters in a way intended to improve the performance of the control system. *See:* **control system, feedback.**   219,206,56,329

control system, automatic. A control system that operates without human intervention. *See:* **control system, feedback.**   56,329

control system, automatic feedback. A feedback control system that operates without human intervention. *See:* **control system, feedback.**   56,329

control system, cascade. A control system in which the output of one subsystem is the input for another subsystem. *See:* **control system, feedback.**   56,329

control system, closed-loop. A control system in which the controlled quantity is measured and compared with a standard representing the desired performance. Any deviation from the standard is fed back into the control system in such a sense that it will reduce the deviation of the controlled quantity from the standard. *Note:* In automatic generation control, the controlled quantities are frequency, unit generation, and net interchange.   94

control system, coarse-fine (industrial control). A control system that uses some elements to reduce the difference between the directly controlled variable and its ideal value to a small value and that uses other elements to reduce the remaining difference to a smaller value.   206

control system, dual-mode. A control system in which control alternates between two predetermined modes. *Notes:* The condition for change from one mode to the other is often a function of the actuating signal. One use of dual-mode action is to provide rapid recovery from large deviations without incurring large overshoot. *See:* **control system, feedback.**   56

control system, duty factor (automatic control). A control system in which the signal to the final controlling element consists of periodic pulses whose duration is varied to relate, in some prescribed manner, the time average of the signal to the actuating signal. *Note:* This mode of control differs from two-step control in that the period of the pulses in duty-factor control is predetermined.   56

control system, feedback (1) (general). A control system that operates to achieve prescribed relationships between selected system variables by comparing functions of these variables and using the comparison to effect control. See the diagram above.

(2) (speed governing of hydraulic turbines). A closed-loop or feedback control system is a control system in which the controlled quantity is measured and compared with a standard representing the desired value of the controlled quantity. In hydraulic governors, any deviation from the standard is fed back into the control system in such a sense that it will reduce the deviation between the controlled quantity and the standard providing negative feedback.   8

control system, floating (automatic control). A control system in which the rate of change of the manipulated variable is a continuous (or at least a piecewise continuous) function of the actuating signal. *Note:* The manipulated variable can remain at any value in its operating range when the actuating signal is zero and constant. Hence the manipulated variable is said to "float." When the forward elements in a control loop have integral control action only, the mode of control has been called "proportional-speed floating." The use of the term integral control action is recommended as a replacement for "proportional-speed floating control." *Syn:* **floating control.** *See:* **control system, single-speed floating; control system, multiple-speed floating; control action, integral; neutral zone.**   56

control system, multiple-speed floating (automatic control). A form of floating control system in which the manipulated variable may change at two or more rates each corresponding to a definite range of values of the actuating signal.   56

control system, multi-step (automatic control). *See:* **control system, step.**

control system, on-off. A two-step control system in which a supply of energy to the controlled system is either on or off. *See:* **control system, feedback.**   56,329

control system, positioning (automatic control). A control system in which there is a predetermined relation between the actuating signal and the position of a final controlling element. *Note:* In a "proportional-position control system" there is a continuous linear relation between the value of the actuating signal and the position of a final controlling element.   56

control system, ratio (automatic control). A system that maintains two or more variables at a predetermined ratio. *Note:* Frequently some function of the value of an uncontrolled variable is the command to a system controlling another variable.   56

control system, sampling. Control using intermittently observed values of signals such as the feedback signal or the actuating signal. *Note:* The sampling is often done periodically. *See:* **control system, feedback.**   56,329

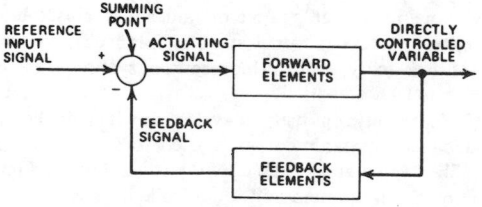

Simplified block diagram indicating essential elements of an automatic control system.

control system, single-speed floating (automatic control). A floating control system in which the manipulated variable changes at a fixed rate, increasing or decreasing depending on the sign of the actuating signal. *Note:* A neutral zone of values of the actuating signal, in which no action occurs, may be used.

56

control system, step (automatic control). A system in which the manipulated variable assumes discrete predetermined values. *Note:* The condition for change from one predetermined value to another is often a function of the value of the actuating signal. When the number of values of the manipulated variable is two, it is called a two-step control system; when more than two, a multi-step control system.   56

control system, two-step. A control system in which the manipulated variable alternates between two predetermined values. *Note:* A control system in which the manipulated variable changes to other predetermined value whenever the actuating signal passes through zero is called a two-step single-point control system. A two-step neutral-zone control system is one in which the manipulated variable changes to the other predetermined value when the actuating signal passes through a range of values known as the neutral zone. The neutral zone may be produced by a mechanical differential gap. The neutral zone is also called overlap, and two-step neutral-zone control overlap control. *See:* **control system, feedback.**

56,329

control system, two-step neutral zone (automatic control). *See:* **control system, two-step.**

control system, two-step single-point (automatic control). *See:* **control system, two-step.**

control terminal (base station) (mobile communication). Equipment for manually or automatically supervising a multiplicity of mobile and/or radio stations including means for calling or receiving calls from said stations. *See:* **mobile communication system.**   181

control track (electroacoustics). A supplementary track usually placed on the same medium with the record carrying the program material. *Note:* Its purpose is to control, in some respect, the reproduction of the program, or some related phenomenon. Ordinarily, the control track contains one or more tones, each of which may be modulated either as to amplitude, frequency, or both. *See:* **phonograph pickup.**   176

control transformers (power and distribution transformer). Step-down transformers generally used in circuits which are characterized by low power levels and which contribute to a control function, such as in heating and air conditioning, printing, and general industrial controls.   53

control unit (1) (digital computer). The parts that effect the retrieval of instructions in proper sequence, the interpretation of each instruction, and the application of the proper signals to the arithmetic unit and other parts in accordance with this interpretation.

255,77

(2) (mobile station) (mobile communication). Equipment including a microphone and/or handset and loudspeaker together with such other devices as may be necessary for controlling a mobile station. *See:* **mobile communication system.**   181

control valve (control systems for steam turbine-generator units). Those valves that control the energy input to the turbine and that are actuated by a controller through the control mechanism.   522

control voltage (power switchgear). The voltage applied to the operating mechanism of a device to actuate it, usually measured at the control power terminals of the mechanism.   103

control winding (1) (rotating machinery). An excitation winding that carries a current controlling the performance of a machine. *See:* **asynchronous machines.**

63

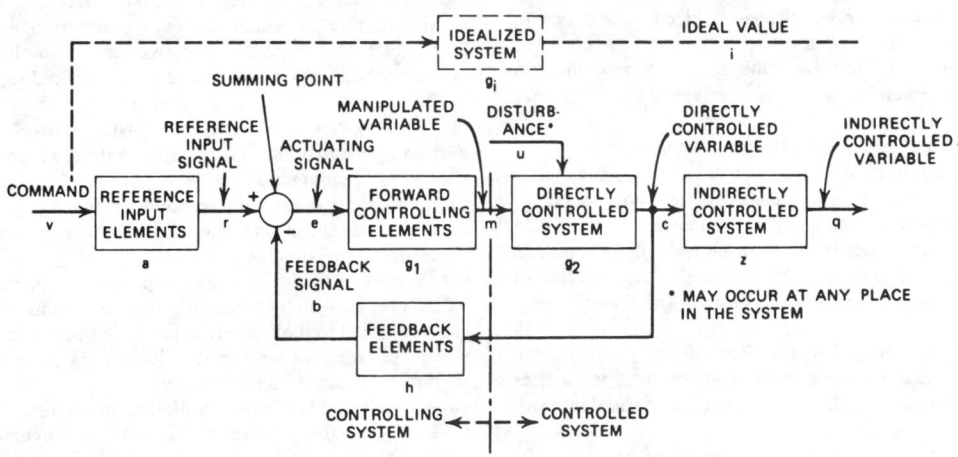

Block diagram of an automatic control system illustrating expansion of the simplified block diagram to a more complex system.

(2) (saturable reactor). A winding by means of which a controlling magnetomotive force is applied to the core. *See:* **magnetic amplifier.**   328

convection current. In an electron stream, the time rate at which charge is transported through a given surface. *See:* **electron emission.**   125,190

convection-current modulation. The time variation in the magnitude of the convection current passing through a surface, or the process of directly producing such a variation. *See:* **electron emission.**   125

convection heater. A heater that dissipates its heat mainly by convection and conduction. *See:* **appliances (including portable).**   328

convective discharge (effluve) (electrical wind) (static breeze) (medical electronics). The movement of a visible or invisible stream of particles carrying away charges from a body that has been charged to a sufficiently high voltage.   192

convenience outlet. *See:* **receptacle (electric distribution).**

conventional-electrode coaxial detector (jargon)(germanium gamma-ray detectors). A coaxial detector in which the outer contact is an n-type layer.   528

conventionally cooled (rotating machinery). A term referring to windings in which the heat generated within the principal portion of the windings must flow through the major ground insulation before reaching the cooling medium.   298,63

conventions (software quality assurance). Requirements employed to prescribe a disciplined, uniform approach to providing consistency in a software product, that is, uniform patterns or forms for arranging data.   481

convergence (multibeam cathode-ray tubes). A condition in which the electron beams intersect at a specified point.   125

convergence, dynamic (multibeam cathode-ray tubes). The process whereby the locus of the point of convergence of electron beams is made to fall on a specified surface during scanning.   125,190

convergence electrode (multibeam cathode-ray tubes). An electrode whose electric field converges two or more electron beams.   125

convergence magnet (multibeam cathode-ray tubes). A magnet assembly whose magnetic field converges two or more electron beams.   125

convergence plane (multibeam cathode-ray tubes). A plane containing the points at which the electron beams appear to experience a deflection applied for the purpose of obtaining convergence.   125

convergence surface (multibeam cathode-ray tubes). The surface generated by the point of intersection of two or more electron beams during the scanning process.   125

conversational (software). Pertaining to an inactive system that provides for interaction between a user and a system similar to a human dialog. *See:* **interactive; system.**   434

conversion (software). Modification of existing software to enable it to operate with similar functional capability in a different environment, for example,

converting a program from FORTRAN to Ada, converting a program that runs on one computer to run on another computer. *See:* **program; modification; software.**   434

conversion efficiency (1) (electrical conversion). In alternating-current to direct-current conversion equipment, the ratio of the product of output direct-current and voltage to input watts expressed in percent. *Note:* It reflects alternating-current power capacity required for a given voltage and current output and does not necessarily reflect watts lost.

Conversion Efficiency

$$= \frac{(E_{dc})(I_{dc})}{P} \, (100 \text{ percent})$$

See: **electric conversion.**   186

(2) (overall) (photoelectric converter). The ratio of available power output to total incident radiant power in the active area for photovoltaic operation. *Note:* This depends on the spectral distribution of the source and junction temperature. *See:* **semiconductor.**   186

(3) (klystron oscillator). The ratio of the high-frequency output power to the direct-current power supplied to the beam. *See:* **velocity-modulated tube.**   244

(4) (solar cells). The ratio of the solar cell's available power output (at a specified voltage) to the total incident radiant power. The cell active area shall be used in this calculation; that is, ohmic contact (but no grid lines) areas on the irradiated side shall be deducted from the total irradiated cell area to determine active area. The spectral distribution of the source and the junction temperature must be specified.   113

conversion loss (nonlinear, active, and nonreciprocal waveguide components). In a frequency converter (mixer), the ratio of the output power at the converted frequency to the available input power at the signal frequency; often expressed in decibels.   530

conversion rate (1) (analog-to-digital converter). The maximum rate at which the start conversion commands can be applied to the converter, to which the converter will respond by providing the desired signal at the output to within a given accuracy.   10

(2) (analog-to-digital converter with multiplexor with sample and hold. The maximum rate at which the start sample commands can be applied to the system to which the system will respond by providing the desired signal at the output to within a given accuracy. (Pre-selected channel).   10

(3) (hybrid computer linkage components). The maximum rate at which the start conversion commands can be applied to the converter, to which the converter will respond by providing the desired signal at the output to within a given accuracy.   10

conversion time (1) (analog-to-digital converter). That time required from the instant at which a conversion command is received and a final digital representation is available for external output to within a given accuracy.   10

(2) (analog-to-digital converter with multiplexor with sample and hold). That time required from the time at which a sample command is received and a final digital representation is available for external output to within a given accuracy. (Pre-selected channel). 10

(3) (hybrid computer linkage components). That time required from the instant at which a conversion command is received and a final digital representation is available for external output to within a given accuracy. 10

(4) (hybrid computer linkage components) (ADC with multiplexor with sample and hold). That time required from the time at which a sample command is received and a final digital representation is available for external output to within a given accuracy (preselected channel). 10

conversion transconductance (heterodyne conversion transducer). The quotient of (1) the magnitude of the desired output-frequency component of currents by (2) the magnitude of the input-frequency (signal) component of voltage when the impedance of the output external termination is negligible for all of the frequencies that may affect the result. *Note:* Unless otherwise stated, the term refers to the cases in which the input-frequency voltage is of infinitesimal magnitude. All direct electrode voltages, and the magnitude of the local-oscillator voltage, must remain constant. *See:* **modulation; transducer.** 125

conversion transducer (1) (general). A transducer in which the signal undergoes frequency conversion. *Note:* The gain or loss of a conversion transducer is specified in terms of the useful signal. *See:* **transducer.** 328

(2) An electric transducer in which the input and the output frequencies are different. *Note:* If the frequency-changing property of a conversion transducer depends upon a generator of frequency different from that of the input or output frequencies, the frequency and voltage or power of this generator are parameters of the conversion transducer. *See:* **heterodyne conversion transducer (converter).** 125

conversion voltage gain (conversion transducer). The ratio of (1) the magnitude of the output-frequency voltage across the output termination, with the transducer inserted between the input-frequency generator and the output termination, to (2) the magnitude of the input-frequency voltage across the input termination of the transducer. 190

convert (data processing). To change the representation of data from one form to another, for example, to change numerical data from binary to decimal or from cards to tape. 255,77

converter (1)(harmonic control and reactive compensation of static power converters). An equipment that changes electrical energy from one form to another. A semiconductor converter is a converter that uses thyristors or diodes as the active elements in the conversion process. 533

(2) (general). A machine or device for changing alternating-current power to direct-current power or vice versa, or from one frequency to another. 63

(3) (frequency converter) (heterodyne reception). The portion of the receiver that converts the incoming signal to the intermediate frequency. 59

(4) (facsimile). A device that changes the type of modulation. *See:* **facsimile (electrical communication).** 12

(5) (industrial control). A network or device for changing the form of information or energy. 219,206

(6) (data transmission). A device for changing one form of information language to another, so as to render the language acceptable to a different machine (that is, card to tape conversion). 59

(7) (test, measurement and diagnostic equipment). A device which changes the manner of representing information from one form to another. 54

(8) (National Electrical Code). A device which changes electrical energy from one form to another, as from alternating current to direct current. 256

converter, analog-to-digital (ADC) (hybrid computer linkage components). *See:* **analog to digital converter.**

converter, digital-to-analog (DAC) (hybrid computer linkage components). *See:* **digital-to-analog converter.**

converter or converter equipment (self-commutated converters). An operative unit for electronic power conversion, comprising one or more electronic switching devices and any associated components, such as transformers, filters, commutation aids, controls, and auxiliaries. 584

converter, reversible power. *See:* **reversible power converter.**

converters, semiconductor. *See:* **semiconductor converters.**

converter, static solid state. *See:* **static solid state converter.**

converter switching element (converter circuit elements)(self-commutated converters). A part of the converter circuit, bounded by two principal terminals, containing one or more semiconductor devices having the property of controllable or noncontrollable conduction in at least one direction. 584

converter tube. An electron tube that combines the mixer and local-oscillator functions of a heterodyne conversion transducer. *See:* **heterodyne conversion transducer (converter).** 125,190

converting station (power operations). A station where machinery is used for changing alternating-current (ac) power to direct-current power or vice versa, or from one frequency to another. 516

conveyor. A mechanical contrivance, generally electrically driven, that extends from a receiving point to a discharge point and conveys, transports, or transfers material between those points. *See:* **conveyor, belt-type; conveyor, chain-type; conveyor, shaker-type; conveyor, vibrating-type.** 328

conveyor, belt-type. A conveyor consisting of an endless belt used to transport material from one place to another. *See:* **conveyor.** 328

conveyor, chain-type. A conveyor using a driven endless chain or chains, equipped with flights that operate in a trough and move material along the trough. *See:* conveyor. 328

conveyor, shaker-type. A conveyor designed to transport material along a line of troughs by means of a reciprocating or shaking motion. *See:* conveyor. 328

conveyor, vibrating-type. A conveyor consisting of a movable bed mounted at an angle to the horizontal, that vibrates in such a way that the material advances. *See:* conveyor. 328

convolution function (burst measurements). The integral of the function $x(\tau)$ multiplied by another function $y(-\tau)$ shifted in time by t

$$\int_{-\infty}^{\infty} x(\tau)y(t - \tau)\mathrm{d}\tau$$

See: burst. 292

convolution integral (automatic control). A mathematical integral operation which is used to describe the time response of a linear element to an input function in terms of the weighting function of the element. The integral generally takes the form $\int_{0}^{t} f(x)g(t - x)\,dx$ where $f(x)$ is an arbitrary input, and $g(t - x)$ is a weighting function which extends backward from instant t through x as far as zero. 56

cooking unit, counter-mounted. An assembly of one or more domestic surface heating elements for cooking purposes, designed for flush mounting in, or supported by, a counter, and which assembly is complete with inherent or separately mountable controls and internal wiring. 256

coolant. *See:* cooling medium.

cooled-input FET preamplifier (germanium gamma-ray detectors). A preamplifier in which the input field-effect transistor (FET) is cooled to achieve a reduction in noise. 528

cooler (heat exchanger) (rotating machinery). A device used to transfer heat between two fluids without direct contact between them. 63

Coolidge tube. An X-ray tube in which the needed electrons are produced by a hot cathode. 190

cooling (power supplies). The cooling of regulator elements refers to the method used for removing heat generated in the regulating process. *Note:* Methods include radiation, convection, and conduction or combination thereof. 186

cooling coil (rotating machinery). A tube through whose wall, heat is transferred between two fluids without direct contact between them. 63

cooling, convection (power supplies). A method of heat transfer that uses the natural upward motion of air warmed by the heat dissipators. 228,186

cooling, duct. *See:* ventilating duct (rotating machinery).

cooling fin (electron device). A metallic part of fin extending the cooling area to facilitate the dissipation of the heat generated in the device. *See:* electron device. 190

cooling, lateral force-air (power supplies). An efficient method of heat transfer by means of side-to-side circulation that employs blower movement of air through or across the heat dissipators. 228,186

cooling medium (rotating machinery) (coolant). A fluid, usually air, hydrogen, or water, used to remove heat from a machine or from certain of its components. 63

cooling system (1) (rectifier). Equipment, that is, parts and their interconnections, used for cooling a rectifier. *Note:* It includes all or some of the following: rectifier water jacket, cooling oils or fins, heat exchanger, blower, water pump, expansion tank, insulating pipes, etcetera. *See:* rectification. 208
(2) (thyristor). Any equipment, that is, parts and their interconnections, used for cooling a thyristor controller. It includes all or some of the following; thyristor heat sink, cooling coils or fins, heat exchanger, fan or blower, water pump, expansion tank, insulating pipes, equipment enclosure, etcetera. 445
(3) (thyristor converter). Equipment, that is, parts and their interconnections, used for cooling a thyristor converter. *Note:* It includes all or some of the following: thyristor heat sink, cooling coils or fins, heat exchanger, fan or blower, water pump, expansion tank, insulating pipes, etcetera. 121

cooling system, direct raw-water (thyristor converter). A cooling system in which water, received from a constantly available supply, such as a well or water system, is passed directly over the cooling surfaces of the thyristor converter and discharged. 121

cooling system, direct raw-water, with recirculation (thyristor converter). A direct raw-water cooling system in which part of the water passing over the cooling surfaces of the thyristor converter is recirculated and raw water is added as needed to maintain the required temperature, the excess being discharged. 121

cooling system, forced-air (thyristor converter). An air cooling system in which heat is removed from the cooling surfaces of the thyristor converter by means of a flow of air produced by a fan or blower. 121

cooling system, heat-exchanger (thyristor converter). A cooling system in which the coolant, after passing over the cooling surfaces of the thyristor converter, is cooled in heat exchanger and recirculated. *Note:* Heat may be removed from the thyristor converter cooling surfaces by liquid or air using the following types of heat exchangers: (1) water-to-water, (2) water-to-air, (3) air-to-water, (4) air-to-air, (5) refrigeration cycle. The liquid in the closed system may be other than water, and the gas in the closed system may be other than air. 121

cooling system, natural-air (thyristor converter). An air cooling system in which heat is removed from the cooling surfaces of the thyristor converter only by the natural action of the ambient air. 121

cooling system regulating equipment (thyristor). Any equipment used for heating and cooling the thyristor controller, together with the devices for controlling and indicating its temperature. 445

cooling-water system (rotating machinery). All parts that are provided for the flow, treatment, or storage of cooling water. 63

coordinate dimension word (numerically controlled machines). A word defining an absolute dimension. 224,207

coordinated operation (power operations). Operation of generation and transmission facilities of two or more interconnected electric systems to achieve greater reliability and economy. 516

coordinated operation of hydroplants (power operations). Operation of a group of hydroplants and storage reservoirs so as to obtain optimum power benefits with due consideration to all other uses. 516

coordinated transpositions (electric supply or communication circuits). Transpositions that are installed for the purpose of reducing inductive coupling, and that are located effectively with respect to the discontinuities in both the electric supply and communication circuits. See: inductive coordination. 328

coordinate system (pulse terms). Throughout the following, a rectangular Cartesian coordinate system is assumed in which, unless otherwise specified: (1) Time (t) is the independent variable taking alone the horizontal axis, increasing in the positive sense from left to right. (2) Magnitude (m) is the dependent variable taken along the vertical axis, increasing the positive sense or polarity from bottom to top. (3) The following additional symbols are used:
(a) e–The base of natural logarithms.
(b) a, b, c, etcetera–Real constants which, unless otherwise specified, may have any value and either sign.
(c) n–A positive integer. 254

coordination of insulation (1) (lightning insulation strength). The steps taken to prevent damage to electric equipment due to overvoltages and to localize flashovers to points where they will not cause damage. Note: In practice, coordination consists of the process of correlating the insulating strengths of electric equipment with expected overvoltages and with the characteristics of protective devices. See: basic impulse insulation level (insulation strength). 224,62

(2) (insulation coordination) (power and distribution transformer). The process of correlating the insulation strengths of electrical equipment with expected overvoltages and with the characteristics of surge protective devices. 53

copolar (radiation) pattern (antennas). A radiation pattern corresponding to the copolarization. See: copolarization. 111

copolarization (antennas). The polarization which the antenna is intended to radiate or receive. See: polarization pattern, Notes 1 and 2. 111

copolar side lobe level, (relative) (antennas). The maximum relative partial directivity (corresponding to the copolarization) of a side lobe with respect to the maximum partial directivity (corresponding to the copolarization) of the antenna. Note: Unless otherwise specified the copolar side lobe level is taken to be that of the highest side lobe of the copolar radiation pattern. 111

copper brush (rotating machinery). A brush composed principally of copper. See: brush. 63

copper-clad aluminum conductors (National Electrical Code). Conductors drawn from a copper-clad aluminum rod with the copper metallurgically bonded to an aluminum core. The copper forms a minimum of 10 percent of the cross-sectional area of a solid conductor or each strand of a stranded conductor. 256

copper-clad steel. Steel with a coating of copper welded to it, as distinguished from copper-plated or copper-sheathed material. 328

copper-covered steel wire. A wire having a steel core to which is bounded a continuous outer layer of copper. See: conductor. 64,57

copper losses. See: load losses.

copy (1)(computer applications). (A) To read data from a source, leaving the source data unchanged, and to write the same data elsewhere in a physical form that may differ from that of the source. For example, to copy data from a magnetic disk onto a magnetic tape. (B) The result of a copy process. For example, a copy of a data file. See: display. 571

(2)(electronic data processing). (A) To reproduce data leaving the original data unchanged. (B) To produce a sequence of character events equivalent, character by character, to another sequence of character events. (C) The sequence of character events produced in (B). See: transfer. 235

cord. One or a group of flexible insulated conductors, enclosed in a flexible insulating covering and equipped with terminals. 328

cord adjuster. A device for altering the pendant length of the flexible cord of pendant. Note: This device may be a rachet reel, a pulley and counterweight, a tent-rope stick, etcetera. 328

cord circuit (telephone switching systems). A connecting circuit, usually terminating in a plug at one or both ends, used at switchboard positions in establishing telephone connections. 55

cord-circuit repeater. A repeater associated with a cord circuit so that it may be inserted in a circuit by an operator. See: repeater. 328

cord connector (cord connector body*) (table tap*). A plug receptacle provided with means for attachment to flexible cord. 328
*Deprecated

cord grip (strain relief). A device by means of which the flexible cord entering a device or equipment is gripped in order to relieve the terminals from tension in the cord. 328

cordless switchboard (telephone switching systems). A telecommunications switchboard in which manually operated keys are used to make connections. 55

core (1) (power and distribution transformer). An element made of magnetic material, serving as part of a path for magnetic flux. 53

(2) (electronic information storage). See: digital computer.

(3) (mechanical recording). The central layer or basic support of certain types of laminated media.   176

(4) (electromagnet). The part of the magnetic structure around which the magnetizing winding is place.   210

(5) (fiber optics). The central region of an optical waveguide through which light is transmitted. *See:* **cladding; normalized frequency; optical waveguide.**   433

(6) (composite insulators). The axially aligned glass fiber reinforced resin rod that forms the mechanically load-bearing component of the insulator.   483

core area (fiber optics). The cross sectional area enclosed by the curve that connects all points nearest the axis on the periphery of the core where the refractive index of the core exceeds that of the homogeneous cladding by k times the difference between the maximum refractive index in the core and the refractive index of the homogeneous cladding, where k is a specified positive or negative constant $k1$. *See:* **cladding; core; homogeneous cladding; tolerance field.**   433

core center (fiber optics). A point on the fiber axis. *See:* **fiber axis; optical axis.**   433

core diameter (fiber optics). The diameter of the circle that circumscribes the core area. *See:* **cladding; core; core area; tolerance field.**   433

core duct (rotating machinery). The space between or through core laminations provided to permit the radial or axial flow of coolant gas. *See:* **rotor (rotating machinery).**   63

core end plate (end plate) (flange) (rotating machinery). A plate or structure at the end of a laminated core to maintain axial pressure on the laminations.   63

core-form transformer (power and distribution transformer). A transformer in which those parts of the magnetic circuit surrounded by the windings have the form of legs with two common yokes.   53

core length (rotating machinery). The dimension of the stator, or rotor, core measured in the axial direction. *See:* **rotor (rotating machinery); stator.**   63

core loss (1). The power dissipated in a magnetic core subjected to a time-varying magnetizing force. *Note:* Core loss includes hysteresis and eddy-current losses of the core.   53

(2) (synchronous machine). The difference in power required to drive the machine at normal speed, when excited to produce a voltage at the terminals on open circuit corresponding to the calculated internal voltage, and the power required to drive the unexcited machine at the same speed. *Note:* The internal voltage shall be determined by correcting the rated terminal voltage for the resistance drop only.   244,63

(3) (electronics power transformer). The measured power loss, expressed in watts, attributable to the material in the core and associated clamping structure, of a transformer that is excited, with no connected load, at a core flux density and frequency equal to that in the core when rated voltage and frequency is applied and rated load current is supplied.   95

(4) (power and distribution transformer). The power dissipated in a magnetic core subjected to a time-varying magnetizing force. Core loss includes hysteresis and eddy-current losses of the core.   53

core-loss current. The in-phase component (with respect to the induced voltage) of the exciting current supplied to a coil. *Note:* It may be regarded as a hypothetical current, assumed to flow through the equivalent core-loss resistance.   197

core loss, open-circuit (rotating machinery). The difference in power required to drive a machine at normal speed, when excited to produce a specified voltage at the open-circuited armature terminals, and the power required to drive the unexcited machine at the same speed.   244

core-loss test (rotating machinery). A test taken on a built-up (usually unwound) core of a machine to determine its loss characteristic. *See:* **stator.**   63

core, relay. *See:* **relay core.**

core test (rotating machinery). FRA test taken on a built-up (usually unwound) core of a machine to determine its loss characteristics or its magnetomotive force characteristics, or to locate short-circuited laminations. *See:* **rotor (rotating machinery); stator.**   63

core-type transformer. A transformer in which those parts of the magnetic-circuit surrounded by the windings have the form of legs with two common yokes.   53

Coriolis acceleration (inertial sensor). That increment of acceleration relative to inertial space which arises from the velocity of a particle relative to a rotating coordinate system. The term Coriolis acceleration is also sometimes used to describe the apparent acceleration relative to a rotating coordinate system of a force-free moving particle.   46

coriolis correction (navigation)(navigation aid terms). An acceleration correction which must be applied to measurements of acceleration with respect to a coordinate system relative to inertial space.   526

cornea (laser-maser). The transparent outer coat of the human eye which covers the iris and the crystalline lens. It is the main refracting element of the eye.   363

corner (waveguide technique). An abrupt change in the direction of the axis of a waveguide. *Note:* Also termed **elbow.** *See:* **waveguide.**   244, 179

corner reflector (1) (antenna). A reflecting object consisting of two or three mutually intersecting conducting flat surfaces. *Note:* Dihedral forms of corner reflectors are frequently used in antennas: trihedral forms are more often used as radar targets. *See:* **antenna; radar.**   111

(2) (radar). Two (dihedral) or three (trihedral) conducting surfaces, mutually intersecting at right angles, designed to return electromagnetic radiations toward their sources and used to render a target more conspicuous to radar observations.   13

corner-reflector antenna. An antenna consisting of a feed and a corner reflector. *See:* **antenna.**   111

corner, waveguide. *See:* **bend, waveguide.**

cornice lighting (illuminating engineering). Lighting comprising sources shielded by a panel parallel to the wall and attached to the ceiling, and distributing light over the wall.   167

corona (1) (air). A luminous discharge due to ionization of the air surrounding a conductor caused by a voltage gradient exceeding a certain critical value. *See:* **tower.**   64
(2) (gas). A discharge with slight luminosity produced in the neighborhood of a conductor, without greatly heating it, and limited to the region surrounding the conductor in which the electric field exceeds a certain value. *See:* **discharge (gas); partial discharge.**
                                                              244, 190
(3) (non-preferred term) (power and distribution transformer). *See:* **partial discharge.**   53
(4) (overhead-power-line corona and radio noise). A luminous discharge due to ionization of the air surrounding an electrode caused by a voltage gradient exceeding a certain critical value. *Note:* For the purpose of this standard, electrodes may be line conductors, hardware, accessories, or insulators.   411
(5) (partial discharge) (corona measurement). A type of localized discharge resulting from transient gaseous ionization in an insulation system when the voltage stress exceeds a critical value. The ionization is usually localized over a portion of the distance between the electrodes of the system.   375

corona charging (electrostatography). Sensitizing by means of gaseous ions of a corona. *See:* **electrostatography.**   191
corona-discharge tube. A low-current gas-filled tube utilizing the corona-discharge properties.   190
corona-extinction gradient (overhead-power-line corona and radio noise). The gradient on that part of an electrode surface at which continuous corona last persists as the applied voltage is gradually decreased.
                                                              411
corona extinction voltage (1) (CEV) (corona measurement). The highest voltage at which continuous corona of specified pulse amplitude no longer occurs as the applied voltage is gradually decreased from above the corona inception value. Where the applied voltage is sinusoidal, the CEV is expressed as $1/\sqrt{2}$ of the peak voltage.   375
(2) (overhead-power-line corona and radio noise). The voltage applied to the electrode to produce the corona-extinction gradient.   411
corona-inception gradient (overhead-power-line corona and radio noise). The gradient on that part of an electrode surface at which continuous corona first occurs as the applied voltage is gradually increased. *See:* **continuous corona.**   411
corona inception test. *See:* **discharge inception test.**
corona inception voltage (1) (CIV) (corona measurement). The lowest voltage at which continuous corona of specified pulse amplitude occurs as the applied voltage is gradually increased. Where the applied voltage is sinusoidal, the CIV is expressed as $1/\sqrt{2}$ of the peak voltage.   375
(2) (overhead-power-line corona and radio noise).

The voltage applied to the electrode to produce the corona-inception gradient.   411
corona level (power distribution, underground cables). *See:* **ionization extinction voltage.**
corona modes (overhead-power-line corona and radio noise). Two principal modes can be distinguished, namely, glow and streamer. Their characteristics and occurrence depend on the polarity of the electrode, the basic ionization characteristics of the ambient air, and the intensity as well as the distribution of the electric field. Thus, the geometry of the electrodes, the ambient weather conditions, and the magnitude as well as the polarity of the applied voltage are the main factors determining corona modes. Corona modes that are possible during alternating half-cycles of the alternating-current wave are essentially similar to those of corresponding direct-current corona modes when effects of space charges left behind from each preceding half-cycle are taken into account. Corona modes listed according to polarity and voltage level and defined in the order of increasing voltage applied to the electrode are given in Table.   411
corona pulse (1) (corona measurement). A voltage or current pulse which occurs at some location in a transformer as a result of a corona discharge.   375
(2) (overhead-power-line corona and radio noise). A voltage or current pulse which occurs at some designated location in a circuit as a result of a corona discharge.   411
corona shielding (corona grading) (rotating machinery). A means adapted to reduce potential gradients along the surface of coils. *See:* **asynchronous machine.**
                                                              63
corona voltmeter. A voltmeter in which the crest value of voltage is indicated by the inception of corona. *See:* **instrument.**   328
coroutines (software). Two or more modules that can call each other, but are not in a superior to subordinate relationship. *See:* **module.**   434
corrected-compass course. *See:* **magnetic course.**
corrected-compass heading. *See:* **magnetic heading.**
correcting signal. *See:* **synchronizing signal.**
correction (1)(digital computer). A quantity (equal in absolute value to the error) added to a calculated or observed value to obtain the true value. *See:* **accuracy rating (instrument); error.**   235
(2) (analog computers). *See:* **error.**   9
correction angle* (navigation aid terms). The angular difference between heading and course of a vehicle. Preferably called drift-correction angle.
*Deprecated   526
correction factor (instrument transformer) (metering). The factor by which the reading of a wattmeter or the registration of a watthour meter must be multiplied to correct for the effects of the error in ratio and the phase angle of the instrument transformer. This factor is the product of the ratio and phase-angle correction factors for the existing conditions of operation.
                                                              212
correction rate (industrial control). The velocity at which the control system functions to correct error in register.   206

Corona Modes

Positive (Anode) Corona		Negative (Cathode) Corona	
Mode	Characteristic	Mode	Characteristic
Burst corona, onset streamer[1]	Moderate amplitude, moderate repitition rate	Trichel streamer (pulse)	Small amplitude, high repetition rate
Glow[2]	Essentially pulseless	Glow[3]	Essentially pulseless
Pre-breakdown streamer	High amplitude, low repetition rate	Pre-breakdown streamer[4]	Moderate amplitude, moderate repetition rate

NOTES:

(1) With alternating voltage, positive onset streamers may be suppressed by space charge created during the negative half-cycles.

(2) With alternating voltage, when onset streamers are suppressed, the positive glow will be the first corona mode as the applied voltage is raised.

(3) With alternating voltage, negative glow may be difficult to observe because of the predominance of Trichel streamers.

(4) With alternating voltage, breakdown usually occurs during the positive half-cycle before the developement of any negative pre-breakdown streamers.

correction time. *See:* **time, settling.**

corrective action (nuclear power quality assurance). Measures taken to rectify conditions adverse to quality and, where necessary, to preclude repetition. 417

corrective maintenance (1) (availability, reliability, and maintainability). The maintenance carried out after a failure has occurred and intended to restore an item to a state in which it can perform its required function. 164

(2) (test, measurement and diagnostic equipment). Actions performed to restore a failed or degraded equipment. It includes fault isolation, repair or replacement of defective units, alignment and checkout. 54

(3) (software). Maintenance performed specifically to overcome existing faults. *See:* **fault; software maintenance.** 434

corrective network. An electric network designed to be inserted in a circuit to improve its transmission properties, its impedance properties, or both. *See:* **network analysis.** 328

correctness (software). (1) The extent to which software is free from design defects and from coding defects, that is, fault free. (2) The extent to which software meets its specified requirements. (3) The extent to which software meets user expectations. *See:* **design; fault; requirements; software.** 434

correctness proof. *See:* **proof of correctness.**

correct relaying-system performance (power switchgear). The satisfactory operation of all equipment associated with the protective-relaying function in a protective-relaying system. It includes the satisfactory presentation of system input quantities to the relaying equipment, the correct operation of the relays in response to these input quantities, and the successful operation of the assigned switching device or devices. 127, 103

correct relay operation (power switchgear). An output response by the relay which agrees with the operating characteristic for the input quantities applied to the relay. *See:* **correct relaying-system performance.** 127, 103

correlated color temperature of a light source (illuminating engineering). The absolute temperature of a blackbody whose chromaticity most nearly resembles that of the light source. 167

correlated photon summing (1) (germanium detectors). The simultaneous detection of two or more photons originating from a single nuclear disintegration. 397

(2) (sodium iodide detector). The simultaneous detection of two or more photons originating from a single disintegration. 423

correlation (test, measurement and diagnostic equipment). That portion of certification which establishes the mutual relationships between similar or identical support test systems by comparing test data collected on specimen hardware or simulators. 54

correlation coefficient function (seismic qualification of Class 1E equipment for nuclear power generating stations). Defines a comparative relationship between two time histories. It provides a statistical estimate of how much two motions are related, as a function of time delay. The numerical range is from zero for unrelated, to $+1.0$ for related motions. 581

correlation detection (modulation systems). Detection based on the averaged product of the received signal and a locally generated function possessing some known characteristic of the transmitted wave. *Notes:* (1) The averaged product can be formed, for example, by multiplying and integrating, or by the use of a matched filter whose impulse response, when reversed in time, is the locally generated function. (2) Strictly, the foregoing definition applies to detection based on cross correlation. The term correlation detection may also apply to detection involving autocorrelation, in which case the locally generated function is merely a delayed form of the received signal.   415

corrosion. The deterioration of a substance (usually a metal) because of a reaction with its environment.   205

corrosion fatigue. Reduction in fatigue life in a corrosive environment.   205

corrosion fatigue limit. The maximum repeated stress endured by a metal without failure in a stated number of stress applications under defined conditions of corrosion and stressing.   205

corrosion rate. The rate at which corrosion proceeds.   205

corrosion-resistant (power and distribution transformer). So constructed, protected, or treated that corrosion will not exceed specified limits under specified test conditions.   53

corrosion-resistant parts (electric installations on shipboard. (1) *General.* Where essential to minimize deterioration due to marine atmospheric corrosion, corrosion-resisting materials, or other materials treated in a satisfactory manner to render them adequately resistant to corrosion should be used. (2) *Corrosion-resisting materials.* Silver, corrosion-resisting steel, copper, brass, bronze, copper-nickel, certain nickel-copper alloys, and certain aluminum alloys are considered satisfactory corrosion-resisting materials within the intent of the foregoing. (3) *Corrosion-resistant treatments.* The following treatments, when properly done and of a sufficiently heavy coating, are considered satisfactory corrosion-resistant treatments within the intent of the foregoing. Electroplating of: cadmium, chromium, copper, nickel, silver, and zinc, sheradizing, galvanizing dipping and painting. (Phosphate or suitable cleaning, followed by the application of zinc chromate primer or equivalent.) (4) *Application.* These provisions should apply to the following components: (A) *Parts.* Interior small parts which are normally expected to be removed in service, such as bolts, nuts, pins, screws, cap screws, terminals, brush-holder studs, springs, etcetera. (B)*Assemblies, subassemblies, and other units.* Where necessary due to the unit function, or for interior protection, such as shafts within a motor or generator enclosure, and surface of stator and rotor. (C) *Enclosures and their fastenings and fittings.* Enclosing cases for control apparatus, outer cases for signal and communication systems (both outside and inside), and similar items together with all their fastenings and fittings which would be seriously damaged or rendered ineffective by corrosion.   3

corrugated horn (antenna). A hybrid-mode horn antenna produced by cutting narrow transverse grooves of specified depth in the interior walls of the horn. *See:* **hybrid-mode horn.**   111

cosecant-squared antenna. A shaped-beam antenna in which the radiation intensity over a part of its pattern in some specified plane (usually the vertical) is proportional to the square of the cosecant of the angle measured from a specified direction in that plane (usually the horizontal). *Note:* Its purpose is to lay down a uniform field along a line that is parallel to the specified direction but that does not pass through the antenna. *See:* **antenna.**   179

cosecant-squared beam antenna. A shaped-beam antenna whose pattern in one principal plane consists of a main beam with well-defined sidelobes on one side, but with the absence of nulls over an extended angular region adjacent to the peak of the main beam on the other side, with the radiation intensity in this region designed to vary as the cosecant-squared of the angle variable. *Note:* The most common applications of this antenna are for use in ground-mapping radars and target acquisition radars, since the cosecant-squared coverage provides constant signal return for targets with the same radar cross section at different ranges but a common height.   111

cosecant-squared pattern (radar). A vertical-plane antenna pattern in which the power varies as the square of the cosecant of the elevation angle. The unique property of this pattern is that it causes the echo strength of a target having constant radar cross section, moving at constant altitude, to be independent of range.   13

cosine-cubed law (illuminating engineering). An extension of the cosine law in which the distance d between the source and surface is replaced by h/cos θ, where h is the perpendicular distance of the source from the plane in which the point is located. It is expressed by $E = (I\cos^3\theta)/h^2$. (See Figure below.) *See:* **cosine law.**   167

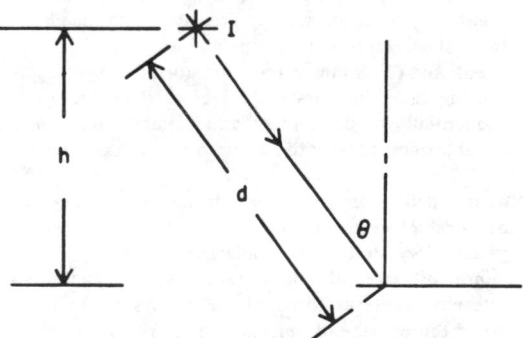

cosine emission law. *See:* **Lambert's cosine law.**

cosine law (illuminating engineering). A law stating that the illuminance on any surface varies as the cosine of the angle of incidence. The angle of incidence θ is the angle between the normal to the surface and the direction of the incident light. The inverse-square law and the cosine law can be combined as $E = (I\cos\theta) / d_2$. *See:* **inverse-square law.**   167

cosmic radio waves (radio wave propagation). Radio waves originating from beyond the solar system.   146 costate (control system). The state of the adjoint system. *See:* control system.   329

cost of incremental fuel (electric power systems) (usually expressed in cents per million British thermal units). The ultimate replacement cost of the fuel that would be consumed to supply an additional increment of generation. *See:* power system, low-frequency and surge testing.   94

costs (power operations). Monies associated with investment or use of electrical plant. *See:* fixed investment costs; fixed opeation costs; variable operating costs.   516

coulomb. The unit of electric charge in SI units (International System of Units). The coulomb is the quantity of electric charge that passes any cross section of a conductor in one second when the current is maintained constant at one ampere.   210

Coulomb's law (electrostatic attraction). The force of repulsion between two like charges of electricity concentrated at two points in an isotropic medium is proportional to the product of their magnitudes and inversely proportional to the square of the distance between them and to the dielectric constant of the medium. *Note:* The force between unlike charges is an attraction.   210

coulometer (voltameter). An electrolytic cell arranged for the measurement of a quantity of electricity by the chemical action produced. *See:* electricity meter (meter).   328

count (radiation counters). A single response of the counting system. *See:* scintillation counter; tube count.   125, 117, 96

count-down (transponder). The ratio of the number of interrogation pulses not answered to the total number of interrogation pulses received.   13, 187

counter (test, measurement and diagnostic equipment). (1) A device such as a register or storage location used to represent the number of occurrences of an event: and (2) An instrument for storing integers, permitting these integers to be increased or decreased sequentially by unity or by an arbitrary integer, and capable of being reset to zero or to an arbitrary integer.   54

counter cells. *See:* counter-electromotive-force cells.

counterclockwise polarized wave (radio wave propagation). *See:* left-handed polarized wave.

counter electromotive force (any system). The effective electromotive force within the system that opposes the passage of current in a specified direction.   328

counter-electromotive-force cells (counter cells). Cells of practically no ampere-hour capability used to oppose the battery voltage. *See:* battery (primary or secondary).   328

counter-mounted cooking unit (National Electrical Code). A cooking appliance designed for mounting in or on a counter and consisting of one or more heating elements, internal wiring, and build-in or separately mountable controls. *See:* wall-mounted oven.   256

counterpoise (1)(overhead lines)(lightning protec-

tion). A conductor or system of conductors, arranged beneath the transmission line, located on, above or most frequently below the surface of the earth, and connected to the footings of the towers or poles supporting the line.   313

(2) (antenna). A system of conductors, elevated above and insulated from the ground, forming a lower system of conductors of an antenna. *Note:* The purpose of a counterpoise is to provide a relatively high capacitance and thus a relatively low impedance path to earth. The counterpoise is sometimes used in medium- and low-frequency applications where it would be more difficult to provide an effective ground connection.   111

counter, radiation. *See:* radiation counter.

counter tube (radiation counters). A device that reacts to individual ionizing events, thus enabling them to be counted. (1) (externally quenched). A radiation-counter tube that requires the use of an external quenching circuit to inhibit reignition. (2) (gas-filled, radiation). A gas tube used for detection of radiation by means of gas ionization. (3) (gas-flow). A radiation-counter tube in which an appropriate atmosphere is maintained by a flow of gas through the tube. (4) (Geiger-Mueller). A radiation-counter tube operated in the Geiger-Mueller region. (5) (proportional). A radiation-counter tube operated in the proportional region. (6) (self-quenched) A radiation-counter tube in which reignition of the discharge is inhibited by internal processes. *See:* anticoincidence (radiation counters).
                                                        96,125

counting channel (liquid-scintillation counting). A region of the pulse-height spectrum which is defined by upper and lower boundaries set by discriminators.
                                                             422

counting efficiency (1) (radiation-counter tubes). The average fraction of the number of ionizing particles or quanta incident on the sensitive area that produce tube counts. *Note:* The operating conditions of the counter and the condition of irradiation must be specified.
                                                             125

(2) (scintillation counters). The ratio of (A) the average number of photons or particles of ionizing radiation that produce counts to (B) the average number incident on the sensitive area. *Note:* The operating conditions of the counter and the conditions of irradiation must be specified. *See:* scintillation counter.
                                                             117

(3) (liquid-scintillation counting). The ratio of the count rate to the disintegration rate, usually expressed as a perentage: $E = (R/A) \times 100$.

E = counting system efficiency

R = net count rate in an individual measurement, counts per minute

A = activity of the radionuclide contained in the check source.

In this standard, activity is expressed in disintegrations per minute, although the recommended unit for activity is the becquerel, where 1 becquerel equals 1 transition per second.   422

counting mechanism (of an automatic line sectionaliz-

er or automatic circuit recloser) (power switchgear).
A device that counts the number of electrical impulses
and, following a predetermined number of successive
electrical impulses, actuates a releasing mechanism. It
resets if the total predetermined number of successive
impulses do not occur in a predetermined time.
                                                    103

**counting operation (of an automatic line sectionalizer
or automatic circuit recloser) (power switchgear).**
Each advance of the counting mechanism towards an
opening operation.   103

**counting operation time (of an automatic line section-
alizer) (power switchgear).** The time between the
cessation of a current above the minimum actuating
current value and the completion of a counting operation.   103

counting rate. Number of counts per unit time. *See:*
anticoincidence (radiation counters).   190

counting-rate meter (pulse techniques). A device that
indicates the time rate of occurrence of input pulses
averaged over a time interval. *See:* **scintillation counter.**   117

**counting rate versus voltage characteristic (gas-filled
radiation-counter tube).** The counting rate as a function of applied voltage for a given constant average
intensity of radiation.   125

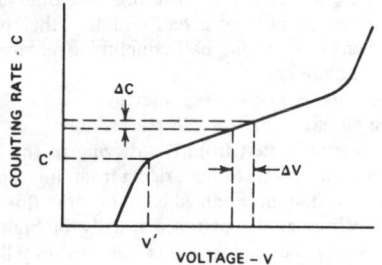

Counting rate-voltage characteristic in which

$$\text{Relative plateau slope} = 100\,\frac{\Delta C/C}{\Delta V}$$

$$\text{Normalized plateau slope} = \frac{\Delta C/\Delta V}{C'/V'} = \frac{\Delta C/C'}{\Delta V/V'}$$

country beam. *See:* **upper beams.**

country code (telephone switching systems). The one-,
two-, or three-digit number that, in the world numbering plan, identifies each country or integrated numbering plan area in the world. The initial digit is always
the world-zone number. Any subsequent digits in the
code further define the designated geographical area
normally identifying a specific country. On an international call, this code is dialed ahead of the national
number.   55

counts, tube, multiple (radiation-counter tubes). *See:*
multiple tube counts.

counts, tube, spurious (radiation-counter tubes). *See:*
spurious tube counts.

couple (1) (storage cell). An element of a storage cell

consisting of two plates, one positive and one negative.
Note: The term couple is also applied to a positive and
a negative plate connected together as one unit for
installation in adjacent cells. *See:* **battery (primary or
secondary); galvanic cell.**   328

(2) (thermoelectric). A thermoelectric device having
two arms of dissimilar composition. *Note:* The term
thermoelement is ambiguously used to refer to either
a thermoelectric arm or to a thermoelectric couple,
and its use is therefore not recommended. *See:* **thermoelectric device.**   191

coupled modes (fiber optics). Modes whose energies
are shared. *See:* **mode.**   433

coupler (1)(navigation aid terms). That portion of a
navigational system which receives signals of one type
from a sensor and transmits signals of a different type
to an actuator. *See:* **autopilot coupler.**   526

**(2)(surge testing for equipment connected to low-
voltage ac power circuits).** A device, or combination
of devices, used to feed a surge from a generator to
powered equipment while limiting the flow of current
from the power source into the generator.   578

(3) (fiber optics). *See:* **optical waveguide coupler.**
                                                    433

coupler, 3-decibel. *See:* **hybrid control.**

coupling (1)(ground system). The association of two or
more circuits or systems in such a way that power or
signal information may be transferred from one to
another. *Note:* Coupling is described as close or loose.
A close-coupled process has elements with small
phase shift between specified variables; close-coupled
systems have large mutual effect shown mathematically by cross-products in the system matrix.
                                                    313

(2) (rotating machinery). A part or combination of
parts that connects two shafts for the purpose of transmitting torque or maintaining alignment of the two
shafts.   63

(3) (data transmission). The association of two or
more circuits or systems in such a way that power or
signal information may be transferred from one to
another.   59

(4) (software). A measure of the interdependence
among modules in a computer program. *See:* **cohesion;
computer program; module.**   434

(5) (waveguide). The power transfer from one transmission path to a particular mode or form in another.
Note: Small, undesired coupling is sometimes called
isolation, decoupling, or cross coupling.   267

**coupling aperture (coupling hole, coupling slot)
(waveguide components).** An aperture in the bounding surface of a cavity resonator, waveguide, transmission line, or waveguide component which permits the
flow of energy to or from an external circuit.
                                                    166

coupling capacitance.(1)(ground systems) The association of two or more circuits with one another by
means of capacitance mutual to the circuits.   313

(2) (interference terminology). The type of coupling
in which the mechanism is capacitance between the
interference source and the signal system, that is, the

interference is induced in the signal system by an electric field produced by the interference source. *See:* **interference.**   188

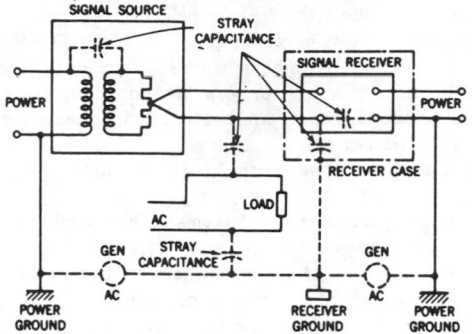

Capacitance coupling (interference).

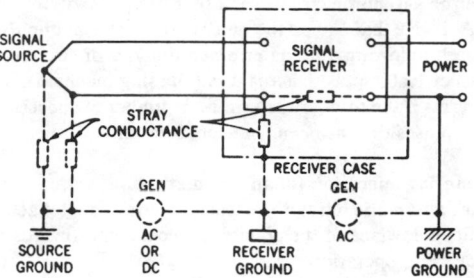

Conductance coupling (interference).

coupling-capacitor voltage transformer (CCVT) (metering). A voltage transformer comprised of a capacitor divider and an electromagnetic unit so designed and interconnected that the secondary voltage of the electromagnetic unit is substantially proportional to, and in phase with, the primary voltage applied to the capacitor divider for all values of secondary burdens within the rating of the coupling-capacitor voltage transformer.   212

coupling coefficient (coefficient of coupling). The ratio of impedance of the coupling to the square root of the product of the total impedances of similar elements in the two meshes. *Notes:* (1) Used only in the case of resistance, capacitance, self-inductance, and inductance coupling. (2) Unless otherwise specified, coefficient of coupling refers to inductance coupling, in which case it is equal to $M / (L_1 L_2)^{1/2}$, where M is the mutual inductance, L_1 the total inductance of one mesh, and L_2 the total inductance of the other. *See:* **network analysis.**   185

coupling coefficient, small-signal (electron stream). The ratio of (1) the maximum change in energy of an electron traversing the interaction space to (2) the product of the peak alternating gap voltage by the electronic charge. *See:* **coupling; coupling coefficient; electron emission.**   125

coupling, conductance (interference terminology). The type of coupling in which the mechanism is conductance between the interference source and the signal system. *See:* **interference; raceway.**   175

coupling efficiency (fiber optics). The efficiency of optical power transfer between two optical components. *See:* **coupling loss.**   433

coupling, electric (rotating machinery). (1) A device for transmitting torque by means of electromagnetic force in which there is no mechanical torque contact between the driving and driven members. *Note:* The slip-type electric coupling has poles excited by direct current on one rotating member, and an armature winding, usually of the double-squirrel cage type, on the other rotating member. (2) A rotating machine

that transmits torque by electric or magnetic means or in which the torque is controlled by electric or magnetic means.   3,63

coupling factor (1) (lightning). The ratio of the induced voltage to the inducing voltage on parallel conductors. *See:* **direct-stroke protection (lightning).**   64
(2) (directional coupler). The ratio of the incident power fed into the main port, and propagating in the preferred direction, to the power output at an auxiliary port, all ports being terminated by reflectionless terminations. *See:* **waveguide.**   185

coupling flange (flange) (rotating machinery). The disc-shaped element of a half coupling that permits attachment to a mating half coupling. *See:* **rotor (rotating machinery).**   63

coupling hole. *See:* **coupling aperture.**

coupling, hysteresis. An electric coupling in which torque is transmitted from the driving to the driven member by magnetic forces arising from the resistance to reorientation of established magnetic flux fields within ferromagnetic material usually of high coercivity. *Note:* The magnetic flux field is normally produced by current in the excitation winding, provided by an external source.   63

coupling, inductance (interference terminology). The

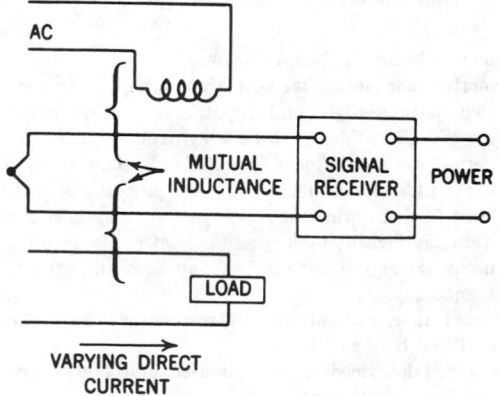

Inductance coupling (interference).

type of coupling in which the mechanism is mutual inductance between the interference is induced in the signal system by a magnetic field produced by the interference source. See the accompanying figure. *See:* **interference.**   188

coupling, induction. An electric coupling in which torque is transmitted by the interaction of the magnetic field produced by magnetic poles on one rotating member and due to an induced voltage in the other rotating member. *Note:* The magnetic poles may be produced by direct-current excitation, permanent-magnet excitation, or alternating-current excitation. Currents due to the induced voltages may be carried in a wound armature, cylindrical cage, or may be present as eddy currents in an electrically conductive disc or cylinder. Couplings utilizing a wound armature or a cylindrical cage are known as slip or magnetic couplings. Couplings utilizing eddy-current effects are known as eddy-current couplings.   63

coupling loop (waveguide components). A conducting loop which permits the flow of energy between a cavity resonator, waveguide, transmission line, or waveguide component and an an external circuit.   166

coupling loss (fiber optics). The power loss suffered when coupling light from one optical device to another. *See:* **angular misalignment loss; extrinsic joint loss; gap loss; insertion loss; intrinsic joint loss; lateral offset loss.**   433

coupling, magnetic friction. An electric coupling in which torque is transmitted by means of mechanical friction. Pressure normal to the rubbing surfaces is controlled by means of an electromagnet and a return spring. *Note:* Couplings may be either magnetically engaged or magnetically released depending upon application.   63

coupling, magnetic-particle. A type of electric coupling in which torque is transmitted by means of a fluid whose viscosity is adjustable by virtue of suspended magnetic particles. *Note:* The coupling fluid is incorporated in a magnetic circuit in which the flux path includes the two rotating members, the fluid, and a magnetic yoke. flux density, and hence the fluid viscosity, are controlled through adjustment of current in a magnetic coil linking the flux path.   63

coupling probe (waveguide components). A probe which permits the flow of energy between a cavity resonator, waveguide, transmission line, or waveguide component and an external circuit.   166

coupling, radiation (interference terminology). The type of coupling in which the interference is induced in the signal system by electromagnetic radiation produced by the interference source. *See:* **interference.**   188

couplings (pothead). Entrance fittings which may be provided with a rubber gland to provide a hermetic seal at the point where the cable enters the box and may have, in addition, a threaded portion to accommodate the conduit used with the cable or have an armor clamp to clamp and ground the armored sheath on armor-covered cable.   289

coupling slat. *See:* **coupling aperture.**

coupling, synchronous (rotating machinery). A type of electric coupling in which torque is transmitted at zero slip, either between two electromagnetic members or like number of poles, or between one electromagnetic member and a reluctance member containing a number of saliencies equal to the number of poles. *Note:* Synchronous couplings may have induction members or other means for providing torque during nonsynchronous operation such as starting. *See:* **electric coupling.**   63

course (navigation aid terms). (1) The intended direction of travel, expressed as an angle in the horizontal plane between a reference line and the course line, usually measured clockwise from the reference line. (2) The intended direction of travel as defined by a navigational facility. (3) Common usage for 'course line'.   526

course-deviation indicator. *See:* **course-line deviation indicator.**

course line (navigation aid terms). The projection in the horizontal plane of a path (proposed path of travel).   526

course linearity (ILS [instrument landing system]) (navigation aid terms). A term used to describe the change in DDM (difference in depth of modulation) of the two modulation signals with respect to displacement of the measuring position from the course line but within the course sector. *Syn:* **desired track; flight path.**   526

course-line computer (navigation aid terms). A device, usually carried aboard a vehicle, to convert navigational signals such as VOR/DME (very high-frequency omnidirectional range/distance measuring equipment) into course extending between any desired points regardless of their orientation with respect to the source of the signals.   526

course-line deviation (navigation aid terms). The amount by which the track of a vehicle differs from its course line, expressed in terms of either an angular or linear measurement.   526

course-line deviation indicator (course deviation indicator)(navigation aid terms). A device providing a visual display of the direction and amount of deviation from the intended course.*Syn:* **flight path deviation indicator; course indicator.**   526

course made good (navigation aid terms). The direction from the point of departure to the position of the vehicle on the horizontal plane.   526

course push (pull)(navigation aid terms). An erroneous deflection of the indicator of a navigational aid, produced by altering the attitude of the receiving antenna. *Note:* This effect is a manifestation of polarization error and results in an apparent displacement of the course line.   526

course roughness (navigation aid terms). A term used to describe the imperfections in a visually indicated course when such imperfections cause the course indicator to make rapid erratic movements. *See:* **scalloping.** 526

course scalloping. *See:* **scalloping.**

course section width (instrument landing systems). The transverse dimension at a specified distance, or the angle in degrees between the sides of the course sector. *See:* **navigation.** 187,13

course sector (ILS [instrument landing system]) (navigation aid terms). A wedge-shaped section of airspace containing the course line and spreading with distance from the ground station; it is bounded on both sides by the loci of points at which the DDM (difference in depth of modulation) is a specified amount, usually the DDM giving full-scale deflection of the course-deviation indicator. 526

course-sector width (ILS [instrument landing system]) (navigation aid terms). The transverse dimension at a specified distance, or the angle in degrees, between the sides of the course sector. 526

course sensitivity (navigation systems)(navigation aid terms). The relative response of a course-line deviation indicator to the actual or simulated departure of the vehicle from the course line. In VOR (very high-frequency) omnidirectional range), Tacan (tactical air navigation), or similar omnirange systems, course sensitivity is often taken as the number of degrees through which the omnibearing selector must be moved to change the deflection of the course-line deviation indicator from full scale on one side to full scale on the other, while the receiver omnibearing-input signal is held constant. 526

course softening (navigation aid terms). The intentional decrease in course sensitivity upon approaching a navigational aid such than the ratio of indicator deflection to linear displacement from the course line tends to remain constant. 526

course width (navigation aid terms). Twice the displacement (of the vehicle), in degrees, to either side of a course line, which produces a specifed indication on the course deviation indicator (usually the specified indication is full scale). 526

cove lighting (illuminating engineering). Lighting comprising light sources shielded by a ledge or horizontal recess, and distributing light over the ceiling and upper wall. 167

cover (power system communication equipment). A protective covering used to enclose or partially enclose equipment that may be mounted in a rack. 453

coverage area (mobile communication). The area surrounding the base station that is within the signal-strength contour that provides a reliable communication service 90 percent of the time. *See:* **mobile communication system.** 181

covered conductor (1)(NESC). A conductor covered with a dielectric having no rated insulating strength or having a rated insulating strength less than the voltage of the circuit in which the conductor is used. 494

(2) (transmission and distribution). A conductor covered with a dielectric having no rated insulating strength or having a rated insulating strength less than the voltage of the circuit in which the conductor is used. 262

covered plate (storage cell). A plate bearing a layer of oxide between perforated sheets. *See:* **battery (primary or secondary).** 328

cover flange (waveguide components). A flat flange used in conjunction with a choke flange to provide a choke joint. 166

cover-up equipment (power line maintenance). Equipment designed to protect people from energized parts in a specific work area. Many different types are available to cover conductors, insulators, dead-end assemblies, structures, and apparatus. Cover-up material may be either flexible or rigid. 458

CPE. *See:* **circular probable error.**

crab angle.* *See:* **drift correction angle; drift angle.** *Deprecated

cracking (composite insulators). rupture of the weathershed material to depths greater than 0.1 millimeter. 483

cradle base (rotating machinery). A device that supports the machine at the bearing housings. 63

crane. A machine for lifting or lowering a load and moving it horizontally, in which the hoisting mechanism is an integral part of the machine. *Note:* It may be driven manually or by power and may be fixed or a mobile machine. *See:* **elevators.** 328

crate (CAMAC). *See:* **CAMAC crate.**

crate (FASTBUS crate)(FASTBUS acquisition and control). The mechanical modules for FASTBUS modules in a crate segment. 480

crate number (c) (subroutines in CAMAC). The symbol c represents an integer which is the crate number component of a CAMAC address. Crate number in this context can be either the physical crate number or it can be an integer symbol which is interpreted by the computer system software to produce appropriate hardware access information. 410

crate segment (FASTBUS acquisition and control). A FASTBUS segment that consists of a backplane mounted on a FASTBUS crate and having connectors to mate with a multiplicity of FASTBUS modules. 480

crawler tractor (conductor stringing equipment). A tracked unit employed to pull pulling lines, sag conductor, level or clear pull and tension sites, and miscellaneous other work. It is also frequently used as a temporary anchor. Sagging winches on this unit are usually arranged in a vertical configuration. *Syn:* **cat; crawler; tractor.** 431

crawling (rotating machinery). The stable but abnormal running of a synchronous or asynchronous machine at a speed near to a submultiple of the synchronous speed. *See:* **asynchronous machine.** 63

crazing (composite insulators). Surface microfractures of the weathershed material to depths less than 0.1 millimeter resulting from ultraviolet exposure. 483

credit-card call (telephone switching systems). A call

in which a credit-card identity is used for billing purposes. 55

creep (watthour meter). A continuous motion of the rotor of a meter with normal operating voltage applied and the load terminals open-circuited. 212

creepage. The travel of electrolyte up the surface of electrode or other parts of the cell above the level of the main body of electrolyte. *See:* **electrolytic cell.** 328

creepage distance (power switchgear) (power and distribution transformer). The shortest distance between two conducting parts measured along the surface or joints of the insulating material between them. 53, 103

creepage surface (rotating machinery). An insulating-material surface extending across the separating space between components at different electric potential, where the physical separation provides the electrical insulation. *See:* **asynchronous machine.** 63

creep distance (outdoor apparatus bushings). The distance measured along the external contour of the weather casing separating the metal parts which have the operating line-to-ground voltage between them. 168

creeping stimulus. *See:* **accumulating stimulus.**

crest factor (of an average reading or root-mean-square voltmeter)(1)(charged-particle detectors)(germanium gamma-ray detectors)(X-ray energy spectrometers)(semiconductor radiation detectors). The ratio of the peak voltage value that an average reading or root-mean-square voltmeter will accept without overloading to the full scale value of the range being used for measurement. 119, 528, 23, 471

(2) (of a periodic function). The ratio of its crest (peak, maximum) value to its root-mean-square (rms) value. 53

(3) (pulse carrier). The ratio of the peak pulse amplitude to the root-mean-square amplitude. *See:* **carrier.** 111,254

crest value (peak value) (1) (power and distribution transformer). The maximum absolute value of a function when such a maximum exists. 53

(2) (surge arrester) (wave, surge, or impulse). The maximum value that it attains. 2

(3)(of a wave, surge, or impulse)(metal-oxide surge arresters for ac power circuits)(surge arrester). The maximum value that it attains. 583,430

crest voltmeter. A voltmeter depending for its indications upon the crest or maximum value of the voltage applied to its terminals. *Note:* Crest voltmeters should have clearly marked on the instrument whether readings are in equivalent root-mean-square values or in true crest volts. It is preferred that the marking should be root-mean-square values of the sinusoidal wave having the same crest value as that of the wave measured. *See:* **instrument.** 328

crest working line voltage (V_{LWM}) (thyristor). The highest instantaneous value of the line voltage excluding all repetitive and nonrepetitive transient voltages, but including voltage variations. 445

crest working voltage (between two points) (semicon- ductor rectifiers). The maximum instantaneous difference of voltage, excluding oscillatory and transient overvoltages, that exists during normal operation. *See:* **crest; rectification; semiconductor rectifier stack.** 291,208

crevice corrosion. Localized corrosion as a result of the formation of a crevice between a metal and a nonmetal, or between two metal surface. 221,205

critical angle (fiber optics). When light propagates in a homogeneous medium of relatively high refractive index (n_{high}) onto a planar interface with a homogeneous material of lower index (n_{low}), the critical angle is defined by

$$\arcsin(n_{low}/n_{high}).$$

Note: When the angle of incidence exceeds the critical angle, the light is totally reflected by the interface. This is termed "total internal reflection". *See:* **acceptance angle; angle of incidence; reflection; refractive index (of a medium); step index profile; total internal reflection.** 433

critical anode voltage (multielectrode gas tubes). Synonymous with anode breakdown voltage (gas tubes). *See:* **gas tube.** 125

critical branch (health care facilities) (National Electrical Code). A subsystem of the Emergency System consisting of feeders and branch circuits supplying energy to task illumination, special power circuits, and selected receptacles serving areas and functions related to patient care, and which can be connected to alternate power sources by one or more transfer switches during interruption of normal power source. 256, 192

critical build-up resistance (rotating machinery). The highest resistance of the shunt winding circuit supplied from the primary winding for which the machine voltage builds up under specified conditions. 63

critical build-up speed (rotating machinery). The limiting speed below which the machine voltage will not build up under specified condition of field-circuit resistance. 63

critical characteristics (1)(equipment)(replacement parts for Class 1E equipment in nuclear power generating stations). Those properties or attributes that are essential for performance of an equipment's safety function. 582

(2)(parts)(replacement parts for Class 1E equipment in nuclear power generating stations). Those properties or attributes of the part that are essential to the safety function of the equipment in which the part is installed. *Note:* Typical critical characteristics are attributes such as dimensions, materials, electrical and temperature parameters, output tolerances, and fluid viscosity. 582

critical controlling current (cryotron). The current in the control that just causes direct-current resistance to appear in the gate, in the absence of gate current and at a specified temperature. *See:* **superconductivity.** 91

critical coupling. That degree of coupling between two circuits, independently resonant to the same frequency, that results in maximum transfer of energy at the resonance frequency. *See:* **coupling.**   328

critical current (superconductor). The current in a superconductive material above which the material is normal and below which the material is superconducting, at a specified temperature and in the absence of external magnetic fields. *See:* **superconductivity.**   191

critical damping. *See:* **damped harmonic system (2) (critical damping).**

critical dimension (waveguide). The dimension of the cross-section that determines the cutoff frequency. *See:* **waveguide.**   328

critical failure. *See:* **failure, critical.**

critical field (1)(magnetrons). The smallest theoretical value of steady magnetic flux density, at a steady anode voltage, that would prevent an electron emitted from the cathode at zero velocity from reaching the anode. *See:* **magnetrons.**   125
(2)(nonlinear, active, and nonreciprocal waveguide components). In a gyromagnetic material, that radiofrequency (rf) magnetic field levelabove which transfer of energy occurs from the uniform precession mode to spin waves; that is, the field corresponding to nonlineal loss threshold.   530

critical freeze protection (electric pipe heating systems). The use of electric pipe heating systems to prevent the temperature of fluids from dropping below the freezing pointof the fluid in important or critical outdoor (usually) piping systems at nuclear generating stations. An example of a critical freeze protection system is the heating for the nuclear service water system.   448

critical frequency (1) (data transmission). In radio propagation (by way of the ionosphere) the limiting frequency below which a wave component is reflected by, and above which it penetrates through, an ionospheric layer of vertical incidence. *Note:* The existence of the critical frequency is the result of electron limitation, that is, the inadequacy of the existing number of free electrons to support reflection at higher frequencies.   59
(2) (radio wave propagation). The limiting frequency below which a magneto-ionic wave component is reflected by, and above which it penetrates through, an ionospheric layer when the waves are incident normal to the layer.   146
(3) (circuits and systems) (network or system). A pole or zero of a transfer or driving-point function.   67

critical grid voltage (multielectrode gas tubes). The grid voltage at which anode breakdown occurs. *Note:* The critical grid voltage is a function of the other electrode voltages or currents and of the environment. *See:* **breakdown voltage (electrode of a gas tube).**   125

critical head (power operations). The head at which the full-gate output of the hydroturbine equals the nameplate generator capacity.   516

critical high-power level (attenuator tubes). The radio-frequency power level at which ionization is produced in the absence of a control-electrode discharge.   125

critical humidity (corrosion). The relative humidity above which the atmospheric corrosion rate of a given metal increases sharply.   205

critical hydroperiod (power operations). Period when the limitations of hydroelectric energy supply due to water conditions are most critical with respect to system load requirements.   516

critical impulse (of a relay) (power switchgear). The maximum impulse in terms of duration and input magnitude which can be applied suddenly to a relay without causing pickup.   103

critical impulse flashover voltage (insulator). The crest value of the impulse wave that, under specified conditions, causes flashover through the surrounding medium on 50 percent of the applications. *See:* **impulse flashover voltage.**   91

critical impulse time (of a relay) (power switchgear). The duration of a critical impulse under specified conditions.   127, 103

criticality (1)(power operations). The state of an assembly of fissionable material in which a stable, self-sustaining chain reaction exists. At this condition a nuclear reactor will produce energy at a constant rate and the effective multiplication factor k_{eff} is exactly equal to 1.   516
(2)(software). A classification of a software error or fault based upon an evaluation of the degree of impact of that error or fault on the development or operation of a system. Often used to determine whether or when a fault will be corrected. *See:* **fault; software error; system.**   434

critical magnetic field (superconductor). The field below which a superconductor material is superconducting and above which the material is normal, at a specified temperature and in the absence of current. *See:* **superconductivity.**   191

critical overtravel time (of a relay) (power switchgear). The time following a critical impulse until movement of the responsive element ceases just short of pickup.   127, 103

critical piece first (software). Pertaining to an approach to software development that focuses on implementing the most critical aspects of a software system first. The critical piece may be defined in terms of services provided, degree of risk, difficulty, or some other criterion. *See:* **software; system.**   434

critical point (1)(feedback control system) (Nichols chart). The bound of stability for the GH $(j\omega)$ plot; the intersection of $|GH| = 1$ with ang $GH = -180$ degrees.
(2) (Nyquist diagram). The bound of stability for the locus of the loop transfer function GH $(j\omega)$; the $(-1. j0)$ point.   56, 329

critical process control (electric pipe heating systems). The use of electric pipe heating systems to increase or maintain, or both, the temperature of fluids (or processes) in important or critical mechanical piping systems including pipes, pumps, valves, tanks, instrumentation, etcetera, in nuclear power generating stations. An example of an important or critical mechanical piping system is the safety injection system.   448

critical processing control (electric pipe heating systems). The use of electric pipe heating systems to increase or maintain, or both, the temperature of fluids (or processes) in important or critical mathematical piping systems including pipes, pumps, valves; tanks, instrumentation, etcetera. An example of an important or critical mechanical piping system would be the safety injection system.   405

critical rate-of-rise of OFF-state voltage (thyristor). The minimum value of the rate of rise of principal voltage which will cause switching from the OFF-state to the ON-state.   445

critical rate-of-rise of ON-state current (thyristor). The maximum value of the rate-of-rise of ON-state current that a thyristor can withstand without deleterious effect. *See:* **principal current.**   191

critical section (software). A segment of code to be executed mutually exclusively with some other segment of code which is also called a critical section. Segments of code are required to be executed mutually exclusively if they make competing uses of a computer resource or data item. *See:* **code; computer; data; execute; segment.**   434

critical software (software verification and validation plans). Software whose failure could have an impact on safety, or could cause large financial or social loss.   511

critical speed (rotating machinery). A speed at which the amplitude of the vibration of a rotor due to shaft transverse vibration reaches a maximum value. *See:* **rotor (rotating machinery).**   63

critical system (health care facilities). A system of feeders and branch circuits in nursing homes and residential custodial care facilities arranged for connection to the alternate power source to restore service to certain critical receptacles, task illumination and equipment.   192

critical temperature (superconductor). The temperature below which a superconductive material is superconducting and above which the material is normal, in the absence of current and external magnetic fields. *See:* **superconductivity.**   191

critical torsional speed (rotating machinery). A speed at which the amplitude of the vibration of a rotor due to shaft torsional vibration reaches a maximum value. *See:* **rotor (rotating machinery).**   63

critical travel (of a relay) (power switchgear). The amount of movement of the responsive element of a relay during a critical impulse, but not subsequent to the impulse.   127, 103

critical voltage (1) (magnetron). The highest theoretical value of steady anode voltage, at a given steady magnetic flux density, at which electrons emitted from the cathode at zero velocity would fail to reach the anode. *See:* **magnetrons.**   125

(2) (relay). *See:* **relay critical voltage.**

critical-voltage parabola (cutoff parabola) (magnetrons). The curve representing in Cartesian coordinates the variation of the critical voltage as a function

of the magnetic induction. *See:* **magnetron.**   244

critical withstand current (impulse)(surge). The highest crest value of a surge of given waveshape and polarity that can be applied without causing disruptive discharge on the test specimen.   62

Crookes dark space. *See:* **cathode dark space.**

cross acceleration (accelerometer). The acceleration applied in a plane normal to an accelerometer input reference axis.   46

crossarm. A horizontal member (usually wood or steel) attached to a pole, post, tower or other structure and equipped with means for supporting the conductors. *Note:* The crossarm is placed at right angles to conductors on straight line poles, but splits the angle on light corners. *See:* **tower.**   64

crossarm guy. A tensional support for a crossarm used to offset unbalanced conductor stress.   64

cross-assembler (software). An assembler that executes on one computer but generates object code for a different computer. *See:* **assembler; code; computer.**   434

cross-axis sensitivity (accelerometer) (inertial sensor). The proportionality constant that relates a variation of accelerometer output to cross acceleration. This sensitivity can vary depending on the direction of cross acceleration.   46

crossband transponder (navigation)(navigation aid terms). A transponder which replies in a different frequency band from that of the received interrogation.   526

crossbar switch. A switch having a plurality of vertical paths, a plurality of horizontal paths, and electromagnetically-operated mechanical means for interconnecting any one of the vertical paths with any one of the horizontal paths.   56

crossbar system. An automatic telephone switching system that is generally characterized by the following features. (1) The selecting mechanisms are crossbar switches. (2) Common circuits select and test the switching paths and control the operation of the selecting mechanisms. (3) The method of operation is one in which the switching information is received and stored by controlling mechanisms that determine the operations necessary in establishing a telephone connection.   328

cross cable bond. A cable bond used for bonding between the armor or lead sheath of adjacent cables. *See:* **cable bond; continuity cable bond.**   64

cross-compiler (software). A compiler that executes on one computer but generates assembly code or object code for a different computer. *See:* **code; compiler; computer.**   434

crossconnection (telephone switching system). Easily changed or removed wire that is run loosely between equipment terminals to establish an electrical association.   55

cross-correlation (excitation control system). The cross-correlation of two random signals $x_1(t)$ and $x_2(t)$ is $R_{12}(t)$ defined by

$$R_{12}(t) = \int_0^t x_1(t - \tau)\, x_2(\tau)\, d\tau.$$

If $x_1(t)$ is a random input to a linear stationary system and $x_2(t)$ is the response, then $R_{12}(t)$ is the inverse Laplace transform of the transfer function of the system. 353

cross coupling (transmission medium). A measure of the undesired power transferred from one channel to another. *See:* **coupling; transmission line.** 267

cross-coupling coefficient (accelerometer) (inertial sensor). The proportionality constant that relates a variation of accelerometer output to the product of acceleration applied normal and parallel to an input reference axis. This coefficient can vary depending on the direction of cross acceleration. 46

cross-coupling errors (gyro) (inertial sensor). The errors in the gyro output resulting from gyro sensitivity to inputs about axes normal to an input reference axis. 46

crossed-field amplifier (microwave tubes). A crossed-field tube or valve, with a nonreentrant slow-wave structure, used as an amplifier. *See:* **microwave tube or valve.** 190

crossed-field tube (microwave). A high-vacuum electron tube in which a direct, alternating, or pulsed voltage is applied to produce an electric field perpendicular both to a static magnetic field and to the direction of propagation of a radio-frequency delay line. *Note:* The electron beam interacts synchronously with a slow wave on the delay line. *See:* **microwave tube (or valve).** 190

cross fire (data transmission). An interfering current in one telegraph or signaling channel resulting from telegraph or signaling current in another channel. 59

crossing angle. *See:* **angle of cut.**

crossing structure (conductor stringing equipment). A structure built of poles and, sometimes, rope nets. It is used whenever conductors are being strung over roads, power lines, communications circuits, highways or railroads and normally constructed in such a way that it will prevent the conductor from falling onto or into any of these facilities in the event of equipment failure, broken pulling lines, loss of tension, etcetera. *Syn:* **guard structure; H-frame; rider structure; temporary structure.** 431

cross light (illuminating engineering). Equal illumination in front of the subject from two directions at substantially equal and opposite angles with the optical axis of the camera and a horizontal plane. 167

cross modulation. A type of intermodulation due to the modulation of the carrier of the desired signal by an undesired signal wave. 111,59

cross modulation distortion (nonlinear, active, and nonreciprocal waveguide components). A third-order distortion product that can occur when nonlineal devices are exposed simultaneously to two carriers of

different frequency where modulation is present. It is measured by using two separate carriers and providing 100% modulation on one. The cross modulation is defined as the power in one sideband of the unmodulated carrier below the power of the carrier. 530

cross neutralization. A method of neutralization used in push-pull amplifiers whereby a portion of the plate-cathode alternating voltage of each tube is applied to the grid-cathode circuit of the other tube. *See:* **amplifier; feedback.** 111,211

crossover (cathode-ray tube). The first focusing of the beam that takes place in the electron gun. *See:* **cathode-ray tubes.** 244,190

crossover, automatic voltage-current (power supplies). The characteristic of a power supply that automatically changes the method of regulation from constant voltage to constant current (or vice versa) as dictated by varying load conditions (see the following figure). *Note:* The constant-voltage and constant-current levels can be independently adjusted within the specified voltage and current limits of the power supply. The intersection of constant-voltage and constant-current lines is called the crossover point E, I and may be located anywhere within the volt-ampere range of the power supply. 186

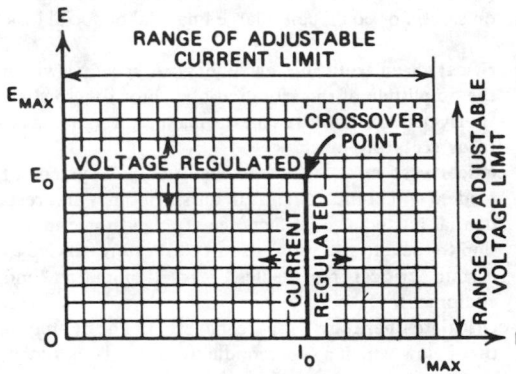

Automatic voltage-current crossover.

crossover characteristic curve (navigation systems such as VOR [very high-frequency omnidirectional range] and ILS [instrument landing system])(navigation aid terms). The graphical representation of the indicator current variation with change of position in the crossover region. 526

crossover frequency (1) (frequency-dividing networks). The frequency at which equal power is delivered to each of two adjacent channels when all channels are properly terminated. *See:* **loudspeaker; transition frequency.** 176

(2) (automatic control). *See:* **gain crossover frequency; phase crossover frequency.**

crossover loss (radar). The reduction in signal-to-noise ratio for a target on the tracking axis relative to that for a target on the peak two-way antenna gain of the beam, for a tracker which uses an offset beam, such as a conical scan tracker. The crossover loss factor is the ratio of the signal-to-noise for a target on the peak two-way antenna gain to that for a target on the tracking axis. 13

crossover network. *See:* **dividing network.**

crossover region (navigation systems). A loosely defined region in space containing the course line and within which a transverse flight yields information useful in determining course sensitivity and flyability. 526

crossover spiral. *See:* **lead-over groove.**

crossover time (charged-particle detectors)(germanium gamma-ray detectors)(semiconductor radiation detectors). The instant at which the waveform of a bipolar pulse passes through a designated level.
119,528,23,118

crossover voltage, secondary-emission (charge-storage tubes). The voltage of a secondary-emitting surface, with respect to cathode voltage, at which the secondary-emission ratio is unity. The crossovers are numbered in progression with increasing voltage. See the following figure. *Note:* The qualifying phrase **secondary-emission** is frequently dropped in general usage. *See:* **charge-storage tube.** 174

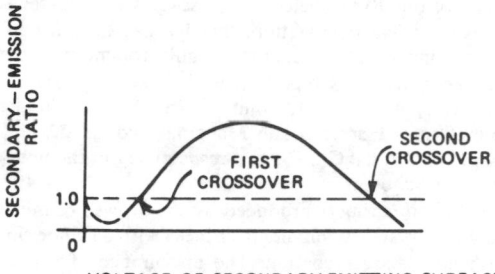

Typical secondary-emission curve.

crossover walk (of a pulse)(1)(charged-particle detectors)(germanium gamma-ray detectors)(semiconductor radiation detectors). The deviation of the crossover time for some variable, such as (pulse) amplitude.
119,528,23,118

crosspoint (telephone switching systems). A controlled device used in extending a transmission or control path. 55

cross-polar (radiation) pattern (antennas). A radiation pattern corresponding to the polarization orthogonal to the copolarization. *See:* **copolarization.** 111

cross polarization (1) (antennas). In a specified plane containing the reference polarization ellipse, the polarization orthogonal to a specified reference polarization. *Note:* The reference polarization is usually the copolarization. 111

(2) (waveguide). The polarization orthogonal to a reference polarization. *Note:* Two fields have orthogonal

polarization if their polarization ellipses have the same axial ratio, major axes at right angles, and opposite senses of rotation. 267

cross-polar side lobe level (relative) (antennas). The maximum relative partial directivity (corresponding to the cross-polarization) of a side lobe with respect to the maximum partial directivity (corresponding to the copolarization) of the antenna. *Note:* Unless otherwise specified the cross-polar side lobe level is taken to be that of the highest side lobe of the cross polar radiation pattern. 111

cross product. *See:* **vector product.**

cross protection. An arrangement to prevent the improper operation of devices from the effect of a cross in electric circuits. 328

cross rectifier circuit. A circuit that employs four or more rectifying elements with a conducting period of 90 electrical degrees plus the commutating angle. *See:* **rectification.** 328

cross section (1) (radar). Often used as a shortened form of "radar cross section"; to be avoided when there is a possibility of confusion with geometric cross section. 13

(2) (antenna). *See:* **back-scattering cross section; radar cross-section; scattering cross section.**

cross-sectional area (conductor) (cross section of a conductor). The sum of the cross-sectional areas of its component wires, that of each wire being measured perpendicular to its individual axis. 59

crosstalk (1) (cable systems in power generating stations). The noise or extraneous signal caused by ac or pulse-type signals in adjacent circuits. 35

(2) (data transmission). (A) Undesired energy appearing in one signal path as a result of coupling from other signal paths. *Note:* Path implies wires, waveguides, or other localized or constrained transmission systems. (B) (electroacoustics). The unwanted sound reproduced by an electroacoustic receiver associated with a given transmission channel resulting from cross coupling to another transmission channel carrying sound-controlled electric waves or, by extension, the electric waves in the disturbed channel that result in such sound. *Note:* In practice, crosstalk may be measured either by the volume of the overheard sounds or by the magnitude of the coupling between the disturbed and the disturbing channels. In the latter case, to specify the volume of the overheard sounds, the volume in the disturbing channel must also be given. 59

crosstalk coupling (crosstalk loss). Cross coupling between speech communication channels or their component parts. *Note:* Crosstalk coupling is measured between specified points of the disturbing and disturbed circuits and is preferably expressed in decibels. *See:* **coupling.** 328

crosstalk, electron beam (charge-storage tubes). Any spurious output signal that arises from scanning or from the input of information. *See:* **charge-storage tube.** 174

crosstalk loss. *See:* **crosstalk coupling.**

crosstalk unit. Crosstalk coupling is sometimes expressed in crosstalk units through the relation

$$\text{Crosstalk units} = 10^{[6-(L/20)]}$$

where L = crosstalk coupling in decibels. *Note:* For two circuits of equal impedance, the number of crosstalk units expresses the current in the disturbed circuit as millionths of the current in the disturbing circuit. *See:* **coupling.** 328

CR-RC shaping (germanium gamma-ray detectors)(X-ray energy spectrometers)(charged-particle detectors). The pulse shaping present in an amplifier that has a simple high-pass filter consisting of a capacitor and a resistor together with a simple low-pass filter, separated by impedance isolation. (Pulse shaping in such an amplifier cuts off at 6 dB (decibels) per octave at both ends of the band.)

119,528,471,23,118

crude metal. Metal that contains impurities in sufficient quantities to make it unsuitable for specified purposes or that contains more valuable metals in sufficient quantities to justify their recovery. *See:* **electrorefining.** 328

crust. A layer of solidified electrolyte. *See:* **fused electrolyte.** 328

cryogenics (1) (general). The study and use of devices utilizing properties of materials near absolute-zero temperature. 255,77

(2) (laser-maser). The branch of physics dealing with very low temperatures. 363

cryotron. A superconductive device in which current in one or more input circuits magnetically controls the superconducting-to-normal transition in one or more output circuits, provided the current in each output circuit is less than its critical value. *See:* **superconductivity.** 191

crystal (communication practice). (1) A piezoelectric crystal. (2) A piezoelectric crystal plate. (3) A crystal rectifier. 328

crystal-controlled oscillator. *See:* **crystal oscillator.**

crystal diode. A rectifying element comprising a semiconducting crystal having two terminals designed for use in circuits in a manner analogous to that of electron-tube diodes. *See:* **rectifier.** 328

crystal loudspeaker (piezoelectric loudspeaker). A loudspeaker in which the mechanical displacements are produced by piezoelectric action. *See:* **loudspeaker.** 328

crystal microphone (piezoelectric microphone). A microphone that depends for its operation of the generation of an electric charge by the deformation of a body (usually crystalline) having piezoelectric properties. *See:* **microphone.** 328

crystal mixer (mixer). A crystal receiver that can be fed simultaneously from a local oscillator and signal source, for the purpose of frequency changing. *See:* **waveguide.** 244

crystal oscillator (crystal-controlled oscillator). An oscillator in which the principal frequency-determining factor is the mechanical resonance of a piezoelectric crystal. *See:* **oscillatory circuit.** 211

crystal pickup (piezoelectric pickup). A phonograph pickup that depends for its operation on the generation of an electric charge by the deformation of a body (usually crystalline) having piezoelectric properties. *See:* **phonograph pickup.** 328

crystal pulling. A method of crystal growing in which the developing crystal is gradually withdrawn from a melt. 66

crystal receiver. A waveguide incorporating a crystal detector for the purpose of rectifying received electromagnetic signals. *See:* **waveguide.** 244,179

crystals hierarchy (piezoelectric, pyroelectric, ferroelectric)(primary ferroelectric terms). Depending on their geometry, crystals are commonly classified into seven systems: triclinic (the least symmetrical), monoclinic, orthorhombic, tetragonal, trigonal, hexagonal, and cubic. The seven systems in turn are divided into point groups (crystal classes) according to their symmetry with respect to a point. There are 32 such crystal classes; 20 are piezoelectric. Piezoelectric crystals have the following property: if stress is applied along certain directions in the crystals, they develop an electric polarization whose magnitude is (within limits) proportional to the applied stress. Conversely, when an electric field is applied along certain directions in a piezoelectric crystal, the crystal is strained by an amount proportional to the applied field. Each crystal system contains at least one piezoelectric class. Ten of the 20 piezoelectric classes possess spontaneous electrical polarization; that is, they have a non-vanishing dipole moment per unit volume and are called *polar*. The ten polar crystal classes are designated 1, 2, m, 2mm, 4, 4mm, 3, 3m, 6, 6mm in the notation of Hermann and Maugin, and C_1, C_2, C_{1h}, C_{2v}, C_4, C_{4v}, C_3, C_{3v}, C_6, C_{6v}, respectively in the notation of Schoenflies. 497

crystal spots. Spots produced by the growth of metal sulfide crystals upon metal surfaces with a sulfide finish and lacquer coating. The appearance of crystal spots is called spotting in. *See:* **electroplating.** 328

crystal-stabilized transmitter. A transmitter employing automatic frequency control, in which the reference frequency is that of a crystal oscillator. *See:* **radio transmitter.** 111

crystal systems (piezoelectricity). The term "crystal" is applied to a solid in which the atoms are arranged in a single pattern repeated throughout the body. In a crystal the atoms may be thought of as occurring in small groups, all groups being exactly alike, similarly oriented, and regularly aligned in all three dimensions. Each group can be regarded as bounded by a parallelepiped and each parallelepiped regarded as one of the ultimate building blocks of the crystal. The crystal is formed by stacking together in all three dimensions replicas of the basic parallelepiped without any spaces between them. Such a building block is called a *unit cell*. Since the choice of a particular set of atoms to form a unit cell is arbitrary, it is evident that there is a wide range of choices in the shapes and dimensions of the unit cell. In practice, that unit cell is selected which is most simply related to the actual crystal faces and X-ray reflections, and which has the symmetry of the crystal itself. Except in a few special cases, the unit cell has the smallest possible size. In crystallography the properties of a crystal are described in terms of the natural coordinate system provided by the crystal itself. The axes of this natural system, indicated by the letters a, b, and, c, are the edges of the unit cell. In a cubic crystal, these axes are of equal length and are mutually perpendicular; in a triclinic crystal they are of unequal lengths and no two are mutually perpendicular. The faces of any crystal are all parallel to planes whose intercepts on the a, b, c axes are small multiples of unit distances or else infinity, in order that their reciprocals, when multiplied by a small common factor, are all small integers or zero. These are the indices of the planes. In this nomenclature we have, for example, faces (100), (010), (001), also called the a, b, c faces, respectively. In the orthorhombic, tetragonal, and cubic systems, these faces are normal to the a, b, c axes, 100, etcetera. Even in the monoclinic and triclinic systems, these faces contain respectively, the b and c, a and c, and, a and b axes. As referred to the

Summary of Crystal Systems

Crystal System		International Point Groups Short	Full	Axis Identification, Crystallographic	Axis Identification, Rectangular (X Y Z)	+/− Axes (Note 3)	Schoenflies Symbol	Example	Formula
Triclinic $c_o < a_o < b_o; \alpha, \beta > 90°$	p	1	1		1(010)	Z X	C_1	Aminoethyl ethanolamine hydrogen d-tartrate (AET)	$C_8H_{17}O_7N_2$
		$\bar{1}$	$\bar{1}$		1(010)		$C_1(S_2)$	Copper sulfate pentahydrate	$CuSO_4 \cdot 5H_2O$
Monoclinic $c_o < a_o, \beta > 90°; \alpha = \gamma = 90°$	p	2	2	2 (b)	1(100) b c	Y	C_2	Ethylene diamine tartrate (EDT)	$C_6H_{14}N_2O_6$
	p	m	m	/m	1(100) b c	Z X	$C_s(C_{1h})$	Lithium trihydrogen selenite	$LiH_3(SeO_3)_2$
		2/m	$\frac{2}{m}$		1(100) b c		C_{2h}	Gypsum	$CaSO_4 \cdot 2H_2O$
Orthorhombic $c_o < a_o < b_o; \alpha = \beta = \gamma = 90°$	p	222	222	2 2 2	a b c		$D_2(V)$	Rochelle salt, except between Curie points	$NaKC_4H_4O_6 \cdot 4H_2O$
	p	mm2	(See Note 1)		(See Note 1)	Z	C_{2v}	Barium sodium niobate	$Ba_2NaNb_5O_{15}$
		mmm	$\frac{2}{m}\frac{2}{m}\frac{2}{m}$	2 2 2	a b c		$D_{2h}(V_h)$	Barite	$BaSO_4$
Tetragonal $a_o = b_o; \alpha = \beta = \gamma = 90°$ (axes: c a1 a2)	p	4	4	4 †	(a_1) (a_2) c	Z	C_4	Potassium strontium niobate	$KSr_2Nb_5O_{15}$
	p	$\bar{4}$	$\bar{4}$	$\bar{4}$ †	(a_1) (a_2) c	Z	S_4	Anorthite	$Ca_2Al_2SiO_7$
		4/m	$\frac{4}{m}$	4 †	(a_1) (a_2) c		C_{4h}	Scheelite	$CaWO_4$
	p	422	422	4 2 2	(a_1) (a_2) c	*	D_4	Nickel sulfate hexahydrate, Paratellurite	$NiSO_4 \cdot 6H_2O, TeO_2$
	p	4mm	4mm	4 /m /m	(a_1) (a_2) c	Z	C_{4v}	Barium titanate	$BaTiO_3$
	p	$\bar{4}2m$	$\bar{4}2m$	$\bar{4}$ 2 2	(See Note 2)	Z	$D_{2d}(V_d)$	Ammonium dihydrogen phosphate (ADP)	$NH_4H_2PO_4$
		4/mmm	$\frac{4}{m}\frac{2}{m}\frac{2}{m}$	4 2 2	(a_1) (a_2) c	*	D_{4h}	Zircon	$ZrSiO_4$
Trigonal $(a_o)_1 = (a_o)_2 = (a_o)_3$ (See 2.2.5) (axes: c a1 a2 a3)	p	3	3	3 †	(a_1) · · · c	Any two	C_3	Sodium periodate trihydrate	$NaIO_4 \cdot 3H_2O$
		$\bar{3}$	$\bar{3}$	$\bar{3}$ †	(a_1) · · · c		$C_{3i}(S_6)$	Dolomite	$CaCO_3MgCO_3$
	p	32	32	3 2 2 2	(a_1) · · · c	X	D_3	α-quartz	SiO_2
	p	3m	3m	3 /m /m /m	(a_1) · · · c	Z Y	C_{3v}	Lithium niobate	$LiNbO_3$
		$\bar{3}m$	$\bar{3}\frac{2}{m}$	$\bar{3}$ 2 2 2	(a_1) · · · c		D_{3d}	Calcite	$CaCO_3$
Hexagonal $(a_o)_1 = (a_o)_2 = (a_o)_3$ (See 2.2.5)	p	6	6	6 †	(a_1) c	Z	C_6	Lithium iodate	$LiIO_3$
	p	$\bar{6}$	$\bar{6}$	$\bar{6}$ †	(a_1) c	X Y	C_{3h}	Lithium peroxide	Li_2O_3
		6/m	$\frac{6}{m}$	6	(a_1) c		C_{6h}	Apatite	$CaF_2 \cdot 3Ca_3P_2O_8$
	p	622	622	6 2 2 2	(a_1) c		D_6	β-quartz	SiO_2
	p	6mm	6mm	6 /m /m /m	(a_1) c	Z	C_{6v}	Cadmium sulfide	CdS
	p	$\bar{6}m2$	$\bar{6}m2$	$\bar{6}$ 2 2 2	(a_1) c	X	D_{3h}	Benitoite	$BaTiSi_3O_9$
		6/mmm	$\frac{6}{m}\frac{2}{m}\frac{2}{m}$	6 2 2 2	(a_1) c		D_{6h}	Beryl	$3BeO \cdot Al_2O_3 \cdot 6SiO_2$
Cubic $a_o = b_o = c_o; \alpha = \beta = \gamma = 90°$ (axes: a1 a2 a3)	p	23	23	2 2 2	(a_1) (a_2) a_3	Z	T	Bismuth germanium oxide	$Bi_{12}GeO_{20}$
		m3	$\frac{2}{m}3$	2 2 2	(a_1) (a_2) a_3	*	T_h	Pyrite	FeS_2
		432	432	4 4 4	(a_1) (a_2) a_3	*	O	Cadmium fluoride	CdF_2
	p	$\bar{4}3m$	$\bar{4}3m$	$\bar{4}$ $\bar{4}$ $\bar{4}$	(a_1) (a_2) a_3	Z	T_d	Gallium arsenide	$GaAs$
		m3m	$\frac{4}{m}\bar{3}\frac{2}{m}$	4 4 4	(a_1) (a_2) a_3	*	O_h	Sodium chloride	$NaCl$

NOTES:

(1) Z is the polar axis, which may be a, b, or c. Depending on whether a, b, or c is polar, the full international point group symbol is $2mm$, $m2m$, or $mm2$, respectively. X is parallel to the smaller of the nonpolar axes. Thus in classes $2mm$, $m2m$, and $mm2$, X is chosen parallel to c, c, and a, respectively.

(2) In class $42m$ the axial choice is as listed here for six of the space groups. For the other six the a axis is chosen at 45 degrees to the twofold axes in order to have the smallest unit cell. In call cases X and Y are chosen parallel to the twofold axes. See 2.2.4.

(3) Axes whose sense is determined by the sign of a piezoelectric constant. It does not necessarily have a polar axis.

For † and *, see text.

set of rectangular axes X, Y, Z, these indices are in general irrational except for cubic crystals. Depending on their degrees of symmetry, crystals are commonly classified into seven systems: triclinic (the least symmetrical), monoclinic, orthorhombic, tetragonal, trigonal, hexagonal, and cubic. The seven systems, in turn, are divided into point groups (classes) according to their symmetry with respect to a point. There are 32 such classes, eleven of which contain enantiomorphous forms. Twelve classes are of too high a degree of symmetry to show piezoelectric properties. Thus twenty classes can be piezoelectric. Every system contains at least one piezoelectric class.

371

crystal-video receiver. A receiver consisting of a crystal detector and a video amplifier. 328

C **scan (electronic navigation).** *See: C* **display.**

C-scope (radar). A cathode-ray oscilloscope arranged to present a C-display. 13

CSR (control and status register) space (FASTBUS acquisition and control). A FASTBUS primary address cycle may specify with a code on the mode select (MS) control lines one of two separate address spaces in a device; CSR space and data space. CSR space contains registers for control of and status reporting registers for the device. Its allocation and usage is part of the FASTBUS specification. *See:* **data space.**

480

CTS. *See:* **carrier test switch.**

cube tap (plural tap). *See:* **multiple plug.**

cubic natural spline (pulse terms). A catenated piecewise sequence of cubic polynominal functions p(1, 2), p(2, 3), . . ., p(n-1, n) between knots t_1m_1 and t_2m_2, t $_2m_2$ and t_3m_3, . . ., t_{n-1} m_{n-1} and t_nm_n, respectively, wherein: (1) At all knots the first and second derivatives of the adjacent polynominal functions are equal, and (2) For all values of t less than t_1 and greater than t_n the function is linear. *See:* Pulse burst envelopes diagram below the **knot** entry. *See:* **waveforms produced by operations on waveforms.** 254

cumulative amplitude probability distribution (control of system electromagnetic compatibility). A cumulative distribution showing the probability that all amplitudes equal to, or above, a stated value are exceeded as a function of that value. 495

cumulative compound (rotating machinery). Applied to a compound machine to denote that the magnetomotive forces of the series and the shunt field windings are in the same direction. *See:* **magnetomotive force.**

63

cumulative demand meter (or register). An indicating demand meter in which the accumulated total of maximum demands during the preceding periods is indicated during the period after the meter has been reset and before it is reset again. *Note:* The maximum demand for any one period is equal or proportional to the difference between the accumulated readings before and after reset. *See:* **electricity meter (meter).**

212

cumulative demand register (metering). A register that

indicates the sum of the previous maximum demand readings prior to reset. When reset, the present reading is added to the previous accumulated readings. The maximum demand for the present reading period is the difference between the present and previous readings. 212

cumulative detection probability (radar). The probability that a target is detected on at least one of N successive scans of a surveillance radar. 13

cuprous chloride cell. A primary cell in which depolarization is accomplished by cuprous chloride. *See:* **electrochemistry.** 328

Curie temperature (electrical heating systems). The temperature at which the magnetic properties of a substance change from ferromagnetic to paramagnetic. 476

Curie-Weiss temperature (ferroelectric material)(primary ferroelectric terms). The intercept θ of the linear portion of the plot of $1/\kappa$ versus T, in the region above the ferroelectric Curie point, where κ is the small-signal relative dielectric permittivity measured at zero bias field along the polar axis, and T is the absolute temperature. *Note:* In many ferroelectrics, κ follows the Curie-Weiss relation

$$\kappa = \epsilon / \epsilon_0 = C/(T - \theta),$$

where

ϵ = small-signal absolute permittivity

ϵ_0 = permittivity of free space (8.854 $\cdot$ 10^{12} coulomb/voltmeter)

C = Curie constant

θ = Curie-Weiss temperature

The Curies-Weiss temperature, θ is always less than or equal to the Curie point T_C and generally within a few degrees of T_C. 497

curl (vector field). A vector that has a magnitude equal to the limit of the quotient of the circulation around a surface element on which the point is located by the area of the surface, as the area approaches zero, provided the surface is oriented to give a maximum value of the circulation: the positive direction of this vector is that traveled by a right hand screw turning about an axis normal to the surface element when an integration around the element in the direction of the turning of the screw gives a positive value to the circulation. If the vector **A** of a vector field is expressed in terms of its three rectangular components A_x, A_y, and A_z, so that the values of A_x, A_y, and A_z are each given as a function of x, y, and z, the curl of the vector field (abbreviated curl **A** or $\nabla \times$ **A**) is the vector sum of the partial derivatives of each perpendicular to it, or

$$\text{curl } \mathbf{A} = \nabla \times \mathbf{A} = \begin{vmatrix} \mathbf{i} & \mathbf{j} & \mathbf{k} \\ \dfrac{\delta}{\delta x} & \dfrac{\delta}{\delta y} & \dfrac{\delta}{\delta z} \\ A_x & A_y & A_z \end{vmatrix}$$

$$= \mathbf{i} \left(\dfrac{\delta A_z}{\delta y} - \dfrac{\delta A_y}{\delta z} \right) + \mathbf{j} \left(\dfrac{\delta A_x}{\delta z} - \dfrac{\delta A_z}{\delta x} \right)$$

$$+ \, \mathbf{k} \left(\frac{\delta A_y}{\delta x} - \frac{\delta A_x}{\delta y} \right)$$

$$= \mathbf{i}(D_y A_z - D_z A_y)$$

$$+ \, \mathbf{j}(D_z A_x - D_x A_z) + \mathbf{k}(D_x A_y - D_y A_z)$$

where **i, j,** and **k** are unit vectors along the x, y, and z axes, respectively. Example: The curl of the linear velocity of points in a rotating body is equal to twice the angular velocity. The curl of the magnetic field strength at a point within an electric conductor is equal to k times the current density at the point where k is a constant depending on the system of units. 210

current (electric) (1) (general). A generic term used when there is no danger of ambiguity to refer to any one or more of the currents specifically described. *Notes:* (A) For example, in the expression "the current in a simple series circuit," the word current refers to the conduction current in the wire of the inductor and the displacement current between the plates of the capacitor. (B) A direct current is a unidirectional current in which the changes in value are either zero or so small that they may be neglected. A given current would be considered a direct current in some applications, but would not necessarily be so considered in other applications. 210
(2) (modified by an adjective). The use of certain adjectives before "current" is often convenient, as in convection current, anode current, electrode current, emission current, etcetera. The definition of conducting current usually applies in such cases and the meaning of adjectives should be defined in connection with the specific applications. 210
(3) (cable insulation materials). Sum of the polarization and conductance currents. 97
current amplification (1)(multiplier phototube). The ratio of the output current to the cathode current due to photoelectric emission at constant electrode voltages. *Notes:* (A) The term output current and photocathode current as here used does not include the dark current. (B) This characteristic is to be measured at levels of operation that will not cause saturation. *See:* **phototube.** 125,117
(2) (magnetic amplifier). The ratio of differential output current to differential control current. 171
(3) (electron multipliers). The ratio of the signal output current to the current applied to the input. *See:* **amplifier.** 190
current, anode. *See:* **electrode current.**
current attenuation. *See:* **attenuation, current.**
current, average discharge (dielectric tests). The sum of the rectified charge quantities passing through the terminals of the test object due to partial discharges during a time interval, divided by this interval. The average discharge current is expressed in coulombs per second (amperes). 139
current balance ratio. The ratio of the metallic-circuit current or noise-metallic (arising as a result of the

action of the longitudinal-circuit induction from an exposure on unbalances outside the exposure) to the longitudinal circuit current or noise-longitudinal in sigma at the exposure terminals. It is expressed in microamperes per milliampere or the equivalent. *See:* **inductive coordination.** 328
current-balance relay (power switchgear). A balance relay that operates by comparing the magnitudes of two current inputs. 127, 103
current-balancing device (thyristor). Device used to achieve satisfactory division of current among parallel connected semiconductor devices, for example, reactor, resistor, impedance. 445
current-balancing reactor. *See:* **reactor, current-balancing.**
current balancing transformer. *See:* **sharing transformer.**
current carrier. In a semiconductor, a mobile conduction electron or hole. 113
current-carrying capacity (contacts). The maximum current that a contact is able to carry continuously or for a specified period of time. *See:* **contactor.** 244, 206
current-carrying part (NESC)(power line maintenance). A conducting part intended to be connected in an electric circuit to a source of voltage. Noncurrent-carrying parts are those not intended to be so connected. 27,103,443,458,494
current circuit (ac high-voltage circuit breaker). That part of the synthetic test circuit from which the major part of the power frequency current is obtained. 426
current clamp (converter circuit elements)(self-commutated converters). A clamp that limits the current through a semiconductor device. 584
current comparator (metering). A device by which the ratio and phase angle between two currents can be measured precisely. *Note:* A common form of current comparator relies on a balance of ampere-turns produced by currents in two or more windings on one or more magnetic cores. 212
current compensator (excitation systems for synchronous machines). An element of the excitation system that acts to compensate for synchronous machine load current effects. *Notes:* (1) Examples are reactive current compensator and active current compensator. A reactive current compensator is a compensator that acts to modify the regulated voltage in accordance with reactive current. An active current compensator is a compensator that acts to modify the regulated voltage in accordance with active current. (2) Historically, terms such as equalizing reactor and cross current compensator have been used to describe the function of a reactive compensator. These terms are deprecated. (3) Reactive compensators are generally applied with synchronous machine voltage regulators to obtain reactive current sharing among synchronous machines operating in parallel. They function in the following two ways: (A) Reactive droop compensation is the more common mathod. It creates a droop in synchronous machine terminal voltage proportional to

reactive current and equivalent to that which would be produced by the insertion of a reactor between the synchronous machine terminals and the paralleling point. (B) Reactive differential compensation is used where droop in synchronous machine voltage is not wanted. It is obtained by a series differential connection of the various synchronous machine, current transformer secondaries, and reactive compensators. The difference current for any synchronous machine from the common series current creates a compensating voltage in the input to the particular synchronous machine voltage regulator which acts to modify the synchronous machine excitation to reduce to minimum (zero) its differential reactive current. (4) Line drop compensators modify synchronous machine terminal voltage by regulator action to compensate for the impedance drop from the machine terminals to a fixed point in the external circuit. Action is accomplished by insertion within the regulator input circuit, [of] a voltage equivalent to the impedance drop. The voltage drops of the resistance and reactance portions of the impedance are obtained, respectively, by an active compensator and a reactive compensator. 507

current, conduction (cable insulation materials). Current in the specimen under steady-state conditions. *Notes:* (1) This is sometimes called "leakage" current. (2) Absorption and capacitive effects are assumed to have been made negligible under steady-state conditions. (3) Surface leakage current is assumed excluded from the measured current. *Syn:* **conductance current.** 97

current crest factor (mercury-lamp ballast combination). The ratio of the peak value of lamp current to the root-mean-square value of lamp current. 271

current cutoff (power supplies). An overload protective mechanism designed into certain regulated power supplies to reduce the load current automatically as the load resistance is reduced. This negative resistance characteristic reduces overload dissipation to negligible proportions and protects sensitive loads. The accompanying figure shows the *E-I* characteristic of a power supply equipped with a current cutoff overload protector. 179

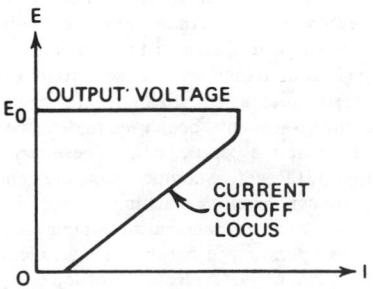

Output characteristics of a power supply equipped with a current cutoff overload protector.

current cycle loop (substation grounding). The combination of conductors and connectors that carries the current of the circuit under test. 475

current delay angle (thyristor). The interval in electrical angular measure by which the starting instant of conduction is delayed in relation to operation that would occur with continuously gated control elements. See α_I of diagram. 445

current density A generic term used where there is no danger of ambiguity to refer either to conduction-current density or to displacement-current density, or to both. 210

current derived voltage (protective relaying). A voltage produced by a combination of currents. *Notes:* (1) The element used to create this voltage in a pilot system is popularly referred to as a filter. A typical example is a filter that is supplied three-phase currents and produces an output voltage proportional to the symmetrical component content of these currents. (For example, $V_F = K_1 IA_1 + K_2 IA_2 + K_0 IA_0$ where IA_1, IA_2, and IA_0 are the symmetrical components of the A phase current and the K are weighting factors.) 128

current efficiency (specified electrochemical process). The proportion of the current that is effective in carrying out that process in accordance with Faraday's law. *See:* **electrochemistry.** 328

current generator (signal-transmission system). A two-terminal circuit element with a terminal current substantially independent of the voltage between its terminals. *Note:* An ideal current generator has zero internal admittance. *See:* **network analysis; signal.** 125

current injection method (ac high-voltage circuit breaker). A synthetic test method in which the voltage circuit is applied to the test circuit breaker before power frequency current zero. 426

current-limit (control) (industrial control). A control function that prevents a current from exceeding its prescribed limits. *Note:* Current-limit values are usually expressed as percent of rated-load value. If the current-limit circuit permits the limit value to increase somewhat instead of being a single value, it is desirable to provide either a curve of the limit value of current as a function of some variable such as speed or to give limit values at two or more conditions of operation. *See:* **control system, feedback.** 219, 206

current-limit acceleration (electric drive) (industrial control). A system of control in which acceleration is so governed that the motor armature current does not exceed an adjustable maximum value. *See:* **electric drive.** 206

current-limit control (electric drive). A system of control in which acceleration, or retardation, or both, are so governed that the armature current during speed changes does not exceed a predetermined value. *See:* **electric drive.** 328

current limiter (protection and coordination of industrial and commercial power systems). A device intended to function only on fault currents of high magnitude and that may not successfully open on lesser

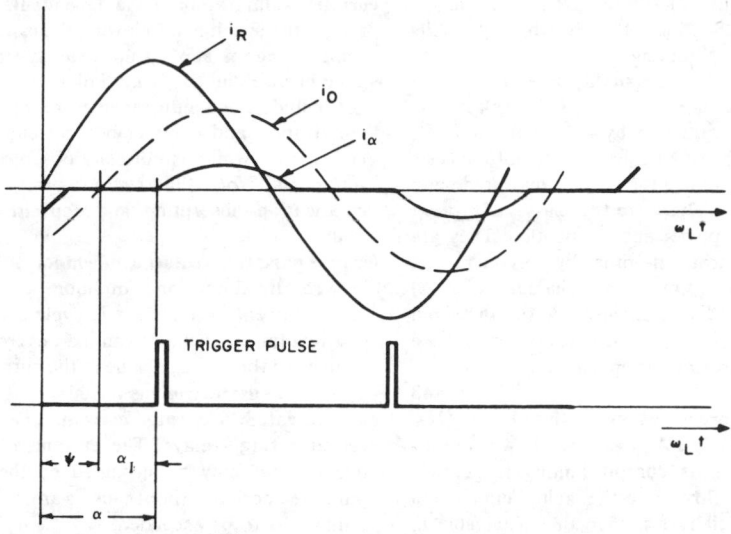

i_R = current in a control element with all control elements continuously gated and a resistive load

> NOTE: In the case of a single phase controller i_R is in phase with the line voltage. The latter may be used as a convenient reference voltage to measure α

i_O = current in a control element with all control elements continuously gated and at the specified load

i_α = current in a control element with a trigger delay angle of α and at the specified load
α = angle of retard
α_I = current delay angle
Ψ = angle between i_R and i_O

> NOTE: In the case of a single phase controller, Ψ is identical with the load power factor angle.

overcurrents regardless of time. Such a device should always be used in series with a fuse, contactor, or circuit breaker to protect against overloads and low-level short circuits. Current limiters are typically added to molded-case circuit breakers, power circuit breakers, or instantaneous circuit protectors.   **504**

current-limiting (peak let-through or cut-off) characteristic curve (of a current-limiting fuse) (power switchgear). A curve showing the relationship between the maximum peak current passed by a fuse and the correlated root-mean-square available current magnitudes under specified voltage and circuit impedance conditions. *Note:* The rms available current may be symmetrical or asymmetrical.   **103, 443**

current limiting, automatic (power supplies). An overload protection mechanism that limits the maximum output current to a preset value, and automatically restores the output when the overload is removed. See the accompanying figure. *See:* **short-circuit protection.**   **179**

current-limiting characteristic curve (current-limiting fuse) (peak let-through characteristic curve) (cutoff characteristic curve). A curve showing the relationship between the maximum peak current passed by a fuse and the correlated root-mean-square available current magnitudes under specified voltage and circuit impedance conditions. *Note:* The root-mean-square available current may be symmetrical or asymmetrical.   **103**

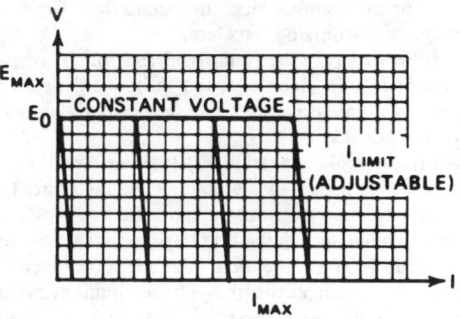

Plot of typical current-limiting curves.

current limiting fuse (protection and coordination of industrial and commercial power systems). A fuse that will interrupt all available currents above its threshold current and below its maximum interrupting rating, limit the clearing time at rated voltage to an interval equal to or less than the first major or symmetrical loop duration, and limit peak let-through current to a value less than the peak current that would be possible with the fuse replaced by a solid conductor of the same impedance. Note that current-limiting action only becomes effective at a specific value of current. Underwriters Laboratories (UL) only recognizes and permits labeling of Classes G, J, L, R, CC, and T as current limiting, although Class K fuses are, in fact, current limiting. Refer to the National Electrical Code (NEC), Section 240-60b, which prohibits fuse clips for current-limiting fuses accepting noncurrent-limiting fuses. *See:* **threshold current.** 504

current limiting fuse unit (power switchgear). A fuse unit that when it is melted by a current within its specified current limiting range, abruptly introduces a high resistance to reduce the current magnitude and duration. *Notes:* (1) There are two classes of current limiting fuse units–power and distribution. They are differentiated one from the other by current ratings and minimum melting time current characteristics. (2) The values specified in standards for the threshold ratio, peak let-thru current, and I^T characteristics are used as the measures of current limiting ability. 103, 443

current-limiting overcurrent protective device (National Electrical Code). A device which, when interrupting currents in its current-limiting range, will reduce the current flowing in the faulted circuit to a magnitude substantially less than that obtainable in the same circuit if the device were replaced with a solid conductor having comparable impedance. 256

current-limiting range (of a current-limiting fuse) (power switchgear). That specified range of currents between the threshold current and the rated interrupting current within which current limitation occurs. 103

current-limiting reactor (power and distribution transformer). A reactor intended for limiting the current that can flow in a circuit under short-circuit conditions, or under other operating conditions such as starting, synchronizing, etcetera. 53

current-limiting resistor (industrial control). A resistor inserted in an electric circuit to limit the flow of current to some predetermined value. *See:* **control system, feedback.** 328

current loss (electric instrument) (voltage circuit current drain) (parallel loss of an electric instrument). In a voltage-measuring instrument, the value of the current when the applied voltage corresponds to nominal end-scale indication. *Note:* In other instruments it is the current in the voltage circuit at rated voltage. *See:* **accuracy rating (instrument).** 280

current margin (neutral direct-current telegraph system). The difference between the steady-state currents flowing through a receiving instrument, corresponding, respectively, to the two positions of the telegraph transmitter. *See:* **telegraphy.** 328

current of traffic. The movement of trains on a main track in one direction specified by the rules. *See:* **railway signal and equipment.** 328

current, peak (low voltage dc power circuit breakers). The instantaneous value of current at the time of its maximum value. 360

current phase-balance protection (power switchgear). A method of protection in which an abnormal condition within the protected equipment is detected by the current unbalance between the phases of a normally balanced polyphase system. 127, 103

current, polarization. Time-dependent, decaying current in the specimen, following the instant that a constant voltage is applied until steady-state conditions have been obtained. *Note:* Polarization current does not include the conductance current. The sum of the polarization and conductance currents in the specimen is that which is normally observed during measurements. *Note:* Polarization current includes both polarization absorption and capacitive-charge currents. 97

current pulsation (rotating machinery). The difference between maximum and minimum amplitudes of the motor current during a single cycle corresponding to one revolution of the driven load expressed as a percentage of the average value of the current during this cycle. *See:* **asynchronous machine.** 63

current, rated. *See:* **rated current.**

current rating (relay). The current at specified frequency that may be sustained by the relay for an unlimited period without causing any of the prescribed limitations to be exceeded. 127

current rating of a relay (power switchgear). The limiting current at specified frequency that may be sustained by the relay for an unlimited period without causing any of the prescribed limitations to be exceeded. 103

current rating, 60-hertz (arrester). A designation of the range of the symmetrical root-mean-square fault currents of the system for which the arrester is designed to operate. *Notes:* (1) An expulsion arrester is given a maximum current rating and may also have a minimum current rating. (2) The designation of the maximum and minimum current ratings of an expulsion arrester not only specifies the useful operating range of the arrester between those extreme values for symmetrical root-mean-square short-circuit current, but indicates that at the point of application of the arrester the root-mean-square short-circuit current for the system should neither be greater than the maximum nor less than the minimum current rating. 328

current ratio (series transformer) (mercury lamp). The ratio of the (root-mean-square) primary current to the root-mean-square secondary current under specified conditions of load. 274

current-recovery ratio (arc-welding apparatus). With a welding power supply delivering current through a short-circuited resistor whose resistance is equivalent to the load setting on the power supply, and with the short-circuit suddenly removed, the ratio of (1) the minimum transient value of current upon the removal of the short-circuit to (2) the final steady-state value is the current-recovery ratio. 264

current regulation (1) (constant-current transformer) (power and distribution transformer). The maximum departure of the secondary current from its rated value, with rated primary voltage at rated frequency applied, and at rated secondary power factor, and with the current variation taken between the limits of a short-circuit and rated load. *Note:* This regulation may be expressed in per unit, or percent, on the basis of the rated secondary current. 53

(2) (thyristor). The method whereby the current is controlled to a specified value. 445

current relay (1) (general). A relay that functions at a predetermined value of current. *Note:* It may be an overcurrent, undercurrent, or reverse-current relay. *See:* **relay.** 1

(2) (power systems relaying) (power switchgear). A relay which responds to current. 103,127

current-responsive element (of a fuse). That part with predetermined characteristics, the melting and severance or severances of which initiate the interrupting function of the fuse. *Note:* The current-response element may consist of one or more fusible elements combined with a strain element or other component(s), or both, which affect(s) the current-responsive characteristics. 103, 443

current ripple (rotating electric machinery). Current ripple, for the purposes of this standard, is defined as

$$[(I_{max}-I_{min})/(I_{max}+I_{min})]\times 100,$$

expressed in percent where I_{max} and I_{min} are the maximum and minimum values of the current waveform, provided that the current is continuous. 424

current-sensing resistor (power supplies). A resistor placed in series with the load to develop a voltage proportional to load current. A current-regulated direct-current power supply regulates the current in the load by regulating the voltage across the sensing resistor. 186

current sensitivity (diode)(nonlinear, active, and non-reciprocal waveguide components). The output current developed by a diode detector for a specified load resistance per unit available radio-frequency (rf) input power. It is expressed in mA/mW. 530

current, short-circuit (transformer-rectifier system). The steady-state value of the input alternating current that flows when the output direct current terminals are short-circuited and rated line alternating voltage is applied to the line terminals. This current is normally of interest when using current limiting transformers or checking current limiting devices. 95

current tap. *See:* **multiple lampholder; plug adapter lampholder.**

current terminals (direct current instrument shunts). Those terminals that are connected into the line whose current is to be measured and that will carry the current of the shunt. 527

current transformer (power and distribution transformer)(instrument transformer). An instrument transformer intended to have its primary winding connected in series with the conductor carrying the current to be measured or controlled. (In window type current transformers, the primary winding is provided by the line conductor and is not an integral part of the transformer.) 53,394

current turn-off time (low-voltage air gap surge-protective devices)(gas-tube surge-protective devices). The time required for the arrester to restore itself to a nonconducting state following a period of conduction. This definition applies only to a condition where the arrester is exposed to a continuous specified direct current (dc) potential under a specified circuit condition. 62,370,556,490

current-type telemeter (power switchgear). A telemeter which employs the magnitude of a single current as the translating means. 103

current withstand rating (emergency and standby power). The maximum allowable current, either instantaneous or for a specified period of time, that a device can withstand without damage. 512

cursor (computer applications). A moveable icon or spot of light on the screen of a display device that indicates the currently selected object or character. 571

curvature (coaxial transmission line). The radial departure from a straight line between any two points on the external surface of a conductor. *See:* **waveguide.** 265

curvature loss. *See:* **macrobend loss.**

curve follower. *See:* **curve-follower function generator.**

curve-follower function generator (analog computers). A function generator that operates by automatically following a curve f(x) drawn or constructed on a surface, as the input x varies over its range. 9

curve, integrated energy (electric power utilization). A curve of demand versus energy showing the amount of energy represented under a load curve, or a load duration curve above any point of demand. *Syn:* **peak percent curve.** *See:* **generating station.** 112

curve, load duration (electric power utilization). A curve of loads, plotted in descending order of magnitude, against time intervals for a specified period. *See:* **generating station.** 2

curve, monthly peak duration (electric power utilization). A curve showing the total number of days within the month during which the net 60 minute clock-hour integrated peak demand equals or exceeds the percent of monthly peak values shown. *See:* **generating station.** 112

curve, reservoir operating rule (electric power supply). A curve, or family of curves (reservoir capability versus time), indicating how a reservoir is to be operated under specified conditions to obtain best or predetermined results. 112

cushion clamp (rotating machinery). A device for securing the cushion to the supporting member. 328

cushioning time (speed governing of hydraulic turbines). The elapsed time during which the (closing)

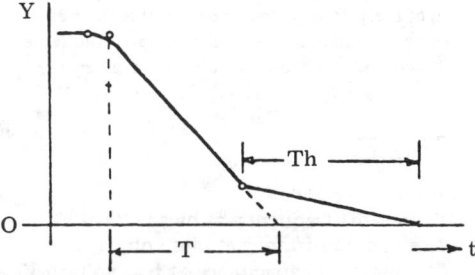

rate of servomotor travel is retarded by the slow closure device. *See:* **slow closure device.** 8

customer (software requirements specifications). The person, or persons, who pay for the product and usually (but not necessarily) decides the requirements. In the context of IEEE Std 830-1984 the customer and the supplier may be members of the same organization. 449

customer alert (watthour meters). A switching output used to indicate a time-of-use period. 485

customer charge (power operations). The amount paid periodically by a customer without regard to demand or energy consumption. 516

customer generation reserve (power operations). The operating reserve available through startup of customer generation. 516

customer load (power operations). Total of loads including distribution system load and losses but excluding station service, transmission losses, and pumping load. 516

customer premises equipment (telephone loop performance). Any equipment connected by customer premises wiring to the customer side of the network interface. 473

cutback technique (fiber optics). A technique for measuring fiber attenuation or distortion by performing two transmission measurements. One is at the output end of the full length of the fiber. The other is within 1 to 3 meters of the input end, access being had by "cutting back" the test fiber. *See:* **attenuation.** 433

cut-in loop. A circuit on the roadway energized to automatically cut in the train control or cab signal apparatus on a passing vehicle. 328

cutoff. *See:* **cutoff frequency.**

cut-off angle (of a luminaire) (illuminating engineering). The angle, measured up from nadir, between the vertical axis and the first line of sight at which the bare source is not visible. 167

cut-off characteristic (of a current-limiting fuse). *See:* **current-limiting characteristic curve.** 103

cutoff frequency (seismic qualification of Class 1E equipment for nuclear power generating stations). The frequency in the response spectrum where the zero period acceleration asymptote begins. This is the frequency beyond which the single-degree-of-freedom oscillators exhibit no amplification of motion, and indicate the upper limit of the frequency content of the waveform being analyzed. 581

cutoff frequency (1)(varactor diodes)(nonlinear, active, and nonreciprocal waveguide components). A figure of merit for a varactor diode. It is the frequency at which Q equal 1. Its relationship to diode series resistance and junction capacitance is given by

$$f_c = \frac{1}{2\pi R_s C_j}.$$ 530

where

f_c = cutoff frequency at bias voltage V_R
R_s = series resistance at bias voltage V_R
C_j = junction capacitance at bias voltage V_R

(2) (of a waveguide). For a given transmission mode in a nondissipative waveguide, the frequency at which the propagation constant is zero. 267

cutoff frequency, effective (electric circuit). A frequency at which its insertion loss between specified terminating impedances exceeds by some specified amount the loss at some reference point in the transmission band. *Note:* The specified insertion loss is usually 3 decibels. *See:* **cutoff frequency; network analysis.** 328

cutoff mode (evanescent mode) (waveguide). A nonpropagating waveguide mode such that the variation of phase along the direction of the guide is negligible. 267

cutoff relay (telephony). A relay associated with a subscriber line, that disconnects the line relay from the line when the line is called or answered. 328

cutoff voltage (1) (electron tube). The electrode voltage that reduces the value of the dependent variable of an electron-tube characteristic to a specified low value. *Note:* A specific cutoff characteristic should be identified as follows: current versus grid cutoff voltage, spot brightness versus grid cutoff voltage, etcetera. *See:* **electrode voltage (electron tube).** 125
(2) (magnetrons). *See:* **critical voltage (magnetrons).**

cutoff waveguide (evanescent waveguide). A waveguide used as a frequency below its cutoff frequency. *See:* **waveguide.** 244, 179

cutoff wavelength (1) (mode in a waveguide) (critical wavelength). That wavelength, in free space or in the unbounded guide medium, as specified, above which a traveling wave in that mode cannot be maintained in the guide. *Note:* For $TE_{m,n}$ or $TM_{m,n}$ waves in hollow rectangular cylinders

$$\lambda_c = 2/[(m/a)^2 + (n/b)^2]^{1/2}$$

where a is the width of the waveguide along the x coordinate and b is the height of the waveguide along the y coordinate.

The following table gives the ratio of cutoff wavelengths to diameters for $TM_{n,m}$ waves in hollow circular metal cylinders:

$n =$	0	1	2	3	4	5
$m = \begin{cases} 1 \\ 2 \\ 3 \\ 4 \\ 5 \end{cases}$	1.307	0.820	0.613	0.483	0.414	0.358
	0.569	0.448	0.373	0.322	0.284	0.2547
	0.363	0.309	0.270	0.241	0.219	0.200
	0.267	0.2375	0.2127	0.1938		
	0.2106	0.1910	0.1750			

The following table gives the ratio of cutoff wavelengths to diameters for $TE_{m,n}$

$n =$		0	1	2	3	4
	1	0.820	1.708	1.030	0.748	0.590
	2	0.448	0.59	0.468	0.382	0.338
$m =$	3	0.309	0.368	0.315	0.277	0.247
	4	0.2375	0.27	0.24	0.22	0.198
	5	0.1910	0.21	0.194	0.18	

See: **guided wave; waveguide.** 244, 179
(2) (uniconductor waveguide). The ratio of the velocity of electromagnetic waves in free space to the cutoff frequency. *See:* **waveguide.** 267
(3) (fiber optics). That wavelength greater than which a particular waveguide mode ceases to be a bound mode. *Note:* In a single mode waveguide, concern is with the cutoff wavelength of the second order mode. *See:* **mode.** 433
(4) (of a waveguide). The free-space wavelength corresponding to the cutoff frequency of the waveguide. 267

cutout (electric distribution systems). An assembly of a fuse support with either a fuseholder, fuse carrier, or disconnecting blade. *Notes:* (1) The fuseholder or fuse carrier may include a conducting element (fuse link), or may act as a disconnecting blade by the inclusion of a nonfusible member. (2) The term **cutout** as defined here is restricted in practice to equipment used on distribution systems. *See:* **distribution** and **distribution cutout.** For fuses having similar components used on power systems. *See:* **power; power fuse.** 103

cutout base. *See:* **fuse holder.**

cutout box (interior wiring) (National Electrical Code). An enclosure designed for surface mounting and having swinging doors or covers secured directly to and telescoping with the walls of the box proper. *See:* **cabinet.** 256

cutout loop (railway practice). A circuit in the roadway that cooperates with vehicle-carried apparatus to cut out the vehicle train control or cab signal apparatus. 328

cut paraboloidal reflector. A reflector that is not symmetrical with respect to its axis. *See:* **antenna.** 111

cut-section. A location within a block other than a signal location where two adjacent track circuits end. 328

cutter (audio and electroacoustics) (mechanical recording head). An electromechanical transducer that transforms an electric input into a mechanical output, that is typified by mechanical motions that may be inscribed into a recording medium by a cutting stylus. *See:* **phonograph pickup.** 176

cutter compensation (numerically controlled machines). Displacement, normal to the cutter path, to adjust for the difference between actual and programmed cutter radii or diameters. 224, 207
cutting down (metal or electrodeposit) (electroplating). Polishing for the purpose of removing roughness or irregularities. *See:* **electroplating.** 328
cutting stylus (electroacoustics). A recording stylus with a sharpened tip that, by removing material, cuts a groove into the recording medium. *See:* **phonograph pickup.** 176
CVD. *See:* **chemical vapor deposition.**
CW radar. A radar which transmits a continuous-wave transmitted signal. 13
cybernetics. *See:* **system science.**
cycle (1)(test pattern language). A complete operation, such as writing or reading, performed by a memory. Also called a period. 463
(2). (A) An interval of space or time in which one set of events or phenomena is completed. (B) Any set of operations that is repeated regularly in the same sequence. The operations may be subject to variations on each repetition. 77
(3) (pulse terms). The complete range of states or magnitudes through which a periodic waveform or a periodic feature passes before repeating itself identically. 254
(4) (data transmission). (A) An interval of space or time in which one set of events or phenomena is completed. Any set of operations that is related regularly in the same sequence. The operations may be subject to variations on each each repetition. (B) The complete series of values of a periodic quantity that occurs during a period. *Note:* It is one complete set of positive and negative values of an alternating current. 59
(5) (electric installations on shipboard)(power and distribution transformer). The complete series of values of a periodic quantity which occurs during a period. (It is one complete set of positive and negative values of an alternating current.) 3,53
cycle of operation (storage cell or battery). The discharge and subsequent recharge of the cell or battery to restore the initial conditions. *See:* **charge.** 328
cycle time. *See:* **pulse period.**
cyclic duration factor (rotating machinery). The ratio between the period of loading including starting and electric braking, and the duration of the duty cycle, expressed as a percentage. *See:* **asynchronous machine.** 63
cyclic irregularity (rotating machinery). The periodic fluctuation of speed caused by irregularity of the prime-mover torque. *See:* **asynchronous machine.** 63
cyclic shift (digital computers). An operation that produces a word whose characters are obtained by a cyclic permutation of the characters of a given word. 235
cyclometer register. A set of 4 or 5 wheels numbered from zero to 9 inclusive on their edges, and so enclosed and connected by gearing that the register

reading appears as a series of adjacent digits. *See:* **watthour meter.** 328

cyclotron. A device for accelerating positively charged particles (for example, protons, deuterons, etcetera) to high energies. The particles in an evacuated tank are guided in spiral paths by a static magnetic field while they are accelerated many times by an electric field of mixed frequency. 190

cyclotron frequency (1). The frequency at which an electron traverses an orbit in a steady uniform magnetic field and zero electric field. *Note:* It is given by the product of the electron charge and the magnetic flux density, divided by 2π times the electron mass. **(2) (radio wave propagation).** Gyrofrequency. *See:* **gyrofrequency; magnetrons.** 125

cyclotron-frequency magnetron oscillations. Those oscillations whose frequency is substantially the cyclotron frequency. 125

cyclotron, frequency-modulated. *See:* **frequency-modulated cyclotron.**

cylindrical antenna. An antenna having the shape of a circular cylinder. *Note:* This term usually describes a linear antenna. 111

cylindrical array (antennas). A two-dimensional array of elements whose corresponding points lie on a cylindrical surface. 111

cylindrical dipole (antennas). A dipole, all of whose transverse cross sections are the same, the shape being circular. 111

cylindrical reflector (antennas). A reflector which is a portion of a cylindrical surface. *Note:* The cylindrical surface is usually parabolic, although other shapes may be employed. 111

cylindrical-rotor generator (solid-iron cylindrical-rotor generator) (turbine generator*). An alternating-current generator driven by a high-speed turbine (usually steam) and having an exciting winding embedded in a cylindrical steel rotor. *See:* **generating station.**

*Deprecated. Favor steam-turbine or gas-turbine generator as distinguished from hydraulic turbine generator. 63

cylindrical wave (radio wave propagation). A wave whose equiphase surfaces form a family of coaxial or confocal cylinders. 146

cytac (navigation aid terms). The designation of loran C in an earlier stage of development. 526

D

D* (pronounced "D-star") (fiber optics). A figure of merit often used to characterize detector performance, defined as the reciprocal of noise equivalent power (NEP), normalized to unit area and unit bandwidth.

$$D* = [\sqrt{A(\Delta f)}]/NEP$$

where A is the area of the photosensitive region of the detector and (Δf)is the effective noise bandwidth. *Syn:* **specific detectivity.** *See:* **detectivity; noise equivalent power.** 433

daisy chain (FASTBUS acquisition and control). A backplane connection between adjacent module stations that allows information to flow between modules independent of the FASTBUS protocol. 480

damped filter (harmonic control and reactive compensation of static power converters). A filter generally consisting of combinations of capacitors, reactors, and resistors which have been selected in such a wayas to present a low impedance over a broad range of frequencies. The filter usually has a relatively low Q(X/R). 533

damped harmonic system (1) (linear system with one degree of freedom). A physical system in which the internal forces, when the system is in motion, can be represented by the terms of a linear differential equation with constant coefficients, the order of the equation being higher than the first. Example: The differential equation of a damped system is often of the form

$$M\,\frac{d^2x}{dt^2} + F\,\frac{dx}{dt} + Sx = f(t)$$

where M, F, and S are positive constants of the system: x is the dependent variable of the system (displacement in mechanics, quantity of electricity, etcetera): and $f(t)$ is the applied force. Examples: A tuning fork is a damped harmonic system in which M represents a mass: F a coefficient of damping: S a coefficient of restitution: and x a displacement. Also an electric circuit containing constant inductance, resistance, and capacitance is a damped harmonic system, in which case M represents the self-inductance of the circuit: F the resistance of the circuit: S the reciprocal of the capacitance: and x the charge that has passed through a cross section of the circuit.

(2) (critical damping). The name given to that special case of damping that is the boundary between underdamping and overdamping. A damped harmonic system is critically damped if F^2 , $4Ms$. *See:* **network analysis.**

(3) (overdamping) (aperiodic damping). The special case of a damping in which the free oscillation does not change sign. A damped harmonic system is overdamped if $F^2 > 4$ MS. *See:* **network analysis.**

(4) (underdamping) (periodic damping). The special case of damping in which the free oscillation changes sign at least once. A damped harmonic system is underdamped if $F^2 < 4$ MS. 210

(5) (underdamped). Damped insufficiently to prevent oscillation of the output following an abrupt input stimulus. *Note:* In an underdamped linear second-order system, the roots of the characteristic equation have complex values. *See:* **control; critical damping; damping; control system, feedback.**   219, 206

(6) (anode) (anticathode) (X-ray tube). An electrode, or part of an electrode, on which a beam of electrons is focused and from which X-rays are emitted. *See:* **electrode (electron tube); radar.**   190

damper bar. *See:* **amortisseur bar.**

damper segment (rotating machinery). One portion of a short-circuiting end ring (of an amortisseur winding) that can be separated into parts for mounting or removal without access to a shaft end. *See:* **rotor (rotating machinery).**   63

damper winding (amortisseur winding) (rotating machinery). A winding consisting of a number of conducting bars short-circuited at the ends by conducting rings or plates and distributed on the field poles of a synchronous machine to suppress pulsating changes in magnitude or position of the magnetic field linking the poles.   63

damping (1)(seismic design of substations). A dynamic property which indicates the ability of a structure to dissipate energy. *Note:* The phenomenon of damping is represented by a quantity called the damping ratio which is a percent of critical damping. After being forced to deflect and allowed to vibrate freely, structures with zero damping vibrate indefinitely. Structures with critical damping return to their static or neutral position in the shortest time without oscillation.   465

(2)(seismic qualification of Class 1E equipment for nuclear power generating stations). An energy dissipation mechanism that reduces the amplification and broadens the vibratory response in the region of resonance. Damping is usually expressed as a percentage of critical damping. Critical damping is defined as the least amount of viscous damping that causes a single-degree-of-freedom system to return to its original position without oscillation after initial disturbance.   581

(3) (noun). The temporal decay of the amplitude of a free oscillation of a system, associated with energy loss from the system.

(4) (adjective). Pertaining to or productive of damping. *Note:* The damping of many physical systems is conveniently approximated by a viscous damping coefficient in a second-order linear differential equation (or a quadratic factor in a transfer function). In this case the system is said to be critically damped when the time response to an abrupt stimulus is as fast as possible without overshoot: underdamped (oscillatory) when overshoot occurs: overdamped (aperiodic) when response is slower than critical. The roots of the quadratic are, respectively, real and equal: complex, and real and unequal. *See:* **control system, feedback.**

(5) (relative underdamped system). A number expressing the quotient of the actual damping of a second-order linear system or element by its critical damping. *Note:* For any system whose transfer function includes a quadratic factor $s^2 + 2z\omega_n s + \omega_n^2$, relative damping is the value of z, since $z + 1$ for

critical damping. Such a factor has a root $- \sigma + j\omega$ in the complex s plane, from which $z = \sigma / \omega_n = \sigma / (\sigma^2 + \omega^2)^{1/2}$. *See:* **control system, feedback.**

(6) (Coulomb). That due to Coulomb friction. *See:* **control system, feedback.**

(7) (instrument). Term applied to the performance of an instrument to denote the manner in which the pointer settles to its steady indication after a change in the value of the measured quantity. Two general classes of damped motion are distinguished as follows: (A) periodic (underdamped) in which the pointer oscillates about the final position before coming to rest, and (B) aperiodic (overdamped) in which the pointer comes to rest without overshooting the rest position. The point of change between periodic and aperiodic damping is called critical damping. *Note:* An instrument is considered for practical purposes to be critically damped when overshoot is present but does not exceed an amount equal to one-half the rated accuracy of the instrument. *See:* **accuracy rating (instrument); moving element (instrument).**   206

damping amortisseur. An amortisseur the primary function of which is to oppose rotation or pulsation of the magnetic field with respect to the pole shoes.   328

damping fluid (gyro, accelerometer) (inertial sensor). A fluid which provides vsicous damping forces or torques to the inertial sensing element.   46

damping magnet. A permanent magnet so arranged in conjunction with a movable conductor such as a sector or disk as to produce a torque (or force) tending to oppose any relative motion between them. *See:* **moving element (instrument).**   328

damping torque (synchronous machine). The torque produced, such as by action of the amortisseur winding, that opposes the relative rotation, or changes in magnitude, of the magnetic field with respect to the rotor poles.   63

damping torque coefficient (synchronous machine). A proportionality constant that, when multiplied by the angular velocity of the rotor poles with respect to the magnetic field, for specified operating conditions, results in the damping torque.   63

damp location (National Electrical Code). Partially protected locations under canopies, marquees, roofed open porches, and like locations, and interior locations subject to moderate degrees of moisture, such as some basements, some barns, and some cold-storage warehouses.   256

danger or hazard buoy (navigation aid terms). Classified as: obstruction, wreck, telegraph, cable, fish net, dredging. *See:* **buoy.**   526

dark adaptation (illuminating engineering). The process by which the retina becomes adapted to a luminance less than about 0.034 cd/m^2 , (2.2×10^{-5}cd/in 2), (0.01 fL).   167

dark current (1) (diode-type camera tube). The current that flows in the output lead of the target in the absence of any external irradiation. Units: amperes (A).   380

(2) (fiber optics). The external current that, under

specified biasing conditions, flows in a photosensitive detector when there is no incident radiation.   433
(3) (photoelectric device) (electron device). The current flowing in the absence of irradiation. *See:* **photoelectric effect; dark current; electrode; dark-current pulses (phototubes).**   244,117, 125
dark-current pulses (phototubes). Dark-current excursions that can be resolved by the system employing the phototube. *See:* **phototube.**   335
darkening (electroplating). The production by chemical action, usually oxidation, of a dark colored film (usually a sulfide) on a metal surface. *See:* **electroplating.**   328
dark pulses. Pulses observed at the output electrode when the photomultiplier is operated in total darkness. These pulses are due primarily to electrons originating at the photocathode.   117
dark space, cathode. *See:* **cathode dark space.**
dark space, Crookes. *See:* **cathode dark space.**
dark-trace screen (cathode-ray tube). A screen giving a spot darker than the remainder of the surface. *See:* **cathode-ray tubes.**   244
dark-trace tube (skiatron) (1) (electronic navigation). A cathode-ray tube having a special screen that changes color but does not necessarily luminesce under electron impact, showing, for example, a dark trace on a bright background. *See:* **cathode-ray tubes; navigation.**   244
(2) (radar). A type of cathode-ray tube having a bright face, on which signals are displayed as dark traces or dark blips; sometimes used as a storage tube or long-persistence display because the dark traces remain on the screen until erased by heat or electron bombardment. This device is now obsolete. *Syn:* **skiatron.**   13
D'Arsonval current (solenoid current) (medical electronics). The current of intermittent and isolated trains of heavily damped oscillations of high frequency, high voltage, and relatively low amperage. *See:* Note under **D'Arsonvalization.**   192
D'Arsonvalization (medical electronics). The therapeutic use of intermittent and isolated trains of heavily damped oscillations of high frequency, high voltage, and relatively low amperage. *Note:* This term is deprecated because it was initially ill-defined and because the technique is not of contemporary interest.   192
data (1)(programmable digital computer systems in safety systems of nuclear power generating stations). A representation of facts, concepts, or instructions in a formalized manner suitable for communication, interpretation or processing by a programmable digital computer.   554
(2)(station control and data acquisition)(supervisory control, data acquisition, and automatic control). Any representation of a digital or analog quantity to which meaning has been assigned.   403,570
(3)(test pattern language). The binary information that is stored in or read out of a memory array.   463
(4)(software). A representation of facts, concepts, or

instructions in a formalized manner suitable for communication, interpretation, or processing by human or automatic means. *See:* **computer data; control data; error data; instructions; reliability data; software experience data.**   434
data abstraction (software). The result of extracting and retaining only the essential characteristic properties of data by defining specific data types and their associated functional characteristics, thus separating and hiding the representation details. *See:* **data; data types; information hiding.**   434
data acquisition (station control and data acquisition)-(supervisory control, data acquisition, and automatic control). The collection of data.   570,403
data acquisition system (supervisory control, data acquisition, and automatic control)(station control and data acquisition). A centralized system which receives data from one or more remote points. A telemetering system. Data may be transported by either analog or digital telemetering. *See:* **telemetering.**   570,403
database (software). (1) A set of data, part or the whole of another set of data, and consisting of at least one file that is sufficient for a given purpose or for a given data processing system. (2) A collection of data fundamental to a system. (3) A collection of data fundamental to an enterprise. *See:* **data; file; system.**   434
data channels (test pattern language). All memory devices have one or more (up to 16) independent data inputs or outputs. Each of these is called a data channel.   463
data characteristic (software unit testing). An inherent, possibly accidental trait, quality, or property of data (for example, arrival rates, formats, value ranges, or relationships between field values).   519
data communication equipment. The equipment that provides the functions required to establish, maintain, and terminate a connection, as well as the signal conversion, and coding required for communication between data terminal equipment and data circuit.   12
data communications (data transmission). The movement of encoded information by means of communications techniques.   59
data cycle (FASTBUS acquisition and control). The portion of a FASTBUS operation in which a master either sends data to or receives data from an attached slave. It begins with the master causing a data sync transition and terminates with the master receiving a data acknowledge transition from the slave.   480
data dictionary (software). (1) A collection of the names of all data items used in a software system, together with relevant properties of those items, for example, length of data item, representation, etcetera. (2) A set of definitions of data flows, data elements, files, data bases, and processes referred to in a leveled data flow diagram set. *See:* **data; data base; data flow diagram; file; processes; software system.**   434
data flow chart. *See:* **data flow diagram.**
data flow diagram (software). A graphic representation of a system, showing data sources, data sinks, storage,

and processes performed on data as nodes, and logical flow of data as links between the nodes. *Syn:* **data flow graph; data flow chart.** *See:* **data; data flow chart; data flow graph; node; processes; system.**   434

data flow graph. *See:* **data flow diagram.**

data-hold (data processing). A device that converts a sampled function into a function of a continuous variable. The output between sampling instants is determined by an extrapolation rule or formula from a set of past inputs.   198

data input (D) (semiconductor memory). The inputs whose states determine the data to be written into the memory.   441

data input/output (DQ) (semiconductor memory). The ports that function as data input during write operations and as data output during read operations.   441

data interchange format (DIF)(computer applications). A standarized data file format allowing data interchange between software packages on personal computers. For example, data interchange between an electronic spread sheet and a word processorcould be accomplished by converting the spread sheet data to data interchange format, then to the format required for the word processor.   571

data link (1)(logical link control). An assembly of two or more terminal installations and the interconnecting communications channel operating according to a particular method that permits information to be exchanged; in this context the term 'terminal installation' does not included the data source and the data sink.   585

(2) (data communication). An assembly of data terminals and the interconnecting circuits operating according to a particular method that permits information to be exchanged between the terminals.   12

(3) (test, measurement and diagnostic equipment). Any information channel used for connecting data processing equipment to any input, output, display device, or other data processing equipment, usually at a remote location.   54

data link layer (logical link control). The conceptual layer of control or processing logic existing in the hierarchical structure of a station that is responsible for maintaining control of the data link. The data link layer functions provide an interface between the station higher layer logic and the data link. These functions include address/control field interpretation, channel access and command protocol data unit (PDU3)/response PDU generation, transmission and interpretation.   585

data logger (power-system communication). A system to measure a number of variables and make a written tabulation and.or record in a form suitable for computer input. *See:* **digital.**   59

data logging (1)(supervisory control, data acquisition, and automatic control). The recording of selected data on suitable media.   570

(2) (power switchgear). An arrangement for the alphanumerical representation of selected quantities on log sheets, papers, magnetic tape, or the like, by means

of an electric typewriter or other suitable devices.   103

(3) (station control and data acquisition). The recording of selected data on suitable media.   403

data-logging equipment. Equipment for numerical recording of selected quantities on log sheets or paper or magnetic tape or the like, by means of an electric typewriter or other suitable device.   103, 202

data output (Q) (semiconductor memory). The outputs whose states represent the data read from the memory.   441

data processing (DP)(1)(computer applications). The systematic performance of operations upon data, such as data manipulation, merging, sorting, and computing.   571

(2)(emergency and standby power). Pertaining to any operation or combination of operations on data.   512

data processing system (computer applications). A system, including computer systems and associated personnel, that performs input, processing, storage, output, and control functions to accomplish a sequence of operations on data.   571

data processor (emergency and standby power). Any device capable of being used to perform operations on data, for example, a desk calculator, tape recorder, analog computer, or digital computer.   512

data rate (supervisory control, data acquisition, and automatic control)(station control and data acquisition). The rate at which a data path (for example, channel) carries data, measured in bits per second (b/s).   570,403

data reconstruction (date processing). The conversion of a signal defined on a discrete-time argument to one defined on a continuous-time argument.   198

data reduction. The transformation of raw data into a more useful form, for example, smoothing to reduce noise.   54

data set (data transmission). A modem serving as a conversion element and interface between a data machine and communication facilities. *See:* **modem.**   59

data sink (data transmission). The equipment which accepts data signals after transmission.   59

data source (data transmission). The equipment which supplies data signals that enter into a data link.   59

data space (FASTBUS acquisition and control). That address space which devices may have which is recommended for use in data operations. There are few constraint applied to data space uses. *See:* **control and status register (CSR) space.**   480

data stabilization (vehicle-borne navigation systems) (navigation aid terms). The stabilization of the output signals with respect to a selected reference invariant with vehicle orientation.   526

data structure (software). A formalized representation of the ordering and accessibility relationships among data items, without regard to their actual storage configuration. *See:* **data.**   434

data terminal (data transmission). A device which

modulates or demodulates data between one input-output device and a data transmission link, or both.
59

data terminal equipment (data communication). The equipment comprising the data source, the data sink, or both.
12

data transmission (data link). The sending of data from one place to another or from one part of a system to another.
59

data type (software). A class of data characterized by the members of the class and the operations that can be applied to them, for example, integer, real, logical. *See:* **data.**
434

dataway (CAMAC). *See:* **CAMAC dataway.**

davit (power line maintenance). An assembly attached to a support or assembled on a structure to provide a rigging point for rope blocks, chains, or hoists so as to manipulate various pieces of apparatus. The davit is a rigid assembly and does not swivel.
458

daylight factor (illuminating engineering). A measure of daylight illuminance at a point on a given plane expressed as a ratio of the illuminance on the given plane at that point to the simultaneous exterior illuminance on a horizontal plane from the whole of an unobstructed sky of assumed or known luminance distribution. Direct sunlight is excluded from both interior and exterior values of illuminance.
167

day-night sound level L_{dn} (audible noise measurement). The day-night sound level L_{dn} rating of a noise is intended to improve upon the L_{eq} rating (see energy-equivalent sound level, L_{eq}) by adding a compensation to nighttime noise intrusion, because people are more sensitive to such nocturnal intrusions. By definition, an upward adjustment of 10 dB(A)'S applied to all sounds occuring between 10 p.m. and 7 a.m. (Local regulations, not based upon L_{dn}, may utilize other nighttime adjustments.) The L_{dn} can be computed from weighted daytime L_d and nighttime L_n values as follows:

$$L_{dn} = 10 \log_{10} \frac{1}{24} \left[15 \text{ antilog } \frac{L_d}{10} + 9 \text{ antilog} \right.$$

$$\left. \left(\frac{L_n + 10}{10} \right) \right] \text{ dB} \qquad \text{(Eq 1)}$$

where

$L_d = L_{eq}$ for the 15 daytime hours

$L_n = L_{eq}$ for the 9 nighttime hours

462

dB. *See:* **decibel.**

dBm (data transmission). A unit for expression of power level in decibels with reference to a power of one milliwatt.
59

dBV. *See:* **voltage level.**

dc. *See:* **direct current.**

D cable. A two-conductor cable, each conductor having the shape of the capital letter D with insulation

between the conductors themselves and between conductors and sheath.
64

dc-ac transfer standards (metering). Instruments used to establish the equality of a root-mean-square (rms) current or voltage (or the average values of alternating power) with the corresponding steady-state direct-current (dc) quantity.
212

dc analog computer (analog computers). An analog computer in which computer variables are represented by the instantaneous values of voltages.
9

dc breakdown voltage (gas tube surge-protective device). The minimum slowly rising dc voltage that will cause breakdown or sparkover when applied across the terminals of an arrester.
370

dc holdover voltage (gas tube surge-protective device). The maximum dc voltage across the terminals of an arrester under which it may be expected to clear and to return to the high impedance state after the passage of a surge, under specified circuit conditions.
370

dc interface (terrestrial photovoltaic power systems). The connections between the array subsystem, the dc auxiliary power source, and the input of the power conditioning subsystem at the input terminals of the power conditioning system (PCS). *See:* **array control**
496

dc level. *See:* **baseline.**

DDA (computing systems). *See:* **digital differential analyzer.**

D dimension (motor). The standard designation of the distance from the centerline of the shaft to the plane through the mounting surface bottom of the feet, in National Electrical Manufacturers Association approved designations.
63

D-display* (radar). Similar to a C-display, but composed of a series of horizontal stripes representing successive elevation angles. Each stripe is a miniature B-display with compressed vertical scale. Horizontal position of a blip represents azimuth, the gross vertical

*Obsolete or rare.

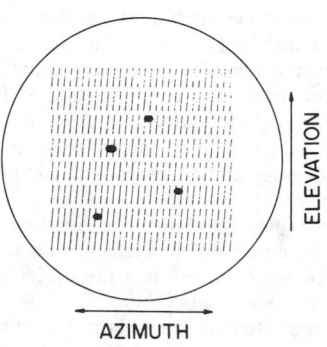

D-Display

scale (the stripe in which the blip appears) represents elevation, and vertical position within the stripe represents range. 13

DDM. *See:* **difference in depth of modulation.**

deactivate (signals and paths)(696 interface devices). To cause a signal to transition from its logically true (active) state to its logically false (inactive) state. Opposite of assert. 538

deactivated shutdown (electric generating unit reliability, availability, and productivity)(power system measurement). The state in which a unit is unavailable for service for an extended period of time because of its removal for economy or reasons not related to the equipment. Under this condition, a unit generally requires weeks of preparation to make it available.
567,432

deactivated shutdown hours (DSH)(electric generating unit reliability, availability, and productivity). The number of hours a unit was in a deactivated shutdown state. 567

deactivation (corrosion). The process of removing active constituents from a corroding liquid (as removal of oxygen from water). 205

deactivation date (electric generating unit reliability, availability, and productivity). The date a unit was placed into the deactivated shutdown state. 567

dead (power line maintenance). A circuit that has been de-energized means that the circuit has been disconnected from all intended electrical sources. However, it could be electrically charged through induction from energized circuits in proximity to it, particularly if the circuits are parallel. *See:* **de-energized.** 458

dead band (1)(supervisory control, data acquisition, and automatic control). The range through which an input can be varied without initiating response.
570

(2) (nuclear power generating station)station control and data acquisition). The range through which an input can be varied without initiating output response.
355,403

(3) (gyro, accelerometer) (inertial sensor). A region between the input limits within which variations in the input produce no detectable change in the output (the electrical analog of backlash). 46

dead band differential (electric pipe heating systems). The difference in degrees between the OFF and the ON stage of temperature controllers. 448

dead-band rating. The limit that the dead band will not exceed when the instrument is used under rated operating conditions. *See:* **accuracy rating (instrument).**
295

dead-break connector (1)(power and distribution transformer). A separable insulated connector designed to be separated and engaged on de-energized circuits only. 53

(2)(separable insulated connectors). A connector designed to be separated and engaged on de-energized circuits only. 454

dead-end (power line maintenance). The point at which mechanical force (primarily) and longitudinal strain is applied to a reliable support. *Syn:* **termination; anchor point; strain attachment.** 458

dead-end guy. An installation of line or anchor guys to hold the pole at the end of a line. *See:* **pole guy; tower.**
64

dead-end tower. A tower designed to withstand unbalanced pull from all of the conductors in one direction together with wind and vertical loads. *See:* **tower.**
64

dead front (1) (National Electrical Code). (A) Without live parts exposed to a person on the operating side of the equipment. (B) **(As applied to switches, circuit breakers, switchboards, and distribution panelboards.)** So designed, constructed and installed that no current-carrying parts are normally exposed on the front. 256

(2) (power and distribution transformer). So constructed that there are no exposed live parts on the front of the assembly. 53

dead-front mounting (of a switching device) (power switchgear). A method of mounting in which a protective barrier is interposed between all live parts and the operator, and all exposed operating parts are insulated or grounded. *Note:* The barrier is usually grounded metal. 103

dead-front switchboard (power switchgear). One which has no exposed live parts on the front. *Note:* The switchboard panel is normally grounded metal and provides a barrier between the operator and all live parts. 103

dead layer(1) (of a semiconductor detector)(charged-particle detectors). An inactive region (layer) in which the energy absorbed from the passage of monoenergetic charged particles does not significantly contribute to the resulting full energy peak. 119

(2)(of a semiconductor radiation conductor)(germanium gamma-ray detectors). A layer of a semiconductor detector in which no significant part of the energy lost by particles can contribute to the resulting signal. *Note:* The semiconductor detector may have an entrance and an exit dead layer. 528

dead layer thickness (of a semiconductor radiation detector)(X-ray energy spectrometers). The thickness of an inactive region (in the form of a layer) through which the incident radiation must pass to reach the sensitive volume. *Syn:* **window.** ,23471

dead leg (1)(electrical heating systems). A portion of a piping system, used to simulate the overall system conditions for control sensing. 476

(2)(electrical heat tracing for industrial applications). A segment of process piping that is not in the normal flow pattern. 523

dead man (overhead construction). *See:* **anchor log.**

deadman's handle (industrial control). A handle of controller or master switch that is designed to cause the controller to assume a preassigned operating condition if the force of the operator's hand on the handle is released. *See:* **electric controller.** 206

deadman's release (industrial control). The effect of that feature of a semiautomatic or nonautomatic control system that acts to cause the controlled apparatus to assume a preassigned operating condition if the operator becomes incapacitated. *See:* **control.**
206

dead-metal part (power and distribution transformer). A part, accessible or inaccessible, which is conductively connected to the grounded circuit under conditions of normal use of the equipment.   53

dead reckoning (DR)(navigation aid terms). The determining of the position of a vehicle at one time with respect to its position at a different time by the application of vectors representing courses and distances.   526

dead room (audio and electroacoustics) (acoustics). Those locations completely within the coverage area where the signal strength is below the level needed for reliable communication. *See:* **mobile communicating system.**   181

dead tank switching device (power switchgear). A switching device in which a vessel(s) at ground potential surrounds and contains the interrupter(s) and the insulating medium.   103

dead time (1)of a circuit breaker on a reclosing operation) (power switchgear). The interval between interruption in all poles on the opening stroke and reestablishment of the circuit on the reclosing stroke. *Notes:* (A)In breakers using arc-shunting resistors, the following intervals are recognized and the one referred to should be stated:

(1) Dead time from interruption on the primary arcing contacts to reestablishment through the primary arcing contacts.

(2) Dead time from interruption on the primary arcing contacts to reestablishment through the secondary arcing contacts.

(3) Dead time from interruption on the secondary arcing contacts to reestablishment on the primary arcing contacts.

(4) Dead time from interruption on the secondary arcing contacts to reestablishment on the secondary arcing contacts.

(B) The dead time of an arcing fault on a reclosing operation is not necessarily the same as the dead time of the circuit breakers involved, since the dead time of the fault is the interval during which the faulted conductor is de-energized from all terminals.   103

(2) **(navigation).** The time interval in an equipment's cycle of operating during which the equipment is prevented from providing normal response. For example, in a radar display, the portion of the interpulse interval which is not displayed; or, in secondary radar, the interval immediately following the transmission of a pulse relay during which the transponder is insensitive to interrogations.   13

(3) **(recovery time) (radiation counter).** The time from the start of a counted pulse until an observable succeeding pulse can occur.   96,125

dead time t_t (sodium iodide detector). The time after a triggering pulse during which the system is unable to retrigger.   423

dead-time correction (radiation counters). A correction to the observed counting rate to allow for the probability of the occurrence of events within the dead time of the system. *See:* **anticoindcidence (radiation counters).**   190

dead zone (industrial control). The period(s) in the operating cycle of a machine during which corrective functions cannot be initiated.   206

debug (1). To examine or test a procedure, routine, or equipment for the purpose of detecting and correcting errors.   194

(2) **(computing systems).** To detect, locate, and remove mistakes from a routine or malfunctions from a computer. *See:* **troubleshoot.**   255,77

debugging (1) (reliability). The operation of an equipment or complex item prior to use to detect and replace parts that are defective or expected to fail, and to correct errors in fabrication or assembly. *See:* **reliability.**   182,54

(2) **(software).** The process of locating, analyzing, and correcting suspected faults. *See:* **faults; testing.**   434

debugging model. *See:* **error model.**

Debye length (radio wave propagation). That distance in a plasma over which a free electron may move under its own kinetic energy before it is pulled back by the electrostatic restoring forces of the polarization cloud surrounding it. Over this distance a net charge density can exist in an ionized gas. The Debye length is given by

$$l_D = \left(\frac{\epsilon_0 k T_e}{e^2 N_e} \right)^{1/2}$$

where ϵ_0 is the permittivity of vacuum, k is Boltzmann's constant, e is the charge of the electron, T_e is the electron temperature, and N_e is the electron number density. *See:* **radio wave propagation.**   146

decade (radiation protection). Synonymous with power of ten.   399

decalescent point (induction heating usage). The temperature at which there is a sudden absorption of heat when metals are raised in temperature. *See:* **dielectric heating; induction heating; coupling; recalescent point.**   125

decay (storage tubes). A decrease in stored information by any cause other than erasing or writing. *Note:* Decay may be caused by an increase, a decrease, or a spreading of stored charge. *See:* **storage tube.**   174,125

decay characteristic (luminescent screen). *See:* **persistence characteristic.**

decay, dynamic (storage tubes). Decay caused by an action such as that of the reading beam, ion currents, field emission, or holding beam. *See:* **storage tube.**   174

decay, static (charge-storage tubes). Decay that is a function only of the target properties, such as lateral and transverse leakage. *See:* **charge-storage tube.**   174

decay time (storage tubes). The time interval during which the stored information decays to a stated fraction of its initial value. *Note:* Information may not decay exponentially. *See:* **storage tube.**   174,125

decay time constant (1)(charged-particle detectors).

The time for a true single-exponential waveform to decay to 36.79 percent of the original step height.

119

(2)(semiconductor radiation detector)germanium gamma-ray detectors)(X-ray energy spectrometers). The time for a true single-exponential waveform to decay to a value of $1/e$ of the original step height.

23,119,528,471

Decca (navigation aid terms). A radio navigation system transmitting on several related frequencies near 100 kHz (kilohertz), useful to about 200 nmi (nautical miles)(370 km [kilometers]) in which sets of hyperbolic lines of position are determined by comparison of the phase of (1) one reference continuous wave signal from a centrally located master with (2) each of several continuous wave signals from slave transmitters located in a star pattern, each about 70 nmi (130 km) from the master.

526

decelerating electrode (electron-beam tubes). An electrode the potential of which provides an electric field to decrease the velocity of the beam electrons.

117,125

decelerating relay (industrial control). A relay that functions automatically to maintain the armature current or voltage within limits, when decelerating from speeds above base speed, by controlling the excitation of the motor field. *See:* **relay.**   225,206

decelerating time (industrial control). The time in seconds for a change of speed from one specified speed to a lower specified speed while decelerating under specified conditions. *See:* **control system, feedback.**

206

deceleration (industrial control). *See:* **retardation.**

deceleration, programmed (numerically controlled machines). A controlled velocity decrease to a fixed percent of the programmed rate.   207

deceleration, timed (industrial control). A control function that decelerates the drive by automatically controlling the speed change as a function time. *See:* **control system, feedback.**   219,206

decentralized computer network (data communication). A computer network, where some of the computing power and network control functions are distributed over several network nodes. *See:* **centralized network.**   12

decibel (1)(power station noise control). Ten times the logarithm to base 10 of a ratio of two powers.

500

(2) (general). One-tenth of a bel, the number of decibels denoting the ratio of the two amounts of power being ten times the logarithm to the base 10 of this ratio. *Note:* The abbreviation dB is commonly used for the term decibel. With P_1 and P_2 designating two amounts of power and n the number of decibels denoting their ratio,

$$n=10 \log_{10}(P_1/P_2)\text{decibel}$$

When the conditions are such that ratios of currents or ratios of voltages (or analogous quantities in other fields) are the square roots of the corresponding power ratios, the number of decibels by which the corresponding powers differ is expressed by the following equations:

$$n=20 \log_{10}(I_1/I_2)\text{decibel}$$
$$n=20 \log_{10}(V_1/V_2)\text{decibel}$$

where I_1/I_2 and V_1/V_2 are the given current and voltage ratios, respectively. By extension, these relations between numbers of decibels and ratios of currents or voltages are sometimes applied where these ratios are not the square roots of the corresponding power ratios; to avoid confusion, such usage should be accompanied by a specific statement of this application. Such extensions of the term described should preferably be avoided.   111

(3) (automatic control). A logarithmetic scale unit relating a variable x (e.g., angular displacement) to a specified reference level x_0;dB $= 20 \log x/x_0$. *Note:* The relation is strictly applicable only where the ratio x/x_0 is the square root of the power ratio P/P_0, as is true for voltage or current ratios. The value dB $= 10 \log_{10} P/P_0$ originated in telephone engineering, and is approximately equivalent to the old "transmission unit."   56

(4) (excitation control systems). In control usage, a logarithmic scale unit relating a variable x to a specified reference level x_0; dB $= 20 \log x/x_0$. *Note:* The relation is strictly applicable only where the ratio x/x_o is the square root of the power ratio P/P_0, as is true for voltage or current ratios.   353

(5) (dB) (data transmission). One-tenth of a bel, the number of decibels denoting the ratio of the two amounts of power being ten times the logarithm to the base 10 of this ratio. *Note:* The abbreviation dB is commonly used for the term decibel. With P_1 and P_2 designating two amounts of power and n the number of decibels denoting their ratio.

$$n=10 \log_{10}(P_1/P_2) \text{ dB}$$

When the conditions are such that ratios of currents or ratios of voltages (or analogous quantities in other fields) are the square roots of the corresponding power ratios, the number of decibels by which the corresponding powers differ is expressed by the following equations:

$$n=20 \log_{10}(V_1/V_2)\text{dB}$$
$$n=20 \log_{10}(I_1/I_2)\text{dB}$$
$$n=20 \log_{10}(V_1/V_2)\text{dBl}$$

where $I_1/I_2, V_1/V_2 =$ given current By extension, these relations between numbers of decibels and ratios of current or voltages are sometimes applied where these ratios are not the square roots of the corresponding power ratios; to avoid confusion, such usage should be accompanied by a specific statement of this application. Such extensions of the term described should preferably be avoided.   59

decibel meter. An instrument for measuring electric power level in decibels above or below an arbitrary reference level. *See:* **instrument.**   328

decile, D_u (electromagnetic site survey). The ratio of the upper decile value (the value of x exceeded 10 percent of the time) of the random variable x to its median value, expressed in decibels.   457

decile, D_l (electromagnetic site survey). The ratio of the lower decile value (the value of x exceeded by 90 percent of the time) of the random variable x to its median value, expressed in decibels.   457

decilog (data transmission). A division of the logarithmic scale used for measuring the logarithm of the ratio of two values of any quantity. Its value is such that the number of decilogs is equal to 10 times the logarithm to the base 10 of the ratio. One decilog, therefore, corresponds to a ratio of $10^{0.1}$ (that is, 1.25892 +).   59

decimal (mathematics of computing). (1) Pertaining to a selection in which there are ten possible outcomes. (2) Pertaining to the numeration system with a radix of ten.   564

decimal code. A code in which each allowable position has one of 10 possible states. The conventional decimal number system is a decimal code.   244,207

decimal number system. *See:* **positional notation.**

decimal point. *See:* **point.**

decineper. One-tenth of a neper.   111

decision gate (navigation aid terms). A specified point near the lower end of an ILS (instrument landing system) approach at which a pilot must make a decision either to complete the landing or to execute a missed-approach procedure.   526

decision instruction (computing systems). An instruction that effects the selection of a branch of a program, for example, a conditional jump instruction.   255,77

decision support services (DSS)(computer applications). (1) The services provided by a decision support system. For example, software components for model building, forecasting, statistical analysis, ad hoc model interrogation,report generation, and graphics.   571

decision support system (DSS)(computer applications). A computer system that supports decision making by performing such functions as modeling, forecasting, and statistical analysis. *See:* **computer-aided management; management and information system.**   571

decision table (1) (ATLAS). A matrix-providing program branching which may be a complex function of a number of variables.   400
(2) **(software).** (A) A table of all contingencies that are to be considered in the description of a problem together with the actions to be taken for each set of contingencies. (B) A presentation in either matrix or tabular form of a set of conditions and their corresponding actions. *See:* **table.**   434

deck (computing systems). A collection of punched cards.   255,77

declared curve (rotating electric machinery). A characteristic curve of the machine type, as obtained by averaging the results of testing four to ten machines, of which at least two shall have had a type test.   424

declination rate of ON-state current (thyristor). Average rate of declination or fall of ON-state current measured from 50 percent I_F to 0.   445

declinometer (navigation aid terms). An instrument for measuring magnetic declination.   526

decode. To produce a single output signal from each combination of a group of input signals. *See:* **translate; matrix, encode.**   235

decoder (electronic computation and control). A matrix of logic elements that selects one or more output channels according to the combination of input signals present.   255,77

decomposition potential (decomposition voltage). The minimum potential (excluding *IR* drop) at which an electrochemical process can take place continuously at an appreciable rate. *See:* **electrochemistry.**   328

decoupling. The reduction of coupling. *See:* **coupling.**   328

decrement (test pattern language). The action of reducing the arithmetic value of a counter by one.   463

decrement factor (safety in ac substation grounding). An adjustment factor used in conjunction with the initial symmetrical ground fault current parameter in safety-oriented grounding calculations. It allows us to obtain a root-mean-square (rms) equivalent of the symmetrical current wave for a given fault duration, accounting for the effect of initial direct-current (dc) offset and its attenuation during the fault.   563

dectra (navigation aid terms). An adaptation of the Decca low frequency (lf) radio navigation system in which two pairs of continuous wave (cw) transmitters are oriented so that the center lines of both pairs are along and at opposite ends of the same great circle path, to provide course guidance along and adjacent to the great circle path. Distance along track may be indicated by synchronized signals from one transmitter from each pair.   526

dedicated word processing (computer applications). Word processing performed on a system used exclusively for that purpose. Contrast with clustered word processing, shared-logic word processing, shared-resource word processing, stand-alone word processing.   571

de-emphasis (1) (data transmission). The use of an amplitude-frequency characteristic complementary to that used for pre-emphasis earlier in the system.   59
(2) **(post emphasis) (post equalization).** The use of an amplitude-frequency characteristic complementary to that used for pre-emphasis earlier in the system. *See:* **pre-emphasis.**   328

de-emphasis network. A network inserted in a system in order to restore the pre-emphasized frequency spectrum to its original form.   111

de-energize (relay). To disconnect the relay from its power source. 259

de-energized (1)(NESC)(power line maintenance). Free from any electrical connection to a source of potential difference and from electric charge; not having a potential different from that of earth. *Note:* The term is used only with reference to current-carrying parts which are sometimes energized (alive). *Syn:* **dead.** 494,458

(2) (conductor stringing equipment). Free from any electrical connection to a source of potential difference and from electric charge; not having a potential different from that of the ground. The term is used only with reference to current-carrying parts that are sometimes alive (energized). To state that a circuit has been deenergized means that the circuit has been disconnected from all intended electrical sources. However, it could be electrically charged through induction from energized circuits in proximity to it, particularly if the circuits are parallel. *Syn:* **dead.** 431

(3) (transmission and distribution) (dead). Free from any electrical connection to a source of potential difference and from electric charge; not having a potential different from that of the earth. *Note:* The term is used only with reference to current-carrying parts which are sometimes energized (alive). 262

deep-bar rotor. A squirrel-cage induction-motor rotor having a winding that is narrow and deep giving the effect of varying secondary resistance, large at standstill and decreasing as the speed rises. *See:* **rotor (rotating machinery).** 63

deep space (communication satellite). Space at distances from the earth approximately equal to or greater than the distance between the earth and the moon. 74

deep space instrumentation facility (DSIF) (communication satellite). A ground network of worldwide communication stations (earth terminals) maintained for providing communications to and from lunar and inter-planetary spacecraft and deep space probes. Each earth terminal utilizes large antennas, low-noise receiving systems and high-power transmitters. 83

deepwell pump. An electrically-driven pump located at the low point in the mine to discharge the water accumulation to the surface. 328

de-excitation (excitation systems for synchronous machines). The removal of an excitation of a synchronous machine, main exciter, or pilot exciter. *Note:* De-excitation may be accomplished by various means, such as a dc field breaker, alternating-current (ac) supply breaker, static switches, phase-back control of controlled rectifiers, or a combination of these. 507

defeater (industrial control). *See:* **interlocking deactivating means.**

defect (solar cells). A localized deviation of any type from the regular structure of the atomic lattice of a single crystal.*See:* **fault.** 113

defined pulse width (semiconductor radiation detectors). The time elapsed between the first and final crossings of the defined zero level for the maximum rated output pulse amplitude. 23

defined reference pulse waveform (pulse measurement). A reference pulse waveform which is defined without reference to any practical or derived pulse waveform. Typically, a defined reference pulse waveform is an ideal pulse waveform. 15

defined zero (semiconductor radiation detectors). An arbitrarily chosen voltage level at the amplifier output resolvable from zero by the measuring apparatus. 23

definite-minimum-time relay (power switchgear). An inverse-time relay in which the operating time becomes substantially constant at high values of input. 127, 103

definite purpose circuit breakers (capacitance current switching). A definite purpose circuit breaker is one that is designed specifically for capacitance current switching. 130

definite-purpose controller (industrial control). Any controller having ratings, operating characteristics, or mechanical construction for use under service conditions other than usual or for use on a definite type of application. *See:* **electric controller.** 206

definite-purpose motor. Any motor designated, listed, and offered in standard ratings with standard operating characteristics or mechanical construction for use under service conditions other than usual or for use on a particular type of application. *Note:* Examples: crane, elevator, and oil-burner motors. *See:* **asynchronous machine.** 328

definite time (relays). A qualifying term indicating that there is purposely introduced a delay action, which delay remains substantially constant regardless of the magnitude of the quantity that causes the action. *See:* **relay.** 103

definite-time acceleration (electric drive) (industrial control). A system of control in which acceleration proceeds on a definite-time schedule. *See:* **electric drive.** 206

definite-time delay (power switchgear). A qualifying term indicating that there is purposely introduced a delay in action, which delay remains substantially constant regardless of the magnitude of the quantity that causes the action. 103

definite-time relay (power switchgear). A relay in which the operating time is substantially constant regardless of the magnitude of the input quantity. *See:* **relay.** 103,127

definition (facsimile). Distinctness or clarity of detail or outline in a record sheet, or other reproduction. 12

definition phase. *See:* **requirements phase.**

definitions of classes (insulation systems of synchronous machines). Insulation systems are those which by service experience or accepted comparative tests with service proven systems can be shown to be capable of continuous operation with the limiting observable temperature rise or hottest spot total temperature as specified in the appropriate American National Standard, C50.13-1977 or C50.14-1977. Insulation systems of synchronous machines shall be classified as Class A, Class B, Class F, or Class H.   298

deflecting electrode (electron-beam tubes). An electrode the potential of which provides an electric field to produce deflection of an electron beam. *See:* **electrode (electron tube).**   125

deflecting force (direct-current recording instrument). At any part of the scale (particularly full scale), the force for that position, measured at the marking device, and produced by the electrical quantity to be measured, acting through the mechanism. *See:* **accuracy rating (instrument).**   328

deflecting voltage (cathode-ray tube). Voltage applied between the deflector plates to create the deflecting electric field. *See:* **cathode-ray tubes.**   244

deflecting yoke. An assembly of one or more coils that provide a magnetic field to produce deflection of an electron beam.   190

deflection axis, horizontal (oscilloscope). The horizontal trace obtained when there is a horizontal deflection signal but no vertical deflection signal.   106

deflection axis, vertical (oscilloscope). The vertical trace obtained when there is a vertical deflection signal and no horizontal deflection signal.   106

deflection blanking (oscilloscopes). Blanking by means of a deflection structure in the cathode-ray tube electron gun that traps the electron beam inside the gun to extinguish the spot, permitting blanking during retrace and between sweeps regardless of intensity setting. *See:* **oscillograph.**   185

deflection center (electron-beam tube). The intersection of the forward projection of the electron path prior to deflection and backward projection of the electron path in the field-free space after deflection.   125

deflection coefficient. *See:* **deflection factor.**

deflection defocusing (cathode-ray tube). A fault of a cathode-ray tube characterized by the enlargement, usually nonuniform, of the deflected spot which becomes progressively greater as the deflection is increased. *See:* **cathode-ray tube.**   244

deflection factor (1) (inverse sensitivity) (general). The reciprocal of sensitivity. *Note:* It is, for example, often used to describe the performance of a galvanometer by expressing this in microvolts per millimeter (or per division) and for a mirror galvanomter at a specified scale distance, usually 1 meter. *See:* **accuracy rating (instrument).**   125
(2) (oscilloscopes). The ratio of the input signal amplitude to the resultant displacement of the indicating spot, for example, volts per division. *See:* **oscillograph.**   125
(3) (spectrum analyzer). The ratio of the input signal amplitude to the resultant output indication. The ratio may be in terms of volts root-mean-square (rms) per division, describes decibels (dB) per division, watts per division, or any other specified factor.   390

deflection plane (cathode-ray tubes). A plane perpendicular to the tube axis containing the deflection center.   190

deflection polarity (oscilloscopes). The relation between the polarity of the applied signal and the direction of the resultant displacement of the indicating spot. *Note:* Conventionally a positive-going voltage causes upward deflection or deflection from left to right. *See:* **oscillograph.**   185

deflection sensibility (oscilloscopes). The number of trace widths per volt that can be simultaneously resolved anywhere within the quality area. *See:* **oscillograph.**   185

deflection sensitivity (magnetic-deflection cathode-ray tube and yoke assembly). The quotient of the spot displacement by the change in deflecting-coil current.   125

deflection yoke (television). An assembly of one or more coils whose magnetic field deflects an electron beam.   125, 18

deflection-yoke pull-back (cathode-ray tubes). (1) color. The distance between the maximum possible forward position of the yoke and the position of the yoke to obtain optimum color purity. **(2) monochrome.** The maximum distance the yoke can be moved along the tube axis without producing neck shadow.   125

deflector (1)(metal-oxide surge arresters for ac power circuits)(surge arrester). A means for directing the flow of the gas discharge from the vent of the arrester.   583,430
(2) (acousto-optics). A device which directs a light beam to an angular position in space upon application of an acoustic frequency.   72

deflector plates. *See:* **deflecting electrode.**

defruiter (radar). Equipment which deletes random nonsynchronous unintentional returns in a beacon system.   13

degassing (electron tube) (rectifier). The process of driving out and exhausting occluded and remanent gases within the vacuum tank or tube, anodes, cathode, etcetera, that are not removed by evacuation alone. *See:* **rectification.**   328

degaussing (navigation aid terms). Neutralization of the strength of the magnetic field of a vessel.   526

degaussing coil. A single conductor or a multiple-conductor cable so disposed that passage of current through it will neutralize or bias the magnetic polarity of a ship or portion of a ship. *Note:* Continuous application of current, with adjustment to suit changes of position or heading of the ship, is required to maintain a degaussed condition.   328

degaussing generator. An electric generator provided for the purpose of supplying current to a degaussing coil or coils.   328

degeneracy (resonant device). The condition where two or more modes have the same resonance frequency. *See:* **waveguide.**   179

degenerate gas. A gas formed by a system of particles whose concentration is very great, with the result of the Maxwell-Boltzmann law does not apply. *Example:* An electronic gas made up of free electrons in the interior of the crystal lattice of a conductor. *See:* **discharge (gas).**   244

degenerate mode (waveguide). In a uniform waveguide, one of a set of modes having the same exponential variation along the direction of the guide, but having different configurations in the transverse plane. In a cavity, one of a set of modes having the same natural frequency. 267

degenerate parametric amplifier (nonlinear, active, and nonreciprocal waveguide components). An inverting parametric device for which the two signal frequencies are identical and equal to one-half the frequency of the pump. *Note:* This exact but restrictive definition is often relaxed to include cases where the signals occupy frequency bands that overlap. *See:* **parametric device.** 530

degeneration. Same as negative feedback. *See:* **feedback.** 111

degradation failure (reliability). See: **failure, degradation.**

degraded backup (station control and data acquisition)supervisory control, data acquisition, and automatic control). A backup capability that does not perform all of the functions of the primary system. 403,570

degraded voltage condition (Class 1E power systems and equipment). A voltage deviation above or below normal to a level which, if sustained, could result in unacceptable performance of, or damage to, the connected loads or their control circuitry. 443

degree Celsius (metric practice). It is equal to the kelvin and is used in place of the kelvin for expressing Celsius temperature (symbol t) defined by the equation $t = T - T_0$ where T is the thermodynamic temperature and $T_0 = 273.15\ K$ by definition. 21

degree of asymmetry (of a current at any time) (power switchgear). The ratio of the direct-current component to the peak value of the symmetrical component determined from the envelope of the current wave at that time. *Note:* This value is 100 percent when the direct-current component equals the peak value of the symmetrical component. 103

degree of coherence (fiber optics). A measure of the coherence of a light source; the magnitude of the degree of coherence is equal to the visibility, V, of the fringes of a two-beam interference experiment, where

$$V = \frac{I_{max} - I_{min}}{I_{max} + I_{min}}$$

I_{max} is the intensity at a maximum of the interference pattern, and I_{min} is the intensity at a minimum. *Note:* Light is considered highly coherent when the degree of coherence exceeds 0.88, partially coherent for values less than 0.88, and incoherent for "very small" values. *See:* **coherence area; coherence length; coherent; interference.** 433

degree of distortion (data transmission). At the digital interface for binary signals, a measure of the time displacement of the transitions between signal states from their ideal instants. The degree of distortion is generally expressed as a percentage of the unit interval. 59

degree of freedom (1) (mesh basis). *See:* **nullity.** 46

(2) (node basis) (inertial sensor) (gyro). An allowable mode of angular motion of the spin axis with respect to the case. The number of degrees-of-freedom is the number of orthogonal axes about which the spin axis is free to rotate. 46

degree of gross start-stop distortion (data transmission). The degree of start-stop distortion determined using the unit interval which corresponds to the actual mean modulation rate of the signal involved. 59

degree of individual distortion (of a particular signal transition) (data transmission). The ratio to the unit interval of the displacement, expressed algebraically, of this transition from its ideal instant. This displacement is considered positive when the transition occurs after its ideal instant (late). 59

degree of isochronous distortion (data transmission). (1) The ratio to the unit interval of the maximum measured difference, irrespective of sign, between the actual and the theoretical intervals separating any two transitions of modulation (or of restitution), these transitions not being necessarily consecutive. (2) The algebraical difference between the highest and lowest value of individual distortion affecting the transitions of an isochronous modulation. (This difference is independent of the choice of the reference ideal instant.) The degree of distortion (of an isochronous modulation or restitution) is usually expressed as a percentage. 59

degree of longitudinal balance (measuring longitudinal balance of telephone equipment operating in the voice band). The ratio of the disturbing longitudinal voltage V_s and the resulting metallic voltage V_m of the network under test expressed in decibels, as follows:

$$\text{longitudinal balance} = 20 \log \mid V_s\ /\ V_m \mid (\text{dB})^1$$

where V_s and V_m are of the same frequency. *Note:* Here and throughout ANSI/IEEE Std 455-1985, log is assumed to mean log to the base 10. 529

degree of start-stop distortion (data transmission). The ratio to the unit interval of the maximum measured difference (irrespective of sign) between the actual interval and the theoretical interval (the appropriate integral multiple of unit intervals) separating any transition from the start transition preceding it. 59

degrees of freedom (1) (mesh basis). *See:* **nullity.**
(2) (node basis). *See:* **rank**
(3) (gyro). An allowable mode of angular motion of the spin axis with respect to the case. The number of degrees-of-freedom is the number of orthogonal axes about which the spin axis is free to rotate. All gyros are either single-degree-of-freedom (SDF) or two-degree-of-freedom (TDF). 46

deionization time (gas tube). The time required for the grid to regain control after anode-current interruption. *Note:* To be exact the dionization time of a gas tube should be presented as a family of curves relating such factors as condensed-mercury temperature, anode and grid currents, anode and grid voltages, and regulation of the rid current. *See:* **gas tube.**   125

deionizing grid (gas tubes). A grid accelerating deionization in its vicinity in a gas-filled valve or tube, and forming a screen between two regions within the envelope.   244

delay (1)(protection and coordination of industrial and commercial power systems). This term is usually applied to the opening time of a fuse when in excess of one cycle, where the time may vary considerably between types and makes and still be within established standards. This word, in itself, has no specific meaning other than in manufacturers' claims unless published standards specify delay characteristics. *See:* **time delay.**

(2) (phase) (data transmission). The time interval by which a pulse is time retarded with respect to a reference time.   59

(3) (time delay) (data transmission). (General). The time interval between the manifestation of a signal at one point and the manifestation or detection of the same signal at another point. *Notes:* (A) Generally, the term **time delay** is used to describe a process whereby an output signal has the same form as an input signal causing it, but is delayed in time; that is, the amplification of all frequency components of the output are related by a single constant to those of corresponding input frequency components but each output component lags behind the corresponding input component by a phase angle proportional to the frequency of the component. (B) Transport delay is synonymous with time delay but usually is reserved for applications that involve the flow of material.   59

delay, absolute (loran). *See:* **absolute delay.**

delay angle (α) (thyristor converter). The time, expressed in degrees (1 cycle of the ac waveform = 360 °), by the starting point of commutation is delayed by phase control in relation to rectifier operationing without phase control, including possible inherent delay angle.   121

delay angle, inherent (α) (thyristor converter). The delay angle which occurs in some connections (for example, 12-pulse connections) in certain operating conditions even where no phase control is applied.   121

delay circuit (pulse techniques). A circuit that produces an output signal that is delayed intentionally with respect to the input signal. *See:* **pulse.**   117

delay coincidence circuit (pulse techniques). A coincidence circuit that is actuated by two pulses, one of which is delayed by a specified time interval with respect to the other. *See:* **pulse.**   117

delay dispersion (dispersive delay line). The change in phase delay over a specified operating frequency range.   81

delay distortion (1)(data transmission). (A) Phase delay distortion (also called phase distortion) which is eitherdeparture from flatness in the phase delay of a circuit, or system over the frequency range required for transmission or the effect of such departure on a transmitted signal, or (B) Envelope delay distortion, which is either departure from flatness in the envelope delay of a circuit or system over the frequency range required for transmission, or the effect the effect of such departure on a transmitted signal.   59

(2) (facsimile). *See:* **envelope delay distortion.**

delayed application (railway practice). The application of the brakes by the automatic train control equipment after the lapse of a predetermined interval of time following its initiation by the roadway apparatus. *See:* **automatic train control.**   328

delayed overcurrent trip. *See:* **delayed release (trip); overcurrent-release (trip).**

delayed plan-position indicator (radar). A plan-position indicator in which the initiation of the time base is delayed.   13

delayed release (trip) (power switchgear). A release with intentional delay introduced between the instant when the activating quantity reaches the release setting and the instant when the release operates.   103

delayed sweep (oscilloscopes). (1) A sweep that has been delayed either by a predetermined period or by a period determined by an additional independent variable. (2) A mode of operation of a sweep, as defined above. *See:* **radar.**   185

delayed test. A service test of a battery made after a specified period of time, which is usually made for comparison with an initial test to determine shelf depreciation. *See:* **battery (primary or secondary).**   328

delay electric blasting cap. An electric blasting cap with a delay element between the priming and detonating composition to permit firing of explosive charges in sequence with but one application of the electric current. *See:* **blasting unit.**   328

delay, envelope. *See:* **envelope delay.**

delay equalizer (data transmission). A corrective network which is designed to make the phase delay or envelope delay of a circuit, or system, substantially constant over a desired frequency range.   59

delaying (operations on a pulse) (pulse terms). A process in which a pulse is delayed in time by active circuitry or by propagation.   254

delaying sweep (oscilloscopes). A sweep used to delay another sweep. *See:* **delayed sweep; oscillograph.**   185

delay line (1) (data transmission). (A) Originally a device utilizing wave propagation for producing a time delay of a signal. (B) Commonly, any real or artificial transmission line or equivalent device designed to introduce delay. (C) A sequential logic element or device with one input channel in which the output channel state at a given instant **t** is the same as the input-channel state at the instant **t-n**, that is, the input sequence undergoes a delay of **n** units. There may be additional taps yielding output channels with smaller values of **n**.   59

(2) (scintillation counting). Commonly, a real or artificial transmission line or equivalent device designed to introduce delay.   117

(3) (digital computer). A sequential logic element or device with one input channel in which the output-channel state at a given instant t is the same as the input-channel state at the instant t-n, that is, the input sequence undergoes a delay of n units. There may be additional taps yielding output channels with smaller values of n. *See:* **acoustic delay line; electromagnetic delay line; magnetic delay line; sonic delay line; pulse.** 235

(4) (sonics and ultrasonics). A device which operates over some defined range of electrical and environmental conditions as a linear passive circuit element. The transfer characteristic has a modulus and argument (phase) which can be constant or a function of frequency. 81

delay line, digital. A delay line designed specifically to accept digital (video) electrical signals. The signals are specified usually as bipolar, RZ, or NRZ. The definitions are based on the output signal being a doublet generated by an input step function. 81

delay line, dispersive. A delay line which has a transfer characteristic with a constant modulus and an argument (phase) which is a nonlinear function of frequency. The phase characteristic of devices of common interest is a quadratic function of frequency, but in general may be represented by higher order polynominals and/or other nonlinear functions. 81

delay-line memory. *See:* **delay-line storage.**

delay line, nondispersive. A delay line which nominally has constant delay over a specified frequency band. The argument (phase) of the transfer function is a linear function of frequency. 81

delay-line storage (delay-line memory) (electronic computation). A storage or memory device consisting of a delay line and means for regenerating and reinserting information into the delay line. 54

delay, phase. The ratio of the total phase shift (radians) experienced by a sinusoidal signal in transmission through a system or transducer, to the frequency (radians/second) of the signal. *Note:* The unit of phase delay is the second. 239

delay pickoff (oscilloscopes). A means of providing an output signal when a ramp has reached an amplitude corresponding to a certain length of time (delay interval) since the start of the ramp. The output signal may be in the form of a pulse, a gate, or simply amplification of that part of the ramp following the pickoff time. 185

delay, pulse (transducer). The interval of time between a specified point on the input pulse and a specified point on the related output pulse. *Note:* (1) This is a general term which applies to the pulse delay in any transducer, such as receiver, transmitter, amplifier, oscillator, etcetera. (2) Specifications may require illustrations. *See:* **transducer.** 254,59

delay pulsing (telephone switching system). A method of pulsing control and trunk integrity check wherein the sender delays the sending of the address pulses until it receives from the far end an off-hook signal (terminating register not yet attached), followed by a steady on-hook signal (terminating register attached). 55

delay relay. A relay having an assured time interval between energization and pickup or between de-energization and dropout. *See:* **relay.** 341

delay, signal. *see:* **signal delay.**

delay slope (dispersive delay line). The ratio of the delay dispersion to the dispersive bandwidth. 81

delay switching system (telephone switching systems). A switching system in which a call is permitted to wait until a path becomes available. 55

delay time (1) (railway practice). The period or interval after the initiation of an automatic train-control application by the roadway apparatus and before the application of the brakes becomes effective. *See:* **automatic train control.** 328

(2) (nondispersive delay line). The transit time of the envelope of an RF tone burst. 81

(3) (thyristor). *See:* **gate controlled delay time (thyristor).**

delimiter A character that provides punctuation in an ATLAS statement. 400

delivery (software). (1) The point in the software development cycle at which a product is released to its intended user for operational use. (2) The point in the software development cycle at which a product is accepted by its intended user. *See:* **operational; software development cycle.** 434

Dellinger effect. *See:* **radio fadeout.**

DeLoach measurement (nonlinear, active, and nonreciprocal waveguide components). A method used in the characterization of varactor diodes that involves the measurement of device resonance parameters in a reduced-height waveguide transmission line. Parameters such as cutoff frequency, Q, capacitance, and inductance can be realized from this measurement. 530

delta. The difference between a partial-select output of a magnetic cell in a ONE state and a partial-select output of the same cell in ZERO state. *See:* **coincident-current selection.** 331

delta B (ΔB). *See:* **delta induction.**

delta connection (power and distribution transformer). So connected that the windings of a three-phase transformer (or the windings for the same rated voltage of single-phase transformers associated in a three-phase bank) are connected in series to for a closed circuit. 53

delta-function light source (scintillation counting). A light source whose rise time, fall time, and FWHM (full width at half maximum) are not more than one third of the corresponding parameters of the output pulse of the photomultiplier. 117

delta induction (toroidal magnetic amplifier cores). The change in induction (flux density) when a core is in a cyclically magnetized condition. *Syn:* **delta flux density.** 170

delta, minimum (power supplies). A qualifier, often appended to a percentage specification to describe that specification when the parameter in question is a variable, and particularly when that variable may approach zero. The qualifier is often known as the minimum delta V, or minimum delta I, as the case may be. 186

delta network (1) (circuits and systems). A network or that part of a network that consists of three branches connected among three terminals. 67
(2) (electromagnetic compatibility). An artificial mains network of specified symmetric and asymmetric impedance used for two-wire mains operation and comprising resistors connected in delta formation between the two conductors, and each conductor and earth. *See:* **electromagnetic compatibility.** 197

delta N_J (ΔNJ)(parametric mode)(nonlinear, active, and nonreciprocal waveguide components). A figure of merit for parametric amplifier varactors that relates to capacitance nonlinearity. 530

delta tan delta (Δ tan δ). The increment in the dielectric dissipation factor (tan) of the insulation measured at two designated voltages. *Note:* When the values of power factors or dissipation factors are in the 0-0·10 range (see dielectric dissipation factor), the value of delta tan delta may be used as the equivalent of the power-factor tip-up value. *See:* **power factor tip-up.** 22

demagnetization (magnetic tape pulse recorders for electricity meters). The removal of the residual magnetization by application of a demagnetizing alternating current (ac) field of sufficient initial magnitude. 551

demand (1)(power operations). Load integrated over a specified interval of time. *See:* **billing demand; rachet demand clause; coincident demand; contract demand; demand factor; instantaneous demand; integrated demand; demand interval; maximum demand; native system demand; noncoincident demand; rachet demand.** 516
(2)(installation or system). The load at the receiving terminals averaged over a specified interval of time. *Note:* Demand is expressed in kilowatts, kilovolt-amperes, kilovars, amperes, or other suitable units. *See:* **alternating-current distribution.** 212

demand, billing (electric power utilization). The demand which is used to determine the demand charges in accordance with the provisions of a rate schedule or contract. 112

demand charge (power operations). That portion of the charge for electric service based upon a customer's demand. 516

demand clause, ratchet (electric power utilization). A clause in a rate schedule which provides that maximum past or present demands be taken into account to establish billings for previous or subsequent periods. *See:* **alternating-current distribution.** 112

demand, coincident (electric power utilization). Any demand that occurs simultaneously with any other demand; also the sum of any set of coincident demands. *See:* **alternating-current distribution.** 112

demand constant (pulse receiver) (metering). The value of the measured quantity for each received pulse, divided by the demand interval, expressed in kilowatts per pulse, kilovars per pulse, or other suitable units. 212

demand, contract (electric power utilization). The demand that the supplier of electric service agrees to have available for delivery. *See:* **alternating-current distribution.** 112

demand deviation (metering). The difference between the indicated or recorded demand and the true demand, expressed as a percentage of the full-scale value of the demand meter or demand register. *See:* **pulse-count deviation.** 212

demand factor (power operations). The ratio of the maximum coincident demand of a system, or part of a system, to the total connected load of the system, or part of the system, under consideration. 516

demand failure rate (reliability data for pumps and drivers, valve actuators, and valves). The probability (per demand) of failure that a component will fail to operate upon demand when required to start, change state, or function. 502

demand, instantaneous (electric power utilization). The load at any instant. 112

demand, integrated (electric power utilization). The demand integrated over a specified period. 112

demand interval (1)(power operations). The period of time during which the electric energy flow is integrated in determining demand. 516
(2) (demand meter or register). The length of the interval of time upon which the demand measurement is based. *Note:* The demand interval of a block-interval demand meter is a specific period of time such as 15, 30, or 60 minutes during which the electric energy flow is average. The demand interval of a lagged-demand meter is the time required to indicate 90 percent of the full value of a constant load suddenly applied. Some meters record the highest instantaneous load. *See:* **demand meter; electricity meter (meter).** 212,200

demand-interval deviation (metering). The difference between the measured demand interval and the specified demand interval, expressed as a percentage of the specified demand interval. 212

demand, maximum. *See:* **maximum demand.** 212

demand meter (metering). A metering device that indicates or records the demand, maximum demand, or both. *Note:* Since demand involves both an electrical factor and a time factor, mechanisms responsive to each of these factors are required, as well as an indicating or recording mechanism. These mechanisms may be separate from or structurally combined with one another. *See:* **contact-making clock demand meter; indicating demand meter; integrating demand meter; lagged demand meter; time characteristic demand meter; timing deviation demand meter.** 212

demand, native system (electric power utilization). The net 60 minute clock-hour peak integrated demand within the system less interruptible loads. *See:* **alternating-current distribution.** 112

demand, noncoincident (electric power utilization). The sum of the individual maximum demands regardless of time of occurrence within a specified period. *See:* **alternating-current distribution.** 112

demand register (metering). A mechanism, for use with an integrating electricity meter, that indicates maximum demand and also registers electric energy (or other integrated quantity). *See:* **cumulative demand register; multiple-pointer form demand register; single-pointer form demand register.** 212

demand-totalizing relay. A device designed to receive and totalize electric pulses from two or more sources for transmission to a demand meter or to another relay. *See:* **demand meter.** 212

demarcation potential (demarcation current*) (current of injury*) (electrobiology). *See:* **injury potential.**
*Deprecated

demarcation strip (data transmission). The terminals at which the telephone company's service ends and the customer's equipment is connected. 59

demineralization. The process of removing dissolved minerals (usually by chemical means). 221

demodulation (1) (information theory). A modulation process wherein a wave resulting from previous modulation is employed to derive a wave having substantially the characteristics of the original modulating wave. *Note:* The term is sometimes used to describe the action of a frequency converter or mixer, but this practice is deprecated except in the case of shifting a single-sideband signal to baseband. 415

(2) (data transmission). A modulation process wherein a wave resulting from previous modulation is employed to derive a wave substantial to the characteristics of the original modulating wave. *Note:* The term is sometimes used to describe the action of a frequency converter or mixer, but this practice is deprecated. 59

demodulator. A device to effect the process of demodulation. *See:* **demodulation.** 242

demonstration (1)(safety systems equipment in nuclear power generating stations). The provision of evidence to support the conclusion derived from assumed premises. 535

(2) (class 1E static battery chargers and inverters). A course of reasoning showing that a certain result is a consequence of assumed premises; an explanation or illustration, as in teaching by use of examples. 408

denormalized number (binary floating-point arithmetic). A nonzero floating-point number whose exponent has a reserved value, usually the format's minimum, and whose explicit or implicit leading significand bit is zero. 469

densitometer (illuminating engineering). A photometer for measuring the optical density (common logarithm of the reciprocal of the transmittance or reflectance) of materials. 167

density (1) (facsimile). A measure of the light-transmitting or reflecting properties of an area. *Notes:* (A) It is expressed by the common logarithm of the ratio of incident to transmitted or reflected light flux. (B) There are many types of density that will usually have different numerical values for a given material; for example, diffuse density, double diffuse density,

specular density. The relevant type of density depends upon the geometry of the optical system in which the material is used. *See:* **scanning.** 12

(2) (electron or ion beam). The density of the electron or ion current of the beam at any given point. 244

(3) (computing systems). *See:* **packing density.**

(4). *See:* **optical density.**

density coefficient. *See:* **environmental coefficient.**

density-modulated tube (space-charge-control tube) (microwave tubes). Microwave tubes or valves characterized by the density modulation of the electron stream by a gating electrode. *Note:* The electron stream is collected on those electrodes that form a part of the microwave circuit, principally the anode. These electrodes are often small compared to operating wavelength so that for this reason space-charge-control tubes or valves are often not considered to be microwave tubes even though they are used at microwave frequencies. 190

density modulation (electron beam). The process whereby a desired time variation in density is impressed on the electrons of a beam. *See:* **velocity-modulated tube.** 244

density-tapered array antenna. *See:* **space-tapered array antenna.**

denuder. That portion of a mercury cell in which the metal is separated from the mercury. 328

dependability (of a relay or relay system) (power switchgear). The facet of reliability that relates to the degree of certainty that a relay or relay system will operate correctly. 127, 103

dependable capability (power operations). The maximum generation, expressed in kilowatt-hours per hour (kWh/h) which a generating unit, station, power source, or system can be depended upon to supply on the basis of average operating conditions. 516

dependable capacity (electric generating unit reliability, availability, and productivity). The maximum capacity, modified for ambient limitations for a specified period of time, such as a month or a season. 567

dependency notation (graphic symbols for logic functions). A means of obtaining simplified symbols for complex elements by denoting the relationships between inputs, outputs, or inputs and outputs, without actually showing all the elements and interconnections involved. *See:* **ANSI/IEEE Std 91-1984, Section 4.** 451

dependent biaxial test (seismic testing of relays). The horizontal and the vertical acceleration components are derived from a single-input signal. 392

dependent contact. A contacting member designed to complete any one of two or three circuits, depending on whether a two- or a three-position device is considered. 328

dependent manual operation (power switchgear). An operation solely by means of directly applied manual energy, such that the speed and force of the operation are dependent upon the action of the attendant. 103

dependent node (network analysis). A node having one or more incoming branches. 282

dependent power operation (power switchgear). An operation by means of energy other than manual, where the completion of the operation is dependent upon the continuity of the power supply (to solenoids, electric or pneumatic motors, etcetera).   103

deperm. To remove, as far as practicable, the permanent magnetic characteristic of a ship's hull by powerful external demangetizing coils.   328

depletion (metal-nitride-oxide field-effect transistor). The state of the silicon surface in the insulated-gate field-effect transistor (IGFET) structure when a gate voltage of such polarity has been applied that all majority carriers have been repelled. The space charge region so formed is depleted of all mobile majority carriers.   386

depletion layer (region)(in a semiconductor)(X-ray energy spectrometers). A region in which the charge-carrier charge density is insufficient to neutralize the net fixed charge density of donors and acceptors. In a diode-type semiconductor radiation detector the depletion region is the sensitive region of the device.   471

depletion mode transistor (metal-nitride-oxide field-effect transistor). An insulated-gate field-effect transistor (IGFET) where the channel connecting source and drain is a preexisting thin layer of the same conductivity type as the source and drain.   386

depletion region (in a semiconductor)(1)(charged-particle conductors). A region in which the charge-carrier charge density is insufficient to neutralize the net fixed charge density of ionized donors and acceptors. In a diode-type semiconductor radiation detector the depletion region is the sensitive region of the device.   119,528

(2)(germanium gamma-ray detectors). A region in which the mobile charge-carrier charge density is insufficient to neutralize the net fixed charge density of donors and acceptors. In a diode-type semiconductor radiation detector the depletion region is the sensitive region of the device.   528

(3) (semiconductor radiation detectors). A region in which the charge-carrier charge density is insufficient to neutralize the net fixed charge density of donors and acceptors. In a diode-type semiconductor radiation detector the depletion region is the sensitive region of the device.   23

depletion voltage (of a semiconductor radiation detector)(charged-particle detectors)(germanium gamma-ray detectors). The voltage at which a junction detector becomes fully depleted.   119, 528

depolarization (1) (electrochemistry). A decrease in the polarization of an electrode at a specified current density. *See:* **electrochemistry.**   328

(2) (medical electronic biology). A reduction of the voltage between two sides of a membrane or interface below an initial value. *See:* **contact potential; electrochemistry.**   192

(3) (antenna). The conversion of power from a reference polarization into the cross polarization.   111

depolarization field (primary ferroelectric terms). A self-generated electric field that opposes the spon-

taneous polarization. In a crystal of finite dimensions there will be a discontinuity of P_s at the crystal surfaces, which gives rise to a bound polarization charge $(Q=P_s a)$ of surface density P_s. This charge gives rise to an electric field called the depolarizing field, which opposes the spontaneous polarization. In real materials the depolarizing field is neutralized by the flow of free charge through the crystal or from the environment. For electrically insulating crystals in an insulating environment, this process of neutralization may be very slow.   497

depolarization front (medical electronics). The border of a wave of electric depolarization, traversing an excitable tissue that has appreciable width and thickness as well as length. *See:* **contact potential.**   192

depolarizer (1). A substance or a means that produces depolarization.   328

(2) (primary cell). A cathodic depolarizer that is adjacent to or a part of the positive electrode. *See:* **electrochemistry; electrolytic cell.**   328

depolarizing mix (primary cell). A mixture containing a depolarizer and a material to improve conductivity. *See:* **electrolytic cell.**   328

deposit attack (deposition corrosion). Pitting corrosion resulting from deposits on a metallic surface.   205

deposited-carbon resistor. A resistor containing a thin coating of carbon deposited on a supporting material.   341

deposition corrosion. *See:* **deposit attack.**

depot maintenance. *See:* **maintenance, depot.**

depth control (cable plowing). The means used to maintain a predetermined plowing depth.   52

depth-finder (navigation aid terms). An instrument for determining the depth of water, particularly an echo sounder.   526

depth of current penetration (induction-heating usage). The thickness of a layer extending inward from the surface of a conductor which has the same resistance to direct current as the conductor as a whole has to alternating current of a given frequency. *Note:* About 87 percent of the heating energy of an alternating current is dissipated in the so-called **depth of penetration.** *See:* **dielectric heating; induction heating.**   14,114

depth of heating (dielectric-heating usage). The depth below the surface of a material in which effective dielectric heating can be confined when the applicator electrodes are applied adjacent to one surface only. *See:* **dielectric heating.**   14,114

depth of modulation (amplitude, frequency, and pulse modulation). The ratio of the maximum minus minimum light intensity to the sum of the maximum and minimum light intensity, namely: $m = (I_{max} - I_{min}) / (I_{max} + I_{min})$. This applies to either the diffracted or the zero order.   72

depth of penetration (induction-heating usage). *See:* **depth of current penetration.**

depth of velocity modulation (electron beams). The ratio of (1) the amplitude of a stated frequency component of the varying velocity of an electron beam, to (2) the average beam velocity. 190,125

derated generation (DG)(electric generating unit reliability, availability, and productivity). The generation that was not available due to unit deratings.DG = equivalent unit derated hours • maximum capacity = EUNDH • MC 567

derating (reliability). The intentional reduction of stress/strength ratio in the application of an item, usually for the purpose of reducing the occurrence of stress-related failures. *See:* **reliability.** 182

derivative control action (electric power systems). *See:* **control action, derivative.**

derivative time (control systems) (speed governing of hydraulic turbines). The derivative time, T_n of a derivative element is also the derivative gain, G_n. The derivative gain is the ratio of the element's percent output to the time derivative of the element's percent input (input slope with respect to time). 8

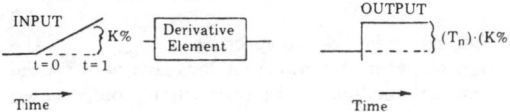

derived envelope (loran C)(navigation aid terms). The waveform equivalent to the summation of the video envelope of the rf (radio frequency) received pulse and the negative of its derivative, in proper proportion; the resulting envelope has a zero crossing at a standard point (for example, 25 μs) from the pulse beginning, serving as an accurate reference point for envelope time-difference measurements, and as a gating point in rejecting the latter part of the received pulse which may be contaminated by skywave transmissions. 526

derived pulse (loran C)(navigation aid terms). A pulse derived by summing the received rf (radio frequency) pulse and an oppositely phased rf pulse so that it has an envelope which is the derivative of the received rf pulseenvelope; the resultant envelope has a zero point and an rf phase reversal at a standard interval (for example, 25 μs) from the pulse beginning and it serves as an accurate reference for cycle time-difference measurements and as a gating point in rejecting the latter part of the received pulse which may be contaminated by sky-wave transmissions. 526

derived reference pulse waveform (pulse measurement). A reference pulse waveform which is derived by a specified procedure or algorithm from the pulse waveform which is being analyzed in a pulse measurement process. 15

derrick. An apparatus consisting of a mast or equivalent members held at the top by guys or braces, with or without a boom, for use with a hoisting mechanism and operating ropes. *See:* **elevators.** 328

descending node (communication satellite). The point of the line of nodes that the satellite passes through as the satellite travels from above to below the equatorial plane. 74

descriptive adjectives (pulse terms). (1) major (minor). Having or pertaining to greater (lesser) importance, magnitude, time, extent, or the like, than another similar feature(s). (2) ideal. Of or pertaining to perfection in, or existing as a perfect exemplar or, a waveform or a feature. (3) reference. Of or pertaining to a time, magnitude, waveform, feature, or the like which is used for comparison with, or evaluation of, other times, magnitudes, waveforms, features, or the like. A reference entity may, or may not, be an ideal entity. 254

design (1) (software). (A) The process of defining the software architecture, components, modules, interfaces, test approach, and data for a software system to satisfy specified requirements. (B) The results of the design process. *See:* **component; data; interface; requirement; software architecture; system.** 434

(2) (pulse terms). *See:* **logic design.** 254

design *A* motor. An integral-horsepower polyphase squirrel-cage induction motor designed for full-voltage starting with normal values of locked-rotor torque and breakdown torque, and with locked-rotor current higher than that specified for design *B, C,* and *D* motors. 63

design analysis (software). (1) The evaluation of a design to determine correctness with respect to stated requirements, conformance to design standards, system efficiency, and other criteria. (2) The evaluation of alternative design approaches. *See:* **correctness; design; preliminary design; requirement; system efficiency.** 434

design analyzer (software). An automated design tool that accepts information about a program's design and produces such outputs as module hierarchy diagrams, graphical representations of control and data structure, and lists of access data blocks. *See:* **automated design tool; data block; data structure; module hierarchy; design.** 434

designated person (NESC). A qualified person designated to perform specific duties under the conditions existing. *Syn:* **designated employee.** 494

designated representative (nuclear power quality assurance). An individual or organization authorized by the purchaser to perform functions in the procurement process. 417

designation (radar). Selection of a particular target and transmission of its particular coordinates from some external source to a radar, usually to initiate tracking. 13

design basis earthquake (DBE) (nuclear power generating stations). That earthquake producing the maximum vibratory ground motion that the nuclear power generating station is designed to withstand without functional impairment to those features necessary to shut down the reactor, maintain the station in a safe condition, and prevent undue risk to the health and safety of the public. 13

design basis event conditions (nuclear power generating stations). Conditions calculated to occur as a result of the design basis events. 31,104

design basis events (DBE)(1)(design and installation

of cable systems for Class 1E circuits in nuclear power generating stations). Postulated events specified by the safety analysis for the station to establish the acceptable performance requirements of the structures and systems. 536

(2)(electric penetration assemblies). Postulated events used in the design to establish the acceptable performance requirements of the structures, systems, and components. 493

(3) (diesel-generator unit) (standby power supplies). Postulated events used in the design to establish the performance requirments of the structures and systems. 99

(4)(safety systems equipment in nuclear power generating stations)valve actuators)(Class 1E motor control ceneters). Postulated events (specified by the safety analysis of the station) used in the design to establish the acceptable performance requirements of the structures and systems. 535,492,440

(5) (single-failure criterion). (A) Postulated (abnormal) events (specified by the safety analysis of the station) used in the design to establish performance requirements of structures, systems, and components.120,26 (B) Postulated abnormal events used in the design to establish the acceptable performance requirements of the structures and systems, and components. 109

(6) (class 1E static battery chargers and inverters). Postulated events, specified by the safety analysis of the station, used in the design to establish the acceptable performance requirements of the structures and systems. 408

(7) (criteria for safety systems). Postulated abnormal events used in the design to establish the acceptable performance requirements of the structures, systems, and components. 387

(8) (class 1E power systems). Postulated abnormal events used in the design to establish the acceptable performance requirements of the structures, systems, and components. 102

(9) (criteria for independence of class 1E equipment and circuits). Postulated abnormal events specified by the safety analysis of the station used in the design to establish the acceptable performance requirements of the structures and systems. 131

design *B* motor. An integral-horsepower polyphase squirrel-cage induction motor, designed for full-voltage starting, with normal locked-rotor and breakdown torque and with locked-rotor current not exceeding specified values. 63

design *C* motor. An integral-horsepower polyphase squirrel-cage induction motor, designed for full-voltage starting with high locked-rotor torque and with locked-rotor current not exceeding specified values. 63

design current (glow lamp). The value of current flow through the lamp upon which rated-life values are based. 283

design *D* motor. An integral-horsepower polyphase squirrel-cage induction motor with rated-load slip of at least 5 percent and designed for full-voltage starting with locked-rotor torque at least 275 percent of rated-load torque and with locked-rotor currents not exceeding specified values. 63

design earthquake (seismic design of substations). The greatest earthquake postulated during the life of the substation for which the user wishes the substation to remain operational. 465

designed availability (nuclear power generating stations). The probability that an item will be operable when needed as determined through the design analyses. 31,75

design element (Ada as a program design language). A basic component or building block in a design. 518

design entity (software design descriptions). An element (component) of a design that is structurally and functionally distinct from other elements and that is separately named and referenced. 568

design input (nuclear power quality assurance). Those criteria, parameters, bases, or other design requirements upon which detailed final design is based. 417

design inspection. See: inspection. 434

design language (software). A language with special constructs and, sometimes, verification protocols used to develop, analyze and document a design. See: design; document; verification. 434

design letters. Terminology established by the National Electrical Manufacturers Association to describe a standard range of characteristics. The characteristics covered are slip at rated load, locked-rotor and breakdown torque, and locked-rotor current. 63

design level (software test documentation). The design decomposition of the software item (for example, subsystem, program, or module). 436

design life (1)(nuclear power generating station). The time during which satisfactory performance can be expected for a specific set of service conditions. Note: The life may be specified in calendar time. However, operating time, number of operating cycles or other performance interval, as appropriate may be used to determine the time. 120

(2)(power generating station)(cable systems). The time during which satisfactory station performance can be expected for a specific set of operating conditions. 477

(3)(safety systems equipment in nuclear power generating stations)(valve actuators). The time during which satisfactory performance can be expected for a specific set of service conditions. Note: The time may be specified in real time, operating time, number of operating cycles or other performance interval, as appropriate. 535,492

(4) (class 1E static battery chargers and inverters). The time during which satisfactory performance can be expected for a specific set of service conditions, based upon component selection and applications. 408

(5) (class 1E motor control centers). The time during which satisfactory performance can be expected for a specific set of service conditions. (See IEEE Std 323-1974.) 440

design limits. Design aspects of the instrument in terms of certain limiting conditions to which the instrument may be subjected without permanent physical damage or impairment of operating characteristics. *See:* **instrument.**   295

design L motor. An integral-horsepower single-phase motor, designed for full-voltage starting with locked-rotor current not exceeding specified values, which are higher than those for design *M* motors.   63

design load (diesel-generator unit). That combination of electric loads, having the most severe power demand characteristic, which is provided with electric energy from a diesel-generator unit for the operation of engineered safety features and other systems required during and following shutdown of the reactor   99

design methodology (software). A systematic approach to creating a design, consisting of the ordered application of a specific collection of tools, techniques, and guidelines. *See:* **design; tools.**   434

design M motor. An integral-horsepower single-phase motor, designed for full-voltage starting with locked-rotor current not exceeding specified values, which are lower than those for design *L* motors.   63

design N motor. A fractional-horsepower single-phase motor, designed for full-voltage starting with locked-rotor current not exceeding specified values, which are lower than those for design *O* motors.   63

design O motor. A fractional-horsepower single-phase motor, designed for full-voltage starting with locked-rotor current not exceeding specified values, which are higher than those for design *N* motors.   63

design output (nuclear power quality assurance). Documents, such as drawings, specifications, and other documents, defining technical requirements of structures, systems, and components.   417

design phase (software)(software verification and validation plans). The period of time in the software life cycle during which the designs for architecture, software components, interfaces, and data are created, documented, and verified to satisfy requirements. *See:* **architecture; component; data; design; interface; requirement; software life cycle.**   434,511

design process (nuclear power quality assurance). Technical and management processes that commence with identification of design input and that lead to and include the issuance of design output documents.   417

design qualification (safety systems equipment in nuclear power generating stations). The generation and maintenance of evidence to demonstrate that equipment can perform within its specification requirements. *Note:* In the context of IEEE Std 627-1980, design qualification is synonymous with equipment qualification or qualification. Normal production testing and preoperational testing performed after installation and acceptance of the equipment is outside the scope of this definition and standard.   535

design requirement (software). Any requirement that impacts or constrains the design of a software system or software system component, for example, function-al requirements, physical requirements, performance requirements, software development standards, software quality assurance standards. *See:* **component; design; functional requirement; performance requirement; requirement; requirements specification; software quality assurance; software system; physical requirement.**   434

design review (software). (1) A formal meeting at which the preliminary or detailed design of a system is presented to the user, customer, or other interested parties for comment and approval. (2) The formal review of an existing or proposed design for the purpose of detection and remedy of design deficiencies which could affect fitness-for-use and environmental aspects of the product, process or service, or for identification of potential improvements of performance, safety, and economic aspects, or both. *See:* **design; detailed design; performance; system.**   434

design service conditions (electric penetration assemblies). The service conditions used as the basis for ratings and for the design qualification of electric penetration assemblies.   493

design specification (software). A specification that documents the design of a system or system component, for example, a software configuration item. Typical contents include system or component algorithms, control logic, data structures, data set-use information, input/output formats, and interface descriptions. *Syn:* **product specification (DOD usage).** *See:* **algorithm; data; data structure; design; document; interface; requirements specification; software configuration item; specification; system.**   434

design tests (1)(electric penetration assemblies). Tests performed to verify that an electric penetration assembly meets design requirements.   493

(2)(metal-enclosed bus and calculating losses in isolated-phase bus). Those tests made to determine the adequacy of a particular type, style, or model of metal-enclosed bus or its component parts to meet its assigned ratings and to operate satisfactorily under normal service conditions or under special conditions if specified. *Note:* Design tests are made only on representative apparatus to substantiate the ratings assigned to all other apparatus of basically the same design. These tests are not intended to be used as a part of normal production. The applicable portion of these design tests may also be used to evaluate modifications of a previous design and to assure that performance has not been adversely affected. Test data from previous similar designs may be used for current designs, where appropriate.   574

(3)(metal-enclosed low-voltage power circuit-breaker switchgear)(metal-clad and station-type cubicle switchgear). Tests made by the manufacturer to determine the adequacy of the design of a particular type, style, or model of equipment or its component parts to meet its assigned ratings and to operate satisfactorily under normal service conditions or under special conditions if specified, and may be used to demonstrate compliance with applicable standards of the industry. *Notes:* (A) Design tests are made on representative apparatus or prototypes to verify the validity of design analysis and calculation methods and to substantiate the ratings assigned to all other apparatus of basically the same design. These tests are not intended to be made on every design variation or to be used as part of normal production. The applicable portion of these design tests may also be used to evaluate modifications of a previous design and to ensure that performance has not been adversely affected. These data from previous similar designs may also be used for current designs, where appropriate. Once made, the tests need not be repeated unless the design is changed so as to modify performance. (B) Design tests are sometimes called type tests.

579,572,573

(4)(metal-oxide surge arresters for ac power circuits). Tests made on each design to establish the performance characteristics and to demonstrate compliance with the appropriate standards of the industry. Once made they need not be repeated unless the design is changed so as to modify performance. 583

(5)(general). Those tests made to determine the adequacy of the design of a particular type, style or model of equipment or its component parts to meet its assigned ratings and to operated satisfactorily under normal service conditions or under special conditions if specified. *Note:* Design tests are made only on representative apparatus to substantiate the ratings assigned to all other apparatus of basically the same design. These tests are not intended to be used as a part of normal production. The applicable portion of these design tests may also be used to evaluate modifications of a previous design and to assure that performance has not been adversely affected. Test data from previous similar designs may be used for current designs, where appropriate. 103,53

(6) (surge arrester). Tests made by the manufacturer on each design to establish the performance characteristics and to demonstrate compliance with the appropriate standards of the industry. Once made they need not be repeated unless the design is changed so as to modify performance. 430

(7) (cable termination). Tests made by the manufacturer to obtain data for design or application, or to obtain information on the performance of each type of high-voltage cable termination. 4

(8)(metal enclosed bus). Those tests made to determine the adequacy of a particular type, style, or model of metal-enclosed bus or its component parts to meet its assigned ratings and to operate satisfactorily under normal service conditions or under special conditions is specified. *See:* General note above. 78

(9) (power and distribution transformer). Those tests made to determine the adequacy of the design of a particular type, style, or model of equipment or its component parts to meet its assigned ratings and to operate satisfactorily under normal service conditions or under special conditions if specified; and to demonstrate compliance with appropriate standards of the industry. *Note:* Design tests are made only on representative apparatus to substantiate the ratings as-

signed to all other apparatus of basically the same design. These tests are not intended to be used as a part of normal production. The applicable portion of these design tests may also be used to evaluate modifications of a previous design and to assure that performance has not been adversely affected. Test data from previous similar designs may be used for current designs, where appropriate. Once made, the tests need not be repeated unless the design is changed so as to modify performance. 53

(10) (power cable joint). Tests made on typical joint designs to obtain data to substantiate the design. These tests are of such nature that after they have once been made, they need not be repeated unless significant changes are made in the material or design which may change the performance of the joint. 34

(11) (power switchgear). Those tests made to determine the adequacy of a particular type, style, or model of equipment with its component parts to meet its assigned ratings and to operate satisfactorily under normal service conditions or under special conditions if specified. *Note:* Design tests are made only on representative apparatus to substantiate the ratings assigned to all other apparatus of basically the same design. These tests are not intended to be used as a part of normal production. The applicable portion (part) of these design tests may also be used to evaluate modifications of a previous design and to assure that performance has not been adversely affected. Test data from previous similar designs may be used for current designs, where appropriate. 103, 443

design unit (Ada as a program design language). A logically related collection of design elements. In an Ada program design language (PDL), a design unit is represented by an Ada compilation unit. 518

design verification. *See:* **verification.** 434

design view (software design descriptions). A subset of design entity attribute information that is specifically suited to the needs of a software project activity. 568

design voltage. The voltage at which the device is designed to draw rated watts input. *See:* **appliance outlet.** 263

design walk-through. *See:* **walk-through.** 434

desired polarization (navigation aid terms). The polarization of the radio wave for which an antenna system is designed. 526

desired track. *See:* **course line.**

desired value. *See:* **value, ideal.**

desk checking (software). The manual simulation of program execution to detect faults through step-by-step examination of the source code for errors in logic or syntax. *See:* **code; error; fault; program execution; simulation; static analysis; syntax.** 434

deskstand. A movable pedestal or stand (adapted to rest on a desk or table) that serves as a mounting for the transmitter of a telephone set and that ordinarily includes a hook for supporting the associated receiver when not in use. *See:* **telephone station.** 328

deskstand telephone set. A telephone set having a deskstand. *See:* **telephone station.** 328

despun antenna. On a rotating vehicle, an antenna whose beam is scanned such that, with respect to fixed reference axes, the beam is stationary.   111

destination (1)(binary floating-point arithmetic). The location for the result of a binary or unary operation. A destination may be either explicitly designated by the user or implicitly supplied by the system (for example, intermediate results in subexpressions or arguments for procedures). Some languages place the results of intermediate calculations in destinations beyond the user's control. Nonetheless, ANSI/IEEE Std 754-1985 defines the result of an operation in terms of that destination's format and the operands' values.   469

(2)(radix-independent floating-point arithmetic). The location for the result of a binary or unary operation. A destination may be either designated by the user or implicitly supplied by the system (that is, intermediate results in subexpressions or arguments for procedures). Some languages place the results of intermediate calculations in destinations beyond the user's control. Nevertheless, ANSI/IEEE Std 854-1987 defines the result of an operation in terms of that destination's precision as well as the operand's values.   588

destination code (telephone switching system). A combination of digits providing a unique termination address in a communication network.   55

destination-code routing (telephone switching systems). The means of using the area and office codes to direct a call to a particular destination regardless of its point of origin.   55

destructive read (computing systems). A read process that also erases the data in the source.   255,77

destructive reading (charge-storage tubes). Reading that partially or completely erases the information as it is being read. *See:* **charge-storage tube.**   174

destructive testing (test, measurement and diagnostic equipment). (1) Prolonged endurance testing of equipment or a specimen until it fails in order to determine service life or design weakness; and (2) Testing in which the preparation of the test specimen of the test itself may adversely affect the life expectancy of the unit under test or render the sample unfit for its intended use.   54

detail contrast (television). See: resolution response.

detailed-billed call (telephone switching systems). A call for which there is a record including the calling and called line identities which will appear in a customer's billing statement.   55

detailed design (software). (1) The process of refining and expanding the preliminary design to contain more detailed descriptions of the processing logic, data structures, and data definitions, to the extent that the design is sufficiently complete to be implemented. (2) The results of the detailed design process. *See:* **data; data structures; design; preliminary design.**   434

detailed-record call (telephone switching systems). A call for which there is a record including the calling

and called line identities that may be used in the billing process as well as for other purposes.   55

detectability factor (1) (radar). In pulsed radar, the ratio of single-pulse signal energy to noise power per unit bandwidth that provides stated probabilities of detection and false alarm, measured in the intermediate-frequency amplifier and using an intermediate-frequency filter matched to the single pulse, followed by optimum video integration.   13

(2) (continuous-wave radar). The ratio of single-look signal energy to noise power per unit bandwidth, using a filter matched to the time on target.   187

detectable failures (1) (nuclear power generating station). Failures that can be identified through periodic testing or can be revealed by alarm or anomalous indication. Component failures which are detected at the channel, division, or system level are detectable failures. *Note:* Identifiable but nondetectable failures are failures identifiable by analysis that cannot be detected through periodic testing or cannot be revealed by alarm or anomalous indication.   102

(2) (single-failure criterion). Detectable failures are those that will be identified through periodic testing or will be revealed by alarm or anamolous indication. Component failures which are detected at the channel or system level are detectable failures.   356

(3) (criteria for safety systems). Failures that will be identified through periodic testing or will be revealed by alarm or anomalous indication. Component failures which are detected at the channel or system level are detectable failures. *Note:* Identifiable but nondetectable failures are failures identified by analysis that cannot be detected through periodic testing or cannot be revealed by alarm or anomalous indication. Refer to IEEE Std 379-1977, Application of the Single Failure Criterion to Nuclear Power Generating Station Class 1E Systems, for application guidance.   387, 428

detecting element. *See:* **element, primary detecting.**

detecting means. The first system element or group of elements that responds quantitatively to the measured variable and performs the initial measurement operation. The detecting means performs the initial conversion or control of measurement energy. *See:* **instrument.**   295

detection. (1) Determination of the presence of a signal. (2) Demodulation. The process by which a wave corresponding to the modulating wave is obtained in response to a modulated wave. *See:* **linear detection; power detection; square-law detection.**   111

detection limit (radiation protection). The extreme of detection or quantification for the radiation of interest by the instrument as a whole or an individual readout scale. The lower detection limit is the minimum quantifiable instrument response or reading. The upper detection limit is the maximum quantifiable instrument response or reading.   399

detection probability (radar). The probability that the signal, when present, will be detected, when a decision is made as to whether signal plus noise was present, or noise alone.   13

detectivity. The reciprocal of noise equivalent power (NEP). *See:* **noise equivalent power (NEP).** 433

detector (1)(monitoring radioactivity in effluents). Any device for converting radiation flux to a signal suitable for observation and measurement. 559

(2) (electromagnetic energy). A device for the indication of the presence of electromagnetic fields. *Note:* In combination with an instrument, a detector may be employed for the determination of the complex field amplitudes. *See:* **auxiliary device to an instrument.** 185

(3) (overhead-power-line corona and radio noise). For purposes of this standard a detector is defined as a device which combines the function of detector (extraction of signal or noise from a modulated input) and weighting (extraction of a particular characteristic of the signal or noise). 411

(4) (radiation protection). A device or component which produces an electronically measurable quantity in response to ionizing radiation. 399

detector, average. *See:* **average detector.**

detector figure of merit (nonlinear, active, and nonreciprocal waveguide components). A measure of the performance of a diode detector. It can be expressed quantitatively as the ratio of the open-circuit voltage sensitivity to the square root of the video resistance. 530

detector geometry (detector jargon)(X-ray energy spectrometers)(charged-particle.detectors)(semiconductor radiation detectors). The physical configuration of a solid-state detector. 471,119,23,118

detector, totally depleted. *See:* **totally depleted detector.**

detector, transmission. *See:* **transmission detector.**

determinant (circuits and systems). A square array of numbers or elements bordered on either side by a straight line. The value of the determinant is a function of its elements. 67

developer (electrostatography). A material or materials that may be used in development. *See:* **electrostatography.** 236,191

development (electrostatography). The act of rendering an electrostatic image viewable. *See:* **electrostatography.** 191

development cycle. *See:* **software development cycle.**

development life cycle. *See:* **software development cycle.** 434

development methodology (software). A systematic approach to the creation of software that defines development phases and specifies the activities, products, verification procedures, and completion criteria for each phase. *See:* **software; verification procedures.** 434

development specification. Synonymous with requirements specification in DOD usage. 434

deviation (1)(navigation aid terms). The angle between the magnetic meridian and the axis of a compass card. Indicates the offset of the compass card from magnetic north. 526

(2) (automatic control). Any departure from a desired or expected value or pattern. 56,219,206

(3) (metric practice). Variation from a specified dimension or design requirement, usually defining upper and lower limits. *See:* **tolerance.** 21

(4) (nuclear power quality assurance). A departure from specified requirements. 417

deviation distortion (data transmission). Distortion in an FM receiver due to inadequate bandwidth and inadequate amplitude modulation rejection, or inadquate discriminator linearity. 59

deviation factor (wave) (rotating machinery). The ratio of the maximum difference between corresponding ordinates of the wave and of the equivalent sine wave when the waves are superposed in such a way as to make this maximum difference as small as possible. *Note:* The equivalent sine wave is defined as having the same frequency and the same root-mean-square value as the wave being tested. *See:* **direct-axis synchronous impedance (rotating machinery).** 63

deviation, frequency. *See:* **frequency deviation.**

deviation from a sine wave (converter characteristics)-(self-commutated converters)(harmonic control and reactive compensation of static power converters). A single number measure of the distortion of a sinusoid due to harmonic components. It is equal to the ratio of the absolute value of the maximum difference between the distorted wave and the fundamental to the crest value of the fundamental. *See:* **maximum theoretical deviation from a sine wave.** 584,533

deviation from a sine wave, maximum theoretical. *See:* **maximum theoretical deviation from a sine wave.**

deviation integral, absolute (automatic control). The time integral of the absolute value of the system deviation following a stimulus specified as to location, magnitude, and time pattern. *Note:* The stimulus commonly employed is a step input. 56

deviation ratio (frequency-modulation system) (data transmission). The ratio of the maximum frequency deviation to the maximum modulating frequency of the system. 59

deviation sensitivity (1)(navigation aid terms). The rate of change of course indication with respect to the change of displacement from the course line. 526

(2) (frequency-modulation receivers). The least frequency deviation that produces a specified output power. 339

deviation, steady-state (control). The system deviation after transients have expired. *Note:* For the purpose of this definition, drift is not considered to be a transient. *See:* **deviation.** 206,329

deviation system (control). The instantaneous value of the ultimately controlled variable minus the command. *Note:* The use of system error to mean a system deviation with its sign changed is deprecated. *Syn:* **system overshoot.** *See:* **deviation.** 206,105

deviation, transient (control). The instantaneous value of the ultimately controlled variable minus its steady-state value. *Syn:* **transient overshoot.** *See:* **deviation.** 206,105

device (1)(electrical equipment)(supervisory control, data acquisition, and automatic control). An operating element such as a relay, contactor, circuit breaker, switch, valve, or governor used to perform a given function in the operation of electrical equipment. 570

(2)(FASTBUS device)(FASTBUS acquisition and control). Any equipment capable of connecting to a segment and responding to the mandatory features of the FASTBUS protocol. 480

(3)(general-system term)(696 interface devices). A circuit or logical group of circuits resident on one or more boards capable of interacting with other such devices through the bus. 538

(4)(measuring longitudinal balance of telephone equipment operating in the voice band). An item of electric equipment that is used in connection with, or as an auxiliary to, other items of electric equipment. 529

(5) (nuclear power generating stations). An item of electric equipment that is used in connection with, or as an auxiliary to, other items of electric equipment. (For example, as used in this document (IEEE Std 649-1980), a device is a starter, contactor, circuit breaker, relay, etcetera. 28, 440

(6) (electrical equipment) (station control and data acquisition). An operating element such as a relay, contractor, circuit breaker, switch, valve, or governor used to perform a given function in the operation of electrical equipment. 403

(7) (packaging machinery). A unit of an electrical system which is intended to carry but not consume electrical energy. 429

device address (DA)(FASTBUS acquisition and control). The (32-m)-bit identifying number assigned to a FASTBUS device that is compared with the signals on the address/data (AD) lines during a logical primary address cycle of a FASTBUS operation. The device address is formed by the group and module address fields. The (remaining) low-order m-bits are assigned to the internal address field. 480

device class--broadcast (FASTBUS acquisition and control). Selective broadcast-class specified by CSR#7 (control and status register #7). Controls device response to subsequent cycles within the broadcast. 480

device rise time (DRT) (photomultipliers for scintillation counting). The mean time difference between the 10- and 90-percent amplitude points on the output waveform for full cathode illumination and delta-function excitation. DRT is measured with a repetitive delta-function light source and a sampling oscilloscope. The trigger signal for the oscilloscope may be derived from the device output pulse, so that light sources such as the the scintillator light source may be employed. 117

dew withstand voltage test (metal-enclosed bus and calculating losses in isolated-phase bus). A test to determine the ability of the insulating system to withstand specified overvoltages for a specified time without flashover or puncture while completely covered with dew. 574

dezincification. Parting of zinc from an alloy (parting

is the preferred term). Note: Other terms in this category, such as denickelification, dealuminification, demolybdenization, etcetera, should be replaced by the term parting. See: electrometallurgy. 205

DF (direction finder). See: radio direction finder.

DF (direction finder) antenna (navigation aid terms). Any antenna used for radio direction finding. 526

DF (direction finder) antenna system (navigation aid terms). One or more DF antennas, their combining circuits and feeder systems, together with the shielding and all electrical and mechanical items up to the termination at the receiver-input terminals. 526

DF (direction finder) noise level (navigation aid terms). In the absence of the desired signals, the average power or rms (root-mean-square) voltage at any specified point in a direction finder system circuit. Note: In rf (radio frequency) and audio channels, The DF noise level is usually measured in terms of the power dissipated in suitable termination. In a video channel, it is customarily measured in terms of voltage across a given impedance, or of the cathode-ray deflection. 526

DF (direction finder) sensitivity (navigation aid terms). That field strength at the DF antenna, in microvolts per meter, which produces a ratio of signal-plus-noise to noise, equal to 20 dB (decibels) in the receiver output, the direction of arrival of the signal being such as to produce maximum pickup in the DF antenna system. 526

DF. See: direction-finder; radio direction-finder.

DF noise level. In the absence of the desired signals, the average power or rms voltage at any specified point in a direction finder system circuit. Note: In radio-frequency and audio channels, the direction finding noise level is usually measured in terms of the power dissipated in suitable termination. In a video channel, it is customarily measured in terms of voltage across a given impedance, or of the cathode-ray deflection. 13

DF sensitivity. That field strength at the DF antenna, in microvolts per meter, which produces a ratio of signal-plus-noise to noise equal to 20 decibels in the receiver output, the direction of arrival of the signal being such as to produce maximum pickup in the direction finding antenna system. 13

dg. See: decilog.

diagnostic (software). (1) Pertaining to the detection and isolation of faults or failures. (2) A message generated by a computer program indicating possible faults in another system component, for example, a syntax fault flagged by a compiler. See: compiler; computer program; failure; fault; system component; syntax. 434

diagnostic factor (evaluation of thermal capability)(-thermal classification of electric equipment and electrical insulation). A variable or fixed stress, which can be applied periodically or continuously during an accelerated test, to measure the degree of aging without in itself influencing the aging process. 506

diagnostic routine (1) (computer). A routine designed

to locate either a malfunction in the computer or a mistake in coding. *See:* **programmed check.**   210

(2) (test, measurement and diagnostic equipment). A logical sequence of tests designed to locate a malfunction in the unit under test.   54

diagnostic test (1) (ATLAS). A test applied to a unit under test (UUT) with the purpose of isolating a fault to a lower level of assembly.   400

(2) (test, measurement and diagnostic equipment). A test performed for the purpose of isolating a malfunction in the unit under test or confirming that there actually is a malfunction.   54

dial (1) (industrial control). A plate or disk, suitably marked, that served to indicate angular position, as for example the position of a handwheel.   206

(2) (automatic control). A type of calling device used in automatic switching that, when wound up and released, generates pulses required for establishing connections.   192

dialing (telephone switching systems). The act of using a calling device.   55

dialing pattern (telephone switching systems). The implementation of a numbering plan with reference to an individual automatic exchange.   55

dial-mobile telephone system (mobile communication). A mobile communication system that can be interconnected with a telephone network by dialing, or a mobile communication system connected on a dial basis with a telephone network. *See:* **mobile communication system.**   181

dial pulse (telephony)(dial-pulse address signaling systems). A momentary interruption or change in the direct-current path of a signalling system to provide address information.   540

dial pulsing (telephony)(dial-pulse address signaling systems). A means of transmitting the address telephone number over a direct-current path. The current is interrupted, at the transmitting end, in a regular, momentary pattern. The number of interruptions corresponds to the digit being transmitted.   540

dial pushing (telephone switching systems). A means of pulsing consisting of regular, momentary interruptions of a direct or alternating current path at the sending end in which the number of interruptions corresponds to the value of the digit or character.   55

dial tone (telephone switching systems). The tone that indicates that the switching equipment is ready to receive signals from a calling device.   55

dial train (register). All the gear wheels and pinions used to interconnect the dial pointers. *See:* **watt-hour meter.**   328

diametric rectifier circuit. A circuit that employs two or more rectifying elements with a conducting period of 180 electrical degrees plus the commutating angle. *See:* **rectifiction.**   328

diamond winding (rotating machinery). A distributed winding in which the individual coils have the same shape and coil pitch.   63

diaphragm (electrolytic cells). A porous or permeable membrane separating anode and cathode compartments of an electrolytic cell from each other or from an intermediate compartments for the purpose of preventing admixture of anolyte and catholyte. *See:* **electrolytic cell.**   328

diathermy (medical electronics). The therapeutic use of alternating currents to generate heat within some part of the body, the frequency being greater than the maximum frequency for neuromuscular response.   192

dibit (data transmission). Two bits; two binary digits.   59

dichotomizing search. *See:* **binary search.**

dichroic filter (fiber optics). An optical filter designed to transmit light selectively according to wavelength (most often, a high-pass or low-pass filter). *See:* **optical filter.**   433

dichroic mirror (fiber optics). A mirror designed to reflect light selectively according to wavelength. *See:* **dichroic filter.**   433

dichromate cell. A cell having an electrolyte consisting of a solution of sulphuric acid and a dichromate. *See:* **electrochemistry.**   328

die (semiconductor). *See:* **chip; semiconductor.**

dielectric (surge arresters). A medium in which it is possible to maintain an electric field with little or no supply of energy from outside sources.   62

dielectric constant (1) (dielectric). That property which determines the electrostatic energy stored per unit volume for unit potential gradient. *Note:* This numerical value usually is given relative to a vacuum. *See:* **dielectric heating.**   14

(2) (antennas). The real part of the complex dielectric constant.   111

dielectric dissipation factor. (1) The cotangent of the dielectric phase angle of a dielectric material or the tangent of the dielectric loss angle. *See:* **dielectric heating.** (2) The ratio of the loss index ϵn to the relative dielectric constant ϵ. *See:* **relative complex dielectric constant.**   22

dielectric filter. *See:* **interference filter.**

dielectric guide. A waveguide in which the waves travel through solid dielectric material. *See:* **waveguide.**   328

dielectric heater. A device for heating normally insulating material by applying an alternating-current field to cause internal losses in the material. *Note:* The normal frequency range is above 10 megahertz. *See:* **interference.**   188

dielectric lens. A lens made of dielectric material and used for refraction of radio-frequency energy. *See:* **antenna; waveguide.**   244

dielectric loss angle (rotating machinery). δ The angle whose tangent is the dissipation factor.   22

dielectric loss factor*. *See:* **loss factor.**
*Deprecated

dielectric phase angle. (1) The angular difference in phase between the sinusoidal alternating voltage applied to a dielectric and the component of the resulting alternating current having the same period as the voltage. *See:* **dielectric heating.** (2) The angle whose contangent is the dissipation factor, or arc cot ϵ''/ϵ'. *See:*

dielectric dissipation factor; relative complex dielectric constant; dielectric heating.   22

dielectric power factor. The cosine of the dielectric phase angle (or the sine of the dielectric loss angle). *See:* **dielectric heating.**   22

dielectric rod antenna. An antenna which employes a shaped dielectric rod as the electrically significant part of a radiating element. *Note:* The polyrod rod antenna is a notable example of the dielectric rod antenna when constructed of polystyrene.   111

dielectric strength (general) (material) (electric strength) (breakdown strength). The potential gradient at which electric failure or breakdown occurs. To obtain the true dielectric strength the actual maximum gradient must be considered, or the test piece and electrodes must be designed so that uniform gradient is obtained. The value obtained for the dielectric strength in practical tests will usually depend on the thickness of the material and on the method and conditions of test.   62

dielectric tests (1) (transformer) (general). Tests which consist of the application of a voltage higher than the rated voltage, for a specified time to assure the withstand strength of insulation materials and spacing. These various types of dielectric tests have been developed to allow selectivity testing the various insulation components of a transformer, without overstressing other components; or to simulate transient voltages which transformers may encounter in service. *See:* **applied voltage tests; induced voltage tests; impulse tests.**   53

(2) (high voltage air switches). Tests that consist of the application of a standard test voltage for a specific time and are designed to determine the adequacy of insulating materials and spacing.   144

(3) (neutral grounding device). Tests that consists of the application of a voltage, higher than the rated voltage, for a specified time to prove compliance with the required voltage class of the device.   91

dielectric waveguide. A waveguide consisting of a dielectric structure.   267

dielectric withstand voltage tests (1)(X-radiation limits for ac high-voltage power vacuum interrupters used in power switchgear)(power switchgear). Tests made to determine the ability of insulating materials and spacings to withstand specified overvoltages for a specified time without flashover or puncture.   103,443,553

dielectric withstand-voltage tests (2) (power and distribution transformer). Tests made to determine the ability of insulating materials and spacings to withstand specified overvoltages for a specified time without flashover or puncture. *Note:* The purpose of the tests is to determine the adequacy against breakdown of insulating materials and spacings under normal or transient conditions.   53

diesel-electric drive (oil-electric drive). A self-contained system of power generation and application in which the power generated by a diesel engine is transmitted electrically by means of a generator and a motor (or multiples of these) for propulsion purposes.

Note: The prefix diesel-electric is applied to ships, locomotives, cars, buses, etcetera, that are equipped with this drive. *See:* **electric locomotive.**   328

diesel-generator unit. An independent source of standby electrical power that consists of diesel-fueled internal combustion engine (or engines) coupled directly to an electrical generator(s), the associated mechanical and electrical auxiliary systems and the control, protection, and surveillance systems.   99

difference amplifier. *See:* **differential amplifier.**

difference channel (monopulse radar). (1) A part of a monopulse receiver dedicated to the amplification, filtering, and other processing of a 'difference' signal, which is generated by comparison of signals received by two (or two sets of) antenna beams, and indicating the departure of the target from the boresight axis. (2) A signal path through a monopulse receiver for processing a "difference" signal which is commonly generated by comparison of two or more signals received by two antenna beams or two sets of antenna beams (that is, by simultaneous lobing).   13

difference detector. A detector circuit in which the output is a function of the difference of the peak amplitudes or root-mean-square amplitudes of the input waveforms. *See:* **navigation.**   187

difference frequency (parametric device). The absolute magnitude of the difference between a harmonic nfp of the pump frequency fp and the signal frequency fs, where n is a positive integer. *Note:* Usually n is equal to one. *See:* **parametric device.**   277,191

difference-frequency parametric amplifier*. *See:* **inverting parametric device.**

*Deprecated

difference in depth of modulation (DDM)(directive systems employing overlapping lobes with modulated signals such as ILS [instrument landing system]) (navigation aid terms). A fraction obtained by subtracting from the percentage of modulation of the smaller signal and dividing by 100.   526

difference limen (differential threshold) (just noticeable difference). The increment in a stimulus that is just noticeable in a specified fraction of trials. *Note:* The relative difference limen is the ratio of the difference limen to the absolute magnitude of the stimulus to which it is related. *See:* **phonograph pickup.**   176

difference pattern (1) (antennas). A radiation pattern characterized by a pair of main lobes of opposite phase, separated by a single null, plus a family of side lobes, the latter usually desired to be at a low level. *Note:* Antennas used in many radar applications are capable of producing a sum pattern and two orthogonal difference patterns. The difference patterns are capable of determining the position of a target in a right/left and up/down sense by antenna pattern pointing, which places the target in the null between the twin lobes of each difference pattern. *See:* **sum pattern.**   111

(2) (radar). A description, often given graphically, of the variation in the strength of the difference signal in a monopulse radar as a function of the difference in angle between the source of the received signal and the boresight axis. *See:* **difference channel.** 13

difference signal. *See:* **differential signal.**

difference slope (radar). In a monopulse radar, the slope of the difference-pattern voltage (normalized with respect to the sum-pattern voltage) as a function of target angle from the boresight axis. The slope is usually specified at the point on the curve where the difference-pattern voltage is zero, which corresponds to the boresight axis. 13

differential (photoelectric lighting control). The difference in foot-candles between the light levels for turn-on and turn-off operation. *See:* **photoelectric control.** 206

differential aeration cell. An oxygen concentration cell. *See:* **electrolytic cell.** 221,205

differential amplifier (1). An amplifier whose output signal is proportional to the algebraic difference between two input signals. *See:* **amplifier.** 185
(2) (signal-transmission system). An amplifier that produces an output only in response to a potential difference between its input terminals (differential-mode signal) and in which outputs from common-mode interference voltages on its input terminals are suppressed. *Note:* An ideal differential amplifier produces neither a differential-mode nor a common-mode output in response to a common-mode interference input. *See:* **amplifier; signal.** 188

differential analyzer (analog computer). A computer designed primarily for the convenient solution of differential equations. 9

differential capacitance (nonlinear capacitor). The derivative with respect to voltage of a charge characteristics, such as an alternating charge characteristic or a mean charge characteristic, at a given point on the characteristic. *See:* **nonlinear capacitance.** 191

differential-capacitance characteristic (nonlinear capac itor). The function relating differential capacitance to voltages. *See:* **nonlinear capacitor.** 191

differential capacitance voltage (power switchgear). The difference in magnitudes of the root-mean-square (rms) system normal frequency line-to-neutral voltage multiplied by the square root of two, with and without the capacitance connected. *Note:* This can be calculated from the equations:

$$\Delta V = \sqrt{2}\, E_S \frac{X_L}{X_C - X_L}$$

$$\Delta V = \sqrt{2}\, E_S \frac{kVAR}{kVA_{SC} - kVAR}$$

where
$kVA_{SC} = \Delta V$ =differential capacitance voltage in volts
E_S =system phase-to-neutral voltage in volts rms
X_L =source inductive reactance to point of application, in ohms per phase

X_C =capacitance reactance of bank being switched in ohms per phase
$kVAR$ =size of bank being switched (three phase)
kVA_{SC} =system short-circuit kVA at point of capacitor application (symmetrical three phase) 27, 103

differential compounded (rotating machinery). Applied to a compound machine to denote that the magnetomotive forces of the series field winding is opposed to that of the shunt field winding. 63

differential control. A system of load control for self-propelled electrically driven vehicles wherein the action of a differential field wound on the field poles of a main generator (or of an exciter) and connected in circuit between the main generator and the traction motors, serves to limit the power demand from the prime mover. *See:* **multiple-unit control.** 328

differential control current (magnetic amplifier). The total absolute change in current in a specified control winding necessary to obtain differential output voltage when the control current is varied very slowly (a quasistatic characteristic). 171

differential control voltage (magnetic amplifier). The total absolute change in voltage across the specified control terminals necessary to obtain differential output voltage when the control voltage is varied very slowly (a quasistatic characteristic). 171

differential *dE/dx* detector(1)(charged particle detectors). A transmission detector whose thickness is small compared to the range of the incident particle. 119
(2)(germanium gamma-ray detectors). A transmission detector whose thickness is small compared to the range of the incident particle, and whose entrance and exit dead layers are small compared to the thickness of the detector. 528

differential Doppler frequency (radio wave propagation). The time rate of change of difference in phase path at two frequencies in a dispersive medium. *Note:* Sometimes called the dispersive Doppler frequency. 146

differential duplex system. A duplex system in which the sent currents divide through two mutually inductive sections of the receiving apparatus, connected respectively to the line and to a balancing artificial line, in opposite directions so that there is substantially no net effect on the receiving apparatus; whereas the received currents pass mainly through one section; or though the two sections in the same direction, and operate the apparatus. *See:* **telegraphy.** 328

differential gain (video transmission system). The difference between (1) the ratio of the output amplitudes of a small high-frequency sine-wave signal at two stated levels of a low frequency signal on which it is superimposed, and (2) unity. *Note:* (A) Differential gain may be expressed in percent by multiplying the above difference by 100. (B) Differential gain may be expressed in decibels by multiplying the common logarithm of the ratio described in (1) above by 20. (C) In this definition, level means a specified position on an amplitude scale applied to a signal wave-form. (D)

The low- and high-frequency signals must be specified. *See:* **television.**   306,178

differential-gain control (gain sensitivity control). A device for altering the gain of a radio receiver in accordance with an expected change of signal level, to reduce the amplitude differential between the signals at the output of the receiver. *See:* **radio receiver.**   328

differential-gain-control circuit (electronic navigation). The circuit of a receiving system that adjusts the gain of a single radio receiver to obtain desired relative output levels from two alternately applied or sequentially unequal input signals. *Note:* This may be accomplished automatically or or manually; if automatic, it is referred to as automatic differential-gain control. Example: Loran circuits that adjust gain between successive pulses from different ground stations.   187,13

differential-gain-control range. The maximum ratio of signal amplitudes (usually expressed in decibels), at the input of a single receiver, over which the differential-gain-control circuit can exercise proper control and maintain the desired output levels. *See:* **navigation.**   187,13

differential gap. *See:* **neutral zone.**

differential Manchester encoding (token ring access method). A signalling method used to encode clock and data bit information into bit symbols. Each bit symbol is split into two halves, where the second half is the inverse symbol of the first half. A 0 bit is represented by a polarity change at the start of the bit time. A 1 bit is represented by no polarity change at the start of the bit time. Differential Manchester encoding is polarity independent.   472

differential mode attenuation (fiber optics). The variation in attenuation among the propagating modes of an optical fiber.   433

differential mode delay (fiber optics). The variation in propagation delay that occurs because of the different group velocities of the modes of an optical fiber. *Syn:* **multimode group delay.** *See:* **group velocity; mode; multimode distortion.**   433

differential-mode-interference (interference terminology). *See:* **interference, differential mode; interference, normal-mode; interference; accuracy rating (instrument).**

differential-mode radio noise. Conducted radio noise that causes the potential of one side of the signal transmission path to be changed relative to another side.   418

differential nonlinearity (percent) (semiconductor radiation detectors). The percentage departure of the slope of the plot of output versus input from the slope of a reference line.   23

differential nonreversible output voltage. *See:* **differential output voltage.**

differential output current (magnetic amplifier). The ratio of differential output voltage to rated load impedance.   171

differential output voltage (magnetic amplifier) (1) (nonreversible output). The voltage equivalent to the algebraic difference between maximum test output voltage and minimum test output voltage.   171

(2) (reversible output). The voltage equivalent to the algebraic difference between positive maximum test output voltage and negative maximum test output voltage.   171

differential permeability (magnetic core testing). The rate of change of the induction with respect to the magnetic field strength.

$$\mu_{\text{dif}} = \frac{1}{\mu_0} \frac{dB}{dH}$$

μ_{dif} = relative differential permeability
dH = infinitely small change in field strength
dB = corresponding change induction   165

differential permittivity (ferroelectric material)(primary ferroelectric terms). The slope of the hysteresis loop (electric displacement versus electric field) at any point. Differential permittivity is usually measured at low frequency (60 Hz) due to the self-heating produced on cycling through a hysteresis loop. The value of differential permittivity is often different from the small-signal permittivity measured under equivalent bias conditions.   497

differential phase (video transmission system). The difference in output phase of a small high-frequency sine-wave signal at the two stated levels of a low-frequency signal on which it is superimposed. *Note:* Notes C and D appended to **differential gain** apply also to **differential phase.** *See:* **television.**   306

differential phase shift (nonlinear, active, and nonreciprocal waveguide components). (1) The difference between insertion phase changes , resulting from a change of configuration or material state, in the two opposite directions of propagation between two ports of a junction. (2)* Differential insertion phase.
*Deprecated.   530

differential-phase-shift keying (modulation systems). A form of phase-shift keying in which the reference phase for a given keying interval is the phase of the signal during the preceding keying interval.   242

differential position (loran, omega, GPS [global positioning system])(navigation aid terms). The difference between position axis determined by separated receivers or antennas. Close proximity error sources are minimized, thereby greatly enhancing the accuracy of this parameter.   526

differential protection (power switchgear). A method of apparatus protection in which an internal fault is identified by comparing electrical conditions at all terminals of the apparatus.   127, 103

differential protective relay (87) (power system device function numbers). A protective relay that functions on a percentage or phase angle or other quantitative difference of two currents or of some other electrical quantities.   402

differential quantum efficiency (fiber optics). In an optical source or detector, the slope of the curve relating output quanta to input quanta.   433

differential relay (1) (power switchgear). A relay that by its design or application is intended to respond to the difference between incoming and outgoing electrical quantities associated with the protected apparatus. 127, 103

(2). A relay with multiple windings that functions when the power developed by the individual windings is such that pickup or dropout results from the algebraic summation of the fluxes produced by the effective windings. 341

differential resistance (semiconductor rectifiers). The differential change of forward voltage divided by a stated increment of forward current producing this change. *See:* **semiconductor rectifier stack.** 208

differential reversible output voltage. *See:* **differential output voltage.**

differential signal. The instantaneous, algebraic difference between two signals. *See:* **oscillograph.** 185

differential threshold. *See:* **difference limen.**

differential trip signal (magnetic amplifier). The absolute magnitude of the difference between trip OFF and trip ON control signal. *See:* **rating and testing magnetic amplifiers.** 171

differentiated (pulse)(pulse amplifier jargon)(germanium gamma-ray detectors)(charged-particle detectors)(semiconductor radiation detectors)(X-ray energy spectrometers). A pulse is differentiated when it is passed through a high-pass network, such a a CR filter. 528,119,23,118,471

differentiating network. *See:* **differentiator.**

differentiator (1) (analog computer). A device producing an output proportional to the derivative of one variable with respect to another, usually time. *See:* **electronic analog computer.** 9

(2) (electronic circuits). A device whose output function is reasonably proportional to the derivative of the input function with respect to one or more variables, for example, a resistance-capacitance network used to select the leading and trailing edges of a pulse signal. 255

(3) (modulation circuits and industrial control) (differentiating circuit) (differentiating network). A transducer whose output waveform is substantially the time derivative of its input waveform. *Note:* Such a transducer preceding a frequency modulator makes the combination a phase modulator; or following a phase detector makes the combination a frequency detector. Its ratio of output amplitude to input amplitude is proportional to frequency and its output phase leads its input phase by 90 degrees. 206,111

(4) (relaying) (power switchgear). A transducer whose output wave form is substantially the time derivative of its input wave form. 103

diffracted wave (1). When a wave in a medium of certain propagation characteristics is incident upon a discontinuity or a second medium, the diffracted wave is the wave component that results in the first medium in addition to the incident wave and the waves corresponding to the reflected rays of geometrical optics. *See:* **radiation.** 328

(2) (audio and electroacoustics). A wave whose front

has been changed in direction by an obstacle or other nonhomogeneity in a medium, rather than by reflection or refraction. *See:* **radiation.** 176

diffraction (1) (general). A process that produces a diffracted wave. *See:* **radiation.** 176

(2) (laser-maser). Deviation of part of a beam, determined by the wave nature of radiation, and occurring when the radiation passes the edge of an opaque obstacle. 363

(3) (fiber optics). The deviation of a wavefront from the path predicted by geometric optics when a wavefront is restricted by an opening or an edge of an object. *Note:* Diffraction is usually most noticeable for openings of the order of a wavelength. However, diffraction may still be important for apertures many orders of magnitude larger than the wavelength. *See:* **far-field diffraction pattern; near-field diffraction pattern.** 433

(4) (radio wave propagation). The deviation of the direction of energy flow of a wave when it passes an obstacle, a restricted aperture, or other inhomogeneities in a medium. *Note:* This deviation does not include effects of reflection. 146

diffraction angle (acousto-optic device). The angle between the Nth order diffraction beam and the zeroth order beam. It is given by the ratio of the optical wavelength λ_0 to the acoustic wavelength, times the order of the diffracted beam $N = \pm 1, \pm 2, \pm 3, \ldots$ so that $\theta_N = N\lambda_0 / \wedge$. 82

diffraction efficiency (acousto-optic device). For the Nth order, the percent ratio of the light intensity diffracted into the Nth order divided by the light intensity in the zeroth order with the acoustic drive power off, thus

$$\eta_N = (I_N/I_O) \times 100$$

For a device of fixed design, the diffraction efficiency will depend on the optical wavelength, beam diameter, angle of incidence, and acoustic drive power. 82

diffraction grating (fiber optics). An array of fine, parallel, equally spaced reflecting or transmitting lines that mutually enhance the effects of diffraction to concentrate the diffracted light in a few directions determined by the spacing of the lines and the wavelength of the light. *See:* **diffraction.** 433

diffraction limited (fiber optics). A beam of light is diffraction limited if: (1) the far-field beam divergence is equal to that predicted by diffraction theory, or (2) in focusing optics, the impulse response or resolution limit is equal to that predicted by diffraction theory. *See:* **beam divergence angle; diffraction.** 433

diffraction loss (laser-maser). That portion of the loss of power in a propagating wave (beam) which is due to diffraction. 363

diffused junction. *See:* **junction, diffused.**

diffused junction detector (charged-particle detectors)(germanium gamma-ray detectors). A semiconductor detector in which the p-n or n-p junction is produced by diffusion of donor or acceptor impurities. 119,528

diffused lighting (illuminating engineering). Lighting provided on the work-plane or on an object light that is not incident predominantly from any particular direction.   167

diffuse ferroelectrics (diffuse ferroelectric single crystals or polycrystalline solid solutions)(primary ferroelectric terms). Materials whose small-signal permittivity, when measured as a function of temperature, indicates by its width a broad or diffuse phase transition between the ferroelectric and nonferroelectric phases. These materials exhibit weak ferroelectric properties in the temperature range of the diffuse phase transition. Diffuse ferroelectrics are noted for their almost anhysteretic P versus E behavior and have been called weak or dilute ferroelectrics, slim-loop materials, penferroelectric, and quasiferroelectric or alpha-phase materials.   497

diffuser (illuminating engineering). A device to redirect or scatter the light from a source, primarily by the process of diffuse transmission.   167

diffuse reflectance (illuminating engineering). The ratio of the flux leaving a surface or medium by diffuse reflection to the incident flux.   167

diffuse reflection (1) (illuminating engineering). That process by which incident flux is redirected over a range of angles. *See:* **diffusing surfaces and media.**
                                        167
(2) (laser-maser). Change of the spatial distribution of a beam of radiation when it is reflected in many directions by a surface or by a medium.   363
(3)(fiber optics). *See:* **reflection.**

diffuse sound field. A sound field in which the time average of the mean-square sound pressure is everywhere the same and the flow of energy in all directions is equally probable. *See:* **loudspeaker.**   176

diffuse transmission (illuminating engineering). That process by which the incident flux passing through a surface or medium is scattered.   167

diffuse transmission density. The value of the photographic transmission density obtained when the light flux impinges normally on the sample and all the transmitted flux is collected and measured.   176

diffuse transmittance (illuminating engineering). The ratio of the diffusely transmitted flux leaving a surface or medium to the incident flux. *Note:* Provision for exclusion of regularly transmitted flux must be clearly described.   167

diffusing panel (illuminating engineering). A translucent material covering the lamps in a luminaire in order to reduce the brightness by distributing the flux over an extended area.   167

diffusing surfaces and media (illuminating engineering). Those surfaces and media that redistribute at least some of the incident flux by scattering.   167

diffusion (laser-maser). Change of the spatial distribution of a beam of radiation when it is deviated in many directions by a surface or by a medium.   363

diffusion capacitance (semiconductor)(nonlinear, active, and nonreciprocal waveguide components). A capacitance enhancement effect associated with p-n junctions. Because the diffusion of electrons and holes takes time, there is a storage effect that is equivalent to adding additional capacitance in shunt with the junction. In the forward bias state, the diffusion capacitance becomes predominant over the space-charge capacitance to the point that the latter can be neglected at lower microwave frequencies. Diffusion capacitance varies inversely with frequency, but increases as minority carrier lifetime increases and must be dealt with in many frequency multiplier and parametric amplifier designs.   530

diffusion constant (charge carrier) (homogeneous semiconductor). The quotient of diffusion current density by the charge-carrier concentration gradient. It is equal to the product of the drift mobility and the average thermal energy per unit charge of carriers. *See:* **semiconductor.**   245

diffusion depth (of a semiconductor radiation detector). *See:* **junction depth.**

diffusion length, charge-carrier (homoegeneous semiconductor). The average distance to which minority carriers diffusion length is equal to the square root of the product of the charge-carrier diffusion constant and the volume lifetime. *See:* **semiconductor.**
                              186,245,66,210

digit (1) (metric practice). One of the ten Arabic numerals (0 to 9).   21
(2) (positional notation) (notation). (A) (1) A character that stands for an integer. (2) Loosely, the integer that the digit stands for. (3) Loosely, any character. (B) A character used to represent one of the non-negative integers smaller than the radix, for example, in decimal notation one of the characters 0 to 9.
                                        255

digit absorption (telephone switching systems). The interpretation and rejection of those digits received, but not required, in the setting of automatic direct control system crosspoints.   55

digital. Pertaining to data in the form of digits. *See:* **analog.**   235

digital coefficient attenuator (1)(DCA) (hybrid computer linkage components). Essentially the same as a digital-to-analog multiplier (DAM). This term is generally reserved for those components that are used as the high speed hybrid replacement for manual and servo potentiometers. *Syn:* **digital potentiometer.**
                                        10
(2) (analog computer). *See:* **hybrid computer linkage components.**   9

digital coefficient potentiometer. *See:* **digital coefficient potentiometer-hybrid computer linkage component.**   9

digital computer (1) (information processing). A computer that operates on discrete data by performing arithmetic and logic processes on these data. Contrasts with analog computer.   255,77
(2) (test, measurement and diagnostic equipment). A computer in which discrete quantities are represented in digital form and which generally is made to solve mathematical problems by iterative use of the fundamental processes of addition, subtraction, multiplication, and division.   54

digital controller (data processing). A controller that accepts an input sequence of numbers and processes them to produce an output sequence of numbers.   198

digital converter (code translator) (power switchgear). A device, or group of devices, that converts an input numerical signal or code of one type into an output numerical signal or code of another type.   103

digital data (data transmission). Pertaining to data in the form of digits or interval quantities. Contrast with **analog data.**   59

digital device (control equipment). A device that operates on the basis of discrete numerical techniques in which the variables are represented by coded pulses or states.   94

digital differential analyzer (DDA) (analog computers). A special-purpose digital computer consisting of many computing elements, all operating in parallel, that performs integration by means of a suitable integration code on incremental quantities and that can be programmed for the solution of differential equations in a manner similar to an analog computer.   9

digital logic elements (analog computers). In an analog computer, a number of digital functional modules, consisting of logic gates, registers, flip-flops, timers, etcetera, all operating in parallel, either synchronously or asynchronously, and whose inputs and outputs are interconnected, according to a "logic program," via patch cards, on a patch board.   9

digitally controlled function generator (analog computers). A hybrid component using DAC's and DAM's to insert the linear segment approximation values to the desired arbitrary function. The values are stored in a self-contained digital core memory, which is accessed by the DAC's and DAM's at digital-computer speeds (microseconds).   9

digital phase lock loop (communication satellite). A circuit for synchronizing the received waveform, by means of discrete corrections.   83

digital potentiometer. *See:* **digital coefficient attenuator.**   10

digital quantity (1)(station control and data acquisition). A variable represented by coded pulses (for example, bits) or states.   403

(2)(supervisory control, data acquisition, and automatic control). A variable represented by a number of discrete units.   570

digital readout clock (power switchgear). A clock that gives (usually with visual indication) a voltage or contact closure pattern of electrical circuitry for a readout of time. A digital readout calendar clock also includes a readout of day, month, and year, usually, also with indication.   103

digital switch (telephone loop performance). A switch that, internally, performs switching only of digital signals of a set format. It is inherently a four-wire entity, requiring a two-wire to four-wire hybrid at the channel interface when accepting analog signals from two-wire channels.   473

digital switching (telephone switching systems). Switching of discrete-level information signals.   55

digital telemeter indicating receiver (power switchgear). A device that receives the numerical signal transmitted from a digital telemeter transmitter and gives a visual numerical display of the quantity measured.   103

digital telemetering (1) (power switchgear). Telemetering in which a numerical representation, as for example some form of pulse code, is generated and transmitted; the number being representative of the quantity being measured.   103

(2) (station control and data acquisition). Telemetering in which a numerical representation is generated and transmitted; the number being representative of the quantity being measured.   403

digital telemeter receiver (power switchgear). A device that receives the numerical signal transmitted by a digital telemeter transmitter and stores it or converts it to a usable form, or both, for such purposes as recording, indication, or control.   103

digital telemeter transmitter (power switchgear). A device that converts its input signal to a numerical form for transmission to a digital telemeter receiver over an interconnecting channel.   103

digital-to-analog (D-A) conversion (1)(station control and data acquisition). Production of an analog signal whose instantaneous magnitude is proportional to the value of a digital input.   403

(2)(supervisory control, data acquisition, and automatic control). Production of an analog signal whose magnitude is proportional to the value of a digital input.   570

digital-to-analog converter (1) (power switchgear). A device, or group of devices, that converts a numerical input signal or code into an output signal some characteristic of which is proportional to the input.   103

(2) (data processing). A device that converts an input number sequence into a function of a continuous variable.   198

(3) (DAC) (hybrid computer linkage components). A circuit or device whose input is information in digital form and whose output is the same information in analog form. In a hybrid computer, the input is a number sequence (or word) coming from the digital computer, while the output is an analog voltage proportional to the digital number.   10

digital-to-analog multiplier (DAM) (hybrid computer linkage components). A device which provides the means of obtaining the continuous multiplication of a specific digital value with a changing analog variable. The product is represented by a varying analog voltage. *Syn:* **multiplying-digital-to-analog converter (MDAC).**   10

digit deletion (telephone switching systems). In the processing of a call, the elimination of a portion of the destination code.   55

digitize (mathematics of computing). To express analog data in digital form.   564

digraph. *See:* **directed graph.**

dimension (metric practice). A geometric element in a design, such as length or angle, or the magnitude of such a quantity.   21 dimension, critical mating (standard connector). Those longitudinal and transverse dimensions assuring nondestructive mating with a corresponding standard connector.   110

diminished-radix complement (mathematics of computing). The complement obtained by subtracting each digit of a given numeral from the largest digit in the numeration system. For example, ones complement in binary notation, nines complement in decimal notation. See: radix complement.   564

dimming reactor (thyristor). A reactor that may be inserted in a lamp circuit at will for reducing the luminous intensity of the lamp. Note: Dimming reactors are normally used to dim headlamps, but may be applied to other circuits, such as gauge lamp circuits.
                                                         328

diode (1) (electron tube). A two-electrode electron tube containing an anode and a cathode. See: equivalent diode.   125
(2) (semiconductor). A semiconductor device having two terminals and exhibiting a nonlinear voltage-current characteristic; in more-restricted usage, a semicondutor device that has the asymmetrical voltage-current characteristic exemplified by a single p-n junction. See: semiconductor.   245

diode characteristic (multielectrode tube). The composite electrode characteristic taken with all electrodes except the cathode connected together.
                                                         125

diode equivalent. The imaginary diode consisting of the cathode of a triode or multigrid tube and a virtual anode to which is applied a composite controlling voltage such that the cathode current is the same as in the triode or multigrid tube.   125

diode function generator (analog computers). A function generator that uses the transfer characteristics of resistive networks containing biased diodes. The desired function is approximated by linear segments whose values are manually inserted by means of potentiometers and switches.   9

diode fuses (semiconductor rectifiers). Fuses of special characteristics connected in series with one or more semiconductor rectifier diodes to disconnect the semiconductor rectifier diode in case of failure and protect the other components of the rectifier. Note: Diode fuses may also be employed to provide coordinated protection in case of overload or short-circuit. See: semiconductor rectifier stack.   208

diode laser. See: injection laser diode (ILD).

dip (electroplating). A solution used for the purpose of producing a chemical reaction upon the surface of a metal. See: electroplating.   328

diplex operation (data transmission). The simultaneous transmission or reception of two signals using a specified common feature, such as a single antenna or a single carrier.   59

diplex radio transmission. The simultaneous transmission of two signals using a common carrier wave. See: radio transmission.   111

dip needle. A device for indicating the angle between the magnetic field and the horizontal. See: magnetometer.   328

dipole. See: dipole antenna; folded dipole antenna; electricdipole; magnetic dipole.

dipole antenna (1) (antennas). Any one of a class of antennas producing a radiation pattern approximating that of an elementary electric dipole. Note: Common usage considers the dipole antenna to be a metal radiating structure which supports a line current distribution similar to that of a thin straight wire so energized that the current has a node only at each end. Syn: doublet antenna.   111
(2) (data transmission). Any one of a class of antennas producing the radiation pattern approximating that of an elementary electric dipole. Note: Common usage considers a dipole antenna to be a metal radiating structure which supports a line current distribution similar to that of a thin straight wire a ½ wavelength long so energized that the current has two nodes, one at each of the far ends.   59
(3) (overhead-power-line corona and radio noise). Any one of a class of antennas having a radiation pattern approximating that of an elementary electric dipole. Note: Common usage considers the dipole antenna to be a metal radiating or receiving structure which supports a line-current distribution similar to that of a thin straight wire, a half wavelength long, so that the current has a node at each end of the antenna.
                                                         411

dipole molecule. A molecule that possesses a dipole moment as a result of the permanent separation of the centroid of positive charge from the centroid of negative charge for the molecule as a whole.   210

dip plating. See: immersion plating.

dip soldering (soldered connections). The process whereby assemblies are brought in contact with the surface of molten solder for the purpose of making soldered connections.   284

direct ac converter (cycloconverter)(self-commutated converters). The alternating current (ac) conversion is accomplished directly, without an intermediate link having different power characteristics, such as direct current (dc) or high-frequency ac.   584

direct-acting machine voltage regulator (power switchgear). A machine voltage regulator having a voltage-sensitive element which acts directly without interposing power-operated means to control the excitation of an electric machine.   103

direct-acting overcurrent trip device. See: direct release (series trip); indirect release (trip); overcurrent release (trip).

direct-acting overcurrent trip device current rating (trip devices for ac and general-purpose dc low-voltage power circuit breakers). The value of current designated by the manufacturer on which trip element calibration marks are based.   560

direct-acting recording instrument. A recording instrument in which the marking device is mechanically connected to, or directly operated by, the primary detector. See: instrument.   328

direct address (computing systems). An address that specifies the location of an operand. See: one-level address.   255, 77

direct addressing (microprocessor assembly language). An addressing mode in which the address is treated as a constant. It is to be distinguished from relative addressing. *See:* **relative addressing.** 466

direct-arc furnace. An arc furnace in which the arc is formed between the electrodes and the charge. *See:* **electrothermis.** 328

direct axis (synchronous machine). The axis that represents the direction of the plane of symmetry of the no-load magnetic-flux density, produced by the main field winding current, normally coinciding with the radial plane of symmetry of a field pole. *See:* **direct-axis synchronous reactance.** 63

direct-axis armature to field transfer function. *See:* **armature to field transfer function.** 63

direct-axis component of armature current. That component of the armature current that produces a magnetomotive force distribution that is symmetrical about the direct axis. 328

direct-axis component of armature voltage. That component of the armature voltage of any phase that is in time phase with the direct-axis component of current in the same phase. *Note:* A direct-axis component of voltage may be produced by: (1) Rotation of the quadrature-axis component of magnetic flux, (2) Variation (if any) of the direct-axis component of magnetic flux, (3) Resistance drop caused by flow of the direct-axis component of armature current. As shown in the phasor diagram, the direct-axis component of terminal voltage, assuming no field magnetization in the quadrature-axis, is given by

$$\mathbf{E}_{ad} = -R\mathbf{I}_{ad} - jX_q\mathbf{I}_{aq}$$

328

direct-axis component of magnetomotive force (rotating machinery). The component of magnetomotive force that is directed along the direct axis. *See:* **asynchronous machine; direct-axis synchronous impedance (rotating machinery).** 63

direct-axis current (rotating machinery). The current that produces direct-axis magnetomotive force. *See:* **direct-axis synchronous reactance.** 63

direct-axis magnetic-flux component (rotating machinery). The magnetic-flux component directed along the direct axis. *See:* **direct-axis synchronous reactance.** 63

direct-axis operational inductance (Ld(s))(synchronous machine parameters by standstill frequency testing)(standstill frequency response testing). The ratio of the Laplace transform of the direct-axis armature flux linkages to the Laplace transform of the direct-axis current, with the field winding short-circuited. 521,565

direct-axis subtransient impedance (rotating machinery). The magnitude obtained by the vector addition of the value for armature resistance and the value for direct-axis subtransient reactance. *Note:* The resist-

ance value to be applied in this case will be a function of frequency depending on rotor iron losses. 63

direct-axis subtransient open-circuit time constant. The time in seconds required for the rapidly decreasing component (negative) present during the first few cycles in the quadrature-axis component of symmetrical armature voltage under suddenly removed symmetrical shot-circuit condition, with the machine running at rated speed to decrease to $1/e$ Δ 0.368 of its initial value. *Note:* If the rotor is made of slid steel no single subtransient time constant exists but a spectrum of time constants will appear in the subtransient region. *See:* **direct-axis synchronous impedance (rotating machinery).** 63

direct-axis subtransient reactance (rotating machinery). The quotient of the initial value of a sudden change in that fundamenatal alternating-current component of armature voltage, which is produced by the total direct-axis primary flux, and the value of this simultaneous change in fundamental alternating-current component of direct-axis armature current, the machine running at rated speed. *Note:* The rated current value is obtained from the tests for the rated current value of direct-axis transient reactance. The rated voltage value is that obtained from a sudden short-circuit test at the terminals of the machine at rated armature voltage, no load. *See:* **direct-axis synchronous reactance (rotating machinery).** 63

direct-axis subtransient short-circuit time constant. The time required for the rapidly changing component, present during the first few cycles in the direct-axis alternating component of a shot-circuit armature current, following a sudden change in operating conditions, to decrease to $1/e$ approx 0.368 of its initial value, the machine running at rated sped. *Note:* The rated current value is obtained from the test for the rated current value of the direct-axis transient reactance. The rated voltage value is obtained from the test for the rated voltage value of the direct-axis transient reactance. *See:* **direct-axis synchronous reactance.** 63

direct-axis subtransient voltage (rotating machinery). The direct-axis component of the terminal voltage which appears immediately after the sudden opening of the external circuit when the machine is running at a specified load, before ay flux variation in the excitation and damping circuits has taken place. 63

direct-axis synchronous impedance (synchronous machine) rotating machinery). The magnitude obtained by the vector addition of the value for armature resistance and the value for direct-axis synchronous reactance. 63

direct-axis synchronous reactance. The quotient of a sustained value of that fundamental alternating-current component of armature voltage that is produced by the total direct-axis flux due to direct-axis armature current and the value of the fundamental alternating-current component of this current, the machine running at rated speed. Unless otherwise specified, the value of synchronous reactance will be that corresponding to rated armature current. For most machines, the armature resistance is negligibly small compared to the synchronous reactance. Hence the synchronous reactance may be taken also as the synchronous impedance.   63

direct-axis transient impedance (rotating machinery). The magnitude obtained by the vector addition of the value for armature resistance and value for direct-axis transient reactance.   63

direct-axis transient open-circuit time constant. The time in seconds required for the root-mean-square alternating-current value of the slowly decreasing component present in the quadrature axis (T'do) component of symmetrical armature voltage on open-circuit to decrease to $1/e \approx 0.368$ of its initial value when the field winding is suddenly short-circuited with the machine running at rated speed. *See:* **direct-axis synchronous impedance (rotating machinery).**   63

direct-axis transient reactance (rotating machinery). The quotient of the initial value of a sudden change in that fundamental alternating-current component of armature voltage, which is produced by the total direct-axis flux, and the value of the simultaneous change in fundamental alternating-current component of direct-axis armature current, the machine running at rated speed and the high-decrement components during the first cycles being excluded. *Note:* The rated current value is that obtained from a three-phase sudden short-circuit test at the terminals of the machine at no load, operating at a voltage such as to give an initial value of the alternating component of current, neglecting the rapidly decaying component of the first few cycles, equal to the rated current. This requirement means that the pr-unit test voltage is equal to the rated current value of transient reactance (per unit). In actual practice, the test voltage will seldom result in initial transient current of exactly rated value, and it will usualy be necessary to determine the reactance from a curve of reactance plotted against voltage. The rated voltage value is that obtained from a three-phase sudden short-circuit test at the terminals of the machine at rated voltage, no load. *See:* **direct-axis synchronous reactance; direct-axis synchronous impedance (rotating machinery).**   63

direct-axis transient short-circuit time constant. The time in seconds required for the root-mean-square value of the slowly decreasing component present in the direct-axis component of the alternating-current component of the armature current under suddenly applied symmetrical short-circuit conditions with the machine running at rated speed, to decrease to $1/e \approx 0.368$ of its initial value.

direct-axis transient voltage (rotating machinery). The direct-axis component of the armature voltage that appears immediately after the sudden opening of the external circuit when running at a specified load, the components that decay very fast during the first few cycles, if any, being neglected. *See:* **direct-axis synchronous reactance.**   63

direct-axis voltage (rotating machinery). The component of voltage that would produce direct-axis current when resistance-limited. *See:* **direct-axis synchronous reactance.**   63

direct-buried transformer (power and distribution transformer). A transformer designed to be buried in the earth with connecting cables.   53

direct capacitances (system of conductors). The direct capacitances of a system of n conductors such as that considered in **coefficients of capacitance (system of conductors)** are the coefficients in the array of linear equations that express the charges on the conductors in terms of their differences in potential, instead of potentials relative to ground.

$$Q_1 = O + C_{12}(V_1 - V_2) + C_{13}(V_1 - V_3)$$
$$+ \cdots + C_{1(n-1)}(V_1 - V_{n-1}) + C_{10}V_1$$

$$Q_2 = C_{21}(V_2 - V_1) + O + C_{23}(V_2 - V_3)$$
$$+ \cdots + C_{2(n-1)}(V_2 - V_{n-1}) + C_{20}V_2$$

$$Q_{n-1} = C_{(n-1)1}(V_{n-1} - V_1, + C_{(n-1)2}(V_{n-1}$$
$$- V_2) + \cdots + O + C_{(n-1)0}V_{n-1}$$

with $C_{rp} = C_{rp}$ and C_{re} not involved but defined as zero. *Note:* The coefficients of capacitance c are related to the direct capacitances C as follows

$$c_{rp} = -C_{rp}, \text{ for } r \neq p$$

and

$$c_{rr} = \sum_{p=1}^{p=n} C_{rp}$$

Note: The relationships of the direct capacitances in a four-conductor system to the other defined capacitances are given in the table below. As in the diagram, the direct capacitances between the conductors of the

Conductor	Self Capacitance
1	$C_{10} + C_{12} + C_{13}$
2	$C_{20} + C_{21} + C_{23}$
3	$C_{30} + C_{31} + C_{32}$

Capacitance diagram showing the equivalent direct capacitance network of a 4-conductor system.

Pair of Conductors: 1 and 2

$$\text{Plenary Capacitance} = C_{12} + \frac{C_{30}(C_{10} + C_{13})(C_{20} + C_{23}) + C_{10}C_{20}(C_{31}C_{32}) + C_{31}C_{32}(C_{10} + C_{20})}{C_{30}(C_{10} + C_{13} + C_{20} + C_{23}) + (C_{10} + C_{20})(C_{31} + C_{32})}.$$

$$\text{Balanced Capacitance} = C_{12} + \frac{(C_{13} + C_{10})(C_{23} + C_{20})}{C_{13} + C_{10} + C_{23} + C_{20}}.$$

Pair of Conductors: 2 and 3

$$\text{Plenary Capacitance} = C_{23} + \frac{C_{10}(C_{20} + C_{21})(C_{30} + C_{31}) + C_{20}C_{30}(C_{31} + C_{21}) + C_{31}C_{21}(C_{20} + C_{30})}{C_{10}(C_{20} + C_{21} + C_{30} + C_{31}) + (C_{20} + C_{30})(C_{31} + C_{21})}.$$

$$\text{Balanced Capacitance} = C_{23} + \frac{(C_{21} + C_{20})(C_{31} + C_{30})}{C_{21} + C_{20} + C_{31} + C_{30}}.$$

Pair of Conductors: 3 and 1

$$\text{Plenary Capacitance} = C_{31} + \frac{C_{20}(C_{10} + C_{12})(C_{30} + C_{32}) + C_{10}C_{30}(C_{32} + C_{12}) + C_{12}C_{32}(C_{10} + C_{30})}{C_{20}(C_{10} + C_{12} + C_{30} + C_{32}) + (C_{10} + C_{30})(C_{12} + C_{32})}.$$

$$\text{Balanced Capacitance} = C_{31} + \frac{(C_{32} + C_{30})(C_{12} + C_{10})}{C_{32} + C_{30} + C_{12} + C_{10}}.$$

system, including ground, are indicated as C_{10}, C_{12}, C_{13}, etcetera. Here the subscript 0 is used to denote the ground (nth)conductor.   210

direct commutation (circuit properties)(self-commutated converters). A commutation between two principal switching branches without the involvement of other switching branches. In converters using devices such as power transistors or gate turn-off thyristors, this is accomplished by turning off the switch in the outgoing branch and turning on the switch in the incoming branch. In converters using circuit-commutated thyristors, commutating capacitors coupled to the switching branches turn off the outgoing switch when the incoming switch is turned on.   584

direct component (illuminating engineering). That portion of the light from a luminaire which arrives at the work-plane without being reflected by room surfaces.   167

direct-connected exciter (rotating machinery). An exciter mounted on or coupled to the main machine shaft so that both machines operate at the same speed. *See:* **asynchronous machine.**   63

direct-connected system. *See:* **headquarters system.**

direct coupled (electrical heating applications to melting furnaces and foreheaths in the glass industry). The power modulation device is conductively connected directly to the electrodes carrying current into the molten glass.   520

direct-coupled amplifier (signal-transmission system). A direct-current amplifier in which all signal connections between active channels are conductive. *See:* **signal.**   188

direct-coupled attenuation (transmit-receive, pre-transmit-receive, and attenuator tubes). The insertion loss measured with the resonant gaps, or their functional equivalent, short-circuited.   333

direct coupling. The association of two or more circuits by means of self-inductance, capacitance, resistance, or a combination of these that is common to the circuits.   313

direct current (dc) (1) (electric installations on shipboard). A unidirectional current in which the changes in value are either zero or so small that they may be neglected. (As ordinarily used, the term designates a practically nonpulsating current.)   3

(2) (power cable systems). Unidirectional current; as used in this guide, the term denotes a practically nonpulsating current.   437

direct-current amplifier (1). An amplifier capable of amplifying waves of infinitesimal frequency. *See:* **amplifier**   111

(2) (signal-transmission system). An amplifier capable of producing a sustained single-valued, unidirectional output in response to a similar but smaller input. *Note:* It generally employs between stages either resistance coupling alone or resistance coupling combined with other forms of coupling. *See:* **amplifier.**   59,188

direct-current balance (amplifiers). An adjustment to avoid a change in direct-current level when changing gain. *See:* **amplifier.**   185

direct-current balancer (electric installations on shipboard). A machine which comprises two or more similar dc machines (usually with shunt or compound excitation) directly coupled to each other and connected in series across the outer conductors of a multiple-wire system of distribution, for the purpose of maintaining the potentials of the intermediate conductors of the system, which are connected to the junction points between the machines.   3

direct-current blocking voltage rating (rectifier circuit element). The maximum continuous direct-current reverse voltage permitted by the manufacturer under stated conditions. *See:* **rectifier circuit element.**   237,66,208

direct-current (dc) breakdown voltage (gas-tube surge-protective devices)(low-voltage air gap surge-protective devices). The minimum slowly rising direct-current (dc) voltage that will cause breakdown or sparkover when applied across the terminals of an arrester.   490, 556

direct-current (dc) circuit breaker (72) (power system device function numbers). A circuit breaker that is used to close and interrupt a dc power circuit under normal conditions or to interrupt this circuit under fault or emergency conditions.   402

direct-current commutating machine (electric installations on shipboard). A machine that comprises a magnetic field excited from a dc source or formed of permanent magnets, and armature and a commutator-connected therewith. Specific types of dc commutating machines are dc generators, motors, synchronous converters, boosters, balancers, and dynamotors.   3

direct-current compensator. *See:* **direct-current balancer.**

direct-current component (of a total current) (power switchgear). That portion of the total current which constitutes the asymmetry.   103

direct-current component of a composite picture signal, blanked picture signal, or picture signal (television). The difference in level between the average value, taken over a specified time interval, and the peak value in the black direction, which is taken as zero. *Note:* The averaging period is usually one line interval or greater. *See:* **television.**   178

direct current (dc) converter (self-commutated converters). A converter for changing dc power at a given voltage to dc power at a higher or lower voltage.   584

direct-current distribution. The supply, to points of utilization, of electric energy by direct current from its point of generation or conversion.   64

direct-current dynamic short-circuit ratio. The ratio of the maximum transient value of a current, after a suddenly applied short circuit, to the final steady-state value.   264

direct-current electric locomotive. An electric locomotive that collects propulsion power from a direct-current distribution system. *See:* **electric locomotive.**   328

direct-current electron-stream resistance (electron tubes). The quotient of electron-stream potential and the direct-current component of stream current. *See:* **electrode current (electron tube).**   125

direct-current erasing head (magnetic recording). One that uses direct current to produce the magnetic field necessary for erasing. *Note:* Direct-current erasing is achieved by subjecting the medium to a unidirectional field. Such a medium is, therefore, in a different magnetic state than one erased by alternating current. *See:* **photograph pickup.**   176

direct current (dc) form factor (converter characteristics)(self-commutated converters). Of a periodic function, the ratio of the rms (root-mean-square) value to the mean value, averaged over a full period of the function.   584

direct-current generator. A generator for production of direct-current power.   63

direct-current (dc) generator-commutator exciter (excitation systems for synchronous machines). An exciter whose energy is derived from a dc generator. The exciter includes a dc generator with its commutator and brushes. It is exclusive of input control elements. The exciter may be driven by a motor, prime mover, or by the shaft of the synchronous machine.   507

direct-current (dc) holdover (gas-tube surge-protective devices)(low-voltage air gap surge-protective devices). In applications where direct-current (dc) voltage exists on a line, a holdover condition is one in which a surge-protective device continues to conduct after it is subjected to an impulse large enough to cause breakdown. Factors which affect the time required to recover from the conducting state include the dc voltage and the current.   490, 556

direct-current (dc) holdover voltage (low-voltage air gap surge-protective devices)(gas-tube surge-protective devices). The maximum dc voltage across the terminals of an arrester under which it may be expected to clear and return to the high impedance state after the passage of a surge, under specified circuit conditions.   556, 490

direct current (dc) input current (converters having dc input)(self-commutated converters). The mean value of the direct current into the input terminals, taken over one period of the ripple current into those terminals.   584

direct current (dc) input power (converters having dc input)(self-commutated converters). The product of the dc supply voltage and the dc input current (mean values of both as defined in dc supply voltage and dc input current).   584

direct-current leakage. *See:* **controlled overvoltage test.**

direct current (dc)-linked ac (alternating current) converter (self-commutated converters). A converter comprising a rectifier and an inverter, with an inmtermediate dc link. *Note:* This definition is intended to include only those circuits in which the dc link is readily identified or explicit, and not those circuits having an implicit dc link but no single pair of conductors that can be identified as the dc link.   584

direct-current magnetic biasing (magnetic recording) Magnetic biasing accomplished by the used of direct current. *See:* **phonograph pickup.**   176

direct-current neutral grid. A network of neutral conductors, usualy grounded, formed by connecting together within a given area of all the neutral conductors of a low-voltage direct-current supply system. *See:* **center of distribution.**   64

direct-current offset (amplifiers). A direct-current level that may be added to the input signal, referred to the input terminals. *See:* **amplifier**   185

direct-current (dc) offset (power fault effects). The difference between the symmetrical current wave and the actual current wave during a power system transient condition. Mathematically, the actual fault current can be broken into two parts, a symmetrical alternating component and a unidirectional (dc) component, either or both when decreasing magnitudes (usually both). The unidirectional component can be of either polarity, but will not change polarity, and will disappear at some predetermined rate. 404

direct-current (dc) offset factor (power fault effects). The ratio of the peak fault current to the peak symmetrical value. 404

direct-current (dc) overcurrent relay (76) (power system device function numbers). A relay that functions when the current in a dc circuit exceeds a given value. 402

direct-current quadruplex system. A direct-current telegraph system that affords simultaneous transmission of two messages in each direction over the same line, operation being obtained by superposing neutral telegraph upon polar telegraph. *See:* **telegraphy.** 328

direct-current rated (thyristor converter). The current in terms of which all test and service current ratings are specified (for example, the per-unit base), except in the case of high-peak loads which are specified in erms of peak load duty. 121

direct-current relay. *See:* **relay, direct-current**

direct-current (dc) reclosing relay (82) (power system device function numbers). A relay that controls the automatic closing and reclosing of a dc circuit interrupter, generally in response to load circuit conditions. 402

direct-current (dc) restoration (television). The reestablishment of the dc and low-frequency components of a video signal that have been lost by ac transmission. 18

direct-current self-synchronous system. A system for transmitting angular position or motion, comprising a transmitter and one or more receivers. The transmitter is an arrangement of resistors that furnishes the receiver with two or more voltages that are functions of transmitter shaft position. The receiver has two or more stationary coils that set up a magnetic field causing a rotor to take up an angular position corresponding to the angular position of the transmitter shaft. *See:* **synchro system.** 328

direct current (dc) supply voltage (converters having dc input)(self-commutated converters). The mean value of the direct voltage between the input terminals, taken over one period of the ripple voltage appearing between the input terminals. 584

direct-current telegraphy. That form of telegraphy in which, in order to form the transmitted signals, direct current is supplied to the line under the control of the transmitting apparatus. *See:* **telegraphy.** 328

direct-current (dc) transmission (television). A form of transmission in which the dc component of the video signal is transmitted. *Note:* In an amplitude-modulated signal with dc transmission, the black level is represented always by the same value of envelope. In a frequency-modulated signal with dc transmission, the black level is represented always by the same value of the instantaneous frequency. 18

direct-current transmission (1) (electric energy). The transfer of electric energy by direct current from its source to one or more main receiving stations. *Note:* For transmitting large blocks of power, high voltage may be used such as obtained with generators in series, rectifiers, etcetera. *See:* **direct-current distribution.** 64

(2) (television). A from of transmission in which the direct-current component of the video signal is transmitted. *Note:* In an amplitude-modulated signal with direct-current transmission, the black level is represented always by the same value of envelope. In a frequency-modulated signal with direct-current transmission, the black level is represented always by the same value of the instantaneous frequency. 163

direct-current winding of rectifier transformer (power and distribution transformer). The secondary winding that is conductively connected to the main electrodes of the rectifier, and that conducts the direct current of the rectifier. 53

direct dc (direct current) converter (dc chopper)(self-commutated converters). The dc conversion is accomplished directly, without an intermediate ac (alternating current) link. 584

direct digital control (electric power systems). A mode of control wherein digital computer outputs are used to directly control a process. *See:* **power system.** 200,94

direct distance dialing (DDD) telephone switching systems). The automatic establishing of toll calls in response to signals from the calling device of a customer. 55

direct-drive machine. An electric driving machine the motor of which is directly connected mechanically to the driving sheave, drum, or shaft without the use of belts or chains, either with or without intermediate gears. 328

directed branch (network analysis). A branch having an assigned direction. *Note:* In identifying the branch direction, the branch jk may be thought of as outgoing from node j and incoming at node k. Alternatively, branch jk may be thought of as originating or having its input at node j and terminating or having its output at node k. The assigned direction is conveniently indicated by an arrow pointing from node j toward node k. 282

directed graph (software). A graph whose edges are unidirectional. *See:* **graph.** 434

directed reference flight (navigation aid terms). That type of stabilized flight which obtains control information from external signals which may be varied as necessary to direct the flight; for example, flight of a guided missile or a target aircraft. 526

direct feeder. A feeder that connects a generating station, substation, or other supply point to one point of utilization. *See:* **radial feeder; center of distribution; auxiliary device to an instrument.** 328

direct glare (illuminating engineering). Glare resulting from high luminances or insufficiently shielded light sources in the field of view. It usually is associated with bright areas, such as luminaires, ceilings, and windows which are outside the visual task or region being viewed.   167

direct grid bias. The direct component of grid voltage. *See:* **electrode voltage (electron tube).**   125

direct-indirect lighting (illuminating engineering). A variant of general diffuse lighting in which the luminaires emit little or no light at angles near the horizontal.   167

direct inductance coupling. *See:* **inductance coupling (communication circuits).**

direct interelectrode capacitance (electron tubes). The direct capacitance between ay two electrodes excluding all capacitance between either electrode and any other electrode or adjacent body.   244

direct inward dialing (DID) (telephone switching systems). A private automatic branch exchange or centrex service feature that permits outside calls to be dialed directly to the stations.   55

direction (navigation aid terms). The position of one point in space relative to another without reference to the distance between them; direction may be either three dimensional or two dimensional, and it is not an angle, but is often indicated in terms of its angular difference from a reference direction. *Note:* Five terms used in navigation--azimuth, bearing, course, heading, and track--involve measurement of angles from reference directions. To specify the reference directions, certain modifiers are used. These are: true, magnetic, compass, relative, grid and gyro.   526

directional antenna. An antenna having the property of radiating or receiving radio waves more effectively in some directions than others. *Note:* This term is usually applied to an antenna whose maximum directivity is significantly greater than that of a halfway dipole.
                                                                111

directional comparison protection (power switchgear). A form of pilot protection in which the relative operating conditions of the directional units at the line terminals are compared to determine whether a fault is in the protected line section.   127, 103

directional-conical reflectance, $\rho(\theta_r,\phi_r;\omega_r)$**(illuminating engineering).** Ratio of reflected flux collected through a conical solid angle to essentially collimated incident flux. *Note:* The direction of incidence must be specified, and the direction and extent of the cone must be specified.   167

directional-conical transmittance, $\tau(\theta_r,\phi_r;\omega_r)$**(illuminating engineering).** Ratio of transmitted flux collected through a conical solid angle to essentially collimated incident flux. *Note:* The direction of incidence must be specified, and the direction and extent of the cone must be specified.   167

directional control (as applied to a protective relay or relay scheme) (power switchgear). A qualifying term that indicates a means of controlling the operating force in a nondirectional relay so that it will not operate until the two or more phasor quantities used to actuate the controlling means (directional relay) are in a predetermined band of phase relations with a reference input.   103

directional coupler (1) (transmission lines). A transmission coupling device for separately (ideally) sampling (through a knowncoupling loss for measuring purposes) either the forward (incident) or the backward (reflected) wave in a transmission line. *Notes:* (A) Similarly, it may be used to excite in the transmission line either a forward or backward wave. (B) A unidirectional coupler has available terminals or connections for sampling only one direction of transmission; a bidirectional coupler has available terminals for sampling both directions. *See:* **auxiliary device to an instrument.**   328

(2) (waveguide components). A four port junction consisting of two waveguides coupled together in such a manner that a single traveling wave in either guide will induce a single traveling wave in the other, the direction of the latter wave being determined by the direction of the former.   166

(3) (fiber optics). *See:* **tee coupler.**

directional-current tripping. *See:* **directional-overcurrent protection; directional-overcurrent relay.**

directional gain directivity index (transducer) (audio and electroacoustics). In decibels, 10 times the logarithm to the base 10 of the directivity factor.
                                                                176

directional-ground relay (power switchgear). A directional relay used primarily to detect single-phase-to-ground faults, but also sensitive to double-phase-to-ground faults. *Note:* This type of relay is usually operated from the zero-sequence components of voltage and current, but is sometimes operated from negative-sequence quantities.   127, 103

directional gyro. A two-degree-of-freedom gyro with provision for maintaining the spin axis approximately horizontal. In this gyro, an output signal is produced by gimbal angular displacement of the case about an axis which is nominally vertical.   46

directional gyro electric indicator. An electrically driven device for use in aircraft for measuring deviation from a fixed heading.   328

directional-hemispherical reflectance, $\rho(\theta_r,\phi_r;2\pi)$**(illuminating engineering).** Ratio of reflected flux collected over the entire hemisphere to essentially collimated incident flux. (See Figure below.) *Note:* The direction of incidence must be specified.   167

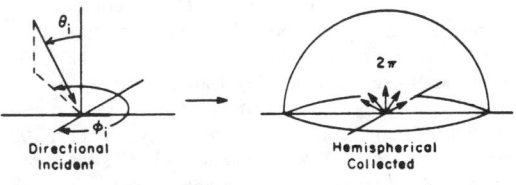

Directional   Hemispherical
Incident   Collected

directional-hemispherical transmittance, $\tau(\theta_r,\phi_r;2\pi)$ **(illuminating engineering).** Ratio of transmitted flux collected over the entire hemisphere to essentially collimated incident flux. (See Figure below.) *Note:* The direction of incidence must be specified.
                                                                167

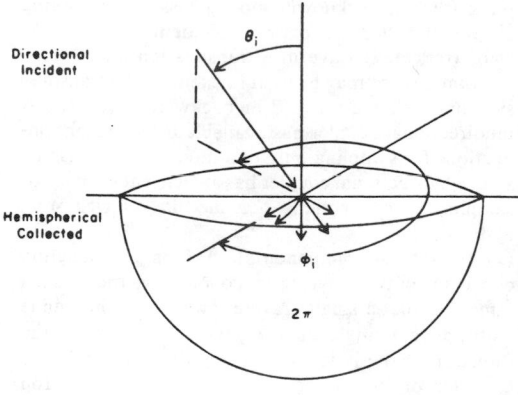

Directional Incident

Hemispherical Collected

θ_i

ϕ_i

2π

directional homing (navigation aid terms). The process of homing wherein the navigational quantity maintained constant is the bearing.   526

directional lighting (illuminating engineering). Lighting provided on the work plane or on an object. Light that is predominantly from a preferred direction.   167

directional localizer (ILS [instrument landing system])(navigation aid terms). A localizer in which maximum energy is directed close to the runway centerline, thus minimizing extraneous reflections.   526

directional microphone. A microphone the response of which varies significantly with the direction of sound incidence. *See:* **microphone.**   328

directional null (antennas). A sharp minimum in the radiation pattern of an antenna which has been produced for the purpose of direction finding or the suppression of unwanted radiation in a specified direction.   111

directional-null antenna. An antenna whose radiation pattern contains one or more directional nulls. *See:* **null steering antenna.**   111

directional-overcurrent protection (power switchgear). A method of protection in which an abnormal condition within the protected equipment is detected by the current being in excess of a predetermined amount and in a predetermined band of phase relations with a reference input.   127, 103

directional-overcurrent relay (power switchgear). A relay consisting of an overcurrent unit and a directional unit combined to operate jointly.   127, 103

directional pattern (radiation pattern). The directional pattern of an antenna is a graphical representation of the radiation or reception of the antenna as a function of direction. *Note:* Cross sections in which directional patterns are frequently given are vertical planes and the horizontal planes or the principal electric and magnetic polarization planes. *See:* **antenna.**   328

directional phase shifter (directional phase changer) (non-reciprocal phase shifter). A passive changer in which the phase change for transmission in one direc-

tion differs from that for transmission in the opposite direction. *See:* **transmission line.**   179

directional-power relay (1) (power switchgear). A relay that operates in conformance with the direction of power.   103

(2) (32) (power system device function numbers). A relay which operates on a predetermined value of power flow in a given direction, or upon reverse power such as that resulting from the monitoring of a generator upon loss of its prime mover.   402

directional-power tripping. *See:* **directional-power relay.**   103

directional relay (power switchgear). A relay that responds to the relative phase position of a current with respect to another current or voltage reference. *Note:* The above definition which applies basically to a single phase directional relay may be extended to cover a polyphase directional relay.   127, 103

directional response pattern (beam pattern) (electroacoustics) (transducer used for sound emission or reception). A description, often presented graphically, of the response of the transducer as a function of the direction of the transmitted or incident sound waves in a specified plane and at a specified frequency. *Notes:* (1) A complete description of the directional response pattern of a transducer would require a three-dimensional presentation. (2) The directional response pattern is often shown as the response relative to the maximum response. *See:* **loudspeaker.**   176

direction finder (DF). *See:* **radio direction finder.**

direction-finder antenna. Any antenna used for radio direction finding. *See:* **navigation.**   187,13

direction finder deviation (navigation aid terms). The amount by which an observed radio bearing differs from the corrected bearing.   526

direction-finder noise level (in the absence of the desired signals). The average power or root-mean-square voltage at any specified point in a direction-finder system circuit. *Note:* In radio-frequency and audio channels, the direction-finder noise level is usually measured in terms of the power dissipated in suitable termination. In a video channel, it is customarily measured in terms of voltage across a given impedance or of the cathode-ray deflection. *See:* **navigation.**   278,187,13

direction-finder sensitivity. The field strength at the direction-finder antenna, in microvolts per meter, that produces a ratio of signal-plus-noise to noise equal to 20 decibels in the receiver output, the direction of arrival of the signal being such as to produce maximum pickup in the direction-finder antenna system. *See:* **navigation.**   278,187,13

direction finding (DF) antenna. Any antenna used for radio direction finding.   13

direction finding (DF) antenna system. One or more DF antennas, their combining circuits and feeder systems, together with the shielding and all electrical and mechanical items up to the termination at the receiver input terminals.   13

See: **radio direction finder.**

direction of lay (cables). The lateral direction, designated as left-hand or right-hand, in which the elements of a cable run over the top of the cable as they recede from an observer looking along the axis of the cable. *See:*   57,64

direction of polarization (radio wave propagation) (1) (linearly polarized wave). The direction of the electric intensity.

(2) (elliptically polarized wave). The direction of the major axis of the electric vector ellipse. *See:* **radiation; radio wave propagation.**   146

direction of propagation (1)(point in a homogeneous isotropic medium). (A) The normal to an equiphase surface taken in the direction of increasing phase lag. *See:* **radiation.**   328
(B) The direction of time-average energy flow. *Notes:* (1) In a uniform waveguide the direction of propagation is often taken along the axis. (2) In the case of a uniform lossless waveguide the direction of propagation at every point is parallel to the axis and in the direction of time-average energy flow. *See:* **radiation; waveguide.**   267,146

(2) (radio wave propagation). At any point in a medium, the direction of time-average energy flow.   146

(3) (waveguide). The direction of time average energy flow in a given mode. *Note:* In the case of a uniform lossless waveguide, the direction of propagation at every point is parallel to the axis.   267

direction of rotation of phasors (power and distribution transformer). Phasor diagrams should be drawn so that an advance in phase of one phasor with respect to another is in the counterclockwise direction. In the following figure, phasor 1 is 120 degrees in advance of phasor 2, and the phase sequence is 1, 2, 3.   53

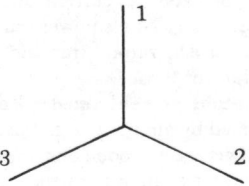

direction operation. Operation by means of a mechanism connected directly to the main operating shaft or an extension of the same.   103,202,27

directive gain *. *See:* **directivity.**
*Deprecated

directive gain in physical media. In a given direction and at a given point in the far field, the ratio of the power flux per unit area from an antenna to thepower flux per unit area from an isotropic radiator at a specified location delivering the same power from the antenna to the medium. *Notes:* (1) The isotropic radiator must be within the smallest sphere containing the antenna. Suggested locations are antenna terminals and points of symmetry if such exist. (2) See IEEE Std 270-1966. Definitions of General (Fundamental and Derived) Electrical and Electronics Terms, and IEEE

Std 211-1969, Definitions of Terms for Radio Wave Propagation, referring to the use of power flux and power flux density, respectively.   111

directivity (1) (antenna) (gain). The value of the directive gain in the direction of its maximum value. *See:* **antenna.**   111,246

(2) (of an antenna) (in a given direction). The ratio of the radiation intensity in a given direction from the antenna to the radiation intensity averaged over all directions. *Notes:* (1) The average radiation intensity is equal to the total power radiated by the antenna divided by 4π. (2) If the direction is not specified, the direction of maximum radiation intensity is implied.   111

(3) (directional coupler). The ratio of the power output at an auxiliary port, when power is fed into the main waveguide or transmission line in the preferred direction, to the power output at the same auxiliary port when power is fed into the main guide or line in the opposite direction, the incident power fed into the main guide or line being the same in each case, and reflectionless terminations being connected to all ports. *Note:* The ratio is usually expressed in decibels.   185

directivity factor (audio and electrocoustics) (1) (transducer used for sound emission). The ratio of the sound pressure squared, at some fixed distance and specified direction, to the mean-square sound pressure at the same distance averaged over all directions from the transducer. *Note:* The distance must be great enough so that the sound pressure appears to diverge spherically from the effective acoustic center of the transducer. Unless otherwise specified, the reference direction is understood to be that of a maximum response. The frequency must be stated. *See:* **loudpeaker.**   176

(2) (transducer used for sound reception). The ratio of the square of the open-circuit voltage produced in response to sound waves arriving in a specified direction to the mean-square voltage that would be produced in a perfectly diffused sound field of the same frequency and mean-square sound pressure. *Notes:* (A) This definition may be extended to cover the case of finite frequency bands whose spectrum may be specified. (B) The average free-field response may be obtained in various ways, such as (1) by the use of a spherical integrator, (2) by numerical integration of a sufficient number of directivity patterns corresponding to different planes, or (3) by integration of one or two directional patterns whenever the pattern of the transducer is known to possess adequate symmetry. *See:* **microphone.**   176

direct lighting (illuminating engineering). Lighting involving luminaires which distribute 90 to 100 percent of the emitted light in the general direction of the surface to be illuminated. The term usually refers to light emitted in a downward direction.   167

direct liquid cooling system (semiconductor rectifiers). A cooling system in which a liquid, received from a constantlyavailable supply, is passed directly over the cooling surfaces of the semiconductor power convert-

er and discharged. *See:* **semiconductor rectifier stack.**
208

direct liquid cooling system with recirculation (semiconductor rectifiers). A direct liquid cooling system in which part of the liquid passing over the cooling surfaces of the semiconductor power converter is recirculated and additional liquid is added as needed to maintain the required temperature, the excess being discharged. *See:* **semiconductor rectifier stack.**   208

directly controlled variable (industrial control) (automatic control). The variable in a feedback control system whose value is sensed to originate the primary feedback signal. *See:* **control system, feedback.**
105

direct on-line starting (rotating machinery). The process of starting a motor by connecting it directly to the supply at a rated voltage. *See:* **asynchronous machine.**
63

direct operation (power switchgear). Operation by means of a mechanism connected directly to the main operating shaft or an extension of the same.   103

direct orbit (communication satellite). An inclined orbit with an inclination between zero and ninety degrees.   74

director element (antennas) (data transmission). A parasitic element located forward of the driven element of an antenna, intended to increase the directive gain of the antenna in the forward direction.
111, 59

directory-assistance call (telephone switching systems). A call placed to request the directory number of a customer.   55

directory number (telephone switching systems). The full complement of digits required to designate a customer in a directory.   55

directory-numbering plan (telephone switching systems). The arrangement whereby each customer is identified by an office and main-station code.
55

direct outward dialing (DOD) (telephone switching systems). A private automatic branch exchange or centrex service feature that permits stations to dial outside numbers without intervention of an attendant.
55

direct-plunger driving machine (elevators). A machine in which the energy is applied by a plunger or piston directly attached to the car frame or platform and that operates in a cylinder under hydraulic pressure. *Note:* **It includes the cylinder and plunger or piston.** *See:* **driving machine (elevators).**   328

direct-plunger elevator. A hydraulic elevator having a plunger or piston directly attached to the car frame or platform. *See:* **elevators.**   328

direct polarity indication (graphic symbols for logic functions). The designation of the internal state produced by the external level of an input, or producing the external level of an output, by the presence or absence of the polarity symbol. *See:* **ANSI/IEEE Std 91-1984, Symbols 3.1-4 through 3.1-9.**   451
See: **mixed logic.**

direct ratio (illuminating engineering). The ratio of the

luminous flux which reaches the work-plane directly to the downward component from the luminaire.
167

direct raw-water cooling system (1) (rectifier). A cooling system in which water, received from a constantly available supply, such as a well or water system, is passed directly over the cooling surfaces of the rectifier and discharged. *See:* **direct liquid cooling system; rectification.**   208

(2) (thyristor controller). A cooling system in which water, received from a constantly available supply, such as a well or water system, is passed directly over the cooling surfaces of the thyristor controller components and discharged.   445

direct raw-water cooling system with recirculation. A direct raw-water cooling system in which part of the water passing over the cooling surfaces of the rectifier is recirculated and raw water is added as needed to maintain the required temperature, the excess being discharged. *See:* **direct liquid cooling system with recirculation; rectification.**   208

direct-recording (facsimile). That type of recording in which a visible record is produced, without subsequent processing, in response to the received signals. *See:* **recording (facsimile).**   12

direct release (series trip) (power switchgear). A release directly energized by the current in the main circuit of a switching device.   103

direct stroke. A lightning stroke direct to any part of a network or electric installation.   244,62

direct-stroke protection (lightning). Lightning protection designed to protect a network or electric installation against direct strokes.   64

direct support maintenance. *See:* **intermediate maintenance.**

direct test (ac high-voltage circuit breaker). A test in which the applied voltage, current, and recovery voltage is obtained from a single power source, which may be comprised of generators, transformers, networks, or combinations of these.   426

direct vacuum-tube current (medical electronics). A current obtained by applying to the part to be treated an evacuated glass electrode connected to one terminal of a generator of high-frequency current (100 to 1000 kilohertz), the other terminal being grounded. *Note:* Deprecated as confusing and as representing an ill-defined and obsolescent procedure.   192

direct-voltage high-potential test (rotating machinery). A test that consists of the application of a specified unidirectional voltage higher than the rated root-mean-square value for a specified time for the purpose of determining (1) the adequacy against breakdown of the insulation system under normal conditions, or (2) the resistance characteristic of the insulation system. *See:* **asynchronous machines.**   63

direct voltage tests definitions (high voltage testing). *See:* **ripple, value of the test voltage.**   150

direct wave (radio wave propagation). A wave propagated directly from a source to a point.   146

direct-wire circuit (one-wire circuit). A supervised circuit, usually consisting of one metallic conductor and a ground return, and having signal receiving equipment responsive to either an increase or a decrease in current. *See:* **protective signaling.** 328

disability glare (illuminating engineering). Glare which reduces visual performance and visibility and often is accompanied by discomfort. *See:* **veiling brightness.** 167

disability glare factor (DGF) (illuminating engineering). A measure of the visibility of a task in a given lighting installation in comparison with its visibility under reference lighting conditions, expressed in terms of the ratio of luminance contrasts having an equivalent effect upon task visibility. The value of DGF takes account of the equivalent veiling luminance produced in the eye by the pattern of luminances in the task surround. 167

disable (supervisory control, data acquisition, and automatic control)station control and data acquisition). A command or condition which prohibits some specific event from proceeding. 570,403

disaster, major storm (transmission and distribution). Designates weather which exceeds design limits of plant and which satisfies all of the following: (1) Extensive mechanical damage to plant. (2) More than a specified percentage of customers out of service. (3) Service restoration times longer than a specified time. *Note:* It is suggested that the specified percentage of customers out of service and restoration times be 10 percent and 24 hours. Percentage of customers out of service may be related to a company operating area rather than to an entire company. Examples of major storm disasters are hurricanes and major ice storms. 112

disc. *See:* **disk; magnetic disk; disk recorder.**

discharge (1) (storage cell). The conversion of the chemical energy of the battery into electric energy. *See:* **charge.** 328
(2) (gas). The passage of electricity through a gas. 244,190

discharge capacity (arrester). *See:* **arrester discharge capacity.**

discharge circuit (surge generator). That portion of the surge-generator connections in which exist the current and voltage variations constituting the surge generated. 64,62

discharge counter (metal-oxide surge arresters for ac power circuits)(surge arrester). A means for recording the number of arrester discharge operations. 583, 430

discharge current (1)(low-voltage air gap surge-protective devices)(gas-tube surge-protective devices). The current that flows through an arrester when sparkover occurs. 556, 490
(2)(metal-oxide surge arresters for ac power circuits). The surge current that flows through an arrester. 583
(3)(surge arrester). The surge current that flows through an arrester when sparkover occurs. 370, 430

discharge current limiting device (series capacitors). A reactor or equivalent device to limit the current

magnitude and frequency of the discharge of the capacitors upon operation of the bypass switch or gap. *See:* **Figure, capacitor group.** 474

discharge detector (ionization or corona detector) (rotating machinery). An instrument that can be connected in or across an energized insulation circuit to detect current or voltage pulses produced by electric discharges within the circuit. *See:* **instrument.** 63

discharge device (1)(capacitor). An internal or external device intentionally connected in shunt with the terminals of a capacitor for the purpose of reducing the residual voltage after the capacitor is disconnected from an energized line. 138
(2)(series capacitors). An internal or external device intentionally connected in parallel with the terminals of a capacitor for the purpose of reducing the residual voltage after the capacitor is disconnected from an energized line. 474

discharge-energy test (rotating machinery). A test for determining the magnitude of the energy dissipated by a discharge or discharges within the insulation. *See:* **asynchronous machine.** 63

discharge extinction voltage (ionization or corona extinction voltage) (rotating machinery). The voltage at which discharge pulses that have been observed in an insulation system, using a discharge detector of specified sensitivity, cease to be detectable as the voltage applied to the system is decreased. *See:* **asynchronous machines.** 63

discharge inception test (corona inception test) (rotating machinery). A test for measuring the lowest voltage at which discharges of the specified magnitude recur in successive cycles when an increasing alternating voltage is applied to insulation. *See:* **asynchronous machine.** 63

discharge inception voltage (ionization or corona inception voltage) (1) (rotating machinery). The voltage at which discharge pulses in an insulation system become observable with a discharge detector of specified sensitivity, as the voltage applied to the system is raised. *See:* **asynchronous machine.** 63
(2) (surge arrester). The root-mean-square value of the power-frequency voltage at which discharges start, the measurement of their intensity being made under specified conditions. 244,62

discharge indicator (surge arrester)(metal-oxide surge arresters for ac power circuits). A means for/of indicating that the arrester has discharged. 430, 583

discharge opening (rotating machinery). A port for the exit of ventilation air. 63

discharge oscillations (laser gyro). Periodic variations in voltage and current at the terminals of a direct current discharge tube which are supported by the negative resistance of the discharge tube itself. 46

discharge probe (ionization or corona probe) (rotating machinery). A portable antenna, safely insulated, and designed to be used with a discharge detector for locating sites of discharges in an energized insulation system. *See:* **instrument.** 63

discharge resistor (1)(excitation systems for synchronous machines). A resistor that, upon interruption of excitation source current, is connected across the field windings of a synchronous machine or an exciter to limit the transient voltage in the field circuit and to hasten the decay of field current of the machine. 507

(2)(power switchgear). A resistor that, upon interruption of excitation source current, is connected across the field windings of a generator, motor, synchronous condenser, or an exciter to limit the transient voltage in the field circuit and to hasten the decay of field current of these machines. 103

discharge tube. An evacuated enclosure containing a gas at low pressure that permits the passage of electricity through the gas upon application of sufficient voltage. *Note:* The tube is usually provided with metal electrodes, but one form permits an electrodeless discharge with induced voltage. 328

discharge voltage (low-voltage air gap surge-protective devices)(gas-tube surge-protective devices)(metal-oxide surge arresters for ac power circuits)(surge arrester). The voltage that appears across the terminals of an arrester during the passage of discharge current. 556, 490, 583, 370, 430

discharge voltage-current characteristic (metal-oxide surge arresters for ac power circuits)(gas-tube surge protective devices)(surge arrester). The variation of the crest values of discharge voltage with respect to discharge current. *Note:* This characteristic is normally shown as a graph based on three or more current-surge measurements of the same wave shape but of different crest values. 370, 430, 490, 583

discharge voltage-time curve (arrester). *See:* arrester discharge voltage-time curve.

discharge withstand current (surge arrester)(metal-oxide surge arresters for ac power circuits). The specified magnitude and wave shape of a discharge current that can be applied to an arrester a specified number of times without causing damage to it. 430, 583

discomfort glare (illuminating engineering). Glare which produces discomfort. It does not necessarily interfere with visual performance or visibility. 167

discomfort glare factor* (illuminating engineering). The numerical assessment of the capacity of a single source of brightness, such as a luminaire, in a given visual environment for producing discomfort. (This term is obsolete and is retained for reference and literature searches). 167
*Obsolete

discomfort glare rating, (DGR) (illuminating engineering). A numerical assessment of the capacity of a number of sources of luminance, such as luminaires, in a given visual environment for producing discomfort. It is the net effect of the individual values of index of sensation M, for all luminous areas in the field of view

$$DGR = \Sigma(M)^a$$

where $a = n^{-0.914}$ n = number of sources (ceiling elements) in the field of view. 167

discone antenna. A biconical antenna with one cone having a vertex angle of 180°. 111

disconnect (1) (release) (telephony). To disengage the apparatus used in a telephone connection and to restore it to its condition when not in use. 193

(2) (watthour meter). A conductor, bar, or nut used to open an electrical circuit for isolation purposes. 419

disconnectable device. A grounding device that can be disconnected from ground by the operation of a disconnecting switch, circuit breaker, or other switching device. 91

disconnected position (of a switchgear-assembly removable element) (power switchgear). That position in which the primary and secondary disconnecting devices of the removable element are separated by a safe distance from the stationary element contacts. *Note:* Safe distance, as used here, is a distance at which the equipment will meet its withstand ratings, both power frequency and impulse, between line and load stationary terminals and phase-to-phase and phase-to-ground on both line and load stationary terminals with the switching device in the closed position. 103

disconnecting blade. *See:* blade (disconnecting blade) (of a switching device).

disconnecting cutout (power switchgear). A cutout having a disconnecting blade for use as a disconnecting or isolating switch. 103

disconnecting device (packaging machinery). A device whereby the conductors of a circuit can be disconnected from their source of supply. 429

disconnecting fuse. *See:* fuse disconnecting switch.

disconnecting means (National Electrical Code) (1). A device, or group of devices, or other means by which the conductors of a circuit can be disconnected from their source of supply. 256

(2) (recreational vehicles). The necessary equipment usually consisting of a circuit breaker or switch and fuses, and their accessories, located near the point of entrance of supply conductors in a recreational vehicle and intended to constitute the means of cutoff for the supply to that recreational vehicle. 256

disconnecting or isolating switch (1) (NESC). A mechanical switching device used for changing the connections in a circuit, or for isolating a circuit or equipment from a (the) source of power. *Note:* It is required to carry normal load current continuously, and also abnormal or short-circuit current for short intervals as specified. It is also required to open or close circuits either when negligible current is broken or made, or when no significant change in the voltage across the terminals of each of the switch poles occurs. *Syn:* disconnector; isolator. 494, 391, 445, 103

(2) (high-power switchgear). A switch used for changing the connections in a circuit, or for isolating a circuit or equipment from the source of power. *Note:* It is required to carry normal load current continuously and also abnormal or short-circuit currents for short intervals as specified. It is also required to open or close circuits either when negligible current is broken

or made, or when no significant change in the voltage across the terminals of each of the switch poles occurs. Some disconnecting switches have some inherent load-break ability which can best be evaluated by the user, based on experience under operating conditions. 443

disconnection (control) (industrial control). Connotes the opening of a sufficient number of conductors to prevent current flow. 206

disconnector. A switch that is intended to open a circuit only after the load has been thrown off by some other means. *Note:* Manual switches designed for opening loaded circuits are usually installed in circuit with disconnectors, to provide a safe means for opening the circuit under load. 178

disconnect signal (telephony). A signal transmitted from one end of a subscriber line or trunk to indicate that the relevant party has released. 193

disconnect-type pothead. A pothead in which the electric continuity of the circuit may be broken by physical separation of the pothead parts, part of the pothead being on each conductor end after the separation. *See:* **pothead.** 4

discontinuity (1). An abrupt nonuniformity in a uniform waveguide or transmission line that causes reflected waves. *See:* **capacitance, discontinuity; waveguide.** 185

(2) (inductive coordination). An abrupt change at a point in the physical relations of electric supply and communication circuits or in electrical parameters of eithe circuit, that would materially affect the coupling. *Note:* Although technically included in the definition, transpositions are not rated as discontinuities because of their application to coordination. *See:* **inductive coordination.** 328

discrete sentence intelligibility. The percent intelligibility obtained when the speech units considered are sentences (usually of simple form and content). *See:* **volume equivalent.** 328

discrete word intelligibility. The percent intelligibility obtained when the speech units considered are words (usually presented so as to minimize the contextual relation between them). *See:* **volume equivalent.** 328

discrimination (1) (any system of transducer). The difference between the losses at specified frequencies, with the system or transducer terminated in specified impedances. *See:* **transmission loss.** 328

(2) (radar). Separation or identification of the differences between nonsimilar signals. 13

discriminator (radar). A circuit in which the output is dependent upon how an input signal differs in some aspect from a standard or from another signal. 13

discriminator, amplitude (pulse techniques). *See:* **discriminator, pulse-height.**

discriminator, constant-fraction pulse-height. A pulse-height discriminator in which the threshold changes with input amplitude in such a way that the triggering point corresponds to a constant fraction of the input pulse height. 117

discriminator, pulse-height (pulse techniques). A circuit that produces a specified output pulse if and only if it receives an input pulse whose amplitude exceeds an assigned value. *See:* **pulse.** 335

dish (1) (radio practice) (data transmission). A reflector, the surface of which is concave, as, for example, a part of a sphere or of a paraboloid of revolution. 59

(2) (radar). A colloquial term for a parabolic microwave antenna reflecting surface. 13

disk recorder (phonograph techniques). A mechanical recorder in which the recording medium has the geometry of a disk. *See:* **phonograph pickup.** 176

dispatching system (mining practice). A system employing radio, telephone, and/or signals (audible or light) for orderly and efficient control of the movements of trains of cars in mines. *See:* **mine fan signal system; mine radio telephone system.** 328

dispatch operation (radio-communication circuit). A method for permitting a maximum number of terminal devices to have access to the same two-way radio communication circuit. *See:* **channel spacing.** 181

dispenser cathode (electron tubes). A cathode that is not coated but is continuously supplied with suitable emission material from a separate element associated with it. *See:* **electron tube.** 244

dispersed magnetic power tape (impregnated tape). *See:* **magnetic powder-impregnated tape.**

dispersion (fiber optics). A term used to describe the chromatic or wavelength dependence of a parameter as opposed to the temporal dependence which is referred to as distortion. The term is used, for example, to describe the process by which an electromagnetic signal is distorted because the various wavelength components of that signal have different propagation characteristics. The term is also used to describe the relationship between refractive index and wavelength. *Note:* Signal distortion in an optical waveguide is caused by several dispersive mechanisms: waveguide dispersion, material dispersion, and profile dispersion. In addition, the signal suffers degradation from multimode "distortion" , which is often (erroneously) referred to as multimode "dispersion". *Syn:* **chromatic dispersion (redundant).** *See:* **distortion; intramodal distortion; material dispersion; material dispersion parameter; multimode distortion; profile dispersion; profile dispersion parameter; waveguide dispersion.** 433

dispersion relation (radio wave propagation). In a source-free region, the functional relation between angular frequency ω and wave vector k for plane waves with the exponential factor exp $[j(\omega t - k \cdot r)]$. 146

dispersive bandwidth (dispersive delay line). The operating frequency range over which the delay dispersion is defined. 81

dispersive Doppler frequency. *See:* **differential Doppler frequency.** 146

dispersive medium (radio wave propagation). A medium whose constitutive relations for plane-wave fields depend on frequency (temporal dispersion) or wave vector (spatial dispersion). 146

displacement power factor (1)(converter characteristics)(self-commutated converters). The displacement component of power factor; the ratio of the active power of the fundamental wave, in watts, to the apparent power of the fundamental wave, in volt-amperes. It is also equal to $\cos\phi_1$, the cosine of the phase displacement angle between the fundamental component of the voltage and current on the ac (alternating current) side of a converter. 584

(2)(harmonic control and reactive compensation of static power converters). The displacement component of power factor; the ratio of the active power of the fundamental wave, in watts, to the apparent power of the fundamental wave. in volt-amperes (including the exciting current of the thyristor converter transformer). 533

(3) (thyristor). The ratio of the active power of the fundamental wave, in watts, to the apparent power of the fundamental wave in voltamperes. This is the cosine of the phase angle by which the fundamental current lags the fundamental voltage. This is the power factor as seen in ut ility metering by watthour and varhour meters assuming that the ac voltages are sinusoidal. 445

display (1)(computer applications). (1) To present data visually. (2) The result of a display process. See: copy. 571

(2)(navigation aid terms). The visual representation of output data. 526

(3)(watthour meters). A means for visually identifying and presenting measured or calculated quantities and other information. 485

(4) (radar). (A) The visual representation of output data. See individual definitions and illustrations of various radar display formats. Note: The letter designations from A to P were devised in the years during and following World War II in an effort to standardize nomenclature. Several of these letter designations are now rarely if ever used, as noted in the individual definitions, but they are still found in some technical literature. The additional designations "plan position indication (PPI)" and "range-height indication (RHI)" are also defined. The standardized type designations do not cover all possible display formats.
(B) A colloquial term for the reflecting surface of a paraboloidal-reflector antenna. 13

(5) (test, measurement and diagnostic equipment). A mechanical, optical, electro-mechanical, or electronic device for presenting information to the operator or maintenance technician about the state or condition of the unit under test or the checkout equipment itself. 54

(6) (oscilloscopes). The visual presentation on the indicating device of an oscilloscope. 106

display flatness (spectrum analyzer). The total variation in displayed amplitude over a specified span, decibel (dB). Note: Display flatness is closely related to frequency response. The main difference is that the tuning control of the spectrum analyzer is not readjusted to center the display. 390

display frequency (spectrum analyzer). The input frequency as indicated by the spectrum analyzer (Hz). 390

display law (spectrum analyzer). The mathematical law that defines the input-output function of the instrument. (1) linear. A display in which the scale divisions are a linear function of the input voltage. (2) square law (power). A display in which the scale divisions are a linear function of the input power. (3) logarithmic. A display in which the scale divisions are a logarithmic function of the input signal. 390

display primaries (color television reception). The colors of constant chromaticity and variable luminance produced by the receiver or any other display device that, when mixed in proper proportions, are used to produce other colors. Note: Usually the three primaries used are red, green, and blue. 18

display reference level (spectrum analyzer). A designated vertical position representing specified input levels. The level may be expressed in dBm, volts, or any other units. 390

displays (nuclear power generating station). Devices which convey information to the operator. 358

display storage tube. A storage tube into which the information is introduced as an electric signal and read at a later time as a visible output. See: storage tube. 174

display tube. A tube, usually a cathode-ray tube, used to display data. 255,77

disposal (handling and disposal of transformer grade insulating liquids containing PCBs). Intentionally or accidentally to discard, throw away, or otherwise complete or terminate the useful life of polychlorinated biphenyls (PCBs) and PCB items. Disposal includes spills, leaks, and other uncontrolled discharges of PCBs as well as actions related to containing, transporting, destroying, degrading, decontaminating, or confining PCBs and PCB items. Examples: Spill, chemical dechlorination, landfill, incineration. 586

disruptive discharge (1)(metal-oxide surge arresters for ac power circuits)(surge arresters). The sudden and large increase in current through an insulating medium, due to the complete failure of the medium under electrostatic stress. 583, 430

(2) (high voltage testing). The term "disruptive discharge" relates to phenomena associated with the failure of insulation under electrical stress, in which the discharge completely bridges the insulation under test, reducing the voltage between the electrodes to zero or nearly to zero. It applies to electrical breakdown in solid, liquid, and gaseous dielectrics and to combinations of these. Discharges between intermediate electrodes or conductors, as well as momentary flashovers (snap-overs), may occur with alternating voltage without permanent reduction of the test voltage to zero. The occurrence of snap-overs may depend on the characteristics of the test circuit. Whether such an occurrence constitutes a failure or not, and whether it requires a repetition of the test, should be established by the appropriate apparatus standard. 150

(3) (high voltage testing). For direct voltage, alternat-

ing voltage, and impulse voltages chopped at or after the crest, the disruptive discharge voltage is the value of the test voltage causing disruptive discharge. For definition of this value, *see* **value of test voltage.** For impulse voltages chopped on the front, the disruptive discharge voltage is defined as the voltage at the instant when disruptive discharge occurs. Disruptive discharge voltages are subject to random variations and, usually a number of observations must be made in order to obtain a statistically significant value of the voltage. The text procedures are generally based on statistical considerations. The same document also gives information on the statistical evaluation of test results.   150

disruptive discharge probability p (high voltage testing). The disruptive discharge probability is the probability that the application of a certain prospective voltage value of a given shape will cause disruptive discharge.   150

disruptive discharge voltage (1). The value of the test voltage for which disruptive discharge takes place. *Note:* The disruptive discharge voltage is subject to statistical variation which can be expressed in different ways as, for example, by the mean, the maximu, and the minimum values of series of observations, or by the mean and standard deviation from the mean, or by a relation between voltage and probability of a disruptive discharge. *See:* **test voltage and current.**   307

(2) (50 percent) (surge arresters). The voltage that has a 50-percent probability of producing a disruptive discharge. *Note:* The term applies mostly to impulse tests and has significance only in cases when the loss of electric strength resulting from a disruptive discharge is temporary.   62,308

(3) (100 percent) (surge arresters). The specified voltage that is to be applied to a test object in a 100-percent disruptive discharge test under specified conditions. *Note:* The term applies mostly to impulse tests and has significance only in cases when the loss of electric strength resulting from a disruptive discharge is temporary. During the test, in general, all voltage applications should cause disruptive discharge.   308,62,307

dissector. *See:* **image dissector.**

dissector tube. A camera tube having a continuous photocathode on which is formed a photoelectric-emission pattern that is scanned by moving its electron optical image over an aperture. *See:* **image dissector tube.**   328

dissipation (1) (electrical energy). Loss of electric energy as heat.   197

(2) (waveguide). The power reduction in a transmission path caused by resistive or conductive loss, or both.   267

dissipation, electrode. *See:* **electrode dissipation.**

dissipation factor (circuits and systems). (1) The ratio of energy dissipated to the energy stored in an element for one cycle, (2) the loss tangent of an element and (3) the inverse of Q. *See:* **dielectric dissipation factor.**   67

dissipation-factor test (rotating machinery). *See:* **loss-tangent test.**

dissymmetrical transducer. Dissymmetrical with respect to a specified pair of terminations when the interchange of that pair of terminations will affect the transmission. *See:* **transducer.**   328

distal stimuli (illuminating engineering). In the physical space in front of the eye one can identify points, lines, surfaces, and three dimensional arrays of scattering particles which constitute the distal physical stimuli which form optical images on the retina. Each element of a surface or volume to which an eye is exposed subtends a solid angle at the entrance pupil. Such elements of solid angle make up the field of view and each has a specifiable luminance and chromaticity. Points and lines are specific cases which have to be dealt with in terms of total candlepower and candlepower per unit length. Distal stimuli are sometimes referred to simply as lights or colors.   167

distance. *See:* **Hamming distance; signal distance.**

distance clearance (power line maintenance). The minimum separation between two conductors, between conductors and supports, or other objects, or between conductors and ground, or the clear space between any objects.   458

distance dialing (telephone switching systems). The automatic establishing of toll calls by means of signals from the calling device of either a customer or an operator.   55

distance mark (range mark) (on a radar display). A calibration marker used on a cathode-ray screen in determining target distance. *See:* **radar.**   187

distance measuring equipment (DME)(navigation aid terms). A radio aid to navigation which provides distance information by measuring total round-trip time of transmission from an interrogator to a transponder and return.   526

distance protection (power switchgear). A method of line protection in which an abnormal condition within a predetermined electrical distance of a line terminal on the protected circuit is detected by measurement of system conditions at that terminal.   127, 103

distance relay (1) (power switchgear). A generic term covering those forms of protective relays in which the response to the input quantities is primarily a function of the electrical circuit distance between the relay location and the point of fault.   127

(2) (21) (power system device function numbers). A relay that functions when the circuit admittance, impedance, or reactance increases or decreases beyond a predetermined value.   402

distance resolution (radar). The ability to distinguish between two targets solely by the measurement of distances; generally expressed in terms of the minimum distance by which two targets of equal strength at the same azimuth and elevation angles must be spaced to be separately distinguishable and measureable.   13

distorted current (ac high-voltage circuit breaker). The current through the test circuit breaker which is influenced by the arc voltage of both the test and auxiliary circuit breakers during the high current interval.   426

distortion (1) (data transmission). An undesired change in waveform. The principal sources of distortion are: (A) A nonlinear relation between input and output at a given frequency, (B) Nonuniform transmission at different frequencies (C) Phase shift not proportional to frequency.   59

(2) (fiber optics). A change of signal waveform shape. *Note:* In a multimode fiber, the signal can suffer degradation from multimode distortion. In addition, several dispersive mechanisms can cause signal distortion in an optical waveguide: waveguide dispersion, material dispersion, and profile dispersion. *See:* **dispersion; profile dispersion.**   433

distortion, amplitude-frequency (electroacoustics). *See:* amplitude-frequency distortion; distortion.

distortion, barrel (camera tubes or image tube). A distortion that results in a progressive decrease in radial magnification in the reproduced image away from the axis of symmetry of the electron optical system. *Note:* For a camera tube, the reproducer is assumed to have no geometric distortion.   125

distortion, envelope delay (1) (general). Of a system or transducer, the difference between the envelope delay at one frequency and the envelope delay at at a reference frequency.   239

(2) (facsimile). That form of distortion which occurs when the rate of change of phase shift with frequency of a circuit or system is not constant over the frequency range required for transmission. *Note:* In facsimile, envelope delay distortion is usually expressed as one-half the difference in microseconds between the maximum and the minimum envelope delays existing between the two extremes of frequency defining the channel used. *See:* **facsimile transmission.**   12

distortion factor (1)(harmonic factor)(harmonic control and reactive compensation of static power converters). The ratio of the root-mean-square of the harmonic content to the root-mean-square value of the fundamental quantity, expressed as a percent of the fundamental.   533

$$DF = \left(\frac{\text{sum of squares of amplitudes of all harmonics}}{\text{square of non-sinusoidal quantity}} \right)^{\frac{1}{2}} \cdot 100\%$$

(2)(power system communication equipment). The ratio of the root-mean-square value of the residue of a voltage wave after the elimination of the fundamental to the root-mean-square value of the original wave.   453

(3) (wave) (rotating machinery). The ratio of the root-mean-square value of the residue of a voltage wave after the elimination of the fundamental to the root-mean-square value of the original wave.   63

distortion, field-time waveform (FD). The linear TV waveform distortion of time components from 64 μs

Distortion Factor =

$$\left[\frac{\text{(sum of squares of amplitudes of all harmonics)}}{\text{(square of amplitude of fundamental)}} \right]^{1/2} (100\%)$$

to 16 μs, that is, time components of the field-time domain.   42

distortion, frequency. *See:* **amplitude-frequency distortion.**

distortion, intermodulation. Nonlinear distortion of a system or transducer characterized by the appearance in the output of frequencies equal to the sums and differences of integral multiples of the two or more component frequencies present in the input wave. *Note:* Harmonic components also present in the output are usually not included as part of the intermodulation distortion. When harmonics are included, a statement to that effect should be made. *See:* **distortion.**   239

distortion, keystone (camera tubes). A distortion such thatthe slope or the length of a horizontal line trace or scan line is linearly related to its vertical displacement. *Note:* A system having keystone distortion distorts a rectangular pattern into a trapezoidal pattern.
125

distortion-limited operation (fiber optics). The condition prevailing when the distortion of the received signal, rather than its amplitude (or power), limits performance. The condition is reached when the system distorts the shape of the waveform beyond specified limits. For linear systems, distortion-limited operation is equivalent to bandwidth-limited operation. *See:* **attenuation-limited operation; bandwidth-limited operation; distortion; multimode distortion.**
433

distortion, linear. That distortion of an electrical signal which is independent of the signal amplitude. *Note:* A small-signal nonuniform frequency response is an example of linear distortion. By contrast, nonlinear distortions of an electrical signal are those distortions that are dependent on the signal amplitude, for example, compression, expansion, and harmonic distortion, etc.   42

distortion, linear TV waveform. The distortion of the shape of a waveform signal where this distortion is independent of the amplitude of the signal. *Notes:* (1) A TV video signal may contain time components with durations from as long as a TV field to as short as a picture element. The shapes of all these time components are subject to distortions. For ease of measurement it is convenient to group these distortions in three separate time domains; short-time waveform distortion, line-time waveform distortion, and field-time waveform distortion. (2) The waveform distortions for times from one field to tens of seconds is not within the scope of this standard.   42

distortion, line-time waveform (LD). The linear TV waveform distortion of time components from 1 μs to 64 μs, that is, time is, time components of the line-time domain.   42

distortion, pattern (oscilloscopes). Any deformation of the pattern from its intended form. (IEC151-14.) *Notes:* (1) In an oscilloscope the intended form is rectilinear and rectangular. (2) An oscilloscope control that affects pattern distortion may be labeled "pattern" or "geometry."   106

distortion, percent harmonic (electroacoustics). A measure of the harmonic distortion in a system or transducer, numerically equal to 100 times the ratio of the square root of the sum of the squares of the root-mean-square voltages (or currents) of each of the individual harmonic frequencies, to the root-mean-square voltage (or current) of the fundamental. *Note:* It is practical to measure the ratio of the root-mean-square amplitude of the residual harmonic voltages (or currents), after the elimination of the fundamental, to the root-mean-square amplitude of the fundamental and harmonic voltages (or currents) combined. This measurement will indicate percent harmonic distortion with an error of less than 5 percent if the magnitude of the distortion does not exceed 30 percent. *See:* **distortion.**   239

distortion, phase delay (system or transducer). The difference between the phase delay at one frequency and the phase delay at a reference frequency.
                                                       239

distortion, pincushion (camera tubes or image tubes). A distortion that results in a progressive increase in radial magnification in the reproduced image away from the axis of symmetry of the electron optical system. *Note:* For a camera tube, the reproducer is assumed to have no geometric distortion. *See:* **distortion, amplitude-frequency (electroacoustics); distortion factor; distortion, percent harmonic (electroacoustics); hiss (in an electron device).**   16, 125

distortion power (polyphase circuit). At the terminals of entry of a polyphase circuit, equal to the sum of the distortion powers for the individual terminals of enry. *Notes:* (1) The distortion power for each terminal of entry is determined by considering each phase conductor, in turn, with the common reference point as a single-phase, two-wire circuit and finding the distortion power for each in accordance with the definition of **distortion power (single-phase two-wire circuit).** The common reference terminal shall be taken as the neutral terminal of entry, if one exists, otherwise as the true neutral point. The sign given to the distortion power for each single-phase current, and therefore to the total for the polyphase circuit, shall be the same as that of the total active power. (2) Distortion power is expressed in volt-amperes when the voltages are in volts and the current in amperes. (3) The distortion power is zero if each voltage has the same wave form as the corresponding current. This condition is fulfilled, of course, when all the currents and voltages are sinusoidal.   210

distortion power (single-phase two-wire circuit). At the two terminals of entry of a single-phase two-wire circuit into a delimited region, a scalar quantity having an amplitude equal to the square root of the difference of the squares of the apparent power and the amplitude of the phasor power. *Note:* Mathematically the amplitude of the distortion power D is given by the equation

$$
\begin{aligned}
D &= (U^2 - S^2)^{1/2} \\
&= (U^2 - P^2 - Q^2)^{1/2} \\
&= \left(\sum_{r=1}^{r=\infty} \sum_{q=1}^{q=\infty} \{E_r^2 I_q^2 \right. \\
&\quad E_r E_q I_r I_q \; \cos \; [(\alpha_r - \beta_r) \\
&\quad \left. - (\alpha_q - \beta_q)]\} \right)^{1/2}
\end{aligned}
$$

where the symbols are as in **power, apparent (single-phase two-wire circuit).** If the voltage and current are quasi-periodic and the amplitudes are slowly varying, the distortion power at any instant may be taken as the value derived from the amplitude of the apparent power and phasor power at that instant. By this definition the sign of distortion power is not definitely determined, and it may be given either sign. In the absence of other definite information, it is to be taken the same as for the active power. Distortion power is expressed in volt-amperes when the voltage is in volts and the current in amperes. The distortion power is zero if the voltage and the current have the same waveform. This condition is fulfilled when the voltage and current are sinusoidal and have the same period, or when the circuit consists entirely of noninductive resistors.   210

distortion, pulse (pulse techniques). The unwanted deviation of a pulse waveform from a reference waveform. *Note:* Some specific forms of pulse distortion have specific names. They include, but are not exclusive to, the following: **overshoot, ringing, preshoot, tilt (droop*), rounding (undershoot and dribble-up*), glitch, bump, spike,** and **backswing.** For further explanation of the forms of pulse distortion, see the following illustrations and IEEE Standard 194 (1977). *See:* **pulse.**   254

distortion, short-time waveform (SD) (TV). The linear TV waveform distortion of time components from 125 ns to 1 μs, that is, time components of the short-time domain.   42

distortion, spiral (camera tubes or image tubes using magnetic focusing). A distortion in which image rotation varies with distance from the axis of symmetry of the electron optical system.   125

distortion tolerance (telegraph receiver). The maximum signal distortion that can be tolerated without error in reception. *See:* **telegraphy.**   194

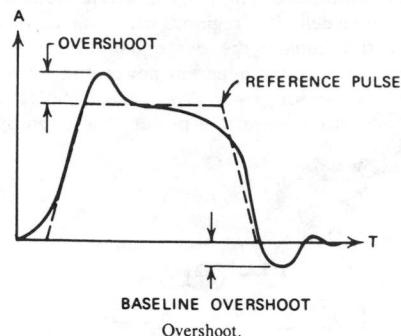

Overshoot.

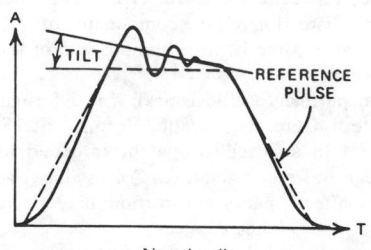

Negative tilt.

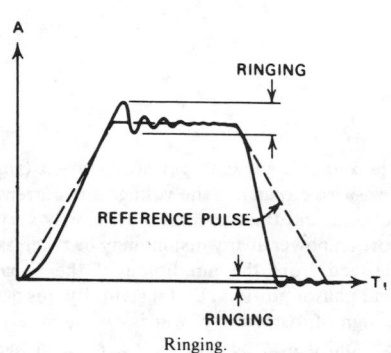

Ringing.

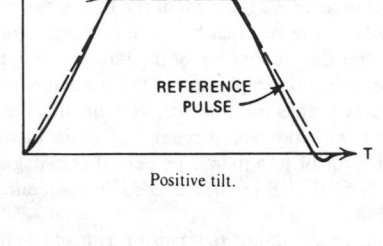

Positive tilt.

distortion, short-time waveform

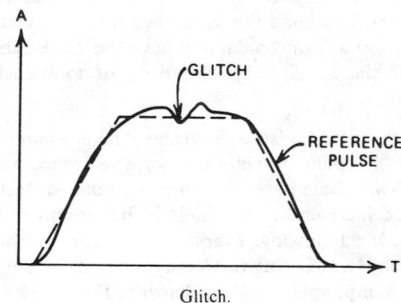

Glitch.

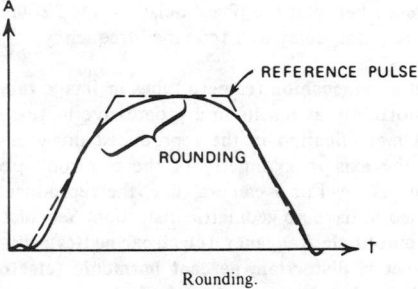

Rounding.

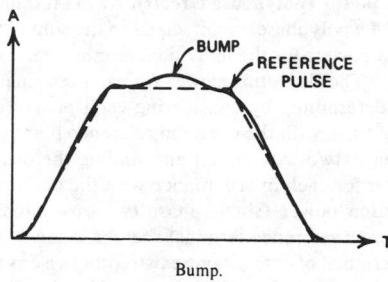

Bump.

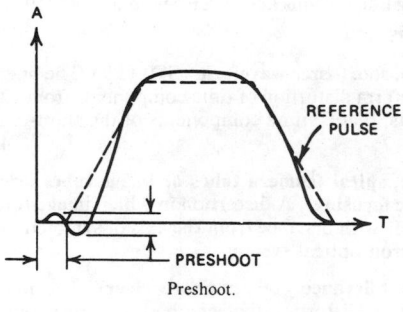

Preshoot.

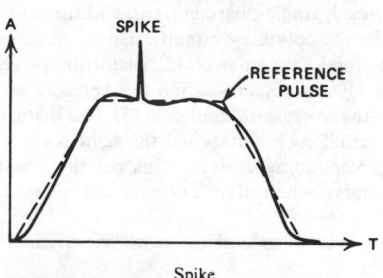

Spike.

distortion, waveform (oscilloscopes). A displayed deviation from the correct representation of the input reference signal. *See:* **oscillograph.**   106

distributed. Spread out over an electrically significant length or area.   328

distributed constant (waveguide). A circuit parameter that exists along the length of a waveguide or transmission line. *Note:* For a transverse electromagnetic wave on a two-conductor transmission line, the distributed constants are series resistance, series inductance, shunt conductance, and shunt capacitance per unit length of line. *See:* **waveguide.**   267

distributed data processing (DDP)(computer applications). The use of computers for processing information within a distributed system.   571

distributed element (for a transmission line) (waveguide). A circuit element that exists continuously along a transmission line. *Note:* For a transverse electromagnetic wave on a two-conductor transmission line, the distributed elements are series resistance, series inductance, shunt conductance, and shunt capacitance per unit length of line.   267

distributed element circuit (microwave tubes). A circuit whose inductance and capacitance are distributed over a physical distance that is comparable to a wavelength. *See:* **microwave tube or valve.**   190

distributed function (dot logic, wired logic)(graphic symbols for logic functions). A logic function (either AND or OR) implemented by connecting together outputs of the appropriate type; these outputs are the inputs of the logic function thus formed; the joined connection is the output. See:ANSI/IEEE Std 91-1984, Symbols 3.3-3 through 3.3-7.   451

distributed-network type. A unit substation which has a single stepdown transformer having its outgoing side connected to a bus through a circuit breaker equipped with relays which are arranged to trip the circuit breaker on reverse power flow to the transformer and to reclose the circuit breaker upon the restoration of the correct voltage, phase angle, and phase sequence at the transformer secondary. The bus has one or more outgoing radial (stub-end) feeders and one or more tie connections to a similar unit substation.   53

distributed processing (supervisory control, data acquisition, and automatic control). A design in which all data is not processed in one processor. Multiple processors in the master station or in the remote stations, or both, share the functions.   570

distributed target (radar). A target composed of a number of scatterers, where the target extent in any dimension is greater than the radar resolution in that dimension. *See:* **complex target.**   13

distributed winding (rotating machinery). A winding, the coils of which occupy several slots per pole. *See:* **armature; rotor (rotating machinery); stator.**   63

distributing cable. *See:* **distribution cable.**

distributing frame. A structure for terminating permanent wires of a central office, private branch exchange, or private exchange and for permitting the easy change of connections between them by means of cross-connecting wires.   193

distributing valve (speed governing systems, hydraulic turbines). The element of the governor-control actuator which controls the flow of hydraulic fluid to the turbine-control servomotor(s).   8

distribution (used as an adjective) (power switchgear). A general term used, by reason of specific physical or electrical characteristics, to denote application or restriction of the modified term, or both, to that part of an electric system used for conveying energy to the point of utilization from a source or from one or more main receiving stations. *Notes:* (1) From the standpoint of a utility system, the area described is between the generating source, or intervening substations, and the customer's entrance equipment. (2) From the standpoint of the customer's internal system, the area described is between a source or receiving station within the customer's plant and the points of utilization.   103

distribution amplifier. *See:* **amplifier, distribution.**

distribution box (mine type). A portable piece of apparatus with enclosure by means of which an electric circuit is carried to one or more machine trailing cables from a single incoming feed line, each trailing cable circuit being connected through individual overcurrent protective devices. *See:* **distributor box; mine feeder circuit.**   328

distribution cable (distributing cable) (communication practice). A cable extending from a feeder cable into a specific area for the purpose of providing service to that area. *See:* **cable.**   328

distribution center (secondary distribution). A point at which is located equipment consisting generally of automatic overload protective devices connected to buses, the principal functions of which are subdivision of supply and the control and protection of feeders, subfeeders, or branch circuits, or any combination of feeders, subfeeders, or branch circuits.   3

distribution coefficients (color) (television). The tristimulus values of monochromatic radiations of equal power. *Note:* Generally represented by overscored, lowercase letters such as $\bar{x}, \bar{y}, \bar{z}$ in the CIE system.   18

distribution current limiting fuse (high-voltage switchgear). A fuse consisting of a fuse support and a current limiting fuse unit. *Note:* In addition, the distribution current limiting fuse is identified by the following characteristics: (1) Dielectric withstand (bil) (basic impulse insulation level) strengths at distribution levels. (2) Application primarily on distribution feeders and circuits. (3) Operating voltage limits correspond to distribution system voltage.   443

distribution cutout (power switchgear). A fuse or disconnecting device consisting of any one of the following assemblies: (1) A fuse support and fuseholder which may or may not include the conducting element (or fuse link). (2) A fuse support and disconnecting blade. (3) A fuse support and fuse carrier which may or may not include the conducting element (fuse link) or disconnecting blade. *Note:* In addition to the distribution cutout is identified by the following characteristics: (1) Dielectric withstand (BIL) strengths at dis-

tribution levels. (2) Application primarily on distribution feeders and circuits. (3) Mechanical construction basically adapted to pole or crossarm mounting except for the distribution oil cutout. (4) Operating voltage limits corresponding to distribution systems voltage.
103, 443

distribution disconnecting cutout (1) (high-voltage switchgear). A distribution cutout, having a disconnecting blade, which is used for closing, opening, or changing the connections in a circuit or system, or for isolating purposes. *Note:* Some load-break ability is inherent in the device but it has no load-break rating. This ability can best be evaluated by the user, based on experience under operating conditions.   443
(2) (power switchgear). *See:* **distribution cutout; disconnecting cutout.**   103

distribution enclosed single-pole air switch (distribution enclosed air switches) (power switchgear). A single-pole disconnecting switch in which the contacts and blade are mounted completely within an insulated enclosure. (Cannot be converted into a distribution cutout or disconnecting fuse.) *Notes:* The distribution enclosed air switch is identified by the following characteristics:
(1) Dielectric withstand (BIL) strengths at distribution level.
(2) Application primarily on distribution feeders and circuits.
(3) Mechanical construction basically adapted to crossarm mounting.
(4) Operating voltage limits correspond to distribution voltage.
(5) Unless incorporating load-break means, it has no interrupting (load-break current) rating. (Some load-break ability is inherent in the device. This ability can be best evaluated only by the user, based on experience under operating conditions.)   103, 443

distribution enter (electric installations on shipboard). A point at which is located equipment consisting generally of automatic overload protective devices connected to buses, the principal functions of which are subdivision of supply and the control and protection of feeders, subfeeders, or branch circuits, or any combination of feeders, subfeeders, or branch circuits.
3

distribution factor (rotating machinery). A factor related to a distributed winding, taking into account the spatial distribution of the slots in which the winding considered is laid, that is the decrease in the generated voltage, as a result of a geometrical addition of the corresponding representative vectors.   63

distribution feeder. *See:* **primary distribution feeder; secondary distribution feeder; distribution center**

distribution fuse cutout (1) (high-voltage switchgear). A distribution cutout having a fuse holder or fuse carrier and fuse link or a fuse unit. *Note:* A fuse cutout is a fuse disconnecting switch. It has some inherent load-break ability but does not have a load-break rating. The load-break ability can best be evaluated by the user, based on experience under operating conditions.   443

(2) (power switchgear). *See:* **distribution cutout; fuse cutout.**   103

distribution main. *See:* **primary distribution mains; secondary distribution mains; center of distribution.**

distribution network. *See:* **primary distribution network; secondary distribution network; center of distribution.**

distribution oil cutout. *See:* **distribution; oil cutout.**

distribution open cutout. *See:* **distribution; open cutout.**

distribution open-link cutout. *See:* **distribution; open-link cutout.**

distribution panelboard (National Electrical Code). A single panel or group of panel units designed for assembly in the form of a single panel; including buses, and with or without switches and/or automatic overcurrent protective devices for the control of light, heat or power circuits of small individual as well as aggregate capacity; designed to be placed in a cabinet or cutout box placed in or against a wall or partition and accessible only from the front.   256

distribution station (power operations). A transforming station where the transmission is linked to the distribution system.   516

distribution switchboard (power switchgear). A power switchboard used for the distribution of electric energy at the voltages common for such distribution within a building. *Note:* Knife switches, air circuit breakers, and fuses are generally used for circuit interruption on distribution switchboards, and voltages seldom exceed 600. However, such switchboards often include switchboard equipment for a high-tension incoming supply circuit and a stepdown transformer.   103

distribution system (power operations). That portion of an electric system which delivers electric energy from transformation points on the transmission or bulk power system to the consumers.   516

distribution temperature of a light source (illuminating engineering). The absolute temperature of a blackbody whose relative spectral distribution is most nearly the same in the visible region of the spectrum as that of the light source.   167

distribution transformer (power and distribution transformer). A transformer for transferring electrical energy from a primary distribution circuit to a secondary distribution circuit or consumer's service circuit. *Note:* Distribution transformers are usually rated in the order of 5-500 kVA.   53

distribution trunk line. *See:* **primary distribution trunk line; center of distribution.**

distributor box. A box or pit through which cables are inserted or removed in a draw-in system of mains. It contains no links, fuses, or switches and its usual function is to facilitate tapping into a consumer's premises. *See:* **distribution box; tower.**   64

distributor duct. A duct installed for occupancy of distribution mains. *See:* **service pipe.**   64

distributor suppressor (internal-combustion engine terminology). A suppressor designed for direct connection to the high-voltage terminals of a distributor cap. *See:* **electromagnetic compatibility.**   220,199

disturbance (1)(communication practice). Any irregular phenomenon associated with transmission that tends to limit or interfere with the interchange of intelligence.   328
(2) (interference terminology). *See:* interference.
(3) (industrial control). An undesired input variable that may occur at any point within a feedback control system. *See:* control system, feedback   206
(4) (storage tubes). That type of spurious signal generated within a tube that appears as abrupt variations in the amplitude of the output signal. *Notes:* (A) These variations are spatially fixed with reference to the target area. (B) The distinction between this and shading. (C) A blemish, a mesh pattern, and moire present in the output are forms of disturbance. Random noise is not a form of disturbance. *See:* storage tube.
                                                  174
disturbed-ONE output (magnetic cell). A ONE output to which partial-read pulses have been applied since that cell was last selected for writing. *See:* coincident-current selection; one output.   331
disturbed-ZERO output (magnetic cell). A ZERO output to which partial-write pulses have been applied since that cell was last selected for reading. *See:* coincident-current selection.   331
dit. *See:* hartley.   61
dither (control circuits). A useful oscillation of small amplitude, introduced to overcome the effects of friction, hysteresis, or clogging. *See:* control system, feedback.   206
dither spillover (laser gyro) (inertial sensor). The residual angle error (expressed in the form of pulses) per dither cycle.   46
divergence (1) (vector field at a point). A scalar equal to the limit of the quotient of the outward flux through a closed surface that surrounds the point by the volume within the surface, as the volume approaches zero. *Note:* If the vector **A** of a vector field is expressed in terms of its three rectangular components A_x, A_y, A_z, so that the values of A_x, A_y, A_z, are each given as a function of x, y, and z, the divergence of the vector field **A** (abbreviated $\nabla\cdot$**A**) is the sum of the three scalars obtained by taking the derivatives of each component in the direction of its axis, or

$$\mathrm{div}\mathbf{A} \equiv \nabla\cdot\mathbf{A} = \frac{\partial A_x}{\partial x} + \frac{\partial A_y}{\partial y} + \frac{\partial A_z}{\partial z}$$

Examples: If a vector field **A** represents velocity of flow such as the material flow of water or a gas or the imagined flow of heat or electricity, the divergence of **A** at any point is the net outward rate of flow per unit volume and pr unit time. It is the time rate of decrease in density of the fluid at that point. Because the density of an incompressible fluid is always zero. The divergence of the flow of heat at a point in a body is equal to the rate of generation of heat per unit volume at the point The divergence of the electric field strength at a point is proportional to the volume density of charge at the point.   210

(2) (fiber optics). *See:* beam divergence.   433
divergence loss (acoustic wave). The part of the transmission loss that is due to the divergence or spreading of the sound rays in accordance with the geometry of the system (for example, spherical waves emitted by a point source). *See:* loudspeaker.   176
diversity. *See:* load diversity.
diversity factor (system diversity factor). The ratio of the sum of the individual maximum demands of the various subdivisions of a system to the maximum demand of the whole system. *Note:* The diversity factor of a part of the system may be similarly defined as the ratio of the sum of the individual maximum demands of the various subdivisions of the part of the system under consideration. *See:* alternating-current distribution; direct-current distribution.   64
diversity gain (radar). The reduction in predetection signal-to-interference energy ratio required to achieve a given level of performance, relative to that of a nondiversity radar, resulting from the use of diversity in frequency, polarization, space, or other characteristics.   13
diversity, seasonal (electric power utilization). Load diversity between two (or more) electric systems which occurs when their peak loads are in different seasons of the year.   112
diversity, time zone (electric power utilization). Load diversity between two (or more) electric systems which occurs when their loads are in different time zones.   112
divided code ringing (divided ringing). A method of code ringing that provides partial ringing selectivity by connecting one-half of the ringers from one side of the line to ground and the other half from the other side of the line to the ground. This term is not ordinarily applied to selective and semiselective ringing systems.
                                                  328
divided ringing. *See:* divided code ringing.
divider (analog computers). (A) A device capable of dividing one variable by another. (B) A device capable of attenuating a variable by a constant or adjustable amount, as an attenuator.   9
dividing network (crossover network) (loadspeaker dividing network). A frequency selective network that divides the spectrum into two or more frequency bands for distribution to different loads *See:* loudspeaker.   176
division (1) (nuclear power generating stations)(accident monitoring instrumentation). The designation applied to a given system or set of components that enables the establishment and maintenance of physical, electrical, and functional independence from other redundant sets of components.   102, 421
(2) (Class 1E equipment and circuits). The designation applied to a given system or set of components that enables the establishment and maintenance of physical, electrical, and functional independence from other redundant sets of components. *Note:* The terms division, train, channel, separation group, safety group or load group, when used in this context, are interchangeable.   131

D layer (radio wave propagation). An ionized layer in the D region. 146

DME. *See:* **distance measuring equipment.**

document (1) (information processing) (computer). (A) A medium and the data recorded on it for human use, for example, a report sheet, a book (B) By extension, any record that has permanence and that can be read by man or machine. 255,77

(2) (nuclear power quality assurance). Any written or pictorial information describing, defining, specifying, reporting, or certifying activities, requirements, procedures, or results. A document is not considered to be a Quality Assurance Record until it satisfies the definition of a Quality Assurance Record as defined in this Supplement. 417

(3) (software). (A) A data medium and the data recorded on it, that generally has permanence and that can be read by man or machine. Often used to describe human readable items only, for example, technical documents, design documents, version description documents. (B) To create a document. *See:* **design.** 434

documentation (1)(design and installation of cable systems for Class 1E circuits in nuclear power generating stations). Any written or pictorial information describing, defining, specifying, reporting or certifying activities, requirements, procedures, or results. 536

(2)(software). (A) A collection of documents on a given subject. (B) The management of documents which may include the actions of identifying, acquiring, processing, storing, and disseminating them. (C) The process of generating a document. (D) Any written or pictorial information describing, defining, specifying, reporting or certifying activities, requirements, procedures, or results. *See:* **document; procedure; requirement; software documentation; system documentation; user documentation.** 434

documentation level. *See:* **level of documentation.**

documents (nuclear power generating stations). Drawings and other records significant to the design, construction, testing, maintenance, and operation of Class 1E equipment and systems for nuclear power generating stations. *Note:* Documents include: (1) Drawings such as instrument diagrams, functional control diagrams, one line diagrams, schematic diagrams, equipment arrangements, cable and tray lists, wiring diagrams. (2) Instrument data sheets. (3) Design specifications. (4) Instruction manuals. (5) Test specifications, procedures, and reports. (6) Device lists. Not to be included as documents are: project schedules, financial reports, meeting minutes, correspondence such as letters and memoranda, and equipment procurement documentation covered by quality assurance programs. 31

Doherty amplifier. A particular arrangement of a radio-frequency linear power amplifier wherein the amplifier is divided into two sections whose inputs and outputs are connected by quarter-wave (90-degree) networks and whose operating parameters are so ad-

justed that, for all values of input signal voltage up to one-half maximum amplitude, Section No. 2 is inoperative and Section No. 1 delivers all the power to the load, which presents an impedance at the output of Section No. 1 that is twice the optimum for maximum output. At one-half maximum input level, Section No. 1 is operating at peak efficiency, but is beginning to saturate. Above this level, Section No. 2 comes into operation, thereby decreasing the impedance presented to Section No. 1, which causes it to deliver additional power into the load until, at maximum signal input, both sections are operating at peak efficiency and each section is delivering one-half the total output power to the load. *See:* **amplifier.** 111

Dolph-Chebyshev array antenna *. *See:* **Dolph-Chebyshev distribution.**

*Deprecated

Dolph-Chebyshev distribution (antennas). A set of excitation coefficients for an equispaced linear array antenna such that the array factor can be expressed as a Chebyshev polynomial. 111

dominant mode (waveguide). In a uniform waveguide, the propagating mode with the lowest cutoff frequency. 267

dominant wave (uniconductor waveguide). The guided wave having the lowest cutoff frequency. *Note:* It is the only wave that will carry energy when the excitation frequency is between the lowest cutoff frequency and the next higher cutoff frequency. *See:* **guided waves; waveguide.**

dominant wavelength (colored light, not purple) (television). The wavelength of the spectrum light that, when combined in suitable proportions with the specified achromatic light, yields a match with the light considered. *Note:* When the dominant wavelength cannot be given (this applies to purples), its place is taken by the complementary wavelength. *See:* **complementary wavelength.** 18

dominant wavelength of a light, λ_d (illuminating engineering). The wavelength of radiant energy of a single frequency that, when combined in suitable proportion with the radiant energy of a reference standard, matches the color of the light. 167

Donnan potential (electrobiology). The potential difference across an inert semipermeable membrane separating mixtures of ions, attributed to differential diffusion. *See:* **electrobiology.** 192

donor (semiconductor). *See:* **impurity, donor; semiconductor.**

door (gate closer). A device that closes a manually opened hoistway door, a car door, or gate by means of a spring or by gravity. *See:* **hoistway (elevator or dumbwaiter).** 328

door contact (burglar-alarm system). An electric contacting device attached to a door frame and operated by opening or closing the door. *See:* **protective signaling.** 328

door or gate power operator. A device, or assembly of devices, that opens a hoistway door and/or a car door or gate by power other than by hand, gravity, springs, or the movement of the car; and that closes them by

power other than by hand, gravity, or the movement of the car. *See:* **hoistway (elevator or dumbwaiter).**
                                                                    328
dopant (1) (acceptor) (semiconductor). An impurity that may induce hole conduction. *See:* **semiconductor device.** *Syn:* **impurity.**
(2) (donor) (semiconductor). An impurity that may induce electron conduction. *See:* **semiconductor device.** *Syn:* **impurity.**   66
*D***open-loop phase angle.** *See:* **loop phase angle.**
doping (semiconductor). Addition of impurities to a semiconductor or production of a deviation from stoichiometric composition, to achieve a desired characteristic. *See:* **semiconductor.**   245
doping compensation (semiconductor). Addition of donor impurities to a p-type semiconductor or of acceptor impurities to an n-type semiconductor. *See:* **semiconductor.**   66
Doppler effect (1) (communication satellite). The effective change of frequency of a received signal due to the relative velocity of a transmitter with respect to receiver. In space communications the frequency shifts due to the Doppler effect may be significant when the velocity of the spacecraft relative to earth is high; the frequency shifts are used to determine the velocity of vehicles.   85
(2) (data transmission). The phenomenon changing the observed frequency of a wave in a transmission system caused by a time rate of change in the effective length of the path of travel between the source and the point of observation.   59
Doppler filter (radar). An equipment designed to enhance the radar response to targets at a selected Doppler frequency relative to targets, clutter, or noise at other frequencies.   13
Doppler frequency (radio wave propagation). Of a wave traveling between a source and a point, the shift in frequency of the wave caused by the change of phase path with time. *Note:* The change may be due to variations in the separation of source and the point or in the refractive index of the intervening medium.
                                                                    146
Doppler-inertial navigation equipment (navigation aid terms). Hybrid navigation equipment which employs both Doppler navigation radar and inertial sensors.
                                                                    526
Doppler navigator (navigation aid terms). A self-contained dead reckoning navigation aid transmitting two or more beams of electromagnetic or acoustic energy outward and downward from the vehicle and utilizing the Doppler effect of the reflected energy, a reference direction, and the relationship of the beams to the vehicle to determine speed and direction of motion over the reflecting surface.   526
Doppler radar (navigation aid terms). A radar which utilizes the Doppler effect to determine the radial component of relative radar target velocity or to select targets having particular radial velocities.   526
Doppler shift. The magnitude of the change in the observed frequency of a wave due to the Doppler effect. The unit is the hertz.   176

Doppler system, pulsed (electronic navigation). *See:* **pulsed Doppler radar system.**
Doppler tracking (communication satellite). A method of determining the position of an observer on earth using the known [exact] satellite transmission frequency and the known satellite ephemeris and measuring the Doppler frequency shift of the signal received from the satellite.   84
Doppler VOR (very high-frequency omnidirectional range)(navigation aid terms). A vhf (very high frequency) radio range, operationally compatible with conventional VOR, less susceptible to siting difficulties because of its increased aperture. In it the variable signal (the signal producing azimuthal information) is developed by sequentially feeding an rf (radio frequency) signal to a multiplicity of antennas disposed in a ring-shaped array; the array usually surrounds the central source of reference signal.   526
dose (dosage) (photovoltaic power system). The radiation delivered to a specified area of the whole body. *Note:* Units of dose are rads or roentgens for X or gamma rays and rads for beta rays and protons. *See:* **photovoltaic power system; solar cells (in a photovoltaic power system).**   186
dose rate (1) (metal-nitride-oxide field-effect transistor). The time rate of deposition of radiation expressed in rads in SI per second. The integration of dose rate over a specified period of time represents the total dose.   386
(2) (photovoltaic power system). Radiation dose delivered per unit time.
                                                                    186
dot cycle (data transmission). One cycle of a periodic alternation between two signaling conditions, with each condition having unit duration. *Note:* Thus, in two-condition signaling, it consists of a dot, or marking element followed by a spacing element.   59
dot product. *See:* **scalar product.**
dot-product line integral. *See:* **line integral.**
dot sequential (color television). Sampling of primary colors in sequence with successive picture elements.
                                                                    18
dot signal (data transmission). A series of binary digits having equal and opposite states, such as a series of alternate 1 and 0 states.   59
double aperture seal (electric penetration assemblies). Two single aperture seals in series.   493
double-break switch (power switchgear). One that opens a conductor of a circuit at two points.   103
double bridge (Thomson bridge). *See:* **Kelvin bridge.**
double-buffered DAC (DAM) (hybrid computer linkage components). A digital-to-analog converter (DAC) or a digital-to-analog multiplier (DAM) with two registers in cascade, one a holding register, and the other the dynamic register.   10
double-circuit system (protective signaling). A system of protectivie wiring in which both the positive and the negative sides of the battery circuit are employed, and that utilizes either an open or a short circuit in the wiring to initiate an alarm. *See:* **protective signaling.**
                                                                    328

double connection (telephone switching systems). A fault condition whereby two separate calls are connected together.   55

double crucible method (fiber optics). A method of fabricating an optical waveguide by melting core and clad glasses in two suitably joined concentric crucibles and then drawing a fiber from the combined melted glass. *See:* **chemical vapor deposition technique.**
                                                                433

double-current generator. A machine that supplies both direct and alternating currents from the same armature winding.   63

double diode (electron device). An electron tube or valve containing two diode systems. *See:* **multiple tube (valve).**   190

double electric conductor seal (electric penetration assemblies). Two single electric conductor seals in series.   493

double-end control. A control system in which provision is made for operating a vehicle from either end. *See:* **multiple-unit control.**   328

double-faced tape. Fabric tape finished on both sides with a rubber or synthetic compound.   64

double-fed asynchronous machine (rotating machinery). An asynchronous machine of which the stator winding and the rotor winding are fed by supply frequencies each of which may be either constant or variable. *See:* **asynchronous machine.**   63

double-gun cathode-ray tube. A cathode-ray tube containing two separate electron-gun systems.   190

double-integrating gyro. A single-degree-of-freedom gyro having no intentional elastic or viscous restraint of the gimbal about the output axis so that the dynamic behavior is primarily established by the inertial properties of the gimbal. In this gyro, an output signal is produced by gimbal angular displacement, relative to the case, which is proportional to the double integral of the angular rate of the case about the input axis.
                                                                46

double length. Pertaining to twice the normal length of a unit of data or a storage device in a given computing system. *Note:* For example, a double-length register would have the capacity to store twice as much data as a single-length or normal register; a double-length word would have twice the number of characters or digits as a normal or single-length word. *See:* **double precision.**   235

double modulation. The process of modulation in which a carrier wave of one frequency is first modulated by a signal wave and a resultant wave is then made to modulate a second carrier wave of another frequency.
                                                                328

double pole-piece magnetic head (electroacoustics). A magnetic head having two separate pole pieces in which pole faces of opposite polarity are on opposite sides of the medium. *Note:* One or both of these pole pieces may be provided with an energizing winding. *See:* **phonograph pickup.**   176

double-pole relay. *See:* **relay, double-pole.**

double precision (mathematics of computing). Pertaining to the use of two computer words to represent a number in order to preserve or gain precision.
                                                                564

double-precision number. *See:* **double-length number.**

double pulse (pulse terms). Two pulse waveforms of the same polarity which are adjacent in time and which are considered or treated as a single feature.
                                                                254

double-secondary current transformer. One that has two secondaries, each on a separate magnetic circuit, with both magnetic circuits excited by the same primary.   203

double-secondary current transformer (instrument transformer)power and distribution transformer). One that has two secondary coils each on a separate magnetic circuit with both magnetic circuits excited by the same primary winding.   53, 394

double-secondary voltage transformer (instrument transformer)(power and distribution transformer). One that has two secondary windings on the same magnetic circuit insulated from each other and the primary.   53, 394

double sideband transmitter (data transmission). A transmitter that transmits the carrier frequency and both sidebands resulting from the modulation of the carrier by the modulating signal.   59

double squirrel cage (rotating machinery). A combination of two squirrel-cage windings mounted on the same induction-motor rotor, one at a smaller diameter than the other. *Note:* It is common but not essential for the two windings to have thesame number of slots. In any case, each bar of the lower (smaller-diameter) winding is located at the bottom of a slot containing a bar of the upper winding. A narrow portion of the slot (called the leakage slot) is provided in the radial separation between the two bars. *See:* **asynchronous machine.**   63

double-superheterodyne reception (triple detection). The method of reception in which two frequency converters are employed before final detection. *See:* **radio receiver.**   328

doublet antenna. *See:* **dipole antenna.**   111

double-throw (as applied to a mechanical switching device) (power switchgear). A qualifying term indicating that the device can change the circuit connections by utilizing one or the other of its two operating positions. *Note:* A double-throw air switch changes circuit connections by moving the switchblade from one of two sets of contact clips into the other.
                                                                103

double-tuned amplifier. An amplifier of one or more stages in which each stage utilizes coupled circuits having two frequencies of resonance, for the purpose of obtaining wider bands than those obtainable with single tuning. *See:* **amplifier.**   328

double-tuned circuit. A circuit whose response is the same as that of two single-tuned circuits coupled together.   328

double-way rectifier (power semiconductor). A rectifier unit which makes use of a double-way rectifier circuit.   66

double-way rectifier circuit. A rectifier circuit in which the current between each terminal of the alternating-voltage circuit and the rectifier circuit elements conductively connected to it flows in both directions. *Note:* The terms single-way and double-way provide a means for describing the effect of the rectifier circuit on current flow in transformer windings connect to rectifier circuit elements. Most rectifier circuits may be classified into these two general types. Double-way rectifier circuits are also referred to as bridge rectifier circuits. *See:* **rectification; rectifier circuit element; power rectifier; single-way rectifier circuit; bridge rectifier circuit.** 208

double-winding synchronous generator. A generator that has two similar windings, in phase with one another, mounted on the same magnetic structure but not connected electrically, designed to supply power to two independent external circuits. 63

doughnut (electronic device). *See:* **toroid.**

dovetail projection. A tenon, commonly flared; used for example, to fasten a pole to the spider. *See:* **stator.** 63

dovetail slot. (1) A recess along the side of a coil slot into which a coil-slot wedge is inserted. (2) A flaring slot into which a dovetail projection is engaged; used for example, to fasten a pole to the spider. *See:* **stator.** 63

dowel (dowel pin). A pin fitting with close tolerance into a hole in abutting pieces to establish and maintain accurate alignment of parts. Frequently designed to resist a shear load at the interface of the abutting pieces. 63

downconverter (nonlinear, active, and nonreciprocal waveguide components). A heterodyne frequency conversion device that converts an input signal to a lower frequency output signal. 530

down lead (lightning protection). The conductor connecting an overhead ground wire or lightning conductor with the grounding system. *See:* **direct-stroke protection (lightning).** 64

downlight (illuminating engineering). A small direct lighting unit which directs the light downward and can be recessed, surface mounted, or suspended. 167

down link (communication satellite). A transmission link carrying information from a satellite or spacecraft to earth. Typically down links carry telemetry, data and voice. 83

down time. (1) **(station control and data acquisition).** The time during which a device or system is not capable of meeting performance requirements. 403
(2)(supervisory control, data acquisition, and automatic control). The time during which a device or system is not capable of meeting performance requirements. 570

downward component (illuminating engineering). That portion of the luminous flux from a luminaire which is emitted at angles below the horizontal. 167

downward modulation. Modulation in which the instantaneous amplitude of the modulated wave is never greater than the amplitude of the unmodulated carrier. 339

DR. *See:* **dead reckoning.**

draft gauge (navigation aid terms). A hydrostatic in-strument installed in vessels to indicate the depth to which a vessel is submerged. 526

drag-in (electroplating). The quantity of solution that adheres to cathodes when they are introduced into a bath. *See:* **electroplating.** 328

drag magnet. *See:* **retarding magnet.**

drag-out (electroplating). The quantity of solution that adheres to cathodes when they are removed from a bath. *See:* **electroplating.** 328

drain (1) (general). The current supplied by a cell or battery when in service. *See:* **battery (primary or secondary).** 328
(2) (metal-nitride-oxide field-effect transistor). Region in the device structure of an insulated-gate field-effect transistor (IGFET) which contains the terminal into which charge carriers flow from the source through the channel. It has the potential which is more attractive than the source for the carriers in the channel. 386

drainage (corrosion). Conduction of current (positive electricity) from an underground metallic structure by means of a metallic conductor. 205

drainage unit (wire-line communication facilities). Center-tapped inductive device designed to relieve conductor-to-conductor and conductor-to-ground voltage stress by draining extraneous currents to ground. It is also designed to serve the purpose of a mutual drainage reactor forcing simultaneous protector-gap operation. 414

drain line (rotating machinery) (bearing oil system). A return pipe line using gravity flow. *See:* **oil cup (rotating machinery).** 63

drawbar pull (cable plowing). The effective pulling force delivered. 52

drawbridge coupler. *See:* **movable-bridge coupler.**

drawdown (power operations). The distance that the water surface of a reservoir is lowered from a given elevation as the result of the withdrawal of water. 516

drawout-mounted device (power switchgear). One having disconnecting devices and in which the removable portion may be removed from the stationary portion without the necessity of unbolting connections or mounting supports. *See:* **stationary-mounted device.** 103

D region (radio wave propagation). The region of the terrestrial ionosphere between about 40 and 90 km altitude responsible for most of the attenuation of radio waves in the range 1 to 100 MHz. 146

drift (1)(navigation aid terms). (A) Drift angle, (B) component of a vehicle's ground speed perpendicular to heading and (C) distance a craft is moved by current and wind. 526
(2) (rotating machinery). A long-time change in synchronous-machine resulting system error resulting from causes such as aging of components, self-induced temperature changes, and random phenomena. *Note:* Maximum acceptable drift is normally a specified change for a specified period of time, for specified conditions. 63
(3) (industrial control). An undesired but relatively

slow change in output over a period of time, with a fixed reference input. *Note:* Drift is usually expressed in percent of the maximum rated value of the variable being measured. *See:* **control system, feedback.**
219,206

(4) (sound recording and reproducing). Frequency modulation of the signal in the range below approximately 0.5 Hz resulting in distortion which may be perceived as a slow changing of the average pitch. *Note:* Measurement of drift is not covered by this definition. 145

(5) (analog computers). In an analog computer, a slowly varying error in an integrator, caused by the integration of offset errors at the inputs, capacitor leakage, or both. Also, any slowly varying error in a computer component. 9

(6) (electronic navigation). *See:* **G drift; G2 drift.**

(7) (oscilloscopes). *See:* **stability.**

drift angle (navigation aid terms). The angular difference between the heading and the track. 526

drift band of amplification (magnetic amplifier). The maximum change in amplification due to uncontrollable causes for a specified period of time during which all controllable quantities have been held constant. *Note:* The units of this drift band are the amplification units per the time period over which the drift band was determined. 171

drift compensation (industrial control). The effect of a control function, device, or means to decrease overall systems drift by minimizing the drift in one or more of the control elements. *Note:* Drift compensation may apply to feedback elements, reference input, or other portions of a system. *See:* **control system, feedback.** 206

drift correction angle (navigation aid terms). The angular difference between the course and the heading. *Syn:* crab angle; correction angle. 526

drift-correction angle. *See:* **correction angle.**

drift, direct-current. *See:* **drift.**

drift, G2. *See* **G2 drift.**

drift, kinematic (radar). *See:* **misalignment drift.**

drift mobility (homogeneous semiconductor). The ensemble average of the drift velocities of the charge-carriers per unit electric field. *Note:* In general, the mobilities of electrons and holes are different. *See:* **semiconductor device.** 245

drift offset (magnetic amplifier). The change in quiescent operating point due to uncontrollable causes over a specified period of time when all controllable quantities are held constant. 171

drift rate (1) (voltage regulators or reference tubes). The slope at a stated time of the smoothed curve of tube voltage drop with time at constant operating conditions. 125

(2) (gyro) (inertial sensor). The time rate of output deviation from the desired output. It consists of random and systematic components and is expressed as an equivalent input angular displacement per unit time with respect to inertial space. 46

drift space (electron tube). A region substantially free of externally applied alternating fields, in which a

relative repositioning of the electrons takes place. 125

drift, stability (electric conversion). Gradual shift or change in the output over a period of time due to change or aging of circuit components. (All other variables held constant). *See:* **electric conversion equipment.** 86

drift stabilization (analog computers). Any automatic method used to minimize the drift of a dc amplifier. 9

drift tunnel (velocity-modulated tube). A piece of metal tubing, held at a fixed potential, that forms the drift space. *Note:* The drift tunnel may be divided into several parts, which constitute the drift electrodes. *See:* **velocity-modulated tube.** 244,290

drift velocity (semiconductor)(nonlinear, active, and nonreciprocal waveguide components). A velocity component in a doped semiconductor that occurs when an electric field exists within the semiconductor material. This component is superimposed upon the thermal carrier's random motion. For electrons, the drift velocity will have a direction opposite to the electric field. 530

drift, zero. Drift with zero input. *See:* **electronic analog computer.** 9

dripproof. So constructed or protected that successful operation is not interfered with when faling drops of liquid or solid particles strike or enter the enclosure at any angle from 0 to 15 degrees from the downward vertical unless otherwise specified. *See:* **asynchronous machine.** 3, 63, 103

dripproof enclosure (1)(metal-enclosed bus and calculating losses in isolated-phase bus). An enclosure usually for indoor application, so constructed or protected that falling drops of liquid or solid particles that strike the enclosure at any angle not greater than 15 degrees from the vertical will not be able to interfere with the successful operation of the metal-enclosed bus. 574

(2) (electric installations on shipboard). An enclosure in which the openings are so constructed that-drops of liquid or solid particles falling on the enclosureat any angle not greater than 15 degrees from the verticaleither cannot enter the enclosure, or if they do enter the enclosure,they will not prevent the successful operation of,or cause damage to,the enclosed equipment. 3

(3) (power and distribution transformer). An enclosure, usually for indoor application, so constructed or protected that falling drops of liquid or solid particles which strike the enclosure at any angle within a specified variation from the vertical shall not interfere with the successful operation of the enclosed equipment. 53

dripproof machine. An open machine in which the ventilating openings are so constructed that drops of liquid or solid particles falling on the machine at any angle not greater than 15 degrees from the vertical cannot enter the machine either directly or by striking and running along a horizontal or inwardly inclined surface. *See:* **asynchronous machine.** 328

driptight (transformer). So constructed or protected as to exclude falling dirt or drops of liquid, under specified test conditions. 53

driptight enclosure. An enclosure so constructed that falling drops of liquid or solid particles striking the enclosure at any angle within a specified variation from the vertical cannot enter the enclosure either directly or by striking and running along a horizontal or inwardly inclined surface. 103,27,53

drive (1) (industrial control). The equipment used for converting available power into mechanical power suitable for the operation of a machine. *See:* **electric drive.** 225,26

(2) (electronic computation and recording). *See:* **tape drive.**

driven element (antenna). A radiating element coupled directly to the feed line. *See:* **antenna.** 179

drive pattern (facsimile). Density variation caused by periodic errors in the position of the recording spot. When caused by gears this is called gear pattern. *See:* **recording (facsimile)** 11

drive pin (disk recording). A pin similar to the center pin, but located to one side thereof, that is used to prevent a disc record from slipping on the turntable. *See:* **phonograph pickup.** 176

drive-pin hole (disk recording). A hole in a disc record that accommodates the turntable drive pin. 176

drive pulse (static magnetic storage). A pulsed magnetomotive force applied to a magnetic cell from one or more sources. 331

driver (1) (communication practice). An electronic circuit that supplies input to another electronic circuit. 328

(2) (software). A program that exercises a system or system component by simulating the activity of a higher level component. *See:* **component; program; system; test driver.** 434

drive strip (chafing strip) (rotating machinery). An insulating strip located in the coil slots between the wedge and the top of the slot armor or the top coil side, to provide protection during assembly of the wedges. *See:* **stator.** 63

driving machine. The power unit that applies the energy necessary to raise and lower an elevator or dumbwaiter car or to drive an escalator or a private-residence inclined lift. 328

driving-point admittance (between the jth terminal and the reference terminal of an n-terminal network). The quotient of (1) the complex alternating component I_j of the current flowing to the jth terminal from its external termination by (2) the complex alternating component V_j of the voltage applied to the jth terminal with respect to the reference point when all other terminals have arbitrary external terminations. *Note:* In specifying the driving-point admittance of a given pair of terminals of a network or transducer having two or more pairs of terminals, no two pairs of which contain a common terminal, all other pairs of terminals are connected to arbitrary admittances. *See:* **electron-tube admittances; network analysis.** 125

driving-point function (linear passive network). A re-

sponse function for which the variables are measured at the same port (terminal pair). 238

driving-point impedance (network). At any pair of terminals the ratio of an applied potential difference to the resultant current at these terminals, all terminals being terminated in any specified manner. *See:* **self-impedance.** 328

driving power, grid. *See:* **grid driving power.**

driving signals (television). Signals that time the scanning at the pickup point. *Note:* Two kinds of driving signals are usually available from a central synchronizing generator. One is composed of pulses at line frequency and the other is composed of pulses at field frequency. *See:* **television.** 178

driving test circuit (measuring longitudinal balance of telephone equipment operating in the voice band). A test circuit used to convert an exciting test voltage into balanced longitudinal voltages on tip and ring leads. 529

droop* (pulse techniques). *See:* **distortion, pulse.**
*Deprecated

droop, frequency (power systems). The absolute change in frequency between steady state no load and steady state full load. 89

drop (data transmission). A connection made between a through transmission circuit and a local terminal unit. 59

drop-away. The electrical value at which the movable member of an electromagnetic device will move ot its de-energized position. 328

dropout (of a relay) (1) (power switchgear). A term for contact operation (opening or closing) as a relay just departs from pickup. Also identifies the maximum value of an input quantity which will allow the relay to depart from pickup. 103

(2) (protective relaying of utility-consumer interconnections). Contact operation (opening or closing) as a relay just departs from pickup. The value at which dropout occurs is usually stated as a percentage of pickup. For example, dropout ratio of a typical instantaneous overvoltage relay is 90 percent. 128

dropout fuse (power switchgear). A fuse in which the fuseholder or fuse unit automatically drops into an open position after the fuse has interrupted the circuit. 103, 443

dropout ratio (of a relay) (power switchgear). The ratio of dropout to pickup of an input quantity. *Note:* This term has been used mostly with relays for which reset is not differentiated from dropout. Hence a similar term, reset ratio, the ratio of reset to pickup, is not generally used, though technically correct. 103

dropouts (data transmission). A loss of discrete data signals due to noise or attenuation hits. 59

dropout time (of a relay) (power switchgear). The time interval to dropout following a specified change of input conditions. *Note:* When the change of input conditions is not specified it is intended to be a sudden change from pickup value of input to zero input. 103

dropout voltage (or current) (emergency and standby power). The voltage (or current) at which a magnetically operated device will release to its deenergized position. It is a level of voltage (or current) that is insufficient to maintain the device in an energized state.   512

drop, voltage, anode (glow-discharge cold-cathode tube). See: anode voltage drop.

drop, voltage, starter (glow-discharge cold-cathode). See: starter voltage drop.

drop, voltage, tube (glow-discharge cold-cathode tube). See: tube voltage drop.

drop wire (drop) (data transmission). A wire suitable for extending an open wire or cable pair from a pole or cable terminal to a building.   59

drum controller (electric installations on shipboard) (industrial control). An electric contoller which utilizes a drum switch as the main switching element. A drum drum controller usually consists of a drum switch and a resistor.   3, 206

drum factor (facsimile). The ratio of usable drum length to drum diameter. *Note:* Before a picture is transmitted, it is necessary to verify that the ratio of used transmitter drum length to transmitter drum diameter is not greater than the receiver drum factor if the receiver is of the drum type. See: **facsimile (electrical communication).**   194

drum puller (conductor stringing equipment). A device designed to pull a conductor during stringing operations. It is normally equipped with its own engine which drives the drum mechanically, hydraulically or through a combination of both. It may be equipped with synthetic fiber rope or wire rope to be used as the pulling line. The pulling line is payed out from the unit, pulled through the travelers in the sag section and attached to the conductor. The conductor is then pulled in by winding the pulling line back onto the drum. This unit is sometimes used with synthetic fiber rope acting as the pilot line to pull heavier pulling lines across canyons, rivers, etcetera. *Syn:* **hoist; single drum hoist; single drum winch; tugger.**   431

drum speed (facsimile). The angular speed of the transmitter of recorder drum. *Note:* This speed is measured in revolutions per minute. See: **recording (facsimile); scanning.**   12

drum switch (industrial control). A switch in which the electric contacts are made of segments or surfaces on the periphery of a rotating cylinder or sector, or by the operation of a rotating cam. See: **control switch; switch.**   206

dry-arcing distance (insulator). The shortest distance through the surrounding medium between terminal electrodes, or the sum of the distances between intermediate electrodes, whichever is the shorter, with the insulator mounted for dry flashover test. See: **insulator.**   261

dry cell. A cell in which the electrolyte is immobilized. See: **electrochemistry.**   328

dry-charged battery (lead-acid batteries for photovoltaic (PV) systems). A battery in which the electrolyte has been removed for ease in shipping or storage or both.   515

dry contact. One through which no direct current flows.   193,55

dry friction. See: **Coulomb friction.**

dry location (National Electrical Code). A location not normally subject to dampness or wetness. A location classified as dry may be temporarily subject to dampness or wetness, as in the case of a building under construction.   256

dry-niche lighting fixtures (National Electrical Code). A lighting fixture intended for installation in the wall of the pool in a niche that is sealed against the entry of pool water by a fixed lens.   256

dry reed relay. A reed relay with dry (nonmercury-wetted) contacts.   341

dry-type (1) (current-limiting reactor) (grounding device). Having the coils immersed in an insulating gas. See: **reactor.**   309,91

(2) (regulator). Having the core and coils not immersed in an insulting liquid. See: **voltage regulator.**   257

(3) (transformer). Having the core and coils neither impregnated with an insulating fluid nor immersed in an insulating oil. See: **dry-type transformer.**   203

dry-type encapsulated water-cooled transformer (electrical heating applications to melting furnaces and forehearths in the glass industry). Dry-type (nonoil-cooled) transformer in which the windings are made from hollow conducting tubes that are cooled by transferring heat to water flowing in the tubes. Since the cooling is internal, the windings are usually totally encapsulated with epoxy resin.   520

dry-type forced-air-cooled transformer (class AFA) (power and distribution transformer). A dry-type transformer which derives its cooling by the forced circulation of air.   53

dry-type nonventilated self-cooled transformer (class ANV) (power and distribution transformer). A dry-type self-cooled transformer which is so constructed as to provide no intentional circulation of external air through the transformer, and operating air through the transformer, and operating at zero gauge pressure.   53

dry-type self-cooled/forced-air-cooled transformer (class AA/FA) (power and distribution transformer). A dry-type transformer which has a self-cooled rating with cooling obtained by the natural circulation of air and a forced-air-cooled rating with cooling obtained by the forced circulation of air.   53

dry-type self-cooled shunt reactor (Class AA)(shunt reactors over 500 kVA). A dry-type shunt reactor which is cooled by the natural circulation of the cooling air.   562

dry-type self-cooled transformer (class AA) (power and distribution transformer). A dry-type transformer which is cooled by the natural circulation of air.   53

dry-type shunt reactor (shunt reactors over 500 kVA). One in which the coils and magnetic circuit are neither impregnated with an insulating fluid nor immersed in an insulating oil.   562

dry-type transformer (1) (electrical heating applications to melting furnaces and forehearths in the glass industry). A transformer that relies on convection or forced air rather than liquid, such as oil or water, for cooling. 520

(2) (power and distribution transformer). A transformer in which the core and coils are in a gaseous or dry compound insulating medium. *See:* **ventilated dry-type transformer; nonventilated dry-type transformer; sealed transformer; gas-filler transformer; compound-filled transformer.** 53

dry vault (power switchgear). A ventilated, enclosed area not subject to flooding. 103

D-scan (radar). *See:* **D-display.**

D-scope (radar). A cathode-ray oscilloscope arranged to present a D-display. *See:* **D-display.** 13

D-star. *See:* **D*.** 433

dual alternate routing (data transmission). (When applied to the routing of two circuits). The transmission facility assigned to one circuit is geographically separated from the transmission facility assigned to the other circuit throughout their entire length. To meet the aforementioned criteria, dual alternate routes are constructed in such a manner that an interruption on one route will not result in an interruption on the other. 59

dual-beam oscilloscope. A multibeam oscilloscope in which the cathode-ray tube produces two separate electron beams that may be individually or jointly controlled. *See:* **multibeam oscilloscope; oscillograph.** 15

dual benchboard (power switchgear). A combination assembly of a benchboard and a vertical hinged panel switchboard placed back to back (no aisle) and enclosed with a top and ends. 103

dual coding (software). A development technique in which two functionally identical versions of a program are developed by different programmers or different programming teams from the same specification. The resulting source code may be in the same or different languages. The purpose of dual coding is to provide for error detection, increase reliability, provide additional documentation, or reduce the probability of systematic programming errors or compiler errors influencing the end result. *See:* **code; compiler; documentation; error; program; reliability; specification.** 434

dual control. A term applied to signal appliances provided with two authorized methods of operation. 328

dual diversity (data transmission). The term applied to the simultaneous combining of four signals and their detection through the use of space, frequency or polarization characteristics. 59

dual-element bolometer unit. An assembly consisting of two bolometer elements and a bolometer mount in which they are supported. *Note:* The bolometer elements are effectively in series to the bias power and in parallel to the RF power. 115

dual-element electrothermic unit. An assembly consisting of two thermopile elements and an electrothermic mount in which they are supported. The thermopile elements are effectively in series to the output voltage and in parallel to the RF power. The termopiles also serve as the power absorber. 47

dual-element fuse (1)(power switchgear). A fuse having current-responsive elements of two different fusing characteristics in series in a single fuse. 103

(2)(protection and coordination of industrial and commercial power systems). A cartridge fuse having two or more current-responsive of different fusing characteristics in series in a single cartridge. This is a construction/design technique frequently used to obtain a time-delay response characteristic. Labeling a fuse as dual element means this fuse meets UL time delay requirements (can carry five times rated current for a minimum of 10 seconds (s) for Class H, K, J, and R fuses) and in this case defines a time-current response characteristic and not necessarily a dual-element construction technique. 504

dual-element substitution effect (error). A component of substitution error, peculiar to dual-element bolometer units, that can cause the effective efficiency to vary with RF input power level. *Note:* This component, usually very small, is included in the effective efficiency correction for substitution error only with reference conditions for input RF power level and frequency. It results from a different division of RF and bias powers between the two elements. 115

dual headlighting system (illuminating engineering). Two double headlighting units, one mounted on each side of the front end of a vehicle. Each unit consists of two sealed-beam lamps mounted in a single housing. The upper or outer lamps have two filaments which supply the lower beam and part of the upper beam, respectively. The lower or inner lamps have one filament which provides the primary source of light for the upper beam. 167

dual mode (semiconductor)(nonlinear, active, and nonreciprocal waveguide components). A class of semiconductor devices used in frequency multipliers. Multiplication results not only from the diode's voltage variable reactive properties, but also from its ability to recover rapidly after storing charge. 530

dual modulation (facsimile). The process of modulting a common carrier wave or subcarrier by two different types of modulation. For example, amplitude and frequency modulation, each conveying separate information. *See:* **facsimile transmission.** 12

dual networks. *See:* **structurally dual networks.**

dual overcurrent trip. *See:* **dual release (trip); overcurrent release (trip).**

dual race (displays). A multitrace operation in which a single beam in a cathode-ray tube is shared by two signal channels. *See:* **alternate display; chopped display; multitrace; oscillograph.** 15,106

dual-range single-pointer-form demand register (mechanical demand registers). An indicating demand register having an arrangement for changing the full-scale capacity from one value to another, usually by reversing the scale plate. For example, Scale Class 1/2; Scale Class 2/6. An interlock assures proper scale and scale-class relation. 548

dual release (trip) (power switchgear). A release that combines the function of a delayed and an instantaneous release. 103

dual service (plural service). Two separate services, usually of different characteristics, supplying one consumer. *Note:* A dual service might consist of an alternating-current and direct-current service, or of 208Y/120 volt 3-phase, 4-wire service for light and some power and a 13.2-kilovolt service for power, etcetera. *See:* service; duplicate service; emergency service; loop service. 64

dual switchboard (power switchgear). A control switchboard with front and rear panels separated a comparatively short distance and enclosed at both ends and top. The panels on at least one side are hinged for access to the panel wiring. 103

dual tone multifrequency (DTMF) signaling(telephone loop performance). Voiceband signaling by simultaneous transmission of two tones, one from a low-frequency and one from a high-frequency group. Each of these groups consists of four voiceband frequency tones no two of which are harmonically related. Only 12 of the 16 combinations are currently in use for customer address signaling. 473

dual-tone multifrequency pulsing (telephone switching systems). A means of pulsing utilizing a simultaneous combination of one of a lower group of frequencies and one of a higher group of frequencies to represent each digit or character. 55

dubbing (electroacoustics). A term used to describe the combining of two or more sources of sound into a complete recording at least one of the sources being a recording. *See:* phonograph pickup; re-recording. 176

duck tape. Tape of heavy cotton fabric, such as duck or drill, which may be impregnated with an asphalt, rubber, or synthetic compound. 64

duct (1)(NESC)(transmission and distribution). A single enclosed raceway for conductors or cable. 494, 262

(2) (underground electric systems). A single enclosed runway for conductors or cables. 64

duct bank (conduit run). An arrangement of conduit providing one or more continuous ducts between two points. *Note:* An underground runway for conductors or cables, large enough for workmen to pass through, is termed a gallery or tunnel. 64

duct edge fair-lead (cable shield). A collar or thimble, usually flared, inserted at the duct entrance in a manhole for the purpose of protecting the cable sheath or insulation from being worn away by the duct edge. 64,57

duct entrance. The opening of a duct at a manhole, distributor box, or other accessible space. 64

ductility factor (seismic design of substations). The ratio of the maximum displacement (ultimate) to the displacement which corresponds to initiation of the yielding. 465

ducting (radar). Confinement of electromagnetic wave propagation to a restricted atmospheric layer by steep gradients in the index of refraction with altitude. 13

duct rodding (rodding a duct). The threading of a duct by means of a jointed rod of suitable design for the purpose of pulling in the cable-pulling rope, mandrel, or the cable itself. 64

duct sealing. The closing of the duct entrance for the purpose of excluding water, gas, or other undesirable substances. 64

duct spacer (vent finger) (rotating machinery). A spacer between adjacent packets of laminations to provide a radial ventilating duct. 63

duct system (National Electrical Code). A continuous passageway for the transmission of air which, in addition to ducts, may include duct fittings, dampers, plemums, fans, and accessory air handling equipment. 256

duct ventilated (pipe ventilated) (rotating machinery). A term applied to apparatus that is so constructed that a cooling gas can be conveyed to or from it through ducts. 63

dumbwaiter. A hoisting and lowering mechanism equipped with a car that moves in guides in a substantially vertical direction, the floor area of which does not exceed 9 square feet, whose total inside height whether or not provided with fixed or removable shelves does not exceed 4 feet, the capacity of which does not exceed 500 pounds, and which is used exclusively for carrying materials. 328

dummy antenna. A device that has the necessary impedance characteristics of an antenna and the necessary power-handling capabilities, but that does not radiate or receive radio waves. *Note:* In receiver practice, that portion of the impedance not included in the signal generator is often called **dummy antenna**. *See:* radio receiver. 339

dummy-antenna system. An electric network that simulates the impedance characteristics of an antenna system. *See:* navigation. 278

dummy coil (rotating machinery). A coil that is not required electrically in a winding, but that is installed for mechanical reasons and left unconnected. *See:* rotor (rotating machinery); stator. 63

dummy load (radio transmission). A dissipative but essentially nonradiating substitute device having impedance characteristics simulating those of the substituted device. *See:* artificial load; radio transmission. 185

dummy parameter. *See:* formal parameter. 434

dump (1) (computing systems). (A) To copy the contents of all or part of a storage, usually from an internal storage into an external storage. (B) A process as in (A). (C) The data resulting from the process as in (A). *See:* dynamic dump; postmortem; selective dump; snapshot dump; static dump. 54

(2) (software). (A) Data that have been dumped. (B) To write the contents of a storage, or of part of a storage, usually from an internal storage to an external medium, for a specific purpose such as to allow other use of the storage, as a safeguard against faults or errors, or in connection with debugging. *See:* data; debugging; error; fault. 434

dump energy (power operations). Energy generated from water, gas, wind, or other source which cannot be stored and which is beyond the immediate needs of the electric system producing the energy. 516

dump power. Power generated from water, gas, wind, or other source that cannot be stored or conserved and that is beyond the immediate needs of the electric system producing the power. *See:* **generating station.**
64

duodecimal. (1) Pertaining to a characteristic or property involving a selection, choice, or condition in which there are twelve possibilities. (2) Pertaining to the numeration system with a radix of twelve.
255,77

duolater coil. *See:* **honeycomb coil.**

duplex (data transmission). Pertaining to a simultaneous two-way independent transmission in both directions.
59

duplex artificial line (balancing network). A balancing network, simulating the impedance of the real line and distant terminal apparatus, that is employed in a duplex circuit for the purpose of making the receiving device unresponsive to outgoing signal currents. *See:* **telegraphy.**
328

duplex benchboard (power switchgear). A combination assembly of a benchboard and a vertical control switchboard placed back to back and enclosed with a top and ends (not grille). Access space with entry doors is provided between the benchboard and vertical control switchboard.
103

duplex cable. (1) A cable composed of two insulated stranded conductors twisted together. *Note:* They may or may not have a common insulating covering. *See:* power distribution, underground construction.
345

(2) A cable composed of two insulated single-conductor cables twisted together. *Note:* The assembled conductors may or may not have a common covering of binding or protecting material.
64

duplex cavity (radar). *See:* **receiver protector.**

duplex channel. *See:* **duplex operation.**
59

duplexer (radar)(nonlinear, active, and nonreciprocal waveguide components). A device that utilizes the finite delay between the transmission of a pulse and the echo thereof so as to permit the connection of the transmitter and receiver to a common antenna. A duplexer commonly employs either a circulator and receiver protector or a balanced network of transmit-receive switches and a receiver protector. *See:* **receiver protector.**
530

duplexing assembly, radar. *See:* **transmit-receive switch.**

duplex lap winding (rotating machinery). A lap winding in which the number of parallel circuits is equal to twice the number of poles.
63

duplex operation (1) (data transmission). (A) (general). The operation of transmitting and receiving apparatus at one location in conjunction with associated transmitting and receiving equipment at another location; the processes of transmission and reception being concurrent. (B) (radio communication) (Two-way radio communication circuit). The operation utilizing two radio-frequency channels, one for each direction of transmission, in such manner that intelligence may be transmitted concurrently in both directions.
111, 59

(2) (radio communication) (two-way radio communication circuit). The operation utilizing two radio-frequency channels, one for each each direction of transmission, in such manner that intelligence may be transmitted concurrently in both directions.
181

duplex signaling (telephone switching systems). A form of polar-duplex signaling for a single physical circuit.
55

duplex switchboard (power switchgear). A control switchboard consisting of panels placed back to back and enclosed with a top and ends (not grille). Access space with entry doors is provided between the rows of panels.
103

duplex system. A telegraph system that affords simultaneous independent operation in opposite directions over the same channel. *See:* **telegraphy**
328

duplex type (breaker-and-a-half arrangement). A unit substation which has two stepdown transformers, each connected to an incoming high-voltage circuit. The outgoing side of each transformer is connected to a radial (stub-end) feeder. These feeders are joined on the feeder side of the power circuit breakers by a normally open-tie circuit breaker.
53

duplex wave winding (rotating machinery). A wave winding in which the number of parallel circuits is four, whatever the number of poles.
63

duplicate. *See:* **copy**

duplicate lines (power transmission). Lines of substantially the same capacity and characteristics, normally operated in parallel, connecting the same supply point with the same distribution point. *See:* **center of distribution.**
64

duplicate service (power transmission). Two services, usually supplied from separate sources, of substantially the same capacity and characteristics. *Note:* The two services may be operated in parallel on the consumer's premises, but either one alone is of sufficient capacity to carry the entire load. *See:* **service; dual service; emergency service; loop service.**
64

duplication check. A check based on the consistency of two independent performances of the same task.
167,77

duration (pulse terms). The absolute value of the interval during which a specified waveform or feature exists or continues.
254

dust-ignition-proof (class II locations) (National Electrical Code). Enclosed in a manner that will exclude ignitable amounts of dusts or amounts that might affect performance or rating and that, where installed and protected in accordance with this Code, will not permit arcs, sparks, or heat otherwise generated or liberated inside of the enclosure to cause ignition of exterior accumulations or atmospheric suspensions of a specified dust on or in the vicinity of the enclosure.
256

dust-ignition proof machine. A totally enclosed machine whose enclosure is designed and constructed in a manner that will exclude ignitable amounts of dusts or amounts that might affect performance or rating, and that, when installation and protection are in conformance with the National Electrical Code (ANSI

CI-1975; section 502-1), will not permit arcs, sparks, or heat otherwise generated or liberated inside of the enclosure to cause ignition of exterior accumulations or atmospheric suspensions of a specific dust on or in the vicinity of the enclosure. *See:* **asynchronous machine.** 232,63

dustproof (1) (general). So constructed or protected that the accumulation of dust will not interfere with successful operation. 206,102,27

(2) (enclosure). An enclosure so constructed or protected that any accumulation of dust that may occur within the enclosure will not prevent the successful operation of, or cause damage to, the enclosed equipment. 3

(3) (luminaire). Luminaire so constructed or protected that dust will not interfere with its successful operation. *See:* **luminaire.** 167

(4) (National Electrical Code). So constructed or protected that dust will not interfere with its successful operation. 256

dust seal (rotating machinery). A sealing arrangement intended to prevent the entry of a specified dust into a bearing. *See:* **asynchronous machine; direct-current commutating machine.** 63

dust-tight (National Electrical Code). So constructed that dust will not enter the enclosing case under specified test conditions. *See:* **ANSI C19.4-1973, for test conditions.** 256

dusttight (1) (enclosure). An enclosure so constructed that dust will not enter the enclosing case. 3, 27

(2) (luminaire) (transformer). An enclosure so constructed that dust will not enter the enclosing case under specified conditions. *See:* **luminaire.** 103,167,53

dust-tight luminaire (illuminating engineering). A luminaire so constructed that dust will not enter the enclosing case. 167

duty (1) (general). A statement of loads including no-load and rest and de-energized periods, to which the machine or apparatus is subjected including their duration and sequence in time. 63

(2) (rating of electric equipment). A statement of the operating conditions to which the machine or apparatus is subjected, their respective durations, and their sequence in time. 310,233

(3) (industrial control) (power and distribution transformer). A requirement of service that defines the degree of regularity of the load. 53

(4) (excitation systems). Those voltage and current loadings imposed by the synchronous machine upon the excitation system including short circuits and al conditions of loading. *Note:* The duty will include the action of limiting devices to maintain synchronous machine loading at or below that defined by American National Standard Requirements for Cylindrical Rotor Synchronous Generators, C50.13-1965. 15

duty continuous (thyristor converter). A duty where the converter equipment carries a direct current of fixed value for an interval sufficiently long for the components of the converter to reach equilibrum temperatures corresponding to the said value of current. 121

duty cycle (1) (general). The time interval occupied by a device on intermittent duty in starting, running, stopping, and idling. 111

(2) (rotating machinery). A variation of load with time which may or may not be repeated, and in which the cycle time is too short for thermal equilibrium to be attained. *See:* **asynchronous machine.** 63

(3) (pulse systems). The ratio of the sm of all pulse durations to the total period, during a specified period of continuous operation. *See:* **navigation.** 187

(4) (welding) (National Electrical Code). The percentage of the time during which the welder is loaded. For instance, a spot welder supplied by a 60-Hertz system (216,000 cycles per hour) making four hundred 15-cycle welds per hour would have a duty cycle of 2.8 percent (400 multiplied by 15, divided by 216,000, multiplied by 100). A seam welder operating 2 cycles "on" and 2 cycles "off" would have a duty cycle of 50 percent. 256

(5) (radar). *See:* **duty factor.**

duty-cycle rating (rotating machinery). The statement of the loads and conditions assigned to the machine by the manufacturer, at which the machine may be operated on duty cycles. *See:* **asynchronous machine.** 63

duty-cycle voltage rating (metal-oxide surge arresters for ac power circuits). The designated maximum permissible voltage between its terminals at which an arrester is designed to perform its duty cycle. 583

duty factor (1) (pulse techniques). The ratio of the pulse duration to the pulse period of a periodic pulse train. *See:* **pulse; pulse carrier.** 85

(2) (electron tubes). The ratio of the ON period to the total period during which an electronic valve or tube is operating. *See:* **ON period (electron tubes); pulse.** 244

(3) (radar). In any system with intermittent operation, the ratio of the active or "ON" ("ON DUTY") time within a specified period to the duration of the specified period. *Note:* In systems with a constant repetition cycle, the **duty factor** usually specified (but not always) is the ratio of one ON period to the total period of one cycle. 13

(4) (automatic control). The ratio of working time to the time taken for the complete sequence of a duty cycle. 56

(5) (waveguide). The ratio of the average power to the peak pulse power passing through the transverse section of the waveguide. *Syn:* **duty cycle.** 267

duty factor control system. *See:* **control system, duty factor.**

duty, peak load (thyristor converter). A type of duty where the rating of the converter is specified in terms of the magnitude and duration of the peak load together with the time of no-load between peaks. 121

duty ratio (1) (pulse system). The ratio of average to peak pulse power. *See:* **navigation.** 187

(2) (radar). (A) In a pulsed radar, the ratio of average to peak pulse power. (B) *Syn:* **duty factor.** *See:* **duty factor.** 13

dwell (numerically controlled machines). A timed delay of programmed or established duration, not cyclic or sequential, that is, not an interlock or hold. 207

dwelling unit (National Electrical Code). One or more rooms for the use of one or more persons as a housekeeping unit with space for eating, living, and sleeping, and permanent provisions for cooking and sanitation. 256

dyadic (mathematics of computing). Pertaining to an operation involving two operands. *See:* **monadic.** 564

dynamic (industrial control) (excitation control systems). A state in which one or more quanitities exhibit appreciable change within an arbitrarily short time interval. *Note:* For excitation control systems, this time interval encompasses up to 15-20 sec., that is, sufficient time to ascertain whether oscillations are decaying or building up with time. *See:* **control system, feedback.** 353

dynamic accuracy (1). Accuracy determined with a time-varying output. Contrast with **static accuracy.** *See:* **electronic analog computer.** 16
(2) (analog computers). Accuracy determined with a time-varying output. 9

dynamic allocation (software). The allocation of addressable storage and other resources to a program while the program is executing. 434

dynamically tuned gyro (DTG) (inertial sensor). A two-degree-of-freedom gyro in which a dynamically tuned flexure and gimbal mechanism both supports the rotor and provides angular freedom about axes perpendicular to the spin axis. *See:* **dynamic tuning.** 46

dynamic analysis (software). The process of evaluating a program based on execution of the program. *See:* **execution; static analysis.** 434

dynamic analyzer (software). A software tool that aids in the evaluation of a computer program by monitoring execution of the program. Examples include instrumentation tools, software monitors, and tracers. *See:* **computer program; execution; instrumentation tools; program; software monitor; software tool; static analyzer; tracer.** 434

dynamic binding (software). Binding performed during execution of a program. *See:* **binding; execution; program; static binding.** 434

dynamic braking (rotating machinery). A system of electric braking in which the excited machine is disconnected from the supply system and connected as a generator, the energy being dissipated in the winding and, if necessary, in a separate resistor. 63

dynamic braking envelope. A curve that defines the dynamic braking limits in terms of speed and tractive force as restricted by such factors asmaximum current flow, maximum permissible voltage, minimum field strength, etcetera. *See:* **dynamic braking.** 328

dynamic characteristic (electron tube) (operating characteristic). *See:* **load (dynamic) characteristic (electron tube).**

dynamic check. *See:* **problem check.**

dynamic computer check. *See:* **problem check.**

dynamic cutoff frequency (semiconductor)(nonlinear, active, and nonreciprocal waveguide components). A figure of merit used for varactor diodes. Unlike fixed cutoff frequency measurements at specific bias voltages, dynamic cutoff frequency is a measure of the varactor's total change in Q from a slight forward bias current to reverse breakdown voltage. This dynamic or total figure of merit is useful in evaluating the frequency multiplier performance of fully driven multipliers. 530

dynamic dump (computing systems). A dump that is performed during the execution of a program. 255,77

dynamic dumping (test, measurement and diagnostic equipment). The printing of diagnostic information without stopping the program being tested. 54

dynamic electrode potential. An electrode potential when current is passing between the electrode and the electrolyte. *See:* **electrochemistry.** 328

dynamic energy sensitivity (photoelectric devices). *See:* **sensitivity dynamic.**

dynamic error (analog computers). An error in a time-varying signal resulting from imperfect dynamic response of a transducer. 9

dynamic holding brake. A braking system designed for the purpose of exerting maximum braking force at a fixed speed only and used primarily to assist in maintaining this fixed speed when a train is descending a grade, but not to effect a deceleration. *See:* **dynamic braking.** 328

dynamic impedance (low voltage varistor surge arresters). A measure of small signal impedance at a given operating point, described as the rate of change of varistor voltage with respect to varistor current at the operating point. 62

dynamic load line (electron device). The locus of all simultaneous values of total instantaneous output electrode current and voltage for a fixed value of load impedance. 190

dynamic loudspeaker. *See:* **moving-coil speaker.**

dynamic microphone. *See:* **moving-coil microphone.**

dynamic problem check. *See:* **problem check.**

dynamic radiation test (metal-nitride-oxide field-effect transistor). Test of the instantaneous effects of radiation obtained by monitoring electrical properties of interest continuously during and immediately after exposure. 386

dynamic range (1)(parametric amplifier)(nonlinear, active, and nonreciprocal waveguide components). The ratio, usually expressed in decibels, of the maximum to the minimum signal input power levels over which the amplifier can operate within some specified range of performance. The minimum level is usually determined by the noise level of the amplifier, while the maximum level is usually set by the maximum tolerable nonlinear effects. 530
(2) (general). The difference, in decibels, between the overload level and the minimum acceptable signal level in a system or transducer. *Note:* The minimum acceptable signal level of a system or transducer is

ordinarily fixed by one or more of the following: noise level, low-level distortion, interference, or resolution level. *See:* **electronic analog computer; signal.**
9

(3) (control system or element). The ratio of two instantaneous signal magnitudes, one being the maximum value consistent with specified criteria of performance, the other the maximum value of noise.
56

(4) (gyro; accelerometer). The ratio of the input range to the threshold. *See:* **input-output characteristics.**
46

(5) (spectrum analyzer). The maximum ratio of two signals simultaneously present at the input which can be measured to a specified accuracy. (A) harmonic dynamic range. The maximum ratio of two harmonically related sinusoidal signals simultaneously present at the input which can be measured with a specified accuracy. (B) nonharmonic dynamic range. The maximum ratio of two nonharmonically related sinusoidal signals simultaneously present at the input which can be measured with a specified accuracy. (C) display dynamic range. The maximum ratio of two nonharmonically related sinusoids each of which can be simultaneously measured on the screen to a specified accuracy.
390

(6) (analog computers). The ratio of the specified maximum signal level capability of a system or component to its noise or resolution level, usually expressed in decibels. Also, the ratio of the maximum to minimum amplitudes of a variable during a computer solution.
9

dynamic range, reading (storage tubes). The range of output levels that can be read, from saturation level to the level of the minimum discernible output signal. *See:* **storage tube.**
174

dynamic range, writing (storage tubes). The range of input levels that can be written under any stated condition of scanning, from the input that will write the minimum usable signal. *See:* **storage tube.**
174

dynamic register (hybrid computer linkage components). The register that produces the analog equivalent voltage or coefficient.
10

dynamic regulation. Expresses the maximum or minimum output variations occurring during transient conditons, as a percentage of the final value. *Note:* Typical transient conditions are instantaneous or permanent input or load changes.

$$\text{Dynamic Regulation} = \frac{E_{max} - E_{final}}{E_{final}} (100)\%$$

$$= \frac{E_{final} - E_{min}}{E_{final}} (100\%)$$

176

dynamic regulator. A transmission regulator in which the adjusting mechanism is in self-equilibrium at only one or a few settings and requires control power to maintain it at any other setting.
328

dynamic response. *See:* **time response; control system, feedback.**

dynamic restructuring (software). (1) The process of changing software components or structure while a system is running. (2) The process of restructuring a data base or data structure during program execution. *See:* **data base; data structure; program execution; software components; system.**
434

dynamic short-circuit output current (converters having ac output)(self-commutated converters). The transient current that flows from the converter into a short-circuit across the output terminals.
584

dynamic signal to noise ratio (digital delay line). The ratio of the minimum peak output signal to the maximum peak noise output when operated with a random bit sequence at a specified clock frequency.
81

dynamic slowdown (industrial control). Dynamic braking applied for slowing down, rather than stopping, a drive. *See:* **electric drive.**
206

dynamic test (test, measurement and diagnostic equipment). A test of one or more of the signal properties or characteristics of the equipment or of any of its constituent items performed while the equipment is energized.
54

dynamic time constant (dynamically tuned gyro) (inertial sensor). The time required for the rotor to move through an angle equal to 63 percent of its final value following a step change in case angular position about an axis normal to the spin axis with the gyro operating open loop. The value depends on the gimbal and rotor damping and drag forces, and is inversely proportional to quadrature spring rate.
46

dynamic torque (electric coupling). That torque of an electric coupling developed or transmitted at a specified value or range of speed differential between input and output members and at specified excitation and other applicable conditions.
416

dynamic transmission (acoustically tunable optical filter). The ratio of the intensity of the light transmitted by the device at the wavelength to be filtered to the light intensity at this this wavelength incident on the device, namely: $T(\lambda) = [I (\lambda) / I_o (\lambda)]$. It includes all static losses as well as the diffraction efficiency of the interaction. For a given design, the dynamic transmission is a function of acoustic drive power.
72

dynamic tuning (dynamically tuned gyro) (inertial sensor). The adjustment of the gimbal inertia or flexure spring rate, or both, of a rotor suspension system or the adjustment of the rotor speed to achieve a condition where the dynamically induced (negative) spring rate cancels the spring rate of the flexure suspension.
46

dynamic variable brake. A dynamic braking system designed to allow the operator to select (within the limits of the electric equipment) the braking force best suited to the operation of a train descending a grade and to increase or decrease this braking force for the purpose of reducing or increasing train speed. *See:* **dynamic braking.**
328

dynamic vertical. *See:* **apparent vertical.**

dynamometer (conductor stringing equipment)(power line maintenance). A device designed to measure loads or tension on conductors. Various models of these devices are used to tension guys or sag conductors. *Syn:* **clock; load cell.**   431, 458

dynamometer, electric (rotating machinery). An electric generator, motor or eddy-current load absorber equipped with means for indicating torque. *Note:* When used for determining power input or output of a coupled machine, means for indicating speed are also provided.   63

dynamometer test (rotating machinery). A braking or motoring test in which a dynamometer is used. *See:* **braking test; asynchronous machine.**   63

dynamotor (electric installations on shipboard) (rotating machinery). A form of converter which combines both motor and generator action, with one magnetic field and with two armatures or with one armature having separate windings.   63, 3

dynatron effect (electron tubes) (dynatron characteristic). An effect equivalent to a negative resistance, which results when the electrode characteristic (or transfer characteristic) has a negative slope. *Example:*

Anode characteristic of a tetrode, or tetrode-connected valve or tube. *See:* **electronic tube.**   244,190

dynatron oscillation. Oscillation produced by negative resistance due to secondary emission. *See:* **oscillatory circuit.**   111

dynatron oscillator. A negative-resistance oscillator in which negative resistance is derived between plate and cathode of a screen-grid tube operating so that secondary electrons produced at the plate are attracted to the higher potential screen grid. *See:* **oscillatory circuit.**   111

dyne. The unit of force in the cgs (centimeter-gram-second) systems. The dyne is 10^{-5} newton.   210

dynode (electron tubes). An electrode that performs a useful function, such as current amplification, by means of secondary emission. *See:* **electrode (electron tube); electron tube.**   117, 125

dynode spots (image orthicons). A spurious signal caused by variations in the secondary-emission ratio across the surface of a dynode that is scanned by the electron beam.   125

E

early failure period (reliability). That possible early period, beginning at a stated time and during which the failure rate decreases rapidly in comparison with that of subsequent period. *See:* **constant failure rate period.**

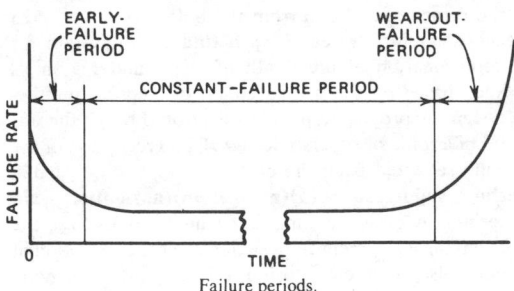

Failure periods.

164,182

early warning radar. Radar employed to search for distant enemy aircraft.   13

earphone (receiver). An electroacoustic transducer intended to be closely coupled acoustically to the ear. *Note:* The term receiver should be avoided when there is risk of ambiguity. *See:* **loudspeaker.**   176

earphone coupler. A cavity of predetermined size and shape that is used for the testing of earphones. The coupler is provided with a microphone for the measurement of pressures developed in the cavity. *Note:* Couplers generally have a volume of 6 cubic centimeters for testing regular earphones and a volume of 2

cubic centimeters for testing insert earphones. *See:* **loudspeaker.**   176

earth, effective radius (radio wave propagation). A value for the radius of the earth that is used in place of the geometrical radius to correct approximately for atmospheric refraction when the index of refraction in the atmosphere changes linearly with height. *Note:* Under conditions of standard refraction the effective radius of the earth is 8.5×10^6 meters, or 4/3 the geometrical radius. *See:* **radiation; radio wave propagation.**   180

earth-fault protection. *See:* **ground protection.**   103

earth inductor. *See:* **generating magnetometer.**

earth rate (vertical, horizontal) (inertial sensor). The angular velocity of the earth with respect to inertial space. Its magnitude is 7.292×10^{-5} rad/s (15.041°/h). This vector quantity is usually expressed as two components in local level coordinates, north (or horizontal) and up (or vertical).   46

earth resistivity (power fault effects). A measurement of the electrical resistance of a unit volume of soil. The commonly used unit of measure is the ohm-meter which refers to the resistance measured between opposite faces of a cubic meter of soil.   404

earth's rate correction (navigation aid terms)(gyro). A rate applied to a gyroscope to compensate for the apparent precession of the spin axis caused by the rotation of the earth.   526, 46

earth station (communication satellite). A ground station designed to transmit to and receive transmission rom communication satellites.   83

earth terminal. *See:* **ground terminal.**

E bend (E-plane bend) (waveguide technique). A smooth change in the direction of the axis of a waveguide, throughout which the axis remains in a plane parallel to the direction of polarization. *See:* **waveguide.** 328

eccentric groove (eccentric circle) (disc recording). A locked groove whose center is other than that of the disc record (generally used in connection with mechanical control of phonographs). *See:* **phonograph pickup.** 176

eccentricity (power distribution, underground cables) (1) (general). The ratio of the difference between the minimum and average thickness to the average thickness of an annular element, expressed in percent. 57

(2) (disk recording). The displacement of the center of the recording groove spiral, with respect to the record center hole. *See:* **phonograph pickup.** 176

Eccles-Jordan circuit. A flip-flop circuit consisting of a two-stage resistance-coupled electron-tube amplifier with its output similarly coupled back to its input, the two conditions of permanent stability being provided by the alternate biasing of the two stages beyond cutoff. *See:* **trigger circuit.** 328

ECCM. *See:* **electronic counter-countermeasures.**

ECCM (electronic counter-countermeasures) improvement factor (radar). The power ratio of the ECM (electronic counter measures) signal level required to produce a given output signal from a receiver using an ECCM technique to the ECM signal level producing the same output from the same receiver without the ECCM technique. 13

echelon (calibration). A specific level of accuracy of calibration in a series of levels, the highest of which is represented by an accepted national standard. *Note:* There may be one or more auxiliary levels between two successive echelons. *See:* **measurement system.** 293,183,155

echo (1)(supervisory control, data acquisition, and automatic control). A communication technique assuring that a word received at the termination point in a system is the same as the word originally transmitted. The received word is retransmitted to the sending device and matched to ensure that the original message was received properly. 570

(2) (data transmission) (general). A wave which has been reflected or otherwise returned with sufficient magnitude and delay to be perceived in some manner as a wave distinct from that directly transmitted. *Note:* Echoes are frequently measured in decibels relative to the directly transmitted wave. 59

(3) (radar). The portion of energy of the transmitted pulse that is reflected to a receiver. 13

(4) (facsimile). A wave which has been reflected at one or more points with sufficient magnitude and time difference to be perceived in some manner as a wave distinct from that of the main transmission. 12

echo area, effective (radar). The area of a fictitious perfect electromagnetic reflector that would reflect the same amount of energy back to the radar as the target. *See:* **navigation.** 187

echo attenuation (data transmission). In a 4-wire or 2-wire circuit in which the two directions of transmission can be separated from each other, the attenuation of the echo currents (which return to the input of the circuit under consideration) is determined by the ratio of the transmitted power to the echo power received expressed in decibels. 59

echo box (radar). A calibrated resonant cavity which stores part of the transmitted pulse power and feeds this exponentially decaying power into the receiving system after completion of the pulse transmission. 13

echo check. A method of checking the accuracy of transmission of data in which the received data are returned to the sending end for comparison with the original data. 54

echo radar (navigation aid terms). The portion of energy of the transmitted pulse which is reflected to a receiver. 526

echo ranging (navigation aid terms). The process of determination of distance by measuring the time interval between transmission of a radiant energy source, usually sound, and the return of its echo. *See:* **radio acoustic ranging.**

echo return loss (ERL)(analog voice frequency circuits). The return loss of a circuit measured with a transmitted signal with a flat spectral distribution between 3 dB frequencies of 560 hertz (Hz) and 1965 Hz. 468

echo, second-time-around (radar). *See:* **second-time-around echo.**

echo sounder (navigation aid terms). An instrument used for echo sounding. *Syn:* **depth finder.** 526

echo sounding (navigation aid terms). Determination of the depth of water by measuring the time interval between emissions of a sonic or ultrasonic signal and the return of its echo from the bottom. 526

echo sounding system (depth finder). A system for determination of the depth of water under a ship's keel, based on the measurement of elapsed time between the propagation and projection through the water of a sonic or supersonic signal, and reception of the echo reflected from the bottom. 328

echo suppressor (1)(navigation)(navigation aid terms). A circuit component which desensitizes the receiving equipment for a period after the reception of one pulse, for the purpose of rejecting pulses arriving later over indirect reflection paths. 526

(2) (data transmission). A voice-operated device for connection to a two-way telephone circuit to attenuate echo current in one direction caused by telephone current in the other direction. 59

eclipsing (radar). The loss of information on radar echoes during intervals when the receiver is blanked because of the occurrence of a transmitter pulse. Numerous such blankings can occur in radars having high pulse repetition frequencies. 13

ECM. *See:* **electronic countermeasures.** 13

economic dispatch (electric power systems). The distribution of total generation requirements among alternative sources for optimum system economy with due consideration of both incremental generating costs and incremental transmission losses. 94

economy energy (power operations). Energy produced in one system and substituted for less economical energy in another system.   516

economy power. Power produced from a more economical source in one system and substituted for less economical power in another system. *See:* **generating station.**   64

eddy current (electrical heating systems). Current that circulates in a metallic material as a result of electromotive forces induced by a variation of magnetic flux.   476

eddy-current braking (rotating machinery). A form of electric braking in which the energy to be dissipated is converted into heat by eddy currents produced in a metallic mass. *See:* **asynchronous machine.**   63

eddy-current loss (1) (parts, hybrids, and packaging). Power dissipated due to eddy currents. *Note:* The eddy-current loss of a magnetic device includes the eddy-current losses in the core, windings, case, and associated hardware.   197

(2) (power and distribution transformer). The energy loss resulting from the flow of eddy currents in a metallic material.   53

eddy currents (power and distribution transformer). The currents that are induced in the body of a conducting mass by the time variation of magnetic flux.   53

Edison distribution system. A three-wire direct-current system, usually about 120-240 volts, for combined light and power service from a single set of mains. *See:* **direct-current distribution.**   64

Edison effect. *See:* **thermionic emission.**

Edison storage battery. An alkaline storage battery in which the positive active material is nickel oxide and the negative an iron alloy. *See:* **battery (primary or secondary).**   328

E-display (radar). A rectangular display in which targets appear as intensity-modulated blips with range indicated by the horizontal coordinate and elevation angle by the vertical coordinate. *Note:* The term E-display has also been applied to a display in which height or altitude is the vertical coordinate, but this

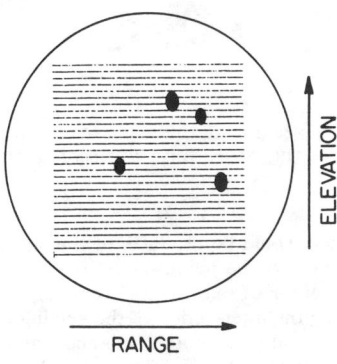

E-Display

usage is deprecated because of ambiguity. The preferred term for such a display is "range-height indication (RHI)".   13

edit (computing systems). To modify the form or format of data, for example, to insert or delete characters such as page numbers or decimal points.   54

editor (software). A computer program that permits selective revision of computer-stored data. *See:* **computer; computer program; data.**   434

EDR. *See:* **electrodermal reaction.**

effective address (microprocessor assembly language). The result of evaluating an address in accordance with its addressing mode.   466

effective aperture (EM-radiation collection device) (radar). Synonymous with effective area for an antenna (IEEE Std 145-1973, ANSI CI6.38-1976), Definitions of Terms for Antennas); also, the effective area of other EM-radiation collecting devices, such as lenses.   13

effective area (of an antenna) (in a given direction). In a given direction, the ratio of the available power at the terminals of a receiving antenna to the power flux density of a plane wave incident on the antenna from that direction, the wave being polarization matched to the antenna. *Notes:* (1) If the direction is not specified, the direction of maximum radiation intensity is implied. (2) The effective area of an antenna in a given direction is equal to the square of the operating wavelength times its gain in that direction divided by 4π. *See:* **polarization match.**   111

effective area antenna (data transmission). The ratio of the power available at the terminals of an antenna to the incident power density of a plane wave from that direction polarized, coincident with the polarization that the antenna would radiate.   59

effective asymmetrical fault current (safety in ac substation grounding). The root-mean-square (rms) value of asymmetrical current wave, integrated over the entire interval of fault duration. See figure below.   563

effective energy (radiation survey instruments). The energy of monochromatic photons which undergoes the same percentage attenuation in a specified filter as the heterogeneous beam under consideration. Aluminum is the filter specified for photon energies less than, or equal to, 100 kiloelectronvolts (keV), copper for photon energies between 100 keV and 1.5 megaelectronvolts (MeV), and lead for photons with energies greater than 1.5 MeV.   558

effective band (facsimile). The frequency band of a facsimile signal wave equal in width to that between zero frequency and maximum keying frequency. *Note:* The frequency band occupied in the transmission medium will in general be greater than the effective band.   12

effective bunching angle (reflex klystrons). In a given drift space, the transit angle that would be required in a hypothetical drift space in which the potentials vary linearly over the same range as in the given space and in which the bunching action is the same as in the given space.   125

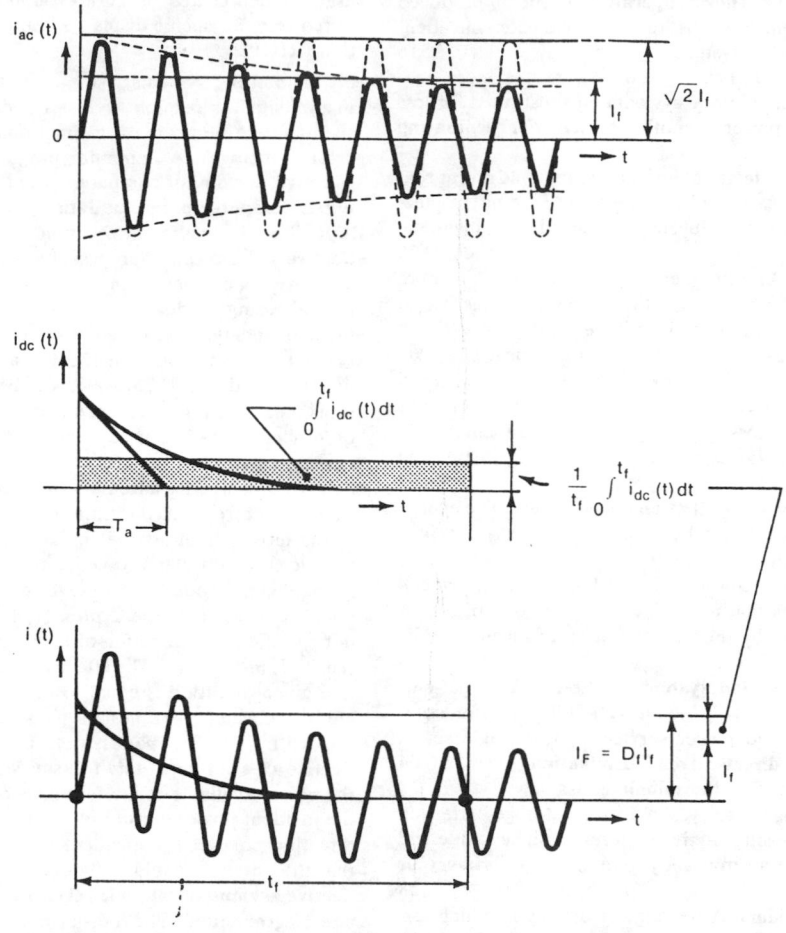

Relationship Between Actual Values of Fault Current and Values of I_F, I_f, and D_f for Fault Duration t_f

NOTE: In terms of this guide, it can be expressed as

$$I_F = D_f(t_f) I_f \qquad\qquad\qquad\qquad\qquad \text{(Eq 2)}$$

where

I_F = effective asymmetrical fault current in A
I_f = (initial) symmetrical ground fault current in A
$D_f(t_f)$ = decrement factor accounting for the effect of a dc offset during the subtransient period of fault current wave on an equivalent time basis of the entire fault duration, t_f, for t_f given in s

effective ceiling cavity reflectance, (illuminating engineering). A number giving the combined reflectance effect of the wall and ceiling reflectance of the ceiling cavity. *See:* **ceiling cavity ratio.** 167

effective center (radiation protection). The point within a detector that produces, for a given set of irradiation conditions, an instrument response equivalent to that which would be produced if the entire detector were located at the point. 399

effective center of mass (accelerometer). That point defined by the intersection of the pendulous axis and an axis parallel to the output axis about which angular acceleration results in minimum accelerometer output. 46

effective center-of-mass for angular acceleration (accelerometer) (inertial sensor). That point defined by the intersection of the pendulous axis and an axis parallel to the output axis about which angular acceleration results in a minimum accelerometer output.
46

effective center-of-mass for angular velocity (accelerometer) (inertial sensor). That point defined by the intersection of the pendulous axis and an axis of constant speed rotation approximately parallel to the input axis, for which the offset due to spin becomes independent of orientation. *See:* **spin offset coefficient.**
46

effective echo area. *See:* **radar cross section.**

effective echoing area (radar). *Syn:* **radar cross section.**
13

effective efficiency (1) (bolometer units). The ratio of the substitution power to the total RF power dissipated within the bolometer unit.
115

(2) (electrothermic unit). The ratio of the substituted reference power (direct current, audio or radio frequency) in the electrothermic unit to the power dissipated within the electrothermic unit for the same direct current output voltage from the electrothermic unit at a prescribed frequency, power level, and temperate. *Notes:* (A) Calibration factor and effective efficiency are related as follows:

$$\frac{K_b}{\eta_c} = 1 - |\Gamma|^2$$

where K_b, η_c and Γ are the calibration factor, effective efficiency, and reflection coefficient of the electrothermic unit, respectively. (B) The reference frequency is to be supplied with the calibration factor.
47

effective floor cavity reflectance, ρfc (illuminating engineering). A number giving the combined reflectance effect of the floor cavity. *See:* **floor cavity ratio.**
167

effective flux penetration (P)(electrical heating systems). The distance into a pipeline or vessel wall that the value of current induced by the magnetic field at the surface would have to penetrate in order to generate the same heat as is generated by the actual induced current distribution in the wall.
476

effective height (antenna). (1) High-frequency usage. The height of the antenna center of radiation above the ground level. *Note:* For an antenna with symmetrical current distribution, the center of radiation is the center of distribution. For an antenna with asymmetrical current distribution, the center of radiation is the center of current moments when viewed from directions near the direction of maximum radiation. **(2) (Low-frequency usage).** *See:* **effective length of an antenna.**
111

effective height antenna (data transmission). (1) The effective height of an antenna is the height of its center of radiation above the effective ground level. **(2)** In low-frequency applications the term *effective height* is applied to loaded or nonloaded vertical antennas and is equal to the moment of the current distribution in the vertical section, divided by the input current. *Note:* For an antenna with symmetrical current distribution, the center of radiation is the center of distribution. For an antenna with asymmetrical current distribution, the center of radiation is the center of current moments when viewed from directions near the direction of maximum radiation.
59

effective induction area of the control current loop (Hall-effect devices). The effective area of the loop enclosed by the control current leads and the relevant conductive path through the Hall element.
107

effective induction area of the output loop (Hall-effect devices). The effective induction area of the loop enclosed by the leads to the Hall terminals and the relevant conductive path through the Hall plate.
107

effective isotropically radiated power (EIRP). *See:* **equivalent isotropically radiated power.**

effective length of linearly polarized antenna. For a linearly polarized antenna receiving a plane wave from a given direction, the ratio of the magnitude of the open circuit voltage developed at the terminals of the antenna to the magnitude of the electric field strength in the direction of the antenna polarization. *Notes:* (1) Alternatively, the effective length is the length of a thin straight conductor oriented perpendicular to the given direction and parallel to the antenna polarization, having a uniform current equal to that at the antenna terminals and producing the same far-field strength as the antenna in that direction. (2) In low-frequency usage the effective length of a vertically polarized ground based antenna is frequently referred to as effective height. Such usage should not be confused with **effective height of an antenna (high frequency usage).**
111

effectively grounded (1)(NESC). Intentionally connected to earth through a ground connection or connections of sufficiently low impedance and having sufficient current-carrying capacity to prevent the build-up of voltages which may result in undue hazard to connected equipment or to persons.
494

(2) (power and distribution transformer). An expression that means grounded through a grounding connection of sufficiently low impedance (inherent or intentionally added, or both) that fault grounds that may occur cannot build up voltages in excess of limits established for apparatus, circuits, or systems so grounded. *Note:* An alternating-current system or portion thereof may be said to be effectively grounded when, for all points on the system or specified portion thereof, the ratio of zero-sequence reactance to positive-sequence reactance is less than three and the ratio of zero-sequence resistance to positive-sequence reactance is less than one for any condition of operation and for any amount of connected generator capacity.
53

effective mode volume (fiber optics). The square of the product of the diameter of the near-field pattern and the sine of the radiation angle of the far-field pattern. The diameter of the near-field radiation pattern is defined here as the full width at half maximum and the radiation angle at half maximum intensity. *Note:* Effective mode volume is proportional to the breadth of the relative distribution of power amongst modes in a multimode fiber. It is not truly a spatial volume but rather an "optical volume" equal to the product of area and solid angle. *See:* **mode volume; radiation pattern.**   433

effective multiplication factor (k_{eff}). The ratio of the average number of neutrons produced by nuclear fission in each generation to the total number of corresponding neutrons absorbed or leaking out of the system. If $k_{eff}=1$, that is, the number of neutrons produced is equal to the number being absorbed or leaking out of the system, a stable, self-sustaining chain reaction exists and the assembly is said to be critical. If $k_{eff} > 1$, the chain reaction is not self-sustaining and will terminate, such a system is said to be subcritical. If $k_{eff} < 1$, the chain reaction is divergent and the system is supercritical.   516

effective radiated power (antennas). In a given direction, the relative gain of a transmitting antenna with respect to the maximum directivity of a half-wave dipole multiplied by the net power accepted by the antenna from the connected transmitter. *Syn:* **equivalent radiated power (ERP).**   111

effective radius of the earth (1)(radio wave propagation) An effective value for the radius of the earth, which is used in place of the geometrical radius to correct approximately for atmospheric refraction when the index of refraction in the atmosphere changes linearly with height. *Note:* Under conditions of standard refraction the effective radius of the earth is 8.5×10^6 meters, or 4/3 the geometrical radius.   146

(2)(data transmission). In radio transmission, a value which is used in place of the geometrical radius to correct for atmospheric refraction when the index of refraction in the atmosphere changes linearly with height. *Note:* Under conditions of standard refraction, the effective radius is 1.33 the geometrical radius.   59

effective resistivity. A factor such that the conduction current density is equal to the electric field in the material divided by the resistivity.   313

effective temperature (laser-maser). The temperature which must be used in the Boltzmann formula to describe the relative populations of two energy levels that may or may not be in thermal equilibrium.   363

efficiency (1)(of a semiconductor radiation conductor for a monoenergetic radiation source)(X-ray energy spectrometers). The ratio of the number of events in the spectral distribution to the total number of photons incident on the active detector volume during the same time interval.   471

(2)(of power conversion)(converter characteristics). The ratio of active (real) output power and active (real) input power. *Note:* Both powers are to be taken as the total average power as given by the formula:

$$P = \frac{1}{T} \int_0^T ei\ dt$$

where T = the period for ac (alternating current) and the ripple period for dc (direct current).   584

(3) (from total loss) (rotating machinery). The method of indirect calculation of efficiency from the measurement of total loss. *See:* **asynchronous machine.**   63

(4) (power and distribution transformer). The ratio of the useful power output to the total power input.   53

(5) (rectification). Ratio of the direct-current component of the rectified voltage at the input terminals of the apparatus to the maximum amplitude of the applied sinusoidal voltage in the specified conditions.   190

(6) (radiation-counter tube). The probability that a tube count will take place with a specified particle or quantum incident in a specified manner.   96,125

(7) (antenna). *See:* **antenna efficiency; aperture illumination efficiency; radiation efficiency.**

(8) (sodium iodide detector). The net number of counts registered by the detector system per unit of time divided by the number of photons of interest originating in the radioactive source that is being measured during the same unit of time.   423

(9) (software). The extent to which software performs its intended functions with a minimum consumption of computing resources. *See:* **functions; software.**   434

(10) (by direct calculation) (rotating machinery). The method by which the efficiency is calculated from the input and output, these having been measured directly. *See:* **asynchronous machine.**   63

(11) (thyristor). The ratio of the load power to the total line power including the contribution of all harmonics.   445

efficiency, effective (bolometer units). The ratio of the substitution power to the total radio-frequency power dissipated within the bolometer unit. *Note:* Effective efficiency includes the combined effect of the direct-current-radio-frequency substitution error and bolometer unit efficiency. *See:* **bolometric power meter.**   183

efficiency, generator (thermoelectric device). *See:* **generator efficiency (thermoelectric couple).**

efficiency, generator, overall (thermoelectric device). *See:* **overall generator efficiency (thermoelectric couple.**

efficiency, generator, reduced (thermoelectric device). *See:* **reduced generator efficiency.**

efficiency, load circuit. *See:* **load circuit efficiency.**

efficiency, overall electrical. *See:* **overall electrical efficiency.**

efficiency, quantum (phototubes). *See:* **quantum efficiency.**

effluent (1)(monitoring radioactivity in effluents). The liquid or gaseous waste streams released to the environment.   559

(2) (radiological monitoring instrumentation). Liquid or airborne radioactive materials released to the environs.   398

effluve. *See:* **convective discharge.**

egoless programming (software). An approach to software development based upon the concept of team responsibility for program development. Its purpose is to prevent the programmer from identifying so closely with his or her output that objective evaluation is impaired. *See:* **program; software.**   434

EHF. *See:* **radio spectrum.**

E-H tee (waveguide components). A junction composed of E- and H-plane tee junctions wherein the axes of the arms intersect at a common point in the main guide. *Note:* Compare hybrid tee.   166

E-H tuner (waveguide components). An E-H tee having E and H arms terminated in movable open- or short-circuit terminations.   166

EI. *See:* **end injection.**

eight-hour rating (magnetic contactor). The rating based on its current-carrying capacity for eight hours, starting with new clean contact surfaces, under conditions of free ventilation, with full-rated voltage on the operating coil, and without causing any of the established limitations to be exceeded.   206

einschleichender stimulus. *See:* **accumulating stimulus.**

Einstein's law (photoelectric device). The law according to which the absorption of a photon frees a photoelectron with a kinetic energy equal to that of the photon less the work function

$$\frac{1}{2}\, mv^2 = h\upsilon - p \;(\text{if } h\upsilon > p).$$

See: **photoelectric effect.**   244

elapsed time printout (sequential events recording systems). The recording of time interval between first and successive detected events.   48

elastances (system of conductors) (coefficients of potential–Maxwell). A set of n conductors of any shape that are insulated from each other and that are mounted on insulating supports within a conducting shell, or on one side of a conducting sheet of infinite extent or above the surface of the earth constitutes a system of n capacitors having mutual elastances and capacitances. *Note:* If the shell (or the earth) is regarded as the electrode common to all n capacitors and the transfers of charge as taking place between shell and the individual electrodes, the sum of the charges on the conductors will be equal and opposite in sign to the charge on the common electrode. The shell (or the earth) is taken to be at zero potential. Let Q_r represent the value of the charge that has been transferred from the shell to the other electrode of the rth capacitor, and let V_r represent the algebraic value of the potential of this electrode resulting from the charges in all n capacitors. If the charges are known the values of the potentials can be computed from the equations:

$$V_1 = S_{11}Q_1 + S_{12}Q_2 + S_{13}Q_3 + \cdots$$
$$V_2 = S_{21}Q_1 + S_{22}Q_2 + S_{23}Q_3 + \cdots$$
$$V_3 = S_{31}Q_1 + S_{32}Q_3 + S_{33}Q_3 + \cdots$$

.
.
.

$$V_r = \sum_{c=1}^{c=n} S_{r,c}Q_c.$$

The multiplying operators $S_{r,r}$ are the self-elastances and the multipliers $S_{r,c}$ are the mutual elastances of the system. Maxwell termed them the coefficients of potential of the system. Their values can be measured by noting that the defining equation for the mutual elastance $S_{r,c}$ is

$$S_{r,c}(\text{reciprocal farad*}) = \frac{V_r\,(\text{volt})}{Q_c\,(\text{coulomb})}$$

(every Q except Q_c being zero).

It can be shown that $S_{r,c} = S_{c,r}$ and that under the conventions stated al the elastances have positive values.   210

*formerly sometimes called the daraf.

elastic-restraint coefficient (gyro) (inertial sensor). The ratio of gimbal restraining torque about an output axis to the output angle.   46

elastic-restraint drift rate (gyro) (inertial sensor). That component of systematic drift rate which is proportional to the angular displacement of a gyro gimbal about an output axis. The relationship of this component of drift rate to gimbal angle can be stated by means of a coefficient having dimensions of angular displacement per unit time per unit angle. This coefficient is equal to the elastic-restraint coefficient divided by angular momentum.   46

elastomer (rotating machinery). Macromolecular material that returns rapidly to approximately the initial dimensions and shape after substantial deformation by a weak stress and release of the stress. *See:* **asynchronous machine.**   63

E layer (radio wave propagation). An ionized layer in the E region.   146

elbow (separable insulated connectors). A connector component for connecting a power cable to a bushing, so designed that when assembled with the bushing, the axes of the cable and bushing are perpendicular. *See:* **bushing insert, Fig 1.**   454

electric. Containing, producing, arising from, actuated by, or carrying electricity, or designed to carry electricity and capable of so doing. Examples: Electric eel, energy, motor, vehicle, wave. *Note:* Some dictionaries indicate electric and electrical as synonymous but usage in the electrical engineering field has in general been restricted to the meaning given in the definitions above. It is recognized that there are borderline cases wherein the usage determines the selection. *See:* **electrical.**   202

electric (or magnetic) field strength (waveguide). The magnitude of the electric (or magnetic) field vector.   267

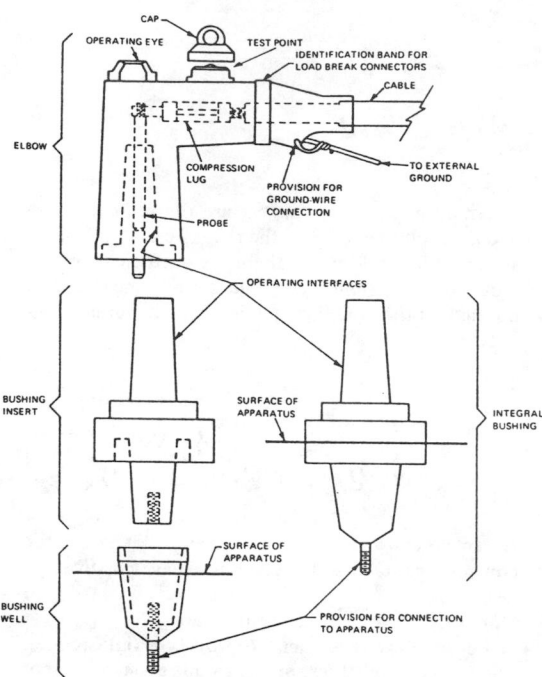

Fig 1
Typical Components of 200 A Separable
Insulated Connector System

electric (power switchgear). Containing, producing, arising from, actuated by, or carrying electricity and capable of doing so. Examples: electric eel, energy, motor, vehicle, wave.   103

electric air-compressor governor. A device responsive to variations in air pressure that automatically starts or stops the operation of a compressor for the purpose of maintaining air pressure in a reservoir between predetermined limits.   328

electrical (1) (power switchgear). Related to, pertaining to, or associated with electricity but not having its properties or characteristics. Examples: electrical engineer, handbook, insulator, rating, school, unit.   103

(2) (general). Related to, pertaining to, or associated with electricity, but not having its properties or characteristics. *Examples:* Electrical engineer, handbook, insulator, rating, school, unit. *Note:* Some dictionaries indicate electric and electrical as synonymous but usage in the electrical engineering field has in general been restricted to the meaning given in the definitions above. It is recognized that there are borderline cases wherein the usage determines the selection. *See:* electric.   202

electrical anesthesia (medical electronics). More or less complete suspension of general or local sensibility produced by electric means.   192

electrical arc (gas). A discharge characterized by a cathode drop that is small compared with that in a glow discharge. *Note:* The electron emission of a cathode is due to various causes (thermionic emission, high-field emission, etcetera) acting simultaneously or separately, but secondary emission plays only a small part. *See:* discharge (gas).   244

electrical back-to-back test (rotating machinery). *See:* **pump-back test.**

electrical boresight (antenna). *See:* electric boresight.

electrical center. *See:* electric center.

electrical codes (1) (general). A compilation of rules and regulations covering electric installations.

(2) official electrical code. One issued by a municipality, state, or other political division, and which may be enforced by legal means.

(3) unofficial electrical code. One issued by other than political entities such as engineering societies, and the enforcement of which depends on other than legal means.

(4) National Electrical Code (NEC). The code of rules and regulations as recommended by the National Fire Protection Association (NFPA) and approved by the American National Standards Institute (ANSI). *Note:* This code is the accepted minimum standard for electric installations and has been accepted by many political entities as their official code, or has been incorporated in whole or in part in their official codes.   256

(5) National Electrical Safety Code(NESC). A set of rules, prepared by the National Electrical Safety Code committee (secretariat held by the Institute of Electrical and Electronics Engineers) and approved by the American National Standards Institute governing: (A) Methods of grounding. (B) Installation and maintenance of electric-supply stations and equipment. (C) Installation and maintenance of overhead supply and communication lines. (D) Installation and maintenance of underground and electric-supply and communications lines. (E) Operation of electric-supply and communication lines and equipment (Work Rules).   391

electrical conductor seal, double (nuclear power generating stations). An assembly of two single electrical conductor seals in series and arranged in such a way that there is a double pressure barrier seal between the inside and the outside of the containment structure along the axis of the conductors.   31

electrical conductor seal, single (nuclear power generating stations). A mechanical assembly providing a single pressure barrier between the electrical conductors and the electrical penetration assembly.   31

electrical degree (rotating machinery). The 360th part of the angle subtended, at the axis of a machine, by two consecutive field poles of like polarity. One mechanical degree is thus equal to as many electrical degrees as there are pairs of poles in the machine.   63

electrical distance (navigation aid terms). The distance between two points expressed in terms of the duration of travel of an electromagnetic wave in free space between the two points. *Note:* An often used unit of electrical distance is the light-microsecond, approximately 300 m (meters) (983 ft).   526

electrical installation, insulation (cable) (electric pipe heating systems). A part that is relied upon to insulate the conductor from other conductors or conducting parts or from ground. Electrical insulation as related to electric pipe heating systems includes that part of a heater that electrically insulates the current carrying conductor(s) from the sheath material.   405

electrical insulating material (thermal classification of electric equipment and electrical insulation). A substance in which the electrical conductivity is very small (approaching zero) and provides electric isolation.   506

electrical insulation system (thermal classification of electric equipment and electrical insulation). An insulating material or a suitable combination of insulating materials specifically designed to perform the functions needed in electric and electronic equipment.   506

electrical interchangeability (of fuse links or fuse units) (power switchgear). The characteristic which permits the designs of various manufacturers to be used interchangeably so as to provide a uniform degree of overcurrent protection and fuse coordination.   103, 443

electrical length (1) (general). The physical length expressed in wavelengths, radians, or degrees. *Note:* When expressed in angular units, it is distance in wavelengths multiplied by 2π to give radians or by 360 to give degrees. *See:* **radio wave propagation; signal wave; wave guide.**   146

(2) (two-port network at a specified frequency). The length of an equivalent lossless reference waveguide or reference air line (which in the ideal case would be evacuated) introducing the same total phase shift as the two-port when each is terminated in a reflectionless termination. *Note:* It is usually expressed in fractions or multiples of waveguide wavelength. When expressed in radians or degrees it is equal to the phase angle of the transmission coefficient $+ 2n\pi$. *See:* **waveguide.**   267

(3) (waveguide). For a traveling wave of a given frequency, a distance in a transmission or guiding medium expressed in wavelengths of the wave in the medium. *Note:* Electrical length is sometimes expressed in radians or degrees.   267

electrical load (power operations). Electric power used by devices connected to an electrical generating system.   516

electrically connected. Connected by means of a conducting path or through a capacitor, as distinguished from connection merely through electromagnetic induction. *See:* **inductive coordination.**   328

electrically heated airspeed tube. A Pitot-static or Pitot-Venturi tube utilizing a heating element for deicing purposes.   328

electrically heated flying suit. A garment that utilizes sewn-in heating elements energized by electric means designed to cover the torso and all or part of the limbs.

Note: It may be a one-piece garment or consist of a coat, trousers, and the like. The lower portion of the one-piece suit is in trouser form.   328

electrically interlocked manual release of brakes (control) (industrial control). A manual release provided with a limit switch that is operated when the braking surfaces are disengaged manually. *Note:* The limit switch may operate a signal, open the control circuit, or perform other safety functions. *See:* **switch.**   206

electrically operated valve (20) (power system device function numbers). An electrically operated, controlled or monitored valve used in a fluid line. *Note:* The functions of the valve may be indicated by the use of the suffixes in 3.3 of ANSI/IEEE C37.2-1979.   402

electrically release-free (trip-free) (as applied to an electrically operated switching device) (power switchgear). A term indicating that the release can open the device even though the closing control circuit is energized. *Note:* Electrically release-free switching devices are usually arranged so that they are also anti-pump. With such an arrangement the closing mechanism will not reclose the switching device after opening until the closing control circuit is opened and again closed.   103

electrically reset relay (power switchgear). A relay that is so constructed that it remains in the picked-up condition even after the input quantity is removed; an independent electrical input is required to reset the relay.   103

electrically short dipole (antennas). A dipole whose total length is small compared to the wavelength. *Note:* For the common case that the two arms are collinear, the radiation pattern approximates that of a hertzian dipole.   111

electrically small antenna. An antenna whose dimensions are such that it can be contained within a sphere whose diameter is small compared to a wavelength at the frequency of operation.   111

electrically suspended gyro (ESG) (inertial sensor). A free gyro in which the main rotating element, the inertial member, is suspended by an electrostatic or an electromagnetic field within an evacuated enclosure.   46

electrically trip-free. *See:* **electrically release-free.**   103

electrical metallic tubing. A thin-walled metal raceway of circular cross section constructed for the purpose of the pulling in or the withdrawing of wires or cables after it is installed in place. *See:* **raceways.**   328

electrical noise (control systems). Unwanted electrical signals, which produce undesirable effects in the circuits of the control systems in which they occur.   43

electrical null (gyro, accelerometer) (inertial sensor). The minimum electrical output. It may be specified in terms of root-mean-square, peak-to-peak, in phase component, or other electrical measurements.   46

electrical null position (gyro; accelerometer). The angular or linear position of a pickoff corresponding to electrical null.   46

electrical objective loudness rating (EOLR) (loudness ratings of telephone connections). For a network

$$\text{EOLR} = -20 \log_{10} \frac{V_T}{\frac{1}{2} V_W} \qquad \text{(Eq 5)}$$

where

V_W = open-circuit voltage of the electric source (in millivolts)

V_T = output voltage of the network (in millivolts)

electrical operation (power switchgear). Power operation by electric energy.   103

electrical penetration assembly (nuclear power generating stations). An electrical penetration assembly provides the means to allow passage of one or more electrical circuits through a single aperture (nozzle or other opening) in the containment pressure barrier, while maintaining the integrity of the pressure barrier.   31

electrical penetration assembly current capacity (nuclear power generating stations). The maximum current that each conductor in the assembly is specified to carry for its duty cycle in the design service environment without causing stabilized temperatures of the conductors or the penetration nozzle-concrete interface (if applicable) to exceed their design limits.   31

electrical penetration assembly short-time overload rating (nuclear power generating stations). The limiting overload current that any one third of the conductors (but in no case less than three of the conductors) in the assembly can carry, for a specified time, in the design service environment, while all remaining conductors carry rated continuous current, without causing the conductor temperatures to exceed those values recommended by the insulated conductor manufacturer as the short-time overload conductor temperature and without causing the stabilized temperature of the penetration nozzle-concrete interface (if applicable) to exceed its design limit.   31

electrical range. The range expressed in equivalent electrical units. *See:* **electrical distance; instrument.**   295

electrical reserve (power operations). The capacity in excess of that required to carry the system load.   516

electrical utility (terrestrial photovoltaic power systems). An organization that provides and distributes electric energy to consumers. In the utility interconnected configuration, solar photovoltaic (PV) systems may be interactive with the utility distribution network to permit the interchange of electric power and energy. *See:* **array control**   496

electrical zero*. *See:* **electrical null position.**   46

*Deprecated

electric back-to-back test. *See:* **pump-back test.**

electric bell. An audible signal device consisting of one or more gongs and an electromagnetically actuated striking mechanism. *Note:* The gong is the resonant metallic member that produces an audible sound when struck. However, the term going is frequently applied to the complete electric bell.   328

electric blasting cap. A device for detonating charges of explosives electrically. *See:* **blasting unit.**   328

electric boresight (antenna). The tracking axis as determined by an electric indication, such as the null direction of a conical-scanning or monopulse antenna system, or the beam-maximum direction of a highly directive antenna. *See:* **reference boresight.**   111

electric braking. A system of braking wherein electric energy, either converted from the kinetic energy of vehicle movement or obtained from a separate source, is one of the principal agents for the braking of the vehicle or train. *See:* **electromagnetic braking (magnetic braking); electropneumatic brake; magnetic track braking; regenerative braking.**   328

electric bus. A passenger vehicle operating without track rails, the propulsion of which is effected by electric motors mounted on the vehicle. *Note:* A prefix diesel-electric, gas-electric, etcetera, may replace the word electric. *See:* **trolley coach (trolley bus) (trackless trolley coach).**   328

electric-cable-reel mine locomotive. An electric mine locomotive equipped with a reel for carrying an electric conductor cable that is used to conduct power to the locomotive when operating beyond the trolley wire. *See:* **electric mine locomotive.**   328

electric capacitance altimeter. An altimeter, the indications of which depend on the variation of an electric capacitance with distance from the earth's surface.   328

electric center (of a power system out of synchronism) (power switchgear). A point at which the voltage is zero when a machine is 180 degrees out of phase with the rest of the system. *Note:* There may be one or more electrical centers depending on the number of the machines and the interconnections among them.   103

electric charge time constant (detector). The time required, after the instantaneous application of a sinusoidal input voltage of constant amplitude, for the output voltage across the load capacitor of a detector circuit to reach 63 percent of its steady-state value. *See:* **electromagnetic compatibility.**   220,199

electric coal drill. An electric motor-driven drill designed for drilling holes in coal for placing blasting charges.   328

electric components (generating stations electric power system). The electric equipment, assemblies, and conductors that together form the electric power systems.   381

electric conduction and convection current density. At any point at which there is a motion of electric charge, a vector quantity whose direction is that of the flow of positive charge at this point, and whose magnitude is the limit of the time rate of flow of net (positive) charge across a small plane area perpendicular to the motion, divided by this area as the area taken approaches zero in a macroscopic sense, so as to always include this point. *Note:* The flow of charge may result from the movement of free electrons or ions but is not, in general, except in microscopic studies, taken to include motions of charges resulting from the polarization of the dielectric. 210

electric console lift. An electrically driven mechanism for raising and lowering an organ console and the organist. *See:* **elevators.** 328

electric constant (permittivity or capacitivity of free space) pertinent to any system of units). The scalar ϵ_0 that in that system relates the electric flux density D, in empty space, to the electric field strength E ($D = \epsilon_0 E$). *Note:* It also relates the mechanical force between two charges in empty space to their magnitudes and separation. Thus in the equation

$$F = Q_1 Q_2 / (n \epsilon_0 r^2)$$

for the force F between charges Q_1 and Q_2 separated by a distance r, ϵ_0 is the electric constant, and n is a dimensionless factor that is unity in unrationalized systems and 4π in a rationalized system. *Note:* In the International System of Units (SI) the magnitude of ϵ_0 is that of $10^7/(4\pi c^2)$ and the dimensions is [$L^{-3} M^{-1} T^4 I^2$]. Here c is the speed of light expressed in the appropriate system of units. 210

electric contact. The junction of conducting parts permitting current to flow. 328

electric controller (electric installations on shipboard). A device, or group of devices, which serves to govern, in some predetermined manner, the electric power delivered to the apparatus to which it is connected. 3

electric-controller rail car. A trail car used in a multiple-unit train, provided at one or both ends with a master controller and other apparatus necessary for controlling the train. *See:* **electric motor car; electric trail car.** 328

electric coupler. A group of devices (plugs, receptacles, cable, etcetera) that provides for readily connecting or disconnecting electric circuits. 328

electric coupler plug. The removable portion of an electric coupler. 328

electric coupler receptacle (electric coupler socket). The fixed portion of an electric coupler. 328

electric coupler socket. *See:* **electric coupler receptacle.**

electric coupling (electric installations on shipboard) A device for transmitting torque by means of electromagnetic force in which there is no mechanical torque contact between the driving and driven members. The slip type electric coupling has poles excited by direct current on one rotating member, and an armature winding, usually of the double squirrel cage type, on the other rotating member. 3

electric course recorder. A device that operates, under control of signals from a master compass, to make a continuous record of a ship's heading with respect to time. 328

electric crab-reel mine locomotive. An electric mine

locomotive equipped with an electrically driven winch, or crab reel, for the purpose of hauling cars by means of a wire rope from places beyond the trolley wire. *See:* **electric mine locomotive.** 328

electric depth recorder. A device for continuously recording, with respect to time, the depth of water determined by an echo sounding system. 328

electric dipole (1) (general). An elementary radiator consisting of a pair of equal and opposite oscillating electric charges an infinitesimal distance apart. *Note:* It is equivalent to a linear current element.

246,179,111

(2) (antenna). The limit of an electric doublet as the separation approaches zero while the moment remains constant. 210

(3). *See:* **hertzian electric dipole.** 111

electric dipole moment (two point charges, q and -q, a distance a apart). A vector at the midpoint between them, whose magnitude is the product $qa1$ and whose direction is along the line between the charges from the negative toward the positive charge. 210

electric-discharge lamp (gas discharge) (illuminating engineering). A lamp in which light (or radiant energy near the visible spectrum) is produced by the passage of an electric currrent through a vapor or a gas. *Note:* Electric-discharge lamps may be named after the filling gas or vapor which is responsible for the major portion of the radiation; for example, mercury lamps, sodium lamps, neon lamps, argon lamps, etcetera. A second method of designating electric-discharge lamps is by physical dimensions or operating parameters; for example, short-arc lamps, high-pressure lamps, low-pressure lamps, etcetera. A third method of designating electric-discharge lamps is by their application; in addition to lamps for illumination there are photochemical lamps, bactericidal lamps, blacklight lamps, sun lamps, etcetera. 167

electric-discharge time constant (detector). The time required, after the instantaneous removal of a sinusoidal input voltage of constant amplitude, for the output voltage across the load capacitor of the detector circuit to fall to 37 percent of its initial value. *See:* **electromagnetic compatibility.** 220,199

electric displacement. *See:* **electric flux density.**

electric displacement density. *See:* **electric flux density.**

electric drive (industrial control). A system consisting of one or several electric motors and of the entire electric control equipment designed to govern the performance of these motors. The control equipment may or may not include various rotating electric machines. 206

electric driving machine. A machine where the energy is applied by an electric motor. *Note:* It includes the motor and brake and the driving sheave or drum together with its connecting gearing, belt, or chain, if any. *See:* **driving machine (elevators).** 328

electric elevator. A power elevator where the energy is applied by means of an electric motor. *See:* **elevators.** 328

electric energy (energy). The electric energy delivered by an electric circuit during a time interval is the integral with respect to time of the instaneous power at the terminals of entry of the circuit to a delimited region. *Note:* If the reference direction for energy flow is selected as into the region when the sign of the energy is positive and out of the region when the sign is negative. If the reference direction is selected as out of the region, the reverse will apply. Mathematically where

$$W = \int_{t_0}^{t+t_0} p \, dt$$

W = electric energy
p = instantaneous power
t = time during which energy is determined.

When the voltages and currents are periodic, the electric energy is the product of the active power and the time interval, provided the time interval is one or more complete periods or is quite long in comparison with the time of one period. The energy is expressed by

$$W = Pt$$

where
P = active power
t = time interval.

If the instantaneous power is constant, as is true when th voltages and currents form polyphase symmetrical sets, there is no restriction regarding the relation of the time interval to the period. If the voltages and currents are quasi-periodic and amplitudes of the voltages and currents are slowly varying, the electric energy is the integral with respect to time of the active power, provided the integration is for a time that is one or more complete periods or that is quite long in comparison with the time of one period.

$$W = \int_{t_0}^{t_0+t} P \, dt$$

where P = active power determined for the condition of voltages and currents having slowly varying amplitudes. Electric energy is expressed in joules (watt-seconds) or watthours when the voltages are in volts and the currents in amperes, and the time interval is in seconds or hours, respectively.   210

electric explosion-tested mine locomotive. An electric mine locomotive equipped with explosion-tested equipment. *See:* **electric mine locomotive.**   328

electric field (radio wave propagation). A state of the region in which stationary charged bodies are subject to forces by virtue of their charges.   146

electric field (1) (general). A vector field of electric field strength or of electric flux density. *Note:* The term is also used to denote a region in which such vector fields have a significant magnitude. *See:* **vector field.**   210

(2) (static). A state of the region in which stationary charged bodies are subject to forces by virtue of their charges.   146

(3) (signal-transmission system). A state of a medium characterized by spatial potential gradients (electric field vectors) caused by conductors at different potentials, that is, the field between conductors at different potentials that have capacitance between them. *See:* **signal.**   188

electric field strength (1)(electric field)(measurement of power frequency electric and magnetic fields from ac power lines). At a given point in space, the ratio of force on a positive test charge placed at the point to the magnitude of the test charge, in the limit that the magnitude of the test charge goes to zero. The electric field strength (E-field) at a point in space is a vector defined by its space components along three orthogonal axes. For steady-state sinusoidal fields, each space component is a complex number or phasor. The magnitudes of the components, expressed by their root-mean-square (rms) values in volts per meter (V/m),and the phases need not be the same. *Note:* The space components (phasors) are not vectors. The space components have a time dependent angle, while vectors have space angles. For example, the sinusoidal electric field E can be expressed in rectangular coordinates as

$$\vec{E} = \hat{a}_x E_x + \hat{a}_y E_y + \hat{a}_z E_z \qquad \text{(Eq 1)}$$

The space component in the x-direction is

$$E_x = \text{Re} \, (E_{x0} \, e^{j\phi x} e^{j\omega t}) = E_{x0} \cos(\phi_x + \omega t)$$

The magnitude, phase angle, and time dependent angle are given by E_{x0}, ϕ_x, and ($\phi_x + \omega t$), respectively. In this representation the space angle of the x-component is specified by the unit vector $\hat{a}_x$. An alternative general representation of a steady-state sinusoidal E-field, derivable algebraically from Eq 1 and perhaps more useful in characterizing power line fields, is a vector rotating in a plane where it describes an ellipse whose semimajor axis represents the magnitude and direction of the maximum value of the electric field, and whose semiminor axis represents the magnitude and direction of the field a quarter cycle later. The electric field in the direction perpendicular to the plane of the ellipse is zero. *See:* **phasor; single-phase ac field; polyphase ac field.**   514

(2) (radio wave propagation). The magnitude of the electric field vector. *Note:* This term has sometimes been called the **electric field intensity,** but such use of the word intensity is deprecated in favor of field strength, since intensity connotes power in optics and radiation. *See:* **radio wave propagation.**

146

(3) (kV/cm) (fly ash resistivity). The ratio of the applied voltage to the ash layer thickness in a test cell used for the laboratory measurement of electrical resistivity of fly ash. bulk density(g/cm^3)(fly ash resistivity). The ratio of ash layer is the ratio of the mass of the particulate in the test cell to the cell volume in a test cell used for the laboratory measurement of electrical resistivity of fly ash.   427

electric field vector (1) (at a point in an electric field). The force on a stationary positive charge per unit charge. *Note:* This may be measured either in newtons per coulomb or in volts per meter. This term is sometimes called the **electric field intensity,** but such use of the word intensity is deprecated since intensity connotes power in optics and radiation. *See:* **radio wave propagation; waveguide.**   267, 146
(2) (radio wave propagation). At a point in an electric field, the force per unit charge acting on a stationary positive charge. *Syn:* **electric field strength; electric vector.** *Note:* This may be expressed either in newtons coulomb or in volts/meter. This term has sometimes been called the electric field intensity, but such use of the word "intensity" is deprecated in favor of field strength since intensity connotes power in optics and radiation. *Syn:* **electric field strength.** *Note:* This term has sometimes been called the electric-field intensity, but such use of the word "intensity" is deprecated in favor of field strength, since intensity connotes power in optics and radiation.   146

electric flux density (1) (electric displacement density)(electric induction*). A quantity related to the charge displaced within the dielectric by application of an electric field. *Notes:* (1) Electric flux density at any point in an isotropic dielectric is a vector that has the same direction as the electric field strength and a magnitude equal to the product of the electric field strength and the absolute capacitivity. The electric flux density is that vector point function whose divergence is the charge density, and that is proportional to the electric field in region free of polarized matter. The electric flux density is given by

$$\mathbf{D} = \epsilon_0 \epsilon \mathbf{E}$$

where $\mathbf{D}$ is the electric flux density, $\epsilon_0 \epsilon$ is the absolute capacitivity, and $\mathbf{E}$ is the electric field strength. (2) In a nonisotropic medium, ϵ becomes a tensor represented by a matrix and $\mathbf{D}$ is not necessarily parallel to $\mathbf{E}$. (3) The concept of a disk-like (Kelvin) cavity, properly oriented normal to $\mathbf{D}$, is frequently employed to visualize and compute the $\mathbf{D}$ vector in material media. (4) The electric flux density at a point is equal to the charge per unit area that would appear on one face of a small thin metal plate introduced in the electric field at the point and so oriented that this charge is a maximum. (5) The symbol Γ_ϵ is often used in modern practice in place of ϵ_0; the symbol ϵ_v has occasionally been used.   146
*Deprecated
(2) (radio wave propagation). A vector quantity relat-

ed to the charge displaced within the medium by application of an electric field. The electric flux density is that vector point function whose divergence is the charge density, and which is proportional to the electric field in regions free of polarized matter. In the medium without dispersion, the electric flux density is given by $\mathbf{D} = \epsilon\mathbf{E}$ where $\mathbf{D}$ is the electric flux density, ϵ is the permittivity, and $\mathbf{E}$ is the electric field vector. *Notes:* (1) In an anisotropic medium, ϵ is a tensor and $\mathbf{D}$ is not necessarily parallel to $\mathbf{E}$. In an isotropic medium, ϵ is a scalar and $\mathbf{D}$ is parallel to $\mathbf{E}$. (2) The concept of a disk-like (Kelvin) cavity, properly oriented normal to $\mathbf{D}$, is frequently used to visualize and compute the $\mathbf{D}$ vector in material media. (3) The electric flux density at a point is equal to the charge per unit area which would appear on one face of a small, thin metal plate introduced in the electric field at the point and so oriented that this charge is a maximum.   146

electric focusing (microwave tubes). The combination of electric fields that acts upon the electron beam in addition to the forces derived from momentum and space charge. *See:* **microwave tube or valve.**   190

electric freight locomotive. An electric locomotive, commonly used for hauling freight trains and generally designed to operate at higher tractive force values and lower speeds than a passenger locomotive of equal horsepower capacity. *Note:* A prefix diesel-electric, gas-electric, turbine-electric, etcetera, may replace the word electric. *See:* **electric locomotive.**   328

electric gathering mine locomotive. An electric mine locomotive, the chief function of which is to move empty cars into, and remove loaded cars from, the working places. *See:* **electric mine locomotive.**   328

electric generator (electric installations on shipboard) A machine which transforms mechanical power into electric power.   3

electric gun heater. An electrically heated element attached to the gun breech to prevent the oil from congealing or the gun mechanism from freezing.   328

electric haulage mine locomotive. An electric mine locomotive used for hauling trains of cars, that have been gathered from the working faces of the mine, to the point of delivery of the cars. *See:* **electric mine locomotive.**   328

electric horn. A horn having a diaphragm that is vibrated electrically. *See:* **protective signaling.**   328

electric-hydraulic governor (hydraulic turbines). A governor in which the control signal is proportional to speed error and the stabilizing and auxiliary signals are developed electrically, summed by appropriate electrical networks, and are then hydraulically amplified. Electrical signals may be analog or digitally derived.   8

electric hygrometer. An instrument for indicating by electric means the humidity of the ambient atmosphere. *Note:* Electric hygrometers usually depend for their operation on the relation between the electric conductance of a film of hygroscopic material and its moisture content. *See:* **instrument.**   328

electric incline railway. A railway consisting of an electric hoist operating a single car with or without counterweights, or two cars in balance, which car or cars travel on inclined tracks. *See:* **elevators.**
                                                             328
electric indication lock. An electric lock connected to a lever of an interlocking machine to prevent the release of the level or latch until the signals, switches, or other units operated, or directly affected by such lever, are in the proper position. *See:* **interlocking (interlocking plant).**   328
electric indication locking. Electric locking adapted to prevent manipulation of levers that would bring about an unsafe condition for a train movement in case a signal, switch, or other operated unit fails to make a movement corresponding with that of its controlling lever; or adapted directly to prevent the operation of one unit in case another unit to be operated first, fails to make the required movement. *See:* **interlocking (interlocking plant).**   328
electric interlocking machine. An interlocking machine designed for the control of electrically operated functions. *See:* **interlocking (interlocking plant).**
                                                             328
electricity meter. A device that measures and registers the integral of an electrical quantity with respect to time.   212
electric larry car. A burden-bearing car for operation on track rails used for short movements of materials, the propulsion of which is effected by electric motors mounted on the vehicle. *Note:* A prefix diesel-electric, gas-electric, etcetera, may replace the word electric. *See:* **electric motor car.**   328
electric length (radio wave propagation). For a wave of a given frequency, a distance in a medium expressed in wavelengths of the wave in the medium. *Note:* The electrical length is sometimes expressed in radians or degrees.   146
electric loading (rotating machinery). The average ampere-conductors of the primary winding per unit length of the air-gap periphery. *See:* **rotor (rotating machinery); stator.**   63
electric lock. A device to prevent or restrict the movement of a lever, a switch, or a movable bridge unless the locking member is withdrawn by an electric device such as an electromagnet, solenoid, or motor. *See:* **interlocking (interlocking plant).**   238
electric locking. The combination of one or more electric locks and controlling circuits by means of which levers of an interlocking machine, or switches, or other units operated in connection with signaling and interlocking, are secured against operation under certain conditions, as follows: (1) approach locking, (2) indication locking, (3) switch-lever locking, (4) time locking, (5) traffic locking. *See:* **interlocking (interlocking plant).**   328
electric locomotive. A vehicle on wheels, designed to operate on a railway for haulage purposes only, the propulsion of which is effected by electric motors mounted on the vehicle. *Note:* While this is a generic term covering any type of locomotive driven by elec-

tric motors, it is usually applied to locomotives receiving electric power from a source external to the locomotive. The prefix electric may also be applied to cars, buses, etcetera, driven by electric motors. A prefix diesel-electric, etcetera, may replace the word electric.
                                                             328
electric-machine regulating system (rotating machinery). A feedback control system that includes one or more electric machines and the associated control.
                                                              63
electric-machine regulator (rotating machinery). A specified element or a group of elements that is used within an electric-machine regulating system to perform a regulating function by acting to maintain a designated variable (or variables) at a predetermined value, or to vary it according to a predetermined plan.
                                                              63
electric mechanism (demand meter). That portion, the action of which, in response to the electric quantity to be measured, gives a measurement of that quantity. *Note:* For example, the electric mechanism of certain demand meters is similar to the ordinary ammeter of wattmeter of the deflection type; in others it is a watt-hour meter or other integrating meter; and in still others it comprises an electric circuit that heats temperature-responsive elements, such as bimetallic spirals, that deflect to move the indicating means. The electrical quantity may be measured in kilowatts, kilowatt-hours, kilovolt-amperes, kilovolt-ampere-hours, amperes, ampere-hours, kilovars, kilovar-hours, or other suitable units. *See:* **demand meter.**   328
electric mine locomotive (1) (general). An electric locomotive designed for use underground; for example, in such places as coal, metal, gypsum, and salt mines, tunnels, and in subway construction.
(2) (storage-battery type). An electric locomotive that receives its power supply from a storage battery mounted on the chassis of the locomotive.
(3) (trolley type). An electric locomotive that receives its power supply from a trolley-wire distribution system.
(4) (combination type). An electric locomotive that receives power either from a trolley-wire distribution system or from a storage battery carried on the locomotive.
(5) (separate tandem). An electric mine locomotive consisting of two locomotive units that can be coupled together or operated from one controller as a single unit, or else separated and operated as two independent units.
(6) (permanent tandem). A locomotive consisting of two locomotive units permanently connected together and provided with one set of controls so that both units can be operated by a single operator.   328
electric motive power unit. A self-contained electric traction unit, comprising wheels and a superstructure capable of independent propulsion from a power supply system, but not necessarily equipped with an independent control system. *Note:* While this is a generic term covering any type of motive power driven by electric motors, it is usually applied to locomotives

receiving electric power from an external source. A prefix diesel-electric, gas-electric, turbine-electric, etcetera, may replace the word electric. *See:* **electric locomotive.**   328

electric motor (1) (electric installations on shipboard). A machine which transforms electric power into mechanical power.   3
(2) (packaging machinery). A device that converts electrical energy into rotating mechanical energy.   429

electric motor car. A vehicle for operating on track rails, used for the transport of passengers or materials, the propulsion of which is effected by electric motors, mounted on the vehicle. *Note:* A prefix diesel-electric, gas-electric, etcetera, may replace the word electric.   328

electric motor controller. A device or group of devices that serve to govern, in some predetermined manner, the electric power delivered to the motor. *Note:* An electric motor controller is distinct functionally from a simple disconnecting means whose principal purpose in a motor circuit is to disconnect the circuit, together with the motor and its controller, from the source of power. *See:* **electric controller.**   206

electric movable-bridge (drawbridge) lock. A device used to prevent the operation of a movable bridge until the device is released. *See:* **interlocking (interlocking plant).**   328

electric network. *See:* **network.**

electric noise (1) (general). Unwanted electrical energy other than crosstalk present in a transmission system.
(2) (interface terminology). A form of interference introduced into a signal system by natural sources that constitutes for that system an irreducible limit on its signal-resolving capability. *Note:* Noise is characterized by randomness of amplitude and frequency distribution and therefore cannot be eliminated by band-rejection filters tuned to preselected frequencies. *See:* **distortion; interference.**   111,63

electric operation. Power operation by electric energy.   103

electric orchestra lift. An electrically driven mechanism for raising and lowering the musicians' platform and the musicians. *See:* **elevators.**   328

electric parachute-flare-launching tube. A tube mounted on an aircraft through which a metal container carrying a parachute flare is launched, the tube being so designed that as the parachute-flare container passes through the tube, an electric circuit is completed that ignites a slow-burning fuse in the container, the fuse being so designed as to permit the container to clear the aircraft before it ignites the parachute flare.   328

electric passenger locomotive. An electric locomotive, commonly used for hauling passenger trains and generally designed to operate at higher speeds and lower tractive-force values than a freight locomotive of equal horsepower capacity. *Note:* A prefix diesel-electric, gas-electric, turbine-electric, etcetera, may replace the word electric. *See:* **electric locomotive.**   328

electric penetration assembly (electric penetration assemblies). An assembly of insulated electric conductors, conductor seals, module seals (if any), and aperture seals that provides the passage of the electric conductors through a single aperture in the nuclear containment structure, while providing a pressure barrier between the inside and the outside of the containment structure. The electric penetration assembly includes terminal (junction) boxes, terminal blocks, connectors and cable supports, and splices which are designed and furnished as an integral part of the assembly.   493

electric permissible mine locomotive. An electric locomotive carrying the official approval plate of the United States Bureau of Mines. *See:* **electric mine locomotive.**   328

electric pin-and-socket coupler (connector). A readily disconnective assembly used to connect electric circuits between components of an aircraft electric system by means of mating pins and sockets.   328

electric pipe heating system (electric pipe heating systems). A system of components and devices consisting of electric heaters, controllers, sensors, dedicated power system components such as transformers, panelboards, cables and systems alarm devices (as required), which, when taken together as a system, is used to increase or maintain the temperature of fluids in mechanical pipes, valves, pumps, tanks, instrumentation, etcetera. *Syn:* **heat tracing system; trace heating system.**   405

electric polarizability. Of an isotropic medium for which the direction of electric polarization and electric field strength are the same at any point in the medium, the magnitude P of the electric polarization at that point divided by the electric field strength there, E. *Note:* In a rationalized system, the electric polarizability $P_e = P/E = \epsilon_0(\epsilon - 1)$.   210

electric polarization (electric field). At any point, the vector difference between the electric flux density at that point and the electric flux density that would exist at that point for the same electric field strength there, if the medium were a vacuum there. *Note:* Electric polarization is the vector limit of the quotient of the vector sum of electric dipole moments in a small volume surrounding a given point, and this volume, as the volume approaches zero in a microscopic sense.   210

electric port (optoelectronic device). A port where the energy is electric. *Note:* A designated pair of terminals may serve as one or more electric ports. *See:* **optoelectronic device.**   191

electric potential. The potential difference between the point and some equipotential surface, usually the surface of the earth, which is arbitrarily chosen as having zero potential (remote earth). *Note:* A point which has a higher potential than a zero surface is said to have a positive potential; one having a lower potential has a negative potential.   313

electric power distribution panel. A metallic or nonmetallic, open or enclosed, unit of an electric system. The operable and the indicating components of an electric system, such as switches, circuit breakers, fuses, indicators, etcetera, usually are mounted on the face of the panel. Other components, such as terminal strips, relays, capacitors, etcetera, usually are mounted behind the panel. 328

electric propulsion apparatus. Electric apparatus (generators, motors, control apparatus, etcetera) provided primarily for ship's propulsion. *Note:* For certain applications, and under certain conditions, auxiliary power may be supplied by propulsion apparatus. *See:* **electric propulsion system.** 328

electric propulsion system. A system providing transmission of power by electric means from a prime mover to a propeller shaft with provision for control, partly or wholly by electric means, of speed and direction. *Note:* An electric coupling (which see) does not provide electric propulsion. 328

electric rate schedule (power operations). A statement of an electric rate and the terms and conditions governing its application. 516

electric reset relay. A relay that is so constructed that it remains in the picked-up condition even after the input quantity is removed: an independent electric input is required to reset the relay. 103, 202, 60

electric resistance-type temperature indicator. A device that indicates temperature by means of a resistance bridge circuit. 328

electric road locomotive. An electric locomotive designed primarily for hauling dispatched trains over the main or secondary lines of a railroad. *Note:* A prefix diesel-electric, gas-electric, turbine-electric, etcetera, may replace the word electric. *See:* **electric locomotive.** 328

electric road-transfer locomotive. An electric locomotive designed primarily so that it may be used either for hauling dispatched trains over the main or secondary lines of a railroad or for transferring relatively heavy cuts of cars for short distances within a switching area. *Note:* A prefix diesel-electric, gas-electric, turbine-electric, etcetera, may replace the word electric. *See:* **electric locomotive.** 328

electric sign (National Electrical Code). A fixed, stationary, or portable self-contained, electrically illuminated utilization equipment with words or symbols designed to convey information or attract attention. 256

electric-signal storage tube. A storage tube into which the information is introduced as an electric signal and read at a later time as an electric signal. *See:* **storage tube.** 174

electric sounding machine. A motor-driven reel with wire line and weight for determination of depth of water by mechanical sounding. 328

electric squib. A device similar to an electric blasting cap but containing a gunpowder composition that simply ignites but does not detonate an explosive charge. *See:* **blasting unit.** 328

electric stage lift. An electrically driven mechanism for raising and lowering various sections of a stage. 328

electric storage subsystem (terrestrial photovoltaic

power systems). The subsystem that stores electric energy. *See:* **array control** 496

electric strength (dielectric strength) (rotating machinery). The maximum potential gradient that the material can withstand without rupture. 6

electric stroboscope. An instrument for observing rotating or vibrating objects or for measuring rotational speed or vibration frequency, or similar periodic quantities, by electrically produced periodic changes in illumination. *See:* **instrument.** 328

electric submersible pump (ESP)(electric submersible pump cable). Deep-well electric submersible pumps as commonly used to lift fluids from subsurface formations. See figure. 484

electric submersible pump (ESP) cable. Three-conductor power cable installed in the well for the purpose of transmitting power from the surface to the motor lead extension cable. 484

electric supply equipment (NESC)(transmission and distribution). Equipment which produces, modifies, regulates, controls, or safeguards a supply of electric energy. *Syn:* **supply equipment.** 494, 262

electric supply lines (NESC)(transmission and distribution). Those conductors used to transmit electric energy and their necessary supporting or containing structures. Signal lines of more than 400 volts are always supply lines within the meaning of the rules, and those of less than 400 volts may be considered as supply lines, if so run and operated throughout. *Syn:* **supply lines.** 494, 262

electric supply station (NESC). Any building, room, or separate space within which electric supply equipment is located and the interior of which is accessible, as a rule, only to qualified persons. This includes generating stations and substations, including their associated generator, storage battery, transformer and switchgear rooms or enclosures but does not include facilities such as pad mounted equipment and installations in manholes and vaults. 494

electric surges (nuclear power generating station). Any spurious voltage or current pulses conducted into the module from external sources. 355

electric susceptibility. Of an isotropic medium, for which the direction of electric polarization and electric field strength are the same, at any point in the medium, the magnitude of the electric polarization at that point of the medium, divided by the electric flux density that would exist at that point for the same electric field strength, if the medium there were a vacuum. *Note:* In a rationalized system the electric susceptibility $\chi_e = P/D(\epsilon - 1)$. 210

electric switching locomotive. An electric locomotive designed for yard movements of freight or passenger cars, its speed and continuous electrical capacity usually being relatively low. *Note:* A prefix diesel-electric, gas-electric, turbine-electric, etcetera, may replace the word electric. *See:* **electric locomotive.** 328

electric switch-lever lock. An electric lock used to prevent the movement of a switch lever or latch in an

315

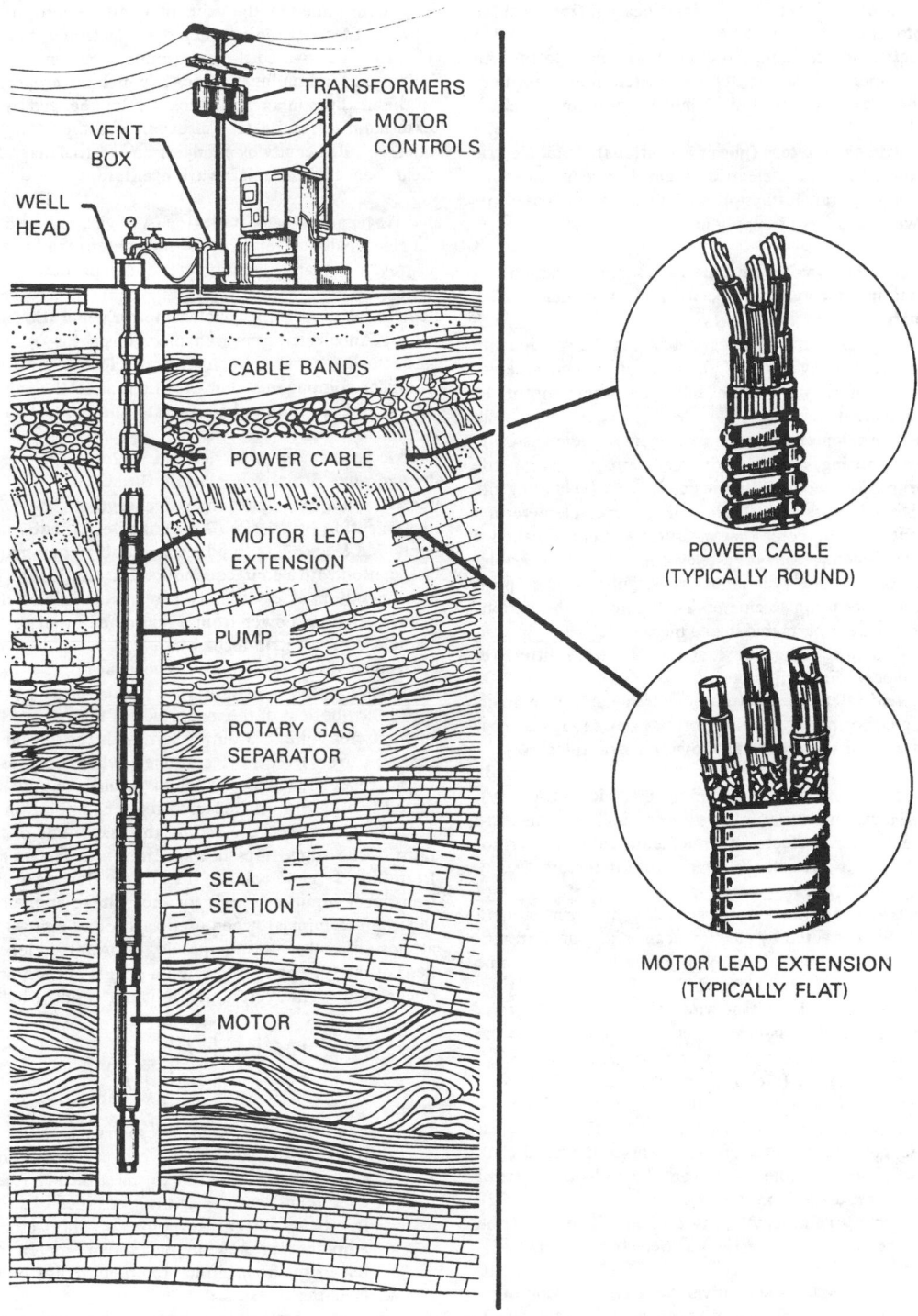

TRANSFORMERS

MOTOR
CONTROLS

VENT
BOX

WELL
HEAD

CABLE BANDS

POWER CABLE

MOTOR LEAD
EXTENSION

PUMP

ROTARY GAS
SEPARATOR

SEAL
SECTION

MOTOR

POWER CABLE
(TYPICALLY ROUND)

MOTOR LEAD EXTENSION
(TYPICALLY FLAT)

Typical Electric Submersible Pump Installation

interlocking machine until the lock is released. *See:* **interlocking (interlocking plant).** 328

electric switch-lever locking. A general term for route or section locking. *See:* **interlocking (interlocking plant).** 328

electric switch lock. An electric lock used to prevent the operation of a switch or a switch movement until the lock is released. *See:* **interlocking (interlocking plant).** 328

electric system loss (power operations). Total electric power loss in the electric system. It consists of transmission, transformation, and distribution losses between sources of supply and points of delivery. 516

electric tachometer (marine usage). An instrument for measuring rotational speed by electric means. *See:* **instrument.** 328

electric telegraph. A telegraph having the relationship of the moving parts of the transmitter and receiver maintained by the use of self-synchronous motors or equivalent devices. 328

electric telemeter (power switchgear). The measuring, transmitting, and receiving apparatus, including the primary detector, intermediate means (excluding the channel) and end devices for electric telemetering. *Note:* A telemeter that measures current is called a teleammeter; voltage, a televoltmeter; power, a telewattmeter; one which measures angular or linear position, a position telemeter. The names of the various component parts making up the telemeter are, in general, self-defining; for example, the transmitter, receiver, indicator, etcetera. 103

electric telemetering (power switchgear). Telemetering performed by an electrical translating means separate from the measurand. *Syn:* **electric telemetry.** 103

electric thermometer (rotating electric machinery). An instrument that utilizes electric means to measure temperature. Electric thermometers include thermocouples and resistance temperature detectors. 424

electric tower car. A rail vehicle, the propulsion of which is effected by electric means and that is provided with an elevated platform, generally arranged to be raised and lowered, for the installation, inspection, and repair of a contact wire system. *Note:* A prefix diesel-electric, gas-electric, etcetera, may replace the word electric. *See:* **electric motor car.** 328

electric trail car (electric trailer). A car not provided with motive power that is used in a train with one or more electric motor cars. *Note:* A prefix diesel-electric, gas-electric, etcetera, may replace the word electric to identify the motor cars. *See:* **electric-control trail car; electric motor car.** 328

electric transducer. A transducer in which all of the waves concerned are electric. *See:* **transducer.** 252, 210

electric transfer locomotive. An electric locomotive designed primarily for transferring relatively heavy cuts of cars for short distances within a switching area. *Note:* A prefix diesel-electric, gas-electric, turbine-

electric, etcetera, may replace the word electric. *See:* **electric locomotive.** 328

electric-tuned oscillator. An oscillator whose frequency is determined by the value of a voltage, current, or power. Electric tuning includes electronic tuning, electrically activated thermal tuning, electromechanical tuning, and tuning methods in which the properties of the medium in a resonant cavity are changed by an external electric means. An example is the tuning of a ferrite-filled cavity by changing an external magnetic field. *See:* **tunable microwave oscillators.** 174, 190

electric turn-and-bank indicator. A device that utilizes an electrically driven gyro for turn determination and a gravity-actuated inclinometer for bank determination. 328

electric valve operator (nuclear power generating stations) An electric powered mechanism for opening and closing a valve, including all electric and mechanical components that are integral to the mechanism and are required to operate and control valve action. 31, 142

electric vector. *See:* **electric field vector.**

electric wave filter. *See:* **electric filter.**

electric wind. *See:* **convective discharge.**

electrification by friction. *See:* **triboelectrification.**

electrified track. A railroad track suitably equipped in association with a contact conductor or conductors for the operation of electrically propelled vehicles that receive electric power from a source external to the vehicle. *See:* **electric locomotive.** 328

electroacoustical reciprocity theorem. For an electroacoustic transducer satisfying the reciprocity principle, the quotient of the magnitude of the ratio of the open-circuit voltage at output terminals (or the short-circuit current) of the transducer, when used as a sound receiver, to the free-field sound pressure referred to an arbitrarily selected reference point on or near the transducer, divided by the magnitude of the ratio of the sound pressure apparent at a distance δ from the reference point to the current flowing at the transducer input terminals (or the voltage applied at the input terminals), when used as a sound emitter, is a constant, called the reciprocity constant, independent of the type or constructional details of the transducer. *Note:* The reciprocity constant is given by

$$\left|\frac{M_o}{S_s}\right| = \left|\frac{M_s}{S_s}\right| = \frac{2\delta}{\rho f}$$

where

M_O = open free-field voltage response, as a sound receiver, in open-circuit volts per newton per square meter, referred to the arbitrary reference point on or near the transducer.

M_s = free-field current response in short-circuit amperes per newton per square meter, referred to the arbitrary reference point on or near the transducer

S_O = sound pressure in newtons per square meter per ampere of input current produced at a distance δ meters from the arbitrary reference point

S_s = sound pressure in newtons per square meter per volt applied at the input terminals produced at a distance δ meters from the arbitrary reference point

f = frequency in hertz

ρ = density of the medium in kilograms per cubic meter

δ = distance in meters from the arbitrary reference point on or near the transducer to the point in which the sound pressure established by the transducer when emitting is evaluated. *See:* **loudspeaker.** 176

electroacoustic transducer (electric system). A transducer for receiving waves and delivering waves to an acoustic system, or vice versa. *See:* **loudspeaker; transducer.** 176

electrobiology. The study of electrical phenomena in relation to biological systems. 192

electrocardiogram. The graphic record of the variation with time of the voltage associated with cardiac activity. *See:* **electrocorticogram (electrobiology); electrodermogram (electrobiology); Galvani's experiment (electrobiology); spindle wave (electrobiology); vector electrocardiogram (electrobiology).** 192

electrocardiographic waves, P, Q, R, S, and, T (medical electronics)(in electrocardiograms obtained from differential electrodes placed on the right arm and left leg). The characteristic tracing consists of five consecutive waves: *P,* a prolonged, low, positive wave: *Q,* brief, low, negative: *R,* brief, high, positive: *S,* brief, low, negative, and *T,* prolonged, low, positive. 192

electrocautery (electrotherapy). An instrument for cauterizing the tissues by means of a conductor brought to a high temperature by an electric current. *See:* **electrotherapy.** 192

electrochemical cell. A system consisting of an anode, cathode, and an electrolyte plus such connections (electric and mechanical) as may be needed to allow the cell to deliver or receive electric energy. 223, 186

electrochemical equivalent (element, compound, radical, or ion) (1) (general). The weight of that substance involved in a specified electrochemical reaction during the passage of a specified quantity of electricity, such as a faraday, ampere-hour, or coulomb. 328 **(2) (oxidation).** The weight of an element or group of elements oxidized or reduced at 100-percent efficiency by a unit quantity of electricity. *See:* **electrochemistry.** 205

electrochemical recording (facsimile). Recording by means of a chemical reaction brought about by the passage of signal-controlled current through the sensitized portion of the record sheet. *See:* **recording (facsimile).** 12

electrochemical series. *See:* **electromotive series.**

electrochemical valve. An electric valve consisting of a metal in contact with a solution or compound across the boundary of which current flows more readily in one direction than in the other direction and in which the valve action is accompanied by chemical changes. 328

electrochemical valve metal. A metal or alloy having properties suitable for use in an electrochemical valve. *See:* **electrochemical valve.** 328

electrochemistry. That branch of science and technology that deals with interrelated transformations of chemical and electric energy. 328

electrocoagulation (medical electronics). The clotting of tissue by heat generated within the tissue by impressed electric currents. 192

electrocorticogram (medical electronics). A graphic record of the variation with time of voltage taken from exposed cortex cerebra. 192

electroculture (medical electronics). The stimulation of growth, flowering, or seeding by electric means. 192

electrocution. The destruction of life by means of electric current. 192

electrode (1) (electrochemistry). An electric conductor for the transfer of charge between the external circuit and the electroactive species in the electrolyte. *Note:* Specifically, in an electrolytic cell, an electrode is a conductor at the surface of which a change occurs from conduction by electrons to conduction by ions or colloidal ions. *See:* **electrolytic cell; electrochemical cell.** 186 **(2) (electron tube).** A conducting element that performs one or more of the functions of emitting, collecting, or controlling by an electric field the movements of electrons or ions. 125 **(3) (biological electronics) (reference, inactive, diffuse, dispersive, indifferent electrode).** (A) A pickup electrode that, because of averaging, shunting, or other aspects of the tissue-current pattern to which it connects, shows potentials not characteristic of the region near the active electrode. (B) Any electrode, in a system of stimulating electrodes, at which due to its dispersive action, excitation is not produced. (C) An electrode of relatively large area applied to some inexcitable or distant tissue in order to complete the circuit with the active electrode that is used for stimulation. 192

electrode, accelerating (electron-beam tube). *See:* **accelerating electrode.**

electrode admittance (*jth* electrode of an *n*-electrode electron tube). The short-circuit driving-point admittance between the *jth* electrode and the reference point measured directly at the *jth* electrode. *Note:* To be able to determine the intrinsic electronic merit of an electron tube, the driving-point and transfer admittances must be defined as if measured directly at the electrodes inside the tube. The definitions of electrode admittance and electrode impedance are included for this reason. *See:* **electron-tube admittances.** 125

electrode alternating-current resistance (electron device). The real component of the electrode impedance. *See:* **self-impedance.** 190

electrode bias (electron tubes). The voltage at which an electrode is stabilized under operating conditions with no incoming signal, but taking into account the voltage drops in the connected circuits. *See:* **electrode voltage.** 244

electrode capacitance (*n*-terminal electron tube). The capacitance determined from the short-circuit driving-point admittance at that electrode. *See:* **electron-tube admittances.** 125

electrode characteristic. A relation, usually shown by a graph, between the electrode voltage and the current of an electrode, all other electrode voltages being maintained constant. 125

electrode conductance. The real part of the electrode admittance. 125

electrode, control. *See:* **control electrode.**

electrode current (electron tube). The current passing to or from an electrode through the interelectrode space. *Note:* The terms cathode current, grid current, anode current, plate current, etcetera, are used to designate electrode currents for these specific electrodes. Unless otherwise stated, an electrode current is measured at the available terminal. 190

electrode current, average. *See:* **average electrode current.**

electrode-current averaging time (electron tubes). The time interval over which the current is averaged in defining the operating capabilities of the electrode. *See:* **electrode current (electron tube).** 190

electrode dark current (1) (phototubes). The component of electrode current remaining when ionizing radiation and optical photons are absent. *Notes:* (A) Optical photons are photons with energies corresponding to wavelengths between 2000 and 1500 angstroms. (B) Since the dark current may change considerably with temperature, the temperature should be specified. *See:* **phototube.** 190
(2) (camera tubes). The current from an electrode in a photoelectric tube under stated conditions of radiation shielding. *See:* **camera tube.** 178, 190

electrode dissipation. The power dissipated in the form of heat by an electrode as a result of electron or ion bombardment, or both, and radiation from other electrodes. *See:* **grid driving power; modes, degenerate.** 190

electrode drop (arc-welding apparatus). The voltage drop in the electrode due to its resistance (or impedance). 264

electrode impedance. The reciprocal of the electrode admittance. *See:* **electron-tube admittances.** 190

electrode impedance, biological. The ratio between two vectors, the numerator being the vector that represents the potential difference between the electrode and biological material, and the denominator being the vector that represents the current between the electrode and the biological material. *See:* **loss angle (biological); polarization capacitance (biological); polarization reactance (biological); polarization resistance (biological).** 192

electrode, pad. *See:* **pad electrode.**

electrode potential, biological. The potential between an electrode and biological material. 192

electrode radiator (cooling fin) (electron tubes). A metallic piece, often of large area, extending the electrode to facilitate the dissipation of the heat generated in the electrode. *See:* **electron tube.** 244,190

electrode reactance (electron device). The imaginary component of the electrode impedance. *See:* **self-impedance.** 190

electrode resistance (1) (general). The reciprocal of the electrode conductance. *Note:* This is the effective parallel resistance and is not the real component of the electrode impedance. 190
(2) (at a stated operating point) (electron device). The quotient of the direct electrode voltage by the direct electrode current. *See:* **self-impedance.** 190

electrodermal reaction (EDR)(medical electronics). The change in electric resistance of the skin during emotional stress. 192

electrodermogram (electromyogram) (electroretinogram) (electrobiology). A graphic record of the variation with time of voltage taken from the given and anatomical structure (skin, muscle, and retina, respectively). *See:* **electrocardiogram.** 192

electrodesiccation (fulguration). The superficial destruction of tissue by electric sparks from a movable electrode. *See:* **electrotherapy.** 192

electrode, signal (camera tube). *See:* **signal electrode.**

electrodes or applicators. *See:* **applicators or electrodes.**

electrode susceptance (electron device). The imaginary component of the electrode admittance. *See:* **self-impedance.** 190

electrode voltage. The voltage between an electrode and the cathode or a specified point of a filamentary cathode. *Note:* The terms grid voltage, anode voltage, plate voltage, etcetera, are used to designate the voltage between these specific electrodes and the cathode. Unless otherwise stated, electrode voltages are understood to be measured at the available terminals. 190,125

electrodiagnosis. The study of functional states of parts of the body either by studying their responses to electric stimulation or by studying the electric potentials (or currents) that they spontaneously produce. 192

electroencephalogram (medical electronics). A graphic record of the changes with time of the voltage obtained by means of electrodes applied to the scalp over the cerebrum. 192

electrographic recording (electrostatography). The branch of electrostatic electrography that employs a charge transfer between two or more electrodes to form directly electrostatic-charge patterns on an insulating medium for producing a viewable record. *See:* **electrostatography.** 191

electrographitic brush (rotating machinery). A brush composed of selected amorphous carbon that, in the process of manufacturer, is carried to a temperature high enough to convert the carbon to the graphitized form. *Note:* This type of brush is exceedingly versatile in that it can be made soft or very hard, also nonabrasive or slightly abrasive. Grades of brushes of this type have a high current-carrying capacity, but differ great-

ly in operating speed from low to high. *See:* **brush.**
63

electrohydraulic elevator. A direct-plunger elevator where liquid is pumped under pressure directly into the cylinder by a pump driven by an electric motor. *See:* **elevators.**
328

electrokinetic potential (zeta potential) (medical electronics). A set of four electric or velocity potentials that accompany relative motion between solids and liquids.
192

electroluminescence (1) (illuminating engineering). The emission of light from a phosphor excited by an electromagnetic field.
167

(2) (light emitting diodes). The emission of light from a material (phosphor or semiconductor) where the exciting mechanism is the application of an electromagnetic field.
162

(3) (fiber optics). Nonthermal conversion of electrical energy into light. One example is the photon emission resulting from electron-hole recombination in a pn junction such as in a light emitting diode. *See:* **injection laser diode.**
433

electroluminescent display device. An optoelectronic device with a multiplicity of electric ports, each capable of independently producing an optic output from an associated electroluminator element. *See:* **optoelectronic device.**
191

electroluminescent display panel. A thin, usually flat, electroluminescent display device. *See:* **optoelectronic device.**
191

electrolysis (underground structures). The destructive chemical action caused by stray or local electric currents to pipes, cables, and other metalwork. *See:* **corrosion.**
64

electrolyte. A conducting medium in which the flow of electric current takes place by migration of ions. *Note:* Many physical chemists define electrolyte as a substance that when dissolved in a specified solvent, usually water, produces an ionically conducting solution. *See:* **electrolytic cell.**
210

electrolyte cells. *See:* **cascade.**

electrolytic. *See:* **cathode.**

electrolytic cell (1) (electrolytic cell line working zone). A receptacle or vessel in which electrochemical reactions are caused by applying electrical energy for the purpose of refining or producing usable materials from other sources.
133

(2) (National Electrical Code). A receptacle or vessel in which electrochemical reactions are caused by applying electrical energy for the purpose of refining or producing usable materials.
256

electrolytic cell line working zone (National Electrical Code). The cell line working zone is the space envelope wherein operation or maintenance is normally performed on or in the vicinity of exposed energized surfaces of electrolytic cell lines or their attachments.
256

electrolytic cleaning. The process of degreasing or descaling a metal by making it an electrode in a suitable bath.
221

electrolytic recording (facsimile). That type of electro-

chemical recording in which the chemical change is made possible by the presence of an electrolyte. *See:* **recording (facsimile).**
12

electrolytic tank. A vessel containing a poorly conducting liquid, in which are inserted conductors that are scale models of an electrode system. *Note:* It is used to obtain potential diagrams. *See:* **electron optics.**
244,190

electrolyzer. An electrolytic cell for the production of chemical products.

electromagmetic interference (emi)(supervisory control, data acquisition, and automatic control). A measure of electromagnetic radiation from equipment.
570

electromagnet. A device consisting of a ferromagnetic core and a coil, that produces appreciable magnetic effects only when an electric current exists in the coil.
210

electromagnetic compatibility (1) (EMC) (station control and data acquisition)(supervisory control, data acquisition, and automatic control). A measure of equipment tolerance to external electromagnetic fields.
403, 570

(2)(control of system electromagnetic compatibility). The ability of a device to function satisfactorily in its electromagnetic environment without introducing intolerable disturbance to that environment (or to other devices).
495

electromagnetic delay line. A delay line whose operation is based on the time of propagation of electromagnetic waves through distributed or lumped capacitance and inductance.
255,77

electromagnetic disturbance (electromagnetic compatibility). An electromagnetic phenomenon that may be superimposed on a wanted signal. *See:* **electromagnetic compatibility.**
199

electromagnetic environment. The electromagnetic field(s) and or signals existing in a transmission medium. *See:* **electromagnetic compatibility.**
199

electromagnetic induction. The production of an electromotive force in a circuit by a change in the magnetic flux linking with that circuit.
197

electromagnetic interference (EMI) (station control and data acquisition). A measure of electromagnetic radiation from equipment.
403

electromagnetic lens (antennas). A three-dimensional structure through which electromagnetic wave can pass, possessing an index of refraction which may be a function of position and a shape which is chosen so as to control the exiting aperture illumination.
111

electromagnetic noise (electromagnetic compatibility). An unwanted electromagnetic disturbance that is not of a sinusoidal character. *See:* **electromagnetic compatibility.**
220,199

electromagnetic radiation (1)(radio frequency radiation hazard warning symbol). The term is restricted to that part of the spectrum commonly defined as the radio frequency region, which for the purpose of this standard includes microwave frequencies.
557

(2) (antennas). The emission of electromagnetic ener-

gy from a finite region in the form of unguided waves.
                                                    111

(3) (laser-maser). The flow of energy consisting of orthogonally vibrating electric and magnetic fields lying transverse to the direction of propagation. X-rays, ultraviolet, visible, infrared, and radio waves occupy various portions of the electromagnetic spectrum and differ only in frequency and wavelength.   363

electromagnetic relay (power switchgear). An electromechanical relay that operates principally by action of an electromagnetic element which is energized by the input quantity.   103

electromagnetic spectrum (radio wave propagation). The spectrum of electromagnetic radiation: in wavelengths, gamma ray, shorter than 0.006 nm; X-ray, 0.006 to 5 nm; ultraviolet, 5nm to 0.4 μm; visible light, 0.4 to 0.7 μm; infrared, 0.7 μm to 1 mm; radio frequency, 1 mm.   146

electromagnetic waves. A wave characterized by variations of electric and magnetic fields. *Note:* Electromagnetic waves are known as radio waves, heat rays, light rays, etcetera, depending on the frequency. *See:* **radio wave propagation; waveguide.**   267,146

electromechanical device (control equipment). A device that is electrically operated and has mechanical motion such as relays, servos, etcetera.   94

electromechanical recording (facsimile). Recording by means of a signal-actuated mechanical device. *See:* **recording (facsimile).**   12

electromechanical relay (power switchgear). A relay that operates by physical movement of parts resulting from electromagnetic, electrostatic, or electrothermic forces created by the input quantities.   103

electromechanical switching system (telephone switching systems). An automatic switching system in which the control functions are performed principally by electromechanical devices.   55

electromechanical transducer. A transducer for receiving waves from an electric system and delivering waves to a mechanical system, or vice versa. *See:* **transducer.**   52, 176

electrometer tube. A vacuum tube having a very low control-electrode conductance to facilitate the measurement of extremely small direct current or voltage.
                                              190,125

electromotive force. *See:* **voltage**

electromotive force series. A list of elements arranged according to their standard electrode potentials.
                                              221,205

electromyograph (medical electronics). An instrument for recording action potentials or physical movements of muscles.   192

electron (ion) beam. A beam of electrons (ions) emitted from a single source and moving in neighboring paths that are confined to a desired region.   244

electron (proton) damage coefficient. The change in a stated quantity (such as minority carrier inverse squared diffusion length) of a given material per unit particle fluence of a stated energy spectrum.   113

electron (1) (noun). An elementary particle containing the smallest negative electric charge. *Note:* The mass

of the electron is approximately equal to 1/1837 of the mass of the hydrogen atom.   244

(2) (adjective). Operated by, containing, or producing electrons. *Examples:* Electron tube, electron emission, and electron gun. *See:* **electronic; electronics.**
                                                    328

electron accelerator, linear. *See:* **linear electron accelerator.**

electronarcosis. The production of transient insensibility by means of electric current applied to the cranium at intensities insufficient to cause generalized convulsions. *See:* **electrotherapy.**   192

electron-beam tube. An electron tube, the performance of which depends upon the formation and control of one or more electron beams.   125

electron collector (microwave tube). The electrode that receives the electron beam at the end of its path. *Note:* The power of the beam is used to produce some desired effect before it reaches the collector. *See:* **velocity-modulated tube.**   244

electron-coupled oscillator. An oscillator employing a multigrid tube with the cathode and two grids operating as an oscillator in any conventional manner, and in which the plate circuit load is coupled to the oscillator through the electron system. *See:* **oscillatory circuit.**   111,240

electron device. A device in which conduction is principally by electrons moving through a vacuum, gas, or semiconductor.   125

electron-device transducer. *See:* **admittance, short-circuit forward.**

electron emission. The liberation of electrons from an electrode into the surrounding space. *Note:* Quantitatively, it is the rate at which electrons are emitted from an electrode.   125

electron gun (electron tubes). An electrode structure that produces and may control, focus, deflect, and converge one or more electron beams. *See:* **electrode (electron tube.)**   125

electron-gun density multiplication (electron tubes). The ratio of the average current density at any specified aperature through which the stream passes to the average current density at the cathode surface.
                                                    125

electronic. Of, or pertaining to, devices, circuits, or systems utilizing electron devices. *Examples:* Electronic control, electronic equipment, electronic instrument, and electronic circuit. *See:* **electron device; electronics.**   125

electronically de-spun antenna (communication satellite). A directional antenna, mounted to a rotating object (namely spin stabilized communication satellite), with beam switching and phasing such that the antenna beam points into the same direction in space regardless of its mechanical rotation.   83

electronic analog computer. An automatic computing device that operates in terms of continuous variation of some physical quantities, such as electric voltages and currents, mechanical shaft rotations, or displacements, and that is used primarily to solve differential equations. *Note:* The equations governing the variation of the physical quantities have the same or very nearly the same form as the mathematical equations under investigation and therefore yield a solution analogous to the desired solution of the problem. Results are measured on meters, dials, oscillograph recorders, or oscilloscopes.   9

electronic contactor (industrial control). A contactor whose function is performed by electron tubes. *See:* **contactor.**   206

electronic controller (industrial control). An electric controller in which the major portion or all of the basic functions are performed by electron tubes.   206

electronic counter-countermeasures (ECCM) (radar). Any electronic technique designed to make a radar less vulnerable to electronic countermeasures (ECM).
                                                                          13

electronic counter-countermeasures (ECCM) improvement factor (radar). The power ratio of the electronic countermeasures (ECM) signal level required to produce a given output signal from a receiver using an ECCM technique to the ECM signal level producing the same output from the same receiver without the ECCM technique.   13

electronic countermeasures (ECM) (radar). Any electronic technique designed to deny detection or accurate information to a radar. Screening with noise, confusion with false targets, and deception by affecting tracking circuits are typical types.   13

electronic direct-current motor controller (industrial control). A phase-controlled rectifying system using tubes of the vapor- or gas-filled variety for power conversion to supply the armature circuit or the armature and shunt-field circuits of a direct-current motor, to provide adjustable-speed, adjustable- and regulated-speed characteristics. *See:* **electronic controller.**
                                                                          206

electronic direct-current motor drive (industrial control). The combination of an electronic direct-current motor controller with its associated motor or motors. *See:* **electronic controller.**   206

electronic efficiency (electron tubes). The ratio of (1) the power at the desired frequency delivered by the electron stream to the circuit in an oscillator or amplifier to (2) the average power supplied to the stream.
                                                                          125

electronic keying. A method of keying whereby the control is accomplished solely by electronic means. *See:* **telegraphy.**   111

electronic line scanning (facsimile). That method of scanning that provides motion of the scanning spot along the scanning line by electronic means. *See:* **scanning.**   12

electronic mail (computer applications). (1) The generation, transmission, and display of correspondence and documents by electronic means. (2) The concepts and technologies employed for the electronic communication of textual material.   571

electronic microphone. A microphone that depends for its operation on a change in the terminal electrical characteristic of an active device when a force is applied to some part of the device. *See:* **microphone.**
                                                                          176

electronic navigation. *See:* **navigation.**

electronic position indicator (EPI)(navigation aid terms). A radio navigation system used in hydrographic surveying which provides circular lines of position.   526

electronic power converter. Electronic devices for transforming electric power. *See:* **rectification.**
                                                                          328

electronic raster scanning (facsimile). That method of scanning in which motion of the scanning spot in both dimensions is accomplished by electronic means. *See:* **scanning.**   12

electronic rectifier (industrial control). A rectifier in which electron tubes are used as rectifying elements. *See:* **electronic controller; rectification of an alternating current.**   206

electronics (1) (adjective). Of, or pertaining to, the field of electronics. *Examples:* Electronics engineer, electronics course, electronics laboratory, and electronics committee. *See:* **electron; electronic.**
(2) (noun). That field of science and engineering that deals with electron devices and their utilization. *See:* **electron device.**   190
(3) (electric installations on shipboard). That branch of science and technology which relates to devices in which conductance is principally by electrons moving through a vacuum, gas, or semiconductor.   3

electronic scanning (antenna). Scanning an antenna beam by electronic or electric means without moving parts. *Syn:* **inertialess scanning.** *See:* **radiation**
                                                                          111

electronic storage register (watthour meters). An electronic circuit, which is an integral part of the time-of-use register, where data are stored for display or retrieval, or both.   485

electronic switching system (telephone switching systems.) An automatic switching system in which the control functions are performed principally by electronic devices.   55

electronic transformer (power and distribution transformer). Any transformer intended for use in a circuit or system utilizing electron or solid-state devices. *Note:* Mercury-arc rectifier transformers and luminous-tube transformers are normally excluded from this classification.   53

electronic trigger circuit (industrial control). A network containing electron tubes in which the output changes abruptly with an infinitesimal change in input at one or more points in the operating range.
                                                                          206

electronic warfare support measures (ESM) (radar). Actions taken to search for, intercept, locate, record, and analyze radiated electromagnetic energy for the purpose of exploiting such radiations in support of military operations.   13

electron injector. The electron gun of a betatron.
                                                                          190

electron lens. A device for the purpose of focusing an electron (ion) beam. *See:* **electron optics.**   244

electron microscope. An electron-optical device that produces a magnified image of an object. *Note:* Detail may be revealed by virtue of selective transmission, reflection, or emission of electrons by the object. 190

electron mirror. An electronic device causing the total reflection of an electron beam. *See:* **electron optics.** 190

electron multiplier. A structure, within an electron tube, that employs secondary electron emission from solids to produce current amplification. *See:* **amplifier; electron emission.** 117,125

electron multiplier transit time. That portion of photo-multiplier transit time corresponding to the time delay between an electron packet leaving the first dynode and the multiplier packet striking the anode. 117

electron optics. The branch of electronics that deals with the operation of certain electronic devices, based on the analogy between the path of electron (ion) beams in magnetic or electric fields and that of light rays in refractive media. 244,190

electron-ray indicator tube. An elementary form of cathode-ray tube used to indicate a change of voltage. *Note:* Such a tube used to indicate the tuning of a circuit is sometimes called a **magic eye.** *See:* **cathode-ray tubes.** 244,190

electron resolution. The ability of the electron multiplier section of the photomultiplier to resolve inputs consisting of n and $n + 1$ electrons. This may be expressed as a fractional full width at half maximum of the nth peak, as the peak to valley ratio of the nth peak to the valley between the nth and $n \times 1$th peaks. 117

electrons, conduction (semiconductor). The electrons in the conduction band of a solid that are free to move under the influence of an electric field. *See:* **semiconductor.** 186

electron sheath (gas) (ion sheath). A film of electrons (or of ions) that has formed on or near a surface that is held at a potential different from that of the discharge. *See:* **discharge (in a gas).** 244,190

electron-stream potential (electron tubes)(any point in an electron stream). The time average of the potential differential difference between that point and the electron-emitting surface. *See:* **electron emission.** 125

electron-stream transmission efficiency (electron tubes) (electrode through which the electron stream passes). The ratio of (1) the average stream current through the electrode to (2) the average stream current approaching the electrode. *Note:* In connection with multitransit tubes, the term electron stream should be taken to include only electrons approaching the electrode for the first time. *See:* **electron emission.** 125

electron telescope. An optical instrument for astronomy including an electronic image transformer associated with an optical telescope. *See:* **electron optics.** 244,190

electron tube. An electron device in which conduction by electrons takes place through a vacuum or gaseous medium within a gastight envelope. *Note:* The envelope may be either pumped during operation or sealed off. 125

electron-tube admittances. The cross-referenced terms generalize the familiar electron-tube coefficients so that they apply to all types of electron devices operated at any frequency as linear transducers. *Note:* The generalizations include the familiar low-frequency tube concepts. In the case of a riode, for example, at relatively low frequencies the short-circuit input admittance reduces to substantially the grid admittance, the short-circuit output admittance reduces to substantially the plate admittance, the short-circuit forward admittance reduces to substantially the grid-plate transconductance, and the short-circuit feedback admittance reduces to substantially the admittance of the grid-plate capacitance. When reference is made to alternating-voltage or -current components, the components are understood to be small enough so that linear relations hold between the various alternating voltages and currents. Consider a generalized network or transducer having n available terminals to each of which is flowing a complex alternating component I_j of the current and between each of which and a a reference point (which may or may not be one of the n network terminals) is applied a complex alternating voltage V_j. This network represents an n-terminal electron device in which each one of the terminals is connected to an electrode. 328

electron-tube amplifier. An amplifier that obtains its amplifying properties by means of electron tubes. 206

electronvolt (eV). The kinetic energy acquired by an electron in passing through a potential difference of 1V in vacuum; 1 eV = 1.602 19 $\times$ 10^{-19} J approximately. 21

electron-wave tube. An electron tube in which mutually interacting streams of electrons having different velocities cause a signal modulation to change progressively along their length. 125

electro-optic effect (fiber optics). A change in the refractive index of a material under the influence of an electric field. *Notes:* (1) Pockels and Kerr effects are electro-optic effects that are respectively linear and quadratic in the electric field strength. (2) "Electro-optic" is often erroneously used as a synonym for "optoelectronic". *See:* **optoelectronic.** 433

electroosmosis. The movement of fluids through diaphragms that is as a result of the application of an electric current. 328

electroosmotic potential (electrobiology). The electrokinetic potential gradient producing unit velocity of liquid flow through a porous structure. *See:* **electrobiology.** 192

electrophonic effect. The sensation of hearing produced when an alternating current of suitable frequency and magnitude from an external source is passed through an animal. 176

electrophoresis. A movement of colloidal ions as a result of the application of an electric potential. *See:* **ion.** 328

electrophoretic potential (electrobiology). The electrokinetic potential gradient required to produce unit velocity of a colloidal or suspended material through a liquid electrolyte. *See:* **electrobiology.**   192

electroplating. The electrodeposition of an adherent coating upon an object for such purposes as surface protection or decoration.   328

electropneumatic brake. An air brake that is provided with electrically controlled valves for control of the application and release of the brakes. *Note:* The electric control is usually in addition to a complete air brake equipment to provide a more prompt and synchronized operation of the brakes on two or more vehicles. *See:* **electric braking.**   328

electropneumatic contactor (1) (industrial control). A contactor actuated by air pressure. *See:* **contactor.**   206

(2) (electropneumatic unit switch). A contactor or switch controlled electrically and actuated by air pressure. *See:* **contactor; control switch.**   1

electropneumatic controller. An electrically supervised controller having some or all of its basic functions performed by air pressure. *See:* **electric controller; multiple-unit control.**   206

electropneumatic interlocking machine. An interlocking machine designed for electric control of electropneumatically operated functions. *See:* **centralized traffic control system.**   328

electropneumatic valve. An electrically operated valve that controls the passage of air.   328

electropolishing (electroplating). The smoothing or brightening of a metal surface by making it anodic in an appropriate solution. *See:* **electroplating.**   328

electrorefining. The process of electrodissolving a metal from an impure anode and depositing it in a more pure state.   328

electroretinogram. *See:* **electrodermogram.**

electroscope. An electrostatic device for indicating a potential difference or an electric charge. *See:* **instrument.**   328

electroshock therapy. The production of a reaction in the central nervous system by means of electric current applied to the cranium. *See:* **electrotherapy.**   192

electrostatic actuator. An apparatus constituting an auxiliary external electrode that permits the application of known electrostatic forces to the diaphragm of a microphone for the purpose of obtaining a primary calibration. *See:* **microphone.**   176

electrostatic coupling (interface terminology). *See:* **coupling, capacitive; signal**

electrostatic deflection (cathode-ray tube). Deflecting an electron beam by the action of an electric field. *See:* **cathode-ray tubes.**   190

electrostatic electrography. The branch of electrostatography that employs an insulating medium to form, without the aid of electromagnetic radiation, latent electrostatic-charge patterns for producing a viewable record. *See:* **electrostatography.**   191

electrostatic electron microscope. An electron microscope with electrostatic lenses. *See:* **electron optics.**   190

electrostatic electrophotography. The branch of electrostatography that employs a photoresponsive medi-

um to form, with the aid of electromagnetic radiation, latent electrostatic-charge patterns for producing a viewable record. *See:* **electrostatography.**   191

electrostatic focusing (electron beam). *See:* **focusing, electrostatic.**

electrostatic instrument. An instrument that depends for its operation on the forces of attraction and repulsion between bodies charged with electricity. *See:* **instrument.**   328

electrostatic lens. An electron lens in which the result is obtained by an electrostatic field. *See:* **electron optics.**

electrostatic loudspeaker (capacitor loudspeaker) (condenser loudspeaker*). A loudspeaker in which the mechanical forces are produced by the action of electrostatic fields. *See:* **loudspeaker.**   176
*Deprecated

electrostatic microphone (capacitor microphone) (condenser microphone*). A microphone that depends for its operation upon variations of its electrostatic capacitance. *See:* **microphone.**   176
*Deprecated

electrostatic recording (facsimile). Recording by means of a signal-controlled electrostatic field. *See:* **recording (facsimile).**   12

electrostatic relay. A relay in which operation depends upon the application or removal of electrostatic charge.   341

electrostatics. The branch of science that treats of the electric phenomena associated with electric charges at rest in the frame of reference.   210

electrostatic storage. A storage device that stores data as electrostatically charged areas on a dielectric surface.   255,77

electrostatic voltmeter. A voltmeter depending for its action upon electric forces. An electrostatic voltmeter is provided with a scale, usually graduated in volts or kilovolts. *See:* **instrument.**   328

electrostatic waves (radio wave propagation). In a plasma, the type of waves whose restoring forces are electrostatic. The associated electric field and particle velocity are in the direction of propagation and there are charge-density fluctuations associated with the waves.   146

electrostatography. The formation and utilization of latent electrostatic-charge patterns for the purpose of recording and reproducing patterns in viewable form.   236,191

electrostenolysis. The discharge of ions or colloidal ions in capillaries through the application of an electric potential. *See:* **ion.**   328

electrotaxis (electrobiology) (galvanotaxis). The act of a living organism in arranging itself in a medium in such a way that its axis bears a certain relation to the direction of the electric current in the medium. *See:* **electrobiology.**   192

electrotherapy. The use of electric energy in the treatment of disease.   192

electrothermal efficiency. The ratio of energy usefully employed in a furnace to the total energy supplied. *See:* **electrothermics.**   328

electrothermal recording (facsimile). That type of recording that is produced principally by signal-controlled thermal action. *See:* **recording (facsimile).**
12

electrothermic-coupler unit (electrothermic power meter). A three-port directional coupler with an electrothermic unit attached to either the side arm or the main arm which is normally used as a feed-through power measuring system. Typically, an electrothermic unit is attached to the side arm of the coupler so that the power at the output port of the main arm can be determined from a measurement of the power in the side arm. This system also can be used as a terminating powermeter by terminating the output port of the directional coupler.
47

electrothermic element (electrothermic power meter). A power absorber and a thermocouple (or thermopile) which are either two separate units or where the thermocouple (or thermopile) is also the power absorber.
47

electrothermic instrument. An instrument that depends for its operation on the heating effect of a current or currents. *Note:* Among the several possible types are (1) the expansion type, including the hot-wire and hot-strip instruments; (2) the thermocouple type; and (3) the bolometric type. *See:* **instrument.**
328

electrothermic mount (electrothermic power meter). A waveguide or transmission line structure which is designed to accept the electrothermic element. It normally contains internal matching devices or other reactive elements to obtain specified impedance conditions at its input terminal when an electrothermic element is installed. It usually contains a means of protecting the electrothermic element and the immediate environment from thermal gradients which would cause an undesirable thermoelectric output.
47

electrothermic power indicator (electrothermic power meter). An instrument which may or may not amplify the low level dc output voltage from the electrothermic unit and provides a display, usually in the form of the D'Arsonval type indication or a digital readout.
47

electrothermic power indicator error (electrothermic power meter). Ability of the metering circuitry to indicate exactly the substituted power within an electrothermic unit. Included are such factors as meter calibration, open loop gain, meter linearity, tracking errors, range switching errors, line voltage errors, and temperature compensation errors.
47

electrothermic power meter. This consists of an electrothermic unit and an electrothermic power indicator.
47

electrothermics. The branch of science and technology that deals with the direct transformation of electric energy and heat.
328

electrothermic substitution power (electrothermic power meter). The power at a reference frequency which, when dissipated in the electrothermic element, produces the same dc electrothermic output voltage that the element produces when subjected to radio frequency power.
47

electrothermic technique of power measurement (electrothermic power meter). A technique wherein the heating effect of power dissipated in an electrothermic element (which consists of an energy absorber and a thermocouple or thermopile) is used to generate a dc voltage. The power is dissipated either in a separate absorber or in the resistance of the electrothermic element. The resultant heat causes a temperature rise in a portion of the element. This temperature rise is sensed by the thermocouple which generates a dc output voltage proportional to the power.
47

electrothermic unit (electrothermic power meter). An assembly consisting of the electrothermic element installed in the electrothermic mount.
47

electrotonic wave (electrobiology). A brief nonpropagated change of potential on an excitable membrane in the vicinity of an applied stimulus; it is often accompanied by a propagated response and always by electrotonus. *See:* **excitability (electrobiology).**
192

electrotonus (1) (physical). The change in distribution of membrane potentials in nerve and muscle during or after the passage of an electric current. *See:* **excitability (electrobiology).**
192

(2) (physiological). The change in the excitability of a nerve or muscle during the passage of an electric current. *See:* **excitability (electrobiology).**
192

electrotyping. The production or reproduction of printing plates by electroforming. *See:* **electroforming.**
328

electrowinning. The electrodeposition of metals or compounds from solutions derived from ores or other materials using insoluable anodes.
328

element (1)(graphic symbols for logic functions). As used within ANSI/IEEE Std 91-1984, a representation of all or part of a logic function within a single outline, which may, in turn, be subdivided into smaller elements representing subfunctions of the overall function. Alternatively, the function so represented.
451

(2)(measuring longitudinal balance of telephone equipment operating in the voice band). Any electric device (such as inductor, resistor, capacitor, generator, or line) with terminals at which it may be directly connected to other devices, elements, or apparatus.
529

(3) (electron tubes). A constituent part of the tube that contributes directly to its electrical operation.
125

(4) (semiconductor device). Any integral part that contributes to its operation. *See:* **semiconductor device.**
245

(5) (integrated circuit). A constituent part of the integrated circuit that contributes directly to its operation. *See:* **integrated circuit.**
312,191

(6) (computing systems). *See:* **combinational logic element; logic element; sequential logic element; threshold element.**

(7) (storage cell). Consists of the positive and negative groups with separators, or separators and retainers, assembled for one cell. *See:* **battery (primary or secondary).**
328

(8) (data) (data transmission). Synonymous with bit as the minimum subdivision within a code grouping representing a character.   59

(9) (primary detecting). That portion of the feedback elements which first either utilizes or transforms energy from the controlled medium to produce a signal that is a function of the value of the directly controlled variable.   105

(10) (antenna). *See:* **array element; director element; driven element; parasitic element; radiating element; reflector element.**   111

(11) (power switchgear) (of a relay). *See:* **relay element.**   103

elemental area (facsimile). Any segment of the scanning line of the subject copy the dimension of which along the line is exactly equal to the nominal line width. *Note:* Elemental area is not necessarily the same as the scanning spot. *See:* **scanning.**   12

elementary diagram (packaging machinery). A diagram using symbols and a plan of connections to illustrate, in simple form, the scheme of control.   429

element cell (array antenna). In an array having a regular arrangement of elements which can be made congruent by translation, an element and a region surrounding it which when repeated by translation covers the entire array without gaps or overlays between cells. *Note:* There are many possible choices for such a cell. Some may be more convenient than others for analytic purposes.   111

element conduction interval (thyristor). That part of an operating cycle in which ON-state current flows in the basic control element.   445

element linear. *See:* **linear system or element.**

element, measuring (automatic control). That portion of the feedback elements which converts the signal from the primary detecting element to a form compatible with the reference input.   56

element nonconduction interval (thyristor). That part of an operating cycle during which no ON-state current flows in the basic control element.   445

element of a fix (navigation aid terms). The specific values of the navigation coordinates necessary to define a position.   526

element, primary detecting (automatic control). That portion of the feedback elements which first either utilizes or transforms energy from the controlled medium to produce a signal which is a function of the value of the directly controlled variable.   56

elements, feedback (control system). The elements in the controlling system that change the feedback signal in response to the directly controlled variable. *See:* **control system, feedback.**   105

elements, forward (automatic control). Those elements situated between the actuating signal and the controlled variable in the closed loop being considered. *See:* **control system, feedback.**   105

elements, loop (control system, feedback) (a closed loop). All elements in the signal path that begins with the loop error signal and ends with the loop return signal. *See:* **control system, feedback.**   56

elements, reference-input (automatic control). The portion of the controlling system that changes the reference input signal in response to the command. See Figure C. *See:* **control system, feedback.**   206

elevated-zero range. A range where the zero value of the measured variable, measured signal, etcetera, is greater than the lower range value. *Note:* The zero may be between the lower and upper range values, at the upper range value, or above the upper range value. For example: (a) -20 to 100, (b) -40 to 0, and (c) -50 to -10. *See:* **instrument.**   295

elevation (illuminating engineering). The angle between the axis of a searchlight drum and the horizontal. For angles above the horizontal, elevation is positive, and below the horizontal negative.   167

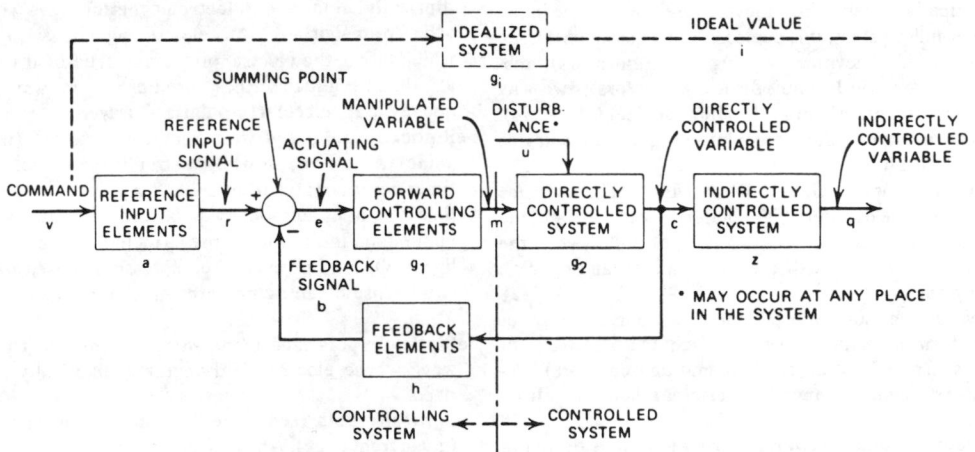

Figure C. Block diagram of an automatic control system illustrating expansion of the simplified block diagram to a more complex system.

elevation rod (lightning protection). The vertical portion of conductor in an air terminal by means of which it is elevated above the object to be protected.
328

elevator. A hoisting and lowering mechanism equipped with a car or platform that moves in guides in a substantially vertical direction, and that serves two or more floors of a building or structure.
328

elevator automatic dispatching device. A device, the principal function of which is to either: (1) operate a signal in the car to indicate when the car should leave a designated landing; or (2) actuate its starting mechanism when the car is at a designated landing. *See:* control (elevators).
328

elevator automatic signal transfer device. A device by means of which a signal registered in a car is automatically transferred to the next car following, in case the first car passes a floor, for which a signal has been registered, without making a stop. *See:* control.
328

elevator car. The load-carrying unit including its platform, car frame, enclosure, and car door or gate. *See:* elevator.
328

elevator car bottom runby (elevator car). The distance between the car buffer striker plate and the striking surface of the car buffer when the car floor is level with the bottom terminal landing. *See:* elevator.
328

elevator-car flash signal device. One providing a signal light, in the car, that is illuminated when the car approaches the landings at which a landing-signal-registering device has been actuated. *See:* control.
328

elevator car-leveling device. Any mechanism that will, either automatically or under the control of the operator, move the car within the leveling zone toward the landing only, and automatically stop it at the landing. *Notes:* (1) Where controlled by the operator by means of up-and-down continuous-pressure switches in the car, this device is known as an **inching device.** (2) Where used with a hydraulic elevator to correct automatically a change in car level caused by leakage in the hydraulic system, this device is known as an **anticreep device.** *See:* elevator; leveling zone (elevators); one-way automatic leveling device; (elevators); two-way automatic maintaining leveling device (elevators); two-way automatic nonmaintaining leveling device (elevators).
328

elevator counterweight bottom runby. The distance between the counterweight buffer striker plate and the striking surface of the counterweight buffer when the car floor is level with the top terminal landing. *See:* elevator.
328

elevator landing. That portion of a floor, balcony, or platform used to receive and discharge passengers or freight. *See:* bottom terminal landing (elevators); elevators; landing zone; top terminal landing (elevators).
328

elevator-landing signal registering device. A button or other device, located at the elevator landing that, when actuated by a waiting passenger, causes a stop signal to be registered in the car. *See:* control.
328

elevator-landing stopping device. A button or other device, located at an elevator landing that, when actuated, causes the elevator to stop at that floor. *See:* control.
328

elevator parking device. An electric or mechanical device, the function of which is to permit the opening, from the landing side, of the hoistway door at any landing when the car is within the landing zone of that landing. The device may also be used to close the door. *See:* control.
328

elevator pit. That portion of a hoistway extending from the threshold level of the lowest landing door to the floor at the bottom of the hoistway. *See:* elevators.
328

elevator separate-signal system. A system consisting of buttons or other devices located at the landings that, when actuated by a waiting passenger, illuminate a flash signal or operate an annunciator in the car, indicating floors at which stops are to be made. *See:* control (elevators).
328

elevator signal-transfer switch. A manually operated switch, located in the car, by means of which the operator can transfer a signal to the next car approaching in the same direction, when he desires to pass a floor at which a signal has been registered in the car. *See:* control.
328

elevator starter's control panel. An assembly of devices by means of which the starter may control the manner in which an elevator, or group of elevators, functions. *See:* control.
328

elevator truck zone. The limited distance above an elevator landing within which the truck-zoning device permits movement of the elevator car. *See:* control.
328

elevator truck-zoning device. A device that will permit the operator in the car to move a freight elevator, within the truck zone, with the car door or gate and a hoistway door open. *See:* control.
328

ELF. *See:* radio spectrum.
146

elliptically polarized (electromagnetic) wave (radio wave propagation). An electromagnetic wave for which either the electric or the magnetic field vector at a fixed point describes an ellipse at the wave frequency. *See:* circularly polarized wave.
146

elliptically polarized field vector (antennas) (waveguide). A field vector whose extremity as a function of times describes an ellipse. *Note:* Any single-frequency field vector is elliptically polarized if "elliptical" is understood in the wide sense as including circular and linear. Often, however, the expression is used in the strict sense meaning noncircular and nonlinear.
111, 267

elliptically polarized plane wave (antennas). A plane wave whose electric field vector is elliptically polarized.
111

elliptically polarized wave (radiowave propagation)-(given frequency). An electromagnetic wave for which the component of the electric vector in a plane normal to the direction of propagation describes an ellipse. *See:* electromagnetic wave; radiation; radio transmitter; waveguide.
267,146

elliptical orbit (communication satellite). An orbit of a satellite in which the distance between the centers of mass of the satellite and of the primary body is not constant. The general type of orbit is a special case. 74

elliptic filter (circuits and systems). A filter having an equiripple pass band and an equiminima stop band. 67

ellipticity. *See note under* **axial ratio.** *See:* **waveguide.**

embedded coil side (rotating machinery). That part of a coil side which lies in a slot between the ends of the core. 63

embedded computer system (software). A computer system that is integral to a larger system whose primary purpose is not computational, for example, a computer system in a weapon, aircraft, command and control, or rapid transit system. *See:* **computer system.** 434

embedded software. Software for an embedded computer system. *See:* **embedded computer system.** 434

embedded temperature detector (1) (electric installations on shipboard). A resistance thermometer or thermocouple built into a machine for the purpose of measuring the temperature. 3
(2) (rotating machinery). An element, usually a resistance thermometer or thermocouple, built into apparatus for the purpose of measuring temperature. *Note:* (A) This is ordinarily installed in a stator slot between coil sides at a location at which the highest temperature is anticipated. (B) Examination or replacement of an embedded detector after the apparatus is placed in service is usually not feasible. 63,27

embossing stylus. A recording stylus with a rounded tip that displaces the material in the recording medium to form a groove. *See:* **phonograph pickup.** 176

embrittlement. Severe loss of ductility of a metal or alloy. 205

emergency announcing system. A system of microphones, amplifier, and loud speakers (similar to a public address system) to permit instructions and orders from a ship's officers to passengers and crew in an emergency and particularly during abandon-ship operations. 328

emergency cells (storage cell). End cells that are held available for use exclusively during emergency discharges. *See:* **battery (primary or secondary).** 328

emergency egress (nuclear security systems). A path or route that provides an immediate exit path or way out of an area in the event of a sudden, unexpected, or dangerous occurrence. 464

emergency electric system (marine). All electric apparatus and circuits the operation of which, independent of ship's service supply, may be required under casualty conditions for preservation of a ship or personnel. 328

emergency generator (marine). An internal-combustion-engine-driven generator so located in the upper part of a vessel as to permit operation as long as the ship can remain afloat, and capable of operation, independent of any other apparatus on the ship, for supply of power to the emergency electric system upon failure of a ship's service power. *See:* **emergency electric system.** 328

emergency lighting (illuminating engineering). Lighting designed to supply illumination essential to the safety of life and property in the event of failure of the normal supply. 167

emergency lighting storage battery (marine transportation). A storage battery for instant supply of emergency power, upon failure of a ship's service supply, to certain circuits of special urgency principally temporary emergency lighting. *See:* **emergency electric system.** 328

emergency operations area(s) (nuclear power generating station). Functional area(s) allocated for the displays used to assess the status of safety systems and the controls for manual operations required during emergency situations. 358

emergency power (power operations). Power required by a system to make up a deficiency between the current firm power demand and the immediately available generating capability. 516

emergency power feedback. An arrangement permitting feedback of emergency-generator power to a ship's service system for supply of any apparatus on the ship within the limit of the emergency-generator rating. *See:* **emergency electric system.** 328

emergency power system (emergency and standby power). An independent reserve source of electric energy that, upon failure or outage of the normal source, automatically provides reliable electric power within a specified time to critical and equipment whose failure to operate satisfactorily would jeopardize the health and safety of personnel or result in damage to property. 512

emergency rating (power operations). The level of power flow that facility can carry for the time sufficient for adjustment of transfer schedules or generation dispatch in an orderly manner, with acceptable loss of life due to the facility involved. *See:* **normal rating.** 516

emergency service. An additional service intended only for use under emergency conditions. *See:* **dual service; duplicate service; loop service; service.** 64

emergency stop switch (elevators). A device located in the car that, when manually operated, causes the electric power to be removed from the driving-machine motor and brake of an electric elevator or from the electrically operated valves and/or pump motor of a hydraulic elevator. *See:* **control.** 328

emergency switchboard. A switchboard for control of sources of emergency power and for distribution to all emergency circuits. *See:* **emergency electric system.** 328

emergency system (1) (health care facilities). A system of feeders and branch circuits meeting the requirements of Article 700 of NFPA 70-1978, National Electrical Code, and intended to supply alternate power to a limited number of prescribed functions vital to the protection of life and safety, with automatic restoration of electrical power within 10 seconds of power interruption.   192

(2) (National Electrical Code). A system of feeders and branch circuits meeting the requirements of Article 700, connected to alternate power sources by a transfer switch and supplying energy to an extremely limited number of prescribed functions vital to the protection of life and patient safety, with automatic restoration of electrical power within 10 seconds of power interruption.   256

emergency-terminal stopping device (elevators). A device that automatically causes the power to be removed from an electric elevator driving-machine motor and brake, or from a hydraulic elevator machine, at a predetermined distance from the terminal landing, and independently of the functioning of the operating device and the normal-terminal stopping device, if the normal-terminal stopping device does not slow down the car as intended. *See:* **control.**   328

emergency transfer capability (electric power supply). The maximum amount of power that can be transmitted following a loss of transmission or generation capacity without causing additional transmission outages. *Syn:* **maximum transmission transfer capability.**   112

emergency voltage limit (power operations). The voltage range that is acceptable without serious system consequences, for the time sufficient for system adjustments to be made. *See:* **normal voltage limit.**   516

E **meter.** *See:* **electricity meter.**

EMF. *See:* **voltage (electromotive force).**

emission (1) (laser-maser). The transfer energy from matter to a radiation field.   363

(2) (radio-noise emission). An act of throwing out or giving off, generally used here in reference to electromagnetic energy.   418

emission characteristic. A relation, usually shown by a graph, between the emission and a factor controlling the emission (such as temperature, voltage, or current of the filament or heater). *See:* **electron emission.**   125

emission current. The current resulting from electron emission.   244,190

emission current, field-free (cathode). *See:* **field-free emission current (cathode).**

emission efficiency (thermionics). The quotient of the saturation current by the heating power absorbed by the cathode. *See:* **electron emission.**   244,190

emissivity (1) (fiber optics). The ratio of power radiated by a substance to the power radiated by a blackbody at the same temperature. Emissivity is a function of wavelength and temperature.   433

(2) (photovoltaic power system). The emittance of a specimen of material with an optically smooth, clean surface and sufficient thickness to be opaque. *See:* **photovoltaic power system; solar cells (photovoltaic power system).**   186

emittance (1) ϵ (illuminating engineering). The ratio of radiance in a given direction (for directional emittance) or radiant exitance (for hemispherical emittance) of a sample of a thermal radiator to that of a blackbody radiator at the same temperature. Formerly, exitance. The use of exitance with this meaning is deprecated.   167

(2) (photovoltaic power system). The ratio of the radiant flux-intensity from a given body to that of a black body at the same temperature.   186

emitter (transistor). A region from which charge carriers that are minority carriers in the base are injected into the base.   210,245,66

emitter, majority (transistor). An electrode from which a flow of majority carriers enters the interelectrode region. *See:* **transistor.**   66

emitter, minority (transistor). An electrode from which a flow of minority carriers enters the interelectrode region. *See:* **semiconductor; transistor.**   328

emitting sole (microwave tubes). An electron source in crossed-field amplifiers that is extensive and parallel to the slow-wave circuit and that may be a hot or cold electron-emitter. *See:* **microwave tube or valve.**   190

empirical propagation model (electromagnetic compatibility). A propagation model that is based solely on measured path-loss data. *See:* **electromagnetic compatibility.**   199

E&M signaling (telephone switching systems). A technique for transferring information between a trunk circuit and a separate signaling circuit over leads designated "E" and "M." The "M" lead transmits to the signaling circuit and the "E" lead transmits to the trunk circuit.   55

EMT (cable system in power generating stations). Electrical metallic tubing.   10

emulation (software). The imitation of all or part of one computer system by another, primarily by hardware, so that the imitating computer system accepts the same data, executes the same programs, and achieves the same results as the imitated system. *See:* **computer system; data; hardware; program.**   434

emulator (software). Hardware, software, or firmware that performs emulation. *See:* **emulation; firmware; hardware; software.**   434

enable (supervisory control, data acquisition, and automatic control)(station control and data acquisition). A command or condition which permits some specific event to proceed.   570,403

enabling pulse (1) (navigation). A pulse which prepares a circuit for some subsequent action.   13,187

(2). A pulse that opens an electric gate normally closed, or otherwise permits an operation for which it is a necessary but not a sufficient condition. *See:* **pulse.**   328

enamel (1) (general). A paint that is characterized by an ability to form an especially smooth film.   215,63

(2) (wire) (rotating machinery). A smooth film applied to wire usually by a coating process. *See:* **rotor (rotating machinery); stator.**   63

encapsulated (rotating machinery). A machine in which one or more of the windings is completely encased by molded insulation. *See:* **asynchronous machine.** 63

encapsulation (1)(of a semiconductor radiation detector)(germanium gamma-ray detectors). The packaging of a detector for protective or mounting purposes, or both. 528

(2) (software). The technique of isolating a system function within a module and providing a precise specification for the module. *See:* **information hiding; module; specification; system function.** 434

enclosed (NESC). Surrounded by case, cage, or fence designed to protect the contained equipment and minimize the possibility under normal conditions of dangerous approach or accidental contact by persons or objects. 494

enclosed brake (industrial control). A brake that is provided with an enclosure that covers the entire brake, including the brake actuator, the brake shoes, and the brake wheel. *See:* **electric drive.** 225,206

enclosed capacitor (shunt power capacitors). A capacitor having enclosed terminals. The enclosure is provided with means for connection to a rigid or flexible conduit. 138

enclosed cutout (power switchgear). A cutout in which the fuse clips and fuseholder or disconnecting blade are mounted completely within an insulating enclosure. 103, 443

enclosed relay. A relay that has both coil and contacts protected from the surrounding medium. *See:* **relay.** 259

enclosed self-ventilated machine (electric installations on shipboard). A machine having openings for the admission and discharge of the ventilating air, which is circulated by means integral with the machine, the machine being otherwise totally enclosed. These openings are so arranged that inlet and outlet ducts or pipes may be connected to them. Such ducts or pipes, if used, must have ample section and be so arranged as to furnish the specified volume of air to the machine, otherwise the ventilation will not be sufficient. 3

enclosed separately ventilated machine (electric installations on shipboard). A machine having openings for the admission and discharge of the ventilating air, which is circulated by means external to and not a part of the machine, the machine being otherwise totally enclosed. These openings are so arranged that inlet and outlet duct pipes may be connected to them. 3

enclosed switch (industrial control) (safety switch). A switch either with or without fuse holders, meter-testing equipment, or accommodation for meters, having all current-carrying parts completely enclosed in metal, and operable without opening the enclosure. *See:* **switch.** 206

enclosed switchboard (power switchgear). A dead-front switchboard that has an overall sheet metal enclosure (not grille) covering back and ends of the entire assembly. *Note:* Access to the interior of the en-closure is usually provided by doors or removable covers. The top may or may not be covered. 103

enclosed switches (indoor or outdoor) (power switchgear). Switches designed for service within a housing restricting heat transfer to the external medium. 103

enclosed switchgear assembly (power switchgear). One that is enclosed on all sides and top. 103

enclosed ventilated (rotating machinery). A term applied to an apparatus with a substantially complete enclosure in which openings are provided for ventilation only. *See:* **asynchronous machine.** 63

enclosed ventilated apparatus. Apparatus totally enclosed except that openings are provided for the admission and discharge of the cooling air. *Note:* These openings may be so arranged that inlet and outlet ducts or pipes may be connected to them. An enclosed ventilated apparatus or machine may be separately ventilated or self-ventilated. 328

enclosure (1)(metal-enclosed bus and calculating losses in isolated-phase bus). A surrounding case or housing used to protect the contained conductor and prevent personnel from accidentally contacting live parts. 574

(2)(power and distribution transformer). A surrounding case or housing used to protect the contained equipment and prevent personnel from accidentally contacting live parts. 53

(3)(power system communication equipment). A surrounding case or housing to protect the contained equipment against external conditions and to prevent personnel from accidentally contacting live parts. 453

(4) (general). A surrounding case or housing used to protect the contained conductor or equipment and prevent personnel from accidentally contacting live parts. *Note:* Material and finish shall conform to the standards for the switchgear enclosed. 202,78

(5) (Class 1E equipment and circuits)(nuclear power generating stations). An identifiable housing such as a cubicle, compartment, terminal box, panel, or enclosed raceway used for electrical equipment or cables. 131

(6) (National Electrical Code). The case or housing of apparatus, or the fence or walls surrounding an installation to prevent personnel from accidentally contacting energized parts, or to protect the equipment from physical damage. 256

(7) (power switchgear). A surrounding case or housing used to protect the contained equipment and to prevent personnel from accidentally contacting live parts. *Note:* Material and finish conform to the standard for the switchgear enclosed. 103

encode (1) (general). To express a single character or a message in terms of a code. 235

(2) (electronic control). To produce a unique combination of a group of output signals in response to each of a group of input signals. 235

(3) (computing systems). To apply the rules of a code. *See:* code; decode; matrix; translate. 255,77

encoder (electronic computation). A network or system in which only one input is excited at a time and each input produces a combination of outputs. *Note:* Sometimes called matrix. 210

end (Class 5) office (EO)(telephone loop performance). A switching system to which customer premises equipment is directly connected by loops. The switch connects loops to loops and loops to trunks. 473

end-around carry (computing systems). A carry generated in the most significant place and forwarded directly to the least significant place, for example, when adding two negative numbers, using nines complement. *See:* carry. 235

end bell. *See:* cable terminal.

end bracket (rotating machinery). A beam or bracket attached to the frame of a machine and intended for supporting a bearing. 63

end capacitor (antennas). A conducting element or group of conducting elements, connected at the end of a radiating element of an antenna, to modify the current distribution on the antenna, thus changing its input impedance. 111

end cells (storage battery) (storage cell). Cells that may be cut in or cut out of the circuit for the purpose of adjusting the battery voltage. *See:* battery (primary or secondary). 328

end device (of a telemeter) (power switchgear). The final system element that responds quantitatively to the measurand through the translating means and performs the final measurement operation. *Note:* An end device performs the final conversion of measurement energy to an indication, record, or the initiation of control. 103

end distortion (data transmission). The shifting of the end of all marking pulses from their proper positions in relation to the beginning of the start pulse, of telegraph signals. 59

end finger (outside space-block) (rotating machinery). A radially extending finger piece at the end of a laminated core to transfer pressure from an end clamping plate or flange to a tooth. *See:* rotor (rotating machinery); stator. 63

end-fire array antenna. A linear array antenna whose direction of maximum radiation lies along the line of the array. 111

end fittings (composite insulators). The insulator attachment hardware that is connected to the core. 483

end injection (Charles or Kino gun (EI)) (microwave tubes). A gun used in the presence of crossed electric and magnetic fields to inject an electron beam into the end of a slow-wave structure. *See:* microwave tube. 190

end-of-block signal (numerically controlled machines). A symbol or indicator that defines the end of one block of data. 224,297

end-of-copy signal (facsimile). A signal indicating termination of the transmission of a complete subject copy. *See:* facsimile signal (picture signal). 12

end office (telephone switching systems). A local office that is part of the toll hierarchy of World Zone 1. An end office is classified as a Class 5 office. *See:* office class. 55

end of program (numerically controlled machines). A miscellaneous function indicating completion of workpiece. *Note:* Stops spindle, coolant, and feed after completion of all commands in the block. Used to reset control and/or machine. Resetting control may include rewinding of tape or progressing a loop tape through the splicing leader. The choice for a particular case must be defined in the format classification sheet. 224,207

end of tape (numerically controlled machines). A miscellaneous function that stops spindle, coolant, and feed after completion of all commands in the block. *Note:* Used to reset control and/or machine. Resetting control will include rewinding of tape, progressing a loop tape through the splicing leader, or transferring to a second tape reader. The choice for a particular case must be defined in the format classification sheet. 54

end-on armature relay. *See:* armature, end-on; relay.

end plate, rotor (rotating machinery). An annular disk (ring) fitted at the outer end of the retaining ring. 63

end-play washers (rotating machinery). Washers of various thicknesses and materials used to control axial position of the shaft. 63

end-point criterion (evaluation of thermal capability) (thermal classification of electric equipment and electrical insulation). A value of property or property degradation (either absolute or percentage change) which defines failure in a functional test. 506

end rail (rotating machinery). A rail on which a bearing pedestal can be mounted. *See:* bearing. 63

end ring, rotor (rotating machinery). The conducting structure of a squirrel-cage or amortisseur (damper) winding that short-circuits all of the rotor bars at one end. *See:* rotor (rotating machinery). 63

end-scale value (electric instrument). The value of the actuating electrical quantity that corresponds to end-scale indication. *Notes:* (1) When zero is not at the end or at the electrical center of the scale, the higher value is taken. (2) Certain instruments such as power-factor meters, ohmmeters, etcetera, are necessarily excepted from this definition. (3) In the specification of the range of multiple-range instruments, it is preferable to list the ranges in descending order, as 750/300/150. *See:* accuracy rating (instrument); instrument. 280

end shield (1) (rotating machinery). A solid or skeletal structure, mounted at one end of a machine, for the purpose of providing a specified degree of protection for the winding and rotating parts or to direct the flow of ventilating air. *Note:* Ordinarily a machine has an end shield at each end. For certain types of machine, one of the end shields may be constructed as an integral part of the stator frame. The end shields may be used to align and support the bearings, oil deflectors, and, for a hydrogen-cooled machine, the hydrogen seals. 63

(2) (magnetrons). A shield for the purpose of confining the space charge to the interaction space. *See:* **magnetrons.** 125

end-shift frame (rotating machinery). A stator frame so constructed that it can be moved along the axis of the machine shaft for purposes of inspection. *See:* **stator.** 63

endurance (metal-nitride-oxide field-effect transistor). The number of write-high–write-low cycles accumulated before any defined unacceptable changes in device properties occur. 386

endurance limit. The maximum stress a metal can withstand without failure during a specified large number (usually 10 million) cycles of stress. 221,205

endurance test (reliability). An experiment carried out to investigate how the properties of an item are affected by the application of stresses and the elapse of time. 164

end user. *See:* **user.**

end user computing (computer applications). The performance of system development and data processing tasks by the user of a computer system. 571

end winding (rotating machinery). That portion of a winding extending beyond the slots. *Note:* It is outside the major flux path and its purpose is to provide connections between parts of the winding within the slots of the magnetic circuit. *See:* **asynchronous machine; rotor (rotating machinery); stator.** 63

end-winding cover (winding shield) (rotating machinery). A cover to protect an end winding against mechanical damage and/or to prevent inadvertent contact with the end winding. 63

end-winding support (rotating machinery). The structure by which coil ends are braced against gravity and electromagnetic forces during start-up (for motors), running, and abnormal conditions such as sudden short-circuit, for example, by blocking and lashings between coils and to brackets or rings. *See:* **stator.** 63

end-window counter tube (radiation). A counter tube designed for the radiation to enter at one end. *See:* **anticoincidence (radiation counters).** 190

end-wire insulation (rotating machinery). Insulation members placed between the end wires of individual coils such as between main and auxiliary windings. *See:* **rotor (rotating machinery); stator.** 63

end wire, winding (rotating machinery). The portion of a random-wound winding that is not inside the core. *See:* **rotor (rotating machinery); stator.** 63

energized (1)(NESC). Electrically connected to a source of potential difference, or electrically charged so as to have a potential significantly different from that of earth in the vicinity. *Syn:* **alive; live.** 494

(2)(conductor stringing equipment) (power line maintenance). Electrically connected to a source of potential difference, or electrically charged so as to have a potential different from that of the ground. *Syn:* **alive; current carrying; hot; live.** 45, 458

(3) (transmission and distribution) (alive or live). Electrically connected to a source of potential difference, or electrically charged so as to have a potential significantly different from that of earth in the vicinity. The term **live** is sometimes used in place of the term **current-carrying**, where the intent is clear, to avoid repetitions of the longer term. 262

energized background noise level (partial discharge measurement in liquid-filled power transformers and shunt reactors). Stated in pC (One pC = 10^{-12} Coulombs), the residual response of the partial discharge measurement system to background noise of any nature after the test circuit has been calibrated and the test object is energized at 50% of its nominal operating voltage. 580

energy (1)(power operations). That which does work or is capable of doing work. As used by electric utilities, it is generally a reference to electrical energy and is measured in kilowatt hours (kWh). *See:* **byproduct energy; energy control center; incremental energy cost; dump energy; economy energy; fuel replacement energy; interchange energy; energy loss; net system energy; off-peak energy; on-peak energy; potential hydro energy.** 516

(2) (metering). The integral of active power with respect to time. 212

(3) (Q)(laser-maser). The capacity for doing work. Energy content is commonly used to characterize the output from pulsed lasers and is generally expressed in joules. 363

(4) (system). The available energy is the amount of work that the system is capable of doing. *See:* **electric energy.** 210

energy and torque (International System of Units) (SI). The vector product of force and moment arm is widely designated by the unit newton meter. This unit for bending moment of torque results in confusion with the unit for energy, which is also newton meter. If torque is expressed as newton meter per radian, the relationship to energy is clarified, since the product of torque and angular rotation is energy:

$$(N \cdot m/rad) \cdot rad = N \cdot m.$$

See: **units and letter symbols.** 21

energy calibration (sodium iodide detector). The relationship between the height of the amplifier output pulse and the energy of the photons originating in the radioactive source. 423

energy charge (power operations). That portion of the charge for electric service based upon the electric energy consumed or billed. 516

energy control center. *See:* **power control center.**

energy costs, incremental (electric power supply). The additional cost of producing or transmitting electric energy above some base cost. 112

energy density (point in a field) (audio and electroacoustics). The energy contained in a given infinitesimal part of the medium divided by the volume of that part of the medium. *Notes:* (1) The term energy density may be used with prefatory modifiers such as instantaneous, maximum, and peak. (2) In speaking of average energy density in general, it is necessary to distinguish between the space average (at a given instant) and the time average (at a given point). 176

energy-density spectrum (burst measurements) (finite energy signal). The square of the magnitude of the Fourier transform of a burst. *See:* **burst; network analysis.** 61

energy dependence (radiation protection). A change in instrument response with respect to radiation energy for a constant exposure or exposure rate. 399

energy distribution (solar cells). The distribution of the flux or fluence of particles with respect to particle energy. 113

energy, dump (electric power supply). Energy generated from water, as, wind, or other source which cannot be stored and which is beyond the immediate needs of the electric system producing the energy. 112

energy, economy (electric power supply). Energy produced in one system and substituted for less economical energy in another system. *See:* **generating station.** 112

energy efficiency (specified electrochemical process). The product of the current efficiency and the voltage efficiency. *See:* **electrochemistry.** 328

energy-equivalent sound level L_{eq} (audible noise measurement). The equivalent sound level L_{eq} is the energy average of the level (usually A-weighted) of a varying sound over a specified period of time. The term *equivalent* signifies that the average of the fluctuating sound would have the same sound-energy level as a steady sound having the same level. The term energy is used because the sound amplitude is averaged on a root-mean-square (rms)-pressure-squared basis, and pressure-squared is proportional to energy. Mathematically, the equivalent sound level is defined as:

$$L_{eq} = 10 \log_{10} \left[\frac{1}{(t_2 - t_1)} \int_{t_1}^{t_2} \frac{p^2(t)}{p^2_{ref}} \, dt \right] \text{ dB}$$

(Eq 2)

where

$p(t)$ = the time varying A-weighted sound level

P_{ref} = the reference pressure, 20 μPa

$(t_2 - t_1)$ = the time period of interest

If the cumulative probability distribution of a noise is known, then the L_{eq} can be estimated by :

$$L_{eq} = 10 \log_{10} \left[\frac{1}{100} \sum_{o}^{n} (P_x - P_{x-1}) \text{ antilog } \frac{L_x}{10} \right] \text{ dB}$$

(Eq 3)

where

L_x = highest noise level in each step

P_x, P_{x-1} are selected adjacent steps along the probability scale expressed in percent probability.

energy flux (audio and electroacoustics). The average rate of flow of energy per unit time through any specified area. *Note:* For a sound wave in a medium of density ρ and for a plane or spherical free wave having a velocity of propagation c, the sound-energy flux through the area S corresponding to an effective sound pressure p is

$$\mathbf{J} = \frac{p^2 S}{\rho c} \cos \theta$$

where θ is the angle between the direction of propagation of the sound and the normal to the area S. 176

energy, fuel replacement (transmission and distribution). Energy generated at a hydroelectric plant as a substitute for energy which would otherwise have been generated by a thermal-electric plant. 112

energy gap (semiconductor). The energy range between the bottom of the conduction band and the top of the valence band. *See:* **semiconductor device.** 245

energy, interchange (transmission and distribution). Energy delivered to or received by one electric system from another. 112

energy-limiting transformer (power and distribution transformer). A transformer that is intended for use on an approximately constant-voltage supply circuit and that has sufficient inherent impedance to limit the output current to a thermally safe maximum value. 53

energy loss (transmission and distribution)(power operations). The difference between energy input and output as a result of transfer of energy between two points. 112, 516

energy metering point (tie line) (electric power systems). The location of the integrating metering equipment used to measure energy transfer on the tie line. *See:* **center of distribution.** 94

energy, net system (transmission and distribution). Energy requirements of a system, including losses, defined as (1) net generation of the system, plus (2) energy received from others, less (3) energy delivered to other systems. 112

energy, nuclear (transmission and distribution). The energy with which nucleons are bound together to form nuclei. When a nucleus is changed or rearranged in a nuclear reaction, (fission, fusion, etc.) or by radioactive decay nuclear energy may be released or absorbed in the form of kinetic energy of the reactants or products. 112

energy, off-peak (transmission and distribution). Energy supplied during designated periods of relatively low system demands. 112

energy, on-peak (transmission and distribution). Energy supplied during designated periods of relatively high system demands. 112

energy, partial discharge (dielectric tests). The energy dissipated by an individual discharge. The partial discharge energy is expressed in joules. 139

energy, potential hydro (electric power supply). The possible aggregate energy obtainable over a specified period by practical use of the available stream flow and river gradient. 112

energy, Q (laser-maser). The capacity for doing work. Energy content is commonly used to characterize the output from pulsed lasers, and is generally expressed in joules. 363

energy ratio (radar). The ratio of signal energy to noise power spectral density in the receiver, which also equals the maximum output signal-to-noise power ratio for a matched-filter system. 13

energy resolution (full width at half maximum)(FWHM)(of a semiconductor radiation detector)(X-ray energy spectrometers). The detector's contribution (including detector leakage current noise), expressed in units of energy, to the FWHM of a pulse-height distribution corresponding to an energy spectrum. 471

energy resolution (FWHM [full width at half maximum]) of a semiconductor radiation conductor/preamplifier combination (germanium gamma-ray detectors). The FWHM, measured at a specific energy and expressed in energy units, of a spectral peak within a pulse-height distribution corresponding to an energy spectrum. 528

energy resolution (FWHM [full width at half maximum]) of a semiconductor radiation detector (germanium gamma-ray detectors). The detector's contribution (including detector leakage current noise), expressed in units of energy, to the FWHM of a pulse-height distribution corresponding to an energy spectrum. 528

energy resolution (percent)(of a semiconductor)(X-ray energy spectrometers). One hundred times the energy resolution divided by the energy for which the resolution is specified. 471

energy resolution (semiconductor radiation detector). (1) (FWHM) (full width at half maximum). The detector's contribution (including detector leakage current noise), expressed in units of energy, to the FWHM of a pulse-height distribution corresponding to an energy spectrum. (2) (percent). One hundred times the energy resolution divided by the energy for which the resolution is specified. 23

energy straggling (germanium gamma-ray detectors)(charged-particle detectors). The random fluctuations in energy loss whereby those particles having the same initial energy lose different amounts of energy when traversing a given thickness of matter. (This process leads to the broadening of spectral lines.) 528

energy straggling. *See:* **straggling, energy.**

engine-driven generator for aircraft. A generator mechanically, hydraulically, or pneumatically coupled to an aircraft propulsion engine to provide power for the electric and electronic systems of an aircraft. It may be classified as follows: (1) Engine-mounted; (2) Remote-driven; (A) Flexible-shaft-driven; (B) Variable-ratio-driven; (C) Air-turbine-driven. *See:* **air transportation electric equipment.** 328

engine equilibrium temperature (periodic testing of diesel-generator units applied as standby power supplies in nuclear power generating stations). The condition at which the jacket water and lube oil temperatures are both within ± 10 °F (5.5 °C) of their normal operating temperatures established by the engine manufacturer. 539

engineered safety features (1)(nuclear power generating station). Features of a unit, other than reactor trip or those used only for normal operation, that are provided to prevent, limit, or mitigate the release of radioactive material. 102

(2)(safety systems equipment in nuclear power generating stations). Features of a unit other than reactor trip or those used only for normal operation, that are provided to prevent, limit, or mitigate the release of radioactive material. 535

engineering units (supervisory control, data acquisition, and automatic control). A unit of measure for use by operating/maintenance personnel usually provided by scaling the input quantity for display (meter, stripchart, or crt). 570

engine-generator system (electric power supply). A system in which electric power for the requirements of a railway vehicle (other than propulsion) is supplied by an engine-driven generator carried on the vehicle, either as an independent source of electric power or supplemented by a storage battery. *See:* **axle generator system.** 328

engine-room control. Apparatus and arrangement providing for control in the engine room, on order from the bridge, of the speed and direction of a vessel. 328

engine synchronism indicator. A device that provides a remote indication of the relative speeds between two or more engines. 328

engine-temperature thermocouple-type indicator. A device that indicates temperature of an aircraft engine cylinder by measuring the electromotive force of a thermocouple. 328

engine-torque indicator. A device that indicates engine torque in pound-feet. *Note:* It is usually converted to horsepower with reference to engine revolutions per minute. 328

English (USA) unit of luminance (illuminating engineering). Candela per square foot(cd/ft^2)also lumen per steradian, square foot ($lm/(sr \cdot ft^2)$). Another unit is candela per square inch (cd/in^2); also, lumen per steradian, square inch ($lm/(sr \cdot in^2)$). *See:* **stilb; Lambertian units of luminance; apostilb; footlambert; lambert.** 167

English language programming (test, measurement and diagnostic equipment). A technique of programming which allows the programmer to write programs and routines in English language statements.   54

enhanced solar radiation (radio wave propagation). The electromagnetic radiation of the sun under other than quiet conditions. *See:* **quiet sun.**   146

enhancement mode transistor (metal-nitride-oxide field-effect transistor) An insulated-gate field effect transistor (IGFET) where the channel connecting source and drain was formed by the effects of an applied gate voltage.   386

enterprise service (telephone switching systems). A service in which calls from certain designated exchanges are completed and billed to a number in another exchange.   55

entity attribute (software design descriptions). A named characteristic or property of a design entity. It provides a statement of fact about the entity.

        568

entrance terminal (distribution oil cutouts) (power switchgear). A terminal with an electrical connection to the fuse contact and suitable insulation where the connection passes through the housing.   103, 443

entry point (routine) (1) (computing systems). Any place to which control can be passed.   255,77

(2) (test, measurement and diagnostic equipment). One of a set of points in an automatic test equipment program where the test conditions are completely stated and are not dependent on previous tests or setups in any way. Such points are the only ones at which it is permissible to begin part of the complete test program. *See:* **re-run point.**   54

entry point identifier (label) (subroutines for CAMAC). The symbol *label* represents an entry point into a programmed procedure. Such a procedure will typically be executed in response to the recognition of a LAM, and it may interrupt the process being executed at the time of recogniton of the LAM. Under these circumstances the procedure must be capable of saving and restoring the state of the computer so that the interrupted process can be resumed. At least one value of labels should identify a system error procedure which deals with LAMs not linked to user processes.

        410

envelope (1) (general) (wave). The boundary of the family of curves obtained by varying a parameter of the wave. For the special case

$$y = E(t) \sin (\omega t + \theta),$$

variation of the parameter θ yields $E(t)$ as the envelope.   210

(2) (automatic control) (wave). Another wave composed of the instantaneous peak values of the original wave of an alternating quantity, and which indicates the variation in amplitude undergone by that quantity.

        56

envelope delay (1) (television radio wave propagation)

(facsimile). The time of propagation, between two points, of the envelope of a wave. *Note:* It is equal to the rate of change with angular frequency of the difference in phase between these two points. It has significance over the band of frequencies occupied by the wave only if this rate is approximately constant over that band. If the system distorts the envelope, the envelope delay at a specified frequency is defined with reference to a modulated wave that occupies a frequency bandwidth approaching zero. *See:* **television; facsimile transmission; radio wave propagation.**

        12,146

(2) (circuits and systems). The time that the envelope of a modulated signal takes to pass from one point in a network (or transmission system) to a second point in the network. *Note:* Envelope delay is often defined the same as group delay, that is, as the rate of change, with angular frequency, of the phase shift between two points in a network. *See:* **group delay time; time delay.**

        67

(3) (facsimile). The time of propagation, between two points of the envelope of a wave. *Note:* The envelope delay is measured by the slope of the phase shift in cycles plotted against the frequency in cycles per second. If the system distorts the envelope the envelope delay at a specified frequency is defined with reference to a modulated wave which occupies a frequency bandwidth approaching zero.   12

(4) (non-real time spectrum analyzer). The display produced on a spectrum analyzer when the resolution bandwidth is greater than the spacing of the individual frequency components.   68

(5) (radio wave propagation). The time of propagation between two points, of the envelope of a wave. It is equal to the rate of change with angular frequency of the difference in phase between these two points. It has significance over the band of frequencies occupied by the wave only if this rate is approximately constant over that band and the envelope retains its shape. *Syn:* **group delay.**   146

(6) (relative delay) (data transmission). The time of propagation, between two points, of the envelope of a wave. *Note:* It is equal to the rate of change with angular frequency of the difference in phase between these two points. It has significance over the band of frequencies occupied by the wave only if this rate is approximately constant over that band. If the system distorts the envelope, the envelope delay at a specified frequency is defined with reference to a modulated wave that occupies a frequency bandwidth approaching zero.   59

envelope delay distortion. *See:* **distortion, envelope delay.**

envelope display (spectrum analyzer). The display produced on a spectrum analyzer when the resolution bandwidth is greater than the spacing of the individual frequency components.   390

envelope, vacuum (electron tubes). *See:* **bulb.**

envelope voltage (electromagnetic site survey). The magnitude of the complex representation of the observed instantaneous voltage. *Note:* Envelope voltage is always a positive quantity permitting the logarithmic operation to be performed upon the value. *Syn: voltage envelope.*   457

environment (1). The universe within which the system must operate. All the elements over which the designer has no control and that affect the system or its inputs and outputs. *See:* system.   209

(2) (class 1E static battery chargers and inverters). The external conditions and influences such as temperature, humidity, altitude, shock and vibration which may affect the life and function of the components or equipment.   408

environmental application factor (reliability data) (reliability data for pumps and drivers, valve actuators, and valves). A multiplicative constant used to modify a failure rate to incorporate the effects of other normal or abnormal environments. *Note:* When available these factors are included in Appendix D of ANSI/IEEE Std 500-1984 P&V in the appropriate chapter prefaces.   447, 502

environmental change of amplification (magnetic amplifier). The change in amplification due to a specified change in one environmental quantity while all other environmental quantities are held constant. *Note:* Use of a coefficient implies a reasonable degree of linearity of the considered quantity with respect to the specified environmental quantity. If significant deviations from linearity exist within the environmental range over which the amplifier is expected to operate, particularly if the amplification, for example, is not a monotonic function of the environmental quantity, the existence of such deviations should be noted.   171

environmental coefficient (control systems). (1) (output from a control system or element having a specified input). The ratio of a change of output to the change in the specified environment (temperature, pressure, humidity, vibration, etc.), measured from a specified reference level, which causes it; in a linear system, it includes the "coefficient of sensitivity," and the "coefficient of zero error." **(2) (sensitivity).** The ratio of a change in sensitivity to the change in the specified environment (measured from a specified reference level) which causes it. **(3) (zero error).** The ratio of a change in zero error to the change in the specified environment (measured from a specified reference level) which causes it.   56

environmental coefficient of amplification (magnetic amplifier). The ratio of the change in amplification to the change in the specified environmental quantity when all other environmental quantities are held constant. *Note:* The units of this coefficient are the amplification units per unit of environmental quantity.   171

environmental coefficient of offset (magnetic amplifier). The ratio of the change in quiescent operating point to the change in the specified environmental quantity when all other environmental quantities are held constant. *Note:* The units of this coefficient are the output units per unit of environmental quantity.   171

environmental coefficient of trip-point stability (magnetic amplifier). The ratio of the change in trip point to the change in the specified environmental quantity when all other environmental quantities are held constant. *Notes:* (1) The units of this coefficient are the control signal units per unit of environmental quantity. (2) Use of a coefficient implies a reasonable degree of linearity of the considered quantity with respect to the specified environmental quantity. If significant deviations from linearity exist within the environmental range over which the amplifier is expected to operate, particularly if the amplification, for example, is not a monotonic function of the environmental quantity, the existence of such deviations should be noted.   171

environmental conditions (1)(electric penetration assemblies). Physical service conditions external to the electric penetration assembly such as ambient temperature, pressure, radiation, humidity, vibration, chemical or demineralized water spray and submergence expected as a result of normal operating requirements, and postulated conditions appropriate for the design basis events applicable to the electric penetration assembly.   493

(2)(nuclear power generating stations). Physical conditions external to the electric penetration assembly including but not limited to ambient temperature, pressure, radiation, humidity, and chemical spray expected as a result of normal operating requirements, and postulated conditions appropriate for the design basis events of the station applicable to the electric penetration assembly. *See:* design basis events; electric penetration assembly.   26

environmental offset (magnetic amplifier). The change in quiescent operating point due to a specified change in one environmental quantity (such as line voltage) while all other environmental quantities are held constant.   171

environmental radio noise (control of system electromagnetic compatibility). The total electromagnetic complex in which an equipment, subsystem, or system may be immersed, exclusive of its own electromagnetic contribution.   495

environmental seal (Class 1E connection assemblies). A device or system that restricts the passage of a gas or liquid through a boundary in conjunction with related cables or wires as an assembly. This does not include fire stops, in-line splices, or containment electric penetrations.   459

environmental temperature (separable insulated connectors). The temperature of the surrounding medium, such as air, water, and earth, into which the heat of the connector is dissipated directly, including the effect of heat dissipation from associated cables and apparatus.   454

environmental trip-point stability (magnetic amplifier). The change in the magnitude of the trip point (either trip OFF or trip ON, as specified) control signal due to a specified change in one environmental quantity (such as line voltage) while all other environmental quantities are held constant.   171

environs (radiological monitoring instrumentation). The uncontrolled area at or near the site boundary. 398

ephapse. The electric junction of two parallel or crossing nerve fibers at which there may occur phenomena similar to those occurring at a synapse. 192

ephemeris (communication satellite). The position vector of a satellite or spacecraft in space with respect to time. 74

E-plane bend (corner) (waveguide components). A waveguide bend (corner) in which the longitudinal axis of the guide remains in a plane parallel to the electric field vector throughout the bend (corner). 166

E-plane, principal (linearly polarized antenna). The plane containing the electric field vector and the direction of maximum radiation. *See:* **antenna; radiation.** 179

E-plane tee junction (series tee) (waveguide components). A waveguide tee junction in which the electric field vector of the dominant mode in each arm is parallel to the plane of the longitudinal axes of the guides. 166

equal-energy source (light) (television). A light source from which the emitted power per unit of wavelength is constant throughout the visible spectrum. 18

equal interval (isophase) light (illuminating engineering). A rhythmic light in which the light and dark periods are equal. 167

equalization (1)(transmission performance of telephone sets). The function a telephone set performs when it automatically adjusts transmitting or receiving, or both, so as to compensate for loop loss. 491

(2) (data transmission). The process of reducing frequency or phase distortion, or both, of a circuit by the introduction of networks to compensate for the difference in attenuation or time delay, or both, at the various frequencies in the transmission band. 59

(3) (feedback control system). Any form of compensation used to secure a closed-loop gain characteristic which is approximately constant over a desired range of frequencies. *See:* **compensation.** 56

(4) (electroacoustics). *See:* **frequency response equalization.**

equalizer (1)(substation grounding). A device to provide equipotential planes for resistance measurements. 475

(2) (rotating machinery). A connection made between points on a winding to minimize any undesirable potential voltage between these points. *See:* **asynchronous machine.** 63

equalizer circuit breaker (22) (power system device function numbers). A breaker that serves to control or to make and break the equalizer or the current-balancing connections for a machine field, or for regulating equipment, in a multiple-unit installation. 402

equalizing charge (storage battery) (storage cell). An extended charge to a measured end point that is given to a storage battery to insure the complete restoration of the active materials in all the plantes of all the cells. *See:* **charge.** 328

equalizing pulses (pulse terms). Pulse trains in which the pulse-repetition frequency is twice the line frequency and that occur just before and just after a vertical synchronizing pulse. *Note:* The equalizing pulses minimize the effect of line-frequency pulse on the interlace. 254

equally tempered scale. A series of notes selected from a division of the octave (usually) into 12 equal intervals, with a frequency ratio between any two adjacent notes equal to the twelfth root of two. 176

Equally tempered intervals

Name of Interval	Frequency Ratio*	Cents
Unison	1:1	0
Minor second or semitone	1.059463:1	100
Major second or whole tone	1.122462:1	200
Minor third	1.189207:1	300
Major third	1.259921:1	400
Perfect fourth	1.334840:1	500
Augmented fourth } Diminished fifth	1.414214:1	600
Perfect fifth	1.498307:1	700
Minor sixth	1.587401:1	800
Major sixth	1.681793:1	900
Minor seventh	1.781797:1	1000
Major seventh	1.887749:1	1100
Octave	2:1	1200

* The frequency ratio is $[(2)^{1/12}]^n$ where n equals the number of the interval. (The number of the interval is its value in cents divided by 100.)

equal vectors. Two vectors are equal when they have the same magnitude and the same direction. 210

equation. *See:* **computer equation.**

equational format (pulse measurement). One or more algebraic equations which specify a waveform wherein, typically, a first equation specifies the waveform from t_0 to t_1, a second equation specifies the waveform from t_1 to t_2, etc. The equational format is typically used to specify hypothetical, ideal, or reference waveforms. 15

equatorial orbit (communication satellite). An inclined orbit with an inclination of zero degrees. The plane of an equatorial orbit contains the equator of the primary body. 74

equilibrium coupling length. *See:* **equilibrium length.**

equilibrium electrode potential. A state electrode potential when the electrode and electrolyte are in equilibrium with respect to a specified electrochemical reaction. *See:* **electrochemistry.** 328

equilibrium length (fiber optics). For a specific excitation condition, the length of multimode optical waveguide necessary to attain equilibrium mode distribution. *Note:* The term is sometimes used to refer to the longest such length, as would result from a worst-case, but undefined excitation. *Syn:* **equilibrium coupling length; equilibrium mode distribution length.** *See:*

equilibrium mode distribution; mode coupling. 433

equilibrium mode distribution (fiber optics). The condition in a multimode optical waveguide in which the relative power distribution among the propagating modes is independent of length. *Syn:* **steady-state condition.** *See:* **equilibrium length; mode; mode coupling.** 433

equilibrium mode distribution length. *See:* **equilibrium length.** 433

equilibrium mode simulator (fiber optics). A device or optical system used to create an approximation of the equilibrium mode distribution. *See:* **equilibrium mode distribution; mode filter.** 433

equilibrium point (control system). A point in state space of a system where the time derivative of the state vector is identically zero. *See:* **control system.** 56

equilibrium potential. The electrode potential at equilibrium. 205

equilibrium reaction potential. The minimum voltage at which an electrochemical reaction can take place. *Note:* It is equal to the algebraic difference of the equilibrium potentials of the anode and cathode with respect to the specified reaction. It can be computed from the free energy of the reaction. Thus

$$\Delta F = -nFE$$

where ΔF is the free energy of the reaction, n is the number of chemical equivalents involved in the reaction, F is the value of the Faraday expressed in calories per volt gram-equivalent (23 060.5) and E is the equilibrium reaction potential (in volts). *See:* **electrochemistry.** 328

equilibrium voltage. *See:* **storage-element equilibrium voltage; storage tube.**

equiphase surface (radio wave propagation). Any surface in a wave over which the field vectors at the same instant are in the same phase or 180° out of phase. 146

equiphase zone (navigation aid terms). The region in space within which difference in phase of two radio signals is indistinguishable. 526

equipment (1)(NESC). A general term including fittings, devices, appliances, fixtures, apparatus, and similar terms used as part of or in connection with an electric supply or communication system. 494

(2)(nuclear power generating station). An assembly of components designed and manufactured to perform specific functions. *Note:* Examples of equipment are motors, transformers, valve operators, and instrumentation and control devices. 120

(3)(safety systems equipment in nuclear power generating stations). An assembly of components designed and manufactured to perform specific functions. *Note:* Certain items which satisfy the definition of the term equipment as used in IEEE Std 627-1980 are those referred to as components in the ASME Boiler and Pressure Vessel Code and its latest addenda, Section III (ANSI/IEEE BPV-III), for example, pumps

and valves. Other examples of equipment are motors, transformers, and instrumentation and control devices. Structures and structural support items are not included in the definition of equipment. 535

(4) (electrical engineering). A general term including materials, fittings, devices, appliances, fixtures, apparatus, machines, etcetera, used as a part of, or in connection with, an electrical installation. 256

(5) (power and distribution transformer) (National Electrical Code) A general term including material, fittings, devices, appliances, fixtures, apparatus, and the like, as a part of, or in connection with, an electrical installation. 53

equipment bonding jumper (National Electrical Code). The connection between two or more portions of the equipment grounding conductor. 256

equipment ground (1) (general). A ground connection to noncurrent-carrying metal parts of a wiring installation or of electric equipment, or both. *See:* **ground.** 64

(2) (surge arresters). A grounding system connected to parts of an installation, normally not alive, with which persons may come into contact. 244,62

equipment-grounding conductor (generating station grounding). The conductor used to connect the non-current-carrying metal parts of equipment, raceways, and other enclosures to the service equipment, the service power source(s) ground, or both. 569

equipment noise (sound recording and reproducing system). The noise output that is contributed by the elements of the equipment during recording and reproducing, excluding the recording medium, when the equipment is in normal operation. *Note:* Equipment noise usually comprises hum, rumble, tube noise, and component noise. *See:* **noise (sound recording and reproducing system).** 350

equipment number (telephone switching systems). A unique, physical or other identification of an input or output termination of a switching network. 193

equipment qualification (1) (class 1E static battery chargers and inverters). The generation and maintenance of evidence to assure that the equipment will meet the system performance requirements. 408

(2) (nuclear power generating stations). (A) The generation and maintenance of evidence to assure that the equipment will operate on demand to meet the system performance requirements. (See IEEE Std 323-1974.)(B)The generation and maintenance of evidence to assure that the equipment will operate on demand, to meet the system performance requirements. 440, 120

equipment signature (test, measurement and diagnostic equipment). The special characteristics of an equipment's response to, or reflection of, impinging impulsive energy, or of its electromagnetic, infrared or acoustical emissions. 54

equipment system (1) (health care facilities). A system of feeders and branch circuits arranged for automatic or manual connection to the alternate power source and which serves primarily three-phase power equipment. 192

(2) (National Electrical Code) (health care facilities). A system of feeders and branch circuits arranged for delayed, automatic or manual connection to the alternate power source and which serves primarily three-phase power equipment. 256

equipment under test (EUT) (radio-noise emission). A representative component, unit or system to be used for evaluation purposes. 418

equipotential (power line maintenance)(conductor stringing equipment). An identical state of electrical potential for two or more items. 458, 45

equipotential cathode. *See:* **cathode, indirectly heated.**

equipotential line or contour. The locus of points having the same potential at a given time. 313

equisignal localizer (navigation aid terms). A localizer in which the localizer on-course line is established as an equality of the amplitudes of two signals. 526

equisignal zone (radio navigation). The region in space within which the difference in amplitude of two radio signals (usually emitted by a single station) is indistinguishable. *See:* **radio navigation.** 328

equivalence (mathematics of computing). A dyadic Boolean operator having the property that if P is a statement and Q is a statement, then the equivalence of P and Q is true if and only is both statements are true or both statements are false. *Note:* The equivalence of P and Q is often represented by $P \equiv Q$. 564

equivalent binary digits (computing systems). The number of binary places required to count the elements of a given set. 255,77

equivalent circuit (1) (general). An arrangement of circuit elements that has characteristics, over a range of interest, electrically equivalent to those of a different circuit or device. *Note:* In many useful applications, the equivalent circuit replaces (for convenience of analysis) a more-complicated circuit or device. *See:* **network analysis.** 210

(2) (piezoelectric crystal unit). An electric circuit that has the same impedance as the unit in the frequency region of resonance. *Note:* It is usually represented by an inductance, capacitance, and resistance in series, shunted by the direct capacitance between the terminals of the crystal unit. *See:* **crystal.** 328

equivalent concentration (ion type). The concentration equal to the ion concentration divided by the valency of the ion considered. *See:* **ion.** 328

equivalent conductance (1) (acid, base, or salt). The conductance of the amount of solution that contains one gram equivalent of the solute when measured between parallel electrodes that are one centimeter apart and large enough in area to include the necessary volume of solution. *Note:* Equivalent conductance is numerically equal to the conductivity multiplied by the volume in cubic centimeters containing one gram equivalent of the acid, base, or salt. *See:* **electrochemistry.** 328

(2) (microwave gas tube). The normalized conductance of the tube in its mount measured as its resonance frequency. *Note:* Normalization is with respect to the characteristic impedance of the transmission line at its junction with the tube mount. *See:* **electron-tube admittances; element (electron tube).** 125

equivalent continuous rating (rotating machinery). The statement of the load and conditions assigned to the machine for test purposes, by the manufacturer, at which the machine may be operated until thermal equilibrium is reached, and which is considered to be equivalent to the duty or duty type. 63

equivalent contrast, $\bar{C}$ (illuminating engineering). A numerical description of the relative visibility of a task. It is the contrast of the standard visibility reference task giving the same visibility as that of a task whose contrast has been reduced to threshold when the background luminances are the same. *See:* **visual task evaluator.** 167

equivalent contrast, $\bar{C}_e$ (illuminating engineering). The actual equivalent contrast in a real luminous environment with nondiffuse illumination. The actual equivalent contrast $\bar{C}_e$ is less than the equivalent contrast due to veiling reflection. $\bar{C}_e = \bar{C} \times CRF$. *See:* **contrast rendition factor.** 167

equivalent core-loss resistance. A hypothetical resistance, assumed to be in parallel with the magnetizing inductance, that would dissipate the same power as that dissipated in the core of the transformer winding for a specified value of excitation. 197

equivalent dark-current input (phototubes). The incident luminous (or radiant) flux required to give a signal output current equal to the output electrode dark current. *Note:* Since the dark current may change considerably with temperature, the temperature should be specified. *See:* **phototubes.** 190

equivalent diode (triode or a multielectrode tube). *See:* **diode, equivalent.**

equivalent diode voltage. *See:* **composite controlling voltage.**

equivalent flat plate area (scattering object). Equal to the wavelength times the square root of the ratio of the back-scattering cross section to 4π. *Note:* A perfectly reflecting plate parallel to the incident wavefront and having this area, if it is large compared to the wavelength, will have approximately the same back-scattering cross section as the object. 111

equivalent 4-wire (data transmission). Use of different frequency bands to form a "high group" and "low group" for the two directions of transmission, thereby permitting operation over a single pair of conductors. 59

equivalent hours (E)(electric generating unit reliability, availability, and productivity). The number of hours a unit was in a time category involving unit derating, expressed as equivalent hours of full outage at maximum capacity. Both unit derating and maximum capacity shall be expressed on a consistent basis, gross or net. Equivalent hours can be calculated for each of the time categories --unit derated hours, in-service unit derated hours, reserve shutdown unit derated hours, planned derated hours, in-service planned derated hours, reserve shutdown planned derated hours, unplanned derated hours, in-service unplanned derated hours, reserve shutdown un-planned derated hours, forced derated hours, in-service forced derated hours, reserve shutdown forced derated hours, maintenance derated hours, in-service maintenance derated hours, reserve shutdown maintenance derated hours, and seasonal derated hours. The symbol designation for the equivalent hours is formed by adding an E in front of the symbol for the corresponding time designation (for example, equivalent unit derated hours is designated EUNDH). Equivalent hours can be calculated from the following equation:

$$E(\ ) = \frac{\Sigma\ D(\ )_i T_i}{MC}$$

where $E(\ )$ = equivalent hours in the time category represented by the parentheses, which can be any one of the time categories in sections 5.11 through 5.16 in ANSI/IEEE Std 762-1987. D = the derating for the time category shown in parentheses, after the ith change in either available capacity (unit deratings) or dependable capacity (seasonal deratings). Note: In order to apportion equivalent hours among the various time categories, appropriate ground rules are established in the reporting system so that after each change in either available capacity or dependable capacity, the sum of all subcategories of unit derating is equal to the unit derating. T_i = the number of hours accumulated in the time category of interest between the ith and the $(i + 1)$th change in either available capacity (unit deratings) or dependable capacity (seasonal deratings). MC = maximum capacity. 567

equivalent input noise sensitivity (spectrum analyzer). The average level of a spectrum analyzer's internally generated noise referenced to the input. *See:* **sensitivity; input signal sensitivity.** 390

equivalent isotropically radiated power (antennas). In a given direction, the gain of a transmitting antenna multiplied by the net power accepted by the antenna from the connected transmitter. *Syn:* **effective isotropically radiated power (EIRP).** 111

equivalent load reflection coefficient. *See:* **reflection coefficient.**

equivalent luminous intensity of an extended source at a specified distance (illuminating engineering). The intensity of a point source which would produce the same illuminance at that distance. Formerly, apparent luminous intensity of an extended source. 167

equivalent network. A network that, under certain conditions of use, may replace another network without substantial effect on electrical performance. *Note:* If one network can replace another network in any system whatsoever without altering in any way the electrical operation of that portion of the system external to the networks, the networks are said to be **networks of general equivalence.** If one network can replace another network only in some particular sys-

tem without altering in any way the electrical operation of that portion of the system external to the networks, the networks are said to be **networks of limited equivalence.** Examples of the latter are networks that are equivalent only at a single frequency, over a single band, in one direction only, or only with certain terminal conditions (such as H and T networks). *See:* **network analysis.** 210

equivalent noise bandwidth (interference terminology) (signal system). The frequency interval, determined by the response-frequency characteristics of the system, that defines the noise power transmitted from a noise source of specified characteristics. *Note:* For Gaussian noise

$$\Delta f = \int_0^{\infty} y(f)^2\, df$$

where $y(f) = Y(0)/Y(f)$ is the relative frequency dependent response characteristic. *See:* **interference.** 188

equivalent noise conductance (interference terminology). A quantitative representation in conductance units of the spectral density of a noise-current generator at a specified frequency. *Notes:* (1) The relation between the equivalent noise noise conductance G_n and the spectral density W_i of the noise-current generator is

$$G_n = \pi W_i / (kT_0)$$

where k is Boltzmann's constant and T_0 is the standard noise temperature (290 kelvins) and $kT_0 = 4.00 \times 10$ 21 watt-seconds. (2) The equivalent noise conductance in terms of the mean-square noise-generator current i^2 within a frequency increment δf is

$$G_n = i^2 / (4kT_0 \Delta f).$$

See: **electron-tube admittances; signal-to-noise signal.** 188,190

equivalent noise current (electron tubes) (interference terminology). A quantitative representation in current units of the spectral density of a noise current generator at a specified frequency. *Notes:* (1) The relation between the equivalent noise current I_n and the spectral density W_i of the noise-current generator is

$$I_n = (2\pi W_i)/e$$

where e is the magnitude of the electron charge. (2) The equivalent noise current in terms of the mean-square noise-generator current $\overline{I^2}$ within a frequency increment δf is

$$I_n = i^2/(2e\Delta f).$$

See: circuit characteristics of electrodes; interference; signal-to-noise ratio.   188,190

equivalent noise input (phototube). The value of incident luminous (or radiant) flux that, when modulated in a stated manner, produces a root-mean-square signal output current equal to the root-mean-square dark-current noise both in the same specified bandwidth (usually 1 hertz). *See:* phototube.   174

equivalent noise referred to input (of a linear amplifier)(germanium gamma-ray detectors)(X-ray energy spectrometers). The value of noise at the input that would produce the same value of noise at the output as does the actual noise source.

      528, 471, 23

equivalent noise resistance. A quantitative representation in resistance units of the spectral density of a noise voltage generator at a specified frequency. *Notes:* (1) The relation between the equivalent noise resistance R_n and the spectral density W_e of the noise-voltage generator is

$$R_n = (\pi W_e)/(k T_0)$$

where k is Boltzmann's constant and T_0 is the standard noise temperature (290 kelvins) and $kT_0 = 4.00 \times 10^{21}$ watt-seconds. (2) The equivalent noise resistance in terms of the mean-square noise-generator voltage $\overline{e^2}$ within a frequency increment δf is

$$R_n = \overline{e^2}/(4kT_0\Delta f).$$

See: interference; signal-to-noise ratio.   125

equivalent noise resistance referred to input (of a linear amplifier)(charged-particle detectors)(germanium gamma-ray detectors). That value of resistor which when applied to the input of a hypothetical noiseless amplifier with the same gain and bandwidth would produce the same output noise.   119, 528

equivalent noise resistance referred to input (semiconductor radiation detectors) (linear amplifier). That value of resistor which when applied to the input of a hypothetical noiseless amplifier with the same gain and bandwidth would produce the same output noise.

      23

equivalent parallel circuit elements (magnetic core testing). Under stated conditions of excitation and coil configuration, the values of inductance and resistance connected parallel so that they give representation to the real permeability of the core (μ'_s) and the total losses in the core (μ''_s)

$$L_p , \mu'_p L_0$$
$$R_p = \omega\mu''_p L_0$$
$$\frac{1}{Z} = \frac{1}{j\omega L_p} + \frac{1}{R_p} = \frac{1}{j\omega\overline{\mu}L_0}$$

$\overline{\mu}$ = complex relative permeability
μ'_p = real component of $\overline{\mu}$ parallel representation
μ''_p = imaginary component of $\overline{\mu}$, parallel representation
L_0 = self inductance of coil with a core of unit relative permeability, but with the same flux distribution as with a ferromagnetic core.
L_p = parallel equivalent self inductance of the coil with a core of $\overline{\mu}$ permeability.
R_p = parallel equivalent loss resistance of the core
ω = angular frequency in radians/sec.

      165

equivalent periodic line (uniform line). A periodic line having the same electrical behavior, at a given frequency, as the uniform line when measured at its terminals or at corresponding section junctions. *See:* transmission line.   210

equivalent radiated power (ERP). *See:* effective radiated power.

equivalent series circuit elements (magnetic core testing). Under stated conditions of excitation and coil configuration, values of a reactance and a resistance connected in series so that they give representation to the real permeability of the core (μ'_s) and to the total losses in the core (μ''_s)

$$L_s = \mu's L_0$$
$$R_s = \omega\mu''_s L_0$$
$$Z = R_s + j\omega L_s = j\omega\overline{\mu}L_0$$

L_s = self inductance of oil with a core of $\overline{\mu}$ permeability; series equivalent inductance.
R_s = equivalent series resistance of coil in ohms with a core of $\overline{\mu}$ permeability.
ω = angular frequency in radians/sec.

      165

equivalent source reflection coefficient (network analyzers). The reflection coefficient equal to that caused by the source impedance Z_s

$$\Gamma_s = \frac{Z_s - Z_o}{Z_s + Z_o}$$

where the source impedance Z_s is the Thevenin impedance and is only considered in the linear range of the source. The Thevenin impedance is the impedance in Thevenin's Theorem. The impedance, Z_o, is the characteristic impedance of the transmission system.

Notes:(1) In order to approximate a Z_o source impedance, that is ,$\Gamma_s = 0$, a directional coupler or suitable power splitter can be used as part of a feedback control circuit to maintain a constant incident power at its main-arm output port independent of the source impedance of the radio-frequency source connected to the main-arm input port of the coupler. (2) At lower frequencies in order to approximate a Z_o source impedance, a Z_o impedance can be put in series with a constant voltage source that is maintained at zero impedance by means of a feedback control circuit independent of the source impedance of the radio-frequency source.   505

equivalent sources. *See:* **Huygens' sources.**   111

equivalent sphere illumination, (ESI) (illuminating engineering). The level of sphere illumination which would produce task visibility equivalent to that produced by a specific lighting environment.   167

equivalent test alternating voltage (charging inductors). A sinusoidal root-mean-square test voltage equal to 0.707 times the power-supply voltage of the network-charging circuit and having a frequency equal to the resonance frequency of charging. *Note:* This is the alternating component of the voltage that appears across the charging inductor in a resonance-charging circuit of the pulse forming network.   137

equivalent two-winding kVA rating (power and distribution transformer). The equivalent two-winding rating of multi-one-half the sum of the kVA ratings of all windings. *Note:* It is customary to base this equivalent two-winding kVA rating on the self-cooled rating of the transformer.   53

equivalent 1-megaelectronvolt electron flux. The flux of electrons of 1-megaelectronvolt energy that changes a stated physical quantity (such as minority carrier diffusion length) of a given material or device to the same value as would the flux of penetrating particles of another stated energy spectrum.   113

equivalent 4-wire (data transmission). Use of different frequency bands to form a "high group" and "low group" for the two directions of transmission, thereby permitting operation over a single pair of conductors.   59

equivocation (information theory). The conditional information content of an input symbol given an output symbol, averaged over all input-output pairs. *See:* **information theory.**   415

erase (charge-storage tubes). To reduce by a controlled operation the amount of stored information. *See:* **storage tube.**   174,125

erasing head (electroacoustics). A device for obliterating any previous magnetic recordings. *See:* **alternating-current erasing head; direct-current erasing head;**

permanent-magnet erasing head; phonograph pickup.   176

erasing rate (charge-storage tubes). The time rate of erasing a storage element line or area, from one specified level to another. Note the distinction between this and erasing speed. *See:* **storage tube.**   174

erasing, selective (storage tubes). Erasing of selected storage elements without disturbing the information stored on other storage elements. *See:* **storage tube.**   174

erasing speed (charge-storage tubes). The linear scanning rate of the beam across the storage surface in erasing. Note the distinction between this and erasing rate. *See:* **storage tube.**   174,125

erasing time, minimum usable (storage tubes). The time required to erase stored information from one specified level to another under stated conditions of operation and without rewriting. *Note:* The qualifying adjectives **minimum usable** are frequently omitted in general usage when it is clear that the minimum usable erasing time is implied. *See:* **storage tube.**   174

erection cut-out (gyro). The feature wherein the signal supplying the erection torque is disconnected in order to minimize vehicle maneuver effects.   46

erection or slaving rate (gyro). The angular rate at which the spin axis is precessed to a reference position. It is expressed as angular displacement per unit time.   46

E region (radio wave propagation). The region of the terrestrial ionosphere between about 90 and 160 km altitude.   146

erg. The unit of work and of energy in the centimeter-gram-second systems. The erg is 10^{-7} joule.   210

erlang (1) (telephone switching systems). Unit of traffic intensity, measured in number of arrivals per mean service time. For carried traffic measurements, the number of erlangs is the average number of simultaneous connections observed during a measurement period.   55

(2) (data transmission). A term used in message loading of telephone leased facilities. One erlange is equal to the number of call-seconds divided by 3600 and is equal to a fully loaded circuit over a one-hour period.   59

erosion (1)(composite insulators). The loss of material by leakage current or corona discharge.   483

(2)(corrosion). Deterioration by the abrasive action of fluids, usually accelerated by the presence of solid particles of matter in suspension. When deterioration is further increased by corrosion, the term erosion-corrosion is often used.   205

error (1) (mathematics). Any discrepancy between a computed, observed, or measured quantity and the true, specified, or theoretically correct value or condition. *Note:* A positive error denotes that the indication of the instrument is greater than the true value. Error = Indication - True. *See:* **absolute error; correction; inherited error; accuracy rating of an instrument; low-frequency testing; power-system testing; surge testing.**   105

(2) (computer or data processing system). Any incorrect step, process, or result. *Note:* In the computer field the term commonly is used to refer to a machine malfunction as a machine error (or computer error) and to a human mistake as a human error (or operator error). Frequently it is helpful to distinguish between these errors as follows; an **error** results from incorrect programming, coding, data transcription, manual operation, etcetera, a **malfunction** results from a failure in the operation of a machine component such as a gate, a flip-flop, or an amplifier. *See:* **dynamic error; linearity error; loading error; resolution error; static error; electronic analog computer.**   210,77,54

(3) (analog computers). (A) In science, the difference between the true value and a calculated or observed value. A quantity (equal in absolute magnitude to the error) added to a calculated, indicated, or observed value to obtain the true value is called a **correction.** (B) In a computer or data processing system, any incorrect step, process, or result. In the computer field, the following terms are commonly used: a machine malfunction is a "machine error" (or "computer error"); an incorrect program is a "program error"; and a human mistake is a "human error" (or "operator error"). Frequently it is helpful to distinguish among these errors as follows: an **error** results from approximations used in numerical methods or imperfections in analog components; a **mistake** results from incorrect programming, coding, data transcription, manual operation, etc; a **malfunction** results from a failure in the operation of a machine component such as a gate, flip-flop, or an amplifier.   9

(4) (automatic control). An indicated value minus an accepted standard value, or true value. *Note:* ASA C85 deprecates use of the term as the negative of deviation. *See:* **accuracy, precision.**   56

(5) (unbalanced transmission-line impedance). 'In any measurement of a particular quantity, the difference between the measurement concerned and the true value of the magnitude of this quantity, taken positive or negative accordingly as the measurement is greater or less than the true value" (Churchill Eisenhart, "Realistic Evaluation of the Precision and Accuracy of Instrument Calibration Systems," *Journal of Research of the National Bureau of Standards,* Vol. 67C, No. 2, April-June 1963).   147

(6) (measurement). The algebraic difference between a value that results from measurement and a corresponding true value.   94

(7) (pascal computer programming language). A violation by a program of the requirements of ANSI/IEEE 770X3.97-1983 that a processor is permitted to leave undetected. *Notes:* (1) If it is possible to construct a program in which the violation or non-violation of this standard requires knowledge of the data read by the program or the implementation definition of implementation-defined features, then violation of that requirement is classified as an error. Processors may report on such violations of the requirement without such knowledge, but there always remain some cases that require execution or simulated execution, or proof procedures with the required knowl-

edge. Requirements that can be verified without such knowledge are not classified as errors. (2) Processors should attempt the detection of as many errors as possible. Permission to omit detection is provided for implementations in which the detection would be an excessive burden.   433

(8) (software). (A) A discrepancy between a computed, observed, or measured value or condition and the true, specified, or theoretically correct value or condition. (B) Human action which results in software containing a fault. Examples include omission or misinterpretation of user requirements in a software specification, incorrect translation or omission of a requirement in the design specification. This is not a preferred usage. *See:* **design specification; failure; fault; requirement; software; specification.**   434

error analysis (software). (1) The process of investigating an observed software fault with the purpose of tracing the fault to its source. (2) The process of investigating an observed software fault to identify such information as the cause of the fault, the phase of the development process during which the fault was introduced, methods by which the fault could have been prevented or detected earlier, and the method by which the fault was detected. (3) The process of investigating software errors, failures, and faults to determine quantitative rates and trends. *See:* **failure; fault; software error; software fault.**   434

error and correction. The difference between the indicated value and the true value of the quantity being measured. *Note:* It is the quantity that algebraically subtracted from the indicated value gives the true value. A positive error denotes theat the indicated value of the instrument is greater than the true value. The correction has the same numerical value as the error of the indicated value, but the opposite sign. It is the quantity that algebraically added to the indicated value gives the true value. If T, I, E and C represent, respectively, the true value, the indicated value, the error, and the correction, the following equations hold:

$$E = I - T; C = T - I.$$

Example: a voltmeter reads 112 volts when the voltage applied to its terminals is actually 110 volts.*See:* **accuracy rating (instrument).**   328

error band (gyro; accelerometer). A band about the specified output function which contains the output data. The error band contains the composite effects of non-linearity, resolution, non-repeatability, hysteresis and other uncertainties in the output data. *See:* **input-output characteristics.**   46

error burst (data transmission). A group of bits in which two successive erroneous bits are always separated by less than a given number x of correct bits. The last erroneous bit in the burst and the first erroneous bit in the following burst are accordingly separated by x correct bits or more. Number x should be specified when describing an error burst.   59

error category (software). One of a set of classes into

which an error, fault, or failure might fall. Categories may be defined for the cause, criticality, effect, life cycle phase when introduced or detected, or other characteristics of the error, fault, or failure. *See:* **criticality; failure; fault; software.**

error coefficient (control system, feedback). The real number C_n by which the nth derivative of the reference input signal is multiplied to give the resulting nth component of the actuating signal. *Note:* The error coefficients may be obtained by expanding in a Maclaurin series the error transfer function as follows:

$$\frac{1}{1 + GH(s)} = C_0 + C_1 s + C_2 s^2 + \ldots + C_n s^n.$$

See: **control system, feedback.**   329

error constant (control system, feedback). The real number K_n by which the nth derivative of the reference input signal is divided to give the resulting nth component of the actuating signal. *Note:* $K_n = 1/C_n$; $K_0 = 1 + Kp$, where K_p is position constant; $K_1 = K_v$ velocity constant; $K_2 = K_a$ acceleration constant; $K_3 = K_j$ jerk constant. In some systems these constants may equal infinity. *See:* **control system, feedback.**   56

error-correcting code. A code in which each telegraph or data signal conforms to specific rules of construction so that departures from this construction in the received signals can be automatically detected, and permits the automatic correction, at the received terminal, or some or all of the errors. *Note:* Such codes require more signal elements than are necessary to convey the basic information. *See:* **error-detecting code; error-detecting and feedback system; error-detecting system.**   194

error data (software). A term commonly (but not precisely) used to denote information describing software problems, faults, failures, and changes, their characteristics, and the conditions under which they are encountered or corrected. *See:* **failure; fault; software.**   434

error-detecting and feedback system. A system employing an error-detecting code and so arranged that a character or block detected as being in error automatically initiates a request for retransmission of the signal detected as being in error.   194

error-detecting code. A code in which each expression conforms to specific rules of construction, so that if certain errors occur in an expression the resulting expression will not conform to the rules of construction and thus the presence of the errors is detected. *Note:* Such codes require more signal elements than are necessary to convey the fundamental information. *See:* **check, forbidden-combination; error correcting code.**   194, 255

error-detecting system (data transmission). A system employing an error-detecting code and so arranged that any signal detected as being in error is either (1)

deleted from the data delivered to the receiver, in some cases with an indication that such deletion has taken place, or (2) delivered to the receiver together with an indication that it has been detected as being in error.   194

error, dynamic (analog computer). An error in a time-varying signal resulting from imperfect dynamic response of a transducer.   9

error, fractional (measurement). The magnitude of the ratio of the error to the true value.   147

error, linearity (analog computer). An error which is the deviation of the output quantity, from a specified linear reference curve.   9

error, matching (analog computer). An error resulting from inaccuracy in matching (two resistors) or mating (a resistor and a capacitor) passive elements. *See:* **electronic analog computer.**   167

error model (software). A mathematical model used to predict or estimate the number of remaining faults, reliability, required test time, or similar characteristics of a software system. *See:* **error prediction; fault; model; reliability; software system.**   434

error prediction (software). A quantitative statement about the expected number or nature of software problems, faults, or failures in a software system. *See:* **error model; failure; fault; software; system.**   434

error prediction model. *See:* **error model.**

error, random (measurement). A component of error whose magnitude and direction vary in a random manner in a sequence of measurements made under nominally identical conditions.   147

error range. The difference between the highest and lowest error values.   255,77

error rate (data transmission). Ratio of the number of characters of a message incorrectly received to the number of characters of the message received.   59

error recovery. *See:* **failure recovery.**

error, resolution (analog computer). The error due to the inability of a transducer to manifest changes of a variable smaller than a given increment.   9

error seeding. *See:* **fault seeding.**

error signal (excitation systems for synchronous machines). In a control system the error signal is the difference between a sensing signal and a constant reference signal. *Note:* In excitation control systems sensing signals may be proportional to synchronous machine terminal voltage, the ratio of terminal voltage to frequency, active or reactive armature current, active or reactive power, power factor, terminal frequency, shaft speed, generator field voltage or field current, and exciter field voltage or field current.   507

error, static (analog computer). An error independent of the time-varying nature of a variable. *Syn:* **D.C. error.**   9

error, systematic (measurement). The inherent bias (offset) of a measurement process or of one of its components.   147

erythema (illuminating engineering). The temporary reddening of the skin produced by exposure to ultraviolet energy. *Note:* The degree of erythema is used as a guide to dosages applied in ultraviolet therapy.   167

erythemal effectiveness (illuminating engineering). The capacity of various portions of the ultraviolet spectrum to produce erythema.   167

erythemal efficiency of radiant flux (for a particular wavelength) (illuminating engineering). The ratio of the erythemal effectiveness of that wavelength to that of wavelength 296.7 nm (nanometers) which is rated as unity. The erythemal efficiency of various wavelengths of radiant flux for producing a minimum perceptible erythema (MPE) is given in the table below. These values have been accepted for evaluating the erythemal effectiveness of sun lamps. *Note:* This term formerly was called "relative erythemal factor".

Erythemal and bactericidal efficiency of ultraviolet radiation

Wavelength (nanometers)	Erythemal Efficiency	Tentative Bactericidal Efficiency
*235.3	—	0.35
240.0	0.56	—
*244.6	0.57	0.58
*248.2	0.57	0.70
250.0	0.57	—
*253.7	0.55	0.85
*257.6	0.49	0.94
260.0	0.42	—
265.0	—	1.00
*265.4	0.25	0.99
*267.5	0.20	0.98
*270.0	0.14	0.95
*275.3	0.07	0.81
*280.4	0.06	0.68
285.0	0.09	—
*285.7	0.10	0.55
*289.4	0.25	0.46
290.0	0.31	—
*292.5	0.70	0.38
295.0	0.98	—
*296.7	1.00	0.27
300.0	0.83	—
*302.2	0.55	0.13
305.0	0.33	—
310.0	0.11	—
*313.0	0.03	0.01
315.0	0.01	—
320.0	0.005	—
325.0	0.003	—
330.0	0.000	—

* Emission lines in the mercury spectrum; other values interpolated.

   167

erythemal exposure (illuminating engineering). The product of erythemal flux density on a surface and time. It usually is measured in erythemal microwatt-minutes per square centimeter. *Note:* For average untanned skin a minimum perceptible erythema requires about 300 microwatt-minutes per square centimeter of radiation at 296.7 nm (nanometers).   167

erythemal flux (illuminating engineering). Radiant flux evaluated according to its capacity to produce erythema of the untanned human skin. It usually is measured in microwatts of ultraviolet radiation weighted in accordance with its erythemal efficiency. Such quantities of erythemal flux would be in erythemal microwatts. *Note:* A commonly used practical unit of erythemal flux is the erythemal unit (EU) or E-viton (erytheme) which is equal to the amount of radiant flux which will produce the same erythemal effect as 10 microwatts of radiant flux at wavelength 296.7 nm (nanometers).   167

erythemal flux density (illuminating engineering). The erythemal flux per unit area of the surface being irradiated. It is equal to the quotient of the incident erythemal flux divided by the area of the surface when the flux is uniformly distributed. It usually is measured in microwatts per square centimeter of erythemally weighed ultraviolet radiation (erythemal microwatts per square centimeter). *Note:* A suggested practical unit of erythemal flux density is the Finsen which is equal to one E-viton per square centimeter.   167

Esaki diode. *See:* **tunnel mode.**

escalator. A power-driven, inclined, continuous stairway used for raising or lowering passengers. *See:* **elevators.**   328

E scan (electronic navigation). *See:* **E display.**

escape character (computing systems). A character used to indicate that the succeeding one or more characters are expressed in a code different from the code currently in use.   255,77

escape ratio (charge-storage tubes). The average number of secondary and reflected primary electrons leaving the vicinity of a storage element per primary electron entering that vicinity. *Note:* The escape ratio is less than the secondary-emission ratio when, for example, some secondary electrons are returned to the secondary-emitting surface by a retarding field. *See:* **charge-storage tube.**   174

E-scope (radar). A cathode-ray oscilloscope arranged to present an E-display.   13

ESM. *See:* **electronic warfare support measures.**

ESONE (FASTBUS acquisition and control). A multinational committee representing European nuclear laboratories. It produced the initial CAMAC specification and collaborates with NIM in the maintenance and extension of CAMAC and in the development of FASTBUS. *See:* **NIM.**   480

ESP cable. *See:* **electric submersible pump cable.**

essential electrical systems (National Electrical Code) (health care facilities). Systems comprised of alternate sources of power, transfer switches, overcurrent protective devices, distribution cabinets, feeders, branch circuits, motor controls, and all connected electrical equipment, designed to provide designated areas with continuity of electrical service during disruption of normal power sources and also designed to minimize the interruptive effects of disruption within the internal wiring system.   256

essentially zero source impedance (electronic power transformer). Implies that the source impedance is low enough so that the test currents under consideration would cause less than five (5) percent distortion (instantaneous) in the voltage amplitude or waveshape at the load terminals. 95

essential performance requirements (nuclear power generating stations). Requirements that must be met if a component, module, or channel is to carry out its part in the implementation of a protective function. 109

essential process control (electric pipe heating systems). The use of electric pipe heating systems to increase or maintain or both, the temperature of fluids (or processes) in desirably available or essential mechanical piping systems including pipes, pumps, valves, tanks, instrumentation, etcetera, in fossil fueled generating stations. An example of an essential process control system is the heating for the fuel oil system. 448

estimated life (performance)(thermal classification of electric equipment and electrical insulation). The expected useful service life based upon service experience or the results of tests performed in accordance with appropriate evaluation procedures established by the responsible technical committee, or both. 506

estimated position (navigation aid terms). The most probable position of a craft determined from incomplete data or data of questionable accuracy. 526

EU. See: **erythemal flux.**

evacuating equipment. The assembly of vacuum pumps, instruments, and other parts for maintaining and indicating the vacuum. See: **rectification.** 328

evanescent field (fiber optics). A time varying electromagnetic field whose amplitude decreases monotonically, but without an accompanying phase shift, in a particular direction is said to be evanescent in that direction. 433

evanescent mode (cutoff mode) (waveguide). A field configuration in a waveguide such that the amplitude of the field diminishes along the waveguide, but the phase is unchanged. The frequency of this mode is less than the critical frequency. See: **waveguide.** 179

evanescent mode. See: **cutoff mode.**

evanescent waveguide. See: **cutoff waveguide.**

event (1)(supervisory control, data acquisition, and automatic control)(station control and data acquisition). A discrete change of state (status) of a system or device. 570, 403

(2) (sequential events recording systems). A change in a process or a change in operation of equipment which is detected by bistable sensors. 48

event recognition (sequential events recording systems). The capability to detect and process changes of state of one or more inputs. 48

everyday load (composite insulators). The bare conductor weight and wind load that predominates for the greatest period of time over the life of a line. 483

evh (power line maintenance). See: extra high voltage. 458

E-viton. See: **erythemal flux.**

evolving fault (power switchgear). A change in the current during interruption whereby the magnitude of current increases to a fault current or to a higher value of fault current in one or more phases. 103

EW (radar). (1) Abbreviation for early warning. (2) Abbreviation for electronic warfare. 13

(2) (radar). Refers to the signal after envelope or phase detection, which in early radar was the displayed signal. Contains the relevant radar information after removal of the carrier frequency. 13

exalted carrier reception. See: **reconditioned carrier reception.**

exception (software). An event that causes suspension of normal program execution. See: **program execution.** 434

exception condition (logical link control). The condition assumed by a logical link control (LLC) upon receipt of a command protocol data unit (PDU) which it cannot execute due to either a transmission error or an internal processing malfunction. 585

excess insertion loss (fiber optics). In an optical waveguide coupler, the optical loss associated with that portion of the light which does not emerge from the nominally operational ports of the device. See: **optical waveguide coupler.** 433

excess meter. An electricity meter that measures and registers the integral, with respect to time, of those portions of the active power in excess of the predetermined value. See: **electricity meter (meter).** 328

excess reactivity (power operations). More reactivity than that needed to achieve criticality. In order to avoid frequent reactor shutdowns to replace fuel that has been consumed and to compensate for the accumulation of fission products which have high neutron absorption cross sections and negative temperature coefficients, excess reactivity is provided in a reactor by including additional fuel in the core at startup. See: **reactivity.** 516

excess-three code (electronic computation). Number code in which the decimal digit n is represented by the four-bit binary equivalent of $n + 3$. Specifically:

decimal digit	excess-three code
0	0011
1	0100
2	0101
3	0110
4	0111
5	1000
6	1001
7	1010
8	1011
9	1100

See: **binary-coded-decimal system.** 235

exchange. See: **central office exchange; private auto-**

matic exchange; private automatic branch exchange; private branch exchange; exchange service. 59

exchangeable power (per unit bandwidth, at a port). The extreme value of the power flow per unit bandwidth from or to a port under arbitrary variations of its terminating impedance. *Notes:* (1) The exchangeable power p_e at a port with a mean-square open-circuit voltage spectral density e^2 and an internal impedance with a real part R is given by the relation

$$p_e = \frac{\overline{e^2}}{4R}$$

(2) The exchangeable power is equal to the available power when the internal impedance of the port has a positive real part. *See:* **signal-to-noise ratio; waveguide.** 190

exchangeable power gain (two-port linear transducer). At a pair of selected input and output frequencies, the ratio of (1) the exchangeable signal power of the output port of the transducer to (2) the exchangeable signal power of the source connected to the input port. *Note:* The exchangeable power gain is equal to the available power gain when the internal impedances of the source and the output port of the transducer have positive real parts. *See:* **signal-to-noise ratio; waveguide.** 190

exchange area (telephone switching systems). The territory included within the boundaries of a telecommunications exchange. 55

exchange, central office. *See:* **central office exchange.**

exchange service (data transmission). A service permitting interconnection of any two customers' telephones through the use of a switching equipment. 59

excitability (electrobiology) (irritability). The inherent ability of a tissue to start its specific reaction in response to an electric current. 192

excitability curve (medical electronics). A graph of the excitability of a given tissue as a function of time, where excitability is expressed either as the reciprocal of the intensity of an electric current just sufficient at a given instant to start the specific reaction of the tissue, or as the quotient of the initial (or conditioning) threshold intensity for the tissue by subsequent threshold intensities. 192

excitation (1) (drive). A signal voltage applied to the control electrode of an electron tube. *See:* **composite controlling voltage (electron tube).** 111

(2) (array antenna). For an array of radiating elements, the specification, in amplitude and phase, of either the voltage applied to each element or the input current to each element. 111

excitation anode (pool-cathode rectifier tube). An electrode that is used to maintain an auxiliary arc in the vacuum tank. *See:* **electrode (electron tube); rectification.** 328

excitation coefficients (antennas). The relative values, in amplitude and phase, of the excitation currents or voltages of the radiating elements of an array antenna. *Syn:* **feeding coefficients.** 111

excitation control system. A feedback control system that includes the synchronous machine and its excitation system. 105

excitation current (1) (power switchgear). The current supplied to unloaded transformers or similar equipment. 103

(2) (no-load current) (power and distribution transformer). The current which flows in any winding used to excite the transformer when all other windings are open-circuited. It is usually expressed in percent of the rated current of the winding in which it is measured. 53

(3) (voltage regulator). The current that maintains the excitation of the regulator. *Note:* It is usually expressed in per unit or in percent of the rated series-winding current of the regulator. *See:* **efficiency; voltage regulator.** 257

excitation equipment (rectifier). The equipment for starting, maintaining, and controlling the arc. *See:* **rectification.** 328

excitation losses (1) (series transformer). The losses in the transformer with the secondary winding open-circuited when the primary winding is excited at rated frequency and at a voltage that corresponds to the primary voltage obtained when the transformer is operating at nominal rated load. *Note:* The measurement should be made with a constant voltage source of supply with not more than 3-percent harmonic deviation from sine wave. 274

(2) (instrument transformer). The watts required to supply the energy necessary to excite the transformer which include the dielectric watts, the core watts, and the watts in the excited winding due to the excitation current. 53, 394

excitation power current transformer (excitation systems for synchronous machines). The elements in a compound source-rectifier excitation system which transfer electrical energy from the synchronous machine armature current to the excitation system at a magnitude and phase relationship required by the excitation system. 507

excitation power potential transformer (excitation systems for synchronous machines). The element or elements in a compound source-rectifier excitation system which transfer electrical energy from the synchronous machine armature terminals to the excitation system at a magnitude and phase relationship required in the excitation system. Also, the element or elements in a potential source-rectifier excitation system which transfer electrical energy either from the machine terminals or from an auxiliary bus to the excitation system at a magnitude level required by the excitation system. 507

excitation purity (light) (television). The ratio of (1) the distance from the reference point to the point representing the sample to (2) the distance along the same straight line from the reference point to the spectrum locus or to the purple boundary, both distances being measured (in the same direction from the reference point) on the CIE chromaticity diagram. *Note:* When giving excitation purity and dominant (or complimentary) wavelength as a pair of values to determine the chromaticity coordinates, the reference point must be the same in all cases, and it must represent the reference standard light (specified achromatic light) mentioned in the definitions of dominant wavelength.   18

excitation purity of a light, p_e (illuminating engineering). The ratio of the distance on the CIE chromaticity diagram between the reference point and the light point to the distance in the same direction between the reference point and the spectrum locus or the purple boundary.   167

excitation-regulating winding (power and distribution transformer)(two-core regulating transformer). In some designs, the main unit will have one winding operating as an autotransformer which performs both functions listed under excitation and regulating windings. Such a winding is called the excitation-regulating winding.   53

excitation response. *See:* voltage response, exciter.

excitation system (1)(excitation systems for synchronous machines). The equipment providing field current for a synchronous machine, including all power, regulating control, and protective elements.   507 **(2)(rotating machinery).** The source of field current for the excitation of a principal electric machine, including means for its control.   63,105

excitation system ceiling voltage (excitation control system). The maximum dc component of system output voltage that may be attained by an excitation system under specified conditions. *Note:* In some excitation systems, ceiling voltage may have both positive and negative values. Also, in some special applications, the excitation system is capable of supplying both positive and negative field current to the synchronous machine.   374

excitation system duty cycle (excitation systems for synchronous machines). An initial operating condition and a subsequent sequence of events of specified duration to which the excitation system will be exposed. *Note:* The duty cycle usually involves a three-phase fault of specified duration which is located electrically close to the synchronous machine. Its primary purpose is to specify the duty that the excitation system components can withstand without incurring maloperation or damage.   507

excitation system, high initial response. An excitation system having an excitation system voltage response time of 0.1 second or less.   105

excitation system nominal ceiling voltage (excitation control system). The ceiling voltage attained by an excitation system under the following conditions: (1) The exciter loaded with a resistor having an ohmic value equal to the resistance of the filed winding to be excited and with this field winding at a temperature of 75 °C for field windings designed to operate at rating with a temperature rise of 60 °C or less; or 100 °C for field windings designed to operate at rating with a temperature rise greater than 60 °C. For rectifier exciters nominal ceiling voltage should be determined with the exciter loaded with a load having resistance

as specified above and sufficient inductance so that regulation effects and voltage and current waveforms can be properly duplicated. For test purposes, providing such a load may often be impractical. In such cases, analytical means may be used to predict performance under actual loading and conditions. (2) For excitation systems employing a rotating exciter, the ceiling should be determined at rated speed. (3) For potential-source rectifier excitation systems, the ceiling should be determined with rated (100 percent) potential applied unless otherwise specified. (4) In compound-rectifier excitation systems both generator voltage and current inputs are utilized as the source of power for the excitation system. The nominal ceiling voltage will be determined under specified reduced generator terminal voltage and increased generator terminal current conditions as would be encountered during power system faults and other disturbances. For some applications where relay coordination is a consideration, the ceiling voltage will be determined by a requirement that the generator produce a specific value of steady-state three-phase short circuit current.   374

excitation system nominal response (excitation systems for synchronous machines). The rate of increase of the excitation system output voltage determined from the excitation system voltage response curve, divided by the rated field voltage. This rate, if maintained constant, would develop the same voltage-time area as obtained from the actual curve over the first half-second interval (unless a different time interval is specified). *See:* **ANSI/IEEE Std 421.1-1986 for notes and required practice.**   507

excitation system output terminals (excitation systems for synchronous machines). The place of output from the equipmant comprising the excitation system. These terminals may be identical with the field winding terminals.   507

excitation system rated current (excitation systems for synchronous machines). The direct current at the excitation system output terminals which the excitation system can supply under defined conditions of its operation. This current is at least that value required by the synchronous machine under the most demanding continuous operating conditions (generally resulting from synchronous machine voltage frequency variations and power factor variations).   507

excitation system rated voltage (excitation systems for synchronous machines). The direct voltage at the excitation system output terminals which the excitation system can provide when delivering excitation system rated current under rated continuous load conditions of the synchronous machine with its field winding at (1) 75 °C for field windings designed to operate at rating with a temperature rise of 60 °C or less; or (2) 100 °C for field windings designed to operate at rating with a temperature rise greater than 60 °C.   507

excitation-system stability (rotating machinery). The ability of the excitation system to control the field voltage of the principal electric machine so that transient changes in the regulated voltage are effectively suppressed and sustained oscillations in the regulated voltage are not produced by the excitation system during steady-load conditions following a change to a new steady-load condition.   63

excitation system stabilizer (excitation systems for synchronous machines). An element or group of elements that modify the forward signal by either series or feedback compensation to improve the dynamic performance of the excitation control system.

    507

excitation system voltage response (1)(excitation control system). The rate of increase or decrease of the excitation system output voltage determined from the excitation system voltage-time response curve, which rate if maintained constant, would develop the same voltage-time area as obtained from the curve for a specified period. The starting point for determining the rate of voltage change is the initial value of the excitation system voltage time response curve. Referring to Fig. 1, (see **exciter or excitation system voltage** diagram) the excitation system voltage response is illustrated by line *ac*. This line is determined by establishing the area *acd* equal to area *abd*. *Notes:* (1) The starting point for determining the rate of voltage change is the initiation of the disturbance, that is, the excitation system voltage time response should include any delay time that may be present. (2) A system having an excitation system voltage response time of 0.1 s or less is defined as a high response excitation system.

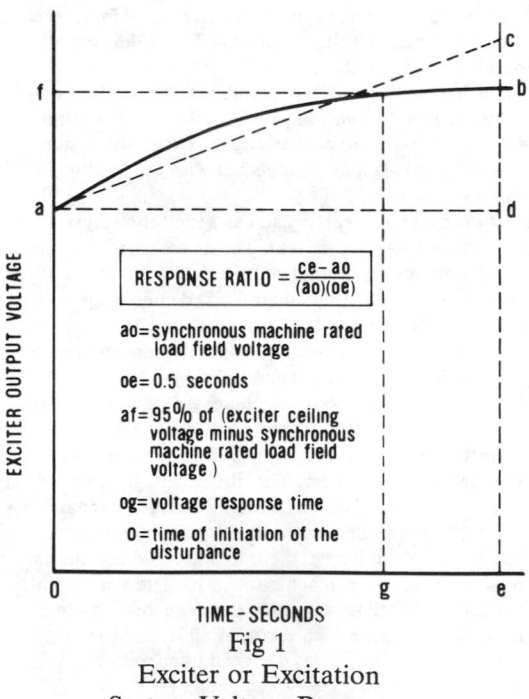

RESPONSE RATIO $= \dfrac{ce-ao}{(ao)(oe)}$

ao = synchronous machine rated load field voltage

oe = 0.5 seconds

af = 95% of (exciter ceiling voltage minus synchronous machine rated load field voltage)

og = voltage response time

O = time of initiation of the disturbance

EXCITER OUTPUT VOLTAGE

TIME - SECONDS

Fig 1
Exciter or Excitation
System Voltage Response

    374

(2)(synchronous machines). The rate of increase or decrease of the excitation system output voltage de-

termined from the excitation system voltage-time response curve, that if maintained constant would develop the same voltage-time area as obtained from the curve for a specified period. The starting point for determining the rate of voltage change is the initial value of the excitation system voltage-time response curve. *Notes:* (1) Similar definitions can be applied to the excitation system major components such as the exciter and regulator. (2) A system having an excitation system voltage response time of 0.1 second or less is defined as a high initial response excitation system.

    105

excitation system voltage response ratio (1)(excitation control system). The numerical value that is obtained when the excitation system voltage response, in volts per second measured over the first 1/2 S interval, unless otherwise specified, is divided by the rated load field voltage of the synchronous machine. *Notes:* (A) Referring to Fig 1, the excitation system response ratio, unless otherwise specified, applies apply only to the increase in excitation system voltage. (B) Response ratio is determined with the exciter voltage initially equal to the rated load field voltage of the synchronous machine to which the exciter is applied, and then suddenly establishing circuit conditions required to obtain nominal exciter ceiling voltage. Excitation system response ratio is determined by suddenly reducing the voltage sensed by the synchronous machine voltage regulator from 100 percent to 80 percent unless otherwise specified. (C) Unless otherwise specified, excitation system response ratio should be determined with the exciter loaded as specified in 3.2.2. If, for practical considerations, the test is performed at no load, analytical means may be utilized to predict the performance under load. (D) For excitation systems employing a rotating exciter, the response ratio should be determined at rated speed. (E) For potential-source rectifier excitation systems, the nature of a power system disturbance greatly affects the available power supply voltages. The ceiling voltage available and the voltage respose time are more meaningful parameters. To specify a response ratio implies equivalence with other systems whose output is not adversely affected by such depressed voltage conditions. Therefore, response ratio is not recommended as a specification parameter for these excitation systems. (F) For compound-rectifier excitation systems, the nature of the power system disturbance and the specific design prameters of the exciter and the synchronous generator influence the performance of the exciter output voltage. For equivalence with rotating exciters, the response ratio should be based on performance under specified reduced generator terminal voltage and increased generator stator current conditions as would be encountered during power system faults and disturbances.   374

(2)(synchronous machine). The numerical value that is obtained when the excitation system voltage response, in volts per second, measured over the first half-second interval, unless otherwise specified, is divided by the rated-load field voltage of the synchronous machine. *Note:* Unless otherwise specified, the excitation system voltage response ratio shall apply only to the increase in excitation system voltage.   105

excitation system voltage response time (excitation systems for synchronous machines)(excitation control system). The time in seconds for the excitation voltage to attain 95% of the difference between ceiling voltage and rated load field voltage under specific conditions.   507, 374, 105

excitation system voltage time response (excitation systems for synchronous machines). The excitation system output voltage expressed as a function of time, under specified conditions. *Note:* A similar definition can be applied to the excitation system major components, the exciter and regulator, separately.   507, 105

excitation voltage. The nominal voltage of the excitation circuit.   164

excitation winding (power and distribution transformer)(two-core regulating transformer). The winding of the main unit which draws power from the system to operate the two-core transformer. *See:* **field winding.**   53

excite (rotating machinery). To initiate or develop a magnetic field in (such as in an electric machine). *See:* **asynchronous machine.**   63

excited-field loudspeaker. A loudspeaker in which the steady magnetic field is produced by an electromagnet. *See:* **loudspeaker.**   328

excited-state maser (laser-maser). A maser in which the terminal level of the amplifying transition is not appreciably populated at thermal equilibrium for the ambient temperature.   363

excited winding (power and distribution transformer)(two-core regulating transformer). The winding of the series unit which is excited from the regulating winding of the main unit.   53

exciter (1)(excitation systems for synchronous machines). The equipment providing the field current for the excitation of a synchronous machine.   507

(2) (rotating machinery). The source of all or part of the field current for the excitation of an electric machine. *Note:* Familar sources include direct-current commutator machines; alternating-current generators whose output is rectified; and batteries. *See:* synchronous machine; direct-current commutating machine.   63,105

(3) (communications)(data transmission). In antenna practice, the portion of a transmitting array, (of the type which includes a reflector or director), which is directly connected with the source of power.   59

(4) (electric installations on shipboard). The source of all or part of the field current for the excitation of an electric machine.   3

exciter, alternator-rectifier (synchronous machines). An exciter whose energy is derived from an alternator and converted to direct current by rectifiers. *Notes:* (1) The exciter includes an alternator and power rectifiers which may be either noncontrolled or controlled, including gate circuitry. (2) It is exclusive of

input control elements. (3) The alternator may be driven by a motor, prime mover, or by the shaft of the synchronous machine. (4) The rectifiers may be stationary or rotating with the alternator shaft.   105

exciter-ceiling voltage (field discharge circuit breakers)(power switchgear)(rotating machinery). The maximum voltage that may be attained by an exciter under specified conditions.   412, 103

exciter ceiling voltage, nominal (rotating machinery). The ceiling voltage of an exciter loaded with a resistor having an ohmic value equal to the resistance of the field winding to be excited and with this field winding at a temperature of: (1) 75 degrees Celsius for field windings designed to operate at rating with temperature rise of 60 degrees Celsius or less. (2) 100 degrees Celsius for field windings designed to operate at rating with a temperature rise greater than 60 degrees Celsius.   328

exciter, compound-rectifier (synchronous machines). An exciter whose energy is derived from the currents and potentials of the alternating current terminals of the synchronous machine and converted to direct current by rectifiers. *Notes:* (1) The exciter includes the power transformers (current and potential), power reactors, and power rectifiers which may be either noncontrolled or controlled, including gate circuitry. (2) It is exclusive of input control elements.   105

exciter, direct current generator-commutator (synchronous machines). An exciter whose energy is derived from a direct current generator. *Notes:* (1) The exciter includes a direct current generator with its commutator and brushes. It is exclusive of input control elements. (2) The exciter may be driven by a motor, prime mover, or by the shaft of the synchronous machine.   105

exciter dome (rotating machinery). Exciter housing for a vertical machine.   63

exciter losses (synchronous machine). The total of the electric and mechanical losses in the equipment supply excitation.   244,63

exciter, main (synchronous machines). The source of all or part of the field current for the excitation of an electric machine, exclusive of another exciter.   105

exciter or direct-current (dc) generator relay (53) (power system device function numbers). A relay that forces the dc machine field excitation to build up during starting or which functions when the machine voltage has built up to a given value.   402

exciter, pilot (synchronous machines). The source of all or part of the field current for the excitation of another exciter.   105

exciter platform (rotating machinery). A deck on which to stand while inspecting the exciter.   63

exciter, potential source-rectifier (synchronous machines). An exciter whose energy is derived from a stationary alternating current potential source and converted to direct current by rectifiers. *Notes:* (1) The exciter includes the power potential transformers, where used, and power rectifiers which may be either noncontrolled or controlled, including gate circuitry. (2) It is exclusive of input control elements.   105

exciter response. *See:* **voltage response, exciter.**

exciter response ratio, main (synchronous machines). The numerical value obtained when the response, in volts per second, is divided by the rated-load field voltage, which response, if maintained constant, would develop, in one half-second, the same excitation voltage-time area as attained by the actual exciter. *Note:* The response is determined with no load on the exciter voltage initially equal to the the rated-load field voltage, and then suddenly establishing circuit conditions which would be used to obtain nominal exciter ceiling voltage. For a rotating exciter, response should be determined at rated speed. This definition does not apply to main exciters having one or more series field, except a light differential series field, or to electronic exciters. 105

exciter voltage response ratio (rotating machinery). *See:* **voltage response ratio.**

exciter voltage-time response (rotating machinery). *See:* **voltage-time response, synchronous-machine excitation system.**

excitron. A single-anode pool tube provided with means for maintaining a continuous cathode spot. 190

exclusion (mathematics of computing). A dyadic Boolean operator having the property that if P and Q is a statement, then the expression P exclusion Q is true if and only if P is true and Q is false. *Note:* P exclusion Q is often represented by a combination of AND and NOT symbols such as $P \wedge \sim Q$. 564

P	Q	$P \wedge \sim Q$
0	0	0
0	1	0
1	0	1
1	1	0

Exclusion truth table

exclusive OR (XOR)(mathematics of computing). A dyadic Boolean operator having the property that if P is a statement and Q is a statement, then P exclusive-OR Q is true if and only if either, but not both, is true. *Note:* P exclusive OR Q is often represented by $P \oplus$ Q or $P \not\vee Q$. *See:* **OR.** 564

P	Q	$P \oplus Q$
0	0	0
0	1	1
1	0	1
1	1	1

Exclusive OR truth table

excursion (computing system). *See:* **reference excursion.** 9,77

execute features (1) (Class 1E power systems). The electrical and mechanical equipment and interconnections that perform a function, associated directly or indirectly with a safety function, upon receipt of a signal from the sense and command features. The scope of the execute features extends from the sense and command features output to and including the actuated equipment-to-process coupling. 102
(2) (safety systems). The electrical and mechanical equipment and interconnections that perform a function, associated directly or indirectly with a safety function, upon receipt of a signal from the sense and command features. The scope of the execute features extends from the sense and command features output to and including the actuated equipment-to-process coupling. *Note:* In some instances protective actions may be performed by execute features that respond directly to the process conditions (for example, check valves, self-actuating relief valves). 428

execution (software). The process of carrying out an instruction or the instructions of a computer program by a computer. *See:* **computer program; instruction.** 434

execution time (software). (1) The amount of actual or central processor time used in executing a program. (2) The period of time during which a program is executing. *See:* **program; run time.** 434

execution time theory (software). A theory that uses cumulative execution time as the basis for estimating software reliability. *See:* **execution time; software reliability.** 434

executive program. *See:* **supervisory program.** 434

executive routine (computing systems). A routine that controls the execution of other routines. *See:* **supervisory routine.** 255,77,54

exercise (test, measurement and diagnostic equipment). To operate an equipment in such a manner that it performs all its intended functions to allow observation, testing, measurement and diagnosis of its operational condition. 54

exit (software). (1) Any instruction in a computer program, in a routine, or in a subroutine, after the execution of which control is no longer exercised by that computer program, that routine, or that subroutine. (2) The point beyond which control is no longer exercised by a routine. *See:* **computer program; instruction; routine; subroutine.** 434

expandability (1)(station control and data acquisition). The capability of a system to be increased in capacity or provided with additional functions. The measurement of expandability of equipments governed by this standard is the ease with which new points or functions, or both, can be added to the system, and the amount of downtime required to expand station equipments. Expandability categories are defined as follows: (A) spare point. Point equipment that is not being utilized but is fully wired and equipped (B) wired point. Point for which all common equipment, wiring and space are provided, but no plug-in point hardware is provided. (C) space only point. Point for

which cabinet space only is provided for future addition of wiring and other necessary equipments. Expandability limits may include but are not restricted to the following: (1) A limit for master or remote station point or memory capacity (addresses or size, or both) preventing the addition of more main memory or point equipment (2) A limit relating to the use of routines, addresses, labels, or buffers such that a modification reduces system capabilities (3) A data rate (for example, communication channel) limit such that the scan or polling cycle is extended when additions are made at the remote stations (4) Design and environmental limits on components (for example, analog to digital converters) such that equipment operation is compromised if the interface is modified or the device relocated.   403
(2)(supervisory control, data acquisition, and automatic control). The capability of a system to be increased in capacity or provided with additional functions.   570
expanded sweep. A sweep of the electron beam of a cathode-ray tube in which the movement of the beam is speeded up during a part of the sweep. *See:* **magnified sweep; radar.**   328
expander (data transmission). A transducer which, for a given amplitude range of input voltages, produces a larger range of output voltages. One important type of expander employs the envelope of speech signals to expand their volume range.   59
expandor (telephone switching systems). A switching entity for connecting a number of inlets to a greater number of outlets.   55
expansion (1) (modulation systems). A process in which the effective gain applied to a signal is varied as a function of the signal magnitude, the effective gain being greater for large than for small signals.   57
(2) (oscillograph). A decrease in the deflection factor, usually as the limits of the quality area are exceeded.   415
(3) (data transmission). A process in which the effective gain applied to a signal is varied as a function of the signal magnitude, the effective gain being greater for large than for small signals; (in a switching stage), a switching stage in which the number of inputs is smaller than the number of outputs.   59
expansion chamber (for an oil cutout) (power switchgear). A sealed chamber separately attachable to the vent opening to provide additional air space into which the gases developed during circuit interruption can expand and cool.   103, 443
expansion orbit (electronic device). The last part of the electron path that terminates at the target. It is outside the equilibrium orbit. *See:* **electron device.**   190
expected data (test pattern language). The binary data that is expected to be read out of a memory array. It is defined by the symbol "Q".   463
expected interruption duration (reliable industrial and commercial power systems planning and design). The expected, or average, duration of a single load interruption event.   561
expendable cap (of an expendable-cap cutout) (power

switchgear). A replacement part or assembly for clamping the button head of a fuse link and closing one end of the fuseholder. It includes a pressure-responsive section which opens to relieve the pressure within the fuseholder, when a predetermined value is exceeded during a circuit interruption.   103, 443
expendable-cap cutout (power switchgear). An open cutout having a fuse support designed for, and equipped with, a fuseholder having an expendable cap.   103, 443
experience or accepted test (insulation systems of synchronous machines). In accordance with IEEE Std 1-1969: "Experience," as used in this standard, means successful operation for a long time under actual operating conditions of machines designed with temperatures at or near the temperature limits. "Accepted test" as used in this standard means a test on a system or model system which simulates the electrical, thermal, and mechanical stresses occurring in service.   298
exploder. *See:* **blasting unit.**
explosionproof apparatus (1)(NESC). Apparatus enclosed in a case that is capable of withstanding an explosion of a specified gas or vapor which may occur within it and of preventing the ignition of a specified gas or vapor surrounding the enclosure by sparks, flashes, or explosion of the gas or vapor within, and which operates at such an external temperature that a surrounding flammable atmosphere will not be ignited thereby.   494
(2) (explosionproof) (mine apparatus). Apparatus capable of withstanding explosion tests as established by the United State Bureau of Mines, namely, internal explosions of methane-air mixtures, with or without coal dust present, without ignition of surrounding explosive methane-air mixtures and without damage to the enclosure or discharge of flame. *See:* **hazardous area groups; hazardous area classes; distribution center; luminaire.**   3,256,178,103,202,232,65
explosionproof enclosure (electric installations on shipboard). An enclosure designed and constructed to withstand an explosion of a specified gas or vapor which may occur within it, and to prevent the ignition of the specified gas or vapor surrounding the enclosure by sparks, flashes, or explosions of the specified gas or vapor which may occur within the enclosure. Explosionproof apparatus should bear Underwriters' Laboratories approval ratings of the proper class and group consonant with the spaces in which flammable volatile liquids, highly flammable gases, mixtures, or highly flammable substances may be present.   3
explosionproof fuse (power switchgear). A fuse, so constructed or protected, that for all current interruptions within its rating shall not be damaged nor transmit flame to the outside of the fuse.   103, 443
explosionproof luminaire (illuminating engineering). A luminaire which is completely enclosed and capable of withstanding an explosion of a specific gas or vapor which may occur within it, and preventing the ignition of a specific gas or vapor surrounding the enclosure by sparks, flashes or explosion of the gas or vapor within.

It must operate at such an external temperature that a surrounding flammable atmosphere will not be ignited thereby.'   167

explosion-tested equipment. Equipment in which the housings for the electric parts are designed to withstand internal explosions of methane-air mixtures without causing ignition of such mixtures surrounding the housings.   328

explosives (conductor stringing equipment). Mixtures of solids, liquids or a combination of the two which, upon detonation, transform almost instantaneously into other products which are mostly gaseous and which occupy much greater volume than the original mixtures. This transformation generates heat which rapidly expands the gases, causing them to exert enormous pressure. Dynamite and Primacord are explosives as manufactured. Aerex, Triex and Quadrex are manufactured in two components and are not true explosives until mixed. Explosives are commonly used to build construction roads, blast holes for anchors, structure footings, etcetera. *Syn:* **Aerex; dynamite; fertilizer; powder; Primacord; Quadrex; Triex.**
     431

exponent (1)(binary floating-point arithmetic). The component of a binary floating-point number that normally signifies the integer power to which two is raised in determining the value of the represented number. Occasionally the exponent is called the signed or unbiased exponent.   469

(2)(mathematics of computing). (A) A superscript indicating the number of times a number is to be used as a factor. (B) The component of a floating-point number that normally signifies the integer power to which the radix is raised in determining the value of the represented number. *Syn:* **characteristic.** *See:* **significand.**   564

(3)(radix-independent floating-point arithmetic). The component of a floating-point number that normally signifies the integer power to which the radix is raised in determining the value of the represented number. Occasionally, the exponent is called the signed or unbiased exponent.   588

exponential (ex) envelope (of a transient recovery voltage) (power switchgear). A voltage-versus-time curve of the general exponential form $e_1 = E_1[1 - ex(t/T)]$ in which e_1 represents the transient voltage across a switching device pole unit, reaching its crest at E_1 at infinite time. *Note:* In practice this envelope curve is derived from a circuit in which a voltage E_1 charges, by means of a switch, a circuit with inductance L in series with impedance Z and capacitance C in parallel. The voltage of e_1 is measured across Z •E_1 represents the ac driving or ceiling voltage which is considered at its peak at the time of a circuit zero and remains practically constant during that portion of the transient defined by the first curve. Hence, it can be considered as dc source during this time. The voltage application is simulated by the closing of the switch. e_1 represents the transient voltage across the circuit breaker pole unit. L represents the equivalent effective inductance on the source side of the circuit

breaker. Z represents the equivalent surge impedance of associated transmission lines. C represents the equivalent lumped capacitance on the source side of the breaker and modifies the exponential envelope by what may be considered as a slight initial time delay, $T_1 \cdot R$ is the transient recovery voltage rate, corresponding to the initial slope of the exponential envelope.   103

exponential-cosine (exponential minus cosine) (ex-cos) envelope (of a transient recovery voltage) (1) (power switchgear). A voltage-versus-time curve which represents the maximum at any time of the 1-cosine (1 minus cosine) envelope and the exponential envelope.   103

(2) (transient recovery voltage). The greater at any instant of: (A) The curve traced by the multiple exponential, transient voltage across Z when a switch is closed on the circuit shown below. It reaches its crest E_1 at $t = \infty$. (B). The 1-cosine curve with its initial crest at P

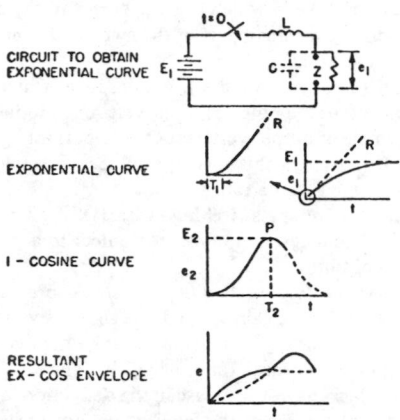

CIRCUIT TO OBTAIN EXPONENTIAL CURVE E_1

EXPONENTIAL CURVE

1 - COSINE CURVE

RESULTANT EX - COS ENVELOPE

E_1 represents the alternating current driving or ceiling voltage which is considered at its peak at the time of a current zero and remains practically constant during that portion of the transient defined by the first curve. Hence, it can be considered as a direct current source during this time. The voltage application is simulated by the closing of the switch. e represents the transient voltage across the circuit breaker pole unit. L represents the equivalent effective inductance on the source side of the circuit breaker. Z represents the equivalent surge impedance of associated transmission lines. C represents the equivalent lumped capacitance on the source side of the breaker and modifies the ex-cos envelope by what may be considered as a slight initial time delay, $T_1 \cdot R$ is the transient recovery voltage rate. Besides forming a basis of rating the above definition is also useful in discussing the changes of transient voltage caused by varying the parameters. *Note:* The ex-cos curve is the standard envelope for rating circuit breaker transient recovery voltage performance for circuit breakers rated 121 kV and above.

exponential function. One of the form $y = ae^{bx}$, where *a* and *b* are constants and may be real or complex. An exponential function has the property that its rate of change with respect to the independent variable is proportional to the function, or $dy/dx = by$.

210

exponential horn. A horn the cross-sectional area of which increases exponentially with axial distance. *Note:* If *S* is the area of a plane section normal to the axis of the horn at a distance *x* from the throat of the horn, S_0 is the area of a plane section normal to the axis of the horn at the throat, and *m* is a constant that determines the rate of taper or flare of the horn, then

$$S = S_0 e^{mx}$$

See: **loudspeaker.** 176

exponential lag. *See:* **lag (first order).**

exponentially damped sine function. A generalized sine function of the form $Ae^{-bx} \sin (x + a$ where $b > 0$. 210

exponential transmission line. A tapered transmission line whose characteristic impedance varies exponentially with electrical length along the line. *See:* **transmission line; waveguide.** 267

exposed (1)(NESC)(transmission and distribution). Not isolated or guarded. 494, 262

(2) (wiring methods). Not concealed. 328

(3) (communication circuits) The circuit is in such a position that in case of failure of supports or insulation, contact with another circuit may result.

256

exposed conductive surfaces (National Electrical Code) (health care facilities). Those surfaces which are capable of carrying electric current and which are unprotected, unenclosed or unguarded, permitting personal contact. Paint, anodizing and similar coatings are not considered suitable insulation, unless they are approved for the purpose. 256

exposed installation (lightning). An installation in which the apparatus is subject to overvoltages of atmospheric origin. *Note:* Such installations are usually connected to overhead transmission lines either directly or through a short length of cable. *See:* **surge arrester (surge diverter).** 244,62

exposed lamp. *See:* **bare lamp.** 167

exposure (laser-maser). The product of an irradiance and its duration. 363

exposure fire (Class 1E equipment and circuits). Fire initiated by other than electrical means or supported by fuel other than cable insulation. 131

exposure time (reliable industrial and commercial power systems planning and design). The time during which a component is performing its intended function and is subject to failure. 561

expression. An ordered set of one or more characters.
235,77

expulsion arrester (surge arresters). An arrester that includes an expulsion element. 430

expulsion fuse or fuse unit (power switchgear)(high voltage switchgear). A vented fuse or fuse unit in which the expulsion effect of gases produced by the

arc and lining of the fuseholder, either alone or aided by a spring, extinguishes the arc. 103, 443

expulsion fuse unit (expulsion fuse) (National Electrical Code) (installations and equipment operating at over 600 volts, nominal). A vented fuse unit in which the expulsion effect of gases produced by the arc and lining of the fuse-holder, either alone or aided by a spring, extinguishes the arc. 256

expulsion-type surge arrester (expulsion-type arrester). *See:* **arrester, expulsion-type.**

extended capability (power operations). The generating capability increment in excess of dependable capability which can be obtained under emergency operating procedures. 516

extended delta connection (power and distribution transformer). A connection similar to a delta, but with a winding extension at each corner of the delta, each of which is 120 degrees apart in phase relationship. *Note:* The connection may be used as an autotransformer to obtain a voltage change or a phase shift, or a combination of both. 53

extended planned derating (electric generating unit reliability, availability, and productivity). The planned derating that is the extension of the basic planned derating beyond its predetermined duration.
567

extended planned outage (electric generating unit reliability, availability, and productivity). The planned outage state that is the extension of the basic planned outage beyond its predetermined duration. *Note:* Extended planned outage applies only when planned work exceed predetermined duration. The extension, due to a condition discovered during the planned outage, is to be classified as Class 1 unplanned outage (see Class 1 unplanned outage [immediate]). Start-up failure would result in Class 0 unplanned outage (see Class 0 unplanned outage [starting failure]). 567

extended return to bias (magnetic tape pulse recorders for electricity meters). A method whereby a recording head current, which results in a magnetic field polarity opposite that of the bias magnet, is applied to the magbnetic tape for a portion of the interval in order to record a pulse. 551

extended segment (FASTBUS acquisition and control). A multiplicity of crate segments accessed by the same group address. Unlike operations on segments linked by segment interconnects, independent operations on each of the segments that are part of an extended segment never proceed concurrently. Depending on the method of implementation, some restrictions may exist as to the placement of masters. Depending on the disposition of modules on the extended segment, some broadcast operations may not be usable or may require special interpretation.
480

extended-service area (telephone switching systems). That part of the local service area that is outside of the boundaries of the exchange area of the calling customer. 55

extended source (laser-maser). An extended source of radiation can be resolved by the eye into a geometrical image, in contrast to a point source of radiation which cannot be resolved into a geometrical image. *See:* **8.1 of ANSI Z136.1-1976 for criteria.**   363

extended-time rating (grounding device). A rated time in which the period of time is greater than the time required for the temperature rise to become constant but is limited to a specified average number of days operation per year.   91

extension (pascal computer programming language). A modification to Section 6 of ANSI/IEEE 770X3.97-1983 that does not invalidate any program complying with this standard, as defined by Section 5.2, except by prohibiting the use of one or more particular spellings of identifiers.   433

extension cord. An assembly of a flexible cord with an attachment plug on one end and a cord connector on the other.   328

extension station. A telephone station associated with a main station through connection to the same subscriber line and having the same call number designation as the associated main station. *See:* **telephone station.**   328

external address *(ext)* (subroutines for CAMAC). The symbol *ext* represents an integer which is used an identifier on an external CAMAC address. The address may represent a register which can be read or written, a complete CAMAC address which can be accessed by control or test functions, or a crate address. The value of *ext* is explicitly defined to be an integer. Normally it can be expected to be an encoded version of the address components, in which the coding has been selected for the most efficient execution of CAMAC actions on the interface to which the implementation applies. Other possibilities are allowed, however. For example, *ext* may be an index or a point into a data structure in which the actual CAMAC address components are stored.   410

external addresses (extb) (subroutines for CAMAC). The symbol *extb* represents an array of integers containing external CAMAC addresses. The array has two elements: (1) The starting address for an Address Scan multiple action; (2) The final address which can be permitted to participate in the Address Scan sequence. Each element has the same form and information content as the parameter *ext*. *See:* **external address.**   410

external audit (nuclear power quality assurance). An audit of those portions of another organization's quality assurance program not under the direct control or within the organizational structure of the auditing organization.   417

external connector (aerial lug). A connector that joins the external conductor to the other current-carrying parts of a cable termination.   4

external field influence (electric instrument). The percentage change (of full-scale value) in indication caused solely by a specified external field. Such a field is produced by a standard method with a current of the same kind and frequency as that which actuates the mechanism. This influence is determined with the most unfavorable phase and position of the field in

relation to the instrument. *Note:* The coil used in the standard method shall be approximately 40 inches in diameter not over 5 inches long, and carrying sufficient current to produce the required field. the current to produce a field to an accuracy of ± 1 percent in air shall be calculated without the instrument in terms of the specific dimensions and turns of the coil. In this coil, 400 ampere-turns will produce a field of approximately 5 oersteds. The instrument under test shall be placed in the center of the coil. *See:* **accuracy rating (instrument).**   280,294

external insulation (1) (apparatus) (power and distribution transformer) (surge arrestors). The external insulating surfaces and the surrounding air. *Note:* The dielectric strength of external insulation is dependent on atmospheric conditions.   62,103,53

(2) (power switchgear) (outdoor ac high-voltage circuit breakers). Insulation that is designed for use outside of buildings and for exposure to the weather.   103

(3) (high voltage testing). The air insulation and the exposed surfaces of solid insulation of equipment, which are subject both to dielectric stresses and to the effects of atmospheric and other external conditions such as contamination, humidity, vermin, etcetera.   150

external logic state (graphic symbols for logic functions). A logic state assumed to exist outside a symbol outline (1) on an input line prior to any external qualifying symbol at that input, or (2) on an output line beyond any external qualifying symbol at that output.   451

externally commutated converter (self-commutated converters). A converter in which the commutating voltages are supplied by the ac supply lines, the ac load, or some other ac source outside the converter. *Note:* Externally commutated converters are excluded from the scope of ANSI/IEEE Std 936-1987, except where they may be linked with a self-commutated converter.   584

externally commutated inverters. An inverter in which the means of commutation is not included within the power inverter. *See:* **self-commutated inverters.**   208

externally operable (National Electrical Code). Capable of being operated without exposing the operator to contact with live parts.   256

externally programmed automatic test equipment (test, measurement and diagnostic equipment). An automatic tester using any programming technique in which the programming instructions are not read directly from within the ATE (automatic test equipment), but from a medium which is added to the equipment such as punched tape, punched cards, and magnetic tape.   54

externally quenched counter tube. A radiation counter tube that requires the use of an external quenching circuit to inhibit reignition.   190

externally ventilated machine (rotating machinery). A machine that is ventilated by means of a separate motor-driven blower. The blower is usually mounted on the machine enclosure but may be separately mounted on the foundation for large machines. *See:* **open-pipe ventilated machine; separately ventilated machine.** 63

external remanent residual voltage (Hall-effect devices). That portion of the zero field residual voltage which is due to remanent magnetic flux density in the external electromagnetic core. 107

external series gap (expulsion-type arrester). An intentional gap between spaced electrodes, in series with the gap or gaps in the arcing chamber. 308,62

external storage (test, measurement and diagnostic equipment). Information storage off-line in media such as magnetic tape, punched tape, and punched cards. 54

external temperature influence (direct current instrument shunts). The percentage change in the output voltage of a shunt (expressed in terms of rated output and measured with low current) when the ambient temperature is changed from 25 °C to 100 °C. 527

external termination (*j*th terminal of an *n*-terminal network). The passive or active two-terminal network that is attached externally between the *j*th terminal and the reference point. *See:* **electron-tube admittances.** 190,125

extinction voltage (gas tube). The anode voltage at which the discharge ceases when the supply voltage is decreasing. 244,190

extinguishing voltage (drop-out voltage) (glow lamp). Dependent upon the impedance in series with the lamp, the voltage across the lamp at which an abrupt decrease in current between operating electrodes occurs and is accompanied by the disappearance of the negative glow. *Note:* In recording or specifying extinguishing voltage, the impedance must be specified. 283

extraband spurious transmitter output (land-mobile communication transmitters). Spurious output of a transmitter outside of its specified band of transmission. 444

extracameral (radiation protection) (radiological monitoring instrumentation). Pertaining to that portion of the instrument exclusive of the detector. 398, 399

extracameral effect (monitoring radioactivity in effluents). Apparent response of an instrument caused by radiation on any other portion of the system than the detector. 559

extracameral response (plutonium monitoring). An instrument response arising from the action of the radiation field on parts of the instrument other than the intended radiosensitive element. 413

extract (electronic computation). To form a new word by juxtaposing selected segments of given words. 210

extract instruction (electronic digital computation). An instruction that requests the formation of a new expression from selected parts of given expressions. 235,255,77

extraction liquor. The solvent used in hydrometallurgi-

cal processes for extraction of the desired constituents from ores or other products. *See:* **electrowinning.** 328

extragalactic radio waves (radio wave propagation). Radio waves from beyond our galaxy. 146

extra high voltage (ehv) (power operations). A term applied to voltage levels which are higher than 230 000 volts (V). 516

extra-high voltage aluminum-sheathed power cable (aluminum sheaths for power cables). Cable used in an electric system having a maximum phase-to-phase root-mean-square (rms) alternating-current (ac) voltage above 242 000 volts (V), the cable having an aluminum sheath as a major component in the construction. 406

extra-high-voltage system (electric power). An electric system having a maximum root-mean-square alternating-current voltage above 240 000 volts to 800 000 volts. *See:* **low-voltage system; medium voltage system; high voltage system.** 49

extraordinary load (composite insulators). The ice or wind load, or both, that may last for as long as one week, recurring as often as once per year. 483

extraordinary wave (radio wave propagation). The magneto-ionic wave component in which the electric vector rotates in the opposite sense to that for the ordinary-wave component. *See:* **ordinary-wave component; radiation.** 146

extraordinary-wave component (radio wave propagation). The magnetoionic wave component in which the electric vector rotates in the opposite sense to that for the ordinary component. *Syn:* **X wave.** *See:* **ordinary-wave component.** 146

extrapolated range for electrons (solar cells). The distance of travel in a material by electrons of a given energy, at which the flux of primary electrons extrapolates to zero. 113

extreme load (composite insulators). The greatest load to occur on the line in a 50-year period. It may last as long as one day. 483

extreme operating conditions (automatic null-balancing electrical instrument). The range of operating conditions within which a device is designed to operate and under which operating influences are usually stated. *See:* **measurement system.** 295

extrinsic joint loss (fiber optics). That portion of joint loss that is not intrinsic to the fibers (that is, loss caused by imperfect jointing). *See:* **angular misalignment loss; gap loss; intrinsic joint loss; lateral offset loss.** 433

extrinsic properties (semiconductor). The properties of a semiconductor as modified by impurities or imperfections within the crystal. *See:* **semiconductor; semiconductor device.** 245

extrinsic semiconductor. A semiconductor whose charge-carrier concentration is dependent upon impurities. *See:* **semiconductor.** 237,66

eye bolt (rotating machinery). A bolt with a looped head used to engage a lifting hook. 63

eyelet (soldered connections). A hollow tube inserted

in a printed circuit or terminal board to provide electric connection or mechanical support for component leads. 284

eye light (illuminating engineering). Illumination on a person to provide a specular reflection from the eyes (and teeth) without adding a significant increase in light on the subject. 167

F

FA. *See:* **transformer, oil-immersed.**

fabric (rotating machinery). A planar structure comprising two or more sets of fiber yarns interlaced in such a way that the elements pass each other essentially at right angles and one set of elements is parallel to the fabric axis. 63

faceplate (cathode-ray tube). The large transparent end of the envelope through which the image is viewed or projected. *See:* **cathode-ray tubes.** 244,190

faceplate controller (industrial control). An electric controller consisting of a resistor and a faceplate switch in which the electric contacts are made between flat segments, arranged on a plane surface, and a contact arm. *See:* **electric controller.** 206

faceplate rheostat (industrial control). A rheostat consisting of a tapped resistor and a panel with fixed contact members connected to the taps, and a lever carrying a contact rider over the fixed members for adjustment of the resistance. 206

facilitation. The brief rise of excitability above normal either after a response of after a series of subthreshold stimuli. *See:* **biological.** 192

facilities charge (power operations). The amount paid by the customer as lump sum, or periodically, as reimbursement for facilities furnished. The charge may include operation and maintenance as well as fixed costs. 516

facility, communication. *See:* **communication facility.** 59

facing (planar structure) (rotating machinery). A fabric, mat, film, or other material used in intimate conjunction with a prime material and forming a relatively minor part of the composite for the purpose of protection, handling, or processing. *See:* **asynchronous machine.** 63

facsimile (fax) (electrical communication) (data transmission). The process, or the result of the process, by which fixed graphic material--including pictures or images--is scanned and the information converted into signal waves which are used either locally or remotely to produce, in record form, a likeness (facsimile) of the subject copy. 59

facsimile signal (picture signal). A signal resulting from the scanning process. 12

facsimile-signal level. The maximum facsimile signal power or voltage (root-mean-square or direct-current) measured at any point in a facsimile system. *Note:* It may be expressed in decibels with respect to some

eye pattern (data transmission). An oscilloscope display of the detector voltage waveform in a data modem. This pattern gives a convenient representation of cross over distortion which is indicated by a closing of the center of the eye. 59

standard value such as 1 milliwatt. *See:* **facsimile signal (picture signal).** 12

facsimile system. An integrated assembly of the elements used for facsimile. *See:* **facsimile (electrical communication).** 12

facsimile transient. A damped oscillatory transient occurring in the output of the system as a result of a sudden change in input. *See:* **facsimile transmission.** 12

facsimile transmission. The transmission of signal waves produced by the scanning of fixed graphic material, including pictures, for reproduction in record form. 111,12

factor of assurance (wire or cable insulation). The ratio of the voltage at which completed lengths are tested to that at which they are used. 64

factor of influence (thermal classification of electric equipment and electrical insulation). A specific physical stress imposed by operation, environment, or test that influences the performance of an insulating material, insulation system, or electric equipment. 506

factory fabricated (electrical heat tracing for industrial applications). A heating cable assembled by the manufacturer, including hot- and cold-end terminations and cold lead. 523

factory renewable fuse unit (power switchgear). A fuse unit that after circuit interruption must be returned to the manufacturer to be restored for service. 103, 443

fade in. To increase signal strength gradually in a sound or television channel. 328

fade out. To decrease signal strength gradually in a sound or television channel. 328

fading (1) (data transmission). (A) (flat). That type of fading in which all frequency components of the received radio signal fluctuate in the same proportions simultaneously. (B) (radio). The variation of radio field intensity caused by changes in the transmission medium, and transmission path, with time. *See:* **selective fading.** 59

(2) (radio wave propagation). The variation of radio field strength caused by changes in the transmission path with time. 146

fail safe (reliability). A designed property of an item which prevents its failures being critical failures. 164

fail-safe* (low-voltage air gap surge-protective devices). *Deprecated 556

failure(1) (raceway)(raceway systems for Class 1E circuits for nuclear power generating stations). The termination of the ability of the raceway system to perform its function. The level of damage done to the raceway system is such that either collapse is imminent or an electrical circuit is interrupted or degraded to an unacceptable level, or both.   513

(2)(reliability data). The termination of the ability of an item or equipment to perform its required function. *Note:* Failures may be unannounced and not detected until the next test or demand (unannounced failure), or they may be announced and detected by any number of methods at the instant of occurrence (announced failure).   447

(3)(reliability data for pumps and drivers, valve actuators, and valves). A subset of a fault and represents an irreversible state of a component such that it must be repaired in order for it to provide its design function. A component failure is generally defined in terms of the system in which it resides. For example, any leak might be considered a failure in a system where fission products are to be contained, and yet leaks may be considered as normal or even required states of other systems (for example, pump packing gland leakage). Failures are sometimes classified as either primary or secondary. (A) A primary failure is the so-called random failure found in literature. It results from no external cause. (B) A secondary failure results when the component is subject to conditions which exceed its design envelope (for example, excessive voltage, pressure, shock, vibration, temperature).   502

(4)(reliability data for pumps and drivers, valve actuators, and valves). The termination of the ability of an item or equipment to perform its required function. *Note:* Failures may be unannounced and not detected until the next test or demand (unannounced failure), or they may be announced and detected by any number of methods at the instant of occurrence (announced failure).   502

(5)(reliable industrial and commercial power systems planning and design). Any trouble with a power system component that causes any of the following to occur: (A) Partial or complete plant shutdown, or below-standard plant operation; (B) Unacceptable performance of user's equipment; (C) Operation of the electrical relaying or emergency operation of the plant electrical system; (D) Deenergization of any electric circuit or equipment. A failure on a public utility supply system may cause the user to have either of the following: (1) A power interruption or loss of service; (2) A deviation from normal voltage or frequency of sufficient magnitude or duration. A failure on an in-plant component causes a forced outage of the component, that is, the component is unable to perform its intended function until it is repaired or replaced. *Syn:* **forced outage.**   561

(6)(safety systems equipment in nuclear power generating stations). The loss of ability of a component, equipment or system to perform a required function.   535

(7)(supervisory control, data acquisition, and automatic control). An event that may limit the capability of an equipment or system to perform its function(s). (A) Critical. Causes a false or undesired operation of apparatus under control. (B) Major. Loss of control or apparatus which does not involve a false operation. (C) Minor. Loss of data relative to power flow or equipment status.   570

(8) (independent) (test, measurement and diagnostic equipment). A failure which occurs without being related to the failure of associated items, distinguished from dependent failure.   54

(9) (nuclear power generating stations). The termination of the ability of an item to perform its required function. Failures may be unannounced and not detected until the next test (unannounced failure), or they may be announced and detected by any number of methods at the instant of occurrence (announced failure).   29

(10) (software). (A) The termination of the ability of a functional unit to perform its required function. (B) The inability of a system or system component to perform a required function within specified limits. A failure may be produced when a fault is encountered. (C) A departure of program operation from program requirements. *See:* **fault; function; functional unit; system.**   434

(11) (station control and data acquisition). An event that may limit the capability of an equipment or system to perform its function(s). (A) Critical. Causes a false or undesired operation of apparatus under control. (B) Major. Loss of control or apparatus which does not involve a false operation. (C) Minor. Loss of data relative to power flow or equipment status.   403

(12) (nuclear power generating stations). The termination of the ability of an item to perform its required function.   102

failure analysis (test, measurement and diagnostic equipment). The logical, systematic examination of an item or its diagram(s) to identify and analyze the probability, causes, and consequences of potential and real failures.   54

failure category. *See:* **error category.**

failure cause (reliability). The circumstances during design, manufacture or use which have led to failure.   164

failure commutation (thyristor converter). A failure to commutate the direct current from the conducting arm to the succeeding arm of a thyristor connection. *Note:* In inverter operation, a commutation failure results in a conduction-through.   121

failure criteria (reliability). Rules for failure relevancy such as specified limits for the acceptability of an item. *See:* **reliability.**   182

failure data. *See:* **error data.**

failure distribution (1)(station control and data acquisition). Is the manner in which failures occur as a function of time; generally expressed in the form of a curve with the abscissa being time.   403

(2)(supervisory control, data acquisition, and auto-

matic control). The manner in which failures occur as a function of time; generally expressed in the form of a curve with the abscissa being time.   570

failure mechanism (reliability data for pumps and drivers, valve actuators, and valves)(reliabilty data). The physical. chemical, or other process which results in failure. *Note:* The circumstance that induces or activates the process is termed the root cause of the failure.   502, 447

failure mode (reliability data for pumps and drivers, valve actuators, and valves)(reliability data). The effect by which a failure is observed to occur.
                502, 447

failure modes and effects analysis (FMEA) (class 1E static battery chargers and inverters). The identification of significant failures, irrespective of cause, and their consequences. This includes electrical and mechanical failures which could conceivably occur under specified service conditions and their effect, if any, on adjoining circuitry or mechanical interfaces displayed in a table, chart, fault tree or other format.   408

failure mode types (reliability data for pumps and drivers, valve actuators, and valves)(reliability data). (1) Catastrophic. A failure mode which is both sudden and complete. *Note:* This failure causes cessation of one or more fundamental functions. This refers to system related failure modes. See Appendix A of ANSI/IEEE Std 500-1984 P&V. (2) Degraded. A failure which is gradual, partial, or both. *Note:* Such a failure does not cease all function but compromises a function. The function may be compromised by any combination of reduced, increased, or erratic outputs. In time, such a failure may develop into catastrophic failure. (3) Incipient. An imperfection in the state or condition of an item or equipment so that a degraded or catastrophic failure can be expected to result if corrective action is not taken.   502, 447

failure rate (forced outage rate) (1)(reliable industrial and commercial power systems planning and design). The mean number of failures of a component per unit exposure time. Usually exposure time is expressed in years and failure rate is given in failures per year.
                561

(2)(reliability data for pumps and drivers, valve actuators, and valves)(reliability data). The expected number of failures of a given type, per item, in a given time interval (for example, valve failures per million valve hours). *Note:* For cyclic items or equipment insert 'in a given number of operating cycles.'
                502, 447

(3) (nuclear power generating stations). The expected number of failures of a given type, per item, in a given time interval (for example, capacitor short-circuit failures per million capacitor hours).   29

(4) (software). (A) The ratio of the number of failures to a given unit of measure, for example, failures per unit of time, failures per number of transactions, failures per number of computer runs. (B) In reliability modeling, the ratio of the number of failures of a given category or severity to a given period of time, for example, failures per second of execution time, fail-

ures per month. *Syn:* **failure ratio.** *See:* **computer; execution time; failure; reliability.**   434

failure-rate acceleration factor (reliability). The ratio of the accelerated testing failure rate to the failure rate under stated reference test conditions and time period. *See:* **reliability.**   182

failure rate, assessed (reliability). The failure rate of an item determined by a limiting value or values of the confidence interval associated with a stated confidence level, based on the same data as the observed failure rate of nominally identical items. *Note:* (1) The source of the data shall be stated. (2) Results can be accumulated (combined) only when all conditions are similar. (3) The assumed underlying distribution of failures against time shall be stated. (4) It should be stated whether a one-sided or a two-sided interval is being used. (5) Where one limiting value is given this is usually the upper limit.   164

failure rate, extrapolated (reliability). Extension by a defined extrapolation or interpolation of the observed or assessed failure rate for durations and/or conditions different from those applying to the observed or assessed failure rate. *Note:* The validity of the extrapolation shall be justified.   164

failure rate, instantaneous. *See:* **instantaneous failure rate.**

failure rate level (reliability). For the assessed failure rate, a value chosen from a specific series of failure rate values and used for stating requirements or for the presentation of test results. *Note:* In a requirement, it denotes the highest permissible assessed failure rate.
                164

failure rate, observed (reliability). For a stated period in the life of an item, the ratio of the total number of failures in a sample to the cumulative observed time on that sample. The observed failure rate is to be associated with particular, and stated time intervals (or summation of intervals) in the life of the items, and with stated conditions. *Note:* (1) The criteria for what constitutes a failure shall be stated. (2) Cumulative time is the sum of the times during which each individual item has been performing its required function under stated conditions.   164

failure rate, predicted (reliability). For the stated conditions of use, and taking into account the design of an item, the failure rate computed from the observed, assessed, or extrapolated failure rates of its parts. *Note:* Engineering and statistical assumptions shall be stated, as well as the bases used for the computation (observed or assessed).   164

failure ratio. *See:* **failure rate.**

failure recovery (software). The return of a system to a reliable operating state after failure. *See:* **failure; system.**   434

failures (supervisory control, data acquisition, and automatic control). (1) Infant mortality. A characteristic pattern of failure, sometimes experienced with new equipment which may contain marginal components, wherein the number of failures decreases rapidly as the number of operating hours increases. A burn-in period period may be utilized to age (mature) an equipment to reduce the number of marginal components. (2) Random. The pattern of failures for equipment that has passed out of its infant mortality period and has not reached the wear-out phase of its operating lifetime. The reliability of an equipment in this period may be computed by the equation

$$R = e^{-\lambda t}$$

where

λ = failure rate

t = time period of interest

(3) Wear out. The pattern of failures experienced when equipment reaches its period of deterioration. Wear-out failure profiles may be approximated by a Gaussian (bell curve) distribution centered on the nominal life of the equipment.   570

fall time (last transition duration), tf,(pulse transformers). The time interval of the pulse trailing edge between the instants at which the instantaneous value first reaches specified upper and lower limits of 90% and 10% of AT.   589

failure to trip (1) (power switchgear). In the performance of a relay or relay system, the lack of tripping which should have occurred considering the objectives of the relay system design.   103

(2) (relay or relay system). In the performance the lack of tripping that should have occurred, considering the objectives of the relay system design. See: **relay.**
                                127

fairlead (aircraft). A tube through which a trailing wire antenna is fed from an aircraft, with particular care in the design as to voltage breakdown and corona characteristics. *Note:* An antenna reel and counter are frequently a part of the assembly.   328

fall time (high- and low-power pulse transformers) (last transition duration) (Tf). The time interval of the pulse trailing edge between the instants at which the instantaneous value first reaches specified upper and lower limits of 90 percent and 10 percent of AT. See: **input pulse shape.**   32,33

fall time of a pulse. The time interval of the trailing edge of a pulse between stated limits. See: **pulse.**
                                185

false alarm (1) (radar). An erroneous radar target detection decision caused by noise or other interfering signals exceeding the detection threshold.   13

(2) (test, measurement and diagnostic equipment). An indicated fault where no fault exists.   54

false-alarm number (radar). The number of possible independent detection decisions during the false-alarm time.   13

false-alarm probability (radar). The probability that noise or other interfering signals will erroneously cause a target detection decision.   13

false-alarm time (radar). The average time between false alarms; that is, the average time between crossings of the target decision threshold by signals not representing targets. In the early work of Marcum, it is the time in which the probability of one or more

false alarms is one-half, but the usage is now deprecated.   13

false course (navigation systems normally providing one or more course lines)(navigation aid terms). A spurious additional course line indication due to undesired reflections or to a maladjustment of equipment.   526

false-proceed operation. The creation or continuance of a condition of the vehicle apparatus in an automatic train control or cab signal installation that is less restrictive than is required by the condition of the track of the controlling section, when the vehicle is at a point where the apparatus, is or should be, in operative relation with the controlling track elements. See: **automatic train control.**   328

false-restrictive operation. The creation or continuance of a condition of the automatic train control or cab signal vehicle apparatus that is more restrictive than is required by the condition of the track of the controlling section when the vehicle apparatus is in operative relation with the controlling track elements, or which is caused by failure or derangement of some part of the apparatus. See: **automatic train control.**
                                328

false tripping (1) (power switchgear). In the performance of a relay or relay system, the tripping which should not have occurred considering the objectives of the relay system design.   103

(2) (relay or relay system). In the performance the tripping that should not have occurred, considering the objectives of the relay system design. See: **relay.**
                                127,103

fan (blower) (rotating machinery). The part that provides an air stream for ventilating the machine.
                                63

fan-beam antenna. An antenna producing a major lobe whose transverse cross section has a large ratio of major to minor dimensions.   111

fan cover (rotating machinery). An enclosure for the fan that directs the flow of air. See: **fan (rotating machinery).**   63

fan duty resistor (electric installations on shipboard). A resistor for use in the armature or rotor circuit of a motor in which the current is approximately proportional to the speed of the motor.   3

fan housing (rotating machinery). The structure surrounding a fan and which forms the outer boundary of the coolant gas passing through the fan.   63

fan-in network (power-system communication). A logic network whose output is a binary code in parallel form of n bits and having up to 2^n inputs with each input producing one of the output codes. See: **digital.**
                                59

fan marker (navigation aid terms). A vhf (very high frequency) radio facility having a vertically-directed fan beam intersecting an airway to provide a fix.
                                526

fan-marker beacon (navigation aid terms). A beacon that transmits vertical beam-horizontal cross section in the shape of a double convex lens.   526

fan shroud (rotating machinery). A structure, either

stationary or rotating, that restricts leakage of gas past the blades of a fan.   63

farad (metric practice). The capacitance of a capacitor between the plates of which there appears a difference of potential of one volt when it is charged by a quantity of electricity equal to one coulomb.   21

faraday. The number of coulombs (96 485) required for an electrochemical reaction involving one chemical equivalent. *See:* **electrochemistry.**   328

Faraday cell (laser gyro). A biasing device consisting of an optical material with a Verdet constant, such as quartz, which is placed between two quarter wave plates and surrounded by a magnetic field in such a fashion that a differential phase change is produced for oppositely directed plane polarized waves.   46

Faraday dark space (gas tube). The relatively nonluminous region in a glow-discharge cold-cathode tube between the negative flow and the positive column. *See:* **gas tube.**   190

Faraday effect. *See:* **Faraday rotation.**

Faraday rotation (1)(nonreciprocal wave rotation)(nonlinear, active, and nonreciprocal waveguide components). A nonreciprocal phenomenon in which the plane of polarization of a linearly polarized electromagnetic plane is rotated clockwise for one direction of propagation, and counterclockwise for the other direction (viewed from the source in each direction), when passing through a gyromagnetic material having a magnetostatic field component along the direction of propagation.   530

(2) (radio wave propagation). The process of rotation of the polarization ellipse of an electromagnetic wave as it propagates in a magneto-ionic medium. *See:* **elliptically polarized wave.**   146

Faraday rotator (nonreciprocal wave rotator)(nonlinear, active, and nonreciprocal waveguide components). A nonreciprocal device providing Faraday rotation, usually in waveguide of circular or square cross section.   530

Faraday's law (electromagnetic induction; circuit). The electromotive force induced is proportional to the time rate of change of magnetic flux linked with the circuit.   210

faradic current (electrotherapy). An asymmetrical alternating current obtained from or similar to that obtained from the secondary winding of an induction coil operated by repeatedly interrupting a direct current in the primary. *See:* **electrotherapy.**   192

faradization (faradism) (electrotherapy). The use of a faradic current to stimulate muscles and nerves. *See:* **faradic current; electrotherapy.**   192

far-end crosstalk. Crosstalk that is propagated in a disturbed channel in the same direction as the direction of propagation of the current in the disturbing channel. The terminal of the disturbed channel at which the far-end crosstalk is present and the energized terminals of the disturbing channel are ordinarily remote from each other. *See:* **coupling.**   328

far-field diffraction pattern (fiber optics). The diffraction pattern of a source (such as a light emitting diode (LED), injection laser diode (ILD), or the output end

of an optical waveguide) observed at an infinite distance from the source. Theoretically, a far-field pattern exists at distances that are large compared with $[s^2]/\lambda$, where s is a characteristic dimension of the source and λ is the wavelength. Example: If the source is a uniformly illuminated circle, then s is the radius of the circle. *Note:* The far-field diffraction pattern of a source may be observed at infinity or (except for scale) in the focal plane of a well-corrected lens. The far-field pattern of a diffracting screen illuminated by a point source may be observed in the image plane of the source. *Syn:* **Fraunhofer diffraction pattern.** *See:* **diffraction; diffraction limited.**   433

far-field pattern (1) (fiber optics). *See:* **far-field radiation pattern.**

(2) (radiation) pattern (antennas). Any radiation pattern obtained in the far field of an antenna. *Note:* Far-field patterns are usually taken over paths on a spherical surface. *See:* **radiation sphere; radiation pattern cut.**   111

far-field radiation pattern. *See:* **radiation pattern.**   433

far-field region (1) (antennas). That region of the field of an antenna where the angular field distribution is essentially independent of the distance from a specified point in the antenna region. *Notes:* (A) In the free space, if the antenna has a maximum overall dimension, D, which is large compared to the wavelength, the far-field region is commonly taken to exist at distances greater than $2D^2/\lambda$ from the antenna, λ being the wavelength. The far-field pattern of certain antennas, such as multibeam reflector antennas, are sensitive to variations in phase over their apertures. For these antennas $2d^2/\lambda$ may be inadequate. (B) In physical media, if the antenna has a maximum overall dimension, D, which is large compared to π/γ, the far-field region can be taken to begin approximately at a distance equal to $\gamma D^2/\pi$ from the antenna, γ being the propagation constant in the medium.   111

(2) (fiber optics). The region, far from a source, where the diffraction pattern is substantially the same as that at infinity. *See:* **far-field diffraction pattern.**   433

(3) (land-mobile communication transmitters). The region of the field of an antenna where the angular field distribution is essentially independent of the distance from the antenna. *Notes:* (1) If the antenna has a maximum overall dimension (D) that is large compared to the wavelength (λ), the far field region is commonly taken to exist at distances greater than $2D_2/\lambda$ from the antenna. (2) For an antenna focused at infinity, the far field region is sometimes referred to as the Fraunhofer region on the basis of analogy to optical terminology.   444

far-field region in physical media (antennas). *See:* **far-field region, Note (B).**   111

far field region, radiating. *See:* **radiating far field region**

FASTBUS (acquisition and control). *Syn:* **primary address cycle.**   480

FASTBUS (FASTBUS acquisition and control). The standard modular high-speed data acquisition and

control systemdefined by ANSI/IEEE Std 960-1986.
480

FASTBUS protocol (FBP)(FASTBUS acquisition and control). The format and sequence of control and data messages in FASTBUS. Formats are specified by the FASTBUS signal line assignments. Sequences are specified by operations. 480

fastener (lightning protection). A device used to secure the conductor to the structure that supports it. *See:* **lighting protection and equipment.** 297

fast groove (disk recording) (fast spiral). An unmodulated spiral groove having a pitch that is much greater than that of the recorded grooves. *See:* **phonograph pickup.** 176

fast-operate, fast-release relay. A high-speed relay specifically designed for both short operate and short release time. 341

fast-operate relay. A high-speed relay specifically designed for short operate time but not necessarily short release time. 341

fast-operate, slow-release relay. A relay specifically designed for short operate time and long release time. 341

fast spiral. *See:* **fast groove.**

fast-time-constant circuit (radar). A circuit with short time-constant used to emphasize signals of short duration to produce discrimination against extended clutter, long-pulse jamming, or noise. 13

fast writing devices (metal-nitride-oxide field-effect transistor). Metal-nitride-oxide semiconductor (MNOS) memory transistors whose threshold window Δv_{HL} is sufficiently large after a writing pulse width of t_w near 1 µs. A write cycle time of about 1 µs makes these devices useful for random access memory (RAM) applications. 386

fatigue. The tendency for a metal to fracture in brittle manner under conditions of repeated cyclic stressing at stress levels below its tensile strength. 103

fault (1)(reliability data for pumps and drivers, valve actuators, and valves). Any undesired state of a component or system. A fault does not necessarily require failure (for example, a pump may not start when required because its feeder breaker was inadvertently left open--a command block). 502

(2) (wire or cable). A partial or total local failure in the insulation or continuity of a conductor. *See:* **center of distribution.** 64

(3) (components). A physical condition that causes a device, a component, or an element to fail to perform in a required manner, for example, a short-circuit, a broken wire, an intermittent connection. *See:* **pattern-sensitive fault; program-sensitive fault.**
255,77

(4) (surge arresters). A disturbance that impairs normal operation, for example, insulation failure or conductor breakage. 244,62

(5) (thyristor power converter). A condition existing when the conduction cycles of some semiconductors are abnormal. *Note:* This usually results in fault currents of substantial magnitude. 121

(6) (power switchgear). *See:* **short circuit.**

(7) (test, measurement and diagnostic equipment). A degradation in performance due to detuning, maladjustment, misalignment, failure of parts, and so forth.
54

(8) (software). (A) An accidental condition that causes a functional unit to fail to perform its required function. (B) A manifestation of an error in software. A fault, if encountered, may cause a failure. *Syn:* **bug.** *See:* **error; failure; function; functional unit; software.**
434

fault bus (fault ground bus) (power switchgear). A bus connected to normally grounded parts of electric equipment, so insulated that all of the ground current passes to ground through fault detecting means.
103

fault bus protection (relaying) (power switchgear). A method of ground fault protection which makes use of a fault bus. 103

fault category. *See:* **error category.**

fault, circulating current (thyristor converter). A circulating current in excess of the design value. *Note:* In a double converter precaution must be taken to control circulating direct current between the forward and reverse sections. 121

fault-closure current rating (separable insulated connectors). The designated rms fault current which a load-break connector can close under specified conditions. 134

fault-closure current rating (separable insulated connectors). The designated root-mean-square fault current which a load-break connector can close under specified conditions. 454

fault current (1)(faulted circuit indicators). Any current through the sensor equal to or in excess of the trip current of the faulted circuit indicator (FCI).
482

(2)(metal-oxide surge arresters for ac power circuits)(surge arrester). The current from the connected power system that flows in a short circuit.
583, 430

(3) (general). A current that flows from one conductor to ground or to another conductor owing to an abnormal connection (including an arc) between the two. *Note:* A fault current flowing to ground may be called a ground fault current. 64,244,62

(4) (health care facilities). A current in an accidental connection between an energized and a grounded or other conductive element resulting from a failure of insulation, spacing, or containment of conductors.
192

fault current withstand (surge arrester). The maximum root-mean-square (rms) symmetrical fault current of a specified duration that a failed distribution class arrester will withstand without an explosive fracture of the arrester housing. 430

fault detection (test, measurement and diagnostic equipment). One or more tests performed to determine if any malfunctions or faults are present in a unit.
54

fault-detector relay (power switchgear). A monitoring relay whose function is to limit the operation of associated protective relays to specific system conditions. 103

faulted circuit indicator. A single or multiphase device designed to sense fault current and provide an indication that the fault current has passed through the power conductor(s) at the point where the FCI (faulted circuit indicator) sensor is installed. 482

fault electrode current (electron tubes) (surge electrode current*). The peak current that flows through an electrode under fault conditions, such as arc-backs and load short-circuits. *See:* **electrode current (electron tube).** 125
*Deprecated

fault ground bus. *See:* **fault bus.**

fault hazard current (health care facilities). The hazard current of a given isolated power system with all devices connected except the line isolation monitor. *See:* **hazard current.** 192

fault incidence angle (power switchgear). The phase angle as measured between the instant of fault inception and a selected reference, such as the zero point on a current or voltage wave. 103

fault indicator (test, measurement and diagnostic equipment). A device which presents a visual display, audible alarm, and so forth, when a failure or marginal condition exists. 54

fault-initiating switch (power switchgear). A mechanical switching device used in applied-fault protection to place a short circuit on an energized circuit and to carry the resulting current until the circuit has been deenergized by protective operation. *Notes:* (1) This switch is operated by a stored-energy mechanism capable of closing the switch within a specified rated closing time at its rated making current. The switch may be opened either manually or by a power-operated mechanism. (2) The applied short circuit may be intentionally limited to avoid excessive system disturbance. 103

fault insertion. *See:* **fault seeding.**

fault interrupter (power switchgear). A self controlled mechanical switching device capable of making, carrying and automatically interrupting an alternating current, but without automatic reclosing capability. It includes an assembly of control elements to detect overcurrents and control the fault interrupter. 103

fault isolation (test, measurement and diagnostic equipment). Tests performed to isolate within the unit under test. 54

fault resistance (surge arresters). The resistance of that part of the fault path associated with the fault itself. 244,62

fault seeding (software). The process of intentionally adding a known number of faults to those already in a computer program for the purpose of estimating the number of indigenous faults in the program. *Syn:* **bug seeding.** *See:* **computer program; fault; indigenous fault; program.** 434

fault symptom (test, measurement and diagnostic equipment). A measurable or visible abnormality in an equipment parameters. 54

fault tolerance (software). The built-in capability of a system to provide continued correct execution in the presence of a limited number of hardware or software faults. *See:* **execution; hardware; software fault; system.** 434

fault withstandability. The ability of electrical apparatus to withstand the effects of prescribed electrical fault current conditions without exceeding specified damage criteria. 70

Faure plate (storage cell) (pasted plate). A plate consisting of electroconductive material, which usually consists of lead-antimony alloy covered with oxides or salts of lead, that is subsequently transformed into active material. *See:* **battery (primary or secondary).** 328

FC (a flat cable) assembly (National Electrical Code). An assembly of parallel conductors formed integrally with an insulating material web specifically designed for field installation in metal surface raceway approved for the purpose. 256

FDHM (fiber optics). Abbreviation for "full duration at half maximum". *See:* **full width (duration) half maximum.** 433

F-display (radar). A rectangular display in which a target appears as a centralized blip when the radar antenna is aimed at it. Horizontal and vertical aiming errors are respectively indicated by horizontal and vertical displacement of the blip. 13

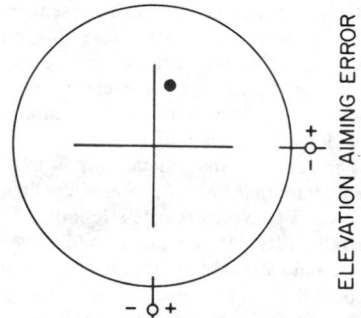

AZIMUTH AIMING ERROR

F-Display

FDNR. *See:* **frequency dependent negative resistor.** 67

feature (metric practice). An individual characteristic of a part, such as screw-thread, taper, or slot. 21

feed (1) (machines). (A) To supply the material to be operated upon to a machine. (B) A device capable of feeding as in (A). 235

(2) (antenna). That portion of an antenna coupled to the terminals which functions to produce the aperture illumination. *Note:* A feed may consist of a distribution network or a primary radiator. 111

feedback (transmission system or section thereof). The returning of a fraction of the output of the input. 111,240

feedback admittance, short-circuit (electron-device transducer). *See:* **admittance, short-circuit feedback.**

feedback control system (hydraulic turbines). A closed-loop or feedback control system is a control system in which the controlled quantity is measured and compared with a standard representing the desired value of the controlled quantity. In hydraulic governors, any deviation from the standard is fed back into the control system in such a sense that it will reduce the deviation between the controlled quantity and the standard providing negative feedback.
    8

feedback impedance (analog computers). In an analog computer, a passive network connected between the output terminal of an operational amplifier and its summing junction.
    9

feedback limit. *See:* **limiter circuit.**

feedback limiter. *See:* **limiter circuit; analog computer.**

feedback loop (numerically controlled machines). The part of a closed-loop system that provides controlled response information allowing comparison with a referenced command.
    224,207

feedback node (branch) (network analysis). A node (branch) contained in a loop. *See:* **linear signal flow graphs.**
    282

feedback oscillator. An oscillating circuit, including an amplifier, in which the output is coupled in phase with the input, the oscillation being maintained at a frequency determined by the parameters of the amplifier and the feedback circuits such as inductance-capacitance, resistance-capacitance, and other frequency-selective element. *See:* **oscillatory circuit.**   111

feedback signal (control) (industrial control). *See:* **signal, feedback.**

feedback winding (saturable reactor). A control winding to which a feedback connection is made.   328

feed circuit (transmission performance of telephone sets). an arrangement for supplying direct- current (dc) power to a telephone set and a terminating circuit.
    491

feeder (1) (electric installations on shipboard). A set ofconductors originating at a main distribution centerand supplying one or more secondary distribution centers,one or more branch-circuit distribution centers,or any combination of these two types of equipment. *Note:* Bus tie circuits between generator and distribution switchboards, including those between main and emergency switchboards, are not considered as feeders.
    3

(2) (packaging machinery). The circuit conductors between the service equipment, or the generator switchboard of an isolated plant, and the branch circuit overcurrent device.
    429

(3) (power distribution). A set of conductors originating at a main distribution center and supplying one or more secondary distribution centers, one or more branch-circuit distribution centers, or any combination of these two types of equipment. *Notes:* (A) Bus the circuits between generator and distribution switchboards, including those between main and emergency switchboards, are not considered as feed-

ers. (B) Feeders terminate at the overcurrent device that protects the distribution center of lesser order.
    3

feeder assembly (National Electrical Code). The overhead or under-chassis feeder conductors, including the grounding conductor, together with the necessary fittings and equipment or a power-supply cord approved for mobile home use, designed for the purpose of delivering energy from the source of electrical supply to the distribution panelboard within the mobile home.
    256

feeder cable (communication practice). A cable extending from the central office along a primary route (main feeder cable) or from a main feeder cable along a secondary route (branch feeder cable) and providing connections to one or more distribution cables. *See:* **cable.**
    328

feeder distribution center. A distribution center at which feeders or subfeeders are supplied. *See:* **distribution center.**
    328

feeder reactor (power and distribution transformer). A current-limiting reactor for connection in series with an alternating-current feeder circuit for the purpose of limiting and localizing the disturbance due to faults on the feeder.
    53

feed function (numerically controlled machines). The relative velocity between the tool or instrument and the work due to motion of the programmed axis (axes).
    224,207

feed groove (rotating machinery). A groove provided to direct the flow of oil in a bearing. *See:* **bearing.**
    63

feeding coefficients. *See:* **excitation coefficients.**

feeding point. The point of junction of a distribution feeder with a distribution main or service connection. *See:* **center of distribution.**
    64

feed line (1) (antennas). A transmission line interconnecting an antenna and a transmitter or receiver or both.
    111

(2) (rotating machinery). A supply pipe line. *See:* **oil cup (rotating machinery).**
    63

feed of an antenna. (1) (continuous aperture antennas). The primary radiator, for example, a horn feeding a reflector. (2) (array antennas). That portion of the antenna which functions to produce the excitation coefficients.
    111

feed rate bypass (numerically controlled machines). A function directing the control system to ignore programmed feed rate and substitute a selected operational rate.
    224,207

feed rate override (numerically controlled machines). A manual function directing the control system to modify the programmed feed rate by a selected multiplier.
    224,207

feedthrough power meter (1) (bolometric power meters). A power-measuring system in which the detector structure is inserted or incorporated in a waveguide or coaxial transmission line to provide a means for measuring (monitoring) the power flow through or beyond the system.
    115

(2) (electrothermic power meters) (measuring sys-

tem). A device which is inserted or incorporated in a waveguide or transmission line and provides a means for measuring (monitoring) the power flow through or beyond the system.   47

feedthrough signal (dispersive and nondispersive delay lines). The undelayed signal resulting from direct coupling between the input and the output of the device.   81

feed tube (cable plowing). A tube attached to the blade of a plow which guides and protects the cable as it enters the earth. **See: fixed feed tube; hinged removable feed tube; floating removable feed tube.**
                    52

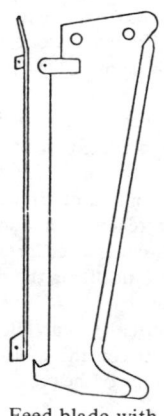

Feed blade with fixed feed tube.

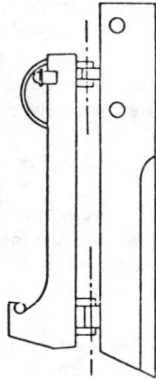

Feed blade and hinged feed tube.

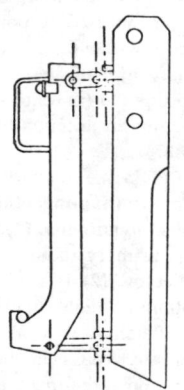

Feed blade with floating feed tube.

fenestra method (illuminating engineering). A procedure for predicting the interior illuminance received from daylight through windows.   167

fenestration (illuminating engineering). Any opening or arrangement of openings (normally filled with media for control) for the admission of daylight.
                    167

ferreed relay. Coined name (Bell Telephone Laboratories) for a special form of dry reed switch having a return magnetic path of high remanence material that provides a bistable, or latching, transfer contact.
                    341

ferri-diode limiter (nonlinear, active, and nonreciprocal waveguide components). A hybrid power limiting device incorporating a ferrite power limiter in cascade with a p-i-n diode or varactor limiter. *See:* **ferrite limiter; receiver protector.**   530

ferrite devices figure of merit (nonlinear, active, and nonreciprocal waveguide components). A measure of performance of the device. It is usually expressed as the ratio of the quantity of interest to the insertion loss in decibels (dB).

ferrite limiter (nonlinear, active, and nonreciprocal waveguide components). A power limiter utilizing the nonlinear characteristics of ferrimagnetic material above a critical or threshold radio-frequency (rf) power level. *See:* **ferridiode limiter; receiver protector.**
                    530

ferritic. The body-centered cubic crystal structure of ferrous metals.   205

ferrodynamic instrument. An electrodynamic instrument in which the forces are materially augmented by the presence of ferromagnetic material. *See:* **instrument.**   328

ferroelastic crystal (primary ferroelectric terms). One that has two or more orientation states in the absence of mechanical stress and electric field, and can be shifted from one to another of these states by a mechanical stress.   497

ferroelectric axis. The crystallograph direction is parallel to the spontaneous polarization vector. *Note:* In some materials the ferroelectric axis may have several possible orientations with respect to the macroscopic crystal. *See:* **ferroelectric domain.**   247

ferroelectric ceramic (primary ferroelectric terms). Typically a sintered polycrystalline material comprising an aggregate of ferroelectric single crystal grains (or crystallites). Each ceramic grain has properties similar to a ferroelectric single crystal, with the possible exception of grains with major dimensions $\ll 1\,\mu m$. *Notes:* (1) A ceramic is, in general, any inorganic, nonmetallic, ordered or disordered material. A ceramic is commonly typified by polycrystallinity and the unique properties associated with grain boundaries. (2) Both single crystal and sintered materials are, strictly speaking, ferroelectric ceramics even though they are often separated in common usage of the terms.   497

ferroelectric Curie point, T_C (primary ferroelectric terms). Temperature at which a ferroelectric material undergoes a structural phase transition to a state where spontaneous polarization vanishes. *Note:* The Curie point is determined at zero applied field.
                    497

ferroelectric Curie temperature. The temperature above which ferroelectric materials do not exhibit reversible spontaneous polarization. *Note:* As the temperature is lowered from above the ferroelectric Curie temperature spontaneous polarization is detected by the onset of a hysteresis loop. The ferroelectric Curie temperature should be determined only with unstrained crystals, at atmospheric pressure, and with no externally applied direct-current fields. (In some ferroelectric multiple hysteresis-loop patterns may be observed at temperatures slightly higher than the ferroelectric Curie temperature under alternating-current fields.) *See:* ferroelectric domains. 247

ferroelectric domains (primary ferroelectric terms). A region of a ferroelectric crystal exhibiting homogeneous and uniform spontaneous polarization. In the close vicinity of domain walls, P_s is different from that in the bulk of the domain, due to energy associated with the domain wall. The equilibrium domain structure is determined by minimization of the domain wall energy and the depolarizing energy. In a conducting ferroelectric crystal, the depolarizing fields can be neutralized by free charge so that the depolarizing energy vanishes and a single domain structure is energetically the most favorable in a perfect crystal. The formation of a domain wall is affected by the local electric field and mechanical stresses. In crystals with many small domains, the field and stress gradients can be large enough to change the measured apparent spontaneous polarization. In small domains P_s can vary considerably across a domain, and even at the domain center may differ from that measured in a large single domain. *Note:* An unpoled ferroelectric material may exhibit a complex domain structure consisting of many domains, each with a different polarization orientation. The direction of the spontaneous polarization within each domain is constrained to a small number of equivalent directions dictated by the symmetry of the prototype. The boundary region between two ferroelectric domains is called a domain wall. Domains can usually be observed by pyroelectric, optical, powder decoration, or electrooptic means. 497

ferroelectric glass-ceramics (primary ferroelectric terms). A multiphase solid containing ferroelectric crystal grains and other phases, one of which must be a glass. Typical grain sizes vary from $10^{-2} - 50 \ \mu m$, depending on chemical composition and nucleation-crystallization conditions. The principal fabrication process of a glass-ceramic has three steps: (1) A molten solution is formed of the ferroelectric and the glass-forming constituents. (2) The melt is rapidly quenched to form a vitreous body. (3) Controlled devitrification is accomplished by annealing. A chief advantage of a glass-ceramic is a wide range of formability, depending upon the viscosity-temperature behavior prior to devitrification. 497

ferroelectric material (primary ferroelectric terms). A material that exhibits, over some range of temperature, a spontaneous electric polarization that can be reversed or reoriented by application of an electric field. The requirement of a nonvanishing spontaneous polarization P_s is a necessary criterion, and the requirement of reversibility or reorientability of P_s is a suficient criterion for a ferroelectric phase. Materials belonging to nonpolar crystal classes at all temperatures, and in which a metastable polar state can be induced by an applied electric field, can also show

reversible pyroelectric behavior, but are not included in the definition of ferroelectrics. The various possible stable orientations of P_s for a given ferroelectric phase are designated as orientation states. A ferroelectric crystal has two or more such orientation states in the absence of an electric field, and it can be switched from one to another of these states by a realizable electric field. Any two of the orientation states are identical (or enantiomorphous) in crystal structure but different in their P_s orientation of zero electric field. 497

ferroelectric polymers (primary ferroelectric terms). Typically, a semicrystalline polymer with a large net dipole moment per unit volume. These materials exhibit a spontaneous electric polarization that can be reversed by the application of a strong electric field (0.1-0.5 MV/cm) and also exhibit piezoelectric and pyroelectric behavior. Examples of ferroelectric polymers include polyvinyl fluoride (PVF), polyvinylidene fluoride (PVF$_2$), and copolymers of vinylidene fluoride with vinyl trifluoroethylene or tetrafluoroethylene. The copolymers of vinylidene fluoride with trifluoroethylene exhibit Curie temperatures in the range of 50-160 °C. The higher trifluoroethylene content materials (greater than 55 mol% VF$_3$) have a second-order transition at lower Curie points; while the materials with higher vinylidene fluoride content (greater than 65 mol% VF$_2$) have a first-order transition at higher Curie points. 497

ferromagnetic material (electrical heating systems). A material that, in general, exhibits hysteresis phenomena and whose permeability is dependent on the magnetizing force. 476

ferroresonance (1)(ferroresonant voltage regulators). The steady-state mode of operation that exists when an alternating voltage of sufficient magnitude is applied to a circuit consisting of capacitance and ferromagnetic inductance causing changes in the ferromagnetic inductance which are repeated each half cycle. *Note:* When certain critical relations exist among circuit parameters, self-sustaining subharmonic or harmonic oscillations may also be excited in the circuit. 456

(2)(power and distribution transformer). A phenomenon usually characterized by overvoltages and very irregular wave shapes and associated with the excitation of one or more saturable inductors through capacitance in series with the inductor. 53

(3)(transformer). A phenomenon usually characterized by overvoltages and very irregular wave shapes and associated with the excitation of one or more saturable inductors through capacitance in series with the inductor. 53

ferroresonant voltage regulation (ferroresonant voltage regulators). The effect obtained by the limiting action of the saturation characteristic of the magnetic material in a ferroresonant circuit, which regulates the output voltage over a specified range of input voltages and a specified frequency of excitation. *Note:* This effect regulates the half-cycle average value of the output voltage. 456

ferroresonant voltage regulator provided with a frequency compensating network (ferroresonant voltage regulators). Output voltage of a ferroresonant voltage regulator changes considerably with the change of the input frequency. An LC network can be added to the regulator output, in series with the load, to compensate this voltage change. *See:* **Figure 'Ferroresonant Transformer Circuit with Frequency Compensation Network' below.** *Note:* Frequency compensating networks, of this series type, are effective in cases where regulators are operated with constant loads but they produce only limited improvement of regulation when loads are variable. 456

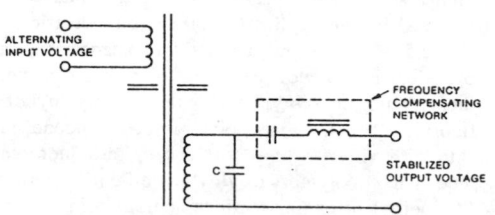

Ferroresonant Transformer Circuit with
Frequency Compensating Network

ferroresonant voltage regulator transformer (ferroresonant voltage regulators). A high-reactance transformer employing magnetic shunts that allow the magnetic functions of the basic series parallel ferroresonant regulator circuits to be combined into a single magnetic component. *Syn:* **ferroresonant transformer.** *See:* **Figure 'Ferroresonant Transformer Voltage Regulator' and Figure 'Schematic of Ferroresonant Transformer Voltage Regulator'.** 456

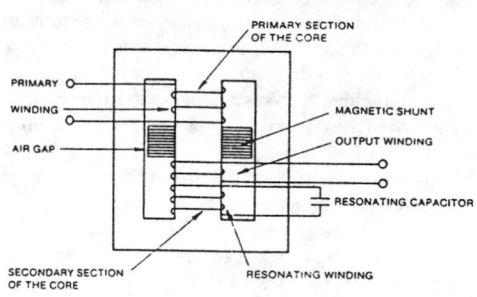

A Common Form of the Ferroresonant
Transformer Voltage Regulator

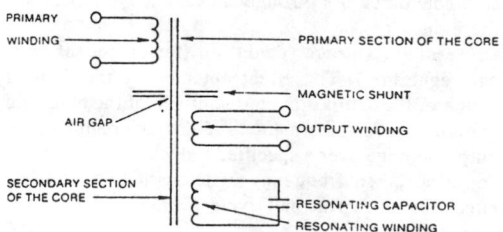

ferroresonant voltage regulator with compensating winding (ferroresonant voltage regulators). A ferroresonant voltage regulator having a compensating winding connected in series with the output winding to attain improved load and line regulation. See figure 'Two-Core Ferroresonant Circuit with Compensating Winding' and Figure 'Ferroresonant Transformer Circuit with Compensating Winding', below. 456

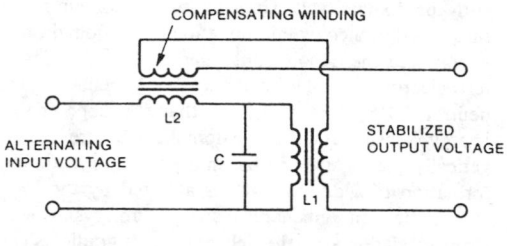

Two-Core Ferroresonant Circuit with
Compensating Winding

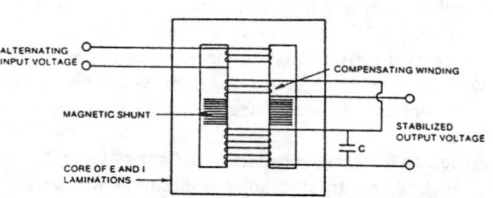

Ferroresonant Transformer Circuit with
Compensating Winding

ferroresonant voltage regulator with compensation for varying load power factor (ferroresonant voltage regulators). Reduction of the amount of output voltage change caused by other than resistive loading and by large changes of load power factor is obtained by providing a capacitive impedance, inserted in series with the output, that essentially matches the output reactance of the regulator. The power factor compensation circuit is usually a capacitive reactance obtained by capacitors alone, as shown in Figure 'Power Factor Compensation Using Series Capacitance' and Figure 'Power Factor Compensation Using Transformer Coupled Capacitance' below **ferroresonant voltage regulator with harmonic filter.** 456

ferroresonant voltage regulator with harmonic filter (harmonic neutralized)(ferroresonant voltage regulators). (1) Magnetically coupled type. Reduction of output harmonics is obtained by effectively filtering the odd harmonics through use of a neutralizing winding that is magnetically coupled to the resonating winding as shown in Figure 'Magnetically Coupled Tuned-Cancellation Type Harmonic Filter' below. (2)Electrically coupled tuned. Cancellation type reduction of output harmonics is obtained by effectively filtering the odd harmonics through use of an inductance in series with the resonating capacitor which effectively filters the major harmonic (the third harmonic) and a saturating inductor to produce odd harmonics which are induced back into the circuit of the regulator to cancel out the remaining odd harmonics. This type of filtering is shown in Figure 'Electrically Connected Tuned-Cancellation Type Harmonic Filter', below. (3) Tuned type. Reduction of output harmonics is obtained by dividing the resonating capacitance into several sections and connecting them to filter the various odd harmonics that exist in the output of the basic regulator. Usually, proper LC filtering of the 3rd, 5th, and 7th harmonics, as indicated in Figure 'Tuned-Type Harmonic Filter', below, will

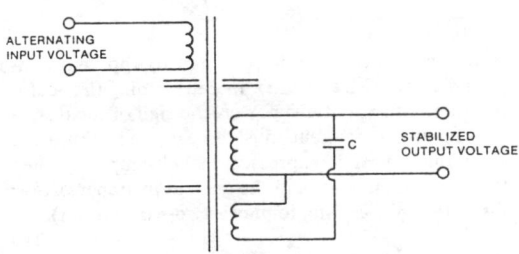

ALTERNATING
INPUT VOLTAGE

C

STABILIZED
OUTPUT VOLTAGE

Magnetically Coupled Tuned-Cancellation Type
Harmonic Filter

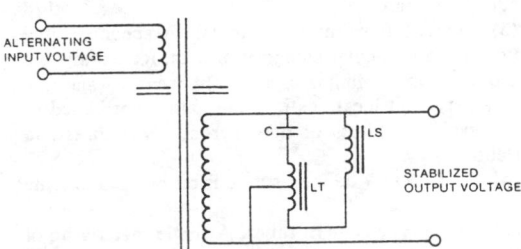

ALTERNATING
INPUT VOLTAGE

C

LS

LT

STABILIZED
OUTPUT VOLTAGE

Electrically Connected Tuned-Cancellation Type
Harmonic Filter

reduce the harmonic content to the same low level as accomplished by the methods shown in Figure 'Magnetically Coupled Tuned-Cancellation Type Harmonic Filter' and Figure 'Electrically Connected Tuned-Cancellation Type Harmonic Filter', above.   456

ferrule (protection and coordination of industrial and commercial power systems). The cylindrical-shaped fuse terminal that also encloses the end of the fuse. In low-voltage fuses, the design is only used in fuses rated up to and including 60 amperes (A). The ferrule may be made of brass or copper, and may be plated with various materials.   504

ferrule (1) (fiber optics). A mechanical fixture, generally a rigid tube, used to confine the stripped end of a fiber bundle or a fiber. *Notes:* (A) Typically, individual fibers of a bundle are cemented together within a ferrule of a diameter designed to yield a maximum packing fraction. (B) Nonrigid materials such as shrink tubing may also be used for ferrules for special applications. *See:* **fiber bundle; packing fraction; reference surface.**   433

(2) (of a cartridge fuse) (power switchgear). A fuse terminal of cylindrical shape at the end of a cartridge fuse.   103

festoon lighting (National Electrical Code). A string of outdoor lights suspended between two points more than 15 feet apart.   256

FET photodetector (fiber optics). A photodetector employing photogeneration of carriers in the channel region of a field-effect transistor (FET) structure to provide photodetection with current gain. *See:* **photocurrent; photodiode.**   433

fiber. *See:* **optical fiber.**

fiber axis (fiber optics). The line connecting the centers of the circles that circumscribe the core, as defined under *bold tolerance field. Syn:* **optical axis.** *See:* **tolerance field..**   433

fiber bandwidth (fiber optics). The lowest frequency at which the magnitude of the fiber transfer function decreases to a specified fraction of the zero frequency value. Often, the specified value is one-half the optical power at zero frequency. *See:* **transfer function.**
     433

fiber buffer (fiber optics). A material that may be used to protect an optical fiber waveguide from physical damage, providing mechanical isolation or protection or both. *Note:* Cable fabrication techniques vary, some resulting in firm contact between fiber and protective buffering, others resulting in a loose fit, permitting the fiber to slide in the buffer tube. Multiple buffer layers may be used for added fiber protection. *See:* **fiber bundle.**   433

fiber bundle (fiber optics). An assembly of unbuffered optical fibers. Usually used as a single transmission channel, as opposed to multifiber cables, which contain optically and mechanically isolated fibers, each of which provides a separate channel. *Notes:* (1) Bundles used only to transmit light, as in optical communications, are flexible and are typically unaligned. (2) Bundles used to transmit optical images may be either flexible or rigid, but must contain aligned fibers. *See:* **aligned bundle; ferrule; fiber optics; multifiber cable; optical cable; optical fiber; packing fraction.**

433

fiber-optic plate (camera tubes). An array of fibers, individually clad with a lower index-of-refraction material, that transfers an optical image from one surface of the plate to the other. *See:* **camera tube.** 190

fiber optics (1). The branch of optical technology concerned with the transmission of radiant power through fibers made of transparent materials such as glass, fused silica, or plastic. *Notes:* (A) Telecommunication applications of fiber optics employ flexible fibers. Either a single discrete fiber or a nonspatially aligned fiber bundle may be used for each information channel. Such fibers are often referred to as "optical waveguides" to differentiate from fibers employed in noncommunications applications. (B) Various industrial and medical applications (typically high-loss) flexible fiber bundles in which individual fibers are spatially aligned, permitting optical relay of an image. An example is the endoscope. (C) Some specialized industrial applications employ rigid (fused) aligned fiber bundles for image transfer. An example is the fiber optics faceplate used on some high-speed oscilloscopes. 433

(2)(data transmission) (FO). The branch of optical technology concerned with the transmission of radiant power through fibers made of transparent materials, such as glass, or fused silica plastic. *Notes:*(A) Communications applications of fiber optics employ flexible fibers. Either a single discrete fiber or a nonspatially aligned fiber bundle may be used for each information channel. Such fibers are generally referred to as "optical waveguides" to differentiate from fibers employed in noncommunications applications. (B) Various industrial and medical applications employ typically high-loss flexible fiber bundles in which individual fibers are spatially aligned, permitting optical relay of an image. An example is the endoscope. (C) Some specialized industrial applications employ rigid (fused) aligned fiber bundles for image transfer. An example is the fiber optics cathode-ray tube (CRT) faceplate used on some high-speed oscilloscopes.

59

fibrillation (medical electronics). A continued, uncoordinated activity in the fibers of the heart, diaphragm, or other muscles consisting of rhythmical but asynchronous contraction and relaxation of individual fibers. 192

fictitious power (polyphase circuit). At the terminals of entry, a vector equal to the (vector) sum of the ficititious powers for the individual terminals of entry. *Note:* The fictitious power for each terminal of entry is determined by considering each phase conductor and the common reference point as a single-phase circuit, as described for distortion power. The sign given to the distortion power in determining the fictitious power for each single-phase circuit shall be the same as that of the total active power. Fictitious power for a polyphase circuit has as its two rectangular com-

ponents the reactive power and the distortion power. If the voltages have the same waveform as the corresponding currents, the magnitude of the fictitious power becomes the same as the reactive power. Fictitious power is expressed in volt-amperes when the voltages are in volts and the currents in amperes.

210

fictitious power (single-phase two-wire circuit). At the two terminals of entry into a delimited region, a vector quantity having as its rectangular components the reactive power and the distortion power. *Note:* Its magnitude is equal to the square root of the difference of the squares of the apparent power and the amplitude of the active power. Its magnitude is also equal to the square root of the sum of the squares of the amplitudes of reactive power and distortion power. If voltage and current have the same waveform, the magnitude of the fictitious power is equal to the reactive power. The magnitude of the ficititious power is given by the equation

$$
\begin{aligned}
F &= (U^2 - p^2)^{1/2} \\
&= (Q^2 + D^2)^{1/2} \\
&= \left\{ \sum_{r=1}^{r=\infty} \sum_{q=1}^{q=\infty} \left[E_r^2 I_q^2 \right. \right. \\
&\quad \left. \left. - E_r E_q I_r I_q \cos(\alpha_r - \beta_r) \cos(\alpha_q - \beta_q) \right] \right\}^{1/2}
\end{aligned}
$$

where the symbols are those of **power, apparent (single-phase two-wire circuit).** In determining the vector position of the fictitious power, the sign of the distortion power component must be assigned arbitrarily. Fictitious power is expressed in volt-amperes when the voltage is in volts and the current in amperes. *See:* **distortion power (single-phase two-wire circuit).**

210

fidelity. The degree with which a system, or a portion of a system, accurately reproduces at its output the essential characteristics of the signal that is impressed upon its input. 111

field (1)(electric submersible pump cable). The term field or in the field may include cable not yet installed or cable that has been removed from its operating environment. 484

(2) (television). One of the two (or more) equal parts into which a frame is divided in interlaced scanning. *See:* **television.** 178

(3) (record) (computing systems). A specified area used for a particular category of data, for example, a group of card columns used to represent a wage rate or a set of bit locations in a computer word used to express the address of the operand. *See:* **threshold field.** 255,77

(4) (ATLAS). A defined subdivision of a statement.

400

(5) (diode-type camera tube). A single raster scan of the target. In the usual 2:1 interlace scan, two fields are required to completely scan the raster frame.

380

(6) (in the field) (power cable systems). Refers generally to apparatus installed in operating position. However, it may include material not yet installed or material that has been removed from its operating environment.   437

field accelerating relay (industrial control). A relay that functions automatically to maintain the armature current within limits, when accelerating to speeds above base speed, by controlling the excitation of the motor field.   206

field application relay (1) (power switchgear). A relay that initiates the application of field excitation to a synchronous machine under specified conditions. *Note:* It is usually a polarized relay sensitive to the slip frequency of the induced field current. It may also remove excitation during an out-of-step condition.
                                                                                    103

(2) (56) (power system device function numbers). A relay that automatically controls the application of the field excitation to an alternating-current (ac) motor at some predetermined point in the slip cycle.   402

field assembled (electrical heat tracing for industrial applications). Heating cable supplied in bulk form with terminating components to be assembled (terminated) by field personnel.   523

field bar (line waveform distortion). A composite pulse, nominally of 8 ms duration, of reference-white amplitude. The field bar is composed of line bars as defined. This signal when displayed on a picture monitor has the form of the window signal shown in the figure below.   42

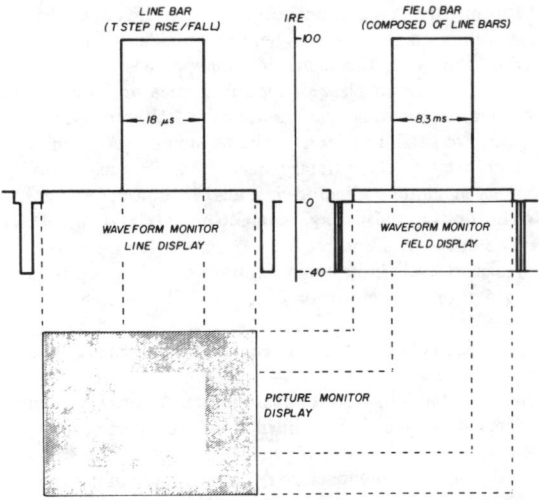

The Window Signal

field-changing contactor (93) (power system device function numbers). A contactor that functions to increase or decrease, in one step, the value of field excitation on a machine.   402

field circuit breaker (41) (power system device function numbers). A device that functions to apply or remove the field excitation of a machine.   402

field coil (rotating machinery) (1) (direct-current and salient-pole alternating-current machines). A suitably insulated winding to be mounted on a field pole to magnetize it.   63

(2) (cylindrical-rotor synchronous machines). A group of turns in the field winding; occupying one pair of slots. *See:* **asynchronous machine.**   63

field-coil flange (rotating machinery). Insulation between the field coil and the pole shoe, and between the field coil and the member carrying the pole body, in a salient-pole machine. *See:* **rotor (rotating machinery); stator.**   63

field contacts (sequential events recording systems). Electricalcontacts which define the state of monitored equipment or a process.   48,58

field contact voltage (sequential events recording systems). The voltage applied to field contacts for the purpose of sensing contact status.   48

field control (motor) (industrial control). A method of controlling a motor by means of a change in the magnitude of the field current. *See:* **control.**   206

field, critical (magnetrons). *See:* **critical field.**

field, cutoff (magnetrons). *See:* **critical field.**

field data (reliability). Data from observations during field use. *Note:* The time stress conditions, and failure or success criteria should be stated in detail.   164

field decelerating relay (industrial control). A relay that functions automatically to maintain the armature current or voltage within limits, when decelerating from speeds above base speed, by controlling the excitation of the motor field. *See:* **relay.**   206

field discharge (as applied to a switching device) (power switchgear). A qualifying term indicating that the switching device has main contacts for energizing and de-energizing the field of a generator, motor, synchronous condenser or exciter; and has auxiliary contacts for short-circuiting the field through a discharge resistor at the instant preceding the opening of the main contacts. The auxiliary contacts also disconnect the field from the discharge resistor at the instant following the closing of the main contacts. *Note:* For dc generator operation the auxiliary contacts may open before the main contacts close.   103

field discharge circuit breaker (1)(rotating electric machinery). A circuit breaker having main contacts for energizing and deenergizing the field of a generator, motor, synchronous condenser, or rotating exciter, and having discharge contacts for short-circuiting the field through the discharge resistor at the instant preceding the opening of the circuit breaker main contacts. The discharge contacts also disconnect the field from the discharge resistor at the instant following the closing of the main contacts. For direct-current generator operation, the discharge contacts may open before the main contacts close. *Note:* When used in the main field circuit of an alternating or direct-current generator, motor, or synchronous condenser, the circuit breaker is designated as a main field discharge circuit breaker. When used in the field circuit of the rotating exciter of the main machine, the circuit breaker is designated as an exciter field discharge circuit breaker. 412

(2)(excitation systems for synchronous machines). A circuit breaker having main contacts for energizing and deenergizing the field of a synchronous machine or rotating exciter and having discharge contacts for short-circuiting the field through the discharge resistor prior to the opening of the circuit breaker main contacts. The discharge contacts also disconnect the field from the discharge resistor following the closing of the main contacts. *Notes:* (A) When used in the main field of a synchronous machine the circuit breaker is designated as a main field discharge circuit breaker. (B) When used in the field circuit of a rotating exciter of the main machine, the circuit breaker is designated as an exciter field discharge circuit breaker. 507

field discharge protection (industrial control). A control function or device to limit the induced voltage in the field when the field current is disrupted or when an attempt is made to change the field current suddenly. *See:* **control.** 219,206

field displacement (nonlinear, active, and nonreciprocal waveguide components). The condition, in a uniform single-mode waveguide, in which the presence of magnetized gyromagnetic material causes the transverse-plane field distributions to be significantly different in the two directions of propagation. 530

field-disturbance sensor (measurement procedure for field-disturbance sensors). A device that employs a point source of radio-frequency (rf) energy to detect motion in the vicinity of the source, and in which the emitter and the receiver (or detector) are essentially at the same point, that is, a space-protected system. 531

field emission. Electron emission from a surface due directly to high-voltage gradients at the emitting surface. *See:* **electron emission.** 125

field-enhanced photoelectric emission. The increased photoelectric emission resulting from the action of a strong electric field on the emitter. *See:* **phototube.** 125

field-enhanced secondary emission. The increased secondary emission resulting from the action of a strong electric field on the emitter. *See:* **electron emission.** 125

field excitation current (Hall-effect devices). The current producing the magnetic flux density in a Hall multiplier. 107

field-failure protection (industrial control). The effect of a device, operative on the loss of field excitation, to cause and maintain the interruption of power in the motor armature circuit. 206

field-failure relay (industrial control). A relay that functions to disconnect the motor armature from the line in the event of loss of field excitation. *See:* **relay.** 206

field forcing (1)(excitation systems for synchronous machines). A control function that rapidly drives the field current of a synchronous machine in the positive or in the negative direction. 507

(2)(industrial control). A control function that temporarily overexcites or underexcites the field of a rotating machine to increase the rate of change of flux. *See:* **control.** 225,206

field forcing relay (industrial control). A relay that functions to increase the rate of change of field flux by underexciting the field of a rotating machine. *See:* **relay.** 206

field frame. *See:* **frame yoke.**

field-free emission current (1) (general). The emission current from an emitter when the electric gradient at the surface is zero. 244,190

(2) (cathode). The electron current drawn from the cathode when the electric gradient at the surface of the cathode is zero. *See:* **electron emission.** 125

field frequency (television). The product of frame frequency multiplied by the number of fields contained in one frequency. *See:* **television.** 178

field-intensity meter* (electric process heating). A calibrated radio receiver for measuring field intensity. *See:* **interference; interference measurement; field-strength meter.** 14

*Deprecated

field I^2R loss. The product of the measured resistance, in ohms, of the field winding, corrected to a specified temperature, and the square of the field current in amperes. 244,63

field-lead insulation (rotating machinery). The dielectric material applied to insulate the enclosed conductor connecting the collector rings to the coil end windings. *Note:* Field leads also include the pole jumpers forming the series connection between the concentric windings on each pole. Where rectangular strap leads are employed, the insulation may consist of either taped mica and glass or moduled mica and glass or moduled mica and glass composites. Where circular rods are used, moulded laminate tubing is frequently employed as the primary insulation. *See:* **asynchronous machine; collector-ring lead insulator.** 63

field-limiting adjusting means (industrial control). The effect of a control function or device (such as a resistor) that limits the maximum or minimum field excitation of a motor or generator. *See:* **control.** 225,206

field loss relay (industrial control). *See:* **motor-field failure relay.**

field molded joint (power cable joint). A joint in which the solid-dielectric joint insulation is fused and curved thermally at the job site. 34

field pole (rotating machinery). A structure of magnetic material on which a field coil may be mounted. *Note:* There are two types of field poles: main and commutating. *See:* **asynchronous machine.** 328

field protection (industrial control). The effect of a control function or device to prevent overheating of the field excitation winding by reducing or interrupting the excitation of the shunt field while the machine is at rest. *See:* **control.** 225,206

field protective relay (industrial control). A relay that functions to prevent overheating of the field excitation winding by reducing or interrupting the excitation of the shunt field. *See:* **relay.**   206

field relay (40) (power system device function numbers). A relay that functions on a given or abnormally low value or failure of machine field current, or on an excessive value of the reactive component of armature current in an alternating-current (ac) machine indicating abnormally low field excitation.   402

field-reliability test. A reliability compliance or determination test made in the field where the operating and environmental conditions are recorded and the degree of control is stated.   164

field-renewable fuse. *See:* **renewable fuse.**

field renewable fuse or fuse unit. *See:* **renewable fuse; fuse unit.**   103

field renewable fuse unit (high-voltage switchgear). A fuse unit that, after circuit interruption, may be readily restored for service by the replacement of the fuse link or refill unit.   443

field-reversal permanent-magnet focusing (microwave tubes). Magnetic focusing by a limited series of field reversals, not periodic, whose location is usually related to breaks in the slow-wave circuit. *See:* **magnetrons.**   190

field rheostat. A rheostat designed to control the exiting current of an electric machine.   206

field sequential (color television). Sampling of primary colors in sequence with successive television fields.   18

field shunting control (shunted-field control). A system of regulating the tractive force of an electrically driven vehicle by shunting, and thus weakening, the traction motor series fields by means of a resistor. *See:* **multiple-unit control.**   328

field splice (cable, field splice, and connection qualification)(design and installation of cable systems for Class 1E circuits in nuclear power generating stations). A permanent joining and reinsulating of conductors in the field to meet the service conditions required.   536, 141

field spool (rotating machinery). A structure for the support of a field coil in a salient-pole machine, either constructed of insulating material or carrying field-spool insulation. *See:* **asynchronous machine.**   63

field-spool insulation (rotating machinery). Insulation between the field spool and the field coil in a salient-pole machine. *See:* **asynchronous machine.**   63

field strength (electromagnetic wave). A general term that usually means the magnitude of the electric field vector, commonly expressed in volts per meter, but that may also mean the magnitude of the magnetic field vector, commonly expressed in amperes (or ampere-turns) per meter. *Note:* At frequencies above about 100 megahertz, and particularly above 1000 megahertz, field strength in the far zone is sometimes identified with power flux density P. For a linearly polarized wave in free space $P = E^{2/(\mu_v/\epsilon_v)}$, where E is the electric field strength, and μ_v and ϵ_v are the magnetic and electric constants of free space, respec-

tively. When P is expressed in watts per square meter and E in volts per meter, the denominator is often rounded off to 120π. See: **electric field strength; magnetic field strength; measurement system.**   213

field-strength meter. A calibrated radio receiver for measuring field strength. *See:* **induction heater; intensity meter.**   14,114

field system (rotating machinery). The portion of a direct-current or synchronous machine that produces the excitation flux. *See:* **asynchronous machine.**   63

field terminal (rotating machinery). A terminations for the field winding. *See:* **rotor (rotating machinery); stator.**   63

field tests (1)(for switchgear)(metal-clad and station-type cubicle switchgear)(metal-enclosed interrupter switchgear)(metal-enclosed bus and calculating losses in isolated-phase bus)(metal-enclosed low-voltage power circuit-breaker switchgear). Tests made after the assembly has been installed at its place of utilization.   103, 572, 573, 574, 579

(2) (cable systems). Tests which may be made on a cable system [including the high-voltage cable termination(s)] by the user after installation, as an acceptance or proof test.   4

(3) (power cable joint). Tests which may be made on the cable and accessories after installation.   34

(4) (power switchgear). Tests made on operating systems usually for the purpose of investigating the performance of switchgear or its component parts under conditions that cannot be duplicated in the factory. *Note:* Field tests are usually supplementary to factory tests and therefore may not provide a complete investigation of capabilities.   103

field-time waveform distortion (FD) (video signal transmission measurement). The linear TV waveform distortion of time components from 64 μ to 16 ms, that is, time components of field-time domain.   42

field-turn insulation (rotating machinery). Insulation in the form of strip or tape separating the individual turns of a field winding. *See:* **asynchronous machine.**   63

field voltage, base. The synchronous machine field voltage required to produce rated voltage on the air-gap line of the synchronous machine at field temperatures of (1) 75°C for field windings designed to operate at rating with a temperature rise of 60 °C or less; or (2) 100 °C for field windings designed to operate at rating with a temperature rise greater than 60 °C. *Note:* This defines one per unit excitation system voltage for use in computer representation of excitation systems.   105

field voltage, no-load. The voltage required across the terminals of the field winding of an electric machine under conditions of no load, rated speed and terminal voltage, and with the field winding at 25 °C.   105

field voltage, rated-load. The voltage required across the terminals of the field winding of an electric machine under rated continuous-load conditions with the field winding at (1) 75 °C for field windings designed to operate at rating with a temperature rise of 60 °C or less; or (2) 100 °C for field windings designed to operate at rating with a temperature rise greater than 60 °C. 105

field winding (excitation systems for synchronous machines)(rotating machinery). A winding on either the stationary or the rotating part of a synchronous machine whose sole purpose is the production of the main electromagnetic field of the machine. 507, 63

field winding terminals (excitation systems for synchronous machines). The place of input to the field winding of the synchronous machine. If there are brushes and sliprings these are to be considered to be part of the field winding. 507

fifth voltage range. *See:* **voltage range.**

fifty percent disruptive-discharge voltage (1) (dielectric tests). The voltage that has a 50 percent probability of producing a disruptive discharge. The term applies mostly to impulse tests and has significance only in cases where the loss of dielectric strength resulting from a disruptive discharge is temporary. 150

(2) (high voltage testing). The fifty percent disruptive discharge voltage is the prospective voltage value which has a fifty percent probability of producing a disruptive discharge. 150

figure (numerical) (metric practice). An arithmetic value expressed by one or more digit. 21

figure of merit (1) (magnetic amplifier). The ratio of power amplification to time constant in seconds. *See:* **beta figure of merit; detector figure of merit; ferrite devices figure of merit; pumped figure of merit.** 171

(2) (thermoelectric couple).

$$\alpha^2[(\rho_1\kappa_1)^{1/2} + (\rho_2\kappa_2)^{1/2}]^{-2}$$

where α is the Seebeck coefficient of the couple and ρ_1, ρ_2, κ_1, and κ_2 are the respective electric resistivities and thermal conductivities of materials 1 and 2. *Note:* This figure of merit applies to materials for the thermoelectric devices whose operation is based on the Seebeck effect or the Peltier effect. *See:* **thermoelectric device.**

(3) (thermoelectric couple, ideal).

$$\overline{\alpha}^{-2}[(\overline{\rho_1\kappa_1})^{1/2} + (\overline{\rho_2\kappa_2})^{1/2}]^{-2}$$

where $\overline{\alpha}$ is the average value of the Seebeck coefficient of the coupled and $\rho_1\kappa_1$ and $\rho_2\kappa_2$ are the average values of the products of the respective electric resistivities and thermal conductivities of materials 1 and 2, where the averages are found by integrating the parameters over the specified temperature range of the couple. *See:* **thermoelectric device.**

(4) (thermoelectric material). The quotient of (A) the square of the absolute Seebeck coefficient α by (B) the product of the electric resistivity ρ and the thermal conductivity κ.

$$\alpha^2/\rho\kappa.$$

Note: This figure of merit applies to materials for thermoelectric devices whose operation is based on the Seebeck effect or the Peltier effect. *See:* **thermoelectric device.** 248,191

(5) (antennas). *See:* **G/T ratio.** 111

(6) (dynamically tuned gyro) (inertial sensor). A design constant which relates the rotor polar moment of inertia and the principal moments of inertia of the gimbal(s). A simplified expression for the figure of merit is:

$$FOM = \cfrac{C}{\overset{n}{\underset{1}{\Sigma}}(A_n + B_n - C_n)}$$

where:
C = rotor polar moment of inertia
A_n, B_n = transverse moments of inertia of the nth gimbal
C_n = polar moment of inertia of the nth gimbal 46

filament (electron tube). A hot cathode, usually in the form of a wire or ribbon, to which heat may be supplied by passing current through it. *Note:* This is also known as a filamentary cathode. *See:* **electrode (electron tube).** 190,125

filament current. The current supplied to a filament to heat it. *See:* **electronic controller; heat current.** 190

filament power supply (electron tube). The means for supplying power to the filament. *See:* **power pack.** 190

filament voltage. The voltage between the terminals of a filament. *See:* **electrode voltage (electron tube); electronic controller.** 328

file (1)(information transfer). One named collection of data. 479

(2)(microprocessor operating systems). A set of related records usually treated as a named unit of storage. 478

(3) (computing systems). A collection of related records treated as a unit. *Note:* Thus in inventory control, one line of an invoice forms an item, a complete invoice forms a record, and the complete set of such records forms a file. 255,77

(4) (software). A set of related records treated as a unit. *See:* **logical file; records.** 434

file gap (computing system). An area on a storage medium, such as tape, used to indicate the end of a file. 255,77

file maintenance (computing systems). The activity of keeping a file up to date by adding, changing, or deleting data. 255,77

file processing (computer applications). The periodic updating of one or more master files to reflect the effects of current data, often from a transaction file. For example, a monthly run updating the inventory file. 571

filiform corrosion. *See:* **underfilm corrosion.**

fill (token ring access method). A bit sequence that may be either 0 bits , 1 bits, or any combination thereof.   472

filled-core annular conductor. A conductor composed of a plurality of conducting elements disposed around a nonconducting supporting material that substantially fills the space enclosed by the conducting elements. *See:* **conductor.**   64

filled-system thermometer. An all-metal assembly consisting of a bulb, capillary tube, and Bourdon tube (bellows and diaphragms are also used) containing a temperature-responsive fill. A mechanical device associated with the Bourdon is designed to provide an indication or record of temperature. *See:* **Bourdon.**   7

filled tape. Fabric tape that has been thoroughly filled with a rubber or synthetic compound, but not necessarily finished on either side with this compound. *See:* **conductor.**   64

filler (filler strip) (1) (rotating machinery). Additional insulating material used to insure a tight depth-wise fit in the slot. *See:* **rotor (rotating machinery); stator.**   63

(2) (mechanical recording). The inert material of a record compound as distinguished from the binder. *See:* **phonograph pickup.**   176

filler strip. *See:* **filler.**

filling compound (power cable joints). A dielectric material poured or otherwise injected into the joint housing. Filling compounds may require heating or mixing prior to filling. Some filling compounds may also serve as the insulation.   34

fill light (illuminating engineering). Supplementary illumination used to reduce shadow or contrast.   167

film (rotating machinery). Sheeting having a nominal thickness not greater than 0.030 centimeters and being substantially homogeneous in nature. *See:* **asynchronous machine; electrochemical valve.**   63

film integrated circuit. An integrated circuit whose elements are films formed in situ upon an insulating substrate. *Note:* To further define the nature of a film integrated circuit, additional modifiers may be prefixed. Examples are (1) thin-film integrated circuit; (2) thick-film integrated circuit. *See:* **magnetic thin film; thin film; electrochemical valve; integrated circuit.**   312,191

filter (1)(harmonic control and reactive compensation of static power converters). A generic term used in describing those types of equipment whose purpose is to reduce the harmonic currents or the voltage flowing in or being impressed upon specific parts of an electrical power system , or both. *See:* **damped filter; high-pass filter; series filter; shunt filter; tuned filter.**   533

(2) (wave filter). A transducer for separating waves on the basis of their frequency. *Note:* A **filter** introduced relatively small insertion loss to waves in one or more frequency bands and relatively large insertion loss to waves of other frequencies.   239

(3) (computing system). (A) A device or program that separates data, signals, or material in accordance with specified criteria. (B) A mask.   255,77

(4) (illuminating engineering). A device for changing, by transmission or reflection, the magnitude or the spectral composition, or both, of the flux incident upon it. Filters are called selective (or colored) or neutral, according to whether or not they alter the spectral distribution of the incident flux.   167

(5). *See:* **band-pass filter; high-pass filter; low-pass filter.**   59

filter, active (circuits and systems). (1) A filter network containing one or more voltage-dependent or current-dependent sources in addition to passive elements. (2) A filter containing energy generating elements.   67

filter, all-pass (circuits and systems). A filter designed to introduce phase shift or delay over a band of frequencies without introducing appreciable attenuation distortion over those frequencies.   67

filter attenuation band (filter stop band) (circuits and systems). A continuous range of frequencies over which the filter introduces an insertion loss whose minimum value is greater than a specific value.   67

filter, band-elimination (1) (signal-transmission system). A filter that has a single attenuation band, neither of the cutoff frequencies being zero or infinite. *See:* **rejection filter; filter.**   239

(2) (circuits and systems). A network designed to eliminate a band or frequencies. Its frequency response has a single pass band bounded by two attenuation bands.   67

filter, Butterworth (circuit and systems). A filter whose pass-band frequency response has a maximally flat shape brought about by the use of Butterworth polynomials as the approximating function.   67

filter capacitor. A capacitor used as an element of an electric wave filter. *See:* **electronic controller.**   206

filter, Chebyshev (circuits and systems). A filter whose pass-band frequency response has an equal-ripple shape brought about by the use of Chebyshev cosine polynomials as the approximating function.   67

filter, comb (circuits and systems). A filter whose insertion loss forms a sequence of narrow pass bands or narrow stop bands centered at multiples of some specified frequency.   67

filter effectiveness (shunt)(harmonic control and reactive compensation of static power converters). Defined by two terms: $\rho_1 \zeta =$ the impedance ratio which determines the per unit current which will flow into the shunt filter $\rho_1 \upsilon =$ the impedance ratio which determines the per unit current which will flow into the power source. $\rho_1 \zeta$ should approach unity and should be very small at the tuned frequency.   533

filter factor (illuminating engineering). The transmittance of "black light" by a filter. *Note:* The relationship among these terms is illustrated by the following formula for determining the luminance of fluorescent materials exposed to "black light":

$$\text{candelas per square meter} = \frac{1}{\pi^*}\ \frac{\text{fluorens}}{\text{square meter}}$$

$$\times \text{ glow factor} \times \text{filter factor}$$

$*\pi$ is omitted when luminance is in footlamberts and the area is in square feet.

When integral-filter "black light" lamps are used, the filter factor is dropped from the formula because it already has been applied in assigning fluoren ratings to these lamps.   167

filter, high-pass. A filter having a single transmission band extending from some cutoff frequency, not zero, up to infinite frequency. *See:* **filter.**   239

filter impedance compensator. An impedance compensator that is connected across the common terminals of electric wave filters when the latter are used in parallel in order to compensate for the effects of the filters on each other. *See:* **network analysis; filter.**   210

filter inductor. An inductor used as an element of an electric wave filter. *See:* **electronic controller.**   206

filter, low-pass. A filter having a single transmission band extending from zero to some cutoff frequency, not infinite. *See:* **filter.**   239

filter matching loss (radar). The loss in output signal-to-noise ratio relative to a matched filter, caused by using a filter whose response is not matched to the transmitted signal. *Syn:* **mismatch loss.**   13

filter mismatch loss (radar). *Syn:* **filter matching loss.**   13

filter pass band (circuits and systems). A frequency band of low attenuation (low relative to other regions termed stop bands). *See:* **filter transmission band.**   67

filter, passive (circuits and systems). A filter network containing only passive elements, such as inductors, capacitors, resistors and transformers.   67

filter reactor (power and distribution transformer). A reactor used to reduce harmonic voltage in alternating-current or direct-current circuits.   53

filter, rejection (signal-transmission system). *See:* **rejection filter; filter.**

filters (power supplies). Resistance-capacitance or inductance-capacitance networks arranged as low-pass devices to attenuate the varying component that remains when alternating-current voltage is rectified. *Note:* In power supplies without subsequent active series regulators, the filters determine the amount of ripple that will remain in the direct-current output. In supplies with active feedback series regulators, the regulator mainly controls the ripple, with output filtering serving chiefly for phase-gain control as a lag element.   186

filter, sound effects (electroacoustics). A filter used to adjust the frequency response of a system for the purpose of achieving special aural effects. *See:* **filter.**   239

filter stop band (circuits and systems). A frequency band of high attenuation (high relative to other regions termed pass bands). *See:* **filter attenuation band.**   67

filter transmission band (circuits and systems). A continuous range of frequencies over which the filter introduces an insertion loss whose maximum value does not exceed a specified value. *See:* **filter pass band.**   67

final approach path. *See:* **approach path.**

final contact pressure. *See:* **contact pressure, final.**

final controlling element (electric power systems). The controlling element that directly changes the value of the manipulated variable.   94

final design (nuclear power quality assurance). Approved design output documents and approved changes thereto.   417

final emergency circuits. All circuits (including temporary emergency circuits) that, after failure of a ship's service supply, may be supplied by the emergency generator. *See:* **emergency electric system.**   328

final emergency lighting. Temporary emergency lighting plus manually controlled lighting of the boat deck and overside to facilitate lifeboat loading and launching. *See:* **emergency electric system.**   328

final sag (transmission and distribution). The sag of a conductor under specified conditions of loading and temperature applied after it has been subjected for an appreciable period to the loading prescribed for the loading district in which it is situated, or equivalent loading, and the loading removed. Final sag includes the effect of inelastic deformation (creep).   262

final-terminal stopping device (elevators). A device that automatically causes the power to be removed from an electric elevator or dumbwaiter driving-machine motor and brake or from a hydraulic elevator or dumbwaiter machine independent of the functioning of the normal-terminal stopping device, the operating device, or any emergency terminal stopping device, after the car has passed a terminal landing. *See:* **control.**   328

final trunk (data transmission). A group of trunks to the higher class office which has no alternate route, and in which the number of trunks provided results in a low probability of calls encountering 'all trunks busy'.   59

final unloaded conductor tension (electric system). The longitudinal tension in a conductor after the conductor has been stretched by the application for an appreciable period, and subsequent release, of the loadings of ice and wind, and temperature decrease, assumed for the loading district in which the conductor is strung (or equivalent loading). *See:* **conductor; initial conductor tension.**   178

final unloaded sag (1) (general). The sag of a conductor after it has been subjected for an appreciable period to the loading prescribed for the loading district in which it is situated, or equivalent loading, and the loading removed. *See:* **sag.**   178

(2) (transmission and distribution). The sag of a conductor after it has been subjected for an appreciable period to the loading prescribed for the loading district in which it is situated, or equivalent loading, and the loading removed. Final unloaded sag includes the effect of inelastic deformation (creep). 262

final value (industrial control). The steady-state value of a specified variable. *See:* **control.** 206

final voltage. *See:* **cutoff voltage.**

finder switch. An automatic switch for finding a calling subscriber line or trunk and connecting it to the switching apparatus. 328

finding (telephone switching systems). Locating a circuit requesting service. 55

fine chrominance primary (National Television System Committee (NTSC) color television). An obsolete term. Use the preferred term, **I chrominance signal.** 18

fines (cable plowing). Particles of earth or rock smaller than 1/8 in greatest dimension. 52

F_1 layer (radio wave propagation). The lower of the two ionized layers normally existing in the F region in the day hemisphere. 146

F_2 layer (radio wave propagation). The single ionized layer normally existing in the F region in the night hemisphere and the higher of the two layers normally existing in the F region in the day hemisphere. 146

finger (rotating machinery). See: **end finger.**

finger line (conductor stringing equipment). A lightweight line, normally sisal, manila or synthetic fiber rope, which is placed over the traveler when it is hung. It usually extends from the ground, passes through the traveler and back to the ground. It is used to thread the end of the pilot line or pulling line over the traveler and eliminates the need for workmen on the structure. These lines are not required if pilot lines are installed then the travelers are hung. 431

finishing (electrotype). The operation of bringing all parts of the printing surface into the same plane, or, more strictly speaking, into positions having equal printing values. *See:* **electroforming.** 328

finishing rate (storage battery) (storage cell). The rate of charge expressed in amperes to which the charging current for some types of lead batteries is reduced near the end of charge to prevent excessive gassing and temperature rise. *See:* **charge.** 328

finite energy signal. *See:* **energy density signal.**

finite state machine (software). A computational model consisting of a finite number of states and transitions between these states. *See:* **model.** 434

finite-time stability. *See:* **stability, finite-time.**

finsen. The recommended practical unit of erythermal flux or intensity of radiation. It is equal to one unit of erythermal flux per square centimeter. 328

fire-alarm system. An alarm system signaling the presence of fire. *See:* **protective signaling.** 328

fire-control radar (navigation aid terms). A radar whose prime function is to provide information for the manual or automatic control of artillery or other weapons. 526

fire detection and fire protection systems (nuclear power generating station). Definitions of terms relat-

ing to fire detection and protection systems and equipment may be found in the National Fire Protection Association (NFPA) Handbook. 439

fire-door magnet. An electromagnet for holding open a self-closing fire door. 328

fire-door release system. A system providing remotely controlled release of self-closing doors in fire-resisting bulkheads to check the spread of fire. *See:* **marine electric apparatus.** 328

fired tube (microwave gas tubes)(nonlinear, active, and nonreciprocal waveguide components). The condition of the tube during which a radio frequency glow exists at either the resonant gap, the resonant window, or both. *See:* **gas tube; limiting threshold.** 125, 530

fireproofing (of cables)(NESC). The application of a fire-resistant covering. 494

fire rating (cable penetration fire stop qualification test).. The term applied to cable penetration fire stops to indicate the endurance in time (hours and minutes to the standard time-temperature curve in ANSI/ASTM E119-76, while satisfying the acceptance criteria specified in this standard. 368

fire-resistance rating. The measured time, in hours or fractions thereof, that the material or construction will withstand fire exposure as determined by fire tests conducted in conformity to recognize standards. 328

fire-resistant. So constructed or treated that it will not be injured readily by exposure to fire. *Syn:* **fire-resistive.** 328

fire-resistive barrier (1) (cable penetration fire stop qualification test). A wall, floor, or floor-ceiling assembly erected to prevent the spread of fire. (To be effective, fire barriers must have sufficient fire resistance to withstand the effects of the most severe fire that may be expected to occur in the area adjacent to the fire barrier and must provide a complete barrier to the spread of fire.) 368

(2)(cable-penetration fire stops, fire breaks, and system enclosures)(design and installation of cable systems for Class 1E circuits in nuclear power generating stations). A wall, floor, or floor-ceiling assembly erected to prevent the spread of fire. 536

fire-resistive barrier rating (cable-penetration fire stops, fire breaks, and system enclosures)(design and installation of cable systems for Class 1E circuits in nuclear power generating stations). This is expressed in time (hours and minutes) and indicates that the wall, floor, or floor-ceiling assembly can withstand, without failure, exposure to a standard fire for that period of time. *Note:* The test fire procedure and acceptance criteria are defined in ASTM E 119-1981. 536

fire-resistive construction. A method of construction that prevents or retards the passage of hot gases or flames as defined by the fire-resistance rating. 328

firing angle (semiconductor rectifier operating with phase control). *See:* **angle of retard (semiconductor rectifier operating with phase control).**

firing power (nonlinear, active, and nonreciprocal waveguide components). The radio-frequency (rf) power level above which a gas tube becomes nonlinear. *See:* **gas tube; receiver protector.**   530

firm capacity (power operations). Firm capacity which is purchased, or sold, in transactions with other systems and which is not from designated units, but is from the overall system of the seller. It is understood that the seller treats this type of transaction as a load obligation.   516

firm power (1)(emergency and standby power). Power intended to be always available, even under emergency conditions.   512

(2)(power operations). Power intended to be available at all times during the period covered by a commitment, even under adverse conditions.   516

firm transfer capability (transmission) (electric power supply). The maximum amount of power that can be interchanged continuously, over an extended period of time. *See:* **generating station.**   200

firmware (1)(supervisory control, data acquisition, and automatic control)(station control and data acquisition). Hardware used for the nonvolatile storage of instructions or data that can be read only by the computer. Stored information is not alterable by any computer program. *See:* **station (2) remote.**
                    570, 403

(2)(watthour meters). A register control program stored in read-only memory and considered to be an integral part of the register.   485

(3) (software). (A) Computer programs and data loaded in a class of memory that cannot be dynamically modified by the computer during processing. (B) Hardware that contains a computer program and data that cannot be changed in its user environment. The computer programs and data contained in firmware are classified as software; the circuitry containing the computer program and data is classified as hardware. (C) Program instructions stored in a read-only storage. (D) An assembly composed of a hardware unit and a computer program integrated to form a functional entity whose configuration cannot be altered during normal operation. The computer program is stored in the hardware unit as an integrated circuit with a fixed logic configuration that will satisfy a specific application or operational requirement. *See:* **computer; computer program; configuration; data; hardware; microcode; microprogram; program instruction; requirement; software.**   434

first (last) transition duration (pulse terms). The transition duration of the first (last) transition waveform in a pulse waveform. *See:* The single phase diagram below the **waveform epoch** entry.   254

first contingency incremental transfer capability (power operations). The amount of power, incremental above normal base power transfers, that can be transferred over the transmission network in a reliable manner, based on the following conditions: (1) With all transmission facilities in service, all facility loadings are within normal ratings, and all voltages are within normal limits; (2) The bulk power system is

capable of absorbing the dynamic power swings and remaining stable following a disturbance resulting in the loss of any single generating unit, transmission circuit or transformer; (3) After the dynamic power swings following a disturbance resulting in the loss of any single generating unit, transmission circuit, or transformer, but before operator-directed system adjustments are made, all transmission facility loadings are within emergency ratings, and all voltages are within emergency limits.   516

first dial (register). That graduated circle or cyclometer wheel, the reading on which changes most rapidly. The test dial or dials, if any, are not considered. *See:* **watthour meter.**   328

first Fresnel zone (data transmission). In optics and radio communication, the circular portion of a wave front transverse to the line between an emitter and a more distant point where the resultant disturbance is being observed, whose center is the intersection of the front with the direct ray and whose radius is such that shortest path from the emitter through the periphery to the receiving point is one-half wave longer than the ray. *Note:* A second zone, a third, etcetera, are defined by successive increases of the path by half-wave increments.   59

first-line release (telephone switching systems). Release under the control of the first line that goes-on-hook.   55

first Townsend discharge (gas). A semi-self- maintained discharge in which the additional ions that appear are due solely to the ionization of the gas by electron collisions. *See:* **discharge (gas).**   244,190

first transition (pulse terms). The major transition waveform of a pulse waveform between the base and the top.   254

first voltage range. *See:* voltage range.

fishbone antenna. An end-fire, traveling wave antenna consisting of a balanced transmission line to which is coupled, usually through lumped circuit elements, an array of closely-spaced, coplanar dipoles.   111

fish tape (fishing wire) (snake). A tempered steel wire, usually of rectangular cross section, that is pushed through a run of conduit or through an inaccessible space, such as a partition, and that is used for drawing in the wires.   328

fission (power operations). The splitting of a nucleus into parts (which are nuclei of lighter elements), accompanied by the release of a relatively large amount of energy (about 200 million electron volts per fission in the case of 235U fission) and frequently one or more neutrons.   516

fitting (National Electrical Code). An accessory such as a locknut, bushing, or other part of a wiring system that is intended primarily to perform a mechanical rather than an electrical function.   256

fittings (raceway)(raceway systems for Class 1E circuits for nuclear power generating stations). Raceway sections that are joined to other raceway sections for the purpose of coupling together or changing the size or direction of the raceway system. These include such items as couplings, elbows, tees, wyes, pulling sleeves, and pull boxes.   513

fix (1)(navigation aid terms). A position determined without reference to any former position. 526
(2) (interference) (electromagnetic compatibility). A device or equipment modification to prevent interference or to reduce an equipment's susceptibility to interference. *See:* **electromagnetic compatibility.** 199

fixed block format (numerically controlled machines). A format in which the number and sequence of words and characters appearing in successive blocks is constant. 224,207
fixed-called-address line (telephone switching systems). A line for originating calls to a fixed called address. 55
fixed cycle (numerically controlled machines). A preset series of operations that direct machine axis movement and/or cause spindle operation to complete such actions as boring, drilling, tapping, or combinations thereof. 244,207
fixed-cycle operation. An operation that is completed in a specified number of regularly timed execution cycles. 255,77
fixed feed tube (cable plowing). A feed tube permanently attached to a blade. It may have removable back plate. *See:* **feed tube.** 52
fixed fraction discriminator. *See:* **constant fraction discriminator.**
fixed-frequency transmitter. A transmitter designed for operation on a single carrier frequency. *See:* **radio transmitter.** 111,240
fixed impedance-type ballast. A reference ballast designed for use with one specific type of lamp and, after adjustment during the original calibration, is expected to hold its established impedance throughout normal use. 271
fixed investment costs (power operations). Monies associated with investment in plant. 516
fixed light (illuminating engineering). A light having a constant luminous intensity when observed from a fixed point. 167
fixed motor connections. A method of connecting electric traction motors wherein there is no change in the motor interconnections throughout the operating range. *Note:* This term is used to indicate that a transition from series to parallel relation is not provided. *See:* **traction motor.** 328
fixed operation costs (power operations). Monies other than those associated with investment in plant, which do not vary or fluctuate with changes in operation or utilization of plant. 516
fixed point (1(electronic computation). Pertaining to a numeration system in which the position of the point is fixed with respect to one end of the numerals, according to some convention. *See:* **floating point, variable point.** 235
(2)(mathematics of computing). Pertaining to a numeration system in which the position of the radix point is fixed with respect to one end of the numerals, according to some convention. *See:* **floating point; variable point.** 564
fixed-point system (electronic computation). *See:* **point.**

fixed rack (power switchgear). An assembly enclosed at top and sides, either open or with door(s) for access, with a top to bottom front panel opening for equipment mounting (for example, nominal 19 inch wide chassis and subpanel assemblies). 103
fixed sequential format. A means of identifying a word by its location in the block. *Note:* Words must be presented in a specific order and all possible words preceding the last desired word must be present in the block. 224,207
fixed signal. The signal of fixed location indicating a condition affecting the movement of a train or engine. 328
fixed storage (computing systems). A storage device that stores data not alterable by computer instructions, for example, magnetic core storage with a lockout feature or punched paper tape. *See:* **nonerasable storage; permanent storage, read-only storage.** 255,77
fixed temperature heat detector (fire protection devices). A device which will respond when its operating element becomes heated to a predetermined level. 71
fixed threshold transistor (metal-nitride-oxide field-effect transistor). Another name for a metal-oxide semiconductor (MOS) type transistor, used in contradistinction to the metal-nitride-oxide semiconductor (MNOS) transistor, which has a variable threshold voltage. 386
fixed transmitter. A transmitter that is operated in a fixed or permanent location. *See:* **radio transmitter.** 111,240
fixed word length (test, measurement and diagnostic equipment). Property of a storage device in which the capacity for bits in each storage word is fixed. 54
fixing (electrostatography). The act of making a developed image permanent. *See:* **electrostatography.** 236,191
fixture stud (stud). A threaded fitting used to mount a lighting fixture to an outlet box. *See:* **cabinet.** 328
flag (1)(microprocessor operating systems parameter types). A yes/no or true/false value. 478
(2) (ATLAS). The first character of an ATLAS statement used to mark the statement as having a special purpose or capability. 400
(3) (computing systems). (A) Any of various types of indicators used for identification, for example, a wordmark. (B) A character that signals the occurrence of some condition, such as the end of a word. 255,77
(4) (software). (A) An indicator that signals the occurrence of an error, state, or other specified condition. (B) Any of various types of indicators used for identification, for example, a word mark. (C) A character that signals the occurrence of some condition, such as the end of a word. (D) To indicate an error, state, or other specified condition in a program. *See:* **error; program; word.** 434
flag alarm (navigation aid terms). An indicator in cer-

tain types of navigation instruments used to warn when the readings are unreliable. 526

flame detector (1) (fire protection devices). A device which detects the infrared, or ultraviolet, or visible radiation produced by a fire. 71

(2) (28) (power system device function numbers). A device that monitors the presence of the pilot or main flame in such apparatus as a gas turbine or a steam boiler. 402

flame flicker detector (fire protection devices). A photoelectric flame detector including means to prevent response to visible light unless the observed light is modulated at a frequency characteristic of the flicker of a flame. 71

flameproof. *See:* **explosionproof.**

flameproof apparatus. Apparatus so treated that it will not maintain a flame or will not be injured readily when subjected to flame. 328

flameproof terminal box. A terminal box so designed that it may form part of a flameproof enclosure. 63

flame protection of vapor openings. Self-closing gauge hatches, vapor seals, pressure-vacuum breather valves, flame arresters, or other reasonably effective means to minimize the possibility of flame entering the vapor space of a tank. *Note:* Where such a device is used, the tank is said to be flameproofed. 297

flame-resistant cable. A portable cable that will meet the flame test requirements of the United States Bureau of Mines. *See:* **mine feeder circuit.** 328

flame resisting. *See:* **flame retarding.**

flame retardant (1) (Class 1E equipment and circuits) (nuclear power generating stations). Capable of limiting the propagation of a fire beyond the area of influence of the energy source that initiated the fire. 131

(2) (power switchgear). So constructed or treated that it will not support flame. 103

flame-retarding (electric installations on shipboard). Flame-retarding materials and structures should have such fire-resisting properties that they will not convey flame nor continue to burn for longer times than specified in the appropriate flame test. Compliance with the requirements of the preceding paragraph should be determined with the apparatus and according to the methods described in the Underwriters' Laboratories Standards for the materials and structures unless specific applicable tests are invoked in these recommendations. 3

flammable air-vapor mixtures. When flammable vapors are mixed with air in certain proportions, the mixture will burn rapidly when ignited. *Note:* The combustion range for ordinary petroleum products, such as gasoline, is from 1 1/2 to 6 percent of vapor by volume, the remainder being air. 297

flammable anesthetics (National Electrical Code) (health care facilities). Gases or vapors such as fluroxene, cyclopropane, divinyl ether, ethyl chloride, ethyl ether, and ethylene, which may form flammable or explosive mixtures with air, oxygen, or reducing gases such as nitrous oxide. 256

flammable anesthetizing location (National Electrical Code) (health care facilities). Any operating room, delivery room, anesthetizing room, corridor, utility room, or any other area if intended for the application of flammable anesthetics. 256

flammable vapors. The vapors given off from a flammable liquid at and above its flash point. 297

flange. *See:* **coupling flange.**

flange, choke (waveguide components). A flange designed with auxiliary transmission line elements to form a choke joint when used with a cover flange. 166

flange, contact (waveguide components). A flat flange used in conjunction with another flat flange to provide a contact joint. 166

flange, flat. *See:* **flange, cover.**

flange, plane. *See:* **flange, cover.**

flare-out (navigation aid terms). That portion of the approach path of an aircraft in which the slope is modified to provide the appropriate rate of descent at touchdown. 526

flarescan (navigation aid terms). A ground-based navigation system used in conjunction with an instrument approach system to provide flare-out vertical guidance to an aircraft by the use of a pulse-space-coded vertically-scanning fan beam that provides elevation-angle data. 526

flash barrier (rotating machinery). A screen of fire-resistant material to prevent the formation of an arc or to minimize the damage caused thereby. 63

flash current (primary cell). The maximum electric current indicated by an ammeter of the dead-beat type when connected directly to the terminals of the cell or battery by wires that together with the meter have a resistance of 0.01 ohm. *See:* **electrolytic cell.** 328

flasher. A device for alternately and automatically lighting and extinguishing electric lamps. *See:* **appliances (including portable).** 328

flasher relay. A relay that is so designed that when energized its contacts open and close at predetermined intervals. *See:* **appliances (including portable).** 328

flashing light (illuminating engineering). A rhythmic light in which the periods of light are of equal duration and are clearly shorter than the periods of darkness. 167

flashing-light signal. A railroad-highway crossing signal the indication of which is given by two red lights spaced horizontally and flashed alternately at predetermined intervals to give warning of the approach of trains, or a fixed signal in which the indications are given by color and flashing of one or more of the signal lights. 328

flashing signal (telephone switching system). A signal for indicating a change or series of changes of state, such as on-hook/off-hook, used for supervisory purposes. 55

flashlight battery. A battery designed or employed to light a lamp of an electric hand lantern or flashlight. *See:* **battery (primary or secondary).** 328

flashover (1) (general). A disruptive discharge through air around or over the surface of solid or liquid insulation, between parts of different potential or polarity, produced by the application of voltage wherein the breakdown path becomes sufficiently ionized to maintain an electric arc. *See:* **test voltage and current.**
299,62

(2) (surge arrester)(metal-oxide surge arresters for ac power circuits). A disruptive discharge around or over the surface of a solid or liquid insulator.
430, 583

(3) (high-voltage ac cable termination). A disruptive discharge around or over the surface of an insulating member, between parts of different potential or polarity, produced by the application of voltage wherein the breakdown path becomes sufficiently ionized to maintain an electric arc.
4

(4) (high voltage testing). Term used when a disruptive discharge occurs over the surface of a solid dielectric in a gaseous or liquid medium.
150

flash plate. A thin electrodeposited coating produced in a short time.
328

flash point. The minimum temperature at which a liquid will give off vapor in sufficient amount to form a flammable air-vapor mixture that can be ignited under specified conditions.
297

flashtube (illuminating engineering). A tube of glass or quartz with electrodes at the ends and filled with a gas, usually xenon. It is designed to produce high intensity flashes of light of extremely short duration.
167

flat-band voltage (metal-nitride-oxide field-effect transistor). Gate voltage that results in zero field at the surface of the silicon. It is related to the threshold voltage by a constant, generally small, voltage increment.
386

flat-compound. A qualifying term applied to a compound-wound generator to denote that the series winding is so proportioned that the terminal voltage at rated load is the same as at no load.
328

flat flange. *See:* **cover flange.**

flat leakage power (microwave gas tubes)(nonlinear, active, and nonreciprocal waveguide components). The peak radio-frequency power transmitted through the tube after the establishment of the steady-state radio-frequency discharge. *See:* **gas tube.**
125, 530

flat loss (gain) (circuits and systems). The frequency independent contribution to the total transfer-function loss (or gain) or a four-terminal network.
67

flat-rate call (telephone switching systems). A call for which no billing is required.
55

flat-rate service (telephone switching systems). Service in which a fixed charge is made for all answered local calls during the billing interval.
55

flat-strip conductor. *See:* **strip (-type) transmission line.**

flat-top antenna. A short vertical monopole antenna with an end capacitor whose elements are all in the same horizontal plane. *See:* **top-loaded vertical antenna; end capacitor.**
111

flat weighting (data transmission). A noise measuring set measuring amplitude frequency characteristics which are flat over a specified frequency range. The frequency range is stated. Flat noise power may be expressed in dBrn (F1-F2) or in dBm (F1-F2). The terms '3 kHz flat weighting' and '15 kHz flat weighting' from 30 Hz mean to the upper frequency indicated.
59

flection-point emission current. That value of current on the diode characteristic for which the second derivative of the current with respect to the voltage has its maximum negative value. *Note:* This current corresponds to the upper flection point of the diode characteristic. *See:* **electron emission.**
125,190

flexible connector (rotating machinery). An electric connection that permits expansion, contraction, or relative motion of the connected parts.
63

flexible coupling (rotating machinery). A coupling having relatively high transverse or torsional compliance. *Notes:* (1) May be used to reduce or eliminate transverse loads or deflections of one shaft from being carried, or felt by the other coupled shaft. (2) May be used to reduce the torsional stiffness between two rotating masses in order to change torsional natural frequencies of the shaft system or to limit transient or pulsating torques carried by the shafts. *See:* **rotor (rotating machinery).**
63

flexible equipment (seismic qualification of Class 1E equipment for nuclear power generating stations). Equipment, structures, and components whose lowest resonant frequency is less than the cutoff frequency on the response spectrum.
581

flexible metal conduit. A flexible raceway of circular cross section specially constructed for the purpose of the pulling in or the withdrawing of wires or cables after the conduit and its fittings are in place. *See:* **raceway.**
328

flexible mounting (rotating machinery). A flexible structure between the core and foundation used to reduce the transmission of vibration.
63

flexible nonmetallic tubing (loom). A mechanical protection for electric conductors that consists of a flexible cylindrical tube having a smooth interior and a single or double wall of nonconducting fibrous material. *See:* **raceways.**
328

flexible tower (frame). A tower that is dependent on the line conductors for longitudinal stability but is designed to resist transverse and vertical loads. *See:* **tower.**
64

flexible waveguide. A waveguide constructed to permit limited bending and twisting or stretching, or both, without appreciable change in its electrical properties.
166, 267

flexure (dynamically tuned gyro) (inertial sensor). An elastic element in a dynamically tuned gyro rotor suspension system, which permits limited angular freedom about axes perpendicular to the spin axis.
46

flicker (television). (1) **general.** Impression of fluctuating brightness or color, occurring when the frequency of the observed variation lies between a few hertz and the fusion frequencies of the images. (2) **television.** A repetitive variation in luminance of a given area in a monochromatic or color display, the visibility of which is a function of repetition rate, duty cycle, luminance, and the decay characteristic.   18

flicker effect (electron tubes). The random variations of the output-current in a valve or tube with an oxide-coated cathode. *Note:* Its value varies inversely with the frequency. *See:* **electron tube.**   244,190

flicker fusion frequency, fff (illuminating engineering). The frequency of intermittent stimulation of the eye at which flicker disappears. *Syn:* **critical fusion frequency (cff); critical flicker frequency (cff).**   167

flicker noise (nonlinear, active, and nonreciprocal waveguide components). One of the sources of noise associated with solid-state devices such as mixers or diode detectors, the amplitude of which varies inversely with frequency. It is also referred to as $1/f$ noise. In the audio-frequency region this noise becomes more significant than either thermal or shot noise.   530

flicker threshold (television). The luminance at which flicker is just perceptible at a given repetition rate, with other variables held constant.   18

flight instrument (navigation aid terms). A vehicle instrument used in the control of the direction of flight, attitude, altitude, or speed of a vehicle.   526

flight path (navigation aid terms). A proposed route in three dimensions. *See:* **course line.**   526

flight-path computer (navigation aid terms). Equipment providing outletsfor the control of the motion of a vehicle along a flight path.   526

flight-path deviation (electronic navigation). The amount by which the flight track of a vehicle differs from its flight path expressed in terms of either angular or linear measurement. *See:* **navigation.**   187,13

flight-path-deviation indicator (electronic navigation). A device providing a visual display of flight-path deviation. *See:* **navigation.**   187,13

flight track (electronic navigation). The path in space actually traced by a vehicle. *See:* **track.**   13,187

flip-flop (electronic computation or control). (1) A circuit or device, containing active elements, capable of assuming either one or two stable states at a given time, the particular state being dependent upon (A) the nature of an input signal, for example, its polarity, amplitude, and duration, and (B) which of two input terminals last received the signal. *Note:* The input and output coupling networks, and indicators, may be considered as an integral part of the flip-flop. (2) A device, as in (1) above, that is capable of counting modulo 2, in which case it might have only one input terminal. (3) A sequential logic element having properties similar to (1) or (2) above. *See:* **control system, feedback; toggle.**   235

flip-flop circuit. A trigger circuit having two conditions of permanent stability, with means for passing from one to the other by an external stimulus. *See:* **trigger circuit.**   59

float (gyro). An enclosed gimbal assembly housing the spin motor and other components such as the pickoff and torquer. This assembly is immersed in a fluid usually at the condition of neutral buoyancy.   46

float-displacement hysteresis (gyro, accelerometer) (inertial sensor). The difference in rebalance torque or equivalent input after displacing the float about the output axis from its null position in successive clockwise and counterclockwise directions by equal amounts (up to its full range of angular freedom unless otherwise specified). The torque may be displaced by applying torques to the float through a torquer or through gyroscopic or acceleration torques in either open- or closed-loop mode. The amount of float-displacement hysteresis may depend on the methods of applying torques, on the mode of operation (open or closed loop), and on the amount and duration of float displacement.   46

floating. A method of operation for storage batteries in which a constant voltage is applied to the battery terminals sufficient to maintain an approximately constant stage of charge. *See:* **trickle charge; charge.**   328

floating battery. A storage battery that is kept in operating condition by a continuous charge at a low rate.   328

floating carrier. *See:* **controlled carrier.**

floating control. *See:* **control system, floating.**

floating grid (electron tubes). An insulated gird, the potential of which is not fixed. *See:* **electronic tube.**   244,190

floating network or component (circuits and systems). A network or component having no terminal at ground potential.   67

floating neutral. One whose voltage to ground is free to vary when circuit conditions change. *See:* **center of distribution.**   159

floating point (1)(mathematics of computing). Pertaining to a numeration system in which each number is represented as a fractional quantity multiplied by an integral power of the radix. *See:* **fixed point; variable point.**   564

(2)(electronic digital computers). Pertaining to a system in which the location of the point does not remain fixed with respect to one end of numerical expressions, but is regularly recalculated. The location of the point is usually given by expressing a power of the base. *See:* **fixed point; variable point.**   235

floating-point number (radix-independent floating-point arithmetic). A digit string characterized by three components: a sign, a signed exponent, and a significand. Its numerical value, if any, is the signed product of its significand and the radix raised to the power of its exponent. In ANSI/IEEE 854-1987 a digit string is not always distinguished from a number it may represent.   588

floating-point system (electronic computation). *See:* **point.**

floating removable feed tube (cable plowing). A feed tube removably attached to a blade so relative motion may occur between the feed tube and the blade around

axis that are essentially vertical and horizontal (perpendicular to direction of travel). *See:* **feed tube.**
52

floating speed (process control). In single-speed or multiple-speed floating control systems, the rate of change of the manipulated variable.
56

floating zero (numerically controlled machines). A characteristic of a numerical machine control permitting the zero reference point on an axis to be established readily at any point in the travel. *Note:* The control retains no information on the location of any previously established zeros. *See:* **zero offset.**
224,207

float storage (gyro). The sum of attitude storage and the torque command storage in a rate integrating gyro. *See:* **attitude storage; torque command storage.**
46

float switch (liquid-level switch) (industrial control). A switch in which actuation of the contacts is effected when a a float reaches a predetermined level. *See:* **switch.**
244,206

flood (charge-storage tubes) (verb). To direct a large-area flow of electrons, containing no spatially distributed information, toward a storage assembly. *Note:* A large-area flow of electrons with spatially distributed information is used in image-converter tubes. *See:* **charge-storage tube.**
174

floodlight (illuminating engineering). A projector designed for lighting a scene or object to a brightness considerably greater than its surroundings. It usually is capable of being pointed in any direction and is of weatherproof construction. *Note:* The beam spread of floodlights may range from relatively narrow (10 degrees) to wide (more than 100 degrees).
167

floodlighting (illuminating engineering). A system designed for lighting a scene or object to a brightness greater than its surroundings. It may be for utility, advertising, or decorative purposes.
167

flood-lubricated bearing (rotating machinery). A bearing in which a continuous flow of lubricant is poured over the top of the bearing or journal at about normal atmospheric pressure. *See:* **bearing.**
63

flood projection (facsimile). The optical method of scanning in which the subject copy is floodlighted and the scanning spot is defined in the path of the reflected or transmitted light. *See:* **scanning (facsimile).**
12

floor acceleration (seismic qualification of Class 1E equipment for nuclear power generating stations). The acceleration of a particular building floor (or equipment mounting) resulting from the motion of a given earthquake. The maximum floor acceleration is the zero period acceleration (ZPA) of the floor response spectrum.
581

floor bushing. A bushing intended primarily to be operated entirely indoors in a substantially vertical position to carry a circuit through a floor or horizontal grounded barrier. Both ends must be suitable for operating in air. *See:* **bushing.**
348

floor cavity ratio (FCR) (illuminating engineering). For a cavity formed by the work-plane, the floor, and the wall surfaces between these two planes, the FCR is computed by using the distance from the floor to the work plane (h_f) as the cavity height in the equations given in the definition for **cavity ratio** .
167

floor lamp (illuminating engineering). A portable luminaire on a high stand suitable for standing on the floor.
167

floor trap (burglar-alarm system). A device designed to indicate an alarm condition in an electric protective circuit whenever an intruder breaks or moves a thread or conductor extending across a floor space. *See:* **protective signaling.**
328

flotation fluid (gyro, accelerometer) (inertial sensor). The fluid that suspends the float inside the instrument case. The float may be fully or partially floated within the fluid. The degree of flotation varies with temperature because the specific gravity of the fluid varies with temperature. In addition, the fluid provides damping. *See:* **damping fluid.**
46

flowchart (software). A graphical representation of the definition, analysis, or solution of a problem, in which symbols are used to represent operations, data, flow, and equipment. *See:* **block diagram; data.**
434

flow diagram (electronic computers). Graphic representation of a program or a routine. *See:* **algorithm; execution.**
210

flow of control (software). The sequence of operations performed in the execution of an algorithm. *See:* **algorithm; execution.**
434

flow relay (power switchgear). A relay that responds to a rate of fluid flow.
103

flow soldering. *See:* **dip soldering.**

flow switch (80) (power system device function numbers). A switch which operates on given values, or on a given rate of change, of flow.
402

fluctuating power (rotating machinery). A phasor quantity of which the vector represents the alternating part of the power, and that rotates at a speed equal to double the angular velocity of the current. *See:* **asynchronous machine.**
63

fluctuating target. A radar target whose echo amplitude varies as a function of time. *See:* **target fluctuation.**
13

fluctuation (1) (pulse terms). Dispersion of the pulse amplitude or other magnitude parameter of the pulse waveforms in a pulse train with respect to a reference pulse amplitude or a reference magnitude. Unless otherwise specified by a mathematical adjective, peak-to-peak fluctuation is assumed. *See:* **mathematical adjectives.**
254

(2) (radar). *See:* **target fluctuation.**
13

fluctuation loss (radar). The apparent loss in radar detectability or measurement accuracy for a target of given average echo return power due to target fluctuation. It may be measured as the increase in required average echo return power of a fluctuating target as compared to a target of constant echo return, to achieve the same detectability or measuement accuracy.
13

fluctuation noise. *See:* **random noise.**

fluence (solar cells). The total time-integrated number of particles that cross a plane unit area from either side.
113

fluid filled joints (power cable joints). Joints in which the joint housing is filled with an insulating material that is fluid at all operating temperatures.   34

fluid loss (rotating machinery). That part of the mechanical losses ina machine having liquid in its air gap that is caused by fluid friction. *See:* **asynchronous machine.**   63

fluidly delayed overcurrent trip. *See:* **fluidly delayed release (trip); overcurrent release (trip).**

fluidly delayed release (trip) (power switchgear). A release delayed by fluid displacement or adhesion.   103

fluid pressure supply system (hydraulic turbines). The pumps, means for driving them, pressure and sump tanks, valves and piping connecting the various parts of the governing system and associated and accessory devices.   8

fluids from essential freeze protection (electric pipe heating systems). The use of electric pipe heating systems to prevent the temperature of fluids from dropping below the freezing point of the fluid in desirably available or essential outdoor (usually) piping systems at fossil fueled generating stations. An example of an essential freeze protection system is the heating for the feedwater system.   448

fluorescence (illuminating engineering). The emission of light as the result of, and only during, the absorption of radiation of shorter wavelengths.   167

fluorescent lamp (illuminating engineering). A low-pressure mercury electric-discharge lamp in which a fluorescing coating (phosphor) transforms some of the ultraviolet energy generated by the discharge into light.   167

flush antenna (aircraft). An antenna having no projections outside the streamlined surface of the aircraft. In general, flush antennas may be considered as slot antennas.   328

flush-mounted antenna. An antenna constructed into the surface of a mechanism, or of a vehicle, without affecting the shape of that surface. *See:* **conformal antenna.**   111

flush-mounted device (1) (power and distribution transformer). A device in which the body projects only a small specified distance in front of the mounting surface.   53

(2) (power switchgear). One in which the body of the device projects only a small specified distance in front of the mounting surface.   103

flush mounted or recessed (illuminating engineering). A luminaire which is mounted above the ceiling (or behind a wall or other surface) with the opening of the luminaire level with the surface.   167

flush mounting (transformers). So designed as to have a minimal front projection when set into and secured to a flat surface.   53

flutter (sound recording and reproducing equipment). Frequency modulation of the signal in the range of approximately 6 Hz to 100 Hz resulting in distortion which may be perceived as a roughening of the sound quality of a tone or program.   145

flutter echo. A rapid succession of reflected pulses resulting from a single initial pulse.   176

flutter rate (sound recording and reproducing). The number of frequency excursions in hertz, in a tone that is frequency-modulated by flutter. *Notes:* (1) Each cyclical variation is a complete cycle of deviation, for example, from maximum-frequency to minimum-frequency and back to maximum-frequency at the rate indicated. (2) If the over-all flutter is the resultant of several components having different repetition rates, the rates and magnitudes of the individual components are of primary importance. *See:* **sound recording and reproducing.**   145

flux (1) (photovoltaic power system). The rate of flow of energy through a surface. *See:* **photovoltaic power system.**   186

(2) (soldering) (connections). A liquid or solid which when heated exercises a cleaning and protective action upon the surfaces to which it is applied.   281

(3) (solar cells). The number of particles that cross a plane unit area per unit time from either side.   113

(4)* (fiber optics). Synonym for **radiant power.**   433

*Obsolete

flux guide (induction heating usage). Magnetic material used to guide electromagnetic flux in desired channels. *Note:* The guides may be used either to direct flux to preferred locations or to prevent the flux from spreading beyond definite regions. *See:* **induction heater.**   14,114

flux linkages. The sum of the fluxes linking the turns forming the coil, that is, in a coil having N turns the flux linkage is

$$\lambda = \phi_1 + \phi_2 + \phi_3 \cdots \phi_N$$

where $\phi_1 =$ flux linking turn 1, $\phi_2 =$ flux linking turn 2, etcetera, and $\phi_N =$ flux linking the Nth turn.   197

fluxmeter. An instrument for use with a test coil to measure magnetic flux. It usually consists of a moving-coil galvanometer in which the torsional control is either negligible or compensated. *See:* **magnetometer.**   328

flux method. *See:* **lumen method.**

flux transfer theory (illuminating engineering). A method of calculating the illuminance in a room by taking into account the interreflection of the light flux from the room surfaces based on the average flux transfer between surfaces.   167

fly ash. The finely divided particles of ash entrained in flue gases arising from the combustion of fuel. The particles of ash may contain incompletely burned fuel. The term has been applied predominantly to the gas-borne ash from boilers with spreader stoker, underfeed stoker, and pulverized fuel (coal) firing. *Note:* The above definition is consistent with the generic concept of the word ash. However, all the particulates (including unburned carbon) in suspension in the flue gases are generally called fly ash and the term herein is used in this sense. 427

flyback (television). The rapid return of the beam in a cathode-ray tube in the direction opposite to that used for scanning. 18

flying spot scanner (optical character recognition). A device employing a moving spot of light to scan a sample space, the intensity of the transmitted or reflected light being sensed by a photoelectric transducer. 255,77

flywheel ring (rotating machinery). A heavy ring mounted on the spider for the purpose of increasing the rotor moment of inertia. *See:* **rotor (rotating machinery).** 63

FM. *See:* **frequency modulation.**

FMCW (radar). *See:* **frequency-modulated continuous waves.**

FM-FM telemetry. *See:* **frequency-modulation-frequency modulation (FM-FM) telemetry.**

FM radio broadcast band (overhead-power-line corona and radio noise). A band of frequencies assigned for frequency-modulated transmission of communication intended to entertain or enlighten the general public. *Note:* In the United States and Canada the frequency range is between 88 and 108 MHz. 411

FOA. *See:* **transformer, oil-immersed.**

focal length (laser-maser). The distance from the secondary nodal point of a lens to the primary focal point. In a thin lens, the focal length is the distance between the lens and the focal point. 363

focal point (laser-maser). The point toward which radiation converges or from which radiation diverges or appears to diverge. 363

focus (oscillograph). Maximum convergence of the electron beam manifested by minimum spot size on the phosphor screen. *See:* **astigmatism; oscillograph.** 185

focusing (1) (defocusing) (radio wave propagation). The convergence (divergence) of energy of a wave. 146

(2) (electron tubes). The process of controlling the convergence of the electron beam. 125

focusing and switching grille (color picture tubes). A color-selecting-electrode system in the form of an array of wires including at least two mutually-insulated sets of conductors in which the switching function is performed by varying the potential difference between them, and focusing is accomplished by maintaining the proper average potentials on the array and on the phosphor screen. 125

focusing coil. *See:* **focusing magnet.**

focusing device. An instrument used to locate the filament of an electric lamp at the proper focal point of lens or reflector optical systems. 328

focusing, dynamic (picture tubes). The process of focusing in accordance with a specified signal in synchronism with scanning. 125,178,190

focusing electrode (beam tube). An electrode the potential of which is adjusted to focus an electron beam. *See:* **electrode (electron tube).** 125,117

focusing, electrostatic (electron beam). A method of

focusing an electron beam by the action of an electric field. 125

focusing grid (pulse techniques). *See:* **focusing electrode.**

focusing magnet. An assembly producing a magnetic field for focusing an electron beam. 125

focusing, magnetic (electron beam). A method of focusing an electron beam by the action of a magnetic field. 190,178

fog (adverse-weather) lamps. Lamps that may be used in lieu of headlamps to provide road illumination under conditions of rain, snow, dust, or fog. *See:* **headlamp.** 167

fog-bell operator. A device to provide automatically the periodic bell signals required when a ship is anchored in fog. 328

fog lamps (illuminating engineering). Units which may be used in lieu of headlamps or in connection with the lower beam headlights to provide road illumination under conditions of rain, snow, dust or fog. *Syn:* **adverse-weather lamps.** 167

foil (foil tape) (burglar-alarm system). A fragile strip of conducting material suitable for fastening with an adhesive to glass, wood, or other insulating material in order to carry the alarm circuit and to initiate an alarm when severed. *See:* **protective signaling.** 328

folded dipole (antenna). An antenna composed of two or more parallel, closely-space dipole antennas connected together at their ends with one of the dipole antennas fed at its center and the others short circuited at their centers. 111

folded monopole antenna. A monopole antenna formed from half of a folded dipole with the unfed element(s) directly connected to the imaging plane. 111

follow current (gas tube surge-protective device) (power) (surge arresters)low-voltage air gap surge-protective devices)(gas-tube protective devices). The current from the connected power source that flows through an arrester during and following the passage of discharge current. 430, 370, 62, 2, 556. 490

follower drive (slave drive) (industrial control). A drive in which the reference input and operation are direct functions of another drive, called the master drive. *See:* **control system, feedback.** 206

font (computing systems). A family or assortment of characters of a given size and style. *See:* **type font.** 255,77

foot (rotating machinery). The part of the stator structure, end shield, or base, that provides means for mounting and fastening a machine to its foundation. *See:* **stator.** 63

footcandle (fc) (1) (illuminating engineering). A unit of illuminance. One footcandle is one lumen per square foot (lm/ft^2). 167

(2) (television). *See:* **illumination.** 18

footings (foundations). Structures set in the ground to support the bases of towers, poles, or other overhead structures. *Note:* Footings are usually skeleton steel pyramids, grills, or piers of concrete. *See:* **tower.** 64

footlambert* (1) (fL) (illuminating engineering). A

lambertian unit of luminance equal to $(1/\pi)$ candela per square foot. This term is obsolete.   167
*Deprecated

(2) (television). A unit of luminance (photometric brightness) equal to $1/\pi$ candela per square foot ($10.7639/\pi$ candelas per square meter), or to the uniform luminance of a perfectly diffusing surface emitting or reflecting light at the rate of 1 lumen per square foot (10.7639 lumens per square meter), or to the average luminance of any surface emitting or reflecting light at that rate. *Notes:* (A) A footcandle is a unit of incident light, and a footlambert is a unit of emitted or reflected light. For a perfectly reflecting or perfectly diffusing surface, the numbers of footcandles is equal to the number of footlamberts. (B) The average luminance of any reflecting surface in footlamberts is, therefore, the product of the illumination in footcandles by the luminous reflectance of the surface.
18
*Deprecated

(3) (light emitting diodes) (fL*). A unit of luminance (photometric brightness) equal to $1/pi$ candela per square foot, or to the uniform luminance of a perfectly diffusing surface emitting or reflecting light at the rate of one lumen per square foot.   162
*Deprecated

footprint (of an antenna beam on a specified surface). An area bounded by a contour on a specified surface formed by the intersection of the surface and that portion of the beam of an antenna above a specified minimum gain level, the orientation of the beam with respect to the surface being specified.   111

foot switch (industrial control). A switch that is suitable for operation by an operator's foot. *See:* **switch.**
206

forbidden combination. A code expression that is defined to be nonpermissible and whose occurrence indicates a mistake or malfunction.   125

forbidden-combination check (electronic computation). *See:* **check, forbidden-combination.**

force. Any physical cause that is capable of modifying the motion of a body. The vector sum of the forces acting on a body at rest or in uniform rectilinear motion is zero.   210

forced-air cooling system (1) (rectifier). An air cooling system in which heat is removed from the cooling surfaces of the rectifier by means of a flow of air produced by a fan or blower. *See:* **rectification.**
291,208

(2) (thyristor controller). A cooling system in which the heat is removed from the cooling surfaces of the thyristor controller components by means of a flow of air produced by a fan or blower.   445

forced derated hours (FDH)(electric generating unit reliability, availability, and productivity). The available hours during which a Class 1, 2, or 3 unplanned derating was in effect.   567

forced drainage (underground metallic structures). A method of controlling electrolytic corrosion whereby an external source of direct-current potential is employed to force current to flow to the structure through the earth, thereby maintaining it in a cathode condition. *See:* **inductive coordination.**   328

forced interruption (electric power systems). An interruption caused by a forced outage. *See:* **outage.**
200

forced-lubricated bearing (rotating machinery). A bearing in which a continuous flow of lubricant is forced between the bearing and journal.   63

forced oscillation (linear constant-parameter system). The response to an applied driving force. *See:* **network analysis.**   210

forced outage (emergency and standby power). A power outage that results from the failure of a system component, requiring that it be taken out of service immediately, either automatically or by manual switching operations, or an outage caused by improper operation of equipment or human error. This type of power outage is not directly controllable and is usually unexpected.   512
See: **failure.**

forced outage duration. *See:* **repair time.**

forced outage hours (FOH)(electric generating unit reliability, availability, and productivity). The number of hours a unit was in a Class 1, 2, or 3 unplanned outage state.   567

forced release (telephone switching systems). Release initiated from sources other than the calling or called line.   55

forced response (circuits and systems). The response of a system resulting from the application of an energy source with the system initially free of stored energy.
67

forced unavailability (reliable industrial and commercial power systems planning and design). The long-term average fraction of time that a component or system is out of service due to failures.   561

forced-ventilated machine. *See:* **open-pipe-ventilated machine.**

force factor (1) (electroacoustic transducer). (A) The complex quotient of the pressure required to block the acoustic system divided by the corresponding current in the electric system; (B) the complex quotient of the resulting open-circuit voltage in the acoustic system. *Note:* Force factors (A) and (B) have the same magnitude when consistent units are used and the transducer satisfies the principle of reciprocity.   176

(2) (electromechanical transducer). (A) The complex quotient of the force required to block the mechanical system divided by the corresponding current in the electric system; (B) the complex quotient of the resulting open-circuit voltage in the electric system divided by the velocity in the mechanical system. *Notes:* (1) Force factors (A) and (B) have the same magnitude when consistent units are used and the transducer satisfies the principle of reciprocity. (2) It is sometimes convenient in an electrostatic or piezoelectric transducer to use the ratios between force and charge or electric displacement, or between voltage and mechanical displacement.   328

forcing (industrial control). The application of control impulses to initiate a speed adjustment, the magnitude of which is greater than warranted by the desired controlled speed in order to bring about a greater rate of speed change. *Note:* Forcing may be obtained by directing the control impulse so as to effect a change in the field or armature circuit of the motor, or both. *See:* **electric drive.** 206

foreign area (telephone switching systems). A numbering plan area other than the one in which the calling customer is located. 55

foreign exchange line (1) (data transmission). A subscriber line by means of which service is furnished to a subscriber at the subscriber's request from an exchange other than the one from which service would normally be furnished. 59

(2) (telephone switching systems). A loop form an exchange other than the one from which service would normally be furnished. 55

forestalling switch. *See:* **acknowledger; forestaller.**

fork beat (facsimile). See: **carrier beat.**

form. Any article such as a printing plate, that is used as a pattern to be reproduced. *See:* **electroforming.** 328

formal language (software). A language whose rules are explicitly established prior to its use. Examples include programming language, such as FORTRAN and Ada, and mathematical or logical languages, such as predicate calculus. *Syn:* **artificial language.** *See:* **natural language; programming language.** 434

formal logic. The study of the structure and form of valid argument without regard to the meaning of the terms in the argument. 255,77

formal parameter (software). A variable used in a subprogram to represent data or program elements to be transmitted to the subprogram by a calling routine. *Syn:* **dummy parameter.** *See:* **actual parameter; data; program; routine; subprogram; variable.** 434

formal specification (software). (1) A specification written and approved in accordance with established standards. (2) In proof of correctness, a description in a formal language of the externally visible behavior of a system or system component. *See:* **component; formal language; proof of correctness; specification; system.** 434

formal testing (software). The process of conducting testing activities and reporting results in accordance with an approved test plan. *See:* **testing; test plan.** 434

format (1) (computing systems). The general order in which information appears on the input medium.

(2) (data transmission). Arrangement of code characters within a group, such as a block or message. 194

(3) Physical arrangement of possible locations of holes or magnetized areas. *See:* **address format.** 107

format classification (numerically controlled machines). A means, usually in an abbreviated notation, by which the motions, dimensional data, type of control system, number of digits, auxiliary functions, etcetera, for a particular system can be denoted. 107

format detail (numerically controlled machines). De-

scribes specifically which words and of what length are used by a specific system in the format classification. 244,207

formation lights (illuminating engineering). A navigation light especially provided to facilitate formation flying. 167

formation voltage. The final impressed voltage at which the film is formed on the valve metal in an electrochemical valve. *See:* **electrochemical valve.** 328

formatted (computer applications). (1) Pertaining to magnetic media, such as tapes or diskettes, that have been initialized and prepared to accept and store data. (2) Pertaining to text that has been organized into a particular arrangement for output or display. 571

form designation (watthour meter). An alphanumeric designation denoting the circuit arrangement for which the meter is applicable and its specific terminal arrangement. The same designation is applicable to equivalent meters of all manufacturers. 212

formette (rotating machinery). *See:* **form-wound motorette.**

form factor (1) (periodic function). The ratio of the root-mean-square value to the average absolute value, averaged over a full period of the function. *See:* **power rectifier; rectification.** 208

(2) (electric process heating). Coil ratio of conductor width to turn to turn space. *See:* **coil shape factor.** 14

(3) f_{1-2} (illuminating engineering). The ratio of the flux directly received by surface 2 (and due to lambertian surface 1) to the total flux emitted by surface 1. It is used in flux transfer theory. 167

forming (1) (electrical) (semiconductor devices). The process of applying electric energy to a semiconductor device in order to modify permanently the electric characteristics. *See:* **semiconductor.** 66

(2) (semiconductor rectifier). The electrical or thermal treatment, or both, of a semiconductor rectifier cell for the purpose of increasing the effectiveness of the rectifier junction. *See:* **rectification.** 237,66

(3) (electrochemical). The process that results in a change in impedance at the surface of a valve metal to the passage of current from metal to electrolyte, when the voltage is first applied. *See:* **electrochemical valve.** 328

forming shell (National Eleetrical Code). A metal structure designed to support a wet-niche lighting fixture assembly and intended for mounting in a swimming pool structure. 256

form-wound (performed winding) (rotating machinery). Applied to a winding whose coils are formed essentially to their final shape prior to assembly into the machine. *See:* **rotor (rotating machinery); stator.** 63

form-wound motorette (formette) (rotating machinery). A motorette for form-wound coils. *See:* **asynchronous machine.** 63

fortuitous distortion (data transmission). A random distortion of telegraph signals such as that commonly produced by interference. 59

fortuitous telegraph distortion. Distortion that includes those effects that cannot be classified as bias or characteristic distortion and is defined as the departure, for one occurrence of a particular signal pulse, from the average combined effects of bias and characteristic distortion. *Note:* Fortuitous distortion varies from one signal to another and is measured by a process of elimination over a long period. It is expressed in percent of unit pulse. *See:* **distortion.**   111

forward (reverse) direction in isolator (nonlinear, active, and nonreciprocal waveguide components). That direction of propagation between two ports of an isolator for which attenuation of waves is lower (higher) than in the opposite direction.   530

forward-acting regulator. A transmission regulator in which the adjustment made by the regulator does not affect the quantity that caused the adjustment. *See:* **transmission regulator.**   328

forward admittance, short-circuit (electron-device transducer). *See:* **admittance, short-circuit forward.**

forward bias (forward voltage) (VF) (light emitting diodes). The bias voltage which tends to produce current flow in the forward direction.   162

forward breakover (thyristor). The failure of the forward blocking action of the thyristor during a normal OFF-state period.   445

forward current (1) (metallic rectifier). The current that flows through a metallic rectifier cell in the forward direction. *See:* **rectification.**   328

(2) (semiconductor rectifier device). The current that flows through a semiconductor rectifier device in the forward direction. *See:* **rectification.**   237,66

(3) (reverse-blocking or reverse-conducting thyristor). The principal current for a positive anode-to-cathode voltage. *See:* **principal current.**

243,66,208,191

(4) (light emitting diodes) (I_F). The current that flows through a semiconductor junction in the forward direction.   162

forward current, average, rating. *See:* **average forward current rating.**

forward direction (1) (metallic rectifier). The direction of lesser resistance to current flow through the cell; that is, from the negative electrode to the positive electrode. *See:* **rectification.**   328

(2) (semiconductor rectifier device). The direction of lesser resistance to steady direct-current flow through the device; for example, from the anode to the cathode. *See:* **semiconductor; semiconductor rectifier stack.**   66

(3) (semiconductor rectifier diode). The direction of lower resistance to steady-state direct-current; that is, from the anode to the cathode.   66

forward error-correcting system. A system employing an error-correcting code and so arranged that some or all signals detected as being in error are automatically corrected at the receiving terminal before delivery to the data sink or to the telegraph receiver.   194

forward gate current (thyristor). The gate current when the junction between the gate region and the adjacent anode or cathode region is forward biased. *See:* **principal current.**   243,66,208,190

forward gate voltage (thyristor). The voltage between the gate terminal and the terminal of the adjacent anode or cathode region resulting from forward gate current. *See:* **principal voltage-current characteristic (principal characteristic).**   243,66,208,190

forward path (signal-transmission system) (feedback -control loop). The transmission path from the loop-error signal to the loop-output signal. *See:* **feedback.**

188

forward period (rectifier circuit element) (rectifier circuit). The part of an alternating-voltage cycle during which forward voltage appears across the rectifier circuit element. *Note:* The forward period is not necessarily the same as the conducting period because of the effect of circuit parameters and semiconductor rectifier cell characteristics. *See:* **rectifier circuit element.**   237,66

forward power dissipation (semiconductor). The power dissipation resulting from forward current.

66

forward power loss (semiconductor device). The power loss within a semiconductor rectifier device resulting from the flow of forward current. *See:* **rectification; semiconductor rectifier stack.**   237,66,208

forward recovery time (semiconductor diode). The time required for the current or voltage to recover to a specified value after instantaneous switching from a stated reverse voltage condition to a stated forward current or voltage condition in a given circuit. *See:* **rectification.**   237,66

forward resistance (metallic rectifier). The resistance measured at a specified forward voltage drop or a specified forward current. *See:* **rectification.**   328

forward-scattering cross section. *See:* **radar cross section.**

forward voltage (1) (rectifiers). Voltage of the polarity that produces the larger current, hence, the voltage across a semiconductor rectifier diode resulting from forward current. *See:* **on-state voltage; forward voltage drop.**

(2) (reverse blocking or reverse conducting thyristor). A positive anode-to-cathode voltage. *See:* **principal characteristic (principal voltage-current characteristic).**   243,66,208,191

forward voltage drop (1) (metallic rectifier). The voltage drop in the metallic rectifying cell resulting from the flow of current through a metallic rectifier cell in the forward direction.

(2) (semiconductor rectifier). *See:* **forward voltage; on-state voltage.**   237,66,108

forward voltage overshoot (thyristor). The difference between the maximum forward OFF-state voltage following turn-off and the instantaneous ac voltage.

445

forward wave (traveling-wave tubes). A wave whose group velocity is in the same direction as the electron steam motion.   190

forward-wave structure (microwave tubes). A slow-wave structure whose propagation is characterized on a ω/β diagram (ω versus phase shift/section) by a positive slope in the region $0 < \beta < \pi$ (in which the group and phase velocity therefore have the same sign). *See:* **microwave tube or valve.**   190

Foster's reactance theorem (circuits and systems). States that the driving-point impedance of a network composed of purely capacitive and inductive reactances is an odd rational function of frequency (ω) which has the following characteristics; (1) a positive slope (2) the poles and zeros of the function are on the $j\omega$ axis, they are simple, they occur in complex conjugate pairs and they alternate.   67

FOT (radio wave propagation). Initials for French *Frequence Optimum de Travail* corresponding to Optimum Working Frequency.   146

Foucault currents. *See:* **eddy currents.**

foul electrolyte. An electrolyte in which the amount of impurities is sufficient to cause an undesirable effect on the operation of the electrolytic cells in which it is employed.

fouling. The accumulation and growth of marine organisms on a submerged metal surface.   205

fouling point (railway practice). The location in a turnout back of a frog at or beyond the clearance point at which insulated joints or details are placed.   328

foundation (rotating machinery). The structure on which the feet or base of a machine rest and are fastened.   63

foundation bolt (rotating machinery). A bolt used to fasten a machine to a foundation.   63

foundation-bolt cone (rotating machinery). A cone placed around a foundation bolt when imbedded in a concrete foundation to provide clearance for adjustment during erection.   63

four-address. Pertaining to an instruction code in which each instruction has four address parts. *Note:* In a typical four-address instruction the address specify the location of two operands, the destination of the result, and the location of the next instruction to be interpreted. *See:* **three-plus-one address.**   125

four-address code (electric computation). *See:* **instruction code.**

four conductor bundle. *See:* **bundle.**

Fourier series. A single-valued periodic function (that fulfills certain mathematical conditions) may be represented by a Fourier series as follows

$$f(x) = 0.5A_0 + \sum_{n=1}^{n=\infty} [A_n \cos nx + B_n \sin nx]$$

$$= 0.5A_0 + \sum_{n=1}^{n=\infty} C_n \sin(nx + \theta_n)$$

where

$$A_n = \frac{1}{\pi} \int_0^{2\pi} f(x) \cos nx \, dx$$

$$n = 0,1,2,3,\cdots$$

$$B_n = \frac{1}{\pi} \int_0^{2} f(x) \sin nx \, dx$$

$$C_n = +(A_n^2 + B_n^2)^{1/2}$$

and

$$\theta_n = \text{arc tan } A_n/B_n$$

Note: $0.5A_0$ is the average of a periodic function $f(x)$ over one primitive period.   210

Fourier spectrum (seismic qualification of Class 1E equipment for nuclear power generating stations). A complex valued function that provides amplitude and phase information as a function of frequency for a time domain waveform.   581

four-plus-one address (computing systems). Pertaining to an instruction that contains four operand addresses and a control address.   255,77

four-pole. *See:* **two-terminal pair network.**

four quadrant DAM (hybrid computer linkage components). A digital-to-analog multiplier (DAM) that accepts both signs of the digital value, giving correct sign output in all four quadrants.   10

four-quadrant multiplier (analog computers). A multiplier in which operation is unrestricted as to the sign of both of the input variables.   9

four-terminal network. A network with four accessible terminals. *Note:* See **two-terminal-pair network** for an important special case. *See:* **two-terminal-pair network; quandripole.**   210

fourth voltage range (railway signal and interlocking). *See:* **voltage range.**

fourth-wire control (telephone switching systems). The wire (in addition to the tip, ring, and sleeve wires) used for transmission of special signals necessary in the establishment or supervision of a call.   55

four-wire channel(1)(data transmission). *See:* **four-wire circuit.**   59

(2)(telephone loop performance). Consists of two unidirectional channels carrying signals in opposite directions.   473

four-wire circuit (data transmission). A two-way circuit using two paths so arranged that the electric waves are transmitted in one direction only by one path and in the other direction only by the other path. *Note:* The transmission paths may or may not employ four wires.   59

four-wire repeater (data transmission). A telephone repeater for use in a four-wire circuit and in which there are two currents in one side of the four-wire circuit and the other serving to amplify the telephone current in the other side of the four-wire circuit.   59

four-wire switching (telephone switching systems). Switching using a separate path, frequency, or time interval for each direction of transmission.   55

four-wire systems (generating station grounding). A three-phase system consisting of three phase conductors and a neutral conductor.   569

four-wire terminating set (data transmission). A hybrid set for interconnecting a four-wire and two-wire circuit.   59

fovea (illuminating engineering). A small region at the center of the retina, subtending about 2 degrees which contains cones but no rods, and forms the site of most distinct vision.   167

FOW. *See:* **transformer, oil-immersed.**

fraction (1)(binary floating-point arithmetic). The field of the significand that lies to the right of its implied binary point.   469

(2)(radix-independent floating-point arithmetic). The component of the significand that lies to the right of its implied radix point.   588

fractional-horsepower brush (rotating machinery). A brush with a cross-sectional area of 1/4 square inch (thickness x width) or less and not exceeding 1 1/2 inches in length, but larger than a miniature brush and smaller than an industrial brush. *See:* **brush (1) (electric machines).**   63

fractional-horsepower motor. A motor built in a frame smaller than that of a motor of open construction having a continuous rating of 1 horsepower at 1700-1800 revolutions per minute. *See:* **asynchronous machine.**   63

fractional-slot winding (rotating machinery). A distributed winding in which the average number of slots per pole per phase is not integral, for example *3 2/7* slots per pole per phase. *See:* **asynchronous machine.**   63

fragility (nuclear power generating stations) (seismic qualification of class 1E equipment) (seismic testing of relays). Susceptibility of equipment to malfunction as the result of structural or operational limitations, or both.   392,28

fragility level (nuclear power generating stations) (seismic qualification of class 1E equipment) (seismic testing of relays). The highest level of input excitation, expressed as a function of input frequency, that an equipment can withstand and still perform the required Class 1E functions.   392,28

fragility response spectrum (FRS) (nuclear power generating stations) (seismic qualification of class 1E equipment) (seismic testing of relays). A TRS (test response spectrum) obtained from tests to determine the fragility level of equipment. *See:* **test response spectrum.**   392,28

frame (1)(token ring access method). A transmission unit that carries a protocol data unit (PDU) on the ring. .   472

(2) (facsimile). A rectangular area, the width of which is the available line and the length of which is determined by the service requirements.   12

(3) (test, measurement and diagnostic equipment). A cross section of tape containing one bit in each channel and possibly a parity bit. *Syn:* **tape line.**   54

(4) (data) (data transmission). A set of consecutive digit time slots in which the position of each digit time slot can be identified by reference to a framing signal.   59

framed plate (storage cell). A plate consisting of a frame supporting active material. *See:* **battery (primary or secondary).**   328

frame frequency (television). The number of times per second that the frame is scanned. *See:* **television.**   328

frame, intermediate distributing. *See:* **intermediate distributing frame.**

frame, main distributing. *See:* **main distributing frame.**

frame rate (data transmission). The repetition rate of the frame.   59

frame ring (rotating machinery). A plate or assembly of flat plates forming an annulus in a radial plane and serving as a part of the frame to stiffen it.   63

frame size (as applied to a low-voltage circuit breaker) (power switchgear). A term which denotes the maximum continuous current rating in amperes for all parts except the coils of the direct-acting trip device.   103

frame split (rotating machinery). A joint at which a frame may be separated into parts.   63

frame synchronization (data transmission). The process whereby a given channel at the receiving end is aligned with the corresponding channel at the transmitting end.   59

framework (rotating machinery). A stationary supporting structure.   63

frame yoke (field frame) (rotating machinery). The annular support for the poles of a direct-current machine. *Note:* It may be laminated or of solid metal and forms part of the magnetic circuit.   63

framing (facsimile). The adjustment of the picture to a desired position in the direction of line progression. *See:* **recording (facsimile).**   12

framing signal (facsimile). A signal used for adjustment of the picture to a desired position in the direction of line progression. *See:* **facsimile signal (picture signal).**   12

Fraunhofer diffraction pattern. *See:* **far-field diffraction pattern.**

Fraunhofer pattern (antennas). A radiation pattern obtained in the Fraunhofer region of an antenna. *Note:* For an antenna focused at infinity a Fraunhofer pattern is a far-field pattern.   111

Fraunhofer region (1) (data transmission). That region of the field in which the energy flow from an antenna proceeds as though coming from a point source located in the vicinity of the antenna. *Note:* If the antenna has a well-defined aperture D in a given aspect, the Fraunhofer region in that aspect is commonly used to exist at distances greater than $2D^2$ from the aperture, being the wavelength.   59

(2) (antenna). The region in which the field of antenna is focused. *Note:* In the Fraunhofer region of an antenna focused at infinity, the values of the fields, when calculated from knowledge of the source distribution of an antenna, are sufficiently accurate when the quadratic phase terms (and higher order terms) are neglected. *See:* **antenna; far-field region.**   111

free capacitance (1) (conductor). The limiting value of its self-capacitance when all other conductors, including isolated ones, are infinitely removed.   210

(2) (between two conductors). The limiting value of the plenary capacitance as all other, including isolated, conductors are infinitely removed.   210

free-code call (telephone switching systems). A call to a service or office code for which no charge is made.   55

free cyanide (electrodepositing solution) (electroplat-

ing). The excess of alkali cyanide above the minimum required to give a clear solution, or above that required to form specified soluble double cyanides. *See:* **electroplating.**   328

free field. A field (wave or potential) in a homogeneous, isotropic medium free from boundaries. In practice, a field in which the effects of the boundaries are negligible over the region of interest. *Note:* The actual pressure impinging on an object (for example, electroacoustic transducer) placed in an otherwise free sound field will differ from the pressure that would exist at that point with the object removed, unless the acoustic impedance of the object matches the acoustic impedance of the medium.   176

free-field current response (receiving current sensitivity) (electroacoustic transducer used for sound reception). The ratio of the current in the output circuit of the transducer when the output terminals are short-circuited to the free-field sound pressure existing at the transducer location prior to the introduction of the transducer in the sound field. *Notes:* (1) The free-field response is defined for a plane progressive sound wave whose direction of propagation has a specified orientation with respect to the principal axis of the transducer. (2) The free-field current response is usually expressed in decibels, namely, 20 times the logarithm to the base 10 of the quotient of the observed ratio divided by the reference ratio, usually 1 ampere per newton per square meter. *See:* **loudspeaker.**   176

free-field microphone (audible noise measurement). A microphone which has been designed to have a flat frequency response for sound waves propagating from a direction perpendicular to the plane of the diaphragm of the microphone.   462

free-field voltage response (receiving voltage sensitivity) (electroacoustic transducer used for sound reception). The ratio of the voltage appearing at the output terminals of the transducer when the output terminals are open-circuited to the free-field sound pressure existing at the transducer location prior to the introduction of the transducer in the sound field. *Notes:* (1) The free-field response is determined for a plane progresive sound wave whose direction of propagation has a specified orientation with respect to the principal axis of the transducer. (2) The free-field voltage response is usually expressed in decibels, namely, 20 times the logarithm to the base 10 of the quotient of the observed ratio divided by the reference ratio, usually 1 volt per newton per square meter. *See:* **loudspeaker.**   176

free gyro. A two-degree-of-freedom gyro in which the spin axis may be oriented in any specific attitude. In this gyro, output signals are produced by an angular displacement of the case about an axis other than the spin axis.   46

free impedance (transducer). The impedance at the input of the transducer when the impedance of its load is made zero. *Note:* The approximation is often made that the free electric impedance of an electroacoustic transducer designed for use in water is that measured with the transducer in air. *See:* **loudspeaker; self-impedance.**   176

free-line call (telephone switching systems). A call to a directory number for which no charge is made.   55

free motion (automatic control). One whose nature is determined only by parameters and initial conditions for the system itself, and not by external stimuli. *Note:* For a linear system, this motion is described by the complementary function of the associated homogeneous differential equation. *Syn:* **free oscillation.**   56

free motional impedance (transducer) (electroacoustics). The complex remainder after the blocked impedance has been subtracted from the free impedance. *See:* **self-impedance.**   176

free oscillation. The response of a system when no external driving force is applied and energy previously stored in the system produces the response. *Note:* The frequency of such oscillations is determined by the parameters in the system or circuit. The term shock-excited oscillation is commonly used. *See:* **oscillatory circuit.**   111

free progressive wave (free wave) (acoustics). A wave in a medium free from boundary effect. A free wave in a steady state can only be approximated in practice.   176

free-radiation frequencies for industrial, scientific, or medical (ISM) apparatus (electromagnetic compatibility). Center of a band of frequencies assigned to industrial, scientific, or medical equipment either nationally or internationally for which no power limit is specified. *See:* **ISM apparatus; electromagnetic compatibility.**   220,167

free-running frequency. The frequency at which a normally synchronized oscillator operates in the absence of a synchronizing signal.   178

free-running sweep (oscilloscopes) (non-real time spectrum analyzer) (spectrum analyzer). A sweep that recycles without being triggered and is not synchronized by any applied signal. *See:* **oscillograph.**   390, 186,68

free-space field intensity. The radio field intensity that would exist at a point in a uniform medium in the absence of waves reflected from the earth or other objects. *See:* **radiation.**   328

free-space loss (antennas). The loss between two isotropic radiators in free space, expressed as a power ratio. *Note:* The free-space loss is not due to dissipation, but rather due to the fact that the power flux density decreases with the square of the separation distance. It is usually expressed in decibels and is given by the formula $20 \log (4\pi D/\lambda)$, where D is the separation of the two antennas and λ is the wavelength.   111

free-space transmission (mobile communication). Electromagnetic radiation that propagates unhindered by the presence of obstructions, and whose power or field intensity decreases as a function of distance squared. *See:* **mobile communication system.**   181

free time (availability). The period of time during which an item is in a condition to perform its required function but is not required to do so.   164 free wave (acoustics). *See:* free progressive wave.

freeze-out (telephone circuit). A short-time denial to a subscriber by a speech-interpolation system.   328

freeze protection (1)(electric pipe heating systems). The use of electric pipe heating systems to prevent the temperature of fluids from dropping below the freezing point of the fluid. Freeze protection is usually associated with outdoor piping, pumps, valves, tanks, instrumentation, etcetera such as water lines.   405

(2). The use of electric pipe heating systems to prevent the temperature of fluids from dropping below the freezing point of the fluid. Freeze protection is usually associated with piping, pumps, valves, tanks, instrumentation, etcetera, such as water lines, that are located outdoors, or in unheated buildings.   448

F region (radio wave propagation). The region of the terrestrial ionosphere above about 160 km altitude.   146

freight elevator. An elevator primarily used for carrying freight on which only the operator and the persons necessary for loading and unloading the freight are permitted to ride. *See:* elevators.   328

Frenkel defect (solar cells). A defect consisting of the displacement of a single atom from its place in the atomic lattice of a crystal, the atom then occupying an interstitial position.   113

frequency (1)(measurement of power frequency electric and magnetic fields from ac power lines). The number of complete cycles of sinusoidal variation per unit time. *Notes:* (A) Electric and magnetic field components have a fundamental frequency equal to that of the power line voltages and currents. (B) For ac power lines, the most widely used frequencies are 60 and 50 hertz (Hz).   514

(2) (general). The number of periods per unit time.   3

(3) (automatic control). The number of periods, or specified fractions of periods, per unit time. *Note:* The frequency may be stated in cycles per second, or in radians per second, where 1 cycle = 2 π radians.   56

(4) (data transmission). (periodic function--wherein time is the independent variable). (A) (General). The number of periods per unit time. (B) (Automatic control). The number of periods, or specified fractions of periods, per unit time. *Note:* The frequency may be stated in cycles per second, or in radians per second, where 1 cycle = two radians. (C) (Transformer). The number of periods occurring per unit time. (D) (Pulse terms). The reciprocal of period.   59

(5) (electric installations on shipboard). The frequency of a periodic quantity, in which time is the independent variable,is the number of periods occurring in unit time.   3

(6) (pulse terms). A pulse radar in which the transmitter carrier frequency is changed between pulses in a random or pseudo-random way by an amount comparable to the reciprocal of the pulsewidth, or a multiple thereof.   13

(7) (power and distribution transformer). The number of periods occurring per unit time.   53

(8) (radio wave propagation). Of a periodic wave, the number of identical cycles per second.   146

frequency-agile radar. A pulse radar in which the transmitter carrier frequency is changed between pulses or between groups of pulses by an amount comparable to or greater than the pulse bandwidth.   13

frequency allocation (table) (electromagnetic compatibility). (1) The process of designating radio-frequency bands for use by specific radio services; or (2) The resulting table (of frequency allocations). *See:* electromagnetic compatibility.   199

frequency allotment (plan) (electromagnetic compatibility). (1) The process of designating radio frequencies within an allocated band for use within specific geographic areas. (2) The resulting plan (of frequency allotment). *See:* electromagnetic compatibility.   199

frequency assignment (list) (electromagnetic compatibility). (1) The process of designating radio frequency for use by a specific station under specified conditions of operations. (2) The resulting list of frequency assignments. *See:* electromagnetic compatibility.   199

frequency band (1). A continuous range of frequencies extending between two limiting frequencies. *Note:* The term frequency band or band is also used in the sense of the term bandwidth. *See:* channel; signal; signal wave.   111,240

(2) (overhead-power-line corona and radio noise). A continuous range of frequencies extending between two limiting frequencies. *Note:* A band of frequencies is also called a channel.   411

(3) (spectrum analyzer). A continuous range of frequencies extending between two limiting frequencies.   390

frequency-band number. The number N in the expression 0.3×10^{N} that defines the range of band N. Frequency band N extends from 0.3×10^{N} hertz, the lower limit exclusive, the upper limit inclusive.   210

frequency band of emission (communication band). The band of frequencies effectively occupied by that emission, or the type of transmission and the speed of signaling used. *See:* radio transmission.   111

frequency bands (mobile communication). The frequency allocations that have been made available for land mobile communications by the Federal Communications Commission, including the spectral bands: 25.0 to 50.0 megahertz, 150.8 to 173.4 megahertz, and 450.0 to 470.0 megahertz. *See:* mobile communication.   181

frequency bias (electric power systems). An offset in the scheduled net interchange power of a control area that varies in proportion to the frequency deviation. *Note:* This offset is in a direction to assist in restoring the frequency to schedule. *See:* power system.   94

frequency changer (1)(self-commutated converters). An alternating current (ac) converter for changing frequency.   584

(2) (rotating machinery). A motor-generator set or other equipment which changes power of an alternating-current system from one frequency to another.   63

frequency-changer set (rotating machinery). A motor-generator set that changes the power of an alternating-current system from one frequency to another.   63

frequency-change signaling (telecommunication). A method in which one or more particular frequencies correspond to each desired signaling condition. *Note:* The transition from one set of frequencies to the other may be either a continuous or a discontinuous change in frequency or in phase. *See:* **frequency modulation.**   194

frequency, chopped. *See:* **rate, chopping.**

frequency control. The regulation of frequency within a narrow range. *See:* **generating station.**   64

frequency-conversion transducer. *See:* **conversion transducer.**

frequency converter. *See:* **frequency changer.**

frequency converter, commutator type (rotating machinery). A polyphase machine the rotor of which has one or two windings connected to slip rings and to a commutator. *Note:* By feeding one set of terminals with a voltage of given frequency, a voltage of another frequency may be obtained from the other set of terminals. *See:* **asynchronous machine.**   63

frequency, corner (asymptotic form of Bode diagram) (control system, feedback). The frequency indicated by a breakpoint, that is, the junction of two confluent straight lines asymptotic to the log gain curve. *Note:* One breakpoint is associated with each distinct real root of the characteristic equation, one with each set of repeated roots, and one with each pair of complex roots. For a single real root, corner frequency (in radians per second) is the reciprocal of the corresponding time constant (in seconds), and the corresponding phase angle is halfway between the phase angles belonging to the asymptotes extended to infinity. *See:* **control system, feedback.**   56,329

frequency, cyclotron. *See:* **cyclotron frequency.**

frequency, damped (automatic control). The apparent frequency of a damped oscillatory time response of a system resulting from a nonoscillatory stimulus. *Note:* The value of the frequency in a particular system depends somewhat on the subsidence ratio. *See:* **control system, feedback.**   56,329

frequency departure (telecommunication). The amount of variation of a carrier frequency or center frequency from its assigned value. *Note:* The term frequency deviation, which has been used for this meaning, is in conflict with this essential term as applied to phase and frequency modulation and is therefore deprecated for future use in the above sense. *See:* **radio transmission.**   111

frequency-dependent negative resistor (circuits and systems). An impedance of the form $1 / (Ks^2)$, where K is a real positive constant and s is the complex frequency variable.   67

frequency deviation (1) (power system). System frequency minus the scheduled frequency. *See:* **frequency modulation; frequency departure.**   94

(2) (telecommunication; frequency modulation). The peak difference between the instantaneous frequency of the modulated wave and the carrier frequency.   111,415

(3) (frequency modulation broadcast receivers). The difference between the instantaneous frequency of the modulated wave and the carrier frequency.   16

(4) (data transmission). In frequency modulation, the peak difference between the instantaneous frequency of the modulated wave and the carrier frequency.   59

frequency distortion. A term commonly used for that form of distortion in which the relative magnitude of the different frequency components of a complex wave are changed in transmission. *Note:* When referring to the distortion of the phase-versus-frequency characteristic, it is recommended that a more specific term such as phase-frequency distortion or delay distortion be used. *See:* **amplitude distortion; distortion; distortion, amplitude-frequency.**   111

frequency diversity (telecommunication). *See:* **frequency diversity reception.**

frequency diversity reception (data transmission). That form of diversity reception that utilizes transmission at different frequencies.   59

frequency divider (1)(antennas). A device for delivering an output wave whose frequency is a proper function, usually a submultiple, of the input frequency. *Note:* Usually the output frequency is an integral submultiple or an integral proper fraction of the input frequency. *See:* **harmonic conversion transducer.**   111

(2)(nonlinear, active, and nonreciprocal waveguide components). A device for delivering output power at a frequency that is usually an integral proper fraction or integral submultiple of the input frequency.   530

frequency division multiple access (communication satellite). A method of providing multiple access to a communication satellite in which the transmissions from a particular earth station occupy a particular assigned frequency band. In the satellite the signals are simultaneously amplified and transposed to a different frequency band and retransmitted. The earth station identifies its receiving channel according to its assigned frequency band in the satellite signal.   84

frequency-division multiplex (telecommunication) (data transmission). The process or device in which each modulating wave modulates a separate subcarrier and the subcarriers are spaced in frequency. *Note:* Frequency division permits the transmission of two or more signals over a common path by using different frequency bands for the transmission of the intelligence of each message signal.   111, 415, 59

frequency-division switching (telephone switching systems). A method of switching that provides a common path with a separate frequency band for each of the simultaneous calls.   55

frequency doubler (1)(antennas). A device delivering

output voltage at a frequency that is twice the input frequency. 111

(2)(nonlinear, active, and nonreciprocal waveguide components). A device for delivering output power at a frequency that is twice the input frequency. 530

frequency drift (1) (nonreal time spectrum analyzer). Gradual shift or change in displayed frequency over a period of time due to change in components (Hz/sec), (Hz/°C), etcetera. 68

(2) (spectrum analyzer). Gradual shift or change in displayed frequency over a period of time due to internal changes in the spectrum analyzer (Hz/s, Hz/°C, etcetera). 390

frequency droop (emergency and standby power). The absolute change in frequency between steady-state no load and steady-state full load. 512

frequency hopping (communication satellite). A modulation technique used for multiple access; frequency-hopping systems employ switching of the transmitted frequencies at a rate equal to or lower than the sampling rate of the information transmitted. Selection of the particular frequency to be transmitted can be made from a fixed sequence or can be selected in pseudo-random manner from a set of frequencies covering a wide bandwidth. The intended receiver would frequency-hop in the same manner as the transmitter in order to retrieve the desired information. 84

frequency, image (heterodyne frequency converters in which one of the two sidebands produced by beating is selected). An undesired input frequency capable of producing the selected frequency by the same process. *Note:* The word **image** implies the mirrorlike symmetry of signal and image frequencies about the beating-oscillator frequency or the intermediate frequency, whichever is the higher. *See:* **radio receiver.** 328

frequency influence (electric instrument) (instruments other than frequency meters). The percentage change (of full-scale value) in the indication of an instrument that is caused solely by a frequency departure from a specified reference frequency. *Note:* Because of the dominance of 60 hertz as the common frequency standard in the United States, alternating-current (power-frequency) instruments are always supplied for that frequency unless otherwise specified. *See:* **accuracy rating (instrument).** 280,294

frequency, instantaneous. *See:* **instantaneous frequency.**

frequency interlace (color television). The effect of intermeshing of the frequency spectrum of a modulated color subcarrier and the harmonics of the horizontal scanning frequency for the purpose of minimizing the visibility of the modulated color subcarrier. 18

frequency linearity (spectrum analyzer). The linearity of the relationship between the frequency of an input signal and the displayed frequency. 390

frequency lock (power-system communication). A means of recovering in a single-sideband suppressed-carrier receiver the exact modulating frequency that is applied to a single-sideband transmitter. *See:* **power-line carrier.** 59

frequency locus (control systems). For a nonlinear system or element whose describing function is both frequency-dependent and amplitude-dependent, a plot of the describing function, in any convenient coordinate system. 56

frequency meter. An instrument for measuring the frequency of an alternating current. *See:* **instrument.** 328

frequency meter, cavity resonator (waveguide components). A cavity resonator used to determine frequency. *See:* **cavity resonator.** 166

frequency-modulated cyclotron. A cyclotron in which the frequency of the accelerating electric field is modulated in order to hold the positively charged particles in synchronism with the accelerating field despite their increase in mass at very high energies. 190

frequency-modulated radar (FM radar). A form of radar in which the radiated wave is frequency modulated and the returning echo beats with the wave being radiated, thus enabling the range to be measured. *See:* **radar.** 328

frequency-modulated transmitter. A transmitter that transmits a frequency-modulated wave. *See:* **radio transmitter.** 111,240

frequency modulation (1) (electrical conversion). The cyclic or random dynamic variation, or both, of instantaneous frequency about a mean frequency during steady-state electric system operation. 186

(2) (FM) (telecommunication) (overhead-power-line corona and radio noise) (data transmission). Angle modulation in which the instantaneous frequency of a sine-wave carrier is caused to depart from the carrier frequency by an amount proportional to the instantaneous value of the modulating wave. *Note:* Combinations of phase and frequency modulation are commonly referred to as **frequency modulation.** 411, 59, 111, 415

frequency modulation-frequency modulation (FM-FM) telemetry (communication satellite). A method of multiplexing many telemetry channels by first frequency modulating subcarriers, combining the modulated subcarriers and finally frequency modulating the radio carrier. This method is widely used for satellite transmissions and follows standards set by Inter Range Instrumentation Group (IRIG). 84

frequency-modulation (friction) noise ("scrape flutter"). Frequency modulation of the signal in the range above approximately 100 Hz resulting in distortion which may be perceived as a noise added to the signal (that is, a noise not present in the absence of a signal). 145

frequency monitor. An instrument for indicating the amount of deviation of a frequency from its assigned value. *See:* **instrument.** 328

frequency multiplier(1)(antennas). A device for delivering an output wave whose frequency is an exact integral multiple of the input frequency. *Note:* Frequency doublers and triplers are common special cases of frequency multipliers. *See:* **harmonic conversion transducer.** 111

(2)(nonlinear, active, and nonreciprocal waveguide

components). A device for delivering output power at a frequency that is an exact positive integer (except for 0 and 1) multiple of an input frequency. Frequency doublers, triplers, quadruplers, etcetera, are all special cases of frequency multipliers. 530

frequency of charging, resonance. The frequency at which resonance occurs in the charging circuit of a pulse-forming network. 137

frequency pulling (oscillator). A change of the generated frequency of an oscillator caused by a change in load impedance. *See:* **oscillatory circuit; waveguide.** 125

frequency, pulse repetition. The number of pulses per unit time of a periodic pulse train or the reciprocal of the pulse period. *Note:* This term also includes the average number of pulses per unit time of aperiodic pulse trains where the periods are of random duration. *See:* **pulse.** 185

frequency quadrupler (nonlinear, active, and nonreciprocal waveguide components). A device for delivering output power at a frequency that is four times the input frequency. 530

frequency range (1) (general). A specifically designated part of the frequency spectrum.

(2) (transmission system). The frequency band in which the system is able to transmit power without attenuating or distorting it more than a specified amount.

(3) (device). The range of frequencies over which the device may be considered useful with various circuit and operating conditions. *Note:* Frequency range should be distinguished from bandwidth, which is a measure of useful range with fixed circuits and operating conditions. *See:* **signal wave.** 210,190,125

(4) (acousto-optic deflector). The frequency range, Δ f, over which the diffraction efficiency is greater than some specified minimum. 72

(5) (spectrum analyzer). That range of frequency over which the instrument performance is specified (hertz to hertz). 390, 68

frequency record (electroacoustics). A recording a various known frequencies at known amplitudes, usually for the purpose of testing or measuring. *See:* **phonograph pickup.** 176

frequency regulation (1)(emergency and standby power). The percentage change in emergency or standby power frequency from steady-state no load to steady-state full load. 512

$$\%R \ = \ \frac{F_{n1} - F_{f1}}{F_{n1}} \ \cdot \ 100$$

(2)(ferroresonant voltage regulators). The maximum amount that the output voltage or current will change as the result of a specified change in line frequency. (Regulation is given either as a percentage of the rated output voltage or current, or as an absolute change, ΔE or ΔI. 456

frequency relay (1) (power switchgear). A relay that

responds to the frequency of an alternating electrical input quantity. 103

(2) (81) (power system device function numbers). A relay that responds to the frequency of an electrical quantity, operating when the frequency of an electrical quantity, operating when the frequency or rate of change of frequency exceeds or is less than a predetermined value. 402

frequency resolution (radar). The ability of a receiver or signal processing system to detect or measure separately two or more signals which differ only in frequency. The classic measure of frequency resolution is the minimum frequency separation of two otherwise identical signals which permits the given system to distinguish that two frequencies are present and to extract the desired information from both of them. When the separation is done by means of a tunable bandpass filter system, the resolution is often specified as the width of the frequency response lobe measured at a specific value (such as 3 decibels) below the peak response. 13

frequency response (1)(transmission performance of telephone sets). Electric or acoustic output level as a function of frequency. 491

(2) (power supplies). The measure of an amplifier or power supply's ability to respond to a sinusoidal program. *Notes:* (A) The frequency response measures the maximum frequency for full-output voltage excursion. (B) Frequency response connotes amplitude-frequency response, which should be used in full, particularly if phase-frequency response is significant. This frequency is a function of the slewing rate and unity-gain bandwidth. *See:* **amplitude-frequency response; power supply.** 186

(3) (spectrum analyzer). The peak-to-peak variation of the displayed amplitude over a specified center frequency range, measured at the center frequency, (dB). *Note:* Frequency response is closely related to display flatness. The main difference is that the tuning control of the spectrum analyzer is readjusted so as to center the display. 390

(4) (speed governing of hydraulic turbines). A characteristic, expressed by formula or graph, which describes the dynamic and steady-state response of a physical system in terms of the magnitude ratio and the phase displacement between a sinusoidally varying input quantity and the fundamental of the corresponding output quantity as a function of the fundamental frequency. 8

(5) (data transmission). The measure of an amplifier or power supply's ability to respond to a sinusoidal program. *Notes:* (A) The frequency response measures the maximum frequency for full-output voltage excursion. (B) Frequency response connotes amplitude-frequency response, which should be used in full, particularly if phase-frequency response is significant. This frequency is a function of the slewing rate and unity-gain bandwidth. 59

(6) (Pascal). *See:* **transfer function (of a device).** 433

frequency-response characteristic (linear system). (1)

(signal-transmission system, industrial control). The frequency-dependent relation, in both gain and phase difference, between steady-state sinusoidal outputs. *Notes:* (A) With nonlinearity, as evidenced by distortion of a sinusoidal input of specified amplitudes, the relation is based on that sinusoidal component of the output having the frequency of the input. (B) Mathematically, the frequency-response characteristic is the complex function of $S = j\omega$:

$$A_0(j\omega)/A_i(j\omega) \exp \{j[\theta_0(j\omega) - \theta_i(j\omega)]\}$$

where

A_i = input amplitude
A_o = output amplitude
θ_i = input phase angle
    (relative to fixed reference)
θ_o = output phase angle
    (relative to same reference)

—

See: **control system, feedback; signal.** 56, 206
(2) (excitation control systems). In a linear system, the frequency-dependent relation, in both gain and phase difference between steady-state sinusoidal inputs and the resultant steady-state sinusoidal outputs. *Note:* A plot of gain in logarithmic terms and phase angle in degrees vs logarithmic frequency is commonly called a Bode diagram. See figure below for identification of the principal characteristics of interest.

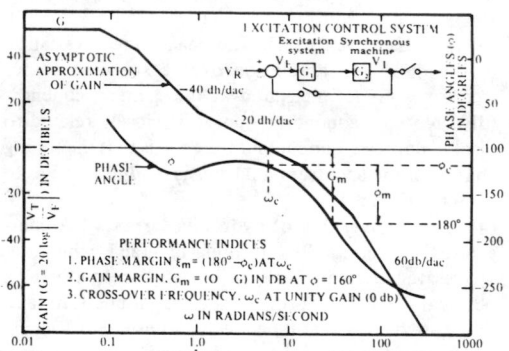

—

Typical open-loop frequency response of an excitation control system with the synchronous machine open circuited. 353
frequency-response equalization (1) (acoustics). The effect of all frequency discriminative means employed in a transmission system to obtain a desired over-all frequency response. 176
(2) (circuits and systems). The process of modifying a frequency response of one network by introducing a frequency response of another network so that, within the band of interest, the combined response follows a specified characteristic. 67
frequency-selective ringing (telephone switching sys-

tems). Selective ringing that employs currents of several frequencies to activate ringers, each of which is tuned mechanically or electrically, or both, to one of the frequencies so that only the desired ringer responds. 55
frequency-selective voltmeter. A selective radio receiver, with provisions for output indication.
314, 199
frequency selectivity (1) (selectivity). (A) A characteristic of an electric circuit or apparatus in virtue of which electric currents or voltages of different frequencies are transmitted with different attenuation. (B) The degree to which a transducer is capable of differentiating between the desired signal and signals or interference at other frequencies. *See:* **transducer.**
328
(2) (characteristic insertion loss) (attenuator). Peak-to-peak variation in decibels through the specified frequency range. 110
frequency-sensitive relay. A relay that operates when energized with voltage, current, or power within specific frequency bands. *See:* **relay.** 259
frequency shift keying (1) (data transmission). That form of frequency modulation in which the modulating signal shifts the output frequency between predetermined values, and the output wave has no phase discontinuity. 59
(2) (FSK) (telecommunication). The form of frequency modulation in which the modulating wave shifts the output frequency between or among predetermined values, and the output wave has no phase discontinuity. *Note:* Commonly, the instantaneous frequency is shifted between two discrete values termed the mark and space frequencies. 61
frequency-shift pulsing (telephone switching systems). A means of transmitting digital information in which a sequence of two frequencies is used. 55
frequency span (nonreal time spectrum analyzer) (spectrum analyzer). The magnitude of the frequency segment displayed (Hz, Hz/div). 390
frequency spectrum (audible noise measurement). A table or graph showing the amplitudes of the frequency components contained in a sound is called the frequency spectrum of that sound. (Spectrum analyzers normally use filters of specified bandwidth to obtain an amplitude measure.) *See:* **octave band, one-third octave band.** 462
frequency stability (1)(data transmission). The measure of the ability to remain on its assigned channel as determined on both a short term (1-second) and a long term (24-hour) basis. 59
(2)(network analyzers). A measure of the amount that a signal source can be expected to vary from its nominal value in a specified time. *Notes:* (A) This can be separated into (1) short-term stability of limited excursion such as phase-jitter and noise, and (2) long-term stability such as drift. (B) (1) May cause inaccuracies in measuring narrow band networks; (2) may cause errors in stored corrections. 505
frequency stabilization. The process of controlling the center or carrier frequency so that it differs from that

of a reference source by not more than a prescribed amount. *See:* **frequency modulation.** 111

frequency standard (1) (electric power systems). A device that produces a standard frequency. *See:* **standard frequency; speed-governing system.** 94 **(2) (facsimile).** A local precision source supplying a stable frequency which is used, among other things, for control of synchronous scanning and recording devices. 11

frequency swing (data transmission). In frequency modulation, the peak difference betwee n the maximum and the minimum values of the instantaneous frequency. *Note:* The term 'frequency swing' is sometimes used to describe the maximum swing permissible under specified conditions. Preferably, such usage should include a specific statement of the conditions. 59

frequency time matrix (communication satellite). A modulation technique used for multiple access: frequency-time matrix systems require the simultaneous presence of energy in more than one time and frequency assignment to produce an output signal. The requirement for presence in several time and/or frequency slots reduces the probability of mutual interference when a number of users are simultaneously transmitting. 84

frequency tolerance (radio transmitter). The extent to which a characteristic frequency of the emission, for example, the carrier frequency itself or a particular frequency in the sideband, may be permitted to depart from a specified reference frequency within the assigned band. *Note:* The frequency tolerance may be expressed in hertz or as a percentage of the reference frequency. *See:* **radio transmitter.** 328

frequency transformation (circuits and systems). The replacing of the frequency variable s in a function $f(s)$ with a new variable z implicitly defined by s, $g(z)$. This may be done, as examples, to convert a low-pass function into a band-pass function or to make calculations less affected by rounding errors. 67

frequency translation (data transmission). The amount of frequency difference between the received audio signals and the original audio signals after passing through a communication channel. 59

frequency tripler(1)(antennas). A device delivering output voltage at a frequency that is three times the input frequency. 111 **(2)(nonlinear, active, and nonreciprocal waveguide components).** A device for delivering output power at a frequency that is three times the input frequency. 530

frequency-type telemeter (power switchgear). A telemeter that employs the frequency of a periodically recurrent electric signal as the translating means. 103

frequency, undamped (frequency, natural). (1) Of a second-order linear system without damping, the frequency of free oscillation in radians per unit time or in hertz. (2) Of any system whose transfer function contains the quadratic factor $s^2 + 2\zeta\omega_n s + \omega_n^2$ in the denominator, the value $\omega_n (0 \leq \zeta < 1)$. (3) Of a closed

loop control system or controlled system, a frequency at which continuous oscillation (hunting) can occur without periodic stimuli. *Note:* In linear systems, the undamped frequency is the phase crossover frequency. With proportional control action only, the undamped frequency of a linear system may be obtained by raising (in most cases) the proportional gain until hunting occurs. This value of gain has been called the "ultimate gain" and the undamped period the "ultimate period." *Syn:* **natural frequency.** 56, 329

frequently-repeated overload rating (power converter). The maximum direct current that can be supplied by the converter on a repetitive basis under normal operating conditions. *See:* **power rectifier.** 208

freshening charge (nuclear power generating stations) (lead storage batteries). The charge given to a storage battery following nonuse or storage. 76

Fresnel contour (antennas). The locus of points on a surface for which the sum of the distances to a source point and an observation point is a constant, differing by a multiple of a half-wavelength from the minimum value of the sum of the distances. *Note:* This definition applies to media which are isotropic and homogeneous. For the general case, the distances along optical paths must be employed. 111

Fresnel diffraction pattern. *See:* **near-field diffraction pattern.**

Fresnel lens antenna. An antenna consisting of a feed and a lens, usually planar, which transmits the radiating power from the feed through the central zone and alternate Fresnel zones of the illuminating field on the lens. *Syn:* **zone-plate lens antenna.** 111

Fresnel pattern (antennas). A radiation pattern obtained in the Fresnel region. 111

Fresnel reflection (fiber optics). The reflection of a portion of the light incident on a planar interface between two homogeneous media having different refractive indices. *Notes:* (1) Fresnel reflection occurs at the air-glass interfaces at entrance and exit ends of an optical waveguide. Resultant transmission losses (on the order of 4 percent per interface) can be virtually eliminated by use of antireflection coatings or index matching materials. (2) Fresnel reflection depends upon the index difference and the angle of incidence; it is zero at Brewster's angle for one polarization. In optical elements, a thin transparent film is sometimes used to give an additional Fresnel reflection that cancels the original one by interference. This is called an antireflection coating. *See:* **antireflection coating; Brewster's angle; index matching material; reflectance; reflection; refractive index.** 433

Fresnel reflection method (fiber optics). The method for measuring the index profile of an optical fiber by measuring the reflectance as a function of position on the end face. *See:* **Fresnel reflection; index profile; reflectance.** 433

Fresnel region (1) (antennas). The region (or regions) adjacent to the region in which the field of an antenna is focused (that is, just outside the Fraunhofer region). *Note:* In the Fresnel region in space, the values of the fields, when calculated from knowledge of the source distribution of an antenna, are insufficiently accurate unless the quadratic phase terms are taken into account, but are sufficiently accurate if the quadratic phase terms are included. *See:* **antenna; radiating near-field region.** 111

(2) (data transmission). (A) (General). The region (or regions) adjacent to the region in which the field of an antenna is focused (that is, just outside the Fraunhofer region). (B) (Data transmission). The region between the antenna and the Fraunhofer region. *Note:* If the antenna has a well-defined aperture D in a given aspect, the Fresnel region in that respect is commonly taken to extend a distance of $2D^2$ in that aspect, being the wavelength. 59

(3) (array element). The input impedance of a radiating element of an array antenna with all other elements in the array open-circuited. *Note:* In general, the self-impedance of a radiating element in an array is not equal to its isolated impedance. *Note:* Strictly speaking, the self-impedance of an array element is not equal to its isolated impedance. However, in many arrays the element spacing is such that it is approximated quite well by the isolated impedance. 111

Fresnel zone (antennas). The region on a surface between successive contours. *Note:* Fresnel zones are usually numbered consecutively with the first zone containing the minimum path length. 111

fretting (corrosion). Deterioration resulting from repetitive slip at the interface between two surfaces. *Note:* When deterioration is further increased by corrosion, the term fretting-corrosion is used. 205

friction and windage loss (rotating machinery). The power required to drive the unexcited machine at rated speed with the brushes in contact, deducting that portion of the loss that results from: (1) Forcing the gas through any part of the ventilating system that is external to the machine and cooler (if used). (2) The driving of direct-connected flywheels or other direct-connected apparatus. *See:* **asynchronous machine.** 244, 63

friction electrification. *See:* **triboelectrification.**

friction tape. A fibrous tape impregnated with a sticky moisture-resistant compound that provides a protective covering for insulation.

fringing capacitance (semiconductor)(nonlinear, active, and nonreciprocal waveguide components). The fixed capacitance between the connecting devices (wires and straps) and the pedestal of a diode enclosure. 530

fritting, relay. *See:* **relay fritting.**

frogging (measuring longitudinal balance of telephone equipment operating in the voice band). A switching technique whereby the tip and ring leads of the test specimen are reversed relative to the driving or terminating test circuits, or both. 529

frog-leg winding (rotating machinery). A composite winding consisting of one lap winding and one wave winding placed on the same armature and connected to the same commutator. 63

front (motor or generator). The front of a normal motor or generator is the end opposite the largest cou-

pling or driving pulley. *See:* **asynchronous machine.** 63

front and back connected fuse (high-voltage switchgear). A fuse in which one or more current carrying conductors are connected directly to the fixed terminals located at the front of the mounting base, the remaining conductors being connected to the studs on the back of the mounting base. 443

front-and-back connected switch. A switch having provisions for some of the circuit connections to be made in front of, and others in back of, the mounting base. 27

front-connected fuse (high-voltage switchgear). A fuse in which the current-carrying conductors are fastened to the fixed terminals in front of the mounting base. 443

front-connected switch. A switch in which the current-carrying conductors are connected to the fixed terminal blocks in front of the mounting base. 27

front contact (1) (general). A part of a relay against which, when the relay is energized, the current-carrying portion of the movable neutral member is held so as to form a continuous path for current. *See:* **a contact.**

(2) (relay systems). A contact which is closed when the relay is picked up. *Syn:* "a" contact. 127

(3) (utility-consumer interconnections). A contact that is open when the relay is deenergized. 128

(4) (power switchgear). *See:* **a contact.** 103

front end (communication satellite). The first stage of amplification or frequency conversion immediately following the antenna in a receiving system. 83

front end computer (data communication). A communications computer associated with a host computer. It may perform line control, message handling, code conversion, error control and applications functions such as control and operation of special purpose terminals. *See:* **communications computer.** 12

front-of-wave impulse sparkover voltage (metal-oxide surge arresters for ac power circuits). The impulse sparkover voltage with a wavefront that rises at a uniform rate and causes sparkover on the wavefront. 583

front-of-wave lightning impulse test (power and distribution transformer). A voltage impulse, with a specified rate-of-rise, that is terminated intentionally by sparkover of a gap which occurs on the rising front of the voltage wave with a specified time to sparkover, and a specified minimum crest voltage. Complete front-of-wave tests (transformer) involve application of the following sequence of impulse waves: (1) one reduced full wave, (2) two front-of-waves, (3) two chopped waves, (4) one full wave. 53

front pitch (rotating machinery). The coil pitch at the connection end of a winding (usually in reference to a wave winding). 63

front porch (television). The portion of a composite picture signal that lies between the leading edge of the horizontal blanking pulse and the leading edge of the corresponding synchronizing pulse. *See:* **television.** 178

front-to-back ratio (1) (antennas). The ratio of the maximum directivity of an antenna to its directivity in a specified rearward direction. *Notes:* (A) This definition is usually applied to beam-type patterns. (B) If the rearward direction is not specified it is taken to be that of the maximum directivity in the rearward hemisphere relative to the antenna's orientation.   111
(2) (data transmission). For a directional antenna, the ratio of its effectiveness toward the front to its effectiveness toward the back.   59
fruit. *See: fruit pulse.*
fruit pulse (fruit). A pulse reply received as the result of interrogation of a transponder by interrogators not associated with the responsor in question.   254
F scan. *See: F display.*
F-scope (radar). A cathode-ray oscilloscope arranged to present an F-display. *See: F-display.*   13
FSK. *See: frequency shift keying.*
FS to AM converter (facsimile). *See: receiving converter, facsimile.*
FTC (radar). Abbreviation for fast time constant. *See: fast-time-constant circuit.*   13
fuel (fuel cells). A chemical element or compound that is capable of being oxidized. *See: electrochemical cell.*   223, 186
fuel adjustment clause (power operations). A clause in a rate schedule that provides for adjustment of the amount of the bill as the cost of fuel varies from a specified base amount per unit.   516
fuel-and-oil quantity electric gauge. A device that measures, by means of bridge circuits and an indicator with separate pointers and scales, the quantity of fuel and oil in the aircraft tanks.   328
fuel battery. An energy-conversion device consisting of more than one fuel cell connected in series, parallel, or both. *See: fuel cell.*   223, 186
fuel-battery power-to-volume ratio. The kilowatt output per envelope volume of the fuel battery (exclusive of the fuel, oxidant, storage, and auxiliaries). *See: fuel cell.*   223, 186
fuel-battery power-to-weight ratio. The kilowatt output per unit weight of the fuel battery (exclusive of the fuel, oxidant, storage, and auxiliaries). *See: fuel cell.*   223, 186
fuel cell. An electrochemical cell that can continuously change the chemical energy of a fuel and oxidant to electric energy by an isothermal process involving an essentially invariant electrode-electrolyte system.   223, 186
fuel-cell Coulomb efficiency. The ratio of the number of electrons obtained from the consumption of a mole of the fuel to the electrons theoretically available from the stated reaction.

$$\text{Coulomb Efficiency} = \frac{\int_0^{t_m} i\,dt}{nF} \times 100$$

t_m = time required to consume a mole of fuel
i = instantaneous current

n = number of electrons furnished in the stated reaction by the fuel molecule
F = Faraday's constant = 96485.3 ± 10.0 absolute joules per absolute volt gram equivalent.

See: fuel cell.   223, 186
fuel-cell standard voltage (at 25 degrees Celsius). The voltage associated with the stated reaction and determined from the equation

$$E^0 = \frac{-J\Delta G^0}{nF}$$

E^0 = fuel-cell standard voltage
J = Joule's equivalent = 4.1840 absolute joules per calorie
ΔG^0 = standard free energy changes in kilo-calories/mole of fuel
n = number of electrons furnished in the stated reaction by the fuel molecule
F = Faraday's constant = 96485.3 ± 10.0 absolute joules per absolute volt gram equivalent.

See: fuel cell.   223, 186
fuel-cell system. An energy conversion device consisting of one or more fuel cells and necessary auxiliaries. *See: fuel cell.*   223, 186
fuel-cell-system energy-to-volume ratio. The kilowatt-hour output per displaced volume of the fuel-cell system (including the fuel, oxidant, and storage). *See: fuel cell.*   223, 186
fuel-cell-system energy-to-weight ratio. The kilowatt-hour output per unit weight of the fuel-cell system (including the fuel, oxidant, and storage). *See: fuel cell.*   223, 186
fuel-cell-system power-to-volume ratio. The kilowatt output per displaced volume of the fuel-cell system (exclusive of the fuel, oxidant, and storage). *See: fuel cell.*   223, 186
fuel-cell-system power-to-weight ratio. The kilowatt output per unit weight of the fuel-cell system (exclusive of the fuel, oxidant, and storage). *See: fuel cell.*   223, 186
fuel-cell-system standard thermal efficiency. The efficiency of a system made up of a fuel cell and auxiliary equipment. *Note:* This efficiency is expressed as the ratio of (1) the electric energy delivered to the load circuit to (2) the enthalpy change for the stated cell reaction.

$$\text{Thermal Efficiency} = \frac{\int_0^{t_m} (E_{IL} \times i_L)\,dt}{\Delta H^0}$$

t_m = time required to consume a mole of fuel
E_{IL} = fuel-cell-system working voltage
i_L = instantaneous current into the load
ΔH^0 = enthalpy change for the stated cell reaction at standard conditions.

See: **fuel cell.**   223, 186

fuel-cell working voltage. The voltage at the terminals of a single fuel-cell delivering current into system auxiliaries and load. *See:* **fuel cell.**   223, 186

fuel-control mechanism (gas turbines). All devices, such as power-amplifying relays, servomotors, and interconnections required between the speed governor and the fuel-control valve. *See:* **speed-governing system.**   58

fuel-control system (gas turbines). Devices that include the fuel-control valve and all supplementary fuel-control devices and interconnections necessary for adequate control of the fuel entering the combustion system of the gas turbine. *Note:* The supplementary fuel-control devices may or may not be directly actuated by the fuel-control mechanism. *See:* **speed-governing system.**   58

fuel-control valve (gas turbines). A valve or any other device operating as a final fuel-metering element controlling fuel input to the gas turbine. *Notes:* (1) This valve or device may be directly or indirectly controlled by the fuel-control mechanism. (2) Variable-displacement pumps, or other devices that operate as the final fuel-control element in the fuel-control system, and that control fuel entering the combustion system are fuel-control valves. *See:* **speed-governing system.**   58

fuel economy. The ratio of the chemical energy input to a generating station to its net electric output. *Note:* Fuel economy is usually expressed in British thermal units per kilowatthour. *See:* **generating station.**   64

fuel elements, nuclear (nuclear power generating station). An assembly of rods, tubes, plates, or other geometrical forms into which nuclear fuel is contained for use in a reactor.   112

fuel-pressure electric gauge. A device that measures the fuel pressure (usually in pounds per square inch) at the carburetor of an aircraft engine. *Note:* It provides remote indication by means of a self-synchronous generator and motor.   328

fuel replacement energy (power operations). Energy generated to substitute for energy which would otherwise have been generated by a different fuel source.   516

fuel reprocessing, nuclear (nuclear power generating station). The processing of irradiated reactor fuel to recover the unused fissionable material, or fission products, or both.   112

fuel stop valve (gas turbines). A device that, when actuated, shuts off all fuel flow to the combustion system, including that provided by the minimum fuel limiter. *See:* **speed-governing system.**   58

fulguration. *See:* **electrodesiccation.**

full automatic plating. Mechanical plating in which the cathodes are automatically conveyed through successive cleaning and plating tanks.   328

full availability (telephone switching systems). Availability that is equal to the number of outlets in the desired group.   55

full-direct trunk group (telephone switching systems). A full trunk group between end offices.   55

full duplex (data transmission) (communication circuit) (telecommunication). Method of operation where each end can simultaneously transmit and receive. *Note:* Full duplex refers to a communications system or equipment capable of simultaneous transmission in two directions.   59

full energy peak (for a monoenergetic photon spectrum for a semiconductor spectrometer system)(X-ray energy spectrometers). The distribution of events within the peak of the pulse-height distribution spectrum representing response to the monoenergetic photon source. *Note:* Notwithstanding other definitions or procedures for subtracting background and other distortions, the full energy peak intensity is defined as not including any events which exceed a Gaussian distribution by more than a factor of two σ.   471

full energy peak efficiency (of a semiconductor radiation detector)(X-ray energy spectrometers). The ratio of the number of events in the full energy peak of the spectral distribution to the total number of photons incident on the active detector volume during the same time interval.   471

full-field relay (industrial control). A relay that functions to maintain full field excitation of a motor while accelerating on reduced armature voltage. *See:* **relay.**   206

full float operation (large lead storage batteries). Operation of a dc system with the battery, battery charger. and load all connected in parallel and with the battery charger supplying the normal dc load plus any self-discharge or charging current, or both, required by the battery. (The battery will deliver current only when the load exceeds the charger output.)   377

full impulse voltage. An aperiodic transient voltage that rises rapidly to a maximum value and falls, usually less rapidly, to zero. See the following figure. *See:* **test voltage and current; full-wave voltage impulse.**

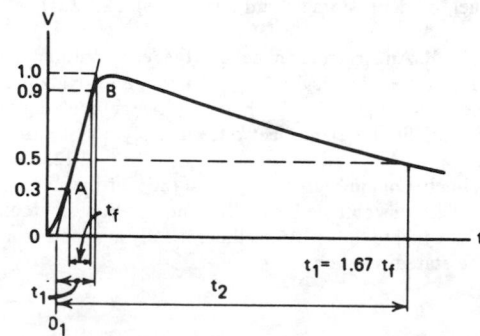

Full impulse voltage.

      307

full-impulse wave (surge arresters). An impulse wave in which there is no sudden collapse.   244, 62

full lightning impulse (high voltage testing) (lightning

impulse tests, general applicability). A full lightning impulse is a lightning impulse that is not interrupted by a disruptive discharge. See Section 1 ANSI/IEEE Std 4-1978 for definition of impulse and for distinction between lightning and switching impulses.   150

full load (test, measurement and diagnostic equipment). The greatest load that a circuit is designed to carry under specific conditions: any additional load is overload.   54

full-load speed (electric drive) (industrial control). The speed that the output shaft of the drive attains with rated load connected and with the drive adjusted to deliver rated output at rated speed. *Note:* In referring to the speed with full load connected and with the drive adjusted for a specified condition other than for rated output at rated speed, it is customary to speak of the full-load speed under the (stated) conditions. *See:* **electric drive.**   206

full magnetic controller (1) (Electric installations on shipboard). An electric controller having all of its basic functions performed by devices, which are operated by electromagnets.   3
(2) (industrial control). An electric controller having all of its basic functions performed by devices that are operated by electromagnets. *See:* **electric controller.**   206, 3

full-pitch winding (rotating machinery). A winding in which the coil pitch is 100 percent, that is, equal to the pole pitch. *See:* **asynchronous machine.**   63

full scale (analog computers). In an analog computer, the nominal maximum value of a computer variable or the nominal maximum value at the output of a computing element. Also sometimes used to indicate the entire computing voltage range, such as 20V is full scale for a compute whose voltage ranges from $+10$V to -10V. The latter definition is generally used in manufacturers' specifications, that is, 0.01 percent of full scale.   9

full-scale value (mechanical demand registers). The maximum scale capacity of the register. If a multiplier exists, the full-scale value will be the product of the maximum scale marking and the multiplying constant.   548

full-screen editor (computer applications). A text editor that allows the user to view a full display screen of data at one time and to enter or alter text by using either commands or cursor control. Scrolling functions allow the user to move up and down within the document. Contrast with line-editor.   571

full span–max span (spectrum analyzer). A mode of operation in which the spectrum analyzer scans an entire selected frequency band.   390

full speed (data transmission). Referring to transmission of data in teleprinter systems at the full rated speed of the equipment.   59

full-trunk group (telephone switching systems). A trunk group, other than a final trunk group, that does not overflow calls to another trunk group.   55

full-voltage starter (industrial control). A starter that connects the motor to the power supply without reducing the voltage applied to the motor. *Note:* Full-

voltage starters are also designated as across-the-line starters. *See:* **starter.**   206

full-wave lightning impulse test (power and distribution transformer). Application of the "standard lightning impulse" wave, a full wave having a front time of 1.2 microseconds and a time to half value of 50 microseconds, described as a 1.2/50 impulse.   53

full-wave rectification (rectifying process) (power supplies). Full-wave rectification inverts the negative half-cycle of the input sinusoid so that the output contains two half-sine pulses for each input cycle. A pair of rectifiers arranged as shown with a center-tapped transformer or a bridge arrangement of four rectifiers and no center tap are both methods of obtaining full-wave rectification. *See:* **rectification; rectifier.**   208

full-wave rectifier. *See:* **full-wave rectification.**

full-wave rectifier circuit. A circuit that changes single-phase alternating current into pulsating unidirectional current, utilizing both halves of each cycle. *See:* **rectification.**   237, 66, 118

full-wave voltage impulse (surge arresters). A voltage impulse that is not interrupted by sparkover, flashover, or puncture. *See:* **full-impulse voltage.**   62

full width (duration) half maximum (fiber optics). A measure of the extent of a function. Given by the difference between the two extreme values of the independent variable at which the dependent variable is equal to half of its maximum value. The term "duration" is preferred when the in dependent variable is time. *Note:* Commonly applied to the duration of pulse waveforms, the spectral extent of emission or absorption lines, and the angular or spatial extent of radiation patterns.   433

full width at fiftieth maximum (FWFM)(X-ray energy spectrometers). Same as full width at half maximum (FWHM) except that measurement is made at one fiftieth of the maximum ordinate rather than one half.   471

full width at half maximum (FWHM)(1)(charged-particle detectors)(germanium gamma-ray detectors)(X-ray energy spectrometers). The full width of a distribution measured at half the maximum ordinate. For a normal distribution, it is equal to

$$2(2 \ln 2)^{1/2}$$

times the standard deviation σ.   119, 528, 471
(2) (germanium detectors). The full width of a gamma-ray peak distribution measured at half the maximum ordinate above the continuum.   397
(3) (scintillation counters). The full width of a distribution measured at half the maximum ordinate. For a normal distribution, it is equal to $2(2 \ln 2)^{1/2}$ times the standard deviation (σ).

Note: The expression **full width at half maximum,** given either as an absolute value or as a percentage of the value of the argument at the maximum of the distribution curve, is frequently used in nuclear physics as an approximate description of a distribution curve. Its significance can best be made clear by refer-

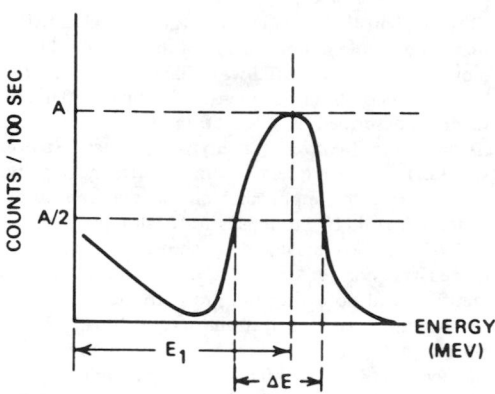

Full width at half maximum (in this case, ΔE).

ence to a typical distribution curve, shown in the figure, of the measurement of the energy of the gamma rays from Cs^{137} with a scintillation counter spectrometer. The measurement is made by determining the number of gamma-ray photons detected in a prescribed interval of time, having measured energies falling within a fixed energy interval (channel width) about the values of energy (channel position) taken as argument of the distribution function. The abscissa of the curve shown is energy in megaelectronvolts (MeV) units and the ordinate is counts per given time interval per megaelectronvolt energy interval. The maximum of the distribution curve shown has an energy E_1 megaelectronvolts. The height of the peak is A_1 counts/100 seconds/megaelectronvolts. The full width at half maximum ΔE is measured at a value of the ordinate equal to $A_1/2$. The percentage full width at half maximum is $100 \Delta E/E_1$. It is an indication of the width of the distribution curve, and where (as in the example cited) the gamma-ray photons are monoenergetic, it is a measure of the resolution of the detecting instrument. When the distribution curve is a Gaussian curve, the percentage full width at half maximum is related to the standard deviation σ by

$$100 \frac{\Delta E}{E_1} = 100 \times 2(2 \ln 2)^{1/2} \times \sigma.$$

See: **scintillation counter.** 117
(4) (sodium iodide detector). The full width of a gamma-ray peak distribution measured at half the maximum ordinate above the continuum. 423
full width at tenth maximum (FWTM)(X-ray energy spectrometers). Same as full width at half maximum (FWHM) except measurement is made at one tenth of the maximum ordinate rather than one half. 471
fully connected network (data communication). A network in which each node is directly connected with every other node. 12
fume-resistant (industrial control). So constructed that it will not be injured readily by exposure to the specified fume. 103, 202, 206

function (1)(microprocessor operating systems). A primitive operation on system-controlled resources. IEEE Std 855 (Trial Use) defines a collection of functions together with suitable input and output parameters. 478
(2) (test, measurement and diagnostic equipment). The action or purpose which a specific item is intended to perform or serve. 54
(3) (software). (A) A specific purpose of an entity or its characteristic action. (B) A subprogram that is invoked during the evaluation of an expression in which its name appears and that returns a value to the point of invocation. See: **subprogram; subroutine.** 434
functional adjectives (pulse terms) (1) linear. Pertaining to a feature whose magnitude varies as a function of time in accordance with the following relation or its equivalent:

$$m = a + bt$$

254
(2) exponential. Pertaining to a feature whose magnitude varies as a function of time in accordance with either of the following relations or their equivalents:

$$m = ae^{-bt}$$
$$m = a (1 - e^{-bt})$$

254
(3) Gaussian. Pertaining to a waveform or feature whose magnitude varies as a function of time in accordance with the following relation or its equivalent:

$$m = ae^{-b(t - c)^2}, \quad b > 0$$

254
(4) trigonometric. Pertaining to a waveform or feature whose magnitude varies as a function of time in accordance with a specified trigonometric function or by a specified relationship based on trigonometric functions (for example, cosine squared). 254
functional area(s) (nuclear power generating station). Location(s) designated within the control room to which displays and controls relating to specific function(s) are assigned. 358
functional component (power switchgear). A device which performs a necessary function for the proper operation and application of a unit of equipment. 103
functional decomposition (software). A method of designing a system by breaking it down into its components in such a way that the components correspond directly to system functions and subfunctions. See: **components; functions; hierarchical decomposition; system.** 434
functional design (software). The specification of the working relationships among the parts of a data processing system. See: **data processing system; preliminary design; specification.** 434
functional designation (abbreviation) (1) (general).

Letters, numbers, words, or combinations thereof, used to indicate the function of an item or a circuit, or of the position or state of a control of adjustment. Compare with: letter combination, reference designation, symbol for a quantity. *See:* **abbreviation.**

173

(2) (electric and electronics parts and equipments). Words, abbreviations, or meaningful number or letter combinations, usually derived from the function of an item (for example: slew, yaw), used on drawings, instructional material, and equipment to identify an item in terms of its function. *Note:* A functional designation is not a reference designation nor a substitute for it.

17

functional diagram (test, measurement and diagnostic equipment). A diagram that represents the functional relationships among the parts of a system.

54

functional nomenclature (generating stations electric power system). Words or terms which define the purpose, equipment, or system for which the component is required.

381

functional requirement (software). A requirement that specifies a function that a system or system component must be capable of performing. *See:* **component; function; requirement; system.**

434

functional specification (software). A specification that defines the functions that a system or system component must perform. *See:* **component; function; performance specification; specification; system.**

434

functional test (1)(ATLAS). A sequence of tests applied to a unit under test (UUT) to establish whether it is functioning correctly.

400

(2)(evaluation of thermal capability)(thermal classification of electric equipment and electrical insulation). A means of evaluation in which an insulating material, insulation system, or electric equipment is exposed to factors of influence, which simulate or are characteristic of actual service conditions.

506

(3)(test pattern language). A test in which the cells of a memory are accessed in a specific order and at a specific rate, while data is being written into them, or read from them.

463

functional test pattern. *See:* **pattern.**

functional unit (1). A system element that performs a task required for the successful operation of the system. *See:* **system.**

209

(2) (software). An entity of hardware, software, or both capable of accomplishing a specified purpose. *See:* **hardware; software.**

434

function check (station control and data acquisition)-(supervisory control, data acquisition, and automatic control). A check of master and remote station equipment by exercising a predefined component or capability. (1) Analog. Monitor a reference quantity (2) Control. Control and indication from a control-check relay (3) Scan. Accomplished when control function check has been performed with all remotes (4) Poll. Accomplished when analog function is performed with all remotes (5) Logging. Accomplished when results of the control function check are logged.

403, 570

function Class-A (back-up) current-limiting fuse. A fuse capable of interrupting all currents from the rated maximum interrupting current down to the rated minimum interrupting current. *Note:* The rated minimum interrupting current for such fuses is higher than the minimum melting current that causes melting of the fusible element in one hour.

103

function Class-G (general purpose) current-limiting fuse (as applied to a high-voltage current-limiting fuse). A fuse capable of interrupting all currents from the rated maximum interrupting current down to the current that causes melting of the fusible element in one hour.

103

function code (f) (subroutines for CAMAC). The symbol f represents an integer which is the function code for a CAMAC action.

410

function codes (fa) (subroutines for CAMAC). The symbol fa represents an array of integers, each of which is the function code for a CAMAC action. The length of fa is given by the value of the first element of cb at the time the subroutine is executed. *See:* **control block.**

410

function, coupling (control systems). A mathematical, graphical, or tabular statement of the influence which one element or subsystem has on another element or subsystem, expressed as the effect.cause ratio of related variables or their transforms. *Note:* For a multi-terminal system described by m differential equations and having m input transforms $R_1 \ldots R_m$ and m output transforms $C_1 \ldots C_m$, the coupling functions consist of all effect.cause ratios which can be formed from transforms bearing unlike-numbered subscripts.

56

function, describing (nonlinear element under periodic input) (control system, feedback). A transfer function based solely on the fundamental, ignoring other frequencies. *Note:* This equivalent linearization implies amplitude dependence with or without frequency dependence. *See:* **control system, feedback.**

56, 329

function, error transfer (closed loop) (control system, feedback). The transfer function obtained by taking the ratio of the Laplace transform of the error signal to the Laplace transform of its corresponding input signal. *See:* **control system, feedback.**   56, 329

function generator (1) (analog computer). A computing element whose output is a specified nonlinear function of its input or inputs. Normal usage excludes multipliers and resolvers.

9

(2) (electric power systems). A device in which a mathematical function such as y, $f(x)$ can be stored so that for any input equal to x, an output equal to $f(x)$ will be obtained. *See:* **speed-governing system.**

94

function generator, bivariant. A function generator having two input variables. *See:* **electronic analog computer.**

9

function generator, card set (analog computer). A diode function generator whose values are stored and set by means of a punched card and mechanical card reading device.

9

function generator, curve-follower (analog computer).

A function generator that operates by automatically following a curve *f(x)* drawn or constructed on a surface, as the input *x* varies over its range. 9

function generator, digitally controlled (analog computer). A hybrid component using digital-to-analog converters and digital-to-analog multipliers to insert the linear segment approximation values to the desired arbitrary function. The values are stored in a self-contained digital core memory, which is accessed by the digital-to-analog converters and digital-to-analog multipliers at digital computer speeds (microseconds). 9

function generator, diode. A function generator that uses the transfer characteristics of resistive networks containing biased diodes. *Note:* The desired function is approximated by linear segments whose values are manually inserted by means of potentiometers and switches. 9

function generator, map-reader. A variant function generator using a probe to detect the voltage at a point on a conducting surface and having coordinates proportional to the inputs. *See:* **electronic analog computer.** 9

function generator, servo. A function generator consisting of a position servo driving a function potentiometer. *See:* **electronic analog computer.** 9

function generator, switch-type. A function generator using a multitap switch rotated in accordance with the input and having its taps connected to suitable voltage sources. *See:* **electronic analog computer.** 9

function, loop-transfer (closed loop) (control system, feedback). The transfer function obtained by taking the ratio of the Laplace transform of the return signal to the Laplace transform of its corresponding error signal. *See:* **control system, feedback.** 56, 329

function, output-transfer (closed loop) (control system, feedback). The transfer function obtained by taking the ratio of the Laplace transform of the output signal to the Laplace transform of the input signal. *See:* **control system, feedback.** 56, 329

function, probability density (control systems). Pertaining to a real random variable x, the derivative with respect to an arbitrary value X of the variable x, of the probability distribution function of X, if a derivative exists. *Note:* The mathematical expression for this function is

$$g(X) = \frac{d}{dX} [f(X)] = \frac{d}{dX} [P \, (x \leqslant X)]$$

56

function, probability distribution (control systems). Pertaining to a real random variable x, the function of an arbitrary value X- X of this variable, whose value is the probability, P, that the random variable is less than or equal to X. *Note:* The mathematical expression for this function is

$$f(X) = P(x \leqslant X)$$

56

function relay (analog computers). In an analog computer, a relay used as a computing element, generally driven by a comparator. 9

function, return-transfer (closed loop) (control system, feedback). The transfer function obtained by taking the ratio of the Laplace transform of the return signal to the Laplace transform of its corresponding input signal. *See:* **control system, feedback.** 56

function switch (analog computers). In an analog computer, a manually operated switch used as a computing element; for example, to modify a circuit, to add or delete an input function or constant, etcetera. 9

function, system-transfer (automatic control). The transfer function obtained by taking the ratio of the Laplace transform of the signal corresponding to the ultimately controlled variable to the Laplace transform of the signal corresponding to the command. *See:* **control system, feedback.** 56, 329

function, transfer (1) (control system, feedback). A mathematical, graphic, or tabular statement of the influence that a system or element has on a signal or action compared at input and at output terminals. *Note:* For a linear system, general usage limits the transfer function to mean the ratio of the Laplace transform of the output to the Laplace transform of the input in the absence of all other signals, and with all initial conditions zero. *See:* **control system, feedback; transfer function.** 56, 329
(2) (antenna). The complex ratio of the output of the device to its input. It is also the combined phase and frequency responses. 151

function, weighting (control system, feedback). A function representing the time response of a linear system, or element to a unit-impulse forcing function: the derivative of the time response to a unit-step forcing function. *Notes:* (1) The Laplace transform of the weighting function is the transfer function of the system or element. (2) The time response of a linear system or element to an arbitrary input is described in terms of the weighting function by means of the convolution integral. *See:* **control system, feedback.** 56, 329

function, work. *See:* **work function.**

fundamental component. The fundamental frequency component in the harmonic analysis of a wave. *See:* **signal wave.** 349

fundamental efficiency (thyristor). The ratio of the fundamental load power to the fundamental line power. 445

fundamental frequency (data transmission). (1) (Signal-transmission system). The reciprocal of the period of a wave. (2) (Mathematically). The lowest frequency component in the Fourier representation of a periodic quantity. (3) (Data transmission) (periodic quantity). The frequency of a sinusoidal quantity having the same period as the periodic quantity. 59

fundamental mode (fiber optics). The lowest order mode of a waveguide. In fibers, the mode designated LP_{01} or HE_{11}. *See:* **mode.** 433

fundamental mode of propagation (laser-maser). The mode in a **beamguide** or **beam resonator** which has a single maximum for the transverse field intensity over the cross-section of the beam. 363

fundamental power (thyristor). The product of the root-mean-square (rms) value of the fundamental current and the rms value of the fundamental voltage multiplied by the cosine of the phase angle by which the fundamental current lags the fundamental voltage. 445

fundamental-type piezoelectric crystal unit. A unit designed to utilize the lowest frequency of resonance for a particular mode of vibration. *See:* **crystal.** 328

furnace transformer (power and distribution transformer). A transformer that is designed to be connected to an electric arc furnace. 53

fuse (1)(protection and coordination of industrial and commercial power systems). A device that protects a circuit by fusing open its current-responsive element when an overcurrent or short-circuit current passes through it. 504

(2) (power switchgear). An overcurrent protective device with a circuit-opening fusible part that is heated and severed by the passage of the overcurrent through it. *Note:* A fuse comprises all the parts that form a unit capable of performing the prescribed functions. It may or may not be the complete device necessary to connect it into an electric circuit. 103, 443

fuse arcing time. *See:* **arcing time (of a fuse).**

fuse blade (of a cartridge fuse) (power switchgear). A cartridge-fuse terminal having a substantially rectangular cross-section. 103

fuse carrier (of an oil cutout) (power switchgear). An assembly of a cap which closes the top opening of an oil-cutout housing, an insulating member, and fuse contacts with means for making contact with the conducting element and for insertion into the fixed contacts of the fuse support. *Note:* The fuse carrier does not include the conducting element (fuse link). 103, 443

fuse clearing time. *See:* **clearing time (total clearing time) (of a fuse).**

fuse clips (power switchgear). The current carrying parts of a fuse support which engage the fuse carrier, fuseholder, fuse unit, or blade. *Syn:* **contact clips; fuse contacts.** 103

fuse condenser (power switchgear). A device that, added to a vented fuse, converts it to a nonvented fuse by providing a sealed chamber for condensation of the gases developed during circuit interruption. 103, 443

fuse contact. *See:* **fuse terminal.**

fuse contacts. *See:* **fuse clips.**

fuse cutout (power switchgear). A cutout having a fuse link or fuse unit. *Note:* A fuse cutout is a fuse disconnecting switch. 103

fused capacitor (1)(series capacitors). A capacitor in combination with a fuse, either external or internal to the case. 474

(2)(shunt power capacitors). A capacitor having fuses mounted on its terminals, or inside a terminal enclosure, or inside the capacitor case for the purpose of interrupting a failed capacitor. 138

fused capacitor unit (series capacitor). A capacitor unit in combination with a fuse, either external or internal to the case, intended to isolate a failed unit from the associated units. 86

fused electrolyte (bath) (fused salt) (electrolyte). A molten anhydrous electrolyte. 328

fused-electrolyte cell. A cell for the production of electric energy when the electrolyte is in a molten state. *See:* **electrochemistry.** 328

fuse disconnecting switch (power switchgear). A disconnecting switch in which a fuse unit or fuseholder and fuse link forms all or a part of the blade. 103, 443

fused quartz (fiber optics). Glass made by melting natural quartz crystals; not as pure as vitreous silica. *See:* **vitreous silica.** 433

fused salt (bath). *See:* **fused electrolyte.**

fused silica. *See:* **vitreous silica; fused quartz.**

fused trolley tap (mining). A specially designed holder with enclosed fuse for connecting a conductor of a portable cable to the trolley system or other circuit supplying electric power to equipment in mines. *See:* **mine feeder circuit.** 328

fused-type voltage transformer (1) (instrument transformer). One that is provided with means for mounting one or more fuses as integral parts of the transformer in series with the primary winding. 394

(2) (power and distribution transformer). One which is provided with the means for mounting a fuse, or fuses, as an integral part of the transformer in series with the primary winding. 53

fuse enclosure package (FEP) (power switchgear). An enclosure supplied with one or more current limiting fuses as a package for which application data covering the specific fuse(s) and enclosure are supplied. 103, 443

fuse filler. *See:* **arc-extinguishing medium (of a fuse).**

fuseholder (1) (of a high-voltage fuse) (power switchgear). An assembly of a fuse tube or tubes together with parts necessary to enclose the contacting element and provide means of making contact with the conducting element and the fuse clips. The fuseholder does not include the contacting element (fuse link or refill unit). 103, 443

(2) (of a low-voltage fuse) (power switchgear). An assembly of base, fuse clips, and necessary insulation for mounting and connecting into the circuit the current-responsive element, with its holding means if used for making a complete device. *Notes:* (A) For low-voltage fuses the current responsive element and holding means are called a fuse. (B) For high-voltage fuses the general type of assembly described above is called a fuse support or fuse mounting. The holding means (fuseholder) and the current-responsive or conducting element are called a fuse unit. 103

fuse hook (power switchgear). A hook provided with an insulating handle for opening and closing fuses or switches and for inserting the fuseholder, fuse unit, or disconnecting blade into and removing it from the fuse support.   103, 443

fuselage lights (illuminating engineering). Aircraft aeronautical lights, mounted on the top and bottom of the fuselage, used to supplement the navigation lights.   167

fuse link (power switchgear). A replaceable part or assembly, comprised entirely or principally of the conducting element, required to be replaced after each circuit interruption to restore the fuse to operating conditions.   103, 443

fuse-link (protection and coordination of industrial and commercial power systems). In British terminology only, a complete enclosed cartridge fuse; in such cases the addition of the carrier, or holder, completes the fuse. In the USA, a renewable, fusible element for fuse cutouts.   504

fuse melting time. *See:* **melting time (of a fuse).**

fuse mounting. *See:* **fuse support (of a high-voltage fuse).**

fuse muffler (1) (high-voltage switchgear). An attachment for the vent of a fuse, that confines the arc and substantially reduces the venting from the fuse.   443

(2) (power switchgear). *See:* **muffler (of a fuse).**   103

fuses (protection and coordination of industrial and commercial power systems). Class H. Cartridge fuses were formerly known as 'NEC-dimensioned fuses.' Class H fuses are tested and listed by Underwriters Laboratories, Inc, under their standard ANSI/UL 198B-1982 in 250 and 600 volt (V) ratings with interrupting capabilities of 10 000 amperes (A). Class H fuses are not marked as current limiting. UL standards for Class H fuses have a time-delay requirement of at least 10 seconds (s) opening time at five times rating in order to have the words 'time delay' on the label.- Class J. These fuses are rated to interrupt 200 000 A alternating current (ac) and meet the standards of Underwriters Laboarories, Inc, for Class J fuses. They are UL-labeled as 'current limiting', are rated only for 600 V (or less) ac, and are of dimensions not interchangeable with other classes. Class J fuses that have a time delay of at least 10 s opening time at five times rated current may have the words 'time delay' on the label. Class K. These fuses meet ANSI/Ul 198D-1982 of Underwriters Laboratories, Inc, for Class K as either K-1, K-5, or K-9. These standards have prescribed values for maximum peak let-through currents and I2t for each subclass, with K-1 having the lowest (most restrictive) values and Class K-9 having the highest (least restrictive) values. Dimensionally the same as Class H fuses, these fuses have no UL-recognized 'rejection feature'. Their ac interrupting rating appears on their labels as 50 000, 100 000, or 200 000 A. They are not labeled as 'current limiting'. The words 'time delay', 'dual element', letter 'D', or phrase of similar significance on the label will indicate the manufacturer has met UL's optional testing for this feature. The use of Class K fuses permits equipment and circuits to be

applied on systems having potential fault currents in excess of 10 000 A. Some hazards may exist in that they can be replaced with Class H fuses by uninformed personnel under present standards. Class L. These fuses meet ANSI/UL 198C-1981 of Underwriters Laboratories, Inc, for Class L fuses, have ratings in the range of 601-6000 A, are rated to interrupt 200 000 A ac, are rated only for 600 V or less ac, and are of specified dimensions larger than those of other fuses rated 600 V (or less). They are intended to be bolted to bus bars and are not used in clips. UL has no definition of time delay for Class L fuses; however, many Class L fuses have substantial overload time-current carrying capability. Class L fuse standards do not include 250 V ratings, dc testing, or direct current (dc) ratings.Class R. These fuses meet ANSI/UL 198E-1982 of Underwriters Laboratories, Inc, for Class R as either RK1 or RK5.Their interrupting ratings are 200 A ac. The standard has prescribed values for maximum peak let-through currents, I2t and threshold current, with subclass RK1 having the lowest (most restrictive) values as compared to subclass RK5. These fuses have dimensions that provide a one-way physical rejection feature, that is, no other-class of fuse will fit into equipment designed to employ Class R fuses; however, Class R fuses can be installed in older Class H or Class K equipment as replacement to upgrade these systems to the maximum allowed by other devices in the system. Class R fuses are available with or without time delay. If marked 'time delay' or similar phrase, they are required to have a minimum opening time of 10 seconds (s) when subjected to a load of five times rated current.   504

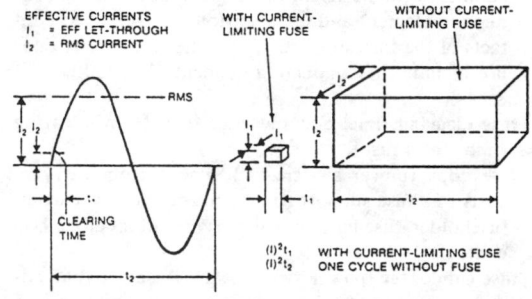

Graphic Representation of $I^2 t$

fuse support (1) (high-voltage switchgear). An assembly of base or mounting support or oil cutout housing, insulator(s) or insulator unit(s), and fuse clips for mounting a fuse carrier, fuse holder, fuse unit, or blade and connecting it into the circuit. *Syn:* **fuse mounting.**   443

(2) (of a high-voltage fuse). An assembly of base, mounting support or oil-cutout housing, fuse clips, and necessary insulation for mounting and connecting into the circuit the current responsive element with its holding means if such means are used for making a complete device. *Notes:* (A) For high-voltage fuses the holding means is called a fuse carrier or fuseholder, and in combination with the current-responsive or conducting element is called a fuse unit. (B) For low-voltage fuses the general type of assembly defined above is called a fuseholder. 103

fuse terminal (power switchgear). The means for connecting the current-responsive element or its holding means, if such means is used for making a complete device, to the fuse clips. 103

fuse time-current characteristic (power switchgear). The correlated values of time and current that designate the performance of all or a stated portion of the functions of a fuse. *Note:* The time-current characteristics of a fuse are usually shown as a curve. 103, 443

fuse time-current tests (power switchgear). Tests that consist of the application of current to determine the relation between the root-mean-square (rms) alternating current or direct current and the time for the fuse to perform the whole or some specified part of its interrupting function. 103, 443

fuse tongs (power switchgear). Tongs provided with an insulating handle and jaws. Fuse tongs are used to insert the fuseholder or fuse unit into the fuse support or to remove it from the support. 103, 443

fuse tube (1) (high-voltage switchgear). A tube of insulating material which encloses the conducting element. 443
(2) (power switchgear). A tube of insulating material that surrounds the current-responsive element, the conducting element or the fuse link. 103

fuse unit (power switchgear). An assembly comprising a conducting element mounted in a fuseholder with parts and materials in the fuseholder essential to the operation of the fuse. 103, 443

fusible element (of a fuse) (power switchgear). That part, having predetermined current-responsive melting characteristics, which may be all or part of the current-responsive element. 103, 443

fusible enclosed (safety) switch (industrial control). A switch complete with fuse holders and either with or without meter-testing equipment or accommodation for meters, having all current-carrying parts completely enclosed in metal, and operable without opening the enclosure. *See:* **switch.** 206

fusion (power operations). The formation of a heavier nucleus from two lighter ones with the attendant release of energy. 516

fusion frequency (television). Frequency of succession of retinal images above which their differences of luminosity or color are no longer perceptible. *Note:* The fusion frequency is a function of the decay characteristic of the display. 18

fusion splice (fiber optics). A splice accomplished by the application of localized heat sufficient to fuse or melt the ends of two lengths of optical fiber, forming a continuous, single fiber. 433

future point (for supervisory control or indication or telemeter selection) (power switchgear). Provision for the future installation of equipment required for a point. *Note:* A future point may be provided with (1) space only, (2) drilling, or other mounting provisions, and wiring only. 103

FW. *See:* **forward wave.**

FW.02M (germanium gamma-ray detectors). Same as FWHM (full width at half maximum) except that the width measurement is made at one fiftieth the maximum ordinate rather than at one half. 528

FW.1M (germanium gamma-ray detectors). Same as FWHM (full width at half maximum) except that measurement is made at one tenth the maximum ordinate rather than at one half. 528

FWFM. *See:* **full width at fiftieth maximum.**

FWHM. (1) *See:* **full width at half maximum.**
(2) (fiber optics). *See:* **full width (duration) half maximum.** 433

FWTM. *See:* **full width at tenth maximum.**

F1A line weighting (data transmission). A noise weighting used in a noise measuring set to measure noise on a line that is terminated by a subset with a number 302 receiver or a similar subset. The meter-scale readings are in dBa (F1 A). 59

F1 layer (radio wave propagation). The lower of the two ionized layers normally existing in the F region in the day hemisphere. *See:* **radiation; radio wave propagation.** 146

F2 layer (radio wave propagation). The single ionized layer normally existing in the F region in the night hemisphere and the higher of the two layers normally existing in the F region in the day hemisphere. *See:* **radiation; radio wave propagation.** 146

G

G (s). *See:* **armature to field transfer function.**
gain (1) (of an antenna)(in a given direction). The ratio of the radiation intensity, in a given direction, to the radiation intensity that would be obtained if the power

accepted by the antenna were radiated isotropically. *Notes:* (A) Gain does not include losses arising from impedance and polarization mismatches. (B) The radiation intensity corresponding to the isotropically

radiated power is equal to the power accepted by the antenna divided by 4π. (C) If an antenna is without dissipative loss, then, in any given direction, its gain is equal to its directivity. (D) If the direction is not specified, the direction of maximum radiation intensity is implied. (E) The term "absolute gain" is used in those instances where added emphasis is required to distinguish gain from relative gain; for example, 'absolute gain measurements'. *Syn: absolute gain.*   111
(2) (waveguide). The power increase in a transmission path in the mode or form under consideration. It is usually expressed as a positive ratio, in decibels.   267

gain-crossover frequency (hydraulic turbines). The frequency at which the gain becomes unity and its decibel value zero.   8

gain integrator (analog computers). For each input, the ratio of the input to the corresponding time rate of change of the output. For fixed input resistors, the "time constant" is determined by the integrating feedback capacitor.   9

gain margin (hydraulic turbines). The reciprocal of the gain at the frequency at which the open-loop phase angle reaches 180 degrees.   8

gain time control (radar). *Syn:* **sensitivity time control.**   13

galactic radio waves (radio wave propagation). Radio waves originating in our galaxy beyond the solar system.   146

gamma (television). The exponent of that power law that is used to approximate the curve of output magnitude versus input magnitude over the region of interest.   18

gamma correction (television). The insertion of a nonlinear output-input characteristic for the purpose of changing the system transfer characteristic.   18

gamma ray resolution (1) (germanium detectors). The measured full width at half maximum (FWHM), after background subtraction, of a gamma-ray peak distribution, expressed in units of energy.   397
(2) (sodium iodide detector). The measured full width at half maximum (FWHM), after background subtraction, of a gamma-ray peak distribution, expressed as a percentage of the energy corresponding to the centroid of the distribution.   423

gapless (metal-oxide surge arresters for ac power circuits). Not possessing gaps, series or parallel, as in gapless arrester .   583

gap loss (fiber optics). That optical power loss caused by a space between axially aligned fibers. *Note:* For waveguide-to-waveguide coupling, it is commonly called "longitudinal offset loss." *See:* **coupling loss.**   433

garage (National Electrical Code). A building or portion of a building in which one or more self-propelled vehicles carrying volatile flammable liquid for fuel power are kept for use, sale, storage, rental, repair, exhibition, or demonstrating purposes, and all that portion of a building which is on or below the floor or floors in which such in which such vehicles are kept and which is not separated therefrom by suitable cutoffs.   256

gas-accumulator relay (power switchgear). A relay that is so constructed that it accumulates all or a fixed proportion of gas released by the protected equipment and operates by measuring the volume of gas so accumulated.   103

gas admixture ratio (nonlinear, active, and nonreciprocal waveguide components). The ratio of partial pressures of the separate constituent gases of the total gas composition used in gas tubes.   530

gas cleanup (nonlinear, active, and nonreciprocal waveguide components). The phenomenon that causes gas atoms or molecules to be absorbed into a solid medium during a gas discharge.   530

gaseous discharge (illuminating engineering). The emission of light from gas atoms excited by an electric current.   167

gas fill (nonlinear, active, and nonreciprocal waveguide components). The process by which a plasma limiter or gas tube is evacuated and an admixture of gases is inserted.   530

gas filled joint (power cable joint). Joints in which the fluid filling the joint housing is in the form of a gas.   34

gas filled protectors (wire-line communication facilities). A discharge gap between two or more electrodes hermetically sealed in a ceramic or glass envelope. These gaps provide protection against excessive voltage in the same manner as carbon block protectors.   414

gas-filled transformer (power and distribution transformer). A sealed transformer, except that the windings are immersed in a dry gas which is other than air or nitrogen.   53

gas-flow counter tube. A radiation-counter tube in which an appropriate atmosphere is maintained by a flow of gas through the tube.   190

gas flow error (laser gyro). The error resulting from the flow of gas in direct current discharge tubes. It is caused by complex interactions among atoms, ions, electrons and tube walls.   46

gas focusing (electron-beam tubes). A method of concentrating an electron beam by gas ionization within the beam. *See:* **focusing, gas; gas tubes.**   190

gas grooves (electrometallurgy). The hills and valleys in metallic deposits caused by streams of hydrogen or other gas rising continuously along the surface of the deposit while it is forming. *See:* **electrowinning.**   328

gasket-sealed relay. A relay in an enclosure sealed with a gasket. *See:* **relay.**   259

gasket, waveguide (waveguide components). A resilient insert usually between flanges intended to serve one or more of the following primary purposes: (1) to reduce gas leakage affecting internal waveguide pressure, (2) to prevent intrusion of foreign material into the waveguide, or (3) to reduce power leakage and arcing.   166

gas multiplication factor (radiation-counter tubes). The ratio of (1) the charge collected from the sensitive volume to (2) the charge produced in this volume by the initial ionizing event.   96 gas-oil sealed system (power and distribution transformer). A system in which the interior of the tank is sealed from the atmosphere, over the temperature range specified, by means of an auxiliary tank or tanks to form a gas-oil seal operating on the manometer principle.   53

gasoline dispensing and service station (National Electrical Code). A location where gasoline or other volatile flammable liquids or liquified flammable gases are transferred to the fuel tanks (including auxiliary fuel tanks) of self-propelled vehicles.   256

gasoline-electric drive. See: gas-electric drive.

gas-pressure relay (power switchgear). A relay so constructed that it operates by the gas pressure in the protected equipment.   103

gasproof. So constructed or protected that the specified gas will not interfere with successful operation.
                                                                           103

gasproof or vaporproof (rotating machinery). So constructed that the entry of a specified gas or vapor under prescribed conditions cannot interfere with satisfactory operating of the machine. See: asynchronous machine.   63

gas ratio. The ratio of the ion current in a tube to the electron current that produces it. See: electrode current.   190

gas seal (rotating machinery). A sealing arrangement intended to minimize the leakage of gas to or from a machine along a shaft. Note: It may be incorporated into a ball or roller bearing assembly.   63

gassing. The evolution of gases from one or more of the electrodes during electrolysis. See: electrolytic cell.
                                                                           328

gas system (rotating machinery). The combination of parts used to ventilate a machine with any gas other than air, including facilities for charging and purging the gas in the machine.   63

gastight (1) (lightning protection). So constructed that gas or air can neither enter nor leave the structure except through vents or piping provided for the purpose.   297

(2) (power switchgear). So constructed that the specified gas will not enter the enclosing case under specified pressure conditions.   103

gas tube. An electron tube in which the pressure of the contained gas or vapor is such as to affect substantially the electrical characteristics of the tube.   190

gas-tube relaxation oscillator (arc-tube relaxation oscillator). A relaxation oscillator in which the abrupt discharge is provided by the breakdown of a gas tube. See: oscillatory circuit.   328

gas-tube surge arrester (gas-tube protective devices). A gap, or gaps, in an enclosed discharge medium, other than air at atmospheric pressure, designed to protect apparatus or personnel, or both, from high transient voltages.   490

gas-turbine-electric drive. A self-contained system of power generation and application in which the power generated by a gas turbine is transmitted electrically by means of a generator and a motor (or multiples of these) for propulsion purposes. Note: The prefix gas-turbine-electric is applied to ships, locomotives, cars, buses, etcetera, that are equipped with this drive. See: electric locomotive.   328

gate (1)(microwave)(nonlinear, active, and nonreciprocal waveguide components). In elementary form, a two-port switch having a single-pole, single-throw function. See: bang snuffer.   530

(2)(X-ray energy spectrometers). A device or element that, depending upon one or more specified inputs, has the ability to permit or inhibit the passage of a signal.   471

(3) (electronic computers). (A) A device having one output channel and one or more input channels, such that the output channel state is completely determined by the contemporaneous input channel states, except during switching transients. (B) A combinational logic element having at least one input channel. (C) An AND gate. (D) An OR gate.   235

(4) (cryotron). An output element of a cryotron. See: superconductivity.   191

(5) (navigation systems). (A) An interval of time during which some portion of the circuit or display is allowed to be operative, or (B) the circuit which provides gating. See: navigation.   187, 13

(6) (metal-nitride-oxide field-effect transistor). This structural element of an insulated-gate field-effect transistor (IGFET) controls the current between source and drain by a voltage applied to its terminal.
                                                                           386

gate-controlled delay time (thyristor). The time interval, between a specified point at the beginning of the gate pulse and the instant when the principal voltage (current) has dropped (risen) to a specified value near its initial value during switching of a thyristor from the OFF state to the ON state by a gate pulse. See: principal voltage-current characteristic.
                                                             243, 66, 208, 191

gate-controlled rise time (thyristor). The time interval between the instants at which the principal voltage (current) has dropped (risen) from a specified value near its initial value to a specified low (high) value, during switching of a thyristor from the OFF state to the ON state by a gate pulse. Note: This time interval will be equal to the rise time of the ON state current only for pure resistive loads. See: principal voltage-current characteristic.   243, 66, 208, 191

gate-controlled turn-off time (turn-off thyristor). The time interval, between a specified point at the beginning of the gate pulse and the instant when the principal current has decreased to a specified value, during switching from the ON state to the OFF state by a gate pulse. See: principal voltage-current characteristic.
                                                             243, 66, 208, 191

gate-controlled turn-on time (thyristor). The time interval, between a specified point at the beginning of the gate pulse and the instant when the principal voltage (current) has dropped (risen) to a specified low (high) value during switching of a thyristor from the OFF state to the ON state by a gate pulse. Turn-on time is the sum of delay time and rise time. See: principal voltage-current characteristic; delay time; rise time.
                                                             243, 204, 208, 191

gate current (semiconductor). The current that results from the gate voltage. *Notes:* (1) Positive gate current refers to conventional current entering the gate terminal. (2) Negative gate current refers to conventional current leaving the gate terminal.   66

gated integrator (germanium gamma-ray detectors). A circuit for obtaining a pulse with an amplitude proportional to the integral of a signal pulse over a definite time interval.   528

gated sweep (oscilloscopes). A sweep controlled by a gate waveform. Also, a sweep that will operate recurrently (free-running, synchronized, or triggered) during the application of a gating signal. *See:* **oscillograph.**   185

gate limit (speed governing system, hydraulic turbines). A device which acts on the governor system to prevent the turbine-control servomotor from opening beyond the position for which the device is set.   58

gate nontrigger current (thyristor). The maximum gate current that will not cause the thyristor to switch from the OFF state to the ON state. *See:* **principal current; gate trigger current.**   243, 204, 208, 191

gate nontrigger voltage (thyristor). The maximum gate voltage that will not cause the thyristor to switch from the OFF state to the ON state. *See:* **principal voltage-current characteristic (principal characteristic); gate trigger voltage.**   243, 204, 208, 191

gate protective action (thyristor converter). Protective action that takes advantage of the switching property in the converter protection network.   121

gate suppression (thyristor power converter). Removal of gating pulses.   121

gate terminal (thyristor). A terminal that is connected to a gate. *See:* **anode.**   243, 204, 191

gate trigger current (thyristor). The minimum gate current required to switch a thyristor from the OFF state to the ON state. *See:* **principal current.**   243, 204, 208, 191

gate trigger voltage (thyristor). The gate voltage required to produce the gate-trigger current. *See:* **principal voltage-current characteristic.**   243, 204, 208, 191

gate turn-off current (gate turn-off thyristor). The minimum gate current required to switch a thyristor from the ON state to the OFF state. *See:* **principal current.**   243, 204, 208, 191

gate turn-off voltage (gate turn-off thyristor). The gate voltage required to produce the gate turn-off current. *See:* **principal voltage-current characteristic.**   243, 204, 208, 191.

gate voltage (thyristor). The voltage between a gate terminal and a specified main terminal. *See:* **principal voltage-current characteristic.**   243, 204, 208, 191

gating (1) (antennas). (A) The process of selecting those portions of a wave that exist during one or more selected time intervals or (B) that have magnitudes between selected limits. *See:* **modulation; wave front.**   111

(2) (radar). The application of enabling or inhibiting pulses during part of a cycle of equipment operation.   13

gating signal (keying signal). A signal that activates or deactivates a circuit during selected time intervals.   328

gating techniques (thyristor). Those techniques employed to provide controller (thyristor) gating signals.   445

gauss (centimeter-gram-second electromagnetic system). The gauss is 10^{-4} webers per square meter or 1 maxwell per square centimeter.   210

Gaussian beam (fiber optics). A beam of light whose electric field amplitude distribution is gaussian. When such a beam is circular in cross section, the amplitude is

$$E(r) = E(0) \ \exp \ [-(r/w)^2],$$

where r is the distance from beam center and w is the radius at which the amplitude is $1/e$ of its value on the axis; w is called the beamwidth. *See:* **beam diameter.**   433

Gaussian beam (laser-maser). A beam of radiation having an approximately spherical wave front at any point along the beam and having transverse field intensity over any wave front which is a Gaussian function of the distance from the axis of the beam.   363

Gaussian density function (radar). Sometimes referred to as normal probability distribution, the Gaussian probability-density function is given by

$$f(X) = \frac{1}{\sigma\sqrt{2\pi}} \exp - \left(\frac{x^2}{2\sigma^2}\right).$$

Often used to describe statistical nature of random noise, where σ = standard deviation.   13

Gaussian distribution (radar). A probability distribution characterized by the probability density function

$$f(x) = \frac{1}{\sqrt{2\pi}\sigma} \exp \left[- \frac{(x-m)^2}{2\sigma^2} \right]$$

where x is the random variable, m is the mean, and σ is the standard deviation; often used for analytical modeling of radar noise and various measurement errors. *Syn:* **normal distribution.**   13

Gaussian filter (circuits and systems). A polynomial filter whose magnitude-frequency response approximates the ideal Gaussian response, the degree of approximation depending on the complexity of the filter. The ideal Gaussian response is given by

$$|H(j\omega)| = \exp \ [-0.3466 \ (\omega/\omega_c)^2]$$

where ω_c 3 dB frequency. Gaussian filters, because of their good transient characteristics (small overshoot

and ringing), find applications in pulse systems.
67

Gaussian pulse (fiber optics). A pulse that has the waveform of a gaussian distribution. In the time domain, the waveform is

$$f(t) = A \ \exp \ [-(t/a)^2],$$

where A is a constant, and a is the pulse half duration at the 1/e points. *See:* **full width (duration) half maximum.**
433

Gaussian response (amplifiers). A particular frequency-response characteristic following the curve $y(f) = e^{-af^2}$. *Note:* Typically, the frequency response approached by an amplifier having good transient response characteristics. *See:* **amplifier.**
185

Gaussian system (units). A system in which centimeter-gram-second electrostatic units are used for electric quantities and centimeter-gram-second electromagnetic units are used for magnetic quantities. *Note:* When this system is used, the factor c (the speed of light) must be inserted at appropriate places in the electromagnetic equations.
210

Gauss law (electrostatics). States that the integral over any closed surface of the normal component of the electric flux density is equal in a rationalized system to the electric charge Q_0 within the surface. Thus

$$\underset{\text{closed surface}}{\int (\mathbf{D} \cdot \mathbf{n}) \, dA} = \underset{\text{volume enclosed}}{\int \rho_0 \, dV} = Q_0.$$

Here **D** is the electric flux density, **n** is a unit normal to the surface, dA the element of area, ρ_0 is the space charge density in the volume V enclosed by the surface.
210

gaussmeter. A magnetometer provided with a scale graduated in gauss or kilogauss. *See:* **magnetometer.**
328

GCA. *See:* **ground-controlled approach.**
GCI. *See:* **ground-controlled intercept.**
G-display. A type of radar display format. *See:* **display.**
13

G **display (radar).** A rectangular display in which a target appears as a laterally centralized blip when the radar antenna is aimed at it in azimuth, and wings appear to grow on the blip as the distance to the target is diminished: horizontal and vertical aiming errors are respectively indicated by horizontal and vertical displacement of the blip. *See:* **navigation.**
13, 187

GDOP. *See:* **geometric dilution of precision.**
G **drift (electronic navigation).** A drift component in gyros (sometimes in accelerometers) proportional to the nongravitational acceleration and caused by torques resulting from mass unbalance. Jargon. *See:* **navigation.**
13, 187

G^2 **drift (electronic navigation).** A drift component in gyros (sometimes in accelerometers) proportional to the square of the nongravitational acceleration and caused by anisoelasticity of the rotor supports. Jargon. *See:* **navigation.**
187, 13

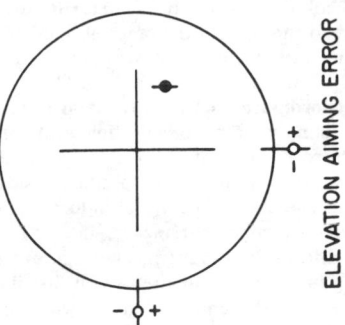

AZIMUTH AIMING ERROR
G-Display

geared-drive machine. A direct-drive machine in which the energy is transmitted from the motor to the driving sheave, drum, or shaft through gearing. *See:* **driving machine (elevators).**
328

geared traction machine (elevators). A geared-drive traction machine. *See:* **driving machine (elevators).**
328

gearless motor. A traction motor in which the armature is mounted concentrically on the driving axle, or is carried by a sleeve or quill that surrounds the axle, and drives the axle directly without gearing. *See:* **traction motor.**
328

gearless traction machine (elevators). A traction machine, without intermediate gearing, that has the traction sheave and the brake drum mounted directly on the motor shaft. *See:* **driving machine (elevators).**
328

gear pattern (facsimile). *See:* **drive pattern.**
gear ratio (watthour meter). The number of revolutions of the rotor of the first dial pointer, commonly denoted by the symbol R_g.
212

Geiger-Mueller counter tube. A radiation-counter tube designed to operate in the Geiger-Mueller region.
328

Geiger-Mueller region (radiation counter tube). The range of applied voltage in which the charge collected per isolated count is independent of the charge liberated by the initial ionizing event.
190, 96, 125

Geiger-Mueller threshold (radiation-counter tube). The lowest applied voltage at which the charge collected per isolated tube count is substantially independent of the nature of the initial ionizing event.
190, 96, 125

Geissler tube. A special form of gas-filled tube for showing the luminous effects of discharges through rarefied gases. *Note:* The density of the gas is roughly one-thousandth of that of the atmosphere. *See:* **gas tube.**
244, 190

general color rendering index (R_a) (illuminating engineering). Measure of the average shift of eight standardized colors chosen to be of intermediate saturation and spread throughout the range of hues. If the color rendering index is not qualified as to the color samples used, R_a is assumed. 167

general coordinated methods (general application to electric supply or communication systems). Those methods reasonably available that contribute to inductive coordination without specific consideration of the requirements for individual inductive exposures. *See:* inductive coordination. 328

general diffuse lighting (illuminating engineering). Lighting involving luminaires which distribute 40 to 60 percent of the emitted light downward and the balance upward, sometimes with a strong component at 90 degrees (horizontal). 167

general insertion loss (or gain) (waveguide). A loss (or gain) resulting from placing two ports of a network between arbitrary generator and load impedances. It is the ratio of the power absorbed in the load when connected to the generator (reference power) to that when the network is inserted. 267

generalized entity. *See:* generalized property.

generalized impedance converter (circuits and systems). A two-port active network characterized by the conversion factor $f(s)$ of the complex frequency variable s and satisfying the following property: when port B is terminated with impedance $Z(s)$ the impedance at port A is given by $Z(s)f(s)$: when port A is terminated with impedance $Z(s)$ the impedance at port B is given by $Z(s)/f(s)$. 67

generalized property (generalized quantity) (generalized entity). Any of the physical concepts in terms of examples of which observable physical systems and phenomena are described quantitatively. *Notes:* (1) Examples are the abstract concepts of length, electric current, energy, etcetera. (2) A generalized property is characterized by the qualitative attribute of physical nature, or dimensionality, but not by a quantitative magnitude. 210

generalized quantity. *See:* generalized property.

general lighting (illuminating engineering). Lighting designed to provide a substantially uniform level of illuminance throughout an area, exclusive of any provision for special local requirements. 167

general purpose branch circuit (National Electrical Code). A branch circuit that supplies a number of outlets for lighting and appliances. 256

general-purpose circuit breaker (alternating current high voltage circuit breakers). A circuit breaker that is not specifically designed for capacitance current switching. 130

general-purpose computer. A computer that is designed to solve a wide class of problems. 255, 77

general-purpose controller (industrial control). Any controller having ratings, characteristics, and mechanical construction for use under usual service conditions. *See:* electric controller. 225, 206

general-purpose current-limiting fuse (power switchgear). A fuse capable of interrupting all currents from the rated maximum interrupting current down to the current that causes melting of the fusible element in one hour. 103, 443

general purpose digital computer. *See:* digital computer.

general-purpose enclosure (1) (power switchgear). An enclosure used for usual service applications where special types of enclosures are not required. 103 (2) (electric installations on shipboard). An enclosure which primarily protects against accidental contact and slight indirect splashing but is neither dripproof nor splashproof. 3

general purpose floodlight (GP) (illuminating engineering). A weatherproof unit so constructed that the housing forms the reflecting surface. The assembly is enclosed by a cover glass. 167

general-purpose induction motor (rotating machinery). Any open motor having a continuous rating of 50-degrees Celsius rise by resistance for Class A insulation, or of 80-degrees Celsius rise for Class B, a service factor as listed in the following tabulation, and designed, listed, and offered in standard ratings with standard operating characteristics and mechanical construction, for use under usual service conditions without restrictions to a particular application or type of application.

Service Factor

Horse-power	Synchronous Speed, revolutions per minute			
	3600	1800	1200	900
1/20	1.4	1.4	1.4	1.4
1/12	1.4	1.4	1.4	1.4
1/8	1.4	1.4	1.4	1.4
1/6	1.35	1.35	1.35	1.35
1/4	1.35	1.35	1.35	1.35
1/3	1.35	1.35	1.35	1.35
1/2	1.25	1.25	1.25	
3/8	1.25	1.25		
1	1.25			

See: asynchronous machine. 63

general-purpose low-voltage dc power circuit breaker (power switchgear). A low-voltage dc power circuit breaker which, during interruption, does not usually prevent the fault current from rising to its E/R value. 103

general purpose low-voltage power circuit breaker (low voltage dc power circuit breakers used in enclosures). A circuit breaker which during interruption does not usually prevent the fault current from rising to its sustained value. 401

general-purpose motor (rotating machinery). Any motor designed, listed and offered in standard ratings with operating characteristics and mechanical construction suitable for use under usual service conditions without restrictions to a particular application or type of application. 63

general-purpose relay. A relay that is adaptable to a variety of applications. *See:* relay. 259

general purpose test equipment (GPTE) (test, measurement and diagnostic equipment). Test equipment which is used for the measurement of a range of parameters common to two or more equipments or systems of basically different design. 54

general-purpose transformers (power and distribution transformer). Step-up or step-down transformers or autotransformers generally used in secondary distribution circuits of 600 V or less in connection with power and lighting service. 53

general support maintenance. *See:* **maintenance, depot.**

general-use snap switch (National Electrical Code). A form of general-use switch so constructed that it can be installed in flush device boxes or on outlet box covers, or otherwise used in conjunction with wiring systems recognized by this Code. 256

general-use switch (National Electrical Code). A switch intended for use in general distribution and branch circuits. It is rated in amperes and it is capable of interrupting its rated current at its rated voltage. 256

generate (computing systems). To produce a program by selection of subsets from a set of skeletal coding under the control of parameters. 255, 77

generated voltage (rotating machinery). A voltage produced in a closed path or circuit by the relative motion of the circuit or its parts with respect to magnetic flux. *See:* **Faraday's law; induced voltage; asynchronous machine, synchronous machine.** 63

generating electric field meter (gradient meter). A device in which a flat conductor is alternately exposed to the electric field to be measured and then shielded from it. *Note:* The resulting current to the conductor is rectified and used as a measure of the potential gradient at the conductor surface. *See:* **instrument.** 328

generating magnetometer (earth inductor). A magnetometer that depends for its operation on the electromotive force generated in a coil that is rotated in the field to be measured. *See:* **magnetometer.** 328

generating station (power operations). A plant wherein electric energy is produced from some other form of energy (for example, chemical, mechanical, or hydraulic) by means of suitable apparatus. 516

generating-station auxiliaries. Accessory units of equipment necessary for the operation of the plant. *Example:* Pumps, stokers, fans, etcetera. *Note:* Auxiliaries may be classified as essential auxiliaries or those that must not sustain service interruptions of more than 15 seconds to 1 minute, such as boiler feed pumps, forced draft fans, pulverized fuel feeders, etcetera: and nonessential auxiliaries that may, without serious effect, sustain service interruptions of 1 to 3 minutes or more, such as air pumps, clinker grinders, coal crushers, etcetera. *See:* **generating station.** 64

generating-station auxiliary power. The power required for operation of the generating station auxiliaries. *See:* **generating station.** 64

generating-station efficiency. *See:* **efficiency.**

generating-station reserve. *See:* **reserve equipment.**

generating unit (unique identification in power plants). The generator, or generators, associated prime mover or movers, auxiliaries and energy supply or supplies that are normally operated together as a single source of electric power. 544

generation rate (semiconductor). The time rate of creation of electron-hole pairs. *See:* **semiconductor device.** 245

generator (1) (rotating machinery). A machine that converts mechanical power into electric power. *See:* **asynchronous machine.** 63
(2) (computing systems). A controlling routine that performs a generate function, for example, report generator, input-output generator. *See:* **function generator; noise generator.** 255, 77

generator, alternating-current. *See:* **alternating-current generator.**

generator, arc welder (1) (generator, alternating-current arc welder). An alternating-current generator with associated reactors, regulators, control, and indicating devices required to produce alternating current suitable for arc welding.
(2) (generator-rectifier, alternating-current direct current arc welder). A combination of static rectifiers and the associated alternating-current generator, reactors, regulators, control, and indicating devices required to produce either direct or alternating current suitable for arc welding.
(3) (generator, direct-current arc welder). A direct-current generator with associated reactors, regulators, control, and indicating devices required to produce direct current suitable for arc welding. 264
(4) (generator-rectifier, direct-current arc welder). A combination of static rectifiers and the associated alternating-current generator, reactors, regulators, controls, and indicating devices required to produce direct current suitable for arc welding. 264

generator efficiency (thermoelectric couple). The ratio of (1) the electric power output of a thermoelectric couple to (2) its thermal power input. *Note:* This is an idealized efficiency assuming perfect thermal insulation of the thermoelectric arms. *See:* **thermoelectric device.** 191

generator-field accelerating relay (industrial control). A relay that functions automatically to maintain the armature current within prescribed limits when a motor supplied by a generator is accelerated to any speed, up to base speed, by controlling the generator field current. *Note:* This definition applies to adjustable-voltage direct-current drives. *See:* **relay.** 225, 206

generator-field control. A system of control that is accomplished by the use of an individual generator for each elevator or dumbwaiter wherein the voltage applied to the driving-machine motor is adjusted by varying the strength and direction of the generator field. *See:* **control (elevators).**

generator field decelerating relay (industrial control). A relay that functions automatically to maintain the armature current within prescribed limits when a motor, supplied by a generator, is decelerated from base speed, or less, by controlling the generator field-current. *Note:* This definition applies to adjustable-voltage direct-current drives. *See:* **relay.** 225, 206

generator/motor. A machine that may be used as either a generator or a motor usually by changing rotational direction. *Note:* (1) This type of machine has particular application in a pumped-storage operation, in which water is pumped into a reservoir during off-peak periods and released to provide generation for peaking loads. (2) This definition eliminates the confusion of terminology for this type of machine. A slant is used between the terms to indicate their equality, and also the machine serves one function or the other and not both at the same time. The word **generator** is placed first to provide a distinction in speech between this term and the commonly used term **motor-generator,** which has an entirely different meaning. *See:* **asynchronous machine.** 63

generator set. A unit consisting of one or more generators driven by a prime mover. *See:* **asynchronous machine.** 63

generette (rotating machinery). A test jig designed on the principle of a motorette, for endurance tests on sample lengths of coils or bars for large generators. *See:* **asynchronous machine.** 63

generic actuator group (valve actuators). An actuator or family of actuators within a range of sizes with similar design principles, materials, manufacturing processes, limiting stresses, operating principles, and design margins. 492

generic connection assembly (Class 1E connection assemblies). A connection assembly that represents a family of connection assemblies having similar materials, manufacturing processes, aasembly techniques, limiting stresses, design, and operating principles. 459

generic design (electric penetration assemblies). A family of equipment units having similar materials, manufacturing processes, limiting stresses, design, and operating principles, that can be represented for qualification purposes by a representative unit(s). 493

generic environment (nuclear power generating stations). A set of environmental conditions intended to envelope the range of expected environments. 440

generic equipment (nuclear power generating stations). A family of equipment units having similar materials, manufacturing processes, limiting stresses, design, and operating principles that can be represented for qualification purposes by representative units. 440

generic qualification (Class 1E connection assemblies). Qualification to a set of requirements designed to envelop the service conditions plus margin of a number of specific applications. 459

geocentric latitude (navigation). The acute angle between (1) a line joining a point with the earth's geometric center and (2) the earth's equatorial plane. 187, 13

geocentric vertical. *See:* **geometric vertical.**

geodesic. The shortest line between two points measured on any mathematically derived surface that includes the points. *See:* **navigation.** 187, 13

geodesic lens antenna. A two-dimensional lens, with uniform index of refraction, disposed on a surface such that the rays in the lens follow geodesic (minimal) paths of the surface. 111

geodetic latitude (navigation). The angle between the normal to the spheroid and the earth's equatorial plane: the latitude generally used in maps and charts. *Syn:* **geographic latitude.** *See:* **navigation.** 187, 13

geographic (map) vertical. The direction of a line normal to the surface of the geoid. *See:* **navigation.** 187, 13

geographical address (GA)(FASTBUS acquisition and control). The primary address of a device based on the physical (geographical) location of the module, and determined by coded backplane pins, or (on a cable segment) by switches. For a crate segment geographical address zero is for the rightmost position when the crate is viewed from the front and the address increases by one for each module position moved to the left. 480

geographical address control (GAC)(FASTBUS acquisition and control). Logic associated with each segment for supervising and generating signals for geographical addressing. 480

geographic latitude. *See:* **geodetic latitude.**

geoid. The shape of the earth as defined by the hypothetical extension of mean sea level continuously through all land masses. *See:* **navigation.** 187, 13

geometric (geocentric) vertical (navigation). The direction of the radius vector drawn from the center of the earth through the location of the observer. *See:* **navigation.** 187

geometrical adjectives (pulse terms). (1) trapezoidal. Having or approaching the shape of a trapezoid. (2) rectangular. Having or approaching the shape of a rectangle. (3) triangular. Having or approaching the shape of a triangle. (4) sawtooth. Having or approaching the shape of a right angle. (See: **composite waveform,** waveform D.) (5) rounded. Having a curved shape characterized by a relatively gradual change in slope. 254

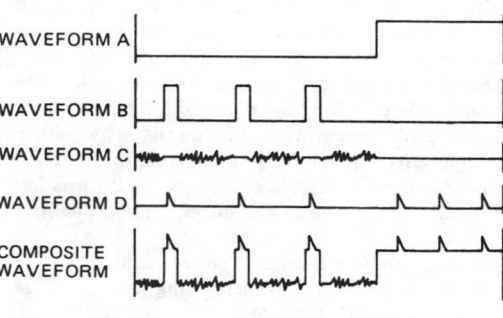

Composite waveform

geometrical factor (navigation). The ratio of the change in a navigational coordinate to the change in distance, taken in the direction of maximum navigational coordinate change: the magnitude of the gradient of the navigational coordinate. *See:* **navigation.**
13, 187

geometric dilution of position (GDOP) (radar). An expression which refers to increased measurement errors in certain regions of coverage of the measurement system. It applies to systems which combine several surface of position measurements such as range only, angle only, or hyperbolic (range difference) to locate the object of interest. When two lines of position cross at a small acute angle, the measurement accuracy is reduced along the axis of the acute angle. 13

geometric dilution of precision (radar). An increase in measurement errors in certain regions of coverage of a measurement system that combines several surface-of-position measurements such as range only, angle only, or range difference (hyperbolic) to locate the object of interest. When two lines of position cross at a small acute angle, the measurement accuracy is reduced along the axis of the acute angle. 13

geometric distortion (television). The displacement of elements in the reproduced picture from the correct relative positions in the perspective plane projection of the original scene. 18

geometric factor (cable calculations) (power distribution, underground cables). A parameter used and determined solely by the relative dimensions and geometric configuration of the conductors and insulation of a cable. 57

geometric inertial navigation equipment. The class of inertial navigation equipment in which the geographic navigational quantities are obtained by computations (generally automatic) based upon the outputs of accelerometers whose vertical axes are maintained parallel to the local vertical, and whose azimuthal orientations are maintained in alignment with a predetermined geographic direction (for example, north). *See:* **navigation.** 187, 13

geometric optics (fiber optics). The treatment of propagation of light as rays. *Note:* Rays are bent at the interface between two dissimilar media or may be curved in a medium in which refractive index is a function of position. *See:* **axial ray; meridional ray; optical axis; paraxial ray; physical optics; skew ray.** 433

geometric rectification error (accelerometer). The error caused by an angular motion of a linear accelerometer input reference axis when this angular motion is coherent with a vibratory cross acceleration input. This is an error that can occur in the application of a linear accelerometer and is not caused by imperfections in the accelerometer. This error is proportional to the square of the cross acceleration and varies with the frequency. 46

geometry (oscilloscopes). The degree to which a cathode-ray tube can accurately display a rectilinear pattern. *Note:* Generally associated with properties of a cathode-ray tube: the name may be given to a cathode-ray-tube electrode or its associated control. *See:* **oscillograph.** 185

geometry, detector. *See:* **detector geometry.**

geotropism (radiation protection). A change in instrument response with a change in instrument orientation as a result of gravitational effects. 399

germicidal efficiency of radiant flux. See bactericidal efficiency of radiant flux.

germicidal exposure. *See:* **bactericidal exposure.**

germicidal flux. *See:* **bactericidal flux.**

germicidal flux density. *See:* **bactericidal flux density.**

getter (electron tubes). a substance introduced into an electron tube to increase the degree of vacuum by chemical or physical action on the residual gases. *See:* **electrode (of an electron tube); electronic tube.**
244, 190

ghost (television). A spurious image resulting from an echo. *See:* **television.** 328

ghost pulse. *See:* **ghost signals.**

ghost signals (loran). (1) Identification pulses that appear on the display at less than the desired loran station full pulse repetition frequency. (2) Signals appearing on the display that have a basic repetition frequency other than that desired. *See:* **navigation.**
187

ghost target (radar). An apparent target in a radar which does not correspond in position or frequency or both, to any real target, but which results from distortion or misinterpretation by the radar circuitry of other real target signals which are present. For example, it may result from range-Doppler ambiguities in the radar waveform used, from intermodulation distortion due to circuit amplitude nonlinearities, or from combining data from two antenna systems or waveforms.
13

Gibb's phenomenon (circuits and systems). Overshoot phenomenon obtained near a discontinuity point of a signal when the spectrum of that signal is truncated abruptly. 67

GIC. *See:* **generalized impedance converter.**

giga (G)(mathematics of computing). A prefix indicating 10^9. 564

gilbert (centimeter-gram-second electromagnetic system). The unit of magnetomotive force. The gilbert is one oersted-centimeter. 210

Gill-Morrell oscillator. An oscillator of the retarding-field type in which the frequency of oscillation is dependent not only on electron transit time within the tube, but also on associated circuit parameters. *See:* **oscillatory circuits.** 111

gimbal (gyro). A device which permits the spin axis to have one or two angular degrees of freedom. 46

gimbal error (gyro). The error resulting from angular displacements of gimbals from their reference positions such that gimbal pickoffs do not measure the true angular-motion of the case about the input reference axis. 46

gimbal freedom (gyro). The maximum angular displacement of a gimbal about its axis. 46

gimbal lock (gyro). A condition of a two-degree-of-

freedom gyro wherein the alignment of the spin axis with an axis of freedom deprives the gyro of a degree-of-freedom and, therefore, of its useful properties.
46

gimbal retardation (gyro). A measure of output axis friction torque when the gimbal is rotated about the output axis. It is expressed as an equivalent input.
46

gimbal-unbalance torque (dynamically tuned gyro) (inertial sensor). The acceleration sensitive torque caused by gimbal unbalance along the spin axis due to nonintersection of the flexure axes. Under constant acceleration, it appears as a second harmonic of the rotor spin frequency because of the single degree of freedom of the gimbal relative to the support shaft. When the gyro is subjected to vibratory acceleration applied normal to the spin axis at twice the rotor spin frequency, this torque results in a rectified unbalance drift rate. *See:* **synchronous vibration sensitivity; two-N (2N) translational sensitivity.**
46

gin (power line maintenance). An assembly which, when attached to a support or assembled on a structure, provides a rigging point for rope blocks, blocks, etcetera, so as to manipulate various pieces of apparatus. The gin unlike the davit, is not rigid and the boom swivels, affording greater maneuverability.
458

gland seal (rotating machinery). A seal used to prevent leakage between a moveable and a fixed part.
63

glare (illuminating engineering). The sensation produced by luminances within the visual field that are sufficiently greater than the luminance to which the eyes are adapted to cause annoyance, discomfort, or loss in visual performance, or visibility. *Note:* The magnitude of the sensation of glare depends upon such factors as the size, position, and luminance of a source, the number of sources and the luminance to which the eyes are adapted.
167

glass half cell (glass electrode). A half cell in which the potential measurements are made through a glass membrane. *See:* **electrolytic cell.**
328

GLC circuit. *See:* **simple parallel circuit.**

glide path (electronic navigation). The path used by an aircraft in approach procedures as defined by an instrument landing facility. *See:* **navigation.**
13, 187

glide-path receiver. An airborne radio receiver used to detect the transmissions of a ground-installed glide-path transmitter. *Note:* It furnishes a visual, audible, or electric signal for the purpose of vertically guiding an aircraft using an instrument landing system.
328

glide slope (electronic navigation). An inclined surface generated by the radiation of electromagnetic waves and used with a localizer in an instrument landing system to create a glide path. *See:* **navigation.**
186, 13

glide-slope angle (electronic navigation). The angle in the vertical plane between the glide slope and the horizontal. *See:* **navigation.**
186, 13

glide-slope deviation (electronic navigation). The ver-

tical location of an aircraft relative to a glide slope, expressed in terms of the angle measured at the intersection of the glide slope with the runway: or the linear distance above or below the glide slope. *See:* **navigation.**
186, 13

glide slope facility (navigation). The ground station of an ILS (instrument landing system) which generates the glide slope.
13

glide-slope sector (instrument landing system). A vertical sector containing the glide slope and within which the pilot's indicator gives a quantitative measure of the deviation above and below the glide slope: the sector is bounded above and below by a specified difference in depth of modulation, usually that which gives full-scale deflection of the glide-slope deviation indicator. *See:* **navigation.**
186, 13

glint (radar). The inherent random component of error in measurement of position or Doppler frequency of a complex target due to interference of the reflections from different elements of the target. Glint may have peak values beyond the target extent in the measured coordinate. *Note:* Not to be confused with "scintillation error".
13

glitch. A perturbation of the pulse waveform of relatively short duration and of uncertain origin. *See:* **distortion, pulse.**
185

global (broadcast) destination service access point (DSAP) address (logical link control). The predefined logical link control (LLC) DSAP address (all ones) used as a broadcast (all parties) address. It can never be the address of a single LLC on the data link.
585

global. Relating to the whole of an ATLAS test procedure.
400

global broadcast (FASTBUS acquisition and control). A broadcast to slaves on all segments of a multisegment system that can be reached from the originating segment.
480

global stability. *See:* **stability, global.**

globe (illuminating engineering). A transparent or diffusing enclosure intended to protect a lamp, to diffuse and redirect its light, or to change the color of the light.
167

glossmeter (illuminating engineering). An instrument for measuring gloss as a function of the directionally selective reflecting properties of a material in angles near to and including the direction giving specular reflection.
167

glow (mode) current (gas tube surge arrester). The current which flows after breakdown when circuit impedance limits the follow current to a value less than the glow-to-arc transition current.
62

glow corona (overhead-power-line corona and radio noise). Glow corona is a stable, essentially steady discharge of constant luminosity occurring at either positive or negative electrodes.
411

glow current (gas-tube surge-protective devices). The current that flows after breakdown when circuit impedance limits the follow current to a value less than the glow-to-arc transition current. It is sometimes called the glow mode current.
490

glow discharge (1) (electron tubes). A discharge of electricity through gas characterized by (A) a change of space potential, in the immediate vicinity of the cathode, that is much higher than the ionization potential of the gas. (B) a low, approximately constant, current density at the cathode, and a low cathode temperature. (C) the presence of a cathode glow. *See:* **gas tubes; lamp.** 125
(2) (illuminating engineering). An electric discharge characterized by a low, approximately constant, current density at the cathode, low cathode temperature, and a high, approximately constant, voltage drop. 167

glow-discharge tube. A gas tube that depends for its operation on the properties of a glow discharge. 125

glow factor (illuminating engineering). A measure of the visible light response of a fluorescent material to "black light". It is equal to π times the luminance in candelas per square meter produced on the material divided by the incident "black light" flux density in milliwatts per square meter. π is omitted when luminance is in footlamberts and the area is in square feet. It may be measured in lumens per milliwatt. *Note:* See note under **filter factor.** 167

glow lamp (illuminating engineering). An electric-discharge lamp whose mode of operation is that of a glow discharge, and in which light is generated in the space close to the electrodes. 167

glow mode current. *See:* **glow current**
glow, negative. *See:* **negative glow.**
glow-switch. An electron tube containing contacts operated thermally by means of a glow discharge. 190

glow-to-arc transition current (gas-tube surge-protective devices). The current required for the arrester to pass from the glow mode into the arc mode. 490
glow-tube. *See:* **glow discharge tube.**
glow voltage (gas-tube surge-protective devices). The voltage drop across the arrester during glow-current flow. It is sometimes called the glow mode voltage. 490

glue-line heating (dielectric heating usage). An arrangement of electrodes designed to give preferential heating to a thin film of material of relatively high loss factor between alternate layers of relatively low loss factor material. *See:* **dielectric heating.** 14, 114
go. *See:* **go/no-go.**
goniometer (electronic navigation). A combining device used with a plurality of antennas so that the direction of maximum radiation or of greatest response may be rotated in azimuth without physically moving the antenna array. 187, 13
goniophotometer (illuminating engineering). A photometer for measuring the directional light distribution characteristics of sources, luminaires, media, and surfaces. 167
go/no-go (test, measurement and diagnostic equipment). A set of terms (in colloquial usage) referring to the condition or state of operability of a unit which can only have two parameters: (1) go, functioning properly, or (2) no-go, not functioning properly. 54

governing system (hydraulic turbines). The combination of devices and mechanisms which detect speed deviation and convert it into a change in servomotor position. It includes the speed sensing elements, the governor control actuator, the hydraulic pressure supply system, and the turbine control servomotor. 8, 58
governor (65) (power system device function numbers). The assembly of fluid, electrical, or mechanical control equipment used for regulating the flow of water, steam, or other medium to the prime mover for such purposes as starting, holding speed or load, or stopping. 402
governor actuator rating (speed governing systems, hydraulic turbines). The governor actuator rating is the flow rate in volume per unit time which the governor actuator can deliver at a specified pressure drop. The pressure drop shall be measured across the terminating pipe connections to the turbine control servomotors at the actuator. This pressure drop is measured with the specified minimum normal working pressure of the pressure supply system delivered to the supply port of the actuator distributing valve. 58
governor control actuator (hydraulic turbines). The combination of devices and mechanisms which detects a speed error and develops a corresponding hydraulic control output to the turbine control servomotors but does not include the turbine control servomotors. Includes gate, blade, deflector, or needle control, or all equipment as appropriate. 8
governor control actuator rating (hydraulic turbines). The governor actuator rating is the flow rate in volume per unit time which the governor actuator can deliver at a specified pressure drop. The pressure drop shall be measured across the terminating pipe connections to the turbine control servomotors at the actuator. This pressure drop is measured with the specified minimum normal working pressure of the fluid pressure supply system delivered to the supply port of the actuator distributing valve. 8
governor-controlled gates (hydro-turbine). Include the gates that control the energy input to the turbine and that are normally actuated by the speed governor directly or through the medium of the speed-control mechanism. *See:* **speed-governing system.** 94
governor-controlled valves (steam turbine). Include the valves that control the energy input to the turbine and that are normally actuated by the speed governor directly or through the medium of the speed-control mechanism. *See:* **speed-governing system.** 94
governor dead band (automatic generation control). The magnitude of the total change in steady-rate speed within which there is no resulting measurable change in the position of the governor-controlled valves. *Note:* Dead band is the measure of the insensitivity of the speed-governing system and is expressed in percent of rated speed. 94
governor dead time (hydraulic turbines). Dead time is the time interval between the initiation of a specified change in steady-state speed and the first detectable movement of the turbine control servomotor. 8

governor pin. *See:* **centrifugal-mechanism pin.**

governor speed-changer. A device by means of which the speed-governing system may be adjusted to change the speed or power output of the turbine while the turbine is in operation. *See:* **speed-governing system.**   94

governor speed-changer position. The position of the speed changer indicated by the fraction of its travel from the position corresponding to minimum turbine speed to the position corresponding to maximum speed and energy input. It is usually expressed in percent. *See:* **speed-governing system.**   94

governor spring. *See:* **centrifugal-mechanism spring.**

governor weights. *See:* **centrifugal-mechanism weights.**

graded index optical waveguide (fiber optics). A waveguide having a graded index profile in the core. *See:* **graded index profile; step index optical waveguide.**
                                                   433

graded index profile (fiber optics). Any refractive index profile that varies with radius in the core. Distinguished from a step index profile. *See:* **dispersion; mode volume; multimode optical waveguide; normalized frequency; optical waveguide; parabolic profile; profile dispersion; profile parameter; refractive index; step index profile; power-law index profile.**
                                                   433

graded insulation (electronics power transformer). The selective arrangement of the insulation components of a composite insulation system to more nearly equalize the voltage stresses throughout the insulation system.   95

graded junction (semiconductor)(nonlinear, active, and nonreciprocal waveguide components). A specially designed p-n junction with a p+ -type of region and an n-type of region whose doping levels increase linearly with distance from the junction.   530

graded-time step-voltage test (rotating machinery). A controlled overvoltage test in which calculated voltage increments are applied at calculated time intervals. *Note:* Usually, a direct-voltage test with the increments and intervals so calculated that dielectric absorption appears as a constant shunt-conductance: to simplify interpretation. *See:* **asynchronous machine.**
                                                    63

grade-of-service (telephone switching systems). The proportion of total calls, usually during the busy hour, that cannot be completed immediately or served within a prescribed time.   55

gradient (scalar field). At a point, a vector (denoted by ∇u) equal to, and in the direction of, the maximum space rate of change of the field. It is obtained as a vector field by applying the operator nabl to a scalar function. Thus if $u = f(x,y,z)$

$$\nabla u = \text{grad } u$$
$$= \mathbf{i}\,\frac{\partial u}{\partial x} + \mathbf{j}\,\frac{\partial u}{\partial y} + \mathbf{k}\,\frac{\partial u}{\partial z}.$$

                                                   210

gradient meter. *See:* **generating electric field meter.**

gradient microphone. A microphone the output of which corresponds to a gradient of the sound pressure. *Note:* Gradient microphones may be of any order as, for example, zero, first, second, etcetera. A pressure microphone is a gradient microphone of zero order. A velocity microphone is a gradient microphone of order one. Mathematically, from a directivity standpoint for plane waves the root-mean-square response is proportional to $\cos^n \theta$, where θ is the angle of incidence and n is the order of the microphone. *See:* **microphone.**
                                                   328

grading (telephone switching systems). Partial commoning or multiplying of the outlets of connecting networks where there is limited availability to the outgoing group or subgroup of outlets.   55

grading device (composite insulators). A device for controlling the potential gradient at the end fittings, such as a grading ring or various semiconductive polymeric devices.   483

grading group (telephone switching systems). That part of a grading in which all inlets have access to the same outlets.   55

grading or control ring (surge arresters)(metal-oxide surge arresters for ac power circuits). A metal part, usually circular or oval in shape, mounted to modify electrostatically the voltage gradient or distribution.
                                              430, 583

grading ring (arrester) (control ring). A metal part, usually circular or oval in shape, mounted to modify electrostatically the voltage gradient or distribution.
                                              2, 299, 62

gradual failure. *See:* **failure, gradual.**

graduated (control) (industrial control). Marked to indicate a number of operating positions.   206

grain (photographic material). A small particle of metallic silver remaining in a photographic emulsion after development and fixing. *Note:* In the agglomerate, these grains form the dark area of a photographic image.   176

graininess (photographic material). The visible coarseness under specified conditions due to silver grains in a developed photographic film.   176

granular-filled fuse unit (power switchgear). A fuse unit in which the arc is drawn through powdered, granular, or fibrous material.   103, 443

graph (software). A model consisting of a finite set of nodes having connections called edges or arcs. *See:* **model; nodes.**   434

graph determinant (network analysis). One plus the sum of the loop set transmittances of all nontouching loop sets contained in the graph. *Notes:* (1)The graph determinant is conveniently expressed in the form

$$\Delta = (1 - \Sigma\, L_i + \Sigma\, L_i L_j - \Sigma\, L_i L_j L_k + \cdots)$$

where L_i is the loop transmittance of the ith loop of the graph, the second is over all of the different pairs of nontouching loops, and the third is over all the different triplets of nontouching loops, etcetera. (2) The graph determinant may be written alternatively as

$$\Delta = [(1 - L_1)(1 - L_2) \cdots (1 - L_n)]\dagger$$

where $L_1, L_2 ..., L_n$, are the loop transmittances of the n different loops in the graph, and where the dagger indicates that, after carrying out the multiplications within the brackets, a term will be dropped if it contains the transmittance product of two touching loops. (3)The graph determinant reduces to the return difference for a graph having only one loop. (4)The graph determinant is equal to the determinant of the coefficient equations.   282

graphic (computer applications). A symbol produced by a process such as handwriting, drawing, or printing.   571

graphic character (computer applications). A character, other than a control character, that is normally represented by a graphic.   571

graphic display (supervisory control, data acquisition, and automatic control)(station control and data acquisition). A hardware device (crt, plasma panel, arrays of lamps, or light emitting diodes) used to present pictorial information.   570, 403

graphic symbol (1)(abbreviation). A geometric representation used to depict graphically the generic function of an item as it normally is used in a circuit. *See:* **abbreviation.**   173

(2)(electrical engineering). A shorthand used to show graphically the functioning or interconnections of a circuit. A graphic symbol represents the functions of a part in the circuit. For example, when a lamp is employed as a nonlinear resistor, the nonlinear resistor symbol is used. Graphic symbols are used on single-line (one-line) diagrams, on schematic or elementary diagrams, or, as applicable, on connection or wiring diagrams. Graphic symbols are correlated with parts lists, descriptions, or instructions by means of designations.   25

graphite brush. A brush composed principally of graphite. *Note:* This type of brush is soft. Grades of brushes of this type differ greatly in current-carrying capacity and in operating speed from low to high. *See:* **brush (rotating machinery).**   63, 279

graph transmittance (network analysis). The ratio of signal at some specified dependent node, to the signal applied at some specified source node. *Note:* The graph transmittance is the weighted sum of the path transmittances of the different open paths from the designated source node to the designated dependent node, where the weight for each path is the path factor divided bt the graph determinant.   282

grass (radar). A descriptive colloquialism referring to the appearance of noise on certain displays, such as an A-display.   13

graticule (oscilloscopes). A scale for measurement of quantities displayed on the cathode-ray tube of an oscilloscope. *See:* **oscilloscope.**   185

graticule area (oscilloscopes). The area enclosed by the continuous outer graticule lines. *Notes:* Unless other-

wise stated the graticule area shall be equal to or less than the viewing area. *See:* **oscillograph; quality-area; viewing area.**   185

graticule, internal (oscilloscopes). A graticule whose rulings are a permanent part of the inner surface of the cathode-ray tube faceplate.   106

grating. *See:* **ultrasonic space grating.**

grating lobe (antennas). A lobe, other than the main lobe, produced by an array antenna when the interelement spacing is sufficiently large to permit the in-phase addition of radiated fields in more than one direction.   111

gravitational acceleration unit (g). The symbol g denotes a unit of acceleration equal in magnitude to the local value of gravity, unless otherwise specified. *Note:* (1) In some applications, a standard value of g may be specified. (2) For an earthbound accelerometer, the attractive force of gravity acting on the proof mass must be treated as an applied upward acceleration of one g.   46

gravity gradient stabilization (communication satellite). The use of the gravity gradient along a satellite structure for controlling its attitude. This method usually requires long booms to create the necessary mass distribution.   74

gravity vertical. *See:* **mass-attraction vertical.**

gravity wave (radio wave propagation). A wave in a fluid whose restoring force arises from the stratification of the medium in the presence of gravity. It may appear as either a surface wave or an internal wave.   146

gray (metric practice). The absorbed dose when the energy per unit mass imparted to matter by ionizing radiation is one joule per kilogram. *Note:* The gray is also used for the ionizing radiation quantities: specific energy imparted, kerma, and absorbed dose index, which have the SI unit joule per kilogram.   21

graybody (illuminating engineering). A temperature radiator whose spectral emissivity is less than unity and the same at all wavelengths.   167

Gray code (mathematics of computing). A binary code in which sequential numbers are represented by binary expressions, each of which differs from the preceding expression in one place only.   564

gray scale (television). An optical pattern in discrete steps between light and dark. *Note:* A gray scale with ten steps is usually included in resolution test charts.   18

Gregorian reflector antenna. A paraboloidal reflector antenna with a concave subreflector, usually ellipsoidal in shape, located at a distance from the vertex of the main reflector which is greater than the prime focal length of the main reflector. *Note:* To improve the aperture efficiency of the antenna, the shapes of the main reflector and subreflector are sometimes modified.   111

grid (circuit) resistor (industrial control). A resistor used to limit grid current. *See:* **electronic controller.**   206

grid control (industrial control). Control of anode current of an electron tube by means of proper variation

(control) of the control-grid potential with respect to the cathode of the tube. *See:* **electronic controller.**
206

grid-controlled mercury-arc rectifier. A mercury-arc rectifier in which one or more electrodes are employed exclusively to control the starting of the discharge. *See:* **rectifier.**
111

grid course (navigation). Course relative to grid north. *See:* **navigation.**
187, 13

grid current (analog computer). The current flowing between the summing junction and the grid of the first amplifying stage of an operational amplifier. *Note:* Grid current results in an error voltage at the amplifier output. *See:* **electronic analog computer; electronic controller.**
9

grid-drive characteristic (electron tubes). A relation, usually shown by a graph, between electric or light output and control-electrode voltage measured from cutoff.
125

grid driving power (electron tubes). The average of the product of the instantaneous values of the alternating components of the grid current and the grid voltage over a complete cycle. *Note:* This power comprises the power supplied to the biasing device and to the grid. *See:* **electrode dissipation.**
125

grid emission. Electron or ion emission from a grid. *See:* **electron emission.**
125

grid emission, primary (thermionic). Current produced by electrons or ions thermionically emitted from a grid. *See:* **electron emission.**
190

grid emission, secondary. Electron emission from a grid due directly to bombardment of its surface by electrons or other charged particles. *See:* **electron emission.**
190

grid-glow tube. A glow-discharge cold-cathode tube in which one or more control electrodes initiate but do not limit the anode current, except under certain operating conditions. *Note:* This term is used chiefly in the industrial field.
190

grid, ground. *See:* **ground grid.**

grid heading (navigation). Heading relative to grid north. *See:* **navigation.**
187, 13

grid-leak detector. A triode or multielectrode tube in which rectification occurs because of electron current to the grid. *Note:* The voltage associated with this flow through a high resistance in the grid circuit appears in amplified form in the plate circuit.
328

grid mesh (generating station grounding). Any one of the open spaces enclosed by the grid conductors.
569

grid modulation (electron tubes). Modulation produced by the application of the modulating voltage to the control grid of any tube in which the carrier is present. *Note:* Modulation in which the grid voltage contains externally generated pulses is called **grid pulse modulation.**
111

grid neutralization (electron tubes). The method of neutralizing an amplifier in which a portion of the grid-cathode alternating-current voltage is shifted 180 degrees and applied to the plate-cathode circuit through a neutralizing capacitor. *See:* **amplifier; feedback.**
111

grid north (navigation). An arbitrary reference direction used in connection with a system of rectangular coordinates superimposed over a chart. *See:* **navigation.**
187, 13

grid number *n* (electron tubes). A grid occupying the *n*th position counting from the cathode. *See:* **electron tube.**
244, 190

grid pitch (electron tubes). The pitch of the helix of a helical grid. *See:* **electron tube.**
244, 190

grid pulse modulation. Modulation produced in an amplifier or oscillator by application of one or more pulses to a grid circuit.
111

grids (high-power rectifier). Electrodes that are placed in the arc stream and to which a control voltage may be applied. *See:* **rectification.**
328

grid system (substation grounding). A system consisting of interconnected bare conductors buried in the earth or in concrete to provide a common ground for electrical devices and metallic structures.
475

grid transformer (industrial control). Supplies an alternating voltage to a grid circuit or circuits.
206

grid voltage. *See:* **electrode voltage; electronic controller.**

grid voltage, critical. *See:* **control-grid voltage.**

grid voltage supply (electron tube). The means for supplying to the grid of the tube a potential that is usually negative with respect to the cathode. *See:* **power pack.**
328

grip, conductor (conductor stringing equipment). A device designed to permit the pulling of conductor without splicing on fittings, eyes, etc. It also permits the pulling of a "continuous" conductor where threading is not possible. The designs of these grips vary considerably. Grips such as the Klein (chicago) and Crescent utilize an open sided rigid body with opposing jaws and swing latch. In addition to pulling conductors, this type is commonly used to tension guys and in some cases, pull wire rope. The design of the Comealong (Pocketbook, Suitcase, "4" Bolt, etc.) incorporates a bail attached to the body of a clamp which folds to completely surround and envelop the conductor. "Bolts" are then used to close the clamp and obtain a grip. *Syn:* **buffalo; chicago grip; comealong; crescent; grip; klein; pocketbook; slip-grip; suit case; "4" bolt; "6" bolt; "7" bolt.**
45

grip, woven wire (conductor stringing equipment). A device designed to permit the temporary joining or pulling of conductors without the need of special eyes, links or grips. *Syn:* **basket; chinese finger; kellem; sock; wire mesh grip.**
45

groove (mechanical recording). The track inscribed in the record by the cutting or embossing stylus. *See:* **phonograph pickup.**
176

groove angle (disk recording). The angle between the two walls of an unmodulated groove in a radial plane perpendicular to the surface of the recording medium. *See:* **phonograph pickup.**
176

groove diameter. *See:* **tape-wound core.**

groove shape (disk recording). The contour of the groove in a radial plane perpendicular to the surface of the recording medium. *See:* **phonograph pickup.**
176

groove speed (disk recording). The linear speed of the groove with respect to the stylus. *See:* **phonograph pickup.**   176

groove width. *See:* **tape-wound core.**

gross actual generation (GAAG) (power system measurement). The energy that was generated by a unit in a given period.   432

gross available capacity (GAC) (power system measurement). The gross dependable capacity, modified for equipment limitation at any time.   432

gross available generation (GAG) (power system measurement). The gross energy that could have been generated in a given period if operated continuously at its gross available capacity.   432

gross demonstrated capacity. The gross steady output that a generating unit or station has produced while demonstrating its maximum performance under stipulated conditions. *See:* **generating station.**   64

gross dependable capacity (GDC) (power system measurement). The gross maximum capacity, modified for ambient limitations for a specified period of time, such as a month or a season.   432

gross generation (electric power systems). The generated output power at the terminals of the generator.   94

gross head (power operations). The difference of elevations between water surfaces of the forebay and tailrace under specified conditions.   516

gross heat rate (power operations). A measure of generating station thermal efficiency, generally expressed as British thermal unit per kilowatt-hour (Btu/kWh). *Note:* It is computed by dividing the total Btu content of the fuel burned (or of heat released from a nuclear reactor) by the resulting kilowatt-hours (kWh) generated.   516

gross information content. A measure of the total information, redundant or otherwise, contained in a message. *Note:* It is expressed as the number of bits or hartleys required to transmit the message with specified accuracy over a noiseless medium without coding. *See:* **bit.**   328

gross maximum capacity (GMC) (power system measurement). The maximum capacity that a unit can sustain over a specified period of time. To establish this capacity, formal demonstration is required. The test should be repeated periodically. This demonstrated capacity level shall be corrected to generating conditions for which there would be minimum ambient restriction. When a demonstration test has not been conducted, the estimated maximum capacity of the unit shall be used.   432

gross maximum generation (GMG) (power system measurement). The energy that could have been produced by a unit in a given period of time if operated continuously at gross maximum capacity.   432

gross rated capacity. The gross steady output that a generating unit or station can produce for at least two hours under specified operating conditions. *See:* **generating station.**   64

gross reserve generation (GRG) (power system measurement). The energy that a unit could have produced in a given period but did not, because it was not required by the system. This is the difference between gross available generation and gross actual generation: GRG = GAG + GAAG   432

gross seasonal unavailable generation (GSUG) (power system measurement). The difference between the energy that would have been generated if operating continuously at gross maximum capacity and the energy that would have been generated if operating continuously at gross dependable capacity, calculated only during the time the unit was in the available state. See Appendix D, ANSI/IEEE Std 762-1980.   432

gross unit unavailable generation (GUUG) (power system measurement). The difference between the energy that would have been generated if operating continuously at gross dependable capacity and the energy that would have been generated if operating continuously at available capacity. This is the energy that could not be generated by a unit due to planned and unplanned outages and unit deratings.   432

ground(1)(ground system). A conducting connection, whether intentional or accidental, by which an electric circuit or equipment is connected to the earth, or to some conducting body of relatively large extent that serves in place of the earth.*Note:* It is used for establishing and maintaining the potential of the earth (or of the conducting body) or approximately that potential, on conductors connected to it, and for conducting body current to and from the earth (or the conducting body).   313, 256, 91

(2)(safety in ac substation grounding). A conducting connection, whether intentional or accidental, by which an electric circuit or equipment is connected to the earth or to some conducting body of relatively large extent that serves in place of the earth.   563

(3) (transmission path). (A) A direct conducting connection to the earth or body of water that is a part thereof. (B) A conducting connection to a structure that serves a function similar to that of an earth ground (that is, a structure such as a frame of an air, space, or land vehicle that is not conductively connected to earth).   25

(4) (electrolytic cell line working zone). A conducting connection, whether intentional or accidental, by which an electric circuit or equipment is connected to earth, or to some conducting body that serves in place of earth.   133

(5) (National Electrical Code). A conducting connection whether intentional or accidental, between an electrical circuit or equipment and the earth, or to some conducting body that serves in place of the earth.   256

groundable parts (power switchgear). Those parts that may be connected to ground without affecting operation of the device.   103, 443

ground absorption (data transmission). The loss of energy in transmission of radio waves, due to dissipation in the ground.   59

ground acceleration (seismic qualification of Class 1E

equipment for nuclear power generating stations). The acceleration of the ground resulting from the motion of a given earthquake. The maximum ground acceleration is the zero period acceleration (ZPA) of the ground response spectrum. 28,581

ground-area open floodlight (O) (illuminating engineering). A unit providing a weatherproof enclosure for the lamp socket and housing. No cover glass is required. 167

ground-area open floodlight with reflector insert (OI) (illuminating engineering). A weatherproof unit so constructed that the housing forms only part of the reflecting surface. An auxiliary reflector is used to modify the distribution of light. No cover glass is required. 167

ground bar (lightning). A conductor forming a common junction for a number of ground conductors. 244, 62

ground-based navigation aid. An aid that requires facilities located upon land or sea. *See:* **navigation.** 187, 13

ground bus (power switchgear). A bus to which the grounds from individual pieces of equipment are connected, and that, in turn, is connected to ground at one or more points. 103

ground bushing (separable insulated connectors). An accessory device designed to electrically ground and mechanically seal a de-energized power cable terminated with an elbow. 454

ground cable bond. A cable bond used for grounding the armor or sheaths of cables or both. *See:* **ground.** 64

ground clamp (grounding clamp). A clamp used in connecting a grounding conductor to a grounding electrode or to a thing grounded. *See:* **ground.** 64

ground clutter (radar). Clutter resulting from the ground or objects on the ground. 13

ground conductor (lightning). A conductor providing an electric connection between part of a system, or the frame of a machine or piece of apparatus, and a ground electrode or a ground bar. *See:* **grounded conductor.** 244, 62

ground conduit. A conduit used solely to contain one or more grounding conductors. *See:* **ground.** 64

ground connection. *See:* **grounding connection.**

ground contact (of a switchgear assembly). A self-coupling separable contact provided to connect and disconnect the ground connection between the removable element and the ground bus of the housing and so constructed that it remains in contact at all times except when the primary disconnecting devices are separated by a safe distance. *Note:* Safe distance, as used here, is a distance at which the equipment will meet its withstand-voltage ratings, both low-frequency and impulse, between line and load terminals with the switching device in the closed position. 103

ground contact indicator *See:* **line isolation monitor.**

ground-controlled approach (GCA) (radar). A ground radar system providing information by which aircraft approaches to landing may be directed via radio communications; the system consists of a precision approach radar (PAR) and an airport surveillance radar (ASR). 13

ground-controlled intercept (GCI)(radar). A radar system by means of which a controller on the ground may direct an aircraft to make an interception of another aircraft. 13

ground controlled interception. *See:* **GCI.**

ground current(1)(ground system). Current flowing in the earth or in a grounding connection. 313

(2)(safety in ac substation grounding). A current flowing into or out of the earth or its equivalent serving as a ground. 563

ground-derived navigation data (air navigation). Data obtained from measurements made on land or sea at locations external to the vehicle. *See:* **navigation.** 187, 13

ground detection rings (rotating machinery). Collector rings connected to a winding and its core to facilitate the measurement of insulation resistance on a rotor winding. *See:* **rotor (rotating machinery).** 63

ground detector. An instrument or an equipment used for indicating the presence of a ground on an ungrounded system. *See:* **ground.** 64

ground detector relay (64) (power system device function numbers). A relay that operates on failure of machine or other apparatus insulation to ground. *Note:* This function is not applied to a device connected in the secondary circuit of current transformers in a normally grounded power system, where other device numbers with a suffix G or N should be used, that is, 51N for an ac time overcurrent relay connected in the secondary neutral of the current transformers. *See:* **ANSI/IEEE C37.2-1979.** 402

grounded(1)(ground system). A system, circuit, or apparatus referred to is provided with a ground. 313

(2)(NESC). Connected to or in contact with earth or connected to some extended conductive body which serves instead of the earth. 494

(3)(power and distribution transformer)(electric systems)(power line maintenance)(conductor stringing equipment). Connected to earth or to some extended conducting body that serves instead of the earth, whether the connection is intentional or accidental. 53, 91, 458, 431

(4)(safety in ac substation grounding). A system, circuit, or apparatus referred to is provided with ground for the purposes of establishing a ground return circuit and for maintaining its potential at approximately the potential of earth. 563

(5) (transmission and distribution). Connected to or in contact with earth or connected to some conductive body which serves instead of the earth. 262

(6) (effectively grounded communication system) (National Electrical Code). Permanently connected to earth through a ground connection of sufficiently low impedance and having sufficient ampacity to prevent the building up of voltages that may result in undue hazard to connected equipment or to persons. 256

grounded capacitance. *See:* **self-capacitance of a conductor; ground.**

grounded-cathode amplifier. An electron-tube amplifier with the cathode at ground potential at the operating frequency, with input applied between the control grid and ground, and the output load connected between plate and ground. *Note:* This is the conventional amplifier circuit. *See:* **amplifier.** 111

grounded circuit. A circuit in which one conductor or point (usually the neutral conductor or neutral point of transformer or generator windings) is intentionally grounded, either solidly or through a noninterrupting current limiting grounding device. *See:* **ground; grounded conductor; grounded system.** 64, 91

grounded concentric wiring system. A grounded system in which the external (outer) conductor is solidly grounded and completely surrounds the internal (inner) conductor through its length. The external conductor is usually uninsulated. *See:* **ground.** 64, 91

grounded conductor (1)(NESC)(transmission and distribution). A conductor which is intentionally grounded, either solidly or through a noninterrupting current-limiting device. 494

(2) (electric system). A conductor that is intentionally grounded, either solidly or through a current limiting device. *See:* **ground.** 91, 262

(3) (National Electrical Code). A system or circuit conductor that is intentionally grounded. 256

grounded, directly. *See:* **grounded, solidly.**

grounded, effectively (1)(grounding of industrial and commercial power systems). An expression that means grounded through a grounding connection of sufficiently low impedance (inherent or intentionally added or both) that ground fault that may occur cannot build up voltages in excess of limits established for apparatus, circuits, or systems so grounded. *Notes:*(A) An alternating-current system or portion thereof may be said to be effectively grounded when, for all points on the system or specified portion thereof, the ratio of zero-sequence reactance to positive-sequence reactance is not greater than three and the ratio of zero-sequence resistance to positive-sequence reactance is not greater than one for any condition of operation and for any amount of connected generator capacity. (B) This definition is basically used in the application of line-to-neutral surge arresters. Surge arresters with less than line-to-line voltage ratings are applicable on effectively grounded systems. *See:* **ground; grounded.** 152

(2)(National Electrical Safety Code). *See:* **effectively grounded.** 391

(3)(system grounding). Grounded through a sufficiently low impedance such that for all system conditions the ratio of zero-sequence reactance (X_0/X_1) is positive and less than 3, and the ratio of zero-sequence resistance to positive-sequence reactance (R_0/X_1) is positive and less than 1. *Note:* The effectively grounded system permits the application of surge arresters with less than line-to-line voltage ratings. Ground fault currents will be approximately of the same magnitude as three-phase fault currents. 152

grounded-grid amplifier. An electron-tube amplifier circuit in which the control grid is at ground potential at the operating frequency, with input applied between cathode and ground, and output load connected between plate and ground. *Note:* The grid-to-plate impedance of the tube is in parallel with the load instead of acting as a feedback path. *See:* **amplifier.** 111

grounded, impedance. Grounded through impedance. *Note:* The components of the impedance need not be at the same location. 91

grounded neutral system (surge arresters). A system in which the neutral is connected to ground, either solidly or through a resistance or reactance of low value. 244, 62

grounded-neutral terminal type voltage transformer. One that has the neutral end of the high-voltage winding connected to the case or mounting base. 203

grounded parts (power switchgear). Parts that are intentionally connected to ground. 103, 443

grounded-plate amplifier (cathode-follower). An electron-tube amplifier circuit in which the plate is at ground potential at the operating frequency, with input applied between control grid and ground, and the output load connected between cathode and ground. *See:* **amplifier.** 111

grounded potentiometer (analog computers). A potentiometer with one end terminal attached directly to ground. 9

grounded solidly (system grounding). Connected directly through an adequate ground connection in which no impedance has been intentionally inserted. *Note:* This term, though commonly used, is somewhat confusing since a transformer may have its neutral solidly connected to ground, and yet the connection may be so small in capacity as to furnish only a very-high-impedance ground to the system to which it is connected. In order to define grounding positively and logically as to degree, the term effective grounding has come into use. The term solidly grounded will therefore be used in this standard only in referring to a solid metallic connection from system neutral to ground; that is, with no impedance intentionally added in the grounding circuit. 152

grounded system (1)(NESC). A system of conductors in which at least one conductor or point is intentionally grounded, either solidly or through a noninterrupting current-limiting device. 494

(2)(system grounding). A system of conductors in which at least one conductor or point (usually the middle wire or neutral point of transformer or generator windings) is intentionally grounded, either solidly or through an impedance. *Note:* Various degrees of groundings are used, from solid or effective grounding to the high-impedance grounding obtained from a small grounding transformer used only to secure enough ground current for relaying, to the high-resistance grounding which secures control of transient overvoltage but may not furnish sufficient current for ground-fault relaying. Fig 1(b) and Fig 1(c) show two points at which a system may be grounded and the corresponding voltage relationships. Note that ac-

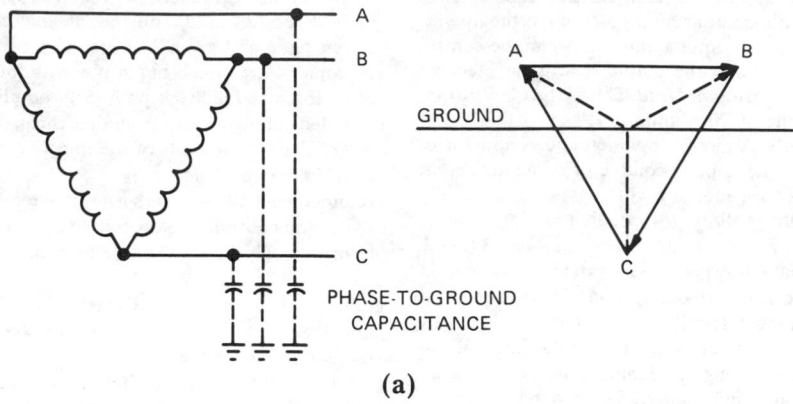

PHASE-TO-GROUND
CAPACITANCE

(a)

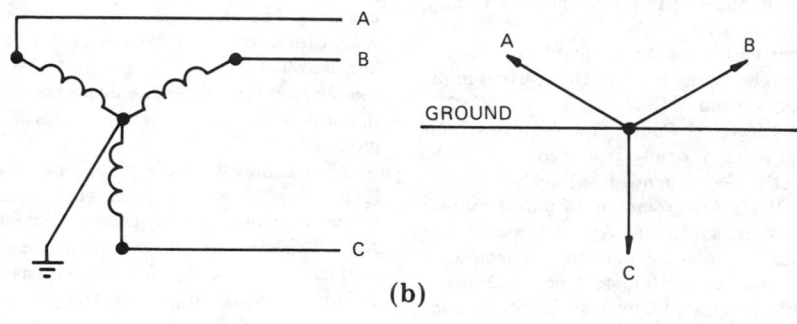

(b)

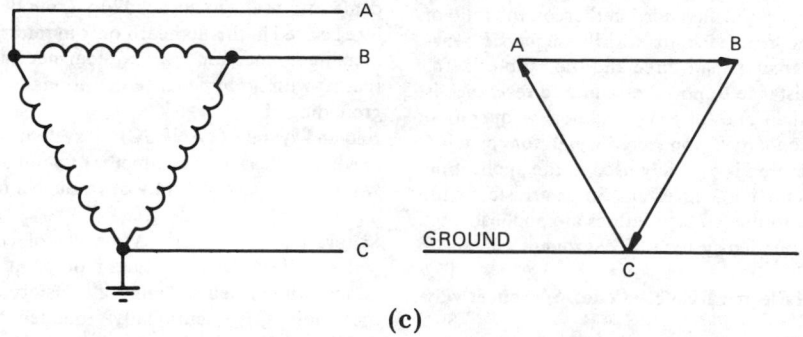

(c)

Voltages to Ground under Steady-State Conditions
(a) Ungrounded System (b) Grounded Wye-Connected System
(c) Corner Grounded Delta-Connected System

cording to NEMA SG 4-1975 [6] there are system voltage limitations for corner grounding.   152

(3) (transmission and distribution). A system of conductors in which at least one conductor or point is intentionally grounded, either solidly or through a noninterrupting current-limiting device.   262

(4) (power and distribution transformer). A system of conductors in which at least one conductor or point (usually the middle wire or neutral point of transformer or generator windings) is intentionally grounded, either solidly or through a current-limiting device.   53

(5) (surge arrester). An electric system in which at least one conductor or point (usually the neutral conductor or neutral point of transformer or generator windings) is intentionally grounded, either solidly or through a grounding device.   430

ground electrode (surge arresters). A conductor or group of conductors in intimate contact with the ground for the purpose of providing a connection with the ground.   244, 62

ground end (grounding device). The end or terminal of the device that is grounded directly or through another device. *See:* **grounding device.**   91

ground equalizer inductors (antenna). Coils of relatively low inductance, placed in the circuit connected to one or more of the grounding points of an antenna to distribute the current to the various points in any desired manner. *Note:* Broadcast usage only and now in disuse. *See:* **antenna.**   111

ground fault (surge arresters). An insulation fault between a conductor and ground or frame.   244, 62

ground-fault circuit-interrupter (1) (health care facilities). A device whose function is to interrupt the electric circuit to the load when a fault current to ground exceeds some predetermined value that is less than that required to operate the overcurrent protective device of the supply circuit.   192

(2) (National Electrical Code). A device intended for the protection of personnel that functions to interrupt the electric current to the load within an established period of time when a fault current to ground exceeds some predetermined value that is less than that required to operate the overcurrent protective device of the supply circuit.   256

ground-fault neutralizer. A grounding device that provides an inductive component of current in a ground fault that is substantially equal to and therefore neutralizes the rated-frequency capacitive component of the ground-fault current, thus rendering the system resonant grounded.   91

ground-fault neutralizer grounded (resonant grounded) (power and distribution transformer). Reactance grounded through such values of reactance that, during a fault between one of the conductors and earth, the rated-frequency current flowing in the grounding reactances and the rated-frequency capacitance current flowing between the unfaulted conductors and earth shall be substantially equal. *Notes:* (1) In the fault these two components of current will be substantially 180 degrees out of phase. (2) When a system is ground-fault neutralizer grounded, it is expected that the quadrature component of the rated-frequency single-phase-to-ground fault current will be so small that an arc fault in air will be self-extinguishing.   53

ground-fault protection of equipment (National Electric Code). A system intended to provide protection of equipment from damaging line-to-ground arcing fault currents by operating to cause a disconnecting means to open all ungrounded conductors of the faulted circuit. This protection is provided at current levels less than that required to protect conductors from damage through the operation of a supply circuit overcurrent device.   256

ground grid (1)(ground system). A system of grounding electrodes consisting of interconnected bare cables buried in the earth to provide a common ground for electrical devices and metallic structures. *Note:* It may be connected to auxiliary grounding electrodes to lower its resistance.   313

(2)(generating station grounding). A buried geometric network of interconnected bare conductors.   569

(3) (conductor stringing equipment). A system of interconnected bare conductors arranged in a pattern over a specified area and on or buried below the surface of the earth. Normally, it is bonded to ground rods driven around and within its perimeter to increase its grounding capabilities and provide convenient connection points for grounding devices. The primary purpose of the grid is to provide safety for workmen by limiting potential differences within its perimeter to safe levels in case of high currents which could flow if the circuit being worked became energized for any reason or if an adjacent energized circuit faulted. Metallic surface mats and gratings are sometimes utilized for this same purpose. When used, these grids are employed at pull, tension and midspan splice sites. *Syn:* **counterpoise; ground gradient mat; ground mat.**   431

(4) (ground resistance). A system of grounding electrodes consisting of interconnected bare cables buried in the earth to provide a common ground for electric devices and metallic structures. *Note:* It may be connected to auxiliary grounding electrodes to lower its resistance. *See:* **grounding device.**   313

ground indication. An indication of the presence of a ground on one or more of the normally ungrounded conductors of a system. *See:* **ground.**   64

grounding cable. A cable used to make a connection to ground. *See:* **grounding conductor.**   63

grounding-cable connector. The terminal mounted on the end of a grounding cable.   63

grounding clamp. *See:* **ground clamp.**

grounding, coefficient of. The ratio E_{LG}/E_{LL}, expressed as a percentage, of the highest root-mean-square line-to-ground power-frequency voltage E_{LG} on a sound phase, at a selected location, during a fault to ground affecting one or more phases to the line-to-line power-frequency voltage E_{LL} which would be obtained, at the selected location, with the fault removed. *Notes:*

(1) Coefficients of grounding for three-phase systems are calculated from the phase-sequence impedance components as viewed from the selected location. For machines use the subtransient reactance. (2) The coefficient of grounding is useful in the determination of an arrester rating for a selected location. (3) A value not exceeding 80 percent is obtained approximately when for all system conditions the ratio of zero-sequence reactance to positive-sequence reactance is positive and less than 3, and the ratio of zero-sequence resistance to positive-sequence reactance is positive and less than 1.   2

grounding conductor(1)(ground system). The conductor that is used to establish a ground and that connects an equipment, device, wiring system, or another conductor (usually the neutral conductor) with the grounding electrode or electrodes.   256, 313
(2) (mining) (safety ground conductor) (safety ground) (frame ground). A metallic conductor used to connect the metal frame or enclosure of an equipment, device, or wiring system with a mine track or other effective grounding medium. *See:* **mine feeder circuit.**   328
(3)(NESC)(transmission and distribution). A conductor which is used to connect the equipment or the wiring system with a grounding electrode or electrodes.   262, 494
(4) (National Electrical Code). A conductor used to connect equipment or the grounded circuit of a wiring system to a grounding electrode or electrodes.   256

grounding connection.(ground system) A connection used in establishing a ground and consists of a grounding conductor, a grounding electrode and the earth (soil) that surrounds the electrode or some conductive body which serves instead of the earth.   313

grounding device (electric power). An impedance device used to connect conductors of an electric system to ground for the purpose of controlling the ground current or voltages to ground or a nonimpedance device used to temporarily ground conductors for the purpose of the safety of workmen. *Note:* The grounding device may consist of a grounding transformer or a neutral grounding device, or a combination of these. Protective devices, such as surge arresters, may also be included as an integral part of the device.   91

grounding elbow (separable insulated connectors). An accessory device designed to electrically ground and mechanically seal a bushing insert, or integral bushing.   454

grounding electrode (ground electrode)(ground system). A conductor used to establish a ground.   64, 313

grounding electrode conductor (National Electrical Code). The conductor used to connect the grounding electrode to the equipment grounding conductor and/or to the grounded conductor of the circuit at the service equipment or at the source of a separately derived system.   256

grounding jumper (electric appliances). A strap or wire to connect the frame of the range to the neutral conductor of the supply circuit. *See:* **appliance outlet.**   263

grounding outlet (safety outlet). An outlet equipped with a receptacle of the polarity type having, in addition to the current-carrying contacts, one grounded contact that can be used for the connection of an equipment grounding conductor. *Note:* This type of outlet is used for connection of portable appliances. *See:* **ground.**   64

grounding pad (rotating machinery). A contact area, usually on the stator frame, provided to permit the connection of a grounding terminal. *See:* **stator.**   63

grounding switch (general). A mechanical switching device by means of which a circuit or piece of apparatus may be electrically connected to ground. *Note:* A grounding switch is used to ground a piece of equipment to permit maintenance personnel to work with safety.   103, 27

grounding system (1)(ground system). Consists of all interconnected grounding connections in a specific area.   313
(2) (health care facilities). A system of conductors which provides a low impedance return path for leakage and fault currents. It coordinates with, but may be locally more extensive than, the grounding system described in Article 250 of NFPA 70, National Electrical Code.   192
(3) (surge arresters). A complete installation comprising one or more ground electrodes, ground conductors, and ground bars as required.   244, 62

grounding terminal (rotating machinery). A terminal used to make a connection to a ground. *See:* **stator.**   63

grounding transformer (power and distribution transformer). A transformer intended primarily to provide a neutral point for grounding purposes. *Note:* It may be provided with a δ winding in which resistors or reactors are connected. *See:* **stabilizing windings; voltage rating of a grounding transformer; rated kVA of a grounding transformer.**   53

ground insulation (rotating machinery). Insulation used to insure the electric isolation of a winding from the core and mechanical parts of a machine. *See:* **asynchronous machine; coil insulation.**   63

ground isolation (sequential events recording systems). The disconnection of selected field contact circuits from the contact voltage supply to allow identification of the grounded field contact wires. *See:* **field contacts.**   48

ground level (mobile communication). The elevation of the ground above mean sea level at the antenna site or other point of interest. *See:* **mobile communication system.**   181

ground light (illuminating engineering). Visible radiation from the sun and sky reflected by surfaces below the plane of the horizon.   167

ground loop (analog computer). A potentially detrimental loop formed when two or more points in an electrical system that are nominally at ground potential are connected by a conducting path such that either or both points are not at the same ground potential.   9

ground mat (ground system). A system of bare connectors, on or below the surface of the earth, connected to a ground or a ground grid to provide protection from dangerous touch voltages. *Note:* Plates and gratings of suitable area are common forms of ground mats.   313

ground overcurrent (1)(power switchgear). The net (phasor sum) current flowing in the phase and neutral conductors or the total current flowing in the normal neutral to ground connection which exceeds a predetermined value.   103

(2) (antenna). A conducting or reflecting plane functioning to image a radiating structure. *Syn:* **imaging plane.**   111

(3) (electromagnetic compatibility)(radio-noise emission). A conducting surface or plate used as a common reference point for circuit returns and electric or signal potentials.   199, 418

(4) (transmission and distribution). An assumed plane of true ground or zero potential. *See:* **direct-stroke protection (lightning).**   64

ground plane, effective (mobile communication). The height of the average terrain above mean sea level as measured for a distance of 100 meters out from the base of the antenna in the desired direction of communication. It may be considered the same as ground level only in open flat country. *See:* **mobile communication system.**   181

ground plate (grounding plate). A plate of conducting material buried in the earth to serve as a grounding electrode. *See:* **ground.**   64

ground-position indicator (GPI) (electronic navigation). A dead-reckoning tracer or computer similar to an air position indicator (API) with provision for taking account of drift. *See:* **navigation; radio navigation.**
                                                                    187, 13

ground protection (1) (ground-fault protection). A method of protection in which faults to ground within the protected equipment are detected irrespective of system phase conditions.   202, 60, 127

(2) (power switchgear). A method of protection in which faults to ground within the protected equipment are detected.   103

ground-referenced navigation data. Data in terms of a coordinate system referenced to the earth or to some specified portion thereof. *See:* **navigation.**
                                                                    187, 13

ground reflected wave (data transmission). The component of the ground wave that is reflected from the ground.   59

ground relay (power switchgear). A relay that by its design or application is intended to respond primarily to system ground faults.   103

ground resistance (grounding electrode). The ohmic resistance between the grounding electrode and a remote grounding electrode of zero resistance. *Note:* By 'remote' is meant at a distance such that the mutual resistance of the two electrodes is essentially zero.
                                                                        313

ground return. *See:* **ground clutter.**

ground-return circuit (1)(ground system). A circuit in which the earth is utilized to complete the circuit.
                                                                        313

(2)(safety in ac substation grounding). A circuit in which the earth or an equivalent conducting body is utilized to complete the circuit and allow current circulation from or to its current source.   563

(3) (data transmission). A circuit which has a conductor (or two or more in parallel) between two points and which is completed through the ground or earth.
                                                                        59

(4)(earth-return circuit) (transmission and distribution). A circuit in which the earth is utilized to complete the circuit. *See:* **ground; telegraphy; transmission line.**   64

ground-return current (line residual current) (electric supply line). The vector sum of the currents in all conductors on the electric supply line. *Note:* Actually the ground-return current in this sense may include components returning to the source in wires on other pole lines, but from the inductive coordination standpoint these components are substantially equivalent to components in the ground. *See:* **inductive coordination.**   328

ground-return system. A system in which one of the conductors is replaced by ground.   244, 62

ground rod (conductor stringing equipment). A rod that is driven into the ground to serve as a ground terminal, such as a copper-clad rod, galvanized iron rod or galvanized iron pipe. Copper-clad steel rods are commonly used during conductor stringing operations to provide a means of obtaining an electrical ground using portable grounding devices. *Syn:* **ground electrode.**   431

ground speed (navigation). The speed of a vehicle along its track. *See:* **navigation.**   13, 187

ground-start signaling (telephone switching systems). A method of signaling using direct current in a ground return path to indicate a service request.   55

ground-state maser (laser-maser). A maser in which the terminal level of the amplifying transition is appreciably populated at thermal equilibrium for the ambient temperature.   363

ground support equipment (GSE) (test, measurement and diagnostic equipment). All equipment (implements, tools, test equipment devices--mobile or fixed--and so forth) required on the ground to make an aerospace system (aircraft, missile, and so forth) operational in its intended environment.   54

ground surveillance radar. A radar set operated at a fixed point for observation and control of the position of aircraft or other vehicles in the vicinity. *See:* **navigation.**   187

ground system (antenna). That portion of an antenna consisting of a system of conductors or a conducting surface in or on the ground.   111

ground system of an antenna. That portion of an antenna closely associated with and including an extensive conducting surface which may be the earth itself. *See:* **antenna.**   179

ground terminal (1)(metal-oxide surge arresters for ac power circuits)(surge arrester). The conducting part provided for connecting the arrester to ground.
430, 583

(2) (lightning protection system). The portion extending into the ground, such as a ground rod, ground plate, or the conductor itself, serving to bring the lightning protection system into electric contact with the ground.
244, 62

ground transformer. *See:* **grounding transformer.**

ground wave (1)(antennas)(data transmission). A radio wave that is propagated over the earth and is ordinarily affected by the presence of the ground and troposphere. *Notes:* (A) The ground wave includes all components of a radio wave over the earth except ionospheric and tropospheric waves. (B) The ground wave is refracted because of variations in the dielectric constant of the troposphere including the condition known as a surface duct. *See:* **radiation; radio wave propagation.**
246, 59

(2) (radio wave propagation). From a source in the vicinity of a planetary surface, that wave which would exist in the vicinity of that surface in the absence of an ionosphere.
146

ground wire (1) (data transmission) (telecommunication). A conductor leading to an electric connection with the ground.
59

(2) (overhead power line). A conductor having grounding connections at intervals, that is suspended usually above but not necessarily over the line conductor to provide a degree of protection against lightning discharges. *See:* **ground; overhead ground wire.**
64

group (1) (communications). *See:* **channel group.**

(2) (storage cell). An assembly of plates of the same polarity burned to a connecting strap. *See:* **battery (primary or secondary).**
328

(3) (electric and electronics parts and equipments). A collection of units, assemblies, or subassemblies which is a subdivision of a set or system, but which is not capable of performing a complete operational function. Typical examples: antenna group, indicator group.
17

group address (GP)(FASTBUS acquisition and control). The high order (left justified) bits assigned in the device address field of a FASTBUS address which are used to identify the segment on which a device is located; more than one group address may be assigned to a given segment. *See:* **base group address.**
480

group alerting (telephone switching systems). A central office feature for simultaneously signaling a group of customers from a control station providing an oral or recorded announcement.
55

group ambient temperature (cable or duct) (power distribution, underground cables). The no-load temperature in a group with all other cables or ducts in the group loaded.
57

group-busy tone (telephone switching systems). A tone that indicates to operators that all trunks in a group are busy.
55

group, commutating. A group of thyristor converter circuit elements and the alternating-voltage supply elements conductively connected to them in which the direct current of the group is commutated between individual elements that conduct in succession.
121

group delay (1)(dispersive and nondispersive delay line). The derivative of radian phase with respect to radian frequency, $\partial\phi/\partial\omega$. It is equal to the phase delay for an ideal nondispersive delay device, but may differ greatly in actual devices where there is ripple in the phase vs. frequency characteristic. *See:* **envelope delay.**
146

(2)(network analyzers). In practice $\Delta\omega$ must be sufficiently greater than zero to permit adequate measurement resolution. If $\Delta\omega$ is too large, however, the limit in the defining equation for group delay will not be reached, and the measured group delay will depend upon $\Delta\omega$. Therefore, the value of $\Delta\omega$ used in a measurement should be specified.
505

group delay time. The rate of change, with angular frequency, of the total phase shift through a network. *Notes:* (1) Group delay time is the time interval required for the crest of a group of interfering waves to travel through a 2-port network, where the component wave trains have slightly different individual frequencies. (2) Group delay time is usually very close in value to envelope delay and transmission time delay, and in the case of vanishing spectrum bandwidth of the signal these quantities become identical. *See:* **measurement system.**
293

group (multi-cast) destination service access point (DSAP) address (logical link control). A destination address assigned to a colection of logical link controls (LLC) to facilitate their being addressed collectively. The least significant bit is set equal to 1.
585

group flashing light (illuminating engineering). A flashing light in which the flashes are combined in groups, each including the same number of flashes, and in which the groups are repeated at regular intervals. The duration of each flash is clearly less than the duration of the dark periods between flashes, and the duration of the dark periods between flashes is clearly less than the duration of the dark periods between groups.
167

group index (denoted N) (fiber optics). For a given mode propagating in a medium of refractive index n, the velocity of light in vacuum, c, divided by the group velocity of the mode. For a plane wave of wavelength λ, it is related thus to the refractive index:

$$N = n - \lambda(dn/d\lambda)$$

See: **group velocity; material dispersion parameter.**
433

grouping (1) (facsimile). Periodic error in the spacing of recorded lines. *See:* **facsimile signal (picture signal).**
12

(2) (electroacoustics). Nonuniform spacing between the grooves of a disk recording.
176

group loop (analog computers). A potentially detrimental loop formed when two or more points in an electrical system that are nominally at group potential are connected by a conducting path such that either or both points are not at the same ground potential. 9

group operation (power switchgear). The operation of all poles of a multipole switching device by one operating mechanism. 103

group path (radio wave propagation). For a pulsed signal traveling between two points in a medium, the product of the speed of light in vacuum and the travel time of the pulse between the two points, provided the shape of the pulse is not significantly changed. *Note:* Sometimes called apparent path or apparent range. 146

groups, commutating, set of (thyristor converter). Two or more commutating groups that have simultaneous commutations. 121

group-series loop insulating transformer (power and distribution transformer). An insulating transformer whose secondary is arranged to operate a group of series lamps or a series group of individual-lamp transformers. 53

group velocity (1) (fiber optics). (A) For a particular mode, the reciprocal of the rate of change of the phase constant with respect to angular frequency. *Note:* The group velocity equals the phase velocity if the phase constant is a linear function of the angular frequency. (B) Velocity of the signal modulating a propagating electromagnetic wave. *See:* **differential mode delay; group index; phase velocity.** 433

(2) (radio wave propagation). Of a traveling wave, the velocity of propagation of the envelope, provided that this moves without significant change of shape. The magnitude of the group velocity is equal to the reciprocal of the rate of change of phase constant with angular frequency. *Note:* Group velocity differs in magnitude from phase velocity if the phase velocity varies with frequency, and differs in direction from phase velocity if the phase velocity varies with direction. 146

(3) (traveling wave). The velocity of propagation of the envelope, provided that this moves without significant change of shape. *Notes:* (1) The magnitude of the group velocity is equal to the reciprocal of the change of phase constant with angular frequency. (2) Group velocity differs in magnitude from phase velocity if the phase velocity varies with frequency and differs in direction from phase velocity if the phase velocity varies with direction. *See:* **radio wave propagation.** 146

(4) (waveguide). Of a traveling wave at a single frequency, and for a given mode, the velocity at which the energy is transported in the direction of propagation. 267

grout (rotating machinery). A very rich concrete used to bond the feet, sole plates, bedplate, or rail of a machine to its foundation. 63

grown junction. *See:* **junction, grown.**

G scan (radar). *See:* **G display.**

G-scope (radar). A cathode-ray oscilloscope arranged to present a G-display. 13

G/T ratio (of an antenna). The ratio of the gain to the noise temperature of an antenna. *Syn:* **figure of merit.** *Notes:* (1) Usually the antenna-receiver system figure of merit is specified. For this case the figure of merit is the gain of the antenna divided by the system noise temperature referred to the antenna terminals. (2) The system figure of merit at any reference plane in the RF system is the same as that taken at the antenna terminals since both the gain and system noise temperature are referred to the same reference plane at the antenna terminals. 111

guard (interference terminology). A conductor situated between a source of interference and a signal path in such a way that interference currents are conducted to the return terminal of the interference source without entering the signal path. *See:* **interference.** 60

guard band (data transmission). A frequency band between two channels which gives a margin of safety against mutual interference. 59

guard circle (disk recording). An inner concentric groove inscribed, on disk records, to prevent the pickup from being damaged by being thrown to the center of the record. 176

guarded (1)(NESC). Covered, fenced, enclosed, or otherwise protected, by means of suitable covers or casings, barrier rails or screens, mats or platforms, designed to minimize the possibility, under normal conditions, of dangerous approach or accidental contact by persons or objects. *Note:* Wires which are insulated, but not otherwise protected, are not considered as guarded. 494

(2) (transmission and distribution). Covered, fenced, enclosed, or otherwise protected, by means of suitable covers or casings, barrier rails or screens, mats or platforms, designed to prevent dangerous approach or contact by persons or objects. *Note:* Wires which are insulated, but not otherwise protected, are not considered as guarded. 262

(3) (electrolytic cell line working zone). Covered, shielded, fenced, enclosed, or otherwise protected by means of suitable covers or casings, barrier rails or screens, mats or platforms to remove the likelihood of the dangerous contact or approach by persons or objects to a point of danger. 133

(4) (National Electrical Code). Covered, shielded, fenced, enclosed, or otherwise protected by means of suitable covers, casings, barriers, rails, screens, mats, or platforms to remove the likelihood of approach or contact by persons or objects to a point of danger. 256

guarded input (oscilloscopes). A shielded input where the shield is driven by a signal in phase with and equal in amplitude to the input signal. 106

guarded machine (rotating machinery). An open machine in which all openings giving direct access to live or rotating parts (except smooth shafts) are limited in size by the design of the structural parts, or by screens, grilles, expanded metal, etcetera, to prevent accidental contact with such parts. Such openings are of such size as not to permit the passage of a cylindrical rod 1/2 inch in diameter, except where the distance from the guard to the live or rotating parts is more than 4 inches: they are of such size as not to permit the passage of a cylindrical rod 3/4 inch in diameter. *See:* **asynchronous machine.** 63

guarded release (telephone switching systems). A technique for retaining a busy condition during the restoration of a circuit to its idle state. 55

guard electrode (testing of electric power system components). One or more electrically conducting elements, arranged and connected in an electric instrument or measuring circuit so as to divert unwanted conduction or displacement currents from, or confine wanted currents to, the measurement device. 201

guard-ground system (interference terminology). A combination of guard shields and ground connections that protects all or part of a signal transmission system from common-mode interference by eliminating ground loops in the protected part. *Note:* Ideally the guard shield is connected to the source ground. The source is usually grounded also to the source ground by bonding of the transducer to the test body. The filter, signal receiver, etcetera, are floating with respect to their own grounded cases. This necessitates physically isolating the signal receiver and filter chassis from the cases and using isolation transformers in power supplies, or isolating input circuits from cases and using isolating input transformers. This arrangement in effect places the signal receiver and filter electrically at the source. By means of a similar guard, the load can be placed effectively at the source. See the accompanying figures. 60

guard-ground system

Ideal guard shield. The guard shield consists of the signal source shield (if present), the signal shield, the filter shield, and the signal receiver shield. *See:* **interference.**

guard shield (interference terminology). A guard that is in the form of a shielding enclosure surrounding all or part of the signal path. *Note:* A guard shield is effective against both capacitively coupled and conductively coupled interference whereas a simple guard conductor is usually effective only against conductively coupled interference. *See:* **guard-ground system; interference.** 60

guard shield

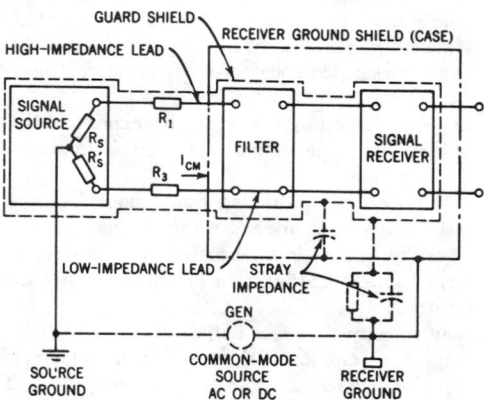

Guard shield connections when connection at source is not convenient. When $(R_3 + R_S') \ll (R_1 + R_s)$, this arrangement causes common-mode current in the low-impedance lead, but protects the more critical high-impedance lead from current flow.

guard signal (power switchgear). A signal sent over a communication channel to make the system secure against false information by preventing or guarding against the relay operation of a circuit breaker or other relay action until the signal is removed and replaced by a tripping or permissive signal. 103

guard wire. A grounded wire erected near a lower-voltage circuit or public crossing in such a position that a high (or higher) voltage overhead conductor cannot come into accidental contact with the lower-voltage circuit, or with persons or objects on the crossing without first becoming grounded by contact with the guard wire. *See:* **ground.** 64

guidance (missile). The process of controlling the flight path through space through the agency of a mechanism within the missile. *See:* **guided missile.** 328

guide (high-voltage switchgear). An attachment used to secure proper alignment when operating a fuse or switch. 443

guide bearing (rotating machinery). A bearing arranged to limit the transverse movement of a vertical shaft. *See:* **bearing.** 63

guided missile. An unmanned device whose flight path through space may be controlled by a self-contained mechanism. *See:* **beam rider guidance; command**

guidance; guidance; homing guidance; preset guidance. 328

guided mode. *See:* **bound mode.**

guided ray (fiber optics). In an optical waveguide, a ray that is completely confined to the core. Specifically, a ray at radial position r having direction such that

$$\sin\theta(r) = [n^2(r) - n^2(a)]^{1/2}$$

where $\theta(r)$ is the angle the ray makes with the waveguide axis, n (r) is the refractive index, and n(a) is the refractive index at the core radius. Guided rays correspond to bound (or guided) modes in the terminology of mode descriptors. *Syn:* **bound ray; trapped ray.** *See:* **bound mode; leaky ray.** 433

guided wave (radio wave propagation). A wave whose energy is concentrated within or near boundaries between materials of different properties, and is propagated along those boundaries. 146

guide flux. *See:* **form factor; shield.**

guideline (nuclear power quality assurance). A suggested practice that is not mandatory in programs intended to comply with a standard. The word "should" denotes a guideline; the word "shall" denotes a requirement. 417

guide wavelength. The wavelength in a waveguide, measured in the longitudinal direction. *See:* **waveguide.** 179

gun-control switch. A switch that closes an electric circuit, thereby actuating the gun-trigger-operating mechanism of an aircraft, usually by means of a solenoid. 328

Gunn oscillator (nonlinear, active, and nonreciprocal waveguide components). A direct dc-to-rf (direct current to radio frequency) conversion device in which the active element of the oscillator is a bulk III-V semiconductor device having a negative dc resistance characteristic. Conduction can occur in either direction, although the substrate contact is considered to be the cathode. The Gunn diode is a transferred electron device with practical output frequencies ranging from approximately 4 GHz (gigahertz) to more than 60 GHz. 530

guy. A tension member having one end secured to a fixed object and the other end attached to a pole, crossarm, or other structural part that it supports. *See:* **guy wire; tower.** 64

guy anchor. The buried element of a guy assembly that provides holding strength or resistance to guy wire pull. *Note:* The anchor may consist of a plate, a screw or expanding device, a log of timber, or a mass of concrete installed at sufficient depth and of such size as to develop strength proportionate to weight of earth or rock it tends to move. The anchor is designed to provide attachment for the anchor rod which extends above surface of ground for convenient guy connection. *See:* **guy; tower.** 64

guy insulator. An insulating element, generally of elongated form with transverse holes or slots for the purpose of insulating two sections of a guy or provide insulation between structure and anchor and also to provide protection in case of broken wires. Porcelain guy insulators are generally designed to stress the porcelain in compression, but wood insulators equipped with suitable hardware are generally used in tension. *See:* **tower.** 64

guy wire. A stranded cable used for a semiflexible tension support between a pole or structure and the anchor rod, or between structures. *See:* **guy; tower.** 64

gyration impedance (circuits and systems). A characteristic of a gyrator that may be expressed in terms of the impedance matrix elements as $\sqrt{z_{12} \times z_{21}}$. *See:* **gyrator.** 67

gyrator (1)(nonlinear, active, and nonreciprocal waveguide components). A two-port nonreciprocal device that provides insertion phases differing by 180 degrees for the two opposite directions of propagation. 530

(2) (antennas and propagation). (A) A directional phase changer in which the phase changes in opposite directions differ by π radians or 180 degrees. (B) Any nonreciprocal passive element employing gyromagnetic properties. *See:* **waveguide.** 244, 179

gyro (gyroscope) (inertial sensor). A device using angular momentum (usually of a spinning rotor) to sense angular motion of its case with respect to inertial space about one or two axes orthogonal to the spin axis. *Notes:* (1) This definition does not include more complex systems such as stable platforms using gyros as components; (2) Certain devices such as laser gyros that perform similar functions but do not use angular momentum may also be classified as gyros. 46

gyrocompass. A compass consisting of a continuously driven Foucault gyroscope whose supporting ring normally confines the spinning axis to a horizontal plane, so that the earth's rotation causes the spinning axis to assume a position in a plane passing through the earth's axis, and thus to point to true north. 328

gyrocompass alignment (inertial systems). A process of self-alignment in azimuth based upon measurements of misalignment drift about the nominal east-west axis of the system. *See:* **navigation.** 187

gyrocompassing. *See:* **gyrocompass alignment.**

gyro flux-gate compass (gyro flux-valve compass). A device that uses saturable reactors in conjunction with a vertical gyroscope, to sense the direction of the magnetic north with respect to the aircraft heading. 328

gyro flux-valve compass. *See:* **gyro flux-gate compass.**

gyrofrequency (radio wave propagation). The lowest natural frequency at which charged particles spiral in a fixed magnetic field. It is a vector quantity expressed by

$$f_h = \frac{1}{2\pi}\frac{q\mathbf{B}}{m}$$

where q is the charge of the particles, B is the magnetic induction and m is the mass of the particles. *Syn:* **cyclotron frequency.** 146

gyro gain. The ratio of the output angle of the gimbal to the input angle of a rate integrating gyro at zero frequency. *Note:* It is numerically equal to the ratio of the rotor angular momentum to the damping coefficient.   46

gyro horizon electric indicator. An electrically driven device for use in aircraft to provide the pilot with a fixed artificial horizon. *Note:* It indicates deviation from level flight.   328

gyromagnetic effect (nonlinear, active, and nonreciprocal waveguide components). The phenomenon by which the magnetization of a material or medium, subjected to a magnetostatic field, upon disturbance relaxes back to equilibrium by damped precessional motion about the direction of that field.   530

gyromagnetic filter (garnet, YIG)(nonlinear, active, and nonreciprocal waveguide components). A filter whose operation depends on the gyromagnetic effect.   530

gyromagnetic limiter (ferrite, garnet, YIG)(nonlinear, active, and nonreciprocal waveguide components). A power limiter whose operation depends on saturation effects in a gyromagnetic material.   530

gyromagnetic material (medium)(nonlinear, active, and nonreciprocal waveguide components). A material (medium), such as ferrite or garnet, capable of exhibiting the gyromagnetic effect.   530

gyromagnetic permeability tensor (nonlinear, active, and nonreciprocal waveguide components). A tensor used to describe the permeability properties exhibited by a gyromagnetic material, appropriate to electromagnetic wave propagation. *Syn:* **Polder tensor.**   530

gyromagnetic resonance absorption (nonlinear, active, and nonreciprocal waveguide components). That amount of power continuously absorbed in a gyromagnetic material subjected to a magnetostatic field when a disturbance causes a steady precession of the magnetization of that material at a rate near the gyromagnetic resonance frequency.   530

gyromagnetic resonance field (nonlinear, active, and nonreciprocal waveguide components). The magnetostatic field which, when applied to a gyromagnetic material, causes gyromagnetic resonance to occur at a particular frequency.   530

gyromagnetic resonance frequency (nonlinear, active, and nonreciprocal waveguide components). The damped natural frequency for precession of the magnetization of a gyromagnetic material subjected to a particular magnetostatic field.   530

gyromagnetic resonance linewidth (nonlinear, active, and nonreciprocal waveguide components). The difference between magnetostatic field levels slightly above and slightly below gyromagnetic resonance for which the gyromagnetic resonance absorption falls to half the peak value. This resonance occurs in a gyromagnetic material with uniformly precessing magnetization at a fixed frequency.   530

H

H (*H* beacon) (electronic navigation). A designation applied to two types of facilities: (1) A nondirectional radio beacon for homing by means of an airborne direction finder. (2) A radar air navigation system using an airborne interrogator to measure the distances from two ground transponders. *See:* **navigation.**   187, 13

halation (cathode-ray tube). An annular area surrounding a spot, that is due to the light emanating from the spot being reflected from the front and rear sides of the face plate. *See:* **cathode-ray tubes.**   244

half-adder. A combinational logic element having two outputs, *S* and *C,* and two inputs, *A* and *B,* such that the outputs are related to the inputs according to the following equations:

$$S = A \text{ OR } B \text{ (exclusive OR)}.$$
$$C = A + B.$$

S denotes sum without carry, *C* denotes carry. Two half-adders may be used for performing binary addition.   77

half-amplitude recovery time (Geiger-Müller counters). The time interval from the start of a full-amplitude pulse to the instant a succeeding pulse can attain an amplitude of 50 percent of the maximum amplitude of a full-amplitude pulse.   96

half cell. An electrode immersed in a suitable electrolyte. *See:* **electrolytic cell.**   328

half coil. *See:* **armature bar.**

half duplex (communications) (data transmission). Pertaining to a transmission over a circuit capable of transmitting in either direction, but only one direction at a time.   59

half-duplex channel (half duplex operation) (data transmission). A channel of a duplex system arranged to permit operation in either direction but not in both directions simultaneously.   59

half-duplex operation (telegraph system). Operation of a duplex system arranged to permit operation in either direction but not in both directions simultaneously. *See:* **telegraphy.**   328

half-duplex repeater. A duplex telegraph repeater provided with interlocking arrangements that restrict the transmission of signals to one direction at a time. *See:* **telegraphy.**   328

half-power beamwidth (antennas). In a radiation pattern cut containing the direction of the maximum of a lobe, the angle between the two directions in which the radiation intensity is one-half the maximum value. *See:* **principal half-power beamwidth.**   111

half section (circuits and systems). A bisected tee or pi section. A basic L-section building block of image-parameter filters.   67

halftone characteristic (facsimile). A relation between the density of the recorded copy and the density of the subject copy. *Note:* The term may also be used to relate the amplitude of the facsimile signal to the density of the subject copy or the record copy when only a portion of the system is under consideration. In a frequency-modulation system an appropriate parameter is to be used instead of the amplitude. *See:* **recording (facsimile).**   12

halftones (storage tubes). *See:* **level; storage tube.**

half-wave dipole. A wire antenna consisting of two straight collinear conductors of equal length, separated by a small feeding gap, with each conductor approximately a quarter-wavelength long. *Note:* This antenna gets its name from the fact that its overall length is approximately a half-wavelength--enough to cause the input impedance to be pure real.   111

half-wave rectification (power supplies). In the rectifying process, half-wave rectification passes only one-half of each incoming sinusoid, and does not pass the opposite half-cycle. The output contains a single half-sine pulse for each input cycle. A single rectifier provides half-wave rectification. Because of its poorer efficiency and larger alternating-current component, half-wave rectification is usually employed in non-critical low-current circumstances. See the accompanying figure. *See:* **rectifier circuit element; rectification; rectifier.**   186

half-wave rectification

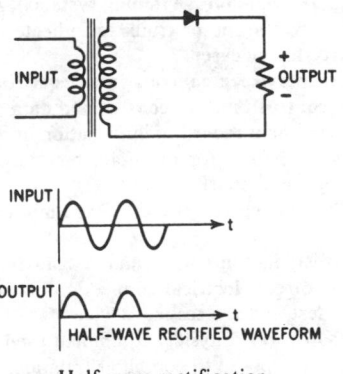

Half-wave rectification

—

Hall analog multiplier. A Hall multiplier specifically designed for analog multiplication purposes.   107

Hall angle. The angle between the electric field vector and the current density vector.   107

Hall coefficient. The coefficient of proportionality R in the relation

$$E_H = R(J \times B)$$

where
E_H is the resulting transverse electric field,
J is the current density,
B is the magnetic flux density.
Note: The sign of the majority carrier charge can usually be inferred from the sign of the Hall coefficient.   107

Hall effect (Hall effect devices). (1) (in conductors and semiconductors). The change of the electric conduction caused by that component of the magnetic field vector normal to the current density vector, which, instead of being parallel to the electric field, forms an angle with it. *Note:* In conductors and semiconductors of noncubic single crystals, the current density and electric field vectors may not be parallel in the absence of an applied magnetic field. For such crystals the more general definition below should be used.(2) (in any material, including ferromagnetic and similar materials). The change of the electric conduction caused by that component of the magnetic field vector applied normal to the current density vector, which causes the angle between the current density vector and the electric field to change from the magnitude that existed prior to the introduction of the magnetic field. *Note:* For ferromagnetic and similar materials there are two effects, the "ordinary" Hall effect due to the applied external magnetic flux as described for conductors and semiconductors and the "extraordinary" Hall effect due to the magnetization in the ferromagnetic or similar material. In the absence of the "extraordinary" Hall effect and the effects outlined in the preceding note, the current density vector and the electric field vector will be parallel when there is no external magnetic flux.   107

Hall effect device. A device in which the Hall effect is utilized.   107

Hall generator. A Hall plate, together with leads, and, where used, encapsulation and ferrous or nonferrous backing plate(s).   107

Hall mobility (electric conductor). The quantity μ_H in the relation $\mu_H = R\sigma$, where R , Hall coefficient and σ = conductivity. *See:* **semiconductor.**   245

Hall modulator. A Hall effect device that is specifically designed for modulation purposes.   107

Hall multiplier. A Hall effect device that contains a Hall generator together with a source of magnetic flux density and that has an output that is a function of the product of the control current and the field excitation current.   107

Hall plate. A three-dimensional configuration of any material in which the Hall effect is utilized.   107

Hall probe. A Hall effect device specifically designed for measurement of magnetic flux density.   107

Hall terminals. The terminals between which the Hall voltage appears.   107

Hall voltage. The voltage generated in a Hall plate due to the Hall effect.   107

halogen-quenched counter tube. A self-quenched

counter tube in which the quenching agent is a halogen, usually bromine or chlorine.   190

halving interval (HIC)(evaluation of thermal capability)(thermal classification of electric equipment and electrical insulation). The number corresponding to the interval in °C determined from the thermal endurance relationship expresses the halving of the time-to-end-point centered on the temperature of the TI or RTI. In case of graphical derivation the times corresponding to the TI or RTI (for example, 20 00 h) and one half that value (for example, 10 000 h) will usually produce an acceptable approximation. *See:* **thermal endurance graph.**   506

Hamming code (mathematics of computing). Any of several error-correcting codes invented by the mathematician Richard Hamming, which use redundant information bits to detect and correct any single error in a transmitted character.   564

Hamming distance (mathematics of computing). The number of digit positions in which two binary numerals, characters, or words of the same length are different. For example, the Hamming distance between 100101 and 101001 is two.   564

hand (head or butt) cable (mining). A flexible cable used principally in making electric connections between a mining machine and a truck carrying a reel of portable cable. *See:* **mine feeder circuit.**   328

hand burnishing (electroplating). Burnishing done by a hand tool, usually of steel or agate. *See:* **electroplating.**   328

H and D curve. *See:* **Hurter and Driffield curve.**

hand elevator. An elevator utilizing manual energy to move the car. *See:* **elevators.**   328

handhole (1)(NESC). An access opening, provided in equipment or in a below-the-surface enclosure in connection with underground lines, into which men reach but do not enter, for the purpose of installing, operating, or maintaining equipment or cable or both.   494

(2) (transmission and distribution). An opening in an underground system containing cable, equipment, or both into which workmen reach but do not enter.   262

handling device (of metal-clad switchgear). That accessory that is used for the removal, replacement, or transportation of the removable element.   103

hand operation (industrial control). Actuation of an apparatus by hand without auxiliary power. *See:* **switch.**   244, 206

hand receiver. An earphone designed to be held to the ear by the hand. *See:* **loudspeaker.**   328

hand-reset relay (power switchgear). A relay that is so constructed that it remains in the picked-up condition even after the input quantity is removed; specific manual action is required to reset the relay.   103

handset (transmission performance of telephone sets). An assembly that includes a handle and a telephone set transmitter and receiver. Other components such as the speech network may also be located in the handset.   491

handset telephone. *See:* **hand telephone set.**

handshake (1)(FASTBUS acquisition and control). An interlocked exchange of signals between a master and a slave, controlling the transfer of data.   480
(2)(test, measurement and diagnostic equipment). A hardware or software sequence of events requiring mutual consent of conditions prior to change.   54

handshake cycle (digital interface for programmable instrumentation). The process whereby digital signals effect the transfer of each data byte across the interface by means of an interlocked sequence of status and control signals. Interlocked denotes a fixed sequence of events in which one event in the sequence must occur before the next event may occur.   40, 378

hand telephone set (handset telephone). A telephone set having a handset and a mounting that serves to support the handset when the latter is not in use. *Note:* The prefix desk, wall, drawer, etcetera, may be applied to the term **hand telephone set** to indicate the type of mounting. *See:* **telephone station.**   328

handwheel (industrial control). A wheel the rim of which serves as a handle for manual operation of a rotary device.   206

hand winding (rotating machinery). A winding placed in slots or around poles by a human operator. *See:* **rotor (rotating machinery); stator.**   63

hang-off (gyro, accelerometer) (inertial sensor). The displacement of an inertial sensing element from its null position which occurs when an input is applied and which is due to the finite compliance of a capture loop or a restoring spring.   46

hangover (facsimile). *See:* **tailing.**

hang-up hand telephone set (suspended-type handset telephone) (bracket-type handset telephone). A hand telephone set in which the mounting is arranged for attachment to a vertical surface and is provided with a switch bracket from which the handset is suspended. *See:* **telephone station.**   328

hang-up signal (telephone switching systems). A signal transmitted over a line or trunk to indicate that the calling party has released.   55

hard copy (supervisory control, data acquisition, and automatic control)(station control and data acquisition). A permanent record of information in readable form for human use, for example, reports, listings, displays, logs, and charts.   570,403

hard limiting (analog computers). *See:* **limiter circuit.**   9

hard line (test, measurement and diagnostic equipment). Any direct electrical connection between the unit under test and the testing device.   54

hardware (software). Physical equipment used in data processing, as opposed to computer programs, procedures, rules, and associated documentation. *See:* **computer programs; data; documentation; procedures; software.**   434

hardwire (test, measurement and diagnostic equipment). Circuitry with the absence of electrical elements, such as resistors, inductors, capacitors: circuits containing only wire and terminal connections with no intervening switching inherent.   54 hardwired (supervisory control, data acquisition, and automatic control)(station control and data acquisition). The implementation of processing steps within a device by way of the placement of conductors between components within the device. The processing steps are not alterable except by modifying the conducting paths between components. 570, 403

harmful interference (electromagnetic compatibility). Any emission, radiation, or induction that endangers the functioning, or seriously degrades, obstructs, or repeatedly interrupts a radiocommunication service or any other equipment or system operating in accordance with regulations. *See:* electromagnetic compatibility. 199

harmonic (1)(converter characteristics)(self-commutated converters)(harmonic control and reactive compensation of static power converters). A sinusoidal component of a periodic wave or quantity having a frequency that is an integral multiple of the fundamental frequency. *Note:* For example, a component, the frequency of which is twice the fundamental frequency, is called a second harmonic. *See:* characteristic harmonic; noncharacteristic harmonic; harmonic content; harmonic components; relative harmonic content. 584, 533

(2)(data transmission). A sinusoidal component of a periodic wave or quantity having a frequency that is an integral multiple of the fundamental frequency. *Note:* For example, a component, the frequency of which is twice the fundamental frequency, is called the second harmonic. 59

harmonic analyzer. A mechanical device for measuring the amplitude and phase of the various harmonic components of a periodic function from its graph. *See:* instrument; signal wave; wave analyzer. 328

harmonic components (converter characteristics)-(self-commutated converters). The components of the harmonic content as expressed in terms of the order and rms (root-mean-square) values of the Fourier series terms describing the periodic function. 584

harmonic conjugate. *See:* Hilbert transform.

harmonic content (1)(converter characteristics)(self-commutated converters). The function obtained by subtracting the dc (direct current) and fundamental components from a nonsinusoidal periodic function. 584

(2)(measurement of power frequency electric and magnetic fields from ac power lines). Distortion of a sinusoidal waveform characterized by indication of the magnitude and order of the Fourier series terms describing the wave. *Note:* For power lines the harmonic content is small and of little concern for the purpose of field measurements, except at points near large industrial loads (saturated power transformers, rectifiers, aluminum and chlorine plants, etcetera) where certain harmonics may reach 10% of the line voltage. Laboratory installations may also have voltage or current sources with significant harmonic content. 514

(3)(nonsinusoidal periodic wave) (static power converters). The deviation from the sinusoidal form, expressed in terms of the order and magnitude of the Fourier series terms describing the wave. *See:* power rectifier; rectification. 208

(4) (emergency and standby power systems). A measure of the presence of harmonics in a voltage or current waveform expressed as a percentage of the amplitude of the fundamental frequency at each harmonic frequency. The total harmonic content is expressed as the square root of the sum of the squares of each of the harmonic amplitudes (expressed as a percentage of the fundamental). 89

harmonic conversion transducer (frequency multiplier, frequency divider). A conversion transducer in which the output-signal frequency is a multiple or submultiple of the input frequency. *Notes:* (1) In general, the output-signal amplitude is a nonlinear function of the input-signal amplitude. (2) Either a frequency multiplier or a frequency divider is a special case of harmonic conversion transducer. *See:* heterodyne conversion transducer (converter); transducer. 125

harmonic distortion (data transmission). Nonlinear distortion of a system or transducer characterized by the appearance in the output of harmonics other than the fundamental component when the input wave is sinusoidal. *Note:* Subharmonic distortion may also occur. 59

harmonic factor (harmonic control and reactive compensation of static power converters). The ratio of the root-mean-square (rms) value of all the harmonics to the rms value of the fundamental. 533

harmonic leakage power (TR and pre-TR tubes). The total radio-frequency power transmitted through the fired tube in its mount at frequencies other than the fundamental frequencies generated by the transmitter. 125

harmonic-restraint relay (power switchgear). A restraint relay that is so constructed that its operation is restrained by harmonic components of one or more separate input quantities. 103

harmonics. *See:* harmonic components.

harmonic series. A series in which each component has a frequency that is an integral multiple of a fundamental frequency. 176

harmonic telephone ringer. A telephone ringer that responds only to alternating current within a very narrow frequency band. *Note:* A number of such ringers, each responding to a different frequency, are used in one type of selective ringing. *See:* telephone station. 328

harmonic test (rotating machinery). A test to determine directly the value of one or more harmonics of the waveform of a quantity associated with a machine, relative to the fundamental of that quantity. *See:* asynchronous machine. 63

harsh environment (nuclear power generating station). An environment expected as a result of the postulated service conditions appropriate for the design basis and post-design basis accidents of the station. (A design basis accident is that subset of a design basis event which requires safety function performance). Harsh environments are the result of a loss of cooling accident (LOCA) / high energy line break (HELB) inside containment and post-LOCA or HELB outside containment. 120

hartley (1) (data transmission). A unit of information content, equal to one decadal decision, or the designation of one of ten possible and equally likely values or states of anything used to store or convey information. *Note:* A hartley may be conveyed by one decadal code element. One hartley equals (log of 10 to base 2) times one bit. 59

(2) (information theory). A unit of information content, equal to one decadal decision, or the designation of one of ten possible and equally likely values or states of anything used to store or convey information. *Notes:* (1) A hartley may be conveyed by one decadal code element. One hartley equals (log of 10 to base 2) times one bits. (2) If, in the definition of information content, the logarithm is taken to the base ten, the result will be expressed in hartleys. *Syn:* **dit.** *See:* **bit.** 415

Hartley oscillator. An electron tube or solid state circuit in which the parallel-tuned tank circuit is connected between grid and plate, the inductive element of the tank having an intermediate tap at cathode potential, and the necessary feedback voltage obtained across the grid-cathode portion of the inductor. *See:* **radio frequency generator.** 14

hauptnutzzeit. *See:* **utilization time.**

Hay bridge. A 4-arm alternating-current bridge in which the arms adjacent to the unknown impedance are nonreactive resistors and the opposite arm comprises a capacitor in series with a resistor. *Note:* Normally used for the measurement of inductance in terms of capacitance, resistance, and frequency. Usually, the bridge is balanced by adjustment of the resis-

Hay bridge

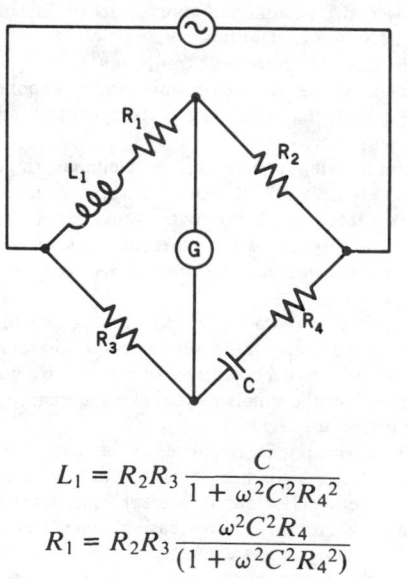

$$L_1 = R_2 R_3 \frac{C}{1 + \omega^2 C^2 R_4^2}$$
$$R_1 = R_2 R_3 \frac{\omega^2 C^2 R_4}{(1 + \omega^2 C^2 R_4^2)}$$

tor that is in series with the capacitor, and of one of the nonreactive arms. The balance depends upon the frequency. It differs from the Maxwell bridge in that in the arm opposite the inductor, the capacitor is in series with the resistor. *See:* **bridge.** 328

hazard (nuclear power generating stations). A specified result of a design basis event that could cause unacceptable damage to systems or components important to safety. 131

hazard beacon (illuminating engineering). An aeronautical beacon used to designate a danger to air navigation. *Syn:* **obstruction beacon.** 167

hazard current (1) (health care facilities). For a given set of connections in an isolated power system, the total current that would flow through a low impedance if it were connected between either isolated conductor and ground. The various hazard currents are: fault hazard current; monitor hazard current; total hazard current. *See:* **fault hazard current; monitor hazard current; total hazard current.** 192

hazard current (2) (National Electrical Code)(health care facilities) For a given set of connections in an isolated system, the total current that would flow through a low impedance if it were connected between either isolated conductor and ground. Fault Hazard Current: The hazard current of a given isolated system with all devices connected except the line isolation monitor. Monitor Hazard Current: The hazard current of the line isolation monitor alone. Total Hazard Current: The hazard current of a given isolated system with all devices, including the line isolation monitor, connected. 256

hazardous (classified) locations (National Electrical Code). Class I Locations. Class I locations are those in which flammable gases or vapors are or may be present in the air in quantities sufficient to produce explosive or ignitible mixtures. Class I locations shall include those specified in (a) and (b) below. (a) Class I, Division 1. A Class I, Division 1 location is a location: (1) in which hazardous concentrations of flammable gases or vapors exist continuously intermittently, or periodically under normal operating conditions; or (2) in which hazardous concentrations of such gases or vapors may exist frequently because of repair or maintenance operations or because of leakage; or (3) in which breakdown or faulty operations of equipment or processes might release hazardous concentrations of flammable gases or vapors, and might also cause simultaneous failure of electric equipment. This classification usually includes locations where volatile flammable liquids or liquefied flammable gases are transferred from one container to another; interiors of spray booths and areas in the vicinity of spraying and painting operations where volatile flammable solvents are used; locations containing open tanks or vats of volatile flammable liquids; drying rooms or compartments for the evaporation of flammable solvents; portions of cleaning and dyeing plants where hazardous liquids are used; gas generator rooms and other portions of gas manufacturing plants where flammable gas may escape; inadequately ventilated pump rooms for flammable gas or for volatile flammable liquids; the interiors of refrigerators and freezers in which volatile flammable materials are stored in open, lightly stoppered, or easily ruptured containers; and all other locations where hazardous concentrations of flammable vapors or gases are likely to occur in the course of normal operations. (b) Class I, Division 2. A Class I, Division 3 location is a location (1) in which volatile flammable liquids or flammable gases are handled, processed, or used, but in which the hazardous liquids, vapors, or gases will normally be confined within closed containers or closed systems from which they can escape only in case of accidental rupture or breakdown of such containers or systems, or in case of abnormal operation of equipment; or (2) in which hazardous concentrations of gases or vapors are normally prevented by positive mechanical ventilation, and which might become hazardous though failure or abnormal operation of the ventilating equipment; or (3) that is adjacent to a Class I, Division I location, and to which hazardous concentrations of gases or vapors might occasionally be communicated unless such communication is prevented by adequate positive-pressure ventilation from a source of clean air, and effective safeguards against ventilation failure are provided. This classification usually includes locations where volatile flammable liquids or flammable gases or vapors are used, but which, in the judgment of the authority having jurisdiction, would become hazardous only in case of an accident or of some unusual operating condition, The quantity of hazardous material that might escape in case of accident, the adequacy of ventilating equipment, the total area involved, and the record of the industry or business with respect to explosions or fires are all factors that merit consideration in determining the classification and extent of each location. Piping without valves, checks, meters and similar devices would not ordinarily introduce a hazardous condition even even though used for hazardous liquids or gases. Locations used for the storage of hazardous liquids or of compressed gases in sealed containers would not normally be considered hazardous unless subject to other hazardous conditions also. Electrical conduits and their associated enclosures separated from process fluids by a single seal or barrier shall be classed as a Division 2 location if the outside of the conduit and enclosures is a nonhazardous location. 500-5 Class II Locations. Class II locations are those that are hazardous because of the presence of combustible dust. Class II locations shall include those specified in (a) and (b) below. (a) Class II, Division 1. A Class II, Division I location is a location (1) in which combustible dust is or may be in suspension in the air continuously, intermittently, or periodically under normal operating conditions, in quantities sufficient to produce explosive or ignitible mixtures; or (2) where mechanical failure or abnormal operation of machinery or equipment might cause such explosive or ignitible mixtures to be produced and might also provide a source of ignition through simultaneous failure of electric equipment, operation of protection devices or

from other causes; or (3) in which combustible dusts of an electrically conductive nature may be present. This classification usually includes the working areas of grain handling and storage plants; rooms containing grinders or pulverizers, cleaners, graders, scalpers, open conveyors or spouts, open bins or hoppers, mixers or blenders, automatic hopper scales, packing machinery, elevator heads and boots, stock distributors, dust and stock collectors (except all-metal collectors vented to the outside), and all similar dust-producing machinery and equipment in grain-processing plants, starch plants, sugar-pulverizing plants, malting plants, hay-grinding plants, and other occupancies of similar nature; coal-pulverizing plants (except where the pulverizing equipment is essentially dust-tight); all working areas where metal dusts and powders are produced, processed, handled, packed, or stored (except in tight containers); and all other similar locations where combustible dust may, under normal operating conditions be present in the air in quantities sufficient to produce explosive or ignitible mixtures. Combustible dusts which are electrically nonconductive include dusts produced in the handling and processing of grain and grain products, pulverized sugar and cocoa, dried egg and milk powders, pulverized spices, starch and pastes, potato and woodflour, oil meal from beans and seed, dried hay, and other oganic materials which may produce combustible dusts when processed or handled. Electrically conductive nonmetallic dusts include dusts from pulverized coal, coke, carbon black, and charcoal. Dusts containing magnesium or aluminum are particularly hazardous and the use of extreme precaution will be necessary to avoid ignition and explosion. (b) Class II, Division 2. A Class II, Division 2 location is a location in which combustible dust will not normally be in suspension in the air or will not be likely to be thrown into suspension by the normal operation of equipment or apparatus in quantities sufficient to produce explosive or ignitible mixtures, but; (1) where deposits or accumulations of such combustible dust may be sufficient to interfere with the safe dissipation of heat from electric equipment or apparatus; or (2) where such deposits or accumulations of combustible dust on, in, or in the vicinity of electric equipment might be ignited by arcs, sparks, or burning material from such equipment. Locations where dangerous concentrations of suspended dust would not be likely but where dust accumulations might form on, or in the vicinity of electric equipment, would include rooms and areas containing only closed spouting and conveyors, closed bins or hoppers, or machines and equipment from which appreciable quantities of dust would escape only under abnormal operating conditions; rooms or areas adjacent to a Class II, Division 1 location as described in (a) above, and into which explosive or ignitible concentrations of suspended dust might be 256

hazardous area class I. The locations in which flammable gases or vapors are or may be present in the air in quantities sufficient to produce explosive or ignitible mixtures. *See:* **explosionproof apparatus.** 256

hazardous electrical condition (electrolytic cell line working zone). Exposure of personnel to surfaces, contact with which may result in the flow of injurious electrical current. 133

hazardous levels of nonionizing electromagnetic radiation (radio frequency radiation hazard warning symbol). Incident electromagnetic energy that may be biologically detrimental or may directly or indirectly cause ignition of explosive materials or vapors. 557

hazardous location (illuminating engineering). An area where ignitable vapors or dust may cause a fire or explosion created by energy emitted from lighting or other electrical equipment or by electrostatic generation. 167

hazardous materials. Those vapors, dusts, fibers or flyings which are explosive under certain conditions. 70

hazardous substance (handling and disposal of transformer grade insulating liquids containing PCBs). A quantity of material offered for transportation in one package or transport vehicle, when the material is not packaged that equals or exceeds the reportable quantity (RQ) specified for the material in Code of Federal Regulations (CFR), Title 40, Parts 116 and 117. 586

HA1 receiver weighting (data transmission). A noise weighting used in a noise measuring set to measure noise across the HA1 receiver of a of a subset with a number 302 receiver or a similar subset. The meter scale readings are in the dBa (HA1). 59

H **bend (waveguide technique).** *See:* ***H*-plane bend.**

HC. *See:* **high conduction.**

HCL. *See:* **relay, high, common, low.**

H **display (radar).** A *B* display modified to include indication of angle of elevation. *Note:* The target appears as two closely spaced blips which approximate a short bright line, the slope of which is in proportion to the sine of the angle of target elevation. 13, 187

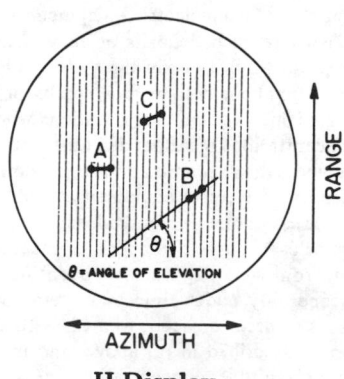

H-Display

head (computing system). A device that reads, records, or erases data on a storage medium. *Note:* For example, a small electromagnet used to read, write, or erase

data on a magnetic drum or tape, or the set of perforating, reading, or marking devices used for punching, reading, or marking on paper tape. 255, 77 *See:* **gross head; critical head; net head; rated head; head water benefits.**

headed brush (rotating machinery). A brush having a top (cylindrical, conical, or rectangular) with a smaller cross section than the cross section of the body of the brush. *Note:* The length of the head shall not exceed 25 percent of the overall length. *See:* **brush.** 279

head-end system (railways). A system in which the electrical requirements of a train are supplied from a generator or generators, located on the locomotive or in one of the cars, customarily at the forward part of the train. *Note:* The generators may be driven by steam turbine, internal-combustion engine, or, if located in one of the cars, by a mechanical drive from a car axle. *See:* **axle generator system.** 328

header (National Electrical Code). A transverse raceway for electric conductors, providing access to predetermined cells of a cellular metal floor, thereby permitting the intallation of electric conductors from a distribution center to the cells. 256

head gap (test, measurement and diagnostic equipment). The space or gap intentionally inserted into the magnetic circuit of the head in order to force or direct the recording flux into or from the recording medium. 54

heading (navigation). The horizontal direction in which a vehicle is pointed, expressed as an angle between a reference line and the line extending in the direction the vehicle is pointed, usually measured clockwise from the reference line. *See:* **navigation.** 187, 13

heading-effect error (navigation). A manifestation of polarization error causing an error in indicated bearing that is dependent upon the heading of a vehicle with respect to the direction of signal propagation. *Note:* Heading-effect error is a special case of attitude-effect error where the vehicle is in a straight level flight: it is sometimes referred to as course push (or pull). *See:* **navigation.** 187, 13

headlamp (illuminating engineering). A major lighting device mounted on a vehicle and used to provide illumination ahead of it. *Syn:* **headlight.** 167

headlight. *See:* **headlamp.** 167

head or butt cable (mining). *See:* **hand cable.**

headquarters system (direct-connected system). A local system to which has been added means of transmitting system signals to and receiving them at an agency maintained by the local government, for example, in a police precinct house, or fire station. *See:* **protective signaling.** 328

head receiver. An earphone designed to be held to the ear by a headband. *Note:* One or a pair (one for each ear) of head receivers with associated headband and connecting cord is known as a headset. *See:* **loudspeaker.** 328

head space (test, measurement and diagnostic equipment). The space between the reading or recording head and the recording medium, such as tape, drum or disk. 54

head water benefits (power operations). The benefits brought about by the storage or release of water by a reservoir project upstream. Application of the term is usually in reference to benefits to a downstream hydroelectric power plant.   **516**

health care facilities (National Electrical Code) (health care facilities). Buildings, portions of buildings, and mobile facilities, that contain but are not necessarily limited solely to premises designed for use as hospitals, nursing homes, residential custodial care facilities, clinics, or medical and dental offices.   **256**

hearing loss (hearing level) (1) (for speech). The difference in decibels between the speech levels at which the average normal ear and the defective ear, respectively, reach the same intelligibility, often arbitrarily set at 50 percent.
(2) (hearing-threshold level) (ear at a specified frequency). The amount, in decibels, by which the threshold of audibility for that ear exceeds a standard audiometric threshold. *Notes:* (A) See: Current issue of American Standard Specification for Audiometers for General Diagnostic Purposes. (B) This concept was at one time called deafness; such usage is now deprecated. (C) Hearing loss and deafness are both legitimate qualitative terms for the medical condition of a moderate or severe impairment of hearing, respectively. Hearing level, however, should only be used to designate a quantitative measure of the deviation of the hearing threshold from a prescribed standard. *See:* **loudspeaker.**   **176**

heat detector (1) (burglar-alarm system). A temperature-sensitive device mounted on the inside surface of a vault to initiate an alarm in the event of an attack by heat or burning. *See:* **protective signaling.**   **328**
(2) (fire alarm system). A device which detects abnormally high temperature or rate-of-temperature rise to initiate a fire alarm.   **71**

heater (1) (electric pipe heating systems). A length of resistance material connected between terminals and used to generate heat electrically. Heaters, as used in this application, can take the form of cables with various sheath materials, blankets, and pads. *Syn:* **electric heater; heating element.**   **405**
(2) (electron tube). An electric heating element for supplying heat to an indirectly heated cathode. *See:* **electrode (electron tube).**   **125**

heater coil. *See:* **load, work or heater coil (induction heating usage).**

heater connector (heater plug). A cord connector designed to engage the male terminal pins of a heating or cooking appliance.   **328**

heater current. The current flowing through a heater. *See:* **cathode preheating time; filament current; electronic controller.**   **190**

heater transformer (industrial control). Supplies power for electron-tube filaments or heaters of indirectly heated cathodes. *See:* **electronic controller.**   **206**

heater voltage. The voltage between the terminals of a heater. *See:* **electronic controller; electrode voltage (electron tube).**   **190**

heater warm-up time. *See:* **cathode heating time.**

heat exchanger. *See:* **cooler.**

heat-exchanger cooling system (1) (rectifier). A cooling system in which the coolant, after passing over the cooling surfaces of the rectifier, is cooled in a heat exchanger and recirculated. *Note:* The coolant is generally water of a suitable purity, or water that has been treated by a corrosion-inhibitive chemical. Antifreeze solutions may also be used where there is exposure to low temperatures. The heat exchanger is usually either: (A) water-to-water where the heat is removed by raw water, (B) water to-air where the heat is removed by air supplied by a blower, (C) air-to-water, (D) air-to-air, (E) refrigeration cycle. The liquid in the closed system may be other than water, and the gas in the closed system may be other than air. *See:* **rectification; rectifier.**   **208**
(2) (thyristor controller). A cooling system in which the coolant, after passing over the cooling surfaces of the thyristor controller components, is cooled in a heat exchanger and recirculated. *Note:* Heat may be removed from the thyristor controller component's cooling surfaces by liquid or air using the following types of heat exchangers: (A) water-to-water (B) water-to-air (C) air-to-water (D)air-to-air (E) refrigeration cycle. The liquid in the closed system may be other than water, and the gas in the closed system may be other than air.   **445**

heating cycle. One complete operation of the thermostat from ON to ON or from OFF to OFF.   **263**

heating, dielectric. *See:* **dielectric heating.**

heating element. A length of resistance material connected between terminals and used to generate heat electrically. *See:* **appliance outlet.**   **263**

heating, glue line. *See:* **glue-line heating.**

heating pattern. The distribution of temperature in a load or charge. *See:* **induction heating; industrial electronics.**   **14, 114**

heating station (dielectric heating usage). The assembly of components, which includes the work coil or applicator and its associated production equipment.   **14**

heating time, tube (mercury-vapor tube). *See:* **preheating time.**

heating unit (electrical appliances). An assembly containing one or more heating elements, electric terminals or leads, electrical insulation, and a frame, casing, or other suitable supporting means. *See:* **appliance outlet; appliances (including portable).**   **263**

heating-up run (rotating electric machinery). A period of operation with current and ventilation designed to bring the machine to approximately its temperature-rise limit.   **424**

heat loss (1)(electrical heating systems). A quantitative value of energy flow from a hot object to a cooler object or cooler surrounding ambient.   **476**
(2)(electrical heat tracing for industrial applications). A quantitative value of energy flow from a pipe, vessel, or equipment to the surrounding ambient.   **523**
(3)(waveguide terms). The part of the transmission

loss due to the conversion of electric energy into heat. *See:* **waveguide.** 267

heat rate (generating station). A measure of generating station thermal efficiency, generally expressed as BTU per kilowatt hour. *Note:* It is computed by dividing the total BTU content of the fuel burned (or of heat released from a nuclear reactor) by the resulting kilowatt hours generated. 112

heat rejection rate (nuclear power generating station). The rate a which a module emits heat energy to its environment (watts/hr or Btu). 355

heat-shield (cathode) (electron tubes). A metallic surface surrounding a hot cathode, in order to reduce the radiation losses. *See:* **electron tube.** 244, 190

heat sink (1)(electrical heating systems)(electrical heat tracing for industrial applications). A part that conducts and dissipates heat away from the pipeline or vessel (the pipe or equipment). Heat sinks, as related to pipe heating systems, can be pipe supports, valve operators, etcetera. 476, 523

(2)(electric pipe heating systems). A part that absorbs heat. Heat sinks, as related to electric pipe heating systems, are those masses of materials that are directly connected to mechanical piping, valves, tanks, etcetera that can absorb the heat generated by heaters, thus reducing the effect of the heater. Typical heat sinks can be pipe hangers, valve operators, etcetera. 405

(3) (semiconductor rectifier diode). A mass of metal generally having much greater thermal capacity than the diode itself and intimately associated with it. It encompasses that part of the cooling system to which heat flows from the diode by thermal conduction only and from which heat may be removed by the cooling medium. *See:* **semiconductor rectifier stack.** 208

(4) (photovoltaic power system). A material capable of absorbing heat: a device utilizing such material for the thermal protection of components or systems. 186

heat tracing (1)(electrical heating systems). A heating system where the externally applied heat source follows (traces) the object to be heated. 476

(2)(electrical heat tracing for industrial applications). The utilization of electric heating cables, other electric heating devices, and support components that are externally applied and used to maintain or raise the temperature of fluids in piping and associated equipment. 523

heat-transfer aids (electrical heat tracing for industrial applications). Thermally conductive materials, such as metallic foils or heat-transfer cements, used to increase the heat-transfer rates from the heating cables to the process piping or equipment. 523

Heaviside-Campbell mutual-inductance bridge. A mutual-inductance bridge of the Heaviside type in which one of the inductive arms contains a separate inductor that is included in the bridge arm during the first of a pair of measurements and is short-circuited during the second. *Note:* The balance is independent of the frequency. *See:* **Heaviside mutual-inductance bridge; bridge.** 328

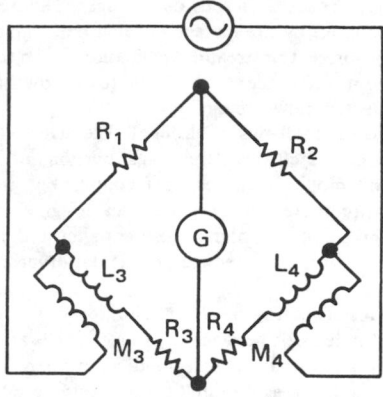

$$R_x = (R_3 - R'_3) \frac{R_2}{R_1}$$

$$L_x = (M - M') \left(1 + \frac{R_2}{R_1}\right)$$

Heaviside-Lorentz system of units. A rationalized system based on the centimeter, gram, and second and is similar to the Gaussian system but differs in that a factor 4π is explicitly inserted to multiply r^2 in each of the formulations of the Coulomb Laws. 210

Heaviside mutual-inductance bridge. An alternating-current bridge in which two adjacent arms contain self-inductance, and one or both of these have mutual inductance to the supply circuit, the other two arms being normally nonreactive resistors. *Note:* Normally used for the comparison of self- and mutual inductances. The balance is independent of the frequency. *See:* **bridge.** 328

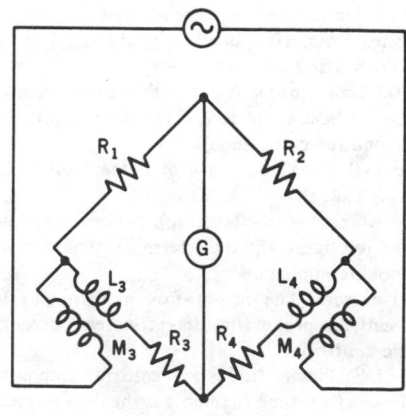

$$R_1 R_4 = R_2 R_3$$

$$L_3 - L_4 \left(\frac{R_1}{R_2}\right) = -(M_3 - M_4) \left(1 + \frac{R_1}{R_2}\right)$$

heavy duty floodlight (HD) (illuminating engineering). A weatherproof unit having a substantially constructed metal housing into which is placed a separate and removable reflector. A weatherproof hinged door with cover glass encloses the assembly but provides an unobstructed light opening at least equal to the effective diameter of the reflector.   167

heavy load (watthour meter). *See:* **test current (TA).**

HE11mode (fiber optics). Designation for the fundamental mode of an optical fiber. *See:* **fundamental mode.**   433

height (antenna). *See:* **effective height.**

height-finding radar (radar). A radar whose function is to measure the range and elevation angle to a target, thus permitting computation of altitude or height; such a radar usually accompanies a surveillance radar which determines other target parameters.   13

height marker (radar). *See:* **calibration marks.**

height marks (radar). *See* **calibration marks.**

height, pulse. *See:* **amplitude, pulse.**

helical antenna (data transmission). An antenna whose configuration is that of a helix. *Note:* The diameter, pitch, and number of turns in relation to the wavelength provide control of the polarization state and directivity of helical antennas.   59

helical plate (storage cell). A plate of large area formed by helically wound ribbed strips of soft lead inserted in supporting pockets or cells of hard lead. *See:* **battery (primary or secondary).**   328

help menu (computer applications). A menu that gives the user a choice of topics for which help information is available on a given computer system.   571

hemispherical-conical   reflectance,$\rho(2\pi;\omega_r)$.(illuminating engineering). Ratio of reflected flux collected over a conical solid angle to the incident flux from the entire hemisphere. (See Figure below). *Note:* The direction and extent of the cone must be specified.   167

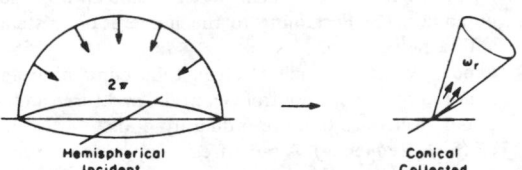

Hemispherical Incident   Conical Collected

hemispherical-conical   transmittance,$\tau(2\pi;\omega_r)$.(illuminating engineering). Ratio of transmitted flux collected over a conical solid angle to the incident flux from the entire hemisphere. (See Figure below). *Note:* The direction and extent of the cone must be specified.   167

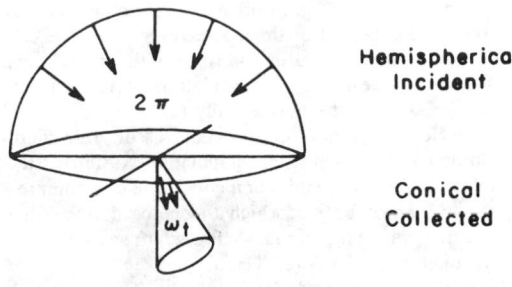

Hemispherical Incident

Conical Collected

hemispherical-directional reflectance,$\rho(2\pi;\theta_r,\phi_r)$.(illuminating engineering). Ratio of reflected flux over an element of solid angle surrounding the given direction to the incident flux from the entire hemisphere. (See Figure below). *Note:* The direction of collection and the size of the solid angle "element" must be specified.   167

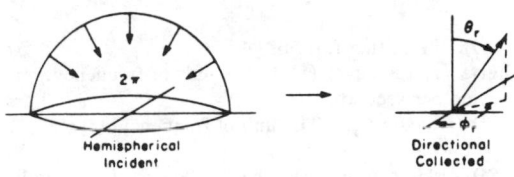

Hemispherical Incident   Directional Collected

hemispherical-directional transmittance,$\tau(2\pi;\theta_r,\phi_r)$. (illuminating engineering). Ratio of transmitted flux collected over an element of solid angle surrounding the given direction to the incident flux from the entire hemisphere. *Note:* The direction of collection and size of the solid angle "element" must be specified.   167

hemispherical reflectance* (illuminating engineering). The ratio of all of the flux leaving a surface or medium by reflection to the incident flux. This term is retained for reference purposes only.   167
*Obsolete

hemispherical transmittance (illuminating engineering). The ratio of the transmitted flux leaving a surface or medium to the incident flux. *Note:* If transmittance is not preceded by an adjective descriptive of the angles of view, hemispherical transmittance is implied. *See:* **transmission (illuminating engineering).**   167

HEM wave. *See:* **hybrid electromagnetic wave.**

henry (metric practice). The inductance of a closed circuit in which an electromotive force of one volt is produced when the electric current in the circuit varies uniformly at a rate of one ampere per second.   21

heptode. A seven-electrode electron tube containing an anode, a cathode, a control electrode, and four additional electrodes that are ordinarily grids.   125

hermetically sealed relay. A relay in a gastight enclosure that has been completely sealed by fusion or other comparable means to insure a low rate of gas leakage over a long period of time. *See:* **relay.** 259

hermetic motor. A stator and rotor without shaft, end shields, or bearings for installation in refrigeration compressors of the hermetically sealed type. 63

hermetic refrigerant motor-compressor (air-conditioning and refrigerating equipment) (National Electrical Code). A combination consisting of a compressor and motor, both of which are enclosed in the same housing, with no external shaft or shaft seals, the motor operating in the refrigerant. 256

hermitian form (circuits and systems). The nxn matrix [A] is hermitian if its conjugate transpose is equal to [A] itself. In terms of a set of complex variables; x_1, x_2, . . . x_n; the quantity

$$[\overline{x}_1 \ \overline{x}_2 \ldots \overline{x}_n][A]\begin{bmatrix} x_1 \\ x_2 \\ \vdots \\ x_n \end{bmatrix}$$

is the hermitian form of [A]. 67

hertz (1) (general) (Hz). The unit of frequency, one cycle per second. 3, 59, 53

(2) (transformer). The unit of frequency, (cycles per second). 53

(3) (metric practice). The frequency of a periodic phenomenon of which the period is one second. 21

(4) (laser-maser). The unit which expresses the frequency of a periodic oscillation in cycles per second. 363

Hertzian electric dipole (antennas). An elementary source consisting of a time-harmonic electric current element of specified direction and infinitesimal length. *Notes:* (1) The continuity equation relating current to charge requires that opposite ends of the current element be terminated by equal and opposite amounts of electric charge, these amounts also varying harmonically with time. (2) As its length approaches zero, the current must approach infinity in such a manner that the product of current and length remains finite. 111

Hertzian magnetic dipole (antennas). A fictitious elementary source consisting of a time-harmonic magnetic current element of specified direction and infinitesimal length. *Notes:* (1) The continuity equation relating current to charge requires that opposite ends of the current element be terminated by equal and opposite amounts of magnetic charge, these amounts also varying harmonically with time. (2) As its length approaches zero, the current must approach infinity in such a manner that the product of current and length remains finite. (3) A magnetic dipole has the same radiation pattern as an infinitesimally small electric current loop. 111

heterodyne (nonlinear, active, and nonreciprocal waveguide components). The process occurring in a frequency converter by which the signal input frequency is changed by superimposing a local oscillation

to produce an output having the same modulation information as the original signal but at a frequency which is either the sum or the difference of the signal and local oscillator frequencies. 530

heterodyne conversion transducer (converter). A conversion transducer in which the useful output frequency is the sum or difference of (1) the input frequency and (2) an integral multiple of the frequency of another wave usually derived from a local oscillator. *Note:* The frequency and voltage or power of the local oscillator are parameters of the conversion transducer. Ordinarily, the output-signal amplitude is a linear function of the input-signal amplitude over its useful operating range. 125

heterodyne frequency. *See:* **beats.**

heterodyne reception (beat reception). The process of reception in which a received high-frequency wave is combined in a nonlinear device with a locally generated wave, with the result that in the output there are frequencies equal to the sum and difference of the combining frequencies. *Note:* If the received waves are continuous waves of constant amplitude, as in telegraphy, it is customary to adjust the locally generated frequency so that the difference frequency is audible. If the received waves are modulated the locally generated frequency is generally such that the difference frequency is superaudible and an additional operation is necessary to reproduce the original signal wave. *See:* **superheterodyne reception.** 328

heterojunction (fiber optics). A junction between semiconductors that differ in their doping level conductivities, and also in their atomic or alloy compositions. *See:* **homojunction.** 433

heteropolar machine (rotating machinery). A machine having an even number of magnetic poles with successive (effective) poles of opposite polarity. *See:* **asynchronous machine.** 63

heuristic. Pertaining to exploratory methods of problem solving in which solutions are discovered by evaluation of the progress made toward the final result. *See:* **algorithm.** 255, 77, 54

Hevea rubber. Rubber from the *Hevea brasiliensis* tree. *See:* **insulation.**

hexadecimal (mathematics of computing). (1) Pertaining to a selection in which there are sixteen possible outcomes. (2) Pertaining to the numeration system with a radix of 16. 564

hexode. A six-electrode electron tube containing an anode, a cathode, a control electrode, and three additional electrodes that are ordinarily grids. 125

HF (high-frequency). A radar frequency band between 3 megahertz and 30 megahertz. 13

HF (high-frequency) radar. A radar operating at frequencies between 3 to 30 megahertz. 13 *See:* **radio spectrum.**

hickey. (1) A fitting used to mount a lighting fixture in an outlet box or on a pipe or stud. *Note:* It has openings through which fixture wires may be brought out of the fixture stem. (2) A pipe-bending tool. 328

hierarchical decomposition (software). A method of designing a system by breaking it down into its components through a series of top-down refinements. *See:* **components; functional decomposition; modular decomposition; stepwise refinement; system; top-down.** 434

hierarchy (software). A structure whose components are ranked into levels of subordination according to a specific set of rules. *See:* **components; levels.** 434

high conduction (HC) threshold voltage v_{HC} (metal-nitride-oxide field-effect transistor). The threshold voltage level resulting from a write-high pulse, which puts the transistor into the HC (high-conduction) state. 386

high direct voltage (1) (power cable systems). A direct voltage above 5000 volts (V) supplied by test equipment of limited capacity. 437

(2) (rotating machinery). A direct voltage above 5000 V supplied by portable test equipment of limited capacity. 6

high energy piping (nuclear power generating station). Piping serving as the pressure boundary for fluid systems that, during normal plant conditions, are either operating or maintaining temperature or pressure when the maximum operating temperature exceeds 200=F or the maximum operating pressure exceeds 275 pounds per square inch gauge (psig). 439

high (H) level (graphic symbols for logic functions). A level within the more positive (less negative) of the two ranges of the logic levels chosen to represent the logic states. 451

higher layer (logical link control). The conceptual layer of control or processing logic existing in the hierarchical structure of a station that is above the data link layer and upon which the performance of data link layer functions are dependent; for example, device control, buffer allocation, logical link control (LLC) station management, etcetera. 585

higher order language (software). A programming language that usually includes features such as nested expressions, user defined data types, and parameter passing not normally found in lower order languages, that does not reflect the structure of any one given computer or class of computers, and that can be used to write machine independent source programs. A single higher order language may represent multiple machine operations. *Syn:* **high level language.** *See:* **assembly language; computer; data types; machine language; programming language; source programs.** 434

higher-order mode (waveguide or transmission line). Any mode of propagation characterized by a field configuration other than that of the fundamental or first-order mode with lowest cutoff frequency. *See:* **waveguide.** 185

higher-order mode of propagation (laser-maser). A mode in a **beamguide** or **beam resonator** which has a plurality of maxima for the transverse field intensity over the cross-section of the beam. 363

high-fidelity signal (speech quality measurements). A signal transmitted over a system comprised of a microphone, amplifier, and loudspeaker or earphones. A

tape recorder may be part of the system. All components should be of the best quality the state of the art permits. 126

high-field-emission arc (gas). An electric arc in which the electron emission is due to the effect of a high electric field in the immediate neighborhood of the cathode, the thermionic emission being negligible. *See:* **discharge (gas).** 244, 190

high frequency (HF) radar (radar). A radar operating at frequencies between 3 to 20 MHz. 13

high-frequency furnace (coreless-type induction furnace). An induction furnace in which the heat is generated within the charge, or within the walls of the containing crucible, or in both, by currents induced by high-frequency flux from a surrounding solenoid. 328

high-frequency induction heater or furnace. A device for causing electric current flow in a charge to be heated, the frequency of the current being higher than that customarily distributed over commercial networks. *See:* **induction heating.** 14, 114

high-frequency stabilized arc welder. A constant-current arc-welding power supply including a high-frequency arc stabilizer and suitable controls required to produce welding current primarily intended for tungsten-inert-gas arc welding. *See:* **constant-current arc-welding power supply.** 264

high-gain dc amplifier (analog computers). An amplifier that is capable of amplification substantially greater than required for a specified operation throughout a frequency band extending from zero to some maximum. Also, an operational amplifier without feedback circuit elements. *See:* **operational amplifier.** 9

high-impedance rotor. An induction-motor rotor having a high-impedance squirrel cage, used to limit starting current. *See:* **rotor (rotating machinery).** 63

high initial response (excitation systems for synchronous machines). An excitation system capable of attaining 95% of the difference between ceiling voltage and rated-load field voltage in 0.1 s or less under specified condition. 507

high intensity discharge (HID) lamp (illuminating engineering). An electric discharge lamp in which the light producing arc is stabilized by wall temperature, and the arc tube has a bulb wall loading in excess 3 W/cm^2. HID lamps include groups of lamps known as mercury, metal halide, and high-pressure sodium. 167

high-key lighting (illuminating engineering). A type of lighting which, applied to a scene, results in a picture having graduations falling primarily between gray and white; dark grays or blacks are present, but in very limited areas. 167

high-level firing time (microwave) (switching tubes). The time required to establish a radio-frequency discharge in the tube after the application of radio-frequency power. *See:* **gas tube.** 190, 125

high-level language (HLL)(high-level microprocessor language). High-level language to be extended by IEEE trial use Std 755-1985. HLLs so extended are sometimes known as implementation languages. 470

high level language. *See:* **higher order language.**
434

high-level modulation. Modulation produced at a point in a system where the power level approximates that at the output of the system. 111, 240

high-level radio-frequency signal (1)(microwave gas tubes). A radio-frequency signal of sufficient power to cause the tube to become fired. *See:* **gas tube.**
125

(2)(microwave gas tubes)(nonlinear, active, and non-reciprocal waveguide components). A radio-frequency signal above the threshold power level necessary to cause the tube to become nonlinear (fired). *See:* **gas tube; limiting threshold; below-threshold firing time.**
530

high-level voltage standing-wave ratio (microwave switching tubes)(nonlinear, active, and nonreciprocal waveguide components). The voltage standing-wave ratio caused by a fired tube located between a generator and matched termination in the waveguide. *See:* **gas tube.** 530, 125

high lights (any metal article). Those portions that are most exposed to buffing or polishing operations, and hence have the highest luster. 328

high-limit temperature (electrical heat tracing for industrial applications). The maximum allowable heat-tracing system temperature. 523

high-low signaling (telephone switching systems). A method of loop signaling in which a high-resistance bridge is used to indicate an on-hook condition and a low resistance bridge is used to indicate an off-hook condition. 55

high-pass filter (1)(data transmission). A filter having a single transmission band extending from some cutoff frequency (not zero) up to infinite frequency.
59

high pass filter (data transmission)(harmonic control and reactive compensation of static power converters). A filter having a single transmission band extending from some cutoff frequency (not zero) up to infinite frequency. 59, 533

high peaking. The introduction of an amplitude-frequency characteristic having a higher relative response at the higher frequencies. *See:* **television.**
178

high pot (hi-pot). *See:* **high-potential test.**

high-potential test (power operations). A test that consists of the application of a voltage higher than the rated voltage for a specified time for the purpose of determining the adequacy against breakdown of insulating materials and spacings under normal conditions. *Note:* The test is used as a proof test of new apparatus, a maintenance test on older equipment, or as one method of evaluating developmental insulation systems. *Syn:* **high pot (hi-pot).** 516

high-power-factor mercury-lamp ballast. A multiple-supply type power-factor-corrected ballast, so designed that the input current is at a power factor of not less than 90 percent when the ballast is operated with center rated voltage impressed upon its input terminals and with a connected load, consisting of the ap-

propriate reference lamp(s), operated in the position for which the ballast is designed. 271

high power factor transformer (power and distribution transformer). A high-reactance transformer that has a power-factor-correcting device, such as a capacitor, so that the input current is at a power factor of not less than 90 percent when the transformer delivers rated current to its intended load device. 53

high-pressure contact (as applied to high-voltage disconnecting switches). One in which the pressure is such that the stress in the material of either of the contact surfaces is near the elastic limit of the material so that conduction is a function of pressure. 103

high-pressure sodium lamp (HPS) (illuminating engineering). A high intensity discharge (HID) lamp in which light is produced by radiation from sodium vapor operating at a partial pressure about $1.33 \times 10^4 \, Pa$ (100 Torr). Includes clear and diffuse-coated lamps.
167

high-pressure vacuum pump. A vacuum pump that discharges at atmospheric pressure. *See:* **rectification.**
328

high-purity germanium (HPGe)(germanium gamma-ray detectors). Germanium whose net concentration of electrically active impurities that are stable at room temperature is such that conventional-sized radiation detectors made from it achieve full depletion at reasonable bias voltages. (The net concentration of electrically active impurities is typically less than 3 · 1010 cm-3.) 528

high-reactance rotor. An induction-motor rotor having a high-reactance squirrel cage, used where low starting current is required and where low locked-rotor and breakdown torques are acceptable. *See:* **rotor (rotating machinery).** 63

high-reactance transformer (power and distribution transformer). An energy-limiting transformer that has sufficient inherent reactance to limit the output current to a maximum value. 53

high-resistance rotor (rotating machinery). An induction motor rotor having a high-resistance squirrel cage, used when reduced locked-rotor current and increased locked-rotor torque are required. 63

high rupturing capacity (HRC)(protection and coordination of industrial and commercial power systems). In British and Canadian terminology, high rupturing capacity, equivalent to USA high interrupting capacity and generally indicating capability of interruption of at least 100 000 root-mean-square (rms) amperes (A) for low-voltage fuses. 504

high-speed carry (electronic computation). A carry process such that if the current sum in a digit place is exactly one less than the base, the carry input is bypassed to the next place. *Note:* The processing necessary to allow the bypass occurs before the carry input arrives. Further processing required in the place as a result of the carry input, occurs after the carry has passed by. Contrasted with **cascaded carry.** *See:* **standing-on-nines carry.** 235

high-speed excitation system. An excitation system capable of changing its voltage rapidly in response to a change in the excited generator field circuit. *See:* **generating station.** 64

high-speed grounding switch. *See:* **fault-initiating switch.** 103

high-speed limit (speed/load reference)(control systems for steam turbine-generator units). A device or input that limits the speed/load reference setting to a predetermined upper limit. This device may establish the upper limit of the synchronizing speed range.
522

high-speed low-voltage dc power circuit breaker (1) (power switchgear). A low-voltage dc power circuit breaker which, during interruption, limits the magnitude of the fault current so that its crest is passed not later than a specified time after the beginning of the fault current transient, where the system fault current, determined without the circuit breaker in the circuit, falls between specified limits of current at a specified time. *Note:* The specified time in present practice is 0.01 second. 103
(2) (low-voltage dc power circuit breakers used in enclosures). A circuit breaker which, when applied in a circuit with the parameter values specified in American National Standard C37.16-1979, Preferred Rating, Related Requirements and Application Recommendations for Low-Voltage Power Circuit Breakers and AC Power Circuit Protectors, Tables 12 and 12A, tests "b" (5 A/μ_S initial rate of rise of current), forces a current crest during interruption within 0.01 s after the current reaches the pickup setting of the instantaneous trip device. *Note:* For total performance characteristics at other than test circuit parameter values, consult the manufacturer. 401

high-speed regulator (power supplies). A power supply regulator circuit that, by the elimination of its output capacitor, has been made capable of much higher slewing rates than are normally possible. *Note:* High-speed regulators are used where rapid step-programming is needed: or as current regulators, for which they are ideally suited. *See:* **slewing rate.** 186

high-speed relay (power switchgear). A relay that operates in less than a specified time. *Note:* The specified time in present practice is fifty milliseconds (three cycles on a 60 Hz basis). 103

high-speed short-circuiting switch. *See:* **fault-initiating switch.**

high state (1)(programmable instrumentation). The relatively more positive signal level used to assert a specific message content associated with one of two binary logic states. 40
(2)(signals and paths)(microcomputer system bus). The more positive voltage level used to represent one of two logical binary states. 542
(3)(signals and paths)(696 interface devices). The electrically more positive signal level used to assert a specific message content associated with one of two binary logic states. 538

high usage trunk (data transmission). A group of trunks for which an engineered alternate route is provided, and for which the number of trunks is deter-

mined on the basis of relative trunk efficiencies and economic considerations. 59

high-usage trunk group (telephone switching systems). A trunk group engineered on the basis of relative trunk efficiencies and economic considerations which will overflow traffic. 55

high-velocity camera tube (anode-voltage stabilized camera tube) (electron device). A camera tube operating with a beam of electrons having velocities such that the average target voltage stabilizes at a value approximately equal to that of the anode. 190

high voltage (system voltage ratings). A class of nominal system voltages equal to or greater than 100,000 volts or less than 230,000 volts. *See:* **low voltage; medium voltage; nominal system voltage.** 260

high-voltage aluminum-sheathed power cable (aluminum sheaths for power cables). Cable used in an electric system having a maximum phase-to-phase root-mean-square (rms) alternating-current (ac) voltage above 72 500 volts (V) to 242 000 V, the cable having an aluminum sheath as a major component in its construction. 406

high-voltage and low-voltage windings (power and distribution transformer). The terms high voltage and low voltage are used to distinguish the winding having the greater from that having the lesser voltage rating.
53

high-voltage cable termination. A device used for terminating alternating current power cables having laminated or extruded insulation rated 2.5 kV and above, which are classified according to the following: Class 1 Termination. Provides electric stress control for the cable insulation shield terminus: provides complete external leakage insulation between the cable conductor(s) and ground: and provides a seal to the end of the cable against the entrance of the external environment and maintains the pressure, if any, of the cable system. Class 2 Termination. Provides electric stress control for the cable insulation shield terminus: and provides complete external leakage insulation between the cable conductor(s) and ground. Class 3 Termination. Provides electric stress control for the cable insulation shield terminus. *Note:* Some cables do not have an insulation shield. Terminations for such cables would not be required to provide electric stress control. In such cases, this requirement would not be part of the definition.

high-voltage disconnect jack (wire-line communication facilities). Jack used to disconnect cable pairs for testing purposes. Helps safeguard personnel from remote ground potentials. 414

high-voltage isolating relays (wire-line communication facilities). A high-voltage isolating relay provides for the repeating of dc on-off signals while maintaining longitudinal isolation. High-voltage isolating relays may be used in conjunction with isolating transformers or may be used as stand-alone devices for dc tripping or dc telemetering. *See:* **isolating transformers with high-voltage isolating relays.** 414

high-voltage power vacuum interrupter (X-radiation limits for ac high-voltage power vacuum interrupters

used in power switchgear). An interrupter in which the separable contacts function within a single evacuated envelope and which is intended for use in power switchgear. 553

high-voltage relay. (1) A relay adjusted to sense and function in a circuit or system at a specific maximum voltage. (2) A relay designed to handle elevated voltages on its contacts, coil, or both. 341

high-voltage system (generating station grounding). An electric system having a maximum root-mean-square ac voltage above 72.5 kilovolts (kV). 569

high-voltage telephone repeater (wire-line communication facilities). A high-voltage telephone repeater provides high-voltage longitudinal isolation, while permitting voice and signalling to pass. This is accomplished by using a short span, carrier transmission system and high-voltage, isolation capacitors or transformers. The repeater is intended to provide ordinary telephone service in a power station environment without interference to other noninterruptible, critical circuits. 414

high-voltage time test. An accelerated life test on a cable sample in which voltage is the factor increased. 64

highway (CAMAC system). An interconnection between CAMAC crate assemblies or between one or more CAMAC crate assemblies and an external controller. 51

highway crossing back light (railway practice). An auxiliary signal light used for indication in a direction opposite to that provided by the main unit of a highway crossing signal. 328

highway crossing bell (railway practice). A bell located at a railroad-highway grade crossing and operated to give a characteristic and arrestive signal to give warning of the approach of trains. 328

highway crossing signal. An electrically operated signal used for the protection of highway traffic at railroad-highway grade crossings. 328

Hilbert transform (harmonic conjugate) (real functional $x(t)$ of the real variable t). The real function $x(t)$ that is the Cauchy principal value of

$$\frac{1}{\pi} \int_{-\infty}^{\infty} \frac{x(\tau)\mathrm{d}\tau}{t - \tau}.$$

This transformation shifts all Fourier components by 90°, $\cos\omega^t$, for example, into $\sin\omega^t$ and $\sin\omega^t$ into $-\cos\omega^t$. *See:* **analytic signal; network analysis.** 61

hinge clip (of a switching device) (power switchgear). The clip to which the blade is movably attached. 103

hinged-iron ammeter. A special form of moving-iron ammeter in which the fixed portion of the magnetic circuit is arranged so that it can be caused to encircle the conductor, the current in which is to be measured. This conductor then constitutes the fixed coil of the instrument. *Note:* The combination of a current transformer of the split-core type with an ammeter is often

used similarly to measure alternating current, but should be distinguished from the hinged-iron ammeter. *See:* **instrument.** 328

hinged removable feed tube (cable plowing). A feed tube removably attached to a blade so relative motion may occur between the feed tube and the blade around an essentially vertical axis. 52

hipot (test, measurement and diagnostic equipment). A colloquialism for **high potential test:** A testing technique whereby a high voltage source is applied to an insulating material to determine the condition of that material. 54

hiss (electron device). Noise in the audio-frequency range, having subjective characteristics analogous to prolonged sibilant sounds. 239, 190

hit (radar). A target echo from one single pulse. 13

H network. A network composed of five branches, two connected in series between an input terminal and an output terminal, two connected in series between another input terminal and output terminal, and the fifth connected from the junction point of the first two branches to the junction point of the second two branches. *See:* **network analysis.**

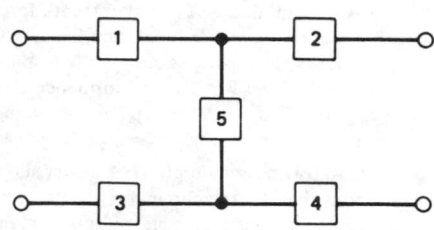

Branches 1 and 2 are the first two branches between an input and an output terminal; branches 3 and 4 are the second two branches; and branch 5 is the branch between the junction points. 210

hodoscope. An apparatus for tracing the path of a charged particle in a magnetic field. *See:* **electron optics.** 244, 190

Hoeppner connection (power and distribution transformer). A three-phase transformer connection involving transformation from a wye winding to the combination of a delta winding and a zigzag winding which are connected permanently in parallel. *Note:* This connection is used when a wye-delta connection is needed, with ground connections on both primary and secondary windings. 53

hoghorn antenna. A reflector antenna consisting of a sectoral horn which physically intersects a reflector in the form of a parabolic cylinder, a part of one of the nonparallel sides of the horn being removed to form the antenna aperture. 111

hoist (power line maintenance). An apparatus for moving a load by the application of pulling force and not including a car or platform running in guides. These devices are normally designed using roller or link chain and built-in leverage to enable heavy loads to be lifted or pulled. They are often used to dead-end a conductor during sagging and clipping-in operations and when tensioning guys. *Syn:* **chain hoist; chain tugger; coffin hoist.** 458

hoist back-out switch (mining). A switch that permits operation of the hoist only in the reverse direction in case of overwind. *See:* **mine hoist.** 328

hoisting-rope equalizer. A device installed on an elevator car or counterweight to equalize automatically the tensions in the hoisting wire ropes. *See:* **elevator.** 328

hoist overspeed device (mining). A device that can be set to prevent the operation of a mine hoist at speeds greater than predetermined values and usually causes an emergency brake application when the predetermined speed is exceeded. *See:* **mine hoist.** 328

hoist overwind device (mining). A device that can be set to cause an emergency break application when a cage or skip travels beyond a predetermined point into a danger zone. *See:* **mine hoist.** 328

hoist signal code (mining). Consists of prescribed signals for indicating to the hoist operator the desired direction of travel and whether men or materials are to be hoisted or lowered in mines. *See:* **mine hoist.** 328

hoist signal system (mining). A system whereby signals can be transmitted to the hoist operator (and in some instances by him to the cager) for control of mine hoisting operations. *See:* **mine hoist.** 328

hoist slack-brake switch (mining). A device for automatically cutting off the power from the hoist motor and causing the brake to be set in case the links in the brake rigging require tightening or the brakes require relining. *See:* **mine hoist.** 328

hoist trip recorder (mining). A device that graphically records information such as the time and number of hoists made as well as the delays or idle periods between hoists. *See:* **mine hoist.** 328

hoistway (National Electrical Code). Any shaftway, hatchway, well hole, or other vertical opening or space in which an elevator or dumbwaiter is designed to operate. 256

hoistway access switch (elevators). A switch, located at a landing, the function of which is to permit operation of the car with the hoistway door at this landing and the car door or gate open, in order to permit access at the top of the car or to the pit. *See:* **control (elevators).** 328

hoistway-door combination mechanical lock and electric contact (elevators). A combination mechanical and electric device, the two related, but entirely independent, functions of which are: (1) to prevent operation of the driving machine by the normal operating device unless the hoistway door is in the closed position, and (2) to lock the hoistway door in the closed position and prevent it from being opened from the

landing side unless the car is within the landing zone. *Note:* As there is no positive mechanical connection between the electric contact and the door-locking mechanism, this device insures only that the door will be closed, but not necessarily locked, when the car leaves the landing. Should the lock mechanism fail to operate as intended when released by a stationary or retiring car-cam device, the door can be opened from the landing side even though the car is not at the landing. If operated by a stationary car-cam device, it does not prevent opening the door from the landing side as the car passes the floor. *See:* **hoistway (elevator or dumbwaiter).** 328

hoistway-door electric contact (elevators). An electric device, the function of which is to prevent operation of the driving machine by the normal operating device unless the hoistway door is in the closed position. *See:* **hoistway (elevator or dumbwaiter).** 328

hoistway-door interlock (elevators). A device having two related and interdependent functions that are (1) to prevent the operation of the driving machine by the normal operating device unless the hoistway door is locked in the closed position: and (2) to prevent the opening of the hoistway door from the landing side unless the car is within the landing zone and is either stopped or being stopped. *See:* **hoistway (elevator or dumbwaiter).** 328

hoistway-door or gate locking device (elevators). A device that secures a hoistway or gate in the closed position and prevents it from being opened from the landing side except under specified conditions. *See:* **hoistway (elevator or dumbwaiter).** 328

hoistway enclosure. The fixed structure, consisting of vertical walls or partitions, that isolates the hoistway from all other parts of the building or from an adjacent hoistway and in which the hoistway doors and door assemblies are installed. *See:* **hoistway (elevator or dumbwaiter).** 328

hoistway-gate separate mechanical lock (elevators). A mechanical device, the function of which is to lock a hoistway gate in the closed position after the car leaves a landing and prevent the gate from being opened from the landing side unless the car is within the landing zone. *See:* **hoistway (elevator or dumbwaiter).** 328

hoistway-unit system (elevators). A series of hoistway-door interlocks, hoistway-door electric contacts, or hoistway-door combination mechanical locks and electric contacts, or a combination thereof, the function of which is to prevent operation of the driving machine by the normal operating device unless all hoistway dobrs are in the closed position and, where so required, are locked in the closed position. *See:* **hoistway (elevator or dumbwaiter).** 328

hold (1) (electronic digital computer). An untimed delay in the program, terminated by an operator or interlock action. 207

(2) (analog computers). In an analog computer, the computer control state in which the problem solution is stopped and held at its last values usually by automatic disconnect of integrator input signals. 9

(3) (industrial control). A control function that arrests the further speed change of a drive during the acceleration or deceleration portion of the operating cycle. *See:* **control system, feedback.** 206

(4) (charge-storage tubes) (verb). To maintain storage elements at an equilibrium voltage by electron bombardment. *See:* **charge-storage tube; data processing.** 174, 125

(5) (test, measurement and diagnostic equipment). (A) The function of retaining information in one storage device after transferring it to another device: and (B) A designed stop in testing. 54

hold-closed mechanism (automatic circuit recloser). A device that holds the contacts in the closed position following the completion of a predetermined sequence of operations as long as current flows in excess of a predetermined value. 103, 202

hold-closed operation (automatic circuit recloser). An opening followed by the number of closing and opening operations that the hold-closed mechanism will permit before holding the contacts in the closed position. 103

hold-down bail (separable insulated connectors). An externally mounted device designed to prevent separation at the operating interface of an elbow and an apparatus bushing. 454

hold down block (conductor stringing equipment). A device designed with one or more single groove sheaves to be placed on the conductor and used as a means of holding it down. This device functions essentially as a traveler used in an inverted position. It is normally used in midspan to control conductor uplift caused by stringing tensions, or at splicing locations to control the conductor as it is allowed to rise after splicing is completed. *Syn:* **hold down roller; hold down traveler; splice release block.** 431

holding current (thyristor). The minimum principal current required to maintain the thyristor in the ON-state, after latching current has been reached and after removal of gate signal. *See:* **latching current.** 445

holding-down bolt. A bolt that fastens a machine to its bedplate, rails, or foundation. 63

holding frequency (take the swings). A condition of operating a generator or station to maintain substantially constant frequency irrespective of variations in load. *Note:* A plant so operated is said to be regulating frequency. *See:* **generating station.** 64, 94

holding load. A condition of operating a generator or station at substantially constant load irrespective of variations in frequency. *Note:* A plant so operated is said to be operating on base load. *See:* **base load; generating station.** 64

holding register (hybrid computer linkage components). The register, in a double-buffered digital-to-analog converter (DAC) or a digital-to-analog multiplier (DAM), that holds the next digital value to be transferred into the dynamic register. 10

holding time (data transmission). The length of time a communication channel is in use for each transmission. Includes both message time and operating time. 59

holding tone (analog voice frequency circuits). A tone, usually 1004 hertz (Hz), transmitted over a communication circuit for performing noise tests on systems using compandors or quantizers or for the measurement of jitter or transients. The tone is transmitted at a predetermined level and filtered out at the noise measuring set. *See:* **C-notch.** 468

holdoff. *See:* **intervals; sweep holdoff.**

hold-off diode (charging inductors). A diode that is placed in series with the charging inductor and connected to the common junction of the switching element and the pulse-forming network in a pulse generator. *Note:* The use of a hold-off diode in the charging circuit of a pulse-forming network allows the capacitors of the network to charge to full voltage and remain at this voltage until the switch conducts. This permits the use of pulse-repetition frequencies of equal to or less than twice the frequency of resonance charging. 137

hold out (power line maintenance). Operating order or operating-order idntification tag, or marker. *See:* **hold card.** 458

holdup-alarm attachment. A general term for the various alarm-initiating devices used with holdup-alarm systems, including holdup buttons, footrails, and others of a secret or unpublished nature. *See:* **protective signaling.** 328

holdup-alarm system. An alarm system signaling a robbery or attempted robbery. *See:* **protective signaling.** 328

hole (semiconductor). A mobile vacancy in the electronic valence structure of a semiconductor that acts like a positive electron charge with a positive mass. *See:* **semiconductor.** 245

hole burning (of an absorption or an emission line) (laser-maser). The frequency dependent saturation of attenuation or gain that occurs in an inhomogeneously broadened transition when the saturating power is confined to a frequency range small compared with the inhomogeneous linewidth. 363

hollow-core annular conductor (hollow-core conductor). A conductor composed of a plurality of conducting elements disposed around a supporting member that does not fill the space enclosed by the elements: alternatively, a plurality of such conducting elements disposed around a central channel and interlocked one with the other or so shaped that they are self-supporting. *See:* **conductor.** 64

hollow-core conductor. *See:* **hollow-core annular conductor.**

home area (telephone switching systems). The numbering plan area in which the calling customer is located. 55

home computer (computer applications). A personal computer designed to be used in the home. 571

home signal (railway practice). A fixed signal at the entrance of a route or block to govern trains or engines entering or using that route or block. 328

homing (1) (navigation). Following a course directed toward a point by maintaining constant some navigational coordinate (other than altitude). *See:* **radio navigation.** 13, 187

(2) (telephone switching systems). Resetting of a sequential switching operation to a fixed starting point. 55

homing beacon (navigation aid terms). A beacon that provides homing guidance. 526

homing guidance. That form of missile guidance wherein the missile steers itself toward a target by means of a mechanism actuated by some distinguishing characteristic of the target. *See:* **guided missile.** 328

homing relay. A stepping relay that returns to a specified starting position prior to each operating cycle. *See:* **relay.** 259

homochromatic gain (optoelectronic device). The radiant gain or luminous gain for specified identical spectral characteristics of both incident and emitted flux. *See:* **optoelectronic device.** 191

homodyne reception (zero-beat reception). A system of reception by the aid of a locally generated voltage of carrier frequency. 328

homogeneous cladding (fiber optics). That part of the cladding wherein the refractive index is constant within a specified tolerance, as a function of radius. *See:* **cladding; tolerance field.** 433

homogeneous line-broadening (laser-maser). An increase of the width of an absorption or emission line, beyond the natural linewidth, produced by a disturbance (for example, collisions, lattice vibrations, etcetera) which is the same for each of the emitters. 363

homogeneous series (of current limiting fuse units) (power switchgear). A series of fuse units, deviating from each other only in such characteristics that, for a given test, the testing of one or a reduced number of particular fuse units of the series may be taken as representative of all the fuse units of the series. 103, 443

homojunction (fiber optics). A junction between semiconductors that differ in their doping level conductivities but not in their atomic or alloy compositions. *See:* **heterojunction.** 433

homopolar machine* (rotating machinery). A machine in which the magnetic flux passes in the same direction from one member to the other over the whole of a single air-gap area. Preferred term is acyclic machine. 63

*Deprecated

honeycomb coil. A coil in which the turns are wound in crisscross fashion to form a self-supporting structure or to reduce distributed capacitance. *Syn:* **duolateral coil.** 341

hook operation. *See:* **stick (hook) operation.**

hook ring (air switch) (power switchgear). A ring provided on the switch blade for operation of the switch with a switch stick. 103

hook stick. *See:* **switch stick (switch hook).**

hopper (computing systems). *See:* **card hopper.**

horizontal amplifier (oscilloscopes). An amplifier for signals intended to produce horizontal deflection. *See:* **oscillograph.** 185

horizontal hold control (television). A synchronizing control that adjusts the free-running period of the horizontal deflection oscillator. 18

horizontally polarized field vector. A linearly polarized field vector whose direction is horizontal. 111

horizontally polarized plane wave (antennas). A plane wave whose electric field vector is horizontally polarized. 111

horizontally polarized wave (1) (general). A linearly polarized wave whose direction of polarization is horizontal. *See:* **radiation.** 328

(2) (radio wave propagation). A linearly polarized wave whose electric field vector is horizontal. *See:* **perpendicular polarization.** 146

horizontal machine. A machine whose axis of rotation is approximately horizontal. 63

horizontal plane of a searchlight (illuminating engineering). The plane which is perpendicular to the vertical plane through the axis of the searchlight drum and in which the train lies. 167

horizontal ring induction furnace. A device for melting metal, comprising an annular horizontal placed open trough or melting channel, a primary inductor winding, and a magnetic core which links the melting channel with the primary winding. 14

horn (1) (acoustic practice). A tube of varying cross-sectional area for radiating or receiving acoustic waves. *Note:* Normally it has different terminal areas that provide a change of acoustic impedance and control of the directional response pattern. *See:* **loudspeaker.** 176

(2) (antenna). An antenna consisting of a waveguide section in which the cross-sectional area increases toward an open end which is the aperture. 111

horn antenna (data transmission). A radiating element having the shape of a horn. 59

horn gaps (wire-line communication facilities). An air gap metal electrode device, consisting of a straight vertical, electrode device, consisting of a straight vertical, round electrode and an angular shaped round electrode. In the case of a telephone pair, there will be one common grounded, center straight, vertical electrode and two angular electrodes, one for each side of the pair. Horn gaps are used usually outdoors on open wire lines exposed to high-voltage power transmission lines exposed to high-voltage power transmission lines and in conjunction with isolating or drainage transformers. They are also frequently used alone out along the open-wire pair. They provide protection both against lightning and power contacts. 414

horn-gap switch (power switchgear). A switch provided with arcing horns. 103

horn mouth. Normally the opening, at the end of a horn, with larger cross-sectional area. *See:* **loudspeaker.** 176

horn reflector antenna (1) (antenna). An antenna consisting of a portion of a paraboloidal reflector fed with an offset horn which physically intersects the reflector, part of the wall of the horn being removed to form the antenna aperture. *Note:* The horn is usually pyramidal or conical, with an axis perpendicular to that of the paraboloid. 111

(2) (communication satellite). A form of reflector

antenna, where the energy coming from the throat of a horn is reflected by a segment of a paraboloid. This type of antenna has a very low backlobe. *See:* **cassegrainian feed.**   83

horn throat (audio and electroacoustics). Normally the opening, at the end of a horn, with the smaller cross-sectional area. *See:* **loudspeaker.**   176

horsepower rating, basis for single-phase motor. A system of rating for single-phase motors, whereby horsepower values are determined, for various synchronous speeds, from the minimum value of breakdown torque that the motor design will provide.   63

hose (liquid cooling) (rotating machinery). The flexible insulated or insulating hydraulic connections applied between the conductors and either a central manifold or coolant passage.   63

hoseproof (rotating machinery). *See:* **waterproof machine.**

hospital (National Electrical Code) (health care facilities). A building or part thereof used for the medical, psychiatric, obstetrical or surgical care, on a 24-hour basis, of 4 or more inpatients. Hospital, wherever used in this Code, shall include general hospitals, mental hospitals, tuberculosis hospitals, children's hospitals, and any such facilities providing inpatient care.   256

host computer (data communication). A computer, attached to a network, providing primary services such as computation, data base access or special programs or programming languages. *See:* **communications computer.**   12

host machine (software). (1) The computer on which a program or file is installed. (2) A computer used to develop software intended for another computer. (3) A computer used to emulate another computer. (4) In a computer network, a computer that provides processing capabilities to users of the network. *See:* **computer; computer network; emulate; file; network; program; software; target machine.**   434

host processor (FASTBUS acquisition and control). The data processing and control processor assigned to exercise overall supervision over a FASTBUS system. Contains detailed knowledge of the system topology.   480

hot. *See:* **energized.**

hot cathode (thermionic cathode). A cathode that functions primarily by the process of thermionic emission.   125

hot-cathode lamp (illuminating engineering). An electric-discharge lamp whose mode of operation is that of an arc discharge. The cathodes may be heated by the discharge or by external means.   167

hot-cathode tube (thermionic tube). An electron tube containing a hot cathode.   190, 125

hot-end termination (HET)(electrical heat tracing for industrial applications). The termination applied to the end of a heating cable, opposite where the power is supplied.   523

hot plate. An appliance fitted with heating elements and arranged to support a flat-bottomed utensil containing the material to be heated. *See:* **appliances (including portable).**   328

hot reserve. The thermal reserve generating capacity maintained at a temperature and in a condition to permit it to be placed into service promptly. *See:* **generating station.**   64

hottest-spot temperature (hot spot)(1)(electric equipment)(thermal classification of electric equipment and electrical insulation). The highest temperature attained in any part of the insulation of electric equipment. (Difficulties in its determination are encountered. See ANSI/IEEE Std 1-1986, Section 4).   506

(2)(power and distribution transformers). The highest temperature inside the transformer winding. It is greater than the measured average temperature (using the resistance change method) of the coil conductors.   53

hottest-spot temperature allowance (electric equipment)(1)(thermal classification of electric equipment and electrical insulation). The designated difference between the hottest-spot temperature and the observable insulation temperature. (The value is arbitrary, difficult to determine, and depends on many factors, such as size and design of the equipment).   506

(2)(equipment rating). A conventional value selected to approximate the degrees of temperature by which the limiting insulation temperature rise exceeds the limiting observable temperature rise. *See:* **limiting insulation temperature.**   233

hot-wire instrument. An electrothermic instrument that depends for its operation on the expansion by heat of a wire carrying a current. *See:* **instrument.**   328

hot-wire microphone. A microphone that depends for its operation on the change in resistance of a hot wire produced by the cooling or heating effects of a sound wave. *See:* **microphone.**   328

hot-wire relay. A relay in which the operating current flows directly through a tension member whose thermal expansion actuates the relay. *See:* **relay.**   259

Houlding measurement (nonlinear, active, and nonreciprocal waveguide components). A method used in the determination of a varactor diode figure of merit (cutoff frequency). The measurement involves matching the device under test at a fixed bias level in a tunable cavity and interpreting the reflection coefficient data when the bias level is changed.   530

house cable (communication practice). A distribution cable within the confines of a single building or a series of related buildings but excluding cable run from the point of entrance to a cross-connecting box, terminal frame, or point of connection to a block cable. *See:* **cable.**   328

house turbine. A turbine installed to provide a source of auxiliary power. *See:* **generating station.**   64

housing (1) (rotating machinery). Enclosing structure, used to confine the internal flow of air or to protect a machine from dirt and other harmful material.   63

(2) (of an oil cutout) (power switchgear). A part of the fuse support that contains the oil and provides means for mounting the fuse carrier, entrance terminals, and fixed contacts. The housing includes the means for mounting the cutout on a supporting structure and openings for attaching accessories such as a vent or an expansion chamber.   103, 443

(3) (power cable joint). A metallic or other enclosure for the insulated splice.   34

H-plane tee junction (shunt tee) (wavegude components). A waveguide tee junction in which the magnetic field vectors of the dominant mode in all arms are parallel to the plane containing the longitudinal axes of the arms.   166

HRC. *See:* **high rupturing capacity.**

H-scope (radar). A cathode-ray oscilloscope arranged to present an H-display.   13

hub (conductor stringing equipment). A reference point established through a land survey. A hub or POT (point on tangent) is a reference point for use during construction of a line. The number of such points established will vary with the job requirements. Monuments, however, are usually associated with state or federal surveys and are intended to be permanent reference points. Any of these points may be used as a reference for transit sagging operations, provided all necessary data pertaining to them is known. It is quite common to establish additional temporary hubs as required for this purpose. *Syn:* **monument; POT.**   431

hue (television). The attribute of visual sensation designated by blue, green, yellow, red, purple, etcetera. *Note:* This attribute is the psychosensorial correlate (or nearly so) of the colorimetric quantity dominant wavelength.   18

hue of a perceived color (illuminating engineering). The attribute which determines whether it is red, yellow, green, blue, or the like.   167

hum sidebands (spectrum analyzer). Undesired responses created within the spectrum analyzer, appearing on the display, that are separated from the desired response by the fundamental or harmonics of the power line frequency.   390

Huygens' sources (antennas). Electric and magnetic sources which, if properly distributed on a closed surface S in substitution for the actual sources inside S, will insure the result that the electromagnetic field at all points outside S is unchanged. *Syn:* **equivalent sources.**   111

HVdc converter station filter system (high-voltage direct- current systems). The harmonic filter system is designed to suppress, at their source, one or more predominant harmonic frequency currents and voltages which appear on the ac and dc transmission lines because of the ac-dc and dc-ac conversion processes. Harmonic filter system components consist of resistors and reactors (capacitive and inductive) of fixed or variable (controlled) values which make up discrete tuned filters (band-pass-to-ground) designed to limit the magnitude of a specific harmonic current and voltage, or high-pass (broad-band-to-ground) filters which can be effective over a wide frequency range. Carrier

frequency noise can be caused by high-frequency current oscillations occurring during solid- state or mercury arc valve commutation. One method of noise suppression is by placing series inductors in the ac supply to the conversion equipment. In addition, precautions should be taken to minimize coupling though the interwinding capacitance of the converter transformers and also through the dc neutral circuit. Methods of reducing power system influence levels in this range (5 k Hz and above) need to be carefully analyzed at the design state to provide adequate filtering in order that the interference to carrier frequency communications systems can be avoided. A high-pass (broad-band-to-ground) filter is used to suppress harmonics over a wide frequency range. It consists of a parallel RL network in series with a capacitor and is not sharply tuned. It can be designed to be effective above any of the harmonics from about the 11th to the 20th harmonic, and can also serve to suppress carrier frequency noise. A discrete frequency (band-pass-to-ground) filter can be tuned to one or more specific lower order harmonics up to the 17th or 18th harmonic (that is, odd multiples of a fundamental on the ac bus, and even multiples on the dc side). On the dc side, series smoothing reactors and surge capacitors should be considered, in addition to band-pass and high-pass filters, as means of suppressing harmonics.   373

HVdc converter station noise (high-voltage direct-current systems). The processes of rectification and inversion create undesirable harmonic voltages and currents on both the ac and dc portions of the power system. Audio-frequency harmonics are low-order multiples of the fundamental ac frequencies which exist at the HVdc converter station line terminals. The audio-frequency range of major concern is up to approximately 5 kHz. The voltage and current wave distortion which result from the ac-dc conversion process produce frequencies in the carrier range of approximately 5 to 100 kHz and above. Currents at these frequencies appear on the ac or dc, or both, systems at excessive levels if they are unfiltered or inadequately reduced by other means at the converter terminals. Harmonic and carrier frequency currents can propagate for up to 160 km (99.4 mi) or more on transmission lines. The actual distance of propagation varies and is dependent on the frequency, wavelength, and attenuation by the power system impedance. The order of characteristic voltage harmonics found on a dc transmission line is given by KP, and the order of characteristic current harmonics found on the ac lines at the HVdc line terminals is given by $KP \pm 1$, where K, the rectifier phase number, is the total number of rectifier conduction pulses per cycle based on the ac system frequency, and $P = 1, 2, 3,...$, any positive integer. The table (see below) gives orders of characteristic harmonics for 6- and 12-pulse converters. The predominant audio- frequency range harmonics and carrier range frequencies, if unfiltered or improperly filtered, will tend to create undesirable longitudinal voltages in communications circuits located in proximity to the HVdc line or associated ac lines, by means

Orders of Characteristic Harmonics

Pulse No	DC Side	AC Side
K	KP	KP±1
6	0, 6, 12, 18, 24	1, 5,7, 11, 13, 17,19, 23, 25
12	0, 12, 24	1, 11, 13, 23, 25

of electromagnetic or elctrostatic, or both (also dc ionic drift) coupling. Filters for these noise sources are usually installed at the converter station, but because of design limitations, may not be completely effective. These longitudinally induced harmonics can, if sufficient in amplitude, be manifested as audible noise in communications systems because they act on inherent communication circuit unbalances (the conversion of common-mode potentials to differential-mode potentials). Data transmission and pulse-type signals can also be adversely affected depending on the situation. 373

HVdc transmission facility (high-voltage direct-current systems). A Facility consisting of converters located at terminal stations connected by a transmission line, bus, or cable systems, which operate at elevated potentials and currents, and transmit electrical energy between ac systems. The converters. when functioning as a rectifier, change alternating current to direct (unidirectional) current; the transmission line transfers the power between terminal stations where the converters, functioning as an inverter, change the direct current back into alternating current. HVdc transmission facilities can also serve as asynchronous ties between ac systems. 373

hybrid balance (data transmission). A measure of the degree of balance between two impedances connected to two conjugate sides of a hybrid set, and is given by the formula for return loss. 59

hybrid coil (bridge transformer) (data transmission). A single transformer having effectively three windings, which is designed to be connected to four branches of a circuit so as to render these branches conjugate in pairs. 59

hybrid computer (analog computers). A computer which consists of two main computers, one a dc analog computer, and the other a digital computer, with appropriate control and signal interface, such that they may simultaneously operate or solve, or both, upon different portions of a single problem. *See:* **hybrid computer linkage components.** 9

hybrid junction (waveguide components). A waveguide or transmission-line arrangement with four ports which, when the ports have reflectionless terminations, has the property that energy entering ay one port is transferred (usually equally) to two of the remaining three. 166

hybrid mode (1)(fiber optics). A mode possessing components of both electric and magnetic field vectors in the direction of propagation. *Note:* Such modes correspond to skew (non-meridional) rays. *See:* **mode; skew ray; transverse electric mode; transverse magnetic mode.** 433

(2)(waveguide). A waveguide mode such that both the electric and magnetic field vectors have components in the direction of propagation of the mode as well as transverse components. 267

hybrid-mode horn (antenna). A horn antenna excited by one or more hybrid waveguide modes in order to produce a specified aperture illumination. 111

hybrid set (data transmission). Two or more transformers interconnected to form a network having four pairs of accessible terminals to which may be connected four impedances so that the branches containing them may be made conjugate in pairs when the impedances have the proper values but not otherwise. 59

hybrid tee (magic tee) (waveguide components). A hybrid junction composed of an E-H tee with internal matching elements, which is reflectionless for a wave propagating into the junction from one pot when the other three ports have reflectionless terminations. 166

hybrid wave (radio wave propagation). An electromagnetic wave in which either the electric or magnetic field vector is linearly polarized normal to the plane of propagation and the other vector is elliptically polarized in this plane. *See:* **transverse-electric hybrid wave; transverse-magnetic hybrid wave.** 146

hydraulically release-free (trip-free) (as applied to a hydraulically operated switching device) (power switchgear). A term indicating that by hydraulic control the switching device is free to open at any position in the closing stroke if the release is energized. *Note:* This release-free feature is operative even though the closing control switch is held closed. 103

hydraulic operation (power switchgear). Power operation by movement of a liquid under pressure. 103

hydro capability (power operations). The capability supplied by hydroelectric sources under specified water conditions. 516

hydroelectric station (power operations). An electric generating station utilizing hydroenergy for the motive force of its prime movers. 516

hydrolysis (composite insulators). The chemical reaction between the ions of water and polymer materials resulting in depolymerization and a change of electrical and mechanical properties. 483

hydrothermal coordination (power operations). Coordinated operation of hydroelectric, pumped-storage hydro, and steam electric stations so as to obtain minimum costs for the system over a predetermined time interval.												516

hyperabrupt junction (semiconductor)(nonlinear, active, and nonreciprocal waveguide components). A specially designed p-n junction that provides a greater capacitance change over a given voltage range than does an abrupt junction. These devices offer a linear frequency versus voltage characteristic over a limited voltage range whenused in a voltage controlled oscillator. The slope of a log-log plot of abrupt-junction capacitance versus voltage is 0.5, whereas a hyperabrupt junction has a slope between 0.5 and 2.0 in the

hyperabrupt voltage region.												530

hysteresis coupling (electric coupling). An electric coupling in which torque is transmitted by forces arising from the resistance to reorientation of established magnetic fields within a ferromagnetic material.												416

hysteresis loss (magnetic) (power and distribution transformer). The energy loss in magnetic material which results from an alternating magnetic field as the elementary magnets within the material seek to align themselves with the reversing magnetic field.												53

I

IA. *See:* **laser gyro axes.**

IAGC. *See:* **instantaneous automatic gain control.**

ice detection light (illuminating engineering). An inspection light designed to illuminate the leading edge of the wing to check for ice formation.												167

iceproof (high voltage air switches, insulators, and bus supports). So constructed or protected that ice will not interfere with successful operation.												575

ice proof (power switchgear). So constructed or protected that ice of a specified composition and thickness will not interfere with successful operation.												103

I chrominance signal (National Television System Committee (NTSC) color television). The sidebands resulting from suppressed-carrier modulation of the chrominance subcarrier by the I video signal. *Note:* The signal is transmitted in vestigial form, the upper sideband being limited to a frequency within the top of the picture transmission channel (approximately 0.6 MHz above the chrominance subcarrier), and the lower sideband extending to approximately 1.5 MHz below the subcarrier. The phase of the signal, for positive I video signals, is 123deg with respect to the (B-Y) axis.												18

ICI*. *See:* **CIE.**

*Deprecated

ICW. *See:* **interrupted continuous wave.**

ideal capacitor (nonlinear capacitor). A capacitor whose transferred charge characteristic is single-valued. *See:* **nonlinear capacitor.**												191

ideal dielectric. *See:* **dielectric, perfect.**

ideal filter (circuits and systems). (1) (frequency domain) A filter that passes, without attenuation, all frequencies inside specified frequency limits while rejecting all other frequencies. (2) (time domain) A filter with a time domain response identical to the excitation except for a constant delay.												67

ideal noise diode. *See:* **noise diode, ideal.**

ideal paralleling (rotating machinery). Paralleling by adjusting the voltage, and frequency and phase angle for alternating-current machines, such that the conditions of the incoming machine are identical with those

of the system with which it is being paralleled. *See:* **asynchronous machine.**												63

ideal transducer. *See:* **transducer, ideal.**

ideal value (1) (control systems: general) (control) (industrial control) (automatic control). The value of a selected variable that would result from a perfect system operating from the same command as the actual system under consideration. *See:* **control system, feedback.**												206

(2) (synchronous-machine regulating system). The value of a controlled variable (for example, generator terminal voltage) that results from a desired or agreed-upon relationship between it and the commands (commands such as voltage regulator setting, limits, and reactive compensators).												63

idenitified (as applied to equipment) (National Electrical Code). Recognizable as suitable for the specific purpose, function use, environment, application, etcetera, where described in a particular Code requirement. *See: equipment.* (FPN) Suitability of equipment for a specific purpose, environment or application may be determined by a qualified testing laboratory, inspection agency, or other organization concerned with product evaluation. Such identification may include labeling or listing; labeled; listed; Section 90-6 of the NEC.												256

identification (radar). The knowledge that a particular radar return signal is from a specific target. This knowledge may be obtained by determining size, shape, timing, position, maneuvers, rate of change of any of these parameters, or by means of coded responses through secondary radar.												13

identification beacon (navigation aid terms). A beacon that transmits coded signals to identify a geographic position.												526

identifier (software). (1) A symbol used to name, indicate, or locate. Identifiers may be associated with such things as data structures, data items,or program locations. (2) A character or group of characters used to identify or name an item of data and possibly to indicate certain properties of that data. *See:* **data; data structure; program.**												434

identity friend or foe (IFF). Equipment used for transmitting radio signals between two stations located on ships, aircraft, or ground, for automatic identification. *Notes:* (1) The usual basic parts of equipment are interrogators, transpondors, and respondors. (2) Usually the initial letters of the name (IFF) are used instead of the full name. *See:* **radio transmission.**

328

I-display* (radar). A display used in a conical-scan radar, in which a target appears as a complete circle when the radar antenna is pointed at it and in which the radius of the circle is proportional to target range. Incorrect aiming of the antenna changes the circle to a segment whose arc length is inversely proportional to the magnitude of the pointing error and the position of the segment indicates the direction in which the antenna should be moved to restore correct aiming.
*Rare 13

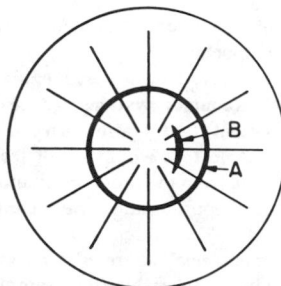

I-Display

idle bar (rotating machinery). An open circuited conductor bar in the rotor of a squirrel-cage motor, used to give low starting current in a moderate torque motor. *See:* **rotor (rotating machinery).** 63

idle circuit (telephone loop performance). The condition of a transmission channel in the talk state when no signal is present. 473

idle period (gas tube). That part of an alternating-voltage cycle during which a certain arc path is not carrying current. 244, 190

idler circuit (parametric device)(nonlinear, active, and nonreciprocal waveguide components). A portion of a parametric device that chiefly determines the behavior of the device at an idler frequency. *See:* **parametric device.** 530

idler frequency (parametric device)(nonlinear, active, and nonreciprocal waveguide components). A sum frequency (or difference frequency) generated within the parametric device other than the input, output, or pump frequencies that requires specific circuit consideration to achieve the desired device performance. *See:* **parametric device.** 530

idle time (electric drive). The portion of the available time during which a system is believed to be in good operating condition but is not in productive use. *See:* **electric drive.** 206

IEC (television). International Electrotechnical Commission. *Note:* The French name is *Commission Electrotechnique Internationale (CEI).* 18

IF. *See:* **intermediate frequency.**
IFF. *See:* **identity friend or foe.**
IGFET. *See:* **insulated-gate field-effect transistor.**
ignition (low voltage surge arresters). *See:* **breakdown; sparkover (in case of a gap).**
ignition control (industrial control). Control of the starting instant of current flow in the anode circuit of a gas tube. *See:* **electronic controller.** 206
ignition switch (industrial control). A manual or automatic switch for closing or interrupting the electric ignition circuit of an internal-combustion engine at the option of the machine operator, or by an automatic function calling for unattended operation of the engine. *Note:* Provisions for checking individual circuits of the ignition system for relative performance may be incorporated in such switches. *See:* **switch.** 206
ignition transformer (power and distribution transformer). Step-up transformer generally used for electrically igniting oil, gas, or gasoline in domestic, commercial, or industrial heating equipment. 53
ignitor. A stationary electrode that is partly immersed in the cathode pool and has the function of initiating a cathode spot. *See:* **electrode (electron tube); electronic controller.** 190
ignitor-current temperature drift (microwave gas tubes). The variation in ignitor electrode current caused by a change in ambient temperature of the tube. *See:* **gas tube.** 125
ignitor discharge (microwave switching tubes). A direct-current glow discharge, between the ignitor electrode and a suitably located electrode, used to facilitate radio-frequency ionization. *See:* **gas tube.** 125
ignitor discharge (microwave switching tubes)(nonlinear, active, and nonreciprocal waveguide components). A direct-current glow discharge between the ignitor electrode and a suitably located electrode, used to facilitate radio-frequency ionization. *See:* **gas tube; radioactive nuclear ignitor.** 530
ignitor electrode (microwave switching tubes). An electrode used to initiate and sustain the ignitor discharge. *See:* **gas tube.** 125
ignitor firing time (microwave switching tubes)(nonlinear, active, and nonreciprocal waveguide components). The time interval between the application of a direct voltage to the ignitor electrode and the establishment of the ignitor discharge. *See:* **gas tube; receiver protector.** 530
ignitor interaction (microwave gas tubes). The difference between the insertion loss measured at a specified ignitor current and that measured at zero ignitor current. *See:* **gas tube.** 125
ignitor leakage resistance (microwave switching tubes). The insulation resistance, measured in the absence of an ignitor discharge, between the ignitor electrode terminal and the adjacent radio-frequency electrode. *See:* **gas tube.** 125
ignitor oscillations (microwave gas tubes). Relaxation oscillations in the ignitor circuit. *Note:* If present, these oscillations may limit the characteristics of the tube. *See:* **gas tube.** 125

ignitor voltage drop (microwave switching tubes). The direct voltage between the cathode and the anode of the ignitor discharge at a specified ignitor current. *See:* **gas tube.** 125

ignitron. A single-anode pool tube in which an ignitor is employed to initiate the cathode spot before each conducting period. *See:* **electronic controller.** 125

ignored conductor. *See:* **isolated conductor.**

ILD. *See:* **injection laser diode.**

illuminance (footcandle or lux) meter (1) (illuminating engineering). An instrument for measuring illuminance on a plane. Instruments which accurately respond to more than one spectral distribution are color corrected, that is, the spectral response is balanced to $V(\lambda)$ or $V'(\lambda)$. Instruments which accurately respond to more than one spatial distribution of incident flux are cosine corrected, that is, the response to a source of unit luminous intensity, illuminating the detector from a fixed distance and from different directions decreases as the cosine of the angle between the incident direction and the normal to the detector surface. The instrument is comprised of some form of photodetector with or without a filter driving a digital or analog readout through appropriate circuitry. 167

(2) (television). *See:* **illumination.** 18

illuminance conversion factors (illuminating engineering). 167

illuminance, E $= d\Phi/d A$ (illuminating engineering). The density of the luminous flux incident at a point on a surface. Average illuminance is the quotient of the luminous flux incident on a surface by the area of the surface. 167

illumination (footcandle) meter. An instrument for measuring the illumination on a surface. *Note:* Most such instruments consist of barrier-layer cells connected to a meter calibrated in footcandles. *See:* **photometry.** 167

illumination (1) (illuminating engineering). An alternate, but deprecated, term for illuminance. It is frequently used since illuminance is subject to confusion with luminance and illuminants, especially when not clearly pronounced. *Note:* The term illumination also

is commonly used in a qualitative or general sense to designate the act of illuminating or the state of being illuminated. Usually the context will indicate which meaning is intended, but occasionally it is desirable to use the expression *level of illumination* to indicate that the quantitative meaning is intended. 167

(2) (television). (A) general. The density of the luminous flux incident on a surface; it is the quotient of the luminous flux by the area of the surface when the latter is uniformly illuminated. **(B) at a point of a surface.** The quotient of the luminous flux incident on an infinitesimal element of surface containing the point under consideration by the area of that element. *Notes:* (1) The term illumination also is commonly used in a qualitative or general sense to designate the act of illuminating or the state of being illuminated. Usually the context will indicate which meaning is intended, but occasionally it is desirable to use the expression level of illumination to indicate that the quantitative meaning is intended. The term illuminance, which sometimes is used in place of illumination, is subject to confusion with luminance and illuminates, especially when not clearly pronounced. (2) The units of measurements are: footcandle (lumen per square foot, lm/ft^2 lux (lumen per square meter, lx or lm/m^2). This unit of illumination is recommended by the IEC phot (lumen per square centimeter, LM / CM^2). 18

(3) (light emitting diodes). (E_v, do_vdA). The density of the luminous flux incident on a surface: illumination is the quotient of the luminous flux by the area of the surface. When the area of the surface is uniformly illuminated, the SI unit for illumination is the lux (lx) that is, lumen per square meter. 162

(4) (antenna). *See:* **aperture illumination.**

illuminator (radar). A system designed to impose electromagnetic radiation on a designated target so that the reflections can be used by another sensor, typically for purposes of homing. *See:* **semiactive guidance.** 13

illustrative diagram (industrial control). A diagram whose principal purpose is to show the operating principle of a device or group of devices without necessari-

1 *lumen* $=1/683$ *lightwatt* (sec 3.7)
1 *lumen-hour* $=60$ *lumen* $-$ *minutes*
1 *footcandle* $=1$ *lumen/ft²*
1 *watt-second* $=1$ *joule* $=10^{7\text{ergs}}$
1 *phot*-1 *lumen/cm²*
1 *lux*-1 *lumen/m²*
Number of $-$ *Multiplied by*

Equals number of	Foot candles	Lux*	Phots	Milliphots
footcandles	1	0.0929	929	0.929
lux*	10.76	1	10,000	10
phot	0.00108	0.0001	1	.001
milliphot	1.076	0.1	1,000	1

*The International System (SI) unit

ly showing actual connections or circuits. Illustrative diagrams may use pictures or symbols to illustrate or represent devices or their elements. Illustrative diagrams may be made of electric, hydraulic, pneumatic, and combination systems. They are applicable chiefly to instruction books, descriptive folders, or other media whose purpose is to explain or instruct. *See:* **control.**   210, 206

ILS (navigation). An internationally adopted instrument landing system for aircraft, consisting of a vhf localizer, a uhf glide slope and 75 MHz markers.   13

ILS marker beacon. *See:* **outer, middle or boundary marker.**

ILS reference point. A point on the centerline of the ILS runway designated as the optimum point of contact for landing: in International Civil Aviation Organization standards this point is from 150 to 300 meters (500 to 1000 feet) from the approach end of the runway.   13

image (1) (optoelectronic device). A spatial distribution of a physical property, such as radiation, electric charge, conductivity, or reflectivity, mapped from another distribution of either the same or another physical property. *Note:* The mapping process may be carried out by a flux of photons, electric charges, or other means. *See:* **optoelectronic device.**   191
(2) (computing systems). *See:* **card image.**

image antenna. The imaginary counterpart of an actual antenna, assumed for mathematical purposes to be located below the surface of the ground and symmetrical with the actual antenna above ground. *See:* **antenna.**   179

image attenuation (circuits and systems). The real part of the image transfer constant. *See:* **image transfer constant.**   67

image burn. *See:* **retained image.**

image camera tube. *See:* **image tube.**

image converter (solid state). An optoelectronic device capable of changing the spectral characteristics of a radiant image. *Note:* Examples of such changes are infrared to visible and X-ray to visible. *See:* **optoelectronic device.**   191

image-converter panel. A thin, usually flat, multicell image converter. *See:* **optoelectronic device.**   191

image-converter tube (camera tubes). An image tube in which an infrared or ultraviolet image input is converted to a visible image output. *See:* **camera tube.**   190

image dissector (optical character recognition) (computing systems). A mechanical or electronic transducer that sequentially detects the level of light in different areas of a completely illuminated sample space.   255, 77

image dissector tube (dissector tube). A camera tube in which an electron image produced by a photoemitting surface is focused in the plane of a defining aperture and is scanned past that aperture. *See:* **television.**   125

image element (optoelectronic device). The smallest portion of an image having a specified correlation with the corresponding portion of the original. *Note:* In some imaging systems the size of the image elements is determined by the structure of the image space, in others by the carrier employed for the mapping process. *See:* **optoelectronic device.**   191

image frequency. *See:* **frequency, image.**

image iconoscope. A camera tube in which an electron image is produced by a photoemitting surface and focused on one side of a separate storage target that is scanned on the same side by an electron beam, usually of high-velocity electrons. *See:* **television.**   125

image impedances (transducer). The impedances that will simultaneously terminate all of its inputs and outputs in such a way that at each of its inputs and outputs the impedances in both directions are equal. *Note:* The image impedances of a four-terminal transducer are in general not equal to each other, but for any symmetrical transducer the image impedances are equal and are the same as the iterative impedances. *See:* **self-impedance; transducer.**   185

image intensifier (solid state). An optoelectronic amplifier capable of increasing the intensity of a radiant image. *See:* **optoelectronic device.**   191

image-intensifier panel. A thin, usually flat, multicell image intensifier. *See:* **optoelectronic device.**   191

image-intensifier tube. An image tube in which the output radiance is (1) in approximately the same spectral region as, and (2) substantially greater than, the photocathode irradiance. *See:* **camera tube; image tube.**   190

image orthicon. A camera tube in which an electron image is produced by a photoemitting surface and focused on a separate storage target, which is scanned on its opposite side by a low-velocity electron beam.   125

image parameters (circuits and systems). Fundamental network functions, namely image impedances and the image transfer function, that are used to design or describe a filter. *See:* **image impedances; image transfer constant.**   67

image phase (circuits and systems). The imaginary part of the image transfer constant. *See:* **image transfer constant.**   67

image phase constant. The imaginary part of the image transfer constant. *See:* **transducer; transfer constant.**   210

image ratio (heterodyne receiver). The ratio of (1) the field strength at the image frequency to (2) the field strength at the desired frequency, each field being applied in turn, under specified conditions, to produce equal outputs. *See:* **radio receiver.**   339

image response. Response of a superheterodyne receiver to the image frequency, as compared to the response to the desired frequency. *See:* **radio receiver.**   328

image storage (diode-type camera tube). The ability of the diode array target to integrate an image for times longer than the conventional frame time.   380

image-storage device. An optoelectronic device capable of retaining an image for a selected length of time. *See:* **optoelectronic device.**   191 image-storage panel (optoelectronic device). A thin, usually flat, multicell image-storage device. *See:* opto-electronic device. 191

image-storage tube. A storage tube into which the information is introduced by means of radiation, usually light, and read at a later time as a visible output. *See:* storage tube. 174

image transfer constant (electric transducer) (transfer constant). The arithmetic mean of the natural logarithm of the ratio of input to output phasor voltages and the natural logarithm of the ratio of the input to output phasor currents when the transducer is terminated in its image impedances. *Note:* For a symmetrical transducer the transfer constant is the same as the propagation constant. *See:* transducer. 210

image tube. An electron tube that reproduces on its fluorescent screen an image of an irradiation pattern incident on its photosensitive surface. *See:* camera tube. 125

imaginary part (circuits and systems). If a complex quantity is represented by 2 components $A + jB$, B is called the imaginary part. 67

imaging plane. *See:* ground plane.

imbedded temperature-detector insulation (rotating machinery). The insulation surrounding a temperature detector, taking the place of a coil separator in its area. 63

immediate address (computing systems). Pertaining to an instruction in which an address part contains the value of an operand rather than its address. *See:* zero-level address. 255, 77

immediate-nonsynchronized ringing (telephone switching systems). An arrangement whereby a pulse of ringing is sent to the called line when the connection is completed, irrespective of the state of the ringing cycle. 55

immediate restoration of service (National Electrical Code) (health care facilities). Automatic restoration of operation with an interruption of not more than 10 seconds as applied to those areas and functions served by the Emergency System, except for areas and functions for which Article 700 [of the National Electrical Code] otherwise makes specific provisions. 256

immediate-synchronized ringing (telephone switching systems). An arrangement whereby the ringing cycle starts with a complete interval of ringing sent to the called line when the connection is completed. 55

immersed gun (microwave tubes). A gun in which essentially all the flux of the confining magnetic field passes perpendicularly through the emitting surface of the cathode. *See:* microwave tube. 190

immersion plating (dip plating). The deposition, without application of an external electromotive force, of a thin metal coating upon a less noble metal by immersing the latter in a solution containing a compound of the metal to be deposited. *See:* electroplating. 328

immittance (linear passive networks). A response function for which one variable is a voltage and the other a current. *Note:* Immittance is a general term for both impedance and admittance, used where the distinction is irrelevant. 238

immittance comparator. An instrument for comparing the impedance or admittance of the two circuits, components, etcetera. *See:* auxiliary device to an instrument. 185

immittance converter (circuits and systems). A two-port circuit capable of making the input immittance of one port (H_{in}) the product of the immittance connected to the other port (H_1) a positive or negative real constant ($\pm 1k$) and some internal immittance (H_i) i.e. $H_{in} = \pm kH_1H_i$. 67

immittance matrix (circuits and systems). A two-dimensional array of immittance quantities that relate currents to voltages at the ports of a network. 67

immunity to interference (electromagnetic compatibility). The property of a receiver or any other equipment or system enabling it to reject a radio disturbance. *See:* electromagnetic compatibility. 199

impact ionization gain (diode-type camera tube). The dimensionless ratio of the target signal current to the photocathode current which produced this signal, both averaged over a frame time or over a time long compared to the frame time. 380

impaired insulation (insulation systems of synchronous machines). The word "impaired" is here used in the sense of causing any change which could disqualify the insulating material for continuously performing its intended function whether creepage spacing, mechanical support, or dielectric barrier action. The electrical and mechanical properties of the insulation must not be impaired by the prolonged application of the hottest spot or limiting observable temperature permitted for the specific insulation class. 298

IMPATT oscillator (nonlinear, active, and nonreciprocal waveguide components). A direct dc-rf (direct current to radio frequency) conversion device in which the active element of the oscillator is a p-n junction diode biased into the avalanche breakdown mode. The term IMPATT is an acronym for IMPact Avalanche Transit Time. 530

impedance (1) (linear constant-parameter system). (A) The corresponding impedance function with p replaced by $j\omega$ in which ω is real. (B) The ratio of the phasor equivalent of a steady-state sine-wave voltage or voltagelike quantity (driving force) to the phasor equivalent of a steady-state sine-wave current or currentlike quantity (response). *Notes:* (1) Definitions (A) and (B) are equivalent. The *real* part of impedance is the resistance. The *imaginary* part is the reactance. (C) A physical device or combination of devices whose impedance as defined in (A) or (B) can be determined. (2) This sentence illustrates the double use of the word impedance, namely for a physical characteristic of a device or system (definitions (A) and (B)) and for a device (definition (C)). In the latter case the word impedor may be used to reduce confusion. Definition (C) is a second use of impedance and is independent of definitions (A) and (B). *Editor's Note:* The ratio Z is commonly expressed in terms of its orthogonal components, thus:

$$Z = R + jX$$

where Z, R, and X are respectively termed the impedance, resistance, and reactance, all being measured in ohms. In a simple circuit consisting of R, L, and C all in series, Z becomes

$$Z = R + j(\omega L - 1/\omega C),$$

where $\omega = 2\pi f$ and f is the frequency. *See: reactance; resistance; network analysis; input impedance; feedback impedance; impedance function.* 210

(2) (electric machine). Linear operator expressing the relation between voltage (increments) and current (increments). Its inverse is called the admittance of an electric machine. *Notes:* (A) If a matrix has as its elements impedances, it is usually referred to as impedance matrix. Frequently the impedance matrix is called impedance for short. (B) Usually such impedances are defined with the mechanical angular velocity of the machine at steady state. *See: asynchronous machine.* 63

(3) (two-conductor transmission line). The ratio of the complex voltage between the conductors to the complex current on one conductor in the same transverse plane. 179

(4) (circular or rectangular waveguide). A nonuniquely defined complex ratio of the voltage and current at a given transverse plane in the waveguide, which depends on the choice of representation of the characteristic impedance. *See: characteristic impedance; waveguide.* 179

(5) (linear system under sinusoidal stimulus) (automatic control). The complex-number ratio of a force-like variable to the resulting velocity-like steady-state variable: a type of transfer function expressed as voltage per unit current, force per unit velocity, pressure difference per unit volume or mass flux, temperature difference per unit heat flux. *See: function, transfer.* 56

(6) (antenna). *See: input impedance; intrinsic impedance; mutual impedance; self-impedance.*

(7) (of a waveguide). A value relating any two of the three quantities, power (P), complex voltage (V), and complex current (I), in a given mode at a specified transverse plane in a waveguide; the value is non-unique, depending on how the voltage and current quantities are defined and on the selected ratio ($V^{2/p}$, P/I^2, or V/I). 267

impedance bond (railway practice). An iron-core coil of low resistance and relatively high reactance used on electrified railroads to provide a continuous direct-current path for the return propulsion current around insulated joints and to confine the alternating-current signaling energy to its own track circuit. 328

impedance, characteristic wave. *See: wave impedance, characteristic.*

impedance compensator. An electric network designed

to be associated with another network or a line with the purpose of giving the impedance of the combination a desired characteristic with frequency over a desired frequency range. *See: network analysis.* 210

impedance, conjugate. An impedance the value of which is the complex conjugate of a given impedance. *Note:* For an impedance associated with an electric network, the complex conjugate is an impedance with the same resistance component and a reactance component the negative of the original. 239, 210, 185

impedance drop (power and distribution transformer). The phasor sum of the resistance voltage drop and the reactance voltage drop. *Note:* For transformers, the resistance drop, the reactance drop, and the impedance drop are, respectively, the sum of the primary and secondary drops reduced to the same terms. They are determined from the load-loss measurements and are usually expressed in per unit, or in percent. 53

impedance, effective input (output) (1) (electron valve or tube). The quotient of the sinusoidal component of the control-electrode voltage (output-electrode voltage) by the corresponding component of the current for the given electrical conditions of all the other electrodes. *See: ON period.* 244, 190

(2) (circuits and systems). The quotient of voltage by current at the input port of a device when it is operating normally (usually steady-state). 67

impedance, essentially zero source (transformer electrical tests). Source impedance low enough so that the test currents under consideration would cause less than five (5) percent distortion (instantaneous) in the voltage amplitude or waveshape at the load terminals. 95

impedance feedback (analog computer). A passive network connected between the output terminal of an operational amplifier and its summing junction. 9, 77

impedance function (defined for linear constant-parameter systems or parts of such systems). That mathematical function of p that is the ratio of a voltage or voltage-like quantity (driving force) to the corresponding current or current-like quantity (response) in the hypothetical case in which the former is e^{pt} (e is the natural log base, p is arbitrary but independent of t, t is an independent variable that physically is usually time) and the latter is a steady-state response of the form $e^{pt}.2(p)$. Note: In electric circuits **voltage** is always the driving force and **current** is the response even though as in nodal analysis the current may be the independent variable: in electromagnetic radiation **electric field strength** is always considered the driving force and **magnetic field strength** the response, and in mechanical systems **mechanical force** is always considered as a driving force and **velocity** as a response. In a general sense the dimension (and unit) of **impedance** in a given application may be whatever results from the ratio of the dimensions of the quantity chosen as the driving force to the dimensions of the quan-

tity chosen as the response. However, in the types of systems cited above any deviation from the usual convention should be noted. *See:* **network analysis.**
210

impedance grounded (power and distribution transformer). Grounded through impedance. *Note:* The components of impedance and the device to be grounded need not be at the same location.
62, 53, 64, 91

impedance heating (electrical heating systems). An electrical heating system where the object to be heated generates heat as a result of an alternating current (ac) passing through it.
476

impedance, image. *See:* **image impedances.**

impedance, input. *See:* **input impedance.**

impedance inverter (circuits and systems). (1) network possessing an input (output) impedance that is proportional to the reciprocal of the load (source) impedance. (2) A symmetrical four-terminal network having the impedance inverting and phase characteristics of a quarter-wave length transmission line at its specified frequency or a chain matrix where A, D, O, B, jK and C, $j.K$ (K is a constant relating the input impedance Z to the load impedance Z_L by the relationship Z, $K^2.Z_L$).
67

impedance irregularity (data transmission). A term used to denote impedance mismatch in a transmission medium. For example, a section of cable in an open-wire line constitutes an impedance irregularity.
59

impedance, iterative (transducer or a 2-port network). The impedance that, when connected to one pair of terminals, produces a like impedance at the other pair of terminals. *Notes:* (1) It follows that the iterative impedance of a transducer or a network is the same as the impedance measured at the input terminals when an infinite number of identically similar units are formed into an iterative or recurrent structure by connecting the output terminals of the first unit to the input terminals of the second, the output terminals of the second to the input terminals of the third, etcetera. (2) The iterative impedances of a four-terminal transducer or network are, in general, not equal to each other but for any symmetrical unit the iterative impedances are equal and are the same as the image impedances. The iterative impedance of a uniform line is the same as its characteristic impedance.
239, 185

impedance kilovolt-amperes (1) (regulator). The kilovolt-amperes (kVA) measured in the shunt winding with the series winding short-circuited and with sufficient voltage applied to the shunt winding to cause rated current to flow in the windings. *See:* **voltage regulator.**
257

(2) (rated) (power and distribution transformer). The kilovolt-amperes (kVA) measured in the excited winding with the other winding short-circuited and with sufficient voltage applied to the excited winding to cause rated current to flow in the winding.
53

impedance, load (1) (general). The impedance presented by the load to a source or network.
185

(2) (semiconductor radiation detectors). The impedance shunting the detector, and across which the detector output voltage signal is developed.
23

impedance, loaded applicator. *See:* **loaded applicator impedance.**

impedance matching (electrical heating applications to melting furnaces and forehearths in the glass industry). The use of a transformer to match line-supply voltage levels to the voltage levels required by the molten-glass load.
520

impedance, matching. *See:* **load matching.**

impedance matrix (multiport network). A matrix operator that interrelates the voltages at the various ports to the currents at the same and other ports.
185

impedance mismatch factor (antennas). The ratio of the power accepted by an antenna to the power incident at the antenna terminals from the transmitter. *Note:* The impedance mismatch factor is equal to one minus the magnitude squared of the input reflection coefficient of the antenna.
111

impedance, normalized. The ratio of an impedance to a specified reference impedance. *Note:* For a transmission line or a waveguide, the reference impedance is usually a characteristic impedance.
185

impedance of a shunt reactor (shunt reactors over 500 kVA). The phasor sum of the reactance and resistance, expressed in ohms per phase, it may be derived from the rated kilovoltampere (kVA) and rated voltage.
562

impedance, output (1) (device, transducer, or network). The impedance presented by the output terminals to a load. *Notes:* (A) Output impedance is sometimes incorrectly used to designate load impedance. (B) This is a frequency-dependent function, and is used to help describe the performance of the power supply and the degree of coupling between loads. *See:* **electrical conversion; self-impedance.**
186

(2) (electron device). The output electrode impedance at the output electrodes. *See:* **self-impedance.**
190

(3) (power supplies). The effective dynamic output impedance of a power supply is derived from the ratio of the measured peak-to-peak change in output voltage to a measured peak-to-peak change in load alternating current. Output impedance is usually specified throughout the frequency range from direct current to 100 kilohertz. *See:* **self-impedance.**
186

(4) (analog computer). The impedance presented by the transducer to a load.
9

(5) (transformer-rectifier system). Internal impedance in ohms measured at the direct current terminals when the rectifier is continuously providing direct current to a load. This impedance is perferably expressed as a curve of impedance in ohms versus frequency, over the frequency range of interest to the application.
95

(6) (Hall generator). The impedance between the Hall terminals.
107

impedance permeability (magnetic core testing). An ac permeability related to the total rms exciting current, including harmonics.

$$\mu_z = \frac{B_i}{H_z \, \mu_0}$$

where

$H_z = \sqrt{2}\, NI/1 =$ equivalent peak field strength, amperes/meters

$B_i =$ maximum intrinsic flux density, tesla

$I =$ rms exciting current, amperes

$N =$ exciting coil turns.   165

impedance ratio (divider). The ratio of the impedance of the two arms connected in series to the impedance of the low-voltage arm. *Note:* In determining the ratio, account should be taken of the impedance of the measuring cable and the instrument. The impedance ratio is usually given for the frequency range within which it is approximately independent of frequency. For resistive dividers the impedance ratio is generally derived from a direct-current measurement such as by means of a Wheatstone bridge.   307, 201

impedance ratio factor (harmonic control and reactive compensation of static power converters). The ratio of the source impedance at the point in the system under consideration to the equivalent total impedance from the source to the converter circuit elements which commutate simultaneously.   533

impedance relay (power switchgear). A distance relay in which the threshold value of operation depends only on the magnitude of the ratio of voltage to current applied to the relay, and is substantially independent of the phase angle of the impedance. *See:* **distance relay, figure (a).**   103

impedance, source. The impedance presented by a source of energy to the input terminals of a device, or network. *See:* **impedance, input; network analysis; self impedance.**   239, 185

impedance, unloaded applicator. *See:* **transformer, matching.**

impedance voltage (1) (transformer). The voltage required to circulate rated current through one of two specified windings of a transformer when the other winding is short-circuited, with the windings connected as for rated voltage operation. *Note:* It is usually expressed in per unit, or percent, of the rated voltage of the winding in which the voltage is measured.   53

(2) (constant-current transformer). The measured primary voltage required to circulate rated secondary current through the short-circuited secondary coil for a particular coil separation. *Note:* It is usually expressed in per unit, or percent, of the rated primary voltage. *See:* **constant-current transformer.**   203

(3) (current-limiting reactor). The product of its rated ohms impedance and rated current. *See:* **reactor.**   309

(4) (regulator). The voltage required to circulate rated current through one winding of the regulator when another winding is short circuited, with the respective windings connected as for rated voltage operation. *Note:* It is usually referred to the series winding, and then expressed in per unit, or percent, of the rated voltage of the regulator. *See:* **voltage regulator.**   257

(5) (neutral grounding device). An effective resistance component corresponding to the impedance losses, and a reactance component corresponding to the flux linkages of the winding.   91

impedance voltage of a constant-current transformer (power and distribution transformer). The measured primary voltage required to circulate rated secondary current through the short-circuited secondary coil for a particular coil separation. *Note:* It is usually expressed in per-unit or percent of the rated primary voltage.   53

impedance voltage of a transformer (power and distribution transformer). The voltage required to circulate rated current through one of two specified windings of a transformer when the other winding is short-circuited, with the windings connected as for rated voltage operation. *Note:* It is usually expressed in per unit, or percent, of the rated voltage of the winding in which the voltage is measured.   53

impedance, wave. *See:* **wave impedance.**

impedor. A device, the purpose of which is to introduce impedance into an electric circuit. *See:* **network analysis.**   210

imperfect debugging (software). In reliability modeling, the assumption that attempts to correct or remove a detected fault are not always successful. *See:* **fault; reliability.**   434

imperfect dielectric. *See:* **dielectric, imperfect.**

imperfection (crystalline solid). Any deviation in structure from that of an ideal crystal. *Note:* An ideal crystal is perfectly periodic in structure and contains no foreign atoms. *See:* **semiconductor.**   245

impingement attack (corrosion). Localized erosion-corrosion resulting from turbulent or impinging flow of liquids.   205

implementation (software). (1) A realization of an abstraction in more concrete terms, in particular, in terms of hardware, software, or both. (2) A machine executable form of a program, or a form of a program that can be translated automatically to a machine executable form. (3) The process of translating a design into code and debugging the code. *See:* **abstraction; code; debugging; design; hardware; program; software.**   434

implementation-defined (pascal computer programming language). Possibly differing between processors, but defined for any particular processor.   433

implementation-dependent (pascal computer programming language). Possibly differing between processors and not necessarily defined for any particular processor.   433

implementation phase (software)software verification and validation). The period of time in the software life cycle during which a software product is created from design documentation and debugged. *See:* **design documentation; installation and checkout phase; software life cycle; software product; test phase.**   434, 511

implementation requirement (software). Any requirement that impacts or constrains the implementation of a software design, for example, design descriptions, software development standards, programming language requirements, software quality assurance standards. *See:* **implementation; programming language; quality assurance; requirement; software design.**
434

implication (mathematics of computing). A dyadic Boolean operator having the property that if P is a statement and Q is a statement, then the expression 'P implies Q' is true in all cases except when P is true and Q is false. *Note:* P implies Q is often represented as P → Q.
564

implicit computation (analog computers). Computation using a self-nulling principle in which, for example, the variable sought is first assumed to exist, after which a synthetic variable is produced according to an equation and compared with a corresponding known variable and the difference between the synthetic and the known variable driven to zero by adjusting the assumed variable. Although the term applies to most analog circuits, even to a single operational amplifier, it is restricted usually to computation performed by (A) circuits in which a function is generated at the output of a single high-gain dc amplifier by inserting an element generating the inverse function in the feedback path, (B) circuits in which combinations of computing elements are interconnected in closed loops to satisfy implicit equations, or (C) circuits in which linear or nonlinear differential equations yield the solutions to a system of algebraic or transcental equations in the steady-state.
9

imprecision. *See:* **precision.**

impregnant (rotating machinery). A solid, liquid, or semiliquid material that, under conditions of application, is sufficiently fluid to penetrate and completely or partially fill or coat interstices and elements of porous or semiporous substances or composites.
63

impregnate (rotating machinery). The act of adding impregnant (bond or binder material) to insulation or a winding. *Note:* The impregnant, if thermosetting, is usually cured in the process. *See:* **vacuum-pressure impregnation.**
63

impregnated (fibrous insulation). A suitable substance replaces the air between the fibers, even though this substance does not fill completely the spaces between the insulated conductors. *Note:* To be considered suitable, the impregnating substance must have good insulating properties and must cover the fibers and render them adherent to each other and to the conductor.
328

impregnated insulation (insulation systems of synchronous machines). Insulating is considered to be "impregnated" when a suitable substance provides a bond between components of the structure and also a suitable degree of filling and surface coverage sufficient to give adequate performance under the extremes of temperature, surface contamination (moisture, dirt, etcetera), and electrical and mechanical stress expected in service. The impregnant must not flow or deteriorate enough at operating temperature so as to seriously affect performance in service.
298

impregnated tape (dispersed magnetic powder tape). *See:* **magnetic powder-impregnated tape.**

impregnation, winding (rotating machinery). The process of applying an insulating varnish to a winding and, when required, baking to cure the varnish.
63

improper ferroelectric (primary ferroelectric terms). A ferroelectric in which the polarization is not the primary order parameter.
497

improvement threshold (angle-modulation systems). The condition of unity for the ratio of peak carrier voltage to peak noise voltage after selection and before any nonlinear process such as amplitude limiting and detection. *See:* **amplitude modulation.**
111

impulse (1)(metal-oxide surge arresters for ac power circuits). A surge of unidirectional polarity.
583
(2) (power systems) (surge arresters). A surge of unidirectional polarity.
2, 436
(3) (mathematics). (automatic control). A pulse that begins and ends within a time so short that it may be regarded mathematically as infinitesimal although the area remains finite. *See:* **pulse.**
210

impulse bandwidth (1)(electromagnetic site survey). The ratio of the maximum value of the voltage at the output of a network (when properly corrected for network sinewave gain at the stated reference frequency) to the spectrum amplitude of the pulse applied at the input. In networks with a single-humped response the reference frequency is taken as that at which the gain is maximum.
457
(2)(radio noise from overhead power lines and substations). The peak value of the response envelope divided by the spectrum amplitude of an applied impulse.
509
(3) (general). When an inpulse is passed through a network with a restricted passband, the output generally consists of a wave train, the envelope of which builds up to a maximum value and then decays approximately exponentially. The impulse bandwidth of such a network is defined as the ratio of that maximum value (when properly corrected for network sine wave gain at a stated reference frequency) to the spectrum amplitude of the pulse applied at the input. In networks with a single humped response, the reference frequency is taken as that at which the gain is maximum. (Overcoupled or stagger-tuned networks should not be used for measurement of spectrum amplitude of impulses.) *See:* **impulse strength.**
30
(4) (radio noise from overhead power lines). The peak value of the response envelope divided by the spectrum amplitude of an applied impulse. *See:* **electromagnetic compatibility.**
36
(5) (spectrum analyzer) (non-real time spectrum analyzer). The peak value of the time response envelope divided by the spectrum amplitude (assumed flat within the bandpass) of an applied pulse.
390
(6) (overhead-power-line corona and radio noise). The peak value of the response envelope divided by

the frequency spectrum amplitude of an applied impulse.   411

impulse circuitry (nonlinear, active, and nonreciprocal waveguide components). A term given to the circuitry associated with either a step recovery or a dual mode varactor frequency multiplier. As charge is stored in the multiplier diode during each positive cycle of the input frequency and released during each negative cycle, a magnetic field is built up in an impulse inductor that stores all the circuit energy as charge approaches zero. Multiplication occurs when the inductor releases its energy in the form of an impulse voltage across the diode at the time of switching or high-impedance recovery.   530

impulse current (current testing). Ideally, an aperiodic transient current that rises rapidly to a maximum value and falls usually less rapidly to zero. A rectangular impulse current rises rapidly to a maximum value, remains substantially constant for a specified time and then falls rapidly to zero. *See:* **test voltage and current.**   307

impulse currents (high voltage testing). Two types of impulse currents are dealt with. The first type has a shape which increases from zero to a crest value in a short time, and thereafter decreases to zero, either approximately exponentially or in the manner of a heavily damped sine curve. This type is defined by the virtual front time T_1 and the virtual time to half-value T_2. The second type has an approximately rectangular shape and is defined by the virtual duration of the peak and the virtual total duration.   150

impulse current tests (high voltage testing). *See:* **impulse currents, value of the test current, virtual front time T_1, virtual origin O_1, virtual time to half-value T_2, virtual duration of peak of a rectangular impulse current, virtual total duration of a rectangular impulse current.**   150

impulse, doublet (automatic control). An impulse having equal positive and negative peaks.   56

impulse (shock) excitation. A method of producing oscillator current in a circuit in which the duration of the impressed voltage is relatively short compared with the duration of the current produced. *See:* **oscillatory circuit.**   111, 240

impulse flashover voltage (1) (insulator). The crest value of the impulse wave that, under specified conditions, causes flashover through the surrounding medium.   261
(2) (surge arresters). The crest voltage of an impulse causing a complete disruptive discharge through the air between electrodes of a test specimen. *See:* **critical impulse flashover voltage; insulator;**   64, 62

impulse flashover volt-time characteristic. A curve plotted between flashover voltage for an impulse and time to impulse flashover, or time lag of impulse flashover. *See:* **insulator.**   64

impulse generator (electromagnetic compatibility). A standard reference source of broadband impulse energy.   314

impulse inertia (surge arresters). The property of insulation whereby more voltage must be applied to pro-

duce disruptive discharge, the shorter the time of voltage application.   64, 62

impulse insulation level. An insulation strength expressed in terms of the crest value of an impulse withstand voltage. *See:* **basic impulse insulation level (BIL) (insulation strength).**   276

impulse noise (antennas) (data transmission) (overhead-power-line corona and radio noise). Noise characterized by transient disturbances separated in time by quiescent intervals. *Notes:* (1) The frequency spectrum of these disturbances are substantially uniform over the useful pass band of the transmission system. (2) The same source may produce impulse noise in one system and random noise in a different system. *See:* **signal-to-noise ratio.**   111, 59, 411

impulse-noise selectivity (receiver) (receiver performance). A measure of the ability to discriminate against impulse noise. *See:* **receiver performance.**   181

impulse protective level (metal-oxide surge arresters for ac power circuits). For a defined waveshape, the higher of the maximum sparkover value or the corresponding discharge-voltage value.   583

impulse protective volt-time characteristic (metal-oxide surge arresters for ac power circuits). The discharge-voltage time response of the device to impulses of a designated wave shape and polarity, but of varying magnitudes.   583

impulse ratio (surge arresters). The ratio of the flashover, sparkover, or breakdown voltage of an impulse to the crest value of the power-frequency, sparkover, or breakdown voltage.   64, 62

impulse relay. (1) A relay that follows and repeats current pulses, as from a telephone dial. (2) A relay that operates on stored energy of a short pulse after the pulse ends. (3) A relay that discriminates between length and strength of pulses, operating on long or strong pulses and not operating on short or weak ones. (4) A relay that alternately assumes one of two positions as pulsed. (5) Erroneously used to describe an integrating relay. *See:* **relay.**   341

impulse response (linear network) (1) (circuits and systems). The response, as a function of time, of a network when the excitation is a unit impulse. Hence, the impulse response of a network is the inverse Laplace transform of the network function in the frequency domain.   67
(2) (automatic control). *See:* **response, impulse-forced.**
(3) (fiber optics). The function h(t) describing the response of an initially relaxed system to an impulse (Dirac-delta) function applied at time t = 0. The root-mean-square (rms) duration, o_{rms}, of the impulse response is often used to characterize a component or system through a single parameter rather than a function:

$$o_{rms} = [1/M_0 \int_{-\infty}^{\infty} (T-t)^2 h(t) dt]^{1/2}$$

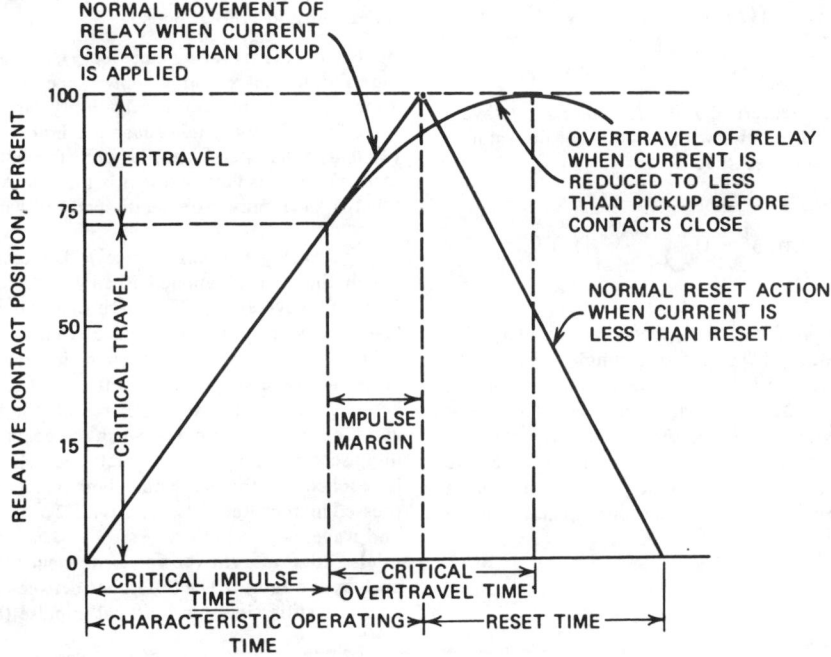

Relationship of relay operating times.

where $M_0 = \int^\infty_{-\infty h} (t)dt$
$T = 1/M_0 \int^\infty_{-\infty th} (t)dt.$

Note: The impulse response may be obtained by deconvolving the input waveform from the output waveform, or as the inverse Fourier transform of the transfer function. *See:* **root-mean-square (rms) pulse duration; transfer function.**   433

impulse root-mean-square (rms) sound level (measurement of sound pressure levels of ac power circuit breakers). The maximum rms value reached by a sound wave, with the mean (or average) taken over a short, specified time interval. Unit: decibel (dB A, B, or C). For the purposes of ANSI/IEEE Std C37.082-1982, the averaging time is that given by a resistance-capacitance charging circuit with a 35 millisecond (ms) time constant.   552

impulses (high voltage testing). An intentionally applied aperiodic transient voltage or current which usually rises rapidly to a peak value and then falls more slowly to zero. Such an impulse is in general well represented by the sum of two exponentials. For special purposes, impulses having approximately linearly rising fronts or of oscillating or approximately rectangular form are used. The term "impulse" is to be distinguished from the term "surge" which refers to transients occurring in electrical equipment or networks in service.   150

impulse sparkover voltage (1)(gas-tube surge-protective devices)(low-voltage air gap surge-protective de-

vices). The highest value of voltage attained by an impulse of a designated wave shape and polarity applied across the terminals of an arrester prior to the flow of discharge current. Sometimes referred to as surge or impulse breakdown voltage.
                   370, 490, 556

(2)(metal-oxide surge arresters for ac power circuits). The highest value of voltage attained by an impulse of a designated wave shape and polarity applied across the terminals of an arrester that will cause gap sparkover prior to the flow of discharge current.*Note:* Sometimes referred to as surge breakdown voltage or impulse breakdown voltage inconnection with low voltage surge arresters.   2, 430, 583

impulse sparkover voltage-time curve (arrester)(gas-tube surge-protective devices). A curve that relates the impulse sparkover voltage to the time to sparkover.   370, 490

impulse sparkover volt-time characteristic (metal-oxide surge arresters for ac power circuits). The gap sparkover response of the device to impulses of a designated wave shape and polarity, but of varying magnitudes. *Note:* For an arrester, this characteristic is shown by a graph of values of crest voltage plotted against time to sparkover.   2, 430, 583

impulse strength. The area under the amplitude-time relation for the impulse. *Note:* This definition can be clarified with the aid of Fig 1. Let $A(t)$ be some function of time having a value other than zero only between the times t_1 and $t_1 + \delta$. Then let the area under the curve $A(t)$ be designated by σ:

$$\sigma = \int_{\infty}^{\infty} A(t)dt = \int_{t_1}^{t_1+\delta} + A(t)dt$$

To define the theoretical or ideal impulse, let $A(t)$ vary in a reciprocal manner with δ such that the value σ remains constant, so that

$$\sigma = \lim \delta \to 0 \int_{t_1}^{t_1+\delta} A(t)dt$$

In the limit the function $A(t)$ becomes an ideal "impulse" of "strength" σ. As an example, consider the function shown in Fig 2. Here a rectangular pulse of finite duration Δt and height A is shown. Now let $A = \sigma/\Delta t$ where σ is (for the present argument) an arbitrary constant, and let $\Delta t \to 0$. In the limit we have an impulse of strength σ. When $\sigma = 1$, one has a "unit impulse." In many conventional applications the amplitude $A(t)$ has the dimension volts and σ then has the dimension volt-seconds.

impulse test (1) (rotating machinery). A test for applying to an insulated component an aperiodic transient voltage having predetermined polarity, amplitude, and wave-form. *See:* **asynchronous machine.**   91
(2) (surge arresters) (power and distribution transformer). An insulation test in which the voltage applied is an impulse voltage of specified wave shape.   53
(3) (neutral grounding devices). Dielectric test in which the voltage applied is an impulse voltage of specified wave shape. The wave shape of an impulse test wave is the graph of the wave as a function of time or distance. *Note:* It is customary in practice to express the wave shape by a combination of two numbers, the first part of which represents the wave front and the second the time between the beginning of the impulse and the instant at which one-half crest value is reached on the wave tail, both values being expressed in microseconds, such as a 1.2×50 microsecond wave.   91
impulse time margin (power switchgear) (in the operation of a relay). The difference between characteristic operating times and critical impulse times.   127, 101
impulse transmission. That form of signaling, used principally to reduce the effects of low-frequency interference, that employs impulses of either or both polarities for transmission to indicate the occurrence of transitions in the signals. *Note:* The impulses are generally formed by suppressing the low-frequency components, including direct current, of the signals. *See:* **telegraphy.**   328
impulse transmitting relay. A relay that closes a set of contacts briefly while going from the energized to the de-energized position or vice versa. *See:* **relay.**   259
impulse-type telemeter. A telemeter that employs characteristics of intermittent electric signals, other than their frequency, as the translating means. *See:* **telemetering.**   328
impulse voltage (current) (surge arresters). Synonymous with voltage of an impulse wave (current of an impulse wave).   244, 62
impulse wave (surge arresters). A unidirectional wave of current or voltage of very short duration containing no appreciable oscillatory components. *See:* **insulator.**   244, 62
impulse withstand voltage (1)(metal-oxide surge arresters for ac power circuits). The crest value of an impulse that, under specified conditions, can be applied without causing a disruptive discharge.   2, 430, 583
(2) (general). The crest value of an applied impulse voltage that, under specified conditions, does not cause a flashover, puncture, or disruptive discharge on the test specimen. *See:* **insulator; surge arrester (surge diverter).**   91
(3) (power switchgear). The crest voltage of an impulse that, under specified conditions, can be applied without causing flashover or puncture.   103

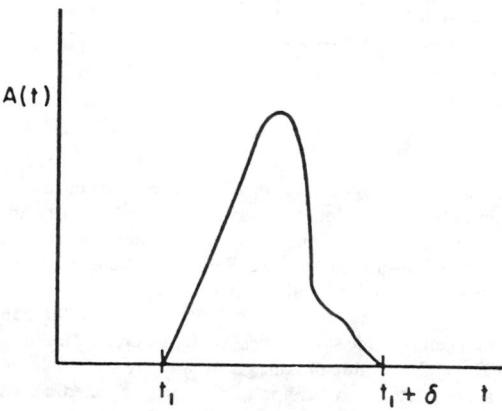

Fig. 1. A pulse of arbitrary shape.

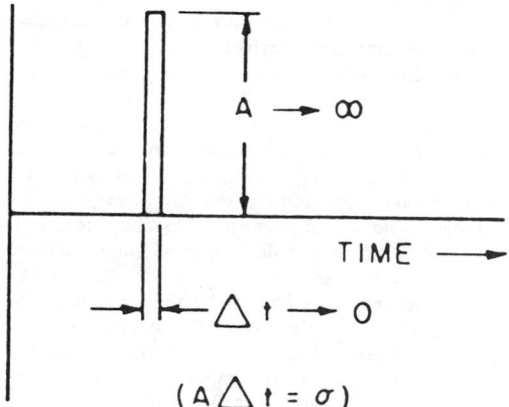

$$(A \triangle t = \sigma)$$

Fig. 2. A rectangular pulse.

impulsive noise (1)(control of system electromagnetic compatibility). Noise characterized by transient disturbances separated in time by quiescent intervals. *Notes:* (A) The frequency spectrum of these disturbances must be substantially uniform over the useful pass band of the transmission system. (B) The same source may produce an output characteristic of impulsive noise in one system and of random noise in a different system.   495
(2)(measurement of sound pressure levels of ac power circuit breakers). A noise characterized by brief excursions of sound pressure (acoustic impulses) which significantly exceed the ambient noise. The duration of a single impulse is usually less than one second (See ANSI S1.13-1971 . For the purpose of ANSI/IEEE Std C37.082-1982, the noise produced by the closing or opening of a circuit breaker, or their combination, is classified as impulsive noise. Other components, such as compressor unloader exhausts, may be sources of impulsive noise.   552
(3) (electromagnetic compatibility). Noise, the effect of which is resolvable into a succession of discrete impulses in the normal operation of the particular system concerned. *See:* **electromagnetic compatibility.**   220, 199
impurity (1) (acceptor) (semiconductor). *See:* **dopant; semiconductor device.**   66
(2) (donor) (semiconductor). *See:* **dopant; semiconductor device.**   66
(3) (crystalline solid). An imperfection that is chemically foreign to the perfect crystal. *See:* **semiconductor.**   245
(4) (chemical) (semiconductor). An atom within a crystal, that is foreign to the crystal.   66
impurity, stoichiometric. A crystalline imperfection arising from a deviation from stoichiometric composition.   245
inactive (signals and paths)(696 interface devices). A signal in its logically false state,   538
inactive region (of a semiconductor radiation detector)(germanium gamma-ray detectors)(charged particle detectors))X-ray energy spectrometers). A region of a detector in which charge created by ionizing radiation does not contribute significantly to the signal charge.   528, 119, 471, 23
inadvertent interchange (electric power systems) (control area). The time integral of the net interchange minus the time integral of the scheduled net interchange. *Note:* This includes the intentional interchange energy resulting from the use of frequency and.or other bias as well as the unscheduled interchange energy resulting from human or equipment errors.   94, 200
in band signaling (1) (data transmission). Signaling which utilizes frequencies within the voice or intelligence band of a channel.   59
(2) (telephone switching systems). Analog generated signaling that uses the same path as a message and in which the signaling frequencies are in the same band used for the message.   55
incandescence (illuminating engineering). The self-

emission of radiant energy in the visible spectrum, due to the thermal excitation of atoms or molecules.   167
incandescent filament lamp (illuminating engineering). A lamp in which light is produced by a filament heated to incandescence by an electric current.   167
incandescent-filament-lamp transformer (series type). A transformer that receives power from a current-regulated series circuit and that transforms the power to another circuit at the same or different current from that in the primary circuit. *Note:* If of the insulating type, it also provides protection to the secondary circuit, casing, lamp, and associated luminaire from the high voltage of the primary circuit.   271
inching (rotating machinery). Electrically actuated angular movement or slow rotation of a machine, usually for maintenance or inspection. *See:* **asynchronous machine; direct-current commutating machine.**   63
inch-pound units (metric practice). Units based upon the yard and the pound commonly used in the United States of America and defined by the National Bureau of Standards. *Note:* Units having the same names in other countries may differ in magnitude.   21
incident. *See:* **software test incident.**
incidental amplitude-modulation factor (signal generator). That modulation factor resulting unintentionally from the process of frequency modulation and.or phase modulation. *See:* **signal generator.**   185
incidental and restricted radiation. Radiation in the radio-frequency spectrum from all devices excluding licensed devices. *See:* **mobile communication system.**   181
incidental frequency modulation (signal generator). The ratio of the peak frequency deviation to the carrier frequency, resulting unintentionally from the process of amplitude modulation. *See:* **signal generator.**   185
incidental phase modulation (signal generator). The peak phase deviation of the carrier, in radians, resulting unintentionally from the process of amplitude modulation. *See:* **signal generator.**   185
incidental radiation of conducted power (frequency-modulated mobile communications receivers). Radio-frequency energy generated or amplified by the receiver, which is detectable outside the receiver.   123
incident wave (1) (radio wave propagation). In a medium of certain propagation characteristics, a wave which impinges on a discontinuity or a medium of different propagation characteristics.   146
(2) (forward wave) (uniform guiding systems). A wave traveling along a waveguide or transmission line in a specified direction toward a discontinuity, terminal plane, or reference plane. *See:* **reflected wave; waveguide.**   185
(3) (surge arresters). A traveling wave before it reaches a transition point.   244, 62
(4) (waveguide). At a transverse plane in a transmission line or waveguide, a wave traveling in a reference direction. *See:* **reflected wave; transmitted wave.**   267

incipient failure detection (nuclear power generating station). Tests designed to monitor performance characteristics and detect degradation prior to failure(s) which would prevent performance of the Class 1E functions. *Note:* Incipient failure testing requires module test checks, inspection, etcetera, on a sufficient time basis to establish performance trends. At the outset, the test cycle and corresponding limits of deviation of module performance or status must be established. Specific parameter trend patterns, exceeding of performance limits shall require that the module be removed and adjusted, replaced or serviced. As used here "sufficient time basis" would never be less than periodic surveillance test interval. In any event the internal must be justified technically based upon such things as manufacturer recommended, periodic preventive maintenance procedures, past operating experience, etcetera. These module tests require testing on line and.or removal from service. 355

inclined-blade blower (rotating machinery). A fan made with flat blades mounted so that the plane of the blades is parallel to and displaced from the axis of rotation of the rotor. *See:* **fan (rotating machinery).** 63

inclined orbit (communication satellite). An orbit of a satellite which is not equatorial, and not polar. 74

inclosed. *See:* **enclosed.**

inclusion (fiber optics). Denoting the presence of extraneous or foreign material. 433

inclusive OR **(computing systems).** *See:* OR.

incoherent (fiber optics). Characterized by a degree of coherence significantly less than 0.88. *See:* **coherent; degree of coherence.** 433

incoherent scattering (radio wave propagation). The phenomenon of generating waves with random variations in phase, amplitude, polarization, and direction of propagation when an incident wave encounters matter. 146

incoming traffic (telephone switching systems). Traffic received directly from trunks by a switching entity. 55

incomplete diffusion (illuminating engineering). That in which the diffusing medium partially redirects the incident flux by scattering while the remaining fraction of incident flux is redirected without scattering, that is, a fraction of the incident flux can remain in an image-forming state. *Syn:* **partial diffusion.** 167

incomplete sequence relay (48) (power system device function numbers). A relay that generally returns the equipment to the normal, of off, position and locks it out if the normal starting, operating, or stopping sequence is not properly completed within a predetermined time. If the device is used for alarm purposes only, it should preferably be designated as 48A (alarm). 402

incorrect relaying-system performance (power switchgear). Any operation or lack of operation of the relays or associated equipment that, under existing conditions, does not conform to correct relaying-systems performance. 103, 127

incorrect relay operation (power switchgear). Any output response or lack of output response by the relay that, for the applied input quantities, is not correct. 103, 127

increment (test pattern language). The action of increasing the arithmetic value of a counter by one. 463

incremental computer. A special-purpose computer that is specifically designed to process changes in the variables as well as the absolute value of the variables themselves, for example, digital differential analyzer. 225, 77

incremental cost of delivered power (source). The additional per unit cost that would be incurred in supplying another increment of power from that source to the composite system load. 94

incremental cost of reference power (source). The additional per-unit cost that would be incurred in supplying another increment of power from that source to a designated reference point on a transmission system. 94

incremental delivered power (electric power systems). The fraction of an increment in power from a particular source that is delivered to any specified point such as the composite system load, usually expressed in percent. 94

incremental dimension (numerically controlled machines). A dimension expressed with respect to the preceding point in a sequence of points. *See:* **dimension; long dimension; normal dimension; short dimension.** 224, 207

incremental energy cost (power operations). The additional cost of producing or transmitting electric energy above some base cost. 516

incremental feed (numerically controlled machines). A manual or automatic input of preset motion command for a machine axis. 244, 207

incremental fuel cost of generation (any particular source). The cost, usually expressed in mill.kilowatt-hour, that would be expended for fuel in order to produce an additional increment of generation at any particular source. 94

incremental generating cost (source at any particular value of generation) (electric power systems). The ratio of the additional cost incurred in producing an increment of generation to the magnitude of that increment of generation. *Note:* All variable costs should be taken into account including maintenance. 94

incremental heat rate (steam turbo-generator unit at any particular output). The ratio of a small change in heat input per unit time to the corresponding change in power output. *Note:* Usually, it is expressed in British thermal unit/kilowatt-hour. 94

incremental hysteresis loss (magnetic material). The hysteresis loss in a magnetic material when it is subjected simultaneously to a biasing and an incremental magnetizing force. 210

incremental induction. At a point in a material that is subjected simultaneously to a polarizing magnetizing force and a symmetrical cyclically varying magnetizing force, one-half the algebraic difference of the maximum and minimum values of the magnetic induction at that point. 210

incremental loading (electric power systems). The assignment of loads to generators so that the additional cost of producing a small increment of additional generation is identical for all generators in the variable range. 94

incremental magnetizing force (magnetic material). At a point that is subjected simultaneously to a biasing magnetizing force and a symmetrical cyclic magnetizing force, one-half of the algebraic difference of the maximum and minimum values of the magnetizing force at the point. 210

incremental maintenance cost (any particular source) (electric power systems). The additional cost for maintenance that will ultimately be incurred as a result of increasing generation by an additional increment. 94

incremental permeability (magnetic induction) (1) (general). The ratio of the cyclic change in the magnetic induction to the corresponding cyclic change in magnetizing force when the mean induction differs from zero. *Note:* In anisotropic media, incremental permeability becomes a matrix. 210
(2) (magnetic core testing). The permeability with stated alternating magnetic field conditions in the presence of a stated static magnetic field.

$$\mu_\Delta = \frac{1}{\mu_0} \frac{\Delta B}{\Delta H}$$

μ_Δ = relative incremental permeability
ΔH = total cyclic variation of the magnetic field strength
ΔB = corresponding total cyclic change in induction. 165

incremental resistance (forward or reverse of a semiconductor rectifier diode) (semiconductor). The quotient of a small incremental voltage by a small incremental current at a stated point on the static characteristic curve. 66

incremental sweep (oscilloscopes). A sweep that is not a continuous function, but that represents the independent variable in discrete steps. *See:* **oscillograph; stairstep sweep.** 185

incremental time constant (electric coupling). The time constant applicable for a small incremental change of excitation voltage about a specified operating value. 416

incremental transmission loss (electric power systems). The fraction of power loss incurred by transmitting a small increment of power from a point to another designated point. *Note:* One of these points may be mathematical (rather than physical), such as the composite system load. 94

incremental worth of power (designated point on a transmission system). The additional per-unit cost that would be incurred in supplying another increment of power from any variable source of a system in economic balance to such designated point. *Note:* When the designated point is the composite system

load, the incremental worth of power is commonly called **lambda** or **Lagrangian multiplier.** 94

increment (network) starter (industrial control). A starter that applies starting current to a motor in a series of increments of predetermined value and at predetermined time intervals in closed-circuit transition for the purpose of minimizing line disturbance. One or more increments may be applied before the motor starts. *See:* **starter.** 225, 206

indefinite admittance matrix (network analysis) (circuits and systems). A matrix associated with an n-node network whose elements have the dimension of admittance and, when multiplied into the vector of node voltages, gives the vector of currents entering the n nodes. 67

independence (1) (Class 1E equipment and circuits) (Class 1E power systems for nuclear power generating stations) The state in which there is no mechanism by which any single design basis event, such as a flood, can cause redundant equipment to be inoperable. 131, 102

(2) (nuclear power generating stations). No common failure mode for any design basis event. 102

independent auxiliary (generating stations electric power system). An item capable of performing its function without dependence on a similar item or the component it serves. 58, 381

independent ballast (mercury lamp). A ballast that can be mounted separately outside a lighting fitting or fixture. 269

independent biaxial test (seismic testing of relays). The horizontal and the vertical acceleration components are derived from two different input signals, which are phase incoherent. 392

independent conformity. *See:* **conformity.**

independent contact. A contacting member designed to close one circuit only. 328

independent firing (industrial control). The method of initiating conduction of an ignitron by obtaining power for the firing pulse in the ignitor from a circuit independent of the anode circuit of the ignitron. *See also:* **electronic controller.** 206

independent ground electrode (surge arresters). A ground electrode or system such that its voltage to ground is not appreciably affected by currents flowing to ground in other electrodes or systems. 244, 62

independent linearity. *See:* **linearity of a signal.**

independent manual operation (of a switching device) (power switchgear). A stored-energy operation where manual energy is stored and released, such that the speed and force of this operation are independent of the action of the attendant. 103

independent pole tripping (power switchgear). The application of multipole circuit breakers in such a manner that a malfunction of one or more poles or associated control circuits will not prevent successful tripping of the remaining pole(s). *Notes:* (1) Circuit breakers used for independent pole tripping must inherently be capable of individual pole opening. (2) Independent pole tripping is applied on ac power systems to enhance system stability by maximizing the probability of clearing at least some phases of a multi-phase fault. 103

independent power operation (power switchgear). An operation by means of energy other than manual where the completion of the operation is independent of the continuity of the power supply. 103

independent transformer. A transformer that can be mounted separately outside a luminaire. 274

independent verification and validation (software). (1) Verification and validation of a software product by an organization that is both technically and managerially separate from the organization responsible for developing the product. (2) Verification and validation of a software product by individuals or groups other than those who performed the original design, but who may be from the same organization. The degree of independence must be a function of the importance of the software. *See:* **design; software; software product; validation; verification.** 434

index (electronic computation). (1) An ordered reference list of the contents of a file or document, together with keys or reference notations for identification or location of those contents. (2) A symbol or a number used to identify a particular quantity in an array of similar quantities. For example, the terms of an array represented by X1, X2, . . . , X100 have the indexes 1, 2, . . . , 100, respectively. (3) Pertaining to an index register. 255, 77

index dip (fiber optics). A decrease in the refractive index at the center of the core, caused by certain fabrication techniques. Sometimes called profile dip. *See:* **refractive index profile.** 433

index matching material (fiber optics). A material, often a liquid or cement, whose refractive index is nearly equal to the core index, used to reduce Fresnel reflections from a fiber end face. *See:* **Fresnel reflection; mechanical splice; refractive index.** 433

index of cooperation, international (facsimile in rectilinear scanning). The product of the total length of a scanning or recording line by the number of scanning or recording lines per unit length divided by pi. *Notes:* (1) For rotating devices the Index of Cooperation is the product of the drum diameter times the number of lines per unit length. (2) The prior IEEE Index of Cooperation was defined for rectilinear scanning or recording as the product of the total line length by the number of lines per unit length. This has been changed to agree with international standards. 12

index of illuminant metamerism (of two objects that are metameric when illuminated by a reference source) (illuminating engineering). Measure of the degree of color difference between the two objects when a specified test source is substituted for the reference source. 167

index of observer metamerism (of two objects that are metameric when viewed by a reference observer) (illuminating engineering). Measure of the degree of color difference between the two objects when a specfied test observer is substituted for the reference observer. 167

index of refraction. *See:* **refractive index (of a medium).** 433

index of sensation (M) (of a source) (illuminating engineering). A number which expresses the effects of source luminance (L_s), solid angle factor (Q), position index (P), and the field luminance (F) on discomfort glare rating.

$$M = L_{sQ} \text{ over } PF^{0.44}$$

(See solid angle factor for an equation defining Q). *Note:* A restatement of this formula lends itself more directly to computer applications. *See:* **discomfort glare rating.** 167

index profile (fiber optics). In an optical waveguide, the refractive index as a function of radius. *See:* **graded index profile; parabolic profile; power-law index profile; profile dispersion; profile dispersion parameter; profile parameter; step index profile.** 433

index register (computing systems). A register whose content is added to or subtracted from the operand address prior to or during the execution of an instruction. 255, 54

indicated bearing (direction finding). A bearing from a direction-finder site to a target transmitter obtained by averaging several readings: the indicated bearing is compared to the apparent bearing to determine accuracy of the equipment. *See:* **navigation.** 187, 13

indicated bearing offset (direction finder [DF] installations)(navigation aid terms). The mean different between the indicated and apparent bearings of a number of signal sources, the sources being, for the most part, uniformly distributed in azimuth. 526

indicated value (power meters). The uncorrected value determined by observing the indicating display of the instrument. 47, 115

indicating circuit (electric installations on shipboard). That portion of the control circuit of a control apparatus or system which carries intelligence to visual or audible devices which indicate the state of the apparatus controlled. 3

indicating control switch (power switchgear). One that indicates its last control operation. 103

indicating demand meter (metering). A demand meter equipped with a readout that indicates demand, maximum demand, or both. 212

indicating fuse (power switchgear). A fuse that automatically indicates that the fuse has interrupted the circuit. 103, 443

indicating instrument (electrical heating applications to melting furnaces and forehearths in the glass industry). An instrument in which only the present value of the quantity measured is visually indicated. 520

indicating or recording mechanism (demand meter). That mechanism that indicates or records the measurement of the electrical quantity as related to the demand interval. *Note:* This mechanism may be operated directly by and be a component part of the electric mechanism, or may be structurally separate from it. The demand may be indicated or recorded in kilowatts, kilovolt-amperes, amperes, kilovars, or other suitable units. This mechanism may be of an indicating type, indicating by means of a pointer related to its position on a scale or by means of the cumulative reading of a number of dial or cyclometer indicators: or a graphic type, recording on a circular or strip chart: or of a printing type, recording on a tape. It may record the demand for each demand interval or may indicate only the maximum demand. *See:* **demand meter.**   328

indicating scale (recording instrument). A scale attached to the recording instrument for the purpose of affording an easily readable value of the recorded quantity at the time of observation. *Note:* For recording instruments in which the production of the graphic record is the primary function, the chart scale should be considered the primary basis for accuracy ratings. For instruments in which the graphic record is secondary to a control function the indicating scale may be more accurate and more closely related to the control than is the chart scale. *See:* **moving element (instrument).**   328

indication (station control and data acquisition)-(supervisory control, data acquisition, and automatic control). A light or other signal (audio or visual) provided by the man/machine interface that signifies a particular condition.   403, 570

indication (status) function. *See:* **supervisory control functions**

indication point (railway practice). The point at which the train control or cab signal impulse is transmitted to the locomotive or vehicle apparatus from the roadway element.   328

indication with memory. *See:* **supervisory control functions**

indicator (faulted circuit indicators). That portion of the FCI (faulted circuit indicator) which indicates that fault current has been sensed.   482
See: **display.**

indicator light. A light that indicates whether or not a circuit is energized. *See:* **appliances outlet.**   263

indicators (class 1E power systems for nuclear power generating stations). Devices that display information to the operator.   31, 102

indicator symbol (logic diagrams). A symbol that identifies the state or level of an input or output of a logic symbol with respect to the logic symbol definition.   88

indicator travel. The length of the path described by the indicating means or the tip of the pointer in moving from one end of the scale to the other. *Notes:* (1) The path may be an arc or a straight line. (2) In the case of knife-edge pointers and others extending beyond the scale division marks, the pointer shall be considered as ending at the outer end of the shortest scale division marks. *See:* **moving element (instrument).**   295

indicator tube. An electron-beam tube in which useful information is conveyed by the variation in cross section of the beam at a luminescent target.   190, 125

indicial admittance. The instantaneous response to unit step driving force. *Note:* This is a time function that is not an admittance of the type defined under **admittance.** *See:* **network analysis.**   210

indigenous fault (software). A fault existing in a computer program that has not been inserted as part of a fault seeding process. *See:* **computer program; fault; fault seeding.**   434

indirect-acting machine voltage regulator (power switchgear). A machine voltage regulator having a voltage-sensitive element that acts indirectly, through the medium of an interposing device such as contactors or a motor, to control the excitation of an electric machine. *Note:* A regulator is called a generator voltage regulator when it acts in the field circuit of a generator and is called an exciter voltage regulator when it acts in the field circuit of the main exciter.   103

indirect-acting recording instrument. A recording instrument in which the level of measurement energy of the primary detector is raised through intermediate means to actuate the marking device. *Note:* The intermediate means are commonly either mechanical, electric, electronic, or photoelectric. *See:* **instrument.**   328

indirect address (computing systems). An address that specifies a storage location that contains either a direct address or another indirect address. *See:* **multilevel address.**   255, 77

indirect-arc furnace. An arc furnace in which the arc is formed between two or more electrodes.   328

indirect commutation (circuit properties)(auxiliary commutation)(self-commutated converters). A commutation between a principal switching branch and an auxiliary switching branch succeeded by a commutation to the next principal switching branch. Indirect commutation is employed in some types of converters using circuit-commutated thyristors, where the auxiliary branch includes a commutating capacitor(s) to turn off the outgoing principal switch when the auxiliary switch is turned on. *Note:* In some converter circuits, several auxiliary branches may be involved consecutively.   584

indirect component (illuminating engineering). That portion of the luminous flux from a luminaire which arrives at the work-plane after being reflected by room surfaces.   167

indirect-drive machine (elevators). An electric driving machine, the motor of which is connected indirectly to the driving sheave, drum, or shaft by means of a belt or chain through intermediate gears. *See:* **driving machine (elevators).**   328

indirect lighting (illuminating engineering). Lighting involving luminaires which distribute 90 to 100 percent of the emitted light upward.   167

indirectly controlled variable (control) (industrial control) (automatic control). A variable that is not directly measured for control but that is related to, and influenced by, the directly controlled variable. *See also:* **control system, feedback.**   206

indirectly heated cathode (equipotential cathode) (unipotential cathode). A hot cathode to which heat is supplied by an independent heater. *See:* **electrode (electron tube).**   125

indirect manual operation (of a switching device) (power switchgear). Operation by hand through an operating handle mounted at a distance from, and connected to the switching device by, mechanical linkage.   103

indirect operation (of a switching device) (power switchgear). Operating by means of an operating mechanism connected to the main operating shaft or an extension of it, through offset linkages and bearings.   103

indirect release (trip) (of a switching device) (power switchgear). A release energized by the current in the main circuit through a current transformer, shunt, or other transducing device.   103

indirect stroke (surge arresters). A lightning stroke that does not strike directly any part of a network but that induces an overvoltage in it.   244, 62

indirect-stroke protection (lightning). Lightning protection designed to protect a network or electric installation against indirect strokes. *See:* **direct-stroke protection (lightning).**   64

individual branch circuit (National Electrical Code). A branch circuit that supplies only one utilization equipment.   256

individual-equipment test requirements (radio-noise emission). The set of explicit requirements specifying the test conditions, instrumentation, equipment under test (EUT) operation, etcetera, to be used in testing a specific EUT for conducted or radiated radio noise. Such requirements should take precedence over the requirements of this standard (ANSI C63.4 -1981).   418

individual-lamp autotransformer (power and distribution transformer). A series autotransformer that transforms the primary current to a higher or lower current as required for the operation of an individual street light.   53

individual-lamp insulating transformer (power and distribution transformer). An insulating transformer used to protect the secondary circuit, casing, lamp, and associated luminaire of an individual street light from the high-voltage hazard of the primary circuit.   53

individual line (1) (telephone switching systems). A line arranged to serve one main station.   55
(2) (data transmission). A subscriber line arranged to serve only one main station although additional stations may be connected to the line as extensions. An individual line is not arranged for discriminatory ringing with respect to the stations on that line.   59

individual pole operation (of a multipole circuit breaker or switching device) (power switchgear). A descriptive term indicating that any pole(s) of the device can be caused to change state (open or close) without changing the state of the remaining pole(s). Devices may have capability for individual pole opening, individual pole closing, or both.   103

individual trunk (telephone switching systems). A trunk, link, or junctor that serves only one input group of a grading.   55

indoor (prefix) (1) (power and distribution transformer). Not suitable for exposure to the weather. *Note:* For example, indoor equipment designed for indoor service or for use in a weatherproof housing.   53
(2) (prefix)(shunt power capacitors). Not suitable for exposure to the weather. *Note:* For example, an indoor capacitor unit is designed for indoor service or for use in a weatherproof housing. *See:* **outdoor.**   138
(3) (power switchgear). Designed for use only inside buildings, or weatherproof (weather-resistant) enclosures.   103, 443

indoor arrester (surge arrester)(metal-oxide surge arresters for ac power circuits). An arrester that, because of its construction, is protected from the weather.   430, 583

indoor enclosure (power system communication equipment). An enclosure for use where another housing provides protection against exposure to the weather.   453

indoor shunt reactor (shunt reactors over 500 kVA). One which, because of its construction, must be protected from the weather.   562

indoor termination (cable termination). A termination intended for use where it is protected from direct exposure to both solar radiation and precipitation. Terminations designed for use in sealed enclosures where the external dielectric strength is dependent upon liquid or special gaseous dielectrics are also included in this category.   4

indoor transformer (power and distribution transformer). A transformer which, because of its construction, must be protected from the weather.   53

indoor wall bushing. A wall bushing of which both ends are suitable for operating only where protection from the weather is provided. *See:* **bushing.**   348

induced charge (ferroelectric device). The charge that flows when the condition of the device is changed from that of zero applied voltage (after having previously been saturated with either a positive or negative voltage) to at least that voltage necessary to saturate in the same sense. *Note:* The induced charge is dependent on the magnitude of the applied voltage, which should be specified in describing this characteristic of ferroelectric devices. *See:* **ferroelectric domain.**   247

induced control voltage (Hall-effect devices). The electromotive force induced in the loop formed by the control current leads and the current path through the Hall plate by a varying magnetic flux density, when there is no control current.   107

induced current (1) (general). Current in a conductor due to the application of a time-varying electromagnetic field. *See:* **induction heating.**   14
(2) (interference terminology). The interference current flowing in a signal path as a result of coupling of the signal path with an interference field, that is, a field produced by an interference source. *See:* **interference.**   188
(3) (lightning strokes). The current induced in a net-

work or electric installation by an indirect stroke. *See:* **direct-stroke protection (lightning).**   64

induced electrification. The separation of charges of opposite sign onto parts of a conductor as a result of the proximity of charges on other objects. *Note:* The charge on a portion of such a conductor is often called an induced charge or a bound charge.   210

induced emission (laser-maser). *See:* **stimulated emission.**

induced field current (synchronous machine). The current that will circulate in the field winding (assuming the circuit is closed) due to transformer action when an alternating voltage is applied to the armature winding, for example, during starting of a synchronous motor.   63

induced-potential tests (electric power). Dielectric tests in which the test voltages are suitable-frequency alternating voltages, applied or induced between the terminals.   91

induced voltage (1) (general). A voltage produced around a closed path or circuit by a change in magnetic flux linking that path. *See:* **Faraday's law.** *Notes:* (A) Sometimes more narrowly interpreted as a voltage produced around a closed path or circuit by a time rate of change in magnetic flux linking that path when there is no relative motion between the path or circuit and the magnetic flux. (B) A single-phase stator winding energized with alternating current, produces a pulsating magnetic field which causes a voltage to be induced in a blocked rotor circuit, and the same magnetic field may be interpreted in terms of two magnetic fields of constant amplitude traveling in opposite directions around the air gap, causing two voltages to be generated in a blocked rotor circuit. (C) Whether a voltage is defined as being induced or generated is often simply a matter of point of view. *See:* **Faraday's law; generated voltage; induction motor.**   63
(2) **(lightning strokes).** The voltage induced on a network or electric installation by an indirect stroke. *See:* **direct-stroke protection (lightning).**   64
(3) **(corona measurement).** Voltage which is induced in a winding. Induced voltage also includes voltage applied across a winding.   375

induced voltage tests (power and distribution transformer). Induced voltage tests are dielectric tests on transformer windings in which the appropriate test voltages are developed in the windings by magnetic induction. *Note:* Power for induced voltage tests is usually supplied at higher-than-rated frequency to avoid core saturation and excessive excitation current.   53

inducing current (power fault effects). The current that would flow in a single conductor electric supply line with ground return to give the same value of induced voltage in a telecommunication line (at a particular separation) as the vectorial sum of all voltages induced by the various currents in in the inductive exposure as a result of a line-to-ground fault.   404

inductance. The property of an electric circuit by virtue of which a varying current induces an electromotive force in that circuit or in a neighboring circuit.   197

inductance coil. *See:* **inductor.**

inductance, effective (1) (general). The imaginary part of an inductive impedance divided by the angular frequency.   185
(2) **(winding).** The self-inductance at a specified frequency and voltage level, determined in such a manner as to exclude the effects of distributed capacitance and other parasitic elements of the winding but not the parasitic elements of the core.   197

inductance grounded (system grounding). Grounded through impedance, the principal element of which is inductance. *Note:* The conditions of an inductance-grounded system are that

$$X_0 / X_1$$

lie within the range of 3-10 and

$$R_0 / X_0 \leq 1$$

The ground-fault current becomes 25 percent or more of the three-phase fault current. Inductance grounding becomes 'effective grounding' if

$$X_0 / X_1$$

is reduced to 3 or less.   152

induction coil. (1) A transformer used in a telephone set for interconnecting the transmitter, receiver, and line terminals. (2) A transformer for converting interrupted direct current into high-voltage alternating current. *See:* **telephone station.**   341

induction compass. A device that indicates an aircraft's heading, in azimuth. Its indications depend on the current generated in a coil revolving in the earth's magnetic field.   328

induction-conduction heater. A heating device in which electric current is conducted through but is restricted by induction to a preferred path in a charge. *See:* **induction heating.**   14

induction coupling (electric coupling). An electric coupling in which torque is transmitted by the interaction of the magnetic field produced by magnetic poles on one rotating member and induced currents in the other rotating member. *Note:* The magnetic poles may be produced by direct current excitation, permanent magnet excitation, or alternating current excitation. The induced currents may be carried in a wound armature or squirrel cage, or may appear as eddy currents.   416

induction cup (of a relay) (power switchgear). A form of relay armature in the shape of a cylinder with a closed end that develops operating torque by its location within the fields of electromagnets that are excited by the input quantities.   127, 103

induction cylinder (of a relay) (power switchgear). A form of relay armature in the shape of an open-ended cylinder that develops operating torque by its location within the fields of electromagnets that are excited by the input quantities.   127, 103

induction disk (of a relay) (1) (power switchgear). A form of relay armature in the shape of a disk that usually serves the combined function of providing an operating torque by its location within the fields of an electromagnet excited by the input quantities and a restraining force by motion within the field of a permanent magnet. 127, 103

(2) (utility consumer interconnections). A thin circular (or spiraled) disk of nonmagnetic conducting material in which eddy currents are produced to create torque about an axis of rotation. 128

induction factor, A_L (magnetic core testing). Under stated conditions, the self inductance that a coil of specified shape and dimensions placed on the core in a given position should have, if it consisted of one turn.

$$A_L = \frac{L}{N^2}$$

A_L = Induction factor (henrys/turns^2)
L = Self inductance of the coil on the core, in henrys
N = Number of turns of the coil. 165

induction frequency converter. A wound-rotor induction machine in which the frequency conversion is obtained by induction between a primary winding and a secondary winding rotating with respect to each other. *Notes:* (1) The secondary winding delivers power at a frequency proportional to the relative speed of the primary magnetic field and the secondary member. (2) In case the machine is separately driven, this relative speed is maintained by an external source of mechanical power. (3) In case the machine is self-driven, this relative speed is maintained by motor action within the machine obtained by means of additional primary and secondary windings with number of poles differing from the number of poles of the frequency-conversion windings. In special cases one secondary winding performs the function of two windings, being short-circuited with respect to the poles of the driving primary winding and open-circuited with respect to the poles of the primary-conversion winding. *See:* **converter.** 63

induction furnace. A transformer of electric energy to heat by electromagnetic induction. 328

induction generator (1) (rotating machinery). An induction machine, when driven above synchronous speed by an external source of mechanical power, used to convert mechanical power to electric power. *See:* **asynchronous machine.** 63

(2) (electric installations on shipboard). An induction machine driven above synchronous speed by an external source of mechanical power. 3

induction heater (interference terminology). A device for causing electric current to flow in a charge of material to be heated. Types of induction heaters can be classified on the basis of frequency of the induced current, for example, a low-frequency induction heater usually induces power-frequency current in the charge: a medium-frequency induction heater induces currents of frequencies between 180 and 540 hertz: a high-frequency induction heater induces currents having frequencies from 1000 hertz upward. 60

induction heater or furnace, core type. *See:* **core type induction heater or furnace.**

induction heating (electrical heating systems). The generation of heat in any conducting material by means of magnetic flux-induced currents. 476

induction instrument. An instrument that depends for its operation on the reaction between a magnetic flux (or fluxes) set up by one or more currents in fixed windings and electric currents set up by electromagnetic induction in movable conducting parts. *See:* **instrument.** 328

induction loop (of a relay) (power switchgear). A form of relay armature consisting of a single turn or loop that develops operating torque by its location within the fields of electromagnets that are excited by the input quantities. 127, 103

induction loudspeaker. A loudspeaker in which the current that reacts with the steady magnetic field is induced in the moving member. *See:* **loudspeaker.** 176

induction machine (electric installations on shipboard). An asynchronous ac machine which comprises a magnetic circuit interlinked with two electric circuits, or sets of circuits, rotating with respect to each other and in which power is transferred from one circuit to another by electromagnetic induction. Examples of induction machines are induction generators, induction motors, and certain types of frequency converters and phase converters. 3

induction motor (electric installations on shipboard). An ac motor in which a primary winding on one member (usually the stator) is connected to the power source and a polyphase secondary winding or a squirrel-cage secondary winding on the other member (usually the rotor) carries induced current. 3

induction-motor meter. A motor-type meter in which the rotor moves under the reaction between the currents induced in it and a magnetic field. *See:* **electricity meter (meter).** 212

induction regulator (electrical heating applications to melting furnaces and forehearths in the glass industry). A regulating transformer, having a primary winding in shunt and a secondary winding in series with a circuit for gradually adjusting the voltage, phase relation, or both, of the circuit by changing the relative magnetic coupling of the existing (primary) and series (secondary) windings. 520

induction ring heater. A form of core-type induction heater adapted principally for heating electrically conducting charges of ring or loop form, the core being open or separable to facilitate linking the charge. *See:* **induction heater.** 14, 114

induction vibrator. A device momentarily connected between the airplane direct-current supply and the primary winding of the magneto, thus converting the magneto to an induction coil. *Note:* It provides energy to the spark plugs of an aircraft engine during its starting period. 328

induction voltage regulator (power and distribution transformer). A regulating transformer having a primary winding in shunt and a secondary winding in series with a circuit, for gradually adjusting the voltage or the phase relation, or both, of the circuit by changing the relative magnetic coupling of the exciting (primary) and series (secondary) windings. 53

induction watthour meter. A motor-type meter in which currents induced in the rotor interact with a magnetic field to produce the driving torque.
212

inductive assertion method (software). A proof of correctness technique in which assertions are written describing program inputs, outputs, and intermediate conditions, a set of theorems is developed relating satisfaction of the input assertions to satisfaction of the output assertions, and the theorems are proved to be true. *See:* **assertions; input assertions; output assertions; program; proof of correctness.** 434

inductive coordination (electric supply and communication systems). The location, design, construction, operation, and maintenance in conformity with harmoniously adjusted methods that will prevent inductive interference. 328

inductive coupling (ground system) (communication circuits). The association of two or more circuits with one another by means of inductance mutual to the circuits or the mutual inductance that associates the circuits. *Note:* This term, when used without modifying words, is commonly used for coupling by means of mutual inductance, whereas coupling by means of self-inductance common to the circuits is called direct inductive coupling. **(inductive coordination practice).** The interrelation of neighboring electric supply and communication circuits by electric or magnetic induction, or both. 313

inductive exposure. A situation of proximity between electric supply and communication circuits under such conditions that inductive interference must be considered. *See:* **inductive coordination.** 328

inductive gap (microwave receiver protector)(nonlinear, active, and nonreciprocal waveguide components). In cell-type waveguide receiver protectors, this is the slot width or distance between iris plates. *See:* **resonant gap.** 530

inductive influence (voice-frequency electrical-noise test). (1) (electric supply circuit with is associated apparatus.) Those characteristics that determine the character and the intensity of the inductive field that it produces. (2) One of the three factors that may be present when there is noise on a telephone communication circuit. It is a measure of the interfering effect of the power system. 376

inductive interference (electric supply and communication systems). An effect, arising from the characteristics and inductive relations of electric supply and communication systems, of such character and magnitude as would prevent the communication circuits from rendering service satisfactorily and economically if methods of inductive coordination were not applied. *See:* **inductive coordination.** 328

inductively coupled circuit. A coupled circuit in which the common element is mutual inductance. *See:* **network analysis.** 210

inductive microphone. *See:* **inductor microphone.**

inductive neutralization (shunt neutralization) (coil

neutralization). A method of neutralizing an amplifier whereby the feedback susceptance due to an interelement capacitance is canceled by the equal and opposite susceptance of an inductor. *See:* **amplifier; feedback.** 111

inductive residual voltage (Hall-effect devices). The electromotive force induced in the loop formed by the Hall voltage leads and the conductive path through the Hall plate by a varying magnetic flux density, when there is no control current. 107

inductive susceptiveness (communication circuit with its associated apparatus). Those characteristics that determine, so far as such characteristics are able to determine, the extent to which the service rendered by the circuit can be adversely affected by a given inductive field. *See:* **inductive coordination.** 328

inductor (1) (general). A device consisting of one or more associated windings, with or without a magnetic core, for introducing inductance into an electric circuit.

(2) (railway practice). A roadway element consisting of a mass of iron with or without a winding, that acts inductively on the vehicle apparatus of the train control, train stop, or cab signal system. 341

inductor alternator. An inductor machine for use as a generator, the voltage being produced by a variation of flux linking the armature winding without relative displacement of field magnet or winding and armature winding. 63

inductor, charging. An inductive component used in the charging circuit of a pulse-forming network.
137

inductor circuit (railway practice). A circuit including the inductor coil and the two lead wires leading therefrom taken through roadway signal apparatus as required. 328

inductor dynamotor (rotating machinery). A dynamotor inverter having toothed field poles and an associated stationary secondary winding for conversion of direct current to high-frequency alternating current by inductor-generator action. *See:* **convertor.** 63

inductor frequency-converter (rotating machinery). An inductor machine having a stationary input alternating-current winding, which supplies the excitation, and a stationary output winding of a different number of poles in which the output frequency is induced through change in field reluctance by means of a toothed rotor. *Note:* If the machine is separately driven, the rotor speed is maintained by an external source of mechanical power. If the machine is self-driven, the primary winding and rotor function as in a squirrel-cage induction motor or a reluctance motor. *See:* **convertor.** 63

inductor machine (rotating machine). A synchronous machine in which one member, usually stationary, carries main and exciting windings effectively disposed relative to each other, and in which the other member, usually rotating, is without windings but carries a number of regular projections. (Permanent magnets may be used instead of the exciting winding). 63

inductor microphone (inductive microphone). A moving-conductor microphone in which the moving element is in the form of a straight-line conductor. *See:* **microphone.** 328

inductor type synchronous generator (electric installations onshipboard). A generator in which the field coils are fixed in magnetic position relative to the armature conductors, the electromotive forces being produced by the movement of masses of magnetic material. 3

inductor-type synchronous motor. An inductor machine for use as a motor, the torques being produced by forces between armature magnetomotive force and salient rotor teeth. *Note:* Such motors usually have permanent-magnet field excitation, are built in fractional-horsepower ratings, frames and operate at low speeds, 300 revolutions per minute or less. 63

industrial brush (rotating machinery). A brush having a cross-sectional area (width x thickness) of more than 1.4 square inch or a length of more than 1 1.2 inches, but larger than a fractional-horsepower brush. *See:* **brush (1) (electric machines).** 63

industrial control. Broadly, the methods and means of governing the performance of an electric device, apparatus, equipment, or system used in industry. 206

industrial electric locomotive. An electric locomotive, used for industrial purposes, that does not necessarily conform to government safety regulations as applied to railroads. *Note:* This term is generally applied to locomotives operating in surface transportation and does not include mining locomotives. A prefix diesel-electric, gas-electric, etcetera, may replace the word electric. *See:* **electric locomotive; industrial electronics.** 328

industrial process supervisory system. A supervisory system that initiates signal transmission automatically upon the occurrence of an abnormal or hazardous condition in the elements supervised, which include heating, air-conditioning, and ventilating systems, and machinery associated with industrial processes. *See:* **protective signaling.** 328

inertance (automatic control). A property expressible by the quotient of a potential difference (temperature, sound pressure, liquid level) divided by the related rate of change of flow: the thermal or fluid equivalent of electrical inductance or mechanical moment of inertia. *See:* **control system, feedback.** 56

inert gas-pressure system (power and distribution transformer). A system in which the interior of the tank is sealed from the atmosphere, over the temperature range specified, by means of a positive pressure of inert gas maintained from a separate inert gas source and reducing valve system. 53

inertia compensation (industrial control). The effect of a control function during acceleration or deceleration to cause a change in motor torque to compensate for the driven-load inertia. *See:* **control system, feedback.** 225, 206

inertia constant (machine). The energy stored in the rotor when operating at rated speed expressed as kilo-watt-seconds per kilovolt-ampere rating of the machine. *Note:* The inertia constant is

$$h = \frac{0.231 \times Wk^2 \times n^2 \times 10^{-6}}{kVA}$$

where

h = inertia constant in kilowatt-seconds per kilovolt-ampere

Wk^2 = moment of inertia in pound-feet^2

n = speed in revolutions per minute

kVA = rating of machine in kilovolt-amperes

See: **asynchronous machine.** 63

inertialess scanning. *See:* **electronic scanning.**

inertial navigation equipment. A type of dead-reckoning navigation equipment whose operation is based upon the measurement of accelerations: accelerations are sensed dynamically by devices stabilized with respect to inertial space, and the navigational quantities (such as vehicle velocity, angular orientation, or positional information) are determined by computers and/or other instrumentation. *See:* **navigation.** 187, 13

inertial navigator. A self-contained, dead-reckoning navigation aid using inertial sensors, a reference direction, and initial or subsequent fixes to determine direction, distance, and speed: single integration of acceleration provides speed information and a double integration provides distance information. *See:* **navigation.** 187, 13

inertial sensor. A position, attitude or motion sensor whose references are completely internal, except possibly for initialization. 46

inertial space (navigation). A frame of reference defined with respect to the fixed stars. *See:* **navigation.** 187, 13

inertia relay. A relay with added weights or other modifications that increase its moment of inertia in order either to slow it or to cause it to continue in motion after the energizing force ends. *See:* **relay.** 259

infant mortality failures (station control and data acquisition). A characteristic pattern of failure, sometimes experienced with new equipments which may contain marginal components, wherein the number of failures per unit of time decrease rapidly as the number of operating hours increase. A burn-in period may be utilized to age (or mature) an equipment to reduce the number of marginal components. 403

infinite multiplication factor (k) (power operations). The ratio of the average number of neutrons produced in each generation of nuclear fissions to the average number of corresponding neutrons absorbed. Since neutron leakage out of the system is ignored, k is the effective multiplication factor for an infinitely large assembly. 516

infix notation (mathematics of computing). A method of forming mathematical expressions in which each operator is written between its operands and the expression is interpreted subject to rules of operator precedence and grouping symbols. For example, A added to B and the result multiplied by C is represent-

ed as $(A+B)\cdot C$. *See:* **postfix notation; prefix notation.**
564

inflection point (tunnel-diode characteristic). The point on the forward current-voltage characteristic at which the slope of the characteristic reaches its most negative value. *See:* **peak point (tunnel-diode characteristic).**
315, 191

inflection-point current (tunnel-diode characteristic). The current at the inflection point. *See:* **peak point (tunnel-diode characteristic).**
315, 191

inflection-point emission current (electron tubes). That value of current on the diode characteristic for which the second derivative of the current with respect to the voltage is zero. *Note:* This current corresponds to the inflection point of the diode characteristic and is, under suitable conditions, an approximate measure of the maximum space-charge-limited emission current.
125

inflection-point voltage (tunnel-diode characteristic). The voltage at which the inflection point occurs. *See:* **peak point (tunnel-diode characteristic).**
315, 191

influence (1) (upon an instrument) (specified variable or condition). The change in the indication of the instrument caused solely by a departure of the specified variable or condition from its reference value, all other variables being held constant. *See:* **accuracy rating of an instrument.**
280

(2) (upon a recording instrument). The change in the recorded value caused solely by a departure of the specified variable or condition from its reference value, all other variables being held constant. *Note:* If the influences in any direction from reference conditions are not equal, the greater value applies.
294

information (1) (general). The meaning assigned to data by known conventions.
255, 77

(2) (nuclear power generating station). Data describing the status and performance of the plant.
358

information center (IC)(computer applications). (1) A user-oriented computer system that provides nontechnical users direct access to data and software for information processing tasks such as report generation, data modeling and manipulation, and word processing. *See:* **decision-support services.** (2) Support personnel for a computer system as in (1).
571

information content (message or a symbol from a source). The negative of the logarithm of the probability that this particular message or symbol will be emitted by the source. *Notes:* (1) The choice of logarithmic base determines the unit of information content. (2) The probability of a given message or symbol's being emitted may depend on one or more preceding messages or symbols. (3) The quantity has been called self-information. *See:* **bit; hartley; information theory.**
160

information content, average. *See:* **average information content.**

information display channel (accident monitoring instrumentation). An arrangement of electrical and mechanical components or modules, or both, from measured process variable to display device as required to

sense and display conditions within the generating stations.
421, 361

information display channel failure (accident monitoring instrumentation). A situation where the display disagrees, in a substantive manner, (that is, the maximum error within which the information must be conveyed to the operator has been exceeded), with the conditions or status of the plant.
421

information field (I)(logical link control). The sequence of octets occurring between the control field and the end of the logical link control (LLC) protocol data unit (PDU). The information field contents of I, TEST, and unnumbered information (UI) PDUs are not interpreted at the LLC sublayer.
585

information hiding (software). The technique of encapsulating software design decisions in modules in such a way that the module's interfaces reveal as little as possible about the module's inner workings; thus, each module is a 'black box' to the other modules in the system. The discipline of information hiding forbids the use of information about a module that is not in the module's interface specification. *See:* **encapsulation; encapsulating software design; interfaces; interface specification; module; system.**
434

information, mutual. *See:* **mutual information.**

information theory. (1) In the narrowest sense, is used to describe a body of work, largely about communciation problems but not entirely about electrical communication, in which the information measures are central. (2) In a broader sense it is taken to include all statistical aspects of communication problems, including the theory of noise, statistical decision theory as applied to detection problems, and so forth. *Note:* This broader field is sometimes called 'statistical communication theory'. (3)In a still broader sense its use includes theories of measurement and observation that use other measures of information, or none at all, and indeed work on any problem in which information, in one of its colloquial senses, is important.
415

information transfer (data transmission). The final result of data transmission from a data source to a data sink. The information transfer rate may or may not be equal to the transmission modulation rate.
59

infrared (IR) (fiber optics). The region of the electromagnetic spectrum between the long-wavelength extreme of the visible spectrum (about 0.7 μm) and the shortest microwaves (about 1 mm).
433

infrared radiation (illuminating engineering). For practical purposes any radiant energy within the wavelength range 770 to 10^6 nm (nanometers) is considered infrared energy. *See:* **regions of electromagnetic spectrum.**
167

infrared radiation (laser-maser). Electromagnetic radiation with wavelengths which lie within the range 0.7 μm to 1 mm.
363

inherent transient recovery voltage (transient recovery voltage)power switchgear). The TRV (transient recovery voltage) produced by the circuit with no modifying effect of the switching device. *Note:* The magnitude of the TRV for a given circuit and voltage

is affected by the degree of current asymmetry. Symmetrical current usually produces the highest TRV magnitudes and is used as the basis for TRV-rated values. An asymmetrical current normally reduces the TRV magnitude from the symmetrical current case.
486, 487, 488, 577

inhibit (supervisory control, data acquisition, and automatic control)(station control and data acquisition). To prevent a specific event from occurring.
403, 570

inhibited oil (power and distribution transformer). Mineral transformer oil to which a synthetic oxidation inhibitor has been added. 53

inhibitor (insulating oil). Any substance that when added to an electrical insulating fluid retards or prevents undesirable reactions. 461

inhomogeneous line-broadening (laser-maser). An increase of the width of an absorption or emission line, beyond the natural linewidth, produced by a disturbance (for example, strain, imperfections, etcetera) which is not the same for all of the source emitters.
363

initial condition (analog computers). The value of a variable at the start of computation. A more restricted definition refers solely to the initial value of an integrator. Also used as a synonym for the computer-control state "reset." *See:* **reset.** 9

initial luminous exitance (illuminating engineering). The density of luminous flux leaving a surface within an enclosure before interreflections occur. *Note:* For light sources this is the luminous exitance as defined herein. For nonself-luminous surfaces it is the reflected luminous exitance of the flux received directly from sources within the enclosure or from daylight. *See:* **luminous exitance.** 167

initial symmetrical ground fault current (safety in ac substation grounding). The maximum root-mean-square (rms) value of symmetrical fault current after the instant of a ground fault initiation. As such, it represents the rms value of the symmetrical component in the first half-cycle of a current wave that develops after the instant of fault at time zero. Generally,

$$I_{f(0+)} = 3I_0''$$

where

$I_{f(0+)}$ = initial symmetrical ground fault current
I_0'' = rms value of zero-sequence symmetrical current that develops immediately after the instant of fault initiation, that is, reflecting the subtransient reactances of rotating machines contributing to the fault.

Note: Elsewhere in the guide, this initial symmetrical fault current is shown in an abbreviated notation, as If, or is referred to only as 3i0. The underlying reason for this latter notation is that, for purposes of this guide, the initial symmetrical fault current is assumed to remain constant for the entire duration of the fault.
563

initiating relay (power switchgear). A programming relay whose function is to constrain the action of dependent relays until after it has operated. 103

injected current (ac high-voltage circuit breakers). The current which flows through the test circuit breaker from the voltage source of a current injection circuit when this circuit is applied to the test circuit breaker. 426

injected-current frequency (ac high-voltage circuit breaker). The frequency of the injected current.
426

injection fiber. *See:* **launching fiber.** 433

injection laser diode (ILD) (fiber optics). A laser employing a forward-biased semiconductor junction as the active medium. *Syn:* **diode laser; semiconductor laser.** *See:* **active laser medium; chirping; laser; superradiance.** 433

injection time (ac high-voltage circuit breaker). The time with respect to the power frequency current zero when the voltage circuit is applied. 426

in line (monitoring radioactivity in effluents). A system where the detector assembly is adjacent to or immersed in the total effluent stream. 559

inner product. *See:* **polarization vector, Note 2.**

in-phase spring rate (dynamically tuned gyro) (inertial sensor). The residual difference, in a dynamically tuned gyro, between the dynamically induced spring rate and the flexure spring rate. 46

input (1) (data transmission). (A) The data to be processed. (B) The state or sequence of states occurring on a specified input channel. (C) The device or collective set of devices used for bringing data into another device. (D) A channel for impressing a state on a device or logic element. (E) The process of transferring data from an external storage to an internal storage. 59
(2) (to a relay) (power switchgear). A physical quantity or quantities to which the relay is designed to respond. *Notes:* (A) A physical quantity that is not directly related to the prescribed response of a relay, (though necessary, to or in some way affecting the relay operation,) is not considered part of input. (B) Time is not considered a relay input, but it is a factor in performance. 103

input assertion (software). A logical expression specifying one or more conditions that program inputs must satisfy in order to be valid. *See:* **program.** 434

input-axis misalignment (gyro, accelerometer) (inertial sensor). The angle between an input axis and its associated input reference axis when the device is at a null condition. (The magnitude of this angle is unambiguous, but when components are reported, the convention should always be identified. IEEE standards use both direction cosines and right-handed Euler angles, depending on the principal field of application. Other conventions, differing both in signs and designation of axes, are sometimes used. 46

input data (test pattern language). The binary data that is written into a memory array. It is identified by the symbol 'D'. 463

input impedance (1) (analog computers). In an analog computer, a passive network connected between the input terminal or terminals of an operational amplifier and its summing junction. 9

(2) (at a transmission line port) (waveguide). The impedance at the transverse plane of the port. *Note:* This impedance is independent of the generator impedance.   267

input limiter. *See:* **limiter circuit.**

input-output characteristic (transmission performance of telephone sets). Electric or acoustic output level as a function of the input level.   491

input power (total)(converters having dc input)(self-commutated converters). The mean value of the instantaneous power into the input terminals, taken over one period of the ripple component. *Note:* If either the voltage or the current is ripple-free, the dc power is the total power.   584

input pulse shape (pulse transformers). Current pulse or source voltage pulse applied through associated impedance. The shape of the input pulse is described by a current- or voltage-time relationship and is defined with the aid of the Input Pulse Shape figure below. *Note:* A general amplitude quantity is designated by A, which may be current I or voltage V. *See:* **pulse amplitude; rise time; pulse duration; fall time; trailing edge amplitude; tilt; overshoot; backswing; return swing; rolloff; ringing; leading edge linearity; quiescent value; leading edge; pulse top; trailing edge; output pulse shape; voltage-time product; voltage-time product rating.**   589

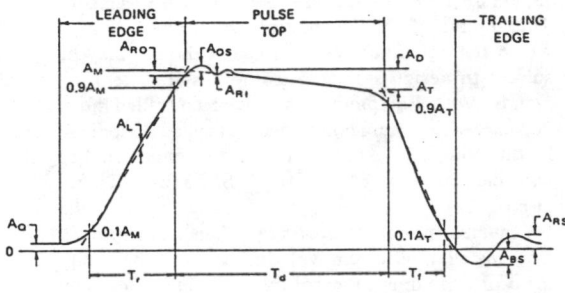

Input Pulse Shape

input signal (hydraulic turbines). A control sign injected at any point into a control system.   8

input signal level sensitivity (spectrum analyzer). The input signal level that produces an output equal to twice the value of the average noise alone. This may be power or voltage relationship, but must be stated so. *See:* **equivalent input noise sensitivity; sensitivity.**   390

inrush current (1) (electronics power transformer). The maximum root-mean-square or average current value, determined for a specified interval, resulting from the excitation of the transformer with no con-

nected load, and with essentially zero source impedance, and using the minimum primary turns tap available and its rated voltage.   95

(2) (packaging machinery). Of a solenoid or coil, the steady-state current taken from the line with the armature blocked in the rated maximum open position.   429

insertion (series capacitors). The opening of the capacitor bypass device to place the series capacitor in-service with or without load current flowing.   474

insertion current (series capacitors). The steady-state root-mean-square current that flows through the series capacitor after the bypass device has opened.   474

insertion gain (data transmission). Resulting from the insertion of a transducer in a transmission system, the ratio of power delivered to that part of the system following the transducer to the power delivered to that same part before insertion.   59

insertion loss (1)(audible noise measurement). In the context of ANSI/IEEE Std 656-1985, the insertion loss of a component (for example, a microphone windscreen) is the difference in decibels between the sound-pressure level measured before the insertion of the component and the sound-pressure level measured after the insertion of the component, provided that the source of the sound and all other conditions remain unchanged. The effect of the added component on the frequency response of a sound-measurement system should be considered and recorded.   462

(2) (data transmission). Resulting from the insertion of a transducer in a transmission system, the ratio of (A) the power delivered to that part of the system following the transducer, before insertion of the transducer, to (B) the power delivered to that same part of the system after insertion of the transducer. *Note:* If the input or output power, or both, consist of more than one component, such as multifrequency signal or noise, then the particular components used and their weighting are specified.   59

(3) (fiber optics). The total optical power loss caused by the insertion of an optical component such as a connector, splice, or coupler.   433

insertion loss (or gain). *See:* **general insertion loss (or gain); matched insertion loss (or gain); matched generator insertion loss (or gain); matched load insertion loss (or gain).**

insertion voltage (series capacitors). The steady-state root-mean-square voltage appearing across the series capacitor upon the interruption of the bypass current with the opening of the bypass device.   474

in service (electric generating unit reliability, availability, and productivity)(power system measurement). The state in which a unit is electrically connected to the system.   567, 432

in-service forced derated hours (IFDH)(electric generating unit reliability, availability, and productivity). The in-service hours during which a Class 1, 2, or 3 unplanned derating was in effect.   567

in-service maintenance derated hours (IMDH)(elec-

tric generating unit reliability, availability, and productivity). The in-service hours during which a Class 4 unplanned derating was in effect.   567

in-service planned derated hours (IPDH)(electric generating unit reliability, availability, and productivity). The in-service hours during which a basic or extended planned derating was in effect.   567

in-service test (metering). A test made during the period that the meter is in service. It may be made on the customer's premises without removing the meter from its mounting, or by removing the meter for test, either on the premises or in a laboratory or meter shop.
   212

in-service unit derated hours (IUNDH)(electric generating unit reliability, availability, and productivity). The in-service hours during which a unit derating was in effect.   567

in-service unplanned derated hours (IUDH)(electric generating unit reliability, availability, and productivity). The in-service hours during which an unplanned derating was in effect.   567

inside air temperature. *See:* **average inside air temperature.**

inside top air temperature (power and distribution transformer). The temperature of the air inside a dry-type transformer enclosure, measured in the space above the core and coils.   53

in sight from (within sight from, within sight) (National Electrical Code). Where this Code specifies that one equipment shall be "in sight from," "within sight from", or "within sight", etcetera, of another equipment, one of the equipments specified shall be visible and not more than 50 ft distant from the other.
   256

inspection (1) (nuclear power quality assurance). Examination or measurement to verify whether an item or activity conforms to specified requirements.
   417

(2) (software). (1) A formal evaluation technique in which software requirements, design, or code are examined in detail by a person or group other than the author to detect faults, violations of development standards, and other problems. (2) A phase of quality control which by means of examination, observation or measurement determines the conformance of materials, supplies, components, parts, appurtenances, systems, processes or structures to predetermined quality requirements. *See:* **code; code audit; design; fault; process; system; walk-through.**   434

inspection, meter installation. *See:* **meter installation inspection.**

inspector (nuclear power quality assurance). A person who performs inspection activities to verify conformance to specific requirements.   417

installation and checkout phase (software). The period of time in the software life cycle during which a software product is integrated into its operational environment and tested in this environment to ensure that it performs as required. *See:* **operational; software life cycle; software product.**   434

installation and checkout phase (software verification and validation plans). The period of time in the software life cycle during which a software product is integrated into its operational environment and tested in this environment to ensure that it performs as required.   511

installed incremental transfer capability (power operations). The amount of power, incremental above normal base power transfers, that can be transferred over the transmission network without giving consideration to the effect of transmission facility outages. All facility loadings are within normal ratings and all voltages are within normal limits.   516

installed life (1)(electric penetration assemblies). The interval of time from installation to permanent removal from service, during which the electric penetration assembly is expected to perform its required function(s). *Note:* Components of the assembly may require periodic replacement; thus, the installed life of such components is less than the installed life of the assembly.   493

(2)(safety systems equipment in nuclear power generating stations)(valve actuators). The interval from installation to removal, during which the equipment or component thereof may be subject to design service conditions and system demands. *Note:* Equipment may have an installed life of 40 years with certain components changed periodically; thus, the installed life of the changed components would be less than 40 years.   535, 120, 492

(3) (class 1E static battery chargers and inverters). The interval from installation to removal, during which the equipment or component thereof may be subject to design service conditions and system demands. *Note:* Equipment may have an installed life of 20 years with certain components changed periodically; thus, the installed life of the components would be less than 20 years (ANSI/IEEE Std 382-1972) (revised 1980).   408

(4) (nuclear power generating stations). The interval from installation to removal, during which the equipment or component thereof may be subject to design service conditions and system demands. (See IEEE Std 323-1974.) (Revised 1983). *Note:* Equipment may have an installed life of 40 years with certain components changed periodically; thus, the installed life of the components would be less.   440

(5) (valve actuators). The interval, from installation to removal, during which the equipment or component thereof may be subject to service conditions and system demands.   142

installed nameplate capacity (electric generating unit reliability, availability, and productivity). The full-load continuous gross capacity of a unit under specified conditions, as calculated from the electric generator nameplate based on the rated power factor. *Note:* The nameplate rating of the electric generator may not be indicative of the unit maximum or dependable capacity, since some other item or equipment (such as the turbine) may limit unit output.   567

installed reserve (power operations). The reserve capacity installed on a system.   516

instantaneous (power switchgear). A qualifying term indicating that no delay is purposely introduced in the action of the device.   103

instantaneous automatic gain control (1)(radar). A quick-acting automatic gain control that responds to variations of mean clutter level, or jamming over different range or angular regions, avoiding receiver saturation.   13

(2) (IAGC) (radar) (nonlinear, active, and nonreciprocal waveguide components). A fast-acting automatic gain control that responds to variations of received signal, avoiding receiver saturation.   530

instantaneous demand (power operations). The load at any instant.   516

instantaneous frequency (1) (data transmission). The time rate of change of the angle of an angle-modulated wave. *Note:* If the angle is measured in radians, the frequency in hertz is the time rate of change of the angle divided by 2.   59

(2) (radio wave propagation). Of a traveling wave,

$$\frac{1}{2\pi}$$

times the time rate of change of phase of the wave.   146

instantaneous overcurrent or rate-of-rise relay (50) (power system device function numbers). A relay that functions instantaneously on an excessive value of current or on an excessive rate of current rise.   402

instantaneous peak power (waveguide). The maximum instantaneous power passing through the transverse section of a waveguide during the interval of interest.   267

instantaneous phase or ground trip element. *See:* instantaneous; direct-acting overcurrent trip device.   103

instantaneous Poynting vector (radio wave propagation). Of an electro-magnetic wave, the vector product of the electric and magnetic field vectors. Its integral over a surface is the instantaneous electro-magnetic power flow through the surface.   146

instantaneous suppression with automatic current regulation (thyristor). A combination of instantaneous trip or suppression and current regulation in which suppression is followed immediately by a regulated current.   445

instantaneous trip (National Electrical Code). (As applied to Circuit Breakers.) A qualifying term indicating that no delay is purposely introduced in the tripping action of the circuit breaker.   256

instantaneous trip or suppression (thyristor). The means to sense an overload and reduce the output current to zero, as fast as practicable.   445

instant of chopping (high voltage testing) (chopped impulses) The instant of chopping is the instant when the initial discontinuity occurs.   150

instant start fluorescent lamp (illuminating engineer-

ing). A fluorescent lamp designed for starting by a high voltage without preheating of the electrodes.   167

instruction (1)(programmable digital computer systems in safety systems of nuclear power generating stations). A meaningful expression in a computer programming language that specifies an operation to a digital computer.'   554

(2)(software). (A) A program statement that causes a computer to perform a particular operation or set of operations. (B) In a programming language, a meaningful expression that specifies one operation and identifies its operands. *See:* operand; program; programming language.   434

instruction set (software). The set of instructions of a computer, of a programming language, or of the programming languages in a programming system. *See:* computer; instruction; programming language; system.   434

instruction set architecture (software). An abstract machine characterized by an instruction set. *See:* abstract machine; instruction set.   434

instruction trace. *See:* trace.

instrument (1) (plutonium monitoring). A complete system designed to quantify a particular type of ionizing radiation.   413

(2) (radiation protection). A complete system designed to quantify one or more particular ionizing radiation or radiations.   399

instrumentation. *See:* program instrumentation.

instrumentation tool (software). A software tool that generates and inserts counters or other probes at strategic points in another program to provide statistics about program execution such as how thoroughly the program's code is exercised. *See:* code; execution; program; software tool.   434

instrument landing system (1) (general). A generic term for a system which provides the necessary lateral, longitudinal and vertical guidance in an aircraft for a low approach or landing. *See:* ILS.   13

(2) (ILS). An internationally adopted instrument landing system for aircraft, consisting of a very-high frequency localizer, an ultra-high-frequency glide slope, and 75-megahertz markers. *See:* instrument landing system reference point.   187

instrument landing system marker beacon (electronic navigation). *See:* boundary marker; navigation.

instrument landing system reference point (electronic navigation). A point on the centerline of the instrument landing system runway designated as the optimum point of contact for landing: in standards of the International Civil Aviation Organization this point is from 500 to 1000 feet from the approach end of the runway. *See:* navigation.   187

instrument multiplier. A particular type of series resistor that is used to extend the voltage range beyond some particular value for which the instrument is already complete. *See:* auxiliary device to an instrument; voltage-range multiplier (recording instrument).   280

instrument relay. A relay whose operation depends

upon principles employed in measuring instruments such as the electrodynamometer, iron vane, D'Arsonval galvanometer, and moving magnet. *See: relay.*
259

instrument shunt (direct current instrument shunts). A particular type of resistor designed to be connected in parallel with the measuring device to extend the current range beyond some particular value for which the instrument is already complete. 527

instrument switch (power switchgear). One used to connect or disconnect an instrument, or to transfer it from one circuit or phase to another. Examples: ammeter switch; voltmeter switch. 103

instrument terminals (direct current instrument shunts). Those terminals which provide a voltage drop proportional to the current in the shunt and to which the instrument or other measuring device is connected. 527

instrument transformer (1) (instrument transformer). A transformer that is intended to reproduce in its secondary circuit, in a definite and known proportion, the current or voltage of its primary circuit with the phase relations substantially preserved. 394

(2) (power and distribution transformer). A transformer which is intended to reproduce in its secondary circuit, in a definite and known proportion, the current or voltage of its primary circuit, with the phase relations and waveform substantially preserved. *See: continuous-thermal-current rating factor (RF); transformer correction factor; true ratio; marked ratio; ratio correction factor (RCF); percent ratio; percent ratio correction of an instrument transformer; phase angle of an instrument transformer; phase-angle correction factor; polarity; secondary winding of an instrument transformer; excitation losses for an instrument transformer; voltage transformer; cascade-type voltage transformer; insulated-neutral terminal type voltage transformer; double-secondary voltage transformer; fused-type voltage transformer; turn ratio of a voltage transformer; thermal burden rating of a voltage transformer; rater voltage of a voltage transformer; rated secondary voltage; current transformer; bushing-type current transformer; double-secondary current transformer; multiple-secondary current transformer; multi-ratio current transformer; window-type current transformer; wound-type current transformer; three-wire type current transformer; rated current; rated secondary current; turn ratio of a current transformer.* 53

instrument-transformer correction factor (watt meter or watthour meter). *See: transformer correction factor.*

instrument transformer, dry-type. *See: dry-type.*

instrument transformer, liquid-immersed. *See: liquid-immersed.*

instrument transformer, low-voltage winding. Winding that is intended to be connected to the measuring or control devices. 203

insulated (1)(NESC). Separated from other conducting surfaces by a dielectric (including air space) offering a high resistance to the passage of current. *Note:*

When any object is said to be insulated, it is understood to be insulated for the conditions to which it is normally subjected. Otherwise, it is, within the purpose of these rules, uninsulated. 494

(2) (electrolytic cell line working zone). Separated from other conducting surfaces by a dielectric substance or air space permanently offering a high resistance to the passage of current and to disruptive discharge through the substance or space. *Note:* When any object is said to beA insulated, it is understood to be insulated in a manner suitable for the conditions to which it is subjected. Otherwise, within th purpose of this definition, it is uninsulated. Insulating covering of conductors is one means for making the conductors insulated. 133

(3) (transmission and distribution). Separated from other conducting surfaces by a dielectric (including air space) offering a high resistance to the passage of current. *Note:* When any object is said to be insulated, it is understood to be insulated in a suitable manner for the conditions to which it is subjected. Otherwise, it is, within the purpose of these rules, uninsulated.
262

insulated bearing (rotating machinery). A bearing that is insulated to prevent the circulation of stray currents. *See: bearing.* 63

insulated bearing housing (rotating machinery). A bearing housing that is electrically insulated from its supporting structure to prevent the circulation of stray currents. *See: bearing.* 63

insulated bearing pedestal (rotating machinery). A bearing pedestal that is electrically insulated from its supporting structure to prevent the circulation of stray currents. *See: bearing.* 63

insulated bolt. A bolt provided with insulation. 328

insulated cap (separable insulated connectors). An accessory device designed to electrically insulate and shield and mechanically seal a bushing insert or integral bushing. 454

insulated conductor (1)(NESC). A conductor covered with a dielectric (other than air) having a rated insulating strength equal to or greater than the voltage of the circuit in which it is used. 494

(2)) (National Electrical Code). A conductor encased within material of composition and thickness that is recognized by this Code as electrical insulation.
256

(3)(transmission and distribution). A conductor covered with a dielectric having a rated insulating strength equal to or greater than the voltage of the circuit in which it is used 262

insulated coupling (rotating machinery). A coupling whose halves are insulated from each other to prevent the circulation of stray current between shafts. *See: rotor (rotating machinery).* 63

insulated flange (piping). Element of a flange-type coupling, insulated to interrupt the electrically conducting path normally provided by metallic piping. *See: rotor (rotating machinery).* 63

insulated-gate field-effect transistor (IGFET) (metal-

nitride-oxide field-effect transistor). A four-terminal device consisting of two separate areas of one conductivity type called source and drain with a terminal each, separated from each other by a substrate of opposite conductivity type with its terminal and straddled by an electrode with terminal called gate, which is insulated from the silicon by a layer of insulator material, frequently silicon dioxide, called gate.

386

insulated-gate field-effect transistor (IGFET) symbols (metal-nitride-oxide field-effect transistor). IGFET types may be categorized as memory–nonmemory, enhancement mode–depletion mode, and n-channel –p-channel. Standard symbols for memory transistors do not exist. The diagram below presents the standard electrical symbols for the nonmemory transistors and the symbols used in this standard for memory transistors. The symbols used for the memory transistors must be considered provisional until specific standards have been finalized. 386

insulated joint (1) (conduit). A coupling or joint used to insulate adjacent pieces of conduits, pipes, rods, or bars. 1

(2) (cable). A device that mechanically couples and electrically insulates the sheath and armor of contiguous lengths of cable. *See:* **tower.** 64

insulated-neutral terminal type voltage transformer (instrument transformer). A voltage transformer that has the neutral end of the high voltage winding insulated from the case or base and connected to a terminal that provides insulation for a lower voltage than required for the line terminal. 394

insulated parking bushing (separable insulated connectors). An accessory device designed to electrically insulate and shield and mechanically seal a power cable terminated with an elbow. 454

insulated rail joint. A joint used to insulate abutting rail ends electrically from one another. 328

insulated splice (power cable joint). A splice with a dielectric medium applied over the connected conductors and adjacent cable insulation. 34

insulated static wire. An insulated conductor on a power transmission line whose primary function is protection of the transmission line from lightning and one of whose secondary function is communications. 59

insulated supply system. *See:* **ungrounded system.**

insulated tool or device (power line maintenance). A tool or device which has conductive parts and is either coated or covered with a dielectric material. 458

insulated turnbuckle. An insulated turnbuckle is one so constructed as to constitute an insulator as well as a turnbuckle. *See:* **tower.** 64

insulating (covering of a conductor, or clothing, guards, rods, and other safety devices). A device that, when interposed between a person and current-carrying parts, protects the person making use of it against electric shock from the current-carrying parts with which the device is intended to be used: the opposite of conducting. *See:* **insulated; insulation.** 262

insulating (isolating) joint (power cable joint). A cable joint which mechanically couples and electrically separates the sheath, shield, and armor on contiguous lengths of cable. 34

insulating cell (rotating machinery). An insulating liner, usually to separate a coil-side from the grounded surface at a slot. *See:* **rotor (rotating machinery); stator.** 63

insulating material (insulant) (rotating machinery). A substance or body, the conductivity of which is zero or, in practice, very small. *See:* **asynchronous machine.** 244, 63

insulating materials (classes of) (high-power switchgear). For the purpose of establishing temperature limits, insulating materials are classified as follows:

Class 90. Materials or combinations of materials such as cotton, silk, or paper without impregnation. Other materials, or combinations of materials may be included in this class if, by experience or accepted tests, they can be shown to be capable of operation at 90 °C.

Class 105. Materials or combinations of materials such as cotton, silk, and paper when suitable impregnated or coated or when immersed in a dielectric liquid such as oil. Other materials or combinations of materials may be included in this class if, by experience or accepted tests, they can be shown to be capable of operation at 105 °C.

Class 130. Materials or combinations of materials such as mica, glass fiber, asbestos, etcetera, with suitable bonding substances. Other materials or combinations of materials, not necessarily inorganic, may be included in this class if, by experience or accepted tests, they can be shown to be capable of operation at 130 °C.

Class 155. Materials or combinations of materials such as mica, glass fiber, asbestos, etcetera, with suitable bonding substances. Other materials or combinations of materials, not necessarily inorganic, may be included in this class if, by experience or accepted tests, they can be shown to be capable of operation at 155 °C.

Class 180. Materials or combinations of materials such as silicone elastomer, mica, glass fiber, asbestos, etcetera, with suitable bonding substances such as appropriate silicone resins. Other materials or combinations of materials may be included in this class if, by experience or accepted tests, they can be shown to be capable of operation at 180 °C.

Class 220. Materials or combinations of materials that by experience or accepted tests can be shown to be capable of operation at 220 °C.

Over Class 220. Insulation that consists entirely of mica, porcelain, glass, quartz, and similar inorganic materials. Other materials or combinations of materials may be included in this class if, by experience or accepted tests, they can be shown to be capable of operation at temperatures over 220 °C. *Notes:* (1) Insulation is considered to be *impregnated* when a suitable substance provides a bond between components of the structure and also a degree of filling and surface coverage sufficient to give adequate performance under the extremes of temperature, surface contamination (moisture, dirt, etcetera), and mechanical stress expected in service. The impregnant shall not flow or deteriorate enough at operating temperature so as to seriously affect performance in service. (2) The electrical and mechanical properties of the insulation shall not be impaired by the prolonged application of the limiting insulation temperature permitted for the specific insulation class. The word *impaired* is here used in the sense of causing any change that could disqualify the insulating material for continuously performing its intended function, whether creepage, spacing, mechanical support, or dielectric barrier action. (3) In the above descriptions the words *accepted tests* are intended to refer to recognized test procedures established for the thermal evaluation of materials by themselves or in simple combinations. Experience or test data, used in classifying insulating materials, are distinct from the experience or test data derived for the use of materials in complete insulation systems. The thermal endurance of complete systems may be determined by test procedures specified by the responsible technical committees. A material that is classified as suitable for a given temperature in the above tabulation may be found suitable for a different temperature, either higher or lower, by an insulation system test procedure. For example, it has been found that some materials suitable for operation at one temperature in air may be suitable for a higher temperature when used in a system operated in an inert gas atmosphere. (4) It is important to recognize that other characteristics, in addition to thermal endurance, such as mechanical strength, moisture resistance, and corona endurance, are required in varying degrees in different applications for the successful use of insulating materials. 443

insulating materials. *See:* **insulation, class ratings.**

insulating spacer. Insulating material used to separate parts. *See:* **rotor (rotating machinery); stator.** 63

insulating tool or device (power line maintenance). (1) A tool or device designed primarily to provide insulation from an energized part or conductor. It can be composed entirely of insulating materials. Examples: conductor cover; stick; insulating tape. (2) A tool or device which has conductive parts separated by a dielectric. 458

insulating transformer. A transformer used to insulate one circuit from another. 53, 133

insulation (as applied to cable)(1)(NESC). That which is relied upon to insulate the conductor from other conductors or conducting parts or from ground. 494

(2) (rotating machinery) (electric system). Material or a combination of suitable nonconducting materials that provide electric isolation of two parts at different voltages. 63

(3) (cable). That part that is relied upon to insulate the conductor from other conductors or conducting parts or from ground. *See:* **cable; insulation, class ratings.** 64

(4) (power cable joint). A material of suitable dielectric properties, capable of being field-applied, and used to provide and maintain continuity of insulation across the splice. The material need not be identical to the cable insulation, but should be electrically and physically compatible, including factor-molded insulating components that are field-installed. 34

(5) (high-voltage switchgear). A material having the property of an insulator used to separate parts of the same or different potential. 443

insulation breakdown (electrical insulation tests). A rupture of insulation that results in a substantial transient or steady increase in leakage current at the specified test voltage. 116

insulation breakdown current (electrical insulation tests). The current delivered from the test apparatus when a dielectric breakdown occurs. 116

insulation class. *See:* **insulation level.**

insulation class (1) (outdoor apparatus bushings). The voltage by which the bushing is identified and which designates the level on which the electrical performance requirements are based. 316, 168

(2) (grounding device). A number that defines the insulation levels of the device. 91

(3) (transformer). Deprecated. *See:* **insulation level.** 53

insulation, class ratings (electric-machine-windings and electric cables) (1) (temperature endurance). These temperatures are, and have been in most cases over a long period of time, benchmarks descriptive of the various classes of insulating materials, and various accepted test procedures have been or are being developed for use in their identification. They should not be confused with the actual temperatures at which these same classes of insulating materials may be used in the various specific types of equipment, nor with the temperatures on which specified temperature rise in equipment standards are based. (1) In the following definitions the words **accepted tests** are intended to refer to recognized test procedures established for the thermal evaluation of materials by themselves or in simple combinations. Experience or test data, used in classifying insulating materials, are distinct from the experience or test data derived for the use of materials in complete insulation systems. The thermal endurance of complete systems may be determined by test procedures specified by the responsible technical committees. A material that is classified as suitable for a given temperature may be found suitable for a different temperature, either higher or lower, by an insulation system test procedure. For example, it has been found that some materials suitable for operation at one temperature in air may be suitable for a higher temperature when used in a system operated in an inert gas atmosphere. Likewise some insulating materials when operated in dielectric liquids will have lower or higher thermal endurance than in air. (2) It is important to recognize that other characteristics, in addition to thermal endurance, such as mechanical strength, moisture resistance, and corona endurance, are required in varying degrees in different applications for the successful use of insulating materials. *See also:* **insulation.(A) class 90 insulation.** Materials or combinations of materials such as cotton, silk, and paper without impregnation. *Note:* Other materials or combinations of materials may be included in this class if by experience or accepted tests they can be shown to have comparable thermal life at 90 degrees Celsius. (B) class 105 insulation. Materials or combinations of materials such as cotton, silk, and paper when suitably impregnated or coated or when immersed in a dielectric liquid. *Note:* Other materials or combinations may be included in this class if by experience or accepted tests they can be shown to have comparable thermal life at 105 degrees Celsius. (C) class 130 insulation. Materials or combinations of materials such as mica, glass fiber, asbestos, etcetera, with suitable bonding substances. *Note:* Other materials or combinations of materials may be included in this class if by experience or accepted tests they can be shown to have comparable thermal life at 130 degrees Celsius.(D) class 155 insulation. Materials or combinations of materials such as mica, glass fiber, asbestos, etcetera, with suitable bonding substances. *Note:* Other materials or combinations of materials may be included in this class if by experience or accepted tests they can be shown to have comparable thermal life at 155 degrees Celsius.(E) class 180 insulation. Materials or combinations of materials such as silicone elastomer, mica, glass fiber, asbestos, etcetera, with suitable bonding substances such as appropriate silicone resins. *Note:* Other materials or combinations of materials may be included in this class if by experience or accepted tests they can be shown to have comparable thermal life at 180 degrees Celsius.(F) class 220 insulation. Materials or combinations of materials which by experience or accepted tests can be shown to have the required thermal life at 220 degrees Celsius.(G) class over-220 insulation. Materials consisting entirely of mica, porcelain, glass, quartz, and similar inorganic materials. *Note:* Other materials or combinations of materials may be included in this class if by experience or accepted tests they can be shown to have the required thermal life at temperatures over 220 degrees Celsius. **(2) (letter symbols).(A) class O insulation.** *See:* **class 90 insulation.(B) class A insulation.** (1) Cotton, silk, paper, and similar organic materials when either impregnated or immersed in a liquid dielectric. (2) Molded and laminated materials with cellulose filler, phenolic resins, and other resins of similar properties. (3) Films and sheets of cellulose acetate and other cellulose derivatives of similar properties. (4) Varnishes (enamel) as applied to conductors. *Note:* An insulation is considered to be impregnated when a suitable substance replaces the air between its fibers, even if this substance does not completely fill the spaces between the insulated conductors. The impregnating substances, in order to be considered suitable, must have good insulating properties; must entirely cover the fibers and render them adherent to each other and to the conductor; must not produce interstices within itself as a consequence of evaporation of the solvent or through any other cause; must not flow during the operation of the machine at full working load or at the temperature limit specified; and must

not unduly deteriorate under prolonged action of heat.(C) class B insulation. Mica, asbestos, glass fiber, and similar inorganic materials in built-up form with organic binding substances. *Note:* A small proportion of class A materials may be used for structural purposes only. Glass fiber or asbestos magnet-wire insulations are included in this temperature class. These may include supplementary organic materials, such as polyvinyl acetal or polyamide films. The electrical and mechanical properties of the insulated winding must not be impaired by application of the temperature permitted for class B material. (The word **impaired** is here used in the sense of causing any change that could disqualify the insulating material for continuous service.) The temperature endurance of different class B insulation assemblies varies over a considerable range in accordance with the percentage of class A materials employed, and the degree of dependence placed on the organic binder for maintaining the structural integrity of the insulation.(D) class H insulation. Insulation consisting of; (1) mica, asbestos, glass fiber, and similar inorganic materials in built-up form with binding substances composed of silicone compounds or materials with equivalent properties; (2) silicone compounds in rubbery or resinous forms or materials with equivalent properties. *Note:* A minute proportion of class A materials may be used only where essential for structural purposes during manufacture. The electrical and mechanical properties of the insulated winding must not be impaired by the application of the hottest-spot temperature permitted for the specific insulation class. The word **impaired** is here used in the sense of causing any change that could disqualify the insulating materials for continuously performing its intended function, whether creepage spacing, mechanical support, or dielectric barrier action.(E) class C insulation. Insulation consisting entirely of mica, porcelain, glass, quartz, and similar inorganic materials. (F) class F insulation. Materials or combinations of materials such as mica, glass fiber, asbestos, etc., with suitable bonding substances. Other materials or combinations of materials, not necessarily inorganic, may be included in this class if by experience or accepted tests they can be shown to be capable of operation at 155 °C.   3

insulation coordination (insulation strength). *See:* **coordination of insulation (insulation strength).**

insulation failure or device breakdown (thyristor). The failure of a semiconductor or an insulator to support its rated voltage.   445

insulation fault (surge arresters). Accidental reduction or disappearance of the insulation resistance between conductor and ground or between conductors.
    244, 62

insulation, graded. The selective arrangement of the insulation components of a composite insulation system to more nearly equalize the voltage stresses throughout the insulation system.   95

insulation level (1)(series capacitors). The combination of power frequency and impulse-test voltage values which characterize the insulation of the capacitor bank with regard to its capability of withstanding the electric stresses between platform and earth.   474

(2) (power and distribution transformer). An insulation strength expressed in terms of a withstand voltage.   53

(3) (surge arresters). A combination of voltage values (both power-frequency and impulse) that characterize the insulation of an equipment with regard to its capability of withstanding dielectric stresses. *See:* **basic impulse insulation level (BIL) (insulation strength).**   276, 244, 62

insulation power factor (1) (power and distribution transformer). The ratio of the power dissipated in the insulation, in watts, to the product of the effective voltage and current in volt-amperes, when tested under a sinusoidal voltage and prescribed conditions. *Note:* If the current is also sinusoidal, the insulation power factor is equal to the cosine of the phase angle between the applied voltage and the resulting current.   53

(2) (rotating machinery). The ratio of dielectric loss in an insulation system to the applied apparent power, when measured at power frequency under designated conditions of voltage, temperature, and humidity. *Note:* Being the sine of an angle normally small, it is practically equal to loss tangent or dissipation factor, the tangent of the same angle. The angle is the complement of the angle whose cosine is the power factor; asynchronous machine.*See:* **loss tangent.**   287, 63

insulation resistance (aircraft, missile, and space equipment). The electrical resistance measured at specified direct-current potentials between any electrically insulated parts, such as a winding and other parts of the machine.   116

insulation resistance, direct current (1) (insulated conductor). The resistance offered by its insulation to the flow of current resulting from an impressed direct voltage. *See:* **conductor.**   59

(2) (between two electrodes in contact with or embedded in a specimen). The ratio of the direct voltage applied to the electrodes to the total current between them. *Note:* It is dependent upon both the volume and surface resistances of the specimen.   210

(3) (rotating machinery). The quotient of a specified direct voltage maintained on an insulation system divided by the resulting current at a specified time after the application of the voltage under designated conditions of temperature, humidity, and previous charge. *Note:* If steady state has not been reached the apparent resistance will be affected by the rate of absorption by the insulation of electric charge. *See:* **asynchronous machine.**   63

insulation-resistance test. A test for measuring the resistance of insulation under specified conditions.   63

insulation-resistance versus voltage test (rotating machinery). A series of insulation-resistance measurements, made at increasing direct voltages applied at successive intervals and maintained for designated periods of time, with the object of detecting insulation system defects by departures of the measured charac-

teristic from a typical form. Usually this is a controlled overvoltage test. *See:* **asynchronous machine.**   63

insulation shielding (1)(NESC). An envelope which encloses the insulation of a cable and provides an equipotential surface in contact with the cable insulation.   494

(2) (power distribution). Conducting and.or semiconducting elements applied directly over and in intimate contact with the outer surface of the insulation. Its function is to eliminate ionizable voids at the surface of the insulation and confine the dielectric stress to the underlying insulation. *See:* **power distribution.**   57

(3) (cable systems). A nonmagnetic, metallic material applied over the insulation of the conductor or conductors, to confine the electric field of the cable to the insulation of the conductor or conductors.   35

insulations, laminar (cable-insulation materials). Dielectric materials, either fibrous or film, or composite, comprising two or more layers of insulation arranged in series, and normally impregnated or flooded with an insulating liquid, or both.   97

insulation sleeving (tubing). A varnish-treated or resin-coated flexible braided tube providing insulation when placed over conductors, usually at connections or crossovers.   63

insulations, solid (cable-insulation materials). Firm, essentially homogeneous, dielectric materials comprising virtually complete solid-phase structures and having no liquid phase.   97

insulation system (1) (electric installations on shipboard). An assembly of insulating materials in a particular type of equipment. The class of the insulation system may be designated by letters, numbers, or other symbols. An insulation system class utilizes material having an appropriate temperature index and operates at such temperatures above stated ambient temperatures as the equipment standard specifies based on experience or accepted test data. The system may alternatively contain materials of ay class, provided that experience or a recognized test procedure for the equipment has demonstrated equivalent life expectancy.   3

(2) (power and distribution transformer). An assembly of insulating materials in a particular type, and sometimes size, of equipment.   53

(3) (random-wound ac electric machinery) class A. A system utilizing materials having a preferred temperature index of 105 and operating at such temperature rises above stated ambient temperature as the equipment standard specifies based on experience or accepted test data. This system may alternatively contain materials of any class, provided that experience or a recognized system test procedure for the equipment has demonstrated equivalent life expectancy. The preferred temperature classification for a Class A insulation system is 105 °C. **class B.** A system utilizing materials having a preferred temperature index of 130 and operating at such temperature rises above stated ambient temperature as the equipment standard specifies based on experience or accepted test data. This system may alternatively contain materials of any class, provided that experience or a recognized system test procedure for the equipment has demonstrated equivalent life expectancy. The preferred temperature classification for a Class B insulation system is 130 ° C. **class C.** A system utilizing materials having a preferred temperature index of over 240 and operating at such temperatures above stated ambient temperatures as the equipment standard specifies based on experience or accepted test data. This system may alternatively contain materials of any class, provided that experience or a recognized test procedure for the equipment has demonstrated equivalent life expectancy. The preferred temperature classification for a Class C insulation system is over 240 °C. **class F.** A system utilizing materials having a preferred temperature index of 155 and operating at such temperature rises above stated ambient temperatures as the equipment standard specifies based on experience or accepted test data. This system may alternatively contain materials of any class, provided that experience or a recognized test procedure for the equipment has demonstrated equivalent life expectancy. The preferred temperature classification for a Class F insulation system is 155 °C. **class H.** A system utilizing materials having a preferred temperature index of 180 and operating at such temperature rises above stated ambient temperature as the equipment standard specifies based on experience or accepted test data. This system may alternatively contain materials of any class, provided that experience or a recognized test procedure for the equipment has demonstrated equivalent life expectancy. The preferred temperature classification for a Class H insulation system is 180 °C. **class N.** A system utilizing materials having a preferred temperature index of 200 and operating at such temperature rises above stated ambient temperatures as the equipment standard specifies based on experience or accepted test data. This system may alternatively contain materials of any class, provided that experience or a recognized test procedure for the equipment has demonstrated equivalent life expectancy. The preferred temperature classification for a Class N insulation system is 200 °C. **class R.** A system utilizing materials have a preferred temperature index of 220 and operating at such temperatures above stated ambient temperatures as the equipment standard specifies based on experience or accepted test data. This system may alternatively contain materials of any class, provided that experience or a recognized test procedure for the equipment has demonstrated equivalent life expectancy. The preferred temperature classification for a Class R insulation system is 220 °C. **class S.** A system utilizing materials having a preferred temperature index of 240 and operating at such temperatures above stated ambient temperatures as the equipment standard specifies based on experience or accepted test data. This system may alternatively contain materials of any class, provided that experience or a recognized test procedure for the equipment has demonstrated equiv-

alent life expectancy. The preferred temperature classification for a Class S insulation system is 240 °C.
154

insulation systems defined (insulation systems of synchronous machines). An insulation system is an assembly of insulating materials. For definition purposes, the insulation systems of synchronous machine windings (either field or armature) are divided into three components. These components are the coil insulation with its accessories, the connection and winding support insulation, and the associated structural parts.
298

insulation, temperature class ratings (transformer). These temperatures are and have been, in most cases over a long period of time, benchmarks descriptive of the various classes of insulating materials, and various accepted test procedures have been or are being developed for use in their identification. They should not be confused with the actual temperatures at which these same classes of insulating materials may be used in the various specific types of equipment, nor with the temperatures on which specified temperature rise in equipment standards are based. (1) In the following definitions the words "accepted tests" are intended to refer to recognized test procedures established for the thermal evaluation of materials by themselves or in simple combinations. Experience or test data, used in classifying insulating materials, are distinct from the experience or test data derived for the use of materials in complete insulation systems. The thermal endurance of complete systems may be determined by test procedures specified by the responsible technical committees. A material that is classified as suitable for a given temperature may be found suitable for a different temperature, either higher or lower, by an insulation system test procedure. For example, it has been found that some materials suitable for operation at one temperature in air may be suitable for a higher temperature when used in a system operated in an inert gas atmosphere. Likewise some insulating materials when operated in dielectric liquids will have lower or higher thermal endurance than in air. (2) It is important to recognize that other characteristics, in addition to thermal endurance, such as mechanical strength, moisture resistance, and corona endurance, are required in varying degrees in different applications for the successful use of insulating materials. **class 105 insulation system.** Materials or combinations of materials such as cotton, silk, and paper when suitably impregnated or coated or when immersed in a dielectric liquid. *Note:* Other materials or combinations may be included in this class if by experience or accepted tests the insulation system can be shown to have comparable thermal life at 105 °C. **class 120 insulation system.** Materials or combinations of materials such as cotton, silk, and paper when suitably impregnated or coated or when immersed in a dielectric liquid: and which possess a degree of thermal stability which allows them to be operated at a temperature 15 °C higher than Class 105 insulation materials. *Note:* Other materials or combinations may be included in this class if by experience or accepted tests the insulation system can be shown to have comparable thermal life at 120 °C. **class 150 insulation system.** Materials or combinations of materials such as mica, glass fiber, asbestos, etcetera, with suitable bonding substances. *Note:* Other materials or combinations of materials may be included in this class if by experience or accepted tests the insulation system can be shown to have comparable life at 150 degrees Celsius. **class 185 insulation system.** Materials or combinations of materials such as silicone elastomer, mica, glass fiber, asbestos, etcetera, with suitable bonding substances such as appropriate silicone resins. *Note:* Other materials or combinations of materials may be included in this class if by experience or accepted tests the insulation system can be shown to have comparable thermal life at 185 °C. **class 220 insulation system.** Materials or combinations of materials such as silicone elastomer, mica, glass fiber, asbestos, etcetera, with suitable bonding substances such as appropriate silicone resins. *Note:* Other materials or combinations of materials may be included in this class if by experience or accepted tests, the system can be shown to have comparable thermal life at 220 degrees Celsius. **class over-220 insulation system.** Materials consisting entirely of mica, porcelain, glass quartz, and similar inorganic materials. *Note:* Other materials or combinations of materials may be included in this class if by experience or accepted tests the insulation system can be shown to have the required thermal life at temperatures over 220 °C. **class O insulation** (nonpreferred term). **class A insulation** (nonpreferred term). *See:* **class 105 insulation system. class B insulation** (nonpreferred term). **class C insulation** (nonpreferred term). **class F insulation** (nonpreferred term). **class H insulation** (nonpreferred term). 53

insulator (1) (power switchgear). A device intended to give flexible or rigid support to electrical conductors or equipment and to insulate these conductors or equipment from ground or from other conductors or equipment. An insulator comprises one or more insulating parts to which connecting devices (metal fittings) are often permanently attached. 70
(2) (transmission and distribution)(NESC). Insulating material in a form designed to support a conductor physically and electrically separate it from another conductor or object. 391, 262, 494
insulator arcing horn. A metal part, usually shaped like a horn, placed at one or both ends of an insulator or of a string of insulators to establish an arcover path, thereby reducing or eliminating damage by arcover to the insulator or conductor or both. *See:* **tower.** 64
insulator arcing ring. A metal part, usually circular or oval in shape, placed at one or both ends of an insulator or of a string of insulators to establish an arcover path, thereby reducing or eliminating damage by arcover to the insulator or conductor or both. *See:* **tower.** 64
insulator arcing shield (insulator grading shield). An arcing ring so shaped and located as to improve the voltage distribution across or along the insulator or insulator string. *See:* **tower.** 64

insulator arcover. A discharge of power current in the form of an arc, following a surface discharge over an insulator. *See:* **tower.** 64
insulator cover (power line maintenance). Electrical protection equipment designed specifically to cover insulators. Examples: dead-end cover; pole-top cover; ridge-pin cover; *Syn:* **hood; pocketbook.** *See:* **cover-up equipment.** 458
insulator grading shield. *See:* **insulator arcing shield.**
insulator lifter (conductor stringing equipment). A device designed to permit insulators to be lifted in a string to their position on a structure. *Syn:* **insulator saddle; potty seat.** 431
insulator string. Two or more suspension insulators connected in series. *See:* **tower.** 64
insulator unit (power switchgear). An insulator assembled with such metal parts as may be necessary for attaching it to other insulating units or device parts. 103, 443
intake opening (rotating machinery). A port for the entrance of ventilation air. 63
intake port (rotating machinery). An opening provided for the entrance of a fluid. 63
integer (microprocessor operating systems parameter types). A whole number that may be positive, negative, or zero and has a range of at least -32767 to + 32767. The qualifier 'long' is used to qualify the size of an integer. A long integer has a range of at least $(- 2**31 + 1)$ to $(+ 2**31 - 1)$. 478
integer adjectives (pulse terms). The ordinal integers (that is, first, second, . . . *nth,* last) or the cardinal integers (that is, 1, 2, . . . *n*) may be used as adjectives to identify or distinguish between similar or identical features. The assignment of integer modifiers should be sequential as a function of time within a waveform epoch or within features thereof. 254
integer array (inta) (subroutines for CAMAC). The symbol *inta* represents an integer array, the length and contents of which are not defined in this standard. It is intended to contain system-dependent or implementation-dependent information associated with the definition of a LAM. If no such information is required, the array need not be used. This information can include parameters necessary for interrupt linkage, event specification, etc. The doucumentation for an implementation must describe the requirements for any parameters contained in this array. 410
integral action rate (process control) (proportional plus integral control action devices). For a step input, the ratio of the initial rate of change of output due to integral control action to the change in steady-state output due to proportional control action. *Note:* Integral action rate is often expressed as the number of repeats per minute because it is equal to the number of times per minute that the proportional response to a step input is repeated by the initial integral response. 56
integral bushing (separable insulated connectors). An apparatus bushing designed for use with another connector component such as an elbow. *See:* **bushing insert, See figure.** 454

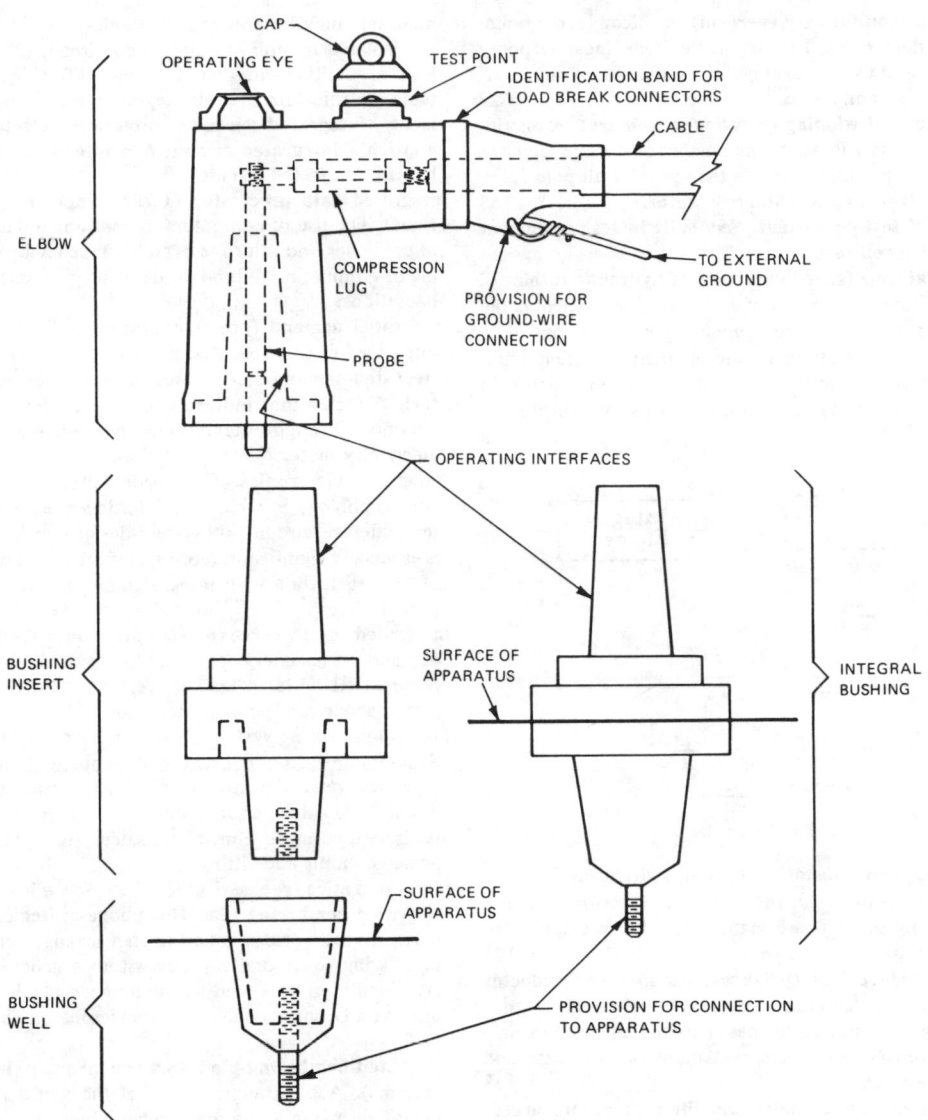

Typical Components of 200 A Separable
Insulated Connector System

integral control action (electric power systems). *See:* control action, integral.

integral coupling (rotating machinery). A coupling flange that is a part of a shaft and not a separate piece. *See:* **rotor (rotating machinery).**   63

integral-horsepower motor. A motor built in a frame as large as or larger than that of a motor of open construction having a continuous rating of 1 horsepower at 1700 1800 revolutions per minute. *See:* **asynchronous machine.**   328

integral nonlinearity (INL)(of a pulse amplifying system)(X-ray energy spectrometers). The maximum nonlinearity (deviation) over the specified operating range of a system, usually expressed as a percentage of the maximum of the specified range.   471

integral nonlinearity (percent) (semiconductor radiation detectors). The departure from linear response expressed as a percentage of the maximum rated output pulse amplitude.   23

integral-slot winding (rotating machinery). A distributed winding in which the number of slots per pole per phase is an integer and is the same for all poles. *See:* **rotor (rotating machinery); stator.**   63

integral test equipment. *See:* **built in test equipment (BITE); self-test.**

integral time (speed governing of hydraulic turbines). The integral time, T_x, of an integrating element is the time required for the element's percent output to be equal in magnitude to the element's percent input where that input is a step function. The integral gain of an element is the reciprocal of its integral time.   8

INPUT

$K\%$ — | Integral Element |

$t = 0$

Time

OUTPUT

$K\%$

$t = 0$ $t = T_x$

Time

integral unit substation (power switchgear). A unit substation in which the incoming, transforming, and outgoing sections are manufactured as a single compact unit.   53, 103

integrated (pulse) (pulse amplifier) (semiconductor radiation detectors). A pulse is "integrated" when it is passed through a lowpass network, such as a single RC (resistance-capacitance) network or a cascaded RC network.   23

integrated (pulse)(pulse amplifier jargon)(charged-particle detectors)(germanium gamma-ray detectors)(X-ray energy spectrometers). A pulse is integrated when it is passed through a low-pass network, such as a single resistance-capacitance RC network or a cascaded RC network.   528, 471

integrated alarm system (alarm monitoring and reporting systems for fossil-fueled power generating stations). An alarm display system consisting of window annunciators combined with cathode-ray tube (CRT), printer, or mimic display.   501

integrated antenna system. A radiator with an active or nonlinear circuit element or network incorporated physically within the structure of the radiator.   111

integrated circuit (solid state). A combination of interconnected circuit elements inseparably associated on or within a continuous substrate. *Note:* To further define the nature of an integrated circuit, additional

modifiers may be prefixed. Examples are: (1) dielectric-isolated monolithic integrated circuit, (2) beam-lead monolithic integrated circuit, (3) silicon-chip tantalum thin-film hybrid integrated circuit. *See:* **element (integrated circuit); integrated electronics; multichip integrated circuit; film integrated circuit; hybrid integrated circuit.**   312, 191

integrated data processing (IDP)(computer applications). The use of computers to coordinate a number of processes and improve overall efficiency by reducing or eliminating redundant data entry or processing operations.   571

integrated demand (power operations). The demand integrated over a specified period.   516

integrated-demand meter (block-interval demand meter). A meter that indicates or records the demand obtained through integration. *See:* **demand meter; electricity meter.**   212

integrated electronics. The portion of electronic art and technology in which the interdependence of material, device, circuit, and system-design consideration is especially significant: more specifically, that portion of the art dealing with integrated circuits. *See:* **integrated circuit.**   312, 191

integrated energy curve (power operations). A curve of demand versus energy showing the amount of energy represented under a load curve, or a load duration curve, above any point of demand.   516

integrated heating system (National Electrical Code). A complete system consisting of components such as pipelines, vessels, heating elements, heat transfer medium, thermal insulation, moisture barrier, nonheating leads, temperature controller, safety signs, junction boxes, conduit and fittings.   256

integrated mica (reconstituted mica). *See:* **mica paper.**

integrated-numbering plan (telephone switching systems). In the world-numbering plan, arrangements for identifying telephone stations within a geographical area identified by a world-zone number which is also used as a country code. *See:* **world zone number.**   55

integrated numbering-plan area (telephone switching systems). A geographical area of the world that is identified by a world-zone number which is also used as a country code. *See:* **world zone number.**   55

integrated optical circuit (IOC) (fiber optics). An optical circuit, either monolithic or hybrid, composed of active and passive components, used for coupling between optoelectronic devices and providing signal processing functions.   433

integrated plow (static or vibratory plows) (cable plowing). A self-contained or integral plow-prime mover unit.

integrated radiation, (L) (laser-maser). The integral of the radiance over the exposure duration. *Syn:* **pulsed radiation** $(W \cdot s \cdot cm^{-2} \cdot sr^{-1})$.   363

integrating (block interval) demand meter (metering). A meter that integrates power or a related quantity over a fixed time interval, and indicates or records the average.   212

integrating accelerometer (inertial sensor). A device

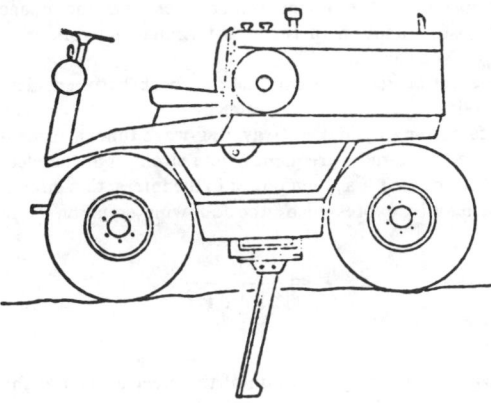

Integrated plow.

which produces an output that is proportional to the time integral of an input acceleration. 46

integrating amplifier (analog computers). An operational amplifier that produces an output signal equal to the time integral of a weighted sum of the input signals. *Note:* In an analog computer, the term integrator is synonymous with integrating amplifier. 9, 10

integrating circuit (integrator) (integrating network). *See:* **integrator.**

integrating network. *See:* **integrator.**

integrating photometer (illuminating engineering). One which enables total luminous flux to be determined by a single measurement. The usual type is the Ulbricht sphere with associated photometric equipment for measuring the indirect illuminance of the inner surface of the sphere. (The measuring device is shielded from the source under measurement.) 167

integrating preamplifier (germanium gamma-ray detectors)(X-ray energy spectrometers). A pulse preamplifier in which individual pulses are intentionally integrated by passive or active circuits. 528, 471

integrating relay. A relay that operates on the energy stored from a long pulse or a series of pulses of the same or varying magnitude, for example, a thermal relay. *See:* **relay.**

integration (software). The process of combining software elements, hardware elements, or both into an overall system. *See:* **hardware; software; system.** 434

integration loss (radar). The loss incurred by integrating a signal noncoherently instead of coherently. 13

integration testing (software verification and validation plans)(software). An orderly progression of testing in which software elements, hardware elements, or both are combined and tested until the entire system has been integrated. 511, 434

integration tests (programmable digital computer systems in safety systems of nuclear power generating stations). Tests performed during the hardware-software integration process prior to computer system validation to verify compatibility of the software and the computer system hardware. 554

integrator (1) (analog computers). A device producing an output proportional to the integral of one variable or of a sum of variables, with respect to another variable, usually time. *See:* **integrating amplifier.** 9

(2) (as applied to relaying) (power switchgear). A transducer whose output wave form is substantially the time integral of its input wave form. 103

(3) (digital differential analyzer). A device using an accumulator for numerically accomplishing an approximation to the mathematical process of integration. 235

integrator, gain (analog computer). For each input, the ratio of the input to the corresponding time rate of change of the output. For fixed input resistors, the "time constant" is determined by the integrating feedback capacitor. 9

integrity (software). The extent to which unauthorized access to or modification of software or data can be controlled in a computer system. *See:* **computer system; data; modification; security; software.** 434

intelligence bandwidth. The sum of the audio- (or video-) frequency bandwidths of the one or more channels. 111

intelligibility. *See:* **articulation.**

intensifier (baseline clipper) (non-real time spectrum analyzer). A means of changing the relative brightness between the signal and baseline portion of the display. 68

intensifier electrode. An electrode causing post acceleration. *See:* **post-accelerating electrode; electrode (electron tube).** 125

intensity (1) (fiber optics). The square of the electric field amplitude of a light wave. Intensity is proportional to irradiance and may be used in place of the term "irradiance" when only relative values are important. *See:* **irradiance; radiant intensity; radiometry.** 433

(2) (oscilloscopes). A term used to designate brightness or luminance of the spot. *See:* **oscillograph.** 185

intensity amplifier (oscilloscopes). An amplifier for signals controlling the intensity of the spot. *See:* **oscillograph.** 185

intensity (candlepower) distribution curve (illuminating engineering). A curve, often polar, which represents the variation of luminous intensity of a lamp or luminaire in a plane through the light center. *Note:* A vertical candlepower distribution curve is obtained by taking measurements at various angles of elevation in a vertical plane through the light center; unless the plane is specified, the vertical curve is assumed to represent an average such as would be obtained by rotating the lamp or luminaire about its vertical axis. A horizontal intensity distribution curve represents measurements made at various angles of azimuth in a horizontal plane through the light center. 167

intensity level (specific sound-energy flux level)

(sound-energy flux density level) (acoustics). In decibels, of a sound is 10 times the logarithm to the base 10 of the ratio of the intensity of this sound to the reference intensity. The reference intensity shall be stated explicitly. *Note:* In discussing sound measurements made with pressure or velocity microphones, especially in enclosures involving normal modes of vibration or in sound fields containing standing waves, caution must be observed in using the terms intensity and intensity level. Under such conditions it is more desirable to use the terms pressure level or velocity level since the relationship between the intensity and the pressure or velocity is generally unknown. 176

intensity modulation (1) (general). The process, or effect, of varying the electron-beam current in a cathode-ray tube resulting in varying brightness or luminance of the trace. *See:* **oscillograph; television.** 185

(2) (radar). A process used in certain displays whereby the luminance of the signal indication is a function of the received signal strength. 13, 187

intensity of magnetization. *See:* **magnetization.**

interaction (nuclear power generating station). A direct or indirect effect of one device or system upon another. 357

interaction-circuit phase velocity (traveling-wave tubes). The phase velocity of a wave traveling on the circuit in the absence of electron flow. *See:* **magnetrons.** 190

interaction crosstalk coupling (between a disturbing and a disturbed circuit in any given section). The vector summation of all possible combinations of crosstalk coupling, within one arbitrary short length, between the disturbing circuit and all circuits other than the disturbed circuit (including phantom and ground-return circuits) with crosstalk coupling, within another arbitrary short length, between the disturbed circuit and all circuits other than the disturbing circuit. *See:* **coupling.** 328

interaction factor (1) (transducer). The factor in the equation for the received current that takes into consideration the effect of multiple reflections at its terminals. *Note:* For a transducer having a transfer constant θ, image impedances Z_{I_1} and Z_{I_2}, and terminating impedances Z_S and Z_R, this factor is

$$\frac{1}{1 - \dfrac{Z_{I_2} - Z_R}{Z_{I_2} + Z_R} \times \dfrac{Z_{I_1} - Z_S}{Z_{I_1} + Z_S} \times e^{-2\theta}}.$$

(2) (electrothermic power meters). The ratio of power incident from an rf source to the power delivered by the source to a nonreflecting load: mathematically, $|1 - \Gamma_g|^2$ where Γ_g is the complex reflection coefficient of the source. 47

(3) (circuits and systems). A factor in the equation for the insertion voltage ratio that takes into account the impedance mismatch variation at one end of the network due to an impedance mismatch at the other end. The factor is written in terms of the source and load

impedance, the image impedances and the image transfer function of the four-terminal network. *See:* (1) *Note* above. 67

interaction gap. An interaction space between electrodes. 125

interaction impedance (traveling-wave tubes). A measure of the radio-frequency field strength at the electron stream for a given power in the interaction circuit. It may be expressed by the following equation

$$K = \frac{E^2}{2(\omega/v)^2 P}$$

where E is the peak value of the electric field at the position of electron flow, ω is the angular frequency, v is the interaction-circuit phase velocity, and P is the propagating power. If the field strength is not uniform over the beam, an effective interaction impedance may be defined. 125

interaction loss (transducer). The interaction loss expressed in decibels is 20 times the logarithm to base 10 of the scalar value of the reciprocal of the interaction factor. *See:* **attenuation.**

interaction space (traveling-wave tubes). A region of an electron tube in which electrons interact with an alternating electromagnetic field. 125

interactive (software). Pertaining to a system in which each user entry causes a response from the system. *See:* **conversational; system.** 434

intercalated tapes (insulation). Two or more tapes, generally of different composition, applied simultaneously in such a manner that a portion of each tape overlies a portion of the other tape. 64

intercardinal plane (antennas). Any plane that contains the intersection of two successive cardinal planes and is at an intermediate angular position. *Note:* In practice the intercardinal planes are located by dividing the angle between successive cardinal plane into equal parts. Often it is sufficient to bisect the angle so that there is only one intercardinal plane between successive cardinal planes. 111

intercarrier sound system. A television receiving system in which use of the picture carrier and the associated sound-channel carrier produces an intermediate frequency equal to the difference between the two carrier frequencies. *Note:* This intermediate frequency is frequency modulated in accordance with the sound signal. *See:* **television.** 328

intercept call (telephone switching systems). A call to a line or an unassigned code that reaches an operator, a recorded announcement, or a vacant-code tone. 55

intercept trunk (telephone switching systems). A central office termination that may be reached by a call to a vacant number, changed number, or line out-of-order. 55

interchangeable bushing (outdoor apparatus bushing). A bushing designed for use in both power transformers and circuit breakers. 168

interchange circuit (data transmission). The length of

cable used for signaling between the digital subset and the customer's equipment.   59

interchange energy (power operations). Energy delivered to or received by one electric system from another.   516

interchannel interference (modulation system). In a given channel, the interference resulting from signals in one or more other channels.   242

interclutter visibility (radar). The ability of a radar to detect moving targets which occur in resolution cells between points of strong clutter; usually applied to moving-target indication or pulse-Doppler radars. The higher the radar range or angle resolution, the better the interclutter visibility, since a smaller fraction of the cells contain strong point clutter.   13

intercom (interphone). The interference resulting from signals in one or more other channels.*See:* **intercommunicating system.**   242

intercommunicating system (intercom) (interphone). A privately owned two-way communication system without a central switchboard, usually limited to a single vehicle, building, or plant area. Stations may or may not be equipped to originate a call but can answer any call.   328

interconnected delta connection (power and distribution transformer). A three-phase connection using six windings (two per phase) connected in a six-sided circuit with six bushings to provide a fixed phase-shift between two three-phase circuits without change in voltage magnitude. *Note:* The interconnected delta connection is sometimes described as a "hexagon autotransformer," or a "squashed delta."   53

interconnected star connection of polyphase circuits. *See:* **zig-zag connection of polyphase circuits.**

interconnected system (electric power systems). A system consisting of two or more individual power systems normally operating with connecting tie lines. *See:* **power system.**   94, 200

interconnecting channel (of a supervisory system) (power switchgear). The transmission link, such as the direct wire, carrier, or microwave channel (including the direct current, tones, etcetera) by which supervisory control or indication signals or selected telemeter readings are transmitted between the master station and the remote station or stations, in a single supervisory system.   103

interconnection device. *See:* **adapter.**

interconnection diagram (packaging machinery). A diagram showing the connections between the terminals in the control panel and outside points, such as connections to motors and auxiliary devices.   429

interconnection tie. A feeder interconnecting two electric supply systems. *Note:* The normal flow of energy in such a feeder may be in either direction. *See:* **center of distribution.**   64

interconnect *See:* **segment interconnect; buffered interconnect.**

interdendritic corrosion. Corrosive attack that progresses preferentially along interdendritic paths. *Note:* This type of attack results from local differences in composition, that is, coring, commonly encountered in alloy castings.   205

interdigit (interdigital) time (measuring the performance of tone address signaling systems). The time interval between successive signal present intervals during which no signal present condition exists. This time includes the signal off condition and transition intervals between signal off condition and signal present condition on both state transitions.   508

interdigital magnetron. A magnetron having axial anode segments around the cathode, alternate segments being connected together at one end, remaining segments connected together at the opposite end.   190, 125

interdigit interval (telephony)(dial-pulse address signaling systems). In dial-pulse signaling, an extended make interval used to separate and distinguish successive dial-pulse address digits.   540

interelectrode capacitance (j-I interelectrode capacitance C_{jl} of an n-terminal electrode tube). The capacitance determined from the short-circuit transfer admittance between the jth and the lth terminals. *Note:* This quantity is often referred to as direct interelectrode capacitance. *See:* **electron-tube admittance.**   125

interelectrode transadmittance (j-I interelectrode transadmittance of an n-electrode electron tube). The short-circuit transfer admittance from the jth electrode to the lth electrode. *See:* **electron-tube admittances.**   125

interelectrode transconductance (j-I interelectrode transconductance). The real part of the j-l interelectrode transadmittance. *See:* **electron-tube admittances.**   125

interelement influences (polyphase wattmeters). The percentage change in the recorded value that is caused solely by the action of the stray field of one element upon the other element. *Note:* This influence is determined at the specified frequency of calibration with rated current and rated voltage in phase on both elements or such lesser value of equal currents in both elements as gives end-scale deflection. Both current and voltage in one element shall then be reversed, and, for rating purposes, one-half the difference in the readings in percent is the interelement influence. *See:* **accuracy rating (instrument).**   280, 294

interface (general-system term)(1)(696 interface devices). A shared electrical boundary between parts of a computer system, through which information is conveyed.   538
(2)(microcomputer system bus). A shared boundary between two systems, or between parts of systems, through which information is conveyed.   542
(3)(microprocessor operating systems). A shared boundary between two layers or modules of software.   478
(4)(watthour meters). The means for transmitting information between a solid-state time-of-use register and peripheral equipment.   485
(5)(general). A shared boundary.   255, 77
(6) (nuclear power generating stations) (class 1E equipment). A junction or junctions between a Class 1E equipment and another equipment or device. (Ex-

amples: connection boxes, splices, terminal boards, electrical connections, grommets, gaskets, cables, conduits, enclosures, etcetera.) 120, 31

(7)(programmable instrumentation). A shared boundary between a considered system and another system, or between parts of a system, through which information is conveyed. 40

(8) (test, measurement and diagnostic equipment). A shared boundary involving the specification of the interconnection between two equipments or systems. The specification includes the type, quantity and function of the interconnection circuits and the type and form of signals to be interchanged via those circuits. See: adapter. 54

(9)(data transmission). (A) A common boundary–for example, a physical connection between two systems or two devices. The boundary may be mechanical such as the physical surfaces and spacings in mating parts, modules, components, or subsystems, or electrical, such as matching signal levels, impedances, or power levels of two or more subsystems.(B) A concept involving the specification of the interconnection between two equipments or systems. The specification includes the type, quantity, and function of the interconnection circuits and the type and form of signals to be interchanged by these circuits. 59

(10) (software). (1) A shared boundary. An interface might be a hardware component to link two devices or it might be a portion of storage or registers accessed by two or more computer programs. (2) To interact or communicate with another system component. See: computer program; hardware component; system component. 434

(11) (programmable instrumentation). A common boundary between a considered system and another system, or between parts of a system, through which information is conveyed. 378

interface-CCITT (data transmission). The present European and possible world standard for interface requirements between data processing terminal equipment and data communication equipment. The CCITT standard resembles very closely the American EIA, Standard RS-232-C. This standard is considered mandatory in Europe and on the other continents. 59

interface control (software configuration management plans). The process of (1) identifying all functional and physical characteristics relevant to the interfacing of two or more configuration items provided by one or more organizations; (2) ensuring that proposed changes to these characteristics are evaluated and approved prior to implementation. See: baseline; configuration item; configuration management; configuration control; configuration control board; configuration audit; configuration identification; configuration status accounting; software library. 546

interface–EIA standard RS-232-C (data transmission). A standardized method adopted by the Electronic Industries Association to ensure uniformity of interface between data communication equipment and data processing terminal equipment. The standard in-

terface has been generally accepted by a great majority of the manufacturers of data transmission and business equipment. 59

interface--MIL STD 188B (data transmission). The standard method of interface established by the Department of Defense and is presently mandatory for use by the departments and agencies of the Department of Defense for the installation of all new equipment. This standard provides the interface requirements for interconnection between data communication security devices, data processing equipment, or other special military terminal devices. 59

interface, operating (connector). The surfaces at which a connector is normally separated. 134

interface requirement (software). A requirement that specifies a hardware, software, or data base element with which a system or system component must interface, or that sets forth constraints on formats, timing, or other factors caused by such an interface. See: data base; hardware; interface; requirement; software; system. 434

interface specification (software). A specification that sets forth the interface requirements for a system or system component. See: component; interface requirement; specification; system. 434

interface system (general-system term)(1)(696 interface devices). The device independent functional, electrical, and mechanical elements of an interface necessary to effect unambiguous communication among a set of devices. Driver and receiver circuits, signal line descriptions, timing and control conventions, data transfer protocols, and functional logic circuits are typical system elements kilobyte $1024 = 2$ 10megabyte $1\,048\,576 = 2^{20}$ 538

(2)(microcomputer system bus). The device-dependent electrical and functional interface elements necessary for communication between devices. Typical elements are: driver and receiver circuits, signal line descriptions, timing and control conventions, and functional logic circuits. 542

(3)(programmable instrumentation). The device-independent mechanical, electrical, and functional elements of an interface necessary to effect communication among a set of devices. Cables, connector, driver and receiver circuits, signal line descriptions, timing and control conventions, and functional logic circuits are typical interface system elements. 40

interface testing (software). Testing conducted to ensure that program or system components pass information or control correctly to one another. See: program; system component; testing. 434

interfacial connection (soldered connections). A conductor that connects conductive patterns on opposite sides of the base material. 284

interference (1) (data transmission). In a signal transmission path, either extraneous power which tends to interfere with the reception of the desired signals or the disturbance of signal which results. 59

(2) (electric-power-system measurements). Any spurious voltage or current appearing in the circuits of the instrument. Note: The source of each type of in-

terference may be within the instrument case or external. The instrument design should be such that the effects of interference arising internally are negligible.   295

(3) (induction or dielectric-heating usage). The disturbance of any electric circuit carrying intelligence caused by the transfer of energy from induction- or dielectric-heating equipment.   14, 114

(4) (fiber optics). In optics, the interaction of two or more beams of coherent or partially coherent light. *See:* **coherent; degree of coherence; diffraction.**
     433

(5) (overhead-power-line corona and radio noise). Impairment to a useful signal produced by natural or man-made sources. *Note:* Distortions caused by reflections, shielding, or extraneous power in a signal's frequency range are all examples of interference.   411

interference, common-mode (signal-transmission system). *See:* **common-mode interference.**

interference coupling ratio (signal-transmission system). The ratio of the interference produced in a signal circuit to the actual strength of the interfering source (in the same units). *See:* **interference.**   188

interference, differential-mode (signal-transmission system). Interference that causes the potential of one side of the signal transmission path to be changed relative to the other side. *Note:* That type of interference in which the interference current path is wholly in the signal transmission path. *See:* **interference.**
     188

interference field strength (electromagnetic compatibility). Field strength produced by a radio disturbance. *Note:* Such a field strength has only a precise value when measured under specified conditions. Normally, it should be measured according to publications of the International Special Committee on Radio Interference. *See:* **electromagnetic compatibility.**
     220, 199

interference filter (fiber optics). An optical filter consisting of one or more thin layers of dielectric or metallic material. *See:* **dichroic filter; interference; optical filter.**   433

interference guard bands. The two bands of frequencies additional to, and on either side of, the communication band and frequency tolerance, which may be provided in order to minimize the possibility of interference. *See:* **channel.**   111

interference, longitudinal (signal-transmission system). *See:* **interference, common-mode.**

interference measurement (induction or dielectric-heating). A measurement usually of field intensity to evaluate the probability of interference with sensitive receiving apparatus. *See:* **dielectric heating; induction heating.**   14

interference, normal-mode (signal-transmission system). A form of interference that appears between measuring circuit terminals. *See:* **interference, differential-mode; accuracy rating (instrument).**   295

interference pattern. The resulting space distribution when progressive waves of the same frequency and kind are superposed. *See:* **wave front.**   328

interference power (electromagnetic compatibility). Power produced by a radio disturbance. *Note:* Such a power has only a precise value when measured under specified conditions. *See:* **electromagnetic compatibility.**   220, 199

interference, series-mode (signal-transmission system). *See:* **interference, differential-mode.**

interference susceptibility (mobile communication). A measure of the capability of a system to withstand the effects of spurious signals and noise that tend to interfere with reception of the desired intelligence. *See:* **mobile communication system.**   181

interference testing (test, measurement and diagnostic equipment). A type of on-line testing that requires disruption of the normal operation of the unit under test. *See:* **noninterference testing.**   54

interference-test input (amplitude-modulation broadcast receivers). The least interfering-signal or signal field, of specified carrier frequency, which results in interference test output. It is expressed in decibels below 1 volt, or in microvolts, or in the case of loop measurements in decibels below 1 volt per meter or microvolts per meter.   524

interference-test output (amplitude-modulation broadcast receivers). Output that is 30 decibels less than, or 0.001 of the power of, the normal test output.   524

interference, transverse. *See:* **interference, differential-mode.**

interference voltage (electromagnetic compatibility). Voltage produced by a radio disturbance. *Note:* Such a voltage has a precise value only when measured under specified conditions. Normally, it should be measured according to recommendations of the International Special Committee on Radio Interference. *See:* **electromagnetic compatibility.**   220, 199

interferometer (1) (fiber optics). An instrument that employs the interference of light waves for purposes of measurement. *See:* **interference.**   433

(2) (radar). A receiving system which determines the angle of arrival of a wave by phase comparison of the signals received at separate antennas or separate points on the same antenna.   13

interferometer antenna. An array antenna in which the interelement spacings are large compared to wavelength and element size so as to produce grating lobes.   111

interflectance method* (illuminating engineering). A lighting design procedure for predetermining the luminances of walls, ceiling, and floor and the average illuminance on the work-plane based on integral equations. It takes into account both direct and reflected flux. (This term is retained for reference and literature searches).

* Obsolete   167

interflected component (illuminating engineering). That portion of luminous flux from a luminaire which arrives at the work-plane after being reflected one or more times from room surfaces, as determined by the flux transfer theory. *See:* **flux transfer theory.**
     167

interflection (illuminating engineering). The multiple reflection of light by the various room surfaces before it reaches the work-plane or other specified surface of a room. *Syn: interreflection.*   167

intergranular corrosion. Corrosion that occurs preferentially at grain boundaries.   205

interior communication systems (marine). Those systems providing audible or visual signals or transmission of information within or on a vessel.   328

interior wiring system ground. A ground connection to one of the current-carrying conductors of an interior wiring system. *See: ground.*   64

interlaboratory standards. Those standards that are used for comparing reference standards of one laboratory with those of another, when the reference standards are of such nature that they should not be shipped. *See: measurement system.*   293, 183

interlaced scanning (television). A scanning process in which the distance from center to center of successively scanned lines is two or more times the nominal line width, and in which the adjacent lines belong to different fields. *See: television.*   178

interlace factor (television). A measure of the degree of interlace of nominally interlaced fields. *Note:* In a two-to-one interlaced raster, the interlace factor is the ratio of the smaller of two distances between the centers of adjacent scanned lines to one-half the distance between the centers of sequentially scanned lines at a specified point.   18

interlacing impedance voltage of a Scott-connected transformer (power and distribution transformer). The interlacing impedance voltage of Scott-connected transformers is the single-phase voltage applied from the midtap of the main transformer winding to both ends, connected together, which is sufficient to circulate in the supply lines a current equal to the rated three-phase line current. The current in each half of the winding is 50% of this value. The per-unit or percent interlacing resistance is the measured watts expressed on the base of the rated kVA of the teaser winding. The per-unit or percent interlacing impedance is the measured voltage expressed on the base of the teaser-voltage.   53

interleave. To arrange parts of one sequence of things or events so that they alternate with parts of one or more other sequences of things or events and so that each sequence retains its identity.   255, 77

interleaved windings (of a transformer) (power and distribution transformer). An arrangement of transformer windings where the the primary and secondary windings, and the tertiary windings, if any, are subdivided into disks (or pancakes) or layers and interleaved on the same core.   53

interlock (1) (power and distribution transformer). A device actuated by the operation of some other device with which it is directly associated, to govern succeeding operations of the same or allied devices. *Note:* Interlocks may be either electric or mechanical.   53

(2) (power switchgear). A device actuated by the operation of some other device with which it is directly

associated, to govern succeeding operations of the same or allied devices. *Note:* An interlock system is a series of interlocks applied to associated equipment in such a manner as to prevent or allow operation of the equipment only in a prearranged sequence. Interlocks are classified into three main divisions: mechanical interlocks, electrical interlocks, and key interlocks, based on the type of interconnection between the associated devices.   103

interlock bypass. A command to temporarily circumvent a normally provided interlock.   224, 207

interlocking (interlocking plant) (railway). An arrangement of apparatus in which various devices for controlling track switches, signals, and related appliances are so interconnected that their movements must succeed one another in a predetermined order, and for which interlocking rules are in effect. *Note:* It may be operated manually or automatically.   328

interlocking deactivating means (defeater) (industrial control). A manually actuated provision for temporarily rendering an interlocking device ineffective, thus permitting an operation that would otherwise be prevented. For example, when applied to apparatus such as combination controllers or control centers, it refers to voiding of the mechanical interlocking mechanism between the externally operable disconnect device and the enclosure doors to permit entry into the enclosure while the disconnect device is closed. *See: electric controller.*   206

interlocking limits (interlocking territory) (railway). An expression used to designate the trackage between the opposing home signals of an interlocking. *See: interlocking.*   328

interlocking machine (railway). An assemblage of manually operated levers or equivalent devices, for the control of signals, switches, or other units, and including mechanical or circuit locking, or both, to establish proper sequence of movements. *See: interlocking (interlocking plant).*   328

interlocking plant. *See: interlocking.*

interlocking relay (railway). A relay that has two independent magnetic circuits with their respective armatures so arranged that the dropping away of either armature prevents the other armature from dropping away to its full stroke.   328

interlocking signals (railway). The fixed signals of an interlocking.   328

interlocking station (railway). A place from which an interlocking is operated.   328

interlocking territory. *See: interlocking limits.*

interlock relay. A relay with two or more armatures having a mechanical linkage, or an electric interconnection, or both, whereby the position of one armature permits, prevents, or causes motion of another armature. *See: relay.*   259

intermediate contacts (of a switching device) (power switchgear). Contacts in the main circuit which part after the main contacts and before the arcing contacts have parted.   103

intermediate current ratings of distribution fuse links

(high-voltage switchgear). A series of distribution fuse link ratings chosen from a series of preferred numbers which are spaced between the preferred current ratings, but which may not provide coordination therewith. Coordination between adjacent intermediate ratings may be secured to the same degree as between adjacent preferred current ratings. 443

intermediate distributing frame (IDF) (telephone switching systems). A frame where crossconnections are made only between units of central office equipment. 55

intermediate frequency (1)(nonlinear, active, and non-reciprocal waveguide components). (A) (general). A frequency to which a signal wave is shifted locally as an intermediate step in transmission or reception. (B) (superheterodyne reception). The difference frequency resulting from a frequency conversion before demodulation. 530
(2) (data transmission). A frequency to which a signal wave is shifted locally as an intermediate step in transmission or reception. 59
(3) (superheterodyne reception). The frequency resulting from a frequency conversion before demodulation. See: radio transmission. 339
(4) (IF) (overhead-power-line corona and radio noise). The frequency resulting from a frequency conversion before modulation. 411

intermediate-frequency-harmonic interference (super-heterodyne receivers). Interference due to radio-frequency-circuit acceptance of harmonics of an intermediate-frequency signal. 339

intermediate-frequency interference ratio. See: intermediate-frequency response ratio; radio receiver.

intermediate-frequency response ratio (super-heterodyne receivers). The ratio of (1) the field strength at a specified frequency in the intermediate frequency band to (2) the field strength at the desired frequency, each field being applied in turn, under specified conditions, to produce equal outputs. See: radio receiver. 339

intermediate-frequency transformer. A transformer used in the intermediate-frequency portion of a heterodyne system. Note: Intermediate-frequency transformers are frequently narrow-band devices. 197

intermediate layer (solar cells). The material on the solar cell surface that provides improved spectral match between the cell and the medium in contact with this surface. 113

intermediate maintenance (test, measurement and diagnostic equipment). Maintenance which is the responsibility of and performed by designated maintenance activities for direct support of using organizations. Its phases normally consist of calibration, repair or replacement of damaged or unserviceable parts, components or assemblies: the emergency manufacture of nonavailable parts and providing technical assistance to using organizations. 54

intermediate means (measurement sequence). All system elements that are used to perform necessary and distinct operations in the measurement sequence be-

tween the primary detector and the end device. Note: The intermediate means, where necessary, adapts the operational results of the primary detector to the input requirements of the end device. See: measurement system. 295

intermediate metal conduit (National Electrical Code). A metal raceway of circular cross section with integral or associated couplings, connectors and fittings approved for the installation of electrical conductors. 256

intermediate office (telephone switching systems). A switching entity where trunks are terminated for purposes of interconnection to other offices. 55

intermediate repeater (data transmission). A repeater for use in a trunk of line at a point other than an end. 59

intermediate subcarrier. A carrier that may be modulated by one or more subcarriers and that is used as a modulating wave to modulate a carrier or another intermediate subcarrier. See: carrier; subcarrier. 111

intermittent duty (1) (general). A requirement of service that demands operation for alternate periods of (A) load and no load: or (B) load and rest: or (C) load, no load, and rest, such alternate intervals being definitely specified. 310, 233,257
(2) (rotating machinery). A duty in which the load changes regularly or irregularly with time. See: asynchronous machine; voltage regulator. 63
(3) (National Electrical Code). Operation for alternate intervals of (A) load and no load; or (B) load and rest; or (C) load, no load and rest. 256
(4) (electric installations on shipboard). A requirement of service that demands operation for alternate periods of (A) load and no load; or (B) load and rest; or (C) load, no load and rest; such alternate intervals being definitely specified. 3
(5) (packaging machinery). A requirement of service that demands operation for alternate intervals of load and no-load; or load and rest; or load, no-load, and rest; such alternate intervals being definitely specified. 429
(6) (power and distribution transformer). A requirement of service that demands operation for alternate periods of (a) load and no load; or (b) load and rest; or (c) load, no load, and rest; such alternate intervals being definitely specified. 53

intermittent-duty rating. The specified output rating of a device when operated for specified intervals of time other than continuous duty. 111

intermittent failure. See: failure, intermittent.

intermittent fault (surge arresters). A fault that recurs in the same place and due to the same cause within a short period of time. 244, 62

intermittent inductive train control. Intermittent train control in which the impulses are communicated to the vehicle-carried apparatus inductively. See: automatic train control. 328

intermittent rating. See: periodic rating.

intermittent test (battery). A service test in which the battery is subjected to alternate discharges and periods of recuperation according to a specified program until the cutoff voltage is reached. *See:* **battery (primary or secondary).**   328

intermittent train control. A system of automatic train control in which impulses are communicated to the locomotive or vehicle at fixed points only. *See:* **automatic train control; intermittent inductive train control.**   328

intermodal distortion. *See:* **multimode distortion.**

intermodulation (nonlinear transducer element). The modulation of the components of a complex wave by each other, as a result of which waves are produced that have frequencies equal to the sums and differences of integral multiples of those of the components of the original complex wave. *See:* **modulation.**   111, 199

intermodulation distortion (1)(data transmission). Nonlinear distortion of a system or transducer, characterized by the appearance in the output of frequencies equal to the sums and differences of integral multiples of the two or more component frequencies present in the input wave. Harmonic components also present in the output are usually not included as part of the intermodulation distortion. When harmonics are included, a statement to that effect should be made.   59
(2)(nonlinear, active, and nonreciprocal waveguide components). Distortion produced by undesired intermodulation.   530

intermodulation interference (mobile communication). The modulation products attributable to the components of a complex wave that on injection into a nonlinear circuit produces interference on the desired signal. *See:* **mobile communication system.**   181

intermodulation product intercept point (nonlinear, active, and nonreciprocal waveguide components). Intermodulation products have an output-versus-input characteristic which, when graphically displayed, would theoretically intercept the plot of the desired output-versus-input if the nonlinear device continued to operate linearly without compression. The signal input level at which this theoretical point would occur is called the intercept point and is usually defined in dBm (decibel referred to one milliwatt). The figure below is a graphical representation of the intercept points for a single-tone second order and a two-tone third-order intermodulation product.   530

intermodulation products (nonlinear, active, and nonreciprocal waveguide components). The undesired responses in a nonlinear device that result from harmonics of two or more signals.   530

intermodulation rejection (spectrum analyzer). The ratio of the sensitivity level and the level of either of two equal amplitude signals which produce any intermodulation product at the sensitivity level.   390

intermodulation spurious emission (land-mobile communication transmitters). External radio frequency (RF) emission of a transmitter which is a product of the nonlinear mixing process in the final stage of the

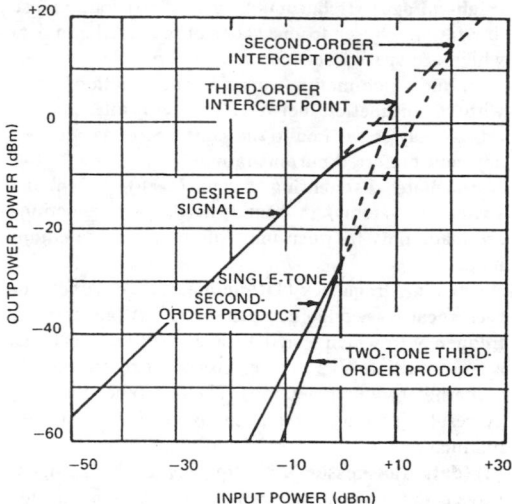

Intermodulation IM Product Intercept Point

transmitter which occurs when external RF power is coupled through the antenna output.   444

intermodulation spurious response (1) (receiver performance). The receiver audio output resulting from the mixing of *n*th-order frequencies, in the nonlinear elements of the receiver, in which the resultant carrier frequency is equivalent to the assigned frequency. *See:* **spurious response; receiver performance.**   181
(2) (nonreal time spectrum analyzer). The spectrum analyzer response resulting from the mixing of the *n*th order frequencies, in the nonlinear elements of the spectrum analyzer, in which the resultant response is equivalent to the tuned frequency. *Syn:* **intermodulation distortion.**   68
(3) (frequency-modulated mobile communications receivers). The response resulting from the mixing of two or more undesired frequencies in the nonlinear elements of the receiver in which a resultant frequency is generated that falls within the receiver passband.   123

(4) (spectrum analyzer). The spectrum analyzer response resulting from the mixing of the nth order frequencies of the input signal in the nonlinear elements of the spectrum analyzer, in which the resultant response is equivalent to the tuned frequency. *Syn:* **intermodulation distortion.**   390

internal address field (IA)(FASTBUS acquisition and control). The group of low-order bits (right justified and contiguous to the device address field on the left) assigned in the address of a FASTBUS device which is used to identify internal locations within a FASTBUS device. Secondary address cycles allow the number of different locations accessed to exceed that available in the internal address field.   480

internal audit (nuclear power quality assurance). An audit of those portions of an organization's quality assurance program retained under its direct control and within its organizational structure.   417

internal bias (teletypewriter). Bias, either marking or spacing, that may occur within a start-stop printer receiving mechanism and that will have an effect on the margins of operation.   194

internal blocking (telephone switching systems). The unavailability of paths in a switching network between a given inlet and any suitable idle outlet.   55

internal connector (pothead). A connector that joins the end of the cable to the other current-carrying parts of a pothead. *See:* **pothead; transformer.**
                                                    4, 288, 289

internal correction voltage (electron tubes). The voltage that is added to the composite controlling voltage and is the voltage equivalent of such effects as those produced by initial electron velocity and contact potential. *See:* **composite controlling voltage (electron tube).**   125

internal heating (electrolysis). The electrolysis of fused electrolytes is the method of maintaining the electrolyte in a molten condition by the heat generated by the passage of current through the electrolyte. *See:* **fused electrolyte.**   328

internal impedance (rotating machinery). The total self-impedance of the primary winding under steady conditions. *Note:* For a three-phase machine, the primary current is considered to have only a positive-sequence component when evaluating this quantity. *See:* **asynchronous machine.**   63

internal impedance drop (rotating machinery). The product of the current and the internal impedance. *Note:* This is the phasor difference between the generated internal voltage and the terminal voltage of a machine. *See:* **asynchronous machine.**   63

internal insulation (1) (apparatus) (surge arresters). The insulation that is not directly exposed to atmospheric conditions.   308, 62, 53

(2) (high voltage testing). Internal insulation comprises the internal solid, liquid, or gaseous elements of the insulation of equipment, which are protected from the effects of atmospheric and other external conditions such as contamination, humidity, vermin, etcetera.   150

(3) (power and distribution transformer). The insulation that is not directly exposed to atmospheric conditions.   53

internal load (power operations). Equal to customer load plus the station service load plus the transmission losses.   516

internal logic state (graphic symbols for logic functions). A logic state assumed to exist inside a symbol outline at an input or an output. See figure below.

EXTERNAL LOGIC   EXTERNAL LOGIC
STATES OR   STATE OR
LOGIC LEVELS   LOGIC LEVEL

                                                    451

internally-programmed automatic test equipment (test, measurement and diagnostic equipment). An automated tester using any programming technique in which a substantial amount of programming information is stored within the equipment, although it may originate from external media.   54

internal oxidation. *See:* **subsurface corrosion.**

internal remanent residual voltage (Hall-effect devices). That portion of the zero field residual voltage which is due to the remanent magnetic flux density in the ferromagnetic encapsulation of the Hall generator.   107

internal resistance (battery). The resistance to the flow of an electric current within a cell or battery. *See:* **battery (primary or secondary).**   328

internal storage (test, measurement and diagnostic equipment). Storage facilities forming an integral part of the machine.   54

internal traffic (telephone switching systems). Traffic originating and terminating within the network being considered.   55

internal triggering (1) (oscilloscopes). The use of a portion of a deflection signal (usually the vertical deflection signal) as a triggering-signal source. *See:* **oscillograph.**   185

(2) (non-real time spectrum analyzer). The use of a deflection signal (usually the vertical deflection signal) as a triggering source.   68

internal trigger–video trigger (spectrum analyzer). The use of a deflection signal (usually the vertical deflection signal) as a triggering source.   390

international call (telephone switching systems). A call to a destination outside of the national boundaries of the calling customer.   55

International Commission on Illumination. *See:* **CIE.**

international direct distance dialing (IDDD) (telephone switching systems). The automatic establishing of international calls by means of signals from the calling device of a customer.   55

international distance dialing (IDD) (telephone switching systems). The automatic establishing of international calls by means of signals from the calling device of either a customer or an operator.   55

international interzone call (telephone switching systems). A call to a destination outside of a national or integrated numbering-plan area.   55

international intrazone call (telephone switching systems). A call to a destination within the boundaries of an integrated numbering-plan area, but outside the national boundaries of the calling customer.   55

International Morse code (Continental code). A system of dot and dash signals, differing from the American Morse code only in certain code combinations, used chiefly in international radio and wire telegraphy. *See:* **telegraphy.**   328

international number (telephone switching systems). The combination of digits representing a country code plus a national number.   55

international operating center (telephone switching systems). In World Zone 1, a center where telephone operators handle originating and terminating international interzone calls and may also handle international intrazone calls. *See:* **world zone number.**   55

international originating toll center (telephone switching systems). In World Zone 1, a toll center where telephone operators handle originating international interzone calls. *See:* **world zone number.**   55

international switching center (telephone switching systems). A toll office that normally serves as a point of entry or exit for international interzone calls.   55

International System of Electrical Units. A system that uses the **international ampere** and the **international ohm.** *Notes:* (1) The international ampere was defined as the current that will deposit silver at the rate of 0.00111800 gram per second: and the international ohm was defined as the resistance at 0 degrees Celsius of a column of mercury having a length of 106.300 centimeters and a mass of 14.4521 grams. (2) The International System of Electrical Units was in use between 1893 and 1947 inclusive. By international agreement it was discarded, effective January 1, 1948 in favor of the MKSA system. (3) Experiments have shown that as these units were maintained in the United States of America, 1 international ohm equalled 1.000495 ohm and that 1 international ampere equalled 0.999835 ampere. *See:* **International System of Units.**   210

International System of Units (SI). A universal coherent system of units in which the following six units are considered basic: meter, kilogram, second, ampere, Kelvin degree, and candela. *Notes:* (1) The MKSA system of electrical units (MKSA System of Units) is a constituent part of this system adequate for mechanics and electromagnetism. (2) The electrical units of this system should not be confused with the units of the earlier International System of Electrical Units which was discarded January 1, 1948. (3) The International System of Units (abbreviated SI) was promulgated in 1960 by the Eleventh General Conference on Weights and Measures. *See:* **units and letter symbols.**   210

interoffice call (telephone switching systems). A call between lines connected to different central offices.   55

interoffice trunk (telephony). A direct trunk between local central offices in the same exchange.   328

interoperability (software). The ability of two or more systems to exchange information and to mutually use the information that has been exchanged. *See:* **compatibility; system.**   434

interphase rod or shaft (high voltage air switches, insulators, and bus supports). A component of a switch operating mechanism designed to connect two or more poles of a multipole switch for group operation.   27

interphase transformer. An autotransformer, or a set of mutually coupled reactors, used to obtain parallel operation between two or more simple rectifiers that have ripple voltages that are out of phase. *See:* **autotransformer; rectifier transformer.**   203, 53

interphase transformer (power and distribution transformer). An autotransformer, or a set of mutually coupled reactors, used to obtain parallel operation between two or more simple rectifiers that have ripple voltages that are out of phase.   53

interphase-transformer loss (rectifier transformer). The losses in the interphase transformer that are incident to the carrying of rectifier load. *Note:* They include both magnetic core loss and conductor loss. *See:* **rectifier transformer.**   258

interphase-transformer rating. Consists of the root-mean-square current, root-mean-square voltage, and frequency, at the terminals of each winding, for the rated load of the rectifier unit, and a designated amount of phase control, as assigned to it by the manufacturer. *See:* **duty; rectifier transformer.**   203

interphone (intercom). *See:* **intercommunicating system.**

interphone equipment (aircraft). Equipment used to provide telephone communications between personnel at various locations in an aircraft.   328

interpolation (signal interpolation) (submarine cable telegraphy). A method of reception characterized by synchronous restoration of unit-length signal elements which are weak or missing in the received signals as a result of one or more of such factors as suppression at the transmitter, attenuation in transmission, or discrimination in the receiving networks. *Note:* This is sometimes referred to as local correction. *See:* **telegraphy.**   328

interpolation function (burst measurements). A function that may be used to obtain additional values between sampled values. *See:* **burst.**   292

interpole. *See:* **commutating pole.**

interposing relay (1)(supervisory control, data acquisition, and automatic control)(station control and data acquisition). A device which enables the energy in a high-power circuit to be switched by a low-power control signal.   570, 403
(2) (of a supervisory system) (power switchgear). An auxiliary relay at the master or remote station, the contacts of which serve: (A) to energize a circuit (for closing, opening, or other purpose) of an element of remote station equipment when the selection of a desired point has been completed and when suitable operating signals are received through the supervisory equipment from the master station; or (B) to connect in the circuit the telemeter transmitting and receiving equipments, respectively, at the remote and master stations. *Note:* The interposing relays are considered part of a supervisory system.   103

interpret (software). To translate and to execute each source language statement of a computer program before translating and executing the next statement. *See:* **assemble; compile; computer program; source language.**   434

interpreter (software). (1) Software, hardware, or firmware used to interpret computer programs. (2) A computer program used to interpret. *See:* **assembler; compiler; computer program; firmware; hardware; interpret; software.**   434

inter-record gap (test, measurement and diagnostic

equipment). An interval of space or time deliberately left between recording portions of data or records. Such spacing is used to prevent errors through loss of data or overwriting and permits tape stop-start operations.   54

interreflection. *See:* **interflection.**

interrogation (radar). In a transponder system, the signal or combination of signals intended to trigger a response.   13

interrogative supervisory system (power switchgear). A system whereby the master station controls all operations of the system, and whereby all indications are obtained on a master station request or interrogation basis. *Note:* The normal state is usually one of continuous interrogation or polling of the remote stations for changes in status.   103

interrogator (radar). (1) Same as interrogator-responsor; (2) The transmitting part of an interrogator-responsor.   13

interrogator-responsor (IR) (radar). A combined radio transmitter and receiver for interrogating a transponder and reporting the resulting replies independently of a radar echo display. *See:* **interrogator.**
                                                                 13

interrupt (software). A suspension of a process such as the execution of a computer program, caused by an event external to that process, and performed in such a way that the process can be resumed. *Syn:* **interruption.** *See:* **computer program; execution; process.**
                                                                434

interrupted continuous wave (ICW). A continuous wave that is interrupted at a constant audio-frequency rate. *See:* **radio transmission.**   111, 211

interrupted quick-flashing light (illuminating engineering). A quick flashing light in which the rapid alternations are interrupted by periods of darkness at regular intervals.   167

interrupter (power switchgear). An element designed to interrupt specified currents under specified conditions.   103

interrupter blade (of an interrupter switch) (power switchgear). A blade used in the interrupter for breaking the circuit.   103

interrupter switch (power switchgear). An air switch, equipped with an interrupter, for making or breaking specified currents, or both. *Note:* The nature of the current made or broken, or both, may be indicated by suitable prefix: that is, load interrupter switch, fault interrupter switch, capacitor current interrupter switch, etcetera.   103

interruptible load (power operations). A load which can be interrupted as defined by contract.   516

interruptible load reserve (power operations). The operating reserve available through disconnection of interruptible loads.   516

interruptible power (power operations). Power which can be interrupted as defined by contract.   516

interrupting capacity (packaging machinery). The highest current at rated voltage that the device can interrupt.   429

interrupting current (power switchgear). The current

in a pole of a switching device at the instant of the initiation of the arc. *Syn:* **breaking current.**   103

interrupting rating (protection and coordination of industrial and commercial power systems). A rating based on the highest root-mean-square (rms) alternating current that the fuse is required to interrupt under the conditions specified. The interrupting rating, in itself, has no direct bearing on any current-limiting effect of the fuse.   504

interrupting tests (power switchgear). Tests that are made to determine or check the interrupting performance of a (switching) device.   103, 443

interrupting (total break) time (of a mechanical switching device) (power switchgear). The interval between the time when the actuating quantity of the release circuit reaches the operating value, the switching device being in a closed position, and the instant of arc extinction on the primary arcing contacts. *Notes:* (1) Interrupting time is numerically equal to the sum of opening time and arcing time. (2) In multipole devices interrupting time may be measured for each pole or for the device as a whole, in which latter case the interval is measured to the instant of arc extinction in the last pole to clear.   103

interruption (1)(electric power systems). The loss of service to one or more consumers or other facilities. *Note:* It is the result of one or more component outages, depending on system configuration. *See:* **outage.**
                                                                290

(2)(reliable industrial and commercial power systems planning and design). The loss of electric power supply to one or more loads.   561

interruption duration (electron power systems). The period from the initiation of an interruption to a consumer or other facility until service has been restored to that consumer or facility. *See:* **outage.**   200

interruption duration index (electric power systems). The average interruption duration for consumers interrupted during a specified time period. It is estimated from operating history by dividing the sum of all consumer interruption durations during the specified period by the number of consumer interruptions during that period. *See:* **outage.**   200

interruption, forced (electric power systems). An interruption caused by a forced outage.   112

interruption frequency (reliable industrial and commercial power systems planning and design). The expected (average) number of power interruptions to a load per unit time, usually expressed as interruptions per year.   561

interruption frequency index (electric power systems). The average number of interruptions per consumer served per time unit. *Note:* It is estimated from operating history by dividing the number of consumer interruptions observed in a time unit by the number of consumers served. A consumer interruption is considered to be one interruption of one consumer. *See:* **outage.**   200

interruption, momentary (electric power systems). An interruption of duration limited to the period required to restore service by automatic or supervisory-controlled switching operations or by manual switching at locations where an operator is immediately available. *Note:* Such switching operations must be completed in a specified time not to exceed 5 minutes. 112

interruption, scheduled (electric power systems). An interruption caused by a scheduled outage. *See:* **outages.** 112

interruption, sustained (electric power systems). Any interruption not classified as a momentary interruption. 112

interruption to service (power switchgear). The isolation of an electrical load from the system supplying that load, resulting from an abnormality in the system. 103

interrupt operation (FASTBUS acquisition and control). A FASTBUS write operation to an interrupt service device, notifying it that the sender requires attention. 480

interrupt service device (ISD)(FASTBUS acquisition and control). A processor or other device which can respond to interrupt operations. 480

interspersing (rotating machinery). Interchanging the coils at the edges of adjacent phase belts. *Note:* The purpose of interspersing depends on the type of machine in which it is done. In asynchronous motors it is used to reduce harmonics that can cause crawling. *See:* **asynchronous machine.** 63

intersymbol interference (transmission system) (modulation systems). Extraneous energy from the signal in one or more keying intervals that tends to interfere with the reception of the signal in another keying interval, or the disturbance that results. 242

intersystem electromagnetic compatibility (control of system electromagnetic compatibility). The condition that enables a system to function without perceptible degradation due to electromagnetic sources in another system. 495

intertoll dialing (telephony). Dialing over intertoll trunks. 328

intertoll trunk (telephone switching systems). A trunk between two toll offices. 55

interturn insulation (rotating machinery). The insulation between adjacent turns, often in the form of strips. 63

interturn test. *See:* **turn-to-turn test (rotating machine-ry).**

interval (1) (acoustics). The spacing between two sounds in pitch or frequency, whichever is indicated by the context. *Note:* The frequency interval is expressed by the ratio of the frequencies or by a logarithm of this ratio. 176

(2) (pulse terms). The algebraic time difference calculated by subtracting the time of a first specified instant from the time of a second specified instant. 254

interval, sweep holdoff (oscilloscopes). The interval between sweeps during which the sweep and/or trigger cicuits are inhibited. 106

intrabeam viewing (laser-maser). The viewing condition whereby the eye is exposed to all or part of a laser beam. 363

intramodal distortion (fiber optics). That distortion

resulting from dispersion of group velocity of a propagating mode. It is the only distortion occurring in single mode waveguides. *See:* **dispersion; distortion.** 433

intraoffice call (telephone switching systems). A call between lines connected to the same central office. 55

intrasystem electromagnetic compatibility (control of system electromagnetic compatibility). The condition that enables the various portions of a system to function without perceptible degradation due to electromagnetic sources in other portions of the same system. 495

intrinsically safe equipment and wiring (National Electrical Code). Equipment and wiring that are incapable of releasing sufficient electrical or thermal energy under normal or abnormal conditions to cause ignition of a specific hazardous atmospheric mixture in its most easily ignited concentration. This equipment is suitable for use in division 1 locations. Division 2 Equipment and Wiring are equipment and wiring which in normal operation would not ignite a specific hazardous atmosphere in its most easily ignited concentration. The circuits may include sliding or make-and-break contacts releasing insufficient energy to cause ignition. Circuits not containing sliding or make-and-break contacts may have higher energy levels potentially capable of causing ignition under fault conditions. 65

intrinsic coercive force. The magnetizing force at which the intrinsic induction is zero when the material is in a symmetrically cyclically magnetized condition. 210

intrinsic induction (magnetic polarization). At a point in a magnetized body, the vector difference between the magnetic induction at that point and the magnetic induction that would exist in a vacuum under the influence of the same magnetizing force. This is expressed by the equation $\mathbf{B}_i = \mathbf{B} - \mu_0\mathbf{H}$. *Note:* In the centimeter-gram-second electromagnetic-unit system, $\mathbf{B}_i/4\pi$ is often called magnetic polarization. 210

intrinsic joint loss (fiber optics). That loss, intrinsic to the fiber, caused by fiber parameter (for example, core dimensions, profile parameter) mismatches when two nonidentical fibers are joined. *See:* **angular misalignment loss; extrinsic joint loss; gap loss; lateral offset loss.** 433

intrinsic loss (or gain) (waveguide). A loss (or gain) resulting from placing two ports of a network between generator and load impedance whose values are adjusted for maximum power absorbed in the load. It is the ratio of the available power from the generator (without generator adjustment) to the power delivered to the load with the network present. 267

intrinsic permeability. The ratio of intrinsic normal induction to the corresponding magnetizing force. *Note:* In anisotropic media, intrinsic permeability becomes a matrix. 210

intrinsic properties (semiconductor). The properties of a semiconductor that are characteristic of the pure, ideal crystal. *See:* **semiconductor.**

245, 66, 210, 186

intrinsic semiconductor (germanium gamma-ray detectors)(charged-particle detectors)(X-ray energy spectrometers). A semiconductor containing an equal number of free holes and electrons throughout its volume. (The term 'intrinsic germanium' is often used incorrectly for 'high purity germanium'.

528, 119, 471

intrinsic semiconductor. *See:* **semiconductor, intrinsic.**

intrinsic temperature range (semiconductor). The temperature range in which the charge-carrier concentration of a semiconductor is substantially the same as that of an ideal crystal. *See:* **semiconductor.**

214, 66, 210

introspective testing. *See:* **self-test.**

intrusion detection (nuclear security systems). Sensing the presence of an intruder or object within specific confines.

464

invalid frame (logical link control). A protocol data unit (PDU) that either: (1) Does not contain an integral number of octets, (2) Does not contain at least two address octets and a control octet, (3) Is identified by the physical layer or medium access control (MAC) sublayer as containing data bit errors.

585

inverse electrode current. The current flowing through an electrode in the direction opposite to that for which the tube is designed. *See:* **electrode current (electron tube).**

190, 125

inverse magnitude contours. *See:* **magnitude contours.**

inverse networks. Two two-terminal networks are said to be inverse when the product of their impedances is independent of frequency within the range of interest. *See:* **network analysis.**

210

inverse neutral telegraph transmission. That form of transmission employing zero current during marking intervals and current during spacing intervals. *See:* **telegraphy.**

194

inverse Nyquist diagram. *See:* **Nyquist diagram.**

inverse-parallel connection (industrial control). An electric connection of two rectifying elements such that the cathode of the first is connected to the anode of the second, and the anode of the first is connected to the cathode of the second.

206

inverse period (rectifier element). The nonconducting part of an alternating-voltage cycle during which the anode has a negative potential with respect to the cathode. *See:* **rectification.**

328

inverse-square law (illuminating engineering). A law stating that the illuminance E at a point on a surface varies directly with the intensity I of a point source, and inversely as the square of the distance d between the source and the point. If the surface at the point is normal to the direction of the incident light, the law is expressed by $E = I / d^2$. *Note:* For sources of finite size having uniform luminance this gives results which are accurate within one percent when d is at least five times the maximum dimension of the source as viewed from the point on the surface. Even though practical interior luminaires do not have uniform luminance,

this distance, d, is frequently used as the minimum for photometry of such luminaires, when the magnitude of the measurement error is not critical.

167

inverse time (1) (National Electrical Code). (As applied to Circuit Breakers.) A qualifying term indicating there is purposely introduced a delay in the tripping action of the circuit breaker, which delay decreases as the magnitude of the current increases.

256

(2) (industrial control). *See:* **inverse-time relay.**

inverse-time delay (power switchgear). A qualifying term indicating that there is purposely introduced a delaying action, the delay decreasing as the operating force increases.

103

inverse-time relay (power switchgear). A relay in which the input quantity and operating time are inversely related throughout at least a substantial portion of the performance range. *Note:* Types of inverse-time relays are frequently identified by such modifying adjectives as "definite minimum time," "moderately," "very," and "extremely" to identify relative degree of inverseness of the operating characteristics of a given manufacturer's line of such relays. *See:* **definite-minimum-time relay; figure.**

103

inverse transfer function (control systems). The reciprocal of a transfer function.

56

inverse transfer locus (control systems). The locus of the inverse transfer function.

56

inverse voltage (rectifier). The voltage applied between the two terminals in the direction opposite to the forward direction. This direction is called the backward direction. *See:* **rectification of an alternating current.**

244

inversion (metal-nitride-oxide field-effect transistor). The state of the silicon surface in the insulated-gate field-effect transistor (IGFET) structure when the voltage leading to depletion has been further increased such that a thin layer of minority carriers becomes stable at the surface.

386

inversion efficiency. The ratio of output fundamental power to input direct power expressed in percent. *See:* **self-commutated inverters.**

208

inversion ratio (laser-maser). In a maser medium, the negative of the ratio of (1) the population difference between two nondegenerate energy states under a condition of population inversion to (2) the population difference at equilibrium.

363

inverted (rotating machinery). Applied to a machine in which the usual functions of the stationary and revolving members are interchanged. *Example:* an induction motor in which the primary winding is on the rotor and is connected to the supply through sliprings, and the secondary is on the stator. *See:* **asynchronous machine.**

63

inverted input (oscilloscopes). An input such that the applied polarity causes a deflection polarity opposite from conventional deflection polarity.

106

inverted-turn transposition (rotating machinery). A form of transposition used on multiturn coils in which one or more turns are given a 180-degree twist in the end winding or at the coil nose or series loop. *See:* **rotor (rotating machinery); stator.**

63

inverter (1)(**electric power**). A machine, device, or system that changes direct-current power to alternating-current power. *See:* **inverting amplifier; electronic analog computer.**   9, 10
(2)(**self-commutated converters**). A converter for conversion from direct current (dc) to alternating current (ac).   584
inverting amplifier (analog computers). An operational amplifier that produces an output signal of nominally equal magnitude and opposite algebraic sign to the input signal. *Note:* In an analog computer, the term inverter is synonymous with inverting amplifier.
                    9
inverting parametric device. A parametric device whose operation depends essentially upon three frequencies, a harmonic of the pump frequency and two signal frequencies, of which the higher signal-frequency is the difference between the pump harmonic and the lower signal frequency. *Note:* Such a device can exhibit gain at either of the signal frequencies provided power is suitably dissipated at the other signal frequency. It is said to be inverting because if one of the two signals is moved upward in frequency, the other will move downward in frequency. *See:* **parametric device.**   191
invisible range. *See:* **visible range.**
inward-wats service (telephone switching systems). A reverse-charge, flat-rate, or measured-time direct distance dialing service to a specific directory number.
                    55
I/O (input/output). Input or output or both.
                    255, 77
IOC. *See:* **integrated optical circuit.**
ion. An electrically charged atom or radical.   205
ion activity (ion species). The thermodynamic concentration, that is, the ion concentration corrected for the deviation from the law of ideal solutions. *Note:* The activity of a single ion species cannot, however, be measured thermodynamically. *See:* **ion.**   328
ion burn. *See:* **ion spot.**
ion charging (charge-storage tubes). Dynamic decay caused by ions striking the storage surface. *See:* **charge-storage tube.**   174
ion concentration (species of ion). The concentration equal to the number of those ions, or of moles or equivalent of those ions, contained in a unit volume of an electrolyte.   328
ion exchange technique (fiber optics). A method of fabricating a graded index optical waveguide by an ion exchange process. *See:* **chemical vapor disposition technique; double crucible method; graded index profile.**   433
ion gun. A device similar to an electron gun but in which the charged particles are ions. *Example:* proton gun. *See:* **electron optics.**   244, 190
ionic-heated cathode (electron tube). A hot cathode that is heated primarily by ionic bombardment of the emitting surface.   125
ionic-heated-cathode tube. An electron tube containing an ionic-heated cathode.   125
ion implantation (germanium gamma-ray detectors)(-charged-particle detectors). A process in which a beam of energetic ions incident upon a solid results in the imbedding of those ions into the material.
                    528, 119
ion-implanted contact (germanium gamma-ray detectors)(charged-particle detectors). A detector contact consisting of a junction produced by ion implantation.
                    528, 119
ionization (1) (**power lines**). A breakdown that occurs in parts of a dielectric when the electric stress in those parts exceeds a critical value without initiating a complete breakdown of the insulation system. *Note:* Ionization can occur both on internal and external parts of a device. It is a source of radio noise and can damage insulation.   229, 62
(2) (**outdoor apparatus bushings**). (A) The formation of limited avalanches of electrons developed in insulation due to an electric field. (B) **Ionization current** is the result of capacitive discharges in an insulating medium due to electron avalanches under the influence of an electric field. *Note:* The occurrence of such currents may cause: (1) radio noise. (2) damage to insulation.   168
(3) (**corona measurement**). Any process by which neutral molecules or atoms dissociate to form positively and negatively charged particles.   375
(4) (**overhead-power-line corona and radio noise**). The process or the result of any process by which a neutral atom or molecule acquires either a positive or a negative charge.   411
ionization current (1)(**metal-oxide surge arresters for ac power circuits**)(**surge arrester**). The electric current resulting from the movement of electric charges in an ionized medium, under the influence of an applied electric field.   583, 430
(2) (**vacuum tubes**). *See:* **gas current.**
ionization extinction voltage (corona level) (cables). The minimum value of falling root-mean-square voltage that sustains electric discharge within the vacuous or gas-filled spaces in the cable construction or insulation.   57
ionization factor (power distribution, underground cables) (dielectric). The difference between percent power factors at two specified values of electric stress. The lower of the two stresses is usually so selected that the effect of the ionization on power factor at this stress is negligible.   57
ionization-gauge tube. An electron tube designed for the measurement of low gas pressure and utilizing the relationship between gas pressure and ionization current.   190
ionization measurement. The measurement of the electric current resulting from the movement of electric charges in an ionized medium under the influence of the prescribed electric field.   287
ionization or corona detector. *See:* **discharge detector.**
ionization or corona inception voltage. *See:* **discharge inception voltage.**
ionization or corona probe. *See:* **discharge probe.**
ionization smoke detector (fire protection devices). A device which has a small amount of radioactive mate-

rial which ionizes the air in the sensing chamber, thus rendering it conductive and permitting a current flow through the air between two charged electrodes. This gives the sensing chamber an effective electrical conductance. When smoke particles enter the ionization area, they decrease the conductance of the air by attaching themselves to the ions, causing a reduction in mobility. When the conductance is less than the predetermined level, the detector circuit responds.
71

ionization time (gas tube). The time interval between the initiation of conditions for and the establishment of conduction at some stated value of tube voltage drop. *Note:* To be exact the ionization time of a gas tube should be presented as a family of curves relating such factors as condensed-mercury temperature, anode and grid currents, anode and grid voltages, and regulation of the grid current.
125

ionization vacuum gauge. A vacuum gauge that depends for its operation on the current of positive ions produced in the gas by electrons that are accelerated between a hot cathode and another electrode in the evacuated space. *Note:* It is ordinarily used to cover a pressure range of 10^{-4} to 10^{-10} conventional millimeters of mercury. *See:* **instrument.**
328

ionization voltage (metal-oxide surge arresters for ac power circuits)(surge arrester). A high-frequency voltage appearing at the terminals of an arrester, generated by all sources, but particularly by ionization current within the arrester, when a power-frequency voltage is applied across the terminals.   583, 430

ionizing event (gas-filled radiation-counter tube). Any interaction by which one or more ions are produced.
125

ionizing radiation (1) (air). (A) Particles or photons of sufficient energy to produce ionization in their passage through air. (B) Particles that are capable of nuclear interactions with the release of sufficient energy to produce ionization in air.   335

(2) (scintillation counting). Particles or photons of sufficient energy to produce ionization in interactions with matter.   117

ion migration. A movement of ions in an electrolyte as a result of the application of an electric potential. *See:* **ion.**   328

ionogram (radio wave propagation). A record showing the group paths of ionospherically returned echoes as function of frequency.   146

ionosonde (radio wave propagation). A radar-like device that transmits radio waves vertically or obliquely to the ionosphere and uses the received reflected waves to form an ionogram. *Note:* The transmission technique may involve simple or complex waveforms.
146

ionosphere (1) (data transmission). That part of the earth's outer atmosphere where ions and free electrons are normally present in quantities sufficient to affect propagation of radio waves.   59

(2) (radio wave propagation). That part of a planetary atmosphere where ions and electrons are present in quantities sufficient to affect the propagation of radio waves.   146

ionosphere disturbance. A variation in the state of ionization of the ionosphere beyond the normally observed random day-to-day variation from average values for the location, date, and time of day under consideration. *Note:* Since it is difficult to draw the line between normal and abnormal variations, this definition must be understood in a qualitative sense. *See:* **radiation.**   328

ionosphere-height error (electronic navigation). The systematic component of the total ionospheric error due to the difference in geometrical configuration between ground paths and ionospheric paths. *See:* **navigation.**   13, 187

ionospheric error (electronic navigation). The total systematic and random error resulting from the reception of the navigational signal via ionospheric reflections: this error may be due to (1) variations in transmission paths, (2) nonuniform height of the ionosphere, and (3) nonuniform propagation within the ionosphere. *See:* **navigation.**   13, 187

ionospheric storm. An ionospheric disturbance characterized by wide variations from normal in the state of the ionosphere, including effects such as turbulence in the F region, increases in absorption, and often decreases in ionization density and increases in virtual height. *Note:* The effects are most marked in high magnetic latitudes and are associated with abnormal solar activity. *See:* **radiation.**   328

ionospheric tilt error (electronic navigation). The component of the ionospheric error due to nonuniform height of the ionosphere. *See:* **navigation.**
187, 13

ionospheric wave. *See:* **sky wave.**

ion repeller (charge-storage tubes). An electrode that produces a potential barrier against ions. *See:* **charge-storage tube.**   174

ion sheath. *See:* **electron sheath.**

ion spot (1) (camera tubes or image tubes). The spurious signal resulting from the bombardment or alteration of the target or photocathode by ions. *See:* **television.**   178, 125

(2) (cathode-ray-tube screen). An area of localized deterioration of luminescence caused by bombardment with negative ions.   178, 125

ion transfer (electrotherapy) (ionic medication) (iontophoresis) (medical ionization) (ion therapy) (ionotherapy). The forcing of ions through biological interfaces by means of an electric field. *See:* **electrotherapy.**   192

ion trap (cathode-ray tube). A device to prevent ion burn by removing the ions from the beam. *See:* **isolation transformer; transformer, isolating.**

IR. *See:* **infrared; interrogator-respondor.**

IRA. *See:* **laser gyro axes.**

IR drop (electrolytic cell). The drop equal to the product of the current passing through the cell and the resistance of the cell.   328

IR-drop compensation transformer (power and distribution transformer). A provision in the transformer by which the voltage drop due to transformer load current and internal transformer resistance is partially or completely neutralized. Such transformers are suitable only for one-way transformation, that is, not interchangeable for step-up and step-down transformations.   53

iris (1) (waveguide technique). A metallic plate, usually of small thickness compared with the wavelength, perpendicular to the axis of a waveguide and partially blocking it. *Notes:* (A) An iris acts like a shunt element in a transmission line: it may be inductive, capacitive, or resonant. (B) When only a single mode can be supported an iris acts substantially as a shunt admittance.   179

(2) (waveguide components). A partial obstruction at a transverse cross-section formed by one or more metal plates of small thickness compared with the wavelength.   166

(3) (laser-maser). The circular pigmented membrane which lies behind the cornea of the human eye. The iris is perforated by the pupil.   363

ironclad plate (storage cell). A plate consisting of an assembly of perforated tubes of insulating material and of a centrally placed conductor. *Note:* "Ironclad" is a registered trademark of ESB Incorporated. *See:* **battery (primary or secondary).**   328

irradiance (1) (at a point of a surface) [E] (laser-maser). Quotient of the radiant flux incident on an element of the surface containing the point by the area of that element. Unit: $W \cdot cm^{-2}$.   363

(2) (fiber optics). Radiant power incident per unit area upon a surface, expressed in watts per square meter. "Power density" is colloquially used as a synonym. *See:* **radiometry.**   433

irreversible dark current increase (diode-type camera tube). That dark current increase which results from irradiation of the target by soft X-rays.   380

irreversible process. An electrochemical reaction in which polarization occurs. *See:* **electrochemistry.**
                                                                    328

irreversible target dark current increase (diode-type camera tube). That dark current increase which is permanent and increases with hours of operation.
                                                                    380

irrigation machines (National Electrical Code). An irrigation machine is an electrically driven or controlled machine, with one or more motors, not hand portable, and used primarily to transport and distribute water for agricultural purposes.   256

I scan. *See:* **I display**

I-scope (radar). A cathode-ray oscilloscope arranged to present an I-display. *See:* **I-display.**   13

island effect (electron tubes). The restriction of the emission from the cathode to certain small areas of it (islands) when the grid voltage is lower than a certain value. *See:* **electronic tube.**   244

ISM apparatus (industrial, scientific, and medical apparatus: electromagnetic compatibility). Apparatus intended for generating radio-frequency energy for industrial, scientific or medical purposes. *See:* **electromagnetic compatibility; industrial electronics.**
                                                               220, 199

isocandela line (illuminating engineering). A line plot-

ted on any appropriate set of coordinates to show directions in space, about a source of light, in which the intensity is the same. A series of such curves, usually for equal increments of intensity, is called an isocandela diagram.   167

isochronous speed governing (gas turbines). Governing with steady-state speed regulation of essentially zero magnitude.   98, 58

isocon mode (camera tube). A low-noise return-beam mode of operation utilizing only back-scattered electrons from the target to derive the signal, with the beam electrons specularly reflected by the electrostatic field near the target being separated and rejected. *See:* **camera tube.**   190

isoelectric point. A condition of net electric neutrality of a colloid, with respect to its surrounding medium. *See:* **ion.**   328

isokeraunic level (lightning). The average annual number of thunderstorm days. *See:* **direct-stroke protection (lightning).**   64

isolated (1) (power line maintenance). (A) Physically separated, electrically and mechanically, from all sources of electrical energy. Such separation may not eliminate the effects of electrical induction. (B) Not readily accessible to persons unless special means for access are used.   458

(2) (transmission and distribution) (National Electrical Code)(NESC). Not readily accessible to persons unless special means for access are used.
                                                            262,494, 256

(3) (conductor stringing equipment). (A) Physically separated, electrically and mechanically, from all sources of electrical energy. Such separation may not eliminate the effects of electrical induction. (B) An object not readily accessible to persons unless special means for access are used.   431

(4) (electrolytic cell line working zone). An object not readily accessible to persons unless special means for access are used.   133

isolated by elevation (NESC). Elevated sufficiently so that persons may safely walk underneath.   494

isolated conductor (ignored conductor). In a multiple-conductor system, a conductor either accessible or inaccessible, the charge of which is not changed and to which no connection is made in the course of the determination of any one of the capacitances of the remaining conductors of the system.   210

isolated impedance of an array element (antennas). The input impedance of a radiating element of an array antenna with all other elements of the array absent.   111

isolated-neutral system. A system that has no intentional connection to ground except through indicating, measuring, or protective devices of very-high impedance. *See:* **grounded system.**   244, 62

isolated patient lead (health care facilities). A patient lead whose impedance to ground or the power line is sufficiently high that connecting the lead to ground, or to either conductor of the power line, results in current flow in the lead which is below a hazardous limit.
                                                                    192

isolated-phase bus (1)(generating station grounding). A metal-enclosed bus in which each phase conductor is enclosed by an individual metal housing separated from adjacent conductor housings by an air space. 569

(2)(power switchgear). One in which each phase conductor is enclosed by an individual metal housing separated from adjacent conductor housings by an air space. *Note:* The bus may be self-cooled or may be forced-cooled by means of circulating a gas or liquid. 103

isolated plant (electric power). An electric installation deriving energy from its own generator driven by a prime mover and not serving the purpose of a public utility. 328

isolated power system (1) (health care facilities). A system comprising an isolating transformer or its equivalent, a line isolation monitor and its ungrounded circuit conductors. *See:* **NFPA 70.** 192

(2) (National Electrical Code). A system comprising an isolating transformer or its equivalent, a line isolation monitor, and its ungrounded circuit conductors. 256

isolating amplifier (signal-transmission system). *See:* **amplifier, isolating.**

isolating contactor (29) (power system device function numbers). A device that is used expressly for disconnecting one circuit from another for the purposes of emergency operation, maintenance, or test. 402

isolating device (nuclear power generating station). A device in a circuit which prevents malfunctions in one section of a circuit from causing unacceptable influences in other sections of the circuit or other circuits. 102

isolating switch (National Electrical Code). A switch intended for isolating an electric circuit from the source of power. It has no interrupting rating, and it is intended to be operated only after the circuit has been opened by some other means. 256

isolating time (of a sectionalizer) (power switchgear). The time between the cessation of a current above the minimum actuating current value which caused the final counting and opening operation and the maximum separation of the contacts. 103

isolating transformer (signal-transmission system). *See:* **transformer, isolating.**

isolating transformers (wire-line communication facilities). Isolating (insulating) transformers provide longitudinal (common mode) isolation of the communication facility. They can also be designed for use in a combined isolating-drainage transformer configuration. 414

isolating transformers with high-voltage isolating relays (wire-line communication facilities). This assembly provides protection for standard telephone service, and consists basically of an isolating transformer and a high-voltage isolating relay. The transformer provides a path for voice and ringing frequencies while the relay provides a means for repeating dc signals. A locally supplied battery or dc power supply is required for operating of the telephone and relay. 414

isolation (1)(nonlinear, active, and nonreciprocal waveguide components). (A) Circulator. The ratio of insertion loss to an isolated port relative to insertion loss to the coupled port in a circulator. (B) Ferrite isolator. The ratio of insertion loss in the reverse direction to insertion loss in the forward direction in an isolator. (C) Mixer. The degree to which the amplitude of an undesired wave is suppressed relative to the amplitude of the desired wave. (D) Switch. The ratio of insertion loss in the OFF (open) state to the insertion loss in the ON (closed) state in a switch. 530

(2) (antennas). A measure of power transfer from one antenna to another. *Note:* The isolation between antennas is the ratio of power input to one antenna to the power received by the other, usually expressed in decibels. *See:* **radiation.** 111

isolation amplifier (buffer). *See:* **amplifier, isolation.**

isolation boundary (periodic testing of diesel-generator units applied as standby power supplies in nuclear power generating stations). A supporting system, subsystem, or device (valve, control power circuit breaker, switch, etcetera) which provides a boundary with the diesel-generator unit. Failures of the device or the supporting system, or subsystem are not considered diesel-generator unit failures. 539

isolation by elevation. *See:* **isolated by elevation.**

isolation device (nuclear power generating stations) (class 1E equipment and circuits). A device in a circuit which prevents malfunctions in one section of a circuit from causing unacceptable influences in other sections of the circuit or other circuits. 31, 131

isolation transformer (1) (health care facilities). A transformer of the multiple-winding type, with the primary and secondary windings physically separated, which inductively couples its ungrounded secondary winding to the grounded feeder system that energizes its primary winding. 192

(2) (National Electrical Code) (health care facilities). A transformer of the multiple-winding type, with the primary and secondary windings physically separated, which inductively couples its secondary winding to the grounded feeder systems that energize its primary winding, thereby preventing primary circuit potential from being impressed on the secondary circuits. 256

isolation voltage (power supplies). A rating for a power supply that specifies the amount of external voltage that can be connected between any output terminal and ground (the chassis). This rating is important when power supplies are connected in series. *See:* **power supply.** 186

isolation zones (nuclear security systems). Any area, adjacent to a perimeter physical barrier, cleared of all objects that conceal or shield an individual. 464

isolator (1) (switchgear). *See:* **disconnecting or isolating switch.**

(2) (waveguide). A passive attenuator in which the loss in one direction is much greater than that in the opposite direction. *See:* **waveguide.** 244

(3) (fiber optics). A device intended to prevent return

reflections along a transmission path. *Note:* The Faraday isolator uses the magneto-optic effect.   433

isolux (isofootcandle) line (illuminating engineering). A line plotted on any appropriate set of coordinates to show all the points on a surface where the illuminance is the same. A series of such lines for various illuminance values is called an isolux (isofootcandle) diagram.   167

isophase. *See* **equal interval.**

isopreference (speech quality measurements). Two speech signals are isopreferent when the votes averaged over all listeners show an equal preference for the speech test and speech reference signals.   126

isotropic (fiber optics). Pertaining to a material whose electrical or optical properties are independent of direction of propagation and of polarization of a traveling wave. *See:* **anisotropic; birefringent medium.**
     433

isotropic radiator (antennas). A hypothetical, lossless antenna having equal radiation intensity in all directions. *Note:* An isotropic radiator represents a convenient reference for expressing the directive properties of actual antennas.   111

I^2t **characteristic (of a fuse) (power switchgear).** The amount of ampere-squared seconds passed by the fuse during a specified period and under specified conditions. *Notes:* (1) The specified period may be the melting, arcing, or total clearing time. The sum of melting and arcing I^2t is the clearing I^2t. (2) The melting characteristic is related to a specified current wave shape, and the arcing I^2t to specified voltage and circuit-impedance conditions.   103

I^2t **(protection and coordination of industrial and commercial power systems).** The measure of heat energy developed within a circuit during the fuses melting or clearing. Generally stated as *melting* I^2t or *clearing* I^2t.   504

item (1) (nuclear power generating station). Any level of unit assembly, including structure, system, subsystem, subassembly, module, component, part, equipment or material. *Note:* This term applies specifically to the subject matter of IEEE Std 467-1980.   438
(2) (nuclear power quality assurance). An all inclusive term used in place of any of the following: appurtenance, assembly, component, equipment, material, module, part, structure, subassembly, subsystem, system, or unit.   417
(3) (computing systems). A collection of related characters, treated as a unit. *See:* **file.**   255
(4) (reliability). An all-inclusive term to denote any level of hardware assembly: that is, system, segment of a system, subsystem, equipment, component, part, etcetera. *Note:* Item includes items, population of items, sample, etcetera, where the context of its use so justifies. *See:* **reliability.**   182

item, nonrepaired (reliability). An item that is not repaired after a failure. *See:* **reliability.**   182

item or equipment hazard rate (reliability data for pumps and drivers, valve actuators, and valves). The instantaneous failure rate of an item or equipment or its conditional probability of failure versus time.
     447, 502

item, repaired (reliability). An item that is repaired after a failure. *See:* **reliability.**   182

iteration (software). (1) The process of repeatedly executing a given sequence of programming language statements until a given condition is met or while a given condition is true. (2) A single execution of a loop. *See:* **loop; programming language.**   434

iterative (test, measurement and diagnostic equipment). Describing a procedure or process which repeatedly executes a series of operations until some condition is satisfied. An iterative procedure may be implemented by a loop in a routine.   54

iterative impedance. *See:* **impedance, iterative; self-impedance.**

iterative operation (analog computer). Similar in many respects to repetitive operation, except that the automatic recycling of the computer is controlled by programmed logic circuits, which generally include a program change for a parameter(s), variable(s), or combinations of these between successive solutions, resulting in an iterative process which tends to converge on desired values of the parameter(s) or variables(s) that have been changed. *See:* **repetitive operation.**
     9

I · T product (harmonic control and reactive compensation of static power converters). The inductive influence expressed in terms of the product of its root-mean-square magnitude in amperes (I) times its telephone influence factor (TIF).   533

I Video Signal (National Television System Committee (NTSC) color television). One of the two video signal (E'_I and E'_Q) controlling the chrominance in the NTSC system. *Note:* It is a linear combination of gamma-corrected primary color signals, E'_R, E'_G, and E'_B as follows:

$$E'_I = -0.27\,(E'_B - E'_Y) + 0.74\,(E'_R - E'_Y)$$
$$= 0.60E'_R - 0.28E'_G - 0.32E'_B$$
     18

J

jack (electric circuits). A connecting device, ordinarily designed for use in a fixed location, to which a wire or wires of a circuit may be attached and that is arranged for the insertion of a plug.   341

jack bolt (rotating machinery). A bolt used to position or load an object.   63

jacket (1)cable)(electrical heat tracing for industrial applications). A thermoplastic or thermosetting plastic covering, sometimes fabric reinforced, applied over the insulation, core, metallic sheath, or armor of a cable.   57, 523
(2)(NESC). A protective covering over the insulation, core, or sheath of a cable.   494
(3) (primary dry cell). An external covering of insulating material, closed at the bottom. *See:* **electrolytic cell.**   328
jack shaft (rotating machinery). A separate shaft carried on its own bearings and connected to the shaft of a machine. *See:* **rotor (rotating machinery).**   63
jack system (rotating machinery). A system design to raise the rotor of a machine. *See:* **rotor (rotating machinery).**   63
jamming (radar). A form of electronic countermeasures (ECM) in which interfering signals, typically noise-like, are transmitted at frequencies in the receiving band of a radar to obscure or distort the radar signal.   13
jam transfer (hybrid computer linkage components). The transfer operation, in a double-buffered digital-to-analog converter (DAC) or digital-to-analog multiplier (DAM), in which the digital value is simultaneously loaded into both the holding and dynamic registers.   10
Jansky (radio wave propagation). A unit of spectral power flux density: 10^{-26} times one watt per square meter per hertz (often called one flux unit).   146
jar (storage cell). The container for the element and electrolyte of a lead-acid storage cell and unattacked by the electrolyte. *See:* **battery (primary or secondary).**   328
J-display (radar). A modified A-display in which the time base is a circle and targets appear as radial deflections from the time base.

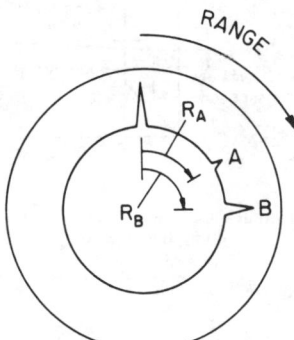

NOTE: Two targets A and
B at different ranges.

J-Display

                                                             13
jerk (inertial sensor). A vector that specifies the time rate of change of the acceleration; the third derivative of displacement with respect to time.   46

jitter (1) (data transmission) (repetitive wave). (A) Time-related, abrupt, spurious variations in the duration of any specified, related interval. (B) Amplitude-related, abrupt, spurious variations in the magnitude of successive cycles. (C) Frequency-related, abrupt, spurious variations in the frequency of successive pulses. (D) Phase-related, abrupt, spurious variations in the phase of the frequency modulation of successive pulses referenced to the phase of a continuous oscillator. *Note:* Qualitative use of jitter requires the use of a generic derivation of one of the categories to identify whether the jitter is time, amplitude, frequency, or phase related and to specify which form within the category, for example, pulse delay-time jitter, pulse-duration jitter, pulse-separation jitter. Quantitative use of jitter requires that a specified measure of the time or amplitude related variation, (for example, average, root-mean-square, or peak-to-peak) be included in addition to the generic term that specifies whether the jitter is time, amplitude, frequency, or phase related.   59
(2) (oscilloscopes, electronic navigation, and television). Small, rapid aberrations in the size or position of a repetitive display, indicating spurious deviations of the signal or instability of the display circuit. *Note:* Frequently caused by mechanical or electronic switching systems or faulty components. It is generally continuous, but may be random or periodic.
                                                          18, 59
(3) (radar). (1) Small, rapid and generally continuous variations in the size, shape, or position of observable information, frequently caused by mechanical and electronic switching systems of faulty components. Also refers to zero-mean random errors in successive target positions measurements due to target echo characteristics, propagation, or receiver thermal noise. (2) Intentional variation of a radar parameter, for example, pulse interval.   13
(4) (facsimile). Raggedness in the received copy caused by erroneous displacement of recorded spots in the direction of scanning. *See:* **recording (facsimile).**
                                                             12
(5) (pulse terms). Dispersion of a time parameter of the pulse waveforms in a pulse train with respect to a reference time, interval, or duration. Unless otherwise specified by a mathematical adjective, peak-to-peak jitter is assumed. *See:* **mathematical adjectives.**
                                                            254
jnd. *See:* **just noticeable difference.**
jog (inch) (control) (industrial control). A control function that provides for the momentary operation of a drive for the purpose of accomplishing a small movement of the driven machine. *See:* **electric drive.**
                                                          225, 206
jog control point, supervisory control. *See:* **supervisory control point, jog control.**
jogging (packaging machinery). The quickly repeated closure of the circuit to start a motor from rest for the purpose of accomplishing small movements of the driven machine. *Syn:* **inching.**   429
jogging speed (industrial control). The steady-state

speed that would be attained if the jogging pilot device contacts were maintained closed. *Note:* It may be expressed either as an absolute magnitude of speed or a percentage of maximum rated speed. *See:* **control system, feedback.**   206

Johnson noise (interference terminology). The noise caused by thermal agitation (of electron charge) in a dissipative body. *Notes:* (1) The available thermal (Johnson) noise power N from a resistor at temperature T is $N = kT\Delta f$, where k is Boltzmann's constant and Δf is the frequency increment. (2) The noise power distribution is equal throughout the radio frequency spectrum, that is, the noise power is equal in all equal frequency increments. *See:* **signal.**   188

joint (interior wiring). A connection between two or more conductors.   328

joint, compression (conductor stringing equipment). A tubular compression fitting designed and fabricated from aluminum, copper or steel to join conductors or overhead ground wires. It is usually applied through the use of hydraulic or mechanical presses. However, in some cases, automatic, wedge type joints are utilized. *Syn:* **sleeve: splice.**   45

joint insulation. *See:* **connection insulation.**

jointly owned generation (power operations). Generation facility owned jointly by several electric utilities each entitled to a share of the capability. *Note:* One of the participating utilities operates the facility.   516

joint use (NESC). Simultaneous use by two or more kinds of utilities.   494, 262

Jordan bearing. A sleeve bearing and thrust bearing combined in a single unit. *See:* **bearing.**   63

joule (1) (metric practice). The work done when the point of application of a force of one newton is displaced a distance of one meter in the direction of the force.   21

(2) (laser-maser) (J). A unit of energy: one (1) joule = 1 watt·second.   363

Joule effect. The evolution of thermal energy produced by an electric current in a conductor as a consequence of the electric resistance of the conductor. *See:* **thermoelectric device; Joule's law.**   191

Joule heat. The thermal energy resulting from the Joule effect. *See:* **thermoelectric device.**   191

Joule's law (heating effect of a current). The rate at which heat is produced in an electric circuit of constant resistance is equal to the product of the resistance and the square of the current.   210

journal (shaft). A cylindrical section of a shaft that is intended to rotate inside a bearing. *See:* **bearing; armature.**   63

journal bearing (rotating machinery). A bearing that supports the cylindrical journal of a shaft. *See:* **bearing.**   63

jpd. *See:* **just perceptible difference.**   18

***J* scan (electronic navigation).** *See: J* **display.**

J-scope (radar). A cathode-ray oscilloscope arranged to present a J-display.   13

jump (electronic computation). (1) To (conditionally or unconditionally) cause the next instruction to be obtained from a storage location specified by an address part of the current instruction when otherwise it would be specified by some convention. (2) An instruction that specifies a jump. *Note:* If every instruction in the instruction code specifies the location of the next instruction (for example, in a three-plus-one-address code), then each one is not called a jump instruction unless it has two or more address parts that are conditionally selected for the jump. *See:* **conditional jump; transfer; unconditional jump.**   235

jumper (1)(power line maintenance). A conductive tool used to maintain electrical continuity across equipment or a conductor which can be opened mechanically to enable various operations of live-line work to be performed. *Syn:* **bypass.**   458

(2) (telephone switching systems). Crossconnection wire(s).   55

(3) (conductor stringing equipment). (A) The conductor that connects the conductors on opposite sides of a deadend structure. (B) A conductor placed across the clear space between the ends of two conductors or metal pulling lines which are being spliced together. Its purpose then is to act as a shunt to prevent workmen from accidentally placing themselves in series between the two conductors.   45

jump resonance (ferroresonant voltage regulators). A phenomenon associated with ferroresonant regulators where the output voltage suddenly changes to the regulating mode of operation at some value of the ascending input voltage (See Figure 'Output Versus Input Voltage' and Figure 'Reversal Point With Jump Resonance' below), or suddenly drops out of the regulating mode of operation with descending input voltage.   456

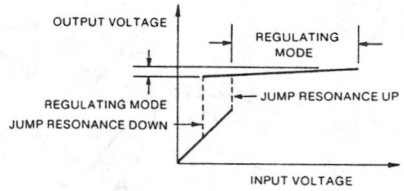

Output Versus Input Voltage with Jump Resonance

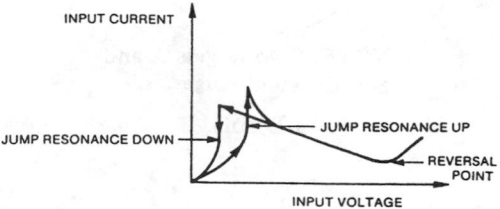

Reversal Point with Jump Resonance

junction (of a semiconductor radiation detector)(germanium gamma-ray detectors)(charged-particle detectors)(X-ray energy spectrometers). A region of transition between semiconductor regions of different electrical properties (for example, n-n$^+$, p-n, p-p$^+$ semiconductors) or between a metal and a semiconductor. 528, 119, 471

junction, alloy (semiconductor). See: alloy or fused junction.

junction, n-n (semiconductor). A region of transition between two regions having different properties in n-type semiconducting material. See: semiconductor device. 328

junction, p-n (semiconductor). A region of transition between p- and n-type semiconducting material. See: semiconductor device. 328

junction, p-p (semiconductor). A region of transition between two regions having different properties in p-type semiconducting material. See: semiconductor device. 328

junction box An enclosed distribution panel for connecting or branching one or more corresponding electric circuits without the use of permanent splices. See: cabinet. 64

junction, collector (semiconductor device). A junction normally biased in the high-resistance direction, the current through which can be controlled by the introduction of minority carriers. Note: The polarity of the voltage across the junction reverses when a switching occurs. See: semiconductor; semiconductor device; transistor. 210, 245, 66

junction depth (1)(of a p-n semiconductor radiation detector)(X-ray energy spectrometers)(germanium gamma-ray detectors)(charged-particle detectors). The distance below the crystal surface at which the conductivity type changes. 471, 528, 119

(2) (solar cells). The distance from the illuminated surface to the center line of the junction in a solar cell. 113

junction, diffused (semiconductor). A junction that has been formed by the diffusion of an impurity within a semiconductor crystal. See: semiconductor. 245

junction, doped (semiconductor). A junction produced by the addition of an impurity to the melt during crystal growth. See: semiconductor. 245

junction, emitter (semiconductor device). A junction normally biased in the low-resistance direction to inject minority carriers into an interelectrode region. See: semiconductor; transistor. 210

junction, fused. See: alloy or fused junction.

junction, grown (semiconductor). A junction produced during growth of a crystal from a melt. See: semiconductor device. 245

junction loss (wire communication). That part of the repetition equivalent assignable to interaction effects arising at trunk terminals. See: transmission loss. 328

junction point. See: node.

junction pole (wire communication). A pole at the end of a transposition section of an open wire line or the pole common to two adjacent transposition sections. See: open wire. 328

junction, rate-grown (semiconductor). A grown junction produced by varying the rate of crystal growth. See: semiconductor device. 245

junction resistance (thermoelectric device). The difference between the resistance of two joined materials and the sum of the resistances of the unjoined materials. See: thermoelectric device. 191

junction temperature (T_j) (light emitting diodes). The temperature of the semiconductor junction. 162

junction transposition (s-pole transposition) (wire communication). A transposition located at the junction pole (s pole) between two transposition sections of an open wire line. See: open wire. 328

junctor (1) (crossbar systems) (wire communication). A circuit extending between frames of a switching unit and terminating in a switching device on each frame. 328

(2) (telephone switching systems). Within a switching system, a connection or circuit between inlets and outlets of the same or different switching networks. 55

just noticeable difference (jnd) (visual) (television). The smallest difference between luminances or colors, occurring either alone or together, of (usually) adjacent areas that is easily discernible or obvious in the course of ordinary observation. 18

just operate value, relay. See: relay just operate value.

just perceptible difference (jpd) (visual) (television). The smallest difference between luminances or colors, occurring either alone or together, of (usually) adjacent areas that is discernible in the course of careful observation under the most favorable conditions. 18

just scale (acoustics). A musical scale formed by taking three consecutive triads each having the ratio 4:5:6, or 10:12:15. Note: Consecutive triads are triads such that the highest note of one is the lowest note of the next. 176

K

Karnaugh map (mathematics of computing). A rectangular diagram of a logical expression drawn with overlapping rectangles representing a unique combination of the logic variables and such that an intersection is shown for all combinations. The rows and columns are headed with combinations of the variables in a Gray code sequence. See: Mahoney map. 564

K-band. A radar frequency band between 18 gigahertz

(GHz) and 27 GHz, usually in the International Telecommunications Union (ITU) assigned band 23 GHz to 24.2 GHz.   13

K-display (radar). A modified A display used with a lobe-switching antenna in which a target appears as a pair of vertical deflections. When the radar antenna is correctly pointed at the target the deflections (blips) are of equal height, and when not so pointed, the difference in blip height is an indication of the direction and magnitude of pointing error.   13

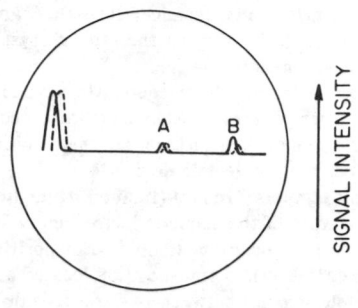

RANGE

NOTE: Two targets A and B at different ranges; radar aimed at target A.

K-Display

keep-alive circuit (transmit-receive, receiver protector) (nonlinear, active, and nonreciprocal waveguide components). A circuit for providing residual ionization for the purpose of reducing the initiation time of the main discharge. *See:* **duplexer; ignitor discharge; radioactive nuclear ignitor.**   530

kelvin (metric practice). Unit of thermodynamic temperature, it is the fraction 1/273.16 of the thermodynamic temperature of the triple point of water (adopted by 13th General Conference on Weights and Measures).

The International Practical Temperature Scale of 1968 (IPTS-68) (amended edition 1975) is defined by a set of interpolation equations and the following reference temperatures:[a]

	K	°C
Hydrogen, solid-liquid-gas equilibrium	13.81	−259.34
Hydrogen, liquid-gas equilibrium at 33.330 6 kPa (25/76 standard atmosphere)	17.042	−256.108
Hydrogen, liquid-gas equilibrium	20.28	−252.87
Neon, liquid-gas equilibrium	27.102	−246.048
Oxygen, solid-liquid-gas equilibrium	54.361	−218.789
Oxygen, liquid-gas equilibrium	90.188	−182.962
Water, solid-liquid-gas equilibrium	273.16	0.01
Water, liquid-gas equilibrium	373.15	100.00
Zinc, solid-liquid equilibrium	692.73	419.58
Silver, solid-liquid equilibrium	1235.08	961.93
Gold, solid-liquid equilibrium	1337.58	1064.43

The 1976 provisional 0.5 to 30 K Temperature Scale (EPT-76) is defined by a set of reference temperatures which includes the four lowest defining points of the IPTS-68 in addition to the following points:

	K	°C
Cadmium, superconducting transition	0.519	−272.631
Zinc, superconducting transition	0.851	−272.299
Aluminum, superconducting transition	1.179_6*	-271.970_4*
Indium, superconducting transition	3.414_5*	-269.735_5*
Helium-four, liquid-gas equilibrium	4.222_1*	-268.927_9*
Lead, superconducting transition	7.199_9*	-265.950_1*
Neon, solid-liquid-gas equilibrium	24.5591	−248.5909

*Subscript digits are uncertain.

[a] Except for the triple points and one equilibrium hydrogen point (17.042 K) the assigned values of temperature are for equilibrium states at a pressure

$$p_o = 1 \text{ standard atmosphere (101.325 kPa)}.$$

The degree Celsius and the kelvin are units of identical size and the scales using these two units are related by a difference of exactly 273.15 kelvins. It is seldom necessary to recognize the differences between the thermodynamic and practical scales, but the General Conference on Weights and Measures in 1968 established the following quantity symbols for use if desired:

Quantity[3]	Quantity Symbol	Unit
Thermodynamic temperature	T	kelvin
Celsius temperature	t	degree Celsius
Practical Kelvin temperature	T_{68}	kelvin
Practical Celsius temperature	t_{68}	degree Celsius

Kelvin bridge (double bridge) (Thomson bridge). A 7-arm bridge intended for comparing the 4-terminal resistances of two 4-terminal resistors or networks, and characterized by the use of a pair of auxiliary resistance arms of known ratio that span the adjacent potential terminals of the two 4-terminal resistors that are connected in series by a conductor joining their adjacent current terminals. *See:* **bridge.**   328, 21

$$R_x = R_s \frac{R_2}{R_1} - \frac{R_c R_3 (R_4/R_3 - R_2/R_1)}{R_3 + R_4 + R_c}$$

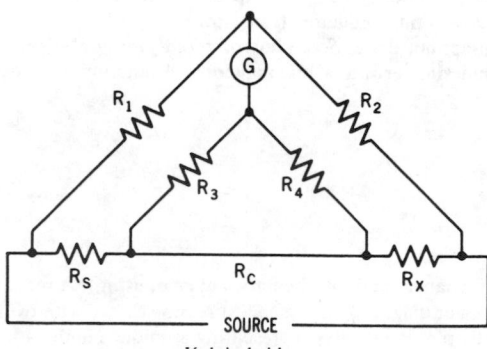

Kelvin bridge

Kendall effect (facsimile). A spurious pattern or other distortion in a facsimile record, caused by unwanted modulation products arising from the transmission of a carrier signal, appearing in the form of a rectified baseband that interferes with the lower sideband of the carrier. *Note:* This occurs principally when the single sideband width is greater than half the facsimile carrier frequency. *See:* **recording (facsimile).** 12

kenotron (tube or valve). A hot-cathode vacuum diode. *Note:* This term is used primarily in the industrial and X-ray fields. 190

kernel (software). (1) A nucleus or a core, as in the kernel of an operating system. (2) An encapsulation of an elementary function. Kernels can be combined to form some or all of an operating system or set of firmware. (3) A model used in computer selection studies to evaluate computer performance. *See:* **computer; encapsulation; firmware; function; model; operating system; performance.** 434

Kerr electrostatic effect. *See:* **electrooptical effect in dielectrics.**

key (1) (telephone switching system). A hand-operated switching device ordinarily comprising concealed spring contacts with an exposed handle or pushbutton, capable of closing or opening one or more parts of a circuit. 55
(2) (rotating machinery). A bar that by being recessed partly in each of two adjacent members serves to transmit a force from one to the other. *See:* **rotor (rotating machinery).** 63
(3) (software). One or more characters, within a set of data, that contains information about the set, including its identification. *See:* **data.** 434

keyboard (test, measurement and diagnostic equipment). A device for the encoding of data by key depression which causes the generation of the selected code element. 54

keyer. A device that changes the output of a transmitter from one value of amplitude or frequency to another in accordance with the intelligence to be transmitted. *Note:* This applies generally to telegraph keying. *See:* **radio transmission.** 111

keying (1) (modulating systems). Modulation involving a sequence of selections from a finite set of discrete states. *See:* **telegraphy.** 61
(2) (telegraph). The forming of signals, such as those employed in telegraph transmission, by an abrupt modulation of the output of a direct-current or an alternating-current source as, for example, by interrupting it or by suddenly changing its amplitude or frequency or some other characteristic. *See:* **telegraphy.** 111
(3) (television). A signal that enables or disables a network during selected time intervals. *See:* **television.** 178

keying interval (periodically keyed transmission system) (modulation systems). One of the set of intervals starting from a change in state and equal in length to the shortest time between changes of state. *Note:* The keying interval equals the symbol duration. 61

keying rate (modulation systems). The reciprocal of the duration of the keying interval. 242

keying wave (telegraphic communication). *See:* **marking wave.**

keyless ringing (telephony). A form of machine ringing on manual switchboards that is started automatically by the insertion of the calling plug into the jack of the called line. 328

key light (illuminating engineering). The apparent principal source of directional illumination falling upon a subject or area. 167

key pulsing (telephone switching systems). A switchboard arrangement using a nonlocking keyset and providing for the transmission of a signal corresponding to each of the keys depressed. 55

key-pulsing signal (telephone switching systems). In multifrequency and key pulsing, a signal used to prepare the equipment for receiving digits. 55

keyshelf (telephone switching systems). The shelf on which are mounted control keys for use by operators or other personnel. 55

keystone distortion (television). A form of geometric distortion that results in a trapezoidal display of a nominally rectangular raster or picture. *See:* **television.** 178

keyway (rotating machinery). A recess provided for a key. *See:* **key (rotating machinery).** 63

kick-sorter (British) (pulse techniques). *See:* **pulse-height analyzer; pulse.**

kilo (mathematics of computing). (1) (k) A prefix indicating one thousand. (2) (K) In statements involving size of computer storage, a prefix indicating 2^{10} or 1024. 564

kilogram (metric practice). The unit of mass; it is equal to the mass of the international prototype of the kilogram. (adopted by 1st and 3rd General Conference on Weights and Measures 1889 and 1901). 21

kilovar (1000 vars)(shunt power capacitors). The practical unit of reactive power, equal to the product of the root-mean-square (rms) voltage in kilovolts (kV), the rms current in amperes (A), and the sine of the angle between them. 138

kilovolt-ampere rating (voltage regulator). The product of the rated load amperes and the rated range of regulation in kilovolts. *Note:* The kilovolt-ampere rating of a three-phase voltage regulator is the product of the rated load amperes and the rated range of regulation in kilovolts multiplied by 1.732. *See:* **voltage regulator.** 257

kinematic drift (electronic navigation). *See:* **misalignment drift.**

kinescope. *See:* **picture tube.**

kinetic energy. The energy that a mechanical system possesses by virtue of its motion. *Note:* The kinetic energy of a particle at any instant is $(1.2)mv^2$, where m is the mass of the particle and v is its velocity at that instant. The kinetic energy of a body at any instant is the sum of the kinetic energies of its several particles. 210

Kingsbury bearing. *See:* **tilting-pad bearing.**
kino gun. *See:* **end injection.**

Kirchhoff's laws (electric networks). (1) The algebraic sum of the currents toward any point in a network is zero. (2) The algebraic sum of the products of the current and resistance in each of the conductors in any closed path in a network is equal to the algebraic sum of the electromotive forces in that path. *Note:* These laws apply to the instantaneous values of currents and electromotive forces, but may be extended to the phasor equivalents of sinusoidal currents and electromotive forces by replacing algebraic sum by phasor sum and by replacing resistance by impedance. *See also:* **network analysis.**   210

klydonograph (surge voltage recorder). *See:* **Lichtenberg figure camera.**

klystron. A velocity-modulated tube comprising, in principle, an input resonator, a drift space, and an output resonator. *See:* **reflex klystron; power klystron.**   244

knife switch (power switchgear). A form of switch in which the moving element, usually a hinged blade, enters or embraces the stationary contact clips. *Note:* In some cases, however, the blade is not hinged and is removable.   103

knockout (power and distribution transformer). A portion of the wall of a box or cabinet so fashioned that it may be readily removed by a hammer, screwdriver, and pliers at the time of installation in order to provide a hole for the attachment of a raceway cable or fitting.   53

knot (pulse terms). A point $t_j k$ m_k (k = 1,2,3, ..., n) in a sequence of points wherein $t_k \leq t_{k_{\pi A}} 1$ through which a spline function passes. *See:* **Pulse burst envelopes diagram below.**   254

$K_a$band radar. A radar operating at frequencies between 27 and 40 gigahertz usually in the International Telecommunications Union assigned band 33.4 to 36 gigahertz.   13

K_u-band. A radar frequency band between 12 gigahertz (GHz) and 18 GHz, usually in one of the International Telecommunications Union (ITU) assigned bands 13.4 GHz to 14.4 GHz or 15.7 GHz to 17.7 GHz.   13

K-scope (radar). A cathode-ray oscilloscope arranged to present a K-display.   13

kV · T product (harmonic control and reactive compensation of static power converters). Inductive influence expressed in terms of the product of its root-mean-square magnitude in kilovolts (kV) times its telephone influence factor (TIF).   533

kVA or volt-ampere short-circuit input rating of a high-reactance transformer (power and distribution transformer). One that designates the input kVA or volt-amperes at rated primary voltage with the secondary terminals short-circuited.   53

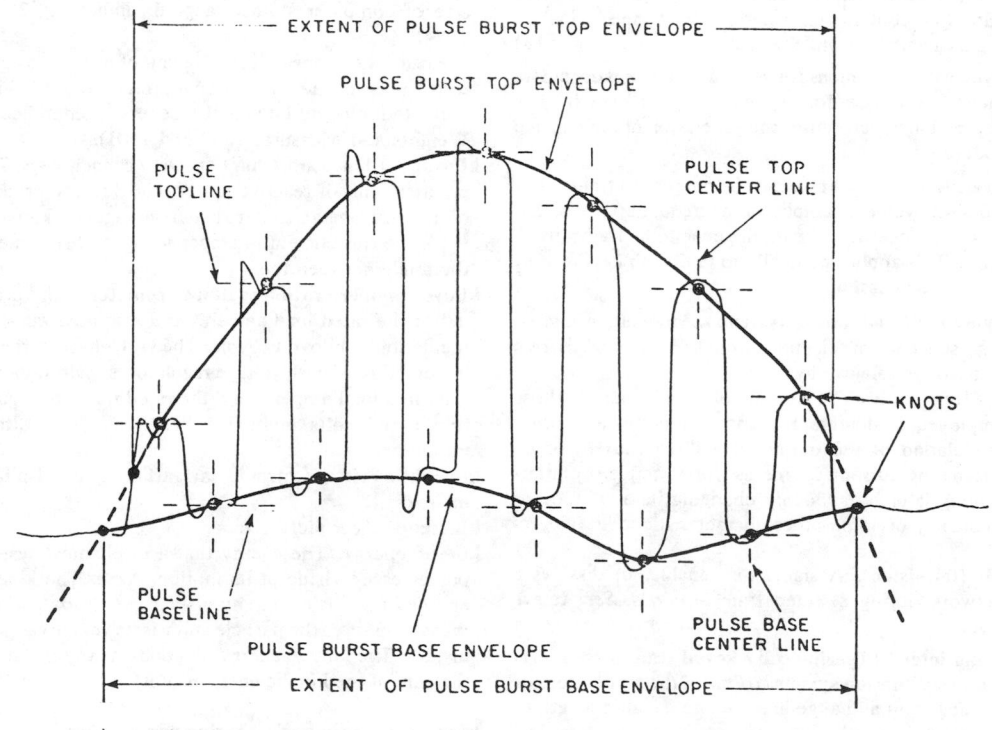

Pulse burst envelopes.

L

L(s). *See:* **operational inductance.**

LA. *See:* **laser gyro axes.**

label (software). (1) One or more characters, within or attached to a set of data, that contain information about the set, including its identification. (2) In computer programming, an identifier of an instruction. (3) An identification record for a tape or disk file. *See:* **computer; data; file; identifier; instruction; record.**
434

labeled (National Electrical Code). Equipment or materials to which has been attached a label, symbol, or other identifying mark of an organization acceptable to the authority having jurisdiction and concerned with product evaluation, that maintains periodic inspection of production of labeled equipment or materials and by whose labeling the manufacturer indicates compliance with appropriate standards or performance in a specified manner.
256

laboratory (meter). A laboratory responsible for maintaining reference standards and assigning values to the working standards used for the testing of electricity meters and auxiliary devices. (2) (independent standards). A standards laboratory maintained by, and responsible to, a company or authority that is not under the same administrative control as the laboratories or companies submitting instruments for calibration.
212

laboratory reference standards (metering). Standards that are used to assign and check the values of laboratory secondary standards.
212

laboratory resistivity ($\Omega \cdot cm$). Of a fly ash deposit is the ratio of the applied electric potential across the layer to the induced current density. The value of resistivity is specific for a given ash sample and depends upon the test variables or conditions: temperature, composition of the gaseous environment (especially water and sulfuric acid vapor content), the magnitude of the applied electric field strength, as well as the porosity of the ash layer. Measured resistivity also is a function of test procedure details, including items such as initial test temperature, the rate of heating or cooling, and the time that the voltage is applied prior to reading current.
427

laboratory secondary standards (metering). Standards that are used in the routine calibration tasks of the laboratory.
212

laboratory test. *See:* **shop test.**

laboratory working standards. Those standards that are used for the ordinary calibration work of the standardizing laboratory. *Note:* Laboratory working standards are calibrated by comparison with the reference standards of that laboratory. *See:* **measurement system.**
183

labyrinth seal ring (rotating machinery). Multiple oil

catcher ring surrounding a shaft with small clearance. *See:* **bearing.**
63

lacing, stator-winding end-wire (rotating machinery). Cord or other lacing material used to bind the stator-winding end wire, to hold in place labels and devices placed on or in the end wire, and to position lead cables at their take-off points on the end wire. *See:* **stator.**
63

lacquer-film capacitor. A capacitor in which the dielectric is primarily a solid lacquer film and the electrodes are thin metallic coatings deposited thereon.
341

lacquer master*. *See:* **lacquer original.**
*Deprecated

lacquer original (electroacoustics) (lacquer master*). An original recording on a lacquer surface for the purpose of making a master. *See:* **phonograph pickup.**
176
*Deprecated

lacquer recording (electroacoustics). Any recording made on a lacquer recording medium. *See:* **phonograph pickup.**
176

ladder network (1) (general). A network composed of a sequence of H, L, T, or pi networks connected in tandem. *See:* **network analysis.**
210
(2) (circuits and systems). A cascade or tandem connection of alternating series and shunt arms.
67

ladder, rope (conductor stringing equipment). A ladder having vertical synthetic or manila suspension members and wood, fiberglass or metal rungs. The ladder is suspended from the arm or bridge of a structure to enable workmen to work at the conductor level, hang travelers, perform clipping-in operations, etcetera. *Syn:* **jacobs ladder.**
45

ladder, tower (conductor stringing equipment). A ladder complete with hooks and safety chains attached to one end of the side rails. These units are normally fabricated from fiberglass, wood or metal. The ladder is suspended from the "arm" or bridge of a structure to enable workmen to work at the conductor level, to hang travelers, perform clipping-in operations, etc. In some cases, these ladders are also used as linemen's platforms. *Syn:* **hook ladder.**
45

ladder-winding insulation (rotating machinery). An element of winding insulation in the form of a single sheet precut to fit into one or more slots and with a broad area at each end to provide end-wire insulation. *See:* **rotor (rotating machinery); stator.**
63

lag (1) (general). The delay between two events.
255, 77
(2) (control circuits). Any retardation of an output with respect to the casual input. *See:* **telegraphy; control system; feedback.**
56, 206
(3) (telegraph system). The time elapsing between the operation of the transmitting device and the response of the receiving device. *See:* **telegraphy.**
328

(4) (distance/velocity) (automatic control). A delay attributable to the transport of material or the finite rate of propagation of a signal or condition. *Syn:* **lag, transportation; lag transport.** *See:* **control system; feedback.**   56

(5) (first-order). The change in phase due to a linear element of transfer function. $1/(1 + Ts)$. *Syn:* **linear lag.**

(6) (second order) (automatic control). In a linear system or element, lag which results from changes of energy storage at two separate points in the system, or from effects such as acceleration. *Note:* It is representable by a second-order differential equation, or by a quadratic factor such as $s^2 + 2z\omega_n s + \omega_n^2$ in the denominator of a transfer function. *Syn:* **quadratic lag.**   56

(7) multi-order (automatic control). In a linear system or element, lag of energy storage in two or more separate elements of the system. *Note:* It is evidenced by a differential equation of order higher than one, or by more than one time-constant. It may sometimes be approximated by a delay followed by a first-order or second-order lag.   56

(8) transfer (automatic control). *See:* **lag, first-order; lag, second-order; lag, multi-order.**   56

(9) (camera tubes). A persistence of the electrical-charge image for a small number of frames.   125 ·

lagged-demand meter. A meter in which the indication of the maximum demand is subject to a characteristic time lag by either mechanical or thermal means. *See:* **electricity meter.**   212

lag networks (power supplies). Resistance-reactance components, arranged to control phase-gain roll-off versus frequency. *Note:* Used to assure the dynamic stability of a power-supply's comparison amplifier. The main effect of a lag network is a reduction of gain at relatively low frequencies so that the slope of the remaining rolloff can be relatively more gentle. 186

LAM access specifier (m) (subroutines for CAMAC). The symbol m represents an integer which is used to indicate the mode of access of a LAM and the lowest-order address component for LAM addressing. If m is zero or positive, it is interpreted as subaddress and the LAM is assumed to be accessed via dataless functions at this subaddress. If m is negative, it is interpreted as the negative of a bit position for a LAM which is accessed via reading, setting, or clearing bits in the group 2 registers at subaddresses 12, 13, or 14.

                    410

lambda (power operations). The incremental operating cost at the load center, commonly expressed in mils per kilowatt-hour (mil/kWh).   516

lambert (television). A unit of luminance (photometric brightness) equal to $1/\pi$ candela per square centimeter and, therefore, equal to the uniform luminance of a perfectly diffusing surface emitting or reflecting light at the rate of 1 lumen per square centimeter. *Note:* The lambert is also the average luminance of any surface emitting or reflecting light at the rate of 1 lumen per square centimeter. For the general case, the average must take account of variation of luminance with

angle of observation, and also of its variation from point to point on the surface considered.   18
*Deprecated

Lambertian radiator. *See:* **Lambert's cosine law.** 433

Lambertian surface (1) (illuminating engineering). A surface that emits or reflects light in accordance with Lambert's cosine law. A lambertian surface has the same luminance regardless of viewing angle.   167

(2) (laser-maser). An ideal surface whose emitted or reflected radiance is independent of the viewing angle.

                    363

lambertian units of luminance (illuminating engineering). The luminance of a surface in a specified direction also has been expressed in terms of the luminous exitance the surface would have if the luminances in all direction, within the hemisphere on the side of the surface being considered, were the same as the luminance in the specified direction. In other words, luminance has been expressed as the luminous exitance of a lambertian surface whose luminance equals the luminance of the surface in question and in the specified direction.

Lambertian Luminance Unit		Lambertian Luminous Exitance		Equivalent Luminance
$1\ asb$	$=$	$1\ lm/m^2$	$=$	$(1/\pi cd/m^2$
$1\ fL$	$=$	$1\ lm/ft^2$	$=$	$(1/\pi)cd/ft^2$
			$=$	$(1/144\pi)cd/in^2$
$1\ L$	$=$	$1\ lm/cm^2$	$=$	$(1/\pi)cd/m^2$

Note: The lambertian units of luminance are numerically equal to the corresponding units of luminous exitance. Thus, the luminance (in lambertian units) of a surface could be determined, and the numerical value directly used in the equation for a lambertian reflecting surface relating illuminance and luminous exitance, that is, $M = \rho E$ where ρ is the reflectance. In practice no surface follows exactly the cosine formula of emission or reflection and many do not even approximate it; hence the luminance is not uniform but varies with the angle from which it is viewed. Since the raison d'etre for this system was the use of a relation of generally unknown and variable accuracy when applied to real surfaces, the use of lambertian units of luminance has not been acceptable since 1967. *See:* **Lambert's cosine law.**   167

Lambert's cosine law (1) (fiber optics). The statement that the radiance of certain idealized surfaces, known as Lambertian radiators, Lambertian sources, or Lambertian reflectors, is independent of the angle from which the surface is viewed. *Note:* The radiant intensity of such a surface is maximum normal to the surface and decreases in proportion to the cosine of the angle from the normal. *Syn:* **cosine emission law.**

                    433

(2) ($I_\theta = I_0 \cos\theta$) (illuminating engineering). The luminous intensity in any direction from a plane perfectly diffusing surface element varies as the cosine of the angle between that direction and the perpendicular to the surface element.   167

LAM identifier lam (subroutines for CAMAC). The

symbol *lam* represents an integer which is used as the identifier of a CAMAC LAM signal. The information associated with the identifier must include not only the CAMAC address but also information about the means of accessing and controlling the LAM, that is, whether it is accessed via dataless functions at a subaddress or via read/write functions in group 2 registers. The value of *lam* is explicitly defined to be a non-zero integer; whether it is an encoded representation of the information required to describe the LAM or simply provides a key for accessing the information in a system-data structure is an implementation decision. The value 0 is used to indicate, where appropriate, that no LAM is being specified.   410

lamp (illuminating engineering). A generic term for a man-made source of light. By extension, the term is also used to denote sources that radiate in regions of the spectrum adjacent to the visible. *Note:* A lighting unit consisting of a lamp with shade, reflector, enclosing globe, housing, or other accessories is also called a 'lamp'. In such cases, in order to distinguish between the assembled unit and the light source within it, the latter is often called a 'bulb' or 'tube', if it is electrically powered. *See:* **luminaire.**   167

lamp burnout factor (illuminating engineering). The fractional loss of task illuminance due to burned out lamps left in place for long periods.   167

lamp post (illuminating engineering). A standard support provided with the necessary internal attachments for wiring and the external attachments for the bracket and luminaire.   167

lamp shielding angle, Φ (illuminating engineering). The angle between the plane of the baffles or louver grid and the plane most nearly horizontal which is tangent to both the lamps and the louver blades. *Note:* The lamp shielding angle frequently is larger than the louver shielding angle, but never smaller.   167

land (electroacoustics). The record surface between two adjacent grooves of a mechanical recording. *See also:* **phonograph pickup.**   176

landing beacon (navigation aid terms). A beacon used to guide aircraft in landing.   526

landing direction indicator (illuminating engineering). A device to indicate visually the direction currently designated for landing and takeoff.   167

landing light (illuminating engineering). An aircraft aeronautical light designed to illuminate a ground area from the aircraft.   167

landing zone (elevator). A zone extending from a point eighteen inches below a landing to a point eighteen inches above the landing. *See:* **elevator landing.**   328

landmark beacon (illuminating engineering). An aeronautical beacon used to indicate the location of a landmark used by pilots as an aid to enroute navigation.   167

lane (navigation system) (electronic navigation). The projection of a corridor of airspace on a navigation chart, the right and left sides of the corridor being defined by the same (ambiguous) values of the navigation coordinate (phase or amplitude), but within which

lateral position information is provided (for example, a Decca lane in which there is a 360-degree change of phase). *See:* **navigation.**   187, 13

language (1)(software requirements specifications). A means of communication, with syntax and semantics, consisting of a set of representations, conventions and associated rules used to convey information.   449
(2) (electronic computers). (A) A system consisting of: (1) a well defined, usually finite, set of characters: (2) rules for combining characters with one another to form words or other expressions: (3) a specific assignment of meaning to some of the words or expressions, usually for communicating information or data among a group of people, machines, etcetera. (B) A system similar to the above but without any specific assignment of meanings. Such systems may be distinguished from (A) above, when necessary, by referring to them as formal or uninterpreted languages. Although it is sometimes convenient to study a language independently of any meanings, in all practical cases at least one set of meanings is eventually assigned. *See:* **code; machine language.**   235, 54

language code (telephone switching systems). On an international call, an address digit that permits an originating operator to obtain assistance in a desired language.   55

language printout (dedicated-type sequential events recording systems). A word description composed of alphanumeric characters used to further identify inputs and their status. *See:* **language.**   48

language processor (software). (1) A computer program that performs such functions as translating, interpreting, and other tasks required for processing a specified programming language, for example, a FORTRAN processor, a COBOL processor. (2) A software tool that performs such functions as translating, interpreting, and other tasks required for processing a specific language, such as a requirements specification language, a design language, or a programming language. *See:* **computer program; design language; function; interpreting; programming language; requirements specification language; software tool.**   434

lapel microphone. A microphone adapted to positioning on the clothing of the user. *See:* **microphone.**   328

Laplace's equation. The special form taken by Poisson's equation when the volume density of charge is zero throughout the isotropic medium. It is $\Delta^2 v = 0$.   210

Laplace transform (unilateral) (function *f(t)*). The quantity obtained by performing the operation

$$F(s) = \int_0^\infty f(t)\mathrm{e}^{-st}\mathrm{d}t$$

where $s = \sigma + j\omega$. *See:* **control system, feedback.**   56

laptop computer (computer applications). A portable computer designed for use on one's lap.   571

lap winding. A winding that completes all its turns under a given pair of main poles before proceeding to the next pair of poles. *Note:* In commutator machines the ends of the individual coils of a simplex lap winding are connected to adjacent commutator bars: those of a duplex lap winding are connected to alternate commutator bars etcetera. *See:* **asynchronous machine.** 328

large signal performance (excitation systems for synchronous machines). Response of an excitation control system, excitation system, or elements of an excitation system to signals which are large enough that nonlinearities must be included in the analysis of the response to obtain realistic results. 507

larry (mine). A motor-driven burden-bearing track-mounted car designed for side or end dumping and used for hauling material such as coal, coke, or mine refuse. 328

laser (1) (fiber optics). A device that produces optical radiation using a population inversion to provide Light Amplification by Stimulated Emission of Radiation and (generally) an optical resonant cavity to provide positive feedback. Laser radiation may be highly coherent temporally, or spatially, or both. *See:* **active laser medium; injection laser diode; optical cavity.** 433

(2) (illuminating engineering). An acronym for Light Amplification by Stimulated Emission of Radiation. The laser produces a highly monochromatic and coherent (spatial and temporal) beam of radiation. A steady oscillation of nearly a single electromagnetic mode is maintained in a volume of an active material bounded by highly reflecting surfaces called a resonator. The frequency of oscillation varies according to the material used and by the methods of initially exciting or pumping the material. 167

(3) (laser-maser). A device which produces an intense, coherent, directional beam of light by stimulating electronic, ionic, or molecular transitions to lower energy levels. Also, an acronym for light amplification by stimulated emission of radiation. *Syn:* **optical maser.** 363

laser diode. *See:* **injection laser diode.**

laser gyro. A device which measures angular rotation by optical heterodyning of counter rotating optical beams. 46

laser gyro axes (inertial sensor). LA and NA are two perpendicular axes in the plane of the laser beams, and are normal to the IA. The IRA, LRA and NRA are reference axes defined with respect to the mounting provisions. These axes are nominally parallel to IA, LA and NA respectively. Generally, the laser gyro will have two electrodes of one sign and one of the other. Plane axis LA is the center line of the laser leg containing the single electrode. Plane axis NA is the line in the laser beam plane perpendicular to LA and bisecting the leg containing LA. Generally, plane axis NA can be thought of as the axis of symmetry. 46

laser medium. *See:* **active laser medium.**

laser safety officer (laser-maser). One who is knowledgeable in the evaluation and control of laser hazards and has authority for supervision of the control of laser hazards. 363

laser system (laser-maser). An assembly of electrical, mechanical, and optical components which includes a laser. 363

lasing medium (laser-maser). A material emitting coherent radiation by virtue of stimulated electronic or molecular transitions to lower energy levels. 363

lasing threshold (1) (fiber optics). The lowest excitation level at which a laser's output is dominated by stimulated emission rather than spontaneous emission. *See:* **laser; spontaneous emission; stimulated emission.** 433

(2) (laser gyro). The discharge current at which the gain of the laser just exceeds the losses. 46

last-line release (telephone switching systems). Release under control of the last line that goes on-hook. 55

last transition (pulse terms). The major transition waveform of a pulse waveform between the top and the base. 254

latch (power switchgear). An attachment used to hold a fuse or switch in a closed position. 103, 443

latching current (1) (of a switching device) (power switchgear). The making current during a closing operation in which the device latches or the equivalent. 103

(2) (thyristor). The minimum principal current required to maintain the thyristor in the ON-state immediately after switching from the OFF-state to the ON-state has occurred and the triggering signal has been removed. *See:* **holding current.** 445

latching relay (power switchgear). A relay that is so constructed that it maintains a given position by means of a mechanical latch until released mechanically or electrically. 103

latch-in relay. A relay that maintains its contacts in the last position assumed without the need of maintaining coil energization. 341

latency (1) (biological electronics). The condition in an excitable tissue during the interval between the application of a stimulus and the first indication of a response. 192

(2) (electronic computation). The time between the completion of the interpretation of an address and the start of the actual transfer from the addressed location. 255, 77

latent period (electrobiology). The time elapsing between the application of a stimulus and the first indication of a response. *See:* **excitability (electrobiology).** 192

lateral conductor (transmission and distribution)-(NESC). A wire or cable extending in a general horizontal direction at an angle to the general direction of the line conductors. 262, 494

lateral critical speeds (rotating machinery). The speeds at which the amplitudes of the lateral vibrations of a machine rotor due to shaft rotation reach their maximum values. *See:* **rotor (rotating machinery).** 63

lateral-cut recording. *See:* **lateral recording.**

lateral insulator (storage cell). An insulator placed between the plates and the side wall of the container in which the element is housed. *See:* **battery (primary or secondary).**   328

lateral offset loss (fiber optics). A power loss caused by transverse or lateral deviation from optimum alignment of source to optical waveguide, waveguide to waveguide, or waveguide to detector. *Syn:* **transverse offset loss.**   433

lateral profile (1)(radio noise from overhead power lines and substations)(overhead-power-line corona and radio noise). The radio noise field strength at ground level plotted as a function of the horizontal distance from, and at a right angle to, the power line conductors.   411, 509

(2) (radio noise). The electric field strength at ground level plotted as a function of the horizontal distance from, and at a right angle to, the line conductors.   36

lateral recording (lateral-cut recording). A mechanical recording in which the groove modulation is perpendicular to the motion of the recording medium and parallel to the surface of the recording medium.   176

lateral width (light distribution). The lateral angle between the reference line and the width line, measured in the cone of maximum candlepower. *Note:* This angular width includes the line of maximum candlepower. *See:* **street-lighting luminaire.**   167

lateral working space (electric power distribution). The space reserved for working between conductor levels outside the climbing space, and to its right and left. *See:* **conflict (wiring system); tower.**   178

lattice (navigation). A pattern of identifiable intersecting lines of position, which lines are laid down by a navigation aid. *See:* **navigation.**   187,13

lattice channeling (in a semiconductor radiation conductor)(germanium gamma-ray detectors)(charged-particle detectors). A phenomenon that results in a crystallographic directional dependence of the rate of energy loss of ionizing particles.   528, 119

lattice network. A network composed of four branches connected in series to form a mesh, two nonadjacent junction points serving as input terminals, while the

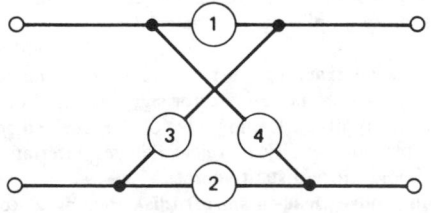

Lattice network. The junction points between branches 4 and 1 and between 3 and 2 are the input terminals; the junction points between branches 1 and 3 and between branches 2 and 4 are the output terminals.

remaining two junction points serve as output terminals. *See:* **network analysis.**   210

launch angle (fiber optics). The angle between the light input propagation vector and the optical axis of an optical fiber or fiber bundle. *See:* **launch numerical aperture.**   433

launcher (waveguide components). An adapter used to provide a waveguide or transmission line port for a wave propagating structure.   166

launching fiber (fiber optics). A fiber used in conjunction with a source to excite the modes of another fiber in a particular fashion. *Note:* Launching fibers are most often used in test systems to improve the precision of measurements. *Syn:* **injection fiber.** *See:* **mode; pigtail.**   433

launch numerical aperture (LNA) (fiber optics). The numerical aperture of an optical system used to couple (launch) power into an optical waveguide. *Notes:* (1) LNA may differ from the stated NA of a final focusing element if, for example, that element is underfilled or the focus is other than that for which the element is specified. (2) LNA is one of the parameters that determine the initial distribution of power among the modes of an optical waveguide. *See:* **acceptance angle; launch angle.**   433

laundry area (National Electrical Code). An area containing or designed to contain either a laundry tray, clothes washer, and/or a clothes dryer.   256

lay (1) (cables). The helical arrangement formed by twisting together the individual elements of a cable.   64, 57

(2) (helical element of a cable). The axial length of a turn of the helix of that element. *Notes:* (1) Among the helical elements of a cable may be each strand in a concentric-lay cable, or each insulated conductor in a multiple-conductor cable. (2) Also termed pitch.   345

lay-up (nuclear power generating stations). Idle condition of equipment and systems during and after installation, with protective measures applied as appropriate.   31, 143

L-band. A radar frequency band between 1 gigahertz (GHz) and 2 GHz, usually in the International Telecommunications Union (ITU) assigned band 1.215 GHz to 1.4 GHz; it may refer also to the 0.89 GHz to 0.94 GHz ITU assignment.   13

L-band radar (radar). A radar operating at frequencies between 1 and 2 GHz, usually in the ITU assigned band 1.215 to 1.4 GHz; may refer also to the 0.89 to 0.94 GHz ITU assignment.   13

LC. *See:* **low conduction threshold voltage.**

LC auxiliary switch. *See:* **auxiliary switch; LC contact.**

LC contact (power switchgear). A latch-checking contact that is closed which the operating mechanism linkage is relatched after an opening operation of the switching device.   103

L_d(s). *See:* **direct-axis operational inductance.**

L-display (radar). Similar to a K-display, but signals from the two lobes are placed back to back. A target appears as a pair of deflections, one on each side of a central time base representing range. Both deflections are of equal amplitude when the radar antenna is pointed directly at the target, any inequality representing relative pointing error. The time base (range scale) can be vertical, as in the illustration, or horizontal. *Syn:* **bearing deviation indicator.** 13

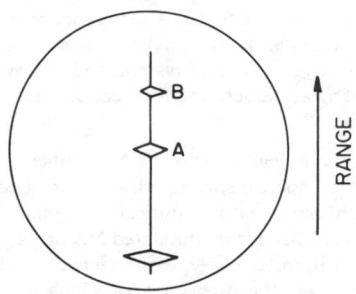

POINTING ERROR
NOTE: Two targets A and B at different ranges; radar aimed at target A.

L-Display

lead (power and distribution transformer). A conductor that connects a winding to its termination (that is, terminal, bushing, terminal board, or connection to another winding). 53

lead box (terminal housing) (rotating machinery). A box through which the leads are passed in emerging from the housing. 63

lead cable (rotating machinery). A cable type of conductor connected to the stator winding, used for making connections to the supply line or among circuits of the stator winding. *See:* **stator.** 63

lead clamp (salient-pole construction) (rotating machinery). A device used to retain and support the field leads between the hub and rotor rim along the rotor spider arms. *See:* **rotor (rotating machinery).** 63

lead collar (salient-pole construction) (rotating machinery). A bushing used to insulate field leads at a point of support between a collector ring and the rotor rim. *See:* **rotor (rotating machinery).** 63

lead-covered cable (lead-sheathed cable). A cable provided with a sheath of lead for the purpose of excluding moisture and affording mechanical protection. *See:* **armored cable.** 64

leader (computing systems). The blank section of tape at the beginning and end of a reel of tape. *Note:* The otherwise blank section may, however, include a parity check. 244, 207

leader cable. A navigational aid consisting of a cable around which a magnetic field is established, marking the path to be followed. *See:* **radio navigation.** 328

leader-cable system. A navigational aid in which a path to be followed is defined by the detection and comparison of magnetic fields emanating from a cable system that is installed on the ground or under water. *See:* **navigation.** 187, 13

leader cone (conductor stringing equipment). A tapered cone made of rubber, neoprene or polyurethane that is used to lead a conductor splice through the travelers, thus making a smooth transition from the smaller diameter conductor to the larger diameter splice. It is also used at the connection point of the pulling line and running board to assist in a smooth transition of the running board over the travelers, thus significantly reducing the shock loads. *Syn:* **tapered hose; nose cone.** 45

lead, first-order (control system, feedback). The change in phase due to a factor $(1 + Ts)$ in the numerator of a transfer function. *See:* **control system, feedback.** 329

lead-in. That portion of an antenna system that connects the elevated conductor portion to the radio equipment. *See:* **antenna.** 179

leading-edge (first transition)(1)(pulse transformers). That portion of the pulse occurring between the time the instantaneous value first becomes greater than A Q to the time of the intersection of straight-line segments used to determine A_M. 32, 33, 589
(2) (radar). A radar range tracking technique in which the range error signal is based on the range delay of the leading edge of the received echo. This provides ability to reject delayed interference, chaff, and more distant sources. 13
(3) (television). The major portion of the rise of a pulse. *See:* **television.** 337

leading edge (first transition) linearity (A_L) (high- and low-power pulse transformers). The maximum amount by which the instantaneous pulse value deviates during the rise time interval from a straight line intersecting the 10 percent and 90 percent A_M amplitude points used in determining rise time. It is expressed in amplitude units or in percent of 0.8 A_M. See also: **input pulse shape.** 32, 33, 589

leading-edge pulse. The first major transition away from the pulse baseline occurring after a reference time. *See:* **pulse; television.** 185

leading-edge pulse time. The time at which the instantaneous amplitude first reaches a stated fraction of the peak pulse amplitude. *See:* **television.** 254, 210, 162

leading-edge tracking. A radar range tracking technique in which the range error signal is based on the range delay of the leading edge of the received echo. This provides ability to reject delayed interference, chaff, and more distant sources. 13

lead-in groove (lead-in spiral) (disk recording) (electroacoustics). A blank spiral groove at the beginning of a recording generally having a pitch that is much greater than that of the recorded grooves. *See:* **phonograph pickup.** 334

leading wire (mining). An insulated wire strung separately or as a twisted pair, used for connecting the two free ends of the circuit of the blasting caps to the blasting unit. *See:* **blasting unit.** 328

lead-in spiral (disk recording). *See:* **lead-in groove.**

lead-in wire. A conductor connecting an electrode to an external circuit. *See:* **electron tube.** 244, 190

lead networks (power supplies). Resistance-reactance components arranged to control phase-gain roll-off versus frequency. *Note:* Used to assure the dynamic stability of a power-supply's comparison amplifier. The main effect of a lead network is to introduce a phase lead at the higher frequencies, near the unity-gain frequency. 228, 186

lead-out groove (throw-out spiral) (disk recording). A blank spiral groove at the end of a recording generally of a pitch that is much greater than that of the recorded grooves and that terminates in either a locked or an eccentric groove. *See:* **phonograph pickup.** 176

lead-over groove (crossover spiral). In disk recording, a groove cut between successive short-duration recordings on the same disk, to enable the pickup stylus to travel from one cut to the next. *See:* **phonograph pickup.** 176

lead polarity (power and distribution transformer). A designation of the relative instantaneous direction of the currents in the leads of a transformer. Primary and secondary leads are said to have the same polarity when, at a given instant, the current enters the primary lead in question and leaves the secondary lead in question in the same direction as though the two leads formed a continuous circuit. The lead polarity of a single-phase distribution or power transformer may be either additive or subtractive. If adjacent leads from each of the two windings in question are connected together and voltage applied to one of the windings: (1) the lead polarity is additive if the voltage across the other two leads of the windings in question is greater than that of the higher voltage winding alone; (2) the lead polarity is subtractive if the voltage across the other two leads of the windings in question is less than that of the higher voltage winding alone. The polarity of a polyphase transformer is fixed by the internal connections between phases; it is usually designated by means of a phasor diagram showing the angular displacements of the voltages in the windings and a sketch showing the marking of the leads. The phasors of the phasor diagrams represent induced voltages. The standard rotation of phasors is counterclockwise. 53

leads, load. *See:* **load leads** or **transmission line.**

lead storage battery. A storage battery the electrodes of which are made of lead and the electrolyte consists of a solution of sulfuric acid. *See:* **battery (primary or secondary).** 328

leak (handling and disposal of transformer grade insulating liquids containing PCBs). Any instance in which a polychlorinated biphenyl (PCB) unit (PCB article, PCB container, PCB equipment) has any PCBs on any portion of its external surface. 586

leakage (1) (analog computers). (A) Undesirable conductive paths in certain components, specifically, in capacitors, a path through which a slow discharge may take effect: in problem boards, interaction effects between electrical signals through insufficient insulation between patch bay terminals. (B) Current flowing through such paths. 9

(2) (signal-transmission system.) Undesired current flow between parts of a signal-transmission system or between the signal-transmission system and point(s) outside the system. *See:* **signal.** 188

(3) (health care facilities). This is any current, including capacitively coupled current, not intended to be applied to a patient but which may be conveyed from exposed metal parts of an appliance to ground or to other accessible parts of an appliance. 192

(4) (transmission lines and waveguides). Radiation or conduction of signal power out of or into an imperfectly closed and shielded system. The leakage is usually expressed in decibels below a specified reference power. 185

(5) (insulation). The current that flows through or across the surface of insulation and defines the insulation resistance at the specified direct-current potential. 116

(6) (semiconductor radiation detector). The total detector current flowing at the operating bias in the absence of radiation. 23

leakage (conduction) current (1) (rotating machinery). The nonreversible constant current component of measured current which remains after the capacitive current and absorption current have disappeared. *Note:* Leakage current passes through the insulation volume, through any defects in the insulation, and across the insulation surface. 6

leakage current (of a semiconductor radiation conductor)(germanium gamma-ray detectors)(X-ray energy spectrometers)(charged-particle detectors). The total detector current flowing at the operating bias in the absence of radiation. 528, 471, 119

leakage current, input (amplifiers). A direct current (of either polarity) that would flow in a short circuit connecting the input terminals of an amplifier. 106

leakage distance (insulator). The sum of the shortest distances measured along the insulating surfaces between the conductive parts, as arranged for dry flashover test. *Note:* Surfaces coated with semiconducting glaze shall be considered as effective leakage surfaces, and leakage distance over such surfaces shall be included in the leakage distance. *See:* **insulator.** 261

leakage distance of external insulation. *See:* **creepage distance.**

leakage flux. Any magnetic flux, produced by current in an instrument transformer winding, which does not link all turns of all windings. 203

leakage flux, relay. *See:* **relay leakage flux.**

leakage inductance (one winding with respect to a second winding). A portion of the inductance of a winding that is related to a difference in flux linkages in the two windings: quantitatively, the leakage inductance of winding 1 with respect to winding 2

$$L_{if} = \frac{\partial\left(\lambda_{11} - \dfrac{N_1}{N_2}\lambda_{21}\right)}{\partial i_1}$$

where λ_{11} and λ_{21} are the flux linkages of windings 1 and 2, respectively, resulting from current i_1 in winding 1: and N_1 and N_2 are the number of turns of windings 1 and 2, respectively.   197

leakage power (TR and Pre-TR tubes). The radio-frequency power transmitted through a fired tube. *See:* **flat leakage power; (TR and Pre-TR tubes); harmonic leakage power; (TR and Pre-TR tubes).**   103

leakage power (TR [transmit-receive] tubes and power limiters)(nonlinear, active, and nonreciprocal waveguide components). The radio-frequency power transmitted through a fired tube. *See:* **flat leakage power (TR and pre-TR tubes); harmonic leakage power (TR and pre-TR tubes).**   530

leakage radiation (radio transmitting system). Radiation from anything other than the intended radiating system. *See:* **radio transmission.**   111

leaky mode (fiber optics). In an optical waveguide, a mode whose field decays monotonically for a finite distance in the transverse direction but which becomes oscillatory everywhere beyond that finite distance. Specifically, a mode for which

$$[n^2(a)k^2 - (script\ el/a)^2]^{1/2} = \beta = n(a)k$$

where β is the imaginary part (phase term) of the axial propagation constant, *script el* is the azimuthal index of the mode, n(a) is the refractive index at r=a, the core radius, and k is the free-space wavenumber, $2\pi/\lambda$, and λ is the wavelength. Leaky modes correspond to leaky rays in the terminology of geometric optics. *Note:* Leaky modes experience attenuation, even if the waveguide is perfect in every respect. *Syn:* **tunneling mode.** *See:* **bound mode; cladding mode; leaky ray; mode; unbound mode.**   433

leaky ray (fiber optics). In an optical waveguide, a ray for which geometric optics would predict total internal reflection at the core boundary, but which suffers loss by virtue of the curved core boundary. Specifically, a ray at radial position r having direction such that

$$n^2(r) - n^2(a) = \sin^2\theta$$
$$(r)\text{and}\ \sin^2\theta(r) = [n^2$$
$$(r) - n^2(a)]/[1 - (r/a)^2\cos^2\phi(r)]$$

where $\theta(r)$ is the angle the ray makes with the waveguide axis, n (r) is the refractive index, a is the core radius, and $\phi(r)$ is the azimuthal angle of the projection of the ray on the transverse plane. Leaky rays correspond to leaky (or tunnelling) modes in the terminology of mode descriptors. *Syn:* **tunnelling ray.** *See:* **bound mode; cladding ray; guided ray; leaky mode.**   433

leaky-wave antenna. An antenna that couples power in

small increments per unit length, either continuously or discretely, from a traveling wave structure to free space.   111

learning system. An adaptive system with memory. *See:* **system science.**   209

leased channel (data transmission). A point-to-point channel reserved for sole use of a single leasing customer.   59

leased line or private wire network (data transmission). A series of points interconnected by telegraph or telephone channels, and reserved for the exclusive use of one customer.   59

least significant bit (LSB)(1)(mathematics of computing). The bit having the smallest effect on the value of a binary numeral; usually the rightmost bit.   564
(2)(station control and data acquisition). In an n bit binary word its contribution is (0 or 1) toward the maximum word value of $2^n - 1$.   403

leave-word call (telephone switching systems). A person-to-person call on which the designated called person was not available and the operator left instructions for its later establishment.   55

Lecher wires (data transmission). Two parallel wires on which standing waves are set up, frequently for the measurement of wavelength.   59

LED. *See:* **light emitting diode.**

Leduc current (electrotherapy). A pulsed direct current commonly having a duty cycle of 1:10. *See:* **electrotherapy.**   192

left-handed (counterclockwise) polarized wave (radio wave propagation). An elliptically polarized electromagnetic wave in which the rotation of the electric field vector with time is counterclockwise for a stationary observer looking in the direction of the wave normal. *Note:* For an observer looking from a receiver toward the apparent source of the wave, the direction of rotation is reversed.   146

leg (circuit). Any one of the conductors of an electric supply circuit between which is maintained the maximum supply voltage. *See:* **center of distribution.**   64

legitimate access (nuclear security systems). The proper and correct access authorization.   464

leg, thermoelectric. Alternative term for thermoelectric arm. *See:* **thermoelectric device.**   248, 191

leg wire (mining). One of the two wires attached to and forming a part of an electric blasting cap or squib. *See:* **blasting unit.**   328

Lenard tube. An electron beam tube in which the beam can be taken through a section of the wall of the evacuated enclosure.   190

length (1) (electronic computers). (A) A measure of the magnitude of a unit of data, usually expressed as a number of subunits, for example, the length of a record is 32 blocks, the length of a word is 40 binary digits, etcetera. (B) The number of subunits of data, usually digits or characters, that can be simultaneously stored linearly in a given device, for example, the length of the register is 12 decimal digits or the length of the counter is 40 binary digits. (C) A measure of the amount of time that data are delayed when being

transmitted from point to point, for example, the length of the delay line is 384 microseconds. *See:* **double length; word length; storage capacity.**
                                                                                    235

(2) (antenna). *See:* **effective length.**

length of lay (cable) (power distribution underground cables). The axial length of one turn of the helix of any helical element.   57

lens and aperture (phototube housing) (industrial control). The cooperating arrangement of a light-refracting member and an opening in an opaque diaphragm through which all light reaching the phototube cathode must pass. *See:* **electronic controller.**   206

lens antenna. An antenna consisting of an electromagnetic lens and a feed (antenna) which illuminates it.
                                                                                    111

lens distance relay (1) (power switchgear). A distance relay that has an operating characteristic comprising the common area of two intersecting circular relay characteristics. *See:* **distance relay; figure b.**   103

(2) (relaying). A distance relay that has an operating characteristic comprising the common area of two intersecting mho relay characteristics (Figure 1b).
                                                                                    79

lens, electromagnetic. A three-dimensional structure propagating electromagnetic waves, with an effective index of refraction differing from unity, employed to control the aperture illumination.   111

lens multiplication factor (phototube housing) (industrial control). The maximum ratio of the light flux reaching the phototube cathode with the lens and aperture in place to the light flux with the lens and aperture removed. *See:* **photoelectric control.**   206

Lenz's law (induced current). The current in a conductor as a result of an induced voltage is such that the change in magnetic flux due to it is opposite to the change in flux that caused the induced voltage.
                                                                                    210

letter. An alphabetic character used for the representation of sounds in a spoken language.   255, 77

letter combination (abbreviation). One or more letters that form a part of the graphic symbol for an item and denote its distinguishing characteristic. *Compare with:* **functional designation: reference designation: symbol for a quantity.** *See:* **abbreviation.**   173

let-through sparkover (surge arrester). A measure of the highest lightning surge an arrester is likely to withstand without sparkover in $3\,\mu s$ or less. The value determined by a 1.2 x 50-μs impulse sparkover test. *See:* **ANSI/IEEE C62.1-1981.**   62

level (1) (data transmission). The magnitude of a quantity, especially when considered in relation to an arbitrary reference value. Level may be stated in the units in which the quantity itself is measured (for example, dB) expressing the ratio to a reference value.
                                                                                    59

(2) (software). (A) The degree of subordination of an item in a hierarchical arrangement. (B) A rank within a hierarchy. An item is of the lowest level if it has no subordinates and of the highest level if it has no superiors. *See:* **hierarchy.**   434

(3) (general) (quantity). Magnitude, especially when considered in relation to an arbitrary reference value. Level may be stated in the units in which the quantity itself is measured (for example, volts, ohms, etcetera) or in units (for example, decibels) expressing the ratio to a reference value. *Notes:* (A) Examples of kinds of levels in common use are electric power level, sound-pressure level, voltage level. (B) The level as here defined is measured in two common units: in decibels when the logarithmic base is 10, or in nepers when the logarithmic base is e. The decibel requires that k be 10 for ratios of power, or 20 for quantities proportional to the square root of power. The neper is used to repre-

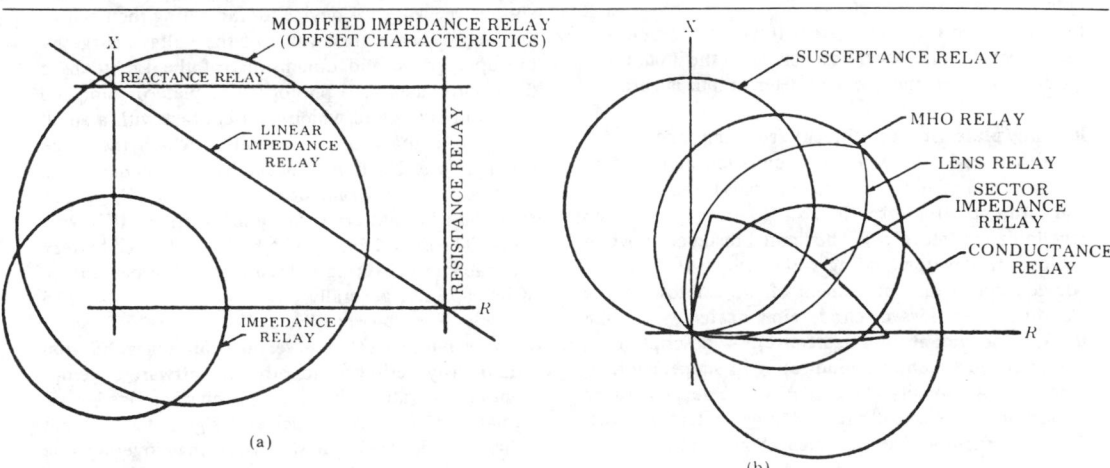

Fig. 1. (a) Operating characteristic of distance relays. (b) Operating characteristics of distance relays that are inherently directional.

sent ratios of voltage, current, sound pressure, and particle velocity. The neper requires that k be 1. (C) In symbols $L = k \log_r(q.q_0)$ where L = level of kind determined by the kind of quantity under consideration. r = base of the logarithm of the reference ratio. q = the quantity under consideration. q_0 = reference quantity of the same kind. k = a multiplier that depends upon the base of the logarithm and the nature of the reference quantity. See: blanking level; transmission level; reference black level; reference white level; signal level (electroacoustics).   176

(4) (charge-storage tubes). A charge value which can be stored in a given storage element and distinguished in the output from other charge values. See: channel.   125

level above threshold (sensation level) (sound) (acoustics). The pressure level of the sound in decibels above its threshold of audibility for the individual observer.   176

level band pressure (for a specified frequency band) (electroacoustics). The sound-pressure level for the sound contained within the restricted band. Notes: (1) The reference pressure must be specified. (2) The band may be specified by its lower and upper cutoff frequencies or by its geometric or arithmetic center frequency and bandwidth. The width of the band may be indicated by a prefatory modifier, for example, octave band (sound pressure) level, half-octave band level, third-octave band level, 50-hertz band level. See: loudspeaker.   176

level compensator (signal transmission). An automatic transmission-regulating feature or device to minimize the effects of variations in amplitude of the received signal. See: telegraphy.   328

level detector (as applied to relaying) (power switchgear). A device that produces a change in output at a prescribed input level.   103

leveling (electronic navigation). See: platform erection.

leveling block (rotating machinery). See: leveling plate.

leveling error (accelerometer) (inertial sensor). The angle between the local horizontal and the input reference axis when the accelerometer output is zero.   46

leveling plate (leveling block) (rotating machinery). A heavy pad built into the foundation and used to support and align the bed plate or rails using shims for adjustment before grouting.   63

leveling zone (elevator). The limited distance above or below an elevator landing within which the leveling device may cause movement of the car toward the landing. See: elevator-car leveling device.   328

level of documentation (software). A description of required documentation indicating its scope, content, format, and quality. Selection of the level may be based on project cost, intended usage, extent of effort, or other factors. See: documentation; quality.   434

level, relay. See: relay level.

levels, usable (storage tubes). The output levels, each related to a different input, that can be distinguished from one another regardless of location on the storage surface. Note: The number of usable levels is normally limited by shading and disturbance. See: storage tube.   174

level switch (71) (power system device function numbers). A switch which operates on given values, or on a given rate of change, of level.   402

lever blocking device (railway signaling). A device for blocking a lever so that it cannot be operated.   328

lever indication (railway signaling). The information conveyed by means of an indication lock that the movement of an operated unit has been completed.   328

LF. See: radio spectrum.

liberator tank (electrorefining). Sometimes known as a depositing-out tank, an electrolytic cell equipped with insoluble anodes for the purpose of either decreasing, or totally removing the metal content of the electrolyte by plating it out on cathodes. See: electrorefining.   328

librarian. See: software librarian.

library See: software library; system library.

library automation (computer applications). The application of automated techniques to library operations such as processing of documents, interlibrary communication, and on-line catalog access.   571

library routine (high-level microprocessor language). A function (which returns a value) or a procedure (which does not return a value) supplied with the implementation of the high-level language (HLL). For the purposes of IEEE trial use Std 755-1985, a short routine coded entirely in the language supported by the implementation, but which the user is obliged to include with the application program, is also said to be a library routine if its complete specification (that is, source listing) is included in the user documentation.   470

Lichtenberg figure camera (klydonograph) (surge-voltage recorder). A device for indicating the polarity and approximate crest value of the voltage surge by the appearance and dimensions of the Lichtenberg figure produced on a photographic plate or film, the emulsion coating of which is in contact with a small electrode coupled to the circuit in which the surge occurs. Note: The film is backed by an extended plane electrode. See: instrument.   328

life (lead-acid batteries for photovoltaic (PV) systems). The period during which a fully charged battery is capable of delivering at least a specified percentage of its capacity, generally 80%.   515

life cycle. See: software life cycle.

life-cycle phase (software verification and validation plans). Any period of time during software development or operation that may be characterized by a primary type of activity (such as design or testing) that is being conducted. These phases may overlap one another; for verification and validation (V&V) purposes, no phase is concluded until its development products are fully verified.   511 life performance curve (illuminating engineering). A curve which represents the variation of a particular characteristic of a light source (luminous flux, intensity, etcetera) throughout the life of the source. *Note:* Life performance curves sometimes are called maintenance curves as, for example, lumen maintenance curves.   167

life safety branch (1) (health care facilities). A subsystem of the Emergency System consisting of feeders and branch circuits, meeting the requirements of Article 700 of NFPA 70-1978, National Electrical Code, and intended to provide adequate power needs to ensure safety to patients and personnel, and which can be automatically connected to alternate power sources during interruption of the normal power source.   192

life support equipment (nuclear power generating station). The breathing apparatus, medical supplies, sanitary facilities, and food and water supplies required to sustain operators for an extended period of time during abnormal operating conditions.   439

life test. *See:* accelerated test (reliability).

life test of lamps (illuminating engineering). A test in which lamps are operated under specified conditions for a specified length of time, for the purpose of obtaining information on lamp life. Measurements of photometric and electric characteristics may be made at specified intervals of time during this test.   167

lifetime rated pulse currents (low voltage varistor surge arresters). Derated values of rated peak single pulse transient current for impulse durations exceeding that of an 8 x 20-μs waveshape, and for multiple pulses which may be applied over the device rated lifetime.   62

lifetime, volume (semiconductor). The average time interval between the generation and recombination of minority carriers in a homogeneous semiconductor. *See:* semiconductor; semiconductor device.   245

lifting eye (fuseholder, fuse unit, or disconnecting blade) (high-voltage switchgear) (power switchgear). An eye provided for receiving a fuse hook or switch hook for inserting the fuse or disconnecting blade into and for removing it from the fuse support.   443, 103

lifting-insulator switch (power switchgear). One in which one or more insulators remain attached to the blade, move with it, and lift it to the open position.   103

light (1) (fiber optics). (1) In a strict sense, the region of the electromagnetic spectrum that can be perceived by human vision, designated the visible spectrum and nominally covering the wavelength range of 0.4μm to 0.7μm. (2)In the laser and optical communication fields, custom and practice have extended usage of the term to include the much broader portion of the electromagnetic spectrum that can be handled by the basic optical techniques used for the visible spectrum. This region has not been clearly defined but, as employed by most workers in the field, may be considered to extend from the near-ultraviolet region of approxi-

mately 0.3μm, through the visible region, and into the mid-infrared region to 30μm. *See:* infrared (IR); optical spectrum; ultraviolet (UV).   433

(2) (illuminating engineering). Radiant energy that is capable of exciting the retina and producing a visual sensation. The visible portion of the electromagnetic spectrum extends from about 380 to 770 nm. *Note:* The subjective impression produced by stimulating the retina is sometimes designated as light. Visual sensations are sometimes arbitrarily defined as sensations of light, and in line with this concept it is sometimes said that light cannot exist until an eye has been stimulated. Electrical stimulation of the retina or the visual cortex is described as producing flashes of light. In illuminating engineering, however, light is a physical entity--radiant energy weighted by the luminous efficiency function. It is a physical stimulus which can be applied to the retina. *See:* spectral luminous efficiency of radiant flux; values of spectral luminous efficiency for photopic vision.   167

light adaptation (illuminating engineering). The process by which the retina becomes adapted to a luminance greater than about 3.4 cd/m^2,(2.2$\times 10^{-3}$ cd/in 2)(1.0fL).   167

light center (illuminating engineering). The center of the smallest sphere that would completely contain the light-emitting element of the lamp.   167

light center length (illuminating engineering). The distance from the light center to a specified reference point on the lamp.   167

light current. *See:* photocurrent.

lighted beacon (navigation aid terms). A beacon that transmits signals by light waves (for example, light house).   526

lighted buoy (navigation aid terms). A buoy with a light that has characteristics for detection and identification. *See:* buoy.   526

light emitting diode (LED) (1) (fiber optics). A pn junction semiconductor device that emits incoherent optical radiation when biased in the forward direction. *See:* incoherent.   433

(2) (illuminating engineering). A p-n junction solid-state diode whose radiated output is a function of its physical construction, material used, and exciting current. The output may be in the infrared or in the visible region.   167

lightguide. *See:* optical waveguide.

lighting branch circuit (electric installations on shipboard). A circuit supplying energy to lighting outlets only. (Lighting branch circuits also may supply portable desk or bracket fans, small heating appliances, motors of 1/4 hp (186 1/2 W) and less, and other portable apparatus of not over 600 W each.)   3

lighting effectiveness factor (LEF) (illuminating engineering). The ratio of equivalent sphere illumination to measured or calculated task illuminance.   167

lighting outlet (1) (electric installations on shipboard). An outlet intended for the direct connection of a lampholder or a lighting fixture.   3

(2) (National Electrical Code). An outlet intended for the direct connection of a lampholder, a lighting fix-

ture, or a pendant cord terminating in a lampholder. 256

light load (watthour meter). The current at which the meter is adjusted to bring its response near the lower end of the load range to the desired value. It is usually 10 percent of the test current for a revenue meter and 25 percent for a standard meter. 212

light loss factor (LLF) (illuminating engineering). The ratio of the illuminance on a given area after a period of time to the initial illuminance on the same area. *Note:* The light loss factor is used in lighting calculations as an allowance for the depreciation of lamps, light control elements, and room surfaces to values below the initial or design conditions, so that a minimum desired level of illuminance may be maintained in service. The light loss factor had formerly been widely interpreted as the ratio of average to initial illuminance. This term was formerly called **maintenance factor.** 167

lightness (of a perceived patch of surface color) (illuminating engineering). The brightness of an area judged relative to the brightness of a similarly illuminated area that appears to be white or highly transmitting. 167

lightning (surge arrester)(metal-oxide surge arresters for ac power circuits). An electric discharge that occurs in the atmosphere between clouds or between clouds and ground. 430, 583

lightning and switching impulses (high voltage testing). A distinction is made between lightning impulses and switching impulses on the basis of the duration of the front. Impulses with front durations from less than one up to a few tens of a microsecond are, in general, considered as lightning impulses and those having front durations of some tens up to thousands of microseconds as switching impulses. Generally, switching impulses are also characterized by considerably longer total durations than those of lightning impulses. *Note:* If requires, a more rigorous distinction may be made in an apparatus standard. Such a distinction is made in IEEE Std 28-1974 (ANSI C62.1-1975), Surge Arresters for Alternating-Current Power Circuits, and is based on a virtual front time of 30 microseconds. 150

lightning impulse insulation level (power and distribution transformer). An insulation level expressed in kilovolts of the crest value of a lightning impulse withstand voltage. 53

lightning impulse protection level (of a protective device) (power and distribution transformer). The maximum lightning impulse voltage expected at the terminals of a surge protective device under specified conditions of operation. 53

lightning impulse test (transformer) (power and distribution transformer). Application of the following sequence of impulse waves: (1) one reduced full wave, (2) two chopped waves, (3) one full wave. 53

lightning overvoltage (surge arrester). The crest voltage appearing across an arrester or insulation caused by a lightning surge. 62

lightning surge (surge arrester)(metal-oxide surge ar- resters for ac power circuits). A transient electric disturbance in an electric circuit caused by lightning. 430, 583

light pattern (optical pattern) (Buchmann-Meyer pattern) (mechanical recording). A pattern that is observed when the surface of the record is illuminated by a light beam of essentially parallel rays. *Note:* The width of the observed pattern is approximately proportional to the signal velocity of the recorded groove. 176

light pipe. An optical transmission element that utilizes unfocused transmission and reflection to reduce photon losses. *Note:* Light pipes have been used to distribute the light more uniformly over a photocathode. *See:* **phototube.** 335

light ray (fiber optics). The path of a point on a wavefront. The direction of a light ray is generally normal to the wavefront. *See:* **geometric optics.** 433

light source (industrial control). A device to supply radiant energy capable of exciting a phototube or photocell. *See:* **photoelectric control.** 206

light-source color (illuminating engineering). The color of the light emitted by the source. *Note:* The color of a point source may be defined by its luminous intensity and chromaticity coordinates; the color of an extended source may be defined by its luminance and chromaticity coordinates. 167

light transition load (rectifier circuit). The transition load that occurs at light load, usually at less than 5 percent of rated load. *Note:* Light transition load is important in multiple rectifier circuits. A similar effect occurs in rectifier units using saturable-reactor control. *See:* **rectification; rectifier circuit element.** 237, 66

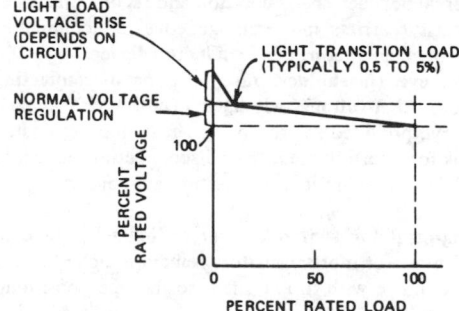

Voltage regulation characteristic showing light transition load.

light valve (electroacoustics). A device whose light transmission can be made to vary in accordance with an externally applied electrical quantity, such as voltage, current, electric field, magnetic field, or an electron beam. *See:* **phonograph pickup.** 176, 178

lightwatt. *See:* **spectral luminous efficiency.**

limit (1) (mathematical). A boundary of a controlled variable.

(2) (synchronous machine regulating systems). The boundary at or beyond which a limiter functions. 63

(3) (industrial control). The designated quantity is controlled so as not to exceed a prescribed boundary condition. *See:* **control system, feedback.** 206

(4) (test, measurement and diagnostic equipment). The extreme of the designated range through which the measured value of a characteristic may vary and still be considered acceptable. 54

limit cycle (control systems). A closed curve in the state space of a particular control system, from which state trajectories may recede, or which they may approach, for all initial states sufficiently close to the curve. 56

limit cycle, stable (control systems). One which is approached asymptotically by a state trajectory for all initial states sufficiently close. 56

limit cycle, unstable (control systems). One from which state trajectories recede for all initial states sufficiently close. 56

limited availability (telephone switching systems). Availability that is less than the number of outlets in the desired group. 55

limited proportionality, region of. *See:* **region of limited proportionality.**

limited signal (radar). A signal that is limited in amplitude by the dynamic range of the system. 13, 187

limited stability. A property of a system characterized by stability when the input signal falls within a particular range and by instability when the signal falls outside this range. 349

limiter (1) (excitation systems for synchronous machines). An element of the excitation system which acts to limit a variable by modifying or replacing the functions of the primary detector element when predetermined conditions have been reached. *Notes:* Examples: (A) An under excitation limiter prevents the voltage regulator from lowering the excitation of the synchronous machine below a prescribed level. (B) An over excitation limiter prevents the voltage regulator from raising the excitation of the synchronous machine above a level which would cause a thermal overload in the machine field; refer to ANSI C50.13-1977. (C) A volts per hertz limiter acts, through the voltage regulator to correct for a machine terminal voltage to frequency ratio that is considered abnormal. (D) Other types of limiters may be used to control various quantities, such as, rotor angle, excitation output, etcetera. *See:* **ferri-diode limiter; ferrite limiter; gyromagnetic limiter; multipactor limiter; passive limiter; p-i-n diode limiter; plasma limiter; quasiactive limiter.** 507

(2) (data transmission). (A) A device in which some characteristic of the output is automatically prevented from exceeding a predetermined value. (B) More specifically, a transducer in which the output amplitude is substantially linear with regard to the input up to a predetermined value and substantially constant thereafter. *Note:* For waves having both positive and negative values, the predetermined value is usually independent of sign. 59

(3) (rotating machinery). An element or group of elements that acts to limit by modifying or replacing the functioning of a regulator when predetermined conditions have been reached. *Note:* Examples are minimum excitation limiter, maximum excitation limiter, maximum armature-current limiter. 63

(4) (radio receivers). A transducer whose output is constant for all inputs above a critical value. *Note:* A limiter may be used to remove amplitude modulation while transmitting angle modulation. *See:* **radio receiver; transducer.** 111

(5) (excitation systems). A feedback element of the excitation system which acts to limit a variable by modifying or replacing the function of the primary detector element when predetermined conditions have been reached. 105

limiter circuits (analog computers). A circuit of nonlinear elements that restrict the electrical excursion of a variable in accordance with some specified criteria. **Hard limiting** is a limiting action with negligible variation in output in the range where the output is limited. **Soft limiting** is a limiting action with appreciable variation in output in the range where the output is limited. A **bridge limiter** is a bridge circuit used as a limiter circuit. In an analog computer, a **feedback limiter** is a limiter circuit usually employing biased diodes shunting the feedback component of an operational amplifier; an input limiter is a limiter circuit usually employing biased diodes in the amplifier input channel that operates by limiting the current entering the summing junction. **linear system or element.** A system with the properties: if y_1 is the response to x_1 and y_2 is the response to x_2, then (i) $(y_1 + y_2)$ is the response to $(x_1 + x_2)$ and (ii) ky_1 is the response to kx_1. *See:* **stop.** 9

limiting (automatic control). The intentional imposition or inherent existence of a boundary on the range of a variable, for example, on the speed of a motor. 56, 105

limiting ambient temperature (1) (electric equipment) (thermal classification of electric equipment and electrical insulation). The highest (or lowest) ambient temperature at which electric equipment is expected to give specified performance under specified conditions, for example, rated load. 506

(2)(equipment rating). An upper or lower limit of a range of ambient temperatures within which equipment is suitable for operation at its rating. Where the term is used without an adjective the upper limit is meant. *See:* **limiting insulation system temperature.** 233

limiting angular subtense (α_{min}) (laser-maser). The apparent visual angle which divides intrabeam viewing from extended source viewing. 363

limiting aperture (laser-maser). The maximum circular area over which radiance and radiant exposure can be averaged. 363

limiting hottest-spot temperature (electric equipment) (thermal classification of electric equipment and electrical insulation). The highest temperature attained in any part of the insulation of electric equipment, which is operating under specified conditions, usually at maximum rating and the upper limiting ambient temperature. 506

limiting hottest-spot temperature. *See:* limiting insulation temperature.

limiting insulation system temperature (limiting hottest-spot temperature) (power and distribution transformer). The maximum temperature selected for correlation with a specified test condition of the equipment with the object of attaining a desired service life of the insulation system. 53

limiting insulation temperature rise (equipment rating). The difference between the limiting insulation temperature and the limiting ambient temperature. *Syn:* limiting hottest-spot temperature. *See:* limiting insulation system temperature. 233

limiting observable temperature rise (equipment rating). The limit of observable temperature rise specified in equipment standards. *See:* limiting insulation system temperature. 233

limiting polarization (radio wave propagation). The resultant polarization of a wave after it has emerged from a magnetoionic medium. 146

limiting resolution (1) (diode-type camera tube). The high frequency bar pattern which can be visually distinguished as separate bars on a display. Units: Television lines per raster height (TVL/RH). 380
(2) (television). A measure of overall system resolution usually expressed in terms of the maximum number of lines per picture height discriminated on a television test chart. *Note:* For a number of lines N (alternate black and white lines), the width of each line is 1/N times the picture height. 18

limiting temperature (power and distribution transformer). The maximum temperature at which a component or material may be operated continuously with no sacrifice in normal life expectancy. 3

limit, lower. *See:* lower limit.

limit of error. *See:* uncertainty.

limit of temperature rise (1) (contacts). The temperature rise of contacts, above the temperature of the cooling air, when tested in accordance with the rating shall not exceed the following values. All temperatures shall be measured by the thermometer method. Laminated contacts: 50 degrees Celsius; solid contacts: 75 degrees Celsius.
(2) (resistors). The temperature rise of resistors above the temperature of the cooling air, when test is made in accordance with the rating, shall not exceed the following temperatures for the several classes of resistors: Class A, cast resistors, 450 degrees Celsius: Class B, imbedded resistors, outside of imbedding material, 250 degrees Celsius: Class C, strap or ribbon wound on Class C insulation, 600 degrees Celsius continuous and 800 degrees Celsius intermittent: class D, enameled wire or strap wound resistance, 350 degrees Celsius. Temperatures to be measured by thermocouple method. 1

limits of interference (electromagnetic compatibility). Maximum permissible values of radio interference as specified in International Special Committee on Radio Interference recommendations or by other competent authorities or organizations. *See:* electromagnetic compatibility 220, 199

limit switch. A switch that is operated by some part or motion of a power-driven machine or equipment to alter the electric circuit associated with the machine or equipment. *See:* switch. 206

limit, upper. *See:* upper limit.

line (1) (electric power). A component part of a system extending between adjacent stations or from a station to an adjacent interconnection point. A line may consist of one or more circuits. *See:* system. 91
(2) (trace) (cathode-ray tube). The path of a moving spot. 190
(3) (electromagnetic theory). *See:* maxwell.
(4) (acoustics). *See:* acoustic delay line; delay line; electromagnetic delay line; magnetic delay line; sonic delay line.
(5) (data transmission). *See:* channel. 59

lineal electric current element. *See:* linear electric current element.

line amplifier. *See:* amplifier, line.

linear accelerometer. A device which measures translational acceleration along an input axis. In this accelerometer, an output signal is produced by the reaction of the proof mass to a translatory acceleration input. The output is usually an electrical signal proportional to applied translational acceleration. 46

linear amplifier (magnetic). An amplifier in which the output quantity is essentially proportional to the input quantity. *Note:* This may be interpreted as an amplifier that has no intentional discontinuities in the output characteristic over the useful input range of the amplifier. 171

linear antenna. An antenna consisting of one or more segments of straight conducting cylinders. *Note:* This term has restricted usage, and applies to straight cylindrical wire antennas. This term should not be confused with the conventional usage of linear in circuit theory. *See:* linear array antenna. 111

linear array. *See:* linear array antenna.

linear array antenna. A one-dimensional array of elements whose corresponding points lie along a straight line. 111

linear broadcast (FASTBUS acquisition and control). A broadcast to a subset of the segments affected by a global broadcast. The subset can be either a specified segment or up to a specified segment or beyond a specified segment. 480

linear charging (direct current) (charging inductors). A special case of resonance charging of the capacitance in an oscillatory series resistance-inductance-capacitance (RLC) circuit where the capacitor is repetitively discharged at a predetermined voltage at a rate much greater than twice the natural resonance of the RLC circuit. *Note:* The inductance of the charging inductor for linear charging is much greater than that for resonance charging for a given pulse-repetition frequency. Under the above conditions, the current through the charging inductor at the time the capacitance is discharged is not zero and the voltage across the capacitance is still rising. 137

linear cross-field amplifier (microwave tubes). A crossed-field amplifier in which a nonre-entrant beam interacts with a forward wave. *See:* **microwave tube or valve.** 190

linear detection (information theory). The form of detection of an amplitude-modulated signal in which the output voltage is a linear function of the envelope of the input wave. 61

linear distortion (linear waveform distortion). That distortion of an electric signal which is independent of the signal amplitude. *Note:* A small-signal nonuniform frequency response is an example of linear distortion. By contrast, nonlinear distortions of an electrical signal are those distortions that are dependent on the signal amplitude, for example, compression, expansion, and harmonic distortion, etcetera. 42

linear electrical parameters (uniform line). The series resistance, series inductance, shunt conductance, and shunt capacitance per unit length of line. *Note:* The term **constant** is frequently used instead of **parameter.** *See:* **transmission line.** 179

linear electric current element. *See:* **Hertzian electric dipole.**

linear electron accelerator. An evacuated metal tube in which electrons are accelerated through a series of small gaps (usually in the form of cavity resonators in the high-frequency range) so arranged and spaced that at a specific excitation frequency, the stream of electrons on passing through successive gaps gains additional energy from the electric field in each gap. 190

linear element (fiber optics). A device for which the output electric field is linearly proportional to the input electric field and no new wavelengths or modulation frequencies are generated. A linear element can be described in terms of a transfer function or an impulse response function. 433

linear gate (X-ray energy spectrometers). A gate whose presence does not affect the linearity of the gated signal. 471

linear-impedance relay (power switchgear). A distance relay for which the operating characteristic on an R-X diagram is a straight line. *Note:* It may be described by the equation $Z \gg K/\cos(\theta - \alpha)$ where K and α are constants and θ is the phase angle by which the input voltage leads the input current. *See:* **distance relay; figure (a).** 103

linear interpolation (numerically controlled machines). A mode of contouring control that uses the information contained in a block to produce velocities proportioned to the distance moved in two or more axes simultaneously. 224, 207

linearity (1) (analog computers). A property of a component describing a constant ratio of incremental cause and effect. **Proportionality** is a special case of linearity in which the straight line passes through the origin. **Zero-error reference** of a linear transducer is a selected straight-line function of the input from which output errors are measured. **Zero-based linearity** is transducer linear defined in terms of a zero-error reference where zero input coincides with zero output. 9

(2) (industrial control) (nuclear power generating station). The closeness with which a curve of a function approximates a straight line. *See:* **control system, feedback.** 219, 206, 355

(3) (test, measurement and diagnostic equipment). The condition wherein the change in the value of one quantity is directly proportional to the change in the value of another quantity. 54

linearity control (television). A control to adjust the variation of scanning speed during the trace interval to minimize geometric distortion. 18

linearity error (1) (analog computers). An error which is the deviation of the output quantity, from a specified linear reference curve. 9

(2) (Hall-effect devices). The deviation of the actual characteristic curve of a Hall generator from the linear approximation to this curve. 107

linearity error, percent of full scale (Hall-effect devices). The maximum deviation, expressed as a percent of full scale, of the actual characteristic curve of a Hall effect device from the straight-line approximation to the characteristic curve derived by minimizing and equalizing the positive and negative deviations of the curve from the straight line. 107

linearity error, percent of reading (Hall-effect devices). The maximum percent deviation of the actual characteristic curve of a Hall effect device from the straight-line approximation to the curve derived by minimizing and equalizing the positive and negative percent deviations of the characteristic curve from the straight line. 107

linearity of a multiplier (analog computers). (A) The ability of an electromechanical or electronic multiplier to generate an output voltage that varies linearly with either one of its two inputs, provided the other input is held constant. (B) The accuracy with which the above requirement is met. Linearity of a potentiometer is the accuracy with which a potentiometer yields a linear but not necessarily a proportional relationship between the angle of rotation of its shaft and the voltage appearing at the output arm, in the absence of loading errors. 9

linearity of a potentiometer. The accuracy with which a potentiometer yields a linear but not necessarily a proportional relationship between the angle of rotation of its shaft and the voltage appearing at the output arm, in the absence of loading errors. *See:* **normal linearity.** 9

linearity of a signal (automatic control). The closeness with which its plot against the variable it represents approximates a straight line. *Note:* The property is usually expressed as a "non-linearity," for example, a maximum deviation. The straight line should be specified as drawn to give limited absolute deviation (independent linearity), to give minimum rms deviation (dependent linearity), to pass through the zero point, or to pass through both end points. 56

linearity, programming (power supplies). The linearity of a programming function refers to the correspondence between incremental changes in the input signal (resistance, voltage, or current) and the consequent incremental changes in power-supply output. *Note:* Direct programming functions are inherently linear for the bridge regulator and are accurate to within a percentage equal to the supply's regulating ability.
186

linearity region (instrument approach system and similar guidance systems). The region in which the deviation sensitivity remains constant within specified values. *See:* **navigation.** 187

linearity, sweep (oscilloscopes). Maximum displacement error of the independent variable between specified points on the display area. 106

linearity, vertical (oscilloscopes). The change in deflection factor of an oscilloscope as the display is positioned vertically within the graticule area. *See:* **compression; expansion.** 106

linear lag. *See:* **lag (5) (first-order).**

linear light (illuminating engineering). A luminous signal having a perceptible physical length. 167

linearly polarized (LP) mode (fiber optics). A mode for which the field components in the direction of propagation are small compared to components perpendicular to that direction. *Note:* The LP description is an approximation which is valid for weakly guiding waveguides, including typical telecommunication grade fibers. *See:* **mode; weakly guiding fiber.**
433

linearly polarized field vector (antennas). At a point in space, a field vector whose extremity describes a straight line segment as a function of time. *Note:* Linear polarization may be viewed as a special case of elliptical polarization where the axial ratio has become infinite. 111

linearly polarized plane wave (antennas). A plane wave whose electric field vector is linearly polarized.
111

linearly polarized wave (1) (data transmission) (plane-polarized wave). (A)(General). A transverse wave in which the displacements at all points along a line in the direction of propagation lie in a plane passing through this line.(B) (radio wave propagation). An electromagnetic wave whose electric and magnetic field vectors always lie along fixed lines at a given point. 146

linearly rising front-chopped impulse (high voltage testing) (chopped impulses). A voltage rising with approximately constant steepness, until it is chopped by a disruptive discharge, as described as a linearly rising front-chopped impulse. To define such an impulse, the best fitting straight line is drawn through the part of the front at the impulse between 50- and 90-percent of the amplitude at the instant of chopping; the intersections of this line with the 50- and 90-percent amplitudes then being designated E and F, respectively (see figure below). The impulse is defined by: (1) the voltage at the instant of chopping; (2) the rise time T_r (this is the time interval between E and F

multiplied by 2.5); and (3) the virtual steepness S (this is the slope of the straight line E-F, usually expressed in kilovolts per microsecond). The impulse is considered to be approximately linear if the front, from 50 percent amplitude up to the instant of chopping, is entirely enclosed between two lines parallel to the line E-F, but displaced from it in time by $+ - 0.05\ T_r$.
150

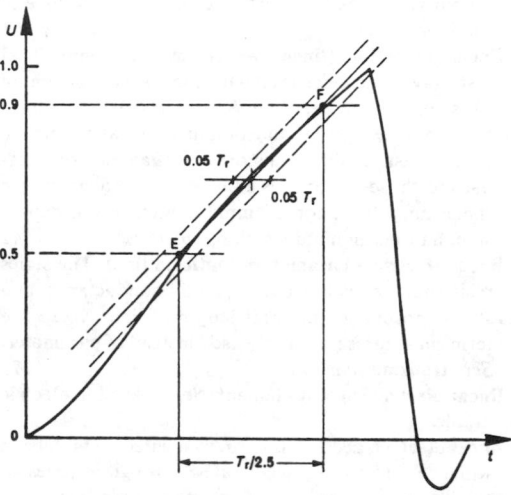

Fig 2.5
Linearly Rising Front-Chopped Impulse

linearly rising switching impulse. An impulse in which the impulse voltage rises at an approximately constant rate. Its amplitude is limited by the occurrence of a disruptive discharge that chops the impulse. 108

linear modulator. A modulator in which, for a given magnitude of carrier, the modulated characteristic of the output wave bears a substantially linear relation to the modulating wave. *See:* **network analysis.** 328

linear network (or system) (circuits and systems). A network (or system) that has both the proportionality and the superposition properties. For example $H(\alpha x_1 + \beta x_2) = \alpha H(x_1) + \beta H(x_2)$. 67

linear polarization (illuminating engineering). The process by which the transverse vibrations of light waves are oriented in or parallel to the same plane. Polarization may be obtained by using either transmitting or reflecting media. 167

linear potentiometer (analog computers). A potentiometer in which the voltage at a movable contact is a linear function of the displacement of the contact.
9

linear power amplifier. A power amplifier in which the signal output voltage is directly proportional to the signal input voltage. *See:* **amplifier.** 111

linear programming (computing systems). *See:* **programming, linear.**

linear pulse amplifier (pulse techniques). A pulse amplifier in which the peak amplitude of the output pulses is directly proportional to the peak amplitude of the corresponding input pulses, if the input pulses are alike in shape. *See:* **pulse.** 335

linear rectifier. A rectifier, the output current or voltage of which contains a wave having a form identical with that of the envelope of an impressed signal wave. *See:* **rectifier.** 111, 240

linear system (hydraulic turbines). A system with the properties that if y_1 is the response to x_1, and y_2 is the response to x_2, then $(y_1 + y_2)$ is the response to $(x_1 + x_2)$, and ky_1 is the response to kx_1. 8

linear system or element (1) (analog computer). A system or element with the properties that if y_1 is the response to x_1, and y_2 is the response to x_2, then $(y_1 + y_2)$ is the response to $(x_1 + x_2)$, and ky_1 is the response to kx_1. *See:* **control system, feedback.** 9

(2) (circuits and systems). An electrical element whose cause and effect relationship follows the proportionality rule. 67

(3) (automatic control). One whose time response to several simultaneous inputs is the sum of their independent time responses. *Note:* It is representable by a linear differential equation, and has a transfer function which is constant for any value of input within a specified range. A system or element not meeting these conditions is described as "nonlinear." 56

(4) (excitation control systems). *Note:* The ANSI C.85 standard definition given above does not permit handling of systems with zero input response. Therefore, the following alternate definition is recommended: "Let a system have zero input response Z(t) and response to two independent inputs $R_1(t)$ and $R_2(t)$ is $C_1(t) + Z(t)$ and $C_2(t) + Z(t)$ respectively. Then the system is linear if the input $aR_1(t) + bR_2(t)$ produces output $aC_1(t) + bC_2(t) + Z(t)$. Otherwise, the system is nonlinear. A linear system is both superimposable and homogeneous. It is mathematically modeled by a linear differential equation. If the system is stationary, the transfer function of the system is constant and not a function of time." 353

linear system with one degree of freedom. *See:* **damped harmonic system.**

linear transducer. A transducer for which the pertinent measures of all the waves concerned are linearly related. *Notes:* (1) By linearly related is meant any relation of linear character whether by linear algebraic equation, by linear differential equation, by linear integral equation, or by other linear connection. (2) The term **waves concerned** connotes actuating waves and related output waves, the relation of which is of primary interest in the problem at hand. *See:* **transducer.** 252, 210

linear TV waveform distortion (linear waveform distortion). The distortion of the shape of a waveform signal where this distortion is independent of the amplitude of the signal. *Notes:* (1) A TV video signal may contain time components with duration from as long as a TV field to as short as a picture element. The shapes of all these time components are subject to distortions. For ease of measurement it is convenient to group these distortions in three separate time domains: short-time waveform distortion, line-time waveform distortion, and field-time waveform distortion. (2) The waveform distortions for times from one field to tens of seconds is not within the scope of this standard. 42

linear varying parameter (varying parameter). A parameter that varies with time or position or both, but not with any dependent variable. *Note:* Unless otherwise specified, **varying parameter** refers to a linear-varying parameter, not to a nonlinear parameter. 210

linear-varying-parameter network. A linear network in which one or more parameters vary with time. 349

line bar (linear waveform distortion). A pulse, nominally of 18 microseconds duration, of reference-white amplitude (see figure below). The rise and fall portions of the line bar are T steps as defined. 42

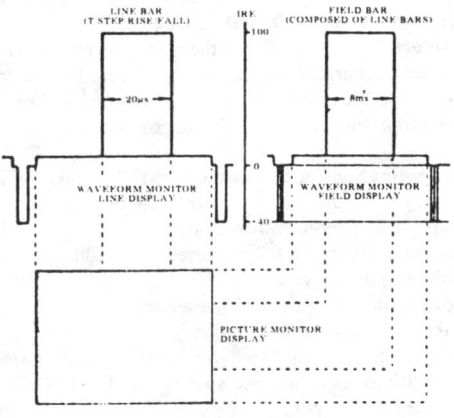

The window signal.

line breaker (line switch) (electrically driven vehicles). A device that combines the functions of a contactor and of a circuit breaker. *Note:* This term is also used for circuit breakers that function to interrupt circuit faults and do not combine the function of a contactor. *See:* **multiple-control unit.** 328

line, bull (conductor stringing equipment). A high strength line, normally manila or synthetic fiber rope, used for pulling and hoisting large loads. 45

line-busy tone (telephone switching systems). A tone that indicates that a station termination is not available. 55

line-charging capacity (synchronous machine). The reactive power when the machine is operating synchronously at zero power factor, rated voltage, and with the field current reduced to zero. *Note:* This quantity has no inherent significance in the actual capability.

line charging current. The current supplied to an unloaded line or cable. 103

line circuit (1) (railway signaling). A signal circuit on an overhead or underground line. 328

(2) (telephone switching systems). An interface circuit between a line and a switching system. 55

line-closing switching-surge factor (power switchgear). The ratio of the line-closing switching-surge maximum voltage to the crest of the normal-frequency line-to-ground voltage at the source side of the closing switching device immediately prior to closing.
103

line-closing switching-surge maximum voltage (power switchgear). The maximum transient crest voltage to ground measured on a transmission line during a switching surge which results from energizing that line. 103

line concentrator (telephone switching systems). A concentrator in which the inlets are lines. 55

line conductor (overhead supply or communication lines)(1)(NESC). A wire or cable intended to carry electric currents, extending along the route of the line, supported by poles, towers or other structures, but not including vertical or lateral conductors. 494

(2) (electric power). One of the wires or cables carrying electric current, supported by poles, towers, or other structures, but not including vertical or lateral connecting wires. See: conductor; tower.
178, 64

line coordination (data transmission). The process of ensuring that equipment at both ends of a circuit are set up for a specific transmission. 59

line current (thyristor). The current in the lines of the supplying power system. 445

line discipline. See: control procedure.

line display (spectrum analyzer). The display produced on a spectrum analyzer when the resolution bandwidth is less than the spacing of the individual frequency components. 390

line-drop compensator (voltage regulator). A device which causes the voltage-regulating relay to increase the output voltage by an amount that compensates for the impedance drop in the circuit between the regulator and a predetermined location on the circuit (sometimes referred to as the load center). See: voltage regulator. 53

line-drop signal (manual switchboard). A drop signal associated with a subscriber line. 328

line-drop voltmeter compensator. A device used in connection with a voltmeter that causes the latter to indicate the voltage at some distant point of the circuit. See: auxiliary device to an instrument. 328

line editor (computer applications). A text editor that allows the user to change text, with cursor control, on only one line at a time. Multiple lines may be viewed or changed through editing commands. See: full-screen editor. 571

line end and ground end (electric power). (1) Line end is that end of a neutral grounding device that is connected to the line circuit directly or through another device. (2) Ground end is that end that is connected to ground directly or through another device. See: grounding device. 91, 64

line fill. The ratio of the number of connected main telephone stations on a line to the nominal main-station capacity of that line. See: cable. 328

line, finger (conductor stringing equipment). A lightweight line, normally sisal, manila or synthetic fiber rope, which is placed over the traveler when it is hung. It usually extends from the ground, passes through the traveler and back to the ground. It is used to thread the end of the pilot line or pulling line over the traveler and eliminates the need for workmen on the structure. These lines are not required if pilot lines are installed when the travelers are hung. 45

line-focus tube (X-ray tubes). An X-ray tube in which the focal spot is roughly a line. 190

line frequency (television). The number of times per second that a fixed vertical line in the picture is crossed in one direction by the scanning spot. Note: Scanning during vertical return intervals is counted. See: television. 178

line-frequency line current (thyristor). The root-mean-square (rms) value of the fundamental component of the line current, the frequency of which is the line frequency. 445

line-frequency line voltage (thyristor). That sine wave component of the line voltage, whose frequency is the line frequency. The root-mean-square (rms) value of that component. 445

line group (telephone switching systems). A multiplicity of lines served by a common set of links.
55

line guy. Tensional support for poles or structures by attachment to adjacent poles or structures. See: tower.
64

line hit (data transmission). An electric interference causing the introduction of spurious signals on a circuit. 59

line-impedance stabilization network (1) (LISN) (radio-noise emission). A network inserted in the supply-mains lead of the equipment under test (EUT) that provides a specified measuring impedance for radio-noise voltage/current measurements and isolates the EUT from the supply mains at radio frequencies.
418

(2) (electromagnetic compatibility). See: artificial mains network.

line-insulation resistance (telephone switching systems). The resistance between the loop conductors and ground or between each other. 55

line insulator (pin, post). An assembly of one or more shells, having means for semirigidly supporting line conductors. See: insulator. 261

line integral (dot-product line integral). The line integral between two points on a given path in the region occupied by a vector field is the definite integral of the dot product of a path element and the vector. Thus

$$I = \int_a^b V \cos \theta \, ds$$

$$= \int_a^b \mathbf{V} \cdot ds$$

$$= \int_a^b (V_x dx + V_y dy + V_z dz)$$

where $\mathbf{V}$ is the vector having a magnitude V, ds **the vector element of the path,** θ the angle between $\mathbf{V}$ and ds. *Example:* The magnetomotive force between two points on a line connecting two points in a magnetic field is the line integral of the magnetic field strength, that is, the definite integral between the two points of the dot product of a vector element of the length of the line and the magnetic strength at the element.
210

line isolation monitor (1) (health care facilities). An instrument which continually checks the hazard current from an isolated circuit to ground. 192 **(2) (National Electrical Code) (health care facilities).** A test instrument designed to continually check the balanced and unbalanced impedance from each line of an isolated circuit to ground and equipped with a built-in test circuit to exercise the alarm without adding to the leakage current hazard. Line isolation monitor was formerly known as ground contact indicator. 256

line lamp (wire communication). A switchboard lamp for indicating an incoming line signal. 328

line lightning performance. The performance of a line expressed as the annual number of lightning flashovers on a circuit mile or tower-line mile basis. *See:* **direct-stroke protection (lightning).** 64

line-load control (telephone switching systems). A means of selectively restricting call attempts during emergencies so as to permit the handling of essential traffic. 55

line lockout (telephone switching systems). A means for the handling of permanent line signals to prevent further recognition as a call attempt. 55

line loss (power operations). Power loss on a transmission or distribution line. 516

lineman (power line maintenance). A person qualified to perform various line-work operations including aerial and ground work. 458

lineman's platform (conductor stringing equipment). A device designed to be attached to a wood pole or metal structure, or both, to serve as a supporting surface for workmen engaged in deadending operations, clipping-in, insulator work, etcetera. The designs of these devices vary considerably. Some resemble short cantilever beams, others resemble swimming pool diving boards, and still others as long as 40 ft (12 m) are truss structures resembling bridges. Materials commonly used for fabrication are wood, fiberglass and metal. *Syn:* **Baker board; D-board; deadend board; deadend platform; diving board.** 431

line microphone. A directional microphone consisting of a single straight line element, or an array of contiguous or spaced electroacoustic transducing elements disposed on a straight line, or the acoustical equivalent of such an array. *See:* **microphone; line transducer.** 328

line number, television (measuring resolution). *See:* **television line number.**

line of nodes (communication satellite). The line which is common to both the orbital plane and the equatorial plane and which passes through the ascending and descending nodes. 74

line of position (LOP) (navigation). The intersection of two surfaces of position, normally plotted as lines on the earth's surface, each line representing the locus of constant indication of the navigational information. *See:* **navigation.** 187, 13

line or input regulation (electrical conversion). Static regulation caused by a change in input. *See:* **electrical conversion.** 186

line or trace (cathode-ray tubes). The path of the moving spot on the screen or target. 125

line pairs per raster height (LP/RH) (diode-type camera tube). The spatial frequency of a uniform periodic array referred to a unit length equal to the raster height. The array may be sinusoidal or comprised of equal width alternating light and dark bars (lines). Each period or cycle includes one light and one dark region or line, hence the term line pairs. 380

line parameters. A sufficient set of parameters to specify the transmission characteristics of the line. 210

line power (thyristor). The total power delivered from the line to the controller. 445

line printing. The printing of an entire line of characters as a unit. 77

liner (1) (rotating machinery). (A) A separate insulating member that is placed against a grounded surface. (B) A layer of insulating material that is deposited on a grounded surface. *See:* **slot liner; pole cell insulation.** 63 **(2) (dry cell) (primary cell).** Usually a paper or pulpboard sheet covering the inner surface of the negative electrode and serving to separate it from the depolarizing mix. *See:* **electrolytic cell.** 328

line regulation (ferroresonant-voltage-regulators). The maximum amount that the output voltage or current will change as the result of a specified change in line voltage. (Regulation is given either as a percentage of the rated output voltage or current, or as an absolute change ΔE or ΔI.) 456

line regulation (power supplies). The maximum steady-state amount that the output voltage or current will change as the result of a specified change in input line voltage (usually for a step change between 105 125 or 210 250 volts, unless otherwise specified). Regulation is given either as a percentage of the output voltage or current, or as an absolute change ΔE or ΔI. 228

line relay (railway practice). c relay that receives its operating energy over a circuit that does not include the track rails. 328

line replaceable unit (LRU) (test, measurement and diagnostic equipment). A unit which is designated by the plan for maintenance to be removed upon failure from a larger entity (equipment, system) in the latter's operational environment. 54

line residual current. *See:* **ground-return current.**

lines (1) (National Electrical Safety Code). (A) communication lines. The conductors and their supporting or containing structures which are used for public or private signal or communication service, and which operate at potentials not exceeding 400 volts to ground or 750 volts between any two points of the circuit, and the transmitted power of which does not exceed 150 watts. When operating at less than 150 volts, no limit is placed on the transmitted power of the system. Under specified conditions, communication cables may include communication circuits exceeding the preceding limitation where such circuits are also used to supply power solely to communication equipment. *Note:* Telephone, telegraph, railroad-signal, data, clock, fire, police-alarms, cable television and other systems conforming with the above are included. Lines used for signaling purposes, but not included under the above definition, are considered as supply lines of the same voltage and are to be insulated. (B) electric supply lines. Those conductors used to transmit electric energy and their necessary supporting or containing structures. Signal lines of more than 400 volts are always supply lines within the meaning of the rules, and those of less than 400 volts may be considered as supply lines, if so run and operated throughout. *Syn:* **supply lines.** 391
(2). *See:* **communication lines; electric-supply lines.**

line side (data transmission). Data terminal connections to a communications circuit between two data terminals. 59

line source (1) (antennas). A continuous distribution of sources of electromagnetic radiation, lying along a line segment. *Note:* Most often in practice the line segment is straight. 111
(2) (fiber optics). (1) In the spectral sense, an optical source that emits one or more spectrally narrow lines as opposed to a continuous spectrum. (2) In the geometric sense, an optical source whose active (emitting) area forms a spatially narrow line. *See:* **monochromatic.** 433

line source corrector. A linear array antenna feed with radiating element locations and excitations chosen to correct for aberrations present in the focal region fields of a reflector. 111

line spectrum (1) (fiber optics). An emission or absorption spectrum consisting of one or more narrow spectral lines, as opposed to a continuous spectrum. *See:* **monochromatic; spectral line; spectral width.** 433
(2) (spectrum analyzer). A spectrum composed of discrete frequency components. 390

line speed (data transmission). The maximum rate at which signals may be transmitted over a given channel; usually in baud or b/s (bits/sec). *See:* **baud.** 59

line spread function (diode-type camera tube). The spatial distribution of the signal, or the time distribution of the signal resulting from the scanning process, produced when an image of an extremely narrow line is formed on the photosurface. 380

line stretcher (waveguide components). A section of waveguide or transmission line having an adjustable physical length. 166

line switch (90) (power system device function numbers). A switch used as a disconnecting, load-interrupter, or isolating switch in an ac or dc power circuit, when this device is electrically operated or has electrical accessories, such as an auxiliary switch, magnetic lock, etcetera. 402

line switching (data transmission). The switching technique of temporarily connecting two lines together so that the stations directly exchange information. 59

line tap (in respect to system protection) (power switchgear). A connection to a line with equipment that does not feed energy into a fault on the line in sufficient magnitude to require consideration in the relay plan. 103

line terminal (1) (metal-oxide surge arresters for ac power circuits). The conducting part provided for connecting the arrester to the circuit conductor. 583

(2) (in respect to system protection) (power switchgear). A connection to a line with equipment that can feed energy into a fault on the line in sufficient magnitude to require consideration in the relay plan and which has means for automatic disconnection. 103

(3) (surge arrester). The conducting part provided for connecting the arrester to the circuit conductor. *Note:* When a line terminal is not supplied as an integral part of the arrester, and the series gap is obtained by providing a specified air clearance between the line end of the arrester and a conductor, or arcing electrode, etcetera, the words line terminal used in the definition refer to the conducting part that is at line potential and that is used as the line electrode of the series gap. 430

(4) (rotating machinery). A termination of the primary winding for connection to a line (not neutral or ground) of the power supply or load. *See:* **asynchronous machine; rotor (rotating machine); stator.** 63

line-time waveform distortion (LD) (linear waveform distortion). The linear TV waveform distortion of time components from 1 microsecond to 64 microseconds, that is, time components of the line-time domain. 42

line-to-background ratio (of a spectral line) (X-ray energy spectrometers). The ratio of the intensity of a monoenergetic line to the intensity of the background immediately adjacent to the line. 471

line-to-line voltage. *See:* **voltage sets.**

line-to-neutral voltage. *See:* **voltage sets.**

line transducer. A directional transducer consisting of a single straight-line element, or an array of contiguous or spaced electroacoustic transducing elements disposed on a straight line, or the acoustical equivalent of such an array. *See:* **line microphone.** 176

line transformer. A transformer for interconnecting a transmission line and terminal equipment for such purposes as isolation, line balance, impedance matching, or for additional circuit connections. 239

line trap. *See:* **carrier-current line trap.**

line trigger (spectrum analyzer). Triggering from the power line frequency. 390

line triggering. Triggering from the power line frequency. *See:* **oscillograph.** 185, 68

line-type fire detector (fire protection devices). A device in which detection is continuous along a path. 71

line voltage (thyristor). The voltage between the lines of the supplying power system. 445

line voltage notch (harmonic control and reactive compensation of static power converters). The dip in the supply voltage to a converter due to the momentary short circuit of the ac lines during a commutation interval. Alternatively, the momentary dip in supply voltage caused by the reactive drops in the supply circuit during the high rates of change in currents occurring in the ac lines during commutation. 533

line weighting (data transmission). A noise weighting used in a noise measuring set to measure noise on a line that is terminated by a subset with a number 144 receiver or a similar subset. The meter-scale readings are in dBrn (144 line). *See:* **noise definitions.** 59

linewidth (1) (laser-maser). The interval in frequency or wavelength units between the points at which the absorbed power (or emitted power) of an absorption (or emission) line is half of its maximum value when measured under specified conditions. 363

(2) (fiber optics). *See:* **spectral width.** 433

line work (power line maintenance). Various operations performed by a person on electrical facilities, including ground work, aerial work, and associated maintenance. 458

link (1)(protection and coordination of industrial and commercial power systems). The current-responsive element in a fuse that is designed to melt under over-current conditions and so interrupt the circuit. A renewal link is one intended for use in Class H low-voltage renewable fuses. 504

(2) (data transmission). A channel or circuit designed to be connected in tandem with other channels or circuits. In automatic switching, a link is a path between two units of switching apparatus within a central office. 59

(3) (communication satellite). A complete facility over which a certain type of information is transmitted: including all elements from source transducer to output transducer. 83

(4) (telephone switching systems). A connection between switching stages within the same switching system. 55

linkage (programming) (computing systems). Coding that connects two separately coded routines. 255

linkage editor (software). A computer program used to create one load module from one or more independent translated object modules or load modules by resolving cross-references among the object modules, and

possibly by relocating elements. Note that not all object modules require linking prior to execution. *See:* **computer program; load module; module.** 434

linkage voltage test, direct-current test (rotating machinery). A series of current measurements, made at increasing direct voltages, applied at successive intervals, and maintained for designated periods of time. *Note:* This may be a controlled overvoltage test. *See:* **asynchronous machine.** 63

link-break cutout (power switchgear). A load-break fuse cutout that is operated by breaking the fuse link to interrupt the load current. 103, 443

linked list. *See:* **chained list.**

link, swivel (conductor stringing equipment). A swivel device designed to connect pulling lines and conductors together in series or connect one pulling line to the drawbar of a pulling vehicle. The device will spin and help relieve the torsional forces which build up in the line or conductor under tension. *Syn:* **swivel.** 45

lin-log receiver (radar). A receiver having a linear amplitude response for small-amplitude signals and a logarithmic response for large-amplitude signals. 13

lip microphone. A microphone adapted for use in contact with the lip. *See:* **microphone.** 328

liquid controller (industrial control). An electric controller in which the resistor is a liquid. *See:* **electric controller.** 206

liquid cooling (rotating machinery). *See:* **manifold insulation.**

liquid counter tube (radiation counters). A counter tube suitable for the assay of liquid samples. It often consists of a thin glass-walled Geiger-Mueller tube sealed into a test tube providing an annular space for the sample. *See:* **anticoincidence (radiation counters).** 190

liquid crystal display (LCD) (illuminating engineering). A display made of material whose reflectance or transmittance changes when an electric field is applied. 167

liquid development (electrostatography). Development in which the image-forming material is carried to the field of the electrostatic image by means of a liquid. *See:* **electrostatography.** 236, 191

liquid-filled fuse unit (power switchgear). A fuse unit in which the arc is drawn through a liquid. 103, 443

liquid-filled, or liquid-cooled transformer (electrical heating applications to melting furnaces and fore-hearths in the glass industry). A transformer that is immersed in a tank of insulating fluid that removes heat from the windings by conduction. The liquid is then cooled by running through external radiators. 520

liquid-flow counter tube (radiation counters). A counter tube specially constructed for measuring the radioactivity of a flowing liquid. *See:* **anticoincidence (radiation counters).** 190

liquid-function potential. The contact potential between two electrolytes. It is not susceptible of direct measurement. 328

liquid-immersed transformer (power and distribution transformer). A transformer in which the core and coils are immersed in an insulating liquid.   53

liquid-in-glass thermometer. A thin-walled glass bulb attached to a glass capillary stem closed at the opposite end, with the bulb and a portion of the stem filled with an expansive liquid, the remaining part of the stem being filled with the vapor of the liquid or a mixture of this vapor and an inert gas. Associated with the stem is a scale in temperature degrees so arranged that when calibrated the reading corresponding to the end of the liquid column indicates the temperature of the bulb.   7

liquid-level switch. *See:* **float switch.**

liquid resistor (industrial control). A resistor comprising electrodes immersed in a liquid.   244, 206

liquid-scintillation (liquid-scintillation counters). A solution consisting of an organic solvent (or mixture of solvents) and one or more organic scintillator solutes.   498

liquid-scintillation solution. A solution consisting of an organic solvent (or mixture of solvents) and one or more organic scintillator solutes.   422

liquidtight flexible metal conduit (National Electrical Code). A raceway of circular cross section having an outer liquidtight, nonmetallic, sunlight-resistant jacket over an inner flexible metal core with associated couplings, connectors, and fittings and approved for the installation of electric conductors.   256

Lissajous figure (oscilloscopes). A special case of an *X–Y* plot in which the signals applied to both axes are sinusoidal functions. For a stable display the signals must be harmonics. Lissajous figures are useful for determining phase and harmonic relationships.   106

list (1) (ATLAS). An ordered single-dimensioned set of data.   400
(2) (computing systems). *See:* **push-down list; pushup list.**
(3) (software). (A) An ordered set of items of data. (B) To print or otherwise display items of data that meet specified criteria. *See:* **chained list; data.**   434

listed (National Electrical Code). Equipment or materials included in a list published by an organization acceptable to the authority having jurisdiction and concerned with product evaluation, that maintains periodic inspection of production of listed equipment or materials and whose listing states either that the equipment or material meets appropriate standards or has been tested and found suitable for use in a specified manner. The means for identifying listed equipment may vary for each organization concerned with product evaluation, some of which do not recognize equipment as listed unless it is also labeled. The authority having jurisdiction should utilize the system employed by the listing organization to identify a listed product.   256

listener echo. Echo that reaches the ear of the listener.   328

listening group (speech quality measurements). A group of persons assembled for the purpose of speech quality testing. Number, selection, characteristics, and training of the listeners depend upon the purpose of the test.   126

listing (software). (1) A computer output in the form of a human-readable list. (2) A human-readable, textual computer output. *See:* **computer; list.**   434

list processing (software). A method of processing data in the form of lists. Usually, chained lists are used so that the logical order of items can be changed without altering their physical locations. *See:* **chained list; data; list.**   434

lithium-drifted detector (semiconductor radiation detectors)(charged-particle detectors)(germanium gamma-ray detectors)(X-ray energy spectrometers). A detector made by the lithium compensation process.   23, 119, 528, 471

lithium drifting (semiconductor radiation detectors)(charged-particle detectors)(germanium gamma-ray detectors)(X-ray energy spectrometers). A technique for compensating p-type material by causing lithium ions to move through a crystal under an applied electric field in such a way as to compensate the charge of the bound acceptor impurities.   23, 119, 528, 471

litz wire (litz). A conductor composed of a number of fine, separately insulated strands, usually fabricated in such a way that each strand assumes, to substantially the same extent, the different possible positions in the cross section for the conductor. Litz is an abbreviation for "litzendraht".   328

live (electric system). *See:* **alive; energized.**

live cable test cap. A protective structure at the end of a cable that insulates the conductors and seals the cable sheath.   64

live-front (industrial control) (transformer). So constructed that there are exposed live parts on the front of the assembly. *See:* **live-front switchboard.**   206, 53

live-front switchboard (power switchgear). One having exposed live parts on the front.   103

live-metal part (power and distribution transformer). A part consisting of electrically conductive material which can be energized under conditions of normal use of the equipment.   53

live parts (power switchgear). Those parts which are designed to operate at voltage different from that of the earth.   103, 443

live room (audio and electroacoustics). A room that has an unusually small amount of sound absorption. *See:* **loudspeaker.**   176

live tank switching device (power switchgear). A switching device in which the vessel(s) housing the interrupter(s) is at a potential above ground.   103

live zone (industrial control). The period(s) in the operating cycle of a machine during which corrective functions can be initiated.   206

LNA. *See:* **launch numerical aperture.**

***L* network (1) (general).** A network composed of two branches in series, the free ends being connected to one pair of terminals and the junction point and one free end being connected to another pair of terminals. *See:* **network analysis.**   210

(2) (circuits and systems). An unbalanced ladder network composed of a series arm and a shunt arm.
67

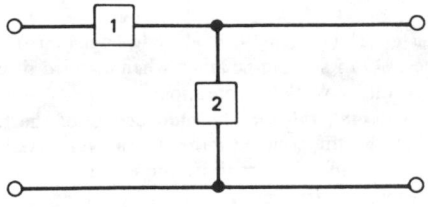

L network. The free ends are left-hand terminal pair; the junction point and one free end are the right-hand terminal pair.

load (1)(power operations). *See:* **base load; center load; customer load; electrical load; internal load; interruptible load; pumping load; system load; system maximum hourly load; load variation; overload.**
516

(2) (induction and dielectric heating usage) (charge). The material to be heated. *See:* **dielectric heating; induction heating.**
14

(3) (output) (power and distribution transformer). The apparent power in megavolt-amperes, kilovolt-amperes, or volt-amperes that may be transferred by the transformer.
53

(4) (rotating machinery). All the numerical values of the electrical and mechanical quantities that signify the demand to be made on a rotating machine by an electric circuit or a mechanism at a given instant.
63

(5) (programming). To place data into internal storage.
255, 77

(6) (electric) (electric utilization). The electric power used by devices connected to an electrical generating system. *See:* **generating station.**
200

(7) (automatic control). (A) An energy-absorbing device. (B) The material, force, torque, energy, or power applied to or removed from a system or element.
56

(8) (data transmission). A power-consuming device connected to a circuit. One use of the word 'load' is to denote a resistor or impedance which replaces some circuit element temporarily or permanently removed. For example, if a filter is disconnected from a line, the line may be artificially terminated in an impedance which simulates the filter that was removed. The artificial termination is then called a load or a dummy load.
59

(9) (test, measurement and diagnostic equipment). (A) To read information from cards or tape into memory: (B) Building block or adapter providing a simulation of the normal termination characteristics of a unit under test: and (C) The effect that the test equipment has on the unit under test or vice versa.
54

(10) (power switchgear). The true or apparent power consumed by power utilization equipment performing its normal function.
103

load-and-go (computing system). An operating technique in which there are no stops between the loading and execution phases of a program, and which may include assembling or compiling.
255, 77

load angle (synchronous machine). The angular displacement, at a specified load, of the center line of a field pole from the axis of the armature magnetomotive force pattern.
63

load-angle curve (load-angle characteristic) (synchronous machine). A characteristic curve giving the relationship between the rotor displacement angle and the load, for constant values of armature voltage, field current, and power factor.
63

load-band of regulated voltage (rotating machinery). The band or zone, expressed in percent of the rated value of the regulated voltage, within which the synchronous-machine regulating system will hold the regulated voltage of a synchronous machine during steady or gradually changing conditions over a specified range of load.
63

load, base (electric power utilization). The minimum load over a given period of time. *See:* **generating station.**
112

load binder (conductor stringing equipment). A toggle device designed to secure loads in a desired position. Normally used to secure loads on mobile equipment. *Syn:* **binder; chain binder.**
45

load-break connector (1)(separable insulated connectors). A connector designed to close and interrupt current on energized circuits.
454

(2) (power and distribution transformer). A separable insulated connector designed to close and interrupt current on energized circuits.
53

load-break cutout (power switchgear). A cutout with means for interrupting load currents.
103, 443

load-break tests (load-interrupting tests) (high-voltage switchgear). Tests that consist of manual or remote control opening of a device, which is provided with a means for breaking load, while the device is carrying a prescribed current under specified conditions.
443

load center (electric power utilization). A point at which the load of a given area is assumed to be concentrated. *See:* **generating station.**
200

load center (power operations). A point at which the load of a given area is assumed to be concentrated.
516

load (dynamic) characteristic (electron tube). For an electron tube connected in a specified operating circuit at a specified frequency, a relation, usually represented by a graph, between the instantaneous values of a pair of variables such as an electrode voltage and an electrode current, when all direct electrode supply voltages are maintained constant.
125

load circuit (induction and dielectric heating usage). The network including leads connected to the output terminals of the generator. *Note:* The load circuit consists of the coupling network and the load material at the proper position for heating.
14

load-circuit efficiency (induction and dielectric heating usage). The ratio of the power absorbed by the load to the power delivered at the generator output terminals. *See:* **dielectric heating; induction heating; network analysis; overall electrical efficiency.** 14

load-circuit power input. The power delivered to the load circuit. *Note:* It is the product of the alternating component of the voltage across the load circuit, the alternating component of the current passing through it (both root-mean-square values), and the power factor associated with these two quantities. *See:* **network analysis.** 111

load coil (induction heating usage). An electric conductor that, when energized with alternating current, is adapted to deliver energy by induction to a charge to be heated. *See:* **induction heating load, work, or heater coil.** 114

load control (watthour meters). A switching output for external load management. 485

load controller (control systems for steam turbine-generator units). The load controller includes only those components and control elements that are responsive to energy output and load reference, that furnish an input signal to the control mechanism for the purpose of controlling the load. 522

load current (1) (thyristor). The current in the load. 445

(2) (tube) (electron tubes). The current output utilized in an external load circuit. *See:* **electron tube.** 244

(3) (watthour meter). *See:* **test current (TA).** 212

load current output (arc-welding apparatus). The current in the welding circuit under load conditions. 264

load curve. A curve of power versus time showing the value of a specific load for each unit of the period covered. *See:* **generating station.** 64

load curves, daily (electric power supply). Curves of net 60-minute integrated demand for each clock hour of a 24-hour day. *See:* **generating station.** 200

load diversity (power operations). The difference between the sum of the maxima of two or more individual loads and the coincident or combined maximum load usually measured in kilowatts (kW) over a specified period of time. 516

load diversity power. The rate of transfer of energy necessary for the realization of a saving of system capacity brought about by load diversity. *See:* **generating station.** 64

load division (load balance) (industrial control). A control function that divides the load in a prescribed manner between two or more power devices connected to the same load. *See:* **control system, feedback.** 206

load, dummy (artificial load). See: **dummy load.**

load duration curve (power operations). A curve of loads, plotted in descending order of magnitude, against time intervals for a specified period. 516

load duty repetitive (thyristor converter). A type of load duty where overloads appear intermittent but cyclic and so frequent that thermal equilibrium is not obtained between all overloads. 121

loaded applicator impedance (dielectric heating). The complex impedance measured at the point of application, with the load material at the proper position for heating, at a specified frequency. *See:* **dielectric heating.** 14, 114

loaded Q (1) (working Q) (electric impedance). The value of Q of such impedance when coupled or connected under working conditions. 328

(2) (switching tubes). The unloaded Q of the tube modified by the coupled impedances. *Note:* As here used, Q is equal to 2π times the energy stored at the resonance frequency divided by the energy dissipated per cycle in the tube, or for cell-type tubes, in the tube and its external resonant circuit. *See:* **gas tube.** 125

loaded impedance (transducer). The impedance at the input of the transducer when the output is connected to its normal load. *See:* **self-impedance.** 176

loaded loop (telephone loop performance). A loop into which lumped inductance (loading coil) is introduced at fixed intervals to compensate for the distributed cable capacitance. The addition of loading coils, properly placed, reduces mid-voiceband loss, flattens the frequency response over most of the voiceband, but creates a sharp cut-off at the high-frequency band edge. 473

loaded voltage ratio (electronics power transformer). Equal to the secondary voltage divided by the primary voltage. For linear loads the ratio is stated for a specified load current and power factor. For rectifier loads, the ratio should be given for the specified circuit configuration, including the filters, and the rated direct-current load. Unless otherwise stated, the ratio is given for rated conditions, line voltage, frequency, load, and stabilized temperature. Primary voltages is given as line to line and secondary voltages as leg values (terminal to neutral or center tap if used) unless otherwise indicated. *See:* **turns ratio.** 95

loader (software). (1) A routine that reads an object program into main storage prior to its execution. (2) A routine, commonly a computer program, that reads data into main storage. *See:* **computer program; data; execution; object program; routine.** 434

load factor. The ratio of the average load over a designated period of time to the peak load occurring in that period. *See:* **generating station.** 64, 200

load-frequency control. The regulation of the power output of electric generators within a prescribed area in response to changes in system frequency, tie line loading, or the relation of these to each other, so as to maintain the scheduled system frequency or the established interchange with other areas within predetermined limits or both. *See:* **generating station.** 64

load ground (signal-transmission system). The potential reference plane at the physical location of the load. *See:* **signal.** 188

load group (nuclear power generating station). An arrangement of buses, transformers, switching equipment, and loads fed from a common power supply within a division. 102

load impedance (of a semiconductor radiation detector)(charged-particle detectors)(germanium gamma-ray detectors)(X-ray energy spectrometers). The impedance shunting the detector, and across which the detector output voltage signal is developed.
119, 528, 471

load-impedance diagram (oscillators). A chart showing performance of the oscillator with respect to variations in the load impedance. *Note:* Ordinarily, contours of constant power and of constant frequency are drawn on a chart whose coordinates are the components of either the complex load impedance or of the reflection coefficient. *See:* **Rieke diagram; oscillatory circuit.** 125

load-indicating automatic reclosing equipment (power switchgear). An automatic reclosing equipment that provides for reclosing the circuit interrupter automatically in response to sensing of predetermined conditions of the load circuit. *Note:* This type of automatic reclosing equipment is generally used for direct-current load circuits. 103

load-indicating resistor (power switchgear). A resistor used, in conjunction with suitable relays or instruments, in an electric circuit, for the purpose of determining or indicating the magnitude of the connected load. 103

loading (1) (antennas). The modification of a basic antenna such as a dipole or monopole caused by the addition of conductors or circuit elements that change the input impedance or current distribution or both.
111

(2) (data transmission) (communication practice). The insertion of reactance in a circuit for the purpose of improving its transmission characteristics in a given frequency band. *Note:* The term is commonly applied, in wire communication practice, to the insertion of loading coils in series in a transmission line at uniform intervals, and in radio practice, to the insertion of one or more loading coils anywhere in a transmission circuit. 59

(3) (H88) (data transmission). A commonly used type of reactive load used on leased lines to produce a specific bandwidth characteristic. 59

(4) (automatic control). Act of transferring energy into or out of a system. 56

loading coil. An inductor inserted in a circuit to increase its inductance for the purpose of improving its transmission characteristics in a given frequency band. *See:* **loading.** 341

loading-coil spacing. The line distance between the successive loading coils of a coil-loaded line. *See:* **loading.** 328

loading error (1) (analog computers). An error due to the effect of a load impedance upon the transducer or signal source driving it. 9

(2) (test, measurement and diagnostic equipment). The error introduced when data are incorrectly transferred from one medium to another. 54

loading machine (mining). A machine for loading ma-

terials such as coal, ore, or rock into cars or other means of conveyance for transportation to the surface of the mine. 328

load, internal. *See:* **load, system.**

load interrupter switch (power switchgear). An interrupter switch designed to interrupt currents not in excess of the continuous-current rating of the switch. *Notes:* (1) It may be designed to close and carry abnormal or short-circuit currents as specified. (2) In international (IEC) practice a device with such performance characteristics is called a switch. 103

load, interruptible (electric power utilization). A load which can be interrupted as defined by contract. *See:* **generating station.** 112

load leads or transmission lines (induction and dielectric heating usage). The connections or transmission line between the power source or generator and load, load coil or applicator. *See:* **dielectric heating; induction heating.** 14

load-limit changer (speed-governing system). A device that acts on the speed-governing system to prevent the governor-controlled fuel valves from opening beyond the position for which the device is set. *See:* **speed-governing system.** 94, 98, 58

load losses (1) (power and distribution transformer). Those losses which are incident to the carrying of a specified load. Load losses include I^2R loss in the windings due to load and eddy currents; stray loss due to leakage fluxes in the windings, core clamps, and other parts, and the loss due to circulating currents (if any) in parallel windings, or in parallel winding strands. 53

(2) (copper losses) (series transformer). The load losses of a series transformer are the I^2R losses, computed from the rated currents for the windings and the measured direct-current resistances of the windings corrected to 75 degrees Celsius. 274

load map (software). A computer generated list that identifies the location or size of all or selected parts of a memory-resident computer program or of memory-resident data. *See:* **computer; computer program; data.** 434

load matching (1) (induction and dielectric heating). The process of adjustment of the load-circuit impedance to produce the desired energy transfer from the power source to the load. 14

(2) (circuits and systems). The technique of either adjusting the load-circuit impedance or inserting a network between two parts of a system to produce the desired power transfer or signal transmission.
67

load-matching network. An electric network for accomplishing load matching. *See:* **induction heating; dielectric heating.** 14

load-matching switch (induction and dielectric heating). A switch in the load-matching network to alter its characteristics to compensate for some sudden change in the load characteristics, such as passing through the Curie point. *See:* **dielectric heating; induction heating.** 14

load module (software). A program unit that is suitable

for loading into main storage for execution; it is usually the output of a linkage editor. *See:* execution; linkage editor; program. 434

load power (thyristor). The total power delivered from the controller to the load. 445

load power factor (converters having ac output)(self-commutated converters). Characteristic of an ac (alternating current) load in terms expressed by the ratio of active power to apparent power assuming an ideal sinusoidal voltage. 584

load profile (diesel-generator unit). The magnitude and duration of loads applied in a prescribed time sequence, including the transient and steady state characteristics of the individual loads. 99

load range (watthour meter). The range in amperes over which the meter is designed to operate continuously with specified accuracy. 212

load regulation (ferroresonant voltage regulators). The maximum amount that the output voltage will change as the result of a specified change in load current. (Regulation is given either as a percentage of the rated output voltage or as an absolute change, Δ E.) 456

load resistance (of a semiconductor radiation detector)(semiconductor radiation detector)(charged-particle detectors)(germanium gamma-ray detectors)(X-ray energy spectrometers). The resistive component of the load impedance.
23, 119, 528, 471

load-resistor contactor (73) (power system device function numbers). A contactor that is used to shunt or insert a step of load limiting, shifting, or indicating resistance in a power circuit, or to switch a space heater in circuit, or to switch a light or regenerative load resistor of a power rectifier or other machine in and out of circuit. 402

load restoration (power switchgear). The process of scheduled load restoration when the abnormality causing load shedding has been corrected. 103

load rheostat. A rheostat whose sole purpose is to dissipate electric energy. *Note:* Frequently used for load tests of generators. *See:* control. 244, 206

load saturation curve (load characteristic) (synchronous machine). The saturation curve of a machine on a specified constant load current. 63

load shedding (power switchgear)(emergency and standby power). The process of deliberately removing preselected loads from a power system in response to an abnormal condition in order to maintain the integrity of the system. 103, 512

load-shifting resistor (power switchgear). One used in an electric circuit to shift load from one circuit to another. 103

load, sliding. A load sliding inside or along a fixed length of waveguide or transmission line. *See:* waveguide. 185

load switch or contactor (induction heating). The switch or contactor in an induction heating circuit which connects the high-frequency generator or power source to the heater coil or load circuit. *See:* induction heating. 14

load, system (electric power systems). Total loads within the system including transmission and distribution losses. 112

load, system maximum hourly (electric power systems). The maximum hourly integrated system load. This is an energy quantity usually expressed in kilowatt hours per hour. 112

load tap-changer (LTC) (power and distribution transformer). A selector switch device, which may include current interrupting contractors, used to change transformer taps with the transformer energized and carrying full load. 53

load-tap-changing transformer (power and distribution transformer). A transformer used to vary the voltage, or the phase angle, or both, of a regulated circuit in steps by means of a device that connects different taps of tapped winding(s) without interrupting the load. 53

load time (hybrid computer linkage components). The time required to read in the digital value to a register of the digital-to-analog converter, measured from the instant that the digital computer commands a digital-to-analog converter or a digital-to-analog multiplier "load." 10

load transfer switch. A switch used to connect a generator or power source optionally to one or another load circuit. *See:* dielectric heating; induction heating. 14

load, transition. The load at which a thyristor converter changes from one mode of operation to another. *Note:* The load current corresponding to a transition load is determined by the intersection of extensions of successive portions of the direct-voltage regulation curve where the curve changes shape or slope. 121

load variation (power operations). Maximum system load over a time interval minus the integrated load over the same time interval. 516

load variation within the hour (electric power utilization). The short-time (three minutes) net peak demand minus the net 60-minute clock-hour integrated peak demand of a supplying system. *See:* generating station. 200

load voltage (1) (arc-welding apparatus). The voltage between the output terminals of the welding power supply when current is flowing in the welding circuit. 264

(2) (thyristor). The voltage across the load. 445

load voltage unbalance (thyristor). If the voltages measured between the pairs of load terminals are not equal, a voltage unbalance exists. 445

load, work, or heater coil (induction heating). An electric conductor which when energized with alternating current is adapted to deliver energy by induction to a charge to be heated. *See:* induction heating. 14

lobe (directional lobe) (radiation lobe) (antenna lobe) (data transmission). A lobe is a portion of the directional pattern bounded by one or two cones of nulls. 59

lobe switching (radar). A means of direction finding in which a directive radiation pattern is periodically shifted in position so as to produce a variation of the signal at the target. The signal variation provides information on the amount and direction of displacement of the target from the pattern mean position. *Syn:* **sequential lobing.** 13

local. Relating to a bounded part of an ATLAS test procedure. 400

local backup (power switchgear). A form of backup protection in which the backup protective relays are at the same station as the primary protective relays. 103

local broadcast (FASTBUS acquisition and control). A broadcast which is effective only on the originating segment. 480

local channel (data transmission). A channel connecting a communications subscriber to a central office. 59

local control (programmable instrumentation). A method whereby a device is programmable by means of its local (front or rear panel) controls in order to enable the device to perform different tasks. *Syn:* **manual control.** 378

local control/alarm (electric pipe heating systems). The locations where control, alarm, or both signal or function take place. With respect to electric pipe heating systems, these are usually mounted in close proximity to the individual heating circuits that they operate. 448

localized general lighting (illuminating engineering). Lighting utilizing luminaires above the visual task and also contributing to the illuminance of the surround. 167

local lighting (illuminating engineering). Lighting providing illuminance over a relatively small area or confined space without providing any significant general surrounding lighting. 167

local loop (data transmission). That part of a communication circuit between the subscriber's equipment and the line terminating equipment in the exchange (either 2-wire or 4-wire). 59

local oscillator (beating oscillator) (data transmission). An oscillator in a superheterodyne circuit whose output is mixed with the received signal to produce a sum or difference frequency equal to the intermediate frequency of the receiver. 59

locked-rotor current (packaging machinery). The steady-state current taken from the line with the rotor locked and with rated voltage (and rated frequency in the case of alternating-current motors) applied to the motor. 429

locked-rotor torque (electric installations on shipboard). The minimum torque of a motor which it will develop at rest for all angular positions of the rotor,- with rated voltage applied at rated frequency. 3

locking-in (data transmission). The shifting and automatic holding of one or both of the frequencies of two oscillating systems which are coupled together, so that the two frequencies have the ratio of two integral numbers. 59

locking ring (rotating machinery). A ring used to prevent motion of a second part. 63

lock-in rate (laser gyro). One half of the absolute value of the algebraic difference between the two rates defining the region over which lock-in occurs. 46

lockout (telephone circuit controlled by two voice-operated devices). The inability of one or both subscribers to get through, either because of excessive local circuit noise or continuous speech from either or both subscribers. *See:* **sweep lockout.** 328, 185

lockout-free (as applied to a recloser or sectionalizer) (power switchgear). A general term denoting that the lockout mechanism can operate even though the manual operating level is held in the closed position. *Note:* When used as an adjective modifying a device, the device has this operating capability. 103

lockout mechanism (of an automatic circuit recloser) (power switchgear). A device that locks the contacts in the open position following the completion of a predetermined sequence of operations. 103

lockout operation (of a recloser) (power switchgear). An opening operation followed by the number of closing and opening operations that the mechanism will permit before locking the contacts in the open position. 103

lockout protection device (series capacitors). A device to block the opening of the bypass device and insertion of the switching step following the closure of the bypass device from a cause which warrants inspection or maintenance. 474

lockout relay (1) (power switchgear). An electrically reset or hand-reset auxiliary relay whose function is to hold associated devices inoperative until it is reset. 103

(2) (86) (power system device function numbers). A hand or electrically reset auxiliary relay that is operated upon the occurrence of abnormal conditions to maintain associated equipment or devices inoperative until it is reset. 402

lockplate (locking plate) (rotating machinery). A plate used to prevent motion of a second part (for example, to prevent a bolt or nut from turning). 63

lock-up relay. A relay that locks in the energized position by means of permanent magnetic bias (requiring a reverse pulse for releasing) or by means of a set of auxiliary contacts that keep its coil energized until the circuit is interrupted. *Note:* Differs from a latching relay in that locking is accomplished magnetically or electrically rather than mechanically. Sometimes used for latching relay. *See:* **relay.** 259

LOF. *See:* **lowest observed frequency.**

log (supervisory control, data acquisition, and automatic control). A printed record of data. 570

logarithmic decrement (1) (underdamped harmonic system). The natural logarithm of the ratio of a maximum of the free oscillation to the next following maximum. The logarithmic decrement of an underdamped harmonic system is

$$\ln\left(\frac{X_1}{X_2}\right) = \frac{2\pi F}{(4MS - F^2)^{1/2}}$$

where X_1 and X_2 are the two maxima.   210

(2) (automatic control). A measure of damping of a second-order linear system, expressed as the Napierian logarithm (with negative sign) of the ratio of the greater to the lesser of a pair of consecutive excursions of the variable (in opposite directions) about an ultimate steady-state value. *Note:* For the system characterized by a quadratic factor

$$1/(s^2 + 2z\omega_n s + \omega_n{}^2)$$ its value is
$$-\pi z/(1 - z^2)^{1/2}$$

Twice this value defines the envelopes of the damping, but C85 prefers the above definition for reasons of convenience noted under damping factor. *See:* **damping, relative; damping coefficient, viscous; subsidence ratio.**   56

logic. (1) The result of planning a data-processing system or of synthesizing a network of logic elements to perform a specified function. (2) Pertaining to the type or physical realization of logic elements used, for example, diode logic, AND logic. *See:* **formal logic; logic design; symbolic logic.**   235, 54

logical address (FASTBUS acquisition and control). A primary address of 32 bits consisting of the device address and internal address. It is independent of the location of the device on a segment.   480

logical file (software). A file independent of its physical environment. Portions of the same logical file may be located in different physical files , or several logical files or parts of logical files may be located in one physical file. *See:* **file.**   434

logical link control (LLC)(1)(logical link control). That part of a data station that supports the logical link control functions of one or more logical links. The LLC generates protocol data units (PDU) and response PDUs for transmission, and interprets command PDUs and response PDUs. Specific responsibilities assigned to a LLC include: (A) Initiation of control signal interchange. (B) Organization of data flow. (C) Interpretation of received command PDUs and generation of appropriate response PDUs. (4) Actions regarding error control and error recovery functions in the LLC sublayer.   585

(2)(token ring access method). That part of the data link layer that supports media independent data link functions, and uses the services of the medium access control sublayer to provide services to the network layer.   472

logical operations with pulses (pulse terms). This section considers the pulse as a logical operator. Some operations defined in **operations on a pulse, operations by a pulse,** and **operations involving the interaction of pulses,** frequently are logical operations in the sense of this section. (1) general. AND, NAND, OR, NOR, EXCLUSIVE OR, INVERSION, inhibiting, enabling, disabling, counting, or other logical operations may be

performed. (2) slivering. A process in which a (typically, unwanted) pulse of relatively short duration is produced by a logical operation. Typically, slivering is a result of partial pulse coincidence. (3) gating. A process in which a first pulse enables or diables portions of a second pulse or other event for the duration of the first pulse. (4) shifting. A process in which logical states in a specified sequence are transferred without alteration of the sequence from one storage element to another by the action of a pulse.   254

logical record (software). A record independent of its physical environment. Portions of the same logical record may be located in different physical records, or several logical records or parts of logical records may be located in one physical record. *See:* **record.**
    434

logical shift (mathematics of computing). A shift that affects all positions in a register, word, or numeral, including the sign position. For example, $+231.702$ shifted two places to the left becomes 3170.200. *Note:* A logical shift may be applied to the multiple-precision representation of a number. *See:* **arithmetic shift.**
    564

logical truth value (l) (subroutines for CAMAC). The symbol *l* represents a logical truth value which can be either true or false.   410

logic board (power-system communication). An assembly of decision-making circuits on a printed-circuit mounting board. *See:* **digital.**   59

logic design (electronic computation). (1) The planning of a computer or data-processing system prior to its detailed engineering design. (2) The synthesizing of a network of logic elements to perform a specified function. (3) The result of (1) and (2) above, frequently called the logic of the system, machine, or network.   235

logic diagram (1) (digital computers). A diagram representing the logic elements and their interconnections without necessarily expressing construction or engineering details.   235,54

(2) (graphic symbols for logic diagrams). A diagram that depicts the two-state device implementation of logic functions with logic symbols and supplementary notations, showing details of signal flow and control, but not necessarily the point-to-point wiring.   88

logic element (electronic digital computers). A combinational logic element or sequential logic element.
    235

logic function (graphic symbols for logic functions). A definition of the relationships that hold among a set of input and output logic variables.   451

logic instruction (computing systems). An instruction that executes an operation that is defined in symbolic logic, such as AND, OR, NOR.   255

logic level (graphic symbols for logic functions). Any level within one or two overlapping ranges of values of a physical quantity used to represent the logic states. *Note:* A logic variable may be equated to any physical quantity for which two distinct ranges of values can be defined. In ANSI/IEEE Std 91-1984, these distinct ranges of values are referred to as logic levels and are denoted H and L. H is used to denote the logic level with the more positive algebraic value, and L is used to denote the logic level with the less positive algebraic value. In the case of systems in which logic states are equated with other physical properties (for example, positive or negative pulses, presence or absence of a pulse), H and L may be used to represent these properties or may be replaced by more suitable designations.   451

logic operation (electronic computation) (1) (general). Any nonarithmetical operation. *Note:* Examples are: extract, logical (bit-wise) multiplication, jump, data transfer, shift, compare, etcetera.
(2) (sometimes). Only those nonarithmetical operations that are expressible bit-wise in terms of the propositional calculus or two-valued Boolean algebra.   235

logic operator. *See:* AND; exclusive OR; NAND; NOT; OR.

logic shift (computing systems). A shift that affects all positions.   255

logic state (graphic symbols for logic functions). One of two possible abstract states that may be taken on by a logic (binary) variable.   451

logic symbol (electronics computation). (1) A symbol used to represent a logic element graphically. (2) A symbol used to represent a logic connective.
                                                        235, 255

logic, 0 1. The representation of information by two states termed 0 and 1. *See:* **bit; dot cycle; mark or space; digital.**   59

log-normal distribution (radar). A probability distribution characterized by the probability density function

$$f(x) = \frac{1}{x\sigma\sqrt{2\pi}} \exp\left[-\frac{(\ln x - \ln x_m)^2}{2\sigma^2}\right], x \geqslant 0$$

$$= 0, \qquad\qquad\qquad x < 0$$

where x is the random variable, σ is the standard deviation of ln x, and x_m is the median value of x. This function is often used for statistical modeling of the radar cross section of certain types of radar targets and clutter.   13

log periodic antenna (1) (antennas). Any one of a class of antennas having a structural geometry such that its impedance and radiation characteristics repeat periodically as the logarithm of frequency.   111
(2) (overhead-power-line corona and radio noise). Any of a class of antenna having a structural geometry such that its electrical characteristics repeat periodically as the logarithm of frequency.   411

long dimension (numerically controlled machines). Incremental dimensions whose number of digits is one more to the left of the decimal point than for a normal dimension, and the last digit shall be zero, that is, XX.XXX0 for the example under normal dimension.   207

long-distance navigation. Navigation utilizing self-contained or external reference aids or methods usable at comparatively great distances. *Note:* Examples of long-distance aids are loran, Doppler, inertial, and celestial navigation. *See:* **short-distance navigation; approach navigation; navigation.**   187, 13

long distance trunk (data transmission). That type of trunk, which permits trunk-to-trunk connection and which interconnects local, secondary, primary, and zone centers.   59

longitudinal (common mode) signal (telephone loop performance). The longitudinal voltage is half the algebraic sum of the voltages to ground in the two conductors (tip and ring). The longitudinal current is the algebraic sum of the current in these conductors.
                                                              473

longitudinal (common) mode voltage (low voltage surge protective devices)gas-tube surge-protective devices). The voltage common to all conductors of a group as measured between that group at a given location and an arbitrary reference (usually earth).
                                                        370, 62, 490

longitudinal attenuation (overhead-power-line corona and radio noise). The decrease in radio noise field strength caused by the propagation of radio frequency energy along an overhead power line and through the earth. *Notes:* (1) In North American practice units are decibels per mile. (2) For multiconductor systems, such as normally found in electric power systems, it is convenient to describe wave propagation as made up of a set of noninteracting modes, each with its own attenuation constant. (3) In the context of this standard, the radio frequency energy is the result of corona.   411

longitudinal balance (1)(analog voice frequency circuits). The electrical symmetry of the two wires comprising a pair with respect to ground. *See:* **longitudinal circuit (telephony).**   468
(2)(data transmission). A measure of the similarity of impedance to ground (or common) for the two or more conductors of a balanced circuit. This term is used to express the degree of susceptibility to common mode interference.   59

longitudinal balance, degree of. *See:* **degree of longitudinal balance.**

longitudinal circuit (measuring longitudinal balance of telephone equipment operating in the voice band). A circuit formed by one communication conductor (or by two or more communication conductors in parallel) with a return through ground or through any other conductors except those which are taken with the original conductor or conductors to form a metallic circuit.   529

longitudinal circuit port (measuring longitudinal balance of telephone equipment operating in the voice band). A place of access in the longitudinal transmission path of a device or network where energy may be supplied or withdrawn, or where the device or network variables may be measured.   529

longitudinal impedance (measuring longitudinal balance of telephone equipment operating in the voice band). Impedance presented by a longitudinal circuit at any given single frequency.   529 longitudinal interference. *See:* common-mode interference; accuracy rating (instrument); signal.

longitudinal magnetization (magnetic recording). Magnetization of the recording medium in a direction essentially parallel to the line of travel.   176

longitudinal mode (laser-maser). Refers to modes that have the same field distributions transverse to the beam, but a different number of half period field variations along the axis of the beam. *See:* longitudinal resonances.   363

longitudinal offset loss. *See:* gap loss.

longitudinal profile (1)(overhead-power-line corona and radio noise). The radio noise field strength at ground level measured at constant lateral distance from the power line and plotted as a function of distance along the line.   411
(2)(radio noise from overhead power lines and substations). The radio noise field strength at ground level measured at constant horizontal distance from the power line and plotted as a function of distance along the line.   509

longitudinal redundancy check (LRC) (data transmission). A system of error control based on the formation of a block check following preset rules. *Note:* The check formation rule is applied in the same manner to each character.   194

longitudinal resonances (in a beam resonator) (laser-maser). Resonances corresponding to modes having the same field distribution transverse to the beam, but differing in the number of half period field variations along the axis of the beam. *Note:* Such resonances are separated in frequency by approximately $v/2L$ where v is the speed of light in the resonator and $2L$ is the round trip length of the beam in the resonator.
                                                            363

longitudinal voltage (power fault effects). A voltage acting in series with the longitudinal circuit.   404

longitudinal wave. A wave in which the direction of displacement at each point of the medium is the same as the direction of the propagation.   210

long-line adapter (telephone switching systems). Equipment inserted between a line circuit and the associated station(s) to allow conductor loop resistances greater than the maximum for which a system is designed.   55

long-line current (corrosion). Current (positive electricity) flowing through the earth from an anodic to a cathodic area that returns along an underground metallic structure. *Note:* Usually used only where the areas are separated by considerable distance and where the current results from concentration cell action. *See:* stray-current corrosion.   205

long-pitch winding (rotating machinery). A winding in which the coil pitch is greater than the pole pitch.
                                                             63

long term stability (LTS)(power supplies)(ferroresonant voltage regulators). The change in output voltage or current as a function of time, at constant line voltage, load, and ambient temperature (sometimes referred to as 'drift').   456

long-time-delay phase trip element (power switch-

gear). A direct-acting trip device element that functions with a purposely delayed action (seconds).
                                                            103

long-time rating (National Electrical Code). A rating based on an operating interval of five minutes or longer.   256

longwall machine (mining). A power-driven machine used for undercutting coal on relatively long faces.
                                                            328

long-wire antenna. A wire antenna that, by virtue of its considerable length in comparison with the operating wavelength, provides a directional pattern. *See:* antenna.   111

look (radar). A colloquial expression for a single attempt at detection of a target.   13

look-up (computing systems). *See:* table look-up.

loom. *See:* flexible nonmetallic tubing.

loop (1)(telephone loop performance). The transmission and signaling channel, with or without gain, between the center of the end office switch and the network interface. It also extends direct current (dc) power to the network interface.   473
(2) (signal-transmission system and network analysis). A set of branches forming a closed current path, provided that the omission of any branch eliminates the closed path. *See:* signal; ground loop; mesh.
                                                         282, 188
(3) (software). A set of instructions that may be executed repeatedly while a certain condition prevails. *See:* instruction; iteration.   434
(4) (telephone circuit) (data transmission). In communications, loop signifies a type of facility, normally the circuit between the subscriber and central office. (Usually a metallic circuit).   59

loop antenna (1) (antennas). An antenna whose configuration is that of a loop. *Note:* If the current in the loop, or in multiple parallel turns of the loop, is essentially uniform and the loop circumference is small compared with the wavelength, the radiation pattern approximates that of a magnetic dipole.   111
(2) (data transmission). An antenna consisting of one or more complete turns of conductor, excited so as to provide an essentially uniform circulatory current, and a radiation pattern approximating that of an elementary magnetic dipole.   59
(3) (overhead-power-line corona and radio noise). An antenna consisting of one or more turns of conductor. If the circulatory current is essentially uniform, the antenna will have a radiation pattern approximating that of an elementary magnetic dipole. *Note:* The loop antenna measures the magnetic field component of the electromagnetic wave.   411

loop circuit (railway signaling). A circuit that includes a source of electric energy, a line wire that conducts current in one direction, and connections to the track rails at both ends of the line to complete the circuit through the two rails in parallel in the other direction.
                                                            328

loop control (industrial control). The effect of a control function or a device to maintain a specified loop of material between two machine sections by automatic speed adjustment of at least one of the driven sections. *See:* **control system, feedback.**   225, 206

loop converter (data transmission). A device used for conversion of dc (direct-current) loop current pulses to relay contact closures and thereby provide circuit isolation.   59

loop (leakage) current (power supplies). A direct current flowing in the feedback loop (voltage control) independent of the control current generated by the reference Zener diode source and reference resistor. *Note:* The loop (leakage) current remains when the reference current is made zero. It may be compensated for, or nulled, in special applications to achieve a very-high impedance (zero current) at the feedback (voltage control) terminals.   186

loop equations. *See:* **mesh or loop equations.**

loop factor (network analysis). *See:* **path factor.**

loop feeder (power distribution). A number of tie feeders in series, forming a closed loop. *Note:* There are two routes by which any point on a loop feeder can receive electric energy, so that the flow can be in either direction. *See:* **center of distribution.**   64

loop gain (communications) (data transmission). The sum of the gains which are given to a signal of a particular frequency in passing around a closed loop. The loop may be a repeater, carrier terminal, or a complete system. The loop gain may be less than the sum of the individual amplifier gain because singing may occur if full amplification is used. The maximum usable gain is determined by, and may not exceed, the losses in the closed path.   59

loop gain characteristic. *See:* **gain characteristic, loop.**

loop graph (network analysis). A signal flow graph each of whose branches is contained in at least one loop. *Note:* Any loop graph embedded in a general graph can be found by removing the cascade branches.   282

looping-in (interior wiring). A method of avoiding splices by carrying the conductor or cable to and from the outlet to be supplied.   328

loop noise bandwidth (communication satellite). One of the fundamental parameters of a phase lock loop. It is the equivalent bandwidth of a square cut-off lowpass filter, which, when multiplied by a flat input noise spectral density, produces the loop noise variance.   85

loop phase angle. *See:* **phase angle, loop.**

loop pulsing (telephone switching systems). Dial pulsing using loop signaling.   55

loop sensitivity-test input (amplitude-modulation broadcast receivers). The least signal field of a specified carrier frequency, modulated 30 percent at 400 cycles, and applied as induced pickup in the loop of the receiver, which results in normal test output when all controls are adjusted for greatest sensitivity. It is expressed in decibels below 1 volt per meter, or in microvolts per meter.   524

loop service (power distribution). Two services of substantially the same capacity and characteristics supplied from adjacent sections of a loop feeder. *Note:* The two sections of the loop feeder are normally tied

together on the consumer's bus through switching devices. *See:* **service.**   64

loop-service (ring) feeder (power switchgear). A feeder that supplies a number of separate loads distributed along its length and that terminates at the same bus from which it originated.   103

loop-set transmittance (network analysis). The product of the negatives of the loop transmittances of the loops in a set.   282

loop signaling (telephone switching systems). A method of signaling over direct current circuit paths that utilize the metallic loop formed by the trunk conductors and terminating bridges.   55

loop stick antenna. A loop receiving antenna with a ferrite rod core used for increasing its radiation efficiency.   111

loop test. A method of testing employed to locate a fault in the insulation of a conductor when the conductor can be arranged to form part of a closed circuit or loop.   328

loop transmittance (1)(network analysis). The product of the branch transmittances in a loop.

(2) (branch). The loop transmittance of an interior node inserted in that branch. *Note:* A branch may always be replaced by an equivalent sequence of branches, thereby creating interior nodes.

(3) (node). The graph transmittance from the source node to the sink node created by splitting the designated node.   282

loose coupling. Any degree of coupling less than the critical coupling. *See:* **coupling.**   328

loose leads (rotating machinery). A form of termination in which the machine terminals are loose cable leads.   63

LOP (electronic navigation). *See:* **line of position.**

loran (1). A long-distance radio-navigation system in which hyperbolic lines of position are determined by measuring arrival-time differences of pulses transmitted in fixed time relationship from two fixed-base transmitters. *Note:* **Loran A,** generally useful to distances of 500 to 1500 nautical miles (900 to 2800 kilometers) over water, depending upon the availability of sky wave, uses a baseline of about 300 nautical miles (550 kilometers), operated at approximately 2 megahertz and gives time-difference measurement by the matching of the leading edges of the pulses, usually with the aid of an oscilloscope. **Loran C,** generally useful to distances of 1000 to 1500 nautical miles (1850 to 2800 kilometers) over water, uses a baseline of about 500 nautical miles, operates at approximately 100 kilohertz: it provides a coarse measurement of time-difference through the matching of pulse envelopes, and a fine measurement by the comparing of phase between the carrier waves. **Loran D** is a shorter-baseline and lower-power adaptation of loran C for tactical applications. *See:* **radio direction-finder (radio compass); radio navigation.**   13, 187, 3

(2) (electric installations on shipboard). A long-range radio navigational aid of the hyperbolic type whose position lines are determined by the measurement of the difference in the time of arrival of synchronized pulses.   3

loran repetition rate (electronic navigation). *See:* **pulse repetition frequency.**

Lorenz number. The quotient of (1) the electronic thermal conductivity by (2) the product of the absolute temperature and the component of the electric conductivity due to electrons and holes. *See:* **thermoelectric device.**   248, 191

lorhumb line (navigation system chart, such as a loran chart with its overlapping families of hyperbolic lines). A line drawn so that it represents a path along which the change in values of one of the families of lines retains a constant relation to the change in values of another of the families of lines. *See:* **navigation.**
                                                      187, 13

loss (1) (power). (A) Power expended without accomplishing useful work. Such loss is usually expressed in watts. (B) **(communications).** The ratio of the signal power that could be delivered to the load under specified reference conditions to the signal power delivered to the load under actual operating conditions. Such loss is usually expressed in decibels. *Note:* Loss is generally due to dissipation or reflection due to an impedance mismatch or both. *See:* **transmission loss.**
                                                          197

(2) (network analysis). *See:* **related transmission terms.**

(3) (waveguide). The power reduction in a transmission path in the mode or modes under consideration. It is usually expressed as a positive ratio, in decibels.
                                                          267

(4) (fiber optics). *See:* **absorption; angular misalignment loss; attenuation; backscattering; differential mode attenuation; extrinsic joint loss; gap loss; insertion loss; intrinsic joint loss; lateral offset loss; macrobend loss; material scattering; microbend loss; nonlinear scattering; Rayleigh scattering; reflection; transmission loss; waveguide scattering.**   433

loss angle (1) (biological). The complement Φ of the phase angle θ (between the electrode potential vector and the current vector).*See:* **electrode impedance (biological).**   192

(2) (magnetic core testing). (A) Dissipation factor. The angle by which the fundamental component of the magnetizing current lags the fundamental component of the exciting current in a coil with a ferromagnetic core. The tangent of this angle is defined as the ratio of the in-phase and quadrature components of the impedance of the coil.

$$\tan \delta_n = \frac{R_s}{\omega L_s} = \frac{\omega L_p}{R_p} = \frac{\mu''_s}{\mu'_s} = \frac{\mu'_p}{\mu''_p}$$

$$\delta_n = \text{loss angle}$$

(B) Relative dissipation factor.
Defined as:

$$\frac{\tan \delta_n}{\mu_i} = \frac{\mu''_s}{(\mu'_s)^2} = \frac{R_s}{\mu_i \omega L_s} = \frac{\omega L_p}{\mu_i R_p} = \frac{\omega L_0}{R_p}$$

Quality factor Q. *See:* General definition.
For inductive devices it is defined as the inverse of the tangent of the loss angle.

$$Q = 1/(\tan \delta_n)$$

Q = quality factor
δ_n = loss angle.   165

loss compensator. *See:* **transformer-loss compensator.**

loss, electric system. Total electric energy loss in the electric system. It consists of transmission, transformation, and distribution losses between sources of supply and points of delivery.   112

losses (grounding device) (electric power). I^2R loss in the windings, core loss, dielectric loss (for capacitors), losses due to stray magnetic fluxes in the windings and other metallic parts of the device, and in cases involving parallel windings, losses due to circulating currents. *Note:* The losses as here defined do not include any losses produced by the grounding device in adjacent apparatus or materials not part of the device. Losses will normally be considered at the maximum rated neutral current but may in some cases be required at other current ratings, if more than one rating is specified, or at no load, as for grounding transformers. *Note:* The losses may be given at 25 degrees Celsius or at 75 degrees Celsius. *See:* **grounding device.**   91

losses of a current-limiting reactor. Losses that are incident to the carrying of current. They include: (1) The resistance and eddy-current in the winding due to load current. (2) Losses caused by circulating current in parallel windings. (3) Stray losses caused by magnetic flux in other metallic parts of the reactor and in the reactor enclosure when the enclosure is supplied as an integral part of the reactor installation. (4) The losses produced by magnetic flux in adjacent apparatus or material not an integral part of the reactor or its enclosure are not included. *See:* **reactor.**   309

loss factor (1) (electric power generation). The ratio of the average power loss to the peak-load power loss during a specified period of time. *See:* **generating station.**   64

(2) (dielectric heater material). The product of its dielectric constant and the tangent of its dielectric loss angle. *See:* **dielectric heating; depth of current penetration; electric constant.**   14

loss function (control system). An instantaneous measure of the cost of being in state x and of using control u at time t. *See:* **performance index.**   329

loss, insertion. *See:* **insertion loss.**

loss of control power protection (series capacitors). A means to initiate the closing of the bypass device upon the loss of normal control power.   474

loss-of-excitation relay (power switchgear). A relay that compares the alternating voltages and currents at the terminal of a synchronous machine and operates to produce an output if the relationship between these quantities indicates that the machine has substantially lost its field excitation.   103

loss of forming (semiconductor rectifier). A partial loss in the effectiveness of the rectifier junction. *See:* **rectification.** 237, 66

loss of voltage condition (Class 1E power systems and equipment). A voltage reduction to a level which results in the immediate loss of equipment capability to perform its intended function. 443

loss on ignition (LOI) (fly ash resistivity). In a fly ash sample a measure of the completeness of the combustion process. It is calculated as the ratio of weight loss upon ignition at $750 = C$ of a previously dried ($100 = C$) sample to the weight of the dry sample, expressed in a percentage. 427

loss, return. *See:* **return loss.**

loss tangent (1) (general). The ratio of the imaginary part of the complex dielectric constant of a material to its real part. 185

(2) (tan δ) (rotating machinery). The ratio of dielectric loss in an insulation system, to the apparent power required to establish an alternating voltage across it of a specified amplitude and frequency, the insulation being at a specified temperature. *Note:* It is the cotangent of the power-factor angle. *See:* **asynchronous machine.** 63

loss-tangent test (dissipation-factor test) (rotating machinery). A test for measuring the dielectric loss of insulation at predetermined values of temperature, frequency, and voltage or dielectric stress, in which the dielectric loss is expressed in terms of the tangent of the complement of the insulation power-factor angle. *See:* **asynchronous machine.** 63

loss, total (rotating machinery). The difference between the active electrical power (mechanical power) input and the active electrical power (mechanical power) output. 63

loss, transmission. *See:* **transmission loss.**

lossy medium (laser-maser). A medium which absorbs or scatters radiation passing through it. 363

lost call (telephone switching systems). A call that cannot be completed due to blocking. 55

loudness equation (loudness ratings of telephone connections). Loudness voltages (in millivolts) and pressures (in pascals) are determined in accordance with Eq 6.

$$S_E, S_M, V_W \text{ or } V_T =$$

$$\left\{ \sum_{j=2}^{N} \left(\log_{10} \frac{f_j}{f_{j-1}} \right) \left[\frac{\left(10^{\frac{x_j}{20}}\right)^{\frac{1}{2.2}} + \left(10^{\frac{x_{j-1}}{20}}\right)^{\frac{1}{2.2}}}{2} \right]^{2.2} \right\}$$
$$\log_{10} f_N/f_1$$

(Eq 6)

where

f_j = specific frequencies of the N frequencies selected for analysis

x_j = the signal level (in dBPa or dBmV)[3] at frequency f_j

Loudness voltages and pressures are expressed in decibel-like form using Eq 7.

$$S'_E, S'_M, V'_W \text{ or } V'_T = 20 \log_{10} X$$

$$\left\{ \sum_{j=2}^{N} \left(\log_{10} \frac{f_j}{f_{j-1}} \right) \left[\frac{\left(10^{\frac{x_j}{20}}\right)^{\frac{1}{2.2}} + \left(10^{\frac{x_{j-1}}{20}}\right)^{\frac{1}{2.2}}}{2} \right]^{2.2} \right\}$$
$$\log_{10} f_N/f_1$$

(Eq 7)
409

loudness rating (loudness ratings of telephone connections). The amount of frequency independent gain that must be inserted into a system under test so that speech sounds from the system under test and a reference system are equal in loudness. 409

louver (illuminating engineering). A series of baffles used to shield a source from view at certain angles or to absorb unwanted light. The baffles usually are arranged in a geometric pattern. *Syn:* **louver grid.** 167

louvered ceiling (illuminating engineering). A ceiling area lighting system comprising a wall-to-wall installation of multicell louvers shielding the light sources mounted above it. 167

louver grid. *See:* **louver.**

louver shielding angle, θ (illuminating engineering). The angle between the horizontal plane of the baffles or louver grid and the plane at which the louver conceals all objects above. *Note:* The planes usually are so chosen that their intersection is parallel with the louvered blade. 167

low (L) level (graphic symbols for logic functions). A level with the more negative (less positive) of the two ranges of logic levels chosen to represent the logic states. 451

low (normal) power-factor mercury lamp ballast. A ballast of the multiple-supply type that does not have a means for power-factor correction. 271

low conduction (LC) threshold voltage $_v$LC (metal-nitride-oxide field-effect transistor). The threshold voltage level resulting from a write-low pulse, which puts the transistor into the LC state. 386

low-energy power circuit (National Electrical Code). A circuit that is not a remote-control or signaling circuit but has its power supply limited in accordance with the requirements of Class 2 and Class 3 circuits. 256

lower (passing) beams (illuminating engineering). One or more beams directed low enough on the left to avoid glare in the eyes of oncoming drivers, and intended for use in congested areas and on highways when meeting other vehicles within a distance of 300 m (1000 ft). Formerly "traffic beam". 167

lower bracket (rotating machinery). A bearing bracket mounted below the level of the core of a vertical machine.   63

lower burst reference (audio and electroacoustics). A selected multiple of the long-time average magnitude of a quantity, smaller than the upper burst reference. See the figure under **burst duration.** *See:* **burst (audio and electroacoustics).**   176

lower coil support (rotating machinery). A support to restrain field-coil motion in the direction away from the air gap. *See:* **rotor (rotating machinery); stator.**   63

lower guide bearing (rotating machinery). A guide bearing mounted below the level of the core of a vertical machine.   63

lower-half bearing bracket (rotating machinery). The bottom half of a bracket that can be separated into halves for mounting or removal without access to a shaft end.   63

lower limit (test, measurement and diagnostic equipment). The minimum acceptable value of the characteristic being measured.   54

lower-range value. The lowest quantity that a device is adjusted to measure. *Note:* The following compound terms are used with suitable modifications in the units: **measured-variable lower-range value, measured signal lower-range value,** etcetera. *See:* **instrument.**   295

lower-sideband parametric down-converter. An inverting parametric device used as a parametric down-converter. *See:* **parametric device.**   191

lowest observed frequency (radio wave propagation). In ionospheric sounding, the lowest radio frequency at which echo signals can be detected or observed with a particular equipment.   146

lowest useful frequency (radio wave propagation). For sky-wave signals in the MF/HF spectrum, the lowest frequency effective under specified conditions for ionospheric propagation of radio waves between two specified points on a planetary surface. *Note:* The lowest useful frequency is determined by factors such as absorption, transmitter power, antenna gain, receiver characteristics, type of service, and noise conditions.   146

lowest useful high frequency (radio wave propagation). The lowest high frequency effective for ionospheric propagation of radio waves between two specified points, under specified ionospheric conditions, and under specified factors such as absorption, transmitter power, antenna gain, receiver characteristics, type of service, and noise conditions. *See:* **radiation; radio wave propagation.**   146

low-frequency dry-flashover voltage. The root-mean-square voltage causing a sustained disruptive discharge through the air between electrodes of a clean dry test specimen under specified conditions.   64

low-frequency flashover voltage (insulator). The root-mean-square value of the low-frequency voltage that, under specified conditions, causes a sustained disruptive discharge through the surrounding medium. *See:* **insulator.**   261

low-frequency furnace (core-type induction furnace). An induction furnace that includes a primary winding, a core of magnetic material, and a secondary winding of one short-circuited turn of the material to be heated.   328

low-frequency high-potential test (rotating machinery). A high-potential test which applies a low-frequency voltage, between 0.1 hertz and 1.0 hertz, to a winding.   63

low-frequency impedance corrector. An electric network designed to be connected to a basic network, or to a basic network and a building-out network, so that the combination will simulate at low frequencies the sending-end impedance, including dissipation, of a line. *See:* **network analysis.**   328

low-frequency induction heater or furnace. A device for inducing current flow of commercial power-line frequency in a charge to be heated. *See:* **dielectric heating, induction heating.**   14, 114

low-frequency puncture voltage (insulator). The root-mean-square value of the low-frequency voltage that, under specified conditions, causes disruptive discharge through any part of the insulator. *See:* **insulator.**   261

low-frequency wet-flashover voltage. The root-mean-square voltage causing a sustained disruptive discharge through the air between electrodes of a clean test specimen on which water of specified resistivity is being sprayed at a specified rate.   64

low-frequency withstand voltage (insulator). The root-mean-square value of the low-frequency voltage that, under specified conditions, can be applied without causing flashover or puncture. *See:* **insulator.**   261

low-key lighting (illuminating engineering). A type of lighting which, applied to a scene, results in a picture having graduations falling primarily between middle gray and black, with comparatively limited areas of light grays and whites.   167

low-level analog signal cable (cable systems in power generating stations). Cable used for transmitting variable current or voltage signals for the control or instrumentation of plant equipment and systems, or both.   35

low-level digital signal circuit cable (cable systems in power generating stations). Cable used for transmitting coded information signals, such as those derived from the output of an analog-to-digital converter, or the coded output from a digital computer or other digital transmission terminals.   35

low-level modulation (communication). Modulation produced at a point in a system where the power level is low compared with the power level at the output of the system.   111, 240

low-level radio-frequency signal (TR, ATR, and Pre-TR tubes). A radio-frequency signal with insufficient power to cause the tube to become fired.   125

low-pass filter (data transmission). A filter having a single transmission band extending from zero frequency up to some cutoff frequency, not infinite.   59 low power factor transformer (power and distribution transformer). A high-reactance transformer that does not have means for power-factor correction. 53

low-pressure contact (area contact) (power switchgear). One in which the pressure is such that stress in the material is well below the elastic limit of both contact surface materials, such that conduction is a function of area. 103

low-pressure sodium lamp (illuminating engineering). A discharge lamp in which light is produced by radiation from sodium vapor operating at a partial pressure of 0.13 to 1.3 Pa (10^{-3} to 10^{-2} $Torr$) 167

low-pressure vacuum pump. A vacuum pump that compresses the gases received directly from the evacuated system. See: rectification. 328

low-priority effort (electric generating unit reliability, availability, and productivity). Repairs were carried out with less than a normal effort. See: repair urgency. 567

low remanence current transformer. One with a remanence not exceeding 10 percent of maximum flux. 203

low-speed limit (speed/load reference)(control systems for steam turbine-generator units). A device or input that limits the speed/load reference setting to a predetermined lower limit. This device may establish the lower limit of the synchronizing speed range. 522

low state (1)(programmable instrumentation). The relatively less positive signal level used to assert a specific message content associated with one of two binary logic states. 40
(2)(signals and paths)(microcomputer system bus). The more negative voltage level used to represent one of two logical binary states. 542
(3)(signals and paths)(696 interface devices). The electrically less positive signal level used to assert a specified message content associated with one or two binary logic states. 538

low-velocity camera tube (electron device) (cathode-voltage-stabilized camera tube). A camera tube operating with a beam of electrons having velocities such that the average target voltage stabilizes at a value approximately equal to that of the electron-gun cathode. 190

low-voltage (1) (National Electrical Code). An electromotive force rated nominal 24 volts, nominal or less, supplied from a transformer, converter, or battery. 256
(2) (system voltage ratings). A class of nominal system voltages 1000 or less. See: medium voltage; high voltage; nominal system voltage. 260

low-voltage ac power circuit breaker (power switchgear). See: circuit breaker (Note) (power switchgear).

low-voltage aluminum-sheathed power cable (aluminum sheaths for power cables). Cable used in an electric system having a maximum phase-to-phase root-mean-square (rms) voltage 1000 volts (V) or less, the cable having an aluminum sheath as a major component in its construction. 406

low-voltage electrical and electronic equipment (radio-noise emission). Electrical and electronic equipment with operating input voltages of up to 600 volts direct current or root-mean-square alternating current. 418

low-voltage integrally fused power circuit breaker (power switchgear). An assembly of a general-purpose ac low-voltage power circuit breaker and integrally mounted current-limiting fuses which together function as a coordinated protective device. 103

low-voltage power cable (cable systems in power generating stations). Cable designed to supply power to utilization devices of the plant auxiliary system, operated at 600 V or less. 35

low-voltage power cables (cable systems in substations). Those cables used on systems operating at 1000 V or less. They are designed to supply operating power to utilization devices. 382

low voltage protection (1)(NESC). The effect of a device operative on the reduction or failure of voltage so as to cause and maintain the interruption of power supply to the equipment protected. 494
(2) (power switchgear). See: undervoltage protection. 103

low-voltage release. The effect of a device, operative on the reduction or failure of voltage, to cause the interruption of power supply to the equipment, but not preventing the reestablishment of the power supply on return of voltage. See: generating system. 262

low-voltage system (electric power). An electric system having a maximum root-mean-square alternating-current voltage of 1000 volts or less. See: voltage classes. 49

LP mode. See: linearly polarized mode.

LP^{01} mode (fiber optics). Designation of the fundamental linearly polarized (LP) mode. See: fundamental mode. 433

$L_q(s)$. See: quadrature operational inductance.

LRA. See: laser gyro axes.

LRC (data transmission). See: longitudinal redundancy check.

LRU. See: line replaceable unit.

L scan (electronic navigation). See: L display.

L scope (electronic navigation). See: L display.

L-scope (radar). A cathode-ray oscillosope arranged to present an L-display. 13

LS dividing network. See: dividing network.

LTS (long-term stability). See: stability, long-term.

LUF. See: lowest useful frequency.

lug (electric installations on shipboard). A wire connector device to which the electrical conductor is attached by mechanical pressure or solder. 3

lug, stator mounting (rotating machinery). A part attached to the outer surface of stator core or a stator shell to provide a means for the bolting or equivalent attachment to the appliance, machine, or other foundation. See: stator. 63

Lukasiewicz notation. See: prefix notation.

lumen (or flux) method (lighting calculation). A lighting design procedure used for predetermining the number and types of lamps or luminaires that will provide a desired average level of illumination on the work plane. *Note:* It takes into account both direct and reflected flux. *See:* **inverse-square law (illuminating engineering).** 167

lumen (1) (color terms) (television). The unit of luminous flux. The luminous flux emitted within unit solid angle (one steradian) by a point source having a uniform intensity of one candela. 18

(2) (lm) (illuminating engineering). SI unit of luminous flux. Radio-metrically, it is determined from the radiant power. Photometrically, it is the luminous flux emitted within a unit solid angle (one steradian) by a point source having a uniform luminous intensity of one candela. *See:* **luminous flux.** 167

lumen hour. A unit of quantity of light (luminous energy). It is the quantity of light delivered in one hour by a flux of one lumen. *See:* **light.** 167

lumen method (illuminating engineering). A lighting design procedure used for determining the relation between the number and types of lamps or luminaires, the room characteristics, and the average level of illuminance on the work-plane. It takes into account both direct and reflected flux. *Syn:* **flux method.** 167

lumen-second (lm/s) (illuminating engineering). A unit of quantity of light, the SI unit of luminous energy (also called a talbot). It is the quantity of light delivered in one second by a luminous flux of one lumen. 167

luminaire (illuminating engineering). A complete lighting unit consisting of a lamp or lamps together with the parts designed to distribute the light, to position and protect the lamps and to connect the lamps to the power supply. *Syn:* **light fixture.** 167

luminaire ambient temperature factor (illuminating engineering). The fractional loss of task illuminance due to improper operating temperature of a gas discharge lamp. 167

luminaire dirt depreciation (LDD) (illuminating engineering). The fractional loss of task illuminance due to luminaire dirt accumulation. 167

luminaire efficiency (illuminating engineering). The ratio of luminous flux (lumens) emitted by a luminaire to that emitted by the lamp or lamps used therein. 167

luminaire surface depreciation factor (illuminating engineering). The loss of task illuminance due to permanent deterioration of luminaire surfaces. 167

luminance (1) $L = d^2\Phi/(d\omega dA\cos\theta)$ **(in a direction and at a point of a real or imaginary surface) (illuminating engineering).** The quotient of the luminous flux at an element of the surface surrounding the point and propagated in directions defined by an elementary cone containing the given direction, by the product of the solid angle of the cone and the area of the orthogonal projection of the element of the surface on a plane perpendicular to the given direction. The luminous flux may be leaving, passing through, and arriving at the surface or both. Formerly, photometric brightness. By introducing the concept of luminous intensity, luminance may be expressed as $L = dE/(d\omega\cos\theta)$.

Here, luminance at a point of a surface in a direction, is interpreted as the quotient of luminous intensity in the given direction, produced by an element of the surface surrounding the point, by the area of the orthogonal projection of the element of surface on a plane, perpendicular to the given direction. (Luminance may be measured at a receiving surface by using $L = dE/(d\omega\cos\theta)$. This value may be less than the the the luminance of the emitting surface due to the attenuation of the transmitting media.) *Note:* In common usage the term brightness usually refers to the strength of sensation which results from viewing surfaces or spaces from which light comes to the eye. In much of the literature the term brightness, used alone, refers to both luminance and sensation. The context usually indicates which meaning is intended. Previous usage notwithstanding, neither the term brightness, not the term photometric brightness should be used to denote the concept of luminance. 167

(2) (average luminance) (average photometric brightness). The total lumens actually leaving the surface per unit area. *Notes:* (A) Average luminance specified in this way is identical in magnitude with **luminous exitance,** which is the preferred term. (B) In general, the concept of average luminance is useful only when the luminance is reasonably uniform throughout a very wide angle of observation and over a large area of the surface considered. It has the advantage that it can be computed readily for reflecting surfaces by multiplying the incident luminous flux density (illumination) by the luminous reflectance of the surface. For a transmitting body it can be computed by multiplying the incident luminous flux density by the luminous transmittance of the body. 18

luminance channel (color television). Any path that is intended to carry the luminance signal. 18

luminance channel bandwidth (color television). The bandwidth of the path intended to carry the luminance signal. 18

luminance coefficient, LC (illuminating engineering). The ratio of average initial wall or ceiling cavity luminance times π to the total lamp flux (lumens) divided by the floor area. *Notes:* (1) If the luminance is in cd/in^2, the floor area must be in square inches. (2) If the luminance is in cd/ft^2 (or in cd/m^2), the floor area must be in square feet (or in square meters). (3)* If the luminance is in footlamberts, the "π" is omitted and the floor area must be in square feet. 167

*Deprecated

luminance contrast (illuminating engineering). The relationship between the luminances of an object and its immediate background. It is equal to $(L_1 - L_2)/L_1$ or $(L_2 - L_1)/L_1 = \Delta L/L_1$, where L_1 and L_2 are the luminances of the background and object, respectively. The form of the equation must be specified. The ratio $\Delta L/L$ is known as Weber's fraction. *Note:* See note under luminance. Because of the relationship among luminance, illumination, and reflectance, contrast often is expressed in terms of reflectance when only reflecting surfaces are involved. Thus, contrast is equal to $(\rho_1 - \rho_2)/\rho_1$ or $(\rho_2 - \rho_1)/\rho_1$ where ρ_1 and ρ_2

are the reflectances of the background and object, respectively. This method of computing contrast holds only for perfectly diffusing surfaces; for other surfaces it is only an approximation unless the angles of incidence and view are taken into consideration. *See:* **reflectance.** 167

luminance difference threshold (illuminating engineering). This can apply to the difference between two separated objects on a common background, or the difference between two juxtaposed patches separated by a contrast border, or the difference between a uniform small object and its background. In the latter case, a contrast border, separates the object from its background. 167

luminance factor. The ratio of the luminance (photometric brightness) of a surface or medium under specified conditions of incidence, observation, and light source, to the luminance (photometric brightness) of a perfectly reflecting or transmitting, perfectly diffusing surface or medium under the same conditions. *Note:* Reflectance or transmittance cannot exceed unity, but luminance factor may have any value from zero to values approaching infinity. *See:* **lamp.** 167

luminance factor, β (illuminating engineering). The ratio of the luminance of a surface or medium under specified conditions of incidence, observation, and light source, to the luminance of a completely reflecting surface or medium under the same conditions. *Note:* Reflectance or transmittance cannot exceed unity, but luminance factor may have any value from zero to values approaching infinity. 167

luminance factor of room surfaces (illuminating engineering). Factors by which the average work-plane illuminance is multiplied to obtain the average luminances of walls, ceilings and floors. 167

luminance flicker (color television). The flicker that results from fluctuation of luminance only. 18

luminance primary (color television). One of a set of three transmission primaries whose amount determines the luminance of a color. *Note:* This is an obsolete term because it is useful only in a linear system. 18

luminance ratio (illuminating engineering). The ratio between the luminances of any two areas in the visual field. *Note:* See last paragraph of the note under **luminance.** 167

luminance signal (NTSC color television). A signal that has major control of the luminance. *Note:* It is a linear combination of gamma-corrected primary color signals, E'_R, E'_G, and E'_B as follows:

$$E'_Y = 0.30E'_R + 0.59E'_G + 0.11E'_B$$

The proportions expresses are strictly true only for television systems using the NTSC original standard receiver primaries having the CIE color points listed below, when they are mixed to produce white light having the same appearance as standard illuminant C.

Color	x	y
Red (R)	0.67	0.33
Green (G)	0.21	0.71
Blue (B)	0.14	0.08

18

luminance (photometric brightness) threshold. The minimum perceptible difference in luminance for a given state of adaptation of the eye. See the note under **luminance.** *See:* **visual field.** 167

luminescence (illuminating engineering). Any emission of light not ascribable directly to incandescence. 167

luminescent-screen tube. A cathode-ray tube in which the image on the screen is more luminous than the background. *See:* **cathode-ray tubes.** 244, 190

luminosity (television). Ratio of luminous flux to the corresponding radiant flux at a particular wavelength. It is expressed in lumens per watt. 18

luminosity coefficients (television). The constant multipliers for the respective tristimulus values of any color, such that the sum of the three products is the luminance of the color. 18

luminous ceiling (illuminating engineering). A ceiling area lighting system comprising a continuous surface of transmitting material of a diffusing or light controlling character with light sources mounted above it. 167

luminous density, $w = dQ/dV$ (illuminating engineering). Quantity of light (luminous energy) per unit volume. 167

luminous efficacy of a source of light (illuminating engineering). The quotient of the total luminous flux emitted by the total lamp power input. It is expressed in lm/W. *Note:* The term luminous efficiency has in the past been extensively used for this concept. 167

luminous efficacy of radiant flux (illuminating engineering). The quotient of the total luminous flux by the total radiant flux. It is expressed in lm/W. 167

luminous efficiency (television). The ratio of the luminous flux to the radiant flux. *Note:* Luminous efficiency is usually expressed in lumens per watt of radiant flux. It should not be confused with the term efficiency as applied to a practical source of light, since the latter is based on the power supplied to the source instead of the radiant flux from the source. For energy radiated at a single wavelength, luminous efficiency is synonymous with luminosity. 18

luminous exitance, $M = d\Phi/dA$ (illuminating engineering). The density of luminous flux leaving a surface at a point. Formerly luminous emittance. *Note:* This is the total luminous flux emitted, reflected, and transmitted from the surface and is independent of direction. 167

luminous flux (1)(television). The time rate of flow of light. 18

Φ **(2)(illuminating engineering).** Radiant flux (radiant power), the time rate of flow of radiant energy, evaluated in terms of a standardized visual response.

$$\Phi_v = K_m \int \Phi_e, \lambda V(\lambda) d\lambda$$

where Φ_v is in lumens, Φ_e, λ is in watts per nanometer, λ is in nanometers, $V(\lambda)$ is the spectral luminous efficiency, in lm/W. Unless otherwise indicated, the luminous flux is defined for photopic vision. For scotopic vision, the corresponding spectral luminous efficiency $V'(\lambda)$ and the corresponding maximum spectral luminous efficiency K'_m are submitted in the above equation. K_m and K'_m are derived from the basic SI definition of luminous intensity and have the values 683 lm/W and 1754 lm/W respectively. *Note:* The value of $K_m = 683$ lm/W was recommended by the International Committee for Weights and Measures in 1977. *See:* **candela; spectral luminous efficiency of radiant flux; values of spectral luminous efficiency for photopic vision; values of spectral luminous efficiency for scotopic vision.** 167

luminous flux density at a surface, $d\Phi/dA$ (illuminating engineering). The luminous flux per unit area at a point on a surface. *Note:* This need not be a physical surface; it may also be a mathematical plane. *See:* **illuminance; luminous exitance.** 167

luminous gain (optoelectronic device). The ratio of the emitted luminous flux to the incident luminous flux. *Note:* The emitted and incident luminous flux are both determined at specified ports. *See:* **optoelectronic device.** 191

luminous intensity (of a source of light in a given direction) (television). The luminous flux per unit solid angle in the direction in question. Hence, it is the luminous flux on a small surface normal to that direction, divided by the solid angle (in steradians) that the surface subtends at the source. *Note:* Mathematically, a solid angle must have a point as its apex; the definition of luminous intensity, therefore, applies strictly only to a point source. In practice, however, light emanating from a source whose dimensions are negligible in comparison with the distance over which it is observed may be considered as coming from a point. 18

luminous intensity, $I = d\Phi/d\omega$ (of a point source of light in a given direction). The luminous flux per unit solid angle in the direction in question. Hence, it is the luminous flux on a small surface centered on and normal to that direction divided by the solid angle (in steradians) which the surface subtends at the source. Luminous intensity may be expressed in candelas or in lumens per steradian (lm/sr). *Note:* Mathematically a solid angle must have a point as its apex; the definition of luminous intensity, therefore, applies strictly only to a point source. In practice, however, light emanating from a source whose dimensions are negligible in comparison with the distance from which it is observed may be considered as coming from a point. Specifically, this implies that with change of distance (1) the variation in solid angle subtended by the source at the receiving point approaches $1/(\text{distance})^2$ and that (2) the average luminance of the projected source area as seen from the receiving point does not vary

appreciably. For extended sources see **equivalent luminous intensity of an extended source at a specified distance.** The word intensity as defined above is used to designate luminous intensity (or candlepower). It is also widely used in other ways either informally or formally in other disciplines. Stimulus intensity may be used to designate the retinal illuminance of a proximal stimulus or the luminance of a distal stimulus. Intensity is used in the same sense with other modalities such as audition. Intensity has been used to designate the level of illuminance on a surface or the flux density in the cross section of a beam of light. In physical optics, intensity usually refers to the square of the wave amplitude. *See:* **distal stimuli; proximal stimuli.** 167

luminous reflectance (illuminating engineering). Any of the geometric aspects of reflectance in which both the incident and transmitted flux are weighed by the luminous efficiency of radiant flux [V (λ)]. *Note:* Unless otherwise qualified, the term "luminous reflectance" is meant by the term "reflectance." 167

luminous sensitivity (phototube). *See:* **sensitivity (camera tube or phototube).**

luminous transmittance (illuminating engineering). Any of the geometric aspects of transmittance in which the incident and transmitted flux are weighed by the luminous efficiency of radiant flux[$V(\lambda)$]. *Note:* unless otherwise qualified, the term luminous transmittance is meant by the term transmittance. 167

luminous-tube transformer (power and distribution transformer). Transformers, autotransformers, or reactors (having a secondary open-circuit rms of 1000 V or more) for operation of cold-cathode and hot-cathode luminous tubing generally used for signs, illumination, and decoration purposes. 53

lumped. Effectively concentrated at a single point. 328

lumped capacitive load (power switchgear). A lumped capacitance which is switched as a unit. 103

lumped element circuit (microwave tubes). A circuit consisting of discrete inductors and capacitors. *See:* **microwave tube or valve.** 190

Luneburg lens antenna. A lens antenna with a circular cross section having an index of refraction varying only in the radial direction such that a feed located on or near a surface or edge of the lens produces a major lobe diametrically opposite the feed. 111

lux (1) (lx) (illuminating engineering). The SI unit of illuminance. One lux is one lumen per square meter (lm/m^2). *See:* **illuminance conversion factors.** 167

(2) (metric practice). The illuminance produced by a luminous flux of one lumen uniformly distributed over a surface of one square meter. 21

Luxemburg effect. A nonlinear effect in the ionosphere as a result of which the modulation on a strong carrier wave is transferred to another carrier passing through the same region. *See:* **radiation.** 328

Lyapunov function (control system) (equilibrium point x_e of a system). A scalar differentiable function $V(x)$

defined in some open region including $\mathbf{x}_e$ such that in that region

$$V(\mathbf{x}) > 0 \text{ for } \mathbf{x} \pm \mathbf{x}_e \qquad (1)$$
$$V(\mathbf{x}_e) = 0 \qquad (2)$$

$$V'(\mathbf{x}) \le 0 \qquad (3)$$

Notes: (1) The open region may be defined by (norm of $\mathbf{x} - \mathbf{x}_e$) < constant. (2) For the system $\mathbf{x}' = \mathbf{f}(\mathbf{x})$, $V'(\mathbf{x}) \equiv [\text{grad } V(\mathbf{x}) \cdot \mathbf{f}(\mathbf{x})]$. *See:* **control system, feedback.** 329

M

machine (1) (general). An article of equipment consisting of two or more resistant, relatively constrained parts that, by a certain predetermined intermotion, may serve to transmit and modify force, motion, or electricity so as to produce some given effect or transformation or to do some desired kind of work. *Notes:* (A) If a matrix has as its elements impedances, it is usually referred to as impedance matrix. Frequently the impedance matrix is called impedance for short. (B) Usually such impedances are defined with the mechanical angular velocity of the machine at steady state. *See:* **asynchronous machine.** 210, 63
(2) (computing systems). *See:* **Turing machine; universal Turing machine.**
(3) (sound measurements). A machine is any rotating electrical device of which the acoustical characteristics are to be measured. (A) A **small machine** has a maximum linear dimension of 250 mm. This dimension is over major surfaces, excluding minor surface protuberances as well as shaft extension, and is measured either parallel to the shaft, or at right angles to it, according to which dimension gives the greater measurement. (B) A **medium machine** has a maximum linear dimension from 250 mm to 1 m as measured for **small machine.** (C) A **large machine** has a maximum linear dimension in excess of 1 m as measured for **small machine.**
machine address (computing systems). *See:* **absolute address.**
machine-aided (computer applications). Pertaining to a process or function performed with the assistance of one or more computers. 571
machine, aircraft electric. An electric machine designed for operation aboard aircraft. *Note:* Minimum weight and size and extreme reliability for a specified (usually short) life are required while operating under specified conditions of coolant temperature and flow, and for air-cooled machines, pressure and humidity. 63
machine check. (1) An automatic check. (2) A programmed check of machine functions. *See:* **check; automatic.** 235
machine code (computing systems). An operation code that a machine is designed to recognize. 255, 77, 54
machine, electric. An electric apparatus depending on electromagnetic induction for its operation and having

one or more component members capable of rotary and/or linear movement. *See:* **asynchronous machine.** 63
machine equation. *See:* **computer equation.**
machine final-terminal stopping device (stop-motion switch) (elevators). A final-terminal stopping device operated directly by the driving machine. *See:* **control.** 328
machine instruction (computing systems). An instruction that a machine can recognize and execute. 255, 77, 54
machine language (1) (electronic digital computers). (A) A language, occurring within a machine, ordinarily not perceptible or intelligible to persons without special equipment or training. (B) A translation or transliteration of (A) above into more conventional characters but frequently still not intelligible to persons without special training. 235
(2) (software). A representation of instructions and data that is directly executable by a computer. *See:* **assembly language; data; higher order language; instruction.** 434
machine oriented language (test, measurement and diagnostic equipment). (1) A language designed for interpretation and use by a machine without translation: (2) a system for expressing information which is intelligible to a specific machine: for example, a computer or class of computers. Such a language may include instructions which define and direct machine operations, and information to be recorded by or acted upon by these machine operations: and (3) the set of instructions expressed in the number system basic to a computer, together with symbolic operation codes with absolute addresses, relative addresses, or symbolic addresses. *Syn:* **machine language** and **assembly language.** 54
machine or transformer thermal relay (49) (power system device function numbers). A relay that functions when the temperature of a machine armature or other load-carrying winding or element of a machine or the temperature of a power rectifier or power transformer (including a power rectifier transformer) exceeds a predetermined value. 402
machine positioning accuracy, precision, or reproducibility (numerically controlled machines). Accuracy, precision, or reproducibility of position sensor or transducer and interpeting system, the machine elements, and the machine positioning servo. *Note:* Cut-

ter, spindle, and work deflection, and cutter wear are not included. (May be the same as control positioning accuracy, precision, or reproducibility in some systems.)   224, 207

machine ringing (telephone switching systems). Ringing that once started continues automatically, rhythmically until the call is answered or abandoned.
        55

machinery spaces (electric installations on shipboard). Where used in these recommendations, those spaces which are primarily used for machinery of any type or equipment for the control of such machinery, as boiler, engine, ship's service generator, and evaporator rooms.   3

machine time. *See:* **time.**   9

machine-tool control transformers (power and distribution transformer). Step-down transformers which may be equipped with fuse or other overcurrent protection device, generally used for the operation of solenoids, contactors, relays, portable tools, and localized lighting.   53

machine winding (rotating machinery). A winding placed in slots or around poles directly by a machine. *See:* **rotor (rotating machinery); stator.**   63

machine word (computing systems). *See:* **computer word.**

machining accuracy, precision, or reproducibility. Accuracy, precision, or reproducibility obtainable on completed parts under normal operating conditions.
        224, 207

macro (1) (ATLAS). A defined procedure or sequence of operations or characters which is inserted in the procedure each time its name is invoked.   400
(2) (software). A predefined sequence of instructions that is inserted into a program during assembly or compilation at each place that its corresponding macroinstruction appears in the program. *Syn:* **macroinstruction.** *See:* **instruction; macroinstruction; program.**   434

macrobending (fiber optics). In an optical waveguide, all macroscopic deviations of the axis from a straight line; distinguished from microbending. *See:* **macrobend loss; microbend loss; microbending.**   433

macrobend loss (fiber optics). In an optical waveguide, that loss attributable to macrobending. Macrobending usually causes little or no radiative loss. *Syn:* **curvature loss.** *See:* **macrobending; microbend loss.**   433

macro instruction (computing systems). An instruction in a source language that is equivalent to a specified sequence of machine instructions.   255, 77

macroinstruction (software). An instruction in a source language that is to be replaced by a defined sequence of instructions in the same source language. The macroinstruction may also specify values for parameters in the instructions that are to replace it. *See:* **instruction; parameter; source language.**   434

macroprocessor (software). The portion of some assemblers and compilers that allows a programmer to define and use macros. *See:* **assembler; compiler; macro.**   434

magazine (computing systems). *See:* **input magazine.**

magic tee. *See:* **hybrid tee.**

magner (1) (polyphase circuit). At the terminals of entry into a delimited region, the algebraic sum of the reactive power for the individual terminals of entry when the the voltages are all determined with respect to the same arbitrarily selected common reference point in the boundary surface (which may be the neutral terminal of entry). The reference direction for the currents and the reference polarity for the voltages must be the same as for the instantaneous power and the active power. The reactive power for each terminal entry is determined by considering each conductor and the common reference point as a single-phase two-wire circuit and finding the reactive power for each in accordance with the definition of **magner (single-phase two-wire circuit).** If the voltages and currents are sinusoidal and of the same period, the reactive power Q for a three-phase circuit is given by

$$Q = E_a I_a \sin(\alpha_a - \beta_a) + E_b I_b \sin(\alpha_b - \beta_b) + E_c I_c \sin(\alpha_c - \beta_c)$$

where the symbols have the same meaning as in **power, instantaneous (polyphase circuit).** If there is no neutral conductor and the common point for voltage measurement is selected as one of the phase terminals of entry, the expression will be changed in the same way as that for **power, instantaneous (polyphase circuit).** If both the voltages and currents in the preceeding equations constitute symmetrical polyphase set of the same phase sequence

$$Q = 3 E_a I_a \sin(\alpha_a - \beta_a).$$

In general the reactive power Q at the $(m + 1)$ terminals of entry of a polyphase circuit of m phases to a delimited region, when one of the terminals is the neutral terminal of entry, is expressed by the equation

$$Q = \sum_{s=1}^{s=m} \sum_{r=1}^{r=\infty} E_{sr} I_{sr} \sin(\alpha_{sr} - \beta_{sr})$$

where the symbols have the same meaning as in **power, active (polyphase circuit).** The reactive power can also be stated in terms of the root-mean-square amplitude of the symmetrical components of the voltages and currents as

$$Q = m \sum_{k=0}^{k=m-1} \sum_{r=1}^{r=\infty} E_{kr} I_{kr} \sin(\alpha_{kr} - \beta_{kr})$$

where the symbols have the same meaning as in **power, active (polyphase circuit).** When the voltages and currents are quasi-periodic and the amplitudes of the voltages and currents are slowly varying, the reactive power for the circuit of each conductor may be determined for this condition as in **power, reactive (mag-**

ner) (single-phase two-wire circuit). The reactive power for the polyphase circuit is the sum of the reactive power values for the individual conductors. Reactive power is expressed in vars when the voltages are in volts and the currents in amperes. *Note:* The sign of reactive power resulting from the above definition is the opposite of that given by the definition in the 1941 edition of the American Standard Definitions of Electrical Terms. The change has been made in accordance with a recommendation approved by the Standards Committee of the Institute of Electrical and Electronics Engineers, by the American National Standard Institute, and by the International Electrotechnical Commission.   210

(2) (single-phase two-wire circuit). At the two terminals of entry of a single-phase two-wire circuit into a delimited region, for the special case of a sinusoidal voltage and a sinusoidal current of the same period, is equal to the product obtained by multiplying the root-mean-square value of the voltage between one terminal of entry and the second terminal of entry, considered as the reference terminal, by the root-mean-square value of the current through the first terminal and by the sine of the angular phase difference by which the voltage leads the current. The reference direction for the current and the reference polarity for the voltage must be the same as for active power at the same two terminals. Mathematically the reactive power Q, for the case of sinusoidal voltage and current, is given by

$$Q = EI \sin(\alpha - \beta)$$

in which the symbols have the same meaning as in **power, instantaneous (two-wire circuit).** For the same conditions, the reactive power Q is also equal to the imaginary part of the product of the phasor voltage and the conjugate of the phasor current, or to the negative of the imaginary part of the product of the conjugate of the phasor voltage and the phasor current. Thus

$$Q = \mathrm{Im} EI^*$$
$$= -\mathrm{Im} E^*I$$
$$= \frac{1}{2j} [EI^* - E^*I]$$

in which E and I are the phasor voltage and phasor current, respectively, and * denotes the conjugate of the phasor to which it is applied. If the voltage is an alternating voltage and the current is an alternating current, the reactive power for each harmonic component is equal to the product obtained by multiplying the root-mean-square amplitude of that harmonic component of the voltage by the root-mean-square amplitude of the same harmonic component of the current and by the sine of the angular phase difference by which that harmonic component of the voltage leads the same harmonic component of the current.

Mathematically the reactive power of the rth harmonic component of Q_r is given by

$$Q_r = E_r I_r \sin(\alpha_r - \beta_r)$$
$$= \mathrm{Im} E_r I_r^*$$
$$= \frac{1}{2j} [E_r I_r^* - E_r^* I_r]$$
$$= -\mathrm{Im} E_r^* I_r$$

in which the symbols have the same meaning as in **power, instantaneous (two-wire circuit) and power, active (single-phase two-wire circuit) (average power) (power).** The reactive power at the two terminals of entry of a single-phase two-wire circuit into a delimited region, for an alternating voltage and current, is equal to the sum of the values of reactive power for every harmonic component. Mathematically the reactive power Q for an alternating voltage and current, is given by

$$Q = Q_1 + Q_2 + Q_3 + Q_4 + \cdots + Q_r + \cdots$$
$$= E_1 I_1 \sin(\alpha_1 - \beta_1)$$
$$+ E_2 I_2 \sin(\alpha_2 - \beta_2) + \cdots$$
$$= \sum_{r=1}^{r=\infty} Q_r$$
$$= \sum_{r=1}^{r=\infty} E_r I_r \sin(\alpha_r - \beta_r)$$

in which the symbols have the same meaning as in **power, instantaneous (two-wire circuit).** If the voltage and current are quasi-periodic functions of the form given in **power, instantaneous (two-wire circuit),** and the amplitudes are slowly varying, so that each may be considered to be constant during any one period, but to have slightly different values in successive periods, the reactive power at any time t may be taken as

$$Q = \sum_{r=1}^{r=\infty} E_r(t) I_r(t) \sin(\alpha_r - \beta_r)$$

by analogy with the expression for active power. When the reactive power is positive, the direction of flow of quadergy is in the reference direction of energy flow. Because the reactive power for each harmonic may have either sign, the direction of the reactive power for a harmonic component may be the same as or opposite to the direction of the total reactive power. The value of reactive power is expressed in vars when the voltage is in volts and the current in amperes. *Notes:* (A) The sign of reactive power resulting from the above definition is the opposite of that given by the definition in the 1941 edition of the American Standard Definitions of Electrical Terms. The change has been made in accordance with a recommendation approved by the Standards Committee of the Institute of Electrical and Electronics Engineers, by the American National Standards Institute, and by the Electrotechnical Commission. (B) Any designation of positive reactive power as inductive reactive power is deprecated. If the reference direction is from the generator toward the load, reactive power is positive if the load is predominantly inductive and negative if the load is predominantly capacitive. Thus a capacitor is a source of quadergy and an inductor is a consumer of quadergy. Designations of two kinds of reactive power are unnecessary and undesirable. 210

(3) (circuits and systems). The product of voltage and the component of alternating current that is 90° out of phase with it. In a passive network reactive power represents the energy that is exchanged alternatively between a capacitive and an inductive storage medium. 67

magnesium cell. A primary cell with the negative electrode made of magnesium or its alloy. *See:* **electrochemistry.** 328

magnet. A body that produces a magnetic field external to itself. 210

magnet, focusing. *See:* **focusing magnet.**

magnetic (as applied to a switching device) (power switchgear). A term indicating that interruption of the circuit takes place between contacts separable in an intense magnetic field. *Note:* With respect to contactors, this term indicates the means of operation. 103

magnetic (electron) microscope. An electron microscope with magnetic lenses. *See:* **electron optics.** 244, 190

magnetic air circuit breaker. *See:* **magnetic; air circuit breaker.**

magnetically shielded type instrument. An instrument in which the effect of external magnetic fields is limited to a stated value. The protection against this influence may be obtained either through the use of a physical magnetic shield or through the instrument's inherent construction. *See:* **instrument.** 280

magnetic amplifier. A device using one or more saturable reactors, either alone or in combination with other circuit elements, to secure power gain. Frequency conversion may or may not be included. 341

magnetic area moment. *See:* **magnetic moment.**

magnetic-armature loudspeaker. A magnetic loudspeaker whose operation involves the vibration of a ferromagnetic armature. *See:* **loudspeaker.** 328

magnetic axis (coil or winding) (rotating machinery). The line of symmetry of the magnetic-flux density produced by current in a coil or winding, this being the location of approximately maximum flux density, with the air gap assumed to be uniform. *See:* **rotor (rotating machinery); stator.** 63

magnetic bearing (navigation). Bearing relative to magnetic north. *See:* **navigation.** 187, 13

magnetic biasing (magnetic recording). The simultaneous conditioning of the magnetic recording medium during recording by the superposing of an additional magnetic field upon the signal magnetic field. *Note:* In general, magnetic biasing is used to obtain a substantially linear relationship between the amplitude of the signal and the remanent flux density in the recording medium. *See:* **alternating-current magnetic biasing; direct-current magnetic biasing; phonograph pickup.** 176

magnetic bias, relay. *See:* **relay magnetic bias.**

magnetic blowout (industrial control). A magnet, often electrically excited, whose field is used to aid the interruption of an arc drawn between contacts. *See:* **contactor.** 206

magnetic brake (industrial control). A friction brake controlled by electromagnetic means. 206

magnetic-brush development (electrostatography). Development in which the image-forming material is carried to the field of the electrostatic image by means of ferromagnetic particles acting as carriers under the influence of a magnetic field. *See:* **electrostatography.** 191

magnetic card (computing systems). A card with a magnetic surface on which data can be stored by selective magnetization of portions of the flat surface. 255, 77

magnetic circuit. A region at whose surface the magnetic induction is tangential. *Note:* The term is also applied to the minimal region containing essentially all the flux, such as the core of a transformer. 210

magnetic compass. A device for indicating the direction of the horizontal component of a magnetic field. *See:* **magnetometer.** 328

magnetic-compass repeater indicator. A device that repeats the reading of a master direction indicator, through a self-synchronous coupling means. 328

magnetic constant (pertinent to any system of units) (permeability of free space). The magnetic constant is the scalar dimensional factor that in that system relates the mechanical force between two currents to their magnitudes and geometrical configurations. More specifically, μ_0 is the magnetic constant when the element of force $d\mathbf{F}$ of a current element $I_1 d\mathbf{I}_1$ on another current element $I_2 d\mathbf{I}_2$ at a distance r is given by

$$d\mathbf{F} = \mu_0 I_1 I_2 d\mathbf{I}_1 \times (d\mathbf{I}_2 \times \mathbf{r}_1)/nr^2$$

where $\mathbf{r}_1$ is a unit vector in the direction from $d\mathbf{I}_1$ to $d\mathbf{I}_2$, and n is a dimensionless factor which is unity in unrationalized systems and 4π in a rationalized system. *Note:* In the centimeter-gram-second (cgs) electromagnetic system μ_0 is assigned the magnitude unity and the dimension numeric. In the centimeter-gram-second (cgs) electrostatic system the magnitude of μ is that of $1/c^2$ and the dimension is $[L^{-2}T^2]$. In the International System of Units (SI) μ_0 is assigned the magnitude $4\pi \cdot 10^{-7}$ and has the dimension $[LMT^{-2}I^{-2}]$. 210

magnetic contactor (industrial control). A contactor actuated by electromagnetic means. *See:* **contactor.** 206

magnetic control relay. A relay that is actuated by electromagnetic means. *Note:* When not otherwise qualified, the term refers to a relay intended to be operated by the opening and closing of its coil circuit and having contacts designed for energizing and/or de-energizing the coils of magnetic contactors or other magnetically operated device. *See:* **relay.** 206

magnetic core. A configuration of magnetic material that is, or is intended to be, placed in a rigid spatial relationship to current-carrying conductors and whose magnetic properties are essential to its use. *Note:* For example, it may be used: (1) to concentrate an induced magnetic field as in a transformer, induction coil, or armature: (2) to retain a magnetic polarization for the purpose of storing data: or (3) for its nonlinear properties as in a logic element. It may be made of iron wires, iron oxide, coils of magnetic tape, ferrite, thin film, etcetera. 235, 255, 77, 54

magnetic course (navigation). Course relative to magnetic north. *See:* **navigation.** 187, 13

magnetic deflection (cathode-ray tube). Deflecting an electron beam by the action of a magnetic field. *See:* **cathode-ray tubes.** 244, 190

magnetic delay line (computing systems). A delay line whose operation is based on the time or propagation of magnetic waves. 255, 77

magnetic deviation. Angular difference between compass north and magnetic north caused by magnetic effects in the vehicle. *See:* **navigation.** 187

magnetic device (packaging machinery). A device actuated by electromagnetic means. 429

magnetic dipole. *See:* **Hertzian magnetic dipole.**
111

magnetic dipole moment (centimeter-gram-second electromagnetic-unit system). The volume integral of **magnetic polarization** is often called magnetic dipole moment. 210

magnetic direction indicator (MDI). An instrument providing compass indication obtained electrically from a remote gyro-stabilized magnetic compass or equivalent. *See:* **radio navigation.** 328

magnetic disk. A flat circular plate with a magnetic surface on which data can be stored by selective polarization of portions of the flat surface.
235, 255, 77, 54

magnetic dissipation factor (magnetic material). The cotangent of its loss angle or the tangent of its hysteretic angle. 210

magnetic drum. A right circular cylinder with a magnetic surface on which data can be stored by selective polarization of portions of the curved surface.
235, 255, 77, 54

magnetic field (radio wave propagation). A state of a region such that a moving charged body in the region is subject to a force in proportion to its charge and to its velocity. 146

magnetic field intensity. *See:* **magnetic field strength.**

magnetic field interference. A form of interference induced in the circuits of a device due to the presence of a magnetic field. *Note:* It may appear as common-mode or normal-mode interference in the measuring circuit. *See:* **accuracy rating (instrument).** 294

magnetic field strength (1) (radio wave propagation). The magnitude of the magnetic field vector. 146 **(2) (magnetizing force).** That vector point function whose curl is the current density, and that is proportional to magnetic flux density in regions free of magnetized matter. *Note:* A consequence of this definition is that the familiar formula

$$\mathbf{H} = \frac{1}{4\pi} \int \mathbf{J} \times \nabla(1/r) \, d\nu$$
$$- \frac{1}{4\pi} \nabla \int \mathbf{M} \cdot \nabla(1/r) \, d\nu$$

(where **H** is the magnetizing force, **J** is current density, and **M** is magnetization) is a mathematical identity. 210

magnetic field strength produced by an electric current (Biot-Savart law) (Ampere's law). The magnetic field strength, at any point in the neighborhood of a circuit in which there is an electric current *i,* can be computed on the assumption that every infinitesimal length of circuit produces at the point an infinitesimal magnetizing force and the resulting magnetizing force at the point is the vector sum of the contributions of all the elements of the circuit. The contribution, d**H**, to the magnetizing force at a point *P* caused by the current *i* in an element d*s* of a circuit that is at a distance *r* from *P,* has a direction t at is perpendicular to both d*s* and *r* and a magnitude equal to

$$\frac{i \, ds \sin \theta}{r^2}$$

where θ is the angle between the element d*s* and the line *r*. In vector notation

$$dH = \frac{i\,[\mathbf{r} \times \mathbf{ds}]}{r^2}$$

This law is sometimes attributed to Biot and Savart, sometimes to Ampere, and sometimes to Laplace, but no one of them gave it in its differential form.
210

magnetic field vector (radio wave propagation). (1)(any point in a magnetic field) The magnetic induction divided by the permeability of the medium. (2)In a medium with linear and isotropic magnetic property, the magnetic induction divided by the permeability of the medium. *Syn:* **magnetic vector.** *See:* **radio wave propagation.** 146

magnetic figure of merit. The ratio of the real part of complex apparent permeability to magnetic dissipation factor. *Note:* The magnetic figure of merit is a useful index of the magnetic efficiency of a material in various electromagnetic devices. 210

magnetic flux (electrical heating systems). A condition in a medium produced by a magnetomotive force, such that when altered in magnitude, a voltage is induced in an electrical circuit linked with the flux.
476

magnetic flux (through an area). The surface integral

of the normal component of the magnetic induction over the area. Thus

$$\phi_A = \int_A (\mathbf{B} \cdot \mathrm{d}A)$$

where ϕ_A is the flux through the area A, and $\mathbf{B}$ is the magnetic induction at the element $\mathrm{d}A$ of this area. *Note:* The net magnetic flux through any closed surface is zero. 210

magnetic flux density (1)(electrical heating systems). Flux per unit area through an element normal to the direction of flux. 476
(2)(magnetic field)(measurement of power frequency electric and magnetic fields from ac power lines). The vector quantity (B-field) of divergence zero at all points, which determines the component of the Coulomb-Lorentz force, that is proportional to the velocity of the carrier. *Note:* In a zero electric field, the force yyy is the velocity of the electric charge q. The vector properties of the field produced by currents in power lines are the same as those given above for the electric field. The magnitudes of the field components are expressed by their root-mean-square (rms) values in tesla (1 T = 10^4 G). 514
See: **magnetic induction.**
magnetic flux leakage (electrical heating systems). That portion of the total magnetic flux in a circuit that does not intercept the material that contains the magnetic flux doing heating. 476
magnetic focusing. *See:* **focusing, magnetic.**
magnetic freezing. *See:* **relay magnetic freezing.**
magnetic friction clutch (coupling) (electric coupling). A friction clutch (coupling) in which the pressure between the friction surfaces is produced by magnetic attraction. 416
magnetic head (magnetic recording). A transducer for converting electric variations into magnetic variations for storage on magnetic media, or for reconverting energy so stored into electric energy, or for erasing such stored energy. 176
magnetic heading (navigation). Heading relative to magnetic north. *See:* **navigation.** 187, 13
magnetic hysteresis (electrical heating systems). The property of a magnetic material to convert electric energy to heat by virtue of the fact that the magnetic induction for a given magnetizing force depends upon the previous conditions of magnetization. 476
magnetic hysteresis loss (magnetic material). (1) The power expended as a result of magnetic hysteresis when the magnetic induction is periodic. (2) The energy loss per cycle in a magnetic material as a result of magnetic hysteresis when the induction is cyclic (not necessarily periodic). *Note:* Definitions (1) and (2) are not equivalent: both are in common use. 210
magnetic hysteretic angle. The mean angle by which the exciting current leads the magnetizing current. *Note:* Because of hysteresis, the instantaneous value of the hysteretic angle will vary during the cycle: the hysteretic angle is taken to be the mean value. 210

magnetic induction (1) (radio wave propagation). The vector **B** producing a torque on a plane current-loop carrying current **I** in accordance with the relation Torque $\mathbf{IA_n} \times \mathbf{B}$ where n is the unit vector along the positive normal to the loop and A is its area. *Syn:* **magnetic flux density.** 146
(2) (signal-transmission system). The process of generating currents or voltages in a conductor by means of a magnetic field. *See:* **magnetic flux density; signal.** 188
magnetic ink. An ink that contains particles of a magnetic substance whose presence can be detected by magnetic sensors. 255, 77
magnetic ink character recognition (MICR)(computer applications). The automatic recognition of magnetic ink characters. Contrast with optical character recognition. 571
magnetic latching relay. (1) A relay that remains operated from remanent magnetism until reset electrically. (2) A bistable polarized (magnetically latched) relay. 341
magnetic loading (rotating machinery). The average flux per unit area of the air-gap surface. *See:* **asynchronous machine.** 63
magnetic loss angle (core). The angle by which the fundamental component of the core-loss current leads the fundamental component of the exciting current in an inductor having a ferromagnetic core. *Note:* The loss angle is the complement of the hysteretic angle. 210
magnetic loss factor, initial (material). The product of the real component of its complex permeability and the tangent of its magnetic loss angle, both measured when the magnetizing force and the induction are vanishingly small. *Note:* In anisotropic media, magnetic loss factor becomes a matrix. 210
magnetic loudspeaker. A loudspeaker in which acoustic waves are produced by mechanical forces resulting from magnetic reactions. *See:* **loudspeaker.** 328
magnetic microphone. *See:* **variable-reluctance microphone.**
magnetic mine. A submersible explosive device with a detonator actuated by the distortion of the earth's magnetic field caused by the approach of a mass of magnetic material such as the hull of a ship. *See:* **degauss.** 328
magnetic moment (1) (magnetized body). The volume integral of the magnetization

$$\mathbf{m} = \int \mathbf{M}\, dv$$

(2) (current loop).

$$\mathbf{m} = I \int \mathbf{n}\, da$$
$$= (I/2) \int \mathbf{r} \times d\mathbf{r}$$

where **n** is the positive normal to a surface spanning the loop, and **r** is the radius vector from an arbitrary origin to a point on the loop. *Notes:* (A) The numerical value of the moment of a plane current loop is IA, where A is the area of the loop. (B) The reference direction for the current in the loop indicates a clockwise rotation, when the observer is looking through the loop in the direction of the positive normal. 210

magnetic north. The direction of the horizontal component of the earth's magnetic field toward the north magnetic pole. *See:* **navigation.** 187, 13

magnetic overload relay. An overcurrent relay the electric contacts of which are actuated by the electromagnetic force produced by the load current or a measure of it. *See:* **relay.** 206

magnetic particle coupling (electric coupling). An electric coupling which transmits torque through the medium of magnetic particles in a magnetic field between coupling members. 416

magnetic pickup. *See:* **variable-reluctance pickup.**

magnetic-plated wire. A magnetic wire having a core of nonmagnetic material and a plated surface of ferromagnetic material. 176

magnetic-platform influence (electric instrument). The change in indication caused solely by the presence of a magnetic platform on which the instrument is placed. *See:* **accuracy rating (instrument).** 280

magnetic polarization. In the centimeter-gram-second electromagnetic-unit system, the intrinsic induction divided by 4π is sometimes called **magnetic polarization** or **magnetic dipole moment per unit volume.** *See:* **intrinsic induction.** 210

magnetic poles (magnet). Those portions of the magnet toward which or from which the external magnetic induction appears to converge or diverge, respectively. *Notes:* (1) By convention, the north-seeking pole is marked with N, or plus, or is colored red. (2) The term is also sometimes applied to a fictitious magnetic charge. 210

magnetic pole strength (magnet). The magnetic moment divided by the distance between its poles. *Note:* Many authors use the above quantity multiplied by the magnetic constant: the two choices are numerically equal in the centimeter-gram-second electromagnetic-unit system. 210

magnetic-powder-impregnated tape (impregnated tape) (dispersed-magnetic-powder tape). A magnetic tape that consists of magnetic particles uniformly dispersed in a nonmagnetic material. 176

magnetic power factor. The cosine of the magnetic hysteretic angle (the sine of the magnetic loss angle). 210

magnetic recorder. Equipment incorporating an electromagnetic transducer and means for moving a magnetic recording medium relative to the transducer for recording electric signals as magnetic variations in the medium. *Note:* The generic term **magnetic recorder** can also be applied to an instrument that has not only facilities for recording electric signals as magnetic variations, but also for converting such magnetic variations back into electric variations. *See:* **phonograph pickup.** 176

magnetic recording (facsimile). Recording by means of a signal-controlled magnetic field. *See:* **recording (facsimile).** 12

magnetic recording head. In magnetic recording, a transducer for converting electric currents into magnetic fields, in order to store the electric signal as a magnetic polarization of the magnetic medium. 176

magnetic recording medium. A material usually in the form of a wire, tape, cylinder, disk, etcetera, on which a magnetic signal may be recorded in the form of a pattern of magnetic polarization. 176

magnetic reproducer. Equipment incorporating an electromagnetic transducer and means for moving a magnetic recording medium relative to the transducer, for reproducing magnetic signals as electric signals. 176

magnetic reproducing head. In magnetic recording, a transducer for collecting the flux due to stored magnetic polarization (the recorded signal) and converting it into an electric voltage. 176

magnetic rotation (polarized light) (Faraday effect). When a plane polarized beam of light passes through certain transparent substances along the lines of a strong magnetic field, the plane of polarization of the emergent light is different from that of the incident light. 210

magnetic sensitivity (Hall-effect devices). The ratio of the voltage across the Hall terminals to the magnetic flux density for a given magnitude of control current. 107

magnetic shunt (ferroresonant voltage regulators). The section of the core of the ferroresonant transformer taht provides the major path for flux generated by the primary winding current that does not link the secondary winding. In addition, the shunts provide a major path for the flux resulting from the output and resonating winding currents that do not link the primary winding. 456

magnetic spectrograph. An electronic device based on the action of a constant magnetic field on the paths of electrons, and used to separate electrons with different velocities. *See:* **electron device.** 244, 191

magnetic starter (packaging machinery). A starter actuated by electromagnetic means. 429

magnetic storage. A method of storage that uses the magnetic properties of matter to store data by magnetization of materials such as cores, films, or plates, or of material located on the surface of tapes, disks, or drums, etcetera. *See:* **magnetic-core; magnetic drum; magnetic tape.** 235

magnetic storm. A disturbance in the earth's magnetic field, associated with abnormal solar activity, and capable of seriously affecting both radio and wire transmission. *See:* **radio transmitter.** 328

magnetic susceptibility (isotropic medium). In rationalized systems, the relative permeability minus unity.

$$k = \mu_r - 1 = B_i/\mu_0 H$$

Notes: (1) In unrationalized systems, $k = (\mu_r - 1)4\pi$. (2) The susceptibility divided by the density of a body is called the susceptibility per unit mass, or simply the mass susceptibility. The symbol is χ. Thus

$$\chi = k/\rho$$

where ρ is the density. χ multiplied by the atomic weight is called the atomic susceptibility. The symbol is χ_A. (3) In anisotropic media, susceptibility becomes a matrix. 210

magnetic tape (homogeneous or coated). (1) A tape with a magnetic surface on which data can be stored by selective polarization of portions of the surface. (2) A tape of magnetic material used as the constituent in some forms of magnetic cores. *See:* **coated magnetic tape.** 235, 255,54

magnetic tape handler (test, measurement and diagnostic equipment). A device which handles magnetic tape and usually consists of a tape transport and magnetic tape reader with associated electrical and electronic equipments. Most units provide for tape to be wound and stored on reels: however, some units provide for the tape to be stored loosely in closed bins. 54

magnetic tape reader (test, measurement and diagnostic equipment). A device capable of converting information from magnetic tape where it has been stored as variations in magnetizations into a series of electrical impulses. 54

magnetic test coil (search coil) (exploring coil). A coil that, when connected to a suitable device, can be used to measure a change in the value of magnetic flux linked with it. *Note:* The change in the flux linkage may be produced by a movement of the coil or by a variation in the magnitude of the flux. Test coils used to measure magnetic induction B are often called B coils: those used to determine magnetizing force H may be called H coils. A coil arranged to rotate through an angle of 180 degrees about an axis of symmetry perpendicular to its magnetic axis is sometimes called a flip coil. *See:* **magnetometer.** 328

magnetic thin film. A layer of magnetic material, usually less than 10 000 angstroms thick. *Note:* In electronic computers, magnetic thin films may be used for logic or storage elements. *See:* **coated magnetic tape; magnetic core; magnetic tape.** 235

magnetic track braking. A system of braking in which a shoe or slipper is applied to the running rails by magnetic means. *See:* **electric braking.** 328

magnetic variometer. An instrument for measuring differences in a magnetic field with respect to space or time. *Note:* The use of variometer to designate a continuously adjustable inductor is deprecated. *See:* **magnetometer.** 328

magnetic vector (radio wave propagation). *See:* **magnetic field vector.** 146

magnetic vector potential. An auxiliary solenoidal vector point function characterized by the relation that its curl is equal to the magnetic induction and its divergence vanishes.

$$\text{Curl } \mathbf{A} = \mathbf{B} \qquad \text{Divergence } \mathbf{A} = 0$$

Note: These relations are satisfied identically by

$$A = (\mu_0/4\pi)[\int \mathbf{M} \times \nabla(1/r)\mathrm{d}v + \int (\mathbf{J}/r)\mathrm{d}v]$$

where v is the volume. 210

magnetization (intensity of magnetization) (at a point of a body). The intrinsic induction at that point divided by the magnetic constant of the system of units employed:

$$M = \mathbf{B}_i/\mu_0 = (\mathbf{B} - \mu_0\mathbf{H})/\mu_0.$$

Note: The magnetization can be interpreted as the volume density of magnetic moment. 210

magnetizing current (1) (transformers). A hypothetical current assumed to flow through the magnetizing inductance of a transformer. 197

(2) (rotating machinery). The quadrature (leading) component (with respect to the induced voltage) of the exciting current supplied to a coil. 210, 63

magnetizing force. *See:* **magnetic field strength.**

magnetizing inductance. A hypothetical inductance, assumed to be in parallel with the core-loss resistance, that would store the same amount of energy as that stored in the core for a specified value of excitation. 197

magnet meter (magnet tester). An instrument for measuring the magnetic flux produced by a permanent magnet under specified conditions of use. It usually comprises a torque-coil or a moving-magnet magnetometer with a particular arrangement of pole-pieces. *See:* **magnetometer.** 328

magneto. *See:* **magnetoelectric generator.**

magneto central office. A telephone central office for serving magneto telephone sets. 193

magnetoelectric generator (electric installations on shipboard). An electric generator, in which the magnetic flux is provided by one or more permanent magnets. 3

magneto-hydrodynamic wave (radio wave propagation). A low-frequency wave in an electrically highly conducting fluid (such as a plasma) permeated by a static magnetic field. The restoring forces of the waves are, in general, the combination of a magnetic tensile stress along the magnetic field lines and the comprehensive stress between the field lines and the fluid pressure (for example, Alfven wave). *Note:* This is also called a hydromagnetic wave. 146

magneto-ionic medium (radio wave propagation). An ionized gas which is permeated by a fixed magnetic field. 146

magneto-ionic mode (radio wave propagation). *See:* **magneto-ionic wave component.**

magneto-ionic wave component (radio wave component). At a given frequency, either of the two plane electromagnetic waves which can travel in a homogeneous magneto-ionic medium without change of polarization. *Syn:* **magneto-ionic mode.** 146

magnetometer. An instrument for measuring the intensity or direction (or both) of a magnetic field or of a component of a magnetic field in a particular direction. *Note:* The term is more usually applied to instruments that measure the intensity of a component of a magnetic field, such as horizontal-intensity magnetometers, vertical-intensity magnetometers, and total-intensity magnetometers.   328

magnetomotive force (acting in any closed path in a magnetic field). The line integral of the magnetizing force around the path.   210

magneto-optic (fiber optics). Pertaining to a change in a material's refractive index under the influence of a magnetic field. Magneto-optic materials generally are used to rotate the plane of polarization.   433

magnetopause (radio wave propagation). The outer boundary of a planetary magnetic field.   146

magnetoresistive coefficient (Hall generator). The ratio at a specified magnetic flux density B of the rate of change of resistance with magnetic flux density to the resistance R_B at the specified magnetic flux density B: defined by the equation

$$\alpha_B = \frac{1}{R_B}\frac{dR_B}{dB}.$$

magnetoresistive effect. The change in the resistance of a current-carrying Hall plate when acted upon by a magnetic field. *Notes: (1)* An increase in magnetic field may cause either an increase or a decrease in ferromagnetic and similar Hall plates, whereas there is usually an increase with Hall plates made of other material. *(2)* There are two factors affecting the changes in resistance: first, a bulk effect due to the characteristics of the Hall plate, and second, a geometric effect due to the shape of the Hall plate and to the presence or absence of shorting bars made of conducting material deliberately, as in the shorting bars plated on some magnetoresistors or the microconductors dispersed in other magnetoresistors or inadvertently, as in the case of the control current electrodes in a Hall generator, added to the current-carrying Hall plate.   107

magnetoresistive ratio (Hall generator). The ratio of the resistance R_B, at a magnetic flux density B, to the resistance R_0, at zero magnetic flux density: defined by the equation

$$\alpha_M = \frac{R_B}{R_0}.$$

107

magnetosphere (radio wave propagation). The region of a planetary atmosphere where a planet's magnetic field, as modified by the solar wind, controls the motions of charged particles. This includes part of the E-region of the terrestrial ionosphere up to the magnetopause.   146

magnetostriction. The phenomenon of elastic deformation that accompanies magnetization.   210

magnetostriction loudspeaker. A loudspeaker in which the mechanical displacement is derived from the deformation of a material having magnetostrictive properties. *See:* **loudspeaker.**   328

magnetostriction microphone. A microphone that depends for its operation on the generation of an electromotive force by the deformation of a material having magnetostrictive properties. *See:* **microphone.**
                    328

magnetostriction oscillator. An oscillator with the plate circuit inductively coupled to the grid circuit through a magnetostrictive element, the frequency of oscillation being determined by the magnetomechanical characteristics of the coupling element. *See:* **oscillatory circuit.**   111

magnetostrictive relay. A relay in which operation depends upon dimensional changes of a magnetic material in a magnetic field. *See:* **relay.**   259

magneto switchboard (telephone switching systems). A telecommunication switchboard for serving magneto telephone sets.   55

magneto telephone set. A local-battery telephone set in which current for signaling by the telephone station is supplied from a local hand generator, usually called a magneto. *See:* **telephone station.**   328

magneto-telluric (M-T). An adjective denoting natural magnetic and electric fields, and effects produced by them.   132

magneto-telluric current (radio wave propagation). A current induced in the earth by time-varying magnetic fields of external origin.   146

magneto-telluric fields (radio wave propagation). Electric and magnetic fields induced in the earth by external time-varying sources.   146

magnetron (induction and dielectric heating). An electron tube characterized by the interaction of electrons with the electric field of a circuit element in crossed steady electric and magnetic fields to produce alternating-current power output.   14

magnetron injection gun (microwave tubes). A gun that produces a hollow beam of high total permeance that flows parallel to the axis of a magnetic field. *See:* **magnetrons; microwave tube (or valve).**   190

magnetron oscillator. An electron tube in which electrons are accelerated by a radial electric field between the cathode and one or more anodes and by an axial magnetic field that provides a high-energy electron stream to excite the tank circuits. *See:* **magnetron.**
                    111

magnet valve (electric controller). A valve controlling a fluid, usually air, operated by an electromagnet. *See:* **multiple-unit control.**   1

magnet wire (rotating machinery). Single-strand wire with a thin flexible insulation, suitable for winding coils. *See:* **rotor (rotating machinery); stator.**
                    63

magnified sweep (oscilloscopes). A sweep whose time per division has been decreased by amplification of the sweep waveform rather than by changing the time constants used to generate it. *See:* **oscillograph.**
                    185

magnitude. The quantitative attribute of size, intensity, extent, etcetera, that allows a particular entity to be placed in order with other entities having the same attribute. *Notes:* (1) The magnitude of the length of a given bar is the same whether the length is measured in feet or in centimeters. (2) The word magnitude is used in other senses. The definition given here is the basic one needed for the logical buildup of later definitions. 210

magnitude characteristic (linear passive networks). The absolute value of a response function evaluated on the imaginary axis of the complex-frequency plane. 238

magnitude contours (control system, feedback). Loci of selected constant values of the magnitude of the return transfer function drawn on a plot of the loop transfer function for real frequencies. *Note:* Such loci may be drawn on the Nyquist or inverse Nyquist diagrams, or Nichols chart. *See:* **control system, feedback.** 56

magnitude origin line (pulse terms). A line of specified magnitude which, unless otherwise specified, has a magnitude equal to zero and extends through the waveform epoch. *See:* The single pulse diagram below the **waveform epoch** entry. 254

magnitude parameters and references (pulse terms). (Unless otherwise specified, derived from data within the waveform epoch.) *See:* **base magnitude; top magnitude; pulse amplitude; magnitude reference lines; magnitude reference points.** 254

magnitude ratio (hydraulic turbines). The ratio of the peak magnitude of the output signal to the peak magnitude of a constant-frequency constant-amplitude sinusoidal input signal. 8

magnitude referenced point (pulse terms). A point at the intersection of a magnitude reference line and a waveform. 254

magnitude reference line (pulse terms). A line parallel to the magnitude origin line at a specified magnitude. 254

magnitude reference lines (pulse terms). (1) baseline (topline). The magnitude reference line at the base (top) magnitude. *See:* The single pulse diagram below the **waveform epoch** entry. 254

(2) percent reference magnitude. A reference magnitude specified by:

$$(x)\%M_r = M_b + \frac{x}{100}(M_t - M_b)$$

where
$0 < x < 100$
$(x)\%M_r$ = percent reference magnitude
M_b = base magnitude
M_t = top magnitude
M_b, M_t and $(x)\%M_r$ are all in the same unit of measurement

(3) proximal (distal) line. A magnitude reference line at a specified magnitude in the proximal (distal) region of a pulse waveform. Unless otherwise specified, the proximal (distal) line is at the 10 (90) percent reference magnitude. *See:* The single pulse diagram below the **waveform epoch** entry.

(4) mesial line. A magnitude reference line at a specified magnitude in the mesial region of a pulse waveform. Unless otherwise specified, the mesial line is at the 50 percent reference magnitude. *See:* The single pulse diagram below the **waveform epoch** entry.

magnitude reference points (pulse terms). *See:* proximal (distal) point; mesial point. 254

magnitude-related adjectives (pulse terms). (1) proximal (distal). Of or pertaining to a region near to (remote from) a first state or region of origin.

(2) mesial. Of or pertaining to region between the proximal and distal regions. 254

Mahoney map (mathematics of computing). A diagram used in logic design, simplification, or optimization; invented by Matthew V. Mahoney. *See:* **Karnaugh map.** 564

main (interior wiring). A feeder extending from the service switch, generator bus, or converter bus to the main distribution center. 328

main (primary) switchgear connections (power switchgear). Those that electrically connect together devices in the main circuit, or connect them to the bus, or both. 103

main anode (pool-cathode tube). An anode that conducts load current. *Note:* The word main is used only when it is desired to distinguish the anode to which it is applied from an auxiliary electrode such as an excitation anode. It is used only in connection with pool-tube terms. *See:* **electrode (electron tube).** 190

main bang (radar). A transmitted pulse. 254

main bonding jumper (National Electrical Code). The connection between the grounded circuit conductor and the equipment grounding conductor at the service. 256

main capacitance (capacitance potential device). The capacitance between the network connection and line. *See:* **outdoor coupling capacitor.** 351

main contacts (of a switching device) (power switchgear). Contacts which carry all or most of the current of the main current. 103

main distributing frame, (MDF) (telephone switching systems). A frame where crossconnections are made between the outside plant and central office equipment. 55

main distribution center. A distribution center supplied directly by mains. *See:* **distribution center.** 328

main exciter (rotating machinery). An exciter that supplies all or part of the power required for the excitation of the principal electric machine or machines. *See:* **asynchronous machine.** 63

main exciter response ratio (nominal exciter respnse). The numerical value obtained when the response, in volts per second, is divided by the rated-load field voltage: which response, if maintained constant, would develop, in one-half second, the same excitation voltage-time area as attained by the actual exciter. *Note:* The response is determined with no load on the exciter, with the exciter voltage initially equal to the rate-load field voltage, and then suddenly establishing circuit conditions that would be used to obtain nominal exciter ceiling voltage. For a rotating exciter, the response should be determined at the rated speed. This definition does not apply to main exciters having one or more series fields, except a light differential series field, or to electronic exciters.   328

main gap (glow-discharge tubes). The conduction path between a principal cathode and a principal anode.   125

main lead (rotating machinery). A conductor joining a main terminal to the primary winding. *See:* **asynchronous machine.**   63

main lobe. *See* **major lobe.**

main protection. *See:* **primary protection.**

main reflector (antennas). The largest reflector of a multiple reflector antenna.   111

mains. *See:* **primary distribution mains; secondary distribution mains; center of distribution.**

mains coupling coefficient (electromagnetic compatibility). *See:* **mains decoupling factor.**

mains decoupling factor (mains coupling coefficient) (electromagnetic compatibility). The ratio of the radio-frequency voltage at the mains terminal to the interfering apparatus to the radio-frequency voltage at the aerial terminals of the receiver. *Note:* Generally expressed in logarithmic units. *See:* **electromagnetic compatibility.**   220, 199

main secondary terminals. The main secondary terminals provide the connections to the main secondary winding. *See:* **main secondary winding.**   351

main secondary winding (capacitance potential device). Provides the secondary voltage or voltages on which the potential device ratings are based. *See:* **main secondary terminals.**   351

mains-interference immunity (mains-interference ratio). The degree of protection against interference conducted by its supply mains as measured under specified conditions. *Note:* see International Special Committee on Radio Interference recommendation 25.1 and International Electrotechnical Commission publication 69 or subsequent publications where the term mains-interference ratio is used. *See:* **electromagnetic compatibility.**   220, 199

main station. A telephone station with a distinct call number designation, directly connected to a central office. *See:* **telephone station.**   328

main-station code (telephone switching systems). The digits designating a main station: these usually follow an office code.   55

main switchgear connections (primary switchgear connection). Those that electrically connect together devices in the main circuit, or connect them to the bus, or both.   103

maintain (maintenance) temperature (electrical heating systems). Specified temperature of the fluid or process material that the heating system is designed to hold at equilibrium under specified design conditions.   476

maintainability (1) (software). (A) The ease with which software can be maintained. (B) The ease with which maintenance of a functional unit can be performed in accordance with prescribed requirements. (C) ability of an item under stated conditions of use to be retained in, or restored to, within a given period of time, a specified state in which it can perform its required functions when maintenance is performed under stated conditions and while using prescribed procedures and resources. *See:* **function; functional unit; maintenance; procedure; requirement; software.**   434

(2) (reliability). Ability of an item, under stated conditions of use, to be retained in or restored to a state in which it can perform its required functions, when maintenance is performed under stated conditions and using prescribed procedures and resources. *Notes* (A): Maintainability can, depending on the particular analysis situation, be stated by one or several maintainability characteristics, such as discrete probability distribution, mean active maintenance time, etcetera. (B) The value of the maintainability characteristic may differ for different maintenance situations. (C) When the term maintainability is used as a maintainability characteristic, it always denotes the probability that the active maintenance is carried out within a given period of time. (D) The required function may be defined as a stated condition.   164

maintaining voltage (glow lamp) (operating voltage). The voltage measured across the lamp electrodes when the lamp is operating.   283

maintain temperature (electrical heat tracing for industrial applications). Specified temperature of the fluid or process material which the heat tracing is designed to hold at equilibrium under specified design conditions.   523

maintenance (1) (computing systems). Any activity intended to keep equipment, programs or a data base in satisfactory working condition, including tests, measurements, replacements, adjustments, and repairs. *See:* **file maintenance; software maintenance.**   255, 77, 434

(2) (test, measurement and diagnostic equipment). Activity intended to keep equipment (hardware) or programs (software) in satisfactory working condition, including tests, measurements, replacements, adjustments, repairs, program copying, and program improvement. Maintenance is either preventive or corrective.   54

(3) (reliability). The combination of all technical and corresponding administrative actions intended to retain an item in, or restore it to, a state in which it can perform its required function. *Note:* The required function may be defined as a stated condition.   164

maintenance bypass (nuclear power generating stations). Removal of the capability of a channel, component, or piece of equipment to perform a protective action due to a requirement for replacement, repair, test, or calibration. *Note:* A maintenance bypass is not the same as an operating bypass. A maintenance bypass may reduce the degree of redundancy of equipment but it will not result in the loss of a safety function.   428

maintenance concept (test, measurement and diagnostic equipment). A description of the general scheme for maintenance and support of an item in the operational environment. 54

maintenance, depot (test, measurement and diagnostic equipment). Maintenance performed on material requiring major overhaul or a complete rebuild of parts, subassemblies, and end items, including the manufacture of parts, modification, testing, and reclamation as required. Depot maintenance serves to support lower categories of maintenance by providing technical assistance and performing that maintenance beyond their responsibility. Depot maintenance provides stocks of serviceable equipment by using more extensive facilities for repair than are available in lower level maintenance activities. 54

maintenance derated hours (MDH)(electric generating unit reliability, availability, and productivity). The available hours during which a Class 4 unplanned derating was in effect. 567

maintenance engineering analysis (test, measurement and diagnostic equipment). A process performed during the development stage to derive the required maintenance resources such as personnel, technical data, support equipment, repair parts, and facilities. 54

maintenance factor. *See:* **light loss factor.**

maintenance, intermediate. *See:* **intermediate maintenance.**

maintenance interval (1)(class 1E battery chargers and inverters). The period, defined in terms of real time, operating time, number of operating cycles, or a combination of these, during which satisfactory performance is required without maintenance or adjustments. 408
(2)(switchgear assemblies for Class 1E applications in nuclear power generating stations). The period, defined in terms of real time, operating time, number of operating cycles, or a combination of these, during which satisfactory performance is expected without maintenance or adjustments. 576

maintenance level (test, measurement and diagnostic equipment). The level at which maintenance is to be accomplished, that is, organizational, intermediate, and depot. 54

maintenance operation device (power switchgear). A removable device for use with power operated circuit breakers which is used for manual operation of a deenergized circuit breaker during maintenance only. *Note:* This device is not to be used for closing the circuit breaker on an energized circuit. 103

maintenance, organizational. *See:* **organizational maintenance.**

maintenance outage hours (MOH)(electric generating unit reliability, availability, and productivity). The number of hours a unit was in a Class 4 unplanned outage state. 567

maintenance phase. *See:* **operation and maintenance phase.**

maintenance plan (software). A document that identifies the management and technical approach that will be used to maintain software products. Typically included are topics such as tools, resources, facilities, and schedules. *See:* **document; software product; tool.** 434

maintenance proof test (rotating machinery). A test applied to an armature winding after being in service to determine that it is suitable for continued service. It is usually made at a lower voltage than the acceptance proof test. 6

maintenance, scheduled (generation). Capability which has been scheduled to be out of service for maintenance. 112

maintenance test (1)(electric submersible pump cable). Test made after removal of the cable from the well. It is intended to detect deterioration of the cable to determine suitability for reuse. 484
(2)(power cable systems). A test made during the operating life of a cable. It is intended to detect deterioration of the system and to check the entire workmanship so that suitable maintenance procedures can be initiated. 437

main-terminal (1)(bidirectional thyristor). (A) The main terminal that is named 1 by the device manufacturer. (B) The main terminal that is named 2 by the device manufacturer. *See:* **anode.** 243, 66, 191

main terminal (2)(rotating machinery). A termination for the primary winding. *See:* **asynchronous machine.** 63

main terminals (thyristor). The terminals through which the principal current flows. *See:* **anode.** 243, 66, 191

main transformer (power and distribution transformer). The term "main transformer" as applied to two single-phase Scott-connected units for three-phase to two-phase or two-phase to three-phase operation, designates the transformer that is connected directly between two of the phase wires of the three-phase lines. *Note:* A tap is provided at the midpoint for connection to the teaser transformer. 53

main unit (power and distribution transformer). The core and coil unit which furnishes excitation to the series unit. 53

main winding, single-phase induction motor. A system of coils acting together, connected to the supply line, that determines the poles of the primary winding, and that serves as the principal winding for transfer of energy from the primary to the secondary of the motor. *Note:* In some multispeed motors, the same main winding will not be used for both starting operation and running operation. *See:* **asynchronous machine.** 63

major alarm (telephone switching systems). An alarm indicating trouble or the presence of hazardous conditions needing immediate attention in order to restore or maintain the system capability. 55

major cycle (electronic computation). In a storage device that provides serial access to storage positions, the time interval between successive appearances of a given storage position. 235

major failure. *See:* **failure, major.**

major insulation (outdoor apparatus bushings). Insulating material internal to the bushing between the line potential conductor and ground. 168

majority (computing systems). A logic operator having the property that if P is a statement, Q is a statement, R is a statement,...,then the majority of $P, Q, R,$...,is true if more than half the statements are true, false if half or less are true. 255, 77

majority carrier (semiconductor). The type of charge carrier constituting more than one half the total charge-carrier concentration. *See:* **semiconductor; semiconductor device.** 245

major lobe. The radiation lobe containing the direction of maximum radiation. *Note:* In certain antennas, such as multilobed or split-beam antennas, there may exist more than one major lobe. *Syn:* **main lobe.** *See:* **antenna.** 179

major loop (control). A continuous network consisting of all of the forward elements and the primary feedback elements of the feedback control system. *See:* **control system, feedback.** 206

major pulse waveform features (pulse terms). *See:* **base; top; first transition; last transition.** 254

major scheduled generation station shutdown. Periodic shutdowns of the generating station for an extended time scheduled for major reconditioning of the station, for example, fuel reloading. 31

make-break operation (pulse operation) (data transmission). Used to describe a method of data transmission by means of opening and closing a circuit to produce a series of current pulses. 59

make busy (telephone switching systems). Conditioning a circuit to be unavailable for service. 55

make-busy signal (telephone switching systems). A signal transmitted from the terminating end of a trunk to prevent the seizure of the originating end. 55

make, % make (telephony)(dial-pulse address signaling systems). In dial-pulse signaling, make is that portion of the signal in which the dialing contacts are closed (make). % is the ratio of make time to the total pulse period (make + break) time. 540

making capacity (industrial control). The maximum current or power that a contact is able to make under specified conditions. *See:* **contactor.** 244, 206

making current (of a switching device) (power switchgear). The value of the available current at the time the device closes. *Notes:* (1) Its root-mean-square (rms) value is measured from the envelope of the current wave at the time of the first major current peak. (2) The making current may also be expressed in terms of instantaneous value of current in which case it is measured at the first major peak of the current wave. This is designated peak making current. 27, 103

making current, rated (switching device). The maximum root-mean-square current against which the recloser is required to close under specified conditions. *Note:* The root-mean-square value is measured from the envelope of the current wave at the time of the first major current peak. [See ANSI C37.05-1964 (R1969), Methods for Determining the Values of a Sinusoidal Current Wave and Normal-Frequency Recovery Voltage for AC High-Voltage Circuit Breakers.] 92

making-current tests (high-voltage switchgear). Tests that consist of manual or remote control closing of the device against a prescribed current. 443

malfunction (1) (electronic digital computer). An error that results from failure in the hardware. *See:* **error; mistake; fault.** 235, 54, 9

(2) (test, measurement, and diagnostic equipment). *See:* **fault.** 54

(3) (analog computer). *See:* **error.** 10

(4) (class 1E static battery chargers and inverters)-seismic qualification of class 1E equipment)(nuclear power generating stations). The loss of capability of Class 1E equipment to initiate or sustain a required function, or the initiation of undesired spurious action which might result in consequences adverse to safety. 408, 28, 440

malicious call, (telephone switching systems). A call of an harassing, abusive, obscene, or threatening nature. 55

management information system (MIS)(computer applications). An automated system designed to provide managers with the information required to make basic decisions. *See:* **computer-aided management; decision support system.** 571

manhole (electric systems) (1) (More accurately termed **splicing chamber** or **cable vault**). A subsurface chamber, large enough for a man to enter, in the route of one or more conduit runs, and affording facilities for placing and maintaining in the runs, conductors, cables, and any associated apparatus. *See:* **cable vault; splicing chamber.** 64

(2)(NESC). A subsurface enclosure which personnel may enter and which is used for the purpose of installing, operating, and maintaining submersible equipment and cable. 494

(3) An opening in an underground system that workmen or others may enter for the purpose of installing cables, transformers, junction boxes, and other devices, and for making connections and tests. *See:* **cable vault; distribution center; splicing chamber.** 178

(4) (transmission and distribution). A subsurface enclosure which personnel may enter and which is used for the purpose of installing, operating, and maintaining submersible equipment, cable, or both. 262

manhole chimney. A vertical passageway for workmen and equipment between the roof of the manhole and the street level. 64

manhole cover (NESC). A removable lid which closes the opening to a manhole or similar subsurface enclosure. 494

manhole cover frame. The structure that caps the manhole chimney at ground level and supports the cover. 64

manhole grating (NESC). A grid which provides ventilation and a protective cover for a manhole opening. 494

manifold insulation (liquid cooling) (rotating machinery). The insulation applied between ground and a manifold connecting several parallel liquid-cooling paths in a winding. *See:* **stator.** 63

manifold-pressure electric gauge. A device that measures the pressure of fuel vapors entering the cylinders of an aircraft engine. *Note:* The gauge is provided with a scale, usually graduated in inches of mercury, absolute. It provides remote indication by means of a self-synchronous generator and motor. 328

manipulated variable (control) (industrial control). A quantity or condition that is varied as a function of the actuating signal so as to change the value of the directly controlled variable. *Note:* In any practical control system, there may be more than one manipulated variable. Accordingly, when using the term it is necessary to state which manipulated variable is being discussed. In process control work, the one immediately preceding the directly controlled system is usually intended. *See:* **control system, feedback.** 206

man/machine interface (MMI) (station control and data acquisition). The operator contact with equipments governed by ANSI/IEEE C37, 1-1979, Mil-Std-1472 [26] is recommended as a reference for use in the design and evaluation of the man/machine interface to equipments governed by this standard. Alternative human engineering data may be specified by the user. The man/machine interface for operation concerns standards and recommendations for information displays, control capabilities, colors and man/machine interaction of equipments governed by this standard. 403

man-made noise (electromagnetic compatibility). Noise generated in machines or other technical devices. *See:* **electromagnetic compatibility.** 199

manned space flight network (MSFN) (communication satellite). A network of ground communication and tracking facilities maintained for the support of manned space flight programs. 83

mantissa (mathematics of computing). (1) The fractional part of a logarithm. Contrast with 'characteristic'. (2) For floating-point arithmetic, see 'significand'. 564

manual (1) (electric systems). Operated by mechanical force, applied directly by personal intervention. *See:* **distribution center.** 206
(2)(NESC). Capable of being operated by personal intervention. 494

manual block-signal system. A block or a series of consecutive blocks governed by block signals operated manually upon information by telegraph, telephone, or other means of communication. *See:* **block signal system.** 328

manual central office. A central office of a manual telephone system. 193

manual checkout (test, measurement and diagnostic equipment). A checkout system which relies completely on manual operation, operator decision and evaluation of results. 54

manual control (1)(excitation systems for synchronous machines). In excitation control system usage, manual control refers to maintaining synchronous machine terminal voltage by operator action. *Note:* Manual control means may include an exciter field rheostat, controlled rectifiers, or a direct-current (dc) regulator controlling either exciter field current or

exciter output voltage, or other means that do not include regulation of synchronous machine terminal voltage. 507
(2) (power switchgear). Control in which the main devices under control, whether manually or power operated are controlled by an attendant. 103
(3) (station control and data acquisition). Control in which the system or main device, whether direct or power-aided in operation, is directly controlled by an attendant. 403
(4) (programmable instrumentation). *See:* **local control.** 378

manual controller (electric installations on shipboard) An electric controller having all of its basic functions performed by devices which are operated by hand. 3

manual data input (numerically controlled machines). A means for the manual insertion of numerical control commands. 207

manual fire-alarm system. A fire-alarm system in which the signal transmission is initiated by manipulation of a device provided for the purpose. *See:* **protective signaling.** 328

manual holdup-alarm system. An alarm system in which the signal transmission is initiated by the direct action of the person attacked or of an observer of the attack. *See:* **protective signaling.** 328

manual input (computing systems). (1) The entry of data by hand into a device at the time of processing. (2) The data entered as in (1). 255, 77

manual load (armature current) division (industrial control). The effect of a manually operated device to adjust the division of armature currents between two or more motors or two or more generators connected to the same load. *See:* **control system, feedback.** 206

manual lockout device (power switchgear). A device that holds the associated device inoperative unless a predetermined manual function is performed to release the locking feature. 103

manually operated door or gate. A door or gate that is opened and closed by hand. *See:* **hoistway (elevator or dumbwaiter).** 328

manually release- (trip-) free. *See:* **mechanically release-(trip-) free** (for a manually operated switching device). 103

manually trip-free. *See:* **mechanically release-free.**

manual mobile telephone system. A mobile communication system manually interconnected with any telephone network, or a mobile communication system manually interconnected with a telephone network. 181

manual operation (power switchgear). Operation by hand without using any other source of power. 103

manual potentiometer (analog computers). A potentiometer which is set by hand, also known as a hand-set potentiometer. 9

manual release (electromagnetic brake) (industrial control). A device by which the braking surfaces may be manually disengaged without disturbing the torque adjustment. *See:* **electric drive.** 225, 206

manual-reset, manual-release (control-brakes). A manual release that requires an additional manual action to re-engage the braking surfaces.   225, 206

manual-reset relay. *See:* **relay, manual-reset.**

manual-reset thermal protector (rotating machinery). A thermal protector designed to perform the function by opening the circuit to or within the protected machine, but requiring manual resetting to close the circuit. *See:* **starting-switch assembly.**   63

manual ringing (telephone switching systems). Ringing that is started by the manual operation of a key and continues only while the key is held operated.
   55

manual speed adjustment (industrial control). A speed adjustment accomplished manually. *See:* **electric drive.**   206

manual switchboard (telephone switching systems). A telecommunication switchboard for making interconnections manually by plugs and jacks or keys.
   55

manual telecommunications exchange (telephone switching systems). A telecommunications exchange in which connections between stations are manually set by means of plugs and jacks or keys.   55

manual telecommunication system (telephone switching systems). A telecommunications system in which connections between customers are ordinarily established manually by operators in accordance with orders given orally by the calling parties.   55

manual test equipment (test, measurement and diagnostic equipment). Test equipment that requires separate manipulations for each task (for example, connection to signal to be measured, selection of suitable range, and insertion of stimuli).   54

manual transfer or selector device (43) (power system device function numbers). A manually operated device that transfers the control circuits in order to modify the plan of operation of the switching equipment or of some of the devices.   402

manual trip device (power switchgear). A device which is connected to the tripping linkage and which can be operated manually to trip a switching device.
   103

manufactured building (National Electrical Code). Any building which is of closed construction and which is made or assembled in manufacturing facilities on or off the building site, other than mobile homes or recreational vehicles.   256

manufacturer (rotating electric machinery). The organization supplying the electric machinery to the purchaser. For the purpose of this standard it may include a repair contractor.   424

manuscript (numerically controlled machines). An ordered list of numerical control instructions. *See:* **programming.**   207

map. To establish a correspondence between the elements of one set and the elements of another set.
   225, 77

map program (software). A compiler or assembler feature that generates a load map. *See:* **assemble; compiler; load map.**   434

map vertical (navigation). *See:* **geographic map vertical.**

margin (1)(electric penetration assemblies). The difference between the most severe design service conditions and the conditions used in the design qualification to account for normal variations in commercial production of equipment and reasonable errors in defining satisfactory performance.   493

(2)(safety systems equipment in nuclear power generating stations)(nuclear power generating station)(valve actuators). The difference between service conditions and the conditions used for equipment qualification.   120, 535, 492, 440

(3)(switchgear assemblies for Class 1E applications in nuclear power generating stations). The difference between the demonstrated capability of the equipment and that required in service for specific conditions.
   576

(4) (data transmission). (A) Digital. Of a receiving equipment, the maximum degree of distortion of the received signal which is compatible with the correct translation of all of the signals which it may possibly receive. *Note:* This maximum degree of distortion applies without reference to the form of distortion effecting the signals. In other words, it is the maximum degree of the most unfavorable distortion acceptable, beyond which incorrect translation occurs, which determines the value of the margin. The condition of the measurements of the margin are to be specified in accordance with the requirements of the system. (B) Analog. The excess of receive level beyond that needed for proper operation.   59

(5) (orientation margin) (printing telegraphy) (teletypewriter). That fraction of a perfect signal element through which the time of selection may be varied in one direction from the normal time of selection, without causing errors while signals are being received. *Note:* There are two distinct margins, determined by varying the time of selection in either direction from normal. *See:* **telegraphy.**   194

(6) (electric penetration assemblies). The difference between the most severe specified service conditions of the plant and the conditions used in type testing to account for normal variations in commercial production of equipment and reasonable errors in defining satisfactory performance.   26

(7) (valve actuators). The difference between the most severe specified service conditions of the plant and the conditions used in type testing.   142

marginal check (electronic computation). A preventive maintenance procedure in which certain operating conditions (for example, supply voltage or frequency) are varied about their nominal values in order to detect and locate incipient defective parts. *See:* **check.**   235

marginal checking (test, measurement and diagnostic equipment). A system or method of determining circuit weaknesses and incipient malfunctions by varying the operating conditions of the circuitry.   54

marginal relay. A relay that functions in response to predetermined changes in the value of the coil current or voltage. *See:* **relay.**   259

marginal testing (test, measurement and diagnostic equipment). Testing that presents results on an indicator that has tolerance bands for evaluating the signal or characteristic being tested. (For example: a green band might indicate an acceptable tolerance range: a yellow band, a tolerance range representing marginal operation: and a red band, a tolerance that is unsatisfactory for operation of the item). 54

margin of commutation γ (margin angle). The time, expressed in degrees, (1 cycle of the ac waveform, 360°) from the termination of commutation in inverter operation to the next point of intersection between the two halfwaves of the voltage phases which have just commutated. *Note:* At this point of intersection, the converter circuit element which has just terminated conduction changes from reverse blocking state to OFF state. 121

marine distribution panel. A panel receiving energy from a distribution or subdistribution switchboard and distributing energy to energy-consuming devices or other distribution panels or panelboards of a ship. *See:* **marine electric apparatus.** 328

marine electric apparatus. Electric apparatus designed especially for use on shipboard to withstand the conditions peculiar to such application. 328

marine generator and distribution switchboard. Re-ceives energy from the generating plant and distributes directly or indirectly to all equipment of a ship supplied by the generating plant. *See:* **marine electric apparatus.** 328

Marinelli beaker (1)(germanium gamma-ray detectors). A reentrant (inverted well) beaker that will hold a radioactive sample and fit over a detector endcap such that the detector is essentially surrounded by the sample. 528

(2)(germanium semiconductor detector). Reentrant (inverted well) beaker. It is available in a variety of sizes for use in large volume, low level, measurements. The beaker specified herein is shown in Fig. 1. Fig. 2 shows a schematic of a typical sample-detector geometry. The specified beaker is considered to be of 450 mL capacity. The actual volume is greater than this, but, for purposes of this standard, the beaker is to be filled to 450 mL ± 2 mL. The beaker specified was selected because of: (A) High counting efficiency for the sample material used (B) Commercial availability at low cost (C) Common usage in many laboratories (D) Physical convenience 388

Marinelli beaker standard source (MBSS) (germanium semiconductor detector). A standard Marinelli beaker containing a carrier with radioactive material.

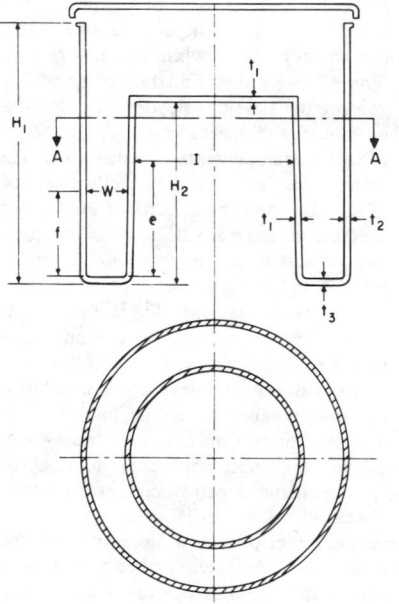

	mm	inches
H_1	104.1 ± 1.3	4.10 ± 0.05
H_2	68.33 ± 0.15	2.690 ± 0.006
I	$\left[77.40 - 0.008\,e\right]$ ± 0.10 avg., ± 0.25 max.	$\left[3.048 - 0.008\,e\right]$ ± 0.004 avg., ± 0.010 max.
W	$\left[14.83 + 0.008\,f\right]$ ± 0.10 avg., ± 0.25 max.	$\left[0.584 + 0.008\,f\right]$ ± 0.004 avg., ± 0.010 max.
t_1	1.90 ± 0.1	0.075 ± 0.004
t_2	2.00 ± 0.25	0.079 ± 0.010
t_3	3.60 ± 0.15	0.142 ± 0.006

MATERIAL : PLASTIC OF DENSITY 1.1 ± 0.1.

SECTION A-A

Fig 1
Standard Marinelli Beaker

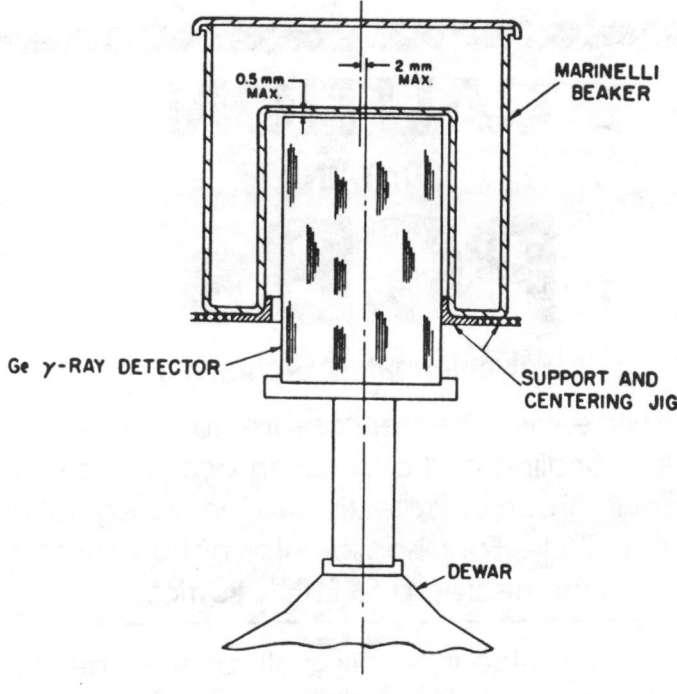

Fig 2
Marinelli Beaker with Solid State Detector

An MBSS may be a certified MBSS, a calibrated MBSS, a certified solution MBSS, or a calibrated solution MBSS. The calibration uncertainty of the photon emission rate for the filled beaker shall be not more than 3 percent unless otherwise stated. *Note:* The photon emission rate as used in this standard is the number of photons per second resulting from the decay of radionuclides in the source and is thus higher than the detected rate at the surface. 388

marine panelboard. A single panel or a group of panel units assembled as a single panel, usually with automatic overcurrent circuit breakers or fused switches, in a cabinet for flush or surface mounting in or on a bulkhead and accessible only from the front, serving lighting branch circuits or small power branch circuits of a ship. *See:* **marine electric apparatus.** 328

mariner's compass. A magnetic compass used in navigation consisting of two or more parallel polarized needles secured to a circular compass card that is delicately pivoted and enclosed in a glass-covered bowl filled with alcohol to support by flotation the weight of the moving parts. *Note:* The compass bowl is supported in gimbals mounted in the binnacle. The compass card is graduated to show the 32 points of the compass in addition to degrees. 328

marine subdistribution switchboard. Essentially a section of the marine generator and distribution switchboard (connected thereto by a bus feeder and remotely located) that distributes energy in a certain section of a vessel. *See:* **marine electric apparatus.** 328

mark (1)(handling and disposal of transformer grade insulating liquids containing PCBs). The descriptive name, instructions, cautions, or other information applied to polychlorinated biphenyls (PCBs) and PCB items or other objects subject to these regulations. See figures below. 586

(2)(computing systems). *See:* **flag.**

marked (handling and disposal of transformer grade insulating liquids containing PCBs). The marking of polychlorinated biphenyl (PCB) items and PCB storage areas and transport vehicles by means of applying a legible mark by painting, fixation of an adhesive label, or by any other method that meets the requirement of these regulations. See figures below **mark**. 586

marked ratio (instrument transformer) (metering) (power and distribution transformer). The ratio of the rated primary value to the rated secondary value as stated on the nameplate. 53

marker (telephone switching systems). A wired-logic

566

CAUTION

CONTAINS

PCBs

(Polychlorinated Biphenyls)

A toxic environmental contaminant requiring
special handling and disposal in accordance with
U.S. Environmental Protection Agency Regulations
40 CFR 761—For Disposal Information contact
the nearest U.S. E.P.A. Office.

In case of accident or spill, call toll free the U.S.
Coast Guard National Response Center:
800:424-8802

Also Contact
Tel. No.

PC-6 LABELMASTER CHICAGO, IL 60660

Fig 1
Large PCB Mark—M_L

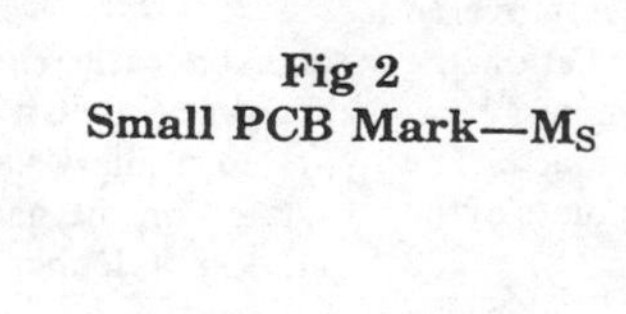

CAUTION CONTAINS **PCBs**
(Polychlorinated Biphenyls)
FOR PROPER DISPOSAL INFORMATION
CONTACT U.S. ENVIRONMENTAL
PROTECTION AGENCY
PC-1 LABELMASTER CHICAGO, IL 60660

Fig 2
Small PCB Mark—M_S

control circuit that, among other functions, tests, selects, and establishes paths through a switching stage or stages.   55

marker-beacon receiver. A receiver used in aircraft to receive marker-beacon signals that identify the position of the aircraft when over the marker-beacon station.   328

marker lamp (railway practice). A signal lamp placed at the side of the rear end of a train or vehicle, displaying light of a particular color to indicate the rear end and to serve for identification purposes.   328

marker light (railway practice). A light that by its color or position, or both, is used to qualify the signal aspect.   328

marker or marker beacon (navigation aid terms). A radio beacon to designate a small area.   526

marker radio beacon (navigation aid terms). A beacon that indicates a specific location.   526

marker signal (oscilloscopes). A signal introduced into the presentation for the purpose of identification, calibration, or comparison.   185

marking (marking and spacing intervals) (telegraph communication) (data transmission). Intervals that correspond, according to convention, to one condition or position of the originating transmitting contacts, usually a closed condition; spacing intervals are the intervals that correspond to another condition of the originating transmitting contacts, usually an open condition. *Note:* The terms "mark" and "space" frequently used for the corresponding conditions. The waves corresponding to the marking and spacing intervals are frequently designated as marking and spacing waves, respectively.   59

marking pulse (teletypewriter). The signal pulse that, in direct current, neutral, operation, corresponds to a circuit-closed or current-on condition.   194

marking wave (keying wave) (telegraph communication). The emission that takes place while the active portions of the code characters are being transmitted. *See:* **radio transmitter.**   111

M-array glide slope (instrument landing systems). A modified null-reference glide-slope antenna system in which the modification is primarily an additional antenna used to obtain a high degree of energy cancellation at the low elevation angles. *Note:* Called M because it was 13th in a series of designs. This system is used at locations where higher terrain exists in front of the approach end of the runway, in order to reduce unwanted reflections of energy into the glide-slope sector. *See:* **navigation.**   187, 13

maser (1) (data transmission)(microwave amplification by stimulated emission of radiation). The general class of microwave amplifiers based on molecular interaction with electromagnetic radiation. The non-electronic nature of the maser principle results in very low noise.   59

(2) (laser-maser). A device for amplifying or generating radiation by induced transitions of electrons, atoms, molecules, or ions between two energy levels having a **population inversion**. Also, an acronym for **microwave amplification by stimulated emission of radiation.**   363

mask (computing systems). (1) A pattern of characters that is used to control the retention or elimination of portions of another pattern of characters. (2) A filter.   255, 77

masking (1) (acoustics). (A) The process by which the threshold of audibility for one sound is raised by the presence of another (masking) sound. (B) The amount by which the threshold of audibility of a sound is raised by the presence of another (masking) sound. The unit customarily used is the decibel. *See:* **loudspeaker.**

(2) (color television). A process to alter color rendition in which the appropriate color signals are used to modify each other. *Note:* The modification is usually accomplished by suitable cross coupling between primary color-signal channels. *See:* **television.**
    176, 178

masking audiogram. A graphic presentation of the masking due to a stated noise. *Note:* This is plotted in decibels as a function of the frequency of the masked tone. *See:* **loudspeaker.**   176

mass (International System of Units) (SI). The SI unit of mass is the kilogram. This unit, or one of the multiples formed by attaching an SI prefix to gram, is preferred for all applications. Among the base and derived units of SI, the unit of mass is the only one whose name, for historical reasons, contains a prefix. Names of decimal multiples and submultiples of the unit of mass are formed by attaching prefixes to the word gram. The megagram (Mg) is the appropriate unit for measuring large masses such as have been expressed in tons. However, the name ton has been given to several large mass units that are widely used in commerce and technology–the long ton of 2240 lb, the short ton of 2000 pounds, and metric ton of 1000 kilograms (also called the tonne). None of these terms are SI. The term metric ton should be restricted to commercial usage, and no prefixes should be used with it. Use of the term tonne is deprecated. *See:* **units and letter symbols.**   21

mass-attraction vertical. The normal to any surface of constant geopotential: it is the direction that would be indicated by a plumb bob if the earth were not rotating. *See:* **navigation.**   13, 187

mass spectrograph. An electronic device based on the action of a constant magnetic field on the paths of ions, used to separate ions of different masses. *See:* **electron device.**   244

mass unbalance (gyro). The characteristic of the gyro resulting from lack of coincidence of the center of supporting forces and the center of mass. It gives rise to torques caused by linear accelerations which lead to acceleration-sensitive drift rates.   46

mast (power transmission and distribution). A column or narrow-base structure of wood, steel, or other material, supporting overhead conductors, usually by means of arms or brackets, span wires, or bridges. *Note:* Broad-base lattice steel supports are often known as towers: narrow-base steel supports are often known as masts. *See:* **pole; tower.**   64

mast arm. *See:* **bracket.**

master (FASTBUS acquisition and control). A device which is capable of asserting or controlling an operation on a segment according to the FASTBUS protocol. A master may, in addition, contain slave logic.   480

master clock. *See:* **clock.**

master compass. A magnetic or gyro compass arranged to actuate repeaters, course recorders, automatic pilots, or other devices.   328

master contactor (4) (power system device function numbers). A device, generally controlled by [a master element] device function 1 or the equivalent and the required permissive and protective devices, that serves to make and break the necessary control circuits to place an equipment into operation under the desired conditions and to take it out of operation under other abnormal conditions.   402

master controller (load-frequency control system) (1) (electric power generators). The central device that develops corrective action, in response to the area control error, for execution at one or more generating units.   94
(2) (car retarders). A controller that governs the operation of one or more magnetic or electropneumatic controllers. *Note:* It is designed to coordinate the movement or the pressure of the retarder with the movement of the retarder level. *See:* **car retarder; multiple-unit control.**   1
(3) (land transportation vehicles). A device which generates local and trainlike control signals to the propulsion and/or brake systems.   1

master direction indicator. A device that provides a remote reading of magnetic heading. It receives a signal from a magnetic sensing element.   28

master drive (industrial control). A drive that sets the reference input for one or more follower drives. *See:* **control system, feedback.**   206

master element (power system device function numbers). The initiating device, such as a control switch, etcetera which serves either directly or through such permissive devices as protective and time-delay relays to place an equipment in or out of operation. *Note:* This number is normally used for a hand-operated device, although it may also be used for an electrical or mechanical device for which no other function number is suitable.   402

master file (computer applications). In data processing, an organized collection of records that is relatively permanent. For example, a file of employee names, addresses, and salary information. Contrast with transaction file. *See:* **main file.**   571

master form. An original form from which, directly or indirectly, other forms may be prepared. *See:* **electroforming.**   328

master ground (conductor stringing equipment). A portable device designed to short circuit and connect (bond) a deenergized circuit or piece of equipment, or both, to an electrical ground. Normally located remote from, and on both sides of, the immediate work site. Primarily used to provide safety for personnel during construction, reconstruction or maintenance operations. *Syn:* **ground set; ground stick.**   431

master library (software). A software library containing formally released versions of software and documentation. *See:* **documentation; production library; software; software library.**   434

master oscillator (data transmission). An oscillator so arranged as to establish the carrier frequency of the output of an amplifier.   59

master reference system for telephone transmission. Adopted by the International Advisory Committee for Long Distance Telephony (CCIF), a primary reference telephone system for determining, by comparison, the performance of other telephone systems and components with respect to the loudness, articulation, or other transmission qualities of received speech. *Note:* The determination is made by adjusting the loss of a distortionless trunk in the master reference system for equal performance with respect to the quality under consideration.   328

master routine (electronic computation). *See:* **subroutine.**

master sequence device (34) (power system device function numbers). A device such as a motor-operated multicontact switch, or the equivalent, or a programming device, such as a computer, that establishes or determines the operating sequence of the major devices in an equipment during starting and stopping or during other sequential switching operations.   402

mastership (FASTBUS acquisition and control). A master is asserting mastership when it has control of the segment to which it is attached and is asserting grant acknowledge (GK) or address sync (AS).   480

master/slave operation (power supplies). A system of interconnection of two regulated power supplies in which one (the master) operates to control the other (the slave). *Note:* Specialized forms of the master.-slave configuration are used in (1) complementary tracking (plus and minus tracking around a common point), (2) parallel operation to obtain increased current output for voltage regulation, (3) compliance extension to obtain increased voltage output for current regulation.   186

master station (1) (power switchgear)(data transmission). (A) (supervisory system). The station from which remotely located units of switchgear or other equipment are controlled by supervisory control or that receives supervisory indications or selected telemeter readings. (B) (electronic navigation). One station of a group of stations, as in LORAN, that is used to control or synchronize the emission of the other stations.   103, 59
(2) (of a supervisory system) (station control and data acquisition). The entire complement of devices, functional modules, and assemblies which are electrically interconnected to effect the master station supervisory functions. The equipment includes the interface with the communication channel but does not include the interconnecting channel.   403
(3) (electronic navigation). One station of a group of stations, as in loran, that is used to control or synchro-

nize the emission of the other stations. *See:* **radio navigation.**   187, 13

master-station supervisory equipment (data transmission) (power switchgear). That part of a (single) supervisory system that includes all necessary supervisory control relays, keys, lamps, and associated devices located at the master station for selection, control, indication, and other functions to be performed.   103, 59

master switch (industrial control) (electric installations on shipboard). A switch that dominates the operation of contactors, relays, or other remotely operated devices. *See:* **switch.**   206, 3

master terminal unit (1)(station control and data acquisition). Refers to the master station of a supervisory control system. *See:* **master station.**   403
(2)(supervisory control, data acquisition, and automatic control). The master station of a supervisory control system. *See:* **station (1) Master.**   570

mast-type antenna for aircraft. A rigid antenna of streamlined cross section consisting essentially of a formed conductor or conductor and supporting body.   328

mat (rotating machinery). A randomly distributed unwoven felt of fibers in a sheetlike configuration having relatively uniform density and thickness. *See:* **rotor (rotating machinery); stator.**   63

matched condition. *See:* **matched termination**

matched filter (radar). A filter which maximizes the output ratio of peak signal power to mean noise power. For white Gaussian noise, it has a frequency response that is the complex conjugate of the transmitted spectrum, or equivalently, has an impulse response that is the time inverse of the transmitted waveform.   13

matched generator insertion loss (or gain)(waveguide). A loss (or gain) resulting from placing two ports of a network between a load having an arbitrary impedance and a matched generator. It is the ratio of the power absorbed in the load when connected to the generator (reference power) to that when the network is inserted.   267

matched impedances. Two impedances are matched when they are equal. *Note:* Two impedances associated with an electric network are matched when their resistance components are equal and when their reactance components are equal. *See:* **network analysis.**   210

matched insertion loss (or gain) (waveguide). A loss (or gain) resulting from placing two ports of a network between a matched generator and a matched load. It is the ratio of the power absorbed in the load when connected to the generator (reference power) to that when the network is inserted.   267

matched load insertion loss (or gain) (waveguide). A loss (or gain) resulting from placing two ports of a network between a generator having an arbitrary impedance and a matched load. It is the ratio of the power absorbed in the load when connected to the generator (reference power) to that when the network is inserted.   267

matched terminated line (waveguide). A transmission line having no reflected wave at any transverse section.   267

matched termination (waveguide components). A termination matched with regard to the impedance in a prescribed way; for example, (A) a reflectionless termination or (B) a conjugate termination.   166

matched transmission line (data transmission). A transmission line is said to be matched at any transverse section if there is no wave reflection at that section.   59

matched waveguide. *See:* **matched terminated line.**

matching, impedance. *See:* **load, matching.**

matching, load. *See:* **load matching.**

matching loss (radar). The loss in S/N (signal-to-noise) output relative to a matched filter, caused by using a filter of other than matched response to the transmitted signal. *Syn:* **mismatch loss.**   13

matching section (transforming section) (waveguide transformer) (waveguide). A length of waveguide of modified cross section, or with a metal or dielectric insert, used for impedance transformation. *See:* **waveguide.**   179

matching transformer (induction heater). A transformer for matching the impedance of the load to the optimum output characteristic of the power source.   14

material (nuclear power generating station). A substance or combination of substances used as constituents in the manufacture of components, modules, or items. *Note:* This term applies specifically to the subject matter of IEEE Std 467-1980.   438

material absorption. *See:* **absorption.**

material dispersion (fiber optics). That dispersion attributable to the wavelength dependence of the refractive index of material used to form the waveguide. Material dispersion is characterized by the material dispersion parameter M. *See:* **dispersion; distortion; material dispersion parameter; profile dispersion parameter; waveguide dispersion.**   433

material dispersion parameter (M) (fiber optics).

$$M(\lambda) = -1/c(dN/d\lambda) = \lambda/c(d^2n/d\lambda^2)$$

where n is the refractive index, N is the group index: $N = n - \lambda(dn/d\lambda)$, λ is the wavelength, and c is the velocity of light in vacuum. *Notes:* (1) For many optical waveguide materials, M is zero at a specific wavelength λ_0, usually found in the 1.2 to 1.5 μm range. The sign convention is such that M is positive for wavelengths shorter than λ_0 and negative for wavelengths longer than λ_0. (2) Pulse broadening caused by material dispersion in a unit length of optical fiber is given by M times spectral linewidth ($\Delta\lambda$), except at $\lambda = \lambda_0$, where terms proportional to $(\Delta\lambda)^2$ are important. (See Note 1). *See:* **group index; material dispersion.**   433

material scattering (fiber optics). In an optical waveguide, that part of the total scattering attributable to the properties of the materials used for waveguide fabrication. *See:* **Rayleigh scattering; scattering; waveguide scattering.**   433

material temperature class (evaluation of thermal capability)(thermal classification of electric equipment and electrical insulation). The lowest value of a range of temperature indices for insulating materials.
506

mathematical adjectives (pulse terms). All definitions in this section are stated in terms of time (the independent variable) and magnitude (the dependent variable). Unless otherwise specified, the following terms apply only to waveform data within a waveform epoch. These adjectives may be used to describe the relation(s) between other specified variable pairs (for example, time and power, time and voltage, etcetera). (1) instantaneous. Pertaining to the magnitude at a specified time. (2) positive (negative) peak. Pertaining to the maximum (minimum) magnitude. (3) peak-to-peak. Pertaining to the absolute value of the algebraic difference between the positive peak magnitude and the negative peak magnitude. (4) root-mean-square (rms). Pertaining to the square root of the average of the square of the magnitude. If the magnitude takes on n discrete values m_j, the root-mean-square magnitude is

$$M_{rss} = \left[\sum_{j=1}^{j=n} m_j^2\right]^{1/2}$$

If the magnitude is a continuous function of time m(t),

$$M_{rss} = \left[\int_{t_1}^{t_2} m^2(t)dt\right]^{1/2}$$

The summation or the integral extends over the interval of time for which the rms magnitude is desired or, if the function is periodic, over any integral number of periodic repetitions of the function. (5) average. Pertaining to the mean of the magnitude. If the magnitude takes on n discrete values m_j, the average magnitude is

$$M_{rms} = \left[\left(\frac{1}{n}\right)\sum_{j=1}^{j=n} m_j^2\right]^{1/2}$$

If the magnitude is a continuous function of time m(t)

$$M_{rms} = \left[\left(\frac{1}{t_2 - t_1}\right)\int_{t_1}^{t_2} m^2(t)dt\right]^{1/2}$$

The summation or the integral extends over the interval of time for which the average magnitude is desired or, if the function is periodic, over any integral number of periodic repetitions of the function. (6) average absolute. Pertaining to the mean of the absolute magnitude. If the magnitude takes on n discrete values m_j, the average absolute magnitude is

$$\overline{M} = \left(\frac{1}{n}\right)\sum_{j=1}^{j=n} m_j$$

If the magnitude is a continuous function of time m(t)

$$\overline{M} = \left(\frac{1}{t_2 - t_1}\right)\int_{t_1}^{t_2} m(t)dt$$

The summation or the integral extends over the interval of time for which the average absolute magnitude is desired or, if the function is periodic, over any integral number of periodic repetitions of the function. (7) root sum of squares (rss). Pertaining to the square root of the arithmetic sum of the squares of the magnitude. If the magnitude takes on n discrete values m_j, the root sum of squares magnitude is

$$|\overline{M}| = \left(\frac{1}{n}\right)\sum_{j=1}^{j=n} |m_j|$$

If the magnitude is a continuous function of time m(t),

$$|\overline{M}| = \left(\frac{1}{t_2 - t_1}\right)\int_{t_1}^{t_2} |m(t)|dt$$

The summation or the integral extends over the interval of time for which the root sum of squares magnitude is desired r, if the function is periodic, over any integral number of periodic repetitions of the function.
254

mathematical check. A programmed check of a sequence of operations that makes use of the mathematical properties of the sequence. Sometimes called a **control.** *See:* **programmed check.** 25

mathematical model (analog computers). A set of equations used to represent a physical system.
9

mathematical quantity. *See:* **mathematico-physical quantity.**

mathematical simulation (analog computers). The use of a model of mathematical equations generally solved by computers to represent an actual or proposed system.
9

mathematical symbol (abbreviation). A graphic sign, a letter or letters (which may have letters or numbers, or both, as subscripts or superscripts, or both), used to denote the performance of a specific mathematical operation, or the result of such operation, or to indicate a mathematical relationship. *Compare with:* **symbol for a quantity, symbol for a unit.** *See:* **abbreviation.**   173

mathematico-physical quantity (symbolic quantity) (mathematical quantity) (abstract quantity). A concept, amenable to the operations of mathematics, that is directly related on one (or more) physical quantity and is represented by a letter symbol in equations that are statements about that quantity. *Note:* Each mathematical quantity used in physics is related to a corresponding physical quantity in a way that depends on its defining equation. It is characterized by both a qualitative and a quantitative attribute (that is, dimensionality and magnitude).   210

matrix (1) (color television). (A) (noun). An array of coefficients symbolic of a color coordinate transformation. *Note:* This definition is consistent with mathematical usage. (B) (verb). To perform a color coordinate transformation by computation or by electrical, optical, or other means.   18

(2) (mathematics). (A) A two-dimensional rectangular array of quantities. Matrices are manipulated in accordance with the rules of matrix algebra. (B) By extension, an array of any number of dimensions.   255, 77

(3) (electronic computers). A logic network whose configuration is an array of intersections of its input-output leads, with elements connected at some of these intersections. The network usually functions as an encoder or decoder. *Note:* A translating matrix develops several output signals in response to several input signals: a decoder develops a single output signal in response to several input signals (therefore sometimes called an *and* matrix): an encoder develops several output signals in response to a single input signal and a given output signal may be generated by a number of different input signals (therefore sometimes called an OR matrix). *See:* **decode; encode; translate.**   235

(4) Loosely, any encoder, decoder, or translator.   210

(5) (electrochemistry). A form used as a cathode in electroforming. *See:* **electroforming.**   328

matrix circuit (color television). *See:* **matrix unit.**

matrix, fundamental. *See:* **matrix, transition.**

matrix, system (control systems). A matrix of transfer functions which relate the Laplace transforms of the system outputs and of the system inputs.   56

matrix, transition (control systems). A matrix which maps the state of a linear system at one instant of time into another state at a later instant of time, provided that the system inputs are zero over the closed time interval between the two instants of time. *Note:* This is also the matrix of solutions of the homogeneous equations. *Syn:* **matrix, fundamental.**   56

matrix unit (matrix circuit) (color television). A device that performs a color coordinate transformation by electrical, optical, or other means.   18

matte dip (electroplating). A dip used to produce a matte surface on a metal. *See:* **electroplating.**   328

matte surface (illuminating engineering). A surface from which the reflection is predominantly diffuse, with or without a negligible specular component. *See:* **diffuse reflection.**   167

maximum asymmetric short-circuit current (rotating machinery). The instantaneous peak value reached by the current in the armature winding within a half of a cycle after the winding has been suddenly short-circuited, when conditions are such that the initial value of any aperiodic component of current is the maximum possible.   63

maximum average power (attenuator). That maximum specified input power applied for a minimum of one hour (unless specified for a longer period) at the maximum operating temperature with output terminated in the characteristic impedance which will not permanently change the specified properties of the attenuator after return to ambient temperature at a power level 20 dB below maximum specified input power.   110

maximum average power output (television). The maximum radio-frequency output power that can occur under any combination of signals transmitted, averaged over the longest repetitive modulation cycle. *See:* **television.**   328

maximum bundle gradient (overhead-power-line corona and radio noise). For a bundle of two or more subconductors, the highest value among the maximum gradients of the individual subconductors. For example, for a three-conductor bundle with individual maximum subconductor gradients of 16.5, 16.9, and 17.0 kV/cm, the maximum bundle gradient would be 17.0 kV/cm.   411

maximum capability (power operations). The maximum generation expressed in kilowatt-hours per hour (kWh/h) which a generating unit, station, power source, or system can be expected to supply under optimum operating conditions.   516

maximum capacity (MC)(electric generating unit reliability, availability, and productivity). The maximum capacity that a unit can sustain over a specified period of time. The maximum capacity can be expressed as gross maximum capacity (GMC) or net maximum capacity (NMC). To establish this capacity, formal demonstration is required.The test is repeated periodically. This demonstrated capacity level is corrected to generating conditions for which there is minimum ambient restriction. When a demonstration test has not been conducted, the estimated maximum capacity of the unit is used.   567

maximum common-mode signal. *See:* **common-mode signal maximum.**

maximum continuous operating voltage (MCOV) (metal-oxide surge arresters for ac power circuits). The maximum designated root-mean-square (rms) value of power frequency voltage that may be applied continuously between the terminals of the arrester.   583

maximum continuous rating (rotating machinery). The maximum values of electric and mechanical loads at which a machine will operate successfully and continuously. *Note:* An overload may be implied, along with temperature rises higher than normal standards for the machine. *See:* **asynchronous machine.**   63

maximum control current (magnetic amplifier). The maximum current permissible in each control winding either continuously or for designated operating intervals as specified by the manufacturer and shall be specified as either root-mean-square or average.   171

maximum current (instrument) (wattmeter or power-factor meter). A stated current that, if applied continuously at maximum stated operating temperature and with any other circuits in the instrument energized at rated values, will not cause electric breakdown or any observable physical degradation. *See:* **instrument.**   280

maximum demand (power operations). The largest of a particular type of demand occurring within a specified period.   516

maximum-demand pointer (demand meter) (friction pointer of a demand meter). A means used to indicate the maximum demand that has occurred since its previous resetting. The maximum-demand pointer is advanced up the scale of an indicating demand meter by the pointer pusher. When not being advanced, it is held stationary, usually by friction, and it is reset manually when the meter is read for billing purposes. *See:* **demand meter.**   328

maximum design rating (MDR)(composite insulators). The maximum mechanical load that the insulator is designed to withstand continuously for the life of the insulator.   483

maximum design voltage (1) (device). The highest voltage at which the device is designed to operate. *Note:* When expressed as a rating this voltage is termed rated maximum voltage.   127
(2) (to ground) (outdoor electric apparatus). The maximum voltage at which the bushing is designed to operate continuously.   287
(3) (power switchgear). (A) of a device. The highest voltage at which the device is designed to operate. (B) of a relay. The highest root-mean-square or direct voltage at which a relay is designed to be energized continuously.   103
(4) (power and distribution transformer). The highest rms phase-to-phase voltage that equipment components are designed to withstand continuously, and to operate in a satisfactory manner without derating of any kind.   53

maximum-deviation sensitivity (in frequency-modulation receivers). Under maximum system deviation, the least signal input for which the output distortion does not exceed a specified limit. *See:* **frequency modulation.**   339

maximum effort (electric generating unit reliability,

availability, and productivity). Repairs were accomplished in the shortest possible time. *See:* **repair urgency.**   567

maximum excursion (electric conversion). The maximum positive or negative deviation from the initial or steady value caused by a transient condition.   186

maximum exposure temperature(1)(electrical heating systems). The highest temperature to which an object may be exposed for a finite period of time.   476
(2)(electrical heat tracing for industrial applications). The highest temperature to which a device in the heat-tracing system may be exposed for a given period of time.   523

maximum generation (MG)(electric generating unit reliability, availability, and productivity). The energy that could have been produced by a unit in a given period of time if operated continuously at maximum capacity. Maximum generation can be expressed as gross maximum generation (GMG) or net maximum generation (NMG).

$$\begin{aligned} MG &= \text{period hours} \cdot \text{maximum capacity} \\ &= PH \cdot MC \\ GMG &= PH \cdot GMC \\ NMG &= PH \cdot NMC \end{aligned}$$

   567

maximum ground acceleration (seismic design of substations). The maximum value of acceleration input to the equipment during a given earthquake for a particular site.   465

maximum instantaneous fuel change (gas turbines). The fuel change allowable for an instantaneous or sudden increased or decreased load or speed demand. *Note:* It is expressed in terms of equivalent load change in percent of rated load.   98, 58

maximum keying frequency (fundamental scanning frequency) (facsimile). The frequency in hertz numerically equal to the spot speed divided by twice the scanning spot X dimension. *See:* **scanning (facsimile).**   12

maximum limiting resolution (diode-type camera tube). The highest value of limiting resolution obtained under optimum irradiance conditions using a stationary bar pattern. Units: LP/RH.   380

maximum modulating frequency (facsimile). The highest picture frequency required for the facsimile transmission system. *Note:* The maximum modulating frequency and the maximum keying frequency are not necessarily equal. *See:* **facsimile transmission.**   12

maximum momentary speed variation (hydraulic turbines). The maximum momentary change of speed when the load is suddenly changed a specified amount.   8

maximum observed frequency (radio wave propagation). In ionospheric sounding, the highest radio frequency at which echo signals can be detected or observed with a particular equipment.   146

maximum operating voltage (household electric ranges). The maximum voltage to which the electric parts of the range may be subjected in normal operation. *See:* **appliances outlet.** 263

maximum output (receivers). The greatest average output power into the rated load regardless of distortion. *See:* **radio receiver.** 339

maximum output voltage (magnetic amplifier). The voltage across the rated load impedance with maximum control current flowing through each winding simultaneously in a direction that increases the output voltage. *Notes:* (1) Maximum output voltage shall be specified either as root-mean-square or average. (2) While specification may be either root-mean-square or average, it remains fixed for a given amplifier. 171

maximum peak power (attenuator). That maximum peak power at the maximum specified pulse-length and average power which, when applied for a minimum of one hour (unless specified for a longer period) at the maximum operating temperature, while the output is terminated in the characteristic impedance, will not permanently change the specified properties of the attenuator when returned to ambient temperature at a power level 20 dB below the maximum specified input power or lower. 110

maximum power output (hydraulic turbines). The maximum output which the turbine-generator unit is capable of developing at rated speed with maximum head and maximum gate. 8

maximum pulse rate (metering). The number of pulses per second at which a pulse device is nominally rated. 212

maximum pulse repetition rate (digital delay line). The maximum pulse repetition rate shall be equal to $1/2 \Delta t$, where Δt is the time spacing between the peaks of the output doublet. 81

maximum rate of fuel change (gas turbines). The rate of fuel change that is allowable after the maximum instantaneous fuel change, when an instantaneous speed or load demand upon the turbine is greater than that corresponding to the maximum instantaneous fuel change. *Note:* It is expressed in percent of equivalent load change per second. 98, 58

maximum safe input power (1) (spectrum analyzer). The power applied at the input which will not cause degradation of the instrument characteristics. *Note:* Input signal conditions, for example, peak or average power, should be specified. 390

(2) (non-real time spectrum analyzer). The power applied at the input which will not cause degradation of the instrument characteristics. 68

(3) (electrothermic unit). The maximum peak pulse or cw input power which will cause no permanent change in the calibration or characteristics of the electrothermic unit. Specify in watt-microseconds the maximum (safe) input energy per pulse and the applicable pulse repetition frequency in hertz or in kilohertz. Specify in watts or milliwatts the maximum (safe) input peak pulse power. 47

maximum OFF voltage (magnetic amplifier). The maximum output voltage existing before trip ON control signal is reached as the control signal is varied from trip OFF to trip ON. 171

maximum sensitivity (frequency-modulation systems). The least signal input that produces a specified output power. 339

maximum sine-current differential permeability (toroidal magnetic amplifier cores). The maximum value of sine-current differential permeability obtained with a specified sine-current magnetizing force. 170

maximum single-conductor (or subconductor) gradient (overhead- power-line corona and radio noise). The maximum value attained by the gradient $E\theta$ as θ varies over the range of 0 to 2π, where $E\theta$ is the gradient on the surface of the power-line conductor (or subconductor) expressed as a function of angular position θ. Unless otherwise stated, the gradient is a nominal gradient. 411

maximum sound pressure (for any given cycle of a periodic wave). The maximum absolute value of the instantaneous sound pressure occurring during that cycle. *Note:* In the case of a sinusoidal sound wave this maximum sound pressure is also called the pressure amplitude. 176

maximum speed (industrial control). The highest speed within the operating speed range of the drive. *See:* **electric drive.** 206

maximum surge current rating (rectifier circuit) (non-repetitive) (semiconductor rectifier). The maximum forward current having a specified waveform and short specified time interval permitted by the manufacturer under stated conditions. *See:* **average forward current rating (rectifier circuit element).** 208

maximum system deviation (frequency-modulation systems). The greatest frequency deviation specified in the operation of the system. *Note:* Maximum system deviation is expressed in kilohertz. In the case of FCC authorized frequency modulation broadcast systems in the range from 88 to 108 megahertz, the maximum system deviation is ± 75 kHz. 16

maximum system voltage (1)(metal-oxide surge arresters for ac power circuits). The highest voltage at which a system is operated. *Note:* This is generally considered to be the maximum system voltage as prescribed in ANSI C84.1-1982. 583

(2) (power and distribution transformer). The highest rms phase-to-phase voltage which occurs on the system under normal operating conditions, and the highest rms phase-to-phase voltage for which equipment and other system components are designed for satisfactory continuous operation without derating of any kind. (This voltage excludes voltage transients and temporary overvoltages caused by abnormal system conditions such as faults, load rejection, etcetera.) 53

maximum test output voltage (magnetic amplifier) (1) (nonreversible output). The output voltage equivalent to the summation of the minimum output voltage plus 66 2/3 percent of the difference between the rated and minimum output voltages. 171

(2) (reversible output). (A) Positive maximum test output voltage is the output voltage equivalent to 66 2/3 percent of the rated output voltage in the positive direction. (B) Negative maximum test output voltage is the output voltage equivalent to $66\,2/3$ percent of the rated output voltage in the negative direction.

171

maximum theoretical deviation from a sine wave (converter characteristics) (self-commutated converters) (harmonic control and reactive compensation of static power converters). For a nonsinusoidal wave, the ratio of the arithmetic sum of the amplitudes of all harmonics in the wave to the amplitude of the fundamental.

584, 533

maximum total sag (electric systems) (transmission and distribution). The sag at the midpoint of the straight line joining the two points of support of the conductor. *See:* **sag.**

178, 262

maximum undistorted output (amplitude-modulation broadcast receivers). The so-called maximum undistorted output is arbitrarily taken as the least power output which contains, under given operating conditions, a total power at harmonic frequencies equal to 1 percent of the apparent power at the fundamental frequency. This corresponds to a root-sum-square total voltage at harmonic frequencies equal to 10 percent of the root-sum-square voltage at the fundamental frequency, if measured across a pure resistance. (The root-sum-square voltage of a complex wave is the square root of the sum of the squares of the component voltages.)

524

maximum usable frequency (radio wave propagation). The highest frequency of radio waves that can be used between two points under specified conditions for reliable transmission by reflection from the regular layers of the ionosphere.

146

maximum useful output. *See:* **maximum undistorted output.**

maximum value of the electric field strength (measurement of power frequency electric and magnetic fields from ac power lines). At a given point, the root-mean-square (rms) value of the semimajor axis magnitude of the electric field ellipse. *See:* **electric field strength.**

514

maximum value of the magnetic field (measurement of power frequency electric and magnetic fields from ac power lines). At a given point, the root-mean-square (rms) value of the semimajor axis magnitude of the magnetic field ellipse.

514

maximum voltage (instrument) (wattmeter, power-factor meter or frequency meter). A stated voltage that, if applied continuously at the maximum stated operating temperature and with any other circuits in the instrument energized at rated values, will not cause electric breakdown or any observable physical degradation. *See:* **accuracy rating (instrument); instrument.**

280

maximum voltage rating (separable insulated connectors). The highest phase-to-ground or phase-to-ground and phase-to-phase voltage (root-mean-square) at which a connector is designed to operate.

454

maxwell (line). The unit of magnetic flux in the centimeter-gram-second electromagnetic system. *Note:* The maxwell is 10^{-8} weber.

210

Maxwell bridge (general). A 4-arm alternating-current bridge characterized by having in one arm an inductor in series with a resistor and in the opposite arm a capacitor in parallel with a resistor, the other two arms being normally nonreactive resistors. *Note:* Normally used for the measurement of inductance (or capacitance) in terms of resistance and capacitance (or inductance). The balance is independent of the frequency, and at balance the ratio of the inductance to the capacitance is equal to the product of the resistances of either pair of opposite arms. It differs from the Hay bridge in that in the arm opposite the inductor, the capacitor is shunted by the resistor. *See:* **bridge.**

328

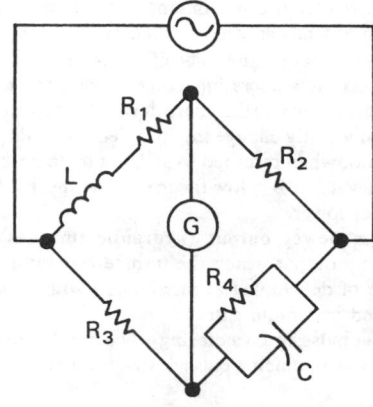

$$R_1 R_4 = R_2 R_3 = L/C$$

Maxwell bridge

Maxwell direct-current commutator bridge. A 4-arm bridge characterized by the presence in one arm of a commutator, or 2-way contactor, that, with a known periodicity, alternately connects the unknown capacitor in series with the bridge arm and then opens the bridge arm while short-circuiting the capacitor, the other three arms being nonreactive resistors. *Note:* Normally used for the measurement of capacitance in terms of resistance and time. The bridge is normally supplied from a battery and the detector is a direct-current galvanometer. *See:* **bridge.**

328

Maxwell inductance bridge. A 4-arm alternating-current bridge characterized by having inductors in two adjacent arms and usually, nonreactive resistors in the other two arms. *Note:* Normally used for the comparison of inductances. The balance is independent of the frequency. *See:* **bridge.**

328

Maxwell mutual-inductance bridge. An alternating-current bridge characterized by the presence of mutual inductance between the supply circuit and that arm of the network that includes one coil of the mutual inductor, the other three arms being normally non-reactive resistors. *Note:* Normally used for the measurement of mutual inductance in terms of self-inductance. The balance is independent of the frequency. *See:* **bridge.**

328

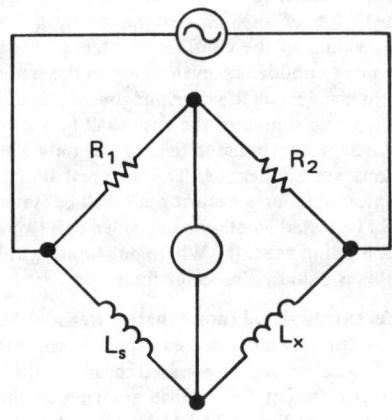

$$C = \frac{R_1}{nR_2R_3} \cdot \times \frac{\left[1 - \frac{R_1{}^2}{(R_1 + R_2 + R_B)(R_1 + R_3 + R_G)} \right]}{\left[1 + \frac{R_1R_B}{R_3(R_1 + R_2 + R_B)} \right]\left[1 + \frac{R_1R_G}{R_2(R_1 + R_3R_G)} \right]}$$

Maxwell direct-current commutator bridge

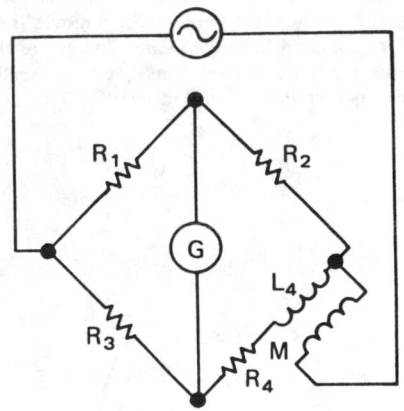

$$R_x = R_s \frac{R_2}{R_1} \qquad L_x = L_s \frac{R_2}{R_1}$$

Maxwell inductance bridge

$$R_1R_4 = R_2R_3 \qquad L_4 = -M\left(1 + \frac{R_2}{R_1} \right)$$

Maxwell mutual-inductance bridge

Maxwell's equations (Maxwell's laws). The fundamental equations of macroscopic electrmagnetic field theory. All real (physical) electric and magnetic fields satisfy Maxwell's equations, namely

$$\nabla \times \mathbf{E} = \frac{\partial \mathbf{B}}{\partial t} \qquad \nabla \cdot \mathbf{B} = 0$$

$$\nabla \times \mathbf{H} = \frac{\partial \mathbf{D}}{\partial t} + \mathbf{J} \qquad \nabla \cdot \mathbf{D} = q_r$$

where **E** is electric field strength, **D** is electric flux density, **H** is magnetic field strength, **B** is magnetic flux density, **J** is current density, and q_r is volume charge density.   210

May Day. *See:* **radio distress signal.**

MBWO (reentrant field type microwave backward-wave oscillator) (microwave tubes). *See:* **carcinotron.**

MCA. *See:* **multichannel analyzer.**

MC cable (metal-clad cable) (National Electrical Code). A factory assembly of one or more conductors, each individually insulated and enclosed in a metallic sheath of interlocking tape, or a smooth or corrugated tube.   256

McCulloh circuit. A supervised, metallic loop circuit having manually or automatically operated switching equipment at the receiving end, that, in the event of a break, a ground, or a combination of a break and a ground at any point in the metallic circuit, conditions the circuit, by utilizing a ground return, for the receipt of signals from suitable signal transmitters on both sides of the point of trouble. *See:* **protective signalling.**   328

MCW. *See:* **modulated continuous wave.**

MDI. *See:* **magnetic direction indicator.**

M-display (radar). A type of A display in which one target range is determined by moving an adjustable pedestal, notch, or step along the baseline until it coincides with the horizontal position of the target-signal deflection; the control which moves the pedestal is calibrated in range. *Note:* The use of the term M-display is uncommon. This display is usually identified as a variant of an A-display.   13

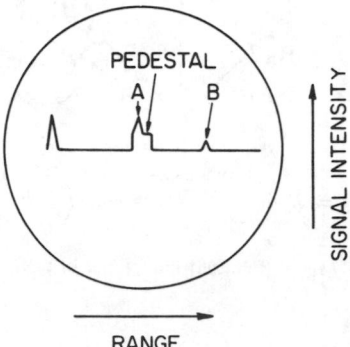

M-Display
(with pedestal)

MDS. *See:* **minimum detectable signal.**

MEA (electronic navigation). *See:* **minimum en-route altitude.**

mean charge (nonlinear capacitor). The arithmetic mean of the transferred charges corresponding to a particular capacitor voltage, as determined from a specified alternating charge characteristic. *See:* **nonlinear capacitor.**   191

mean-charge characteristic (nonlinear capacitor). The function relating mean charge to capacitor voltage. *Note:* Mean-charge characteristic is always single-valued. *See:* **nonlinear capacitor.**   191

mean first slip time (communication satellite). The mean time for a phase-lock loop starting to lock to a slip one or more cycles.   85

mean free path (acoustics). For sound waves in an enclosure, the average distance sound travels between successive reflections in the enclosure.   191, 176

mean Hall plate temperature (Hall effect devices). The value of the temperature averaged over the volume of the Hall plate.   107

mean horizontal intensity (candlepower) (illuminating engineering). The average intensity (candelas) of a lamp in a plane perpendicular to the axis of the lamp and which passes through the luminous center of the lamp.   167

mean life (reliability). The arithmetic mean of the times to failure of a group of nominally identical items. *See:* **reliability.**   182

mean life, assessed (non-repaired items) (reliability). The mean life of an item determined by a limiting value or values of the confidence interval associated with a stated confidence level, based on the same data as the observed mean life of nominally identical items. *Notes: (1)* The source of the data shall be stated. *(2)* Results can be accumulated (combined) only when all conditions are similar. *(3)* The assumed underlying distribution of failures against time shall be stated. *(4)* It should be stated whether a one-sided or a two-sided interval is being used. *(5)* Where one limiting value is given this is usually the lower limit.   164

mean life, extrapolated (non-repaired items) (reliability). Extension by a defined extrapolation or interpolation of the observed or assessed mean life for stress conditions different from those applying in the observed or assessed mean life. *Note:* The validity of the extrapolation shall be justified.   164

mean life, observed (non-repaired items) (reliability). The mean value of the lengths of observed times to failure of all items in a sample under stated conditions. *Note:* The criteria for what constitutes a failure shall be stated.   164

mean life, predicted (non-repaired items) (reliability). For the stated conditions of use, and taking into account the design of an item, the mean life computed from the observed, assessed or extrapolated mean life of its parts. *Note:* Engineering and statistical assumptions shall be stated, as well as the bases used for the computation (observed or assessed).   164

mean of reversed direct-current values (alternating-current instruments). The simple average of the indications when direct current is applied in one direction and then reversed and applied in the other direction. *See:* **accuracy rating (instrument).**   280

mean pulse time. The arithmetic mean of the leading-edge pulse time and the trailing-edge pulse time. *Note:* For some purposes, the importance of a pulse is that it exists (or is large enough) at a particular instant of time. For such applications the important quantity is the mean pulse time. The leading-edge pulse time and trailing-edge pulse time are significant primarily in that they may allow a certain tolerance in timing.   254

mean side lobe level (antennas). The average value of the relative power pattern of an antenna taken over a specified angular region, which excludes the main beam, relative to the peak of the main beam.   111

means of grounding (neutral grounding in electrical utility systems). The generic agent by which various degrees of grounding are achieved; for example, inductance grounding, resistance grounding, and resonant grounding.   591

mean spherical luminous intensity (illuminating engineering). Average value of the luminous intensity in all directions for a source. Also, the quotient of the total emitted luminous flux of the source by 4π.

$$I_{ms}=(1 \text{ over } 4\pi)\int_0^{\pi}I \; d\omega=\Phi_{total}/(4\pi).$$

164

mean temperature coefficient of output voltage (Hall effect devices). The arithmetic average of the percentage changes in output voltage per degree Celsius taken over a given temperature range for a given control current magnitude and a given magnetic flux density.   107

mean time between (or before) failures (MTBF) (power supplies). A measure of reliability giving either the time before first failure or, for repairable equipment, the average time between repairs. Mean time between (before) failures may be approximated or predicted by summing the reciprocal failure rates of individual components in an assembly.   186

mean time between failure (MTBF)(supervisory control, data acquisition, and automatic control)(station control and data acquisition). The time interval (hours) that may be expected between failures of an operating equipment.   570, 403

mean time between failures (1) (power system communications). (MTF). The average time (preferably expressed in hours) between failures of a continuously operating device, circuit, or system.   59

(2) (nuclear power generating stations) (MTBF). The arithmetic average of operating times between failures of an item.   29

(3) (repairable items) (reliability). The product of the number of items and their operating time divided by the total number of failures. *See:* **reliability.**   182

mean time between failures, assessed (repaired items) (reliability). The mean time between failures of an item determined by a limiting value or values of the confidence interval associated with a stated confidence level, based on the same data as the observed mean time between failures of nominally identical items. *Notes:* (1) The source of the data shall be stated. (2) Results can be accumulated (combined) only when all conditions are similar. (3) The assumed underlying distribution of failures against time shall be stated. (4) It should be stated whether a one-sided or a two-sided interval is being used. (5) Where one limiting value is given this is usually the lower limit.   164

mean time between failures, extrapolated (repaired items) (reliability). Extension by a defined extrapolation or interpolation of the observed or assessed mean time between failures for duration and.or conditions different from those applying to the observed or assessed mean time between failures. *Note:* The validity of the extrapolation shall be justified.   164

mean time between failures, observed (repaired items) (reliability). For a stated period in the life of an item, the mean value of the length of time between consecutive failures, computed as the ratio of the cumulative observed time to the number of failures under stated conditions. *Notes:* (1) The criteria for what constitutes a failure shall be stated. (2) Cumulative time is the sum of the times during which each individual item has been performing its required function under stated conditions. (3) This is the reciprocal of the observed failure rate during the period.   164

mean time between failures, predicted (repaired items) (reliability). For the stated conditions of use, and taking into account the design of an item, the mean time between failures computed from the observed, assessed, or extrapolated failure rates of its parts. *Note:* Engineering and statistical assumptions shall be stated, as well as the bases used for the computation (observed or assessed).   164

mean time to failure (nonrepaired items) (reliability). The total operating time of a number of items divided by the total number of failures. *See:* **reliability.**   182

mean time to failure, assessed (non-repaired items) (reliability). The mean time to failure of an item determined by a limiting value or values of the confidence interval associated with a stated confidence level, based on the same data as the observed mean time to failure of nominally identical items. *Notes:* (1) The source of the data shall be stated. (2) Results can be accumulated (combined) only when all conditions are similar. (3) The assumed underlying distribution of failures against time shall be stated. (4) It should be stated whether a one-sided or a two-sided interval is being used. (5) Where one limiting value is given this is usually the lower limit.   164

mean time to failure, extrapolated (non-repaired items) (reliability). Extension by a defined extrapolation or interpolation of the observed or assessed mean time to failure for durations and.or conditions different from those applying to the observed or assessed mean time to failure. *Note:* The validity of the extrapolation shall be justified. 164

mean time to failure, observed (non-repaired items) (reliability). For a stated period in the life of an item, the ratio of the cumulative time for a sample to the total number of failures in the sample during the period, under stated conditions. *Notes:* (1) The criteria for what constitutes a failure shall be stated. (2) Cumulative time is the sum of the times during which each individual item has been performing its required function under stated conditions. (3) This is the reciprocal of the observed failure rate during the period. 164

mean time to failure, predicted (non-repaired items) (reliability). For the stated conditions of use, and taking into account the design of an item the mean time to failure computed from the observed, assessed or extrapolated mean times to failure of its parts. *Note:* Engineering and statistical assumptions shall be stated, as well as the bases used for the computation (observed or assessed). 164

mean time to repair (MTTR) (1) (station control and data acquisition) (supervisory control, data acquisition, and automatic control). The time interval (hours) that may be expected to return a failed equipment to proper operation. 403,570
(2) (nuclear power generating stations). The arithmetic average of time required to complete a repair activity. 31, 29

mean zonal candlepower (illuminating engineering). The average intensity (candelas) of a symmetrical luminaire or lamp at an angle to the luminaire or lamp axis which is in the middle of the zone under consideration. 167

measurand (power switchgear). A physical or electrical quantity, property or condition that is to be measured. 103

measure. The number (real, complex, vector, etcetera) that expresses the ratio of the quantity to the unit used in measuring it. 210

measured current (rotating machinery). The total direct current resulting from the application of direct voltage to insulation and including the leakage current, the absorption current, and, theoretically, the capacitive current. Measured current is the value read on the microammeter during a direct high voltage test of insulation. 6

measured service (telephone switching systems). Service in which charges are assessed in terms of the number of message units during the billing interval. 55

measured signal (automatic null-balancing electric instrument). The electrical quantity applied to the measuring-circuit terminals of the instrument. *Note:* It is the electrical analog of the measured variable. *See:* **measurement system.** 295

measured value (power meters). An estimate of the value of a quantity obtained as a result of a measurement. *Note:* Indicated values may be corrected to give measured values. 115, 47

measured variable (measurand) (automatic null-balancing electric instrument). The physical quantity, property, or condition that is to be measured. *Note:* Common measured variables are temperature, pressure, thickness, speed, etcetera. *See:* **measurement system.** 295

measure equations. Equations in which the quantity symbols represent pure numbers, the measures of the physical quantities corresponding to the symbols. 210

measurement. The determination of the magnitude or amount of a quantity by comparison (direct or indirect) with the prototype standards of the system of units employed. *See:* **absolute measurement.** 210

measurement component. A general term applied to parts or subassemblies that are primarily used for the construction of measurement apparatus. *Note:* It is used to denote those parts made or selected specifically for measurement purposes and does not include standard screws, nuts, insulated wire, or other standard materials. *See:* **measurement system.** 328

measurement device. An assembly of one or more basic elements with other components and necessary parts to form a separate self-contained unit for performing one or more measurement operations. *Note:* It includes the protecting, supporting, and connecting, as well as the functioning, parts, all of which are necessary to fulfill the application requirements of the device. It should be noted that end devices (which see) are frequently but not always complete measurement devices in themselves, since they often are built-in with all or part of the intermediate means or primary detectors to form separate self-contained units. *See:* **measurement system.** 328

measurement energy. The energy required to operate a measurement device or system. *Note:* Measurement energy is normally obtained from the measurand or from the primary detector. *See:* **measurement system.** 328

measurement equipment. A general term applied to any assemblage of measurement components, devices, apparatus, or systems. *See:* **measurement system.** 328

measurement inverter (chopper). *See:* **measuring modulator.**

measurement mechanism. An assembly of basic elements and intermediate supporting parts for performing a mechanical operation in the sequence of measurement. *Note:* For example, it may be a group of components required to effect the proper motion of an indicating or recording means and does not include such parts as bases, covers, scales, and accessories. It may also be applied to a specific group of elements by substituting a suitable qualifying term: such as time-switch mechanism or chart-drive mechanism. *See:* **measurement system.** 328

measurement range (instrument). That part of the total range within which the requirements for accuracy are to be met. *See:* **instrument.** 328

measurement system. One or more measurement devices and any other necessary system elements interconnected to perform a complete measurement from the first operation to the end result. *Note:* A measurement system can be divided into general functional groupings, each of which consists of one or more specific functional steps or basic elements. 328

measurement uncertainty (test, measurement and diagnostic equipment). The limits of error about a measured value between which the true value will lie with the confidence stated. 54

measurement voltage divider (voltage ratio box) (volt box). A combination of two or more resistors, capacitors, or other circuit elements so arranged in series that the voltage across one of them is a definite and known fraction of the voltage applied to the combination, provided the current drain at the tap point is negligible or taken into account. *Note:* The term volt box is usually limited to resistance voltage dividers intended to extend the range of direct-current potentiometers. *See:* **auxiliary device to an instrument.** 328

measuring accuracy, precision, or reproducibility (numerically controlled machines). Accuracy, precision, or reproducibility of position sensor or transducer and interpreting system. 224, 207

measuring and test equipment (1)(nuclear power generating stations). Devices or systems used to calibrate, measure, gage, test or inspect in order to control or to acquire data to verify to specified requirements. 41

(2) (test, measurement and diagnostic equipment). *See:* **test, measurement and diagnostic equipment.** 54

measuring mechanism (recording instrument). The parts that produce and control the motion of the marking device. *See:* **moving element (instrument).** 294

measuring modulator (measurement inverter) (chopper). An intermediate means in a measurement system by which a direct-current or low-frequency alternating-current input is modulated to give a quantitatively related alternating-current output usually as a preliminary to amplification. *Note:* The modulator may be of any suitable type such as mechanical, magnetic, or varistor. The mechanical types, which are actuated by vibrating or rotating members, may be classified as (1) contacting, (2) microphonic, or (3) generating (capacitive or inductive). *See:* **auxiliary device to an instrument.** 328

measuring or test equipment (nuclear power quality assurance) (M&TE). Devices or systems used to calibrate, measure, gage, test, or inspect in order to control or to acquire data to verify conformance to specified requirements. 417

measuring units and relay logic (surge withstand capability tests). Analog or digital devices which analyze the input currents and voltages to determine the immediate status of that part of the power system that they were installed to protect and to provide the control signal to trip circuit breakers. 90

mechanical back-to-back test (rotating machinery). A test in which two identical machines are mechanically coupled together and the total losses of both machines are calculated from the difference between the electrical input to one machine and the electrical output of the other machine. *See:* **efficiency.** 63

mechanical bias. *See:* **relay mechanical bias.**

mechanical braking (industrial control). The kinetic energy of the drive motor and the driven machinery is dissipated by the friction of a mechanical brake. *See:* **electric drive.** 206

mechanical condition monitor (39) (power system device function numbers). A device that functions upon the occurrence of an abnormal mechanical condition (except that associated with bearings as covered under device function 38) [bearing protective device] , such as excessive vibration, eccentricity, expansion, shock, tilting, or seal failure. 402

mechanical current rating (neutral grounding device) (electric power). The symmetrical root-mean-square alternating-current component of the completely offset current wave that the device can withstand without mechanical failure. *Note:* The mechanical forces depend upon the maximum crest value of the current wave. However, for convenience, the mechanical current rating is expressed in terms of the root-mean-square value of the alternating-current component only of a completely displaced current wave. Specifically, the crest value of the completely offset wave will then be 2.82 times the mechanical current rating. *See:* **grounding device.** 91

mechanical cutter (mechanical recorder). An equipment for transforming electric or acoustic signals into mechanical motion of approximately like form and inscribing such motion in an appropriate medium by cutting or embossing. 176

mechanical data processing (computer applications). A method of data processing that involves the use of small, simple, mechanical machines. 571

mechanical fatigue test (rotating machinery). A test designed to determine the effect of a specific repeated mechanical load on the life of a component. *See:* **asynchronous machine.** 63

mechanical filter. *See:* **mechanical wave filter.**

mechanical freedom (accelerometer). The maximum linear or angular displacement of the accelerometer's proof mass relative to its case. 46

mechanical-hydraulic governor (hydraulic turbines). A governor in which the control signal proportional to speed error and necessary stabilizing and auxiliary signals are developed mechanically, summed by a mechanical system, and are then hydraulically amplified. 8

mechanical-impact strength (insulator). The impact that, under specified conditions, the insulator can withstand without damage. *See:* **insulator.** 261

mechanical impedance. The complex quotient of the effective alternating force applied to the system, by the resulting effective velocity. 176

mechanical inertia time (hydraulic turbines). A characteristic of the machine due to the inertia of the rotating components of the machine defined as:

$$T_M = \frac{(Wk^2)(N)^2(10^{-6})}{(1.61)(HP)}$$

where: W = weight of machine rotating parts, pounds; k = radius of gyration, feet; N = rated speed, rev/min; HP = rated output of turbine, horsepower. *Notes:* (1) T_M is also approximately equal to $2H$ where H is the inertia constant. (2) To calculate T_M using International SI units:

T_M = $J\omega^2_0/P_0$ where
J = the polar moment of inertia in kgm^2 calculated by dividing Wk^2 in newton-meters by acceleration of gravity 9.81 m/second^2.

$$= Mk^2 = GD^2$$

ω_0 = $\pi 2N/60$, rad/second
P_0 = rated ouput of turbine, watts.   8

mechanical interchangeability (of fuse links) (power switchgear). The characteristic that permits the designs of various manufacturers to be interchanged physically so they fit into and withstand the tensile stresses imposed by various types of prescribed cutouts made by different manufacturers.   103, 443

mechanical interlocking machine. An interlocking machine designed to operate the units mechanically. *See:* **interlocking (interlocking plant).**   328

mechanical latching relay. A relay in which the armature or contacts may be latched mechanically in the operated or unoperated position until reset manually or electrically.   341

mechanical limit (neutral grounding devices). The rated maximum instantaneous value of current, in amperes, that the device will withstand without mechanical failure.   91

mechanically delayed overcurrent trip. *See:* **mechanically delayed release (trip); overcurrent release (trip).**

mechanically delayed release (trip) (power switchgear). A release delayed by a mechanical device.   103

mechanically de-spun antenna (communication satellite). A rotating directional antenna, mounted to a rotating object (namely spin stabilized communication satellite): the rotation of the antenna is counter to the rotation of the body it is mounted to, such that the antenna beam points into the same direction of space.   83

mechanically release-free (trip-free)(as applied to a switching device) (power switchgear). A term indicating that the release can open the device even though (1) in a manually operated switching device the operating lever is being moved toward the closed position; or (2) in a power-operated switching device,

such as solenoid- or spring-actuated types, the operating mechanism is being moved toward the closed position either by continued application of closing power or by means of a maintenance closing lever.   103

mechanically reset relay. *See:* **hand-reset relay.**

mechanically timed relays. Relays that are timed mechanically by such features as clockwork, escapement, bellows, or dashpot. *See:* **relay.**   259

mechanical modulator (electronic navigation). (1) A device that varies some characteristic of a carrier wave so as to transmit information, the variation being accomplished by physically moving or changing a circuit element. (2) In instrument landing systems, a particular arrangement of radio-frequency transmission lines and bridges with resonant sections coupled to the lines and motor-driven capacitor plates that alter the resonance so as to produce 90- and 150-hertz modulations. *See:* **navigation.**   187, 13

mechanical operation (of a switch) (power switchgear). Operation by means of an operating mechanism connected to the switch by mechanical linkage. *Note:* Mechanically operated switches may be actuated either by manual, electrical, or other suitable means.   103

mechanical part (electric machine). Any part having no electric or magnetic function.   63

mechanical plating. Any plating operation in which the cathodes are moved mechanically during the deposition. *See:* **electroplating.**   328

mechanical recorder. *See:* **mechanical cutter.**

mechanical rectifier. A rectifier in which rectification is accomplished by mechanical action. *See:* **rectification.**   328

mechanical reproducer. *See:* **phonograph pickup.**

mechanical shock. A significant change in the position of a system in a nonperiodic manner in a relatively short time. *Note:* It is characterized by suddenness and large displacements and develops significant internal forces in the system. *See:* **shock motion.**   176

mechanical short-time current rating (current transformer). The root-mean-square value of the alternating-current component of a completely displaced primary current wave that the transformer is capable of withstanding, with secondary short-circuited. *Note:* Capable of withstanding means that after a test the current transformer shows no visible sign of distortion and is capable of meeting the other specified applicable requirements. *See:* **instrument transformer.**   203

mechanical splice (fiber optics). A fiber splice accomplished by fixtures or materials, rather than by thermal fusion. Index matching material may be applied between the two fiber ends. *See:* **fusion splice; index matching material; optical waveguide splice.**   433

mechanical switching device (power switchgear). A switching device designed to close and open one or more electric circuits by means of guided separable contacts. *Note:* The medium in which the contacts separate may be designated by suitable prefix; that is, air, gas, oil, etcetera.   103

mechanical terminal load (high voltage air switches, insulators, and bus supports). The external mechanical load at each terminal equivalent to the combined mechanical forces to which the air switch may be subjected. 575

mechanical time constant (critically damped indicating instrument). The period of free oscillation divided by 2π. *See:* **electromagnetic compatibility.** 220, 199

mechanical transmission system. An assembly of elements adapted for the transmission of mechanical power. *See:* **phonograph pickup.** 176

mechanical trip (trip arm) (railway practice). A roadway element consisting in part of a movable arm that in operative position engages apparatus on the vehicle to effect an application of the brakes by the train-control system. 328

mechanical wave filter (mechanical filter). A filter designed to separate mechanical waves of different frequencies. *Note:* Through electromechanical transducers, such a filter may be associated with electric circuits. *See:* **filter.** 328

mechanical wrap or connection (soldered connections). The securing of a wire or lead prior to soldering. 284

mechanism (1) (indicating instrument). The arrangement of parts for producing and controlling the motion of the indicating means. *Note:* It includes all the essential parts necessary to produce these results but does not include the base, cover, dial, or any parts, such as series resistors or shunts, whose function is to adapt the instrument to the quantity to be measured. *See:* **instrument; moving element (instrument).** 280

(2) (recording instrument). Includes (A) the arrangement for producing and controlling the motion of the marking device: (B) the marking device: (C) the device (clockwork, constant-speed motor, or equivalent) for driving the chart: (D) the parts necessary to carry the chart. *Note:* It includes all the essential parts necessary to produce these results but does not include the base, cover, indicating scale, chart, or any parts, such as series resistors or shunts, whose function is to make the recorded value of the measured quantity correspond to the actual value. *See:* **moving element (instrument).** 328

(3) (of a switching device) (power switchgear). The complete assembly of levers and other parts that actuates the moving contacts of a switching device. 103

media-independent information transfer (information transfer). Used as a general term to refer to any volume conforming to IEEE Trial Use Std 949. 479

median water conditions (power operations). Precipitation and runoff conditions which provide water for hydroelectric energy development approximating the median amount and distribution available over a long time period, usually the period of record. 516

medium (1) (computing systems). The material, or configuration thereof, on which data are recorded, for example, paper tape, cards, magnetic tape. 255, 77

(2) (information transfer). A vehicle capable of transferring data. 479

(3) (token ring access method). The material on which the data may be represented. Twisted pairs, coaxial cables, and optical fibers are examples of media. 472

medium access control (MAC) (1) (logical link control). That part of a data station that supports the medium access control functions that reside just below the logical link control sublayer. The MAC procedures included framing/deframing data units, performing error checking, and acquiring the right to use the underlying physical medium. 585

(2) (token ring access method). The portion of the IEEE 802 data station that controls and mediates the access to the ring. 472

medium attachment unit (MAU) (medium attachment units and repeater units). The portion of the physical layer between the medium dependent interface (MDI) and attachment unit interface (AUI) that interconnects the trunk coaxial cable to the branch cable and contains the electronics which send, receive, and manage the encoded signals impressed on, and recovered from the trunk coaxial cable. 543

medium dependent interface (medium attachment units and repeater units). The mechanical and electrical interface between the trunk cable medium and the medium attachment unit (MAU). *See:* **coaxial cable interface.** 543

medium interface connector (MIC) (token ring access method). The connector between the station and trunk coupling unit (TCU) at which all transmitted and received signals are specified. 472

medium noise (sound recording and reproducing system). The noise that can be specifically ascribed to the medium. *See:* **noise (sound recording and reproducing system).** 350

medium voltage (1) (cable systems in power generating stations). 601 to 15 000 V. 35

(2) (system voltage ratings). A class of nominal system voltages greater than 1000 and less than 100 000 volts. *See:* **low voltage; high voltage; nominal system voltage.** 260

medium-voltage aluminum-sheathed power cable (aluminum sheaths for power cables). Cable used in an electric system having a maximum phase-to-phase root-mean-square (rms) alternating-current (ac) voltage above 1000 volts (V) to 72 500 V, the cable having an aluminum sheath as a major component in its construction. 406

medium-voltage power cable (cable systems in power generating stations). Cables designed to supply power to utilization devices of the plant auxiliary system, operated at 601 to 15 000 V. 35

medium-voltage system (electric power for industrial and commercial systems only). An electric system having a maximum root-mean-square alternating-current voltage above 1000 volts to 72 500 volts. *See:* **voltage classes.** 49

mega (M) (mathematics of computing). (1) A prefix indicating one million. (2) In statements involving size of computer storage, a prefix indicating 2^{20}, or 1 048 576. 564

Meissner oscillator. An oscillator that includes an isolated tank circuit inductively coupled to the input and output circuits of an amplifying device to obtain the proper feedback and frequency. *See:* **oscillatory circuit.** 111

mel. A unit of pitch. By definition, a simple tone of frequency 1000 hertz, 40 decibels above a listener's threshold, produces a pitch of 1000 mels. *Note:* The pitch of any sound that is judged by the listener to be n times that of the 1-mel tone is n mels. 176

melting channel. The restricted portion of the charge in a submerged resistor or horizontal-ring induction furnace in which the induced currents are concentrated to effect high energy absorption and melting of the charge. *See:* **induction heating.** 14, 114

melting-speed ratio (of a fuse) (power switchgear). A ratio of the current magnitudes required to melt the current responsive element at two specified melting times. *Notes:* (1) Specification of the current wave shape is required for time less than one-tenth of a second. (2) The lower melting time in present use is 0.1 second, and the higher minimum melting current times are 100 seconds for low-voltage fuses and 300 or 600 seconds, whichever specified, for high-voltage fuses. 103

melting time (1)(of a fuse) (power switchgear). The time required for overcurrent to sever the current-responsive element. 103 **(2)(protection and coordination of industrial and commercial power systems).** The time required to melt the current-responsive element on a specified overcurrent. Where the fuse is current limiting in less than half-cycle, the melting time may be approximately half or less of the clearing time. *Syn:* **pre-arcing. See figure under arcing time.** 504

membrane potential. The potential difference, of whatever origin, between the two sides of a membrane. *See:* **electrobiology.** 192

memory (electronic computation). *See:* **storage; storage medium.**

memory action (of a relay). A method of retaining an effect of an input after the input ceases or is greatly reduced, so that this input can still be used in producing the typical response of the relay. *Note:* For example, memory action in a high-speed directional relay permits correct response for a brief period after the source of voltage input necessary to such a response is short-circuited. 103

memory capacity (electronic computation). *See:* **storage capacity.**

memory cycle (test, measurement and diagnostic equipment). The time required to read information from memory and replace it. 54

memory relay. (1) A relay having two or more coils, each of which may operate independent sets of contacts, and another set of contacts that remain in a position determined by the coil last energized. (2) Sometimes erroneously used for polarized relay. *See:* **relay.** 259

menu (computer applications). A list of options available for selection by the user of a computer system. *See:* **help menu; menu selection.** 571

menu selection (computer applications). (1) The process of choosing an item from a menu. (2) The item chosen from a menu. 571

mercury-arc converter, pool-cathode. *See:* **pool-cathode mercury-arc converter; oscillatory circuit.**

mercury-arc rectifier. A gas-filled rectifier tube in which the gas is mercury vapor. *See:* **rectification.** 244, 190

mercury cells. Electrolytic cells having mercury cathodes with which deposited metals form amalgams. 328

mercury-contact relays. (1) **mercury plunger relay:** A relay in which the magnetic attraction of a floating plunger by a field surrounding a sealed capsule displaces mercury in a pool to effect contacting between fixed electrodes. (2) **mercury-wetted-contact relay:** A form of reed relay in which the reeds and contacts are glass enclosed and are wetted by a film of mercury obtained by capillary action from a mercury pool in the base of a glass capsule vertically mounted. (3) **mercury-contact relay:** A relay mechanism in which mercury establishes contact between electrodes in a sealed capsule as a result of the capsule's being tilted by an electromagnetically actuated armature, either on pick-up or dropout or both. *See:* **mercury relay.** 341

mercury fluorescent lamp (illuminating engineering). An electric discharge lamp having a high-pressure mercury arc in an arc tube, and an outer envelope coated with a fluorescing substance (phosphor) which transforms some of the ultraviolet energy generated by the arc into light. 167

mercury-hydrogen spark-gap converter (dielectric heater usage). A spark-gap generator or power source which utilizes the oscillatory discharge of a capacitor through an inductor and a spark gap as a source of radio-frequency power. The spark gap comprises a solid electrode and a pool of mercury in a hydrogen atmosphere. *See:* **induction heating.** 14, 114

mercury lamp (illuminating engineering). A high intensity discharge (HID) lamp in which the major portion of the light is produced by radiation from mercury operating at a partial pressure in excess of $1.013 \times 10^5 Pa$(1 atmosphere). Includes clear, phosphor-coated (mercury-fluorescent), and self-ballasted lamps. 167

mercury-lamp ballast. *See:* **ballast.**

mercury-lamp transformer (series type). *See:* **constant-current (series) mercury-lamp transformer.**

mercury motor meter. A motor-type meter in which a portion of the rotor is immersed in mercury, which serves to direct the current through conducting portions of the rotor. *See:* **electricity meter.** 341

mercury oxide cell. A primary cell in which depolarization is accomplished by oxide of mercury. *See:* **electrochemistry.** 341

mercury-pool cathode (gas tube). A pool cathode consisting of mercury. 244, 190

mercury relay. A relay in which the movement of mercury opens and closes contacts. *See:* **mercury-contact relay.**   259

mercury vapor lamp transformers (multiple-supply type) (power and distribution transformer). Transformers, autotransformers, or reactors for operating mercury or metallic iodide vapor lamps for all types of lighting applications, including indoor, outdoor area, roadway, uviarc, and other process and specialized lighting.   53

mercury-vapor tube. A gas tube in which the active gas is mercury vapor.   125

merge (computing systems). To combine two or more sets of items into one, usually in a specified sequence.
                                                                255, 77

meridional ray (fiber optics). A ray that passes through the optical axis of an optical waveguide (in contrast with a skew ray, which does not). *See:* **axial ray; geometric optics; numerical aperture; optical axis; paraxial ray; skew ray.**   433

mesh. A set of branches forming a closed path in a network, provided that if any one branch is omitted from the set, the remaining branches of the set do not form a closed path. *Note:* The term **loop** is sometimes used in the sense of mesh. *See:* **network analysis.**
                                                                210

mesh-connected circuit. A polyphase circuit in which all the current paths of the circuit extend directly from the terminal of entry of one phase conductor to the terminal of entry of another phase conductor, without any intermediate interconnections among such paths and without any connection to the neutral conductor, if one exists. *Note:* In a three-phase system this is called the delta (or Δ) connection. *See:* **network analysis.**   210

mesh current. A current assumed to exist over all cross sections of a given closed path in a network. *Note:* A mesh current may be the total current in a branch included in the path, or it may be a partial current such that when combined with others the total current is obtained. *See:* **network analysis.**   210

mesh or loop equations. Any set of equations (of minimum number) such that the independent mesh or loop currents of a specified network may be determined from the impressed voltages. *Notes:* (1) For a given network, different sets of equations, equivalent to one another, may be obtained by different choices of mesh or loop currents. (2) The equations may be differential equations, or algebraic equations when impedances and phasor equivalents of steady-state single-frequency sine-wave quantities are used. *See:* **network analysis.**   210

mesial point (pulse terms). A magnitude referenced point at the intersection of a waveform and a mesial line. *See:* The single pulse diagram below the **waveform epoch** entry.   254

mesopic vision (illuminating engineering). Vision with fully adapted eyes at luminance conditions between those of photopic and scotopic vision, that is, between about 3.4 cd/m^2,(2.2 × 10^{-3} cd/in^2) (1.0 fL) and 0.034 cd/m^2, (2.2 × 10^{-5} cd/in^2) (0.01 fL).   167

message (telephone switching systems). An answered call or the information content thereof.   55

message source. That part of a communication system where messages are assumed to originate. *See:* **information theory.**   160

message switch (data transmission). A technique whereby messages are routed to the appropriate receiver by way of message address codes rather than by switching of the communication channel itself.
                                                                59

message switching (data communications). A method of handling messages over communications networks. The entire message is transmitted to an intermediate point (that is, a switching computer), stored for a period of time, perhaps very short, and then transmitted again towards its destination. The destination of each message is indicated by an address integral to the message. *See:* **circuit switching.**   12

message telecommunication network (telephone switching systems). An arrangement of switching and transmission facilities to provide telecommunication services to the public.   55

message-timed release (telephone switching systems). Release effected automatically after a measured interval of communication.   55

message unit (telephone switching systems). A basic chargeable unit based on the duration and destination of a call.   55

message-unit call (telephone switching systems). A call for which billing is in terms of accumulated message units.   55

metacompiler. *See:* **compiler generator.**

metalanguage (1)(ATLAS). A form of notation used to rigorously define the syntax and sometimes the semantics of another language.   400

(2)(software). A language used to specify a language or languages.   434

metal clad. The conducting parts are entirely enclosed in a metal casing.   328

metal-clad switchgear (1)(electric power distribution for industrial plants). Metal-enclosed power switchgear characterized by the following necessary features. (A) The main circuit switching and interrupting device is of the removable type arranged with a mechanism for moving it physically between connected and disconnected positions and equipped with self-aligning and self-coupling primary and secondary disconnecting devices. (B) Major parts of the primary circuit, such as the circuit switching or interrupting devices, buses, potential transformers, and control power transformers, are enclosed by grounded metal barriers. Specifically included is an inner barrier in front of or a part of the circuit interrupting device to ensure that no energized primary circuit components are exposed when the unit door is opened. (C) All live parts are enclosed within grounded metal compartments. Automatic shutters prevent exposure of primary circuit elements when the removable element is in the test, disconnected, or fully withdrawn position. (D) Primary bus conductors and connections are covered with insulating material throughout. For special configurations, insulated barriers between phases and between phase and ground may be specified. (E) Mechanical interlocks are provided to ensure a proper and safe operating sequence. (F) Instruments, meters, relays, secondary control devices, and their wiring are isolated by grounded metal barriers from all primary circuit elements with the exception of short lengths of wire, such as at instrument transformer terminals. (G) The door through which the circuit interrupting device is inserted into the housing may serve as an instrument or relay panel and may also provide access to a secondary or control compartment within the housing. *Notes:* (1) Auxiliary frames may be required for mounting associated auxiliary equipment, such as potential transformers, control power transformers, etcetera. (2) The term metal-clad switchgear can be properly used only if metal-enclosed switchgear conforms to the foregoing definition. All metal-clad switchgear is metal-enclosed, but not all metal-enclosed switchgear can be correctly designated as metal-clad. The most prevalent type of switching and interrupting device used in metal-clad switchgear is the air-magnetic power circuit breaker over 1000 volts (V).   510

(2)(metal-clad and station-type cubicle switchgear). Metal-enclosed power switchgear characterized by the following necessary features: (A) The main switching and interrupting device is of the removable (drawout) type arranged with a mechanism for moving it physically between connected and disconnected positions and equipped with self-aligning and self-coupling primary disconnecting devices and disconnectable control wiring connections. (B) Major parts of the primary circuit, that is the circuit switching or interrupting devices, buses, voltage transformers, and control power transformers, are completely enclosed by grounded metal barriers, that have no intentional openings between compartments. Specifically included is a metal barrier in front of or a part of the circuit interrupting device to ensure that, when in the connected position, no primary circuit components are exposed by the opening of a door. (C) All live parts are enclosed within grounded metal compartments. (D) Automatic shutters that cover primary circuit elements when the removable element is in the disconnected, test, or removed position. (E) Primary bus conductors and connections are covered with insulating materials throughout. (F) Mechanical interlocks are provided for proper operating sequence under normal operating conditions. (G) Instruments, meters, relays, secondary control devices and their wiring are isolated by grounded metal barriers from all primary circuit elements with the exception of short lengths of wire such as at instrument transformer terminals. (H) The door through which the circuit interrupting device is inserted into the housing may serve as an instrument or relay panel and may also provide access to a secondary or control compartment within the housing. *Notes:* (1) Auxiliary vertical sections may be required for mounting devices or for use as bus transition. (2) The term metal-clad (as applied to switchgear

assemblies) is correctly used only in connection with switchgear conforming fully to the definition of metal-clad switchgear given here. Metal-clad switchgear is metal-enclosed but not all metal-enclosed switchgear can be correctly designated as metal-clad.   572

(3)(power switchgear). Metal-enclosed power switchgear characterized by the following necessary features: (A) The main switching and interrupting device is of the moveable type arranged with a mechanism for moving it physically between connected and disconnected positions and equipped with self-aligning and self-coupling primary and secondary disconnecting devices. (B) Major parts of the primary circuit, that is, the circuit switching or interrupting devices, buses, voltage transformers, and control power transformers are completely enclosed by grounded metal barriers which have no intentional openings between compartments. Specifically included is a metal barrier in front of or a part of the circuit interrupting device to insure that, when in the connected position, no primary circuit components are exposed by the opening of a door. (C) All live parts are enclosed within grounded metal compartments. Automatic shutters prevent exposure of primary circuit elements when the removable element is in the disconnected, test, or removed position. (D) Primary bus conductors and connections are covered with insulating material throughout. (E) Mechanical interlocks are provided to ensure a proper and safe operating sequence. (F) Instruments, meters, relays, secondary control devices and their wiring are isolated by metal barriers from all primary circuit elements with the exception of short lengths of wire such as at instrument transformer terminals. (G) The door through which the circuit-interrupting device is inserted into the housing may serve as an instrument or relay panel and may also provide access to a secondary or control compartment within the housing. *Notes:* (1)Auxiliary vertical sections may be required for mounting devices or for use as bus transition. (2) The term metal-clad (as applied to switchgear assemblies) is correctly used only in connection with switchgear conforming fully to the definition above. Metal-clad switchgear is metal-enclosed but not all metal-enclosed switchgear can be correctly designated as metal-clad.   103

metal distribution ratio (electroplating). The ratio of the thicknesses (weights per unit areas) of metal upon two specified parts of a cathode. *See:* **electroplating.**   328

metal-enclosed (1)(as applied to a switchgear assembly or components thereof) (power switchgear). Surrounded by a metal case or housing, usually grounded.   103

(2)(as applied to metal-enclosed bus)(metal-enclosed bus and calculating losses in isolated-phase bus). Surrounded by a metal case or housing, with provisions for grounding.   574

metal-enclosed bus (1)(electric power distribution for industrial plants). An assembly of rigid electrical buses with associated connections, joints, and insulating supports, all housed within a grounded metal enclosure. Three basic types of metal-enclosed bus construction are recognized: nonsegregated phase, segregated phase, and isolated phase. The most prevalent type used in industrial power systems is the nonsegregated phase, which is defined as one in which all phase conductors are in a common metal enclosure without barriers between the phases. When metal-enclosed buses over 100 volts (V) are used with metal-clad switchgear, the bus conductors and connections are covered with insulating material throughout. When metal-enclosed buses are associated with metal-enclosed 1000 V and below power circuit breaker switchgear or metal-enclosed interrupter switchgear, the primary bus conductors and connections are usually bare. 510

(2)(generating station grounding). An assembly of rigid conductors with associated connections, joints, and insulating supports within a grounded metal enclosure. 569

(3)(metal-enclosed bus and calculating losses in isolated-phase bus). An assembly of conductors with associated connections, joints, and insulating supports within a grounded metal enclosure. The conductors may be either rigid or flexible. 574

(4) (power switchgear)(power systems relaying). An assembly of conductors with associated connections, joints, and insulating supports within a grounded metal enclosure. The conductors may be either rigid or flexible. *Note:* In general, three basic types of construction are used: nonsegregated-phase, segregated-phase, and isolated-phase. 103

metal-enclosed equipment (shunt power capacitors). A capacitor equipment assembly enclosed in a metal enclosure or metal house, usually grounded, to prevent accidental contact with live parts. *Syn:* **housed equipment.** 138

metal-enclosed interrupter switchgear (1)(electric power distribution for industrial plants). Metal-enclosed power switchgear including the following equipment as required: (A) Interrupter switches. (B) Power fuses. (C) Bare bus and connections. (D) Instrument and control power transformers. (E) Control wiring and accessory devices. The interrupter switches and power fuses may be of the stationary or removable type. For the removable type, mechanical interlocks are provided to ensure a proper and safe operating sequence. 510

(2)(metal-enclosed interrupter switchgear). Metal-enclosed power switchgear including the following equipment as required: (A) Interrupter switches. (B) Power fuses (current limiting or noncurrent limiting. (C) Bare bus and connections. (D) Instrument transformers. (E) Control wiring and accessory devices. The interrupter switches and power fuses may be stationary or removable (drawout) type. When removable type, automatic shutters that cover primary circuit elements when the removable element is in the disconnected, test, or removed position, and mechanical interlocks are to be provided for proper operating sequence. 573

(3)(power switchgear). Metal-enclosed power switchgear including the following equipment as required: (A) interrupter switches, (B) power fuses, (C) bare bus and connections, (D) instrument transformers, and (E) control wiring and accessory devices. The interrupter switches and power fuses may be stationary or removable type. When removable type, mechanical interlocks are provided to insure a proper and safe operating sequence. 103

metal-enclosed low-voltage power circuit-breaker switchgear (1)(metal-enclosed low-voltage power circuit-breaker switchgear). Low-voltage (LV) switchgear of multiple or individual enclosures, including the following equipment as required: (A) Low-voltage power circuit breakers (fused or unfused) in accordance with ANSI/IEEE C37.13-1981. (B) Bare bus and connections. (C) Instrument and control power transformers. (D) Instruments, meters, and relays. (E) Control wiring and accessory devices. The low-voltage power circuit breakers are contained in individual grounded metal compartments and controlled either remotely or from the front of the enclosure. The circuit breakers may be stationary or removable (drawout) type; when of removable type, mechanical interlocks are provided for proper operating sequence. 579

(2)(power switchgear). Metal-enclosed power switchgear of multiple or individual enclosure including the following equipment as required: (A) low-voltage power circuit breaker (fused or unfused), (B) bare bus and connections, (C) instrument and control power transformers, (D) instruments, meters, and relays, and (E) control wiring and accessory devices. The low-voltage power circuit breakers are contained in individual grounded metal compartments and controlled either remotely or from the front of the panels. The circuit breakers may be stationary or removable type. When removable type, mechanical interlocks are provided to insure a proper and safe operating sequence. 103

metal-enclosed power switchgear(1)(power switchgear)(metal-enclosed interrupter switchgear). A switchgear assembly completely enclosed on all sides and top with sheet metal (except for ventilating openings and inspection windows) containing primary power circuit switching or interrupting devices, or both, with buses and connections and may include control and auxiliary devices. Access to the interior of the enclosure is provided by doors or removable covers. *Note:* Metal-clad switchgear, station-type cubicle switchgear, metal-enclosed interrupter switchgear, and low-voltage power circuit breaker switchgear are specific types of metal-enclosed power switchgear. 103,573

(2)(metal-enclosed low-voltage power circuit-breaker switchgear)(metal-clad and station-type cubicle switchgear). A switchgear assembly completely enclosed on all sides and top with sheet metal (except for ventilating openings and inspection windows) containing primary power-circuit switching or interrupting devices or both, with buses and connections. The assembly may include control and auxiliary devices. Access to the interior of the enclosure is provided by doors or removable covers or both.   579, 572

metal-enclosed 1000 volts (V) and below power circuit breaker switchgear (electric power distribution for industrial plants). Metal-enclosed power switchgear, including the following equipment as required: (1) 1000 V and below power circuit breakers (fused or unfused. (2) Bare bus and connections. (3) Instrument and control power transformers. (4) Instruments, meters, relays. (5) Control wiring and accessory devices. (6) Cable and busway termination facilities. The 1000 V and below power circuit breakers are contained in individual grounded metal compartments and controlled either remotely or from the front of the panels. The circuit breakers are usually of the drawout type, but may be nondrawout. When drawout-type circuit breakers are used, mechanical interlocks must be provided to ensure a proper and safe operating sequence.   510

metal fog (metal mist) (electrolysis). A fine dispersion of metal in a fused electrolyte. *See:* **fused electrolyte.**   328

metal-graphite brush (rotating machinery). A brush composed of varying percentages of metal and graphite, copper or silver being the metal generally used. *Note:* This type of brush is soft. Grades of brushes of this type have extremely high current-carrying capacities, but differ greatly in operating speed from low to high. *See:* **brush (rotating machinery).**   63, 279

metal halide lamp (illuminating engineering). A high intensity discharge (HID) lamp in which the major portion of the light is produced by radiation of metal halides and their products of dissociation–possibly in combination with metallic vapors such as mercury. Includes clear and phosphor-coated lamps.   167

metallic (differential) signal (telephone loop performance). The metallic voltage is the algebraic difference between the voltages to ground in the two conductors (tip and ring). The metallic current is half the algebraic difference between the current in these conductors.   473

metallic circuit (1)(data transmission). A circuit of which the ground or earth forms no part.   59 **(2)(measuring longitudinal balance of telephone equipment operating in the voice band).** A circuit of which the ground (earth) forms no part.   529

metallic impedance (telephone equipment)(measuring longitudinal balance of telephone equipment operating in the voice band). Impedance presented by a metallic circuit at any given single frequency, at or across the terminals of one of its transmission ports.   39, 529

metallic longitudinal induction ratio. (M-L ratio). The ratio of the metallic-circuit current or noise-metallic arising in an exposed section of open wire telephone line, to the longitudinal-circuit current or noise-longitudinal in sigma. It is expressed in microamperes per milliampere or the equivalent. *See:* **noise-metallic; inductive coordination.**   328

metallic outer covering (electrical heat tracing for in-dustrial applications). A metal sheath or braid used to provide physical protection for heating cable and. in some cases, to provide an electrical ground path.   523

metallic rectifier (electric installations on shipboard). A metallic rectifier cell is a device consisting of a conductor and semiconductor forming a junction. The junction exhibits a difference in resistance to current flow in the two directions through the junction. This results in effective current flow in one direction only. A metallic rectifier stack is a single columnar structure of one or more metallic rectifier cells.   3

metallic rectifier cell. A device consisting of a conductor and a semiconductor forming a junction. *Notes:* (1) Synonymous with **metallic rectifying cells.** (2) Such cells conduct current in each direction but provide a rectifying action because of the large difference in resistance to current flow in the two directions. (3) A metallic rectifier stack is a single columnar structure of one or more metallic rectifier cells. *See:* **rectification.**   3

metallic rectifier stack assembly. The combination of one or more stacks consisting of all the rectifying elements used in one rectifying circuit. *See:* **rectification.**   328

metallic rectifier unit. An operative assembly of a metallic rectifier, or rectifiers, together with the rectifier auxiliaries, the rectifier transformers, and the essential switchgear. *See:* **rectification.**   328

metallic transmission port (telephone equipment) (measuring longitudinal balance of telephone equipment operating in the voice band). A place of access in the metallic transmission path of a device or network where energy may be supplied or withdrawn, or where the device or network variables may be measured. The terminals of such a port are sometimes referred to as the tip and ring terminals. *Note:* In any particular case, the transmission ports are determined by the way the device is used, and not by its structure alone.   39, 529

metallic transmission port (telephone equipment). A place of access in the metallic transmission path of a device or network where energy may be supplied or withdrawn, or where the device or network variables may be measured. The terminals of such a port are sometimes referred to as the tip and ring terminals. *Note:* In any particular case, the transmission ports are determined by the way the device is used, and not by its structure alone. *See:* **device; network.**   39

metallic voltage (telephone equipment)(measuring longitudinal balance of telephone equipment operating in the voice band). The voltage across a metallic circuit.   39, 529

metallized brush. *See:* **metal-graphite brush.**

metallized paper capacitor. A capacitor in which the dielectric is primarily paper and the electrodes are thin metallic coatings deposited thereon.   341

metallized screen (cathode-ray tube). A screen covered on its rear side (with respect to the electron gun) with metallic film, usually aluminized, transparent to electrons and with a high optical reflection factor, which passes on to the viewer a large part of the light emitted by the screen on the electron-gun side. *See:* **cathode-ray tubes.**   244, 190

metal master (metal negative) (no. 1 master) (disk recording) (electroacoustics). *See:* **original master.**

metal mist (electrolysis). *See:* **metal fog.**

metal negative (metal master) (no. 1 master) (disk recording) (electroacoustics). *See:* **original master.**

metal-nitride-oxide-semiconductor transistor (MNOS transistor). In analogy with the metal-oxide-semiconductor (MOS) transistor, this acronym derives from the layer sequence in the gate region of the IGFET, namely, Metal-Nitride-Oxide-Semiconductor: MNOS Memory Transistor. Usually it has a variable threshold voltage. Some devices with this layer sequence have fixed threshold voltages.   386

metal-oxide-semiconductor (MOS) transistor (metal-nitride-oxide field-effect transistor). A type of IGFET, referring specifically to the layer sequence in the gate region of the IGFET, namely, Metal-Oxide-Semiconductor.   386

metal-oxide surge arrester (MOSA)(metal-oxide surge arresters for ac power circuits). A surge arrester utilizing valve elements fabricated from nonlinear resistance metal-oxide materials.   583

metalworking machine tool (National Electrical Code). A power-driven machine not portable by hand, used to shape or form metal by cutting, impact, pressure, electrical techniques, or a combination of these processes.   256

metamers (illuminating engineering). Lights of the same color but of different spectral energy distribution. *Note:* The term 'metamers' is also used to denote objects which when illuminated by a given source and viewed by a given observer produce metameric lights.   167

meter (m) (1) (laser-maser) (m). A unit of length in the international systems of units: currently defined as a fixed number of wavelengths, in vacuum, of the orange-red line of the spectrum of krypton 86. Typically, the meter is sub-divided into the following units: Centimeter 10^{-2} m(cm) Millimeter 10^{-3} m(mm) Micrometer 10^{-6} m(grkmm) Nanometer 10^{-9} m(nm)   363

See: **demand meter; electricity meter; watthour meter.**

(2)(metric practice) meter (m). A unit of length in the international system of units; currently defined as a fixed number of wavelengths, in vacuum, of the orange-red line of the spectrum of krypton 86. Typically, the meter is subdivided into the following units:

$$\text{centimeter} = 10^{-2}\text{m(cm)}$$
$$\text{millimeter} = 10^{-3}\text{m(mm)}$$
$$\text{micrometer} = 10^{-6}\text{m}(\mu\text{m})$$
$$\text{nanometer} = 10^{-9}\text{m(nm)}$$

21

meter installation inspection (metering). Examination of the meter, auxiliary devices, connections, and surrounding conditions, for the purpose of discovering mechanical defects or conditions that are likely to be detrimental to the accuracy of the installation. Such an examination may or may not include an approximate determination of the percentage registration of the meter.   212

meter laboratory. *See:* **laboratory (1) meter.**

meter relay. Sometimes used for instrument relay. *See:* **relay.**   259

meter shop. A place where meters are inspected, repaired, tested, and adjusted.   212

meter socket (socket)(watthour meter sockets). An enclosure which has matching jaws to accommodate the bayonet-type (blade) terminals of a detachable watthour meter and has a means of connections for the termination of the circuit conductors. It may be a single-position socket for one meter or a multiposition trough socket for two or more meters.   549

meter support (watthour meter sockets). That part of a ringless-type meter socket which positions and supports a detachable watthour meter.   549

method of pulse measurement. A method of making a pulse measurement comprises: the complete specification of the functional characteristics of the devices, apparatus, instruments, and auxiliary equipment to be used: the essential adjustments required: the procedure to be used in making essential adjustments: the operations to be performed and their sequence: the corrections that will ordinarily need to be made: the procedures for making such corrections: the conditions under which all operations are to be carried out. *See:* **pulse measurement.**   15

methods or types of grounding (neutral grounding in electrical utility systems). The equipment, procedure, or scheme used for attaining the particular means.   591

micro (μ)(mathematics of computing). A prefix indicating one millionth.   564

metrology (test, measurement and diagnostic equipment). The science of measurement for determination of conformance to technical requirements including the development of standards and systems for absolute and relative measurements.   54

MEW. *See:* **microwave early warning.**

MF. *See:* **radio spectrum.**

mho (siemens). The unit of conductance (and of admittance) in the International System of Units (SI). The mho is the conductance of a conductor such that a constant voltage of 1 volt between its ends produces a current of 1 ampere in it.   210

mho relay (power switchgear). A distance relay for which the inherent operating characteristic on an R-X diagram is a circle which passes through the origin. *Note:* The operating characteristics may be described by the equation $Z = K\cos(\theta - \alpha)$ where K and α are constants and θ is the phase angle by which the input voltage leads the input current. *See:* **distance relay; figure (b).**   103

MIC (electromagnetic compatibility). *See:* **mutual interference chart.**

mica flake (rotating machinery). Mica lamina in thickness not over approximately 0.0028 centimeter having a surface area parallel to the cleavage plane under 1.0 centimeter square. *See:* **rotor (rotating machinery); stator.** 63

mica folium (rotating machinery). A relatively thin flexible bonded sheet material composed of overlapping mica splittings with or without backing or facing. *See:* **rotor (rotating machinery); stator.** 63

mica paper (integrated mica) (reconstituted mica) (rotating machinery). Mica flakes having an area under approximately 0.200 centimeter square combined laminarly into a substantial sheet-like configuration with or without binder, backing, or facing. *See:* **rotor (rotating machinery); stator.** 63

mica sheet (rotating machinery). A composite of overlapping mica splittings bonded into a planar structure with or without backing or facing. *See:* **rotor (rotating machinery); stator.** 63

mica splitting (rotating machinery). Mica lamina in thickness approximately 0.0015 centimeter to 0.0028 centimeter having a surface area parallel to the basal cleavage plane of at least 1.0 centimeter square. *See:* **rotor (rotating machinery); stator.** 63

mica tape (rotating machinery). A composite tape composed of overlapping mica splittings bonded together with or without backing or facing. *See:* **rotor (rotating machinery); stator.** 63

MICR. *See:* **magnetic ink character recognition.**

microbar. A unit of pressure formerly in common usage in acoustics. One microbar is equal to 1 dyne per square centimeter and equals 0.1 newton per square meter. The newton per square meter is now the preferred unit. *Note:* The term bar properly denotes a pressure of 10^6 dynes per square centimeter. Unfortunately the bar was once used in acoustics to mean 1 dyne per square centimeter, but this is no longer correct. 176

microbending (fiber optics). In an optical waveguide, sharp curvatures involving local axial displacements of a few micrometers and spatial wavelengths of a few millimeters. Such bends may result from waveguide coating, cabling, packaging, installation, etcetera. *Note:* Microbending can cause significant radiative losses and mode coupling. *See:* **macrobending.** 433

microbend loss (fiber optics). In an optical waveguide, that loss attributable to microbending. *See:* **microbend loss.** 433

microchannel plate (electron image tube). An array of small aligned channel multipliers usually used for intensification. *See:* **amplifier; camera tube.** 190

microcode (software). (1) A symbolic representation of a microprogram. (2) The internal representation of a microprogram in its storage medium. *See:* **firmware; microprogram.** 434

microelectronic device (electric and electronics parts and equipments). An item of inseparable parts and hybrid circuits, usually produced by integrated circuit techniques. Typical examples are microcircuit, integrated circuit package, micromodule. 17

micrographics (computer applications). That branch of science and technology concerned with methods and techniques for converting information to or from microform. *See:* **office automation.** 571

micrometer (μm) (laser-maser). A unit of length equal to 10^{-6} m. In common practice a micrometer is a micron. 363

micron (metric system). The millionth part of a meter. *Note:* According to the set of submultiple prefixes now established in the International System of Units, the preferred term would be micrometer. However, use of the same word to denote a small length, and also to denote an instrument for measuring a small length, could occasionally invite confusion. Therefore it seems unwise to deprecate, at this time, the continued use of the word micron. 415

microphone. An electroacoustic transducer that responds to sound waves and delivers essentially equivalent electric waves. 176

microphonics (1) (general). The noise caused by mechanical shock or vibration of elements in a system. 239

(2) (interference terminology). Electrical interference caused by mechanical vibration of elements in a signal transmission system. *See:* **signal.** 188

(3) (microphonic effect) (microphonism) (electron device) (electron tubes). The undesired modulation of one or more of the electrode currents resulting from the mechanical vibration of one or more of the valve or tube elements. *See:* **electron tube.** 125

microphonism. *See:* **microphonics (electron tubes).**

microprogram (software). A sequence of elementary instructions that corresponds to a computer operation, that is maintained in special storage, and whose execution is initiated by the introduction of a computer instruction into an instruction register of a computer. Microprograms are often used in place of hard-wired logic. *See:* **computer; execution; firmware; instruction.** 434

micropulsation (radio wave propagation). Small amplitude fluctuations (usually less than 10^{-7} T) of the earth's magnetic field with periods usually on the order of seconds or minutes. These fluctuations usually result from magneto-hydrodynamic waves in the magnetosphere. 146

microradiometer (radio-micrometer). A thermosensitive detector of radiant power in which a thermopile is supported on and connected directly to the moving coil of a galvanometer. *Note:* This construction minimizes lead losses and stray electric pickup. *See:* **electric thermometer (temperature meter).** 328

microspark (overhead-power-line corona and radio noise). A spark breakdown occurring in the miniature air gap formed by two conducting or insulating surfaces. (This is sometimes called a gap discharge). 411

microstrip. *See:* **strip-type transmission line; (waveguides).**

microstrip antenna. An antenna which consists of a thin metallic conductor bonded to a thin grounded dielectric substrate. *Note:* The metallic conductor typically has some regular shape, for example, rectangular, circular, or elliptical. Feeding is often by means of a coaxial probe or a microstrip transmission line. 111

microstrip array (antennas). An array of microstrip antennas. 111

microstrip dipole (antennas). A microstrip antenna of rectangular shape with its width much smaller than its length. 111

microsyn (gyro, accelerometer) (inertial sensor). An electromagnetic device used as a pickoff or torquer in single-degree-of-freedom gyros and accelerometers. It has a stator, fastened to the sensor case, containing primary and secondary sets of windings, and a rotor, without windings, which is attached to the float. The operational mode is determined by the configuration of the stator windings and the method of alternating-current (ac) excitation. The device produces either an output voltage proportional to the rotational displacement of the rotor from its null or reference position; for example, movement of float with respect to case (pickoff), or torque on the rotor that is a function of the currents through the primary and secondary windings of the stator (torquer). 46

microwave early warning (MEW). A United States high-power, long-distance radar of the World War II era with a number of separate displays giving high resolution and large traffic-handling capacity in detecting and tracking targets. 13

microwave landing system (MLS) (radar). An airfield approach radar generating a guideline for target landing. 13

microwave-pilot protection (power switchgear). A form of pilot protection in which the communication means between relays is a beamed microwave radio channel. 103

microwaves (data transmission). A term used rather loosely to signify radio waves in the frequency range from about 1000 megahertz (mHz) upwards. 59

microwave therapy. The therapeutic use of electromagnetic energy to generate heat within the body, the frequency being greater than 100 megahertz. *See:* **electrotherapy.** 192

middle marker. A marker facility in an ILS (instrument landing system) which is installed approximately 1000 meters (3500 feet) from the approach end of the runway on the localizer course line to provide a fix. 13

mid-peak period (watthour meters). The period of time during which the specified mid-peak rate applies. 485

MIG. *See:* **magnetic injection gun.**

MIIT. *See:* **media-independent information transfer.**

mild environment (nuclear power generating station). An environment expected as a result of normal service conditions and extremes (abnormal) in service conditions where seismic is the only design basis event (DBE) of consequence. 120

mile of standard cable (MSC). Two units, both loosely designated as a mile of standard cable, were formerly used as measures of transmission efficiency. One, correctly known as an 800-hertz mile of standard cable, signified an attenuation constant, independent of frequency, of 0.109. The other signified the effect upon speech volume of an actual mile of standard cable, equivalent to an attenuation constant of approximately 0.122. Both units are now obsolete, having been replaced by the decibel. One 800-hertz mile of standard cable is equal to approximately 0.95 decibel. One standard cable mile is equivalent in effect on speech volume to approximately 1.06 decibels. 328

milestone (software). A scheduled event for which some project member or manager is held accountable and which is used to measure progress, for example, a formal review, issuance of a specification, product delivery. *See:* **delivery; specification.** 434

milli (m)(mathematics of computing). A prefix indicating one thousandth. 564

milliroentgen (mR)(X-radiation limits for ac high-voltage power vacuum interrupters used in power switchgear). The amount of X-radiation which produces 2.58×10^{-7} coulomb per kilogram of air. 553

mill scale (corrosion). The heavy oxide layer formed during hot fabrication or heat treatment of metals. Especially applied to iron and steel. 205

Mills cross antenna system. A multiplicative array antenna system consisting of two linear receiving arrays positioned at right angles to one another and connected together by a phase modulator or switch such that the effective angular response of the output is related to the product of the radiation patterns of the two arrays. 111

mimic bus (power switchgear). A single-line diagram of the main connections of a system constructed on the face of a switchgear or control panel, or assembly. 103

mine-fan signal system. A system that indicates by electric light or electric audible signal, or both, the slowing down or stopping of a mine ventilating fan. *See:* **dispatching system.** 328

mine feeder circuit. A conductor or group of conductors, including feeder and sectionalizing switches or circuit breakers, installed in mine entries or gangways and extending to the limits set for permanent mine wiring beyond which limits portable cables are used. 328

mine hoist. A device for raising or lowering ore, rock, or coal from a mine and for lowering and raising men and supplies.

mine jeep. a special electrically driven car for underground transportation of officials, inspectors, repair, maintenance, surveying crews, and rescue workers. 328

mine radio telephone system. A means to provide communication between the dispatcher and the operators on the locomotives where the radio impulses pass along the trolley wire and down the trolley pole to the radio telephone set. *See:* **dispatching system.** 328

mineral-insulated, metal sheathed cable (National Electrical Code). A factory assembly of one or more conductors insulated with a highly compressed refractory mineral insulation and enclosed in a liquidtight and gastight continuous copper sheath. 256

miner's electric cap lamp. A lamp for mounting on the miner's cap and receiving electric energy through a cord that connects the lamp with a small battery.
328

miner's hand lamp. A self-contained mine lamp with handle for convenience in carrying. 328

mine tractor. A trackless, self-propelled vehicle used to transport equipment and supplies and for general service work. 328

mine ventilating fan. A motor-driven disk, propeller, or wheel for blowing (or exhausting) air to provide ventilation of a mine. 328

miniature brush (electric machines). A brush having a cross-sectional area of less than 1/64 square inch with the thickness and width thereof less than 1/8 inch or, in the case of a cylindrical brush, a diameter less than 1/8 inch. *See:* **brush.** 279

minimal perceptible erythema. The erythemal threshold. *See:* **ultraviolet radiation.** 167

minimum access code (test, measurement and diagnostic equipment). A system of coding which minimizes the effect of delays for transfer of data or instructions between storage and other machine units. 54

minimum clearance between poles (phases) (power switchgear). The shortest distance between any live parts of adjacent poles (phases). *Note:* Cautionary differentiation should be made between clearance and spacing or center-to-center distance. 103

minimum clearance to ground (power switchgear). The shortest distance between any live part and adjacent grounded parts. 103

minimum conductance function (linear passive networks). *See:* **minimum resistance (conductance) function.**

minimum detectable signal (MDS) (radar). The minimum signal level which gives reliable detection in the presence of white Gaussian noise. Being a statistical quantity, it must be described in terms of a probability of detection and a probability of false alarm. 13

minimum discernible signal (radar). The minimum detectable signal for a system using an operator and display or aural device for detection. 13

minimum-distance code (computing systems). A binary code in which the signal distance does not fall below a specified minimum value. 255, 77

minimum-driving-point function (linear passive networks). A driving-point function that is a minimum-resistance, minimum-conductance, minimum-reactance, and minimum-susceptance function. 238

minimum en-route altitude (MEA) (electronic navigation). The lowest altitude between radio fixes that assures acceptable navigational signal coverage and meets obstruction clearance requirements for instrument flight. *See:* **navigation.** 187, 13

minimum firing power (microwave switching tubes). The minimum radio-frequency power required to initiate a radio-frequency discharge in the tube at a specified ignitor current. *See:* **gas tube.** 125

minimum flashover voltage (impulse). The crest value of the lowest voltage impulse, of a given wave shape and polarity that causes flashover. 64, 62

minimum fuel limiter (gas turbines). A device by means of which the speed-governing system can be prevented from reducing the fuel flow below the minimum for which the device is set as required to prevent unstable combustion or blowout of the flame.
58

minimum illumination (sensitivity) (industrial control). The minimum level, in footcandles of a photoelectric lighting control, at which it will operate. *See:* **photoelectric control.** 206

minimum impulse flashover voltage (neutral grounding devices). The crest value of the lowest voltage impulse at a given wave shape and polarity that causes flashover. 91

minimum input shaft torque (electric coupling). The minimum input torque required to drive an electric coupling with zero output torque load, either with or without rated excitation as specified. 416

minimum melting current (power switchgear). (1) (of a fuse). The smallest current at which a current-responsive fuse element will melt. (2) (high-voltage switchgear). The smallest current at which a current responsive fuse element will melt at any specified time. 103

minimum ON-state voltage (thyristor). The minimum positive principal voltage for which the differential resistance is zero with the gate open-circuited. *See:* **principal voltage-current characteristic (principal characteristic).** 191

minimum ON voltage (magnetic amplifier). The minimum output voltage existing before the trip OFF control signal is reached as the control signal is varied from trip ON to trip OFF. 171

minimum output voltage (magnetic amplifier). The minimum voltage attained across the rated load impedance as the control ampere-turns are varied between the limits established by (1) positive maximum control currents flowing through all the corresponding control windings simultaneously and (2) negative maximum control currents flowing through all the corresponding control windings simultaneously.
171

minimum perceptible erythema (MPE) (illuminating engineering). The erythemal threshold. 167

minimum-phase function (linear passive networks). A transmittance from which a nontrivial realizable all-pass function cannot be factored without leaving a nonrealizable remainder. *Note:* For lumped-parameter networks, this is equivalent to specifying that the function has no zeros in the interior of the right half of the complex-frequency plane. 238

minimum phase network (1) (data transmission). A network for which the phase shift at each frequency equals the minimum value which is determined uniquely by the attenuation-frequency characteristic in accordance with the following equation:

$$B_c = \frac{1}{\pi} \int_{-\infty}^{+\infty} \frac{\mathrm{d}A}{\mathrm{d}u} \log \coth \frac{|u|}{2} \, \mathrm{d}u$$

where B_c is phase shift (radians) at a particular frequency f_c, A is attenuation (nepers) as a function of frequency f, and u is $\log(f/f_c)$. *Note:* A ladder network employing lumped impedances, with no coupling between the branches, is an example of a minimum phase network. A bridged T or lattice network of the all-pass type is a nonminimum phase network.   59

(2) (excitation control systems). See definition above. *Notes:* (A) A network for which the transfer function expressed as a function of s has neither poles nor zeros in the right-hand s plane. Networks (elements) or systems having either poles or zeros in the right half s-plane do not have minimum phase characteristics, assuming there is no right-hand S-plane pole-zero cancellation. (B) Elements whose response is described by transfer functions having transport lags also exhibit non-minimum phase characteristics. The frequency response characteristics of a typical element with transport lag is given in figure below.   353

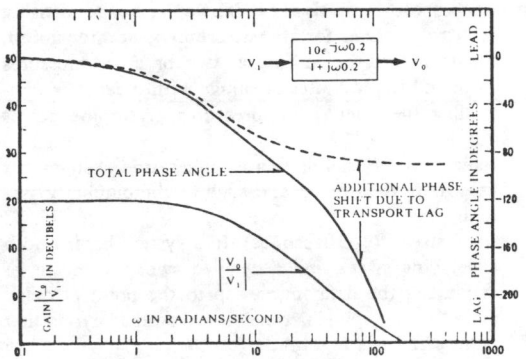

Element with one time constant and a transport lag.

(3) (two specified terminals or two branches). A network for which the transfer admittance expressed as a function of p (*See:* **impedance function**) has neither poles nor zeros in the right-hand p plane. *Note:* A simple T section of real lumped constant parameters without coupling between branches is a minimum-phase network, whereas a bridged-T or lattice section of all-pass type may not be. *See:* **network analysis; impedance function.**   210

minimum pulse down time (FASTBUS acquisition and control). In order for all devices to detect the zero state of a signal between the occurence of two successive one states of the signal, the zero state must last for at least a bus dependent minimum pulse down time.   480

minimum-reactance function (linear passive networks). A driving-point impedance from which a reactance function cannot be subtracted without leaving a nonrealizable remainder. *Notes:* (1) For lumped-parameter networks, this is equivalent to specifying that the impedance function has no poles on the

imaginary axis of the complex-frequency plane, including the point at infinity. (2) A driving-point impedance (admittance) having neither poles nor zeros on the imaginary axis is both a minimum-reactance and a minimum-susceptance function.   238

minimum reception altitude (MRA) (electronic navigation). The lowest en-route altitude at which adequate signals can be received to determine specific radio-navigation fixes. *See:* **navigation.**   187, 13

minimum-resistance (conductance) function (linear passive networks). A driving-point impedance (admittance) from which a positive constant cannot be subtracted without leaving a nonrealizable remainder.   238

minimum single-conductor (or subconductor) gradient (overhead-power-line corona and radio noise). The minimum value attained by the gradient $E\theta$ as given in maximum single-conductor (or subconductor) gradient as θ varies over the range 0 to 2π.   411

minimum speed (adjustable-speed drive) (industrial control). The lowest speed within the operating speed range of the drive. *See:* **electric drive.**   206

minimum-susceptance function (linear passive networks). A driving-point admittance from which a susceptance function cannot be subtracted without leaving a nonrealizable remainder. *Notes:* (1) For lumped-parameter networks, this is equivalent to specifying that the admittance function has no poles on the imaginary axis of the complex-frequency plane, including the point at infinity. (2) A driving-point immitance having neither poles nor zeros on the imaginary axis is both a minimum-susceptance and a minimum-reactance function.   238

minimum tasks (software verification and validation plans). Those verification and validation (V&V) tasks applicable to all projects. V&V planning for critical software includes all such tasks; these tasks are recommended for the V&V of noncritical software.   511

minimum test output voltage (nonreversible output) (magnetic amplifier). The output voltage equivalent to the summation of the minimum output voltage plus 33 1/3 percent of the difference between the rated and minimum output voltages.   171

minitrack (communication satellite). A ground based tracking system for satellites using interferometers. It requires a minimum satellite instrumentation, hence the name.   83

minor alarm, (telephone switching systems). An alarm indicating trouble which does not seriously impair the system capability.   55

minor cycle (electronic computation). In a storage device that provides serial access to storage positions, the time interval between the appearance of corresponding parts of successive words.   235, 210

minor failure. *See:* **failure, minor.**

minority carrier (semiconductor). The type of charge carrier constituting less than one half the total charge-carrier concentration. *See:* **semiconductor.**   245

minor lobe (antennas). Any radiation lobe except a major lobe.   111

minor loop (industrial control). A continuous network consisting of both forward elements and feedback elements and is only a portion of the feedback control system. *See:* **control system, feedback.**   206

minor railway tracks. Railway tracks included in the following list. (1) Spurs less than 2000 feet long and not exceeding two tracks in the same span. (2) Branches on which no regular service is maintained or which are not operated during the winter season. (3) Narrow-gauge tracks or other tracks on which standard rolling stock cannot, for physical reasons, be operated. (4) Tracks used only temporarily for a period not exceeding 1 year. (5) Tracks not operated as a common carrier, such as industrial railways used in logging, mining, etcetera.   262

minus input. *See:* **inverted input.**

misalignment drift (gyros). The part of the total apparent drift component due to uncertainty of orientation of the gyro input axis with respect to the coordinate system in which the gyro is being used. *See:* **navigation.**   187, 13

misalignment loss. *See:* **angular misalignment loss; gap loss; lateral offset loss.**

miscellaneous function (numerically controlled machines). An on-off function of a machine such as spindle stop, coolant on, clamp.   224, 207

misfire (1) (gas tube). A failure to establish an arc between the main anode and cathode during a scheduled conducting period. *See:* **gas tube; rectification.**   190, 190

(2) (mining). The failure of a blasting charge to explode when expected. *Note:* In electric firing, this usually is the result of a broken blasting circuit or insufficient current through the electric blasting cap. *See:* **blasting unit.**   328

mismatch. The condition in which the impedance of a load does not match the impedance of the source to which it is connected. *See:* **self-impedance.**   179

mismatch factors (power meters). Resulting from a combination of interaction factor and reflection factor resulting from reflective source and load impedances which relate incident, absorbed, and delivered power to a nonreflection load.   47

mismatch loss. *See:* **matching loss.**   13

mismatch uncertainty (power meters). Uncertainty in an assigned value that is caused by uncorrected or uncertain values for one or both of the mismatch factors.   115, 47

missing (misgating) (thyristor converter). A condition where the onset of conduction of an arm is substantially delayed from its correct instant of time. *Note:* If an arm fails to turn on during inverter service, there is a commutation failure resulting in a conduction-through.   121

mission (1) (systems, man, and cybernetics). The operating objective for which the system was intended. *See:* **system.**   209

(2) (nuclear power generating stations). The singular objective, task, or purpose of an item or system.   29

mission time (nuclear power generating stations). The time during which the mission must be performed without interruption.   29

mistake (1) (electronic computation). A human action that produces an unintended result. *Note:* Common mistakes include incorrect programming, coding, manual operation, etcetera.   255, 77, 235

(2) (analog computer). *See:* **error.**   9

mistrigger (misfire) (thyristor). The failure of a thyristor to conduct at the correct instant of time.   445

mixed highs (color television). Those high-frequency components of the picture signal that are intended to be reproduced achromatically in a color picture.   18

mixed logic (logic diagram). The defining of the 1-state of the variables as the more positive or less positive of the two possible levels, depending upon the absence or presence of the polarity indicator symbol. *Syn:* **direct polarity indication.**   88

mixed-loop series street-lighting system. A street-lighting system that comprises both open loops and closed loops. *See:* **alternate-current distribution; direct-current distribution.**   64

mixed-pressure turbine, condensing or noncondensing (control systems for steam turbine-generator units). Steam enters the turbine at two or more pressures through separate inlet openings with means for controlling the inlet steam pressures or turbine power output.   522

mixed radix. Pertaining to a numeration system that uses more than one radix, such as the biquinary system.   255, 77

mixed sweep (oscilloscopes). In a system having both a delaying sweep and a delayed sweep, a means of displaying the delaying sweep to the point of delay pickoff and displaying the delayed sweep beyond that point. *See:* **oscillograph.**   185

mixer (data transmission). (1) In a sound transmission, recording or reproducing system, a device having two or more inputs, usually adjustable, and a common output, which operates to combine linearly in a desired proportion the separate input signals to produce an output signal; (2) the stage in a heterodyne receiver in which the incoming signal is modulated with the signal from the local oscillator to produce the intermediate-frequency signal; (3) a process of intermingling of data traffic flowing between concentration and expansion stages.   59

mixer tube. An electron tube that performs only the frequency-conversion function of a heterodyne conversion transducer when it is supplied with voltage or power from an external oscillator.   125

MKSA system of units. A system in which the basic units are the meter, kilogram, and second, and the ampere is a derived unit defined by assigning the magnitude $4\pi \times 10^{-7}$ to the rationalized magnetic constant (sometimes called the permeability of space). *Notes:* (1) At its meeting in 1950 the International Electrotechnical Commission recommended that the MKSA system be used only in the rationalized form. (2) The electrical units of this system were formerly called the *practical* electrical units. (3) If the MKSA system is used in the unrationalized form the magnetic constant is 10^{-7} henry/meter and the electric constant is $10^7/c^2$ farads/meter. Here c, the speed of light, is approximately 3×10^8 meters.second. (4) In this system, dimensional analysis is customarily used with the four independent (basic) dimensions: mass, length, time, current.   210

M-L ratio. *See:* **metallic-longitudinal induction ratio.**

MLS. *See:* **microwave landing system.**

mnemonic (test, measurement and diagnostic equipment). Assisting or intending to assist a human memory and understanding. Thus a mnemonic term is usually an abbreviation, that is easy to remember: for example, mpy for multiply and acc for accumulator.
                                                         54

mnemonic code (test, measurement and diagnostic equipment). A pseudo code in which information, usually instructions, is represented by symbols or characters which are readily identified with the information.   54

mnemonic symbol (software). A symbol chosen to assist the human memory, for example, an abbreviation such as 'mpy' for 'multiply'.   434

MNOS transistor. *See:* **metal-nitride-oxide semiconductor transistor.**

mobile (X-ray) (National Electrical Code). Equipment mounted on a permanent base with wheels and/or casters for moving while completely assembled.
                                                        256

mobile communication system. Combinations of interrelated devices capable of transmitting intelligence between two or more spatially separated radio stations, one or more of which shall be mobile.   181

mobile home (National Electrical Code). A factory-assembled structure or structures equipped with the necessary service connections and made so as to be readily movable as a unit or unit(s) without a permanent foundation. The phrase "without a permanent foundation" indicates that the support system is constructed with the intent that the mobile home placed thereon will be moved from time to time at the convenience of the owner.   256

mobile home accessory building or structure (National Electrical Code). Any awning, cabana, ramada, storage cabinet, carport, fence, windbreak or porch established for the use of the occupant of the mobile home upon a mobile home lot.   256

mobile home lot (National Electrical Code). A designated portion of a mobile home park designed for the accommodation of one mobile home and its accessory buildings or structures for the exclusive use of its occupants.   256

mobile home park (National Electrical Code). A contiguous parcel of land which is used for the accommodation of occupied mobile homes.   256

mobile home service equipment (National Electrical Code). The equipment containing the disconnecting means, overcurrent protective devices, and receptacles or other means for connecting a mobile home feeder assembly.   256

mobile radio service. Radio service between a radio station at a fixed location and one or more mobile stations, or between mobile stations. *See:* **radio transmission.**   328

mobile station (communication). A radio station designed for installation in a vehicle and normally operated when in motion. *See:* **mobile communication system.**   181

mobile telemetering. Electric telemetering between points that may have relative motion, where the use of interconnecting wires is precluded. *Note:* Space radio is usually employed as an intermediate means for mobile telemetering, but radio may also be used for telemetering between fixed points. *See:* **telemetering.**
                                                        328

mobile telephone system (automatic channel access). A mobile telephone system capable of operation on a plurality of frequency channels with automatic selection at either the base station or any mobile station of an idle channel when communication is desired. *See:* **mobile communication system.**   181

mobile transmitter. A radio transmitter designed for installation in a vessel, vehicle, or aircraft, and normally operated while in motion. *See:* **radio transmitter.**   111

mobile unit substation (power switchgear). A unit substation mounted and readily movable as a unit on a transportable device.   103

mobility (semiconductor). *See:* **drift mobility.**

mobility, Hall (electric conductor). *See:* **Hall mobility.**

modal analysis (power-system communication). A method of computing the propagation of a wave on a multiconductor power line.   59

modal channel (X-ray energy spectrometers). That channel in the distribution containing the largest number of counts.   471

modal direction (transmission performance of telephone sets). The assumed direction of speech transmission on a modal head. Also, the axis of an artificial mouth,   491

modal distance (telephony). The distance between the center of the grid of a telephone-handset transmitter cap and the center of the lips of a human talker (or the reference point of an artificial mouth), when the handset is in the modal position.   122

modal head (transmission performance of telephone sets). Head dimensions that are modal for a human population. The modal head is the same as that adopted by the *Comité Consultatif International Télégraphique et Téléphonique* (CCITT) for the measurement of *Affaiblissement équivalent pour netteté* (equivalent articulation loss). The applicable dimensions are shown in the figure below.
                                                        491

modal noise (fiber optics). Noise generated in an optical fiber system by the combination of mode dependent optical losses and fluctuation in the distribution of optical energy among the guided modes or in the relative phases of the guided modes. *Syn:* **speckle noise.** *See:* **mode.**   433

modal point (transmission performance of telephone sets). The position of the center of the lips of a modal head. Also, the corresponding reference point of an

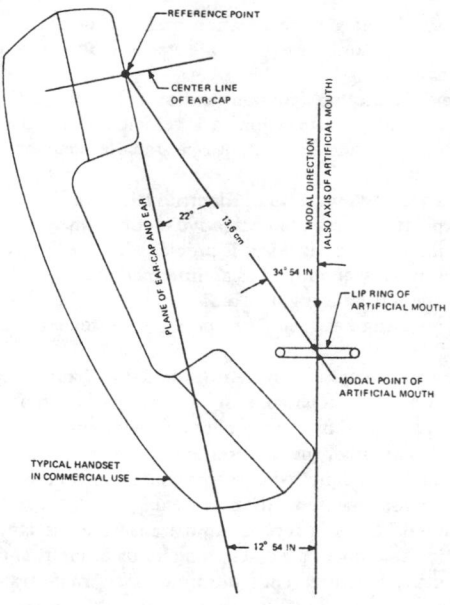

REFERENCE POINT

CENTER LINE
OF EAR·CAP

PLANE OF EAR·CAP AND EAR

22°

13.6 cm

MODAL DIRECTION
(ALSO AXIS OF ARTIFICIAL MOUTH)

34° 54 IN

LIP RING OF
ARTIFICIAL MOUTH

MODAL POINT OF
ARTIFICIAL MOUTH

TYPICAL HANDSET
IN COMMERCIAL USE

12° 54 IN

artificial mouth, the center of the external plane of the lip ring. 491

modal position (1)(telephony). The position a telephone handset assumes when the receiver of the handset is held in close contact with the ear of a person with head dimensions that are modal for a population. For this standard, the modal position is defined, by the modal head adopted by the CCITT (*Comité Consulatif International Télégraphique et Téléphonique*) Laboratory for the measurement of AEN. The point of reference for the handset and the head is the center of the circular plane of contact of the handset earcap and the ear. If the handset earcap is not circular or has no external plane of contact, an effective center and an effective plane of contact must be determined. The modal point is the position of the center of the lips with respect to the center and plane of the earcap point of reference. 122

(2)(transmission performance of telephone sets). The position a telephone-set handset assumes when the ear-cap of the handset is held in close contact with the ear of a modal head and the modal direction is in the plane defined by the axes of the transmitter cap and ear-cap. 491

mode (1)(binary floating point arithmetic)(radix-independent floating-point arithmetic). A variable that a user may set, sense, save, and restore to control the execution of subsequent arithmetic operations. The default mode is the mode that a program can assume to be in effect unless an explicitly contrary statement is included in either the program or its specification. The following mode is implemented: (A) Rounding, to control the direction of rounding errors. (B) In certain implementations, rounding precision, to shorten the precision of results. (C) The implementor may, at his option, implement the following modes: traps disabled or enabled, to handle exceptions. 469, 588

(2) (electron tubes). A state of a vibrating system to which corresponds one of the possible resonance frequencies (or propagation constants). *Note:* Not all dissipative systems have modes. *See:* **modes, degenerate; oscillatory circuit.** 190, 125

(3) (fiber optics). In any cavity or transmission line, one of those electromagnetic field distributions that satisfies Maxwell's equations and the boundary conditions. The field pattern of a mode depends on the wavelength, refractive index, and cavity or waveguide geometry. *See:* **bound mode; cladding mode; differential mode attenuation; differential mode delay; equilibrium mode distribution; equilibrium mode simulator; fundamental mode; hybrid mode; leaky modes; linearly polarized mode; mode volume; multimode distortion; multimode laser; multimode optical waveguide; single mode optical waveguide; transverse electric mode; transverse magnetic mode; unbound mode.** 433

mode conversion (waveguide). The transformation of an electromagnetic wave from one mode of propagation to one or more other modes. 267

mode conversion loss (or gain) (waveguide). The loss (or gain) due to the conversion of power from one waveguide mode to another. 267

mode coupler (waveguides). A coupler that provides preferential coupling to a specific wave mode. *See:* **waveguide.** 185

mode coupling (fiber optics). In an optical waveguide, the exchange of power among modes. The exchange of power may reach statistical equilibrium after propagation over a finite distance that is designated the equilibrium length. *See:* **equilibrium length; equilibrium mode distribution; mode; mode scrambler.** 433

mode dispersion. *See:* **multimode distortion.**

mode (or modal) distortion. *See:* **multimode distortion.**

mode filter (1) (fiber optics). A device used to select, reject, or attenuate a certain mode or modes. 433

(2) (waveguide components). A device designed to pass energy along a waveguide in one or more selected modes of propagation, and substantially to reject energy carried in other modes. 166

mode, higher-order (waveguide or transmission line). Any mode of propagation characterized by a field configuration other than that of the fundamental or first-order mode with lowest cutoff frequency. 185

model (1). A mathematical or physical representation of the system relationships. *See:* **mathematical model; system.** 209

(2) (software). A representation of a real world process, device, or concept. *See:* **analytical model; availability model; debugging model; error model; process; reliability model; simulation; statistical test model.** 434

modeling. Technique of system analysis and design using mathematical or physical idealizations of all or a portion of the system. Completeness and reality of the model are dependent on the questions to be answered, the state of knowledge of the system, and its environment. *See:* system.   209

modem (1)(data transmission). A contraction of MOdulator-DEModulator, an equipment that connects data terminal equipment to a communication line.   59
(2)(supervisory control, data acquisition, and automatic control). A MOdulator/DEModulator device which converts serial binary digital data to and from the signal form appropriate for the respective communication channel.   570

mode mixer. *See:* mode scrambler.   433

mode of operation (rectifier circuit). The characteristic pattern of operation determined by the sequence and duration of commutation and conduction. *Note:* Most thyristor converters and rectifier circuits have several modes of operation, which may be identified by the shape of the current wave. The particular mode obtained at a given load depends upon the circuit constants. *See:* rectification; rectifier circuit element.
121

mode of propagation (waveguides). A form of propagation of guided waves that is characterized by a particular field pattern in a plane transverse to the direction of propagation, which field pattern is independent of position along the axis of the waveguide. *Note:* In the case of uniconductor waveguides the field pattern of a particular mode of propagation is also independent of frequency. *See:* waveguide.   267, 319

mode of resonance (waveguide). A form of natural electromagnetic oscillation in a resonator, characterized by a particular field pattern.   267

mode of vibration (vibratory body, such as a piezoelectric crystal unit). A pattern of motion of the individual particles due to (1) stresses applied to the body: (2) its properties: and (3) the boundary conditions. Three common modes of vibration are (A) flexural, (B) extensional, and (C) shear. *See:* crystal.   328

mode scrambler (fiber optics). (1) A device for inducing mode coupling in an optical fiber. (2) A device composed of one or more optical fibers in which strong mode coupling occurs. *Note:* Frequently used to provide a mode distribution that is independent of source characteristics or that meets other specifications. *Syn:* mode mixer. *See:* mode coupling.   433

mode stripper. *See:* cladding mode stripper.

mode transducer (waveguide components). A device for transforming an electromagnetic wave from one mode of propagation to another.   166

mode transformer. *See:* mode transducer.

mode voltage. *See:* glow voltage.

mode volume (fiber optics). The number of bound modes that an optical waveguide is capable of supporting; for $V5$, approximately given by $V^2/2$ and $(V_2/2[g/(g+2)]$, respectively, for step index and power-law profile waveguides, where g is the profile parameter, and V is normalized frequency. *See:* effective mode volume; mode; normalized frequency; power-law index profile; step index profile; V number.
433

modification (software). (1) A change made to soft-

ware. (2) The process of changing software. *See:* software.   434

modified circuit transient recovery voltage (power switchgear). The circuit transient recovery voltage modified in accordance with the normal-frequency recovery voltage and the assymetry of the current wave obtained on a particular interruption. *Note:* This voltage indicates the severity of the particular interruption with respect to recovery-voltage phenomena.
103

modified impedance relay (power switchgear). An impedance form of distance relay for which the operating characteristic of the distance unit on an R-X diagram is a circle having its center displaced from the origin. *Note:* It may be described by the equation:

$$Z^2 - 2K_1 Z \cos(\theta - \alpha) = K_2^2 - K_1^2$$

where K_1, K_2 and α are constants and θ is the phase angle by which the input voltage leads the input current.   103

modified index of refraction (radio wave propagation). In the troposphere, the sum of the refractive index at a given height above the mean geometrical surface and the ratio of this height to the mean geometrical radius.
146

modified inherent transient recovery voltage (power switchgear). The TRV (transient recovery voltage) that results from the interaction of a circuit (that produces the inherent transient recovery voltage) and the impedance (capacitors, resistors, etcetera) of an interrupting device without the modifying effects of an arc and its voltage. Modifying impedances, such as capacitors and resistors, are sometimes included as part of a switching device to modify the TRV.
577

modified inherent transient recovery voltage (transient recovery voltage). The TRV (transient recovery voltage) that results from the interaction of a circuit (that produces the inherent TRV) and the impedance (capacitors, resistors, etcetera) of an interrupting device without the modifying effects of an arc and its voltage. Modifying impedances, such as capacitors and resistors, are sometimes included as part of a switching device to modify the TRV.
486, 487, 488

modular decomposition (software). A method of designing a system by breaking it down into modules. *See:* hierarchical decomposition; module; system.
434

modularity (software). The extent to which software is composed of discrete components such that a change to one component has minimal impact on other components. *See:* component; software.   434

modular programming (software). A technique for developing a system or program as a collection of modules. *See:* module; program; system.   434

modulated 12.5T pulse (MOD 12.5T) (linear waveform distortion). A burst of color subcarrier frequency of nominally 3.58 MHz. The envelope of the burst is $\sin^2$ shaped with a HAD of nominally 1.56 μs. The MOD 12.5T pulse consists of a luminance and a chrominance component. The envelope of the frequency spectrum consists of two parts, namely signal energy concentrated in the luminance region below 0.6 MHz and

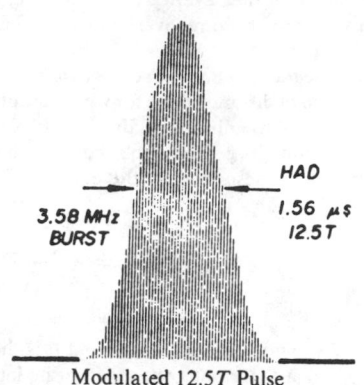

Modulated 12.5T Pulse

Envelope of Frequency Spectrum of Modulated 12.5T Pulse

in the chrominance region from roughly 3 MHz to 4.2 MHz.   42

modulation (1) (data transmission). (1) (Carrier) (A) The process by which some characteristic of a carrier is varied in accordance with a modulating wave. (B) The variation of some characteristic of a carrier. (2) (Signal transmission system) (A) A process whereby certain characteristics of a wave, often called the carrier, are varied or selected in accordance with a modulating function. (B) The result of such a process. *See: angle modulation; modulation index.*   59

(2) (diode-type camera tube). The ratio of the difference between the maximum and minimum signal currents divided by the sum. To avoid ambiguity, the optical input image intensity shall be assumed to be sinusoidal in the direction of scan.   380

(3) (fiber optics). A controlled variation with time of any property of a wave for the purpose of transferring information.   433

(4) (overhead-power-line corona and radio noise). The process by which some characteristic of a carrier is varied in accordance with a modulating wave.
                                                                411

modulation contrast (C_m) (diode-type camera tube). The ratio of the difference between the peak and the minimum values of irradiance to the sum of the peak and the minimum value of irradiance of an image or specified portion of an image. equation   380

modulation index (angle modulation with a sinusoidal modulating function) (data transmission). The ratio of the frequency deviation of the modulated wave to the frequency of the modulating function. *Note:* The modulation index is numerically equal to the phase deviation expressed in radians.   59

modulation threshold (illuminating engineering). In the case of a square wave or sine wave grating, manipulation of luminance differences can be specified in terms of modulation and the threshold may be called the modulation threshold.

$$\text{modulation} = \frac{L_{\text{max}} - L_{\text{min}}}{L_{\text{max}} + L_{\text{min}}}$$

Periodic patterns that are not sine wave can be specified in terms of the modulation of the fundamental sine wave component. The number of periods or cycles per degree of visual angle represents the spatial frequency.   167

modulation transfer function (MTF) or $R_o(N)$, (diode-type camera tube). $R_o(N)$, the modulus of the optical transfer function (OTF), is synonymous with the sine amplitude response. That is, the response of the imaging sensor to sinewave images. When the modulation transfer functions or MTFs of a linear sensor's components are known, the overall system MTF can be found by multiplying the individual component MTFs together. equation where $R_{01}(N)$, $R_{02}(N)$,...,$R_{0j}(N)$ are the component MTF.   380

module (1)(FASTBUS module)(FASTBUS acquisition and control). Any FASTBUS device that can be housed in a FASTBUS crate, that can connect to a crate segment and that conforms with the mandatory specifications in ANSI/IEEE Std 960-1986 for a FASTBUS module.   480

(2) (accident monitoring instrumentation). Any assembly of interconnected components which constitutes an identifiable device, instrument, or piece of equipment. A module can be removed as a unit, and replaced with a spare. It has definable performance characteristics which permit it to be tested as a unit. A module could be a card, a drawout circuit breaker, or other subassembly of a larger device, provided it meets the requirements of this definition.   421

(3) (cable penetration fire stop qualification test). An opening in a fire resistive barrier so located and spaced from adjacent modules (openings) that its respective cable penetration fire stop's performance will not affect the performance of cable penetration fire stops in any adjacent module. A module may take on any shape to permit the passage of cables from one or any number of raceways.   386

(4) (nuclear power generating station). Any assembly of interconnected components which constitutes an identifiable device, instrument, or piece of equipment. A module can be disconnected, removed as a unit, and replaced with a spare unit. It has definable performance characteristics which permit it to be tested as a unit. A module could be a card, a drawout circuit breaker, or other subassembly of a larger device, provided it meets the requirements of this definition.
                                                          387, 102

(5) (software). (A) A program unit that is discrete and

identifiable with respect to compiling, combining with other units, and loading, for example, the input to, or output from, an assembler, compiler, linkage editor, or executive routine. (B) A logically separable part of a program. *See:* **assembler; compiler; linkage editor; program; routine.** 434

module accelerated aging (advanced life conditioning) (nuclear power generating station). The acceleration process designed to achieve an advanced life condition in a short period of time. It is the process of subjecting a module or component to stress conditions in accordance with known measurable physical or chemical laws of degradation in order to render its physical and electrical properties similar to those it would have at an advanced age operating under expected service conditions. In addition, when operations of a device are cyclical, acceleration is achieved by subjecting the device to the number of cycles anticipated during its qualified life. 355

module accuracy (nuclear power generating station). Conformity of a measurement value to an accepted standard value or true value. *Note:* For further information, see Process Measurement and Control Terminology SAMA PMC-20.1-1973. 355

module address (MA)(FASTBUS acquisition and control). The group of bits assigned in the device address field of a FASTBUS address which identifies the module on its segment. The module address may partially overlap the group address. 480

module aging (natural) (nuclear power generating station). The change with passage of time of physical chemical, or electrical properties of a component or module under design range operating conditions which may result in degradation of significant performance characteristics. 355

module auxiliary connector (FASTBUS acquisition and control). The standard connector that mounts above the module segment connector on a module circuit board. 480

module calibration (nuclear power generating station). Adjustment of a device, to bring the module's output to a desired value or series of values, within a specified tolerance, for a particular value or series of values of the input or measurements used to establish the input-output function of the module. 355

module circuit board (FASTBUS acquisition and control). The printed board that is the circuit part of a FASTBUS module. 480

module common mode rejection (nuclear power generating station). The ability of a module with a differential input stage to cancel or reject a signal applied equally to both inputs. 355

module components (nuclear power generating station). Items from which the module is assembled (for example, resistors, capacitors, wires, connectors, transistors, springs, etcetera.) 355

module conformity (nuclear power generating station). The closeness with which the curve of a function approximates a specified curve. 355

module contact rating (nuclear power generating station). The electrical power-handling capability of relay or switch contacts. This should be specified as continuous or interrupting, resistive or inductive, ac or dc. 355

module design range operating conditions (nuclear power generating station). The range or environmental and energy supply operating conditions within which a module is designed to operate. 355

module drift (nuclear power generating station). A change in output-input relationship over a period of time, normally determined as the change in output over a specified period of time for one or more input values which are held constant under specified reference operating conditions. 355

module electromagnetic interference (nuclear power generating station). Any unwanted electromagnetically transmitted energy appearing in the circuitry of a module. 355

module energy supply (nuclear power generating station). Electrical energy, compressed fluid, manual force or other such input to the module which will establish the power for its operation. 355

module failure trending (nuclear power generating station). Systematic documentation and analysis of the frequency of a particular failure mode. 355

module frequency response (nuclear power generating station). The frequency-dependent relation, in both amplitude and phase, between steady-state sinusoidal inputs and the resulting fundamental sinusoidal outputs. 355

module input overrange constraints (nuclear power generating station.) The upper and/or lower values of the input signal which may be applied to a module without causing damage or otherwise altering permanent characteristics of the module or causing undesired saturation effects. 355

module input signal range (nuclear power generating station). The region between the limits within which a quantity is measured or received, expressed by stating the lower and upper values of the input signal. 355

module isolation characteristics (nuclear power generating station). provisions for electrical isolation of particular sections of a module from each other; such as input and output circuitry, control and protection circuitry, and redundant protection circuitry. 355

module load capability (nuclear power generating station). The range of load values within which a module will perform to its specified performance characteristics. 355

module output impedance (nuclear power generating station). The internal impedance presented by a module at its output terminals to a load. 355

module output ripple (nuclear power generating station). The ac component of a dc output signal harmonically related in frequency to either the supply voltage or a voltage generated within the module (for example, carrier demodulation). 355

module output signal range (nuclear power generating station). The region between the limits within which a quantity is transmitted, expressed by stating the lower and upper values of the output signal.   355

module pulse characteristics (nuclear power generating station). Information such as pulse duration, amplitude, rise time, decay time, separation and shape.
355

module qualified life (nuclear power generating station). The life expectancy in years (or cycles of operation, if applicable) over which the module has been demonstrated to be qualified for use, as established by type tests, analysis or other qualification method.
355

module range and characteristics of adjustments (nuclear power generating station). Such information as upper and lower range-limits of calibration capability and where applicable, their relationship to the calibrated range of the module.   355

module reference operating conditions (nuclear power genberating station). The range of environmental operating conditions of a module within which environmental influences are negligible.   355

module reproducibility (nuclear power generating station). The closeness of agreement among repeated measurements of the output for the same value of input made under the same operating conditions over a period of time, approaching from both directions.
355

module response time (nuclear power generating station.) The time required for an output change from an initial value to a specified percentage of the final steady-state value, resulting from the application of a specified input change under specified conditions. For digital equipment (that is, relays, solid state logic, delay networks, etcetera). Response time is the time required for a change from an initial state to a specified final state resulting from application of specified input under specified conditions.   355

modules (electric pipe heating systems). Any assembly of interconnected components which constitutes an identifiable device, instrument, or piece of equipment which can be disconnected, removed as a unit and replaced with a spare, and has definable performance characteristics which permit it to be tested as a unit. A module can be a card or other subassembly.
448

module segment connector (FASTBUS acquisition and control). The standard connector that mounts on a FASTBUS module and mates with the crate segment connector for connection of the module to the segment.   480

module signal to noise rato (nuclear power generating station). The output signal with input signal applied minus the output signal with no input signal applied divided by the output signal with no input signal applied.   355

module strength. *See:* cohesion.

module supplementary board (FASTBUS acquisition and control). Any board in a FASTBUS module that does not make direct connection with the crate segment.   480

module type tests (nuclear power generating station). Tests made on one or more production units to dem-

onstrate that the performance characteristics of the module(s) conform to the module's specifications.
355

modulo N check (1) (data transmission). A form of check digits, such that the number of ones in each number A operated upon is compared with a check number B, carried along with A and equal to the remainder of A when divided by N; for example, in a modulo 4 check, the check number will be 0, 1, 2, or 3 and the remainder of A when divided by 4 must equal the reported check number B, or else an error or malfunction has occurred; a method of verification by congruences; for example, casting out nines. *See:* **residue check.**   194

(2) (computing systems). *See:* **residue check**

modulus (phasor). Its absolute value. The modulus of a phasor is sometimes called its amplitude.   210

MOF. *See:* **maximum observed frequency.**   146

Moho (Mohorovičić discontinuity). Seismic discontinuity situated about 35 kilometers below the continents and about 10 kilometers below the oceans. Crudely speaking, its separates the earth's crust and mantle.   132

moiré (television). The spurious pattern in the reproduced picture resulting from interference beats between two sets of periodic structures in the image. *Note:* The most common cause of moiré is the interference between scanning lines and some other periodic structure such as a line pattern in the original scene, a mesh or dot pattern in the camera sensor (for example, the target mesh in an image orthicon), or the phosphor dots or other structure in a shadow-mask picture tube. Moiré may result from the interference between the subcarrier elements of the chrominance signal and another periodic structure. In systems using an fm carrier, such as magnetic or video-disc record-playback systems, moire may also be caused by interference between the upper sidebands of the fm carrier and lower sidebands of harmonics of the fm carrier. In general, moiré may be caused by interference beats between any two periodic structures that are not perfectly aligned and not of the same frequency.
18

moisture-resistant (1) (packaging machinery). So constructed or treated that exposure to a moist atmosphere will not readily cause damage.   429

(2) (power switchgear). Not readily injured by exposure to a moist atmosphere.   103

molded-case circuit breaker (1) (electric installations on shipboard). A circuit breaker assembled as an integral unit in a supporting and enclosing housing of insulating material; the overcurrent and tripping means being of the thermal type, the magnetic type, or a combination of both.   3

(2) (power switchgear). One that is assembled as an integral unit in a supporting and enclosing housing of molded insulating material.   103

mole (metric practice). The amount of substance of a system which contains as many elementary entities as there are atoms in 0.012 kilogram of carbon-12 (adopted by 14th General Conference on Weights and

Measures). *Note:* When the mole is used, the elementary entities must be specified and may be atoms, molecules, ions, electrons, other particles, or specified groups of such particles.   21

momentary current (power switchgear). The current flowing in a device, an assembly, or a bus at the major peak of the maximum cycle as determined from the envelope of the current wave. *Note:* The current is expressed as the root-mean-square (rms) value, including the direct-current component, and may be determined by the method shown in American National Standard Methods for Determining the Values of a Sinusoidal Current Wave and Normal-Frequency Recovery Voltage for AC High-Voltage Circuit Breakers, ANSI/IEEE C37.09-1979.   103

momentary interruption. *See:* **interruption, momentary.**

momentary rating (X-ray) (National Electrical Code). A rating based on an operating interval that does not exceed five seconds.   256

monadic (mathematics of computing). Pertaining to an operation involving a single operand. *See:* **dyadic.**
                    564

monitor (token ring access method). That function that recovers from various error situations. It is contained in each ring station; however, only the monitor in one of the stations on a ring is the active monitor at any point in time. The monitor function in all other stations on the ring is in standby mode.   472

monitor hazard current (health care facilities). The hazard current of the line isolation monitor alone. *See:* **hazard current.**   192

monitoring (1)(data transmission). In communication, an observation of the characteristics of transmitted signals.   59
(2)(electric pipe heating systems). To check the operation and performance of an equipment or system by sampling the results of the operation. Monitoring with respect to electric pipe heating systems usually consists of checking system temperatures or operation of the heater circuits; voltage, current, etcetera.
                    448

monitoring relay (power switchgear). A relay which has as its function to verify that system or control-circuit conditions conform to prescribed limits.
                    103

monochromatic (1) (color) (television). Having spectral emission over an extremely small region of the visible spectrum.   18
(2) (fiber optics). Consisting of a single wavelength or color. In practice, radiation is never perfectly monochromatic but, at best, displays a narrow band of wavelengths. *See:* **coherent; line source; spectral width.**   433

monochromator (fiber optics). An instrument for isolating narrow portions of the spectrum.   433

monochrome (television). Having only one chromaticity, usually achromatic.   18

monochrome channel (television). Any path that is intended to carry the monochrome signal.   18

monochrome channel bandwidth (television). The

bandwidth of the path intended to carry the monochrome signal.   18

monochrome signal (television). (1) **monochrome television.** A signal wave for controlling the luminance values in the picture. (2) **color television*** *See:* **luminance signal.**   18
*Deprecated

monochrome television. The electric transmission and reception of transient visual images in only one chromaticity, usually achromatic. *Note:* Also termed black-and-white television.   18

monochrome transmission (television). The transmission of a signal wave for controlling the luminance values in the picture, but not the chromaticity values. *Note:* Also termed black-and-white transmission.
                    18

monoclinic system (piezoelectricity). A monoclinic crystal has either a single axis of twofold symmetry or a single plane of reflection symmetry, or both. Either the twofold axis or the normal to the plane of symmetry (they are the same if both exist, and this direction is called the unique axis in any case) is taken as the b or Y axis. Of the two remaining axes, the smaller is the c axis. In class 2, $+Y$ is chosen so that d_{22} is positive; $+Z$ is chosen parallel to c (sense trivial), and $+X$ such that it forms a right-handed system with $+Z$ and $+Y$. In class m, $+Z$ is chosen so that d_{33} is positive, and $+X$ so that d_{11} is positive, and $+Y$ to form a right-handed system. *Note:* "Positive" and "negative" may be checked using a carbon-zinc flashlight battery. The carbon anode connection will have the same effect on meter deflection as the $+$ end of the crystal axis upon *release* of compression. *See:* **crystal systems.**   371

monocular visual field (illuminating engineering). The field for a single eye.   167

monomode optical waveguide. *See:* **single mode optical waveguide.**   433

monopulse (radar). A radar technique in which information concerning the angular location of a source or target is obtained by comparison of signals received in two or more simultaneous antenna beams, as distinguished from techniques such as lobe switching or conical scanning in which the beams are generated sequentially. The simultaneity of the beams makes it possible to obtain a two-dimensional angle estimate from a single pulse (hence the name monopulse), although multiple pulses are usually employed to improve the accuracy of the estimate or to provide Doppler resolution. The monopulse principle can be used with continuous wave as well as pulsed radar. *Syn:* **simultaneous lobing.**   13

monostatic cross section (antennas). The scattering cross section in the direction toward the source. *Syn:* **back scattering cross section.** *See:* **bistatic cross section.**   111

monthly peak duration curve (power operations). A curve showing the total number of days within the month during which the net 60 min clock-hour integrated peak demand equals or exceeds the percent of monthly peak values shown.   516

MOPA (radar). Abbreviation for a master oscillator/power amplifier type of coherent transmitter.
13

MOS transistor. *See:* **metal-oxide-semiconductor transistor.**
386

most significant bit (MSB)(mathematics of computing). The bit having the greatest effect on the value of a binary numeral; usually the leftmost bit.
564

motive power (valve actuators). The electric, fluid, air, nitrogen, or mechanical energy required to operate the actuator.
492

motor branch circuit (electric installations on shipboard). A branch circuit supplying energy only to one or more motors and associated motor controllers.
3

motor-circuit switch (1) (National Electrical Code). A switch, rated in horsepower, capable of interrupting the maximum operating overload current of a motor of the same horsepower rating as the switch at the rated voltage.
256
(2) (packaging machinery). A switch intended for use in a motor branch circuit. It is rated in horsepower and is capable of interrupting the maximum operating overload current of a motor of the same rating at the rated voltage.
429

motor conduit box (packaging machinery). An enclosure on a motor for the purpose of terminating a conduit run and joining motor to power conductors.
429

motor control center (nuclear power generating stations). A floor mounted assembly of one or more enclosed vertical sections having a common horizontal power bus and principally containing combination motor starting units. These units are mounted one above the other in the vertical sections. The sections may incorporate vertical buses connected to the common power bus, thus extending the common power supply to the individual units. Units may also connect directly to the common power bus by suitable connections.
440

motor control circuit (National Electrical Code). The circuit of a control apparatus or system that carries the electric signals directing the performance of the controller but does not carry the main power current. Motor control circuits tapped from the load side of the motor branch-circuits, short-circuit protective devices shall not be considered to be branch circuits and shall be permitted to be protected by either supplementary or branch-circuit overcurrent protective devices.
256

motor-generator set (electric installations on shipboard). A machine which consists of one or more motors mechanically coupled to one or more generators.
3

motor home (National Electrical Code). A vehicular unit designed to provide temporary living quarters for recreational, camping or travel use built on or permanently attached to a self-propelled vehicle chassis or on a chassis cab or van which is an integral part of the completed vehicle. *See:* **recreational vehicle.** 256

motor lead extension cable (electric submersible pump cable). Three-conductor cable running from above the pump to the motor including motor connecting plug.
484

motor meter. A meter comprising a rotor, one or more stators, and a retarding element by which the resultant speed of the rotor is made proportional to the quantity being integrated (for example, power or current) and a register connected to the rotor by suitable gearing so as to count the revolutions of the rotor in terms of the accumulated integral (for example, energy or charge). *See:* **electricity meter (meter).**
212

motor parts (electric). A term applied to a set of parts of an electric motor. Rotor shaft, conventional stator-frame (or shell), end shields, or bearings may not be included, depending on the requirements of the end product into which the motor parts are to be assembled.
63

motor reduction unit (electric installations on shipboard). A motor, with an integral mechanical means of obtaining a speed different from the speed of the motor. Motor reduction units are usually designed to obtain a speed lower than that of the motor, but may also be built to obtain a speed higher than that of the motor.
3

motor synchronizing. Synchronizing by means of applying excitation to a machine running at slightly below synchronous speed. *See:* **asynchronous machine.**
63

motor type watthour meter. A motor in which the speed of the rotor is proportional to the power, with a readout device that counts the revolutions of the rotor.
212

mount (switching tubes). The flange or other means by which the tube, or tube and cavity, are connected to a waveguide. *See:* **gas tube.** 190, 125

mounted plow (static or vibratory plows) (cable plowing). A unit which, to be operable, is semipermanently attached to and dependent upon a prime mover.

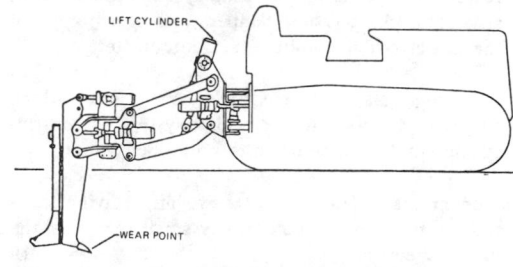

LIFT CYLINDER

WEAR POINT

Mounted plow.

52

mounting lug, stator. *See:* **lug, stator mounting.**

mounting position (of a switch or fuse support) (power switchgear). The position determined by and corresponding to the position of the base of the device. *Note:* The usual positions are: (1) horizontal upright, (2) horizontal overhung, (3) vertical, and (4) angle. Modifications of these notes appear in C37.40-1981.
103, 443

mounting ring (rotating machinery). A ring of resilient or nonresilient material used for mounting an electric machine into a base at the end shield hub.   63

mounting structure (substation rigid-bus structures). A structure for mounting an insulating support.   566

movable bridge (drawbridge) rail lock. A mechanical device used to insure that the movable bridge rails are in proper position for the movement of trains. *See:* **interlocking.**   328

movable bridge coupler (drawbridge coupler). A device for engaging and disengaging signal or interlocking connections between the shore and a movable bridge span.   328

moving-base-derived navigation data. Data obtained from measurements made at moving cooperative facilities located external to the navigated vehicle. *See:* **navigation.**   187, 13

moving-base navigation aid. An aid that requires cooperative facilities located upon a moving vehicle other than the one being navigated. *Notes:* (1) The cooperative facilities may move along a predictable path that is referenced to a specified coordinate system such as in the case of a nongeostationary navigation satellite. (2) Such an aid may also be designed solely to permit one moving vehicle to home upon another. *See:* **navigation.**   187, 13

moving-base-referenced navigation data. Data in terms of a coordinate system referenced to a moving vehicle other than the one being navigated. *See:* **navigation.**   187, 13

moving-coil loudspeaker (dynamic loudspeaker). A moving-conductor loudspeaker in which the moving conductor is in the form of a coil conductively connected to the source of electric energy. *See:* **loudspeaker.**   328

moving-coil microphone (dynamic microphone). A moving-conductor microphone in which the movable conductor is in the form of a coil. *See:* **microphone.**   328

moving-conductor loudspeaker (moving conductor). A loudspeaker in which the mechanical forces result from magnetic reactions between the field of the current and a steady magnetic field. *See:* **loudspeaker.**   328

moving-conductor microphone. A microphone the electric output of which results from the motion of a conductor in a magnetic field. *See:* **microphone.**   328

moving contact (power switchgear). A conducting part which bears a contact surface arranged for movement to and from the stationary contact.   103

moving-contact assembly (rotating machinery). That part of the starting switch assembly that is actuated by the centrifugal mechanism. *See:* **centrifugal starting switch.**   328

moving element (instrument). Those parts that move as a direct result of a variation in the quantity that the instrument is measuring. *Notes:* (1) The weight of the moving element includes one-half the weight of the springs, if any. (2) The use of the term movement is deprecated.   280, 294

moving-iron instrument. An instrument that depends for its operation on the reactions resulting from the current in one or more fixed coils acting upon one or more pieces of soft iron or magnetically similar material at least one of which is movable. *Note:* Various forms of this instrument (plunger, vane, repulsion, attraction, repulsion-attraction) are distinguished chiefly by mechanical features of construction. *See:* **instrument.**   328

moving-magnet instrument. An instrument that depends for its operation on the action of a movable permanent magnet in aligning itself in the resultant field produced either by another permanent magnet and by an adjacent coil or coils carrying current, or by two or more current-carrying coils, the axes of which are displaced by a fixed angle. *See:* **instrument.**   328

moving-magnet magnetometer. A magnetometer that depends for its operation on the torques acting on a system of one or more permanent magnets that can turn in the field to be measured. *Note:* Some types involve the use of auxiliary magnets (Gaussian magnetometer), others electric coils (sine or tangent galvanometer). *See:* **magnetometer.**   328

moving target indication (MTI) (radar). A technique that enhances the detection and display of moving radar targets by suppressing fixed targets. Doppler processing is one method of implementation.   13

moving-target indicator improvement factor. *See:* **MTI improvement factor.**

MPE (laser-maser). Maximum permissible exposure.   363

M **peak (closed loop) (1) (control system, feedback).** The maximum value of the magnitude of the return transfer function for real frequencies, the value at zero frequency being normalized to unity. *See:* **control system, feedback.**   56

(2) (excitation control systems) (Mp). See definition above. *Note:* See Figure under **bandwidth.** Mp is the maximum value of the closed-loop amplitude response.   353

m**-phase circuit.** A polyphase circuit consisting of m distinct phase conductors, with or without the addition of a neutral conductor. *Note:* In this definition it is understood that m may be assigned the integral value of three or more. For a two-phase circuit see: **two-phase circuit: two-phase, three-wire circuit: two-phase, four-wire circuit: two-phase, five-wire circuit.** *See:* **network analysis.**   210

MRA. *See:* **minimum reception altitude.**

MSC. *See:* **mile of standard cable.**

M **scan (radar).** *See:* M **display.**

M-scope (radar). A cathode-ray oscilloscope arranged to present an M-display.   13

MSFN. *See:* **manned space flight network.**

M-T. *See:* **magneto-telluric.**

MTBF (reliability). *See:* **mean time between failures.**

MTE. *See:* **mean time between errors.**

MTF. *See:* **mean time between failures.**

MTI. *See:* **moving-target indication.**

MTI improvement factor (radar MTI). The signal-to-clutter ratio at the output of the clutter filter divided by the signal-to-clutter ratio at the input of the clutter filter, averaged uniformly over all target radial velocities of interest.   13

mu (μ) circuit (feedback amplifier). That part that amplifies the vector sum of the input signal and the fed-back portion of the output signal in order to generate the output signal. *See:* **feedback.**   328

MUF. *See:* **maximum usable frequency.**

mu factor (μ factor) (n-terminal electron tubes). The ratio of the magnitude of infinitesimal change in the voltage at the *j*th electrode to the magnitude of an infinitesimal change in the voltage at the *l*th electrode under the conditions that the current to the *m*th electrode remain unchanged and the voltages of all other electrodes be maintained constant. *See:* **electron-tube admittances.**   190, 125

muffler (of a fuse) (power switchgear). An attachment for the vent of a fuse, or a vented fuse, that confines the arc and substantially reduces the venting from the fuse.   103

multiaddress (computers). *See:* **multiple-address.**

multianode tank (multianode tube). An electron tube having two or more main anodes and a single cathode. *Note:* This term is used chiefly for pool-cathode tubes.
                                                    190

multibeam antenna. An antenna capable of creating a family of major lobes from a single nonmoving aperture, through use of a multiport feed, with one-to-one correspondence between input ports and member lobes, the latter characterized by having unique main beam pointing directions. *Note:* Often the multiple main beam angular positions are arranged to provide complete coverage of a solid angle region of space.
                                                    111

multibeam oscilloscopes. An oscilloscope in which the cathode-ray tube produces two or more separate electron beams that may be individually, or jointly, controlled. *See:* **dual-beam oscilloscope; oscillograph.**
                                                    185

multicable penetrator (electric installations on shipboard). A device consisting of multiple nonmetallic cable seals assembled in a surrounding metal frame, for insertion in openings in decks, bulkheads, or equipment enclosures and through which cables may be passed to penetrate decks or bulkheads or to enter equipment without impairing their original fire or watertight integrity.   3

multicavity magnetron. A magnetron in which the circuit includes a plurality of cavities. *See:* **magnetron.**
                                                    125

multicellular horn (electroacoustics). A cluster of horns with juxtaposed mouths that lie in a common surface. *Note:* The purpose of the cluster is to control the directional pattern of the radiated energy. *See:* **loudspeaker.**   176

multichannel analyzer (MCA)(X-ray energy spectrometers). An instrument which digitizes analog amplitude signal pulses and stores them in a memory as a function of their analog amplitude.   471

multichannel radio transmitter. A radio transmitter having two or more complete radio-frequency portions capable of operating on different frequencies, either individually or simultaneously. *See:* **radio transmitter.**   111

multichip integrated circuit. An integrated circuit whose elements are formed on or within two or more semiconductor chips that are separately attached to a substrate. *See:* **integrated circuit.**   312

multiconductor bundle. *See:* **bundle.**

multi-constant speed motor (rotating machinery). A multi-speed motor whose two or more definite speeds are constant or substantially constant over its normal range of loads: for example A synchronous or an induction motor with windings capable of various pole groupings.   63

multidimensional system (control system). A system whose state vector has more than one element. *See:* **control system.**   56

multielectrode tube. An electron tube containing more than three electrodes associated with a single electron stream.   125

multi-element conduction interval (thyristor). That part of the conduction interval when ON-state current flows in more than one basic control element simultaneously.   445

multifamily dwelling (National Electrical Code). A building containing three or more dwelling units.
                                                    256

multifiber cable (fiber optics). An optical cable that contains two or more fibers, each of which provides a separate information channel. *See:* **fiber bundle; optical cable assembly.**   433

multifiber joint (fiber optics). An optical splice or connector designed to mate two multifiber cables, providing simultaneous optical alignment of all individual waveguides. *Note:* Optical coupling between aligned waveguides may be achieved by various techniques including proximity butting (with or without index matching materials), and the use of lenses.   433

multifrequency transmitter. A radio transmitter capable of operating on two or more selectable frequencies, one at a time, using present adjustments of a single radio-frequency portion. *See:* **radio transmitter.**
                                                    111, 240

multigrounded neutral system (power and distribution transformer). A distribution system of the four-wire type where all transformer neutrals are grounded, and neutral conductors are directly grounded at frequent points along the circuit.   53

multilateration (radar). The location of an object by means of two or more range measurements from different reference points. It is a useful technique with radar because of the inherent accuracy of radar range measurement. The use of three reference points, **trilateration,** is common practice.   13

multilayer filter. *See:* **interference filter.**

multilevel address (computing systems). *See:* **indirect address.**

multilevel security (software). A mode of operation permitting data at various security levels to be concurrently stored and processed in a computer system when at least some users have neither the clearance nor the need-to-know for all data contained in the system. *See:* **computer system; data; security.** 434

multimeter. *See:* **circuit analyzer.**

multimode distortion (fiber optics). In an optical waveguide, that distortion resulting from differential mode delay. *Note:* The term 'multimode dispersion' is often used as a synonym; such usage, however, is erroneous since the mechanism is not dispersive in nature. *Syn:* **intermodal distortion; mode (or modal) distortion.** *See:* **distortion.** 433

multimode group delay. *See:* **differential mode delay.**

multimode laser (fiber optics). A laser that produces emission in two or more transverse or longitudinal modes. *See:* **laser; mode.** 433

multimode optical waveguide (fiber optics). An optical waveguide that will allow more than one bound mode to propagate. *Note:* May be either a graded index or step index waveguide. *See:* **bound mode; mode; mode volume; multimode distortion; normalized frequency; power-law index profile; single mode optical waveguide; step index optical waveguide.** 433

multimode waveguide. A waveguide used to propagate power in more than one mode at a frequency of interest. 267

multioffice exchange (telephone switching systems). A telecommunications exchange served by more than one local central office. 55

multioutlet assembly (National Electrical Code). A type of surface or flush raceway designed to hold conductors and receptacles, assembled in the field or at the factory. 256

multipactor limiter (nonlinear, active, and nonreciprocal waveguide components). A high-vacuum device that uses the multipacting phenomenon to limit high microwave power levels. *See:* **multipacting.** 530

multiparty ringing (telephone switching systems). By custom, any arrangement that provides for the individual ringing of more than four parties. 55

multipath (1)(facsimile). *See:* **multipath transmission.**
(2) (radar). The propagation of a wave from one point to another by more than one path. When multipath occurs in radar, it usually consists of a direct path and one or more indirect paths by reflection from the surface of the earth or sea or from large man-made structures. At frequencies below approximately 40 megahertz (MHz) it may also include more than one path through the ionosphere. 13

multipath error (radar). The error (for example, in the measurement of the angle of arrival) caused by multipath. 13

multipath transmission (radio wave propagation). The propagation phenomenon that results in signals reaching the receiving antenna by two or more paths. 146

multiple (1) (noun). A group of terminals arranged to make a circuit or group of circuits accessible at a number of points at any one of which connection can be made.
(2) (verb). To connect in parallel, or to render a circuit

accessible at a number of points at any one of which connection can be made. 328
(3) (analog computers). A junction into which patch cords may be plugged to form a common connection. 9

multiple access (communication satellite). The capability of having simultaneous access to one communication satellite from a number of ground stations. 84

multiple-address (multiaddress) (computers). Pertaining to an instruction that has more than one address part. 235

multiple-address code (electronic computation). *See:* **instruction code; electronic computation.**

multiple-beam headlamp (illuminating engineering). A headlamp so designed as to permit the driver of a vehicle to use any one of two or more distributions of light on the road. 167

multiple-beam klystron (microwave tubes). An *O*-type tube having more than one electron beam, and resonators coupled laterally but not axially. *See:* **microwave tube or valve.** 190

multiple circuit. Two or more circuits connected in parallel. *See:* **center of distribution.** 64

multiple-conductor cable. A combination of two or more conductors cabled together and insulated from one another and from sheath or armor where used. *Note:* Specific cables are referred to as 3-conductor cable, 7-conductor cable, 50-conductor cable, etcetera. 64

multiple-conductor concentric cable. A cable composed of an insulated central conductor with one or more tubular stranded conductors laid over it concentrically and insulated from one another. *Note:* This cable usually has only two or three conductors. Specific cables are referred to as 2-conductor concentric cable, 3-conductor concentric cable, etcetera. 64

multiple-current generator. A generator capable of producing simultaneously currents or voltages of different values, either alternating-current or direct-current. 63

multiple feeder (power switchgear). One that is connected to a common load in multiple with one or more feeders from independent sources. 103

multiple frame transmission (token ring access method). A transmission where more than one frame is transmitted when a token is captured. 472

multiple-gun cathode-ray tube. A cathode-ray tube containing two or more separate electron-gun systems. 190

multiple hoistway (elevators). A hoistway for more than one elevator or dumbwaiter. *See:* **hoistway (elevator or dumbwaiter).** 328

multiple lampholder (current tap). A device that by insertion in a lampholder, serves as more than one lampholder. 328

multiple lightning stroke. A lightning stroke having two or more components. *See:* **direct-stroke protection (lightning).** 64

multiple metallic rectifying cell. An elementary metallic rectifier having one common electrode and two or more separate electrodes of the opposite polarity. *See:* **rectification.**   328

multiple modulation. A succession of processes of modulation in which the modulated wave from one process becomes the modulating wave for the next. *Note:* In designating multiple-modulation systems by their letter symbols, the processes are listed in the order in which the modulating function encounters them. For example, PPM-AM means a system in which one or more signals are used to position-modulate their respective pulse subcarriers which are spaced in time and are used to amplitude-modulate a carrier.   111

multiple plug (cube tap) (plural tap). A device that, by insertion in a receptacle, serves as more than one receptacle.   328

multiple-pointer form demand register (metering). An indicating demand register from which the demand is obtained by reading the position of the multiple pointers relative to their scale markings. The multiple pointers are resettable to zero.   212

multiple precision (mathematics of computing). Pertaining to the use of two or more computer words to represent a number in order to preserve or gain precision. *See:* **double precision.**   564

multiple rectifier circuit. A rectifier circuit in which two or more simple rectifier circuits are connected in such a way that their direct currents add, but their commutations do not coincide. *See:* **rectification; rectifier circuit element.**   208

multiple rho (electronic navigation). A generic term referring to navigation systems based on two or more distance measurements for determination of position. *See:* **navigation.**   187, 13

multiple-secondary current transformer (1) (instrument transformer). One that has three or more secondary coils each on a separate magnetic circuit with all magnetic circuits excited by the same winding.   394

(2) (power and distribution transformer). One which has three or more secondary coils each on a separate magnetic circuit with all magnetic circuits excited by the same primary winding.   53

multiple-shot blasting unit. A unit designed for firing a number of explosive charges simultaneously in mines, quarries, and tunnels. *See:* **blasting unit.**   328

multiple sound tract. Consists of a group of sound tracks, printed adjacently on a common base, independent in character but in a common time relationship, for example, two or more have been used for stereophonic sound recording. *See:* **multitrack recording system; phonograph pickup.**   176

multiple speed floating. *See:* **control system, multiple-speed floating.**

multiple spot scanning (facsimile). The method in which scanning is carried on simultaneously by two or more scanning spots, each one analyzing its fraction of the total scanned area of the subject copy. *See:* **scanning (facsimile).**   12

multiple street-lighting system. A street-lighting system in which street lights, connected in multiple, are supplied from a low-voltage distribution system. *See:* **alternating-current distribution; direct-current distribution.**   64

multiple-supply-type ballast. A ballast designed specifically to receive its power from an approximately constant-voltage supply circuit and that may be operated in multiple (parallel) with other loads supplied from the same source. *Note:* The deviation in source voltage ordinarily does not exceed plus or minus 5 percent, but in the case of ballasts designed for a stated input voltage range, the deviation may by greater as long as it is within the stated range.   271

multiple switchboard (telephone switching systems). A telecommunications switchboard having each line connected to two or more jacks so that the line is within the reach of several operators.   55

multiple system (electrochemistry). The arrangement in a multielectrode electrolytic cell whereby in each cell all of the anodes are connected to the positive bus bar and all of the cathodes to the negative bus bar. *See:* **electrorefining.**   328

multiple transit signals (dispersive and nondispersive delay lines). Spurious signals having delay time related to the main signal delay by small odd integers. *Notes:* (1) Specific multiple transit signals may be labeled the third transit (triple transit), fifth transit, etc. (2) There is often a tradeoff available between multiple transit signal levels and bandwidth, delay time, insertion loss, and VSWR (voltage standing-wave ratio).   81

multiple tube (or valve). A space-charge-controlled tube or valve containing within one envelope two or more units or groups of electrodes associated with independent electron streams, through sometimes with one or more common electrodes. Examples: Double diode, double triode, triode-heptode, etcetera. *See:* **multiple-unit tube.**   125

multiple tube counts (radiation counter tubes). Spurious counts induced by previous tube counts.   125, 96

multiple-tuned antenna. An antenna designed to operate, without modification, in any of a number of pre-set frequency bands.   111

multiple twin quad (telephony). A quad in which the four conductors are arranged in two twisted pairs, and the two pairs twisted together. *See:* **cable.**   328

multiple-unit control (electric traction). A control system in which each motive-power unit is provided with its own controlling apparatus and arranged so that all such units operating together may be controlled from any one of a number of points on the units by means of a master controller.   328

multiple-unit electric car. An electric car arranged either for independent operation or for simultaneous operation with other similar cars (when connected to form a train of such cars) from a single control station. *Note:* A prefix diesel-electric, gas-electric, etcetera, may replace the word electric. *See:* **electric motor car.**   328

multiple-unit electric locomotive. A locomotive composed of two or more multiple-unit electric motive-power units connected for simultaneous operation of all such units from a single control station. *Note:* A prefix diesel-electric, turbine-electric, etcetera, may replace the word electric. *See:* **electric locomotive.**
                                                              328
multiple-unit electric motive-power unit. An electric motive-power unit arranged either for independent operation or for simultaneous operation with other similar units (when connected to form a single locomotive) from a single control station. *Note:* A prefix diesel-electric, gas-electric, turbine-electric, etcetera, may replace the word electric. *See:* **electric locomotive.**
                                                              328
multiple-unit electric train. A train composed of multiple-unit electric cars. *See:* **electric motor car.**
                                                              328
multiple-unit tube. *See:* **multiple tube (or valve).**
multiplex (communication) (data transmission). To interleave or simultaneously transmit two or more messages on a single channel.   59
multiplexer (supervisory control, data acquisition, and automatic control). (1) A device that allows the interleaving of two or more signals to a single line or terminal. (2) A device for selecting one of a number of inputs and switching its information to the output.
                                                              570
multiplexing (modulation systems) (data transmission). The combining of two or more signals into a single wave (the multiplex wave) from which the signals can be individually recovered.   59
multiplex lap winding (rotating machinery). A lap winding in which the number of parallel circuits is equal to a multiple of the number of poles.   63
multiplexor (hybrid computer linkage components). An electronic multiposition switch under the control of a digital computer, generally used in conjunction with an analog-to-digital converter (ADC), that allows for the selection of any one of a number of analog signals (up to the maximum capacity of the multiplexor), as the input to the ADC. A device that allows the interleaving of two or more signals to a single line or terminus.   10
multiplex printing telegraphy. That form of printing telegraphy in which a line circuit is employed to transmit in turn one character (or one or more pulses of a character) for each of two or more independent channels. *See:* **frequency-division multiplexing; time-division multiplexing; telegraphy.**   328
multiplex radio transmission. The simultaneous transmission of two or more signals using a common carrier wave. *See:* **radio transmission.**   111
multiplex wave winding (rotating machinery). A wave winding in which the number of parallel circuits is equal to a multiple of two, whatever the number of poles.   163
multiplication factor (k)(1)(power operations). A measure of the change in the neutron population in a reactor core from one generation to the subsequent generation. *See:* **effective multiplication factor; infinite multiplication factor.**   516

(2) (multiplier type of valve or tube) (thermionics). The ratio of the output current to the primary emission current. *See:* **electron emission.**   244, 190
multiplicative array antenna system. A signal-processing antenna system consisting of two or more receiving antennas and circuitry in which the effective angular response of the output of the system is related to the product of the radiation patterns of the separate antennas.   111
multiplier (1) (general). A device that has two or more inputs and whose output is a representation of the product of the quantities represented by the input signals.   210
(2) (analog computers). In an analog computer, a device capable of multiplying one variable by another.
                                                              9
(3) (linearity). *See:* **constant multiplier; normal linearity; servo multiplier.**
multiplier, constant (computing systems). A computing element that multiplies a variable by a constant factor. *See:* **electronic analog computer; multiplier (linearity).**   9, 10
multiplier, electronic. An all-electronic device capable of forming the product of two variables. *Note:* Examples are a time-division multiplier, a square-law multiplier, an amplitude-modulation-frequency-modulation (AM-FM) multiplier, and a triangular-wave multiplier. *See:* **electronic analog computer.**   9
multiplier, four-quadrant (analog computer). A multiplier in which operation is unrestricted as to the sign of both of the input variables.   10
multiplier, one-quadrant. A multiplier in which operation is restricted to a single sign of both input variables. *See:* **electronic analog computer.**   9
multiplier phototube. A phototube with one or more dynodes between its photocathode and output electrode. *See:* **amplifier; photocathode.**   125, 117
multiplier potentiometer (analog computers). Any of the ganged potentiometers of a servo multiplier that permit the multiplication of one variable by a second variable.   9
multiplier section, electron (electron tubes). *See:* **electron multiplier.**
multiplier servo. An electromechanical multiplier in which one variable is used to position one or more ganged potentiometers across which the other variable voltages are applied. *See:* **electronic analog computer; multiplier (linearity).**   9, 10
multiplier, two-quadrant. A multiplier in which operation is restricted to a single sign of one input variable only. *See:* **electronic analog computer.**   9, 10
multiplying-digital-to-analog converter (MDAC) (hybrid computer linkage components). *See:* **digital-to-analog multiplier (DAM).**
multipoint circuit (data transmission). A circuit interconnecting several stations.   59
multipoint connection (data communication). A configuration in which more than two stations are connected to a shared communications channel.   12
multipole fuse (1) (power switchgear). *See:* **pole (pole unit) (of a switching device or fuse) Second note.**   103

(2) (high-voltage switchgear). An assembly of two or more single-pole fuses. 443

multipole operation (of a circuit breaker or switching device) (power switchgear). A descriptive term indicating that all poles of the device are linked mechanically, electrically, or by other means such that they change state (open or close) substantially simultaneously. Devices may have capability for multipole opening, multipole closing, or both. 103

multiposition relay. A relay that has more than one operate or nonoperate position, for example, a stepping relay. *See:* **relay.** 259

multiposition switches (industrial control). (1) **self-returning switch.** A switch that returns to a stated position when it is released from any one of a stated set of other positions. (2) **spring return switch.** A switch in which the self-returning function is effected by the action of a spring. (3) **gravity-return switch.** A switch in which the self-returning function is effected by the action of weight. (4) **self-positioning switch.** A switch that assumes a certain operating position when it is placed in the neighborhood of the position. *See:* **switch.** 206

multipressure zone pothead (electric power distribution). A pressure-type pothead intended to be operated with two or more separate pressure zones that may be at different pressures. *See:* **pressure-type pothead; single-pressure zone potheads.** 4

multiprocessing (computing systems). Pertaining to the simultaneous execution of two or more programs or sequences of instructions by a computer network consisting of two or more processors. *See:* **parallel processing; multiprogramming.** 77

multiprocessor (computing systems). A computer capable of multiprocessing. 255, 77

multiprogramming (1) (computing systems). Pertaining to the interleaved execution of two or more programs by a computer. *See:* **parallel processing.** 255, 77

(2) (software). (A) A mode of operation that provides for the interleaved execution of two or more computer programs by a single processor. (B) Pertaining to the concurrent execution of two or more computer programs by a computer. (C) The concurrent execution of two or more functions as though each function operates alone. *See:* **computer program; execution.** 434

multi-radio-frequency-channel transmitter. *See:* **multichannel radio transmitter.**

multirate meter. A meter that registers at different rates or on different dials at different hours of the day. *See:* **electricity meter.** 328

multi-ratio current transformer (power and distribution transformer)(instrument transformer). One from which more than one ratio can be obtained by the use of taps on the secondary winding. 53, 394

multirestraint relay (power switchgear). A restraint relay that is so constructed that its operation may be restrained by more than one input quantity. 103

multisection coil (rotating machinery). A coil consisting of two or more coil sections or a group of turns,

each section or group being individually insulated. 63

multisegment magnetron. A magnetron with an anode divided into more than two segments, usually by slots parallel to its axis. *See:* **magnetrons.** 125

multispeed motor (1)(rotating machinery). One that can be operated at any one of two or more definite speeds, each being practically independent of the load. *Note:* For example, a direct-current motor with two armature windings or an induction motor with windings capable of various pole groupings. *See:* **asynchronous machine.** 63

(2) (electric installations on shipboard). A motor which can be operated at any one of two or more definite speeds, each being practically independent of the load. For example, a dc motor with two armature windings, or an induction motor with windings capable of various pole groupings. 3

multistage tube (X-ray tubes). An X-ray tube in which the cathode rays are accelerated by multiple ring-shaped anodes at progressively higher potential. 190

multistate indication. *See:* **supervisory control functions**

multi-step control. *See:* **control system, step.**

multitrace (oscilloscopes). A mode of operation in which a single beam in a cathode-ray tube is shared by two or more signal channels: *See:* **alternate display; chopped display; dual trace; oscillograph.** 185

multitrack recording system. A system that provides two or more recording tracks on a medium, resulting in either related or unrelated recordings in common time relationship. *See:* **multiple sound track; phonograph pickup.** 176

multivalent function. If to any value of u there corresponds more than one value of x (or more than one set of values of x_1, x_2, ..., x_n) then u is a multivalent function. Thus $u = \sin x$, $u = x^2$ are multivalent. 210

multivalued function. If to any value of x (or any set of values of x_1, x_2, ..., x_n) there corresponds more than one value of u, then u is a multivalued function. Thus $u = \cos^{-1} x$ is multivalued. 210

multi-variable function generator (analog computers). A function generator with more than one input. 9

multivariable system (control system). A system whose input vector and.or output vector has more than one element. 56

multivibrator. A relaxation oscillator employing two electron tubes to obtain the in-phase feedback voltage by coupling the output of each to the input of the other through, typically, resistance-capacitance elements. *Notes:* (1) The fundamental frequency is determined by the time constants of the coupling elements and may be further controlled by an external voltage. (2) A multivibrator is termed free-running or driven, according to whether its frequency is determined by its own circuit constants or by an external synchronizing voltage. The name multivibrator was originally given to the free-running multivibrator, having been suggested by the large number of harmonics produced. (3) When such circuits are normally in a nonoscillating state and a trigger signal is required to start a single cycle of operation, the circuit is commonly called a one-shot, a flip-flop, or a start-stop multivibrator. *See:* **oscillatory circuit.**   111

multivoltage control (elevators). A system of control that is accomplished by impressing successively on the armature of the driving-machine motor a number of substantially fixed voltages such as may be obtained from multicommutator generators common to a group of elevators. *See:* **control (elevators).**   328

multiwire branch circuit (National Electrical Code). A branch circuit consisting of two or more ungrounded conductors having a potential difference between them, and a grounded conductor having equal potential difference between it and each ungrounded conductor of the circuit and which is connected to the neutral conductor of the system.   256

multiwire element (antennas). A radiating element composed of several wires connected in parallel, the assemblage being the electrical equivalent of a single conductor larger than any one of the individual wires.   111

municipal fire-alarm system. A manual fire-alarm system in which the stations are accessibly located for operation by the public, and the signals of which register at a central station maintained and operated by the municipality. *See:* **protective signaling.**   328

municipal police report system. A system of strategically located stations from any one of which a patrolling policeman may report his presence to a supervisor in a central office maintained and operated by the municipality. *See:* **protective signaling.**   328

Munsell chroma (1) (illuminating engineering). The index of perceived (Y) and chromaticity coordinates (x,y) for CIE Standard Illuminant C and the CIE 1931 Standard Observer.   167
(2) (television). The dimension of the Munsell system of color that corresponds most closely to saturation. *Note:* Chroma is frequently used, particularly in English works, as the equivalent of saturation.   18

Munsell color system (1) (illuminating engineering). A system of surface- color specification based on perceptually uniform color scales for the three variables: Munsell hue, Munsell value, and Munsell chroma. For an observer of normal color vision, adapted to daylight, and viewing the specimen when illuminated by daylight and surrounded with a middle gray to white background, the Munsell hue, value, and chroma of the color correlate well with the hue, lightness, and perceived chroma. *Note:* A number of other color specification systems have been developed, usually for specific commercial purposes. These are described and discussed in considerable detail in several of the references listed in ANSI/IES RP-16-1980.   167
(2) (television). A system of surface-color specification based on perceptually uniform color scales for the three variables: Munsell hue, Munsell value, and Munsell chroma. *Notes:* (1) For an observer of normal color vision adapted to daylight, viewing the specimen when illuminated by daylight viewing and surrounding by a middle-gray-to-white background, the Munsell hue, value, and chroma of the color correlate well with the hue, lightness, and saturation of the perceived color. (2) A number of other color specification systems have been developed, usually for specific commercial purposes.   18

Munsell hue, (1) H (illuminating engineering). The index of the hue of the perceived object color defined in terms of the luminance factor (V) and chromaticity coordinates (x,y) for CIE Standard Illuminant C and the CIE 1931 Standard Observer.   167
(2) (television). The index of the hue of the perceived object color defined in terms of the Y value and chromaticity coordinates (x,y) of the color of the light reflected or transmitted by the object.   18

Munsell value (1) V (illuminating engineering). The index of the lightness of the perceived object color defined in terms of the luminance factor (Y) for CIE Standard Illuminant C and the CIE 1931 Standard Observer. *Note:* The exact definition gives Y as a 5th power function of V so that tabular or iterative methods are needed to find V as a function of Y. However, V can be estimated within $+-0.1$ by $V = 11.6 \, (Y/100)^{1/3} - 1.6$ or within $+-0.6$ by $V = Y^{1/2}$ where Y is the luminance factor expressed in percent.   167
(2) (television. The index of the lightness of the perceived object color defined in terms of the Y value. *Note:* Munsell value is approximately equal to the square root of the reflectance expressed in percent.   18

musa antenna (multiple-unit steerable antenna). *See:* **electronically scanned antenna; antenna.**

must operate value. *See:* **relay must operate value.**

mutation. *See:* **program mutation.**

mutual capacitance between two conductors. *See:* **balanced capacitance between two conductors.**

mutual conductance. The control-grid-to-anode transconductance. *See:* ON **period (electron tubes).**   244, 190

mutual coupling effect (1)(on input impedance of an array element) (antennas). The change in input impedance of an array element from the case when all other elements are present but open circuited to the case when all other elements are present and excited.   111
(2) (on the radiation pattern of an array antenna). The change in antenna pattern from the case when a particular feeding structure is attached to the array and mutual impedances among elements are ignored in deducing the excitation, to the case when the same feeding structure is attached to the array and mutual impedances among elements are included in deducing the excitation.   111

mutual impedance (1)(antenna). The mutual impedance between any two terminal pairs in a multielement array antenna is equal to the open-circuit voltage produced at the first terminal pair divided by the current supplied to the second when all other terminal pairs are open circuited. *See:* **antenna.**   246, 179

(2)(power fault effects). The ratio of the total induced open-circuit voltage on the disturbed circuit to the disturbing electric supply system phase current with the effects of all conductors taken into account. 404

mutual inductance. The common property of two electric circuits whereby an electromotive force is induced in one circuit by a change of current in the other circuit. *Notes:* (1) The coefficient of mutual inductance M between two windings is given by the following equation

$$M = \frac{\partial i}{\partial \lambda}$$

where λ is the total flux linkage of one winding and i is the current in the other winding. (2) The voltage e induced in one winding by a current i in the other winding is given by the following equation

$$e = -\left[M\frac{di}{dt} + i\frac{dM}{dt} \right]$$

If M is constant $e = -M\dfrac{di}{dt}$

197

mutual inductor. An inductor for changing the mutual inductance between two circuits. 210

mutual information (information theory). The amount of information about one event, say x = a, provided by another event, say y = b, for example, $I_{x,y}^{(a;b)} = \log$ $P_{x\ y}^{(a\ b)}$ over $P_{x}^{(a)}$, where $P_{x}^{(a)}$ is the probability that x = a and $P_{x\ y}^{(a\ b)}$ is the probability that x = a and $P_{x\ y}^{(a\ b)}$ is the conditional probability that x = a given that y = b. *Note:* This quantity is symmetric in its two arguments, that is, $I_{y;x}^{(b;a)} = I_{x;y}^{(a;b)}$. *See:* information. 415

mutual interference chart (MIC) (electromagnetic compatibility). A plot or matrix, with ordinate and abscissa representing the tuned frequencies of a single transmitter-receiver combination, that indicates potential interference to normal receiver operation by reason of interaction of the two equipments under consideration at any combination of tuned transmit.-receive frequencies. *Note:* This interaction includes transmitter harmonics and other spurious emissions, and receiver spurious responses and images. *See:* electromagnetic compatibility. 199

mutually exclusive events (nuclear power generating stations)(reliability analysis of nuclear power generating station safety systems). Events that cannot exist simultaneously. 29, 587

mutual resistance of grounding electrodes. Equal to the voltage change in one of them produced by a change of one ampere of direct current in the other, and is expressed in ohms. 313

mutual surge impedance (surge arresters). The apparent mutual impedance between two lines, both of infinite length. *Note:* It determines the relationship between the surge voltage induced into one line by a surge current of short duration in the other. 244, 62

N

NA (1) (inertia sensor). *See:* laser gyro axes. 46
(2) (fiber optics). *See:* numerical aperture. 433
nameplate (rating plate) (rotating machinery). A plaque giving the manufacturer's name and the rating of the machine. 63
NaN (radix-independent floating-point arithmetic)-(binary floating-point arithmetic). Not a number, a symbolic entity encoded in floating-point format. There are two types of NaNs. Signaling NaNs signal the invalid (operation) exception whenever they appear as operands. Quiet NaNs propagate through almost every arithmetic operation without signaling exceptions. 558, 469
NAND (mathematics of computing). A Boolean operator having the property that if P is a statement, Q is a statement, R is a statement,..., then the NAND of P, Q, R, ... is true if and only if at least one statement is false. *Note:* The NAND of P and Q is often represented by $\overline{PQ}$. 564
nano (mathematics of computing). A prefix indicating 10^{-9}.
narrow-angle diffusion (illuminating engineering).

That in which flux is scattered at angles near the direction which the flux would take by regular reflection or transmission. 167
narrow angle luminaire (illuminating engineering). A luminaire which concentrates the light within a cone of comparatively small solid angle. 167
narrow-band axis (color television) (phasor representation of the chrominance signal). The direction of the phasor representing the coarse chrominance primary. 18, 178
narrow-band interference (electromagnetic compatibility). For purposes of measurement, a disturbance of spectral energy lying within the bandpass of the measuring receiver in use. *See:* electromagnetic compatibility. 199
narrowband radio noise. Radio noise having a spectrum exhibiting one or more sharp peaks, narrow in width compared to the nominal bandwidth of, and far enough apart to be resolvable by, the measuring instrument (or the communication receiver to be protected). 418
narrowband response spectrum (seismic qualification

of Class 1E equipment for nuclear power generating stations). A response spectrum that describes the motion in which amplified response occurs over a limited (narrow) range of frequencies.   581

narrow band spurious emission (land-mobile communication transmitters). Any spurious output emitted from a radio transmitter, other than on its assigned frequency, which produces a disturbance of spectral energy lying within the bypass of the measuring receiver in use.   444

N-ary (1) (software). (A) Characterized by a selection, choice or condition that has n possible different values or states. (B) Of a fixed radix numeration system, having a radix of n.   434

(2) (information theory). A code whose output alphabet consists of N symbols.   415

nat. *See: nit.*

national call (telephone switching systems). A toll call to a destination outside of the local service area of the calling customer but within the boundaries of the country in which he is located.   55

national distance dialing (telephone switching systems). The automatic establishing of a national call by means of signals from the calling device of either a customer or an operator.   55

National Electrical Code (NEC) dimensions (protection and coordination of industrial and commercial power systems). Dimensions once stated in the National Electrical Code, but now found in ANSI/UL 198B-1982 and in ANSI/UL 198D-1982. These dimensions are common to Class H and K fuses and provide interchangeability between manufacturers for fuses and fusible equipment of a given ampere and voltage range.   504

national number (telephone switching systems). The combination of digits representing an area code and a directory number that, for the purpose of distance dialing, uniquely identifies each main station within each of the world's geographical areas that is identified by a country code.   55

national-numbering plan (telephone switching systems). Any plan for identifying telephone stations within a geographical area identified by a unique country code.   55

national numbering-plan area (telephone switching systems). A geographical area of the world where a country code and the national boundaries are related uniquely.   55

national radioactivity standard source (germanium detectors). A calibrated radioactive source prepared and distributed as a standard reference material by the U.S. National Bureau of Standards.   397

national standard (plutonium monitoring). An instrument source or other system or device maintained and promulgated as such by the United States National Bureau of Standards.   413

national standards (metering). Those standards of electrical measurements that are maintained by the National Bureau of Standards.   212

nationwide toll dialing (telephony). A system of automatic switching whereby an outward toll operator can complete calls to any basic-numbering-plan area in the country covered by the system.   328

native system demand (power operations). The net 60 min clock-hour peak integrated demand within the system less interruptible loads.   516

natural air cooling system (thyristor controller). A cooling system in which the heat is removed from the cooling surfaces of the thyristor controller components only by the natural action of the ambient air.   445

natural bandwidth (laser-maser). The linewidth of an absorption or emission line when spontaneous emission is the dominant process determining spectral distribution.   363

natural frequency (1)(seismic design of substations). The frequency(s) at which a body vibrates due to its own physical characteristics (mass, shape, boundary conditions, and elastic restoring forces brought into play) when the body is distorted and then released, while restrained or supported at specified points.   465

(2)(seismic qualification of Class 1E equipment for nuclear power generating stations). The frequency(s) at which a body vibrates due to its own individual characteristics (mass and stiffness) when the body is distorted in a specific direction and then released.   581

(3) (surge arresters). The frequency or frequencies at which the circuit will oscillate if it is free to do so. *See: frequency, undamped.*   308, 62

(4) (gyro; accelerometer). That frequency at which the output lags the input by ninety degrees. It is generally applied only to inertial sensors with approximate second order response.   46

(5) (nuclear power generating stations) (seismic qualification of class 1E equipment). The frequency or frequencies at which a body vibrates due to its own physical characteristics (mass, shape) and elastic restoring forces brought into play when the body is distorted in a specific direction and then released, while restrained or supported at specified points.   28

(6) (automatic control). *See: frequency, undamped.*

natural language (software). A language whose rules are based on current usage without being explicitly prescribed. Examples include English, Chinese, French, and Swahili. *See: formal language.*   434

natural linewidth (laser-maser). The linewidth of an absorption or emission line when spontaneous emission is the dominant process determining spectral distribution.   363

natural noise (electromagnetic compatibility). Noise having its source in natural phenomena and not generated in machines or other technical devices. *See: electromagnetic compatibility.*   220, 199

natural period. The period of the periodic part of a free oscillation of the body or system. *Notes:* (1) When the period varies with amplitude, the natural period is the period as the amplitude approaches zero. (2) A body or system may have several modes of free oscillation, and the period may be different for each.   210

Naveam (navigation aid terms). A radio navigational warning of dangers to shipping in the Eastern, Atlantic, Mediterranean, and Red Seas.   526

navigation (navigation aid terms). The process of directing a vehicle so as to reach the intended destination.   526

navigational aid (navigation aid terms). An instrument, system, device, chart, or method intended to assist in the navigation of a vehicle.   526

navigational astronomy (navigation aid terms). That part of astronomy of direct use to a navigator, comprised principally of celestial coordinates, time, and apparent motions of celestial bodies.   526

navigational radar (surface search radar). A high-frequency radio transmitter-receiver for the detection, by means of transmitted and reflected signals, of any object (within range) projecting above the surface of the water and for visual indication of its bearing and distance. *See:* **radio direction-finder (radio compass).**

navigational satellite (navigation aid terms). An artificial earth orbiting satellite designed for navigational purposes. *See:* **satellite navigation; global positioning system.**   526

navigation coordinate (navigation aid terms). Any one of a set of quantities; the set serving to define a position.   526

navigation light system (illuminating engineering). A set of aircraft aeronautical lights provided to indicate the position and direction of motion of an aircraft to pilots of other aircraft or to ground observers.   167

navigation parameter (navigation aid terms). A measurable characteristic of motion or position used in the process of navigation.   526

navigation quantity (navigation aid terms). A measured value of a navigation parameter.   526

n-channel device (metal-nitride-oxide field-effect transistor). Insulated-gate field-effect transistor (IGFET) where source and drain are regions of n-type conductivity.   386

***n*-conductor cable (electric power distribution).** *See:* **multiple-conductor cable.**

***n*-conductor concentric cable (electric power distribution).** *See:* **multiple-conductor concentric cable.**

***N*-contours.** *See:* **phase contours.**

NDB (electronic navigation). *See:* **nondirectional beacon.**

N-display* (radar). A K-display having an adjustable pedestal, notch, or step, as in the M-display, for the measurement of range. *Note:* This display is usually regarded as a variant of an A-display rather than as a separate type.

*Rare.   13

near-end crosstalk. Crosstalk that is propagated in a disturbed channel in the direction opposite to the direction of propagation of the current in the disturbing channel. *Note:* The terminal of the disturbed channel at which the near-end crosstalk is present is ordinarily near to or coincides with the energized terminal of the disturbing channel. *See:* **coupling.**   328

near-field (radiation) pattern (antennas). Any radia-

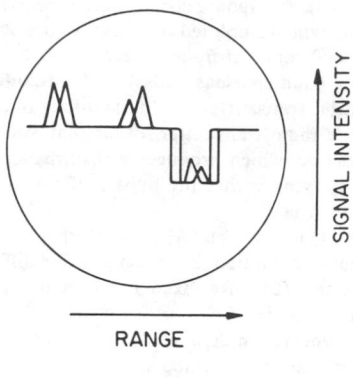

N-Display
(with notch)

tion pattern obtained in the near-field of an antenna. *Note:* Near-field patterns are usually taken over paths on planar, cylindrical or spherical surfaces. *See:* **Fresnel pattern; radiation pattern cut.**   111

near-field diffraction pattern (fiber optics). The diffraction pattern observed close to a source or aperture, as distinguished from far-field diffraction pattern. *Note:* The pattern in the output plane of a fiber is called the near-field radiation pattern. *Syn:* **Fresnel diffraction pattern.** *See:* **diffraction; far-field diffraction pattern; far-field region.**   433

near-field pattern. *See:* **near-field radiation pattern; radiation pattern.**

near-field radiation pattern. *See:* **radiation pattern.**

near-field region (1) (antennas). That part of space between the antenna and far-field region. *Note:* In lossless media, the near-field may be further subdivided into reactive and radiating near-field regions.   111

(2) (fiber optics). The region close to a source, or aperture. The diffraction pattern in this region typically differs significantly from that observed at infinity and varies with distance from the source. *See:* **far-field diffraction pattern; far-field region.**   433

(3) (land-mobile communication transmitters). The region of the field of an antenna between the reactive near field region and the far field region wherein radiation fields predominate and wherein the angular field distribution is dependent upon distance from the antenna. *Notes:* (1) If the antenna has a maximum overall dimension which is not large compared to the wavelength, this field region may not exist. (2) For an antenna focused at infinity, the radiating near field region is sometimes referred to as the Fresnel region on the basis of analogy to optical terminology.   444

near-field region in physical media. *   167
*Deprecated

near-field region, radiating. The region of the field of an antenna between the reactive near-field region and the far-field region wherein radiation fields predominate and wherein the angular field distribution is dependent upon distance from the antenna. *Notes:* (1) If the antenna has a maximum over-all dimension which is not large compared to the wavelength, this field region may not exist. (2) For an antenna focused at infinity, the radiating near-field region is sometimes referred to as the Fresnel region on the basis of analogy to optical terminology. *See:* **radiating near-field region.**   111

near-field region, reactive. *See:* **reactive near-field region.**

near-field scanning (fiber optics). The technique for measuring the index profile of an optical fiber by illuminating the entrance face with an extended source and measuring the point-by-point radiance of the exit face. *See:* **refracted ray method.**   433

near-side (of a segment interconnect (SI) or a buffered interconnect (BI))(FASTBUS acquisition and control). That port of an SI or BI that is electrically closer to the originating master.   480

neck (cathode-ray tube). The small tubular part of the envelope near the base. *See:* **cathode-ray tube.**   244, 190

negate. To perform the logic operation NOT.   255, 77

negative after-potential (electrobiology). Relatively prolonged negativity that follows the action spike in a homogeneous fiber group. *See:* **contact potential.**   192

negative conductance (circuits and systems). The conductance of a negative-resistance device.   67

negative conductor. A conductor connected to the negative terminal of a source of supply. *Note:* A negative conductor is frequently used as an auxiliary return circuit in a system of electric traction. *See:* **center of distribution.**   64

negative-differential-resistance region (thyristor). Any portion of the principal voltage-current characteristic in the switching quadrant(s) within which the differential resistance is negative. *See:* **principal voltage-current characteristic (principal characteristic).**   243, 66, 191

negative electrode (1) (primary cell). The anode when the cell is discharging. *Note:* The negative terminal is connected to the negative electrode. *See:* **electrolytic cell.**
(2) (metallic rectifier). The electrode from which the forward current flows within the cell. *See:* **rectification.**   328

negative feedback (1) (circuits and systems). The process by which part of the signal in the output circuit of an amplifying device reacts upon the input circuit in such a manner as to counteract the initial power, thereby decreasing the amplification.   67

(2) (control) (industrial control). A feedback signal in a direction to reduce the variable that the feedback represents. *See:* **control system, feedback; feedback.**   219, 206

(3) (degeneration) (stabilized feedback) (data transmission). The process by which a part of the power in the output circuit of an amplifying device reacts upon the input circuit in such a manner as to reduce the initial power, thereby decreasing the amplification.   59

negative glow (1) (gas tube). The luminous glow in a glow-discharge cold-cathode tube between the cathode dark space and the Faraday dark space. *See:* **gas tube.**   190
(2) (overhead-power-line corona and radio noise). Negative glow occurs at field strengths above those required for Trichel streamers. Negative glow is confined to a small portion of the electrode and appears as a small stationary luminous bluish fan. The corona current of negative glow is essentially pulseless. *See:* **Trichel streamers.**   411

negative logic convention (graphic symbols for logic functions). The representation of the 1-state and the 0-state by the low (L) and high (H) levels, respectively.   451

negative matrix (negative). A matrix the surface of which is the reverse of the surface to be ultimately produced by electroforming. *See:* **electroforming.**   328

negative modulation (amplitude-modulation television system). That form of modulation in which an increase in brightness corresponds to a decrease in transmitted power. *See:* **television.**   328

negative-phase-sequence (phase-reversal) relay. A relay that responds to the negative-phase-sequence component of a polyphase input quantity. *Note:* Frequently employed in three-phase systems. *See:* **relay.**   103, 127

negative-phase-sequence impedance (rotating machinery). The quotient of the negative-sequence rated-frequency component of the voltage, assumed to be sinusoidal, at the terminals of a machine rotating at synchronous speed, and the negative-sequence component of the current at the same frequency. *Note:* It is equal to the asynchronous impedance for a slip equal to 2. *See:* **asynchronous machine.**   63

negative-phase-sequence reactance (rotating machinery). The quotient of the reactive fundamental component of negative-sequence primary voltage due to sinusoidal negative-sequence primary current of rated frequency, and the value of this current, the machine running at rated speed. *See:* **asynchronous machine.**   58

negative-phase-sequence relay (power switchgear). A relay that responds to the negative-phase-sequence component of a polyphase input quantity.   103

negative phase-sequence resistance (rotating machin-

ery). The quotient of the in-phase fundamental component of negative-sequence primary voltage, due to sinusoidal negative-sequence primary current of rated frequency, and the value of this current, the machine running at rated speed. *See:* **asynchronous machine.**

63

negative-phase-sequence symmetrical components. Of an unsymmetrical set of polyphase voltages or currents of M phases, that set of symmetrical components that have the $(m - 1)$st phase sequence. That is, the angular phase lag from the first member of the set to the second, from every other member of the set to the succeeding one, and from the last member to the first, is equal to $(m - 1)$ times the characteristic angular phase difference, or $(m - 1)2\pi/m$ radians. The members of this set will reach their positive maxima uniformly but in the reverse order of their designations. *Note:* The negative-phase-sequence symmetrical components for a three-phase set of unbalanced sinusoidal voltages ($m = 3$), having the primitive period, are represented by the equations

$$e_{a2} = (2)^{1/2}E_{a2}\cos{(\omega t + \alpha_{a2})}$$

$$e_{b2} = (2)^{1/2}E_{a2}\cos{\left(\omega t + \alpha_{a2} - \frac{4\pi}{3}\right)}$$

$$e_{c2} = (2)^{1/2}E_{a2}\cos{\left(\omega t + \alpha_{a2} - \frac{2\pi}{3}\right)}$$

derived from the equation of symmetrical components of a set of polyphase (alternating) voltages. Since in this case $r = 1$ for every component (of first harmonic), the third subscript is omitted. Then k is 2 for $(m -1)$st sequence, and s takes on the algebraic values 1, 2, and 3 corresponding to phases a, b, and c. The sequence of maxima occurs in the order, a, c, b, which is the reverse or negative of the order for $k = 1$. *See:* **polyphase alternating currents; symmetrical components.**

210

negative-polarity lightning stroke. A stroke resulting from a negatively charged cloud that lowers negative charge to the earth. *See:* **direct-stroke protection (lightning).**

64

negative pre-breakdown streamers (overhead-power-line corona and radio noise). Streamers occurring at field strengths close to breakdown. The discharge appears as bright filament with very little branching and extends far into the gap. The associated current pulse has high magnitude, long duration, and low repetition rate.

411

negative-resistance device. A resistance in which an increase in current is accompanied by a decrease in voltage over the working range.

210

negative-resistance oscillator. An oscillator produced by connecting a parallel-tuned resonant circuit to a two-terminal negative-resistance device. (One in which an increase in voltage results in a decrease in current.) *Note:* Dynatron and transitron oscillators,

arc converters, and oscillators of the semiconductor type are examples. *See:* **oscillatory circuit.**

111, 211

negative-resistance repeater (data transmission). A repeater in which gain is provided by a series or a shunt negative resistance, or both.

59

negative-sequence reactance. The ratio of the fundamental reactive component of negative-sequence armature voltage, resulting from the presence of fundamental negative-sequence armature current of rated frequency, to this current, the machine being operated at rated speed. *Notes:* (1) The rated current value of negative-sequence reactance is the value obtained from a test with a fundamental negative-sequence current equal to rated armature current. The rated voltage value of negative-sequence reactance is the value obtained from a line-to-line short-circuit test at two terminals of the machine at rated speed, applied from no load at rated voltage, the resulting value being corrected when necessary for the effect of harmonic components in the current. (2) For any unbalanced short-circuits, certain harmonic components of current, if present, may produce fundamental reactive components of negative-sequence voltage that modify the ratio of the total fundamental reactive component of negative-sequence voltage to the fundamental component of negative-sequence current. This effect can be included by multiplying the negative-sequence reactance, before it is used for short-circuit calculations, by a wave distortion factor, equal to or less than unity, that depends primarily upon the type of short-circuit, and upon the characteristics of the machine and the external circuit, if any, between the machine and the point of short-circuit.

328

negative-sequence resistance. The ratio of the fundamental component of in-phase armature voltage, due to the fundamental negative-sequence component of armature current to this component of current at rated frequency. *Note:* This resistance, which forms a part of the negative-sequence impedance for use in circuit calculations to establish relationships between voltages and currents, is not directly applicable in the calculations of the total loss in the machine caused by negative-sequence currents. This loss is the product of the square of the fundamental component of the negative-sequence current and the difference between twice the negative-sequence resistance and the positive-sequence resistance, that is, $I_2^2 (2R_2 \text{ mins } R_1)$.

328

negative temperature (laser-maser) An **effective temperature** used in the Boltzmann factor to describe a **population inversion.** *Note:* If n_2 particles populate the higher of two states and n_1 particles populate the lower state, their ratio is conventionally expressed by the Boltzmann factor

$$\frac{n_2}{n_1} = \exp(-kT)$$

where k is the Boltzmann constant and T the (effective) absolute temperature.   363

negative terminal (of a battery). The terminal toward which positive electric charge flows in the external circuit from the positive terminal. *Note:* The flow of electrons in the external circuit is to the positive terminal and from the negative terminal. *See:* **battery (primary or secondary).**   328

negative-transconductance oscillator. Oscillator in which the output of the device is coupled back to the input without phase shift, the condition for oscillation being satisfied by the negative conductance of the device. *See:* **oscillatory circuit.**   111

negative vectors. Two vectors are mutually negative if their magnitudes are the same and their directions opposite.   210

negentropy (information theory). *See:* **average information content.**

neighborhood. Of any point u_0 in a three-dimensional space, the volume enclosed by a sphere drawn with u_0 as center. Of any point u_0 in a two-dimensional space, the area enclosed by a circle drawn with u_0 as center.   210

neon indicator (tube). A cold-cathode gas-filled tube containing neon, used as a visual indicator of a potential difference or a field. *See:* **gas tube.**   244, 190

neper (data transmission). A division of the logarithmic scale so that the number of nepers is equal to the natural logarithm of the scalar ratio of two currents or two voltages. *Notes:* (1) With I_1 and I_2 designating the scalar value of two currents, and n the number of nepers denoting their scalar ratio: $n = \log_e(I_1/I_2)$. (2) One neper equals 0.8686 bel. (3) The neper is a dimensionless unit.   59

nerve-block (electrobiology). The application of a current to a nerve so as to prevent the passage of a propagated potential. *See:* **excitability (electrobiology).**   192

nest (software). (1) To incorporate a structure or structures of some kind into a structure of the same kind. For example, to nest one loop (the nested loop) within another loop (the nesting loop); to nest one subroutine (the nested subroutine) within another subroutine (the nesting subroutine). (2) To place subroutines or data in other subroutines or data at a different hierarchical level so that subroutines can be executed as recursive subroutines or so that the data can be accessed recursively. *See:* **data; level; loop; subroutine.**   434

nest or section (multiple system). A group of electrolytic cells placed close together and electrically connected in series for convenience and economy of operation. *See:* **electrorefining.**   328

net assured capability (electric power supply). The net dependable capability of all power sources available to a system, including firm power contracts and applicable emergency interchange agreements, less that reserve assigned to provide for scheduled maintenance outages, equipment and operating limitations, and forced outages of power sources.   112

net capability (electric power supply). The maximum generation expressed in kilowatt hours per hour which

a generating unit, station, power source, or system can be expected to supply under optimum operating conditions.   112

net dependable capability (1) (electric power supply). The maximum generation, expressed in kilowatt hours per hour which a generating unit, station, power source, or system can be depended upon to supply on the basis of average operating conditions.   112
(2) (power system engineering). The maximum system load, expressed in kilowatthours per hour that a generating unit, station, or power source can be depended upon to supply on the basis of average operating conditions. *Note:* This capability takes into account average conditions of weather, quality of fuel, degree of maintenance and other operating factors. It does not include provision for maintenance outages. *See:* **generating station.**   64, 200

net generation (electric power systems). Gross generation less station or unit power requirements.   94

net head (power operations). The gross head less all hydraulic losses except those chargeable to the turbine.   516

net information content. A measure of the essential information contained in a message. *Note:* It is expressed as the minimum number of bits or hartleys required to transmit the message with specified accuracy over a noiseless medium. *See:* **bit.**   328

net interchange (power and/or energy) (control area). The algebraic sum of the power and/or energy on the area tie lines. *Note:* Positive net interchange is due to excess generation out of the area.   94

net interchange deviation (control area) (electric power systems). The net interchange minus the scheduled net interchange.   94

net interchange schedule programmer (speed-governing system). A means of automatically changing the net interchange schedule from one level to another at a predetermined time and during a predetermined period or at a predetermined rate. *See:* **speed-governing system.**   94

net load capability. The maximum system load expressed in kilowatt hours per hour that a generating unit, station, or power source can be expected to supply under good operating conditions. *Notes:* (1) This capability provides for variations of load within the hour that it is assumed are to be spread among all of the power sources. (2) This is sometimes called net rated capability. *See:* **generating station.**   64, 200

net loss (circuit equivalent) (data transmission). The sum of all the transmission losses occurring between the two ends of the circuit, minus the sum of all the transmission gains.   59

net system energy (power operations). Energy requirements of a system, including losses, defined as (1) net generation of the system, plus (2) energy delivered to other systems.   516

network (1)(measuring longitudinal balance of telephone equipment operating in the voice band). A combination of elements or devices.   529
(2) (communication) (data transmission). (A) A se-

ries of points interconnected by communication channels. (B) The switched telephone network is the network of telephone lines normally used for dialed telephone calls. (C) A private network is a network of communications channels confined to the use of one customer.   59
(3) (software). (A) An interconnected or interrelated group of nodes. (B) In connection with a disciplinary or problem oriented qualifier, the combination of material, documentation, and human resources that are united by design to achieve certain objectives, for example, a social science network, a science information network. *See:* **documentation; node.**   434
(4) (distribution of electric energy). An aggregation of interconnected conductors consisting of feeders, mains, and services. *See:* **alternating-current distribution; network analysis.**   64
network analysis (network). The derivation of the electrical properties, given its configuration and element values.   210
network analyzer (network analyzers). A system that measures the two-port transmission and one-port reflection characteristics of a multiport in its linear range at a common input and output frequency. *Note:* This includes systems that (1) Measure magnitude of reflection and transmission coefficient only as well as those systems that measure magnitude and phase (2) Operate manually or automatically (3) Cover a frequency range either continuously or in small enough steps to make coverage practically continuous (4) Vary frequency manually or automatically, stepped, or swept.   505
network capacitance (charging inductors). The effective capacitance of the pulse-forming circuit.   137
network control (telephone switching systems). The means of determining and establishing the required connections in response to information received from the system control.   55
network feeder. A feeder that supplies energy to a network. *See:* **center of distribution.**   64
network function. Any impedance function, admittance function, or other function of *p* that can be expressed in terms of or derived from the determinant of a network and its cofactors. *Notes:* (1) This includes not only impedance and admittance functions as previously defined, but also voltage ratios, current ratios, and numerous other quantities. (2) Certain network functions are sometimes classified together for a given purpose (for example, those with common zeros or common poles). These represent subgroups that should be specifically defined in each special case. (3) In the case of distributed-parameter networks, the determinant may be an infinite one: the definition still holds. *See:* **network analysis.**   210
network interface (telephone loop performance). The interface between the public switched telephone network and the customer premises wiring.   473
network leased line or private wire (data transmission). A series of points interconnected by telegraph or telephone channels, and reserved for the exclusive use of one customer.   59

network limiter (power switchgear). An enclosed fuse for disconnecting a faulted cable from a low-voltage network distribution system and for protecting the unfaulted portions of that cable against serious thermal damage.   103
network, load matching. *See:* **load matching network.**
network management (NMT)(token ring access method). The conceptual control element of a station which interfaces with all of the layers of the station and is responsible for the setting and resetting of control parameters, obtaining reports of error conditions, and determining if the station should be connected to or disconnected from the medium.   472
network master relay (power switchgear). A relay that functions as a protective relay by opening a network protector when power is back-fed into the supply system and as a programming relay by closing the protector in conjunction with the network phasing relay when polyphase voltage phasors are within prescribed limits.   103
network phasing relay (power switchgear). A monitoring relay which has as its function to limit the operation of a network master relay so that the network protector may close only when the voltages on the two sides of the protector are in a predetermined phasor relationship.   103
network primary distribution system. A system of alternating-current distribution in which the primaries of the distribution transformers are connected to a common network supplied from the generating station or substation distribution buses. *See:* **alternating-current distribution.**   64
network, private line telegraph. *See:* **private line telegraph network.**
network, private line telephone. *See:* **private line telephone network.**
network protector (power switchgear) (power and distribution transformer). An assembly comprising a circuit breaker and its complete control equipment for automatically disconnecting a transformer from a secondary network in response to predetermined electrical conditions on the primary feeder or transformer, and for connecting a transformer to a secondary network either through manual control or automatic control responsive to predetermined electrical conditions on the feeder and the secondary network. *Note:* The network protector is usually arranged to connect automatically its associated transformer to the network when conditions are such that the transformer, when connected, will supply power to the network and to automatically disconnect the transformer from the network when power flows from the network to the transformer.   53, 103
network restraint mechanism (power switchgear). A device that prevents opening of a network protector on transient power reversals which either do not exceed a predetermined value or persist for a predetermined time.   103
network secondary distribution system. A system of alternating-current distribution in which the secondaries of the distribution transformers are connect-

ed to a common network for supplying light and power directly to consumers' services. *See:* **alternating-current distribution.**   64

network synthesis (network). The derivation of the configuration and element values, with given electrical properties. *See:* **network analysis.**   210

network transformer (power and distribution transformer). A transformer designed for use in a vault to feed a variable capacity system of interconnected secondaries. *Note:* A network transformer may be of the submersible or of the vault type. It usually, but not always, has provision for attaching a network protector.   53

network tripping and reclosing (networking) equipment) (power switchgear). An equipment that automatically connects its associated power transformer to an ac network when conditions are such that the transformer, when connected, will supply power to the network and that automatically disconnects the transformer from the network when power flows from the network to the transformer.   103

neuroelectricity. Any electric potential maintained or current produced in the nervous system. *See:* **electrobiology.**   192

neutral (rotating machinery). The point along an insulated winding where the voltage is the instantaneous average of the line terminal voltages during normal operation. *See:* **asynchronous machine.**   63

neutral conductor (when one exists) (circuit consisting of three or more conductors). The conductor that is intended to be so energized, that, in the normal steady state, the voltages from every other conductor to the neutral conductor, at the terminals of entry of the circuit into a delimited region, are definitely related and usually equal in amplitude. *Note:* If the circuit is an alternating-current circuit, it is intended also that the voltages have the same period and the phase difference between any two successive voltages, from each of the other conductors to the neutral conductor, selected in a prescribed order, have a predetermined value usually equal to 2π radians divided by the number of phase conductors m. *See:* **center of distribution; network analysis.**   210

neutral direct-current telegraph system (single-current system) (single Morse system). A telegraph system employing current during marking intervals and zero current during spacing intervals for transmission of signals over the line. *See:* **telegraphy.**   328

neutral ground (power and distribution transformer). An intentional ground applied to the neutral conductor or neutral point of a circuit, transformer, machine, apparatus, or system.   53

neutral grounding capacitor (electric power). A neutral grounding device the principal element of which is capacitance. *Note:* A neutral grounding capacitor is normally used in combination with other elements, such as reactors and resistors. *See:* **grounding device.**   91, 64

neutral grounding device (electric power). A grounding device used to connect the neutral point of a system of electric conductors to earth. *Note:* The device may consist of a resistance, inductance, or capacitance element, or a combination of them. *See:* **grounding device; neutral grounding impedor.**   91, 64

neutral grounding impedor (electric power). A neutral grounding device comprising an assembly of at least two of the elements resistance, inductance, or capacitance. *See:* **grounding device; neutral grounding device.**   91, 64

neutral grounding reactor (1) (transformer). A current-limiting inductive reactor for connection in the neutral for the purpose of limiting and neutralizing disturbances due to ground faults.   53
(2) (neutral grounding devices). One in which the principal element of which is inductive reactance.   91

neutral grounding resistor (electric power). A neutral grounding device, the principal element of which is resistance. *See:* **grounding device.**   91, 64

neutral grounding wave trap. A neutral grounding device comprising a combination of inductance and capacitance designed to offer a very high impedance to a specified frequency or frequencies. *Note:* The inductances used in neutral grounding wave traps should meet the same requirements as neutral grounding reactors.   91

neutralization. A method of nullifying the voltage feedback from the output to the input circuits of an amplifier through the tube interelectrode impedances. *Note:* Its principal use is in preventing oscillation in an amplifier by introducing a voltage into the input equal in magnitude but opposite in phase to the feedback through the interelectrode capacitance. *See:* **amplifier; feedback.**   111, 211

neutralizing indicator. An auxiliary device for indicating the degree of neutralization of an amplifier. (For example, a lamp or detector coupled to the plate tank circuit of an amplifier.) *See:* **amplifier.**   111

neutralizing transformers (wire-line communication facilities). A device which introduces a voltage into a circuit pair to oppose an unwanted voltage. It neutralizes extraneous longitudinal voltages resulting from ground potential rise, or longitudinal induction, or both, which simultaneously allowing ac or dc metallic signals to pass. These transformers are primarily used to protect communication circuits at power stations, or along routes where exposure to power line induction is a problem, or both.   414

neutralizing voltage. The alternating-current voltage specifically fed from the grid circuit to the plate circuit (or vice versa), deliberately made 180 degrees out of phase with, and equal in amplitude to, the alternating-current voltage similarly transferred through undesired paths, usually the grid-to-plate tube capacitance. *See:* **amplifier.**   111

neutral keying (data transmission). A form of telegraph signal which has current either on or off in the circuit with "on" as mark, "off" as space.   59

neutral lead (rotating machinery). A main lead connected to the common point of a star-connected winding. *See:* **asynchronous machine.**   63

neutral point (1) (system). The point that has the same

potential as the point of junction of a group of equal nonreactive resistances if connected at their free ends to the appropriate main terminals or lines of the system. *Note:* The number of such resistances is 2 for direct-current or single-phase alternating-current: 4 for two-phase (applicable to 4-wire systems only) and 3 for three-, six- or twelve-phase systems. *See:* **alternating-current distribution; center of distribution; direct-current distribution.**   64

(2) (transformer)(power and distribution transformer). (A) The common point of a Y connection in a polyphase system. (B) The point of a symmetrical system which is normally at zero voltage.   53

neutral relay (1) (sometimes called nonpolarized relay). A relay in which the movement of the armature does not depend upon the direction of the current in the circuit controlling the armature.   259

(2) (power switchgear). A relay that responds to quantities in the neutral of a power circuit.   103

neutral terminal. The terminal connected to the neutral of a machine or apparatus. *See:* **asynchronous machine.**   63

neutral wave trap (electric power). A neutral grounding device comprising a combination of inductance and capacitance designed to offer a very high impedance to a specified frequency or frequencies. *See:* **grounding device.**   91

neutral zone (control element). The range of values of input for which no change in output occurs. *Note:* The neutral zone is an adjustable parameter in many two-step and floating control systems. *See:* **dead band.**   56

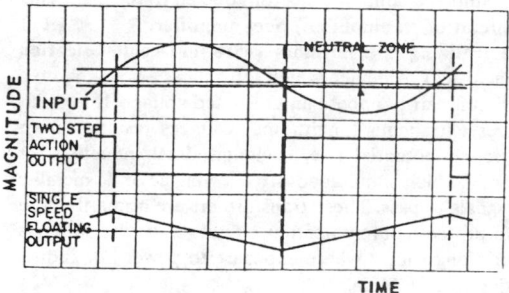

Neutral zone

new installation (elevators). Any installation not classified as an existing installation by definition, or an existing elevator, dumbwaiter, or escalator moved to a new location subsequent to the effective date of a code. *See:* **elevators.**   328

new sync (data transmission). Allows for a rapid transition from one transmitter to another on multipoint private line data networks.   59

newton (metric practice). That force which, when applied to a body having a mass of one kilogram, gives it an acceleration of one meter per second squared.   21

next transfer address (NTA)(FASTBUS acquisition and control). A pointer in a slave to the module register which will be accessed during the next data transfer. The NTA register may be written during a primary address cycle and may be read or written during a secondary address cycle. During block or pipeline transfers the NTA may be automatically modified between data cycles.   480

n gate thyristor (anode gate SCR). A thyristor in which the gate terminal is connected to the *n* region adjacent to the region to which the anode terminal is connected and which is normally switched to the ON state by applying a negative signal between gate and anode terminals.   445

Nichols chart (Nichols diagram) (control system, feedback). A plot showing magnitude contours and phase contours of the return transfer function referred to ordinates of logarithmic loop gain and to abscissae of loop phase angle. *See:* **control system, feedback.**   56

nickel-cadmium storage battery. An alkaline storage battery in which the positive active material is nickel oxide and the negative contains cadmium. *See:* **battery (primary or secondary).**   328

night (illuminating engineering). The hours between the end of evening civil twilight and the beginning of morning civil twilight. *Note:* Civil twilight ends in the evening when the center of the sun's disk is six degrees below the horizon and begins in the morning when the center of the sun's disk is six degrees below the horizon.   167

night alarm. An electric bell or buzzer for attracting the attention of an operator to a signal when the switchboard is partially attended.   328

night effect (radio navigation systems)(navigation aid terms). A special case of error occurring predominantly at night when sky-wave propagation is at the maximum.   526

NIM (FASTBUS acquisition and control). (1) A committee sponsored by the U.S. Department of Energy and associated with the U.S. National Bureau of Standards. It produced the NIM instrumentation system specifications, endorsed the use of CAMAC, and collaborates with ESONE in the maintenance and extension of CAMAC. (2) A standardized modular instrumentation system consisting of NIM modules and NIM BINS as defined in U.S. Department of Energy Report TID-20893, Standard Nuclear Instrument Modules.   480

911 call (telephone switching systems). A call to an emergency service bureau.   55

nines complement. The radix-minus-one complement of a numeral whose radix is ten.   255, 77

ninety-percent response time of a thermal converter (electric instrument). The time required for 90 percent of the change in output electromotive force to occur after an abrupt change in the input quantity to a new constant value. See note 1 of **response time of**

a thermal converter. *See:* thermal converter.
                                                    280
nipple (rigid metal conduit). A straight piece of rigid metal conduit not more than two feet in length and threaded on each end. *See:* raceways.   101
nit (1)(television). The unit of luminance (photometric brightness) equal to one candela per square meter. *Note:* Candela per square meter is the unit of luminance in the International System of Units (SI). Nit is the name recommended by the International Commission on Illumination (CEI).   18
  (2) (information theory). Also called 'nat'.   415
N-layer (logical link control). A subdivision of the architecture, constituted by subsystems of the same rank (N).   585
n-level address (computing systems). A multilevel address that specifies *n* levels of addressing.
                                                 255, 77
noble potential. A potential substantially cathodic to the standard hydrogen potential. *See:* stray current corrosion.   205
no-busy test call (telephone switching systems). A call in which busy testing is inhibited.   55
node (1) (network analysis). One of the set of discrete points in a flow graph.   282
  (2) (software). (A) An end point of any branch of a network or graph, or a junction common to two or more branches. (B) In a tree structure, a point at which subordinate items of data originate. (C) In a network, a point where one or more functional units interconnect transmission lines. (D) The representation of a state or an event by means of a point on a diagram. *See:* data; functional unit; graph; network.   434
node absorption (network analysis). A flow-graph transformation whereby one or more dependent nodes disappear and the resulting graph is equivalent with respect to the remaining node signals. *Note:* For example, a circuit analog of node absorption is the star-delta transformation.   282
node equations (network). Any set of equations (of minimum number) such that the independent node voltages of a specified network may be determined from the impressed currents. *Notes:* (1) The number of node equations for a given network is not necessarily the same as the number of mesh or loop equations for that network. (2) Notes for mesh or loop equations, with appropriate changes, apply here. *See:* network analysis.   210
node signal (network analysis). A variable X_k associated with node k.   282
node voltage (network). The voltage from a reference point to any junction point (node) in a network. *Note:* The assumptions of lumped-network theory are such that the path of integration is immaterial.   210
no-go. *See:* go/no-go.
noise (1) (analog computers). Unwanted disturbances superimposed upon a useful signal, which tend to obscure its information content. Random noise is the part of the noise that is unpredictable, except in a statistical sense.   9
  (2) (data transmission) (general). (1) Unwanted dis-

turbances superposed upon a useful signal that tend to obscure its information content. (2) An undesired disturbance within the useful frequency band. *Note:* Undesired disturbances within the useful frequency band produced by other services may be called interference. *See:* background noise; circuit noise; circuit noise level; noise-to-ground; impulse noise; noise level; noise-longitudinal; noise-metallic; random noise; reference noise; thermal noise; white noise; noise figure; noise measurement units.   59
  (3) An undesired disturbance within the useful frequency band. *Note:* Undesired disturbances within the useful frequency band produced by other services may be called interference.   111
  (4) (phototubes). The random output that limits the minimum observable signal from the phototube. *See:* phototube.   335, 117
  (5) (facsimile). Any extraneous electric disturbance tending to interfere with the normal reception of a transmitted signal. *See:* facsimile transmission.
                                                     12
  (6) (hybrid computer linkage components). Unwanted disturbances superimposed upon a useful signal that tends to obscure its information content expressed in millivolts peak and referred to the input voltage.   10
  (7) (electrical noise). Electrical noise is unwanted electrical signals, which produce undesirable effects in the circuits of the control systems in which they occur.
                                                     43
  (8) (overhead-power-line corona and radio noise). An undesired disturbance within the useful frequency band.   411
  (9) (oscilloscopes). Any extraneous electric disturbance tending to interfere with the normal display.
                                                    106
noise amplitude distribution (NAD)(control of system electromagnetic compatibility). A distribution showing the pulse amplitude that is obtained or exceeded as a function of pulse repetition rate.   495
noise, audio-frequency. Any unwanted disturbance in the audio-frequency range.   239
noise-current generator. A current generator, the output of which is described by a random function of time. *Note:* At a specified frequency, a noise-current generator can often be adequately characterized by its mean-square current within the frequency increment Δf or by its spectral density. If the circuit contains more than one noise-voltage generator or noise-current generator, the correlation coefficients among the generators must also be specified. *See:* network analysis; signal; signal-to-noise ratio.   188, 190
noise definitions (data transmission).
  dBa. For F1 A weighted noise measurement, usually obtained with a WECO 2B noise meter. 0 dBa equivalent to 1000 Hz tone with a power of -85 dBm. Or, a 3 kHz white noise band of -82 dBm. Filter produces a 3 dB loss over flat indication.
  dBmp. Filter produces a 2.5 dB loss compared to no weighting. For psophometrically weighted noise according to CCITT; 0 dBmp is equivalent to a 1000 Hz

tone with a power of -1 dBm (or an 800 Hz tone with a power of 0 dBm). Or, a 3 kHz band of white noise with a power of +2.5 dBm. Filter produces a 2.5 dB loss compared to no weighting.

dBrn 144. Obsolete unit for Type 144 weighted-noise measurement.

dBrnC. For C-message weighted noise; 0 dBrnc is equivalent to a 1000 Hz tone with a power of -90 dBm (10^{-12}W (watt) or 90 dB below 1 mW (milliwatt). Or, 3 kHz white-noise band of a power of 88 dBm (actually 88.5 dBm). Filter produces a 2 dB loss compared to no weighting.

dBrnCO, dBaO, pWpO, dBmpO. Noise units as measured in dBrnC, dBa, pWp, and dBmp at (or referred to) a 0 transmission level point.

pWp. For psophometrically weighted noise; 1 pWp is equivalent to a 1000 Hz tone with a power of -91 dBm (or an 800 Hz tone of -90 dBm). Or, a 3 kHz band of white noise with a power of 88 dBm.

$$1 \text{ pWp} = \frac{1 \text{ pW}}{1.78} = 0.56 \text{ pW}$$

59

noise diode, ideal. A diode that has an infinite internal impedance and in which the current exhibits full shot noise fluctuations. *See:* **signal-to-noise ratio.**   190

noise, electrical. *See:* **electrical noise.**

noise equivalent power (NEP) (fiber optics). At a given modulation frequency, wavelength, and for a given effective noise bandwidth, the radiant power that produces a signal-to-noise ratio of 1 at the output of a given detector. *Notes:* (1) Some manufacturers and authors define NEP as the minimum detectable power per root unit bandwidth; when defined in this way, NEP has the units of watts/(hertz)$^{1/2}$. Therefore, the term is a misnomer, because the units of power are watts. *See:* **D*; detectivity.** (2) Some manufacturers define NEP as the radiant power that produces a signal-to-dark-current noise ratio of unity. This is misleading when dark-current noise does not dominate, as is often true in fiber systems.   433

noise factor (two-port transducer). At a specified input frequency the ratio of (1) the total noise power per unit bandwidth at a corresponding output frequency available at the output port when the noise temperature of its input termination is standard (290° K) at all frequencies (Reference: Definition for Average Noise Factor to (2) that portion of (1) engendered at the input frequency by the input termination at the Standard Noise Temperature 290° K). *Notes:* (A) For heterodyne systems there will be, in principle, more than one output frequency corresponding to a single input frequency, and vice versa: for each pair of corresponding frequencies a noise factor is defined. (2) includes only that noise from the input termination which appears in the output via the principal-frequency transformation of the system, that is, via the signal-frequency transformation(s), and does not include spurious contributions such as those from an unused image-frequency or an unused idler-frequency transformation. (B) The phrase "available at the output

port" may be replaced by "delivered by system into an output termination." (C) To characterize a system by a noise factor is meaningful only when the admittance (or impedance) of the input termination is specified. *Syn:* **noise figure.**   125

noise figure (noise factor) (interference terminology, linear system) (data transmission). (1) At a selected input frequency. The ratio of (A) the total noise power per unit bandwidth (at a corresponding output frequency) delivered by the system into an output termination to (B) the portion thereof engendered at the input frequency by the input termination, whose noise temperature is standard (290 K (kelvins) at all frequencies). *Notes:* (1) Numerically, the noise factor F at frequency f can be expressed as

$$\overline{F}(f) = \frac{P_{\text{noise out}}}{GP_{\text{noise in}}}$$

where $P_{noise\ out}$ and $P_{noise\ in}$ are taken at frequency f and 290 kelvins and G is the gain of the system at frequency f. (b) For heterodyne systems there will be, in principle, more than one output frequency corresponding to a single input frequency and vice versa; for each pair of corresponding frequencies a noise factor is defined. (c) The phrase "available at the output terminals" may be replaced by "delivered by the system into an output termination" without changing the sense of the definition. (d) The term "noise factor" is used where it is desired to emphasize that the noise figure is a function of input frequency. (2) (Average). The ratio of (A) the total noise power delivered by the system into its output termination when the noise temperature of its input termination is standard (290 kelvins) at all frequencies to (B) the portion thereof engendered by the input termination. *Notes:* (a) For heterodyne systems, portion (A) includes only that noise from the input termination that appears in the output via the principal frequency transformation of the system, and does not include spurious contributions such as those from image frequency transformations. (b) A quantitative relation between average noise factor F and spot noise factor F (f) is

$$\overline{F} = \frac{\displaystyle\int_0^\infty F(f)G(f)\mathrm{d}f}{\displaystyle\int_0^\infty G(f)\mathrm{d}f}$$

where f is the input frequency and F (f) is the ratio of (a) the signal power delivered by the system into its output termination to (b) the corresponding signal power available from the input termination at the input frequency. For heterodyne systems, (a) comprises only power appearing in the output via the principal frequency transformation of the system. (c) To characterize a system by an average noise factor is meaningful only when the admittance (or impedance) of the input termination is specified.   59

noise figure, average (communication satellite). Of a two-port transducer the ratio of the total noise power to the input noise power, when the input termination is at the standard temperature of 290° kelvin. *See:* **noise factor.**   85

noise-free equivalent amplifier (signal-transmission system). An ideal amplifier having no internally generated noise that has the same gain and input.output characteristics as the actual amplifier. *See:* **signal.**   188

noise generator (analog computers). In an analog computer, a computing element used purposely to introduce noise of specified amplitude distribution, spectral density, or root-mean square value, or appropriate combination therefore into other computing elements.   9

noise generator diode (electron device). A diode in which the noise is generated by shot effect and the noise power of which is a definite function of the direct current.   190

noise killer (telegraph circuit). An electric network inserted usually at the sending end, for the purpose of reducing interference with other communication circuits. *See:* **telegraphy.**   328

noise level (1) (audio and electroacoustics). (A) The noise power density spectrum in the frequency range of interest, (B) the average noise power in the frequency range of interest, or (C) the indication on a specified instrument. *Notes:* (1) In (C), the characteristics of the instrument are determined by the type of noise to be measured and the application of the results thereof. (2) Noise level is usually expressed in decibels relative to a reference value. *See:* **signal-to-noise ratio.**   239

(2) (speech quality measurements). The A-weighted sound level of the noise.   126

(3) (data transmission). (A) The noise power density spectrum in the frequency range of interest. (B) The average noise power in the frequency range of interest. (C) The indication on a specified instrument. *Notes:* (1) In (C), the characteristics of the instrument are determined by the type of noise to be measured and the application of the results thereof. (2) Noise level is usually expressed in decibels relative to a reference value.   59

noise-level test (rotating machinery). A test taken to determine the noise level produced by a machine under specified conditions of operation and measurement. *See:* **asynchronous machine.**   63

noise linewidth (charged-particle detectors)(X-ray energy spectrometers)(germanium gamma-ray dectectors)(semiconductor radiation detectors). The contribution of noise to the width of a spectral peak.   119, 528, 471, 23

noise, longitudinal (data transmission). In telephone practice, the 1/1000th part of the total longitudinal-circuit noise current at any given point in one or more telephone wires.   59

noise measurement units (data transmission). The following units are used to express weighted and unweighted circuit noise. They include terms used in American and International practice.

dBa-dBrn adjusted. Weighted circuit noise power, in dB, referred to 3.16 pW (picowatts) (-85 dBm) which is 0 dBa. Use of F1 A line or HA1 receiver weighting is indicated in parenthesis as required.

dBrn (Describes above reference noise). Weighted circuit noise power in dB referred to 1.0 pW (-90 dBm) which is 0 dBrn. Use of 144 line, 144 receiver, or C message weighting, parentheses, as required. With C-message weighting, as 1 mW (milliwatt), 1000 Hz tone will read $+90$ dBrn, but the same power as white noise distributed over a $3kH_3$ band (nominally 300 to 3300 Hz) will read approximately $+88.5$ dBrn (rounded off to $+88$ dBrn, due to frequency weighting). With 144 weighting, as 1 mW, 1000 Hz tone will also read $+90$ dBrn, but the same $3kH_3$ white-noise power would read only $+82$ dBrn, due to the different frequency weighting.   59

noise measuring set. *See:* **circuit noise meter.**

noise, metallic (data transmission). The weighted noise current in a metallic circuit at a given point when the circuit is terminated at that point in the nominal characteristic impedance of the circuit.   59

noise power (Hall effect devices). The power generated by a random electromagnetic process.   107

noise pressure equivalent (electroacoustic transducer or system used for sound reception). The root-mean-square sound pressure of a sinusoidal plane progressive wave that, if propagated parallel to the principal axis of the transducer, would produce an open-circuit signal voltage equal to the root-mean-square of the inherent open-circuit noise voltage of the transducer in a transmission band having a bandwidth of 1 hertz and centered on the frequency of the plane sound wave. *Note:* If the equivalent noise pressure of the transducer is a function of secondary variables, such as ambient temperature or pressure, the applicable values of these quantities should be stated explicitly. *See:* **phonograph pickup.**   176

noise quieting (receiver) (receiver performance). A measure of the quantity of radio-frequency energy, at a specified deviation from the receiver center frequency, required to reduce the noise output by a specified amount. *See:* **receiver performance.**   181

noise reduction (photographic recording and reproducing). A process whereby the average transmission of the sound track of the print (averaged across the track) is decreased for signals of low level and increased for signals of high level. *Note:* Since the noise introduced by the sound track is less at low transmission, this process reduces film noise during soft passages. The effect is normally accomplished automatically. *See:* **phonograph pickup.**   176

noise sidebands (non-real time spectrum analyzer). Undesired response caused by noise internal to the spectrum analyzer appearing on the display around a desired response.   390

noise temperature (1) (general) (at a pair of terminals and at a specific frequency). The temperature of a passive system having an available noise power per unit bandwidth equal to that of the actual terminals. *Note:* Thus, the noise temperature of a simple resistor is the actual temperature of the resistor, while the noise temperature of a diode may be many times the observed absolute temperature. *See:* **signal; signal-to-noise ratio.**   188

(2) (standard). The standard reference temperature T for noise measurements is 290°K. *Note:* $kT_0/e = 0.0250$ volt, where e is the magnitude of the electronic charge and k is Boltzmann's constant.   125

(3) (antenna). The temperature of a resistor having an available thermal noise power per unit bandwidth equal to that at the antenna output at a specified frequency. *Note:* Noise temperature of an antenna depends on its coupling to all noise sources in its environment as well as noise generated within the antenna. *See:* **antenna.**   111

(4) (at a port and at a selected frequency). A temperature given by the exchangeable noise-power density divided by Boltzmann's constant, at a given port and at a stated frequency. *Notes:* (A) When expressed in units of kelvins, the noise temperature T is given by the relation

$$T = N/k$$

where N is the exchangeable noise-power density in watts per hertz at the port at the stated frequency and k is Boltzmann's constant expressed as joules per kelvin ($k \simeq 1.38 \times 10^{-23}$ joules per kelvin). (B) Both N and T are negative for a port with an internal impedance having a negative real part. *See:* **signal-to-noise ratio; waveguide.**   190

(5) (at a port). The temperature of a passive system having an available noise power per unit bandwidth equal to that of the actual port, at a specified frequency. A uniform temperature throughout the passive system is implied. *See:* **thermal noise.**   125

noise temperature, average operating (communication satellite). Equivalent temperature of passive system having an available noise power equal to that of the operating system. In space communication systems the noise temperature is generally composed of contributions from the background, atmospheric and receiver front end noise.   85

noise-to-ground (data transmission). In telephone practice, the weighted noise current through the 100 000 Ω (ohm) circuit of a circuit noise meter, connected between one or more telephone wires and ground.   59

noise transmission impairment (NTI). The noise transmission impairment that corresponds to a given amount of noise is the increase in distortionless transmission loss that would impair the telephone transmission over a substantially noise-free circuit by an amount equal to the impairment caused by the noise. Equal impairments are usually determined by judgment tests or intelligibility tests.   328

noise-voltage generator (interference terminology). A voltage generator the output of which is described by a random function of time. *Note:* At a specified frequency, a noise-voltage generator can often be adequately characterized by its mean-square voltage with

the frequency increment Δf or by its spectral density. If the circuit contains more than one noise-current generator or noise-voltage generator, the correlation coefficients among the generators must be specified. *See:* **network analysis; signal; signal-to-noise ratio.**   125

noise weighting (data transmission). In measurement of circuit noise, a specific amplitude frequency characteristic of a noise measuring set. It is designed to give numerical readings which approximate the amount of transmission impairment due to the noise, to an average listener, using a particular class of telephone subset. The noise weightings generally used were established by the agencies concerned with public telephone service and are based on characteristics of specific commercial telephone subsets, representing successive stages of technological development. The coding of commercial apparatus appears in the nomenclature of certain weightings. The same weighting nomenclature and units are used in the military versions and in the commercial noise measuring sets.   59

no-load (adjective) (rotating machinery). The state of a machine rotating at normal speed under rated conditions, but when no output is required of it. *See:* **asynchronous machine.**   63

no-load (excitation) losses (power and distribution transformer). Those losses which are incident to the excitation of the transformer. No-load (excitation) losses include core loss, dielectric loss, conductor loss in the winding due to exciting current, and conductor loss due to circulating current in parallel windings. These losses change with the excitation voltage.   53

no-load field current (excitation systems for synchronous machines). The direct current in the field winding of the synchronous machine required to produce rated voltage at no-load and rated speed.   507

no-load field voltage (excitation systems for synchronous machines). The voltage required across the terminals of the field winding of the synchronous machine under conditions of no-load, rated speed, and rated terminal voltage, and with the field winding at 25 °C.   507

no-load loss (1)(electronics power transformer). The input power, expressed in watts, to a completely assembled transformer that is excited at rated terminal voltage and frequency, but not supplying load current. *Syn:* **excitation loss.**   95

(2)(power operations). Power losses in an electric facility when energized at rated voltage and frequency, but not carrying load.   516

no-load saturation curve (no-load characteristic) (of a synchronous machine). The saturation curve of a machine on no-load.   63

no-load speed (industrial control) (of an electric drive). The speed that the output shaft of the drive attains with no external load connected and with the drive adjusted to deliver rated output at rated speed. *Note:* In referring to the speed with no external load connected and with the drive adjusted for a specified condition other than for rated output at rated speed, it is customary to speak of the no-load speed under the (stated) conditions. *See:* electric drive.   206

no-load test (synchronous machine). A test in which the machine is run as a motor providing no useful mechanical output from the shaft.   63

nomenclature (electric power system) (generating stations electric power system). The words and terms used to identify electric power systems.   381

nominal band of regulated voltage. The band of regulated voltage for a load range between any load requiring no-load field voltage and any load requiring rated-load field voltage with any compensating means used to produce a deliberate change in regulated voltage inoperative.   328

nominal battery voltage (National Electrical Code). The voltage computed on the basis of 2.0 volts per cell for the lead-acid type and 1.2 volts per cell for the alkali type.   256

nominal collector ring voltage (rotating machinery). *See:* rated-load field voltage.

nominal conductor gradient (overhead-power-line corona and radio noise). The gradient determined for a smooth cylindrical conductor whose diameter is equal to the outside diameter of the actual (stranded) conductor.   411

nominal control current (Hall generator). That value of control current that, if exceeded, will cause the linearity error of the device to exceed a rated magnitude.   107

nominal discharge current (arrester). The discharge current having a designated peak value and waveshape, that is used to classify an arrester with respect to protective characteristics. *Note:* It is also the discharge current that is used to initiate follow current in the operating duty test.   62

nominal input power (loudspeaker measurements). Nominal input power is equal to the square of the true rms voltage at the input terminals of the loudspeaker, divided by the rated impedance. *See:* rated impedance.   19

nominal line pitch (television). The average separation between centers of adjacent lines forming a raster.   18

nominal line width (facsimile). The average separation between centers of adjacent scanning or recording lines. *See:* recording (facsimile); scanning (facsimile).   12

nominal metallic impedance (measuring longitudinal balance of telephone equipment operating in the voice band). Impedance based on lumped constants of a metallic circuit at a single given frequency.   529

nominal pull-in torque (synchronous motor). The torque it develops as an induction motor when operating at 95 percent of synchronous speed with rated voltage applied at rated frequency. *Note:* This quantity is useful for comparative purposes when the inertia of the load is not known.   63

nominal rate of rise (of an impulse wave front)(metal-oxide surge arresters for ac power circuits)(electric power)(transformer)(surge arrester). The slope of the line that determines the virtual zero. It is usually expressed in volts or amperes per microsecond.   583, 2, 62, 53, 430

nominal ratio. *See:* marked ratio.   203

nominal synchronous-machine excitation-system ceiling voltage. The ceiling voltage of the excitation system with: (1) The exciter and all rotating elements at rated speed. (2) The auxiliary supply voltages and frequencies at rated values. (3) The excitation system loaded with a resistor having a value equal to the resistance of the field winding to be excited at a temperature of: (A) 75 °C for field windings designed to operate at rating with a temperature rise of 60 °C or less, and (B) 100 °C for field windings designed to operate at rating with a temperature rise greater than 60 °C. (4) The manual control means adjusted as it would be to produce the rated voltage of the excitation system, if this manual control means is not under the control of the voltage regulator when the regulator is in service, unless otherwise specified. (The means used for controlling the exciter voltage with the voltage regulator out of service is normally called the manual control means.) (5) The voltage sensed by the synchronous machine voltage regulator reduced to give the maximum output from the regulator. Note that the regulator action may be simulated in test by applying to the field of the exciter under regulator control the maximum output developed by the regulator.   63

nominal system voltage (1)(metal-oxide surge arresters for ac power circuits)(power switchgear). A nominal value assigned to designate a system of a given voltage class. *Note:* See ANSI C84.1-1982.   583, 103

(2)(power and distribution transformer). The system voltage by which the system is designated and to which certain operating characteristics of the system are related. (The nominal voltage of a system is near the voltage level at which the system normally operates and provides a per-unit base voltage for system study purposes. To allow for operating contingencies, systems generally operate at voltage levels about 5 to 10 percent below the maximum system voltage for which system components are designed.)   53

(3) (rotating electric machinery). A number used to denote the general level of voltage of the system described and, in the case of an externally powered system, selected from the list of preferred values below.   424

(4) (surge arrester). A nominal value assigned to designate a system of a given voltage class. *Note:* See ANSI C84.1-1977. maximum (highest) system voltage. The highest voltage at which a system is operated. *Note:* This is generally considered to be the maximum system voltage as prescribed in ANSI C84.1-1977.   430

(5) (system voltage ratings). The root-mean-square phase-to-phase voltage by which the system is designated and to which certain operating characteristics of the system are related. *Note:* The nominal system voltage is near the voltage level at which the system normally operates. To allow for operating contingencies, systems generally operate at voltage levels about five to ten percent below the maximum system voltage for which system components are designed. *See:* system voltage; maximum system voltage; service voltage; utilization voltage; low voltage; medium voltage high voltage. 260

nominal thickness (cable element) (power distribution, underground cables). The specified, indicated, or named thickness. *Note:* In general, measured thicknesses will approximate but will not necessarily be identical with nominal thicknesses. 57

nominal value (metric practice). A value assigned for the purpose of convenient designation; existing in name only. 21

nominal voltage (National Electrical Code). A nominal value assigned to a circuit or system for the purpose of conveniently designating its voltage class (as 120/240, 480Y/277,600 etcetera.) The actual voltage at which a circuit operates can vary from the nominal within a range that permits satisfactory operation of equipment. *See:* "Voltage Ratings for Electric Power Systems and Equipment (60 Hz.)," ANSI C84.1-1970 and supplement C84.la-1973. 256

nonadjustable (National Electrical Code). (As applied to Circuit Breakers.) A qualifying term indicating that the circuit breaker does not have any adjustment to alter the value of current at which it will trip or the time required for its operation. 256

nonatmospheric paths (atmospheric correction factors to dielectric tests). Paths, such as through a gas or vacuum sealed from the atmosphere, through a liquid such as oil, or through a solid, or a combination thereof. 50

nonautomatic (National Electrical Code). Action requiring personal intervention for its control. As applied to an electric controller, nonautomatic control does not necessarily imply a manual controller, but only that personal intervention is necessary. *See:* automatic. 256

nonautomatic extraction turbine (condensing or noncondensing)(control systems for steam turbine-generator units). Steam is extracted from one or more stages, but without means of controlling the pressure(s) of the extracted steam. 522

nonautomatic opening (nonautomatic tripping). The opening of a switching device only in response to an act of an attendant. 103

nonautomatic operation (power switchgear). Operation controlled by an attendant. 103

nonautomatic tripping. *See:* nonautomatic opening.

nonblocking switching network (telephone switching systems). A switching network in which any idle outlet can always be reached from any given inlet under all traffic conditions. 55

noncharacteristic harmonic (1)(converter characteristics)(self-commutated converters). Those harmonics which are not produced by semiconductor converter equipment in the course of normal operation. These may be a result of beat frequencies, a demodulation of

characteristic harmonics and the fundamental component, unbalance in the ac (alternating current) power system, or unsymmetrical control of the converter. 584

(2)(harmonic control and reactive compensation of static power converters). Those harmonics which are not produced by semiconductor converter equipment in the course of normal operation. These may be the result of beat frequencies, a demodulation of characteristic harmonics and the fundamental, or unbalance in the ac power system or unsymmetrical delay angle. 533

noncode fire-alarm system. A local fire-alarm system in which the alarm signal is continuous and is usually sounded by vibrating bells. *See:* protective signaling. 328

noncoherent MTI (moving-target indication). A form of MTI in which a moving target is detected without use of an internal reference phase signal. 13

noncoincident demand (power operations). The sum of the individual maximum demands regardless of time of occurrence within a specified period. 516

noncomposite color picture signal (National Television System Committee (NTSC) color television). The electric signal that represents complete color picture information but excludes line and field sync signals. 18

nonconformance (nuclear power quality assurance). A deficiency in characteristic, documentation, or procedure that renders the quality of an item or activity unacceptable or indeterminate. 417

nonconforming load (electric power systems). A customer load, the characteristics of which are such as to require special treatment in deriving incremental transmission losses. 99

noncontact plunger. *See:* choke piston.

noncontinuous electrode. A furnace electrode the residual length of which is discarded when too short for further effective use. *See:* electrothermics. 328

noncontinuous enclosures (generating station grounding). Refers to a type of isolated-phase bus in which the enclosure is sectionalized with insulation between sections to block the longitudinal flow of current in the enclosure. 569

noncontinuously acting regulator (excitation systems for synchronous machines). A regulator that requires a sustained finite charge in the controlled variable to initiate corrective action. 507

noncritical failure (test, measurement and diagnostic equipment). Any failure which degrades performance or results in degraded operation requiring special operating techniques or alternative modes of operation which could be tolerated throughout a mission but should be corrected immediately upon completion of mission. 54

nondestructive read (computing systems) (1) (general). A read process that does not erase the data in the source. 255, 77

(2) (magnetic cores). A method of reading the magnetic state of a core without changing its state. 331

nondestructive reading (charge-storage tubes). Reading that does not erase the stored information. *See:* **charge-storage tube.** 174, 54

nondestructive testing (test, measurement and diagnostic equipment). Testing of a nature which does not impair the usability of the item. 54

nondirectional beacon (NDB)(air navigation)(navigation aid terms). A radio facility which can be used with an airborne DF (direction finder) to provide a line of position. *Syn:* **compass locator; H; H-beacon; locator.** 526

nondirectional microphone. *See:* **omnidirectional microphone.**

nondisconnecting fuse (power switchgear). An assembly consisting of a fuse unit or fuseholder and a fuse support having clips for directly receiving the associated fuse unit or fuseholder, which has no provision for guided operation as a disconnecting switch. 103, 443

nonenclosed (transformer). Not surrounded by a medium which will prevent a person accidentally contacting live parts. 53

nonenclosed switches, indoor or outdoor (power switchgear). Switches designed for service without a housing restricting heat transfer to the external medium. 103

nonenergy-limiting transformer (power and distribution transformer). A constant-potential transformer that does not have sufficient inherent impedance to limit the output to a thermally safe maximum value. 53

nonerasable storage (computing systems). *See:* **fixed storage.**

nonexposed installation (lightning). An installation in which the apparatus is not subject to overvoltages of atmospheric origin. *Note:* Such installations are usually connected to cable networks. 244, 62

nonfirm power (power operations). Power supplied or available under an arrangement which does not have the availability feature of firm power. 516

nongyroscopic angular sensor (inertial sensor). An angular sensor whose functions do not depend on the angular momentum of a spinning rotor. 46

nonhoming (telephone switching systems). Resumption of a sequential switching operation from its last setting. 55

noninjecting contact (of a semiconductor radiation detector)(1)(charged-particle detectors)(X-ray energy spectrometers). A contact at which the carrier density in the adjacent semiconductor material is not changed from its equilibrium value. 119, 471
(2)(germanium gamma-ray detectors). A purely resistive contact, that is, one that has a linear voltage-current characteristic throughout its entire operating range. 528

noninterference testing (test, measurement and diagnostic equipment). A type of on-line testing that may be carried out during normal operation of the unit under test without affecting the operation. *See:* **interference testing.** 54

noninverting parametric device. A parametric device whose operation depends essentially upon three frequencies, a harmonic of the pump frequency and two signal frequencies, of which one is the sum of the other plus the pump harmonic. *Note:* Such a device can never provide gain at either of the signal frequencies. It is said to be noninverting because if either of the two signals is moved upward in frequency, the other will move upward in frequency. *See:* **parametric device.** 191

nonlinear capacitor. A capacitor having a mean-charge characteristic or a peak-charge characteristic that is not linear, or a reversible capacitance that varies with bias voltage. 191

nonlinear circuit. *See:* **nonlinear network.**

nonlinear distortion. Distortion caused by a deviation from a desired linear relationship between specified measures of the output and input of a system. *Note:* The related measures need not to be output and input values of the same quantity: for example, in a linear detector the desired relation is between the output signal voltage and the input modulation envelope: or the modulation of the input carrier and the resultant detected signal. *See:* **close-talking pressure-type microphone; distortion.** 239

nonlinear exponent (low voltage varistor surge arresters). A measure of varistor voltage nonlinearity between two given operating currents, expressed as $I = KV^{\alpha}$; where I is any current between the two operating currents I_1 and I_2; K is a device constant; V is the varistor voltage; and α is the logarithm of the ratio of the two operating currents (I_1 / I_2) divided by the logarithm of the ratio of the varistor voltages at the two operating currents (V_2 / V_1). 62

nonlinear ideal capacitor. An ideal capacitor whose transferred-charge characteristic is not linear. *See:* **nonlinear capacitor.** 191

nonlinearity (1)(gyro: accelerometer). The systematic deviations from the least squares straight line for input-output relationships which nominally can be represented by a linear equation. 46
(2)(of a pulse amplifying system)(X-ray energy spectrometers). Distortion caused by a deviation from a desired linear relationship between specified measures of the output and input pulse amplitudes of a system or device. 471

nonlinear network (signal-transmission system). (1) A network (circuit) not specifiable by linear differential equations with time and/or position coordinates as the independent variable. *Note:* It will not operate in accordance with the superposition theorem. (2) A network (circuit) in which the signal transmission characteristics depend on the input signal magnitude. *See:* **network analysis; signal.** 210, 188

nonlinear parameter. A parameter dependent on the magnitude of one or more of the dependent variables or driving forces of the system. *Note:* Examples of dependent variables are current, voltage, and analogous quantities. 210

nonlinear resistor-type arrester (valve type). An arrester having a single or a multiple spark gap connected in series with nonlinear resistance. *Note:* If the arrester has no series gap, the characteristic element limits the follow current to a magnitude that does not interfere with the operation of the system. *See:* **arrester, valve type.**   62

nonlinear scattering (fiber optics). Direct conversion of a photon from one wavelength to one or more other wavelengths. In an optical waveguide, nonlinear scattering is usually not important below the threshold irradiance for stimulated nonlinear scattering. *Note:* Examples are Raman and Brillouin scattering. *See:* **photon.**   433

nonlinear series resistor (arrester). The part of the lightning arrester that, by its nonlinear voltage-current characteristics, acts as a low resistance to the flow of high discharge currents thus limiting the voltage across the arrester terminals, and as a high resistance at normal power-frequency voltage thus limiting the magnitude of follow current.   62

nonlinear system or element. *See:* **linear system or element.**

nonlined construction (primary cell) (dry cell). A type of construction in which a layer of paste forms the only medium between the depolarizing mix and the negative electrode. *See:* **electrolytic cell.**   328

nonline frequency components (thyristor). Expressed by the frequency and the root-mean-square (rms) value of the components having a different frequency than the line frequency.   445

nonline frequency content (thyristor). The function obtained by subtracting the line frequency component from a nonsinusoidal periodic function. *Note:* For the case of asynchronous ON-OFF control, this function varies with time.   445

nonload-break connector. A connector designed to be separated and engaged on de-energized circuits.   134

nonloaded *Q* (basic *Q*) (of an electric impedance). The value of *Q* of such impedance without external coupling or connection.   328

nonmagnetic relay armature stop. *See:* **relay armature stop, nonmagnetic.**

nonmagnetic ship. A ship constructed with an amount of magnetic material so small that it causes negligible distortion of the earth's magnetic field. *See:* **degauss.**   328

nonmechanical switching device (power switchgear). A switching device designed to close or open, or both, one or more electric circuits by means other than by separable mechanical contacts.   103

nonmetallic extensions (National Electrical Code). An assembly of two insulated conductors within a nonmetallic jacket or an extruded thermoplastic covering. The classification includes both surface extensions, intended for mounting directly on the surface of walls or ceilings, and aerial cable, containing a supporting messenger cable as an integral part of the cable assembly.   256

nonmetallic-sheathed cable (National Electrical Code). A factory assembly of two or more insulated conductors having an outer sheath of moisture-resistant, flame-retardant, nonmetallic material.   256

nonminimum phase function (linear networks) (circuits and systems). A network function that is not minimum-phase. *See:* **minimum phase function.**   67

nonmultiple switchboard (telephone switching systems). A telecommunications switchboard having each line connected to only one jack.   55

nonmultiple transit spurious signals (dispersive and nondispersive delay lines). Signals not related to the main signal delay by a simple integer may be labeled by the delay time of that signal.   81

nonnumerical action (switch). That action that does not depend on the called number (such as hunting an idle trunk).   328

nonoperate value. *See:* **relay nonoperate value.**

nonphantomed circuit. A two-wire or four-wire circuit that is not arranged to form part of a phantom circuit. *See:* **transmission line.**   328

nonphysical primary color (color) (television). A primary represented by a point order outside the area of the chromaticity diagram enclosed by the spectrum locus and the purple boundary. *Note:* Nonphysical primaries cannot be produced because they require negative power at some wavelengths. However, they have properties that facilitate colorimetric calculation.   18

nonplanar network. A network that cannot be drawn on a plane without crossing of branches. *See:* **network analysis.**   210

nonpolarized electrolytic capacitor. An electrolytic capacitor in which the dielectric film is formed adjacent to both metal electrodes and in which the impedance to the flow of current is substantially the same in both directions. *See:* **electrolytic capacitor.**   328

non-polychlorinated biphenyl (PCB) transformer (handling and disposal of transformer grade insulating liquids containing PCBs). Transformers that contain less than 50 parts per million (ppm) PCB. No transformer may ever be considered to be a non-PCB transformer unless its dielectric fluid has been tested or otherwise verified to contain less than 50 ppm PCB. Examples: Transformers so identified.   586

nonpreferred ATLAS. An ATLAS structure which is found to be superfluous or troublesome in real use, but which is retained in the standard for upward compatibility. *See:* **ATLAS.**   400

nonprocedural programming language (software unit testing). A computer programming language used to express the parameters of a problem rather than the steps in a solution (for example, report writer or sort specification languages). *See:* **procedural programming language.**   519

nonprofessional projector (National Electrical Code). Those types other than described under professional projector.   256

n-on-p solar cells (photovoltaic power system). Photovoltaic energy-conversion cells in which a base of *p*-type silicon (having fixed electrons in a silicon lattice and positive holes that are free to move) is overlaid with a surface layer of *n*-type silicon (having fixed positive holes in a silicon lattice with electrons that are free to move).   186

nonquadded cable. *See:* **paired cable.**

nonreciprocal (circuits and systems). A device or network that does not have the property of reciprocity.   67

nonreciprocal differential insertion phase (nonlinear, active, and nonreciprocal waveguide components) The difference between insertion phases in the two opposite directions of propagation between two ports of a junction.   530

nonrelevant failure. *See:* **failure, nonrelevant.**

nonrenewable (one-time) fuse or fuse unit (power switchgear). A fuse or fuse unit not intended to be restored for service after circuit interruption.   103

nonrenewable fuse unit (high-voltage switchgear). A fuse unit that, after circuit interruption, cannot readily be restored for service.   103, 443

nonrepaired item (reliability). *See:* **item, nonrepaired.**

nonrepetitive peak line voltage(V_{LSM})(thyristor). The highest instantaneous value of any nonrepetitive transient line voltage. *Note:* The voltage L_{LSM} may originate from operating circuit breakers, atmospheric disturbances, etcetera. This kind of voltage may be minimized by the provision of surge suppression components.   445

nonrepetitive peak reverse voltage rating (rectifier circuit) (semiconductor rectifier). The maximum value of non-repetitive peak reverse voltage permitted by the manufacturer under stated conditions. *See:* **average forward-current rating (of a rectifier circuit).**   237

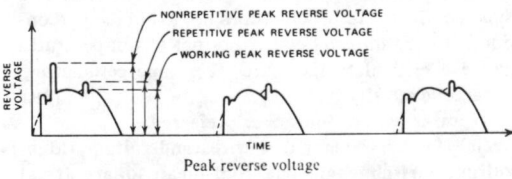

Peak reverse voltage

nonrepetitive peak OFF-state voltage (thyristor). The maximum instantaneous value of any nonrepetitive transient OFF-state voltage that occurs across the thyristor. *See:* **principal voltage-current characteristic (principal characteristic).**   243, 66, 208, 191

nonrepetitive transient reverse voltage (1) (reverse-blocking thyristor). The maximum instantaneous value of any unrepetitive transient reverse voltage that occurs across a thyristor. *See:* **principal voltage-current characteristic (principal characteristic).**   208

(2) (semiconductor rectifier). The maximum instantaneous value of the reverse voltage, including all non-repetitive transient voltages but excluding all repetitive transient voltages, that occurs across a semiconductor rectifier cell, rectifier diode, or rectifier stack. *See:* **rectification; semiconductor rectifier stack.**   237

nonrequired time (availability). The period of time during which the user does not require the item to be in a condition to perform its required function.   164

nonrestorable fire detector (fire protection devices). A device whose sensing element is designed to be destroyed by the process of detecting a fire.   71

nonreturn to zero (magnetic tape pulse recorders for electricity meters). A method whereby a pulse is recorded on the magnetic tape by a polarity reversal of the recording head current.   551

nonreturn to zero (NRZ) code (power-system communication). A code form having two states termed zero and one, and no neutral or rest condition. *See:* **digital.**   59

nonreversible output (magnetic amplifier). *See:* **maximum test output voltage.**

nonreversible power converter (semiconverter). An equipment containing assemblies of mixed power thyristor and diode devices that is capable of transferring energy in only one direction (that is, from the alternating-current side to the direct-current side). *See:* **power rectifier.**   208

nonreversing (industrial control). A control function that provides for operation in one direction only. *See:* **control system, feedback.**   225, 206

nonsalient pole (rotating machinery). The part of a core, usually circular, that by virtue of direct-current excitation of a winding embedded in slots and distributed over the interpolar (and possibly over some or all of the polar) space, acts as a pole. *See:* **asynchronous machine.**   63

nonsalient-pole machine. *See:* **cylindrical-rotor machine.**

nonsaturation region of an insulated-gate field-effect transistor (IGFET) (metal-nitride-oxide field-effect transistor). A portion of the I_{DS} versus V_{DS} characteristic where I_{DS} is strongly dependent on V_{DS}. This is true when $0 < |V_{DS}| < |V_{GS} - V_T|$.   386

nonsegregated-phase bus (1)(generating station grounding). A metal-enclosed bus in which all phase conductors are in a common metal enclosure without barriers between the phases.   569

(2)(power switchgear). One in which all phase conductors are in a common metal enclosure without barriers between the phases. *Note:* When associated with metal-clad switchgear, the primary bus conductors and connections are covered with insulating material throughout.   103

nonselective collective automatic operation (elevators). Automatic operation by means of one button in the car for each landing level served and one button at each landing, wherein all stops registered by the momentary actuation of landing or car buttons are made irrespective of the number of buttons actuated or of the sequence in which the buttons are actuated. *Note:* With this type of operation the car stops at all landings for which buttons have been actuated, making the stops in the order in which the landings are reached after the buttons have been actuated, but irrespective of its direction of travel. *See:* **control (elevators).**   328

nonself-maintained discharge (gas). A discharge characterized by the fact that the charged particles are produced solely by the action of an external ionizing agent. *See:* **discharge (gas).**   244, 190

nonself-restoring insulation (1) (high voltage testing). Insulation which loses its insulating properties, or does not recover them completely, after a disruptive discharge caused by the application of a test voltage. In insulation of this kind, disruptive discharges generally, but not necessarily, occur in the internal part of the insulation.   150

(2) (power and distribution transformer). An insulation which loses its insulating properties or does not recover them completely, after a disruptive discharge caused by the application of a test voltage; insulation of this kind is generally, but not necessarily, internal insulation.   53

nonshield insulated splice (power cable joint). An insulated splice in which no conducting material is employed over the insulation for electric stress control.   34

nonspinning reserve (power operations). That operating reserve capable of being connected to the bus and loaded within a specified time.   516

nonstop switch (elevators). A switch that, when operated, will prevent the elevator from making registered landing stops. *See:* **control.**   328

nonstorage display (display storage tubes). Display of nonstored information in the storage tube without appreciably affecting the stored information. *See:* **storage tube.**   174

nonsynchronous transmission (data transmission). A transmission process so that between any two significant instants in the same group, there is always an integral number of unit intervals. Between two significant instants located in different groups, there is not always an integral number of unit intervals. *Note:* In data transmission, this group is a block or a character. In telegraphy, this group is a character.   59

nontouching loop set (network analysis). A set of loops no two of which have a common node.   282

nonuniformity (transmission lines and waveguides). The degree with which a characteristic quantity, for example, impedance, deviates from a constant value along a given path. *Note:* It may be defined as the maximum amount of deviation from a selected nominal value. For example, the nonuniformity of the characteristic impedance of a slotted coaxial line may be 0.05 ohm due to dimensional variations.   185

nonvented fuse (power switchgear). A fuse without intentional provision for the escape of arc gases, liquids, or solid particles to the atmosphere during circuit interruption.   103, 443

nonvented power fuse (National Electrical Code) (installations and equipment operating at over 600 volts, nominal). A fuse without intentional provision for the escape of arc gases, liquids, or solid particles to the atmosphere during circuit interruption.   256

nonventilated (power and distribution transformer). So constructed as to provide no intentional circulation of external air through the enclosure.   53

nonventilated dry-type transformer (dry-type general purpose distribution and power transformers)(power and distribution transformer). A dry-type transformer which is so constructed as to provide no intentional circulation of external air through the transformer, and operating at zero gauge pressure.   555, 53

nonventilated enclosure (1)(metal-enclosed bus and calculating losses in isolated-phase bus)(metal enclosed bus). An enclosure so constructed as to provide no intentional circulation of external air through the enclosure.   574, 78

(2) (power switchgear). An enclosure so constructed as to provide no intentional circulation of external air through the enclosure. *Note:* Doors or removable covers are usually gasketed and humidity control may be provided by filtered breathers.   103

nonvolatile storage (test, measurement and diagnostic equipment). A storage device which can retain information in the absence of power. Contrast to volatile storage.   54

no op (computing systems). An instruction that specifically instructs the computer to do nothing, except to proceed to the next instruction in sequence.

    255, 77

NOR (mathematics of computing). A Boolean operator having the property that if P is a statement, Q is a statement, R is a statement, ..., then the NOR of P,Q,R, ..., is true if and only if all statements are false. *Note:* P NOR Q is often represented by P ↓ Q.

    564

norator (circuits and systems). A two-terminal ideal element the current through which and the voltage across which can each be arbitrary.   67

normal (1) (state of a superconductor). The state of a superconductor in which it does not exhibit superconductivity. *Example:* Lead is normal at temperatures above a critical temperature. *See:* **superconducting; superconductivity.**   191

(2) (power generation). *See:* **preferred.**

normal (low) frequency dew-withstand voltage (high-voltage switchgear). The root-mean-square (rms) voltage that can be applied to an insulator or a device, completely covered with condensed moisture, under specified conditions for a specified time without causing flashover or puncture.   443

normal (low) frequency dry-withstand voltage (high-voltage switchgear). The root-mean-square (rms) voltage that can be applied to a dry device under specified conditions for a specified time without causing flashover or puncture.   443

normal (low) frequency wet-withstand voltage (high-voltage switchgear). The root-mean-square (rms) voltage that can be applied to a wetted device under specified conditions for a specified time without causing flashover or puncture.   443

normal (through) dielectric heating applications. The metallic electrodes are arranged on opposite sides of the material so that the electric field is established through it. *Note:* The electrodes may be classified as plate electrodes, roller electrodes, or concentric electrodes. (1) Plate electrodes may have plane surfaces or surfaces of any desired shape to meet a particular condition and the spacing between them may be uniform or varied. (2) Roller electrodes are rollers separated by the material which moves between them. (3) Concentric electrodes consist of an enclosed and a surrounding electrode, with the material placed between the two. *See:* **dielectric heating.** 14

normal base power transfers (power operations). Those power transfers that are considered to be a part of normal base system loadings for the condition being analyzed. Other trasfers, such as emergency power or opportunistic economy energy transfers, are excluded even though they may be provided for in contractual arrangements. 516

normal clear. A term used to express the normal indication of the signals in an automatic block system in which an indication to proceed is displayed except when the block is occupied. 328

normal clear system. A term describing the normal indication of the signals in an automatic block signal system in which an indication to proceed is displayed except when the block is occupied. *See:* **centralized traffic control system.** 328

normal contact. A contact that is closed when the operating unit is in the normal position. 328

normal dimension. Incremental dimensions whose number of digits is specified in the format classification. For example, the format classification would be plus 14 for a normal dimension: X.XXXX. *See:* **dimension; incremental dimension; long dimension; short dimension.** 207

normal distribution. *See:* **Gaussian distribution.**

normal effort (electric generating unit reliability, availability, and productivity). Repairs were carried out with normal repair crews working normal shifts. *See:* **repair urgency.** 567

normal frequency (power switchgear). The frequency at which a device or system is designed to operate. 103

normal frequency (low frequency) dew withstand voltage (power switchgear). The normal-frequency withstand voltage applied to insulation completely covered with condensed moisture. *See:* **normal-frequency withstand voltage.** 103

normal frequency (low-frequency) dry withstand voltage (power switchgear). The normal-frequency withstand voltage applied to dry insulation. *See:* **normal-frequency withstand voltage.** 103

normal-frequency (low-frequency) wet withstand voltage (power switchgear). The normal-frequency withstand voltage applied to wetted insulation. *See:* **normal-frequency withstand voltage.** 103

normal-frequency (low-frequency) withstand voltage (power switchgear). The normal-frequency voltage that can be applied to insulation under specified conditions for a specified time without causing flashover or puncture. *Note:* This value is usually expressed as a root-mean-square (rms) value. *See:* **normal-frequency dew withstand voltage; normal-frequency dry withstand voltage; normal-frequency wet withstand voltage.** 103

normal-frequency line-to-line recovery voltage (power switchgear). The normal-frequency recovery voltage, stated on a line-to-line basis, that occurs on the source side of a three-phase circuit-interrupting device after interruption is complete in all three poles. 103

normal-frequency pole-unit recovery voltage (power switchgear). The normal-frequency recovery voltage that occurs across a pole unit of a circuit-interrupting device upon circuit interruption. 103

normal-frequency recovery voltage (power switchgear). The normal-frequency root-mean-square (rms) voltage that occurs across the terminals of an ac circuit-interrupting device after the interruption of the current and after the high-frequency transients have subsided. *Note:* For determination of the normal-frequency recovery voltage, see ANSI/IEEE C37.09-1979. 103, 443

normal-glow discharge (gas). The glow discharge characterized by the fact that the working voltage decreases or remains constant as the current increases. *See:* **discharge (gas).** 244, 190

normal induction (magnetic material). The maximum induction in a magnetic material that is in a symmetrically cyclically magnetized condition. 210

normalize (1)(mathematics of computing). To shift the fixed point part of a floating point number, and make the corresponding adjustment to the exponent, to ensure that the fixed-point part lies within some prescribed range. The number represented remains unchanged. 564

(2) (test, measurement and diagnostic equipment). (A) To adjust the characteristic and fraction of a floating decimal point number thus eliminating leading zeros in the fraction; (B) To adjust a measured parameter to a value acceptable to an instrument or measurement technique. 54

(3) (circuits and systems). To divide an impedance or frequency by a reference quantity thereby making the result dimensionless. 67

normalized admittance (waveguide). The reciprocal of the normalized impedance. 267

normalized directivity. *See:* **antenna (aperture) illumination efficiency.**

normalized frequency (fiber optics). A dimensionless quantity (denoted by V), given by

$$V = \frac{2\pi a}{\lambda} \sqrt{n_1^2 - n_2^2}$$

where a is waveguide core radius, λ is wavelength in vacuum, and n_1 and n_2 are the maximum refractive index in the core and refractive index of the homogeneous cladding, respectively. In a fiber having a power-law profile, the approximate number of bound modes is $(V^2/2)[g/(g + 2)]$, where g is the profile parameter. *Syn:* **V number.** *See:* **bound mode; mode volume; parabolic profile; power-law index profile; single mode optical waveguide.** 433

normalized impedance (waveguide). The ratio of an impedance and the corresponding characteristic impedance. *Note:* The normalized impedance is independent of the convention used to define the characteristic impedance, provided that the same convention is also taken for the impedance to be normalized.   267

normalized response (automatic control). One obtained by dividing a measured value and dimension by some convenient reference value and dimension: usually the quotient is nondimensional. *See:* **indicial response; gain, static.**   56

normalized transimpedance (magnetic amplifier). The ratio of differential output voltage to the product of differential control current and control winding turns.   171

normal joint (power cable joint). A joint which is designed not to restrict movement of dielectric fluid between cables being joined.   34

normal linearity (computing systems). Transducer linearity defined in terms of a zero-error reference, that is chosen so as to minimize the linearity error. *Note:* In this case, the zero input does not have to yield zero output and full-scale input does not have to yield full-scale output. The specification of normal linearity, therefore, is less stringent than zero-based linearity. *See:* **electronic analog computer; linearity of a multiplier (linearity).**   9

normally closed. *See:* **normally open and normally closed.**

normally closed contact. A contact, the current-carrying members of which are in engagement when the operating unit is in its normal position.   328

normally open and normally closed (industrial control). When applied to a magnetically operated switching device, such as a contactor or relay, or to the contacts thereof, these terms signify the position taken when the operating magnet is deenergized. Applicable only to nonlatching types of devices. *See:* **contacter.**   206

normally open contact (open contact). A contact, the current-carrying members of which are not in engagement when the operating unit is in the normal position.   328

normal-mode interference. *See:* **interference, normal-mode.**

normal mode noise (transverse or differential) (cable systems in power generating stations). The noise voltage which appears differentially between two signal wires and which acts on the signal sensing circuit in the same manner as the desired signal. Normal mode noise may be caused by one or more of the following: (1) Electrostatic induction and differences in distributed capacitance between the signal wires and the surroundings (2) Electromagnetic induction and magnetic fields linking unequally with the signal wires (3) Junction or thermal potentials due to the use of dissimilar metals in the connection system (4) Common mode to normal mode noise conversion.   10

normal number (radix-independent floating-point arithmetic). A nonzero number that is finite and not subnormal.   588

normal operations area (nuclear power generating sta-

Maximum Allowable Shield Length With Only One Ground

Size Conductor	One Cable Per Duct (ft)	Three Cables Per Duct (ft)
1/0 AWG	1250	4500
4/0 AWG	865	3000
350 kcmil	710	2260
500 kcmil	580	1870
750 kcmil	510	1500
1000 kcmil	450	——
2000 kcmil	340	——

tion). A functional area allocated for those displays and controls necessary for the tasks routinely performed during plant startup, shutdown and power operation modes.   358

normal permeability. The ratio of normal induction to the corresponding maximum magnetizing force. *Note:* In anisotropic media, normal permeability becomes a matrix.   210

normal position (of a device). A predetermined position that serves as a starting point for all operations.   328

normal rating (power operations). The level of power flow that facilities can carry through a series of daily load cycles without loss of life to the facility involved. *See:* **emergency rating.**   516

normal stop system. A term used to describe the normal indication of the signals in an automatic block signal system in which the indication to proceed is given only upon the approach of a train to an unoccupied block. *See:* **centralized traffic-control system.**   328

normal-terminal stopping device (elevators). A device, or devices, to slow down and stop an elevator or dumbwaiter car automatically at or near a terminal landing independently of the functioning of the operating device. *See:* **control.**   328

normal test output (amplitude-modulation broadcast receivers). (1) For receivers capable of delivering at least 1 watt maximum undistorted output, the normal test output is an audio-frequency power of 0.5 watt delivered to a standard dummy load. (2) For receivers capable of delivering 0.1 but less than 1 watt maximum undistorted output, the normal test output0.05 watt audio-frequency power delivered to a standard dummy load. When this value is used, it should be so specified. Otherwise, the 0.5-watt value is assumed. (3)For receivers capable of delivering less than 0.1 watt maximum undistorted output, the normal test output is 0.005 watt audio-frequency power delivered to a standard dummy load. When this value is used, it should be so specified. (4) For automobile receivers, normal test output is 1.0 watt audio-frequency power delivered to a standard dummy load.   524

normal transfer capability (power operations). The maximum amount of power that can be transmitted continuously over the transmission network.   516 normal velocity storage (digital acceleromter). The velocity information that is stored in the accelerometer during the application of an acceleration within its input range.   383

normal voltage limit (power operations). The voltage range that is acceptable on a sustained basis. *See: emergency voltage limit.*   516

normal weather (electric power systems). All weather not designated as adverse. *See: outage.*   200

normal weather persistent-cause forced-outage rate (for a particular type of component) (electric power systems). The mean number of outages per unit of normal weather time per component. *See: outage.*   200

north (navigation aid terms). The primary reference direction relative to the earth. True north is the direction of the north geographical pole. Magnetic north is the direction north as determined by the earth's magnetic line of force.   526

north-stabilized plan-position indicator (radar). A special case of azimuth-stabilized plan-position indicator in which the reference direction is magnetic north. *See: navigation.*   187, 13

Norton's theorem (circuits and systems). States that a linear time-invariant one-port is equivalent to a circuit which consists of the driving-point admittance of the one-port shunted by the short-circuit current of the one-port.   67

Norton transformation (circuits and systems). A four-terminal network transformation of a series (shunt) ladder element to an equivalent pi (tee) network in cascade with an ideal transformer. The pi (tee) network arms are identical to the series (shunt) arm except for multiplying factors related to the ideal transformer turns-ratio.   67

nose suspension. A method of mounting an axle-hung motor or a generator to give three points of support, consisting of two axle bearings (or equivalent) and a lug or nose projecting from the opposite side of the motor frame, the latter supported by a truck or vehicle frame. *See: traction motor.*   328

NOT (mathematics of computing). A monadic Boolean operator having the property that if P is a statement, then the expression 'NOT P' is true if P is false, and false if P is true. *Note:* NOT P is often represented by ~ P, P̄, P'.   564

notation. *See: positional notation.*

notch filter (circuits and systems). A band-elimination filter, sometimes used to eliminate a single frequency for example, 60 Hz.   67

notching (relays). A qualifying term applied to a relay indicating that a predetermined number of separate impulses is required to complete operation. *See: relay.*   328

notching or jogging device (66) (power system device function numbers). A device that functions to allow only a specified number of operations of a given device, or equipment, or a specified number of successive operations within a given time of each other. It is also a device that functions to energize a circuit periodically or for fractions of specified time intervals, or that is used to permit intermittent acceleration or jogging of a machine at low speeds for mechanical positioning.   402

notching relay (power switchgear). A programming relay in which the response is dependent upon successive impulses of the input quantity.   103

not connected (NC) (semiconductor memory). The inputs/outputs that are not connected to any active part of the circuit or any other pin or any conductive surface of the package.   441

note. A conventional sign used to indicate the pitch, or the duration, or both, of a tone. It is also the tone sensation itself or the oscillation causing the sensation. *Note:* The word serves when no distinction is desired among the symbol, the sensation, and the physical stimulus.   176

NPR (data transmission)(Noise-Power Ratio, expressed in dB [decibels]). This is a term commonly used in noise loading technique. Usually an uncalibrated receiver is used to measure the noise power in a channel of a system loaded with noise, first with full noise in the channel and the noise source. The ratio of these readings is the NPR. An NPR reading is independent of the noise bandwidth of the receiver; provided the same bandwidth is used in both noise measurements and the band stop filters are wide enough.   59

NRA. *See: laser gyro axes.*

N scan (electronic navigation). *See: N display.*

N scope. *See: N display.*

N-scope (radar). A cathode-ray oscilloscope arranged to present an N-display.   13

n-terminal electron tubes. *See: capacitance, input.*

N-terminal network. A network with N accessible terminals. *See: network analysis.*   210

N-terminal pair network. A network with $2N$ accessible terminals grouped in pairs. *Note:* In such a network one terminal of each pair may coincide with a network node. *See: network analysis.*   210

Nth field lag (diode-type camera tube). The fraction of the output signal which is read out in the N th field after the initial reading scan of an input signal which has been completely extinguished just before the scanning beam reaches that portion of the target irradiated by the input signal. This can be readily understood with reference to the decay characteristics diagram below. A similar definition can be made for the buildup of a signal as illustrated in the buildup characteristics diagram below.   380

Nth harmonic. The harmonic of frequency N times that of the fundamental component.   349

NTI. *See: noise transmission impairment.*

n-type crystal rectifier. A crystal rectifier in which forward current flows when the semiconductor is negative with respect to the metal. *See: rectifier.*   328

n-type semiconductor. *See: semiconductor, n-type.*

nuclear burnup (power operations). (1) A measure of nuclear reactor fuel consumption, usually expressed as energy produced per unit weight of fuel exposed (megawatt-days per metric ton of fuel). (2) Percentage of fuel atoms that have undergone fission (atom percent burnup). 516

nuclear energy (power operations). The energy with which nucleons are bound together to form nuclei. When a nucleus is changed or rearranged in a nuclear reaction (fission, fusion, etcetera) or by radioactive decay, nuclear energy may be released or absorbed in the form of kinetic energy of the reactants or products. *See:* **nuclear fuel elements; nuclear fuel reprocessing; nuclear reactor.** 516

nuclear fuel elements (power operations). An assembly of rods, tubes, plates, or other geometrical forms into which nuclear fuel is contained for use in a reactor. 516

nuclear fuel reprocessing (power operations). The processing of irradiated reactor fuel to recover the unused fissionable material, or fission products, or both. 516

nuclear power generating station (station). A plant wherein electric energy is produced from nuclear energy by means of suitable apparatus. The station may consist of one or more generating units. 102

nuclear power generating stations, class ratings. (1) Class 1 electric equipment. The electric equipment that is essential to the safe shutdown and isolation of the reactor or whose failure or damage could result in significant release of radioactive material. **(2) Class-1 structures and equipment.** Structures and equipment that are essential to the safe shutdown and isolation of the reactor or whose failure or damage could result in significant release of radioactive material. **(3) Class 1E.** The safety classification of the electric equipment and systems that are essential to emergency reactor shutdown, containment isolation, reactor core cooling, and containment and reactor heat removal, or are otherwise essential in preventing significant release of radioactive material to the environment. **(4) Class 1E control switchboard.** A rack panel, switchboard, or similar type structure fitted with any Class 1E equipment. **(5) Class 1E electric systems.** The systems that provide the electric power used to shut down the reactor and limit the release of radioactive material following a design basis event. **(6) Class II structures and equipment.** Structures and equipment that are important to reactor operation but are not essential to the safe shutdown and isolation of the reactor and whose failure cannot result in a significant release of radioactive material. **(7) Class III structures and equipment.** Structures and equipment that are not essential to the operation, safe shutdown, or isolation of the reactor and whose failure cannot result in the release of radioactive material. 28, 143, 31, 102, 142, 140, 120, 141

nuclear reactor (power operations). An apparatus by means of which a fission chain reaction can be initiated, maintained, and controlled. 516

nuclear safety related (nuclear power generating stations). That term used to call attention to safety classifications incorporated in the body of the document so marked. *Note:* As used in IEEE Std 494-1974, the term calls attention to the safety classification Class 1E. 156

null (1)(direction finding systems)(navigation aid terms). The condition of minimum output as a function of the direction of arrival of the signal, or of the rotation of the response pattern of the DF (direction finder) antenna system. 526 **(2)(microprocessor operating systems parameter types).** A value whose definition is to be supplied within the context of a specific operating system. This value is a representation of the set of no numbers or no value for the operating system in use. 478 **(3) (gyro, accelerometer) (inertial sensor).** The condition of minimum output. *See:* **electrical null.** 46 **(4) (signal-transmission system).** The condition of zero error-signal achieved by equality at a summing junction between an input signal and an automatically or manually adjusted balancing signal of phase or polarity opposite to the input signal. *See:* **signal.** 188

nullator (circuits and systems). An idealized one-port that is simultaneously an open and short circuit, that is, $V = I = O$. The nullator is a bilateral, loss-less one-port. 67

null balance (automatic null-balancing electric instrument) (instruments). The condition that exists in the circuits of an instrument when the difference between an opposing electrical quantity within the instrument and the measured signal does not exceed the dead band. *Note:* The value of the opposing electrical quantity produced within the instrument is related to the position of the end device. *See:* **control system, feedback; measurement system.** 295

null-balance system. A system in which the input is measured by producing a null with a calibrated balancing voltage or current. *See:* **signal.** 188

nullity (degrees of freedom on mesh basis) (network). The number of independent meshes that can be selected in a network. The nullity N is equal to the number of branches B minus the number of nodes V plus the number of separate parts P. $N = B - V + P$. *See:* **network analysis.** 210

null junction (power supplies). The point on the Kepco bridge at which the reference resistor, the voltage-control resistance, and one side of the comparison amplifier coincide. *Note:* The null junction is maintained at almost zero potential and is a virtual ground. *See:* **summing point.** 228, 185

null offsetting (gyro, accelerometer) (inertial sensor). A calibration or test technique by which the electrical null position is intentionally shifted, resulting in a rotation of the input axis relative to the input reference axis. 46

null operation (FASTBUS acquisition and control). A primary address cycle followed by no data cycles. It determines if the system contains a device capable of responding to the primary address used. It can be used to reserve segment interconnects for an arbitration locked sequence. 480

null-reference glide slope (navigation aid terms). A glide-slope system using a two-element array in which the slope angle is defined by the first null above the horizontal in the field pattern of the upper antenna.
526

null steering (antennas). To control, usually electronically, the direction at which a directional null appears in the radiation pattern of an operational antenna.
111

null-steering antenna. An antenna having in its radiation pattern one or more directional nulls that can be steered, usually electronically.
111

null triggering/zero-crossing triggering (thyristor). A method of triggering the controller circuit elements such that the associated angle of retard is zero.
445

number (electronic computation). (1) Formally, an abstract mathematical entity that is a generalization of a concept used to indicate quantity, direction, etcetera. In this sense a number is independent of the manner of its representation. (2) Commonly: A representation of a number as defined above (for example, the binary number 10110, the decimal number 3695, or a sequence of pulses). (3) An expression, composed wholly or partly of digits, that does not necessarily represent the abstract entity mentioned in the first meaning. *Note:* Whenever there is a possibility of confusion between meaning (A) and meaning (B) or (C), it is usually possible to make an unambiguous statement by using **number** for meaning (A) and **numerical expression** for meaning (B) or (C).
235

number crunching (computer applications). Computer processing that relies heavily on the arithmetic and logical capabilities of the central processing unit, as contrasted with processing that entails extensive input/output or data movement.
571

number group (telephone switching systems). An arrangement for associating equipment numbers with mainstation codes.
55

numbering plan (telephone switching systems). A plan employing codes and directory numbers for identifying main stations and other terminations within a telecommunication system.
55

numbering-plan area (NPA) (telephone switching systems). A geographical subdivision of the territory covered by a national or integrated numbering plan. An NPA is identified by a distinctive area code (NPA code).
55

numbering-plan area code, (NPA) (telephone switching systems). A one, two-, or three-digit number that, for the purpose of distance dialing, designates one of the geographical areas within a country (and in some instances neighboring territories) that is covered by a separate numbering plan.
55

number of loops (magnetically focused electron beam). The number of maxima in the beam diameter between the electron gun and the target, or between a point on the photocathode and the target.
125

number of rectifier phases (rectifier circuit). The total number of successive, nonsimultaneous commutations occuring within that rectifier circuit during each cycle when operating without phase control. *Note:* It is also equal to the order of the principal harmonic in the direct-current potential wave shape. The number of rectifier phases influences both alternating-current and direct-current waveforms. In a simple single-way rectifier the number of rectifier phases is equal to the number of rectifying elements. *See:* **rectification; rectifier circuit element.**
237, 66

number of scanning lines (television) (numerically). The total number of lines, both active and blanked, in a frame. *Note:* In any specified scanning system, this number is inherently the ratio of the line frequency to the frame frequency and is always a whole number. In a two-to-one odd-line interlaced system, it is always an odd whole number.
18

number system (electronic computation). Loosely, a numeration system. *See:* **positional notation.**
255, 77

number 1 master (metal master) (metal negative) (disk recording) (electroacoustics). *See:* **original master.**

number 1 mold (mother) (metal positive). A mold derived by electroforming from the original master. *See:* **phonograph pickup.**
176

number 2, number 3, etcetera master. A master produced by electroforming from a number 1, number 2, etcetera mold. *See:* **phonograph pickup.**
176

number 2, number 3, etcetera mold. A mold derived by electroforming from a number 2, number 3, etcetera master. *See:* **phonograph pickup.**
176

numeral. A representation of a number. *See:* **binary numeral.**
255, 77

numeration system (numeral system). A system for the representation of numbers, for example, the decimal system, the Roman numeral system, the binary system.
255, 77

numerical action (switch). That action that depends on at least part of the called number.
328

numerical analysis. The study of methods of obtaining useful quantitative solutions to problems that have been expressed mathematically, including the study of the errors and bounds on errors in obtaining such solutions.
255, 77

numerical aperture (NA) (fiber optics). (1) The sine of the vertex angle of the largest cone of meridional rays that can enter or leave an optical system or element, multiplied by the refractive index of the medium in which the vertex of the cone is located. Generally measured with respect to an object or image point and will vary as that point is moved. (2) For an optical fiber in which the refractive index decreases monotonically from n sub 1 on axis to n sub 2 in the cladding the numerical aperture is given by

$$NA = \sqrt{n_1^2 - n_2^1}$$

(3) Colloquially, the sine of the radiation or acceptance angle of an optical fiber, multiplied by the refractive index of the material in contact with the exit or

entrance face. This usage is approximate and imprecise, but is often encountered. *See:* **acceptance angle; launch numerical aperture; meridional ray; radiation angle; radiation pattern.**   433

numerical control. Pertaining to the automatic control of processes by the proper interpretation of numerical data.   255, 77

numerical control system (numerically controlled machines). A system in which actions are controlled by the direct insertion of numerical data at some point. *Note:* The system must automatically interpret at least some portion of these data.   224, 207

numerical data. Data in which information is expressed by a set of numbers or symbols that can only assume discrete values or configurations.   207

numerical display (illuminating engineering). An electrically operated display of digits. Tungsten filaments, gas discharges, light-emitting diodes, liquid crystals, projected numerals, illuminated numbers and other principles of operation may be used. *Syn:* **digital display.**   167

numerical reliability (software). The probability that an item will perform a required function under stated conditions for a stated period of time. *See:* **function.**   434

numeric printout (dedicated-type sequential events recording systems). A brief coded method of identifying inputs using numeric characters only.   48, 58

nurses' stations (National Electrical Code) (health care facilities). Areas intended to provide a center of nursing activity for a group of nurses working under one nurse supervisor and serving bed patients, where the patient calls are received, nurses are dispatched, nurses' notes written, inpatient charts prepared, and medications prepared for distribution to patients. Where such activities are carried on in more than one location within a nursing unit, all such separate areas are considered a part of the nurses' station.   256

nursing home (National Electrical Code) (health care facilities). A building or part thereof used for the lodging, boarding and nursing care, on a 24-hour basis, of 4 or more persons who, because of mental or physical incapacity, may be unable to provide for their own needs and safety without the assistance of another person. Nursing home, wherever used in the National Electrical Code shall include nursing and convalescent homes, skilled nursing facilities, intermediate care facilities, and infirmaries of homes for the aged.   256

N-user (logical link control). An $N+1$ entity that uses the services of the N-layer, and below, to communicate with another $N+1$ entity.   585

nutating feed (radar). A technique of conical scanning in which the polarization remains unchanged.   13

nutation (gyro). The oscillation of the spin axis of a two-degree-of-freedom gyro about two orthogonal axes normal to the mean position of the spin axis.   46

nutation field (radar). *See:* **conical scan.**

nutation frequency (gyro) (inertial sensor). The frequency of the coning or periodic wobbling motion of the rotor spin axis which results from a transient input. For an undamped rotor, the nutation frequency equals the product of the rotor spin frequency and the ratio of the rotor polar moment of inertia to the effective rotor transverse moment of inertia.   46

Nyquist diagram (data transmission). The Nyquist diagram of a feedback amplifier is a plot, in rectangular coordinates, of the real and imaginary parts of the factor μ/β for frequencies from zero to infinity, where μ is the amplification in the absence of feedback, and β is the fraction of the output voltage that is superimposed on the amplifier input. *Note:* The criterion for stability of a feedback amplifier is that the curve of the Nyquist diagram shall not enclose the point $X = -1$, $Y = 0$, where μ/β equals $X + jY$.   59

Nyquist interval (data transmission). The maximum separation in time which can be given to regularly spaced instantaneous samples of a wave of bandwidth W for complete determination of the wave form of the signal. Numerically, it is equal to $1/2W$ seconds.   59

Nyquist rate (channel) (information theory). The reciprocal of the Nyquist interval.   415

O

OA. *See:* **transformer, oil-immersed.**

OBI (omnibearing indicator). An instrument which presents an automatic and continuous indication of an omnibearing. *See:* **omnibearing indicator.**   13

object color. The color of the light reflected or transmitted by the object when illuminated by a standard light source, such as source *A, B,* or *C* of the *Commission Internationale de l'Eclairage* (CIE). *See:* **standard source; color.**   167

objective evidence (nuclear power quality assurance). Any documented statement of fact, other information, or record, either quantitative or qualitative, pertaining to the quality of an item or activity, based on observations, measurements, or tests which can be verified.   417

objective loudness rating (OLR) (loudness ratings of telephone connections). The rating of a connection or its components when measured according to this standard.   409

object language. *See:* **target language.**

object program (software). A fully compiled or assembled program that is ready to be loaded into the computer. *See:* **assembled program; compile; computer; source program.**   434 observable (control system). A property of a component of a state whereby its value at a given time can be computed from measurements on the output over a finite past interval. *See:* control system.   329

observable, completely (control system). The property of a plant whereby all components of the state are observable. *See:* control system; observable; plant.   329

observable insulation temperature (electric equipment) (thermal classification of electric equipment and electrical insulation). The temperature of the insulation in electric equipment, which is measured in a specified way, for example, with a thermometer, embedded thermocouple, resistance detector, or by winding resistance or other suitable procedure.   506

observable temperature (equipment rating). The temperature of equipment obtained on test or in operation. *See:* limiting insulation temperature.   320

observable temperature rise (electric equipment) (thermal classification of electric equipment and electrical insulation). The difference between the observable insulation temperature and the ambient temperature.   506

observation time (radar). The time interval over which a radar echo signal may be integrated for detection or measurement.   13

observed instantaneous availability. At a stated instant of time the proportion of occasions when an item can perform a required function. *Notes:* (1) Occasions can refer to either a number of items at a single instant of time or to one or more items at instants repeated in time. (2) The run-up time is counted in down-time after repair and is counted in the up-time when the equipment is brought into use for the first time. (3) The observed instantaneous availability is to be associated with a period of time and with stated conditions of use and maintenance.   164

observed mean active maintenance time. The ratio of the sum of the active maintenance times to the total number of maintenance actions. *Note:* The maintenance conditions applied shall be stated.   164

observed mean availability. The ratio of the cumulative time for which an item can perform a required function to the cumulative time under observation, or at instants of time (chosen by a sampling technique), the mean of the proportion of a number of nominally identical items which can perform their required function. *Notes:* (1) When one limiting value is given, this is usually the lower limit. (2) The observed mean availability is to be associated with a stated period of time and with stated conditions of use and maintenance.   164

observed reliability (reliability) (1) non-repaired items. For a stated period of time, the ratio of the number of items which performed their functions satisfactorily at the end of the period to the total number of items in the sample at the beginning of the period.   164

(2) repaired item or items. The ratio of the number of occasions on which an item or items performed their functions satisfactorily for a stated period of time to the total number of occasions the item or items were required to perform for the same period. *Note:* The criteria for what constitutes satisfactory function shall be stated.   164

obstruction beacon. *See:* hazard beacon.

obstruction lights (illuminating engineering). Aeronautical ground lights provided to indicate obstructions.   167

occulting light (illuminating engineering). A rhythmic light in which the periods of light are clearly longer than the periods of darkness.   167

occupied bandwidth (radio-noise emission). The frequency bandwidth such that, below its lower and above its upper frequency limits, the mean powers radiated are each equal to 0.5 percent of the total mean power radiated by a given emission. In some cases, for example multichannel frequency division systems, the percentage of 0.5 percent may lead to certain difficulties in the practical application of the definition of occupied bandwidth; in such cases a different percentage may be useful.   418

OCR. *See:* optical character recognition.

octad (octade)(mathematics of computing). A group of three bits used to represent one octal digit.   564

octal (mathematics of computing). (1) Pertaining to a selection in which there are eight possible outcomes. (2). Pertaining to the numeration system with a radix of eight.   564

octant. *See:* sextant.

octantal error (navigation)(navigation aid terms). An error in measured bearing caused by the finite spacing of the antenna elements in systems using spaced antennas to provide bearing information (such as VOR [very high-frequency omnidirectional range]): this error varies in a sinusoidal manner throughout the 360° and has four positive and four negative maximums.   526

octave (1)(data transmission). In electric communication, the interval between two frequencies having a ratio of 2 to 1.   59

(2)(seismic testing of relays). The interval between two frequencies which have a frequency ratio of two. For example, 1 to 2, 2 to 4, 4 to 8 Hz, etcetera.   392

octave band, one-third octave band (audible noise measurement). Many sounds, including audible noise fro a transmission line, are broad band, having components which are continuously distributed over a range of frequencies. The spectrum of such a sound can be approximated in terms of a series of octave band or one-third octave band pressure levels. A band is designated by its center frequency, f_0, which is the geometric mean of the upper and lower frequencies of the band. An octave band extends from a lower frequency $(f_0/\sqrt{2})$ to twice the lower frequency $(\sqrt{2} f_0)$. A one-third octave band extends from a lower frequency $(f_0/{}^6\sqrt{2})$ to ${}^3\sqrt{2}$ times the lower frequency $({}^6\sqrt{2} f_0)$. The octave (one-third octave) band sound-pressure level is the integrated sound-pressure level of all spectral components in the specified octave or one-third octave band. For example, see ANSI/ASC S1.6-1984 [4]. 462

octave-band pressure level (acoustics) (octave pressure level) (sound). The band pressure level for a frequency band corresponding to a specified octave. *Note:* The location of an octave-band pressure level on a frequency scale is usually specified as the geometric mean of the upper and lower frequencies of the octave. 176

octet (1)(mathematics of computing). A group of eight adjacent digits operated upon as a unit. 564 **(2)(logical link control).** A bit-oriented element that consists of eight contiguous binary bits. 585

octode. An eight-electrode electron tube containing an anode, a cathode, a control electrode, and five additional electrodes that are ordinarily grids. 125

octonary (electronic computation). *See:* octal.

odd-even check (computing systems). *See:* parity check.

O-display (radar). An A-display modified by the inclusion of an adjustable notch for measuring range. 13

odolite (navigation aid terms). An optical instrument for accurately measuring horizontal and vertical angles. 526

odometer (navigation aid terms). A device attached to a vehicle for counting the number of revolutions of a drive shaft or wheel. 526

oersted. The unit of magnetic field strength in the unrationalized centimeter-gram-second (cgs) electromagnetic system. The oersted is the magnetic field strength in the interior of an elongated uniformly wound solenoid that is excited with a linear current density in its winding of one abampere per 4π centimeters of axial length. 210

off-axis mode (laser-maser). *See:* higher order mode of propagation. An off-axis mode will incorporate one or more of the maxima which lie off the axis of a beam. 363

off-center display. A plan-position-indicator display, the center of which does not correspond to the position of the radar antenna. *See:* radar. 328

off-center PPI (radar). A plan position indicator (PPI) which has the zero position of the time base at a point other than at the center of the display, thus providing the equivalent of a larger display for a selected portion of the service area. *Syn:* offset PPI. 13

offered traffic (telephone switching systems). A measure of the calls requesting service during a given period of time. 55

off-hook (telephone switching systems). A closed station line or any supervisory or pulsing condition is indicative of this state. 55

office automation (computer applications). The automation of information traffic through the use of any or all of the following: voice processing; word and data processing; reprographics; records processing or micrographics; telecommunications. *See:* electronic mail. 571

office class (telephone switching systems). A designation (Class 1, 2, 3, 4, 5) given to each office in World Zone 1 involved in the completion of toll calls. The class is determined according to the office's switching function, its interrelation with other switching offices, and its transmission requirements. The class designation given to the switching points in the network determines the routing pattern for all calls. Class 1 is higher in rank than Class 2: Class 2 is higher than Class 3: and so on. *See:* world zone number. 55

office code (telephone switching systems). The digits that designate a block of main-station codes within a numbering-plan area. 55

office test (meter). A test made at the request or suggestion of some department of the company to determine the cause of seemingly abnormal registration. *See:* service test (field test). 328

OFF-impedance (thyristor). The differential impedance between the terminals through which the principal current flows, when the thyristor is in the OFF state at a stated operating point. *See:* principal voltage-current characteristic (principal characteristic). 243, 66, 191

off line (1)(monitoring radioactivity in effluents). A system where an aliquot is withdrawn from the effluent stream and conveyed to the detector assembly. 559 **(2) (computing systems).** Pertaining to equipment or devices not under direct control of the central processing unit. 255, 77 **(3) (test, measurement and diagnostic equipment).** (A) Operation of input/output and other devices not under direct control of a device: and (B) Peripheral equipment operated outside of, and not under control of the system, for example, the off-line printer. 54

off-line operation (emergency and standby power). (1) Pertaining to computer systems not under direct control of the central processing unit. (2) Pertaining to uninterruptible power supply systems whereby an inverter is off during normal operating conditions. 512

off-line testing (test, measurement and diagnostic equipment). Testing of the unit under test removed from its operational environment or its operational equipment. Shop testing. 54

off-net call (telephone switching systems). A call from a switched-service network to a station outside that network. 55

off-peak energy (power operations). Energy supplied during designated periods of relatively low system demands. 516

off-peak period (watthour meters). The period of time during which the specified off-peak rate applies. 485

off-peak power. Power supplied during designated periods of relatively low system demands. *See:* generating station. 64

OFF period: (1) (electron tubes). The time during an operating cycle in which the electronic tube is nonconducting. *See:* ON period (electron tubes). 244 **(2) (circuit switching element) (inverters).** The part of an operating cycle during which essentially no current flows in the circuit switching element. *See:* **self-commutated inverters.** 208

off road vehicle (conductor stringing equipment). A vehicle specifically designed and equipped to traverse sand, swamps, muddy tundra or rough mountainous terrain. Vehicles falling into this category are usually all wheel drive or tracked units. In some cases, units equipped with special air bag rollers having a soft footprint are utilized. *Syn:* **all terrain vehicle (ATV); swamp buggy.** 431

offset (1)(supervisory control, data acquisition, and automatic control). A predetermined value modifying the actual value so as to improve the integrity of the system, for example, the use of a 4 mA (milliampere) signal to represent zero in a 4 mA to 20 mA system. 570

(2) (transducer). The component of error that is constant and independent of the inputs, often used to denote bias. 10

(3) (distance relay). The displacement of the operating characteristic on an R-X diagram from the position inherent to the basic performance class of the relay. *Note:* A relay with this characteristic is called an offset relay. 103, 127

(4) (course computer) (electronic navigation). An automatic computer that translates reference navigational coordinates into those required for a predetermined course. *See:* **navigation.** 187, 13

(5) (pulse terms). The algebraic difference between two specified magnitude reference lines. Unless otherwise specified, the two magnitude reference lines are the waveform baseline and the magnitude origin line. *See:* The single pulse diagram below the **waveform epoch** entry. 254

(6) (analog computers). In a transducer, the component of error that is constant and independent of the inputs, often used to denote bias. 9

(7) (station control and data acquisition). A predetermined value modifying the actual value so as to improve the integrity of the system, for example, the use of a 4 mA signal to represent zero in a 4 to 20mA system. 403

offset angle (electroacoustics) (lateral disk reproduction). The offset angle is the smaller of the two angles between the projections into the plane of the disk of the vibration axis of the pickup stylus and the line connecting the vertical pivot (assuming a horizontal disk) of the pickup arm with the stylus point. *See:* **phonograph pickup.** 176

offset (outboard) bearing (air switch) (power switchgear). A component of a switch operating mechanism designed to provide support for a torsional operating member and a crank which provides reciprocating motion for switch operation. 103

offset, clipping. *See:* **clipping offset.**

offset course computer (navigation aid terms). An automatic computer which translates reference navigational coordinates into those required for a predetermined course. 526

offset paraboloidal reflector antenna. A reflector antenna whose main reflector is a portion of a paraboloid which is not symmetrical with respect to its focal axis, and does not include the vertex so that aperture blockage by the feed is reduced or eliminated. 111

offset voltage (power supplies). A direct-current potential remaining across the comparison amplifier's input terminals (from the null junction to the common terminal) when the output voltage is zero. The polarity of the offset voltage is such as to allow the output to pass through zero and the polarity to be reversed. It is often deliberately introduced into the design of power supplies to reach and even pass zero-output volts. 186

offset waveform (pulse terms). A waveform whose baseline is offset from, unless otherwise specified, the magnitude origin line. 254

OFF state (thyristor). The condition of the thyristor corresponding to the high-resistance low-current portion of the principal voltage-current characteristic between the origin and the breakover point(s) in the switching quadrant(s). *See:* **principal voltage-current characteristic (principal characteristic).** 66

OFF-state current (thyristor). The principal current when the thyristor is in the OFF state. *See:* **principal current.** 66

OFF-state power dissipation (thyristor). The power dissipation resulting from OFF-state current. 66

OFF-state voltage (thyristor). The principal voltage when the thyristor is in the OFF state. *See:* **principal voltage-current characteristic (principal characteristic).** 66

ohm (1) (general). The unit of resistance (and of impedance) in the International System of Units (SI). The ohm is the resistance of a conductor such that a constant current of one ampere in it produces a voltage of one volt between its ends. 210

(2) (metric practice). The electric resistance between two points of a conductor when a constant difference of potential of one volt, applied between these two points, produces in this conductor a current of one ampere, this conductor not being the source of any electromotive force. 21

ohmic contact (1) (of a semiconductor radiation detector)(charged-particle detectors)(X-ray energy spectrometers). A purely resistive contact, that is, one that has a linear voltage-current characteristic throughout its entire operating range. 23, 245, 471

(2) (semiconductor). A contact between two materials, possessing the property that the potential difference across it is proportional to the current passing through. *See:* **semiconductor.** 186

ohmic resistance test (rotating machinery). A test to measure the ohmic resistance of a winding, using direct current. *See:* **asynchronous machine.** 63

ohmmeter. A direct-reading instrument for measuring electric resistance. It is provided with a scale, usually graduated in either ohms, megohms, or both. If the scale is graduated in megohms, the instrument is usually called a megohmmeter. *See:* **instrument.** 328

Ohm's law. The current in an electric circuit is inversely proportional to the resistance of the circuit and is directly proportional to the electromotive force in the circuit. *Note:* Ohm's law applies strictly only to linear constant-current circuits.   210

OHR. *See:* **over-the-horizon radar.**

oil (1) (packaging machinery). Used as a prefix and applied to a device that interrupts an electric circuit; indicates that the interruption occurs in oil.   429
(2) (power and distribution transformer). The term "oil" includes the following insulating and cooling liquids: Type I Mineral Oil (uninhibited oil), Type II Mineral Oil (inhibited oil), and Askarel.   53
(3) (outdoor apparatus bushings). Mineral transformer oil.   168

oil buffer (elevators). A buffer using oil as a medium that absorbs and dissipates the kinetic energy of the descending car or counterweight. *See:* **elevators.**   328

oil-buffer stroke (oil buffer) (elevators). The oil-displacing movement of the buffer plunger or piston, excluding the travel of the buffer-plunger accelerating device. *See:* **elevators.**   328

oil catcher (rotating machinery). A recess to carry off oil. *See:* **oil cup (rotating machinery).**   63

oil cup (rotating machinery). An attachment to the oil reservoir for adding oil and controlling its upper level.   63

oil cutout (oil-filled cutout) (power switchgear). A cutout in which all or part of the fuse support and its fuse link or disconnecting blade are mounted in oil with complete immersion of the contacts and the fusible portion of the conducting element (fuse link), so that arc interruption by severing of the fuse link or by opening of the contacts will occur under oil.   103, 443

oil feeding reservoirs. Oil storage tanks situated at intervals along the route of an oil-filled cable or at oil-filled joints of solid cable for the purpose of keeping the cable constantly filled with oil under pressure.   64

oil-filled (designated liquid-filled) (prefix). The prefix oil-filled or designated liquid-filled as applied to equipment indicates that oil or the designated liquid is the surrounding medium.   328

oil-filled bushing (outdoor electric apparatus). A bushing in which the space between the inside surface of the weather casing and the major insulation (or conductor where no major insulation is used) is filled with oil.   168

oil-filled cable. A self-contained pressure cable in which the pressure medium is low-viscosity oil having access to the insulation. *See:* **pressure cable; self-contained pressure cable.**   64

oil-filled pipe cable. A pipe cable in which the pressure medium is oil having access to the insulation. *See:* **pressure cable; pipe cable; power distribution, underground construction.**   64

oil-fill stand pipe (rotating machinery). *See:* **oil overflow plug.**

oil groove (rotating machinery). A groove cut in the surface of the bearing lining or sometimes in the journal to help to distribute the oil over the bearing surface. *See:* **oil cup (rotating machinery).**   63

oil-immersed (1) (transformers, reactors, regulators, and similar components). Having the coils immersed in an insulating liquid. *Note:* The insulating liquid is usually (though not necessarily) oil. *See:* **transformer, oil-immersed.**   257
(2) (grounding device). Means that the windings are immersed in an insulating oil.   91

oil-immersed forced-air-cooled shunt reactor (Class OFA)(shunt reactors over 500 kVA). An oil-immersed shunt reactor which is cooled by forced circulation of the cooling air over the cooling surface.   562

oil-immersed forced-oil-cooled transformer with forced-air cooler (class FOA). A transformer having its core and coils immersed in oil and cooled by the forced circulation of this oil through external oil-to-air heat-exchanger equipment utilizing forced circulation of air over its cooling surface.   53

oil-immersed forced-oil-cooled transformer with forced-water cooler (Class FOW). A transformer having its core and coils immersed in oil and cooled by the forced circulation of this oil through external oil-to-water heat-exchanger equipment utilizing forced circulation of water over its cooling surface.   53

oil-immersed forced-oil-cooled with forced-air cooler shunt reactor (Class FOA)(shunt reactors over 500 kVA). An oil-immersed shunt reactor cooled by the forced circulation of oil through external oil-to-air heat-exchanger equipment utilizing forced circulation of air over its cooling surface.   562

oil-immersed forced-oil-cooled with forced-water cooler shunt reactor (Class FOW)(shunt reactors over 500 kVA). An oil-immersed shunt reactor cooled by the forced circulation of the oil through external oil-to-water heat-exchanger equipment utilizing forced circulation od water over its cooling surface.   562

oil-immersed self-cooled / forced-air-cooled / forced-air-cooled transformer (class OA/FA/FA) (power and distribution transformer). A transformer having its core and coils immersed in oil and having a self-cooled rating obtained by the natural circulation of air over the cooling surface, a forced-air-cooled rating obtained by the forced circulation of air over a portion of the cooling surface, and an increased forced-air-cooled rating obtained by the increased forced circulation of air over a portion of the cooling surface.   53

oil-immersed self-cooled / forced-air-cooled / forced-oil-cooled transformer (class OA/FA/FOA) (power and distribution transformer). A transformer having its core and coils immersed in oil and having a self-cooled rating with cooling obtained by the natural circulation of air over the cooling surface, a forced-air-cooled rating with cooling obtained by the forced circulation of air over this same cooling surface, and a forced-oil-cooled rating with cooling surface, and a forced-oil-cooled rating with cooling obtained by the forced circulation of oil over the core and coils and adjacent to this same cooling surface over which the air is being forced circulated.   53 oil-immersed self-cooled/forced-air-cooled transformer (class OA/FA) (power and distribution transformer). A transformer having a self-cooled rating with cooling obtained by the natural circulation of air over the cooling surface, and a forced-air-cooled rating with cooling obtained by the forced circulation of air over this same cooling surface.   53

oil-immersed self-cooled / forced-air, forced-oil-cooled / forced-air, forced-oil-cooled transformer (class OA/ FOA/FOA) (power and distribution transformer). A transformer similar to class OA/FA/FOA transformer except that its auxiliary cooling controls are arranged to start a portion of the oil pumps and a portion of the fans for the first auxiliary rating and the remainder of the pumps and fans for the second auxiliary rating.   53

oil-immersed self-cooled shunt reactor (Class OA)(shunt reactors over 500 kVA). An oil-immersed shunt reactor which is cooled by natural circulation of the cooling air over the cooling surface.   562

oil-immersed self-cooled transformer (class OA) (power and distribution transformer). A transformer having its core and coils immersed in oil, the cooling being effected by the natural circulation of air over the cooling surface.   53

oil-immersed shunt reactor (shunt reactors over 500 kVA). One in which the coils and magnetic current are immersed in an insulating oil.   562

oil-immersed transformer. A transformer in which the core and coils are immersed in an insulating oil.   53

oil-immersed water-cooled/self-cooled transformer (Class OW/A). A transformer having its core and coils immersed in oil and having a water-cooled rating with cooling obtained by the natural circulation of oil over the water-cooled surface, and a self-cooled rating with cooling obtained by the natural circulation of air over the cooling surface.   53

oil-immersed water-cooled shunt reactor (Class OW)(shunt reactors over 500 kVA). An oil-immersed shunt reactor which is cooled by the natural circulation of the cooling oil over the water-cooled surface.   562

oil-immersed water-cooled transformer (class OW) (power and distribution transformer). A transformer having its core and coils immersed in oil, the cooling being effected by the natural circulation of oil over the water-cooled surface.   53

oil immersible current limiting fuse (power switchgear). A current limiting fuse unit suitable for application requiring total or partial immersion directly in oil or other dielectric liquid of a transformer or switchgear.   103

oil-impregnated paper-insulated bushing. A bushing in which the major insulation is provided by paper impregnated with oil.   168

oilless circuit breaker. See: circuit breaker. Note.

oil level gauge (rotating machinery). An indicating device showing oil level in the oil reservoir.   63

oil-lift bearing (rotating machinery). A journal bearing in which high-pressure oil is forced under the shaft journal or thrust runner to establish a lubricating film. See: bearing.   63

oil-lift system (rotating machinery). A system that lubricates a bearing before starting by forcing oil between the journal or thrust runner and bearing surfaces. See: oil cup (rotating machinery).   63

oil-overflow plug (oil-fill stand-pipe) (rotating machinery). An attachment to the oil reservoir that can be opened to allow excess oil to escape, to inspect the oil level, or to add oil.   63

oil pot (oil reservoir) (rotating machinery). A bearing reservoir for a vertical-shaft bearing. See: oil cup (rotating machinery).   63

oil-pressure electric gauge. A device that measures the pressure of oil in the line between the oil pump and the bearings of an aircraft engine. The gauge is provided with a scale, usually graduated in pounds per square inch. It provides remote indication by means of self-synchronous generator and motor.   328

oilproof enclosure (electric installations on shipboard). An enclosure constructed so that oil vapors, or free oil not under pressure, which may accumulate within the enclosure will not prevent successful operation of, or cause damage to, the enclosed equipment.   3

oil reservoir (rotating machinery). See: oil pot.

oil-resistant gaskets (power and distribution transformer). Those made of material which is resistant to oil or oil fumes.   53

oil ring (rotating machinery). A ring encircling the shaft in such a manner as to bring oil from the oil reservoir to the sleeve bearing and shaft. See: bearing.   63

oil-ring guide (rotating machinery). A part whose main purpose is the restriction of the motion of the oil ring. See: oil cup (rotating machinery).   63

oil-ring lubricated bearing. A bearing in which a ring, encircling the journal, and rotated by it, raises oil to lubricate the bearing from a reservoir into which the ring dips.   63

oil-ring retainer (rotating machinery). A guard to keep the oil ring in position on the shaft.   63

oil seal (bearing seal) (bearing oil seal) (rotating machinery). A part or combination of parts in a bearing assembly intended to prevent leakage of oil from the bearing.   63

oil switch (high-voltage switchgear). A switch with contacts that separate in oil.   443

oil thrower (oil slinger) (rotating machinery). A peripheral ring or ridge on a shaft adjacent to the journal and which is intended to prevent any flow of oil along the shaft. See: oil cup (rotating machinery).   63

oiltight (power and distribution transformer). So constructed as to exclude oils, coolants, and similar liquids under specified test conditions.   53

oiltight enclosure (electric installations on shipboard). An enclosure constructed so that oil vapors, or free oil, not under pressure, which may be present in the surrounding atmosphere, cannot enter the enclosure.   3

oiltight pilot devices (industrial control). Devices such as push-button switches, pilot lights, and selector switches that are so designed that, when properly installed, they will prevent oil and coolant from entering around the operating or mounting means. *See:* **switch.** 206

oil, uninhibited. Mineral transformer oil to which no synthetic oxidation inhibitor has been added. *See:* **oil-immersed transformer.** 203

oil-well cover (rotating machinery). A cover for an oil reservoir. *See:* **oil cup (rotating machinery).** 63

oil wick (rotating machinery). Wool, cotton, or similar material used to bring oil to the journal surface by capillary action. *See:* **oil cup (rotating machinery).** 63

omega (navigation aid terms). A very long distance navigation system operating at approximately 10 kHz (kilohertz), in which hyperbolic lines of position are determined by measurement of the difference in travel time of continuous wave signals from two transmitters separated by 5000 nmi (nautical miles) to 6000 nmi (9000 km [kilometers] to 11 000 km) or in which changes in distances from the transmitters are measured by counting rf (radio frequency) wavelengths in space of lanes as the vehicle moves from a known position, the lanes being counted by phase comparison with a stable oscillator aboard the vehicle, 526

omnibearing (navigation aid terms). A magnetic bearing indicated by a navigational receiver on transmission from an omnirange. 526

omnibearing converter (navigation aid terms). A device which combines the omnibearing signal with vehicle heading information to furnish electrical signals for the operation of the pointer of a radio magnetic indicator. 526

omnibearing-distance facility (navigation aid terms). A combination of an omnirange and a distance measuring facility, so that both bearing and distance information may be obtained; tacan and VOR/DME are omnibearing distance facilities. 526

omnibearing-distance navigation (navigation aid terms). Radio navigation utilizing a polar coordinate system as a reference, making use of omnibearing-distance facilities. *Syn:* **rho-theta navigation.** 526

omnibearing indicator (OBI)(navigation aid terms). An instrument which presents an automatic and continuous indication of an omnibearing. 526

omnibearing indicator. *See:* **OBI.**

omnibearing line. *See:* **radial.**

omnibearing selector (navigation aid terms). A control used with an omnirange receiver so that any desired omnibearing may be selected; deviation from on-course for any selected bearing is displayed on the course line deviation indicator. 526

omnidirectional antenna. An antenna having an essentially nondirectional pattern in a given plane of the antenna and a directional pattern in any orthogonal plane. *Note:* For ground-based antennas the omnidirectional plane is usually horizontal. 111

omnidirectional microphone (nondirectional microphone). A microphone the response of which is essen-

tially independent of the direction of sound incidence. *See:* **microphone.** 328

omnidirectional range (omnirange)(navigation aid terms). A radio facility providing bearing information at or from such facilities at all azimuths within its service area and providing direct indication of the magnetic bearing (omnibearing) of that station from any direction. 526

omnirange. *See:* **omnidirectional range.**

on-course curvature (navigation)(navigation aid terms). The rate of change of the indicated course with respect to distance along the course line or path. 526

one-address. Pertaining to an instruction code in which each instruction has one address part. Also called single address. In a typical one-address instruction the address may specify either the location of an operand to be taken from storage, the destination of a previously prepared result, or the location of the next instruction to be interpreted. In a one-address machine, the arithmetic unit usually contains at least two storage locations, one of which is an accumulator. For example, operations requiring two operands may obtain one operand from the main storage and the other from the storage location in the arithmetic unit that is specified by the operation part. *See:* **single-address.** 235

one-address code (electronic computation). *See:* **instruction code.**

one-family dwelling (National Electrical Code). A building consisting solely of one dwelling unit. 256

one-fluid cell. A cell having the same electrolyte in contact with both electrodes. *See:* **electrochemistry.** 328

one hour rating (rotating electric machinery). The output that the machine can sustain for 1 hour starting cold under the conditions of Section 4 of IEEE Std 11-1980 without exceeding the limits of temperature rise of Section 5. 424

one hundred percent disruptive-discharge voltage (dielectric tests). A specified minimum voltage that is to be applied to a test object in a 100 percent disruptive-discharge test under specified conditions. The term applies mostly to impulse tests and has significance only in cases where the loss of dielectric strength resulting from a disruptive discharge is temporary. 150

100 percent insulation level (National Electrical Code). Cables in this category shall be permitted to be applied where the system is provided with relay protection such that ground faults will be cleared as rapidly as possible, but in any case within 1 minute. While these cables are applicable to the great majority of cable installations which are on grounded systems, they shall be permitted to be used also on other systems for which the application of cables is acceptable provided the above clearing requirements are met in completely de-energizing the faulted section. 256

133 percent insulation level (National Electrical Code). This insulation level corresponds to that formerly designated for ungrounded systems. Cables in this category shall be permitted to be applied in situations where the clearing time requirements of the 100 percent level category cannot be met, and yet there is adequate assurance that the faulted section will be de-energized in a time not exceeding 1 hour. Also they shall be permitted to be used when additional insulation strength over the 100 percent level category is desirable.   256

one-level address (computing systems). *See:* **direct address.**

one-line diagram (single-line) (industrial control). A diagram which shows, by means of single lines and graphic symbols, the course of an electric circuit or system of circuits and the component devices or parts used therein.   25

one minus cosine (high voltage circuit breakers). The 1-cosine curve starting at zero and reaching a peak of E_2 at time T_2. The crest is denoted by P. *Note:* The 1-cosine curve is the standard envelope for rating circuit breaker transient recovery voltage performance for circuit breakers rated 72.5 kV and below.
                                                          148

one minus cosine (1-cosine) envelope (transient recovery voltage) (1) (power switchgear). A voltage-versus-time curve of the general form e_2, E_2 (1 mins cos Kt) in which e_2 represents the transient voltage across a switching device pole unit, reaching its crest E_2 at a time T_2.   103

1 minus cosine (1 - cosine) envelope (of a transient recovery voltage) (power switchgear). A voltage-versus-time curve of the general form $e_2E_2(1-\cos Kt)$ in which e_2 represents the transient voltage across a switching device pole unit, reaching its crest at E_2 at a time T_2.   103

one-N (1N) modulation (dynamically tuned gyro) (inertial sensor). The modulation of the pickoff output at spin frequency.   46

one-N (1N) translational sensitivity (dynamically tuned gyro). *See:* **radial unbalance torque.**

ONE output (magnetic cell). (1) The voltage response obtained from a magnetic cell in a ONE state by a reading or resetting process. (2) The integrated voltage response obtained from a magnetic cell in a ONE state by a reading or resetting process. *See:* **distributed-ONE output; ONE state.**   331

one-plus (1+) call (telephone switching systems). A type of station-to-station call in which the digit one is dialed as an access code.   55

one-plus-one address. Pertaining to an instruction that contains one operand address and a control address.
                                                      255, 77

ones complement (mathematics of computing). The diminished-radix complement of a binary numeral, which is formed by subtracting each digit from 1. For example, the ones complement of 1101 is 0010.
                                                          564

one-sided switching array (telephone switching systems). A switching array where each crosspoint interconnects multiples within one group.   55

ONE state. A state of a magnetic cell wherein the magnetic flux through a specified cross-sectional area

has a positive value, when determined from an arbitrarily specified direction of positive normal to that area. A state wherein the magnetic flux has a negative value, when similarly determined, is a ZERO state. A ONE output is (1) the voltage response obtained from a magnetic cell in a ONE state by a reading or resetting process, or (2) the integrated voltage response obtained from a magnetic cell in a ONE state by a reading or resetting process. A ZERO output is (1) the voltage response obtained from a magnetic cell in a ZERO state by a reading or resetting process, or (2) the integrated voltage response obtained from a magnetic cell in a ZERO state by a reading or resetting process. A ratio of a ONE output to a ZERO output is a ONE-to-ZERO ratio. A pulse, for example a drive pulse, is a **write pulse** if it causes information to be introduced into a magnetic cell or cells, or is a **read pulse** if it causes information to be acquired from a magnetic cell or cells.   331

one-third octave (seismic testing of relays). The interval between two frequencies which have a frequency ratio of 2 1/3. For example, 1 to 1.26, 1.26 to 1.59, 159 to 2.0 Hz, etcetera.   392

1-state (logic). The logic state represented by the binary number 1 and usually standing for an active or true logic condition.   88

one-time fuse (protection and coordination of industrial and commercial power systems). Strictly speaking, any nonrenewable fuse, but generally accepted and used to describe any Class H nonrenewable cartridge fuse, sith a single (as opposed to dual) fusing element and intended to interrupt not over 10 000 amperes (A).   504
See: **nonrenewable fuse.**

O **net loss (circuit equivalent) (circuit).** The net loss is the sum of all the transmission losses occurring between the two ends of the circuit, minus the sum of all the transmission gains. *See:* **transmission loss.**
                                                          328

ONE-to-partial-select ratio. The ratio of a ONE output to a partial-select output. *See:* **coincident-current selection.**   331

ONE-to-ZERO ratio. A ratio of a ONE output to a ZERO output. *See:* **ONE state.**   331

O **network.** A network composed of four impedance branches connected in series to form a closed circuit, two adjacent junction points serving as input terminals while the remaining two junction points serve as output terminals. *See:* **network analysis.**   328

one-unit call (telephone switching systems). A call for which there is a single-unit charge for an initial minimum interval.   55

one-way automatic leveling device. A device that corrects the car level only in case of under-run of the car, but will not maintain the level during loading and unloading. *See:* **elevator-car leveling device.**   328

one-way correction (industrial control). A method of register control that effects a correction in register in one direction only.   206

one-way trunk (telephone switching systems). A trunk between two switching entities accessible by calls from one end only. At the originating end, the one-way trunk is known as an outgoing trunk: at the terminating end, it is known as an incoming trunk. 55

one-wire circuit. *See:* **direct-wire circuit.**

one-wire line (open-wire lead). *See:* **open-wire pole line.**

on-hook (telephone switching systems). An open station line or any supervisory or pulsing condition is indicative of this state. 55

ON impedance (thyristor). The differential impedance between the terminals through which the principal current flows, when the thyristor is in the ON state at a stated operating point. *See:* **principal voltage-current characteristic (principal characteristic).** 66

online. (1) Pertaining to equipment or devices under direct control of the central processing unit. (2) Pertaining to a user's ability to interact with a computer. 77, 54

on-line operation (emergency and standby power). (1) Pertaining to equipment or devices under direct control of the central processing unit. (2) Pertaining to uninterruptible power supply systems whereby an inverter is on during normal operation conditions. 512

on-line testing (test, measurement and diagnostic equipment). Testing of the unit under test in its operational environment. *See:* **interference testing; non-interference testing.** 54

on-load factor (thyristor). The ratio of the controller ON-state interval to the operating period in the ON-OFF control mode, often expressed as a percentage. 445

on-net call (telephone switching systems). A call within a switched-service network. 55

ON-OFF control (thyristor). The starting instant may be synchronous or asynchronous with respect to the line voltage. The controller ON-state interval is equal to or greater than half a line period. *See:* **operation modes (Typical Examples of Operation Modes) E, (limiting case.)** 445

ON-OFF keying (modulation systems). A binary form of amplitude modulation in which one of the states of the modulated wave is the absence of energy in the keying interval. *Note:* The terms mark and space are often used to designate, respectively, the presence and absence of energy in the keying interval. *See:* **telegraphy.** 242

ON-OFF test (test, measurement and diagnostic equipment). A test conducted by repeatedly switching on and off either the signal, power, or load connected to the unit under test while observing the reaction or performance of some parameter of that unit under test. A test frequently used to isolate offending equipment while conducting compatibility, interference, or system performance evaluations. 54

O **noise unit.** An amount of noise judged to be equal in interfering effect to the one-millionth part of the current output of a particular type of standard generator of artificial noise, used under specified conditions. *Note:* This term was formerly used in connection with ear balance measurements, but has been largely su-

perseded by dBa employed with indicating noise meter. Approximately seven noise units of noise on a telephone line are frequently taken as equivalent to reference noise. *See:* **signal-to-noise ratio.** 328

on-peak energy (power operations). Energy supplied during designated periods of relatively high system demands. 516

on-peak period (watthour meters). The period of time during which the specified on-peak rate applies. 485

on-peak power. Power supplied during designated periods of relatively high system demands. *See:* **generating station.** 64

ON period (electron tube or valve). The time during an operating cycle in which the electron tube or valve is conducting. 244, 190

on site (monitoring radioactivity in effluents). Location within a facility that is controlled with respect to access by the general public. 559

ON state (thyristor). The condition of the thyristor corresponding to the low-resistance low-voltage portion of the principal voltage-current characteristic in the switching quadrant(s). *Note:* In the case of reverse-conducting thyristors, this definition is applicable only for a positive anode-to-cathode voltage. *See:* **principal voltage-current characteristic (principal characteristic).** 66

ON-state current (thyristor). The principal current when the thyristor is in the ON state. 66

ON-state voltage (thyristor). The principal voltage when the thyristor is in the ON state. *See:* **principal voltage-current characteristic (principal characteristic).** 66

opacity (electroacoustics) (optical path). The reciprocal of transmission. *See:* **transmission (transmittance).** 176

open amortisseur. An amortisseur that has no connections between poles. 328

open area (electromagnetic compatibility). *See:* **test site.**

open-center display. A plan-position-indicator display on which zero range corresponds to a ring around the center of the display. *See:* **radar.** 328

open-center PPI (plan-position indicator) (radar). A PPI in which the display of the initiation of the time base precedes that of the transmitted pulse. 13

open-circuit characteristic (synchronous machine). *See:* **open-circuit saturation curve.**

open-circuit control. A method of controlling motors employing the open-circuit method of transition from series to parallel connections of the motors. *See:* **multiple-unit control.** 328

open circuit cooling (rotating machinery). A method of cooling in which the coolant is drawn from the medium surrounding the machine, passes through the machine and then returns to the surrounding medium. 63

open-circuit impedance (1) (general). An impedance of a network that has a specified pair or group of terminals open circuited. **(2) (four-terminal network or line).** The input-output- or transfer-impedance parameters z_{11}, z_{22}, z_{12} and z_{21} of a four-terminal network when the far-end is open circuited. 67

open-circuit inductance (transformer). The apparent inductance of a winding with all other windings open-circuited. 197

open-circuit potential. The measured potential of a cell from which no current flows in the external circuit. 108

open-circuit saturation curve (open-circuit characteristic) (synchronous machine). The saturation curve of a machine with an open-circuited armature winding. 63

open circuit signaling (data transmission). That type of signaling in which no current flows while the circuit is in the idle condition. 59

open-circuit test (synchronous machine). A test in which the machine is run as a generator with its terminals open-circuited. 63

open-circuit transition (1) (multiple-unit control). A method of changing the connection of motors from series to parallel in which the circuits of all motors are open during the transfer. *See:* **multiple-unit control.** 328

(2) (industrial control) (reduced-voltage controllers, including star-delta controllers). A method of starting in which the power to the motor is interrupted during normal starting sequence. *See:* **electric controller.** 206

open circuit transition auto-transformer starting (rotating machinery). The process of auto-transformer starting whereby the motor is disconnected from the supply during the transition from reduced to rated voltage. 63

open-circuit voltage (1) (battery). The voltage at its terminals when no appreciable current is flowing. *See:* **battery (primary or secondary).** 328

(2) (arc-welding apparatus). The voltage between the output terminals of the welding power supply when no current is flowing in the welding circuit. 264

open conductor (NESC). A type of electric supply or communication line construction in which the conductors are bare, covered or insulated and without grounded shielding, individually supported at the structure either directly or with insulators. *Syn:* **open wire.** 494

open cutout (power switchgear). A cutout in which the fuse clips and fuseholder, fuse unit, or disconnecting blade are exposed. 103, 443

open-delta connection (power and distribution transformer). A connection similar to a delta-delta connection utilizing three single-phase transformer, but with one single-phase transformer removed. *Note:* The two remaining transformers of an open-delta bank will carry 57.7 percent of the load carried by the bank using three identical transformers connected delta-delta. 53

open ended. Pertaining to a process or system that can be augmented. 255, 77

open-ended coil (rotating machinery). A partly preformed coil the turns of which are left open at one end to facilitate their winding into the machine. *See:* **asynchronous machine.** 63

open fuse trip device (ac power circuit breakers)(low-voltage ac power circuit protectors). A device that operates to open (trip) all poles of a circuit breaker (protector) in response to the opening, or absence, of one or more fuses integral to the circuit protector on which the device is mounted. After operating, the device shall prevent closing of the circuit breaker (protector) until reset operation is performed. *Note:* Since some open-fuse trip devices may operate by sensing the voltage across the fuses, they may not prevent closing of the circuit breaker (protector) with an open or missing fuse, but in most cases will cause an immediate trip if such an operation is performed. There is a practical limit of load impedance above which the device (sensing voltage across an open or missing fuse) will not function as described. 425, 158

opening eye (of a fuse holder, fuse unit, or disconnecting blade) (high-voltage switchgear). An eye provided for receiving a fuse hook or switch hook for opening and closing the fuse. 443

opening operating time (of a switch) (power switchgear). The interval during which a switch is being operated to move from the fully closed to the fully open position. 103

opening operation (of a switching device). *See:* **open operation (of a switching device).** 103

opening time (1) (of a mechanical switching device) (power switchgear). The time interval between the time when the actuating quantity of the release circuit reaches the operating value, and the instant when the primary arcing contacts have parted. Any time delay device forming an integral part of the switching device is adjusted to its minimum setting or, if possible, is cut out entirely for the determination of opening time. *Note:* The opening time includes the operating time of an auxiliary relay in the release circuit when such a relay is required and supplied as part of the switching device. 103

(2) (of a sectionalizer). *See:* **isolating time (of a sectionalizer).** 103

open line wire charging current (power switchgear). Current supplied to an unloaded open-wire line. *Note:* Current is expressed in root-mean-square (rms) amperes. 103

open-link cutout (power switchgear). A cutout that does not employ a fuseholder and in which the fuse support directly receives an open-link fuse link or a disconnecting blade. 103, 443

open-link fuse link (power switchgear). A replaceable part or assembly comprised of the conducting element and fuse tube, together with the parts necessary to confine and aid in extinguishing the arc and to connect it directly into the fuse clip of the open-link fuse support. 103, 443

open-link fuse support (power switchgear). An assembly of base or mounting support, insulators or insulator unit and fuse clips for directly mounting an open-link fuse link and for connecting it into the circuit. 103, 443

open loop (automatic control). A signal path without feedback. *See:* **control system; feedback.** 56

open loop control (station control and data acquisition). A form of control without feedback. 403

open-loop control system (1) (general). A system in which the controlled quantity is permitted to vary in accordance with the inherent characteristics of the control system and the controlled power apparatus for any given adjustment of the controller. *Note:* No function of the controlled variable is used for automatic control of the system. It is not a feedback control system. *See:* control; control system; network analysis. 151, 321
(2) (hydraulic turbines). A control system that has no means for comparing the output with the input for control purposes. 8

open-loop gain (power supplies). The gain, measured without feedback, is the ratio of the voltage appearing across the output terminal pair to the causative voltage required at the (input) null junction. The open-loop gain is denoted by the symbol A in diagrams and equations. *See:* closed loop; loop gain; power supply. 228, 186

open loop measurement (data transmission). A measurement made in which a circuit has at least one of two hybrid sets disconnected and thereby opening the loop. 59

open-loop series street-lighting system. A street-lighting system in which the circuits each consist of a single line wire that is connected from lamp to lamp and returned by a separate route to the source of supply. *See:* alternating-current distribution; direct-current distribution. 64

open-loop system. A control system that has no means for comparing the output with input for control purposes. 224

open machine (1) (rotating machinery). A machine in which no mechanical protection as such is embodied and there is no restriction to ventilation other than that necessitated by good mechanical construction. *See:* asynchronous machine. 63
(2) (electric installations on shipboard). A machine having ventilating openings which permit passage of external cooling air over and around the windings. 3

open-numbering plan (telephone switching systems). A numbering plan in which the number of digits dialed varies according to the requirements of the telecommunications message network. 55

open operation (of a switching device) (power switchgear). The movement of the contacts from the normally closed to the normally open position. *Note:* The letter O signifies this operation: Open. 103

open path (network analysis). A path along which no node appears more than once. 282

open-phase protection (power switchgear). A form of protection that operates to disconnect the protected equipment on the loss of current in one phase conductor of a polyphase circuit, or to prevent the application of power to the protected equipment on the absence of one or more phase voltages of a polyphase system. 103

open-phase relay (power switchgear). A polyphase relay designed to operate when one or more input phases of a polyphase circuit are open. 103

open pipe-ventilated machine. An open machine except that openings for the admission of the ventilating air are so arranged that inlet ducts or pipes can be connected to them. This air may be circulated by means integral with the machine or by means external to and not a part of the machine. In the latter case, this machine is sometimes known as a **separately ventilated machine** or a **forced-ventilated machine**. Mechanical protection may be defined as under **dripproof machine, splashproof machine, guarded machine,** or **semiguarded machine.** *See:* asynchronous machine. 328

open region (1) (three-dimensional space). A volume that satisfies the following conditions: (A) any point of the region has a neighborhood that lies within the region: (B) any two points of the region may be connected by a continuous space curve that lies entirely in the region.
(2) (two-dimensional space). An area that satisfies the following conditions: (A) any point of the region has a neighborhood that lies within the region: (B) any two points of the region may be connected by a continuous curve that lies entirely in the region. 210

open relay. An unenclosed relay. *See:* relay. 259

open resonator (laser-maser). An open resonator and a beam resonator are identical. 363

open subroutine (computing systems). A subroutine that must be relocated and inserted into a routine at each place it is used. *See:* closed subroutine. 255, 77

open switchgear assembly (power switchgear). One that does not have enclosures as part of the structure. 103

open terminal box (rotating machinery). A terminal box that is, normally, open only to the interior of the machine. 63

open-wire circuit. A circuit made up of conductors separately supported on insulators. *Note:* The conductors are usually bare wire, but they may be insulated by some form of continuous insulation. The insulators are usually supported by crossarms or brackets on poles. *See:* open wire. 328

open-wire lead (open-wire line). *See:* open-wire pole line.

open-wire line charging current (high voltage circuit breakers). Current supplied to an unloaded open wire line. *Note:* Current is expressed in root-mean-square amperes. 130

open-wire pole line (open-wire line) (open-wire lead). A pole line whose conductors are principally in the form of open wire. 328

open wiring (on insulators) (National Electrical Code). An exposed wiring method using cleats, knobs, tubes, and flexible tubing for the protection and support of single insulated conductors run in or on buildings, and not concealed by the building structure. 256

operable equipment (test, measurement and diagnostic equipment). An equipment which, from its most re-

cent performance history and a cursory electrical and mechanical examination, displays an indication of operational performance for all required functions.
54

operand (1)(**mathematics of computing**). A variable, constant, or function upon which an operation is to be performed. For example, in the expression $A = B + 3$, B and 3 are the operands.
564

(2)(**microprocessor assembly language**). Data which is to be operated on; also, an address denoting data which is to be operated on.
466

(3)(**software**). (A) An entity on which an operation is performed. (B) That which is operated upon. An operand is usually identified by an address part of an instruction. *See:* **address; instruction; operator.**
434

operate (**analog computers**). In an analog computer, the computer-control state in which input signals are connected to all appropriate computing elements, including integrators, for the generation of the solution.
9

operated unit. A switch, signal, lock, or other device that is operated by a lever or other operating means.
328

operating basis earthquake (OBE)(1)(seismic qualification of Class 1E equipment for nuclear power generating stations). An earthquake that could reasonably be expected to occur at the plant site during the operating life of the plant considering the regional and local geology and seismology and specific characteristics of local subsurface material. It is that earthquake that produces the vibratory ground motion for which those features of the nuclear power plant, necessary for continued operation without undue risk to the health and safety of the public, are designed to remain functional.
581

(2)(**valve actuators**). The earthquake which could easily be expected to affect the plant site during the operating life of the plant. It is that earthquake which produces the vibratory ground motion for which those features of the nuclear power plant necessary for continued operation without undue risk to the health and safety of the public are designed to remain functional.
492

(3) (**class 1E static battery chargers and inverters)-(nuclear power generating stations)(seismic qualification of Class 1E equipment)(seismic testing of relays).** That earthquake which could reasonably be expected to affect the plant site during the operating life of the plant. It is that earthquake which produces the vibratory ground motion for which those features of the nuclear power plant necessary for continued operation, without undue risk to the health and safety of the public, are designed to remain functional.
408, 440, 28, 392

operating bypass (nuclear power generating station). (1) Inhibition of the capability to accomplish a protective function that could otherwise occur in response to a particular set of generating station conditions. *Note:* An operating bypass is not the same as a maintenance bypass. Different modes of plant operation may ne-

cessitate an automatic or manual bypass of a protective function. Operating bypasses are used to permit mode changes (for example, prevention of initiation of safety injection during the cold shutdown mode).
387

(2) Normal and permissive removal of the capability to accomplish a protective function that could otherwise occur in response to a particular set of generating station conditions. *Note:* Typically, operating bypasses are used to permit a change to a different mode of generating station operation (for example, prevention of initiation of safety injection during cold shutdown conditions).
20

operating characteristic (of a relay) (power switchgear). The response of the relay to the input quantities which result in relay operation.
103, 127

operating conditions (1)(reliability data)(reliability data for pumps and drivers, valve actuators, and valves). The loading or demand cyclic operation, or both, of an item between zero and 100% of its related capability(ies).
379, 502

(2) (**general**). The whole of the electrical and mechanical quantities that characterize the work of a machine, apparatus, or supply network, at a given time.
310

operating cycle (nuclear power generating stations). The complete sequence of operations that occur during a response to a demand function.
142, 31

operating device (elevators). The car switch, pushbutton, lever, or other manual device used to actuate the control. *See:* **control.**
328

operating duty (of a switching device). A specified number and kind of operations at stated intervals.
103

operating duty cycle (surge arrester)(metal-oxide surge arresters for ac power circuits). One or more unit operations, as specified.
2, 430, 583

operating-duty test (surge arresters). A test in which working conditions are simulated by the application to the arrester of a specified number of impulses while it is connected to a power supply of rated frequency and specified voltage.
308, 62

operating experience (1)(safety systems equipment in nuclear power generating stations). Verifiable service data for equipment.
535

(2) (**class 1E static battery chargers and inverters**). Accumulation of verifiable service data for conditions equivalent to those for which particular equipment is to be qualified.
408

(3) (**nuclear power generating stations**). Accumulation of verifiable service data for conditions equivalent to those for the equipment to be qualified or for which particular equipment is to be qualified.
120

operating failure rate (reliability data for pumps and drivers, valve actuators, and valves). The probability (per hour) of failure for those operating components required to operate or function for a period of time.
502

operating floor (packaging machinery). A floor or platform used by the operator under normal operating conditions.
429

operating frequency (thyristor). The operating frequency is the reciprocal value of the operating period.
445

operating frequency line current (thyristor). The root-mean-square (rms) value of the fundamental component of the line current, whose frequency is the operating frequency.
445

operating frequency load voltage (thyristor). The root-mean-square (rms) value of the fundamental component of the load voltage, whose frequency is the operating frequency.
445

operating influence. The change in a designated performance characteristic caused solely by a prescribed change in a specified operating variable from its reference operating condition to its extreme operating condition, all other operating variables being held within the limits of reference operating conditions. *Notes:* (1) It is usually expressed as a percentage of span. (2) If the magnitude of the influence is affected by direction, polarity, or phase, the greater value shall be taken.
295

operating interface (separable insulated connectors). The surfaces at which a connector is normally separat-ed. See figure below **bushing insert** and the figure **operating interface** below.
454

operating life (gyro; accelerometer). The accumulated timeof operation throughout which a gyro or accelerometer exhibits specified performance when maintained and calibrated in accordance with a specified schedule.
46

operating line; operating curve (electron device). the locus of all simultaneous values of total instantaneous electrode voltage and current for given external circuit conditions.
190

operating mechanism (1)(of a switching device) (power switchgear). That part of the mechanism that actuates all the main-circuit contacts of the switching device either directly or by the use of pole-unit mechanisms.
103

(2)(84) (power system device function numbers). The complete electrical mechanism or servomechanism, including the operating motor, solenoids, position switches, etcetera for a tap changer, induction regulator, or any similar piece of apparatus which otherwise has no device function number.
402

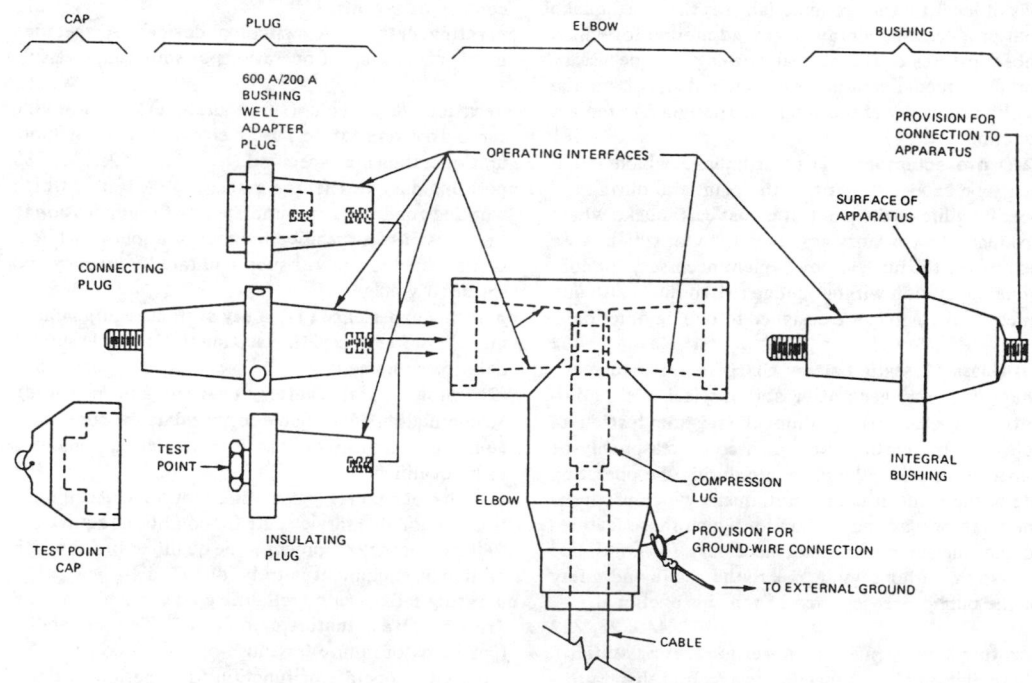

Typical Components of 600 A Separable Insulated Connector System

operating modes (nuclear power generating station). The nuclear power plant modes as defined by the technical specifications for the plant.   358

operating noise temperature. The temperature in kelvins given by

$$T_{\mathrm{op}} = \frac{N_0}{kG_s}$$

where N_0 is the output noise power per unit bandwidth at a specified output frequency flowing into the output circuit (under operating conditions), k is Boltzmann's constant, and G_s is the ratio of (1) the signal power delivered at the specified output frequency into the output circuit (under operating conditions) to (2) the signal power available at the corresponding input frequency or frequencies to the system (under operating conditions) at its accessible input terminations. *Notes:* (A) In a nonlinear system T_{op} may be a function of the signal level. (B) In a linear two-port transducer with a single input and a single output frequency, if the noise power originating in the output termination and reflected at the output port can be neglected, T_{op} is related to the noise temperature of the input termination T_i and the effective input noise temperature T_e by the equation

$$T_{\mathrm{op}} = T_i + T_e.$$

See: **transducer.**   125

operating overload (packaging machinery). The overcurrent to which electric apparatus is subjected in the course of the normal operating conditions that it may encounter. *Notes:* (1) The maximum operating overload is to be considered six times normal full-load current for alternating-current industrial motors and control apparatus; four times normal full-load current for direct-current industrial motors and control apparatus used for reduced-voltage starting; and ten times normal full-load current for direct-current industrial motors and control apparatus used for full-voltage starting. (2) It should be understood that these overloads are currents that may persist for a very short time only, usually a matter of seconds.   429

operating period (thyristor). The time between starting instants of successive controller ON-state intervals in the ON-OFF control mode.   445

operating point (working point) (electron device). The point on the family of characteristic curves corresponding to the average voltages or currents of the electrodes in the absence of a signal. *See:* **quiescent point.**   190

operating range (1)(navigation aid terms). The maximum distance at which reliable service is provided by an aid to navigation.   526

(2)(plutonium monitoring). The region between the limits within which a quantity is measured.   413

operating reserve (power operations). That reserve above firm system load necessary to provide for: (1) regulation within the hour to cover minute to minute variations; (2) load forecasting error; (3) loss of equipment; (4) local area protection.   516

operating speed range. The range between the lowest

and highest rated speeds at which the drive may perform at full load. *See:* **electric drive.**   206

operating system (software). Software that controls the execution of programs. An operating system may provide services such as resource allocation scheduling, input/output control, and data management. Although operating systems are predominantly software, partial or complete hardware implementations are possible. An operating system provides support in a single spot rather than forcing each program to be concerned with controlling hardware. *See:* **data; execution; hardware; implementation; program; system software.**   434

operating tap voltage (capacitance potential devices). Indicates the root-mean-square voltage to ground at the point of connection (potential tap) of the device network to the coupling capacitor or bushing. This is the voltage on which certain insulation tests are based. *See:* **outdoor coupling capacitor.**   351

operating temperature (1) (power supplies). The range of environmental temperatures in which a power supply can be safely operated (typically, 20 to 50 degrees Celsius). *See:* **ambient operating temperature (range); power supply.**   186

(2) (gyro; accelerometer). The temperature at one or more gyro or accelerometer elements when the device is in the specified operating environment. These elements may include the spin motor winding, the flange, the pickoff, the torquer, the temperature sensor, etcetera.   46

operating temperature limits (attenuator). Maximum temperature in degrees Celsius at which attenuator will operate with full input power. *Note:* Derating function for maximum power versus temperature must be specified to show maximum temperature in degrees Celsius at which attenuator will operate 10 dB below full input power.   110

operating temperature, maximum, (electrical insulation tests). The stabilized temperature obtained from operation of the equipment at rated load and duty cycle in the maximum ambient temperature specified for the device under test.   116

operating temperature range (Hall-effect devices). The range of ambient temperature over which the Hall effect device may be operated with nominal control current and a specified maximum magnetic flux density.   107

operating temperature, room (electrical insulation tests). The temperature of the equipment expected at rated load and duty cycle in an ambient temperature of 20 °C $\pm$ 5° (68 °F $\pm$ 9°). An equipment item that has been operated through its normal duty cycle or has stabilized to the approximate normal running temperature may be assumed to be at room ambient operating temperature.   116

operating time (1)(reliability data for pumps and drivers, valve actuators, and valves). The period of time that an active item or equipment is functioning effectively.   502

(2) (of a relay) (power switchgear). The time interval from occurrence of specified input conditions to a specified operation.   103

(3) (availability)(reliability data). The period of time during which an item performs its intended function.
      164, 379

(4) (seismic testing of relays). The time interval from occurrence of specified input conditions to a specified operation.   392

operating voltage (power switchgear). The voltage of the system on which a device is operated. *Note:* This voltage, if alternating, is usually expressed as a root-mean-square (rms) value.   103

operating voltage range (hybrid computer linkage components). The minimum and maximum values of the analog input voltage which can be represented by the output to within a given accuracy.   10

operation (1)(FASTBUS acquisition and control). A primary address cycle followed by zero or more data cycles and a termination sequence.   480

(2)(mathematics of computing). The action specified by an operator on one or more operands. For example, in the expression $A = B + 3$, the process of adding B to 3 to obtain the result A.   564

(3) (of a switching device) (power switchgear). Action of the parts of the device to perform its normal function.   103

(4) (train control). The functioning of the automatic train-control or cab-signaling system that results from the movement of an equipped vehicle over a track element or elements for a block with the automatic train-control apparatus in service, or which results from the failure of some part of the apparatus. *See:* **automatic train control.**   328

(5) (elevators). The method of actuating the control. *See:* **control.**

(6) (electronic digital computers). (A) A defined action, namely, the act of obtaining a result from one or more operands in accordance with a rule that completely specifies the result for any permissible combination of operands. (B) The set of such acts specified by such a rule, or the rule itself. (C) The act specified by a single computer instruction. (D) A program step undertaken or executed by a computer, for example, addition, multiplication, extraction, comparison, shift, transfer. The operation is usually specified by the operator part of an instruction. (E) The event of specific action performed by a logic element. (F) Loosely: command.   235, 255

operational (software). Pertaining to the status given to a software product once it has entered the operation and maintenance phase. *See:* **operation and maintenance phase; software product.**   434

operational amplifier (analog computers). (1) An amplifier, usually a high-gain dc amplifier, designed to be used with external circuit elements to perform a specified computing operation or to provide a specified transfer function. (2) An amplifier, usually a high-gain dc amplifier, with external circuit elements, used for performing a specified computing operation. (*See:* **integrating amplifier; summing amplifier,** and **inverting amplifier.** *Notes:* (1) The gain and phase characteristics are generally designed to permit large variations in the feedback circuit without instability. (2) The

input terminal of an operational amplifier (1) is the summing junction of an operational amplifier (2) and is generally designed to draw current that is negligibly small relative to signal currents in the feedback impedance.   9

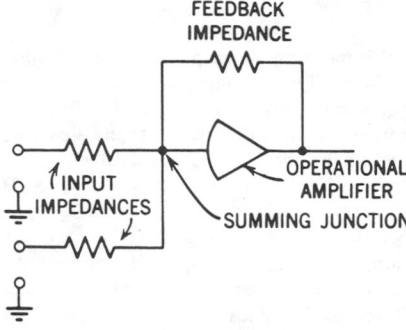

Operational amplifier, typical arrangement

operational availability (nuclear power generating stations). The measured characteristic of an item expressed by the probability that it will be operable when needed as determined by periodic test and resultant analyses.   75

operational gain. *See:* **closed-loop gain.**

operational impedance (Z (s)) (rotating machinery). Defined by the equation

$$Z(s) = V(s) \text{ over } I(s)$$

where V(s) is the Laplace transform of the voltage and I(s) is the Laplace transform of the current.
    63

operational inductance (L (s)) (rotating machinery). Defined by the equation

$$L(s) = \Lambda(s) \text{ over } I(s)$$

where $\Lambda(s)$ is the Laplace transform of the flux linkages and I(s) is the Laplace transform of the current.
    63

operational maintenance. *See:* **non-interference testing.**

operational maintenance influence (instruments). The effect of routine operations that involve opening the case, such as to inspect or mark records, change charts, add ink, alter control settings, etcetera. *See:* **accuracy rating (instrument).**   295

operational power supply. A power supply whose control amplifier has been optimized for signal-processing applications rather than the supply of steady-state power to a load. A self-contained combination of operational amplifier, power amplifier, and power supplies for higher-level operations.   186

operational programming. The process of controlling the output voltage of a regulated power supply by means of signals (which may be voltage, current, resistance, or conductance) that are operated on by the power supply in a predetermined fashion. Operations may include algebraic manipulations, multiplication, summing, integration, scaling, and differentiation.

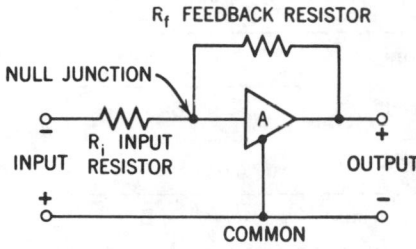

Operational programming

186

operational relay (analog computers). A relay that may be driven from one position or state to another by an operational amplifier or a relay amplifier. *See:* **function relay.**   9

operational reliability (software). The reliability of a system or software subsystem in its actual use environment. Operational reliability may differ considerably from reliability in the specified or test environment. *See:* **reliability; software subsystem; system.**   434

operational testing (software). Testing performed by the end user on software in its normal operating environment. *See:* **software; testing.**   434

operational tests(nuclear power generating stations). Tests conducted in a qualification program to demonstrate operational capability.   440

operation and maintenance phase (software). The period of time in the software life cycle during which a software product is employed in its operational environment, monitored for satisfactory performance, and modified as necessary to correct problems or to respond to changing requirements. *See:* **operational; performance; requirement; software life cycle; software product.**   434

operation and maintenance phase (software verification and validation plans). The period of time in the software life cycle during which a software product is employed in its operational environment, monitored for satisfactory performance, and modified as necessary to correct problems or to respond to changing requirements.   511

operation code. (1) The operations that a computing system is capable of executing, each correlated with its equivalent in another language; for example, the binary or alphanumeric codes in machine language along with their English equivalents; the English description of operations along with statements in a programming language such as **Cobol, Algol,** or **Fortran.** (2) The code that represents or describes a specific operation. The operation code is usually the operation part of the instruction.   235

operation, coordinated (1) (electric power supply). Operation of generation and transmission facilities of two or more interconnected electrical systems to achieve greater reliability and economy.

(2) (hydro plants). Operation of a group of hydro plants and storage reservoirs so as to obtain optimum power benefits with due consideration to all other uses.   112

operation factor. The ratio of the duration of actual service of a machine or equipment to the total duration of the period of time considered. *See:* **generating station.**   64

operation indicator (of a relay). *See:* **target (of a relay).**

operation influence (electrical influence). The maximum variation in the reading of an instrument from the initial reading, when continuously energized at a prescribed point on the scale under reference conditions over a stated interval of time, expressed as a percentage of full-scale value. *See:* **accuracy rating (instrument).**   280

operation modes (thyristor). In thyristor ac power controllers different operation modes are possible. These operation modes may be periodic or nonperiodic. See tables below.   445

operation part (instruction) (electronic computation). The part that usually specifies the kind of operation to be performed, but not the location of the operands. *See:* **instruction code.**   235, 210

operation, quantizing. *See:* **quantizing operation.**

operations by a pulse (pulse terms) (general). Activation, blanking, clearing, deactivation, deflection, reading, resetting, selection, sequencing, setting, starting, stopping, storing, switching, and writing may occur or be performed.   254

operations involving the interaction of pulses (pulse terms). Addition, chopping, coding, comparison, decoding, encoding, mixing, modulation, subtraction, summation, and superposition may occur or be performed. *See:* **complex waveforms.**   254

operations on a pulse (pulse terms) (general). Amplification, attenuation, conditioning, conversion, coupling demodulation, detection, discrimination, filtering, inversion, reception, reflection, and transmission may occur or be performed.   254

operations research (OR)(computer applications). The design of models for complex problems concerning the optimal allocation of available resources, and the application of of mathematical methods for the solution of these problems.   571

operation, synchronized (power operations). An operation wherein power facilities are electrically connected and controlled to operate at the same frequency.   112

operation time (electron tubes). The time after simultaneous application of all electrode voltages for a current to reach a stated fraction of its final value. Conventionally the final value is taken as that reached after a specified length of time. *Note:* All electrode voltages are to remain constant during measurement. The tube elements must all be at room temperature at the start of the test.   125

operator (1)(mathematics of computing). A mathematical or logical symbol that represents an action to be performed in an operation. For example, in the

Operation Modes

Operation Mode		Starting Instant of the Controller On-State Interval	Example Figure
Switch	Non-Periodic	Random	A
		Selected	B
Control	ON-OFF Control Periodic	Asynchronous*	C
		Synchronous*	D, E
	Phase Control Periodic	Synchronous*	F

*With respect to the Line Voltage.

expression A = B + 3, + is the operator, representing addition.   564

(2)(microprocessor assembly language). A symbol denoting an operation to be performed.   466

(3) (telephone switching systems). A person who handles switching and signaling operations needed to establish connections between stations or who performs various auxiliary functions associated therewith.   55

(4) (nuclear power generating station). A person licensed to operate the plant.   358

(5) (software). (A) In symbol manipulation, a symbol that represents the action to be performed in an operation. Examples of operators are +, −, *, /. (B) In the description of a process, that which indicates the action to be performed on operands. (C) A person who operates a machine. *See:* **operand; process.**   434

operator code (telephone switching systems). The digits dialed by operators to reach other operators.

   55

operator-handled call (telephone switching systems). A call in which information necessary for its completion, other than the number of the calling station, is verbally given by or to an operator.   55

operator loss (radar). A loss in effective signal-to-noise ratio manifested by reduced detection probability or increased false alarm rate, when detection is performed by a human operator rather than an ideal thresholding device.   13

operator number identification (ONI) (telephone switching systems). An arrangement in which the operator requests the identity of the calling station and enters it into the system for automatic message accounting.   55

operator's telephone set (operator's set). A set consisting of a telephone transmitter, a head receiver, and associated cord and plug, arranged to be worn so as to leave the operator's hands free. *See:* **telephone station.**   328

opposition (electrical engineering). The relation between two periodic functions when the phase difference between them is one-half of a period.   210

optical (optoelectronic device) coupling coefficient (between two designated ports). The fraction of the radiant or luminous flux leaving one port that enters the other port. *See:* **optoelectronic device.**   191

optical ammeter. An electrothermic instrument in which the current in the filament of a suitable incandescent lamp is measured by comparing the resulting illumination with that produced when a current of known magnitude is used in the same filament. The comparison is commonly made by using a photoelectric cell and indicating instrument. *See:* **instrument.**

   328

optical axis (fiber optics). In an optical waveguide, synonymous with "fiber axis".   433

optical bandwidth (acoustically tunable optical filter). The width at the 50 percent (mins3 decibel) points of the optical intensity versus optical wavelength response curve of the device, measured under the conditions of white light input and fixed acoustic frequency.

   72

optical blank (fiber optics). A casting consisting of an optical material molded into the desired geometry for grinding, polishing, or (in the case of optical waveguides) drawing to the final optical/mechanical specifications. *See:* **preform.**   433

optical cable (fiber optics). A fiber, multiple fibers, or fiber bundle in a structure fabricated to meet optical, mechanical, and environmental specifications. *Syn:* **opticalfiber cable.** *See:* **fiber bundle; optical cable assembly.**   433

optical cable assembly (fiber optics). An optical cable that is connector terminated. Generally, an optical cable that has been terminated by a manufacturer and is ready for installation. *See:* **fiber bundle; optical cable.** 433

optical cavity (fiber optics). A region bounded by two or more reflecting surfaces, referred to as mirrors, end mirrors, or cavity mirrors, whose elements are aligned to provide multiple reflections. The resonator in a laser is an optical cavity. *Syn:* **resonant cavity.** *See:* **active laser medium; laser.** 433

optical character recognition (OCR)(computer applications). The automatic recognition of graphic characters using light-sensitive devices such as optical mark readers. Contrast with magnetic ink character recognition. 579

optical combiner (fiber optics). A passive device in which power from several input fibers is distributed among a smaller number (one or more) of input fibers. *See:* **star coupler.** 433

optical conductor*. *See:* **optical waveguide.**
*Deprecated

optical connector. *See:* **optical waveguide connector.**

optical coupler (wire-line communication facilities). An optical coupler provides isolation using a short length, optical path. 414

optical data bus (fiber optics). An optical fiber network, interconnecting terminals, in which any terminal can communicate with any other terminal. *See:* **optical link.** 433

optical density, Dλ (1) (laser-maser). Logarithm to the base ten of the reciprocal of the transmittance: $D\lambda = -\log_{10\tau\mu}$, where τ is transmittance. *See:* **photographic transmission density.** 363
(2) (fiber optics). A measure of the transmittance of an optical element expressed by: $\log_{10}(1/T)$ or $-\log_{10}T$, where T is transmittance. The analogous term log $_{10}(1/R)$ is called reflection density. *Note:* The higher the optical density, the lower the transmittance. Optical density times 10 is equal to transmission loss expressed in decibels (dB); for example, an optical density of 0.3 corresponds to a transmission loss of 3 dB. *See:* **transmission loss; transmittance.** 433

optical detector (fiber optics). A transducer that generates an output signal when irradiated with optical power. *See:* **optoelectronic.** 433

optical fiber (fiber optics). Any filament or fiber, made of dielectric materials, that guides light, whether or not it is used to transmit signals. *See:* **fiber bundle; fiber optics; optical waveguide.** 433

optical fiber cable. *See:* **optical cable.**

optical fiber waveguide. *See:* **optical waveguide.**

optical filter (fiber optics). An element that selectively transmits or blocks a range of wavelengths. 433

optical landing system (navigation aid terms). A shipboard gyro stabilized or shore-based device which indicates to the pilot his displacement from a preselected glide path. 526

optical link (fiber optics). Any optical transmission channel designed to connect two end terminals or to be connected in series with other channels. *Note:* Sometimes terminal hardware (for example, transmitter/receiver modules) is included in the definition. *See:* **optical data bus.** 433

optically active material (fiber optics). A material that can rotate the polarization of light that passes through it. *Note:* An optically active material exhibits different refractive indices for left and right circular polarizations (circular birefringence). *See:* **birefringent medium.** 433

optically pumped laser (laser-maser). A laser in which the electrons are excited into an upper energy state by the absorption of light from an auxiliary light source. 363

optical path length (fiber optics). In a medium of constant refractive index n, the product of the geometrical distance and the refractive index. If n is a function of position,

$$\text{optical path length} = \int n\, ds,$$

where ds is an element of length along the path. *Note:* Optical path length is proportional to the phase shift a light wave undergoes along a path. *See:* **optical thickness.** 433

optical pattern (mechanical recording). *See:* **light pattern.**

optical photons (scintillation counting). Photons with energies corresponding to wavelengths between approximately 120=1800 m. 117

optical power. *See:* **radiant power.**

optical pyrometer. A temperature-measuring device comprising a standardized comparison source of illumination and source convenient arrangement for matching this source, either in brightness or in color, against the source whose temperature is to be measured. The comparison is usually made by the eye. *See:* **electric thermometer (temperature meter).** 328

optical repeater (fiber optics). In an optical waveguide communication system, an optoelectronic device or module that receives a signal, amplifies it (or, in the case of a digital signal, reshapes, retimes, or otherwise reconstructs it) and retransmits it. *See:* **modulation.** 433

optical scanner (character recognition). (1) A device that scans optically and usually generates an analog or digital signal. (2) A device that optically scans printed or written data and generates their digital representations. *See:* **visual scanner; electronic analog computer.** 255, 77

optical sound recorder. *See:* **photographic sound recorder.**

optical sound reproducer. *See:* **photographic sound reproducer.**

optical spectrum (fiber optics). Generally, the electromagnetic spectrum within the wavelength region extending from the vacuum ultraviolet at 40 nanometers (nm) to the far infrared at 1 millimeter (mm). *See:* **infrared; light.** 433

optical thickness (fiber optics). The physical thickness of an isotropic optical element, times its refractive index. *See:* **optical path length.** 433

optical time domain reflectometry (fiber optics). A method for characterizing a fiber wherein an optical pulse is transmitted through the fiber and the resulting light scattered and reflected back to the input is measured as a function of time. Useful in estimating attenuation coefficient as a function of distance and identifying defects and other localized losses. *See:* **Rayleigh scattering; scattering.** 433

optical tracker (navigation aid terms). A device for determining the direction of a luminous body relative to a set of reference axes using visible light vice, infrared, or radio frequencies. 526

optical transfer function (OTF) (diode-type camera tube). The spatial frequency response of an imaging sensor to a point source input. That is, the Fourier transform of the output image waveshape when the input image is a point, is known as the two-dimensional optical transfer function. In the one- dimensional case, the optical transfer function is the Fourier transform of the output image when the input image is a line. In the most common one-dimensional form, the OTF, designated $R_o(N)$, is written

$$R_o(N) = |R_o(N)| \exp[j\Phi(N)].$$

The OTF contains both amplitude and phase information. 380

optical waveguide (fiber optics). (1) Any structure capable of guiding optical power. (2) In optical communications, generally a fiber designed to transmit optical signals. *Syn:* **lightguide; optical conductor (Deprecated); optical fiber waveguide.** *See:* **cladding; core; fiber bundle; fiber optics; multimode optical waveguide; optical fiber; single mode waveguide; tapered fiber waveguide.** 433

optical waveguide connector (fiber optics). A device whose purpose is to transfer optical power between two optical waveguides or bundles, and that is designed to be connected and disconnected repeatedly. *See:* **multifiber joint; optical waveguide coupler.** 433

optical waveguide coupler (fiber optics). (1) A device whose purpose is to distribute optical power among two or more ports. *See:* **star coupler; tee coupler.** (2) A device whose purpose is to couple optical power between a waveguide and a source or detector. 433

optical waveguide preform. *See:* **preform.**

optical waveguide splice (fiber optics). A permanent joint whose purpose is to couple optical power between two waveguides. 433

optical waveguide termination (fiber optics). A configuration or a device mounted at the end of a fiber or cable which is intended to prevent reflection. *See:* **index matching material.** 433

optic amplifier. An optoelectronic amplifier whose signal input and output ports are electric. *Note:* This is in accord with the accepted terminologies of other electric-signal input and output amplifiers such as dielectric, magnetic, and thermionic amplifiers. *See:* **optoelectronic device.** 191

optic axis (fiber optics). In an anisotropic medium, a direction of propagation in which orthogonal polarizations have the same phase velocity. Distinguished from "optical axis". *See:* **anisotropic.** 433

optic port. A port where the energy is electromagnetic radiation, that is, photons. *See:* **optoelectronic device.** 191

optimal control. An admissible control law that gives a performance index an extremal value. *See:* **control system.** 329

optimization. The procedure used in the design of a system to maximize or minimize some performance index. May entail the selection of a component, a principle of operation, or a technique. *See:* **system.** 209

optimum bunching (electron tubes) (traveling-wave tube). The bunching condition that produces maximum power at the desired frequency in an output gap. 190, 125

optimum linearizing load resistance (Hall generator). The load resistance that produces the least linearity error. 107

optimum working frequency (radio wave propagation). The most effective frequency for ionospheric propagation of radio waves between two points under specified conditions. 146

optional stop (numerically controlled machines). A miscellaneous function command similar to a program stop except that the control ignores the command unless the operator has previously pushed a button to validate the command. 207

optional tasks (software verification and validation plans). Those verification and validation (V&V) tasks that are applicable to some, but not all software, or that may require the use of specific tools or techniques. 511

optoelectronic (fiber optics). Pertaining to a device that responds to optical power, emits or modifies optical radiation, or utilizes optical radiation for its internal operation. Any device that functions as an electrical-to-optical or optical-to-electrical transducer. *Notes:* (1) Photodiodes, light emitting diodes (LED), injection lasers and integrated optical elements are examples of optoelectronic devices commonly used in optical waveguide communications. (2) "Electro-optical" is often erroneously used as a synonym. *See:* **electro-optic effect; optical detector.** 433

optoelectronic amplifier. An optoelectronic device capable of power gain, in which the signal ports are either all electric ports or all optic ports. *See:* **optoelectronic device.** 191

optoelectronic cell. The smallest portion of an optoelectronic device capable of independently performing all the specified input and output functions. *Note:* An optoelectronic cell may consist of one or more optoelectronic elements. *See:* **optoelectronic device.** 191

optoelectronic device. An electronic device combining optic and electric ports. 191

optoelectronic element. A distinct constituent of an optoelectronic cell, such as an electroluminor, photoconductor, diode, optical filter, etcetera. *See:* **optoelectronic device.** 191

OR (mathematics of computing). A Boolean operator having the property that if P is a statement, Q is a statement, R is a statement, ..., then the OR of P,Q,R, ... is true if and only if at least one statement is true. *Note:* P OR Q is often represented by P $\vee$ Q or P + Q. *See:* **exclusive OR.** 564

orbit (navigation aid terms). The path of a celestial body relative to another body around which it revolves.   526

orbital inclination (communication satellite). The angle between the plane of the orbit and the plane of the equator measured at the ascending node.   74

orbital plane (communication satellite). The plane containing the radius vector and the velocity vector of a satellite, the system of reference being that specified for defining the orbital elements. In the idealized case of the unperturbed orbit, the orbital plane is fixed relative to the equatorial plane of the primary body.   74

orbital stability. *See:* **stability of a limit cycle.**

OR circuit (electronic computation). *See:* **OR gate.**

order (1) (general). To put items in a given sequence.   255, 77

(2) (in electronic computation). (A) Synonym for instruction. (B) Synonym for command. (C) Loosely, synonym for **operation part.** *Note:* The use of **order** in the computer field as a synonym for terms similar to the above is losing favor owing to the ambiguity between these meanings and the more-common meanings in mathematics and business. *See:* **instruction.**   235

order parameter (ξ)(primary ferroelectric terms). A parameter, or functionally related set of parameters, that can describe the reduction in symmetry occurring at a phase transition from a nonferroic phase to a ferroic phase.   497

order tone (telephone switching systems). A tone that indicates to an operator that verbal information can be transferred to another operator.   55

order wire (communication practice). An auxiliary circuit for use in the line-up and maintenance of communication facilities.   193

ordinary wave (O wave). The magnetoionic wave component that, when viewed below the ionosphere in the direction of propagation, has counterclockwise or clockwise elliptical polarization, respectively, according as the earth's magnetic field has a positive or negative component in the same direction. *See:* **radiation.**   328

ordinary-wave component (radio wave propagation). That magneto-ionic wave component deviating the least, in most of its propagation characteristics, relative to those expected for a wave in the absence of a fixed magnetic field. More exactly, if at fixed electron density, the direction of the fixed magnetic field were rotated until its direction was transverse to the direction of phase propagation, the wave component whose propagation would then be independent of the magnitude of the fixed magnetic field. *Syn:* **O wave.**   146

organic scintillator solute material (liquid-scintillation counters). An organic compound that can absorb radiant energy and immediately (typically within 10-9 s) reemit this energy as photons in the visible or ultraviolet range. This material is sometimes referred to as the scintillator or the fluor.   498

organizational maintenance (test, measurement and diagnostic equipment). Maintenance which is the responsibility of and performed by using organizations on its assigned equipment. Its phases normally consist of inspecting, servicing, lubricating, adjusting and the replacing of parts, minor assemblies and subassemblies.   54

organizing. *See:* **self-organizing.**

OR gate (OR circuit) (electronic computation). A gate whose output is energized when any one or more of the inputs is in its prescribed state. An OR gate performs the function of the logical OR.   210

orientation (illuminating engineering). The relation of a building with respect to compass directions.   167

orifice. An opening or window in a side or end wall of a waveguide or cavity resonator, through which energy is transmitted. *See:* **waveguide.**   328

orifice plate (rotating machinery). A restrictive opening in a passage to limit flow.   63

original master (electroacoustics) (metal master) (metal negative) (number 1 master) (disk recording). The master produced by electroforming from the face of a wax or lacquer recording.   176

original supplier (replacement parts for Class 1E equipment in nuclear power generating stations). Supplier of the original Class 1E equipment, as opposed to the original manufacturer of the part.   582

originating traffic (telephone switching systems). Traffic received from lines.   55

orthicon. A camera tube in which a beam of low-velocity electrons scans a photoemissive mosaic capable of storing an electric-charge pattern. *See:* **television.**   125

orthogonality (oscilloscopes). The extent to which traces parallel to the vertical axis of a cathode-ray-tube display are at right angles to the horizontal axis. *See:* **oscillograph.**   185

orthogonal polarization (with respect to a specified polarization) (antennas). In a common plane of polarization, the polarization for which the inner product of the corresponding polarization vector and that of the specified polarization is equal to zero. *Notes:* (1) The two orthogonal polarizations can be represented as two diametrical points on the Poincaré sphere. (2) Two elliptically polarized fields having the same plane of polarization have orthogonal polarizations if their polarization ellipses have the same axial ratio, major axes at right angles and opposite senses of polarization. *See:* **polarization vector, Note 2, for definition of the inner product; cross polarization.**   111, 267

orthorhombic system (piezoelectricity). An orthorhombic crystal has three mutually perpendicular twofold axes or two mutually perpendicular planes of reflection symmetry, or both. The *a, b, c* axes are of unequal length. For classes 222 and $2/m\ 2/m\ 2/m$ unit distances are chosen such that $c_0 < a_0 < b_0$. For the remaining class, which is polar, Z will always be the polar axis regardless of whether it is *a, b,* or *c* in the crystallographer's notation. Axes X and Y will then be chosen so that X is parallel to the remaining axis that is smallest. This class therefore may be properly designated $mm2$, $2mm$, or $m2m$, depending on whether c, a, or b is the polar axis. Axis sense is trivial except for the polar class for which $+Z$ is chosen such that d_{33} is positive. *Note:* "Positive" and "negative" may be checked using a carbon-zinc flashlight battery. The carbon anode connection will have the same effect on meter deflection as the + end of the crystal axis upon release of compression. *See:* **crystal systems.**

                                       371

_O_ scan (radar). *See:* **_O_ display.**

oscillating current. A current that alternately increases and decreases but is not necessarily periodic.

                                       210

oscillation (1) (general). The variation, usually with time, of the magnitude of a quantity with respect to a specified reference when the magnitude is alternately greater and smaller than the reference. *See:* **vibration.**

                                       176

(2) (vibration). A generic term referring to any type of a response that may appear in a system or in part of a system. *Note:* Vibration is sometimes used synonymously with oscillation, but it is more properly applied to the motion of a mechanical system in which the motion is in part determined by the elastic properties of the body.   210

(3) (gas turbines). The periodic variation of a function between limits above or below a mean value, for example, the periodic increase and decrease of position, speed, power output, temperature, rate of fuel input, etcetera within finite limits.   98, 58

oscillator (1) (general). Apparatus intended to produce or capable of maintaining electric or mechanical oscillations.   244, 210

(2) (electronics). A nonrotating device for producing alternating current, the output frequency of which is determined by the characteristics of the device. *See:* **industrial electronics.**   111

oscillator starting time, pulsed. *See:* **pulsed oscillator starting time.**   254

oscillator tube, positive grid. *See:* **positive-grid oscillator tube.**

oscillatory circuit. A circuit containing inductance and capacitance so arranged that when shock excited it will produce a current or a voltage that reverses at least once. If the losses exceed a critical value, the oscillating properties will be lost.   328

oscillatory surge (metal-oxide surge arresters for ac power circuits)(surge arresters). A surge that includes both positive and negative polarity values.

                       583, 430, 299, 62, 64, 2

oscillogram. A record of the display presented by an oscillograph or an oscilloscope. *See:* **oscillograph.**

                                       185

oscillograph. An instrument primarily for producing a record of the instantaneous values of one or more rapidly varying electrical quantities as a function of time or of another electrical or mechanical quantity. *Note:* (1) Incidental to the recording of instantaneous values of electrical quantities, these values may be-

come visible, in which case the oscillograph performs the function of an oscilloscope. (2) An oscilloscope does not have inherently associated means for producing records. (3) The term includes mechanical recorders.

oscillograph tube (oscilloscope tube). A cathode-ray tube used to produce a visible pattern that is the graphic representation of electric signals, by variations of the position of the focused spot or spots in accordance with these signals.   125

oscillography. The art and practice of utilizing the oscillograph. *See:* **oscillograph.**   185

oscilloscope. An instrument primarily for making visible the instantaneous value of one or more rapidly varying electrical quantities as a function of time or of another electrical or mechanical quantity. *See:* **cathode-ray oscilloscope; oscillograph.**   328

oscilloscope, dual-beam. An oscilloscope in which the cathode-ray tube produces two separate electron beams that may be individually or jointly controlled. *See:* **oscilloscope, multibeam; oscillograph.**   106

oscilloscope, multibeam. An oscilloscope in which the cathode-ray tube produces two or more separate electron beams that may be individually or jointly controlled. *See:* **oscilloscope, dual-beam.**   106

_O_-scope (radar). A cathode-ray oscilloscope arranged to present an O-display.   13

Ostwald color system (illuminating engineering). A system of describing colors in terms of color content, white content, and black content. It is usually exemplified by color charts in triangular form with Full Color, White, and Black at the apexes providing a gray scale of White and Black mixtures, and parallel scales of Constant White Content as these grays are mixed with varying proportions of the Full Color. Each chart represents a constant dominant wavelength (called hue), and the colors lying on a line parallel to the gray scale represent constant purity (called Shadow Series).

                                       167

OTH. *See:* **over-the-horizon radar.**

other insulation characteristics (insulation systems of synchronous machines). It is important to recognize that other characteristics, in addition to thermal endurance, such as mechanical strength, moisture resistance, and corona endurance are required in varying degrees in different applications for the successful use of insulating materials.   298

other tests (power and distribution transformer). Tests so identified in individual product standards which may be specified by the purchaser in addition to routine tests. (Examples: impulse, insulation power factor, audible sound.) *Note:* Transformer "General Requirements" Standards (such as ANSI C57.12.00-1973, IEEE Std 462-1973), General Requirements for distribution, Power and Regulating Transformers classify various tests as "routine," "design," or "other" depending on the size, voltage, and type of transformer involved.   53

OTH radar. *See:* **over-the-horizon radar.**

_O_-type tube or valve (microwave tubes). A microwave tube in which the beam, the circuit, and the focusing field have symmetry about a common axis. The interaction between the beam and the circuit is dependent upon velocity modulation, the suitably focused beam being launched by a gun structure outside one end of the microwave circuit and principally collected outside the other end of the microwave structure. *See:* **microwave tube.**   190

Oudin current (desiccating current) (resonator current) (medical electronics). A brush discharge produced by a high-frequency generator that has an output range of 2 to 10 kilovolts and a current sufficient to evaporate tissue water without charring. It is usually applied through a small needlelike electrode with the reference or ground electrode being relatively large and diffuse.   192

Oudin resonator. A coil of wire often with an adjustable number of turns, designed to be connected to a source of high-frequency current, such as a spark gap and induction coil, for the purpose of applying an effluve (convective discharge) to a patient.   192

outage (1)(electric power systems). The state of a component when it is not available to perform its intended function due to some event directly associated with that component. *Notes:* (A) An outage may or may not cause an interruption of service to consumers: depending on system configuration. (B) This definition derives from transmission and distribution applications and does not necessarily apply to generation outages.   200, 112
(2)(reliable industrial and commercial power systems planning and design). The state of a component or system when it is not available to properly perform its intended function.   561

outage duration (electric power systems). The period from the initiation of an outage until the affected component once again becomes available to perform its intended function. *Note:* Outage durations may be defined for specific types of outages, for example, permanent forced outage duration, transient forced outage duration, and scheduled outage duration. *See:* **outage.**   112, 200

outage duration, permanent forced (electric power systems). The period from the initiation of the outage until the component is replaced or repaired.   112

outage duration, scheduled (electric power systems). The period from the initiation of the outage until construction, preventive maintenance, or repair work is completed.   112

outage duration, transient forced (electric power systems). The period from the initiation of the outage until the component is restored to service by switching or fuse replacement. *Note:* Thus transient forced outage duration is really switching time.   112

outage, equipment (relay systems). The electrical isolation of equipment from the electric system such that it can no longer perform usefully for the duration of such isolation. *Note:* Since the term "outage" can also refer to service as well as equipment, it should always carry the appropriate modifier.   127

outage, forced (electric power systems). An outage that results from conditions directly associated with a component requiring that it be taken out of service immediately, either automatically or as soon as switching operations can be performed, or an outage caused by improper operation of equipment or human error. *Notes:* (1) This definition derives from transmission and distribution applications and does not necessarily apply to generation outages. (2) The key test to determine if an outage should be classified as forced or scheduled is as follows. If it is possible to defer the outage when such deferment is desirable, the outage is a scheduled outage: otherwise, the outage is a forced outage. Deferring an outage may be desirable, for example, to prevent overload of facilities or an interruption of service to consumers.   89, 112

outage, permanent forced (electric power systems). An outage whose cause is not immediately self-clearing, but must be corrected by eliminating the hazard or by repairing or replacing the component before it can be returned to service. An example of a permanent forced outage is a lightning flashover which shatters an insulator thereby disabling the component until repair or replacement can be made. *Note:* This definition derives from transmission and distribution applications and does not necessarily apply to generation outages.   112

outage, power. Complete absence of power at the point of use.   89

outage rate (electric power systems). For a particular classification of outage and type of component, the mean number of outages per unit exposure time per component. *Note:* Outage rates may be defined for specific weather conditions and types of outages. For example, permanent forced outage rates may be separated into adverse weather permanent forced outage rate and normal weather permanent forced outage rate.   112

outage rate, adverse weather permanent forced (electric power systems). For a particular type of component, the mean number of outages per unit of adverse weather exposure time per component.   112

outage rate, normal weather permanent forced (electric power supplies). For a particular type of component, the mean number of outages per unit of normal weather exposure time per component.   112

outage, scheduled (electric power systems). A loss of electric power that results when a component is deliberately taken out of service at a selected time, usually for purposes of construction, preventive maintenance, or repair. *Notes:* (1) This derives from transmission and distribution applications and does not necessarily apply to generation outages. (2) The key test to determine if an outage should be classified as forced or scheduled is as follows. If it is possible to defer the outage when such deferment is desirable, the outage is a scheduled outage: otherwise, the outage is a forced outage. Deferring an outage may be desirable, for example, to prevent overload of facilities or an interruption of service to consumers.   112

outage times (reliability data)(reliability data for pumps and drivers, valve actuators, and valves). (1) Out of Service. The average time required to get the failure, analyze it, obtain spare parts, repair and return the item or equipment to service, including planned delays. (2) Restoration. The average time required to get to the failure, analyze it, obtain spare parts, repair, and return the item or equipment to service, excluding planned delays. (3) Repair. The average time required to analyze the failure, repair, and return the item or equipment to service. This excludes planned delays and waiting for spares or tools.   447, 502

outage, transient forced (electric power systems). An outage whose cause is immediately self-clearing so that the affected component can be restored to service either automatically or as soon as a switch or circuit breaker can be reclosed or a fuse replaced. *Notes:* (1) An example of a transient forced outage is a lightning flashover which does not permanently disable the flashed component. (2) This definition derives from transmission and distribution applications and does not necessarily apply to generation outages.   112

outdoor (1) (power and distribution transformer). Suitable for installation where exposed to the weather.   53

(2) (power switchgear). Designed for use outside buildings.   103, 443

(3) (shunt power capacitors) (prefix). Designed for use outside buildings and exposure to the weather.   138

outdoor arrester (surge arrester)(metal-oxide surge arresters for ac power circuits). An arrester that is designed for outdoor use.   430, 583, 2

outdoor coupling capacitor. A capacitor designed for outdoor service that provides, as its primary function, capacitance coupling to a high-voltage line. *Note:* It is used in this manner to provide a circuit for carrier-frequency energy to and from a high-voltage line and to provide a circuit for power-frequency energy from a high-voltage line to a capacitance potential device or other voltage-responsive device.   351

outdoor enclosure (1)(power switchgear). An enclosure for outdoor application designed to protect against weather hazards such as rain, snow, or sleet. *Note:* Condensation is minimized by use of space heaters.   103

(2)(power system communication equipment). An enclosure constructed to protect equipment therein from the weather and accidental contact that would interfere with the successful operation.   453

outdoor shunt reactor (shunt reactors over 500 kVA). One of weather-resistant construction.   562

outdoor termination (cable termination). A termination intended for use where it is not protected from direct exposure to either solar radiation or precipitation.   4

outdoor transformer (power and distribution transformer). A transformer of weather-resistant construction suitable for service without additional protection from the weather.   53

outdoor wall bushing. A wall bushing on which one or both ends (as specified) are suitable for operating continuously outdoors. *See:* **bushing.**   348

outdoor weatherproof enclosure (series capacitor). An

enclosure so constructed or protected that exposure to the weather will not interfere with the successful operation of the equipment contained therein.   86

outer frame (rotating machinery). The portion of a frame into which the inner frame with its assembled core and winding is installed.   63

outer marker (navigation aid terms). A marker facility in an ILS (instrument landing system) which is installed at approximately 5 nmi (nautical miles)(9 km [kilometers]) from the approach end of the runway on the localizer course line to provide height, distance, and equipment functioning checks to aircraft on intermediate and final approach.   526

outgoing traffic (telephone switching systems). Traffic delivered directly to trunks from a switching entity.   55

outlet (electric installations on shipboard) (National Electrical Code). A point on the wiring system at which current is taken to supply utilization equipment.   3, 256

outlet box. A box used on a wiring system at an outlet. *See:* **cabinet.**   328

outline lighting (National Electrical Code). An arrangement of incandescent lamps or electric discharge tubing to outline or call attention to certain features such as the shape of a building or the decoration of a window.   256

out of band signaling (1) (data transmission). Signaling which utilizes frequencies within the guard band between channels. This term is also used to indicate the use of a portion of a channel bandwidth provided by the medium such as a carrier channel, but denied to the speech or intelligence path by filters. It resultsin a reduction of the effective available bandwidth.   59

(2) (telephone switching systems). Analog generated signaling that uses the same path as a message and in which the signaling frequencies are lower or higher than those used for the message.   55

out-of-phase (power switchgear) (as prefix to a characteristic quantity). A qualifying term indicating that the characteristic quantity applies to operation of the circuit breaker in out-of-phase conditions. *See:* **out-of-step.**   157

out-of-phase conditions (power switchgear). Abnormal circuit conditions of loss or lack of synchronism between the parts of an electrical system on either side of a circuit breaker in which, at the instant of operation of the circuit breaker, the phase angle between rotating phasors representing the generated voltages on either side exceeds the normal value and may be as much as 180° (phase opposition).   157

out-of-roundness (conductor). The difference between the major and minor diameters at any one cross section. *See:* **waveguide.**   265

out of step. A system condition in which two or more synchronous machines have lost synchronism with respect to one another and are operating at different average frequencies.   103, 127

out-of-step protection (power switchgear)(power system). A form of protection that separates the appropriate parts of a power system, or prevents separation that might otherwise occur, in the event of loss of synchronism. 103, 127

outpulsing (telephone switching systems). Pulsing from a sender. 55

output (1) (electronic digital computer) (data transmission). (A) Data that have been processed. (B) The state or sequence of states occurring on a specified output channel. (C) The device or collective set of devices used for taking data out of a device. (D) A channel for expressing a state of a device or logic element. (E) The process of transferring data from an internal storage to an external storage device. 59

(2) (rotating machinery). (A) (generator). The power (active, reactive, or apparent) supplied from its terminals. **(B) (motor).** The power supplied by its shaft. *See:* **asynchronous machine.** 63

output, acoustic (telephony). The sound pressure level developed in an artificial ear, measured in dB referred to an rms sound pressure of 2×10^{-5} newtons per square meter (N/m^2). 122

output angle (1) (gyro). The angular displacement of a gimbal about its output axis with respect to its support. 46

(2) (Pascal). *See:* **radiation angle.**

output assertion (software). A logical expression specifying one or more conditions that program outputs must satisfy in order for the program to be correct. *See:* **program.** 434

output attenuation (signal generator). The ratio, expressed in decibels (dB), of any selected output, relative to the output obtained when the generator is set to its calibration level. *Note:* It may be necessary to eliminate the effect of carrier distortion and/or modulation feedthrough by the use of suitable filters. *See:* **signal generator.** 185

output axis (OA) (gyro; accelerometer). An axis of freedom provided with a pickoff which generates an output signal as a function of the output angle. 46

output-axis-angular-acceleration drift rate (gyro) (inertial sensor). A drift rate that is proportional to the angular acceleration with respect to inertial space of the gyro case about the output axis. The relationship of this component of drift rate to angular acceleration can be stated by means of a coefficient having dimensions of angular displacement per unit time divided by angular displacement per unit time-squared. 46

output capacitance (*n*-terminal electron tube). The short-circuit transfer capacitance between the output terminal and all other terminals, except the input terminal, connected together. *See:* **electron-tube admittances.** 125

output-capacitor discharge time (power supply). The interval between the time at which the input power is disconnected and the time when the output voltage of the unloaded regulated power supply has decreased to a specified safe value. *See:* **regulated power supply.** 347

output circuit (protective relay system). An output

from a relay system which exercises direct or indirect control of a power circuit breaker, such as trip or close. 90

output control characteristics (thyristor). Output operating characteristics which can be deliberately selected or controlled, or both. 445

output control range (thyristor). The continuous range over which the output of a power controller can be changed by control signal input. 445

output current (converters having ac output)(self-commutated converters). The total rms (root-mean-square) current from the output terminals. 584

output-dependent overshoot and undershoot. Dynamic regulation for load changes. *See:* **electric conversion equipment.** 186

output electrode (electron tubes). The electrode from which is received the amplified, modulated, detected, etcetera, voltage. *See:* **electron tube.** 244, 190

output enable (G) (semiconductor memory). The inputs that when false cause the output to be in the OFF or high impedance state. This pin must be true for the output to be in any other state. 441

output factor. The ratio of the actual energy output, in the period of time considered, to the energy output that would have occurred if the machine or equipment had been operating at its full rating throughout its actual hours of service during the period. *See:* **generating station.** 64

output frequency stability (inverters). The deviation of the output frequency from a given set value. *See:* **self-commutated inverters.** 208

output gap (electron tubes) (traveling-wave tubes). An interaction gap by means of which usable power can be abstracted from an electron stream. 190

output impedance (1)(analog computers). The impedance presented by the transducer to a load. 9

(2)(converters having ac output)(self-commutated converters). The impedance presented by the converter to the load for specified frequencies. 584

output phase displacement (power inverters that have polyphase output) (inverters). The angular displacement between fundamental phasors. *See:* **self-commutated inverters.** 208

output power (1) (general). The power delivered by a system or transducer to its load. 239

(2) (electron tube or valve). The power supplied to the load by the electron tube or valve at the output electrode. *See:* **ON period (electron tube or valve).** 244, 190

output pulse (digital accelerometer). A pulse which represents the minimum unit of velocity increment (g·s, m/s). 383

output pulse amplitude (digital delay line). Peak amplitude of the output doublet which is obtained across the specified output load for a given amplitude of input step. 81

output pulse duration (digital delay line). Time spacing between the 10 percent amplitude point of the rise of the first peak to the 10 percent amplitude point of the fall of the second peak. 81

output pulse shape (low and high power pulse trans-

formers). Load current pulse flowing in a winding or voltage pulse developed across a winding in response to application of an input pulse. *Notes:* (1) The shape of the output pulse is described by a current- or voltage-time relationship. (2) Typically, a prominent feature of the output pulse is an accentuated backswing (last transition overshoot) (*ABS*). *See:* **pulse amplitude; rise time; pulse duration; fall time; trailing edge; tilt; overshoot; backswing; return swing; roll off; ringing; leading edge; quiescent value; leading edge; pulse top.** 32, 33

output pulse shape (pulse transformers). Load current pulse flowing in a winding or voltage pulse developed across a winding in response to application of an input pulse. The shape of the output pulse is described by a current- or voltage-time relationship. The following definitions for the input pulse shape apply to the output pulse shape: pulse amplitude; rise time; pulse duration; fall time; trailing edge; tilt; overshoot; backswing; return swing; rolloff; ringing; leading edge linearity; quiescent value; leading edge; pulse top; trailing edge. Typically, a prominent feature of the output pulse is an accentuated backswing (last transition overshoot), ABS. 589

output range (gyro; accelerometer). The product of input range and scale factor. *See:* **input-output characteristics.** 46

output reference axis (ORA) (gyro; accelerometer). The direction of an axis defined by the case mounting surfaces and/or external case markings. It is nominally parallel to the output axis. 46

output resonator (catcher) (electron tubes). A resonant cavity, excited by density modulation of the electron beam, that supplies useful energy to an external circuit. *See:* **velocity-modulated tube.** 244, 190

output ripple voltage (regulated power supply). The portion of the output voltage harmonically related in frequency to the input voltage and arising solely from the input voltage. *Note:* Unless otherwise specified, percent ripple is the ratio of root-mean-square value of the ripple voltage to the average value of the total voltage expressed in percent. In television, ripple voltage is usually expressed explicitly in peak-to-peak volts to avoid ambiguity. *See:* **regulated power supply.** 347

output signal (hydraulic turbines). The physical reaction of of any element of a control system to an input signal. 8

output span (gyro; accelerometer). The algebraic difference between the upper and lower values of the output range. *See:* **input-output characteristics.** 46

output-structure transit time. That portion of the photomultiplier transit time occurring with the output structure. 117

output torque without excitation (electric coupling). The torque an electric coupling will transmit or develop with zero excitation. 416

output voltage (converters having ac output)(self-commutated converters). The fundamental rms (root-mean-square) voltage (unless otherwise speci-

fied for a particular load) between the output terminals. 584

output voltage regulation (power supply). The change in output voltage, at a specified constant input voltage, resulting from a change of load current between two specified values. *See:* **regulated power supply.** 347

output voltage stabilization (power supply). The change in output voltage, at a specified constant load current, resulting from a change of input voltage between two specified values. *See:* **regulated power supply.** 347

output voltage versus input voltage characteristics (ferroresonant voltage regulators). Ferroresonant regulators may have output versus input characteristics as shown in Figure 'Output Versus Input Voltage with Jump Resonance' (*See:* **jump resonance** and figure 'Output Versus Input Voltage without Jump Resonance'), below.

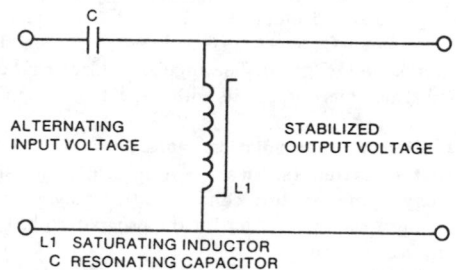

L1 SATURATING INDUCTOR
C RESONATING CAPACITOR

**Basic Series Ferroresonant
Voltage Regulator**

456

output winding (1)(ferroresonant voltage regulators). The winding of the ferroresonant transformer used to provide the regulated output voltage. *Note:* It is wound on the secondary section of the core and separated from the primary by a magnetic shunt. 456 **(2) (secondary winding(s)).** The winding(s) from which the output is obtained. *See:* **magnetic amplifier.** 415

outrigger (of a switching-device terminal) (power switchgear). An attachment that is fastened to or adjacent to the terminal pad of a switching device to maintain electrical clearance between the conductor and other parts or, when fastened adjacent, to relieve mechanical strain on the terminal, or both. 103

outside plant (communication practice). That part of the plant extending from the line side of the main distributing frame to the line side of the station or private-branch-exchange protector or connecting block, or to the line side of the main distributing frame in another central office building. 328

outside space block. *See:* **end finger.**

outward-wats service (telephone switching systems). A flat-rate or measured-time direct distance dialing service for defined geographical groups of numbering plan areas. 55

oven (analog computers). An enclosure and associated

sensors and heaters for maintaining components at a controlled and usually constant temperature.

10, 9

oven, wall-mounted. A domestic oven for cooking purposes designed for mounting in or on a wall or other surface. *See:* **appliances.**   256

overall (transmission performance of telephone sets). The acoustic output level of a telephone set due to an acoustic input to another telephone set to which it is connected by a test circuit. The acoustic input may be varied either in frequency or level. The output is measured in an artificial ear and the input is measured at the calibration position of an artificial mouth.

491

over-all electrical efficiency (dielectric and induction heater). The ratio of the power absorbed by the load material to the total power drawn from the supply lines. *See:* **load circuit efficiency; dielectric heating; induction heating.**   14,114

over-all generator efficiency (thermoelectric device). The ratio of (1) electric power output to (2) thermal power input. *See:* **thermoelectric device.**   191

overall objective loudness rating (OOLR) (loudness ratings of telephone connections).

$$\text{OOLR} = -20 \log_{10} \frac{S_E}{S_M}$$

where

S_M = sound pressure at the mouth reference point (in pascals)
S_E = sound pressure at the ear reference point (in pascals)   409

overall power efficiency (laser-maser). The ratio of the useful power output of the device to the total input power.   363

overall regulation (power supplies)(ferroresonant voltage regulators). The maximum amount that the output will change as a result of the specified change in line voltage, output load, input frequency, temperature, or time. *Note:* Line regulation, load regulation, effect of frequency variation,stability, and temperature coefficient are defined and usually specified separately.   456

overbunching (electron tubes). The bunching condition produced by the continuation of the bunching process beyond the optimum condition.   125

overburden (earth conductivity). The surface layers or regions of the earth that are water bearing and are subject to weathering. They comprise predominantly sand, gravel, clays, and poorly consolidated rocks.

132

overcast sky (illuminating engineering). A sky which has 100 percent cloud cover; the sun is not visible.

167

overcompounded. A qualifying term applied to a compound-wound generator to denote that the series winding is so proportioned that the terminal voltage at rated load is greater than at no load.   328

overcurrent (1)(National Electrical Code). Any cur-

rent in excess of the rated current of equipment or the ampacity of a conductor. It may result from overload (see definition) short circuit, or ground fault. A current in excess of rating may be accommodated by certain equipment and conductors for a given set of conditions. Hence the rules for overcurrent protection are specific for particular situations.   256

(2)(packaging machinery). An abnormal current greater than the full-load value.   429

overcurrent protection (1) (overload protection) (electric installations on shipboard). The effect of a device operative on excessive current (but not necessarily on short circuit) to cause and maintain the interruption of current flow to the device governed.

3

(2) (power and distribution transformer) (thyristor) (power switchgear). A form of protection(s) that operates when current exceeds a predetermined value.

53, 445, 103

overcurrent protective device (packaging machinery). A device operative on excessive current that causes and maintains the interruption of power in the circuit.

429

overcurrent relay (power switchgear). A relay that operates when its input current exceeds a predetermined value.   127, 103

overcurrent release (trip) (power switchgear). A release that operates when the current in the main circuit is equal to or exceeds the release setting.

103

overcutting (disk recording). The effect of excessive level characterized by one groove cutting through into an adjacent one. *See:* **phonograph pickup.**   176

overdamped (industrial control). A degree of damping that is more than sufficient to prevent the oscillation of the output following an abrupt stimulus. *Note:* For a linear second order system the roots of the characteristic equation must then be real and unequal. *See:* **control system, feedback.**   67, 206

overdamping (aperiodic damping). The special case of damping in which the free oscillation does not change sign. A damped harmonic system is overdamped if $F^2 > 4MS$. *See:* **damped harmonic system** for equation, definitions of letter symbols, and referenced terms.   210

overflow (mathematics of computing). (1) The condition that arises when the result of an arithmetic operation exceeds the capacity of the number representation system used in a digital computer. (2) The carry digit arising from this condition.   564

overflow traffic (telephone switching systems). That part of the offered traffic that cannot be carried by a group of servers.   55

overhang packing (rotating machinery). Insulation inserted in the end region of the winding to provide spacing and bracing. *See:* **rotor (rotating machinery); stator.**   63

overhead electric hoist. A motor-driven hoist having one or more drums or sheaves for rope or chain, and supported overhead. It may be fixed or traveling. *See:* **hoist.**   328

overhead ground wire (1)(**generating station grounding**). A grounded, bare conductor suspended horizontally between supporting rods or masts to provide protection from lightning strokes for structures, equipment, or suspended conductors within the zone of protection created by the masts and overhead ground wire combination.　　　　　　　　569

(2) (**lightning protection**). Grounded wire or wires placed above phase conductors for the purpose of intercepting direct strokes in order to protect the phase conductors from the direct strokes. They may be grounded directly or indirectly through short gaps. *See:* **direct-stroke protection (lightning)**.　　45

(3) (**conductor stringing equipment**) (**lightning protection**). Multiple grounded wire or wires placed above phase conductors for the purpose of intercepting direct strokes in order to protect the phase conductors from the direct strokes. *Syn:* **earth wire; shield wire; skywire; static wire**.　　45

overhead insulated ground (static or sky) wire-coupling protector (wire-line communication facilities). A device for protecting carrier terminals which are used in conjunction with overhead, insulated, ground wires (static wire) of a power transmission line.

414

overhead line charging current (power switchgear). Current supplied to an unloaded overhead line. *Note:* Current is expressed in root-mean-square (rms) amperes.　　　　103

overhead power line, corona (overhead-power-lines corona and radio noise). Corona occurring at the surfaces of electrodes during the positive or negative polarity of the power-line voltage. *Notes:* (1) Surfaces irregularities such as stranding, nicks, scratches, and semiconducting or insulating protrusions are usual corona sites. (2) Dry or wet airborne particles in proximity of electrodes may cause corona discharges. (3) Weather has a pronounced influence on the occurrence and characteristics of overhead-power-line corona.　　　　　　　411

overhead structure (elevators). All of the structural members, platforms, etcetera, supporting the elevator machinery, sheaves, and equipment at the top of the hoistway. *See:* **elevators.**　　　　328

overhead system service-entrance conductors (National Electrical Code). The service conductors between the terminals of the service equipment and a point usually outside the building, clear of building walls, where joined by tap or splice to the service drop.

256

overinterrogation control (electronic navigation). *See:* **gain turn-down.**

overlap. The distance the control of one signal extends into the territory that is governed by another signal or other signals. *See:* **neutral zone.**　　　328

overlap angle (1) (gas tube). The time interval, in angular measure, during which two consecutive arc paths carry current simultaneously.　　244, 190

(2) (**rectifier circuits**). *See:* **commutating angle.**

overlap X (**facsimile**). The amount by which the recorded spot X dimension exceeds that necessary to form a most nearly constant density line. *Note:* This effect arises in that type of equipment which responds to a constant density in the subject copy by a succession of discrete recorded spots. *See:* **recording (facsimile).**　　　　12

overlap control. *See:* **control system, two-step.**

overlap Y (**facsimile**). The amount by which the recorded spot Y dimension exceeds the nominal line width. *See:* **recording (facsimile).**　　12

overlap interval (circuit properties)(self-commutated converters). The time interval during which two commutating converter branches are carrying principal current simultaneously.　　　　584

overlap testing (nuclear power generating stations). Overlap testing consists of channel, train or load group verification by performing individual tests on the various components and subsystems of the channel, train, or load group. The individual component and subsystem tests shall check parts of adjacent subsystems, such that the entire channel, train, or load group will be verified by testing of individual components or subsystems.　　　　159

overlay (software). (1) In a computer program, a segment that is not permanently maintained in internal storage. (2) The technique of repeatedly using the same areas of internal storage during different stages of a program. (3) In the execution of a computer program, to load a segment of the computer program in a storage area hitherto occupied by parts of the computer program that are not currently needed. *See:* **computer program; execution; program; segment.**

434

overload (1)(**power operations**). Loading in excess of normal rating of equipment.　　　516

(2)(**protection and coordination of industrial and commercial power systems**). Generally used in reference to an overcurrent that is not of sufficient magnitude to be termed a short circuit. An overload is normally that overcurrent value from 100 percent of fuse rating up to ten times fuse rating. *See:* **short circuit.**

504

(3)(**power and distribution transformer**). Output of current power, or torque, by a device, in excess of the rated output of the device on a specified rating basis.

53

(4)(**analog computers**). In an analog computer, a condition existing within or at the output of a computing element that causes a substantial computing error because of the voltage or current saturation of one or more of the parts of the computing element. Similar to an overflow of an accumulator in a digital computer.

9

(5) (**National Electrical Code**). Operation of equipment in excess of normal, full load rating, or of a conductor in excess of rated ampacity which, when it persists for a sufficient length of time, would cause damage or dangerous overheating. A fault, such as a short circuit or ground fault, is not an overload. *See:* **overcurrent.**　　　　256

(6) (**radiation protection**). Response of less than full scale (that is, maximum scale reading) when exposed

to radiation intensities greater than the upper detection limit.   399

(7) (test, measurement and diagnostic equipment). To exceed the rated capacity of.   54

(8) (thyristor power computer). A condition existing when the load current exceeds the continuous rating of the converter unit in magnitude or time, or both, but the conduction cycles and waveforms remain essentially normal.   121

overload capacity (1) (antennas) (industrial control). The current, voltage, or power level beyond which permanent damage occurs to the device considered. This is usually higher than the rated load capacity. *Note:* To carry load greater than the continuous rating, may be acceptable for limited use.   111

(2) (accelerometer). The maximum acceleration to which an accelerometer may be subjected beyond the normal operating range without causing a permanent change in the specified performance characteristics.   46

overload characteristic (ferroresonant voltage regulator). That portion of the output voltage versus output current characteristic of ferroresonant regulators from rated current to short-circuit current. Figure "Overload Characteristic with Unsaturated Series Inductance" below, shows the typical overload characteristic when inductor L2 (See Figure "Basic Series Parallel Ferroresonant Voltage Regulator" in the definition of the same name) does not saturate. Figure "Overload Characteristic with Saturated Series Inductanc", below, shows the effect of L2 saturation in the overload condition where the short-circuit current is less than maximum overload current.   456

overload factor (electromagnetic compatibility). The ratio of the maximum value of a signal for which the operation of the predetector circuits of the receiver does not depart from linearity by more than one decibel, to the value corresponding to full-scale deflection of the indicating instrument. *See:* **electromagnetic compatibility.**   220, 199

overload level (system or component). That level above which operation ceases to be satisfactory as a result of signal distortion, overheating, or damage. *See:* **level.**   239

overload ON-state current (thyristor). An ON-state current of substantially the same wave shape as the normal ON-state current and having a greater value than the normal ON-state current. *See:* **principal current.**   208

overload point, signal (electronic navigation). The maximum input signal amplitude at which the ratio of output to input is observed to remain within a prescribed linear operating range. *See:* **navigation.**   187, 13

overload protection (industrial control). The effect of a device operative on excessive current, but not necessarily on short circuit, to cause and maintain the interruption of current flow to the device governed. *See:* **overcurrent protection.**   206

overload pulse (X-ray energy spectrometers). An signal that drives a section of the amplifying chain into saturation.   471

overload recovery time (diode-type camera tube). A measure of the ability of the camera tube to recover from a specified overload signal, defined as the increased time required for the readout process to reach

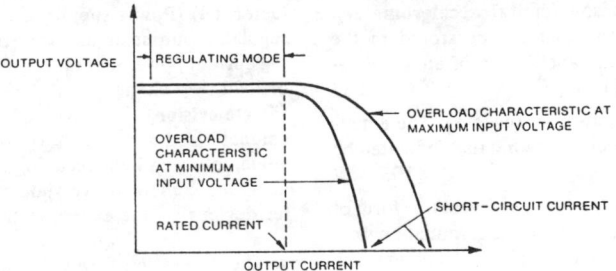

Fig 10
Overload Characteristic with Unsaturated
Series Inductance

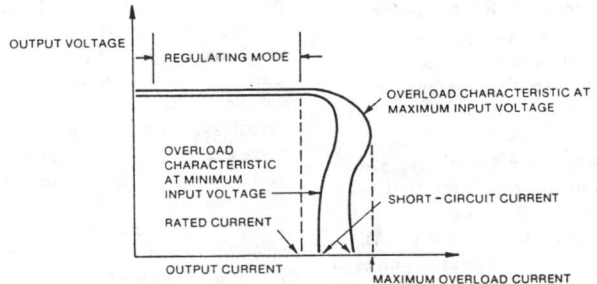

Fig 11
Overload Characteristic with Saturated
Series Inductance

its nonoverload third-field value. Units: seconds or numbers of fields.   380

overload relay (1) (general). A relay that responds to electric load and operates at a preset value of overload. *Note:* Overload relays are usually current relays but they may be power, temperature, or other relays.   103

(2) (overcurrent). *Note:* An overload relay is intended to protect the load (for example, motor armature) or its controller, and does not necessarily protect itself. *See:* **overcurrent relay; undercurrent relay.**   206

(3) (electric installations on shipboard). An overcurrent relay which functions at a predetermined value of overcurrent to cause the disconnection of the power supply.An overload relay is intended to protect the motor or controller and does not necessarily protect itself.   3

(4) (packaging machinery). An overcurrent relay that functions at a predetermined value of overcurrent to cause disconnection of the load from the power supply.   429

overmoded waveguide. A waveguide used to propagate a single mode, but capable of propagating more than one mode at the frequency of interest. *See:* **waveguide.**   267

overpotential. *See:* **overvoltage.**

overrange (noun) (1) (system or element). Any excess value of the response above its nominal full-scale value, or deficiency below the nominal minimum value.   56

(2) (test, measurement and diagnostic equipment). An input to a measuring device which exceeds in magnitude the capability of a given range.   54

overrange velocity storage (digital accelerometer). The velocity information that can be stored in the accelerometer during the application of an acceleration exceeding its input range.   383

overreach (of a relay) (power switchgear). The extension of the zone of protection beyond that indicated by the relay setting.   103, 127

overreaching protection (power switchgear). A form of protection in which the relays at one terminal operate for faults beyond the next terminal. They may be constrained from tripping until an incoming signal from a remote terminal has indicated whether the fault is beyond the protected line section.   103, 127

override (general system terms)(microcomputer system bus). A bus master overrides the bus control logic when it is necessary to guarantee itself back-to-back bus cycles. This is called overriding the bus, temporarily preventing other masters from using the bus.   542

overshoot (1)(first transition overshoot), A_{os} **(pulse transformers).** The amount by which the first maximum occurring in the pulse-top region exceeds the straight-line segment fitted to the top of the pulse in determining A_M. It is expressed in amplitude units or as a percentage of A_M.   589

(2) (pulse terms). A distortion which follows a major transition. *See:* Note in **preshoot** entry.   254

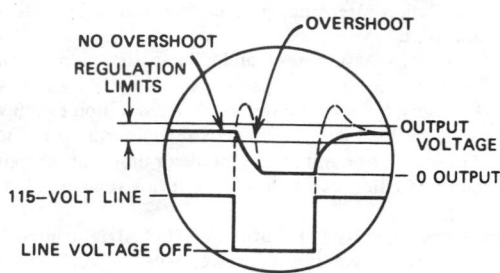

Scope view of turn-off/turn-on effects on a power supply, showing overshoot.

(3) (oscilloscopes). In the display of a step function (usually of time), that portion of the waveform which, immediately following the step, exceeds its nominal or final amplitude.   106

(4) (data transmission). (A) (Instrument). The amount of the overtravel of the indicator beyond its final steady deflection when a new constant value of the measured quantity is suddenly applied to the instrument. The overtravel and deflection are determined in angular measure and the overshoot is expressed as a percentage of the change in steady deflection. *Notes:* (1) Since in some instruments the percentage depends on the magnitude of the deflection, a value corresponding to an initial swing from zero to end scale is used in determining the overshoot for rating purposes. (2) Overshoot and damping factor have a reciprocal relationship. The percentage overshoot may be obtained by dividing 100 by the damping factor. **(B) (Power supplies).** A transient rise beyond regulated output limits, occurring when the alternating current power input is turned on or off, and for line or load step changes.   59

(5) (television). That part of a distorted wave front characterized by a rise above (or a fall below) the final value, followed by a decaying return to that final value. *Note:* Generally overshoots are produced in transfer devices having excessive transient response.   18

overshoot duration (low voltage varistor surge arresters). The time between the point at which the wave exceeds the clamping voltage and the point at which the voltage overshoot has decayed to 50 percent of its peak. For the purpose of this definition, clamping voltage is defined with an 8 x 29 µs current waveform of the same peak current amplitude as the waveform used for the overshoot duration.   62

overshoot response time (low voltage varistor surge arresters). The time between the point at which the wave exceeds the clamping voltage and the peak of the voltage overshoot. For the purpose of this definition, clamping voltage is defined with an 8 x 20 µ current waveform of the same peak current amplitude as the waveform used for the response time.   62

overshoot switch-off (transformer-rectifier system). The transient output voltage pulse occurring as the result of deenergization of the core on switch-off.   95

overshoot switch-on (transformer rectifier system) is the transient voltage on the output direct voltage following the completion of capacitor charging in the direct current circuit. It may be expressed as a percentage of excess over the steady-state direct voltage.
95

overshoot, system. *See:* **deviation, system.**

overshoot transient. *See:* **deviation transient.**

oversized waveguide. A waveguide operated in its dominant mode, but far above cutoff; sometimes termed quasioptical waveguide. 267

over-size insulation (electrical heat tracing for industrial applications). A term applied to thermal insulation when the inside diameter of the thermal insulation must be larger than the nominal outside diameter of a particular pipe so as to accommodate the heating cable. 523

overslung car frame. A car frame to which the hoisting-rope fastenings or hoisting-rope sheaves are attached to the crosshead or top member of the car frame. *See:* **hoistway (elevator or dumbwaiter).**
328

overspeed (hydraulic turbines). Any speed in excess of rated speed expressed as a percent of rated speed.
8

overspeed and overtemperature protection system (gas turbines). The overspeed governor, overtemperature detector, fuel stop valve(s), blow-off valve, other protective devices and their interconnections to the fuel stop valve, and to the blow-off valve, if used, that are required to shut off all fuel flow and shut down the gas turbine. 98

overspeed device (12) (power system device function numbers). Usually a direct-connected speed switch which functions on machine overspeed. 402

overspeed governor (gas turbines). A control element that is directly responsive to speed and that actuates the overspeed and overtemperature protection system when the turbine reaches the speed for which the device is set. 98

overspeed protection (1) (industrial control). The effect of a device operative whenever the speed rises above a preset value to cause and maintain an interruption of power to the protected equipment or a reduction of its speed. *See:* **relay.** 206

(2) (relay systems). A form of protection that operates when the speed of rotation exceeds a predetermined value. 127

(3) (power switchgear). A form of protection that operates when the speed of rotation exceeds a predetermined value. 103

overspeed test (rotating machinery). A test on a machine rotor to demonstrate that it complies with specified overspeed requirements. *See:* **rotor (rotating machinery).**
63

overtemperature (rotating machinery). Unusually high temperature from causes such as overload, high ambient temperature, restricted ventilation, etcetera.
37

overtemperature detector (gas turbines). The primary sensing element that is directly responsive to temperature and that actuates the overspeed and overtemperature protection system when the turbine temperature reaches the value for which the device is set.
98

overtemperature protection (power supplies). A thermal relay circuit that turns off the power automatically should an overtemperature condition occur. 186

over-the-horizon radar (navigation aid terms). Radar using sufficiently low carrier frequencies, usually in the high-frequency (hf) band, so that ground-wave or ionospherically refracted sky-wave propagation can allow detection far beyond the ranges allowed by line-of-sight propagation. 526

overtone. *See:* **harmonic.**

overtone-type piezoelectric-crystal unit (circuits and systems). (1) An overtone driven by the action of the piezoelectric effect: (2) **(crystal unit)** A resonator constructed from a piezoelectric crystal material and designed to operate in the vicinity of an overtone of that device. 67

overtravel (of a relay) (power switchgear). The amount of continued movement of the responsive element after the input is changed to a value below pickup. *See:* **impedance relay.** 103

Relationship of Relay Operating Time for
Electromechanical Relays
(PSRC)

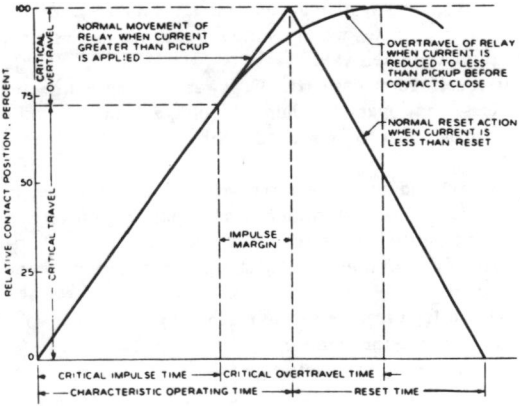

overvoltage (overpotential) (1) (general). A voltage above the normal rated voltage or the maximum operating voltage of a device or circuit. A direct test overvoltage is a voltage above the peak of the line alternating voltage. *See:* **insulation testing (large alternating-current rotating machinery).** 6

(2) (rotating machinery). An abnormal voltage higher than the normal service voltage, such as might be caused from switching or lightning surges. 37

(3) (surge arresters) (system voltage). Abnormal voltage between two points of a system that is greater than the highest value appearing between the same two points under normal service conditions. *Note:* Overvoltages may be low frequency, temporary, and transient—meaning a lightning or switching surge overvoltage.   53, 244, 62

(4) (radiation-counter tubes). The amount by which the applied voltage exceeds the Geiger-Mueller threshold. *See:* **gas-filled radiation-counter tube.**
190

(5) (electrochemistry). The displacement of an electrode potential from its equilibrium (reversible) value because of flow of current. *Note:* This is the irreversible excess of potential required for an electrochemical reaction to proceed actively at a specified electrode, over and above the reversible potential characteristic of that reaction. *See:* **electrochemistry.**   205

(6) (overpotential) (rotating machinery). A voltage above the normal rated voltage or the maximum operating voltage of a device or circuit. A direct test overvoltage is a voltage above the peak of the alternating line voltage.   6

overvoltage due to resonance (surge arresters). Overvoltage at the fundamental frequency of the installation, or of a harmonic frequency, resulting from oscillation of circuits.   244, 62

overvoltage protection. The effect of a device operative on excessive voltage to cause and maintain the interruption of power in the circuit or reduction of voltage to the equipment governed.   206

overvoltage relay (power switchgear). A relay that operates when its input voltage exceeds a predetermined value.   103, 127

(59) (power system device function numbers). A relay which operates when its input voltage is more than a predetermined value.   402

overvoltage release (trip) (power switchgear). A release that operates when the voltage of the main circuit is equal to or exceeds the release setting.
103

overvoltage suppressors (thyristor). Devices used in the ac power controller to attenuate repetitive and nonrepetitive overvoltages of internal or external origin, for example, snubbers, surge arresters, limiters, etcetera.   445

overvoltage test (rotating machinery). A test at voltages above the rated operating voltage.   63

overwriting (charge-storage tubes). Writing in excess of that which produces write saturation. *See:* **charge-storage tube.**   174

OW. *See:* **oil-immersed water-cooled transformer.**

O **wave (radio wave propagation).** *See:* **ordinary-wave component; radio wave propagation.**

Owen bridge. A 4-arm alternating-current bridge in which one arm, adjacent to the unknown inductor, comprises a capacitor and resistor in series: the arm opposite the unknown consists of a second capacitor, and the fourth arm of a resistor. *Note:* Normally used for the measurement of self-inductance in terms of capacitance and resistance. Usually, the bridge is balanced by adjustment of the resistor that is in series with the first capacitor and of another resistor that is inserted in series with the unknown inductor. The balance is independent of frequency. *See:* **bridge.**
328

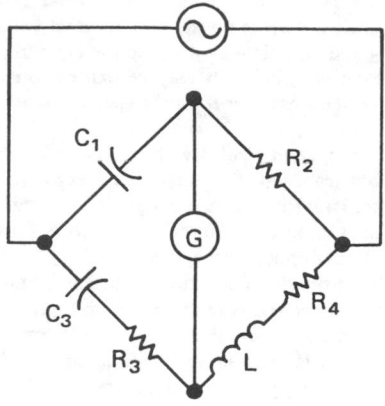

$$C_3 R_4 = C_1 R_2 \qquad L = C_1 R_3 R_2$$

Owen bridge

owner (nuclear power quality assurance). The person, group, company, agency, or corporation who has or will have title to the nuclear power plant.   417

oxidant. A chemical element or compound that is capable of being reduced. *See:* **electrochemical cell.**
223

oxidation (electrochemical cells and corrosion). Loss of electrons by a constituent of a chemical reaction. *See:* **electrochemical cell.**   185

oxidation inhibitor (insulating oil). Any substance added to an insulating fluid to improve its resistance to deleterious attack in an oxidizing environment. For example, 2, 6-ditertiary-butyl para-cresol or 2, 6-ditertiary-butyl phenol, or both, are sometimes added to petroleum insulating oil to improve its oxidation stability.   461

oxide-cathode (thermionics). *See:* **oxide-coated cathode.**

oxide-coated cathode (oxide-cathode) (thermionics). A cathode whose active surface is a coating of oxides of alkaline earths on a metal. *See:* **electron emission.**
244, 190

oxidizing (electrotyping). The treatment of a graphited wax surface with copper sulfate and iron filings to produce a conducting copper coating. *See:* **electroforming.**   328

oxygen-concentration cell. A galvanic cell resulting primarily from differences in oxygen concentration. *See:* **electrolytic cell.**   205

ozone-producing radiation (illuminating engineering). Ultraviolet energy shorter than 220 nm (nanometers) which decomposes oxygen O_2 thereby producing ozone O_3. Some ultraviolet sources generate energy at 184.9 nm which is particularly effective in producing ozone.   167

P

P.I.D. *See:* **control action, proportional plus integral plus derivative.**

P.I. *See:* **control action, proportional plus integral.**

PABX. *See:* **private automatic branch exchange.**

pace voltage (surge arresters). A voltage generated by ground current between two points on the surface of the ground at a distance apart corresponding to the conventional length of an ordinary pace. 244, 62

pacing (Class 1E connection assemblies). A method of ongoing qualification by parallel age conditioning. 459

pack. To compress several items of data in a storage medium in such a way that the individual items can later be recovered. 255, 77

package, core (rotating machinery). The portion of core lying between two adjacent vent ducts or between an end plate and the nearest vent duct. 63

packaged magnetron. An integral structure comprising a magnetron, its magnetic circuit, and its output matching device. *See:* **magnetron.** 190, 125

packaging machine. Any automatic, semiautomatic, or hand-operated apparatus that performs one or more packaging functions, such as, but not limited to, the fabrication, preparation, filling, closing, labeling, or preparing, or both, for final distribution of any type of package or container used to protect or display, or both, any product. 429

packet (data communication). A group of binary digits including data and control elements which is switched and transmitted as a composite whole. The data and control elements and possibly error control information are arranged in a specified format. 12

packet switching (data communication). A data transmission process, utilizing addressed packets, whereby a channel is occupied only for the duration of transmission of the packet. *See:* **circuit switching; message switching; store-and-forward switching.** 12

packing density (computing systems). The number of useful storage cells per unit of dimension, for example, the number of bits per inch stored on a magnetic tape or drum track. 255, 77, 54

packing fraction (fiber optics). In a fiber bundle, the ratio of the aggregate fiber cross-sectional core area to the total cross-sectional area (usually within the ferrule) including cladding and interstitial areas. *See:* **ferrule; fiber bundle.** 433

packing gland. An explosionproof entrance for conductors through the wall of an explosionproof enclosure, to provide compressed packing completely surrounding the wire or cable, for not less than π% inch measured along the length of the cable. *See:* **mine feeder circuit.** 328

pad (attenuating pad) (data transmission). A nonadjustable passive network that reduces the power level of a signal without introducing appreciable distortion. *Note:* A pad may also provide impedance matching. 59

pad electrode (dielectric heater usage). One of a pair of electrode plates between which a load is placed for dielectric heating. *See:* **dielectric heating.** 14, 114

pad mounted (1) (National Electrical Safety Code). A method of supporting equipment, generally at ground level. 391

(2) (power switchgear). A general term describing equipment positioned on a surface mounted pad located outdoors. The equipment is usually enclosed with all exposed surfaces at ground potential. 103

pad-mounted equipment (NESC). A general term describing enclosed equipment, the exterior of which enclosure is at ground potential, positioned on a surface-mounted pad. 494

pad-mounted transformer (power and distribution transformer). An outdoor transformer utilized as part of an underground distribution system, with enclosed compartment(s) for high-voltage and low-voltage cables entering from below, and mounted on a foundation pad. 53

pad-type bearing (rotating machinery). A journal or thrust-type bearing in which the bearing surface is not continuous but consists of separate pads. *See:* **bearing.** 63

pair. A term applied in electric transmission to two like conductors employed to form an electric circuit. *See:* **cable.** 328

paired brushes (pair of brushes) (rotating machinery). Two individual brushes that are joined together by a common shunt or terminal. *Note:* They are not to be confused with a split brush. *See:* **brush (rotating machinery).** 63

paired cable (nonquadded cable). A cable in which all of the conductors are arranged in the form of twisted pairs, none of which is arranged with others to form quads. *See:* **cable.** 328

pairing (scanning) (television). The condition in which lines appear in groups of two instead of being equally spaced. 18

pair of brushes (rotating machinery). *See:* **paired brushes.**

PAM. *See:* **pulse-amplitude modulation.**

pancake coil. A coil having the shape of a pancake, usually with the turns arranged in the form of a flat spiral. 328

pancake motor. A motor that is specially designed to have an axial length that is shorter than normal. 63

panel (1) (power switchgear)(Class 1E control boards). A unit of one or more sections of flat material suitable for mounting electric devices. 103, 140

(2) (industrial control)(packaging machinery). An element of an electric controller consisting of a slab or plate on which various component parts of the controller are mounted and wired. 206, 124, 429

(3) (photovoltaic converter). Combination of shingles or subpanels as a mechanical and electric unit required to meet performance specifications. *See:* semiconductor.

(4) (computing systems). *See:* control panel; problem board.

(5) (solar cells). The largest unit combination of solar cells or subpanels that is mechanically designed to facilitate manufacture and handling and that will establish a basis for electrical performance by test. 113

panelboard (National Electrical Code). A single panel or group of panel units designed for assembly in the form of a single panel; including buses, automatic overcurrent devices, and with or without switches for the control of light, heat, or power circuits; designed to be placed in a cabinet or cutout box placed in or against a wall or partition and accessible only from the front. See: **switchboard.** 256

panel efficiency (1) (photoelectric converter). The ratio of available power output to incident radiant power intercepted by a panel composed of photoelectric converters. *Note:* This is less than the efficiency of the individual photoelectric converters because of area not covered by photoelectric converters, Joule heating, and photoelectric-converter mismatches. *See:* semiconductor. 186

(2) (solar cells). The ratio of available electric power output to total incident radiant power intercepted by the area of a panel composed of solar cells. *Note:* This depends on the spectral distribution of the radiant power source and junction temperature(s), requires uniform normal illumination on the intercepting area, and results in an efficiency less than the efficiency of the individual solar cells because of area not covered by solar cells, incident energy heating, solar cell mismatch, optical losses, and wiring losses. 113

panel-frame mounting (of a switching device). Mounting on a panel frame in the rear of a panel with the operating mechanism on the front of the panel. 103

panel system. An automatic telephone switching system that is generally characterized by the following features: (1) The contacts of the multiple banks over which selection occurs are mounted vertically in flat rectangular panels. (2) The brushes of the selecting mechanism are raised and lowered by a motor that is common to a number of these selecting mechanisms. (3) The switching pulses are received and stored by controlling mechanisms that govern the subsequent operations necessary in establishing a telephone connection. 328

paper-lined construction (dry cell) (primary cell). A type of construction in which a paper liner, wet with electrolyte, forms the principal medium between the negative electrode, usually zinc, and the depolarizing mix. (A layer of paste may lie between the paper liner

and the negative electrode.) *See:* electrolytic cell. 328

PAR. *See:* precision approach radar.

parabolic profile (fiber optics). A power-law index profile with the profile parameter, g, equal to 2. *Syn:* quadratic profile. *See:* graded index profile; multimode optical waveguide; power-law index profile; profile parameter. 433

parabolic torus reflector (antennas). A toroidal reflector formed by rotating a segment of a parabola about a nonintersecting coplanar line. 111

paraboloidal reflector (antennas). An axially symmetric reflector which is a portion of a paraboloid. *Note:* This term may be applied to any reflector which is a portion of a paraboloid provided the term is appropriately qualified. For example, if the reflector is a portion of a paraboloid but does not include its vertex then it may be called an "off-set paraboloidal reflector". 111

paraelastic crystal (primary ferroelectric terms). By analogy with paraelectric crystals, a crystal in which mechanical strain S is a single-valued function of mechanical stress T, whose elastic compliance exhibits an obvious Curie-Weiss behavior with temperature over some given temperature range, and which at some critical temperature T_c undergoes a phase transition to a ferroelastic phase. Crystals that clearly have a paraelastic phase include the metallic alloys Nb_3Sn, In-Th, Au-Zn-Sn, and lithium ammonium tartrate. 497

paraelastic phase (primary ferroelectric terms). A phase that encompasses the range of temperature in which the elastic compliance exhibits Curie-Weiss behavior. 497

paraelectric Curie temperature (of a ferroelectric material). The intercept of the linear portion of the plot of $1/\epsilon$ versus T, where ϵ is the small signal dielectric permittivity measured at zero bias field and T is the absolute temperature in the region above the ferroelectric Curie temperature where ϵ generally follows the Curie-Weiss relation. *See:* ferroelectric domain. 247

paraelectric phase (primary ferroelectric terms). Encompasses the range of temperature or pressure over which the permittivity exhibits Curie-Weiss behavior. 497

paraelectric region (ferroelectric material). The region above the Curie point where the small signal permittivity increases with decreasing temperature. *See:* ferroelectric Curie point; small signal permittivity. 80

parallel (parallel elements) (1) (network). (A) Two-terminal elements are connected in parallel when they are connected between the same pair of nodes. (B) Two-terminal elements are connected in parallel when any cut-set including one must include the others. *See:* network analysis. 210

(2) (electronic computers). (A) Pertaining to the simultaneity of two or more processes. (B) Pertaining to the simultaneity of two or more similar or identical processes. (C) Pertaining to simultaneous processing of the individual parts of a whole, such as the bits of

a character and the characters of a word, using separate facilities for the various parts. *See:* **serial-parallel.**
235, 210, 255, 77

(3) (radio wave propagation). Of a propagating wave for which the electric field vector lies parallel to the plane of incidence. *Note:* Sometimes called vertical polarization. 146

parallel-connected capacitance (as applied to interrupter switches) (power switchgear). Capacitances are defined to be parallel-connected when the crest value of inrush current to the capacitance being switched exceeds the switch inrush current capability for single capacitance. 103

parallel connection. The arrangement of cells in battery made by connecting all positive terminals together and all negative terminals together, the voltage of the group being only that of one cell and the current drain through the battery being divided among the several cells. *See:* **battery (primary or secondary).** 328

parallel digital computer. One in which the digits are handled in parallel. Mixed serial and parallel machines are frequently called serial or parallel according to the way arithmetic processes are performed. An example of a parallel digital computer is one that handles decimal digits in parallel although it might handle the bits that comprise a digit either serially or in parallel. *See:* **serial digital computer.** 210

parallel feeder (power switchgear). One that operates in parallel with one or more feeders of the same type form the same source. *Note:* These feeders may be of the stub-, multiple-, or tie-feeder type. 103

parallel heating cable (electrical heat tracing for industrial applications). Heating elements that are electrically connected in parallel, either continuously or in zones, so that watt density per lineal length is maintained irrespective of any change in length for the continuous type or for any number of discrete zones. 523

paralleling (rotating machinery). The process by which a generator is adjusted and connected to run in parallel with another generator or system. *See:* **asynchronous machine.** 63

paralleling reactor (power and distribution transformer). A current-limiting reactor for correcting the division of load between parallel-connected transformers which have unequal impedance voltages. 53

parallel-mode interference (signal-transmission system). *See:* **interference, common-mode.**

parallel operation (power supplies). Voltage regulators, connected together so that their individual output currents are added and flow in a common load. Several methods for parallel connection are used: spoiler resistors, master/slave connection, parallel programming, and parallel padding. Current regulators can be paralleled without special precaution. 186

parallel padding (power supplies). A method of parallel operation for two or more power supplies in which their current limiting or automatic crossover output characteristic is employed so that each supply regulates a portion of the total current, each parallel supply adding to the total and padding the output only when the load current demand exceeds the capability or limit setting-of the first supply. 186

parallel (or perpendicular) polarization (facsimile). A linear polarization for which the field vector is parallel (or perpendicular) to some reference plane. *Note:* These terms are applied mainly to uniform plane waves incident upon a plane of discontinuity (surface of the earth, surface of a dielectric or a conductor). Then the convention is to take as reference the plane of incidence, that is, the plane containing the direction of propagation and the normal to the surface of discontinuity. If these two directions coincide, the reference plane must be specified by some other convention. 11

parallel processing (computing systems). Pertaining to the simultaneous execution of two or more sequences of instructions or one sequence of instructions operating on two or more sets of data, by a computer having multiple arithmetic and/or logic units. *See:* **multiprocessing; multiprogramming.** 255, 77

parallel programming (power supplies). A method of parallel operation for two or more power supplies in which their feedback terminals (voltage-control terminals) are also paralleled. These terminals are often connected to a separate programming source. 186

parallel rectifier. A rectifier in which two or more similar rectifiers are connected in such a way that their direct currents add and their commutations coincide. *See:* **power rectifier.** 208

parallel rectifier circuit. A rectifier circuit in which two or more simple rectifier circuits are connected in such a way that their direct currents add and their commutations coincide. *See:* **rectification; rectifier circuit element.** 66

parallel resonance (circuits and systems). The sinusoidal steady state condition that exists in a circuit composed of an inductor and a capacitor connected in parallel when the applied frequency is such that (1) the driving-point impedance is a maximum, or (2) the susceptance of the two parallel arms are equal in magnitude, or (3) the phase-angle of the driving-point impedance is zero. Sometimes defined as above for more general RLC (resistance-inductance-capacitance) networks. 67

parallel search storage (computing systems). *See:* **associative storage.** 255, 77

parallel storage (computing systems). A storage device in which characters, words, or digits are dealt with simultaneously. 255, 77

parallel-T network (twin-T network). A network composed of separate T networks with their terminals connected in parallel. *See:* **network analysis.** 328

parallel transmission (data transmission). Simultaneous transmission of the bits making up a character, either over separate channels or on different carrier frequencies on one channel. 194

parallel two-terminal pair networks. Two-terminal pair networks are connected in parallel at the input or at the output terminals when their respective input or

output terminals are in parallel. *See:* **network analysis.**
332

paralyzable system (X-ray energy spectrometers).
Any system or device whose response characteristics
contain a region where the ratio of output-to-input
count rate decreases with increasing input count rate.
471

paramagnetic material. Material whose relative permeability is slightly greater than unity and practically
independent of the magnetizing force.   210

parameter (1) (mathematical). A variable that is given
a constant value for a specific purpose or process.
255, 77

(2) (physical). One of the constants entering into a
functional equation and corresponding to some characteristic property, dimension, or degree of freedom.

(3) (electrical). One of the resistance, inductance,
mutual inductance, capacitance, or other element values included in a circuit or network. Also called **network constant.**   210

(4) (control systems). A quantity of property treated
as a constant but which may sometimes vary or be
adjusted.   56

(5) (test, measurement and diagnostic equipment).
(A) Any specific quantity or value affecting or describing the theoretical or measurable characteristics
of a unit being considered which behaves as an independent variable or which depends upon some functional interaction of other quantities in a theoretically
determinable manner: and (B) In programming, a
variable that is given a constant value for a specific
purpose or process.   54

(6) (software). (A) A variable that is given a constant
value for a specified application and that may denote
the application. (B) A variable that is used to pass
values between program routines. *See:* **actual parameter; formal parameter; variable.**   434

**parameter potentiometer (analog computers) (scale-
factor potentiometer or coefficient potentiometer).** A
potentiometer used in an analog computer to represent a problem parameter such as a coefficient or a
scale factor.   9

parametric amplifier. An inverting parametric device
used to amplify a signal without frequency translation
from input to output. *Note:* In common usage, this
term is a synonym for reactance amplifer. *See:* **parametric device.**   277, 191

parametric converter. An inverting parametric device
or noninverting parametric device used to convert an
input signal at one frequency into an output signal at
a different frequency. *See:* **parametric device.**
191

parametric device. A device whose operation depends
essentially upon the time variation of a characteristic
parameter usually understood to be a reactance.
191

parametric down-converter. A parametric converter in
which the output signal is at a lower frequency than
the input signal. *See:* **parametric device.**   191

parametric up-converter. A parametric converter in
which the output signal is at a higher frequency than

the input signal. *See:* **parametric device.**   191

parametric variation (automatic control). A change in
those system properties generally regarded as constants which affect the dependent variables describing
system operation.   56

parasitic element (1) (data transmission). A radiating
element that is not coupled directly to the feed lines
of an antenna and that materially affects the radiation
pattern or impedance of an antenna, or both.
59

(2) (antennas). A radiating element that is not coupled directly to the feed lines of an antenna and that
materially affects the radiation pattern and.or impedance of an antenna. *Note:* Compare with driven element. *See:* **antenna.**   111

(3) (circuits and systems). An unwanted circuit element that is an unavoidable adjunct of a wanted circuit
element.   67

parasitic oscillations. Unintended self-sustaining oscillations, or transient impulses. *See:* **oscillatory circuit.**
111, 211

paraxial ray (fiber optics). A ray that is close to and
nearly parallel with the optical axis. *Note:* For purposes of computation, the angle, θ, between the ray
and the optical axis is small enough for $\sin \theta$ or $\tan$
θ to be replaced by θ (radians). *See:* **light ray.**
433

parcel plating. Electroplating upon only a part of the
surface of a cathode. *See:* **electroplating.**   328

**parity (1)(for FASTBUS)(FASTBUS acquisition and
control).** A bit, optionally appended to a FASTBUS
word, whose value is chosen to make the total number
of one bits (including the parity bit) odd. It is used for
error checking since receipt of an even number of one
bits implies a transmission error.   480

(2)(mathematics of computing). (A) An error detection method in which the total number of ones in a
binary word, byte, character, or message is set to an
odd or even number by appending a redundant bit.
This number is subsequently checked to ensure that it
remains odd or even. (B) The property of oddness or
evenness possessed by a word, byte, character, or
message. This property is determined by the total
number of ones.   564

parity bit (computing systems). A binary digit appended to an array of bits to make the sum of all the bits
always odd or always even.   255, 77, 54

parity check (electronic computation). A summation
check in which the bits in a character or block are
added (modulo 2) and the sum checked against a single, previously computed parity digit: that is, a check
that tests whether the number of ones is odd or even.
255, 77, 194, 54

**park electrical wiring systems (National Electrical
Code).** All of the electrical wiring fixtures, equipment
and appurtenances related to electrical installations
within a mobile home park, including the mobile home
service equipment.   256

parking lamp (illuminating engineering). A lighting
device placed on a vehicle to indicate its presence
when parked.   167

parking stand (separable insulated connectors). A bracket, designed for installation on an apparatus, suitable for holding accessory devices, such as insulated parking bushing and grounding bushing. 454

parse (software). To determine the syntactic structure of an artificial or natural language unit by decomposing the unit into more elementary subunits and establishing the relationships among the subunits, for example, blocks, statements, and expressions may be decomposed into statements, expressions, and operators and operands. *See:* **artificial language; block; natural language; operand; operator.** 434

parsec (pc). The distance at which 1 astronomical unit subtends an angle of 1 second of arc: approximately, 1 pc = 206 265 AU = 30857 × 10^{12}m. 21

part (unique identification in power plants and related facilities). An element of a component not amenable to further disassembly for maintenance purposes. 544, 545

partial (audio and electroacoustics). (1) A physical component of a complex tone. (2) A component of a sound sensation that may be distinguished as a simple tone that cannot be further analyzed by the ear and that contributes to the timbre of the complex sound. *Notes:* (A) The frequency of a partial may be either higher or lower than the basic frequency and may or may not be an integral multiple or submultiple of the basic frequency. If the frequency is not a multiple or submultiple, the partial is inharmonic. (B) When a system is maintained in steady forced vibration at a basic frequency equal to one of the frequencies of the normal modes of vibration of the system, the partials in the resulting complex tone are not necessarily identical in frequency with those of the other normal modes of vibration. 176

partial automatic control (station control and data acquisition). Control which is a combination of manual and automatic control. 403

partial-automatic station (power switchgear). A station that includes protection against the usual operating emergencies, but in which some or all of the steps in the normal starting or stopping sequence, or in the maintenance of the required character of service, must be performed by a station attendant or by supervisory control. 103

partial-automatic transfer (or throwover) equipment (power switchgear). An equipment that automatically transfers load to another (emergency) source of power when the original (preferred) source to which it has been connected fails, but that will not automatically retransfer the load to the original source under any conditions. *Note:* The restoration of the load to the preferred source from the emergency source upon the reenergization of the preferred source after an outage may be of the continuous-circuit restoration type or the interrupted-circuit restoration type. 103

partial body irradiation (electrobiology). Pertains to the case in which part of the body is exposed to the incident electromagnetic energy. *See:* **electrobiology.** 322

partial carry (parallel addition). A technique in which some or all of the carries are stored temporarily instead of being allowed to propagate immediately. *See:* **carry.** 255, 77

partial correctness (software). In proof of correctness, a designation indicating that a program's output assertions follow logically from its input assertions and processing steps. *See:* **input assertion; output assertion; program; proof of correctness; total correctness.** 434

partial directivity (of an antenna for a given polarization). In a given direction, that part of the radiation intensity corresponding to a given polarization divided by the total radiation intensity averaged over all directions. *Note:* The (total) directivity of an antenna, in a specified direction, is the sum of the partial directivities for any two orthogonal polarizations. 111

partial discharge (1)(partial discharge measurement in liquid-filled power transformers and shunt reactors). An electric discharge that only partially bridges the insulation between conductors. *See:* **corona.** 580

(2) **(high voltage testing).** Discharges which do not completely bridge the insulation between electrodes. 150

(3) **(PD) (power and distribution transformer).** An electric discharge which only partially bridges the insulation between conductors, and which may or may not occur adjacent to a conductor. *Notes:* (A) Partial discharges occur when the local electric-field intensity exceeds the dielectric strength of the dielectric involved, resulting in local ionization and breakdown. Depending on intensity, partial discharges are often accompanied by emission of light, heat, sound, and radio influence voltage (with a wide frequency range). (B) The relative intensity of partial discharge can be observed at the transformer terminals by measurement of the apparent charge (coulombs). However, the apparent charge (terminal charge) should not be confused with the actual charge transferred across the discharging element in the dielectric which in most cases cannot be ascertained. Partial discharges tests using the radio influence voltage techniques which are responsive to the apparent terminal charges are generally used for measurement of relative discharge intensity. (C) Partial discharges can also be detected and located using sonic techniques. (D) "Corona" has also been used to describe partial discharges. This is a non-preferred term since it has other unrelated meanings. 53

(4) **(dielectric tests).** An electrical discharge that only partially bridges the insulation between conductors. *Note:* The term corona has also frequently been used with this connotation. It is recommended that such usage be discontinued in favor of the term partial discharge. 139

(5) **(power switchgear).** A localized electric discharge resulting from ionization in an insulation system when the voltage stress exceeds the critical value. This discharge partially bridges the insulation between electrodes. 103

partial discharge (corona) extinction voltage (cable

termination). The voltage at which partial discharge (corona) is no longer detectable on instrumentation adjusted to a specified sensitivity, following the application of a specified higher voltage. 4

partial discharge-free test voltage (partial discharge measurement in liquid-filled power transformers and shunt reactors). A specified voltage applied in accordance with a specified test procedure, at which the test object should not exhibit partial discharges above the acceptable energized background noise level. 580

partial effective area (of an antenna for a given polarization and direction). In a given direction, the ratio of the available power at the terminals of a receiving antenna to the power flux density of a plane wave incident on the antenna from that direction and with a specified polarization differing from the receiving polarization of the antenna. 111

partial failure. *See:* **failure, partial.**

partial-fraction expansion (circuits and systems). A sum of fractions that is used to represent a function which is a ratio of polynomials. The denominators of the fractions are the poles of the function. 67

partial gain (of an antenna for a given polarization). In a given direction, that part of the radiation intensity corresponding to a given polarization divided by the radiation intensity that would be obtained if the power accepted by the antenna were radiated isotropically. *Note:* The (total) gain of an antenna, in a specified direction, is the sum of the partial gains for any two orthogonal polarizations. 111

partially dead region or layer (of a semiconductor detector)(X-ray energy spectrometers). Any region or layer on or in the detector which contributes an output pulse which is less than the full energy peak for that incident radiation. 471

partially shielded insulated splice (power cable joint). An insulated splice in which a conducting material is employed over a portion of the insulation for electric stress control. 34

partial-read pulse. Any one of the currents applied that cause selection of a cell for reading. *See:* **coincident-current selection.** 331

partial reference designation (electric and electronics parts and equipments). A reference designation that consists of a basic reference designation and which may include, as prefixes, some but not all of the reference designations that apply to the subassemblies or assemblies within which the item is located. 17

partial-select output. (1) The voltage response of an unselected magnetic cell produced by the application of partial-read pulses or partial-write pulses. (2) The integrated voltage response of an unselected magnetic cell produced by the application of partial-read pulses or partial-write pulses. *See:* **coincident-current selection.** 331

partial system test. *See:* **stimulation, physical.**

partial-write pulse. Any one of the currents applied that cause selection of a cell for writing. *See:* **coincident-current selection.** 331

particle accelerator. Any device for accelerating charged particles to high energies, for example, cyclotron, betatron, Van der Graaff generator, linear accelerator, etcetera. 190

particle velocity (sound field). The velocity of a given infinitesimal part of the medium, with reference to the medium as a whole, due to the sound wave. *Note:* The terms **instantaneous particle velocity, effective particle velocity, maximum particle velocity,** and **peak particle velocity** have meanings that correspond with those of the related terms used for sound pressure. 176

parting (corrosion). The selective corrosion of one or more components of a solid solution alloy. 205

parting limit (corrosion). The maximum concentration of a more-noble component in an alloy, above which parting does not occur within a specific environment. 205

partitioning (software requirements specifications). Decomposition; the separation of the whole into its parts. 449

partition noise (electron device). Noise caused by random fluctuations in the distribution of current between the various electrodes. 190

partly cloudy sky (illuminating engineering). A sky which has 30 to 70 percent cloud cover. 167

part programming, computer (numerically controlled machines). The preparation of a manuscript in computer language and format required to accomplish a given task. The necessary calculations are to be performed by the computer. 224, 207

part programming, manual (numerically controlled machines). The preparation of a manuscript in machine control language and format required to accomplish a given task. The necessary calculations are to be performed manually. 224, 207

parts (replacement parts for Class 1E equipment in nuclear power generating stations). Items from which the equipment is assembled (for example, resistors, capacitors, wires, connectors, transistors, tubes, lubricants, O-rings, and springs). 582

part-winding starter (industrial control). A starter that applies voltage successively to the partial sections of the primary winding of an alternating-current motor. *See:* **starter.** 206

part-winding starting (rotating machinery). A method of starting a polyphase induction or synchronous motor, by which certain specially designed circuits of each phase of the primary winding are initially connected to the supply line. The remaining circuit or circuits of each phase are connected to the supply in parallel with initially connected circuits, at a predetermined point in the starting operation. 63

party line (1) (data transmission). A subscriber line arranged to serve more than one main station, with discriminatory ringing for each station. 59

(2) (telephone switching systems). A line arranged to serve more than one main station, with distinctive ringing for each station. 55

Paschen's law (gas). The law stating that, at a constant temperature, the breakdown voltage is a function only of the product of the gas pressure by the distance between parallel plane electrodes. *See:* **discharge (gas).**   244, 190

pass band (1) (data transmission). A range of frequency spectrum which can be passed at low attenuation. *See:* **bandpass filter.**   59

(2) (circuits and systems). A band of frequencies that pass through a filter with little loss (relative to other frequency bands such as a stop band).   67

pass-band ripple (circuits and systems). The difference between maxima and minima of loss in a filter pass-band. If the differences are of constant amplitude then the filter is said to be equiripple.   67

pass element (power supplies). A controlled variable-resistance device, either a vacuum tube or power transistor, in series with the source of direct-current power. The pass element is driven by the amplified error signal to increase its resistance when the output needs to be lowered or to decrease its resistance when the output must be raised. *See:* **series regulator.**   186

passenger elevator. An elevator used primarily to carry persons other than the operator and persons necessary for loading and unloading. *See:* **elevators.**   328

pass/fail criteria (software test documentation). Decision rules used to determine whether a software item or a software feature passes or fails a test.   436

passivation (corrosion). The process or processes (physical or chemical) by means of which a metal becomes passive.   205

passivator (corrosion). An inhibitor that changes the potential of a metal appreciably to a more cathodic or noble value.   205

passive-active cell (corrosion). A cell composed of passive and active areas. *See:* **electrolytic cell.**   205

passive electric network. An electric network containing no source of energy. *See:* **network analysis.**
                                                       328

passive homing guidance (navigation aid terms). Guidance in which a craft or missile is directed toward a destination by means of natural radiation from the destination.   526

passive limiter (nonlinear, active, and nonreciprocal waveguide components). A nonlinear device that suppresses input radio-frequency (rf) power without the aid of an external bias.   530

passive satellite (communication satellite). A communication satellite which is a reflector and performs no active signal processing.   83

passive station (data transmission). All stations on a multipoint network, other than the master and slave(s), which, during the information message transfer state, monitor the line for supervisory sequences, ending characters, etcetera.   59

passive test (test, measurement and diagnostic equipment). A test conducted upon an equipment or any part thereof when the equipment is not energized. *Syn:* **cold test.**   54

passive transducer. *See:* **transducer, passive.**

passivity (1) (chemical). The condition of a surface that retards a specified chemical reaction at that surface. *See:* **electrochemistry.**   328

(2) (electrolytic or anodic). Such a condition of an anode that the normal anodic reaction is retarded. *See:* **electrochemistry.**   328

paste (dry cell) (primary cell). A gelatinized layer containing electrolyte that lies adjacent to the negative electrode. *See:* **electrolytic cell.**   328

pasted sintered plate (alkaline storage battery). A plate consisting of fritted metal powder in which the active material is impregnated. *See:* **battery (primary or secondary).**   328

patch (1) (in general). To connect circuits together temporarily by means of a cord, known as a patch cord.   328

(2) (computing systems). To modify a routine in a rough or expedient way.   255, 77

(3) (software). (A) A modification to an object program made by replacing a portion of existing machine code with modified machine code. (B) To modify an object program without recompiling the source program. *See:* **code; modification; object program; source program.**   434

patch bay (analog computers). In an analog computer, a concentrated assembly of the inputs and outputs of computing elements, control elements, tie points, reference voltages, and ground points that offers a means of electrical connection.   9

patch board. *See:* **problem board.**

patchcord (test, measurement and diagnostic equipment). An interconnecting cable for plugging or patching between terminals: commonly employed on patchboard, plugboard, and in maintenance operations.   54

patch panel. *See:* **problem board; electronic analog computer.**

path (1)(navigation)(navigation aid terms). A line connecting a series of points in space and constituting a proposed or traveled route. *See:* **flight path; flight track; course line.**   526

(2) (network analysis). Any continuous succession of branches, traversed in the indicated branch directions.
                                                       282

(3) (data transmission). *See:* **channel.**   59

(4) (telephone switching systems). The set of links and junctors joined in series to establish a connection.

                                                       55

path analysis (software). Program analysis performed to identify all possible paths through a program, to detect incomplete paths, or to discover portions of the program that are not on any path. *See:* **program.**
                                                       434

path condition (software). A set of conditions that must be met in order for a particular program path to be executed. *See:* **program.**   434

path (loop) factor (network analysis). The graph determinant of that part of the graph not touching the specified path (loop). *Notes:* (1) A path (loop) factor is obtainable from the graph determinant by striking out all terms containing transmittance products of loops that touch that path (loop). (2) For loop L_k, the loop factor is

$$-\partial\Delta/\partial L_k$$

282

path length (1) (general). The length of a magnetic flux line in a core. *Note:* In a toroidal core with nearly equal inside and outside diameters, the value

$$l_m = \frac{\pi}{2}(\text{O.D.} + \text{I.D.})$$

where O.D. and I.D. are the outside and inside diameters of the core, is commonly used.   331
(2) (laser gyro). The length of the optical path traversed in a single pass by the laser beams.   46
pathocathode radiant sensitivity. *See:* **sensitivity, cathode radiant.**
path transmittance (network analysis). The product of the branch transmittances in that path.   282
patient care-related electrical appliance (health care facilities). An electrical appliance that is intended to be used for diagnostic, therapeutic or monitoring purposes in a patient care area.   192
patient equipment grounding point (health care facilities). A jack or terminal which serves as the collection point for redundant grounding of electric appliances serving a patient vicinity or for grounding other items in order to eliminate electromagnetic interference problems.   192
patient grounding point (National Electrical Code) (health care facilities). A jack or terminal bus which serves as the collection point for redundant grounding of electric appliances serving a patient vicinity.   256
patient lead (health care facilities). Any deliberate electrical connection which may carry current between an appliance and a patient. This may be a surface contact (for example, an electrocardiogram (ECG) electrode); an invasive connection (for example, implanted wire or catheter); or an incidental long-term connection (for example, conductive tubing). It is not intended to include adventitious or casual contacts such as pushbutton, bed surface, lamp, hand-held appliance, etcetera.   192
patient vicinity (National Electrical Code) (health care facilities). In an area which patients are normally cared for the patient vicinity is the space with surfaces likely to be contacted by the patient or an attendant who can touch the patient. This encloses a space within the room 6 ft. beyond the perimeter of the bed in its nominal location, and extending vertically 7 1/2 ft. above the floor.   256
patina (corrosion). A green coating consisting principally of basic sulfate and occasionally containing small amounts of carbonate or chloride, that forms on the surface of copper or copper alloys exposed to the atmosphere a long time.   205
patrol tours (nuclear security systems). An inspection by a member of the security organization along a predetermined route to observe the route area's security conditions.   464

patrol tour stations (nuclear security systems). Points along patrol tour routes where security force member progress is acknowledged.   464
pattern. *See:* **radiation pattern.**
pattern (test pattern language). The sequence of addresses and data used to test a semiconductor memory.   463
pattern, heating. *See:* **heating station.**
pattern-propagation factor (radar). Ratio of the strength that is actually present at a point in space to that which would have been present if free space propagation had occurred with the antenna beam directed toward the point in question. This factor is used in the radar equation to modify the strength of the transmitted or received signal to account for the effect of multipath propagation, diffraction, refraction, and pattern of an antenna.   13
pattern recognition. The identification of shapes, forms, or configurations by automatic means.
     255, 77
pattern select (FASTBUS acquisition and control). A broadcast address specifying that all devices seeing the broadcast remain attached to the master only if their timing (T) pins are asserted during the immediately ensuing write data cycle.   480
pattern-sensitive fault. A fault that appears in response to some particular pattern of data.   255, 77
PAX (telephony). *See:* **private automatic exchange.**
pay station. *See:* **public telephone station.**
P-band radar. Occasionally used to denote the 420 megahertz (MHz) to 450 MHz International Telecommunications Union assigned band, more generally described as the ultrahigh frequency (UHF) band.
     13
PBX. *See:* **private branch exchange.**
PBX trunk. *See:* **private-branch-exchange trunk.**
PCB. *See:* **polychlorinated biphenyl; askarel.**
p-channel device (metal-nitride-oxide field-effect transistor). Insulated-gate field-effect transformer (IGFET) where source and drain are regions of p-type conductivity.   386
PCS. *See:* **plastic clad silica.**
P.D. *See:* **control action, proportional plus derivative.**
P-display (radar). A type of radar display format, more commonly called plan position indication (PPI). *See:* **display.**   13
PDM. *See:* **pulse-duration modulation.**
PEAK. Channel number corresponding to the peak of a distribution. *See:* **crest.**   117
peak (crest) restriking voltage (surge arresters). The maximum instantaneous voltage that is attained by the re-striking voltage.   308, 62
peak alternating gap voltage (electron tube) (traveling-wave tubes). The negative of the line integral of the peak alternating electric field taken along a specified path across the gap. *Note:* The path of integration must be stated.   125
peak anode current. The maximum instantaneous value of the anode current. *See:* **electronic controller.**
     206
peak burst magnitude (audio and electroacoustics).

The maximum absolute peak value of voltage, current, or power for a burstlike excursion. *See:* The figure attached to the definition of **burst duration; burst (audio and electroacoustics).**   176

peak cathode current (steady-state). The maximum instantaneous value of a periodically recurring cathode current.   125

peak-charge characteristic (nonlinear capacitor). The function relating one-half the peak-to-peak value of transferred charge in the steady state to one-half the peak-to-peak value of a specified applied symmetrical alternating capacitor voltage. *Note:* Peak-charge characteristic is always single-valued. *See:* **nonlinear capacitor.**   191

peak current (low-voltage dc power circuit breakers used in enclosures). The instantaneous value of current at the time of its maximum value.   401

peak detector (overhead-power-line corona and radio noise). A detector, the output voltage of which approximates the true peak value of an applied signal or noise.   411

peak distortion (data transmission). The largest total distortion of telegraph signals noted during a period of observation.   59

peak electrode current (electron tube). The maximum instantaneous current that flows through an electrode. *See:* electrode current (electron tube).   190

peak flux density. The maximum flux density in a magnetic material in a specified cyclically magnetized condition.   331

peak forward anode voltage (electron tube). The maximum instantaneous anode voltage in the direction in which the tube is designed to pass current. *See:* **electrode voltage (electron tube); electronic controller.**   190

peak forward current rating (repetitive) (rectifier circuit element). The maximum repetitive instantaneous forward current permitted by the manufacturer under stated conditions. *See:* **average forward current rating (rectifier circuit element).**   208

peak forward voltage (of a rectifying element). The maximum instantaneous voltage between the anode and cathode during the positive nonconducting period. *See:* **rectification.**   328

peak induction (of toroidal magnetic amplifier cores). The magnetic induction corresponding to the peak applied magnetizing force specified. *Note:* It will usually be slightly less than the true saturation induction. *Syn:* **peak flux density.**   170

peaking circuit. A circuit capable of converting an input wave into a peaked waveform.   328

peaking network. A type of interstage coupling network in which an inductance is effectively in series (series peaking network) or in shunt (shunt peaking network) with the parasitic capacitance to increase the amplification at the upper end of the frequency range. *See:* **network analysis.**   328

peaking station (power operations). A generating station which is normally operated to provide power only during maximum load periods.   516

peaking time (1)(semiconductor radiation detectors). The time elapsed from the first zero crossing of the defined zero level to the departure from peak amplitude of a pulse equal to the maximum rated amplifier output.   23

(2) (tp)(of an amplifier output pulse)(germanium gamma-ray detectors)(charged-particle detectors. The time between the 1% amplitude point on the leading edge and the 100% amplitude point of a pulse (provided that the pulse does not have a flat top). For flat-top pulses, the peaking time is defined as the time between the 1% amplitude point and the midpoint of the flat top.   528, 119

peak inrush current (electronics power transformer). The peak instantaneous current value resulting from the excitation of the transformer with no connected load, and with essentially zero source impedance, and using the minimum turns primary tap and rated voltage.   95

peak instantaneous sound pressure level (measurement of sound pressure levels of ac power circuit breakers). Maximum unweighted positive or negative pressure peak value reached by an impulsive sound wave at any time during the period of observation. Unit: decibel (dB). For the purpose of ANSI/IEEE Std C37.082-1982, readings can be considered as peak instantaneous sound pressure level if the C-weighting is used and the response time of the instrument is 50 μs or less. Peak instantaneous sound pressure level is sometimes referred to as impact noise.   552

peak inverse anode voltage (electron tube). The maximum instantaneous anode voltage in the direction opposite to that in which the tube is designed to pass current. *See:* **electrode voltage (electron tube); electronic controller.**   190

peak inverse voltage (PIV) (semiconductor diode). The maximum instantaneous anode-to-cathode voltage in the reverse direction that is actually applied to the diode in an operating circuit. *Notes:* (A) This is an applications term not to be confused with **breakdown voltage,** which is a property of the device. (B) In semiconductor work the preferred term is **peak reverse voltage.** *See:* **semiconductor; peak reverse voltage.**   245

peak inverse voltage, maximum rated (semiconductor diode). The recommended maximum instantaneous anode-to-cathode voltage that may be applied in the reverse direction. *See:* **semiconductor.**   245

peak jitter. *See:* **jitter.**

peak let-through characteristic curve (of a current-limiting fuse). *See:* **current-limiting characteristic curve (of a current limiting fuse).**

peak let-through current (1)(current-limiting fuse) (peak cutoff current). The highest instantaneous current passed by the fuse during the interruption of the circuit.   103

(2)(protection and coordination of industrial and commercial power systems). The maximum instantaneous current through a current-limiting fuse during the total clearing time. Since this is an instantaneous value, it may well exceed the root-mean-square (rms) available current, but will be less than the peak current available without a fuse in the circuit if the fault level is high enough for it to operate in its current-limiting mode. *See:* **Figure under arcing time.** 504

peak let-through cut-off current (of a current-limiting fuse) (power switchgear). The highest instantaneous current passed by the fuse during the interruption of the circuit. 103, 443

peak limiter. A device that automatically limits the magnitude of a signal to a predetermined maximum value by changing its amplification. *Notes:* (1) The term is frequently applied to a device whose gain is quickly reduced and slowly restored when the instantaneous magnitude of the input exceeds a predetermined value. (2) In this context, the terms **instantaneous magnitude** and **instantaneous peak power** are used interchangeably. 178

peak load (1) (general). The maximum load consumed or produced by a unit or group of units in a stated period of time. It may be the maximum instantaneous load or the maximum average load over a designated interval of time. *Note:* Maximum average load is ordinarily used. In commercial transactions involving peak load (peak power) it is taken as the average load (power) during a time interval of specified duration occurring within a given period of time, that time interval being selected during which the average power is greatest. *See:* **generating station.** 64
(2) (motor) (rotating machinery). The largest momentary or short-time load expected to be delivered by a motor. It is expressed in percent of normal power or normal torque. *See:* **asynchronous machine.** 63

peak load station (electric power supply). A generating station that is normally operated to provide power during maximum load periods. *See:* **generating station.** 200

peak magnetizing force (1) (toroidal magnetic amplifier cores). The maximum value of applied magnetomotive force per mean length of path of the core. 165
(2) (peak field strength). The upper or lower limiting value of magnetizing force associated with a cyclically magnetized condition. 331

peak nominal varistor voltage (low voltage varistor surge arresters). Voltage across the varistor measured at a specified pulsed direct-current (dc) current of specific duration coincident with a specified alternating-current (ac) current crest. 62

peak overvoltages (for current-limiting fuses) (power switchgear). The peak value of the voltage that can exist across a current-limiting fuse during its arcing interval. 103, 443

peak percent curve. *See:* **integrated energy curve.**

peak point (tunnel-diode characteristic). The point on the forward current-voltage characteristic corresponding to the lowest positive (forward) voltage at which $dI/dV = 0$. 315, 191

peak-point current (tunnel-diode characteristic). The current at the peak point. *See:* **peak point (tunnel-diode characteristic).** 315, 191

peak-point voltage (tunnel-diode characteristic). The

voltage at which the peak point occurs. *See:* **peak point (tunnel-diode characteristic).** 315, 191

peak power, instantaneous. *See:* **instantaneous peak power.**

peak power output (modulated carrier system). The output power, averaged over a carrier cycle, at the maximum amplitude that can occur with any combination of signals to be transmitted. *See:* **radio transmitter; television.** 111

peak power pulse (waveguide). The root-mean-square (rms) value of rectangular pulse of radio frequency (RF) power passing through the transverse section of a waveguide. 267

peak pulse amplitude (television). The maximum absolute peak value of the pulse excluding those portions considered to be unwanted, such as spikes. *Note:* Where such exclusions are made, it is desirable that the amplitude chosen be illustrated pictorially. *See:* **pulse.** 59

peak pulse power, carrier-frequency. The power averaged over that carrier-frequency cycle that occurs at the maximum of the pulse of power (usually one half the maximum instantaneous power). 254

peak radiant responsivity (diode-type camera tube). The peak value of the spectral response of the tube usually specified together with the wavelength at which it occurs. Units: amperes watt^{-1} (AW^{-1}). 380

peak repetitive ON-state current (thyristor). The peak value of the ON-state current including all repetitive transient currents. *See:* **principal current.** 415, 66, 208, 191

peak responsibility. The load of a customer, a group of customers, or a part of the system at the time of occurrence of the system peak load. *See:* **generating station.** 64

peak reverse voltage (semiconductor rectifier). The maximum instantaneous value of the reverse voltage that occurs across a semiconductor rectifier device, or rectifier stack. *See:* **rectification.** 66

peak sound pressure (for any specified time interval). The maximum absolute value of the instantaneous sound pressure in that interval. *Note:* In the case of a periodic wave, if the time interval considered is a complete period, the peak sound pressure becomes identical with the maximum sound pressure. 176

peak switching current (rotating machinery). The maximum peak transient current attained following a switching operation on a machine. *See:* **asynchronous machine.** 63

peak torque (electric coupling). The maximum torque an electric coupling will transmit or develop for any speed relation on input and output members, with rated excitation and at specified operating conditions. 416

peak value (1) (alternating voltage). The maximum value excluding small high-frequency oscillations arising, for instance, from partial discharges in the circuit. *See:* **test voltage and current.** 307
(2) (crest value) (voltage or current) (surge arresters). The maximum value of an impulse. If there are

small oscillations superimposed at the peak, the peak value is defined by the maximum value and not the mean curve drawn through the oscillations.

308, 62

(3) (impulse current) (virtual peak value). Normally the maximum value. With some test circuits, overshoot or oscillations may be present on the current. The maximum value of the smooth curve drawn through the oscillations is defined as the virtual peak value. It will depend on the type of test whether the value of the test current shall be defined by the actual peak or a virtual peak value. *Note:* The term **peak value** is to be understood as including the term **virtual peak value** unless otherwise stated. *See:* **test voltage and current.**   307

(4) (impulse voltage) (virtual peak value). Normally the maximum value. With some test circuits, oscillations or overshoot may be present on the voltage. If the amplitude of the oscillations is not greater than 5 percent of the peak value and the frequency is at least 0.5 megahertz or, alternatively, if the amplitude of the Construction for derivation of virtual peak values overshoot is not greater than 5 percent of the peak value and the duration not longer than 1 microsecond, when for the purpose of measurement a mean curve may be drawn, the maximum amplitude of which is defined as the virtual peak value. (See the accompanying figure.) If the frequency is less than or if the duration is greater than described above, the peak of the oscillation may be used as the peak value. *Note:* The term **peak value** is to be understood as including the term **virtual peak value** unless otherwise stated. *See:* **test voltage and current.**   307

(5) (high voltage testing). The peak value of an alternating voltage is the maximum value, except that small high-frequency oscillations arising, for instance, from partial discharges are disregarded.   150

peak wavelength (1) (fiber optics). The wavelength at which the radiant intensity of a source is maximum. *See:* **spectral line; spectral width.**   433

(2) (λp) (light emitting diodes). The wavelength at which the spectral radiant intensity is a maximum.

162

peak working voltage (1)(pulse transformers). The maximum instantaneous voltage stress that may appear under operation across the insulation being considered, including abnormal and transient conditions.

589

(2) (charging inductors). The algebraic sum of the maximum alternating crest voltage and the direct voltage of the same polarity appearing between the terminals of the inductor winding or between the inductor winding and the grounded elements.   137

(3) (PWV) (corona measurement). The maximum instantaneous voltage that may appear under normal rated conditions across the insulation being considered. This insulation may be within a winding, between windings, or between windings and ground.

375

pedestal. A substantially flat-topped pulse that elevates the base level for another wave. *See:* **pulse.**   328

pedestal bearing (rotating machinery). A complete assembly of a bearing with its supporting pedestal.

63

pedestal bearing insulation (rotating machinery). The insulation applied either below the bearing liner shell and the adjacent pedestal support or between the base of the pedestal and the machine bed plate, to break the current path that may be formed through the shaft to the outboard bearing to the frame to the drive-end bearing and thence back to the shaft. *Note:* The volt-

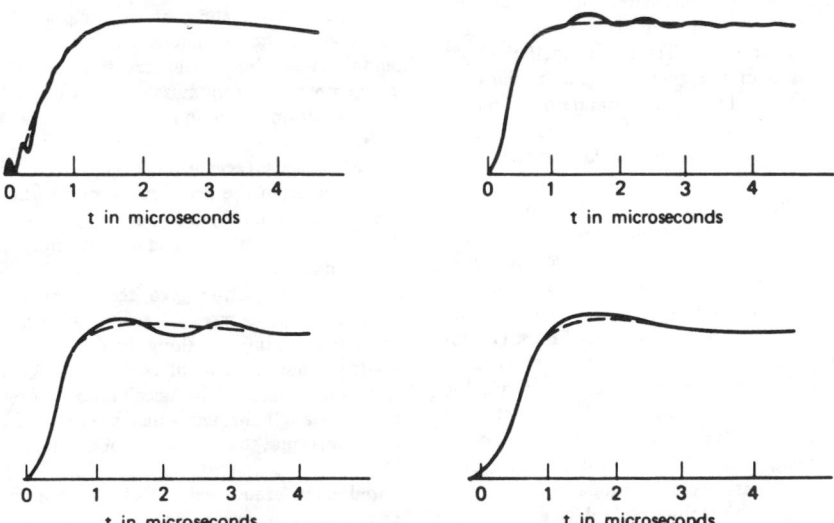

Construction for derivation of virtual peak values

age is usually very low. However, very destructive bearing currents can flow in this path if some insulating break is not provided. High-pressure moulded laminates are usually employed for this type of insulation.   63

pedestal delay time (amplitude, frequency, and pulse modulation). The time elapsed between the application of an electronic command signal to the electronic driver and the time the diffracted light reaches the 10 percent intensity point.   72

peeling. The unwanted detachment of a plated metal coating from the base metal. *See:* **electroplating.**
                                                                           328

peer protocol (logical link control). The sequence of message exchanges between two entities in the same layer that utilize the services of the underlying layers to effect the successful transfer of data or control information, or both, from one location to another location.   585

peg count (telephone switching systems). The notation of the number of occurrences of an event.   55

Peltier coefficient, absolute. The product of the absolute temperature and the absolute Seebeck coefficient of the material: the sign of the Peltier coefficient is the same as that of the Seebeck coefficient. *Note:* The opposite sign convention has also been used in the technical literature. *See:* **thermoelectric device.**
                                                                           191

Peltier coefficient of a couple. The quotient of (1) the rate of Peltier heat absorption by the junction of the two conductors by (2) the electric current through the junction: the Peltier coefficient is positive if Peltier heat is absorbed by the junction when the electric current flows from the second-named conductor to the first conductor. *Notes:* (A) The opposite sign convention has also been used in the technical literature. (B) The Peltier coefficient of a couple is the algebraic difference of either the relative or absolute Peltier coefficients of the two conductors. *See:* **thermoelectric device.**   191

Peltier coefficient, relative. The Peltier coefficient of a couple composed of the given material as the first-named conductor and a specified standard conductor. *Note:* Common standard conductors are platinum, lead, and copper. *See:* **thermoelectric device.**
                                                                           191

Peltier effect. The absorption or evolution of thermal energy, in addition to the Joule heat, at a junction through which an electric current flows: and in a non-homogeneous, isothermal conductor, the absorption or evolution of thermal energy, in addition to the Joule heat, produced by an electric current. *Notes:* (1) For the case of a nonhomogeneous, nonisothermal conductor, the Peltier effect cannot be separated from the Thomson effect. (2) A current through the junction of two dissimilar materials causes either an absorption or liberation of heat, depending on the sense of the current, and at a rate directly proportional to it to a first approximation. *See:* **thermoelectric device.**
                                                                       210, 191

Peltier heat. The thermal energy absorbed or evolved

as a result of the Peltier effect. *See:* **thermoelectric device.**   191

penalty factor (electric power system). A factor that, when multiplied by the incremental cost of power at a particular source, produces the incremental cost of delivered power from that source. Mathematically, it is

$$\frac{1}{(1 - \text{Incremental Transmission Loss})^*}$$

*Expressed as a decimal.   94

pencil beam (1) (radar). Antenna pattern having a narrow major lobe with approximately circular contours of equal radiation intensity.   13
(2) (antenna). A unidirectional antenna having a narrow major lobe with approximately circular contours of equal radiation intensity in the region of the major lobe.   111

pencil-beam antenna. An antenna whose radiation pattern consists of a single main lobe with narrow principal half-power beamwidths and side lobes having relatively low levels. *Note:* The main lobe usually has approximately elliptical contours of equal radiation intensity in the angular region around the peak of the main lobe. This type of pattern is diffraction limited in practice. It is often called a sum pattern in radar applications.   111

pendant. A device or equipment that is suspended from overhead either by means of the flexible cord carrying the current or otherwise.   328

pending master (FASTBUS acquisition and control). The master which participated in and won the most recent arbitration cycle. As a result it will assume bus mastership when the current master releases the bus.
                                                                           480

pendulosity (gyro, accelerometer)(inertial sensor). The product of the mass and the distance from the center of mass to the center of support or pivot measured along the pendulous axis.   46

pendulous accelerometer (inertial sensor). A device that employs a proof mass which is suspended to permit a rotation about an axis perpendicular to an input axis.   46

pendulous axis (accelerometer). A line through the mass center of the proof mass, perpendicular to and intersecting the output axis in pendulous devices. The positive direction is defined from the output axis to the proof mass.   46

pendulous integrating gyro accelerometer. A device using a single-degree-of-freedom gyro having an intentional pendulosity along the spin axis which is servo driven about the input axis at a rate which balances the torque induced by acceleration along the input axis. The angle through which the servoed axis rotates is proportional to the integral of applied acceleration.
                                                                           46

pendulous reference axis (PRA) (accelerometer). The direction of an axis as defined by the case mounting surfaces and.or external case markings. It is nominally parallel to the pendulous axis.   46

penetration, depth of. *See:* **depth of penetration.**

penetration frequency (radio wave propagation). Same as critical frequency. 146

pentode. A five-electrode electron tube containing an anode, a cathode, a control electrode, and two additional electrodes that are ordinarily grids. 125

pen travel. The length of the path described by the pen in moving from one end of the chart scale to the other. The path may be an arc or a straight line. *See:* **moving element (instrument).** 295

perceived chroma of an area of surface color (illuminating engineering). The attribute according to which it appears to exhibit more or less chromatic color judged in proportion to the brightness of a similarly illuminated area that appears to be white or highly transmitting. In a given set of viewing conditions, and at luminance levels that result in photopic vision, a stimulus of a given chromaticity and luminance factor exhibits approximately constant perceived chroma for all levels of illumination; but for a stimulus of a given chromaticity viewed at a given level of illumination, the perceived chroma generally increases if the luminance factor is increased. 167

perceived color (illuminating engineering). The proximal stimulus applied to the retina initiates color which is perceived as a substance occupying the space in front of the observer's eyes. Color may be perceived as self-luminous or as being reflected or transmitted light. It may be perceived as being confined to a point or line or arrayed as a surface or film or distributed in three dimensions as in the case of the perceived image of a patch of fog. A perceived image may be perceived as composed of volume color as in the case of fog or as covered by surface color as in the case of a piece of chalk. In the case of the sky or a patch of color seen through an aperture where it cannot be identified as belonging to a specific object, it is called aperture color and judgement is suspended as to whether the color is self-luminous or perceived by reflected or transmitted light. The color of a point source of light may be perceived and described as such. This is a special case of a self-luminous color. 167

perceived light-source color (illuminating engineering). The color perceived to belong to a light source. 167

perceived object color. The color perceived to belong to an object, resulting from characteristics of the object, of the incident light, and of the surround, the viewing direction, and observer adaptation. *See:* **color.** 167

percentage differential relay (power switchgear). A differential relay in which the designed response to the phasor difference between incoming and outgoing electrical quantities is modified by a restraining action of one or more of the input quantities. *Note:* The relay operates when the magnitude of the phasor difference exceeds the specified percentage of one or more of the input quantities. 103

percentage error (watthour meter). The difference between its percentage registration and 100 percent. A meter whose percentage registration is 95 percent is said to be 5 percent slow, or its error is −5 percent. A meter whose percentage registration is 105 percent is 5 percent fast, or its error is +5 percent. 212

percentage immediate appreciation (telephone transmission system). The percentage of the total number of spoken sentences that are immediately understood without conscious deductive effort when each sentence conveys a simple and easily understandable idea. *See:* **volume equivalent.** 328

percentage modulation. (1) In angle modulation, the fraction of a specified reference modulation, expressed in percent. (2) In amplitude modulation, the modulation factor expressed in percent. *Note:* It is sometimes convenient to express percentage modulation in decibels below 100-percent modulation. *See:* **radio transmission.** 211

percentage modulation, effective (single, sinusoidal input component). The ratio of the peak value of the fundamental component of the envelope to the direct-current component in the modulated conditions, expressed in percent. *Note:* It is sometimes convenient to express percentage modulation in decibels below 100-percent modulation. 111

percentage registration (watthour meter) (accuracy) (percentage accuracy). The ratio of the actual registration of the meter to the true value of the quantity measured in a given time, expressed as a percentage. *See:* **electricity meter (meter).** 212

percent articulation. *See:* **articulation.**

percent energy resolution. *See:* **energy resolution (percent).**

percent flutter (reproduced tone) (sound recording and reproducing). The root-mean-square deviation from the average frequency, expressed as a percentage of average frequency. *See:* **sound recording and reproducing.** 145

percent harmonic distortion. *See:* **distortion, percent harmonic.**

percent impairment of hearing (percent hearing loss). An estimate of a person's ability to hear correctly. It is usually based, by means of an arbitrary rule, on the pure-tone audiogram. The specific rule for calculating this quantity from the audiogram now varies from state to state according to a rule or law. *Note:* The term disability of hearing is sometimes used for impairment of hearing. Impairment refers specifically to a person's illness or injury that affects his personal efficiency in the activities of daily living. Disability has the additional medicolegal connotation that an impairment reduces a person's ability to engage in gainful activity. Impairment is only a contributing factor to the disability. 176

percent impedance (rectifier transformer). The percent of rated alternating-current winding voltage required to circulate current equivalent to rated line kilovolt-amperes in the alternating-current winding with all direct-current winding terminals short-circuited. *See:* **rectifier transformer.** 258

percent intelligibility. *See:* **articulation.**

percent-make-and-break (telephone switching systems). The proportions of a dial pulse cycle during which the circuit is closed (make) and opened (break) respectively.   55

percent pulse waveform distortion (pulse terms). Pulse waveform distortion expressed as a percentage of, unless otherwise specified, the pulse amplitude of the reference pulse waveform.   254

percent pulse waveform feature distortion (pulse terms). Pulse waveform feature distortion expressed as a percentage of, unless otherwise specified, the pulse amplitude of the reference pulse waveform.

254

percent ratio (1) (instrument transformer). The true ratio expressed in percent of the marked ratio.

394

(2) (power and distribution transformer). The true ratio expressed in percent of the marked ratio.

53

percent ratio correction (instrument transformer). The difference between the ratio correction factor and unity, expressed in percent. [(RCF-1) X 100].

203

percent ripple (1)(power system communication equipment). The ratio of the effective (root-mean-square) value of the ripple voltage or current to the average value of the total (direct current) voltage or current, expressed in percent.   453

(2) (electrical conversion). The percent ripple voltage is defined as the ratio of the root-mean-square(RMS) value of the voltage pulsations (E_{max} to E_{min}) to the average value of the total voltage.

$$\text{Percent Ripple} = \frac{\text{RMS Ripple}}{E_{\text{nominal}}}(100\%)$$

Note: In most applications the definition has been revised to simplify the calculations by defining percent ripple as the ratio of the root-mean-square (RMS) value of the voltage pulsations to the nominal no-load output voltage of the converter E_{nominal}

$$\text{Percent Ripple} = \frac{\text{RMS Ripple}}{E_{\text{av}}}(100\%)$$

See: **electrical conversion.**   186

percent ripple voltage or current. *See:* **percent ripple.**

percent steady-state deviation (control). The difference between the ideal value and the final value, expressed as a percentage of the maximum rated value of the directly controlled variable (or another variable if specified). *See:* **control system, feedback.**   206

percent syllable articulation. *See:* **syllable articulation.**

percent system deviation (control). At any given point on the time response, the difference between the ideal value and the instantaneous value, expressed as a percentage of the maximum rated value of the directly controlled variable (or another variable if specified). *See:* **deviation (control); control system, feedback.**

206

percent total flutter (sound recording and reproducing). The value of flutter indicated by an instrument that responds uniformly to flutter of all rates from 0.5 up to 200 hertz. *Note:* Except for the most critical tests, instruments that respond uniformly to flutter of all rates up to 120 hertz are adequate, and their indications may be accepted as showing percent total flutter. *See:* **sound recording and reproducing.**   145

percent transformer correction (instrument transformer). Difference between the transformer correction factor and unity expressed in percent. *Note:* The percent transformer correction-factor error is positive if the transformer correction factor is greater than unity. If the percent transformer correction-factor error is positive, the measured watts or watthours will be less than the true value.   203

percent transformer correction-factor error. Difference between the transformer correction factor and unity expressed in percent. *Note:* The percent transformer correction-factor error is positive if the transformer correction factor is greater than unity. If the percent transformer correction-factor error is positive, the measured watts or watthours will be less than the true value.   305

percent transient deviation (control). The difference between the instantaneous value and the final value, expressed as a percentage of the maximum rated value of the directly controlled variable (or another variable if specified). *See:* **control system, feedback.**   206

percent unbalance of phase voltages (electrical conversion). The ratio of the maximum deviation of a phase voltage from the average of the total phases to the average of the phase voltages, expressed in percent.

186

$$\% \text{ Unbalance} = \frac{\text{RMS Phase Voltage} - \text{RMS Average Phase Voltages}}{\text{RMS Average Phase Voltage}} \times 100\%$$

perfect dielectric (ideal dielectric). A dielectric in which all of the energy required to establish an electric field in the dielectric is recoverable when the field or impressed voltage is removed. Therefore, a perfect dielectric has zero conductivity and all absorption phenomena are absent. A complete vacuum is the only known perfect dielectric.   210

perfect diffusion (illuminating engineering). That in which flux is uniformly scattered in accordance with Lambert's cosine law. *See:* **Lambert's cosine law.**

167

perfective maintenance (software). Maintenance performed to improve performance, maintainability, or other software attributes. *See:* **adaptive maintenance; corrective maintenance; maintainability; maintenance; performance; software.** 434

perfect transformer. *See:* **ideal transformer.**

perforated tape. Tape in which a code hole(s) and a tape-feed hole have been punched in a row. 207

perforator (1) (telegraph practice). A device for punching code signals in paper tape for application to a tape transmitter. *Note:* A perforating device that is automatically controlled by incoming signals is called a reperforator. *See:* **telegraphy.** 255, 77, 194

(2) (test, measurement and diagnostic equipment). A device for punching digital information into tape for application to a tape transmitter or tape reader. Sometimes called tape punch. 54

performance (software). (1) The ability of a computer system or subsystem to perform its functions. (2) A measure of a computer system or subsystem to perform its functions, for example, response time, throughput, number of transactions. *See:* **computer system; function; performance requirement; subsystem; system; throughput.** 434

performance characteristic (device) (1) (power switchgear). An operating characteristic, the limit or limits of which are given in the design test specifications. 103, 202

(2) (transformer). Those characteristics (such as impedance, losses, dielectric test levels, temperature rise, sound level, etcetera) which describe the performance of the equipment under specified conditions of operation. 53

performance chart (magnetron oscillators). A plot on coordinates of applied anode voltage and current showing contours of constant magnetic field, power output, and over-all efficiency. *See:* **magnetron.** 125

performance evaluation (software). The technical assessment of a system or system component to determine how effectively operating objectives have been achieved. *See:* **component; system.** 434

performance index (excitation control systems). A scalar measure of the quality of system behavior. It is frequently a function of system output and control input over some specified time interval and.or frequency range. A quadratic performance index is a quadratic function of system states and this form finds wide applications to linear systems. 353

performance monitor (test, measurement and diagnostic equipment). A device which continuously or periodically scans a selected number of test points to determine if the unit is operating within specified limits. The device may include provisions for insertion of stimuli 54

performance requirement (software). A requirement that specifies a performance characteristic that a system or system component must possess, for example, speed, accuracy, frequency. *See:* **accuracy; component; performance; requirement; system.** 434

performance specification (software). (1) A specification that sets forth the performance requirements for a system or system component. (2) Synonymous with 'requirements specification'. (U.S.Navy usage). *See:* **component; functional specification; performance requirement; requirements specification; specification; system.** 434

performance test (lead storage batteries). A constant current capacity test made on a battery normally in the 'as found' condition, after being in service, to detect any change in the capacity determined by the acceptance test. 38

performance tests (rotating machinery). The tests required to determine the characteristics of a machine and to determine whether the machine complies with its specified performance. *See:* **asynchronous machine.** 63

performance verification (test, measurement and diagnostic equipment). A short, precise check to verify that the unit under test is operational and performing its intended function. 54

periapsis (communication satellite). The least distant point from the center of a primary body (or planet) to an orbit around it. 74

perigee (navigation aid terms). That orbital point nearest the earth when the earth is the center of attraction. 526

perimeter lights (illuminating engineering). Aeronautical ground lights provided to indicate the perimeter of a landing pad for helicopters. 167

period (1) (large lead storage batteries). An interval of time in the battery duty cycle during which the current is assumed to be constant for purposes of cell sizing calculations. 377

(2) (pulse terms). The absolute value of the minimum interval after which the same characteristics of a periodic waveform or a periodic feature recur. 254

period, critical hydro (electric power supply). Period when the limitations of hydroelectric energy supply due to water conditions are most critical with respect to system load requirements. 112

period hours (PH)(electric generating unit reliability, availability, and productivity). The number of hours a unit was in the active state. 567

periodically sampled equivalent time format (pulse measurement). A format which is identical to the periodically sampled real time format, below, except that the time coordinate is equivalent to and convertible to real time. Typically, each datum point is derived from a different measurement on a different wave in a sequence of waves. *See:* **sampled format.** 15

periodically sampled real time format (pulse measurement). A finite sequence of magnitudes m_0, m_1, m_2, ..., m_n each of which represents the magnitude of the wave at times t_0, $t_0 + \Delta t$, $t_0 + 2\Delta t$, ..., $t_0 + n\Delta t$, respectively, wherein ... the data may exist in a pictorial format or as a list of numbers. *See:* **sampled format.** 15

periodic-automatic-reclosing equipment (power switchgear). An equipment that provides for automatically reclosing a circuit-switching device a specified number of times at specified intervals between reclosures. *Note:* This type of automatic reclosing equipment is generally used for alternating-current circuits.   103

periodic check (test, measurement and diagnostic equipment). A test or series of tests performed at designated intervals to determine if all elements of the unit under test are operating within their designated limits.   54

periodic damping. *See:* **underdamping.**

periodic duty (National Electrical Code) (packaging machinery) (power and distribution transformer). Intermittent operation in which the load conditions are regularly recurrent.   256, 53, 429

periodic electromagnetic wave (radio wave propagation). A wave in which the electric field vector is repeated in detail in either of two ways: (1) at a fixed point, after the lapse of a time known as the period, or (2) at a fixed time, after the addition of a distance known as the wavelength.   146

periodic frequency modulation (converters having ac output)(self-commutated converters). The periodic variation of the output frequency from its rated value.   584

periodic function. A function that satisfies $f(x) = f(x + nk)$ for all x and for all integers n, k being a constant. For example,

$$\sin (x + a) = \sin (x + a + 2n\pi).$$

210

periodic line (transmission lines). A line consisting of successive identically similar sections, similarly oriented, the electrical properties of each section not being uniform throughout. *Note:* The periodicity is in space and not in time. An example of a periodic line is the loaded line with loading coils uniformly spaced. *See:* **transmission line.**   210

periodic output voltage modulation (converters having ac output)(self-commutated converters). The periodic variation of output voltage amplitude at frequencies less than the fundamental output frequency.   584

periodic permanent-magnet focusing (PPM) (microwave tubes). Magnetic focusing derived from a periodic array of permanent magnets. *See:* **magnetron.**   190

periodic pulse train (automatic control). A pulse train made up of identical groups of pulses, the groups repeating at regular intervals.   56

periodic rating (1) (electric power sources). The load that can be carried for the alternate periods of load and rest specified in the rating, the apparatus starting at approximately room temperature, and for the total time specified in the rating, without causing any of the specified limitations to be exceeded. *See:* **asynchronous machine.**   206
(2) (relay). A rating that defines the current or voltage that may be sustained by the relay during intermittent periods of energization as specified, starting cold and operating for the total time specified without causing

any of the prescribed limitations to be exceeded.   103

periodic slow-wave circuit (microwave tubes). A circuit whose structure is periodically recurring in the direction of propagation. *See:* **microwave (tube or valve).**   190

periodic tests (periodic testing). (nuclear power generating stations). Tests performed at scheduled intervals to detect failures and verify operability.   102, 109, 355

periodic wave. A wave in which the displacement at each point of the medium is a periodic function of the time. Periodic waves are classified in the same manner as periodic quantities.   210

periodic waveguide. A waveguide in which propagation is obtained by periodically arranged discontinuities or periodic modulations of the material boundaries.   267

peripheral air-gap leakage flux (rotating machinery). The component of air-gap magnetic flux emanating from the rotor or stator, that flows from pole to pole without entering the radially opposite surface of the air gap. *See:* **rotor (rotating machinery); stator.**   63

peripheral equipment (test, measurement and diagnostic equipment). Equipment external to a basic unit. A tape unit, for example, is peripheral equipment to a computer.   54

peripheral vision (illuminating engineering). The seeing of objects displaced from the primary line of sight and outside the central visual field.   167

peripheral visual field (illuminating engineering). That portion of the visual field which falls outside the region corresponding to the foveal portion of the retina.   167

periscope (navigation aid terms). An optical instrument which displaces the line of sight parallel to permit a view which otherwise may be obstructed itself.   526

periscope antenna. An antenna consisting of a very directive feed located close to ground level and oriented so that its beam illuminates an elevated reflector which is oriented so as to produce a horizontal beam.   111

periscopic sextant (navigation aid terms). A sextant designed to be mounted inside a vehicle, with a tube extending vertically upward through the skin of the vehicle.   526

permanent connection (substation grounding). A grounding connector that will retain its electrical and mechanical integrity for the design life of the conductor within limits established by IEEE Std 837-1984.   475

permanent echo (radar). A signal reflected from an object fixed with respect to the radar site.   13

permanent fault (surge arresters). A fault that can be cleared only by action taken at the point of fault.   244, 62

permanent-field synchronous motor. A synchronous motor similar in construction to an induction motor in which the member carrying the secondary laminations and windings carries also permanent-magnet field poles that are shielded from the alternating flux by the laminations. It starts as an induction motor but operates normally at synchronous speed. *See:* **permanent-magnet synchronous motor.** 63

permanently grounded device. A grounding device designed to be permanently connected to ground, either solidly or through current transformers and/or another grounding device. 91

permanently installed decorative fountains and reflection pools (National Electrical Code). Those that are constructed in the ground, on the ground, or in a building in such a manner that the pool cannot be readily disassembled for storage and are served by electrical circuits of any nature. These units are primarily constructed for their aesthetic value and not intended for swimming or wading. 256

permanently installed swimming, wading and therapeutic pools (National Electrical Code). Those that are constructed in the ground, on the ground, or in a building in such a manner that the pool cannot be readily disassembled for storage, whether or not served by electrical circuits of any nature. 256

permanent magnet (PM). A ferromagnetic body that maintains a magnetic field without the aid of external electric current. 244

permanent-magnet erasing head (electroacoustics). A head that uses the fields of one or more permanent magnets for erasing. *See:* **phonograph pickup.** 176

permanent-magnet focusing (microwave tubes). Magnetic focusing derived from the use of a permanent magnet. *See:* **magnetrons.** 190

permanent-magnet generator (magneto). A generator in which the open-circuit magnetic flux field is provided by one or more permanent magnets. 63

permanent-magnet loudspeaker. A moving-conductor loudspeaker in which the steady field is produced by means of a permanent magnet.
See: **loudspeaker.** 328

permanent-magnet moving-coil instrument (d'Arsonval instrument). An instrument that depends for its operation on the reaction between the current in a movable coil or coils and the field of a fixed permanent magnet. *See:* **instrument.** 328

permanent-magnet moving-iron instrument (polarized-vane instrument). An instrument that depends for its operation on the action of an iron vane in aligning itself in the resultant magnetic field produced by a permanent magnet and the current in an adjacent coil of the instrument. *See:* **instrument.** 328

permanent-magnet, second-harmonic, self-synchronous system. A remote-indicating arrangement consisting of a transmitter unit and one or more receiver units. All units have permanent-magnet rotors and toroidal stators using saturable ferromagnetic cores and excited with alternating current from a common external source. The coils are tapped at three or more equally spaced intervals, and the corresponding taps are connected together to transmit voltages that consist principally of the second harmonic of the excitation voltage. The rotors of the receiver units will assume the same angular position as that of the transmitter rotor. *See:* **synchro system.** 328

permanent-magnet synchronous motor (rotating machinery). A synchronous motor in which the field system consists of one or more permanent magnets. *See:* **permanent-field synchronous motor.** 63

permanent signal (telephone switching systems). A sustained off-hook supervisory signal originating outside a switching system. 55

permanent-signal alarm, (telephone switching systems). An alarm resulting from the simultaneous accumulation of a predetermined number of permanent signals. 55

permanent-signal tone, (telephone switching systems). A tone that indicates to an operator or other employee that a line is in a permanent-signal state. 55

permanent-split capacitor motor. A capacitor motor with the same value of effective capacitance for both starting and running operations. *See:* **asynchronous machine.** 63

permanent storage (computing systems). *See:* **fixed storage.**

permeability (1)(general). A general term used to express various relationships between magnetic induction and magnetizing force. These relationships are either (A) absolute permeability, that in general is the quotient of a change in magnetic induction divided by the corresponding change in magnetizing force: or (B) specific (relative) permeability, which is the ratio of the absolute permeability to the magnetic constant. *Notes:* (1) Relative permeability is a pure number that is the same in all unit systems: the value and dimension of absolute permeability depend upon the system of units employed. (2) In anisotropic media, permeability becomes a matrix. 210

(2)(mu)(electrical heating systems). Ratio of the magnetic flux density to the corresponding magnetizing force. 476

permeability of free space. *See:* **magnetic constant permeameter.** An apparatus for determining corresponding values of magnetizing force and flux density in a test specimen. From such values of magnetizing force and flux density, normal induction curves or hysteresis loops can be plotted and magnetic permeability can be computed. *See:* **magnetic constant.** 328

permeameter. An apparatus for determining corresponding values of magnetizing force and flux density in a test specimen. From such values of magnetizing force and flux density, normal induction curves or hysteresis loops can be plotted and magnetic permeability can be computed. *See:* **magnetometer.** 328

permeance. The reciprocal of reluctance. 210

permissible mine equipment. Equipment that complies with the requirements of and is formally approved by the United States Bureau of Mines after having passed the inspections and the explosion and/or other tests specified by that Bureau. *Note:* All equipment so approved must carry the official approval plate required as identification for permissible equipment. 328

permissible response rate (steam generating unit). The maximum assigned rate of change in generation for load-control purposes based on estimated and known limitations in the turbine, boiler, combustion control, or auxiliary equipment. The permissible response rate for a hydro-generating unit is the maximum assigned rate of change in generation for load-control purposes based on estimated and known limitations of the water column, associated piping, turbine, or auxiliary equipment. *See:* **speed-governing system.** 94

permissive (1) (as applied to a relay system) (power switchgear). A general term indicating that functional cooperation of two or more relays is required before control action can become effective. 103
(2) (relay system). A general term indicating that functional cooperation of two or more relays is required before control action can become effective. 127

permissive block. A block in manual or controlled manual territory, governed by the principle that a train other than a passenger train may be permitted to follow a train other than a passenger train in the block. *See:* **block signal system; controlled manual block signal system.** 328

permissive control (electric power systems). A control mode in which generating units are allowed to be controlled only when the change will be in the direction to reduce area-control error. *See:* **speed-governing system.** 94

permissive control device (69) (power system device function numbers). Generally a two-position device that in one position permits the closing of a circuit breaker, or the placing of an equipment into operation, and in the other position prevents the circuit breaker or the equipment from being operated. 402

permittivity (small-signal, ferroelectric material)(primary ferroelectric terms). The incremental change in electric displacement per unit electric field when the magnitude of the measuring field is very small compared to the coercive electric field. The small signal relative permittivity, κ, is equal to the ratio of the absolute permittivity ϵ to the permittivity of free space ϵ_0, that is $\kappa = \epsilon/\epsilon_0$. Macroscopically, ϵ is found by measuring the capacitance. The units of permittivity are coulombs/volt-meter or farads/meter. In a ferroelectric, the measuring field or voltage must be sufficiently small in order to prevent ferroelectric domain reorientation from contributing to the permittivity. *Note:* The value of the small-signal permittivity may depend on the remanent polarization, electric field, mechanical stress, sample history, or frequency of the measuring field. (Measurements are usually made at a frequency of 1 kHz or higher.) 497
permittivity. *See* **absolute capacitivity.**
permittivity in physical media (antennas). The real part of the complex permittivity. *See:* **complex dielectric constant.** 111
permittivity of free space. *See:* **electric constant.**
perpendicular magnetization (magnetic recording). Magnetization of the recording medium in a direction perpendicular to the line of travel and parallel to the

smallest cross-sectional dimension of the medium. *Note:* In this type of magnetization, either single pole-piece or double pole-piece magnetic heads may be used. *See:* **phonograph pickup.** 176
perpendicular polarization (radio wave propagation). Of a propagating wave for which the electric field vector lies perpendicular to the plane of incidence. *Note:* Sometimes called horizontal polarization. 146
persistence (oscilloscopes). The decaying luminosity of the luminescent screen 1phosphor.screen0 after the stimulus has been reduced or removed. *See:* **phosphor decay.** 106
persistence characteristic (1) (camera tubes). The temporal step response of a camera tube to illumination. *See:* **methods of measurement.** 125
(2) (decay characteristic) (luminescent screen). A relation, usually shown by a graph, between luminance (or emitted radiant power) and time after excitation is removed. 125
persistent-cause forced outage (electric power systems). A component outage whose cause is not immediately self-clearing but must be corrected by eliminating the hazard or by repairing or replacing the affected component before it can be returned to service. *Note:* An example of a persistent-cause forced outage is a lightning flashover that shatters an insulator thereby disabling the component until repair or replacement can be made. *See:* **outage.** 200
persistent-cause forced-outage duration (electric power systems). The period from the initiation of a persistent-cause forced outage until the affected component is replaced or repaired and made available to perform its intended function. *See:* **outage.** 200
persistent current (superconducting material). A magnetically induced current that flows undiminished in a superconducting material or circuit. *See:* **superconductivity.** 191
persistent-image device. An optoelectronic amplifier capable of retaining a radiation image for a length of time determined by the characteristics of the device. *See:* **optoelectronic device.** 191
persistent-image panel (optoelectronic device). A thin, usually flat, multicell persistent-image device. *See:* **optoelectronic device.** 191
personal computer (computer applications). A single-user microcomputer designed for personally controllable applications. *See:* **desktop computer; laptop computer; home computer.** 571
personal computing (computer applications). (1) Computing performed using a personal computer. (2) Computing performed in an environment in which the user has complete control over the data and access to software with which the data may be manipulated. 571
personal ground (conductor stringing equipment). A portable device designed to connect (bond) a deenergized conductor or piece of equipment, or both, to an electrical ground. Distinguished from a master ground in that it is utilized at the immediate site when work is to be performed on a conductor or piece of equip-

ment which could accidentally become energized. *See:* **ground stick; red head; working head.** 431

person-to-person call, (telephone switching systems). A call intended for a designated person. 55

perturbed electric or magnetic field (electric and magnetic fields from ac power lines). (1) weakly perturbed field. The field at a point will be regarded as weakly perturbed if the magnitude does not change by more than 5 percent or the direction does not vary by more than 5 degrees, or both, when an object is introduced into the vicinity. The electric field at the surface of the object is in general strongly perturbed (see (3) below) by the presence of the object. At power frequencies the magnetic field is not in general perturbed by the presence of objects which are free of magnetic materials. Exceptions to this are regions near the surface of nonmagnetic electric conductors which develop eddy currents because of the B-field time variation. (2) moderately perturbed field. The field at a point will be regarded as moderately perturbed if the magnitude varies between 5 percent and 30 percent or the direction varies between 5 degrees and 30 degrees, or both, when an object is introduced into the vicinity. (3) strongly perturbed field. The field at a point will be regarded as strongly perturbed if the magnitude varies in excess of 30 percent or the direction varies in excess of 30 degrees, or both, when an object is introduced into the vicinity. 407

perturbed field (measurement of power frequency electric and magnetic fields from ac power lines). A field that is changed in magnitude or direction ,or both, by the introduction of an object. *Note:* The electric field at the surface of the object is, in general, strongly perturbed by the presence of the object. At power frequencies the magnetic field is not, in general, greatly perturbed by the presence of objects that are free of magnetic materials. Exceptions to this are regions near the surface of thick electric conductors where eddy currents alter time-varying magnetic fields. 514

per unit (PU) (power fault effects). The reference unit, established as a calculating convenience, for expressing all power system electrical parameters on a common reference base. One per unit (PU) is 100 percent of the base chosen. The PU system in power systems engineering is used to obtain a better comparison of the performance of the power system elements of different ratings, similar to the decibel system used for equating the losses and levels of different telecommunications systems. 404

per-unit quantity (rotating machinery). The ratio of the actual value of a quantity to the base value of the same quantity. The base value is always a magnitude, or in mathematical terms, a positive, real number. The actual value of the quantity in question (current, voltage, power, torque, frequency, etcetera) can be of any kind: root-mean-square, instantaneous, phasor, complex, vector, envelope, etcetera. *Note:* The base values, though arbitrary, are usually related to characteristic values, for example, in case of a machine, the base power is usually chosen to be the rated power (active

or apparent), the base voltage to be the rated root-mean-square voltage, the base frequency, the rated frequency. Despite the fact that the choice of base values is rather arbitrary, it is of advantage to choose base values in a consistent manner. The use of a consistent per-unit system becomes a practical necessity when a complicated system is analyzed. *See:* **asynchronous machine.** 63

per-unit resistance. The measured watts expressed in per-unit on the base of the rated kilovolt-amperes of the teaser winding. *See:* **efficiency.** 328

per-unit system (rotating machinery). The system of base values chosen in a consistent manner to facilitate analysis of a device or system, when per-unit quantities are used. Its importance becomes paramount when analog facilities (network analyzer, analog and hybrid computers) are utilized. *Note:* In electric network analysis and electromechanical system studies, usually four independent fundamental base values are chosen. The rest of the base values are derived from the fundamental ones. In most cases power, voltage, frequency, and time are chosen as fundamental base values. The base power must be the same for all types: apparent, active, reactive, instantaneous. The base time is usually 1 second. From the above, all other base values can be found, for example, base power times base time equals base energy, etcetera. The per-unit system can cover extensive networks because the base voltages of network sections connected by transformers can differ, in which case an **ideal per-unit transformer** is usually introduced having a turns ratio equal to the quotient of the effective turns ratio of the actual transformer and the ratio of base voltage values. By keeping the power, frequency, and time bases the same, only those base quantities will differ for different network sections that are directly or indirectly related to voltage (for example, current, impedance, reactance, inductance, capacitance, etcetera) but those related to power, frequency, and time only (for example, energy, torque, etcetera) will remain unchanged. *See:* **asynchronous machine.** 63

per-unit value (ac rotating machinery)(basic per-unit quantities for ac rotating machines). The actual value divided by the value of the base quantity when both actual and base values are expressed in the same units. 517

perveance. The quotient of the space-charge-limited cathode current by the three-halves power of the anode voltage in a diode. *Note:* Perveance is the constant G appearing in the Child-Langmuir-Schottky equation

$$i_k = Ge_b{}^{3/2}.$$

When the term perveance is applied to triode or multigrid tube, the anode voltage e_b is replaced by the composite controlling voltage e of the equivalent diode. 190

Petri net (software). An abstract, formal model of information flow, showing static and dynamic properties of a system. A Petri net is usually represented as a graph having two types of nodes (called places and transitions) connected by arcs, and markings (called tokens) indicating dynamic properties. *See:* **graph; model; node; state diagram; system.** 434

PFM telemetry. *See:* pulse frequency modulation (PFM) telemetry.

p gate thyristor (cathode date SCR). A thyristor in which the gate terminal is connected to the *p* region adjacent to the region to which the cathode terminal is connected and that is normally switched to the ON state by applying a positive signal between gate and cathode terminals. 445

pH (of a solution A). The pH is obtained from the measurements of the potentials *E* of a galvanic cell of the form H_2: solution A: saturated potassium chloride (KCl); reference electrode with the aid of the equation

$$\text{pH} = \frac{E - E_0}{(RT/F) \ln 10} = \frac{E - E_0}{2.303\ RT/F}$$

in which E_0 is a constant depending upon the nature of the reference electrode, *R* is the gas constant in joules per mole per degree, *T* is the absolute temperature in kelvins, and *F* is the Faraday constant in coulombs per gram equivalent. Historically pH was defined by

$$\text{pH} = \log \frac{1}{[H^+]}$$

in which [H+] is the hydrogen ion concentration. According to present knowledge there is no simple relation between hydrogen ion concentration or activity and pH. *See:* **ion; ion activity.** Values of pH may be regarded as a convenient scale of acidities. 328

phanotron. A hot-cathode gas diode. *Note:* This term is used primarily in the industrial field. 190

phantom circuit (data transmission). A third circuit derived from two physical circuits by means of repeating coils installed at the terminals of the physical (side) circuits. A phantom circuit is a superimposed circuit derived from two suitably arranged pairs of wires, called side circuits, the two wires of each pair being effectively in parallel. 59

phantom-circuit loading coil. A loading coil for introducing a desired amount of inductance in a phantom circuit and a minimum amount of inductance in the constituent side circuits. *See:* **loading.** 328

phantom-circuit repeating coil (phantom-circuit repeat coil). A repeating coil used at a terminal of a phantom circuit, in the terminal circuit extending from the midpoints of the associated side-circuit repeating coils. 328

phantom group. A group of four open wire conductors suitable for the derivation of a phantom circuit. *See:* **open wire.** 328

phantom target (radar). (1) An echo box, or other reflection device, that produces a particular blip on the radar indicator. (2) A condition, maladjustment, or phenomenon (such as a temperature inversion) that produces a blip on the radar indicator resembling blips of targets for which the system is being operated. *See:* **echo box; navigation.** 13

phase (of a periodic phenomenon *f(t)*, for a particular value of *t*). The fractional part *t*/*P* of the period *P* through which *t* has advanced relative to an arbitrary origin. *Note:* The origin is usually taken at the last previous passage through zero from the negative to the positive direction. *See:* **simple sine-wave quantity; control system, feedback.** 329

phase advance. *See:* **phase lead.**

phase advancer. A phase modifier that supplies leading reactive volt-amperes to the system to which it is connected. Phase advancers may be either synchronous or asynchronous. *See:* **converter.** 328

phase angle (1) (general). The measure of the progression of a periodic wave in time or space from a chosen instant or position. *Notes:* (A) The phase angle of a field quantity, or of voltage or current, at a given instant of time at any given plane in a waveguide is ($wt - \beta z + \theta$), when the wave has a sinusoidal time variation. The term **waveguide** is used here in its most general sense and includes all transmission lines: for example, **rectangular waveguide, coaxial line, strip line,** etcetera. The symbol β is the imaginary part of the propagation constant for that waveguide, propagation is in the + z direction, and θ is the phase angle when $z = t = 0$. At a reference time $t = 0$ and at the plane *z*, the phase angle $(-\beta z + \Theta)$ will be represented by Φ. (B) Phase angle is obtained by multiplying the phase by 360 degrees or by 2π radians. *See:* **sinewave quantity.** 183

(2) (speed governing of hydraulic turbines). Referring to a simultaneous phasor diagram of the input and output, the angle by which the output signal lags or leads the input signal. 8

phase angle correction factor (instrument transformer). The ratio of the true power factor to the measured power factor. It is a function of both the phase angles of the instrument transformer and the power factor of the primary circuit being measured. *Note:* The phase angle correction factor is the factor that corrects for the phase displacement of the secondary current or voltage, or both, due to the instrument transformer phase angles. The measured watts or watt-hours in the secondary circuits of instrument transformers must be multiplied by the phase angle correction factor and the true ratio to obtain the true primary watts or watt-hours. 53, 394

phase angle, dielectric. *See:* **dielectric phase angle.**

phase angle, loop (automatic control) (closed loop). The value of the loop phase characteristic at a specified frequency. *See:* **phase characteristic; control system, feedback.** 56

phase-angle measuring (78) (power system device function numbers). A relay that functions at a predetermined phase angle between two voltages or between two currents or between voltage and current. *Syn:* **out-of-step protective relay.** 402

phase angle of an instrument transformer (instrument transformer) (power and distribution transformer).

The phase displacement, in minutes, between the primary and secondary values. *Note:* The phase angle of a current transformer is designated by the Greek letter beta β and is positive when the current leaving the identified secondary terminal leads the current entering the identified primary terminal. The phase angle of a voltage transformer is designated by the Greek letter gamma γ and is positive when the secondary voltage from the identified to the unidentified terminal leads the corresponding primary voltage.   53, 394

phase back (electrical heating systems). The amount of retardation (expressed in percent or as an angle) during which the controlling element is prevented from conducting.   476

phase-balance relay (power switchgear). A relay that responds to differences between quantities of the same nature associated with different phases of a normally balance polyphase circuit.   103

phase belt (coil group). A group of adjacent coils in a distributed polyphase winding of an alternating-current machine that are ordinarily connected in series to form one section of a phase winding of the machine. Usually, there are as many such phase belts per phase as there are poles in the machine. *Note:* The adjacent coils of a phase belt do not necessarily occupy adjacent slots: the intervening slots may be occupied by coils of another winding on the same core. Such may be the case in a two-speed machine. *See:* **rotor (rotating machinery); stator.**   63

phase center (antennas). The location of a point associated with an antenna such that, if it is taken as the center of a sphere whose radius extends into the far-field, the phase of a given field component over the surface of the radiation sphere is essentially constant, at least over that portion of the surface where the radiation is significant. *Note:* Some antennas do not have a unique phase center.   111

phase characteristic (1). The variation with frequency of the phase angle of a phasor quantity.   210
(2) (linear passive networks). The angle of a response function evaluated on the imaginary axis of the complex-frequency plane.   238

phase characteristic, loop (automatic control) (closed loop). The phase angle of the loop transfer function for real frequencies. *See:* **control system, feedback.**   56

phase coherence. *See:* **coherent.**

phase-coil insulation (rotating machinery). Additional insulation between adjacent coils that are in different phases. *See:* **asynchronous machine**   63

phase-comparison monopulse (radar). A form of monopulse employing receiving beams with different phase centers as obtained, for example, from side-by-side antennas or separate portions of an array. If the amplitude-versus-angle patterns of the beams are identical, the monopulse is pure phase-comparison, otherwise, it is a combination of phase-comparison and amplitude-comparison. *See:* **monopulse; amplitude-comparison monopulse.**   13

phase-comparison protection (power switchgear). A form of pilot protection that compares the relative phase-angle position of specified currents at the terminals of a circuit.   103

phase conductor (alternating-current circuit). The conductors other than the neutral conductor. *Note:* If an alternating-current circuit does not have a neutral conductor, all the conductors are phase conductors.   210

phase connections (rotating machinery). The insulated conductors (usually arranged in peripheral rings) that make the necessary connections between appropriate phase belts in an alternating-current winding. *See:* **rotor (rotating machinery); stator.**   63

phase constant (1) (fiber optics). The imaginary part of the axial propagation constant for a particular mode, usually expressed in radians per unit length. *See:* **axial propagation constant.**   433
(2) (waveguide). Of a traveling wave, the space rate of change of phase of a field component (or of the voltage or current) in the direction of propagation, in radians per unit length.   267
(3) (wavelength constant) (radio wave propagation). The imaginary part of the propagation constant.   146

phase contours. Loci of the return transfer function at constant values of the phase angle. *Note:* Such loci may be drawn on the Nyquist, inverse Nyquist, or Nichols diagrams for estimating performance of the closed loop with unity feedback. In the complex plane plot of $KG(j\omega)$, these loci are circles with centers at $-1/2, j/2\,N$ and radiuses such that each circle passes through the origin and the point $-1, j0$. In the inverse Nyquist diagram they are straight lines $\gamma = -\,N(x + 1)$ radiating from the point $-1,0$. *See:* **Nichols chart; Nyquist diagram; inverse Nyquist diagram.**   329

phase control (1) (rectifier circuits). The process of varying the point within the cycle at which forward conduction is permitted to begin through the rectifier circuit element. *Note:* The amount of phase control may be expressed in two ways: (1) the reduction in direct-current voltage obtained by phase control or (2) the angle of retard or advance.   208
(2) (thyristor). The starting instant is synchronous with respect to the line voltage. The controller ON-state interval is equal to or less than half the line period.   445

phase control range (thyristor). The range over which it is possible to adjust the angle of retard expressed in electrical degrees.   445

phase converter (rotating machinery). A converter that changes alternating-current power of one or more phases to alternating-current power of a different number of phases but of the same frequency. *See:* **converter.**   63

phase-corrected horn. A horn designed to make the emergent electromagnetic wave front substantially plane at the mouth. *Note:* Usually this is achieved by means of a lens at the mouth. *See:* **circular scanning; waveguide.**   244, 179

phase correction (telegraph transmission). The process of keeping synchronous telegraph mechanisms in substantially correct phase relationship. *See:* **telegraphy.** 328

phase corrector. A network that is designed to correct for phase distortion. *See:* **network analysis.** 210

phase-crossover frequency (hydraulic turbines). The frequency at which the phase angle reaches 180 degrees. 8

phased-array antenna. An array antenna whose beam directon or radiation pattern is controlled primarily by the relative phases of the excitation coefficients of the radiating elements. *See:* **antenna.** 111

phase delay (1) (facsimile) (in the transfer of a single-frequency wave from one point to another in a system). The time delay of a part of the wave identifying its phase. *Note:* The phase delay is measured by the ratio of the total phase shift in cycles to the frequency in hertz. *See:* **facsimile transmission.** 12
(2) (relaying). An equal delay of both the leading and trailing edges of a locally generated block. 103
(3) (dispersive and nondispersive delay lines). The ratio of total radian phase shift, to the specified radian frequency,w. Phase delay is nominally constant over the frequency band of operation for nondispersive delay devices. *See:* **phase lag.** 81

phase delay time. In the transfer of a single-frequency wave from one point to another in a system, the time delay of a part of the wave identifying its phase. *Note:* The phase delay time is measured by the ratio of the total phase delay through the network, in cycles, to the frequency, in hertz. *See:* **measurement system.** 293, 183

phase deviation (1) (angle modulation) (phase modulation). The peak difference between the instantaneous angle of the modulated wave and the angle of the carrier. *Note:* In the case of a sinusoidal modulating function, the value of the phase deviation, expressed in radians, is equal to the modulation index. *See:* **angle or phase (3); phase modulation.** 415
(2) (data transmission). The lack of direct proportionality of phase shift to frequency over the frequency range required for transmission, or the effect of such departure on a transmitted signal. 59

phase difference (1) (general). The difference in phase between two sinusoidal functions having the same periods. 210
(2) (automatic control). (A) Between sinusoidal input and output of the same frequency, phase angle of the output minus phase angle of the input: it is called "phase lead" if the input angle is the smaller, "phase lag" if the larger. (B) Of two periodic phenomena (for example, in nonlinear systems) the difference between the phase angles of their two fundamental waveforms. *Note:* Regarded as part of the transfer function which relates output to input at a specified frequency, phase difference is simply the phase angle $\theta(j\omega)$ in $A(j\omega)$ exp $j\theta(j\omega)$. Measurement of phase difference in the complex case is sometimes made in terms of the angular interval between respective crossings of a mean reference line, but values so measured will generally differ from those made in terms of the fundamental waveforms. *See:* **phase shift.** 56

phase distortion (1) (data transmission). Either the lack of direct proportionality of phase shift to frequency over the frequency range required for transmission, or the effect of such departure on a transmitted signal. 59
(2) (facsimile). *See:* **phase-frequency distortion.**

phased satellite (communication satellite). A satellite, the center of mass of which is maintained in a desired relation relative to other satellites, to a point on earth or to some other point of reference such as the subsolar point. *Note:* If it is necessary to identify those satellites that are not phased satellites, the term "unphased satellites" may be used. 74

phase-failure protection. *See:* **open-phase protection; phase-undervoltage protection.**

phase-frequency distortion (facsimile). Distortion due to lack of direct proportionality of phase shift to frequency over the frequency range required for transmission. *Notes:* (1) **delay distortion** is a special case. (2) This definition includes the case of a linear phase-frequency relation with the zero frequency intercept differing from an integral multiple of p. *See:* **distortion; distortion, phase delay; facsimile transmission; phase distortion.** 12

phase-insulated terminal box (rotating machinery). A terminal box so designed that the protection of phase conductors against electric failure within the terminal box is by insulation only. 63

phase jitter (data transmission). *See:* **jitter.**

phase lag (phase delay) (2-port network). The phase angle of the input wave relative to the output wave ($\Phi_{in} - \Phi_{out}$), or the initial phase angle of the output wave relative to the final phase angle of the output wave ($\Phi_i - \Phi_f$). *Note:* Under matched conditions, **phase lag** is the negative of the angle of the transmission coefficient of the scattering matrix for a 2-port network. *See:* **phase difference.** 293, 183

phase lead (2-port network). The phase angle of the output wave relative to the input wave (infcout minsinfcin), or the final phase angle of the output wave relative to the initial phase angle of the output wave (infnf minsinfni). *Note:* Under matched conditions, **phase lead** is the angle of the transmission coefficient of the scattering matrix for a 2-port network. *Syn:* **phase advance.** *See:* **phase difference.** 293, 183

phase localizer (navigation aid terms). A localizer in which the on-course line is defined by the phase reversal of energy radiated by the sideband antenna system, a reference carrier signal being radiated and used for the detection of phase. 526

phase lock loop (communication satellite). A circuit for synchronizing a variable local oscillator with the phase of a transmitted signal. Widely used in space communication for coherent carrier tracking, and threshold extension, bit synchronization and symbol synchronization. 83

phase locus (for a loop transfer function, say $G(s)$ $H(s)$. A plot in the s plane of those points for which the phase angle, ang GH, has some specified constant value. *Note:* The phase loci for 180 degrees plus or minus n 360 degrees are also root loci. *See:* **control system, feedback.** 56

phase margin (1) (loop transfer function for a stable feedback control system) (excitation control systems). 180 degrees minus the absolute value of the loop phase angle at a frequency where the loop gain is unity. *Note:* Phase margin is a convenient way of expressing relative stability of a linear system under parameter changes, in Nyquist, Bode, or Nichols diagrams. In a conditionally stable feedback control system where the loop gain becomes unity at several frequencies, the term is understood to apply to the value of phase margin at the highest of these frequencies. *See:* **control system, feedback.** 353
(2) (speed governing of hydraulic turbines). 180 degrees minus the absolute value of the open-loop phase angle at a frequency where the open-loop gain is unity. 8

phase meter (phase-angle meter). An instrument for measuring the difference in phase between two alternating quantities of the same frequency. *See:* **instrument.** 328

phase modifier (rotating machinery). An electric machine, the chief purpose of which is to supply leading or lagging reactive power to the system to which it is connected. Phase modifiers may be either synchronous or asynchronous. *See:* **converter.** 63, 3

phase-modulated transmitter. A transmitter that transmits a phase-modulated wave. 111, 240

phase modulation (pm) (1) (data transmission) (information theory)(antennas). Angle modulation in which the angle of a carrier is caused to depart from its reference value by an amount proportional to the instantaneous value of the modulating function. *Notes:* (A) A wave phase modulated by a given function can be regarded as a wave frequency modulated by the time derivative of that function. (B) Combinations of phase and frequency modulation are commonly referred to as frequency modulation. *See:* **angle or phase (C); phase deviation; pulse duration; reactance modulator.** 59, 415, 111
(2) (overhead-power-line corona and radio noise). Angle modulation in which the angle of a carrier is caused to depart from its reference value by an amount proportional to the instantaneous value of the modulating wave. 411

phase-modulation telemetering (electric power systems). A type of telemetering in which the phase difference between the transmitted voltage and a reference voltage varies as a function of the magnitude of the measured quantity. *See:* **telemetering.** 94

phase of a circularly polarized field vector (antennas). In the plane of polarization, the angle that the field vector makes, at a time taken as the origin, with a reference direction and with the angle counted as positive if it is in the same direction as the sense of polarization and negative if it is in the opposite direction to the sense of polarization. 111

phase overcurrent (power switchgear). The current flowing in a phase conductor which exceeds a predetermined value. 103

phase path (radio wave propagation). For a time-harmonic electromagnetic wave traveling between two points in a medium, the product of the speed of light in vacuum and the travel time of an equiphase surface between the two points. 146

phase pattern (of an antenna). The spatial distribution of the relative phase of a field vector excited by an antenna. *Notes:* (1) The phase may be referred to any arbitrary reference. (2) The distribution of phase over any path, surface, or any radiation pattern cut is also called a phase pattern. 111

phase recovery time (microwave gas tubes). The time required for a fired tube to deionize to such a level that a specified phase shift is produced in the low-level radio-frequency signal transmitted through the tube. *See:* **gas tube.** 125

phase relay (power switchgear). A relay that by its design or application is intended to respond primarily to phase conditions of the power system. 103

phase resolution. The minimum change of phase that can be distinguished by a system. *See:* **measurement system.** 183

phase-reversal protection. *See:* **phase-sequence reversal protection.**

phase-reversal relay. *See:* **negative-phase-sequence relay.**

phase-segregated terminal box. A terminal box so designed that the protection of phase conductors against electric failure within the terminal box is by insulation, and additionally by grounded metallic barriers forming completely isolated individual phase compartments so as to restrict any electric breakdown to a ground fault. 63

phase-selector relay (power switchgear). A programming relay whose function is to select the faulted phase or phases thereby controlling the operation of other relays or control devices. 103

phase-separated terminal box. *See:* **phase-segregated terminal box.**

phase separator (rotating machinery). Additional insulation between adjacent coils that are in different phases. *See:* **rotor (rotating machinery); stator.** 63

phase sequence (1) (set of polyphase voltages or currents). The order in which the successive members of the set reach their positive maximum values. *Note:* The phase sequence may be designated in several ways. If the set of polyphase voltages or currents is a symmetrical set, one method is to designate the phase sequence by specifying the integer that denotes the number of times that the angular phase lag between successive members of the set contains the characteristic angular phase difference for the number of phases m. If the integer is zero, the set is of zero phase sequence: if the integer is one, the set is of first phase sequence: and so on. Since angles of lag greater than 2p produce the same phase position for alternating quantities as the same angle decreased by the largest integral multiple of 2p contained in the angle of lag, it may be shown that there are only m distinct symmetrical sets normally designated from 0 to m mins 1 phase sequence. It can be shown that only for the first phase sequence do all the members of the set reach their positive maximum in the order of identification at uniform intervals of time.   210, 63

(2) (transformer) (power and distribution transformer). The order in which the voltages successively reach their positive maximum values. *See:* **direction of rotation of phasors.**   53

phase-sequence indicator. A device designed to indicate the sequence in which the fundamental components of a polyphase set of potential differences, or currents, successively reach some particular value, such as their maximum positive value. *See:* **instrument.**   328

phase-sequence relay (power switchgear). A relay that responds to the order in which the phase voltages or currents successively reach their maximum positive values.   103

phase-sequence reversal (industrial control). A reversal of the normal phase sequence of the power supply. For example, the interchange of two lines on a three-phase system will give a phase reversal.   206

phase-sequence reversal protection (power switchgear). A form of protection that prevents energization of the protected equipment on the reversal of the phase sequence in a polyphase circuit.   103

phase-sequence test (rotating machinery). A test to determine the phase sequence of the generated voltage of a three-phase generator when rotating in its normal direction. *See:* **asynchronous machine.**   63

phase-sequence voltage relay (47) (power system device function numbers). A relay that functions upon a predetermined value of polyphase voltage in the desired phase sequence.   402

phase shift (1) (general). The absolute magnitude of the difference between two phase angles. *Notes:* (A) The phase shift between two planes of a 2-port network is the absolute magnitude of the difference between the phase angles at those planes. The total phase shift, or absolute phase shift, is expressed as the total number of cycles, including any fractional number, between the two planes, where one complete cycle is 2π radians or 360 degrees. Relative phase shift is the total or absolute phase shift less the largest integral number of 2π radians or 360 degrees. The unit of phase shift is, therefore, the radian or the electrical degree. The term **2-port network** is used in its most general sense to include structures of passive or active elements. This includes the case of a given length of waveguide but may also refer to any two ports of a multiport device, where it is understood that a signal is incident only at one port. (B) A phase shift can be either a phase lead (advance) or a phase lag (delay). *See:* **measurement system.**   293, 183

(2) (electrical conversion). The displacement between corresponding points in similar wave shapes and is expressed in degrees lead or lag. *See:* **electrical conversion.**   186

(3) (transfer function). A change of phase angle with frequency, as between points on a loop phase characteristic. *See:* **control system, feedback.**

(4) (signal). A change of phase angle with transmission.   329

(5) (dispersive and nondispersive delay lines). The total number of degrees or radians that a CW (continuous wave) signal experiences as it is transmitted through the delay device at a given frequency within the band of operation. The phase shift is nominally linearly proportional to frequency within the frequency band of operation for a nondispersive delay device.   81

phase-shift circuit (industrial control). A network that provides a voltage component shifted in phase with respect to a reference voltage. *See:* **electronic controller.**   206

phase shifter (data transmission). A device in which the output voltage (or current) may be adjusted, in use or in its design, to have some desired phase relation with the input voltage (or current).   59

phase shifter, waveguide (waveguide components). An essentially lossless device for adjusting the phase of a forward traveling electromagnetic wave at the output of the device relative to the phase at the input.   166

phase-shifting transformer (1) (general). A transformer that advances or retards the phase-angle relationship of one circuit with respect to another. *Notes:* (A) The terms "advance" and "retard" describe the electrical angular position of the load voltage with respect to the source voltage. (B) If the load voltage reaches its positive maximum sooner than the source voltage, this is an "advance" position. (C) Conversely, if the load voltage reaches its positive maximum later than the source voltage, this is a "retard" position. *See:* **primary circuits; regulated circuit; series unit; main unit; series winding; excited winding; regulating winding; excitation winding; excitation-regulating winding; voltage-regulating relay; line-drop compensator; voltage winding (or transformer) for regulating equipment.**   53

(2) (metering)(watthour meters). An assembly of one or more transformers intended to be connected across the phases of a polyphase circuit as to provide voltages in the proper phase relationships for energizing varmeters, varhour meters, or other measurement equipment. This type of transformer is sometimes referred to as a phasing transformer.   212

(3) (power and distribution transformer). A transformer that advances or retards the voltage phase-angle relationship of one circuit with respect to another. *Notes:* (A) The terms "advance" and "retard" describe the electrical angular position of the load voltage with respect to the source voltage. (B) If the load voltage reaches its positive maximum power sooner than the source voltage, this is an "advance" position. (C) Conversely, if the load voltage reaches its positive maximum later than the source voltage, this is a "retard" position. (D) Additional related terms defined in 7.2. *See:* **regulating transformer; load-tap-changing transformer; phase-shifting transformer; circuits; preventative autotransformer; series transformer series and main units of a two-core regulating transformer; windings of a two-core regulating transformer; controls.**   53

form of phase modulation in which the modulating
function shifts the instantaneous phase of the modulated wave among predetermined discrete values.
242

phase-shift oscillator. An oscillator produced by connecting any network having a phase shift of an odd
multiple of 180 degrees (per stage) at the frequency of
oscillation, between the output and the input of an
amplifier. When the phase shift is obtained by resistance-capacitance elements, the circuit is an *R-C*
phase-shift oscillator. *See:* **oscillatory circuit.**
111

phase space (control system). (1) The state space augmented by the independent time variable. (2) One
used synonymously with the state space, usually with
the state variables being successive time derivatives of
each other. *See:* **control system.** 329

**phase spacing (of a fuse or switching device) (power
switchgear).** The distance between center-lines of the
current-carrying parts of the adjacent poles of the
switching device. 103

**phase splitter (phase splitting circuit) (data transmis-
sion).** A device which produces, from a single input
wave, two or more output waves which differ in phase
from one another. 59

phase-splitting circuit. *See:* **phase splitter.**

phase swinging (rotating machinery). Periodic variations in the speed of a synchronous machine above or
below the normal speed due to power pulsations in the
prime mover or driven load, possibly recurring every
revolution. 63

**phase-to-ground per unit overvoltage (power and dis-
tribution transformer).** The ratio of a phase-to-ground
overvoltage to the phase-to-ground voltage corresponding to the maximum system voltage. 53

**phase-to-phase per unit overvoltage (power and distri-
bution transformer).** The ratio of a phase-to-phase
overvoltage to the phase-to-phase voltage corresponding to the maximum system voltage. 53

**phase-to-phase voltage on an alternating-current elec-
tric system.** *See:* **nominal system voltage; maximum
system voltage; service voltage; utilization voltage;
low voltage; medium voltage; high voltage.** 260

phase transfer function. The argument Φ *(N)* of the
modulation transfer function is designated the phase
transfer function or PTF. 380

**phase transition (ferroelectric material)(primary fer-
roelectric terms).** A change in the crystal structure,
usually occurring at a well-defined temperature, which
alters the orientation or magnitude, or both, of the
electric polarization. 497

phase-tuned tube (microwave gas tubes). A fixed-tuned
broad-band transmit-receive tube, wherein the phase
angle through and the reflection introduced by the
tube are controlled within limits. *See:* **gas tube.**
125

phase-undervoltage protection (power switchgear). A
form of protection that disconnects or inhibits connection of the protected equipment on deficient voltage in one or more phases of a polyphase circuit.
103

phase-undervoltage relay (power switchgear). A relay
that operates when one or more phase voltages in a
normally balanced polyphase circuit is less than a
predetermined value. 127, 103

phase vector (of a wave). The vector in the direction of
the wave normal, whose magnitude is the phase constant. 146

phase vector in physical media (antenna). The imaginary part of the propagation vector. 111

phase velocity (1) (fiber optics). For a particular mode,
the ratio of the angular frequency to the phase constant. *See:* **axial propagation constant; coherence
time; group velocity.** 433

(2) **(of a traveling plane wave at a single frequency).**
The velocity of an equiphase surface along the wave
normal. *See:* **radio wave propagation; waveguide.**
179, 180

(3) **(radio wave propagation).** Of a traveling wave at
a single frequency, the velocity of an equiphase surface along the wave normal. 146

(4) **(waveguide).** Of a traveling wave at a given frequency, and for a given mode, the velocity of an equiphase surface in the direction of propagation.
267

phase-versus-frequency response characteristic. A
graph or tabulation of the phase shifts occurring in an
electric transducer at several frequencies within a
band. *See:* **transducer.** 111

phase voltage of a winding (machine or apparatus).
The potential difference across one phase of the machine or apparatus. *See:* **asynchronous machine.**
244, 63

phasing. The adjustment of picture position along the
scanning line. *See:* **scanning (facsimile).** 12

phasing signal. A signal used for adjustment of the
picture position along the scanning line. *See:* **facsimile
signal (picture signal).** 12

phasing time (facsimile). The time interval during
which the start positions of the scanning and recording
strokes are aligned so as to ensure against a split image
at the recorder. 11

**phasing voltage (of a network protector) (power
switchgear).** The voltage across the open contacts of
a selected phase. *Note:* This voltage is equal to the
phasor difference between the transformer voltage and
the corresponding network voltage. 103

**phasor (1)(measurement of power frequency electric
and magnetic fields from ac power lines).** A complex
number expressing the magnitude and phase of a
time-varying quantity. Unless otherwise specified, it is
used only within the context of steady-state alternating linear systems. In polar coordinates, it can be written as $Ae^{j\theta}$ where A is the amplitude or magnitude
(usually root-mean-square (rms), but sometimes indicated as peakvalue and θ is the phase angle. The phase
angle θ should not be confused with the space angle
of a vector. *See:* **electric field strength.** 514

(2) **(metering).** A complex number, associated with
sinusoidally varying electrical quantities, such that the
absolute value (modulus) of the complex number cor-

responds to either the peak amplitude or root-mean-square (rms) value of the quantity, and the phase (argument) to the phase angle at zero time. By extension, the term "phasor" can also be applied to impedance and related complex quantities that are not time dependent. 212

*Deprecated

phasor diagram (synchronous machine). A diagram showing the relationships of as many of the following phasor quantities as are necessary: armature current, armature voltages, the direct and quadrature axes, armature flux linkages due to armature and field winding currents, magnetomotive forces due to armature and field-winding currents, and the various components of air-gap flux. Figure 1 shows a "complete" phasor diagram for an over-excited generator, using the "generator convention" of current and voltages, shown for a nonsalient-pole machine in Figure 2. Figure 3 shows a phasor diagram of the basic quantities for an underexcited motor, using the "motor convention" of Figure 4. *Note:* It should be noted, that in the following

diagrams, the convention has been adopted where the positive direction of the direct axis is 90° ahead of the positive direction of the quadrature axis. (Counterclockwise direction of rotation of field relative to armature). This convention is opposite to that assumed in the definition for quadrature axis in IEEE Std 100-1977. In this latter definition, which conforms to widespread practice, the quadrature axis positive direction is 90° ahead of the positive direction of the direct axis. (Counterclockwise direction of rotation of field relative to armature). 63

phasor difference. *See:* **phasor sum (difference).**

phasor function. A functional relationship that results in a phasor. 210

phasor notation (radio wave propagation). For time-harmonic fields, the complex notation used in the expressions for the field quantities with the exponential time factor exp $(j\omega t)$. For example, for plane waves,

$$E(r, t) = \mathrm{Re}\ \{E(k, \omega) \exp\ [j(\omega t - k \cdot r)]\}$$

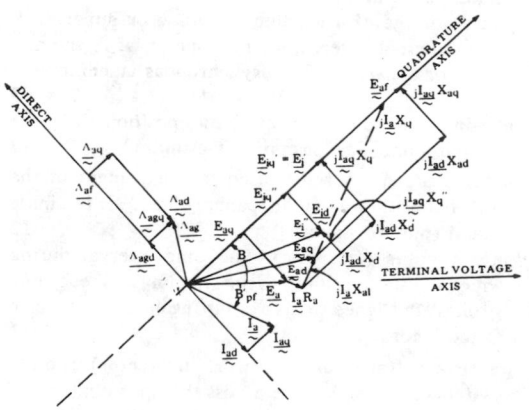

Fig. 1. Phasor diagram of overexcited generator (generator convention).

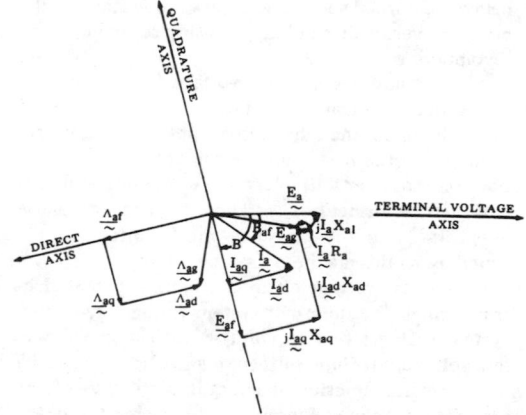

Fig. 3. Phasor diagram of motor operating at underexcited power factor (motor convention).

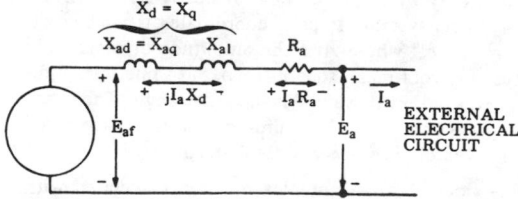

Fig. 2. Steady-state equivalent circuit for nonsalient-pole synchronous machine (generator convention).

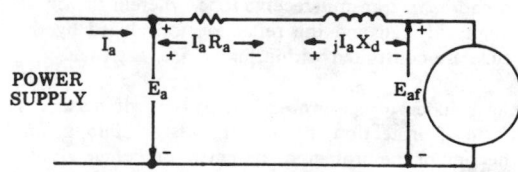

Fig. 4. Steady-state equivalent circuit for nonsalient-pole synchronous machine (motor convention).

Armature Voltages

$\underset{\sim}{E_a}$,	Terminal Voltage.
$\underset{\sim}{\widetilde{E}_{ad}}$	Direct-axis component of terminal voltage.
$\widehat{E}_{aq}$	Quadrature-axis component of terminal voltage.
$\underset{\sim}{\widetilde{E}_{ag}}$,	Voltage behind leakage reactance; this is the voltage due to net air-gap flux, also called "virtual voltage".
$\underset{\sim}{E_{af}}$,	Voltage due to flux produced only by the field-winding current.
$\underset{\sim}{\widetilde{I}_a R_a}$,	The voltage across armature resistance.
$\underset{\sim}{j\widetilde{I}_a X_{al}}$	The voltage across armature leakage reactance.
$\underset{\sim}{j\widetilde{I}_{ad} X_{ad}}$,	The voltage across direct-axis armature magnetizing reactance.
$\underset{\sim}{j\widetilde{I}_{aq} X_{aq}}$,	The voltage across quadrature-axis armature magnetizing reactance.
$\underset{\sim}{j\widetilde{I}_a X_q}$	A voltage based on quadrature-axis synchronous reactance, frequently used to locate the quadrature axis.
$jI_{ad} X_d'$	The voltage across direct-axis transient reactance.
$jI_{aq} X_q'$	The voltage across quadrature axis transient reactance.
$\underset{\sim}{\widetilde{E}_i'}$	Voltage behind transient reactance; this is the transient internal voltage.
$\underset{\sim}{j\widetilde{I}_{ad} \ddot{X}_d''}$	The voltage across direct-axis subtransient reactance.
$\underset{\sim}{j\widetilde{I}_{aq} X_q''}$	The voltage across quadrature-axis subtransient reactance.
$\underset{\sim}{\widetilde{E}_i''}$	Voltage behind subtransient reactance; this is the subtransient internal voltage.
$\underset{\sim}{E_{id}''}$	Direct-axis component of subtransient internal voltage.
$\underset{\sim}{\widetilde{E}_{iq}''}$	Quadrature-axis component of subtransient internal voltage.

Armature Current

$\underset{\sim}{I_a}$,	Armature Current.
$\underset{\sim}{\widetilde{I}_{ad}}$,	Direct-axis component of armature current. The sense of phasor I_{ad} in Figure 1, indicates that flux produced by armature current is opposed to that produced by the field current.
$\underset{\sim}{I_{aq}}$	Quadrature-axis component of armature current.

Flux Linkages with the Armature Winding

$\underset{\sim}{\wedge_{ag}}$,	Flux linkage due to net air-gap flux.
$\underset{\sim}{\wedge_{agd}}$,	Direct-axis component of flux linkage due to net air-gap flux.
$\underset{\sim}{\wedge_{agq}}$,	Quadrature-axis component of flux linkage due to net air-gap flux.
$\underset{\sim}{\wedge_{ad}}$,	Flux linkage due to the direct-axis component of armature current. The sense of $\wedge_{ad}$ in Fig. 1, indicates that the flux produced is in opposition to that produced by the field winding.
$\underset{\sim}{\wedge_{aq}}$,	Flux linkage due to the quadrature-axis component of armature current.
$\underset{\sim}{\wedge_{af}}$,	Flux linkages due to field-winding current.

Circuit Elements

R_a	Positive-sequence resistance of armature winding.

Reactances

X_{al} Armature leakage reactance.

X_{ad} Direct-axis armature magnetizing reactance.

X_{aq} Quadrature-axis armature magnetizing reactance.

X_q Quadrature-axis synchronous reactance.

X_d' Direct-axis transient reactance.

X_q' Quadrature-axis transient reactance.

X_d'' Direct-axis subtransient reactance.

X_q'' Quadrature-axis subtransient reactance.

General

$\approx$ Symbol for a phasor.

where Re indicates the real part, $E(r, t)$ is the instantaneous electric field vector, and $E(\mathbf{k},\omega)e^{Sjkr}$ is the phasor notation for the electric field vector.   146

phasor power (rotating machinery). The phasor representing the complex power. *See:* **asynchronous machine.**   63

phasor power factor. The ratio of the active power to the amplitude of the phasor power. The phasor power factor is expressed by the equation

$$F_{pp} = \frac{P}{S}$$

where F_{pp} is the phasor power factor, P is the active power, S is the amplitude of phasor power. If the voltages and currents are sinusoidal and, for polyphase circuits, form symmetrical sets,

$$A = |A|e^{j\theta A}$$
$$B = |B|e^{j\theta B}$$

See: **displacement power factor (rectifier).**   210

phasor product (quotient). A phasor whose amplitude is the product (quotient) of the amplitudes of the two phasors and whose phase angle is the sum (difference) of the phase angles of the two phasors. If two phasors are

$$F_{pp} = \cos(\alpha - \beta).$$

the phasor product is

$$AB = |AB|e^{J(\theta A + \theta B)}$$

and the quotient is

$$\frac{A}{B} = \left|\frac{A}{B}\right| e^{j(\theta A - \theta B)}$$

210

phasor quantity. (1) A complex equivalent of a simple sine-wave quantity such that the modulus of the former is the amplitude A of the latter, and the phase angle (in polar form) of the former is the phase angle of the latter. (2) Any quantity (such as impedance) that is expressed in complex form. *Note:* In case (1), sinusoidal variation with t enters: in case (2), no time variation (in constant-parameter circuits) enters. The term **phasor quantity** covers both cases.   210

phasor quotient. *See:* **phasor product (quotient).**

phasor reactive factor. The ratio of the reactive power to the amplitude of the phasor power. The phasor reactive factor is expressed by the equation

$$F_{qp} = \frac{Q}{S}$$

where F_{qp} is the phasor reactive factor, Q is the reactive power, S is the amplitude of the phasor power. If the voltages and currents are sinusoidal and, for polyphase circuits, form symmetrical sets,

phasor reactive factor

$$F_{pp} = \sin(\alpha - \beta).$$

210

phasor sum (difference). A phasor of which the real component is the sum (difference) of the real components of two phasors and the imaginary component is the sum (difference) of the imaginary components of two phasors. If two phasors are

$$A = a_1 + ja_2$$
$$B = b_1 + jb_2$$

phasor sum (difference) is

$$A \pm B = (a_1 \pm b_1) + j(a_2 \pm b_2).$$

210

phi (Φ) polarization. The state of the wave in which the E vector is tangential to the lines of latitude of a given spherical frame of reference. *Note:* The usual frame of reference has the polar axis vertical and the origin at or near the antenna. Under these conditions, a vertical dipole will radiate only theta (j) polarization, and a horizontal loop will radiate only phi (Φ) polarization. *See:* **antenna.**   111, 246

Philips gauge. A vacuum gauge in which the gas pressure is determined by measuring the current in a glow discharge. *See:* **instrument.**   328

phon. The unit of loudness level as specified in the definition of loudness level. *See:* **loudspeaker.**
                    176

phonograph pickup (mechanical reproducer). A mechanoelectrical transducer that is actuated by modulations present in the groove of the recording medium and that transforms this mechanical input into an electric output. *Note:* (1) Where no confusion is likely the term **phonograph pickup** may be shortened to **pickup.** (2) A phonograph pickup generally includes a pivoted mounting arm and the transducer itself (the pickup cartridge).   176

phosphene (electrical) (electrotherapy). A visual sensation experienced by a human subject during the passage of current through the eye. *See:* **electrotherapy.**
                    192

phosphor. A substance capable of luminescence. *See:* **cathode-ray tube; fluorescent lamp; radio navigation; television.**   328

phosphor decay. A phosphorescence curve describing energy emitted versus time. *See:* **oscillograph.**
                    185

phosphorescence (illuminating engineering). The emission of light as the result of the absorption of radiation, and continuing for a noticeable length of time after excitation.   167

phosphor screen. All the visible area of the phosphor on the cathode-ray tube faceplate. *See:* **oscillograph.**
                    185

phot (ph) (illuminating engineering). A unit of illuminance equal to one lumen per square centimeter. The use of this unit is deprecated.   167

photocathode. An electrode used for obtaining a photoelectric emission when irradiated. *See:* **electrode (electron tube); phototube.**   117

photocathode blue response. The photoemission current produced by a specified luminous flux from a tungsten filament lamp at 2854 kelvins color temperature when the flux is filtered by a CS 5-58 blue filter of half stock thickness (1.75—2.25 mm). This parameter is useful in characterizing response to scintillation counting sources.   117

photocathode luminous sensitivity. *See:* **sensitivity, cathode luminous.**

photocathode response (diode-type camera tube). The response of a photocathode is the current emitted into vacuum per incident radiant power of specified spectral distribution. It is expressed in amperes watt^{-1} (AW^{-1}).   380

photocathode, semitransparent. *See:* **semitransparent photocathode.**

photocathode spectral quantum efficiency (diode-type camera tube). The ratio of the average number of electrons emitted to the number of photons in the input signal irradiance on the photocathode face as a function of the photon energy, frequency, or wavelength.   380

photocathode spectral-sensitivity characteristic. *See:* **spectral-sensitivity characteristic photocathode.**

photocathode transit time. That portion of the photomultiplier transit time corresponding to the time for photoelectrons to travel from the photocathode to the first dynode.   117

photocathode transit-time difference. The difference in transit time between electrons leaving the center of the photocathode and electrons leaving the photocathode at some specified point on a designated diameter.   117

photocell (photoelectric cell). (1) A solid-state photosensitive electron device in which use is made of the variation of the current-voltage characteristic as a function of incident radiation. *See:* **phototube.**
                    117

(2) A device exhibiting photovoltaic or photoconductive effects. *See:* **phototube.**   244, 190

photochemical radiation (illuminating engineering). Energy in the ultraviolet, visible and infrared regions to produce chemical changes in materials. *Note:* Examples of photochemical processes are accelerated fading tests, photography, photoreproduction and chemical manufacturing. In many such applications a specific spectral region is of importance.   167

photoconductive cell. A photocell in which the photoconductive effect is utilized. *See:* **phototube.**
                    244, 190

photoconductive effect (photoconductivity). A photoelectric effect manifested as a change in the electric conductivity of a solid or a liquid and in which the charge carriers are not in thermal equilibrium with the lattice. *Note:* Many semiconducting metals and their compounds (notably selenium, selenides, and tellurides) show a marked increase in electric conductance when electromagnetic radiation is incident on them. *See:* **photoelectric effect; phototube; photovoltaic effect; photoemissive effect.**   210, 190

photoconductivity (fiber optics). The conductivity increase exhibited by some nonmetallic materials, resulting from the free carriers generated when photon energy is absorbed in electronic transitions. The rate at which free carriers are generated, the mobility of the carriers, and the length of time they persist in conducting states (their lifetime) are some of the factors that determine the amount of conductivity change. *See:* **photoelectric effect.**   433

photocurrent (fiber optics). The current that flows through a photosensitive device (such as a photodiode) as the result of exposure to radiant power. Internal gain, such as that in an avalanche photodiode, may enhance or increase the current flow but is a distinct mechanism. *See:* **dark current; photodiode.**   433

photodiode (fiber optics). A diode designed to produce photocurrent by absorbing light. Photodiodes are used for the detection of optical power and for the conversion of optical power to electrical power. *See:* **ava-**

lanche photodiode (APD); photocurrent; PIN photodiode.   433

photoelectric beam-type smoke detector (fire protection devices). A device which consists of a light source which is projected across the area to be protected into a photosensing cell. smoke between the light source and the receiving photosensing cell reduces the light reaching the cell, causing actuation.   71

photoelectric cathode. *See:* photocathode.

photoelectric color-register controller. A photoelectric control system used as a longitudinal position regulator for a moving material or web to maintain a preset register relationship between repetitive register marks in the first color and reference positions of the printing cylinders of successive colors. *See:* photoelectric control.   206

photoelectric control (industrial control). Control by means of which a change in incident light effects a control function.   206

photoelectric counter (industrial control). A photoelectrically actuated device used to record the number of times a given light path is intercepted by an object. *See:* photoelectric control.   206

photoelectric current. The current due to a photoelectric effect. *See:* photoelectric effect.   244, 206

photoelectric cutoff register controller (industrial control). A photoelectric control system used as a longitudinal position regulator that maintains the position of the point of cutoff with respect to a repetitively referenced pattern on a moving material. *See:* photoelectric control.   206

photoelectric directional counter. A photoelectrically actuated device used to record the number of times a given light path is intercepted by an object moving in a given direction. *See:* photoelectric control.   204

photoelectric door opener. A photoelectric control system used to effect the opening and closing of a power-operated door. *See:* photoelectric control.   204

photoelectric effect (fiber optics). (1) External photoelectric effect: The emission of electrons from the irradiated surface of a material. *Syn:* photoemissive effect. (2) Internal photoelectric effect: photoconductivity.   433

photoelectric emission (electron tube). The ejection of electrons from a solid or liquid by electromagnetic radiation. *See:* field-enhanced photoelectric emission.   125

photoelectric flame detector (fire protection devices). A device whose sensing element is a photocell which either changes its electrical conductivity or produces an electrical potential when exposed to radiant energy.   71

photoelectric lighting controller. A photoelectric relay actuated by a change in illumination to control the illumination in a given area or at a given point. *See:* photoelectric control.   206

photoelectric loop control (industrial control). A photoelectric control system used as a position regulator for a strip processing line that matches the average linear speed in one section to the speed in an adjacent section to maintain the position of the loop located between the two sections. *See:* photoelectric control.   206

photoelectric pinhole detector. A photoelectric control system that detects the presence of minute holes in an opaque material. *See:* photoelectric control.   204

photoelectric power system. *See:* photovoltaic power system.

photoelectric pyrometer (industrial control). An instrument that measures the temperature of a hot object by means of the intensity of radiant energy exciting a phototube. *See:* electronic control.   206

photoelectric relay. A relay that functions at predetermined values of incident light. *See:* photoelectric control.   206

photoelectric scanner (industrial control). A single-unit combination of a light source and one or more phototubes with a suitable optical system. *See:* photoelectric control.   206

photoelectric side-register controller (industrial control). A photoelectric control system used as a lateral position regulator that maintains the edge of, or a line on, a moving material or web at a fixed position. *See:* photoelectric control.   206

photoelectric smoke-density control. A photoelectric control system used to measure, indicate, and control the density of smoke in a flue or stack. *See:* photoelectric control.   206

photoelectric smoke detector (industrial control). A photoelectric relay and light source arranged to detect the presence of more than a predetermined amount of smoke in air. *See:* photoelectric control.   206

photoelectric spot-type smoke detector (fire protection devices). A device which contains a chamber with either overlapping or porous covers which prevent the entrance of outside sources of light but which allow the entry of smoke. The unit contains a light source and a special photosensitive cell in the darkened chamber. The cell is either placed in the darkened area of the chamber at an angle different from the light path or has the light blocked from it by a light stop or shield placed between the light source and the cell. With the admission of smoke particles, light strikes the particles and is scattered and reflected into the photosensitive cell. This causes the photosensing circuit to respond to the presence of smoke particles in the smoke chamber.   71

photoelectric system (protective signaling). An assemblage of apparatus designed to project a beam of invisible light onto a photoelectric cell and to produce an alarm condition in the protection circuit when the beam is interrupted. *See:* protective signaling.   328

photoelectric tube. An electron tube, the functioning of which is determined by the photoelectric effect. *See:* phototube.   190

photo-electron. An electron liberated by the photoemissive effect. *See:* photoelectric effect.   190

photo-electron irradiation dark current increase (diode-type camera tube). That irreversible dark current increase which is caused by bombardment of the charge storage target by photo- electrons.   380

photo-electron irradiation deterioration (diode-type camera tube). That irreversible dark current increase which is associated with bombardment of the charge storage target by photo-electrons.   380

photoemission spectrum (scintillator material). The relative numbers of optical photons emitted per unit wavelength as a function of wavelength interval. The emission spectrum may also be given in alternative units such as wave number, photon energies, frequency, etcetera. *Note:* Optical photons are photons with energies corresponding to wavelengths between 2000 and 15 000 angstroms.   335

photoemissive effect. *See:* **photoelectric effect (external).**   433

photoflash lamp (illuminating engineering). A lamp in which combustible metal or other solid material is burned in an oxidizing atmosphere to produce light of high intensity and short duration for photographic purposes.   167

photoformer. A function generator that operates by means of a cathode-ray beam optically tracking the edge of a mask placed on a screen. *See:* **electronic analog computer.**   9

photographic emulsion. The light-sensitive coating on photographic film consisting usually of a gelatin containing silver halide.   176

photographic sound recorder (optical sound recorder). Equipment incorporating means for producing a modulated light beam and means for moving a light-sensitive medium relative to the beam for recording signals derived from sound signals.   176

photographic sound reproducer (optical sound reproducer). A combination of light source, optical system, photoelectric cell, or other light-sensitive device such as a photoconductive cell, and a mechanism for moving a medium carrying an optical sound record (usually film), by means of which the recorded variations may be converted into electric signals of approximately like form.   176

photographic transmission density (optical density). The common logarithm of opacity. Hence, film transmitting 100 percent of the light has a density of zero, transmitting 10 percent a density of 1, and so forth. Density may be diffuse, specular, or intermediate. Conditions must be specified.   176

photometer (illuminating engineering). An instrument for measuring photometric quantities such as luminance, luminous intensity, luminous flux or illuminance.   167

photometric brightness. *See:* **luminance.**

photometry (1) (illuminating engineering). The measurement of quantities associated with light. *Note:* Photometry may be visual in which the eye is used to make a comparison, or physical in which measurements are made by means of physical receptors.   167

(2) (television). (A) general. The measurement of quantities referring to radiation evaluated in accordance with the visual effect it produces, as based on certain conventions. (B) visual. That branch of photometry in which the eye is used to make comparison.

(C) physical. That branch of photometry in which the measurement is made by means of physical receptors.   18

photomultiplier. *See:* **multiplier phototube.**

photomultiplier transit time (scintillation counting). The time difference between the incidence of a delta-function light pulse on the photocathode of the photomultiplier and the occurrence of the half-amplitude point on the output-pulse leading edge.   117

photomultiplier tube. *See:* **multiplier phototube.**

photon (1) (fiber optics). A quantum of electromagnetic energy. The energy of a photon is h_ν where h is Planck's constant and ν is the optical frequency. *See:* **nonlinear scattering; Planck's constant.**   433

(2) (range protection). A quantum of electromagnetic radiation irrespective of origin.   399

photon emission spectrum, scintillator material (scintillation counting). The relative numbers of optical photons emitted per unit wavelength as a function of wavelength interval. The emission spectrum may also be given in alternative units such as wavenumber, photon energies, frequency, and so on.   117

photon emitting diode (light emitting diodes). A semiconductor device containing a semiconductor junction in which radiant flux is nonthermally produced when a current flows as a result of an applied voltage.   162

photon noise. *See:* **quantum noise.**

photopic spectral luminous efficiency function (V_λ) (photometric standard observer for photopic vision). The photopic spectral luminous efficiency function gives the ratio of the radiant flux at wavelength λ_m to that at wavelength λ, when the two fluxes produce the same photopic luminous sensations under specified photometric conditions, λ_m being chosen so that the maximum value of this ratio is unity. Unless otherwise indicated, the values used for the spectral luminous efficiency function relate to photopic vision by the photometric standard observer having the characteristics laid down by the International Commission on Illumination (CIE).   162

photopic vision (illuminating engineering). Vision mediated essentially or exclusively by the cones. It is generally associated with adaptation to a luminance of at least $3.4 cd/m^2 (2.2 \times 10^{-3} cd/in^2)(1.0fL)$.   167

photosensitive recording (facsimile). Recording by the exposure of a photosensitive surface to a signal-controlled light beam or spot. *See:* **recording (facsimile).**   12

photosensitive tube. *See:* **photoelectric tube.**

photosensitizers (laser-maser). Substances which increase the sensitivity of a material to irradiation by electromagnetic radiation.   363

phototube (photoelectric tube). An electron tube that contains a photocathode and has an output depending at every instant on the total photoelectric emission from the irradiated area of the photocathode. *See:* **field-enhanced photoelectric emission.**   117

phototube gain (liquid-scintillation counting). The ratio of the signal output current to the photoelectric signal current from the photocathode.   422

phototube, gas. *See:* **gas phototube.**

phototube housing (industrial control). An enclosure containing a phototube and an optical system. *See:* **photoelectric control.** 206

phototube, multiplier. *See:* **multiplier phototube.**

phototube, vacuum. *See:* **vacuum phototube.**

photovaristor. A varistor in which the current-voltage relation may be modified by illumination, for example, cadmium sulphide or lead telluride. *See:* **semiconductor device.** 342

photovoltaic (PV) array (terrestrial photovoltaic power systems). The smallest installed assembly of photovoltaic (PV) panels, support structure, foundation, and other components as required, such as a tracker. *See:* **array control** 496

photovoltaic (PV) array subfield (terrestrial photovoltaic power systems). One or more arrays associated by a distinguishing feature, such as field geometry or electrical interconnection. *See:* **array control** 496

photovoltaic (PV) cell (terrestrial photovoltaic power systems). The basic device that converts sunlight directly into dc electricity. *See:* **array control** 496

photovoltaic (PV) module (terrestrial photovoltaic power systems). The smallest, complete, environmentally protected assembly of photovoltaic (PV) cells (flat plte-type), or receiver(s) and optics (concentrator-type), and related components, such as interconnects and mounting, that accepts unconcentrated sunlight. *See:* **array control** 496

photovoltaic (PV) panel (terrestrial photovoltaic power systems). One or more photovoltaic (PV) modules assembled and wired and designed to provide a field-installable unit. *See:* **array control** 496

photovoltaic (PV) receiver (terrestrial photovoltaic power systems). An assembly of one or more photovoltaic (PV) cells that accepts concentrated sunlight and incorporates means for thermal and electric energy removal. *See:* **array control** 496

photovoltaic (PV) system-utility interface (terrestrial photovoltaic power systems). The interconnection between the power conditioning subsystem, the on-site ac loads, and the utility. *See:* **array control** 496

photovoltaic effect (fiber optics). The production of a voltage difference across a pn junction resulting from the absorption of photon energy. The voltage difference is caused by the internal drift of holes and electrons. *See:* **photon.** 433

photovoltaic power system (terrestrial photovoltaic power systems). A system that converts sunlight directly into electric energy and processes it into a form suitable for use by the intended load. The system will include an array subsystem and may also include the following major subsystems: power conditioning, storage, thermal, and system monitor and control. A photovoltaic (PV) system-utility interface may also be included. *See:* **array control** 496

physical (PHY) layer (token ring accesss method). The layer responsible for interfacing with the medium, detecting and generating signals on the medium, and

converting and processing signals received from the medium access cotrol layer. 472

physical circuit (data transmission). A two-wire metallic circuit that is not arranged for phantom use. 59

physical concept. Anything that has existence or being in the ideas of man pertaining to the physical world. Examples are magnetic fields, electric currents, electrons. 210

physical damage (rotating machinery). This contributes to electrical insulation failure by opening leakage paths through the insulation. Included here are: physical shock, vibration, overspeed, short-circuit forces, erosion by foreign matter, damage by foreign objects, and thermal cycling. 37

physical entity. *See:* **physical quantity.**

physical medium attachment (PMA)(medium attachment units and repeater units). The portion of the medium attachment interface (MAU) that contains the functional circuitry. 543

physical optics (fiber optics). The branch of optics that treats light propagation as a wave phenomenon rather than a ray phenomenon, as in geometric optics. optics. 433

physical photometer (illuminating engineering). An instrument containing a physical receptor and associated filters, which is calibrated so as to read photometric quantities directly. *See:* **physical receptor.** 167

physical property. Any one of the generally recognized characteristics of a physical system by which it can be described. 210

physical quantity (physical entity) (concrete quantity). A particular example of a measurable physical property of a physical system. It is characterized by both a qualitative and a quantitative attribute (that is, kind and magnitude). It is independent of the system of units and equations by which it and its relation to other physical quantities are described quantitatively. 210

physical receptor (illuminating engineering). A device which generates electric current or voltage or undergoes a change of resistance or generates a charge when radiation is incident on it. 167

physical requirement (software). A requirement that specifies a physical characteristic that a system or system component must possess, for example, material, shape, size, weight. *See:* **component; requirement; system.** 434

physical signaling (PLS)(medium attachment units and repeater units). That portion of the physical layer, contained within the data terminal equipment (DTE) that provides the logical and functional coupling between medium attachment unit (MAU) and data link layers. 543

physical stimuli (illuminating engineering). May be either distal or proximal. 167

physical system. A part of the real physical world that is directly or indirectly observed or employed by mankind. 210

physical unit. *See:* **unit.**

pi (π) mode (magnetrons). The mode of operation for which the phases of the fields of successive anode openings facing the interaction space differ by p radians. *See:* **magnetrons.** 125

pi (π) network. A network composed of three branches connected in series with each other to form a mesh, the three junction points forming an input terminal, an output terminal, and a common input and output terminal, respectively. See accompanying figure. *See:* **network analysis.** 210

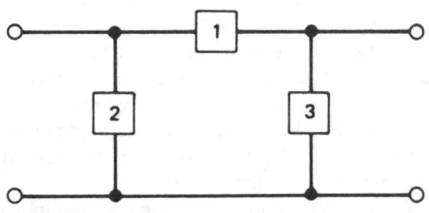

pi network. The junction point between branches 1 and 2 forms an input terminal, that between branches 1 and 3 forms an output terminal, and that between branches 2 and 3 forms a common input and output terminal.

pi (π) point. A frequency at which the insertion phase shift of an electric structure is 180 degrees or an integral multiple thereof. 328

pickle (corrosion) (electroplating). A solution or process used to loosen or remove corrosion products such as oxides, scale, and tarnish from a metal. *See:* **electroplating.** 205

pickling (electroplating) (1) (chemical). The removal of oxides or other compounds from a metal surface by means of a solution that acts chemically upon the compounds.
(2) (electrolytic). Pickling during which a current is passed through the pickling solution to the metal (cathodic pickling) or from the metal (anodic pickling). *See:* **electroplating.** 328

pickoff (1) (gyro; accelerometer). A device which produces a signal output, generally a voltage, as a function of the relative linear or angular displacement between two elements. 46
(2) (test, measurement and diagnostic equipment). A sensing device that responds to movement to create a signal or to effect some type of control. 54

pickoff axis (dynamically tuned gyro) (inertial sensor). The axis of angular displacement between the rotor and the case that results in the maximum signal per unit of rotation from the pickoff. 46

pickup (1) (of a relay) (power switchgear). The action of a relay as it makes designated response to progressive increase of input. As a qualifying term, the state of a relay when all response to progressive increase of input has been completed. Also used to identify the minimum value of an input quantity reached by progressive increases which will cause the relay to reach the pickup state from reset. *Note:* In describing the performance of relays having multiple inputs,

pickup has been used to denote contact operation, in which case pickup value of any input is meaningful only when related to all other inputs. 103
(2) (electronics). A device that converts a sound, scene, or other form of intelligence into corresponding electric signals (for example, a microphone, a television camera, or a phonograph pickup). *See:* **microphone; phonograph pickup; television.** 178, 54

pickup and seal voltage (magnetically operated device) (industrial control). The minimum voltage at which the device moves from its deenergized into its fully energized position. *See:* **initial contact pressure.** 302

pickup current. *See:* **pickup value.**

pick-up factor (DF [direction finder] antenna system)(navigation aid terms). An index of merit expressed as the voltage across the receiver input impedance divided by the signal field strength to which the antenna system is exposed, the direction of arrival and polarization of the wave being such as to give maximum response. 526

pickup factor, direction-finder antenna system. An index of merit expressed as the voltage across the receiver input impedance divided by the signal field strength to which the antenna system is exposed, the direction of arrival and polarization of the wave being such as to give maximum response. *See:* **navigation.** 278, 187, 13

pickup spectral characteristic (color television). The set of spectral responses of the device, including the optical parts, that converts radiation to electric signals, as measured at the output terminals of the pickup tubes. *Note:* Because of nonlinearity, the spectral characteristics of some kinds of pickup tubes depend upon the magnitude of radiance used in the measurement. 18

pickup tube. *See:* **camera tube.**

pickup value. The minimum input that will cause a device to complete contact operation or similar designated action. *Note:* In describing the performance of devices having multiple inputs, the pickup value of an input is meaningful only when related to all other inputs. 103

pickup voltage (or current) (magnetically operated device). The voltage (or current) at which the device starts to operate when its operating coil is energized under conditions of normal operating temperature. *See:* **contactor.** 1, 206

pico (p)(mathematics of computing). A prefix indicating 10^{-12}. 564

pictorial format (pulse measurement). A graph, plot, or display in which a waveform is presented for observation or analysis. Any of the waveform formats defined in the following subsections may be presented in the pictorial format. 15

picture element (pixel). The smallest area of a television picture capable of being delineated by an electric signal passed through the system or part thereof. *Note:* It has three important properties, namely P_v, the vertical height of the picture element: P_h, the horizontal length of the picture element, and P_a, the aspect ratio of the picture element. In addition, N_p, the total number of picture elements in a complete picture, is of interest since this number provides a convenient way of comparing systems. For convenience P_v and P_h are normalized for V, the vertical height of the picture: that is, P_v or P_h must be multiplied by V to obtain the actual dimension in a particular picture. P_v is defined as $P_v = 1/N$, where N is the number of active scanning lines in the raster. P_h is defined as $P_h = t_r A/t_e$, where t_r is the average value of the rise and delay times (10 percent to 90 percent) of the most rapid transition that can pass through the system or part thereof, t_e is the duration of the part of a scanning line that carries picture information, and A is the aspect ratio of the picture. (At present all broadcast television systems have a horizontal to vertical aspect ratio of 4/3.) P_a is defined as $P_a = P_h/P_v = t_r AN/t_e$ and N_p is defined as $N_p = (1/P_v) \times (A/P_h) = Nt_e/T_r$. See: **television.** 177

picture frequencies (facsimile). The frequencies which result solely from scanning subject copy. Note: This does not include frequencies that are part of a modulated carrier signal. See: **scanning (facsimile).** 12

picture inversion (facsimile). A process that causes reversal of the black and white shades of the recorded copy. See: **facsimile transmission.** 12

picture signal (television or facsimile). The signal resulting from the scanning process. See: **television.** 178

picture transmission (telephotography). The electric transmission of a picture having a gradation of shade values. 328

picture tube (kinescope) (television). A cathode-ray tube used to produce an image by variation of the beam current as the beam scans a raster. See: **television.** 18

Pierce gun (microwave tubes). A gun that delivers an initially convergent electron beam. If a magnetic focusing scheme is used, the beam is made to enter the field at the minimum beam diameter or else, if the magnetic field threads through the cathode, the magnetic field must have a shape that is consistent with the desired beam that imparts certain flow characteristics to the electron beam. In Brillouin flow, angular electron velocity about the axis is imparted to the beam on entry into the magnetic field and the resulting inwardly directed force balances both the space charge and centrifugal forces. In practice, values of field up to twice the theoretical equilibrium value may be found necessary. In confined flow there is no overall angular velocity of the beam about the beam axis. Individual electron trajectories are tight helices (of radius small compared to beam radius) whose axis is along a magnetic-field line. The required magnetic field is several times greater than the Brillouin value and the flux must intersect the cathode surface. See: **microwave tube or valve.** 190

Pierce oscillator. An oscillator that includes a piezoelectric crystal connected between the input and the output of a three-terminal amplifying element, the feedback being determined by the internal capacitances of the amplifying elements. Note: This is basically a Colpitts oscillator. See: **Colpitts oscillator; oscillatory circuit.** 111, 211

piezoelectric accelerometer. A device that employs a piezoelectric material as the principal restraint and pickoff. It is generally used as a vibration sensor. 46

piezoelectric crystal cut, type. See: **type of piezoelectric crystal cut.**

piezoelectric-crystal element. A piece of piezoelectric material cut and finished to a specified geometrical shape and orientation with respect to the crystallographic axes of the material. See: **crystal.** 341

piezoelectric-crystal plate. A piece of piezoelectric material cut and finished to specified dimensions and orientation with respect to the crystallographic axes of the material, and having two major surfaces that are essentially parallel. See: **crystal.** 341

piezoelectric-crystal unit. A complete assembly, comprising a piezoelectric-crystal element mounted, housed, and adjusted to the desired frequency, with means provided for connecting it in an electric circuit. Such a device is commonly employed for purposes of frequency control, frequency measurement, electric wave filtering, or interconversion of electric waves and elastic waves. Note: Sometimes a piezoelectric-crystal unit may be an assembly having in it more than one piezoelectric-crystal plate. Such an assembly is called a mutliple-crystal unit. See: **crystal.** 341

piezoelectric effect. Some materials become electrically polarized when they are mechanically strained. The direction and magnitude of the polarization depend upon the nature and amount of the strain, and upon the direction of the strain. In such materials the converse effect is observed, namely, that a strain results from the application of an electric field. 210

piezoelectric loudspeaker. See: **crystal loudspeaker.**

piezoelectric microphone. See: **crystal microphone.**

piezoelectric pickup. See: **crystal pickup.**

piezoelectric transducer. A transducer that depends for its operation on the interaction between electric charge and the deformation of certain materials having piezoelectric properties. Note: Some crystals and specially processed ceramics have piezoelectric properties. 176

pigtail (fiber optics). A short length of optical fiber, permanently fixed to a component, used to couple power between it and the transmission fiber. See: **launching fiber.** 433

pile-up (of signal pulses)(X-ray energy spectometers). Two pulses (signals) are said to be piled-up when a second pulse occurs before the transient response of the preceding pulse has decayed to a negligible value. 471

pileup. See: **relay pileup.**

pile-up rejection (X-ray energy spectrometers). A technique used to identify and reject pulses (signals) that are piled up. 471

pillbox antenna. A reflector antenna having a cylindrical reflector enclosed by two parallel conducting plates perpendicular to the cylinder, spaced less than one wavelength apart. *Syn:* **cheese antenna.**   111

pilot (transmission system). A signal wave, usually a single frequency, transmitted over the system to indicate or control its characteristics.   328

pilotage (navigation aid terms). The process of directing a vehicle by reference to recognizable landmarks or soundings, or to electronic or other aids to navigation. Observations may be by any means including optical, aural, mechanical, or electronic.   526

pilot cell (storage battery). A selected cell whose condition is assumed to indicate the condition of the entire battery. *See:* **battery (primary or secondary).**   328

pilot channel. A channel over which a pilot is transmitted.   328

pilot circuit (industrial control). The portion of a control apparatus or system that carries the controlling signal from the master switch to the controller. *See:* **control.**   206

pilot director indicator. A device that indicates to the pilot information as to whether or not the aircraft has departed from the target track during a bombing run.   328

pilot exciter (1)(excitation systems for synchronous machines). The equipment providing the field current for the excitation of another exciter.   507
(2)(rotating machinery) (electric installations on shipboard). The source of all or part of the field current for the excitation of another exciter.   63, 3

pilot fit (spigot fit) (rotating machinery). A clearance hole and mating projection used to guide parts during assembly.   63

pilot house control (illuminating engineering). A mechanical means for controlling the elevation and train of a searchlight from a position on the other side of the bulkhead or deck on which it is mounted.   167

pilot lamp. A lamp that indicates the condition of an associated circuit. In telephone switching, a pilot lamp is a switchboard lamp that indicates a group of line lamps, one of which is or should be lit.   328

pilot light. A light, associated with a control, that by means of position or color indicates the functioning of the control.   328

pilot line (conductor stringing equipment). A lightweight line, normally synthetic fiber rope, used to pull heavier pulling lines which in turn are used to pull the conductor. Pilot lines may be installed with the aid of finger lines or by helicopter when the insulators and travelers are hung. *Syn:* **lead line; leader; P-line; straw line.**   431

pilot line winder (conductor stringing equipment). A device designed to payout and rewind pilot lines during stringing operations. It is normally equipped with its own engine which drives a drum or a supporting shaft for a reel mechanically, hydraulically or through a combination of both. These units are usually equipped with multiple drums or reels, depending upon the number of pilot lines required. The pilot line is payed out from the drum or reel, pulled through the travelers in the sag section, and attached to the pulling line on the reel stand or drum puller. It is then rewound to pull the pulling line through the travelers.   431

pilot protection (power switchgear). A form of line protection that uses a communication channel as a means to compare electrical conditions at the terminals of a line.   103

pilot streamer (lightning). The initial low-current discharge that begins when the voltage gradient exceeds the breakdown voltage of air. *See:* **direct-stroke protection (lightning).**   64

pilot wire. An auxiliary conductor used in connection with remote measuring devices or for operating apparatus at a distant point. *See:* **center of distribution.**   64

pilot-wire-controlled network. A network whose switching devices are controlled by means of pilot wires. *See:* **alternating-current distribution.**   64

pilot wire protection (power switchgear). Pilot protection in which a metallic circuit is used for the communicating means between relays at the circuit terminals. *See:* **wire-pilot protection.**   103

pilot-wire regulator. An automatic device for controlling adjustable gains or losses associated with transmission circuits to compensate for transmission changes caused by temperature variations, the control usually depending upon the resistance of a conductor or pilot wire having substantially the same temperature conditions as the conductors of the circuits being regulated. *See:* **transmission regulator.**   328

pinboard. A perforated board that accepts manually inserted pins to control the operation of equipment.   255, 77, 54

pinch (electron tubes). The part of the envelope of an electron tube or valve carrying the electrodes and through which pass the connections to the electrodes. *See:* **electron tube.**   244, 190

pinch effect (1) (rheostriction). The phenomenon of transverse contraction and sometimes momentary rupture of a fluid conductor due to the mutual attraction of the different parts carrying currents. *See:* **electrothermics; induction heating.**   210
(2) (disk recording). A pinching of the reproducing stylus tip twice each cycle in the reproduction of lateral recordings due to a decrease of the groove angle cut by the recording stylus when it is moving across the record as it swings from a negative to a positive peak.   176
(3) (induction heating). The result of an electromechanical force that constricts, and sometimes momentarily ruptures, a molten conductor carrying current at high density. *See:* **motor effect; skin effect.**   14

p-i-n detector (charged-particle detectors)(germanium gamma-ray detectors)(semiconductor radiation detectors)(X-ray energy spectrometers). A detector consisting of an intrinsic or nearly intrinsic region between a p and n region.   528, 23, 471

PIN diode (fiber optics). A diode with a large intrinsic region sandwiched between p- and n-doped semiconducting regions. Photons absorbed in this region create electron-hole pairs that are then separated by an electric field, thus generating an electric current in a load circuit.　　433

p-i-n diode attenuator (nonlinear, active, and nonreciprocal waveguide components). A device that provides a predetermined value of attenuation in a transmission line in response to a precise value of bias.　　530

p-i-n diode limiter (nonlinear, active, and nonreciprocal waveguide components). A passive microwave power limiter that utilizes the nonlinear conductivity of p-i-n diodes.　　530

pin insulator. A complete insulator, consisting of one insulating member or an assembly of such members without tie wires, clamps, thimbles, or other accessories, the whole being of such construction that when mounted on an insulator pin it will afford insulation and mechanical support to a conductor that has been properly attached with suitable accessories. *See:* **insulator; tower.**　　64

pin jack. A single-conductor jack having an opening for the insertion of a plug of very small diameter.　　341

pink noise (speech quality measurements). A random noise whose spectrum level has a negative slope of 10 decibels per decade.　　126

pins (electron tube or valve). Metal pins connected to the electrodes that plug into the holder. They ensure the electric connection between the electrodes and the external circuit and also mechanically fix the tube in its holder. *See:* **electron tube.**　　244, 190

pip. A popular term for a sharp deflection in a visible trace. *See:* **radar.**　　328

pipe cable. A pressure cable in which the container for the pressure medium is a loose-fitting rigid metal pipe. *See:* **pressure cable; oil-filled pipe cable; gas-filled pipe cable.**　　64

pipe guide. A component of a switch operating mechanism designed to maintain alignment of a vertical rod or shaft.　　27

pipeline (National Electrical Code)(electrical heating systems). A length of pipe including pumps, valves, flanges, control devices, strainers and/or similar equipment for conveying fluids.　　256, 476

pipelined transfer (FASTBUS acquisition and control). The portion of a FASTBUS operation in which a master either sends data to or causes data to be sent by an attached slave on every transition of data sync. The slave acknowledges receipt of or sends data with every transition of data acknowledge. The master does not wait for an acknowledge signal from the slave before causing another data sync transition.　　480

pipe-ventilated (rotating machinery). *See:* **duct-ventilated.**

pip-matching display (navigation)(navigation aid terms). A display in which the received signal appears as a pair of blips, the comparison of the characteristics of which provides a measure of the desired quantity.　　526

Pirani gauge. A bolometric vacuum gauge that depends for its operation on the thermal conduction of the gas present: pressure being measured as a function of the resistance of a heated filament ordinarily over a pressure range of 10^{-1} to 10^{-4} conventional millimeter of mercury. *See:* **instrument.**　　328

piston (high-frequency communication practice) (plunger). A conducting plate movable along the inside of an enclosed transmission path and acting as a short-circuit for high-frequency currents. *See:* **waveguide.**　　328

piston attenuator (waveguide). A variable cutoff attenuator in which one of the coupling devices is carried on a sliding member like a piston. *See:* **waveguide.**　　244, 179

pistonphone. A small chamber equipped with a reciprocating piston of measurable displacement that permits the establishment of a known sound pressure in the chamber. *See:* **loudspeaker.**　　176

pit (rotating machinery). A depressed area in a foundation under a machine.　　63

pitch (acoustics) (audio and electroacoustics). The attribute of auditory sensation in terms of which sounds may be ordered on a scale extending from low to high, such as a musical scale. *Notes:* (1) Pitch depends primarily upon the frequency of the sound stimulus, but it also depends upon the sound pressure and wave form of the stimulus. (2) The pitch of a sound may be described by the frequency of that simple tone, having a specified sound pressure or loudness level, that seems to the average normal ear to produce the same pitch. (3) The unit of pitch is the mel.　　176

pitch angle. *See:* **pitch attitude.**

pitch attitude (navigation aid terms). The angle between the longitudinal axis of the vehicle and the horizontal. *Syn:* **pitch angle.**　　526

pitch factor (rotating machinery). The ratio of the resultant voltage induced in a coil to the arithmetic sum of the magnitudes of the voltages induced in the two coil sides. *See:* **armature.**　　63

pits. Depressions produced in metal surfaces by nonuniform electrodeposition or from electrodissolution: for example, corrosion. *See:* **electrodeposition.**　　328

pitting (corrosion). Localized corrosion taking the form of cavities at the surface.　　205

pitting factor (corrosion). The depth of the deepest pit resulting from corrosion divided by the average penetration as calculated from weight loss.　　205

PIV. *See:* **peak inverse voltage; peak reverse voltage (semiconductor rectifier).**

pivot-friction error. Error caused by friction between the pivots and the jewels: it is greatest when the instrument is mounted with the pivot axis horizontal. *Note:* This error is included with other errors into a combined error defined in **repeatability.** *See:* **moving element (instrument).**　　280

pixel. *See:* **picture element.**

place. In positional notation, a position corresponding to a given power of the base, a given cumulated product, or a digit cycle of a given length. It can usually be specified as the *nth* character from one end of the numerical expression.　　235

plain conductor. A conductor consisting of one metal only. *See:* **conductors.**　　64

plain flange (plane flange) (plain connector) (waveguide). A coupling flange with a flat face. *See:* waveguide.   179

planar array (antennas). A two-dimensional array of elements whose corresponding points lie in a plane.   111

planar network. A network that can be drawn on a plane without crossing of branches. *See:* network analysis.   210

Planckian locus (television). The locus of chromaticities of Planckian (blackbody or full) radiators having various temperatures. *See:* chromaticity diagram.   18

Planck radiation law (illuminating engineering). An expression representing the spectral radiance of a blackbody as a function of the wavelength and temperature. This law commonly is expressed by the formula

$$L_\lambda = I_\lambda/A' = c_{1L}\lambda^{-5}[e^{(c_2/\lambda T)} - 1]^{-1}$$

in which L_λ is the spectral radiance, I_λ is the spectral radiant intensity, A' is the projected area ($A\cos\theta$) of the aperture of the blackbody, e is the base of natural logarithms (2.71828), T is absolute temperature, c_{1L} and c_2 are constants designated as the first and second radiation constants. *Note:* The designation c_{1L} is used to indicate that the equation in the form given here refers to the radiance L, or to the intensity I per unit projected area A', of the source. Numerical values are commonly given not for c_{1L} but for c_1 which applies to the total flux radiated from a blackbody aperture, that is, in a hemisphere (2π steradians), so that, with the Lambert cosine law taken into account, $c_1 = \pi c_{1L}$. The currently recommended value of c_1 is 3.7415 $\times$ $10^{-16} W\cdot m^2$ or 3.7415 $\times$ $10^{-12} W\cdot cm^2$. Then c_{1L} is $1.1910 \times 10^{-16} W\cdot m^2 sr^{-1}$ or $1.1910 \times 10^{-12} W\cdot cm^2 sr^{-1}$. If, as is more convenient, wavelengths are expressed in micrometers and area in square centimeters, $c_{1L} = 1.1910 \times 10^1 W\cdot \mu m^4\cdot cm^{-2} sr - 1$, L$\lambda$ being given in $W\cdot cm^{-2\cdot sr - 1\cdot \mu - 1}$. The presently recommended value of c_2 is 1.43879 cm kelvin. The Planck law in the following form gives the energy radiated from the blackbody in a given wavelength interval ($\lambda_1 - \lambda_2$):

$$Q = \int_{\lambda_1}^{\lambda_2} Q_\lambda d\lambda$$

$$= Atc_1 \int_{\lambda_1}^{\lambda_2} \lambda^{-5} (e^{(c_2/\lambda T)} - 1)^{-1} d\lambda$$

If A is the area of the radiation aperture or surface in square centimeters, t is time in seconds, λ is wavelength in micrometers, $c_1 = 3.7415 \times 10^4 W\cdot \mu m^4\cdot cm^{-2}$, then Q is the total energy in watt seconds (joules) emitted from this area (that is, in the solid angle 2π) in time t, within the wavelength interval ($\lambda_1 - \lambda_2$). *Note:* It often is convenient, as is done here, to use different units of length in specifying wavelengths and areas, respectively. If both quantities are expressed in centimeters and the corresponding value for c_1

(3.7415×10^{-5} erg $\cdot$ cm $\cdot$ sec^{-1}) is used, this equation gives the total emission of energy in ergs from area A (that is in the solid angle 2π), for time t, and for the interval $\lambda_1 - \lambda_2$ in centimeters. *See:* **radiant energy (in illuminating engineering).**   167

Planck's constant (fiber optics). The number h that relates the energy E of a photon with the frequency ν of the associated wave through the relation $E = h$ $\nu.h = 6.626 \times 10^{-34}$ joule second. *See:* **photon.**   433

plane angle (International System of Units) (SI). The SI unit for plane angle is the radian. Use of the degree and its decimal submultiples is permissible when the radian is not a convenient unit. Use of the minute and second is discouraged except for special fields such as cartography. *See:* **units and letter symbols.**   21

plane bend (corner) (waveguide components). A waveguide bend (corner) in which the longitudinal axis of the guide remains in a plane parallel to the plane of the magnetic field vectors throughout the bend (corner).   166

plane-earth factor (radio wave propagation). The ratio of the electric field strength that would result from propagation over an imperfectly conducting plane earth to that which would result from propagation over a perfectly conducting plane.   146

plane flange (waveguide components). *See:* **cover flange.**

plane of polarization (1) (data transmission). For a plane polarized wave, the plane containing the electric intensity and the direction of propagation.   59
(2) (radio wave propagation). For a plane-polarized wave, the plane containing the electric and magnetic field vectors.   146
(3) (antennas). A plane containing the polarization ellipse. *Notes: (A)* When the ellipse degenerates into a line segment, the plane of polarization is not uniquely defined. In general, any plane containing the segment is acceptable: however, for a plane wave in an isotropic medium, the plane of polarization is taken to be normal to the direction of propagation. *(B)* In optics the expression plane of polarization is associated with a linearly polarized plane wave (sometimes called a plane polarized wave) and is defined as a plane containing the field vector of interest and the direction of propagation. This usage would contradict the above one and is deprecated.   111

plane of propagation (radio wave propagation). Of an electromagnetic wave, the plane containing the attenuation vector and the wave normal; in the common degenerate case where these vectors have the same direction, the plane containing the electric vector and the wave normal.   146

plane-parallel resonator (laser-maser). A beam resonator comprising a pair of plane mirrors oriented perpendicular to the axis of the beam.   363

plane-polarized wave (radio wave propagation) At a point in a homogeneous medium, an electromagnetic wave whose electric and magnetic field vectors at all times lie in a fixed plane.   146

plane wave (1) (antennas). A wave in which the only

spatial dependence of the field vectors is through a common exponential factor whose exponent is a linear function of position. *Notes:* (1) In homogeneous and isotropic space the electric field vector, magnetic field vector, and the propagation vector are mutually perpendicular. The ratio of the magnitude of the electric to the magnetic field vector is equal to the intrinsic impedance of the medium; for free space the intrinsic impedance is equal to 376.731 ohms or approximately 120π ohms. (2) A plane wave can be resolved into two component waves corresponding to two orthogonal polarizations. The total power density of the plane wave at a given point in space is equal to the sum of the power densities in the orthogonal component waves.   111

(2) (fiber optics). A wave whose surfaces of constant phase are infinite parallel planes normal to the direction of propagation.   433

(3) (radio wave propagation). A wave whose equiphase surfaces form a family of parallel planes.   146

plane-wave exponential factor (radio wave propagation). The factor $\exp j(\omega t - k\cdot r)$ in the expression for the field parameters of plane waves, where ω is the angular frequency; t, the time; r, the position vector; and k, the wave vector.   146

planned derated hours (PDH)(electric generating unit reliability, availability, and productivity). The available hours during which a basic or extended planned derating was in effect.   567

planned derating (electric generating unit reliability, availability, and productivity). That portion of the unit derating that is scheduled well in advance. *See:* **basic planned derating; extended planned derating.**   567

planned outage (electric generating unit reliability, availability, and productivity). The state in which a unit is unavailable due to inspection, testing, nuclear refueling, or overhaul. A planned outage is scheduled well in advance.   567

planned outage hours (POH)(electric generating unit reliability, availability, and productivity). The number of hours a unit was in the basic or extended planned outage state.   567

planned stop. *See:* **optional stop.**

plan position indication (PPI) (radar). A display on which target blips are shown in plan position, thus forming a map-like display, with radial distance from the center representing range and with the angle of the radius vector representing azimuth angle. *Syn:* **plan position indicator.** *See:* **display.**   13

plan position indicator (PPI)(navigation aid terms). A type of radar display format.   526

plan-position-indicator scope (radar). A cathode-ray oscilloscope arranged to present a plan-position-indicator display.   13

plant (control system). For a given system, that part which is to be controlled and whose parameters are unalterable. *See:* **process equipment.**   56

plant-capacity factor. *See:* **plant factor.**

plant dynamics (control system). Equations which describe the behavior of the plant. *See:* **control system.**   329

plant factor (plant-capacity factor). The ratio of the average load on the plant for the period of time considered to the aggregate rating of all the generating equipment installed in the plant. *See:* **generating station.**   64

plasma (radio wave propagation). A macroscopically neutral assembly of charged and possibly also uncharged particles. A plasma is said to be cold if the thermal effects of charged particles on dynamic processes in the plasma can be neglected for the particular problem involved. It is said to be hot (or warm) if the thermal effects are not negligible.   146

plasma frequency (radio wave propagation). A natural frequency of oscillation of charged particles in a plasma given by

$$f_N = \frac{1}{2\pi} \left(\frac{Nq^2}{\epsilon_0 m} \right)^{1/2}$$

where q is the charge per particle, m is the particle mass, N is the particle number density, and ϵ_0 is the permittivity of free space, all quantities being expressed in meter-kilogram-second (MKS) units. Note: For electrons, $f_n = 8.979 \, N^{1/2}$ in the International System of Units (SI). *See:* **radio wave propagation.**   146

plasma sheath (radio wave propagation). A layer of charged particles of substantially one sign which accumulates around a body in plasma.   146

plastic (rotating machinery). A material that contains as an essential ingredient an organic substance of large molecular weight, is solid in its finished state, and, at some stage in its manufacture or in its processing into finished articles, can be shaped by flow.   63

plastic clad silica fiber (fiber optics). An optical waveguide having silica core and plastic cladding.   433

plate (electron tubes). A common name for an anode in an electron tube. *See:* **electrode (electron tube).**   125

plate (anode) efficiency. The ratio of load circuit power (alternating current) to the plate power input (direct current). *See:* **network analysis.**   111

plate (anode) load impedance. The total impedance between anode and cathode exclusive of the electron stream. *See:* **network analysis.**   111

plate (anode) modulation. Modulation produced by introducing the modulating signal into the plate circuit of any tube in which the carrier is present. *See:* **modulator.**   111, 211

plate (anode) neutralization. A method of neutralizing an amplifier in which a portion of the plate-to-cathode alternating voltage is shifted 180 degrees and applied to the grid-cathode circuit through a neutralizing capacitor. *See:* **amplifier, feedback.**   111, 211

plate (anode) power input. The power delivered to the plate (anode) of an electron tube by the source of supply. *Note:* The direct-current power delivered to the plate of an electron tube is the product of the mean plate voltage and the mean plate current.   111

plate 701 plug adapter lampholder

plate (anode) pulse modulation. Modulation produced in an amplifier or oscillator by application of externally generated pulses to the plate circuit.   111

plateau (radiation-counter tubes). The portion of the counting-rate-versus-voltage characteristic in which the counting rate is substantially independent of the applied voltage.   125, 96

plateau length (radiation-counter tubes). The range of applied voltage over which the plateau of a radiation-counter tube extends.   96, 125

plateau slope, normalized (radiation counter tubes). The slope of the substantially straight portion of the counting rate versus voltage characteristic divided by the quotient of the counting rate by the voltage at the Geiger-Mueller threshold.   125

plateau slope, relative (radiation-counter tubes). The average percentage change in the counting rate near the midpoint of the plateau per increment of applied voltage. *Note:* Relative plateau slope is usually expressed as the percentage change in counting rate per 100-volt change in applied voltage (see accompanying figure).   96,125

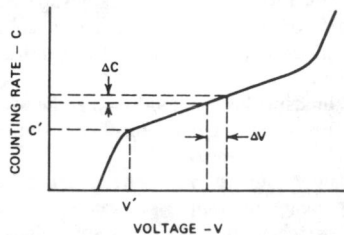

Counting rate–voltage characteristic in which

$$\text{Relative plateau slope} = 100 \frac{\Delta C/C}{\Delta V}$$

$$\text{Normalized plateau slope} = \frac{\Delta C/\Delta V}{C'/V'} = \frac{\Delta C/C'}{\Delta V/V'}$$

plate-circuit detector. A detector functioning by virtue of a nonlinearity in its plate-circuit characteristic.   328

plated-through hole (electronic and electrical applications). Deposition of metal on the side of a hole and on both sides of a base to provide electric connection, and an enlarged portion of conductor material surrounding the hole on both sides of the base.   284

plate keying. Keying effected by interrupting the plate-supply circuit. *See:* **telegraphy.**   111

plate out (1)(nuclear power generating stations). A thermal, electrical, chemical, or mechanical action that results in a loss of material by deposition on surfaces between sampling point and detector.   31

(2)(monitoring radioactivity in effluents). A thermal, electrical, chemical, or mechanical action that results in a loss of material by deposition on surfaces between sampling point and detector.   559

platform control power (series capacitors). Energy source(s) available at platform potential for performing operational and control functions.   474

platform erection (navigation aid terms). In the alignment of inertial systems, the process of bringing the vertical axis of a stable platform system into agreement with the local vertical.   526

platform fault-detection device (series capacitors). A device to detect insulation failure on the platform that results in current flowing from normal current carrying circuit elements to the platform.   474

platform-to-ground signaling devices (series capacitors). Devices to transmit operating, control, and alarm functions to and from the platform.   474

plating rack (electroplating). Any frame used for suspending one or more electrodes and conducting current to them during electrodeposition. *See:* **electroplating.**   328

playback. A term used to denote reproduction of a recording.   328

playback loss. *See:* **translation loss.**

plenary capacitance (between two conductors). The capacitance between two conductors when the changes in the charges on the two are equal in magnitude but opposite in sign and the other $n - 2$ conductors are isolated conductors. *See:* **direct capacitances of a system of conductors.**   210

plenum (National Electrical Code). An air compartment or chamber to which one or more ducts are connected and which forms part of an air-distribution system.   256

pliotron. A hot-cathode vacuum tube having one or more grids. *Note:* This term is used primarily in the industrial field.   190

plotting board. *See:* **recorder, X-Y (plotting board).**

plotting chart (navigation aid terms). A chart designed primarily for plotting dead reckoning lines of position.   526

plow (cable plowing). Equipment capable of laying cable, flexible conduit, etcetera, underground.   52

plow blade (cable plowing). A soil-cutting tool.   52

plow blade amplitude (cable plowing). Maximum displacement of plow blade tip from mean position induced by the vibrator (half the stroke).   52

plow blade frequency (cable plowing). Rate of blade tip vibration in hertz.   52

plowing (cable plowing). A process for installing cable, flexible conduit, etc., by cutting or separating the earth, permitting the cable or flexible conduit to be placed or pulled in behind the blade.   52

plug. A device, usually associated with a cord, that by insertion in a jack or receptacle establishes connection between a conductor or conductors associated with the plug and a conductor or conductors connected to the jack or receptacle.   341

plug adapter (plug body). A device that by insertion in a lampholder serves as a receptacle.   328

plug adapter lampholder (current tap). A device that by insertion in a lampholder serves as one or more receptacles and a lampholder.   328

plugboard (1) (general). A perforated board that accepts manually inserted plugs to control the operation of equipment. *See:* **control panel.**   255, 77

(2) (test, measurement and diagnostic equipment). Patchboard the use of which is restricted to punched card machines. *See:* **patchboard.**   54

plug braking (rotating machinery). A form of electric braking of an induction motor obtained by reversing the phase sequence of its supply. *See:* **asynchronous machine.**   63

plug fuses (protection and coordination of industrial and commercial power systems). Plug fuses are rated 125 volts (V) and are available with current artings up to 30 amperes (A). Their use is limited to circuits rated 125 V or less. However, they may also be used in circuits supplied from a system having a grounded neutral, and in which no conductor operates at more than 150 V to ground. The National Electrical Code (NEC) requires type S fuses in all new installations of plug fuses because they are tamper resistant and size limiting, thus making it difficult to overfuse.   504

plugging (industrial control). A control function that provides braking by reversing the motor line voltage polarity or phase sequence so that the motor develops a counter-torque that exerts a retarding force. *See:* **electric drive.**   206

plug-in. A communication device when it is so designed that connections to the device may be completed through pins, plugs, jacks, sockets, receptacles, or other forms of ready connectors.   328

plug-in-type bearing (rotating machinery). A complete journal bearing assembly, consisting of a bearing liner and bearing housing and any supporting structure that is intended to be inserted into a machine end-shield. *See:* **bearing.**   63

plug-in unit (CAMAC). *See:* **CAMAC plug-in unit.**

plumb-bob vertical (navigation aid terms). The direction indicated by a simple, ideal, frictionless pendulum that is motionless with respect to the earth; it indicates the direction of the vector sum of the gravitational and centrifugal accelerations of the earth at the location of the observer.   526

plumbing (1) (communication) (data transmission). A term employed in communication practice to designate coaxial lines or wave guides and accessory equipment for radio-frequency transmission.   59

(2) (radar). A colloquial expression for pipelike waveguide circuit elements and transmission lines.   13

plumb mark (conductor stringing equipment). A mark placed on the conductor located vertically below the insulator point of support for steel structures and vertically above the pole center line at ground level for wood pole structures used as a reference to locate the center of the suspension clamp.   431

plumb marker pole (conductor stringing equipment). A small diameter, lightweight pole, with a marking device attached to one end, having sufficient length to enable a workman to mark the conductor directly below him from a position on the bridge or arm of the structure. This device is utilized to mark the conduc-

tor immediately after completion of sagging. *Syn:* **marker; offset marker (pole).**   431

plunger relay. A relay operated by a movable core or plunger through solenoid action. *See:* **relay.**   259

plunger, waveguide. *See:* **short circuit, adjustable.**

plural service. *See:* **dual service.**

plural tap (cube tap). *See:* **multiple plug.**

plus input (oscilloscopes). An input such that the applied polarity causes a deflection polarity in agreement with conventional deflection polarity.   106

PM. *See:* **permanent magnet.**

PM. *See:* **phase modulation.**

PM (microwave tubes). *See:* **permanent-magnet focusing.**

pneumatically release-free (trip-free) (as applied to a pneumatically operated switching device) (power switchgear). A term indicating that by pneumatic control the switching device is free to open at any position in the closing stroke if the release is energized. *Note:* This release-free feature is operative even though the closing control switch is held closed.   103

pneumatic bellows, relay. *See:* **relay pneumatic bellows.**

pneumatic controller. A pneumatically supervised device or group of devices operating electric contacts in a predetermined sequence. *See:* **multiple-unit control.**   1

pneumatic loudspeaker. A loudspeaker in which the acoustic output results from controlled variation of an air stream. *See:* **loudspeaker.**   328

pneumatic operation (power switchgear). Power operation by means of compressed gas.   103

pneumatic switch. A pneumatically supervised device opening or closing electric contacts, and differs from a pneumatic controller in being purely an ON and OFF type device. *See:* **control switch.**   1

pneumatic transducer. A unilateral transducer in which the sound output results from a controlled variation of an air stream.   176

pneumatic tubing system (protective signaling). An automatic fire-alarm system in which the rise in pressure of air in a continuous closed tube, upon the application of heat, effects signal transmission. *Note:* Most pneumatic tubing systems contain means for venting slow pressure changes resulting from temperature fluctuations and therefore operate on the so-called rate-of-rise principle. *See:* **protective signaling.**   328

PN sequence. *See:* **pseudonoise sequence.**

pocket-type plate (of a storage cell). A plate of an alkaline storage battery consisting of an assembly of perforated oblong metal pockets containing active material. *See:* **battery (primary or secondary).**   328

poid. The curve traced by the center of a sphere when it rolls or slides over a surface having a sinusoidal profile.   176

Poincare sphere (antennas). A sphere whose points are associated in a one-to-one fashion to all possible polarization states of a plane wave [field vector] according to the following rules: The longitude equals twice the tilt angle and the latitude is twice the angle whose cotangent is the negative of the axial ratio of the polarization ellipse. *Notes:* (1) For this definition the axial ratio carries a sign. (2) The points of the northern hemisphere of a *Poincare* sphere represent polarizations with a right-hand sense. The north pole represents left-hand circular polarization and the south pole right-hand circular polarization. The points of the equator represent all possible linear polarizations.

111

point (1) (for supervisory control or indication or telemeter selection) (power switchgear). All of the supervisory control or indication devices, in a system, exclusive of the common devices, in the master station and in the remote station that are necessary for: (A) Energizing the closing, opening, or other circuits of a unit, or set of units of switchgear or other equipment being controlled, or (B) Automatic indication of the closed or open or other positions of a unit, or set of units of switchgear or other equipment for which indications are being obtained, or (C) Connecting a telemeter transmitting equipment into the circuit to be measured and to transmit the telemeter reading over a channel to a telemeter receiving equipment. *Note:* A point may serve for any two or all three of the purposes described above; for example, when a supervisory system is used for the combined control and indication of remotely operated equipment, point (for supervisory control) and point (for supervisory indication) are combined into a single control and indication point.   103

(2) (positional notation). (A) The character, or the location of an implied symbol, that separates the integral part of a numerical expression from its fractional part. For example, it is called the binary point in binary notation and the decimal point in decimal notation. If the location of the point is assumed to remain fixed with respect to one end of the numerical expressions, a fixed-point system is being used. If the location of the point does not remain fixed with respect to one end of the numerical expression, but is regularly recalculated, then a floating-point system is being used. *Note:* A fixed-point system usually locates the point by some convention, while a floating-point system usually locates the point by expressing a power of the base. *See:* **branchpoint; breakpoint; checkpoint; entry point; fixed point; floating point; rerun point; variable point.**   210, 255, 77

(B) The character, or implied location of such a character, that separates the integral part of a numerical expression from the fractional part. Since the place to the left of the point has unit weight in the most commonly used systems, the point is sometimes called the units point, although it is frequently called the binary point in binary notation and the decimal point in decimal notation. *See:* **breakpoint; fixed point; floating point.**   235

(3) (lightning protection). The pointed piece of metal used at the upper end of the elevation rod to receive a lightning discharge.   328

point contact (semiconductors). A pressure contact between a semiconductor body and a metallic point. *See:* **semiconductor; semiconductor device.**   245

point detector. A device that is a part of a switch-operating mechanism and is operated by a rod connected to a switch, derail, or movable-point frog to indicate that the point is within a specified distance of the stock rail.   328

point equipment (point)(supervisory control, data acquisition, and automatic control). Elements of a supervisory system, exclusive of the basic common equipment, which are peculiar to and required for the performance of a discrete supervisory function. (1) Alarm point. Station (remote or master, or both) equipment(s) that inputs a signal to the alarm function. (2) Accumulator point. Station (remote or master, or both) equipment(s) that accepts a pulsing digital input signal to accumulate a total of pulse counts. (3) Analog point. Station, (remote or master, or both) equipment(s) that inputs an analog quantity to the analog function. (4) Control point. Station (remote or master, or both) equipment(s) that operates to perform the control function. (5) Indication (status) point. Station (remote or master, or both) equipment(s) that accepts a digital input signal for the function of indication. (6) Sequence of events point. Station (remote or master, or both) equipment(s) that accepts a digital input signal to perform the function of registering sequence of events. (7) Telemetering selection point. Station (remote or master, or both) equipment(s) for the selective operation of telemetering transmitting equipment to appropriate telemetering receiving equipment over an interconnecting communication channel. This type of point is more commonly used in electromechanical or stand-alone type of supervisory control. (8) Spare point. Point equipment that is not being utilized but is fully wired and equipped. (9) Wired point. Point for which all common equipment, wiring, and space are provided. To activate the point requires only the addition of plug-in hardware. (10) Space only point. Point for which cabinet space only is provided for future addition or wiring and other necessary plug-in equipment. *Note:* A point may serve for one or more of the purposes described above, for example, when a supervisory system is used for combined control and supervision of remotely operated equipment, a point for supervisory control and point for supervisory indication may be combined into a single control and indication point.   570

pointer (software). (1) An identifier that indicates the location of an item of data. (2) A data item whose value is the location of another data item. *See:* **data; identifier.**   434

pointer pusher (demand meter). The element that advances the maximum demand pointer in accordance with the demand and in integrated-demand meters is reset automatically at the end of each demand interval. *See:* **demand meter.**   328

pointer shift due to tapping. The displacement in the position of a moving element of an instrument that occurs when the instrument is tapped lightly. The

displacement is observed by a change in the indication of the instrument. *See:* **moving element (instrument).**
280

point ID printout (sequential events recording systems). A brief coded method of identifying inputs using alphanumeric characters, usually used in computer based systems.
48

pointing accuracy (communication satellite). The angular difference between the direction in which the main beam of an antenna points and the required pointing direction.
85

point method (formerly called "point-by-point method") (illuminating engineering). A lighting design procedure for predetermining the illuminance at various locations in lighting installations, by use of luminaire photometric data. The direct component of illuminance due to the luminaires and the interreflected component of illuminance due to the room surfaces are calculated separately. The sum is the total illumination at a point.
167

point of fixation (illuminating engineering). A point or object in the visual field at which the eyes look, and upon which they are focused.
167

point of observation (illuminating engineering). For most purposes it may be assumed that the distribution of luminance in the field of view can be described as if there were a single point of observation located at the midpoint of the baseline connecting the centers of the entrance pupils of the two eyes. For many problems it is necessary, however, to regard the centers of the entrance pupils as separate points of observation for the two eyes.
167

point source (laser-maser). A source of radiation whose dimensions are small enough compared with the distance between source and receptor for them to be neglected in calculations.
363

point-to-point control system. *See:* **positioning control system.**

point-to-point radio communication. Radio communication between two fixed stations. *See:* **radio transmission.**
328

point transposition. A transposition, usually in an open wire line, that is executed within a distance comparable to the wire separation, without material distortion of the normal wire configuration outside this distance. *See:* **open wire.**
328

Poisson's equation. In rationalized form:

$$\nabla^2 V = - \frac{\rho}{\epsilon_0 \epsilon}$$

where $\epsilon_0 \epsilon$ is the absolute capacitivity of the medium, V the potential, and ρ the charge density at any point.
210

polar axis (primary ferroelectric terms). A direction that is parallel to the spontaneous polarization vector. When a polar crystal is heated or cooled, the internal or external electrical conduction generally cannot provide enough current to compensate for the change in polarization with temperature, and the crystal develops an electric charge on its surface. For this reason, polar crystals are called pyroelectric. *Note:* For crystal class m, the polar axis is in an arbitrary direction in a plane (the mirror plane). The polar axis for crystal class 1 can be in any arbitrary direction.
497

polar contact. A part of a relay against which the current-carrying portion of the movable polar member is held so as to form a continuous path for current.
328

polar direct-current telegraph system. A system that employs positive and negative currents for transmission of signals over the line. *See:* **telegraphy.**
328

polar-duplex signaling (telephone switching systems). Any method of bidirectional signaling over a line using ground potential compensation and polarity sensing.
55

polarential telegraph system. A direct-current telegraph system employing polar transmission in one direction and a form of differential duplex transmission in the other direction. *Note:* Two kinds of polarential systems, known as types *A* and *B*, are in use. In half-duplex operation of a type-*A* polarential system the direct-current balance is independent of line resistance. In half-duplex operation of a type-*B* polarential system the direct-current balance is substantially independent of the line leakage. *See:* **telegraphy.**
194

Polaris correction (navigation aid terms). A correction to be applied to the corrected sextant altitude of Polaris to obtain latitude.
526

polarity (1) (power and distribution transformer) (instrument transformer). The designation of the relative instantaneous directions of the currents entering the primary terminals and leaving the secondary terminals during most of each half cycle. *Note:* Primary and secondary terminals are said to have the same polarity, when, at a given instant during most of each half cycle, the current enters the identified, similarity marked primary lead and leaves the identified, similarly marked secondary terminal in the same direction as though the two terminals formed a continuous circuit.
53

(2) (battery). An electrical condition determining the direction in which current tends to flow on discharge. By common usage the discharge current is said to flow from the positive electrode through the external circuit. *See:* **battery (primary or secondary).**
328

(3) (television) (picture signal). The sense of the potential of a portion of the signal representing a dark area of a scene relative to the potential of a portion of the signal representing a light area. Polarity is stated as black negative or black positive. *See:* **television.**
178

polarity and angular displacement (regulator). Relative lead polarity of a regulator or a transformer is a designation of the relative instantaneous direction of current in its leads. In addition to its main transformer windings, a regulator commonly has auxiliary transformers or auxiliary windings as an integral part of the regulator. The same principles apply to the polarity of all transformer windings. *Notes:* (1) Primary and secondary leads are said to have the same polarity when at a given instant the current enters an identified secondary lead in the same direction as though the two leads formed a continuous circuit. (2) The relative lead polarity of a single-phase transformer may be either additive or subtractive. If one pair of adjacent leads from the two windings is connected together and voltage applied to one of the windings, then: (A) The relative lead polarity is additive if the voltage across the other two leads of the windings is greater than that of the higher-voltage winding alone. (B) The relative lead polarity is subtractive if the voltage across the other two leads of the winding is less than that of the higher-voltage winding alone. (3) The polarity of a polyphase transformer is fixed by the internal connections between phases as well as by the relative locations of leads: it is usually designated by means of a vector line diagram showing the angular displacement of windings and a sketch showing the marking of leads. The vector lines of the diagram represent induced voltages and the recognized counterclockwise direction of rotation is used. The vector line representing any phase voltage of a given winding is drawn parallel to that representing the corresponding phase voltage of any other winding under consideration. *See:* **voltage regulator.** 257

polarity marks (instrument transformer). The identifications used to indicate the relative instantaneous polarities of the primary and secondary current and voltages. *Notes:* (1) On voltage transformers during most of each half cycle in which the identified primary terminal is positive with respect to the unidentified primary terminal, the identified secondary terminal is also positive with respect to the unidentified secondary terminal. (2) The polarity marks are so placed on current transformers that during most of each half-cycle, when the direction of the instantaneous current is into the identified primary terminal, the direction of the instantaneous secondary current is out of the correspondingly identified secondary terminal. (3) This convention is in accord with that by which standard terminal markings H_1, X_1, etcetera, are correlated. *See:* **instrument transformer.** 203

polarity or polarizing voltage device (36) (power system device function numbers). A device that operates, or permits the operation of, another device on a predetermined polarity only, or verifies the presence of a polarizing voltage in the equipment. 402

polarity-related adjectives (pulse terms). (1) unipolar. Of, having, or pertaining to a single polarity.

(2) bipolar. Of, having, or pertaining to both polarities. 254

polarity test (rotating machinery). A test taken on a machine to demonstrate that the relative polarities of the windings are correct. *See:* **asynchronous machine.** 63

polarizability. The average electric dipole moment produced per molecule per unit of electric field strength. 210

polarization (1)(primary ferroelectric terms). The electric dipole moment per unit volume. Polarization is related to electric displacement D through the linear expression

$$D_i = P_i + \epsilon_0 E_i$$

where the derived constant ϵ_0 (usually called the permittivity of free space) equals $8.854 \cdot 10^{-12}$ coulomb/volt-meter. In ferroelectric materials both D and P are nonlinear functions of E and may depend on previous history of the material. When the electric field is applied along a polar axis that is also a special axis of the prototype phase of the crystal, this expression may then be regarded as a scalar equation, since D, E, and P all point along the same direction. When the term $\epsilon_0 E$ in the above expression is negligible compared to P (as in the case for most ferroelectric materials), D is nearly equal to P; therefore, the D versus E and P versus E plots of the hysteresis loop become, in practice, equivalent. *Note:* The polarization P may be expressed as the bound surface charge per unit area of a free surface normal to the direction of P. 497

(2) (as applied to a relay) (power switchgear). A term identifying the input that provides a reference for establishing the direction of system phenomena such as direction of power or reactive flow, or direction to a fault or other disturbance on a power system. 103

(3) (of a waveguide mode). In some cases, the polarization of the electric field vector on the axis of symmetry of a waveguide. In general, however, the polarization of the mode is not identical to the polarization of the electric field vector in the mode, since the latter varies from point to point in the guide cross-section. *Notes:* (A) Polarization is that property of a degenerate waveguide mode which characterizes a particular mode within a set of degenerate modes. The main application of this concept is to waveguides of square or circular cross-section. (B) When two orthogonal modes can be identified in a square or circular waveguide, a polarization ellipse can be associated with the field vectors and considered in terms of axial ratio, etcetera. 267

(4) (radio wave propagation). Of an electromagnetic wave, a description of the angular variation with time of either the electric or the magnetic field vector at a fixed point. *See:* **elliptically polarized wave.** 146

(5) (radiated wave). That property of a radiated electromagnetic wave describing the time-varying direction and amplitude of the electric field vector: specifically, the figure traced as a function of time by the extremity of the vector at a fixed location in space, as observed along the direction of propagation. *Note:* In general the figure is elliptical and it is traced in a clockwise or counterclockwise sense. The commonly referenced circular and linear polarizations are obtained when the ellipse becomes a circle or a straight line, respectively. Clockwise sense rotation of the electric vector is designated **right-hand polarization** and counterclockwise sense rotation is designated

left-hand polarization. *See:* **radiation.**

179, 111, 146

(6) (desired) (electronic navigation). The polarization of the radio wave for which an antenna system is designed. *See:* **navigation.** 278, 187, 13

(7) (antenna). In a given direction, the polarization of the wave radiated by the antenna. Alternatively, the polarization of a plane wave incident from the given direction which results in maximum available power at the antenna terminals. *Notes:* (A) The polarization of these two waves is the same in the following sense: In the plane perpendicular to the direction considered, their electric fields describe similar ellipses. The sense of rotation on these ellipses is the same if each one is referred to the corresponding direction of propagation, outgoing for the radiated field, incoming for the incident plane wave. (B) When the direction is not stated, the polarization is taken to be the polarization in the direction of maximum gain. 111

(8) (battery). The change in voltage at the terminals of the cell or battery when a specified current is flowing, and is equal to the difference between the actual and the equilibrium (constant open-circuit condition) potentials of the plates, exclusive of the *IR* drop. *See:* **polarization (electrolytic).** 328

polarization capacitance (biological). The reciprocal of the product of electrode capacitive reactance and 2π times the frequency.

$$C_{\mathrm{p}} = \frac{1}{2\pi f X_{\mathrm{p}}}$$

See also: **electrode impedance (biological).** 192

polarization current. *See:* **current, polarization.**

polarization, desired. *See:* **desired polarization.**

polarization diversity reception (data transmission). That form of diversity reception that utilizes separate vertically and horizontally polarized receiving antennas. 59

polarization efficiency (antennas). The ratio of the power received by an antenna from a given plane wave of arbitrary polarization to the power that would be received by the same antenna from a plane wave of the same power density and direction of propagation, whose state of polarization has been adjusted for a maximum received power. *Notes:* (1) The polarization efficiency is equal to the magnitude of the inner product of the polarization vector describing the receiving polarization of the antenna and the polarization vector of the plane wave incident at the antenna. (2) If the receiving polarization of an antenna and the polarization of an incident plane wave are properly located as points on the *Poincare*sphere, then the polarization efficiency is given by the square of the cosine of one-half the angular separation of the two points. *See:* **polarization vector, Note 2.** *Syn:* **polarization mismatch factor.** 111

polarization ellipse (waveguide). The locus of the extremity of a field vector at a fixed point in space. *See:* **polarization of a field vector. Notes (1), (2), (3), and (4).** 267

polarization error (navigation)(navigation aid terms). The error arising from the transmission or reception of an electromagnetic wave having a polarization other than that intended for the system. 526

polarization index (rotating machinery). The ratio of the the insulation resistance of a machine winding measured at 1 min after voltage has been applied divided into the measurement at 10 min. *See:* **ANSI/IEEE Std 43-1974.** 6

polarization index test (rotating machinery). A test for measuring the ohmic resistance of insulation at specified time intervals for the purpose of determining the polarization index. *See:* **asynchronous machine.** 63

polarization match (antennas). The condition that exists when a plane wave, incident upon an antenna from a given direction, has a polarization which is the same as the receiving polarization of the antenna in that direction. *See:* **receiving polarization (of an antenna).** 111

polarization mismatch loss (antennas). The magnitude, expressed in decibels, of the polarization efficiency. 111

polarization of a field vector (1) (antennas). For a field vector at a single frequency at a fixed point in space, the polarization is that property which describes the shape and orientation of the locus of the extremity of the field vector and the sense in which this locus is traversed. *Notes:* (A) For a time harmonic (or single-frequency) vector, the locus is an ellipse with center at the origin. In some cases, this ellipse becomes a circle or a segment of a straight line. The polarization is then called, respectively, circular and linear. (B) The orientation of the ellipse is defined by its plane, called the plane of polarization, and by the direction of its axes. (For a linearly polarized field, any plane containing the segment locus of the field vector is a plane of polarization.) (C) The shape of the ellipse is defined by the axial ratio (major axis)/(minor axis). This ratio varies between infinity and 1 as the polarization changes from linear to circular. Sometimes the ratio is defined as (minor axis)/(major axis). (D) The sense of polarization is indicated by an arrow placed on the ellipse. Alternatively, if the observation is made from a particular side of the plane of polarization the sense can be qualified as clockwise or counterclockwise. It can also be called right hand (or left hand) if, when placing the thumb of the right hand (or left hand) in a specified reference direction normal to the plane of polarization, the sense of travel on the ellipse is indicated by the fingers of the hand. (E) The field vector considered may be the electric field, the magnetic field, or any other field vector, for example, the velocity field in a warm plasma. 111

(2) (waveguide). For a field vector at a single frequency at a fixed point in space, the polarization is that property which describes the shape and orientation of the locus of the extremity of the field vector and the sense in which this locus is traversed. *Notes:* (A) For a time harmonic (or single frequency) vector, the locus is an ellipse with center at the origin. In some cases, this ellipse becomes a circle or a segment of a straight line. The polarization is then called "circular" and "linear", respectively. (B) The orientation of the ellipse is defined by its plane, called the plane of polarization, and by the direction of its axes. (For a linearly polarized field, any plane containing the segment locus of the field vector is a plane of polarization.) (C) The shape of the ellipse is defined by the axial ratio which is the major axis/minor axis. Sometimes the ratio is defined as the reciprocal of the above, that is, minor axis/major axis. Sometimes the ratio is defined as the reciprocal of the above, that is, minor axis. (D) The sense of polarization is indicated by an arrow placed on the ellipse. Alternatively, if the observation is made from a particular side of the plane of polarization, the sense can be qualified as "clockwise" or "counterclockwise". It can also be called "right hand" (or "left hand"), if, when placing the thumb of the right hand (or left hand) in a specified reference direction normal to the plane of polarization, the sense of travel on the ellipse is indicated by the fingers of the hand. (E) The field vector considered may be the electric field, the magnetic field, or any other field vector, for example, the velocity field in a warm plasma.   267

polarization of an antenna. In a given direction from the antenna, the polarization of the wave transmitted by the antenna. *Note:* When the direction is not stated, the polarization is taken to be the polarization in the direction of peak gain. *See:* **polarization of a wave radiated by an antenna.**   111

polarization of a wave (radiated by an antenna in a specified direction). In a specified direction from an antenna and at a point in its far field, the polarization of the (locally) plane wave which is used to represent the radiated wave at that point. *Note:* At any point in the far field of an antenna the radiated wave whose electric field strength is the same as that of the wave and whose direction of propagation is in the radial direction from the antenna. As the radial distance approaches infinity, the radius of curvature of the radiated wave's phase front also approaches infinity and thus in any specified direction the wave appears locally as a plane wave.   111

polarization pattern (of an antenna). (1) The spatial distribution of the polarizations of a field vector excited by an antenna taken over its radiation sphere. (2) The response of a given antenna to a linearly polarized plane wave incident from a given direction and whose direction of polarization is rotating about an axis parallel to its propagation vector; the response being plotted as a function of the angle that the direction of polarization makes with a given reference direction. *Notes:* (A) When describing the polarizations over the radiation sphere (definition 1 above), or a portion of it, reference lines are specified over the sphere, in order to measure the tilt angles of the polarization ellipses and the direction of polarization for linear polarizations. An obvious choice, though by no means the only one, is a family of lines tangent at each point on the sphere to either the θ or ϕ coordinate line

associated with a spherical coordinate system of the radiation sphere. (B) At each point on the radiation sphere the polarization is usually resolved into a pair of orthogonal polarizations, the copolarization and the cross polarization. To accomplish this the copolarization must be specified at each point on the radiation shere. For certain linearly polarized antennas, it is common practice to define the copolarization in the following manner: First specify the orientation of the copolar electric field vector at a pole of the radiation sphere. Then, for all other directions of interest (points on the radiation sphere), require that the angle that the copolar electric field vector makes with each great circle line through the pole remain constant over that circle, the angle being that at the pole. In practice, the axis of the antenna's main beam is directed along the polar axis of the radiation sphere. The antenna is then appropriately oriented about this axis to align the direction of its polarization with that of the defined copolarization at the pole. The manner of defining copolarization can be extended to the case of elliptical polarization by defining the constant angles using the major axes of the polarization ellipses rather than the copolar electric field vector. The sense of polarization must also be specified. (C) The polarization pattern (definition 2 above) generally has the shape of a dumbbell. The polarization ellipse of the antenna in the given direction is similar to one which can be described in the dumbbell shape with points of tangency at the maxima and minima points; thus the axial ratio and tilt angle can be obtained from the polarization pattern. *See:* **copolarization; cross polarization; tilt angle.**   111

polarization-phase vector (for a field vector) (antennas). That polarization vector, among all of those that define the same polarization, which carries the phase information of the field vector whose polarization it represents. *Note:* The polarization-phase vector of the field vector $\bar{E}$ is given by

$$\bar{e} = = \bar{E}/E$$

where E is the magnitude of $\bar{E}$; that is, the positive square root of $\bar{E}^{*} \cdot \bar{E}$.   111

polarization potential (biological). The boundary potential over an interface. *See:* **electrobiology.**   192

polarization ratio (antennas). The magnitude of a complex polarization ratio.   111

polarization reactance (biological). The impedance multiplied by the sine of the angle between the potential vector and the current vector.

$$X_p = Z_p \sin \theta$$

See: **electrode impedance (biological).**   192

polarization receiving factor. The ratio of the power received by an antenna from a given plane wave of arbitrary polarization to the power received by the same antenna from a plane wave of the same power density and direction of propagation, whose state of polarization has been adjusted for the maximum received power. *Note:* It is equal to the square of the absolute value of the scalar product of the polarization unit vector of the given plane wave with that of the radiation field of the antenna along the direction opposite to the direction of propagation of the plane wave. *See:* **waveguide.** 267

polarization resistance (biological). The impedance multiplied by the cosine of the phase angle between the potential vector and the current vector.

$$R_p = Z_p \cos \theta$$

See: **electrode impedance (biological).** 192
polarization state. *See:* **state of polarization.** 111

polarization unit vector (field vector) (at a point). A complex field vector divided by its magnitude. *Notes:* (1) For a field vector of one frequency at a point, the polarization unit vector completely describes the state of polarization, that is, the axial ratio and orientation of the polarization ellipse and the sense of rotation on the ellipse. (2) A complex vector is one each of whose components is a complex number. The magnitude is the positive square root of the scalar product of the vector and its complex conjugate. *See:* **waveguide.** 267

polarization vector (for a field vector) (antennas). A unitary vector which describes the state of polarization of a field vector at a given point in space. *Notes:* (1) Polarization vectors differing only by a unitary factor ($e^{j\alpha}$ where α is real) correspond to the same polarization state. (2) The appropriate inner product, $\hat{e}_1.\hat{e}_2$, for two polarization vectors in the same plane of polarization is given by $\hat{e}_1.\hat{e}_2 = \hat{e}_1{}^*.\hat{e}_2$ where $\hat{e}_1$ and $\hat{e}_2$ represent the polarization vectors corresponding to polarizations 1 and 2. (3) The inner product of polarization vectors representing the same polarization is equal to unity. The inner product of two polarization vectors representing two orthogonal polarizations is zero. (4) The inner product of a polarization vector corresponding to a specified polarization, $\hat{E}_1$, and a complex electric field vector $\vec{E}$, at a point in space will yield the component of the electric field vector corresponding to the specified polarization, $\vec{E}_1$; that is $\vec{E}_1 = \hat{e}_1{}^*.\vec{E}$. (5) The basis vectors for the components of the polarization vector may correspond to any two orthogonal polarizations, the most common being two orthogonal linear polarizations or right-hand and left-hand circular polarizations. 111

polarized electrolytic capacitor. An electrolytic capacitor in which the dielectric film is formed adjacent to only one metal electrode and in which the impedance to the flow of current in one direction is greater than in the other direction. *See:* **electrolytic capacitor.** 328

polarized plug (packaging machinery). A plug so arranged that it may be inserted in its counterpart only in a predetermined position. 429

polarized relay. A relay that consists of two elements, one of which operates as a neutral relay and the other

of which operates as a polar relay. *See:* **neutral relay; polar relay.** 328

polarized snubber (converter circuit elements)(self-commutated converters). A snubber, including a diode, in which the limiting action depends on the direction of voltage or current. 584

polarizer. A substance that when added to an electrolyte increases the polarization. *See:* **electrochemistry.** 328

polar mode (analog computer). *See:* **resolver.**

polar navigation (navigation aid terms). Navigation in polar regions where unique considerations and techniques are applied. 526

polar operation (data transmission). Circuit operation in which mark and space transitions are represented by a current reversal. 59

polar orbit (communication satellite). An inclined orbit with an inclination of 90°. The plane of a polar orbit contains the polar axis of the primary body. 74

polar regions (navigation aid terms). The regions near the geographic poles. Definite limits for these regions are not recognized. 526

polar relay. A relay in which the direction of movement of the armature depends upon the direction of the current in the circuit controlling the armature. *See:* **electromagnetic relay; neutral relay; polarized relay.** 328

pole (1) (illuminating engineering). A standard support generally used where overhead lighting distribution circuits are employed. 167

(2) (pole unit) (of a switching device or fuse) (power switchgear). That portion of the device associated exclusively with one electrically separated conducting path of the main circuit of the device. *Notes:* (A) Those portions which provide a means for mounting and operating all poles together are excluded from the definition of a pole. (B) A switching device or fuse is called single-pole if it has only one pole. If it has more than one pole, it may be called multipole (two-pole, three-pole, etcetera) and provided, in the case of a switching device, that the poles are or can be coupled in such a manner as to operate together. 103

(3) (electric power or communication). A column of wood or steel, or some other material, supporting overhead conductors, usually by means of arms or brackets. *See:* **field pole; pole shoe; tower.** 64

pole body (rotating machinery). The part of a field pole around which the field winding is fitted. *See:* **asynchronous machine.** 63

pole-body insulation (rotating machinery). Insulation between the pole body and the field coil. *See:* **asynchronous machine.** 63

pole bolt (rotating machinery). A bolt used to fasten a pole to the spider. 63

pole-cell insulation (salient pole) (rotating machinery). Insulation that constitutes the liner between the field pole coil and the salient pole body. *See:* **rotor (rotating machinery).** 63

pole-changing winding (rotating machinery). A winding so designed that the number of poles can be changed by simple changes in the coil connections at the winding terminals. *See:* **rotor (rotating machinery); stator.** 63

pole end plate (rotating machinery). A plate or structure at each end of a laminated pole to maintain axial pressure on the laminations. *See:* **asynchronous machine.** 63

pole face (rotating machinery). The surface of the pole shoe or nonsalient pole forming one boundary of the air gap. *See:* **asynchronous machine.** 63

pole-face bevel (rotating machinery). The portion of the pole shoe that is beveled so as to increase the length of the radial air gap. *See:* **asynchronous machine.** 63

pole face, relay. *See:* **relay pole face.**

pole-face shaping (rotating machinery). The contour of the pole shoe that is shaped other than by being beveled, so as to produce nonuniform radial length of the air gap. *See:* **rotor (rotating machinery); stator.** 63

pole fixture. A structure installed in lieu of a single pole to increase the strength of a pole line or to provide better support for attachments than would be provided by a single pole. Examples are *A* fixtures, *H* fixtures, etcetera. 328

pole guy. A tension member having one end securely anchored and the other end attached to a pole or other structure that it supports against overturning. *See:* **tower.** 64

pole line. A series of poles arranged to support conductors above the surface of the ground: and the structures and conductors supported thereon. 328

pole, offset marker (conductor stringing equipment). A small diameter, lightweight pole with a marking device attached to one end, having sufficient length to enable a workman to mark the conductor directly below him from a position on the bridge or arm of the structure. This device is normally utilized when it is necessary to mark the conductor immediately after completion of initial sag for "offset clipping" required to balance the horizontal forces on the structure. *Syn:* **offset marker: marker.** 45

pole piece. A piece or an assembly of pieces of ferromagnetic material forming one end of a magnet and so shaped as to appreciably control the distribution of the magnetic flux in the adjacent medium. 210

pole pitch (rotating machinery). The peripheral distance between corresponding points on two consecutive poles: also expressed as a number of slot positions. *See:* **armature; rotor (rotating machinery); stator.** 63

pole shoe. The portion of a field pole facing the armature that serves to shape the air gap and control its reluctance. *Note:* For round-rotor fields, the effective pole shoe includes the teeth that hold the field coils and wedges in place. *See:* **field pole; rotor (rotating machinery); stator.** 63

pole slipping (rotating machinery). The process of the secondary member of a synchronous machine slipping one pole pitch with respect to the primary magnetic flux. 63

pole steps. Devices attached to the side of a pole, con-

veniently spaced to provide a means for climbing the pole. *See:* **tower.** 64

pole tip (rotating machinery). The leading or trailing extremity of the pole shoe. *See:* **rotor (rotating machinery); stator.** 63

pole-type transformer (power and distribution transformer). A transformer which is suitable for mounting on a pole or similar structure. 53

pole-unit mechanism (of a switching device) (power switchgear). That part of the mechanism that actuates the moving contacts of one pole. 103

pole-zero cancellation (PZ)(X-ray energy spectrometers). A technique used to cancel out the effects of a singularity in an amplifier's transfer function in order to effect a monotonic return of signal pulses to the baseline. 471

policies (safety systems equipment in nuclear power generating stations). Management directives that describe the organization, principles, plans, or courses of action. 534

policy (control system). *See:* **control law.**

poling (ferroelectric material)(primary ferroelectric terms). The process by which a dc electric field exceeding the coercive field is applied to a multidomain ferroelectric to produce a net remanent polarization. 497

poling (1) (general). The adjustment of polarity. Specifically, in wire line practice, it signifies the use of transpositions between transposition sections of open wire or between lengths of cable to cause the residual crosstalk couplings in individual sections or lengths to oppose one another. 328

(2) (ferroelectric material). The process by which a direct-current electric field exceeding the coercive field is applied to a multidomain ferroelectric to produce a net remanent polarization. *See:* **coercive electric field; ferroelectric domain; polarization; remanent polarization.** 80

polishing (electroplating). The smoothing of a metal surface by means of abrasive particles attached by adhesive to the surface of wheels or belts. *See:* **electroplating.** 328

Polish notation. *See:* **prefix notation.**

poll (data transmission). A flexible, systematic method, centrally controlled for permitting stations on a multipoint circuit to transmit without contending for the line. 59

polling (data request)(supervisory control, data acquisition, and automatic control). The process by which a data acquisition system selectively requests data from one or more of its remote terminals. A remote terminal may be requested to respond with all, or a selected portion of, the data available. 570

polling. *See:* **supervisory control.**

polling supervisory system (station control and data acquisition). A system in which the master interrogates each remote to ascertain if there has been a change since the last interrogation. Upon detection of a change the master may request data immediately. 403

polychlorinated biphenyl (PCB)(handling and dispos-

al of transformer grade insulating liquids containing PCBs). Any chemical substance that is limited to the biphenyl molecule that has been chlorinated to varying degrees or any combination of substances that contains such substance $\geq$ 50 parts per million (ppm) dry weight basis. Examples: PCB liquids and nonliquids.   586

polychlorinated biphenyl (PCB) article (handling and disposal of transformer grade insulating liquids containing PCBs). Any manufactured article, other than a PCB container that contains PCBs and whose surface(s) has been in direct contact with PCBs. Examples: PCB large high- and low-voltage capacitors; PCB transformer; PCB cooler motor.   586

polychlorinated biphenyl (PCB) article container (handling and disposal of transformer grade insulating liquids containing PCBs). Any package, can bottle, bag, barrel, drum, tank, or other device used to contain PCB articles or PCB equipment, and whose surface(s) has not been in direct contact with PCBs. Examples: Shipping or storage cartons for capacitors.   586

polychlorinated biphenyl (PCB) container (handling and disposal of transformer grade insulating liquids containing PCBs). Any package, can, bottle, bag, barrel, drum, tank, or other device that contains PCBs or PCB articles and whose surface(s) has been in direct contact with PCBs. Examples: Bottle, barrel, drum, or box.   586

polychlorinated biphenyl (PCB) contaminated electrical equipment (handling and disposal of transformer grade insulating liquids containing PCBs). Any electrical equipment, including but not limited to transformers (including those used in railway locomotives and self-propelled cars), capacitors, circuit breakers, reclosers, voltage regulators, switches (including sectionalizers and motor starters), bushings, electromagnets, and cable, that contain 50 parts per million (ppm) or greater PCBs but less than 500 ppm PCB. Oil-filled electrical equipment other than circuit breakers, reclosers, and cable whose PCB concentration is unknown must be assumed to be PCB-contaminated electrical equipment. Examples: Some oil-filled units; some retrofilled units.   586

polychlorinated biphenyl (PCB) equipment (handling and disposal of transformer grade insulating liquids containing PCBs). Any manufactured item, other than a PCB container, that contains a PCB article or other PCB equipment. Examples: Microwave oven, power-factor-corrected lighting ballast.   586

polychlorinated biphenyl (PCB) item (handling and disposal of transformer grade insulating liquids containing PCBs). Any PCB article, PCB article container, PCB container, or PCB equipment that deliberately or unintentionally contains or has a part of it any PCB or PCBs at a concentration of 50 parts per million (ppm) or greater. Examples: PCB askarel contaminated transformer (mineral) oil, or coolants retrofilled to transformer formerly cooled with askarel.   586

polychlorinated biphenyl (PCB) storage for disposal (handling and disposal of transformer grade insulat-

ing liquids containing PCBs). The facilities meet the following criteria: (1) Adequate roof and walls to prevent rain water from reaching the stored PCBs and PCB items. (2) An adequate floor that has continuous curbing with a minimum 6 inches high curb. The floor and curbing provide a containment volume equal to at least two times the internal volume of the largest PCB article or PCB container stored therein or 25% of the total internal volume of all PCB articles or PCB containers stored therein, whichever is greater. (3) No drain valves, floor drains, expansion joints, sewer lines, or other openings that would permit liquids to flow from the curbed area. (4) Floors and curbing constructed of continuous smooth and impervious materials, such as Portland cement, concrete, or steel, to prevent or minimize penetration of PCBs. (5) Not located at a site that is below the 100-year flood water elevation.   586

polychlorinated biphenyl (PCB) transformer. Any transformer that contains 500 parts per million (ppm) PCB or greater. Examples: PCB askarel-insulated units; some oil-filled units; some retrofilled units.   586

polyphase (as applied to a relay) (power switchgear). A descriptive term indicating that the relay is responsive to polyphase alternating electrical input quantities. *Note:* A multiple-unit relay with individual units responsive to single-phase electrical inputs is not a polyphase relay even though the several single-phase units constitute a polyphase set.   103

polyphase ac fields (measurement of power frequency electric and magnetic fields from ac power lines). Fields whose space components may not be in phase. These field will be produced by polyphase power lines. The field at any point can be described by the field ellipse, that is, by the magnitude and direction of its semiminor axis. *Note:* For polyphase power lines the electric field at large distances ($\geq$ 15 m) away from the outer phases (conductors) can frequently be considered a single-phase field because the minor axis of the electric field ellipse is only a fraction (less than 10%) of the major axis when measured at a height of 1 m above ground level. Similar remarks apply to the magnetic field.   514

polyphase circuit. An alternating-current circuit consisting of more than two intentionally interrelated conductors that enter (or leave) a delimited region at more than two terminals of entry and that are intended to be so energized that in the steady state the alternating voltages between successive pairs of terminals of entry of the phase conductors, selected in a systematic chosen sequence, have: (1) the same period, (2) definitely related and usually equal amplitudes, and (3) definite and usually equal phase differences. If a neutral conductor exists, it is intended also that the voltages from the successive phase conductors to the neutral conductor be equal in amplitude and equally displaced in phase. *Note:* For all polyphase circuits in common use except the two-phase three-wire circuit, it is intended that the voltage amplitudes and the phase differences of the systematically chosen voltages between phase conductors be equal. For a two-phase three-wire circuit it is intended that voltages between two successive pairs of terminals be equal and have a phase difference of p.2 radians, but that the voltage between the third pair of terminals have an amplitude $(2)^{1.2}$ times as great as the other two, and a phase difference from each of the other two of 3 p.4 radians. *See:* **mesh connection of polyphase circuit; star connection of polyphase circuit; zig-zag connection of polyphase circuit.**   210

polyphase machine (rotating machinery). A machine that generates or utilizes polyphase alternating-current power. These are usually three-phase machines with three voltages displaced 120 electrical degrees with respect to each other. *See:* **asynchronous machine.**   63

polyphase symmetrical set (1) (polyphase voltages). A symmetrical set of polyphase voltages in which the angular phase difference between successive members of the set is not zero, π radians, or a multiple thereof. The equations of **symmetrical set of polyphase voltages** represent a polyphase symmetrical set of polyphase voltages if k/m is not zero, 1/2, or a multiple thereof. (The symmetrical set of voltages represented by the equations of **symmetrical set of polyphase voltages** may be said to have polyphase symmetry if k/m is not zero, 1/2, or a multiple of 1/2) *Note:* This definition may be applied to a two-phase four-wire or five-wire circuit if m is considered to be 4 instead of 2. It is not applicable to a two-phase three-wire circuit. *See:* **symmetrical set of polyphase voltages.**   210
(2) **(polyphase currents).** This definition is obtained from the corresponding definitions for voltage by substituting the word current for voltage, the symbol I for E, and β for α wherever they appear. The subscripts are unaltered.   210

polyphase synchronous generator (electric installations on shipboard). A generator whose ac circuits are so arranged that two or more symmetrical alternating electromotive forces with difinite phase relationships are produced at its terminals. Polyphase synchronous generators are usually two-phase, producing two electromotive forces displaced 90 electrical degrees with respect to one another, or three-phase, with three electromotive forces, displaced 120 electrical degrees with respect to each other. (Polyphase generators as used for marine service are generally three-phase. For special cases they may be two-phase.)   3

polyplexer (radar). Equipment combining the functions of duplexing and lobe switching.   13

pondage (power operations). Hydroreserve and limited storage capacity that provides only daily or weekly regulation of streamflow.   516

pondage station. A hydroelectric generating station with storage sufficient only for daily or weekend regulation of flow. *See:* **generating station.**   64

p-on-n solar cells (photovoltaic power system). Photovoltaic energy-conversion cells in which a base of *n-type* silicon (having fixed positive holes in a silicon lattice and electrons that are free to move) is overlaid with a surface layer of *p-type* silicon (having fixed

electrons in a silicon lattice and positive holes that are free to move).   186

pool-cathode mercury-arc converter. A frequency converter using a mercury-arc pool-type discharge device.   14, 114

pool rectifier. A gas-filled rectifier with a pool cathode, usually mercury.   244, 190

pool tube. A gas tube with a pool cathode. *See:* **electronic controller.**   190

population, conceptual (results from a measurement process). The set of measurements that would result from infinite repetition of a measurement process in a state of statistical control.   115

population inversion (laser-maser). A nonequilibrium condition of a system of weakly interacting particles (electronics, atoms, molecules, or ions) which exists when more than one-half of the particles occupy the higher of two energy states.   363

pores (electroplating). Micro discontinuities in a metal coating that extend through to the base metal or underlying coating. *See:* **electroplating.**   363

port (1) (electronic devices or networks). A place of access to a device or network where energy may be supplied or withdrawn or where the device or network variables may be observed or measured. *Notes:* (A) In any particular case, the ports are determined by the way the device is used and not by its structure alone. (B) The terminal pair is a special case of a port. (C) In the case of a waveguide or transmission line, a port is characterized by a specified mode of propagation and a specified reference plane. (D) At each place of access, a separate port is assigned to each significant independent mode of propagation. (E) In frequency changing systems, a separate port is also assigned to each significant independent frequency response. *See:* **network analysis; optoelectronic device; waveguide.**   191, 190, 185
(2) **(rotating machinery).** An opening for the intake or discharge of ventilating air.   63
(3) **(for a waveguide component).** A means of access characterized by a specified reference plane and a specified propagating mode in a waveguide which permits power to be coupled into or out of a waveguide component. *Notes:* (A) At low frequencies the port is synonymous with a terminal pair. (B) To each propagating mode at a specified reference plane there corresponds a distinct port.   267

portability (software). The ease with which software can be transferred from one computer system or environment to another. *See:* **computer system; software.**   434

portable (X-ray) (National Electrical Code). X-ray equipment designed to be hand-carried.   256

portable appliance (National Electrical Code). An appliance which is actually moved or can easily be moved from one place to another in normal use. For the purpose of this article, the following major appliances other than built-in are considered portable if cord-connected; refrigerators, gas range equipment, clothes washers, dishwashers without booster heaters, or other similar appliances.   256

portable battery. A storage battery designed for convenient transportation. *See:* **battery (primary or secondary).** 328

portable computer (computer applications). A personal computer that is designed and configured to permit transportation as a piece of handheld luggage. *Note:* Federal regulations limit use of the term 'portable' to objects weighing no more than 21 pounds. *See:* **laptop computer; transportable computer.** 571

portable concentric mine cable. A double-conductor cable with one conductor located at the center and with the other conductor strands located concentric to the center conductor with rubber or synthetic insulation between conductors and over the outer conductor. *See:* **mine feeder circuit.** 328

portable lighting (illuminating engineering). Lighting involving equipment designed for manual portability. 167

portable luminaire (illuminating engineering). A lighting unit which is not permanently fixed in place. 167

portable mine blower. A motor-driven blower to provide secondary ventilation into spaces inadequately ventilated by the main ventilating system and with the air directed to such spaces through a duct. 328

portable mine cable. An extra-flexible cable, used for connecting mobile or stationary equipment in mines to a source of electric energy when permanent wiring is prohibited or impracticable. *See:* **mine feeder circuit.** 328

portable mining-type rectifier transformer. A rectifier transformer that is suitable for transporting on skids or wheels in the restrictive areas of mines. *See:* **rectifier transformer.** 258

portable parallel duplex mine cable. A double or triple-conductor cable with conductors laid side by side without twisting, with rubber or synthetic insulation between conductors and around the whole. The third conductor, when present, is a safety ground wire. *See:* **mine feeder circuit.** 328

portable shunt (direct current instrument shunts). An instrument shunt with insulating base which may be laid on, or fastened to, any flat surface. It may be used also for switchboard applications where the current is relatively low and connection bars are not used. 527

portable standard watthour meter. A portable meter, principally used as a standard for testing other meters. It is usually provided with several current and voltage ranges and with a readout indicating revolutions and fractions of a revolution of the rotor. *Note:* Electronic portable standards not using a rotor may have a readout indicating equivalent revolutions and fractions of revolutions, or other units such as percentage registration. 212

portable station (mobile communication). A mobile station designed to be carried by or on a person. **Personal** or **pocket** stations are special classes of portable stations. *See:* **mobile communication system.** 181

portable traffic control light (illuminating engineering). A signalling light designed for manual portability

that produces a controllable distinctive signal for purposes of directing aircraft operations in the vicinity of an aerodrome. 167

portable transmitter. A transmitter that can be carried on a person and may or may not be operated while in motion. *Notes:* (1) This has been called a transportable transmitter, but the designation portable is preferred. (2) This includes the class of so-called **walkie-talkies, handy-talkies,** and **personal** transmitters. *See:* **radio transmission; radio transmitter; transportable transmitter.** 211, 111

portable X- or gamma-radiation survey instrument (radiation survey instruments). An instrument with a self-contained energy source (for example, batteries) designed to measure exposure rate while being carried. Such instruments may also have the capability to measure integral exposure, but instruments with the capability of measuring integral exposure only are specifically are specifically excluded from this definition. 558

port difference (hybrid). A port that yields an output proportional to the difference of the electric field quantities existing at two other ports of the hybrid. *See:* **waveguide.** 185

port signal (data transmission). The signal used to telemeter the real time occurrence of polarity reversals of a power voltage or current. The signal may be a pulse train, a square voltage wave, a frequency shift keying (FSK) tone or an FSK carrier wave. Use is generally for frequency or phase-angle telemetering. 59

port sum (hybrid). A port that yields an output proportional to the sum of the electric-field quantities existing at two other ports of the hybrid. *See:* **waveguide.** 185

position (1)(FASTBUS acquisition and control). The location of a module in a crate. The position number corresponds to the geographical address. 480 **(2)(navigation)(navigation aid terms).** The location of a point with respect to a specified or implied coordinate system. 526 *See:* **punch position, sign position.**

positional crosstalk (multibeam cathode-ray tubes). The variation in the path followed by an one electron beam as the result of a change impressed on any other beam in the tube. 125

positional notation (1) (general). A number representation that makes use of an ordered set of digits, such that the value contributed by each digit depends on its position as well as on the digit value. *Note:* The Roman numeral system for example, does not use positional notation. *See:* **binary system; binary-coded-decimal system; biquinary system; decimal system; Gray code.**
(2) One of the schemes for representing numbers, characterized by the arrangement of digits in sequence, with the understanding that successive digits are to be interpreted as coefficients of successive powers of an integer called the base (or radix) of the number system. *Notes:* (A) In the binary number system the successive digits are interpreted as coefficients

of the successive powers of the base two, just as in the decimal number system they relate to successive powers of the base ten. (B) In the ordinary number systems each digit is a character that stands for zero or for a positive integer smaller than the base. (C) The names of the number systems with bases from 2 to 20 are: binary, ternary, quaternary, quinary, senary, septenary, octonary (also octal), novenary, decimal, unidecimal, duodecimal, terdenary, quaterdenary, quindenary, sexadecimal (also hexadecimal), septendecimal, octodenary, novendenary, and vicenary. The sexagenary number system has the base 60. The commonly used alternative of saying **base-3**, **base-4**, etcetera, in place of ternary, quaternary, etcetera, has the advantage of uniformity and clarity. (D) In the most common form of positional notation the expression $\pm$ $a_n a_{n-1} \cdots a_2 a_1 a_0 \cdot a_{-1} a_{-2} \cdots a_{-m}$ is an abbreviation for the sum

$$\pm \sum_{i=-m}^{n} a_i r^i$$

where the point separates the positive powers from the negative powers, the a_i are integers $(0 \le a_i < r)$ called digits, and r is an integer, greater than one, called the base (or radix). *See:* **base; radix.**
(3) A number-representation system having the property that each number is represented by a sequence of characters such that successive characters of the sequence represent integral coefficients of accumulated products of a sequence of integers (or reciprocals of integers) and such that the sum of these products, each multiplied by its coefficient, equals the number. Each occurrence of a given character represents the same coefficient value. *Note:* The biquinary system is an example of (C).
(4) A number-representation system such that if the representations are arranged vertically in order of magnitude with digits of like significance in the same column, then each column of digits consists of recurring identical cycles (for numbers sufficiently large in absolute value) whose length is an integral multiple of the cycle length in the column containing the next-less-significant digits. *Note:* (B), (C), and (d) are not mutually exclusive. The biquinary system is an example of (C) and (D): whereas the Gray code system is an example of (D) only. The binary and decimal systems are examples of (B), (C), and (D).
235, 210, 77

positional response (close-talking pressure microphone). The response-frequency measurements conducted with the principal axis of a microphone collinear with the axis of the artificial voice and the combination of microphone and artificial voice placed at various angles to the horizontal plane. *Note:* Variations in positional response of carbon microphones may be due to gravitational forces. *See:* **close-talking pressure-type microphone.** 249

position changing mechanism (75) (power system device function numbers). A mechanism that is used for moving a main device from one position to another in an equipment; as, for example, shifting a removable circuit breaker unit to and from the connected, disconnected, and test positions. 402

position-control system (industrial control). A control system that attempts to establish and.or maintain an exact correspondence between the reference input and the directly controlled variable, namely physical position. *See:* **control system, feedback.** 206

position index, P (illuminating engineering). A factor which represents the relative average luminance for a sensation at the borderline between comfort and discomfort (BCD), for a source located anywhere within the visual field. 167

position indicator (elevators). A device that indicates the position of the elevator car in the hoistway. It is called a hall position indicator when placed at a landing or a car position indicator when placed in the car. *See:* **control.** 328

position influence (electric instrument). The change in the indication of an instrument that is caused solely by a position departure from the normal operating position. *Note:* Unless otherwise specified, the maximum change in the recorded value caused solely by an inclination in the most unfavorable direction from the normal operating position. *See:* **accuracy rating (instrument).** 280, 294, 295

positioning-control system (numerically controlled machines). A system in which the controlled motion is required only to reach a given end point, with no path control during the transition from one end point to the next. 207

position lights (illuminating engineering). The aircraft aeronautical lights which form the basic or internationally recognized navigation light system. *Note:* The system is composed of a red light showing from dead ahead to 110 degrees to the left, a green light showing from dead ahead to 110 degrees to the right, and a white light showing to the rear through 140 degrees. *Syn:* **navigation lights.** 167

position light signal. A fixed signal in which the indications are given by the position of two or more lights. 328

position of the effective short (microwave switching tube). The distance between a specified reference plane and the apparent short-circuit of the fired tube in its mount. *See:* **gas tube.** 125

position-sensitive detector ((charged-particle detectors). A detector which measures the impact position, in one or more dimensions, of ionizing radiation upon the detector surface. 119

position sensor or position transducer (numerically controlled machines). A device for measuring a position and converting this measurement into a form convenient for transmission. 207

position stopping (industrial control). A control function that provides for stopping the driven equipment at a preselected position. *See:* **electric controller.** 206

position switch (33) (power system device function numbers). A switch that makes or breaks contact

when the main device or piece of apparatus which has no device function number reaches a given position.
402

position-type telemeter. *See:* **ratio-type telemeter.**

positive after-potential (electrobiology). Relatively prolonged positivity that follows the negative after-potential. *See:* **contact potential.** 192

positive column (gas tube). The luminous glow, often striated, in a glow-discharge cold-cathode tube between the Faraday dark space and the anode. *See:* **gas tube.** 190

positive conductor. A conductor connected to the positive terminal of a source of supply. *See:* **center of distribution.** 64

positive creep effect (semiconductor rectifier). The gradual increase in reverse current with time, that may occur when a direct-current reverse voltage is applied to a semiconductor rectifier cell. *See:* **rectification.**
66

positive electrode (1) (primary cell). The cathode when the cell is discharging. The positive terminal is connected to the positive electrode. *See:* **electrolytic cell.** 328
(2) (metallic rectifier). The electrode to which the forward current flows within the metallic rectifying cell. *See:* **rectification.** 328

positive feedback (regeneration) (data transmission). The process by which a part of the power in the output circuit of an amplifying device reacts upon the input circuit in such a manner as to reinforce the initial power, thereby increasing the amplification. 59

positive glow (overhead-power-line corona and radio noise). Positive glow appears at field strength above those required for burst corona and onset streamers. Positive glow is a bright blue discharge appearing as a luminous sheet adhering closely and uniformly to the electrode. The corona current of positive glow is essentially pulseless. *See:* **burst corona; onset streamers.**
411

positive grid. *See:* **retarding field (positive-grid) oscillator.**

positive-grid oscillator tube (Barkhausen tube). A triode operating under oscillating conditions such that the quiescent voltage of the grid is more positive than that of either of the other electrodes. 190

positive logic convention (graphic symbols for logic functions). The representation of the 1-state and the 0-state by the high (H) and low (L) levels, respectively. 451

positive matrix (positive). A matrix with a surface like that which is to be ultimately produced by electroforming. *See:* **electroforming.** 328

positive modulation (in an amplitude-modulation television system). That form of modulation in which an increase in brightness corresponds to an increase in transmitted power. *See:* **television.** 328

positive nonconducting period (rectifier element). The nonconducting part of an alternating-voltage cycle during which the anode has a positive potential with respect to the cathode. *See:* **power rectifier; rectification.** 208

positive noninterfering and successive fire-alarm system. A manual fire-alarm system employing stations and circuits such that, in the event of simultaneous operation of several stations, one of the operated stations will take control of the circuit, transmit its full signal, and then release the circuit for successive transmission by other stations that are held inoperative until they gain circuit control. *See:* **protective signaling.** 328

positive onset streamers (overhead-power-line corona and radio noise). Streamers occurring at field strengths at and slightly above the corona-inception gradient. These appear as bright. blue "brushes" increasing in length to several inches as the gradient is increased. The associated current pulses are of appreciable magnitude, short duration (in the range of hundreds of nanoseconds), and low repetition rate (less than 1 kHz). *Note:* Occurrence of burst corona and positive onset streamers requires the same range of field strength. 411

positive-phase-sequence reactance (rotating machinery). The quotient of the reactive fundamental component of the positive-sequence primary voltage due to the sinusoidal positive-sequence primary current of rated frequency, and the value of this current, the machine running at rated speed. *See:* **asynchronous machine.** 63

positive-phase-sequence relay (power switchgear). A relay that responds to the positive-phase-sequence component of a polyphase input quantity.
127, 103

positive-phase-sequence resistance (rotating machinery). The quotient of the in-phase component of positive-sequence primary voltage corresponding to direct load losses in the primary winding and stray load losses due to sinusoidal positive-sequence primary current, and the value of this current, the machine running at rated speed. 63

positive-phase-sequence symmetrical components (of an unsymmetrical set of polyphase voltages or currents of m phases). The set of symmetrical components that have the first phase sequence. That is, the angular phase lag from the first member of the set to the second, from every other member of the set to the succeeding one, and from the last member to the first, is equal to the characteristic angular phase difference, or $2\pi/m$ radians. The members of this set will reach their positive maxima uniformly in their designated order. The positive-phase-sequence symmetrical components for a three-phase set of unbalanced sinusoidal voltages ($m = 3$), having the primitive period, are represented by the equations

$$e_{a1} = (2)^{1/2} E_{a1} \cos(\omega t + \alpha_{a1})$$

$$e_{b1} = (2)^{1/2} E_{a1} \cos\left(\omega t + \alpha_{a1} - \frac{2\pi}{3}\right)$$

$$e_{c1} = (2)^{1/2} E_{a1} \cos\left(\omega t + \alpha_{a1} - \frac{4\pi}{3}\right)$$

derived from the equation of **symmetrical components of a set of polyphase (alternating) voltages.** Since in this case $r = 1$ for every component (of 1st harmonic) the third subscript is omitted. Then k is 1 for 1st sequence and s takes on the algebraic values 1, 2, and 3 corresponding to phases a, b, and c. The sequence of maxima occurs in the order a, b, c. *See:* **network analysis.**   210

positive plate (storage cell). The grid and active material from which current flows to the external circuit when the battery is discharging. *See:* **battery (primary or secondary).**   328

positive-polarity lightning stroke. A stroke resulting from a positively charged cloud that lowers positive charge to the earth. *See:* **direct-stroke protection (lightning).**   64

positive pre-breakdown streamers (overhead-power-line corona and radio noise). Streamers occurring at field strengths above those required for onset streamers and positive glow. The discharge appears as a light blue filament with branching extending far into the gap. The associated current pulses have high magnitude, short duration (in the range of hundreds of nanoseconds), and low repetition rate (in the range of a few kilohertz). *Note:* When appearing in multiple, these streamers are usually referred to as a plume. When the plume occurs between an electrode and an airborne particle (snow, rain, aerosols, etcetera) coming into near proximity or impacting on the electrode, it is referred to as an impingement plume. When the plume occurs due to the disintegration of water drops resting on the electrode surface, it is referred to as a spray plume.   411

positive-sequence impedance. The quotient of that component of positive-sequence rated-frequency voltage, assumed to be sinusoidal, that is due to the positive-sequence component of current, divided by the positive-sequence component of current. *See:* **asynchronous machine.**   63

positive-sequence resistance. That value of resistance that, when multiplied by the square of the fundamental positive-sequence rated-frequency component of armature current and by the number of phases, is equal to the sum of the copper loss in the armature and the load loss resulting from that current, when the machine is operating at rated speed. Positive-sequence resistance is normally that corresponding to rated armature current. *Note:* Inasmuch as the load loss may not vary as the square of the current, the positive-sequence resistance applies accurately only near the current for which it was determined.   328

positive terminal (battery). The terminal from which the positive electric charge flows through the external circuit to the negative terminal when the cell discharges. *Note:* The flow of electrons in the external circuit is to the positive terminal and from the negative terminal. *See:* **battery (primary or secondary).**   328

post (waveguide). A cylindrical rod placed in a transverse plane of the waveguide and behaving substan-

tially as a shunt susceptance. *See:* **waveguide.**   179

post-accelerating (deflection) electrode (intensifier electrode). An electrode to which a potential is applied to produce post-acceleration. *See:* **electrode (electron tube).**   90

post acceleration (electron-beam tube). Acceleration of the beam electrons after deflection.   125

post-arc current (ac high-voltage circuit breaker). The current which flows through the arc gap of a circuit breaker immediately after current zero, and which has a substantially lower magnitude than the test current.   426

post-deflection acceleration. *See:* **post-acceleration.**

post-dialing delay (telephone switching systems). In an automatic telecommunication system that time interval between the receipt of the last called address digit from the calling station and the application of ringing to the called station.   55

post emphasis. *See:* **deemphasis.**

post equalization. *See:* **deemphasis.**

postfix notation (mathematics of computing). A method of forming mathematical expressions in which each operator is preceded by its operands. For example, A added to B and the result multiplied by C is expressed as $AB + C\times$. *Syn:* reverse **Polish notation.** *See:* **infix notation; prefix notation.**   564

post insulator (composite insulators). Intended to be loaded in tension, bending, or compression. The most common types are a horizontal line post where the post projects nearly horizontally from a pole and is loaded in flexure by the conductor, and a station post insulator used as a bus support in an outdoor substation.   483

post-mortem (computing systems). Pertaining to the analysis of an operation after its completion.   255, 54

post-mortem dump (computing systems). A static dump used for debugging purposes that is performed at the end of a machine run.   255

post processor (numerically controlled machines). A set of computer instructions that transform tool centerline data into machine motion commands using the proper tape code and format required by a specific machine.control system. Instructions such as feedrate calculations, spindle-speed calculations, and auxiliary-function commands may be included.   207

post puller. An electric vehicle having a powered drum handling wire rope used to pull mine props, after coal has been removed, for the recovery of the timber.   328

post, waveguide (waveguide components). A cylindrical rod placed in a transverse plane of the waveguide and behaving substantially as a shunt susceptance.   166

potential diagram (electrode-optical system). A diagram showing the equipotential curves in a plane of symmetry of an electron-optical system. *See:* **electron optics.**   244, 190

potential difference, electrostatic. *See:* **electrostatic potential difference.**

potential, electrostatic. *See:* **electrostatic potential.**

potential energy (of a body or of a system of bodies, in a given configuration with respect to an arbitrarily chosen reference configuration). The work required to bring the system from an arbitrarily chosen reference configuration to the given configuration without change in other energy of the system. 210

potential false-proceed operation. The existence of a condition of vehicle or roadway apparatus in an automatic train control or cab-signal installation under which a false-proceed operation would have occurred had a vehicle approached or entered a section where normally a restrictive operation would occur.

328

potential gradient. A vector of which the direction is normal to the equipotential surface, in the direction of decreasing potential, and of which the magnitude gives the rate of variation of the potential. 244

potential hydro energy (power operations). The possible aggregate energy obtainable over a specified period by practical use of the available streamflow and river gradient. 516

potential profile. A plot of potential as a function of distance along a specified path. 313

potential source-rectifier exciter (excitation systems for synchronous machines). An exciter whose energy is derived from a stationary alternating-current (ac) potential source and converted to direct current by rectifiers. The exciter includes the power potential transformers and power rectifiers which may be either noncontrolled or controlled , including gate circuitry. It is exclusive of input control elements. The source of ac power may come from the machine terminals or from a station auxiliary bus or a separate winding within the synchronous machine. 507

potential transformer (1) (power and distribution transformer). *See:* **voltage transformer; cascade-type voltage transformer; insulated-neutral terminal type voltage transformer; double-secondary voltage transformer; fused-type voltage transformer; turn ratio of a voltage transformer; thermal burden rating of a voltage transformer; rated voltage of a voltage transformer; rated secondary voltage.** 53
(2) (voltage transformer). An instrument transformer that is intended to have its primary winding connected in shunt with a power-supply circuit, the voltage of which is to be measured or controlled. *See:* **instrument transformer.** 212

potential transformer, cascade-type. A single high-voltage line-terminal potential transformer with the primary winding distributed on several cores with the cores electromagnetically coupled by coupling windings and the secondary winding on the core at the neutral end of the high-voltage winding. Each core is insulated from the other cores and is maintained at a fixed potential with respect to ground and the line-to-ground voltage. *See:* **instrument transformer.**

305

potential transformer, double-secondary. One that has two secondary windings on the same magnetic circuit insulated from each other and the primary. Either or both of the secondary windings may be used for measurement or control. *See:* **instrument transformer.**

305

potential transformer, fused-type. One that is provided with the means for mounting a fuse, or fuses, as an integral part of the transformer in series with the primary winding. *See:* **instrument transformer.** 305

potential transformer, grounded-neutral terminal type. One that has the neutral end of the high-voltage winding connected to the case or mounting base. *See:* **instrument transformer.** 305

potential transformer, insulated-neutral terminal type. One that has the neutral end of the high-voltage winding insulated from the case or base and connected to a terminal that provides insulation for a lower-voltage insulation class than required for the rated insulation class of the transformer. *See:* **instrument transformer.** 305

potential transformer, single-high-voltage line terminal. One that has the line end of the primary winding connected to a terminal insulated from ground for the rated insulation class. The neutral end of the winding may be (1) insulated from ground but for a lower insulation class than the line end (insulated neutral) or (2) connected to the case or base (grounded neutral). *See:* **instrument transformer.** 305

potential transformer, two-high-voltage line terminals. One that has both ends of the high-voltage winding connected to separate terminals that are insulated from each other, and from other parts of the transformer, for the rated insulation class of the transformer. *See:* **instrument transformer.** 305

potentiometer (1) (general). A three-terminal rheostat, or a resistor with one or more adjustable sliding contacts, that functions as an adjustable voltage divider. *See:* **potentiometer, function; normal linearity; potentiometer, multiplier; potentiometer, parameter; attenuator; electronic analog computer.** 9
(2) (measurement techniques). An instrument for measuring an unknown electromotive force or potential difference by balancing it, wholly or in part, by a known potential difference produced by the flow of known currents in a network of circuits of known electrical constants. *See:* **instrument.** 328
(3) (analog computers). A resistive element with two end terminals and a movable contact. *See:* **attenuator.**

9

potentiometer, digital coefficient (analog computer). *See:* **digital coefficient potentiometer (hybrid computer linkage component).**

potentiometer, follow-up. A servo potentiometer that generates the signal for comparison with the input signal. *See:* **electronic analog computer.** 9

potentiometer, function. A multiplier potentiometer in which the voltage at the movable contact follows a prescribed functional relationship to the displacement of the contact. *See:* **linearity.** 9

potentiometer granularity (analog computers). The physical inability of a potentiometer to produce an output voltage that varies in other than discrete steps, due either to contacting individual turns of wire in a wire-would potentiometer or to discrete irregularities of the resistance element of composition or film potentiometers.   9, 10

potentiometer, grounded. A potentiometer with one end terminal attached directly to ground. *See:* **electronic analog computer.**   9, 10

potentiometer, linear. A potentiometer in which the voltage at a movable contact is a linear function of the displacement of the contact. *See:* **linearity.**   9

potentiometer, manual (analog computer). A potentiometer which is set by hand. *Syn:* **hand set potentiometer.**   9

potentiometer, multiplier. Any of the ganged potentiometers of a servo multiplier that permit the multiplication of one variable by a second variable. *See:* **electronic analog computer.**   9,10

potentiometer, parameter (scale-factor potentiometer) (coefficient potentiometer). A potentiometer used in an analog computer to represent a problem parameter such as a coefficient or a scale factor. *See:* **electronic analog computer.**   9,10

potentiometer, servo. A potentiometer driven by a positional servomechanism. *See:* **electronic analog computer.**   9, 10

potentiometer set (analog computers). A computer-control state that supplies the same operating potentiometer loading as under computing conditions, and thus allows correct potentiometer adjustment. *See:* **problem check.**   9, 10

potentiometer, sine-cosine. A function potentiometer with movable contacts attached to a rotating shaft so that the voltages appearing at the contacts are proportional to the sine and cosine of the angle of rotation of the shaft, the angle being measured from a fixed referenced position. *See:* **electronic analog computer.**   9

potentiometer, tapered. A function potentiometer that achieves a prescribed functional relationship by means of a nonuniform winding. *See:* **electronic analog computer.**   9

potentiometer, tapped. A potentiometer, usually a servo potentiometer, that has a number of fixed contacts (or taps) to the resistance element in addition to the end and movable contacts. *See:* **electronic analog computer.**   9

potentiometer, ungrounded. A potentiometer with neither end terminal attached directly to ground. *See:* **electronic analog computer.**   9, 10

pothead. A device that seals the end of a cable and provides insulated egress for the conductor or conductors.   288, 289, 323

pothead body. The part of a pothead that joins the entrance fitting to the insulator or to the insulator lid. *See:* **pothead; transformer.**   288, 289, 323

pothead bracket or mounting plate. The part of the pothead used to attach the pothead to the supporting structure. *See:* **pothead; transformer.**   323

pothead bracket or mounting-plate insulator. An insulator used to insulate the pothead from the supporting structure for the purpose of controlling cable sheath currents. *See:* **pothead; transformer.**   323

pothead entrance fitting. A fitting used to seal or attach the cable sheath, armor, or other coverings to the pothead. *See:* **pothead; transformer.**   323

pothead insulator. An insulator used to insulate and protect each conductor passing through the pothead. *See:* **pothead; transformer.**   288, 289, 323

pothead insulator lid. The part of a multi-conductor pothead used to join two or more insulators to the body. *See:* **pothead; transformer.**   288, 289, 323

pothead mounting plate. The part of the pothead used to attach the pothead to the supporting structure. *See:* **transformer.**   288, 289

pothead mounting-plate insulator. An insulator used to insulate the pothead from the supporting structure for the purpose of controlling cable sheath currents. *See:* **transformer.**   288, 289

pothead sheath insulator. An insulator used to insulate an electrically conductive cable sheath or armor from the metallic parts of the pothead in contact with the supporting structure for the purpose of controlling cable sheath currents. *See:* **pothead.**   323

Potier reactance (rotating machinery). An equivalent reactance used in place of the primary leakage reactance to calculate the excitation on load by means of the Potier method. *Note:* It takes into account the additional leakage of the excitation winding on load and in the overexcited region: it is greater than the real value of the primary leakage reactance. It is useful for the calculation of excitation of the machine at other loads and power factors. The height of a Potier reactance triangle determines the reactance drop, and the reactance X_P is equal to the reactance drop divided by the current. The value of Potier reactance is that obtained from the no-load normal-frequency saturation curve: and normally with the excitation for rated voltage and current at zero power factor (overexcited), and at rated frequency. Approximate values of Potier reactance may be obtained from test load excitations at loads differing from rated load, and at power factors other than zero.The excitation results in the range from zero power factor overexcited to unity power factor are close enough to the test values for most practical applications   63

Potter horn (antennas). A circular horn with one or more abrupt changes in diameter which excites two or more waveguide modes in order to produce a specified aperture illumination.   111

poured joint (power cable joint). A joint insulated by the means of a hot or cold poured insulating medium which solidifies.   34

poured joint (power cable joints). A joint insulated by the means of a hot or cold poured insulating medium which solidifies.   34

power (1)(fiber optics). *See:* **irradiance; radiant intensity; radiant power.**   433

(2)(power operations). The rate (in kilowatts)(kW) of generating, transferring, or using energy. *See:* **power control center; emergency power; power factor adjustment clause; firm power; interruptible power; nonfirm power; power pool.**   516

(3)(used as an adjective) (power switchgear). A general term, used, by reason of specific physical or electrical characteristics, to denote application or restriction or both, to generating stations, switching stations, or substations. The term may also denote use or application to energy purposes as contrasted with use for control purposes. 103

(4) (laser-maser) Φ. The time rate at which energy is emitted, transferred, or received: usually expressed in watts (or in joules per second). 363

power, active (polyphase circuit) (power†). At the terminals of entry of a polyphase circuit into a delimited region, the algebraic sum of the active powers for the individual terminals of entry when the voltages are all determined with respect to the same arbitrarily selected common reference point in the boundary surface (which may be the neutral terminal of entry). *Notes:* (1) The active power for each terminal of entry is determined by considering each conductor and the common reference point as a single-phase two-wire circuit and finding the active power for each in accordance with the definition of **power, active (single-phase two-wire circuit).** If the voltages and currents are sinusoidal and of the same period, the active power *P* for a three-phase circuit is given by

$$P = E_a I_a \cos(\alpha_a - \beta_a) + E_b I_b \cos(\alpha_b - \beta_b) \\ + E_c I_c \cos(\alpha_c - \beta_c)$$

where the symbols have the same meaning as in **power, instantaneous (polyphase circuit).** (2) If there is no neutral conductor and the common point for voltage measurement is selected as one of the phase terminals of entry, the expression will be changed in the same way as that for **power, instantaneous (polyphase circuit).** (3) If both the voltages and the currents in the preceding equations constitute symmetrical sets of the same phase sequences

$$P = 3E_a I_a \cos(\alpha_a - \beta_a).$$

(4) In general the active power *P* at the ($m = 1$) terminals of entry of a polyphase circuit of *m* phases to a delimited region, when one of the terminals is the neutral terminal of entry, is expressed by the equation

$$P = \sum_{s=1}^{s=m} \sum_{r=1}^{r=\infty} E_{sr} I_{sr} \cos(\alpha_{sr} - \beta_{sr})$$

where E_{sr} is the root-mean-square amplitude of the *r*th harmonic of the voltage e_s, from phase conductor to neutral. I_{sr} is the root-mean-square amplitude of the *r*th harmonic of the current i_s through terminal *s*. α_{sr} is the phase angle of the *r*th harmonic of e_s with respect to a common reference. β_{sr} is the phase angle of the *r*th harmonic of i_s with respect to the same reference

as the voltages. The indexes *s* and *r* have the same meaning as in **power, instantaneous (polyphase circuit).** (5) The active power can also be stated in terms of the root-mean-square amplitudes of the symmetrical components of the voltages and currents as

$$P = m \sum_{k=0}^{k=m-1} \sum_{r=1}^{r=\infty} E_{kr} I_{kr} \cos(\alpha_{kr} - \beta_{kr})$$

where *m* is the number of phase conductors, *k* denotes the number of the symmetrical component, and *r* denotes the number of the harmonic component. (6) When the voltages and currents are quasi-periodic and the amplitudes of the voltages and currents are slowly varying, the active power for the circuit of each conductor may be determined for this condition as in **power, active (single-phase two-wire circuit).** The active power for the polyphase circuit is the sum of the active power values for the individual conductors. The active power is also the time average of the instantaneous power for the polyphase circuit. (7) Mathematically the active power at any time t_0 is

$$P = \frac{1}{T} \int_{t_0-T/2}^{t_0+T/2} p\,dt$$

where *p* is the instantaneous power and *T* is the period. This formulation may be used when the voltage and current are periodic or quasi-periodic so that the period is defined. The active power is expressed in watts when the voltages are in volts and the currents in amperes.

†When it is clear that average power and not instantaneous power is meant, **power** is often used for **active power** 210

power, active (single-phase two-wire circuit) (power) †. At the terminals of entry of a single-phase, two-wire circuit into a delimited region, when the voltage and current are periodic or quasi-periodic, the time average of the values of the instantaneous power, the average being taken over one period. *Notes:* (1) Mathematically, the active power *P* at a time t_0 is given by the equation

$$P = \frac{1}{T} \int_{t_0-T/2}^{t_0+T/2} p\,dt$$

where *T* is the period, and *p* is the instantaneous power. (2) If both the voltage and current are sinusoidal and of the same period the active power *P* is given by

$$P = EI \cos(\alpha - \beta)$$

in which the symbols have the same meaning as in **power, instantaneous (two-wire circuit).** (3) If both the voltage and current are sinusoidal, the active power P is also equal to the real part of the product of the phasor voltage and the conjugate of the phasor current, or to the real part of the product of the conjugate of the phasor voltage and the phasor current. Thus

$$P = \text{Re } \mathbf{EI^*}$$
$$= \text{Re } \mathbf{E^*I}$$
$$= \frac{1}{2}[\mathbf{EI^*} + \mathbf{E^*I}]$$

in which $\mathbf{E}$ and $\mathbf{I}$ are the root-mean-square phasor voltage and root-mean-square phasor current, respectively (see **phasor quantity**), and the * denotes the conjugate of the phasor to which it is applied. (4) If the voltage is an alternating voltage and the current is an alternating current (see **alternating voltage** and **alternating current**), the active power is given by the equations

$$P = E_1I_1\cos(\alpha_1 - \beta_1)$$
$$+ E_2I_2\cos(\alpha_2 - \beta_2) + \cdots$$
$$= \sum_{r=1}^{r=\infty} \mathbf{E}_r\mathbf{I}_r\cos(\alpha_r - \beta_r)$$
$$= \text{Re} \sum_{r=1}^{r=\infty} E_rI_r$$
$$= \frac{1}{2}\sum_{r=1}^{r=\infty}[E_rI_r + E_rI_r]$$

in which r is the order of the harmonic component of the voltage (see **harmonic components (harmonics)**) and r is also the order of the harmonic component of the current. E_r and I_r are the phasors corresponding to the r th harmonic of the voltage and current, respectively. (5) If the voltage and current are quasi-periodic functions of the form given in **power, instantaneous (two-wire circuit)**, the integral over the period T will not result in the simple expressions that are obtained when E_r and I_r are constant. However, if the relative rates of change of the quantities are so small that each may be considered to be constant during any one period, but to have slightly different values in successive periods, the active power at any time t is very closely approximated by

$$P = \sum_{r=1}^{r=\infty} E_r(t)I_r(t)\cos(\alpha_r - \beta_r)$$

which is analogous to the preceding expression. When the amplitudes of voltage and current are slowly changing, the active power may be represented by this expression. (6) Active power is expressed in watts when the voltage is in volts and the current in amperes.

† When it is clear that average power and not instantaneous power is meant, **power** is often used for **active power**.   210

power amplification (1) (general). The ratio of the power level at the output terminals of an amplifier to that at the input terminals. Also called **power gain.** *See:* **amplifier; power gain.**   111
(2) **(magnetic amplifier).** The product of the voltage amplification and the current amplification.   171
power amplifier. *See:* **amplifier, power.**
power, apparent (1) (rotating machinery). The product of the root-mean-square current and the root-mean-square voltage. *Note:* It is a scalar quantity equal to the magnitude of the phasor power. *See:* **asynchronous machine.**   63
(2) **(polyphase circuit).** At the terminals of entry of a polyphase circuit, a scalar quantity equal to the magnitude of the vector power. *Notes:* (A) In determining the apparent power, the reference terminal for voltage measurement shall be taken as the neutral terminal of entry, if one exists, otherwise as the true neutral point. (B) If the ratios of the components of the vector power, for each of the terminals of entry, to the corresponding apparent power are the same for every terminal of entry, the total apparent power is equal to the arithmetic apparent power for the polyphase circuit: otherwise the apparent power is less than the arithmetic apparent power. (C) If the voltages have the same wave form as the corresponding currents, the apparent power is equal to the amplitude of the phasor power. (D) Apparent power is expressed in volt-amperes when the voltages are in volts and the currents in amperes.   210
(3) **(single-phase two-wire circuit).** At the two terminals of entry of a single-phase two-wire circuit into a delimited region, a scalar equal to the product of the root-mean-square voltage between one terminal of entry and the second terminal of entry, considered as the reference terminal, and the root-mean-square value of the current through the first terminal. *Notes:* (A) Mathematically the apparent power U is given by the equation

$$U = EI$$
$$= (\pm)(E_1{}^2 + E_2{}^2 + \cdots + E_r{}^2 + \cdots)^{1/2}$$
$$\times (I_1{}^2 + I_r{}^2 + \cdots + I_q{}^2 + \cdots)^{1/2}$$

in which E and I are the root-mean-square amplitudes of the voltage and current, respectively. E_r and I_q are the root-mean-square amplitudes of the rth harmonic of voltage and the qth harmonic of current, respectively. (B) If both the voltage and current are sinusoidal and of the same period, so that the distortion power is zero, the apparent power becomes

$$U = EI = E_1I_1$$

in which E_1 and I_1 are the root-mean-square amplitudes of voltage and current of the primitive period. The apparent power is equal to the amplitude of the phasor power. (C) If the voltage and current are quasiperiodic and the amplitude of the voltage and current components are slowly varying, the apparent power at any instant may be taken as the value derived from the amplitudes of the components at that instant. (D) Apparent power is expressed in volt-amperes when the voltage is in volts and the current in amperes. Because apparent power has the property of magnitude only and its sign is ambiguous, it does not have a definite direction of flow. For convenience it is usually treated as positive.   210

power, auxiliary (thyristor power converter). Input power used by the thyristor converter to perform its various auxiliary functions as opposed to the power that may be flowing between the ac supply and the load.   121

power, auxiliary, general (thyristor power converter). The power required for fans or blowers, relays, breaker control, phase loss detection, etcetera.   121

power, available (1) (audio and electroacoustics). The maximum power obtainable from a given source by suitable adjustment of the load. *Note:* For a source this is equivalent to a constant sinusoidal electromotive force in series with an impedance independent of amplitude, the available power is the mean-square value of the electromotive force divided by four times the resistive part of the impedance of the source.
                                239, 252, 210

(2) (at a port). The maximum power that can be transferred from the port to a load. *Note:* At a specified frequency, maximum power transfer will take place when the impedance of the load is the conjugate of that of the source. The source impedance must have a positive real part. *See:* **network analysis.**   190

(3) (of a sound field with a given object placed in it). The power that would be extracted from the acoustic medium by an ideal transducer having the same dimensions and the same orientation as the given object. The dimensions and the orientation with respect to the sound field must be specified. *Note:* The acoustic power available to an electroacoustic transducer, in a plane-wave sound field of given frequency, is the product of the free-field intensity and the effective area of the transducer. For this purpose the effective area of an electroacoustic transducer, for which the surface velocity distribution is independent of the manner of excitation of the transducer, is set 1.4 times the product of the receiving directivity factor and the square of the wavelength of a free progressive wave in the medium. If the physical dimensions of the transducer are small in comparison with the wavelength, the directivity factor is near unity, and the effective area varies inversely as the square of the frequency. If the physical dimensions are large in comparison with the wavelength, the directivity factor is nearly proportional to the square of the frequency, and the effective

area approaches the actual area of the active face of the transducer.   176

(4) (signal generators). The power at the output port supplied by the generator into a specified load impedance. *See:* **signal generator.**   185

power, average phasor (single-phase two-wire, or polyphase circuit). A phasor of which the real component is the average active power and the imaginary component is the average reactive power. The amplitude of the phasor power is

$$S_{av} = [(P_{av})^2 + (Q_{av})^2]^{1/2}$$

which P_{av} and Q_{av} are the active and the reactive power, respectively.   210

power capacitor (shunt power capacitors). An assembly of dielectric and electrodes in a container (case), with terminals brought out, which is intended to introduce capacitance into an electric power circuit. *Note:* The abbreviated term capacitor is used interchangeably with power capacitor throughout this standard.
                                138

power capacity (waveguide). The maximum power which can be carried by the waveguide under a specified set of environmental and circuit conditions with a desired safety factor.   267

power, carrier-frequency, peak pulse. *See:* **peak pulse power, carrier-frequency.**

power-circuit limit switch (industrial control). A limit switch the contacts of which are connected into the power circuit. *See:* **switch.**   206

power circuit protector (power switchgear) (low-voltage ac power circuit protectors). An assembly consisting of a modified low-voltage power circuit breaker, which has no direct-acting tripping devices, with a current-limiting fuse in series with the load terminals of each pole.   103, 158

power-closed car door or gate (elevators). A door or gate that is closed by a car-door or gate power closer or by a door or gate power operator. *See:* **elevators.**
                                328

power coefficient (characteristic insertion loss) (attenuator). Temporary and reversible variation in decibels when input power is varied from 20 dB below full rated power or lower to full rated power after steady-state condition has been reached.   110

power, commercial. Power furnished by an electric power utility company: when available, it is usually the prime power source.   89

power conditioning subsystem (PCS)(terrestrial photovoltaic power systems). The subsystem that converts the dc power from the array subsystem to dc or ac power that is compatible with system requirements. *See:* **array control**   496

power control center (power operations). The location where power system operators monitor, analyze, or control power systems using digital or analog teleprocessing systems.   516

power density (1) (fiber optics). *See:* **irradiance.**
                                433

(2) (radio wave propagation) Of a traveling wave, the

time average of the Poynting vector. *Syn:* **power flux density.** 146

(3)(of an electromagnetic wave)(control of system electromagnetic compatibility). Emitted power per unit cross-sectional area normal to the direction of propagation. 495

power-density spectrum (electromagnetic compatibility). A plot of power density per unit frequency as a function of frequency. 199

power detection. That form of detection in which the power output of the detecting device is used to supply a substantial amount of power directly to a device such as a loudspeaker or recorder. *See:* **detection.** 328

power dissipation, (P) (light emitting diodes). The time average product of current times voltage of the device. 162

power distribution, underground cables. *See:* **cable bedding; cable separator; base ambient temperature; aluminum-covered steel wire.**

power divider (waveguide). A device for producing a desired distribution of power at a branch point. *See:* **waveguide.** 244, 179

power, effective radiated (mobile communication). The product in a given direction of the effective gain of the antenna in that direction over a half-wave dipole antenna, and the antenna power input. *See:* **mobile communication system.** 181

power elevator. An elevator utilizing energy other than gravitational or manual to move the car. *See:* **elevators.** 328

power, emergency (electric power systems). Power required by a system to make up a deficiency between the current firm power demand and the immediately available generating capability. 112

power factor (1)(electrical heating systems). The ratio of the circuit power (watts) to the circuit volt-amperes. 476

(2)(total)(converter characteristics)(self-commutated converters). The ratio of the total active power in watts to the total apparent power in volt-amperes (the product of root-mean-square [rms] voltage and rms current) on the ac (alternating current) side of the converter. *Note:* This definition includes the effect of harmonic components of current and voltage, as well as the effect of phase displacement between current and voltage. If either the voltage or the current is an undistorted sine wave, and d is the distortion factor of the other, then the following relationship is true: 584

(3)(total)(harmonic control and reactive compensation of static power converters). The ratio of the total power input in watts to the total volt-ampere input to the converter. *Notes:* (1) This definition includes the effect of harmonic components of current and voltage, the effect of phase displacement between current and voltage, and the exciting current of the transformer. Volt-amperes is the product of root-mean-square (rms) voltage and rms current. (2) The power factor is determined at the ac line terminals of the converter. 533

(4) (general). The ratio of total watts to the total root-mean-square (RMS) volt-amperes.

$$F_P = \frac{\Sigma \text{ Watts per Phase}}{\Sigma \text{ RMS Volt-Amperes per Phase}}$$
$$= \frac{\text{Active Power}}{\text{Apparent Power}}$$

Note: If the voltages have the same waveform as the corresponding currents, power factor becomes the same as phasor power factor. If the voltages and currents are sinusoidal and, for polyphase circuits, form symmetrical sets, $F_P = \cos(\alpha - \beta)$. *See:* **asynchronous machine.** 210, 244, 186, 63

(5) (rectifier or rectifier unit) (thyristor converter). The ratio of the total watts input (total power input in watts) to the total volt-ampere input to the rectifier, rectifier unit or converter. *Notes:* (A) This definition includes the effect of harmonic components of current and voltage, the effect of phase displacement between the current and voltage, and the exciting current of the transformer. Volt-ampere is the product of root-mean square volts and root-mean-square amperes. (B) It is determined at the alternating-current line terminals of the thyristor converter or rectifier unit. *See:* **power rectifier; rectification.** 121

(6) (dielectric) (rotating machinery). The cosine of the dielectric phase angle or the sine of the dielectric loss angle. 63

(7) (outdoor apparatus bushings) (insulation). The ratio of the power dissipated in the insulation, in watts, to the product of the effective voltage and current in voltamperes, when tested under a sinusoidal voltage and prescribed conditions. *Note:* The insulation power factor is equal to the cosine of the phase angle between the voltage and the resulting current when both the voltage and current are sinusoidal. 168

(8) (metering). The ratio of the active power to the apparent power. 212

(9)(thyristor). The ratio of the total watts to the total voltamperes. *Note:* This definition includes the effect of harmonic components of current and voltage, and the effect of phase displacement between current and voltage. 445

power factor adjustment clause (power operations). A clause in a rate schedule that provides for an adjustment in the billing if the customer's power factor varies from a specified reference. 516

power-factor angle. The angle whose cosine is the power factor. *See:* **asynchronous machine.** 63

power factor, arithmetic. The ratio of the active power to the arithmetic apparent power. The arithmetic power factor is expressed by the equation

$$F_{pa} = \frac{P}{U_a}$$

where F_{pa} = arithmetic power factor
P = active power
U_a = arithmetic apparent power.

Note: Normally power factor, rather than arithmetic power factor, will be specified, but in particular cases, especially when the determination of the apparent power for a polyphase circuit is impracticable with the available instruments, arithmetic power factor may be used. When arithmetic power factor and power factor differ, arithmetic power factor is the smaller.   210

power factor, coil Q. *See:* **coil Q.**

power-factor-corrected mercury-lamp ballast. A ballast of the multiple-supply type that has a power-factor-correcting device, such as a capacitor, so that the input current is at a power factor in excess of that of an otherwise comparable low-power-factor ballast design, but less than 90 percent, when the ballast is operated with center rated voltage impressed upon its input terminals and with a connected load, consisting of the appropriate reference lamp(s), operated in the position for which the ballast is designed. The minimum input power factor of such a ballast should be specifically stated.   271

power factor, dielectric. *See:* **dielectric power factor.**

power factor, displacement. *See:* **displacement power factor.**

power-factor influence (electric instruments). The change in the recorded value that is caused solely by a power-factor departure from a specified reference power factor maintaining constant power (or vars) at rated voltage, and not exceeding 120 percent of rated current. It is to be expressed as a percentage of the full-scale value. *See:* **accuracy rating (instrument).**   280, 294

power-factor meter. A direct-reading instrument for measuring power factor. It is provided with a scale graduated in power factor. *See:* **instrument.**   328

power factor relay (54) (power system device function numbers). A relay that operates when the power factor in an alternating-current (ac) circuit rises above or falls below a predetermined value.   402

power-factor tip-up (rotating-machinery stator-coil insulation). The difference between the power-factors measured at two different designated voltages applied to an insulation system, other conditions being constant. *Notes:* (1) Used mainly as a measure of discharges, and hence of voids, within the system at the higher voltage. (2) The incremental change in power factor divided by incremental change in voltage applied to an insulation system. *See:* **asynchronous machine.** (3) Tip-up tests may be made using dissipation factor (tan δ) instead of power factor. In this case the tip-up is often identified as Δ tan δ or delta tan delta.   22

power-factor tip-up test (rotating machinery). A test applied to insulation to determine the power-factor tip-up. *See:* **asynchronous machine.**   63

power-factor-voltage characteristic (rotating machinery stator-coil insulation). The relation between the magnitude of the applied test voltage and the measured power factor of the insulation. *Note:* The characteristic is usually shown as a curve of power factor plotted against test voltage. *See:* **asynchronous machine.**   22

power failure (emergency and standby power). Any variation in electric power supply that causes unacceptable performance of the user's equipment.   512

power feeder. A feeder supplying principally a power or heating load. *See:* **feeder.**

power flux density. *See:* **power density.**   146

power-frequency current-interrupting rating (of an expulsion arrester) (surge arrester). A designation of the range of the symmetrical root-mean-square fault currents of the system for which the arrester is designed to operate. An expulsion arrester is given a maximum current-interrupting rating and may also have a minimum current-interrupting rating.   430

power-frequency sparkover voltage (surge arrester)(-metal-oxide surge arresters for ac power circuits). The root-mean-square (rms) value of the lowest power-frequency sinusoidal voltage that will cause sparkover when applied across the terminals of an arrester.   430, 583

power-frequency withstand voltage (surge arrester)(-metal-oxide surge arresters for ac power circuits). A specified root-mean-square (rms) test voltage at power frequency that will not cause a disruptive discharge.   430,583

power fuse (power switchgear). A fuse consisting of an assembly of a fuse support and a fuse unit or fuseholder which may or may not include the refill unit or fuse link. *Note:* The power fuse is identified by the following characteristics: (1) Dielectric withstand (BIL) strengths at power levels. (2) Application primarily in stations and substations. (3) Mechanical construction basically adapted to station and substation mountings.   103, 443

power fuse unit (National Electrical Code) (installations and equipment operating at over 600 volts, nominal). A vented, nonvented or controlled vented fuse unit in which the arc is extinguished by being drawn through solid material, granular material, or liquid, either alone or aided by a spring.   256

power gain (1) (data transmission). (General). The ratio of the signal power that a transducer delivers to its load to the signal power absorbed by its input circuit. *Notes:* (A) Power gain is usually expressed in decibels. (B) If more than one component is involved in the input or output, the particular components used are specified. (C) If the output signal power is at a frequency other than the input signal power, the gain is a conversion gain.   59

(2) (antenna) (A) (referred to a specified polarization). The power gain of an antenna, reduced by the ratio of that portion of the radiation intensity corresponding to the specified polarization to the radiation intensity. (B) **(in physical media).** In a given direction and at a given point in the far field, the ratio of the power flux per unit area from an antenna to the power flux per unit area from an isotropic radiator at a specified location with the same power input as the subject antenna. *Note:* The isotropic radiator must be within the smallest sphere containing the antenna. Suggested locations are antenna terminals and points of symmetry if such exist.   111

(3) (two-port linear transducer). At a specified frequency, the ratio of (A) the signal power that the transducer delivers to a specified load, to (B) the signal power delivered to its input Port. *Note:* The power gain is not defined unless the input impedance of the transducer has a positive real part.   125, 111

power influence (telephone loop performance). The power of a longitudinal signal induced in a telephone circuit by an electromagnetic field emanating from a conductor or conductors of a power system. In common usage, power influence is synonymous with longitudinal noise.   473

power, instantaneous (1) (circuit). At the terminals of entry into a delimited region the rate at which electric energy is being transmitted by the circuit into or out of the region. *Note:* Whether power into the region or out of the region is positive is a matter of convention and depends upon the selected reference direction of energy flow. *See:* **sign of power of energy.**   210

(2) (polyphase circuit). At the terminals of entry into a delimited region, the algebraic sum of the products obtained by multiplying the voltage between each terminal of entry and some arbitrarily selected common point in the boundary surface (which may be neutral terminal of entry) by the current through the corresponding terminal of entry. *Notes:* (A) The reference direction of each current must be the same, either into or out of the delimited region. The reference polarity for each voltage must be consistently chosen, either with all the positive terminals at the terminals of entry and all negative terminals at the common reference point, or vice-versa. If the reference direction for currents is into the delimited region and the positive reference terminals for voltage are at the phase terminals of entry, the power will be positive when the energy flow is into the delimited region and negative when the flow is out of the delimited region. Reversal of either the reference direction of the reference polarity will reverse the relation between the sign of the power and the direction of energy flow. (B) When the circuit has a neutral terminal of entry, it is usually to select the neutral terminal as the common point for voltage measurement, because one of the voltages is then always zero, and, when both the currents and voltages form symmetrical polyphase sets of the same phase sequence, the average power for each single-phase circuit consisting of one phase conductor and the neutral conductor, will be the same. When the voltages and currents are sinusoidal and the voltages are measured to the neutral terminal of entry as the common point, the instantaneous power at the four points of entry of a three-phase circuit with neutral is given by

$$p = E_a I_a \left[\cos(\alpha_a - \beta_a) + \cos(2\omega t + \alpha_a + \beta_a)\right]$$
$$+ E_b I_b \left[\cos(\alpha_b - \beta_b) + \cos(2\omega t + \alpha_b + \beta_b)\right]$$
$$+ E_c I_c \left[\cos(\alpha_c - \beta_c) + \cos(2\omega t + \alpha_c + \beta_c)\right]$$

where E_a, E_b, E_c are the root-mean-square amplitudes of the voltages from the phase conductors, a, b, and c, respectively, to the neutral conductor at the terminals of entry, I_a, I_b, I_c are the root-mean-square amplitudes of the currents in the phase conductors a, b, and c. α_a, α_b, α_c are the phase angles of the voltages E_a, E_b, E_c with respect to a common reference, β_a, β_b, β_c are the phase angles of the currents I_a, I_b, I_c with respect to the same reference as the voltages. (C) If there is no neutral conductor, so that there are only three terminals of entry, the point of entry of one of the phase conductors may be chosen as the common voltage point, and the voltages from that conductor to the common point becomes zero. If, in the preceding, the terminal of entry of phase conductor b is chosen as the common point E_a is replaced by E_{ab} in the first line, E_c is replaced by E_{cb} in the third line, and the second line, being zero, is omitted. (D) If both the voltages and currents in the preceding equations constitute symmetrical polyphase sets of the same phase sequence, then $p = 3E_a I_a \cos(\alpha_a - \beta_a)$. Because this expression and similar expressions for m phases are independent of time, it follows that the instantaneous power is constant when the voltages and currents constitute polyphase symmetrical sets of the same phase sequence. (E) However, if the polyphase sets have single-phase symmetry or zero-phase symmetry rather than poly-phase symmetry, the higher frequency terms do not cancel, and the instantaneous power is not a constant. (F) In general, the instantaneous power p at the $(m + 1)$ terminals of entry of a polyphase circuit of m phases to a delimited area, when one of the terminals is that of the neutral conductor, is expressed by the equation

$$p = \sum_{s=1}^{s=m} e_s i_s$$
$$= \sum_{s=1}^{s=m} \sum_{r=1}^{r=\infty} \sum_{q=1}^{q=\infty} E_{sr} I_{sq}$$
$$\left[\cos[(r-q)\omega t + \alpha_{sr} - \beta_{sq}]\right.$$
$$\left. + \cos[(r+q)\omega t + \alpha_{sr} + \beta_{sq}]\right]$$

where e_s is the instantaneous alternating voltage between the sth terminal entry and the terminal of voltage reference, which may be the true neutral point, the neutral conductor, or another point in the boundary surface. i_s is the instantaneous alternating current through the sth terminal of entry. E_{sr} is the root-mean-square amplitude of the rth harmonic of voltage e_s. I_{sq} is the root-mean-square amplitude of the qth harmonic of current i_s. α_{sr} is the phase angle of the rth harmonic of e_s with respect to a common reference. β_{sq} is the phase angle of the qth harmonic of i_s, with

respect to the same reference as the voltages. The index s runs through the phase letters identifying the m-phase conductor of an m-phase system, a,b,c, etcetera, and then concludes with the neutral conductor n if one exists. The indexes r and q identify the order of the harmonic term in each e_s and i_s, respectively, and run through all the harmonics present in the Fourier series representation of each alternating voltage and current. If the terminal voltage reference is that of the neutral conductor, the terms for $s = n$ will vanish. If the voltages and current are quasi-periodic, of the form given in **power, instantaneous (two-wire circuit)**, this expression is still valid but E_{sr} and I_{sq} become a periodic functions of time. (G) Instantaneous power is expressed in watts when the voltages are in volts and the currents in amperes. *See:* **single-phase symmetrical set of polyphase voltages; zero-phase symmetrical set of polyphase voltage; single-phase symmetrical set of polyphase currents; zero-phase symmetrical set of polyphase currents.** 210
(3) **(two-wire circuit).** At the two terminals of entry into a delimited region, the product of the instantaneous voltage between one terminal of entry and the second terminal of entry, considered as the reference terminal, and the current through the first terminal. *Notes:* (A) The entire path selected for the determination of each voltage must lie in the boundary surface of the delimited region or be so selected that the voltage is the same as that analog such a path. (B) Mathematically the instantaneous power p is given by $p = ei$ in which e is the voltage between the first terminal of entry and the second (reference) terminal of entry and i is the current through the first terminal of entry in the reference direction. (C) If both the voltage and current are sinusoidal and of the same period, the instantaneous power at any instant t is given by the equation

$$p = ei = [(2)^{1/2} E \cos(\omega t + \alpha)]$$
$$\times [(2)^{1/2} I \cos(\omega t + \beta)]$$
$$= 2EI \cos(\omega t + \alpha)\cos(\omega t + \beta)$$
$$= EI[\cos(\alpha - \beta) + \cos(2\omega t + \alpha + \beta)]$$

in which E and I are the root-mean-square amplitudes of voltage and current, respectively, and α and β are the phase angles of the voltage and current, respectively, from the same reference. (D) If the voltage is an alternating voltage and the current is an alternating current of the same primitive period (see **alternating voltage, alternating current**, and **(period (primitive period) of a function)**), the instantaneous power is given by the equation

$$p = ei$$
$$= E_1 I_1 [\cos(\alpha_1 - \beta_1)$$
$$+ \cos(2\omega t + \alpha_1 + \beta_1)]$$
$$+ E_2 I_2 [\cos(\omega t + \alpha_2 - \beta_1)$$
$$+ \cos(3\omega t + \alpha_2 + \beta_1)]$$

$$+ E_1 I_2 [\cos(\omega t - \alpha_1 + \beta_2)$$
$$+ \cos(3\omega t + \alpha_1 + \beta_2)]$$
$$+ E_2 I_2 [\cos(\alpha_2 - \beta_2)$$
$$+ \cos(4\omega t + \alpha_2 + \beta_2)]$$
$$+ \cdots$$

This equation can be written conveniently as a double summation

$$p = \sum_{r=1}^{r=\infty} \sum_{q=1}^{q=\infty} E_r I_q \{\cos[(r - q)\omega t + \alpha_r - \beta_q]$$
$$+ \cos[(r + q)\omega t + \alpha_r - \beta_q]\}$$

in which r is the order of the harmonic component of the voltage and q is the order of the harmonic component of the current (see **harmonic components (harmonics)**), and E, I, α, and β apply to the harmonic denoted by the subscript. (E) If the voltage and current are quasi-periodic functions of the form

$$e = (2)^{1/2} \sum_{r=1}^{r=\infty} E_r(t)\cos(r\omega t + \alpha_r)$$

$$i = (2)^{1/2} \sum_{q=1}^{q=\infty} I_q(t)\cos(q\omega + \beta_r)$$

where $E_r(t)$, $I_q(t)$ are aperiodic functions of t, the instantaneous power is given by the equation

$$p = ei$$
$$= E_1(t)I_1(t))[\cos(\alpha_1 - \beta_1)$$
$$+ \cos(2\omega t + \alpha_1 + \beta_1)]$$
$$+ E_2(t)I_1(t)[\cos(\omega t + \alpha_2 - \beta_1)$$
$$+ \cos(3\omega t + \alpha_2 + \beta_1)]$$
$$+ E_1(t)I_2(t)[\cos(\omega t - \alpha_1 + \beta_2)$$
$$+ \cos(3\omega t + \alpha_1 + \beta_2)]$$
$$+ E_2(t)I_2(t)[\cos(\alpha_2 - \beta_2)$$
$$+ \cos(4\omega t + \alpha_2 + \beta_2)]$$
$$+ \cdots$$

(F) Instantaneous power is expressed in watts when the voltage is in volts and the current in amperes. *Note:* See note 3 of **reference direction of energy.** The sign of the energy will be positive if the flow of power is in the reference direction and negative if the flow is in the opposite direction. 210
power inverter (electric installations on shipboard). A converter unit in which the direction of average power flow is from the dc circuit to the ac circuit. 3
power klystron (microwave tubes). A klystron, usually an amplifier, with two or more cavities uncoupled except by the beam, designed primarily for power amplification or generation. *See:* **microwave tube or value.** 190

power-law index profile (fiber optics). A class of graded index profiles characterized by the following equations:

$$n(r)=n_1(1-2\Delta(r/a)^g)^{1/2} \quad r=a$$
$$n(r)=n_2=n_1(1-2\Delta)^{1/2} \quad r=a$$
$$where\; \Delta=n^2_1-n^2_2 \;\; over \;\; 2\;n^2_1$$

where n(r) is the refractive index as a function of radius, n_1 is the refractive index on axis, n_2 is the refractive index of the homogeneous cladding, a is the core radius, and g is a parameter that defines the shape of the profile. *Notes:* (1) α is often used in place of g. Hence, this is sometimes called an alpha profile. (2) For this class of profiles, multimode distortion is smallest when g takes a particular value depending on the material used. For most materials, this optimum value is around 2. When g increases without limit, the profile tends to a step index profile. *See:* **graded index profile; mode volume; profile parameter; step index profile.**   433

power level (data transmission). The magnitude of power averaged over a specified interval of time. *Note:* Power level may be expressed in units in which the power itself is measured or in decibels indicating the ratio to a reference power. This ratio is usually expressed either in decibels, referred to one mW (milliwatt), abbreviated dBm, or in decibels referred to one W (watt), abbreviated dBW.   59

power-line carrier (1)(overhead-power-line corona and radio noise). The use of radio frequency energy, generally below 600 kHz (kilohertz), to transmit information over transmission lines whose primary purpose is the transmission of power.   411

(2) (protective relaying of utility-consumer interconnections). A high-frequency signal superimposed on the normal voltage on a power circuit. It is customarily coupled to the power line by means of a coupling capacitor. A tuning device provides series resonance at the carrier frequency. Prevention of shorting of the carrier signal by a fault external to the protected line is ordinarily provided by a line trap.   128

power loss (data transmission). (1) From a circuit, in the sense that it is converted to another form of power not useful for the purpose at hand (for example I^2R loss). A physical quantity measured in watts in the International System of Units (SI) and having the dimensions of power. For a given R, it will vary with the current in R. (2) Defined as the ratio of two powers. If P_o is the output power and P_i is the input power of a transducer or network under specified conditions, P_o/P_i is a dimensionless quantity that would be unity if $P_o=P_i$. (3) (Logarithmic). Loss may also be defined as the logarithm, or a quantity directly proportional to the logarithm of a power ratio, such as P_o/P_i. Thus if loss $=10\;\log_{10}(P_o/P_i)$, the loss is zero when $P_o=P_i$. This is the standard for measuring loss in decibels. *Notes:* (A) It should be noted that in cases (2) and (3) the loss for a given linear system is the same whatever may be the power levels. Thus (2) and (3) give characteristics of the system, and do not depend, as (1) does

on the value of the current or other dependent quantity. (B) If more than one component is involved in the input or output, the particular components used must be specified. This ratio is usually expressed in decibels. (C) If the output signal power is at a frequency other than the input signal power, the loss is a conversion loss. (4) (Electric instrument) (watt loss). In the circuit of a current-or-voltage- measuring instrument, the active power at its terminals for end-scale indication. *Note:* For other than current or voltage- measuring instruments, for example, wattmeters, the power loss of any circuit is expressed at a stated value of current or of voltage.   59

power, nonfirm (electric power systems). Power supplied or available under an arrangement which does not have the availability feature of firm power.   112

power, nonreactive (polyphase circuit). At the terminals of entry of a polyphase circuit, a vector equal to the (vector) sum of the nonreactive powers for the individual terminals of entry. *Note:* The nonreactive power for each terminal of entry is determined by considering each phase conductor and the common reference point as a single-phase circuit, as described for distortion power. The sign given to the distortion power in determining the nonreactive power for each single-phase circuit shall be the same as that of the total active power. Nonreactive power for a polyphase circuit has as its two rectangular components the active power and the distortion power. If the voltages have the same waveform as the corresponding currents, the magnitude of the nonreactive power becomes the same as the active power. Nonreactive power is expressed in volt-amperes when the voltages are in volts and the currents in amperes.   210

power, nonreactive (single-phase two-wire circuit). At the two terminals of entry of a single-phase two-wire circuit into a delimited region, a vector quantity having as its rectangular components the active power and the distortion power. Its magnitude is equal to the square root of the difference of the squares of the apparent power and the amplitude of the reactive power. Its magnitude is also equal to the square root of the sum of the squares of the amplitudes of the active power and the distortion power. If voltage and current have the same waveform, the magnitude of the nonreactive power is equal to the active power. The amplitude of the nonreactive power is given by the equation

$$N = (U^2 - Q^2)^{1/2} = (P^2 + D^2)^{1/2}$$
$$= \left\{ \sum_{r=1}^{r=\infty} \sum_{q=1}^{q=\infty} [E_r^2 I_q^2 - E_r E_q I_r I_q \right.$$
$$\left. \times \sin(\alpha_r - \beta_r)\sin(\alpha_q - \beta_q)] \right\}^{1/2}$$

where the symbols are those in **power, apparent (single-phase two-wire circuit).** In determining the vector position of the nonreactive power, the sign of the distortion power component must be assigned arbitrarily. Nonreactive power is expressed in volt-amperes when the voltage is in volts and the current in amperes. *See:* **distortion power (single-phase two-wire circuit).** 210

power-operated door or gate (elevators). A hoistway door and/or a car door or gate that is opened and closed by a door or gate power operator. *See:* **hoistway (elevator or dumbwaiter).** 328

power operation (power switchgear). Operation by other than hand power. 103

power outage (emergency and standby power). Complete absence of power at the point of use. 512

power outlet (National Electrical Code). An enclosed assembly which may include receptacles, circuit breakers, fuseholders, fused switches, buses and watt-hour meter mounting means; intended to supply and control power to mobile homes, recreational vehicles or boats, or to serve as a means for distributing power required to operate mobile or temporarily installed equipment. 256

power output (hydraulic turbines). The electrical output of the turbine generator unit as measured at the generator terminals. 8

power output, instantaneous. The rate at which energy is delivered to a load at a particular instant. *See:* **radio transmitter.** 111

power pack. A unit for converting power from an alternating-current or direct-current supply into alternating-current or direct-current power at voltages suitable for supplying an electronic device. 328

power, partial discharge (dielectric tests). The power fed into the terminals of the test object due to partial discharges. The average discharge power is expressed in watts. 139

power, peak pulse. *See:* **peak pulse power.**

power, phase control (thyristor converter). The power used to synchronize the phase control of the thyristor converter to the ac supply input phases. 121

power, phasor (polyphase circuit). At the terminals of entry of a polyphase circuit into a delimited region, a phasor (or plane vector) that is equal to the (phasor) sum of the phasor powers for the individual terminals of entry when the voltages are all determined with respect to the same arbitrarily selected common reference point in the boundary surface (which may be the neutral terminal of entry). The reference direction for the currents and the reference polarity for the voltages must be the same as for instantaneous power, active power, and reactive power. The phasor power for each terminal of entry is determined by considering each conductor and the common reference point as a single-phase, two-wire circuit and finding the phasor power for each in accordance with the definition of **power, phasor (single-phase two-wire circuit).** The phasor power S is given by $S = P + jQ$ where P is the active power for the polyphase circuit and Q is the reactive power for the same terminals of entry. If the voltages and currents are sinusoidal and of the same period, the phasor power S for a three-phase circuit is

given by

$$S = E_a I_a{}^* + E_b I_b{}^* + E_c I_c{}^*$$

where E_a, E_b, and E_c are the phasor voltages from the phase conductors a, b, and c, respectively, to the neutral conductor at the terminals of entry, I_a, I_b, and I_c are the conjugate of the phasor currents in the phase conductor, so that there are only three terminals of entry, the point of entry of one of the phase conductors may be chosen as the common voltage point, and the phasor from that conductor to the common voltage point becomes zero. If the terminal of entry of phase conductor b is chosen as the common point, the phasor power of a three-phase, three-wire circuit becomes

$$S = E_{ab} I_a{}^* + E_{cb} I_c{}^*$$

where E_{ab}, E_{cb} are the phasor voltages from phase conductor a to b and from c to b, respectively. If both the voltages and currents in the preceding equations constitute symmetrical sets of the same phase sequence $S = 3E_a I_a{}^*$. In general the phasor power at the $(m + 1)$ terminals of entry of a polyphase circuit of m phases to a delimited region, when one of the terminals is the neutral terminal of entry, is expressed by the equation

$$S = \sum_{s=1}^{s=m} \sum_{r=1}^{r=\infty} E_{sr} I_{sr}{}^*$$

where E_{sr} is the phasor representing the rth harmonic of the voltage from phase conductor s to neutral at the terminals of entry $I_{sr}{}^*$ is the conjugate of the phasor representing the rth harmonic of the current through the sth terminal of entry. The phasor power can also be stated in terms of the symmetrical components of the voltages and currents as

$$S = m \sum_{k=0}^{k=m-1} \sum_{r=1}^{r=\infty} E_{kr} I_{kr}{}^*$$

where E_{kr} is the phasor representing the symmetrical component of kth sequence of the rth harmonic of the line-to-neutral set of polyphase voltages at the terminals of entry. $I_{kr}{}^*$ is the conjugate of the phasor representing the symmetrical component of the kth sequence of the rth harmonic of the polyphase set of currents through the terminals of entry. Phasor power is expressed in volt-amperes when the voltages are in volts and the currents in amperes. *Notes:* This term was once defined as **vector power.** With the introduction of the term **phasor quantity**, the name of this term has been altered to correspond. The definition has also been altered to agree with the change in the sign of reactive power. *See:* **power, reactive (magner) (single-phase two-wire circuit); power, reactive (magner) (polyphase circuit).** 210

power, phasor (single-phase two-wire circuit). At the two terminals of entry of a single-phase two-wire circuit into a delimited region, a phasor (or plane vector) of which the real component is the active power and the imaginary component is the reactive power at the same two terminals of entry. When either component of phasor power is positive, the direction of that component is in the reference direction. The phasor power S is given by $S = P + jQ$ where P and Q are the active and reactive power, respectively. If both the voltage and current are sinusoidal, the phasor power is equal to the product of the phasor voltage and the conjugate of the phasor current.

$$\mathbf{E} = Ee^{j\alpha}; \qquad \mathbf{I} = Ie^{j\beta};$$

the phasor power is

$$\mathbf{S} = P + jQ = \mathbf{EI}^* = EIe^{j(\alpha-\beta)}$$
$$= EI[\cos(\alpha - \beta) + j\sin(\alpha - \beta)].$$

If the voltage is an alternating voltage and the current is an alternating current, the phasor power for each harmonic component is defined in the same way as for the sinusoidal voltage and sinusoidal current. Mathematically the phasor power of the rth harmonic component $\mathbf{S}_r$ is given by

$$\mathbf{S}_r = P_r + jQ_r = \mathbf{E}_r\mathbf{I}_r^* = E_rI_re^{j(\alpha r-\beta r)}$$
$$= E_rI_r[\cos(\alpha_r - \beta_r) + j\sin(\alpha_r - \beta_r)].$$

The phasor power at the two terminals of entry of a single-phase two-wire circuit into a delimited region, for an alternating voltage and current, is equal to the (phasor) sum of the values of the phasor power for every harmonic. Mathematically, this relation may be expressed

$$\mathbf{S} = \mathbf{S}_1 + \mathbf{S}_2 + \mathbf{S}_3 + \cdots = \Sigma\mathbf{S}_r$$
$$= \mathbf{E}_1\mathbf{I}_1^* + \mathbf{E}_2\mathbf{I}_2^* + \mathbf{E}_3\mathbf{I}_3^* \cdots = \Sigma\,\mathbf{E}_r\mathbf{I}_r^*$$
$$= (P_1 + P_2 + P_3 + \cdots)$$
$$+ j(Q_1 + Q_2 + Q_3 + \cdots) = \Sigma(P_r + jQ_r).$$

The amplitude of the phasor power is equal to the square root of the sum of the squares of the active power and the reactive power. Mathematically, if S is the amplitude of the phasor power and θ is the angle between the phasor power and the real-power axis,

$$\mathbf{S} = Se^{j\theta}$$
$$S = (P^2 + Q^2)^{1/2} = [(P_1 + P_2 + P_3 + \cdots)^2 + (Q_1 + Q_2 + Q_3 + \cdots)^2]^{1/2}$$
$$\theta = \tan^{-1}\frac{Q}{P} = \tan^{-1}\frac{Q_1 + Q_2 + Q_3 + \cdots}{P_1 + P_2 + P_3 + \cdots}.$$

If the voltage and current are quasi-periodic and the amplitude of the voltage and current components are slowly varying, the phasor power may still be taken as the phasor having P and Q as its components, the values of P and Q being determined for these conditions, as specified in **power, active (single-phase two-wire circuit) (average power) (power)** and **power, reactive (magner) (single-phase two-wire circuit)**, respectively. For this condition the phasor power will be a function of time. If the voltage and current have the same waveform, the amplitude of the phasor power is equal to the apparent power, but they are not the same for all other cases. The phasor power is expressed in volt-amperes when the voltage is in volts and the current in amperes. *Note:* This term was once defined as vector power. With the introduction of the term **phasor quantity**, the name of this term has been altered to agree with the change in the sign of reactive power. *See:* **power, reactive (magner) (single-phase two-wire circuit); alternating current.** 210

power pool (power operations). Term referring to a group of power systems operating as an interconnected system and pooling their resources. 516

power primary detector (electric power systems). A power-measuring device for producing an output proportional to power input. *See:* **speed-governing system.** 94

power quantities (single-phase three-wire circuit) and (two-phase circuit). The definitions of the power quantities for a single-phase circuit of more than two wires and of a two-phase circuit are essentially the same as those expressions involve m, the number of phases or phase conductors, the numeral 2 should be used for single-phase, three-wire systems, and the numeral 4 for two-phase, four-wire and five-wire systems. *See:* **polyphase symmetrical sets (polyphase voltages).** 210

power rating (waveguide attenuator). The maximum power that, if applied under specified conditions of environment and duration, will not produce a permanent change that causes any performance characteristics to be outside of specifications. This includes characteristic insertion loss and standing-wave ratio. *See:* **waveguide.** 185

power rating or voltage rating (line and connectors) (coaxial transmission line). That value of transmitted power or voltage that permits satisfactory operation of the line assembly and provides an adequate safety factor below the point where injury or appreciably shortened life will occur. *See:* **transmission line.** 265

power, reactive. The product of voltage and out-of-phase component of alternating current. In a passive network, reactive power represents the alternating exchange of stored energy (inductive or capacitive) between two areas. *See:* **magner (polyphase circuit); magner (single-phase two-wire circuit).** 63

power rectifier. A rectifier unit in which the direction of average energy flow is from the alternating-current circuit to the direct-current circuit. 208

power rectifier transformer (power and distribution transformer). A rectifier transformer connected to mercury-arc or semiconductor rectifiers for electrochemical service, steel processing applications, electric furnace applications, mining applications, transportation applications, and direct-current transmissions. 53

power reflectance of a radome (antennas). At a given point on a radome, the ratio of the power flux density that is internally reflected from the radome to that incident on the radome from an internal radiating source. 111

power relay (power switchgear). A relay that responds to a suitable product of voltage and current in an electric circuit. *See:* **active power relay; reactive power relay.** 103, 127

power response (close-talking pressure-type microphone). The ratio of the power delivered by a microphone to its load, to the applied sound pressure as measured by a Laboratory Standard Microphone placed at a stated distance from the plane of the opening of the artificial voice. *Note:* The power response is usually measured as a function of frequency in decibels (dB) above 1 milliwatt per newton per square centimeter $[mW/(N/m^2)]$ or 1 milliwatt per 10 microbars $[mW/10\mu bar]$. *See:* **close-talking pressure-type microphone.** 249

power selsyn (synchros or selsyns). An inductive type of positioning system having two or more similar mechanically independent slip-ring machines with corresponding slip rings of all machines connected together and the stators fed from a common power source. *See:* **synchro-system.** 63

power sensitivity error. The maximum deviation from linearity over each power range of either the electrothermic unit or the electrothermic power meter. Expressed in percent. 47

power service protector (power switchgear). An assembly consisting of a modified low-voltage power circuit-breaker, which has no direct-acting tripping devices, with a current limiting fuse connected in series with the load terminals of each pole. 103

power, signal electronics (thyristor converter). The power used for the analog or digital system power supplies, or both, required for the thyristor converter control and protection systems. 121

powers of units (International System of Units) (SI). An exponent attached to a symbol containing a prefix indicates that the multiple or submultiple of the unit (the unit with its prefix) is raised to the power expressed by the exponent. For example:

$$1 \text{ cm}^3 = (10^{-2} \text{ m})^3 = 10^{-6} \text{ m}^3$$
$$1 \text{ ns}^{-1} = (10^{-9} \text{ s})^{-1} = 10^9 \text{ s}^{-1}$$
$$1 \text{ mm}^2/\text{s} = (10^{-3} \text{ m})^2/\text{s} = 10^{-6} \text{ m}^2/\text{s}$$

See: **units and letter symbols; prefixes and symbols.** 21

power source isolation. Absence of a direct-current circuit (path) between the power source and the system power supply outputs. 48

power sources (accident monitoring instrumentation)-(nuclear power generating station). The electrical and mechanical equipment and its interconnections necessary to generate or convert power. *Note:* Electric power source and power supply are interchangeable within the context of ANSI/IEEE Std 308-1980. 428, 421, 102

power spectral density (PSD)(seismic qualification of Class 1E equipment for nuclear power generating stations). The mean squared amplitude per unit frequency of a waveform. PSD is expressed in g^12/Hz versus frequency for acceleration waveforms. 581

power storage. That portion of the water stored in a reservoir available for generating electric power. *See:* **generating station.** 64

power-supply assembly (National Electrical Code). The conductors, including the grounding conductors, insulated from one another, the connectors, attachment plug caps, and all other fittings, grommets, or devices installed for the purpose of delivering energy from the source of electrical supply to the distribution panel within the recreational vehicle. 256

power supply circuit (relay system). An input circuit to a relay system which supplies auxiliary power for the proper functioning of the relay system. 90

power supply, direct-current (alternating-current to direct-current). Generally, a device consisting of a transformer, rectifier, and filter for converting alternating current to a prescribed direct voltage or current. 186

power supply, direct-current regulated. A direct-current power supply whose output voltage is automatically controlled to remain within specified limits for specified variations in supply voltage and load current. *See:* **direct-current (power-system communication).** 415

power supply, uninterruptible (UPS). A system designed to provide power, without delay or transients, during any period when the normal power supply is incapable of performing acceptably. 89

power-supply voltage range (transmitter performance). The range of voltages over which there is not significant degradation in the transmitter or receiver performance. *See:* **audio-frequency distortion.** 181

power switchboard (power switchgear). A type of switchboard including primary power-circuit switching and interrupting devices together with their interconnections. *Note:* Knife switches, fuses, and air circuit breakers are the commonly used switching and interrupting devices. 103

power system (1) (generating stations electric power system). The electric power sources, conductors, and equipment required to supply electric power. 381

(2) (electric). A group of one or more generating sources and.or connecting transmission lines operated under common management or supervision to supply load. 94

power system, emergency. An independent reserve

source of electric energy which, upon failure or outage of the normal source, automatically provides reliable electric power within a specified time to critical devices and equipment whose failure to operate satisfactorily would jeopardize the health and safety of personnel or result in damage to property.   89

power system stabilizer (excitation systems for synchronous machines). An element or group of elements that provide an additional input to the regulator to improve power system performance. *Note:* A number of different quantities may be used as input to the power system stabilizer, such as, shaft speed, frequency, synchronous machine electrical power, etcetera.   507

power system, standby. An independent reserve source of electric energy which, upon failure or outage of the normal source, provides electric power of acceptable quality and quantity so that the user's facilities may continue in satisfactory operation.   89

power-temperature coefficient. The change in power required to hold the bolometer element at the desired operating resistance per unit change in ambient temperature. *Note:* This quantity is expressed in microwatts per degree Celsius.   115

power transfer relay. A relay so connected to the normal power supply that the failure of such power supply causes the load to be transferred to another power supply.   328

power transformer (power and distribution transformer). A transformer which transfers electric energy in any part of the circuit between the generator and the distribution primary circuits.   53

power transmittance of a radome (antennas). In a given direction, the ratio of the power flux density emerging from a radome with an internal source to the power flux density that would be obtained if the radome were removed.   111

power type relay. A term for a relay designed to have heavy-duty contacts usually rated 15 amperes or higher. Sometimes called a **contactor.**   341

power, utility. *See:* **power, commercial.**

power, vector (polyphase circuit). At the terminals of entry of a polyphase circuit, a vector of which the three rectangular components are, respectively, the active power, the reactive power, and the distortion power at the same terminals of entry. In determining the components, the reference terminals for voltage measurement are taken as the neutral terminal of entry, if one exists, otherwise as the true neutral point. The vector power is also the (vector) sum of the vector powers for the individual terminals of entry. The vector power for each terminal of entry is determined by considering each phase conductor and the common reference point as a single-phase circuit, as described for distortion power. The sign given to the distortion power in determining the vector power for each single-phase circuit are the same as that of the total active power. The magnitude of the vector power is the apparent power. If the voltages have the same waveform as the corresponding currents, the magnitude of the vector power is equal to the amplitude of the phasor

power. Vector power is expressed in volt-amperes when the voltages are in volts and the currents in amperes. *See:* **network analysis.**   210

power vector (single-phase two-wire circuit). At the two terminals of entry of a single-phase two-wire circuit into a delimited region, a vector whose magnitude is equal to the apparent power, and the three rectangular components of which are, respectively, the active power, the reactive power, and the distortion power at the same two terminals of entry. Mathematically the vector power **U** is given by

$$\mathbf{U} = \mathbf{i}P + \mathbf{j}Q + \mathbf{k}D$$

where **i, j,** and **k** are unit vectors along the three perpendicular axes, respectively. *P, Q,* and *D* are the active power, reactive power, and distortion power, respectively. The direction cosines of the angles between the vector power U and the three rectangular axes are

$$\cos \phi = \frac{P}{U}$$

$$\cos \Psi = \frac{Q}{U}$$

$$\cos \theta = \frac{D}{U}$$

The magnitude of the vector power is the apparent power, or

$$U = (P^2 + Q^2 + D^2)^{1/2}$$
$$= \left(\sum_{r=1}^{r=\infty} \sum_{q=1}^{q=\infty} E_r{}^2 I_q{}^2 \right)^{1/2}$$

where the symbols are those of the preceding definitions. The geometric power diagram shows the relationships among the different types of power. Active power, reactive power, and distortion power are represented in the directions of the three rectangular axes. The accompanying diagram corresponds to a case in which all three are positive.

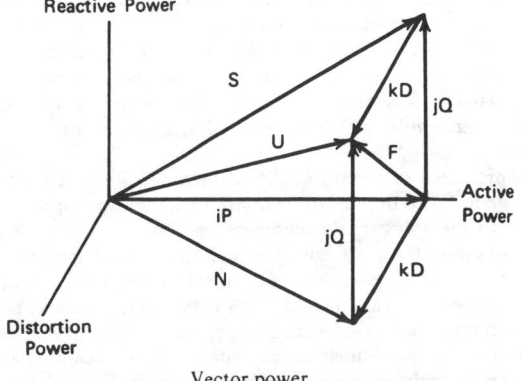

Vector power

Since the sign of D is not definitely determined, D may be drawn in either direction along the axis. The position of U is thus also ambiguous, as it may occupy either of two positions, for D positive or negative. When the sign of D has been assumed, the vector positions of the fictitious power F and the nonreactive power N are determined. They have been shown in the figure, with the assumption that D has the same sign as P. Vector power is expressed in volt-amperes when the voltage is in volts and the current in amperes. *Notes:* (1) The vector power becomes a plane vector having the same magnitude as the phasor power if the voltage and the current have the same wave form. This condition is fulfilled as a special case when the voltage and current are sinusoidal and of the same period. (2) The term **vector power** as defined in the 1941 edition of the American Standard Definitions of Electrical Terms has now been called **phasor power** (*see* **power, phasor (single-phase two-wire circuit)**) and the present definition of vector power is new. *See:* **network analysis.** 210

power winding (saturable reactor). A winding to which is supplied the power to be controlled. Commonly the functions of the output and power windings are accomplished by the same winding, which is then termed the output winding. *See:* **magnetic amplifier.** 328

Poynting's vector. If there is a flow of electromagnetic energy into or out of a closed region, the rate of flow of this energy is, at any instant, proportional to the surface integral of the vector product of the electric field strength and the magnetizing force. This vector product is called Poynting's vector. If the electric field strength is E and the magnetizing force is **H**, then Poynting's vector is given by

$$U = E \times H \text{ and } U = E \times H/4\pi$$

in rationalized and unrationalized systems, respectively. Poynting's vector is often assumed to be the local surface density of energy flow per unit time. 210

PPI. *See:* **plan position indication.**

ppm (circuits and systems). Parts per million, that is, $x.10^6$ is x parts per million. 67

PPM. *See:* **periodic permanent-magnet focusing.**

practical reference pulse waveform (pulse measurement). A reference pulse waveform which is derived from a pulse which is produced by a device or apparatus. 15

practices (software quality assurance). Requirements employed to prescribe a disciplined, uniform approach to the software development process. 481

preamplifier. An amplifier connected to a low-level signal source to present suitable input and output impedances and provide gain so that the signal may be further processed without appreciable degradation in the signal-to-noise ratio. *Notes:* (1) A preamplifier may include provision for equalizing and.or mixing.

(2) Further processing frequently includes further amplification in a main amplifier. *See:* **amplifier.** 239

pre-arcing time. *See:* **melting time.**

preassigned multiple access (communication satellite). A method of providing multiple access in which the satellite channels are preassigned at both ends of the path. 84

precedence call (telephone switching systems). A call on which the calling party has elected to use one of several levels of priority available to him. 55

precession (navigation aid terms). The change in the direction of the axis of rotation of a spinning body, as a gyroscope, when acted upon by a torque. 526

precision (1)(mathematics of computing). The degree of exactness or discrimination with which a quantity is stated. *See:* **accuracy.** 564

(2)(monitoring radioactivity in effluents). The degree of agreement of repeated measurements of the same property expressed in terms of dispersion of test results about the mean result obtained by repetitive testing of a homogeneous sample under specified conditions. The precision of a method is expressed quantitatively as the standard deviation computed from the results of a series of controlled determinations. 599

(3) (general). The quality of being exactly or sharply defined or stated. A measure of the precision of a representation is the number of distinguishable alternatives from which it was selected, which is sometimes indicated by the number of significant digits it contains. *See:* **accuracy; double precision** 235, 9

(4) (measurement process). The quality of coherence or repeatability of measurement data, customarily expressed in terms of the standard deviation of the extended set of measurement results from a well-defined (adequately specified) measurement process in a state of statistical control. The standard deviation of the conceptual population is approximated by the standard deviation of an extended set of actual measurements. *See:* **accuracy; reproducibility.** 13, 187

(5) (pulse measurement). The degree of mutual agreement between the results of independent measurements of a pulse characteristic, property, or attribute yielded by repeated application of a pulse measurement process. 15

(6) (analog computers). Exactly or sharply defined or stated. A measure of the precision of a representation is the number of distinguishable alternatives from which it was selected, which is sometimes indicated by the number of significant digits it contains. *See:* **accuracy.** 9

(7) (metric practice). The degree of mutual agreement between individual measurements, namely repeatability and reproducibility. *See:* **accuracy.** 21

(8) (nuclear power generating stations) (measuring and test equipment). The quality of an instrument scale or readout being exactly or sharply defined or stated. 41

(10) (plutonium monitoring)(radiological monitoring instrumentation). The degree of agreement of repeated measurements of the same property, expressed quantitatively as the standard deviation computed from the results of the series of measurements.

413, 398

(11) (software). (A) A measure of the ability to distinguish between nearly equal values, for example, four-place numerals are less precise than six-place numerals; nevertheless, a properly computed four-place numeral may be more accurate than an improperly computed six-place numeral. (B) A degree of discrimination with which a quantity is stated, for example, a three-digit numeral discriminates among 1000 possibilities. *See:* **accuracy.**   434

precision-approach radar (PAR). A radar system located on an airfield for observation of the position of an aircraft with respect to an approach path and specifically intended to provide guidance to the aircraft in the approach.   526

precision device (packaging machinery). A device that will operate within prescribed limits and will consistently repeat operations within those limits.   429

precision wound (rotating machinery). A coil wound so that maximum nesting of the conductors occurs, usually with all crossovers at one end, with conductor aligned and positioned with respect to each adjacent conductor. *See:* **rotor (rotating machinery); stator.**

63

precompiler (software). A computer program that pre-processes source code, part of which may be unacceptable to a compiler, to generate equivalent code acceptable to the compiler, for example, a preprocessor which converts structured FORTRAN to ANSI standard FORTRAN. *See:* **code; compiler; computer program; preprocessor.**   434

preconditioning (industrial control). A control-function that provides for manually or automatically establishing a desired condition prior to normal operation of the system. *See:* **control system, feedback.**

206

precursor. *See:* **undershoot.**

predicted mean active maintenance time. The mean active maintenance time of an item calculated by taking into account the reliability characteristics and the mean active maintenance time of all of its parts and other relevant factors according to the stated conditions. *Notes:* (1) Maintenance policy, statistical assumptions and computing methods shall be stated. (2) The source of the data shall be stated.   164

predicted reliability. *See:* **reliability, predicted.**

predictive alarming (alarm monitoring and reporting systems for fossil-fueled power generating stations). A method of alerting the operator to a potential problem in time for him to respond and initiate corrective action to mitigate the problem.   501

predissociation. A process by which a molecule that has absorbed energy dissociates before it has had an opportunity to lose energy by radiation. *See:* **gas-filled radiation-counter tubes.**   125

predistortion (pre-emphasis) (system) (transmitter performance). A process that is designed to emphasize or de-emphasize the magnitude of some frequency components with respect to the magnitude of others. *See:* **pre-emphasis.**   181

pre-emphasis (pre-equalization). (1) (General) A process in a system designed to emphasize the magnitude of some frequency components with respect to the magnitude of others, to reduce adverse effects, such as noise, in subsequent parts of the system. *Note:* After transmitting the pre-emphasized signal through the noisy part of the system, de-emphasis may be applied to restore the original signal with a minimum loss of signal-to-noise ratio. (2) (Modulating systems) (recording) An arbitrary change in the frequency response of a recording system from its basic response (such as constant velocity or amplitude) for the purpose of improvement in signal-to-noise ratio, or the reduction of distortion.   59

pre-emphasis network. A network inserted in a system in order to emphasize one range of frequencies with respect to another. *See:* **network analysis.**   111

preemption (telephone switching systems). On a precedence call, the disconnection and subsequent reuse of part of an established connection of lower priority if all the relevant circuits are busy.   55

preemptive control (test, measurement and diagnostic equipment). An action or function which, by reason of preestablished priority, is able to seize or interrupt the process in progress and cause to be performed a process of higher priority.   54

pre-envelope. *See:* **analytic signal.**

preference (power-system communication) (channel supervisory control). An assembly of devices arranged to prevent the transmission of any signals over a channel other than supervisory control signals when supervisory control signals are being transmitted. *See:* **supervisory control system.**   415

preference level (speech quality measurements). The signal-to-noise ratio ($S.N$) of the speech reference signal when it is isopreferent to the speech test signal.   126

preferred (electric power system) (generating stations electric power system). That equipment and system configuration selected to supply the power system loads under normal conditions.   381

preferred basic impulse insulation level (insulation strength). A basic impulse insulation level that has been adopted as a preferred American National Standard voltage value. *See:* **basic impulse insulation level (BIL) (insulation strength).**   276

preferred current ratings of distribution fuse links (high-voltage switchgear). A series of distribution fuse link ratings so chosen from a series of preferred numbers that a specified degree of coordination may be obtained between adjacent sizes.   443

preferred insulations system classification (electric installationson shipboard). The preferred insulation system classifications are classes A,B,F,H,C, or 105,130,155,180, or greater than 220, and as designated by the equipment standard.   3

preferred power supply (PPS)(preferred power supply for nuclear power generating stations). That power supply from the transmission system to the Class 1E distribution system which is preferred to furnish electric power under accident and post-accident conditions. 541

prefix code (telephone switching systems). One or more digits preceding the national or international number to implement direct distance dialing.

55

prefixes and symbols (International System of Units) (SI). Used to form names and symbols of the decimal mutiples and submultiples of the SI units:

Prefix	Abbreviation	Factor
tera-(megamega-*)	T(MM*)	10^{12}
giga-(kilomega-*)	G(kM*)	10^{9}
mega-	M	10^{6}
myria-		10^{4}
kilo-	k	10^{3}
hecto-	h	10^{2}
deka-		10
deci-	d	10^{-1}
centi-	c	10^{-2}
milli-	m	10^{-3}
decimilli-	dm	10^{-4}
micro-	μ	10^{-6}
nano-(millimicro-*	n (mμ*)	0^{-9}
pico-(micromicro-*)	p($\mu\mu$*)	10^{-12}

* Deprecated

These prefixes or their symbols are directly attached to names or symbols of units, forming multiples and submultiples of the units. In strict terms these must be called "multiples and submultiples of SI units," particularly in discussing the coherence of the system. In common parlance, the base units and derived units, along with their multiples and submultiples, are all called SI units. *See:* **units and letter symbols.**

prefix multipliers. The prefixes listed in the following table, when applied to the name of a unit, serve to form the designation of a unit greater or smaller than the original by the factor indicated. 210

Multiplication Factor	Prefix	Symbol
1 000 000 000 000 000 000 = 10^{18}	exa[a]	E
1 000 000 000 000 000 = 10^{15}	peta[a]	P
1 000 000 000 000 = 10^{12}	tera	T
1 000 000 000 = 10^{9}	giga	G
1 000 000 = 10^{6}	mega	M
1 000 = 10^{3}	kilo	k
100 = 10^{2}	hecto[b]	h
10 = 10^{1}	deka[b]	da
0.1 = 10^{-1}	deci[b]	d
0.01 = 10^{-2}	centi[b]	c
0.001 = 10^{-3}	milli	m
0.000 001 = 10^{-6}	micro	μ
0.000 000 001 = 10^{-9}	nano	n
0.000 000 000 001 = 10^{-12}	pico	p
0.000 000 000 000 001 = 10^{-15}	femto	f
0.000 000 000 000 000 001 = 10^{-18}	atto	a

[a] Adopted by the CGPM in 1975.
[b] To be avoided where possible.

prefix notation (mathematics of computing). A parenthesis-free method of forming mathematical expressions devised by the Polish logician Jan Lukasiewicz, in which each operator is immediately followed by its operands. For example, A added to B and the result multiplied by C is expressed as $\times + ABC$. *Syn:* **Lukasiewicz notation; Polish notation.** *See:* **infix notation; postfix notation.** 564

preform (fiber optics). A glass structure from which an optical fiber waveguide may be drawn. *See:* **chemical vapor deposition technique; ion exchange technique; optical blank.** 433

preformed coil or coil side (rotating machinery). An element of a preformed winding, composed of conductor strands, usually insulated and sometimes transposed, cooling ducts in some designs, turn insulation where number of turns exceeds one, and coil insulation. *See:* **rotor (rotating machinery); stator.**

63

preformed winding. A winding consisting of coils which are given their shape before being assembled in the machine. 63

preheat (switch start) fluorescent lamp (illuminating engineering). A fluorescent lamp designed for operation in a circuit requiring a manual or automatic starting switch to preheat the electrodes in order to start the arc. 167

preheating time (mercury-arc valve). The time required for all parts of the valve to attain operating temperature. 244, 190

preheat-starting (fluorescent lamps) (switch-starting systems). The designation given to those systems in which hot-cathode electric discharge lamps are started from preheated cathodes through the use of a starting switch, either manual or automatic in its operation. *Note:* The starting switch, when closed, connects the two cathodes in series in the ballast circuit so that current flows to heat the cathodes to emission temperature. When the switch is opened, a voltage surge is produced that initiates the discharge. Only the arc current flows through the cathodes after the lamp is in operation. *See:* **fluorescent lamp.** 167

preliminary design (software). (1) The process of analyzing design alternatives and defining the software architecture. Preliminary design typically includes definition and structuring of computer program components and data, definition of the interfaces, and preparation of timing and sizing estimates. (2) The results of the preliminary design process. *See:* **computer program component; data; design; design analysis; functional design; interface; software architecture.** 434

premises wiring (system) (National Electrical Code). That interior and exterior wiring, including power, lighting, control, and signal circuit wiring together with all of its associated hardware, fittings, and wiring devices, both permanently and temporarily installed, which extends from the load end of the service drop, or load end of the service lateral conductors to the outlet(s). Such wiring does not include wiring internal to appliances, fixtures, motors, controllers, motor control centers, and similar equipment. 256

pre-molded joint (power cable joints). A joint made of pre-molded components assembled in the field. 34

preoperational system test (Class 1E power systems). A test to confirm that all individual component parts of a system function as a system and the system functions as designed. A preoperational test is performed following significant modifications or additions made to the facility at later dates. 455

preparatory function (numerically controlled machines). A command changing the mode of operation of the control such as from positioning to contouring or calling for a fixed cycle of the machine.

prepatch panel. *See:* **problem board.**

preprocessor (software). A computer program that effects some preliminary computation or organization. *See:* **computer program; precompiler.** 434

preregister operation (elevators). Operation in which signals to stop are registered in advance by buttons in the car and at the landings. At the proper point in the car travel, the operator in the car is notified by a signal, visual, audible, or otherwise, to initiate the stop, after which the landing stop is automatic. *See:* **control.** 328

pre-rip (cable plowing). A process using a plow blade to loosen the earth prior to plowing and installing the cable, flexible tube, etcetera. 52

preselector. (1) A device placed ahead of a frequency converter or other device, that passes signals of desired frequencies and reduces others. (2) In automatic switching, a device that performs its selecting operation before seizing an idle trunk. 328

presence tests (test, measurement and diagnostic equipment). Actions which verify the presence or absence of signals or characteristics. Such signals or characteristics are those which are not tolerance critical to operation of the item. 54

preset. To establish an initial condition, such as the control values of a loop. 255,77

preset guidance (navigation aid terms). Guidance in which a predetermined path is set into the guidance mechanism of a craft and is not altered after launching. 526

preset speed (industrial control). A control function that establishes the desired operating speed of a drive before initiating the speed change. *See:* **electric drive.** 225,206

preshoot (pulse terms). A distortion which precedes a major transition. *Note:* Colloquial term which qualitatively describes a type of distortion. 254

pressing (disk recording). A pressing is a record produced in a record-molding press from a master or stamper. *See:* **phonograph pickup.** 176

pressure (solderless) connector (National Electrical Code). A device that establishes a connection between two or more conductors or between one or more conductors and a terminal by means of mechanical pressure and without the use of solder. 256

pressure altimeter (navigation aid terms). An altime-

ter that measures and indicates altitude above a datum plane by means of an aneroid which responds to the change in atmospheric pressure with height. 526

pressure barrier seal (nuclear power generating stations). Consists of an aperture seal and an electrical conductor seal. 31

pressure cable. An oil-impregnated paper-insulated cable in which positive gauge pressure is maintained on the insulation under all operating conditions. 64

pressure coefficient. *See:* **environmental coefficient.**

pressure connector (packaging machinery). A conductor terminal applied with pressure so as to make the connection mechanically and electrically secure. 429

pressure-containing terminal box (rotating machinery). A terminal box so designed that the products of an electric breakdown within the box are completely contained inside the box. 63

pressure controller (control systems for steam turbine-generator units). Includes only those components and control elements that generate one or more signal(s) for the control mechanism in response to pressure set point and pressure feedback signals for the purpose of controlling pressure. 522

pressure control system (control systems for steam turbine-generator units). A system that controls the pressure at a sensing point in a designated location. Typically, it includes the pressure-sensing element, the controller, the control mechanism, and the controlled valve(s). 522

pressure-lubricated bearing (rotating machinery). A bearing in which a continuous flow of lubricant is forced into the space between the journal and the bearing. *See:* **bearing.** 63

pressure microphone. A microphone in which the electric output substantially corresponds to the instantaneous sound pressure of the impressed sound waves. *Note:* A pressure microphone is a gradient microphone of zero order and is nondirectional when its dimensions are small compared to a wavelength. *See:* **microphone.** 328

pressure reference changer (control systems for steam turbine-generator units). A device for producing the pressure reference signal to the pressure controller in response to a manual or automatic adjustment. 522

pressure relay (power switchgear). A relay that responds to liquid or gas pressure. 103

pressure-relief device (arrester). A means for relieving internal pressure in an arrester and preventing explosive shattering of the housing, following prolonged passage of follow current or internal flashover of the arrester. 62

pressure-relief terminal box (rotating machinery). A terminal box so designed that the products of an electric breakdown within the box are relieved through a pressure-relief diaphragm. 63

pressure-relief test (arresters). A test made to ascertain that an arrester failure will not cause explosive shattering of the housing. 62

pressure retaining boundary (nuclear power generat-

ing stations). The pressure retaining boundary includes those surfaces of the aperture seal, the conductor feed-through plate, the conductor seal (or seals), and the conductor (or conductors) which are exposed to the containment environment.   31

pressure switch (1) (industrial control). A switch in which actuation of the contacts is effected at a predetermined liquid or gas pressure.   308,206

(2) (63) (power system device function numbers). A switch which operates on given values, or on a given rate of change, of pressure.   402

pressure system (protective signaling). A system for protecting a vault by maintaining a predetermined differential in air pressure between the inside and outside of the vault. Equalization of pressure resulting from opening the vault or cutting through the structure initiates an alarm condition in the protection circuit. *See:* **protective signaling.**   328

pressure-type pothead. A pressure-type pothead is a pothead intended for use on positive-pressure cable systems. *See:* **multipressure zone pothead; single pressure zone potheads.**   323

pressure-type termination (cable termination). A Class 1 termination intended for use on positive pressure cable systems. (1) Single-pressure zone termination: a pressure type termination intended to operate with one pressure zone; (2) multipressure zone termination: a pressure type termination intended to be operated with two or more pressure zones.   4

pressure wire connector. A device that establishes the connection between two or more conductors or between one or more conductors and a terminal by means of mechanical pressure and without the use of solder.   328

pressurized (rotating machinery). Applied to a sealed machine in which the internal coolant is kept at a higher pressure than the surrounding medium.   63

prestressed concrete structures (NESC). Concrete structures which include metal tendons that are tensioned and anchored either before or after curing of the concrete.   494

prestrike current (lightning). The current that flows in a lightning stroke prior to the return stroke current. *See:* **direct-stroke protection (lightning).**   64

pretersonic. Ultrasonic and with frequency higher than 500 megahertz.   352

pretransmit-receive tube. A gas-filled radio-frequency switching tube used to protect the transmit-receive tube from excessively high power and the receiver from frequencies other than the fundamental. *See:* **gas tube.**   125

preventative autotransformer (power and distribution transformer). An autotransformer (or center-tapped reactor) used in load-tap-changing and regulating transformers, or step-voltage regulators to limit the circulating current when operating on a position in which two adjacent taps are bridged, or during the change of taps between adjacent positions.   53

preventive maintenance (1) (test, measurement and diagnostic equipment). Tests, measurement, replacements, adjustments, repairs and similar activities, carried out with the intention of preventing faults or malfunctions from occurring during subsequent operation. Preventive maintenance is designed to keep equipment and programs in proper operating condition and is performed on a scheduled basis.   54

(2) (reliability). The maintenance carried out at predetermined intervals or corresponding to prescribed criteria, and intended to reduce the probability of failure or the performance degradation of an item.   164

prf (1)(laser-maser). Abbreviation for pulse-repetition frequency. High prf = more than 1 Hz. *See:* **pulse-repetition frequency.**   363

(2)(radar). (PRF). *See:* **pulse repetition frequency.**   13

primaries (color) (television). The colors of constant chromaticity and variable amount that, when mixed in proper proportions, are used to produce or specify other colors. *Note:* Primaries need not be physically realizable.   18

primary (1)(supervisory control, data acquisition, and automatic control). An equipment or subsystem which normally contributes to system operation. *See:* **backup.**   570

(2) (instrument transformer). The winding intended for connection to the circuit to be measured or controlled.   203

(3) (used as an adjective) (power switchgear). (A) First to operate; for example, primary arcing contacts, primary detector. (B) First in preference; for example, primary. protection. (C) Referring to the main circuit as contrasted to auxiliary or control circuits; for example, primary disconnecting devices. (D) Referring to the energy input side of transformers, or the conditions (voltages) usually encountered at this location; for example, primary unit substation.   103

(4) (electric machines and devices). The part of a machine having windings that are connected to the power supply line (for a motor or transformer) or to the load (for a generator).   63

primary address (FASTBUS acquisition and control). An address assigned to a device by means of which a master is able to establish contact with the device or a subdivision of the device. Primary address types are logical, geographical and broadcast addresses.   480

primary address cycle (FASTBUS acquisition and control). The portion of a FASTBUS operation in which a master addresses a slave on the address/data (A/D) lines. The address type is specified by the enable geographical (EG) and mode select (MS) control lines. It begins with the master asserting the address sync (AS) line and terminates with the master receiving an address acknowledgement on the address acknowledgement (AK) line. Logical, geographical or broadcast addresses are asserted during primary address cycles.   480

primary arcing contacts (of a switching device) (power switchgear). The contacts on which the initial arc is drawn and the final current, except for the arc-shunting-resistor current, is interrupted after the main contacts have parted. 103

primary battery. *See:* **battery (primary or secondary); electrochemistry; primary cell.**

primary calibration (monitoring radioactivity in effluents). The determination of the electronic system accuracy when the detector is exposed in a known geometry to radiation from sources of known energies and activity levels traceable to the National Bureau of Standards (NBS). 559

primary cell. A cell that produces electric current by electrochemical reactions without regard to the reversibility of those reactions.Some primary cells are reversible to a limited extent. *See:* **electrochemistry.** 328

primary center (telephone switching systems). A toll office to which toll centers and toll points may be connected. Primary centers are classified as Class 3 offices. *See:* **office class.** 55

primary circuit (power and distribution transformer). The circuit on the input side at the regulator. 53

primary coating (fiber optics). The material in intimate contact with the cladding surface, applied to preserve the integrity of that surface. *See:* **cladding.** 433

primary-color unit (television). The area within a color cell occupied by one primary color. *See:* **television.** 125

primary current ratio (electroplating). The ratio of the current densities produced on two specified parts of an electrode in the absence of polarization. It is equal to the reciprocal of the ratio of the effective resistances from the anode to the two specified parts of the cathode. *See:* **electroplating.** 328

primary detector (1) (power switchgear). The first system element or group of elements that responds quantitatively to the measurand and performs the initial measurement operation. A primary detector performs the initial conversion or control of measurement energy and does not include transformers, amplifiers, shunts, resistors, etcetera when these are used as auxiliary means. 103 **(2) (power systems).** That portion of the measurement device which either utilizes or transforms energy from the controlled medium to produce a measurable effect which is a function of change in the value of the controlled variable. 94

primary disconnecting devices (of a switchgear assembly) (power switchgear). Self-coupling separable contacts provided to connect and disconnect the main circuits between the removable element and the housing. 103

primary distribution feeder. A feeder operating at primary voltage supplying a distribution circuit. *Note:* A primary feeder is usually considered as that portion of the primary conductors between the substation or point of supply and the center of distribution. 204

primary distribution mains. The conductors that feed from the center of distribution to direct primary loads or to transformers that feed secondary circuits. *See:* **center of distribution.** 204

primary distribution system. A system of alternating-current distribution for supplying the primaries of distribution for supplying the primaries of distribution transformers from the generating station or substation distribution buses. *See:* **alternating-current distribution; center of distribution.** 204

primary distribution trunk line. A line acting as a main source of supply to a distribution system. *See:* **center of distribution.** 204

primary electron (thermionics). An electron in a primary emission. *See:* **electron emission.** 308,190

primary fault. The initial breakdown of the insulation of a conductor, usually followed by a flow of power current. *See:* **center of distribution.** 204

primary flow (carriers). A current flow that is responsible for the major properties of the device. 342

primary line of sight (illuminating engineering). The line connecting the point of observation and the point of fixation. In terms of a single eye, it is the line connecting the point of fixation and the center of the entrance pupil. 167

primary line-to-ground voltage (coupling capacitors and capacitance potential devices). Refers to the high-tension root-mean-square line-to-ground voltage of the phase to which the coupling capacitors or potential device, in combination with its coupling capacitor or bushing, is connected. *See:* **rated primary line-to-ground voltage.** 351

primary network. A network supplying the primaries of transformers whose secondaries may be independent or connected to a secondary network. *See:* **primary distribution network; center of distribution.** 204

primary overcurrent protective device of apparatus (nuclear power generating stations). A device or apparatus which normally performs the function of circuit interruption. 26

primary protection (as applied to a relay system) (power switchgear). First-choice relay protection in contrast with backup relay protection. 127, 103

primary radar (navigation aid terms). A radar system, subsystem, or mode of operation in which the return signals are the echoes obtained by reflection from the target. Since this is the normal method of radar operation, the word primary is omitted unless necessary to distinguish it from secondary. *See:* **secondary radar.** 526

primary radiator (antennas). The radiating element of a reflector or lens antenna which is coupled to the transmitter or receiver directly or through a feed line. *Note:* For some applications an array of radiating elements is employed. 111

primary reactor starter (industrial control). A starter that includes a reactor connected in series with the primary winding of an induction motor to furnish reduced voltage for starting. It includes the necessary switching mechanism for cutting out the reactor and connecting the motor to the line. *See:* **starter.** 225,206

primary resistor starter (industrial control). A starter that includes a resistor connected in series with the primary winding of an induction motor to furnish re-

duced voltage for staring. It includes the necessary switching mechanism for cutting out the resistor and connecting the motor to the line. *See: starter.*
206

primary section of the core (ferroresonant voltage regulators). The section of the core of a ferroresonant transformer on which the primary winding is wound.
456

primary service area (radio broadcast transmitter). The area within which reception is not normally subject to objectionable interference or fading. *See: radio transmitter.*
328

primary standard (luminous standards) (illuminating engineering). A light source by which the unit of light is established and from which the values of other standards are derived. This order of standard also is designated as the national standard. *Note:* A satisfactory primary (national) standard must be reproducible from specifications. Primary (national) standards usually are found in national physical laboratories, such as the National Bureau of Standards. *See: candela.*
167

primary supply voltage (mobile communication). The voltage range over which a radio transmitter, a radio receiver, or selective signaling equipment is designed to operate without degradation in performance. *See: mobile communication system.*
181

primary switchgear connections. *See: main switchgear connections.*
103

primary transmission feeder. A feeder connected to a primary transmission circuit. *See: center of distribution.*
64

primary unit substation (1) (power switchgear). *See: unit substation. Note.*
103

(2) (power and distribution transformer). A substation in which the low-voltage section is rated above 1000 V.
53

primary voltage rating of a general-purpose specialty transformer (power and distribution transformer). The input circuit voltage for which the primary winding is designed, and to which operating and performance characteristics are referred.
53

primary winding (1)(ferroresonant voltage regulators). The winding of the ferroresonant transformer to which the input voltage is applied.
456

(2) (power and distribution transformer). The winding on the energy input side.
53

(3) (voltage (regulator). The shunt winding. *See: voltage regulator.*
257

(4) (rotating machinery) (motor or generator). The winding carrying the current and voltage of incoming power (for a motor) or power output (for a generator). The choice of what constitutes a primary circuit is arbitrary for certain machines having bilateral power flow. In a synchronous or direct-current machine, this is more commonly called the armature winding. *See: armature.*
63

prime (charge-storage tubes). To charge storage elements to a potential suitable for writing. *Note:* This is a form of erasing. *See: charge-storage tube; television.*
125

prime meridian (navigation aid terms). The meridiam of longitude 0° almost universally considered as Greenwich, England.
526

prime mover (emergency and standby power). The machine used to develop mechanical horsepower to drive an emergency or standby generator to produce electrical power.
512

prime power (1)(emergency and standby power). That source of supply of electrical energy that is normally available and used continuously day and night, usually supplied by an electric utility company, but sometimes supplied by base-loaded user-owned generation.
512

(2)(transmission and distribution). The maximum potential power (chemical, mechanical, or hydraulic) constantly available for transformation into electric power. *See: generating station.*
64

priming rate (charge-storage tubes). The time rate of priming a storage element, line, or area from one specified level to another. Note the distinction between this and **priming speed.** *See: charge-storage tube.*
174

priming speed (charge-storage tubes). The lineal scanning rate of the beam across the storage surface in priming. Note the distinction between this and **priming rate.** *See: charge-storage tube.*
174,125

primitive period (function). *See: period (function).*

principal axis (1) (close-talking pressure-type microphone). The axis of a microphone normal to the plate of the principal acoustic entrance of a microphone, and that passes through the center of the entrance. *See: close-talking pressure-type microphone.*
249

(2) (transducer used for sound emission or reception). A reference direction used in describing the directional characteristics of the transducer. It is usually an axis of structural symmetry, or the direction of maximum response, but if one of these does not coincide with the reference direction, it must be described explicitly.
176

principal axis of compliance (gyro; accelerometer). An axis along which an applied force results in a displacement along that axis only. The acceleration squared error due to anisoelectricity is zero when acceleration is along a principal axis of compliance.
46

principal branch (converter circuit elements)(main branch)(self-commutated converters). A branch involved in the major transfer of energy from one side of the converter to the other.
584

principal characteristics. *See: principal voltage-current characteristic.*

principal-city office (telephone switching systems). An intermediate office that has the screening and routing capabilities to accept traffic to all end office within one or more numbering-plan areas.
55

principal current (1)(circuit properties)(of a converter switching element or branch)(self-commutated converters). The on-state current of the semiconductor devices in a switching element or branch flowing between its principal terminals. *Note:* The principal current is often referred to as the 'current' of the switching element or branch.
584

(2)(thyristor). A generic term for the current through the collector junction. *Note:* It is the current though both main terminals. 445

principal half-power beamwidths (antennas). For a pattern whose major lobe has a half-power contour which is essentially elliptical, the half-power beamwidths in the two pattern cuts which contain the major and minor axes of the ellipse respectively. 111

principal power (thyristor). The power which is consumed in the load circuit plus the losses in the power circuit elements including switching losses. 445

principal restraint (accelerometer). The means by which a measurable force or torque is generated to oppose the force or torque produced by an acceleration along or about an input axis. 46

principal terminal (converter circuit elements)(self-commutated converters). A terminal (of a device or circuit element) through which passes the current transmitting the power that is controlled by the device or circuit element. The term is used for distinction fro control terminals, monitoring signal terminals, etcetera. *Note:* Examples of principal terminals are the anode and cathode of thyristor or diode devices, the collector and emitter of bipolar transistor devices, and the source and drain of field-effect transistor devices. 584

principal voltage (thyristor). The voltage between the main terminals. *Notes:* (1) In the case of reverse blocking and reverse conducting thyristors, the principal voltage is called positive when the anode potential is higher than the cathode potential, and called negative when the anode potential is lower than the cathode potential. (2) For bidirectional thyristors, the principal voltage is called positive when the potential of main terminal 2 is higher than the potential of main terminal 1. 445

principal voltage-current characteristic (thyristor). The function, usually represented graphically, relating the principal voltage to the principal current with gate current, where applicable, as a parameter. *Syn:* **principal characteristic.** 243,66,191

printed circuit (soldered connections). A pattern comprising printed wiring formed in a predetermined design in, or attached to, the surface or surfaces of a common base. 284

printed circuit antenna. An antenna of some desired shape bonded onto a dielectric substrate. *Note:* The microstrip antenna is a notable example. *See:* **microstrip antenna.** 111

printed-circuit assembly. A printed-circuit board on which separately manufactured component parts have been added. 284

printed-circuit board (1) (general). A board for mounting of components on which most connections are made by printed circuitry. 415

(2) (double-sided). A board having printed circuits on both sides. 415

(3) (single-sided). A board having printed circuits on one side only. 415

printed wiring (soldered connections). A portion of a printed circuit comprising a conductor pattern for the purpose of providing point-to-point electric connection only. 284

printer (teleprinter) (teletypewriter). A printing telegraph instrument having a signal-actuated mechanism for automatically printing received messages. It may have a keyboard similar to that of a typewriter for sending messages. The term receiving-only is applied to a printer having no keyboard. *See:* **telegraphy.** 194

printing. *See:* **line printing.**

printing demand meter. An integrated demand meter that prints on a paper tape the demand for each demand interval and indicates the time during which the demand occurred. *See:* **electricity meter (meter).** 328

printing recorder (protective signaling). An electromechanical recording device that accepts electric signal impulses from transmitting circuits and converts them to a printed record of the signal received. *See:* **protective signaling.** 328

printout (test, measurement and diagnostic equipment). The output of a device which is printed on some type of printer. 54

print-through. The undesirable transfer of a recorded signal from a section of a magnetic recording medium to another section of the same medium when these sections are brought into proximity. *Note:* The resulting copy usually is distorted. 176

priority string (power-system communication). A series connection of logic circuits such that inputs are accommodated in accordance with their position in the string, one end of the string corresponding to the highest priority. *See:* **digital.** 415

privacy system (radio transmission). A system designed to make unauthorized reception difficult. *See:* **radio transmission.** 328

private automatic branch exchange (PABX) (telephone switching systems) (data transmission). A private branch exchange that is automatic. 55, 59

private automatic exchange (PAX) (telephone switching systems) (data transmission). A private non-branch exchange that is automatic. 55, 59

private branch exchange (PBX) (telephone switching systems) (data transmission). A private telecommunications exchange that includes access to a public telecommunications exchange. 55, 59

private branch exchange hunting (telephone switching systems). An arrangement for searching over a group of trunks at the central office, any one of which would provide a connection to the desired private branch exchange. 55

private-branch-exchange trunk (PBX) (telephone switching systems). A line used as a trunk between a private branch exchange and the central office that serves it. 55

private exchange. A telephone exchange serving a single organization and having no means for connection with a public telephone exchange. 193,101

private line (private wire) (data transmission). A channel or circuit furnished to a subscriber for the subscriber's exclusive use. 59

private line telegraph network (data transmission). A system of points interconnected by leased telegraph channels and providing hard-copy or five-track punched paper tape, or both, at both sending and receiving points.   59

private line telephone network (data transmission). A series of points interconnected by leased voice-grade telephone lines, with switching facilities or exchange operated by the customer.   59

private non-branch exchange (telephone switching systems). A series of points interconnected by leased voice-grade telephone lines, with switching facilities or exchange operated by the customer.   415

private residence. A separate dwelling or a separate apartment in a multiple dwelling that is occupied only by the members of a single family unit.   328

private residence elevator. A power passenger electric elevator, installed in a private residence, and that has a rated load not in excess of 700 pounds, a rated speed not in excess of 50 feet per minute, a net inside platform area not in excess of 12 square feet, and a rise not in excess of 50 feet. *See:* **elevators.**   328

private-residence inclined lift. A power passenger lift, installed on a stairway in a private residence, for raising and lowering persons from one floor to another. *See:* **elevator.**   328

private telecommunication exchange (telephone switching systems). A telecommunications exchange for a single organization.   55

privileged instruction (software). An instruction that may be used only by a supervisory program. *See:* **instruction; supervisory program.**   434

probability density function (control of system electromagnetic compatibility). The first derivative of the probability distribution function; it represents the probability of obtaining a given value.   495

probability distribution (nuclear power generating stations). The mathematical function that relates the probability of an event to an elapsed time or to a number of trials.   29

probability distribution function (1)(control of system electromagnetic compatibility). The function of x whose value is the probability that the amplitude is greater than, or equal to, x. *Note:* The probability distribution function is a nondecreasing function ranging from zero to unity.   495

(2)(reliability analysis of nuclear power generating station safety systems). The mathematical function that gives Prob $(X \leq x)$ where X is a random variable and x is a particular value of X.   587

probe (potential) (gas). An auxiliary electrode of small dimensions compared with the gas volume, that is placed in a gas tube to determine the space potential. *See:* **discharge (gas).**   244,190

probe loading. The effect of a probe on a network, for example, on a slotted line, the loading represented by a shunt admittance or a discontinuity described by a reflection coefficient. *See:* **measurement system.**
                                                            185

probe pickup, residual. *See:* **residual probe pickup.**
problem. *See:* **benchmark problem.**

problem board (analog computers). In an analog computer, a removable frame of receptacles for patch cords and plugs that, through a patch bay, offers a means for interconnecting the inputs and outputs of computing elements, etcetera according to the computer diagram. *Syn:* **patch board, patch panel, prepatch panel.**   9

problem check (analog computers). One or more tests used to assist in obtaining the correct machine solution to a problem. Static check consists of one or more tests of computing elements, their interconnections, or both, performed under static conditions. Dynamic check consists of one or more tests of computing elements, their interconnections, or both, performed under dynamic conditions. Rate test is a test that verifies that the time constants of the integrators are those selected. This term also refers to the computer-control state that implements the rate test previously described. Dynamic problem check is any dynamic check used to ascertain the correct performance of some or all of the computer components. *See:* **computer-control state.**   9

problem oriented language (computing systems). A programming language designed for the convenient expression of a given class of problems.
                                                      255, 77, 54

problem variable. *See:* **scale factor.**

procedural programming language (software unit testing). A computer programming language used to express the sequence of operations to be performed by a computer (for example, COBOL). *See:* **nonprocedural programming language.**   519

procedure (1) (computing systems). The course of action taken for the solution of a problem.   255, 77

(2) (nuclear power quality assurance). A document that specifies or describes how an activity is to be performed.   417

(3) (software). (A) A portion of a computer program which is named and which performs a specific task. (B) The course of action taken for the solution of a problem. (C) The description of the course of action taken for the solution of a problem. (D) A set of manual steps to be followed to accomplish a task each time the task is to be done. *See:* **computer program; function; module; subprogram; subroutine.**   434

procedure-oriented language (computing systems). A programming language designed for the convenient expression of procedures used in the solution of a wide class of problems.   255, 77, 54

process (1)(microprocessor operating systems). A unit of activity characterized by a single sequential thread of execution, a current state, and an associated set of system resources.   478

(2) (automatic control). The collective functions performed in and by the equipment in which a variable is to be controlled. *Syn:* **controlled system.**   56

(3) (software). (A) In a computer system, a unique, finite course of events defined by its purpose or by its effect, achieved under given conditions. (B) To perform operations on data in process. *See:* **computer system; data.**   434

process control (1)(electric pipe heating systems). The use of electric pipe heating systems to increase or maintain, or both, the temperature of fluids (or processes) in mechanical piping systems including pipes, pumps, tanks, instrumentation in nuclear power generating stations.   448

(2) (automatic control). Control imposed upon physical or chemical changes in a material. *See:* **control system, feedback.**   56

(3) (electric pipe heating systems). The use of electric pipe heating systems to increase or maintain, or both, the temperature of fluids (or processes) in mechanical piping systems including pipes, pumps, valves, tanks, instrumentation, etcetera.   405

process equipment (automatic control). Apparatus with which physical or chemical changes in a material are produced. *Syn:* **plant.** *See:* **directly controlled system; indirectly controlled system.**   56

processing. *See:* **data processing; information processing; multiprocessing; parallel processing.**

processing cycle (computer applications). A single, complete execution of data processing that is periodically repeated.   571

processor (1) (computing systems). (A) (hardware). A data processor.(B) (pascal computer programming language). A system or mechanism that accepts a program as input, prepares it for execution, and executes the process so defined with data to produce results. *Note:* A processor may consist of an interpreter, a compiler and run-time system, or other mechanism, together with an associated host computing machine and operating system, or other mechanism for achieving the same effect. A compiler in itself, for example, does not constitute a processor.   433

(2) (software). A computer program that includes the compiling, assembling, translating, and related functions for a specific programming language, for example, **Cobol** processor, **Fortran** processor. *See:* **data processor; multiprocessor.**   255, 77

processor interface (PI)(FASTBUS acquisition and control). The interface device between a processor and a FASTBUS segment.   480

processtag (microprocessor operating systems parameter types). A 'tag' returned by one function for use by another. Its contents may not be examined or changed. Its form is system dependent. A processtag is only valid within a given process and should not be passed between processes.   478

procurement document (nuclear power quality assurance). Purchase requisitions, purchase orders, drawings, contracts, specifications, or instructions used to define requirements for purchase.   417

procurement documents (nuclear power generating station). Those documents such as specifications, contracts, letters of intent, work orders, purchase orders or proposals and their acceptance which authorize the seller to perform services or supply equipment, material or facilities to the purchaser. *Note:* This term applies specifically to the subject matter of IEEE Std 467-1980.   438

product certification. *See:* **certification.**

production (routine) (power cable joint). Tests made on joint components or subassemblies during production for the purpose of quality control.   34

production library (software). A software library containing software approved for current operational use. *See:* **operational; software library.**   434

production tests (1)(metal-enclosed low-voltage power circuit-breaker switchgear)(metal-clad and station-type cubicle switchgear). Tests made for quality control by the manufacturer on every device or representative samples, or on parts or materials as required to verify during production that the product meets the design specifications and applicable standards. *Notes:* (A) Certain quality assurance tests on identified critical parts of repetitive high-production devices may be tested on a planned statistical sampling basis. (B) Production tests are sometimes called routine tests.
579, 572, 573

(2) (for switchgear). Those tests made to check the quality and uniformity of the workmanship and materials used in the manufacture of switchgear or its components.   103

(2) (power cable joints) (routine). Tests made on joint components or sub-assemblies during production for the purpose of quality control.   34

product modulator. A modulator whose modulated output is substantially equal to the product of the carrier and the modulating wave. *Note:* The term implies a device in which intermodulation between components of the modulating wave does not occur. *See:* **modulation.**   328

product relay (power switchgear). A relay that operates in response to a suitable product of two alternating electrical input quantities.   103

product sensitivity. The ratio of Hall voltage to the product of control current and magnetic flux density at any point on the product sensitivity characteristic curve of a Hall generator.   107

product specification. *See:* **design specification.**

professional projector (National Electrical Code). The professional projector is a type using 35- or 70-millimeter film which has a minimum width of 1 3/8 inches and has on each edge 5.4 perforations per inch, or a type using carbon arc, Xenon, or other light source equipment which develops hazardous gases, dust or radiation.   256

profile. *See:* **graded index profile; index profile; parabolic profile; power-law index profile; step index profile.**

profile dispersion (fiber optics). (1) In an optical waveguide, that dispersion attributable to the variation of refractive index contrast with wavelength, where contrast refers to the difference between the maximum refractive index in the core and the refractive index of the homogeneous cladding. Profile dispersion is usually characterized by the profile dispersion parameter, defined by the following entry. (2) In an optical waveguide, that dispersion attributable to the variation of refractive index profile with wavelength. The profile variation has two contributors: (a) variation in refractive index contrast, and (b) variation in profile parameter. *See:* **dispersion; distortion; refractive index profile.** 433

profile dispersion parameter (P) (fiber optics).

$$P(\lambda) = \frac{n_1}{N_1} \frac{\lambda}{\Delta} \frac{d\Delta}{d\lambda}$$

where n_1, N_1 are, respectively, the refractive and group indices of the core, and $n_1\sqrt{1 - 2\Delta}$ is the refractive index of the homogeneous cladding, $N_1 = n_1 - \lambda(dn_1/d\lambda)$, and Δ is the refractive index constant. Sometimes it is defined with the factor (-2) in the numerator. *See:* **dispersion.** 433

profile parameter (fiber optics). The shape-defining parameter, g, for a power-law index profile. *See:* **power-law index profile; refractive index profile.** 433

prognosis (test, measurement and diagnostic equipment). The use of test data in the evaluation of a system or equipment for potential or impending malfunctions. 54

program (1)(microprocessor operating systems). A collection of processes working together to accomplish a common task. 478
(2) (general). A sequence of signals transmitted for entertainment or information. *See:* **communication.** 239
(3) (electronic computation). (A) A plan for solving a problem. (B) Loosely, a routine. (C) To devise a plan for solving a problem. (D) Loosely, to write a routine. *See:* **acceleration, programmed; communication; computer program; object program; source program; target program.** 235
(4) (telephone switching systems). A set of instructions arranged in a predetermined sequence to direct the performance of a planned action or actions. 55
(5) (P) (semiconductor memory). The inputs that when true enable programming, or writing into, a programmable read only memory (PROM). 441
(6) (software). (A) A computer program. (B) A schedule or plan that specifies actions to be taken. (C) To design, write, and test computer programs. *See:* **computer program; design.** 434

program amplifier. *See:* **amplifier, line.**

program architecture (software). The structure and relationships among the components of a computer program. The program architecture may also include the program's interface with its operational environment. *See:* **component; computer program; interface.** 434

program block (software). In problem-oriented languages, a computer program subdivision that serves to group related statements, delimit routines, specify storage allocation, delineate the applicability of labels, or segment paths of the computer program for other purposes. *See:* **computer program; label; routine; segment.** 434

program correctness. *See:* **correctness.**

program design language. *See:* **design language.**

program extension (software). An enhancement made to existing software to increase the scope of its capabilities. *See:* **enhancement; software.** 434

program instrumentation (software). (1) Probes, such as instructions or assertions, inserted into a computer program to facilitate execution monitoring, proof of correctness, resource monitoring, or other activities. (2) The process of preparing and inserting probes into a computer program. *See:* **assertion; computer program; execution; instruction; proof of correctness.** 434

program level. The magnitude of program in an audio system expressed in volume units. 239

program library (software). An organized collection of computer programs. *See:* **computer program; software library; system library.** 434

programmable (programmable instrumentation). That characteristic of a device that makes it capable of accepting data to alter the state of its internal circuitry to perform two or more specific tasks. 40

programmable digital computer (programmable digital computer systems in safety systems of nuclear power generating stations). A device that can store instructions and is capable of the execution of a systematic sequence of operations performed on data that is controlled by internally stored instructions. 554

programmable equipment (supervisory control, data acquisition, and automatic control). A remote or master station having one or more of its operations specified by a program contained in a memory device. 570

programmable measuring apparatus (programmable instrumentation). A measuring apparatus that performs specified operations on command from the system and, if it is a measuring apparatus proper, may transmit the results of the measurement(s) to the system. 40

programmable stimuli (test, measurement and diagnostic equipment). Stimuli that can be controlled in accordance with instructions from a programming device. 54

programmed check. A check procedure designed by the programmer and implemented specifically as a part of his program. *See:* **check, automatic; check problem; mathematical check.** 255, 77

programmed control (industrial control). A control system in which the operations are determined by a predetermined input program from cards, tape, plug boards, cams, etcetera. *See:* **control system, feedback.** 206

programmer (1) (power switchgear). An arrangement of operating elements or devices that initiates, and often controls, one or a series of operations in a given sequence. 103
(2) (test, measurement and diagnostic equipment). (A) A device having the function of controlling the timing and sequencing operations: and (B) a person who prepares sequences of instructions for a programmable machine. 54

programmer-comparator (test, measurement and diagnostic equipment). (1) A device which reads commands and data from a sequential program usually on tape or cards: (2) sets up delays, switching, and stimuli, and performs measurements as directed by the program: and (3) compares the results of each measurement with fixed programmed tolerance limits to arrive at a decision. Often numerous other operations, such as branching on no-go or other conditions, are included. 54

programming (1) (electronic computation). The ordered listing of a sequence of events designed to accomplish a given task. *See:* **linear programming; multiprogramming; automatic programming.**
244, 207, 54

(2) (power supplies). The control of any power-supply functions, such as output voltage or current, by means of an external or remotely located variable control element. Control elements may be variable resistances, conductances, or variable voltage or current sources. 186

programming delay (power switchgear). A relay whose function is to establish or detect electrical sequences. 103

programming language (software). An artificial language designed to generate or express programs. *See:* **artificial language; program.** 434

programming, linear (1) general. Optimization problem characterization in which a set of parameter values are to be determined, subject to given linear constraints, optimizing a cost function that is linear in the parameter. *See:* **system.** 209
(2) (computing systems). The analysis or solution of problems in which linear function of a number of variables is to be maximized or minimized when those variables are subject to a number of constraints in the form of linear inequalities. 255, 77

programming, nonlinear. Optimization problem in which any or all of the following are nonlinear in the variables: (1) The objective functions. (2) The defining interrelationships among the variables, the plant description. (3) The constraints. *See:* **system.** 209
programming, quadratic. Optimization problem in which: (1) The objective function is a quadratic function of the variable. (2) The plant description is linear. *See:* **system.** 209
programming relay. A relay whose function is to establish or detect electrical sequences. 103, 127
programming speed (power supplies). Describes the time requires to change the output voltage of a power supply from one value to another. The output voltage must change across the load and because the supply's filter capacitor forms a resistance-capacitance network with the load and internal source resistance, programming speed can only be described as a function of load. Programming speed is the same as the recovery-time specification for current-regulated operation: it is not related to the recovery-time specification for voltage-regulated operation. 186
programming support environment (software). An integrated collection of tools accessed via a single command language to provide programming support capabilities throughout the software life cycle. The environment typically includes tools for designing, editing, compiling, loading, testing, configuration management, and project management. *See:* **command language; configuration management; software life cycle; testing; tool.** 434
program mutation (software). (1) A program version purposely altered from the intended version to evaluate the ability of program test cases to detect the alteration. (2) The process of creating program mutations in order to evaluate the adequacy of program test

Remote programming connection showing programming of power supplies.

data. *Syn:* **program mutant.** *See:* **program; program test data; test cases.** 434

program protection (software). The application of internal or external controls to preclude any unauthorized access or modification to a computer program. *See:* **computer program; modification.** 434

program-sensitive fault (computing systems). A fault that appears in response to some particular sequence of program steps. 255, 77

program specification (software). (1) Any specification for a computer. program. *Syn:* **design specification.** *See:* **design specification; functional specification; performance specification; requirements specification.** 434

program stop (numerically controlled machines). A miscellaneous function command to stop the spindle, coolant, and feed after completion of other commands in the block. It is necessary for the operator to push a button in order to continue with the remainder of the program. 244, 207

program support library. *See:* **software development library.**

program synthesis (software). The use of software tools to aid in the transformation of a program specification into a program which realizes that specification. *See:* **program; program specification; software tool; specification.** 434

program tracking (communication satellite). A technique for tracking a satellite by pointing a high gain antenna towards the satellite which employs a computer program for antenna pointing: known orbital parameters are used as an input to the computer program. 84

program validation. *Syn:* **computer program validation.** *See:* **validation.**

progressive grading (telephone switching systems). A grading in which the outlets of different grading groups are connected together in such a way that the number of grading groups connected to each outlet is larger for later choice outlets. 55

progressive scanning*. *See:* **sequential scanning.** 18

*Deprecated

progressive scanning (television). *See:* **sequential scanning.**

project (unique identification in power plants). A single- or multiple-unit power plant or major independent related facility. A project is composed of systems and structures and may be defined to include the design, construction, operation, and related activities associated with the project during its life cycle. 544

projected peak point (tunnel-diode characteristic). The point on the forward current-voltage characteristic where the current is equal to the peak-point current and where the voltage is greater than the valley-point voltage. *See:* **peak point (tunnel-diode characteristic).** 191

projected peak-point voltage (tunnel-diode characteristic). The voltage at which the projected peak point occurs. *See:* **peak point (tunnel-diode characteristic).** 191

project file. *See:* **project notebook.**

projection tube (electron device). A cathode-ray tube specifically designed for use with an optical system to produce a projected image. 190

project notebook (software). A central repository of written material such as memos, plans, technical reports, etcetera, pertaining to a project. *Syn:* **project file.** *See:* **software development notebook.** 434

projector (illuminating engineering). A lighting unit which, by means of mirrors and lenses, concentrates the light to a limited solid angle so as to obtain a high value of luminous intensity. 167

project plan (software). A management document describing the approach that will be taken for a project. The plan typically describes the work to be done, the resources required, the methods to be used, the configuration management and quality assurance procedures to be followed, the schedules to be met, the project organization, etcetera. *See:* **configuration management; document; quality assurance.** 434

prompt (software). (1) A message informing a user that a system is ready for the next command, message, or other user action. (2) To inform a user that a system is ready for the next command, element, or other input. *See:* **system.** 434

proof (1) (packaging machinery). Used as a suffix, indicates that apparatus is so constructed, protected, or treated that its successful operation is not interfered with when subjected to specified material or condition. 429

(2) (suffix)(shunt power capacitors)(power and distribution transformer). Apparatus is designed as splashproof, dustproof, etcetera, when so constructed, protected, or treated that its successful operation is not interfered with when subjected to the specified material or condition. 138, 53

(3) (suffix) (high-voltage switchgear). So constructed, protected or treated that successful operation is not interfered with when the device is subjected to the specified material or condition. *Note:* Explosion-proof requires that the fuse shall not be injured and flame shall not be transmitted to the outside of the fuse for all current interruptions within the rating of the fuse. 443

proof mass (accelerometer). The effective mass whose inertia transforms an acceleration along or about an input axis into a force or torque. The effective mass takes into consideration flotation and contributing parts of the suspension. 46

proof of correctness (software). (1) A formal technique used to prove mathematically that a program satisfies its specifications. (2) A program proof that results from applying this technique. *See:* **partial correctness; program; specification; total correctness.** 434

proof test (1)(evaluation of thermal capability)(thermal classification of electric equipment and electrical insulation). A means of evaluation in which an arbitrary fixed level of a diagnostic factor is applied periodically. In this case, the number of failures among multiple test specimens (rather than the magnitude of the diagnostic factor) defines the end-point of the test. *See:* **diagnostic factor.** 506

(2) (withstand test) (rotating machinery). A "fail" or "no fail" test of the insulation system of a rotating machine made to demonstrate whether the electrical strength of the insulation is above a predetermined minimum value. 6

proof-test load (composite insulators). The routine mechanical load that is applied to an insulator at the time of its manufacture. 483

propagated potential (biological). A change of potential involving depolarization progressing along excitable tissue. 192

propagating mode (waveguide). A waveguide mode such that the variation of phase along the direction of the guide is not negligible. 267

propagation (electrical practice) (data transmission). The travel of waves through or along a medium. 59

propagation constant (1) (fiber optics). For an electromagnetic field mode varying sinusoidally with time at a given frequency, the logarithmic rate of change, with respect to distance in a given direction, of the complex amplitude of any field component. *Note:* The propagation constant is a complex quantity. 433

(2) (overhead-power-line corona and radio noise). The propagation constant of a traveling plane wave at a given frequency is the complex quantity whose real part is the attenuation constant in nepers per unit length and whose imaginary part is the phase constant in radians per unit length. 411

(3) (radio wave propagation). Of a traveling wave in a homogeneous medium, the negative of the partial logarithmic derivative, with respect to distance in the direction of the wave, of the phasor quantity describing the wave. *Note:* In the case of cylindrical or spherical traveling waves, the amplitude factors $1/\sqrt{r}$ and $1/r$, respectively, are not to be included in the phasor quantity. 146

(4) (waveguide). Of a traveling wave at a given frequency and for a given mode, the complex quantity whose real part is the attenuation constant in nepers per unit length and whose imaginary part is the phase constant in radians per unit length. 267

(5) (transmission lines and transducers). (A) (per unit length of a uniform line). The natural logarithm of the ratio of the phasor current at a point of the line, to the phasor current at a second point, at unit distance from the first point along the line in the direction of transmission, when the line is infinite in length or is terminated in its characteristic impedance. (B) (per section of a periodic line). The natural logarithm of the ratio of the phasor current entering a section, to the phasor current leaving the same section, when the periodic line is infinite in length or is terminated in its iterative impedance. (C) (of an electric transducer). The natural logarithm of the ratio of the phasor current entering the transducer, to the phasor current leaving the transducer, when the transducer is terminated in its iterative impedance. 210

(6) (circuits and systems). The image transfer constant for a symmetrical transducer. 67

propagation delay (sequential events recording systems) (power generation). The time interval between the appearance of a signal at any circuit input and the appearance of the associated signal at that circuit output. 48

propagation factor (radio wave propagation). For a time-harmonic wave propagating from one point to another, the ratio of the complex electric field strength at the second point to that value which would exist at the second point if propagation took place in a vacuum. 146

propagation loss. The total reduction in radiant power surface density. The propagation loss for any path traversed by a point on a wave front is the sum of the spreading loss and the attenuation loss for that path. *See:* **radio transmission.** 210

propagation mode (1) (in a periodic beamguide) (laser-maser). A form of propagation characterized by identical field distributions over cross-sections of the beam at positions separated by one period of the guide. 363

(2) (overhead-power-line corona and radio noise). A concept for treating radio noise propagation along a set of overhead-power-line conductors. Modal waves form a complete set of noninteracting components into which the propagated wave may be separated. *Note:* For a three-phase horizontal single-circuit transmission line with one conductor per phase and without ground wires the following modes are defined: Mode 1 - The transmission path is between the center phase and the outside phases. It has lowest attenuation and lowest surge impedance. Mode 2 - The transmission path is between outside phases. It has intermediate attenuation and intermediate surge impedance. Mode 3 - The transmission path is along all three phases and returning through ground. It has highest attenuation and highest surge impedance. 411

propagation model. An empirical or mathematical expression used to compute propagation path loss. *See:* **electromagnetic compatibility.** 199

propagation ratio (radio wave propagation). For a time-harmonic wave propagating from one point to another, the ratio of the complex field strength at the second point to that at the first point. 146

propagation vector (radio wave propagation). For a traveling time-harmonic wave at a given frequency, the complex vector whose real part is the attenuation vector and whose imaginary part is the phase vector. 146

propagation vector in physical media (antennas). The complex vector $\overline{\gamma}$ in plane wave solutions of the form $e^{-\gamma r}$ for an e^{jwt} time variation and $\overline{r}$ the position vector. *See:* **attenuation vector in physical media; phase vector in physical media; wave vector in physical media; propagation constant in physical media.** 111

propeller-type blower (rotating machinery). An axial-flow fan with air-foil-shaped blades. *See:* **fan (rotating machinery).** 63

proper ferroelectric (primary ferroelectric terms). A ferroelectric in which the polarization is the primary order parameter. 497

proper operation. The functioning of the train control

or cab signaling system to create or continue a condition of the vehicle apparatus that corresponds with the condition of the track of the controlling section when the vehicle apparatus is in operative relation with the track elements of the system.   328

proportional amplifier (industrial control). An amplifier in which the output is a single value and an approximately linear function of the input over its operating range. *See:* **control system, feedback.**
    206

proportional control action (electric power systems). *See:* **control action, proportional.**

proportional counter tube. A radiation-counter tube designed to operate in the proportional region.
    9

proportional gain (hydraulic turbines). The proportional gain G_p of a proportional element is the ratio of the element's percent output to its percent input. A linear relationship is assumed.   8

proportionally. *See:* **linearity.**

proportional plus derivative control. *See:* **control action, proportional plus derivative.**

proportional plus integral control. *See:* **control action, proportional plus integral.**

proportional plus integral plus derivative control. *See:* **control action proportional plus integral plus derivative.**

proportional region (radiation-counter tubes). The range of operating voltage for a counter tube in which the gas amplification is greater than unity and is independent of the amount of primary ionization. *Notes:* (1) In this region the pulse size from a counter tube is proportional to the number of ions produced as a result of the initial ionizing event. (2) The proportional region depends on the type and energy of the radiation.   125

proprietary system (protective signaling). A local system sounding and.or recording alarm and supervisory signals at a control center located within the protected premises, the control center being under the supervision of employees of the proprietor of the protected premises. *Note:* According to the United States Underwriters' rules, a proprietary system must be a recording system. *See:* **protective signaling.**   328

propulsion-control transfer switch. Apparatus in the engine room for transfer of control from engine room to bridge and vice versa. *Note:* Engine-room control is provided on all ships. Bridge control with a transfer switch is optional and is used principally on small vessels such as tugs or ferries, usually with a direct-current propulsion system.   328

propulsion set-up switch. Apparatus providing ready means to set up for operation under varying conditions where practicable: for example, cutout of one or more generators when multiple units are provided. *See:* **electric propulsion system.**   328

prorated section (metal-oxide surge arresters for ac power circuits)(surge arresters). A complete, suitably housed part of an arrester, comprising all necessary components, including gaseous medium, in such a proportion as to accurately represent, for a particular test, the characteristics of a complete arrester.
    583, 430

prorated unit (arrester). A completely housed prorated section of an arrester that may be connected in series with other prorated units to construct an arrester of higher voltage rating.   62

prospective characteristics of a test voltage (high voltage testing). The prospective characteristics of a test voltage causing disruptive discharge are the characteristics which would have been obtained if no disruptive discharge had occurred. When a prospective characteristic is referred to, this should always be stated.   150

prospective current (1) (ac high-voltage circuit breaker). The current that would flow if it were not influenced by the circuit breaker.   426
(2) (available current) (1) (surge arresters). The root-mean-square symmetrical short-circuit current that would flow at a given point in a circuit if the arrester(s) at that point were replaced by links of zero impedance.   308, 62

prospective current of a circuit (with respect to a switching device situated therein). *See:* **available (prospective) current (of a circuit with respect to a switching device situated therein).**

prospective peak (crest) value (of a chopped impulse) (surge arresters). The peak (crest) value of the full-wave impulse voltage from which a chopped impulse voltage is derived.   62

prospective peak value of test voltage (switching impulse testing). The voltage that would be obtained if no disruptive discharge occurred before the crest.
    108

prospective short-circuit current (at a given point in a circuit). *See:* **available (prospective) short-circuit current (at a given point in a circuit).**

prospective short-circuit test current (at the point of test). *See:* **available (prospective) short-circuit test current (at the point of test).**

protected area (nuclear security systems). A controlled-access area encompassed by physical barriers.
    464

protected enclosure (electric installations on shipboard). An enclosure in which all openings are protected with wire screen, expanded metal, or perforated covers. A common from of specifications for "protected enclosure" is: 'The openings should not exceed 1/2 sq. in. in area and should be of such shape as not to permit the passage of a rod larger than 1/2 in. in diameter, except where the distance of exposed live parts from the guard is more than 4 in. the openings may be 3/4 sq. in. in area and must be of such shape as not to permit the passage of a rod larger than 3/4 in. in diameter."   3

protected location (computing systems). A storage location reserved for special purposes in which data cannot be stored without undergoing a screening procedure to establish suitability for storage therein.
                                                     255, 77

protected machine. *See:* **guarded machine.**

protected zone. *See:* **cone of protection.**

protection (1) (computing systems). *See:* **storage protection.**

(2) (software). An arrangement for restricting access to or use of all, or part, of a computer system. *See:* **computer system.**   434

protection system (nuclear power generating station).
(1) That part of the sense and command features involved in generating those signals used primarily for the reactor trip system and engineered safety features.
                                                     428, 102
(2) The electrical and mechanical devices (from measured process variables to protective action system input terminals) involved in generating those signals associated with the protective functions. These signals include those that initiate reactor trip, engineered safety features (for example, containment isolation, core spray, safety injection, pressure reduction, and air cleaning), and auxiliary supporting features.   387
(3) The electrical and mechanical devices (from measured process variables to protective action system input terminals) involved in generating those signals associated with the protective functions. These signals include those that actuate reactor trip and actuate engineered safety features (for example, containment isolation, core spray, safety injection, pressure reduction, and air cleaning).   20

protective action (nuclear power generating station).
(1) The initiation of a signal within the sense and command features, or the operation of equipment within the execute features, for the purpose of accomplishing a safety function   102
(2) The initiation of a signal or operation of equipment within the protection system or protective action system for the purpose of accomplishing a protective function in response to a generating station condition having reached a limit specified in the design basis:
(A) Protection System: Protective action at the channel level is the initiation of a signal by a single channel when the sensed variable(s) reaches a specified limit.
(B) Protective Action System: Protective action at the system level is the operation of sufficient actuated equipment, including the appropriate auxiliary supporting features, to accomplish a protective function. Examples of protective actions at the system level are: rapid insertion of control rods, closing of containment isolation valves, operation of safety injection, and core spray.   387
(C) The initiation of a signal within the sense and command features or the operation of equipment within the execute features for the purpose of accomplishing a safety function.   428
(3) (A) At the channel level, the initiation of a signal by a single channel when the variable sensed exceeds a limit. (B) At the system level, initiation of the operation of a sufficient number of actuators to effect a protective function.   109, 159, 31
(4) The initiation of a signal or operation of equipment within the safety system for the purpose of accomplishing a protective function in response to a generating station condition having reached a limit specified in the design basis. *Notes:* (A) Protective action at the channel level is the initiation of a signal by a single channel when the sensed variable(s) reaches a specified limit. (B) Protective action at the system level is the operation of sufficient actuated equipment including the appropriate auxiliary supporting features to accomplish a protective function. Examples of protective actions at the system level are: rapid insertion of control rods, closing of containment isolation valves, and operation of safety injection and core spray.
                                                     20

protective action set point (nuclear power generating stations). The reference value to which the measured variable is compared for the initiation of protective action.   31

protective action system (1) (class 1E power systems for nuclear power generating stations). The electrical and mechanical equipment (from the protection system output to and including the actuated equipment-to-process coupling) that performs a protective action when it receives a signal from the protective system. *Note:* Examples of protective action systems are: control rods and their trip mechanisms; isolation valves, their operators and their contactors; and emergency service water pumps and associated valves, their motors and circuit breakers. In some instances protective actions may be performed by protective action system equipment that responds directly to the process conditions (for example, check valves, self-actuating relief valves).   102
(2) (nuclear power generating station). The electrical and mechanical equipment (from the protection system output to and including the actuated equipment-to-process coupling) that performs a protective action when it receives a signal from the protection system. *Notes:* (A) Examples of protective action systems are: control rods, and their mechanisms; isolation valves, their operators, and their contractors; and emergency service water pumps and associated valves, their motors, and circuit breakers. (B) In some instances protective actions may be performed by protective action system equipment that responds directly to the process conditions (for example, check valves, self-actuating relief valves).   387

(C) An arrangement of equipment that performs a protective action when it receives a signal from the protection system.   20

protective covering (power cable joint). A field-applied material to provide environmental protection over the joint or housing, or both.   34

protective device (series capacitors). A bypass gap or other device which limits the voltage on the capacitor segment to a predetermined level when overcurrent flows through the series capacitor (that is, during system faults, system swings, or other abnormal events), and which is capable of carrying capacitor discharge, system fault, and load current for specified durations.   474

protective function (nuclear power generating station). (1) Any one of the functions necessary to mitigate the consequences of a design basis event (for example, reduce power, isolate containment, or cool the core). *Note:* A protective function is design basis objective that must be accomplished: a successfully completed protective action at the system level, including the sensing of one or more variables, will accomplish the protective function. However, the design may be such that a given protective function may be accomplished by any one of several protective actions at the system level.   356

(2) The completion of those protective actions at the system level required to maintain plant conditions within the allowable limits established for a design basis event (for example, reduce power, isolate containment, or cool the core).   102, 387

(3) The sensing of one or more variables associated with a particular generating station condition, the signal processing and the initiation and completion of the protective action at values of the variables established in the design bases.   109, 20

protective gap (power switchgear). A gap placed between live parts and ground to limit the maximum over-voltage that may occur.   103

protective lighting (illuminating engineering). A system intended to facilitate the nighttime policing of industrial and other properties.   167

protective margin (surge arresters). The value of the protective ratio minus one expressed in percent ((PR mins 1) mult 100).   62

protective power gap (series capacitor). A bypass gap which limits the voltage on the capacitor segment to a predetermined level when system fault occurs on the line, and which is capable of carrying capacitor discharge, system fault, and load currents for specified durations.   86

protective ratio (surge arresters). The ratio of the insulation withstand characteristics of the protected equipment to the arrester protective level, expressed as a multiple of the latter figure.   62

protective relay (1)(power operations). A device whose function is to detect defective lines or apparatus or other power system conditions of an abnormal or dangerous nature and to initiate appropriate control action.   516

(2) **(power switchgear).** A relay whose function is to detect defective lines or apparatus or other power system conditions of an abnormal or dangerous nature and to initiate appropriate control circuit action. *Note:* A protective relay may be classified according to its input quantities, operating principal, or performance characteristics.   127, 103

(3) **(seismic testing of relays).** A relay whose function is to detect defective lines or apparatus or other power system conditions of an abnormal or dangerous nature and to initiate appropriate control circuit action. *Note:* A protective relay may be classified according to its operating quantities, operating principle, or performance characteristics.   392

protective screen (burglar-alarm system). A lightweight barrier of either solid strip or lattice construction, carrying electric protection circuits, and barring access through a normal opening to protected premises. *See:* **protective signaling.**   328

protective signaling. Protective signaling comprises the initiation, transmission, and reception of signals involved in the detection and prevention of property loss or damage due to fire, burglary, robbery, and other destructive conditions, and in the supervision of persons and of equipment concerned with such detection and prevention.   328

protective system (class 1E power systems for nuclear power generating stations). The electrical and mechanical devices (from measured process variables to protective action system input terminals) involved in generating those signals associated with the protective functions. These signals include those that initiate reactor trip, engineered safety features (for example, containment isolation, core spray, safety injection, pressure reduction, and air cleaning) and auxiliary supporting features.   102

protector tube (1) (surge arresters). An expulsion arrester used primarily for the protection of line ard switch insulation.   2, 62

(2) **(electron-tube type).** A glow-discharge cold-cathode tube that employs a low-voltage breakdown between two or more electrodes to protect circuits against overvoltage.   190

protocol (1)(supervisory control, data acquisition, and automatic control). A strict procedure required to initiate and maintain communication.   570

(2) **(data communication).** A formal set of conventions governing the format and relative timing of message exchange between two communications terminals. *See:* **control procedure.**   12

(3) **(software).** (A) A set of conventions or rules that govern the interactions of processes or applications within a computer system or network. (B) A set of rules that govern the operation of functional units to achieve communication. *See:* **computer system; functional unit; network; process.**   434

protocol data unit (PDU)(1)(logical link control). The sequence of contiguous octets delivered as a unit from or to the medium access control (MAC) sublayer. A valid logical link control (LLC) PDU is at least 3 octets in length, and contains two address fields and a control field. A PDU may or may not include an information field in addition.   585

(2)(token ring access method). Information delivered as a unit between peer entities which contains control information and, optionally, data. 472

proton microscope. A device similar to the electron microscope but in which the charged particles are protons. *See:* **electron optics.** 244, 190

proton range (solar cells). The maximum distance traversed through a material by a proton of a given energy. 113

prototype standard. A concrete embodiment of a physical quantity having arbitrarily assigned magnitude, or a replica of such embodiment. *Note:* As an illustration of the distinction between prototype standard and unit, the length of the United States Prototype Meter Bar is not exactly one meter. 210

proximal (distal) point (pulse terms). A magnitude referenced point at the intersection of a waveform and a proximal (distal) line. *See:* The single pulse diagram below the **waveform epoch** entry. 254

proximal stimuli (illuminating engineering). The distribution of illuminance on the retina constitutes the proximal stimulus. 167

proximity-coupled dipole array antenna. An array antenna consisting of a series of coplanar dipoles, loosely coupled to the electromagnetic field of a balanced transmission line, the coupling being a function of the proximity and orientation of the dipole to the transmission line. 111

proximity effect (electric circuits and lines). The phenomenon of non-uniform current distribution over the cross section of a conductor caused by the time variation of the current in a neighboring conductor. *See:* **induction heating.** 14, 210

proximity-effect error (navigation systems)(navigation aid terms). An error in determination of system performance caused by improper use of measurements made in the near field of the antenna system. 526

proximity-effect ratio (power distribution, underground cables). The quotient obtained by dividing the alternating-current resistance of a cable conductor subject to proximity effect, by the alternating-current resistance of an identical conductor free of proximity effect. 57

proximity influence. The percentage change in indication caused solely by the fields produced from two edgewise instruments mounted in the closest possible proximity, one on each side (or above and below for horizontal-scale instruments). *Note:* Proximity influence of alternating-current instruments on either alternating-current or direct-current types is determined by energizing two instruments, one on each side of the test instrument (or above and below) at 90 percent of end-scale value (in phase with the current in the instrument under test, if the latter is alternating current). The current in the two outside instruments only shall be reversed. For rating purposes, the proximity influence shall be taken as one-half the difference in the readings in percentage of full scale. In direct-current permanent-magnet moving-coil instruments the field produced by the current in the instru-

ment is small compared with the field from the permanent magnet. The proximity influence on either an alternating-current or direct-current test instrument will be the difference in reading, expressed as a percentage of full-scale value, of the instrument under test mounted alone on the panel, compared with the reading when two direct-current instruments are mounted in closest possible proximity, each with current applied to give 90-percent end-scale deflection. All three instruments shall be of the same manufacture and size. *See:* **accuracy rating (instrument).** 280

proximity switch (industrial control). A device that reacts to the proximity of an actuating means without physical contact or connection therewith. *See:* **switch.** 225, 206

P scan (electronic navigation). *See:* **plan-position indicator.**

P scope (1) (electronic navigation). *See:* **plan-position indicator.**

(2) (radar). A cathode-ray oscilloscope arranged to present a P-display. 13

pseudo code (1) (software). A combination of programming language and natural language used for computer program design. *See:* **computer program design; natural language; programming language.** 434

(2) (test, measurement and diagnostic equipment). An arbitrary code, independent of the hardware of a computer, which has the same general form as actual computer code but which must be translated into actual computer code if it is to direct the computer. 54

pseudo-coning (strapdown inertial system) (inertial sensor). A system error created when the system computer attempts to cancel a steady coning input term which in actuality does not exist. Because of certain coupling errors in the gyro, a rate input about only one axis can produce outputs on both axes of the gyro. If the coupling error, for example, is angular acceleration sensitivity, the two outputs produced will have the same form as if a true coning motion was applied to the gyro. 46

pseudo-instruction (test, measurement and diagnostic equipment). An instruction which resembles the instructions acceptable to the computer but which must be translated into actual computer instructions in order to control the computer. 54

pseudolatitude (navigation aid terms). A latitude in a coordinate system which has been arbitrarily displaced from the earth's conventional latitude system so as to move the meridian convergence zone (polar region) away from the place of intended operation. 526

pseudolongitude (navigation aid terms). A longitude in a coordinate system which has been arbitrarily dosplace from the earth's conventional longitude system so as to move the meridian convergence zone (polar region) away from the place of intended operation. 526

pseudonoise (PN) sequence (communication satellite). A binary sequence with a very desirable transortho-

gonal auto-correlation property. In space communications commonly used for synchronization and ranging.
84

pseudorandom number sequence. A sequence of numbers, determined by some defined arithmetic process, that is satisfactorily random for a given purpose, such as by satisfying one or more of the standard statistical tests for randomness. Such a sequence may approximate any one of several statistical distributions, such as uniform distribution or normal Gaussian distribution.
255, 77

PSK. *See:* **phase-shift keying.**

psychometric chroma (illuminating engineering). A correlate of perceived chroma defined in terms of CIELUV or CIELAB. Equal scale intervals correspond approximately to equal differences in perceived chroma.
167

psychometric hue-angle (illuminating engineering). A correlate of hue defined in terms of CIELUV or CIE-LAB.
167

psychometric lightness (illuminating engineering). A correlate of lightness defined in terms of CIELUV or CIELAB. Equal scale intervals correspond approximately to equal differences in (perceived) lightness.
167

psychometric saturation (illuminating engineering). A correlate of saturation defined in terms of CIELUV. Equal scale intervals correspond approximately to equal differences of (perceived) saturation. *Note:* Psychometric saturation cannot be calculated in terms of CIELAB.
167

PTM. *See:* **pulse-time modulation.**

p-type crystal rectifier. A crystal rectifier in which forward current flows when the semiconductor is positive with respect to the metal. *See:* **rectifier.**
328

p-type semiconductor. *See:* **semiconductor,** *p-type.*

public-address system. A system designed to pick up and amplify sounds for an assembly of people.
328

public telecommunications exchange (telephone switching systems). A telecommunications exchange that serves the public.
55

public telephone station (pay station). A station available for use by the public, generally on the payment of a fee that is deposited in a coin collector or is paid to an attendant. *See:* **telephone station.**
328

pull blade (cable plowing). A plow blade used to pull direct burial conductors into position by means of a suitable pulling grip attachment at the heel of the blade.
52

pull box. A box with a blank cover that is inserted in one or more runs of raceway to facilitate pulling in the conductors, and may also serve the purpose of distributing the conductors. *See:* **cabinet.**
328

pulley (sheave) (rotating machinery). A shaft-mounted wheel used to transmit power by means of a belt, chain, band, etcetera. *See:* **rotor (rotating machinery).**
63

pulling eye. A device that may be fastened to the conductor or conductors of a cable or formed by or fas-

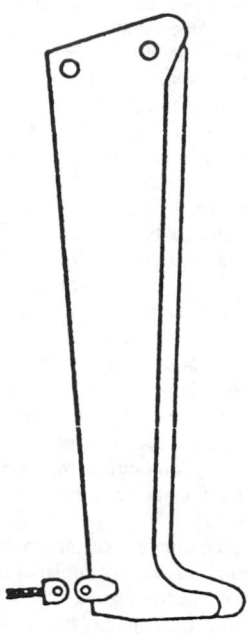

Pull blade.

tened to the wire armor and to which a hook or rope may be directly attached in order to pull the cable into or from a duct. *Note:* Pulling eyes are sometimes equipped, like test caps, with facilities for oil feed or vacuum treatment.
64

pulling figure (oscillator). The difference between the maximum and minimum values of the oscillator frequency when the phase angle of the load-impedance reflection coefficient varies through 360 degrees, while the absolute value of this coefficient is constant and equal to a specified value, usually 0.20. (Voltage standing-wave ratio 1.5.) *See:* **oscillatory circuit; waveguide.**
125

pulling into synchronism (rotating machinery). The process of synchronizing by changing from asynchronous speed to synchronous.
63

pulling iron (NESC). An anchor secured in the wall, ceiling, or floor of a manhole or vault to attach rigging used to pull cable.
494

pulling line (conductor stringing equipment). A high strength line, normally synthetic fiber rope or wire rope, used to pull the conductor. However, on reconstruction jobs where a conductor is being replaced, the old conductor often serves as the pulling line for the new conductor. In such cases, the old conductor must be closely examined for any damage prior to the pulling operation. *Syn:* **bull line; hard line; light line; sock line.**
431

pulling out of synchronism (rotating machines). The process of losing synchronism by changing from synchronous speed to a lower asynchronous speed (for a motor) or higher asynchronous speed (for a generator). 63

pulling tension (NESC). The longitudinal force exerted on a cable during installation. 494

pulling vehicle (conductor stringing equipment). Any piece of mobile ground equipment capable of pulling pilot lines, pulling lines or conductors. However, helicopters may be considered as a "pulling vehicle" when utilized for the same purpose. 45

pull-in test (synchronous machine). A test taken on a machine that is pulling into synchronism from a specified slip. 63

pull-in time (acquisition time) (communication satellite). The time required for achieving synchronization in a phase-lock loop. 85

pull-in torque (synchronous motor). The maximum constant torque under which the motor will pull its connected inertia load into synchronism, at rated voltage and frequency, when its field excitation is applied. *Note:* The speed to which a motor will bring its load depends on the power required to drive it and whether the motor can pull the load into step from this speed depends on the inertia of the revolving parts, so that the pull-in torque cannot be determined without having the Wk^2 as well as the torque of the load. 63

pull or transfer box. A box without a distribution panel, within which one or more corresponding electric circuits are connected or branched. 64

pull-out test (rotating machinery). A test to determine the conditions under which an alternating-current machine develops maximum torque while running at specified voltage and frequency. *See:* **asynchronous machine.** 63

pull-out torque (electric installations on shipboard)-(synchronous machine). The maximum sustained torque which the machine or a synchronous motor will develop at synchronous speed with rated voltage applied at rated frequency and with normal excitation. 3, 63

pull site (conductor stringing equipment). The location on the line where the puller, reel winder and anchors (snubs) are located. This site may also serve as the pull or tension site for the next sag section. *Syn:* **reel setup; tugger setup.** 431

pull-up torque (alternating-current motor). The minimum torque developed by the motor during the period of acceleration from rest to the speed at which breakdown torque occurs with rated voltage applied at rated frequency. 63

pulsating function. A periodic function whose average value over a period is not zero. For example, $f(t) = A + B \sin \omega t$ is a pulsating function where neither A nor B is zero. 210

pulse (1) (impulse) (data transmission). A brief excursion of a quantity from normal. 59

(2) (relaying) (power switchgear). A brief excursion of a quantity from its initial level. 103

(3) (automatic control). A variation of a signal whose value is normally constant: this variation is characterized by a rise and a decay, and has a finite duration. 56

(4) (pulse terms). A wave which departs from a first nominal state, attains a second nominal state, and ultimately returns to the first nominal state. Throughout the remainder of this document the term pulse is included in the term wave. 254

pulse accumulator (or register) (of a telemeter system) (power switchgear). A device that accepts and stores pulses and makes them available for readout on demand. 103

pulse advance (delay) (pulse terms). The occurrence in time of one pulse waveform before (after) another pulse waveform. 254

pulse advance (delay) interval (pulse terms). The interval by which, unless otherwise specified, the pulse start time of one pulse waveform preceded (follows), unless otherwise specified, the pulse start time of another pulse waveform. 254

pulse amplifier (pulse techniques). An amplifier designed specifically for the amplification of electric pulses. *See:* **pulse.** 335

pulse amplifier or relay (metering). A device used to change the amplitude or waveform of a pulse for retransmission to another pulse device. 212

pulse amplitude (1) (pulse terms). The algebraic difference between the top magnitude and the base magnitude. *See:* The single pulse diagram below the **waveform epoch** entry. 254

(2) (light emitting diodes). A general term indicating the magnitude of a pulse measured with respect to the normally constant value unless otherwise stated. 162

pulse amplitude, A_M (pulse transformers). That quantity determined by the intersection of a line passing through the points on the leading edge where the instantaneous value reaches 10% and 90% of A_M and a straight line that is the best least-squares fit to the pulse in the pulse-top region (usually this is fitted visually rather than numerically). For pulses deviating greatly from the ideal trapezoidal pulse shape, a number of successive approximations may be necessary to determine Am. *Note:* The pulse amplitude A_M may be arrived at by applying the following procedure. (1) Visually or numerically determine the best straight line fit to the pulse in the pulse-top region and extend this straight line into the leading-edge region. (2) An initial estimate of A_M is the first intersection of the pulse (in the late leading-edge or early pulse-top regions) with the straight line fitted to the pulse top. (3) Using the estimate of A_M calculate $0.1\,A_M$ and $0.9\,A_M$ and draw a straight line through these two points of the pulse-leading edge. (4) The intersection of the leading-edge straight line and the pulse-top straight line gives an improved estimate of A_M. (5) Repeat steps 3 and 4 until the estimate of A_M does not change. The converged estimate is the pulse amplitude Am. 589

pulse amplitude modulation (PAM) (antennas) (pulse terms) (data transmission). Modulation in which the modulating wave is caused to amplitude modulate a pulse carrier. 111, 254

pulse amplitude, peak. *See:* **peak pulse amplitude.**

pulse amplitude, root-mean-square (effective). *See:* **root-mean-square (effective) pulse amplitude.**

pulse average time (t_w) **(light emitting diodes).** The time interval between the instants at which the instantaneous pulse amplitude first and last reaches a specified fraction of the peak pulse amplitude, namely, 50 percent.   162

pulse bandwidth. The smallest continuous frequency interval outside of which the amplitude (of the spectrum) does not exceed a prescribed fraction of the amplitude at a specified frequency. Caution: This definition permits the spectrum amplitude to be less than the prescribed amplitude within the interval. *Notes:* (1) Unless otherwise stated, the specified frequency is that at which the spectrum has its maximum amplitude. (2) This term should really be pulse spectrum bandwidth because it is the spectrum and not the pulse itself that has a bandwidth. However, usage has caused the contraction and for that reason the term has been accepted. *See:* **signal.**   254

pulse base (pulse waveform) (pulse techniques). That major segment having the lesser displacement in amplitude from the baseline, excluding major transitions. *See:* **pulse.**   185

pulse, bidirectional. *See:* **bidirectional pulse.**

pulse broadening (fiber optics). An increase in pulse duration. *Note:* Pulse broadening may be specified by the impulse response, the root-mean-square pulse broadening, or the full-duration-half-maximum pulse broadening. *See:* **impulse response; root-mean-square pulse broadening; full width (duration) half maximum.**   433

pulse burst (1) (radar). A sequence of pulses, usually generated coherently and batch-processed for Doppler resolution, and often having a total duration less than the radar echo delay time.   13
(2) (pulse terms). A finite sequence of pulse waveforms.   254

pulse burst base envelope (pulse terms). Unless otherwise specified, the waveform defined by a cubic natural spline with knots at (1) that point of intersection of the pulse burst top envelope and the pulse burst waveform which precedes the first pulse waveform in a pulse burst, (2) each point of intersection of the base center line and the baseline between adjacent pulse waveforms in a pulse burst, and (3) that point of intersection of the pulse burst top envelope and the pulse burst waveform which follows the last pulse waveform in a pulse burst. *See:* Pulse burst envelopes diagram, below the **knot** entry.   254

pulse burst time-related definitions (pulse terms). (1) pulse burst duration. The interval between the pulse start time of the first pulse waveform and the pulse stop time of the last pulse waveform in a pulse burst. (2) pulse burst separation. The interval between the pulse stop time of the last pulse waveform in a pulse burst and the pulse start time of the first pulse waveform of the immediately following pulse burst. (3) pulse burst repetition period. The interval between the pulse start time of the first pulse waveform in a pulse burst and the pulse start time of the first pulse wave-

form in the immediately following pulse burst in a sequence of periodic pulse bursts. (4) pulse burst reception frequency. The reciprocal of burst repetition period.   254

pulse burst top envelope (pulse terms). Unless otherwise specified, the waveform defined by a cubic natural spline with knots at (1) the first transition mesial point of the first pulse waveform in a pulse burst, (2) each point of intersection of the top centerline and the topline of each pulse waveform in a pulse burst, and (3) the last transition mesial point of the last pulse waveform in a pulse burst. *See:* Pulse burst envelopes diagram below the **knot** entry.   254

pulse capacity (metering). The number of pulses per demand interval that a pulse receiver can accept and register without loss.   212

pulse capture (gyro, accelerometer) (inertial sensor). A technique which uses discrete quanta of torque-time (force-time) area to generate a restoring torque (force).   46

pulse carrier. A carrier consisting of a series of pulses. *Note:* Usually, pulse carriers are employed as subcarriers. *See:* **carrier.**   111, 254

pulse, carrier-frequency. *See:* **carrier-frequency pulse.**

pulse code. (1) A pulse train modulated so as to represent information. (2) Loosely, a code consisting of pulses, such as Morse code, Baudot code, binary code. *See:* **pulse.**   254, 185

pulse code modulation (PCM) (1) (data transmission). The type of pulse modulation where the magnitude of the signal is sampled and each sample is approximated to a nearest reference level (this process is called quantizing). Then a code, which represents the reference level, is transmitted to the distant location. The figure below is an example of one form of PCM which has eight reference levels. It can be seen that a straight binary code would require a group of three pulses to be transmitted for each sample. The main advantage of PCM is the fact that at the receiving end only the presence or absence of a pulse must be detected.   415

(2) (pulse terms). A modulation process involving the conversion of a waveform from analog to digital form by means of coding. *Notes:* (A) This is a generic term, and additional specification is required for a specific purpose. (B) The term is commonly used to signify that form of pulse modulation in which a code is used to represent quantized values of instantaneous samples of the signal wave.   242, 254

pulse coder (navigation aid terms). A device for varying one or more of the characteristics of a pulse or of a pulse train so as to transmit information.   526

pulse coincidence (noncoincidence) (pulse terms). The occurrence (lack of occurrence) of two or more pulse waveforms in different waveforms either essentially simultaneously or for a specified interval.   254

pulse coincidence (noncoincidence) duration (pulse terms). The interval between specified points on two or more pulse waveforms in different waveforms during which pulse coincidence (noncoincidence) exists.   254

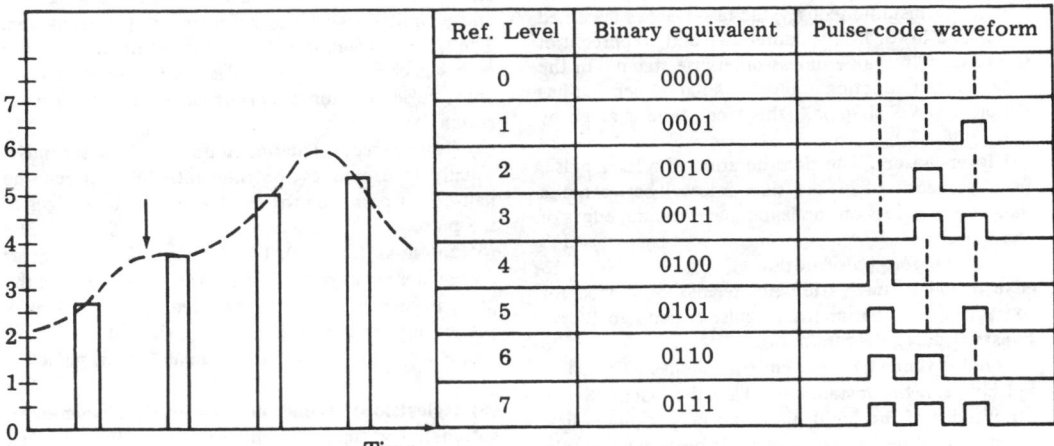

Ref. Level	Binary equivalent	Pulse-code waveform
0	0000	
1	0001	
2	0010	
3	0011	
4	0100	
5	0101	
6	0110	
7	0111	

A form of pulse code modulation

pulse compression (radar). The coding and processing of a signal pulse of long time duration to one of short time duration and high range resolution, while maintaining the benefits of high pulse energy.   13

pulse control (rotating electric machinery). Means by which the voltage applied to a machine circuit departs from being essentially constant if unidirectional, or from being essentially sinusoidal if alternating. Pulse-control devices include, but are not limited to, choppers, inverters, and rectifiers.   424

pulse corner (pulse waveform feature). A continuous pulse waveform feature of specified extent which includes a region of maximum curvature or a point of discontinuity in the waveform slope. Unless otherwise specified, the extent of the corners in a rectangular or trapezoidal pulse waveform are as specified in the following table: 60

Corner	First Point	Last Point
First	first base point	first transition proximal point
Second	first transition distal point	top center point
Third	top center point	last transition distal point
Fourth	last transition proximal point	last base point

pulse corrector (telephone switching systems). Equipment to reestablish, within predetermined limits, the make/break ratio of dial pulses.   55

pulse count (control of system electromagnetic compatibility). The number of pulses in some specified time interval.   495

pulse-count deviation (metering). The difference between the number of recorded pulses and the number of pulses supplied to the input terminals of a pulse recorder (true count), expressed as a percentage of the true count. Pulse-count deviation is applicable to each data channel of a pulse recorder.   212

pulse counter (pulse techniques). A device that indicates or records the total number of pulses that it has received during a time interval. *See:* **pulse.**   117

pulsed Doppler radar. A Doppler radar that uses pulsed transmissions. *Syn:* **pulse-Doppler radar.**   13

pulse decay time (t_d)(1)(X-ray energy spectrometers)-(germanium gamma-ray detectors). The interval between the instants at which the instantaneous value last reaches specified upper and lower limits, namely, 90 percent and 10 percent of the peak pulse value unless otherwise stated. (In the case of a step function applied to an amplifier that has simple capacitance-resistance to resistance-capacitance (CR-RC) shaping, the decay time is given by $t_d = 3.36CR$.)   471, 528

(2) (light emitting diodes) (general) (pulse fall time) (t_f). The interval between the instants at which the instantaneous amplitude last reaches specified upper and lower limits, namely, 90 percent and 10 percent of the peak pulse amplitude unless otherwise stated. *See:* **pulse.**   178, 119, 162, 118, 23

(3) (data transmission). The interval of time required for the trailing edge of a pulse to decay from 90 percent to 10 percent of the peak-pulse amplitude.   59

(4) (semiconductor radiation detectors) (last transition time) (t_d). The interval between the instants at which the instantaneous value last reaches specified upper and lower limits, namely, 90 and 10 percent of the peak pulse value unless otherwise stated. (In the case of a step function applied to an amplifier that has simple CR-RC shaping, the decay time is given by $t_d = 3.36\ CR$). 23

(5)(laser-maser). The time duration of a laser pulse: usually measured as the time interval between the half-power points on the leading and trailing edges of the pulse. 363

(6)(radar). See: **pulsewidth.** 13

pulse decoder (navigation aid terms). A device for extracting information from a pulse-coded signal. Syn: **constant-delay discriminator.** 526

pulse delay time (t_d) (light emitting diodes). The interval between the instants at which the instantaneous amplitudes of the input pulse and output pulses first reach a specified fraction of their peak pulse amplitudes, namely, 10 percent. 162

pulse delay, transducer. See: **delay, pulse.**

pulse delay, transmitter. See: **delay, pulse.**

pulse device (for electricity metering). The functional unit for initiating, transmitting, retransmitting, or receiving electric pulses, representing finite quantities, such as energy, normally transmitted from some form of electricity meter to a receiver unit. 212

pulse distortion. See: **distortion.** 433

pulsed laser (laser-maser). A laser which delivers its energy in the form of a single pulse or train of pulses. The duration of a pulse $\le$ 0.25 s. 363

pulse Doppler (pulse-Doppler radar)(navigation aid terms). A Doppler radar that uses pulsed transmission. 526

pulsed optical feedback preamplifier (germanium gamma-ray detectors). A charge-sensitive preamplifier in which the charge that accumulates on the feedback capacitor is periodically reset by a pulse of light incident on a suitable photosensitive element (for example, the gate of the n-p junction of the input FET [field effect transistor]). 528

pulsed oscillator. An oscillator that is made to operate during recurrent intervals by self-generated or externally applied pulses. See: **oscillatory circuit.** 111

pulsed-oscillator starting time. The interval between the leading-edge pulse time of the pulse at the oscillator control terminals and the leading-edge pulse time of the related output pulse. 254

pulse droop (television). A distortion of an otherwise essentially flat-topped rectangular pulse characterized by a decline of the pulse top. See: **television.** 254

pulse duration (1) (data transmission). (A) (Loosely). The duration of a rectangular pulse whose energy and peak power equal those of the pulse in question. Note: When determining the peak power, any transients of relatively short duration are frequently ignored. (B) television) (radiation counters) (telecommunications). The time interval between the first and last instants at which the instantaneous amplitude reaches a stated fraction of the peak pulse amplitude. 59

(2) (fiber optics). The time between a specified reference point on the first transition of a pulse waveform and a similarly specified point on the last transition. The time between the 10 percent, 50 percent, or 1/e points is commonly used, as is the root-mean-square (rms) pulse duration. See: **root-mean-square pulse duration.** 433

(3) (laser-maser). The time duration of a laser pulse; usually measured as the time interval between the half-power points on the leading and trailing edges of the pulse. 363

(4)(antennas)(loosely). The duration of a rectangular pulse whose energy and peak power equal those of the pulse in question. Note: When determining the peak power, any transients of relatively short duration are frequently ignored. See: **phase modulation; pulse.** 111

(5) (television) (radiation counters) (telecommunications). The time interval between the first and last instants at which the instantaneous amplitude reaches a stated fraction of the peak pulse amplitude. See: **pulse; signal.** 178, 117

(6) (pulse terms). The duration between pulse start time and pulse stop time. See: The single pulse diagram below the **waveform epoch** entry. 254

(7)(of an amplifier output pulse)(charged-particle detectors). The time interval between the 50 percent of maximum amplitude points of a pulse. 119

(8)(90%),t_p(pulse transformers). The time interval between the instants at which the instantaneous value reaches 90% of A_M on the leading edge and 90% of [1]A_T on the trailing edge. Notes: (1) Often the input pulse tilt (droop) is only a few percentages, and in those cases pulse duration may be considered as the time interval between the first and last instants at which the instantaneous value reaches 90% of A_M. (2) Pulse duration may be specified at a value other than 90% A_M and A_T in special cases. 589

pulse-duration discriminator (radar). Circuit in which the output is a function of the deviation of the input pulse duration from a reference. 13

pulse duration distribution (PDD)(electromagnetic site survey). The fraction of pulse duration at level vithat exceeds time T. See figure under **average crossing rate.** 457

pulse-duration modulation (PDM) (1) (antennas) (pulse terms). Pulse-time modulation in which the value of each instantaneous sample of the modulating wave is caused to modulate the duration of a pulse. Notes: (A) The deprecated terms **pulsewidth modulation** and **pulse-length modulation** also have been used to designate this system of modulation. (B) In **pulse-duration modulation,** the modulating wave may vary the time of occurrence of the leading edge, the trailing edge, or both edges of the pulse. 111, 254

(2) (data transmission). Pulse-time modulation in which the value of each instantaneous sample of the modulating wave is caused to modulate the duration of a pulse. Notes: (A) The deprecated terms 'pulsewidth modulation' and 'pulse-length modulation' also have been used to designate this system of modulation. (B) In pulse-duration modulation, the modulating wave may vary the time of occurrence of the leading edge, the trailing edge, or both edges of the pulse. 59

pulse-duration-modulation torquing (gyro, accelerometer) (inertial sensor). A torquing mechanism that provides current to a sensor torquer of fixed amplitude but variable pulse duration proportional to input. The duration may be quantized to enable digital interpretation and readout. *See:* **binary torquing; ternary torquing.** 46

pulse-duration telemetering (pulse-width modulation) (electric power systems). A type of telemetering in which the duration of each transmitted pulse is varied as a function of the magnitude of the measured quantity. *See:* **telemetering.** 94

pulse duration time (t_p) (light emitting diodes). The time interval between the first and last instants at which the instantaneous amplitude reaches a stated fraction of the peak pulse amplitude, namely, 90 percent. 162

pulse duty factor (light emitting diodes). The ratio of the average pulse duration to the average pulse spacing. *Note:* This is equivalent to the product of the average pulse duration and the pulse-repetition rate. 254

pulse energy (pulse terms). The energy transferred or transformed by a pulse(s). Unless otherwise specified by a mathematical adjective, the total energy over a specified interval is assumed. *See:* **mathematical adjectives.** 254

pulse fall time (1) (photomultipliers for scintillation counting). The interval between the instants at which the instantaneous amplitude last reaches specified upper and lower limits, namely, 90 percent and 10 percent of the peak pulse amplitude unless otherwise stated. 117

(2) (t_f) (charged-particle detectors). The interval between the instants at which the instantaneous value last reaches specified upper and lower limits, namely, 90 percent and 10 percent of the peak pulse value unless otherwise stated. (In the case of a step function applied to an amplifier that has simple capacitance-resistance to resistance-capacitance (CR-RC) shaping, the fall time is given by $t_f = 3.36\ CR$). 119

pulse-forming line (radar). A passive electric network in a radar modulator whose propagation delay determines the length of the modulation pulse. 13

pulse frequency modulation (PFM) (data transmission) (pulse terms). A form of pulse-time modulation in which the pulse-repetition rate is the characteristic varied. *Note:* A more precise term for pulse-frequency modulation would be 'pulse-repetition-rate modulation'. 254, 59

pulse frequency modulation (PFM) telemetry (communication satellite). A telemetry system where the information is coded according to subcarrier frequency, pulse duration and pulse repetition rate. Often used for satellite telemetry. 83

pulse, Gaussian. A pulse shape tending to follow the Gaussian curve corresponding to $A(t) = e^{-a(b-t)^2}$. *See:* **pulse.** 185

pulse-height analyzer (radiation counters). An instrument capable of indicating the number or rate of occurrence of pulses falling within each of one or more specified amplitude ranges. *See:* **pulse.** 117

pulse-height discriminator (liquid-scintillation counting). A circuit that produces an output signal if it receives an input pulse whose amplitude exceeds an assigned value. 422

pulse-height resolution constant, electron (photomultipliers). The product of the square of the electron (photomultiplier) pulse-height resolution expressed as the fractional full width at half maximum (FWHM / A_1), and the mean number of electrons per pulse from the photocathode. *See:* **phototube.** 335

pulse-height resolution, electron (photomultiplier). A measure of the smallest change in the number of electrons in a pulse from the photocathode that can be discerned as a change in height of the output pulse. Quanitatively, it is the fractional standard deviation (σ / A_1) of the pulse-height distribution curve for output pulses resulting from a specified number of electrons per pulse from the photocathode. *Note:* The fractional full width at half maximum of the pulse-height distribution curve (FWHM/A_1) is frequently used as a measure of this resolution, where A_1 is the pulse height corresponding to the maximum of the distribution curve. *See:* **pulse.** 117

pulse-height selector (pulse techniques). A circuit that produces a specified output pulse when and only when it receives an input pulse whose amplitude lies between two assigned values. *See:* **pulse.** 335

pulse initiator (metering). Any device, mechanical or electrical, used with a meter to initiate pulses, the number of which are proportional to the quantity being measured. It may include an external amplifier or auxiliary relay or both. 212

pulse-initiator coupling ratio (metering). The number of revolutions of the pulse-initiating shaft for each output pulse. 212

pulse-initiator gear ratio (metering). The ratio of meter rotor revolutions to revolutions of the pulse-initiating shaft. 212

pulse-initiator output constant (metering). The value of the measured quantity for each outgoing pulse of a pulse initiator, expressed in kilowatt hours per pulse, kilovarhours per pulse, or other suitable units. 212

pulse-initiator output ratio (metering). The number of revolutions of the meter rotor per output pulse of the pulse initiator. 212

pulse-initiator ratio (1)(metering). The ratio of revolutions of the first gear of the pulse initiator to revolutions of the pulse-initiating shaft. 212

(2)(watthour meter). The ratio of revolutions of first gear of pulse device to revolutions of pulse initiating shaft. *Note:* It is commonly denoted by the symbol P_r. *See:* **auxiliary device to an instrument.** 212

pulse-initiator shaft reduction (metering). The ratio of revolutions of the meter rotor to the revolutions of the first gear of the pulse initiator. 212

pulse interleaving. A process in which pulses from two or more sources are combined in time-division multiplex for transmission over a common path.   254

pulse interrogation. The triggering of a transponder by a pulse or pulse mode. *Note:* Interrogations by means of pulse modes may be employed to trigger a particular transponder or group of transponders.   254

pulse interval. *See:* pulse spacing.

pulse-interval modulation. A form of pulse-time modulation in which the pulse spacing is varied.   254

pulse jitter. A relatively small variation of the pulse spacing in a pulse train. *Note:* The jitter may be random or systematic, depending on its origin, and is generally not coherent with any pulse modulation imposed.   254

pulse length (fiber optics). Often erroneously used as a synonym for pulse duration.   433

pulse-length modulation. *See:* pulse-duration modulation.

pulse measurement. The assignment of a number and a unit of measurement to a characteristic, property, or attribute of a pulse wherein the number and unit assigned indicate the magnitude of the characteristic which is associated with the pulse. Typically, this assignment is accomplished by comparison of a transform of the pulse, its pulse waveform, with a scale or reference which is calibrated in the unit of measurement. *See:* method of pulse measurement.   15

pulse measurement process. A realization of a method of pulse measurement in terms of specific devices, apparatus, instruments, auxiliary equipment, conditions, operators, and observers. *See:* method of pulse measurement.   15

pulse mode. (1) A finite sequence of pulses in a prearranged pattern used for selecting and isolating a communication channel. (2) The prearranged pattern.   254

pulse-mode multiplex. A process or device for selecting channels by means of pulse modes. *Note:* This process permits two or more channels to use the same carrier frequency.   254

pulse mode, spurious. *See:* spurious pulse mode.

pulse-modulated radar. A form of radar in which the radiation consists of a series of discrete pulses. *See:* radar.   328

pulse modulation (continuous-wave). Modulation of one or more characteristics of a pulse carrier. *Note:* In this sense, the term is used to describe methods of transmitting information on a pulse carrier.   61

pulse modulation, width. *See:* pulse duration modulation.

pulse modulator. A device that applies pulses to the element in which modulation takes place. *See:* modulation.   111

pulse multiplex. *See:* pulse-mode multiplex.

pulse number (1)(circuit properties)(of a group of principal branches or of a complete converter)(self-commutated converters). The number of nonsimultaneous commutations from one principal branch to another during one cycle of operation, considering the group or the complete converter, respectively.   584

(2)(harmonic control and reactive compensation of static power converters). The total number of successive nonsimultaneous commutations occurring within that converter circuit during each cycle when operating without phase control. It is also equal to the order of the principal harmonic in the direct voltage, that is, the number of pulses present in the dc output voltage in one cycle of the supply voltage.   533

pulse operation. The method of operation in which the energy is delivered in pulses. *Note:* Pulse operation is usually described in terms of the pulse shape, the pulse duration, and the pulse-recurrence frequency. *See:* pulse.   59

pulse packet (radar). The volume of space occupied by a single radar pulse. The dimensions of this volume are determined by the angular width of the beam, the duration of the pulse, and the distance from the antenna.   13

pulse-pair resolution (photomultiplier). The time interval between two equal-amplitude delta-function optical pulses such that the valley between the two corresponding anode pulses falls to fifty percent of the peak amplitude.   117

pulse period (1)(measuring the performance of tone address signaling systems). When sending a sequence of signals, the time interval from the start of one signal present condition to the start of the next signal present condition. *Syn:* cycle time.   508

(2)(telephony)(dial-pulse address signaling systems). The time from the start of one break interval of a dial pulse in a train until the start of the next break interval. Milliseconds is the preferred unit of time to express the duration of the pulse period.   540

pulse permeability (magnetic core testing). The value of amplitude permeability when the rate of change of induction (that is, the exciting voltage) is held substantially constant over a period of time during each cycle. The frequency, amplitude, duration of the exciting voltage, and the time interval for which the permeability is measured must be stated.

$$\mu_\pi = \frac{1}{\mu_0} \frac{\Delta B}{\Delta H}$$

μ_π = pulse permeability, relative

ΔB = change in induction during the stated time interval

ΔH = associated change in magnetic field strength

Note: When pulse permeability is to be related to a specific circuit condition, a second subscript may be used, for example, $\mu_{\pi a}$ would represent the relative amplitude permeability determined under pulsed excitation.   165

pulse position modulation (ppm) (data transmission)-(pulse-phase modulation)(antennas). Pulse-time modulation in which the value of each instantaneous sample of a modulating wave is caused to modulate the position in time of a pulse.   59, 111

pulse power (pulse terms). The power transferred or transformed by a pulse(s). Unless otherwise specified by a mathematical adjective average power over a specified interval is assumed. *See:* **mathematical adjectives.** 254

pulse power, carrier-frequency peak. *See:* **peak pulse power, carrier-frequency.**

pulse power, peak. *See:* **peak pulse power.**

pulse quadrant (pulse waveform feature). One of the four continuous and contiguous waveform features of specified extent which include a region of maximum curvature or a point of discontinuity in the waveform slope. Unless otherwise specified, the extent of the quadrants in a rectangular or trapezoidal pulse waveform are as specified in the following table: 60

Quadrant	First Point	Last Point
First	first base point	first transition mesial point
Second	first transition mesial point	top center point
Third	top center point	last transition mesial point
Fourth	last transition mesial point	last base point

pulse, radio-frequency. *See:* **radio-frequency pulse.**

pulse rate (1) (electronic navigation). *See:* **pulse-repetition frequency.**

(2) (watthour meter). The number of pulses per demand interval at which a pulse device is nominally rated. *See:* **auxiliary device to an instrument.** 212

pulse rate, maximum. *See:* **maximum pulse rate.**

pulse rate telemetering (electric power systems). A type of telemetering in which the number of unidirectional pulses per unit time is varied as a function of the magnitude of the measured quantity. *See:* **telemetering.** 94

pulse ratio (telephony)(dial-pulse address signaling systems). The percentage of the total dial-pulse period during which the circuit is in each of the two interval states. This ratio is customarily expressed as % break (preferred) of % make. 540

pulse rebalance. *See:* **pulse capture.**

pulse receiver (metering). The unit that receives and registers the pulses. It may include a periodic resetting mechanism, so that a reading proportional to demand may be obtained. 212

pulse recorder (metering). A device that receives and records pulses over a given demand interval. *Note:* It may record pulses in a machine-translatable form on magnetic tape, paper tape, or other suitable media. 212

pulse-recorder channel (metering). A means of conveying information. It consists of an individual input, output, and intervening circuitry required to record pulse data on the recording media. 212

pulse regeneration (data transmission). The process of restoring a series of pulses to their original timing, form, and relative magnitude. 59

pulse relay, totalizing. *See:* **totalizing pulse relay.**

pulse repeater (transponder). A device used for receiving pulses from one circuit and transmitting corresponding pulses into another circuit. It may also change the frequency and waveforms of the pulses and perform other functions. *See:* **pulse; repeater.** 111

pulse-repetition frequency (1) (charging inductors). The pulse-repetition rate of a periodic pulse train. 137

(2) (PRF)(radar). The number of pulses per unit of time, usually per second. *Syn:* **pulse repetition rate.** 13

pulse repetition interval (PRI)(radar). The time duration between successive pulses. The reciprocal of the pulse repetition frequency. *Syn:* **pulse repetition period.** 13

pulse repetition period. *See:* **pulse repetition interval.**

pulse repetition rate. *See:* **pulse repetition frequency.**

pulse reply. The transmission of a pulse or pulse mode by a transponder as the result of an interrogation. 254

pulse response characteristics (pulse response curve) (electromagnetic compatibility). The relationship between the indication of a quasi-peak voltmeter and the repetition rate of regularly repeated pulses of constant amplitude. *See:* **electromagnetic compatibility.** 220, 199

pulse rise time (t_r) (1)(charged-particle detectors)-(germanium gamma-ray detectors)(X-ray energy spectrometers). The interval between the instants at which the instantaneous value first reaches specified lower and upper limits, namely, 10 percent and 90 percent of the peak pulse value unless otherwise specified. (In the case of a step function applied to an RC (resistance-capacitance) low-pass filter, the rise time is given by $t_r = 2.2\ RC$. In the case of a step function applied to an amplifier that has simple CR-RC (capacitance-resistance to resistance-capacitance)shaping, that is, one high-pass and one low-pass RC filter of equal time constants, the rise time is given by $t_r = 0.57\ RC$.) 119, 528, 471

(2) (data transmission). The interval between the instants at which the instantaneous amplitude first reaches specified lower and upper limits, namely 10 % and 90 % of the peak-pulse amplitude unless otherwise stated. 59

(3) (semiconductor radiation detectors) (first transition time) (t_r). The interval between the instants at which the instantaneous value first reaches specified lower and upper limits, namely, 10 and 90 percent of the peak pulse value unless otherwise specified. *Notes:* (A) In the case of a step function applied to an RC low-pass filter, the rise time is given by $t_r = 2.2\ RC$. (B) In the case of a step function applied to an amplifier that has simple $CR\text{--}RC$ shaping, that is, one high-pass and one low-pass RC filter of equal time constants, the rise time is given by $t_r = 0.57\ RC$. 23

pulse separation (pulse terms). The interval between the trailing-edge pulse time of one pulse and the lead-

ing-edge pulse time of the succeeding pulse. *See:* **leading-edge pulse time; pulse; trailing-edge pulse time.**
254

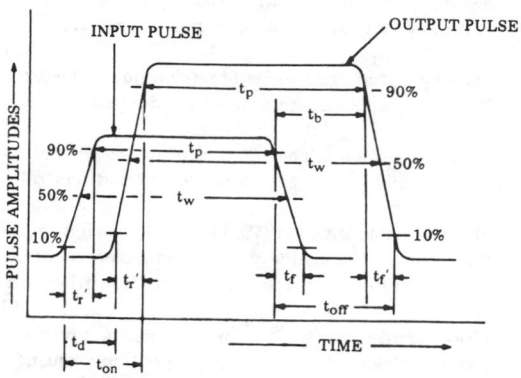

Diagram illustrating pulse time symbology

—

pulses, equalizing. *See:* **equalizing pulses.**

pulse shape (pulse terms). (A) For descriptive purposes a pulse waveform may be imprecisely described by any of the adjectives, or combinations thereof, in **descriptive adjectives, major (minor): polarity related adjectives: geometrical adjectives: and functional adjectives, exponential.** (B) For tutorial purposes a hypothetical pulse waveform may be precisely defined by the further addition of the adjective ideal. *See:* **descriptive adjectives, ideal.** (C) For measurement or comparison purposes a pulse waveform may be precisely defined by the further addition of the adjective reference. *See:* **descriptive adjectives, reference.**
254

pulse shaper (pulse techniques). Any transducer used for changing one or more characteristics of a pulse. *Note:* This term includes pulse regenerators. *See:* **pulse.**
254

pulse shaping. Intentionally changing the shape of a pulse.
254

pulse, single-polarity. *See:* **unidirectional pulse.**

pulse spacing (pulse interval). The interval between the corresponding pulse times of two consecutive pulses. *Note:* The term pulse interval is deprecated because it may be taken to mean the duration of the pulse instead of the space or interval from one pulse to the next. Neither term means the space between pulses.
254

pulse spacing distribution (PSD)(electromagnetic site survey). The fraction of pulse spacing time at level vi that exceeds time T. (See Figure under 'average crossing rate').
457

pulse spectrum (signal-transmission system). The frequency distribution of the sinusoidal components of the pulse in relative amplitude and in relative phase. *See:* **signal.**
254

pulse speed (telephony)(dial-pulse address signaling systems). The number of dial pulses occurring per unit of time per unit of time. Pulses per second (pls/s) is the preferred unit to express pulse speed.
540

pulse spike (automatic control). An unwanted pulse of relatively short duration, superimposed on the main pulse. *See:* **spike.**
56

pulse start (stop) time (pulse terms). The instant specified by a magnitude referenced point on the first (last) transition of a pulse waveform. Unless otherwise specified, the pulse start (stop) time is at the mesial point on the first (last transition). *See:* The single pulse diagram below the **waveform epoch** entry.
254

pulse storage time (t_s) (light emitting diodes). The interval between the instants at which the instantaneous amplitudes of the input and output pulses last reach a specified fraction of their peak pulse amplitudes, namely, 90 percent.
162

pulse stretcher (spectrum analyzer). A pulse shaper that produces an output whose duration is greater than that of the input pulse and whose amplitude is proportional to that of the peak amplitude of that input pulse.
117, 390

pulse techniques. *See:* **burst.**

pulse tilt. A distortion in an otherwise essentially flat-topped rectangular pulse characterized by either a decline or a rise of the pulse top. *See:* **television.**
254, 178

pulse time, leading edge. *See:* **leading-edge pulse time.**

pulse time, mean. *See:* **mean pulse time.**

pulse-time modulation (PTM) (data transmission) (antennas) (pulse terms). Modulation in which the value of instantaneous samples of the modulating wave are caused to modulate the time of occurrence of some characteristic of a pulse carrier. *Note:* Pulse-duration modulation, pulse-position modulation, and pulse-interval modulation, are particular forms of pulse-time modulation.
111, 254, 59

pulse time reference points (pulse terms). (1) top center point. A specified time referenced point or magnitude referenced point on a pulse waveform top. If no point is specified, the top center point is the time referenced point at the intersection of a pulse waveform and the top center line. *See:* The single pulse diagram below the **waveform epoch** entry. (2) first (last) base point. Unless otherwise specified, the first (last) datum point in a pulse epoch. *See:* The single pulse diagram below the **waveform epoch** entry, as well as **pulse train time-related definitions, base center point.**
254

pulse time, trailing-edge. *See:* **trailing-edge pulse time.**

pulse timing of video pulses (television). The determination of an occurrence of a pulse or a specified portion thereof at a particular time. *See:* **pulse width of video pulses (television); television; time of rise (decay) of video pulses (television).**
336

pulse top (1)(pulse transformers). That portion of the pulse occurring between the time of intersection of straight-line segments used to determine A_T and $A_{T'}$
589

(2)(instrumentation and measurement). That major segment of a pulse waveform having the greater displacement in amplitude from the baseline. *See:* **pulse.**
185

pulse train (1)(telephony)(dial-pulse address signaling systems). In dial-pulse signaling, a series of contiguous pulses of undetermined length.   540

(2) (pulse terms). A continuous repetitive sequence of pulse waveforms.   254

(3) (thyristor). A gate signal applied during the desired conducting interval, or parts thereof, made up of a train of pulses of predetermined duration, amplitude, and frequency.   445

(4) (signal-transmission system) (industrial control). A sequence of pulses. *See:* **pulse; signal.**
          210, 56, 178, 188, 206

(5) (radar). A sequence of pulses used to accomplish a function such as MTI or increased effective signal-to-noise ratio. A pulse train of duration less than the radar echo delay time is usually referred to as a pulse burst.   13

pulse train, bidirectional. *See:* **bidirectional pulse train.**

pulse train, periodic. *See:* **periodic pulse train.**

pulse-train spectrum (pulse-train frequency-spectrum). The frequency distribution of the sinusoidal components of the pulse train in amplitude and in phase angle.   254

pulse train time-related definitions (pulse terms). (1) pulse repetition period. The interval between the pulse start time of a first pulse waveform and the pulse start time of the immediately following pulse waveform in a periodic pulse train. (2) pulse repetition frequency. The reciprocal of pulse repetition period. (3) pulse separation. The interval between the pulse stop time of a first pulse waveform and the pulse start time of the immediately following pulse waveform in a pulse train. (4) duty factor. Unless otherwise specified, the ratio of the pulse waveform duration to the pulse repetition period of a periodic pulse train. (5) on-off. ratio. Unless otherwise specified, the ratio of the pulse waveform duration to the pulse separation of a periodic pulse train. (6) base center line. The time reference line at the average of the pulse stop time of a first pulse waveform and the pulse start time of the immediately following pulse waveform in a pulse train. (7) base center point. A specified time referenced point or magnitude referenced point on a pulse train waveform base. If no point is specified, the base center point is the time referenced point at the intersection of a pulse train waveform base and a base center line. *See:* **pulse time reference points, first (last) base point.** (8) pulse train epoch. The span of time in a pulse train for which waveform data are known or knowable and which extends from a first base center point to the immediately following base center point.   254

pulse train top (base) envelope (pulse terms). Unless otherwise specified, the waveform defined by a cubic natural spline with knots at each point of intersection of the top center line and topline (the base center line and the baseline) of each (between adjacent) pulse waveforms(s) in a pulse train.   254

pulse train, unidirectional. *See:* **unidirectional pulse train.**

pulse transmitter (1). A pulse-modulated transmitter whose peak power-output capabilities are usually large with respect to average power-output rating. *See:* **pulse; radio transmitter.**   111

(2) (77) (power system device function numbers). Used to generate and transmit pulses over a telemetering or pilot-wire circuit to the remote indicating or receiving device.   402

pulse turn-off time (t_{off}) (light emitting diodes). The arithmetic sum of the pulse storage time, and the pulse decay time of the output pulse.   162

pulse turn-on time (t_{on}) (light emitting diodes). The arithmetic sum of the pulse delay time, and the pulse rise time of the output pulse.   162

pulse-type telemeter (power switchgear). A telemeter that employs characteristics of intermittent electric signals other than their frequency, as the translating means. *Note:* These pulses may be utilized in any desired manner to obtain the final indications, such as periodically counting the total number of pulses; or measuring their "on" time, their "off" time, or both.   103

pulse unidirectional. *See:* **unidirectional pulse.**

pulse waveform distortion (pulse terms). The algebraic difference in magnitude between all corresponding points in time of a pulse waveform and a reference pulse waveform. Unless otherwise specified by a mathematical adjective, peak-to-peak pulse waveform distortion is assumed. *See:* **mathematical adjectives.** *See:* pulse waveform distortion and pulse waveform feature distortion diagram below.   254

pulse waveform feature distortion (pulse terms). The algebraic difference in magnitude between all corresponding points in time of a pulse waveform and a reference pulse waveform feature. Unless otherwise specified by a mathematical adjective, peak-to-peak pulse waveform feature distortion is assumed. *See:* **mathematical adjectives.** *See:* pulse waveform distortion and pulse waveform feature distortion diagram below the **pulse waveform distortion** entry.   254

pulsewidth (radar). The time interval between the points on the leading and trailing edges at which the instantaneous value bears a specified relation to the maximum instantaneous value of the pulse, usually the time interval between the half-power points of the pulse. *Syn:* **pulse duration.**   13

pulse width (1) (television). *See:* **pulse duration.**

(2) (data transmission). The time interval between the points on the leading and trailing edges at which the instantaneous value bears a specified relation to the maximum instantaneous value of the pulse.   194

(3) (fiber optics). Often erroneously used as a synonym for pulse duration.   433

pulse width at half maximum, $T_{0.5}$ (of an amplifier output pulse)(germanium gamma-ray detectors). The time interval between the 50% of maximum amplitude points of a pulse.   528

pulse-width modulation (PWM)(converter characteristics)(self-commutated converters). Pulse-time modulation in which the value of each instantaneous sample of the modulating wave is caused to modulate the duration of a pulse. The modulating frequency

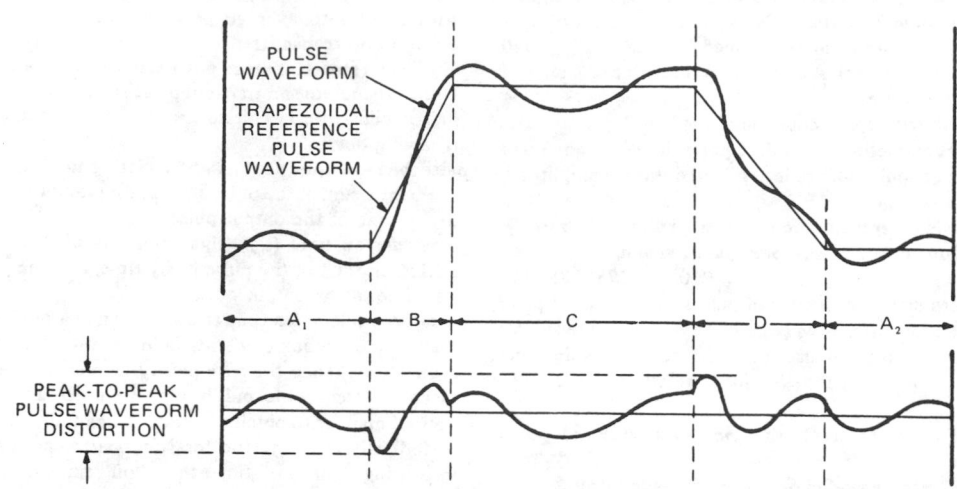

EXTENT OF DATA INCLUDED IN PULSEWAVEFORM FEATURE DISTORTION:
A₁ AND A₂-PULSE BASE DISTORTION   C-PULSE TOP DISTORTION
B-FIRST TRANSITION DISTORTION   D-LAST TRANSITION DISTORTION

Pulse waveform distortion and pulse waveform feature distortion.

may be fixed or variable. Examples of waveforms produced by PWM are shown in the figure below.
584

Waveforms Produced by Pulse-Width Modulation

(a) DC Waveform

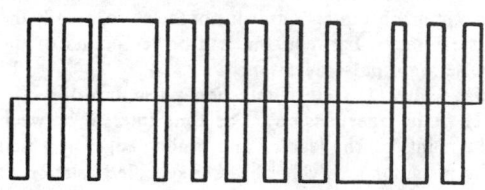

(b) AC Waveform

pulse-width modulation. *See:* **pulse-duration modulation.**

pulse-width-modulation torquing (gyro, accelerometer). *See:* **pulse-duration-modulation torquing.**

pulse width of video pulses (television). The duration of a pulse measured at a specified level. *See:* **pulse timing of video pulses (television).**   336

pulsing (telephone switching systems). The signaling over the communication path of signals representing one or more digits required to set up a call.   55

pulsing circuit (peaking circuit) (industrial control). A circuit designed to provide abrupt changes in voltage or current of some characteristic pattern. *See:* **electronic controller.**   206

pulsing transformer (industrial control). Supplies pulses of voltage or current. *See:* **electronic controller.**
206

pump (parametric device). The source of alternating-current power that causes the nonlinear reactor to behave as a time-varying reactance. *See:* **parametric device.**   191

pump-back test (electrical back-to-back test) (rotating machinery). A test in which two identical machines are mechanically coupled together, and they are both connected electrically to a power system. The total losses of both machines are taken as the power input drawn from the system. *See:* **asynchronous machine.**
63

pumped figure of merit (parametric amplifier)(nonlinear, active, and nonreciprocal waveguide components). The ratio of the half-amplitude fundamental Fourier component of pumped elastance to the series resistance of the varactor diode. This quantity, designated as $m_1\omega_c$, has the dimensions of angular frequency and describes the noise, gain, and impedance characteristics of parametric amplifiers. The parameter m_1 is the normalized fundamental component of elastance and ω_c is the total angular cutoff frequency.
530

pumped-storage hydro capability (power operations). The capability supplied by hydroelectric sources under specified water conditions using a reservoir that is alternately filled by pumping and depleted by generating.   516

pumped storage station (power operations). A hydroelectric generating station at which electric energy is normally generated during periods of relatively high system demand by utilizing water which has been pumped into a storage reservoir usually during periods of relatively low system demand. 516

pumped tube. An electron tube that is continuously connected to evacuating equipment during operation. *Note:* This term is used chiefly for pool-cathode tubes. 190

pump efficiency (laser-maser). The ratio of the power or energy absorbed from the pump to the power or energy available from the pump source. 363

pump-free control. *See:* antipump (pump-free) device.

pumping load (power operations). Totals of loads caused by pumping in pumped-storage stations within the system. 516

punch (computing systems). *See:* keypunch, zone punch.

punched card. A card on which a pattern of holes or cuts is used to represent data. 77, 54

punched tape (computing systems). A tape on which a pattern of holes or cuts is used to represent data. 255, 77, 54

punched tape handler (test, measurement and diagnostic equipment). A device which handles punched tape and usually consists of a tape transport and punched tape reader with associated electrical and electronic equipments. Most units provide for tape to be wound and stored in reels: however, some units provide for the tape to be stored loosely in closed bins. 54

punched tape reader (test, measurement and diagnostic equipment). A device capable of converting information from punched tape, where it has been stored in the form of a series of holes, into a series of electrical impulses. 54

punching (rotating machinery). A lamination made from sheet material using a punch and die. *See:* rotor (rotating machinery); stator. 63

punch position (computing systems). A site on a punched tape or card where holes are to be punched. 255, 77, 54

punch-through voltage (of a semiconductor radiation detector)(X-ray energy spectrometers). The voltage at which a junction detector becomes fully depleted. *See:* depletion voltage. 471

puncture (1) (high voltage testing). Term used when a disruptive discharge occurs through a solid dielectric. A disruptive in a solid dielectric produces permanent loss of dielectric strength; in a liquid or gaseous dielectric, the loss may be only temporary. 150

(2) (voltage testing). A disruptive discharge through the body of a solid dielectric. *See:* test voltage and current. 307, 201

puncture voltage (surge arresters). The voltage at which the test specimen is electrically punctured. 64, 62

pupil (1) (laser-maser). The variable aperture in the iris through which light travels toward the interior of the eye. 363

(2) (pupillary aperture) (illuminating engineering). The opening in the iris which admits light into the eye. 167

purchaser (1) (nuclear power quality assurance). The organization responsible for establishment of procurement requirements and for issuance, administration, or both, of procurement documents. 417

(2) (rotating electric machinery). The organization placing the contract for the machinery or its repair; often called the "user". 424

pure tone. *See:* simple tone.

purification of electrolyte. The treatment of a suitable volume of the electrolyte by which the dissolved impurities are removed in order to keep their content in the electrolyte within desired limits. *See:* electrorefining. 328

purity (excitation purity) (color). *See:* excitation purity (light).

Purkinje phenomenon (illuminating engineering). The reduction in subjective brightness of a red light relative to that of a blue light when the luminances are reduced in the same proportion without changing the respective spectral distributions. In passing from photopic to scotopic vision, the curve of spectral luminous efficiency changes, the wavelength of maximum efficiency being displaced toward the shorter wavelengths. 167

purple boundary (television) (illuminating engineering). The straight line drawn between the ends of the spectrum locus on a chromaticity diagram. 18, 167

push brace. A supporting member, usually of timber, placed between a pole or other structural part of a line and the ground or a fixed object. *See:* tower. 64

pushbutton (industrial control). Part of an electric device, consisting of a button that must be pressed to effect an operation. *See:* switch. 244, 206

pushbutton dial (telephone switching systems). A type of calling device used in automatic switching that has an activator per digit that generates distinctive pulsing. 55

pushbutton station (industrial control). A unit assembly of one or more externally operable pushbutton switches, sometimes including other pilot devices such as indicating lights or selector switches, in a suitable enclosure. *See:* switch. 206

pushbutton switch (pushbutton). A master switch, usually mounted behind an opening in a cover or panel, and having an operating plunger or button extending forward in the opening. Operation of the switch is normally obtained by pressure of the finger against the end of the button. *See:* switch. 328

pushbutton switching. A reperforator switching system in which selection of the outgoing channel is initiated by an operator. 194

pushdown list (computing systems). A list that is constructed and maintained so that the next item to be retrieved is the most recently stored item in the list, that is, last in, first out. *See:* pushup list. 255, 77

pushdown storage (software). A storage device that handles data in such a way that the next item to be retrieved is the most recently stored item still in the storage device, that is, last-in-first-out (LIFO). *See:* **data; stack.** 434

pushing figure (oscillator). The change of oscillator frequency with a specified change in current, excluding thermal effects. *See:* **tuning sensitivity, electronic; oscillatory circuit; television.** 125

push-pull amplifier circuit. *See:* **amplifier, balanced.**

push-pull circuit. A circuit containing two like elements that operate in 180-degree phase relationship to produce additive output components of the desired wave, with cancellation of certain unwanted products. *Note:* Push-pull amplifiers and push-pull oscillators are examples. *See:* **amplifier** 415

push-pull currents. Balanced currents. *See:* **waveguide.** 267

push-pull microphone. A microphone that makes use of two like microphone elements actuated by the same sound waves and operating 180 degrees out of phase. *See:* **microphone.** 328

push-pull operation (electron device). The operation of two similar electron devices or of an equivalent double-unit device, in a circuit such that equal quantities in phase opposition are applied to the input electrodes, and the two outputs are combined in phase. 190

push-pull oscillator. A balanced oscillator employing two similar tubes in phase opposition. *See:* **balanced oscillator; oscillatory circuit.** 111

push-pull voltages. Balanced voltages. *See:* **waveguide.** 267

push-push circuit. A circuit employing two similar tubes with grids connected in phase opposition and plates in parallel to a common load, and usually used as a frequency multiplier to emphasize even-order harmonics. 111

push-push currents. Currents flowing in the two conductors of a balanced line that, at every point along the line, are equal in magnitude and in the same direction. *See:* **waveguide.** 267

push-push voltages. Voltages (relative to ground) on the two conductors of a balanced line that, at every point along the line, are equal in magnitude and have the same polarity. *See:* **waveguide.** 267

push-to-type operation. That form of telegraph operation, employing a one-way reversible circuit, in which the operator must keep a switch operated in order to send from his station. It is generally used in radio transmission where the same frequency is employed for transmission and reception. *See:* **telegraphy.** 328

pushup list (computing systems). A list that is constructed and maintained so that the next item to be retrieved and removed is the oldest item still in the list, that is, first in, first out. *See:* **pushdown list.** 255, 77

PV array. *See:* **photovoltaic (PV) array**

PV array subfield. *See:* **photovoltaic (PV) array subfield**

PVC (cable system in power generating stations). Conduit fabricated from polyvinyl chloride. 35

PV cell. *See:* **photovoltaic (PV) cell**

PV module. *See:* **photovoltaic (PV) module**

PV panel. *See:* **photovoltaic (PV) panel**

PV receiver. *See:* **photovoltaic (PV) receiver.**

PV system-utility interface. *See:* **photovoltaic (PV) system-utility interface**

pyramidal horn antenna. A horn antenna the sides of which form a pyramid. *See:* **antenna.** 179, 111

pyroconductivity. Electric conductivity that develops with rising temperature, and notably upon fusion, in solids that are practically nonconductive at atmospheric temperatures. 328

pyroelectric effect (primary ferroelectric terms). The appearance of an electric charge at the surface of a polar material when uniform heating or cooling changes the polarization. If the polar material is electroded and an external resistance is connected between the electrodes, the current that flows is a pyroelectric current. All pyroelectrics are polar; ferroelectrics are a subgroup of the polar materials and, therefore, they are both pyroelectric and piezoelectric. The differ from the more general pyroelectrics principally by the reversibility or reorientability of their spontaneous polarization P_s. *Note:* Due to the existence of free charge, the pyroelectric charge may be rapidly compensated. 497

pyrometer. A thermometer of any kind usable at relatively high temperatures (above 500 degrees Celsius). *See:* **electric thermometer (temperature meter).** 328

Q

Q chrominance signal (National Television System Committee (NTSC) color television). The sidebands resulting from suppressed-carrier modulation of the chrominance subcarrier by the Q video signal. *Note:* The signal is transmitted in double-sideband form, the sidebands extending approximately 0.6 MHz above and below the chrominance subcarrier. The phase of the signal, for positive Q video signals, is 33° with respect to the (B−Y) axis. 18

Q coil. *See:* **coil Q.**

Q-hour meter. An electricity meter that measures the quantity obtained by effectively lagging the applied voltage to a watthour meter by 60 degrees. This quantity is one of the quantities used in calculating quadergy (varhours). 212

Q-meter (quality-factor meter). An instrument for measuring the quality factor Q of a circuit or circuit element. *See:* **instrument.** 328

Q of an electrically small tuned antenna. An inverse measure of the bandwidth or an antenna as determined by its impedance. It is numerically equal to one half the magnitude of the ratio of the incremental change in impedance to the incremental change in frequency at resonance, divided by the ratio of the antenna resistance to the resonance frequency. *Note:* The Q of an antenna also is a measure of the energy stored to the energy radiated or dissipated per cycle.   111

Q of a resonant antenna. The ratio of 2π times the energy stored in the fields excited by the antenna to the energy radiated and dissipated per cycle. *Note:* For an electrically small antenna, it is numerically equal to one-half the magnitude of the ratio of the incremental change in impedance to the corresponding incremental change in frequency at resonance, divided by the ratio of the antenna resistance to the resonant frequency.   111

Q-percentile life (1) (assessed) (non-repaired items) (reliability). The Q-percentile life determined as limiting value or values of the confidence interval with a stated confidence level, based on the same ata as the observed Q-percentile life of nominally identical items. *Notes:* (A) The source of the data should be stated. (B) Results can be accumulated (combined) only when all conditions are similar. (C) The assumed underlying distribution of failures against time should be stated. (D) It should be stated whether a one-sided or two-sided interval is being used. (E) Where one limiting value is given this is usually the lower limit. **(2) extrapolated.** Extension by a defined extrapolation or interpolation of the observed or assessed Q-percentile life for stress conditions different from those applying to the assessed Q-percentile life and for different percentages. *Note:* The validity of the extrapolation should be justified. **(3) observed.** The length of observed time at which a stated proportion (Q%) of a sample of items has failed. *Notes:* (A) The criteria for what constitutes a failure should be stated. (B) The Q-percentile life is also that life at which (100-Q)% reliability is observed. **(4) predicted.** For the stated conditions of use, and taking into account the design of an item, assessed or extrapolated Q-percentile lives of its parts. *Note:* Engineering and statistical assumptions should be stated, as well as the bases used for the computation (observed or assessed).

Q response (q) (subroutines for CAMAC). The symbol *q* represents a logical truth value which corresponds to the CAMAC Q response. It is set to true if the Q response is 1, to false if the Q response is 0.   410

Q responses (qa) (subroutines for CAMAC). The symbol *qa* represents an array of Q response values. Each element of *qa* has the same form and can have the same values as the parameter *q*. The length of *qa* is given by the value of the first element of *cb* at the time the subroutine is executed. *See:* **Q response; control block.**   410

Q-switch (laser-maser). A device for producing very short (Δ 30 ns), intense laser pulses by enhancing the

storage and dumping of electronic energy in and out of the lasing medium, respectively.   363

Q-switched laser (laser-maser). A laser which emits short (Δ 30 ns), high-power pulses by utilizing a Q-switch.   363

Q-switching (laser-maser). Producing very short (Δ 30 ns), intense pulses by enhancing the storage and dumping of electronic energy in and out of the laser-maser medium, respectively.   363

quad. A structural unit employed in cable, consisting of four separately insulated conductors twisted together. *See:* **cable.**   328

quadded cable. A cable in which at least some of the conductors are arranged in the form of quads. *See:* **cables.**   328

quadergy (1) (general). Delivered by an electric circuit during a time interval when the voltages and currents are periodic, the product of the reactive power and the time interval, provided the time interval is one or more complete periods or is quite long in comparison with the time of one period. If the reference direction for energy flow is selected as into the region the net delivery of quadergy will be into the region when the sign of the quadergy is positive and out of the region when the sign is negative. If the reference direction is selected as out of the region, the reverse will apply. The quadergy is expressed by

$$K = Qt$$

where Q is the reactive power and t is the time interval. If the voltages and currents form polyphase symmetrical sets, there is no restriction regarding the relation of the time interval to the period. If the voltages and currents are quasi-periodic and the amplitudes of the voltages and currents are slowly varying, the quadergy is the integral with respect to time of the reactive power, provided the integration is for a time that is one or more complete periods or that is quite long in comparison with the time of one period. Mathematically

$$K = \int_{t_0}^{t_0+t} Q\,\mathrm{d}t$$

where Q is the reactive power determined for the condition of voltages and current having slowly varying amplitudes. Quadergy is expressed in var-seconds or var-hours when the voltages are in volts and the currents in amperes, and the time is in seconds or hours, respectively. *See:* **network analysis.**   210 **(2) (metering).** The integral of reactive power with respect to time.   212

quadrantal error (navigation)(navigation aid terms). An angular error in measured bearing caused by characteristics of the vehicle or station which adversely affect the direction of signal propagation; the error varies in a sinusoidal manner throughout the 360° and has two positive and two negative maximums.   526 quadratic lag. *See:* lag (6) (second-order).

quadratic profile. *See:* parabolic profile.

quadrature. The relation between two periodic functions when the phase difference between them is one-fourth of a period. *See:* network analysis.   210

quadrature-acceleration drift rate (dynamically tuned gyro) (inertial sensor). A drift rate about an axis normal to both the spin axis and the axis along which an acceleration is applied. This drift rate results from a torque about the axis of applied acceleration and is in quadrature with that due to mass unbalance.   46

quadrature axis (synchronous machine). The axis that represents the direction of the radial plane along which the main field winding produces no magnetization, normally coinciding with the radial plane midway between adjacent poles. *Notes:* (1) The positive direction of the quadrature axis is 90 degrees ahead of the positive direction of the direct axis, in the direction of rotation of the field relative to the armature. (2) The definitions of currents and voltages given in the terms listed below are applicable to balanced load conditions and for sinusoidal currents and voltages. They may also be applied under other conditions to the positive-sequence fundamental-frequency components of currents and voltages. More generalized definitions, applicable under all conditions, have not been agreed upon.   63

quadrature-axis component (1) (armature voltage). That component of the armature voltage of any phase that is in time phase with the quadrature-axis component of current in the same phase. *Notes:* A quadrature-axis component of voltage may be reproduced by: (A) rotation of the direct-axis component magnetic flux; (B) variation (if any) of the quadrature-axis component of magnetic flux; (C) resistance drop caused by flow of the quadrature-axis component of armature current. The quadrature-axis component of terminal voltage is related to the synchronous internal voltage by

$$\mathbf{E}_{aq} = \mathbf{E}_i - R\mathbf{I}_{aq} - jX_d\mathbf{I}_{ad}.$$

See: phasor diagram (synchronous machine).   328

(2) (armature current). That component of the armature current that produces a magnetomotive-force distribution that is symmetrical about the quadrature axis. *See:* quadrature-axis component of armature voltage; quadrature-axis component of magnetomotive force.   328

(3) (magnetomotive force) (rotating machinery). The component of a magnetomotive force that is directed along an axis in quadrature with the axis of the poles. *See:* asynchronous machine; direct-axis synchronous impedance.   63

quadrature-axis current (rotating machinery). The current that produces quadrature-axis magnetomotive force. *See:* direct-axis synchronous impedance.   63

quadrature-axis magnetic flux (rotating machinery).

The magnetic-flux component directed along the quadrature axis. *See:* direct-axis synchronous impedance.   63

quadrature-axis operational inductance $L_{d\,(s)}$(synchronous machine parameters by standstill frequency testing)(standstill frequency response testing). The ratio of the Laplace transform of the quadrature-axis armature flux linkages to the Laplace transform of the quadrature-axis current.   521, 565

quadrature-axis subtransient impedance (rotating machinery). The operator expressing the relation between the initial change in armature voltage and a sudden change in quadrature-axis armature current, with only the fundamental-frequency components considered for both voltaged and current, with no change in the voltage applied to the field winding, and with the rotor running at steady speed. In terms of network theory it corresponds to the quadrature-axis impedance the machine displays against disturbances (modulations) with infinite frequency. *Note:* If no rotor winding is along the quadrature axis and/or the rotor is not made out of solid steel, this impedance equals the quadrature-axis synchronous impedance. *See:* asynchronous machine; direct-axis synchronous impedance.   63

quadrature-axis subtransient open-circuit time constant. The time in seconds required for the rapidly decreasing component (negative) present during the first few cycles in the direct-axis component of symmetrical short-circuit conditions with the machine running at rated speed, to decrease to $1/c$ Δ 0.368 of its initial value.   328

quadrature-axis subtransient reactance. The ratio of the fundamental component of reactive armature voltage due to the initial value of the fundamental quadrature-axis component of alternating-current component of the armature current, to this component of current under suddenly applied balanced load conditions and at rated frequency. Unless otherwise specified, the quadrature-axis subtransient reactance will be that corresponding to rated armature current.   328

quadrature-axis subtransient short-circuit time constant. The time in seconds required for the rapidly decreasing component present during the first few cycles in the quadrature-axis component of the alternating-current component of the armature current under suddenly applied symmetrical short-circuit conditions, with the machine running at rated speed to decrease to $1/e$ Δ 0.368 of its initial value.   328

quadrature-axis subtransient voltage (rotating machinery). The quadrature-axis component of the terminal voltage which appears immediately after the sudden opening of the external circuit when the machine is running at a specified load, before any flux variation in the excitation and damping circuits has taken place.   63

quadrature-axis synchronous impedance (synchronous machine) (rotating machinery). The impedance of the armature winding under steady-state conditions where the axis of the armature current and magnetomotive force coincides with the quadrature axis. In large machines where the armature resistance is negligibly small, the quadrature-axis synchronous impedance is equal to the quadrature-axis synchronous reactance.   63

quadrature-axis synchronous reactance. The ratio of the fundamental component of reactive armature voltage, due to the fundamental quadrature-axis component of armature current, to this component of current under steady-state conditions and at rated frequency. Unless otherwise specified, the value of quadrature-axis synchronous reactance will be that corresponding to rated armature current.   328

quadrature-axis transient impedance (rotating machinery). The operator expressing the relation between the initial change in armature voltage and a sudden change in quadrature-axis armature current component, with only the fundamental frequency components considered for both voltage and current, with no change in the voltage applied to the field winding, with the rotor running at steady speed, and by considering only the slowest decaying component and the steady-state component of the voltage drop. In terms of network theory it corresponds to the quadrature-axis impedance the machine display against disturbances (modulation) with infinite frequency by considering only two poles pairs* (or poles*), namely those with smallest (including zero) real parts, of the impedance function. *Note:* If no rotor winding is along the quadrature axis and/or the rotor is not made out of solid steel, this impedance equals the quadrature-axis synchronous impedance. *See:* **asynchronous machine; direct-axis synchronous impedance.**   63

*The term pole refers here to the roots of the denominator of the impedance function.

quadrature-axis transient open-circuit time constant. The time in seconds required for the root-mean-square alternating-current value of the slowly decreasing component present in the direct-axis component of symmetrical armature voltage on open-circuit to decrease to $1/c \Delta 0.368$ of its initial value when the quadrature field winding (if any) is suddenly short-circuited with the machine running at rated speed. *Note:* This time constant is important only in turbine generators.   328

quadrature-axis transient reactance. The ratio of the fundamental component of reactive armature voltage, due to the fundamental quadrature-axis component of the alternating-current component of the armature current, to this component of current under suddenly applied load conditions and at rated frequency, the value of current to be determined by the extrapolation of the envelope of the alternating-current component of the current wave to the alternating-current component of the current wave to the instant of the sudden application of load, neglecting the high-decrement current during the first few cycles. *Note:* The quadrature-axis transient reactance usually equals the quadrature-axis synchronous reactance except in solid-rotor machines, since in general there is no really effective field current in the quadrature axis.   328

quadrature-axis transient short-circuit time constant. The time in seconds required for the root-mean-square alternating-current value of the slowly decreasing component present in the direct-axis component of the alternating-current component of the armature current under suddenly applied short-circuit conditions with the machine running at rated speed to decrease to $1/e \Delta 0.368$ of its initial value.   328

quadrature-axis transient voltage (rotating machinery). The quadrature-axis component of the terminal voltage that appears immediately after the sudden opening of the external circuit when running at a specified load, neglecting the components with very rapid decay that may exist during the first few cycles. *See:* **asynchronous machine; direct-axis synchronous impedance.**   63

quadrature-axis voltage (rotating machinery). The component of voltage that would produce quadrature-axis current when resistance limited.   63

quadrature hybrid (waveguide components). A hybrid junction which has the property that a wave leaving one output port is in phase quadrature with the wave leaving the other output port.   166

quadrature modulation. Modulation of two carrier components 90 degrees apart in phase by separate modulating functions.   242

quadrature spring rate (dynamically tuned gyro) (inertial sensor). When the case of a dynamically tuned gyro is displaced with respect to the gyro rotor through an angle about an axis perpendicular to the spin axis, a torque proportional to and 90 degrees away from the displacement acts in a direction to reduce this angle and to align the rotor with the case. The torque is usually due to windage, a squeeze-film force, or flexure hysteresis. This spring rate results in a drift rate coefficient having dimensions of angular displacement per unit time per unit angle of displacement about an input axis.   46

quadri pole. *See:* **two-terminal pair network.**

quadrupole parametric amplifier. A beam parametric amplifier having transverse input and output couplers for the signal, separated by a quadrupole structure that is excited by a pump to obtain parametric amplification of a cyclotron wave. *See:* **parametric device.**   277,191

qualification (1)(nuclear power generating station). The generation and maintenance of evidence to ensure that the equipment will operate on demand to meet the system performance requirements.   120

(2) (nuclear power quality assurance)(personnel). The characteristics or abilities gained through education, training, or experience, as measured against established requirements, such as standards or tests, that qualify an individual to perform a required function.   417

(3)(raceway system)(raceway systems for Class 1E circuits for nuclear power generating stations). Demonstration in the form of certificates of compliance, analysis reports, or testing reports that the raceway system meets the design requirements.   513

qualification testing (software). Formal testing, usually conducted by the developer for the consumer, to demonstrate that the software meets its specified requirements. *See:* **acceptance testing; formal testing; requirement; software; system testing.** 434

qualification tests (safety systems equipment in nuclear power generating stations). Tests conducted on safety systems equipment to demonstrate the capability to meet specified functional requirements under the action of specified test levels of environmental and operational parameters. *Note:* These tests subject a sample or samples of equipment to specified conditions designed to simulate normal, abnormal, containment test, design-basis-event, including loss-of-coolant, and post-design-basis-event conditions. 534

qualified (NESC). Having adequate knowledge of the installation, construction or operation of apparatus and the hazards involved. 494

qualified diesel-generator unit. A diesel-generator unit that meets the qualification of IEEE Std 387-1984. 99

qualified life (1)(cable, field splice, and connection qualification)(design and installation of cable systems for Class 1E circuits in nuclear power generating stations). The period of time for which satisfactory performance can be demonstrated for a specific set of service conditions. 536

(2)(valve actuators)(electric penetration assemblies). The period of time, prior to the start of a design basis event, for which the equipment was demonstrated to meet the design requirements for the specified service conditions. *Note:* At the end of the qualified life, the equipment remains capable of performing the safety function(s) required for the postulated design basis events and post design basis events. 492, 493

(3)(nuclear power generating station). The period of time, prior to a design basis event, for which equipment was demonstrated to meet the design requirements for the specified service conditions. *Note:* At the end of the qualified life, the equipment is expected to be capable of performing the safety functions(s) required for the postulated design basis and post-design basis events. 120

(4)(safety systems equipment in nuclear power generating stations). The period of time, prior to the start of a design-basis event, for which the equipment was demonstrated to meet the design requirements for the specified service conditions. *Note:* At the end of the qualified life, the equipment is required to be capable of performing the safety function(s) required for the postulated design basis and post design-basis events. 535

(5)(seismic qualification of Class 1E equipment for nuclear power generating stations). The period of time, prior to the start of a design basis event (DBE), for which the equipment was demonstrated to meet the design requirements for the specified service conditions. 581

(6) (class 1E motor control)(class 1E static battery chargers and inverters). The period of time for which satisfactory performance can be demonstrated for a specific set of service conditions. *Note:* The qualified life of a particular equipment item may be changed during its installed life where justified 120, 141, 440, 408

(7) (nuclear power generating stations). The period of time that can be verified for which the electric penetration assembly will meet all design requirements for the specified service conditions. *See:* **electric penetration assembly.** 26

qualified life test (electric penetration assemblies). Tests performed on preconditioned test specimens to verify that an electric penetration assembly will meet design requirements at the end of its qualified life. 493

qualified module (nuclear power generating station). Module which exhibits performance characteristics which are acceptable for Class 1E service in a nuclear power generating station and which satisfy the aging criteria and other requirements of this document. 355

qualified person (National Electrical Code). One familiar with the construction and operation of the equipment and the hazardous involved. 256

qualified procedures (nuclear power quality assurance). An approved procedure that has been demonstrated to meet the specified requirements for its intended purpose. 417

qualifying symbol (graphic symbols for logic functions). A symbol added to the basic outline of an element to designate the physical or logic characteristics of an input or output of the element or the overall logic characteristics of the element. 451

qualitative adjectives (pulse terms). *See:* **descriptive adjectives; time related adjectives; magnitude related adjectives; polarity related adjectives; geometrical adjectives.**

qualitative distortion terms (pulse terms). See: **preshoot; overshoot; rounding; spike; ringing; tilt; valley.**

quality (software). The totality of features and characteristics of a product or service that bear on its ability to satisfy given needs. *See:* **software quality.** 434

quality area. The area of the cathode-ray-tube phosphor screen that is limited by the cathode-ray tube and instrument specification. *Note:* If the quality area and the graticule area are not equal, this must be specified. *See:* **graticule area; viewing area; oscillograph.** 185

quality assurance (1)(monitoring radioactivity in effluents). All those planned and systematic actions necessary to provide adequate confidence that a system or component will perform satisfactorily in service. 559

(2)(software quality assurance). A planned and systematic pattern of all actions necessary to provide adequate confidence that the item or product conforms to established technical requirements. 434, 442

quality assurance record (nuclear power quality assurance). A completed document that furnishes evidence of the quality of items or activities, or both, affecting quality. 417

quality control (nuclear power generating station).

Those quality assurance actions which provide a means to control and measure the characteristics of an item, process or facility to established requirements. (See ANSI 45.2.10-1973).   438

quality factor (Q) (1) (network, structure, or material). Two pi times the ratio of the maximum stored energy to the energy dissipated per cycle at a given frequency. *Notes:* (A) The Q of an inductor at any frequency is the magnitude of the ratio of its reactance to its effective series resistance at that frequency. (B) The Q of a capacitor at any frequency is the magnitude of the ratio of is susceptance to its effective shunt conductance at that frequency. (C) The Q of a simple resonant circuit comprising an inductor and a capacitor is given by

$$Q = \frac{Q_L Q_C}{Q_L + Q_C}$$

where Q_L and Q_C are the Q's of the inductor and capacitor, respectively, at the resonance frequency. If the resonant circuit comprises an inductance L and a capacitance C in series with an effective R, the value of Q is

$$Q = \frac{1}{R}\left(\frac{L}{C}\right)^{1/2}$$

(D) An approximate equivalent definition, which can be applied to other types of resonant structures, is that the Q is the ratio of the resonance frequency to the bandwidth between the frequencies on opposite sides of the resonance frequency (known as half-power points) where the response of the resonant structure differs by 3 decibels from that at resonance. (E) The Q of a magnetic or dielectic material at any frequency is equal to 2π times the ratio of the maximum stored energy to the energy dissipated in the material per cycle. (F) For networks that contain several elements, and distributed parameter systems, the Q is generally evaluated at a frequency of resonance. (G) The non-loaded Q of a system is the value of Q obtained when only the incidental dissipation of the system element is present. **The loaded** Q of a system is the value of Q obtained when the system is coupled to a device that dissipates energy. (H) The **period** in the expression for Q is that of the driving force, not that of energy storage, which is usually half that of the driving force. *See:* **network analysis.**   210

quality factor (2)(harmonic control and reactive compensation of static power converters). Two π times the ratio of the maximum stored energy to the energy dissipated per cycle at a given frequency. An approximate equivalent definition is that the Q is the ratio of the resonance frequency to the bandwidth between those frequencies on opposite sides of the resonance frequency where the response of the resonant structure differs by 3 decibels (dB) from that at resonance. If the resonant circuit comprises an inductance L and

a capacitance C in series with an effective resistance R, the value of Q is:

$$Q = \frac{1}{R}\left(\frac{L}{C}\right)^{\frac{1}{2}}$$

533

(3) (circuits and systems). In active filters the transfer functions are generally broken down into second-order sections expressed by biquadratic functions as follows:

$$T(s) = \frac{n_2 s^2 + n_1 s + n_0}{d_2 s^2 + d_1 s + d_0}$$

Such transfer functions are generally re-arranged in the following form where the zero Q-factor, Q_z, and the pole Q-factor, Q_P, may be identified.

$$T(s) = K\frac{s^2 + s\dfrac{\omega_{0z}}{Q_z} + \omega_{0z}{}^2}{s^2 + s\dfrac{\omega_{0p}}{Q_p} + \omega_{0p}{}^2}$$

67

quality metric (software). A quantitative measure of the degree to which software possesses a given attribute which affects its quality. *See:* **quality; software.**   434

quality of lighting (illuminating engineering). Pertains to the distribution of luminance in a visual environment. The term is used in a positive sense and implies that all luminances contribute favorably to visual performance, visual comfort, ease of seeing, safety, and esthetics for the specific visual tasks involved.   167

quantitative adjectives (pulse terms). *See:* **integer adjectives; mathematical adjectives; functional adjectives.**

quantitative testing (test, measurement and diagnostic equipment). Testing that monitors or measures the specific quantity, level or amplitude of a characteristic to evaluate the operation of an item. The outputs of such tests are presented as finite or quantitative values of the associated characteristics.   54

quantity equations. Equations in which the quantity symbols represent mathematico-physical quantities possessing both numerical values and dimensions.   210

quantity of electricity. *See:* **electric charge**

quantity of light (luminous energy,$Q = \int \Phi dt$**) (illuminating engineering).** The product of the luminous flux by the time it is maintained. It is the time integral of luminous flux.   167

quantity of light (luminous energy),

$$Q_v = \int \varphi_v dt)$$

(light emitting diodes). The product of the luminous flux by the time it is maintained. It is the time integral of luminous flux.   162

quantization (1) (telecommunication). A process in which the continuous range of values of an input signal is divided into nonoverlapping subranges, and to each subrange a discrete value of the output is uniquely assigned. Whenever the signal value falls within a given subrange, the output has the corresponding discrete value. *Note:* Quantized may be used as an adjacent modifying various forms of modulation, for example, quantized pulse-amplitude modulation. *See:* **quantization distortion (quantization noise); quantization level).**   415,255,77,194
(2) (gyro; accelerometer). The analog-to-digital conversion of a gyro or accelerometer output signal which gives an output that changes in discrete steps as the input varies continuously.   46
(3) (data transmission). In communication, quantization is a process in which the range of values of a wave is divided into a finite number of smaller subranges, each of which is represented by an assigned (or quantized) value within the subrange. *Note:* "Quantized" may be used as an adjective modifying various forms of modulation, for example, quantized pulse amplitude modulation.   59

quantization distortion (quantization noise) (data transmission). The inherent distortion introduced in the process of quantization.   59

quantization error (supervisory control, data acquisition, and automatic control). The amount that the digital quantity differs from the analog quantity. *See:* **analog-to-digital (a/d) conversion.**   570

quantization level (1) (data transmission). A particular subrange of a symbol designating it.   59
(2) (telecommunication). The discrete value of the output designating a particular subrange of the input. *See:* **quantization.**   242

quantization noise. *See:* **quantization distortion.**

quantize. To subdivide the range of values of a variable into a finite number of nonoverlapping subranges or intervals, each of which is represented by an assigned value within the subrange, for example, to represent a person's age as a number of whole years.   255,77

quantized pulse modulation. Pulse modulation that involves quantization.   328

quantized system. One in which at least one quantizing operation is present.   56

quantizing error (radar). An error caused by conversion of a variable having a continuous range of values to a quantized form having only discrete values, as in analog-to-digital conversion. The error is the difference between the original (analog) value and its quantized (digital) representation. *See:* **quantization; quantize.**   13

quantizing loss (radar). (1) In phased arrays, a loss that occurs when the beam is phase steered by digitally controlled phase shifters, due to the quantizing errors in the phase shifts applied to the various elements. (2) In signal processing, a loss that occurs when elements of a composite signal (for example, complex amplitudes of pulses in a pulse train) are quantized (digitized) before being combined. *See:* **quantizing error.**   13

quantizing noise (analog voice frequency circuits). The noise introduced during the process of digitally encoding an analog signal.   468

quantizing operation. One which converts one signal into another having a finite number of predetermined magnitude values.   56

quantum efficiency (1) (photocathodes). The average number of electrons photoelectrically emitted from the photocathode per incident photon of a given wavelength. *Note:* The quantum efficiency varies with the wavelength, angle of incidence, and polarization of the incident radiation. *See:* **photocathodes; photoelectric converter; phototubes; semiconductors.**   125,117
(2) (laser, maser, laser material, or maser material). The ratio of the number of photons or electrons emitted by a material at a given transition to the number of absorbed particles.   363
(3) (fiber optics). In an optical source or detector, the ratio of output quanta to input quanta. Input and output quanta need not both be photons.   433

quantum noise (fiber optics). Noise attributable to the discrete or particle nature of light. *Syn:* **photon noise.**   433

quantum-noise-limited operation (fiber optics). Operation wherein the minimum detectable signal is limited by quantum noise. *See:* **quantum noise.**   433

quarter-phase or two-phase circuit. A combination of circuits energized by alternating electromotive forces that differ in phase by a quarter of a cycle, that is, 90 degrees. *Note:* In practice the phases may vary several degrees from the specified angle. *See:* **center of distribution.**   64

quarters (electric installations on shipboard). Where used in these recommendations, those spaces provided for passengers or crew, as specified, which are actually used for berthing, mess spaces, offices, private baths, toilets and showers, and lounging rooms, smoking rooms, and similar spaces.   3

quarter-thermal-burden ambient-temperature rating. The maximum ambient temperature at which the transformer can be safely operated when the transformer is energized at rated voltage and frequency and is carrying 25 percent of its thermal-burden rating without exceeding the specified temperature limitations.   328

quartet (mathematics of computing). A group of four adjacent digits operated upon as a unit.   564

quasi-analog signal (data transmission). A digital signal after conversion to a form suitable for transmission over a specified analog channel. The specifications of an analog channel includes the frequency of range, frequency of bandwidth, signal-to-noise ratio (snr), and envelope delay distortion. When this form of signaling is used to convey message traffic over the public dial-up network telephone systems, it is often referred to as voice data. 59

quasi-Gaussian shaping (germanium gamma-ray detectors)(charged-particle detectors). Pulse shaping consisting of one differentiation followed by four or more integrations resulting in a pulse shape that is approximated by a Gaussian curve. For n integrations the shaping network is sometime denoted as

$$CR - (RC)^n$$

528, 119

quasi-impulsive noise (electromagnetic compatibility). A superposition of impulsive and continuous noise. See: electromagnetic compatibility. 220,199

quasi-peak detector (1) (electromagnetic compatibility). A detector having specified electrical time constants that, when regularly repeated pulses of constant amplitude are applied to it, delivers an output voltage that is a fraction of the peak value of the pulses, the fraction increasing towards unity as the pulse repetition rate is increased. See: electromagnetic compatibility. 220,199

(2) (overhead-power-line corona and radio noise). A detector having specified electrical time constants which, when regularly repeated pulses of constant amplitude are applied to it, delivers an output voltage that is a fraction of the peak value of the pulses, the fraction increasing toward unity as the pulse repetition rate is increased. Note: According to ANSI C63.2-1979, American National Standard Specification for Electromagnetic Noise and Field Strength Instrumentation, 10 kHz to 1 GHz, the quasi-peak detector has a charging time constant of 1 ms and a discharging time constant of 600 ms. The corresponding values according to International Special Committee on Radio Interference Publications, CISPR 1(1972), Specification for CISPR Radio Interference Measuring Apparatus for the Frequency Range 0.15 MHz to 30 MHz; CISPR 2(1975), Specification for CISPR Radio Interference Measuring Apparatus for the Frequency Range 25 MHz to 300 MHz; and CISPR 4 (1967), CISPR Measuring Set Specification for the Frequency Range 300 MHz to 1000 MHz, are 1 ms and 160 ms for the frequency range of 0.15 to 30 MHz and 1 ms and 550 ms for the frequency range 30 to 1000 MHz. 411

quasi-peak voltmeter (electromagnetic compatibility). A quasi-peak detector coupled to an indicating instrument having a specific mechanical time-constant. See: electromagnetic compatibility. 220,199

quasi-square wave (converter characteristics)(self-commutated converters). The stepped waveform obtained from the difference of two phase-shifted square waves of equal amplitude. An example is shown below. 584

quaternary code (information theory). A code whose output alphabet consists of four symbols. See: ternary code. 415

quenched sample (liquid-scintillation counters). A counting sample (material of interest plus liquid-scin-

quasi-square wave

tillation solution) that contain adulterants that reduce the photon output from the vials. 498

quenched spark gap converter (dielectric heater usage). A spark-gap generator or power source that utilizes the oscillatory discharge of a capacitor through an inductor and a spark gap as a source of radio-frequency power. The spark gap comprises one or more closely spaced gap operating in series. See: induction heating; industrial electronics. 14,114

quenching (gas-filled radiation-counter tube). The process of terminating a discharge in a Geiger-Mueller radiation-counter tube by inhibiting a reignition. Note: This may be effected internally (internal quenching or self-quenching) by use of an appropriate gas or vapor filling, or externally (external quenching) by momentary reduction of the applied potential difference. 125,96

quenching circuit (radiation counters). A circuit that reduces the voltage applied to a Geiger-Mueller tube after an ionizing event, thus preventing the occurrence of subsequent multiple discharges. Usually the original voltage level is restored after a period that is longer than the natural recovery time of the Geiger-Mueller tube. See: anticoincidence (radiation counters). 190

query (data transmission). The process by which a master station asks a slave station to identify itself and to give its status. See: poll. 59

queue (software). A list that is accessed in a first-in, first-out manner. See: stack. 434

quick-break (power switchgear). A term used to describe a device which has a high contact opening speed independent of the operator. 103

quick-break switch (high-voltage switchgear). A switch that has a high contact opening speed independent of the operator. 443

quick charge (storage battery). See: boost charge.

quick-flashing light (illuminating engineering). A rhythmic light exhibiting very rapid regular alternations of light and darkness. There is no restriction on the ratio of the durations of the light to the dark periods. 167

quick-make (power switchgear). A term used to describe a device which has a high contact closing speed independent of the operator. 103

quick release (control brakes). The provision for effecting more rapid release than would inherently be obtained. See: electric drive. 206

quick set (control brakes). The provision for effecting more rapid setting than would inherently be obtained. See: electric drive. 206

quick startup reserve (power operations). The operating reserve available within a specified time through startup and synchronization of quick start internal combustion generation. 516

quiescent. *See:* **supervisory control.**

quiescent-carrier telephony. That form of carrier telephony in which the carrier is suppressed whenever there are no modulating signals to be transmitted. 328

quiescent current (electron tubes). The electrode current corresponding to the electrode bias voltage. *See:* **electronic tube.** 244,190

quiescent operating point (magnetic amplifier). The output obtained under any specified external conditions when the signal is non-time-varying and zero. 171

quiescent point (amplifier). That point on its characteristic that represents the conditions existing when the signal input is zero. *Note:* The quiescent values of the parameters are not in general equal to the average values existing in the presence of the signal unless the characteristic is linear and the signal has no direct-current component. *See:* **operating point.** 328

quiescent supervisory system. *See:* **supervisory system (3) Quiescent.**

quiescent supervisory system (1) (power switchgear). A supervisory system that is normally alert but inactive, and transmits information or control signals only when a change in status occurs at the remote station or when a demand operation is initiated at the master station. 103

(2) (station control and data acquisition). A system which is normally alert but inactive, and transmits information only when a change in indication occurs at the remote station or when a command operation is initiated at the master station. *See:* **supervisory system.** 403

quiescent value (base magnitude), AQ (pulse transformers). The maximum value existing between pulses. 589

quiet automatic volume control. Automatic volume control that is arranged to be operative only for signal strengths exceeding a certain value, so that noise or other weak signals encountered when tuning between strong signals are suppressed. *See:* **radio receiver.** 328

quiet ground (health care facilities). A system of grounding conductors, insulated from portions of the conventional grounding of the power system, which interconnects the grounds of electric appliances for

the purpose of improving immunity to electromagnetic noise. 192

quieting sensitivity (frequency-modulation receivers). The minimum unmodulated signal input for which the output signal-noise ratio does not exceed a specified limit, under specified conditions. *See:* **radio receiver; receiver performance.** 181

quiet sun (radio wave propagation). The sun in the absence of unusual optical or radio frequency activity. 146

quiet tuning. A circuit arrangement for silencing the output of a radio receiver except when the receiver is accurately tuned to an incoming carrier wave. *See:* **radio receiver.** 328

quill drive. A form of drive in which a motor or generator is geared to a hollow cylindrical sleeve, or quill, or the armature is directly mounted on a quill, in either case, the quill being mounted substantially concentrically with the driving axle and flexibly connected to the driving wheels. *See:* **traction motor.** 328

quinary. *See:* **biquinary.**

quinhydrone electrode. *See:* **quinhydrone half cell.**

quinhydrone half cell. A half cell with an electrode of an inert etal (such as platinum gold) in contact with a solution saturated with quinhydrone. *Syn:* **quinhydrone electrode.** *See:* **electrochemistry.** 328

quotation board. A manually or automatically operated panel equipped to display visually the price quotations received by a ticker circuit. Such boards may provide displays of large size or may consist of small automatic units that ordinarily display one item at a time under control of the user. *See:* **telegraphy.** 328

quotient. *See:* **phasor product.**

quotient relay. A relay that operates in response to a suitable quotient of two alternating electrical input quantities. *See:* **relay.** 103

Q video signal (National Television System Committee (NTSC) color television). One of the two video signals (E'_I and E'_Q) controlling the chrominance in the NTSC system. *Note:* It is a linear combination of gamma-corrected primary color signals, E'_R, E'_G, and E'_B, as follows:

$$E'_Q = 0.41(E'_B - E'_Y) + 0.48(E'_R - E'_Y)$$
$$= 0.21E'_R - 0.52E'_G + 0.31E'_B$$

 18

R

rabbet, mounting (1)(packaging machinery). Any channel for holding wires, cables, or bus bars; designed expressly for, and used solely for, this purpose. 429
(2)(rotating machinery). A male or female pilot on a face or flange type of end shield of a machine, used for

mounting the machine with a mating rabbet. The rabbet may be circular, or of other configuration and need not be continuous. 63

raceway (1)(NESC). Any channel designed expressly and used solely for holding conductors. 494

(2)(raceway systems for Class 1E circuits for nuclear power generating stations). Any channel that is designed and used expressly for supporting wires, cables, or bus bars. Raceways consist primarily of, but are not restricted to, cable tray, conduits, and wireways.   513

(3) (electric system). Any channel for enclosing, and for loosely holding wires, cables, or busbars in interior work that is designed expressly for, and used solely for, this purpose. *Note:* Raceways may be of metal or insulating material and the term includes rigid metal conduit, rigid nonmetallic conduit, flexible metal conduit, electrical metallic tubing, underfloor raceways, cellular concrete-floor raceways, cellular metal-floor raceways, surface metal raceways, structural raceways, wireways and busways, and auxiliary gutters or moldings.   256

(4) (cable penetration fire stop qualification test) (Class 1E equipment and circuits). Any channel that is designed and used for supporting or enclosing wires, cable, or bus bars. Raceways consist primarily of, but are not restricted to, cable trays and conduits.
   131, 368

(5) (National Electrical Code). An enclosed channel designed expressly for holding wires, cables or busbars with additional functions as permitted in this Code.
   256

raceway penetration (raceway systems for Class 1E circuits for nuclear power generating stations). An opening for a raceway in a floor or wall to permit passage of cables from one side to the other. The raceway may or may not be continuous through the opening.   513

raceway system (raceway systems for Class 1E circuits for nuclear power generating stations). An integrated assembly of raceways, fittings, supports, accessories, and anchorages.   513

rachet demand (power operations). The maximum past or present demands which are taken into account to establish billings for previous or subsequent periods.
   516

rachet demand clause (power operations). A clause in a rate schedule which provides that maximum past or present demands be taken into account to establish billings for previous or subsequent periods.   516

rack (control boards, panels, and racks). A framework, constructed of rails or steel members, for mounting an assembly of modules for monitoring, measuring, and controlling remotely operated systems.   140

racon (navigation aid terms). A radar beacon which returns a coded signal providing identification of the beacon as well as range and bearing. *Syn:* **radar transponder beacon.** *See:* **radar beacon.**   526

rad (photovoltaic power system). An absorbed radiation unit equivalent to 100 ergs/gram of absorber. *See:* **photovoltaic power system.**   186

radar (navigation aid terms). A device for transmitting electromagnetic signals and receiving echoes from objects of interest (targets) within its volume of coverage. Presence of a target is revealed by its echo or its transponder reply. Additional information about a target provided by a radar includes one or more of the following: distance (range), by the elapsed time between transmissions of the signal and reception of the return signal; direction, by use of directive antenna patterns; rate of change of range, by measurement of Doppler shift; description or classification of target, by analysis of echoes and their variation with time. The name radar was originally an acronym for Radio Detection and Ranging. *Note:* Some radars can operate in a passive mode, in which the transmitter is turned off and information about targets is derived by receiving radiation emanating from the targets themselves or reflected by the targets from external sources.
   526

radar altimeter. *See:* **radio altimeter.**

radar astronomy (radio wave propagation). The study of extraterrestrial objects by transmission of artificially generated radio waves toward such objects and examination of the radio waves scattered or reflected from these objects.   146

radar beacon (navigation aid terms). A transponder used for replying to interrogations from a radar.
   526

radar bearing (navigation aid terms). A bearing obtained by a radar.   526

radar camouflage. The art, means, or result of concealing the presence of the nature of an object from radar detection by the use of coverings or surfaces that considerably reduce the radio energy reflected toward a radar. *See:* **radar.**   328

radar cross section (RCS) (1) (radar). A measure of the reflective strength of a radar target; usually represented by the symbol σ, measured in square meters, and defined as 4π times the ratio of the power per unit solid angle scattered in a specified direction to the power per unit area in a plane wave incident on the scatterer from a specified direction. More precisely, it is the limit of that ratio as the distance from the scatterer to the point where the scattered power is measured approaches infinity. Three cases are distinguished: (A) monostatic or backscattering RCS when the incident or pertinent scattering directions are coincident but opposite in sense, (B) forward-scattering RCS when the two directions and senses are the same, and (C) bistatic RCS when the two directions are different. If not identified, RCS is usually understood to refer to case (A). In all three cases, RCS is a function of frequency, transmitting and receiving polarizations, and target aspect angle (except for a sphere). *Syn:* **effective echoing area; backscattering (or forward-scattering, bistatic-scattering) cross section.**   13

(2) (antennas). For a given scattering object, upon which a plane wave is incident, that portion of the scattering cross section corresponding to a specified polarization component of the scattered wave. *See:* **scattering cross section.**   111

radar duplexing assembly. *See:* **receiver protector; transmit-receive switch; circulator; duplexer (radar).**

radar equation. A mathematical expression for primary radar which, in its basic form, relates radar parameters such as transmitter power, antenna gain, wavelength, effective echo area of the target, distance to the target, and receiver input power. The basic equation may be modified to take into account other factors, such as attenuation caused by a radome, attenuation due to atmospheric losses or precipitation, and various other losses and propagation effects. *Syn:* **radar range equation; range equation.** 13

radar fix (navigation aid terms). A position fix established by means of radar data. 526

radar letter designations. The radar letter designations are consistent with the recommended nomenclature of the International Telecommunications Union (ITU), as shown in Table 2, below. Note that the high frequency (HF) and the very high frequency (VHF) definitions are identical in the two systems. The essence of the radar nomenclature is to subdivide the existing ITU bands, in accordance with radar practice, without conflict or ambiguity. The letter band designations are not to be construed as being a substitute for the specific frequency limits of the frequency bands. The specific frequency limits are to be used when appropriate, but when a letter designation of a radar-frequency band is called for, those of Table 1, below, are to be used. The letter designations described in IEEE Std 521-1984 are designed for radar usage and are used in current practice. They are not meant to be used for other radio or telecommunication purposes, unless they pertain to radar. The letter designations for Electronic Countermeasure operations as described in Air Force Regulation No 55-44, Army Regulation No 105-86, and Navy OPNAV Instruction 3430.9B are not consistent with radar practice and are not used to describe radar-frequency bands. 44

radar performance figure. The ratio of the pulse power of the radar transmitter to the power of the minimum signal detectable by the receiver. 13

radar range equation. *See:* **radar equation.** 13

radar relay. An equipment for relaying the radar video and appropriate synchronizing signals to a remote location. 13

radar responder beacon. *See:* **racon.**

radar shadow. Absence of radar illumination because of an intervening reflecting or absorbing object; the shadow is manifested on the display by the absence of blips from targets in the shadow area. 13

radar transmitter. The transmitter portion of a radio detecting and ranging system. 111,181

radial (navigation)(navigation aid terms). One of a number of lines of position defined by an azimuthal navigation facility; the radial is identified by its bearing (usually the magnetic bearing) from the facility. 526

radial air gap (rotating machinery). See: **air gap (gap) (rotating machinery).**

radial-blade blower (rotating machinery). A fan made with flat blades mounted so that the plane of the blades passes through the axis of rotation of the rotor. *See:* **fan (blower) (rotating machinery).** 63

radial distribution feeder. *See:* **radial feeder.**

radial feeder. A feeder supplying electric energy to a substation or a feeding point that receives energy by no other means. *Note:* The normal flow of energy in such a feeder is in one direction only. *See:* **center of distribution.** 64

radially outer coil side. *See:* **bottom coil slot.**

radial magnetic pull (rotating machinery). The radial force acting between rotor and stator resulting from the radial displacement of the rotor from magnetic center. *Note:* Unless other conditions are specified, the value of radial magnetic pull will be for no load and rated voltage, and for rated no load field current and rated frequency as applicable. 63

radial power factor (paper-insulated power cable). The power factor of individual insulating tapes of a power cable as a function of the radial location of the insulating tapes through the insulation wall. 537

radial probable error (RPE). *See:* **circular probable error.**

radial system. A system in which independent feeders branch out radially from a common source of supply. *See:* **alternating-current distribution; direct-current distribution.** 64

radial-time-base display. *See:* **plan-position indicator.**

radial transmission feeder. *See:* **radial feeder.**

radial transmission line (waveguide). A pair of parallel conducting planes used for propagating waves whose phase fronts are concentric coaxial circular cylinders having their common axis normal to the planes; sometimes applied to tapered versions, such as biconical lines. 267

radial type. A unit substation which has a single step-down transformer and which has an outgoing section for the connection of one or more outgoing radial (stub end) feeders. 53

radial-unbalance torque (dynamically tuned gyro) (inertial sensor). The acceleration-sensitive torque caused by radial unbalance due to noncoincidence of the flexure axis and the center of mass of the rotor. Under constant acceleration, it appears as a rotating torque at the rotor spin frequency. When the gyro is subjected to vibratory acceleration along the spin axis at the spin frequency, this torque results in a rectified unbalanced drift rate. 46

(2) (laser-maser). A unit of angular measure equal to the angle subtended at the center of a circle by an arc whose length is equal to the radius of the circle. One (1) radian Δ 57.3 degrees; 2π radians = 360 degrees. 363

radian (metric practice). The plane angle between two radii of a circle which cut off on the circumference an arc equal in length to the radius. 21

radiance (1) (fiber optics). Radiant power, in a given direction, per unit solid angle per unit of projected area of the source, as viewed from that given direction. Radiance is expressed in watts per steradian per square meter. *See:* **brightness; conservation of radiance; radiometry.** 433

(2) (L) (laser-maser). Radiant flux or power output per unit solid angle unit area ($W \cdot sr^{-1} \cdot cm^{-2}$). 363

Table 1

Standard Radar-Frequency Letter Band Nomenclature

Band Designation	Nominal Frequency Range	Specific Frequency Ranges for Radar Based on ITU Assignments for Region 2, see Note (1)
HF	3 MHz–30 MHz	Note (2)
VHF	30 MHz–300 MHz	138 MHz–144 MHz
		216 MHz–225 MHz
UHF	300 MHz–1000 MHz (Note 3)	420 MHz–450 MHz (Note 4)
		890 MHz–942 MHz (Note 5)
L	1000 MHz–2000 MHz	1215 MHz–1400 MHz
S	2000 MHz–4000 MHz	2300 MHz–2500 MHz
		2700 MHz–3700 MHz
C	4000 MHz–8000 MHz	5250 MHz–5925 MHz
X	8000 MHz–12 000 MHz	8500 MHz–10 680 MHz
K_u	12.0 GHz–18 GHz	13.4 GHz–14.0 GHz
		15.7 GHz–17.7 GHz
K	18 GHz–27 GHz	24.05 GHz–24.25 GHz
K_a	27 GHz–40 GHz	33.4 GHz–36.0 GHz
V	40 GHz–75 GHz	59 GHz–64 GHz
W	75 GHz–110 GHz	76 GHz–81 GHz
		92 GHz–100 GHz
mm (Note 6)	110 GHz–300 GHz	126 GHz–142 GHz
		144 GHz–149 GHz
		231 GHz–235 GHz
		238 GHz–248 GHz (Note 7)

NOTES: (1) These frequency assignments are based on the results of the World Administrative Radio Conference of 1979. The ITU defines no specific service for radar, and the assignments are derived from those radio services which use radar: radiolocation, radio-navigation, meteorological aids, earth exploration satellite, and space research.

(2) There are no official ITU radiolocation bands at HF. So-called HF radars might operate anywhere from just above the broadcast band (1.605 MHz) to 40 MHz or higher.

(3) The official ITU designation for the ultra high frequency band extends to 3000 MHz. In radar practice, however, the upper limit is usually taken as 1000 MHz, L and S bands being used to describe the higher UHF region.

(4) Sometimes called P band, but use is rare.

(5) Sometimes included in L band.

(6) The designation mm is derived from *millimeter* wave radar, and is also used to refer to V and W bands when general information relating to the region above 40 GHz is to be conveyed.

(7) The region from 300 GHz–3000 GHz is called the submillimeter band.

Table 2

Comparison of Radar-Frequency Letter Band Nomenclature with ITU Nomenclature

Radar Nomenclature		International Telecommunications Union Nomenclature			
Radar Letter Designation	Frequency Range	Frequency Range	Band No	Adjectival Band Designation	Corresponding Metric Designation
HF	3 MHz–30 MHz	3 MHz–30 MHz	7	High frequency (HF)	Dekametric waves
VHF	30 MHz–300 MHz	30 MHz–300 MHz	8	Very high frequency (VHF)	Metric waves
UHF	300 MHz–1000 MHz				
L	1 GHz–2 GHz	0.3 GHz–3 GHz	9	Ultra high frequency (UHF)	Decimetric waves
S	2 GHz–4 GHz				
C	4 GHz–8 GHz				
X	8 GHz–12 GHz	3 GHz–30 GHz	10	Super high frequency (SHF)	Centimetric waves
K_u	12 GHz–18 GHz				
K	18 GHz–27 GHz				
K_a	27 GHz–40 GHz				
V	40 GHz–75 GHz	30 GHz–300 GHz	11	Extremely high frequency (EHF)	Millimetric Waves
W	75 GHz–110 GHz				
mm	110 GHz–300 GHz				

(3) (radiant intensity per unit area at a point on a surface and in a given direction (television). The quotient of the radiant intensity in the given direction of an infinitesimal element of the surface containing the point under consideration, by the area of the orthogonally projected area of the element on a plane perpendicular to the given direction. *Note:* The usual unit is the watt per steradian per square meter. This is the radiant analog of luminance.   18

(4) (light emitting diodes).$(L_e = d^2\phi_e/d_\psi(dA\cos\theta) = dI_e/(dA\cos\theta))$. At a point of the surface of a source, of a receiver, or of any other real or virtual surface, the quotient of the radiant flux leaving, passing through or arriving at an element of the surface surrounding the point, and propagated in the direction defined by an elementary cone containing the given direction, by the product of the solid angle of the cone and the area of the orthogonal projection of the element of the surface on a plane perpendicular to the given direction. *Note:* In the defining equation *theta* is the angle between the normal to the element of the surface and the direction of observation.   162

(5) (illuminating engineering)$L = d^2\Phi/[d\omega(dA\cdot\cos\theta)] = dI/(dA\cdot\cos\theta)$(in a direction at a point of the surface of a receiver, or of any other real or virtual surface) (illuminating engineering). Properly, this should be a second partial derivative since area and solid angle are independent variables. However, the symbol "d" is used due to the convenience in printing and typing. For this specific use, no possible errors or confusion are foreseen. This practice is in accord with the International Lighting Vocabulary (CIE No.17 (E-1.1.) 1970) and the practice of the National Bureau of Standards (NBS Technical Note 910-1, 1976). The quotient of the radiant flux leaving, passing through, or arriving at an element of the surface surrounding the point, and propagated in directions defined by an elementary cone containing the given direction, by the product of the solid angle of the cone and the area of the orthogonal projection of the element of the surface on a plane perpendicular to the given direction. *Note:* In the defining equation fI theta fR is the angle between the normal to the element of the surface and the given direction.   167

radian frequency (circuits and systems). The number of radians per unit time. The unit of time is generally the second and the radian frequency *omega* is therefore $2\pi f$ where f is the frequency in hertz.   67

radiant density (we = dQe/dV) (light emitting diodes). Radiant energy per unit volume; joules per m3.   162

radiant efficiency of a source of radiant flux (ξ_e) (light emitting diodes). The ratio of the total radiant flux to the forward power dissipation (total electrical lamp power input).   162

radiant emittance (fiber optics). Radiant power emitted into a full sphere (4π steradians) by a unit area of a source; expressed in watts per square meter. *Syn:* **radiant exitance.** *See:* **radiometry; radiant flux density at a surface.**   433

radiant energy (1) (fiber optics). Energy that is trans-

ferred via electromagnetic waves, that is, the time integral of radiant power; expressed in joules. *See:* **radiometry.**   433

(2) (light emitting diodes) (Qe). Energy traveling in the form of electromagnetic waves. It is measured in units of energy such as joules, ergs or kilowatt-hours.   162

(3) (laser maser). Energy emitted, transferred, or received in the form of radiation. Unit: joule (J).   363

radiant energy, Q (illuminating engineering). Energy traveling in the form of electromagnetic waves. It is measured in units of energy such as joules or kilowatt hours.   167

radiant energy density, $w = dQ/dV$ (illuminating engineering). Radiant energy per unit volume; for example, joules per cubic meter.   167

radiant exitance. *See:* **radiant emittance.**

radiant exposure (H) (laser-maser). Surface density of the radiant energy received. Unit: $J\cdot cm^{-2}$.   363

radiant flux (1) (light emitting diodes) ($\phi_e = dQe/dt$). The time rate of flow of radiant energy. *Note:* It is expressed preferably in watts, or in ergs per second.   18,178,162,167

(2) (television). Power emitted, transferred, or received in the form of radiation. *Note:* It is expressed preferably in watts.   18

(3) (laser-maser) (radiant power). Power emitted, transferred, or received in the form of radiation. Unit: watt (W).   363

radiant flux ϕ, radiant power (laser-maser). Power emitted, transferred, or received in the form of radiation. Unit: W.   363

radiant flux density at a surface ($M_e = d\Phi_e/dA$, $E_e = d\Phi_e/dA$). The quotient of radiant flux at the element of surface of the area of that element: that is, watts per cm2. When referring to radiant flux emitted from a surface, this has been called **radiant emittance*** (symbol: **M**). The preferred term for radiant flux leaving a surface is **radiant exitance** (symbol: M). When referring to radiant flux incident on a surface, it is called **irradiance** (symbol: E).   162

radiant flux density at a surface, $d\Phi/dA$ (illuminating engineering). The quotient of radiant flux at an element of surface to the area of that element; e.g., watts per square meter. When referring to radiant flux emitted from a surface, this has been called radiant emittance (a deprecated term) (symbol: *M*). The preferred term for radiant flux leaving a surface is radiant exitance (symbol: *M*). The radiant exitance per unit wavelength interval is called spectral radiant exitance. When referring to radiant flux incident on a surface it is called irradiance (symbol: *E*).   167

radiant gain (optoelectronic device). The ratio of the emitted radiant flux to the incident radiant flux. *Note:* The emitted and incident radiant flux are both determined at specified ports. *See:* **optoelectronic device.**   191

radiant heater. A heater that dissipates an appreciable part of its heat by radiation rather than by conduction or convection. *See:* **appliances (including portable).**

radiant incidence. *See:* **irradiance.** 433

radiant intensity (1) (fiber optics). Radiant power per unit solid angle, expressed in watts per steradian. *See:* **intensity; radiometry.** 433

(2) (of a source, in a given direction) (television). The quotient of the radiant power emitted by a source, or by an element of source, in an infinitesimal cone containing the given direction, by the solid angle of that cone. *Note:* It is expressed preferably in watts per steradian. 18

(3) (light emitting diodes). ($I_e = d\Phi_e/d\omega$). The radiant flux proceeding from the source per unit solid angle in the direction considered; that is watts per steradian. 162

(4) (laser-maser). Quotient of the radiant flux leaving the source, propagated in an element of solid angle containing the given direction, by the element of solid angle. Unit: watt per steradian ($W \cdot sr^{-1}$). 363

(5) (illuminating engineering). ($I = d\Phi/d\omega$). The radiant flux proceeding from a source per unit solid angle in the direction considered; for example, watts per steradian. *Note:* Mathematically a solid angle must have a point as its apex; the definition of radiant intensity, therefore, applies strictly only to a point source. In practice, however, radiant energy emanating from a source whose dimensions are negligible in comparison with the distance from which it is observed may be considered as coming from a point. Specifically, this implies that with change of distance (A) the variation in solid angle subtended by the source at the receiving point approaches $1/(distance)^2$ and that (B) the average radiance of the projected source area as seen from the receiving point does not vary appreciably. 167

radiant power (fiber optics). The time rate of flow of radiant energy, expressed in watts. The prefix is often dropped and the term "power" is used. Colloquial synonyms: flux; optical power; power; radiant flux. *See:* **radiometry.** 433

radiant sensitivity (phototube). *See:* **sensitivity, radiant.**

radiated interference (electromagnetic compatibility). Radio interference resulting from radiated noise or unwanted signals. *See:* **electromagnetic compatibility.** 199

radiated noise (radio noise from overhead power lines and substations). Radio noise energy in the form of an electromagnetic field including both the radiation and induction components of the field. 509

radiated power output (transmitter performance). The average power output available at the antenna terminals, less the losses of the antenna, for any combination of signals transmitted when averaged over the longest repetitive modulation cycle. *See:* **audiofrequency distortion.** 181

radiated radio noise. Radio-noise energy in the form of an electromagnetic field including both the radiation and induction components of the field. 418

radiated spurious emission power (land-mobile communication transmitters). Any part of the spurious emission power output radiated from the transmission enclosure, independent of any associated transmission lines or antenna, in the form of an electromagnetic field composed of variations of the intensity of electric and magnetic fields. 444

radiating element. A basic subdivision of an antenna that in itself is capable of effectively radiating or receiving radio waves. *Note:* Typical examples of a radiating element are a slot, horn, or dipole antenna. *See:* **antenna.** 111

radiating element (antennas). A basic subdivision of an antenna which in itself is capable of radiating or receiving radio waves. *Note:* Typical examples of a radiating element are a slot, horn, or dipole antenna. 111

radiating far field region (land-mobile communication transmitters). Measurement is performed at or beyond a distance of 3λ, but not less than 1 meter (m). *See:* **far field region.** 444

radiating near-field region (1)(antennas). That portion of the near-field region of an antenna between the far-field and the reactive portion of the near-field region, wherein the angular field distribution is dependent upon distance from the antenna. *Notes:* (A) If the antenna has a maximum overall dimension which is not large compared to the wavelength, this field region may not exist. (B) For an antenna focused at infinity, the radiating near-field region is sometimes referred to as the Fresnel region on the basis of analogy to optical terminology. 111

(2)(land-mobile communication transmitters). Measurement is limited to the region external to the induction field and extending to the outer boundary of the reactive field which is commonly taken to exist at a distance of $\lambda/2\pi$. Either the electric or magnetic component of the radiated energy may be used to determine the magnitude of power present. *See:* **near field region.** 444

radiation (1) (nuclear) (nuclear work). The usual meaning of radiation is extended to include moving nuclear particles, charged or uncharged. 190,125,96

(2) (data transmission). In radio communication, the emission of energy in the form of electromagnetic waves. The term is also used to describe the radiated energy. 59

radiation angle (fiber optics). Half the vertex angle of the cone of light emitted by a fiber. *Note:* The cone is usually defined by the angle at which the far-field irradiance has decreased to a specified fraction of its maximum value or as the cone within which can be found a specified fraction of the total radiated power at any point in the far field. *Syn:* output angle. *See:* **acceptance angle; far-field region; numerical aperture.** 433

radiation counter. An instrument used for detecting or measuring radiation by counting action. 125

radiation detector. Any device whereby radiation produces some physical effect suitable for observation and/or measurement. *See:* **anticoincidence (radiation counters).** 190

radiation efficiency (antenna)(data transmission) .
The ratio of (1) the total power radiated by an antenna
to (2) the net power accepted by the antenna from the
connected transmitter. *See:* **antenna.**   111, 59
radiation, electromagnetic (antenna). The emission of
energy in the form of electromagnetic waves.

111
**radiation induced data loss characteristic (metal-ni-
tride-oxide field-effect transistor).** The collection of
threshold voltage data as a function of total dose or
dose rate after initial high-conduction or low-conduction threshold voltage levels had been written into the
device.   386
radiation intensity (1)(data transmission). The radiation intensity in a given direction is the power radiated
from an antenna per unit solid angle in that direction.

59
(2)(in a given direction). The power radiated from an
antenna per unit solid angle in that direction. *See:*
antenna.   246, 111
radiation lobe (antenna pattern). A portion of the radiation pattern bounded by regions of relatively weak
radiation intensity. *See:* **antenna.**   246,111
radiation loss (1) (transmission system). That part of
the transmission loss due to radiation of radio-frequency power. *See:* **waveguide.**   267
(2) (waveguide). A power loss due to electromagnetic
radiation leaving a network.   267
radiation mode (fiber optics). In an optical waveguide,
a mode whose fields are transversely oscillatory
everywhere external to the waveguide, and which exists even in the limit of zero wavelength. Specifically,
a mode for which

$$\beta = [n^2(a)k^2 - (\text{script } l/a)^2]^{1/2}[zc]$$

where β is the imaginary part (phase term) of the axial
propagation constant, script l is the azimuthal index of
the mode, $n(a)$ is the refractive index at $r=a$, the core
radius, and k is the free-space wavenumber, $2\pi/\lambda$,
where λ is the wavelength. Radiation modes correspond to refracted rays in the terminology of geometric optics. *Syn:* **unbound mode.** *See:* **bound mode;
leaky mode; mode; refracted ray.**   433
radiation pattern (1)(antenna pattern). A graphical
representation of the radiation properties of the antenna as a function of space coordinates. *Notes:* (A) In
the usual case the radiation pattern is determined in
the far-field region and is represented as a function of
directional coordinates. (B) Radiation properties include power flux density, field strength, phase, and
polarization.   111
(2)(antennas). The spatial distribution of a quantity
which characterizes the electromagnetic field generated by an antenna. *Notes:* (1) The distribution can be
expressed as a mathematical function or as a graphical
representation. (2) The quantities which are most oft-
en used to characterize the radiation from an antenna
are proportional to or equal to power flux density,
radiation intensity, directivity, phase, polarization,
and field strength. (3) The spatial distribution over any
surface or path is also an antenna pattern. (4) When

the amplitude or relative amplitude of a specified
component of the electric field vector is plotted
graphically, it is called an amplitude pattern, field
pattern, or voltage pattern. When the square of the
amplitude or relative amplitude is plotted, it is called
a power pattern. (5) When the quantity is not specified, an amplitude or power pattern is implied. *Syn:*
antenna pattern.   111
(3)(fiber optics). Relative power distribution as a
function of position or angle. *Notes:* (1) Near-field
radiation pattern describes the radiant emittance (W
$\cdot m^{-2}$) as a function of position in the plane of the exit
face of an optical fiber. (2) Far-field radiation pattern
describes the irradiance as a function of angle in the
far-field region of the exit face of an optical fiber. (3)
Radiation pattern may be a function of the length of
the waveguide, the manner in which it is excited, and
the wavelength. *See:* **far-field region; near-field region.**   433
radiation pattern cut (antennas). Any path on a surface
over which a radiation pattern is obtained. *Note:* For
far-field patterns the surface is that of the radiation
sphere. For this case the path formed by the locus of
points for which θ is a specified constant and ϕ is a
variable is called a conical cut. The path formed by the
locus of points for which ϕ is a specified constant and
θ is a variable is called a great circle cut. The conical
cut with θ equal to 90° is also a great circle cut. A spiral
path which begins at the north pole ($\theta=0°$) and ends
at the south pole ($\theta=180°$) is called a spiral cut.

111
radiation protection guide (electrobiology). Radiation
level that should not be exceeded without careful consideration of the reasons for doing so. *See:* **electrobiology.**   322
radiation pyrometer (radiation thermometer). A pyrometer in which the radiant power from the object or
source to be measured is utilized in the measurement
of its temperature. The radiant power within wide or
narrow wavelength bands filling a definite solid angle
impinges upon a suitable detector. The detector is
usually a thermocoupler or thermopile or a bolometer
responsive to the heating effect of the radiant power,
or a photosensitive device connected to a sensitive
electric instrument. *See:* **electric thermometer (temperature meter).**   328
radiation resistance (antenna). The ratio of the power
radiated by an antenna to the square of the rms antenna current referred to a specified point. *Note:* This
term is of limited utility in lossy media.   111
radiation sphere (for a given antenna). A large sphere
whose center lies within the volume of the antenna
and whose surface lies in the far field of the antenna,
over which quantities characterizing the radiation
from the antenna are determined. *Notes:* (1) The location of points on the sphere are given in terms of the
θ and ϕ coordinates of a standard spherical coordinate system whose origin coincides with the center of
the radiation sphere. (2) If the antenna has a spherical
coordinate system associated with it, then it is desirable that its coordinate system coincide with that of the
radiation sphere.   111

radiation thermometer. *See:* **radiation pyrometer.**

radiation trapping (laser-maser). The suppression or delay of fluorescence in an optically thick absorbing medium resulting from **absorption** and re-emission.   363

radiative relaxation time (laser-maser). The relaxation time which would be observed if only processes involving the radiation of electromagnetic energy were effective in producing relaxation.   363

radiator (1) (illuminating engineering). An emitter of radiant energy.   167
(2) (telecommunication). Any antenna or radiating element that is a discrete physical and functional entity.   111

radio-acoustic ranging (navigation aid terms). Determining distance by a combination of radio and sound. *Syn:* **echo ranging.**   526

radioactive check source (liquid-scintillation counters). A radioactive sample used to monitor the operational status of an instrument. The approximate activity should be known.   498

radioactivity standard (sodium iodide detector). A radioactivity standard, as used in this text, is either a radioactivity standard that has been certified by a laboratory recognized as a country's National Standardizing Laboratory for radioactivity measurements or a radioactivity standard that has been obtained from a supplier who participates in measurement assurance activities with the National Standardizing Laboratory when such standards are available. In such measurement assurance activities, the supplier's calibration value should agree with the National Standardizing Laboratory value within the overall uncertainty stated by the supplier in its certification of the same batch of sources or in its certification of similar sources.   423

radio altimeter (navigation aid terms). An altimeter using radar principles for height measurement. Height is determined by measurement of propagation time of a radio signal transmitted from the vehicle and reflected back to the vehicle from the terrain below. *Syn:* **radar altimeter.**   526

radio astronomy (radio wave propagation). The branch of astronomy dealing with the passive reception and analysis of electromagnetic radiations of radio wavelength from extraterrestrial sources.   146

radio-autopilot coupler (navigation aid terms). Equipment providing means by which electrical signals from navigation receivers control the vehicle autopilot.   526

radio beacon (navigation aid terms). A facility, usually a nondirectional radio station, emitting identifiable signals intended for radio direction finding observations. *See:* **nondirectional beacon.**   526

radio-beacon buoy (navigation aid terms). A buoy equipped with a marker-radio beacon. *See:* **buoy.**   526

radio broadcasting. Radio transmission intended for general reception. *See:* **radio transmission.**   111,240

radio channel (data transmission) (antennas). A band of frequencies of a width sufficient to permit its use for radio communication. *Note:* The width of the channel depends on the type of transmission and the tolerance for the frequency of emission. Normally allocated for radio transmission in a specified type of service or by a specified transmitter.   111, 59

radio circuit. A means for carrying out one radio communication at a time in either or both directions between two points. *See:* **radio channel; radio transmission.**   328

radio compass. A direction-finder used for navigational purposes. *See:* **radio navigation.**   328

radio compass indicator. A device that, by means of a radio receiver and rotatable loop antenna, provides a remote indication of the relationship between a radio bearing and the heading of the aircraft.   328

radio compass magnetic indicator. A device that provides a remote indication of the relationship between a magnetic bearing, radio bearing, and the aircraft's heading.   328

radio control. The control of mechanism or other apparatus by radio waves. *See:* **radio transmission.**   328

radio detection (radio warning). The detection of the presence of an object by radiolocation without precise determination of its position. *See:* **radio transmission.**   328

radio direction-finder (RDF)(navigation aid terms). A device used to determine the direction of arrival of radio signals. *Syn:* **directional finder (DF).**   526

radio direction finding (navigation aid terms). A procedure for determining the bearing, at a receiving point, of the source of a radio signal by observing the direction of arrival and other properties of the signal.   526

radio distress signal (SOS). Radiotelegraph distress signal consists of the group . . . . - - - . . . . in Morse code, transmitted on prescribed frequencies. The radiotelephone distress signal consists of the spoken words **May Day (m'aidez = help me).** *Note:* By international agreement, the effect of the distress signal is to silence all radio traffic that may interfere with distress calls.   328

radio disturbance (electromagnetic compatibility). An electromagnetic disturbance in the radio-frequency range. *See:* **radio interference; radio noise.**   199

radio Doppler. The direct determination of the radial component of the relative velocity of an object by an observed frequency change due to such velocity. *See:* **radio transmission.**   328

radio fadeout (Dellinger effect). A phenomenon in radio propagation during which substantially all radio waves that are normally reflected by ionospheric layers in or above the E region suffer partial or complete absorption. *See:* **radiation.**   328

radio field strength (radio wave propagation). The electric or magnetic field strength at radio frequency.   146

radio frequency (1) (data transmission). (A) (Loosely). The frequency in the portion of the electromagnetic spectrum that is between the audio-frequency portion

and the infrared portion. (B) A frequency useful for radio transmission. *Note:* The present practicable limits of radio frequency are roughly 10 kHz (kilohertz) to 100 000 MHz (megahertz). Within this frequency range electromagnetic radiation may be detected and amplified as an electric current at the wave frequency.
59

(2) (radio wave propagation). A frequency at which electromagnetic radiation may be detected and amplified as an electric current at the wave frequency.
146

radio-frequency alternator. A rotating-type generator for producing radio-frequency power.   111

radio-frequency attenuator (signal-transmission system). A low-pass filter that substantially reduces the radio-frequency power at its output relative to that at its input, but transmits lower-frequency signals with little or no power loss. *See:* **signal.**   188

radio-frequency converter (industrial electronics). A power source for producing electric power at a frequency of 10 kilohertz and above.   114

radio-frequency generator (1) (signal-transmission system). A source of radio-frequency energy.
188

(2) (induction heating). A power source for producing electric power at a frequency of 10 kilohertz and above.   14

radio-frequency generator, electron tube type (induction and dielectric usage). A power source comprising an electron-tube oscillator, an amplifier if used, a power supply and associated control equipment. *See:* **Colpitts oscillator; Hartley oscillator; magnetron; tuned grid-tuned plate oscillator.**   14

radio frequency protection guides (RFPG)(radio frequency electromagnetic fields). The radio frequency field strengths or equivalent plane wave power densities which should not be exceeded without (1) careful consideration of the reasons for doing so, (2) careful estimation of the increased energy deposition in the human body, and (3) careful consideration of the increased risk of unwanted biological effects.   450

radio-frequency pulse. A radio-frequency carrier amplitude modulated by a pulse. The amplitude of the modulated carrier is zero before and after the pulse. *Note:* Coherence of the carrier (with itself) is not implied.   254

radio-frequency switching relay. A relay designed to switch frequencies that are higher than commercial power frequencies with low loss.   341

radio-frequency system loss (mobile communication). The ratio expressed in decibels of (1) the power delivered by the transmitter to its transmission line to (2) the power required at the receiver-input terminals that is just sufficient to provide a specified signal-to-noise ratio at the audio output of the receiver. *See:* **mobile communication system.**   181

radio-frequency transformer. A transformer for use with radio-frequency currents. *Note:* Radio-frequency transformers used in broadcast receivers are generally shunt-tuned devices that are tunable over a relatively broad range of frequencies. *See:* **radio transmission.**
197

radio gain (radio wave propagation). Of a radio system, the reciprocal of the system loss.   146

radio horizon (radio wave propagation)(data transmission)(of an antenna). Of The locus of the farthest points at which direct rays from the antenna become tangential to the planetary surface. *Note:* On a spherical surface the horizon is a circle. The distance to the horizon is affected by atmospheric reflection.
146, 111

radio-influence field (RIF) (electromagnetic compatibility). Radio-influence field is the radio noise field emanating from an equipment or circuit, as measured using a radio noise meter in accordance with specified methods. *See:* **electromagnetic compatibility.**
197

radio-influence tests (power switchgear). Tests that consist of the application of voltage and the measurement of the corresponding radio-influence voltage produced by the device being tested.   103, 443

radio influence voltage (RIV)(1)(metal-oxide surge arresters for ac power circuits)(surge arrester). A high-frequency voltage, generated by all sources of ionization current, that appears at the terminals of electric-power apparatus or on power circuits.
583, 430

(2) (outdoor apparatus bushings). A high-frequency voltage generated as a result of ionization, which may be a propagated by conduction, induction, radiation or a combined effect of all three.   168

(3) (high voltage ac cable terminations). The radio noise appearing on conductors of electric equipment or circuits, as measured using a radio-noise meter as a two-terminal voltmeter in accordance with specified methods.   4

(4) (overhead-power-line corona and radio noise). The radio frequency voltage appearing on conductors of electrical equipment or circuits, as measured using a radio noise meter as a two-terminal voltmeter in accordance with specified methods (generally termed conducted measurements) in NEMA 107-1964 (R1971,R1976), Methods of Measurement of Radio Influence Voltage (RIV) of High Voltage Apparatus.
411

(5) (power and distribution transformer). A radio frequency voltage generally produced by partial discharge and measured at the equipment terminals for the purpose of determining the electromagnetic interference effect of the discharges. *Notes:* (A) "RIV" can be measured with a coupled radio interference measuring instrument and is commonly measured at approximately 1 MHz, although a wide frequency range is involved. (B) "RIV" values are often used as an "index" of "partial discharge" intensity. (C) The RIV of equipment was historically measured to determined the influence of energized equipment on radio broadcasting, hence - RIV.   53

radio interference (overhead-power-line corona and radio noise). Impairment of the reception of a wanted radio signal caused by an unwanted radio signal or a radio disturbance.   411

radio interferometer (radio wave propagation). A type

of radio telescope that uses methods involving the interference of two or more beams of radiation received by means of physically separated collecting elements to achieve high angular resolution of the brightness temperature distribution of a radio source. 146

radiolocation (navigation aid terms). Position determination by means of radio aids for purposes other than those of navigation. 526

radio magnetic indicator (RMI)(navigation aid terms). A combined indicating instrument which converts omnibearing indications to a display resembling an ADF (automatic direction finder) display, one in which the indicator points toward the omnirange station; it combines omnibearing, vehicle heading, and relative bearing. 526

radiometric sextant (navigation aid terms). An instrument which measures the direction to a celestial body by detecting and tracking the nonvisible natural radiation of the body; such radiation includes radio, infrared, and ultraviolet. 526

radiometry (fiber optics). The science of radiation measurement. The basic quantities of radiometry are listed below. 433

radio navigation (navigation aid terms). Navigation based upon the reception of radio signals. 526

radio noise (1)(radio noise from overhead power lines and substations). Any unwanted disturbance within the radio frequency band, such as undesired electromagnetic waves in any transmission channel or device. 509

(2). An electromagnetic noise that may be superimposed upon a wanted signal and is within the radio-frequency range. For the purposes of this standard, an electromagnetic disturbance of a sinusoidal character is also considered radio noise. 418

(3) (overhead-power-line corona and radio noise). Any unwanted disturbance within the radio frequency band, such as undesired electric waves in any transmission channel or device. *See:* **IEEE Std 430-1976, Procedures for Measurement of Radio Noise from Overhead Power Lines.** 36, 411

radio noise field strength (overhead-power-line corona and radio noise). A measure of the field strength at a point (as a radio receiving station) of electromagnetic waves of an interfering character. *Notes:* (1) In practice the quantity measured is not the field strength of the interfering waves but some quantity that is proportional to, or bears a known relation to, the field strength. (2) It is commonly measured in average microvolts, quasi-peak microvolts, peak microvolts, or peak microvolts in a unit bandwidth per meter, according to which detector function of a radio noise meter is used. 411

radiophare (navigation aid terms). A term often used in international terminology, meaning radio beacon. 526

radio propagation path (mobile communication). For a radio wave propagating from one point to another, the great-circle distance between the transmitter and receiver antenna sites. *See:* **mobile communication system.** 181

radio proximity fuse. A radio device contained in a missile to detonate it within predetermined limits of distance from a target by means of electromagnetic interactions with the target. *See:* **radio transmission.** 111

RADIOMETRIC TERMS

TERM NAME	SYMBOL	QUANTITY	UNIT
Radiant energy	Q	Energy	joule(J)
Radiant power Syn: optical power	ϕ	Power	watt(W)
Irradiance	E	Power incident per unit area (irrespective of angle)	$W \cdot m^{-2}$
Spectral irradiance	E λ	Irradiance per unit wave length interval at a given wave length	$W \cdot m^{-2} \cdot nm^{-1}$
Radiant emittance Syn: radiant excitance	W	Power emitted (into a full sphere) per unit area	$W \cdot m^{-1}$
Radiant intensity	I	Power per unit solid angle	$W \cdot sr^{-1}$
Radiance	L	Power per unit angle per unit projected area	$W \cdot sr^{-1} \cdot m^{-2}$
Spectral radiance	L λ	Radiance per unit wave length interval at a given wave length	$W \cdot sr^{-1} \cdot m^{-2} \cdot nm^{-1}$

radio range (navigation aid terms). A radio facility which provides radial lines of position by having characteristics in its emission which are convertible to bearing information and useful in the lateral guidance of aircraft.   526

radio range-finding. The determination of the range of an object by means of radio waves. *See:* **radio transmission.**   328

radio receiver. A device for converting radio-frequency power into perceptible signals.   181

radio relay system (radio relay). A point-to-point radio transmission system in which the signals are received and retransmitted by one or more intermediate radio stations. *See:* **radio transmission.**   328

radio shielding. A metallic covering in the form of conduit and electrically continuous housings for airplane electric accessories, components, and wiring, to eliminate radio interference from aircraft electronic equipment.   328

radiosonde. An automatic radio transmitter in the meteorological-aids service, usually carried on an aircraft, free balloon, kite, or parachute, that transmits meteorological data. *See:* **radio transmitter.**   111

radio source (radio wave propagation). A celestial object or region that emits radio waves.   146

radio spectrum (radio wave propagation). The radio frequency portion of the electromagnetic spectrum. The frequency ranges are: *E*xtreme *F*requency (EHF), 300 to 30 GHz; *S*uper HF, 30 to 3 GHz; *U*ltra HF, 3 GHz to 300 MHz; *V*ery HF, 300 to 30 MHz; HF, 30 to 3 MHz; *M*edium *F*requency, 3 MHz to 300 kHz; *L*ow F, 300 to 30 kHz; *V*ery LF, 30 to 3 kHz; *E*xtreme LF, 3 kHz to 3 Hz; *U*ltra LF, lower than 3 Hz.   146

radio star (communication satellite). A discrete source in the celestial sphere emitting electrical random noise. *See:* **background noise.**   85

radio station. A complete assemblage of equipment for radio transmission or reception, or both. *See:* **radio transmission.**   328

radio telescope (radio wave propagation). An instrument used to collect and detect radio emission from an object or region at which it is pointed.   146

radio transmission. The transmission of signals by means of radiated electromagnetic waves other than light or heat waves.   328

radio transmitter. A device for producing radio-frequency power, for purposes of radio transmission.   111

radio warning. *See:* **radio detection.**

radio wave (radio wave propagation). An electromagnetic wave of radio frequency. *Note:* Current usage includes frequencies up to 3000 GHz.   146

radio wave propagation (radio wave propagation). The transfer of energy by electromagnetic radiation at radio frequencies.   146

radix (1)(mathematics of computing). A quantity whose successive integer powers are the implicit multipliers of the sequence of digits that represent a number in some positional notation systems. For example, if the radix is 5, then 143.2 means 1 times 5 to the second power, plus 4 times 5 to the first power, plus 3 times 5 to the zero power, plus 2 times 5 to the minus-one power. *Syn:* **base.**   564

(2)(radix-independent floating-point arithmetic). The base for the representation of floating point numbers   588

radix complement (mathematics of computing). The complement obtained by subtracting each digit of a given numeral from the largest digit in the numeration system, then adding 1 to the least significant digit of the result and executing any required carries. For example, twos complement in binary notation, tens complement in decimal notation. *See:* **diminished-radix complement.**   564

radix-minus-one complement. A numeral in radix notation that can be derived from another by subtracting each digit from one less than the radix, for example, nines complement in decimal notation, ones complement in binary notation.   255,77

radix point (mathematics of computing). In positional notation, the character, expressed or implied, that separates the integral part of a numerical expression from the fractional part. For example, binary point, decimal point, hexadecimal point, or octal point.   564

radome (antennas). A cover usually intended for protecting an antenna from the effects of its physical environment without degrading its electrical performance.   111

rads in Si (metal-nitride-oxide field-effect transistor). Amount of radiation measured by its ionizing effect in silicon; 1 rad equals 100 erg of energy deposited in a gram of irradiated solid. This number can be translated into the density of electron-hole pairs (ehp/cm^3) by the following operation:

$$n_{eh} \text{ [ehp/volume]} = \gamma \text{ [energy/mass]} \times \rho \text{ [mass/volume]} \times N_{eh} \text{ [ehp/energy]},$$

where

n_{eh} = volume density of ehp,

γ = total radiation dose as energy dissipated per unit mass, typically expressed in rads,

d = density of solid in mass/volume,

N_{eh} = number of ehp's created per energy dissipated.

In many solids, $N_{eh} \sim 1/bEg$, and in silicon particularly, $b = 3.6[\text{ehp}]^{-1}$. *Note:* [15], and Eg = 1.0eV. This means that one ehp is created per 3.6 electron-volts of energy dissipated. In order to permit the use of the total dose expressed in rads directly, use is made of the identity that 1 erg = 6.2×10^{11} eV. From this $N_{eh} \sim 6.2 \times 10^{11}$ [eV/erg] [ehp]/3.6[eV], and since 1 rad = 100 ergs/g, this makes $N_{eh} \sim 1.7 \times 10^{13}$ [ehp/g rads]. Thus, for silicon, n_{eh} [ehp/cm^3] = $1.7 \ 10^{13} \times \gamma$ [rads] $\times \rho$ [g/cm^3].   386

rail clamp (mining). A device for connecting a conductor or a portable cable to the track rails that serve as the return power circuit in mines. *See:* **mine feeder circuit.**   328

rain clutter (radar). Return from rain which impairs or obscures return from targets.   13

rainproof (National Electrical Code) (power and distribution transformer). So constructed, protected, or treated as to prevent rain from interfering with the successful operation of the apparatus under specified test conditions. 53, 256

raintight (1) (National Electrical Code). So constructed or protected that exposure to a beating rain will not result in the entrance of water under specified test conditions. 256
(2) (power and distribution transformer). So constructed or protected as to exclude rain under specified test conditions. 53

raise-lower control point. *See: supervisory control point, raise-lower.*

Raman-Nath region (acousto-optic device). The region that occurs when the Bragg Region inequality is reversed, that is $L < n\Lambda^2/\lambda_0$. The angle of incidence is generally zero degrees, and light is diffracted into many diffraction orders. 82

ramp (1) (thyristor). A controlled change in output at a predetermined linear rate, from one value to another. 445
(2) (railway control). A roadway element consisting of a metal bar of limited length, with sloping ends, fixed on the roadway, designed to make contact with and raise vertically a member supported on the vehicle. 328
(3) (automatic control). *See: signal, unit-ramp.*
(4) (pulse terms) (single transition). A linear feature. 254

ramp response (1) (null-balancing electric instrument). A criterion of the dynamic response of an instrument when subjected to a measured signal that varies at a constant rate. *See: accuracy rating (instrument).* 295
(2) (automatic control).* *See: response, ramp-forced.* 56

*Deprecated

ramp response time (null-balancing electric instrument). The time lag, expressed in seconds, between the measured signal and the equivalent positioning of the end device when the measured signal is varying at constant rate. *See: accuracy rating (instrument).* 295

ramp response-time rating (null-balancing electric instrument). The maximum ramp response time for all rates of change of measured signal not exceeding the average velocity corresponding to the span step-response-time-rating of the instrument when the instrument is used under rated operating conditions. *Example:* If the span step-response-time-rating is 4 seconds, the ramp response-time rating shall apply to any rate of change of measured signal not exceeding 25 percent of span per second. *See: accuracy rating (instrument).* 295

ramp shoe. *See: shoe.*

random (1) (data transmission). A condition not localized in time or frequency. 59
(2) (automatic control). Describing a variable whose value at a particular future instant cannot be predicted exactly, but can only be estimated by a probability distribution function. 56

random access (computing systems). (1) Pertaining to the process of obtaining data from, or placing data into storage where the time required for such access is independent of the location of the data most recently obtained or placed in storage. (2) Pertaining to a storage device in which the access time is effectively independent of the location of the data. 255,77,54

random access programming (test, measurement and diagnostic equipment). Programming without regard for the sequence required for access to the storage position called for in the program. 54

random drift rate (gyro). The non-systematic time varying component of drift rate under specified operating conditions. It is expressed as an rms value or standard deviation of angular displacement per unit time. 46

random errors (navigation aid terms). Those errors which cannot be predicted except on a statistical basis. 526

random failure (1) (class 1E static battery chargers and inverters). Any failure whose cause or mechanism, or both, make its time of occurrence unpredictable. 408
(2) (reliability). *See: failure, random.*

random failures (station control and data acquisition). The pattern of failures for equipments that have passed out of their infant mortality period and have not reached the wear-out phase of their operating life-time. The reliability of an equipment in this period may be computed by the equation: $R = e^-\mu t$ where λ is the failure rate and t is the time period of interest. 403

random-incidence microphone (audible noise measurement). A microphone which has been designed to have a flat frequency response in a diffuse sound field where sound waves are arriving equally from all directions. 462

random noise (1)(control of system electromagnetic compatibility)(overhead-power-line corona and radio noise)(radio-noise emission). Noise that comprises transient disturbances ocurring at random. *Note:* The part of the noise that is unpredictable except in a statistical sense. The term is most frequently applied to the limiting case where the number of transient disturbances per unit time is large, so that the spectral characteristics are the same as those of thermal noise. Thermal noise and shot noise are special cases of random noise. 495, 411, 418, 9, 24
(2) (fluctuation noise) (data transmission). The noise which comprises transient disturbances occurring at random. The term is more frequently applied to the limiting case where the number of transient disturbances per unit time is large, so that the spectral characteristics are the same as those of thermal noise. Thermal noise and shot noise are special cases of random noise. 59

random noise bandwidth (overhead-power-line corona and radio noise). The width in hertz of a rectangle having the same area and maximum amplitude as the square of the amplifier response to a sinusoidal input. 411

random number. *See:* **pseudorandom number sequence.**

random paralleling (rotating machinery). Paralleling of an alternating-current machine by adjusting its voltage to be equal to that of the system, but without adjusting the frequency and phase angle of the incoming machine to be sensibly equal to those of the system. *See:* **asynchronous machine.**   63

random photon summing (sodium iodide detector). The simultaneous detection of two or more photons originating from the disintegrations of more than one atom.   423

random separation (NESC). Installed with no deliberate separation.   494

random summing (germanium detectors). The simultaneous detection of two or more photons originating from the disintegration of more than one atom.   397

random walk (gyro) (inertial sensor). (1) (angle). The angular error buildup with time due to white noise in angular rate. This error is typically expressed in degrees per square root of hour $(°/\sqrt{h})$ (2) (rate). The drift rate error buildup with time due to white noise in angular acceleration. This error is typically expressed in degrees per hour per square root of hour $[(°/h) / (\sqrt{h})]$.   46

random winding (rotating machinery). A winding in which the individual conductors of a coil side occupy random position in a slot. *Syn:* **mush-wound; pancake-wound.** *See:* **rotor (rotating machinery); stator.**   63

random-wound motorette. A motorette for random-wound coils.   63

range (1)(electric pipe heating systems). The capability span of an instrument, the region between the lower and upper limits of a measured or generated function. With respect to electric pipe heating systems, range is usually defined as the difference between the lowest available set point and the highest available set point.   448

(2) (radiation protection). The set of values lying between the upper and lower detection limits.   399

(3) (radiological monitoring instrumentation). The set of values lying between the upper and lower detection limits.   398

(4) (computing systems). (A) The set of values that a quantity or function may assume. (B) The difference between the highest and lowest value that a quantity or function may assume. *See:* **error range.**   255,77

(5) (electronic navigation). An ambiguous term meaning either: (A) a distance, as in artillery techniques and radar measurements or (B) a line of position, located with respect to ground references, such as a very-high frequency omnidirectional radio range (VOR) station, or a pair of lighthouses, or an aural radio range (A-N) radio beacon. *Note:* In electronic navigation, the reader must be particularly wary, since the two meanings of the word **range** often occur in close proximity. *See:* **radio range.**

range and elevation guidance for approach and landing

(regal). A ground-based navigation system used in conjunction with a localizer to compute vertical guidance for proper glide-slope and flare-out during an instrument approach and landing; it uses a digitally coded vertically scanning fan beam that provides data for both elevation angle and distance. *See:* **navigation.**   187,13

range equation. *See:* **radar equation.**   13

range extender (telephone switching systems). Equipment inserted in a switched connection to allow an increased loop resistance.   55

range-height indication (RHI) (radar). An intensity-modulated display in which horizontal and vertical distances of a blip from an origin in the lower left part of the display represent target ground range and target height respectively. The display is generated by successive range sweeps starting at the origin and inclined at an angle that varies progressively in accordance with the elevation scan of the radar antenna at a selected azimuth. The height scale of the display is usually expanded relative to the range scale. *Syn:* **range-height indicator.**   13

range lights (illuminating engineering). Groups of color-coded boundary lights provided to indicate the direction and limits of a preferred landing path normally on an aerodrome without runways but exceptionally on an aerodrome with runways.   167

range mark (or marker) (radar). A calibration marker used on a display to aid in measuring target range (distance from the radar).   13

range noise (radar). The noise-like variation in the apparent distance of a target, caused by changes in phase and amplitude of the target scattering sources, and including radial components of glint and scintillation error.   13

range resolution (radar). The ability to distinguish between two targets solely by the measurement of their ranges (distances from the radar); usually expressed in terms of the minimum distance by which two targets of equal strength at the same azimuth and elevation angles must be spaced to be separately distinguishable.   13

ranging (communication satellite). The measurement of distance between two points and a precisely known reference point. A multiplicity of tones or a PN (pseudonoise) sequence ranging code is often used.   84

rank (network) (degrees of freedom on a node basis). The number of independent cut-sets that can be selected in a network. The rank R is equal to the number of nodes V minus the number of separate parts P. Thus $R = V - P$. *See:* **network analysis.**   210

rapid access loop (test, measurement and diagnostic equipment). In internal memory machines, a small section of memory which has much faster accessibility than the remainder of the memory.   54

rapid start fluorescent lamp (illuminating engineering). A fluorescent lamp designed for operation with a ballast that provides a low-voltage winding for preheating the electrodes and initiating the arc without a starting switch or the application of high voltage.   167

rapid-starting systems (fluorescent lamps). The designation given to those systems in which hot-cathode electric discharge lamps are operated with cathodes continuously heated through low-voltage heater windings built as part of the ballast, or through separate low-voltage secondary transformers. Sufficient voltage is applied across the lamp and between the lamp and fixture to initiate the discharge when the cathodes reach a temperature high enough for adequate emission. The cathode-heating current is maintained even after the lamp is in full operation. *Note:* In Europe this system is sometimes referred to as an instant-start system.   268

raster (cathode-ray tubes) (television). A predetermined pattern of scanning lines that provides substantially uniform coverage of an area.   125, 18

raster burn (camera tubes). A change in the characteristics of that area of the target that has been scanned, resulting in a spurious signal corresponding to that area when a larger or tilted raster is scanned.   125

raster scan (radar). A method of sweeping a cathode ray tube screen or an antenna beam characterized by a network of parallel sweeps either from side to side or from top to bottom.   13

ratchet demand (electric power utilization). The maximum past or present demands that are taken into account to establish billings for previous or subsequent periods. *See:* **alternating-current distribution.**   200

ratchet demand clause (electric power utilization). A clause in a rate schedule that provides that maximum past or present demands be taken into account to establish billings for previous or subsequent periods. *See:* **alternating-current distribution.**   200

ratchet relay. A stepping relay actuated by an armature-driven ratchet. *See:* **relay.**   259

rate action (process control). That component of proportional plus rate control action or of proportional plus reset plus rate control action for which there is a continuous linear relation between the rate of change of the directly controlled variable and the position of a final control element. *See:* **control action; derivative.**   56

rate base (power operations). The net plant investment or valuation bases specified by a regulatory authority, upon which a utility is permitted to earn a specified rate of return.   516

rate, chopping (cathode-ray oscilloscopes). The rate at which channel switching occurs in chopped display operation.   106

rate compensation heat detector (fire protection devices). A device which will response when the temperature of the air surrounding the device reaches a predetermined level, regardless of the rate of temperature rise.   71

rate control action (electric power systems). Action in which the output of the controller is proportional to the input signal and the first derivative of the input signal. Rate time is the time interval by which the rate action advances the effect of the proportional control action. *Note:* Applies only to a controller with proportional control action plus derivative control action. *See:* **speed-governing system.**   94

rated (power switchgear). A qualifying term that, applied to an operating characteristic, indicates the designated limit or limits of the characteristic for application under specified conditions. *Note:* The specific limit or limits applicable to a given device is specified in the standard for that device, and included in the title of the rated characteristic, that is, rated maximum voltage, rated frequency range, etcetera.   103

rated accuracy (1)(automatic null-balancing electric instrument). The limit that errors will not exceed when the instrument is used under any combination of rated operating conditions. *Notes:* (1) It is usually expressed as a percent of the span. It is preferred that a + sign or − sign or both precede the number or quantity. The absence of a sign infers a $\pm$ sign. (2) Rated accuracy does not include accuracy of sensing elements or intermediate means external to the instrument. *See:* **accuracy rating (instrument).**   295

(2)(direct current instrument shunts). The limit of error, expressed as a percentage of the rated output voltage, with two thirds rated current applied for one half hour to allow for self heating. It represents the expected accuracy of the shunt obtainable under normal conditions of use.   527

rated accuracy of instrument shunts (electric power systems). The limit of error, expressed as a percentage of rated voltage drop, with two-thirds rated current applied for one-half hour to allow for self-heating. *Note:* Practically, it represents the expected accuracy of the shunt obtainable over normal operating current ranges. *See:* **accuracy rating (instrument).**   201

rated alternating voltage (rectifier unit) (rated alternating-current winding voltages) (rectifier). The root-mean-square voltages between the alternating-current line terminals that are specified as the basis for rating. *Note:* When the alternating-current winding of the rectifier transformer is provided with taps, the rated voltage shall refer to a specified tap that is designated as the rated-voltage tap. *See:* **rectification; rectifier transformer.**   291, 208, 258

rated apparent efficiency (thyristor). Rated output volt-amperes divided by rated input power, generally expressed as percent.   445

rated average tube current. The current capacity of a tube, in average amperes, as assigned to it by the manufacturer for specified circuit conditions. *See:* **rectification.**   291

rated burden (capacitance potential device). The maximum unity-power-factor burden, specified in watts at rated secondary voltage, that can be carried for an unlimited period when energized at rated primary line-to-ground voltage, without causing the established limitations to be exceeded. *See:* **outdoor coupling capacitor.**   351

rated capacity (large lead storage batteries). The ampere-hour capacity assigned to a lead storage cell by its manufacturer for a given discharge time, at a specified electrolyte temperature and specify gravity, to a given end-of-discharge voltage.   377

rated circuit voltage. Used to designate the rated, root-mean-square, line-to-line, voltage of the circuit on which coupling capacitors or the capacitance potential device in combination with its coupling capacitor or bushings designed to operate. *See:* **outdoor coupling capacitor.** 351

rated continuous controller current (thyristor). The rated root-mean-square (rms) value of the maximum controller current which can be carried continuously without exceeding established limitations under prescribed conditions of operation. 445

rated continuous current (1) (electric installations on shipboard). The designed limit in rms amperes dc amperes which a switch or circuit breaker will carry continuously without exceeding the limit of observable temperature rise. 3

(2) (high-voltage switchgear). The maximum root-mean-square (rms) current in amperes, at rated frequency which a device will carry continuously without exceeding the allowable temperature rise and total temperature listed in the table below **allowable continuous current.** 443

(3) (neutral grounding devices). The current expressed in amperes, root-mean-square, that the device can carry continuously under specified service conditions without exceeding the allowable temperature rise. 91

rated continuous output current (converters having ac output)(self-commutated converters). The maximum output current that can be carried continuously without exceeding established limitations under prescribed conditions of operation. 584

rated controller current (thyristor). Rated root-mean-square (rms) value of the controller current which is specified by the manufacturer under the prescribed operation mode as a basis of declaring the duty cycles and overcurrent capability. 445

rated current (1)(power and distribution transformer). The primary current selected for the basis of performance specifications of a current transformer. 53

(2) (neutral grounding device) (electric power). The thermal current rating. The rated current of resistors whose rating is based on constant voltage is the initial root-mean-square symmetrical value of the current that will flow when rated voltage is applied. *See:* **grounding device.** 91

rated current of a shunt reactor (shunt reactors over 500 kVA). Derived from the rated voltage and rated kilovoltamperes (kVA). 562

rated direct current. *See:* **direct current, rated.**

rated direct-current winding voltage (rectifier). The root-mean-square voltage of the direct-current winding obtained by turns ratio from the rated alternating-current winding voltage of the rectifier transformer. *See:* **rectifier transformer.** 258

rated direct voltage (power inverter). The nominal direct input voltage. *See:* **self-commutated inverters.** 208

rated duty. That duty that the particular machine or apparatus has been designed to comply with. 210

rated dynamic short circuit load current (thyristor). The maximum permissible peak transient current which can be supplied into a short circuited load. This is stated in terms of I^2t, number of cycles and maximum peak value. In general this places a constraint on the minimum source of impedance. 445

rated efficency (thyristor). Rated output power divided by rated input power, generally expressed as percent. 445

rated excitation-system voltage (rotating machinery). The main exciter rated voltage. 63

rated field current (excitation systems for synchronous machines). The direct current in the field winding of the synchronous machine when operating at rated voltage, current, power factor, and speed. 507

rated field voltage (excitation systems for synchronous machines). The voltage required across the terminals of the field winding of the synchronous machine under rated continuous load conditions of the synchronous machine with its field winding at (1) 75 °C for field windings designed to operate at rating with a temperature rise of 60 °C or less; or (2) 100 °C for field windings designed to operate at rating with a temperature rise greater than 60 °C. 507

rated frequency (1)(frequency range)(converters having ac output)(self-commutated converters). The rated value of the fundamental frequency of the output voltage or the range over which the fundamental frequency may be adjusted. 584

(2) (power system or interconnected system). The normal frequency in hertz for which alternating-current generating equipment operating on such system is designed. 94

(3) (arrester). The frequency, or range of frequencies, of the power systems on which the arrester is designed to be used. 308

(4) (grounding device). The frequency of the alternating current for which it is designed. *Note:* Some devices, such as neutral wave traps, may have two or more rated frequencies; the rated frequency of the circuit and the frequencies of the harmonic or harmonics the devices are designed to control. *See:* **grounding devices.** 91

(5) (frequency rating) (of a fuse) (high-voltage switchgear). The system frequency for which it is designed. 443

rated fundamental output current (converters having ac output)(self-commutated converters). The fundamental output current specified by the manufacturer as a basis for rating. 584

rated fundamental output voltage (converters having ac output)(self-commutated converters). The fundamental output voltage specified by the manufacturer as a basis for rating. 584

rated head (1)(hydraulic turbines). The value stated on the turbine nameplate. 8

(2)(power operations). The head at which a turbine operating at rated speed will deliver rated capacity at specified gate and efficiency. 516

rated ice breaking ability (high voltage air switches,

insulators, and bus supports). The maximum thickness of ice deposited on the device which will not interfere with the successful opening or closing of the device. Suggested ratings: zero 3/8 inches (in)(10 millimeters)(mm), and 3/4 in (20 mm).   575

rated impedance (loudspeaker measurements). The rated impedance of a loudspeaker driver or system is that value of a pure resistance which is to be substituted for the driver of system when measuring the electric power delivered from the source. This should be specified by the manufacturer.   19

rated impulse protective level (arrester). The impulse protective level with the residual voltage referred to the nominal discharge current.   62

rated impulse withstand voltage (apparatus). An assigned crest value of a specified impulse voltage wave that the apparatus must withstand without flashover, disruptive discharge, or other electric failure.
                    62

rated input power (1)(ferroresonant voltage regulators). The input power to the ferroresonant regulator with the rated load and under stated operating conditions.   456
(2)(thyristor). The total real power at the lines of the controller at rated line current and voltage.   445

rated input volt-amperes (1)(ferroresonant voltage regulators). The input volt-amperes to the ferroresonant regulator with the rated load and under stated operating conditions.   456
(2)(thyristor). The product of rated line voltage and current.   445

rated insulation class (neutral grounding device) (electric power). An insulation class expressed in root-mean-square kilovolts, that determines the dielectric tests that the device shall be capable of withstanding. See: grounding device; outdoor coupling capacitor.   92

rated internal pressure (power cable joint). The rated internal pressure of a joint is the nominal internal operating pressure. This will depend on the type of cable being joined and the service conditions.   34

rated interrupting current (rated interrupting capacity) (current interrupting rating) (of a fuse) (high-voltage switchgear). The designated value of the highest available root-mean-square (rms) short-circuit current which the fuse is required to interrupt successfully under stated conditions.   443

rated kilovoltampere (current-limiting reactor). The kilovoltamperes that can be carried for the time specified at rated frequency without exceeding the specified temperature limitations, and within the limitations of established standards. See: reactor.   309

rated kilovoltampere (kVA) of a shunt reactor (shunt reactors over 500 kVA). The apparent power at rated voltage for which the shunt reactor is designed.
                    562

rated kilowatts of a constant-current transformer (power and distribution transformer). The kilowatt output at the secondary terminals with rated primary voltage and frequency, and with rated secondary current and power factor, and within the limitations of established standards.   53

rated kVA (kilovoltampere) of a grounding transformer (power and distribution transformer). The short-time kilovolt-ampere rating is the product of the rated line-to-neutral voltage at rated frequency, and the maximum constant current that can flow in the neutral for the specified time without causing specified temperature-rise limitations to be exceeded, and within the limitations of established standards for such equipment.   53

rated kVA (kilovoltampere) of a transformer (power and distribution transformer). The output that can be delivered for the time specified at rated secondary voltage and rated frequency without exceeding the specified temperature-rise limitations under prescribed conditions.   53

rated kVA tap (in a transformer) (power and distribution transformer). A tap through which the transformer can deliver its rated kVA output without exceeding the specified temperature rise.   53

rated life (glow lamp). The length of operating time, expressed in hours, that produces specified changes in characteristics. Note: In lamps for indicator use the characteristic usually is light output; the end of usual life is considered to be when light output reaches 50 percent of initial, or when the lamp becomes inoperative at line voltage. In lamps used as circuit components, the characteristic is usually voltage; life is determined as the length of time for a specified change from initial.   283

rated line current (thyristor). Rated root-mean-square (rms) value of the current in the lines at rated controller current for the specified controller connection.
                    445

rated line frequency (thyristor). The frequency or range of frequencies at which the controller can operate. Note: Some wide ranges may require a derating curve to express this rating meaningfully.   445

rated line kilovoltampere (kVA) rating (rectifier transformer). The kilovoltampere rating assigned to it by the manufacturer corresponding to the kilovoltampere drawn from the alternating-current system at rated voltage and kilowatt load on the rectifier under the normal mode of operation. See: rectifier transformer.
                    258

rated line voltage (thyristor). Rated root-mean-square (rms) value of the line voltage.   445

rated load (1) (elevator, dumbwaiter, escalator, or private residence inclined lift). The load which the device is designed and installed to lift at the rated speed. See: elevator.   328
(2) (rectifier unit). The kilowatt power output that can be delivered continuously at the rated output voltage. It may also be designated as the one-hundred-percent-load or full-load rating of the unit. Note: Where the rating of a rectifier unit does not designate a continuous load it is considered special. See: continuous rating; rectification.   208

rated load-break current (load break current rating). The designated value of the maximum root-mean-square (rms) current which a device having operable means for interrupting load currents is required to interrupt successfully under stated conditions when opened by manual or remote control means. 443

rated-load current (air-conditioning equipment) (National Electrical Code). The rated-load current for a hermetic refrigerant motor-compressor is the current resulting when the motor-compressor is operated at the rated load, rated voltage and rated frequency of the equipment it serves. 256

rated-load field voltage (nominal collector ring voltage) (rotating machinery). The voltage required across the terminals of the field winding of an electric machine under rated continuous-load conditions with the field winding at: (1) 75 degrees Celsius for field windings designed to operate at rating with a temperature rise of 60 degrees Celsius or less. (2) 100 degrees Celsius for field windings designed to operate at rating with a temperature rise greater than 60 degrees Celsius. 63

rated load power factor (thyristor). A range of load power factors over which a controller may be operated. 445

rated-load torque (rated torque) (rotating machinery). The shaft torque necessary to produce rated power output at rated-load speed. *See:* **asynchronous machine.** 63

rated load voltage (thyristor). The root-mean-square (rms) voltage delivered at the controller load terminals with rated line voltage and rated continuous controller current. 445

rated making current (high-voltage switchgear). The maximum root-mean-square (rms) current against which the device is required to close successfully when switched from the open to the closed position. 443

rated maximum interrupting of main contacts voltage (field discharge circuit breakers). The maximum direct-current (dc) voltage, including voltage induced in the machine field by current in the machine armature, at which the field discharge circuit breaker main contacts are required to interrupt the excitation source current. The magnitude of the dc component of the total voltage across the main contacts is equal to the displacement of the axis. 412

rated maximum voltage (maximum voltage rating) (high-voltage switchgear). The highest root-mean-square (rms) voltage at which the device is designed to operate. *Note:* This voltage corresponds to the maximum tolerable zone primary voltage at distribution transformers for distribution cutouts and single-pole air switches, and at substations and on transmission systems for power fuses given in ANSI C84.1-1977. 443

rated mechanical operations (high voltage air switches, insulators, and bus supports). The minimum number of operating cycles that an air switch can perform without requiring replacement of parts. 575

rated mechanical terminal load (high voltage air switches, insulators, and bus supports). The static force of conductors equivalent to the external mechanical load, applied at each terminal in specified directions, than an air switch can withstand. 575

rated minimum displacement factor (thyristor). The minimum ratio of input power to the input volt-amperes (at fundamental line frequency) at which a controller may be operated. 445

rated minimum interrupting current (high-voltage switchgear). The designated value of the smallest current that a fuse is required to interrupt at a designated voltage under prescribed conditions. 443

rated minimum tripping current (automatic circuit reclosers). The minimum rms current which causes a device to operate. 92

rated momentary current (maximum voltage rating) (high-voltage switchgear). The maximum current measured at the major peak of the maximum cycle, which the device or assembly is required to carry. *Notes:* (1) The current is expressed as the root-mean-square (rms) value including the direct-current component, as determined from the envelope of the current wave by the method shown in Appendix A of ANSI/IEEE Std C37.41-1981. (2) This rating is an index of the ability of the disconnecting device to withstand electromagnetic forces under short-circuit conditions. 443

rated nominal voltage class (field discharge circuit breakers). The voltage to which operating and performance characteristics are referred. 412

rated nonrepetitive peak line voltage (thyristor). The maximum value of the transient peak instantaneous voltage, U_{LSM}, appearing across the lines with the controller disconnected. 445

rated nonrepetitive peak OFF-state voltage (thyristor). The maximum instantaneous value of any nonrepetitive transient OFF-state voltage which may occur across the thyristor without damage. 445

rated operating conditions (automatic null-balancing electric instrument). The limits of specified variables or conditions within which the performance ratings apply. *See:* **measurement system.** 295

rated output (1)(converters having ac output)(self-commutated converters). The apparent output power for specified load conditions. 584

(2)(electrical heat tracing for industrial applications). Total wattage or watt/unit length of heating cable, at rated voltage , temperature and length. 523

rated output capacity (inverters). The kilovoltampere output at specified load power-factor conditions. *See:* **self-commutated inverters.** 208

rated output current (1)(converters having ac output)-(self-commutated converters). The total rms (root-mean-square) output current specified by the manufacturer as a basis of declaring the duty cycles and overcurrent capability, and of selecting the conductor to the load. 584

(2) (magnetic amplifier). Rated output current that the amplifier is capable of supplying to the rated load impedance, either continuously or for designated operating intervals, under nominal conditions of supply voltage, supply frequency, and ambient temperature such that the intended life of the amplifier is not reduced or a specified temperature rise is not exceeded.

Rated output current shall be specified either as root-mean-square or average. *Notes:* (A) When other than rated load impedance is used, the root-mean-square value of the rated output current should not be exceeded. (B) While specification may be either root-mean-square or average, it remains fixed for a given amplifier. *See:* **rating and testing magnetic amplifiers.**
171

rated output frequency (inverters). The fundamental frequency or the frequency range over which the output fundamental frequency may be adjusted. *See:* **self-commutated inverters.** 208

rated output power (thyristor). The total real power available to the controller load at rated controller current and rated load voltage. 445

rated output voltage (magnetic amplifier). The voltage across the rated load impedance when rated output current flows. Rated output voltage shall be specified by the same measure as rated output current (that is, both shall be stated as root-mean-square or average). *Note:* While specification may be either root-mean-square or average, it remains fixed for a given amplifier. *See:* **rating and testing magnetic amplifiers.**
171

rated output voltamperes (1) (magnetic amplifier). The product of the rated output voltage and the rated output current. 171
(2) (thyristor). The product of rated load voltage and current. 445

rated output voltamperes of the ferroresonant regulator (ferroresonant voltage regulators). The sum of the rated output winding voltamperes under stated operating conditions. 456

rated output winding voltamperes (ferroresonant voltage regulators). The product of the output voltage and output current (root-mean-square values) at the rated load and under stated operating conditions. 456

rated peak single pulse transient current (low voltage varistor surge arresters). Maximum peak current which may be applied for a single 8 x 20-μs impulse, with rated line voltage also applied, without causing device failure. 62

rated performance (automatic null-balancing electric instrument). The limits of the values of certain operating characteristics of the instrument that will not be exceeded under an combination of rated operating conditions. 295

rated power output (hydraulic turbines). The value stated on the generator nameplate. 8

rated primary current (current transformer). Current selected for the basis of performance specifications. *See:* **instrument transformer.** 305

rated primary line-to-ground voltage. The root-mean-square line-to-ground voltage for which the potential device, in combination with its coupling capacitor or bushing, is designed to deliver rated burden at rated secondary voltage. The rated primary line-to-ground voltage is equal to the rated circuit voltage (line-to-line) divided by $(3)^{1/2}$. *See:* **primary line-to-ground voltage.** 351

rated primary voltage (1) (constant-voltage trans-former) (power and distribution transformer). The voltage calculated from the rated secondary voltage by turn ratio. *Notes:* (A) See turn ratio of a transformer and its note, for the definition of the turn ratio to be used. (B) In the case of a multiwinding transformer, the rated voltage of any other winding is obtained in a similar manner. 53

(2) (instrument transformer). (A) The rated primary voltage (of a potential (voltage) transformer) is the voltage selected for the basis of performance guarantees. (B) The rated primary voltage (of a current transformer) designates the insulation class of the primary winding. *Note:* A current transformer can be applied on a circuit having a nominal system voltage corresponding to or less than the rated primary voltage of the current transformer. *See:* **instrument transformer.** 305

rated primary voltage of a constant current transformer (power and distribution transformer). The primary voltage for which the transformer is designed, and to which operation and performance characteristics are referred. 53

rated range of regulation (voltage regulator). The amount that the regulator will raise or lower its rated voltage. *Note:* The rated range may be expressed in per unit, or in percent, of rated voltage; or it my be expressed in kilovolts. *See:* **voltage regulator.**
257

rated recurrent peak voltage (low voltage varistor surge arresters). Maximum recurrent peak voltage which may be applied for a specified duty cycle and waveform. 62

rated secondary current (1) (constant voltage transformer) (power and distribution transformer). The secondary current obtained by dividing the rated kVA by the rated secondary voltage, kV. *Note:* The relationship above applies directly for single-phase transformers, but requires additional consideration of the connections involved in three-phase transformers.
53

(2) (power and distribution transformer). The rated current divided by the marked ratio. 53

rated secondary current of a constant-current transformer (power and distribution transformer). The secondary current for which the transformer is designed and to which operation and performance characteristics are referred. 53

rated secondary voltage (1) (constant voltage transformer) (power and distribution transformer). The voltage at which the transformer is designed to deliver rated kVA and to which operating and performance characteristics are referred. 53

(2) (power and distribution transformer)(voltage transformer). The voltage divided by the marked ratio. 53

(3) (capacitance potential device). This is the root-mean-square secondary voltage for which the potential device, in combination with its coupling capacitor or bushing, is designed to deliver its rated burden when energized at rated primary line-to-ground voltage. *See:* **secondary voltage; outdoor coupling capacitor.** 351

rated short-circuit withstand current (of metal-enclosed bus)(metal-enclosed bus and calculating losses in isolated-phase bus). The maximum root-mean-square (rms) total current that it can carry momentarily without electrical, thermal, or mechanical damage or permanent deformation. The current is the rms value, including the direct-current (dc) component, at the major peak of the maximum cycle as determined from the envelope of the current wave during a given test time interval. 574

rated short-term output current (converters having ac output)(self-commutated converters). The maximum output current that can be carried for a specified time without exceeding the established limitations under prescribed conditions of operation. 584

rated short-time current (1)(rated for isolated-phase bus)(metal-enclosed bus and calculating losses in isolated-phase bus). The maximum symmetrical current that the bus must carry without exceeding a specified total temperature in a given time interval. 574 **(2)(short-time current rating) (of a disconnecting device) (high-voltage switchgear).** The maximum root-mean-square (rms) total current (including the direct-current component) which the device is required to carry successfully for a specified short-time interval. *Note:* The ratings recognized the limitations imposed by both thermal and electromagnetic effects. 443

rated short-time of main contacts voltage (field discharge circuit breakers). The highest direct-current (dc) voltage at which the circuit breaker main contacts shall be required to interrupt exciter short-circuit current. 412

rated single pulse transient energy (low voltage varistor surge arresters). Energy which may be dissipated for a single impulse of maximum rated current at a specified waveshape, with rated root-mean-square (rms) voltage or rated direct-current (dc) voltage also applied, without causing device failure. 62

rated source impedance (thyristor). The equivalent impedance of the line voltage source, including the connections to the terminals of the converter. 445

rated speed (hydraulic turbines). The value stated on the unit nameplate. 8

rated supply current (magnetic amplifier). The root-mean-square current drawn from the supply when the amplifier delivers rated output current. 171

rated supply voltage (converters having dc input)-(self-commutated converters). The supply voltage specified by the manufacturer as a basis for rating. 584

rated symmetrical interrupting current (accelerometer). The root-mean-square value of the symmetrical component of the highest current which a device is required to interrupt under the operating duty, rated maximum voltage, and circuit constants specified. 92

rated system deviation. The specified maximum permissible carrier frequency deviation. Nominal values for mobile communications systems are $\pm$ 15 kilohertz or $\pm$ 5 kilohertz. 123

rated system voltage (current-limiting reactor). The voltage to which operations and performance characteristics are referred. It corresponds to the nominal system voltage of the circuit on which the reactor is intended to be used. *See:* **reactor.** 309

rated OFF voltage (magnetic amplifier). The output voltage existing with trip OFF control signal applied. 171

rated ON voltage (magnetic amplifier). The output voltage existing with trip ON control signal applied. Rated ON voltage shall be specified either as root-mean-square or average. *Note:* While specification may be either root-mean-square or average it remains fixed for a given amplifier. 171

rated thermal current (neutral grounding device). The root-mean-square neutral current in amperes which the device is rated to carry under standard operating conditions for rated time without exceeding temperature limits. 91

rated three-second current (three-second current rating) (high-voltage switchgear). The root-mean-square (rms) total current, including the direct-current component which the device, or assembly, is required to carry for 3 seconds (s). *Note:* For practical purposes, this current is measured at the end of the first second. This rating is an index of the ability of the disconnecting device to withstand heat that may be generated under short-circuit conditions. 443

rated time (grounding device) (electric power). The time during which it will carry its rated current, or withstand its rated voltage, or both, under standard conditions without exceeding standard limitations, unless otherwise specified. *See:* **grounding device.** 91

rated-time temperature rise (grounding device). The maximum temperature rise above ambient attained by the winding of a device as the result of the flow of rated thermal current (or, for certain resistors, the maintenance of rated voltage across the terminals) under standard operating conditions, for rated time and with a starting temperature equal to the steady-state temperature. It may be expressed as an average or a hot-spot winding rise. 91

rated torque (rotating machinery). *See:* **rated-load torque.**

rated transient average power dissipation (low voltage varistor surge arresters). Maximum average power which may be dissipated due to a group of pulses occurring within a specified isolated time period, without causing device failure. 62

rated value (converter characteristics)(self-commutated converters). A specified value for the electrical, thermal, mechanical, and environmental quantities assigned by the manufacturer to define the operating conditions under which a converter is expected to give satisfactory service. 584

rated values (thyristor). A specified value for the electrical, thermal, mechanical, and environmental quantities assigned by the manufacturer to define the operating conditions under which a controller is expected to give satisfactory service. The rated values may change if the operating mode is different from that specified. *Note:* For calculating or measuring the root-mean-square (rms) value, several integration intervals are possible depending upon the operation mode. The interval used should be exactly specified.   445

rated voltage (1)(electrical heat tracing for industrial applications). The voltage to which operating and performance characteristics of heating cables are referred.   523

(2)(electric submersible pump cable). The rated voltage is expressed in terms of phase-to-phase voltage of a three-phase system.   484

(3) (power and distribution transformer). The voltage to which operating and performance characteristics of apparatus and equipment are referred. *Note:* Deviation from rated voltage may not impair operation of equipment, but specified performance characteristics are based on operation under rated conditions. However, in many cases apparatus standards specify a range of voltage within which successful performance may be expected.   53

(4) (power cable joint). The rated voltage of a joint is the voltage at which it is designed to operate under usual service conditions. Unless otherwise specified, the voltage rating is assigned with the understanding that the joint will be applied on the three-phase circuits whose nominal phase-to-phase voltage rating does not exceed that of the joint.   34

(5) (power cable systems). For cables, either single-conductor or multiple-conductor, the rated voltage is expressed in terms of phase-to-phase voltage of a three-phase system. For single-phase systems a rated voltage of $\sqrt{3}\cdot$ the voltage to ground should be assumed.   437

(6) (rotating electric machinery). The voltage specified at the terminals of a machine.   424

(7) (step-voltage or induction-voltage regulator). The voltage between terminals of the series winding, with rated voltage applied to the regulator, when the regulator is in the maximum raise position and is delivering rated kilovoltampere output at 80-percent power factor. *See:* **voltage regulator.**   257

(8) (arrester). The designated maximum permissible root-mean-square value of power-frequency voltage between its line and earth terminals at which it is designed to operate correctly.   62

(9) (grounding device) (electric power). The root-mean-square voltage, at rated frequency, that may be impressed between its terminals under standard conditions for its rated time without exceeding standard limitations, unless otherwise specified. *See:* **grounding device.**   91

rated voltage adjustment range (thyristor). The range over which the steady state load voltage can be varied.   445

rated voltage of a shunt reactor (shunt reactors over 500 kVA). The voltage to which operating and performance characteristics are referred.   562

rated voltage of a voltage transformer (instrument transformer)(power and distribution transformer). The primary voltage selected for the basis of performance specifications of a voltage transformer.   394, 53

rated voltage of a winding (power and distribution transformer). The voltage to which operating and performance characteristics are referred.   53

rated watts input (household electric ranges). The power input in watts (or kilowatts) that is marked on the range nameplate, heating units, etcetera. *See:* **appliance outlet.**   263

rated withstand current (surge current) (surge arresters). The crest value of a surge, of given wave shape and polarity, to be applied under specified conditions without causing disruptive discharge on the test specimen.   62

rated withstand voltage (insulation strength). The voltage that electric equipment is required to withstand without failure or disruptive discharge when tested under specified conditions and within the limitations of established standards. *See:* **basic impulse insulation level (insulation strength).**   276

rated 15-cycle current (15-cycle current rating) (of a disconnecting device or assembly) (high-voltage switchgear). The root-mean-square (rms) symmetrical current of an asymmetrical wave produced by a circuit having a prescribed X/R ratio, which the device or assembly is required to carry for 15 cycles. *Note:* This rating is an index of the ability of the disconnecting device to withstand heat that may be generated under short-circuit conditions.   443

rate gyro. Generally, a single-degree-of-freedom gyro having primarily elastic restraint of the spin axis about the output axis. In this gyro, an output signal is produced by precession of the gimbal, the precession angle being proportional to the angular rate of the case about the input axis.   46

rate-integrating gyro. A single-degree-of-freedom gyro having primarily viscous restraint of the spin axis about the output axis. In this gyro, an output signal is produced by precession of the gimbal, the precession angle being proportional to the integral of the angular rate of the case about the input axis.   46

rate-of-change protection (power switchgear). A form of protection in which an abnormal condition causes disconnection or inhibits connection of the protected equipment in accordance with the rate of change of current, voltage, power, frequency, pressure, etcetera.   127, 103

rate-of-change relay (power switchgear). A relay that responds to the rate of change of current, voltage, power, frequency, pressure, etcetera.   127, 103

rate of decay (audio and electroacoustics). The time rate at which the sound pressure level (or other stated characteristic) decreases at a given point and at a given time. *Note:* Rate of decay is frequently expressed in decibels per second.   176

rate of punching (test, measurement and diagnostic equipment). Number of characters, blocks, words, or frames of information placed in the form of holes distributed on cards or tape per unit time. The number of cards punched per unit time.   54

rate of reading (test, measurement and diagnostic equipment). Number of cards, characters, blocks, words, or frames sensed by a sensing device per unit time.   54

rate-of-rise current tripping. *See:* **rate-of-rise release (trip).**   103

rate-of-rise detector (fire protection devices). A device which will response when the temperature rises at a rate exceeding a predetermined amount.   71

rate of rise limiters (thyristor). Devices used to control the rate of rise of current or voltage to the semiconductor device, or both. *Note:* Current rate of rise limiters may include linear or nonlinear devices.
                                                          445

rate of rise of reapplied forward voltage (thyristor). The average slope of the reapplied voltage measured as the slope, of a line from the intersection of the voltage waveform and the axis, to the point where the waveform achieves 63 percent of the maximum forward OFF-state voltage.   445

rate of rise of restriking voltage (transient recovery voltage rate) (usually abbreviated to rrrv) (surge arresters). The rate, expressed in volts per microsecond, that is representative of the increase of the restriking voltage.   308,62

rate-of-rise release (trip) (power switchgear). A release that operates when the rate of rise of the actuating quantity in the main circuit exceeds the release setting.   103

rate-of-rise suppressors (semiconductor rectifiers). Devices used to control the rate of rise of current and/or voltage to the semiconductor devices in a semiconductor power converter. *See:* **semiconductor rectifier stack.**   208

rate-of-rise trip. *See:* **rate-of-rise release.**

rate signal (industrial control). A signal that is the time derivative of a specified variable. *See:* **control system, feedback.**   206

rate-squared sensitivity (nongyroscopic angular sensor) (inertial sensor). A term specific to angular accelerometers which describes an error torque about the input axis proportional to the product of input rates on the other two axes. This is analogous to anisoinertia torque.   46

rate, sweep. *See:* **sweep time/division.**

rate test. *See:* **problem check.**

rating (1)(metal-oxide surge arresters for ac power circuits). The designation of an operating limit for a device.   583

(2)(rating of electric equipment) (general). The whole of the electrical and mechanical quantities assigned to the machine, apparatus, etcetera, by the designer, to define its working in specified conditions indicates on the rating plate. *Note:* The rating of electric apparatus in general is expressed in voltamperes, kilowatts, or other appropriate units. Resistors are generally rated in ohms, amperes, and class of service.
                                                      310, 415

(3) (rotating machinery). The numerical values of electrical quantities (frequency, voltage, current, apparent and active power, power factor) and mechanical quantities (power, torque), with their duration and sequences, that express the capability and limitations of a machine. The rated values are usually associated with a limiting temperature rise of insulation and metallic parts. *See:* **asynchronous machine.**   63

(4) (power switchgear). The designated limit(s) of the rated operating characteristic(s) of a device. *Note:* Such operating characteristics as current, voltage, frequency, etcetera, may be given in the rating.
                                                  103, 127, 443

(5) (current-limiting reactor). The voltamperes that it can carry, together with any other characteristics, such as system voltage, current and frequency assigned to it by the manufacturer. *Note:* It is regarded as a test rating that defines an output that can be carried under prescribed conditions of test, and within the limitations of established standards. *See:* **reactor.**
                                                          309

(6) (interphase transformer). The root-mean-square current, root-mean-square voltage, and frequency at the terminals of each winding, when the rectifier unit is operating at rated load and with a designated amount of phase control. *See:* **duty; rectifier transformer.**   53

(7) (rectifier transformer). The kilovoltampere output, voltage, current, frequency, and number of phases at the terminals of the alternating-current winding; the voltage (based on turn ratio of the transformer), root-mean-square current, and number of phases at the terminals of the direct-current winding, to correspond to the rated load of the rectifier unit. *Notes:* (A) Because of the current wave shapes in the alternating- and direct-current windings of the rectifier transformer, these windings may have individual ratings different from each other and from those of power transformers in other types of service. The ratings are regarded as test ratings that define the output that can be taken from the transformer under prescribed conditions of test without exceeding any of the limitations of the standards. (B) For rectifier transformers covered by established standards, the root-mean-square current ratings and kilovoltampere ratings of the windings are based on values derived from rectangular rectifier circuit element currents without overlap. *See:* **rectifier transformer.**   53

(8) (surge arrester). The designation of an operating limit for a device.   430

(9) (transformer). A voltampere output together with any other characteristics, such as voltage, current, frequency, powerfactor, and temperature rise, assigned to it by the manufacturer.   53

(10) (relay). *See:* **relay rating.**

(11) (rotating electric machinery). The output at the shaft if a motor, or at the terminals if a generator, assigned to a machine under specified conditions of speed, voltage, temperature rise, etcetera.   424

(12)(power and distribution transformer). The rating of a transformer consists of a volt-ampere output together with any other characteristics, such as voltage, current, frequency, power factor, and temperature rise, assigned to it by the manufacturer. It is regarded as a rating associated with an output which can be taken from the transformer under prescribed conditions and limitations of established standards.

53

rating, emergency (generating station). Capability of installed equipment for a short time interval.

112

rating, normal. Capacity of installed equipment.

112

rating of diesel-generator unit (nuclear power generating stations). (1) Continuous rating. The electric power output capability that the diesel-generator unit can maintain in the service environment for 8760 h of operation per (common) year with only scheduled outages for maintenance. (2) Short time rating. The electric power output capability that the diesel-generator unit can maintain in the service environment for 2 h in any 24 h period, without exceeding the manufacturer's design limits and without reducing the maintenance interval established for the continuous rating. *Note:* Operation at this higher rating does not limit the use of the diesel-generator unit at its continuous rating.

99

rating of interphase transformer (power and distribution transformer). The root-mean-square current, root-mean-square voltage, and frequency at the terminals of each winding, when the rectifier unit is operating at rated load and with a designated amount of phase control.

53

rating plate (rotating machinery). *See:* **nameplate.**

ratio (magnetic storage). *See:* **squareness ratio.**

ratio correction factor (power and distribution transformer) (RCF) (instrument transformer). The ratio of the true ratio to the marked ratio. The primary current or voltage is equal to the secondary current or voltage multiplied by the marked ratio times the ratio correction factor.

53

ratio meter. An instrument that measures electrically the quotient or two quantities. A ratio meter generally has no mechanical control means, such as springs, but operates by the balancing of electromagnetic forces that are a function of the position of the moving element. *See:* **instrument.**

328

rationalized system of equations. A rationalized system of electrical equations is one in which the proportionality factors in the equations that relate (1) the surface integral of electric flux density to the enclosed charge, and (2) the line integral of magnetizing force to the linked current, are each unity. *Notes:* (A) By these choices, some formulas applicable to configuration having spherical or circular symmetry contain an explicit factor of 4π or 2π; for example, Coulomb's law is $f = q_1 q_2 / (4\pi \epsilon_0 r^2)$. (B) The differences between the equations of a rationalized system and those of an unrationalized system may be considered to result from either (a) the use of a different set of units to measure the same quantities or (b) the use of the same set of units to measure quantities that are quantitatively different (though of the same physical nature) in the two systems. The latter consideration, which represents a changed relation between certain mathematicophysical quantities and the associated physical quantities, is sometimes called **total rationalization.**

210

ratio set (valve actuators). A set of performance parameter values described by a range of numerical values whoseboundaries have been established by doubling and halving the numerical mean value of a selected physical performance parameter.

492

ratio-type (position-type) telemeter (power switchgear). A telemeter that employs the relative phase position between, or the magnitude relation between, two or more electrical quantities as the translating means. *Note:* Examples of ratio-type telemeters include ac or dc position matching systems.

103

ratproof electric installing. Apparatus and wiring designed and arranged to eliminate harborage and runways for rats. *See:* **marine electric apparatus.**

328

rat race.* *See:* **hybrid ring.**

*Deprecated

ray. *See:* **light ray.**

Ray Dist. (navigation aid terms). A radio navigation system used in hydrographic and geophysical surveying.

526

Rayleigh density function (radar). A probability density function describing the behavior of some variable, given by

$$f(X) = \frac{1}{\sigma_{AVG}} \exp - \left(\frac{X}{\sigma_{AVG}}\right)$$

Often used to describe the signal statistics after envelope detection.

13

Rayleigh disk. A special form of acoustic radiometer that is used for the fundamental measurement of particle velocity.

176

Rayleigh distribution (radar). A probability distribution characterized by the probability density function.

$$f(x) = \frac{x}{\sigma^2} \exp\left(-\frac{x^2}{2\sigma^2}\right), x \geqslant 0$$

$$= 0, \qquad\qquad x < 0$$

where x is the random variable and $2\sigma^2$ is the average value of x^2. This function is often used to model the statistics of the amplitude of noise at i-f or at video after linear envelope detection. It also applies to the signal voltage amplitude of certain models of fluctuating targets.

13

Rayleigh scattering (1) (fiber optics). Light scattering by refractive index fluctuations (inhomogeneities in material density or composition) that are small with respect to wavelength. The scattered field is inversely proportional to the fourth power of the wavelength. *See:* **material scattering; scattering; waveguide scattering.**

433

(2) **(laser-maser).** Scattering of radiation in the course of its passage through a medium containing particles, the sizes of which are small compared with the wavelength of the radiation.

363

RCS. *See:* **radar cross section.**

RDF (navigation aid terms). *Note:* At one time this

term was used by the British to mean radio distance-finding, that is, radar. *See:* **radio direction finding.**
526

R-display (radar). An A display with a segment of the time base expanded near the blip for greater precision in range measurement and visibility of pulse shape. *Note:* Usually regarded as an optional feature of an A-display rather than being identified by the nomenclature "R-display".
13

reach (1) (of a relay) (power switchgear). The extent of the protection afforded by a relay in terms of the impedance or circuit length as measured from the relay location. *Note:* The measurement is usually to a point of fault, but excessive loading or system swings may also come within reach or operating range of the relay.
103

(2) (protective relaying). The maximum distance from the relay location to a fault for which a particular relay will operate. The reach may be stated in terms of miles, primary ohms, or secondary ohms.
128

reactance (1) (electric installations on shipboard). The reactance of a portion of a circuit for a sinusoidal current and potential difference of the same frequency is the product of the sine of the angular phase difference between the current and potential difference times the ratio of the effective potential difference to the effective current, there being no source of power in the portion of the circuit under consideration. The reactance of a circuit is different for each component of an alternating current. If

$$e = E_{1m}\sin(\omega t + \alpha_1) + E_{2m}\sin(2\omega t + \alpha_2) + \ldots$$

and

$$i = I_{1m}\sin(\omega t + \beta_1) + I_{2m}\sin(2\omega t + \beta_2) + \ldots$$

then the reactances, X_1, X_2, etcetera, for the different components are

$$X_1 = \frac{E_{1m}\sin(\alpha_1 - \beta_1)}{I_{1m}} = \frac{E_1\sin(\alpha_1 - \beta_1)}{I_1}$$

$$X_2 = \frac{E_{2m}\sin(\alpha_2 - \beta_2)}{I_{2m}} = \frac{E_2\sin(\alpha_2 - \beta_2)}{I_2}$$

etcetera *Note:* The reactance for the entire periodic current is not the sum of the reactances of the components. A definition of reactance for a nonsinusoidal periodic current has not yet been agreed upon.
3

(2) (general). The imaginary part of impedance. *See:* **reactor.**
210,185

(3) (portion of a circuit for a sinusoidal current and potential difference of the same frequency). The product of the sine of the angular phase difference between the current and potential difference times the ratio of the effective potential different to the effective current, there being no source of power in the portion

of the circuit under consideration. The reactance of a circuit is different for each component of an alternating current. If

$$e = E_{1m}\sin(\omega t + \alpha_1)$$
$$+ E_{2m}\sin(2\omega t + \alpha_2) + \ldots$$

and

$$i = I_{1m}\sin(\omega t + \beta_1)$$
$$+ I_{2m}\sin(2\omega t + \beta_2) + \ldots$$

then the reactances, X_1, X_2, etcetera, for the different components are

$$X_1 = \frac{E_{1m}\sin(\alpha_1 - \beta_1)}{I_{1m}} = \frac{E_1\sin(\alpha_1 - \beta_1)}{I_1}$$

$$X_2 = \frac{E_{2m}\sin(\alpha_2 - \beta_2)}{I_{2m}} = \frac{E_2\sin(\alpha_2 - \beta_2)}{I_2}$$

etcetera. *Note:* The reactance for the entire periodic current is not the sum of the reactance of the components. A definition of reactance for a nonsinusoidal periodic current has not been agreed upon.
3

reactance amplifier. *See:* **parametric amplifier.**

reactance drop (1) (power and distribution transformer). The component of the impedance voltage drop in quadrature with the current.
53

(2) (general). The voltage drop in quadrature with the current.
257

reactance, effective synchronous. An assumed value of synchronous reactance used to represent a machine in a system study calculation for a particular operating condition.
63

reactance frequency multiplier. A frequency multiplier whose essential element is a nonlinear reactor. *Note:* The nonlinearity of the reactor is utilized to generate harmonics of a sinusoidal source. *See:* **parametric device.**
277,191

reactance function (1) (linear passive networks). The driving-point impedance of a lossless network. *Note:* This is an odd function of the complex frequency.
238

(2) (circuits and systems). A function that is realizable as a driving-point impedance with ideal inductors and capacitors. It must meet the conditions described in Foster's reactance theorem.
67

reactance grounded (power and distribution transformer). Grounded through impedance, the principal element of which is reactance. *Note:* The reactance may be inserted either directly, in the connection to ground, or indirectly, by increasing the reactance of the ground return circuit. The latter may be done by intentionally increasing the zero-sequence reactance of apparatus connected to ground, or by omitting some of the possible connections from apparatus neutrals to ground.
91, 53

reactance modulator. A device, used for the purpose of modulation, whose reactance may be varied in accordance with the instantaneous amplitude of the modulating electromotive force applied thereto. *Note:* Such a device is normally an electron-tube circuit and is commonly used to effect phase or frequency modulation. *See:* **frequency modulation (telecommunication); modulation; phase modulation.**   111,211

reactance relay (power switchgear). A linear-impedance form of distance relay for which the operating characteristic of the distance unit on an R-X diagram is a straight line of constant reactance. *Note:* The operating characteristic may be described by either equation X = K, or $Z\sin\theta$ equals K, where K is a constant, and θ is the angle by which the input voltage leads the input current. See figure below.   103

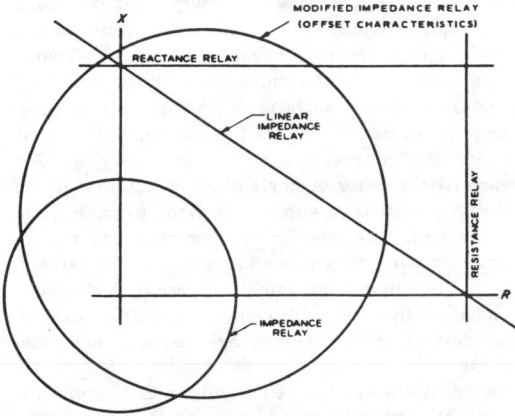

reaction curve (process control). The plot of a time response.   56

reaction frequency meter. *See:* **absorption frequency meter.**

reaction time (illuminating engineering). The interval between the beginning of a stimulus and the beginning of the response of an observer.   167

reaction torque (or force) (gyro; accelerometer). A torque (or force) exerted on a gimbal, gyro rotor or accelerometer proof mass, usually as a result of applied electrical excitations exclusive of torquer (or forcer) command signals.   46

reaction wheels (communication satellite). A set of gyro wheels used for controlling the attitude of a satellite.   74

reactivation date (electric generating unit reliability, availability, and productivity). The date a unit was returned to the active state from the deactivated shutdown state.   567

reactive attenuator (waveguide). An attenuator that absorbs no energy. *See:* **waveguide.**   244,179

reactive current (rotating machinery). The component of a current in quadrature with the voltage. *See:* **asynchronous machine.**   244,63

reactive-current compensator (rotating machinery). A compensator that acts to modify the functioning of a voltage regulator in accordance with reactive current.   63

reactive factor. The ratio of the reactive power to the apparent power. The reactive factor is expressed by the equation

$$F_q = \frac{Q}{U}$$

where F_q = reactive factor
Q = reactive power
U = apparent power.

If the voltages have the same waveform as the corresponding currents, reactive factor becomes the same as phasor reactive factor. If the voltages and currents are sinusoidal and for polyphase circuits form symmetrical sets

$$F_q = \sin(\alpha - \beta).$$

See: **network analysis.**   210

reactive-factor meter. An instrument for measuring reactive factor. It is provided with a scale graduated in reactive factor. *See:* **instrument.**   328

reactive field (of an antenna). Electric and magnetic fields surrounding an antenna and resulting in the storage of electromagnetic energy rather than in the radiation of electromagnetic energy.   111

reactive ignition cable (electromagnetic compatibility). High-tension ignition cable, the core of which is so constructed to give a high reactive impedance at radio frequencies. *See:* **electromagnetic compatibility.**   220,199

reactive near-field region (antennas). That portion of the near-field region immediately surrounding the antenna, wherein the reactive field predominates. *Note:* For a very short dipole, or equivalent radiator, the outer boundary is commonly taken to exist at a distance $\lambda/2\pi$ from the antenna surface, where λ is the wavelength.   111

reactive power (1) (general). *See:* **magner**
(2) (metering). For sinusoidal quantities in a two-wire circuit, reactive power is the product of the voltage, the current, and the sine of the phase angle between them. For nonsinusoidal quantities, it is the sum of all harmonic components, each determined as above. In a polyphase circuit, it is the sum of the reactive powers of the individual phases.   212

reactive power relay (power switchgear). A power relay that responds to reactive power.   103

reactive voltampere-hour meter. *See:* **varhour meter.**

reactive voltampere meter. *See:* **varmeter.**

reactivity (power operations). A measure of the departure of a nuclear reactor from criticality. Mathematically,

$$\rho = (k_{\text{eff}} - 1) \div K_{\text{eff}}$$

If ρ is positive (excess reactivity), the reactor is supercritical and its power level is increasing. If ρ is negative (negative reactivity), the power level of the reactor decreases. For a reactor at criticality, for instance, constant power level, the reactivity is zero. *Note:* Other measures are also used to express reactivity. *See:* **excess reactivity.** 516

reactor (shunt reactors over 500 kVA). A device used for introducing impedance into an electric circuit, the principal element of which is inductive reactance. 562

reactor (1) (electric installations on shipboard). A device, the primary purpose of which is to introduce reactance into a circuit. A reactor is a device used for introducing reactance into a circuit for purposes such as motor starting, paralleling transformers, and control of current. 3
(2) (power and distribution transformer). An electromagnetic device, the primary purpose of which is to introduce inductive reactance into a circuit. 53
(3) (radiological monitoring instrumentation). As used in this standard, reactor means a nuclear reactor designed for and capable of operation at a steady state reactor power level of $\geq R \ 1MW_{th}$. 398

reactor, ac (thyristor converter). An inductive reactor that is inserted between the transformer and the thyristor converter for the purpose of controlling the rate of rise of current in the thyristor and possibly the magnitude of fault current. 121

reactor, amplistat. A reactor conductively connected between the direct-current winding of a rectifier transformer and rectifier circuit elements that when operating in conjunction with other similar reactors, provides a relatively small controlled direct-current voltage range at the rectifier output terminals. *See:* **reactor.** 203

reactor, bus. A current-limiting reactor for connection between two different buses or two sections of the same bus for the purpose of limiting and localizing the disturbance due to a fault in either bus. *See:* **reactor.** 203

reactor, current-balancing. A reactor used in semiconductor rectifiers to achieve satisfactory division of current among parallel-connected semicondutor diodes. *See:* **reactor.** 203

reactor, current-limiting. A reactor intended for limiting the current that can flow in a circuit under short-circuit conditions, or under other operating conditions such as starting, synchronizing, etcetera. *See:* **reactor.** 203

reactor, dc (thyristor converter). An inductive reactor between the dc output of the thyristor converter and the load in order to limit the magnitude of fault current and also, in some cases, to limit the magnitude of ripple current in the load. In this latter case, it is called a ripple reactor. 121

reactor, diode-current-balancing. A reactor with a set of mutually coupled windings that, operating in conjunction with other similar reactors, forces substantially equal division of current among the parallel paths of a rectifier circuit element. *See:* **reactor.** 203

reactor facility (radiological monitoring instrumentation). The structures, systems and components used for the operation of a nuclear reactor. If a site contains more than one nuclear reactor, reactor facility means all structures, systems and components used for operation of the nuclear reactors at the site. 398

reactor, feeder. A current-limiting reactor for connection in series with an alternating-current feeder circuit for the purpose of limiting and localizing the disturbance due to faults on the feeder. *See:* **reactor.** 203

reactor, filter. A reactor used to reduce harmonic voltage in alternating-current or direct-current circuits. *See:* **reactor.** 203

reactor, paralleling. A current-limiting reactor for correcting the division of load between parallel-connected transformers that have unequal impedance voltages. *See:* **reactor.** 203

reactor starting (rotating machinery). The process of starting a motor at reduced voltage by connecting it initially in series with a reactor (inductor) which is short-circuited for the running condition. 63

reactor, starting (transformers). A current-limiting reactor for decreasing the starting current of a machine or device. *See:* **reactor.** 203

reactor-start motor. A single-phase induction motor of the split-phase type with a main winding connected in series with a reactor for starting operation and an auxiliary winding with no added impedance external to it. For running operation, the reactor is short-circuited or otherwise made ineffective, and the auxiliary winding circuit is opened. *See:* **asynchronous machine.** 63

reactor, synchronizing. A current-limiting reactor for connecting momentarily across the open contacts of a circuit-interrupting device for synchronizing purposes. *See:* **reactor.** 203

read (electronic computation). To acquire information usually from some form of storage. *See:* **destructive read; nondestructive read; write.** 235,255,77,54,125

read-around number (storage tubes). The number of times reading operations are performed on storage elements adjacent to any given storage element without more than a specified loss of information from that element. *Note:* The sequence of operations (including priming, writing, or erasing), and the storage elements on which th operations are performed, should be specified. *See:* **storage tube.** 174,125

read-around ratio* (storage tubes). *See:* **read-around number.**
*Deprecated

read cycle (read)(FASTBUS acquisition and control). A cycle in which the direction of data flow is from slave(s) toward a master. 480

read delay t_{rd} (metal-nitride-oxide field-effect transistor). Time period between the end of the writing pulse and the start of the read condition. 386

read disturb (metal-nitride-oxide field-effect transistor). A change in the instantaneous threshold voltage of a metal-nitride-oxide-semiconductor (MNOS) transistor due to the very act of measuring it. 386

read head (test, measurement and diagnostic equip-

ment). A sensor that converts information stored on punched tape, magnetic tape, magnetic drum, and so forth into electrical signals.   54

readily accessible (National Electrical Code)(power and distribution transformer)(packaging machinery). Capable of being reached quickly for operation, renewal, or inspections, without requiring those to whom ready access is requisite to climb over or remove obstacles or to resort to portable ladders, chairs, etcetera. *See:* **accessible.**   256, 53, 429

readily climbable (NESC). Having sufficient handholds and footholds to permit an average person to climb easily without using a ladder or other special equipment.   494

readiness test (test, measurement and diagnostic equipment). A test specifically designed to determine whether an equipment or system is operationally suitable for a mission.   54

reading (recording instrument). The value indicated by the position of the index that moves over the indicating scale. *See:* **accuracy rating (instrument).**   328

reading rate (storage tubes). The rate of reading successive storage elements. *See:* **storage tube.**   174

reading speed (storage tubes). *See:* **data processing; storage tube.**

reading speed, minimum usable (storage tubes). The slowest scanning rate under stated operating conditions before a specified degree of decay occurs. *Note:* The qualifying adjectives **minimum usable** are frequently omitted in general usage when it is clear that the minimum usable reading speed is implied. *See:* **storage tube.**   174

reading time (storage tubes). The time during which stored information is being read. *See:* **storage tube.**   174

reading time, maximum usable (storage tubes). The length of time a storage element, line, or area can be read before a specified degree of decay occurs. *Notes:* (1) This time may be limited by static decay, dynamic decay, or a combination of the two. (2) It is assumed that rewriting is not done. (3) The qualifying adjectives **maximum usable** are frequently omitted in general usage when it is clear that the maximum usable reading time is implied. *See:* **storage tube.**   174

read-in lag (diode-type camera tube). The fraction of the steady-state ON signal which is read out in any field after initiation of irradiance.   380

read-mostly devices (metal-nitride-oxide field-effect transistor). Metal-nitride-oxide semiconductor (MNOS) memory transistors whose retention under constant read condition is in excess of one year. This makes these devices applicable in electrically-alterable read-only memories (EAROMs). A typical writing pulse width is 1 ms.   386

read number, maximum usable (storage tubes). The number of times a storage element, line, or area can be read without rewriting before a specified degree of decay results. *Note:* The qualifying adjectives **maximum usable** are frequently omitted in general usage when it is clear that the maximum usable read number is implied. *See:* **storage tube.**   174,190

read-only storage (computing systems). See: **fixed storage.**

readout (1) (radiological monitoring instrumentation) (radiation protection) (plutonium monitoring). The device that conveys information regarding the measurement to the user.   398, 399, 413
(2) (test, measurement and diagnostic equipment). (A) The device used to present output information to the operator, either in real time or as an output of a storage medium; and (B) the act of reading, transmitting, displaying information either in real time or from an internal storage medium of an operator or an external storage medium or peripheral equipment.   54

readout, command (numerically controlled machines). Display of absolute position as derived from position command. *Note:* In many systems the readout information may be taken directly from the dimension command storage. In others it may result from the summation of command departures.   207

read-out lag (diode-type camera tube). The fraction of the initial signal which is read out in any field after the image illumination is interrupted.   380

readout, position (numerically controlled machines). Display of absolute position as derived from a position transducer.   207

read pulse. A pulse that causes information to be acquired from a magnetic cell or cells. *See:* **ONE state.**   331

ready-to-receive signal (facsimile). A signal sent back to the facsimile transmitter indicating that a facsimilee receiver is ready to accept the transmission. *See:* **facsimile signal (picture signal).**   12

realizable function (linear passive networks). A response function that can be realized by a network containing only positive resistance, inductance, capacitance, and ideal transformers. *Note:* This is the sense of realizability in the theory of linear, passive, reciprocal, time-invariant networks.   238

realized gain (antennas). (1) (General) The gain of an antenna reduced by the losses due to the mismatch of the antenna input impedance to a specified impedance. *Note:* The realized gain does not include losses due to polarization mismatch between two antennas in a complete system. (2) (Partial). The partial gain of an antenna for a given polarization reduced by the loss due to the mismatch of the antenna input impedance to a specified impedance.   111

real time (1)(processing)(emergency and standby power). Pertaining to the actual time during which a physical process transpires or pertaining to the performance of a computation during the actual time of related physical processing in order that results of the computation can be used in guiding the physical process.   512
(2) (analog computers). Using an ordinary clock as a time standard, the number of seconds measured between two events occurring in a physical system. By contrast, computer time is the number of seconds measured, with the same clock, between corresponding events in the simulated system. The ratio of the

time interval between two events in a simulated system to the time interval between the corresponding events in the physical system is the time scale. Computer time is equal to the product of real time and the time scale. Real-time computation is computer operation in which the time scale is unity. Machine time is synonymous with computer time. *See:* **scale factor.**
    9

(3) (software). (A) Pertaining to the processing of data by a computer in connection with another process outside the computer according to time requirements imposed by the outside process. This term is also used to describe systems operating in conversational mode and processes that can be influenced by human intervention while they are in progress. (B) Pertaining to the actual time during which a physical process transpires, for example, the performance of a computation during the actual time that the related physical process transpires, in order that results of the computation can be used in guiding the physical process. *See:* **computer; conversational; data; process; system.**
    434

real time printout (sequential events recording systems). The recording of actual time that an input signal was received as correlated to a time standard.
    48,58

real time testing (test, measurement and diagnostic equipment). The testing of a system or its components at its normal operating frequency or timing. 54

recalescent point (metal). The temperature at which there is a sudden liberation of heat when metals are lowered in temperature. *See:* **dielectric heating; induction heating; coupling; Curie point.** 14,114

receive characteristic (telephony). The acoustic output level of a telephone set as a function of the electrical input level. The output is measured in an **artificial ear,** and the input signal is obtained from an available constant-power source of specified impedance.
    114,415

received power (mobile communication). The root-mean-square value of radio-frequency power that is delivered to a load that correctly terminates an isotropic reference antenna. The reference antenna most commonly used is the half-wave dipole. *See:* **mobile communication system.** 181

receive-only equipment. Data communication equipment capable of receiving signals, but not arranged to transmit signals. 194

receiver (1) (facsimile). The apparatus employed to translate the signal from the communications channel into a facsimile record of the subject copy. *See:* **facsimile (in electrical communication).** 12
(2) (telephone switching systems). A part of an automatic switching system that receives signals from a calling device or other source for interpretation and action. 55

receiver gating (radar). The application of enabling or inhibiting pulses to one or more stages of a receiver only during the part of a cycle of operation when reception is either desired or undesired, respectively. *See:* **gating.** 13

receiver ground (signal-transmission system). The potential reference at the physical location of the signal receiver. *See:* **signal.** 188
receiver linear dynamic range (electromagnetic site survey). The interval between the minimum detectable signal and the 1 decibel (dB) gain compression point within which the receiver gain deviates from a constant value by less than 1 dB. 457
receiver-off-hook tone (telephone switching systems). A tone on a line to indicate an abnormal off-hook condition. 55
receiver operating characteristic curves (ROC curves) (radar). Plots of probability of detection versus probability of false alarm for various input signal-to-noise ratios and detection threshold settings. 13
receiver, power-line carrier. *See:* **power-line carrier receiver.**
receiver primaries (color television). *See:* **display primaries.**
receiver pulse delay. *See:* **transducer pulse delay.**
receiver relay (power switchgear). An auxiliary relay whose function is to respond to the output of a communications set such as an audio, carrier, radio, or microwave receiver. 127, 103
receiver, telephone. An earphone for use in a telephone system. 176
receiver 1 dB gain compression point (electromagnetic site survey). The input signal level to an otherwise linear receiver for which the gain has been decreased 1 dB (decibel) below the value measured for the mimimum detectable input signal (within the linear response range). 457
receiving (1)(nuclear power quality assurance). Taking delivery of an item at a designated location. 417
(2)(transmission performance of telephone sets). The acoustic output level of a telephone set due to an electric input to the telephone set or connecting test circuit. The electric input may be varied either in frequency or level. The output is measured in an artificial ear and the input is measured as the open-circuit voltage from a source of constant available power.
    491
receiving converter, facsimile (frequency-shift to amplitude-modulation converter). A device which changes the type of modulation from frequency shift to amplitude. *See:* **facsimile transmission.** 12
receiving-end crossfire (telegraph channel). The crossfire from one or more adjacent telegraph channels at the end remote from the transmitting end. *See:* **telegraphy.** 328
receiving loop loss. That part of the repetition equivalent assignable to the station set, subscriber line, and battery supply circuit that are on the receiving end. *See:* **transmission loss.** 328
receiving objective loudness rating (ROLR) (loudness ratings of telephone connections).

$$ROLR = -20 \log_{10} \frac{S_E}{\frac{1}{2} V_W}$$

where

V_W = open-circuit voltage of the electric source (in millivolts)

S_E = sound pressure at the ear reference point (in pascals)

Note: Normally occurring ROLRs will be in the 40 to 55 decibel (dB) range. These numbers are a result of the units chosen and have no physical significance. 409

receiving polarization (of an antenna). That polarization of a plane wave, incident from a given direction and of a given intensity, which results in maximum available power at the antenna terminals. *Notes:* (1) The receiving polarization of an antenna is related to the antenna's polarization on transmit in the following way: in the same plane of polarization, the polarization ellipses have the same axial ratio, the same sense of polarization and the same spatial orientation. Since their senses of polarization and spatial orientation are specified by viewing their polarization ellipses in the respective directions into which they are propagating, one should note that (A) although their senses of polarization are the same, they would appear to be opposite if both waves were viewed in the same direction, and (B) their tilt angles are such that they are the negative of one another with respect to a common reference. (2) The receiving polarization may be used to specify the polarization characteristic of a nonreciprocal antenna which may transmit and receive arbitrarily different polarizations. 111

receiving voltage sensitivity. *See:* **free-field voltage response.**

receptacle (1) (National Electrical Code). A receptacle is a contact device installed at the outlet for the connection of a single attachment plug. A single receptacle is a single contact device with no other contact device on the same yoke. A multiple receptacle is a single device containing two or more receptacles. 256

(2) (electric installations on shipboard). A device installed in a receptacle outlet to accommodate an attachment plug. 3

receptacle circuit. A branch circuit to which only receptacle outlets are connected. *See:* **branch circuit.** 328

receptacle outlet (National Electrical Code) (electric installations on shipboard). An outlet where one or more receptacles are installed. 256, 3

reception diversity (data transmission). That method of radio reception whereby, in order to minimize the effects of fading, a resultant signal is obtained by combination or selection, or both, of two or more sources of received-signal energy that carry the same modulation or intelligence, but that may differ in strength or signal-to-noise ratio at any given instant. 59

receptive field (medical electronics). The region in which activity is observed by means of the pickup electrode. 192

reciprocal bearing (navigation aid terms). The opposite direction to a bearing. 526

reciprocal color temperature (illuminating engineer-ing). Color temperature T_c expressed on a reciprocal scale $(1/T_c)$. An important use stems from the fact that a given small increment in reciprocal color temperature is approximately equally perceptible regardless of color temperature. Also, color temperature conversion filters for sources approximating graybody sources change the reciprocal color temperature by nearly the same amount anywhere on the color temperature scale. *Note:* The unit is the reciprocal megakelvin (MK^{-1}). The reciprocal color temperature expressed in this unit has the numerical value $10^6/T_c$ when T_c is expressed in kelvins. The acronym "mirek" (for micro-reciprocal-kelvin) occasionally has been used in the literature. The acronym "mired" (for micro-reciprocal-degree) is now considered obsolete. 167

reciprocal transducer. A transducer in which the principle of reciprocity is satisfied. *Note:* The use of the term **reversible transducer** as a synonym for reciprocal transducer is deprecated. *See:* **transducer.** 210

reciprocating mechanism (high voltage air switches, insulators, and bus supports). An operating mechanism which produces longitudinal motion of the operating means to open or close the switching device. 27

reciprocity (multiport network). The property described by symmetry of the impedance matrix. In the case of a network with identical ports, it is also described by symmetry of the scattering matrix. 185

reciprocity theorem. States that if an electromagnetic force E at one point in a network produces a current I at a second point in the network, then the same voltage E acting at the second point will produce the same current I at the first point. 328

reclamation (insulating oil). The restoration to usefulness by the removal of contaminants and products of degradation such as polar, acidic, or colloidal materials from used electrical insulating liquids by chemical or adsorbent means. *Note:* The methods listed under reconditioning are usually performed in conjunction with reclaiming. Reclaiming typically includes treatment with clay or other adsorbents. 461

reclosing device (power operations). A control device which initiates the reclosing of a circuit after it has been opened by a protective relay. 516

reclosing fuse (power switchgear). A combination of two or more fuseholders, fuse units, of fuse links mounted on a fuse support or supports, mechanically or electrically interlocked, so that one fuse can be connected into the circuit at a time and the functioning of that fuse automatically connects the next fuse into the circuit, with or without intentionally-added time delay, thereby permitting one or more service restorations without replacement of fuse links, refill units, or fuse units. 103, 443

reclosing interval (of an automatic circuit recloser) (power switchgear). The open-circuit time between an automatic opening and the succeeding automatic reclosure. *See:* **clearing time.** 103

reclosing relay (power switchgear). A programming

relay whose function is to initiate the automatic re-closing of a circuit breaker. 127, 103

reclosing time (of a circuit breaker) (power switch-gear). The interval between the time when the actuating quantity of the release (trip) circuit reaches the operating value (the breaker being in the closed position) and the reestablishment of the circuit on the primary arcing contacts on the reclosing stroke.
103

reclosure (relay). The automatic closing of a circuit-interrupting device following automatic tripping. Re-closing may be programmed for any combination of instantaneous, time-delay, single-shot, multiple-shot, synchronism-check, dead-line-live-bus, or dead-bus-live-line operation. 128

recognition. *See:* **character recognition; magnetic-ink character recognition; optical character recognition; pattern recognition.**

recombination center (solar cells). A defect having electrical properties so as to facilitate the recombination of mobile charge carriers (electrons or holes) with one each of the opposite polarity. 137

recombination rate (volume). The time rate at which free electrons and holes recombine within the volume of a semiconductor. *See:* **semiconductor device.**
113

recombination velocity (semiconductor surface). The quotient of the normal component of the electron (hole) current density at surface by the excess electron (hole) charge density at the surface. *See:* **semiconductor device.** 245

reconditioned carrier reception (exalted-carrier reception). The method of reception in which the carrier is separated from the sidebands for the purpose of eliminating amplitude variations and noise, and then added at increased level to the sideband for the purpose of obtaining a relatively undistorted output. This method is frequently employed, for example, when a reduced-carrier single-sideband transmitter is used. *See:* **radio receiver.** 328

reconditioning (insulating oil). The removal of insoluble contaminants, moisture, and dissolved gases from used, electrical insulating liquids by mechanical means. *Note:* The typical means employed are settling, filtering, centrifuging, and vacuum drying or degassing. 461

reconstituted mica (integrated mica) (rotating machinery). *See:* **mica paper.**

reconstruction (National Electrical Safety Code). Replacement of any portion of an existing installation by new equipment or construction. Does not include ordinary maintenance replacements. 391

record (1) (microprocessor operating systems). A collection of related data or words treated as a unit and saved in a position dependent fashion within a file or other such unit. 478
(2) (software). A collection of related data or words treated as a unit. *See:* **data; logical record; word.**
434

recorded announcement (telephone switching systems). A prerecorded oral message received on a call.
55

recorded spot, X dimension (facsimile). The effective recorded-spot dimension measured in the direction of the recorded line. *Notes:* (1) By effective dimension is meant the largest center-to-center spacing between recorded spots which gives minimum peak-to-peak variation of density of the recorded line. (2) This term applies to that type of equipment which responds to a constant density in the subject copy by a succession of discrete recorded spots. *See:* **recording (facsimile).**
12

recorded spot, Y dimension (facsimile). The effective recorded-spot dimension measured perpendicularly to the recorded line. *Note:* By effective dimension is meant the largest center-to-center distance between recorded lines which gives minimum peak-to-peak variation of density across the recorded lines. *See:* **recording (facsimile).** 12

recorded value. The value recorded by the marking device on the chart, with reference to the division lines marked on the chart. *See:* **accuracy rating (instrument).** 328

recorder (1) (analog computers). A device that makes a permanent record, usually graphic, of varying signals. 9
(2) (facsimile). That part of the facsimile receiver which performs the final conversion of electric picture signal to an image of the subject copy on the record medium. *See:* **facsimile (electrical communication); recording (facsimile).** 12

recorder, strip-chart. A recorder in which one or more records are made simultaneously as a function of time. *See:* **electronic analog computer.** 9,10

recorder-warning tone (telephone switching systems). A tone that indicates periodically that the conversion is being electrically recorded. 55

record gap (1) (computing systems) (storage medium). An area used to indicate the end of a record.
255,77
(2) (test, measurement and diagnostic equipment). An interval of space or time associated with a record to indicate or signal the end of the record. 54

recording (facsimile). The process of converting the electrical signal to an image on the record medium. *See:* **direct recording; electrochemical recording; electrolytic recording; electromechanical recording; electrostatic recording; electrothermal recording; ink vapor recording; magnetic recording; photosensitive recording.** 12

recording channel (electroacoustics). The term refers to one of a number of independent recorders in a recording system or to independent recording tracks on a recording medium. *Note:* One or more channels may be used at the same time for covering different ranges of the transmitted frequency band, for multichannel recording, or for control purposes. *See:* **phonograph pickup.** 176

recording-completing trunk (telephone switching systems). A one-way trunk for operator recording, extending, and automatic completing of toll calls.
55

recording demand meter. A demand meter that records

on a chart the demand for each demand interval. *See:* **electricity meter (meter).** 328

recording, instantaneous. A phonograph recording that is intended for direct reproduction without further processing. *See:* **phonograph pickup.** 176

recording instrument (electrical heating applications to melting furnaces and forehearths in the glass industry). An instrument that makes a graphic record of the value of one or more quantities as a function of another variable, usually time. 520

recording loss (mechanical recording). The loss in recorded level whereby the amplitude of the wave in the recording medium differs from the amplitude executed by the recording stylus. *See:* **phonograph pickup.** 176

recording spot (facsimile). The image area found at the record medium by the facsimile recorder. *See:* **recording (facsimile).** 12

recording stylus. A total that inscribes the groove into the recording medium. *See:* **phonograph pickup.** 176

recording trunk (communications). A trunk extending from a local central office or private branch exchange to a toll office, that is used only for communication with toll operators and not for completing toll connections. 328

record medium (facsimile). A physical medium on which the facsimile recorder forms an image of the subject copy. *See:* **recording (facsimile).** 12

record sheet (facsimile). The medium which is used to produce a visible image of the subject copy in record form. The record medium and the record sheet may be identical. *See:* **recording (facsimile).** 12

record spot (facsimile). The image of the recording spot on the record sheet. *See:* **recording (facsimile).** 12

recoverable light loss factors (illuminating engineering). Factors which give the fractional light loss that can be recovered by cleaning or lamp replacement. 167

recovered charge (semiconductor). The charge recovered from a semiconductor device after switching from a forward current condition to a reverse condition. 66

recovery current (semiconductor rectifier). The transient component of reverse current associated with a change from forward conduction to reverse voltage. *See:* **rectification.** 66

recovery cycle (electrobiology). The sequence of states of varying excitability following a conditioning stimulus. The sequence may include periods such as absolute refractoriness, relative refractoriness, supernormality, and subnormality. *See:* **excitability (electrobiology).** 192

recovery time (1) (power supplies). Specifies the time needed for the output voltage or current to return to a value within the regulation specification after a step load or line change. *Notes:* (A) Recovery time, rather than response time, is the more meaningful and therefore preferred way of specifying power-supply performance, since it relates to the regulation specification.

(B) For load change, current will recover at a rate governed by the rate-of-change of the compliance voltage across the load. This is governed by the resistance-capacitance time constant of the output filter capacitance, internal source resistance, and load resistance. *See:* **programming speed; radar.** 186

(2) **(Geiger-Mueller counters).** The minimum time from the start of a counted pulse to the instant a succeeding pulse can attain a specific percentage of the maximum value of the counted pulse. *See:* **radar.** 125,96

(3) **(antitransmit-receive tubes).** The time required for a fired tube to deionize to such a level that the normalized conductance and susceptance of the tube in its mount are within specified ranges. *Note:* Normalization is with respect to the characteristic admittance of the transmission line at its junction with the tube mount. *See:* **gas tube; radar.** 125

(4) **(gas tubes).** The time required for the control electrode to regain control after anode current interruption. *Note:* To be exact, the deionization and recovery time of a gas tube should be presented as families of curves relating such factors as condensed-mercury temperature, anode current, anode and control electrode voltages, and control-circuit impedance. *See:* **radar.** 125

(5) **(gas turbines).** The interval between two conditions of speed occurring with a specified sudden change in the steady-state electric load on the gas-turbine-generator unit. It is the time in seconds from the instant of change from the initial load condition to the instant when the decreasing oscillation of speed finally enters a specified speed band. *Note:* The specified speed band is taken with respect to the midspeed of the steady-state speed band occurring at the subsequent steady-state load condition. The recovery time for a specified load increase and the same specified load decrease may not be identical and will vary with the magnitude of the load change. 58

(6) **(delay).** *See:* **relay recovery time.**

(7) **(TR and Pre-TR tubes).** The time required for a fired tube to deionize to such a level that the attenuation of a low-level radio-frequency signal transmitted through the tube is decreased to a specified value. 125

(8) **(reverse-blocking thyristor or semiconductor diode).** *See:* **reverse recovery time (thyristor or semiconductor diode).**

recovery voltage (1) (power switchgear). The voltage that occurs across the terminals of a pole of a circuit interrupting device upon interruption of the current. 103, 443

(2) **(surge arrester).** The voltage that occurs across the terminals of a pole of circuit-interrupting device upon the interruption of the current. *Note:* For an arrester, this occurs as a result of interruption of the follow current. 430

recreational vehicle (National Electrical Code). A vehicular type unit primarily designed as temporary living quarters for recreational, camping, or travel use,

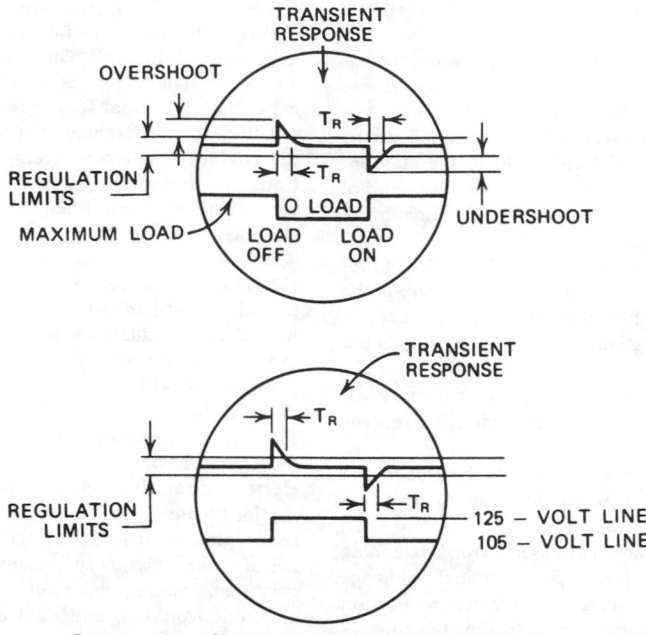

Recovery time. Oscilloscope views showing (top) the effects
of a step load change, and (bottom) the effects of a step
line change. T_R = recovery time.

which either has its own motive power or is mounted
on or drawn by another vehicle, The basic entities are:
travel trailer, camping trailer, truck camper and motor
home.   256
recreational vehicle park (National Electrical Code).
A plot of land upon which two or more recreational
vehicle sites are located, established or maintained for
occupancy by recreational vehicles of the general public as temporary living quarters for recreation or vaca-
tion purposes.   256
recreational vehicle site (National Electrical Code). A
plot of ground within a recreational vehicle park intended for the accommodation of either a recreational
vehicle, tent, or other individual camping unit on a
temporary basis.   256
**recreational vehicle site feeder circuit conductors
(National Electrical Code).** The conductors from the
park service equipment to the recreational vehicle site
supply equipment.   256
**recreational vehicle site supply equipment (National
Electrical Code).** The necessary equipment, usually a
power outlet, consisting of a circuit breaker or switch
and fuse and their accessories, located near the point
of entrance of supply conductors to a recreational
vehicle site and intended to constitute the disconnecting means for the supply to that site.   256
recreational vehicle stand (National Electrical Code).
That area of a recreational vehicle site intended for the
placement of a recreational vehicle.   256
rectangular array. *See:* **rectangular grid array.**

rectangular grid array (antennas). A rectangular arrangement of array elements, in a plane, such that
lines connecting corresponding points of adjacent elements form rectangles.   111
rectangular impulse (surge arresters). An impulse that
rises rapidly to a maximum value, remains substantially constant for a specified period, and then falls rapidly
to zero. The parameters that define a rectangular impulse wave are polarity, peak value, duration of the
peak, total duration.   62
rectangular mode. *See:* **resolver.**
rectangular-shape logic symbol. A logic symbol in
which the logic function is indicated by a qualifying
symbol in its interior.   88
rectangular wave (data transmission). A periodic wave
which alternately assumes one of two fixed values, the
time of transition being negligible in comparison with
the duration of each fixed value.   59
rectification. The term used to designate the process by
which electric energy is transferred from an alternating-current circuit to a direct-current circuit.
                                                      237,66
rectification error (accelerometer). A steady state error in the output while vibratory disturbances are acting on an accelerometer. Anisoelasticity is one source
of rectification error.   46
rectification factor. The quotient of the change in average current of an electrode by the change in amplitude
of the alternating sinusoidal voltage applied to the
same electrode, the direct voltages of this and other
electrodes being maintained constant. *See:* **conduc-**

tance for rectification; transrectification factor.
       125

rectification failure (58) (power system device function numbers). A device that functions if one or more anodes of a power rectifier fail to fire, or to detect an arc bac, or on failure of a diode to conduct or block properly.   402

rectification of an alternating current. Process of converting an alternating current to a unidirectional current. *See:* **electronic rectifier; inverse voltage (rectifier); semiconductor device.**   244,190

rectified unbalance (dynamically tuned gyro). *See:* **gimbal unbalance torque.**

rectified value (alternating quantity). The average of all the positive values of the quantity during an integral number of periods. Since the positive values of a quantity y are represented by the expression

$$\frac{1}{2}\Big[\,y + |y|\,\Big],$$

$$y_\mathrm{r} = \frac{1}{T}\int_0^T \frac{1}{2}\Big[\,y + |y|\,\Big]\,\mathrm{d}t.$$

Note: The word positive and the sign $+$ may be replaced by the word negative and the sign $-$.
       210

rectifier (1)(self-commutated converters). A converter for conversion from alternating current (ac) to direct current (dc).   584
(2) (electric installations on shipboard). A converter in which the direction of average power flow is from the ac circuit to the dc circuit.   3
(3) (generating stations electric power system). A device for converting ac to dc.   381

rectifier anode. An electrode of the rectifier from which the current flows into the arc. *Note:* The direction of current flow is considered in the conventional sense from positive to negative. The cathode is the positive direct-current terminal of the apparatus and is usually a pool of mercury. The neutral of the transformer secondary system is the negative direct-current terminal of the rectifier unit. *See:* **rectification.**   328

rectifier assembly. A complete unit containing rectifying components, wiring, and mounting structure capable of converting alternating-current power to direct-current power. *See:* **converter.**   63

rectifier cathode. The electrode of the rectifier into which the current flows from the arc. *Note:* The direction of current flow is considered in the conventional sense from positive to negative. The cathode is the positive direct-current terminal of the rectifier unit and is usually a pool of mercury. The neutral of the transformer secondary system is the negative direct-current terminal of the rectifier unit. *See:* **rectification.**
       328

rectifier circuit element. A circuit element bounded by two circuit terminals that has the characteristic of conducting current substantially in one direction only. *Note:* The rectifier circuit element may consist of more than one semiconductor rectifier cell, rectifier

diode, or rectifier stack connected in series or parallel or both, to operate as a unit.   66

rectifier electric locomotive. An electric locomotive that collects propulsion power from an alternating-current distribution system and converts this to direct current for application to direct-current traction motors by means of rectifying equipment carried by the locomotive. *Note:* A rectifier electric locomotive may be defined by the type of rectifier used on the locomotive, such as **ignitron electric locomotive.** *See:* **electric locomotive.**   328

rectifier electric motor car. An electric motor car that collects propulsion power from an alternating-current distribution system and converts this to direct current for application to direct-current traction motors by means of rectifying equipment carried by the motor car. *Note:* A rectifier electric motor car may be defined by the type of rectifier used on the motor car, such as **ignitron electric motor car.** *See:* **electric motor car.**   328

rectifier instrument. The combination of an instrument sensitive to direct current and a rectifying device whereby alternating currents or voltages may be measured. *See:* **instrument.**   328

rectifier junction (semiconductor rectifier cell or diode). The junction in a semiconductor rectifier cell that exhibits asymmetrical conductivity. *See:* **semiconductor; semiconductor rectifier stack.**   66

rectifier stack (semiconductor). An integral assembly of one or more rectifier diodes, including its associated mounting and cooling attachments if integral with it. *See:* **semiconductor rectifier stack.**   66

rectifier transformer (power and distribution transformer). A transformer that operates at the fundamental frequency of an alternating-current system and designated to have one or more output windings conductively connected to the main electrodes of a rectifier. *See:* **power rectifier transformer; alternating current winding of a rectifier transformer; direct-current winding of rectifier transformer; rating of rectifier transformer; interphase transformer; rating of interphase transformer; anode paralleling reactor; commutating reactor.**   53

rectifier tube (valve). An electronic tube or valve designed to rectify alternating current.   190

rectifier unit. An operative assembly consisting of the rectifier, or rectifiers, together with the rectifier auxiliaries, the rectifier transformer equipment, and the essential switchgear. *See:* **rectification; rectifier transformer.**   208

rectifying device. An elementary device, consisting of one anode and its cathode, that has the characteristic of conducting current effectively in only one direction. *See:* **rectification.**   328

rectifying element. A circuit element that has the property of conducting current effectively in only one direction. *Note:* When a group of rectifying devices is connected, either in parallel or series arrangement, to operate as one circuit element, the group of rectifying devices should be considered as a rectifying element. *See:* **rectification; rectifying device; rectifier circuit**

element; rectifying junction; metallic rectifying cell.
328

rectifying junction (barrier layer) (blocking layer). The region in a metallic rectifier cell that exhibits the asymmetrical conductivity. *See:* **rectification.**
328

rectilinear scanning (television). The process of scanning an area in a predetermined sequence of straight parallel scanning lines.
18

rector, shunt. *See:* **shunt reactor.**

recurrence rate (pulse techniques). *See:* **pulse repetition frequency.**

recurrent sweep. A sweep that repeats or recurs regularly. It may be free-running or synchronized. *See:* **oscillograph.**
185

recursive routine (software). A routine that may be used as a subroutine of itself, calling itself directly or being called by another subroutine, one that it itself has called. The use of a recursive routine usually requires the keeping of records of the status of its unfinished uses in, for example, a pushdown list. *See:* **list; routine; subroutine.**
434

redirecting surfaces or media (illuminating engineering). Those which change the direction of the flux without scattering the redirected flux.
167

redistribution (storage or camera tubes). The alteration of the charge pattern on an area of a storage surface by secondary electrons from any other part of the storage surface. *See:* **charge-storage tube.**
174,125

reduced full-wave test (power and distribution transformer). A wave similar in shape and duration to that involved in a "full-wave lightning impulse test", but reduced in magnitude. *Note:* The reduced full wave normally has a crest value between 50 and 70 percent of the full-wave value involved, and is used for comparison of oscillograms in failure detection.
53

reduced generator efficiency (thermoelectric device). The ratio of (1) a specified generator efficiency to (2) the corresponding Carnot efficiency. *See:* **thermoelectric device.**
191

reduced kVA (kilovoltampere) tap (in a transformer) (power and distribution transformer). A tap through which the transformer can deliver only an output less than rated kVA without exceeding the specified temperature rise. The current is usually that of the rated kVA tap.
53

reduced-voltage starter (industrial control). A starter, the operation of which is based on the application of a reduced voltage to the motor. *See:* **starter.**
244,206

reducing joint. A joint between two lengths of cable the conductors of which are not the same size. *See:* **branch joint; cable joint; straight joint.**
64

redundancy (1)(emergency and standby power). Duplication of elements in a system or installation for the purpose of enhancing the reliability or continuity of operation of the system or installation.
512

(2) (data transmission). In the transmission of information, the fraction of the gross information content of a message which can be eliminated without loss of

essential information. Numerically, it is one minus the ratio of the net information content to the gross information content, expressed in percent.
59

(3) (software). The inclusion of duplicate or alternate system elements to improve operational reliability by ensuring continued operation in the event that a primary element fails. *See:* **operational reliability; system.**
434

(4) (source) (information theory). The amount by which the logarithm of the number of symbols available at the source exceeds the average information content per symbol of the source. *Note:* The term redundancy has been used loosely in other senses. For example, a source whose output is normally transmitted over a given channel has been called redundant if the channel utilization index is less than unity. *See:* **information theory.**
415

(5) (reliability). In an item, the existence of more than one means for performing a given function.
164

redundancy, active (reliability). That redundancy wherein all means for performing a given function are operating simultaneously.
164

redundancy check (1) (electronic computation). *See:* **check, forbidden-combination.**

(2) (power generation). *See:* **check, redundant.**

redundancy, standby (reliability). That redundancy wherein the alternative means for performing a given function are inoperative until needed.
164

redundant (1). Pertaining to characters that do not contribute to the information content. Redundant characters are often used for checking purposes or to improve reliability. *See:* **parity; self-checking code; error-detecting code; check, forbidden-combination; check digit.**
235

(2) (cable systems in power generating stations). Applied to two or more systems serving the same objective, where they are also either: (A) Systems where personnel or public safety is involved, such as fire pumps. (B) Systems provided with redundancy because of the severity of economic consequences of equipment damage. (Turbine-generator ac and dc bearing oil pumps are examples of redundant equipment under this definition.)
35

(3) (electric pipe heating systems). The introduction of auxiliary elements and components to a system to perform the same function as other elements in the system for the purpose of improving reliability. Redundant electric pipe heating systems consist of two heaters and two controllers, each with its own sensor, supplied from two power systems, all independent of each other but all applied to the same mechanical piping, valves, tanks, etcetera. Redundant electric pipe heating systems are referred to as primary and backup in this recommended practice. *Syn:* **redundancy.**
405

redundant check (checking code)(data transmission). A check which uses extra digits (check bits) short of complete duplication, to help detect the absence of error within the character or block. *Syn:* **checking code.**
59

redundant equipment or system (1)(diesel-generator

unit). An equipment or system that duplicates the essential function of another equipment or system to the extent that either may perform the required function regardless of the state of operation or failure of the other.   99, 131

(2) (accident monitoring instrumentation). A piece of equipment or a system that duplicates the essential function of another piece of equipment or system to the extent that either may perform the required function regardless of the state of operation or failure of the other. *Note:* Redundancy can be accomplished by use of identical equipment, equipment diversity, or functional diversity.   421, 102

(3) (nuclear power generating station). Same as the preceding definition except for the following note. *Note:* The term redundant could include identical equipment, equipment diversity, or functional diversity.   387

redundant systems (cable systems). Two or more systems serving the same objective, where they are also either systems where personnel or public safety is involved, such as fire pumps, or systems provided with redundancy because of the severity of economic consequences of equipment damage. *Note:* Turbine-generator alternating-current and direct-current bearing oil pumps are examples of redundant equipment under this definition.   35

reed relay. A relay using glass-enclosed, magnetically closed reeds as the contact members. Some forms are mercury wetted.   341

reel puller (conductor stringing equipment). A device designed to pull a conductor during stringing operations. It is normally equipped with its own engine which drives the supporting shaft for the reel mechanically, hydraulically or through a combination of both. The shaft, in turn, drives the reel. The application of this unit is essentially the same as that for the drum puller. Some of these devices function as either a puller or tensioner. *See:* **drum puller.**   431

reel stand (conductor stringing equipment). A device designed to support one or more reels and having the possibility of being skid, trailer or truck mounted. These devices may accomodate rope or conductor reels of varying sizes and are usually equipped with reel brakes to prevent the reels from turning when pulling is stopped. They are used for either slack or tension stringing. The designation of reel trailer or reel truck implies that the trailer or truck has been equipped with a reel stand (jacks) and may serve as a reel transport or payout unit, or both, for stringing operations. Depending upon the sizes of the reels to be carried, the transporting vehicles may range from single axle trailers to semitrucks with trailers having multiple axles. *Syn:* **reel trailer; reel transporter; reel truck.**   431

reel tensioner (conductor stringing equipment). a device designed to generate tension against a pulling line or conductor during the stringing phase. Some are equipped with their own engines which retard the supporting shaft for the reel mechanically, hydraulically or through a combination of both. The shaft, in turn, retards the reel. Some of these devices function as either a puller or tensioner. Other tensioners are only equipped with friction type retardation. *Syn:* **retarder; tensioner.**   431

reel winder (conductor stringing equipment). A device designed to serve as a recovery unit for a pulling line. It is normally equipped with its own engine which drives a supporting shaft for a reel mechanically, hydraulically or through a combination of both. The shaft, in turn, drives the reel. It is normally used to rewind a pulling line as it leaves the bullwheel puller during stringing operations. This unit is not intended to serve as a puller, but sometimes serves this function where only low tensions are involved. *Syn:* **takeup reel.**   431

reentrant beam (microwave tubes). An undeterminated recirculating electron beam. *See:* **microwave tube (orvalve).**   190

reentrant-beam crossed-field amplifier (amplitron) (microwave tubes). A crossed-field amplifier in which the beam is reentrant and interacts with either a forward or a backward wave. *See:* **microwave tube (or valve).**   190

reentrant circuit (microwave tubes). A slow-wave structure that closes upon itself. *See:* **microwave tube (or valve).**   190

reentrant switching network (telephone switching systems). A switching network in which outlets (usually last choice) from a given connecting stage are connected to inlets of the same or previous stage.   55

re-entry communication (communication satellite). Communication during re-entry of a space vehicle into the atmosphere. Usually the ionization requires a special system of modulation to overcome the communication blackout.   84

referee test (metering). A test made by or in the presence of one or more representatives of a regulatory body or other impartial agency.   212

reference (computing system). *See:* **linearity.**

reference accuracy (automatic null-balancing electric instrument). A number or quantity that defines the limit of error under reference operating conditions. *Notes:* (1) It is usually expressed as a percent of the span. It is preferred that a + sign or − sign or both precede the number of quantity. The absence of a sign infers a ± sign. (2) Reference accuracy does not include accuracy of sensing elements or intermediate means external to the instrument. *See:* **error and correction; accuracy rating (instrument).**   295

reference air line. A uniform section of air-dielectric transmission line of accurately calculable characteristic impedance used as a standard immittance. *See:* **transmission line.**   185

reference audio noise power output (mobile communications receivers). The average audio noise power present at the output of an unsquelched receiver having no radio-frequency signal input in which the audio gain has been adjusted for the reference audio power output.   123

reference audio power output (mobile communications

receivers). The manufacturer's rated audio-frequency power available at the output of a properly terminated receiver, when responding to a standard test modulated radio-frequency input signal at a -80 dBW level.   123

reference ballast (illuminating engineering). A ballast which is specially constructed, having certain prescribed characteristics and which is used for testing electric-discharge lamps and other ballasts.   167

reference ballasts. Specially constructed series ballasts having certain prescribed characteristics. *Note:* They serve as comparison standards for use in testing ballasts or lamps and are used also in selecting the reference lamps that are necessary for the testing of ballasts. Reference ballasts are characterized by a constant impedance over a wide range of operating current. They also have constant characteristics that are relatively uninfluenced by time and temperature. *See:* **fixed-impedance type (reference ballast); primary standards (illuminating engineering); variable-impedance type (reference ballast).**   167

reference black level (television). The picture-signal level corresponding to a specified maximum limit for black peaks.   178

reference block (numerically controlled machines). A block within the program identified by an *o* (letter o) in place of the word address *n* and containing sufficient data to enable resumption of the program following an interruption. This block should be located at a convenient point in the program that enables the operator to reset and resume operation.   207

reference boresight (antennas). A direction established as a reference for the alignment of an antenna. *Note:* The direction can be established by optical, electrical or mechanical means. *See:* **electrical boresight.**
                               111

reference conditions. The values assigned for the different influence quantities at which or within which the instrument complies with the requirements concerning errors in indication. *See:* **accuracy rating (instrument).**   280

reference current (fluorescent lamp). The value of current specified in a specific lamp standard. *Note:* It is normally the same as the value of current for which the corresponding lamp is rated. Since the reference ballast is a standard that is representative of the impedance of lamp power sources installed, it is not necessary to change this current value unless major changes in lamp standards require modification of the ballast impedance. For this reason, reference ballast characteristics are specified in terms of, and with reference to, reference current.   270

reference deflection (volume measurements of electrical speech and program waves). The deflection to the meter-scale point marked 0 vu, 100, or both. *Note:* This is the deflection at which the meter should be used. *See:* **vu.**   523

reference designation (1) (symbols), (abbreviation). Numbers, or letters and numbers, used to identify and locate units, portions thereof, and basic parts of a specific set. *Compare with:* **functional designation** and **symbol for a quantity.** *See:* **abbreviation.**   173
(2) (electric and electronics parts and equipments). Letters or numbers, or both, used to identify and locate discrete units, portions thereof, and basic parts of a specific set. *Note:* A reference designation is not a letter symbol, abbreviation, or functional designation for an item.   17
reference direction (1)(navigation aid terms). A direction from which other directions are reckoned; for example, true north, grid north, and so on.   526
(2) (energy) (specified circuit). With reference to the boundary of a delimited region, the arbitrarily selected direction in which electric energy is assumed to be transmitted past the boundary, into or out of the region. *Notes:* (A) When the actual direction of energy flow is the same as the reference direction, the sign is negative. (B) Unless specifically stated to the contrary, it shall be assumed that the reference direction for all power, energy, and quadergy quantities associated with the circuit is the same as the reference direction of the energy flow. (C) In these definitions it will be assumed that the reference direction of the current in each conductor of the circuit is the same as the reference direction of energy flow.   210
reference directivity. *See:* **standard directivity.**

reference distance (sound measurement). A standard 1 m distance from the major machine surfaces at which mean sound level data shall be reported.   129

reference excursion (analog computer). The range from zero voltage to nominal full-scale operating voltage. *See:* **electronic analog computer.**   9
reference frequency (information theory). The frequency upon which a phasor or amplitude-or-phase representation of signals is based.   415
reference frequency, upper and lower. *See:* **bandwidth.**
reference grounding point (1) (health care facilities). A terminal bus which is the equipment grounding bus or an extension of the equipment grounding bus and is a convenient collection point for grounding of electrical appliances and equipment, and, when necessary and appropriate, exposed conductive surfaces in a patient vicinity.   192
(2) (National Electrical Code) (health care facilities). A terminal bus which is the equipment grounding bus or an extension of the equipment grounding bus and is a convenient collection point for grounding all electric appliances, equipment, and exposed conductive surfaces in a patient vicinity.   256
reference input signal (industrial control). *See:* **signal, reference input.**
reference lamp (1) (mercury). A seasoned lamp that under stable burning conditions, in the specified operating position (usually vertically, base up), and in conjunction with the reference ballast rated input voltage, operates at values of lamp volts, watts, and amperes, each within $\pm$ 2 percent of the nominal

values.   274,271

(2) (fluorescent). Seasoned lamps that under stable burning conditions, in conjunction with the reference ballast specified for the lamp size and rating, and at the rated reference ballast supply voltage, operate at values of lamp volts, watts, and amperes each within $\pm$ 2 1/2 percent of the values, and under conditions established by present standards. *See:* **reference ballasts.**   268

reference line (1)(navigation aid terms). A line from which angular or linear measurements are reckoned.   526

(2) (illuminating engineering). Either of two radial lines where the surface of the cone of maximum candlepower is intersected by a vertical plane parallel to the curb line and passing through the light-center of the luminaire.   167

reference lines and points (pulse terms). Constructs which are (either actually or figuratively) superimposed on waveforms for descriptive or analytical purposes. Unless otherwise specified, all defined lines and points lie within a waveform epoch. *See:* **time origin line; magnitude origin line; time reference line; time referenced point; magnitude reference line; magnitude referenced point; knot; cubic natural spline.**   254

reference modulation (VOR [very high-frequency omnidirectional range])(navigation aid terms). That modulation of the ground-station radiation which produces a signal in the air-borne receiver whose phase is independent of the bearing of the receiver; the reference signal derived from this modulation is used for comparison with the variable signal.   526

reference noise (data transmission). The magnitude of circuit noise that will produce a circuit-noise-meter reading equal to that produced by 10^{-12} watt of electric power at 1000 Hz (hertz).   59

reference operating conditions (automatic null-balancing electric instrument). The conditions under which reference performance is stated and the base from which the values of operating influences are determined. *See:* **measurement system.**   295

reference performance (1) (watthour meter). Performance at specified reference conditions for each test, used as a basis for comparison with performance under other conditions of the test.   212

(2) (automatic null-balancing electric instrument). The limits of the values of certain operating characteristics of the instrument that will not be exceeded under any combination of reference operating conditions. *See:* **electricity meter (meter); test (instrument or meter).**   295

reference plane. A plane perpendicular to the direction of propagation in a waveguide or transmission line, to which measurement or immittance, electric length, reflection coefficients, scattering coefficients, and other parameters may be referred. *See:* **waveguide.**   185

reference plane, electrical (standard connector). A transverse plane of the waveguide or transmission line on the drawing standardizing the critical mating di-

mensions shown in relation to the mechanical reference plane. *Notes:* (1) The electrical reference planes of two mating standard connectors forming a mated standard connector pair nearly coincide. (2) The electrical and mechanical reference planes of standard connectors do not necessarily coincide except for precision coaxial connectors complying with the IEEE Std 287-1968. Precision Coaxial Connectors and many connectors for uniconductor waveguides.   110

reference plane, mechanical (standard connector). A transverse plane of the waveguide or transmission line to which all critical, longitudinal dimensions are referenced to assure nondestructive mating; it is the only plane where a mated standard connector pair butt against one another. *Note:* Usually a stable, rugged metal surface.   110

reference power supply. A regulated, electronic power supply furnishing the reference voltage. *See:* **electronic analog computer.**   9

reference radius (sound measurement). The sum of the reference distance and one half the maximum linear dimension as defined for **small-, medium-,** or **large** machines. *See:* **machine (3).**   129

reference sensitivity (mobile communications receivers). The level of a radio-frequency signal with standard test modulation which provides a 12-decibel sinad with at least 50 percent reference audio power output.   123

reference standards (measuring and test equipment) (nuclear power generating stations). Standards (that is, primary, secondary and working standards, where appropriate) used in a calibration program. These standards establish the basic accuracy limits for that program.   41

reference standard watthour meter. A meter used to maintain the unit of electric energy. It is usually designed and operated to obtain the highest accuracy and stability in a controlled laboratory environment.   212

reference surface (fiber optics). That surface of an optical fiber which is used to contact the transverse-alignment elements of a component such as a connector. For various fiber types, the reference might be the fiber core, cladding, or buffer layer surface. *Note:* In certain cases the reference surface may not be an integral part of the fiber. *See:* **ferrule; optical waveguide connector.**   433

reference system (loudness ratings of telephone connections). A system that provides 0 decibel (dB) acoustic gain between a mouth reference point at 25 millimeters (mm) in front of a talker's lips and an ear reference point at the entrance to the ear canal of a listener, when the listener is using an earphone. This system is assigned a loudness rating of 0 dB. The frequency characteristic of the system must be flat over the range 300-3300 hertz (Hz) and show infinite attenuation outside of this range. *Notes:* (1) If an actual reference system is constructed for subjective comparison purposes, the system response at 300 and 3300 Hz shall be down 3 ± 1 dB relative to the midband response. The gain of the system shall be adjusted to compensate for the finite slope of the filter skirts and deviation from flatness of the pass band. The amount of this adjustment can be determined by first calculating the objective loudness rating (OLR) (3.3) over a frequency range that includes at least the 50 dB down points of the real response, and next calculating the OLR of the ideal response, over the same frequency range. The difference between the OLRs is the required gain adjustment. (2) To the extent that the artificial mouth and artificial ear replicate their average human counterparts, a virtual reference system is introduced by reason of the calibration of the test system described in Section 5.   409

reference test field (direction-finder testing)(navigation aid terms). That field strength, in microvolts per meter, numerically equal to the DF (direction finder) sensitivity.   526

reference threshold squelch adjustment (mobile communications receivers). The minimum adjustment position of the squelch control required to reduce the reference audio noise power output by at least 40 decibels.   123

reference time (magnetic storage). An instant near the beginning of switching chosen as an origin for time measurements. It is variously taken as the first instant at which the instantaneous value of the drive pulse, the voltage response of the magnetic cell, or the integrated voltage response reaches a specified fraction of its peak pulse amplitude.   77

reference voltage (analog computers). In an analog computer, a voltage used as a standard of reference, usually the nominal full scale of the computer.   9

reference volume (volume measurements of electrical speech and program waves). The level which gives a reading of 0 vu on a standard volume indicator. *Notes:* (1) The methods of reading and calibration are described in Section 3 of ANSI/IEEE Std 152-1953. (2) The 'reading of 0 vu' is the algebraic sum of the meter and attenuator readings on the standard volume indicator. *See:* **vu; standard volume indicator.**   523

reference waveguide. A uniform section of waveguide with accurately fabricated internal cross-sectional dimensions used as a standard of immittance. *See:* **waveguide.**   185

reference white (television). (1) **original scene.** The light from a nonselective diffuse reflector that is lighted by the normal illumination of the scene. *Notes:* (A) Normal illumination is not intended to include lighting for special effects. (B) In the reproduction of recorded material, the word scene refers to the original scene. (2) **color television display.** That white with which the display device simulates reference white of the original scene. *Note:* In general, the reference whites of the original scene and of the display device are not colorimetrically identical.   18

reference white level (television). The picture-signal level corresponding to a specified maximum limit for white peaks. *See:* **television.**   178

refill unit (of a high-voltage fuse unit) (power switch-gear). An assembly comprised of a conducting element, the complete arc-extinguishing medium, and parts normally required to be replaced after each circuit interruption to restore the fuse unit to its original operating condition.   103, 443

reflectance (1) (fiber optics). The ratio of reflected power to incident power. *Note:* In optics, frequently expressed as optical density or as a percent; in communication applications, generally expressed in decibels (dB). Reflectance may be defined as specular or diffuse, depending on the nature of the reflecting surface. Formerly: "reflection". *See:* **reflection.**   433

(2), $\rho = \Phi_r/\Phi_i$ **(of a surface or medium) (illuminating engineering).** The ratio of the reflected flux to the incident flux. Reflectance is a function of:
(1) Geometry (a) of the incident flux (b) of collection for the reflected flux (2) Spectral distribution (a) characteristic of the incident flux (b) weighting function for the collected flux (3) Polarization (a) of the incident flux (b) component defined for the collected flux *Notes:* (1) Unless the state of polarization for the incident flux and the polarized component of the reflected flux are stated, it shall be considered that the incident flux is unpolarized and that the total reflected flux (including all polarizations) is evaluated. (2)Unless qualified by the term 'spectral' (see spectral reflectance) or other modifying adjectives, luminous reflectance (see luminous reflectance) is meant. (3) If no qualifying geometric adjective is used, reflectance for hemispherical collection is meant. *See:* **bihemispherical reflectance; hemispherical-conical reflectance; hemispherical-directional reflectance; conical-hemispherical reflectance; biconical reflectance; conical-directional reflectance; directional-hemispherical reflectance; directional conical reflectance; bidirectional reflectance.** (4) Certain of the reflectance terms are theoretically imperfect and are recognized only as practical concepts to be used when applicable. Physical measurements of the incident and reflected flux are always biconical in nature. Directional reflectances (see above) cannot exist since one component is finite while the other is infinitesimal; here the reflectance-distribution function (see bidirectional reflectance-distribution function is required. However, the concepts of directional and hemispherical reflectances have practical application in instrumentation, measurements, and calculations when including the aspect of the nearly zero or nearly 2 PI conical angle would increase complexity without appreciably affecting the immediate results. (5) In each case of conical incidence or collection, the solid angle is not restricted to a right circular cone, but may be of any cross section including rectangular, a ring, or a combination of two or more solid angles.   167

(3) (laser-maser) (reflectivity ρ**).** The ratio of total reflected radiant power to total incident power.   363

reflectance factor, R (illuminating engineering). The ratio of the flux actually reflected by a sample surface to that which would be reflected into the same reflected-beam geometry by an ideal (lossless), perfectly diffuse (lambertian) standard surface irradiated in exactly the same way as the sample. Note the analogies to reflectance in the fact that nine canonical forms are possible paralleling bihemispherical reflectance, hemispherical- conical reflectance, hemispherical-directional reflectance, conical-hemispherical reflectance, biconical reflectance, conical- directional reflectance, directional-hemispherical reflectance, directional-conical reflectance, and bidirectional reflectance, that spectral may be applied as a modifier, that it may be luminous or radiant reflectance factor, etcetera.   167

reflected binary code. *See:* **Gray code.**

reflected glare (illuminating engineering). Glare resulting from reflections of high luminance in polished or glossy surfaces in the field of view. It usually is associated with reflections from within a visual task or areas in close proximity to the region being viewed. *See:* **veiling reflections.**   167

reflected harmonics (electric conversion). Harmonics produced in the prime source by operation of the conversion equipment. These harmonics are produced by current-impedance (IZ) drop due to nonsinusoidal load currents, and by switching or commutating voltages produced in the conversion equipment. *See:* **electrical conversion.**   186

reflected-light scanning (industrial control). The scanning of changes in the magnitude of reflected light from the surface of an illuminated web.   206

reflected wave (1) (data transmission). When a wave in one medium is incident upon a discontinuity or a different medium, the reflected wave is the wave component that results in the first medium in addition to the incident wave. *Note:* The reflected wave includes both the reflected rays of geometrical optics and the diffracted wave.   59

(2) (overhead-power-line corona and radio noise). Same as the preceding definition but without the note..   411

(3) (waveguide). At a transverse plane in a transmission line or waveguide, a wave returned from a reflecting discontinuity in a direction opposite to the incident wave. *See:* **incident wave.**   267

(4) (surge arresters). A wave, produced by an incident wave, that returns in the opposite direction to the incident wave after reflection at the point of transition.   244,62

reflection (1) (fiber optics). The abrupt change in direction of a light beam at an interface between two dissimilar media so that the light beam returns into the medium from which it originated. Reflection from a smooth surface is termed specular, whereas reflection from a rough surface is termed diffuse. *See:* **critical angle; reflectance; reflectivity; total internal reflection.**   433

(2) (illuminating engineering). A general term for the process by which the incident flux leaves a (stationary) surface or medium from the incident side, without change in frequency. *Note:* Reflection is usually a combination or regular and diffuse reflection. *See:* **regular (specular) reflection; diffuse reflection.**   167

(3) (laser-maser). Deviation of radiation following incidence on a surface.   363

reflection coefficient (waveguide). At a given frequency, at a given point, and for a given mode of propagation, the ratio of some quantity associated with the reflected wave to the corresponding quantity in the incident wave. *Note:* The reflection coefficient may be different for different associated quantities, and the chosen quantity must be specified. The voltage reflection coefficient is most commonly used and is defined as the ratio of the complex electrical field strength (or voltage) of the reflected wave to that of the incident wave. Examples of other quantities are power or current.   267

reflection color tube. A color-picture tube that produces an image by means of electron reflection techniques in the screen region. *See:* **television.**   190,125

reflection error (navigation aid terms). The error due to the fact that some of the total received signal arrives from a reflection rather than all by way of the direct path.   526

reflection factor (1) (data transmission). The reflection factor between two impedances Z_1 and Z_2 is:

$$\frac{(4Z_1Z_2)^{1/2}}{Z_1 + Z_2}$$

Physically, the reflection factor is the ratio of the current delivered to a load, whose impedance is not matched to the source, to the current that would be delivered to a load of matched impedance.   59

(2) (reflex klystron). The ratio of the number of electrons of the reflected beam to the total number of electrons that enter the reflector space in a given time. *See:* **velocity-modulated tube.**   244,190

(3) (electrothermic power meter). The ratio of the power absorbed in, to the power incident upon, a load; mathematically, $1 - |\Gamma_d|^2$, where $|\Gamma_d|$ is the magnitude of the reflection coefficient of the load.   47

reflectionless termination. *See:* **termination, reflectionless.**

reflectionless transmission line. A transmission line having no reflected wave at any transverse section. *See:* **transmission line.**   185

reflectionless waveguide. A waveguide having no reflected wave at any transverse section. *See:* **waveguide.**   185

reflection loss (1) (data transmission). The reflection loss for a given frequency at the junction of a source of power and a load is given by the formula

$$20 \log_{10} \frac{Z_1 + Z_2}{(4Z_1Z_2)^{1/2}} \text{ decibels}$$

where Z_1 is the impedance of the source of power and Z_2 is the impedance of the load. Physically, the reflection loss is the ratio, expressed in decibels (dB), of the scalar values of the volt-amperes delivered to a load of the same impedance as the source. The reflection loss is equal to the number of decibels which corresponds to the scalar value of the reciprocal of the reflection factor. *Note:* When the two impedances have opposite phases and appropriate magnitudes, a reflection gain may be obtained.   59

(2) (or gain) (waveguide). The ratio of incident to transmitted power at a reference plane of a network.   267

reflection mode photocathode (photomultipliers for scintillation counting). A photocathode wherein photoelectrons are emitted from the same surface as that on which the photons are incident.   117

reflection modulation (storage tubes). A change in character of the reflected reading beam as a result of the electrostatic fields associated with the stored signal. A suitable system for collecting electrons is used to extract the information from the reflected beam. *Note:* Typically the beam approaches the target closely at low velocity and is then selectively reflected toward the collection system. *See:* **charge-storage tube.**   174

reflective array antenna. An antenna consisting of a feed and an array of reflecting elements arranged on a surface and adjusted so that the reflected waves from the individual elements combine to produce a prescribed secondary pattern. *Note:* The reflecting elements are usually waveguides containing electrical phase shifters and are terminated by short circuits. *Syn:* **reactive reflector antenna.**   111

reflectivity (1) (fiber optics). The reflectance of the surface of a material so thick that the reflectance does not change with increasing thickness; the intrinsic reflectance of the surface, irrespective of other parameters such as the reflectance of the rear surface. No longer in common usage. *See:* **reflectance.**   433

(2) (photovoltaic power system). The reflectance of an opaque, optically smooth, clean portion of material.   186

reflectometer (1) (illuminating engineering). A photometer for measuring reflectance. *Note:* Reflectometers may be visual or physical instruments.   167

(2). An instrument for the measurement of the ratio of reflected-wave to incident-wave amplitudes in a transmission system. *Note:* Many instruments yield only the magnitude of this ratio. *See:* **instrument.**   185

reflectometer, time-domain. An instrument designed to indicate and to measure reflection characteristics of a transmission system connected to the instrument by monitoring the step-formed signals entering the test object and the superimposed reflected transient signals on an oscilloscope equipped with a suitable time-base sweep. The measuring system, basically, consists of a fast-rise function generator, a tee coupler, and an oscilloscope connected to the probing branch of the coupler. *See:* **instrument.**   185

reflector (1) (data transmission). One or more conductors or conducting surfaces for reflecting radiant energy.   59

(2) (illuminating engineering). A device used to redirect the flux from a source by the process of reflection.   167

(3) (wave propagation). A reflector comprises one or more conductors or conducting surfaces for reflecting radiant energy. *See:* **antenna; reflector element.**   313

reflector antenna. An antenna consisting of one or more reflecting surfaces and a radiating [receiving] feed system. *Note:* Specific reflector antennas often carry the name of the reflector used as part of the term used to specify it; for example, parabolic reflector antenna.   111

reflector element (antenna). A parasitic element located in a direction other than forward of the driven element of an antenna intended to increase the directive gain of the antenna in the forward direction. *See:* **antenna.**   111

reflector space (reflex klystron). The part of the tube following the buncher space, and terminated by the reflector. *See:* **velocity-modulated tube.**   244,190

reflex baffle (audio and electroacoustics). A loudspeaker enclosure in which a portion of the radiation from the rear of the diaphragm is propagated outward after controlled shift of phase or other modification, the purpose being to increase the useful radiation in some portion of the frequency spectrum. *See:* **loudspeaker.**   176

reflex bunching. The bunching that occurs in an electron stream that has been made to reverse its direction in the drift space.   125

reflex circuit. A circuit through which the signal passes for amplification both before and after a change in its frequency.   328

reflex klystron (microwave tubes). A single-resonator oscillator klystron in which the electron beam is reversed by a negative electrode so that it passes twice through the resonator, thus providing feedback. *See:* **microwave tube.**   190

reforming (semiconductor rectifier). The operation of restoring by an electric or thermal treatment, or both, the effectiveness of the rectifier junction after loss of forming. *See:* **rectification.**   237,66

refracted near-field scanning method. *See:* **refracted ray method.**

refracted ray (fiber optics). In an optical waveguide, a ray that is refracted from the core into the cladding. Specifically a ray at radial position r having direction such that

$$\frac{n^2(r) - n^2(a)}{1 - (r/a)^2 \cos^2 \phi(r)} \leq \sin^2 \theta\,(r)$$

where $\phi(r)$ is the azimuthal angle of projection of the ray on the transverse plane, $\theta(r)$ is the angle the ray makes with the waveguide axis, n(r) is the refractive index at the core radius, and a is the core radius. Refracted rays correspond to radiation modes in the terminology of mode descriptors. *See:* **cladding ray; guided ray; leaky ray; radiation mode.**   433

refracted ray method (fiber optics). The technique for measuring the index profile of an optical fiber by scanning the entrance face with the vertex of a high numerical aperture cone and measuring the change in power of refracted (unguided) rays. *Syn:* **refracted near-field scanning method.** *See:* **refraction; refracted ray.**   433

refracted wave (radio wave propagation)(data transmission). That part of an incident wave which travels from one medium into a second medium. *Syn:* **transmitted wave.** 146,59

refraction (1) (fiber optics). The bending of a beam of light in transmission through an interface between two dissimilar media or in a medium whose refractive index is a continuous function of position (graded index medium). *See:* **angle of deviation; refractive index (of a medium).** 433
(2) (radio wave propagation). Of a traveling wave, the change in direction of propagation resulting from the spatial variation of refractive index of the medium. 146

refraction error (navigation aid terms). Error due to the bending of one or more wave paths by the propagation medium. 526

refraction loss. That part of the transmission loss due to refraction resulting from nonuniformity of the medium. 176

refractive index (1) (data transmission). (1) (wave transmission medium). The ratio of the phase velocity in free space to that in the medium. (2) (dielectric for electromagnetic wave). The ratio of the sine of the angle of incidence to the sine of the angle of refraction as the wave passes from a vacuum into the dielectric. The angle of incidence θ_i is the angle between the direction of travel of the wave in vacuum and the normal to the surface of the dielectric. The angle of refraction θ_r is the angle between the direction of travel of the wave after it has entered the dielectric and the normal to the surface. Refractive index is related to the dielectric constant through the following relation:

$$n = \frac{\sin \theta_i}{\sin \theta_r} = (\epsilon')^{1/2}$$

where ϵ' is the real dielectric constant. Since ϵ' and n vary with frequency, the above relation is strictly correct only if all quantities are measured at the same frequency. The refractive index is also equal to the ratio of the velocity of the wave in the vacuum to the velocity in the dielectric medium. 59
(2) (of a medium) (fiber optics). Denoted by n, the ratio of the velocity of light in vacuum to the phase velocity in the medium. *Syn:* **index of refraction.** *See:* **cladding; core; critical angle; dispersion; Fresnel reflection; fused silica; graded index optical waveguide; group index; index matching material; index profile; linearly polarized mode; material dispersion; mode; normalized frequency; numerical aperture; optical path length; power-law index profile; profile dispersion; scattering; step index optical waveguide; weakly guiding fiber.** 433
(3) (radio wave propagation). Of a wave transmission medium, the ratio of the phase velocity in a vacuum to that in the medium. 146

refractive index contrast (fiber optics). Denoted by Δ, a measure of the relative difference in refractive index of the core and cladding of a fiber, given by Δ $= (n^2_1 - n^2_2)/2 \, n^2_1$ where n_1 and n_2 are, respectively, the maximum refractive index in the core and the refractive index of the homogeneous cladding. 433

refractive index profile (fiber optics). The description of the refractive index along a fiber diameter. *See:* **graded index profile; parabolic profile; power-law index profile; profile dispersion; profile dispersion parameter; profile parameter; step index profile.** 433

refractive modulus (1) (excess modified index of refraction) (data transmission). The excess over unity of the modified index of refraction, expressed in millionths. It is represented by M and is given by the equation:

$$M = (n + h/a - 1)10^6$$

where n is the index of refraction at a height h above sea level, and a is the radius of the earth. 59
(2) (radio wave propagation). In the troposphere, the excess over unity of the modified index of refraction, expressed in millionths. It is represented by M and is given by the above equation where a is the mean geometrical radius of the surface and n is the index of refraction at a height h above the local surface. 146

refractivity (radio wave propagation). The amount by which the refractive index exceeds unity. Often measured in parts per million called N-units. 146

refractor (illuminating engineering). A device used to redirect the flux from a source, primarily by the process of refraction. 167

refractory. A nonmetallic material highly resistant to fusion and suitable for furnace roofs and linings. 328

refresh rate (supervisory control, data acquisition, and automatic control)(station control and data acquisition). The number of times in each second that the information displayed on a nonpermanent display, for example, a crt, is rewritten or reenergized. 570, 403

REGAL (range and elevation guidance for approach and landing)(navigation aid terms). A ground-based navigation system used in conjunction with a localizer to compute vertical guidance for proper glide-slope and flare-out during an instrument approach and landing; it uses a digitally-coded vertically-scanning fan beam that provides data for both elevation angle and distance. 526

regenerate (electronic storage devices) (1). To bring something into existence again after decay of its own accord or after intentional destruction.
(2) (storage devices in which physical states used to represent data deteriorate). To restore the device to its latest undeteriorated state. *See:* **rewrite.** 235

regenerated leach liquor (electrometallurgy). The solution that has regained its ability to dissolve desired constituents from the ore by the removal of those constituents in the process of electrowinning. *See:* **electrowinning.** 328

regeneration (1) (storage tubes). The replacing of stored information lost through static decay and dynamic decay. *See:* **storage tube.** 174,125
(2). *See:* **positive feedback.**

regeneration of electrolyte. The treatment of a depleted electrolyte to make it again fit for use in an electrolyte cell. *See:* **electrorefining.** 328

regenerative braking (industrial control). A form of dynamic braking in which the kinetic energy of the motor and driven machinery is returned to the power supply system. *See:* **asynchronous machine; dynamic breaking; electric drive.** 206,63

regenerative branch (converter circuit elements)-(self-commutated converters). An auxiliary branch intended to transfer energy from the load to the supply side of the converter. 584

regenerative divider (regenerative modulator). A frequency divider that employs modulation, amplification, and selective feedback to produce the output wave. 328

regenerative fuel-cell system. A system in which the reactance may be regenerated using an external energy source. *See:* **fuel cell.** 186

regenerative repeater (1) (data transmission). A repeater that performs pulse regeneration. *Note:* The retransmitted signals are practically free from distortion. 59
(2) (fiber optics). A repeater that is designed for digital transmission. *Syn:* **regenerator.** *See:* **optical repeater.** 433

regenerator. *See:* **regenerative repeater.**

regional center (telephone switching systems). A toll office to which a number of sectional enters are connected. Regional centers are classified as Class 1 offices. *See:* **office class.** 55

region, Geiger-Mueller (radiation-counter tubes). See: **Geiger-Mueller region.**

region of limited proportionality (radiation-counter tubes). The range of applied voltage below the Geiger-Mueller threshold, in which the gas amplification depends upon the charge liberated by the initial ionizing event. 125

region, proportional (radiation-counter tubes). *See:* **proportional region.**

regions of electromagnetic spectrum (1) (illuminating engineering). For convenience of reference, the electromagnetic spectrum is arbitrarily divided as follows:
Vacuum ultraviolet
Extreme ultraviolet 10-100 nm
Far ultraviolet 100-200 nm
Middle ultraviolet 200-300 nm
Near ultraviolet 300-380 nm
Visible 380-770 nm
Near (short wavelength) 770-1400 nm infrared
Intermediate infrared 1400-5000 nm
Far (long wavelength) 5000-1 000 000 nm infrared
Note: The spectral limits indicated above have been chosen as a matter of practical convenience. There is a gradual transition from region to region without sharp delineation. Also, the division of the spectrum is not unique. In various fields of science, the classifi-

cations may differ due to the phenomena of interest. Another division of the ultraviolet spectrum often used by photobiologists is given by the International Commission on Illumination (CIE):
UV-A 315 to 400 nm
UV-B 280 to 315 nm
UV-C 100 to 280 nm 167
(2) (light emitting diodes). For convenience of reference the electromagnetic spectrum near the visible spectrum is divided as follows.

Spectrum	Wavelength in nanometers
far ultraviolet	10–280
middle ultraviolet	280–315
near ultraviolet	315–380
visible	380–780
infrared	790–10^5

Note: The spectral limits indicated above should not be construed to represent sharp delineations between the various regions. There is a gradual transition from region to region. The above ranges have been established for practical purposes. *See:* **radiant energy (illuminating engineering).** 167

register (1) (electronic computation). A device capable of retaining information, often that contained in a small subset (for example, one word), of the aggregate information in a digital computer. *See:* **address register; circulating register; index register; instruction register; shift register.** 235
(2) (telephone switching systems). A part of an automatic switching system that receives and stores signals from a calling device or other source for interpretation and action. 55

register constant (meter). The factor by which the register reading must be multiplied in order to provide proper consideration of the register, or gear, ratio and of the instrument transformer ratios to obtain the registration in the desired unit. *Note:* It is commonly denoted by the symbol *Kr.* *See:* **electricity meter (meter); moving element (instrument).** 212

register length (electronic computation). The number of characters that a register can store. 210

register marks or lines (industrial control). Any mark or line printed or otherwise impressed on a web of material and which is used as a reference to maintain register. *See:* **photoelectric control.** 206

register, mechanical (pulse techniques). An electromechanical indicating pulse counter. *See:* **pulse.** 335

register ratio (watthour meter). The number of revolutions of the first gear of the register, for one revolution of the first dial pointer. *Note:* This is commonly denoted by the symbol R_r. 212

register reading. The numerical value indicated by the register. Neither the register constant nor the test dial (or dials), if any exist, is considered. *See:* **electricity meter (meter).** 328

registration (1) (general). Accurate positioning relative to a reference. 255,77

(2) (display device). The condition in which corresponding elements of the primary-color images are in geometric coincidence. *See:* **registration (camera device).** 328

(3) (camera device). The condition in which corresponding elements of the primary-color images are scanned in time sequence. *See:* **registration (display device).** 328

(4) (watthour meter). The registration of a meter is the apparent amount of electric energy (or other quantity being measured) that has passed through the meter, as shown by the register reading. It is equal to the product of the register reading and the register constant. The registration during a given period is equal to the product of the register constant and the difference between the register readings at the beginning and the end of the period. *See:* **percentage registration** 212

regressed (illuminating engineering). A luminaire which is mounted above the ceiling with the opening of the luminaire above the ceiling line. 167

regression testing (software). Selective retesting to detect faults introduced during modification of a system or system component, to verify that modifications have not caused unintended adverse effects, or to verify that a modified system or system component still meets its specified requirements. *See:* **component; fault; modification; requirement; system.** 434

regular (specular) reflectance (illuminating engineering). The ratio of the flux leaving a surface or medium by regular (specular) reflection to the incident flux. 167

regular (specular) reflection (illuminating engineering). That process by which incident flux is redirected at the specular angle. *See:* **specular angle.** 167

regular transmission (illuminating engineering). That process by which incident flux passes through a surface or medium without scattering. 167

regular transmittance (illuminating engineering). The ratio of the regularly transmitted flux leaving a surface or medium to the incident flux. 167

regulated circuit (power and distribution transformer). The circuit on the output side of the regulator, and in which it is desired to control the voltage, or the phase relation, or both. *Note:* The voltage may be held constant at any selected point on the regulated circuit. 53

regulated frequency. Frequency so adjusted that the average value does not differ from a predetermined value by an appreciable amount. *See:* **generating station.** 64

regulated power supply. A power supply that maintains a constant output voltage (or current) for changes in the line voltage, output load, ambient temperature, or time. 186

regulated-power-supply efficiency. The ratio of the regulated output power to the input power. *See:* **regulated power supply.** 347

regulated voltage, band of (synchronous machines). The band or zone, expressed in percent of the rated value of the regulated voltage, within which the exci-

tation system will hold the regulated voltage of an electric machine during steady or gradually changing conditions over a specified range of load. 105

regulated voltage, nominal band of (synchronous machines). The band of regulated voltage for a load range between any load requiring no-load field voltage and any load requiring rated-load field voltage with any compensating means used to produce a deliberate change in regulated voltage inoperative. 105

regulating autotransformer (rectifier). A transformer used to vary the voltage applied to the alternating-current winding of rectifier transformer by means of de-energized autotransformer taps, and with load-tap-changing equipment to vary the voltage over a specified range on any of the autotransformer taps. *See:* **rectifier transformer.** 258

regulating circuit (thyristor). A circuit which together with the power controller and the thyristor trigger equipment forms a system for automatic control of the desired variable. 445

regulating device (90) (power system device function numbers). A device that functions to regulate a quantity, or quantities, such as voltage, current, power, speed, frequency, temperature, and load at a certain value or between certain (generally close) limits for machines, tie lines, or other apparatus. 402

regulating limit setter (speed-governing system). A device in the load-frequency-control system for limiting the regulating range on a station or unit. *See:* **speed-governing system.** 94

regulating range (load-frequency control). A range of power output within which a generating unit is permitted to operate. *See:* **speed-governing system.** 94

regulating relay (power switchgear). A relay whose function is to detect a departure from specified system operating conditions and to restore normal conditions by acting through supplementary equipment. 103

regulating system, synchronous-machine. An electrimachine regulating system consisting of one or more principal synchronous electric machines and the associated excitation system. 63

regulating transformer. A transformer used to vary the voltage, or the phase angle, or both, of an output circuit (referred to as the **regulated circuit**) controlling the output within specified limits, and compensating for fluctuations of load and input voltage (and phase angle, when involved within specified limits. *See:* **primary circuits; regulated circuits; series unit; main unit; series winding; excited winding; regulating winding; excitation winding; excitation-regulating winding; voltage-regulating relay; line-drop compensator; voltage winding (or transformer) for regulating equipment.** 53

regulating winding (power and distribution transformer). The winding of the main unit in which taps are changed to control the voltage or phase angle of the regulated circuit through the series unit. 53

regulation (1)(ferroresonant voltage regulators). The maximum amount that the output will change as a result of the specified change in line voltage, output load, temperature, or time. *Note:* line regulation, load regulation, stability, and temperature coefficient are defined and usually specified separately. 456

(2) (rotating machinery). The amount of change in voltage or speed resulting from a load change. *See:* **asynchronous machine.** 63

(3) (overall) (power supplies). The maximum amount that the output will change as a result of the specified change in line voltage, output load, temperature, or time. *Note:* Line regulation, load regulation, stability, and temperature coefficient are defined and usually specified separately. *See:* **line regulation; load regulation; power supply; stability; temperature coefficient.** 186

(4) (electrical conversion). The change of one of the controlled or regulated output parameters resulting from a change of one or more of the unit's variables within specificaton limits. *See:* **electrical conversion.** 186

(5) (transformer-rectifier system). The change in output voltage as the load current is varied. It is usually expressed as a percentage of the rated load voltage when the load current is changed by its rated value.

$$\text{Percent regulation} = 100\,\frac{(E_1 - E_2)}{E_2}$$

where E_1 is the no-load voltage and E_2 is the voltage at rated load current and the line voltage is held constant at rated value. 95

(6) (automatic control). *See:* **load regulation.**

regulation changer (speed governing systems, hydraulic turbines). A device by means of which the speed regulation may be adjusted while the turbine is operating. 58

regulation curve (generator). A characteristic curve between voltage and load. The speed is either held constant, or varied according to the speed characteristics of the prime mover. The excitation is held constant for separately excited fields, and the rheostat setting is held constant for self-excited machines. 328

regulation, frequency (power systems). The percentage change in frequency from steady state no load to steady state full load, which is a function of the engine and governing system:

$$\%R = \frac{F_{nl} - F_{fl}}{F_{nl}} \times 100$$

89

regulation, load (synchronous machines). The steady-state decrease of the value of the specified variable resulting from a specified increase in load, generally from no-load to full-load unless otherwise specified. 105

regulation pull-out (regulation drop-out) (power supply). The load currents at which the power supply fails to regulate when the load current is gradually increased or decreased. *See:* **regulated power supply.** 347

regulator, continuously acting (synchronous machines). A regulator that initiates a corrective action for a sustained infinitesimal change in the controlled variable. 105

regulator, noncontinuously acting (synchronous machines). A regulator that requires a sustained finite change in the controlled variable to initiate corrective action. 105

regulator, rheostatic-type (synchronous machines). A regulator that accomplishes the regulating function by mechanically varying a resistance. *Note:* Historically, rheostatic-type regulators have been further defined as direct acting and indirect acting. An indirect-acting-type regulator is a rheostatic type that controls the excitation of the exciter by acting on an intermediate device which is not considered part of the regulator or exciter. A direct-acting-type regulator is a rheostatic type that directly controls the excitation of an exciter by varying the input to the exciter field circuit. 105

regulator, synchronous-machine (1) (rotating machinery). An electric-machine regulator that controls the excitation of a synchronous machine. 63

(2) (excitation system). One that couples the output variables of the synchronous machine to the input of the exciter through feedback and forward controlling elements for the purpose of regulating the synchronous machine output variables. *Note:* In general, the regulator is assumed to consist of an error detector, preamplifier, power amplifier, stabilizers, auxiliary inputs, and limiters. 105

reguline. A word descriptive of electrodeposits that are firm and coherent. *See:* **electrodeposition.** 328

reheat turbine, condensing or noncondensing (control systems for steam turbine-generator units). Steam enters the turbine initially at one pressure, then is extracted at a lower pressure and temperature, and reheated. The steam is then readmitted to the turbine. 522

reignition (1) (power switchgear). A resumption of current between the contacts of a switching device during an opening operation after an interval of zero current of less than 1/4 cycle at normal frequency. 103

(2) (radiation-counter tubes). A process by which multiple counts are generated within a counter tube by atoms or molecules excited or ionized in the discharge accompanying a tube count. 96, 125

reinforced plastic (rotating machinery). A plastic with some strength properties greatly superior to those of the base resin, resulting from the presence of high-strength fillers imbedded in the composition. *Note:* The reinforcing fillers are usually fibers, fabrics, or mats made of fibers. The plastic laminates are the most common and strongest. 63

reinsertion (series capacitors). The restoration of load current to the series capacitor from the bypass path. See figure under **capacitor group.** 474

reinsertion current (series capacitors). The transient current, load current, or both, flowing through the series capacitor after the opening of the bypass path.
474

reinsertion voltage (series capacitors). The transient voltage, steady-state voltage, or both, appearing across the series capacitor after the opening of the bypass path.
474

rejection band (1) (uniconductor waveguide). The frequency range below the cutoff frequency. *See:* **waveguide.**
267

(2) (circuits and systems). *See:* **stop band.**

rejection filter (signal-transmission system). A filter that attenuates alternating currents between given upper and lower cutoff frequencies and transmits substantially all others. Also, a filter placed in the signal transmission path to attenuate interference. *See:* **signal.**
188

related transmission terms (loss and gain). The term **loss** used with different modifiers has different meanings, even when applied to one physical quantity such as power. In view of definitions containing the word **loss** (as well as others containing the word **gain**), the following brief explanation is presented. (1) Power loss from a circuit, in the sense that it is converted to another form of power not useful for the purpose at hand (for example, I^2R loss) is a physical quantity measured in watts in the International System of Units (SI) and having the dimensions of power. For a given R, it will vary with the square of the current in R. (2) Loss may be defined as the ratio of two powers, for example: if P_o is the output power and P_i the input power of a network under specified conditions, P_i/P_o is a dimensionless quantity that would be unity if P_o , P_i. Thus, no power loss in the sense of (1) means a loss, defined as the ratio P_i/P_o, of unity. The concept is closely allied to that of efficiency. (3) Loss may also be defined as the logarithm, or as directly proportional to the logarithm, of a power ratio such as P_i/P_o. Thus if loss $= 10 \log_{10} P_i/P_o$ the loss is zero when P_o , P_i. This is the standard for measuring loss in decibels. It should be noted that in cases (2) and (3) the **loss** (for a given linear system) is the same whatever may be the power levels. Thus (2) and (3) give characteristics of the system and do not depend (as (1) does) on the value of the current or other dependent quantity. **Power** refers to average power, not instantaneous power. *See:* **network analysis.**
210

relative address (computing systems). The number that specifies the difference between the absolute address and the base address.
255, 77, 54

relative addressing (microprocessor assembly language). An addressing mode in which the effective address is formed by adding an offset to the program counter (or a portion thereof) during execution.
466

relative bearing (navigation aid terms). Bearing relative to heading.
526

relative capacitivity. *See:* **relative dielectric constant.**

relative chrominance level (RCL) (linear waveform distortion). The difference between the level of the luminance and chrominance signal components. An inaccuracy in RCL will cause saturation inaccuracy of all colors in a color TV picture.
42

relative chrominance time (RCT) (linear waveform distortion). The difference in absolute time between the luminance and chrominance signal components. An inaccuracy in RCT will cause registration inaccuracy of all colors relative to their luminance components in a color TV picture.
42

relative coding (computing systems). Coding that uses machine instructions with relative addresses.
255, 77

relative complex dielectric constant (homogeneous isotropic material) (complex capacitivity) (complex permittivity). The ratio of the admittance between two electrodes of a given configuration of electrodes with the material as a dielectric to the admittance between the same two electrodes of the configuration with vacuum as dielectric or

$$\epsilon^* \equiv \epsilon' - j\epsilon'' = Y/(j\omega C_v)$$

where Y is the admittance with the material and $j\omega C$ v is the admittance with vacuum. Experimentally, vacuum must be replaced by the material at all points where it makes a significant change in the admittance. *Note:* The word **relative** is frequently dropped. *See:* **dielectric loss index; relative dielectric constant.**
210

relative contrast sensitivity (illuminating engineering). The relation between the reciprocal of the luminous contrast of a task at visibility threshold and the background luminance expressed as a percentage of the value obtained under a very high level of diffuse task illumination.
167

relative damping (1) (specific damping) (instrument). Under given conditions, the ratio of the damping torque at a given angular velocity of the moving element to the damping torque that, if present at this angular velocity, would produce the condition of critical damping. *See:* **accuracy rating (instrument).**
328

(2) (automatic control) (under damped system). A number expressing the quotient of the actual damping of a second-order linear system or element by its critical damping. *Note:* For any system whose transfer function includes a quadratic factor $s^2 + 2z\omega_n s + \omega^2_n$, relative damping is the value of z, since z = 1 for critical damping. Such a factor has a root $- \sigma + j\omega$ in the complex s-plane, from which $z = \sigma/\omega_n = \sigma(\sigma^2 + \omega^2)^{1/2}$.
56

relative dielectric constant (homogeneous isotropic material) (relative capacitivity) (relative permittivity). The ratio of the capacitance of a given configuration of electrodes with the material as a dielectric to the capacitance of the same electrode configuration with a vacuum (or air for most practical purposes) as the dielectric or

$$\epsilon' = C_x/C_v$$

where C_x is the capacitance with the material and C_v is the capacitance with vacuum. Experimentally, vacuum must be replaced by the material at all points where it makes a significant change in the capacitance. *See:* **electric flux density; relative capacitivity.**
    210

relative directive gain (physical media). In a given direction and at a given point in the far field, the ratio of the power flux per unit area from an antenna to the power flux per unit area from a reference antenna at a specified location and delivering the same power from the antenna to the medium. *Note:* All or part of the reference antenna must be within the smallest sphere containing the subject antenna.
    111

relative fundamental content (converter characteristics)(self-commutated converters). The ratio of the rms (root-mean-square) value of the fundamental component to the rms value of the total nonsinusoidal periodic function.
    584

relative gain (of an antenna). The ratio of the gain of an antenna in a given direction to the gain of a reference antenna. *Note:* Unless otherwise specified the maximum gains of the antennas are implied. *See:* **relative power gain.**
    111

relative grid (navigation aid terms). Navigation in a relative grid as opposed to an absolute coordinate system (for example, geo-referenced). A relative grid, arbitrarily constructed by designating a point as the origin and constructing a set of axes U, V, W enables members to navigate in this relative grid by virtue of their U, V, W coordinates.
    526

relative harmonic content (converter characteristics)-(self-commutated converters). The ratio of the rms (root-mean-square) value of the harmonic content to the rms value of the total nonsinusoidal periodic function.
    584

relative interfering effect (single-frequency electric wave in an electroacoustic system). The ratio, usually expressed in decibels, of the amplitude of a wave of specified reference frequency to that of the wave in question when the two waves are equal in interfering effect. The frequency of maximum interfering effect is usually taken as the reference frequency. Equal interfering effects are usually determined by judgment tests or intelligibility tests. *Note:* When applied to complex waves, the relative interfering effect is the ratio, usually expressed in decibels, of the power of the reference wave to the power of the wave in question when the two waves are equal in interfering effect.
    328

relative lead polarity (transformer). A designation of the relative instantaneous directions of current in its leads. *Notes:* (1) Primary and secondary leads are said to have the same polarity when at a given instant during most of each half cycle, the current enters an identified, or marked, primary lead and leaves the similarly identified, or marked, secondary lead in the same direction as though the two leads formed a continuous circuit. (2) The relative lead polarity of a single-phase transformer may be either additive or sub-

tractive. If one pair of adjacent leads from the two windings is connected together and voltage applied to one of the windings, then: (A) The relative lead polarity is additive if the voltage across the other two leads of the windings is greater than that of the higher-voltage winding alone. (B) The relative lead polarity is subtractive if the voltage across the other two leads of the windings is less than that of the higher-voltage winding alone. (3) The relative lead polarity is indicated by identification marks on primary and secondary leads of like polarity, or by other appropriate identification. *See:* **routine test; constant-current transformer.**
    203

relative luminosity (television). The ratio of the value of the luminosity at a particular wavelength to the value at the wavelength of maximum luminosity.
    18

relatively refractory state (electrobiology). The portion of the electric recovery cycle during which the excitability is less than normal. *See:* **excitability (electrobiology).**
    192

relative nonline frequency content (thyristor). The ratio of the root-mean-square (rms) value of the nonline frequency content to the total rms value of the nonsinusoidal periodic function. 445 subline frequency components (thyristor). Expressed by the frequency and the root-mean-square (rms) value of the components having a lower frequency than the line frequency (dc included).
    445

relative partial gain (of an antenna with respect to a reference antenna of a given polarization). In a given direction, the ratio of the partial gain of an antenna, corresponding to the polarization of the reference antenna, to the maximum gain of the reference antenna.
    111

relative permeability. The ratio of normal permeability to the magnetic constant. *Note:* In anisotropic media, relative permeability becomes a matrix.
    210

relative permittivity in physical media (antenna). The ratio of the complex permittivity to the permittivity of free space. *See:* **relative dielectric constant.**
    111

relative phase of an elliptically polarized field vector (antennas). The phase angle of the unitary factor by which the polarization-phase vector for the given field vector differs from that of a reference field vector with the same polarization. *Notes:* (1) The relative phase of an elliptically polarized field $\bar{E}_1$ can only be defined with respect to that of another field $\bar{E}_0$ having the same polarization. In that case, the polarization vectors $\hat{e}_1$ and $\hat{e}_0$ have the same direction (in the three dimensional complex space C^3) and, being of unit magnitudes, they differ only by a unitary factor:

$$\hat{e}_1 = e^{j\alpha}\hat{e}_0$$

The angle α is the phase difference between E_1 and E_0. (2) The field vectors $E_1(t) = \text{Re}E_1 e^{j\omega t}$ and $E_0(t) = \text{Re}E_0 e^{j\omega t}$ describe similar ellipses as t varies. The angle α is 2π times the area of the sector shown on the figure divided by the area of the ellipse described by the extremity of $E_0(t)$. For circular polarization, α is the angle between E_0 and E_1 at any instant of time.

(3) The phase of an elliptically polarized field vector can be expressed relative to a spatial direction in it plane of polarization. For example, the phase angle is given by 2π times the area of the sector shown on the figure, which is bounded by $E(0)$ and the reference, divided by the area of the ellipse described by $E(t)$. (4) The angle is positive if it is in the same direction as the sense of polarization and negative if it is in the opposite direction to the sense of polarization.   111

relative power gain (antenna). The ratio of the power gain in a given direction to the power gain of a reference antenna in its reference direction. *Note:* Common reference antennas are half-wave dipoles, electric dipoles, magnetic dipoles, monopoles, and calibrated horn antennas.

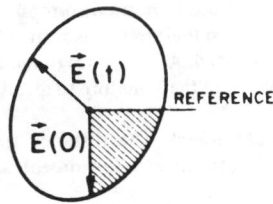

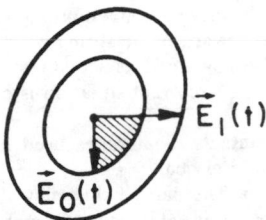

111

relative power gain in physical media (antenna). In a given direction and at a given point in the far field, the ratio of the power flux per unit area from an antenna to the power flux per unit area from a reference antenna at a specified location with the same power input to both antennas. *Note:* All or part of the reference antenna must be within the smallest sphere containing the subject antenna.   111

relative redundancy (of a source) (information theory). The ratio of the redundancy of the source to the logarithm of the number of symbols available at the source. *See:* **information theory.**   415

relative refractive index (radio wave propagation). Of two media, the ratio of their refractive indices.   146

relative response (audio and electroacoustics). The ratio, usually expressed in decibels, of the response under some particular conditions to the response under reference conditions. Both conditions should be stated explicitly.   176

relative temperature index (RTI)(1)(evaluation of thermal capability)(thermal classification of electric equipment and electrical insulation). The tempera-

ture index of a new or candidate insulating material, which corresponds to the accepted temperature index of a reference material for which considerable test and service experience has been obtained. Both new and reference material are subjected to the same aging and diagnostic procedure in a comparative test. *See:* **thermal endurance graph.**   506

(2)(solid electrical insulating materials). Derived at an arbitrary time by comparing the life values from thermal endurance graphs from a new and a referenced material with considerable service experience. *See:* **IEEE Std 98-1984, Section 17.**   452

relaxation (laser-maser). The spontaneous return of a system towards its equilibrium condition.   363

relaxation oscillator. Any oscillator whose fundamental frequency is determined by the time of charging or discharging of a capacitor or inductor through a resistor, producing waveforms that may be rectangular or sawtooth. *Note:* The frequency of a relaxation oscillator may be self-determined or determined by a synchronizing voltage derived from an external source. *See:* **electronic controller; oscillatory circuit.**

111

relaxation time (laser-maser). The time required for the deviation from equilibrium of some system parameter to diminish to $1/e$ of its initial value.   363

relay (1) (general). An electric device that is designed to interpret input conditions in a prescribed manner and after specified conditions are met to respond to cause contact operation or similar abrupt change in associated electric control circuits. *Notes:* (A) Inputs are usually electric, but may be mechanical, thermal, or other quantities. Limit switches and similar simple devices are not relays. (B) A relay may consist of several units, each responsive to specified inputs, the combination providing the desired performance characteristic.   110, 127

(2) (electric and electronics parts and equipments). An electrically controlled, usually two-state, device that opens and closes electrical contacts to effect the operation of other devices in the same or another electric circuit. *Notes:* (A) A relay is a device in which a portion of one or more sets of electrical contacts is moved by an armature and its associated operating coil. (B) This concept is extended to include assembled reed relays in which the armature may act as a contact. For individual magnetic reed switches, see **switch.**   17

(3) (packaging machinery). A device that is operative by a variation in the conditions of one electric circuit to affect the operation of other devices in the same or another electric circuit. *Note:* Where relays operate in response to changes in more than one condition, all functions should be mentioned.   429

(4) (power switchgear). An electrical device designed to respond to input conditions in a prescribed manner and after specified conditions are met to cause contact operation or similar abrupt change in associated electric control circuits. *Notes:* (A) Inputs are usually electrical, but may be mechanical, thermal or other quantities or a combination of quantities. Limit switches or similar simple devices are not relays. (B) A relay may consist of several relay units, each responsive to specified inputs with the combination providing the desired overall performance characteristic of the relay.   103

relay actuation time. The time at which a specified contact functions.   259

relay actuation time, effective. The sum of the initial actuation time and the contact chatter intervals following such actuation.   259

relay actuation time, final. The time of termination of chatter following contact actuation.   259

relay actuation time, initial. The time of the first closing of a previously open contact or the first opening of a previously closed contact.   259

relay actuator. The part of the relay that converts electric energy into mechanical work.   259

relay adjustment. The modification of the shape or position of relay parts to affect one or more of the operating characteristics, that is, armature gap, restoring spring, contact gap.   259

relay air gap. Air space between the armature and the pole piece. This is used in some relays instead of a nonmagnetic separator to provide a break in the magnetic circuit.   259

relay, alternating-current. A relay designed for operation from an alternating-current source. *See:* **relay.**   259

relay amplifier. An amplifier that drives an electromechanical relay. *See:* **electronic analog computer.**   9,10

relay antifreeze pin. Sometimes used for **relay armature stop, nonmagnetic.**   259

relay armature (electromechanical relay). The moving element that contributes to the designed response of the relay and that usually has associated with it a part of the relay contact assembly.   103

relay armature, balanced. An armature that is approximately in equilibrium with respect to both static and dynamic forces.   259

relay armature bounce. *See:* **relay armature rebound.**   259

relay armature card. An insulating member used to link the movable springs to the armature.   259

relay armature contact. (1) A contact mounted directly on the armature. (2) Sometimes used for **relay contact, movable.**   259

relay armature, end-on. An armature whose motion is in the direction of the core axis, with the pole face at the end of the core and perpendicular to this axis.   259

relay armature, flat-type. An armature that rotates about an axis perpendicular to that of the core, with the pole face on a side surface of the core.   259

relay armature gap. The distance between armature and pole face.   259

relay armature hesitation. Delay or momentary reversal of armature motion in either the operate or release stroke.   259

relay armature lifter. *See:* **relay armature stud.**

relay armature, long-lever. An armature with an armature ratio greater than 1:1.   259

relay armature overtravel. The portion of the available stroke occurring after the contacts have touched.   259

relay armature ratio. The ratio of the distance through which the armature stud or card moves to the armature travel.   259

relay armature rebound. Return motion of the armature following impact on the backstop.   259

relay armature, short-lever. An armature with an armature ratio of 1:1 or less.   259

relay armature, side. An armature that rotates about an axis parallel to that of the core, with the pole face on a side surface of the core.   259

relay armature stop. Sometimes used for **relay backstop.**   259

relay armature stop, nonmagnetic. A nonmagnetic member separating the pole faces of core and armature in the operated position, used to reduce and stabilize the pull from residual magnetism in release.   259

relay armature stud. An insulating member that transmits the motion of the armature to an adjacent contact member.   259

relay armature travel. The distance traveled during operation by a specified point on the armature.   259

relay, automatic reset. (1) A stepping relay that returns to its home position either when it reaches a predetermined contact position, or when a pulsing circuit fails to energize the driving coil within a given time. May either pulse forward or be spring reset to the home position. (2) An overload relay that restores the circuit as soon as an overcurrent situation is corrected.   259

relay back contacts. Sometimes used for **relay contacts, normally closed.**   259

relay backstop. The part of the relay that limits the movement of the armature away from the pole face or core. In some relays a normally closed contact may serve as backstop.   259

relay backup. That part of the backup protection that operates in the event of failure of the primary relays.   103, 202

relay bank. *See:* **relay level.**

relay bias winding. An auxiliary winding used to produce an electric bias.   259

relay blades. Sometimes used for **relay contact springs.**   259

relay bracer spring. A supporting member used in conjunction with a contact spring.   259

relay bridging. (1) A result of contact erosion, wherein a metallic protrusion or bridge is built up between opposite contact faces to cause an electric path between them. (2) A form of contact erosion occurring on the break of a low-voltage, low-inductance circuit, at the instant of separation, that results in melting and resolidifying of contact metal in the form of a metallic protrusion or bridge. (3) Make-before-break contact action, as when a wiper touches two successive contacts simultaneously while moving from one to the other.   259

relay brush. *See:* **relay wiper.**

relay bunching time. The time during which all three contacts of a bridging contact combination are electrically connected during the armature stroke. 259

relay bushing. Sometimes used for **relay spring stud.** 259

relay chatter time. The time interval from initial actuation of a contact to the end of chatter. 259

relay coil. One or more windings on a common form. 259

relay coil, concentric-wound. A coil with two or more insulated windings, wound one over the other. 259

relay-coil dissipation. The amount of electric power consumed by a winding. For the most practical purposes, this equals the I^2R loss. 259

relay-coil resistance. The total terminal-to-terminal resistance of a coil at a specified temperature. 259

relay-coil serving. A covering, such as thread or tape, that protects the winding from mechanical damage. 259

relay-coil temperature rise. The increase in temperature of a winding above the ambient temperature when energized under specified conditions for a given period of time, usually the time required to reach a stable temperature. 259

relay-coil terminal. A device, such as a solder lug, binding post, or similar fitting, to which the coil power supply is connected. 259

relay-coil tube. An insulated tube upon which a coil is wound. 259

relay comb. An insulating member used to position a group of contact springs. 259

relay contact actuation time. The time required for any specified contact on the relay to function according to the following subdivisions. When not otherwise specified contact actuation time is **relay initial actuation time.** For some purposes, it is preferable to state the actuation time in terms of final actuation time or effective actuation time. 259

relay contact arrangement. The combination of contact forms that make up the entire relay switching structure. 259

relay contact bounce. Sometimes used for **relay contact chatter,** when internally caused. 259

relay contact chatter. The undesired intermittent closure of open contacts or opening of closed contacts. It may occur either when the relay is operated or released or when the relay is subjected to external shock or vibration. 259

relay contact chatter, armature hesitation. Chatter ascribed to delay or momentary reversal in direction of the armature motion during either the operate or the release stroke. 259

relay contact chatter, armature impact. Chatter ascribed to vibration of the relay structure caused by impact of the armature on the pole piece in operation, or on the backstop in release. 259

relay contact chatter, armature rebound. Chatter ascribed to the partial return of the armature to its operated position as a result of rebound from the backstop in release. 259

relay contact chatter, externally caused. Chatter resulting from shock or vibration imposed on the relay by external action. 259

relay contact chatter, external shock. Chatter ascribed to impact experienced by the relay or by the apparatus of which it forms a part. 259

relay contact chatter, initial. Chatter ascribed to vibration produced by opening or closing the contacts themselves, as by contact impact in closure. 259

relay contact chatter, internally caused. Chatter resulting from the operation or release of the relay. 259

relay contact chatter, transmitted vibration. Chatter ascribed to vibration originating outside the relay and transmitted to it through its mounting. 259

relay contact combination. (1) The total assembly of contacts on a relay. (2) Sometimes used for contact form. 259

relay contact, fixed. *See:* **relay contact, stationary.**

relay contact follow. The displacement of a stated point on the contact-actuating member following initial closure of a contact. 259

relay contact follow, stiffness. The rate of change of contact force per unit contact follow. 259

relay contact form. A single-pole contact assembly. 259

relay contact functioning. The establishment of the specified electrical state of the contacts as a continuous condition. 259

relay contact gap. *See:* **relay contact separation.**

relay contact, movable. The member of a contact pair that is moved directly by the actuating system. 259

relay contact pole. Sometimes used for **relay contact, movable.** 259

relay contact rating. A statement of the conditions under which a contact will perform satisfactorily. 259

relay contacts. The current-carrying parts of a relay that engage or disengage to open or close electric circuits. 259

relay contacts, auxiliary. Contacts of lower current capacity than the main contacts: used to keep the coil energized when the original operating circuit is open, to operate an audible or visual signal indicating the position of the main contacts, or to establish interlocking circuits, etcetera. 259

relay contacts, back. Sometimes used for **relay contacts, normally closed.** 259

relay contacts, break. *See:* **relay contacts, normally closed.**

relay contacts, break-make. A contact form in which one contact opens its connection to another contact and then closes its connection to a third contact. 259

relay contacts, bridging. A contact form in which the moving contact touches two stationary contacts simultaneously during transfer. 259

relay contacts, continuity transfer. Sometimes used for **relay contacts, make-break.** 259

relay contacts, double break. A contact form in which

one contact is normally closed in simultaneous connection with two other contacts. 259

relay contacts, double make. A contact form in which one contact, which is normally open, makes simultaneous connection when closed with two other independent contacts. 259

relay contacts, dry. (1) Contacts which neither break nor make current. (2) Erroneously used for **relay contacts, low level.** 259

relay contacts, early. Sometimes used for **relay contacts, preliminary.** 259

relay contact separation. The distance between mating contacts when the contacts are open. 259

relay contacts, front. Sometimes used for **relay contacts, normally open.** 259

relay contacts, interrupter. An additional set of contacts on a stepping relay, operated directly by the armature. 259

relay contacts, late. Contacts that open or close after other contacts when the relay is operated. 259

relay contacts, low-capacitance. A type of contact construction providing low intercontact capacitance. 259

relay contacts, low-level. Contacts that control only the flow of relatively small currents in relatively low-voltage circuits: for example, alternating currents and voltages encountered in voice or tone circuits, direct currents and voltages of the order of microamperes and microvolts, etcetera. 259

relay contacts, make. *See:* **relay contacts, normally open.**

relay contacts, make-break. A contact form in which one contact closes connection to another contact and then opens its prior connection to a third contact. 259

relay contacts, multiple-break. Contacts that open a circuit in two or more places. 259

relay contacts, nonbridging. A contact arrangement in which the opening contact opens before the closing contact closes. 259

relay contacts, normally closed. A contact pair that is closed when the coil is not energized. 259

relay contacts, normally open. A contact pair that is open when the coil is not energized. 259

relay contacts, off-normal. Contacts on a multiple switch that are in one condition when the relay is in its normal position and in the reverse condition for any other position of the relay. 259

relay contacts, preliminary. Contacts that open or close in advance of other contacts when the relay is operating. 259

relay contact spring. (1) A current-carrying spring to which the contacts are fastened. (2) A non-current-carrying spring that positions and tensions a contact-carrying member 259

relay contacts, sealed. A contact assembly that is sealed in a compartment separate from the rest of the relay. 259

relay contacts, snap-action. A contact assembly having two or more equilibrium positions, in one of which the contacts remain with substantially constant contact pressure during the initial motion of the actuating member, until a condition is reached at which stored energy snaps the contacts to a new position of equilibrium. 259

relay contact, stationary. The member of a contact pair that is not moved directly by the actuating system. 259

relay contact wipe. The sliding or tangential motion between two contact surfaces when they are touching. 259

relay core. The magnetic member about which the coil is wound. 259

relay critical voltage (current). That voltage (current) that will just maintain thermal relay contacts operated. 259

relay cycle timer. A controlling mechanism that opens or closes contacts according to a preset cycle. 259

relay damping ring, mechanical. A loose member mounted on a contact spring to reduce contact chatter. 259

relay, direct-current. A relay designed for operation from a direct-current source. *See:* **relay.** 259

relay, double-pole. A term applied to a contact arrangement to denote that it includes two separate contact forms: that is, two single-pole contact assemblies. 259

relay, double-throw. A term applied to a contact arrangement to denote that each contact form included is a break-make. 259

relay driving spring. The spring that drives the wipers of a stepping relay. 259

relay drop-out. *See:* **relay release.**

relay, dry circuit. Erroneously used for a relay with either dry or low-level contacts. *See:* **relay contacts, low-level.** 259

relay duty cycle. A statement of energized and deenergized time in repetitive operation, as: 2 seconds on, 6 seconds off. 259

relay electric bias. An electrically produced force tending to move the armature towards a given position. 259

relay, electric reset. A relay that may be reset electrically after an operation. 259

relay, electromagnetic. A relay, controlled by electromagnetic means, that opens and closes electric contacts. *See:* **relay.** 259

relay, electrostatic. A relay in which the actuator element consists of nonconducting media separating two or more conductors that change their relative positions because of the mutual attraction or repulsion by electric charges applied to the conductors. *See:* **relay.** 259

relay, electrostrictive. A relay in which an electrostrictive dielectric serves as the actuator. *See:* **relay.** 259

relay electrothermal expansion element. An actuating element in the form of a wire strip or other shape having a high coefficient of thermal expansion. 259

relay element. A subassembly of parts. *Note:* The combination of several relay elements constitutes a relay unit.					103,127

relay finish lead. The outer termination of the coil.					259

relay, flat-type. *See:* **relay armature, flat-type.**

relay frame. The main supporting portion of a relay. This may include parts of the magnetic structure.					259

relay freezing, magnetic. Sticking of the relay armature to the core as a result of residual magnetism.					259

relay fritting. Contact erosion in which the electrical discharge makes a hole through the film and produces molten matter that is drawn into the hole by electrostatic forces and solidifies there to form a conducting bridge.					259

relay front contacts. Sometimes used for **relay contacts, normally open.**					259

relay functioning time. The time between energization and operation or between de-energization and release.					259

relay functioning value. The value of applied voltage, current, or power at which the relay operates or releases.					259

relay header. The subassembly that provides support and insulation to the leads passing through the walls of a sealed relay.					259

relay heater. A resistor that converts electric energy into heat for operating a thermal relay.					259

relay heel piece. The portion of a magnetic circuit of a relay that is attached to the end of the core remote from the armature.					259

relay, high, common, low (HCL). A type of relay control used in such devices as thermostats and in relays operated by them, in which a momentary contact between the common lead and another lead operates the relay, that then remains operated until a momentary contact between the common lead and a third lead causes the relay to return to its original position.					259

relay hinge. The joint that permits movement of the armature relative to the stationary parts of the relay structure.					259

relay hold. A specified functioning value at which no relay meeting the specification may release.					259

relay housing. An enclosure for one or more relays, with or without accessories, usually providing access to the terminals.					259

relay hum. The sound emitted by relays when their coils are energized by alternating current or in some cases by unfiltered rectified current.					259

relay inside lead. *See:* **relay start lead.**

relay inverse time. A qualifying term applied to a relay indicating that its time of operation decreases as the magnitude of the operating quantity increases.					259

relay just-operate value. The measured functioning value at which a particular relay operates.					259

relay just-release value. The measured functioning value for the release of a particular relay.					259

relay leakage flux. The portion of the magnetic flux that does not cross the armature-to-pole-face gap.					259

relay level. A series of contacts served by one wiper in a stepping relay.					259

relay load curves. The static force displacement characteristic of the total load of the relay.					259

relay magnetic bias. A steady magnetic field applied to the magnetic circuit of a relay.					259

relay magnetic gap. Nonmagnetic portion of a magnetic circuit.					259

relay, manual reset. A relay that may be reset manually after an operation.					259

relay mechanical bias. A mechanical force tending to move the armature towards a given position.					259

relay mounting plane. The plane to which the relay mounting surface is fastened.					259

relay must-operate value. A specified functioning value at which all relays meeting the specification must operate.					259

relay must-release value. A specified functioning value, at which all relays meeting the specification must release.					259

relay nonfreeze pin. Sometimes used for **relay armature stop, nonmagnetic.**					259

relay nonoperate value. A specified functioning value at which no relay meeting the specification may operate.					259

relay normal condition. The de-energized condition of the relay.					259

relay operate. The condition attained by a relay when all contacts have functioned. *See:* **relay contact actuation time.**					259

relay operate time. The time interval from coil energization to the functioning time of the last contact to function. Where not otherwise stated the functioning time of the contact in question is taken as its initial functioning time.					259

relay operate time characteristic. The relation between the operate time of an electromagnetic relay and the operate power.					259

relay operating frequency. The rated alternating-current frequency of the supply voltage at which the relay is designed to operate.					259

relay outside lead. *See:* **relay finish lead.**

relay overtravel. Amount of contact wipe. *See:* **relay armature overtravel; relay contact wipe.**					259

relay pickup value. Sometimes used for **relay must-operate value.**					259

relay pileup. A set of contact arms, assemblies, or springs, fastened one on top of the other with insulation between them.					259

relay pneumatic bellows. Gas-filled bellows, sometimes used with plunger-type relays to obtain time delay.					259

relay pole face. The part of the magnetic structure at the end of the core nearest the armature.					259

relay pole piece. The end of an electromagnet, sometimes separable from the main section, and usually shaped so as to distribute the magnetic field in a pattern best suited to the application.					259

relay pull curves. The force-displacement characteristics of the actuating system of the relay.					259

relay pull-in value. Sometimes used for **relay must-operate value.** 259

relay pusher. Sometimes used for **relay armature stud.** *See:* relay. 259

relay rating. A statement of the conditions under which a relay will perform satisfactorily. 259

relay recovery time. A cooling time required from heater de-energization of a thermal time-delay relay to subsequent re-energization that will result in a new operate time equal to 85 percent of that exhibited from a cold start. 259

relay recovery time, instantaneous. Recovery time of a thermal relay measured when the heater is de-energized at the instant of contact operation. 259

relay recovery time, saturated. Recovery time of a thermal relay measured after temperature saturation has been reached. 259

relay release. The condition attained by a relay when all contacts have functioned and the armature (where applicable) has reached a fully opened position. 259

relay release time. The time interval from coil de-energization to the functioning time of the last contact to function. Where not otherwise stated the functioning time of the contact in question is taken as its initial functioning time. 259

relay reoperate time. Release time of a thermal relay. 259

relay reoperate time, instantaneous. Reoperate time of a thermal relay measured when the heater is de-energized at the instant of contact operation. 259

relay reoperate time, saturated. Reoperate time of a thermal relay measured when the relay is de-energized after temperature saturation (equilibrium) has been reached. 259

relay residual gap. Sometimes used for **relay armature stop, nonmagnetic.** 259

relay restoring spring. A spring that moves the armature to the normal position and holds it there when the relay is de-energized. 259

relay retractile spring. Sometimes used for **relay restoring spring.** 259

relay return spring. Sometimes used for **relay restoring spring.** 259

relay saturation. The condition attained in a magnetic material when an increase in field intensity produces no further increase in flux density. 259

relay sealing. Sometimes used for **relay seating.** 259

relay seating. The magnetic positioning of an armature in its final desired location. 259

relay seating time. The elapsed time after the coil has been energized to the time required to seat the armature of the relay. 259

relay shading coil. Sometimes used for **relay shading ring.** 259

relay shading ring. A shorted turn surrounding a portion of the pole of an alternating-current magnet, producing a delay of the change of the magnetic field in that part, thereby tending to prevent chatter and reduce hum. 259

relay shields, electrostatic spring. Grounded conducting members located between two relay springs to minimize electrostatic coupling. 259

relay shim, nonmagnetic. Sometimes used for **relay armature stop, nonmagnetic.** 259

relay, single-pole. A relay in which all contacts connect, in one position or another, to a common contact. 259

relay, single-throw. A relay in which each contact form included is a single contact pair. 259

relay sleeve. A conducting tube placed around the full length of the core as a short-circuited winding to retard the establishment or decay of flux within the magnetic path. 259

relay slow-release time characteristic. The relation between the release time of an electromagnetic relay and the conductance of the winding circuit or of the conductor (sleeve or slug) used to delay release. The conductance in this definition is the quantity $N^2.R$, where N is the number of turns and R is the resistance of the closed winding circuit. (For a sleeve or slug N, 1). 259

relay slug. A conducting tube placed around a portion of the core to retard the establishment or decay of flux within the magnetic path. 259

relay soak. The condition of an electromagnetic relay when its core is approximately saturated. 259

relay soak value. The voltage, current, or power applied to the relay coil to insure a condition approximating magnetic saturation. 259

relay spool. A flanged form upon which a coil is wound. 259

relay spring buffer. Sometimes used for **relay spring stud.** 259

relay spring curve. A plot of spring force on the armature versus armature travel. 259

relay spring stop. A member that controls the position of a pretensioned spring. 259

relay spring stud. An insulating member that transmits the motion of the armature from one movable contact to another in the same pileup. 259

relay stack. Sometimes used for **relay pileup.** 259

relay stagger time. The time interval between the actuation of any two contact sets. 259

relay starting switch (rotating machinery). A relay, actuated by current, voltage, or the combined effect of current and voltage, used to perform a circuit-changing function in the primary winding of a single-phase induction motor within a predetermined range of speed as the rotor accelerates: and to perform the reverse circuit-changing operation when the motor is disconnected from the supply line. One of the circuit changes that is usually performed is to open or disconnect the auxiliary-winding circuit. *See:* **starting-switch assembly.** 63

relay start lead. The inner termination of the coil. 259

relay static characteristic. The static force-displacement characteristic of the spring system or of the actuating system. 259

relay station (mobile communication). A radio station used for the reception and retransmission of the signals from another radio station. *See:* **mobile communication system.**   181

relay system (surge withstand capability). An assembly usually consisting of measuring units, relay logic, communications interfaces, and necessary power supplies. The communications link is not considered as a part of a relay system.   90

relay thermal. A relay that is actuated by the heating effect of an electric current. *See:* **relay**   259

relay, three-position. Sometimes used for a center-stable polar relay. *See:* **relay.**   259

relay transfer contacts. Sometimes used for **relay contacts, break-make.**   259

relay transfer time. The time interval between opening the closed contact and closing the open contact of a break-make contact form.   259

relay unit (1) (general). An assembly of relay elements that in itself can perform a relay function. *Note:* One or more relay units constitutes a relay.   127
(2) (power switchgear). (A) A subassembly of parts. *Note:* The combination of several relay elements constitutes a relay unit. (B) An assembly of relay elements that in itself can perform a relay function. *Note:* One or more relay units constitutes a relay.   103

relay winding. Sometimes used for **relay coil.**
                    257

relay wiper. The moving contact on a rotary stepping switch or relay.   259

release (telephone switching systems). Disengaging the apparatus used in a connection and restoring it to its idle condition upon recognizing a disconnect signal.   55

release (trip) coil (of a mechanical switching device). A coil used in the electromagnet which initiates the action of a release (trip).   103

release (tripping) delay (of a mechanical switching device). Intentional time-delay introduced into contact parting time in addition to opening time. *Note:* In devices employing a shunt release, release delay includes the operating time of protective and auxiliary relays external to the device. In devices employing direct or indirect release, release delay consists of intentional delay introduced into the function of the release.   103

release (tripping mechanism) (of a mechanical switching device). A device, mechanically connected to a mechanical switching device, which releases the holding means and permits the opening or closing of the switching device.   103

release (trip) setting (power switchgear). A calibrated point at which the release is set to operate.   103

release delay (mechanical switching device). Intentional time delay introduced into contact-parting time in addition to opening time. *Note:* In devices employing a shunt release, release delay includes the operating time of protective and auxiliary relays external to the device. In devices employing direct or indirect release, release delay consists of intentional delay introduced into the function of the release. *Syn:* **tripping delay.**   103

release-delay (trip-delay) setting (power switchgear). A calibrated setting of the time interval between the time when the actuating value reaches the release setting and the time when the release operates.
                    103

release-free (trip-free) (as applied to a mechanical switching device). A descriptive term that indicates that the opening operation can prevail over the closing operation during specified parts of the closing operation.   103

release-free (trip-free) in any position (power switchgear). A descriptive term indicating that a switching device is release-free at any part of the closing operation. *Note:* If the release circuit is completed through an auxiliary switch, electrical release will not take place until such auxiliary switch is closed.   103

release-free (trip-free) relay. An auxiliary relay whose function is to open the closing circuit of an electrically operated switching device so that the opening operation can prevail over the closing operation. *See:* **trip-free (release-free) relay.**   103

release mechanism (mechanical switching device). A device, mechanically connected to the mechanical switching device, that releases the holding means and permits the opening or closing of the switching device. *Syn:* **tripping mechanism.**   103

release signal (telephone switching systems). A signal transmitted from one end of a line or trunk to indicate that the called party has disconnected.   55

release time, relay. *See:* **relay release time.**

relevant failure. *See:* **failure, relevant.**

reliability (1) (general). (A) The ability of an item to perform a required function under stated conditions for a stated period of time. *Note:* The term reliability is also used as a reliability characteristic denoting a probability of success, or a success ratio. *See:* **wear-out failure period; observed reliability; assessed reliability; extrapolated reliability.**   182,164
(B) The probability that a device will function without failure over a specified time period or amount of usage. *Notes:* (1) Definition (B) is most commonly used in engineering applications. In any case where confusion may arise, specify the definition being used. (2) The probability that the system will perform its function over the specified time should be equal to or greater than the reliability.   255, 77, 209
(2) (relay or relay system). A measure of the degree of certainty that the relay, or relay system, will perform correctly. *Note:* Reliability denotes certainty of correct operation together with assurance against incorrect operation from all extraneous causes. *See:* **dependability; security.**   103, 202, 60, 127
(3) (reliability analysis of nuclear power generating station safety systems). The characteristic of an item or system expressed by the probability that it will perform a required mission under stated conditions for a stated mission time.   587
(4) (nuclear power generating stations). The characteristic of an item expressed by the probability that it will perform a required (function) (mission) under stated conditions for a stated (mission) (period of) time.   29,357

(5) (of a relay or relay system) (power switchgear). A measure of the degree of certainty that the relay, or relay system, will perform correctly. *Note:* Reliability denotes certainty of correct operation together with assurance against incorrect operation from all extraneous sources. *See:* **dependability; security.**   103
(6) (software). The ability of an item to perform a required function under stated conditions for a stated period of time. *See:* **function; software reliability.**   434

reliability allocation (nuclear power generating stations). The assignment of reliability subgoals to subsystems and elements thereof within a system for the purpose of meeting the overall reliability goal for the system, if each of these subgoals is attained.   31,159

reliability, assessed. The reliability of an item determined within stated confidence limits from tests or failure data on nominally identical items. The source of the data shall be stated. Results can only be accumulated (combined) when all the conditions are similar. *Note:* Alternatively, point estimates may be used, the basis of which shall be defined. *See:* **reliability.**   182

reliability assessment (software). The process of determining the achieved level of reliability of an existing system or system component. *See:* **component; reliability; system.**   434

reliability compliance test (reliability). An experiment used to show whether or not the value of a reliability characteristic of an item complies with its stated reliability requirements.   164

reliability data (software). Information necessary to assess the reliability of software at selected points in the software life cycle. Examples include error data and time data for reliability models, program attributes such as complexity, and programming characteristics such as development techniques employed and programmer experience.   434

reliability determination test (reliability). An experiment used to determine the value of a reliability characteristic of an item. *Note:* Analysis of available data may also be used for reliability determination.   164

reliability evaluation. *See:* **reliability assessment.**

reliability, extrapolated (reliability). Extension by a defined extrapolation or interpolation of the observed or assessed reliability for durations and/or conditions different from those applying to the observed or assessed reliability. *Note:* The validity of the extrapolation shall be justified.   164

reliability goal (nuclear power generating stations). A design objective, stated numerically, applied to reliability or availability.   31

reliability growth (software). The improvement in software reliability that results from correcting faults in the software. *See:* **faults; software; software reliability.**   434

reliability, inherent. The potential reliability of an item present in its design. *See:* **reliability.**   182

reliability model (software). A model used for predict-

ing, estimating, or assessing reliability. *See:* **model; reliability; reliability assessment.**   434

reliability modeling (nuclear power generating stations). A logical display in a block diagram format and a mathematical representation of component functions as they occur in sequence which is required to produce system success.   31,159

reliability, operational. The assessed reliability of an item based on field data. *See:* **reliability.**   182

reliability, predicted. For the stated conditions of use, and taking into account the design of an item, the reliability computed from the observed, assessed, or extrapolated reliabilities of its parts. *Note:* Engineering and statistical assumptions shall be stated, as well as the bases used for the computation (observed or assessed).   164

reliability, test. The assessed reliability of an item based on a particular test with stated stress and stated failure criteria. *See:* **reliability.**   182

relief door (rotating machinery). A pressure-operated door to prevent excessive gas pressure within a housing.   63

relieving (electroplating). The removal of compounds from portions of colored metal surfaces by mechanical means. *See:* **electroplating.**   328

relieving anode (pool-cathode tube). An auxiliary anode that provides an alternative conducting path for reducing the current to another electrode. *See:* **electrode (electron tube).**   328

relocatable machine code (software). Machine language code that requires relative addresses to be translated into absolute addresses prior to computer execution. *See:* **absolute machine code; address; computer execution; machine language code.**   434

relocate (computing systems) (programming). To move a routine from one portion of storage to another and to adjust the necessary address references so that the routine, in its new location, can be executed.   255, 77

reluctance (magnetic circuit). The ratio of the magnetomotive force to the magnetic flux through any cross section of the magnetic circuit.   210

reluctance motor. A synchronous motor similar in construction to an induction motor, in which the member carrying the secondary circuit has salient poles, without permanent magnets or direct-current excitation. It starts as an induction motor, is normally provided with a squirrel-cage winding, but operates normally at synchronous speed.   63

reluctance synchronizing (rotating machinery). Synchronizing by bringing the speed of a salient pole synchronous machine to near-synchronous speed, but without applying excitation to it.   63

reluctance torque (synchronous motor). The torque developed by the motor due to the flux produced in the field poles by action of the armature-reaction magnetomotive force.   63

reluctivity. The reciprocal of permeability. *Note:* In anisotropic media, reluctivity becomes a matrix.   210

remanence. The magnetic flux density that remains in

a magnetic circuit after the removal of an applied magnetomotive force. *Note:* This should not be confused with **residual flux density.** If the magnetic circuit has an air gap, the remanence will be less than the residual flux density.   331

remanent charge (ferroelectric device). The charge remaining when the applied voltage is removed. *Note:* The remanent charge is essentially independent of the previously applied voltage, provided this voltage was sufficient to cause saturation. If the device was not or cannot be saturated, the value of the previously applied voltage should be stated when measurements of remanent charge are reported. See the figure under **total charge.** *See:* **ferroelectric domain.**   247

remanent induction (magnetic material). The induction when the magnetomotive force around the complete magnetic circuit is zero. *Note:* If there are no air gaps or other inhomogeneities in the magnetic circuit, the remanent induction will equal the residual induction: if there are air gaps or other inhomogeneities, the remanent induction will be less than the residual induction.   210

remanent polarization (ferroelectric material)(primary ferroelectric terms). The value of the polarization P_r that remains after an applied electric field is removed. Remanent polarization can be measured by integrating the compensating surface charge released on heating a poled ferroelectric to a temperature above its Curie point. *Note:* When the magnitude of this electric field is sufficient to saturate the polarization (usually $3E_c$ that is, three times the coercive electric field), the polarization remaining after the field is removed is termed the saturation remanent polarization P_r. In a single-domain ferroelectric material, the saturation remanent polarization is equal to the spontaneous polarization.   497

remote access (test, measurement and diagnostic equipment). Pertaining to communication with a data processing facility by one or more stations that are distant from that facility.   54

remote backup (power switchgear). A form of backup protection in which the protection is at a station or stations other than that which has the primary protection.   103

remote concentrator (telephone switching systems). A concentrator located away from a serving system control.   55

remote control (1) (general). Control of an operation from a distance: this involves a link, usually electrical, between the control device and the apparatus to be operated. *Note:* Remote control may be over (A) direct wire, (B) other types of interconnecting channels such as carrier-current or microwave, (C) supervisory control, or (D) mechanical means. *See:* **control.**   244,200

(2) (digital interface for programmable instrumentation). A method whereby a device is programmable via its electrical interface connection in order to enable the device to perform different tasks.   40

(3) (power switchgear). Control of a device from a distant point. *Note:* Remote control may be over (1)

direct wire, or over (2) other types of interconnecting channels such as carrier-current or microwave, or by (3) supervisory control or by (4) mechanical means.   103

(4) (programmable instrumentation). A method whereby a device is programmable via its electrical interface connection in order to enable the device to perform different tasks.   378

remote-control circuit (National Electrical Code). Any electric circuit that controls any other circuit through a relay or an equivalent device.   256

remote-control circuits. Any electric circuit that controls any other circuit through a relay or an equivalent device.   256

remote-cutoff tube. *See:* **variable μ tube.**

remote data logging (power switchgear). An arrangement for the numerical representation of selected telemetered quantities on log sheets or paper or magnetic tape, or the like, by means of an electric typewriter, teletype, or other suitable devices.   103

remote error-sensing (power supplies). A means by which the regulator circuit senses the voltage directly at the load. This connection is used to compensate for voltage drops in the connecting wires.   186

remote indication (power switchgear). Indication of the position or condition of remotely located devices. *Note:* Remote indication may be over (1) direct wire, or over (2) other types of interconnecting channels such as carrier-current or microwave, or by (3) supervisory indication or by (4) mechanical means.   103

remote job entry (data communication). Submission of jobs through an input device that has access to a computer through a communications link.   12

remote line (electroacoustics). A program transmission line between a remote pickup point and the studio or transmitter site. *See:* **transmission line.**   239

remotely controlled operation (power switchgear). Operation of a device by remote control.   103

remotely operated (as applied to equipment)(NESC). Capable of being operated from a position external to the structure in which it is installed or from a protected position within the structure.   494

remote magnetic sensor (navigation aid terms). A magnetic sensor located on a vehicle away from disturbances which provides an electrical signal in synchro format which is proportional to the vehicle heading relative to magnetic north. Often called a flux valve.   526

remote manual operation. *See:* **indirect manual operation.**

remote metering. *See:* **telemetering.**

remote operation. *See:* **remotely controlled operation.**

remote release. *See:* **remote trip (remote release).**

remote station (of a supervisory system)(power switchgear). A remotely located station wherin units of switchgear or other equipment are controlled by supervisory control or from which supervisory indications or selected telemeter readings are obtained.   103

remote-station supervisory equipment (power switch-

gear). That part of a (single) supervisory system that includes all supervisory control relays and associated devices located at the remote station for selection, control, indication, and other functions to be performed.   103

remote switching entity (telephone switching systems). An entity for switching inlets to outlets located away from a serving system control.   55

remote terminal unit (RTU)(supervisory control, data acquisition, and automatic control). A remote station equipment of a supervisory system. *See:* **station (2) Remote.**   570

remote trip (remote release) (power switchgear). A general term applied to a relay installation to indicate that the switching device is located physically at a point remote from the initiating protective relay, device, or source of release power or all these. *Note:* This installation is commonly called transfer trip when a communication channel is used to transmit the signal for remote tripping. *Syn:* **remote release.**
                                                          103, 127

removable element (of a switchgear assembly). That portion that normally carries the circuit-switching and circuit-interrupting devices and the removable part of the primary and secondary disconnecting devices.
                                                              103

rendezvous (software). The interaction that occurs between two parallel tasks when one task has called an entry of the other task, and a corresponding accept statement is being executed by the other task on behalf of the calling task.   434

renewable (field-renewable) fuse or fuse unit (power switchgear). A fuse or fuse unit that, after circuit interruption, may be restored readily for service by the replacement of the renewal element, fuse link, or refill unit.   103

renewable fuse (protection and coordination of industrial and commercial power systems). A fuse in which the element, usually a zinc link, may be replaced after the fuse has opened. Once a very popular item, this fuse is gradually losing popularity due to the possibility of using higher ampere-rated links or multiple links in the field, which can present a hazard.   504

renewable parts (power switchgear). Those parts which it is necessary to replace during maintenance as a result of wear.   103

renewal element (of a low-voltage fuse)(power switchgear). That part of a renewable fuse that is replaced after each interruption to restore the fuse to operating condition.   103

reoperate time, relay. *See:* **relay reoperate time.**

repair (1) (nuclear power quality assurance). The process of restoring a nonconforming characteristic to a condition such that the capability of an item to function reliably and safely is unimpaired, even though that item still does not conform to the original requirement.   417
(2) (test, measurement and diagnostic equipment). The restoration or replacement of parts or components of material as necessitated by wear and tear, damage, failure of parts or the like in order to maintain the specific item of material in efficient operating condition.   54

repairable item (reliability). *See:* **item, repaired.**

repair rate (1)(nuclear power generating stations). The expected number of repair actions of a given type completed per unit of time.   29
(2)(reliability analysis of nuclear power generating station safety systems). The expected number of repair actions of a given type completed on a given item per unit of time.   587

repair time (reliable industrial and commercial power systems planning and design). The repair time of a failed component or the duration of a failure is the clock time from the occurrence of the failure to the time when the component is restored to service, either by repair of the failed component or by substitution of a spare component for the failed component. It is not the time required to restore service to a load by putting alternate circuits into operation. It includes time for diagnosing the trouble, locating the failed component, waiting for parts, repairing or replacing, testing, and restoring the component to service. *Syn:* **forced outage duration.**   561

repair urgency (electric generating unit reliability, availability, and productivity). When a planned or unplanned outage is initiated, the urgency with which repair activities are carried out is classified according to one of three classes as defined in maximum effort, normal effort, and low-priority effort.   567

repeat (token ring access method). The action of a station in receiving a bit stream (for example, frame, token, or fill) from the previous station and placing it on the medium to the next station. The station repeating the bit stream may copy it into a buffer or modify control bits as appropriate.   472

repeatability (1)(electric pipe heating systems). The closeness of agreement among a number of consecutive measurements of the output for the same value of the input under the same operating conditions approaching from the same direction. With respect to electric pipe heating systems, repeatability is usually associated with temperature controllers and is the difference in degrees for repeated operation at a specific temperature setting.   448
(2)(measurement) (control equipment)(general). The closeness of agreement among repeated measurements of the same variable under the same conditions. *See:* **measurement system.**   94
(3)(supervisory control, data acquisition, and automatic control). The measure of agreement among multiple readings of an output for the same value of input, made under the same operating conditions, approaching from the same direction, using full-range traverses.   570
(4) (electrical analog indicating instruments). The ability of an instrument to repeat its readings taken when the pointer is deflected upscale, compared to the readings taken when the pointer is deflected downscale, expressed as a percentage of the full-scale value. *See:* **measurement system; moving element (instrument).**   280

(5) (analog computers). A quantitative measure of the agreement among repeated operations. 9

(6) attenuator, variable in fixed steps. Maximum difference in decibels of residual or incremental characteristic insertion loss for a selected position between the extreme values of a first and second set of ten measurements before and after the specified stepping life. 110

(7) two-port, due to insertion/removal cycle. The maximum difference in decibels between the extreme value of ten measurements before and ten measurements after the number of complete insertion/removal cycles specified in insertion/removal life. 110

(8) (gyro, accelerometer) (inertial sensor). The closeness of agreement among repeated measurements of the same variable under the same conditions when changes in conditions or nonoperating periods occur between measurements. 46

(9) continuously variable attenuator, due to cycling. Maximum difference in decibels between the extreme values of a first set of ten measurements for a selected calibration point, five of which are approached from the opposite direction, and the extreme values of a similar second set after the specified cycling life. 110

(10) (nuclear power generating station). The closeness of agreement among a number of consecutive measurements of the output for the same value of the input under the same operating conditions, approaching from the same direction. 355

repeatability

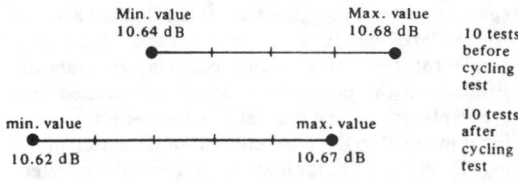

```
     Min. value              Max. value
     10.64 dB                10.68 dB        10 tests
     ●————┼————┼————┼————●                   before
                                             cycling
                                             test
  min. value              max. value         10 tests
  ●————┼————┼————┼————┼————                  after
  10.62 dB              10.67 dB             cycling
                                             test
```

(11) (software). *See:* test repeatability. 434

repeater (1)(medium attachment units and repeater units). A device used to extend the length, topology, or interconnectivity of the physical medium beyond that imposed by a single segment, up to the maximum allowable end-to-end trunk transmission line length. Repeaters perform the basic actions of restoring signal amplitude, waveform, and timing applied to normal data and collision signals. 543

(2)(token ring access method). A device used to extend the length, topology, or interconnectivity of the transmission medium beyond that imposed by a single transmission segment. 472

(3) (data transmission). A combination of apparatus for receiving either one-way or two-way communication signals and delivering corresponding signals which are either amplified, reshaped, or both. A repeater for one-way communication signals is termed a "one-way repeater" and one for two-way communication signals a "two-way repeater". 59

(4) (communication satellite). A receiver-transmitter combination, often aboard a satellite or spacecraft, which receives a signal, performs signal processing (amplification, frequency translation, etcetera) and retransmits it. Used in active communication satellite to relay signals between earth stations. *Syn:* **transponder.** 83

(5) (fiber optics). *See:* **optical repeater.** 433

repeater station (data transmission). An intermediate point in a transmission system where line signals are received, amplified or reshaped, and retransmitted. 59

repeller (reflector) (electron tubes). An electrode whose primary function is to reverse the direction of an electron stream. *See:* **electrode (electron tube).** 125

reperforator. *See:* perforator.

reperforator switching center. A message-relaying center at which incoming messages are received on a reperforator that perforates a storage tape from which the message is retransmitted into the proper outgoing circuit. The reperforator may be of the type that also prints the message on the same tape, and the selection of the outgoing circuit may be manual or under control of selection characters at the head of the message. *See:* **telegraphy.** 194

repertory. *See:* instruction repertory.

repetition equivalent (of a complete telephone connection, including the terminating telephone set). A measure of the grade of transmission experienced by the subscribers using the connection. It includes the combined effects of volume, distortion, noise, and all other subscriber reactions and usages. The repetition equivalent of a complete telephone connection is expressed numerically in terms of the trunk loss of a working reference system when the latter is adjusted to given an equal repetition rate. 328

repetition instruction. An instruction that causes one or more instructions to be executed an indicated number of times. 255, 77

repetition rate (n)(partial discharge measurement in liquid-filled power transformers and shunt reactors). The partial discharge pulse repetition rate n is the average number of partial discharge pulses per second measured over a selected period of time. 580

repetitively pulsed laser (laser-maser). A laser with multiple pulses of radiant energy occurring in a sequence. 363

repetitive operation (analog computers). A condition in which the computer operates as a repetitive device; the solution time may be a small fraction of a second or as long as desired, after which the problem is automatically and repetitively cycled through reset, hold, and operate. 9

repetitive peak forward current (semiconductor). The peak value of the forward current including all repetitive transient currents. 66

repetitive peak line voltage (V_{LRM}) (thyristor). The highest instantaneous value of the line voltage including all repetitive transient voltages, but excluding all nonrepetitive transient voltages. 445

repetitive peak reverse current (semiconductor). The maximum instantaneous value of the reverse current that results from the application of repetitive peak reverse voltage. 66

repetitive peak reverse voltage (1) (semiconductor rectifier). The maximum instantaneous value of the reverse voltage, including all repetitive transient voltages but excluding all nonrepetitive transient voltages, that occurs across a semiconductor rectifier cell, rectifier diode, or rectifier stack. *See:* **principal voltage-current characteristic; rectification; semiconductor rectifier stack.** 237, 243, 66, 208, 191

(2) (reverse blocking thyristor). The maximum instantaneous value of the reverse voltage which occurs across the thyristor, including all repetitive transient voltages, but excluding all non-repetitive transient voltages. 66

repetitive peak reverse-voltage rating (rectifier circuit element). The maximum value of repetitive peak reverse voltage permitted by the manufacturer under stated conditions. *See:* **average forward-current rating (rectifier circuit).** 237, 66, 208

repetitive peak OFF-state current (semiconductor). The maximum instantaneous value of the OFF-state current that results from the application of repetitive peak-OFF-state voltage. 66

repetitive peak OFF-state voltage. The maximum instantaneous value of the OFF-state voltage that occurs across a thyristor, including all repetitive transient voltages, but excluding all nonrepetitive transient voltages. 243, 66, 208, 191

repetitive peak ON-state current (semiconductor). The peak value of the ON-state current including all repetitive transient currents. 66

replacement part. A part for use in place of an existing component of switching equipment. 27

replica temperature relay (power switchgear). A thermal relay whose internal temperature rise is proportional to that of the protected apparatus or conductor, over a range of values and durations of overloads. 103

reply (transponder operation)(navigation aid terms). A radio-frequency signal or combination of signals transmitted as a result of an interrogation. 526

representative sample (nuclear power generating stations). Production/prototype equipment used in a qualification program which is equivalent to that for which qualification is sought in terms of design, function, materials, and manufacturing techniques and processes. 440

reproducibility (1) (general). The ability of a system or element to maintain its output/input precision over a relatively long period of time. *See:* **accuracy; precision.** 207

(2) (transmission lines and waveguides). The degree to which a given set of conditions or observations, using different components or instruments each time, can be reproduced. *See:* **measurement system.** 185

(3) (automatic null-balancing electric instrument). The closeness of agreement among repeated measurements by the instrument for the same value of input made under the same operating conditions, over a long period of time, approaching from either direction. *Notes:* (1) It is expressed as a maximum nonreproducibility in percent of span for a specified time. (2) Reproducibility includes drift, repeatability, and dead band. *See:* **measurement system.** 295

(4) (precision) (radiation protection). The degree of agreement of repeated measurements of the same property expressed quantitatively as the standard deviation computed from the results of the series of measurements. 399

(5)(supervisory control, data acquisition, and automatic control). The measure of agreement among multiple readings of the output for the same value of input, made under the same operating conditions, approaching from either direction, using full-range traverses. 570

reproducing stylus. A mechanical element adapted to following the modulations of a record groove and transmitting the mechanical motion thus derived to the pickup mechanism. *See:* **phonograph pickup.** 176

reproduction speed (facsimile). The area of copy recorded per unit time. *See:* **recording (facsimile).** 12

repulsion-induction motor. A motor with repulsion-motor windings and short-circuited brushes, without an additional device for short-circuiting the commutator segments, and with a squirrel-cage winding in the rotor in addition to the repulsion motor winding. 63

repulsion motor. A single-phase motor that has a stator winding arranged for connection to a source of power and a rotor winding connected to a commutator. Brushes on the commutator are short-circuited and are so placed that the magnetic axis of the rotor winding is inclined to the magnetic axis of the stator winding. This type of motor has a varying-speed characteristic. *See:* **asynchronous machine.** 63

repulsion-start induction motor. A single-phase motor with repulsion-motor windings and brushes, having a commutator-short-circuiting device that operates at a predetermined speed of rotation to convert the motor into the equivalent of a squirrel-cage motor for running operation. For starting operation, this motor performs as a repulsion motor. *See:* **asynchronous machine.** 63

request test (metering). A test made at the request of a customer. 212

required input motion (RIM)(valve actuators). The input motion in terms of acceleration, velocity, and displacement expressed as a function of frequency that a device being tested shall withstand and still perform its intended function. 492

required inputs (software verification and validation plans). The set of items necessary to perform the minimum verification and validation (V&V) tasks mandated within any life-cycle phase. 511

required outputs (software verification and validation plans). The set of items produced as a result of performing the minimum verification and validation (V&V) tasks mandated within any life-cycle phase.   511

required reserve (power operations). The system planned reserve capability needed to ensure a specified standard of service.   516

required response spectrum (RRS) (1) (seismic qualification of Class 1E equipment for nuclear power generating stations). The response spectrum issued by the user or his agent as part of his specifications for qualification or artificially created to cover future applications. The RRS constitutes a requirement to be met.   581

(2) (valve actuators). The required response spectrum issued by the user or his agent as part of his specifications for proof testing, or artificially created to cover future applications. The RRS constitutes a requirement to be met.   492

(3) (nuclear power generating stations) (seismic qualification of class 1E equipment). The response spectrum issued by the user or his agent as part of his specifications for proof testing, or artificially created to cover future applications. The RRS constitutes a requirement to be met.   28

(4) (seismic testing of relays). The response spectrum issued by the user or his agent as part of his specifications for proof testing, or artificially created to cover future applications. The RRS constitutes a requirement to be met.   392

required time (availability). The period of time during which the user requires the item to be in a condition to perform its required function.   164

requirement (software). (1) A condition or capability needed by a user to solve a problem or achieve an objective. (2) A condition or capability that must be met or possessed by a system or system component to satisfy a contract, standard, specification, or other formally imposed document. The set of all requirements forms the basis for subsequent development of the system or system component. See: component; document; requirements analysis; requirements phase; requirements specification; specification; system.   434

requirements analysis (software). (1) The process of studying user needs to arrive at a definition of system or software requirements. (2) The verification of system or software requirements. See: software requirement; system; verification.   434

requirements inspection. See: inspection.

requirements phase (1)(software). The period of time in the software life cycle during which the requirements for a software product, such as the functional and performance capabilities, are defined and documented.   434

(2)(software verification and validation plans). The period of time in the software life cycle during which the requirements, such as functional and performance capabilities for a software product, are defined and documented.   511

requirements specification (software). A specification that sets forth the requirements for a system or system component, for example, a software configuration item. Typically included are functional requirements, performance requirements, interface requirements, design requirements, and development standards.   434

requirements specification language (software). A formal language with special constructs and verification protocols used to specify, verify, and document requirements. See: document requirement; formal language; verification.   434

requirements verification. See: verification.

reradiation. (1) The scattering of incident radiation, or (2) the radiation of signals amplified in a radio receiver. See: radio receiver.   328

rerecording (electroacoustics). The process of making a recording by reproducing a recorded sound source and recording this reproduction. See: dubbing.   176

rerecording system (electroacoustics). An association of reproducers, mixers, amplifiers, and recorders capable of being used for combining or modifying various sound recordings to provide a final sound record. Note: Recording of speech, music, and sound effects may be so combined. See: dubbing; phonograph pickup.   176

re-refining (insulating oil). The use of primary refining processes on used electrical insulating liquids that are suitable for further use as electrical insulating liquids. Note: Techniques may include a combination of distillation and acid, clay or hydrogen treating, and other physical and chemical means.   461

rering signal (telephone switching systems). A signal initiated by an operator at the calling end of an established connection to recall the operator at the called end or the customer at either end.   55

rerun point (computing systems). The location in the sequence of instructions in a computer program at which all information pertinent to the rerunning of the program is available.   255, 77, 54

reseal voltage rating (surge arrester). The maximum arrester recovery voltage permitted for a specified time following one or more unit operation(s) with discharge currents of specified magnitude and duration.   62

reserve (1) (test, measurement and diagnostic equipment). The setting aside of a specific portion of memory for a storage area.   54

(2) (electric power system) (generating stations electric power system). A qualifying term used to identify equipment and capability that is available and is in excess of that required for the load. Note: The reserve may be connected to the system and partially loaded or may be made available by closing switches, contactors, or circuit breakers. Reserve not in operation and requiring switching is sometimes called standby equipment.   381

(3) (power operations). See: customer generation reserve; electrical reserve; installed reserve; interruptible load reserve; nonspinning reserve; operating re-

serve; quick startup; required reserve; spinning reserve; voltage reduction reserve. 516

reserve cell. A cell that is activated by shock or other means immediately prior to use. *See:* **electrochemistry.** 328

reserved (FASTBUS acquisition and control). Bus lines, connector pins, codes, bits, etcetera held for future assignment by the NIM committee. They are not to be used until and except as so assigned. 480

reserved segment interconnect (FASTBUS acquisition and control). A segment interconnect is said to be reserved if it has gained mastership of the far-side segment and is asserting GK=1 (grant acknowledge =1) onto that segment. 480

reserve, electrical (electric power supply). The capability in excess of that required to carry the system load. 112

reserve equipment. The installed equipment in excess of that required to carry peak load. *Note:* Reserve equipment not in operation is sometimes referred to as **standby equipment.** *See:* **generating station.** 64

reserve generation (RG)(electric generating unit reliability, availability, and productivity). The energy that a unit could have produced in a given period but did not, because it was not required by the system. This is the difference between available generation and actual generation. 567

reserve, installed (electric power supply). The reserve capability installed on a system. 112

reserve, nonspinning (electric power supply). That operating reserve capable of being connected to the bus and loaded within a specified time. 112

reserve, operating (electric power supply). That reserve above firm system load required to provide for: (1) Regulation within the hour to cover minute to minute variations (2) Load forecasting error (3) Loss of equipment (4) Local area protection. The operating reserve consists of spinning or nonspinning reserve, or both. 112

reserve, required (electric power supply). The system planned reserve capability needed to insure a specified standard of service. 112

reserve shutdown (electric generating unit reliability, availability, and productivity)(power system measurement). The state in which a unit is available but not in service. *Note:* This is sometimes referred to as economy shutdown. 567, 432

reserve shutdown forced derated hours (RSFDH)(electric generating unit reliability, availability, and productivity). The reserve shutdown hours during which a Class 1, 2, or 3 unplanned derating was in effect. 567

reserve shutdown hours (RSH)(electric generating unit reliability, availability, and productivity). The number of hours a unit was in the reserve shutdown state. 567

reserve shutdown maintenance derated hours (RSMDH)(electric generating unit reliability, availability, and productivity). The reserve shutdown hours during which a Class 4 unplanned derating was in effect. 567

reserve shutdown planned derated hours (RSPDH)(electric generating unit reliability, availability, and productivity). The reserve shutdown hours during which a basic or extended planned derating was in effect. 567

reserve shutdown unit derated hours (RSUNDH)(electric generating unit reliability, availability, and productivity). The reserve shutdown hours during which a unit derating was in effect. 567

reserve shutdown unplanned derated hours (RSUDH)(electric generating unit reliability, availability, and productivity). The reserve shutdown hours during which an unplanned derating was in effect. 567

reservoir operating curve (power operations). A curve, or family of curves (reservoir capability versus time), indicating how a reserve is to be operated under specified conditions to obtain best or predetermined results. 516

reservoir storage (power operations). The volume of water in a reservoir at a given time. 516

reset (1) (electronic digital computation) (industrial control). (A) To restore a storage device to a prescribed state, not necessarily that denoting zero. (B) To place a binary cell in the initial or zero state. *See:* **set.** 235, 210, 255, 77, 206, 54 **(2) (analog computers).** The computer control state in which integrators are held constant and the proper initial condition voltages or charges are applied or reapplied. *See:* **initial condition.** 9 **(3) (of a relay) (power switchgear).** The action of a relay as it makes designated response to decreases in input. Reset as a qualifying term denotes the state of a relay when all response to decrease of input has been completed. Reset is also used to identify the maximum value of an input quantity reached by progressive decreases that will permit the relay to reach the state of complete reset from pickup. *Note:* In defining the designated performance of relays having multiple inputs, reset describes the state when all inputs are zero and also when some input circuits are energized, if the resulting state is not altered from the zero-input condition. 103, 127

reset action (process control). A component of control action in which the final control element is moved at a speed proportional to the extent of proportional-position control action. *Note:* This term applies only to a multiple control action including proportional-position control action. *See:* **control action, proportional plus integral; control system, positioning.** 56

reset, automatic. A function which operates to automatically re-establish specific conditions. 70

reset control action (electric power systems). Action in which the controller output is proportional to the input signal and the time integral of the input signal. The number of times per minute that the integral control action repeats the proportional control action is called the reset rate. *Note:* Applies only to a controller with proportional control action plus integral control action. *See:* **speed-governing system.** 94

reset current or voltage (faulted circuit indicators). The nominal rms (root-mean-square) value of current or voltage that will cause the indicator of the automatic current or voltage reset FCI (faulted circuit indicator) to change from FAULT to NORMAL indication.   482

reset device. A device whereby the brakes may be released after an automatic train-control brake application.   328

reset dwell time. The time spent in reset. In cycling the computer from reset, to operate, to hold, and back to reset, this time must be long enough to permit the computer to recover from any overload and to charge or discharge all integrating capacitors to appropriate initial voltages. *See:* **electronic analog computer.**   9

reset interval (1) (automatic circuit recloser or automatic line sectionalizer). The time required, after a counting operation, for the counting mechanism to return to the starting position of that counting operation.   103
(2) (automatic circuit recloser). The time required for the counting mechanism to return to the starting position.   92

reset, manual. A function which requires a manual operation to re-establish specific conditions.   70

reset on inertial navigation systems (navigation aid terms). Use of external data (for example, position fix) to refine alignment of and to calibrate the inertial navigation system.   526

reset pulse. A drive pulse that tends to reset a magnetic cell.   331

reset rate (proportional plus reset control action or proportional plus reset plus rate control action) (process control). The number of times per minute that the effect of proportional-position control action is repeated. *See:* **integral action rate.**   56

reset switch (industrial control). A machine-operated device that restores normal operation to the control system after a corrective action. *See:* **photoelectric control.**   206

resettability (1)(electric pipe heating systems). The restoring of a mechanism, electrical circuit, or device to the prescribed state. Resettability is usually associated with temperature controllers and is the difference in degrees when returning to original temperature setting.   448
(2)(oscillator). The ability of the tuning element to retune the oscillator to the same operating frequency for the same set of input conditions. *See:* **tunable microwave oscillators.**   174

reset time (1)(faulted circuit indicators). The time required for the FCI (faulted circuit indicator) to return automatically to NORMAL indication after its reset current or voltage has been established, or for the elapsed time automatic reset FCI to reset.   482
(2) (of a relay) (power switchgear). The time interval from occurrence of specified conditions to reset. *Note:* When the conditions are not specified it is intended to apply to a picked-up relay and to be a sudden change from pickup value of input to zero input.   103

(3) (of an automatic circuit recloser or automatic line sectionalizer) (power switchgear). The time required, after one or more counting operations, for the counting mechanism to return to the starting position.   103

residential-custodial care facility (National Electrical Code) (health care facilities). A building, or part thereof, used for the lodging or boarding of 4 or more persons who are incapable of self-preservation because of age, or physical or mental limitation. This includes facilities such as homes for the aged, nurseries (custodial care for children under 6 years of age), and mentally retarded care institutions. Day care facilities that do not provide lodging or boarding for institutional occupants are not classified as residential custodial care facilities.   256

residual-component telephone-influence factor (three-phase synchronous machine). The ratio of the square root of the sum of the squares of the weighted residual harmonic voltages to three times the root-mean-square no-load phase-to-neutral voltage.   63

residual current (protective relaying). The sum of the three-phase currents on a three-phase circuit. The current that flows in the neutral return circuit of three wye-connected current transformers is residual current.   128

residual-current state (thermionics). The state of working of an electronic valve or tube in the absence of an accelerating field from the anode of a diode or equivalent diode, in which the cathode current is due to the nonzero velocity of emission of electrons. *See:* **electron emission; inductive coordination.**   244, 190

residual error (electronic navigation). The sum of the random errors and the uncorrected systematic errors. *See:* **navigation.**   13, 187

residual-error rate. *See:* **undetected-error rate.**

residual flux density. The magnetic flux density at which the magnetizing force is zero when the material is in a symmetrically cyclically magnetized condition. *See:* **remanence.**   331

residual FM (frequency modulation) (spectrum analyzer). Short term displayed frequency instability (jitter) of the spectrum analyzer caused by instability of the local oscillators. Given in terms of peak-to-peak frequency deviation (Hz). *Notes:* (1) Any influencing factors such as phase lock on or off, etcetera should be given. (2) For the purpose of this standard "short term" shall mean measurements made during a specified period of time. The recommended time duration is 20s to 20 μs per division. This will accommodate incidental FM from less than 1 Hz to tens of kHz. The manufacturer shall specify the time to be used. *Syn:* **incidental FM.**   390

residual frequency-modulation (non-real time spectrum analyzer). Short term displayed frequency instability or jitter of the spectrum analyzer local oscillators. Given in terms of peak-to-peak frequency deviation, hertz (Hz). Any influencing factors (for example, phase lock on or off) must be given.   68

residual induction (1) (magnetic material). The magnetic induction corresponding to zero magnetizing force in a material that is in a symmetrically cyclically magnetized condition.   210

(2) (residual flux density) (toroidal magnetic amplifier cores). The magnetic induction at which the magnetizing force is zero when the magnetic core is cyclically magnetized with a half-wave sinusoidal magnetizing force of a specified peak magnitude. *Note:* This use of the term **residual induction** differs from the standard definition that requires symmetrically cyclically magnetized conditions.   165

residual magnetism (ferromagnetic bodies). A property by which they retain a certain magnetization (induction) after the magnetizing force has been removed.   244, 210

residual modulation. *See:* **carrier noise; carrier noise level.**

residual probe pickup (constancy of probe coupling) (slotted line). The noncyclical variation of the amplitude of the probe output over its complete range of travel when reflected waves are eliminated on the slotted section by proper matching at the output and the input, discounting attenuation along the slotted section. It is defined by the ratio of one-half of the total variation to the average value of the probe output, assuming linear amplitude response of the probe, at a specified frequency(ies) within the range of usage. *Note:* This quantity consists of two parts of which one is reproducible and the other is not. The repeatable part can be eliminated by subtraction in repeated measurements, while the nonrepeatable part must cause an error. The residual probe pickup depends to some extent on the insertion depth of the probe. *See:* **measurement system; residual standing-wave ratio.**   185

residual reflected coefficient (reflectometer). The erroneous reflection coefficient indicated when the reflectometer is terminated in reflectionless terminations. *See:* **measurement system.**   185

residual relay (power switchgear). A relay that is so applied that its input, derived from external connections of instrument transformers, is proportional to the zero-phase-sequence component of a polyphase quantity.   103

residual response (1) (non-real time spectrum analyzer). A spurious response in the absence of an input.   68

(2) (spectrum analyzer). A spurious response in the absence of an input, not including noise and zero pip.   390

residual standing-wave ratio (SWR) (slotted line). The standing-wave ratio measured when the slotted line is terminated by a reflectionless termination and fed by a signal source that provides a nonreflecting termination for waves reflected toward the generator. *Note:* Residual standing-wave ratio does not include the residual noncyclical probe pickup or the attenuation encountered as the probe is moved along the line. *See:* **residual probe pickup.**   185

residual voltage (1) (arrester) (discharge voltage). The voltage that appears between the line and ground terminals of an arrester during the passage of discharge current. *See:* **inductive coordination.**   308, 62

(2) (protective relaying). The sum of the three line-to-neutral voltages on a three-phase circuit.   128

residue check (computing systems). A check in which each operand is accompanied by the remainder obtained by dividing this number by *n*, the remainder then being used as a check digit or digits. *See:* **modulo *n* check.**   255, 77

resin (rotating machinery). Any of various hard brittle solid-to-soft semisolid amorphous fusible flammable substances of either natural or synthetic origin: generally of high molecular weight, may be either thermoplastic or thermosetting.   63

resin-bonded paper-insulated bushing (outdoor electric apparatus). A bushing in which the major insulation is provided by paper bonded with resin.   168

resist (electroplating). Any material applied to part of a cathode or plating rack to render the surface nonconducting. *See:* **electroplating.**   328

resistance (1) (network analysis). (A) That physical property of an element, device, branch, network, or system that is the factor by which the mean-square conduction current must be multiplied to give the corresponding power lost by dissipation as heat or as other permanent radiation or loss of electromagnetic energy from the circuit. (B) The real part of impedance. *Note:* Definitions (A) and (B) are not equivalent but are supplementary. In any case where confusion may arise, specify definition being used. *See:* **resistor.**   210, 185, 206

(2) (shunt). The quotient of the voltage developed across the instrument terminals to the current passing between the current terminals. In determining the value, account should be taken of the resistance of the instrument and the measuring cable. The resistance value is generally derived from a direct-current measurement such as by means of a double Kelvin bridge. *See:* **test voltage and current.**   307

(3) (automatic control). A property opposing movement of material, or flow of energy, and involving loss of potential (voltage, temperature, pressure, level).   56

(4) (antenna). *See:* **antenna resistance; radiation resistance.**

resistance, apparent (insulation testing). Ratio of the voltage across the electrodes in contact with the specimen to the current between them as measured under the specified test conditions and specified electrification time.   97

resistance box. A rheostat consisting of an assembly of resistors of definite values so arranged that the resistance of the circuit in which it is connected may be changed by known amounts.   210

resistance braking. A system of dynamic braking in which electric energy generated by the traction motors is dissipated by means of a resistor. *See:* **dynamic braking.**   328

resistance bridge smoke detector (fire protection devices). A device which responds to an increase of smoke particles and moisture, present in products of combustion, which fall on an electrical bridge grid. As these conductive substances fall on the grid they reduce the resistance of the grid and cause the detector to respond. 71

resistance-capacitance (RC) coupling. Coupling between two or more circuits, usually amplifier stages, by means of a combination of resistance and capacitance elements. *See:* **coupling.** 328

resistance-capacitance characteristic, input (oscilloscopes). The direct-current resistance and parallel capacitance to ground present at the input of an oscilloscope. 106

resistance-capacitance oscillator (RC oscillator). Any oscillator in which the frequency is determined principally by resistance-capacitance elements. *See:* **oscillatory circuit.** 328

resistance drop (power and distribution transformer). The component of the impedance voltage drop in phase with the current. 53

resistance furnace. An electrothermic apparatus, the heat energy for which is generated by the flow of electric current against ohmic resistance internal to the furnace. 328

resistance grading (cr corona shielding). A form of corona shielding embodying high resistance material on the surface of the coil. *Syn:* **corona shielding.** 63

resistance grounded (1)(power and distribution transformer). Grounded through impedance, the principal element of which is resistance. *Note:* The resistance may be inserted either directly, in the connection to the ground, or indirectly, as for example, in the secondary of a transformer, the primary of which is connected between neutral and ground, or in series with the delta-connected secondary of a wye-delta grounding transformer. 53

(2)(system grounding). Grounded through impedance, the principal element of which is resistance. *Note:* The high-resistance-grounded system is designed to meet the criterion of $R_0 \leq X_{C0}$ in order to limit transient overvoltages due to arcing ground faults. The ground-fault current is usually limited to less than 10A. X_{C0} is the distributed per-phase capacitive reactance to ground of the system. The low-resistance-grounded system permits a higher ground-fault current (on the order of 25 A to several hundred amperes) to obtain sufficient current for selective relay performance. For the usual system the criterion for limiting transient overvoltages is $R_0/X_0 \geq 2$. 152

resistance lamp. An electric lamp used to prevent the current in a circuit from exceeding a desired limit. 328

resistance magnetometer. A magnetometer that depends for its operation upon the variation of electrical resistance of a material immersed in the field to be measured. *See:* **magnetometer.** 328

resistance method of temperature determination (power and distribution transformer). The determination of the temperature by comparison of the resistance of a winding at the temperature to be determined, with the resistance at a known temperature. 53

resistance modulation (bolometric power meters). A change in resistance of the bolometer resulting from a change in power (RF, ac, or dc) dissipated in the element. *Note:* The resistance modulation sensitivity is the (dc) change in resistance per unit (dc) change in power at normal bias and at a constant ambient temperature. Resistance modulation frequency response is the frequency of repetitive (sinusoidal) power change for which the peak-to-peak resistance change is 3 dB lower than the asymptotic, maximum value at zero frequency. 115

resistance modulation effect (bolometric power meters). A component of substitution error (for dc power substitution) in bolometer units in which both ac and dc bias is used. *Note:* This component is dependent upon the frequency of the ac bias and the frequency response of the element: it is usually very small, and usually not included in the effective efficiency correction for substitution error. It is caused by resistance modulation of the element, and is more pronounced in barretters than in thermistors. 115

resistance relay(1). A linear-impedance form of distance relay for which the operating characteristic on an *R-X* diagram is a straight line of constant resistance. *Note:* The operating characteristic may be described by the equation R , K, or $Z \cos \theta$, K, where K is a constant, and θ is the angle by which the input voltage leads the input current. 127

(2)(power switchgear). A linear-impedance form of distance relay for which the operating characteristic on an R-X diagram is a straight line of constant resistance. *Note:* The operating characteristic may be described by the equation R = K or $Z\cos\theta = K$, where K is a constant, and θ is the angle by which the input voltage leads the input current. *See:* **distance relay; Figure (a)**

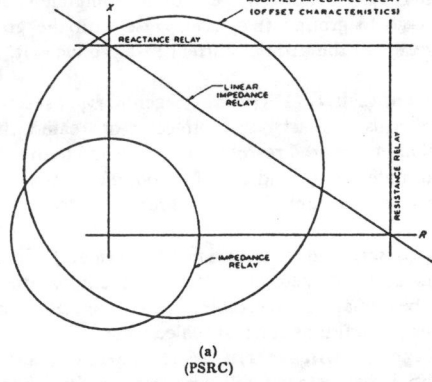

103

resistance starting (industrial control). A form of reduced-voltage starting employing resistances that are short-circuited in one or more steps to complete the starting cycle. *See:* **resistance starting, generator-field; resistance starting, motor-armature.** 225, 206

resistance starting, generator-field. Field resistance starting provided by one or more resistance steps in series with the shunt field of a generator, the output of which is connected to a motor armature. *See:* **resistance starting; resistance starting, motor-armature.**

206

resistance starting, motor-armature. Motor resistance starting provided by one or more resistance steps connected in series with the motor armature. *See:* **resistance starting; resistance starting, generator-field.**

206

resistance-start motor. A form of split-phase motor having a resistance connected in series with the auxiliary winding. The auxiliary circuit is opened when the motor has attained a predetermined speed. *See:* **asynchronous machine.**

328

resistance temperature detector (resistance thermometer detector) (resistance thermometer resistor). A resistor made of some material for which the electrical resistivity is a known function of the temperature and that is intended for use with a resistance thermometer. It is usually in such a form that it can be placed in the region where the temperature is to be determined. *Note:* A resistance temperature detector with its support and enclosing envelope, is often called a resistance thermometer bulb. *See:* **electric thermometer (temperature meter); embedded temperature detector.**

328

resistance thermometer (resistance temperature meter). An electric thermometer that operates by measuring the electric resistance of a resistor, the resistance of which is a known function of its temperature. The temperature-responsive element is usually called a **resistance temperature detector.** *Note:* The resistance thermometer is also frequently used to designate the sensor and its enclosing bulb alone, for example, as in platinum thermometer, copper-constantan thermometer, etcetera. *See:* **electric thermometer (temperature meter); instrument.**

7

resistance to ground (surge arresters). The ratio, at a point in a grounding system, of the component of the voltage to ground that is in phase with the ground current, to the ground current that produces it.

244, 62

resistant (suffix) (1) (rotating machinery). Material or apparatus so constructed, protected or treated, that it will not be injured readily when subjected to the specified material or condition, for example, fire-resistant, moisture-resistant. *See:* **asynchronous machine.**

127, 27

(2) (power and distribution transformers). So constructed, protected, or treated that the apparatus will not be damaged when subjected to the specified material or conditions for a specified time.

53

(3) (power switchgear). So constructed, protected, or treated that damage will not occur readily when the device is subjected to the specified material or condition.

103

resistive attenuator (waveguide). A length of waveguide designed to introduce a transmission loss by the use of some dissipative material. *See:* **absorptive attenuator (waveguide); waveguide.**

244, 179

resistive conductor. A conductor used primarily because it possesses the property of high electric resistance.

64

resistive coupling. The association of two or more circuits with one another by means of resistance mutual to the circuits.

313

resistive distributor brush (electromagnetic compatibility). Resistive pickup brush in an ignition distributor cap. *See:* **electromagnetic compatibility.**

220, 199

resistive feedback preamplifier (germanium gamma-ray detectors). A charge-sensitive preamplifier in which charge that accumulates on the feedback capacitor is continually discharged through a resistor in parallel with the capacitor.

528

resistive ignition cable (electromagnetic compatibility). High-tension ignition cable, the core of which is made of resistive material. *See:* **electromagnetic compatibility.**

220, 199

resistivity (material). A factor such that the conduction-current density is equal to the electric field in the material divided by resistivity.

313

resistivity, volume. The reciprocal of volume conductivity, measured in siemens per centimeter, which is a steady-state parameter.

97

resistor (electric installations on shipboard). A device the primary purpose of which is to introduce resistance into an electric circuit. Note: A resistor as used in electric circuits for purposes of operation, protection, or control, commonly consists of an aggregation of units. Resistors, as commonly supplied, consist of wire, metal ribbon, cast metal, or carbon compounds supported by or embedded in an insulating medium. The insulating medium may enclose and support the resistance material as in the case of the porcelain tube type or the insulation may be provided only at the points of support as in the case of heavy duty ribbon or cast iron grids mounted in metal frames.

3

resistor, bias (semiconductor radiation detector). The resistor through which bias voltage is applied to a detector.

23, 118

resistor furnace. A resistance furnace in which the heat is developed in a resistor that is not a part of the charge.

328

resistor-start motor. A single-phase induction motor with a main winding and an auxiliary winding connected in series with a resistor, with the auxiliary winding circuit opened for running operation.

63

re-solution (electrodeposition). The passing back into solution of metal already deposited on the cathode. *See:* **electrodeposition.**

328

resolution (1)(supervisory control, data acquisition, and automatic control). The least value of the measured quantity which can be distinguished.   570

(2) (data transmission)(communication). (A) The result of deriving from a sound, scene, or other form of intelligence, a series of discrete elements wherefrom the original may subsequently be synthesized. (B) The degree to which nearly equal values of a quantity can be discriminated. (C) The fineness of detail in a reproduced spatial pattern. (D) The degree to which a system or a device distinguishes fineness of detail in a spatial pattern.

59

(3) **(storage tube).** A measure of the quantity of information that may be written into and read out of a storage tube. *Notes:* (A) Resolution can be specified in terms of number of bits, spots, lines, or cycles. (B) Since the relative amplitude of the output may vary with the quantity of information, the true representation of the resolution of a tube is a curve of relative amplitude versus quantity. *See:* **storage tube.**
174

(4) **(television).** A measure of ability to delineate picture detail. *Note:* Resolution is usually expressed in terms of a number of lines N (normally alternate black and white lines) the width of each line is $1/N$ times the picture height. In television practice, where the raster has a 4/3 aspect ratio, resolution, measured in either the horizontal or the vertical direction, is the number of test chart lines observable in a distance equal to the vertical dimension of the raster.
18

(5) **(oscilloscopes).** A measure of the total number of trace lines discernible along the coordinate axes, bounded by the extremities of the graticule or other specific limits. *See:* **oscillograph.**
185

(6) **(transmission lines and waveguides).** The degree to which nearly equal values of a quantity can be discriminated.
185

(7) **(industrial control).** The smallest distinguishable increment into which a quantity is divided in a device or system. *See:* **control system, feedback.**
206

(8) **(digital delay line).** The time spacing between peaks of the doublet.
81

(9) **(acousto-optic deflector).** The ratio of the angular swing to the minimum resolvable angular spread of one spot. The minimum spot size depends on the optical beam amplitude and phase distribution, as well as the criteria used to define minimum spot size. When the Rayleigh criteria is used for minimum spot size, resolution, N, is given by $N = 1/\alpha\ \tau\Delta f$, with $\alpha = 1$ rectangular beam, constant amplitude: 1.22 circular beam, constant amplitude: 1.34 circular beam, Gaussian amplitude. For operation in the scanning mode, the resolution will be reduced as the scan time approaches the access time.
72

(10) **(non-real time spectrum analyzer).** The ability to display adjacent responses discretely (hertz, hertz decibel down). The measure of resolution is the frequency separation of two responses which merge with a 3 decibel notch. (A) Equal Amplitude Signals. As a minimum instruments will be specified and controls labeled on the basis of two equal amplitude responses under the best operation conditions. (B) Unequal Amplitude Signals. The frequency difference between two signals of specified unequal amplitude when the notch formed between them is 3 decibel down from the smaller signal shall be termed Skirt Resolution. (C) Optimum Resolution. For every combination of frequency span and sweep time there exists a minimum obtainable value of resolution (R). This is the Optimum Resolution (Ro), which is defined theoretically as:

$$Ro = K \sqrt{\frac{\text{Frequency Span}}{\text{Sweep Time}}}$$

The factor "K" shall be unity unless otherwise specified.
68

(11) **(pulse measurement).** The smallest change in the pulse characteristic, property, or attribute being measured which can unambiguously be discerned or detected in a pulse measurement process.
15

(12) **(electrothermic power meters).** The smallest discrete or discernible change in power that can be measured. In this standard, resolution includes the estimated uncertainty with which the power changes can be determined on the readout scale.
47

(13) **(gyro: accelerometer).** The largest value of the minimum change in input, for inputs greater than the threshold, which produces a change in output equal to some specified percentage (at least 50 percent) of the change in output expected using the nominal scale factor. *See:* **input-output characteristics.**
46

(14) **(plutonium monitoring).** The minimum detectable change in instrument response.
413

(15) **(spectrum analyzer).** (A) general. The ability to display adjacent responses discretely (Hz, Hz dB down). The measure of resolution is the frequency separation of two responses which merge with a 3 dB notch. (B) resolution. As a minimum, instruments will be specified and controls labeled on the basis of two equal amplitude responses under the best operational conditions. (C) skirt resolution. The frequency difference between two signals of specified unequal amplitude when the notch formed between them is 3 dB down from the smaller signal. (D) optimum resolution. For every combination of frequency span and sweep time there exists a minimum obtainable value of resolution (R). This the the optimum resolution (R_o), which is defined theoretically as: The factor K shall be unity unless otherwise specified.
390

(16) **(station control and data acquisition).** The least value of the measured quantity which can be distinguished.
403

resolution bandwidth (1) **(non-real time spectrum analyzer).** The width in hertz, of the spectrum analyzer's response to a (continuous wave) signal. This width is usually defined as the frequency difference at specified points on the response curve, such as the 3 or 6 decibel down points. The manufacturer will specify the decibel down points to be used.
68

(2) **(spectrum analyzer).** The width in Hz, of the spectrum analyzer's response to a continuous wave (CW) signal. This width is usually defined as the frequency difference at specified points on the response curve, such as the 3 or 6 dB down points. The manufacturer will specify the dB down points to be used.
390

resolution element (radar). A spatial and velocity region contributing echo energy which can be separated from that of adjacent regions by action of the antenna or receiving system. In conventional radar its dimensions are given by the beamwidths of the antenna, the transmitter pulsewidth, and the receiver bandwidth; its dimensions may be increased by the presence of spurious response regions (sidelobes), or decreased by use of specially coded transmissions and appropriate processing techniques. *Syn:* resolution cell.   13

resolution, energy (percent). *See:* energy resolution (percent).

resolution error (analog computers). The error due to the inability of a transducer to manifest changes of a variable smaller than a given increment.   9

resolution of output adjustment (of any output parameter, voltage, frequency, etcetera) (inverters). The minimum increment of change in setting. *See:* self-commutated inverters.   208

resolution response (square-wave) (television) (1) (camera device). The ratio of (A) the peak-to-peak signal amplitude given by a test pattern consisting of alternate black and white vertical bars of equal widths corresponding to a specified line number to (B) the peak-to-peak signal amplitude given by large-area blacks and large-area whites having the same luminances as the black and white bars in the test pattern. *Note:* Horizontal scanning lines are assumed. *See:* television.

(2) (display device). The ratio of (A) the peak-to-peak luminance given by a square-wave test signal whose half-period corresponds to a specified line number to (B) the peak-to-peak luminances given by a test signal producing large-area blacks and large-area whites having the same amplitudes as the signal of (A). *Note:* In a display device, resolution response, at relatively high line numbers, is sometimes called **detail contrast.**   317

resolution, structural (color-picture tubes). The resolution as limited by the size and shape of the screen elements. See also: **television.**   178

resolution time (1) (counter tube or counting system) (radiation counters). The minimum time interval between two distinct events that will permit both to be counted. *See:* **anticoincidence (radiation counters).**   190

(2) (sequential events recording systems). The minimum time interval between any two distinct events that will permit both to be recorded in sequence of occurrence. *See:* event.   48, 58

resolution time correction (radiation counters). Correction to the observed counting rate to allow for the probability of the occurrence of events within the resolution time. *See:* **anticoincidence (radiation counters).**   190

resolution wedge (television). A narrow-angle wedge-shaped pattern calibrated for the measurement of resolution and composed of alternate contrasting strips that gradually converge and taper individually to preserve equal widths along any given line at right angles to the axis of the wedge. *Note:* Alternate strips may be black and white of maximum contrast or strips of different colors.   18

resolver (analog computers). A device or computing element used for vector resolution or composition. The rectangular mode is the mode of operation that produces a transformation from polar to rectangular coordinates or a rotation of rectangular coordinates. The polar mode of operation that produces a transformation from rectangular to polar coordinates.   9

resolving power (illuminating engineering). The ability of the eye to perceive the individual elements of a grating or any other periodic pattern with parallel elements measured by the number of cycles per degree that can be resolved. The resolution threshold is the period of the pattern that can be just resolved. Visual acuity, in such a case, is the reciprocal of one-half of the period expressed in minutes. The resolution threshold for a pair of points or lines is the distance between them when they can just be distinguished as two, not one, expressed in minutes of arc.   167

resolving time (1)(navigation aid terms). The minimum time interval by which two events must be separated, to be distinguishable in a navigation system, by the time measurement alone.   526

(2) (radiation counters). The time from the start of a counted pulse to the instant a succeeding pulse can assume the minimum strength to be detected by the counting circuit. *Note:* This quantity pertains to the combination of tube and recording circuit. *See:* **ionizing radiation.**   117, 96, 125

(3) (liquid-scintillation counting). The minimum time that exists between successive events if they are to be counted as separate events.   422

resonance (1)(seismic design of substations). A dynamic condition which occurs when any input frequency of vibration coincides with one of the natural frequencies of the structure. In a plot of the response of the structure (acceleration, velocity, displacement) versus input frequency for a constant input, as the input frequency approaches one of the natural frequencies of the structure the response increases to a maximum value if damping is less than critical. The response of the structure at resonance may be much greater than the input, if the damping is low.   465

(2) (circuits and systems). The enhancement of the response of a physical system to a periodic excitation when the excitation frequency is equal to a natural frequency of the system.   67

(3) (automatic control). Of a system or element, a condition evidenced by large oscillatory amplitude, which results when a small amplitude of a periodic input has a frequency approaching one of the natural frequencies of the driven system. *Note:* In a feedback control system, this occurs near the stability limit.   56

(4) (data transmission). A condition in a circuit containing inductance and capacitance in which the capacitive reactance is equal to the inductive reactance. This condition occurs at only one frequency in a circuit with fixed constants, and the circuit is said to be 'tuned' to this frequency. The resonance frequency can be changed by varying the value of the capacitance or inductance of the circuit.   59

(5) (radio wave propagation). Of a traveling wave, the

change in amplitude as the frequency of the wave approaches or coincides with a natural frequency of the medium (for example, plasma frequency). 146

resonance bridge. A 4-arm alternating-current bridge in which both an inductor and a capacitor are present in one arm, the other three arms being (usually) nonreactive resistors, and the adjustment for balance includes the establishment of resonance for the applied frequency. *Note:* Normally used for the measurement of inductance, capacitance, or frequency. Two general types can be distinguished according as the inductor and capacitor are effectively in series or in parallel. *See:* **bridge.**

$$R_1R_4 = R_2R_3$$
$$\omega^2 LC = 1$$

Series resonance bridge

328

resonance charging (charging inductors) (1) direct current. The charging of the capacitance (of a pulse-forming network) to the initial peak value of voltage in an oscillatory series resistance- inductance-capacitance (RLC) circuit, when supplied by a direct voltage. *Note:* in order to provide a pulse train, the network capacitance is repetitively discharged by a synchronous switch at the time when the current through the charging inductor is zero and the peak voltage to which the network capacitance is charged approaches two times the power-supply direct voltage. (2) alternating current. The charging of the capacitance (of a pulse-forming network) to the peak value of voltage selected, in an oscillatory resistance-inductance-capacitance (RLC) circuit, when supplied by an alternating voltage. *Note:* In order to provide a pulse train, the network capacitance is repetitively discharged at a time in the charging cycle when the current through the charging inductor is zero. At these times the voltage may be essentially:

$$\frac{\pi E_p}{2}, \pi E_p, \frac{3\pi E_p}{2}, \text{etc.}$$

The value chosen depends upon the pulse-repetition rate and the frequency of the alternating voltage. (E_p = peak alternating voltage supply.) 137

resonance curve, carrier-current line trap (power-system communication). A graphical plot of the ohmic impedance of a carrier current line trap with respect to frequency at frequencies near resonance. *See:* **power-line carrier.** 59

resonance frequency (resonant frequency) (1) network. Any frequency at which resonance occurs. *Note:* For a given network, resonance frequencies may differ for different quantities, and almost always differ from natural frequencies. For example, in a simple series resistance-inductance-capacitance circuit there is a resonance frequency for current, a different resonance frequency for capacitor voltage, and a natural frequency differing from each of these. *See:* **network analysis.** 210

(2) (crystal unit). The frequency for a particular mode of vibration to which, discounting dissipation, the effective impedance of the crystal unit is zero. *See:* **crystal.** 328

resonance frequency of charging (charging inductors). The frequency at which resonance occurs in the charging circuit of a pulse-forming network. *Note:* In this document, it will be assumed to be the frequency determined as follows:

$$f_0 = \frac{1}{2\pi\sqrt{LC_0}}$$

where
f_0 = resonance frequency of charging
C_0 = capacitance of pulse-forming network
L = charging inductance. 137

resonance mode (laser-maser). A natural oscillation in a resonator characterized by a distribution of fields which have the same harmonic time dependence throughout the resonator. 363

resonant cavity. *See:* **optical cavity.**

resonant frequency (1)(seismic qualification of Class 1E equipment for nuclear power generating stations). A frequency at which a response peak occurs in a system subjected to forced vibration. This frequency is accompanied by a phase shift of response relative to the excitation. 581

(2) (antenna). A frequency at which the input impedance of an antenna is nonreactive. 111

(3) (network). *See:* **resonance frequency.**

resonant gap (microwave gas tubes). The small region in a resonant structure interior to the tube, where the electric field is concentrated. 125

resonant grounded. *See:* **ground-fault neutralizer grounded.**

resonant grounded system (arc-suppression coil) (surge arresters). A system grounded through a reactor, the reactance being of such value that during a single line-to-ground fault, the power-frequency inductive current passed by this reactor essentially neutralizes the power-frequency capacitive component of the ground-fault current. *Note:* With resonant grounding of a system, the net current in the fault is limited to such an extent that an arc fault in air would be self-extinguishing. *See:* **ground.**   91, 244, 62

resonant iris (waveguide components). An iris designed to have equal capacitive and inductive susceptances at the resonant frequency.   166

resonant line oscillator. An oscillator in which the principal frequency-determining elements are one or more resonant transmission lines. *See:* **oscillatory circuit.**   111, 211

resonant mode (1) (general). A component of the response of a linear device that is characterized by a certain field pattern, and that when not coupled to other modes is representable as a single-tuned circuit. *Note:* When modes are coupled together, the combined behavior is similar to that of the corresponding single-tuned circuits correspondingly coupled. *See:* **waveguide.**   328

(2) (cylindrical cavities). When a metal cylinder is closed by two metal surfaces perpendicular to its axis a cylindrical cavity is formed. The resonant modes in this cavity are designated by adding a third subscript to indicate the number of half-waves along the axis of the cavity. When the cavity is a rectangular parallelepiped the axis of the cylinder from which the cavity is assumed to be made should be designated since there are three possible cylinders out of which the parallelepiped may be made. *See:* **guided wave.**
    343

resonant wavelengths (cylindrical cavities). Those given by $\lambda_r = 1[(1/\lambda_c)^2 + (1/2c)^2]^{1/2}$ where λ_c is the cutoff wavelength for the transmission mode along the axis, l is the number of half-period variations of the field along the axis, and c is the axial length of the cavity. *See:* **guided wave.**   343

resonating (steady-state quantity or phasor). The maximizing or minimizing of the amplitude or other characteristic provided the maximum or minimum is of interest. *Notes:* (1) Unless otherwise specified, the quantity varied to obtain the maximum or minimum is to be assumed to be frequency. (2) Phase angle is an example of a quantity in which there is usually no interest in a maximum or a minimum. (3) In the case of amplification, transfer ratios, etcetera, the amplitude of the phasor is maximized or minimized: in the case of currents, voltages, charges, etcetera, it is customary to think of the amplitude of the steady-state simple sine-wave quantity as being maximized or minimized. *See:* **network analysis.**   210

resonating capacitor (ferroresonant voltage regulators). Provides the capacitance associated with ferroresonant reulating circuits for the purpose of producing ferroresonance.   456

resonating capacitor volt-amperes (ferroresonant voltage regulators). The product of the voltage across the resonating capacitor and the current through the resonating capacitor (root-mean-square values) under stated operating conditions.   456

resonating winding (ferroresonant voltage regulators). The winding of the ferroresonant transformer used to connect the resonating capacitance to the circuit. *Note:* It is wound on the secondary section of the core and is separated from the primary winding by a magnetic shunt. It may itself be the output winding or a portion of the output winding.   456

resonator (1) (general). A device, the primary purpose of which is to introduce resonance into a system. *See:* **network analysis.**   210

(2) (circuits and systems). (A) A resonating system. (B) A device designed to operate in the vicinity of a natural frequency of that device. (C) (electrical circuit) An electrical network designed to present a given natural frequency at its terminal.   67

resonator grid (electron tubes). An electrode, connected to a resonator, that is traversed by an electron beam and that provides the coupling between the beam and the resonator. *See:* **velocity-modulated tube.**
    244

resonator mode (oscillator). A condition of operation corresponding to a particular field configuration for which the electron stream introduces a negative conductance into the coupled circuit. *See:* **oscillatory circuit.**   125

resonator, waveguide (waveguide components). A waveguide or transmission line structure which can store oscillating electromagnetic energy for time periods that are long compared with the period of the resonant frequency, at or near the resonant frequency.
    166

responder beacon. *See:* **transponder.**

response (1) (logical link control). In data communications, a reply represented in the control field of a response protocol data unit (PDU). It advises the address destination logical link control (LLC) of the action taken by the source LLC to one or more command PDUs.   585

response (2) (of a device or system) (power switchgear). A quantitative expression of the output as a function of the input under conditions that must be explicitly stated. *Note:* The response characteristic, often presented graphically, gives the response as a function of some independent variable such as frequency or time.   103

(3) (radiation protection). The instrument reading.
    399

response, acceleration-forced (automatic control). The total (transient plus steady-state) time response resulting from a sudden increase in the rate of the rate of change of input from zero to some finite value.
    56

response, forced (automatic control). A time response which is produced by a stimulus external to the system or element under consideration. *Note:* The response may be described in terms of the causal variable. *See:* **response, acceleration-forced; response, impulse-forced.**   56

response function (linear passive networks). The ratio of response to excitation, both expressed as functions of the complex frequency, $s = \sigma + j\omega$. *Note:* The response function is the Laplace transform of the response due to unit impulse excitation.   238

response, G (high voltage testing). The response G of a measuring system is the output, as a function of time or frequency, when an input voltage or current is applied to the system.   150

response, Gaussian (amplifiers) (oscilloscopes). A particular frequency response characteristic following the curve $y(f) = e^{-af^2}$. Typically, the frequency response approached by an amplifier having good transient response characteristics.   106

response, impulse-forced (automatic control). The total (transient plus steady-state) time response resulting from an impulse at the input. *Syn:* **impulse response.**   56

response, indicial (process control). The output of a system or element, expressed as a function of time, when forced from initial equilibrium by a unit-step input. *Note:* In the time domain, it is the graphic statement of the characteristic of a system or element analogous to the frequency-response characteristic of the transfer function.   56

response, instrument (1) (dynamic). The behavior of the instrument output as a function of the measured signal, both with respect to time. *See:* **damping characteristic; frequency response; ramp response; step response; accuracy rating (instrument).**   295
(2) (forced). The total steady-state plus transient time response resulting from an external input.   329

response protocol data unit (PDU)(logical link control). All PDUs sent by a logical link control (LLC) in which the command/response (C/R) bit is equal to '1'.   585

response, ramp-forced automatic control. The total (transient plus steady-state) time response resulting from a sudden increase in the rate of change of input from zero to some finite value. *Syn:* **ramp response.**   56

response, sinusoidal (sine-force). The forced response due to a sinusoidal stimulus. *Note:* A set of steady-state sinusoidal responses for sinusoidal inputs at different frequencies is called the frequency-response characteristic. *See:* **control system, feedback.**   329

response spectrum (1)(seismic design of substations). A plot of the maximum response of single-degree-of-freedom bodies of different natural frequencies, at a damping value expressed as a percent of critical damping, rigidly mounted on the surface of interest (that is, on the foundation for the foundation response spectrum or on the floor of a building for that floor's response spectrum) when that surface is subjected to a given earthquake motion as modified by intervening structures.   465
(2)(seismic qualification of Class 1E equipment for nuclear power generating stations)(valve actuators). A plot of the maximum response, as a function of oscillator frequency, of an array of single-degree-of-freedom (SDOF) damped oscillators subjected to the same base excitation.   581, 492
(3) (as applied to relays) (seismic testing of relays).

A plot of the peak acceleration response of damped, single-degree-of-freedom bodies, at a damping value expressed as a percent of critical damping of different natural frequencies, when these bodies are rigidly mounted on the surface of interest.   392

response, steady-state (system or element). The part of the time response remaining after transients have expired. *Note:* The term **steady-state** may also be applied to any of the forced-response terms: for example **steady-state sinusoidal response.** *See:* **control system, feedback; response, sinusoidal.**   329

response, step-forced (automatic control). The total (transient plus steady-state) time response resulting from a sudden change from one constant level of input to another.   56

response time (1)(faulted circuit indicators). The time required for the FCI (faulted circuit indicator) to respond to a specified value of fault current.   482
(2)(monitoring radioactivity in effluents). The time interval from a step change in the input concentration at the instrument inlet to a reading of 90 percent (nominally equivalent to 2.2 time constants) of the ultimate recorded output.   559

response time (1) (data transmission).
(A) (magnetic amplifier). The time (preferably in seconds; may also be in cycles of supply frequency) required for the output quantity to change by some agreed-upon percentage of the differential output quantity in response to a step change in control signal equal to the differential control signal. *Note:* The initial and final output quantities correspond to the test output quantities. The response time is the maximum obtained including differences arising from increasing or decreasing output quantity or time phase of signal application.
(B) (turn-ON response time) (control devices). The time required for the output voltage to change from rated OFF voltage to rated ON amplifiers, one serving to amplify the telephone voltage in response to a step change in control signal equal to 120 percent of the differential trip signal. *Note:* The absolute magnitude of the initial signal condition is the absolute magnitude of the trip OFF control signal plus 10 percent of the differential trip signal.
(C) (Turn OFF response time) (control devices). The time required for the output voltage to change from rated ON voltage to rated OFF voltage in response to a step change in control signal equal to 120 percent of the differential trip signal. *Note:* The absolute magnitude of the initial signal condition is the absolute magnitude of the trip ON control signal minus 10 percent of the differential trip signal.
(D) (electrically tuned oscillator). The time following a change in the input to the tuning element required for a characteristic to reach a predetermined range of values within which it remains.
(E) (instrument). The time required after an abrupt change has occurred in the measured quantity to a new constant value until the pointer, or indicating means, has first come to apparent rest in its new position. *Notes:* (1) Since, in some instruments, the re-

sponse time depends on the magnitude of the deflection, a value corresponding to an initial deflection from zero to end scale is used in determining response time for rating purposes. (2) The pointer is at apparent rest when it remains within a range on either side of its final position equal to one-half the accuracy rating, when determined as specified above.

(F) (bolometric power meter). The time required for the bolometric power indication to reach 90 percent of its final value after a fixed amount of radio-frequency power is applied to the bolometer unit.

(G) (thermal converter). The time required for the output electromotive force to come to its new value after an abrupt change has occurred in the input quantity (current, voltage, or power) to a new constant value. *Notes:* (1) Since, in some thermal converters, the response time depends upon the magnitude and direction of the change, the value obtained for an abrupt change from zero to rated input quantity is used for rating purposes. (2) The output electromotive force is considered to have come to its new value when all but 1 percent of the change in electromotive force

has been indicated.

(H) (industrial control). The time required, following the initiation of a specified stimulus to a system, for an output going in the direction of necessary corrective action to first reach a specified value. *Note:* The response time is expressed in seconds.

(I) (electrical conversion). The elapsed time from the initiation of a transient until the output has recovered to 63 percent of its maximum excursion.

(J) (arcwelding apparatus). The time required to attain conditions within a specified amount of their final value in an automatically regulated welding circuit after a definitely specified disturbance has been initiated.

(K) (photoelectric lighting control) (industrial control). The time required for operation following an abrupt change in illumination from 50 percent above to 50 percent below the minimum illumination sensitivity.

(L) (control system or element) (time of response) (control system, feedback). The time required for an output to make the change from an initial value to a

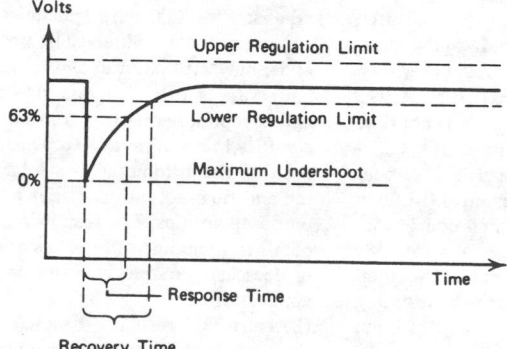

Response and recovery time for a critically damped circuit.

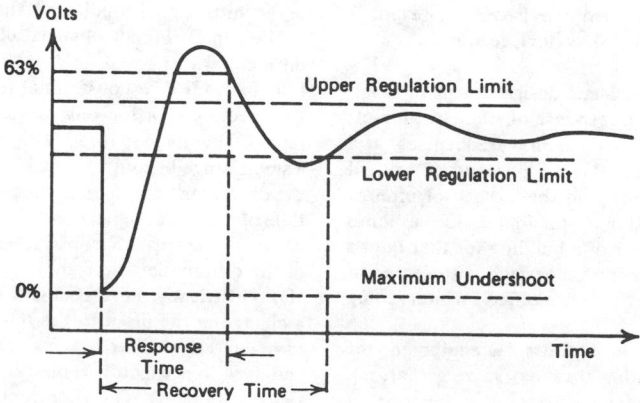

Response and recovery time for an underdamped circuit

large specified percentage of the steady state, either before overshoot or in the absence of overshoot. *Note:* If the term is unqualified, time of response of a first-order system to a unit-step stimulus is generally understood; otherwise the pattern and magnitude of the stimulus should be specified. Usual percentages are 90, 95, or 99.

(M) (data circuit). The amount of time elapsed between generation of an inquiry at a data communications terminal and receipt of a response at that same terminal. Response time, thus defined includes: transmission time to the computer; processing time at the computer, including access time to obtain any file records needed to answer the inquiry; and transmission time back to the terminal. 59

(2) (station control and data acquisition). The time between initiating some operation and obtaining results. 403

(3) (sequential events recording systems). The time interval between receiving a finite input status change and the recognition by the system of the status change. The time interval is usually expressed in milliseconds. 48

(4) (temperature measurement). The time required for the indication of a thermometer, which has been subjected to an essentially instantaneous change in temperature, to traverse 63 percent of the temperature interval involved. Following such a temperature change the indication of the thermometer may be expected to traverse 99 percent of the temperature interval in a period ranging from 5 to 8 time constants so defined, depending on the details of its construction. 7

(5) (lagged demand meter). *See:* **time characteristic (lagged-demand meter).** 212

response timer (FASTBUS acquisition and control). A timing device within a FASTBUS master or segment interconnect used to terminate an operation which has failed to complete within a given (excessive) period of time. 480

response time, ramp-forced (automatic control). The time interval by which an output lags an input, when both are varying at a constant rate. 56

response time, T (high voltage testing). The response time T of a measuring system is indicative of the errors encountered when measuring rapidly changing voltages or currents:

$$T \approx \frac{a_i - a_m}{da_m/dt}$$

where a_1 is the value of a ramp input function at some specific time and a_m is the measured value of that quantity, provided the rates of change of both the input function and the measured value of that function are constant and equal. *Note:* For particulars concerning the response time and related response parameters see Section 4 of IEEE Std 4-1978. 150

response to signal removal (measuring the perfor- mance of tone address signaling systems). The time interval from the end of signal present condition to the time the receiver indication terminates. 508

response to signal start (measuring the performance of tone address signaling systems). The time interval from start of a signal present condition to the time at which the appropriate indication occurs in the receiver. 508

response to tone removal (measuring the performance of tone address signaling systems). The time interval from the end of tone present condition to the time the receiver indication terminates. 508

response to tone start (measuring the performance of tone address signaling systems). The time interval from the start of a tone present condition to the time at which the appropriate indication occurs in the receiver. 508

responsivity (fiber optics). The ratio of an optical detector's electrical output to its optical input, the precise definition depending on the detector type; generally expressed in amperes per watt or volts per watt of incident radiant power. *Note:* 'Sensitivity' is often incorrectly used as a synonym. 433

responsor (navigation aid terms). The receiving component of an interrogator-responsor. 526

rest and de-energized (rotating machinery). The complete absence of all movement and of all electric or mechanical supply. *See:* **asynchronous machine.** 63

restart (computing systems). To reestablish the execution of a routine, using the data recorded at a checkpoint. 255, 77, 54

resting potential (biological). The voltage existing between the two sides of a living membrane or interface in the absence of stimulation. 192

restorable fire detector (fire protection devices). A device whose sensing element is not ordinarily destroyed by the process of detecting a fire. Restoration may be manual or automatic. 71

restoring force gradient (direct-acting recording instrument). The rate of change, with respect to the displacement, of the resultant of the electric, or of the electric and mechanical, forces tending to restore the marking device to any position of equilibrium when displaced from that position. *Note:* The force gradient may be constant throughout the entire travel of the marking device or it may vary greatly over this travel, depending upon the operating principles and the details of construction. *See:* **accuracy rating (instrument).** 328

restoring torque gradient (instrument). The rate of change, with respect to the deflection, of the resultant of the electric, or electric and mechanical, torques tending to restore the moving element to any position of equilibrium when displaced from that position. *See:* **accuracy rating (instrument).** 328

restraint relay (power switchgear). A relay so constructed that its operation in response to one input is restrained or controlled by a second input. 103

restricted radiation frequencies for industrial, scientific, and medical equipment. Center of a band of frequencies assigned to industrial, scientific, and medical equipment either nationally or internationally and for which a power limit is specified. *See:* **electromagnetic compatibility.** 220, 199

restricted-service tone (telephone switching systems). A class-of-service tone that indicates to an operator that certain services are denied the caller. 55

restrike (power switchgear). A resumption of current between the contacts of a switching device during an opening operation after an interval of zero current at 1/4 cycle at normal frequency or longer. 103

restrike time (nuclear security systems). The time period during which a momentary loss or reduction of illumination results from the need to cool down to restrike the arc after a momentary loss or reduction of electrical power to a luminaire. 464

restriking voltage (1) (gas tube). The anode voltage at which the discharge recommences when the supply voltage is increasing before substantial deionization has occurred. 244, 190

(2) (industrial control). The voltage that appears across the terminals of a switching device immediately after the breaking of the circuit. *Note:* This voltage may be considered as composed of two components. One, which subsists in steady-state conditions, is direct current or alternating current at service frequency, according to the system. The other is a transient component that may be oscillatory (single or multifrequency) or nonoscillatory (for example, exponential) or a combination of these depending on the characteristics of the circuit and the switching device. *See:* **switch.** 244, 206

result (1)(binary floating-point arithmetic). The bit string (usually representing a number) that is delivered to the destination. 469

(2)(radix-independent floating-point arithmetic). The digit string (usually representing a number) that is delivered to the destination. 588

resultant color shift (illuminating engineering). The difference between the perceived color of an object illuminated by a test source and of the same object illuminated by a reference source, taking account of the state of chromatic adaptation in each case; that is, the resultant of colorimetric shift and adaptive color shift. *See:* **state of chromatic adaptation.** 167

retained image (image burn). A change produced in or on the target that remains for a large number of frames after the removal of a previously stationary light image and that yields a spurious electric signal corresponding to that light image. *See:* **camera tube.** 125

retainer. *See:* **separator (storage cell).**

retaining ring (rotating machinery) (1) (steel). A mechanical structure surrounding parts of a rotor to restrain radial movement due to centrifugal action. *See:* **rotor (rotating machinery).** 63

(2) (insulation). The insulation forming a dielectric and mechanical barrier between the rotor end windings and the high-strength steel retaining ring. *See:* **rotor (rotating machinery).** 63

retaining ring liner (rotating machinery). Insulating ring between the end winding and the metallic ring

which secures the coil ends against centrifical force. 63

retardation (deceleration) (industrial control). The operation of reducing the motor speed from a high level to a lower level or zero. *See:* **electric drive.** 206

retardation coil. *See:* **inductor.**

retardation test (rotating machinery). A test in which the losses in a machine are deduced from the rate of deceleration of the machine when only these losses are present. *See:* **asynchronous machine.** 63

retard coil. *See:* **inductor.**

retarding-field (positive-grid) oscillator. An oscillator employing an electron tube in which the electrons oscillate back and forth through a grid maintained positive with respect to the cathode and the plate. The frequency depends on the electron-transit time and may also be a function of the associated circuit parameters. The field in the region of the grid exerts a retarding effect that draws electrons back after passing through it in either direction. Barkhausen-Kurz and Gill-Morell oscillators are examples. *See:* **oscillatory circuit.** 111

retarding magnet. A magnet used for the purpose of limiting the speed of the rotor of a motor-type meter to a value proportional to the quantity being integrated. *See:* **watthour meter; braking magnet; drag magnet.** 328

retard transmitter. A transmitter in which a delay period is introduced between the time of actuation and the time of transmission. *See:* **protective signaling.** 328

retention (metal-nitride-oxide field-effect transistor). The time period defined by the time elapsed between the instant of writing a metal-nitride-oxide semiconductor (MNOS) transistor into a given high conduction or low conduction (HC or LC) state, and the instant when either state becomes indistinguishable from the other. 386

retention characteristic (metal-nitride-oxide field-effect transistor). A plot of both high conduction (HC) and low conduction (LC) threshold voltages v_{HC} or v_{LC} as a function (commonly the logarithm) of the time t_{rd} elapsed after the instant of writing. 386

retention time, maximum (storage tubes). The maximum time between writing into a storage tube and obtaining an acceptable output by reading. *See:* **storage tube.** 125

retentivity (magnetic material). That property that is measured by its maximum residual induction. *Note:* The maximum residual induction is usually associated with a hysteresis loop that reaches saturation, but in special cases this is not so. 210

reticle (navigation aid terms). A system of lines, etcetera, placed in the focal plane of an optical instrument to serve as a reference. 526

retina (1) (illuminating engineering). A membrane lining the posterior part of the inside of the eye. It comprises photoreceptors (cones and rods) which are sensitive to light, and nerve cells which transmit to the optic nerve the responses of the receptor elements. 167

(2) (laser-maser). That sensory membrane which receives the incident image formed by the cornea and lens of the human eye. The retina lines the inside portion of the eye.   363

retirement phase (software). The period of time in the software life cycle during which support for a software product is terminated. *See:* software life cycle; software product.   434

retrace (oscillography). Return of the spot on the cathode-ray tube to its starting point after a sweep: also that portion of the sweep waveform that returns the spot to its starting point. *See:* oscillograph.   185

retrace blanking. *See:* blanking.

retrace interval (television). The interval corresponding to the direction of sweep not used for delineation. *See:* flyback.   18

retrace line. The line traced by the electron beam in a cathode-ray tube in going from the end of one line or field to the start of the next line or field.   339

retrieval. *See:* information retrieval.

retrodirective antenna. An antenna whose backscattering cross section is comparable to the product of its maximum directivity and its area projected in the direction toward the source, and is relatively independent of the source direction. *Note:* Active devices can be added to enhance the return signal. For this case the term shall be qualified by the word 'active'; for example, active retrodirective antenna.   111

retrofill (handling and disposal of transformer grade insulating liquids containing PCBs). The process of replacing the dielectric liquid in a transformer.   586

retrograde orbit (communication satellite). An inclined orbit with an inclination between 90° and 180°.   74

retro-reflector (illuminating engineering). A device designed to reflect light in a direction close to that at which it is incident, whatever the angle of incidence.   167

retry period (FASTBUS acquisition and control). The time a master waits after failing to receive a response before trying the operation again. This time is randomized to avoid system deadlocks.   480

return. *See:* carriage return.

return-beam mode (camera tube). A mode of operation in which the output current is derived, usually through an electron multiplier, from that portion of the scanning beam not accepted by the target. *See:* camera tube.   190

return beam multiplier gain (diode-type camera tube). The dimensionless ratio between the output signal current at the final anode of the electron multiplier in a return beam camera tube and the modulated portion of the beam current falling on the first dynode of the multiplier. The output signal current is the value of the output current less the dark current.   380

return difference (network analysis). One minus the loop transmittance.   282

return interval (television). *See:* retrace interval.

return loss (1) (data transmission). (A) At a discontinuity in a transmission system the difference between the power incident upon the discontinuity. (B) The ratio in decibels of the power incident upon the discontinuity to the power reflected from the discontinuity. *Note:* This ratio is also the square of the reciprocal to the magnitude of the reflection coefficient. Return loss $= 20 \log_{10}(1/\Gamma)$. (C) More broadly, the return loss is a measure of the dissimilarity between two impedances, being equal to the number of decibels that corresponds to the scalar value of the reciprocal of the reflection coefficient, and hence being expressed by the following formula:

$$20 \log_{10} \left| \frac{Z_1 + Z_2}{Z_1 - Z_2} \right| \text{ decibel}$$

where Z_1 and $Z_2 = $ the two impedances.   267, 59

(2) (or gain) (waveguide). The ratio of incident to reflected power at a reference plane of a network.   267

return stroke (lightning). The luminescent, high-current discharge that is initiated after the stepped leader and pilot streamer have established a highly ionized path between charge centers. *See:* direct-stroke protection (lightning).   64

return swing (last transition ringing), A_{RS} (pulse transformers). The maximum amount by which the instantaneous pulse value is below the zero axis in the region following the backswing. It is expressed in amplitude units or as a percentage of A_M.   589

return to bias (magnetic tape pulse recorders for electricity meters). A method whereby a recording head current, which results in a magnetic field polarity opposite that of the bias magnet, is applied momentarily in order to record a pulse.   551

return trace (television). The path of the scanning spot during the retrace interval.   18

reusability (software). The extent to which a module can be used in multiple applications. *See:* module.   434

reverberant sound. Sound that has arrived at a given location by a multiplicity of indirect paths as opposed to a single direct path. *Notes:* (1) Reverberation results from multiple reflections of sound energy contained within an enclosed space. (2) Reverberation results from scattering from a large number of inhomogeneities in the medium or reflection from bounding surfaces. (3) Reverberant sound can be produced by a device that introduces time delays that approximate a multiplicity of reflections. *See:* echo.   176

reverberation. The presence of reverberant sound.   176

reverberation chamber. An enclosure especially designed to have a long reverberation time and to produce a sound field as diffuse as possible. *See:* anechoic chamber.   176

reverberation room. *See:* reverberation chamber.

reverberation time. The time required for the mean-square sound pressure level, or electric equivalent, originally in a steady state, to decrease 60 decibels after the source output is stopped.   176

reverberation-time meter. An instrument for measuring the reverberation time of an enclosure. *See:* **instrument.** 328

reversal (storage battery) (storage cell). A change in normal polarity of the cell or battery. *See:* **charge.** 328

reversal point (ferroresonant voltage regulators). That point on the input current versus input voltage characteristics where the input current reaches a minimum value and begins to increase. *See:* **Figure 'Reversal Point with Jump Resonance' (jump resonance) and Figure 'Reversal Point without Jump Resonance' below.** 456,

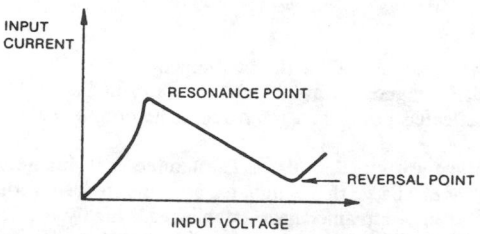

Fig 9
Reversal Point without Jump Resonance

reverse-battery signaling (telephone switching systems). A method of loop signaling in which the direction of current in the loop is changed to convey on-hook and off-hook signals. 55

reverse-battery supervision (telephone switching systems). A form of supervision employing reverse-battery signaling. 55

reverse bias (reverse voltage) (V_R) (light emitting diodes). The bias voltage that is applied to an LED (light emitting diode) in the reverse direction. 162

reverse-blocking current (reverse-blocking thyristor). The reverse current when the thyristor is in the reverse-blocking state. *See:* **principal current.** 243, 66, 208, 191

reverse-blocking diode-thyristor. A two-terminal thyristor that switches only for positive anode-to-cathode voltages and exhibits a reverse-blocking state for negative anode-to-cathode voltages. 445

reverse-blocking impedance (reverse-blocking thyristor). The differential impedance between the two terminals through which the principal current flows, when the thyristor is in the reverse-blocking state at a stated operating point. *See:* **principal voltage-current characteristic (principal characteristic).** 191

reverse-blocking state (reverse-blocking thyristor). The condition of a reverse-blocking thyristor corresponding to the portion of the anode-to-cathode voltage-current characteristic for reverse currents of lower magnitude than the reverse-breakdown current. *See:* **principal voltage-current characteristic (principal characteristic).** 191

reverse-blocking triode-thyristor (SCR). A three-terminal thyristor that switches only for positive anode-to-cathode voltages and exhibits a reverse-blocking state for negative anode-to-cathode voltages. 445

reverse-breakdown current (reverse-blocking thyristor). The principal current at the reverse-breakdown voltage. *See:* **principal current.** 24

reverse-breakdown voltage (reverse-blocking thyristor). The value of negative anode-to-cathode voltage at which the differential resistance between the anode and cathode terminals changes from a high value to a substantially lower value. *See:* **principal voltage-current characteristic (principal characteristic).** 24

reverse-conducting diode-thyristor. A two-terminal thyristor that switches only for positive anode-to-cathode voltages and conducts large currents at negative at negative anode-to-cathode voltages comparable in magnitude to the ON-state voltages. 445

reverse-conducting triode-thyristor. A three-terminal thyristor that switches only for positive anode-to-cathode voltages and conducts large currents at negative anode-to-cathode voltages comparable in magnitude to the ON-state voltage. 445

reverse contact. A contact that is closed when the operating unit is in the reverse position. 328

reverse current (1) (general). Current that flows upon application of reverse voltage. 328
(2) (reverse-blocking or reverse-conducting thyristor). The principal current for negative anode-to-cathode voltage. *See:* **principal current.** 24
(3) (metallic rectifier). The current that flows through a metallic rectifier cell in the reverse direction. *See:* **rectification.** 328
(4) (semiconductor rectifier). The total current that flows through a semiconductor rectifier device in the reverse direction. *See:* **rectification.** 245

reverse-current cleaning. *See:* **anode cleaning.**

reverse-current cutout. A magnetically operated direct-current device that operates to close an electric circuit when a predetermined voltage condition exists and operates to open an electric circuit when more than a predetermined current flows through it in the reverse direction. (1) **Fixed-voltage type:** A reverse-current cutout that closes an electric circuit whenever the voltage at the cutout terminal exceeds a predetermined value and is of the correct polarity. It opens the circuit when more than a predetermined current flows through it in the reverse direction. (2) **Differential-voltage type:** A reverse-current cutout that closes an electric circuit when a predetermined differential voltage appears at the cutout terminal, provided this voltage is of the correct polarity and exceeds a predetermined value. It opens the circuit when more than a predetermined current flows through it in the reverse direction. 328

reverse-current relay (power switchgear). A relay that operates on a current flow in a direct-current circuit in a direction opposite to a predetermined reference direction. 103

reverse-current release (trip) (power switchgear). A release that operates upon reversal of the direct current in the main circuit from a predetermined direction.   103

reverse-current tripping: reverse-power tripping. *See:* reverse-current release (trip) (power switchgear).

reverse direction (1) (semiconductor rectifier diode). The direction of higher resistance to steady-state direct-current: that is, from the cathode to the anode.   66

(2) (light emitting diodes). The direction of higher resistance to steady direct-current flow through the device: that is, from the cathode to the anode.   162

reverse-electrode coaxial detector (germanium gamma-ray detectors). A coaxial detector in which the outer contact is a p-type layer.   528

reverse emission (back emission) (vacuum tubes). The inverse electrode current from an anode during that part of a cycle in which the anode is negative with respect to the cathode. *See:* electron emission.   190

reverse gate current (thyristor). The gate current when the junction between the gate region and the adjacent anode or cathode region is reverse biased. *See:* principal current.   191

reverse gate voltage (thyristor). The voltage between the gate terminal and the terminal of an adjacent region resulting from reverse gate current. *See:* principal voltage-current characteristic (principal characteristic).   191

reverse or OFF-state voltage dividers (thyristor). Devices employed to assure satisfactory division of reverse or OFF-state voltage among series-connected semiconductor devices under transient or steady state conditions , or both.   445

reverse period (rectifier circuit) (rectifier circuit element). The part of an alternating-voltage cycle during which the current flows in the reverse direction. See note under blocking period. *See:* rectifier circuit element.   208

reverse-phase or phase-balance current relay (46) (power system device function numbers). A relay that functions when the polyphase currents are unbalanced or contain negative phase-sequence components above a given amount.   402

reverse Polish notation. *See:* postfix notation.

reverse position (device). The opposite of the normal position.   328

reverse power dissipation (semiconductor). The power dissipation resulting from reverse current.   66

reverse power loss (semiconductor rectifier). The power loss resulting from the flow of reverse current. *See:* rectification; semiconductor rectifier stack.   66

reverse-power tripping. *See:* reverse-current release (trip).

reverser. A switching device for interchanging electric circuits to reverse the direction of motor rotation. *See:* multiple-unit control.   328

reverse recovery current (semiconductor rectifier). The transient component of reverse current associated with a change from ON state conduction to reserve voltage. *See:* rectification.   66

reverse recovery interval (thyristor). The interval between the instant when the principal ON-state current flowing through a semiconductor passes through zero, and the instant when the reverse current has decayed to 10 percent of the peak reverse value.   445

reverse recovery time (reverse-blocking thyristor or semiconductor diode). The time required for the principal current or voltage to recover to a specified value after instantaneous switching from an ON state to a reverse-voltage or current. *See:* principal voltage-current characteristic (principal characteristic); rectification.   191

reverse resistance (metallic rectifier). The resistance measured at a specified reverse voltage or a specified reverse current. *See:* rectification.   328

reverse voltage (1) (rectifier). Voltage of that polarity that produces the smaller current. *See:* principal voltage-current characteristic; rectification.   328

(2) (semiconductor) (reverse blocking or reverse conducting thyristor). A negative anode-to-cathode voltage.   66

reverse voltage dividers (rectifier). Devices employed to assure satisfactory division of reverse voltage among series-connected semiconductor rectifier diodes. Transformers, bleeder resistors, capacitors, or combinations of these may be employed. *See:* power rectifier.   208

reverse wave. *See:* reflected wave.

reversibility (Hall generator). The ratio of the change in absolute magnitude of the Hall voltage to the mean absolute magnitude of the Hall voltage, when the control current is kept constant and the magnetic field is changed from a given magnitude of one polarity to the same magnitude of the opposite polarity.   107

reversible capacitance (nonlinear capacitor). The limit, as the amplitude of an applied sinusoidal capacitor voltage approaches zero, of the ratio of the amplitude of the resulting in-phase fundamental-frequency component of transferred charge to the amplitude of the applied voltage, for a given constant bias voltage superimposed on the sinusoidal voltage. *See:* nonlinear capacitor.   191

reversible-capacitance characteristic (nonlinear capacitor). The function relating the reversible capacitance to the bias voltage. *See:* nonlinear capacitor.   191

reversible dark current increase (diode-type camera tube). That increase of the target dark current which results from electron bombardment of the charge storage target by the scanning electron beam. This is manifested as a dark current increase which is reversible.   380

reversible motor. A motor whose direction of rotation can be selected by change in electric connections or by mechanical means but the motor will run in the selected direction only if it is at a standstill or rotating below a particular speed when the change is initiated. *See:* asynchronous machine.   328

reversible permeability. The limit of the incremental permeability as the incremental change in magnetizing force approaches zero. *Note:* In anisotropic media, reversible permeability becomes a matrix.   210

reversible permittivity (ferroelectric material). The change in displacement per unit field when a very small relatively high-frequency alternating signal is applied to a ferroelectric at any point of a hysteresis loop. *See:* **ferroelectric domain.** 247

reversible potential. *See:* **equilibrium potential.**

reversible power converter (1) (static power converters). An equipment containing thyristor converter assemblies connected in such a way that energy transfer is possible from the alternating-current side to the direct-current side and from the direct-current side to the alternating-current side with or without reversing the current in the direct-current circuit. *See:* **power rectifier.** 208

(2) (thyristor converter). A converter in which the transfer of energy is possible both from the ac side to the dc side and vice versa. 121

reversible process. An electrochemical reaction that takes place reversibly at the equilibrium electrode potential. *See:* **electrochemistry.** 328

reversible target dark current increase (diode-type camera tube). That increase in dark current and dark current nonuniformity and monitoring and is not permanent. It is removable through target operation under special operating procedures or with a nonoperating rest period. 380

reversible turbine (power operations). A hydraulic turbine, normally installed in a pumped-storage station, which can be used alternately as a pump or as a prime mover. 516

reversing (industrial control). The control function of changing motor rotation from one direction to the opposite direction. *See:* **electric drive.** 206

reversing device (9) (power system device function numbers). A device that is used for the purpose of reversing a machine field or for performing any other reversing functions. 402

reversing motor. One the torque and hence direction of rotation of which can be reversed by change in electric connections or by other means. These means may be initiated while the motor is running at full speed, upon which the motor will come to a stop, reverse, and attain full speed in the opposite direction. *See:* **asynchronous machine.** 328

reversing starter (industrial control). An electric controller for accelerating a motor from rest to normal speed in either direction of rotation. *See:* **electric controller; starter.** 206

reversing switch. A switch intended to reverse the connections of one part of a circuit. *See:* **switch.** 206

reverting call (telephone switching systems). A call between two stations on the same party line. 55

reverting-call tone (telephone switching systems). A tone that indicates to a calling customer that the called party is on the same line. 55

revertive pulsing (telephone switching systems). A means of pulsing for controlling distant selections whereby the near end receives signals from the far end. 55

review. *See:* **design review.** 434

rewind (test, measurement and diagnostic equipment). To return a tape to its beginning or a passed location. 54

rework (nuclear power quality assurance). The process by which an item is made to conform to original requirements by completion or correction. 417

rewrite. (1) To write again. (2) In a destructive-read storage device, to return the data to the state it had prior to reading. *See:* **regenerate.** 235

RF (radio frequency) link (test, measurement and diagnostic equipment). A radio frequency channel or channels used to connect the unit under test with the testing device. 54

RF. *See:* **radio frequency.**

rheobase (medical electronics). The intensity of the steady cathodal current of adequate duration that when suddenly applied just suffices to excite a tissue. 192

rheostat (1). An adjustable resistor so constructed that its resistance may be changed without opening the circuit in which it may be connected. 210

(2) (70) (power system device function numbers). A variable resistance device used in an electric circuit, which is electrically operated or has other electrical accessories, such as auxiliary, position, or limit switches. 402

rheostatic braking (industrial control). A form of dynamic braking in which electric energy generated by the traction motors is controlled and dissipated by means of a resistor whose value of resistance may be varied. *See:* **dynamic braking; electric drive.** 206

rheostatic control (elevators). A system of control that is accomplished by varying resistance and.or reactance in the armature and.or field circuit of the driving-machine motor. *See:* **control (elevators).** 328

rheostatic-type voltage regulator (rotating machinery). A regulator that accomplishes the regulating function by mechanically varying a resistance. 63

rheostat loss (synchronous machine). The I^2R loss in the rheostat controlling the field current. 298

rheostriction. *See:* **pinch effect.**

RHI. *See:* **range-height indication.**

rhombic antenna. An antenna composed of long-wire radiators comprising the sides of a rhombus. The antenna usually is terminated in a resistance. *Note:* The length of the sides of the rhombus, the angle between the sides, the elevation above ground, and the value of the termination resistance are proportioned to give the desired radiation properties. 111

rho rho (navigation aid terms). A generic term referring to navigation systems based on the measurement of two distances for determination of position. 526

rho theta (navigation aid terms). A generic term referring to polar coordinate navigation systems for determination of position of a vehicle through measurement of distance and direction. 526

rhumbatron (electron tube) (microwave tube). A resonator, usually in the form of a torus. *See:* **velocity-modulated tube.** 244, 190

rhumbline (navigation aid terms). A line on the surface of the earth making the same oblique angle with all meridians. 526

rhythmic light (illuminating engineering). A light which, when observed from a fixed point, has a luminous intensity which changes periodically. 167

ribbon microphone. A moving-conductor microphone in which the moving conductor is in the form of a ribbon that is directly driven by the sound waves. *See:* **microphone.** 328

ribbon transducer. A moving-conductor transducer in which the movable conductor is in the form of a thin ribbon. 176

Richardson-Dushmann equation (thermionics). An equation representing the saturation current of a metallic thermionic cathode in the saturation-current state:

$$ J = A_0 \left(1 - r\right) T^2 \exp\left(-\frac{b}{T}\right) $$

where
J = density of the saturation current
T = absolute temperature
A_0 = universal constant equal to 120 amperes per centimeter^2 kelvin^2
b = absolute temperature equivalent to the work function
r = reflection coefficient, which allows for the irregularities of the surface.
See: **work function; electron emission.**
244, 190

Richardson effect. *See:* **thermionic emission.**

ridged horn (antenna). A horn antenna in which the waveguide section is ridged. 111

ridge waveguide. A waveguide with interior projections extending along the length and in contact with the boundary wall. *See:* **waveguide.** 179

Rieke diagram (oscillator performance). A chart showing contours of constant power output and constant frequency drawn on a polar diagram whose coordinates represent the components of the complex reflection coefficient at the oscillator load. *See:* **load impedance diagram; oscillatory circuit.** 190, 125

RIF (electromagnetic compatibility). *See:* **radio-influence field.**

rigging (power line maintenance). An assembly of material used to manipulate or support various tools or equipment in both energized and de-energized line work. 458

right-hand (or left-hand) polarization (1) (plane wave). The polarization of a plane wave when the electric field vector is right-hand (or left-hand) polarized, taking as the reference the direction of propagation. 111
(2) (field vector). A polarization such that the sense of rotation of the extremity of the field vector with time is in the direction of the fingers of the right hand (or left hand) when the thumb of that hand is in some reference direction perpendicular to the plane of po-

larization. *Note:* For a linearly polarized field vector the sense of polarization is not defined. 111

right-handed (clockwise) polarized wave (radio wave propagation). An elliptically polarized electromagnetic wave in which the rotation of the electric field vector with time is clockwise for a stationary observer looking in the direction of the wave normal. *Note:* For an observer looking from a receiver toward the apparent source of the wave, the direction of rotation is reversed. 146

right of access (nuclear power quality assurance). The right of a purchaser or designated representative to enter the premises of a supplier for the purpose of inspection, surveillance, or quality assurance audit. 417

rigid-bus structure (substation rigid-bus structures). A bus structure comprised of rigid conductors supported by rigid insulators. 566

rigid equipment (seismic qualification of Class 1E equipment for nuclear power generating stations). Equipment, structures, and components whose lowest resonant frequency is greater than the cutoff frequency on the response spectrum. 581

rigid metal conduit. A raceway specially constructed for the purpose of the pulling in or the withdrawing of wires or cables after the conduit is in place and made of metal pipe of standard weight and thickness permitting the cutting of standard threads. *See:* **raceway.** 328

rigid tower. A tower that depends only upon its own structural members to withstand the load that may be placed upon it. *See:* **angle tower; dead-end-tower; flexible tower; tower.** 64

rim (spider rim) (rotating machinery). The outermost part of a spider. A rotating yoke. *See:* **rotor (rotating machinery).** 63

ring (plug). A ring-shaped contacting part, usually placed in back of the tip but insulated therefrom. 328

ring around (radar). (1) The undesired triggering of a transponder by its own transmitter; (2) The triggering of a transponder at all bearings, causing a ring-type presentation on a plan-position indicator (PPI). 13

ring array. *See:* **circular array.**

ringback signal (telephone switching systems). A signal initiated by an operator at the called end of an established connection to recall the originating operator. 55

ringback tone (telephone switching systems). A tone that indicates to a caller that a ringing signal is being applied to a destination outlet. 55

ring circuit (waveguide practice). A hybrid T having the physical configuration of a ring with radial branches. *See:* **waveguide.** 328

ring counter. A re-entrant multistable circuit consisting of any number of stages arranged in a circle so that a unique condition is present in one stage, and each input pulse causes this condition to transfer one unit around the circle. *See:* **trigger circuit.** 328

ringdown signaling (1)(telephone switching systems).

A method of alerting an operator in which ringing is sent over the line to operate a device or circuit to produce a steady indication (normally a visual signal). 55

(2)(data transmission). The application of a signal to the line for the purpose of bringing in a line signal or supervisory signal at a switchboard or ringing a user's instrument. (Historically, this was a low frequency signal of about 20 Hz from the user on the line for calling the operator or for disconnect). 59

ringer (station ringer). *See:* **telephone ringer.**

ringer box. *See:* **bell box.**

ring feeder. *See:* **loop-service feeder.**

ring head (electroacoustics). A magnetic head in which the magnetic material forms an enclosure with one or more air gaps. The magnetic recording medium bridges one of these gaps and is in contact with or in close proximity to the pole pieces on one side only. 176

ring heater, induction. *See:* **induction ring heater.**

ringing (1)(first transition ringing), A_{RI} (pulse transformers). The maximum amount by which the instantaneous pulse value deviates from the straight-line segment fitted to the top of the pulse in determining A_M in the pulse top region following rolloff, or overshoot, or both. It is expressed in amplitude units or as a percentage of A_M. 589

(2) (data transmission). The production of an audible or visible signal at a station or switchboard by means of an alternating or pulsating current, or a damped oscillation occurring in the output signal of a system, as a result of a sudden change in input signal. 59

(3) (telephone switching systems). An alternating or pulsing current primarily intended to produce a signal at a station or switchboard. 55

(4) (facsimile). *See:* **facsimile transient.**

(5) (pulse terms). A distortion in the form of a superimposed damped oscillatory waveform which, when present, usually follows a major transition. *See:* Note in **preshoot** entry. 254

ringing cycle (telephone switching systems). A recurring sequence made up of ringing signals and the intervals between them. 55

ringing key. A key whose operation sends ringing current over the circuit to which the key is connected. 328

ring latency (token ring access method). In a token ring medium access control system, the time (measured in bit times at the data transmission rate) required for a signal to propagate once around the ring. The ring latency time includes the signal propagation delay through the ring medium plus the sum of the propagation delays through each station connected to the token ring. 472

ringless-type meter socket (watthour meter sockets). A meter socket which has no provision for a socket sealing ring but has other means of holding a detachable watthour meter in place, such as a cover which is secured in place by a latch. 549

ring oscillator. An arrangement of two or more pairs of tubes operating as push-pull oscillators around a ring, usually with alternate successive pairs of grids and plates connected to tank circuits. Adjacent tubes around the ring operate in phase opposition. The load is supplied by coupling to the plate circuits. *See:* **oscillatory circuit.** 111, 240

ring time (radar). The time during which the indicated output of an echo box remains above a specified signal-to-noise ratio. The ring time is used in measuring the performance of radar equipment. 13

ring-type meter socket (watthour meter sockets). A meter socket which has a socket rim. 549

riometer (relative ionospheric opacity meter) (radio wave propagation). A radio-frequency receiving device that measures the ionospheric absorption experienced by cosmic radio noise passing through the ionosphere. 146

ripple (1) (general). The alternating-current component from a direct-current power supply arising from sources within the power supply. *Notes:* (A) Unless specified separately, ripple includes unclassified noise. (B) In electrical-conversion technology, ripple is expressed in peak, peak-to-peak, root-mean-square volts, or as percent root-mean-square. (C) Unless otherwise specified, **percent ripple** is the ratio of the root-mean-square value of the ripple voltage to the absolute value of the total voltage, expressed in percent. *See:* **percent ripple.** 210

(2) (high voltage testing). Ripple is the periodic deviation from the arithmetic mean value of the voltage. The amplitude of the ripple is defined as half the difference between the maximum and minimum values. The ripple factor is the ratio of the ripple amplitude to the arithmetic mean value. 150

ripple amplitude (1) (power converters). The maximum value of the instantaneous difference between the average and instantaneous value of a pulsating unidirectional wave. *Note:* The amplitude is a useful measure of ripple magnitude when a single harmonic is dominant. Ripple amplitude is expressed in percent or per unit referred to the average value of the wave. *See:* **power rectifier; rectification.** 204, 208

(2) (circuits and systems). The fine variations on a frequency plot of an impedance function or of a transfer function are called ripple. The ripple amplitude is the difference between the maximum and the minimum value of the function. 67

ripple content (converter characteristics)(self-commutated converters). The periodic ac (alternating current) function that may be superimposed on a steady zero-frequency (dc)(direct current) voltage or current. 584

ripple current. *See:* **ripple voltage or current.**

ripple factor. The ratio of the ripple magnitude to the arithmetic mean value of the voltage. *See:* **electrical conversion; interference; power pack; radio receiver.** 307, 201

ripple filter. A low-pass filter designed to reduce the ripple current, while freely passing the direct current, from a rectifier or generator. *See:* **filter.** 59

ripple voltage (rectifier or generator). The alternating-voltage component of the unidirectional voltage from a direct-current power supply arising from sources within the power supply. *See:* **interference; rectifier.**   111, 211

ripple voltage or current. The alternating component whose instantaneous values are the difference between the average and instantaneous values of a pulsating unidirectional voltage or current. *See:* **rectification.**
                                                   291, 237

rise. *See:* **travel.**

rise-and-fall pendant. A pendant the height of which can be regulated by means of a cord adjuster.
                                                        328

riser cable (communication practice). The vertical portion of a house cable extending from one floor to another. In addition, the term is sometimes applied to other vertical sections of cable. *See:* **cable.**   328

rise time (1)(first transition duration), t_r **(pulse transformers).** The time interval of the leading edge between the instants at which the instantaneous value first reaches the specified lower and upper limits of 10% and 90% of A_M. Limits other than 10% and 90% may be specified in special cases.   589

(2) (industrial control). The time required for the output of a system (other than first-order) to make the change from a small specified percentage (often 5 or 10) of the steady-state increment to a large specified percentage (often 90 or 95), either before overshoot or in the absence of overshoot. *Note:* If the term is unqualified, response to a step change is understood: otherwise the pattern and magnitude of the stimulus should be specified. *See:* **control system, feedback.**
                                                        206

(3) (radiation-counter tubes). The interval between the instants at which the instantaneous value first reaches specified lower and upper limits, namely, 10 and 90 percent of the peak pulse value.   96

rise time, fall time (amplitude, frequency, and pulse modulation). The time for the light intensity to increase from the 10 to 90 percent intensity points. The fall time is the time for the light intensity to fall from the 90 to 10 percent intensity points.   72

rise time, pulse. *See:* **pulse rise time.**

rising-sun magnetron. A multicavity magnetron in which resonators of two different resonance frequencies are arranged alternately for the purpose of mode separation. *See:* **magnetrons.**   125

risk (1)(nuclear power generating stations). The expected detriment per unit time to a person or a population from a given cause.   29, 357

(2)(reliability analysis of nuclear power generating station safety systems). A measure of the probability and severity of undesired effects. Often taken as the simple product of probability and consequence.
                                                        587

RIV. *See:* **radio influence voltage.**

RLC circuit. *See:* **simple series circuit.**

RMI. *See:* **radio magnetic indicator.**

rms. *See:* **root-mean-square value (of a periodic quantity).**

rms pulse duration. *See:* **root-mean-square (rms) pulse duration.**

roadband interference (measurement) (electromagnetic compatibility). A disturbance that has a spectral energy distribution sufficiently broad, so that the response of the measuring receiver in use does not vary significantly when tuned over a specified number of receiver bandwidths. *See:* **electromagnetic compatibility.**   199

roadway (1)(NESC). The portion of highway, including shoulders, for vehicular use. *Note:* A divided highway has two or more roadways. *See:* **shoulder; traveled way.**   494

(2) (transmission and distribution). The portion of highway, including shoulders, for vehicular use. A divided highway has two or more roadways. *Notes:* (A) the shoulder is the portion of the roadway contiguous with the traveled way for accomodation of stopped vehicles for emergency use and for lateral support of base and surface courses. (B) The traveled way is the portion of the roadway for the movement of vehicles, exclusive of shoulders and full-time parking lanes.
                                                        262

roadway element (track element). That portion of the roadway apparatus associated with automatic train stop, train control, or cab signal systems, such as a ramp, trip arm, magnet, inductor, or electric circuit, to which the locomotive apparatus is directly responsive. *See:* **automatic train control.**   328

robustness (software). The extent to which software can continue to operate correctly despite the introduction of invalid inputs. *See:* **software.**   434

ROC curves. *See:* **receiver operating characteristic curves.**

rock-dust distributor. *See:* **rock duster.**

rock duster (rock-dust distributor). A machine that distributes rock dust over the interior surfaces of a coal mine by means of air from a blower or pipe line or by means of a mechanical contrivance, to prevent coal dust explosions.   328

rodding a duct. *See:* **duct rodding.**

rod, ground. *See:* **ground rod.**

rods (illuminating engineering). Retinal receptors which respond at low levels of luminance even down below the threshold for cones. At these levels there is no basis for perceiving differences in hue and saturation. No rods are found in the center of the fovea.
                                                        167

Roebel transposition (rotating machinery). An arrangement of strands occupying two heightwise tiers in a bar (half coil), wherein at regular intervals through the core length, one top strand and one bottom strand cross over to the other tier in such a way that each strand occupies every vertical position in each tier so as to equalize the voltage induced in each of the strands, thereby eliminating current that would otherwise circulate among the strands. Looking from one end of the slot, the strands are seen to progress in a clockwise direction through the core length through what may be interpreted as an angle of 360 degrees so that the strands occupy the same position at both ends of the core. There are several variations of the Roebel transposition in use. In a bar having four tiers of copper, the two pairs of tiers would each have a Roebel transposition. The uninsulated bar, then, would be assembled as two Roebel-transposed bars, side-by-side. In order to transpose against voltages induced by end-winding flux, various modifications of the transposition in the slot, and extension of the Roebel transposition into the end winding have been used. *See:* **rotor (rotating machinery; stator.**   63

roll angle. *See:* **roll attitude.**

roll attitude (navigation aid terms). The angle between the horizontal and the lateral axis of the craft. *Syn:* **roll angle.** *See:* **bank.**   526

roller bearing (rotating machinery). A bearing incorporating a peripheral assembly of rollers. *See:* **bearing.**   63

roller, uplift. *See:* **uplift roller.**

rolling contacts (industrial control). A contact arrangement in which one cooperating member rolls on the other. *See:* **contactor.**   244, 206

rolling transposition. A transposition in which the conductors of an open wire circuit are physically rotated in a substantially helical manner. With two wires a complete transposition is usually executed in two consecutive spans. *See:* **open wire.**   328

roll-in-jewel error. Error caused by the pivot rolling up the side of the jewel and then falling to a lower position when tapped. This effect is not present when instruments are mounted with the axis of the moving element in a vertical position. (Roll-in-jewel error includes pivot-friction error that is small compared to the roll-in-jewel error.) *See:* **moving element of an instrument.**   280

rolloff (1)(rounding after first transition), A_{RO} **(pulse transformers).** The amount by which the instantaneous pulse value is less than A_M at the point in time of the intersection of straight-line segments used to determine A_M. It is expressed in amplitude units or as a percentage of A_M.   589

roll over angle (conductor stringing equipment). For tangent stringing, the sum of the vertical angles between the conductor and the horizontal on both sides of the traveler. Resultants of these angles must be considered when stringing through line angles. Under some stringing conditions, such as stringing large diameter conductor, excessive roll over angles can cause premature failure of a conductor splice if it is allowed to pass over the travelers.   431

roof bushing. A bushing intended primarily to carry a circuit through the roof, or other grounded barriers of a building, in a substantially vertical position. Both ends must be suitable for operating in air. At least one end must be suitable for outdoor operation. *See:* **bushing.**   348

roof conductor. The portion of the conductor above the eaves running along the ridge, parapet, or other portion of the roof.   328

room ambient temperature (electrical insulation tests). 20°C $\pm$ 5° (68°F $\pm$ 9°).   116

room bonding point (National Electrical Code) (health care facilities). A grounding terminal or group of terminals which serves as a collection point for grounding exposed metal or conductive building surfaces in a room.   256

room cavity ratio (RCR) (illuminating engineering). For a cavity formed by a plane of the luminaires, the work-plane, and the wall surfaces between these two planes, the RCR is computed by using the distance from the work-plane to the plane of the luminaires (h_r) as the cavity height in the equations given in the definition for **cavity ratio.**   167

room coefficient, K* (illuminating engineering). A number computed from wall and floor areas. *Note:* The room coefficient is computed from

$$K_r = \frac{\text{height} \times (\text{length} + \text{width})}{2 \times \text{length} \times \text{width}}$$

(This term is retained for reference and literature searches).   167
*Obsolete

room index* (illuminating engineering). A letter designation for a range of room ratios. (This term is retained for reference and literature searches).   167
*Obsolete

room ratio* (illuminating engineering). A number indicating room proportions, calculated from the length, width, and ceiling height (or luminaire mounting height) above the work plane. It is used to simplify lighting design tables by expressing the equivalence of room shapes with respect to the utilization of direct or interreflected light. (This term is retained for reference and literature searches).   167
*Obsolete

room surface dirt depreciation (RSDD) (illuminating engineering). The fractional loss of task illuminance due to dirt on the room surface.   167

room utilization factor (illuminating engineering). The ratio of the luminous flux (lumens) received on the work-plane to that emitted by the luminaire. *Note:* This ratio sometimes is called interflectance. Room utilization factor is based on the flux emitted by a complete luminaire, whereas *coefficent of utilization* is based on the rated flux generated by the lamps in a luminaire.   167

root compiler (software). A compiler whose output is a machine independent, intermediate-level representation of a program. The root compiler, when combined with a machine dependent code generator, comprises a full compiler. *See:* **code generator; compiler; program.**   434

root locus (1) (control system, feedback) (for a closed loop whose characteristic equation is $KG(s)H(s) + 1 = 0$**).** A plot in the *s* plane of all those values of *s* that make $G(s)H(s)$ a negative real number: those points that make the loop transfer function $KG(s)H(s) = -1$ are roots. *Note:* The locus is conveniently sketched from the factored form of $KG(s)H(s)$: each branch starts at a pole of that function with $K = 0$. With increasing K, the locus proceeds along its several branches toward a zero of that function and, often asymptotic to one of several equiangular radial lines, toward infinity. Roots lie at points on the locus for which (1) the sum of the phase angles of component $G(s)H(s)$ vectors totals 180 degrees, and for which (2) $1/K = |G(s)H(s)|$. Critical damping of the closed loop occurs when the locus breaks away from the real axis: instability when it crosses the imaginary axis. *See:* **control system, feedback.** 56

(2) (excitation control systems). Consider a linear, stationary, system with closed loop transfer function $C(S)/R(S)$ where R(S) is the Laplace Transform of the excitation (input) driving function of the closed loop system and C(S) is the Laplace Transform of the response (output) function of the closed loop system. When $C(S)/R(S)$ is a function of the gain, K, of one element in either the forward or reverse signal path, the poles of $C(S)/R(S)$ in the S-plane will in general be a function of K. A plot in the S-plane of the loci of poles of the closed loop transfer function as K varies is known as a root locus. 353

root-mean-square (effective) burst magnitude (audio and electroacoustics). The square root of the average square of the instantaneous magnitude of the voltage or current taken over the burst duration. See the figure attached to the definition of **burst duration.** *See:* **burst (audio and electroacoustics).** 253, 176

root-mean-square (effective) pulse amplitude. The square root of the average of the square of the instantaneous amplitude taken over the pulse duration. 254

root-mean-square (rms) deviation (fiber optics). A single quantity characterizing a function given, for f(x), by

$$\sigma_{rms} = [1/M_0 \int_{-\infty}^{\infty} (x - M_1)^2 \, f(x) dx]^{1/2}$$
$$where \quad M_0 = \int_{-\infty}^{\infty} f(x) dx$$
$$M_1 = 1/M_0 \int_{-\infty}^{\infty} xf(x) dx$$

Note: The term rms deviation is also used in probability and statistics, where the normalization, M_o, is unity. Here, the term is used in a more general sense. *See:* **impulse response; root-mean-square (rms) pulse broadening; root-mean-square (rms) pulse duration; spectral width.**

root-mean-square (rms) pulse broadening (fiber optics). The temporal rms deviation of the impulse response of a system. *See:* **root-mean-square (rms) deviation; root-mean-square (rms) pulse duration.**

root-mean-square (rms) pulse duration (fiber optics). A special case of root-mean-square deviation where the independent variable is time and f(t) is pulse waveform. *See:* **root-mean-square (rms) deviation.**

root-mean-square detector (overhead-power-line corona and radio noise). A detector, the output voltage of which approximates the root-mean-square value of an applied signal or noise. 411

root-mean-square reverse-voltage rating (rectifier device). The maximum sinusoidal root-mean-square reverse voltage permitted by the manufacturer under stated conditions. *See:* **average forward-current rating (rectifier circuit).** 237, 66

root-mean-square ripple. The effective value of the in-

stantaneous difference between the average and instantaneous values of a pulsating unidirectional wave integrated over a complete cycle. *Note:* The root-mean-square ripple is expressed in percent or per unit referred to the average value of the wave. *See:* **rectification.** 291, 208

root-mean-square sound pressure. *See:* **effective sound pressure.**

root mean square value (high voltage testing). The root mean square value of an alternating voltage is the square root of the mean value of the square of the voltage values during a complete cycle. 150

root-mean-square value (periodic function) (effective value*). The square root of the average of the square of the value of the function taken throughout one period. Thus, if y is a periodic function of t

$$Y_{rms} = \left[\frac{1}{T} \int_a^{a+T} y^2 \, dt \right]^{1/2}$$

where Y_{rms} is the root-mean-square value of y, a is any value of time, and T is the period. If a periodic function is represented by a Fourier series, then:

$$Y_{rms} = \frac{1}{(2)^{1/2}} \left(\frac{1}{2} A_0{}^2 + A_1{}^2 + A_2{}^2 \cdots + B_1{}^2 \right.$$
$$\left. + B_2{}^2 + \cdots \right)^{1/2}$$
$$= \frac{1}{(2)^{1/2}} \left(\frac{1}{2} A_0{}^2 + C_1{}^2 + C_2{}^2 + \cdots + C_n{}^2 \right)^{1/2}$$

210

*Deprecated

root-sum-square. The square root of the sum of the squares. *Note:* Commonly used to express the total harmonic distortion. *See:* **radio receiver.** 339

rope block (power line maintenance). A device designed with one or more sheaves, a wood or steel shell, and an attachment hook or shackle commonly used in pairs with a rope reeved through the sheaves. The primary purpose of this device is to provide mechanical advantage so as to lift or move equipment. *Syn:* **block and tackle.** 458

roped-hydraulic driving machine (elevators). A machine in which the energy is applied by a piston, connected to the car with wire ropes, that operates in a cylinder under hydraulic pressure. It includes the cylinder, the piston, and the multiplying sheaves if any and their guides. *See:* **roped-hydraulic elevator; driving machine (elevators).** 328

roped-hydraulic elevator. A hydraulic elevator having its piston connected to the car with wire ropes. *See:* **roped-hydraulic driving machine (elevators); elevators.** 328

rope ladder (conductor stringing equipment). A ladder having vertical synthetic or manila suspension members and wood, fiberglass or metal rungs. The ladder is suspended from the arm or bridge of a structure to enable workmen to work at the conductor level, hang travelers, perform clipping-in operations, etcetera. *Syn:* **Jacobs ladder.** 431

rope-lay conductor or cable. A cable composed of a central core surrounded by one or more layers of helically laid groups of wires. *Note:* This kind of cable differs from a concentric-lay conductor in that the main strands are themselves stranded. In the most common type of rope-lay conductor or cable, all wires are of the same size and the central core is a concentric-lay conductor. *See:* **conductor.**   64

rosette. An enclosure of porcelain or other insulating material, fitted with terminals and intended for connecting the flexible cord carrying apendant to the permanent wiring. *See:* **cabinet.**   328

rotary attenuator (waveguide). A variable attenuator in circular waveguide having absorbing vanes fixed diametrically across one section: the attenuation is varied by rotation of this section about the common axis. *See:* **waveguide.**   179

rotary converter. A machine that combines both motor and generator action in one armature winding connected to both a commutator and slip rings, and is excited by one magnetic field. It is normally used to change alternating-current power to direct-current power.   63

rotary dial (telephone switching systems). A type of calling device used in automatic switching that generates pulses by manual rotation and release of a dial, the number of pulses being determined by how far the dial is rotated before being released.   55

rotary generator (induction heating). An alternating-current generator adapted to be rotated by a motor or prime mover. *See:* **dielectric heating; industrial electronics.**   14, 114

rotary inverter. A machine that combines both motor and generator action in one armature winding. It is excited by one magnetic field and changes direct-current power to alternating-current power. (Usually it has no amortisseur winding.)   63

rotary joint (waveguide components). A coupling for efficient transmission of electromagnetic energy between two waveguide or transmission line structures designed to permit unlimited mechanical rotation of one structure.   166

rotary phase changer (rotary phase shifter) (waveguide). A phase changer that alters the phase of a transmitted wave in proportion to the rotation of one of its waveguide sections.   244, 179

rotary relay. (1) A relay whose armature moves in rotation to close the gap between two or more pole faces (usually with a balanced armature). (2) Sometimes used for stepping relay. *See:* **relay.**   259

rotary solenoid relay. A relay in which the linear motion of the plunger is converted mechanically into rotary motion. *See:* **relay.**   259

rotary switch. A bank-and-wiper switch whose wipers or brushes move only on the arc of a circle. *See:* **switch.**   328

rotary system. An automatic telephone switching system that is generally characterized by the following features: (1) The selecting mechanisms are rotary switches. (2) The switching pulses are received and stored by controlling mechanisms that govern the subsequent operations necessary in establishing a telephone connection.   328

rotary voltmeter. *See:* **generating voltmeter.**

rotatable frame (rotating machinery). A stator frame that can be rotated by a limited amount about the axis of the machine shaft. *See:* **stator.**   63

rotatable phase-adjusting transformer (phase-shifting transformer). A transformer in which the secondary voltage may be adjusted to have any desired phase relation with the primary voltage by mechanically orienting the secondary winding with respect to the primary. The primary winding of such a transformer usually consists of a distributed symmetrical polyphase winding and is energized from a polyphase circuit. *See:* **auxiliary device to an instrument.**   328

rotating amplifier (excitation systems for synchronous machines). An electric machine in which a small energy change in the field is amplified to a large energy change at the armature terminals.   507

rotating-anode tube (X-ray). An X-ray tube in which the anode rotates. *Note:* The rotation continually brings a fresh area of its surface into the beam of electrons, allowing greater output without melting the target.   190

rotating control assembly (rotating machinery). The complete control circuits for a brushless exciter mounted to permit rotation. *See:* **rotor (rotating machinery).**   63

rotating field. A variable vector field that appears to rotate with time.   210

rotating-insulator switch (power switchgear). One in which the opening and closing travel of the blade is accomplished by the rotation of one or more of insulators supporting the conducting parts of the switch.   103

rotating joint (waveguides). A coupling for transmission of electromagnetic energy between two waveguide structures designed to permit mechanical rotation of one structure. *See:* **waveguide.**   330

rotating machinery. *See:* **machine, electric.**

rotation plate (rotating machinery). A plaque showing the proper direction of rotor rotation. *See:* **rotor (rotating machinery).**   63

rotation test (rotating machinery). A test to determine that the rotor rotates in the specified direction when the voltage applied agrees with the terminal markings. *See:* **asynchronous machine.**   63

rotor (1) (watthour meter). That part of the meter that is directly driven by electromagnetic action.   212

(2) (rotating machinery). The rotating member of a machine, with shaft. *Note:* In a direct-current machine with stationary field poles, universal, alternating-current series, and repulsion-type motors, it is commonly called the armature.   63

rotor angular momentum (gyro). The product of spin angular velocity and rotor moment of inertia about the spin axis.   46

rotor bar (rotating machinery). A solid conductor that constitutes an element of the slot section of a squirrel-cage winding. *See:* **rotor (rotating machinery).**   63

rotor bushing (rotating machinery). A ventilated or nonventilated piece or assembly used for mounting onto a shaft, an assembled rotor core whose inside opening is larger than the shaft. *See:* **rotor (rotating machinery).**   63

rotor coil (rotating machinery). A unit of a rotor winding of a machine. *See:* **rotor (rotating machinery).**   63

rotor core (rotating machinery). That part of the magnetic circuit that is integral with, or mounted on, the rotor shaft. It frequently consists of an assembly of laminations.   63

rotor-core assembly (rotating machinery). The rotor core with a squirrel-cage or insulated-conductor winding, put together as an assembly. *See:* **rotor (rotating machinery).**   63

rotor core lamination (rotating machinery). A sheet of magnetic material, containing teeth, slots, or other perforations dictated by design, which forms the rotor core when assembled with other identical or similar laminations.   63

rotor displacement angle (load angle) (rotating machinery). The displacement caused by load between the terminal voltage and the armature voltage generated by that component of flux produced by the field current. *See:* **rotor (rotating machinery).**   63

rotor end ring (rotating machinery). The conducting structure of a squirrel-cage winding that short-circuits all of the rotor bars at one end. *See:* **rotor (rotating machinery).**   63

rotor moment-of-inertia (gyro) (inertial sensor). The moment of inertia of a gyro rotor about its spin axis.   46

rotor-resistance starting (rotating machinery). The process of starting a wound-rotor induction motor by connecting the rotor initially in series with starting resistors that are short-circuited for the running operation. *See:* **asynchronous machine.**   63

rotor rotation detector (gyro). A device which produces a signal output as a function of the speed of the rotor.   46

rotor slot armor (cylindrical-rotor synchronous machine) (rotating machinery). Main ground insulation surrounding the slot or core portions of a field coil assembled on a slotted rotor. *See:* **rotor (rotating machinery).**   63

rotor speed sensitivity (dynamically tuned gyro) (inertial sensor). The change in in-phase spring rate due to a change in gyro rotor speed.   46

rotor spider. *See:* **spider.**

rotor winding (rotating machinery). A winding on the rotor of a machine. *See:* **rotor (rotating machinery).**   63

roughness (navigational system display)(navigation aid terms). Irregularities resembluing scalloping, but distinguished by their random, noncyclic nature. *Syn:* **course roughness.**   526

round conductor. Either a solid or stranded conductor of which the cross section is substantially circular. *See:* **conductor.**   64

rounding (pulse techniques). *See:* **distortion, pulse.**

rounding error (test, measurement and diagnostic equipment). The error resulting from deleting the less significant digits of a quantity and applying some rule of correction to the part retained. A common round-off rule is to take the quantity to the nearest digit. Thus the value of Pi , 3.14159265 . . . , rounded to four decimals is 3.1416.   54

round off. To delete the least-significant digit or digits of a numeral and to adjust the part retained in accordance with some rule.   255, 77, 54

round rotor (cyclindrical rotor) (rotating machinery). A rotor of cylindrical shape in which the coil sides of the winding are contained in axial slots. *See:* **rotor (rotating machinery).**   63

route (telephone switching systems). A particular order of a set of switching entities through which call connections may be established.   55

route locking. Locking effective when a train passes a signal and adapted to prevent manipulation of levers that would endanger the train while it is within the limits of the route entered. It may be so arranged that a train in clearing each section of the route releases the locking affecting that section. *See:* **interlocking (interlocking plant).**   328

route table (FASTBUS acquisition and control). The list of group addresses recognized by an SI (segment interconnect) for passing operations to its far-side segment.   480

route tracing mode (FASTBUS acquisition and control). A mode of SI (segment interconnect) operation which generates an error diagnostic response instead of the normal passing of an operation.   480

routine (software). (1) A computer program segment that performs a specific task. (2) A program, or a sequence of instructions called by a program, that may have some general or frequent use. *See:* **computer program segment; function; instruction; procedure; program; subprogram; subroutine.**   434

routine test (1)(electrical heat tracing for industrial applications). A test which is carried out by the manufacturer of the heating cable during the production process.   523

(2)(rotating electric machinery). A test showing that each machine has been run and found to be sound electrically and mechanically, and is essentially identical with those that have been type tested. *See:* **type test.**   424

routine tests (1)(metal-oxide surge arresters for ac power circuits)(surge arrester). Tests made by the manufacturer on every device or representative samples, or on parts or materials, as required, to verify that the product meets the design specifications.   583, 430

(2) (general). Tests made for quality control by the manufacturer on every device or representative samples, or on parts or materials as required to verify during production that the product meets the design specifications.   4, 210, 203, 53

(3) (rotating machinery). The tests applied to a machine to show that it has been constructed and assembled correctly, is able to withstand the appropriate high-voltage tests, is in sound working order both electrically and mechanically, and has the proper electrical characteristics. *See:* **asynchronous machine; limiting insulation temperature (limiting hottest-spot temperature).** 63

(4) (cable termination). Tests made on each high-voltage cable termination or upon a representative number of devices, or parts thereof, during production for the purpose of quality control. 4

(5) (for switchgear). *See:* **production tests (for switchgear).**

routing code (telephone switching systems). A digit or combination of digits used to direct a call towards its destination. 55

routing pattern (telephone switching systems). The implementation of a routing plan with reference to an individual automatic exchange. 55

routing plan (telephone switching systems). A plan for directing calls through a configuration of switching entities. 55

roving (rotating machinery). A loose assemblage of fibers drawn or rubbed into a single strand with very little twist. In spun yarn systems, the product of the stage or stages just prior to spinning. 63

row (test pattern language). A group of words or bits in a memory, identified by a common X-address. 463

row binary. Pertaining to the binary representation of data on punched cards in which adjacent positions in a row correspond to adjacent bits of data, for example, each row in an 80-column card may be used to represent 80 consecutive bits of two 40-bit words. 255, 77

row enable (RAS) (RE) (semiconductor memory). The input used to strobe in the row address in multiplexed address random access memories (RAM). 441

RPE. (radial probable error). *See:* **circular probable error.**

RRRV. *See:* **rate of rise of restriking voltage.**

RS (cable systems in power generating stations). Rigid steel conduit. 35

R scan. *See:* **R display.**

R-scope (radar). A cathode-ray oscilloscope arranged to present an R-display. 13

rubber tape. A tape composed of rubber or rubberlike compounds that provides insulation for joints. 328

rudder-angle-indicator system. A system consisting of an indicator (usually in the wheel house) so controlled by a transmitter connected to the rudder stock as to show continually the angle of the rudder relative to the center line of the ship. 328

ruling span (conductor stringing equipment). A calculated deadend span length which will have the same changes in conductor tension due to changes of temperature and conductor loading as will be found in a series of spans of varying lengths between deadends. 431

rumble (electroacoustics). Low-frequency vibration of the recording or reproducing drive mechanism superimposed on the reproduced signal. *See:* **phonograph pickup.** 176

rumble, turntable. *See:* **turntable rumble.**

run (computing systems). A single, continuous performance of a computer routine. 255, 77, 54

runaway pipeline (vessel) temperature (electrical heating systems). The highest equilibrium temperature on the pipeline or vessel that can occur when the heating system is continuously energized in the maximum ambient temperature. 476

runaway pipe temperature (electrical heat tracing for industrial applications). The highest equilibrium pipe temperature that occurs when the heating cable is continuously energized at the maximum ambient. 523

run-down time (gyro). The time interval required for the gyro rotor to reach a specified speed, or during which the gyro exhibits specified performance, after removal of rotor excitation at a specified speed. 46

run/halt switch (RH)(FASTBUS acquisition and control). A switch normally operated by the run/halt switch activator bar on crate segments and on the ATC (arbitration timing control) on cable segments which stops bus traffic so that it may be possible to insert or remove modules without affecting other modules on the segment. 480

running board (conductor stringing equipment). A pulling device designed to permit stringing more than one conductor simultaneously with a single pulling line. For distribution stringing, it is usually made of lightweight tubing with the forward end curved gently upward to provide smooth transition over pole cross-arm rollers. For transmission stringing, the device is either made of sections hinged transversely to the direction of pull or of a hard nose rigid design, both having a flexible pendulum tail suspended from the rear. This configuration stops the conductors from twisting together and permits smooth transition over the sheaves of bundle travelers. *Syn:* **alligator; bird; monkey tail; sled.** 431

running circuit breaker (42) (power system device function numbers). A device whose principal function is to connect a machine to its source or running or operating voltage. This function may also be used for a device, such as a contactor, that is used in series with a circuit breaker or other fault protecting means, primarily for frequent opening and closing of the circuit. 402

running ground (conductor stringing equipment). A portable device designed to connect a moving conductor or wire rope, or both, to an electrical ground. These devices are normally placed on the conductor or wire rope adjacent to the pulling or tensioning equipment located at either end of a sag section. Primarily used to provide safety for personnel during construction or reconstruction operations. *Syn:* **ground roller; moving ground; rolling ground; traveling ground.** 431

running-light-indicator panel (telltale). A panel in the wheelhouse providing audible and visible indication of the failure of any running light connected thereto. 328

running lights (navigation lights). Lanterns constructed and located as required by navigation laws, to permit the heading and approximate course of a vessel to be determined by an observer on a nearby vessel. *Note:* Usual running lights are port side, starboard side, mast-head, range, and stern lights.   328

running open-phase protection (industrial control). The effect of a device operative on the loss of current in one phase of a polyphase circuit to cause and maintain the interruption of power in the circuit.   206

running operation (single-phase motor). (1) For a motor employing a starting switch or relay: operation at speeds above that corresponding to the switching operation. (2) For a motor not employing a starting switch or relay: operation in the range of speed that includes breakdown-torque speed and above. *See:* **asynchronous machine.**   63

running tension control (industrial control). A control function that maintains tension in the material at operating speeds. *See:* **control system, feedback.**
    225, 206

run-of-river station (power operations). A hydroelectric generating station utilizing limited pondage or the flow of the stream as it occurs.   516

runout rate (industrial control). The velocity at which the error in register accumulates.   204

run time (software). (1) A measure of the time expended to execute a program. While it ordinarily reflects the expended central processor time, it may also include peripheral processing and peripheral accessing time, for example, a run time of 5 hours. (2) The instant at which a program begins to execute. *See:* **execution time; program.**   434

run time variable (RTV) (test, measurement and diagnostic equipment). An application program condition in which the stimuli is varied under system control based on a measurement result.   54

run-up time (gyro). The time interval required for the gyro rotor to reach a specified speed from standstill.
    46

runway alignment indicator (illuminating engineering). A group of aeronautical ground lights arranged and located to provide early direction and roll guidance on the approach to a runway.   167

runway centerline lights (illuminating engineering). Runway lights installed in the surface of the runway along the centerline indicating the location and direction of the runway centerline and are of particular value in conditions of very poor visibility.   167

runway-edge lights (illuminating engineering). Lights installed along the edges of a runway marking its lateral limits and indicating its direction.   167

runway-end identification light (illuminating engineering). A pair of flashing aeronautical ground lights symmetrically disposed on each side of the runway at the threshold to provide additional threshold conspicuity.   167

runway-exit lights (illuminating engineering). Lights placed on the surface of a runway to indicate a path of the taxiway centerline.   167

runway lights (illuminating engineering). Aeronautical ground lights arranged along or on a runway.
    167

runway threshold. *See:* **approach-light beacon.**

runway visibility (illuminating engineering). The meteorological visibility along an identified runway. Where a transmissometer is used for measurement, the instrument is calibrated in terms of a human observer; that is, the sighting of dark objects against the horizon sky during daylight and the sighting of moderately intense unfocused lights of the order of 25 candelas at night.   167

runway visual range (RVR)(1)(navigation aid terms). The forward distance a human pilot can see along the runway during an approach to landing; this distance is derived from electro-optical instruments operated on the ground and it is improved (increased) by the use of lights (such as high-intensity runway lights).
    526

(2) (illuminating engineering in the United States). An instrumentally derived value, based on standard calibrations, that represents the horizontal distance a pilot will see down the runway from the approach end; it is based on the sighting of either high intensity runway lights or on the visual contrast of other targets--whichever yields the greater visual range.
    167

rural districts (NESC). All places not urban. This may include thinly settled areas within city limits.
    494

rural line. A line serving one or more subscribers in a rural area.   328

rust (corrosion). A corrosion product consisting primarily of hydrated iron oxide. *Note:* This term is properly applied only to iron and ferrous alloys.
    205

rust-resistant. So constructed, protected or treated that rust will not exceed a specified limit when subjected to a specified rust resistance test.   53

RVR. *See:* **runway visual range.**

R-X diagram (power switchgear). A graphic presentation of the characteristics of a relay unit in terms of the ratio of voltage to current and the phase angle between them. *Note:* For example, if a relay just operates with ten volts and ten amperes in phase, one point on the operating curve of the relay would be plotted as one ohm on the R axis (that is, $R = 1$, $X = 0$ where R is the abscissa and X is the ordinate). *See:* **distance relay, Figures (a) and (b).**   103

R X plot (protective relaying). A graphical method of showing the characteristics of a relay element in terms of the ratio of voltage to current and the angle between them. For example, if a relay barely operates with 10 V and 10 A in phase, one point on the operating curve of the relay would be plotted as 1 Ω on the R axis (that is, $R = 1$, $X = 0$).   128

S

sabin (audio and electroacoustics). A unit of absorption having the dimensions of area. *Notes:* (1) The metric sabin has dimensions of square meters. (2) When used without a modifier, the sabin is the equivalent of one square foot of a perfectly absorptive surface. 176

sacrificial protection (corrosion). Reduction or prevention of corrosion of a metal in an environment acting as an electrolyte by coupling it to another metal that is electrochemically more active in that particular electrolyte. *See:* **stray-current corrosion.** 205

safeguard (electrolytic cell line working zone). A precautionary measure or stipulation, or a technical contrivance to prevent accidents. 133

safe shutdown earthquake (SSE)(1)(seismic qualification of Class 1E equipment for nuclear power generating stations). An earthquake that is based upon an evaluation of the maximum earthquake potential considering the regional and local geology and seismology and specific characteristics of local subsurface material. It is that earthquake that produces the maximum vibratory ground motion for which certain structures, systems, and components are designed to remain functional. These structures, systems, and components are those necessary to ensure (A) The integrity of the reactor coolant pressure boundary (B) The capability to shut down the reactor and maintain it in a safe shutdown condition (C) The capability to prevent or mitigate the consequences of accidents that could result in potential off-site exposures comparable to the guideline exposures of 10 FR, Ch 1, Section 100. 581

(2) (valve actuators). That earthquake which produces the maximum vibratory ground motion for which certain structural systems are designed to remain functional. These structures, systems, and components are those necessary to ensure (A) The integrity of the reactor coolant pressure boundary (B) The capability to shut down the reactor and maintain it in a safe shutdown condition (C) The capability to prevent or mitigate the consequences of an accident which could result in potential offsite exposures comparable to the exposure guideline of CFR 10, Energy--Nuclear Regulatory Commission, Part 100, Dec 5, 1973. 492

(3) (nuclear power generating stations). That earthquake which produces the maximum vibratory ground motion for which certain structures, systems, and components are designed to remain functional. 440

(4) (seismic qualification of class 1E equipment) (seismic testing of relays). That earthquake which produces the maximum vibratory ground motion for which certain structures, systems, and components are designed to remain functional. These structures, systems, and components are those necessary to assure (ensure): (A) the integrity of the reactor coolant pres-

sure boundary, (B) the capability to shutdown the reactor and maintain it in a safe shutdown condition, or (C) the capability to prevent or mitigate the consequences of accidents which could result in potential offsite exposures comparable to the guideline exposures of Code of Federal Regulations, Title 10, Part 100 (December 5, 1973). 392

safety class features (Class 1E equipment and circuits). Structures design to protect Class 1E equipment against the effects of design basis events. *Note:* For the purposes of this standard, separate safety class structures can be separate rooms in the same building. The rooms may share a common wall. 131

safety class structures (nuclear power generating station) (1) (Class 1E equipment and circuits). Structures designed to protect Class 1E equipment against the effects of design basis events. 102

(2) (nuclear power generating stations). Structures designed to protect Class 1E equipment against the effects of the design basis events. *Note:* For the purposes of this document, separate safety class structure can be separate rooms in the same building. The rooms may share a common wall. 131

safety control feature (deadman's feature). That feature of a control system that acts to reduce or cut off the current to the traction motors or to apply the brakes, or both, if the operator relinquishes personal control of the vehicle. *See:* **multiple-unit control.** 328

safety control handle (deadman's handle). A safety attachment to the handle of a controller, or to a brake valve, causing the current to the traction motors to be reduced or cut off, or the brakes to be applied, or both, if the pressure of the operator's hand on the handle is released. *Note:* This function may be applied alternatively to a foot-operated pedal or in combination with attachments to the controller or the brake valve handles, or both. *See:* **multiple-unit control.** 328

safety function (1) (accident monitoring instrumentation). One of the processes or conditions (for example, emergency negative reactivity insertion, post accident heat removal, emergency core cooling, post accident radioactivity removal, and containment isolation) essential to maintain plant parameters within acceptable limits established for a Design Basis Event (DBE). 421

(2) (nuclear power generating station) (Class 1E equipment and circuits). One of the processes or conditions (for example, emergency negative reactivity insertion, postaccident heat removal, emergency core cooling, postaccident radioactivity removal, and containment isolation) essential to maintain plant parameters within acceptable limits established for a design basis event. *Note:* A safety function is achieved by the

completion of all required protective actions by the reactor trip system and the engineered safety features, or both, concurrent with the completion of all required protective actions by the auxiliary supporting features.
102

(3) (nuclear power generating stations) (safety systems). One of the processes or conditions (for example, emergency negative reactivity insertion, post-accident heat removal, emergency core cooling, post-accident radioactivity removal, and containment isolation) essential to maintain plant parameters within acceptable limits established for a design basis event. *Note:* A safety function is achieved by the completion of all required protective actions by the reactor trip system or the engineered safety features concurrent with the completion of all required protective actions by the auxiliary supporting features, or both. (See Appendix A, IEEE Std 603-1980 for an illustrative example.)
428

safety group (1) (nuclear power generating station) (Class 1E equipment and circuits). A given minimal set of interconnected components, modules, and equipment that can accomplish a safety function. *Note:* A safety group may include one or more divisions. In a design where each division can accomplish a safety function, each division is a safety group. However, a design consisting of three 50 percent capacity systems separated into 3 divisions would have 3 safety groups, each safety group requiring that any two out of three divisions be operating to accomplish the safety function.
102

(2) (nuclear power generating stations) (safety systems). A given minimal set of interconnected components, modules, and equipment that can accomplish a safety function. *Note:* A safety group may include one or more divisions. (See Appendix A, IEEE Std 603-1980 for an illustrative example.)
428

safety life line (conductor stringing equipment). A safety device normally constructed from synthetic fiber rope and designed to be connected between a fixed object and the body belt of a workman working in an elevated position when his regular safety strap cannot be utilized. *Syn:* **life line; safety line; scare rope.**
431

safety outlet. *See:* **grounding outlet.**

safety-related (nuclear power generating station). Any Class IE power or protection system device included in the scope of IEEE-279-1971 or IEEE-308-1974.
357

safety system (1) (accident monitoring instrumentation). Those systems (the reactor trip system, and an engineered safety feature, or both, including all their auxiliary supporting features and other auxiliary features) which provide a safety function. A safety system is comprised of more than one safety group of which any one safety group can provide the safety function.
421

(2) (class 1E power systems for nuclear power generating stations). The collection of systems required to minimize the probability and magnitude to release of radioactive material to the environment by maintain-

ing plant conditions within the allowable limits established for each design basis event. *Note:* The safety system is the aggregate of one or more protection systems and one or more protective action systems. It includes the engineered safety features, the reactor trip system, and the auxiliary supporting features.
102

(3) (nuclear power generating station) (Class 1E equipment and circuits). Those systems (the reactor trip system and an engineered safety feature, or both, including all their auxiliary supporting and other auxiliary features) which provide a safety function. A safety system is comprised of more than one safety group of which any one safety group can provide the safety function.
102

(4) (nuclear power generating station) (design of control room complex). The collection of systems required to minimize the probability and magnitude of release of radioactive material to the environment by maintaining plant conditions within the allowable limits established for each design basis event. *Note:* The safety system is the aggregate of one or more protective action systems. It includes, but is not necessarily limited to, the engineered safety features, the reactor trip system, and the auxiliary supporting features. (See IEEE Std 603-1980, Criteria for Safety Systems for Nuclear Power Generating Stations.)
439

(5) (nuclear power generating station) (single-failure criterion). The collection of systems required to mitigate the consequences of design basis events.
356

(6) (nuclear power generating stations) (safety systems). Those systems (the reactor trip system, an engineered safety feature, or both, including all their auxiliary supporting features and other auxiliary features) which provide a safety function. A safety system is comprised of more than one safety group of which any one safety group can provide the safety function.
428

safety systems (safety systems equipment in nuclear power generating stations). Systems which provide a safety function (the reactor trip system, an engineered safety feature, or both, including all their auxiliary supporting features). A safety system is comprised of more than one safety group of which any one safety group can provide the safety function.
534

safe working space (electrolytic cell line working zone). The space required to safeguard personnel from hazarous electrical conditions during the conduct of their work in operating and maintaining cells and their attachments. This space includes space allowance for tools and equipment that may be involved.
133

safe working voltage to ground (electric recording instrument). The highest safe voltage in terms of maximum peak value that should exist between any circuit of the instrument and its case. *See:* **test (instrument or meter).**
294

safe work practices (electrolytic cell line working zone). Those operating and maintenance procedures that are effective in preventing accidents.
133

SAFI. *See:* **semiautomatic flight inspection.**

sag (1)(NESC). (A) The distance measured vertically from a conductor to the straight line joining its two points of support. Unless otherwise stated in the rule, the sag referred to is the sag at the midpoint of the span. (B) initial unloaded sag. The sag of a conductor prior to the application of any external load. (C) final sag. The sag of a conductor under specified conditions of loading and temperature applied, after it has been subjected for an appreciable period to the loading prescribed for the loading district in which it is situated, or equivalent loading, and the loading removed. Final sag includes the effect of inelastic deformation (creep). (D) final unloaded sag, The sag of a conductor after it has been subjected for an appreciable period to the loading prescribed for the loading district in which it is situated, or equivalent loading, and the loading removed. Final unloaded sag includes the effect of inelastic deformation (creep). (E) total sag. The distance measured vertically from the conductor to the straight line joining its two points of support, under conditions of ice loading equivalent to the total resultant loading for the district in which it is located. (F) maximum total sag. The total sag at the midpoint of the straight line joining the two points of support of the conductor. (G) apparent sag of a span. The maximum distance between the wire in a given span, and the straight line between the two points of support of the wire, measured perpendicularly from the straight line. *See:* **Figure below.** (H) sag of a conductor at any point in the span. The distance measured vertically from the particular point in the conductor to a straight line between its two points of support. (I) apparent sag at any point in the span. The distance, at the particular point in the span, between the wire and the straight line between the two points of support of the wire, measured perpendicularly from the straight line.

494

(2) (transmission and distribution). (A) The distance measured vertically from a conductor to the straight line joining its two points of support. Unless otherwise stated, the sag referred to is the sag at the midpoint of the span. *See:* initial unloaded sag; final sag; final unloaded sag; total sag; maximum total sag; apparent sag of a span; sag of a conductor at any point in a span; apparent sag at any point in the span. Sag and apparent sag.(B) (apparent sag at any point).The departure of the wire at the particular point in the span from the straight line between the two points of support of the span, at 60 degrees Fahrenheit, with no wind loading. See Figure 'Sag and Apparent Sag' below.

178

sag of a conductor at any point in a span (transmission and distribution). The distance measured vertically from the particular point in the conductor to a straight line between its two points of support.

sag section (conductor stringing equipment). The section of line between snub structures. More than one sag section may be required in order to sag properly the actual length of conductor which has been strung. *Syn:* **pull; setting; stringing section.**   431

sag span (conductor stringing equipment). A span selected within a sag section and used as a control to

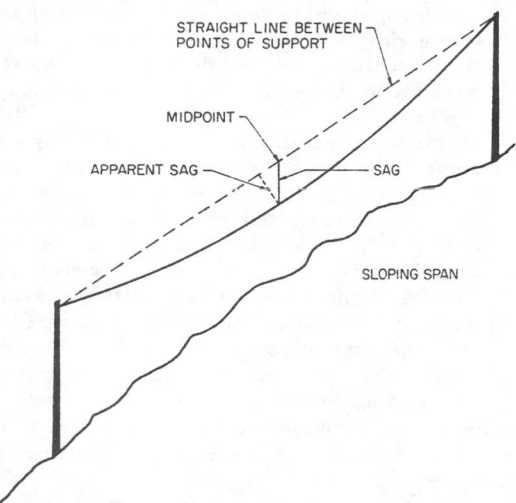

STRAIGHT LINE BETWEEN POINTS OF SUPPORT

MIDPOINT

APPARENT SAG

SAG

SLOPING SPAN

Sag and Apparent Sag

determine the proper sag of the conductor, thus establishing the proper conductor level and tension. A minimum of two, but normally three, sag spans are required within a sag section in order to sag properly. In mountainous terrain or where span lengths vary radically, more than three sag spans could be required within a sag section. *Syn:* **control span.**   431

sag target (conductor stringing equipment). A device used as a reference point to sag conductors. It is placed on one structure of the sag span. The sagger, on the other structure of the sag span, can use it as his reference to determine the proper conductor sag. *Syn:* **sag board; target.**   431

sal ammoniac cell. A cell in which the electrolyte consists primarily of a solution of ammonium chloride. *See:* **electrochemistry.**   328

salient pole (rotating machinery). A field pole that projects from the yoke or hub towards the primary winding core. *See:* **rotor (rotating maxhinery).**

63

salient-pole machine. An alternating-current machine in which the field poles project from the yoke toward the armature and/or the armature winding self-inductance undergoes a significant single cyclic variation for a rotor displacement through one pole pitch. *See:* **asynchronous machine.**   63

salient pole synchronous induction motor (rotating machinery). A salient pole synchronous motor having a coil winding for starting purposes embedded in the pole shoes. The terminal leads of this coil winding are connected to collector rings.   63

salinity indicator system. A system, based on measurement of varying electric resistance of the solution, to indicate the amount of salt in boiler feed water, the output of an evaporator plant, or other fresh water. *Note:* Indication is usually in grains per gallon.

328

sampled data. Data in which the information content can be, or is, ascertained only at discrete intervals of time. *Note:* Sampled data can be analog or digital. *See:* **control system, feedback.**   224, 207

sampled-data control system (industrial control). A system that operates with sampled data. *See:* **control system, feedback.**   206

sampled format (pulse measurement). A waveform which is a series of sample magnitudes taken sequentially or nonsequentially as a function of time. It is assumed that nonsequential samples may be rearranged in time sequence to yield the following samples formats. *See:* **periodically sampled real time format; periodically sampled equivalent time format; aperiodically sampled real time format; aperiodically sampled equivalent time format.**   15

sampled signal. The sequence of values of a signal taken at discrete instants. *See:* **control system.**   56

sample equipment (1)(nuclear power generating stations). Production equipment tested to obtain data that are valid over a range of ratings and for specific services.   120, 408

(2)(safety systems equipment in nuclear power generating stations). Equipment, representative of a design, used to obtain data that are valid over a range of ratings and for specific service conditions.   535

sample valve actuator. A representative unit manufactured in accordance with the manufacturer's quality control system and specifications for production units.   142

sample valve operator (nuclear power generating stations). A production valve operator type tested to obtain data that are valid over a range of sizes and for the specific services. *Note:* All salient factors must be shown to be common to the sample valve operator and to the intended service valve operator. Commonality of factors such as materials of construction, lubrication, mechanical stresses and clearances, manufacturing processes, and dielectric properties may be established by specification, test, or analyses.   142

sampling (pulse terms). A process in which strobing pulses yield signals which are proportional to the magnitude (typically, as a function of time) of a second pulse or other event.   254

sampling circuit (sampler). A circuit whose output is a series of discrete values representative of the values of the input at a series of points in time.   328

sampling control. *See:* **control system, sampling.**

sampling gate (radar). A device that extracts information from the input wave only when activated by a selector pulse or sampling pulse.   13

sampling, instantaneous. The process for obtaining a sequence of instantaneous values of a wave. *Note:* These values are called instantaneous samples.   111

sampling interval (automatic control). The time between samples in a sampling control system. *See:* **control system, feedback.**   329

sampling period (automatic control). The time interval between samples in a periodic sampling control system. *See:* **control system, feedback.**   329

sampling smoke detector (fire protection devices). A device which consists of tubing distributed from the detector unit to the area(s) to be protected. An air pump draws air from the protected area back to the detector through the air sampling ports and piping. At the detector, the air is analyzed for smoke particles.   71

sampling tests. Tests carried out on a few samples taken at random out of one consignment. *See:* **asynchronous machine.**   63

SAR. *See:* **synthetic aperture radar.**   13

satellite (communication satellite). A body which revolves around another body and which has a motion primarily and permanently determined by the force of attraction of this body. *Note:* A body so defined which revolves round the sun is called a planet or planetoid. By extension, a natural satellite of a planet may itself have a satellite.   74

satellite navigation (navigation aid terms). Navigation using artificial earth satellites as an aid. Position is computed by determination of either angles, range and range rate, or range and angle measurements of the vehicle relative to the satellite plus satellite ephemeris data received by the vehicle. Satellite ephemeris data can be determined by tracking stations and transmitted to and stored in the satellite's memory for subsequent transmission to vehicle's receivers.   526

satellite phasing (communication satellite). Maintaining the center of mass of a satellite by propulsion within a prescribed small tolerance in a desired relation with respect to other satellites or a point on the earth or some other point of reference, such as the subsolar point.   74

saturable-core magnetometer. A magnetometer that depends for its operation on the changes in permeability of a ferromagnetic core as a function of the field to be measured. *See:* **magnetometer.**   328

saturable-core reactor. *See:* **saturable reactor.**

saturable reactor (saturable core reactor)(1)(electrical heating applications to melting furnaces and fore-hearths in the glass industry). A device that provides output voltage modulation by variation of its circuit reactance. This reactance is controlled by changing the saturation of its magnetic core through variation of a superimposed unidirectional flux.   520

(2) (power and distribution transformer). (A) A magnetic-core reactor whose reactance is controlled by changing the saturation of the core through variation of a super-imposed unidirection flux. (B) A magnetic-core reactor operating in the region of saturation without independent control means. *Note:* Thus a reactor whose impedance varies cyclically with the alternating current (or voltage).   53

saturated signal. *See:* **saturating signal.**

saturated sleeving. A flexible tubular product made from braided cotton, rayon, nylon, glass, or other fibers, and coated or impregnated with varnish, lacquer, a combination of varnish and lacquer, or other electrical insulating materials. The impregnant or coating need not form a continuous film.   328

saturating reactor. A magnetic-core reactor operating in the region of saturation without independent control means. *See:* **magnetic amplifier.** 328

saturating signal (electronic navigation). A signal of an amplitude greater than can be accommodated by the dynamic range of a circuit. *See:* **navigation.** 187, 13

saturation (1) (signal-transmission system). A natural phenomenon or condition in which any further change of input no longer results in appreciable change of output. 219, 206

(2) (automatic control) A condition caused by the presence of a signal or interference large enough to produce the maximum limit of response, resulting in loss of incremental response. *See:* **control system, feedback; signal.** 56, 188

(3) (visual). The attribute of a visual sensation which permits a judgment to be made of the proportion of pure chromatic color in the total sensation. *Note:* This attribute is the psychosensorial correlate (or nearly so) of the colorimetric quantity "purity." 18

(4) (color television). In a tristimulus reproducer, the degree to which the color lies on the triangle as defined by the three reproducing primaries. *Note:* Full saturation is achieved when one or two of the reproduced primary colors have zero intensity. 18

(5) (perceived light-source color). The attribute used to describe its departure from a light-source color of the same brightness perceived to have no hue. *See:* **color.** 167

(6) (diode-type camera tube). The point on the signal transfer characteristic where an increase in the input irradiance signal does not change the resulting output current signal significantly. 380

(7) (of a maser, laser, maser material, or laser material) (laser-maser). A condition in which the attenuation or gain of a material or a device remains at a fixed level or decreases as the input signal is increased. 363

saturation characteristics (nuclear power generating station). A description of the steady state or dynamic conditions or limitations under which a further change in input produces an output response which no longer conforms to the specified steady-state or dynamic input-output relationship. 355

saturation current (1) (thermionics). The value of the current in the saturation state. *See:* **electron emission.** 190

(2) (semiconductor diode). That portion of the steady-state reverse current that flows as a result of the transport across the junction of minority carriers thermally generated within the regions adjacent to the junction. *See:* **semiconductor device.** 245

(3) (diode-type camera tube). The value of the output current signal saturation. Units: amperes (A). 380

saturation curve (machine or other apparatus). A characteristic curve that expresses the degree of magnetic saturation as a function of some property of the magnetic excitation. *Note:* For a direct-current or synchronous machine the curve usually expresses the relation between armature voltage and field current for no load or some specified load current, and for specified speed. 63

saturation factor (rotating machinery). The ratio of the unsaturated value of a quantity to its saturated value. The reciprocal of this definition is also used. 63

saturation flux density (β_s)**(electrical heating systems).** The maximum possible magnetic flux density in a material. 476

saturation flux density. *See:* **saturation induction.**

saturation induction, (B_s) **(magnetic core testing).** The maximum intrinsic value of induction possible in a material. *Notes:* (1) This term is often used for the maximum value of induction at a stated high value of field strength where further increase in intrinsic magnetization with increasing field strength is negligible. (2) S.I. unit: Tesla: cgs unit: Gauss (1 Tesla), 10^4 Gauss. (3) Peak induction (B_m) is the magnetic induction corresponding to the peak applied magnetizing force specified in a test. 165

saturation level (storage tubes). The output level beyond which no further increase in output is produced by further writing (then called write saturation) or reading (then called read saturation). *Note:* The word saturation is frequently used alone to denote saturation level. *See:* **storage tube.** 174, 190

saturation of a perceived color (illuminating engineering). The attribute according to which it appears to exhibit more or less chromatic color judged in proportion to its brightness. In a given set of viewing conditions, and at luminance levels that result in photopic vision, a stimulus of a given chromaticity exhibits approximately constant saturation for all luminances. 167

saturation region of an insulated-gate field-effect transistor (IGFET) (metal-nitride-oxide field-effect transistor). A portion of the I_{DS} versus V_{DS} characteristic where I_{DS} is nearly constant regardless of the value of V_{DS}. This is true when $|V_{DS}| \geq |V_{GS} - V_T|$. 386

saturation state (thermionics). The state of working of an electron tube or valve in which the current is limited by the emission from the cathode. *See:* **electron emission.** 190

sawtooth. *See:* **sawtooth waveform.**

sawtooth sweep. A sweep generated by the ramp portion of a sawtooth waveform. *See:* **oscillograph.** 185

sawtooth wave (television). A periodic wave whose instantaneous value varies substantially linearly with time between two values, the interval required for one direction of progress being longer than that for the other. *Note:* In television practice, the waveform during the retrace interval is not necessarily linear, since only the trace interval is used for active scanning. 18

sawtooth waveform. A waveform containing a ramp and a return to initial value, the two portions usually of unequal duration. *See:* **oscillograph.** 185

S-band. A radar frequency band between 2 gigahertz (GHz) and 4 GHz, usually in one of the International Telecommunications Union (ITU) assigned bands 2.3 GHz to 2.5 GHz or 2.7 GHz to 3.7 GHz. 13

scada channel (supervisory control, data acquisition, and automatic control). The communication path between master and remote stations. See figure below.   570

scalar. A quantity that is completely specified by a single number.   210

scalar field. The totality of scalars in a given region represented by a scalar function $S(x,y,z)$ of the space coordinates x,y,z.   210

scalar function. A functional relationship that results in a scalar.   210

scalar product (of two vectors) (dot product). The scalar obtained by multiplying the product of the magnitudes of the two vectors by the cosine of the angle between them. The scalar product of the two vectors **A** and **B** may be indicated by means of a dot **A** · **B**. If the two vectors are given in terms of their rectangular components, then

$$\mathbf{A} \cdot \mathbf{B} = A_x B_x + A_y B_y + A_z B_z.$$

Example: Work is the scalar product of force and displacement.   210

scale (1) (acoustics). A musical scale is a series of notes (symbols, sensations, or stimuli) arranged from low to high by a specified scheme of intervals, suitable for musical purposes.   176

(2) (computing systems). To change a quantity by a factor in order to bring its range within prescribed limits.   255, 77

(3) (instrument scale). *See:* full scale.

scale class (mechanical demand registers). Denotes, with respect to single-pointer-form, dual-range single-pointer form, or cumulative-form demand registers, the relationship between the full-scale value of the register and the kilovoltampere (kVA) rating of the meter with which the register is used.   548

scale factor (1) (analog computers). The multiplication factor necessary to transform problem variables to computer variables. A problem variable is a variable appearing in the mathematical model of the problem. A computer variable is a dependent variable as represented on the computer. *See:* time.   9

(2) (gyro; accelerometer). The ratio of a change in output to a change in the input intended to be measured. Scale factor is generally evaluated as the slope of the straight line that can be fitted by the method of least squares to input-output data obtained by varying the input cyclically over the input range. *See:* input-output characteristics.   46

(3) (laser gyro) (inertial sensor). The ratio of the change in angular displacement about the input axis to a change in output (arc-seconds per pulse). The laser gyro scale factor is directly proportional to the total path length and operating wavelength, and inversely proportional to the effective enclosed ring area.   46

(4)(measuring system)(high voltage testing). The scale factor of a measuring system is the factor by which the output indication is multiplied to determine

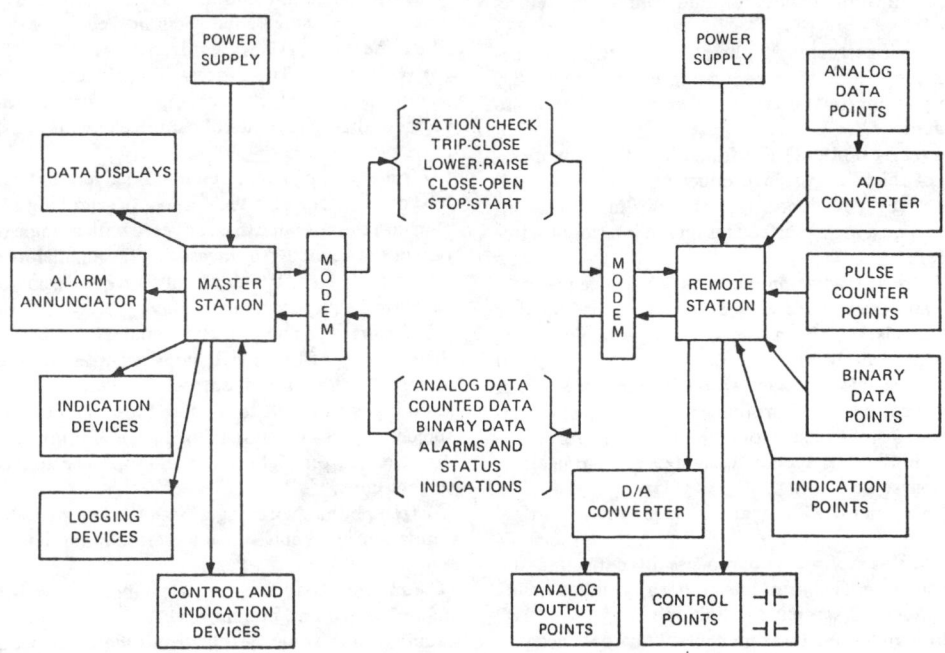

Scada System Data/Control Flow

the measured value of the input quantity or function. It is, in principle, a constant but its validity may be restricted to a specific duration or frequency range in which case the duration or range for which it is valid shall be stated.   150

scale-factor potentiometer. *See:* **parameter potentiometer.**

scale length (electric instrument). The length of the path described by the indicating means or the tip of the pointer in moving from one end of the scale to the other. *Notes:* (1) In the case of knife-edge pointers and others extending beyond the scale division marks, the pointer shall be considered as ending at the outer end of the shortest scale division marks. In multiscale instruments the longest scale shall be used to determine the scale length. (2) In the case of antiparallax instruments of the step-scale type with graduations on a raised step in the plane of and adjacent to the pointer tip, the scale length shall be determined by the end of the scale divisions adjacent to the pointer tip. *See:* **accuracy rating (instrument); instrument.**   280

scale-of-two counter. A flip-flop circuit in which successive similar pulses, applied at a common point, cause the circuit to alternate between its two conditions of permanent stability. *See:* **trigger circuit.**
                                                         328

scaler (radiation counters). An instrument incorporating one or more scaling circuits and used for registering the number of counts received. *See:* **anticoincidence (radiation counters).**   190

scaler, pulse (pulse techniques). A device that produces an output signal whenever a prescribed number of input pulses has been received. It frequently includes indicating devices for interpolation. *See:* **pulse.**
                                                         335

scale span (instrument). The algebraic difference between the values of the actuating electrical quantity corresponding to the two ends of the scale. *See:* **instrument.**   328

scaling (corrosion). (1) The formation at high temperatures of thick corrosion product layer(s) on a metal surface. (2) The deposition of water-insoluble constituents on a metal surface (as on the interior of water boilers).   205

scaling circuit (radiation counters). A device that produces an output pulse whenever a prescribed number of input pulses has been received. *See:* **anticoincidence (radiation counters).**   190

scalloping (navigation aid terms). The irregularities in the field pattern of the ground facility due to unwanted reflections from obstructions or terrain features, exhibited in flight as cyclical variations in bearing error. *Syn:* **course scalloping.**   526

scan (1) (general). To examine sequentially part by part.   255, 77

(2) (oscillography). The process of deflecting the electron beam. *See:* **graticule area; uniform luminance area; phosphor screen; oscillograph.**   185

(3) (interrogation) (station control and data acquisition). The process by which a data acquisition system interrogates remote stations or points for data.
                                                         403

(4) (interrogation) (supervisory control, data acquisition, and automatic control). The process by which a data acquisition system interrogates remote stations of points for data.   570

scan angle (antennas). The angle between the direction of the maximum of the major lobe or a directional null and a reference direction. *Notes:* (1) The term beam angle applies to the case of a pencil beam antenna. (2) The reference boresight is usually chosen as the reference direction. *Syn:* **beam angle.**   111

scan cycle (supervisory control, data acquisition, and automatic control). The time in seconds required to obtain a collection of data (for example, all data from one remote, all data from all remotes, and all data of a particular type from all remotes).   570

scanner (1) (facsimile). That part of the facsimile transmitter which systematically translates the densities of the subject copy into signal waveform. *See:* **scanning (facsimile).**   12

(2) (industrial control). (A) A multiplexing arrangement that sequentially connects one channel to a number of channels. (B) An arrangement that progressively examines a surface for information. *See:* **control system, feedback.**   219, 206

(3) (test, measurement and diagnostic equipment). A device that sequentially samples a number of data points. *See:* **flying-spot scanner; optical scanner; visual scanner.**   54

scanning (1)(navigation aid terms). A programmed motion given to the major lobe of an antenna for the purpose of searching a larger angular region that can be covered with a single direction of the beam, or for measuring angular location of a target; also, the analogous process using range gates or frequency domain filters. *See:* **supervisory control.**   526

(2) (television). The process of analyzing or synthesizing successively, according to a predetermined method, the light values of picture elements constituting a picture area.   372

(3) (radar). A programmed motion given to the major lobe of an antenna for the purpose of searching a larger angular region than can be covered with a single direction of the beam, or for measuring the angular location of a target; also, the analogous process using range gates or frequency-domain filters.   13

(4) (facsimile). The process of analyzing successively the densities of the subject copy according to the elements of a predetermined pattern. *Note:* The normal scanning is from left to right and top to bottom of the subject copy as when reading a page of print. Reverse direction is from right to left and top to bottom of the subject copy.   12

(5) (telephone switching systems). The periodic examination of circuit states under common control.
                                                         55

(6) (antenna beam). A repetitive motion given to the major lobe of an antenna.   111

scanning, high-velocity (electron tube). The scanning of a target with electrons of such velocity that the secondary-emission rate is greater than unity. *See:* **television.**   125

scanning line (television). A single continuous narrow strip that is determined by the process of scanning. *Note:* In most television systems, the scanning lines that occur during the retrace intervals are blanked. The total number of scanning lines is numerically equal to the ratio of line frequency to frame frequency.
                                                                                                                                18

scanning linearity (television). A measure of the uniformity of scanning speed during the unblanked trace interval.   18

scanning line frequency (facsimile). *See:* **stroke speed (scanning or recording line frequency).**

scanning line length (facsimile). The total length of scanning line is equal to the spot speed divided by the scanning line frequency. *Note:* This is generally greater than the length of the available line. *See:* **scanning (facsimile).**   12

scanning loss (1) (radar system employing a scanning antenna). The reduction in sensitivity, usually expressed in decibels, due to scanning across a target, compared with that obtained when the beam is directed constantly at the target. *See:* **antenna.**   111
(2) (radar). (A) The reduction in sensitivity of a scanning radar due to motion of the beam between transmission and reception of the signal. *See:* **beam-shape loss.** (B)(radar). Variations in the signal received from a complex target due to changes in aspect angle or other causes. *Note:* Because this term has been applied variously to target fluctuation and scintillation error, use of one of these more specific terms is recommended to avoid ambiguity.   13

scanning, low-velocity (electron tube). The scanning of a target with electrons of velocity less than the minimum velocity to give a secondary-emission ratio of unity. *See:* **television.**   178, 190, 125

scanning speed (television). The time rate of linear displacement of the scanning spot.   18

scanning spot (1) (television). The area with which the scanned area is being explored at any instant in the scanning process. *See:* **television.**   328
(2) (facsimile). The area on the subject copy viewed instantaneously by the pickup system of the scanner. *See:* **scanning (facsimile).**   12

scanning spot, X dimension (facsimile). The effective scanning-spot dimension measured in the direction of the scanning line on the subject copy. *Note:* The numerical value of this will depend upon the type of system used. *See:* **scanning (facsimile).**   12

scanning spot, Y dimension (facsimile). The effective scanning-spot dimension measured perpendicularly to the scanning line on the subject copy. *Note:* The numerical value of this will depend upon the type of system used. *See:* **scanning (facsimile).**   12

scanning supervisory system (station control and data acquisition). A system in which the master controls all information exchange. The normal state is usually one of repetitive communication with the remote stations.
                                                                                                                                403

scanning velocity (spectrum analyzer). Frequency span divided by sweep time.   390
scan pitch (facsimile). The number of scanning lines

per unit length measured perpendicular to the direction of scanning.   11

scan rate (data transmission). The quantity of remote functions or stations that a master station can poll in a given time period.   59

scan sector (antenna). The angular interval over which the major lobe of an antenna is scanned.   111

scan time (1) (sequential events recording systems). The time required to examine the state of all inputs.
                                                                                                                                48
(2) (acousto-optic deflector). The time for the light beam to be scanned over the angular swing of the deflector.   72

scatterband (interrogation systems)(navigation aid terms). The total bandwidth occupied by the various received signals from interrogators operating with carriers on the same nominal radio frequency; the scatter results from the individual deviations from the nominal frequency.   526

scattering (1) (laser-maser). The angular dispersal of power from a beam of radiation (or the perturbation of the field distribution of a resonance mode) either with or without a change in frequency, caused for example by inhomogeneities or nonlinearities of the medium or by irregularities in the surfaces encountered by the beam.   363
(2) (fiber optics). The change in direction of light rays or photons after striking a small particle or particles. It may also be regarded as the diffusion of a light beam caused by the inhomogeneity of the transmitting medium. *See:* **leaky modes; material scattering; mode; nonlinear scattering; Rayleigh scattering; refractive index (of a medium); unbound mode; waveguide scattering.**   433
(3) (data transmission). The production of waves of changed direction, frequency, or polarization when radio waves encounter matter. *Note:* The term is frequently used in a narrower sense, implying a disordered change in the incident energy.   59
(4) (radio wave propagation). A process in which the energy of a traveling wave is dispersed in direction due to interaction with inhomogeneities of the medium.
                                                                                                                                146

scattering coefficient. Element of the scattering matrix. *See:* **scattering matrix.**   185

scattering cross section (antennas). For a scattering object and an incident plane wave of a given frequency, polarization and direction, an area which, when multiplied by the power flux density of the incident wave, would yield sufficient power that could produce by omnidirectional radiation the same radiation intensity as that in a given direction from the scattering object. *Note:* The scattering cross section is equal to 4π times the ratio of the radiation intensity of the scattered wave in a specified direction to the power flux density of the incident plane wave. *See:* **monostatic cross section; bistatic cross section; radar cross section.**   111

scattering loss (1) (laser-maser). That portion of the loss in received power which is due to scattering.
                                                                                                                                363

(2) (acoustics). That part of the transmission loss that is due to scattering within the medium or due to roughness of the reflecting surface. 176

scattering matrix (1) (waveguide components). A square array of complex numbers consisting of the transmission and reflection coefficients of a waveguide component. As most commonly used, each of these coefficients relates the complex electric field strength (or voltage) of a reflected or transmitted wave to that of an incident wave. The subscripts of a typical coefficient S_{ij} refer to the output and input ports related by the coefficient. These coefficients, which may vary with frequency, apply at a specified set of input and output reference planes. 319, 179

(2) (circuits and systems). An nxn (square) matrix used to relate incident waves and reflected waves for an n-port network. If the incident wave quantities for the ports are denoted by the vector A and the reflected wave quantities by the vector B then the scattering matrix S is defined such that $B = SA$. where:

$$a_i = \frac{1}{\sqrt{R_e Z_i}} (V_i + Z_i I_i)$$

$$b_i = \frac{1}{\sqrt{R_e Z_i}} (V_i - Z_i I_i).$$

Z_i is the port normalization impedance with $R_e Z_i > 0$. One formula for scattering matrix is $S = [Z+R]^{-1}[Z-R]$ where Z is the open circuit impedance matrix that describes the network and R is a diagonal matrix representing the source or load resistances at each port. It should be noted that the scattering matrix is defined with respect to a specific set of port terminations. Physical interpretations can be given to the scattering coefficients for example, $|S_{ij}|^2$ is the fraction of available power that is delivered to the port termination at port i due to a source at port j. 67

scheduled frequency (electric power systems). The frequency that a power system or an interconnected system attempts to maintain. 94

scheduled frequency offset (electric power systems). The amount, usually expressed in hundredths of a hertz, by which the frequency schedule is changed from rated frequency in order to correct a previously accumulated time deviation. 94

scheduled interruption (electric power systems). An interruption caused by a scheduled outage. *See:* **outage.** 200

scheduled maintenance. *See:* **preventive maintenance.**

scheduled net interchange (electric power systems) (control area). The mutually prearranged intended net power and.or energy on the area tie lines. 94

scheduled outage (1)(emergency and standby power). A power outage that results when a component is deliberately taken out of service at a selected time, usually for purposes of construction, preventive maintenance, or repair. This type of outage is directly controllable and usually predictable. 512

(2)(reliable industrial and commercial power systems planning and design). An outage that results when a component is deliberately taken out of service at a selected time, usually for purposes of construction, maintenance, or repair. 561

scheduled outage duration (reliable industrial and commercial power systems planning and design). The period from the initiation of a scheduled outage until construction, preventative maintenance, or repair work is completed and the affected component is made available to perform its intended function. 561

scheduled outage rate (reliable industrial and commercial power systems planning and design). The mean number of scheduled outages of a component per unit exposure time. 561

schedule, electric rate (electric power supply). A statement of an electric rate and the terms and conditions governing its application. 112

schedule setter or set-point device (speed-governing system). A device for establishing or setting the desired value of a controlled variable. *See:* **speed-governing system.** 94

schematic diagram (elementary diagram). A diagram which shows, by means of graphic symbols, the electrical connections and functions of a specific circuit arrangement. The schematic diagram facilitates tracing the circuit and its functions without regard to the actual physical size, shape, or location of the component device or parts. 25

Scherbius machine (rotating machinery). A polyphase alternating-current commutator machine capable of generator or motor action, intended for connection in the secondary circuit of a wound-rotor induction motor supplied from a fixed-frequency polyphase power system, and used for speed and/or power-factor control. The magnetic circuit components are laminated and may be of the salient-pole type or of the cylindrical-rotor uniformly slotted type, either type having a series-connected armature reaction compensating winding as part of the field system. The control field winding may be separately or shunt-excited with or without an additional series-excited field winding. *See:* **asynchronous machine.** 63

Schering bridge. A 4-arm alternating-current bridge in which the unknown capacitor and a standard loss-free capacitor form two adjacent arms, while the arm adjacent to the standard capacitor consists of a resistor and a capacitor in parallel, and the fourth arm is a nonreactive resistor. *Note:* Normally used for the measurement of capacitance and dissipation factor. Usually, one terminal of the source is connected to the junction of the unknown capacitor with the standard capacitor. With this connection, if the impedances of the capacitance arms are large compared to those of the resistance arms, most of the applied voltage appears across the former, the maximum test voltage being limited by the rating of the standard capacitor. If the detector and the source of electromotive force are interchanged the resulting circuit is called a **conjugate Schering bridge.** The balance is independent of frequency. *See:* **bridge.** 328

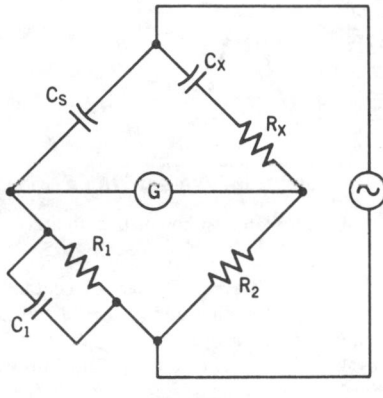

$$C_x R_2 = C_S R_1$$
$$C_x R_x = C_1 R_1$$

Schering bridge.

Schlieren method (acoustics). The technique by which light refracted by the density variations resulting from acoustic waves is used to produce a visible image of a sound field.   176

Schmitt trigger. A solid state element that produces an output when the input exceeds a specified turn-on level, and whose output continues until the input falls below a specified turn-off level.   70

Schottky barrier contact (charged-particle detectors). A metal-semiconductor contact structure whose rectification properties are heavily influenced by the difference in material work functions. These contacts frequently utilize an interfacial metal/semiconductor compound (for example, a silicide).   119

Schottky barrier contact (germanium gamma-ray detectors). A metal semiconductor contact structure whose rectification properties are heavily influenced by the difference in material work functions . These contacts may utilize an interfacial metal / semiconductor compound.   528

Schottky barrier radiation detector (germanium gamma-ray detectors)(charged-particle detectors). A semiconductor radiation detector whose blocking contact is a Schottky barrier contact.   528, 119

Schottky effect. *See:* **Schottky emission.**

Schottky emission (electron tubes). The increased thermionic emission resulting from an electric field at the surface of the cathode. *See:* **electron emission.**
                125

Schottky noise (electron tubes). The variation of the output current resulting from the random emission from the cathode. *See:* **electronic tube.**   244, 190

Schuler tuning (inertial navigation system)(navigation aid terms). The application of parameter values such that accelerations do not deflect the platform system from any vertical to which it has been set; a Schuler-tuned system, if fixed to the mean surface of a nonrotating earth, exhibits a natural period of 84.4 minutes.
                526

scintillation (1) (radar). Variations in the signal received from a complex target due to changes in aspect angle, etcetera. *Note:* Because this term has been applied variously to target fluctuation and scintillation error, use of one of the more specific terms is recommended to avoid ambiguity.   13

(2) (radio wave propagation). The phenomenon of rapid fluctuation of the amplitude and the phase of a wave passing through a medium with small-scale irregularities that cause irregular changes in the transmission path or paths with time. *See:* **fading.**   146

(3) (scintillators). The optical photons emitted as a result of the incidence of a particle or photon of ionizing radiation on a scintillator. *Note:* Optical photons unless otherwise specified are photons with energies corresponding to wavelengths between 2000 and 15 000 angstroms. *See:* **ionizing radiation; radiation.**
                117

(4) (laser-maser). This term is used to describe the rapid changes in irradiance levels in a cross section of a laser beam.   363

scintillation counter. The combination of scintillation-counter heads and associated circuitry for detection and measurement of ionizing radiation.   117

scintillation-counter cesium resolution. The scintillation-counter energy resolution for the gamma ray or conversion electron from cesium-137. *See:* **scintillation counter.**   117

scintillation-counter energy resolution. A measure of the smallest difference in energy between two particles or photons of ionizing radiation that can be discerned by the scintillation counter. Quantitatively it is the fractional standard deviation (σ / E_1) of the energy distribution curve. *Note:* The fractional full width at half maximum of the energy distribution curve (FWHM/E_1) is frequently used as a measure of the scintillation-counter energy resolution where E_1 is the mode of the distribution curve. *See:* **scintillation counter.**   117

scintillation-counter energy-resolution constant. The product of the square of the scintillation-counter energy resolution, expressed as the fractional full width at half maximum (FWHM/E_1), and the specified energy. *See:* **scintillation counter.**   335

scintillation counter head. The combination of scintillators and phototubes or photocells that produces electric pulses or other electric signals in response to ionizing radiation. *See:* **phototube; scintillation counter.**   335

scintillation-counter time discrimination. A measure of the smallest interval of time between two individually discernible events. Quantitatively it is the standard deviation of the time-interval curve. *Note:* The full width at half maximum of the time-interval curve is frequently used as a measure of the time discrimination. *See:* **scintillation counter.**   333

scintillation decay time. The time required for the rate of emission of optical photons of a scintillation to decrease from 90 percent to 10 percent of its maximum value. *Note:* Optical photons, for the purpose of this Standard, are photons with energies corresponding to wavelengths between 2000 and 15 000 angstroms. *See:* **scintillation counter.** 117

scintillation duration. The time interval from the emission of the first optical photon of a scintillation until 90 percent of the optical photons of the scintillation have been emitted. *Note:* Optical photons are photons with energies corresponding to wavelengths between 2000 and 15 000 angstroms. *See:* **scintillation counter.** 117

scintillation error (radar). Error in radar-derived target position or Doppler frequency caused by interaction of the scintillation spectrum with frequencies used in sequential measurement techniques. *Note:* Not to be confused with glint. 13

scintillation rise time. The time required for the rate of emission of optical photons of a scintillation to increase from 10 percent to 90 percent of its maximum value. *Note:* Optical photons are photons with energies corresponding to wavelengths between 2000 and 15 000 angstroms. *See:* **scintillation counter.** 117

scintillator. The body of scintillator material together with its container. *See:* **scintillation counter.** 117

scintillator conversion efficiency. The ratio of the optical photon energy emitted by a scintillator to the incident energy of a particle or photon of ionizing radiation. *Note:* The efficiency is generally a function of the type and energy of ionizing radiation. Optical photons are photons with energies corresponding to wavelengths between 2000 and 15 000 angstroms. *See:* **scintillation counter.** 335

scintillator material. A material that emits optical photons in response to ionizing radiation. *Notes:* (1) There are five major classes of scintillator materials, namely: (A) inorganic crystals such as NaI(Tl) single crystals, ZnS(Ag) screens, (B) organic crystals (such as, anthracene, *trans*-stilbene), (C) solution scintillators: (1) liquid, (2) plastic, (3) glass, (D) gaseous scintillators, (E) Cerenkov scintillators. (2) Optical photons are photons with energies corresponding to wavelengths between 2000 and 15 000 angstroms. *See:* **scintillation counter.** 117

scintillator-material total conversion efficiency. The ratio of the optical photon energy produced to the energy of a particle or photon of ionizing radiation that is totally absorbed in the scintillator material. *Note:* The efficiency is generally a function of the type and energy of the ionizing radiation. Optical photons are photons with energies corresponding to wavelengths between 2000 and 15 000 angstroms. *See:* **scintillation counter.** 117

scintillator photon distribution (in number). The statistical distribution of the number of optical photons produced in the scintillator by total absorption of monoenergetic particles. *Note:* Optical photons are photons with energies corresponding to wavelengths between 2000 and 15 000 angstroms. *See:* **scintillation counter.** 117

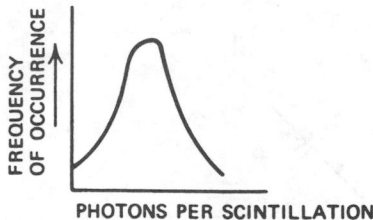

Scintillator photon distribution

scope (navigation aid terms). The face of a cathode-ray tube or a display of similar appearance. A colloquial abbreviation of oscilloscope. 526

scoring system (electroacoustics) (motion-picture production). A recording system used for recording music to be reproduced in timed relationship with a motion picture. 176

scotopic spectral luminous efficiency function, (Vλ'), (photometric standard observer for scotopic vision) (light emitting diodes). The ratio of the radiant flux at wavelength λ_m, to that at wavelength λ, when the two fluxes produce the same scotopic luminous sensations under specified photometric conditions, λ_m, being chosen so that the maximum value of this ratio is unity. Unless otherwise indicated, the values used for the spectral luminous efficiency function relate to scotopic vision by the photometric standard observer having the characteristics laid down by the International Commission on Illumination. 162

scotopic vision (illuminating engineering). Vision mediated essentially or exclusively by the rods. It is generally associated with adaptation to a luminance below about 0.034 cd/m^2, ($2.2 \times 10^{-5} cd/in^2$), (0.01 fL). 167

Scott-connected transformer, interlacing impedance voltage. The single-phase voltage applied from the midtap of the main transformer winding to both ends, connected together, that is sufficient to circulate in the supply lines a current equal to the three-phase line current. The current in each half of the winding is 50 percent of this value. *See:* **efficiency.** 207

Scott-connected transformer per-unit resistance. The measured watts expressed in per-unit on the base of the rated kilovoltampere of the teaser winding. 207

Scott or T-connected transformer (power and distribution transformer). An assembly used to transfer energy from a three-phase circuit to a two-phase circuit, or vice versa; or from a three-phase circuit to another three-phase circuit. The assembly consists of a main transformer with a tap at its midpoint connected directly between of the phase wires of a three-phase circuit, and of a teaser transformer connected between the mid-tap of the main transformer and a third phase wire of the three-phase circuit. The other windings of the transformers may be connected to provide either a two-phase or a three-phase output. Alternatively, this may be accomplished with an assembly utilizing a three-legged core with main and teaser coil assem-

blies located on the two outer legs, and with a center leg which has no coil assembly and provides a common magnetic circuit for the two outer legs. *See:* **main transformer; teaser transformer; interlacing impedance voltage of a Scott-connected transformer.**
53

SCR (1). *See:* **semiconductor controlled rectifier.**
(2) **silicon controlled rectifier.** *See:* Note under **semiconductor controlled rectifier.**
(3) **selenium controlled rectifier.** *See:* Note under **semiconductor controlled rectifier.**

scram (power operations). The rapid shutdown of a nuclear reactor. Usually, a scram is accomplished by rapid insertion of safety or control rods, or both. Emergencies or deviations from normal operation may require scramming the reactor by manual or automatic means.
516

scraper hoist. A power-driven hoist operating a scraper to move material (generally ore or coal) to a loading point.
328

screen (1) (rotating machinery). A port cover with multiple openings used to limit the entry of foreign objects.
263

(2) **(cathode-ray tubes).** The surface of the tube upon which the visible pattern is produced. *See:* **electrode (electron tube).**
125

screened conductor cable (insulated conductors). A cable in which the insulated conductor or conductors is.are enclosed in a conducting envelope or envelopes.
257

screen factor (electron-tube grid). The ratio of the actual area of the grid structure to the total area of the surface containing the grid. *See:* **electron tube.**
244, 190

screen grid. A grid placed between a control grid and an anode, and usually maintained at a fixed positive potential, for the purpose of reducing the electrostatic influence of the anode in the space between the screen grid and the cathode. *See:* **electrode (of an electron tube); grid.**
125

screen-grid modulation. Modulation produced by application of a modulating voltage between the screen grid and the cathode of any multigrid tube in which the carrier is present.
211

screening (telephone switching systems). The ability to accept or reject calls by using trunk or line class or trunk or line number information
255

screening test (reliability). A test, or combination of tests, intended to remove unsatisfactory items or those likely to exhibit early failures. *See:* **reliability.**
164

screen protected. *See:* **guarded.**
screen, viewing. *See:* **viewing area.**
screw machine (elevators). An electric driving machine, the motor of which raises and lowers a vertical screw through a nut with or without suitable gearing, and in which the upper end of the screw is connected directly to the car frame or platform. The machine may be of direct or indirect drive type.
328

sculling error (strapdown inertial system) (inertial sensor). A system error resulting from the combined input of linear vibration along one axis and an angular oscillation, at the same frequency, around a perpendicular axis. In the computer processing, an apparent rectified acceleration is produced along an axis perpendicular to these two axes.
46

seal (window) (in a waveguide). A gastight or watertight membrane or cover designed to present no obstruction to radio-frequency energy. *See:* **waveguide.**
179

sealable equipment (National Electrical Code). Equipment enclosed in a case or cabinet that is provided with a means of sealing or locking so that live parts cannot be made accessible without opening the enclosure. The equipment may or may not be operable without opening the enclosure.
256

seal, aperture. *See:* **aperture seal.**
seal, double electric conductor (nuclear power generating stations). An assembly of two single electric conductor seals in series and arranged in such a way that there is a double pressure barrier seal between the inside and the outside of the containment structure along the axis of the conductors.
226

sealed (1) (power and distribution transformer). So constructed that the enclosure will remain hermetically sealed within specified limits of temperature and pressure.
53

(2) **(rotating machinery).** Provided with special seals to minimize either the leakage of the internal coolant out of the enclosure or the leakage of medium surrounding the enclosure into the machine. *See:* **asynchronous machine.**
263

sealed-beam headlamp (illuminating engineering). An integral optical assembly designed for headlighting purposes, identified by the name 'Sealed Beam' branded on the lens.
167

sealed cell (nuclear power generating stations) (lead storage batteries). A cell in which the only passage for the escape of gases from the interior of the cell is provided by a vent of effective spray-trap design adapted to trap and return to the cell particles of liquid entrained in the escaping gases.
76

sealed cell or battery (National Electrical Code). A sealed cell or battery is one which has no provision for the addition of water or electrolyte or for external measurement of electrolyte specific gravity.
256

sealed dry-type transformer (dry-type general purpose distribution and power transformers). Self-cooled. A dry-type self-cooled transformer with a hermetically sealed tank. *Note:* The insulating gas may be air, nitrogen or other gases, such as fluorocarbons.
555

sealed dry-type transformer, self-cooled (class GA) (power and distribution transformer). A dry-type self-cooled transformer with a hermetically sealed tank. *Note:* The insulating gas may be air, nitrogen, or other gases (such as fluorocarbons) with high dielectric strength.
53

sealed end (cable) (shipping seal). The end fitted with a cap for protection against the loss of compound or the entrance of moisture.
264

sealed refrigeration compressor (hermetic type). A mechanical compressor consisting of a compressor and a motor, both of which are enclosed in the same sealed housing, with no external shaft or shaft seals, the motor operating in the refrigerant atmosphere. *See:* **appliances.** 256

sealed-tank system (power and distribution transformer). A method of oil preservation in which the interior of the tank is sealed from the atmosphere and in which the gas plus the oil volume remains constant over the temperature range. 53

sealed transformer (power and distribution transformer). A dry-type transformer with a hermetically sealed tank. 53

sealed tube. An electron tube that is hermetically sealed. *Note:* This term is used chiefly for pool-cathode tubes. 190

sealing gap (industrial control). The distance between the armature and the center of the core of a magnetic circuit-closing device when the contacts first touch each other. *See:* **electric controller; initial contact pressure.** 302

sealing voltage (or current) (contactors). The voltage (or current) necessary to complete the movement of the armature of a magnetic circuit-closing device from the position at which the contacts first touch each other. *See:* **contactor; control switch.** 21

seal-in relay (power switchgear). An auxiliary relay that remains picked up through one of its own contacts which bypasses the initiating circuit until deenergized by some other device. 103

seal, pressure barrier (nuclear power generating stations). A seal that consists of an aperture seal and an electric conductor seal. 226

seal, single electric conductor (nuclear power generating stations). A mechanical assembly providing a single pressure barrier between the electric conductors and the electric penetration. 226

search (1) (information processing). To examine a set of items for those that have a desired property. *See:* **binary search; dichotomizing search.** 255, 77

(2) (test, measurement and diagnostic equipment). The scanning of information contained on a storage medium by comparing the information of each field with a predetermined standard until an identity is obtained. 254

searchlight (illuminating engineering). A projector designed to produce an approximately parallel beam of light. *Note:* The optical system of a searchlight has an aperture of greater than 20 cm (8 inches). 167

searchlighting (radar). The process of projecting a radar beam continuously at a particular object or in a particular direction as contrasted to scanning. 13

search radar (navigation aid terms). A radar used primarily for the detection of targets in a particular volume of interest. 526

sea return (navigation aid terms). The radar response from the sea surface. 526

seasonal derated hours (SDH)(power system mea-surement)(electric generating unit reliability, availability, and productivity). The available hours during which a seasonal derating was in effect. 432, 567

seasonal derating (1)(electric generating unit reliability, availability, and productivity). The difference between maximum capacity and dependable capacity. 567

(2) (SD) (power system measurement). The difference between gross maximum capacity and gross dependable capacity:

$$SD = GMC - GDC$$

Note: The concept of derating applies only when the unit is in the available state. *See:* **ANSI/IEEE Std 762-1980, Appendix D.** 432

seasonal diversity (power operations). Load diversity between two (or more) electric systems which occurs when their peak loads are in different seasons of the year. 516

seasonal unavailable generation (SUG)(electric generating unit reliability, availability, and productivity). The difference between the energy that would have been generated if operating continuously at maximum capacity and the energy that would have been generated if operating continuously at dependable capacity, calculated only during the time the unit was in the available state.

$$
\begin{aligned}
SUG &= \text{equivalent seasonal derated hours} \\
&\quad \cdot \text{maximum capacity} \\
&= ESDH \cdot MC
\end{aligned}
$$

567

season cracking (corrosion). Cracking resulting from the combined effect of corrosion and internal stress. A term usually applied to stress-corrosion cracking of brass. 205

SEC. *See:* **secondary-electron conduction.**

second (metric practice). The duration of 9 192 631 770 periods of the radiation corresponding to the transition between the two hyperfine levels of the ground state of the cesium-133 atom. (adopted by 13 General Conference on Weights and Measures 1967). *Notes:* This definition supersedes the ephemeris second as the unit of time. 21

secondary (used as an adjective) (power switchgear). (1) Operates after the primary device; for example: secondary arcing contacts. (2) Second in preference. (3) Referring to auxiliary or control circuits as contrasted with the main circuit; for example, secondary disconnecting devices, secondary and control wiring. (4) Referring to the energy output side of transformers or the conditions (voltages) usually encountered at this location; for example, secondary fuse, secondary unit substation. 103

secondary address (FASTBUS acquisition and control). An address for use within a device. It is provided by a secondary address cycle which loads the NTA (next transfer address) register of the device following a primary address cycle or a data cycle. 480

secondary address cycle (FASTBUS acquisition and

control). A data cycle in which a master uses the address/data (AD) lines to load a secondary address into the NTA (next transfer address) register of a device.   480

secondary alarm station (SAS)(nuclear security systems). A continuously manned alarm station that is capable of providing all necessary backup security system functions in the event the central alarm station (CAS) is disabled or compromised.   464

secondary and control wiring (small wiring) (power switchgear). Wire used with switchgear assemblies for control circuits and for connections between instrument transformer secondaries, instruments, meters, relays, or other equipment.   103

secondary arcing contacts (of a switching device). The contacts on which the arc of the arc-shunting-resistor current is drawn and interrupted.   103

secondary calibration (monitoring radioactivity in effluents)(nuclear power generating stations). The determination of the response of a system with an applicable source whose effect on the system was established at the time of a primary calibration.
    559,231

secondary current rating (transformer). The secondary current existing when the transformer is delivering rated kilovoltamperes at rated secondary voltage. *See:* **transformer.**   203

secondary disconnecting devices (of a switchgear assembly). Self-coupling separable contacts provided to connect and disconnect the auxiliary and control circuits between the removable element and the housing.
    103

secondary distribution feeder. A feeder operating at secondary voltage supplying a distribution circuit.
    264

secondary distribution mains. The conductors connected to the secondaries of distribution transformers from which consumers' services are supplied. *See:* **center of distribution.**   264

secondary distribution network. A network consisting of secondary distribution mains. *See:* **center of distribution.**   264

secondary distribution system. A low-voltage alternating-current system that connects the secondaries of distribution transformers to the consumers' services. *See:* **alternating-current distribution; center of distribution.**   264

secondary distribution trunk line. A line acting as a main source of supply to a secondary distribution system. *See:* **center of distribution.**   264

secondary electron (thermionics). An electron detached from a surface during secondary emission by an incident electron. *See:* **electron emission.**
    244, 190

secondary-electron conduction (SEC). The transport of charge under the influence of an externally applied field in low-density structured materials by free secondary electrons traveling in the interparticle spaces (as opposed to solid-state conduction). *See:* **camera tube.**   190

secondary-electron conduction (SEC) camera tube. A camera tube in which an electron image is generated by a photocathode and focused on a target composed of (1) a backplate and (2) a secondary-electron-conduction layer that provides charge amplification and storage. *See:* **camera tube.**   190

secondary emission. Electron emission from solids or liquids due directly to bombardment of their surfaces by electrons or ions. *See:* **electron emission.**   125

secondary-emission characteristic (surface) (thermionics). The relation, generally shown by a graph, between the secondary-emission rate of a surface and the voltage between the source of the primary emission and the surface. *See:* **electron emission.**   244, 190

secondary-emission ratio (electrons). The average number of electrons emitted from a surface per incident primary electron. *Note:* The result of a sufficiently large number of events should be averaged to ensure that statistical fluctuations are negligible.   125

secondary failure (reliability). *See:* **failure, secondary.**

secondary fault. An insulation breakdown occurring as a result of a primary fault. *See:* **center of distribution.**
    264

secondary fuse (power switchgear). A fuse used on the secondary-side circuits of transformers. *Note:* In high-voltage fuse parlance such a fuse is restricted for use on a low-voltage secondary distribution system that connects the secondaries of distribution transformers to consumers' services.   103

secondary grid emission. Electron emission from a grid resulting directly from bombardment of its surface by electrons or other charged particles.   125

secondary neutral grid. A network of neutral conductors, usually grounded, formed by connecting together within a given area all the neutral conductors of individual transformer secondaries of the supply system. *See:* **center of distribution.**   264

secondary power. The excess above firm power to be furnished when, as, and if available. *See:* **generating station.**   264

secondary radar (navigation aid terms). (1) A radar technique or mode of operation in which the return signals are obtained from a beacon, transponder, or repeater carried by the target, as contrasted with primary radar in which the return signals are obtained by reflection from the target. (2) A radar, or that portion of a radar, that operates on this principle. *See:* **primary radar.**   526

secondary radiator. That portion of an antenna having the largest radiating aperture, consisting of a reflecting surface or a lens, as distinguished from its feed.
    111

secondary section of the core (ferroresonant voltage regulators). The section of the ferroresonant transformer on which the output and resonating windings are wound. In steady-state operation, this section of the core is normally driven into magnetic saturation.
    456

secondary-selective type (low voltage-selective type). A unit substation which has two stepdown transformers each connected to an incoming high-voltage circuit. The outgoing side of each transformer is connected to a separate bus through a suitable switching and protective device. The two sections of bus are connected by a normally open switching and protective device. Each bus has one or more outgoing radial (stub-end) feeders. 53

secondary service area (radio broadcast station). The area within which satisfactory reception can be obtained only under favorable conditions. *See:* **radio transmitter.** 328

secondary short-circuit current rating of a high-reactance transformer (power and distribution transformer). One that designates the current in the secondary winding when the primary winding is connected to a circuit of rated primary voltage and frequency and when the secondary terminals are short-circuited. 53

secondary, single-phase induction motor. The rotor or stator member that does not have windings that are connected to the supply line. *See:* **asynchronous machine; induction motor.** 263

secondary standard (luminous standards) (illuminating engineering). A stable light source calibrated directly or indirectly by comparison with a primary standard. This order of standard also is designated as a reference standard. *Note:* National secondary (reference) standards are maintained at national physical laboratories; laboratory secondary (reference) standards are maintained at other photometric laboratories. 167

secondary unit substation (1). *See:* **unit substation. Note.**
(2) (power and distribution transformer). A substation in which the low-voltage section is rated 1000 V (volts) and below. 53

secondary voltage (capacitance potential device). The root-mean-square voltage obtained from the main secondary winding, and when provided, from the auxiliary secondary winding. *See:* **rated secondary voltage; outdoor coupling capacitor.** 351

secondary voltage rating (power and distribution transformer). The load circuit voltage for which the secondary winding is designed. 53

secondary winding (1) (power and distribution transformer). The winding on the energy output side. 53
(2) (rotating machinery). Any winding that is not a primary winding. *See:* **asynchronous machine; voltage regulator.** 263
(3) (voltage regulator). The series winding. *See:* **voltage regulator.** 257
(4)(instrument transformer) (power and distribution transformer). The winding that is intended to be connected to the measuring or control devices. 394, 53

second-channel attenuation. *See:* **selectance.**

second-channel interference. Interference in which the extraneous power originates from a signal of assigned (authorized) type in a channel two channels removed from the desired channel. *See:* **interference; radio receiver.** 339

second contingency incremental transfer capability

(power operations). The amount of power, incremental above normal base power transfers, that can be transferred over the transmission network in a reliable manner, based on the following conditions: (1) With all transmission facilities in service, all facility loadings are within normal ratings and all voltages are within normal limits. (2) The bulk power system is capable of absorbing the dynamic power swings and remaining stable following a disturbance resulting in the sequential and overlapping outage of two facilities, either being a generating unit, transmission circuit, or transformer with system adjustments made between the two outages as required. (3) After the dynamic power swings following a disturbance resulting in the loss of the second facility, either a generating unit , transmission circuit, or transformer, but before further operator-directed system adjustments are made, all transmission facility loadings are within emergency ratings and all voltages are within emergency limits. *Note:* The term second contingency is used to specifically exclude simultaneous outages. Use of the term double contingency has been avoided, since it is often used to include both simultaneous and sequential outages. 516

second-order nonlinearity coefficient (accelerometer). The proportionality constant that relates a variation of the output to the square of the input applied parallel to an input reference axis. 46

second-time-around echo (radar). An echo received after a time delay exceeding one pulse repetition interval but less than two pulse repetition intervals. Third-time-around (etcetera) echoes are defined in a corresponding manner. The generic term multiple-time-around is sometimes used. 13

second Townsend discharge (gas). A semi-self-maintained discharge in which the additional ionization is due to the secondary electrons emitted by the cathode under the action of the bombardment by the positive ions present in the gas. *See:* **discharge (gas).** 190

second voltage range (railway signal). *See:* **voltage range.**

secretary/librarian (software). The software librarian on a chief programmer team. *See:* **chief programmer team; software librarian.** 434

section (1) (rectifier unit). A part of a rectifier unit with its auxiliaries that may be operated independently. *See:* **rectification.** 208
(2) (thyristor converter). Those parts of a thyristor converter unit containing the power thyristors (and when also used, the power diodes) together with their auxiliaries (including individual transformers or cell windings of double converters and circulating current reactors, if any), in which the main direct current when viewed from the converter unit dc terminals always flows in the same direction. A thyristor converter section is supposed to be operated independently. *Note:* A converter equipment may have either only one section or one forward and one reverse section. 121

sectional center, (telephone switching systems). A toll office to which may be connected a number of primary

centers, toll centers, or toll points. Sectional centers are classified as Class 2 offices. *See:* **office class.**
                                                                    255

sectionalized linear antenna. A linear antenna in which reactances are inserted at one or more points along the length of the antenna.   111

sectionalizer. *See:* **automatic line sectionalizer.**

section locking. Locking effective while a train occupies a given section of a route and adapted to prevent manipulation of levers that would endanger the train while it is within that section. *See:* **interlocking (interlocking plant).**   328

section, sag. *See:* **sag section.**

sectoral horn antenna. A horn antenna with two opposite sides of the horn parallel and the two remaining sides diverging.   111

sector cable. A multiple-conductor cable in which the cross section of each conductor is substantially a sector of a circle, an ellipse, or a figure intermediate between them. *Note:* Sector cables are used in order to obtain decreased overall diameter and thus permit the use of larger conductors in a cable of given diameter.   264

sector display (1) (radar). A limited display in which only a sector of the total service area of the radar system is shown; usually the sector to be displayed is selectable.   13

(2) (continuously rotating radar-antenna system). A range-amplitude display used with a radar set, the antenna system of which is continuously rotating. The screen, which is of the long-persistence type, is excited only while the beam of the antenna is within a narrow sector centered on the object. *See:* **radar.**   187

sector impedance relay (power switchgear). A form of distance relay that by application and design has its operating characteristic limited to a sector of its operating circle on the R-X diagram. *See:* **distance relay Figure (b).**   103

sector scanning. A modification of circular scanning in which the direction of the antenna beam generates a portion of a cone or a plane.   111

security (1) (of a relay or relay system) (power switchgear). That facet of reliability that relates to the degree of certainty that a relay or relay system will not operate incorrectly.   103

(2) (software). The protection of computer hardware and software from accidental or malicious access, use, modification, destruction, or disclosure. Security also pertains to personnel, data, communications, and the physical protection of computer installations. *See:* **data; hardware; modification; protection; software.**
                                                                    434

security kernel (software). A small, self-contained collection of key security-related statements that works as a privileged part of an operating system. All criteria specified by the kernel must be met for a program or data to be accessed. *See:* **data; kernel; operating system; program; security.**   434

security system (nuclear security systems). As used in the context of IEEE Std 692-1986, security system is intended to mean the aggregate assemblage of hardware that includes all components, equipment, barriers, etcetera, necessary for the physical protection of nuclear power generating stations against the design basis threat of radiological sabotage.   464

sedimentation potential (electrobiology). The electrokinetic potential gradient resulting from unity velocity of a colloidal or suspended material forced to move by gravitational or centrifugal forces through a liquid electrolyte. *See:* **electrobiology.**   192

sediment separator (rotating machinery). Any device, used to collect foreign material in the lubricating oil. *See:* **oil cup (rotating machinery).**   263

Seebeck coefficient (of a couple) (for homogeneous conductors). The limit of the quotient of (1) the Seebeck electromotive force by (2) the temperature difference between the junctions as the temperature difference approaches zero: by convention, the Seebeck coefficient of a couple is positive if the first-named conductor has a positive potential with respect to the second conductor at the cold junction. *Note:* The Seebeck coefficient of a couple is the algebraic difference of either the relative or absolute Seebeck coefficients of the two conductors. *See:* **thermoelectric device.**
                                                                    191

Seebeck coefficient, absolute. The integral, from absolute zero to the given temperature, of the quotient of (1) the Thomson coefficient of the material by (2) the absolute temperature. *See:* **thermoelectric device.**
                                                                    191

Seebeck coefficient, relative. The Seebeck coefficient of a couple composed of the given material as the first-named conductor and a specified standard conductor. *Note:* Common standards are platinum, lead, and copper. *See:* **thermoelectric device.**   191

Seebeck effect. The generation of an electromotive force by a temperature difference between the junctions in a circuit composed of two homogeneous electric conductors of dissimilar composition: or, in a nonhomogeneous conductor, the electromotive force produced by a temperature gradient in a nonhomogeneous region. *See:* **thermoelectric effect; thermoelectric device.**   191

Seebeck electromotive force. The electromotive force resulting from the Seebeck effect. *See:* **thermoelectric device.**   191

seeding. *See:* **fault seeding.**   434

seek (microprocessor operating systems). An activity that positions a pointer at a specific location within a data file.   478

segment (1)(FASTBUS acquisition and control). A specific transmission medium which supports the FASTBUS protocol and to which FASTBUS devices may attach. A segment is capable of supporting autonomous operation and communicating with other segments via segment interconnects.   480

(2) (software). A self-contained portion of a computer program that may be executed without the entire computer program necessarily being maintained in internal storage at any one time. (A) To divide a computer program into segments. (B) The sequence of computer program statements between two consecutive branch points. *See:* **component; computer program; module; path analysis; subprogram.** 434

segmental conductor. A stranded conductor consisting of three or more stranded conducting elements, each element having approximately the shape of the sector of a circle, assembled to give a substantially circular cross section. The sectors are usually lightly insulated from each other and, in service, are connected in parallel. *Note:* This type of conductor is known as type-*M* conductor in Canada. *See:* **conductor.** 264

segmental-rim rotor (rotating machinery). A rotor in which the rim is composed of interleaved segmental plates bolted together. *See:* **rotor (rotating machinery).** 263

segment extender (SE)(FASTBUS acquisition and control). A device for connecting two segments to form an extended segment or part of an extended segment. 480

segment interconnect (SI)(FASTBUS acquisition and control). A device which implements an intersegment connection such that the FBP (FASTBUS protocol) on the two segments is synchronized. When an operation is passing through an SI, the SI acts as a slave on the near-side and as a master on the far-side. 480

segment shoe (bearing shoe) (rotating machinery). A pad that is part of the bearing surface of a pad-type bearing. *See:* **bearing.** 263

segregated-phase bus (1)(generating station grounding). A metal-enclosed bus in which phase conductors are in a common metal enclosure but are segregated by metal barriers between phases. 569
(2)(power switchgear). One in which all phase conductors are in a common metal enclosure, but are segregated by metal barriers between phases. 103

seismic category I (nuclear power generating station). The classification assigned to those structures, systems, and components of a nuclear power plant, including foundations and supports, which must be designed to withstand the effects of the Safe Shutdown Earthquake (SSE) and remain functional. 439

seizure signal (telephone switching systems). A signal transmitted from the sending end of a trunk to the far end to indicate that its sending end has been selected. 255

selectance (amplitude-modulation broadcast receivers). The ratio of the ordinates of a selectivity graph, described in Section 4.05.03 of ANSI/IEEE Std 186-1948, between the resonant frequency and another frequency differing from the resonant frequency by a specified multiple of the width of one channel. (The width of one broadcast channel is 10 kilocycles.) It is expressed in decibels or voltage ratios. The ratio at a frequency n channels above the resonant frequency is denoted by S_{+n} and at a frequency n channels below the resonant frequency is denoted by S_{-n}. The geometric mean of these ratios is denoted by S_n. Expressed in decibels, the value of S_n is the average value of S_{+n} and S_{-n}. The terms 'adjacent-channel attenuation' (ACA) and 'second-channel attenuation'

(2ACA) are used to refer to S_1 and S_2, respectively. 524

select before operate, supervisory control. *See:* **supervisory control system, select before operate.**

selecting (telephone switching systems). Choosing a particular group of one or more servers in the establishment of a call connection. 255

selection (computing systems). *See:* **amplitude selection; coincident-current selection.**

selection check (electronic computation). A check (usually an automatic check) to verify that the correct register, or other device, is selected in the interpretation of an instruction. 235, 255, 77

selection ratio. The least ratio or a magnetomotive force used to select a cell to the maximum magnetomotive force used that is not intended to select a cell. *See:* **coincident-current selection.** 331

selective collective automatic operation (elevators). Automatic operation by means of one button in the car for each landing level served and by UP and DOWN buttons at the landings, wherein all stops registered by the momentary actuation of the car buttons are made as defined under **nonselective collective automatic operation,** but wherein the stops registered by the momentary actuation of the landing buttons are made in the order in which the landings are reached in each direction of travel after the buttons have been actuated. With this type of operation, all UP landing calls are answered when the car is traveling in the up direction and all DOWN landing calls are answered when the car is traveling in the down direction, except in the case of the uppermost or lowermost calls, which are answered as soon as they are reached irrespective of the direction of travel of the car. *See:* **control (elevators).** 328

selective dump (computing systems). A dump of a selected area of storage. 255, 77

selective opening (tripping) (power switchgear). The application of switching devices in series such that (of the devices carrying fault current) only the device nearest the fault will open and the devices closer to the source will remain closed and carry the remaining load. 103

selective overcurrent trip. *See:* **selective release (trip); overcurrent release (trip).**

selective overcurrent tripping. *See:* **selective opening (tripping); overcurrent release (trip).**

selective pole switching (power switchgear). The practice of tripping and reclosing one or more poles of a multipole circuit breaker without changing the state of the remaining pole(s) with tripping being initiated by protective relays which respond selectively to the faulted phases. *Note:* Circuit breakers applied for selective pole switching must inherently be capable of individual pole opening. 103

selective release (selective trip). A delayed release with selective settings that will automatically reset if the actuating quantity falls and remains below the release setting for a specified time. 103

selective release (trip) (power switchgear). A delayed release with selective settings that will automatically reset if the actuating quantity falls and remains below the release setting for a specified time.   103

selective ringing (telephone switching systems). Ringing in which only the ringer at the desired main station on a party line responds.   255

selective signaling equipment (mobile communication). Arrangements for signaling, selective from a base station, of any one of a plurality of mobile stations associated with the base station for communication purposes. *See:* **mobile communication system.**   181

selectivity (1) (circuits and systems). The characteristic of a filter that determines the extent to which the filter is capable of altering the frequency spectrum of a signal. A highly selective filter has an abrupt transition between a pass-band region and a stop-band region.   267

(2) (of a protective system) (power switchgear). A general term describing the interrelated performance of relays and breakers, and other protective devices; complete selectivity being obtained when a minimum amount of equipment is removed from service for isolation of a fault or other abnormality.   103

selector, amplitude (pulse techniques). *See:* **selector, pulse-height.**

selector pulse (navigation aid terms). A pulse which is used to identify, for selection, one event in a series of events.   526

selector, pulse-height (pulse techniques). *See:* **pulse-height selector.**

selector switch (1) (general) (power switchgear). A switch arranged to permit connecting a conductor to any one of a number of other conductors.   103,27

(2) (industrial control). A manually operated multiposition switch for selecting alternative control circuits.   206

self-adapting. Pertaining to the ability of a system to change its performance characteristics in response to its environment.   255, 77

self-aligning bearing (rotating machinery). A sleeve bearing designed so that it can move in the end shield to align itself with the journal of the shaft. *See:* **bearing.**   263

self-ballasted lamp (illuminating engineering). Any arc discharge lamp of which the current-limiting device is an integral part.   167

self-capacitance (conductor) (grounded capacitance) (total capacitance). In a multiple-conductor system, the capacitance between this conductor and the other $(n - 1)$ conductors connected together. *Note:* The self-capacitance of a conductor equals the sum of its $(n - 1)$ direct capacitances to the other $(n - 1)$ conductors.   210

self-checking code (electronic computation). A code that uses expressions such that one (or more) error(s) in a code expression produces a forbidden combination. Also called an error-detecting code. *See:* **check, forbidden combination; error-detecting code; parity.**   235

self-closing door or gate (elevators). A manually opened hoistway door and/or a car door or gate that closes when released. *See:* **hoistway (elevator or dumbwaiter).**   328

self-commutated converter (forced-commutated converter)(self-commutated converters). A converter in which commutation is accomplished by components within the converter. *Note:* In converters using switching devices that can interrupt or turn off current, such as transistors or gate turn-off thyristors, rejection of the current produces a voltage across the device to commutate the current to another device. In converters using circuit-commutated thyristors, the commutating voltages required to transfer current from one device to another are usually supplied by capacitors.   584

self-commutated inverters. An inverter in which the commutation elements are included within the power inverter.   208

self-contained instrument. An instrument that has all the necessary equipment built into the case or made a corporate part thereof. *See:* **instrument.**   280

self-contained navigation aid (navigation aid terms). An aid which consists only of facilities carried by the vehicle.   526

self-contained pressure cable. A pressure cable in which the container for the pressure medium is an impervious flexible metal sheath, reinforced if necessary, that is factory assembled with the cable core, *See:* **gas-filled cable; oil-filled cable; pressure cable.**   264

self-coupling separable contacts (switchgear assembly disconnecting device). Contacts, mounted on the stationary and removable elements of a switchgear assembly, that align and engage or disengage automatically when the two elements are brought into engagement or disengagement.   202

self-damping (conductor self-damping measurements). Of a conductor subjected to a load T is defined by the power dissipated per unit length of a conductor vibrating in a natural mode, with a loop length l and an antinode displacement amplitude y and a frequency f. The power per unit conductor length P is expressed as a function in the nth mode.   385

$$P = f_n\,(T, l, f, y)$$

self-excited. A qualifying term applied to a machine to denote that the excitation is supplied by the machine itself.   263

self-field (Hall generator). The magnetic field caused by the flow of control current through the loop formed by the control current leads and the relevant conductive path through the Hall plate.   107

self-impedance (array element). The input impedance of a radiating element of an array antenna with all other elements in the array open-circuited. *Note:* In general, the self-impedance of a radiating element in an array is not equal to its isolated impedance.   111

self-inductance. The property of an electric circuit

whereby an electromotive force is induced in that circuit by a change of current in the circuit. *Notes:* (1) The coefficient of self-inductance L of a winding is given by the following expression:

$$L = \frac{\partial \lambda}{\partial i}$$

where λ is the total flux-linkage of the winding and i is the current in the winding. (2) The voltage e induced in the winding is given by the following equation:

$$e = -\left[L\frac{di}{dt} + i\frac{dL}{dt} \right]$$

If L is constant

$$e = -L\frac{di}{dt}$$

(3) The definition of self-inductance L is restricted to relatively slow changes in i that is, to low frequencies, but by analogy with the definitions, equivalent inductances may often be evolved in high-frequency applications such as resonators, waveguide equivalent circuits, etcetera. Such inductances, when used, must be specified. The definition of self-inductance L is also restricted to cases in which the branches are small in physical size compared with a wavelength, whatever the frequency. Thus in the case of a uniform 2-wire transmission line it may be necessary even at low frequencies to consider the parameters as distributed rather than to have one inductance for the entire line.
           210, 177

self-information. *See:* **information content.**

self-lubricating bearing (rotating machinery). A bearing lined with a material containing its own lubricant such that little or no additional lubricating fluid need be added subsequently to ensure satisfactory lubrication of the bearing. *See:* **bearing.**   263

self-maintained discharge (gas). A discharge characterized by the fact that it maintains itself after the external ionizing agent is removed. *See:* **discharge (gas).**   244, 190

self-organizing. Pertaining to the ability of a system to arrange its internal structure.   255, 77

self-phasing array antenna system. A receiving antenna system which introduces a phase distribution among the array elements so as to maximize the received signal regardless of the direction of incidence. *See:* **retrodirective antenna.**   111

self-propelled electric car. An electric car requiring no external source of electric power for its operation. *Note:* Diesel-electric, gas-electric, and storage-battery-electric cars are examples of self-propelled cars. The prefix self-propelled is also applied to buses. *See:* **electric motor car.**   328

self-propelled electric locomotive. An electric locomotive requiring no external source of electric power for its operation. *Note:* Storage-battery, diesel-electric, gas-electric and turbine-electric locomotives are examples of self-propelled electric locomotives. *See:* **electric locomotive.**   328

self-pulse modulation. Modulation effected by means of an internally generated pulse. *See:* **blocking oscillator; oscillatory circuit.**   111

self-quenched counter tube. A radiation counter tube in which reignition of the discharge is inhibited by internal processes.   328

self-rectifying X-ray tube. An X-ray tube operating on alternating anode potential.   190

self regulation. *See:* **inherent regulation.**

self-reset manual release (control) (industrial control). A manual release that is operative only while it is held manually in the release position. *See:* **electric controller.**   206

self-reset relay (automatically reset relay). A relay that is so constructed that it returns to its reset position following an operation after the input quantity is removed.   103, 127

self-restoring fire detector (fire protection devices). A restorable fire detector whose sensing element is designed to be returned to normal automatically.   271

self-restoring insulation (1) (high voltage testing). Self-restoring insulation is insulation which completely recovers its insulating properties after a disruptive discharge caused by the application of a test voltage. In insulation of this kind, disruptive discharges, frequently, but not necessarily, occur in the external part of the insulation.   150

(2) (power and distribution transformer). Insulation which completely recovers its insulating propproperties after a disruptive discharge caused by the application of a test voltage; insulation of this kind is generally, but not necessarily, external insulation.   53

self-rest relay (automatically reset relay). A relay that is so constructed that it returns to its reset position following an operation after the input quantity is removed.   103

self-saturation (magnetic amplifier). The saturation obtained by rectifying the output current of a saturable reactor.   328

self-supporting aerial cable. A cable consisting of one or more insulated conductors factory assembled with a messenger that supports the assemblage, and that may or may not form a part of the electric circuit. *See:* **conductor.**   264

self-surge impedance. *See:* **surge impedance.**

self-test (test, measurement and diagnostic equipment). A test or series of tests, performed by a device upon itself, which shows whether or not it is operating within designed limits. This includes test programs on computers and automatic test equipment which check out their performance status and readiness.   254

self-test capability (test, measurement and diagnostic equipment). The ability of a device to check its own circuitry and operation. The degree of self-test is de-

pendent on the ability to fault detect and isolate.
254

self-ventilated (rotating machinery). Applied to a machine which has its ventilating air circulated by means integral with the machine. *See:* **asynchronous machine.**
23, 63

self-ventilated machine (electric installations on shipboard). A machine which has its ventilating air circulated by means integral with the machine.
3

semantics (1). The relationships between symbols and their meanings.
255, 77

(2) (ATLAS). The connotative meaning of words within an ATLAS statement.
400

(3) (software). (A) The relationships of characters or groups of characters to their meanings, independent of the manner of their interpretation and use. (B) The relationships between symbols and their meanings. (C) The discipline of expressing the meanings of computer language constructs in metalanguages. *See:* **computer; metalanguage; syntax.**
434

semaphore (1)(microprocessor operating systems). A system variable used to synchronize concurrent processes by indicating whether an action has been completed or an event has occurred.
478

(2)(software). A shared variable used to synchronize concurrent processes by indicating whether an action has been completed or an event has occurred. *See:* **concurrent processes.**
434

semiactive guidance (radar). A bistatic-radar homing system in which a receiver in the guided vehicle derives guidance information from electromagnetic signals scattered from the target, which is illuminated by a transmitter at a third location. *See:* **illuminator.**
13

semiactive homing guidance (navigation aid terms). Guidance in which a craft is directed toward a destination by means of information received from the destination in response to transmissions from a source other than the craft.
526

semianalytic inertial navigation equipment. The same as geometric inertial navigation equipment except that the horizontal measuring axes are not maintained in alignment with a geographic direction. *Note:* The azimuthal orientations are automatically computed. *See:* **navigation.**
187, 13

semiautomatic. Combining manual and automatic features so that a manual operation is required to supply to the automatic feature the actuating influence that causes the automatic feature to function.
328

semiautomatic controller. An electric controller in which the influence directing the performance of some of its basic functions is automatic. *See:* **electric controller.**
206

semiautomatic flight inspection (SAFI)(navigation aid terms). A specialized and largely automatic system for evaluating the quality of information in signals from ground-based navigational aids; data from navigational aids along and adjacent to any selected air route are simultaneously received by a specially equipped SAFI aircraft as it proceeds under automatic control along the route, evaluated at once for gross errors, and re-

corded for subsequent processing and detailed analysis at a computer-equipped central ground facility. *Note:* Flight inspection means the evaluation of performance of navigational aids by means of in-flight measurements.
526

semiautomatic gate (elevators). A gate that is opened manually and that closes automatically as the car leaves the landing. *See:* **hoistway (elevator or dumbwaiter).**
328

semiautomatic holdup-alarm system. An alarm system in which the signal transmission is initiated by the indirect and secret action of the person attacked or of an observer of the attack. *See:* **protective signalling.**
328

semiautomatic plating. Mechanical plating in which the cathodes are conveyed automatically through only one plating tank. *See:* **electroplating.**
328

semiautomatic signal. A signal that automatically assumes a stop position in accordance with traffic conditions, and that can be cleared only by cooperation between automatic and manual controls.
328

semiautomatic station (station control and data acquisition). A station that requires both automatic and manual modes to maintain the required character of service.
403

semiautomatic telephone systems. A telephone system in which operators receive orders orally from the calling parties and establish connections by means of automatic apparatus.
328

semiautomatic test equipment (test, measurement and diagnostic equipment). Any automatic testing device which requires human participation in the decision-making, control, or evaluative functions.
254

semiconducting jacket. A jacket of such resistance that its outer surface can be maintained at substantially ground potential by contact at frequent intervals with a grounded metallic conductor, or when buried directly in the earth.
264, 57

semiconducting material. A conducting medium in which the conduction is by electrons, and holes, and whose temperature coefficient of resistivity is negative over some temperature range below the melting point. *See:* **semiconductor; semiconductor device.**
210

semiconducting paint (rotating machinery). A paint in which the pigment or portion of pigment is a conductor of electricity and the composition is such that when converted into a solid film, the electrical conductivity of the film is in the range between metallic substances and electrical insulators.
263

semiconducting tape (power distribution, underground cables). A tape of such resistance that when applied between two elements of a cable the adjacent surfaces of the two elements will maintain substantially the same potential. Such tapes are commonly used for conductor shielding and in conjunction with metallic shielding over the insulation.
257

semiconductive ignition cable (electromagnetic compatibility). High-tension ignition cable, the core of which is made of semiconductive material. *Note:* Semiconductive is understood here as referring to conductivity and no other physical properties. *See:* **electromagnetic compatibility.**
220, 199

semiconductor. An electronic conductor, with resistivity in the range between metals and insulators, in which the electric-charge-carrier concentration increases with increasing temperature over some temperature range. *Note:* Certain semiconductors possess two types of carriers, namely, negative electrons and positive holes.   245

semiconductor, compensated. A semiconductor in which one type of impurity or imperfection (for example, donor) paritally cancels the electric effects of the other type of impurity or imperfection (for example, acceptor). *See:* **semiconductor.**   245, 23

semiconductor controlled rectifier (SCR). An alternative name used for the reverse-blocking triode-thyristor. *Note:* The name of the actual semiconductor material (selenium, silicon, etcetera) may be substituted in place of the word **semiconductor** in the name of the components. *See:* **thyristor.**   245

semiconductor converters, classification. The following designations are intended to describe the functional characteristics of converters, but not necessarily the circuits or components used. *Note:* Forms A through D refer only to the converters. Rotational direction of motors may be changed by field or armature reversal.
(1) form A converter. A single converter unit in which the direct current can flow in one direction only and which is not capable of inverting energy from the load to the ac supply. Operates in quadrant I only (semi-converter).
(2) form B converter. A double converter unit in which the direct current can flow in either direction but which is not capable of inverting energy from the load to the ac supply. Operates in quadrants I and III only.
(3) form C converter. A single converter unit in which the direct current can flow in one direction only and which is capable of inverting energy from the load to the ac supply. Operates in quadrant I and IV.
(4) form D converter. A double converter unit in which the direct current can flow in either direction and which is capable of inverting energy from the load to the ac supply. Operates in quadrant I, II, III, and IV.   121

semiconductor device An electron device in which the characteristic distinguishing electronic conduction takes place within a semiconductor. *See:* **semiconductor.**   210

semiconductor device circuit breaker (thyristor). A circuit breaker of special characteristics used to isolate or protect semiconductor devices from overcurrent.   445

semiconductor device fuse (thyristor). A fuse of special characteristics connected in series with one or more semiconductor devices to isolate or protect the semiconductor.   445

semiconductor device lead inductance (nonlinear, active, and nonreciprocal waveguide components). The inductance of a semiconductor device associated with the strap, mesh, or wire connections used to contact the semiconductor chip. In general, a larger cross-sectional contacting area results in decreased lead inductance.   530

semiconductor diode (circuits and systems). A two-terminal device formed of a semiconductor junction having a nonlinear characteristic which will conduct electric current more in one direction than in the other.   267

semiconductor-diode parametric amplifier. A parametric amplifier using one or more varactors. *See:* **parametric device.**   191

semiconductor, extrinsic (1) (general). A semiconductor with charge-carrier concentration dependent upon impurities. *See:* **semiconductor.**   245
(2) (power semiconductor). A semiconductor in which the concentrations of holes and electrons are unbalanced by the introduction of impurities.   266

semiconductor frequency changer. A complete equipment employing semiconductor devices for changing from one alternating-current frequency to another. *See:* **semiconductor rectifier stack.**   208

semiconductor, intrinsic (1) (general). A semiconductor whose charge-carrier concentration is substantially the same as that of the ideal crystal. *See:* **semiconductor.**   245
(2) (power semiconductor). A semiconductor in which holes and electrons are created solely by thermal excitation across the energy gap. In an intrinsic semiconductor the concentration of holes and electrons must always be the same.   266

semiconductor junction (light emitting diodes). A region of transition between semiconductor regions of different electrical properties.   162

semiconductor laser. *See:* **injection laser diode (ILD).**

semiconductor, n-type. An extrinsic semiconductor in which the conduction electron concentration exceeds the mobile hole concentration.*Note:* It is implied that the net ionized impurity concentration is donor type. *See:* **semiconductor.**   245

semiconductor, n-type. An n-type semiconductor in which the excess conduction electron concentration is very large.*See:* **semiconductor.**   245

semiconductor, n$^+$-type. An n-type semiconductor in which the excess conduction electron concentration is very large. *See:* **semiconductor.**   245

semiconductor, p$^+$-type. A p-type semiconductor in which the excess mobile hole concentration is very large.*See:* **semiconductor.**   245

semiconductor, p-type. An extrinsic semiconductor in which the mobile hole concentration exceeds the conduction electron concentration.*Note:* It is implied that the net ionized impurity concentration is acceptor type. *See:* **semiconductor.**   245

semiconductor power converter. A complete equipment employing semiconductor devices for the transformation of electric power. *See:* **semiconductor rectifier stack.**   208

semiconductor radiation detector (1)(germanium gamma-ray detectors). A semiconductor device that utilizes the production and motion of excess free charge carriers in the semiconductor for the detection and measurement of particles or photons of incident radiation.   528

(2)(X-ray energy spectrometers)(charged-particle detectors). A semiconductor device that utilizes the production and motion of excess free charge carriers for the detection and measurement of incident radiation. 471, 119

semiconductor rectifier (electric installations on shipboard). A semiconductor rectifier cell is a device consisting of a conductor and semiconductor forming a junction. The junction exhibits a difference in resistance to current flow in the two directions through the junction. This results in effective current flow in one direction only. The semiconductor rectifier stack is a single columnar structure of one or more semiconductor rectifier cells. 3

semiconductor rectifier cell. A semiconductor device consisting of one cathode, one anode, and one rectifier junction. *See:* **semiconductor; semiconductor rectifier stack.** 237, 66

semiconductor rectifier cell combination. The arrangement of semiconductor rectifier cells in one rectifier circuit, rectifier diode, or rectifier stack. The semiconductor rectifier cell combination is described by a sequence of four symbols written in the order 1-2-3-4 with the following significances: (1) Number of rectifier circuit elements. (2) Number of semiconductor rectifier cells in series in each rectifier circuit element. (3) Number of semiconductor rectifier cells in parallel in each rectifier circuit element. (4) Symbol designating circuit. If a semiconductor rectifier stack consists of sections of semiconductor rectifier cells insulated from each other, the total semiconductor rectifier cell combination becomes the sum of the semiconductor rectifier cell combinations of the individual insulated sections. If the insulated sections have the same semiconductor rectifier cell combination, the total semiconductor rectifier cell combination may be indicated by the semiconductor rectifier cell combination of one section preceded by a figure showing the number of insulated sections. Example: 4(4-1-1-B) indicates four single-phase full-wave bridges insulated from each other assembled as one semiconductor rectifier stack. *Notes:* (A) The total number of semiconductor rectifier cells in each semiconductor rectifier cell combination is the product of the numbers in the combination. (B) This arrangement can also be applied by analogy to give a semiconductor rectifier diode combination. *See:* **semiconductor rectifier cell.** 237, 66

Symbol	Circuit	Example
H	half wave	1-1-1-H
C	center tap	2-1-1-C
B	bridge	4-1-1-B
		6-1-1-B
Y	wye	3-1-1-Y
S	star	6-1-1-S
D	voltage doubler	2-1-1-D

semiconductor rectifier diode (thyristor). A semiconductor diode having an asymmetrical voltage-current characteristic, used for the purpose of rectification, and including its associated housing, mounting, and cooling attachment if integral with it. 445

semiconductor rectifier stack. An integral assembly, with terminal connections, of one or more semiconductor rectifier diodes, and includes its associated mounting and cooling attachments if integral with it. *Note:* It is a subassembly of, but not a complete semiconductor rectifier. 237, 66, 208

semiconverter, bridge. A bridge in which one commutating group uses thyristors and the other uses diodes. 121

semi-direct lighting (illuminating engineering). Lighting involving luminaires which distribute 60 to 90 percent of the emitted light downward and the balance upward. 167

semienclosed. (1) Having the ventilating openings in the case protected with wire screen, expanded metal, or perforated covers or (2) having a solid enclosure except for a slot for an operating handle or small openings for ventilation, or both. 328

semienclosed brake (industrial control). A brake that is provided with an enclosure that covers the brake shoes and the brake wheel but not the brake actuator. *See:* **control.** 225, 206

semiflush-mounted device (power switchgear). One in which the body of the device projects in front of the mounting surface a specified distance between the distance specified for flush-mounted and surface-mounted devices. 103

semiguarded machine (rotating machinery). One in which part of the ventilating openings, usually in the top half, are guarded as in the case of a guarded machine but the others are left open. 263

semi-high-speed low-voltage dc power circuit breaker (1) (power switchgear). A low-voltage dc power circuit breaker which, during interruption, limits the magnitude of the fault current so that its crest is passed not later than a specified time after the beginning of the fault current transient, where the system fault current, determined without the circuit breaker in the circuit, falls between specified limits of current at a specified time. *Note:* The specified time in present practice is 0.03 second. 103

(2) (low-voltage dc power circuit breakers used in enclosures). A circuit breaker which, when applied in a circuit with the parameter values specified in American National Standard C37.16-1979, Tables 11 and 11A, tests "b" (1.7 A μs initial rate of rise of current), forces a current crest during interruption within 0.030 seconds (s) after the current reaches the pickup setting of the instantaneous trip device. *Note:* For total performance at other than test circuit parameters values, consult the manufacturer. 401

semi-indirect lighting (illuminating engineering). Lighting involving luminaires which distribute 60 to 90 percent of the emitted light upward and the balance downward. 167

semi-magnetic controller (electric installations on shipboard). An electric controller having only part of its basic functions performed by devices which are operated by electromagnets. 3

semioutdoor reactor. A reactor suitable for outdoor use provided that certain precautions in installation (specified by the manufacturer) are observed. For example, protection against rain. 309

semiprotected enclosure (electric installations on shipboard). An enclosure in which all of the openings, usually in the top half, are protected as in the case of a "protected enclosure,"but the others are left open. 3

semiremote control. A system or method of radio-transmitter control whereby the control functions are performed near the transmitter by means of devices connected to but not an integral part of the transmitter. *See:* **radio transmitter.** 111

semiselective ringing (telephone switching systems). Ringing wherein the ringers at two or more of the main stations on a party line respond simultaneously, differentiation being by the number of rings. 255

semistop joint (power cable joint). A joint which is designed to restrict movement of the dielectric fluid between cables being joined. 34

semistrain insulator (semitension assembly). Two insulator strings at right angles, each making an angle of about 45 degrees with the line conductor. *Note:* These assemblies are used at intermediate points where it may be desirable to partially anchor the conductor to prevent too great movement in case of a broken wire. *See:* **tower.** 264

semit (half-step). *See:* **semitone.**

semitone (semit) (half-step). The interval between two sounds having a basic frequency ratio approximately the twelfth root of two. *Note:* In equally tempered semitones, the interval between any two frequencies is 12 times the logarithm to the base 2 (or 39.86 times the logarithm to the base 10) of the frequency ratio. 176

semitransparent photocathode (camera tube or phototube). A photocathode in which radiant flux incident on one side produces photoelectric emission from the opposite side. *See:* **electrode (electron tube); phototubes.** 125

sender (telephone switching systems). Equipment that generates and transmits signals in response to information received from another part of the system. 255

sending-end crossfire. The crossfire in a telegraph channel from one or more adjacent telegraph channels transmitting from the end at which the crossfire is measured. *See:* **telegraphy.** 328

sending-end impedance (line). The ratio of an applied potential difference to the resultant current at the point where the potential difference is applied. The sending-end impedance of a line is synonymous with the driving-point impedance of the line. *Note:* For an infinite uniform line the sending-end impedance and the characteristic impedance are the same: and for an infinite periodic line the sending-end impedance and the iterative impedance are the same. *See:* **self-impedance; waveguide.** 328

send-only equipment. Data communication channel equipment capable of transmitting signals, but not arranged to receive signals. 194

sensation level (sound) (acoustics). *See:* **level above threshold.**

sense (navigation)(navigation aid terms). The pointing direction of a vector representing some navigation parameter. 526

sense and command features (nuclear power generating station) (power systems). The electrical and mechanical components and interconnections involved in generating those signals associated directly or indirectly with the safety functions. The scope of the sense and command features extends from the measured process variables to the execute features input terminals. 102

sense finder. That portion of a direction-finder that permits determination of direction without 180-degree ambiguity. *See:* **radio receiver.** 328

sense of polarization (antennas). For an elliptical or circularly polarized field vector, the sense of rotation of the extremity of the field vector when the origin is fixed. *Note:* When the plane of polarization is viewed from a specified side, if the extremity of the field vector rotates clockwise [counterclockwise] the sense is right-handed [left-handed] . For a plane wave the plane of polarization is viewed looking in the direction of propagation. 111

sensibility, deflection (oscilloscopes). The number of trace widths per volt of input signal that can be simultaneously resolved anywhere within the quality area. 106

sensing (navigation aid terms). The process of finding the sense, as, for example, in direction finding, the resolution of the 180° ambiguity in bearing indication; and, as in phase or amplitude-comparison systems such as ILS (instrument landing system) and VOR (very high-frequency omnidirectional range), the establishment of a relation between course displacement signal and the proper response in the control of the vehicle. 526

sensing element (initial element). *See:* **primary detector.**

sensitive relay. A relay that operates on comparatively low input power, commonly defined as 100 milliwatts or less. *See:* **relay.** 259

sensitive volume (radiation-counter tubes). That portion of the tube responding to specific radiation. 190, 96, 125

sensitivity (1) (general comment). Definitions of sensitivity fall into two contrasting categories. In some fields, sensitivity is the ratio of response to cause. Hence increasing sensitivity is denoted by a progressively larger number. In other fields, sensitivity is the ratio of cause to response. Hence increasing sensitivity is denoted by a progressively smaller number. *See:* **sensitivity coefficient.**

(2)(electric pipe heating systems). The ratio of the magnitude of a device response to the magnitude of the quantity measured. In electric pipe heating systems sensitivity is usually associated with temperature controls and alarms and addresses their response function.   448

(3)(monitoring radioactivity in effluents). The minimum amount of contaminant that can be repeatedly be detected by an instrument.   559

(4) (measuring device). The ratio of the magnitude of its response to the magnitude of the quantity measured. *Notes:* (A) It may be expressed directly in divisions per volt, millimeters per volt, milliradians per microampere, etcetera, or indirectly by stating a property from which sensitivity can be computed (for example, ohm per volt for a stated deflection. (B) In the case of mirror galvanometers it is customary to express sensitivity on the basis of a scale distance of 1 meter. *See:* **accuracy rating (instrument).**   54

(5) (radio receiver or similar device). Taken as the minimum input signal required to produce a specified output signal having a specified signal-to-noise ratio. *Note:* This signal input may be expressed as power or as voltage, with input network impedance stipulated.   59

(6) (transmission lines, waveguides, and nuclear techniques). The least signal input capable of causing an output signal having desired characteristics. *See:* **ionizing radiation.**   185

(7) (camera tube or phototube). The quotient of output current by incident luminous flux at constant electrode voltages. *Note:* (A) The term output current as here used does not include the dark current. (B) Since luminous sensitivity is not an absolute characteristic but depends on the special distribution of the incident flux, the term is commonly used to designate the sensitivity to light from a tungsten-filament lamp operating at a color temperature of 2870 kelvins. *See:* **sensitivity, cathode luminous; phototube.**   328

(8) (gyro; accelerometer). The ratio of a change in a parameter to a change in an undesirable or secondary input. For example: a scale factor temperature sensitivity of a gyro or accelerometer is the ratio of change in scale factor to a change in temperature.   46

(9) (electrothermic unit) (A) dissipated power: The ratio of the dc output voltage of the electrothermic unit to the microwave power dissipated within the electrothermic unit at a prescribed frequency, power level, and temperature. **(B) incident power:** The ratio of the dc output voltage of the electrothermic unit to the microwave power incident upon the electrothermic unit at a prescribed frequency, power level, and temperature.   47

(10) (electrothermic-coupler unit). The ratio of the dc output voltage of the electrothermic unit on the side arm of the directional coupler to the power incident upon a nonreflecting load connected to the output port of the main arm of the directional coupler at a prescribed frequency, power level, and temperature. If the electrothermic unit is attached to the main arm of the directional coupler, the sensitivity is the ratio of

the dc output voltage of the electrothermic unit attached to the main arm of the directional coupler to the microwave power incident upon a nonreflecting load connected to the output port of the side arm of the directional coupler at a prescribed frequency, power level, and temperature.   47

(11) (non-real time spectrum analyzer) (volts, decibels above or below one milliwatt). Measure of a spectrum analyzer's ability to display minimum level signals. IF (intermediate frequency) bandwidth, display mode, and any other influencing factors must be given. *Notes:* (A) equivalent input noise. The average level of a spectrum analyzer's internally generated noise referenced to the input. (B) input signal level. The input signal level that produces an output equal to twice the value of the average noise alone. This may be power or voltage relationship, but must be so stated.   68

(12) (nuclear power generating stations). (A) The minimum amount of contaminant that can repeatedly be detected by an instrument. (B) The ratio of a change in output magnitude to the change in input which causes it, after the steady-state has been reached.   355

(13) (automatic control). Of a control system or element, or combination, the ratio of a change in output magnitude to the change of input which causes it, after the steady state has been reached. *Note:* ASA C85 deprecates use of "sensitivity" to describe smallness of a dead-band. *See:* **amplification; gain.**   56

(14) (fiber optics). Imprecise synonym for responsivity. In optical system receivers, the minimum power required to achieve a specified quality of performance in terms of output signal-to-noise ratio or other measure.   433

(15) (radiation protection). The ratio of a change in response to the corresponding change in the field being measured.   399

(16) (spectrum analyzer). Measure of a spectrum analyzer's ability to display minimum level signals, (V, dBm). Intermediate frequency (IF) bandwidth, display mode, and any other influencing factors must be given. *See:* **equivalent input noise sensitivity; input signal level sensitivity.**   390

sensitivity analysis (nuclear power generating stations). An analysis which determines the variation of a given function caused by changes in one or more parameters about a selected reference value.   29, 31

sensitivity analysis (reliability analysis of nuclear power generating station safety systems). An analysis that assesses the variation in the value of a given function caused by changes in one or more arguments of the function.   587

sensitivity, cathode luminous (photocathodes). The quotient of photoelectric emission current from the photocathode by the incident luminous flux under specified conditions of illumination. *Notes:* (1) Since cathode luminous sensitivity is not an absolute characteristic but depends on the spectral distribution of the incident flux, the term is commonly used to designate the sensitivity to radiation from a tungsten filament lamp operating at a color temperature of 2870 kelvins. (2) Cathode luminous sensitivity is usually measured with a collimated beam at normal incidence. *See:* **phototube.**   117, 125

sensitivity, cathode radiant (photocathodes). The quotient of the photoelectric emission current from the photocathode by the incident radiant flux at a given wavelength under specified conditions of irradiation. *Note:* Cathode radiant sensitivity is usually measured with a collimated beam at normal incidence.
              117, 125

sensitivity coefficient (1) (automatic control) (control system). The partial derivative of a system signal with respect to a system parameter. *See:* **control system.**
                  56

(2) (circuits and systems). A coefficient used to relate the change of a system function F due to the variation of one of its parameters x. In some applications (for example control theory) absolute changes are important and the sensitivity coefficient is defined as the ∂ F/∂x. In other applications (for example, filter theory) relative changes are important and then sensitivity is defined as

$$\partial(\mathrm{Ln}\ F)/\partial(\mathrm{Ln}x) = (\partial F/\partial x)/(F/x).$$

                  67

sensitivity, deflection. (1) (magnetic-deflection cathode ray tube and yoke assembly). The quotient of the spot displacement by the change in deflecting-coil current.   190

(2) (oscilloscopes). The reciprocal of the deflection factor (for example, divisions.volt).   106

sensitivity, dynamic (phototubes). The quotient of the modulated component of the output current by the modulated component of the incident radiation at a stated frequency of modulation. *Note:* Unless otherwise stated the modulation wave shape is sinusoidal. *See:* **phototube.**   174, 190

sensitivity, illumination (camera tubes or phototubes). The quotient of signal output current by the incident illumination, under specified conditions of illumination. *Notes:* (1) Since illumination sensitivity is not an absolute characteristic but depends on the spectral distribution of the incident flux, the term is commonly used to designate the sensitivity to radiation from a tungsten-filament lamp operating at a color temperature of 2870K. (2) Illumination sensitivity is usually measured with a collimated beam at normal incidence. (3) *See:* **transfer characteristic (camera tubes).**
                 125

sensitivity, incremental (instrument) (nuclear techniques). A measure of the smallest change in stimulus that produces a statistically significant change in response. Quantitatively it is usually expressed as the change in the stimulus that produces a change in response equal to the standard deviation of the response. *See:* **ionizing radiation.**   335

sensitivity level (response level) (sensitivity) (response) (in electroacoustics) (of a transducer) (in

decibels). 20 times the logarithm to the base 10 of the ratio of the amplitude sensitivity S_A to the reference sensitivity S_0, where the amplitude is a quantity proportional to the square root of power. The kind of sensitivity and the reference sensitivity must be indicated. *Note:* For a microphone, the free-field voltage/pressure sensitivity is the kind often used and a common reference sensitivity is S_0, 1 volt per newton per square meter. The square of the sensitivity is proportional to a power ratio. The free-field voltage sensitivity-squared level, in decibels, is therefore $S_A = 10 \log (S_A^2/S_0^2) = 20 \log (S_A/S_0)$. Often, **sensitivity-squared level** in decibels can be shortened, without ambiguity, to **sensitivity level** in decibels, or simply **sensitivity** in decibels.   176

sensitivity, luminous (camera tubes or phototubes). The quotient of signal output current by incident luminous flux, under specified conditions of illumination. *Notes:* (1) Since luminous sensitivity is not an absolute characteristic but depends on the spectral distribution of the incident flux, the term is commonly used to designate the sensitivity to radiation from a tungsten-filament lamp operating at a color temperature of 2870K. (2) Luminous sensitivity is usually measured with a collimated beam at normal incidence.
                 125

sensitivity, quieting (test, measurement and diagnostic equipment). The level of a continuous wave (CW) input signal which will reduce the noise output level of a frequency-modulation (FM) receiver by a specified amount, usually 20 decibels (dB). 54

sensitivity, radiant (camera tube or phototube). The quotient of signal output current by incident radiant flux at a given wavelength, under specified conditions of irradiation. *Note:* Radiant sensitivity is usually measured with a collimated beam at normal incidence. *See:* **luminous flux; phototube; radiant flux.** 125

sensitivity, threshold (test, measurement and diagnostic equipment). The smallest quantity that can be detected by a measuring instrument or automatic control system.   54

sensitivity time control (STC) (radar). Programmed variation of the gain (sensitivity) of a radar receiver as a function of time within each pulse repetition interval or observation time in order to prevent overloading of the receiver by strong echoes from targets or clutter at close ranges.   13

sensitizing (electrostatography). The act of establishing an electrostatic surface charge of uniform density on an insulating medium. *See:* **electrostatography.**
              236, 191

sensitometry. The measurement of the light response characteristics of photographic film under specified conditions of exposure and development. 176

sensor (1)(electrical heating applications to melting furnaces and forehearths in the glass industry). A device that responds to a physical stimulus (such as heat and light) and transmits a resulting signal.
                520

(2)(electric pipe heating systems). The first system element that responds quantitatively to the measure and performs the initial measurement operation. Sensors, as used in electric pipe heating systems, respond to the temperature of the system and may be directly connected to controllers, alarms, or both. Sensors can be mechanical (bulb, bimetallic) or electrical (thermocouple, resistance-temperature detector (RTD), thermistor). *Syn:* **sensing element.**   448

(3) (nuclear power generating stations). (A) That portion of a channel which first responds to changes in, and performs the primary measurement of, a plant variable or condition. (B) A device directly responsive to the value of the measured quantity.   355

(4) (temperature measurement). That portion of a temperature-measuring system that responds to the temperature being measured.   7

(5) (test, measurement and diagnostic equipment). A transducer which converts a parameter at a test point to a form suitable for measurement by the test equipment. *See:* **pick-up; pick-off.**   54

(6) (nuclear power generating station) (safety systems). (A) That portion of a channel which responds to changes in a plant variable or condition, and converts the measured process variable into a safety system signal (for example, electric, pneumatic). (B) That portion of a channel which responds to changes in a plant variable or condition and converts the measured process variable into an electric or pneumatic signal.
387

(7) (electric pipe heating systems). The first system element that responds quantitatively to the measurand and performs the initial measurement operation. Sensors, as used in electric pipe heating systems, respond to the temperature of the system and may be directly connected to controllers. Sensors can be mechanical (bulb, bimetallic) or electrical (thermocouple, resistance-temperature detector (RTD), thermistor. *Syn:* **sensing element.**
405

sensor, active (test, measurement and diagnostic equipment). A sensor requiring a source of power other than the signal being measured.   54

sensor, passive (test, measurement and diagnostic equipment). A sensor requiring no source of power other than the signal being measured.   54

sensory saturation (nuclear power generating station). The impairment of effective operator response to an event due to excessive amount of display information which must be evaluated prior to taking action.
358

sentinel (computing systems). *See:* **flag.**

separable insulated connector (1)(separable insulated connectors). A fully insulated and shielded system for terminating and electrically connecting an insulated power cable to electrical apparatus, other power cables, or both, so designed that the electrical connection can be readily established or broken by engaging or separating the connector at the operating interface. See figure 'bushing insert', and figure 'operating interface' below.   454

(2) (power and distribution transformer). A system for terminating and electrically connecting an insulat-

ed power cable to electrical apparatus, other power cables, or both, so designed that the electrical connection can be readily established or broken by engaging or separating mating parts of the connector at the operating interface.   53

separate excitation (emergency and standby power). A source of generator field excitation power derived from a source independent of the generator output power.   512

separate excitation device (power system device function numbers). A device that connects a circuit, such as the shunt field of a synchronous converter, to a source of separate excitation during the starting sequence; or one that energizes the excitation and ignition circuits of a power rectifier.   402

separately excited (rotating machinery). A qualifying term applied to a machine to denote that the excitation is obtained from a source other than the machine itself.   63

separately ventilated machine (electric installations onshipboard). A machine which has its ventilating air supplied by an independent fan or blower external to the machine.   3

separate parts of a network. The parts that are not connected. *See:* **network analysis.**   210

separate terminal enclosure (rotating machinery). A form of termination in which the ends of the machine winding are connected to the incoming supply leads inside a chamber that need not be fully enclosed and may be formed by the foundations beneath the machine.   63

separation (1)(frequency modulation). The process of deriving individual channel signals (for example, for stereophonic systems) from a composite transmitted signal. *Note:* Separation describes the ability of a receiver to produce left and right stereophonic channel signals at its output terminals and is a measured parameter for stereo receivers only. Left-channel signal separation is defined as the ratio in decibels of the output voltage of the left output of the receiver to that of the right output when an "L"-only signal is received. Right-channel separation is similarly defined.
16

(2)(separation and identification)(design and installation of cable systems for Class 1E circuits in nuclear power generating stations). Physical independence of redundant circuits, components, and equipment. (Physical independence may be achieved by space, barriers, shields, etcetera.   536

separation criteria (electromagnetic compatibility). Curves that relate the frequency displacement to the minimum distance between a receiver and an undesired transmitter to insure that the signal-to-interference ratio does not fall below a specified value. *See:* **electromagnetic compatibility.**   199

separation distance (Class 1E equipment and circuits). Space which has no interposing structures, equipment, or materials that could aid in the propagation of fire or that could otherwise disable Class 1E systems or equipment.   131

separator (1) (storage cell). A spacer employed to

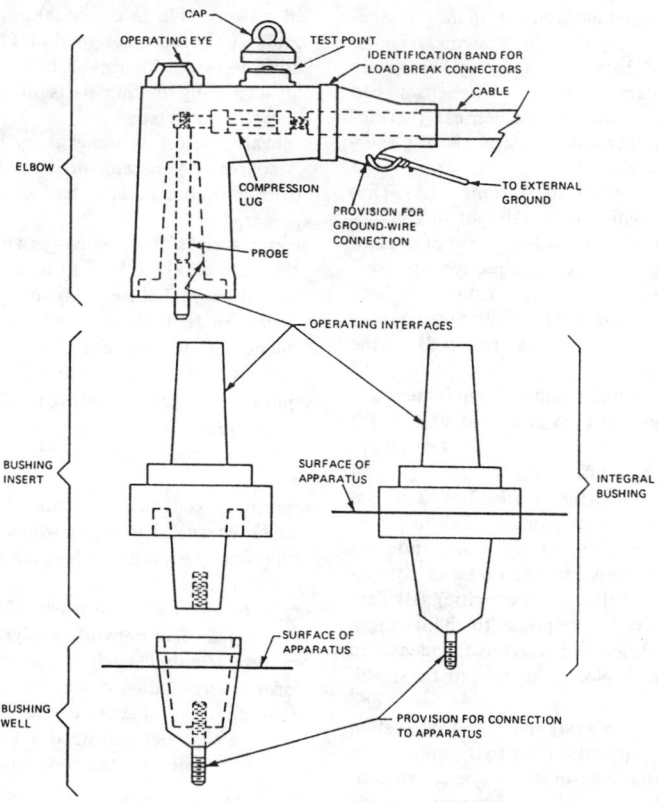

**Typical Components of 200 A Separable
Insulated Connector System**

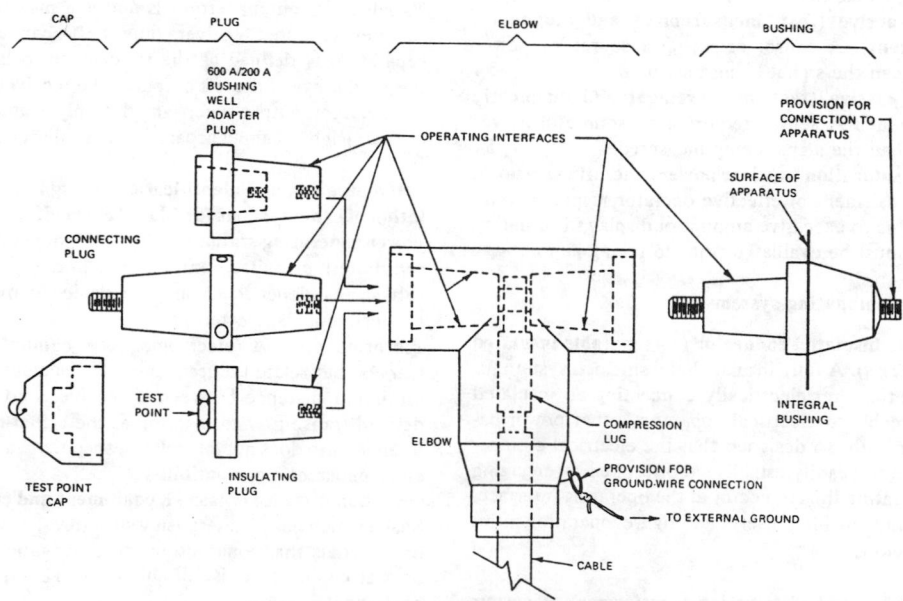

**Typical Components of 600 A Separable
Insulated Connector System**

prevent metallic contact between plates of opposite polarity within the cell. (Perforated sheets are usually called retainers.) *See:* **battery (primary or secondary).**
328

(2) (computing systems). *See:* **delimiter.**

separator, insulation slot (rotating machinery). Insulation member placed in a slot between individual coils, such as between main and auxiliary windings. *See:* **rotor (rotating machinery); stator.**   263

sequence. *See:* **calling sequence; collating sequence; pseudorandom number sequence.**

sequence filter. *See:* **sequence network.**

sequence network (power switchgear). An electrical circuit that produces an output proportional to one or more of the sequence components of a polyphase system of voltages or currents, for example positive-sequence network, or zero-sequence network.   103

sequence number. A number identifying the relative location of blocks or groups of blocks on a tape.
207

sequence-number readout. Display of the sequence number punched on the tape. *See:* **block-count readout.**   207

sequence of events function. *See:* **supervisory control functions**

sequence of operation (packaging machinery). A written detailed description of the order in which electrical devices and other parts of the industrial equipment should function.   429

sequence switch. A remotely controlled power-operated switching device used as a secondary master controller. *See:* **multiple-unit control.**   328

sequence table (electric controller). A table indicating the sequence of operation of contactors, switches, or other control apparatus for each step of the periodic duty. *See:* **multiple-unit control.**   1

sequential (formatted system) (telecommunication). If the signal elements are transmitted successively in time over a channel, the transmission is said to be **sequential.** If the signal elements are transmitted at the same time over a multiwire circuit, the transmission is said to be **coincident.** *See:* **bit.**   194

sequential access (test, measurement and diagnostic equipment). A system in which the information becomes available in a one after the other sequence only, whether all of it is desired or not.   54

sequential commutation (circuit properties)(self-commutated converters). Commutation occurs from one to the next of three or more principal switching branches arranged as a multipulse group that conduct in cyclic sequential order for usually (but not always) equal time intervals. The commutation may be direct or indirect. *Note:* An example of a converter employing a sequential commutation is given in the figure below.
584

sequential control (computing systems). A mode of computer operation in which instructions are executed consecutively unless specified otherwise by a jump.
255, 77, 54

sequential events recording system (SERS). A system which monitors bistable equipment operations and

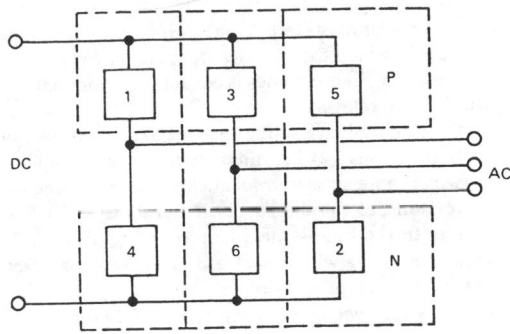

(a) Two 3-Pulse Commutating Groups: P, N

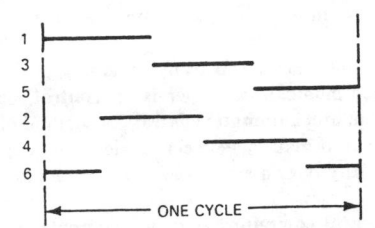

(b) Conducting Intervals of Principal Switching Branches 1–6

process status and records changes of state in the order of detected occurrences. This monitoring may be accomplished using a device dedicated solely to this function, or using a multifunction system such as a data acquisition computer system.   48, 58

sequential lobing. *See:* **lobe switching.**

sequential logic function (graphic symbols for logic functions). A logic function in which there exists at least one combination of input states for which there is more than one possible resulting combination of states at the outputs. *Note:* The outputs are functions of variables in addition to the present states of the inputs, such as time, previous internal states of the element, etcetera.   451

sequential memory (sequential events recording systems). The memory which stores events in the same order in which they were received by the system. The memory capacity can be expressed as the number of events or levels. *See:* **event; level.**   48, 58

sequential operation. Pertaining to the performance of operations one after the other.   255, 77

sequential processes (software). Processes that execute in such a manner that one must finish before the next begins. *See:* **concurrent processes; process.**   434

sequential programming (test, measurement and diagnostic equipment). The programming of a device by which only one arithmetical or logical operation can be executed at one time.   54

sequential relay. A relay that controls two or more sets

of contacts in a predetermined sequence. *See:* **relay.**
259

sequential scanning (television). A rectilinear scanning process in which the distance from center to center of successively scanned lines is equal to the nominal line width. *See:* **television.** 328

serial. (1) Pertaining to the time sequencing of two or more processes. (2) Pertaining to the time sequencing of two or more similar or identical processes, using the same facilities for the successive processes. (3) Pertaining to the time-sequential processing of the individual parts of a whole, such as the bits of a character, the characters of a word, etcetera, using the same facilities for successive parts. *See:* **serial-parallel.**
235

serial access (computing systems). Pertaining to the process of obtaining data from, or placing data into, storage when there is a sequential relation governing the access time to successive storage locations.
255, 77

serial by bit. *See:* **serial transmission.** 59

serial communication (supervisory control, data acquisition, and automatic control). A method of transmitting information between devices by sending all bits serially over a single communication channel.
570

serial digital computer. A digital computer in which the digits are handled serially. Mixed serial and parallel machines are frequently called serial or parallel according to the way arithmetic processes are performed. An example of a serial digital computer is one that handles decimal digits serially although it might handle the bits that comprise a digit either serially or in parallel. *See:* **parallel digital computer.** 210

serial operation (telecommunication) (data transmission). The flow of information in time sequence, using only one digit, word, line, or channel at a time.
59

serial-parallel. Pertaining to processing that includes both serial and parallel processing, such as one that handles decimal digits serially but handles the bits that comprise a digit in parallel. 235

serial transmission (data transmission) (telecommunications). Used to identify a system wherein the bits of a character occur serially in time. Implies only a single transmission channel. *Syn:* **serial by bit.** 59

series capacitor. A device that has the primary purpose of introducing capacitive reactance in series with an electric circuit. 474

series capacitor bank (series capacitors). An assembly of capacitors and associated auxiliaries, such as structures, support insulators, switches, and protective devices, with control equipment required for a complete operating installation. 474

series circuit. A circuit supplying energy to a number of devices connected in series, that is, the same current passes through each device in completing its path to the source of supply. *See:* **center of distribution.**
64

series circuit lighting transformer (power and distribution transformer). Dry-type individual lamp insu-

lating transformer, autotransformer, and group series loop transformers for operation of incandescent or memory lamps on series lighting circuits such as for street and airport lighting. 53

series coil sectionalizer (power switchgear). A sectionalizer in which main circuit current impulses above a specified value, flowing through a solenoid or operating coil, provide the energy required to operate the counting mechanism. 103

series connected starting-motor starting (rotating machinery). The process of starting a motor by connecting its primary winding to the supply in series with the primary windings of a starting motor, this latter being short-circuited for the running condition. 63

series connection. The arrangement of cells in a battery made by connecting the positive terminal of each successive cell to the negative terminal of the next adjacent cell so that their voltages are additive. *See:* **battery (primary or secondary).** 328

series distribution system. A distribution system for supplying energy to units of equipment connected in series. *See:* **alternating-current distribution; direct-current distribution.** 64

series elements (network). (1) Two-terminal elements are connected in series when they form a path between two nodes of a network such that only elements of this path, and no other elements, terminate at intermediate nodes along the path. (2) Two-terminal elements are connected in series when any mesh including one must include the others. *See:* **network analysis.**
210

series-fed vertical antenna. A vertical antenna which is insulated from ground and whose feed line connects between ground and the lower end of the antenna.
111

series filter (harmonic control and reactive compensation of static power converters). That type of filter which reduces harmonics by putting a high series impedance between the harmonic source and the system to be protected. 533

series gap (1)(metal-oxide surge arresters for ac power circuits). An intentional gap(s) between spaced electrodes in series with the valve elements across which all or part of the impressed arrester terminal voltage appears. 583
(2)(surge arrester). An intentional gap(s) between spaced electrodes: it is in series with the valve or expulsion element of the arrester, substantially isolating the element from line or ground, or both, under normal line-voltage conditions. 430

series heater (electrical heat tracing for industrial applications). Heating elements that are designed to have a specific resistance at a given temperature for a given length. 523

series loading. Loading in which reactances are inserted in series with the conductors of a transmission circuit. *See:* **loading.** 328

series-mode interference (signal-transmission system). *See:* **interference, differential-mode.**

series modulation. Modulation in which the plate circuits of a modulating tube and a modulated amplifier tube are in series with the same plate voltage supply.   328

series operation (power supplies). The output of two or more power supplies connected together to obtain a total output voltage equal to the sum of their individual voltages. Load current is equal and common through each supply. The extent of series connection is limited by the maximum specified potential rating between any output terminal and ground. For series connection of current regulators, master/slave (compliance extension) or automatic crossover is used. *See:* **isolation voltage.**   186

series overcurrent tripping. *See:* **direct (series) release (trip); overcurrent release (trip).**

series-parallel connection. The arrangement of cells in a battery made by connecting two or more series-connected groups, each having the same number of cells so that the positive terminals of each group are connected together and the negative terminals are connected together in a corresponding manner. *See:* **battery (primary or secondary).**   328

series-parallel control. A method of controlling motors wherein the motors, or groups of them, may be connected successively in series and in parallel. *See:* **multiple-unit control.**   328

series-parallel network. Any network, containing only two-terminal elements, that can be constructed by successively connecting branches in series and.or in parallel. *Note:* An elementary example is the parallel combination of two branches, one containing resistance and inductance in series, the other containing capacitance. This network is sometimes called a **simple parallel circuit.** *See:* **network analysis.**   210

series-parallel primary current transformer (instrument transformer). One that has two insulated primaries, which are intended for connection in series or parallel to provide different rated currents.   203

series-parallel starting (rotating machinery). The process of starting a motor by connecting it to the supply with the primary winding phase circuits initially in series, and changing them over to a parallel connection for running operation. *See:* **asynchronous machine.**   63

series rectifier circuit. A rectifier circuit in which two or more simple rectifier circuits are connected in such a way that their direct voltages add and their commutations coincide. *See:* **rectification; rectifier circuit element.**   66

series regulator (power supplies). A device placed in series with a source of power that is capable of controlling the voltage or current output by automatically varying its series resistance. *See:* **passive element.**   186

series relay. *See:* **current relay.**

series resistor (electric instrument). A resistor that forms an essential part of the voltage circuit of an instrument and generally is used to adapt the instrument to operate on some designated voltage or voltages. The series resistor may be internal or external to the instrument. *Note:* Inductors, capacitors, or combinations thereof are also used for this purpose. *See:* **auxiliary device to an instrument.**   280

series snubber (converter circuit elements)(self-commutated converters). Circuit elements, usually including an inductor, connected in series with a switching device to limit the rate of rise or fall of current through the device when switching on or off, respectively.   584

series street-lighting transformer (power and distribution transformer). A series transformer that receives energy from a current-regulating series circuit and that transforms the energy to another winding at the same or different current from that in the primary.   53

series system. The arrangement in a multielectrode electrolytic cell whereby in each cell an anode connected to the positive bus bar is placed at one end and a cathode connected to the negative bus bar is placed at the other end, with the intervening unconnected electrodes acting as bipolar electrodes. *See:* **electrorefining.**   328

series tee junction. *See:* **E-plane tee junction.**

series transformer (power and distribution transformer). A transformer with a "series" winding and an "exciting" winding, in which the "series" winding is placed in a series relationship in a circuit to change voltage or phase, or both, in that circuit as a result of input received from the "exciting" winding. *Note:* Applications of series transformers include: (1)Use in a transformer such as a load-tap-changing or regulating transformer to change the voltage or current duty of the load-tap-changing mechanism. (2) Inclusion in a circuit for power factor correction to indirectly insert series capacitance in a circuit by connecting capacitors to the exciting winding.   53

series transformer rating (power and distribution transformer). The lumen rating of the series lamp, or the wattage rating of the multiple lamps, that the transformer is designed to operate.   53

series-trip recloser (power switchgear). A recloser in which main-circuit current above a specified value, flowing through a solenoid or operating coil, provides the energy necessary to open the main contacts.   103

series two-terminal pair networks. Two-terminal pair networks are connected in series at the input or at the output terminals when their respective input or output terminals are in series. *See:* **network analysis.**   332

series undercurrent tripping. *See:* **direct (series) release (trip); undercurrent release (trip).**

series unit (power and distribution transformer). The core and coil unit which has one winding connected in series in the line circuit.   53

series winding (1) (autotransformer) (power and distribution transformer). That portion of the autotransformer winding which is not common to both the primary and the secondary circuits, but is connected in series between the input and output circuits.   53

(2) (power and distribution transformer). The winding of the series unit which is connected in series in the line circuit. *Note:* If the main unit of a two-core transformer is an autotransformer, both units will have a series winding. In such cases, one is referred to as the series winding of the autotransformer and the other, the series winding of the series unit.   53

series-wound (rotating machinery). A qualifying term applied to a machine to denote that the excitation is supplied by a winding or windings connected in series with or carrying a current proportional to that in the armature winding. *See:* **asynchronous machine.**
    63

series-wound motor (1) (rotating machinery). A commutator motor in which the field circuit and armature circuit are connected in series. *See:* **asynchronous machine.**   63
(2) **(electric installations on shipboard).** A commutator motor in which the field circuit and armature circuit are connected in series. (It operates at a much higher speed at light load than at full load.)   3
(3) **(National Electrical Code).** The conductors and equipment for delivering energy from the electricity supply system to the wiring system of the premises served.   256

service (1) (electric systems). The conductors and equipment for delivering electric energy from the secondary distribution or street main, or other distribution feeder, or from the transformer, to the wiring system of the premises served. *Note:* For overhead circuits, it includes the conductors from the last line pole to the service switch or fuse. The portion of an overhead service between the pole and building is designated as service drop.   256
(2) **(controller) (industrial control).** The specific application in which the controller is to be used, for example: (A) general purpose, (B) definite purpose, for example, crane and hoist, elevator, machine tool, etcetera. *See:* **electric controller.**   206

service area (navigation)(navigation aid terms). The area within which a navigational aid provides either generally satisfactory service or a specific quality of service.   526

service area (power operations). Territory in which a utility system is required or has the right to supply or make available electric service to ultimate consumers.   516

service band. A band of frequencies allocated to a given class of radio service. *See:* **radio transmission.**
    111, 240

service bits (telecommunication). Those bits that are neither check nor information bits. *See:* **bit.**   194

service cable (National Electrical Code). Service conductors made up in the form of a cable.   256

service capacity (cell or battery). The electric output (expressed in ampere-hours, watthours, or similar units) on a service test before its working voltage falls to a specified cutoff voltage. *See:* **battery (primary or secondary).**   328

service circuit (telephone switching systems). A circuit used for signaling purposes connected to and disconnected from a communication path during the progress of a call.   55

service class (use in primitives)(logical link control).

A parameter used to convey the type of service required or desired (for example, priority).   585

service code (telephone switching systems). Any of the destination codes for use by customers to obtain directory assistance or repair service, or to reach the business office of the telecommunications company.
    55

service condition (thermal classification of electric equipment and electrical insulation). A combination of factors of influence, which are to be expected in a specific application of electric equipment.   506

service conditions (1)(cable, field splice, and connection qualification)(design and installation of cable systems for Class 1E circuits in nuclear power generating stations). Environmental, power and signal conditions expected as a result of normal operating requirements, extremes in operating requirements, and postulated conditions appropriate for the design basis events of the station.   536
(2)**(electric penetration assemblies).** Environmental, power, and signal conditions expected as a result of normal operating requirements, expected extremes in operating requirements, and postulated conditions appropriate for the design basis events applicable to the electric penetration assembly.   493
(3)**(safety systems equipment in nuclear power generating stations)(nuclear power generating station)-(valve actuators).** Environmental, loading, power and signal conditions expected as a result of normal operating requirements, expected extremes (abnormal) in operating requirements, and postulated conditions appropriate for the design-basis events of the station.
    535, 120, 492
(4) **(Class 1E static battery chargers and inverters).** Environmental, power, and signal conditions expected as a result of normal operation requirements, expected extremes in operating requirements, and postulated conditions appropriate for the design basis events of the station.   408
(5) **(nuclear power generating stations) (Class 1E motor control centers).** Environmental, loadings, power, and signal conditions expected as a result of normal operating requirements, expected extremes (abnormal) in operating requirements, and postulated conditions appropriate for the design basis events of the station.   440

service conductors (National Electrical Code). The supply conductors that extend from the street main or from transformers to the service equipment of the premises supplied.   256

service corrosion (dry cell). The consumption of the negative electrode as a result of useful current delivered by the cell. *See:* **electrolytic cell.**   328

service current, continuous (1) (thyristor converter). The value of direct current which a converter unit or section can supply to its load for unlimited time periods under specified conditions. **(2) long-time.** The rms value and duration (minutes) of direct current which may be applied to the converter unit or section within the service current profile. *Note:* This value establishes point B on the service current profile and it may be

identical to the long-time test current. **(3) profile.** The time-current profile that defines the allowable rms currents the converter section can sustain. *Note:* The profile is defined for times from zero to infinity, and the rms current derived from any current-time diagram must not exceed this profile. **(4) short time.** The peak rms value and duration (seconds) of direct current which may be applied to the converter unit or section within the service current profile. *Note:* This value establishes point C on the service current profile.
       121

service data unit (SDU)(token ring access method). Information delivered as a unit between adjacent entities which may also contain a protocol data unit (PDU) of the upper layer.
       472

service date (power system measurement). The date a unit first enters the active state. On this date the reporting of performance data shall begin. *Note:* The service date is not to be confused with the installation date (the date the unit was first electrically connected to the system) or with the commercial operation date (usually related to the satisfactory completion of acceptance tests as specified in the purchase contract).
       432

service drop (1)(NESC). The overhead conductors between the electric supply or communication line and the building or structure being served.
       494
(2) (National Electrical Code). The overhead service conductors from the last pole or other aerial support to and including the splices, if any, connecting to the service-entrance conductors at the building or other structure.
       256

service-entrance cable (National Electrical Code). A single conductor or multiconductor assembly provided with or without an overall covering, primarily used for services and of the following types: (1) Type SE, having a flame-retardant, moisture-resistant covering, but not required to have inherent protection against mechanical abuse. (2) Type USE, recognized for underground use, having a moisture-resistant covering, but not required to have a flame-retardant covering or inherent protection against mechanical abuse. Single-conductor cables having an insulation specifically approved for the purpose do not require an outer covering. Cabled single-conductor Type USE constructions recognized for underground use may have a bare copper conductor cabled with the assembly. Type USE single, parallel, or cabled conductor assemblies recognized for underground use may have a bare copper concentric conductor applied. These constructions do not require an outer overall covering. (3) If Type SE or USE cable consists of two or more conductors, one shall be permitted to be uninsulated.
       256
service entrance conductors (electric system) (1) (overhead system). The service conductors between the terminals of the service equipment and a point usually outside the building, clear of building walls, where joined by tap or splice to the service drop.
       256
(2) (underground system). The service conductors between the terminals of the service equipment and

the point of connection to the service lateral. *Note:* Where service equipment is located outside the building walls, there may be no service-entrance conductors, or they may be entirely outside the building.
       256, 64

service environment (diesel-generator unit). The aggregate of conditions surrounding the diesel-generator unit in its enclosure, while serving the design load during normal, accident, and post-accident operation.
       99

service equipment (National Electrical Code). The necessary equipment, usually consisting of a circuit breaker or switch and fuses, and their accessories, located near the point of entrance of supply conductors to a building or other structure, or an otherwise defined area, and intended to constitute the main control and means of cutoff of the supply.
       256

service evaluation (telephone switching systems). Determination of the quality of service received by the customer.
       55

service factor (general-purpose alternating-current motor). A multiplier that, when applied to the rated power, indicates a permissible power loading that may be carried under the conditions specified for the service factor. *See:* **asynchronous machine.**
       63

service ground. A ground connection to a service equipment or a service conductor or both. *See:* **ground.**
       64

service hours (SH)(electric generating unit reliability, availability, and productivity)(power system measurement). The number of hours a unit was in the in-service state.
       56, 4327

service lateral (National Electrical Code). The underground service conductors between the street main, including any risers at a pole or other structure or from transformers, and the first point of connection to the service-entrance conductors in a terminal box or meter or other enclosure with adequate space, inside or outside the building wall. Where there is no terminal box, meter, or other enclosure with adequate space, the point of connection shall be considered to be the point of entrance of the service conductors into the building.
       256

service life (1) (primary cell or battery). The period of useful service before its working voltage falls to a specified cutoff voltage.
(2) (storage cell or battery). The period of useful service under specified conditions, usually expressed as the period elapsed before the ampere-hour capacity has fallen to a specified percentage of the rated capacity. *See:* **battery (primary or secondary); charge.**
       328

service life of cable (cable systems). The time during which satisfactory cable performance can be expected for a specific set of service conditions.
       477

service period (illuminating engineering). The number of hours per day for which the daylighting provides a specified illuminance level. It often is stated as a monthly average.
       167

service pipe. The pipe or conduit that contains underground service conductors and extends from the junction with outside supply wires into the customer's premises. *See:* distributor duct; service. 64

service-point (National Electrical Code). The point of connection between the facilities of the serving utility and the premises' wiring. *Note:* For clearances of conductors of over 600 volts, see National Electrical Safety Code (ANSI C2-1977). 256

service raceway (National Electrical Code). The raceway that encloses the service-entrance conductors. 256

service rating (rectifier transformer). The maximum constant load that, after a transformer has carried its continuous rated load until there is no further measurable increase in temperature rise, may be applied for a specified time without injury. *See:* rectifier transformer. 258

service request handler(SRH)(FASTBUS acquisition and control). A master responsible for monitoring the service request line, SR, on a segment or group of segments. When SR=1 the SRH requests bus mastership and after obtaining mastership determines which module(s) is asserting SR, either by polling or by a broadcast operation. The SRH may subsequently service the pending request(s) itself, or may issue interrupt messages to other devices on behalf of the module(s) asserting SR. SR is usually asserted only by modules which lack mastership capability. 480

service requirement (thermal classification of electric equipment and electrical insulation). The specified performance to be expected in a specific application under a specified service condition. 506

service routine (computing systems). A routine in general support of the operation of a computer, for example, an input-output, diagnostic, tracing, or monitoring routine. *See:* utility routine. 255, 77, 54

services (logical link control). The capabilities and features provided by an N-layer to an N-user. 585

service, standby (electric power utilization). Service through a permanent connection not normally used but available in lieu of, or as a supplement to, the usual source of supply. 112

service, station. Facilities which provide energy for station use in a generating, switching, converting, or transforming station. 112

service test (1)(lead storage batteries). A special test of the battery's capability, as found, to satisfy the design requirements (battery duty cycle) of the dc system. 38

(2) (primary battery). A test designed to measure the capacity of a cell or battery under specified conditions comparable with some particular service for which such cells are used. 328

(3) (field test) (meter). A test made during the period that the meter is in service. *Note:* A service test may be made on the consumer's premises without removing the meter from its support, or by removing the meter for test, either on the premises or in a laboratory or meter shop. 212

service voltage (system voltage ratings). The root-mean-square phase-to-phase or phase-to-neutral voltage at the point where the electrical system of the

supplier and the user are connected. *See:* system voltage; nominal system voltage; maximum system voltage; service voltage; utilization voltage; low voltage; medium voltage; high voltage. 260

servicing time (electric drive). The portion of down time that is necessary for servicing due to breakdowns or for preventive servicing measures. *See:* electric drive. 1

serving (cable). A wrapping applied over the core of a cable before the cable is leaded, or over the lead if the cable is armored. *Note:* Materials commonly used for serving are jute or cotton. The serving is for mechanical protection and not for insulating purposes. 64

servo. *See:* servomechanism.

servo amplifier (analog computers). An amplifier, used as part of a servomechanism, that supplies power to the electrical input terminals of a mechanical actuator. 9

servomechanism. (1) A feedback control system in which at least one of the systems signals represents mechanical motion. (2) Any feedback control system. (3) An automatic feedback control system in which the controlled variable is mechanical position or any of its time derivatives. *See:* control system, feedback. 255

servomechanism, positional. A servomechanism in which a mechanical shaft is positioned, usually in the angle of rotation, in accordance with one or more input signals. *Note:* Frequently, the shaft is positioned (excluding transient motion) in a manner linearly related to the value of the input signal. However, the term also applies to any servomechanism in which a loop input signal generated by a transmitting transducer can be compared to a loop feedback signal generated by a compatible or identical receiving transducer to produce a loop error signal that, when reduced to zero by movement of the receiving transducer, results in a shaft position related in a prescribed and repeatable manner to the position of the transmitting transducer. *See:* electronic analog computer; servomechanism, repeater. 9

servomechanism, rate. A servomechanism in which a mechanical shaft is translated or rotated at a rate proportional to an input signal amplitude. *See:* electronic analog computer. 9

servomechanism, repeater. A positional servomechanism in which loop input signals from a transmitting transducer are compared with loop feedback signals from a compatible or identical receiving transducer mechanically coupled to the servomechanism to produce a mechanical shaft motion or position linearly related to motion or position of the transmitting transducer. *See:* electronic analog computer. 9

servomechanism type number. In control systems in which the loop transfer function is

$$\frac{K(1 + a_1 s + a_2 s^2 + \ldots + a_i s^i)}{s^n(1 + b_1 s + b_2 s^2 + \ldots + b_k s^k)}$$

where K, a_1, b_1, b_2, etcetera, are constant coefficients, the value of the integer n. *Note:* The value of n determines the low-frequency characteristic of the transfer function. The log-gain–log-frequency curve (Bode diagram) has a zero-frequency slope of zero for $n = 0$, slope $- 1$ for $n = 1$, etcetera. *See:* **control system, feedback.**   256

servomotor limit (hydraulic turbines). A device which acts on the governor system to prevent the turbine-control servomotor from opening beyond the position for which the device is set.   8

servomotor position (hydraulic turbines). The instantaneous position of the turbine control servomotor expressed as a percent of the servomotor stroke. This is commonly referred to as gate position, needle position, blade position, or deflector position, although the relationship between servomotor stroke and the position of the controlled device may not always be linear.   8

servomotor stroke (1) (hydraulic turbines). Travel of the turbine control servomotor from zero to maximum without overtravel at the maximum position or "squeeze" at the minimum position. (Sometimes referred to as "effective servomotor stroke".)   8
(2) (speed governing systems). Travel of the turbine control servomotor from zero to maximum without overtravel at the maximum position or "squeeze" at the minimum position. *Notes:* (A) For a gate servomotor this shall be established as the change in gate position from no discharge to maximum discharge. (B) For a blade servomotor this shall be established as the change in blade position from "flat" to "steep". (C) For a deflector servomotor this shall be established as the change in deflector position from "no deflection" position to "full flow deflected" position with maximum discharge under maximum specified head including overpressure due to water hammer. *Syn:* **effective servomotor stroke.**   58

servomotor time (hydraulic turbines). The equivalent elapsed time for one servomotor stroke (either opening or closing) corresponding to maximum servomotor velocity. Servomotor time can be qualified as: (A) gate, (B) blade, (C) deflector, (D) needle.   8

servomotor velocity limit (hydraulic turbines). A device which functions to limit the servomotor velocity in either the opening, closing, or both directions exclusive of the operation of the slow closure device (above).   8

servo multiplier (analog computers). An electromechanical multiplier in which one variable is used to position one or more ganged potentiometers across which the other variable voltages are applied.
                                                                     9

servo potentiometer (analog computers). A potentiometer driven by a positional servomechanism.
                                                                     9

set (1) (electronic computation). (A) To place a storage device into a specified state, usually other than that denoting ZERO or BLANK (B) To place a binary cell into the state denoting ONE. *See:* **preset; reset.**   235

(2) (used as a verb) (power switchgear). To position the various adjusting devices so as to secure the desired operating characteristic. *Note:* Typical adjustment devices are taps, dials, levers, and scales suitably marked, rheostats that may be adjusted during tests, and switches with numbered positions that refer to recorded operating characteristics.   103
(3) (polyphase currents) (of m phases). A group of m interrelated alternating currents, each in a separate phase conductor, that have the same primitive period but normally differ in phase. They may or may not differ in amplitude and waveform. The equations for a set of m-phase currents, when each is sinusoidal, and has the primitive period, are

$$i_a = (2)^{1/2} I_a \cos{(\omega t + \beta_{a1})}$$
$$i_b = (2)^{1/2} I_b \cos{(\omega t + \beta_{b1})}$$
$$i_c = (2)^{1/2} I_c \cos{(\omega t + \beta_{c1})}$$
$$\vdots$$
$$i_m = (2)^{1/2} I_m \cos{(\omega t + \beta_{m1})}$$

where the symbols have the same meaning as for the general case given later. The general equations for a set of m-phase alternating currents are

$$i_a = (2)^{1/2}[I_{a1} \cos(\omega t + \beta_{a1}) + I_{a2} \cos{(2\omega t} + \beta_{a2}) + \ldots + I_{aq} \cos{(q\omega t + \beta_{aq})} + \ldots]$$
$$i_b = (2)^{1/2}[I_{b1} \cos{(\omega t + \beta_{b1})} = I_{b2} \cos{(2\omega t} + \beta_{b2}) + \ldots + I_{bq} \cos{(q\omega t + \beta_{bq})} + \ldots]$$
$$\vdots$$
$$i_m = (2)^{1/2}[I_{m1} \cos{(\omega t + \beta_{m1})} + I_{m2} \cos{(2\omega t + \beta_{m2})} + \ldots + I_{mq} \cos{(q\omega t + \beta_{mq})} + \ldots]$$

where i_a, i_b, ..., i_m are the instantaneous values of the currents, and I_{a1}, I_{a2}, ..., I_{aq} are the root-mean-square amplitudes of the harmonic components of the individual currents. The first subscript designates the individual current and the second subscript denotes the number of the harmonic component. If there is no second subscript, the quantity is assumed to be sinusoidal. β_{a1}, β_{a2}, ..., β_q are the phase angles of the components of the same subscript determined with relation to a common reference. *Notes:* (A) If the circuit has a neutral conductor, the current in the neutral conductor is generally not considered as a separate current of the set, but as the negative of the sum of all the other currents (with respect to the same reference direction). (B) See Note (C) of **voltage sets (polyphase circuit).** *See:* **network analysis.**   210
(4) (polyphase voltages) (m phases). A group of m interrelated alternating voltages that have the same primitive period but normally differ in phase. They may or may not differ in amplitude and wave form. The equations for a set of m-phase voltages, when each is sinusoidal and has the primitive period, are

$$e_a = (2)^{1/2} E_a \cos (\omega t + \alpha_{a1})$$
$$e_b = (2)^{1/2} E_b \cos (\omega t + \alpha_{b1})$$
$$e_c = (2)^{1/2} E_c \cos (\omega t + \alpha_{c1})$$
$$\vdots$$
$$e_m = (2)^{1/2} E_m \cos (\omega t + \alpha_{m1})$$

where the symbols have the same meaning as for the general case given below. The general equations for a set of m-phase alternating voltages are

$$e_a = (2)^{1/2} [E_{a1} \cos (\omega t + \alpha_{a1}) + E_{a2} \cos (2\omega t + \alpha_{a2}) + \ldots + E_{ar} \cos (r\omega t + \alpha_{ar}) + \ldots]$$

$$e_b = (2)^{1/2} [E_{b1} \cos (\omega t + \alpha_{b1}) + E_{b2} \cos (2\omega t + \alpha_{b2}) + \ldots + E_{br} \cos (r\omega t + \alpha_{br}) + \ldots]$$

$$\vdots$$

$$e_m = (2)^{1/2} [E_{m1} \cos (\omega t + \alpha_{m1}) + E_{m2} \cos (2\omega t + \alpha_{m2}) + \ldots + E_{mr} \cos (r\omega t + \alpha_{mr}) + \ldots]$$

where e_a, e_b, ..., e_m are the instantaneous values of the voltages, and E_{a1}, E_{a2}, ..., E_{ar} the root-mean-square amplitudes of the harmonic components of the individual voltages. The first subscript designates the individual voltage and the second subscript denotes the number of the harmonic component. If there is no second subscript, the quantity is assumed to be sinusoidal. α_{a1}, α_{a2}, ..., a_{ar} are the phase angles of the components with the same subscript determined in relation to a common reference. *Note:* This definition may be applied to a two-phase four-wire or five-wire circuit if m is considered to be 4 instead of 2. A two-phase three-wire circuit should be treated as a special case. *See:* **network analysis.** 210

(5) (test, measurement and diagnostic equipment). (A) A collection: (B) To place a storage device into a specified state, usually other than that denoting zero or blank: and (C) To place a binary cell into the one state. 54

(6) (electric and electronics parts and equipments). A unit or units and necessary assemblies, subassemblies, and basic parts connected or associated together to perform an operational function. Typical examples: search radar set, radio transmitting set, sound measuring set: these include such parts, assemblies, and units as cables, microphone, and measuring instruments. 17

set light (illuminating engineering). The separate illumination of the background or set, other than that provided for principal subjects or areas. 167

set of commutating groups (rectifier). Two or more commutating groups that have simultaneous commutations. *See:* **rectification; rectifier circuit element.** 66, 208

set point (1)(electric pipe heating systems). A fixed or constant (for relatively long time periods) command. With respect to electric pipe heating systems, set points are usually associated with temerature control-

lers or alarms and are the position of of the dials, taps, levels, scales, etcetera, so as to secure the desired operating characteristics. 448

(2) (nuclear power generating stations). A predetermined point within the range of an instrument where protective or control action is initiated. 143

set pulse. A drive pulse that tends to set a magnetic cell. 331

setting (1) (of circuit breaker) (National Electrical Code). The value of current and/or time at which an adjustable circuit breaker is set to trip. 256

(2) (used as a noun) (power switchgear). The desired characteristic, obtained as a result of having set a device, stated in terms of calibration markings or of actual performance bench marks such as pickup current and operating time at a given value of input. *Note:* When the setting is made by adjusting the device to operate as desired in terms of a measured input quantity, the procedure may be the same as in calibration. However, since it is for the purpose of finding one particular position of an adjusting device, which in the general case may have several marked positions that are not being calibrated, the word setting is to be preferred over the word calibration. 103

setting error (power switchgear). The departure of the actual performance from the desired performance resulting from errors in adjustment or from limitations in testing or measuring techniques. 103

setting limitation (power switchgear). The departure of the actual performance from the desired performance resulting from limitations of adjusting devices. 103

settling time (1) (hybrid computer linkage components). The time required from the instant after the "load" has been completed until the digital-to-analog converter (KDAC) or digital-to-analog multiplier (DAM) output voltage is available within a given accuracy (under the condition of a jam transfer for a double-buffered DAC). 10

(2) (automatic control). The time required, following the initiation of a specified stimulus to a linear system, for the output to enter and remain within a specified narrow band centered on its steady-state value. *Note:* The stimulus may be a step, impulse, ramp, parabola, or sinusoid. For a step or impulse, the band is often specified as ± 2 per cent. For nonlinear behavior, both magnitude and pattern of the stimulus should be specified. *Syn:* **correction time.** 56

setup (television). The ratio between reference black level and reference white level, both measured from blanking level. It is usually expressed in percent. *See:* **television.** 178

severity. *See:* **criticality.** 434

sexadecimal. (1) Pertaining to a characteristic or property involving a selection, choice, or condition in which there are sixteen possibilities. (2) Pertaining to the numeration system with a radix of sixteen. *Note:* More commonly called **hexadecimal.** *See:* **positional notation.** 77

sextant (navigation aid terms). A double-reflecting instrument for measuring angles--primarily altitudes--of the celestial bodies. 526

shade (illuminating engineering). A screen made of opaque or diffusing material which is designed to prevent a light source from being directly visible at normal angles of view.   167

shaded-pole motor (rotating machinery). A single-phase induction motor with a main winding and one or more short-circuited windings (or shading coils) disposed about the air gap. The effect of the winding combination is to produce a rotating magnetic field which in turn induces the desired motor action.   63

shading (1) (storage tubes). The type of spurious signal, generated within a tube, that appears as a gradual variation or a small number of gradual variations in the amplitude of the output signal. These variations are spatially fixed with reference to the target area. Note the distinction between this and **disturbance.** *See:* **storage tube; television.**   174
(2) (audio and electroacoustics). A method of controlling the directional response pattern of a transducer through control of the distribution of phase and amplitude of the transducer action over the active face. *See:* **television.**   176
(3) (camera tubes). A brightness gradient in the reproduced picture, not present in the original scene, but caused by the tube.   125

shading coil (1) (rotating machinery). The short-circuited winding used in a shaded-pole motor, for the purpose of producing a rotating component of magnetic flux.   63
(2) (direct-current motors and generators). A short-circuited winding used on a main (excitation) pole to delay the shift in flux caused by transient armature current. Transient commutation is aided by the use of this coil. *See:* **rotor (rotating machinery); stator.**   63

shading wedge (rotating machinery). A strip of magnetic material placed between adjacent pole tips of a shaded-pole motor to reduce the effective separation between the pole tips. The shading wedge usually has a slot running most of its length to provide some separation effect. *See:* **rotor (rotating machinery); stator.**   63

shadow factor (radio wave propagation). The ratio of the electric field strength which would result from propagation over a convex curved surface to that which would result from propagation over a plane, other factors being the same.   146
shadowing (shielding). The interference of any part of an anode, cathode, rack, or tank with uniform current distribution upon a cathode.   328
shadow loss (mobile communication). The attenuation to a signal caused by obstructions in the radio propagation path. *See:* **mobile communication system.**   181
shadow mask (color-picture tubes). A color-selecting-electrode system in the form of an electrically conductive sheet containing a plurality of holes that uses masking to effect color selection. *See:* **television.**   125
shaft (rotating machinery). That part of a rotor that

carries other rotating members and that is supported by bearings in which it can rotate. *See:* **rotor (rotating machinery).**   63
shaft current (rotating machinery). Electric current that flows from one end of the shaft of a machine through bearings, bearing supports, and machine framework to the other end of the shaft, driven by a voltage between the shaft ends that results from flux linking the shaft caused by irregularities in the magnetic circuit. *See:* **rotor (rotating machinery).**   63
shaft extension (rotating machinery). The portion of a shaft that projects beyond the bearing housing and away from the core. *See:* **armature.**   63
shaft revolution indicator. A system consisting of a transmitter driven by a propeller shaft and one or more remote indicators to show the speed of the shaft in revolutions per minute, the direction of rotation and (usually) the total number of revolutions made by the shaft. *See:* **electric propulsion system.**   328
shaft voltage test (rotating machinery). A test taken on an energized machine to detect the induced voltage that is capable of producing shaft currents. *See:* **rotor (rotating machinery).**   63
shall (1)(binary floating-point arithmetic)(information transfer)(microprocessor assembly language). The use of the word 'shall' signifies that which is obligatory in any conforming implementation.   469, 479, 466
(2)('dose calibrator' ionization chambers). Indicates a recommendation that is necessary or essential to meet requirements of American National Standard N42.13-1986.   499
(3) (ionization chambers). Shall indicates a recommendation that is necessary or essential to meet requirements.   396
(4) (sodium iodide detector). In this text, indicates an action required by American National Standard N42.12-1980.   423
(5)(high-level microprocessor language). In IEEE Trial Use Std 755-1985, the use of the word 'shall' signifies that which is obligatory in any conforming implementation.   470
shank (cable plowing). A portion of the plow blade to which a removable wear point is fastened. *See:* **wear point.**   52
shaped-beam antenna. An antenna which is designed to have a prescribed pattern shape differing significantly from that obtained from a uniform-phase aperture of the same size.   111
shape factor (1) (spectrum analyzer). A measure of the asymptotic shape of the resolution bandwidth response curve of a spectrum analyzer. Shape factor is defined as the ratio between bandwidths at two widely spaced points on the response curve, such as the 3 dB and 60 dB down points. *Syn:* **skirt selectivity.**   390
(2) (induction and dielectric heating equipment). *See:* **coil shape factor.**
shaping (operations on a pulse) (pulse terms). A process in which the shape of a pulse is modified to one

which is ideal or more suitable for the intended application wherein time magnitude parameters may be changed. Typically, some property(ies) of the original pulse is preserved. (1) regeneration. A shaping process in which a pulse with desired reference characteristics is developed from a pulse which lacks certain desired characteristics. (2) stretching. A shaping process in which pulse duration is increased. (3) clipping. A shaping process in which the magnitude of a pulse is constrained at one or more predetermined magnitudes. (4) limiting. A clipping process in which the pulse shape is preserved for all magnitudes between predetermined clipping magnitudes. (5) slicing. A clipping process in which the pulse shape is preserved for all magnitudes less (greater) than a predetermined clipping magnitude. (6) differentiation. A shaping process in which a pulse is converted to a wave whose shape is or approximates the time derivative of the pulse. (7) integration. A shaping process in which a pulse is converted to a wave whose shape is or approximates the time integral of the pulse.   254

shaping pulse. The intentional processing of a pulse waveform to cause deviation from a reference waveform. *See:* **pulse.**   185

shaping time constant (semiconductor radiation detectors). The time constants of the bandwidth defining CR (capacitance-resistance) differentiators and RC (resistance-capacitance) integrators used in pulse amplifiers.   23

shared-logic word processing (computer applications). Word processing performed on a system composed of multiple work stations that share the logic and storage sections of a single central processor. *See:* **clustered word processing; dedicated word processing; shared-resource word processing; stand-alone word processing.**   571

shared-resource word processing (computer applications). Word processing performed on a system compose of multiple work stations, each with its own processor but sharing certain resources such as printers and disk drives. *See:* **clustered word processing; dedicated word processing; share-logic word processing; stand-alone word processing.**   571

sharing. *See:* **time sharing.**

sharing transformer (ST), current balancing transformer (CBT), (current balancing reactor)(electrical heating applications to melting furnaces and fore-hearths in the glass industry). Two-winding, iron core devices used in paralleled current paths, connected so that any difference in current between the paths causes an induced voltage that opposes the current difference.   520

shearing machine. An electrically driven machine for making vertical cuts in coal.   328

shear pin (rotating machinery). A dowel designed to shear at a predetermined load and thereby prevent damage to other parts. *See:* **rotor (rotating machinery); shear section shaft.**   63

shear section shaft (rotating machinery). A section of shaft machined to a controlled diameter, or area, designed to shear at a predetermined load and thereby

prevent damage to connected machinery. *See:* **rotor (rotating machinery); shear pin.**   63

shear wave (acoustics) (rotational wave). A wave in an elastic medium that causes an element of the medium to change its shape without a change of volume. *Notes:* (1) Mathematically, a shear wave is one whose velocity field has zero divergence. (2) A shear plane wave in an isotropic medium is a transverse wave. (3) When shear waves combine to produce standing waves, linear displacements may result.   176

sheath (1)(electrical heat tracing for industrial applications). The outermost continuous covering for a cable.   523

(2)(jacket) (cable systems in power generating stations). The overall protective covering for the insulated cable.   35

sheath temperature (electrical heat tracing for industrial applications). The temperature of the outermost continuous covering that may be exposed to the surrounding atmosphere.   523

sheave (1) (conductor stringing equipment). (A) The grooved wheel of a traveler or rigging block. Travelers are frequently referred to as sheaves. (B) A shaft-mounted wheel used to transmit power by means of a belt, chain, band, etcetera. *Syn:* **pulley; roller; wheel.**   431

(2) (rotating machinery). *See:* **pulley.**

shelf corrosion (dry cell). The consumption of the negative electrode as a result of local action. *See:* **electrolytic cell.**   328

shelf depreciation. The depreciation in service capacity of a primary cell as measured by a shelf test. *See:* **battery (primary or secondary).**   328

shelf test. A storage test designed to measure retention of service ability under specified conditions of temperature and cutoff voltage. *See:* **battery (primary or secondary).**   328

shell (1) (insulators). A single insulating member, having a skirt or skirts without cement or other connecting devices intended to form a part of an insulator or an insulator assembly. *See:* **insulator.**   261

(2) (electrolysis). The external container in which the electrolysis of fused electrolyte is conducted. *See:* **fused electrolyte.**   328

(3) (electrotyping). A layer of metal (usually copper or nickel) deposited upon, and separated from, a mold. *See:* **electroforming.**   328

shell-form transformer (power and distribution transformer). A transformer in which the laminations constituting the iron core surround the windings and usually enclose the greater part of them.   53

shell, stator (rotating machinery). A cylinder in tight assembly around the wound stator core, all or a portion of which is machined or otherwise made to a specific outer dimension so that the stator may be mounted into an appliance, machine, or other end product. *See:* **stator.**   63

shell-type motor. A stator and rotor without shaft, end shields, bearings or conventional frame. *Note:* A shell-type motor is normally supplied by a motor manufacturer to an equipment manufacturer for incorporation as a built-in part of the end product. Separate fans or fans larger than the rotor are not included. *See:* **asynchronous machine.** 63

sheltered equipment (test, measurement and diagnostic equipment). Equipment so housed or otherwise protected that the extreme of natural and induced environments are partially or completely excluded or controlled. Examples are laboratory and shop equipment, equipment shielded from sun by a canopy or roof, and so forth. 54

SHF. *See:* **radio spectrum.**

shield (1)(instrumentation cable)(design and installation of cable systems for Class 1E circuits in nuclear power generating stations). Braid copper, metallic sheath, or metallic coated polyester tape (usually copper or aluminum), applied over the insulation of a conductor or conductors for the purpose of reducing elecrostatic coupling between the shielded conductors and others that may be either susceptible to, or generators of, electrostatic fields (noise). When electromagnetic shielding is intended, the term electromagnetic is usually included To indicate the difference in shielding requirement and material. 536

(2) (cable systems in power generating stations). As normally applied to instrumentation cables, refers to metallic sheath (usually copper or aluminum), applied over the insulation of a conductor or conductors for the purpose of providing means for reducing electrostatic coupling between the conductors so shielded and others which may be susceptible to or which may be generating unwanted (noise) electrostatic fields. When electromagnetic shielding is intended, the term "electromagnetic" is usually included to indicate the difference in shielding requirements as well as material. To be effective at power system frequencies, electromagnetic shields would have to be made of high-permeability steel. Such shielding material is expensive and is not normally applied. Other less expensive means for reducing low-frequency electromagnetic coupling, as described herein, are preferred. 35

(3) (power and distribution transformer). A conductive protective member placed in relationship to apparatus or test components to control the shape of magnitude, or both, of electric or magnetic fields, thereby improving performance of apparatus or test equipment by reducing losses, voltage gradients, or interface. 53

(4) (electromagnetic). A housing, screen, or other object, usually conducting, that substantially reduces the effect of electric or magnetic fields on one side thereof, upon devices or circuits on the other side. *See:* **dielectric heating; induction heating; industrial electronics; signal.** 341

(5) (mechanical protection) (rotating machinery). An internal part used to protect rotating parts or parts of the electric circuit. In general, the word **shield** will be preceded by the name of the part that is being protected. 63

(6) (magnetrons). *See:* **end shield.**

(7) (induction heating). A material used to suppress the effect of an electric or magnetic field within or beyond definite regions. 14

(8) (metallic conductors). A housing or other object that substantially reduces the effect of electric or magnetic fields on one side thereof upon devices or circuits on the other side. 57

shielded conductor cable. A cable in which the insulated conductor or conductors is.are enclosed in a conducting envelope or envelopes. 345

shielded ignition harness. A metallic covering for the ignition system of an aircraft engine, that acts as a shield to eliminate radio interference with aircraft electronic equipment. The term includes such items as ignition wiring and distributors when they are manufactured integral with an ignition shielding assembly. 328

shielded insulated splice (power cable joint). An insulated splice in which a conducting material is employed over the full length of the insulation for electric stress control. 34

shielded joint. A cable joint having its insulation so enveloped by a conducting shield that substantially every point on the surface of the insulation is at ground potential or at some predetermined potential with respect to ground. 64

shielded-loop antenna (probe). An electrically-small antenna consisting of a tubular electrostatic shield formed into a loop with a small gap, and containing one or more wire turns for external coupling. 111

shielded pair (signal-transmission system). A two-wire transmission line surrounded by a sheath of conducting material to protect it from the effects of external fields, or to confine fields produced by the transmission line. *See:* **signal; waveguide.** 267

shielded strip transmission line. A strip conductor between two ground planes. Some common designations are: Stripline (trade mark): Tri-plate (trade mark): slab line (round conductor): balanced strip line. *See:* **strip (-type) transmission line; unshielded strip transmission line.** 179

shielded transmission line (1) (signal-transmission system). A transmission line surrounded by a sheath of conducting material to protect it from the effects of external fields, or to confine fields produced by the transmission line. *See:* **signal; waveguide.** 188

(2) (waveguide). A transmission line whose elements essentially confine propagated electrical energy to a finite space inside a conducting sheath. 267

shielded-type cable. A cable in which each insulated conductor is enclosed in a conducting envelope so constructed that substantially every point on the surface of the insulation is at ground potential or at some predetermined potential with respect to ground under normal operating conditions. 64

shield factor (telephone circuit). The ratio of noise, induced current, or voltage when a source of shielding is present, to the corresponding quantity when the shielding is absent. *See:* **induction coordination.** 328

shield grid (gas tubes). A grid that shields the control electrode in a gas tube from the anode or the cathode, or both, with respect to the radiation of heat and the deposition of thermionic activating material and also reduces the electrostatic influence of the anode. It may be used as a control electrode. *See:* **electrode (electron tube); grid.**   125

shielding (1) (power cable joint)(screening). A conducting layer, applied to control the dielectric stresses within tolerable limits and minimize voids. It may be applied over the entire joint insulation, on the tapered insulation ends only, or over irregular conductor or connector surfaces.   34

(2)(X-radiation limits for ac high-voltage power vacuum interrupters used in power switchgear). Barrier of attenuating material used to reduce radiation hazards.   553

shielding angle (1) (of a luminaire) (illuminating engineering). The angle between a horizontal line through the light center and the line of sight at which the bare source first becomes visible.   167

(2) (lightning protection). The angle between the vertical line through the overhead ground wire and a line connecting the overhead ground wire with the shielded conductor. *See:* **direct stroke protection (lightning).**   64

shielding effectiveness (1)(eelectromagnetic compatibility). (A) A measure of the ability of a shield to exclude or confine electromagnetic waves. Usually expressed as the ratio (in the frequency domain) of the incident to the penetrating signal amplitudes in decibels. (B) (Screen rooms and similar enclosures). The ratio of the signal received (from a transmitter) without the shield to the signal received inside the shield; the insertion loss when the shield is placed between the transmitting antenna and the receiving antenna. (C) (Shielded wires and cables). The ratio of the total cable current (shield current plus core wire current) to the current in the core alone, when both are due to sources outside the shield.   199, 415

(2)(measurement of shielding effectiveness of high-performance shielding enclosures). In general, fields penetrating a shielding enclosure arise from both the electric and magnetic components of the electromagnetic energy impinging upon the enclosure. If the penetrating magnetic or electric fields are measured separately, each can be demonstrated to be a function of both the electric and magnetic components of the impinging waveform. In addition, the applied fields are radically altered in penetrating a high-performance enclosure, and the measured penetrating waveforms are affected by the position of the sensor. As a consequence, measurement results are highly sensitive to the test procedures employed, and simple amplitude ratios between any external and internal electric field vectors (or the magnetic field vectors) alone do not correctly characterize the shielding phenomenon. For similar considerations, simple measurements of power-flow ratios in the conventional sense cannot be employed without reference to a standardized test procedure. As a result a definition of a measure of enclosure performance is set forth and must be associated with a specific test and procedure. This measure is termed 'shielding effectiveness'. In the low-frequency region (200 hertz-20 megahertz) shielding effec-

tiveness is expressed by

$$S_H = 20 \log_{10} \left(\frac{H_1}{H_2} \right) \tag{1}$$

where

H_1 = magnetic field in absence of enclosure
H_2 = magnetic field within the enclosure.

For the ultrahigh frequency range (300–1000 megahertz), the shielding effectiveness* is expressed by

$$S_E = 20 \log_{10} \left(\frac{E_1}{E_2} \right) \tag{2}$$

where

E_1 = electric field in absence of enclosure
E_2 = electric field within the enclosure.

For the microwave region (1.7–12.4 gigahertz), the effectiveness* is defined in terms appropriate to the detector indication and is expressed by one of the following equations:

$$S_P = 10 \log_{10} \left(\frac{P_1}{P_2} \right) \tag{3}$$

where

P_1 = power detected in the absence of the enclosure
P_2 = power detected within the enclosure;

$$S_E = 20 \log_{10} \left(\frac{E_1}{E_2} \right) \tag{4}$$

where E_1 and E_2 are defined as in Equation (2); or

$$S_P = A_1 - A_2 \tag{5}$$

where

A_1 = attenuator setting in decibels for a specified detector output in the absence of the enclosure
A_2 = attenuator setting for the same detector output within the enclosure.

* This definition applies only to a specific measurement procedure described in subsequent sections.
     525

shielding failure (lightning protection). The occurrence of a lightning stroke that bypasses the overhead ground wire and terminates on the phase conductor. *See:* **direct-stroke protection (lightning).**   64

shield wire (electromagnetic fields). A wire employed for the purpose of reducing the effects on electric supply or communication circuits from extraneous sources. *See:* **inductive coordination.**   328

shift (mathematics of computing). A displacement of an ordered set of characters one or more places to the left or right. If the characters are the digits of a numeral, a shift may be equivalent to multiplying by a power of the radix. *See:* **arithmetic shift; logical shift.**   564

shift clock (C) (semiconductor memory). The inputs that when operated in a prescribed manner shift internal data in a serial memory.   441

shift, direct-current (oscilloscopes). A deviation of the displayed response to an input step, occurring over a period of several seconds after the input has reached its final value.   106

shift pulse. A drive pulse that initiates shifting of characters in a register.   331

shift register. A register in which the stored data can be moved to the right or left.   255, 77

shim (rotating machinery) (1) mechanical. A lamination usually machined to close-tolerance thickness, for assembly between two parts to control spacing. (2) magnetic. A lamination added to adjust or change the effective air gap in a magnetic circuit. *See:* **rotor (rotating machinery); stator.**   63

shingle (photoelectric converter). Combination of photoelectric converters in series in a shingle-type structure. *See:* **semiconductor.**   186

ship control telephone system. A system of sound-powered telephones (requiring no external power supply for talking) with call bells, exclusively for communication among officers responsible for control and operation of a ship. *Note:* Call bells are usually energized by hand-cranked magneto generators.   328

shipping brace (rotating machinery). Any structure provided to reduce motion or stress during shipment, that must be removed before operation.   63

shipping seal (cable). *See:* **sealed end.**

ship's service electric system. On any vessel, all electric apparatus and circuits for power and lighting, except apparatus provided primarily either for ship propulsion or for the emergency system. *Note:* Emergency and interior communication circuits are normally supplied with power from the ship's service system, upon failure of which they are switched to an independent emergency generator or other sources of supply. *See:* **marine electric apparatus.**   328

shock excitation (oscillatory systems). The excitation of natural oscillations in an oscillatory system due to a sudden acquisition of energy from an external source or a sudden release of energy stored with the oscillatory system. *See:* **oscillatory circuit.**   328

shock motion (mechanical system). Transient motion that is characterized by suddenness, by significant relative displacements, and by the development of substantial internal forces in the system. *See:* **mechanical shock.**   176

shockproof electric apparatus. Electric apparatus designed to withstand, to a specified degree, shock of specified severity. *Note:* The severity is stated in foot-pounds impact on a special test stand equivalent to shock of gunfire, explosion of mine or torpedo, etcetera. *See:* **marine electric apparatus.**   328

shoe (ramp shoe). Part of a vehicle-carried apparatus that makes contact with a ramp.   328

shop instruments. Instruments and meters that are used in regular routine shop or field operations.   212

shop test (laboratory test). A test made upon the receipt of a meter from a manufacturer, or prior to reinstallation. Such tests are made in a shop or a laboratory of a meter department. *See:* **service test (field test).**   328

shoran (navigation aid terms). A radio navigation system which provides circular lines of position. The term is derived from the words short-range navigation.   526

shore feeder. Permanently installed conductors from a distribution switchboard to a connection box (or boxes) conveniently located for the attachment of portable leads for supply of power to a ship from a source on shore. *See:* **marine electric apparatus.**   328

short circuit (1)(gas-tube surge-protective devices). An abnormal connection of relatively low impedance, whether made accidentally or intentionally, between two points of different potential in a circuit.   490
(2)(low-voltage air gap surge-protective devices). An abnormal connection of relatively low impedance, whether made accidentally or intentionally, between two points of different potential in a circuit. *Note:* The term is often applied to the group of phenomena that accompany a short circuit.   556
(3) (power switchgear). An abnormal connection (including an arc) of relatively low impedance, whether made accidentally or intentionally, between two points of different potential. *Note:* The term fault or short- circuit fault is used to describe a short circuit.   103

short circuit, adjustable, waveguide (waveguide components). A longitudinally movable obstacle which reflects essentially all the incident energy.   166

short-circuit current (protection and coordination of industrial and commercial power systems). An overcurrent usually defined as being in excess of ten times normal continuous rating. *See:* **overload.**   504

short-circuit driving-point admittance. *See:* **admittance, short-circuit driving-point.**

short-circuit duration rating (magnetic amplifier). The length of time that a short circuit may be applied to the load terminals nonrecurrently without reducing the intended life of the amplifier or exceeding the specified temperature rise.   171

short-circuiter. A device designed to short circuit the commutator bars when the motor has attained a predetermined speed in some forms of single-phase commutator-type motors. *See:* **asynchronous machine.**   328

short circuit, external dc (thyristor power converter). A short circuit on the dc side outside the converter. *Note:* External short circuits may require different protecting means, depending on the character of the short circuit. Complete dc short circuit occurs when the short-circuit impedance is negligible compared to internal impedance of the converter. Limited dc short circuit occurs when the short-circuit impedance is large enough to limit the fault current. Feeder dc short circuit is a short circuit in a feeder with a separate protective device with much lower rating than the feeding converter (multimotor drives).   121

short-circuit feedback admittance. *See:* **admittance, short-circuit feedback.**

short-circuit flux (magnetic sound records). That flux from a magnetic record which flows across a plane normal to the recorded medium, through a magnetic short circuit placed in intimate contact with the record.   161

short-circuit flux per unit width (magnetic sound rec-

ords). The measured short-circuit flux divided by the measured width of the recorded track. *Note:* The term **fluxivity** has been proposed for the quantity short-circuit flux per unit width. 161

short-circuit forward admittance. *See:* **admittance, short-circuit forward.**

short-circuit impedance (1) (general). A qualifying adjective indicating that the impedance under consideration is for the network with a specified pair or group of terminals short-circuited. *See:* **network analysis; self-impedance.** 210

(2) (line or four-terminal network). The driving-point impedance when the far-end is short-circuited. *See:* **self-impedance.** 328

short-circuit inductance. The apparent inductance of a winding of a transformer with one or more specified windings short circuited often taken as a means of determining the leakage inductance of a winding. 197, 67

short-circuiting or grounding device (57) (power system device function numbers). A primary circuit switching device that functions to short circuit or to ground a circuit in a response to automatic or manual means. 402

short circuiting or grounding relays (wire-line communication facilities). Relay used to ground an exposed communication or telephone pair, usually on open-wire joint use during periods of severe power system disturbance. 414

short circuiting switch (test switches for transformer-rated meters). A single-pole double-throw transfer switch (make before break) used to transfer current away from the meter. 590

short-circuit input admittance. *See:* **admittance, short-circuit input.**

short-circuit input volt-amperes (ferroresonant voltage regulators). The product of the input voltage and input current (root-mean-square values) with the resonating winding short-circuited). 456

short circuit, internal (thyristor power converter). A short circuit caused by converter faults. *Note:* An internal short circuit may be fed from both ac and dc circuits: for example, in the cases of (1) converters with battery or motor loads, (2) converters in a double converter, (3) converters operating as inverters. 121

short-circuit loss (rotating machinery). The difference in power required to drive a machine at normal speed, when excited to produce a specified balanced short-circuit armature current, and the power required to drive the unexcited machine at the same speed. *See:* **asynchronous machine.** 63

short-circuit output admittance. *See:* **admittance, short-circuit output.**

short-circuit output capacitance. *See:* **capacitance, short-circuit output.**

short-circuit protection (power supplies) (automatic). Any automatic current-limiting system that enables a power supply to continue operating at a limited current, and without damage, into any output overload including short circuits. The output voltage must be restored to normal when the overload is removed, as distinguished from a fuse or circuit-breaker system that opens at overload and must be closed to restore power. *See:* **current limiting.** 186

short-circuit ratio (harmonic control and reactive compensation of static power converters). Of a semiconductor converter is the ratio of the short-circuit capacity of the bus in megavoltamperes (MVA) at the point of converter connection to the rating of the converter in megawatts (MW). 533

short-circuit saturation curve (synchronous machine). The relationship between the current in the short-circuited armature winding and the field current. 63

short-circuit time constant (primary winding) (rotating machinery). The time required for the direct-current component present in the short-circuit primary-winding current following a sudden change in operating conditions to decrease to $1/e \, \Delta \, 0.368$ of its initial value, the machine running at rated speed. *See:* **asynchronous machine.** 63

short-circuit transfer admittance. *See:* **admittance, short-circuit transfer.**

short-circuit transfer capacitance. *See:* **capacitance, short-circuit transfer.**

short dimension. Incremental dimensions whose number of digits is the same as normal dimensions except the first digit is zero, that is 0.XXXX for the example under normal dimension. *See:* **dimension; incremental dimension; long dimension; normal dimension.** 207

short-distance navigation. Navigation utilizing aids usable only at comparatively short distances; this term covers navigation between approach navigation and long-distance navigation, there being no distinct, universally accepted demarcation between them. *See:* **navigation.** 187, 13

short field (tapped field*). Where two field strengths are required for a series machine, short field is the minimum-strength field connection. *See:* **asynchronous machine.**
*Deprecated 328

short-line-fault transient recovery voltage (power switchgear). The transient recovery voltage obtained when a circuit switching device interrupts a nearby fault on the line. *Note:* It differs from terminal fault conditions in that the length of line adds a high-frequency sawtooth component to the transient recovery voltage. As the distance to the fault becomes greater, the amplitude of the sawtooth component increases, the rate of rise of the sawtooth component decreases and the fault current decreases. The increased amplitude adversely affects the interrupting capability of the circuit switching device while the decrease in rate of rise and the decrease in current makes interruption easier. The effects are not proportional and a distance is reached where interruption is most severe even though the current is less than for a terminal fault. The critical value varies considerably with the type of circuit switching device (oil, air-blast, gas-blast, etcetera), and with the particular design. The critical distance may be in the order of a mile at the higher voltages. The critical distance is less as lower voltages are considered. 103

short-time current (power switchgear). The current carried by a device, an assembly, or a bus for a specified short time interval. 103

short-time current rating (separable insulated connectors). The designated root-mean-square current which a connector can carry for a specified time under specified conditions. 454

short-time current tests (high-voltage switchgear). Tests that consist of the application of a current higher than the rated continuous current for specified short periods to determine the adequacy of the device to withstand short-circuit currents for the specified short times. 443

short-time delay phase or ground trip element (power switchgear). A direct-acting trip device element that functions with a purposely delayed action (milliseconds). 103

short time duty (1) (National Electrical Code). Operation at a substantially constant load for a short and definitely specified time. 256

(2) (power and distribution transformer). A duty that demands operation at a substantially constant load for a short and definitely specified time. 53

short-time overload rating (rotating electric machinery). The output that the machine can sustain for a specified time starting hot under the conditions of Section 4 of IEEE Std 11-1980, without exceeding the limits of temperature rise of Section 5. 424

short-time rating (1) (packaging machinery). The rating that defines the load which can be carried for a short and definitely specified time, the machine apparatus, or device being at approximately room temperature at the time the load is applied. 429

(2) (power and distribution transformer). Defines the maximum constant load that can be carried for a specified short time without exceeding established temperature-rise limitations, under prescribed conditions. 53

short-time rating of a relay (power switchgear). The highest value of current or voltage or their product that the relay can stand, without injury, for specified short-time intervals (for alternating-current circuits, root-mean-square total value including the direct-current component is used). The rating recognizes the limitations imposed by both the thermal and electromagnetic effects. 103

short time rating of diesel-generator unit (nuclear power generating stations). The electric power output capability that the diesel-generator unit can maintain in the service environment for 2 hours in any 24 hour period, without exceeding the manufacturer's design limits and without reducing the maintenance interval established for the continuous rating. *Note:* Operation at this rating does not limit the use of the diesel-generator unit at its continuous rating. 99

short-time thermal current rating (current transformer). The 1s thermal current rating of a current transformer is the root-mean-square (rms) symmetrical primary current that can be carried for 1 s with the secondary winding short-circuited without exceeding in any winding the limiting temperature. 203

short-time waveform distortion (SD) (linear waveform distortion). Distortion of time components from 125 ns to 1 μs, that is, time components of the short-time domain. 142

shot noise (fiber optics). Noise caused by current fluctuations due to the discrete nature of charge carriers and random or unpredictable (or both) of charged particles from an emitter. *Note:* There is often a (minor) inconsistency in referring to shot noise in an optical system: many authors refer to shot noise loosely when speaking of the mean square shot noise current (amp^2) rather than noise power (watts). *See:* **quantum noise.** 433

should (1) (binary floating-point arithmetic). The use of the word 'should' signifies that which is strongly recommended as being in keeping with the intent of the standard, although architectural or other constraints beyond the scope of ANSI/IEEE Std 754-1985 may on occasion render the recommendations impractical. 469

(2) ('dose calibrator' ionization chambers). Indicates an advisory recommendation that is to be applied when practicable. 499

(3) (high-level microprocessor language). The use of the word 'should' signifies that which is strongly recommended--that is, that which is in keeping with the intent of the standard, despite architectural or other constraints beyond the scope of IEEE trial use Std 755-1985 that, on occasion, may render the recommendations impractical. 470

(4) (information transfer). The use of the word 'should' signifies that which is strongly recommended as being in keeping with the intent of IEEE Trial Use Std 949, despite architectural or other constraints that may on occasion render the recommendations impractical. 479

(5) (microprocessor assembly language). The use of the word 'should' signifies that which is strongly recommended (advised) as being in keeping with the intent of IEEE Std 694-1985. 466

(6) (sodium iodide detector). In this text, indicates an action that is included when practicable. 423

shoulder (NESC). The portion of the roadway contiguous with the traveled way for accomodation of stopped vehicles for emergency use and for lateral support of base and surface course. 494

shoulder lobe (antennas). A radiation lobe which has merged with the major lobe, thus causing the mjor lobe to have a distortion which is shoulder-like in appearance when displayed graphically. *Syn:* **vestigial lobe.** 111

show window (National Electrical Code). Any window used or designed to be used for the display of goods or advertising material, whether it is fully or partly enclosed or entirely open at the rear and whether or not it has a platform raised higher than the street floor level. 256

shrink link (rotating machinery). A bar with an enlarged head on each end for use like a rivet but slipped into place after expansion by heat. It tightens on cooling by shrinkage only.   263

shunt (1) (air switch). A flexible, electrical conductor comprised of braid, cable, or flat laminations designed to conduct current around the mechanical joint between two conductors.   103

(2) (general). A device having appreciable resistance or impedance connected in parallel across other devices or apparatus, and diverting some (but not all) of the current from it. Appreciable voltage exists across the shunted device or apparatus and an appreciable current may exist in it.   210

shunt capacitor bank current (power switchgear). Current, including harmonics, supplied to a shunt capacitor bank.   103

shunt control. A method of controlling motors employing the shunt method of transition from series to parallel connections of the motors. *See:* **multiple-unit control.**   328

shunt excitation (emergency and standby power). A source of generator field excitation power taken from the generator output, normally through power potential transformers connected directly or indirectly to the generator output terminals.   512

shunt-fed vertical antenna. A vertical antenna which is connected directly to ground at its base and whose feed line connects between ground and a point suitably positioned above the base.   111

shunt filter (harmonic control and reactive compensation of static power converters). That type of filter which reduces harmonics by providing a low impedance path to shunt the harmonics from the source away from the system to be protected.   533

shunt gap (metal-oxide surge arresters for ac power circuits). An intentional gap(s) between spaced electrodes that is electrically in parallel with one or more valve elements.   583

shunting or discharge switch (17) (power system device function numbers). A switch that serves to open or to close a shunting circuit around any piece of apparatus (except a resistor), such as a machine field, a machine armature, a capacitor, or a reactor. *Note:* This excludes devices that perform such shunting operatings as may be necessary in the process of starting a machine by [a starting circuit breaker or a running circuit breaker,] devices 6 or 42, or their equivalent, and also excludes [a load-resistor contactor,] device function 73 that serves for the switching of resistors.   402

shunting transition. *See:* **shunt transition.**

shunt leads (instrument). Those leads that connect a circuit of an instrument to an external shunt. The resistance of these leads is taken into account in the adjustment of the instrument. *See:* **auxiliary device to an instrument; instrument.**   280

shunt loading. Loading in which reactances are applied in shunt across the conductors of a transmission circuit. *See:* **loading.**   328

shunt noninterfering fire-alarm system. A manual fire-alarm system employing stations and circuits such

that, in case two or more stations in the same premises are operated simultaneously, the signal from the operated box electrically closest to the control equipment is transmitted and other signals are shunted out. *See:* **protective signaling.**   328

shunt reactor (1)(power and distribution transformer). A reactor intended for connection in shunt to an electric system for the purpose of drawing inductive current. *Note:* The normal use for shunt reactors is to compensate for capacitive currents from transmission lines, cable, or shunt capacitors. The need for shunt reactors is most apparent at light load.   53

(2)(shunt reactors over 500 kVA). A reactor intended for connection in shunt to an electric system for the purpose of drawing inductive current. *Note:* The normal use for shunt reactors is to compensate for capacitive currents from transmission lines, cable, or shunt capacitors. The need for shunt reactors is most apparent at light load.   562

shunt regulator (power supplies). A device placed across the output that controls the current through a series dropping resistance to maintain a constant voltage or current output.   186

shunt release (trip) (power switchgear). A release energized by a source of voltage. *Note:* The voltage may be derived either from the main circuit or from an independent source.   103

shunt snubber (converter circuit elements)(self-commutated converters). Circuit elements, usually including a capacitor, connected in shunt with a switching device to limit the rate of rise of voltage or the peak voltage across the device (or both) when switching from a conducting to a blocking state, or when subjected to an external voltage transient.   584

shunt tee junction. *See:* **H-plane tee junction.**

shunt transition (shunting transition). A method of changing the connection of motors from series to parallel in which one motor, or group of motors, is first shunted or short circuited, then open circuited, and finally connected in parallel with the other motor or motors. *See:* **multiple-unit control.**   328

shunt trip. *See:* **shunt release.**

shunt-trip recloser (power switchgear). A recloser in which the tripping mechanism, by releasing the holding means, permits the main contacts to open, with both the tripping mechanism and the contact opening mechanism deriving operating energy from other than the main circuit.   103

shunt-wound. A qualifying term applied to a direct-current machine to denote that the excitation is supplied by a winding connected in parallel with the armature in the case of a motor, with the load in the case of a generator, or is connected to a separate source of voltage.   328

shunt-wound generator (electric installations on shipboard). A dc generator in which ordinarily the entire field excitation is derived from one winding consisting of many turns with a relatively high resistance. This one winding is connected in parallel with the armature circuit for a self excited generator, and to the load side of another generator or other source of direct current, for a separately excited generator.   3

shunt-wound motor (electric installations on shipboard). A dc motor in which the field circuit and armature circuit are connected in parallel. 3

shutter (1) (of a switchgear assembly). A device that is automatically operated to completely cover the stationary portion of the primary disconnecting devices when the removable element is either in the disconnected position, test position, or has been removed. 103

(2) (electric machine). A protective covering used to close, or to close partially, an opening in a stator frame or end shield. In general, the word **shutter** will be preceded by the name of the part to which it is attached. As used for an electric machine, a shutter is rigid and hence not adjustable. *See:* **stator.** 263

shuttle car. A vehicle on rubber tires or caterpillar treads and usually propelled by electric motors, electric energy for which is supplied by a diesel-driven generator, by storage batteries, or by a power distribution system through a portable cable. Its chief function is the transfer of raw materials, such as coal and ore, from loading machines in trackless areas of a mine to the main transportation system. 328

shuttle car, explosion-tested. A shuttle car equipped with explosion-tested equipment. 328

SI (International System of Units). *See:* **units and letter symbols.**

side back light (illuminating engineering). Illumination from behind the subject in a direction not parallel to a vertical plane through the optical axis of the camera. 167

sideband attenuation. That form of attenuation in which the transmitted relative amplitude of some component(s) of a modulated signal (excluding the carrier) is smaller than that produced by the modulation process. *See:* **wave front.** 111

sideband null (rectilinear navigation system)(navigation aid terms). The surface of position along which the resultant energy from a particular pair of sideband antennas is zero. 526

sideband-reference glide slope (ILS [instrument landing system])(navigation aid terms). A modified null reference glide-slope antenna system in which the upper (sideband) antenna is replace with two antennas, both at lower heights, and fed out of phase, so that a null is produced at the desired glide-slope angle. *Note:* This system is used to reduce unwanted reflections of energy into the glide-slope sector at locations where rough terrain exists in front of the approach end of the runway, by producing partial cancellation of energy at low elevation angles. 526

sidebands (1) (antennas). (A) The frequency bands on both sides of the carrier frequency within which fall the frequencies of the wave produced by the process of modulation. (B) The wave components lying within such bands. *Note:* In the process of amplitude modulation with a sine-wave carrier, the upper sideband includes the sum (carrier plus modulating) frequencies: the lower sideband includes the difference (carrier minus modulating) frequencies. *See:* **amplitude modulation; radio receiver.** 111

(2) (data transmission). A band of frequencies containing components of either the sum (upper side band) or difference (lower sideband) of the carrier and modulation frequencies. 59

sideband suppression (power-system communication). A process that removes the energy of one of the sidebands from the modulated carrier spectrum. 179

side-break switch (power switchgear). One in which the travel of the blade is in a plane parallel to the base of the switch. 103

side circuit (data transmission). A circuit arranged for deriving a phantom circuit. *Note:* In the case of two-wire side circuits, the conductors of each side circuit are placed in parallel to form a side of the phantom circuit. In the case of four-wire side circuits, the lines of the two side circuits which are arranged for transmission in the same direction provide a one-way phantom channel for transmission in that same direction, the two conductors of each line being placed in parallel to provide a side for that phantom channel. Similarly the conductors of the other two lines provide a phantom channel for transmission in the opposite direction. 59

side-circuit loading coil. A loading coil for introducing a desired amount of inductance in a side circuit and a minimum amount of inductance in the associated phantom circuit. *See:* **loading.** 328

side-circuit repeating coil (side-circuit repeat coil). A repeating coil that functions simultaneously as a transformer at a terminal of a side circuit and as a device for superposing one side of a phantom circuit on that side circuit. 328

side-effect (software). Processing or activities performed, or results obtained, secondary to the primary function of a program, subprogram, or operation. *See:* **function; program; subprogram.** 434

sideflash (lightning). A spark occurring between nearby metallic objects or from such objects to the lightning protection system or to ground. *See:* **direct-stroke protection (lightning).** 297, 64

side flashover (lightning). A flashover of insulation resulting from a direct lightning stroke that bypasses the overhead ground wire and terminates on a phase conductor of a transmission line. *See:* **direct-stroke protection (lightning).** 64

side frequency. One of the frequencies of a sideband. *See:* **amplitude modulation.** 111, 415

side lobe (antennas). A radiation lobe in any direction other than that of the major lobe. *See:* **back lobe; copolar side lobe; cross-polar side lobe; mean side lobe level; minor lobe; side lobe level (maximum relative; relative).** 111

sidelobe blanker (radar). A device which employs an auxiliary wide-angle antenna and receiver to sense whether a received pulse originates in the sidelobe region of the main antenna and to gate it from the output signal if it does. 13

sidelobe canceler (or canceller) (radar). A device which employs one or more auxiliary antennas and receivers to allow linear subtraction of interfering signals from the desired output if they are sensed to originate in the sidelobes of the main antenna. 13

side lobe level (antennas). (1) Maximum relative. The maximum relative directivity of the highest side lobe with respect to the maximum directivity of the antenna. (2) Relative. The maximum relative directivity of a side lobe with respect to the maximum directivity of an antenna, usually expressed in decibels. 111

side lobe suppression (antennas). Any process, action or adjustment to reduce the level of the side lobes or to reduce the degradation of the intended antenna system performance resulting from the presence of side lobes. 111

side-lock. Spurious synchronization in an automatic frequency synchronizing system by a frequency component of the applied signal other than the intended component. *See:* **television.** 178

sidelooking radar (navigation aid terms). A ground mapping radar, used aboard aircraft, involving the use of a fixed antenna beam pointing out of the side of an aircraft either abeam or squinted with respect to the aircraft axis. The beam is usually a vertically-oriented fan beam having a narrow azimuth width. The narrow azimuth resolution can either be obtained with a long aperture mounted along the axis of the aircraft or by the use of synthetic aperture radar processing. 526

side marker lights (illuminating engineering). Lamps indicating the presence of a vehicle when seen from the front and sometimes serving to indicate its width. When seen from the side they may also indicate its length. 167

side panel (rotating machinery). A structure enclosing or partly enclosing one side of a machine. 63

sidereal (navigation aid terms). Of or pertaining to the stars. 526

sidereal period (communication satellite). The time duration of one orbit measured relative to the stars. 74

side thrust (skating force) (disk recording). The radial component of force on a pickup arm caused by the stylus drag. *See:* **phonograph pickup.** 176

sidetone (transmission performance of telephone sets). The acoustic output level of a telephone set due to an acoustic input to the same telephone set. The acoustic input may be varied either in frequency or level. The output is measured in an artificial ear and the input is measured at the calibration position of an artificial mouth. *Note:* Where the handset is mounted on a test fixture which includes the artificial mouth and artificial ear, the definition includes transmission through the handset proper; there may be also some vibration effect which is expected to be insignificant for handsets of modern design. 491

sidetone objective loudness rating (SOLR) (loudness ratings of telephone connections).

$$\text{SOLR} = -20 \log_{10} \frac{S_E}{S_M}$$

where

S_M = sound pressure at the mouth reference point (in pascals)

S_E = sound pressure at the ear reference point (in pascals) 409

sidetone path loss (telephony). The difference in dB of the acoustic output level of the receiver of a given telephone set to the acoustic input level of the transmitter of the same telephone set. 122

sidetone telephone set. A telephone set that does not include a balancing network for the purpose of reducing sidetone. *See:* **telephone station.** 328

sidewalk elevator. A freight elevator that operates between a sidewalk or other area exterior to the building and floor levels inside the building below such area, that has no landing opening into the building at its upper limit of travel, and that is not used to carry automobiles. *See:* **elevators.** 328

side-wall pressure (NESC). The crushing force exerted on a cable during installation. 494

siemens (metric practice). The electric conductance of a conductor in which a current of one ampere is produced by an electric potential difference of one volt. 21

sievert (metric practice). The dose equivalent when the absorbed dose of ionizing radiation multiplied by the dimensionless factors Q (quality factor) and N (product of any other multiplying factors) stipulated by the International Commission on Radiological Protection is one joule per kilogram. 21

sigma (σ)**.** The term **sigma** designates a group of telephone wires, usually the majority or all wires of a line, that is treated as a unit in the computation of noise or in arranging connections to ground for the measurement of noise or current balance ratio. *See:* **induction coordination.** 328

sign (1) (National Electrical Code). *See:* **electric sign** 256

(2) (power or energy). Positive, if the actual direction of energy flow agrees with the stated or implied reference direction: negative, if the actual direction is opposite to the reference direction. *See:* **network analysis.** 210

(3) (test, measurement and diagnostic equipment). The symbol which distinguishes positive from negative numbers. 54

signal (1)(signals and paths)(microcomputer system bus). The physical representation of data. 542

(2)(signals and paths)(696 interface devices). The physical representation which conveys data from one point to another. For the purpose of ANSI/IEEE Std 696-1983, this applies to digital electrical signals only. 538

(3) (data transmission). (A) A visual, audible or other indication used to convey information. (B) The intelligence, message or effect to be conveyed over a communication system. (C) A signal wave; the physical embodiment of a message. 59

(4) (overhead-power-line corona and radio noise). The intelligence, message, or effect to be conveyed over a communication system. 411

(5) (programmable instrumentation). The physical representation of information. *Note:* For the purpose of this standard, this is a restricted definition of what is often called "signal" in more general terms, and is hereinafter referred to digital electrical signals only.
378

(6) (computing systems). The event or phenomenon that conveys data from one point to another.
255, 77

(7) (control) (industrial control). Information about a variable that can be transmitted in a system.
206

(8) (telephone switching systems). An audible, visual or other indication of information. 255

(9) (circuits and systems). A phenomenon (visual, audible, or otherwise) used to convey information. The signal is often coded, such as a modulated waveform, so that it requires decoding to be intelligible.
67

signal, actuating (control system, feedback). *See:* **actuating signal.**

signal aspect. The appearance of a fixed signal conveying an indication as viewed from the direction of an approaching train: the appearance of a cab signal conveying an indication as viewed by an observer in the cab. 328

signal back light. A light showing through a small opening in the back of an electrically lighted signal, used for checking the operation of the signal lamp.
328

signal charge (ferroelectric device). The charge that flows when the condition of the device is changed from that of zero applied voltage (after having previously been saturated with either a positive or negative voltage) to at least that voltage necessary to saturate in the reverse sense. *Note:* The signal charge Q_s equals the sum of Q_t and Q_t, as illustrated in the accompanying figure. It is dependent on the magnitude of the applied voltage, which should be specified in describing this characteristic of ferroelectric devices. *See:* **ferroelectric domain.** 247

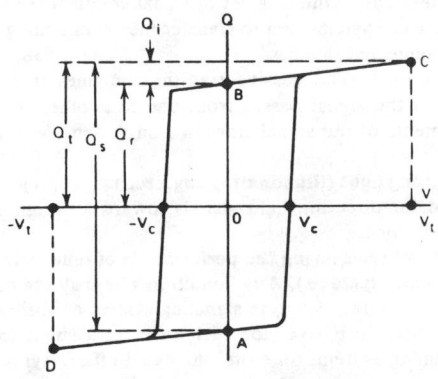

Hysteresis loop for a ferroelectric device

signal circuit (1). Any electric circuit that supplies energy to an appliance that gives a recognizable signal. Such circuits include circuits for door bells, buzzers, code-calling systems, signal lights, and the like. *See:* **appliances.** 256

(2) (protective relay system). Any circuit other than input voltage circuits, input current circuits, power supply circuits, or those circuits which directly or indirectly control power circuit breaker operation.
90

signal contrast (facsimile). The ratio expressed in decibels between white signal and black signal. *See:* **facsimile signal (picture signal).** 12

signal converter (test, measurement and diagnostic equipment). A device for changing a signal from one form or value to another form or value. 54

signal current (diode-type camera tube). The change in target current which occurs when the target is irradiated with photons, or electrons, compared to the case where no radiation is incident on the target.
380

signal decay time (measuring the performance of tone address signaling systems). The time interval between the end of the signal present condition and the beginning of the signal off condition at the end of the signal under consideration. 508

signal delay. The transmission time of a signal through a network. The time is always finite, may be undesired, or may be purposely introduced. *See:* **delay line; oscillograph.** 185

signal, difference. *See:* **differential signal.**

signal distance (computing systems). The number of digit positions in which the corresponding digits of two binary words of the same length are different. *See:* **Hamming distance.** 55, 77

signal distributing (telephone switching systems). Delivering of signals from a common control to other circuits. 55

signal duration (measuring the performance of tone address signaling systems). The time interval during which a signal present condition exists continuously.
508

signal electrode (camera tube). An electrode from which the signal output is obtained. *See:* **electrode (electron tube).** 178

signal element (unit interval) (data transmission). The part of a signal that occupies the shortest interval of signaling code. It is considered to be of unit duration in building up signal combinations. 59

signal, error (1) (automatic control device) (general). A signal whose magnitude and sign are used to correct the alignment between the controlling and the controlled elements. 328

(2) (power supplies). The difference between the output voltage and a fixed reference voltage compared in ratio by the two resistors at the null junction of the comparison bridge. The error signal is amplified to drive the pass elements and correct the output.
228, 186

(3) (closed loop) (control system, feedback). The signal resulting from subtracting a particular return signal from its corresponding input signal. See the accompanying figure. *See:* **control system, feedback.**

56, 105

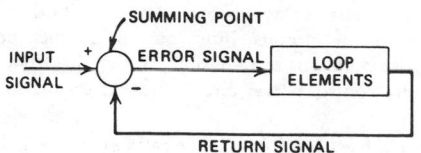

Block diagram of a closed loop

signal, feedback (1) (general). A function of the directly controlled variable in such form as to be used at the summing point. *See:* **control system, feedback.**

206

(2) (control system, feedback). The return signal that results from the reference input signal. See the accompanying figure. *See:* **control system, feedback.**

56, 105

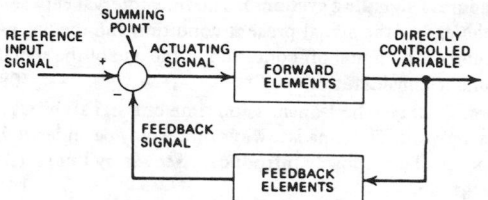

Simplified block diagram indicating essential elements of an automatic control system

signal flow graph (network analysis). A network of directed branches in which each dependent node signal is the algebraic sum of the incoming branch signals at that node. *Note:* Thus,

$$x_1 t_{1k} + x_2 t_{2k} + \ldots + x_n t_{nk} = x_k$$

at each dependent node k, where t_{jk} is the branch transmittance of branch jk.										282

signal frequency shift (frequency-shift facsimile system). The numerical difference between the frequencies corresponding to white signal and black signal at any point in the system. *See:* **facsimile signal (picture signal).**										12

signal generator. A shielded source of voltage or power, the output level and frequency of which are calibrated, and usually variable over a range. *Note:* The output of known waveform is normally subject to one or more forms of calibrated modulation.										185

signal identifier (spectrum analyzer). A means to identify the frequency of the input when spurious responses are possible. A front panel control used to identify the input frequency when spurious responses are present.										390

signal indication. The information conveyed by the aspect of a signal.										328

signaling (1) (data transmission). The production of an audible or visible signal at a station or switchboard by means of an alternating or pulsating current. In a telephone system, any of several methods used to alert subscribers or operators or to establish and control connections.										59

(2) (telephone switching systems). The transmission of address and other switching information between stations and central offices and between switching entities.										55

signaling and doorbell transformers (power and distribution transformer). Step-down transformers (having a secondary of 30 V or less), generally used for the operation of signals, chimes, and doorbells.										53

signaling circuit (National Electrical Code). Any electric circuit that energizes signaling equipment.										256

signaling light. A projector used for directing light signals toward a designated target zone.										167

signal, input (control system, feedback). A signal applied to a system or element. See the figure attached to the definition (3) of **signal, error.** *See:* **control system, feedback.**										56, 105

signal integration. The summation of a succession of signals by writing them at the same location on the storage surface. *See:* **storage tube.**										174

signal level (1)(signals and paths)(microcomputer system bus). The relative magnitude of a signal when compared to an arbitrary reference. Signal levels in ANSI/IEEE Std 796-1983 are specified in volts.										542

(2)(signals and paths)(696 interface devices). The magnitude of a signal when considered in relation to an arbitrary reference magnitude (voltage in the case of ANSI/IEEE Std 696-1983).										538

(3) (programmable instrumentation). The magnitude of signal compared to an arbitrary reference magnitude (voltage in the case of this standard).										378

signal line (programmable instrumentation)(signals and paths)(696 interface devices)(microcomputer system bus). One of a set of signal conductors in an interface system used to transfer messages among interconnected devices.										378, 538, 542

signal lines. The passive transmission lines through which the signal passes from one to another of the elements of the signal transmission system. *See:* **signal.**										188

signalling light (illuminating engineering). A projector used for directing light signals toward a designated target zone.										167

signal off (measuring the performance of tone address signaling systems). Any condition where all the constituent tones of a tone signaling system are below a specified OFF level for each tone. In a single-tone signaling system, tone off and signal off are synonymous terms. *Note:* During a signal off condition, tones that are not used in the signaling system may be at a higher level.										508

signal operation (elevators). Operation by means of single buttons or switches (or both) in the car, and up-or-down direction buttons (or both) at the landings, by which predetermined landing stops may be set up or registered for an elevator or for a group of elevators. The stops set up by the momentary actuation of the car buttons are made automatically in succession as the car reaches those landings, irrespective of its direction of travel or the sequence in which the buttons are actuated. The stops set up by the momentary actuation of the up-and-down buttons at the landing are made automatically by the first available car in the group approaching the landing in the corresponding direction, irrespective of the sequence in which the buttons are actuated. With this type of operation, the car can be started only by means of a starting switch or button in the car. *See:* **control (elevators).**   328

signal, output (control system, feedback). A signal delivered by a system or element. *See:* **control system, feedback.**   256, 105

signal output current (camera tubes or phototubes). The absolute value of the difference between output current and dark current. *See:* **phototube.**
                                                 190, 178, 125

signal parameter (signals and paths)(1)(microcomputer system bus). That element of an electrical quantity whose values or sequence of values convey information.   542
(2)(696 interface devices). That parameter of an electrial quantity whose values or sequence of values convey information.   538

signal present (measuring the performance of tone address signaling systems). Any condition where the presence of tone or tones is sufficient to be recognized as a valid digital or supervisory signal. In a single-tone signaling system, tone present and signal present are synonymous terms; in a two-tone signaling system the signal present state exists where two and only two tones each meet the signal present condition, and the two tones represent a valid combination.   508

signal-processing antenna system. An antenna system having circuit elements associated with its radiating elements that perform functions such as multiplication, storage, correlation, and time modulation of the input signals. *See:* **antenna.**   111

signal purity (network analyzers). A measure of freedom from frequency components other than the desired measurement frequency. It includes harmonics, subharmonics, spurious mixer products, and unwanted components of signal or local oscillator leakage. *Note:* The resulting error in measurement is a function of the detection system and of the frequency response of the network under test, as well as the signal purity.
                                                          505

signal, rate. A signal that is responsive to the rate of change of an input signal.   105

signal, reference input (1) (general). The command expressed in a form directly usable by the system. The reference input signal is in the terms appropriate to the form in which the signal is used, that is, voltage, current, ampere-turns, etcetera. *See:* **control system feedback.**   206
(2) (control system, feedback). A signal external to a control loop that serves as the standard of comparison for the directly controlled variable. See the figure attached to the definition of **signal, feedback.** *See:* **control system, feedback.**   105

signal relay. *See:* **alarm relay.**

signal repeater lights. A group of lights indicating the signal displayed for humping and trimming.   328

signal, return (control system, feedback) (closed loop). The signal resulting from a particular input signal, and transmitted by the loop and to be subtracted from that input signal. See the figure attached to the definition of **signal, error.** *See:* **control system, feedback.**
                                                          105

signal rise time (measuring the performance of tone address signaling systems). The time interval between the end of the signal off condition and the beginning of the signal present condition at the beginning of a tone signal.   508

signal, sampled. *See:* **sampled signal.**

signal-shaping amplifier (telegraph practice). An amplifier and associated electric networks inserted in the circuit, usually at the receiving end of an ocean cable, for amplifying and improving the waveshape of the signals. *See:* **telegraphy.**   328

signal-shaping network (wave-shaping set). An electric network inserted (in a telegraph circuit) for improving the waveshape of the received signals. *See:* **telegraphy.**   328

signal shutter (illuminating engineering). A device which modulates a beam of light by mechanical means for the purpose of transmitting intelligence.   167

signal threshold (telephony)(dial-pulse address signaling systems). The current or voltage value representing a transition between make and break states. The threshold may be expressed either as an absolute value (for example, 15 milliamperes [mA] or 21 volts [V]) or as a percentage deviation between steady-state conditions (for example, the value which is 70% of the difference between make and break values, added to the off-hook value). Signaling detectors having hysteresis will have different signal threshold values for the make-break and break-make transition.   540

signal-to-clutter ratio (radar). The ratio of target echo power to the power received from clutter sources lying within the same resolution element.   13

signal-to-interference ratio. The ratio of the magnitude of the signal to that of the interference or noise. *Note:* The ratio may be in terms of peak values or root-mean-square values and is often expressed in decibels. The ratio may be a function of the bandwidth of the system. *See:* **signal.**   188

signal-to-noise ratio (1)(video magnetic-tape recording systems). The ratio of the peak-to-peak amplitude of the video luminance signal from blanking level to reference white level (100 IRE units), 714 megavolts (mV) to the root-mean-square (rms) amplitude of the random noise, expressed in decibels.

$$SNR = 20 \, Log_{10} \, \frac{E_V}{E_N}$$

where

E_V = peak-to-peak amplitude of the maximum video luminance component (714 mV)

E_N = rms amplitude of random noise.

Note: Unless otherwise specified, the definition for signal-to-noise is as defined here.   460

(2) (data transmission)(snr) (antennas)(overhead-power-line corona and radio noise). (general). The ratio of the value of the signal to that of the noise. *Notes:* (A) This ratio is usually in terms of peak values in the case of impulse noise and in terms of the root-mean-square values in the case of the random noise. (B) Where there is a possibility of ambiguity, suitable definitions of the signal and noise should be associated with the term, as, for example, peak-signal to peak-noise ratio, root-mean-square signal to root-mean-square noise ratio, peak-to-peak signal to peak-to-peak noise ratio, etcetera. (C) This ratio may be often expressed in decibels. (D) This ratio may be a function of the bandwidth of the transmission system.
                         111, 59, 411

(3) (camera tubes). The ratio of peak-to-peak signal output current to root-mean-square noise in the output current. *Note:* Magnitude is usually not measured in tubes where the signal output is taken from target. *See:* **camera tube; television.**

(4) (television transmission). The signal-to-noise ratio at any point is the ratio in decibels of the maximum peak-to-peak voltage of the video television signal, including synchronizing pulse, to the root-mean-square voltage of the noise. *Note:* The signal-to-noise ratio is defined in this way because of the difficulty of defining the root-mean-square value of the video signal or the peak-to-peak value of random noise. *See:* **television.**   328

(5) (mobile communication). The ratio of a specified speech-energy spectrum to the energy of the noise in the same spectrum. *See:* **television.**   181

(6) (sound recording and reproducing system). The ratio of the signal power output to the noise power in the entire pass band.   350

(7) (digital delay line). Ratio of the peak amplitude of the output doublet to the maximum peak of any noise response (or signal) outside of the doublet interval. (Includes overshoot.)   81

(8) (speech quality measurements) *(S/N).* In decibels of a speech signal, the difference between its speech level and the noise level.   126

signal transfer characteristic (diode-type camera tube). The relationship between the input image irradiance incident on the camera tube and the resulting output current signal. It is presented as a plot of the logarithm of the output signal as a function of the logarithm of the input signal.   380

signal-transfer point (telephone switching systems). A switching entity where common channel signaling facilities are interconnected.   55

signal transmission system. *See:* **carrier.**

signal, TV waveform. An electrical signal whose amplitude varies with time in a generally nonsinusoidal manner, whose shape (that is, duration and amplitude) carries the TV signal information.   42

signal, unit-impulse (automatic control). A signal that is an impulse having unity area. *See:* **control system, feedback.**   56

signal, unit-ramp (automatic control). A signal that is zero for all values of time prior to a certain instant and equal to the time measured from that instant. *Note:* The unit-ramp signal is the integral of the unit-step signal. *See:* **control system-feedback.**   56

signal, unit-step (automatic control). A signal that is zero for all values of time prior to a certain instant and unity for all values of time following. *Note:* The unit-step signal is the integral of the unit-impulse signal. *See:* **control system, feedback.**   56

signal wave. A wave whose shape conveys some intelligence, message, or effect.   199

signal winding (input winding) (saturable reactor). A control winding to which the independent variable (signal wave) is applied.   328

signature diagnosis (test, measurement and diagnostic equipment). The examination of signature of an equipment for deviation from known or expected characteristics and consequent determination of the nature and location of malfunctions.   254

sign digit (electronic computation). A character used to designate the algebraic sign of a number.   235

significance (test, measurement and diagnostic equipment). The value or weight given to a position, or to a digit in a position, in a positional numeration system. In most positional numeration systems positions are grouped in sequence of significance, usually more significant towards the left.   54

significand (1)(binary floating-point arithmetic). The component of a binary floating-point number that consists of an explicit or implicit leading bit to the left of its implied binary point and a fraction field to the right.   469

(2)(mathematics of computing). The component of a floating-point number that consists of an explicit or implicit leading digit to the left of its implied radix point and a fraction field to the right. *Syn:* **mantissa.** *See:* **exponent.**   564

(3)(radix-independent floating-point arithmetic). The component of a floating-point number that consists of a leading digit to the left of its implied radix point and a fraction field to the right.   588

significant (nuclear power generating station). Demonstrated to be important by the safety analysis of the station.   102,355

significant digit (1). A digit that contributes to the accuracy or precision of a numeral. The number of significant digits is counted beginning with the digit contributing the most value, called the most-significant digit, and ending with the one contributing the least value, called the least-significant digit.   77

(2) (metric practice). Any digit that is necessary to define a value or quantity.   21

sign position. The position at which the sign of a number is located.   255, 77

silent lobing (radar). A method for scanning an antenna beam to achieve angle tracking without revealing the scanning pattern on the transmitted signal.   13

silicon controlled rectifier (SCR) (thyristor). An alternative name for the reverse blocking triode thyristor. *Note:* Although not an official definition, the term "unidirectional" is sometimes used to describe the single switching class of thyristors consisting of reverse-blocking and reverse-conducting thyristors. This term is useful for comparing or contrasting this class of thyristors with bidirectional thyristors.   445

silicone oil (insulating oil). A generic term for a family of relatively inert liquid organosiloxane polymers used as electrical insulation.   461

silvering (electrotyping). The application of a thin conducting film of silver by chemical reduction upon a plastic or wax matrix. *See:* **electroforming.**   328

silver oxide cell. A cell in which depolarization is accomplished by oxide of silver. *See:* **electrochemistry.**   328

silver storage battery. An alkaline storage battery in which the positive active material is silver oxide and the negative contains zinc. *See:* **battery (primary or secondary).**   328

silver-surfaced or equivalent (power switchgear). The term indicates metallic materials having satisfactory long-term performance and which operate within the temperature rise limits established for silver-surfaced electrical contact parts and conducting mechanical joints.   103

simple *GCL* circuit. *See:* **simple parallel circuit; simple series circuit.**

simple combination of insulating materials (thermal classification of electric equipment and electrical insulation). A number of insulating materials, which together make possible the evaluation of any interaction between them.   506

simple parallel circuit (simple *GCL* circuit). A linear, constant-parameter circuit consisting of resistance, inductance, and capacitance in parallel. *See:* **network analysis.**   210

simple rectifier. A rectifier consisting of one commutating group if single-way or two commutating groups if double-way. *See:* **rectification.**   208

simple rectifier circuit. A rectifier circuit consisting of one commutating group if single-way, or two commutating groups if double-way. *See:* **rectification; rectifier circuit element.**   66

simple scanning (facsimile). Scanning of only one scanning spot at a time during the scanning process.   212

simple series circuit. A resistance, inductance, and capacitance in series. *Syn:* **simple *RLC* circuit.** *See:* **network analysis.**   210

simple sine-wave quantity. A physical quantity that is varying with time t as either $A \sin (\omega t + \theta_A)$ or $A \cos (\omega + \theta_B)$ where $A, \omega, \theta_A, \theta_B$ are constants. (**Simple**

denotes that $A, \omega, \theta_A, \theta_B$ are constants.) *Notes:* (1) It is immaterial whether the sin or cos form is used, so long as no ambiguity or inconsistency is introduced (2) A is the amplitude or maximum value, $\omega t + \theta_A$ (or $\omega t + \theta_B$) the phase, θ_A (or θB) the phase angle. However, when no ambiguity may arise, **phase angle** may be abbreviated **phase.** (3) In certain special applications, for example, modulation, $\omega t + \theta$ is called the angle (of a sine wave), (not phase angle) in order to clarify particular uses of the word "phase." Another permissible term for $(\omega t + \theta)$ is argument (sine wave).   210

simple sound source. A source that radiates sound uniformly in all directions under free-field conditions.   176

simple target (radar). A target having a reflecting surface such that the amplitude of the reflected signal does not vary with the aspect of the target, for example, a metal sphere; distinguished from **complex target.**   13

simple tone (pure tone). (1) A sound wave, the instantaneous sound pressure of which is a simple sinusoidal function of the time. (2) A sound sensation characterized by its singleness of pitch. *Note:* Whether or not a listener hears a tone as simple or complex is dependent upon the ability, experience, and listening attitude. *See:* **complex tone.**   176

simplex channel (simplex operation) (data transmission). A method of operation in which communication between two stations takes place in one direction at a time. *Note:* This includes ordinary transmit-receive operation, press-to-talk operation, voice-operated carrier and other forms of manual or automatic switching from transmit to receive.   59

simplex circuit (data transmission). A circuit derived from a pair of wires by using the wires in parallel with ground return.   59

simplex lap winding (rotating machinery). A lap winding in which the number of parallel circuits is equal to the number of poles.   63

simplex operation (antennas)(radio transmitters). A method of operation in which communication between two stations takes place in one direction at a time. *Note:* This includes ordinary transmit-receive operation, press-to-talk operation, voice-operated carrier and other forms of manual or automatic switching from transmit to receive. *See:* **radio transmission; telegraphy.**   111, 240

simplex signaling (telephone switching systems). A method of signaling over a pair of conductors by producing current flow in the same direction through both of the conductors.   55

simplex supervision (telephone switching systems). A form of supervision employing simplex signaling.   55

simplex wave winding (rotating machinery). A wave winding in which the number of parallel circuits is two, whatever the number of poles.   63

simply connected region (two-dimensional space). A region, such that any closed curve in the region encloses points all of which belong to the region.   210 simply mesh-connected circuit. A circuit in which two, and only two, current paths extend from the terminal of entry of each phase conductor, one to the terminal of entry that precedes and the other to the terminal of entry that follows the first terminal in the normal sequence, and from which the amplitude of the voltages to the first terminal is normally the smallest (when the number of phases is greater than three). *See:* network analysis. 210

simulate (computing systems). To represent the functioning of one system by another, for example, to represent one computer by another, to represent a physical system by the execution of a computer program, to represent a biological system by a mathematical model. *See:* electronic analog computer. 255, 77

simulated fly ash. The entrained ash produced by suspension firing in a small-scale pulverized coal combustor designed and operated with the objective of closely approximating certain selected properties of the fly ash produced in the full-scale steam generator of interest. The combustor should have the capability of providing approximately the same time/ temperature profile for combustion as would occur in a full-scale boiler furnace. This process is applicable particularly when coal from a new source has never been burned in a full-scale boiler. 427

simulated source (sodium iodide detector). A radioactive source consisting of one or more long-lived radionuclides that are chosen to simulate the radiations from a short-lived radionuclide of interest. 423

simulated sources ('dose calibrator' ionization chambers)(ionization chambers). Simulated sources usually contain long-lived radionuclides, alone or in combination, that are chosen to simulate, in terms of photon or particle emission, a short-lived radionuclide of interest. 499, 396

simulation (1) (analog computers). The representation of an actual or proposed system by the analogous characteristics of some device easier to construct, modify, or understand. 9

(2) (software). The representation of selected characteristics of the behavior of one physical or abstract system by another system. In a digital computer system, simulation is done by software, for example, (A) the representation of physical phenomena by means of operations performed by a computer system, (B) the representation of operations of a computer system by those of another computer system. *See:* analytical model; computer system; software; system. 434

(3) (mathematical). The use of a model of mathematical equations generally solved by computers to represent an actual or proposed system. 9

simulator (1) (analog computers). A device used to represent the behavior of a physical system by virtue of its analogous characteristics. In this general sense, all computers are, or can be, simulators. However in a more restricted definition, a simulator is a device used to interact with, or to train, a human operator in the performance of a given task or tasks. 9

(2) (software). A device, data processing system, or computer program that represents certain features of the behavior of a physical or abstract system. *See:* computer program; data; system. 434

(3) (test, measurement and diagnostic equipment). A device or program used for test purposes which simulates a desired system or condition providing proper inputs and terminations for the equipment under test. 54

simultaneous lobing (1) (electronic navigation). A direction-determining technique utilizing the received energy of two concurrent and partially overlapped signal lobes: the relative phase, or the relative power, of the two signals received from a target is a measure of the angular displacement of the target from the equiphase or equisignal direction. Compare with lobe switching. 187, 246, 179

(2) (radar) (antennas). A direction-determining technique utilizing the signals of overlapping lobes existing at the same time. *Syn:* (radar) monopulse. 13, 111

SINAD (land-mobile communication transmitters). An acronym for "signal plus noise plus distortion to noise plus distortion ratio" expressed in decibels (dB), where the "signal plus noise plus distortion" is the audio power recovered from a modulated radio frequency carrier, and the "noise plus distortion" is the residual audio power present after the audio signal is removed. This ratio is a measure of audio output signal quality for a given receiver audio power output level. 444

sinad ratio (mobile communication). A measure expressed in decibels of the ratio of (1) the signal plus noise plus distortion to (2) noise plus distortion produced at the output of a receiver that is the result of a modulated-signal input. *See:* mobile communication system; receiver performance. 123

sinad sensitivity (receiver performance). The minimum standard modulated carrier-signal input required to produce a specified sinad ratio at the receiver output. *See:* receiver performance. 181

sine beats (1)(nuclear power generating stations) (seismic qualification of class 1E equipment). A continuous sinusoid of one frequency, amplitude modulated by a sinusoid of a lower frequency. *Notes:* (A) The amplitudes of the sinusoids represent acceleration and the modulated frequency represents the frequency of the applied seismic stimulus. (B) Beats are usually considered to be the result of the summation of two sinusoids of slightly different frequencies with the frequencies within the beats as the average of the two, and the beat frequency as one-half the difference between the two. However, the sine beats may be an amplitude-modulated sinusoid with pauses between the beats. 28

(2)(seismic qualification of Class 1E equipment for nuclear power generating stations). A continuous sinusoid of one frequency, amplitude modulated by a sinusoid of a lower frequency. 581

sine-current coercive force (toroidal magnetic amplifier cores). The instantaneous value of sine-current magnetizing force at which the dynamic hysteresis loop passes through zero induction. 170

sine-current differential permeability (toroidal magnetic amplifier cores). The slope of the sides of the dynamic hysteresis loop obtained with a sine-current magnetizing force. 170

sine-current magnetizing force (toroidal magnetic amplifier cores). The applied magnetomotive force per unit length for a core symmetrically cyclicly magnetized with sinusoidal current. 170

sine-square ($\sin^2$) pulse (video signal transmission measurement). One cycle of a sine wave, starting and finishing at its negative peaks with an added constant amplitude component of half the peak-to-peak value, thus raising the negative peaks to zero. *Note:* A $\sin^2$ pulse is obtained by squaring a half-cycle of a sine wave. 42

sine-square ($\sin^2$) step (video signal transmission measurement). A step function whose transition from zero to the final value is the sum of a ramp and a negative sinusoid of equal durations, with zero slope at both the zero and the final value of the step. *Notes:* (1) A $\sin^2$ step is obtained by integrating a $\sin^2$ pulse. (2) The attractiveness of both the $\sin^2$ pulse and the $\sin^2$ step lies in the fact that their frequency spectra are limited: that is, they are effectively at zero amplitude beyond a given frequency. For the $\sin^2$, pulse this frequency is a function of the half-amplitude duration (HAD) of the pulse: for the $\sin^2$ step the frequency is a function of the 10 percent to 90 percent rise time. 242

sine wave. A wave that can be expressed as the sine of a linear function of time, or space, or both. 111

sine-wave generator. An alternating-current generator whose output voltage waveform contains a single main frequency with low harmonic content of prescribed maximum level. *See:* **asynchronous machine.** 63

sine-wave response (camera tubes). *See:* **amplitude response.**

sine-wave sweep. A sweep generated by a sine function. *See:* **oscillograph.** 185

singing (1) (antennas). An undesired self-sustained oscillation existing in a transmission system or transducer. *Note:* Very-low-frequency oscillation is sometimes called motor-boating. *See:* 111
(2) **(data transmission).** In a transmission system, an undesired self-sustained oscillation existing in the system. 59

singing margin (gain margin). The excess of loss over gain around a possible singing path at any frequency, or the minimum value of such excess over a range of frequencies. *Note:* Singing margin is usually expressed in decibels. 239

singing point (data transmission). For a circuit which is coupled back to itself, the point at which the gain is just sufficient to make the circuit break into oscillation. 59

singing point margin (data transmission). The amount of additional gain (dB) which can be inserted into a loop without sustained oscillations developing. 59

singing return loss (SRL) (analog voice frequency circuits). The return loss of a circuit measured with two separately transmitted signals with a flat spectral distribution between 3 decibel (dB) frequencies of 260 hertz (Hz) and 500 Hz (SRL Low) and 2200 Hz and 3400 Hz (SRL High). The lower of the two return losses (SRL Low or SRL High) will be the best measure of the margin of the circuit against singing. 468

single-address. Pertaining to an instruction that has one address part. In a typical single-address instruction the address may specify either the location of an operand to be taken from storage, the destination of a previously prepared result, the location of the next instruction to be interpreted, or an immediate address operand. *Syn:* **one-address.** 77

single-address code (electronic computation). *See:* **instruction code.**

single-anode tank (single-anode tube). An electron tube having a single main anode. *Note:* This term is used chiefly for pool-cathode tubes. 190

single aperture seal (electric penetration assemblies)-(nuclear power generating stations). A single seal between the containment aperture and the electric penetration assembly. 493,26

single automatic operation (elevators). Automatic operation by means of one button in the car for each landing level served and one button at each landing so arranged that if any car or landing button has been actuated, the actuation of any other car or landing operating button will have no effect on the operation of the car until the response to the first button has been completed. *See:* **control (elevators).** 328

single-break switch (power switchgear). One that opens each conductor of a circuit at one point only. 103

single-buffered DAC (DAM) (hybrid computer linkage components). A digital-to-analog converter (DAC) or a digital-to-analog multiplier (DAM) with one dynamic register, which also serves as the holding register for the digital value. 10

single capacitance (as applied to interrupter switches). A capacitance is defined to be a single capacitance when the crest of its inrush current does not exceed the switch inrush current capability for single capacitance. 103

single-circuit system (protective signaling). A system of protective wiring that employs only the nongrounded side of the battery circuit, and consequently depends primarily on an open circuit in the wiring to initiate an alarm. *See:* **protective signaling.** 328

single-degree-freedom gyro. A gyro in which the rotor is free to precess (relative to the case) about only the axis orthogonal to the rotor spin axis. *See:* **navigation.** 187,13

single electric conductor seal (electric penetration assemblies)(nuclear power generating stations). A mechanical assembly arranged in such a way that there is a single pressure barrier seal between the inside and the outside of the containment structure along the axis of the electric conductor. 493, 26

single electron distribution (scintillation counting). The pulse-height distribution associated with single electrons originating at the photocathode. 117

single-electron PHR (pulse-height resolution) (scintillation counting). The fractional FWHM (full width at half maximum) of the single-electron distribution of a photomultiplier. 117

single-electron rise time (scintillation counting). The anode-pulse rise time associated with single electrons originating at the photocathode. 117

single-electron transit-time spread (scintillation counting). Transit-time spread measured with single-electron events. 117

single-element fuse (power switchgear). A fuse having a current-responsive element comprising one or more parts with a single fusing characteristic. 103

single-element relay. An alternating-current relay having a set of coils energized by a single circuit. 328

single-end control (single-station control). A control system in which provision is made for operating a vehicle from one end or one location only. *See:* multiple-unit control. 328

single-ended amplifier. An amplifier in which each stage normally employs only one active element (tube, transistor, etcetera) or, if more than one active element is used, in which they are connected in parallel so that operation is asymmetric with respect to ground. *See:* amplifier. 111

single-ended push-pull amplifier circuit (electroacoustics). An amplifier circuit having two transmission paths designed to operate in a complementary manner and connected so as to provide a single unbalanced output without the use of an output transformer. *See:* amplifier. 239

single-faced tape. Fabric tape finished on one side with rubber or synthetic compound. 64

single feeder. A feeder that forms the only connection between two points along the route considered. 64

single-frequency pulsing (telephone switching systems). Dial pulsing using the presence or absence of a single frequency to represent break or make intervals, respectively or vice versa. 55

single-frequency signaling (telephone switching systems). A method for conveying dial pulse and supervisory signals from one end of a trunk to the other using the presence or absence of a single specified frequency. 55

single-frequency signal-to-noise ratio (sound recording and reproducing system). The ratio of the single-frequency signal power output to the noise power in the entire pass band. *See:* noise (sound recording and reproducing system). 350

single-frequency simplex operation (radio communication). The operation of a two-way radio-communication circuit on the same assigned radio-frequency channel, which necessitates that intelligence can be transmitted in only one direction at a time. *See:* channel spacing. 181

single hoistway (elevators). A hoistway for a single elevator or dumbwaiter. *See:* hoistway (elevator or dumbwaiter). 328

single-layer winding (rotating machinery). A winding in which there is only one actual coil side in the depth of the slot. (Also known as one-coil-side-per-slot winding). *See:* asynchronous machine. 63

single-line diagram. *See:* one-line diagram.

single mode optical waveguide (fiber optics). An optical waveguide in which only the lowest order bound mode (which may consist of a pair of orthogonally polarized fields) can propagate at the wavelength of interest. In step index guides, this occurs when the normalized frequency, V, is less than 2.405. For power-law profiles, single mode operation occurs for normalized frequency, V, less than approximately 2.405 $\sqrt{(g+2)/g}$ where g is the profile parameter. *Note:* In practice, the orthogonal polarizations may not be associated with degenerate modes. *Syn:* monomode optical waveguide. *See:* bound mode; mode; multimode optical waveguide; normalized frequency; power-law index profile; profile parameter; step index optical waveguide. 433

single-office exchange (telephone switching systems). A telecommunications exchange served by one central office. 55

single-operator arc welder. An arc-welding power supply designed to deliver current to only one welding arc. 264

single-phase ac fields (measurement of power frequency electric and magnetic fields from ac power lines). Fields whose space components are in phase. These fields will be produced by single-phase power lines. The field at any point can be described in terms of a single direction in space and its time-varying magnitude. 514

single-phase circuit (1) (electric installations on shipboard). A circuit energized by a single alternating electromotive force. A single-phase circuit is usually supplied through two wires. The currents in these two wires, counted outward from the source, differ in phase by 180 degrees or a half cycle. 3
(2) (power and distribution transformer). An alternating-current circuit consisting of two or three intentionally interrelated conductors that enter (or leave) a delimited region at two or three terminals of entry. If the circuit consists of two conductors, it is intended to be so energized that, in the steady state, the voltage between the two terminals of entry is an alternating voltage. If the circuit consists of three conductors, it is intended to be so energized that, in steady state, the alternating voltages between any two terminals of entry have the same period and are in phase or in phase opposition. 53

single-phase electric locomotive. An electric locomotive that collects propulsion power from a single phase of an alternating-current distribution system. *See:* electric locomotive. 328

single-phase machine. A machine that generates or utilizes single-phase alternating-current power. *See:* asynchronous machine. 63

single-phase motor (rotating machinery). A machine that converts single-phase alternating-current electric power into mechanical power, or that provides mechanical force or torque. 63

single-phase symmetrical set (1) (polyphase voltages). A symmetrical set of polyphase voltages in which the angular phase difference between successive members of the set is π radians or odd multiples thereof. The equations of **symmetrical set (polyphase voltages)** represent a single-phase symmetrical set of polyphase voltages if k/m is $1/2$ or an odd multiple thereof. (The symmetrical set of voltages represented by the equations of **symmetrical set (polyphase voltages)** may be said to have single-phase symmetry if k/m is an odd (positive or negative) multiple of $1/2$. *Notes:* (1) A set of polyphase voltages may have single-phase symmetry only if m, the number of members of the set, is an even number. (2) This definition may be applied to a two-phase four-wire or five-wire circuit if m is considered to be 4 instead of 2. It is not applicable to a two-phase three-wire circuit. *See:* **network analysis.**
      210

(2) (polyphase currents). This definition is obtained from the corresponding definitions for voltage by substituting the word **current** for **voltage,** and the symbol I for E and β for α wherever they appear in the equations of **symmetrical set (polyphase voltages).** The subscripts are unaltered. *See:* **network analysis.**
      210

single-phase synchronous generator (electric installations on shipboard). A generator which produces a single alternating electromotive force at its terminals. It delivers electric power which pulsates at double frequency.
      3

single-phase three-wire circuit. A single-phase circuit consisting of three conductors, one of which is identified as the neutral conductor. *See:* **network analysis.**
      210

single-phase two-wire circuit. A single-phase circuit consisting of only two conductors. *See:* **network analysis.**
      210

single-phasing (rotating machinery). An abnormal operation of a polyphase machine when its supply is effectively single-phase. *See:* **asynchronous machine.**
      63

single-pointer form demand register (metering). An indicating demand register from which the demand is obtained by reading the position of a pointer relative to the markings on a scale. The single pointer is resettable to zero.
      212

single-polarity pulse. A pulse in which the sense of the departure from normal is in one direction only. *See:* **unidirectional pulse; pulse.**
      328

single-pole relay. *See:* **relay, single-pole.**

single pole switching (power switchgear). The practice of tripping and reclosing one pole of a multipole circuit breaker without changing the state of the remaining poles, with tripping being initiated by protective relays which respond selectively to the faulted phase. *Notes:* (1) Circuit breakers used for single pole switching must inherently be capable of individual pole opening. (2) In most single pole switching schemes it is the practice to trip all poles for any fault involving more than one phase.
      103

single precision (mathematics of computing). Pertaining to the use of a single computer word to represent a number. *Note:* Single precision is implied in number representation and in computer arithmetic unless multiple precision is specified.
      564

single-pressure-zone potheads. A pressure-type pothead intended to operate with one pressure zone. *See:* **multipressure-zone pothead; pressure-type pothead.**
      323

single pulse (thyristor). A gate signal applied at the commencement of the conducting interval in the form of a single pulse of predetermined duration, amplitude, and frequency.
      445

single service. One service only supplying a consumer. *Note:* Either or both lighting and power load may be connected to the service. *See:* **service.**
      64

single-shot blasting unit. A unit designed for firing only one explosive charge at a time. *See:* **blasting unit.**
      328

single-shot blocking oscillator. A blocking oscillator modified to operate as a single-shot trigger circuit. *See:* **trigger circuit.**
      328

single-shot multivibrator (single-trip multivibrator). A multivibrator modified to operate as a single-shot trigger circuit. *See:* **trigger circuit.**
      328

single-shot trigger circuit (single-trip trigger circuit). A trigger circuit in which a triggering pulse intiates one complete cycle of conditions ending with a stable condition. *See:* **trigger circuit.**
      328

single-sideband modulation (antennas) (data transmission). Modulation whereby the spectrum of the modulating function is translated in frequency by a specified amount either with or without inversion. *See:* **modulation.**
      111, 42

single-sideband transmission (data transmission). The method of operation in which one sideband is transmitted and the other sideband is suppressed. The carrier wave may be either transmitted or suppressed.
      59

single-sideband transmitter. A transmitter in which one sideband is transmitted and the other is effectively eliminated.
      111

single-sided printed-circuit board. *See:* **printed-circuit board.**

single-station control. *See:* **single-end control.**

single step (computing systems). Pertaining to a method of operating a computer in which each step is performed in response to a single manual operation.
      255, 77

single-stroke bell. An electric bell that produces a single stroke on its gong each time its mechanism is actuated. *See:* **protective signaling.**
      328

single sweep (spectrum analyzer). Operating mode for a triggered sweep instrument in which the sweep must be reset for each operation, thus preventing unwanted multiple displays. This mode is useful for trace photography. In the interval after the sweep is reset and before it is triggered, it is said to be an armed sweep.
      390

single throw (switching device). A qualifying term used to indicate that the device has an open and a closed circuit position only.
      103

single-tone keying (modulation systems). That form of keying in which the modulating function causes the carrier to be modulated by a single tone for one condition, which may be either a mark or a space, the carrier being unmodulated for the other condition. *See:* telegraphy.   111, 242, 194

single-track (standard track) (electroacoustics). A variable-density or variable-area sound track in which both positive and negative halves of the signal are linearly recorded. *See:* phonograph pickup.   176

single-trip multivibrator. *See:* single-shot multivibrator.

single-trip trigger circuit. *See:* single-shot trigger circuit.

single-tuned amplifier (circuits and systems). An amplifier characterized by a resonance at a single frequency as indicated by the s-plane representation of its gain which is $A(s) = A_0 s / (s^2 + \omega_0 \xi s + \omega_0^2)$. It rejects low and high frequencies while having a peak gain at a center frequency $s = j\omega_0$. *See:* amplifier.
                    67

single-tuned circuit. A circuit that may be represented by a single inductance and a single capacitance, together with associated resistances.   328

single-valued function. A function u is single valued when to every value of x (or set of values of $x_1, x_2, ..., x_n$) there corresponds one and only one value of u. Thus $u = ax$ is single valued if a is an arbitrary constant.   210

single-way rectifier. A rectifier unit which makes use of a single-way rectifier circuit.   66

single-way rectifier circuit. A rectifier circuit in which the current between each terminal of the alternating voltage circuit and the rectifier circuit element or elements conductively connected to it flows in only one direction.   66

single-winding multispeed motor. A type of multispeed motor having a single winding capable of reconnection in two or more pole groupings. *See:* asynchronous machine.   328

single-wire line (waveguide). A surface-wave transmission line consisting of a single conductor so treated as to confine the propagated energy to the neighborhood of the wire. The treatment may consist of a coating of dielectric.   267

singular point (control system). Synonymous with equilibrium point. *See:* control system.   329

sink (oscillator). The region of a Rieke diagram where the rate of change of frequency with respect to phase of the reflection coefficient is maximum. Operation in this region may lead to unsatisfactory performance by reason of cessation or instability of oscillations. *See:* oscillatory circuit.   125

sink node (network analysis). A node having only incoming branches.   282

sin² pulse. *See:* sine-square pulse.

sin² step. *See:* sine-square step.

sin-square (sin²) (linear waveform distortion). A step function whose transition from zero to the final value is the sum of a ramp and a negative sinusoid of equal durations, with zero slope at both the zero and the final value of the step. *Note:* (1) As shown in detail in Appendix A, a sin² step is obtained by integrating a sin² pulse. (2) The attractiveness of both the sin² pulse and the sin² step lies in the fact that their frequency spectra are limited; that is, they are effectively at zero amplitude beyond a given frequency. For the sin² pulse this frequency is a function of the half-amplitude duration (HAD) of the pulse; for the sin² step the frequency is a function of the 10 percent to 90 percent rise time.   42

sin-square (sin²) pulse (linear waveform distortion). One cycle of a sine wave, starting and finishing at its negative peaks with an added constant amplitude component of half the peak-to-peak value, thus raising the negative peaks to zero. *Note:* As shown in detail in Appendix A, a sin² pulse is obtained by squaring a half-cycle of a sine wave.   42

sinusoidal electromagnetic wave (radio wave propagation). In a homogeneous medium, a wave whose electric field vector is proportional to the sine (or cosine) of an angle that is a linear function of time, or of a distance, or of both.   146

sinusoidal field. A field in which the field quantities vary as a sinusoidal function of an independent variable, such as space or time.   210

sinusoidal function. A function of the form $A \sin (x\ a)$. A is the amplitude, x is the independent variable, and a the phase angle. Note that $\cos (x)$ may be expressed as $\sin 1x$ (6.2)0. *See:* simple sinewave quantity.
                    210

siphon recorder. A telegraph recorder comprising a sensitive moving-coil galvanometer with a siphon pen that is directed by the moving coil across a traveling strip of paper. *See:* telegraphy.   328

site error (navigation)(navigation aid terms). Error due to the distortion in the electromagnetic field by objects in the vicinity of the navigational equipment.
                    526

SI unit of luminance (illuminating engineering). Candela per square meter (cd/m^2); also, lumen per steradian·square meter ($lm/(sr/m^2)$ also, lumen per steradian·square meter ($lm/(sr/m^2)$). This also is called the nit.   167

SI units (ATLAS). *Systeme International;* fundamental physical units and dimensions defined by CIPM (*Comité International des Poids et Mesures),* the International Committee of Weights and Measures.
                    400

six-phase circuit (power and distribution transformer). A combination of circuits energized by alternating electromotive forces which differ in phase by one-sixth of a cycle, that is, 60 degrees. *Note:* In practice, the phases may vary several degrees from the specified angle.   53

size distribution (fly ash resistivity). Size distribution of particulate matter is the cumulative frequency of particle diameter, generally expressed on a mass basis. It describes the probability that a particle diameter x takes a value equal to or less than probability P. Size distribution rather than mean particle size shall be reported.   427

size threshold (illuminating engineering). The minimum perceptible size of an object. It also is defined as the size which can be detected some specific fraction of the times it is presented to an observer, usually 50 percent. It usually is measured in minutes of arc. 167

sizing (software). The process of estimating the amount of computer storage or the number of source lines that will be required for a system or system component. *See:* **component; computer; system.** 434

skate machine. A mechanism, electrically controlled, for placing on, or removing from, the rails a skate that, if allowed to engage with the wheels of a car, provides continuous braking until the car is stopped and that may be electrically or pneumatically operated. 328

skating force. *See:* **side thrust.**

skeleton frame (rotating machinery). A stator frame consisting of a simple structure that clamps the core but does not enclose it. 63

skew (1)(measuring the performance of tone address signaling systems). In a two-tone signal, the time interval from the start of the higher-frequency tone present condition. Skew is negative if the higher-frequency tone starts before the lower-frequency tone. 508

(2) (facsimile). The deviation of the received frame from rectangularity due to asynchronism between scanner and recorder. Skew is expressed numerically as the tangent of the angle of the deviation. *See:* **recording (facsimile).** 12

(3) (magnetic storage). The angular displacement of an individual printed character, group of characters, or other data, from the intended or ideal placement. 255, 77, 54

skewed slot (rotating machinery). A slot of a rotor or stator of an electric machine, placed at an angle to the shaft so that the angular location of the slot at one end of the core is displaced from that at the other end. Slots are commonly skewed in many types of machines to provide more uniform torque, less noise, and better voltage waveform. *See:* **rotor (rotating machinery); stator.** 63

skew ray (fiber optics). A ray that does not intersect the optical axis of a system (in contrast with a meridional ray). *See:* **axial ray; geometric optics; hybrid mode; meridional ray; optical axis; paraxial ray.** 433

skew time (FASTBUS acquisition and control). The minimum time that the assertion of a FASTBUS timing signal must be delayed after the assertion of information and/or control signals to allow for differences in propagation time of signals on a FASTBUS segment. 480

skiatron* (radar) (1) A dark-trace storage-type cathode-ray tube; (2) A display employing an optical system with a dark-trace tube. *See:* **dark-trace tube.**
*Obsolete. 13

skid wire (pipe-type cable) (power distribution, underground cables). Wire or wires, usually D shaped, applied open spiral with curved side outward with a suitable spacing between turns over the outside surface of the cable. Its purpose is to facilitate cable pulling and to provide mechanical protection during installation. 57

skim tape. Filled tape coated on one or both sides with a thin film of uncured rubber or synthetic compound to produce a coating suitable for vulcanization. 64

skin depth (waveguide). Of a conducting material, at a given frequency, the depth at which the surface current density is reduced by one neper. 267

skin effect (induction heating). Tendency of an alternating current to concentrate in the areas of lowest impedance. 14

skin effect heating (electrical heating systems). An electrical heating system where a conductor inside a ferromagnetic material generates heat via I^2R losses in the conductor and ferromagnetic material. 476

skip (computing systems). To ignore one or more instructions in a sequence of instructions. 255, 77

skip distance (1)(navigation aid terms)(data transmission). The minimum separation for which radio waves of a specified frequency can be transmitted at a specified time (interval) between two points on the earth by reflection from the regular ionized layers of the ionosphere. 526, 59

(2) (radio wave propagation). For a given frequency, that distance between two points along the surface of the earth for which this frequency is the maximum usable frequency. It is the minimum separation between the two points for which radio waves of a specified frequency can be transmitted at a specified time via reflection from the regular layers of the ionosphere. 146

sky compass (navigation aid terms). A type of astro compass, designed for use in the arctic during long periods of twilight. 526

sky condition. *See:* **clear sky; partly cloudy sky; cloudy sky; overcast sky.**

sky factor (illuminating engineering). The ratio of the illuminance on a horizontal plane at a given point inside a building due to the light received directly from the sky, to the illuminance due to an unobstructed hemisphere of sky of uniform luminance equal to that of the visible sky. 167

sky light (illuminating engineering). Visible radiation from the sun redirected by the atmosphere. 167

sky noise (communication satellite). Noise contribution of the sky (often the galaxies). *See:* **background noise.** 85

sky wave (radio wave propagation). A radio wave propagated towards, and returning from, an ionosphere. *Note:* This term has sometimes been called ionospheric wave but such use is deprecated since the term "ionospheric wave" is intended to connote wave characteristics in ionospheric plasma. 146

sky-wave contamination (navigation aid terms). Degradation of the received ground-wave signal, or of the desired sky-wave signal, by the presence of delayed ionospheric-wave components of the same transmitted signal. 526

sky-wave correction (navigation)(navigation aid terms). A correction for sky-wave propagation errors applied to measured position data; the amount of the correction is established on the basis of an assumed ionosphere height.   526

sky-wave station-error (sky-wave synchronized loran)(navigation aid terms). The error of station synchronization due to the effect of variations of the ionosphere on the time of transmission of the synchronizing signal from one station to the other.   526

slabbing or arcwall machine. A power-driven mobile-cutting machine that is a single-purpose cutter in that it cuts only a horizontal kerf at variable heights.   328

slab interferometry (fiber optics). The method for measuring the index profile of an optical fiber by preparing a thin sample that has its faces perpendicular to the axis of the fiber, and measuring its index profile by interferometry. *Syn:* **axial slab interferometry.** *See:* **interferometer.**   433

slab line (waveguide). A uniform transmission line consisting of a round conductor between two extended parallel conducting surfaces, so that the propagating wave is essentially confined between the surfaces.   267

slack-rope switch (elevators). A device that automatically causes the electric power to be removed from the elevator driving-machine motor and brake when the hoisting ropes of a winding-drum machine become slack. *See:* **control.**   328

slack stringing (conductor stringing equipment). The method of stringing conductor slack without the use of a tensioner. The conductor is pulled off the reel by a pulling vehicle and dragged along the ground, or the reel is carried along the line on a vehicle and the conductor deposited on the ground. As the conductor is dragged to, or past, each supporting structure, the conductor is placed in the travelers, normally with the aid of finger lines.   431

slant distance (navigation aid terms). The distance between two points that are not at the same elevation. Used in contrast to ground distance.   526

slant range (navigation aid terms). Slant distance between a radar and a target.   526

slave (1)(FASTBUS acquisition and control). A device which responds to masters according to the FBP (FASTBUS protocol).   480

(2)(test, measurement and diagnostic equipment). Device that follows an order given by a master remote control.   54

slave drive. *See:* **electric drive; follower drive.**

slaved tracking (power supplies). A system of interconnection of two or more regulated supplies in which one (the master) operates to control the others (the slaves). The output voltage of the slave units may be equal or proportional to the output voltage of the master unit. (The slave output voltages track the master output voltage in a constant ratio). *See:* **complementary tracking, master/slave.**   228, 186

slave relay. *See:* **auxiliary relay; relay.**

slave station (navigation)(navigation aid terms). A station in which some characteristic of its emission is controlled by a master station.   526

slaving (gyro). The use of a torquer to maintain the orientation of the spin axis relative to an external reference such as a pendulum or magnetic compass.   46

sleet hood (of a switch). A cover for the contacts to prevent sleet from interfering with the successful operation of the switch.   103

sleetproof (1) (general). So constructed or protected that the accumulation of sleet will not interfere with successful operation.   103,27

(2) (power and distribution transformer). So constructed or protected that the accumulation of sleet (ice) under specified test conditions will not interfere with the successful operation of the apparatus.   53

sleeve (1) (plug) (three-wire telephone-switchboard plug). A cylindrically shaped contacting part, usually placed in back of the tip or ring but insulated therefrom.   328

(2) (rotating machinery). A tubular part designed to fit around another part. *Note:* In a sleeve bearing, the sleeve is that component that includes the cylindrical inner surface within which the shaft journal rotates.   63

sleeve bearing (rotating machinery). A bearing with a cylindrical inner surface in which the journal of a rotor (or armature) shaft rotates. *See:* **rotor (rotating machinery).**   63

sleeve conductor. *See:* **sleeve wire.**

sleeve-dipole antenna. A dipole antenna surrounded in its central portion by a coaxial conducting sleeve. *See:* **antenna.**   111

sleeve-monopole antenna. An antenna consisting of half of a sleeve-dipole antenna projecting from a ground plane. *Syn:* **sleeve-stub antenna.**   111

sleeve supervision. The use of the sleeve circuit for transmitting supervisory signals.   328

sleeve-type suppressor (electromagnetic compatibility). A suppressor designed for insertion in a high-tension ignition cable. *See:* **electromagnetic compatibility.**   220, 199

sleeve wire (telephone switching systems). That conductor, usually accompanying the tip and ring leads of a switched connection, which provides for miscellaneous functions necessary to the control and supervision of the connection. In cord-type switchboards, the sleeve wire is that conductor associated with the sleeve contacts of the jacks and plugs.   55

slewing (gyro). The rotation of the spin axis about an axis parallel to that of the applied torque causing the rotation.   46

slewing rate (1) (power supplies). A measure of the programming speed or current-regulator-response timing. The slewing rate measures the maximum rate-of-change of voltage across the output terminals of a power supply. Slewing rate is normally expressed in volts per second ($\Delta E/\Delta T$) and can be converted to a sinusoidal frequency-amplitude product by the equation $f(E_{pp})$ = slewing rate$/\pi$, where E_{pp} is the peak-to-peak sinusoidal volts. Slewing rate $= \pi f(E_{pp})$. *See:* **high-speed regulator.**   186

(2) **(thyristor).** A rate at which the output changes in response to a step change in control signal input.

    445

slewing speed (test, measurement and diagnostic equipment). A continuous speed, usually the maximum at which a tape reader or other rotating device can search for information.   54

slicer (amplitude gate) (clipper-limiter). A transducer that transmits only portions of an input wave lying between two amplitude boundaries. *Note:* The term is used especially when the two amplitude boundaries are close to each other as compared with the amplitude range of the input.   328

slide rail (rotating machinery). A special form of soleplate which is long in the direction of the machine axis to permit sliding the stator frame in the axial direction.

    63

slide-screw tuner (1) (transmission lines and waveguides). An impedance or matching transformer that consists of a slotted waveguide or coaxial-line section and an adjustable screw or post that penetrates into the guide or line and can be moved axially along the slot. *See:* **waveguide.**   185

(2) **(waveguide components).** A waveguide or transmission line tuner employing a post of adjustable penetration, adjustable in position along the longitudinal axis of the waveguide.   166

sliding contact. An electric contact in which one conducting member is maintained in sliding motion over the other conducting member. *See:* **contactor.**

    328

sliding load. *See:* **load, sliding.**

sliding short circuit. A short-circuit termination that consists of a section of waveguide or transmission line fitted with a sliding short-circuiting piston (contacting or noncontacting) that ideally reflects all the energy back toward the source. *See:* **waveguide.**   185

slime, anode. Finely divided insoluble metal or compound forming on the surface of an anode or in the solution during electrolysis. *See:* **electrodeposition.**

    328

slinging wire. A wire used to suspend and carry current to one or more cathodes in a plating tank. *See:* **electroplating.**   328

sling, traveler. *See:* **traveler sling.**

slip (1) (rotating machinery). (A)The quotient of (1) the difference between the synchronous speed and the actual speed of a rotor, to (2) the synchronous speed, expressed as a ratio, or as a percentage.(B)The difference between the speed of a rotating magnetic field and that of a rotor, expressed in revolutions per minute.(C) (electric couplings).The difference between the speeds of the two rotating members. *See:* **asynchronous machine.**   63

(2) **(electric installations on shipboard).** In an induction machine, the difference between its synchronous speed and its operating speed. It may be expressed in the following ways: (A) as a percent of synchronous

speed; (B) as a decimal fraction of synchronous speed; (C) directly in evolutions per minute.   3

slip regulator (rotating machinery). A device arranged to produce a reduction in speed below synchronous speed greater than would be obtained inherently. Such a device is usually in the form of a variable impedance connected in the secondary circuit of a slip ring induction motor.   63

slip relay. A relay arranged to act when one or more pairs of driving wheels increase or decrease in rotational speed with respect to other driving wheels of the same motive power unit. *See:* **multiple-unit control.**

    328

slip ring. *See:* **collector ring.**

slip-ring induction motor. *See:* **wound-rotor induction motor.**

slope angle. *See:* **glide slope angle.**

slope detector (telephony)(dial-pulse address signaling systems). A circuit that provides a means of accurately measuring the open and closed intervals of a contact even though the contact may be shunted by a contact protection network or measured from the far end of a metallic loop, or both.   540

slot (1)(FASTBUS acquisition and control). A module connector position on a crate segment backplane. *See:* **position.**   480

(2)**(rotating machinery).** A channel or tunnel opening onto or near the air gap and passing essentially in an axial direction through the rotor or stator core. A slot usually contains the conductors of a winding, but may be used exclusively for ventilation. *See:* **rotor (rotating machinery); stator.**   63

slot antenna (data transmission). A radiating element formed by a slot in a conducting surface.   59

slot array. An antenna array formed of slot radiators. *See:* **antenna.**   244

slot cell (rotating machinery). A sheet of insulation material used to line a slot before the winding is placed in it. *See:* **rotor (rotating machinery); stator.**

    63

slot coupling factor (slot-antenna array)(navigation aid terms). The ratio of the desired slot current to the available slot current, controlled by changing the depth of penetration of the slot probe into the waveguide.   526

slot current ratio (slot-antenna array)(navigation aid terms). The relative slot currents in the slots of the waveguide reading from its center to its end, with the maximum taken as 1; this ratio is dependent upon the slot spacing factor and the slot coupling factor.

    526

slot discharge (rotating machine). Sparking between the outer surface of coil insulation and the grounded slot surface, caused by capacitive current between conductors and iron. The resulting current pulses have a fundamental frequency of a few kilohertz. *See:* **asynchronous machine.**   63

slot-discharge analyzer (rotating machinery). An instrument designed for connection to an energized winding of a rotating machine, to detect pulses caused by slot discharge, and to discriminate between them and pulses otherwise caused. *See:* **asynchronous machine.** 63

slot insulation (rotating machinery). A sheet or deposit of insulation material used to line a slot before the winding is placed in it. *See:* **asynchronous machine.** 63

slot liner (rotating machinery). Separate insulation between an embedded coil side and the slot which can provide mechanical and electrical protection. 63

slot packing (filler) (rotating machinery). Additional insulation used to pack embedded coil sides to ensure a tight fit in the slots. *See:* **rotor (rotating machinery); stator.** 63

slot pitch (tooth pitch) (rotating machinery). The peripheral distance between fixed points in corresponding positions in two consecutive slots. 63

slot separator insulation. *See:* **separator insulation, slot.**

slot space factor (rotating machinery). The ratio of the cross-sectional area of the conductor metal in a slot to the total cross-sectional area of the slot. *See:* **asynchronous machine.** 63

slot spacing factor (slot-antenna array)(navigation aid terms). A value proportional to the size of the angle between the slot location and the null of the internal standing wave; this factor is dependent upon frequency. 526

slotted armature (rotating machine). An armature with the winding placed in slots. *See:* **armature.** 244, 63

slotted section (slotted line) (slotted waveguide). A section of a waveguide or shielded transmission line the shield of which is slotted to permit the use of a carriage and travelling probe for examination of standing waves. *See:* **auxiliary device to an instrument.** 244, 179, 185

slot-type antenna (aircraft). A slot in the normal streamlined metallic surface of an aircraft, excited electromagnetically by a structure within the aircraft. Radiation is thus obtained without projections that would disturb the aerodynamic characteristics of the aircraft. Radiation from a slot is essentially directive. 328

slot wedge (rotating machinery). The element placed above the turns or coil sides in a stator or rotor slot, and held in place by engagement of wedge (slots) grooves along the sides of the coil slot, or by projections from the sides of the slot tending to close the top of the slot. *Note:* A wedge may be a thin strip of material provided solely as insulation or to provide temporary retention of the coils during the manufacturing process. It may be a piece of structural insulating material or high-strength metal to hold the coils in the slot. Slots in laminated cores are normally wedged with insulating material. *See:* **rotor (rotating machinery); stator.** 63

slow-closure device (hydraulic turbines). A cushioning device which retards the closing velocity of servomotor travel from a predetermined servomotor position to zero servomotor position. 8

slow-operate relay. A slugged relay that has been specifically designed for long operate time but not for long release time. *Caution:* The usual slow-operate relay has a copper slug close to the armature, making it also at least partially slow to release. 341

slow-operating relay. A relay that has an intentional delay between energizing and operation. *See:* **electromagnetic relay.** 328

slow-release relay. A relay that has an intentional delay between de-energizing and release. *Note:* The reverse motion need not have any intentional delay. *See:* **electromagnetic relay.** 341

slow release time characteristic, relay. *See:* **relay slow release time characteristic.**

slow-speed starting (industrial control). A control function that provides for starting an electric drive only at the minimum-speed setting. *See:* **starter.** 206

slow-wave circuit (microwave tubes). A circuit whose phase velocity is much slower than the velocity of light. For example, for suitably chosen helixes the wave can be considered to travel on the wire at the velocity of light but the phase velocity is less than velocity of light by the factor that the pitch is less than the circumference. *See:* **microwave tube or valve.** 190

slug, relay. *See:* **relay slug.**

slug tuning. A means for varying the frequency of a resonant circuit by introducing a slug of material into either the electric or magnetic fields or both. *See:* **network analysis; radio transmission.** 111

slush compound (corrosion). A non-drying oil, grease, or similar organic compound that, when coated over a metal, affords at least temporary protection against corrosion. 205

small-signal (light emitting diodes). A signal which when doubled in magnitude does not produce a change in the parameter being measured that is greater than the required accuracy of the measurement. 162

small-signal forward transadmittance. The value of the forward transadmittance obtained when the input voltage is small compared to the beam voltage. *See:* **electron-tube admittances.** 125

small signal performance (excitation systems for synchronous machines). The response of an excitation cotrol system, excitation system, or elements of an excitation system to signals which are small enough that nonlinearities can be disregarded in the analysis of the response, and operation can be considered to be linear. 507

small-signal permittivity (ferroelectric material). The incremental change in electric displacement per unit electric field when the magnitude of the measuring field is very small compared to the coercive electric field. (Measurements are usually made at a frequency of 1 kilohertz or higher). The small signal relative permittivity κ is equal to the ratio of the absolute permittivity ϵ to the permittivity of free space ϵ_0, that is, $\kappa = \epsilon/\epsilon_0$. *Note:* The value of the small-signal permittivity may depend on the remanent polarization, electric field, mechanical stress, sample history, or frequency of the measuring field. *See:* **Curie-Weiss temperature; paraelectric; remanent polarization; coercive electric field.** 80

small-signal resistance (semiconductor rectifier). The resistive part of the quotient of incremental voltage by incremental current under stated operating conditions. *See:* **rectification.** 66

small wiring. *See:* **secondary and control wiring.**

smashboard signal. A signal so designed that the arm will be broken when passed in the stop position. 328

smoke detector (fire protection devices). A device which detects the visible or invisible particles of combustion. 71

smooth current (rotating electric machinery). Current that remains unidirectional and the ripple of which does not exceed 3 percent. 424

smothered-arc furnace. A furnace in which the arc or arcs is covered by a portion of the charge. 328

snake. *See:* **fish tape.**

snapover. When used in connection with alternating-current testing, a quasi-flashover or quasi-sparkover, characterized by failure of the alternating-current power source to maintain the discharge, thus permitting the dielectric strength of the specimen to recover with the test voltage still applied. *See:* **test voltage and current.** 201

snapshot dump (computing systems). A selective dynamic dump performed at various points in a machine run. 255, 77

snatch block (power line maintenance)(conductor stringing equipment). A device normally designed with a single sheave, a wood or metal shell, and a hook. One side of the shell usually opens to eliminate the need for threading of the line. It is commonly used for lifting loads on a single line, or as a device to control the position or direction, or both, of a fall line or pulling line. *Syn:* **skookum; Washington; Western.** 458, 431

SNM (shielded nonmetallic-sheathed) cable (National Electrical Code). A factory assembly of two or more insulated conductors in an extruded core of a moisture-resistant, flame-resistant nonmetallic material, covered with an overlapping spiral metal tape and wire shield and jacketed with an extruded moisture-, flame-, oil-, corrosion-, fungus-, and sunlight-resistant nonmetallic material. 256

snow (intensity-modulated display). A varying speckled background caused by noise. *See:* **radar; television.** 178

snr psoph (data transmission). Signal-to-noise ratio measured with psophometrically weighted receiver; expressed in dB (decibels). 59

snubber (converter circuit elements)(self-commutated converters). An auxiliary circuit element or combination of elements employed to modify the transient voltage or current of a semiconductor device. *See:* **shunt snubber; series snubber; polarized snubber.** 584

snub structure (conductor stringing equipment). A structure located at one end of a sag section and considered as a zero point for sagging and clipping offset calculations. The section of line between two such structures is the sag section, but more than one sag section may be required in order to sag properly the actual length of conductor which has been strung. *Syn:* **O structure; zero structure.** 431

soak, relay. *See:* **relay soak.**

socket. *See:* **lampholder.**

socket cover (watthour meter sockets). The removal portion of the enclosure that provides access to the meter socket wiring. 549

socket rim (watthour meter sockets). That part of a ring-type meter socket which is required to accommodate the socket sealing ring which holds a detachable watthour meter in place. The socket rim may be a part of the cover, which is secured in place by a fastener, such as a latch or cross-bar. 549

socket sealing ring (watthour meter sockets). A ring used to overlap the socket rim and the detachable watthour meter cover ring to hold and provide means for sealing a detachable watthour meter in place. 549

sodium vapor lamp transformers (multiple-supply type) (power and distribution transformer). Transformers, autotransformers, or reactors for operating sodium vapor lamps for all types of lighting applications, including indoor, outdoor area, roadway, and other process and specialized lighting. 53

sofar (navigation aid terms). A system of navigation providing hyperbolic lines of position determined by shore listening stations. 526

soft limiting. *See:* **limiter circuit.**

soft start (1)(electrical heating systems). The ability of a controlling device to apply power to a load upon energization in a proportional manner irrespective of values of the controlling signals. 476
(2)(thyristor). At turn-on, a gradual increase in output at a predetermined rate from zero or a set minimum to a desired maximum. 445

soft start reset (thyristor). Reset of soft start to initial conditions when ac power is interrupted. 445

software (1)(software engineering terminology). (A) Computer programs, procedures, rules, and possibly associated documentation and data pertaining to the operation of a computer system. (B) Programs, procedures, rules, and any associated documentation pertaining to the operation of a computer system. *See:* **application software; computer program; computer system; data; documentation; hardware; procedure; program; system software.** 434
(2)(programmable digital computer systems in safety systems of nuclear power generating stations). Computer programs and data. 554

software accuracy (programmable digital computer systems in safety systems of nuclear power generating stations). The software attribute that provides a quantitative measure of the magnitude of error. 554

software characteristic (software unit testing). An inherent, possibly accidental, trait, quality, or property of software (for example, functionality, performance, attributes, design constraints, number of states, lines of branches). 519

software configuration management. *See:* **configuration management.** 434

software consistency (programmable digital computer systems in safety systems of nuclear power generating stations). The software attribute that provides uniform design and implementation techniques and notation. 554

software data base. A centralized file of data definitions and present values for data common to, and located internal to, an operational software system. *See:* **data; file; operational software system.** 434

software design description (1)(SDD)(software design descriptions). A representation of a software system created to facilitate analysis, planning, implementation, and decision making. A blueprint or model of the software system. The SDD is used as the primary medium for communicating software design information. 568

(2)(software verification and validation plans). A representation of software created to facilitate analysis, planning, implementation, and decision making. The software design description is used as a medium for communicating software design information, and may be thought of as a blueprint or model of the system. 511

software development cycle. (1) The period of time that begins with the decision to develop a software product and ends when the product is delivered. This cycle typically includes a requirements phase, implementation phase, test phase, and sometimes, installation and checkout phase. (2) The period of time that begins with the decision to develop a software product and ends when the product is no longer being enhanced by the developer. (3) Sometimes used as a synonym for "software life cycle". *See:* **design phase; implementation phase; installation and checkout phase; requirements phase; software life cycle; software product.** 434

software development library. A software library containing computer readable and human readable information relevant to a software development effort. *See:* **computer; software; software library.** 434

software development notebook. A collection of material pertinent to the development of a given software module. Contents typically include the requirements, design, technical reports, code listings, test plans, test results, problem reports, schedules, notes, etcetera, for the module. *See:* **code; design; listing; module; project notebook; requirement; software.** 434

software development plan. A project plan for the development of a software product. *Syn:* **computer program development plan.** *See:* **project plan; software product.** 434

software development process. The process by which user needs are translated into software requirements, software requirements are transformed into design, the design is implemented in code, and the code is tested, documented, and certified for operational use.

See: **code; design; operational; requirement.** 434

software documentation. Technical data or information, including computer listings and printouts, in human-readable form, that describe or specify the design or details, explain the capabilities, or provide operating instructions for using the software to obtain desired results from a software system. *See:* **computer listing; data; design; documentation; software; system; system documentation; user documentation.** 434

software engineering. The systematic approach to the development, operation, maintenance, and retirement of software. *See:* **maintenance; software.** 434

software error tolerance (programmable digital computer systems in safety systems of nuclear power generating stations). The software attribute that provides continuity of operation under postulated nonnominal conditions. 554

software experience data. Data relating to the development or use of software that could be useful in developing models, reliability predictions, or other quantitative descriptions of software. 434

software feature (1)(software test documentation). A distinguishing characteristic of a software item (for example, performance, portability, or functionality). 436

(2)(software unit testing). A software characteristic specified or implied by requirements documentation (for example, functionality, performance, attributes, or design constraints). 519

software item (software test documentation). Source code, object code, job control code, control data, or a collection of these items. 436

software librarian. The person responsible for establishing, controlling, and maintaining a software library. *See:* **software library.** 434

software library. A controlled collection of software and related documentation designed to aid in software development, use, or maintenance. Types include software development library, master library, production library, program library, and software repository. *See:* **documentation; master library; production library; program library; software; software repository; system library.** 434

software life cycle. The period of time that starts when a software product is conceived and ends when the product is no longer available for use. The software life cycle typically includes a requirements phase, design phase, implementation phase, test phase, installation and checkout phase, operation and maintenance phase, and sometimes, retirement phase. *See:* **design phase; implementation phase; installation and checkout phase; operation and maintenance phase; requirements phase; retirement phase; software development cycle; software product.** 434

software maintenance. (1) Modification of a software product after delivery to correct faults. (2) Modification of a software product after delivery to correct faults, to improve performance or other attributes, or to adapt the product to a changed environment. *See:* **adaptive maintenance; corrective maintenance;**

delivery; faults; modification; perfective maintenance; software product.   434

software modularity (programmable digital computer systems in safety systems of nuclear power generating stations). The software attribute that provides a structure of highly independent computer program units that are discrete and identifiable with respect to compiling, combining with other units, and loading.   554

software monitor. A software tool that executes concurrently with another computer program and that provides detailed information about the execution of the other program. *See:* **computer program; execution; program; software tool.**   434

software product. A software entity designated for deliver to a user. *See:* **delivery; user.**   434

software quality. (1) The quality of features aand characterisitics of a software product that bear on its ability to satisfy given needs; for example, conform to specifications. (2) The degree to which software possesses a desired combination of atributes. (3) The degree to which a consumer or user perceives that software meets his or her composite expectations. (4) The composite characteristics of software that determine the degree to which the software in use will meet the expectations of the customer. *See:* **software; software product; specification.**   434

software quality assurance. *See:* **quality assurance.**

software reliability. (1) The probability that software will not cause the failure of a system for a specified time under specified conditions. The probability is a function of the inputs to and use of the system as well as a function of the existence of faults in the softwear. The inputs to the system determine whether existing faults, if any, are encountered. (2) The ability of a program to perform a required function under stated conditions for a stated period of time. *See:* **failure; faults; function; program; software; system.**   434

software repository. A software library providing permanent, archival storage for software and related documentation. *See:* **documentation; software; software library.**   434

software requirements specification (software verification and validation plans). Documentation of the essential requirements (functions, performance, design constraints, and attributes) of the software and its external interfaces.   511

software sneak analysis. A technique applied to software to identify latent (sneak) logic control paths or conditions that could inhibit a desired operation or cause an unwanted operation to occur. *See:* **software.**   434

software test incident (software unit testing). Any event occurring during the execution of a software test that requires investigation.   519

software tool. A computer program used to help develop, test, analyze, or maintain another computer program or its documentation, for example, automated design tool, compiler, test tool, maintenance tool. *See:* **automated design tool; compiler; computer program; documentation; maintenance tool; tool.**   434

software verification and validation plan (software verification and validation plans). A plan for the conduct of software verification and validation.   511

software verification and validation report (software verification and validation plans). Documentation of verification and validation (V&V) results and appropriate software quality assurance results.   511

solar array (photovoltaic power system). A group of electrically interconnected solar cells assembled in a configuration suitable for oriented exposure to solar flux.   186

solar constant (illuminating engineering). The irradiance (averaging 1,353 W/m^2(125.7 W/ft^2), from the sun at its mean distance from the earth 92.9 $X10^6$ miles (1.5 $X10^{11}$m), before modification by the earth's atmosphere.   167

solar induced currents (SIC) (power fault effects). Spurious, quasidirect currents flowing in grounded power systems or telecommunication cables caused by earth potential differences due to geomagnetic storms resulting from the particle emission of solar flares erupting from the surface of the sun. *See:* **auroral.**   404

solar noise (communication satellite). Electrical noise generated by the sun. Exceeds other background noise sources by several orders of magnitude.   85

solar panel (photovoltaic power system). *See:* **solar array.**

solar radiation simulator (illuminating engineering). A device designed to produce a beam of collimated radiation having a spectrum, flux density, and geometric characteristics similar to those of the sun outside the earth's atmosphere.   167

solar wind (communication satellite). Energetic particles emitted by the sun and travelling through space.   74

solderability. That property of a metal surface to be readily wetted by molten solder.   284

soldered joints. The connection of similar or dissimilar metals by applying molten solder, with no fusion of the base metals.   284

solder projections. Icicles, nubs, and spikes are undesirable protrusions from a solder joint.   284

solder splatter. Unwanted fragments of solder.   284

solenoid. An electric conductor wound as a helix with a small pitch, or as two or more coaxial helixes. *See:* **solenoid magnet.**   210

solenoid magnet (solenoid) (industrial control). An electromagnet having an energizing coil approximately cylindrical in form, and an armature whose motion is reciprocating within and along the axis of the coil.   210, 206

solenoid relay. *See:* **plunger relay.**

soleplate (rotating machinery). A support fastened to a foundation on which a stator frame foot or a bracket arm can be mounted. *See:* **slide rail.**   63

solid angle (laser-maser) (ω). The ratio of the area on the surface of a sphere to the square of the radius of that sphere. It is expressed in **steradians.**   363

solid angle factor, Q (illuminating engineering). A function of the solid angle (ω) subtended by a source and is given by

$$Q = 20.4\omega = 1.52\omega^{0.2} - 0.075$$

See: **index of sensation.**   167

solid-beam efficiency (antennas). The ratio of the power received over a specified solid angle when an antenna is illuminated isotropically by uncorrelated and unpolarized waves to the total power received by the antenna. *Note:* This term is sometimes used to mean the power received corresponding to a particular polarization over the solid angle to the total power received. Equivalently, the term is used to mean the power radiated over a specified solid angle by the antenna corresponding to a particular polarization to the total power radiated.   111

solid bushing (outdoor electric apparatus). A bushing in which the major insulation is provided by a ceramic or analogous material.   168

solid conductor. A conductor consisting of a single wire. *See:* **conductor.**   64

solid contact. A contact having relatively little inherent flexibility and whose contact pressure is supplied by another member.   103

solid coupling (rotating machinery). A coupling that makes a rigid connection between two shafts. *See:* **rotor (rotating machinery).**   63

solid electrolytic capacitor. A capacitor in which the dielectric is primarily an anodized coating on one electrode, with the remaining space between the electrodes filled with a solid semiconductor.   341

solid enclosure. An enclosure that will neither admit accumulations of flyings or dust nor transmit sparks or flying particles to the accumulations outside.   328

solid-iron cylindrical-rotor generator. *See:* **cylindrical-rotor generator.**

solidly grounded (power and distribution transformer). Grounded through an adequate ground connection in which no impedance has been inserted intentionally. *Note:* Adequate as used herein means suitable for the purpose intended.   53

solid-material fuse unit (power switchgear). A fuse unit in which the arc is drawn through a hole in solid material.   103, 443

solid-pole synchronous motor. A salient-pole synchronous motor having solid steel pole shoes, and either laminated or solid pole bodies.   63

solid rotor (rotating machinery). (1) A rotor, usually constructed of a high-strength forging, in which slots may be machined to accommodate the rotor winding. (2) A spider-type rotor in which spider hub is not split. *See:* **rotor (rotating machinery).**   63

solid-state component. A component whose operation depends on the control of electric or magnetic phenomena in solids, for example, a transistor, crystal diode, ferrite core.   255, 77

solid-state device (control equipment). A device that may contain electronic components that do not depend on electronic conduction in a vacuum or gas. The electrical function is performed by semiconductors or the use of otherwise completely static components such as resistors, capacitors, etcetera.   94

solid-state relay (or relay unit) (power switchgear). A static relay or relay unit constructed exclusively of solid-state components.   127, 103

solid-state scanning (facsimile). A method in which all or part of the scanning process is due to electronic commutation of a solid-state array of thin-film photosensitive elements. *See:* **facsimile (electrical communication).**   194

solid-type paper-insulated cable. Oil-impregnated, paper-insulated cable, usually lead covered, in which no provision is made for control of internal pressure variations.   64

solution. *See:* **check solution.**

solvent cleaning (electroplating). Cleaning by means of organic solvents. *See:* **electroplating.**   328

solventless (rotating machinery). A term applied to liquid or semiliquid varnishes, paints, impregnants, resins, and similar compounds that have essentially no change in weight or volume when converted into a solid or semisolid.   63

sonar (navigation aid terms). A general name for sonic and ultrasonic ranging, sounding and communication systems.   526

sonic delay line. *See:* **acoustic delay line.**

sonic depth finder (navigation aid terms). A direct reading instrument which determines the depth of water by measuring the time interval between emission of sound and the return of its echo from the bottom.   526

sonne (navigation aid terms). A radio navigation aid that provides a number of characteristic signal zones which rotate in a time sequence; a bearing may be determined by observation (by interpolation) of the instant at which transition occurs from one zone to the following zone. *See:* **consol.**   526

sonobuoy (navigation aid terms). A buoy with equipment for automatically transmitting a radio signal when triggered by an underwater sound signal. *Syn:* **sono-radio buoy.**   526

sort. To arrange data or items in an ordered sequence by applying specific rules.   255, 77, 54

sorter. A person, device, or computer routine that sorts.   255, 77

SOS. *See:* **radio distress signal.**

sound. (1) An oscillation in pressure, stress, particle displacement, particle velocity, etcetera, in a medium with internal forces (for example, elastic, viscous), or the superposition of such propagated oscillations. (2) An auditory sensation evoked by the oscillation described above. *Notes:* (A) In case of possible confusion, the term sound wave or elastic wave may be used for concept (1) and the term sound sensation for concept (2). Not all sound waves can evoke an auditory sensation, for example, an ultrasonic wave. (B) The medium in which the sound exists is often indicated by an appropriate adjective, for example, air-borne, water-borne, structure-borne.   176

sound absorption. (1) The change of sound energy into some other form, usually heat, in passing through a medium or on striking a surface. (2) The property possessed by material and objects, including air, of absorbing sound energy. 176

sound-absorption coefficient (surface). The ratio of sound energy absorbed or otherwise not reflected by the surface, to the sound energy incident upon the surface. Unless otherwise specified, a diffuse sound field is assumed. 176

sound analyzer. A device for measuring the band pressure level, or pressure spectrum level, of a sound at various frequencies. *Notes:* (1) A sound analyzer usually consists of a microphone, an amplifier and wave analyzer, and is used to measure amplitude and frequency of the components of a complex sound. (2) The band pressure level of a sound for a specified frequency band is the effective root-mean-square sound pressure level of the sound energy contained within the bands. *See:* **instrument.** 334

sound articulation (percent sound articulation). The percent articulation obtained when the speech units considered are fundamental sounds (usually combined into meaningless syllables). *See:* **volume equivalent.** 328

sound buoy (navigation aid terms). A buoy equipped with a characteristic sound signal. *See:* **buoy.** 526

sound-detection system (protective signaling). A system for the protection of vaults by the use of sound-detecting devices and relay equipment to pick up and convert noise, caused by burglarious attack on the structure, to electric impulses in a protection circuit. *See:* **protective signaling.** 328

sound-effects filter. *See:* **filter, sound-effects.**

sound energy. Of a given part of a medium, the total energy in this part of the medium minus the energy that would exist in the same part of the medium with no sound waves present. 176

sound field. A region containing sound waves. 176

sound intensity (power station noise control). The average rate of sound energy radiated by a source per unit time. 500

sound intensity (sound-energy flux density) (sound power density) (in a specified direction at a point). The average rate of sound energy transmitted in the specified direction through a unit area normal to this direction at the point considered. *Notes:* (1) The sound intensity in any specified direction a of a sound field is the sound-energy flux through a unit area normal to that direction. This is given by the expression

$$I_a = \frac{1}{T} \int_0^T p v_a \, dt$$

where

T = an integral number of periods or a long time compared to a period

p = the instantaneous sound pressure

v_a = the component of the instantaneous particle velocity in the direction a

t = time

(2) In the case of a free plane or spherical wave having an effective sound pressure, p, the velocity of propagation c, in a medium of density p, the intensity in the direction of propagation is given by

$$I = \frac{p^2}{\rho c}$$

176

sound level (measurement of sound pressure levels of ac power circuit breakers). Weighted sound pressure level obtained by the use of a metering characteristic and the weightings A, B, C (or other) as specified. The weighting used must be indicated. For the purpose of ANSI/IEEE Std C37.082-1982, C weighted sound level is the same as sound pressure level (SPL). Unit: decibel (dB, A, B, or C). 552

sound level, A-weighted (airborne sound measurements on rotating electric machinery). The A-weighted sound level is the weighted sound pressure level, obtained by use of metering characteristics and A-weighting specified in ANSI S1.4-1971, Specification for Sound Level Meters. 129

sound-level meter. An instrument including a microphone, an amplifier, an output meter, and frequency-weighting networks for the measurement of noise and sound levels in a specified manner. *Notes:* (1) The measurements are intended to approximate the loudness level of pure tones that would be obtained by the more-elaborate ear balance method. (2) Loudness level in phons of a sound is numerically equal to the sound pressure level in decibels relative to 0.0002 microbar of a simple tone of frequency 1000 hertz that is judged by the listeners to be equivalent in loudness. (3) Specifications for sound-level meters are given in American National Standard Specification for Sound-Level Meters, S1.4-1971 (or latest revision thereof). *See:* **instrument.** 176

sound power (power station noise control). The total sound energy radiated by a source per unit time. 500

sound power level (airborne sound measurements on rotating electric machinery). The sound power level, in decibels, is equal to 10 times the logarithm to the base 10 of the ratio of a given power to the reference power, 10^{-12} W (see ANSI S1.8-1969). 129

$$L_{\mathrm{w}} = 10 \log_{10} \left(\frac{W}{W_0} \right)$$

where

L_{w} = sound power level

W = measured sound power in watts

W_0 = reference power

sound power level, *A*-weighted (airborne sound measurements on rotating electric machinery). The *A*-weighted sound power level, in decibels, is equal to the sound power level determined by weighting each of the frequency bands.   129

sound pressure (1)(power station noise control). The instantaneous pressure measured in a sound wave, that is, the variation in atmospheric pressure.   500

(2)(transmission performance of telephone sets). The sound pressure at a point, is the total instantaneous pressure at that point, in the presence of a sound wave, minus the static pressure at that point.   491

sound pressure, effective (root-mean-square sound pressure). At a point over a time interval, the root-mean-square value of the instantaneous sound pressure at the point under consideration. In the case of periodic sound pressures, the interval must be an integral number of periods or an interval long compared to a period. In the case of nonperiodic sound pressures, the interval should be long enough to make the value obtained essentially independent of small changes in the length of the interval. *Note:* The term **effective sound pressure** is frequently shortened to **sound pressure.**   176

sound pressure, instantaneous (at a point). The total instantaneous pressure at that point minus the static pressure at that point. *Note:* The commonly used unit is the newton per square meter.   176

sound pressure level (1)(SPL)(measurement of sound pressure levels of ac power circuit breakers). Twenty times the logarithm to the base 10 of the ratio of the pressure of a sound to the reference sound pressure. Unless otherwise specified, the effective root-mean-square (rms) pressure is used. The reference sound pressure is 20 μPa. Unit: decibel (dB).   552

(2)(transmission performance of telephone sets). The sound pressure level, in decibels, of a sound is 20 times the logarithm to the base 10 of the ratio of the pressure of this sound to the reference pressure. The reference is one pascal (Pa).   491

sound probe. A device that responds to some characteristic of an acoustic wave (for example, sound pressure, particle velocity) and that can be used to explore and determine this characteristic in a sound field without appreciably altering the field. *Note:* A sound probe may take the form of a small microphone or a small tubular attachment added to a conventional microphone. *See:* **instrument.**   176

sound recording system. A combination of transducing devices and associated equipment suitable for storing sound in a form capable of subsequent reproduction. *See:* **phonograph pickup.**   176

sound reflection coefficient (surface). The ratio of the sound reflected by the surface to the sound incident upon the surface. Unless otherwise specified, reflection of sound energy in a diffuse sound field is assumed.   176

sound reproducing system. A combination of transducing devices and associated equipment for reproducing recorded sound. *See:* **loudspeaker.**   176

sound spectrum analyzer (sound analyzer). A device or system for measuring the band pressure level of a sound as a function of frequency.   176

sound tract (electroacoustics). A band that carries the sound record. In some cases, a plurality of such bands may be used. In sound film recording, the band is usually along the margin of the film. *See:* **phonograph pickup.**   176

sound transmission coefficient (interface or partition). The ratio of the transmitted to incident sound energy. Unless otherwise specified, transmission of sound energy between two diffuse sound fields is assumed.   176

source (1) (laser-maser). Taken to mean either laser of laser-illuminated reflecting surface.   363

(2) (metal-nitride-oxide field-effect transistor). Region in the device structure of an insulated-gate-field-effect transistor (IGFET) which contains the terminal from which charge carries flow into channel toward the drain. It has the potential which is less attractive than the drain for the carriers in the channel.   386

source efficiency (fiber optics). The ratio of emitted optical power of a source to the input electrical power.   433

source ground (signal-transmission system). Potential reference at the physical location of a source, usually the signal source. *See:* **signal.**   188

source impedance. *See:* **impedance, source; self-impedance.**

source language (software). (A) A language used to write source programs. (B) A language from which statements are translated. *See:* **source programs; target language.**   434

source/load impedance (loudness ratings of telephone connections). For the purposes of IEEE Std 661-1979 the source/load impedance used for determining loudness ratings (see 3.6-3.9) is considered to be 900 Ω resistive. *See:* **impedance matching network; source/ load impedance other than 900 Ω.**   409

source node (network analysis). A node having only outgoing branches.   282

source program (software). (1) A computer program that must be compiled, assembled, or interpreted before being executed by a computer. (2) A computer program expressed in a source language. *See:* **assemble; compile; computer; computer program; interpret; object program; source language.**   434

source resistance. The resistance presented to the input of a device by the source. *See:* **measurement system.**   295

source resistance rating. The value of source resistance that, when injected in an external circuit having essentially zero resistance, will either (1) double the dead band, or (2) shift the dead band by one-half its width. *See:* **measurement system.**   295

space (1) (data transmission). One of the two possible conditions of an element (bit); an open line in a neutral circuit. In Morse code, a duration of two unit intervals between characters and six unit intervals between words.   59

(2) (computing devices). (A) A site intended for the storage of data, for example, a site on a printed page or a location in a storage medium. (B) A basic unit of area, usually the size of a single character. (C) One or more blank characters. (D) To advance the reading or display position according to a prescribed format, for example, to advance the printing or display position horizontally to the right or vertically down.   77

space charge (1) (general). A net excess of charge of one sign distributed throughout a specified volume.

(2) (thermionics). Electric charge in a region of space due to the presence of electrons and/or ions. *See:* **electron emission.**   125

space-charge-control tube. *See:* **density-modulated tube.**

space-charge debunching. Any process in which the mutual interactions between electrons in the stream disperse the electrons of a bunch.   125

space-charge density (thermionics). The space charge per unit volume. *See:* **electron emission.**   125

space-charge generation (semiconductor radiation detector) (germanium gamma-ray detectors) (charged-particle detectors) (X-ray energy spectrometers). The thermal generation of free charge carriers in the space-charge region.   23,528,119,471

space-charge grid. A grid, usually positive, that controls the position, area, and magnitude of a potential minimum or of a virtual cathode in region adjacent to the grid. *See:* **electrode (electron tube); grid.**   125

space-charge-limited current (electron vacuum tubes). The current passing through an interelectrode space when a virtual cathode exists therein. *See:* **electrode current (electron tube).**   125

space-charge region (of a semiconductor radiation detector)(X-ray energy spectrometers)(charged-particle detectors). A region in which the net charge density is significantly different from zero. *See:* **depletion region.**   471,245,23,119

space correction (industrial control). A method of register control that takes the form of a sudden change in the relative position of the web.   206

spacecraft (communication satellite). Any type of space vehicle, including an earth satellite or deep-space probe, whether manned or unmanned, and also rockets and high-altitude balloons which penetrate the earth's outer atmosphere.   74

space current (electron tubes). Synonym in a diode or equivalent diode of cathode current. *See:* **electrode current (electron tube); leakage current (electron tubes); load current (electron tubes); quiescent current (electron tubes).**   244, 190

space diversity. *See:* **space diversity reception.**

space diversity reception (data transmission). That form of diversity reception that utilizes receiving antennas placed in different locations.   59

space-division digital switching (telephone switching systems). Digital switching with separate paths for each call.   55

space-division switching (telephone switching systems). A method of switching that provides a separate path for each of the simultaneous calls.   55

space factor (rotating machinery). The ratio of (1) the sum of the cross-sectional areas of the active or specified material to (2) the cross-sectional area within the confining limits specified. *See:* **asynchronous machine; slot space factor.**   63

space, head. *See:* **head space.**

space heater (1) (general). A heater that warms occupied spaces.

(2) (rotating machinery). A device that warms the ventilating air within a machine and prevents condensation of moisture during shut-down periods. *See:* **appliances (including portable).**   63

space pattern (television). A geometrical pattern on a test chart designed for the measurement of geometric distortion.   372

space probe (communication satellite). A spacecraft with a trajectory extending into deep space.   74

spacer cable (NESC). A type of electric supply line construction consisting of an assembly of one or more covered conductors, separated from each other and supported from a messenger by insulating spacers.   494

space-referenced navigation data (navigation aid terms). Data in terms of a coordinate system referenced to inertial space.   526

spacer shaft (rotating machinery). A separate shaft connecting the shaft ends of two machines. *See:* **armature.**   63

space, state. *See:* **state space.**

space-tapered array antenna. An array antenna whose radiation pattern is shaped by varying the density of driven radiating elements over the array surface. *Syn:* **density-tapered array antenna.**   111

spacing (data transmission). A term which originated with telegraph to indicate an open key condition. Present usage implies the absence of current or carrier on a circuit. It also indicates the binary digit 0 in computer language.   59

spacing pulse (data transmission). A spacing pulse or **space** is the signal pulse that, in direct-current neutral operation, corresponds to a **circuit open** or **no current** condition. *See:* **pulse.**   194

spacing wave (back wave) (telegraph communication). The emission that takes place between the active portions of the code characters or while no code characters are being transmitted. *See:* **radio transmitter.**   111

spalling (corrosion). Spontaneous separation of a surface layer from a metal.   205

span (1) (measuring devices). The algebraic difference between the upper and lower values of a range. *Notes:* (A) For example: (a) Range 0 to 150, span 150: (b) Range −20 to 200, span 220: (c) Range 20 to 150, span 130: (d) Range −100 to −20, span 80. (B) The following compound terms are used with suitable modifications in the units: measured variable span, measured signal span, etcetera. (C) For multirange devices, this definition applies to the particular range that the device is set to measure. *See:* **instrument.**   295

(2) (overhead conductors). (A) The horizontal distance between two adjacent supporting points of a conductor. (B) That part of any conductor, cable, suspension strand, or pole line between two consecutive points of support. *See:* **cable; open wire.** 64

span frequency-response rating. The maximum frequency in cycles per minute of sinusoidal variation of measured signal for which the difference in amplitude between output and input represents an error no greater than five times the accuracy rating when the instrument is used under rated operating conditions. The peak-to-peak amplitude of the sinusoidal variation of measured signal shall be equivalent to full span of the instrument. It must be recognized that the span frequency-response rating is a measure of dynamic behavior under the most adverse conditions of measured signal (that is, the maximum sinusoidal excursion of the measured signal). The frequency response for an amplitude of measured signal less than full span is not proportional to the frequency response for full span. The relationship between the frequency response of different instruments at any particular amplitude of measured signal is not indicative of the relationship that will exist at any other amplitude. *See:* **accuracy rating of an instrument.** 295

span length (NESC). The horizontal distance between two adjacent supporting points of a conductor.
494

span, ruling. *See:* **ruling span.**

span, sag. *See:* **sag span.**

span step-response-time rating. The time that the step-response time will not exceed for a change in measured signal essentially equivalent to full span when the instrument is used under rated operating conditions. The actual span step-response time shall not be less than 2/3 of the span step-response-time rating. (For example, for an instrument of 3-second span step-response-time rating, the span step-response time, under rated operating conditions, will be between 3 and 2 seconds.) It must be recognized that the step-response time for smaller steps is not proportional to the step-response time for full span. *Note:* The end device shall be considered to be at rest when it remains within a band of plus and minus the accuracy rating from its final position. *See:* **accuracy rating (instrument).** 295

span wire (NESC). An auxiliary suspension wire which serves to support one or more trolley contact conductors or a light fixture and the conductors which connect it to a supply system. 494

spare equipment. Equipment complete or in parts, on hand for repair or replacement. *See:* **reserve equipment.** 64

spare point (for supervisory control or indication or telemeter selection) (power switchgear). A point that is not being utilized but is fully equipped with all of the necessary devices for a point. 103

spark (overhead-power-line corona and radio noise). A sudden and irreversible transition from a stable corona discharge to a stable arc discharge. It is a luminous electric discharge of short duration between two electrodes in a insulating medium. It is generally brighter and carries more current than corona, and its color is mainly determined by the type of insulating medium. It generates radio noise of wider frequency spectrum (extending into hundreds of megahertz) and wider magnitude range than corona. A spark is not classified as corona. 411

spark capacitor (spark condenser*). A capacitor connected across a pair of contact points, or across the inductance that causes the spark, for the purpose of diminishing sparking at these points.
*Deprecated 63

spark gap. Any short air space between two conductors electrically insulated from or remotely electrically connected to each other. 297

spark-gap converter, mercury-hydrogen. *See:* **mercury-hydrogen spark-gap converter.**

spark-gap converter, quenched. *See:* **quenched spark-gap converter.**

spark-gap modulation. A modulation process that produces one or more pulses or energy by means of a controlled spark-gap breakdown for application to the element in which modulation takes place. *See:* **oscillatory circuit.** 111

spark gaps (wire-line communication facilities). Spark gaps consist of air dielectric between two electrodes which may be a combination of several basic shapes. Spark gaps are used to protect communication circuits from damage due to voltage stress in excess of their dielectric capabilities. 414

spark killer. An electric network, usually consisting of a capacitor and resistor in series, connected across a pair of contact points, or across the inductance that causes the spark, for the purpose of diminishing sparking at these points. *See:* **network analysis.** 328

sparkover (1)(surge arrester)(gas-tube surge-protective devices)(low-voltage air gap surge-protective devices)gas tube surge-protective device). A disruptive discharge between electrodes of a measuring gap, voltage control gap, or protective device.
430,490,556, 370

(2)(metal-oxide surge arresters for ac power circuits). A disruptive discharge between electrodes of a measuring gap, voltage-control gap, or gap-type protective device. 583

(3) (high voltage testing). Term used when a disruptive discharge occurs in a gaseous or liquid dielectric.
150

spark-plug suppressor (electromagnetic compatibility). A suppressor designed for direct connection to a spark plug. *See:* **electromagnetic compatibility.**
199

spark transmitter. A radio transmitter that utilizes the oscillatory discharge of a capacitor through an inductor and a spark gap as the source of its radio-frequency power. *See:* **radio transmitter.** 111, 240

sparse data scan (SDS)(FASTBUS acquisition and control). A technique by which arrays of modules with low data occupancy may be scanned efficiently, that is, without accessing every potential data site.
480

spatial coherence (1) (fiber optics). *See:* **coherent.**
433

(2) (electromagnetic) (laser-maser). The correlation between electromagnetic fields at points separated in space. *See:* **coherence area.**   363

spatial disturbance (diode-type camera tube). In the output signal from a television camera consists of a broad variety of spurious signals, some of which are observable when no optical input is present, while others are input- level dependent. Spatial disturbances are characterized as either independent of time or as having a temporal variation long with respect to a frame interval, provided the operating conditions, including position and temperature, remain fixed. Tolerance for spatial distrubance covers a broad range, depending upon the application. Cosmetic considerations ultimately reduce to a cost decision. Spurious signals have been classified in the following categories: (1) Fixed pattern. This is a modulation of a uniform background which may be either spatially periodic or random. (2) Shading. This consists of a broad area continuous variation in the background signal, with or without an optical input. The signal corresponding to uniform irradiance is either curved or tilted, causing a brightness variation in the display. (3) Moire. This is a periodic amplitude modulation in the output which is not present in the input, usually due to the interaction of two or more periodic tube elements such as the field mesh, scanning raster, and target. (4) Blemishes. These are bright or dark spots or streaks whose effect is equivalent to viewing the scene through a dirty window. Blemishes affect limited portions of the raster. (5) Geometric distortion. This includes any skewing, bending, displacement or rotation of the image. It can be localized or include the entire raster. *See:* **image storage**   380

spatially aligned bundle. *See:* **aligned bundle.**

spatially coherent radiation. *See:* **coherent.**

speaker. *See:* **loudspeaker.**

special-billing call (telephone switching systems). A call charged to a special number.   55

special character (character set). A character that is neither a numeral, a letter, nor a blank, for example, virgule, asterisk, dollar sign, equals sign, comma, period.   255, 77

special color rendering index (R_i) (illuminating engineering). Measure of the color shift of various standardized special colors including saturated colors, typical foliage, and Caucasian skin. It also can be defined for other color samples when the spectral reflectance distributions are known.   167

special combination protective devices (open-wire or hot-line protectors) (wire-line communication facilities). Combined isolating and drainage transformer type protectors used in conjunction with, but not limited to, horn gaps and grounding relays, are used on open-wire lines to provide protection against lightning, power contacts, or high values or induced voltage.   414

special-dial tone (telephone switching systems). A tone for certain features that indicates that a customer can use his calling device.   55

specialized common carrier (data communication). A company that provides private line communications services, for example, voice, teleprinter, data, facsimile transmission. *See:* **common carrier, value added service.**   12

special permission (National Electrical Code). The written consent of the authority having jurisdiction.   256

special process (nuclear power quality assurance)replacement parts for Class 1E equipment in nuclear power generating stations). A process, the results of which are highly dependent on the control of the process or the skill of the operators, or both, and in which the specified quality cannot be readily determined by inspection or test of the product.   417,582

special-purpose computer. A computer that is designed to solve a restricted class of problems.   255, 77

special purpose electronic test equipment. *See:* **special purpose test equipment.**

special-purpose motor. A motor with special operating characteristics or special mechanical construction, or both designed for a particular application and not falling within the definition of a general-purpose or definite-purpose motor. *See:* **asynchronous machine.**   232, 63

special purpose test equipment (test, measurement and diagnostic equipment). Equipment used for test, repair and maintenance of a specified system, subsystem or module, having application to only one or a very limited number of systems.   54

specialty transformer (power and distribution transformer). A transformer generally intended to supply electric power for control, machine tool, Class 2, signaling, ignition, luminous-tube, cold-cathode lighting series street-lighting, low-voltage general purpose, and similar applications. *See the following types of transformers:* **individual-lamp; series street-lighting; energy-limiting; high-reactance; non-energy-limiting; high power factor; low power factor; insultating; individual-lamp insulating; group-series loop insulating; luminous tube; ignition; series circuit lighting; signaling and doorbell; control; machine tool control; general-purpose; mercury vapor lamp (multiple-supply type); sodium vapor lamp (multiple-supply type); saturable reactor (saturable-core reactor); electronic.** *See:* **series transformer rating; primary voltage rating of a general-purpose speciality transformer; secondary voltage rating; IR-drop compensation transformer; kVA or voltampere short-circuit input rating of a high-reactance transformer; secondary short-cicuit current rating of a high-reactance transformer; class 2 transformer.**   53

special unit capacity purchases (electric power supply). That capacity that is purchased or sold in transactions with other utilities and that is from a designated unit on the system of the seller. It is understood that the seller does not provide reserve capacity for this type of capacity transaction. *See:* **generating station.**   200

specific acoustic impedance (unit area acoustic impedance) (at a point in the medium). The complex ratio of sound pressure to particle velocity. *See:* Note 2 under **acoustic impedance.**   176

specific acoustic reactance. The imaginary component of the specific acoustic impedance.   176

specific acoustic resistance. The real component of the specific acoustic impedance.   176

specification (software). (1) A document that prescribes, in a complete, precise, verifiable manner, the requirements, design, behavior, or other characteristics of a system or system component. (2) The process of developing a specification. (3) A precise statement of a set of requirements to be satisfied by a product, a material or process indicating, wherever appropriate, the procedure by means of which it may be determined whether the requirements given are satisfied. *See:* **component; design; document; design specification; formal specification; functional specification; interface specification; performance specification; procedure; process; requirement; requirements specification.**   434

specification language (software). A language, often a machine-processable combination of natural and formal language, used to specify the requirements, design, behavior, or other characteristics of a system or system component. *See:* **component; design; design language; formal language; natural language; requirement; requirements specification language; system.**   434

specification verification. *See:* **verification.**

specific coordinated methods. Those additional methods applicable to specific situations where general coordinated methods are inadequate. *See:* **inductive coordination.**   328

specific detectivity. *See:* **D*.**

specific emission. The rate of emission per unit area.   244

specific inductive capacitance. *See:* **relative capacitivity.**

specific repetition frequency (loran)(navigation aid terms). One of a set of closely-spaced pulse repetition frequencies derived from the basic repetition frequency and associated with a specific set of synchronized stations.   526

specific repetition rate. *See:* **specific repetition frequency.**

specific unit capacity (power operations). Capacity which is purchased, or sold, in transactions with other systems and which is from a designated unit on the system of the seller.   516

specified achromatic lights. (1) Light of the same chromaticity as that having an equi-energy spectrum. (2) The standard illuminants of colorimetry *A, B,* and *C,* the spectral energy distributions of which were specified by the International Commission on Illumination (CIE) in 1931, with various scientific applications in view. Standard *A:* incandescent electric lamp of color temperature 2854 kelvins. Standard *B:* Standard *A* combined with a specified liquid filter to give a light of color temperature approximately 4800 kelvins. Standard *C:* Standard *A* combined with a specified liquid filter to give a light of color temperature approximately 6500 kelvins. (3) Any other specified white light. *See:* **color.**   244, 178

specified breakaway torque (rotating machinery). The torque which a motor is required to develop to break away its load from rest to rotation.   63

speckle noise. *See:* **modal noise.**

speckle pattern (fiber optics). A power intensity pattern produced by the mutual interference of partially coherent beams that are subject to minute temporal and spatial fluctuations. *Note:* In a multimode fiber, a speckle pattern results from a superposition of mode field patterns. If the relative modal group velocities change with time, the speckle pattern will also change with time. If, in addition, differential mode attenuation is experienced, modal noise results. *See:* **modal noise.**   433

spectral bandwidth (λ_{BW}) (light emitting diodes). The difference between the wavelengths at which the spectral radiant intensity is 50 percent (unless otherwise stated) of the maximum value. The term spectral linewidth is sometimes used.   162

spectral characterisitic (1) (color television). The set of spectral responses of the color separation channels with respect to wavelength. *Notes:* (A) The channel terminals at which the cahracteristics apply must be specified, and an appropriate modifier, such as a pickup spectral characteristic or studio spectral characteristic may be added to the term. (B) Because of nonlinearity, some spectral characteristics depend on the magnitude of radiance used in the measurement. (C) Nonlinearizing and matrixing operations may be performed within the channels. (D) The spectral taking characteristics are uniquely related to the chromaticities of the display primaries. (2) **camera tube.** A relation, usually shown by a graph, between wavelength and sensitivity per unit wavelength interval.(3) **luminescent screen.** The relation, usually shown by a graph, between wavelength and emitted radiant power per unit wavelength interval. *Note:* The radiant power is commonly expressed in arbitrary units. (4) **phototube.** A relation, usually shown by a graph, between the radiant sensitivity and the wavelength of the incident radiant flux.   18

spectral-conversion luminous gain (optoelectronic device). The luminous gain for specified wavelength-intervals of both incident and emitted luminous flux. *See:* **optoelectronic device.**   191

spectral-conversion radiant gain (optoelectronic device). The radiant gain for specified wavelength intervals of both incident and emitted radiant flux. *See:* **optoelectronic device.**   191

spectral-directional emissivity, $\epsilon(\theta,\tau,\lambda,T)$ **(of an element of surface of a temperature radiator at any wavelength and in a given direction).** The ratio of its spectral radiance at that wavelength and in the given direction to that of a black body at the same temperature and wavelength.

$$\epsilon(\lambda,\theta,\phi,T) = = L_\lambda(\lambda,\theta,\phi,T)/L_{blackbody}(\lambda,T)$$

167

spectral emissivity (element of surface of a tempera-

ture radiator at any wavelength). The ratio of its radiant flux density per unit wavelength interval (spectral radiant exitance) at that wavelength to that of a blackbody at the same temperature. *See:* radiant energy (illuminating engineering). 167

spectral-hemispherical emissivity, $\epsilon(\lambda,T)$ (of an element of surface of a temperature radiator). The ratio of its spectral radiant exitance to that of a blackbody at the same temperature. *Note:* Hemispherical emissivity is frequently called "total" emissivity. "Total" by itself is ambiguous, and should be avoided since it may also refer to "spectral-total" (all wavelengths) as well as directional-total (all directions). *See:* spectral-total hemispherical emissivity. 167

spectral irradiance (fiber optics). Irradiance per unit wavelength interval at a given wavelength, expressed in watts per unit area per unit wavelength interval. *See:* irradiance; radiometry. 433

spectral line (1)(charged-particle detectors)(germanium gamma-ray detectors)(X-ray energy spectrometers). A sharply peaked portion of the spectrum that represents a specific feature of the incident radiation, usually the full energy of a monoenergetic radiation. 23,119,528,471

(2) (fiber optics). A narrow range of emitted or absorbed wavelengths. *See:* line source; line spectrum; monochromatic; spectral width. 433

spectral luminous efficacy of radiant flux, $K(\lambda) = \Phi_{v\lambda}$ / $\Phi_{e\lambda}$ (illuminating engineering). The quotient of the luminous flux at a given wavelength by the radiant flux at that wavelength. It is expressed in lumens per watt. *Note:* This term formerly was called "luminosity factor." The reciprocal of the maximum luminous efficacy of radiant flux is sometimes called "mechanical equivalent of light;" that is, the ratio between radiant and luminous flux at the wavelength of maximum luminous efficacy. The most probable value is 0.00146 W/lm, corresponding to 683 lm/W as the maximum possible luminous efficacy. For scotopic vision values (13.7) the maximum luminous efficacy is 1754 "scotopic" lm/W. 167

spectral luminous efficiency of radiant flux (illuminating engineering). The ratio of the luminous efficacy for a given wavelength to the value at the wavelength of maximum luminous efficacy. It is dimensionless. *Note:* The term "spectral luminous efficiency" replaces the previously used terms "relative luminosity" and "relative luminosity factor". 167

spectral luminous flux ($\phi v\lambda$) (light emitting diodes). The luminous flux per unit wavelength interval at wavelength λ that is, lumens per nanometer. 162

spectral luminous gain (optoelectronic device). Luminous gain for a specified wavelength interval of either the incident or the emitted flux. *See:* optoelectronic device. 191

spectral luminous intensity ($I_{v\lambda}$) (light emitting diodes). The luminous intensity per unit wavelength (at wavelength λ), that is, candela per nanometer. 162

spectral-noise density (sound recording and reproducing system). The limit of the ratio of the noise output

within a specified frequency interval to the frequency interval, as that interval approaches zero. *Note:* This is approximately the total noise within a narrow frequency band divided by that bandwidth in hertz. *See:* noise (sound recording and reproducing system). 350

spectral power density (radio wave propagation). The power density per unit bandwidth. *Syn:* spectral power flux density. 146

spectral power flux density. *See:* spectral power density. 146

spectral quantum efficiency ($\eta\lambda$) (diode- type camera tube). The average number of electrons produced in the output signal per photon incident on the camera tube faceplate at a particular photon energy or wavelength. It is a dimensionless quantity which can be conveniently calculated from the spectral response R λ through the relation

$$\eta_\lambda = \frac{1241 \, R\lambda}{\lambda}$$

where $R\lambda$ is in amperes per watt and λ in nanometers. 380

spectral quantum yield (photocathode). The average number of electrons photoelectrically emitted from the photocathode per incident photon of a given wavelength. *Note:* The spectral quantum yield may be a function of the angle of incidence and of the direction of polarization of the incident radiation. *See:* phototube. 335

spectral radiance (1) (fiber optics). Radiance per unit wavelength interval at a given wavelength, expressed in watts per steradian per unit area per wavelength interval. *See:* radiance; radiometry. 433

(2) (laser-maser). The power transmitted in a radiation field per unit frequency (or wavelength) interval unit solid angle unit area normal to a given direction ($W\cdot nm^{-1}\cdot sr^{-1}\cdot m^{-2}$). 363

spectral radiant energy ($Q_\lambda = dQ_e/d_\lambda$) (light emitting diodes). Radiant energy per unit wavelength interval at wavelength λ: that is, joules per nanometer. 162

spectral radiant flux (1) ($\phi_\lambda = d\phi_e/d_\lambda$) (light emitting diodes). Radiant flux per unit wavelength interval at wavelength λ: that is watts per nanometer. 162

(2) $\Phi_\lambda = d\Phi/d\lambda$ (illuminating engineering). Radiant flux per unit wavelength interval at wavelength λ; for example, watts per nanometer. 167

spectral radiant gain (optoelectronic device). Radiant gain for a specified wavelength interval of either the incident or the emitted radiant flux. *See:* optoelectronic device. 190

spectral radiant intensity (1) ($I\lambda = dI_e/d\lambda$) (light emitting diodes). The radiant intensity per unit wavelength interval: for example watts per (steradian-nanometer). 162

(2) $I_\lambda = dI/d\lambda$ (illuminating engineering). Radiant intensity per unit wavelength interval; for example, watts per (steradian-nanometer). 167

Photopic spectral luminous efficiency $V(\lambda)$.
(Unity at wavelength of maximum luminous efficacy.)

λ in nanometers	Standard values	Standard values interpolated at intervals of 1 nanometer								
		1	2	3	4	5	6	7	8	9
380	0.00004	0.000045	0.000049	0.000054	0.000059	0.000064	0.000071	0.000080	0.000090	0.000104
390	0.00012	0.000138	0.000155	0.000173	0.000193	0.000215	0.000241	0.000272	0.000308	0.000350
400	0.0004	0.00045	0.00049	0.00054	0.00059	0.00064	0.00071	0.00080	0.00090	0.00104
410	0.0012	0.00138	0.00156	0.00174	0.00195	0.00218	0.00244	0.00274	0.00310	0.00352
420	0.0040	0.00455	0.00515	0.00581	0.00651	0.00726	0.00806	0.00889	0.00976	0.01066
430	0.0116	0.01257	0.01358	0.01463	0.01571	0.01684	0.01800	0.01920	0.02043	0.02170
440	0.023	0.0243	0.0257	0.0270	0.0284	0.0298	0.0313	0.0329	0.0345	0.0362
450	0.038	0.0399	0.0418	0.0438	0.0459	0.0480	0.0502	0.0525	0.0549	0.0574
460	0.060	0.0627	0.0654	0.0681	0.0709	0.0739	0.0769	0.0802	0.0836	0.0872
470	0.091	0.0950	0.0992	0.1035	0.1080	0.1126	0.1175	0.1225	0.1278	0.1333
480	0.139	0.1448	0.1507	0.1567	0.1629	0.1693	0.1761	0.1833	0.1909	0.1991
490	0.208	0.2173	0.2270	0.2371	0.2476	0.2586	0.2701	0.2823	0.2951	0.3087
500	0.323	0.3382	0.3544	0.3714	0.3890	0.4073	0.4259	0.4450	0.4642	0.4836
510	0.503	0.5229	0.5436	0.5648	0.5865	0.6082	0.6299	0.6511	0.6717	0.6914
520	0.710	0.7277	0.7449	0.7615	0.7776	0.7932	0.8082	0.8225	0.8363	0.8495
530	0.862	0.8739	0.8851	0.8956	0.9056	0.9149	0.9238	0.9320	0.9398	0.9471
540	0.954	0.9604	0.9661	0.9713	0.9760	0.9803	0.9840	0.9873	0.9902	0.9928
550	0.995	0.9969	0.9983	0.9994	1.0000	1.0002	1.0001	0.9995	0.9984	0.9969
560	0.995	0.9926	0.9898	0.9865	0.9828	0.9786	0.9741	0.9691	0.9638	0.9581
570	0.952	0.9455	0.9386	0.9312	0.9235	0.9154	0.9069	0.8981	0.8890	0.8796
580	0.870	0.8600	0.8496	0.8388	0.8277	0.8163	0.8046	0.7928	0.7809	0.7690
590	0.757	0.7449	0.7327	0.7202	0.7076	0.6949	0.6822	0.6694	0.6565	0.6437
600	0.631	0.6182	0.6054	0.5926	0.5797	0.5668	0.5539	0.5410	0.5282	0.5156
610	0.503	0.4905	0.4781	0.4658	0.4535	0.4412	0.4291	0.4170	0.4049	0.3929
620	0.381	0.3690	0.3570	0.3449	0.3329	0.3210	0.3092	0.2977	0.2864	0.2755
630	0.265	0.2548	0.2450	0.2354	0.2261	0.2170	0.2082	0.1996	0.1912	0.1830
640	0.175	0.1672	0.1596	0.1523	0.1452	0.1382	0.1316	0.1251	0.1188	0.1128
650	0.107	0.1014	0.0961	0.0910	0.0862	0.0816	0.0771	0.0729	0.0688	0.0648
660	0.061	0.0574	0.0539	0.0506	0.0475	0.0446	0.0418	0.0391	0.0366	0.0343
670	0.032	0.0299	0.0280	0.0263	0.0247	0.0232	0.0219	0.0206	0.0194	0.0182
680	0.017	0.01585	0.01477	0.01376	0.01281	0.01192	0.01108	0.01030	0.00956	0.00886
690	0.0082	0.00759	0.00705	0.00656	0.00612	0.00572	0.00536	0.00503	0.00471	0.00440
700	0.0041	0.00381	0.00355	0.00332	0.00310	0.00291	0.00273	0.00256	0.00241	0.00225
710	0.0021	0.001954	0.001821	0.001699	0.001587	0.001483	0.001387	0.001297	0.001212	0.001130
720	0.00105	0.000975	0.000907	0.000845	0.000788	0.000736	0.000688	0.000644	0.000601	0.000560
730	0.00052	0.000482	0.000447	0.000415	0.000387	0.000360	0.000335	0.000313	0.000291	0.000270
740	0.00025	0.000231	0.000214	0.000198	0.000185	0.000172	0.000160	0.000149	0.000139	0.000130
750	0.00012	0.000111	0.000103	0.000096	0.000090	0.000084	0.000078	0.000074	0.000069	0.000064
760	0.00006	0.000056	0.000052	0.000048	0.000045	0.000042	0.000039	0.000037	0.000035	0.000032

spectral range (acoustically tunable optical filter). The wavelength region over which the dynamic transmission is greater than some specified minimum value.

72

spectral reflectance, $\rho(\lambda) = \Phi_{r\lambda}/\Phi_{i\lambda}$ (illuminating engineering). The ratio of the reflected flux to the incident flux at a particular wavelength, λ, or within a small band of wavelengths, $\Delta\lambda$, about λ . *Note:* The various geometrical aspects of reflectance may each be considered restricted to a specific region of the spectrum and may be so designated by the addition of the adjective "spectral". 167

spectral response (R_λ) (diode-type camera tube). The spectral response (R_μ) of a camera is the current produced in the output signal per incident radiant power in the input signal as a function of the photon energy frequency, or wavelength. Units: amperes watt^{-1} (AW^{-1}). 380

spectral response characteristic (photoelectric devices). *See:* **spectral sensitivity characteristic.**

spectral responsivity (fiber optics). Responsivity per unit wavelength interval at a given wavelength. *See:* **responsivity.** 433

spectral selectivity (photoelectric device). The change of photoelectric current with the wavelength of the irradiation. *See:* **photoelectric effect.** 244, 190

spectral sensitivity characteristic (camera tubes or pho-totubes). The relation between the radiant sensitivity and the wavelength of the incident radiation, under specified conditions of irradiation. *Note:* Spectral sensitivity characteristic is usually measured with a collimated beam at normal incidence. *See:* **phototube.** 190, 178

spectral temperature (of a radiation field) (laser-maser). The temperature of a black body which produces the same spectral radiance as the radiation field at a given frequency and in a given direction. 363

spectral-total directional emissivity,ϵ,ϕ,T (of an element of surface of a temperature radiator in a given direction) (illuminating engineering). The ratio of its radiance to that of a blackbody at the same temperature.

$$\epsilon(\theta,\phi,T) = = L(\theta,\phi,T)/L_{\text{blackbody}})(T)$$

where θ and ϕ are directional angles and T is temperature. 167

spectral-total hemispherical emissivity, ϵ (of an element of surface of a temperature radiator). The ratio of its radiant exitance to that of a blackbody at the same temperature.

$$\epsilon = \frac{1}{\pi}\int\epsilon(\theta,\phi)\cdot\cos\theta\cdot d\omega = \frac{1}{\pi}\int\int\epsilon$$
$$(\lambda,\theta,\phi)\cdot\cos\theta\cdot d\omega\cdot d\lambda = = M(T)/M_{\text{blackbody}}(T)$$

167

spectral transmittance, $\tau(\lambda) = \Phi_{t\lambda}/\Phi_{i\lambda}$ (illuminating engineering). The ratio of the transmitted flux to the incident flux at a particular wavelength, λ , or within a small band of wavelengths, $\Delta\lambda$, about λ. *Note:* The various geometrical aspects of transmittance may each be considered restricted to a specific region of the spectrum and may be so designated by the addition of the adjective "spectral." 167

spectral tristimulus values. Values per unit wavelength interval and unit spectral radiant flux. *Note:* Spectral tristimulus values have been adopted by the International Commission on Illumination (CIE). They are tabulated as functions of wavelength throughout the spectrum and are the basis for the evaluation of radiant energy as light. 167

spectral width (fiber optics). A measure of the wavelength extent of a spectrum. *Notes:* (1) One method of specifying the spectral linewidth is the full width at half maximum (FWHM), specifically the difference between the wavelengths at which the magnitude drops to one half of its maximum value. This method may be difficult to apply when the line has a complex shape. (2) Another method of specifying spectral width is a special case of root-mean-square (rms) deviation where the independent variable is wavelength (λ), and $f(\lambda)$ is a suitable radiometric quantity. *See:* **root-mean-square (rms) deviation.** (3) The relative spectral width ($\Delta\lambda/\lambda$) is frequently used, where $\Delta\lambda$ is obtained according to Note 1 or Note 2.) *See:* **coherence length; line spectrum; material dispersion.**

433

spectral window (fiber optics). A wavelength region of relatively high transmittance, surrounded by regions of low transmittance. *Syn:* **transmission window.**

433

spectrophotometer (illuminating engineering). An instrument for measuring the transmittance and reflectance of surfaces and media as a function of wavelength. 167

spectroradiometer (illuminating engineering). An instrument for measuring radiant flux as a function of wavelength. 167

spectrum (1)(radiation)(charged-particle detectors)-(germanium gamma-ray detectors)(X-ray energy spectrometers). A distribution of the intensity of radiation as a function of energy or its equivalent electric analog (such as charge or voltage) at the output of a radiation detector. 23,119,528, 471

(2) (fiber optics). *See:* **optical spectrum.** 433

(3) (data transmission). The distribution of the amplitude (and sometimes phase) of the components of a wave as a function of frequency. Spectrum is also used to signify a continuous range of frequencies, usually wide in extent, within which waves have some specified common characteristic. 59

spectrum amplitude (impulse strength and impulse bandwidth). The voltage spectrum of a pulse can be expressed as

$$V(\omega) = R(\omega) + jX(\omega) = \int_{-\infty}^{+\infty} v(t)e^{-jwt}\, dt$$

where

$$R(\omega) = \int_{-\infty}^{+\infty} v(t)\cos \omega t \, dt$$

$$X(\omega) = \int_{-\infty}^{+\infty} v(t)\sin \omega t \, dt$$

and $\omega = 2\pi f$.

Note: See IEEE Std 263-1965, Measurement of Radio Noise Generated by Motor Vehicles and Affecting Mobile Communications Receivers in the Frequency Range 25 to 1000 megahertz. The spectrum then has the amplitude characteristic

$$A(\omega) = \sqrt{R^2(\omega) + X^2(\omega)} \quad \text{(V/rad)/s}$$

and the phase characteristic

$$\varphi(\omega) = \tan^{-1}\frac{X(\omega)}{R(\omega)}$$

The inverse transform can be written

$$v(t) = \frac{1}{\pi}\int_0^{\infty} A(\omega)\cos\,[\omega t + \varphi(\omega)]dw,$$

$$\text{for real } v(t)$$

The spectrum amplitude is also expressible in volts per hertz (volt-seconds) as follows:

$$S(f) = 2A(\omega) \qquad \text{(eq 1)}$$

It is this form that is used as the basis for calibration of commercially available impulse generators. A practical impulse is a function of time duration short compared with the reciprocals of all frequencies of interest. *Note:* For a rectangular pulse, the spectrum is flat within about 1 dB up to a frequency for which the pulse duration is equal to 1/4 cycle. Its spectrum amplitude $S(f)$ is substantially uniform in this frequency range and is equal to twice the area under the impulse time function or 2σ. At frequencies higher than this it is still of interest to define the spectrum amplitude which will usually be less than 2σ.

In most broadband impulse generators a dc voltage is used to charge a calibrated coaxial transmission line. The pulses are produced when the line is discharged into its terminating impedance through mechanically activated contacts. These mechanical contacts may be parts of either a vibrating diaphragm or mercury wetted relay switches. By proper choice of transmission line length and resistive termination, it is possible to produce impulses having a predictable uniform spectrum amplitude range. The advent of solid-state switches has made it possible to switch on a sine wave

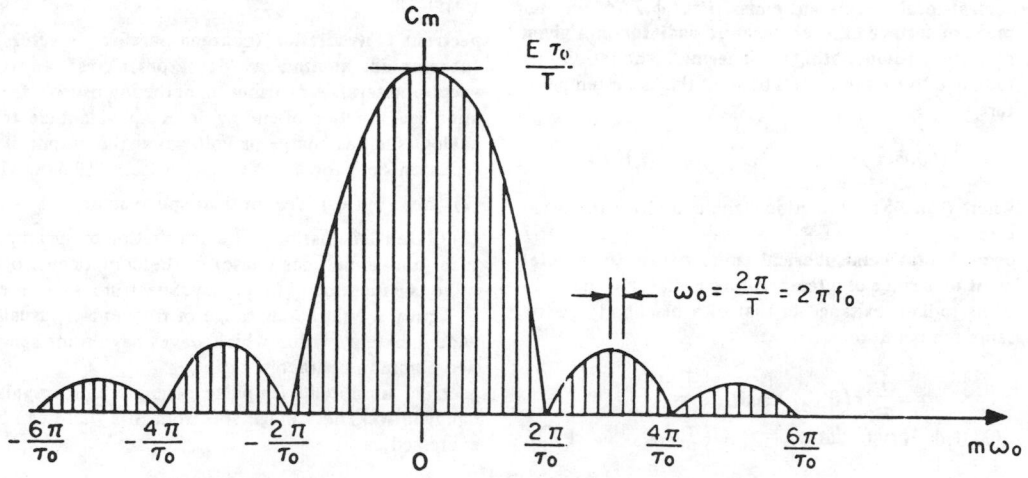

Frequency spectrum of the rectangular pulse train.

for a precisely measurable time interval (τ), producing in the frequency band in the vicinity of the sine wave a spectrum simulating that produced by an impulse. The spectrum amplitude at that particular frequency can be measured in terms of a measurement of the amplitude of the sine wave when not switched, and a measurement of the on time (τ_0) for the switch.
                                                              30

spectrum analyzer. An instrument generally used to display the power distribution of an incoming signal as a function of frequency. *Notes:* (1) Spectrum analyzers are useful in analyzing the characteristics of electrical waveforms in general since, by repetitively sweeping through the frequency range of interest, they display all components of the signal. (2) The display format may be a cathode ray tube or chart recorder.
                                                              68

spectrum intensity (impulse strength and impulse bandwidth). (For spectra which have a continuous distribution of components–components are not discrete–over the frequency range of interest). The spectrum intensity is the ratio of the power contained in a given frequency range to the frequency range as the frequency range approaches zero. It has the dimensions watt-seconds or joules and is usually stated quantitatively in terms of watts per hertz.   30

spectrum level (spectrum density level) (acoustics) (specified signal at a particular frequency). The level of that part of the signal contained within a band 1 hertz wide, centered at the particular frequency. Ordinarily this has significance only for a signal having a continuous distribution of components within the frequency range under consideration. The words **spectrum level** cannot be used alone but must appear in combination with a prefatory modifier: for example, pressure, velocity, voltage. *Note:* For illustration, if L_{ps} be a desired pressure spectrum level, p the effective pressure measured through the filter system, p_0 reference sound pressure, Δf the effective bandwidth of the filter system, and $\Delta_0 f$ the reference bandwidth (1 hertz), then

$$L_{ps} = L_p - 10 \log_{10} \frac{\Delta f}{\Delta_0 f}$$

—

For computational purposes, if L_{ps} is the band pressure level observed through a filter of bandwidth Δf, the above relation reduces to

$$L_{ps} = \log_{10} \frac{p^2/\Delta f}{p_0{}^2/\Delta_0 f}$$

spectrum locus (1) (color). The locus of points representing the chromaticities of spectrally pure stimuli in a chromaticity diagram.   18

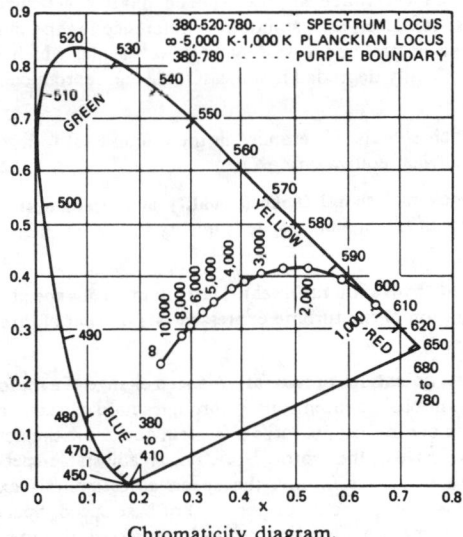

Chromaticity diagram.

(2) (illuminating engineering). The locus of points representing the colors of the visible spectrum in a chromaticity diagram.   167

specular angle (illuminating engineering). That angle between the perpendicular to the surface and the reflected ray that is numerically equal to the angle of incidence and that lies in the same plane as the incident ray and the perpendicular but on the opposite side.   167

specular reflection (1) (fiber optics). *See:* **reflection.**
                                                              433

(2) (laser-maser). A mirrorlike reflection.   363

specular surface (illuminating engineering). A surface from which the reflection is predominantly regular. *See:* **regular reflection.**   167

speech interpolation. The method of obtaining more than one voice channel per voice circuit by giving each subscriber a speech path in the proper direction only at times when his speech requires it.   328

speech level (speech quality measurements). The speech level defined and measured subjectively by comparison of the speech signal with a signal obtained by passing pink noise through a filter with A-weighting characteristics that has been judged to be equal to it in loudness. *Note:* The value of the speech level is defined to be the A-weighted sound pressure level of this noise [dB(A)].   126

speech network (transmission performance of telephone sets). An electrical circuit that connects the transmitter and the receiver to a telephone line or telephone test loop and to each other.   491

speech quality (speech quality measurements). A characteristic of a speech signal that can be described in terms of subjective and objective parameters. Speech quality is evaluated only in terms of the subjective parameter of preference. 126

speech reference signal (speech quality measurements). Used as a standard of reference for the purpose of preference testing, a speech signal which is artificially degraded in a measurable and reproducible way. 126

speech signals. Utterances in their acoustical form or electrical equivalent. 126

speech test signal (speech quality measurements). A speech signal whose speech quality is to be evaluated. 126

speed (hydraulic turbines). The instantaneous speed of rotation of the turbine expressed as a percent of rated speed. 8

speed adjustment (control). A speed change of a motor accomplished intentionally through action of a control element in the apparatus or system governing the performance of the motor. *Note:* For an adjustable-speed direct-current motor, the speed adjustment is expressed in percent (or per unit) of base speed. Speed adjustment of all other motors is expressed in percent (or per unit) of rated full-load speed. *See:* **adjustable-speed motor; base speed of an adjustable-speed motor; electric drive.** 206

speed changer (1) (hydraulic turbines). A device by means of which the governor system may be adjusted to change the speed or power output of the turbine while the turbine is operating. 8

(2) (gas turbines). A device by means of which the speed-governing system is adjusted to change the speed or power output of the turbine during operation. *See:* **asynchronous machine.** 98, 58

speed-changer high-speed stop (gas turbines). A device that prevents the speed changer from moving in the direction to increase speed or power output beyond the position for which the device is set. *See:* **asynchronous machine.** 98, 58

speed controller (control systems for steam turbine-generator units). Includes only those components and control elements that are responsive to speed and speed reference, and that supplies an input signal to the control mechanism for the purpose of controlling speed. 522

speed-control mechanism (electric power systems). Includes all equipment such as relays, servomotors, pressure or power-amplifying devices, levers, and linkages between the speed governor and the governor-controlled valves. *See:* **speed-governing system.** 94

speed deadband (hydraulic turbines). The total magnitude of the change in steady-state speed, expressed in percent of rated speed, required to reverse the direction of travel of the turbine control servomotor.

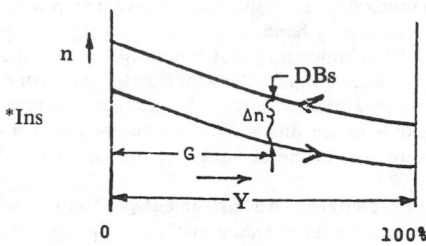

One half of the governor speed deadband is termed the governor speed insensitivity.

$$DB_s = \Delta n$$

8, 58

speed deviation (hydraulic turbines). The instantaneous difference between the actual speed and a reference speed. 8

speed droop (hydraulic turbines). The speed droop and speed regulation graphs may indicate a nonlinear relationship between the two measured variables depending on the adjustment of the governor speed changer and the quantity (servomotor position or generator power output) used to develop the feedback signal used in the governor system. Speed droop and speed regulation are considered positive when speed increases with a decrease in gate position or power output.

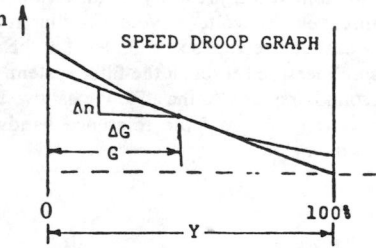

The slope of the speed droop graph at a specified point of operation G. The change in a steady-state speed expressed in percent of rated speed corresponding to the 100 percent turbine servomotor stroke with no change in setting of any governor adjustments and with the turbine supplying power to a load independently of any other power source.

$$D_S = \left(\frac{-\Delta n}{\Delta G}\right) \cdot (100)$$

Speed droop is classified as either permanent or temporary. (1) Permanent speed droop. The speed droop which remains in steady state after the decay action of the damping device has been completed. (2) Temporary speed droop. The speed droop in steady state which would occur if the decay action of the damping device were blocked and the permanent speed droop were made inactive. 8

speed-droop changer (hydraulic turbines). A device by means of which the speed droop may be adjusted while the turbine is operating. 8

speed-governing system. Control elements and devices for the control of the speed or power output of a gas turbine. This includes a speed governor, speed changer, fuel-control mechanism, and other devices and control elements. 94, 98, 57

speed governor (electric power system). Includes only those elements that are directly responsive to speed and that position or influence the action of other elements of the speed-governing system. *See:* **asynchronous machine; gas turbines; speed-governing system.** 94, 98, 58

speed limit (industrial control). A control function that prevents a speed from exceeding prescribed limits. Speed-limit values are expressed as percent of maximum rated speed. If the speed-limit circuit permits the limit value to change somewhat instead of being a single value, it is desirable to provide either a curve of the limit value of speed as a function of some variable, such as load, or to give limit values at two or more conditions of operation. *See:* **control system, feedback.** 219, 206

speed-limit indicator. A series of lights controlled by a relay to indicate the speeds permitted corresponding to the track conditions. 328

speed/load control system (control systems for steam turbine-generator units). A system that controls the speed and load of a steam turbine-generator. The system typically includes the speed and load sensing and referencing elements, the controller(s), the control mechanism(s), and the control valve(s). 522

speed/load reference changer (control systems for steam turbine-generator units). A device or devices by means of which the control system reference may be adjusted to change the speed or load of the turbine while the turbine is in operation. 522

speed of transmission (data transmission). The instantaneous rate of which information is processed by a transmission facility. This quantity is usually expressed in characters per unit time or bits per unit time. **(Rate of transmission** is in more common use.). 59

speed of transmission, effective. Speed, less than rated, of information transfer that can be averaged over a significant period of time and that reflects effects of control codes, timing codes, error detection, retransmission, tabbing, hand keying, etcetera. 194

speed of vision (illuminating engineering). The reciprocal of the duration of the exposure time required for something to be seen. 167

speed or frequency matching device (15) (power system device function numbers). A device that functions to match and hold the speed or the frequency of a machine or of a system equal to, or approximately

equal to, that of another machine, source or system. 402

speed range (industrial control). All the speeds that can be obtained in a stable manner by action of part (or parts) of the control equipment governing the performance of the motor. The speed range is generally expressed as the ratio of the maximum to the minimum operating speed. *See:* **electric drive.** 206

speed ratio (1) (high-voltage switchgear). The ratio between 0.1 second (s) and 300 s or 600 s minimum melting currents, whichever is specified, which designates the relative speed of the fuse link. 443

(2) (of a fuse). *See:* **melting-speed ratio (of a fuse).** 103

speed ratio control (industrial control). A control function that provides for operation of two drives at a preset ratio of speed. *See:* **control system, feedback.** 206

speed-regulating rheostat (industrial control). A rheostat for the regulation of the speed of a motor. *See:* **control.** 244, 206

speed regulation (speed governing of hydraulic turbines). The slope of the speed regulation graph at a specified point P of operation. The change in steady-state speed expressed in percent of rated speed when the power output of the unit is reduced from rated power output to zero power output under rated head with no change in setting of any governor adjustments and with the unit supplying power to a load independently of any other power source.

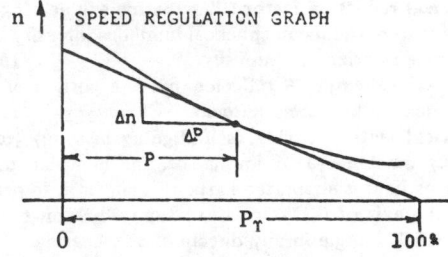

The slope of the speed regulation graph at a specified point, P, of operation. The change in steady state speed expressed in percent of rated speed when the power output of the unit is reduced from rated power output to zero power output under rated head with no change in setting of any governor adjustments and with the unit supplying power to a load independently of any other power source.

$$R_S = \left(\frac{-\Delta n/100}{\Delta P/P_r} \right) \cdot (100)$$

speed regulation changer (hydraulic turbines). A device by means of which the speed regulation may be adjusted while the turbine is operating. 8

speed regulation characteristic (rotating machinery).

The relationship between speed and the load of a motor under specified conditions. *See:* **asynchronous machine.**   63

speed regulation of a constant-speed direct-current motor. The change in speed when the load is reduced gradually from the rated value to zero with constant applied voltage and field-rheostat setting, expressed as a percent of speed at rated load.   328

speed-sensing elements (hydraulic turbines). The speed responsive elements which determine speed and influence the action of other elements of the governing system. Included are the the means used to transmit a signal proportional to the speed of the turbine to the governor.   8

speed variation (industrial control). Any change in speed of a motor resulting from causes independent of the control-system adjustment, such as line-voltage changes, temperature changes, or load changes. *See:* **electric drive.**   206

sphere illumination (illuminating engineering). The illumination on a task from a source providing equal illuminance in all directions about that task, such as a uniformly illuminated sphere with the task located at the center.   167

spherical array (antennas). A two-dimensional array of elements whose corresponding points lie on a spherical surface.   111

spherical hyperbola (navigation aid terms). The locus of the points on the surface of a sphere having a specified constant difference in great circle distances from two fixed points on the sphere.   526

spherical reduction factor (illuminating engineering). The ratio of the mean spherical luminous intensity to the mean horizontal intensity.   167

spherical reflector. A reflector that is a portion of a spherical surface. *See:* **antenna.**   111

spherical-seated bearing (self-aligning bearing) (rotating machinery). A journal bearing in which the bearing liner is supported in such a manner as to permit the axis of the journal to be moved through an appreciable angle in any direction. *See:* **bearing.**
                                                          63

spherical support seat (rotating machinery). A support for a journal bearing in which the inner surface that mates with the bearing shell is spherical in shape, the center of the sphere coinciding approximately with the shaft centerline, permitting the axis of the bearing to be aligned with that of the shaft. *See:* **bearing.**
                                                          63

spherical wave (radio wave propagation). A wave whose equiphase surfaces form a family of concentric spheres.   146

spider (rotor spider) (rotating machinery). A structure supporting the core or poles of a rotor from the shaft, and typically consisting of a hub, spokes, and rim, or some modified arrangement of these.   63

spider rim (rotating machinery). *See:* **rim.**

spider web (rotating machinery). The component of a rotor that provides radial separation between the hub or shaft and the rim or core. *See:* **rotor (rotating machinery).**   63

spike (1) (pulse terms). A distortion in the form of a pulse waveform of relatively short duration superimposed on an otherwise regular or desired pulse waveform. *See:* Note in **preshoot** entry.   254
(2) (as applied to relaying) (power switchgear). An output signal of short duration and limited crest derived from an alternating input of specified polarity. *Note:* The duration of a spike usually does not exceed one millisecond.   103

spike leakage energy (microwave gas tubes). The radio-frequency energy per pulse transmitted through the tube before and during the establishment of the steady-state radio-frequency discharge. *See:* **gas tube.**
                                                         125

spike train (electrotherapy) (courant iteratif). A regular succession of pulses of unspecified shape, frequency, duration, and polarity. *See:* **electrotherapy.**
                                                         192

spill (1)(charge-storage tubes). The loss of information from a storage element by redistribution.   125
(2)(handling and disposal of transformer grade insulating liquids containing PCBs). Spills, leaks, and other uncontrolled discharges of polychlorinated biphenyls (PCBs) constitute the disposal of PCBs. *See:* **disposal.**   586

spillover (antennas). In the transmit mode of a reflector antenna, the power from the feed which is not intercepted by the reflecting elements.   111

spillover loss (radar). In a transmitting antenna having a focusing device such as a reflector or lens illuminated by a feed, spillover loss is the reduction in gain due to the portion of the power radiated by the feed in directions that do not intersect the focusing device. By reciprocity, the same loss occurs when the same antenna is used for reception.   13

spinaxis (navigation aid terms). The axis of rotation of a gyroscope.   526

spin axis (SA) (gyro). The axis of rotation of the rotor.
                                                          46

spin-axis acceleration detuning error (dynamically tuned gyro) (inertial sensor). The error in a dynamically tuned gyro whereby deflection of the flexure, resulting from acceleration along the spin axis, can cause a shift in the tuning frequency. This will result in a change in the gyro output when there also exists a pickoff or capture loop offset.   46

spindle speed (numerically controlled machines). The rate of rotation of the machine spindle usually expressed in terms of revolutions per minute.   207

spindle wave (electrobiology). A sharp, rather large wave considered of diagnostic importance in the electroencephalogram. *See:* **electrocardiogram.**   328

spin-input-rectification drift rate (gyro) (inertial sensor). The drift rate in a single-degree-of-freedom gyro resulting from coherent oscillatory rates about the spin reference axis (SRA) and input reference axis (IRA). It occurs only when gyro and loop dynamics allow the gimbal to move away from null in response to the rate about the input reference axis, resulting in a cross-coupling of the spin reference axis rate. This drift rate is a function of the input rate amplitudes and the phase angle between them.   46

spinner* (radar). Rotating part of a radar antenna, together with directly associated equipment, used to impart any subsidiary motion, such as conical scanning, in addition to the primary slewing of the beam. *Obsolete. 13

spinning reserve (power operations). That operating reserve connected to the bus and ready to take load. 516

spin-offset coefficient (accelerometer) (inertial sensor). The constant of proportionality between bias change and the square of angular rate for an accelerometer which is spun about an axis parallel to its input reference axis and which passes through its effective center-of-mass for angular velocity. 46

spin-output rectification drift rate (gyro). The drift rate in a single-degree-of-freedom gyro resulting from coherent oscillatory rates about the spin reference axis (SRA) and output reference axis (ORA). It occurs only when gyro and loop dynamics allow the float motion to lag case motion when subjected to a rate about the output reference axis, resulting in a cross-coupling of the spin reference axis rate. This drift rate is a function of the input rate amplitudes and the phase angle between them. 46

spin reference axis (SRA) (gyro). An axis normal to the input reference axes and nominally parallel to the spin axis when the gyro outputs have specified values, usually null. 46

spiral antenna. An antenna consisting of one or more conducting wires or tapes arranged as a spiral. *Note:* Spiral antennas are usually classified according to the shape of the surface to which they conform; for example, conical or planar spirals, and according to the mathematical form, for example, equiangular or Archimedean. 111

spiral four (star quad). A quad in which the four conductors are twisted about a common axis, the two sets of opposite conductors being used as pairs. *See:* **cable.** 328

spiral scanning (electronic navigation). Scanning in which the direction of maximum response describes a portion of a spiral. *See:* **antenna.** 179

SPL. *See:* **sound pressure level.**

s-plane (circuits and systems). In the Laplace transform, the notation $s = \sigma + j\omega$ is introduced. The s plane is a coordinate system with σ as the abscissa and ω as the ordinate. The letter "p" is sometimes used instead of "s". 67

splashproof (industrial control) (packaging machinery). So constructed and protected that external splashing will not interfere with successful operation. *See:* **traction motor.** 1,103,27,429

splashproof enclosure. An enclosure in which the openings are so constructed that drops of liquid or solid particles falling on the enclosure or coming towards it in a straight line at any angle not greater than 100 degrees from the vertical cannot enter the enclosure either directly or by striking and running along a surface. 3

splashproof machine. An open machine in which the ventilating openings are so constructed that drops of liquid or solid particles falling on the machine or coming towards it in a straight line at any angle not greater than 100 degrees downward from the vertical cannot enter the machine either directly or by striking and running along a surface. *See:* **asynchronous machine.** 63

splice (1) (optical waveguide). *See:* **optical waveguide splice.** 433
(2) (power cable joint). The physical connection of two or more conductors to provide electrical continuity. 34

splice box (mine type). An enclosed connector permitting short sections of cable to be connected together to obtain a portable cable of the required length. *See:* **mine feeder circuit.**

splice loss. *See:* **insertion loss.** 433

splicing cart (conductor stringing equipment). A unit which is equipped with a hydraulic compressor (press) and all other necessary equipment for performing splicing operations on conductor. *Syn:* **sleeving trailer; splicing trailer; splicing truck.** 431

splicing chamber. *See:* **cable vault; manhole.**

split-anode magnetron. A magnetron with an anode divided into two segments: usually by slots parallel to its axis. *See:* **magnetrons.** 125

split-beam cathode-ray tube (double-beam cathode-ray tube). A cathode-ray tube containing one electron gun producing a beam that is split to produce two traces on the screen. 190

split brush (electric machines). Either an industrial or fractional-horsepower brush consisting of two pieces that are used in place of one brush. The adjacent sides of the split brush are parallel to the commutator bars. *Note:* A split brush is normally mounted so that the plane formed by the adjacent contacting brush sides is parallel to or passes through the rotating axis of the rotor. *See:* **asynchronous machine; brush; brush (rotating machinery).** 279

split collector ring (rotating machinery). A collector ring that can be separated into parts for mounting or removal without access to a shaft end. *See:* **rotor (rotating machinery).** 63

split-conductor cable. A cable in which each conductor is composed of two or more insulated conductors normally connected in parallel. *See:* **segmental conductor.** 64

split-core-type current transformer. *See:* **current transformer.**

split fitting. A conduit split longitudinally so that it can be placed in position after the wires have been drawn into the conduit, the two parts being held together by screws or other means. *See:* **raceway.** 328

split-gate tracker (radar). A form of range tracker using a pair of time gates called an "early gate" and a "late gate", contiguous or partly overlapping in time. When tracking is established, the pair of gates straddles the received pulse that is being tracked. The position of the pair of gates then gives a measure of the time of arrival of the pulse that is, the range of the target from which the echo is received. Deviation of the pair of gates from the proper tracking position increases the signal energy in one gate and decreases it in the other, producing an error signal which moves the pair of gates so as to establish equilibrium. 13

split hub (rotating machinery). A hub that can be separated into parts for ease of mounting on removal from a shaft. *See:* **rotor (rotating machinery).** 63

split hydrophone (audio and electroacoustics). *See:* **split transducer.**

split node (network analysis). A node that has been separated into a source node and a sink node. *Notes:* (1) Splitting a node interrupts all signal transmission through that node. (2) In splitting a node, all incoming branches are associated with the resulting sink node, and all outgoing branches with the resulting source node. 282

split-phase electric locomotive. A single-phase electric locomotive equipped with electric devices to change the single-phase power to polyphase power without complete conversion of the power supply. *See:* **electric locomotive.** 328

split-phase motor. A single-phase induction motor having a main winding and an auxiliary winding, designed to operate with no external impedance in either winding. The auxiliary winding is energized only during the starting operation of the auxiliary-winding circuits and is open-circuited during running operation. *See:* **asynchronous machine.** 63

split projector (audio and electroacoustics). *See:* **split transducer.**

split rotor (rotating machinery). A rotor that can be separated into parts for mounting or removal without access to a shaft end. *See:* **rotor (rotating machinery).** 63

split-sleeve bearing (rotating machinery). A journal bearing having a bearing sleeve that is split for assembly. *See:* **bearing.** 63

split-throw winding (rotating machinery). A winding wherein the conductors that constitute one complete coil side in one slot do not all appear together in another slot. *See:* **asynchronous machine.** 63

split transducer (audio and electroacoustics). A directional transducer in which electroacoustic transducing elements are so divided and arranged that each division is electrically separate. 176

split-winding protection (power switchgear). A form of differential protection in which the current in all or part of the winding is compared to the normally proportional current in another part of the winding. 103

spoiler resistors (power supplies). Resistors used to spoil the load regulation of regulated power supplies to permit parallel operation when not otherwise provided for. 186

s pole. *See:* **junction pole.**

sponge (electrodeposition). A loose cathode deposit that is fluffy and of the nature of a sponge, contrasted with a reguline metal. *See:* **electrodeposition.** 328

spontaneous emission (1) (fiber optics). Radiation emitted when the internal energy of a quantum me-

chanical system drops from an excited level to a lower level without regard to the simultaneous presence of similar radiation. *Note:* Examples of spontaneous emission include: (1) radiation from a light emitting diode (LED), and (2) radiation from an injection laser below the lasing threshold. *See:* **injection laser diode; light emitting diode; stimulated emission; superradiance.** 433

(2) (laser-maser). The emission of radiation from a single electron, atom, molecule, or ion in an excited state at a rate independent of the presence of applied external fields. 363

spontaneous polarization (primary ferroelectric terms). Magnitude of the polarization within a single ferroelectric domain in the absence of an external electric field. A spontaneous polarization P_s is a fundamental property of all pyroelectric crystals, although it is reversible or reorientable only in ferroelectrics. Most ferroelectric phases originate from a nonpolar prototypic phase and all of the polarization is reorientable. However, if the prototypic phase is polar, only a portion of the total spontaneous polarization may be reoriented. This reorientable or reversible portion is commonly called the spontaneous polarization. The figures below illustrate an example of a case where the prototypic phase is tetragonal (4mm) and the ferroelectric phase is monoclinic (m), one of the two special polar groups. The spontaneous polarization, P_s in Fig. 2, is composed of a switchable part, P_1, and a nonswitchable component, P_4; thus P_s is reorientable when P_1 switches between any of its four allowed states. In this example the pyroelectric vector is not collinear with the polar axis P_s, and both P_1 and P_s may independently change direction with temperature. The unit cell is the smallest group of atoms within a crystal whose repetition in space generates the whole crystal. It is electrically neutral in all states of the crystal, but on this microscopic scale, P_s is associated with a polar displacement of the ionic and electronic charges within the crystalline unit cell, and this, in turn, gives rise to a microscopic electric dipole moment. In ferroelectric crystals, dipoles in adjacent unit cells are aligned in the same direction, resulting in a net P_s within a macroscopic volume (much larger than a unit cell) called a *domain*. 497

spontaneous strain (primary ferroelectric terms). The summation of all the strains necessary to convert a ferroelastic crystal from the nonferroelastic prototype state to one of the ferroelastic orientation states. The prototype state, by definition, has zero spontaneous strain. A ferroelastic crystal can be switched from one ferroelastic orientation state to another by mechanical stress. Any two of the states are identical or enantiomorphous in crystal structure but different in mechanical strain tensor at zero mechanical stress (and at zero electrical field). 497

spool insulator. An insulating element of generally cylindrical form having an axial mounting hole and a circumferential groove or grooves for the attachment of a conductor. *See:* **insulator; tower.** 64, 261

spot (oscilloscopes) (cathode-ray tube). The illuminat-

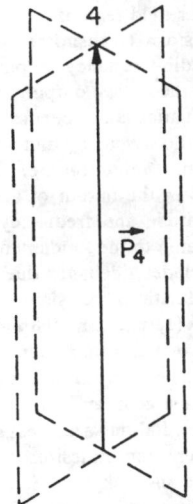

NOTE: The polar axis lies along the four-fold axis of symmetry.

Fig 1
Polar Prototype Phase, Tetragonal (4mm)

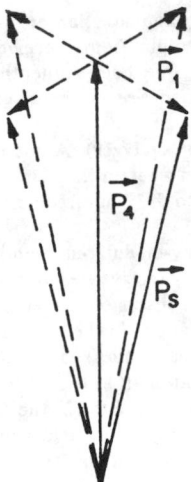

NOTE: The spontaneous polarization is not collinear with the prototype four-fold axis.

Fig 2
Ferroelectric Phase in the
Monoclinic Symmetry (m), Which Was Derived
from the Tetragonal Prototype Phase

ed area that appears where the primary electron beam strikes the phosphor screen of a cathode-ray tube. *Note:* The effect of the impact on this small area of the screen is practically instantaneous. *See:* **cathode-ray tubes; oscillograph.**   244, 190, 185

spotlight (illuminating engineering). A form of floodlight usually equipped with lens and reflectors to give a fixed or adjustable narrow beam.   167

spot-network type. A unit substation which has two stepdown transformers, each connected to an incoming high-voltage circuit. The outgoing side of each transformer is connected to a common bus through circuit breakers equipped with relays which are arranged to trip the circuit breaker on reverse power flow to the transformer and to reclose the circuit breaker upon the restoration of the correct voltage, phase angle and phase sequence at the transformer secondary. The bus has one or more outgoing radial (stub end) feeders.   53

spot noise figure (transducer at a selected frequency) (spot noise factor). The ratio of the output noise power per unit bandwidth to the portion thereof attributable to the thermal noise in the input termination per unit-bandwidth, the noise temperature of the input termination being standard (290 kelvins). The spot noise figure is a point function of input frequency. *See:* **noise figure.** *See:* **signal-to-noise ratio.**   328

spot projection (facsimile). The optical method of scanning or recording in which the scanning or recording spot is defined in the path of the reflected or transmitted light. *See:* **scanning (facsimile); recording (facsimile).**   12

spot size. *See:* **trace width.**

spot speed (facsimile). The speed of the scanning or recording spot within the available line. *Note:* This is generally measured on the subject copy or on the record sheet. *See:* **recording (facsimile); scanning (facsimile).**   12

spotting (electroplating). The appearance of spots on plated or finished metals.   328

spot-type fire detector (fire protection devices). A device whose detecting element is concentrated at a particular location.   71

spot wobble (television). A process wherein a scanning spot is given a small periodic motion transverse to the scanning lines at a frequency above the picture signal spectrum.   18

spread F (radio wave propagation). A phenomenon observed on ionograms displaying a wide range of delays of echo pulses, with pulse durations seemingly increased over that of the transmitted pulse.   146

spreading loss (wave propagation). The reduction in radiant-power surface density due to spreading.   210

spread spectrum (communication satellite). A modulation technique for multiple access, or for increasing immunity to noise and interference. Spread spectrum systems makes use of a sequential noise-like signal structure, for example P.N. (pseudonoise) codes, to spread the normally narrowband information signal over a relatively wide band of frequencies. The receiver correlates these signals to retrieve the original information signal. 84

spring (relay). *See:* **relay spring.**

spring attachment (burglar-alarm system) (spring contact) (trap). A device designed for attachment to a movable section of the protected premises, such as a door, window, or transom, so as to carry the electric protective circuit in or out of such section, and to indicate an open- or short-circuit alarm signal upon opening of the movable section. *See:* **protective signaling.** 328

spring barrel. The part that retains and locates the short-circuiter. *See:* **rotor (rotating machinery).** 328

spring buffer. A buffer that stores in a spring the kinetic energy of the descending car or counterweight. *See:* **elevators.** 328

spring-buffer load rating (spring buffer) (elevators). The load required to compress the spring an amount equal to its stroke. *See:* **elevators.** 328

spring-buffer stroke (elevators). The distance the contact end of the spring can move under a compressive load until all coils are essentially in contact. *See:* **elevators.** 328

spring contact. An electric contact that is actuated by a spring. 328

spring-loaded bearing (rotating machinery). A ball bearing provided with a spring to ensure complete angular contact between the balls and inner and outer races, thereby removing the effect of diametral clearance in both bearings of a machine provided with ball bearing at each end. *See:* **bearing.** 328

spring operation (power switchgear). Stored-energy operation by means of spring-stored energy. 103

sprinkler supervisory system. A supervisory system attached to an automatic sprinkler system that initiates signal transmission automatically upon the occurrence of abnormal conditions in valve positions, air or water pressure, water temperature or level, the operability of power sources necessary to the proper functioning of the automatic sprinkler, etcetera. *See:* **protective signaling.** 328

sprocket hole (test, measurement and diagnostic equipment). The hole in a tape that is used for electrical timing or mechanically driving the tape. 54

spurious count (1) (nuclear techniques). A count from a scintillation counter other than (A) one purposely generated or (B) one due directly to ionizing radiation. *See:* **scintillation counter.** 117

(2) (radiation-counter tubes). A count caused by any event other than the passage into or through the counter tube of the ionizing radiation to which it is sensitive. 96

spurious emission power (land-mobile communication transmitters). Any part of the radio frequency output that is not a component of the theoretical output, as determined by the type of modulation and specified bandwidth limitations. 444

spurious emission power radiation field (land-mobile communications transmitters). That portion of the spurious emission power which may be radiated from

a transmitter enclosure and which can be measured in the near or far field regions. 444

spurious emissions (transmitter performance). Any part of the radio-frequency output that is not a component of the theoretical output, as determined by the type of modulation and specified bandwidth limitations. *See:* **audio-frequency distortion.** 181

spurious output (nonharmonic) (signal generator). Those signals in the output of a source that have a defined amplitude and frequency and are not harmonically related to the fundamental frequency. This definition excludes sidebands due to residual and intentional modulation. *See:* **signal generator.** 185

spurious pulse (1)(nuclear techniques). A pulse in a scintillation counter other than (A) one purposely generated or (B) one due directly to ionizing radiation. *See:* **scintillation counter.** 117

(2)(telephony)(dial-pulse address signaling systems). The intermittent and undesired change of state in a circuit from its on-hook condition (spurious make) or off-hook condition (spurious break) lasting more than 1 millisecond (ms). 540

spurious pulse mode. An unwanted pulse mode, formed by the chance combination of two or more pulse modes, that is indistinguishable from a pulse interrogation or pulse reply. 254

spurious radiation (radio-noise emission). Any emission from an electronic communications equipment at frequencies outside its occupied bandwidth. 418

spurious response (1) (general). Any response, other than the desired response, of an electric transducer or device. 59

(2) (mobile communication or electromagnetic compatibility). Output, from a receiver, due to a signal or signals having frequencies other than that to which the receiver is tuned. *See:* **electromagnetic compatibility.** 181

(3) (spectrum analyzer). A characteristic of a spectrum analyzer wherein the displayed frequency does not conform to the input frequency. *Syn:* **spurii; spur.** 390

(4) (frequency-modulated mobile communications receivers). Any receiver response that occurs because of frequency conversions other than the desired frequency translations in the receiver. 123

spurious-response ratio (radio receiver). The ratio of (1) the field strength at the frequency that produces the spurious response to (2) the field strength at the desired frequency, each field being applied in turn, under specified conditions, to produce equal outputs. *Note:* Image ratio and intermediate-frequency-response ratio are special forms of spurious response ratio. *See:* **radio receiver.** 339

spurious transmitter output.(1)general. Any part of the radio-frequency output that is not implied by the type of modulation (amplitude modulation, frequency modulation, etcetera) and specified bandwidth. *See:* **radio transmitter.** 240

(2) conducted. Any spurious output of a radio transmitter conducted over a tangible transmission path. *Note:* Power lines, control leads, radio-frequency transmission lines and waveguides are all considered as tangible paths in the foregoing definition. Radiation is not considered as tangible path in the foregoing definition. Radiation is not considered a tangible path in this definition. *See:* **radio transmitter.**   240

(3) extraband. *See:* **extraband spurious transmitter output.**   444

(4) inband. Spurious output of a transmitter within its specified band of transmission. *See:* **radio transmitter.**   240

(5) radiated. Any spurious output radiated from a radio transmitter. *Note:* The radio transmitter does not include the associated antenna and transmission lines. *See:* **radio transmitter.**   240

spurious tube counts (radiation-counter tubes). Counts in radiation-counter tubes, other than background counts and those caused by the source measured. *Note:* Spurious counts are caused by failure of the quenching process, electric leakage, and the like. Spurious counts may seriously affect measurement of background counts.   125

sputtering (electroacoustics) (cathode sputtering). A process sometimes used in the production of the metal master wherein the original is coated with an electric conducting layer by means of an electric discharge in a vacuum. *Note:* This is done prior to electroplating a heavier deposit. *See:* **phonograph pickup.**   176

square-law detection (information theory). The form of detection of an amplitude-modulated signal in which the output voltage is a linear function of the square of the envelope of the input wave.   415

squareness ratio (material in a symmetrically cyclically magnetized condition) (magnetic storage). The ratio of (1) the flux density at zero magnetizing force to the maximum flux density. (2) The ratio of the flux density when the magnetizing force has changed halfway from zero toward its negative limiting value, to the maximum flux density. *Note:* Both these ratios are functions of the maximum magnetizing force.
                                            77

square wave (1) (data transmission). A periodic wave that alternately for equal lengths of time assumes one of two fixed values, the time of transition being negligible in comparison.   59

(2) (pulse terms). A periodic rectangular pulse train with a duty factor of 0.5 or an on-off ratio of 1.0.
                                            254

square-wave response (1) (camera tubes). The ratio of (A) the peak-to-peak signal amplitude given by a test pattern consisting of alternate black and white bars of equal widths to (B) the difference in signal between large-area blacks and large-area whites having the same illuminations as the black and white bars in the test pattern. *Note:* Horizontal square-wave response is measured if the bars run perpendicular to the direction of horizontal scan. Vertical square-wave response is measured if the bars run parallel to the direction of horizontal scan. *See:* **amplitude response.**   125

(2) (diode-type camera tube). Square-wave spatial inputs may be used, in which case the response curve is called the contrast transfer function or square-wave

amplitude response. Units: lines per picture height.
                                            380

square-wave response characteristic (camera tubes). The relation between square-wave response and the ratio of (1) a raster dimension to (2) the bar width in the square-wave response test pattern. *Note:* Unless otherwise specified, the raster dimension is the vertical height. *See:* **amplitude response characteristic; television.**   125

squaring amplifier (as applied to relaying) (power switchgear). A circuit which produces a block.
                                            103

squeezable waveguide (radar). A variable-width waveguide for shifting the phase of the radio-frequency wave traveling through it.   13

squeeze section (transmission lines and waveguides). A length of rectangular waveguide so constructed as to permit alteration of the broad dimension with a corresponding alteration in electrical length. *See:* **waveguide.**   185

squeeze trace (electroacoustics). A variable-density sound track wherein, by means of adjustable masking of the recording light beam and simultaneous increase of the electric signal applied to the light modulator, a track having variable width with greater signal-to-noise ratio is obtained. *See:* **phonograph pickup.**
                                            176

squelch (radio receivers) (noun). A circuit function that acts to suppress the audio output of a receiver when noise power that exceeds a predetermined level is present.   123

squelch circuit (1) (data transmission). A circuit for preventing a radio receiver from producing audio-frequency output in the absence of a signal having predetermined characteristics. A squelch circuit may be operated by signal energy in the receiver pass band, by noise quieting, or by a combination of the two (ratio squelch). It may also be operated by a signal having special modulation characteristics (selective squelch).
                                            59

(2) (power switchgear). A circuit for preventing production of an unwanted output in the absence of a signal having predetermined characteristics.   103

squelch clamping (frequency-modulated mobile communications receivers). The characteristic of the receiver, when receiving a normal signal, in which the squelch circuit under certain conditions of modulation will cause suppression of the audio output.   123

squelch selectivity (frequency-modulated mobile communications receivers). The characteristic that permits the receiver to remain squelched when a radio-frequency signal not on the receiver's tuned frequency is present at the input.   123

squelch sensitivity (frequency-modulated mobile communications receivers). The minimum radio-frequency signal input level, with standard test modulation required to increase the audio power output from the reference threshold squelch adjustment condition to within 6 decibels of the reference audio power output.
                                            123

squint (1) (antennas). A condition in which a specified

axis of an antenna, such as the direction of maximum directivity or of a directional null, departs slightly from a specified reference axis. *Notes:* (1) Squint is often the undesired result of a defect in the antenna; but in certain cases squint is intentionally designed in order to satisfy an operational requirement. (2) The reference axis is often taken to be the mechanically defined axis of the antenna, for example, the axis of a paraboloidal reflector.   111

(2) (radar). (A) The angle between the major lobe axis of each lobe and the central axis in a lobe-switching or simultaneous-lobing antenna; (B) The angular difference between the axis of antenna radiation and a selected geometric axis, such as the axis of the reflector, the center of the cone formed by movement of the radiation axis, or the broadside direction of a moving vehicle.   13

squint angle (antennas). The angle between a specified axis of an antenna, such as the direction of maximum directivity or a directional null, and the corresponding reference axis.   111

squirrel-cage induction motor (electric installations on shipboard). A motor in which the secondary circuit consists of a squirrel-cage winding suitably disposed in slots in the secondary core.   3

squirrel-cage rotor (rotating machinery). A rotor core assembly having a squirrel-cage winding. *See:* **rotor (rotating machinery).**   63

squirrel-cage winding (1) (rotating machinery). A winding, usually on the rotor of a machine, consisting of a number of conducting bars having their extremities connected by metal rings or plates at each end.   63

(2) (electric installations on shipboard). A permanently short-circuited winding, usually uninsulated (chiefly used in induction machines) having its conductors uniformly distributed around the periphery of the machine and joined by continuous end rings.   3

squitter (radar). Random output pulses from a transponder caused by ambient noise, or by an intentional random triggering system, but not by the interrogation pulses.   13

SSB. *See:* **single-sideband modulation.**

stability (1) (circuits and systems). An aspect of system behavior associated with systems having the general property that bounded input perturbations result in bounded output perturbations. *Notes:* (A) A stable system will ultimately attain a steady state. (B) Deviations from this steady state due to component aging or environmental changes do not indicate instability, but a change in the system. *See:* **steady-state stability; transient stability.**   67

(2) (perturbations). For convenience in defining various stability concepts, only those parameters or signals that are perturbed are explicitly exhibited, or mentioned, that is, for perturbations in initial states, a perturbed solution is denoted

$$\varphi(\mathbf{x}(t_0) + \Delta\mathbf{x}(t_0);t),$$

where $\Delta\mathbf{x}(t_0)$ represents the perturbation in initial state. Finally, the perturbed-state solution is denoted

$$\Delta\varphi = \varphi(\mathbf{x}(t_0) + \Delta\mathbf{x}(t_0);t) - \varphi(\mathbf{x}(t_0);t).$$

329

(3) (power system stability). In a system of two or more synchronous machines connected through an electric network, the condition in which the difference of the angular positions of the rotors of the machines either remains constant while not subjected to a disturbance, or becomes constant following an aperiodic disturbance. *Note:* If automatic devices are used to aid stability, their use will modify the steady-state and transient stability terms to: **steady-state stability with automatic devices; transient stability with automatic devices.** Automatic devices as defined for this purpose are those devices that are operating to increase stability during the period preceding and following a disturbance as well as during the disturbance. Thus relays and circuit breakers are excluded from this classification and all forms of voltage regulators included. Devices for inserting and removing shunt or series impedance may or may not come within this classification depending upon whether or not they are operating during the periods preceding and following the disturbance. *See:* **steady-state stability; transient stability.**   64

(4) (oscilloscopes). The property of retaining defined electrical characteristics for a prescribed time and environment. *Notes:* (A) Deviations from a stable state may be called drift if it is slow, or jitter or noise if it is fast. In triggered-sweep systems, triggering stability may refer to the ability of the trigger and sweep systems to maintain jitter-free displays of high-frequency waveforms for long (seconds to hours) periods of time. (B) Also, the name of the control used on some oscilloscopes to adjust the sweep for triggered, free-running, or synchronized operation. *See:* **sweep mode control.**   106

(5) (hydraulic turbines). Characteristics of the governing system pertaining to limitation of oscillations of speed or power under sustained conditions, to damping of oscillations of speed following rejection of load, and to damping of speed oscillations under isolated load conditions following sudden load changes.   8

(6) (electrothermic power meters). For a constant input rf power, constant ambient temperature and constant power line voltage, the variation in rf power indication over stated time intervals. *Note:* Long term stability or drift is the maximum acceptable change in 1 hour. Short term stability or fluctuation is the maximum (peak) change in 1 minute.   47

(7) (gyro; accelerometer). A measure of the ability of a specific mechanism or performance coefficient to remain invariant when continuously exposed to a fixed operating condition. (This definition does not refer to dynamic or servo stability).   46

(8) (nuclear power generating station). The ability of a module to attain and maintain a steady state.

355

(9) (software). (A) The ability to continue unchanged despite disturbing or disruptive events. (B) The ability to return to an original state after disturbing or disruptive events. 434

stability, absolute (control system). Global asymptotic stability maintained for all nonlincaritics within a given class. *Note:* A typical problem to which the concept of absolute stability has been applied consists of a system with dynamics described by the vector differential equation

$$\dot{\mathbf{x}} = A\mathbf{x} + \mathbf{b}f(\sigma),$$

$$\sigma = \mathbf{c}^{\tau}\mathbf{x},$$

with a nonlinearity class defined by the conditions

$$f(0) = 0,$$

$$k_1 \leqslant f(\sigma)/\sigma \leqslant k_2.$$

The solution $\mathbf{x}(t) = \mathbf{0}$ is said to be absolutely stable if it is globally asymptotically stable for all nonlinear functions $f(\sigma)$ in the above class. *See:* **control system.**

329

stability, asymptotic (1) (control system) (of a solution $\phi(\mathbf{x}(t_0);t)$). The solution is (1) Lyapunov stable, (2) such that

$$\lim_{t \to \infty} \|\Delta\phi\| = 0,$$

where $\Delta\phi$ is a change in the solution due to an initial state perturbation. See **stability** for explanation of symbols. *Notes:* (A) The solution $\mathbf{x} = 0$ of the system $\mathbf{x} = \mathbf{a}\mathbf{x}$ is asymptotically stable for $\mathbf{a} < 0$, but not for $\mathbf{a} = 0$. In this case

$$\varphi(\mathbf{x}(t_0);t) = \mathbf{x}(t_0)\exp(-\mathbf{a}(t - t_0)).$$

In some cases the rate of convergence to zero depends on both the initial state $\mathbf{x}(t_0)$ and the initial time t_0. See stability, equiasymptotic for stability concepts where the rate of convergence is independent of either $\mathbf{x}(t_0)$ or t_0. *See:* **control system.** 329

stability, bounded-input-bounded-output (1) (control system). Driven stability when the solution of interest is the output solution. *See:* **control system.**(2) **(excitation control systems).** A system exhibits bounded input-bounded output (BIBO) stability if the output is bounded for every bounded input. *Note:* BIBO stability is also known as stability in the sense of Liapunov and it refers to force systems. In nonlinear systems, a bounded limit cycle appearing in the output signal is an example of BIBO stability. 353

stability, conditional (linear feedback control system).

A property such that the system is stable for prescribed operating values of the frequency-invariant factor of the loop gain and becomes unstable not only for higher values, but also for some lower values. *See:* **control system, feedback.** 353

stability, driven (control system) (solution $\phi(\mathbf{u};t)$). For each bounded system input perturbation $\Delta\mathbf{u}(t)$ the output perturbation $\Delta\phi$ is also bounded for $t \geq t_0$. *Note:* A necessary and sufficient condition for a solution of a linear system to be driven-stable is that the solution be uniformly asymptotically stable. See **stability** for explanation of symbols. *See:* **control system.**

329

stability, equiasymptotic (control system). Asymptotic stability where the rate of convergence to zero of the perturbed-state solution is independent of all initial states in some region $\|\Delta\mathbf{x}(t_0)\| \leq v$. *See:* **control system.**

329

stability, excitation-system (synchronous machines). The ability of the excitation system to control the field voltage of the principal electric machine so that transient changes in the regulated voltage are effectively suppressed and sustained oscillations in the regulated voltage are not produced by the excitation system during steady-load conditions or following a change to a new steady-load condition. *Note:* It should be recognized that under some system conditions it may be necessary to use power system stabilizing signals as additional inputs to excitation control systems to achieve stability of the power system including the excitation system. 105

stability factor. The ratio of a stability limit (power limit) to the nominal power flow at the point of the system to which the stability limit is referred. *Note:* In determining stability factors it is essential that the nominal power flow be specified in accordance with the one of several bases of computation, such as rating or capacity of, or average or maximum load carried by, the equipment or the circuits. *See:* **alternating-current distribution.** 64

stability, finite-time (control system) (solutions). For all initial states that originate in a specified region R at time t_0, the resulting solutions remain in another specified region R_ϵ over the given time interval $t_0 \leq t \leq T$. *Notes:* (1) In the definition of finite-time stability the quantities R_π, R_ϵ, and T are prespecified. Obviously, R must be included in R_ϵ. (2) A system may be Lyapunov unstable and still be finite-time stable. For example, a system with dynamics $\mathbf{x} = \mathbf{a}\mathbf{x}, \mathbf{a} > 0$, is Lyapunov unstable, but if

$$R_\delta: |\mathbf{x}| \leqslant \delta,$$

$$R_\epsilon: |\mathbf{x}| \leqslant \epsilon,$$

and $T < \mathbf{a}^{-1} \ln(\epsilon/\delta)$, the system is finite-time stable (relative to the given values of δ, ϵ, and T). *See:* **control system.** 329

stability, global (control system) (solution $\phi(\mathbf{x}(t_0);t)$). Stable for all initial perturbations, no matter how large they may be. *See:* **control system.** 329

stability in-the-whole (control system). Synonymous with **global stability.** *See:* **stability, global.** *See:* **control system.** 329

stability, Lagrange (system) (control system). Every solution that is generated by a finite initial state is bounded. *Note:* An example of a system that is Lagrange stable is a second-order system with a single stable limit cycle. Although this system must contain a point inside the limit cycle that is Lyapunov unstable, the system is still Lagrange stable because every solution remains bounded. *See:* **control system.** 329

stability limit (power limit) (electric systems). The maximum power flow possible through some particular point in the system when the entire system or the part of the system to which the stability limit refers is operating with stability. *See:* **alternating-current distribution.** 64

stability, long-term (LTS) (power supplies). The change in output voltage or current as a function of time, at constant line voltage, load, and ambient temperature (sometimes referred to as drift). 186

stability, Lyapunov (control system) (of a solution ϕ ($\mathbf{x}(t_0)$;t)). For every given $\epsilon > 0$ there exists a $\delta > 0$ (which, in general, may depend on ϵ and on t_0) such that $||\Delta \mathbf{x}(t_0)|| \leq \delta$ implies $||\Delta \phi|| \leq \epsilon$ for $t \geq t_0$. *Notes:* (1) The solution $\mathbf{x} = 0$ of the system $\mathbf{x} = \mathbf{ax}$ is Lyapunov stable if $\mathbf{a} < 0$ and is Lyapunov unstable if $\mathbf{a} > 0$. (2) For a linear system with an irreducible transfer function $T(s)$, Lyapunov stability implies that all the poles of $T(s)$ are in the left-half s plane and that those on the $j\omega$ axis are simple. *See:* **control system.** 329

stability of a limit cycle (control system). Synonymous with orbital stability. *See:* **control system.** 329

stability of the speed-governing system (gas turbines). A characteristic of the system that indicates that the speed-governing system is capable of actuating the turbine fuel-control valve so that sustained oscillations in turbine speed, or rate of energy input to the turbine, are limited to acceptable values by the speed-governing system. 98, 58

stability of the temperature-control system (gas turbines). A characteristic of the system that indicates that the temperature-control system is capable of actuating the turbine fuel-control valve so that sustained oscillations in rate of energy input to the turbine are limited to acceptable values by the temperature-control system during operation under constant system frequency. 98, 58

stability, orbital (control system) (closed solution curve denoted Γ). Implies that for every given $\epsilon > 0$ there exists a $\delta > 0$ (which, in general, may depend on ϵ and on t_0) such that $\rho(\Gamma, \mathbf{x}(t_0)) \leq \delta$ implies $\rho(\Gamma, \phi(\mathbf{x}(t_0); t)) \leq \epsilon$ for $t \geq t_0$, where $\rho(\Gamma, \mathbf{a})$ denotes the minimum distance between the curve Γ and the point $\mathbf{a}$. Here the point $\mathbf{x}(t_0)$ is assumed to be off the curve Γ. *Notes:* (1) Orbital stability does not imply Lyapunov stability of a closed solution curve, since a point on the closed curve may not travel at the same speed as a neighboring point off the curve. (2) Only nonlin-

ear systems can produce the type of solutions for which the concept of orbital stability is applicable. *See:* **control system.** 329

stability, practical (control system). Synonymous with **finite-time stability.** *See:* **stability, finite-time; control system.**

stability, quasi-asymptotic (control system) (solution $\phi\mathbf{x}(\mathbf{x}(t_0);t)$). Implies

$$\lim_{t \to \infty} ||\Delta \phi|| = 0.$$

Notes: (1) Quasi-asymptotic stability is condition (2) in the definition of asymptotic stability and, hence, need not imply Lyapunov stability. (2) An example of a solution that is quasi-asymptotically stable but not asymptotically stable is the solution $\mathbf{x}(t) = 0$ of the system $\mathbf{x} = \mathbf{x}^2$. The solution of the above system for a perturbation $\Delta x(t_0)$ from 0 is

$$\rho(\Delta \mathbf{x}(t_0);t) = \Delta \mathbf{x}(t_0)/[1 - (t - t_0)\Delta \mathbf{x}(t_0)].$$

Obviously, $\phi(\Delta \mathbf{x}(t_0);t)$ approaches zero as t approaches ∞, yet is not Lyapunov stable since it is unbounded at $t = t_0 + (1/\Delta \mathbf{x}(t_0))$. *See:* **control system.** 329

stability, relative (automatic control) (stable underdamped system). The property measured by the relative setting times when parameters are changed. *See:* **control system, feedback.** 329

stability, short-time (control system). Synonymous with **finite-time stability.** *See:* **stability, finite-time.** 329

stability, synchronous-machine regulating-system. The property of a synchronous-machine-regulating system in which a change in the controlled variable, resulting from a stimulus, decays with time if the stimulus is removed. 63

stability, total (control system) (solution, $\phi = \phi(\mathbf{x}(t_0); t)$ of the system $\mathbf{x} = \mathbf{f}(\mathbf{x},t)$). Implies that for every given $\epsilon > 0$ there exist a $\delta_1 > 0$ and a $\delta_2 > 0$ (both of which, in general, may depend on ϵ and t_0) such that $||\Delta \mathbf{x}(t_0)|| \leq \delta_1$ and $||\mathbf{g}(\mathbf{x},t)|| \leq \delta_2$ imply $||\phi - \Psi|| \leq \epsilon$ for $t \geq t_0$, where $\Psi = \Psi(\mathbf{x}(t_0) + \Delta \mathbf{x}(t_0);t)$ is a solution of the system $\mathbf{x} = \mathbf{f}(\mathbf{x},t) + \mathbf{g}(\mathbf{x},t)$. *See:* **control system.** 329

stability, trajectory (control system). Orbital stability where the solution curve is not closed. *See:* **control system.** 329

stability, uniform-asymptotic (control system). Asymptotic stability where the rate of convergence to zero of the perturbed-state solution is independent of the initial time t_0. *Note:* An example of a solution that is asymptotically stable but not uniformly asymptotically stable is the solution

$$\varphi(\mathbf{x}(t_0);t) = \mathbf{x}(t_0)t_0/t$$

of the system $\dot{\mathbf{x}} = -\mathbf{x}/t, t_0 > 0$. Note that the initial rate of decay,

$$\dot{\mathbf{x}}(t_0)/\mathbf{x}(t_0) = -1/t_0,$$

is clearly a function of t_0. Compare with the time-invariant system $\dot{\mathbf{x}} = \mathbf{ax}$ where $\dot{\mathbf{x}}(t_0)/\mathbf{x}(t_0) = \mathbf{a}$ is independent of t_0. The concept of uniformity with respect to the initial time t_0 applies only to time-varying systems. All stable time-invariant systems are uniformly stable. *See:* **control system.** 329

stabilization (1) (control system, feedback). Act of attaining stability or of improving relative stability. *See:* **control system, feedback.** 56

(2)(navigation)(navigation aid terms). Maintenance of a desired orientation of a vehicle or device with respect to one or more reference directions. 526

(3) (direct-current amplifier). *See:* **drift stabilization.**

stabilization, drift. *See:* **drift stabilization.**

stabilization network (analog computers). As applied to operational amplifiers, a network used to shape the transfer characteristics to eliminate or minimize oscillations when feedback is provided. 9

stabilized feedback. Feedback employed in such a manner as to stabilize the gain of a transmission system or section thereof with respect to time or frequency or to reduce noise or distortion arising therein. *Note:* The section of the transmission system may include amplifiers only, or it may include modulators. *See:* **feedback.** 111

stabilized flight (navigation aid terms). That type of flight which obtains control information from devices which sense orientation with respect to external references. 526

stabilized shunt-wound generator (electric installations on shipboard). A stabilized shunt-wound generator is the same as the shunt-wound type, except that a series field winding is added, of such proportion as not to require equalizers for satisfactory parallel operation. The voltage regulation of this type of generator is to comply with that given for shunt-wound generators. 3

stabilized shunt-wound motor (electric installations on shipboard). A shunt-wound motor having a light series winding added to prevent a rise in speed or to obtain a slight reduction in speed, with increase of load. 3

stabilizer, excitation control system (synchronous machines). An element or group of elements that modify the forward signal by either series or feedback compensation to improve the dynamic performance of the excitation control system. 105

stabilizer, power system (synchronous machines). An element or group of elements that provide an additional input to the regulator to improve power system dynamic performance. *Note:* A number of different quantities may be used as input to the power system stabilizer, such as shaft speed, frequency, synchronous machine electrical power, and others. 105

stabilizing winding (power and distribution transformer). A delta connected auxiliary winding used particularly in Y-connected three-phase transformers for such purposes as the following: (1) To stabilize the neutral point of the fundamental frequency voltages, (2) To minimize third-harmonic voltage and the resultant effects on the system, (3) To mitigate telephone influence due to third-harmonic currents and voltages, (4) To minimize the residual direct-current magnetomotive force on the core, (5) To decrease the zero-sequence impedance of transformers with Y-connected windings. *Note:* A winding is regarded as a stabilizing winding if its terminals are not brought out for connection to an external circuit. However, one or two points of the winding which are intended to form the same corner point of the delta may be brought out for grounding, or grounded internally to the tank. For a three-phase transformer, if other points of the winding are brought out, the winding should be regarded as a normal winding as otherwise defined. 53

stable (1) (control system, feedback). Possessing stability. *See:* **control system, feedback.** 56

(2) (excitation control systems). Possessing stability, where, for a feedback control system or element, stability is the property such that its output is asymptotic, that is, will ultimately attain a steady-state, within the linear range and without continuing external stimuli. For certain nonlinear systems or elements, the property such that the output remains bounded, that is, in a limit cycle of continued oscillation, when the input is bounded. *See:* **stability, asymptotic; stability, bounded input–bounded output; stability conditional.** 353

stable element (navigation)(navigation aid terms). An instrument or device which maintains a desired orientation independently of the motion of the vehicle. 526

stable oscillation. A response that does not increase indefinitely with increasing time: an unstable oscillation is the converse. *Note:* The response must be specified or understood: a steady current in a pure resistance network would be stable, although the total charge passing any cross section of a network conductor would be increasing continuously. 210

stable platform (navigation aid terms). A gimbal-mounted platform, usually containing gyros and accelerometers, whose purpose is to maintain a desired orientation in inertial space, independent of the motion of the vehicle. 526

stack (software). A list that is accessed in a last-in, first-out manner. *See:* **list; queue.** 434

stacked-beam radar (radar). A radar that forms two or more simultaneous beams at the same azimuth but at different elevation angles. The beams are usually contiguous or partly overlapping. Each stacked beam feeds an independent receiver. 13

stacker (computing systems). *See:* **card stacker.**

stack, insulator. A rigid assembly of two or more switch and bus insulating units. *See:* **tower.** 261, 64

stage (1) (communication practice). One step, especially if part of a multistep process, or the apparatus employed in such a step. The term is usually applied to an amplifier. *See:* **amplifier.** 328

(2) (thermoelectric device). One thermoelectric couple or two or more similar thermoelectric couples arranged thermally in parallel and electrically connected. *See:* **thermoelectric device.** 248, 191

stage efficiency. The ratio of useful power delivered to the load (alternating current) and the plate power input (direct current). *See:* **network analysis.** 111

stagger (facsimile). Periodic error in the position of the recorded spot along the recorded line. *See:* **recording (facsimile).** 12

staggered-repetition-interval moving-target indicator (radar). A moving-target indicator with multiple interpulse intervals. The interval may vary either from pulse-to-pulse or from scan-to-scan. 13

staggering. The offsetting of two channels of different carrier systems from exact sideband frequency coincidence in order to avoid mutual interference. 328

staggering advantage. The effective reduction, in decibels, of interference between carrier channels, due to staggering. 328

stagger time, relay. *See:* **relay stagger time.**

stagger-tuned amplifier. An amplifier consisting of two or more single-tuned stages that are tuned to different frequencies. *See:* **amplifier.** 328

stain spots (electroplating). Spots produced by exudation, from pores in the metal, of compounds absorbed from cleaning, pickling plating solutions. The appearance of stain spots is called spotting out. 328

staircase (pulse terms). Unless otherwise specified, a periodic and finite sequence of steps of equal magnitude and of the same polarity. 254

staircase signal (television). A waveform consisting of a series of discrete steps resembling a staircase. 18

stairstep sweep (oscilloscopes). An incremental sweep in which each step is equal. The electric deflection waveform producing a stairstep sweep is usually called a staircase or stairstep waveform. *See:* **incremental sweep; oscillograph.** 185

stalled tension control (industrial control). A control function that maintains tension in the material at zero speed. *See:* **electric drive.** 206

stalled torque control (industrial control). A control function that provides for the control of the drive torque at zero speed. *See:* **electric drive.** 206

stalo (STALO) (radar). Acronym for stable local oscillator, a highly stable radio-frequency local oscillator used for heterodyning signals to produce an intermediate frequency. 13

stamper (electroacoustics). A negative (generally made of metal by electroforming) from which finished pressings are molded. *See:* **phonograph pickup.** 176

stand-alone word processing (computer applications). Word processing performed on a system that does not depend on the resources of other equipment to perform word processing activities. *See:* **dedicated word processing; shared-logic word processing; shared-resource word processing.** 571

standard (1) (instrument or source) (radiation protection). (A) national standard. An instrument, source, or other system or device maintained and promulgated

by the U.S. National Bureau of Standards as such. (B) derived or secondary standard. A calibrated instrument, source, or other system or device directly relatable (that is, with no intervening steps) to one or more U.S. National Standards. (C) laboratory standard. A calibrated instrument, source, or other system or device without direct one-step relatability to the U.S. National Bureau of Standards, maintained and used primarily for calibration and standardization. 399

(2) (transmission lines and waveguides). A device having stable, precisely defined characteristics that may be used as a reference. 185

(3) (test, measurement and diagnostic equipment). A laboratory type device which is used to maintain continuity of value in the units of measurement by periodic comparison with higher echelon or national standards. They may be used to calibrate a standard of lesser accuracy or to calibrate test and measurement equipment directly. 54

standard antenna (amplitude-modulation broadcast receivers). An open single-wire antenna (including the lead-in wire) having an effective height of 4 meters. A dummy antenna which closely approximates such an actual antenna over a wide frequency range is shown in Fig 1 below and its impedance characteristics in Fig 2. 524

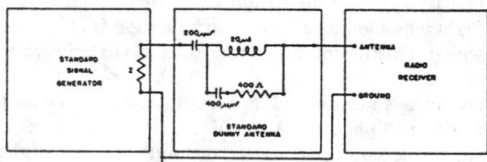

Fig. 1—Standard dummy antenna and method of connections.

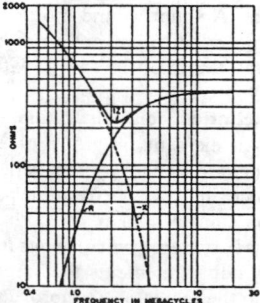

Fig. 2—Impedance characteristics of a dummy antenna.

standard antenna input voltages (amplitude-modulation broadcast receivers). Four standard antenna input voltages are specified for the purpose of certain tests, as follows: (1) A 'distant signal voltage' is taken as 86 decibels below 1 volt, or 50 microvolts. (2) A 'mean-signal voltage' is taken as 46 decibels below 1 volt, or 5 000 microvolts. (3) A 'local signal voltage' is taken as 20 decibels below 1 volt, or 100 000 microvolts. (4) A 'strong-signal voltage' is taken as 1 volt.   524

standard cable. The standard cable formerly used for specifying transmission losses had, in American practice, a linear series resistance and linear shunt capacitance of 88 ohms and 0.054 microfarad, respectively, per loop mile, with no inductance or shunt conductance.   328

standard cell. A cell that serves as a standard of electromotive force. *See:* **Weston normal cell; unsaturated standard cell; auxiliary device to an instrument; electrochemistry.**   328

standard chopped lightning impulse (high voltage testing). A standard chopped lightning impulse is a standard impulse which is chopped by an external gap after 2 to 5 μs. Other times to chopping may be specified by the appropriate apparatus standard. Because of practical difficulties in measurement, the virtual duration of voltage collapse has not been standardized. This characteristic is of importance only for some specific tests.   150

standard code. The operating, block signal, and interlocking rules of the Association of American Railroads.   328

standard compass. A magnetic compass so located that the effect of the magnetic mass of the vessel and other factors that may influence compass indication is the least practicable.   328

standard connector (fixed and variable attenuators). A connector, the critical mating dimensions of which have been standardized to assure nondestructive mating. *Notes:* (1) It butts against its mating standard connector only in the mechanical reference plane. (2) It joins to its waveguide or transmission line with a minimum discontinuity. (3) All its discontinuities are to the maximum extent possible, self-compensated, not within the mating connector.   110

standard connector pair (fixed and variable attenuators). Two standard connectors designed to mate with each other.   110

standard de-emphasis characteristic (frequency modulation). A falling response with modulation frequency, complementary to the standard pre-emphasis characteristic and equivalent to an RC circuit with a time constant of 75 μs. *Note:* The de-emphasis characteristic is usually incorporated in the audio circuits of the receiver.   16

standard directivity (antennas). The maximum directivity from a planar aperture of area A, or from a source line of length L, when excited with a uniform amplitude, equiphase distribution. *Notes:* (1) For planar apertures in which $A\lambda^2$. The value of the standard directivity is $4\pi A/\lambda^2$, with λ the wavelength and with radiation confined to a half space. (2) For line sources with $L\lambda$, the value of the standard directivity is $2L/\lambda$.   111

standard electrode potential. An equilibrium potential for an electrode in contact with an electrolyte, in which all of the components of a specified electrochemical reaction are in their standard states. The standard state for a gas is the pressure of one atmosphere, for an ionic constituent it is unit ion activity, and it is a constant for a solid. *See:* **electrochemistry.**   328

standard frequency (electric power systems). A precise frequency intended to be used for a frequency reference. *Note:* In the U.S. a frequency of 50 Hz is recognized as a standard for all ac lighting and power systems.   3

standard full impulse voltage wave (1) (insulation strength). An impulse that rises to crest value of voltage in 1.2 microseconds (virtual time) and drops to 0.5 crest value of voltage in 50 microseconds (virtual time), both time being measured from the same origin and in accordance with established standards of impulse testing techniques. *Note:* The virtual value for the duration of the wavefront is 1.67 times the time taken by the voltage to increase from 30 percent to 90 percent of its crest value. The origin from which time is measured is the intersection with the zero axis of a straight line drawn through points on the front of the voltage wave at 30-percent and 90-percent crest value.   276

(2) (mercury lamp transformers). An impulse that rises to crest value of voltage in 1.5 microseconds (nominal time) and drops to 0.5 crest value of voltage in 40 microseconds (nominal time), both times being measured from the same time origin and in accordance with established standards of impulse testing techniques. *See:* **basic impulse insulation level (BIL) (insulation strength).**   274

standard illuminant (illuminating engineering). A hypothetical light source of specified relative spectral power distribution. *Note:* The International Commission on Illumination has specified spectral power distributions for standard illuminants A, B, and C, and several D-illuminants.   167

standard illuminant A (illuminating engineering). A blackbody at a temperature of 2856 K. It is defined by its relative spectral power distribution over the range from 300 to 830 nanometers (nm).   167

standard illuminant B (illuminating engineering). A representation of noon sunlight with a correlated color temperature of approximately 4900 K. It is defined by its relative spectral power distribution over the range from 320 to 770 nanometers (nm). *Note:* It is anticipated that at some future date, that is yet to be decided, illuminant B will be dropped from the list of recommended standard illuminants.   167

standard illuminant C (illuminating engineering). A representation of daylight having a correlated color temperature of approximately 6800 K. It is defined by its relative spectral power distribution over the range from 320 to 770 nanometers (nm). *Note:* It is anticipated that at some future date, that is yet to be decided, illuminant C will be dropped from the list of recommended standard illuminants.   167

standard illuminant D65 (illuminating engineering). A

representation of daylight at a correlated color temperature of approximately 6500 K. It is defined by its relative spectral power distribution over the range from 300 to 830 nanometers (nm). *Note:* At present, no artificial source for matching this illuminant has been recommended. 167

standard insulation class (instrument transformer). Denotes the maximum voltage in kilovolts that the insulation of the primary winding is designed to withstand continuously. *See:* **instrument transformer.** 212

standardization. *See:* **echelon; interlaboratory working standards; laboratory reference standards.**

standardize. *See:* **check (instrument or meter)/**

standard lightning impulse (power and distribution transformer). An impulse that rises to crest value of voltage in 1.2 μs (virtual time) and drops to 0.5 crest value of voltage in 50 μs (virtual time), both times being measured from the same origin and in accordance with established standards of impulse testing techniques. It is described as a 1.2/50 μs impulse. *See:* **ANSI C68.1-1968 (IEEE Std 4-1969), Techniques for Dielectric Tests.)** *Note:* The virtual value for the duration of the wavefront is 1.67 times the time taken by the voltage to increase from 30 percent to 90 percent of its crest value. The origin from which time is measured is the intersection with the zero axis of a straight line drawn through points on the front of the voltage wave at 30 percent and 90 percent crest value. 53

standard loop input signals (amplitude-modulation broadcast receivers). (1) A 'distant-signal' loop input is taken as 86 decibels below 1 volt per meter, or 500 microvolts per meter. (2) A 'mean-signal' loop input is taken as 46 decibels below 1 volt per meter, or 5 000 microvolts per meter. (3) A 'local-signal' loop input is taken a 26 decibels below 1 volt per meter, or 50 000 microvolts per meter. (4) A 'strong-signal' loop input is taken as 14 decibels below 1 volt per meter, or 200 000 microvolts per meter. *Note:* The above loop field intensities are not equivalent to the standard antenna input voltages for the corresponding class of service. For example, the 'mean-signal' voltage for antenna operation is 5 000 microvolts. This corresponds to a field intensity if 1 250 microvolts per meter assuming a standard 4-meter antenna, whereas the mean-signal voltage for loop receivers is arbitrarily taken as 5 000 microvolts per meter. 524

standard microphone. A microphone the response of which is accurately known for the condition under which it is to be used. *See:* **instrument.** 328

standard noise temperature (interference terminology). The temperature used in evaluating signal transmission systems for noise factor 290 kelvins (27 degrees Celsius). *See:* **interference.** 188

standard observer (color) (CIE 1931) (television). Receptor of radiation whose colorimetric characteristics correspond to the distribution coefficients $\bar{x}_\lambda$, $\bar{y}_\lambda$, $\bar{z}_\lambda$ adopted by the International Commission on Illumination (CIE) in 1931. 18

standard operating duty. *See:* **operating duty.** 103

standard pitch. *See:* **standard tuning frequency.**

standard potential (standard electrode potential). The reversible potential for an electrode process when all products and reactants are at unit activity on a scale in which the potential for the standard hydrogen half-cell is zero. 221,25

standard propagation (1) (data transmission). The propagation of radio waves over a smooth spherical earth of specialized dielectric constant and conductivity, under conditions of standard refraction in the atmosphere. 59

(2) **(radio wave propagation).** The propagation of radio waves over a smooth spherical earth of uniform dielectric constant and conductivity, under conditions of standard refraction in the atmosphere. 146

standard reference position (of a contact) (power switchgear). The nonoperated or deenergized position of the associated main device to which the contact position is referred. *Note:* Standard reference positions of typical devices are listed in 9.44 of ANSI/IEEE C37.2-1979. 103

Device	Standard Reference Position
Circuit breaker	Main contacts open
Disconnecting switch	Main contacts open
Relay	De-energized position
Contactor	De-energized position
Valve	Closed position

standard refraction (1) (data transmission). The refraction which would occur in an idealized atmosphere in which the index of refraction decreases uniformly with height at rate $39z10^{-6}$ per km (kilometer). *Note:* Standard refraction may be included in ground wave calculations by the use of an effective earth radius of 8.5×10^6 meters, or 4/3 the geometrical earth radius. Refraction exceeding standard refraction is called *superrefraction*, and refraction less than standard refraction is called *subrefraction*. 59

(2) **(radio wave propagation).** The refraction which would occur in an atmosphere in which the refractive index decreased uniformly with height above the earth at the rate of 39×10^{-6} per meter. *Note:* Standard refraction may be included in ground-wave calculations by use of an effective earth radius of 8.5×10^6 m or 4/3 the geometrical radius of the earth. 146

standard register (motor meter) (dial register). A four- or five-dial register, each dial of which is divided into ten equal parts, the division marks being numbered from zero to nine, and the gearing between the dial pointers being such that the relative movements of the adjacent dial pointers are in opposite directions and in a 10-to-1 ratio. *See:* **watthour meter.** 328

standard resistor (resistance standard). A resistor that is adjusted with igh accuracy to a specified value, is but slightly affected by variations in temperature, and is substantially constant over long periods of time. *See:* **auxiliary device to an instrument.** 328

standard rod gap. A gap between the ends of the two

one-half-inch square rods cut off squarely and mounted on suppots so that a length of rod equal to or greater than one-half the gap spacing overhangs the inner edge of each support. It is intended to be used for the approximate measurement of crest voltages. *See:* **instrument.**   328

standards (software quality assurance). Mandatory requirements employed and enforced to prescribe a disciplined, uniform approach to software development, that is, mandatory conventions and practices are in fact standards.   481

standards enforcer (software). A software tool that determines whether prescribed development standards have been followed. The standards may include module size, module structure, commenting conventions, use of certain statement forms, and documentation conventions. *See:* **documentation; module; software tool.**   434

standard source (illuminating engineering). An artificial source having the same spectral distribution as a specified standard illuminant.   167

standard source A (illuminating engineering). A tungsten filament lamp operated at a color temperature of 2856 K (International Practical Temperature Scale, 1968) and approximating the relative spectral power distribution of standard illuminant A.   167

standard source B (illuminating engineering). An approximation of standard illuminant B obtained by a combination of Source A and a special filter.   167

standard source C (illuminating engineering). An approximation of standard illuminant C obtained by a combination of Source A and a special filter.   167

standard source diameter (X-ray energy spectrometers). The diameter of the X-ray emission source which is used to measure the response characteristics of the spectrometer. Unless otherwise specified, this is assumed to be a point source.   471

standard sources ('dose calibrator' ionization chambers). A general term used to refer to the standard sources listed below: (1) national radioactivity standard source. A calibrated radioactive source prepared and distributed as a standard reference material by the US Bureau of Standards. (2) certified radioactivity standard source. A calibrated radioactive source, with stated accuracy, whose calibration is certified by the source supplier as traceable to the National Radioactive Measurements System.   499

standard sphere gap (high voltage testing). A peak-voltage device constructed and arranged in accordance with this document. It consists of two metal spheres of the same diameter, D, with their shanks, operating gear, insulating supports, supporting frame, and leads for connections to the point at which the voltage is to be measured. Standard values of D are 625, 125, 250, 500, 750, 1000, 1500, and 2000 mm. The spacing between the spheres is designated as S.

The points on the two spheres that are closest to each other are called the sparking points. In practice, the disruptive discharge may occur between other neighboring points. Figs 1 and 2 show two arrangements; one of which is typical of sphere gaps with a vertical axis, and the other, of sphere gaps with a horizontal axis.   150

standard switching impulse (power and distribution transformer). A full impulse having a front time of 250 μs and a time to half value of 2500 μs. It is described as a 250/2500 impulse. *Note:* It is recognized that some apparatus standards may have to use a modified wave shape where practical test considerations or particular dielectric strength characteristics make some modification imperative. Transformers, for example, use a modified switching impulse wave with the following characteristics: (1) Time to crest greater than 100 μs. (2) Exceeds 90 percent of crest value for at least 200 μs. (3) Time to first voltage zero on tail not less than 1000 μs, except where core saturation causes the tail to become shorter. *See:* **ANSI C57.12.90-1973 (IEEE Std 262-1973), Test Code for Distribution, Power, and Regulating Transformers.**   53

standard systems (electric installations on shipboard). The following systems of distribution are recognized as standard: (1) Two-wire with single-phase alternating current, or direct current. (2) Three-wire with single-phase alternating current, or direct current. (3) Three-phase three-wire, alternating current. (4) Three-phase, four-wire, alternating current.   3

standard television signal. A signal that conforms to certain accepted specifications. *See:* **television.**   328

standard test frequencies in the broadcast band (amplitude-modulation broadcast receivers). The standard group of seven carrier frequencies for testing is 540, 600, 800, 1 000, 1 200, 1 400, and 1 600 kilohertz. The standard group of three carrier frequencies for testing is 600, 1 000, and 1 400 kilohertz.   524

standard test modulation (frequency-modulated mobile communications receivers). Sixty percent of the rated system deviation at a frequency of 1 kilohertz.   123

standard test problem (test, measurement and diagnostic equipment). An evaluation of the performance of a system, or any part of it, conducted by setting parameters into the system; the parameters are operated on and the results obtained from system read outs.   54

standard test tone (data transmission). A 1 mW (megawatt) (0 dBm) 1000 Hz (hertz) signal applied to the 600 Ω (ohm) audio portion of a circuit at a zero transmission level reference point. If referred to a point with a relative level other than 0, the absolute power of the tone shall be adjusted to suit the relative level at the point of application.   59

standard track (electroacoustics). *See:* **single track electroacoustics).**

standard transmitter test modulation (land-mobile communication transmitters). The standard test

modulation shall be 60 percent of the maximum rated deviation at 1 kilohertz (kHz).   444

standard tuning frequency (standard musical pitch). The frequency for the note A_4, namely, 440 Hz (hertz). See the table below.

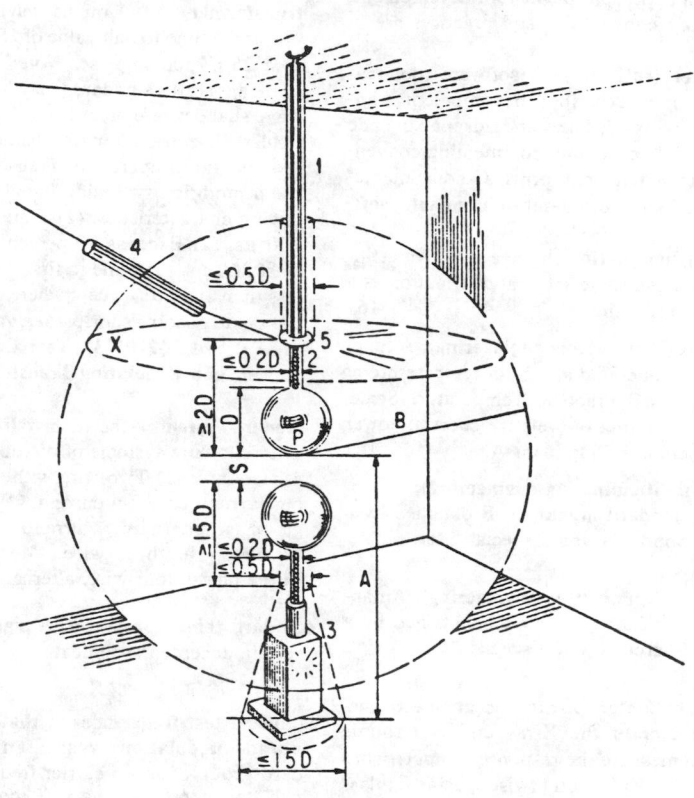

1. Insulating support
2. Sphere shank
3. Operating gear, showing maximum dimensions
4. High-voltage connection with series resistor
5. Stress distributor, showing maximum dimensions
P Sparking point of high-voltage sphere
A Height of P above ground plane
B Radius of space free from external structures
X Item 4 not to pass through this plane within a distance B from P

NOTE: The figure is drawn to scale for a 100 cm sphere gap at radius spacing.

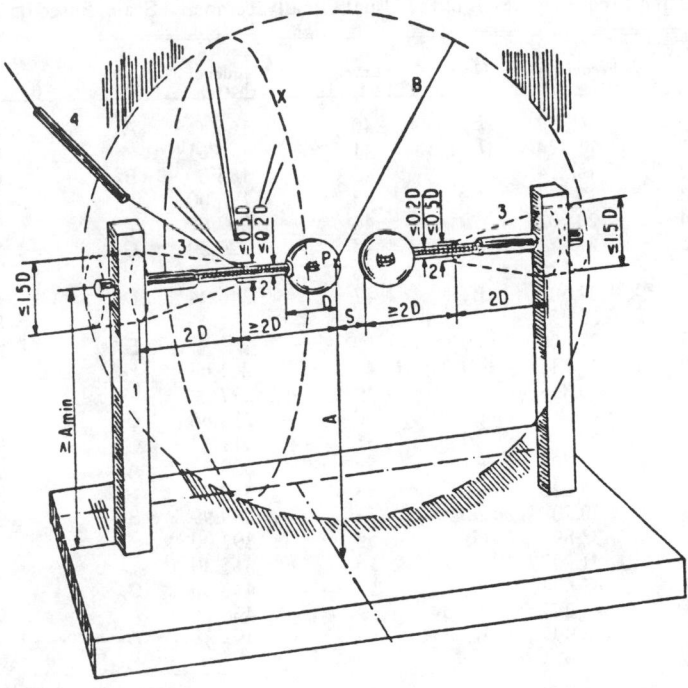

1. Insulating support
2. Sphere shank
3. Operating gear, showing maximum dimensions
4. High-voltage connection with series resistor
P Sparking point of high-voltage sphere
A Height of P above ground plane
B Radius of space free from external structures
X Item 4 not to pass through this plane within a distance B from P

NOTE: The figure is drawn to scale for a 25 cm sphere gap at radius spacing.

standard voltages (electric installations on shipboard). The following voltages are recognized as standard:

	Alternating Current (volts)	Direct Current (volts)
Lighting	115	115
Power	115-200-220-440	115 and 230
Generators	120-208-230-450	120 and 240

Note: Satisfactory to use 120 V lamps.

standard volume indicator (volume measurements of electrical speech and program waves). A device for the indication of volume, and having the characteristic described in Section 3 of ANSI/IEEE Std 152-1953. *Note:* A standard volume indicator consists of at least two parts: (1) A meter (2) An attenuator (adjustable loss) or pad (fixed loss).   523

standard watthour meter. *See:* **portable standard watthour meter; reference standard watthour meter.**
     212

standard-wave error (DF [direction finder] measurements)(navigation aid terms). The bearing error produced by a wave whose vertically and horizontally polarized electric fields are equal and phased so as to give maximum error in the DF, and whose incidence direction is arranged to be 45°.   526

standard working axis (of a semiconductor X-ray energy spectrometer). A straight line drawn between the

Frequencies and Frequency Levels of the Usual Equally Tempered Scale; Based on $A_4 = 440$ Hz

Note Name	Frequency Level (semits)	Frequency (hertz)	Note Name	Frequency Level (semits)	Frequency (hertz)	Note Name	Frequency Level (semits)	Frequency (hertz)
C_0	0	16.352	E_3	40	164.81	A_6	81	1760.0
	1	17.324	F_3	41	174.61		82	1864.7
D_0	2	18.354		42	185.00	B_6	83	1975.5
	3	19.445	G_3	43	196.00			
E_0	4	20.602		44	207.65	C_7	84	2093.0
F_0	5	21.827	A_3	45	220.00		85	2217.5
	6	23.125		46	233.08	D_7	86	2349.3
G_0	7	24.500	B_3	47	246.94		87	2489.0
	8	25.957				E_7	88	2637.0
A_0	9	27.500				F_7	89	2793.8
	10	29.135	C_4	48	261.63		90	2960.0
B_0	11	30.868		49	277.18	G_7	91	3136.0
			D_4	50	293.66		92	3322.4
				51	311.13	A_7	93	3520.0
C_1	12	32.703	E_4	52	329.63		94	3729.3
	13	34.648	F_4	53	349.23	B_7	95	3951.1
D_1	14	36.708		54	369.99			
	15	38.891	G_4	55	392.00			
E_1	16	41.203		56	415.30	C_8	96	4186.0
F_1	17	43.654	A_4	57	440.00		97	4434.9
	18	46.249		58	466.16	D_8	98	4698.6
G_1	19	48.999	B_4	59	493.88		99	4978.0
	20	51.913				E_8	100	5274.0
A_1	21	55.000				F_8	101	5587.7
	22	58.270	C_5	60	523.25		102	5919.9
B_1	23	61.735		61	554.37	G_8	103	6271.9
			D_5	62	587.33		104	6644.9
				63	622.25	A_8	105	7040.0
C_2	24	65.406	E_5	64	659.26		106	7458.6
	25	69.296	F_5	65	698.46	B_8	107	7902.1
D_2	26	73.416		66	739.99			
	27	77.782	G_5	67	783.99			
E_2	28	82.407		68	830.61	C_9	108	8372.0
F_2	29	87.307	A_5	69	880.00		109	8869.8
	30	92.499		70	932.33	D_9	110	9397.3
G_2	31	97.999	B_5	71	987.77		111	9956.1
	32	103.83				E_9	112	10548.
A_2	33	110.00				F_9	113	11175.
	34	116.54	C_6	72	1046.5		114	11840.
B_2	35	123.47		73	1108.7	G_9	115	12544.
			D_6	74	1174.7		116	13290.
				75	1244.5	A_9	117	14080.
C_3	36	130.81	E_6	76	1318.5		118	14917.
	37	138.59	F_6	77	1396.9	B_9	119	15084.
D_3	38	146.83		78	1480.0			
	39	155.56	G_6	79	1568.0			
				80	1661.2	C_{10}	120	16744.

center of the entrance window on the detector and the specified location of the source of X-rays.   471

standby. *See:* **alternative; reserve.**

standby current, dc. *See:* **direct-current (dc) standby current. nominal varistor voltage (low voltage varistor surge arresters).** Voltage across the varistor measured at a specified pulsed direct-current (dc) current of specific duration.   62

standby failure rate (reliability data for pumps and drivers, valve actuators, and valves). The probability (per hour) of failure for those components which are normally dormant or in a standby state until tested or required to operate to perform their function.   502

standby power, ac. *See:* **alternating-current (ac) standby power.**

standby power supply (diesel-generator unit)(nuclear power generating station). The power supply that is selected to furnish electric energy when the preferred power supply is not available.   99,102

standby power system (emergency and standby power). An independent reserve source of electric energy that, upon failure or outage of the normal source, provides electric power of acceptable quality so that the user's facilities may continue in satisfactory operation. 512

standby redundancy (reliability). *See:* **redundancy, standby.**

standby service (power operations). Service through a permanent connection not normally used but available in lieu of, or as a supplement to, the usual source of supply. 516

standing-on-nines carry (parallel addition of decimal numbers). A high-speed carry in which a carry input to a given digit place is bypassed to the next digit place if the current sum in the given place is nine. *See:* **carry.** 255,77

standing wave (1) (overhead-power-line corona and radio noise). A wave in which, for any component of the field, the ratio of its instantaneous value at one point to that at any other point does not vary with time. *Note:* Commonly it is a periodic wave in which the amplitude of the displacement in the medium is a periodic function of the distance in the direction of any line of propagation of the wave. 411
(2) (radio wave propagation). A wave in which, for any component of the field, the ratio of its instantaneous value at one point to that at any other point does not vary with time. 146
(3) (waveguide). A wave in which, for any component in the field, the ratio of its instantaneous amplitude at one point to that at any other point does not vary with time. 267

standing-wave antenna. An antenna whose excitation is essentially equiphase, as the result of two feeding waves which traverse its length from opposite directions, their combined effect being that of a standing wave. 111

standing-wave detector. *See:* **standing-wave meter.**

standing wave dissipation factor (waveguide). The ratio of the transmission loss in an unmatched waveguide to that in the same waveguide when matched. 267

standing-wave indicator. *See:* **standing-wave meter.**

standing-wave loss factor. The ratio of the transmission loss in an unmatched waveguie to that in the same waveguide when matched. *See:* **waveguide.** 267

standing-wave machine. *See:* **standing-wave meter.**

standing-wave meter (standing-wave indicator) (standing-wave detector) (standing-wave machine). An instrument for measuring the standing-wave ratio in a transmission line. In addition a standing-wave meter may include means for finding the location of maximum and minimum amplitudes. See table on facing page. *See:* **instrument.** 328

standing wave ratio (1) (data transmission). The ratio of the amplitude of a standing wave at an antinode to the amplitude of a node. *Note:* The standing wave ratio in a uniform transmission line is

$$\frac{1 + p}{1 - p}$$

where p = the reflection coefficient. 59
(2) (overhead-power-line corona and radio noise). The ratio of the amplitude of a standing wave at an antinode to the amplitude at a node. 411

(3) (waveguide). At a given frequency in a uniform transmission line or waveguide, the ratio of the maximum to the minimum amplitudes of corresponding components of the field (or the voltage or current) along the waveguide in the direction of propagation. *Note:* The standing wave ratio is occasionally expressed as the reciprocal of the ratio defined above. 267

(4) (voltage-standing-wave ratio*) (at a given frequency in a uniform waveguide or transmission line). The ratio of the maximum to the minimum amplitudes of corresponding components of the field (or the voltage or current) appearing along the guide or line in the direction of propagation. *Notation:* SWR (VSWR*). *Note:* Alternatively, the standing-wave ratio may be expressed as the reciprocal of the ratio defined above. 319

*Deprecated
standing-wave-ratio indicator (standing-wave-ratio meter). A device or part thereof used to indicate the standing-wave ratio. *Note:* In common terminology, it is the combination of amplifier and meter as a supplement to the slotted line or bridge, etcetera, when performing impedance or reflection measurements. 15

stand, reel. *See:* **reel stand**

standstill locking (rotating machinery). The occurrence of zero or unusably small torque in an energized polyphase induction motor, at standstill, for certain rotor positions. *See:* **asynchronous machine.** 63

star chain (navigation aid terms). A radio navigation transmitting system comprising a master station about which three (or more) slave stations are symmetrically located. 526

star-connected circuit. A polyphase circuit in which all the current paths of the circuit extend from a terminal of entry to a common terminal or conductor (which may be the neutral conductor). *Note:* In a three-phase system this is sometimes called a Y (or wye) connection. *See:* **star network; network analysis.** 210

star connection (non-preferred term) (power and distribution transformer). *See:* **Y-connection.**

star coupler (fiber optics). A passive device in which power from one or several input waveguides is distributed amongst a larger number of output optical waveguides. *See:* **optical combiner; tee coupler.** 433

star-delta starter (industrial control). A switch for starting a three-phase motor by connecting its windings first in star and then in delta. *See:* **starter.** 244,206

star-delta starting. The process of starting a three-phase motor by connecting it to the supply with the primary winding initially connected in star, then reconnected in delta for running operation. 63

star network. A set of three or more branches with

one-terminal of each connected at a common node. *See:* **network analysis.** 210

star quad. *See:* **spiral four.**

star rectifier circuit. A circuit that employs six or more rectifying elements with a conducting period of 60 electrical degreesplus the commutating angle. *See:* **rectification.** 328

start (1) (industrial control). An electric controller for accelerating a motor from rest to normal speed and to stop the motor. 3,429

(2) (gas tubes). A control electrode, the principal function of which is to establish sufficient ionization to reduce the anode breakdown voltage. *Note:* This has sometimes been referred to as a "trigger electrode." 125

start-dialing signal (semiautomatic or automatic working) (telecommunication). A signal transmitted from the incoming end of a circuit, following the receipt of a seizing signal, to indicate that the necessary circuit conditions have been established for receiving the numerical routing information. 194

start diesel signal (diesel-generator unit). That input signal to the diesel-generator unit start logic which initiates a diesel-generator unit start sequence. 99

start element (data transmission). In a character transmitted in a start-stop system, the first element in each character, which serves to prepare the receiving equipment for the reception and registration of the character. The start element is a spacing signal. 59

starter (1) (electric installations on shipboard). An electric controller for accelerating a motor from rest to normal speed and to stop the motor. (A device designed for starting a motor in either direction of rotation includes the additional function of reversing and should be designated a controller.) 3

(2) (illuminating engineering). A device used in conjunction with a ballast for the purpose of starting an electric-discharge lamp. 167

(3) (packaging machinery). An electric controller for accelerating a motor from rest to normal speed and for stopping the motor. 429

starter gap (gas tube). The conduction path between a starter and the other electrode to which starting voltage is applied. *Note:* Commonly used in the glow-discharge cold-cathode tube. 125

starters (fluorescent lamp). Devices that first connect a fluorescent or similar discharge lamp in a circuit to provide for cathode preheating and then open the circuit so that the starting voltage is applied across the lamp to establish an arc. Starters also include a capacitor for the purpose of assisting the starting operation and for the suppression of radio interference during lamp starting and lamp operation. They may also include a circuit-opening device arranged to disconnect the preheat circuit if the lamp fails to light normally. 326

starter voltage drop (glow-discharge cold-cathode tube). The starter-gap voltage drop after conduction is established in the starter gap. *See:* **electrode voltage (electron tube); gas tube.** 125

starting (rotating machinery). The process of bringing a motor up to speed from rest. *Note:* This includes breaking away, accelerating and if necessary, synchronizing with the supply. 63

starting amortisseur. An amortisseur the primary function of which is to starting of the synchronous machine and its connected load. 328

starting anode. An electrode that is used in establishing the initial arc. *See:* **rectification.** 328

starting attempt (electric generating unit reliability, availability, and productivity). The action to bring a unit from shutdown to the in-service state. Repeated initiations of the starting sequence without accomplishing corrective repairs are counted as a single attempt. 567

starting capacitance (capacitor motor). The total effective capacitance in series with the auxiliary winding for starting operation. *See:* **asynchronous machine.**

starting circuit breaker (6) (power system device function numbers). A device whose principal function is to connect a machine to its source of starting voltage. 402

starting current (1) (rotating machinery). The current drawn by the motor during the starting period. (A function of speed or slip). *See:* **asynchronous machine.** 63

(2) (oscillator). The value of electron-stream current through an oscillator at which self-sustaining oscillations will start under specified conditions of loading. *See:* **magnetrons.** 125

starting failure (electric generating unit reliability, availability, and productivity). The inability to bring a unit from some unavailable state or reserve shutdown state to the in-service state within a specified period. The specified period may be different for individual units. Repeated failures within the specified starting period are counted as a single starting failure. 567

starting motor. An auxiliary motor used to facilitate the starting and accelerating of a main machine to which it is mechanically connected. *See:* **asynchronous machine.** 63

starting open-phase protection. The effect of a device operative to prevent connecting the load to the supply unless all conductors of a polyphase system are energized. 206

starting operation (single-phase motor). (1) The range of operation between locked rotor and switching for a motor employing a starting-switch or relay. (2) The range of operation between locked rotor and a point just below but not including breakdown-torque speed for a motor not employing a starting switch or relay. *See:* **asynchronous machine.** 63

starting reactor (power and distribution transformer). A current-limiting reactor for decreasing the starting current of a machine or device. 53

starting resistor (rotating machinery). A resistor connected in a secondary or field circuit to modify starting performance of an electric machine. *See:* **rotor (rotating machinery); stator.** 63

starting rheostat (industrial control). A rheostat that

controls the current taken by a motor during the period of starting and acceleration, but does not control the speed when the motor is running normally.
  206

starting sheet (electrorefining). A thin sheet of refined metal introduced into an electrolytic cell to serve as a cathode surface for the deposition of the same refined metal. *See:* **electrorefining.** 328

starting-sheet blank (electrorefining). A rigid sheet of conducting material designed for introduction into an electrolytic cell as a cathode for the deposition of a thin temporarily adherent deposit to be stripped off as a starting sheet. *See:* **electrorefining.** 328

starting success (electric generating unit reliability, availability, and productivity). The occurrence of bringing a unit from some unavailable state or the reserve shutdown state to the in-service state within a specified period. The specified period may be different ofr individual units. 567

starting-switch assembly. The make-and-break contacts, mechanical linkage, and mounting parts necessary for starting or running, or both starting and running, split-phase and capacitor motors. *Note:* The starting-switch assembly may consist of a stationary-contact assembly and a contact that moves with the rotor. 328

starting switch, centrifugal (rotating machinery). *See:* **centrifugal starting switch.**

starting switch, relay. *See:* **relay starting switch.**

starting temperature (grounding device). The winding temperature at the start of the flow of thermal current.
  91

starting test (rotating machinery). A test taken on a machine while it is accelerating from standstill under specified conditions. *See:* **asynchronous machine.**
  63

starting torque (1) (electric coupling). The minimum torque of an electric coupling developed with the output member stationary and the input member rotating, with excitation applied. *Note:* Starting torque is usually specified with rated speed of rotation and rated excitation applied. 416
(2) (synchronous motor). The torque exerted by the motor during the starting period. (A function of speed or slip). 63

starting-to-running transition contactor (19) (power system device function numbers). A device that operates to initiate or cause the automatic transfer of a machine from the starting to the running power connection. 402

starting voltage (radiation counters). The voltage applied to a Geiger-Mueller tube at which pulses of 1 volt amplitude appear across the tube when irradiated. *See:* **anticoincidence (radiation counters).** 190

starting winding (rotating machinery). A winding, the sole or main purpose of which is to set up or aid in setting up a magnetic field for producing the torque to start and accelerate a rotating electric machine.
  63

start-pulsing signal (telephone switching systems). A signal transmitted from the receiving end to the sending end of a trunk to indicate that the receiving end is in a condition to receive pulsing. 55

star tracker. *See:* **astrotracker.**

start-record signal. A signal used for starting the process of converting the electric signal to an image on the record sheet. *See:* **facsimile signal (picture signal).**
  12

start signal (1) (start-stop system). Signal serving to prepare the receiving mechanism for the reception and registration of a character, or for the control of a function. 194
(2) (facsimile). A signal that initiates the transfer of a facsimile equipment condition from standby to active. *See:* **facsimile signal (picture signal).** 12
(3) (telephone switching systems). In multifrequency and key pulsing, a signal used to indicate that all digits have been transmitted. 55

start-stop printing telegraphy. That form of printing telegraphy in which the signal-receiving mechanisms are started in operation at the beginning and stopped at the end of each character transmitted over the channel. *See:* **telegraphy.** 328

start-stop system (data transmission). A system in which each group of code elements corresponding to a character is preceded by a start element which serves to prepare the receiving equipment for the reception and registration of a character, and is followed by a stop element during which the receiving equipment comes to rest in preparation for the reception of the next character. 59

start-stop transmission (data transmission). A synchronous transmission in which a group of code elements corresponding to a character signal is preceded by a start signal which serves to prepare the receiving mechanism for the reception and registration of a character and is followed by a stop signal which serves to bring the receiving mechanism to rest in preparation for the reception of the next character. 59

start transition (data transmission). In a character transmitted in a start-stop system, the mark-to-space transition at the beginning of the start element.
  59

startup (of a relay) (power switchgear). The action of a relay as it just departs from complete reset. Startup as a qualifying term is also used to identify the minimum value of the input quantity which will permit this condition. 103

statcoulomb. The unit of charge in the centimeter-gram-second electrostatic system. It is that amount of charge that repels an equal charge with a force of one dyne when they are in a vacuum, stationary, and one centimeter apart. One statcoulomb is approximately 3.335×10^{-10} coulomb. 210

state (high-level microprocessor language). The condition of the target microprocessor, given in terms of the contents of its registers, internal flags, local memory, etcetera. 470

state data (software unit testing). Data that defines an internal state of the test unit and is used to establish that state or compare with existing states. 519

state diagram (software). A directed graph in which

nodes correspond to internal states of a system, and edges correspond to transitions; often used to describe a system in terms of state changes. *See:* **directed graph; node; Petri net; system.** 434

state element (high-level microprocessor language). A microprocessor component containing a distinguishable part of the state information, such as a single register. 470

statement (computer programming). A meaningful expression or generalized instruction in a source language. 255, 77

state of chromatic adaptation (illuminating engineering). The condition of the chromatic properties of the visual system at a specified moment as a result of exposure to the totality of colors of the visual field currently and in the past. *See:* **chromatic adaptation.** 167

state of polarization (of a propagating wave [field vector]) (antennas). At a given point in space, the condition of the polarization of a plane wave [field vector] as described by the axial ratio, tilt angle and sense of polarization. *Syn:* **polarization state.** 111

state of statistical control (pulse measurement process). That state wherein a degree of consistency among repeated measurements of a characteristic, property, or attribute is attained. 15

state space (automatic control). A space which contains the state vectors of a system. *Note:* The number of state variables in the system determines the dimension of the state space. *See:* **control system.** 56

state trajectory (automatic control). The vector function describing the dependence of the state on time and initial state. Note: If ϕ is the state trajectory, then

$$1)\ \phi(t_0, x_0) = x_0$$
$$2)\ \phi(t_2, x_0) = \phi[t_2, \phi(t_1, x_0)]$$

56

state variable formulation, (eigenvalue, eigenvector, characteristic equation) (excitation control systems). A system may be mathematically modeled by assigning variables $x_1, x_2, \ldots, x_n$ to system parameters: when these x's comprise the minimum number of parameters which completely specify the system, they are termed "states" or "state variables." System states arranged in a n-vector form a state vector. The mathematical model of the system may be manipulated into the form

$$dx/dt = \dot{X} = AX + bu$$
$$Y = CX + du$$

where X is the system state vector, u is the input vector, Y is the output vector, and A, b, C, d are matrices of appropriate dimension which specify the system. Such a model is known as a state variable or modern control formulation.

$$\det\,(A - \lambda I) = 0$$

is called the characteristic equation and has n roots which are called eigenvalues (det (·) denotes determinant). When eigenvalues are real, they are the negative inverses of closed loop system time constants. Eigenvalues are also the pole locations of the closed loop transfer function. Any vector e_i such that

$$(A - \lambda_i I)e_i = 0$$
$$\|e_i\| \neq 0$$

is called an eigenvector of the eigenvalue λ_i ($\| \cdot \|$) denotes the square root of the sum of the squares of all entries of a vector. All n eigenvectors of a system form a modal matrix of matrix A when arranged side-by-side in a square matrix. The modal matrix is used in certain analytic procedures in modern control theory whereby large, complex systems are decoupled into many first order systems. 353

state variables (automatic control). Those whose values determine the state. 56

state vector (automatic control). One whose components are the state variables. 56

static (1) (atmospherics). Interference caused by natural electric disturbances in the atmosphere, or the electromagnetic phenomena capable of causing such interference. *See:* **radio transmitter.** 328

(2) (adjective) (automatic control). Referring to a state in which a quantity exhibits no appreciable change within an arbitrarily long time interval.
56

static accuracy (analog computers). Accuracy determined with a constant output. 9

static analysis (software). The process of evaluating a program without executing the program. *See:* **code audit; desk checking; dynamic analysis; inspection; program; static analyzer; walk-through.** 434

static analyzer (software). A software tool that aids in the evaluation of a computer program without executing the program. Examples include syntax checkers, compilers, cross-reference generators, standards enforcers, and flowcharters. *See:* **compiler; computer program; dynamic analyzer; program; software tool; standards enforcer; syntax.** 434

static binding (software). Binding performed prior to execution of a program and not subject to change during execution. *See:* **binding; dynamic binding; execution; program.** 434

static breeze. *See:* **convective discharge.**

static characteristic (electron tubes). A relation, usually represented by a graph, between a pair of variables such as electrode voltage and electrode current, with all other voltages maintained constant. 190

static characteristic, relay. *See:* **relay static characteristic.**

static check. *See:* **problem check.**

static converter (electric installations on shipboard). A unit that employs static rectifier devices such as semiconductor rectifiers or controlled rectifiers (thyristors) transistors, electron tubes, or magnetic amplifiers to change ac power to dc power or vice-versa.   3

static dump (computing systems). A dump that is performed at a particular point in time with respect to a machine run, frequently at the end of a run.   255, 77

static electrode potential. The electrode potential that exists when no current is flowing between the electrode and the electrolyte. *See:* **electrolytic cell.**   328

static error (analog computers). An error independent of the time-varying nature of a variable. Also known as the dc error.   9

static friction. *See:* **stiction.**

static induced current. The charging and discharging current of a pair of Leyden jars or other capacitors, which current is passed through a patient. *See:* **electrotherapy.**   192

staticize (electronic digital computation). (1) To convert serial or time-dependent parallel data into static form. (2) Occasionally, to retrieve an instruction and its operands from storage prior to its execution.   235

staticizer (electronic computation). A storage device for converting time-sequential information into static parallel information.   210

static Kraemer system (rotating machinery). A system of speed control below synchronous speed for wound-rotor induction motors. Slip power is recovered through the medium of a static converter equipment electrically connected between the secondary winding of the induction motor and a power system. *See:* **asynchronous machine.**   63

static load line (electron device). The locus of all simultaneous average values of output electrode current and voltage, for a fixed value of direct-current load resistance.   190

static noise (atmospherics) (telephone practice). Interference caused by natural electric disturbances in the atmosphere, or the electromagnetic phenomena capable of causing such interference. *See:* **static.**   24

static optical transmission (acousto-optic device). The ratio of the transmitted zero order intensity, I_0, to the incident light intensity, I_{in}, when the acoustic drive power is off: thus $T = I_0/I_n$.   82

static overvoltage (surge arresters). An overvoltage due to an electric charge on an isolated conductor or installation.   62

static plow (cable plowing). A plowing unit that depends upon drawbar pull only for its movement through the soil.   52

static pressure (acoustics) (audio and electroacoustics) (at a point in a medium). The pressure that would exist at that point in the absence of sound waves.   176

static radiation test (metal-nitride-oxide field-effect transistor). Test of the more or less permanent changes induced by radiation obtained by a comparison of characteristics before and after exposure.   386

static regulation. Expresses the change from one steady-state condition to another as a percentage of the final steady-state condition.

$$\text{Static Regulation} = \frac{E_{\text{initial}} - E_{\text{final}}}{E_{\text{final}}} \, (100\%).$$

186

static regulator. A transmission regulator in which the adjusting mechanism is in self-equilibrium at any setting and requires control power to change the setting. *See:* **transmission regulator.**   328

static relay (or relay unit) (power switchgear). A relay or relay unit in which the designed response is developed by electronic, solid-state, magnetic or other components without mechanical motion. *Note:* A relay which is composed of both static and electromechanical units in which the designed response is accomplished by static units may be referred to as a static relay.   103

static resistance (forward or reverse) (semiconductor rectifier device). The quotient of the voltage by the current at a stated point on the static characteristic curve. *See:* **rectification.**   66

static short-circuit ratio (arc-welding apparatus). The ratio of the steady-state output short-circuit current of a welding power supply at any setting to the output current at rated load voltage for the same setting.   264

static, solid state converter (induction and dielectric heating equipment). A solid state generator or power source which utilizes semiconductor devices to control the switching of currents through inductive and capacitive circuit elements and thus generate a useable alternating current at a desired output frequency.   14

static test (1) (analog computers). The computer-control state which applies a predetermined set of voltages and conditions to the analog computer, which allows the static check to be executed.   9
(2) (test, measurement and diagnostic equipment). (A) A test of a non-signal property, such as voltage and current, of an equipment or of any of its constituent units, performed while the equipment is energized: and (B) A test of a device in a stationary of helddown position as a means of testing and measuring its dynamic reactions.   54

static torque (electric coupling). The minimum torque an electric coupling will transmit or develop with no relative motion between the input and output members, with excitation applied. *Note:* Static torque is usually specified for rated excitation.   416

static value (light emitting diodes). A non-varying value or quantity of measurement at a specified fixed point, or the slope of the line from the origin to the operating point on the appropriate characteristic curve.   162

static voltampere characteristic (arc-welding apparatus). The curve or family of curves that gives the terminal voltage of a welding power supply as ordinate, plotted against output load current as abscissa, is the static voltampere characteristic of the power supply. 264

static wave current (electrotherapy). The current resulting from the sudden periodic discharging of a patient who has been raised to a high potential by means of an electrostatic generator. *See:* **electrotherapy.** 192

station (1)(generating station grounding). For the purpose of ANSI/IEEE Std 665-1987, 'station' is synonymous with 'generating station'. 569
(2)(or data station)(token ring access method). A physical device that may be attached to a shared medium local area network for the purpose of transmitting and receiving information on that shared medium. A data station is identified by a destination address. 472
(3)(power operations). A facility where several components of a system are colocated. *See:* **generating station; converting station; distribution station; hydroelectric station; peaking station; pumped-storage station; run-of-river station; station service; steam-electric station; storage station; switching station; transforming station.** 516
(4)(supervisory control, data acquisition, and automatic control). (A) Master (of a supervisory system). The entire complement of devices, functional modules, and assmblies which are electrically interconnected to effect the master station supervisory functions. The equipment includes the interface with the communication channel but does not include the interconnecting channel. During communication with one or more remote stations the master station is the superior in the communication hierarchy. (B) Remote (of a supervisory system). The entire complement of devices, functional modules, and assemblies which are electrically interconnected to effect the remote station supervisory functions. The equipment includes the interface with the communication channel but does not include the interconnecting channel. During communication with a master station the remote station is the subordinate in the communication hierarchy. *Notes:* Examples of station equipments include (1) Hardwired. Station supervisory equipment which is comprised entirely of wired-logic elements. (2) Firmware. Station supervisory equipment which uses hardware logic programmed routines in a manner similar to a computer. The routines can only be modified by physically exchanging logic memory elements. (3) Programmable. Station supervisory equipment which uses software routines. (C) Semiautomatic. A station that requires both automatic and manual modes to maintain the required character of service. (D) Submaster. A station that can perform as a master station on one message transaction and as a remote station on another message transaction. 570
stationarity (seismic qualification of Class 1E equipment for nuclear power generating stations). A condition that exists when a waveform is stationary and when its amplitude distribution, frequency content, and other descriptive parameters are statistically constant with time. 581

stationary battery. A storage battery designed for service in a permanent location. *See:* **battery (primary or secondary).** 328
stationary-contact assembly. The fixed part of the starting-switch assembly. *See:* **starting-switch assembly.** 328
stationary contact member (power switchgear). A conducting part having a contact surface which remains substantially stationary. 103
stationary-mounted device (power switchgear). One that cannot be removed except by the unbolting of connections and mounting supports. *Note:* Compare with drawout-mounted device. 103
stationary satellite (communication satellite). A synchronous satellite with an equatorial, circular and direct orbit. A stationary satellite remains fixed in relation to the surface of the primary body. *Note:* A geostationary satellite is a stationary earth satellite. 74
stationary (time invariant) sytem (excitation control systems). Let a system have zero input response $Z(t)$, then the system is stationary (time invariant) if the response to input $R(t)$ is $C(t) + Z(t)$ and the response to input $R(t + T)$ is $C(t + T) + Z(t)$. Otherwise the system is nonstationary. *Note:* A stationary system is modelled mathematically by a stationary differential equation the coefficients of which are not functions of time. 353
stationary wave. *See:* **standing wave.**
station auxiliary (generating station) (generating stations electric power system). An auxiliary at a generating station not assigned to a specific unit. 381
station changing (communication satellite). The changeover of service from one earth station to another, especially in a system using satellites that are not stationary. 84
station check (supervisory check, status update)(supervisory control, data acquisition, and automatic control). The automatic selection, in a definite order, of all the supervisory alarm and indication points associated with one remote station or all remote stations of a system, and the transmission of all the indications to the master station. 570
station-control error (electric power systems). The station generation minus the assigned station generation. *Note:* Refer to note on polarity under **area control error.** 94
station equipment (data transmission). A broad term used to denote equipment located at the customer's premises. The equipment may be owned by the telephone company or the customer. If the equipment is owned by the customer it is referred to as the customer's equipment. 59
station identification (supervisory control, data acquisition, and automatic control). A sequence of signal elements used to identify a station. 570
station line (telephone switching systems). Conductors carrying direct current between a central office and a main station, private branch exchange, or other end equipment. 55
station-loop resistance (telephone switching systems).

The series resistance of the loop conductors, including the resistance of an off-hook station. 55

station number (n) (subroutines for CAMAC). The number *n* represents an integer which is the station number component of a CAMAC address. 410

station, peaking. A generating station which is normally operated to provide power only during maximum load periods. 112

station, pumped storage. A hydroelectric generating station at which electric energy is normally generated during periods of relatively high system demand by utilizing water which has been pumped into a storage reservoir usually during periods of relatively low system demand. 112

station ringer (ringer). *See:* **telephone ringer.**

station, run-of-river. A hydroelectric generation station utilizing limited pondage or the flow of the stream as it occurs. 112

station service (power operations). Facilities which provide power for station use in a generating, switching, converting, or transforming station. 516

station service transformer (generating stations electric power system). A transformer that supplies power from a station high voltage bus to the station auxiliaries and also to the unit auxiliaries during unit startup and shutdown or when the unit auxiliaries transformer is not available, or both. 381

station, steam-electric. An electric generating station utilizing steam for the motive force of its prime movers. 112

station, storage. A hydroelectric generating station associated with a water storage reservoir. 112

station-to-station call (telephone switching systems). A call intended for a designated main station. 55

station-type cubicle switchgear (metal-clad and station-type cubicle switchgear). Metal-enclosed power switchgear characterized by the following required features: (1) The main switching and interrupting device is of the stationary mounted type, composed of a primary circuit compartment and a secondary or mechanism compartment; arranged with gang-operated isolating switches that are mechanically interlocked with main switching or interrupting device. (2) Each phase for the major parts of the primary circuit switching or interrupting devices, buses, and line-to-ground potential transformers are completely enclosed (or segregated) by grounded metal barriers that have no intentional openings between compartments. Specifically included are mechanically interlocked doors in front of or a part of the primary circuit compartment of the circuit switching and interrupting device so that when the group operated isolating switches are closed no primary parts can be exposed by the attempted opening of the interlocked doors. (3) All live parts are enclosed within grounded metal compartments. (4) Primary bus conductors and connections are bare. (5) Mechanical interlocks are provided for proper operating sequence under normal operating conditions. (6) Secondary control devices and their wiring are isolated by grounded metal barriers from all primary circuit elements with the exception of short lengths of wire such as at instrument transformer terminals. (7) The doors to the secondary or mechanism compartment of the primary switching or interrupting device are to provide access to the secondary or control equipment within the housing without danger to exposure to the primary circuit parts. *Note:* Auxiliary vertical sections may be required for mounting devices or for use as bus transition. 572

station-type cubicle switchgear (power switchgear). Metal-enclosed power switchgear composed of the following equipment: (1) primary power equipment for each phase segregated and enclosed by metal, (2) stationary-mounted power circuit breakers, (3) group-operated switches, interlocked with the circuit breakers, for isolating the circuit breakers, (4) bare bus and connections, (5) instrument transformers, (6) control wiring and accessory devices. 103

station-type transformer (power and distribution transformer). A transformer designed for installation in a station or substation. 53

statistical delay (gas tubes) (electron device). The time lag from (1) the application of the specified voltage to initiate the discharge to (2) the beginning of breakdown. *See:* **gas tube.** 190

statistical descriptors (exceedance levels, L-levels) (audible noise measurements). Many sounds have sound-pressure levels that are not constant in time and cannot, without qualification, be adequately characterized by a single value of sound level. One method for dealing with fluctuating or intermittent sounds is to examine the sound level statistically as a function of time. Statistical descriptors are often applied to A-weighted sound levels, and are called exceedance levels or L-levels. For example, the L_{10} is the A-weighted sound level exceeded for 10 percent of the time over a specified time period. The other 90 percent of the time, the sound level is less than L_{10}. Similarly, the L_{50} is the sound level exceeded 50 percent of the time; the L_{90} is the sound level exceeded 90 percent of the time, etcetera. 462

statistical test model (software). A model that relates program faults to the input data set (or sets) which cause them to be encountered. The model also gives the probability that these faults will cause the program to fail. *See:* **data; model; program faults.** 434

stator (1)(watthour meter). An assembly of an induction watthour meter, which consists of a voltage circuit, one or more current circuits, and a combined magnetic circuit so arranged that their joint effect, when energized, is to exert a driving torque on the rotor by the reaction with currents induced in an individual or common conducting disk. 212

(2) (rotating machinery). The portion that includes and supports the stationary active parts. The stator includes the stationary portions of the magnetic circuit and the associated winding and leads. It may, depending on the design, include a frame or shell, winding supports, ventilation circuits, coolers, and temperature detectors. A base, if provided, is not ordinarily considered to be part of the stator. 63

stator coil (rotating machinery). A unit of a winding on the stator of a machine. *See:* **stator.** 63

stator coil pin (rotating machinery). A rod through an opening in the stator core, extending beyond the faces of the core, for the purpose of holding coils of the stator winding to a desired position. *See:* **stator.** 63

stator core (rotating machinery). The stationary magnetic-circuit of an electric machine. It is commonly an assembly of laminations of magnetic steel, ready for winding. *See:* **stator.** 63

stator-core lamination (rotating machinery). A sheet of material usually of magnetic steel, containing teeth and winding slots, or containing pole structures, that forms the stator core when assembled with other identical or similar laminations. *See:* **stator.** 63

stator frame (rotating machinery). The supporting structure holding the stator core or core assembly. *Note:* In certain types of machines, the stator frame may be made integral with one end shield. *See:* **stator.** 63

stator iron (rotating machinery). A term commonly used for the magnetic steel material or core of the stator of a machine. *See:* **stator.** 63

stator resistance starting (rotating machinery). The process of starting a motor at reduced voltage by connecting the primary winding initially in series with starting resistors which are short-circuited for the running condition. 63

stator shell. *See:* **shell, stator.**

stator winding (rotating machinery). A winding on the stator of a machine. *See:* **stator.** 63

stator winding copper (rotating machinery). A term commonly used for the material or conductors of a stator winding. *See:* **stator.** 63

status (supervisory control, data acquisition, and automatic control). Information describing a logical state of a point or equipment. 570

status flag (binary floating point arithmetic)(radix-independent floating-point arithmetic). A variable that may take two states, set and clear. A user may clear a flag, copy it, or restore it to a previous state. When set, a status flag may contain additional system-dependent information, possibly inaccessible to some users. The operations of ANSI/IEEE Std 754-1985 and ANSI/IEEE Std 854-1987 may as a side effect set some of the following flags: inexact result, underflow, overflow, divide by zero, and invalid operation. 469, 588

status memory (sequential events recording systems). The memory which contains the most recently scanned status of all inputs. 48, 58

status point, supervisory control (power-system communication). *See:* **supervisory control point, status.**

STC. *See:* **sensitivity time control.**

steady current. A current that does not change with time. 210

steady state (1) (signal-transmission system). That in which some specified characteristic of a condition, such as value, rate, periodicity, or amplitude, exhibits only negligible change over an arbitrarily long period of time. *See:* **signal.** 105

(2) (industrial control). The condition of a specified variable at a time when no transients are present. *Note:* For the purpose of this definition, drift is not considered to be a transient. *See:* **control system, feedback.** 206

(3) (cable insulation materials). Conditions of current in the material attained when the difference between the maximum and minimum current observed during four consecutive hourly readings is less than 5 percent of the minimum current. 97

(4) (excitation control systems). That in which some specified characteristic of a condition, such as value, rate, periodicity, or amplitude, exhibits only negligible change over an arbitrarily long interval of time. *Note:* It may describe a condition in which some characteristics are static, others dynamic. 353

steady-state condition (fiber optics). *See:* **equilibrium mode distribution.**

steady-state deviation (control). *See:* **deviation, steady-state.**

steady-state governing load band (hydraulic turbines). The magnitude of the envelope of cyclic load variations caused by the governing system, expressed as a percent of rated power output, when the generating unit is operating in parallel with other generators and under steady-state load demand. *Syn:* **power stability index.** 8

steady-state governing speed band (hydraulic turbines). The magnitude of the envelope of the cyclic speed variations caused by the governing system, expressed as a percent of rated speed when the generating unit is operating independently and under steady-state load demand. *Syn:* **speed stability index.** 8

steady-state incremental speed regulation (excluding the effects of deadband) (gas turbines). At a given steady-state speed and power, the rate of change of the steady-state speed with respect to the power output. It is the slope of the tangent to the steady-state speed versus power curve at the point of power output under consideration. It is the difference in steady-state speed, expressed in percent of rated speed, for any two points on the tangent, divided by the corresponding difference in power output, expressed as a fraction of the rated power output. For the basis of comparison, the several points of power output at which the values of steady-state incremental speed regulation are derived are based upon rated speed being obtained at each point of power output. 94, 98, 58

steady-state oscillation. A condition in which some aspect of the oscillation is a continuing periodic function. 176

steady-state short-circuit current (synchronous machine). The steady-state current in the armature winding when short-circuited. 63

steady-state speed regulation (straight condensing and noncondensing steam turbines, nonautomatic extraction turbines, hydro-turbines, and gas turbines). The change in steady-state speed, expressed in percent of rated speed, when the power output of the turbine operating isolated is gradually reduced from rated power output to zero power output with unchanged settings of all adjustments of the speed-governing system. *Note:* Speed regulation is considered positive when the speed increases with a decrease in power output. *See:* **asynchronous machine; speed-governing system.** 94, 98, 58

steady-state stability. A condition that exists in a power system if it operates with stability when not subjected to an aperiodic disturbance. *Note:* In practical systems, a variety of relatively small aperiodic disturbances may be present without any appreciable effect upon the stability, as long as the resultant rate of change in load is relatively slow in comparison with the natural frequency of oscillation of the major parts of the system or with the rate of change in field flux of the rotating machines. 64

steady-state stability factor (system or part of a system). The ratio of the steady-state stability limit to the nominal power flow at the point of the system to which the stability limit is referred. *See:* **stability factor.** 64

steady-state stability limit (steady-state power limit). The maximum power flow possible through some particular point in the system when the entire system or the part of the system to which the stability limit refers is operating with steady-state stability. 64

steady-state temperature rise (grounding device). The maximum temperature rise above ambient which will be attained by the winding of a device as the result of the flow of rated continuous current under standard operating conditions. It may be expressed as an average or a hot-spot winding rise. 91

steady-state value. The value of a current or voltage after all transients have decayed to a negligible value. For an alternating quantity, the root-mean-square value in the steady state does not vary with time. *See:* **asynchronous machine.** 63

steady voltage. *See:* **steady-current.**

steam capability (power operations). The maximum net capability of steam generating units which can be obtained under normal operating practices for a given period of time as calculated based on design or test data or as demonstrated by total plant tests. The limitation on steam capability may be electrical or mechanical in nature. 516

steam-electric station (power operations). An electric generating station utilizing steam for the motive force of its prime movers. 516

steam turbine-electric drive. A self-contained unit of power generation and application in which the power generated by a steam turbine is transmitted electrically by means of a generator and a motor (or multiples of these) for propulsion purposes. *Note:* The prefix steam turbine-electric is applied to ships, locomotives, cars, buses, etcetera, that are equipped with this drive. *See:* **electric locomotive; electric propulsion system.** 328

steel container (storage cell). The container for the element and electrolyte of a nickel-alkaline storage cell. This steel container is sometimes called a can. *See:* **battery (primary or secondary).** 328

steerable-beam antenna. An antenna with a nonmoving aperture for which the direction of the major lobe can be changed by electronically altering the aperture excitation or by mechanically moving a feed of the antenna. 111

steering compass. A compass located within view of a steering stand, by reference to which the helmsman holds a ship on the set course.

Stefan-Boltzmann law (illuminating engineering). The statement that the radiant exitance of a blackbody is proportional to the fourth power of its absolute temperature; that is,

$$M = \sigma T^4$$

Note: The currently recommended value of the Stefan-Boltzmann constant σ is $5.67032 \times 10^{-8} W \cdot m^{-2} \cdot K^{-4}$. 167

stellar guidance (navigation aid terms). Guidance by means of celestial bodies, particularly the stars. 526

stellar-inertial navigation equipment. *See:* **celestial-inertial navigation equipment.**

step (1) (pulse techniques). A waveform that, from the observer's frame of reference, approximates a Heaviside (unit step) function. *See:* **signal, unit step.** 185

(2) (computing systems). (A) One operation in a computer routine. (B) To cause a computer to execute one operation. *See:* **single step.** 255, 77

(3) (pulse terms) (single transition). A transition waveform which has a transition duration which is negligible relative to the duration of the waveform epoch or to the duration of its adjacent first and second nominal states. 254

step-back relay (electric installations on shipboard). A Relay which operates to limit the current peaks of a motor when the armature or line current increases. A step-back relay may, in addition, operate to remove such limitations when the cause of the high current has been removed. 3

step-by-step switch. A bank-and-wiper switch in which the wipers are moved by electromagnet ratchet mechanisms individual to each switch. *Note:* This type of switch may have either one or two types of motion. 328

step-by-step system. An automatic telephone switching system that is generally characterized by the following features. (1) The selecting mechanisms are step-by-step switches. (2) The switching pulses may either actuate the successive selecting mechanisms directly or may be received and stored by controlling mechanisms that, in turn, actuate the selecting mechanisms. 328

step change (control) (industrial control) (step function). An essentially instantaneous change of an input variable from one value to another. *See:* **control system, feedback.** 206

step compensation (correction) (industrial control). The effect of a control function or a device that will cause a step change in an other function when a predetermined operating condition is reached. 206

step-down transformer (power and distribution transformer). A transformer in which the power transfer is from a higher voltage source circuit to a lower voltage circuit. 53

step index optical waveguide (fiber optics). An optical waveguide having a step index profile. *See:* **step index profile.** 433

step index profile (fiber optics). A refractive index profile characterized by a uniform refractive index within the core and a sharp decrease in refractive index at the core-cladding interface. *Note:* This corresponds to a power-law profile with profile parameter, g, approaching infinity. *See:* **critical angle; dispersion; graded index profile; mode volume; multimode optical waveguide; normalized frequency; optical waveguide; refractive index (of a medium); total internal reflection.** 433

stepless (electrical heating applications to melting furnaces and forehearths in the glass industry). Power modulation by means of a device, such as a saturable reactor or thyristor that provides essentially infinite resolution in output voltage, current, or power. 520

step line-voltage change (power supplies). An instantaneous change in line voltage (for example, 105-125 volts alternating current): for measuring line regulation and recovery time. 186

step load change (power supplies). An instantaneous change in load current (for example, zero to full load): for measuring the load regulation and recovery time. 186

stepped (electrical heating applications to melting furnaces and forehearths in the glass industry). Power modulation by means of discrete voltage steps, such as with a tapped transformer. 520

stepped antenna. *See:* **zoned antenna.**

stepped-gate structure (metal-nitride-oxide field-effect transistor). Also source-drain protected structure: a variant of the metal-nitride-oxide semiconductor (MNOS) transistor whose gate dielectric along the channel is divided into two or three parts. One portion has the standard MNOS layer sequence of the memory device. On one or either side of the memory portion, particularly covering the lines where source and drain junction emerge at the silicon surface, is a gate dielectric that is used for the threshold insulated-gate field-effect transistor (IGFET) in a given technology. 386

stepped leader (lightning). A series of discharges emanating from a region of charge concentration at short time intervals. Each discharge proceeds with a luminescent tip over a greater distance than the previous one. *See:* **direct-stroke protection (lightning).** 64

stepped wave (converter characteristics)(self-commutated converters). The waveform obtained from the summation of any number of square waves of the same frequency, each displaced in time from the others. The square waves are often uniformly displaced in time, but are not necessarily of equal amplitudes. An example is shown below.

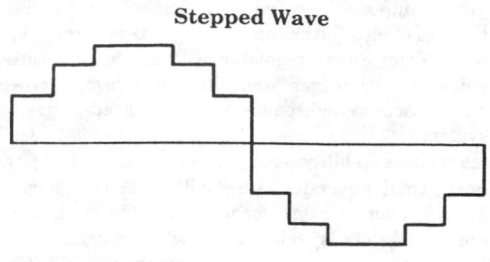

Stepped Wave

584

stepping life (for attenuator variable in fixed steps). Number of times to switch from any selected position to any other selected positions, after which the residual and incremental characteristic insertion loss remain within the specified repeatability. 110

stepping relay (1) (general). A multiposition relay in which moving wiper contacts mate with successive sets of fixed contacts in a series of steps, moving from one step to the next in successive operations of the relay. *See:* **relay.** 259

(2) (rotary type). A relay having many rotary positions, ratchet actuated, moving from one step to the next in successive operations, and usually operating its contacts by means of cams. There are two forms: (A) **directly driven,** where the forward motion occurs on energization, and (B) **indirectly (spring) driven,** where a spring produces the forward motion on pulse cessation. *Note:* The term is also incorrectly used for **stepping switch.** 341

stepping relay, spring-actuated. A stepping relay that is cocked electrically and operated by spring action. *See:* **relay.** 259

step response. A criterion of the dynamic response of an instrument when subjected to an instantaneous change in measured quantity from one value to another. *See:* **accuracy rating of an instrument.** 295

step response, G(t) (high voltage testing). The step response G(t) of a measuring system is the output as a function of time t when the input is a voltage or current step. A convenient form is the "unit step response g(t)" in which the constant value of the output magnitude is denoted as unity when that magnitude, multiplied by the corresponding scale factor, equals the input step. 150

step-response time. The time required for the end device to come to rest in its new position after an abrupt change to a new constant value has occurred in the measured signal. *See:* **accuracy rating of an instrument.** 295

step speed adjustment (industrial control). The speed drive can be adjusted in rather large and definite steps between minimum and maximum speed. *See:* **electric drive.** 206

step-stress test (reliability). A test consisting of several stress levels applied sequentially, for periods of equal duration, to a (one) sample. During each period a stated stress-level is applied and the stress level is increased from one step to the next. *See:* **reliability.** 182, 164

step twist, waveguide (waveguide components). A waveguide twist formed by abruptly rotating about the waveguide longitudinal axis, one or more waveguide sections each nominally a quarter wavelength long.
166

step-up transformer (power and distribution transformer). A transformer in which the power transfer is from a lower voltage source circuit to a higher voltage circuit.
53

step voltage (conductor stringing equipment)(measuring potential of a ground system). The potential difference between two points on the earth's surface separated by a distance of one pace (assumed to be 1 m, one meter) in the direction of maximum potential gradient. This potential difference could be dangerous when current flows through the earth or material upon which a workman is standing, particularly under fault conditions. *Syn:* **step potential.**
431,313

step-voltage regulator (power and distribution transformer). A regulating transformer in which the voltage of the regulated circuit is controlled in steps by means of taps and without interrupting the load. *Note:* Such units are generally 833 kVA (output) and below, single-phase; or 2500 kVA (output) and below, three-phase.
53

step-voltage test (rotating machinery). A controlled overvoltage test in which designated voltage increments are applied at designated times. Time increments may be constant or graded. *See:* **graded-time step-voltage test; asynchronous machine.**
63

step wedge.* *See:* **gray scale.**

*Deprecated

stepwise refinement (software). A system development methodology in which data definitions and processing steps are defined broadly at first and then with increasing detail. *See:* **bottom-up; data; hierarchical decomposition; system development methodology; top-down.**
434

steradian (1) (metric practice). The solid angle which, having its vertex in the center of a sphere, cuts off an area of the surface of the sphere equal to that of a square with sides of length equal to the radius of the sphere.
21

(2) (sr) (laser-maser). The unit of measure for a solid angle. There are 4π sr in a sphere.
363

stereophonic (adjective) (frequency modulation). Pertains to audio information carried by a plurality of channels arranged to afford the listener a sense of the spatial distribution of the sound sources. *Note:* A stereophonic receiver responds to both the L+R main channel and the L−R subcarrier channel of a composite stereophonic signal, so that the one output contains substantially only L information, and the other only R. In addition to the main channel, stereophonic program modulation requires transmission of a 19 kHz pilot signal and the sidebands of a suppressed 38 kHz subcarrier carrying L−R information. This combination is called the composite signal, and it may be used alone or with other subcarrier (SCA) signals to frequency modulate the RF carrier. After pre-emphasis, the left and right channels are added for main channel

information. The right-channel program material is subtracted from the left to derive a difference signal that then amplitude modulates a 38 kHz subcarrier. The subcarrier is suppressed, divided by two, and transmitted as a 19 kHz pilot signal to facilitate demodulation of the suppressed carrier information at the receiver.
16

stick (power line maintenance). A type of insulating tool used in various operations of live-line work. *Syn:* **work-stick; pole; work pole; hot stick.**
458

stick circuit. A circuit used to maintain a relay or similar unit energized through its own contact.
328

stickiness. The condition caused by physical interference with the rotation of the moving element. *See:* **moving element of an instrument.**
280

sticking voltage (luminescent screen). The voltage applied to the electron beam below which the rate of secondary emission from the screen is less than unity. The screen then has a negative charge that repels the primary electrons. *See:* **cathode-ray tubes.**
244, 190

stick (hook) operation (power switchgear). Manual operation of a switching device by means of a switch stick.
103

stiction (1). The force in excess of the coulomb friction required to start relative motion between two surfaces in contact.
329

(2) (static friction) (industrial control). The total friction that opposes the start of relative motion between elements in contact. *See:* **control system, feedback.**
206

stiffness (industrial control). The ability of a system or element to resist deviations resulting from loading at the output. *See:* **control system, feedback.**
56, 206

stiffness coefficient. The factor K (also called spring constant) in the differential equation for oscillatory motion $M\ddot{x} + B\dot{x} + Kx = 0$.
56

stilb (illuminating engineering). A centimeter-gram-second (CGS) unit of luminance. One stilb equals one candela per square centimeter. The use of this term is deprecated.
167

Stiles-Crawford effect (illuminating engineering). The reduced luminous efficiency of rays entering the peripheral portion of the pupil of the eye. This effect applies only to cones and not to rod visual cells. Hence, there is no Stiles-Crawford effect in scotopic vision.
167

stimulated emission (1) (fiber optics). Radiation emitted when the internal energy of a quantum mechanical system drops from an excited level to a lower level when induced by the presence of radiant energy at the same frequency. An example is the radiation from an injection laser diode above lasing threshold. *See:* **spontaneous emission.**
433

(2) (laser-maser). The emission of radiation at a given frequency caused by an applied external radiation field of the same frequency.
363

stimulus (industrial control). Any change in signal that affects the controlled variable: for example, a disturbance or a change in reference input. *See:* **control system, feedback.**   219, 206, 54

stirring effect (induction heater usage). The circulation in a molten charge due to the combined forces of motor and pinch effects. *See:* **induction heating; motor effect; pinch effect.**   14, 114

stop (limit stop). A mechanical or electric device used to limit the excursion of electromechanical equipment. *See:* **limiter circuit.**   9

stop band (circuits and systems). A band of frequencies that pass through a filter with a substantial amount of loss (relative to other frequency bands such as a pass band).   67

stop-band ripple (circuits and systems). The difference between maxima and minima of loss in a filter stop band.   67

stop dowel (rotating machinery). A pin fitted into a hole to limit motion of a second part.   63

stop element (data transmission). In a character transmitted in a start-stop system, the last element in each character, to which is assigned a minimum duration, during which the receiving equipment is returned to its rest condition in preparation for the reception of the next character. The stop element is a marking signal.   59

stop-go pulsing (telephone switching systems). A method of pulsing control wherein the pulsing operation may take place in stages, and the sending end is arranged to pulse the digits continuously unless or until the stop-pulsing signal is received. *Note:* When this occurs, the pulsing of the remaining digits is suspended until the sending end receives a start-pulsing signal.   55

stop joint (power cable joint). A joint which is designed to prevent any transfer of dielectric fluid between the cables being joined.   34

stop lamp (illuminating engineering). A lighting device giving a steady warning light to the rear of a vehicle or train of vehicles, to indicate the intention of the operator to diminish speed or to stop.   167

stop-motion switch (elevators). *See:* **machine final-terminal stopping device.**

stop or throttle valve(s) (control systems for steam turbine-generator units). Those valve(s) that normally provide fast interruption of the main energy input to the turbine. Throttle valves are sometimes used for turbine control during start-up. *Note:* The term stop valve is defined as an open or closed valve. A throttle valve has some portion of its opening through which it can modulate flow.   522

stopping device (5) (power system device function numbers). A control device used primarily to shut down an equipment and hold it out of operation. This device may be manually or electrically actuated, but excludes the function of electrical lockout on abnormal conditions. *See:* **lockout relay, device number 86.**   402

stopping off. The application of a resist to any part of a cathode or plating rack. *See:* **electroplating.**   328

stop-pulsing signal (telephone switching systems). A signal transmitted from the receiving end to the sending end of a trunk to indicate that the receiving end is not in a condition to receive pulsing.   55

stop-record signal (facsimile). A signal used for stopping the process of converting the electrical signal to an image on the record sheet. *See:* **facsimile signal (picture signal).**   12

stop signal (facsimile). A signal which initiates the transfer of a facsimile equipment condition from active to standby. *See:* **facsimile signal (picture signal).**   12

storable swimming or wading pool (National Electrical Code). A pool with a maximum dimension of 15 ft and a maximum wall height of 3 ft and is so constructed that it may be readily disassembled for storage and reassembled to its original integrity.   256

storage (electronic computation). (1) The act of storing information. (2) Any device in which information can be stored, sometimes called a memory device. (3) In a computer, a section used primarily for storing information. Such a section is sometimes called a memory or store (British). *Notes:* (A) The physical means of storing information may be electrostatic, ferroelectric, magnetic, acoustic, optical, chemical, electronic, electric, mechanical, etcetera, in nature. (B) Pertaining to a device in which data can be entered, in which it can be held, and from which it can be retrieved at a later time. *See:* **store.**   255, 77, 54

storage allocation (computing systems). The assignment of sequences of data or instructions to specified blocks of storage.   255, 77

storage assembly (storage tubes). An assembly of electrodes (including meshes) that contains the target together with electrodes used for control of the storage process, those that receive an output signal, and other members used for structural support. *See:* **storage tube.**   174

storage battery (National Electrical Code). A battery comprised of one or more rechargeable cells of the lead-acid, nickel-cadmium, or other rechargeable electrochemical types.   256

storage capacity. The amount of data that can be contained in a storage device. *Notes:* (1) The units of capacity are bits, characters, words, etcetera. For example, capacity might be "32 bits," "10 000 decimal digits," "16 384 words with 10 alphanumeric characters each." (2) When comparisons are made among devices using different character sets and word lengths, it may be convenient to express the capacity in equivalent bits, which is the number obtained by taking the logarithm to the base 2 of the number of usable distinguishable states in which the storage can exist. (3) The storage (or memory) capacity of a computer usually refers only to the internal storage section.   235

storage cell (secondary cell or accumulator) (1) (electric energy). A galvanic cell for the generation of electric energy in which the cell, after being discharged, may be restored to a fully charged condition by an electric current flowing in a direction opposite to the flow of current when the cell discharges. *See:* **electrochemistry.**   328

(2) (computing systems) (information). An elementary unit of storage, for example, a binary cell, a decimal cell. *See:* **electrochemistry.**   255, 77

storage device. A device in which data can be stored and from which it can be copied at a later time. The means of storing data may be chemical, electrical, mechanical, etcetera. *See:* **storage.**   235

storage element (storage tubes). An area of a storage surface that retains information distinguishable from that of adjacent areas. *Note:* The storage element may be a portion of a continuous storage surface or a discrete area such as a dielectric island. *See:* **storage tube.**
                                                                                                          174,125

storage-element equilibrium voltage (storage tubes). A limiting voltage toward which a storage element charges under the action of primary electron bombardment and secondary emission. At equilibrium voltage the escape ratio is unity. *Note:* **Cathode equilibrium voltage, second-crossover equilibrium voltage,** and **gradient-established equilibrium voltage** are typical examples. *See:* **charge-storage tube.**   174

storage-element equilibrium voltage, cathode (storage tubes). The storage element equilibrium voltage near cathode voltage and below first-crossover voltage. *See:* **charge-storage tube.**   174

storage-element equilibrium voltage, collector (storage tubes). *See:* **charge storage tube.**

storage-element equilibrium voltage, gradient established (storage tubes). The storage-element equilibrium voltage, between first- and second-crossover voltages, at which the escape ratio is unity. *See:* **charge-storage tube.**   174

storage-element equilibrium voltage, second-crossover (storage tubes). The storage-element equilibrium voltage at the second-crossover voltage. *See:* **charge-storage tube.**   174

storageid (microprocessor operating systems parameter types). An identifier for a block of data. The identifier is not guaranteed to be valid outside the allocating process and should not be passed between processes.   478

storage integrator (analog computer). An integrator used to store a voltage in the hold condition for future use while the rest of the computer assumes another computer control state. *See:* **electronic analog computer.**   9

storage life (gyro, accelerometer) (inertial sensor). The nonoperating time interval under specified conditions, after which a device will still exhibit a specified operating life and performance. *See:* **operating life.**   46

storage light-amplifier (optoelectronic device). *See:* **image-storage panel.**

storage medium. Any device or recording medium into which data can be stored and held until some later time, and from which the entire original data can be obtained.   207

storage protection (computing systems). An arrangement for preventing access to storage for either reading or writing, or both.   77

storage, reservoir (electric power systems). The volume of water in a reservoir at a given time.   112

storage station (power operations). A hydroelectric generating station associated with a water storage reservoir.   516

storage surface (storage tubes). The surface upon which information is stored. *See:* **storage tube.**
                                                                                                          174

storage temperature (1) (power supply). The range of environmental temperatures in which a power supply can be safely stored (for example, -40 to $+85$ degrees Celsius).   186

(2) (semiconductor device). The range of environmental temperatures in which a semiconductor device can be safely stored.   66

storage temperature (3) (light emitting diodes) (T). The temperature at which the device, without any power applied, is stored.   162

storage temperature range. The range of temperatures over which the Hall generators may be stored without any voltage applied, or without exceeding a specified change in performance.   107

storage time (storage tubes). *See:* **retention time, maximum; decay time; storage time.**

storage tube. An electron tube into which information can be introduced and read at a later time. *Note:* The output may be an electric signal and.or a visible image corresponding to the stored information.   174,125

store. (1) To retain data in a device from which it can be copied at a later time. (2) To put data into a storage device. (3) British synonym for storage. *See:* **storage.**
                                                                                                          235

store-and-forward switching (data communication). A method of switching whereby messages are transferred directly or with interim storage, each in accordance with its own address. *See:* **packet switching.**
                                                                                                          12

store-and-forward switching system (telephone switching systems). A switching system for the transfer of messages, each with its own address or addresses, in which the message can be stored for subsequent transmission.   55

stored energy indicator (power switchgear). An indicator which visibly shows that the stored energy mechanism is in the charged or discharged position.
                                                                                                          103

stored-energy operation (power switchgear). Operation by means of energy stored in the mechanism itself prior to the completion of the operation and sufficient to complete it under predetermined conditions. *Note:* This kind of operation may be subdivided according to: (1) how the energy is stored (spring, weight, etcetera), (2) how the energy originates (manual, electric, etcetera), (3) how the energy is released (manual, electric, etcetera).   103

stored logic (telephone switching systems). Instructions in memory arranged to direct the performance of predetermined functions in response to readout.
                                                                                                          55

stored program (telephone switching systems). A program in memory that a processor can execute.
                                                                                                          55

stored-program computer. A digital computer that, under control of internally stored instructions, can synthesize, alter, and store instructions as though they were data and can subsequently execute these new instructions.   255, 77

stored program control (telephone switching systems). A system control using stored logic.   55

stored-program switching system (telephone switching systems). An automatic switching system having stored program control.   55

storm guys. Anchor guys, usually placed at right angles to direction of line, to provide strength to withstand transverse loading due to wind. *See:* **tower.**   64

storm loading. The mechanical loading imposed upon the components of a pole line by the elements, that is, wind and/or ice, combined with the weight of the components of the line. *Note:* The United States has been divided into three loading districts, light, medium, and heavy, for which the amounts of wind and.or ice have been arbitrarily defined. *See:* **cable; open wire.**   328

straggling, energy (semiconductor radiation detectors). The random fluctuations in energy loss whereby those particles having the same initial energy lose different amounts of energy when traversing a given thickness of matter. (This process leads to the broadening of spectral lines.)   23, 119

straight condensing turbine (control systems for steam turbine-generator units). All the steam enters the turbine at one pressure and all the steam leaves the turbine exhaust at a pressure below atmospheric pressure.   522

straight-cut control system (numerically controlled machines). A system in which the controlled cutting action occurs only along a path parallel to linear, circular, or other machine ways.   207

straightforward trunking (manual telephone switchboard system). That method of operation in which the *A* operator gives the order to the *B* operator over the trunk on which talking later takes place.   328

straight joint (power cable joint). A cable joint used for connecting two lengths of cable, each of which consists of one or more conductors.   34

straight-line coding (computing systems). Coding in which loops are avoided by the repetition of parts of the coding when required.   255, 77

straight noncondensing turbine (control systems for steam turbine-generator units). All the steam enters the turbine at one pressure and all the steam leaves the turbine exhaust at a pressure equal to or greater than atmospheric pressure.   522

straight-seated bearing (rotating machinery) (cylindrical bearing). A journal bearing in which the bearing liner is constrained about a fixed axis determined by the supporting structure. *See:* **bearing.**   63

straight storage system (electric power supply). A system in which the electrical requirements of a car are supplied solely from a storage battery carried on the car. *See:* **axle generator system.**   328

strain-bus structure (substation rigid-bus structures). A bus structure comprised of flexible conductors supported by strain insulators.   566

strain element (strain wire) (of a fuse) (power switchgear). That part of the current-responsive element, connected in parallel with the fusible element in order to relieve it of tensile strain. *Note:* The fusible element melts and severs first and then the strain element melts during circuit interruption.   103, 443

strain insulator. An insulator generally of elongated shape, with two transverse holes or slots.   261

strain stick (power line maintenance). An insulating support tool that is used primarily to relieve mechanical loading at suspension and dead-end configurations so as to replace damaged insulators or hardware.   458

strain wire. *See:* **strain element (fuse).**

strand. (1) One of the wires, or groups of wires, of any stranded conductor. (2) One of a number of paralleled uninsulated conducting elements of a conductor which is stranded to provide flexibility in assembly or in operation. (3) One of a number of paralleled insulated conducting elements which constitute one turn of a coil in rotating machinery. The strands are usually separated electrically through all the turns of a multi-turn coil. Various types of transposition are commonly employed to reduce the circulation of current among the strands. A strand has a solid cross section, or it may be hollow to permit the flow of cooling fluid in intimate contact with the conductor (one form of "conductor cooling"). *See:* **conductor; rotor (rotating machinery); stator.**   63

stranded conductor. A conductor composed of a group of wires or of any combination of groups of wires. *Note:* The wires in a stranded conductor are usually twisted or braided together. *See:* **conductor.**   64

stranded wire. *See:* **stranded conductor.**

strand insulation (rotating machinery). The insulation on a strand or lamination or between adjacent strands or laminations which comprise a conductor.   63

strand restraining clamp (conductor stringing equipment). An adjustable circular clamp commonly used to keep the individual strands of a conductor in place and prevent them from spreading when the conductor is cut. *Syn:* **cable binding block; hose clamp; vise grip plier clamp.**   431

strand-to-strand test (rotating machinery). A test that is designed to apply a voltage of specified amplitude and waveform between the strands of a coil for the purpose of determining the integrity of the strand insulation.   63

strap, anode (magnetron). *See:* **anode strap.**

strapdown (gyro, accelerometer) (inertial sensor). A term which defines the condition when inertial sensors are directly mounted (without gimbals) to a vehicle to sense the linear and angular motion of the vehicle.   46

strap-down inertial navigation equipment (navigation aid terms). Inertial navigation equipment wherein the inertial sensors, (for example, gyros and accelerometers) are directly mounted to the vehicle, (eliminating the stable platform and gimbal system) to sense the linear and angular motion of the vehicle. *Notes:* (1) In this equipment, a computer utilizes gyro information to resolve the accelerations that are sensed along the carrier axes, and to refer these accelerations to an inertial frame of reference. Navigation is then accomplished in the same manner as in systems using a stable platform. (2) Also called strapped down.   526

strap key. A pushbutton circuit controller that is biased by a spring metal strip and is used for opening or closing a circuit momentarily.   328

strapping (multiple-cavity magnetrons). *See:* **anode strap; jumper.**

stray (circuits and systems). An element or occurrence usually not desired in a theoretical design, but unavoidable in a practical realization. For example, the relative proximity of wires can cause stray capacitance. *See:* **parasitic element.**   67

stray-current corrosion. Corrosion caused by current through paths other than the intended circuit or by an extraneous current in the earth. *See:* **cathodic corrosion; long-line current (corrosion); noble potential; sacrificial protection.**   205

stray light (in the eye) (illuminating engineering). Light from a source which is scattered onto parts of the retina lying outside the retinal image of the source.   167

stray load loss (synchronous machine). The losses due to eddy currents in copper and additional core losses in the iron, produced by distortion of the magnetic flux by the load current, not including that portion of the core loss associated with the resistance drop.   244, 63

stray losses (electronics power transformer). Those occurring in the core and case structure that result from the leakage flux of a transformer when supplying rated load current.   95

strays*. Electromagnetic disturbances in radio reception other than those produced by radio transmitting systems. *See:* **radio transmitter.**   328
*Obsolete

streamer (overhead-power-line corona and radio noise). A repetitiive corona discharge characterized by luminous filaments extending into the low electric field intensity region near either a positive or a negative electrode, but not completely bridging the gap.   411

stream flow. The quantity rate of water passing a given point. *See:* **generating station.**   64

streaming (audio and electroacoustics). Unidirectional flow currents in a fluid that are due to the presence of acoustic waves.   176

streaming potential (electrobiology). The electrokinetic potential gradient resulting from unit velocity of liquid forced to flow through a porous structure or past an interface. *See:* **electrobiology.**   192

streetlighting luminaire (illuminating engineering). A complete lighting device consisting of a light source together with its direct appurtenances such as globe, reflector, refractor, housing, and such support as is integral with the housing. The pole, post, or bracket is not considered part of the luminaire. *Note:* Modern streetlighting luminaires contain the ballasts for high intensity discharge lamps where they are used.   167

streetlighting unit (illuminating engineering). The assembly of a pole or lamp post with a bracket and a luminaire.   167

strength-duration (time-intensity) curve (medical electronics). A graph of the intensity curve of applied electrical stimuli as a function of the duration just needed to elicit response in an excitable tissue.   192

strength of a sound source (strength of a simple source). The maximum instantaneous rate of volume displacement produced by the source when emitting a wave with sinusoidal time variation. *Note:* The term is properly applicable only to sources of dimension small with respect to the wavelength.   176

stress-accelerated corrosion. Corrosion that is accelerated by stress.   205

stress analysis (class 1E static battery chargers and inverters). An electrical and thermal design analysis of component applications in specific circuits under the specified range of service conditions.   408

stress corrosion cracking. Spontaneous cracking produced by the combined action of corrosion and static stress (residual or applied).   205

stress relief. A predetermined amount of slack to relieve tension in component or lead wires.   284

stress test (class 1E static battery chargers and inverters). A type test performed on a sample equipment which "stresses" the equipment to the specified range of service conditions.   408

strike deposit (1) (electroplating). A thin film of deposited metal to be followed by other coatings. *See:* **electroplating.**   328

(2) (bath) (electroplating). An electrolyte used to deposit a thin initial film of metal. *See:* **electroplating.**   328

striking (1) (arc) (spark) (gas). The process of establishing an arc or a spark. *See:* **discharge (gas).**   244, 190

(2) (electroplating). The electrode position of a thin initial film of metal, usually at a high current density. *See:* **electroplating.**   328

striking current (gas tube). The starter-gap current required to initiate conduction across the main gap for a specified anode voltage.   244, 190

striking distance (1) (power switchgear). The shortest distance, measured through air, between parts of different polarities.   202

(2) (transformers). The shortest unobstructed distance measured through a dielectric medium such as liquid, gas, or vacuum: between parts of different electric potential.   53

(3) (outdoor apparatus bushings). The shortest tight string distance measured externally over the weather casing between the metal parts which have the operating line to ground voltage between them.   168

string (1)(microprocessor operating systems parameter types). A sequence of characters. In general IEEE Std 855 expects ASCII characters. However, the standard does not preclude other character sets.   478

(2)(software). A linear sequence of entities such as characters or physical elements.   434

stringing (conductor stringing equipment). The pulling of pilot lines, pulling lines and conductors over travelers supported on structures of overhead transmission lines. Quite often, the entire job of stringing conductors is referred to as stringing operations, beginning with the planning phase and terminating after the conductors have been installed in the suspension clamps. 431

string-shadow instrument. An instrument in which the indicating means is the shadow (projected or viewed through an optical system) of a filamentary conductor, the position of which in a magnetic or an electric field depends upon the measured quantity. *See:* **instrument.** 328

strip (electroplating. A solution used for the removal of a metal coating from the base metal. *See:* **electroplating.** 328

strip chart recorder (analog computers). A recorder in which one or more records are made simultaneously as a function of time. 9

strip line (waveguide). A transmission line consisting of a strip conductor above or between extended parallel conducting surfaces. Some common examples of such transmission lines are: (1) Partially-shielded strip transmission line: a strip conductor above a single ground plane. (2) Shielded strip transmission line: a strip conductor between two ground planes. *See:* **slab line.** 267

stripper tank (electrorefining). An electrolytic cell in which the cathode deposit, for the production of starting sheets, is plated on starting-sheet blanks. *See:* **electrorefining.** 328

stripping (1) (electroplating) (mechanical). The removal of a metal coating by mechanical means. 328

(2) (chemical). The removal of a metal coating by dissolving it. 328

(3) (electrolytic). The removal of a metal coating by dissolving it or an underlying coating anodically with the aid of a current. *See:* **electroplating.** 328

stripping compound (electrometallurgy). Any suitable material for coating a cathode surface so that the metal electro deposited on the surface can be conveniently stripped off in sheets. *See:* **electrowinning.** 328

strip terminals (rotating machinery). A form of terminal in which the ends of the machine winding are brought out to terminal strips mounted integral with the machine frame or assembly. 63

strip-type transmission line (waveguides). A transmission line consisting of a conductor above or between extended conducting surfaces. *See:* **shielded strip transmission line; unshielded strip transmission line.** 179

strobing (pulse terms). A process in which a first pulse of relatively short duration interacts with a second pulse or other event of relatively longer duration to yield a signal which is indicative (typically, proportional to) the magnitude of the second pulse during the first pulse. 254

stroboscopic lamp (illuminating engineering). A flashtube designed for repetitive flashing. *Syn:* **strobe light.** 167

stroboscopic tube. A gas tube designed for the periodic production of short light flashes. *See:* **gas tube.** 255, 190

stroke (cable plowing). Peak to peak displacement of the plow blade tip. 52

stroke speed (scanning or recording line frequency) (facsimile). The number of times per minute, unless otherwise stated, that fixed line perpendicular to the direction of scanning is crossed in one direction by a scanning or recording spot. *Note:* In most conventional mechanical systems this is equivalent to drum speed. In systems in which the picture signal is used while scanning in both directions, the stroke speed is twice the above figure. *See:* **recording (facsimile); scanning (facsimile).** 12

strong typing (software). A programming language feature that requires the data type of each data object to be declared, and that precludes the application of operators to inappropriate data objects and, thereby, prevents the interaction of data objects of incompatible types. *See:* **data; data type; operator; programming language; type.** 434

structurally dual networks. A pair of networks such that their branches can be marked in one-ton-one correspondence so that any mesh of one corresponds to a cut-set of the other. Each network of such a pair is said to be the dual of the other. *See:* **network analysis.** 210

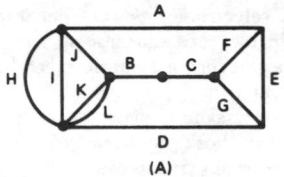
(A)

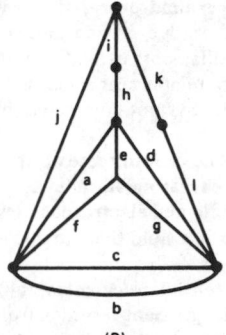
(B)

Structurally dual networks. For example, the mesh EFG in (A) corresponds to the cut-set efg in (B), the mesh bc in (B) to the cut-set BC in (A), and the mesh JA-EGCB in (A) to the cut-set jaegcb in (B).

—

structurally symmetrical network. A network that can be arranged so that a cut through the network produces two parts that are mirror images of each other. *See:* **network analysis.** 210

structure (power line maintenance). Material assembled to support conductors or associated apparatus, or both, used for transmission and distribution of electricity. Examples: service pole; tower.   458

structure base ground (conductor stringing equipment). A portable device designed to connect (bond) a metal structure to an electrical ground. Primarily used to provide safety for personnel during construction, reconstruction or maintenance operations. *Syn:* **butt ground; ground chain; structure ground; tower ground.**   431

structure conflict (1)(NESC). A line is so situated with respect to a second line that the overturning of the first line will result in contact between its supporting structures or conductors and the conductors of the second line, assuming that no conductors are broken in either line.   494

(2) (transmission and distribution). A line is so situated with respect to a second line that the overturning of the first line will result in contact between its supporting structures or conductors and the conductors of the second line, assuming that no conductors are broken in either line. *Exceptions:* Lines are not considered as conflicting under the following conditions: (A) Where one line crosses another. (B) Where two lines are on opposite sides of a highway, street, or alley and are separated by a distance not less than 60% of the height of the taller pole line and not less than 20 feet. *See:* **conductor clearance.**   262

structure, crossing. *See:* **crossing structure.**

structured design (software). A disciplined approach to software design which adheres to a specified set of rules based on principles such as top-down design, stepwise refinement, and data flow analysis. *See:* **data; software design; stepwise refinement; top-down design.**   434

structured program (software). A program constructed of a basic set of control structures, each one having one entry point and one exit. The set of control structures typically includes: sequence of two or more instructions, conditional selection of one of two or more instructions or sequences of instructions, and repetition of an instruction or a sequence of instructons. *See:* **control structure; instruction; program.**   434

structured programming (software). (1) A well-defined software development technique that incorporates top-down design and implementation, and strict use of structured program control constructs. (2) Loosely, any technique for organizing and coding programs that reduces complexity, improves clarity, and facilitates debugging and modification. *See:* **complexity; debugging; implementation; modification; program; software; structured program; top-down design.**   434

structured programming language (software). A programming language that provides the structured program constructs and that facilitates the development of structured programs. *See:* **programming language; structured program.**   434

structure, snub. *See:* **snub structure.**

structures, safety class (Class 1E power systems). Structures designed to protect Class 1E equipment against the effects of design basis events.   102

stub (1) (radial) feeder (power switchgear). A feeder that connects a load to its only source of power.   103

(2) (software). (A) A dummy program module used during the development and testing of a higher-level module. (B) A program statement substituting for the body of a program unit and indicating that the unit is or will be defined elsewhere. *See:* **program; program module; testing.**   434

stub antenna. A short thick monopole.   111

stub feeder (radial feeder). A feeder that connects a load to its only source of power.   103, 202

stub-multiple feeder (power switchgear). A feeder that operates as either a stub or a multiple feeder.   103

stub shaft (rotating machinery). A separate shaft not carried in its own bearings and connected to the shaft of a machine. *See:* **rotor (rotating machinery).**   63

stub-supported coaxial. A coaxial whose inner conductor is supported by means of short-circuited coaxial stubs. *See:* **waveguide.**   328

stub tuner. A stub that is terminated by movable short-circuiting means and used for matching impedance in the line to which it is joined as a branch. *See:* **waveguide.**   179

stub, waveguide (waveguide components). A section of waveguide or transmission line joined to the main guide or transmission line and containing an essentially nondissipative termination.   166

stud (of a switching device) (power switchgear). A rigid conductor between a terminal and a contact.   103

stuffing box (watertight gland). A device for use where a cable passes into a junction box or other piece of apparatus and is so designed as to render the joint watertight.   64

stylus (electroacoustics). A mechanical element that provides the coupling between the recording or the reproducing transducer and the groove of a recording medium. *See:* **phonograph pickup.**   176

stylus drag (needle drag) (electroacoustics). An expression used to denote the force resulting from friction between the surface of the recording medium and the reproducing stylus. *See:* **phonograph pickup.**   176

stylus force (electroacoustics). The vertical force exerted on a stationary recording medium by the stylus when in its operating position. *See:* **phonograph pickup.**   176

subaddress (a) (subroutines for CAMAC). The symbol *a* represents an integer which is the subaddress component of a CAMAC address.   410

subassembly (electric and electronics parts and equipments). Two or more basic parts which form a portion of an assembly or a unit, replaceable as a whole, but having a part or parts which are individually replaceable. *Notes:* (1) The application, size, and construction of an item may be factors in determining whether an item is regarded as a unit, an assembly, a subassembly, or a basic part. A small electric motor might be considered as a part if it is not normally subject to disassembly. (2) The distinction between an assembly and a subassembly is not always exact: an assembly in one instance may be a subassembly in another where it forms a portion of an assembly. Typical examples: filter network, terminal board with mounted parts.

17

subcarrier (facsimile). A carrier which is applied as a modulating wave to modulate another carrier.

12

subclutter visibility (radar). The ratio by which the target echo power may be weaker than the coincident clutter echo power and still be detected with specified detection and false alarm probabilities. Target and clutter powers are measured on a single pulse return and all target radial velocities are assumed equally likely.

13

subdivided capacitor (condenser box). A capacitor in which several capacitors known as sections are so mounted that they may be used individually or in combination.

210

subfeeder. A feeder originating at a distribution center other than the main distribution center and supplying one or more branch-circuit distribution centers. *See:* **feeder.**

328

subharmonic (data transmission). A sinusoidal quantity having a frequency which is an integral submultiple of the fundamental frequency of a periodic quantity to which it is related. For example, a wave the frequency of which is half the fundamental frequency of another wave is called the second subharmonic of that wave.

59

subharmonic detector (series capacitors). A device that detects subharmonic current of specified frequency and duration and initiates an alarm system or corrective action.

474

subharmonic protector (series capacitor). A device to detect subharmonic current of a specified frequency and duration to initiate closing of the capacitor bypass switches.

86

subject copy (facsimile). The material in graphic form which is to be transmitted for facsimile reproduction. *See:* **facsimile (electrical communication).**

12

subjective brightness (illuminating engineering). The subjective attribute of any light sensation giving rise to the percept of luminous magnitude, including the whole scale of qualities of being bright, light, brilliant, dim, or dark. *Note:* The term brightness often is used when referring to the measurable luminance. While the context usually makes it clear as to which meaning is intended, the preferable term for the photometric quality is luminance, thus reserving brightness for the subjective sensation. *See:* **saturation of a perceived color; luminance.**

167

submarine cable. A cable designed for service under water. *Note:* Submarine cable is usually a lead-covered cable with a steel armor applied between layers of jute.

64

submerged-resistor induction furnace. A device for melting metal comprising a melting hearth, a depending melting channel closed through the hearth, a primary induction winding, and a magnetic core which links the melting channel and the primary winding. *See:* **induction heating.**

14

submersible (rotating machinery) (industrial control). So constructed as to be successfully operable when submerged in water under specified conditions of pressure and time. *See:* **constant current transformer.**

103, 63, 3, 53

submersible enclosure (electric installations on shipboard). An enclosure constructed so that the equipment within it will operate successfully when submerged in water under specified conditions of pressure and time.

3

submersible entrance terminals (cableheads) (of distribution oil cutouts) (power switchgear). A hermetically sealable entrance terminal for the connection of cable having a submersible sheathing or jacket.

103, 443

submersible fuse (1) (subway oil cutout). *See:* **submersible; fuse.**

103

(2) (high-voltage switchgear). A fuse that is so constructed that it will operate successfully when submerged in water under specified conditions of pressure and time.

443

submersible transformer (power and distribution transformer). A transformer so constructed as to be successfully operable when submerged in water under predetermined conditions of pressure and time.

53

subnormality (electrical depression) (electrobiology). The state of reduced electrical sensitivity after a response or succession of responses. *See:* **excitability (electrobiology).**

192

subnormal number (radix-independent floating-point arithmetic). A nonzero floating-point number whose exponent is the precision's minimum and whose leading significant digit is zero.

588

subpanel (1) (photoelectric converter). Combination of photoelectric converters in parallel mounted on a flat supporting structure. *See:* **semiconductor.**

186

(2) (solar cells). A combination of solar cells in series/parallel matrix to provide current at array (bus) voltage.

113

subpost car frame (elevators). A car frame all of whose members are located below the car platform. *See:* **hoistway (elevator or dumbwaiter).**

328

subprogram (1) (software). A program unit that may be invoked by one or more other program units. Examples are "procedure", "function", "subroutine". *See:* **function; procedure; program; subroutine.**

434

(2) (test, measurement and diagnostic equipment). A part of a larger program which can be converted into machine language independently.

54

subreflector (antennas). A reflector other than the main reflector of a multiple-reflector antenna.

111

subroutine (1) (electronic computation). (A) In a routine, a portion that causes a computer to carry out a well-defined mathematical or logic operation. (B) A routine that is arranged so that control may be transferred to it from a master routine and so that at the conclusion of the subroutine, control reverts to the master routine. *Note:* Such a subroutine is usually called a closed subroutine. A single routine may simultaneously be both a subroutine with respect to another routine and a master routine with respect to a third. Usually control is transferred to a single subroutine from more than one place in the master routine and the reason for using the subroutine is to avoid having to repeat the same sequence of instructions in different places in the master routine.   210, 54

(2) (software). (A) A sequenced set of statements that may be used in one or more computer programs and at one or more points in a computer program. (B) A routine that can be part of another routine. (C) A subprogram that is invoked by a calling statement, that may or may not receive input values, and that returns any output values through parameter names, program variables, or mechanisms other than the subroutine name itself. *See:* **computer program; function; parameter; procedure; program variable; routine; subprogram.**   434

subroutine, open. A subroutine that must be relocated and inserted into a routine at each place it is used. Contrast with closed subroutine. *See:* **subroutine.**   54

subscriber carrier (telephone loop performance). A system that multiplexes customer signals to achieve pair gain in the loop plant. Usually, it consists of (1) An end office terminal (EOT); it interfaces with the end office (EO) through analog line appearances, one per each integrated loop. If the carrier is a digital system integrated in a digital end office, this terminal and its interfaces are replaced by much simpler all-digital equipment that may be integrated into the switching system. (2) A remote terminal (RT); it interfaces with the cable pairs to the customers' premises through analog interfaces, one per each implemented loop. (3) A transmission medium between the EOT (or EO in a digital integrated system) and the RT; it provides a control channel for internal EOT/EO-to-RT communication and communciation channels for customer traffic. A nonconcentrated system has as many customer channels as implemented loops, with fixed loop-channel assignments. A concentrated system has fewer channels than implemented loops, and changes loop-channel assignments to accomodate changing traffic patterns.   473

subscriber equipment (protective signaling). That portion of a system installed in the protected premises or otherwise supervised. *See:* **protective signaling.**   328

subscriber line (data transmission). A telephone line between a central office and a telephone station, private branch exchange, or other end equipment.   59

subscriber multiple. A bank of jacks in a manual switchboard providing outgoing access to subscriber lines, and usually having more than one appearance across the face of the switchboard.   328

subscriber set (customer set). An assembly of apparatus for use in originating or receiving calls on the premises of a subscriber to a communication or signaling service. *See:* **voice-frequency telephony.**   328

subsidence. *See:* **attenuation; damping.**

subsidence ratio (automatic control). A measure of the damping of a second-order linear oscillation, resulting from step or ramp forcing, expressed as the greater divided by the lesser of two successive excursions in the same direction from an ultimate steady-state value. *Note:* The term is also used loosely to describe the ratio of the first two consecutive peaks of any damped oscillation.   56

subsidiary communications authorization (SCA) subcarrier modulation. The FCC permits broadcasters to transmit privileged information and control signals on subcarriers as specified under the SCA but only when transmitted in conjunction with broadcast programming. *Notes:* (1) With monophonic broadcasting, the SCA service may use from 20 to 75 kHz with no restriction on the number of subcarriers, but the total SCA modulation of the RF (radio frequency) carrier must not exceed 30 percent and the crosstalk into the main channel must be at least 60 dB down. (2) With stereophonic broadcasting, the SCA service is limited to 53 75 kHz, 10 percent modulation of the carrier, and must still comply with the 60 dB crosstalk ratio. A 67 kHz subcarrier with ± 6 kHz modulation is often used.   16

subsidiary conduit (lateral). A terminating branch of an underground conduit run, extending from a manhole or handhole to a nearby building, handhole, or pole. *See:* **cable.**   328

subsonic frequency. *See:* **infrasonic frequency.**

substantial (transmission and distribution). So constructed and arranged as to be of adequate strength and durability for the service to be performed under the prevailing conditions.   262

substation (1)(generating stations electric power system). An area or group of equipment containing switches, circuit breakers, buses, and transformers for switching power circuits and to transform power from one voltage to another or from one system to another.   381

(2) (transmission and distribution). An assemblage of equipment for purposes other than generation or utilization, through which electric energy in bulk is passed for the purpose of switching or modifying its characteristics. Service equipment, distribution transformer installations, or other minor distribution or transmission equipment are not classified as substations. *Note:* A substation is of such size or complexity that it incorporates one or more buses, a multiplicity of circuit breakers, and usually is either the sole receiving point of commonly more than one supply circuit, or it sectionalizes the transmission circuits passing through it by means of circuit breakers. *See:* **alternating-current distribution; direct-current distribution.**   64

substitution error, direct-current-radio-frequency (bolometers). The error arising in the bolometric measurement technique when a quantity of direct-current or audio-frequency power is replaced by a quantity of radio-frequency power with the result that the different current distributions generate different temperature fields that give the bolometer element different values of resistance for the same amounts of power. This error is expressed as

$$\epsilon_s = \frac{\eta_e - \eta}{\eta}$$

where η_e is the effective efficiency of the bolometer units and η is the efficiency of the bolometer unit. *See:* **bolometric power meter.**   185

substitution error, dual-element. A substitution error peculiar to dual-element bolometer units that results from a different division of direct-current (or audio-frequency) and radio-frequency powers between the two elements.   185

substitution power (bolometers). The difference in bias power required to maintain the resistance of a bolometer at the same value before and after radio-frequency power is applied. Commonly, a bolometer is placed in one arm of a Wheatstone bridge that is balanced when the bias current (direct current and/or audio frequency) holds the bolometer at its nominal operating resistance. Following the application of the radio-frequency signal, the reduction in bias power is taken as a measure of the radio-frequency power. This reduction in the bias power is the substitution power and is given by

$$P = I_1^2 R - I_2^2 R$$

where I_1 and I_2 are the bias currents before and after radio-frequency power is applied and R is the nominal operating resistance of the bolometer. *See:* **bolometric power meter.**   115

substrate (1) (integrated circuit). The supporting material upon or within which an integrated circuit is fabricated or to which an integrated circuit is attached.   312

(2) (photovoltaic power system). Supporting material or structure for solar cells in a panel assembly. Solar cells are attached to the substrate.   186

subsurface corrosion. Formation of isolated particles of corrosion product(s) beneath the metal surface. This results from the preferential reaction of certain alloy constituents by inward diffusion of oxygen, nitrogen, sulfur, etcetera (internal oxidation).   205

subsurface switch (power switchgear). A submersible switching assembly suitable for application in a below-grade enclosure that does not allow space for personnel access.   489

sub-surface transformer (power and distribution transformer). A transformer utilized as part of an underground distribution system, connected below ground to high-voltage and low-voltage cables, and located below the surface of the ground.   53

subsynchronous reluctance motor. A form of reluctance motor that has the number of salient poles greater than the number of electrical poles of the primary winding, thus causing the motor to operate at a constant average speed that is a submultiple of its apparent synchronous speed. *See:* **asynchronous machine.**   63

subsynchronous satellite (communication satellite). A satellite, for which the sidereal period of rotation of the primary body about its own axis is an integral multiple of the mean sidereal period of revolution of the satellite about the primary body.   74

subsystem (1)(unique identification in power plants). A portion of a system containing two or more integrated components which, while not completely performing the specific function of a system, may be isolated for design, test, or maintenance.   544

(2)(unique identification in power plants and related facilities). A portion of a system containing two or more interrelated components which may be isolated for design, test, or maintenance.   545

(3) (nuclear power generating station protective systems). That part of the system which effects a particular protective function. These subsystems may include, but are not limited to those actuating: reactor shutdown: safety injection: containment isolation: emergency core cooling: containment pressure and temperature reduction: containment air cleaning.   31

(4) (software). A group of assemblies or components or both combined to perform a single function. *See:* **component; function.**   434

subtransient current (rotating machinery). The initial alternating component of armature current following a sudden short circuit. *See:* **armature.**   63

subtransient internal voltage (synchronous machine) (specified operating condition). The fundamental-frequency component of the voltage of each armature phase that would appear at the terminals immediately following the sudden removal of the load. *Note:* The subtransient internal voltage, as shown in the phasor diagram, is related to the terminal-voltage and phase-current phasors by the equation:

$$\mathbf{E}''_1 = \mathbf{E}_a + R\mathbf{I}_a + jX''_d\mathbf{I}_{ad} + jX''_q\mathbf{I}_{aq}$$

For a machine subject to saturation, the reactances should be determined for the degree of saturation applicable to the specified operating conditions.   63

subtransient reactance (power fault effects). The reactance of a generator at the initiation of a fault. This reactance is used for the initial calculation of the initial symmetrical fault current. The current continuously decreases but it is assumed to be steady at this value as a first step, lasting approximately 0.05 s after a suddenly applied fault.   404

subtrate (metal-nitride-oxide field-effect transistor). This insulated-gate field-effect transistor (IGFET) region separates source from drain and is of opposite conductivity type. The potential on the substrate terminal can only be equally, or less attractive to the carriers in the channel than the source terminal.

386

subway transformer (power and distribution transformer). A submersible-type distribution transformer suitable for installation in an underground vault.

53

sudden failure. *See:* **failure, sudden.**

sudden-pressure relay (power switchgear). A relay that operates by the rate of rise in pressure of a liquid or gas.

103

sudden short-circuit test (synchronous machine). A test in which a short-circuit is suddenly applied to the armature winding of the machine under specified operating conditions.

63

Suez Canal searchlight. A searchlight constructed to the specifications of the Canal Administration that by regulation of the Administration, must be carried by every ship traversing the canal, so located as to illuminate the banks.

328

suicide control (adjustable-speed drive). A control function that reduces and automatically maintains the generator voltage at approximately zero by negative feedback. *See:* **control system, feedback.**

206

suitable test (faulted circuit indicators). Where a condition or a set of conditions are so variable from one utility to another or even within the utility itself that no test can be properly specified for all conditions, it is left to the user to determine their individual test needs. A suitabe test and anticipated service life are mutually agreed to between manufacturer and user.

482

sum frequency (parametric device). The sum of a harmonic (nf_p) of the pump frequency (f_p) and the signal frequency (f_s), where n is a positive integer. *Note:* Usually n is equal to one. *See:* **parametric device.**

191

sum-frequency parametric amplifier. *See:* **noninverting parametric device; parametric device.**

summary punch (test, measurement and diagnostic equipment). A tape or card punch operating in conjunction with another machine to punch data which have been summarized or calculated by the other machine.

54

summation check (computing systems). A check based on the formation of the sum of the digits of a numeral. The sum of the individual digits is usually compared with a previously computed value.

255, 77

summer (computing systems). *See:* **summing amplifier.**

summing amplifier (analog computers). An operational amplifier that produces an output signal equal to a weighted sum of the input signals. *Note:* In an analog computer, the term summer is synonymous with summing amplifier.

9

summing junction (analog computers). The junction common to the input and feedback impedances used with an operational amplifier.

9

summing point (1). Any point at which signals are added algebraically. *Note:* For example the null junction of a power supply is a summing point because, as the input to a high-gain direct-current amplifier, op-

erational summing can be performed at this point. As a virtual ground, the summing point decouples all inputs so that they add linearly in the output, without other interaction. *See:* **operational programming; null junction; power supply.**

56

(2) (industrial control). The point in a feedback control system at which the algebraic sum of two or more signals is obtained. *See:* **control system, feedback.**

206

sum pattern (antennas). A radiation pattern characterized by a single main lobe whose cross-section is essentially elliptical, and a family of side lobes, the latter usually at a relatively low level. *Note:* Antennas which produce sum patterns are often designed to produce a difference pattern and have application in acquisition and tracking radar systems.

111

sun bearing (illuminating engineering). The angle measured in the plane of the horizon between a vertical plane at a right angle to the window wall and the position of this plane after it has been rotated to contain the sun.

167

sun light (illuminating engineering). Direct visible radiation from the sun.

167

superconducting. The state of a superconductor in which it exhibits superconductivity. *Example:* Lead is superconducting below a critical temperature and at sufficiently low operating frequencies. *See:* **normal (state of a superconductor); superconductivity.**

191

superconductive. Pertaining to a material or device that is capable of exhibiting superconductivity. *Example:* Lead is a superconductive metal regardless of temperature. The cryotron is a superconductive computer component. *See:* **superconductivity.**

191

superconductivity. A property of a material that is characterized by zero electric resistivity and, ideally, zero permeability.

191

superconductor. Any material that is capable of exhibiting superconductivity. *Example:* Lead is a superconductor. *See:* **superconductivity.**

191

superdirectivity (antennas). The condition that occurs when the antenna illumination efficiency significantly exceeds one hundred percent. *Note:* Superdirectivity is only obtained at a cost of a large increase in the ratio of average stored energy to energy radiated per cycle.

111

supergroup. *See:* **channel supergroup.**

superheterodyne reception. A method of receiving radio waves in which the process of heterodyne reception is used to convert the voltage of the received wave into a voltage of an intermediate, but usually superaudible, frequency, that is then detected. *See:* **radio receiver.**

328

superimposed ringing (telephone switching systems). Selective ringing that utilizes direct current polarity to obtain selectivity.

55

superluminescent light emitting diode (LED) (fiber optics). An emitter based on stimulated emission with amplification but insufficient feedback for oscillation to build up. *See:* **spontaneous emission; stimulated emission.**

433

superposed circuit. An additional channel obtained from one or more circuits, normally provided for other channels, in such a manner that all the channels can be used simultaneously without mutual interference. *See:* **transmission line.** 328

superposition theorem. States that the current that flows in a linear network, or the potential difference that exists between any two points in such a network, resulting from the simultaneous application of a number of voltages distributed in any manner whatsoever throughout the network is the sum of the component currents at the first point, or the component potential differences between the two points, that would be caused by the individual voltages acting separately. 328

superradiance (fiber optics). Amplification of spontaneously emitted radiation in a gain medium, characterized by moderate line narrowing and moderate directionality. *Note:* This process is generally distinguished from lasing action by the absence of positive feedback and hence the absence of well-defined modes of oscillation. *See:* **laser; spontaneous emission; stimulated emission.** 433

superregeneration. A form of regenerative amplification, frequently used in radio receiver detecting circuits, in which oscillations are alternately allowed to build up and are quenched at a superaudible rate. *See:* **radio receiver.** 328

supersonic frequency. *See:* **ultrasonic frequency.**

supersynchronous satellite (communication satellite). A satellite with mean sidereal period of revolution about the primary body which is an integral multiple of the sidereal period of rotation of the primary body about its axis. 74

supervised circuit (protective signaling). A closed circuit having a current-responsive device to indicate a break in the circuit, and, in some cases, to indicate an accidental ground. *See:* **protective signaling.** 328

supervision (telephone switching systems). The function of indicating and controlling the status of a call. 55

supervisory control (1)(supervisory control, data acquisition, and automatic control). An arrangement for operator control and supervision of remotely located apparatus using multiplexing techniques over a relatively small number of interconnecting channels. 570

(2)(power switchgear). A form of remote control comprising an arrangement for the selective control of remotely located units by electrical means over one or more common interconnecting channels. 103

supervisory control data acquisition system (supervisory control, data acquisition, and automatic control). A system operating with coded signals over communication channels so as to provide control of remote equipment (using typically one communication channel per remote station). The supervisory system may be combined with a data acquisition system, by adding the use of coded signals over communication channels to acquire information about the status of the remote equipment for display or for recording functions. 570

supervisory control functions (supervisory control, data acquisition, and automatic control). Equipment governed by ANSI/IEEE Std C37.1-1987 comprise one or more of the following functions: (1) Alarm function. The capability of a supervisory system to accomplish a predefined action in response to an alarm condition. *See:* **alarm condition.** (2) Analog function. The capability of a supervisory system to accept, record, or display, or do all of these, an analog quantity as presented by a transducer or external device. The transducer may or may not be part of the supervisory control system. (3) Control function. The capability of a supervisory system to selectively perform manual or automatic, or both, operation (singularly or in selected groups) of external devices. Control may be either analog (magnitude or duration) or digital. (4) Indication (status) function. The capability of a supervisory system to accept, record, or display, or do all of these, the status of a device. The status of a device may be derived from one or more inputs giving one or more states of indication. (A) Two-state indication. Only one of the two possible positions of the supervised device is displayed at one time. Such display may be derived from a single set of contacts. (B) Three-state indication. One in which the transitional state or security indication as well as the terminal positions of the supervised device is displayed. Such a display is derived from at least two sets of indicating contacts. (C) Multistate indication. Only one of the predefined states (transitional or discrete, or both) is indicated at a time. Such a display is derived from multiple inputs. (D) Indication with memory. An indication function with the additional capability of storing single or multiple change(s) of status that occur between scans. (5) Accumulator function. The capability of a supervisory system to accept and totalize digital pulses and make them available for display, or recording, or both. (6)Sequence of events function. The capability of a supervisory system to recognize each predefined event, associate a time of occurrence with each event, and present the event data in order of occurrence of the events. 570

supervisory indication (power switchgear). A form of remote indication comprising an arrangement for the automatic indication of the position or condition of remotely located units by electrical means over one or more common interconnecting channels. 103

supervisory program (software). A computer program, usually part of an operating system, that controls the execution of other computer programs and regulates the flow of work in a data processing system. *Syn:* **executive program.** *See:* **computer program; data; execution; executive program; operating system; supervisor; system.** 434

supervisory relay. A relay that, during a call, is generally controlled by the transmitter current supplied to a subscriber line in order to receive, from the associated station, directing signals that control the actions of operators or switching mechanisms with regard to the connection. 328

supervisory routine (computing systems). *See:* **executive routine.**

supervisory scanning cycle (station control and data acquisition). The time interval to start and complete a supervisory scan.　　403

supervisory signal (telephone switching systems). Any signal used to indicate or control the states of the circuits involved in a particular connection.　　55

supervisory station check (power switchgear). The automatic selection in a definite order, by means of a single initiation of the master station, of all of the supervisory points associated with one remote station of a system; and the transmission to the master station of indications of positions or conditions of the individual equipment or device associated with each point.　　103

supervisory system (1)(supervisory control, data acquisition, and automatic control). All control indicating and associated with telemetering equipment at the master station and all of the complementary devices at the remote station, or stations. (A) Continuous update. A system in which the remote station continuously updates indication and telemetering to the master station regardless of action taken by the master station. The remote station may interrupt the continuous data updating to perform a control operation. (B) Polling. A system in which the master interrogates each remote to ascertain if there has been a change since the last interrogation. Upon detection of a change the master may request data immediately. (C) Quiescent. A system which is normally alert but inactive and transmits information only when a change in indication occurs at the remote station or when a command operation is initiated at the master station. (D) Scanning. A system in which the master controls all information exchange. The normal state is usually one of repetitive communication with the remote stations.　　570

(2) (power switchgear). All supervisory control, indicating, and telemeter selection devices in the master station and all of the complementary devices in the remote station, or stations, which utilize a single common interconnecting channel for the transmission of the control or indication signals between these stations.　　103

supervisory system check (power switchgear). The automatic selection in a definite order, by means of a single initiation at the master station, of all supervisory points associated with all of the remote stations in a system; and the transmission to the master station of indications of positions or conditions of the individual equipment or device associated with each point.　　103

supervisory telemeter selection (power switchgear). A form of remote telemeter selection comprising an arrangement for the selective connection of telemeter transmitting equipment to an appropriate telemeter receiving equipment over one or more common interconnecting channels.　　103

supervisory tone (telephone switching systems). A tone that indicates to equipment, an operator or a customer that a particular state in the call has been reached, and which may signify the need for action to be taken. The terms used for the various supervisory tones are usually self-explanatory.　　55

supplementary equipment ground (generating station grounding). A grounding conductor used to connect the equipment frame to local grounding system to minimize potential difference.　　569

supplementary lighting (illuminating engineering). Lighting used to provide an additional quantity and quality of illuminance which cannot readily be obtained by a general lighting system and which supplements the general lighting level, usually for specific work requirements.　　167

supplementary standard illuminant D_{55} (illuminating engineering). A representation of a phase of daylight at a correlated color temperature of approximately 5500 K.　　167

supplementary standard illuminant D_{75} (illuminating engineering). A representation of a phase of daylight at a correlated color temperature of approximately 7500 K.　　167

supplier (1)(nuclear power quality assurance). Any individual or organization who furnishes items or services to a procurement document. An all inclusive term used in place of any of the following: vendor, seller, contractor, subcontractor, fabricator, consultant, and subtier levels.　　417

(2)(software requirements specifications). The person, or persons, who produce a product for a customer. In the context of IEEE Std 830-1984, the customer and the supplier may be members of the same organization.　　449

supply circuit (household electric ranges). The circuit that is the immediate source of the electric energy used by the range. *See:* **appliance outlet.**　　263

supply equipment. *See:* **electric supply equipment.**

supply impedance (1)(converters having dc input) (self-commutated converters). The impedance appearing in the input lines to the converter.　　584

(2)(inverters). The impedance appearing across the input lines to the power inverter with the power inverter disconnected. *See:* **self-commutated inverters.**　　208

supply line, motor (rotating machinery). The source of electric power to which the windings of a motor are connected. *See:* **asynchronous machine.**　　63

supply short-circuit current (converters having dc input)(self-commutated converters). The steady-state current that the dc (direct current) supply system can deliver into a short-circuit across the terminals to which the converter is to be connected.　　584

supply station. *See:* **electric supply station.**

supply transient energy (converters having dc input) (self-commutated converters). The energy that the dc (direct current) supply system, due to a transient, is capable of delivering at the terminals to which the terminal is to be connected.　　584

supply transient overvoltage (converters having dc input)(self-commutated converters). The peak instantaneous voltage that may appear between the input lines to the converter with the converter disconnected.   584

supply transient voltage (inverters). The peak instantaneous voltage appearing across the input lines to the power inverter with the inverter disconnected. *See:* **self-commutated inverters.**   208

supply voltage (electrode) (electron tubes). The voltage, usually direct, applied by an external source to the circuit of an electrode. *See:* **electrode voltage (electron tube).**   244, 190

support (raceway)(raceway systems for Class 1E circuits for nuclear power generating stations). An assembly of structural members whose function is to restrain and provide structural stability for raceways.   513

support components (metal-enclosed bus and calculating losses in isolated-phase bus). The components that add additional strength and rigidity or both to the bus enclosure and are basic subassemblies of the enclosure.   574

support equipment (test, measurement and diagnostic equipment). Equipment required to make an item, system or facility operational in its environment. This includes all equipment required to maintain and operate the item, system or facility and the computer programs related thereto.   54

supporting operations area(s) (nuclear power generating station). Functional area(s) allocated for controls and displays which support plant operation.   358

supporting structure (NESC)(transmission and distribution). The main supporting unit (usually a pole or tower).   494, 262

support ring (rotating machinery). A structure for the support of a winding overhang: either constructed of insulating material, carrying support-ring insulation, or separately insulated before assembly. *See:* **stator.**   63

support-ring insulation (rotating machinery). Insulation between the winding overhang or end winding and the winding support rings. *See:* **rotor (rotating machinery); stator.**   63

support test system (test, measurement and diagnostic equipment). A measurement system used to assess the quality of operational equipments and may include (1) test equipment: (2) ancillary equipment: (3) supporting documentation: (4) operating personnel.   54

suppressed-carrier modulation. Modulation in which the carrier is suppressed. *Note:* By **carrier** is meant that part of the modulated wave that corresponds in a specified manner to the unmodulated wave.   242

suppressed-carrier operation (data transmission). That form of amplitude-modulation carrier transmission in which the carrier wave is suppressed.   59

suppressed time delay (navigation aid terms). A deliberate displacement of the zero of the time scale with respect to the time of emission of a pulse.   526

suppressed-zero instrument. An indicating or recording instrument in which the zero position is below the end of the scale markings. *See:* **instrument.**   328

suppressed-zero range. A range where the zero value of the measured variable, measured signal, etcetera, is less than the lower range value. Zero does not appear on the scale. *Note:* For example: 20 to 100.   295

suppression (computing systems). *See:* **zero suppression.**

suppression characteristic (thyristor). Predicated on a device's ability to block voltage at higher than rated junction temperatures (T_s) when either the voltage or the rate of application of the principal blocking voltage, or both, are below the rated voltage of the silicon controlled rectifier (SCR).   445

suppression distributor rotor. Rotor of an ignition distributor with a built-in suppressor. *See:* **electromagnetic compatibility.**   220, 199

suppression rating (thyristor). Repetitive surge ON-state current. A specified ON-state current of short time duration resulting in a specified junction temperature, above rated, immediately prior to supporting a specified principal voltage without turning on (gate signal removed, gate impedance specified). *Note:* Proper coordination with this rating permits a thyristor power controller to limit fault currents without fuse blowing or circuit breaker action. For a given silicon controlled rectifier (SCR) its suppression characteristic may be defined in one of two ways: (1) T_L and $\hat{I}_f$ together with shape of fault I waveform may be specified together with time $t_2 - t_1$. This then determines maximum V_{line} and shape of reapplied V at time t_2, that is, dv/dt. Alternately, ac frequency and peak V may be given for sinusoidal waveforms. (2) T_s may be specified together with shape and magnitude of reapplied voltage at time t_2. Criteria (2) serves as well as (1) since the magnitude and shape of the fault current determine T_2 together with the time ($t_2 - t_1$).   445

suppression ratio (suppressed-zero range). The ratio of the lower range-value to the span. *Note:* For example: Range 20 to 100

$$\text{Suppression Ratio} = \frac{20}{80} = 0.25$$

295

suppressive wiring techniques (coupling in control systems). Those wiring techniques which result in the reduction of electric or magnetic fields in the vicinity of the wires which carry current without altering the value of the current. Wires which are candidates for suppressive techniques are generally connected to a noise source, may couple noise into a susceptible circuit by induction. Example: twisting or transposing of alternating-current power lines to reduce the intensity of magnetic field produced by current in these lines. *See:* **barrier wiring techniques; compensatory wiring techniques.**   43

suppressor grid. A grid that is interposed between two positive electrodes (usually the screen grid and the plate), primarily to reduce the flow of secondary electrons from one electrode to the other. *See:* **electrode (electron tube); grid.**   125

suppressor spark plug. A spark plug with a built-in

interference suppressor. *See:* **electromagnetic compatibility.**   199

surface active agent. *See:* **wetting agent (electroplating).**

surface barrier contact (1)(germanium gamma-ray detectors)(charged-particle detectors). A metal-insulator-semiconductor (MIS) contact structure whose rectification properties are dominated or heavily influenced by charge trapped at the interfaces and in the insulator.   528, 119

(2)(X-ray energy spectrometers)(semiconductor radiation detectors). A rectifying contact that is characterized by a potential barrier associated with an inversion or accumulation layer; said inversion or accumulation layer being caused by surface charge resulting from the presence of surface states and work function differences, or both.   471,23, 119

surface barrier radiation detector (1)(charged-particle detectors). A radiation detector for which the blocking contact is a surface barrier contact.   119

(2)(X-ray energy spectrometers)(germanium gamma-ray detectors)(semiconductor radiation detectors). A radiation detector for which the principal rectifying junction is a surface barrier contact.   471,528,23

surface channel (of a semiconductor radiation detector) (charged-particle detectors) (germanium gamma-ray detectors). A thin region at a semiconductor surface of p- or n-type conductivity created by the action of an electric field; for example, that due to trapped surface charge (due to charges trapped in surface layers.   119,528

surface connecting cable (electric submersible pump cable). Power cable connecting the ESP (electric submersible pump) cable to surface equipment.   484

surface contamination (plutonium monitoring). Radioactive material deposited on the surface of facilities (floor surfaces, workbench tops, machines, etcetera), equipment, or personnel.   413

surface duct (radio wave propagation). An atmospheric duct for which the lower boundary is the surface bounding the atmosphere.   146

surface leakage. The passage of current over the surface of a material rather than through its volume.   210

surface metal raceway (metal molding). A raceway consisting of an assembly of backing and capping. *See:* **raceways.**   328

surface mounted (illuminating engineering). A luminaire which is mounted directly on the ceiling.   167

surface-mounted device (power switchgear). A device, the entire body of which projects in front of the mounting surface.   103

surface navigation (navigation aid terms). Navigation of a vehicle on the surface of the earth. *See:* **land navigation.**   526

surface noise (mechanical recording). The noise component in the electric output of a pickup due to irregularities in the contact surface of the groove. *See:* **phonograph pickup.**   176

surface of position (navigation aid terms). Any surface defined by a constant value of some navigation quantity.   526

surface operable (power switchgear). A term indicating that the switch and its accessories are operable from above grade.   489

surface-potential gradient. The slope of a potential profile, the path of which intersects equipotential lines at right angles.   313

surface, prescribed (sound measurements). A prescribed surface is a hypothetical surface surrounding the machine on which sound measurements are made.   129

surface search radar. *See:* **navigational radar.**

surface state coefficient (m) (overhead-power-line corona and radio noise). A coefficient (0 m 1) by which the nominal corona inception gradient must be multiplied to obtain the actual corona-inception gradient on overhead-power lines. *Note:* Examples of conditions which affect the surface state are given in overhead power lines corona.   411

surface wave (fiber optics). A wave that is guided by the interface between two different media or by a refractive index gradient in the medium. The field components of the wave may exist (in principle) throughout space (even to infinity) but become negligibly small within a finite distance from the interface. *Note:* All guided modes, but not radiation modes, in an optical waveguide belong to a class known in electromagnetic theory as surface waves.   433

surface wave antenna. An antenna which radiates power from discontinuities in the structure that interrupt a bound wave on the antenna surface.   111

surface-wave transmission line (waveguide). A transmission line in which propagation in other than a TEM mode is constrained to follow the external face of a guiding structure.   267

surge (metal-oxide surge arresters for ac power circuits)(surge arrester). A transient wave of current, potential, or power in an electric circuit.   583, 430

surge arrester (1)(electrical heating applications to melting furnaces and forehearths in the glass industry). A protective device for limiting surge voltages on equipment by discharging or bypassing surge current. It prevents continued flow of current to ground and is capable of repeating these functions, as specified. As surge protective devices, arrester are connected from sensitive circuit points to ground, thus limiting dangerous surge voltage below damaging levels.   520

(2)(metal-oxide surge arresters for ac power circuits). A protective device for limiting surge voltages on equipment by diverting surge current and returning the device to its original status. It is capable of repeating these functions as specified. *Note:* The term 'arrester' as used in ANSI/IEEE Std C62.11-1987 is understood to mean 'surge-arrester'.   583

(3) (ac power circuits). A protective device for limiting surge voltages on equipment by discharging or bypassing surge current: it prevents continued flow of follow current to ground, and is capable of repeating these functions as specified. *Note:* Hereafter, the term "arrester" as used in this standard shall be understood to mean "surge arrester." *Syn:* **lightning arrester***.

2, 430

*Deprecated

(4) (wire-line communication facilities). Devices which guard against dielectric failure of protection apparatus due to lightning or surge voltages in excess of their dielectric capabilities, and serve to interrupt power follow current. Protection of isolating and neutralizing transformers by surge arresters is covered in Annex C. 414

surge (or impulse) breakdown voltage. *See:* **impulse sparkover voltage.** 370

surge capacitor (electrical heating applications to melting furnaces and forehearths in the glass industry). Capacitors used to decrease the slope of the surge voltage wave fronts. They help to reduce the voltage stresses on protected apparatus by spreading the impressed voltage over a greater time span. 520

surge-crest ammeter. A special form of magnetometer intended to be used with magnetizable links to measure the crest value of transient electric currents. *See:* **instrument.** 328

surge diverter. *See:* **surge arrester.**

surge electrode current. *See:* **fault electrode current.**

surge generator (impulse generator) (lightning generator*). An electric apparatus suitable for the production of surges. *Note:* Surge generator types common in the art are: transformer-capacitor; transformer-rectifier; transformer-rectifier-capacitor; parallel charging; series discharging.

*Deprecated. 62, 64

surge impedance (self-surge impedance). The ratio between voltage and current of a wave that travels on a line of infinite length and of the same characteristics as the relevant line. *See:* **characteristic impedance.**

244, 62

surge let-through (surge testing for equipment connected to low-voltage ac power circuits). That part of the surge that passes by a surge protective device with little or no alteration. 578

surge protection. *See:* **rate-of-change protection.**

surge protective level (surge arrester). The highest value of surge voltage that may appear across the terminals under the prescribed conditions. 62

surge protector (gas-tube surge-protective devices). A protective device, consisting of one or more surge arresters, a mounting assembly, optional fuses and short circuiting devices, etcetera, which is used for limiting surge voltages on low-voltage (1000 V (volt) root-mean-square or 1200 V direct current) electrical or electronic equipment or circuits. 490

surge rating (thyristor). Rated values for surge forward current is given for two time regions: (1) For times smaller than one-half cycle (at 50 hertz (Hz) or 60 Hz) down to approximately 1 millisecond (ms), the value is given in terms of maximum rated $\int d\tau_i^2$. They may be given by means of a curve or by single values. No immediate subsequent application of reverse blocking voltage is assumed. (2) Maximum values of surge for-

ward current versus time up to at least 10 cycles. The frequency, the conducting period length, the current waveshape and the reverse blocking voltage capability including the rate-of-rise of voltage for the intervals after and between the surges are specified. In either case a previous application of rate maximum junction temperature is assumed if not otherwise specifically mentioned. 445

surge remnant (surge testing for equipment connected to low-voltage ac power circuits). That part of an applied surge that remains downstream of one or several protective devices. 578

surge suppressor (industrial control). A device operative in conformance with the rate of change of current, voltage, power, etcetera, to prevent the rise of such quantity above a predetermined value. 206

surge, switching. *See:* **switching surge.**

surge voltage recorder (klydonograph). *See:* **Lichtenberg figure camera.**

surveillance (1)(diesel-generator unit). The determination of the state or condition of a system or subsystem. 99

(2) (nuclear power quality assurance). The act of monitoring or observing to verify whether an item or activity conforms to specified requirements. 417

surveillance radar (navigation aid terms). A search radar used to maintain cognizance of selected traffic within a selected area, such as an airport terminal area or air route. 526

survey (plutonium monitoring). The examination of an area for the purpose of detecting the presence of radioactive materials and determining the quantity of that radioactivity. 413

survey contamination control (plutonium monitoring). A survey conducted to determine the presence of unwanted contaminants, normally conducted with alpha or gamma, or both, sensitive instruments. 413

survey dose rate (plutonium monitoring). A survey conducted to determine the dose rate at some specified location or area and usually conducted with gamma exposure rate survey instruments. Neutron surveys may also be required frequently. 413

susceptance. The imaginary part of admittance.

210, 185

susceptance relay (power switchgear). A mho-type distance relay for which the center of the operating characteristic on the R-X diagram is on the X axis. *Note:* The equation that describes such a characteristic is $Z = K \sin \theta$ where K is a constant and θ is the phase angle by which the input voltage leads the input current. *See:* **distance relay; Figure (b).** 103

susceptibility (1)(power fault effects). The property of an equipment which describes its capability to function acceptably when subjected to unwanted electromagnetic energy. 404

(2)(surge testing for equipment connected to low-voltage ac power circuits). The inability of a device, equipment or system to resist an electromagnetic disturbance. *Note:* Susceptibility is the lack of immunity. 578

susceptiveness (NESC)(transmission and distribu-

tion). The characteristics of a communications circuit including its connected apparatus which determine the extent to which it is adversely affected by inductive fields.					494,262

susceptor. Energy absorbing device generally used to transfer heat to another load.					14

suspended (pendant) (illuminating engineering). A luminaire which is hung from a ceiling by supports.					167

suspended-type handset telephone (bracket-type handset telephone). *See:* **hang-up hand telephone.**

suspension (gyro, accelerometer) (inertial sensor). A means of supporting and positioning a float (floated gyro), rotor (dynamically tuned gyro, electrically suspended gyro), or proof mass (accelerometer) with respect to the case.					46

suspension insulator (composite insulators). As used in ANSI/IEEE Std 987-1985, any insulator intended primarily to carry tension loads. It includes tangent, deadend, and vee-string installations.					483

suspension insulator. One or a string of suspension-type insulators assembled with the necessary attaching members and designed to support in a generally vertical direction the weight of the conductor and to afford adequate insulation from tower or other structure. *See:* **insulator; tower.**					64

suspension-insulator unit. An assembly of a shell and hardware, having means for nonrigid coupling to other units or terminal hardware.					261

suspension-insulator weights. Devices, usually cast iron, hung below the conductor on a special spindle supported by the conductor clamp. *Note:* Suspension insulator weights will limit the swing of the insulator string, thus maintaining adequate clearances. In practice, weights of several hundreds of pounds are sometimes used. *See:* **tower.**					64

suspension of reclosing (power line maintenance). To make inoperative automatic reclosing equipment. *Syn:* **hold out; live-line permit; hold order.**					458

suspension strand (messenger). A stranded group of wires supported above the ground at intervals by poles or other structures and employed to furnish within these intervals frequent points of support for conductors or cables. *See:* **cable; open wire.**					328

sustained bypass current detection (series capacitors). A means to detect prolonged current flow through the protective device and to initiate closing of the bypass device.					474

sustained gap-arc protection (series capacitor). A means to detect prolonged arcing of the protective power gap or arcing of the backup gap if included to initiate closing of the capacitor bypass switch.					86

sustained interruption (electric power systems). Any interruption not classified as a momentary interruption. *See:* **outage.**					200

sustained-operation influence. The change in the recorded value, including zero shift, caused solely by energizing the instrument over extended periods of time, as compared to the indication obtained at the end of the first 15 minutes of the application of energy.

It is to be expressed as a percentage of the full-scale value. *Note:* The coil used in the standard method shall be approximately 80 inches in diameter, not over 5 inches long, and shall carry sufficient current to produce the required field. The current to produce a field to an accuracy of ± 1 percent in air shall be calculated without the instrument in terms of specific dimensions and turns of the coil. In this coil, 800 ampere-turns will produce a field of approximately 5 oersteds. The instrument under test shall be placed in the center of the coil. *See:* **accuracy rating (instrument).**					294

sustained oscillation (sustained vibration) (1) (system). The oscillation when forces controlled by the system maintain a periodic oscillation of the system. Example: Pendulum actuated by a clock mechanism.					210

(2) (gas turbines). Those oscillations in which the amplitude does not decrease to zero, or to a negligibly small, final value.					98, 58

sustained overvoltage detection device (series capacitors). A device that detects capacitor voltage above rating but below the operation level of the protective device and initiates an alarm signal or corrective action.					474

sustained overvoltage protection device (series capacitor). A device to detect capacitor voltage that is above rating or predetermined value but is below the sparkover of the protective power gaps, and to initiate the closing of the capacitor bypass switch according to a predetermined voltage-time characteristic.					86

sustained short-circuit test (synchronous machine). A test in which the machine is run as a generator with its terminals short-circuited.					63

sweep. A traversing of a range of values of a quantity for the purpose of delineating, sampling, or controlling another quantity. *Notes:* (1) Examples of swept quantities are (A) the displacement of a scanning spot on the screen of a cathode-ray tube, and (B) the frequency of a wave. (2) Unless otherwise specified, a linear time function is implies; but the sweep may also vary in some other controlled and desirable manner.					178,185

sweep accuracy (oscilloscopes). Accuracy of the horizontal (vertical) displacement of the trace compared with the reference independent variable, usually expressed in terms of average rate error as a percent of full scale. *See:* **oscillograph.**					106

sweep-delay accuracy (oscilloscopes). Accuracy of indicated sweep delay, usually specified in error terms.					106

sweep,delayed. *See:* **delayed sweep.**

sweep duration (sawtooth sweep). The time required for the sweep ramp. *See:* **oscillograph.**					185

sweep duty factor. For repetitive sweeps, the ratio of the sweep duration to the interval between the start of one sweep and the start of the next. *See:* **oscillograph.**					185

sweep, expanded. *See:* **magnified sweep.**

sweep, external (oscilloscopes). A sweep generated external to the instrument.					106

sweep, free-running *See:* **free-running sweep.**

sweep frequency (oscilloscopes). The sweep repetition rate. *See:* **oscillograph.**   185

sweep gate (oscilloscopes). Rectangular waveform used to control the duration of the sweep; usually also used to unblank the cathode-ray tube for the duration of the sweep. *See:* **oscillograph.**   185

sweep, gated. See: **gated sweep.**

sweep generator (oscilloscopes). A circuit that generates a signal used as an independent variable; the signal is usually a ramp, changing value at a constant rate.   106

sweep mode control (oscilloscopes). The control used on some oscilloscopes to set the sweep for triggered, free-running, or synchronized operation.   106

sweep oscillator. An oscillator in which the output frequency varies continuously and periodically between two frequency limits. *See:* **telephone station.**   196

sweep-out time, charge (semiconductor radiation detector). *See:* **charge collection time.**

sweep range (oscilloscopes). The set of sweep-time/division settings provided. *See:* **oscillograph.**   185

sweep recovery time (oscilloscopes). The minimum possible time between the completion of one sweep and the initiation of the next, usually the sweep hold-off interval. *See:* **oscillograph.**   185

sweep, recurrent. *See:* **recurrent sweep.**

sweep reset (oscilloscopes). In oscilloscopes with single-sweep operation, the arming of the sweep generator to allow it to cycle once. *See:* **oscillograph.**   185

sweep, sine-wave. A sweep generated by a sinusoidal function. *See:* **oscillograph.**   106

sweep, stairstep (oscilloscopes). An incremental sweep in which each step is equal. The electrical deflection waveform producing a stairstep sweep is usually called a staircase or stairstep waveform. *See:* **sweep, incremental; oscillograph.**   106

sweep switching (automatic). Alternate display of two or more time bases or other sweeps using a single-beam cathode-ray tube: comparable to dual- or multiple-trace operation of the deflection amplifier.   106

sweep time (acoustically tunable optical filter). The time to continuously tune the filter over its spectral range.   72

sweep time division (spectrum analyzer). The nominal time required for the spot in the reference coordinate to move from one graticule division to the next. Also the name of the control used to select this time.   390

swinging compass (navigation aid terms). An accurate, portable magnetic compass used to indicate magnetic headings during aircraft magnetic compass calibration.   526

swingout panel (packaging machinery). A panel that is hinge-mounted in such a manner that the back of the panel may be made accessible from the front of the enclosure.   429

swing rack cabinet (power switchgear). An assembly enclosed at the top, side and rear with front hinged door for front access having a swing open frame for equipment mounting (for example 19 inch wide chassis and subpanel assemblies).   103

switch (1)(NESC)(transmission and distribution). A device for opening or closing or for changing the connection of a circuit. In these rules, a switch is understood to be manually operable, unless otherwise stated.   494,262

(2)(switching system)(telephone loop performance). A system that establishes communication channels among two or more of its interfaces at customers' demand.   473

(3) (high-voltage switchgear). A device designed to close or open, or both, one or more electric circuits. *See:* **switching device.**   443

(4) (computing systems). A device or programming technique for making a selection, for example, a toggle, a conditional jump.   255, 77

(5) (electric and electronics parts and equipments). A device for making, breaking, or changing the connections in an electric circuit. *Note:* a switch may be operated by manual, mechanical, hydraulic, thermal, barometric, or gravitational means, or by electromechanical means not falling within the definition of "relay."   17

switch base. The main members to which the insulator units are attached.   27

switchboard (1)(NESC). A type of switchgear assembly that consists of one or more panels with electric devices mounted thereon, and associated framework.   494

(2) (National Electrical Code). A large single panel, frame, or assembly of panels on which are mounted, on the face or back or both, switches, overcurrent and other protective devices, buses, and usually instruments. Switchboards are generally accessible from the rear as well as from the front and are not intended to be installed in cabinets. *See:* **panelboard.**   256

(3) (power switchgear). A type of switchgear assembly that consists of one or more panels with electric devices mounted thereon, and associated framework. *Note:* Switchboards may be classified by function; that is, power switchboards or control switchboards. Both power and control switchboards, may be further classified by construction as defined.   103

(4) (electric power systems). A large single panel, frame, or assembly of panels, on which are mounted, on the face or back or both, switches, overcurrent and other protective devices, buses, and usually instruments. *Note:* Switchboards are generally accessible from the rear as well as from the front and are not intended to be installed in cabinets. *See:* **center of distribution; distribution center; panelboard.**   256

(5) (transmission and distribution). When referred to in connection with supply of electricity, a large single panel, frame, or assembly of panels, on which are mounted (on the face, or back, or both) switches, fuses, buses, and usually instruments.   262

switchboard cord. A cord that is used in conjunction

with switchboard apparatus to complete or build up a telephone connection. 328

switchboard lamp (switchboard). A small electric lamp associated with the wiring in such a way as to give a visual indication of the status of a call or to give information concerning the condition of trunks, subscriber lines, and apparatus. 328

switchboard position (telephone switching systems). That portion of a manual switchboard normally provided for the use of one operator. 55

switchboards and panels (electric installations on shipboard). A generator and distribution switchboard receives energy from the generating plant and distributes directly or indirectly to all equipment supplied by the generating plant. A subdistribution switchboard is essentially a section of the generator and distribution switchboard (connected thereto by a bus feeder and remotely located for reasons of convenience or economy) which distributes energy for lighting, heating, and power circuits in a certain section of the vessel. A distribution panel receives energy from a distribution or subdistribution switchboard and distributes energy to energy-consuming devices or other distribution panels or panelboards. A panelboard is a distribution panel enclosed in a metal cabinet. 3

switchboard section (telephone switching systems). A structural unit providing for one or more operator positions. A complete switchboard may consist of one or more sections. 55

switchboard supervisory lamp (cord circuit or trunk circuit). A lamp that is controlled by one or other of the users to attract the attention of the operator. 328

switchboard supervisory relay. A relay that controls a switchboard supervisory lamp. 328

switch compartment (metal-enclosed interrupter switchgear). That portion of the switchgear assembly that contains one switching device, such as an interrupter switch, power fuse interrupter switch combination, etcetera, and the associated primary conductors. 573

switched-service (SSN) network (telephone switching systems). An arrangement of dedicated switching facilities to provide telecommunications services for a specific customer. 55

switched way (power switchgear). A way connected to the bus through a three-pole, group operated switch. 489

switchgear (1)(power switchgear). A general term covering switching and interrupting devices and their combination with associated control, metering, protective and regulating devices, also assemblies of these devices with associated interconnections, accessories, enclosures and supporting structures, used primarily in connection with the generation, transmission, distribution and conversion of electric power. 103

(2)(metal-enclosed low-voltage power circuit-breaker switchgear)(metal-clad and station-type cubicle switchgear). A general term covering switching and interrupting devices and their combination with associated control, instruments, metering, protective and regulating devices, also assemblies of these devices with associated interconnections, accessories, and supporting structures used primarily in connection with the generation, transmission, distribution, and conversion of electric power. 579,572,573

switchgear assembly (1)(metal-enclosed low-voltage power circuit-breaker switchgear)(metal-clad and station-type cubicle switchgear). An assembled equipment (indoor or outdoor) including, but not limited to, one or more of the following categories: switching, interrupting, control, instrumentation, metering, protective and regulating devices, together with their supporting structures, enclosures, conductors, electrical interconnections, and accessories. 579,572,573

(2)(power switchgear). An assembled equipment (indoor or outdoor) including, but not limited to, one or more of the following: switching, interrupting, control, metering, protective and regulating devices, together with their supporting structures, enclosures, conductors, electric interconnections, and accessories. 103

switchgear pothead. A pothead intended for use in a switchgear where the inside ambient air temperature may exceed 40 degrees Celsius. It may be an indoor or outdoor pothead that has been suitably modified by silver surfacing (or the equivalent) the current-carrying parts and incorporates sealing materials suitable for the higher operating temperatures. *See:* **pothead.** 4

switchgear, protective (thyristor converter). The ac circuit devices and the dc circuit devices that may be used in the thyristor converter unit to clear fault conditions. 121

switchhook (hookswitch). A switch on a telephone set, associated with the structure supporting the receiver or handset. It is operated by the removal or replacement of the receiver or handset on the support. *See:* **telephone station; switch stick.** 328

switch indicator. A device used at a noninterlocked switch to indicate the presence of a train in a block. 328

switching (1)(ferroelectric material)(primary ferroelectric terms). The process by which the remanent polarization is reversed (or reoriented) to a new value of P_r (generally equal and opposite). Switching can be produced by electric fields or mechanical stresses. 497

(2) (single-phase motor). The point in the starting operation at which the stator-winding circuits are switched from one connection arrangement to another. *See:* **asynchronous machine.** 63

(3) (test, measurement and diagnostic equipment). The act of manually, mechanically or electrically actuating a device for opening or closing an electrical circuit. 54

switching amplifier (industrial control). An amplifier which is designed to be applied so that its output is sustained at one of two specified states dependent upon the presence of specified inputs. *See:* **control system, feedback.** 206

switching array (telephone switching systems). An assemblage of multipled crosspoints. 55

switching branch (converter circuit elements)(self-commutated converters). A part of the circuit, including at least one switching element, bounded by two principal terminals. *Note:* A switching branch may include one or more simultaneously conducting converter switching elements connected together, commutating reactor windings, and other devices intended to protect the semiconductor devices or to assure their proper function, such as voltage and current dividers, snubbers, etcetera. In the simplest case, a switching branch may consist of only the switching element, which may be a single semiconductor device. The adjective 'switching' may be omitted when the context of converter circuits is clear. 584

switching card (test, measurement and diagnostic equipment). A plug-in device which provides the necessary interconnection to the unit under test.
54

switching circuit (data transmission). Term applied to the method of handling traffic through a switching center, either from a local user or from other switching centers, whereby additional electrical connection is established between the calling and the called station.
59

switching coefficient. The derivative of applied magnetizing force with respect to the reciprocal of the resultant switching time. It is usually determined as the reciprocal of the slope of a curve of reciprocals of switching times versus values of applied magnetizing forces. The magnetizing forces are applied as step functions. 77

switching control center (telephone switching systems). A place where maintenance analysis and control activities are centralized for switching entities situated in different locations. 55

switching current (1) (low-voltage ac power circuit protectors). The value of root-mean-square (rms) symmetrical current expressed in amperes, which the power circuit breaker element of the circuit protector interrupts at the rated maximum voltage and rated frequency under the prescribed test conditions.
158

(2) (power circuit breaker)(power switchgear). The value of current expressed in rms symmetrical amperes which the power circuit breaker element of the circuit protector interrupts at the rated maximum voltage and rated frequency under the prescribed test conditions. 158, 103

switching current rating (separable insulated connectors). The designated root-mean-square current which a load-break connector can connect and disconnect for a specified number of times under specified conditions. 454

switching device (switch) (power switchgear). A device designed to close or open, or both, one or more electric circuits. *Note:* The term 'switch' in International (IEC) practice refers to a mechanical switching device capable of opening and closing rated continuous load current. *See:* **mechanical switching device; nonmechanical switching device.** 103

switching entity (telephone switching systems). A switching network and its control. 55

switching impulse. Ideally, an aperiodic transient impulse voltage that rises rapidly to a maximum value and falls, usually less rapidly, to zero. Switching impulses generally have front times of the order of tens to thousands of microseconds, in contrast to lightning impulses, which have front times from fractions of a microsecond to tens of microseconds. 108

switching impulse insulation level (power and distribution transformer). An insulation level expressed in kilovolts of the crest value of a switching impulse withstand voltage. 53

switching impulse protective level (of a protective device) (power and distribution transformer). The maximum switching impulse expected at the terminals of a surge protective device under specified conditions of operation. 53

switching-impulse sparkover voltage (arrester). The impulse sparkover voltage with an impulse having a virtual duration of wavefront greater than 30 microseconds. 62

switching impulse test (power and distribution transformer). Application of the "standard switching impulse," a full wave having a front time of 250 μ and a time to half value of 2500 μ, described as a 250/2500 impulse. *Note:* It is recognized that some apparatus standards may have to use a modified wave shape where practical test considerations or particular dielectric strength characteristics make some modification imperative. Transformers, for example, use a modified switching impulse wave with the following characteristics: (1) Time to crest greater than 100 μ. (2) Exceeds 90 percent of crest value for at least 200 μ. (3) Time to first voltage zero on tail not less than 1000 μ, except where core saturation causes the tail to become shorter. 53

switching network (telephone switching systems). Switching stages and their interconnections. Within a switching system there may be more than one switching network. 55

switching-network plan (telephone switching systems). The switching stages and their interconnections within a specific switching system. 55

switching overvoltage (surge arrester). Any combination of switching surge(s) and temporary overvoltage(s) associated with a single switching episode.
62

switching plan (telephone switching systems). A plan for the interconnection of switching entities. 55

switching, slave-sweep (oscilloscopes). A combination of sweep switching and multiple-trace operation in which a specific channel is displayed with a specific sweep. 106

switching stage (telephone switching systems). An assemblage of switching arrays within each inlet which can be connected through a single crosspoint to its associated outlet. 55

switching station (power operations). A station where transmission lines are connected without power transformers. 516

switching structure (power switchgear). An open framework supporting the main switching and associated equipment, such as instrument transformers, buses, fuses, and connections. It may be designed for indoor or outdoor use and may be assembled with or without switchboard panels carrying the control equipment.   103

switching surge (1) (conductor stringing equipment). A transient wave of overvoltage in an electrical circuit caused by a switching operation. When this occurs, a momentary voltage surge could be induced in a circuit adjacent and parallel to the switched circuit in excess of the voltage induced normally during steady state conditions. If the adjacent circuit is under construction, switching operations should be minimized to reduce the possibility of hazards to the workmen.   431

(2) (surge arrester). A heavily damped, transient electrical disturbance associated with switching. System insulation flashover may precede or follow the switching in some cases but not all.   62

switching-surge protective level (arrester). The highest value of switching-surge voltage that may appear across the terminals under the prescribed conditions. *Note:* The switching-surge protective levels are given numerically by the maximums of the following quantities: (1) discharge voltage at a given discharge current, and (2) switching-impulse sparkover voltage.   62

switching system (telephone switching systems). A system in which connections are established between inlets and outlets either directly or with intermediate storage.   55

switching-system processor (telephone switching systems). Circuitry to perform a series of switching system operations under control of a program.   55

switching time (1) (electric power circuits). The period from the time a switching operation is required due to a forced outage until that switching operation is performed. Switching operations include reclosing a circuit breaker after a trip-out, opening or closing a sectionalizing switch or circuit breaker, or replacing a fuse link. *See:* **outage.**   200

(2) (magnetic storage cells). (A) T_s, the time interval between the reference time and the last instant at which the instantaneous voltage response of a magnetic cell reaches a stated fraction of its peak value. (B) T_x, the time interval between the reference time and the first instant at which the instantaneous integrated voltage response reaches a stated fraction of its peak value.   77

(3) (settling time) (hybrid computer linkage components). That time required from the time at which a channel is addressed until the selected analog signal is available at the output within a given accuracy.   10

(4) (reliable industrial and commercial power systems planning and design). The period from the time a switching operation is required due to a component failure until that switching operation is completed. Switching operations include such operations as: throwover to an alternate current, opening or closing

a sectionalizing switch or circuit breaker, reclosing a circuit breaker following a trip out due to a temporary fault, etcetera.   561

switching torque (1)(motor having an automatic connection change during the starting period). The minimum external torque developed by the motor as it accelerates through switch operating speed. *Note:* It should be noted that if the torque on the starting connection is never less than the switching torque, the pull-up torque is identical with the switching torque: however, if the torque on the starting connection falls below the switching torque at some speed below switch operating speed, the pull-up and switching torques are not identical. *See:* **asynchronous machine.**   328

(2) single-phase motor. The minimum torque which a motor will provide at switching at normal operating temperature, with rated voltage applied at rated frequency. *See:* **asynchronous machine.**   63

switching transients (radiation survey instruments). Sudden excursions of the meter which occur when the range switch is changed from one position to the next.   558

switch inrush current capability for single capacitance (as applied to interrupter switches). This capability is a function of the rated switching current, for single capacitance, the rated differential capacitance voltage (minimum) and the maximum design voltage of the switch. *Note:* This can be calculated from the equation:

$$\text{Capability, in Peak Amperes} = \sqrt{2}\,I_C\,\sqrt{1 + \frac{0.816 E_m}{\Delta V_{min}}}$$

where

I_C = rated switching current for single capacitance

ΔV_{min} = rated differential capacitance voltage, minimum

E_m = switch rated maximum voltage, in volts, rms.

      103

switch, load matching. *See:* **load-matching switch.**

switch, load transfer. *See:* **load transfer switch.**

switch machine. A quick-acting mechanism, electrically controlled, for positioning track switch points, and so arranged that the accidental trailing of the switch points does not cause damage. A switch machine may be electrically or pneumatically operated. *See:* **car retarder.**   328

switch machine lever lights. A group of lights indicating the position of the switch machine.   328

switch-machine point detector. *See:* **point detector.**

switch mode (thyristor). The starting instant of the controller ON-state interval is nonperiodic. This instant may be random (analogous to contactor operation), or it may be selected, for example, at voltage zero.   445

switch or contactor, load. *See:* **load switch or contactor.**

switch point (watthour meters). The transition from one time-of-use period to another.   485

switchroom (telephone switching systems). That part of a building that houses an assemblage of switching equipment. 55

switch signal. A low two-indication horizontal color light signal with electric lamps for indicating position of switch or derail. 328

switch sleeve. A component of the linkage between the centrifugal mechanism and the starting-switch assembly. *See:* **starting-switch assembly.** 328

switch starting. *See:* **preheat starting.**

switch stick (switch hook) (power switchgear). A device with an insulated handle and a hook or other means for performing stock operation of a switching device. 103

switch train. A series of switches in tandem. 328

swivel link (conductor stringing equipment). A swivel device designed to connect pulling lines and conductors together in series or connect one pulling line to the drawbar of a pulling vehicle. The device will spin and help relieve the torsional forces which build up in the line or conductor under tension. *Syn:* **swivel.** 431

SWR. *See:* **standing-wave-ratio indicator.**

syllabic companding (modulation systems). Companding in which the gain variations occur at a rate comparable to the syllabic rate of speech; but do not respond to individual cycles of the audio-frequency signal wave. 61

syllable articulation (percent syllable articulation). The percent articulation obtained when the speech units considered are syllables (usually meaningless and usually of the consonant-vowel-consonant type). *See:* **articulation (percent articulation); volume equivalent.** 328

symbol (1)(radio frequency radiation hazard warning symbol). The word refers to the overall design, shape and coloring as shown in the figure below.

BAR SCALE (INCHES)

557

(2) (general). A representation of something by reason of relationship, association, or convention. *See:* **logic symbol.** 255, 77

(3) (packaging machinery). A sign, mark, or drawing agreed upon to represent an electrical device of component part thereof. 429

symbol for a quantity (quantity symbol) (abbreviation). A letter (which may have letters or numbers, or both, as subscripts or superscripts, or both), used to represent a physical quantity or a relationship between quantities. *Compare with:* **abbreviation, functional designation, mathematical symbol, reference designation, and symbol for a unit.** *See:* **abbreviation.** 173

symbol for a unit (unit symbol) (abbreviation). A letter, a character, or combinations thereof, that may be used in place of the name of the unit. With few exceptions, the letter is taken from the name of the unit. *Compare with:* **abbreviation, mathematical symbol, symbol for a quantity.** *See:* **abbreviation.** 173

symbolic address (computing systems). An address expressed in symbols convenient to the programmer. 255, 77

symbolic coding (computing systems). Coding that uses machine instructions with symbolic addresses. 255, 77

symbolic execution (software). A verification technique in which program execution is simulated using symbols rather than actual values for input data, and program outputs are expressed as logical or mathematical expressions involving these symbols. *See:* **data; execution; verification.** 434

symbolic logic. The discipline that treats formal logic by means of a formalized artificial language or symbolic calculus whose purpose is to avoid the ambiguities and logical inadequacies of natural languages. 255, 77

symbolic quantity. *See:* **mathematico-physical quantity.**

symmetrical alternating current. A periodic alternating current in which points one-half a period apart are equal and have opposite signs. *See:* **alternating function; network analysis.** 210

symmetrical component (alternating-current component) (of a total current) (power switchgear). That portion of the total current that constitutes the symmetry. 103

symmetrical components (1) (set of polyphase alternating voltages). The symmetrical components of an unsymmetrical set of sinusoidal polyphase alternating voltages of m phases are the m symmetrical sets of polyphase voltages into which the unsymmetrical set can be uniquely resolved, each component set having an angular phase lag between successive members of the set that is a different integral multiple of the characteristic angular phase difference for the number of phases. The successive component sets will have phase differences that increase from zero for the first set to $(m - 1)$ times the characteristic angular phase difference for the last set. The phase sequence of each component set is identified by the integer that denotes the number of times the angle of lag between successive members of the component set contains the characteristic angular phase difference. If the members of an unsymmetrical set of alternating polyphase voltages are not sinusoidal, each voltage is first resolved into its harmonic components, then the harmonic components of the same period are grouped to form unsymmetrical sets of sinusoidal voltages, and finally each harmonic set of sinusoidal voltages is uniquely resolved into its symmetrical components. Because the resolution of a set of polyphase voltages into its harmonic components is also unique, it follows that the resolution of an unsymmetrical set of polyphase voltages into its symmetrical components is unique. There may be a symmetrical-component set of voltages for each of the possible phase sequences from zero to $(m - 1)$ and for each of the harmonics present from 1 to r, where r may approach infinity in particular cases. Each member of a set of symmetrical component voltages of kth phase sequence and rth harmonic may be denoted by

$$e_{ski} = (2)^{1/2} E_{akr} \cos\left(r\omega t + \alpha_{akr} - (s - 1) K \frac{2\pi}{m}\right)$$

where e_{skr} is the instantaneous voltage component of phase sequence k and harmonic r in phase s. E_{akr} is the root-mean-square amplitude of the voltage component of phase sequence k and harmonic r, using phase a as reference, a_{akr} is the phase angle of the first member of the set, selected as phase a, with respect to a common reference. The letter s as the first subscript denotes the phase identification of the individual member, a, b, c, etcetera for successive members, and a denotes that the first phase, a, has been used as a reference from which other members are specified. The second subscript k denotes the phase sequence of the component, and may run from 0 to $m - 1$. The third subscript denotes the order of the harmonic, and may run from 1 to ∞. The letter s as an algebraic quantity denotes the member of the set and runs from 1 for phase a to m for the last phase. Of the m symmetrical component sets for each harmonic, one will be of zero phase sequence, one of positive phase sequence, and one of negative phase sequence. If the number of phases m ($m > 2$) is even, one of the symmetrical component sets for $k = m/2$ will be a single-phase symmetrical set (polyphase voltages). The zero-phase-sequence component set will constitute a zero-phase symmetrical set (polyphase voltages), and the remaining sequence components will constitute polyphase symmetrical sets (polyphase voltages). *See:* **network analysis.**
210

(2) (set of polyphase alternating currents). Obtained from the corresponding definition for **symmetrical components (set of polyphase alternating voltages)** by substituting the word **current** for **voltage** wherever it appears. *See:* **network analysis.** 210

symmetrical fractional-slot winding (rotating machinery). A distributed winding in which the average number of slots per pole per phase is not integral, but in which the winding pattern repeats after every pair of poles, for example, $3\frac{1}{2}$ slots per pole per phase. *See:* **rotor (rotating machinery); stator.** 63

symmetrically cyclically magnetized condition. A condition of a magnetic material when it is in a cyclically magnetized condition and the limits of the applied magnetizing forces are equal and of opposite sign, so that the limits of flux density are equal and of opposite sign. 210

symmetrical network. *See:* **structurally symmetrical network.**

symmetrical periodic function. A function having the period 2π is symmetrical if it satisfies one or more of the following identities.

$$(1)\ f(x) = -f(-x) \qquad (4)\ f(x) = f(-x)$$
$$(2)\ f(x) = -f(\pi + x) \qquad (5)\ f(x) = f(\pi + x)$$
$$(3)\ f(x) = -f(\pi - x) \qquad (6)\ f(x) = f(\pi - x)$$

See: **network analysis.** 210

symmetrical set (1) (polyphase voltages). A symmetrical set of polyphase voltages of m phases is a set of polyphase voltages in which each voltage is sinusoidal and has the same amplitude, and the set is arranged in such a sequence that the angular phase difference between each member of the set and the one following it, and between the last member and the first, can be expressed as the same multiple of the characteristic angular phase difference $2\pi/m$ radians. A symmetrical set of polyphase voltages may be expressed by the equations

$$e_a = (2)^{1/2} E_{ar} \cos(r\omega t + \alpha_{ar})$$
$$e_b = (2)^{1/2} E_{ar} \cos\left(r\omega t + \alpha_{ar} - k\frac{2\pi}{m}\right)$$
$$e_c = (2)^{1/2} E_{ar} \cos\left(r\omega t + \alpha_{ar} - 2k\frac{2\pi}{m}\right)$$
$$e_m = (2)^{1/2} E_{ar} \cos\left(r\omega t + \alpha_{ar} - (m - 1)k\frac{2\pi}{m}\right)$$

where E_{ar} is the root-mean-square amplitude of each member of the set, r is the order of the harmonic of each member, with respect to a specified period. a_{ar} is the phase angle of the first member of the set with respect to a selected reference. k is an integer that denotes the phase sequence. *Notes:* (1) Although sets of polyphase voltages that have the same amplitude and waveform but that are not sinusoidal possess some of the characteristics of a symmetrical set, only in special cases do the several harmonics have the same phase sequence. Since phase sequence is an important feature in the use of symmetrical sets, the definition is limited to sinusoidal quantities. This represents a change from the corresponding definition in the 1941 edition of the American Standard Definitions of Electrical Terms. (2) This definition may be applied to a two-phase four-wire or five-wire circuit if m is considered to be 4 instead of 2. The concept of symmetrical sets is not directly applicable to a two-phase three-wire circuit. 210

(2) (polyphase currents). This definition is obtained from the corresponding definitions for voltage by substituting the word **current** for **voltage,** and the symbol I for E and β for α wherever they appear. The subscripts are unaltered. *See:* **network analysis.**   210

symmetrical terminal voltage (electromagnetic compatibility). Terminal voltage measured in a delta network across the mains lead. *See:* **electromagnetic compatibility.**   199

symmetrical transducer (specified terminations in general). A transducer in which all possible pairs of specified terminations may be interchanged without affecting transmission. *See:* **transducer.**   210

synapse. The junction between two neural elements, which has the property of one-way propagation. *See:* **biological.**   192

sync (television). An abbreviation for synchronizing signal used extensively in speech and writing. *Note:* This abbreviation is so commonly used that it has achieved the status of a word.   18

synchro control transformer (synchro or selsyn devices). A transformer with relatively rotatable primary and secondary windings. The primary inputs is a set of two or more voltages from a synchro transmitter that define an angular position relative to that of the transmitter. The secondary output voltage varies with the relative angular alignment of primary and secondary windings, of the control transformer and the position of the transmitter. The output voltage is substantially zero in value at a position known as correspondence. *See:* **synchro system.**   63

synchro differential receiver (motor) (synchro or selsyn devices). A transformer identical in construction to a synchro differential transmitter but used to develop a torque increasing with the difference in the relative angular displacement (up to about 90 electrical degrees) between the two sets of voltage input signals to its primary and secondary windings, the torque being in a direction to reduce this difference to zero. *See:* **synchro system.**   63

synchro differential transmitter (generator) (rotating machinery). A transformer with relatively rotatable primary and secondary windings. The primary input is a set of two or more voltages that define an angular position. The secondary output is a set of two or more voltages that represent the sum or difference, depending upon connections, of the position defined by the primary input and the relative angular displacement between primary and secondary windings. *See:* **synchro system.**   63

synchronism (rotating machinery). The state where connected alternating-current systems, machines, or a combination operate at the same frequency and where the phase-angle displacements between voltages in them are constant, or vary about a steady and stable average value. *See:* **asynchronous machine.**   63

synchronism-check relay (power switchgear). A verification relay whose function is to operate when two input voltage phasors are within predetermined limits.   103

synchronization (data transmission). A means of ensuring that both transmitting and receiving stations are operating together (equal scanning line frequencies) in a fixed phase relationship.   59

synchronization error (navigation)(navigation aid terms). The error due to imperfect timing of two operations; this may or may not include signal transmission time.   526

synchronization time (gyro) (inertial sensor). The time interval required for the gyro to reach synchronous speed from standstill.   46

synchronized operation (power operations). An operation wherein power facilities are electrically connected and controlled to operate at the same frequency.   516

synchronized sweep (oscilloscopes) (non-real time spectrum analyzer) (spectrum analyzer). A sweep that would free run in the absence of an applied signal but in the presence of the signal is synchronized by it. *See:* **oscillograph.**   390

synchronizing (1) (rotating machinery). The process whereby a synchronous machine, with its voltage and phase suitably adjusted, is paralleled with another synchronous machine or system. *See:* **asynchronous machine.**   63

(2) (facsimile). The maintenance of predetermined speed relations between the scanning spot and the recording spot within each scanning line. *See:* **facsimile (electrical communication).**   12

(3) (television). Maintaining two or more scanning processes in phase.   178

(4) (pulse terms). The process of rendering a first pulse train or other sequence of events synchronous with a second pulse train.   254

synchronizing coefficient (rotating machinery). The quotient of the shaft power and the angular displacement of the rotor. *Note:* It is expressed in kilowatts per electrical radian. Unless otherwise stated, the value will be for rated voltage, load, power-factor, and frequency. *See:* **asynchronous machine.**   63

synchronizing or synchronism-check device (25) (power system device function numbers). A device that operates when two ac circuits are within the desired limits of frequency, phase angle, and voltage, to permit or to cause the paralleling of these two circuits.   402

synchronizing reactor (power and distribution transformer). A current-limiting reactor for connecting momentarily across the open contacts of a circuit-interrupting device for synchronizing purposes.   53

synchronizing relay (power switchgear). A programming relay whose function is to initiate the closing of a circuit breaker between two ac sources when the voltages of these two sources have a predetermined relationship of magnitude, phase angle, and frequency.   103

synchronizing signal (1) (television). The signal employed for the synchronizing of scanning. *Note:* In television, this signal is composed of pulses at rates related to the line and field frequencies. The signal usually originates in a central synchronizing generator and is added to the combination of picture signal and blanking signal, comprising the output signal from the pickup equipment, to form the composite picture signal. In a television receiver, this signal is normally separated from the picture signal and is used to synchronize the deflection generators.   178

(2) (facsimile). A signal used for maintenance of predetermined speed relations between the scanning spot and recording spot within each scanning line. *See:* **facsimile signal (picture signal).**   12

(3) (oscillograph). A signal used to synchronize repetitive functions. *See:* **oscillograph.**   185

(4) (telecommunication). A special signal which may be sent to establish or maintain a fixed relationship in synchronous systems.   194

synchronizing signal compression (television). The reduction in gain applied to the synchronizing signal over any part of its amplitude range with respect to the gain at a specified reference level. *Notes:* (1) The gain referred to in the definition is for a signal amplitude small in comparison with the total peak-to-peak composite picture signal involved. A quantitative evaluation of this effect can be obtained by a measurement of differential gain. (2) Frequently the gain at the level of the peaks of synchronizing pulses is reduced with respect to the gain at the levels near the bases of the synchronizing pulses. Under some conditions, the gain over the entire synchronizing signal region of the composite picture signal may be reduced with respect to the gain in the region of the picture signal. *See:* **television.**   178

synchronizing signal level (television). The level of the peaks of the synchronizing signal. *See:* **television.**   178

synchronizing torque (synchronous machine). The torque produced, primarily through interaction between the armature currents and the flux produced by the field winding, tending to pull the machine into synchronism with a connected power system or with another synchronous machine.   63

synchronous booster converter. A synchronous converter having a mechanically connected alternating-current reversible booster connected in series with the alternating-current supply circuit for the purpose of adjusting the output voltage. *See:* **converter.**   328

synchronous booster inverter. An inverter having a mechanically connected reversible synchronous booster connected in series for the purpose of adjusting the output voltage. *See:* **converter.**   328

synchronous capacitor (synchronous condenser*) (rotating machinery) A synchronous machine running without mechanical load and supplying or absorbing reactive power to or from a power system. *See:* **converter.**

*Deprecated   63

synchronous computer. A computer in which each event, or the performance of each operation, starts as a result of a signal generated by a clock.   255, 77

synchronous condenser (1) (rotating machinery). A synchronous machine running without mechanical load and supplying or absorbing reactive power. *Syn:* **synchronous compensator.**   63

(2) (electric installations on shipboard). A synchronous phase modifier running without mechanical load, the field excitation of which may be varied so as to modify the power factor of the system: or through such modification, to influence the load voltage.   3

synchronous converter (electric installations on shipboard). A converter which combines both motor and generator action in one armature winding and is excited by one magnetic field. It is normally used to change ac power to dc power.   3

synchronous coupling (electric coupling). An electric coupling in which torque is transmitted by attraction between magnetic poles on both rotating members which revolve at the same speed. The magnetic poles may be produced by direct current excitation, permanent magnet excitation, or alternating current excitation, and those on one rotating member may be salient reluctance poles.   416

synchronous device (data transmission). A device whose speed of operation is related to the rest of the system to which the device is connected.   59

synchronous gate. A time gate wherein the output intervals are synchronized with an incoming signal.   111

synchronous generator (electric installations on shipboard). A synchronous alternating-current machine which transforms mechanical power into electric power. (A synchronous machine is one in which the average speed of normal operation is exactly proportional to the frequency of the system to which it is connected.)   3

synchronous impedance (1) (per unit direct-axis). The ratio of the field current at rated armature current on sustained symmetrical short-circuit to the field current at normal open-circuit voltage on the air-gap line. *Note:* This definition of synchronous impedance is used to a great extent in electrical literature and corresponds to the definition of direct-axis synchronous reactance as determined from open-circuit and sustained short-circuit tests. *See:* **positive phase-sequence reactance (rotating machinery).**   328

(2) (rotating machinery). The ratio of the value of the phasor difference between the synchronous internal voltage and the terminal voltage of a synchronous machine to the armature current under a balanced steady-state condition. *Note:* This definition is of rigorous application to turbine type machines only, but it gives a good degree of approximation for salient pole machines. *See:* **synchronous reactance (effective).**   63

synchronous induction motor (rotating machinery). A cylindrical rotor synchronous motor having a secondary coil winding similar to that of a wound rotor induction motor. *Note:* This winding is used for both starting and excitation purposes.   63

synchronous internal voltage (synchronous machine for any specified operating conditions). The fundamental-frequency component of the voltage of each armature phase that would be produced by the steady (or very slowly varying) component of the current in the main field winding (or field windings) acting alone provided the permeance of all parts of the magnetic circuit remained the same as for the specified operating condition. *Note:* The synchronous internal voltage, as shown in the phasor diagram, is related to the terminal-voltage and phase-current phasors by the equation

$$\mathbf{E}_i = \mathbf{E}_a + R\mathbf{I}_a + jX_d\mathbf{I}_{ad} + jX_q\mathbf{I}_{aq}$$

For a machine subject to saturation, the reactances should be determined for the degree of saturation applicable to the specified operating condition. 63

synchronous inverter. An inverter that combines both motor and generator action in one armature winding. It is excited by one magnetic field and changes direct-current power to alternating-current power. *Note:* Usually it has no amortisseur winding. *See:* **converter.** 328

synchronous machine (rotating machinery). A machine in which the average speed of normal operation is exactly proportional to the frequency of the system to which it is connected. 3

synchronous machine, ideal. A hypothetical synchronous machine that has certain idealized characteristics that facilitate analysis. *Note:* The results of the analysis of ideal machines may be applied to similar actual machines by making, when necessary, approximate corrections for the deviations of the actual machine from the ideal machine. The ideal machine has, in general, the following properties; (1) the resitance of each winding is constant throughout the analysis, independent of current magnitude or its rate of change; (2) the permeance of each portion of the magnetic circuit is constant throughout the analysis, regardless of the flux density; (3) the armature circuits are symmetrical with respect to each other; (4) the electric and magnetic circuits of the field structure are symmetrical about the direct axis of the quadrature axis; (5) the self-inductance of the field, and every circuit on the field structure is constant; (6) the self-inductance of each armature circuit is a constant or a constant plus a second-harmonic sinusoidal function of the angular position of the rotor relative to the stator; (7) the mutual inductance between any circuit on the field structure and any armature circuit is a fundamental sinusoidal function of the angular position of the rotor relative to the stator; (8) the mutual inductance between any two armature circuits is a constant or a constant plus a second-harmonic sinusoidal function of the angular position of the rotor relative to the stator; (9) the amplitude of the second-harmonic component of variation of the self-inductance of the armature circuits and of the mutual inductances between any two armature circuits is the same; (10) effects of hysteresis are negligible; (11) effects of eddy currents are negligible or, in the case of solid-rotor machines, may be represented by hypothetical circuits on the field structure symmetrical about the direct axis and the quadrature axis. 63

synchronous machine regulator (excitation systems for synchronous machines). A regulator that couples the output variables of the synchronous machine to the output of the exciter through feedback and forward controlling elements for the purpose of regulating the synchronous machine output variables. 507

synchronous machines (electric installations on shipboard). A machine in which the average speed of normal operation is exactly proportional to the frequency of the system to which it is connected. 3

synchronous motor (1) (rotating machinery). A synchronous machine that transforms electric power into mechanical power. Unless otherwise stated, it is generally understood that it has field poles excited by direct current. *See:* **asynchronous machine.** 63
(2) (electric installations on shipboard). A synchronous machine which transforms electric power into mechanical power. Unless otherwise stated, it is generally understood that a synchronous generator (or motor) has field magnetics excited with direct current. 3

synchronous operation (opening or closing) (power switchgear). Operation of a switching device in such a manner that the contacts are closed or opened at a predetermined point on a reference voltage or current wave. *Note:* Synchronous operation applied on multiphase circuits may require that closing or opening of the contacts of each pole be responsive to a different reference. 103

synchronous reactance (1) (effective) (rotating machinery). An assumed value of synchronous reactance used to represent a machine in a system study calculation for a particular operating condition. *Note:* The synchronous internal voltage, as shown in the phasor diagram, is related to the terminal-voltage and phase-current phasors by the equation

$$\mathbf{E'}_i = \mathbf{E}_a + R\mathbf{I}_a + jX_{eff}\mathbf{I}_a$$

See: **synchronous internal voltage, synchronous impedance.** 63
(2) (power fault effects). The steady state reactance of a generator during fault conditions used to calculate the steady state fault current. The current so calculated excludes the effect of the automatic voltage regulator or governor. 404

synchronous satellite (navigation aid terms). An equatorial satellite orbiting the earth in a west-to-east direction at an altitude of approximately 35 900 kms (kilometers). At this altitude the satellite makes one revolution in 24 h (hours), synchronous with the earth's rotation. 526

synchronous speed (rotating machinery). The speed of rotation of the magnetic flux, produced by or linking the primary winding. 63

synchronous-speed device (13) (power system device function numbers). A device such as a centrifugal-speed switch, a slip-frequency relay, a voltage relay, an undercurrent relay, or any type of device that operates at approximately the synchronous speed of a machine. 402

synchronous system (data transmission). A system in which the sending and receiving instruments are operating continuously at substantially the same rate and are maintained by means of correction if necessary, in a fixed relationship.   59

synchronous transmission (data transmission). A mode of data transmission in which the sending and receiving data processing terminal equipments are operating continuously at substantially the same frequency and are maintained in a desired phase relationship by an appropriate means.   59

synchronous-vibration sensitivity (dynamically tuned gyro) (inertial sensor). The functions that relate drift rates to linear or angular vibrations which are phase coherent with spin frequency or its harmonics. *See:* **one-N (1N) translational sensitivity; two-N (2N) translational sensitivity; two-N (2N) angular sensitivity.**   46

synchronous voltage (traveling-wave tubes). The voltage required to accelerate electrons from rest to a velocity equal to the phase velocity of a wave in the absence of electron flow. *See:* **magnetron.**   190

synchro receiver (or motor) (rotating machinery). A transformer electrically similar to a synchro transmitter and that, when the secondary windings of the two devices are interconnected, develops a torque increasing with the difference in angular alignment of the transmitter and receiver rotors and in a direction to reduce the difference toward zero. *See:* **synchro system.**   63

synchroscope. An instrument for indicating whether two periodic quantities are synchronous. It usually embodies a continuously rotatable element the position of which at any time is a measure of the instantaneous phase difference between the quantities: while its speed of rotation indicates the frequency difference between the quantities: and its direction of rotation indicates which of the quantities is of higher frequency. *Note:* This term is also used to designate a cathode-ray oscilloscope providing either (1) a rotating pattern giving indications similar to that of the conventional synchroscope, or (2) a triggered sweep, giving an indication of synchronism. *See:* **instrument.**   328

synchro system (alternating current). An electric system for transmitting angular position or motion. It consists of one or more synchro transmitters, one or more synchro receivers or synchro control transformers and may include differential synchro machines.   63

synchro transmitter (or generator) (rotating machinery). A transformer with relatively rotatable primary and secondary windings, the output of the secondary winding being two or more voltages that vary with and completely define the relative angular position of the primary and secondary windings. *See:* **synchro system.**   63

synchrotron. A device for accelerating charged particles (for example, electrons) to high energies in a vacuum. The particles are guided by a changing magnetic field while they are accelerated many times in a closed path by a radio-frequency electric field.   190

sync signal. *See:* **synchronizing signal.**

syntax (1) (computing systems). (A) The structure of expressions in a language. (B) The rules governing the structure of a language.   255, 77
(2) The grammatical rules pertaining to the structure of an ATLAS statement.   400
(3) **(software).** (A) The relationship among characters or groups of characters, independent of their meanings or the manner of their interpretation and use. (B) The structure of expressions in a language. (C) The rules governing the structure of a language. *See:* **semantics.**   434

synthetic aperture radar (navigation aid terms). A radar system that generates the effect of a long antenna by signal processing means rather than by the actual use of a long physical antenna.   526

synthetic test (ac high-voltage circuit breaker). A test in which a major part of, or the total current, is obtained from one source (current circuit), and the major part of, or all of the transient recovery voltage from a separate source or sources (voltage circuit).   426

system (1)(general-system term)(696 interface devices). A set of interconnected elements constituted to achieve a given objective by performing specified functions.   538
(2)**(general system terms)(microcomputer system bus).** A set of interconnected elements which achieve a given objective through the performance of a specified function.   542
(3)**(monitoring radioactivity in effluents).** The entire assembled equipment excluding only the sample collecting pipe.   559
(4)**(power operations).** A group of components connected or associated in a fixed configuration to perform a specified function. *See:* **bulk power system; distribution system; system operator; transmission system.**   516
(5)**(reliability data for pumps and drivers, valve actuators, and valves).** A collection of components arranged to provide a desired function (for example, containment spray system, residual heat removal system, high pressure coolant injection system).   502
(6)**(reliable industrial and commercial power systems planning and design).** A group of components connected or associated in a fixed configuration to perform a specified function of distributing power.   561
(7)**(seismic design of substations).** A group of components operating together to perform a function (for example, disconnect switch, support structure and foundation, relay protection system, and telemetering system).   465
(8)**(unique identification in power plants and related facilities).** A combination of two or more interrelated components that perform a specific function related to plant operation and safety. A system may peform a function such as control, monitoring, electrical, mechanical, or structural.   545
(9) **(systems, man, and cybernetics).** An integrated whole even though composed of diverse, interacting, specialized structures and subjunctions. *Notes:* (A)

Any system has a number of objectives and the weights placed on them may differ widely from system to system. (B) A system performs a function not possible with any of the individual parts. Complexity of the combination is implied. 209

(10) (software). (A) A collection of people, machines, and methods organized to accomplish a set of specific functions. (B) An integrated whole that is composed of diverse, interacting, specialized structures and subfunctions. (C) A group or subsystem united by some interaction or interdependence, performing many duties but functioning as a single unit. *See:* **function; subsystem.** 434

(11) (programmable instrumentation). A set of interconnected elements constituted to achieve a given objective by performing a specified function. 378

(12) (controlling). *See:* **controlling system.** 91

(13) (electric and electronics parts and equipments). A combination of two or more sets, generally physically separated when in operation, and such other units, assemblies, and basic parts necessary to perform an operational function or functions. Typical examples: telephone carrier system, ground-controlled approach (GCA) electronic system, telemetering system, facsimile transmission system. 17

system architecture (software). The structure and relationship among the components of a system. The system architecture may also include the system's interface with its operational environment. *See:* **component; interface; system.** 434

system assured capability (power operations). The dependable capability of all power sources available to a system under short range conditions, including firm power contracts, less that reserve assigned to provide for planned outages, equipment and operating limitations, and unplanned outages of power sources. 516

systematic drift rate (gyro). That component of drift rate that is correlated with specific operating conditions. It is composed of acceleration-sensitive drift rate and acceleration-insensitive drift rate. It is expressed as angular displacement per unit time. 46

systematic error (1) (general) (electrothermic power meters). The inherent bias (off-set) of a measurement process or (of) one of its components. 47

(2) (electronic navigation). Error capable of identification due to its orderly character. *See:* **navigation.** 13, 187

system control (telephone switching systems). The means for collecting and processing pulsing and supervisory signals in a switching system. 55

system delay time (mobile communication). The time required for the transmitter associated with the system to provide rated radio-frequency output after activation of the local control (push to talk) plus the time required for the system receiver to provide useful output. *See:* **mobile communication system.** 181

system demand factor. *See:* **demand factor.**

system design (software). (1) The process of defining the hardware and software architectures, components, modules, interfaces, and data for a system to satisfy specified system requirements. (2) The result of the system design process. *See:* **component; data; hardware; interface; module; requirement; software; system.** 434

system deviation (control). *See:* **deviation, system.**

system, directly controlled. That portion of the controlled system that is directly guided or restrained by the final controlling element to achieve a prescribed value of the directly controlled variable. *See:* **control system, feedback.** 329, 206

system, discrete (control system). A system whose signals are inherently discrete. *See:* **control system.** 329

system, discrete-state (control system) (system, finite-state). A system whose state is defined only for discrete values of time and amplitude. *See:* **control system.** 329

system diversity factor. *See:* **diversity factor.**

system documentation (software). Documentation conveying the requirements, design philosophy, design details, capabilities, limitations, and other characteristics of a system. *See:* **design; documentation; requirement; system; user documentation.** 434

system element. One or more basic elements with other components and necessary parts to form all or a significant part of one of the general functional groups into which a measurement system can be classified. While a system element must be functionally distinct from other such elements it is not necessarily a separate measurement device. Typical examples of system elements are: a thermocouple, a measurement amplifier, a millivoltmeter. *See:* **measurement system.** 328

system, finite-state (control system). *See:* **system, discrete state.**

system frequency (electric power system). Frequency in hertz of the power system alternating voltage. 94

system frequency stability (radio system) (mobile communication). The measure of the ability of all stations, including all transmitters and receivers, to remain on an assigned frequency-channel as determined on both a short-term and long-term basis. *See:* **mobile communication system.** 181

system ground (surge arresters). The connection between a grounding system and a point of an electric circuit (for example, a neutral point). 244, 62

system grounding conductor. An auxiliary solidly grounded conductor that connects together the individual grounding conductors in a given area. *Note:* This conductor is not normally a part of any current-carrying circuit including the system neutral. *See:* **ground.** 64

system handshake (FASTBUS acquisition and control). A handshake in a broadcast operation where the handshake signal is from the last segment of the addressed system rather than from individual devices. 480

system, idealized (automatic control). An imaginary system whose ultimately controlled variable has a stipulated relationship to specified commands. *Note:* It is a basis for performance standards. *See:* **control system, feedback.** 56, 329

system, indirectly controlled (industrial control). The portion of the controlled system in which the indirectly controlled variable is changed in response to changes in the directly controlled variable. *See:* **control system, feedback.** 206

system interconnection. The connecting together of two or more power systems. *See:* **alternating-current distribution; direct-current distribution.** 64

system library (software). A controlled collection of system-resident software that can be accessed for use or incorporated into other programs by reference, for example, a group of routines that a linkage editor can incorporate into a program as required. *See:* **linkage editor; program; routine; software library; system-resident software.** 434

system load (power operations). Equal to internal load plus pumping load plus firm sales for resale. 516

system logic (nuclear power generating stations). That equipment that monitors the output of two or more channels and supplies output signals in accordance with a prescribed combination rule (for example, 2 of 3, 2 of 4, etcetera). 109,102

system loss (radio wave propagation). Of a radio system the transmission loss plus the losses in the transmitting and receiving antennas. 146

system margin capability (power operations). The difference between system capability and system load. 516

system matrix (control system). A matrix of transfer functions that relate the Laplace transforms of the system outputs and of the system inputs. *See:* **control system.** 329

system maximum hourly load (power operations). The maximum hourly integrated system load. This is an energy quantity usually expressed in kilowatthours per hour (kWh/h). 516

system monitor and control subsystem (terrestrial photovoltaic power systems). Logic and control circuitry that supervises the overall operation of the system by controlling the interaction between all subsystems. *See:* **array control** 496

system, multidimensional. *See:* **multidimensional system.**

system, multivariable. *See:* **multivariable system.**

system noise (sound recording and reproducing system). The noise output that arises within or is generated by the system or any of its components, including the medium. *See:* **noise (sound recording and reproducing system).**

system of units. A set of interrelated units for expressing the magnitudes of a number of different quantities. 210

system operator (National Electrical Safety Code). A person designated to operate the system or parts thereof. 391

system operator (power operations). Electric utility personnel in charge of system-wide coordination of generation, interchange, and transmission security. 516

system overshoot (control) (industrial control). The largest value of system deviation following the first dynamic crossing of the ideal value in the direction of correction, after the application of a specified stimulus. *See:* **control system, feedback.** 206

system performance testing (nuclear power generating stations). Tests performed on completed systems, including all their electric, instrumentation, controls, fluid, and mechanical subsystems under normal or simulated normal process conditions of temperature, flow, level, pressure, etcetera. *Syn:* **preoperational testing.** 143

system, quantized. *See:* **quantized system.**

system recovery time (mobile communication). The elapsed time from deactivation of the local transmitter control until the local receiver is capable of producing useful output. *See:* **mobile communication system.** 181

system reliability (software). The probability that a system, including all hardware and software subsystems, will perform a required task or mission for a specified time in a specified environment. *See:* **hardware; operational reliability; software reliability; software subsystem; system.** 434

system reserve. The capacity, in equipment and conductors, installed on the system in excess of that required to carry the peak load. *See:* **generating station.** 64

system routing code (telephone switching systems). In World Zone 1, a three-digit code consisting of a country code and two additional numerals that uniquely identifies an international switching center. *See:* **world zone number.** 55

system, sampled-data (control system). One in which at least one sampled signal is present. *See:* **control system.** 329

system science (systems, man, amd cybernetics). The branch of organized knowledge dealing with systems and their properties, the systematized knowledge of systems. *See:* **adaptive system; cybernetics; learning system; system; systems engineering; tradeoff.** 209

systems engineering (systems, man, and cybernetics). The application of the mathematical and physical sciences to develop systems that utilize economically the materials and forces of nature for the benefit of mankind. *See:* **system science.** 209

system software. Software designed for a specific computer system or family of computer systems to facilitate the operation and maintenance of the computer system and associated programs, for example, operating systems, compilers, utilities. *See:* **application software; compiler; computer system; maintenance; operating system; program; software.** 434

system structure (unique identification in power plants). A combination of two or more integrated components, generally physically remote or occupying a large area, interacting to perform a specific function important to plant operation or safety, or both. A system may be civil/structural, that is, a building or structure, mechanical/fluid, or electrical/control. A system, for the purpose of ANSI/IEE Std 803-1983, will not be considered a subsystem of another system. 544

systemtag (microprocessor operating systems parameter types). A 'tag' returned by one function for use by another. Its contents may not be examined or changed. Its form is system dependent. A systemtag is valid system-wide (that is, global) and may be passed between processes. 478

system testing (1)(software). The process of testing an integrated hardware and software system to verify that the system meets its specified requirements. *See:* acceptance testing; hardware; qualification testing; requirement; software system; testing. 434
(2)(software verification and validation plans). The process of testing an integrated hardware and software system to verify that the system meets its specified requirements. 511

system time (supervisory control, data acquisition, and automatic control). A coordinated value of time maintained at stations throughout the power system. 570

system utilization factor. *See:* utilization factor.
system validation. *See:* validation.
system verification. *See:* verification.
system voltage (1)(metal-oxide surge arresters for ac power circuits)(surge arrester). The root-mean-square (rms) power-frequency voltage from line-to-line as distinguished from the voltage from line-to-neutral. 583, 430
(2)(power and distribution transformer). A root-mean-square (rms) phase-to-phase power frequency voltage on a three-phase alternating-current electrical system. 53
system voltage classes. *See:* low voltage; medium voltage; high voltage; voltage classes chart. 260

T

T **(video signal transmission measurement) (linear waveform distortion).** Letter symbol for the duration of a half-period of the nominal upper cut-off frequency of a transmission system. Therefore

$$T = \frac{1}{2f_c}$$

Note: For the TV system M

$$T = \frac{1}{2 \times 4 \text{ (MHz)}} = 125 \text{ (ns)}$$

The duration *T* is commonly referred to as the Nyquist interval. The concept of *T* is employed not only when the frequency cut-off is a physical property of a given system but also when the system is flat and there is no interest in the performance of the system beyond a given frequency. 42

table (software). (1) An array of data, each item of which may be unambiguously identified by means of one or more arguments. (2) A collection of data in which each item is uniquely identified by a label, by its position relative to the other items, or by some other means. *See:* data; label. 434
table lamp (illuminating engineering). A portable luminaire with a short stand suitable for standing on furniture. 167
table look-up. A procedure for obtaining the function value corresponding to an argument from a table of function values. 255, 77
tab sequential format (numerically controlled machines). A means of identifying a word by the number of tab characters preceding the word in the block. The first character in each word is a tab character. Words must be presented in a specific order but all characters in a word, except the tab character, may be omitted

when the command represented by that word is not desired. 207
tabulate. (1) To form data into a table. (2) To print totals. 255, 77
tacan (tactical air navigation)(navigation aid terms). A complete ultra-high frequency (uhf), polar coordinate (rho theta) navigation system using pulse techniques. The distance (rho) function operates as DME (distance measuring equipment) and the bearing function is derived by rotating the ground transponder antenna so as to obtain a rotating multilobe pattern for coarse and fine bearing information. 526
tachometer. A device to measure speed or rotation. *See:* rotor (rotating machinery). 63
tachometer electric indicator. A device that provides an indication of the speed of an aircraft engine, of a helicopter rotor, of a jet engine, and of similar rotating apparatus used in aircraft. Such tachometer indicators may be calibrated directly in revolutions per minute or in percent of some particular speed in revolutions per minute. 328
tachometer generator (rotating machinery). A generator, mechanically coupled to an engine, whose main function is to generate a voltage, the magnitude or frequency of which is used either to determine the speed of rotation of the common shaft or to supply a signal to a control circuit to provide speed regulation. 63
tachometric relay (industrial control). A relay in which actuation of the contacts is effected at a predetermined speed of a moving part. *See:* relay. 206
taffrail log. A device that indicates distance traveled based on the rotation of a screw-type rotor towed behind a ship which drives, through the towing line, a counter mounted on the taffrail. *Note:* An electric contact made (usually) each tenth of a mile causes an

audible signal to permit ready calculation of speed.
                                                                328

tag (1)(NESC). Accident prevention tag (DANGER, PEOPLE AT WORK, etcetera) of a distinctive appearance used for the purpose of personnel protection to indicate that the operation of the device to which it is attached is restricted.   494
(2)(supervisory control, data acquisition, and automatic control). A visual indication, usually at the master station, to indicate that a device has been cleared for field maintenance/construction purposes and is not available for control or data acquisition.
                                                                570
(3) (computing system). (A) Same as flag. (B) Same as label.   77, 54
tag address. *See:* **symbolic address.**
tag line (conductor stringing equipment). A control line, normally manila or synthetic fiber rope, attached to a suspended load to enable a workman to control its movement.   431
tags. Men-at-work tags of distinctive appearance, indicating that the equipment or lines so marked are being worked on.   262
tailing (1) (hangover) (facsimile). The excessive prolongation of the decay of the signal. *See:* **facsimile signal (picture signal).**   12
(2) (electrometallurgy) (hydrometallurgy and ore concentration). The discarded residue after treatment of an ore to remove desirable minerals. *See:* **electrowinning.**   328
tailing time (of an amplifier output pulse)(charged-particle detectors). The time between the 100 percent and 1 percent of maximum amplitude points on the trailing edge of a pulse (provided that the pulse does not have a flat top). For flat topped pulses the tailing time is defined as the time between the midpoint of the flat top and the 1 percent of maximum amplitude point.   119
tail lamp (illuminating engineering). A lighting device used to designate the rear of a vehicle by a warning light.   167
tail-of-wave (chopped wave) impulse test voltage (insulation strength). The crest voltage of a standard impulse wave that is chopped by flashover at or after crest.   287
tail or tailing (on a monoenergetic peak)(X-ray energy spectrometers). Any peak shape distortion that does not comply with the limits defining the full energy peak intensity and that does not come from a source of radiation other than the monoenergetic source in question.   471
talker echo. Echo that reaches the ear of the talker.
                                                                328
talking key. A key whose operation permits conversation over the circuit to which the key is connected.
                                                                328
talk-ringing key (listening and ringing key). A combined talking key and ringing key operated by one handle.   328
tamper-resistant (enclosure). A metal-enclosure for a power switchgear assembly which is designed to resist

damage to or improper operation of the switchgear from willful acts of destruction and which is designed to provide reasonably safe protection against tampering by unauthorized persons who may attempt to gain entry by forcible means, to insert foreign substances into, or otherwise tamper with the assembly.   93
tandem (cascade) (network) (circuits and systems). Networks are in tandem when the output terminals of one network are directly connected to the input terminals of the other network. *See:* **network analysis.**
                                                                67
tandem central office (tandem office). A central office used primarily as a switching point for traffic between other central offices.   93
tandem-completing trunk. A trunk, extending from a tandem office to a central office, used as part of a telephone connection between stations.   328
tandem control (electric power systems). A means of control whereby the area control error of an area or areas A, connected to the interconnected system B only through the facilities of another area C, is included in control of area C generation.   94
tandem drive (industrial control). Two or more drives that are mechanically coupled together. *See:* **electric drive.**   219, 206
tandem office (telephone switching systems). An intermediate office used primarily for interconnecting end offices with each other and with toll connecting trunks.   55
tandem trunk (data transmission). A trunk extending from a central office or a tandem office to a tandem office and used as part of a telephone connection between stations.   59
tangential wave path (data transmission). In radio wave propagation over the earth, a path of propagation of a direct wave, which is tangential to the surface of the earth. The tangential wave path is curved by atmospheric refraction.   59
tank (storage cell). A lead container, supported by wood, for the element and electrolyte of a storage cell. *Note:* This is restricted to some relatively large types of lead-acid cells. *See:* **battery (primary or secondary).**   328
tank circuit (signal-transmission system). A circuit consisting of inductance and capacitance, capable of storing electric energy over a band of frequencies continuously distributed about a single frequency at which the circuit is said to be resonant, or tuned. *Note:* The selectivity of the circuit is proportional to the ratio of the energy stored in the circuit to the energy dissipated. The ratio is often called the Q of the circuit. *See:* **dielectric heating; oscillatory circuit;** Q **quality factor; signal.**   111
tank, single-anode. *See:* **single-anode tank.**
tank voltage. The total potential drop between the anode and cathode bus bars during electrodeposition. *See:* **electroplating.**   328
tap (1) (fiber optics). A device for extracting a portion of the optical signal from a fiber.   433
(2) (in a transformer) (power and distribution transformer). A connection brought out of a winding at

some point between its extremities, to permit changing the voltage, or current, ratio.   53

(3) (general). An available connection that permits changing the active portion of the device in the circuit. *See:* **grounding device.**   91

(4) (reactor). A connection brought out of a winding at some point between its extremities, to permit changing the impedance. *See:* **reactor.**   309

(5) (rotating machinery). A connection made at some intermediate point in a winding. *See:* **rotor (rotating machinery); stator; voltage regulator.**   63

tap-changer, for deenergized operation (power and distribution transformer). A selector switch device used to change transformer taps with the transformer de-energized.   53

tape (1) (rotating machinery). A relatively narrow, long, thin, flexible fabric, mat, or film, or a combination of them with or without binder, not over 20 centimeters in width. *See:* **rotor (rotating machinery); stator.**

(2) (electronic computation). *See:* **magnetic tape.**   63

tape block (test, measurement and diagnostic equipment). A group of frames or tape lines.   54

taped insulation. Insulation of helically wound tapes applied over a conductor or over an assembled group of insulated conductors. (1) When successive convolutions of a tape overlie each other for a fraction of the tape width, the taped insulation is lap wound. This is also called positive lap wound. (2) When a tape is applied so that there is an open space between successive convolutions, this construction is known as open butt or negative lap wound. (3) When a tape is applied so that the space between successive convolutions is too small to measure with the unaided eye, it is a closed butt taping.   64

taped joint (power cable joint). A joint with hand-applied tape insulation.   34

tape drive. A device that moves tape past a head.   255, 77

tape line. *See:* **frame.**

tape preparation. The act of translating command information into punched or magnetic tape.   207

tape punch. *See:* **perforator.**

taper (communication practice). A continuous or gradual change in electrical properties with length, as obtained, for example, by a continuous change of cross-section of a waveguide. *See:* **transmission line.**   210

tape recorder. *See:* **magnetic recorder.**

tapered fiber waveguide (fiber optics). An optical waveguide whose transverse dimensions vary monotonically with length. *Syn:* **tapered transmission line.**   433

tapered key (rotating machinery). A wedge-shaped key to be driven into place, in a matching hole or recess.   63

tapered transmission line (1) (waveguide terms). *See:* **tapered waveguide.**   267

(2) (fiber optics). *See:* **tapered fiber waveguide.**   433

tapered waveguide (1) (waveguide terms). A waveguide in which a physical or electrical characteristic increases or decreases continuously with distance along the axis of the guide.   267

(2) (waveguide components). A waveguide or transmission line in which a physical or electrical characteristic changes progressively with distance along the axis of the guide.   166

taper, waveguide. A section of tapered waveguide. *See:* **waveguide.**   166

tape station. *See:* **tape unit.**

tape thickness. The lesser of the cross-sectional dimensions of a length of ferromagnetic tape. *See:* **tape-wound core.**   331

tape to card. Pertaining to equipment or methods that transmit data from either magnetic tape or punched tape to punched cards.   255, 77

tape transmitter (telegraphy). A machine for keying telegraph code signals previously recorded on tape. *See:* **telegraphy.**   194

tape transport (test, measurement and diagnostic equipment). A device which moves magnetic or punched tape past the tape reader. Reels for storage of the tape are usually provided. *See:* **tape drive.**   54

tape unit. A device containing a tape drive, together with reading and writing heads and associated controls.   255, 77

tape width. The greater of the cross-sectional dimensions of a length of ferromagnetic tape. *See:* **tape-wound core.**   331

tape-wound core. A length of ferromagnetic tape coiled about an axis in such a way that one convolution falls directly upon the preceding convolution. *See:* **wrap thickness.**   331

tapped field. *See:* **short field.**

tapped field control. A system of regulating the tractive force of an electrically driven vehicle by changing the number of effective turns of the traction motor series-field windings by means of an intermediate tap or taps in those windings. *See:* **multiple-unit control.**   328

tapped-secondary current or voltage transformer (instrument transformer). One with two ratios, obtained by use of a tap on the secondary.   203

tapped way (power switchgear). A way solidly connected to the bus.   489

tapper bell. A single-stroke bell having a gong designed to produce a sound of low intensity and relatively high pitch. *See:* **protective signaling.**   328

target (1)(radar)(navigation aid terms). (A) Specifically, an object of radar search or tracking. (B) Broadly, an discrete object which reflects energy back to the radar.   526

(2) (camera tube). A structure employing a storage surface that is scanned by an electron beam to generate a signal output current corresponding to a charge-density pattern stored thereon. *Note:* The structure may include the storage surface that is scanned by an electron beam, the backplate, and the intervening dielectric. *See:* **radar; television.**   190, 178, 125

(3) (storage tubes). The storage surface and its immediate supporting electrodes. *See:* **radar; storage tube.**
174, 190

(4) (operation indicator) (of a relay) (power switchgear). A supplementary device operated either mechanically or electrically, to indicate visibly that the relay has operated or completed its function. *Notes:* (A) A mechanically operated target indicates the physical operation of the relay. (B) An electrically operated target when not further described is actuated by the current in the control circuit associated with the relay and hence indicates not only that the relay has operated but also that it has completed its function by causing current to flow in the associated control circuit. (C) A shunt-energized target only indicates operation of the relay contact and does not necessarily show that current has actually flowed in the associated control circuit.
103

target capacitance (camera tubes). The capacitance between the scanned area of the target and the backplate. *See:* **television.**
178, 190, 125

target cutoff voltage (camera tubes). The lowest target voltage at which any detectable electric signal corresponding to a light image on the sensitive surface of the tube can be obtained. *See:* **television.**
125

target fluctuation (radar). Variation in the amplitude of a target signal, caused by changes in target aspect angle, rotation, or vibration of target scattering sources, or changes in radar wavelength (the amplitude component of target noise). Rapid fluctuation is usually modeled as independent from pulse to pulse, and slow fluctuation is usually modeled as constant from pulse to pulse within a scan and independent from scan to scan. The terms scintillation and amplitude noise have been used in the past as synonyms for target fluctuation and also to denote location errors caused by target fluctuation, and should be avoided because of this ambiguity.
13

target glint (radar). *See:* **glint.**

target language (1). A language that is an output from a given translation process. *See:* **object language.**
255, 77

(2) (software). A language into which source statements are translated. *See:* **source language.**
434

target machine (software). (1) The computer on which a program is intended to operate. (2) The computer being emulated by another computer. *See:* **computer; host machine; program.**
434

target noise (radar). Random variations in observed amplitude, location, and Doppler of a target, caused by changes in target aspect angle, rotation or vibration of target scattering sources, or changes in radar wavelength. *See:* **target fluctuation; scintillation error; glint.**
13

target program. A program written in a target language. *See:* **object program.**
255, 77

target transmitter (electronic navigation). A source of radio-frequency energy suitable for providing test signals at a test site. *See:* **navigation.**
278, 187, 13

target transmitter (navigation aid terms). A source of radio-frequency energy suitable for providing test signals at a test site.
526

target voltage (camera tube with low-velocity scanning). The potential difference between the thermionic cathode and the backplate. *See:* **television.**
178, 190, 125

tariff (1) (data transmission). The published rate for a particular approved commercial service of a common carrier.
59

(2) (electric power utilization)(power operations). A published volume of rate schedules and general terms and conditions.
112,516

tarnish (corrosion). Surface discoloration of a metal caused by formation of a thin film of corrosion product.
205

task illumination (National Electrical Code) (health care facilities). Provision for the minimum lighting required to carry out necessary tasks in the described areas, including safe access to supplies and equipment, and access to exits.
256

taut-band suspension (electric instrument). A mechanical arrangement whereby the moving element of an instrument is suspended by means of ligaments, usually in the form of a thin flat conducting ribbon, at each of its ends. The ligaments normally are in tension sufficient to restrict the lateral motion of the moving element to within limits that permit freedom of useful motion when the instrument is mounted in any position. A restoring torque is produced within the ligaments with rotation of the moving element. *See:* **moving element (instrument).**
280

taxi-channel lights (illuminating engineering). Aeronautical ground lights arranged along a taxi-channel of a water aerodrome to indicate the route to be followed by taxiing aircraft.
167

taxi light (illuminating engineering). An aircraft aeronautical light designed to provide necessary illumination for taxiing.
167

taxiway-centerline lights (illuminating engineering). Taxiway lights placed along the centerline of a taxiway except that on curves or corners having fillets, these lights are placed a distance equal to half the normal width of the taxiway from the outside edge of the curve or corner.
167

taxiway-edge lights (illuminating engineering). Taxiway lights placed along or near the edges of a taxiway.
167

taxiway holding-post light (illuminating engineering). A light or group of lights installed at the edge of a taxiway near an entrance to a runway, or to another taxiway, to indicate the position at which the aircraft should stop and obtain clearance to proceed.
167

taxiway lights (illuminating engineering). Aeronautical ground lights provided to indicate the route to be followed by taxiing aircraft.
167

Taylor distribution (antennas). (1) (circular). A continuous distribution of a circular planar aperture which is equiphase, with the amplitude distribution dependent only on distance from the center of the aperture and such as to produce a pattern with a main beam plus sidelobes. The sidelobe structure is rotationally symmetric, with a specified number of inner sidelobes at a quasi-uniform height, the remainder of the side-lobes decaying in height with their angular separation from the main beam. *Note:* Taylor distributions are often sampled to obtain the excitation for a planar array. (2) (linear). A continuous distribution of a line source which is symmetric in amplitude, has a uniform progressive phase, and yields a pattern with a main beam plus sidelobes. The sidelobe structure is symmetrical, with a specified number of inner sidelobes at a quasi-uniform height, the remainder of the sidelobes decaying in height with their angular separation from the main beam. *Note:* Taylor distributions are often sampled to obtain the excitation for a linear array.   111

TB cell. *See:* **transmitter-blocker cell.**

T-connected (or tee-connected) transformer (power and distribution transformer). A three-phase to three-phase transformer, similar to a Scott-connected transformer. *See:* **Scott or T-connected transformer.**
                                                                53

TE. *See:* **transverse electric.**

tearing (television). An erratic lateral displacement of some scanning lines of a raster caused by disturbance of synchronization. *See:* **television.**   178

teaser transformer (power and distribution transformer). The term "teaser transformer," as applied to two single-phase Scott-connected units for the three-phase to two-phase or two-phase to three-phase operation, designates the transformer that is connected between the midpoint of the main transformer and the third-phase wire of the three-phase system.   53

techniques (software quality assurance). Technical and managerial procedures that aid in the evaluation and improvement of the software development process.
                                                                481

TE$_{mn}$mode (H$_{mn}$mode) (waveguide). (1) In a rectangular waveguide, the subscripts $_m$ and $_n$ denote the number of half-period variations in the electric field parallel to the broad and narrow sides, respectively, of the guide. *Note:* In the United Kingdom, the reverse order is preferred. (2) In a circular waveguide, a mode which has $_m$ diametral planes in which the longitudinal component of the magnetic field is zero, and $_n$ cylindrical surfaces of nonzero radius (including the wall of the guide) at which the tangential component of the electric field is zero. (3) In a resonant cavity consisting of a length of rectangular or circular waveguide, a third subscript is used to indicate the number of half-period variations of the field along the waveguide axis.
                                                                267

tee coupler (fiber optics). A passive coupler that connects three ports. *See:* **star coupler.**   433

teed feeder. A feeder that supplies two or more feeding points. *See:* **center of distribution.**   203

tee junction (waveguide components). A junction of waveguides or transmission lines in which the longitudinal guide axes form a tee.   166

telautograph. A system in which writing movement at the transmitting end causes corresponding movement of a writing instrument at the receiving end. *See:* **telegraphy.**   328

telecommunication (1)(computer applications). The transmission of signals over long distance, such as by telegraph, radio, or television. *See:* **computer conferencing; office automation.**   579
(2)(data transmission). The transmission of information from one point to another.   59

telecommunication loop (telephone switching systems). A channel between a telecommunications station and a switching entity.   55

telecommunications customer (telephone switching systems). One for whom telecommunications service is provided (formerly referred to as a "subscriber").
                                                                55

telecommunications exchange (telephone switching systems). A means of providing telecommunications services to a group of users within a specified geographical area.   55

telecommunications switchboard (telephone switching systems). A manual means of interconnecting telecommunications lines, trunks, and associated circuits, and including signaling facilities.   55

telecommunications switching (telephone switching systems). The function of selectively establishing and releasing connections among telecommunication transmission paths.   55

telecommunications system (telephone switching systems). An assemblage of telecommunications stations, lines, and channels, and switching arrangements for their interconnection, together with all the accessories for providing telecommunications services.   55

telecommuting (computer applications). An employment alternative involving working at home using a computer and telecommunication system instead of commuting between home and workplace.   579

teleconferencing (computer applications). A form of communication that uses telephones, computer networks, and television to allow participants at different geographical locations to confer.   579

telegraph (marine transportation). A mechanized or electric device for the transmission of stereotyped orders or information from one fixed point to another. *Note:* The usual form of telegraph is a transmitter and a receiver, each having a circular dial in sectors upon which are printed standard orders. When the index of the transmitter is placed at any order, the pointer of the receiver designates that order. dual mechanism is generally provided to permit repeat back or acknowledgment of orders.   328

telegraph channel (data transmission). A channel suitable for the transmission of telegraph signals. *Note:* Three basically different kinds of telegraph channels used in multichannel telegraph transmission are (1) One of a number of paths for simultaneous transmission in the same frequency range as in bridge duplex, differential duplex, and quadruplex telegraphy. (2) One of a number of paths for simultaneous transmission in the same frequency range as in bridge duplex, differential duplex, and quadruplex telegraphy. (3) One of a number of paths for successive transmission as in multiplex printing telegraphy. Combinations of these three types may be used on the same circuit.
                                                                59

telegraph concentrator. A switching arrangement by means of which a number of branch or subscriber lines or station sets may be connected to a lesser number of trunk lines or operating positions or instruments through the medium of manual or automatic switching devices in order to obtain more efficient use of facilities. *See:* **telegraphy.** 194

telegraph distortion (data transmission). The condition in which the significant intervals have not all exactly their theoretical durations. The reference point used when measuring telegraph distortion is the initial space-to-mark transition of each character which occurs at the beginning of each "start" element. The slicing level for all measurements is at the 50% point on the rising or falling current waveforms. Percent distortion is expressed by

$$\text{Percent Distortion} = \frac{\Delta t}{t_e} \times 100$$

where

t = time difference between the actual slicing point and the ideal crossover point

t_e = time interval of one signal element. 59

telegraph distributor. A device that effectively associates one direct-current or carrier telegraph channel in rapid succession with the elements of one or more signal sending or receiving devices. *See:* **telegraphy.** 328

telegraph key. A hand-operated telegraph transmitter used primarily in Morse telegraphy. *See:* **telegraphy.** 328

telegraph repeater. An arrangement of apparatus and circuits for receiving telegraph signals from one line and retransmitting corresponding signals into another line. *See:* **telegraphy.** 328

telegraph selector. A device that performs a switching operation in response to a definite signal or group of successive signals received over a controlling circuit. *See:* **telegraphy.** 328

telegraph sender. A transmitting device for forming telegraph signals. Examples are a manually operated telegraph key and a printer keyboard. *See:* **telegraphy.** 328

telegraph signal (telecommunication). The set of conventional elements established by the code to enable the transmission of a written character (letter, figure, punctuation sign, arithmetic sign, etcetera) or the control of a particular function (spacing, shift, line-feed, carriage return, phase correction, etcetera): this set of elements being characterized by the variety, the duration and the relative position of the component elements or by some of these features. 194

telegraph signal distortion. Time displacement of transitions between conditions, such as marking and spacing, with respect to their proper relative positions in perfectly timed signals. *Note:* The total distortion is the algebraic sum of the bias and the characteristic and fortuitous distortions. *See:* **telegraphy.** 328

telegraph sounder. A telegraph receiving instrument by means of which Morse signals are interpreted aurally

(or read) by noting the intervals of time between two diverse sounds. *See:* **telegraphy.** 328

telegraph speed (telecommunication). *See:* **modulation rate.**

telegraph transmission speed. The rate at which signals are transmitted, and may be measured by the equivalent number of dot cycles per second or by the average number of letters or words transmitted, and received per minute. *Note:* A given speed in dot cycles per second (often abbreviated to dots per second) may be converted to **bauds** by multiplying by 2. The baud is the unit of signaling transmission speed recommended by the International Consultative Committee on Telegraph Communication. Where words per minute are used as a measure of transmission speed, five letters and a space per word are assumed. *See:* **telegraphy.** 328

telegraph transmitter. A device for controlling a source of electric power so as to form telegraph signals. *See:* **telegraphy.** 328

telegraph word (conventional). A word comprising five letters together with one letter-space, used in computing telegraph speed in words per minute or traffic capacity. *See:* **telegraphy.** 194

telegraphy (data transmission). A system of telecommunication for the transmission of graphic symbols, usually letters or numerals, by the use of a signal code. It is used primarily for record communication. The term may be extended to include any system of telecommunication for the transmission of graphic symbols or images for reception in record form, usually without gradation of shade values. 59

telemetering (1)(remote metering). Measurement with the aid of intermediate means that permit the measurement to be interpreted at a distance from the primary detector. *Note:* The distinctive feature of telemetering is the nature of the translating means, which includes provision for converting the measurand into a representative quantity of another kind that can be transmitted conveniently for measurement at a distance. The actual distance is irrelevant. 103

(2)(supervisory control, data acquisition, and automatic control). (A) Transmission of measurable quantities using telecommunication techniques. (1) Current-type telemeter. A telemeter that employs the magnitude of a single current as the translating means. (2) Frequency-type telemeter. A telemeter that employs the frequency of a periodically recurring electric signal as the translating means. (3) Pulse-type telemeter. A telemeter that employs characteristics of intermittent electric signals, other than their frequency, as the translating means. (4) Ratio-type telemeter. A telemeter that employs the relative phase position between, or the magnitude relation between, two or more electrical quantities as the translating means. *Note:* Examples of ratio-type telemeters include ac or dc position matching systems. (5) Voltage-type telemeter. A telemeter that employs the magnitude of a single voltage as the translating means. (B) Analog. Telemetering in which some characteristic of the transmitter signal is proportional to the quantity being measured. (C) Digital. Telemetering in which a numerical representation is generated and transmitted, the number being representative of the quantity being measured.   570

(3) (data transmission)(power switchgear). Measurement with the aid of intermediate means that permit the measurement to be interpreted at a distance from the primary detector. *Note:* The distinctive feature of telemetering is the nature of the translating means, which includes provision for converting the measure into a representative quantity of another kind that can be transmitted conveniently for measurement at a distance. The actual distance is irrelevant.   59,103

(4) (station control and data acquisition). Transmission of measurable quantitites using telecommunication techniques. (A) Current-type telemeter. A telemeter which employs the magnitude of a single current as the translating means. (B) Frequency-type telemeter. A telemeter that employs the frequency of a periodically recurring electric signal as the translating means. (C) Pulse-type telemeter. A telemeter that employs characteristics of intermittent electric signals other than their frequency, as the translating means. (D) Ratio-type telemeter. A telemeter that employs the relative phase position between, or the magnitude relation between, two or more electrical quantities as the translating means. *Note:* Examples of ratio-type telemeters include ac or dc position matching system. (E) Voltage-type telemeter. A telemeter that employs the magnitude of a single voltage as the translating means.   403

telemeter service. Metered telegraph transmission between paired telegraph instruments over an intervening circuit adapted to serve a number of such pairs on a shared-time basis. *See:* **electric metering; telegraphy.**   328

telephone air-to-air input-output characteristic. The acoustical output level of a telephone set as a function of the acoustical input level of another telephone set to which it is connected. The output is measured in an artificial ear, and the input is measured free-field at a specified location relative to the reference point of an artificial mouth. *See:* **telephone station.**   196

telephone booth. A booth, closet, or stall for housing a telephone station. *See:* **telephone station.**   328

telephone central office (data transmission). A telephone switching unit, installed in a telephone system providing service to the general public, having the necessary equipment and operating arrangements for terminating and interconnecting lines and trunks. *Note:* There may be more than one central office in the same building.   59

telephone channel (data transmission). A channel suitable for the transmission of telephone signals.   59

telephone connection. A two-way telephone channel completed between two points by means of suitable switching apparatus and arranged for the transmission of telephone currents, together with the associated arrangements for its functioning with the other parts of a telephone system in switching and signaling operations. *Note:* The term is also sometimes used to mean a two-way telephone channel permanently established between two telephone stations.   328

telephone electrical impedance. The complex ratio of the voltage to the current at the line terminals at any given single frequency. *See:* **telephone station.**   196

telephone equalization. A property of a telephone circuit that ideally causes both transmit and receive responses to be inverse functions of current, thus tending to equalize variations in loop loss.   196

telephone exchange (data transmission). A unit of a telephone communication system for the provision of communication service in a specified area which usually embraces a city, town, or village, and its environs. Incoming lines are connected to outgoing lines as required by the individual caller dial code.   59

telephone feed circuit. An arrangement for supplying direct-current power to a telephone set and an alternating-current path between the telephone set and a terminating circuit.   196

telephone frequency characteristics. Electrical and acoustical properties as functions of frequency.   196

telephone handset. A telephone transmitter and receiver combined in a unit with a handle. *See:* **telephone station.**   196

telephone influence factor (TIF)(1)(harmonic control and reactive compensation of static power converters). Of a voltage or current wave in an electric supply circuit, the ratio of the square root of the sum of the squares of the weighted root-mean-square (rms) values of all the sine-wave components (including alternating current (ac) waves both fundamental and harmonic) to the rms value (unweighted) of the entire wave.   533

(2) (high-voltage direct-current systems). A dimensionless quantity which includes C-message weighting and is used to express the effect of the deviation of a voltage or current wave shape from a pure sinusoidal wave on a voice-frequency communication network caused by electromagnetic or electrostatic induction, or both. The frequencies and amplitudes of harmonics present on the power circuit, among other factors, determine a power circuit's inductive influence on a voice communications circuit. TIF expressed in terms of I·T product current and voltage TIF (that is, $kV \cdot T$ product per kilovolt) is a measure of this influence. TIF of a voltage or current wave is the ratio of the square root of the sum of the squares (rss) of the weighted root-mean-square (rms) values of all the sine- wave components (including in ac waves both fundamental and harmonics to the root-mean-square value (unweighted) of the entire wave. C-message weighting is derived from listening tests to indicate the relative annoyance of speech impairment by an interfering signal of frequency f as heard through a modern (since 1960) telephone set. The result, called C-message weighting, is shown in graphical and tubular form in the figure below in terms of relative interfering effect P_f at frequency f.   373

1960 C-Message Weighting Curve

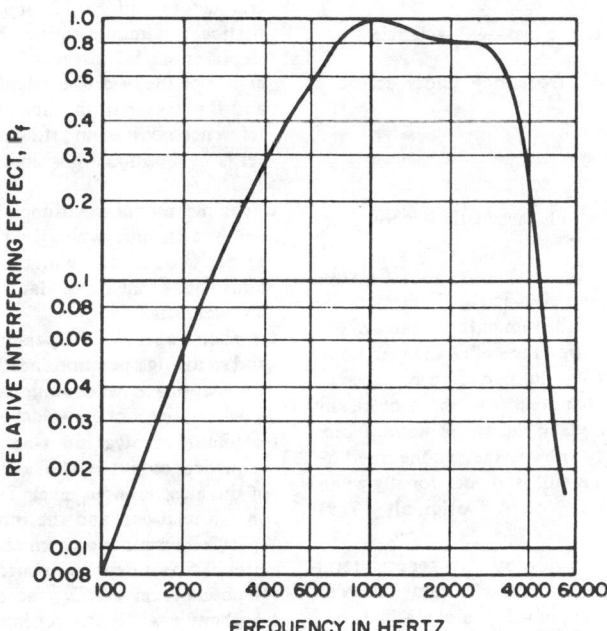

FREQUENCY IN HERTZ

(3) (thyristor). Of a voltage or current wave in an electric supply circuit, the ratio of the square root of the sum of the square of the weighted root-mean-square (rms) values of all sine-wave components (including in alternating waves both the fundamental and harmonics) to the rms (unweighted) values of the entire wave. (The weightings are applied to the individual components of different frequencies according to a prescribed curve).   445

(4) (voice-frequency electrical-noise test). Of a voltage or current wave in an electric supply circuit, the ratio of the square root of the sum of the squares of the weighted root-mean-square values of all the sine-wave components (including, in alternating-current waves, both fundamental and harmonics) to the root-mean-square value (unweighted) of the entire wave. *Note:* (A) The TIF represents the relative interfering effect of voltages and currents at the various harmonic frequencies that appear in power-supply circuits. It is a dimensionless quantity indicative of waveform and not of amplitude. (B) TIF takes into consideration the characteristics of the telephone receiver and the ear (all represented by *c*-message weighting) and the assumption the coupling between the electric-supply circuit and the telephone circuit is directly proportional to the interfering frequency. TIF is also shown as *T* for convenience and is expressed as

$$T = \frac{\sqrt{\Sigma(x_f \cdot w_f)^2}}{x_t} \quad \text{or}$$

$$T = \sqrt{\Sigma\left(\frac{x_t \cdot w_f}{x_t}\right)^2}$$

where
X_t Represents the total effective or rms current (I) or voltage (kV)
X_f Represents the single-frequency effective current (I) or voltage (kV) at frequency f, including the fundamental
W_f Represents the single-frequency TIF weighting at frequency f.
The TIF contribution of power-circuit voltage or current at frequency f may be expressed as follows:

$$T = \frac{X_f \cdot W_f}{X_t}$$

The 1960 TIF weighting characteristic represents the relative interfering effect of a voltage or current in a supply circuit at frequency f. The weighting takes into account the relative subjective effect of frequency f as heard through a telephone set (that is, the *c*-message weighting) and the coupling between the power and telephone circuit, assumed to be directly proportional to frequency. It is defined as

$W_f = 5P_f f$

where

5   = A constant

P_f Represents the c-message weighting at frequency f

f Represents the frequency under consideration

The 1960 TIF weighting characteristic is shown below.   376

(5) (power and distribution transformer). Of a voltage or current wave in an electric supply circuit, the ratio of the square root of the sum of the squares of the weighted root-mean-square values of all the sine-wave components (including in alternating-current waves both fundamental and harmonics) to the root-mean-square value (unweighted) of the entire wave. *Note:* This factor was formerly known as telephone interference factor, which term is still used occasionally when referring to values based on the original (1919) weighting curve.   53

telephone line (data transmission). A general term used in communication practice in several different senses, the more important of which are: (1) The conductor or conductors and supporting or containing structures extending between telephone stations and central offices or between central offices whether they be in the same or in different communities. (2) The conductors and circuit apparatus associated with a particular communication channel.   59

telephone modal distance. The distance between the center of the grid of a telephone handset transmitter and the center of the lips of a human talker (or the reference point of an artificial mouth) when the handset is in the modal position. *See:* **telephone station.**   122, 196

telephone modal position. The position a telephone handset assumes when the receiver of the handset is held in close contact with the ear of a person with head dimensions that are modal for a population. *See:* **telephone station.**   196

telephone operator. A person who handles switching and signaling operations needed to establish telephone connections between stations or who performs various auxiliary functions associated therewith.   193

telephone receive input-output characteristic. The acoustical output level of a telephone set as a function of the electric input level. The output is measured in an artificial ear, and the input is measured across a specified termination connected to the telephone feed circuit. *See:* **telephone station.**   196

telephone receiver. *See:* **receiver, telephone.**

telephone repeater. A repeater for use in a telephone circuit. *See:* **repeater.**   328

1960 SINGLE FREQUENCY TIF VALUES							
FREQ	TIF	FREQ	TIF	FREQ	TIF	FREQ	TIF
60	0.5	1020	5100	1860	7820	3000	9670
180	30	1080	5400	1980	8330	3180	8740
300	225	1140	5630	2100	8830	3300	8090
360	400	1260	6050	2160	9080	3540	6730
420	650	1380	6370	2220	9330	3660	6130
540	1320	1440	6650	2340	9840	3900	4400

1960 SINGLE FREQUENCY TIF VALUES							
FREQ	TIF	FREQ	TIF	FREQ	TIF	FREQ	TIF
660	2260	1500	6680	2460	10340	4020	3700
720	2760	1620	6970	2580	10600	4260	2750
780	3360	1740	7320	2820	10210	4380	2190
900	4350	1800	7570	2940	9820	5000	840
1000	5000						

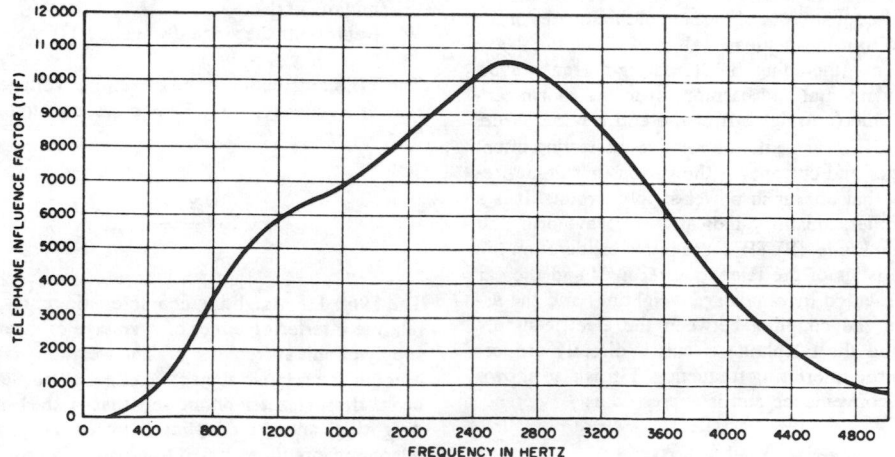

Fig 2
1960 TIF Weighting Curve

telephone ringer (station ringer) (ringer). An electric bell designed to operate on low-frequency alternating or pulsating current and associated with a telephone station for indicating a telephone call to the station. *See:* **telephone station.**   328

telephone set (general) (telephone). An assemblage of apparatus including a telephone transmitter, a telephone receiver, and usually a switch, and the immediately associated wiring and signaling arrangements. *See:* **telephone station.**   328

telephone set (transmission performance of telephone sets). An assembly comprising a handset containing a transmitter and a receiver which are connected to a speech network which may be mounted in the handset or in a separate physical housing.   491

telephone sidetone. The ratio of the acoustical output of the receiver of a given telephone set to the acoustical input of the transmitter of the same telephone set. *See:* **telephone air-to-air input-output characteristic.**   196

telephone speech network. An electric circuit that connects the transmitter and the receiver to a telephone line or telephone test loop and to each other. *See:* **telephone station.**   196

telephone station. An installed telephone set and associated wiring and apparatus, in service for telephone communication. *Note:* As generally applied, this term does not include the telephone sets employed by central-office operators and by certain other personnel in the operation and maintenance of a telephone system.   328

telephone subscriber. A customer of a telephone system who is served by the system under a specific agreement or contract.   193

telephone switchboard. A switchboard for interconnecting telephone lines and associated circuits.   193

telephone system. An assemblage of telephone stations, lines, channels, and switching arrangements for their interconnection, together with all the accessories for providing telephone communication.   193

telephone test circuit (transmission performance of telephone sets). An assembly consisting of a telephone set(s), a test loop(s) and a feed circuit as may be required to realize simulated partial and overall telephone connections.   491

telephone test connection. Two telephone sets connected together by means of telephone test loops and a telephone feed circuit.   196

telephone test loop. A circuit that is interposed between a telephone set and a telephone feed circuit to simulate a real telephone line.   196

telephone transmit input-output characteristic. The electric output level of a telephone set as a function of the acoustical input level. The output is measured across a specified impedance connected to the telephone feed circuit, and the input is measured free-field at a specified location relative to the reference point of an artificial mouth. *See:* **telephone station.**   196

telephone transmitter. A microphone for use in a telephone system. *See:* **telephone station.**   176

telephone-type relay (power switchgear). A type of electromechanical relay in which the significant structural feature is a hinged armature mechanically separate from the contact assembly. This assembly usually consists of a multiplicity of stacked leaf-spring contacts.   103

telephony. *See:* **sleeve conductor; sleeve wire.**

telephotography. *See:* **picture transmission.**

teleprinter (teletypewriter). *See:* **printer.**

teleran (navigation aid terms). A navigation system which employs ground-based search radar equipment along an airway to locate aircraft flying near that airway.   526

teletypewriter (teleprinter). *See:* **printer.**

television (TV). The electric transmission and reception of transient visual images.   328

television broadcast station. A radio station for transmitting visual signals, and usually simultaneous aural signals, for general reception. *See:* **television.**   328

television camera. A pickup unit used in a television system to convert into electric signals the optical image formed by a lens. *See:* **television.**   328

television channel. A channel suitable for the transmission of television signals. The channel for associated sound signals may or may not be considered a part of the television channel. *See:* **channel.**   328

television interference (overhead-power-line corona and radio noise). A radio interference occurring in the frequency range of television signals.   411

television line number. The ratio of the raster height to the half period of a periodic test pattern. *Example:* In a test pattern composed of alternate equal-width black and white bars, the television line number is the ratio of the raster height to the width of each bar. *Note:* Both quantities are measured at the camera-tube sensitive surface. *See:* **television.**   174

television lines per raster height (TVL/RH) (diode-type camera tube). The number of half-cycles of a uniform periodic array referred to a unit length equal to the raster height. The array may be sinusoidal or comprised of equal width alternating light and dark bars (lines). For a given array, the TVL/RH value is numerically twice the spatial frequency in line pairs per raster height (LP/RH) units. *Note:* While the unit TVL/RH has had wide usage throughout the television industry, it is recommended that the more accurately descriptive unit LP/RH be adopted.   380

television picture tube. *See:* **picture tube.**

television receiver. A radio receiver for converting incoming electric signals into television pictures and customarily associated sound. *See:* **television.**   328

television repeater. A repeater for use in a television circuit. *See:* **repeater; television.**   328

television transmitter. The aggregate of such radio-frequency and modulating equipment as is necessary to supply to an antenna system modulated radio-frequency power by means of which all the component parts of a complete television signal (including audio, video, and synchronizing signals) are concurrently transmitted. *See:* **television.**   111 telltale. *See:* **running light indicator panel.**

telluric (power fault effects). Currents circulating in the earth or in conductors connecting two grounded points due to voltages in the earth.					404

TEM. *See:* **transverse electromagnetic.**

TEM mode (1)(waveguide terms). A waveguide mode in which the longitudinal components of the electric and magnetic fields are everywhere zero.					267

(2) (fiber optics). *See:* **transverse electromagnetic mode.**					433

TE mode (1) (fiber optics). *See:* **transverse electric mode.**					433

(2) (H mode) (waveguide). A waveguide mode in which the longitudinal component of the electric field is everywhere zero and the longitudinal component of the magnetic field is not.					267

temperature, ambient air. The temperature of the surrounding air which comes in contact with equipment. *Note:* Ambient air temperature, as applied to enclosed switchgear assemblies, is the average temperature of the surrounding air that comes in contact with the enclosure.					93

temperature class (evaluation of thermal capability) (thermal classification of electric equipment and electrical insulation). A standardization designation of the temperature capability of the insulation in electric equipment, as defined by the appropriate technical committee. It may be determined by experience or test and expressed by letters or numbers.					506

temperature classification (solid electrical insulating materials). The term is reserved for insulation systems as used in specific equipment and is no longer recognized as a description of the temperature capability of individual insulating materials. (*See:* IEEE Std 1-1969, Parts II and III). Historically, the term has been used in reference to insulation systems and to electrical equipment. In the future the term may be reserved for use in rating electrical equipment, while thermal identification may be used in the specification of insulation systems for particular applications.					452

temperature coefficient (1)(power supplies)(ferroresonant voltage regulators). The percent change in the outright voltage or current as a result of a 1 °C change in the ambient operating temperature (percent per degree Celsius).					456

(2) (rotating machinery). The variation of the quantity considered, divided by the difference in temperature producing it. Temperature coefficient may be defined as an average over a temperature range or an incremental value applying to a specified temperature. *See:* **asynchronous machine.**					63

(3) (power supplies). The percent change in the output voltage or current as a result of a 1 degree-Celsius change in the ambient operating temperature (percent per degree Celsius).					186

(4) (automatic control). *See:* **environmental coefficient.**

(5) (variable or fixed attenuator). Maximum temporary and reversible change of insertion loss in decibels per degree Celsius over operating temperature range.					110

temperature coefficient of capacity (storage cell or battery). The change in delivered capacity (ampere-hour or watt hour capacity) per degree Celsius relative to the capacity of the cell or battery at a specified temperature. *See:* **initial test temperature.**					328

temperature coefficient of electromotive force (storage cell or battery). The change in open-circuit voltage per degree Celsius relative to the electromotive force of the cell or battery at a specified temperature. *See:* **initial test temperature.**					328

temperature coefficient of resistance (rotating machinery). The temperature coefficient relating a change in electric resistance to the difference in temperature producing it. *See:* **asynchronous machine.**					63

temperature coefficient of sensitivity (electrothermic power meters). The change in rf sensitivity (microvolts/milliwatts) resulting from a specified temperature change of the electrothermic unit at a specified power level. Expressed in percent per degree celsius.					47

temperature coefficient of voltage drop (glow-discharge tubes). The quotient of the change of tube voltage drop (excluding any voltage jumps) by the change of ambient (or envelope) temperature. *Note:* It must be indicated whether the quotient is taken with respect to ambient or envelope temperature. *See:* **gas tube.**					125

temperature compensated overload relay (electric installations on shipboard). A device that functions at any current in excess of a predetermined value essentially independent of the ambient temperature.					3

temperature control (packaging machinery). A control device responsive to temperature.					429

temperature control device (23) (power system device function numbers). A device that functions to raise or lower the temperature of a machine or other apparatus, or of any medium, when its temperature falls below, or rises above, a predetermined value. *Note:* An example is a thermostat that switches on a space heater in a switchgear assembly when the temperature falls to a desired value as distinguished from a device that is used to provide automatic temperature regulation between close limits and would be designated as device function 90T, [regulating device T.]					402

temperature control system (gas turbines). The devices and elements, including the necessary temperature detectors, relays, or other signal-amplifying devices and control elements, required to actuate directly or indirectly the fuel-control valve, speed of the air compressor, or stator blades of the compressor so as to limit or control the rate of fuel input or air flow inlet to the gas turbine. By this means the temperature in the combustion system or the temperatures in the turbine stages or turbine exhaust may be limited or controlled.					58

temperature conversion (tolerance requirements) (International System of Units). Standard practice for converting tolerances from degrees Fahrenheit to kelvins or degrees Celsius is:

Conversion of temperature tolerance requirements

Tolerance °F	Tolerance K or °C
± 1	± 0.5
± 2	± 1
± 5	± 3
±10	± 5.5
±15	± 8.5
±20	±11
±25	±14

Normally, temperatures expressed in a whole number of degrees Fahrenheit should be converted to the nearest 0.5 kelvin (or degree Celsius). As with other quantities, the number of significant digits to retain will depend upon implied accuracy of the original dimension, for example:

$$100 \pm 5°F \text{ implied accuracy estimated total } 2°F.$$
$$37.7777 \pm 2.7777°C \text{ rounds to } 38 \pm 3°C$$

$$1000 \pm 50°F \text{ implied accuracy estimated total } 20°F.$$
$$537.7777 \pm 27.7777°C \text{ rounds to } 540 \pm 30°C$$

See: **units and letter symbols.** 21

temperature derating (semiconductor device). The reduction in reverse-voltage or forward-current rating, or both, assigned by the manufacturer under stated conditions of higher ambient temperatures. *See:* **average forward-current rating (rectifier circuit); semiconductor rectifier stack.** 208

temperature detectors (gas turbines and rotating electric machinery). The primary temperature-sensing elements that are directly responsive to temperature. *See:* **asynchronous machine; electric thermometer.** 63

temperature, equilibrium (thyristor power converter). The steady-state temperature reached by a component of a thyristor converter under specified conditions of load and cooling. *Note:* The steady-state temperatures are, in general, different for different components. The times necessary to establish steady-state temperatures are also different and proportional to the thermal time constants. 121

temperature index (1)(power and distribution transformer). An index that allows relative comparisons of the temperature capability of insulating materials or insulation systems based on specified controlled test conditions. Preferred values of temperature index numbers are: 53

Number Range	Preferred Temperature Index
90–104	90
105–129	105
130–154	130
155–179	155
180–199	180
200–219	200
220 and above no preferred indices established.	

(2)(TI)(evaluation of thermal capability)(thermal classification of electric equipment and electrical insulation). The number that corresponds to the temperature in °C, derived mathematically or graphically from the thermal endurance relationship at a specified time (often 20 000 h). The temperature index (TI) may be reported for materials and insulation systems. However, for insulation systems it may be preferable to make comparisons at a particular temperature, for example, 130 °C, 155 °C, or over a range of temperatures. (The TI is not used for equipment). *See:* **thermal endurance graph.** 506

(3)(TI)(solid electrical insulating materials). This is the number corresponding to the temperature in degrees Celsius derived from the thermal endurance graph at a given time. See 'Thermal Endurance Graph' below. 452

temperature meter. *See:* **electric thermometer.**

temperature, operating (power supply). *See:* **operating temperature.**

temperature radiator (illuminating engineering). An ideal radiator whose radiant flux density (radiant exitance) is determined by its temperature and the material and character of its surface, and is independent of its previous history. 167

temperature-regulating equipment (rectifier). Any equipment used for heating and cooling a rectifier, together with the devices for controlling and indicating its temperature. *See:* **rectification.** 208

temperature relay (power switchgear). A relay whose operation is caused by specified external temperature. *Note:* Compare with **thermal relay.** 103

temperature relays (gas turbines). Devices by means of which the output signals of the temperature detectors are enabled to control directly or indirectly the rate of fuel energy input, the air flow input, or both, to the combustion system. *Note:* Operation of a temperature relay is caused by a specified external temperature: whereas operation of a thermal relay is caused by the heating of a part of the relay. *See:* **thermal relay.** 103,127

temperature rise (power and distribution transformer). The difference between the temperature of the part under consideration (commonly the "average winding rise" or the "hottest spot winding rise") and the ambient temperature. 53

temperature-rise rate, locked rotor, winding. *See:* **locked-rotor temperature-rise rate, winding.**

temperature-rise test (1) (rotating machinery). A test undertaken to determine the temperature rise above ambient of one or more parts of a machine under specified operating conditions. *Note:* The specified conditions may refer to current, load, etcetera. *See:* **asynchronous machine.** 202, 63

(2) (high voltage air switches). A test in which rated current at rated frequency is applied to equipment to determine its temperature rise. 144

temperature-rise tests (power switchgear). Tests to determine the temperature rise, above ambient, of various parts of the tested device when subjected to specified test quantities. *Note:* The test quantities may be current, load, etcetera. *See:* **allowable continuous**

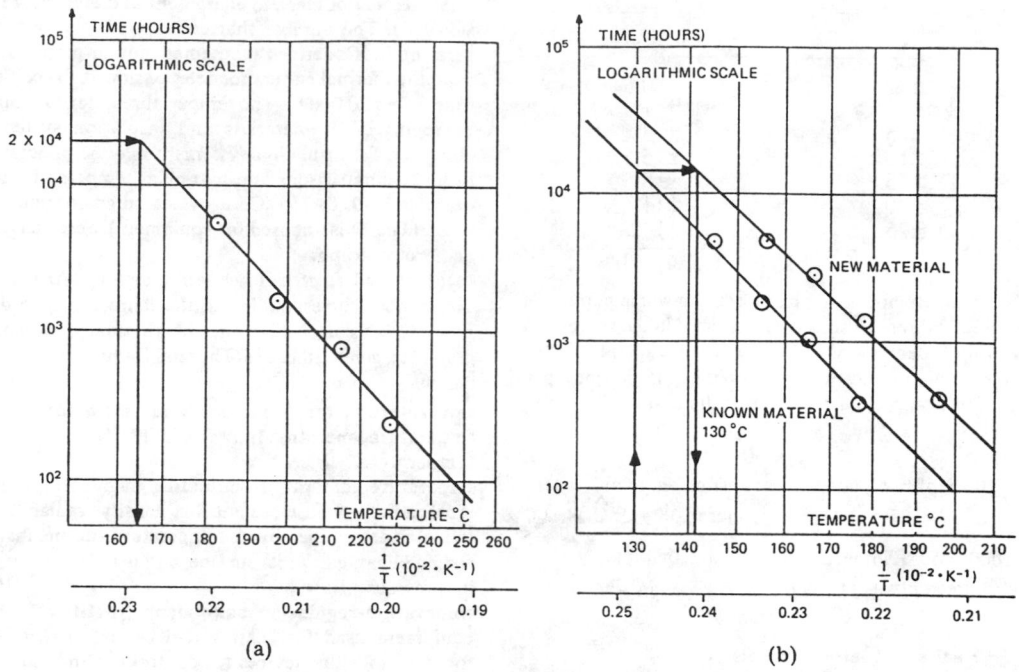

Thermal Endurance Graph
(a) Temperature Index (b) Relative Temperature Index

current for table showing values for various types of devices.   103, 443

temperature sensor (sensing element)(electrical heat tracing for industrial applications). A device that responds to temperature and provides an electrical signal or mechanical operation.   523

temperature stability (electrical conversion). Static regulation caused by a shift or change in output that was caused by temperature variation. This effect may be produced by a change in the ambient or by self-heating. *See:* **electrical conversion.**   186

temporal coherence (1) (fiber optics). *See:* **coherent.**   433

(2) (electromagnetic) (laser-maser). The correlation in time of electromagnetic fields at a point in space.   363

temporally coherent radiation. *See:* **coherent.**

temporal noise (diode-type camera tube). The varying amplitude portion of what should be a fixed amplitude video signal. It is statistical in nature, being random in both time and amplitude.   380

temporary emergency circuits (marine transportation). Circuits arranged for instantaneous automatic transfer to a storage-battery supply upon failure of a ship's service supply. *See:* **emergency electric system.**   328

temporary emergency lighting (marine transportation). The lighting of exits and passages to permit passengers and crew, upon failure of a ship's service lighting, readily to find their way to the lifeboat embarkation deck. *See:* **emergency electric system.**   328

temporary ground. A connection between a grounding system and parts of an installation that are normally alive, applied temporarily so that work may be safely carried out in them.   62

temporary overvoltage (power and distribution transformer). An oscillatory phase-to-ground or phase-to-phase overvoltage at a given location of relatively long duration and which is undamped or only weakly damped. Temporary overvoltages usually originate from switching operations or faults (for example, load rejection, single-phase faults) or from nonlinearities (ferro-resonance effects, harmonics), or both. They may be characterized by the amplitude, their oscillation frequencies, their total duration, or their decrement.   53

temporary storage (programming). Storage locations reserved for intermediate results. *See:* **working storage.** 255, 77, 54

tens complement (mathematics of computing). The radix complement of a decimal numeral, which may be formed by subtracting each digit from 9, then adding 1 to the least significant digit and executing any required carries. For example, the tens complement of 4830 is 5170. 564

tension. *See:* **final unloaded conductor tension; initial conductor tension; conductor.**

tension site (conductor stringing equipment). The location on the line where the tensioner, reel stands and anchors (snubs) are located. This site may also serve as the pull or tension site for the next sag section. *Syn:* **conductor payout station; payout site; reel setup.** 431

tension stringing (conductor stringing equipment). The use of pullers and tensioners to keep the conductor under tension and positive control during the stringing phase, thus keeping it clear of the earth and other obstacles which could cause damage. 431

tension, unloaded (transmission and distribution). (1) Initial: The longitudinal tension in a conductor prior to the application of any external load. (2) Final: The longitudinal tension in a conductor after it has been subjected for an appreciable period to the loading prescribed for the loading district in which it is situated, or equivalent loading, and the loading removed. Final unloaded tension shall include the effect of inelastic deformation (creep).*See:* **unloaded tension.** 262

tenth-power width (in a plane containing the direction of the maximum of a lobe). The full angle between the two directions in that plane about the maximum in which the radiation intensity is one-tenth the maximum value of the lobe. *See:* **antenna.** 111

terminal (1)(supervisory control, data acquisition, and automatic control). (A) A point in a system or communication network at which data can either enter or leave. (B) An input/output device capable of transmitting entries to and obtaining output from the system of which it is a part, for example cathode-ray tube (crt) terminal. 570
(2) (packaging machinery). A point of connection in an electric circuit. 429
(3) (power and distribution transformer). (A) A conducting element of an equipment or a circuit intended for connection to an external conductor. (B) A device attached to a conductor to facilitate connection with another conductor. 53
(4) (terminal connector) (power switchgear). A connector for attaching a conductor to electrical apparatus. 103, 443
(5) (network). A point at which any element may be directly connected to one or more other elements. *See:* **network analysis.** 332
(6) (semiconductor device) (industrial control) (light emitting diodes). An externally available point of connection to one or more electrodes or elements within the device. *See:* **anode; semiconductor; semiconductor rectifier cell.** 245

(7) (communication channels). (A) (general) A point in a system or communication network at which data can either enter or leave. 255, 77
(B) (telegraph circuits). A general term referring to the equipment at the end of a telegraph circuit, modems, input-output and associated equipment. *See:* **telegraph.** 194
(8) (rotating machinery). A conducting element of a winding intended for connection to an external electrical conductor. *See:* **stator.** 63

terminal block (terminal board) (power switchgear). An insulating base equipped with terminals for connecting secondary and control wiring. 103

terminal board (power and distribution transformer). A plate of insulating material that is used to support terminations of winding leads. *Notes:* (1) The terminations, which may be mounted studs or blade connectors, are used for making connections to the supply line, the load, other external circuits, or among the windings of the machine. (2) Small terminal boards may also be termed terminal blocks, or terminal strips. 53

terminal board cover (rotating machinery). A closure for the opening which permits access to the terminal board and prevents accidental contact with the terminals. 62

terminal box (conduit box) (rotating machinery). A form of termination in which the ends of the machine winding are connected to the incoming supply leads inside a box that virtually encloses the connections, and is of minimum size consistent with adequate access and with clearance and creepage-distance requirements. The box is provided with a removable cover plate for access. *See:* **stator.** 63

terminal chamber (power switchgear). A metal enclosed container which includes all necessary mechanical and electrical items to complete the connections to other equipment. 78, 103

terminal conformity. *See:* **conformity.**

terminal connection detail (lead storage batteries). Connections made between rows of cells or at the positive and negative terminals of the battery, which may include lead-plated terminal plates, cables with lead-plated lugs, and lead-plated rigid copper connectors. 38

terminal connector (1) (power switchgear). *See:* **terminal.** 103
(2) (power and distribution transformer). A connector for attaching a conductor to a lead, terminal block, or stud of electric apparatus. 53

terminal corona charge (corona measurement). A charge equal to the product of the capacitance of the insulation system and the terminal corona-pulse voltage. 375

terminal corona-pulse voltage (corona measurement). The pulse voltage resulting from a corona discharge which is represented as a voltage source suddenly applied in series with the capacitance of the insulation system under test, and which would appear at the terminals of the system under open-circuit conditions. 375

terminal guidance (navigation aid terms). Guidance from an arbitrary point, at which midcourse guidance ends, to the destination. 526

terminal interference voltage (electromagnetic compatibility). *See:* **terminal voltage.**

terminal linearity. *See:* **conformity.**

terminal of entry (for a conductor entering a delimited region). That cross section of the conductor that coincides with the boundary surface of the region and that is perpendicular to the direction of the electric field intensity at its every point within the conductor. In a conventional circuit, in which the conductors have a cross section that is uniform and small by comparison of the largest dimension with the length, the terminal of entry is a cross section perpendicular to the axis of the conductor. If the cross section of the conductor is infinitesimal, the terminal of entry becomes the point at which the conductor cuts the surface. *Notes:* (1) It follows from this definition and **delimited region** that the algebraic sum of the currents directed into a delimited region through all the terminals of entry is zero at every instant. (2) The term **terminal of entry** has been introduced because of the need in precise definitions of indicating definitely the terminations of the paths along which voltages are determined. The terms **phase conductor** and **neutral conductor** refer to a portion of a conductor rather than to a particular cross section although they may be considered by a practical engineer as representing a portion along which the integral of the electric intensity is negligibly small. Hence he may treat these terms as synonymous with **terminal of entry** in particular cases. *See:* **network analysis.** 210

terminal pad (power switchgear). A usually flat conducting part of a device to which a terminal connector is fastened. 103, 443

terminal pair (network). An associated pair of accessible terminals, such as input pair, output pair, and the like. *See:* **network analysis.** 153, 210

terminal-per-line system (telephone switching systems). A switching entity having an outlet corresponding to each line. 55

terminal-per-station system (telephone switching systems). A switching entity having an outlet corresponding to each main-station code. 55

terminal repeater (data transmission). A repeater for use at the end of a trunk or line. 59

terminal room (telephone switching systems). That part of a building that contains distributing frames, relays and similar apparatus associated with switching equipment. 55

terminals (1)(metal-oxide surge arresters for ac power circuits)(surge arrester). The conducting parts provided for connecting the arrester across the insulation to be protected. 583,430

(2) (storage battery) (storage cell). The parts to which the external circuit is connected. *See:* **battery (primary or secondary).** 328

terminal screw. *See:* **binding screw.**

terminal, stator winding (rotating machinery). The end of a lead cable or a stud or blade of a terminal

board to which connections are normally made during installation. *See:* **stator.** 63

terminal strip. *See:* **terminal board.**

terminal trunk (data transmission). A trunk circuit connecting switching centers used in conjunction with local switching only in these centers. 59

terminal unit (programmable instrumentation). An apparatus by means of which a connection (and translation, if required) is made between the considered interface system and another external interface system. 40

terminal voltage (terminal interference voltage) (electromagnetic compatibility). Interference voltage measured between two terminals of an artificial mains network. *See:* **electromagnetic compatibility.** 220, 199

terminating (line or transducer). The closing of the circuit at either end by the connection of some device thereto. Terminating does not imply any special condition, such as the elimination of reflection. 328

terminating power meter or measuring system. A device which terminates a waveguide or transmission line in a prescribed manner and contains provisions for measuring the incident of absorbed power. 47, 115

terminating test circuit (measuring longitudinal balance of telephone equipment operating in the voice band). A network connected to a transmission port of a circuit to terminate it in a suitable balanced termination for longitudinal balance testing. This circuit is used when a driving test circuit is connected to one such port and the test specimen has additional transmission ports. 529

terminating toll center code (TTC) (telephone switching systems). In operator distance dialing, the three digits used for identifying the toll center within the area to which a call is routed. 55

terminating traffic (telephone switching systems). traffic delivered directly to lines. 55

termination (1)(terminal chamber)(metal-enclosed bus and calculating losses in isolated-phase bus). A metal enclosure that contains all necessary and mechanical and electrical items to complete the connections to other equipment.*See:* **cable terminal.** 574

(2) (National Electrical Safety Code). *See:* **cable terminal.** 391

(3) (general). A one-port load that terminates a section of a transmission system in a specified manner. *See:* **transmission line.** 185

(4) (rotating machinery). The arrangement for making the connections between the machine terminals and the external conductors. *See:* **stator.** 63

(5) (waveguide components). A one port load in a waveguide or transmission line. 166

termination, conjugate. A termination whose input impedance is the complex conjugate of the output impedance of the source or network to which it is connected. *See:* **transmission line.** 185

termination insulator (cable termination). An insulator used to protect each cable conductor passing through the device and provide complete external leakage insulation between the cable conductor(s) and ground. 4

termination, matched (waveguide components). A termination matched with regard to the impedance in a prescribed way: for example, (A) a reflectionless termination, or (B) a conjugate termination. *See:* **termination, reflectionless; transmission line.** 166

termination proof (software). In proof of correctness, the demonstration that a program will terminate under all specified input conditions. *See:* **program; proof of correctness.** 434

termination, reflectionless. A termination that terminates a waveguide or transmission line without causing a reflected wave at any transverse section. *See:* **transmission line; waveguide.** 267

terminations charge (power operations). The amount paid by a customer when service is terminated at the customer's request. 516

termination sequence (FASTBUS acquisition and control). The process by which the AS/AK (address sync/address acknowledge) lock is broken. 480

ternary. (1) Pertaining to a characteristic or property involving a selection, choice, or condition in which there are three possibilities. (2) Pertaining to the numeration system with a radix of three. *See:* **base; positional notation; radix.** 235, 255, 77

ternary code (information theory). A code whose alphabet consists of three symbols. *See:* **ternary.** 415

ternary torquing (1) (digital accelerometer). System with three stable torquing states (for example, positive, negative and off). 383

(2) (gyro, accelerometer) (inertial sensor). A torquing mechanism that utilizes three levels of torquer current, usually positive and negative, of the same magnitude and a zero current or off condition. The positive and negative torque conditions can be either discrete pulses or pulse-duration-modulated current periods. In both implementations, the case of zero input (acceleration or angular rate) will result in zero torquer current. Ternary torquer power is proportional to the input (acceleration or angular rate), resulting in minimum power as compared to binary torquing. 46

terrain avoidance radar (navigation aid terms). A radar which provides assistance to a pilot for navigation around obstacles by displaying obstacles at or above the pilot's altitude. 526

terrain-clearance indicator (navigation aid terms). An absolute altimeter using the measurement of height above terrain to alert the pilot of danger. 526

terrain echoes. *See:* **ground clutter.**

terrain error (navigation)(navigation aid terms). The error resulting from the use of a wave which has become distorted by the terrain over which it has propagated. 526

terrain following radar (navigation aid terms). A radar which works with the aircraft flight control system to provide low level flight following the contour of the earth's surface at some given altitude. 526

terrestrial-reference flight (navigation aid terms). That type of stabilized flight which obtains control information from terrestrial phenomena, such as earth's magnetic field, atmospheric pressure, etcetera. 526

tertiary winding (power and distribution transformer). An additional winding in a transformer which can be connected to a synchronous condenser, a reactor, an auxiliary circuit, etc. For transformers with Y-connected primary and secondary windings, it may also help: (1) to stabilize voltages to the neutral, when delta connected, (2) to reduce the magnitude of third harmonics when delta connected, (3) to control the value of the zero-sequence impedance, (4) to serve load. 53

tesla. The unit of magnetic induction in the International System of Units (SI). The tesla is a unit of magnetic induction equal to 1 weber per square meter. 210

Tesla current (electrotherapy) (coagulating current). A spark discharge having a drop of 5 to 10 kilovolts in air, from monopolar or bipolar electrodes, generated by a special arrangement of transformers, spark gaps, and capacitors, delivered to a tissue surface, and dense enough to precipitate and oxidize (char) tissue proteins. *Note:* The term **Tesla current** is appropriate if the emphasis is on the method of generation: a **coagulating current,** if the emphasis is on the physiological effects. *See:* **electrotherapy.** 192

test (1)(supervisory control, data acquisition, and automatic control). (A) Certified design. A test performed on a production model specimen of a generic type of equipment to establish a specific performance parameter of that genre of equipment. The condition and results of the test are described in a document that is signed and attested to by the testing engineer and other appropriate responsible individuals. (B) Data. (1) The recorded results of test. (2) A set of data developed specifically to test the adequacy of a computer run or system. They may be actual data taken from previous operations or artificial data created for this purpose. (C) Point. A predefined location within equipment or routines at which a known result should be present if the equipment or routine is operating properly. 570

(2) (ATLAS). An action or group of actions performed on a particular unit under test (UUT) to evaluate a parameter or characteristic. 400

(3) (radiation protection). A procedure whereby the instrument, component, or circuit is evaluated for satisfactory operation. 399

(4) (software test documentation). (A) A set of one or more test cases, or (B) a set of one or more test procedures, or (C) a set of one or more test cases and procedures. 436

(5) (electronic digital computation). (A) To ascertain the state or condition of an element, device, program, etcetera. (B) Sometimes used as a general term to include both check and diagnostic procedures. (C) Loosely, same as check. *See:* **check; check problem.** 235

(6) (instrument or meter). To ascertain its performance characteristics while functioning under controlled conditions.   54

testability (software). (1) The extent to which software facilitates both the establishment of test criteria and the evaluation of the software with respect to those criteria. (2) The extent to which the definition of requirements facilitates analysis of the requirements to establish test criteria. *See:* **requirement; software.**
                                                                    434

test amperes (TA). *See:* **test current (watthour meter).**
                                                                    212

test analysis (test, measurement and diagnostic equipment). The examination of the test results to determine whether the device is in a go or no-go state or to determine the reasons for or location of a malfunction.
                                                                    54

test bed (software). (1) A test environment containing the hardware, instrumentation tools, simulators, and other support software necessary for testing a system or system component. (2) The repertoire of test cases necessary for testing a system or system component. *See:* **component; hardware; instrumentation tool; simulator; software; system; test case; testing.**
                                                                    434

test bench (test, measurement and diagnostic equipment). An equipment specifically designed to provide a suitable work surface for testing a unit in a particular test setup under controlled conditions.   54

test block. *See:* **test switch.**

test block cabinet (watthour meter). An enclosure to house a test block and wiring for a bottom-connected watthour meter.   419

test board. A switchboard equipped with testing apparatus so arranged that connections can be made from it to telephone lines or central-office equipment for testing purposes.   193

test cabinet (for a switchgear assembly). An assembly of a cabinet containing permanent electric connections, with cable connections ions to a contact box arranged to make connection to the secondary contacts on an electrically operated removable element, permitting operation and testing of the removable element when removed from the housing. It includes the necessary control switch and closing relay, if required.
                                                                    103

test call (telephone switching systems). A call made to determine if circuits or equipment are performing properly.   55

test cap. A protective structure that is placed over the exposed end of the cable to seal the sheath or other covering completely against the entrance of dirt, moisture, air, or other foreign substances. *Note:* Test caps are often provided with facilities for vacuum treatment, oil filling, or other special field operations. *See:* **live cable test cap.**   64

test case (1)(software). A specific set of test data and associated procedures developed for a particular objective, such as to exercise a particular program path or to verify compliance with a specific requirement. *See:* **procedure; program; requirement; test data; testing.**
                                                                    434

(2)(software verification and validation plans). Documentation specifying inputs, predicted results, and a set of execution conditions for a test item.
                                                                    511

test case generator. *See:* **automated test generator.**

test case specification (software test documentation). A document specifying inputs, predicted results, and a set of execution conditions for a test item.   436

test circuit breaker (ac high-voltage circuit breaker). The circuit breaker under test.   426

test connection (telephony). Two telephone sets connected together by means of test loops and a feed circuit.   122

test current (TA) (watthour meter). The current specified by the manufacturer for the main adjustment of the meter (heavy- or full-load adjustment). *Notes:* (1) It has been identified as "TA" on revenue meters manufactured since 1960. (2) The main adjustment of a meter used with a current transformer may be made either at the test current or at the rated secondary current of the transformer.   212

test current, continuous (thyristor converter). The value of direct current that a converter unit or section can supply to its load for unlimited time periods under specified conditions.   121

test current in alternating-current circuits (insulation tests). The normal current flowing in the test circuit as the result of insulation leakage and, in alternating-current circuits, is the vector sum of the inphase leakage currents and quadrature capacitive currents.
                                                                    116

test current, long-time (thyristor converter). The specified value of direct current that a converter unit or section shall be capable of carrying for a sustained period (minutes or hours) following continuous operation at a specified lower dc value under specific conditions.   121

test current, short-time (thyristor converter). The value of direct current that may be applied to a unit or section for a short period (seconds) following continuous operation at a specified lower dc value under specific conditions.   121

test data (1) (reliability). Data from observations during tests. *Note:* All conditions should be stated in detail, for example, time, stress conditions and failure or success criteria.   164

(2) (software). Data developed to test a system or system component. *See:* **component; data; system; test case.**   434

(3) (station control and data acquisition). (A) The recorded results of test. (B) A set of data developed specifically to test the adequacy of a computer run or system. They may be actual data taken from previous operations or artificial data created for this purpose.
                                                                    403

test data generator. *See:* **automated test generator.**

test design (software verification and validation plans). Documentation specifying the details of the test approach for a software feature or combination of software features and identifying the associated tests.
                                                                    511

test design specification (software test documentation). A document specifying the details of the test approach for a software feature or combination of software features and identifying the associated tests. 436

test desk (telephone switching systems). A position equipped with testing apparatus so arranged that connections can be made from it to telephone lines or central office equipment for testing purposes. 55

test driver (software). A driver that invokes the item under test and may provide test inputs and report test results. *See: driver.* 434

test duration (nuclear power generating stations). The elapsed time between the test initiation and the test termination. 366

test enclosure (for low-voltage ac power circuit breakers) (power switchgear). A single unit enclosure used for test purposes for a specific frame-size circuit breaker and which conforms to the manufacturer's recommendation for minimum volume, minimum electrical clearances, effective areas and locations of ventilation openings, and configuration of connections to terminals. 103

test frequency (reliability analysis of nuclear power generating station safety systems). The number of tests of the same type per unit time interval; the reciprocal of the test interval. 587

test gas phase (fly ash resistivity). The gaseous environment to which the ash layer being tested is exposed in a test cell used for the laboratory measurement of electrical resistivity of fly ash. 427

test handset (hand test telephone). A handset used for test purposes in a central office or in the outside plant. It may contain in the handle other components in addition to the transducer, as for example a dial, keys, capacitors, and resistors. *See: telephone station.* 328

test incident report (software test documentation). A document reporting on any event that occurs during the testing process which requires investigation. 436

testing (1) (nuclear power quality assurance). An element of verification for the determination of the capability of an item to meet specified requirements by subjecting the item to a set of physical, chemical, environmental, or operating conditions. 417

(2) (software). The process of exercising or evaluating a system or system component by manual or automated means to verify that it satisfies specified requirements or to identify differences between expected and actual results. *See: component; debugging; requirement; system.* 434

(3) (software test documentation). The process of analyzing a software item to detect the differences between existing and required conditions (that is, bugs) and to evaluate features of the software item. 436

test initiation (nuclear power generating stations). The application of a test input or removal of equipment tram service to perform a test. 366

test input (nuclear power generating stations). A real or simulated, but deliberate action which is imposed upon a sensor, channel, train, load group, or other system or device for the purpose of testing. 366

test interval (1)(nuclear power generating stations). The elapsed time between the initiation of identical tests on the same sensor, channel, train, load group, or other specified system or device. 159, 366

(2)(reliability analysis of nuclear power generating station safety systems). The elapsed time between the initiation of identical tests on the same sensor, channel, etcetera. 587

test item (software test documentation). A software item which is an object of testing. 436

test item transmittal report (software test documentation). A document identifying test items. It contains current status and location information. 436

test jack (test switches for transformer-rated meters). A spring-jaw receptacle in the current element of a test switch to provide a bipolar test connection in the metering current circuit without interruption of the current circuit. 550,590

test-jack switch (test switches for transformer-rated meters). A single-pole single-throw disconnect switch used in conjunction with a test jack to provide a parallel current path during normal operating conditions. 550,590

test log (1) (software). A chronological record of all relevant details of a testing activity. *See: testing.* 434

(2) (software test documentation). A chronological record of relevant details about the execution of tests. 436

test logic (test, measurement and diagnostic equipment). The logical, systematic examination of circuits and their diagrams to identify and analyze the probability and consequence of potential malfunctions for determining related maintenance or maintainability design requirements. 54

test loop (transmission performance of telephone sets). A circuit which is interposed between a telephone set and a telephone feed circuit to simulate a telephone line. 491

test, measurement and diagnostic equipment (TMDE). Any system or device used to evaluate the operational condition of a system or equipment to identify and isolate or both any actual or potential malfunction. 54

test model (evaluation of thermal capability)(thermal classification of electric equipment and electrical insulation). A representation of equipment, a component or part of equipment, or the equipment itself, that is suitable for use in a functional test. 506

test objective (software unit testing). An identified set of software features to be measured under specified conditions by comparing actual behavior with the required behavior described in the software documentation. 519

test operating cycle (valve actuators). The movement of an actuator through its required operations travel under specified loading conditions, terminating with a return to the starting position. 492

test oriented language (test, measurement and diagnostic equipment). A computer language utilizing English mnemonics that are commonly used in testing. Examples are measure, apply, connect, disconnect, and so forth. 54

test phase (software)(software verification and validation plans). The period of time in the software life cycle during which the components of a software product are evaluated and integrated, and the software product is evaluated to determine whether or not requirements have been satisfied. *See:* **component; requirement; software life cycle; software product.** 434,511

test plan (1)(safety systems equipment in nuclear power generating stations). A document that identifies the equipment to be qualified, defines the acceptance criteria and the total scope of the testing activities required for qualification to a specified set of conditions. 534

(2)(software verification and validation plans). Documentation specifying the scope, approach, resources, and schedule of intended testing activities. 511

(3) (software). A document prescribing the approach to be taken for intended testing activities. The plan typically identifies the items to be tested, the testing to be performed, test schedules, personnel requirements, reporting requirements, evaluation criteria, and any risks requiring contingency planning. *See:* **document; testing.** 434

(4) (software test documentation). A document describing the scope, approach, resources, and schedule of intended testing activities. It identifies test items, the features to be tested, the testing tasks, who will do each task, and any risks requiring contingency planning. 436

test plug (test switches for transformer-rated meters). A bipolar mating plug to a test jack for inserting instrumentation into the metering current circuit. 590, 550

test point (1)(separable insulated connectors). A capacitively coupled terminal for use with voltage sensing devices. See Fig 'Bushing Insert' and Fig 'Operating Interface' below. 454

(2) (station control and data acquisition). A predefined location within equipments or routines at which known result should be present if the equipment or routine is operating properly. 403

(3) (test, measurement and diagnostic equipment). A convenient, safe access to a circuit or system so that a significant quantity can be measured or introduced to facilitate maintenance, repair, calibration, alignment, and checkout. 54

test point selector (test, measurement and diagnostic equipment). A device capable of selecting test points on an item being tested in accordance with instructions from the programmer. 54

test position (of a switchgear assembly removable element) (power switchgear). That position in which the primary disconnecting devices of the removable element are separated by a safe distance from those in the housing and some or all of the secondary disconnecting devices are in operating contact. *Notes:* (1) A set of test jumpers or mechanical movement of secondary disconnecting devices may be used to complete all secondary connections for test in the test position. This may correspond with the disconnected position. (2) Safe distance, as used here, is a distance at which the equipment will meet its withstand ratings, both power frequency and impulse, between line and load stationary terminals and phase-to-phase and phase-to-ground on both line and load stationary terminals with the switching device in the closed position. 103

test procedure (1)(safety systems equipment in nuclear power generating stations). A document that defines the implementation of the test plan and describes the methodology for performing the specific test. 534

(2)(software verification and validation plans). Documentation specifying a sequence of actions for the execution of a test. 511

(3) (ATLAS). A document that describes the tests, test methods, and test sequences to be performed on a unit under test (UUT) to verify conformance with its test specification with or without fault diagnosis and without reference to specific test equipment. 400

(4) (software). Detailed instructions for the set up, operation, and evaluation of results for a given test. A set of associated procedures is often combined to form a test procedures document. *See:* **document; procedure.** 434

(5) (test, measurement and diagnostic equipment). A document that describes step by step the operation required to test a specific unit with a specific test system. 54

test procedure specification (software test documentation). A document specifying a sequence of actions for the execution of a test. 436

test program (1) (ATLAS). A test program implements the tests, test methods, and test sequences to be performed on a unit under test (UUT) to verify conformance with its test specification with or without fault diagnosis and designed for execution on a specific test system. 400

(2) (test, measurement and diagnostic equipment). A program specifically intended for the testing of a unit under test (UUT). *See:* **program; unit under test.** 54

test program documentation. *See:* **test programming procedures.**

test programming procedures (test, measurement and diagnostic equipment). Documents which explain in detail the composition of test programs including definitions and logic used to compose the program. Provides instructions to implement changes in the program. 54

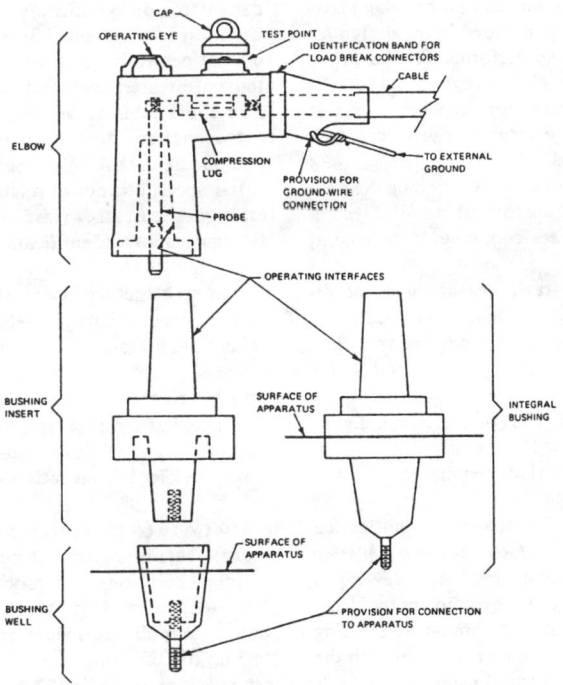

CAP
OPERATING EYE
TEST POINT
IDENTIFICATION BAND FOR
LOAD BREAK CONNECTORS
CABLE

ELBOW

COMPRESSION
LUG
TO EXTERNAL
GROUND
PROVISION FOR
GROUND WIRE
CONNECTION
PROBE

OPERATING INTERFACES

BUSHING
INSERT
SURFACE OF
APPARATUS
INTEGRAL
BUSHING

BUSHING
WELL
SURFACE OF
APPARATUS
PROVISION FOR CONNECTION
TO APPARATUS

**Typical Components of 200 A Separable
Insulated Connector System**

Bushing Insert

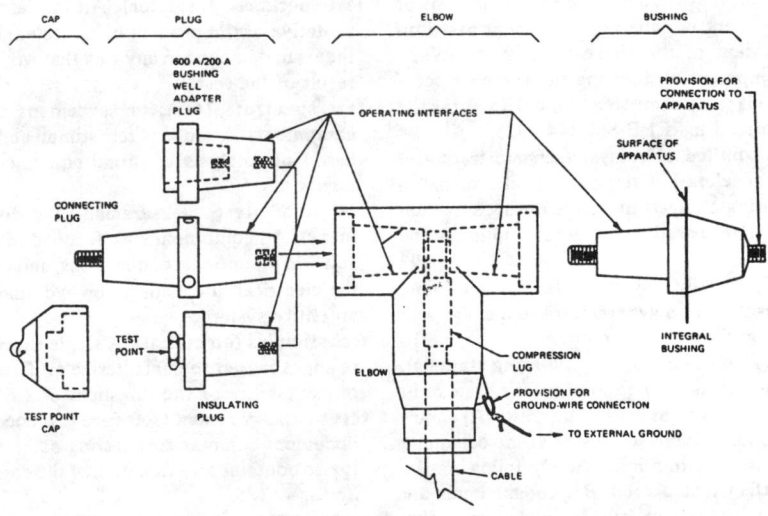

CAP
PLUG
ELBOW
BUSHING

600 A /200 A
BUSHING
WELL
ADAPTER
PLUG
PROVISION FOR
CONNECTION TO
APPARATUS

OPERATING INTERFACES

CONNECTING
PLUG
SURFACE OF
APPARATUS

TEST
POINT

COMPRESSION
LUG
INTEGRAL
BUSHING
TEST POINT
CAP
INSULATING
PLUG
ELBOW
PROVISION FOR
GROUND-WIRE CONNECTION
TO EXTERNAL GROUND

CABLE

**Typical Components of 600 A Separable
Insulated Connector System**

Operating Interface

test provisions (test, measurement and diagnostic equipment). The capability included in the design for conveniently evaluating the performance of a prime equipment, module, assembly, or part.   54

test repeatability (software). An attribute of a test indicating whether the same results are produced each time the test is conducted.   434

test report (software). A document describing the conduct and results of the testing carried out for a system or system component. *See:* component; document; system; testing.   434

test requirement analysis (test, measurement and diagnostic equipment). The examination of documents such as schematics, assembly drawings and specifications for the purpose of deriving test requirements for a unit.   54

test requirement document (TRD) (test, measurement and diagnostic equipment). The document that specifies the tests and test conditions required to test and fault isolate a unit under test.   54

test response spectrum (TRS)(1)(seismic qualification of Class 1E equipment for nuclear power generating stations). The response spectrum that is developed from the actual time history of the motion of the shake table. *Note:* When qualifying equipment bu utilizing response spectra, the TRS is to be compared with the RRS (required response spectrum) using the methods described in 7.6.2 and 7.6.3 of IEEE Std 344-1987.   581

(2) (TRS) (valve actuators). The response spectrum that is constructed using analysis or derived using spectrum analysis equipment based on the actual input test table motion to the device.   492

(3) (TRS) (nuclear power generating stations) (seismic qualification of class 1E equipment). The response spectrum that is constructed using analysis or derived using spectrum analysis equipment based on the actual motion of the shake table. *Note:* When qualifying equipment by utilizing the response spectrum, the TRS is to be compared to the RRS, using the methods described in IEEE Std 344-1975.   28

(4) (TRS, as applied to relays) (seismic testing of relays). The acceleration response spectrum that is constructed using analysis or derived using spectrum analysis equipment based on the actual motion of the shake table.   392

test routine. (1) Usually a synonym for check routine. (2) Sometimes used as a general term to include both check routine and diagnostic routine.   210

test schedule (1)(nuclear power generating stations). The pattern of testing applied to parts of a system. In general, there are two patterns of interest: (A) simultaneous: Redundant items are tested at the beginning of each test interval, one immediately following the other. (B) perfectly staggered: Redundant items are tested such that the test interval is divided into equal subintervals.   29

(2)(reliability analysis of nuclear power generating station safety systems). The pattern of testing applied to systems or the parts of a system. In general, there are two patterns of interest: (A) simultaneous. Redun-

dant items or systems are tested at the beginning of each test interval, one immediately following the other. (B) perfectly staggered. Redundant items or systems are tested such that the test interval is divided into equal subintervals.   587

test sequence (test, measurement and diagnostic equipment). (1) A unique setup of measurements: and (2) A specific order of related tests.   54

test sequence number (test, measurement and diagnostic equipment). Identification of a test sequence.   54

test set architecture (software unit testing). The nested relationships between sets of test cases that directly reflect the hierarchic decomposition of the test objectives.   519

test site (electromagnetic compatibility). A site meeting specified requirements suitable for measuring radio interference fields radiated by an appliance under test. *See:* electromagnetic compatibility.   220, 199

test software (test, measurement and diagnostic equipment). Maintenance instructions which control the testing operations and procedures of the automatic test equipment. This software is used to control the unique stimuli and measurement parameters used in testing the unit under test.   54

test specification (1)(ATLAS). A document that defines the tests to be performed on unit under test (UUT) to verify conformance with its performance specification and without reference to any specific test equipment or test method.   400

(2)(safety systems equipment in nuclear power generating stations). A document that defines the test requirements including test levels and performance requirements.   534

test specimen (insulator). An insulator that is representative of the product being tested: it is a specimen that is undamaged in any way that would influence the result of the test.   261

test spectrum (test, measurement and diagnostic equipment). A range of test stimuli and measurements based on analysis of prime equipment test requirements.   54

test stand (test, measurement and diagnostic equipment). An equipment specifically designed to provide suitable mountings, connections, and controls for testing electrical, mechanical, or hydraulic equipment as an entire system.   54

test stimulus (electrical). A single shock or succession of shocks, used to characterize or determine the state of excitability or the threshold of a tissue.   192

test summary report (software test documentation). A document summarizing testing activities and results. It also contains an evaluation of the corresponding test items.   436

test support software (test, measurement and diagnostic equipment). Computer programs used to prepare, analyze, and maintain test software. Test software includes automatic test equipment (ATE) compilers, translation/analysis programs and punch/print programs.   54

test switch (test block) (power switchgear). A combination of connection studs, jacks, plugs, or switch parts arranged conveniently to connect the necessary devices for testing instruments, meters, relays, etcetera.   103

test termination. The removal of a test input with results of the test being known, or the committal of the equipment for repair based on the results of the test.   31

test testboard (telephone switching systems). A position equipped with testing apparatus so arranged that connections can be made from it to trunks for testing purposes.   55

test unit (software unit testing). A set of one or more computer program modules together with associated control data, (for example, tables), usage procedures, and operating procedures that satisfy the following conditions: (1) All modules are from a single computer program (2) At least one of the new or changed modules in the set has not completed the unit test. (3) The set of modules together with its associated data and procedures are the sole object of a testing process. *Notes:* (A) A test unit may occur at any level of the design hierarchy from a single module to a complete program. Therefore, a test unit may be a module, a few modules, or a complete computer program along with associated data and procedures. (B) A test unit may contain one or more modules that have already been unit tested.   519

test validity (software). The degree to which a test accomplishes its specified goal.   434

test voltage (electrical insulation tests). The voltage applied across the specimen during a test.   116

test voltage, partial discharge-free. A specified voltage applied in a specified test procedure, at which the test object is free from partial discharges exceeding a specified level. This voltage is expressed as a peak value divided by the square root of two. *Note:* The term **corona-free test voltage** has frequently been used with this connotation. It is recommended that such usage be discontinued in favor of the term **partial discharge-free test voltage.**   139

tetanizing current (electrotherapy). The current that, when applied to a muscle or to a motor nerve connected with a muscle stimulates the muscle with sufficient intensity and frequency to produce a smoothly sustained contraction as distinguished from a succession of twitches. *See:* **electrotherapy.**   192

tetragonal system (piezoelectricity). A tetragonal crystal has a single fourfold axis or a fourfold inversion axis. The c axis is taken along this fourfold axis and the Z axis lies along c. The a and b axes are equivalent and are usually called a_1 and a_2. There are seven classes of tetragonal crystals, five of which can be piezoelectric; these are classes $\bar{4}$, 4, $\bar{4}2m$, 422, and $4mm$. Three of these have no twofold axes to guide in a choice of an a axis; however, for all of them except $42m$ there is no alternative to the choice of an a axis in such a way as to make the unit cell of smallest volume. In class 42 m, which has a twofold axis, the smallest cell may not have its a axis parallel to this axis. There are twelve possible arrangements of matter (space groups) that have symmetry $\bar{4}2m$. Of these twelve, six have the smallest cell when the a axis is an axis of twofold symmetry, and six have the smallest cell when a is chosen at 45 degrees to twofold axes (while still perpendicular to the c axis). The international tables for X-ray crystallography [6] now use the smallest cell in all twelve cases. In order for this standard not to be in conflict it is therefore necessary to choose the a axis at 45 degrees to the twofold axes in space groups $P\bar{4}m2$, $P\bar{4}c2$, $P\bar{4}b2$, $P\bar{4}n2$, $I\bar{4}c2$, and $I\bar{4}c2$ of class $\bar{4}2m$. With classes 4 and $4mm$ the $+Z$ axis is chosen so that d_{33} is positive and $+X$ and $+Y$ are parallel to a to form a right-handed system. With class $\bar{4}$, $+Z$ is chosen so that d_{31} is positive and $+X$ and $+Y$ are parallel to a to form a right-handed system. In classes $42m$ and 422 the $+Z$ axis (parallel to c) is chosen arbitrarily. In class $\bar{4}2m$ the $+X$ and $+Y$ axes are chosen parallel to the twofold axes (which are not parallel to the a axis for the space groups listed) such that d_{36} is positive. In class 422 the senses of the $+X$ and $+Y$ axes are trivial but they must form a right-handed system with $+Z$. *Note:* "Positive" and "negative" may be checked using a carbon-zinc flashlight battery. The carbon anode connection will have the same effect on meter deflection as the $+$ end of the crystal axis upon *release* of compression. *See:* **crystal systems.**   371

tetrode. A four-electrode electron tube containing an anode, a cathode, a control electrode, and one additional electrode that is ordinarily a grid.   190

theoretical cutoff frequency (theoretical cutoff) (electric structure). A frequency at which, disregarding the effects of dissipation, the attenuation constant changes from zero to a positive value or vice versa. *See:* **cutoff frequency.**   328

theory. *See:* **information theory.**

therapeutic high-frequency diathermy equipment (National Electrical Code) (health care facilities). Therapeutic high-frequency diathermy equipment is therapeutic induction and dielectric heating equipment.   256

thermal aging (1)(rotating machinery). Normal load/temperature deteriorating influence on insulation.   37

(2)(thermal classification of electric equipment and electrical insulation). The aging that takes place at an elevated temperature.   506

thermal burden rating (potential transformer). The voltamperes that the potential transformer will carry continuously at rated voltage and frequency without causing the specified temperature limitations to be exceeded. *See:* **instrument transformer.**   207

thermal burden rating of a voltage transformer (power and distribution transformer) (instrument transformer). The voltampere output that the transformer will supply continuously at rated secondary voltage without causing the specified temperature limitations to be exceeded.   53, 394

thermal capability (solid electrical insulating materials). Includes the ability to withstand without failure the maximum short time operating temperatures and the long time integrated degradative effect of temperature and time. It constitutes a design limitation on the use of insulating materials in electrical and electronic equipment to the extent that both thermal softening (or other short term effects) and long term aging affect functional properties. 452

thermal cell. A reserve cell that is activated by the application of heat. *See:* **electrochemistry.** 328

thermal conduction. The transport of thermal energy by processes having rates proportional to the temperature gradient and excluding those processes involving a net mass flow. *See:* **thermoelectric device.** 191

thermal conductivity. The quotient of (1) the conducted heat through unit area per unit time by (2) the component of the temperature gradient normal to that area. *See:* **thermoelectric device.** 191

thermal conductivity, electronic. The part of the thermal conductivity resulting from the transport of thermal energy by electrons and holes. *See:* **thermoelectric device.** 191

thermal converter (electric instrument) (thermocouple converter) (thermoelement). A device that consists of one or more thermojunctions in thermal contact with an electric heater or integral therewith, so that the electromotive force developed at its output terminals by thermoelectric action gives a measure of the input current in its heater. *Note:* The combination of two or more thermal converters when connected with appropriate auxiliary equipment so that its combined direct-current output gives a measure of the active power in the circuit is called a thermal watt converter. 328

thermal current converter (electric instrument). A type of thermal converter in which the electromotive force developed at the output terminals gives a measure of the current through the input terminals. *See:* **thermal converter.** 280

thermal current rating (1) (neutral grounding device) (electric power). The root-mean-square neutral current in amperes that it will carry under standard conditions for its rated time without exceeding standard temperature limitations, unless otherwise specified. *See:* **grounding device.** 91

(2) (resistor). The initial root-mean-square symmetrical value of the current that will flow when rated voltage is applied. 91

thermal cutout (National Electrical Code). An overcurrent protective device that contains a heater element in addition to and affecting a renewable fusible member which opens the circuit. It is not designed to interrupt short-circuit currents. 256

thermal duty cycle (nuclear power generating stations). The percentage of time that heat producing electrical current flows in equipment over a specific period of time. 440

thermal electromotive force. Alternative term for Seebeck electromotive force. *See:* **thermoelectric device.** 248, 191

thermal endurance (1)(electric insulation) (rotating machinery). The relationship, between temperature and time spent at that temperature, required to produce such degradation of an electrical insulation that it fails under specified conditions of stress, electric or mechanical, in service or under test. For most of the chemical reactions encountered, this relationship is a straight line when plotted with ordinates of logarithm of time against abscissae of reciprocal of absolute temperature (Arrhenius plot). *See:* **asynchronous machine.** 63

(2)(solid electrical insulating materials). Related to the rate at which important properties deteriorate as a function of temperature and time. It is determined by accelerated testing. 452

thermal endurance graph (evaluation of thermal capability)(thermal classification of electric equipment and electrical insulation). The graphical expression of the thermal endurance relationship in which time to failure is plotted against the reciprocal of the absolute test temperature. 506

thermal endurance relationship (evaluation of thermal capability)(thermal classification of electric equipment and electrical insulation). The expression of aging time to failure as a function of test temperature in an aging test. 506

thermal equilibrium (rotating machinery). The state reached when the observed temperature rise of the several parts of the machine does not vary by more than 2 degrees Celsius over a period of one hour. *See:* **asynchronous machine.** 63

thermal flow switch. *See:* **flow relay.**

thermal insulation (1)(electrical heating systems). Material having air- or gas-filled pockets, void spaces, or heat reflective surfaces that, when properly applied, will reduce the transfer of heat with reasonable effectiveness. 476

(2)(electrical heat tracing for industrial applications). Material having air- or gas-filled pockets, void spaces, or heat-reflective surfaces which, when properly applied, will retard the transfer of heat with reasonable effectiveness under ordinary conditions. 523

(3)(electric pipe heating systems). A material having a relatively high resistance to heat flow and used primarily to retard the flow of heat. 405

thermal limit curves for large squirrel-cage motors. Plots of maximum permissible time versus percent of rated current flowing in the motor winding under specified emergency conditions. These curves can be used in conjunction with the motor time-current curve for a normal start to set protective relays and breakers for motor thermal protection during starting and running conditions. 420

thermally delayed overcurrent trip. *See:* **thermally delayed release; overcurrent release.**

thermally delayed release (trip) (power switchgear). A release delayed by a thermal device. 103

thermally protected (National Electrical Code). (As applied to motors.) The words "Thermally Protected" appearing on the nameplate of a motor or motor-compressor indicate that the motor is provided with a thermal protector. 256

thermal-mechanical cycling (rotating machinery). The

experience undergone by rotating-machine windings, and particularly their insulation, as a result of differential movement between copper and iron on heating and cooling. Also denotes a test in which such actions are simulated for study of the resulting behavior of an insulation system, particularly for machines having a long core length. *See:* **asynchronous machine.**

63

thermal noise (1) (telephone practice). Noise occurring in electric conductors and resistors and resulting from the random movement of free electrons contained in the conducting material. The name derives from the fact that such random motion depends on the temperature of the material. Thermal noise has a flat power spectrum out to extremely high frequencies.

24

(2) (electron tube). The noise caused by thermal agitation in a dissipative body. *Note:* The available thermal noise power N, from a resistor at temperature T, is $N = kT\Delta f$, where k is Boltzmann's constant and Δf is the frequency increment. 125

(3) (resistance noise) (data transmission). Random noise in a circuit associated with the thermodynamic interchange of energy necessary to maintain thermal equilibrium between the circuit and its surroundings. *Note:* The average square of the open-circuit voltage across the terminals of a passive two-terminal network of uniform temperature, due to thermal agitation, is given by:

$$V_T^2 = 4kT \int R(f)\, df$$

where T is the absolute temperature in degrees Celsius, R is the resistance component Ω in ohms of the network impedance at the frequency f measured in hertz, and k is the Boltzmann constant, 1.38×10^{-23}.

59

thermal-overload detection (series capacitors). A means to detect excessive heating of series capacitor bank components and to initiate an alarm signal, or the closing of the associated bypass device, or both.

474

thermal-overload protection (series capacitor). A means to detect excessive heating of capacitor units as a result of a combination of current, ambient temperature, and solar radiations, and to initiate an alarm signal or the closing of the associated capacitor bypass switch, or both. 86

thermal power converter (thermal watt converter) (electric instrument). A complex type of thermal converter having both potential and current input terminals. It usually contains both current and potential transformers or other isolating elements, resistors, and a multiplicity of thermoelements. The electromotive force developed at the output terminals gives a measure of the power at the input terminals. *See:* **thermal converter.** 280

thermal protection (motor). The words **thermal protection** appearing on the nameplate of a motor indicate that the motor is provided with a thermal protector. *See:* **contactor.** 256

thermal protector (1) (National Electrical Code). (As applied to motors) A protective device for assembly as an integral part of a motor or motor-compressor and which, when properly applied, protects the motor against dangerous overheating due to overload and failure to start. The thermal protector may consist of one or more sensing elements integral with the motor or motor-compressor and an external control device.

256

(2) (rotating machinery). A protective device, for assembly as an integral part of a machine, that protects the machine against dangerous overheating due to overload or any other reason. *Notes:* (A) It may consist of one or more temperature-sensing elements integral with the machine and a control device external to the machine. (B) When a thermal protector is designed to perform its function by opening the circuit to the machine and then automatically closing the circuit after the machine cools to a satisfactory operating temperature, it is an automatic-reset thermal protector. (C) When a thermal protector is designed to perform its function by opening the circuit to the machine but must be reset manually to close the circuit, it is a manual-reset thermal protector. *See:* **contactor.**

63

thermal relay (1) (power switchgear). A relay whose operation is caused by heat developed within the relay as a result of specified external conditions. *Note:* Compare with **temperature relay.** 103

(2) (industrial control). A relay in which the displacement of the moving contact member is produced by the heating of a part of the relay under the action of electric currents. *Note:* Compare with **temperature relay.** *See:* **relay.** 244, 206

thermal residual voltage (Hall effect devices). That component of the zero field residual voltage caused by a temperature gradient in the Hall plate. 107

thermal resistance (1) (cable). The resistance offered by the insulation and other coverings to the flow of heat from the conductor or conductors to the outer surface. *Note:* The thermal resistance of the cable is equal to the difference of temperature between the conductor or conductors and the outside surface of the cable divided by rate of flow of heat produced thereby. It is preferably expressed by the number of degrees Celsius per watt per foot of cable. 64

(2) (Hall generator). The difference between the mean Hall plate temperature and the temperature of an external reference point, divided by the power dissipation in the Hall plate. 107

thermal resistance case-to-ambient, ($R_{\theta CA}$) (light emitting diodes). The thermal resistance (steady-state) from the device case to the ambient. 162

thermal resistance, effective (semiconductor rectifier) (semiconductor device). The effective temperature rise per unit power dissipation of a designated junction, above the temperature of a stated external reference point under conditions of thermal equilibrium. *Note:* Thermal impedance is the temperature rise of the junction above a designated point on the case, in degrees Celsius per watt of heat dissipation. *See:* **recti-**

fication; semiconductor; semiconductor rectifier stack.   245

thermal resistance junction-to-ambient, (formerly θ $_{J-A}$) ($R_{\theta JA}$) (light emitting diodes). The thermal resistance (steady-state) from the semiconductor junction (s) to the ambient.   162

thermal resistance junction-to-case, (formerly θ_{J-C}) ($R_{\theta JC}$) (light emitting diodes). The thermal resistance (steady-state) from the semiconductor junction (s) to a stated location on the case.   162

thermal short-time current rating (current transformer). The root-mean-square symmetrical primary current that may be carried for a stated period (five seconds or less) with the secondary winding short-circuited, without exceeding a specified maximum temperature in any winding. See: instrument transformer.   203

thermal subsystem (terrestrial photovoltaic power systems). The subsystem that receives thermal energy from the array subsystem. The thermal energy may be utilized for a thermal load application or dissipated. See: array control   496

thermal telephone receiver (thermophone). A telephone receiver in which the temperature of a conductor is caused to vary in response to the current input, thereby producing sound waves as a result of the expansion and contraction of the adjacent air. See: loudspeaker.   328

thermal tuning. See: tuning, thermal.

thermal voltage converter (electric instrument). A thermoelement of low-current input rating with an associated series impedance or transformer, such that the electromotive force developed at the output terminals gives a measure of the voltage applied to the input terminals. See: thermal converter.   280

thermal watt converter. See: thermal power converter; thermal converter.

thermionic arc (gas). An electric arc characterized by the fact that the thermionic cathode is heated by the arc current itself. See: discharge (gas).   244, 190

thermionic emission (Edison effect) (Richardson effect). The liberation of electrons or ions from a solid or liquid as a result of its thermal energy. See: electron emission.   333,125

thermionic generator. A thermoelectric generator in which a part of the circuit, across which a temperature difference is maintained, is a vacuum or a gas. See: thermoelectric device.   191

thermionic grid emission. Current produced by electrons thermionically emitted from a grid. See: electron emission.   333,125

thermionic tube. An electron tube in which the heating of one or more of the electrodes is for the purpose of causing electron or ion emission. See: hot cathode tube.   125

thermistor (1) (general). An electron device that makes use of the change of resistivity of semiconductor with change in temperature. See: bolometric detector; semiconductor.   245 7

(2) (power semiconductor). A semiconductor device whose electric resistance is dependent upon temperature.   66

(3) (waveguide components). A form of bolometer element having a negative temperature coefficient of resistivity which typically employs a semiconductor bead.   166

thermistor mount (bolometer mount) (waveguide). A waveguide termination in which a thermistor (bolometer) can be incorporated for the purpose of measuring electromagnetic power. See: waveguide.   244, 179

thermocouple. A pair of dissimilar conductors so joined at two points that an electromotive force is developed by the thermoelectric effects when the junctions are at different temperatures. See: electric thermometer (temperature meter); thermoelectric effect.   47, 7

thermocouple converter (thermoelement). See: thermal converter.

thermocouple extension wire. A pair of wires having such electromotive-force-temperature characteristics relative to the thermocouple with which the wires are intended to be used that, when properly connected to the thermocouple, the reference junction is in effect transferred to the other end of the wires.   7

thermocouple instrument. An electrothermic instrument in which one or more thermojunctions are heated directly or indirectly by an electric current or currents and supply a direct current that flows through the coil of a suitable direct-current mechanism, such as one of the permanent-magnet moving-coil type. See: instrument.   328

thermocouple leads. A pair of electrical conductors that connect the thermocouple to the electromotive force measuring device. One or both leads may be simply extensions of the thermoelements themselves or both may be of copper, dependent on the thermoelements in use and upon the physical location of the reference junction or junctions relative to the measuring device.   7

thermocouple thermometer. A temperature-measuring instrument comprising a device for measuring electromotive force, a sensing element called a thermocouple that produces an electromotive force of magnitude directly related to the temperature difference between its junctions, and electrical conductors for operatively connecting the two.   7

thermocouple vacuum gauge. A vacuum gauge that depends for its operation on the thermal conduction of the gas present, pressure being measured as a function of the electromotive force of a thermocouple the measuring junction of which is in thermal contact with a heater that carries a constant current. It is ordinarily used over a pressure range of 10^{-1} to 10^{-3} conventional millimeter of mercury. See: instrument.   328

thermoelectric cooling device. A thermoelectric heat pump that is used to remove thermal energy from a body. See: thermoelectric device.   248, 191

thermoelectric device. A generic term for thermoelectric heat pumps and thermoelectric generators.   191

thermoelectric effect. See: Seebeck effect.

thermoelectric effect error (bolometric power meter).

An error arising in bolometric power meters that employ thermistor elements in which the majority of the bias power is alternating current and the remainder direct current. The error is caused by thermocouples at the contacts of the thermistor leads to the metal oxides of the thermistors. *See:* **bolometric power meter.** 185

thermoelectric generator. A device that converts thermal energy into electric energy by direct interaction of a heat flow and the charge carriers in an electric circuit, and that requires for this process the existence of a temperature difference in the electric circuit. *See:* **thermoelectric device.** 191

thermoelectric heating device. A thermoelectric heat pump that is used to add thermal energy to a body. *See:* **thermoelectric device.** 191

thermoelectric heat pump. A device that transfers thermal energy from one body to another by the direct interaction of an electric current and the heat flow. *See:* **thermoelectric device.** 191

thermoelectric power. *See:* **Seebeck coefficient; thermoelectric device.**

thermoelectric thermometer (thermocouple thermometer). An electric thermometer that employs one or more thermocouples of which the set of measuring junctions is in thermal contact with the body, the temperature of which is to be measured, while the temperature of the reference junctions is either known or otherwise taken into account. *See:* **electric thermometer (temperature meter).** 328

thermoelement (electric instrument). The simplest type of thermal converter. It consists of a thermocouple, the measuring junction of which is in thermal contact with an electric heater or integral therewith. *See:* **thermal converter; thermoelectric arm; thermoelectric couple.** 280

thermogalvanic corrosion. Corrosion resulting from a galvanic cell caused primarily by a thermal gradient. *See:* **electrolytic cell.** 205

thermojunction. One of the surfaces of contact between the two conductors of a thermocouple. The thermojunction that is in thermal contact with the body under measurement is called the measuring junction, and the other thermojunction is called the reference junction. *See:* **electric thermometer (temperature meter).** 328

thermometer. An instrument for determining the temperature of a body or space. 27

thermometer method of temperature determination (power and distribution transformer). The determination of the temperature by mercury, alcohol, resistance, or thermocouple thermometer, any of these instruments being applied to the hottest accessible part of the device. 53

thermophone. An electroacoustic transducer in which sound waves of calculable magnitude result from the expansion and contraction of the air adjacent to a conductor whose temperature varies in response to a current input. *Note:* When used for the calibration of pressure microphones, a thermophone is generally used in a cavity the dimensions of which are small compared to a wavelength. *See:* **microphone.** 176

thermopile. A group of thermocouples connected in series aiding. This term is usually applied to a device used either to measure radiant power or energy or as a source of electric energy. *See:* **electric thermometer (temperature meter).** 47, 7

thermoplastic (1) (noun). A plastic that is thermoplastic in behavior.
(2) (adjective). Having the quality of softening when heated above a certain temperature range and of returning to its original state when cooling below that range. 63

thermoplastic insulating tape. A tape composed of a thermoplastic compound that provides insulation for joints. 328

thermoplastic insulations and jackets (power distribution, underground cables). Insulations and jackets made of materials that are softened by heat for application to the cable and then become firm, tough and resilient upon cooling. Subsequent heating and cooling will reproduce similar changes in the physical properties of the material. 57

thermostatic switch (thermostat). A form of temperature-operated switch that receives its operating energy by thermal conduction or convection from the device being controlled or operated. *See:* **switch.** 206

theta (θ) polarization. The state of the wave in which the E vector is tangential to the meridian lines of a given spherical frame of reference. *Note:* The usual frame of reference has the polar axis vertical and the origin at or near the antenna. Under these conditions, a vertical dipole will radiate only theta (θ) polarization and the horizontal loop will radiate only phi (ϕ) polarization. *See:* **antenna.** 246, 244

Thevenin's theorem. States that the current that will flow through an impedance Z', when connected to any two terminals of a linear network between which there previously existed a voltage E and an impedance Z, is equal to the voltage E divided by the sum of Z and Z'. 328

thickener (hydrometallurgy) (electrometallurgy). A tank in which suspension of solid material can settle so that the solid material emerges from a suitable opening with only a portion of the liquid while the remainder of the liquid overflows in clear condition at another part of the thickener. *See:* **electrowinning.** 328

thick film technology (circuits and systems). A technology in which a thick film (about 1 mil) is screen-printed onto an insulating substrate and then fused to the substrate by firing. *Note:* Resistors, capacitors, and conductors are commonly made by this technology. 67

thin film. Loosely, magnetic thin film. 255, 77

thin film technology (circuits and systems). A technology in which a thin film (a few hundred to a few thousand angstroms in thickness) is applied by vacuum deposition to an insulating substrate. Resistors, capacitors, and conductors are commonly made by this technology. 67

thin film waveguide (fiber optics). A transparent dielectric film, bounded by lower index materials, capable of guiding light. *See:* **optical waveguide.**   433

thinned array antenna. An array antenna which contains substantially fewer driven radiating elements than a conventional uniformly spaced array with the same beamwidth having identical elements. *Note:* Interelement spacings in the thinned array are chosen such that no large grating lobes are formed and side lobes are minimized.   111

thin-wall counter (radiation counters). A counter tube in which part of the envelope is made thin enough to permit the entry of radiation of low penetrating power. *See:* **anticoincidence (radiation counters).**   190

third-order nonlinearity coefficient (accelerometer). The proportionality constant that relates a variation of the output to the cube of the input applied parallel to an input reference axis.   46

third-rail clearance line (railroad). The contour that embraces all cross sections of third rail and its insulators, supports, and guards located at an elevation higher than the top of the running rail. *See:* **electric locomotive.**   328

third-rail electric car. An electric car that collects propulsion power through a third-rail system. *See:* **electric motor car.**   328

third-rail electric locomotive. An electric locomotive that collects propulsion power from a third-rail system. *See:* **electric locomotive.**   328

third voltage range. *See:* **voltage range.**

Thomson bridge (double bridge). *See:* **Kelvin bridge.**

Thomson coefficient (Thomson heat coefficient). The quotient of (1) the rate of Thomson heat absorption per unit volume of conductor by (2) the scalar product of the electric current density and the temperature gradient. The Thomson coefficient is positive if Thomson heat is absorbed by the conductor when the component of the electric current density in the direction of the temperature gradient is positive. *See:* **thermoelectric device.**   191

Thomson effect. The absorption or evolution of thermal energy produced by the interaction of an electric current and a temperature gradient in a homogeneous electric conductor. *Notes:* (1) An electromotive force exists between two points in a single conductor that are at different temperatures. The magnitude and direction of the electromotive force depend on the material of the conductor. A consequence of this effect is that if a current exists in a conductor between two points at different temperatures, heat will be absorbed or liberated depending on the material and on the sense of the current. (2) In a nonhomogeneous conductor, the Peltier effect and the Thomson effect cannot be separated. *See:* **thermoelectric device.**
                                    191

Thomson heat. The thermal energy absorbed or evolved as a result of the Thomson effect. *See:* **thermoelectric device.**   191

thread (control) (industrial control). A control function that provides for maintained operation of a drive at a preset reduced speed such as for setup purposes. *See:* **electric drive.**   206

threaded coupling (rigid steel conduit). An internally threaded steel cylinder for connecting two sections of rigid steel conduit.   101

threading line (conductor stringing equipment). A lightweight flexible line, normally manila or synthetic fiber rope, used to lead a conductor through the bullwheels of a tensioner or pulling line through a bullwheel puller. *Syn:* **bull line.**   431

three-address. Pertaining to an instruction code in which each instruction has three address parts. Also called triple-address. In a typical three-address instruction the addresses specify the location of two operands and the destination of the result, and the instructions are taken from storage in a preassigned order. *See:* **two-plus-one-address.**   235

three-address code. *See:* **instruction code.**

three conductor bundle. *See:* **bundle.**

three-dimensional (3D) radar (navigation aid terms). A radar capable of producing three-dimensional position data on a multiplicity of targets.   526

three-phase ac fields (electric and magnetic fields from ac power lines). Three-phase transmission lines generate a three-phase field whose space components are not in phase. The field at any point can be described by the field ellipse, that is, by the magnitude and direction of the semi-major axis and the magnitude and direction of its semi-major axis. In a three-phase field, the electric field at large distances $\geq$ 15 meters (m) away from the outer phases (conductors) can frequently be considered a single-phase field because the minor axis of the electric field ellipse is only a fraction (less than 10 percent) of the major axis when measured at a height of 1 m. Similar remarks apply for the magnetic field. *See:* **electric field strength.**   407

three-phase circuit (power and distribution transformer) (electric installations on shipboard). A combination of circuits energized by alternating electromotive forces which differ in phase by one-third of a cycle (120 degrees). In practice, the phases may vary several degrees from the specified angle.   53, 3

three-phase electric locomotive. An electric locomotive that collects propulsion power from three phases of an alternating-current distribution system. *See:* **electric locomotive.**   328

three-phase four-wire system. A system of alternating-current supply comprising four conductors, three of which are connected as in a three-phase three-wire system, the fourth being connected to the neutral point of the supply, which may be grounded. *See:* **alternating-current distribution.**   64

three-phase seven-wire system. A system of alternating-current supply from groups of three single-phase transformers connected in Y so as to obtain a three-phase four-wire grounded-neutral system for lighting and a three-phase three-wire grounded-neutral system of a higher voltage for power, the neutral wire being common to both systems. *See:* **alternating-current distribution.**   64

three-phase three-wire system. A system of alternating-current supply comprising three conductors between successive pairs of which are maintained alternating differences of potential successively displaced in phase by one-third of a period. *See:* **alternating-current distribution.** 64

three-plus-one-address. Pertaining to a four-address code in which one address part always specifies the location of the next instruction to be interpreted. 235

three-position relay. A relay that may be operated to three distinct positions. 328

three-state indication. *See:* **supervisory control functions**

three-terminal capacitor. Two conductors (the active electrodes) insulated from each other and from a surrounding third conductor that constitutes the shield. When the capacitor is provided with properly designed terminals and used with shielded leads, the direct capacitance between the active electrodes is independent of the presence of other conductors. (Specialized usage.) 210

three-wire control (industrial control). A control function that utilizes a momentary-contact pilot device and a holding-circuit contact to provide undervoltage protection. *See:* **undervoltage protection; relay.** 206

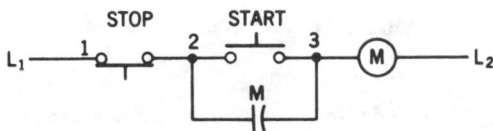

Three-wire control

three-wire system (direct current or single-phase alternating current). A system of electric supply comprising three conductors, one of which (known as the **neutral wire**) is maintained at a potential midway between the potential of the other two (referred to as the outer conductors). *Note:* Part of the load may be connected directly between the outer conductors, the remainder being divided as evenly as possible into two parts each of which is connected between the neutral and one outer conductor. There are thus two distinct voltages of supply, the one being twice the other. *See:* **alternating-current distribution; direct-current distribution.** 64

three-wire type current transformer (power and distribution transformer). One which has two primary windings each completely insulated for the rated insulation level of the transformer. This type of current transformer is for use on a three-wire single-phase service. *Note:* The primary windings and secondary windings are permanently assembled on the core as an integral structure. The secondary current is proportional to the phasor sum of the primary currents. 53

threshold (1)(mathematics of computing). (A) A logic operator having the property that if P is a statement, Q is a statement, R is a statement, ..., then the thresh-

old of P,Q,R, ... is true if at least N statements are true, false if less than N statements are true, where N is a specified nonnegative integer called the threshold condition. (B) The threshold condition as in (A). 564

(2) (illuminating engineering). The value of a variable of a physical stimulus (such as size, luminance, contrast or time) which permits the stimulus to be seen a specific percentage of the time or at a specific accuracy level. In many psychophysical experiments, thresholds are presented in terms of 50 percent accuracy or accurately 50 percent of the time. However, the threshold also is expressed as the value of the physical variable which permits the object to be just barely seen. The threshold may be determined by merely detecting the presence of an object or it may be determined by discriminating certain details of the object. 167

(3) (of a maser or laser). The condition of a maser or laser wherein the gain of its medium is just sufficient to permit the start of oscillation. 363

(4) (radar). As used in communications or radar signal processing, threshold refers to a value of voltage or current equal in value to the smallest signal to be detected by the receiver. 13

threshold audiogram. *See:* **audiogram.**

threshold center voltage V_{TC} The algebraic average of the (HC) and (LC) threshold voltages; that is, $(V_{LC} + V_{HC})/2$.

threshold current (1)(protection and coordination of industrial and commercial power systems). The magnitude of current at which a fuse becomes current limiting, specifically the symmetrical root-mean-square (rms) available current at the threshold of the current-limiting range, where the fuse total clearing time is less than half-cycle at rated voltage and rated frequency, for symmetrical closing, and a power factor of less than 20%. Refer to various peak let-through current curves for each type of fuse. The threshold ratio is the relationship of the threshold current to the fuse's continuous-current rating. 504

(2) (fiber optics). The driving current corresponding to lasing threshold. *See:* **lasing threshold.** 433

(3) (of a current-limiting fuse) (power switchgear). A current magnitude of specified wave shape at which the melting of the current-responsive element occurs at the first instantaneous peak current for that wave shape. *Note:* The current magnitude is usually expressed in root-mean-square (rms) amperes. 103

threshold element. (1) A combinational logic element such that the output channel is in its ONE state if and only if at least *n* input channels are in their ONE states, where *n* is a specified fixed nonnegative integer, called the threshold of the element. (2) By extension, a similar element whose output channel is in its ONE state if and only if at least *n* input channels are in states specified for them, not necessarily the ONE state but a fixed state for each input channel. 235

(3) A device that performs the logic threshold operation but in which the truth of each input statement contributes to the output determination a weight associated with that statement. 255, 77

threshold field. The least magnetizing force in a direction that tends to decrease the remanence, that, when applied either as a steady field of long duration or as a pulsed field appearing many times, will cause a stated fractional change of remanence. 77

threshold frequency (photoelectric device) (photoelectric tubes). The frequency of incident radiant energy below which there is no photoemissive effect.

$$v_0 = p/h.$$

See: **photoelectric effect.** 244, 190

threshold lights (illuminating engineering). Runway lights so placed as to indicate the longitudinal limits of that portion of a runway, channel, or landing path usable for landing. 167

threshold of audibility (threshold of detectability) (specified signal). The minimum effective sound pressure level of the signal that is capable of evoking an auditory sensation in a specified fraction of the trials. The characteristics of the signal, the manner in which it is presented to the listener, and the point at which the sound pressure is measured must be specified. *Notes:* (1) Unless otherwise specified, the ambient noise reaching the ears is assumed to be negligible. (2) The threshold is usually given as a sound pressure level in decibels relative to 20 micronewtons per square meter. (3) Instead of the method of constant stimuli, which is implied by the phrase in a specified fraction of the trials, another psychophysical method (which should be specified) may be employed.
176

threshold of discomfort (for a specified signal) (audio and electroacoustics). The minimum effective sound pressure level at the entrance to the external auditory canal that, in a specified fraction of the trials, will stimulate the ear to a point at which the sensation of feeling becomes uncomfortable. 176

threshold of feeling (tickle) (for a specified signal) (audio and electroacoustics). The minimum effective sound pressure level at the entrance to the external auditory canal that, in a specified fraction of the trials, will stimulate the ear to a point at which there is a sensation of feeling that is different from the sensation of hearing. 176

threshold of pain (for a specified signal) (audio and electroacoustics). The minimum effective sound pressure level at the entrance to the external auditory canal that, in a specified fraction of the trials, will stimulate the ear to a point at which the discomfort gives way to definite pain that is distinct from the mere nonnoxious feeling of discomfort. 176

threshold ratio (of a current-limiting fuse) (power switchgear). The ratio of the threshold current to the fuse current rating. 103

threshold signal (navigation)(navigation aid terms). The smallest signal capable of effecting a recognizable change in navigational information. 526

threshold signal-to-interference ratio (TSI) (electromagnetic compatibility). The minimum signal to interference power, described in a prescribed way, required to provide a specified performance level. *See:* **electromagnetic compatibility.** 199

threshold voltage (1) (metal-nitride-oxide field-effect transistor). Minimum gate voltage necessary for onset of current flow between source and drain of an insulated-gate field-effect transistor (IGFET). This is a serviceable general definition. There are three more specific definitions possible.

$$(1)\ V_T = V_{GS} \text{ at } I_{DS} = 10\ \mu A,$$
$$\text{when } V_{GS} = V_{DS},$$

that is, V_T is the gate voltage necessary to result in a defined low current level;

$$(2)\ V_T = V_{GS} - \left[\frac{I_{DS}\ell}{k'W}\right]^{1/2},$$

that is, V_T is that value that gives the best fit to the I–V relationship of the IGFET;

$$(3)\ V_T = \phi_{ms} + \phi_s - \frac{Q_B}{C_G} - \frac{Q_I}{C_G},$$

that is, V_T is derived from inherent structural parameters only. *Note:* Definition (1) is commonly used in practice, and its use will be implied here unless otherwise specified. While normally I_{DS} is set at 10 μA, a value of I_{DS} independent of lateral geometry is $I_{DS} = 0.5\ (W/l)\ \mu A$. Definition (3) suffers from the fact that in packaged devices, none of the terms can be varified by independent measurement. *See:* **insulated-gate field effect transistor symbols.** 386

(2) (semiconductor rectifiers). The zero-current voltage intercept of a straight-line approximation of the forward current-voltage characteristic over the normal operating range. *See:* **semiconductor rectifier stack.** 208

threshold voltage saturation (metal-nitride-oxide field-effect transistor). For a given gate-to-source voltage, the transistor metal-nitride-oxide semiconductor (MNOS) threshold voltage achieved in either of the two written states for which an order of magnitude increase in pulse width causes less than a 100 mV change in threshold voltage. Thus pulse width can also be achieved by sequence of shorter pulses of the same polarity. 386

threshold voltage window (Δv_{HL}) (metal-nitride-oxide field-effect transistor). The algebraic difference between the two threshold voltages (the threshold voltage after a write-high operation minus the threshold voltage after a write-low operation). 386

threshold wavelength (photoelectric tubes) (photoelectric device). The wavelength of the incident radiant energy above which there is no photoemission effect. *See:* **photoelectric effect.** 244, 190

throat microphone. A microphone normally actuated by mechanical contact with the throat. *See:* **microphone.** 328

through bolt (rotating machinery). A bolt passing axially through a laminated core, that is used to apply pressure to the end plates. 63

throughput (1) (data transmission). The total capability of equipment to process or transmit data during a specified time period. 59

(2) (software). A measure of the amount of work performed by a computer system over a period of time, for example, number of jobs per day. *See:* **computer system.** 434

(3) (automatic control). *See:* **capacity.**

through supervision (1) (communication switching). The automatic transfer of supervisory signals through one or more trunks in a manual telephone switchboard. 193

(2) (telephone switching systems). The capability of apparatus within a switched connection to pass or repeat signaling. 55

throwing power (of a solution) (electroplating). A measure of its adaptability to deposit metal uniformly upon a cathode of irregular shape. In a given solution under specified conditions it is equal to the improvement (in percent) of the metal distribution ratio above the primary-current distribution ratio. *See:* **electroplating.** 328

throw-over equipment. *See:* **automatic transfer equipment.**

thrust bearing (rotating machinery). A bearing designed to carry an axial load so as to prevent or to limit axial movement of a shaft, or to carry the weight of a vertical rotor system. *See:* **bearing.** 63

thrust block (rotating machinery). A support for a thrust-bearing runner. *See:* **bearing.** 63

thrust collar (rotating machinery). The part of a shaft or rotor that contacts the thrust bearing and transmits the axial load. *See:* **rotor (rotating machinery).** 63

thump. A low-frequency transient disturbance in a system or transducer characterized audibly by the onomatopoeic connotation of the word. *Note:* In telephony, thump is the noise in a receiver connected to a telephone circuit on which a direct-current telegraph channel is superposed caused by the telegraph currents. *See:* **signal-to-noise ratio.** 239

Thury transmission system. A system of direct-current transmission with constant current and a variable high voltage. *Note:* High voltage used on this system is obtained by connecting series direct-current generators in series at the generating station: and is utilized by connecting series direct-current motors in series at the substations. *See:* **direct-current distribution.** 64

thyratron. A hot-cathode gas tube in which one or more control electrodes initiate but do not limit the anode current except under certain operating conditions. 190

thyristor (1)(thyristor ac power controllers). A bistable semiconductor device comprising three or more junctions that can be switched from the OFF state to the ON state or vice versa, such switching occurring within at least one quadrant of the principal voltage-current characteristic. 445

(2)(electrical heating applications to melting furnaces and forehearths in the glass industry). A bistable semiconductor device comprising three or more junctions that can be switched from an off (nonconducting) to an on (conducting) condition, or vice versa, by the application of a small electric signal. Such switching occurs within at least one quadrant of the principal voltage-current characteristic. 520

thyristor ac power controller. A power electronic equipment for the control or switching of ac power where switching, multicycle control and phase control are included. The only power controlling element is the thyristor, although other power elements may be included. 445

thyristor assembly. An electrical and mechanical functional assembly of thyristors in combination with diodes, if any, or thyristor stacks, complete with all its connections and auxiliary components, including trigger equipment, together with means for cooling, if any, in its own mechanical structure, but without the controller transformers and other switching devices. *Note:* A thyristor assembly may be combined of several subassemblies, which are made and traded as mechanically combined units, for example, thyristor stacks or other combinations or one or more thyristors with control devices, protective devices, etcetera. 445

thyristor converter (thyristor converter unit, thyristor converter equipment). An operative unit comprising one or more thyristor sections together with converter transformers, essential switching devices, and other auxiliaries, if any of these items exist. System control equipments are optionally included. 121

thyristor converter, bridge (double-way). A bridge thyristor converter in which the current between each terminal of the alternating-voltage circuit and the thyristor converter circuit elements conductively connected to it flows in both directions. *Note:* The terms single-way and double-way (bridge) provide a means for describing the effect of the thyristor converter circuit on current flow in the transformer windings connected to the converter. Most thyristor converters may be classified into these two general types. The term bridge relates back to the single-phase "bridge" which resembles the Wheatstone bridge. 121

thyristor converter, cascade. A thyristor converter in which two or more simple converters are connected in such a way that their direct voltages add, but their commutations do not coincide. 121

thyristor converter circuit element. A group of one or more thyristors, connected in series or parallel or any combination of both, bounded by no more than two circuit terminals and conducting forward current in the same direction between these terminals. *Note:* A circuit element is also referred to as a leg or arm, and in the case of paralleled thyristors each path is referred to as a branch. 121

thyristor converter, multiple. A thyristor converter in which two or more simple thyristor converters are connected in such a way that their direct currents add, but their commutations do not coincide. 121

thyristor converter, parallel. A thyristor converter in which two or more simple converters are connected in such a way that their direct currents add and their commutations coincide. 121

thyristor converter, series. A thyristor converter in which two or more simple converters are connected in such a way that their direct voltages add and their commutations coincide. 121

thyristor converter, simple. A thyristor converter that consists of one commutating group. 121

thyristor converter, single-way. A thyristor converter in which the current between each terminal of the alternating-voltage circuit and the thyristor converter circuit element or elements conductively connected to it flows only in one direction. 121

thyristor converter transformer. A transformer that operates at the fundamental frequency of the ac system and is designed to have one or more output windings conductively connected to the thyristor converter elements. 121

thyristor converter unit, double. Two converters connected to a common dc circuit such that this circuit can accept or give up energy with direct current in both directions. *Note:* The converters may be supplied from separate cell windings on a common transformer, from common cell windings, or from separate transformers, The converter connections may be single way or symmetrical double way. Where two converters are involved, the designated forward converter arbitrarily operates in quadrant I. Quadrant I implies motoring torque in the agreed-upon forward direction. 121

thyristor converter unit, single. A thyristor converter unit connected to a dc circuit such that the direct current supplied by the converter is flowing in only one direction. The single converter section is referred to as a forward converter section. *Note:* When used without a reversing switch a single converter can be used in a reversible power sense only in those cases where single-way thyristor connections or symmetrical double-way thyristor connections are used and where the dc circuit can change from accepting energy to giving up energy without the need for current reversal, for example, a heavily inductive load. When used with a reversing switch, a single converter can be used in a reversible power sense in all cases where single-way thyristor connections or uniform double-way thyristor connections are used. 121

thyristor fuses. Fuses of special characteristics connected in series with one or more thyristors to protect the thyristor or other circuit components, or both. 121

thyristor, reverse-blocking triode. A monocrystalline reverse-blocking semiconductor device with bistable character in the forward direction normally having three pn junctions and a gate electrode at which a suitable electrical signal will cause switching from the off state to the on state within the first quadrant of the anode to cathode voltage–current characteristics. If cooling means are integrated, they are included. *Note:* In this document the word thyristor means a reverse-blocking triode thyristor. *See:* **junction, P-N.**
  121

thyristor stack. A single structure of one or more thyristors with its (their) associated mounting(s), cooling attachments, if any, connections whether electrical or mechanical, and auxiliary components, if any. A thyristor stack may consist of thyristors and semiconductor rectifier diodes in combination and in this case it may be referred to as a non-uniform thyristor stack. Trigger equipments are not included in this definition. 445

thyristor trigger circuit. A circuit for the conversion of a control signal to suitable trigger signals for the thyristors in a thyristor ac power controller including phase shifting circuits, pulse generating circuits, and power supply circuits. 445

ticker. A form of receiving-only printer used in the dissemination of information such as stock quotations and news. *See:* **telegraphy.** 194

tie (rotating machinery). A binding of the end turns used to hold a winding in place or to hold leads to windings for purpose of anchoring. *See:* **rotor (rotating machinery); stator.** 63

tie feeder (power switchgear). A feeder that connects together two or more independent sources of power and has no tapped load between the terminals. *Note:* If a feeder has any tapped load between the two sources, it is designated as a multiple feeder. 103

tie line (electric power systems). A transmission line connecting two or more power systems. *See:* **transmission line.** 94

tie-line bias control (control area). A mode of operation under load-frequency control in which the area control error is determined by the net interchange minus the biased scheduled net interchange. *See:* **speed-governing system.** 94

tie point (tie line) (electric power systems). The location of the switching facilities, when closed, permit power to flow between the two power systems. *See:* **center of distribution.** 94

tier (rotating machinery). A concentric winding is said to have one, two or more tiers according to whether the periphal extremities of the end windings of groups of coils at each end of the machine form one, two or more solids of revolution around the axis of the machine. 63

tie trunk (data transmission). A telephone line or channel directly connecting two private branch exchanges. 55, 59

tie wire. A short piece of wire used to bind an overhead conductor to an insulator or other support. *See:* **conductor; tower.** 64

TIF. *See:* **telephone influence factor.**

tight (1) (packaging machinery). Used as a suffix, indicating that apparatus is so constructed that the enclosing case will exclude the specified material. 429

(2) (suffix) (power and distribution transformer). Apparatus is designed as watertight, dusttight, etc, when so constructed that the enclosing case will exclude the specified material under specified conditions. 53

(3) (used as a suffix) (power switchgear). So constructed that the specific material is excluded under specified conditions. 103, 443

tight coupling. *See:* close coupling.

tilt (1)(directional antenna)(navigation aid terms). The angle which the antenna axis forms with the horizontal. 526

(2)(droop), A_D **(pulse transformers).** The difference between A_M and A_T. It is expressed in amplitude units or in percentage of A_M. 589

(3) (pulse terms). A distortion of a pulse top or pulse base wherein the overall slope over the extent of the pulse top or the pulse base is essentially constant and other than zero. Tilt may be of either polarity. *See:* Note in **preshoot** entry. 254

(4) (high- and low-power pulse transformers). (droop) (A_D). The difference between A_M and A_T. It is expressed in amplitude units or in percent of A_M. *See:* **input pulse shape.** 32, 33

tilt angle (1)(navigation aid terms). The vertical angle between the axis of measurement and a reference axis; the reference is normally horizontal. 526

(2) (of a polarization ellipse) (antennas). When the plane of polarization is viewed from a specified side, the angle measured clockwise from a reference line to the major axis of the ellipse. *Notes:* (A) For a plane wave the plane of polarization is viewed looking in the direction of propagation. (B) The tilt angle is only defined up to multiple of π radians and is usually taken in the range $-\pi/2$, $+\pi/2$ or $0,\pi$. 111

tilt error. *See:* **ionospheric tilt error.**

tilting-insulator switch (power switchgear). One in which the opening and closing travel of the blade is accomplished by a tilting movement of one or more of the insulators supporting the conducting parts of the switch. 103

tilting-pad bearing (Kingsbury bearing) (rotating machinery). A pad-type bearing in which the pads are capable of moving in such a manner as to improve the flow of lubricating fluid between the bearing and the shaft journal or collar (runner). *See:* **bearing.** 63

timbering machine. An electrically driven machine to raise and hold timbers in place while supporting posts are being set after being cut to length by the machine's power-driven saw. 328

timbre. The attribute of auditory sensation in terms of which a listener can judge that two sounds similarly presented and having the same loudness and pitch are dissimilar. *Note:* Timbre depends primarily upon the spectrum of the stimulus, but it also depends upon the waveform, the sound pressure, the frequency location of the spectrum, and the temporal characteristics of the stimulus. *See:* **loudspeaker.** 176

time (1)(supervisory control, data acquisition, and automatic control). (A) Response. The time between initiating some operation and obtaining results. (B) Settling. Time required by channel or terminal equipment to reach an acceptable ooperating condition. 570

(2) (reliability). Any duration of observations of the considered items either in actual operation or in storage, readiness, etcetera, but excluding down time due to a failure. *Note:* In definitions where time is used, this parameter may be replaced by distance, cycles, or other measures of life as may be appropriate. This refers to terms such as acceleration factor, wear-out failure, failure rate, mean life, mean time between failures, mean time to failure, reliability, and useful life. *See:* **reliability.** 182

(3) (electronic computation). *See:* **access time; downtime; real time; reference time; switching time; word time.**

(4) (International System of Units) (SI). The SI unit of time is the second. This unit is preferred and should be used if practical, particularly when technical calculations are involved. In cases where time relates to life customs or calendar cycles, the minute, hour, day, and other calendar units may be necessary. For example, vehicle velocity will normally be expressed in kilometers per hour. *See:* **units and letter symbols.** 21

time above 90 percent (switching impulse testing). The time interval T_d during which the switching impulse exceeds 90 percent of its crest value. 108

time-and-charge-request call, (T&C) (telephone switching systems). A call for which a request is made to be informed of its duration and cost upon its completion. 55

time-averaged Poynting vector (radio wave propagation). Of a periodic electromagnetic wave, the time average of the instantaneous Poynting vector over a period. For time-harmonic waves, it is equal to $(1/2)$ Re [E X H*], where Re indicates real part, E and H are the electric and magnetic field vectors in phasor notations, respectively, and * indicates the complex conjugate. 146

time base (oscilloscopes). The sweep generator in an oscilloscope. *See:* **oscillograph.** 185

time base--primary (watthour meters). A timing system established from the power line-source. 485

time base--secondary (watthour meters). A timing system established from an alternate source when the line source is not available or not used. 485

time bias (electric power systems). An offset in the scheduled net interchange power of a control area that varies in proportion to the time deviation. This offset is in a direction to assist in restoring the time deviation to zero. *See:* **power system.** 200, 94

time bias setting (electric power systems). For a control area, a factor with negative sign that is multiplied by the time deviation to yield the time bias. 200, 94

time, build-up (automatic control). In a continuous step-forced response, the fictitious time interval, which would be required for the output to rise from its initial to its ultimate value, assuming that the entire rise were to take place at the maximum rate. *Note:* It can be evaluated as π/ω_O, where ω_O is the cut-off frequency of an ideal low-pass filter. 56

time characteristic (lagged-demand meter). *See:* **demand meter.**

time characteristic demand meter (lagged-demand meter) (metering). The nominal time required for 90 percent of the final indication, with constant load suddenly applied. *Note:* The time characteristic of lagged-demand meters describes the exponential response of the meter to the applied load. The response of the lagged-demand meter to the load is continuous and independent of the selected discrete time intervals. 212

time coherence. *See:* coherent.

time constant (1) (electrothermic unit). The time required for the dc electrothermic output voltage to reach $1 - (1/e)$, or 63 percent of its final value after a fixed amount of power is applied to the electrothermic unit. 47

(2) (excitation control systems). The value T in an exponential response term A exp $(-t/T)$ or in one of the transform factors $1 + sT$, $1 + jwT$, $1/(1 + sT)$, $1/(1 + jwT)$. *Note:* For the output of a first-order (lag or lead) system forced by a step or an impulse, T is the time required to complete 63.2 percent of the total rise or decay: at any instant during the process, T is the quotient of the instantaneous rate of change divided into the change still to be completed. In higher order systems, there is a time constant for each of the first-order components of the process. In a Bode diagram, breakpoints occur at $w = 1/T$. 353

time constant, derivative action (automatic control). A parameter whose value is equal to $1/2\pi f_d$ where f_d is the frequency (cycles.unit time) on a Bode diagram of the lowest frequency gain corner resulting from derivative control action. 56

time constant of an exponential function (ae^{-bt}). $1/b$, if t represents time and b is real. 210

time constant of fall (pulse) (data transmission). The time required for the pulse to fall from 70.7 to 26.0 percent of its maximum amplitude excluding spike. 59

time constant of integrator (for each input). The ratio of the input to the corresponding time rate of change of the output. *See:* electronic analog computer. 9

time constant of rise (pulse) (data transmission). The time required for the pulse to rise from 26.0 percent to 70.7 percent of its maximum amplitude excluding spike. 59

time constant of the damping device (hydraulic turbines). A time constant which describes the decay of the output signal from the damping device. 8

time constant of the damping device

time-current characteristic (of a fuse). *See:* fuse time-current characteristic.

time-current tests (of a fuse). *See:* fuse time-current tests.

time delay (1)(protection and coordination of industrial and commercial power systems). Meaningless unless defined. This term is now used by National Electrical Manufacturers Association (NEMA), American National Standards Institute (ANSI), and Underwriters Laboratories (UL) to mean, in Classes H, K, J, and R cartridge fuses, a minimum opening time of 10 seconds (s) on an overload current five times the ampere rating of the fuse. Such a delay is particularly useful in allowing the fuse to pass the momentary starting current of a motor, yet not hindering the opening of the fuse should the overload persist. In Class G, CC, and plug fuses, the phrase 'time delay' is required by UL to be a minimum opening time of 12 s on an overload of twice the fuse's ampere rating. The time-delay characteristic does not affect the fuse's short-circuit current clearing ability. 504

(2) (analog computers). The time interval between the manifestation of a signal at one point and the manifestation or detection of the same signal at another point. *Notes:* (A) Generally, the term time delay is used to describe a process whereby an output signal has the same form as an input signal causing it, but is delayed in time; that is, the amplification of all frequency components of the output are related by a single constant to those of corresponding input frequency components but each output component lags behind the corresponding input component by a phase angle proportional to the frequency of the component. (B) Transport delay is synonymous with time delay but usually is reserved for applications that involve the flow of material. 9

(3) (industrial control). A time interval is purposely introduced in the performance of a function. *See:* control system, feedback. 206

time-delay relay. See: delay relay; relay.

time-delay starting or closing relay (2) (power system device function numbers). A device that functions to give a desired amount of time delay before or after any point of operation in a switching sequence or protective relay system, except as specifically provided by incomplete sequence relay, time-delay stopping or opening relay, and alternating current (ac) reclosing relay, device functions 48, 62, and 79. 402

time-delay stopping or opening relay (62) (power system device function numbers). A time-delay relay that serves in conjunction with the device that initiates the shutdown, stopping, or opening operating in an automatic sequence or protective relay system. 402

time deviation (power system). The integrated or accumulated difference in cycles between system frequency and rated frequency. This is usually expressed in seconds by dividing the deviation in cycles by the rated frequency. 94

time dial (time lever) (of a relay) (power switchgear).

An adjustable, graduated element of a relay by which, under fixed input conditions, the prescribed relay operating time can be varied.   103

time difference (loran)(navigation aid terms). The difference in the time of reception of the two signals of a loran rate.   526

time discriminator (electronic navigation). A circuit in which the sense and magnitude of the output is a function of the time difference of the occurrence, and relative time sequence, of two pulses. *See:* **navigation.**
                                                                    13, 187

time distribution analyzer (nuclear techniques). An instrument capable of indicating the number or rate of occurrence of time intervals falling within one or more specified time interval ranges. The time interval is delineated by the separation between pulses of a pulse pair. *See:* **ionizing radiation.**   335

time-division analog switching (telephone switching systems). Analog switching with common time-divided paths for simultaneous calls.   55

time-division digital switching (telephone switching systems). Digital switching with common time-divided paths for simultaneous calls.   55

time division multiple access (TDMA) (communication satellite). A technique whereby earth stations communicate with each other on the basis of non-overlapping time sequenced bursts of transmissions through a common satellite repeater.   84

time-division multiplex (data transmission) (antennas). The process or device in which each modulating wave modulates a separate pulse subcarrier, the pulse subcarriers being spaced in time so that no two pulses occupy the same time interval. *Note:* Time division permits the transmission of two or more signals over a common path by using different time intervals for the transmission of the intelligence of each message signal.   59, 111

time-division switching (telephone switching systems). A method of switching that provides a common path with separate time intervals assigned to each of the simultaneous calls.   55

time-domain reflectometer. *See:* **reflectometer, time domain.**

timed release (telephone switching systems). Release accomplished after a specified delay.   55

time, electrification (cable-insulation material). Time during which a steady direct voltage is applied to electrical insulating materials before the current is measured.   97

time gain control (electronic navigation). *See:* **differential gain-control circuit.**

time gate. A transducer that gives output only during chosen time intervals.   111

time history (seismic design of substations). The record of acceleration, velocity of displacement as a function of time which the floor of a building experiences due to an earthquake.   465

time-interval selector (nuclear techniques). A circuit that produces a specified output pulse when and only when the time interval between two pulses lies between specified limits. *See:* **scintillation counter.**
                                                                    335

time-invariant filtering (germanium gamma-ray detectors). Pulse shaping in which the filter response does not change with respect to time. [CR-(RC)n shaping is an example of time-invariant filtering.]   528

time lag. *See:* **lag.**

time lag of impulse flashover (surge arresters). The time between the instant when the voltage of the impulse wave first exceeds the power-frequency flashover crest voltage and the instant when the impulse flashover causes the abrupt drop in the testing wave.
                                                                    62, 64

time-load withstand strength (of an insulator). The mechanical load that, under specified conditions, can be continuously applied without mechanical or electrical failure. *See:* **insulator.**   261

time locking. A method of locking, either mechanical or electric, that, after a signal has been caused to display an aspect to proceed, prevents, until after the expiration of a predetermined time interval after such signal has been caused to display its most restrictive aspect, the operation of any interlocked or electrically locked switch, movable-point frog, or derail in the route governed by that signal, and that prevents an aspect to proceed from being displayed for any conflicting route. *See:* **interlocking (interlocking plant).**
                                                                    328

time meridian (navigation aid terms). Any meridian used as a reference for reckoning time, particularly a zone.   526

time-of-arrival location (TOA) (radar). A process whereby the position of a radiating transmitter can be located by means of the relative time delay between its signals as received in multiple receivers of known relative position.   13

time of response. *See:* **response time.**

time of rise (decay) of video pulses (television). The duration of the rising (decaying) portion of a pulse measured between specified levels. *See:* **pulse timing of video pulses (television).**   336

time-of-use period (watthour meters). A selected period of time during which a specified rate will apply to the energy usage or demand.   485

time-of-use register (watthour meters). That portion of a watthour meter that, for selected periods of time, accumulates and may display amounts of electric energy, demand, or other quantities measured or calculated.   485

time origin line (pulse terms). A line of constant and specified time which, unless otherwise specified, has a time equal to zero and passes through the first datum time, t_0, of a waveform epoch. *See:* The single pulse diagram below the **waveform epoch** entry.   254

timeout (FASTBUS acquisition and control). A timeout occurs when a protective timer completes its assigned time without the expected event occurring. Timeouts prevent the system from waiting indefinitely in case of error or failure.   480

time-overcurrent relay (power switchgear). An overcurrent relay in which the input current and operating time are inversely related throughout a substantial portion of the performance range.   103

time parameters and references (pulse terms). *See:* pulse start (stop) time; pulse duration; time reference lines; pulse time reference points; transition duration; first (last) transition duration.

time pattern (television). A picture-tube presentation of horizontal and vertical lines or dot rows generated by two stable frequency sources operating at multiples of the line and field frequencies.   372

time per point (multiple-point recorders). The time interval between successive points on printed records. *Note:* For some instruments this interval is variable and depends on the magnitude of change in measured signal. For such instruments, time per point is specified as the minimum and maximum time intervals.   295

time proportioning (electrical heating applications to melting furnaces and forehearths in the glass industry). An operation in which variable length bursts of full cycles of output voltage are alternated with variable length off periods to produce modulation of output.   520

time rate (storage cell). The current in amperes at which a storage battery will be discharged in a specified time, under specified conditions of temperature and final voltage. *See:* **battery (primary or secondary).**   328

time, real (analog computer). *See:* real time.

time referenced point (pulse terms). A point at the intersection of a time reference line and a waveform.   254

time reference line (pulse terms). A line parallel to the time origin line at a specified instant.   254

time reference lines (pulse terms). (1) pulse start (stop) line. The time reference line at pulse start (stop) time. *See:* The single pulse diagram below the **waveform epoch** entry. (2) top center line. The time reference line at the average of pulse start time and pulse stop time. *See:* The single pulse diagram below the **waveform epoch** entry.   254

time-related adjectives (pulse terms). (1) periodic (aperiodic). Of or pertaining to a series of specified waveforms or features which repeat or recur regularly (irregularly in time. (2) coherent (incoherent). Of or pertaining to two or more repetitive waveforms whose constituent features have (lack) time correlation. (3) synchronous (a synchronous). Of or pertaining to two or more repetitive waveforms whose sequential constituent features have (lack) time correlation.   254

time-related definitions (pulse terms). *See:* instant; interval; duration; period; frequency; cycle.   254

time release. A device used to prevent the operation of an operative unit until after the expiration of a predetermined time interval after the device has been actuated.   328

time, response. *See:* response time.

time response (1) (control system, feedback). An output, expressed as a function of time, resulting from the application of a specified input under specified operating conditions. *Syn:* **dynamic response.** *Note:* It consists of a transient component that depends on the initial conditions of the system, and a steady-state component that depends on the time pattern of the input.   206

(2) (synchronous-machine regulator). The output of the synchronous-machine regulator (that is, voltage, current, impedance, or position) expressed as a function of time following the application of prescribed inputs under specified conditions.

(3) (excitation control systems). An output expressed as a function of time, resulting from the application of a specified input under specified operating conditions. See figure below. For a typical time response of a system to step increase of input and for identification of the principal characteristics of interest.

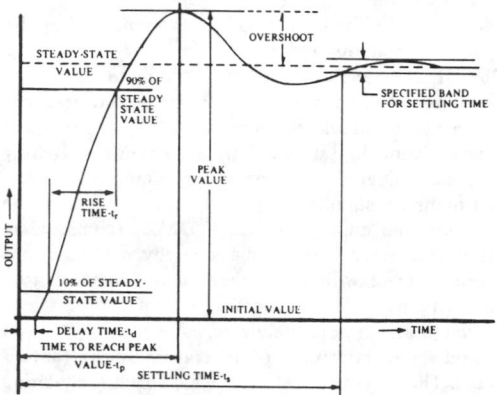

Typical time response of a feedback control system to a step change in input.

time scale (analog computers). *See:* time.   9

time sharing (software). (1) An operating technique of a computer system that provides for the interleaving in time of two or more processes in one processor. (2) Pertaining to the interleaved use of time on a computing system that enables two or more users to execute computer programs concurrently. *See:* **computer program; computer system; process; system.**   434

time signal (navigation aid terms). An accurate signal marking a specified time or time interval.   526

time skew (1) (analog computers). *See:* **analog to digital converter.**   9

(2) (analog-to-digital converter). In an analog to digital conversion process, the time difference between the conversion of one analog channel and any other analog channel, such that the converted (digital) representations of the analog signals do not correspond to values of the analog variables that existed at the same instant of time. Time skew is eliminated, where necessary, by the use of a multiplexor with a sample/hold feature, allowing all input channels to be simultaneously sampled and stored for later conversion. *See:* **switching time; converters, analog to digital.**   9, 10

time sorter (nuclear techniques). *See:* **time distribution analyzer.**

time-to-amplitude converter (scintillation counting). An instrument producing an output pulse whose amplitude is proportional to the time difference between start and stop pulses.   117

time to chopping (switching impulse testing). The time interval T_c between actual zero and the instant when the chopping occurs. 108

time to crest. The time interval T_{cr} between actual zero and the instant when the voltage has reached its crest value. 108

time to half value. The time interval T_h between actual zero and the instant on the tail when the impulse has decreased to half its crest value. 108

time to half-value on the wavetail (virtual duration of an impulse) (surge arresters). *See:* **virtual time to half-value (on the wavetail).**

time to impulse flashover. The time between the initial point of the voltage impulse causing flashover and the point at which the abrupt drop in the voltage impulse takes place. 64

time to impulse sparkover (metal-oxide surge arresters for ac power circuits)(surge arrester). The time between virtual zero of the voltage impulse causing sparkover and the point on the voltage wave at which sparkover occurs. 583,430

time, turn-around (test, measurement and diagnostic equipment). The time needed to service or check out an item for recommitment. 54

time-undervoltage protection (power switchgear). A form of undervoltage protection that disconnects the protected equipment upon a deficiency of voltage after a predetermined time interval. 103

time update (sequential events recording systems). The correction or resetting of a real time clock to match a time standard. *See:* **real time.** 48, 58

time-variant filtering (germanium gamma-ray detectors). Pulse shaping in which the filter response is not constant with respect to time. This is often achieved by a gated integrator following a time-invariant filter. The integrator is gated on for the duration of each pulse. 528

time zone diversity (power operations). Load diversity between two (or more) electric systems which occurs when their loads are in different time zones. 516

timing analyzer (software). A software tool that estimates or measures the execution time of a computer program or portions of a computer program either by summing the execution times of the instructions in each path, or by inserting probes at specific points in the program and measuring the execution times between probes. *See:* **computer program; execution time; instruction; program; software tool.** 434

timing deviation demand meter (metering). The difference between the elapsed time indicated by the timing element and the true elapsed time, expressed as a percentage of the true elapsed time. 212

timing mechanism (1) (demand meter). That mechanism through which the time factor is introduced into the result. The principal function of the timing mechanism of a demand meter is to measure the demand interval, but it has a subsidiary function, in the case of certain types of demand meters, to provide also a record of the time of day at which any demand has occurred. A timing mechanism consists either of a clock or its equivalent, or of a lagging device that delays the indications of the electric mechanism. In thermally lagged meters the time factor is introduced by the thermal time lag of the temperature responsive elements. In the case of curve-drawing meters, the timing element merely provides a continuous record of time on a chart or graph. *See:* **demand meter.** 328

(2) (recording instrument). The time-regulating device usually includes the motive power unit necessary to propel the chart at a controlled rate (linear or angular). *See:* **moving element (instrument).** 294

timing relay (or relay unit) (power switchgear). An auxiliary relay or relay unit whose function is to introduce one or more time delays in the completion of an associated function. 103

timing table. That portion of central-station equipment at which means are provided for operators' supervision of signal reception. *See:* **protective signaling.** 328

tinning (electrotyping). The melting of lead-tin foil or tin plating upon the back of shells. *See:* **electroforming.** 328

tinsel cord. A flexible cord in which the conducting elements are thin metal ribbons wound helically around a thread core. *See:* **transmission line.**

tip (plug) (1). The contacting part at the end of the plug. 328

(2) (electron tubes) (pip). A small protuberance on the envelope resulting from the sealing of the envelope after evacuation. *See:* **electronic tube.** 244, 190

tip and ring wires (telephone switching systems). A pair of conductors associated with the transmission portions of circuits and apparatus. Tip or ring designation of the individual conductors is arbitrary except when applied to cord-type switchboard wiring in which case the conductors are designated according to their association with tip or ring contacts of the jacks and plugs. 55

T junction (waveguide). A junction of waveguides in which the longitudinal guide axes form a T. *Note:* The guide that continues through the junction is the main guide: the guide that terminates at a junction is the branch guide. *See:* **waveguide.** 330, 179

TM. *See:* **transverse magnetic.**

TM$_{mn}$mode (E$_{mn}$mode). (1) In a rectangular waveguide, the subscripts $_m$ and $_n$ denote the number of half-period variation in the magnetic field parallel to the broad and narrow sides, respectively, of the guide. *Note:* In the United Kingdom, the reverse order is preferred. (2) In a circular waveguide, a mode which has $_m$ diametral planes and $_n$ cylindrical surfaces of nonzero radius (including the wall of the guide) at which the longitudinal component of the electric field is zero. (3) In a resonant cavity consisting of a length of rectangular or circular waveguide, a third subscript is used to indicate the number of half-period variations of the field along the waveguide axis. 267

TM mode (1) (fiber optics). *See:* **transverse magnetic mode.** 433

(2) (E mode). A waveguide mode in which the longitudinal component of the magnetic field is everywhere zero and the longitudinal component of the electric field is not. 267

T network. A network composed of three branches with one end of each branch connected to a common junction point, and with the three remaining ends connected to an input terminal, an output terminal, and a common input and output terminal, respectively. *See:* **network analysis.** 332, 210

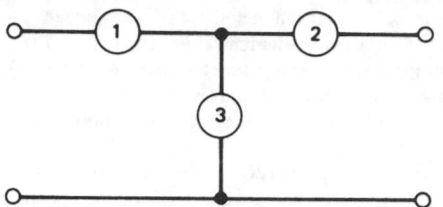

T network. One end of each of the branches 1, 2, and 3 is connected to a common point. The other ends of branches 1 and 2 form, respectively, an input and an output terminal, and the other end of branch 3 forms a common input and output terminal.

TOA (radar). Abbreviation for time of arrival. *See:* **time-of-arrival location.** 13

toe and shoulder (of a Hurter and Driffield (H and D) curve) (photographic techniques). The terms applied to the nonlinear portions of the H and D curve that lie, respectively, below and above the straight portion of this curve. 176

to-from indicator (omnirange receiver)(navigation aid terms). A supplementary device used with an omnibearing selector to resolve the ambiguity of measured omnibearings. 526

toggle. Pertaining to any device having two stable states. *See:* **flip-flop.** 255, 77

token (token ring access method). The symbol of authority that is passed between stations using a token access method to indicate which station is currently in control of the medium. 472

tolerable out-of-service time (nuclear power generating station). The time an information display channel is allowed to be unavailable for use as a post accident monitoring display. 361

tolerable voltage difference (generating station grounding). The maximum potential difference that would cause a body current to flow of such value as not to cause ventricular fibrillation. 569

tolerance (1) (metric practice). The total amount by which a quantity is allowed to vary; thus the tolerance is the algebraic difference between the maximum and minimum limits. 21

(2)(nuclear power generating stations). The allowable deviation from a specified or true value. 41

(3) (software). The ability of a system to provide continuity of operation under various abnormal conditions. *See:* **system.** 434

(4) (test, measurement and diagnostic equipment). The total permissible variation of a quantity from a designated value. 54

tolerance band (1)(converter characteristics)(self-

commutated converters). The range of steady-state values of a stabilized output quantity lying between the limits of operating error. *Notes:* (A) Tolerance band describes the permissible deviation of a stabilized output quantity from a rated or preset value. (B) A statement of tolerance band is useful when a subdivision into output effects and intrinsic errors is not of interest. 584

(2)(thyristor). The range of values specified in terms of permissible deviations of the steady state value of a parameter from a specified nominal value. 445

tolerance field (fiber optics). (1) In general, the region between two curves (frequently two circles) used to specify the tolerance on component size. (2) When used to specify fiber cladding size, the annular region between the two concentric circles of diameter $D + \Delta D$ and $D - \Delta D$. The first circumscribes the outer surface of the homogeneous cladding; the second (smaller) circle is the largest circle that fits within the outer surface of the homogeneous cladding. (3) When used to specify the core size, the annular region between the two concentric circles of diameter $d + \Delta d$ and $d - \Delta d$. The first circumscribes the core area; the second (smaller) circle is the largest circle that fits within the core area. *Note:* The circles of definition 2 need not be concentric with the circles of definition 3. *See:* **cladding; core; concentricity error; homogeneous cladding.** 433

toll board. A switchboard used primarily for establishing connections over toll lines. 328

toll call (telephone switching systems). A call for a destination outside the local-service area of the calling station. 55

toll center (telephone switching systems). A toll office where trunks from end offices are connected to intertoll trunks and where operator's assistance is provided in completing incoming calls and where other traffic operating functions are performed. Toll centers are classified as Class 4C offices. *See:* **office class.** 55

toll connecting trunk (telephone switching systems). A trunk between a local office and a toll office or switchboard. 55

toll line. A telephone line or channel between two central offices in different telephone exchanges. 328

toll office (telephone switching systems). An intermediate office serving toll calls. 55

toll point (telephone switching systems). A toll office where trunks from end offices are connected to the distance dialing network and where operators handle only outward calls or where there are no operators present. Toll points are classified as Class 4P offices. 55

toll restriction (telephone switching systems). A method that prevents private automatic branch exchange stations from completing certain or any toll calls or reaching a toll operator, except through the attendant. 55

toll station. A public telephone station connected directly to a toll telephone switchboard. *See:* **telephone station.** 328

toll switching trunk (telephone switching systems). A trunk for completing calls from a toll office or switchboard to a local office. 55

toll switch train (toll train). A switch train that carries a connection from a toll board to a subscriber line. *See:* **switching system.** 328

toll terminal loss (toll connection). That part of the over-all transmission loss that is attributable to the facilities from the toll center through the tributary office to and including the subscriber's equipment. *Note:* The toll terminal loss at each end of the circuit is ordinarily taken as the average of the transmitting loss and the receiving loss between the subscriber and the toll center. *See:* **transmission loss.** 328

toll train. See: **toll switch train.**

toll transmission selector. A selector in a toll switch train that furnishes toll-grade transmission to the subscriber and controls the ringing. 328

tone (1) (general). (A) A sound wave capable of exciting an auditory sensation having pitch. (B) A sound sensation having pitch. 176

(2) (telephone switching systems). An audible signal transmitted over the telecommunications network. 55

tone, call (telephone switching systems). A tone that indicates to an operator or attendant that a call has reached the position or console. 55

tone control. A means for altering the frequency response at the audio-frequency output of a circuit, particularly of a radio receiver or hearing aid, for the purpose of obtaining a quality more pleasing to the listener. *See:* **amplifier; radio receiver.** 328

tone decay time (measuring the performance of tone address signaling systems). The time interval between the end of the tone present condition and the beginning of the tone off condition at the end of the tone under consideration. 508

tone duration (measuring the performance of tone address signaling systems). The time interval during which a tone present condition exists continuously. 508

tone leak (measuring the performance of tone address signaling systems). The occurrence of any address signaling tone during the signal present or signal off intervals when such tone is not intended. 508

tone localizer. *See:* **equisignal localizer.**

tone-modulated waves. Waves obtained from continuous waves by amplitude modulating them at audio frequency in a substantially periodic manner. *See:* **telegraphy.** 328

tone off (measuring the performance of tone address signaling systems). Any condition where the tone under consideration is below a specified OFF level. 508

tone-operated net-loss adjuster. *See:* **tonlar.**

tone present (measuring the performance of tone address signaling systems). Any condition where the tone under consideration is equal to or greater than a specified threshold value. 508

toner (electrostatography). The image-forming material in a developer that, deposited by the field of an electrostatic-charge pattern, becomes the visible record. *See:* **electrostatography.** 191

tone rise time (measuring the performance of tone address signaling systems). The time interval between the end of the tone off condition and the beginning of the tone present condition at the beginning of the tone under consideration. 508

tongs, fuse. *See:* **fuse tongs.** 103, 443

tonlar. A system for stabilizing the net loss of a telephone circuit by means of a tone transmitted between conversations. The name is derived from the initial letters of the expression **tone-operated net-loss adjuster.** *See:* **tone-frequency telephony.** 328

tool (software). (1) See 'software tool'. (2) A hardware device used to analyze software or its performance. *See:* **hardware; performance; software.** 434

tool function (numerically controlled machines). A command identifying a tool and calling for its selection either automatically or manually. The actual changing of the tool may be initiated by a separate tool-change command. 207

tool offset (numerically controlled machines). A correction for tool position parallel to a controlled axis. 207

tools, access (switchgear assembly). Keys or other special accessories with unique characteristics that make them suitable for gaining access to the tamper-resistant switchgear assembly. 93

tooth (1) (rotating machinery). A projection from a core, separating two adjacent slots, the tip of which forms part of one surface of the air gap. *See:* **rotor (rotating machinery); stator.** 63

(2) (cable plowing). *See:* **wear point.** 52

tooth pitch (rotating machinery). *See:* **slot pitch.**

tooth tip (rotating machinery). That portion of a tooth that forms part of the inner or outer periphery of the air gap. It is frequently considered to be the section of a tooth between the radial location of the wedge and the air gap. *See:* **rotor (rotating machinery); stator.** 63

top (pulse terms). The portion of a pulse waveform which represents the second nominal state of a pulse. 254

top cap (side contact) (electron tubes). A small metal shell on the envelope of an electron tube or valve used to connect one electrode to an external circuit. *See:* **electron tube.** 190

top car clearance (elevators). The shortest vertical distance between the top of the car crosshead, or between the top of the car where no car crosshead is provided, and the nearest part of the overhead structure or any other obstruction when the car floor is level with the top terminal landing. *See:* **hoistway (elevator or dumbwaiter).** 328

top coil side (radially inner coil side) (rotating machinery). The coil side of a stator slot nearest the bore of the stator or nearest the slot wedge. *See:* **stator.** 63

top counterweight clearance (elevator counterweight)

(elevator). The shortest vertical distance between any part of the counterweight structure and the nearest part of the overhead structure or any other obstruction when the car floor is level with the bottom terminal landing. *See:* **hoistway (elevator or dumbwaiter).**
328

top-down (software). Pertaining to an approach that starts with the highest level component of a hierarchy and proceeds through progressively lower levels, for example, top-down design, top-down programming, top-down testing. *See:* **bottom-up; component; hierarchy; level; top-down design; top-down testing.**
434

top-down design (software). The process of designing a system by identifying its major components, decomposing them into their lower level components, and iterating until the desired level of detail is achieved. *See:* **bottom-up design; component; level; system.**
434

top-down testing (software). The process of checking out hierarchically organized programs, progressively, from top to bottom, using simulation of lower level components. *See:* **component; program; simulation.**
434

top half bearing (rotating machinery). The upper half of a split sleeve bearing. *See:* **bearing.**   63

top-loaded vertical antenna. A vertical monopole with additional metallic structure at the top intended to increase the effective height of the antenna and to change its input impedance.   111

top magnitude (pulse terms). The magnitude of the top as obtained by a specified procedure or algorithm. *See:* The single pulse diagram below the **waveform epoch** entry. *See:* IEEE Std 181-1975, Section 4.3, for suitable algorithms.)   254

top side sounding (communication satellite). Ionospheric sounding from medium altitude satellites for measuring ionospheric densities at high altitudes.
85

top terminal landing (elevators). The highest landing served by the elevator that is equipped with a hoistway door and hoisting-door locking device that permits egress from the hoistway side. *See:* **elevator landing.**
328

torchere (illuminating engineering). An indirect floor lamp which sends all or nearly all of its light upward.
167

toroid (doughnut) (electron device). A toroidal-shaped vacuum envelope in which electrons are accelerated. *See:* **electron device.**   190

toroidal coil. A coil wound in the form of a toroidal helix.   329

toroidal reflector (antennas). A reflector formed by rotating a segment of plane curve about a nonintersecting coplanar line. *Note:* The plane curve segment is called the torus cross-section and the coplanar line is called the toroidal axis.   111

torque (1) (instrument). The turning moment on the moving element produced by the quantity to be measured or some quantity dependent thereon acting through the mechanism. This is also termed the deflecting torque and in many instruments is opposed by the controlling torque, which is the turning moment produced by the mechanism of the instrument tending to return it to a fixed position. *Note:* Full-scale torque is the particular value of the torque for the condition of full-scale deflection and as an index of performance should be accompanied by a statement of the angle corresponding to this deflection. *See:* **accuracy rating (instrument).**   328

(2) (International System of Units) (SI). *See:* **energy and torque.**

torque (force) balance accelerometer. A device that measures acceleration by applying a principal restraint through capturing.   46

torque buildup time constant (electric coupling). The time constant applicable when excitation voltage is changed from zero to full value.   416

torque-coil magnetometer. A magnetometer that depends for its operation on the torque developed by a known current in a coil that can turn in the field to be measured. *See:* **magnetometer.**   328

torque-command storage (gyro)(inertial sensor). The transient deviation of the output of a rate integrating gyro from that of an ideal integrator when the gyro is subjected to a torquer command signal. It is a function of the gyro's characteristic time and the torquer time constant. *See:* **attitude storage; float storage.**
46

torque control (1) (of a relay) (power switchgear). A method of constraining the pickup of a relay by preventing the torque-producing element from developing operating torque until another associated relay unit operates.   103

(2) (protective relaying of utility-consumer interconnections). A means of supervising the operation of one relay element with another. For example, an overcurrent relay cannot operate unless the lag coil circuit is closed. It may be closed by the contact of an undervoltage element.   128

torque decay time constant (electric coupling). The time constant applicable when the excitation voltage is changed from full value to zero.   416

torque-generator reaction torque (gyro; accelerometer). *See:* **torquer reaction torque.**

torque margin (electric installations on shipboard). The increase in torque about rated torque to which a propulsion system may be subjected without the motor pulling out of step with the generator.   3

torque motor. A motor designed primarily to exert torque through a limited travel or in a stalled position. *Note:* Such a motor may be capable of being stalled continuously or only for a limited time. *See:* **asynchronous machine.**   63

torquer (gyro; accelerometer). A device which exerts a torque (or force) on a gimbal, a gyro rotor or a proof mass in response to a command signal. *Syn:* **forcer.**
46

torquer axis (gyro; accelerometer) (inertial sensor). The axis about which a force couple is produced by a torquer.   46

torquer-current rectification (gyro; accelerometer)

(inertial sensor). An apparent drift rate (or bias) in an inertial sensor resulting from effects such as torquer nonlinearity or capture loop asymmetry. 46

torquer reaction torque (gyro; accelerometer). The reaction torque which is a function of the frequency and amplitude of the command torque signal.
46

torque time constants (electric coupling). Torque time constants define the time required for the coupling torque to reach 63.2 percent of its total excursion, whenever the magnitude of excitation voltage is instantly changed between specified values. This does not imply that the torque time constant of a given coupling under certain conditions of slip, speed, temperature, and environmental conditions will be the same under other conditions. 416

torquing (gyro; accelerometer). The application of torque to a gimbal or a gyro rotor about an axis-of-freedom for the purpose of precessing, capturing, slaving or slewing. 46

torquing rate (inertial navigation)(navigation aid terms). The angular rate at which the orientation of a gyro, with respect to inertial space, is changed in response to a command. 526

torsional critical speed (rotating machinery). The speed at which the amplitudes of the angular vibrations of a machine rotor due to shaft torsional vibration reach a maximum. *See:* **rotor (rotating machinery).** 63

torsional mechanism. An operating mechanism which transfers rotary motion by torsion through a pipe or shaft from the operating means to open or close the switching device. 27

torsionmeter. A device to indicate the torque transmitted by a propeller shaft based on measurement of the twist of a calibrated length of the shaft. *See:* **electric propulsion system.** 328

total average power dissipation (semiconductor). The sum of the full cycle average forward and full cycle average reverse power dissipation. 66

total break time (mechanical switching device). *See:* **interrupting time.**

total capability for load (electric power supply). The capability available to a system from all sources including purchases. *See:* **generating station.** 200

total capacitance. *See:* **self-capacitance (conductor).**

total charge (ferroelectric device). One-half of the charge that flows as the condition of the device is changed from that of full applied positive voltage to that of full negative voltage (or vice versa). *Note:* Total charge is dependent on the amplitude of the applied voltage which should be stated when measurements of total charge are reported. *See:* **ferroelectric domain.** 247

total clearing time (protection and coordination of industrial and commercial power systems). The total time between the beginning of the specified overcurrent and the final interruption of the circuit, at rated voltage. It is the sum of the minimum melting time plus tolerance and the arcing time. For clearing times in excess of half-cycle, the clearing time is substantial-ly the maximum melting time for low-voltage fuses. *See:* **Figure under arcing time.** 504

total clearing time (of a fuse). *See:* **clearing time (of a fuse).**

total correctness (software). In proof of correctness, a designation indicating that a program's output assertions follow logically from its input assertions and processing steps, and that, in addition, the program terminates under all specified input conditions. *See:* **input assertion; output assertion; partial correctness; proof of correctness.** 434

total (asymmetrical) current (power switchgear). The combination of the symmetrical component and the direct current component of the current. 103

total-current regulation (axle generator). That type of automatic regulation in which the generator regulator controls the total current output of the generator. *See:* **axle generator system.** 328

total cyanide (in a solution for metal deposition) (electroplating). The total content of the cyanide radical (CN), whether present as the simple or complex cyanide of an alkali or other metal. *See:* **electroplating.** 328

total detection efficiency (germanium detectors). The ratio of the total (peak plus Compton) counting rate to the gamma-ray emission rate. *Note:* The terms *standard source* and *radioactivity standard* are general terms used to refer to the sources and standards of National Radioactivity Standard Source and Certified Radioactivity Standard Source. 397

total dose (metal-nitride-oxide field-effect transistor). The total amount of ionizing radiation deposited in the active area of a device over a given period of time. The unit of measure for total dose most commonly used in the present context is rads (SI). 386

total electric current density. At any point, the vector sum of the conduction-current density vector, the convection-current density vector, and the displacement-current density vector at that point. 210

total electrode capacitance (electron tubes). The capacitance of one electrode to all other electrodes connected together. 190

total emissivity (element of surface of a temperature radiator). The ratio of its radiant-flux density (radiant exitance) to that of a blackbody at the same temperature. 167

total for load capability (power operations). The dependable capability available to a system from all sources including purchases. 516

total hazard current (health care facilities). The hazard current of a given isolated system with all devices, including the line isolation monitor, connected. *See:* **hazard current.** 192

total internal reflection (fiber optics). The total reflection that occurs when light strikes an interface at angles of incidence (with respect to the normal) greater than the critical angle. *See:* **critical angle; step index optical waveguide.** 433

totalizing pulse relay (metering). A device used to receive and totalize pulses from two or more sources for proportional transmission to another totalizing relay or to a receiver. 212

totalizing relay. A device used to receive and totalize pulses from two or more sources for proportional transmission to another totalizing relay or to a receiver. *See:* **auxiliary device to an instrument.** 212

total losses (transformer or regulator). The sum of the no-load and load losses, excluding losses due to accessories. 53

total losses of a shunt reactor (shunt reactors over 500 kVA). The sum of the conductor loss, magnetic circuit loss, cooling loss, shielding loss, and any other stray losses in the shunt reactor. 562

totally depleted detector (charged-particle detectors)(germanium gamma-ray detectors)(X-ray energy spectrometers)(semiconductor radiation detectors). A detector in which the thickness of the depletion region is essentially equal to the thickness of the semiconductor material. material.
119,526,471,23

totally enclosed (rotating machinery). A term applied to apparatus with an integral enclosure that is constructed so that while it is not necessarily airtight, the enclosed air has no deliberate connection with the external air except for the provision for draining and breathing. 63

totally enclosed fan-cooled (totally enclosed fan-ventilated). A term applied to a totally enclosed apparatus equipped for exterior cooling by means of a fan or fans, integral with the apparatus but external to the enclosing parts. *See:* **asynchronous machine.**
63, 3

totally enclosed fan-cooled machine (electric installations on shipboard). A totally enclosed machine equipped for exterior cooling by means of a fan or fans, integral with the machine but external to the enclosing parts. 3

totally-enclosed fan-ventilated air-cooled (rotating machinery). Applied to a totally-enclosed machine having an air-to-air heat exchanger in the internal air circuit, the external air being blown through the heat exchanger by a fan mechanically driven by the machine shaft. 63

totally enclosed machine (electric installations on shipboard). A machine so enclosed as to prevent the exchange of air between the inside and outside of the case, but not sufficiently enclosed to be termed airtight. 3

totally enclosed nonventilated (rotating machinery). A term applied to a totally enclosed apparatus that is not equipped for cooling by means external to the enclosing parts. *See:* **asynchronous machine.** 63

totally enclosed pipe-ventilated machine. A totally enclosed machine except for openings so arranged that inlet and outlet ducts or pipes may be connected to them for the admission and discharge of the ventilating air. This air may be circulated by means integral with the machine or by means external to and not a part of the machine. In the latter case, these machines shall be known as separately ventilated or forced ventilated machines. *See:* **asynchronous machine; closed air-circuit.** 328

totally enclosed ventilated apparatus. Apparatus totally enclosed in which the cooling air is carried through the case and apparatus by means of ventilating tubes and the air does not come in direct contact with the windings of the apparatus. 328

totally unbalanced currents (balanced line). Push-push currents. *See:* **waveguide.** 267

total power loss (semiconductor rectifier). The sum of the forward and reverse power losses. *See:* **rectification.** 66

total range (instrument). The region between the limits within which the quantity measured is to be indicated or recorded and is expressed by stating the two end-scale values. *Notes:* (1) If the span passes through zero, the range is stated by inserting zero or 0 between the end-scale values. (2) In specifying the range of multiple-range instruments, it is preferable to list the ranges in descending order, for example, 750/300/150. *See:* **instrument.** 294

total sag. The distance measured vertically from the conductor to the straight line joining its two points of support, under conditions of ice loading equivalent to the total resultant loading for the district in which it is located. 262

total start-stop telegraph distortion. Refers to the time displacement of selecting-pulse transitions from the beginning of the start pulse expressed in percent of unit pulse. 111

total switching time (ferroelectric device). The time required to reverse the signal charge. *Note:* Total switching time is measured from the time of application of the voltage pulse, which must have a rise time much less than and a duration greater than the total switching time. The magnitude of the applied voltage pulse should be specified as part of the description of this characteristic. *See:* **ferroelectric domain.** 247

total telegraph distortion. Telegraph transmisson impairment, expressed in terms of time displacement of mark-space and space-mark transitions from their proper positions relative to one another, in percent of the shortest perfect pulses called the unit pulse. (Time lag affecting all transitions alike does not cause distortion). Telegraph distortion is specified in terms of its effect on code and terminal equipment. Total Morse telegraph distortion for a particular mark or space pulse is expressed as the algebraic sum of time displacements of space-mark and mark-space transitions determining the beginning and end of the pulses, measured in percent of unit pulse. Lengthening of mark is positive, and shortening, negative. *See:* **distortion.**
111

total varactor capacitance. The capacitance between the varactor terminals under specified conditions.
136

total voltage regulation (rectifier). The change in output voltage, expressed in volts, that occurs when the load current is reduced from its rated value to zero or light transition load with rated sinusoidal alternating voltage applied to the alternating-current line terminals, but including the effect of the specified alternating-current system impedance as if it were inserted between the line terminals and the transformer, with the rectifier transformer on the rated tap. *Note:* The measurement shall be made with zero phase control and shall exclude the corrective action of any automatic voltage-regulating means, but not impedance. *See:* **power rectifier; rectification.**   208

touchdown zone lights (illuminating engineering). Barettes of runway lights installed in the surface of the runway between the runway edge lights and the runway centerline lights to provide additional guidance during the touchdown phase of a landing in conditions of very poor visibility.   167

touch voltage (1)(conductor stringing equipment). The potential difference between a grounded metallic structure and a point on the earth's surface separated by a distance equal to the normal maximum reach, approximately 1 meter. This potential difference could be dangerous and could result from induction or fault conditions, or both. *Syn:* **touch potential.**   431

(2)(ground system measuring). The potential difference between a grounded metallic structure and a point on the earth's surface separated by a distance equal to the normal maximum horizontal reach, approximately one meter.   313

tower. A broad-base latticed steel support for line conductors.   64

tower footing resistance (lightning protection). The resistance between the tower grounding system and true ground. *See:* **direct-stroke protection (lightning).**   64

tower ladder (conductor stringing equipment). A ladder complete with hooks and safety chains attached to one end of the side rails. These units are normally fabricated from fiberglass, wood or metal. The ladder is suspended from the arm or bridge of a structure to enable workmen to work at the conductor level, to hang travelers, perform clipping-in operations, etcetera. In some cases, these ladders are also used as linemen's platforms. *Syn:* **hook ladder.**   431

tower loading. The load placed on a tower by its own weight, the weight of the wires with or without ice covering, the insulators, the wind pressure normal to the line acting both on the tower and the wires and the pull from the wires in the direction of the line. *See:* **tower.**   64

towing light. A lantern or lanterns fixed to the mast or hung in the rigging to indicate that a ship is towing another vessel or other objects.   328

Townsend coefficient (gas). The number of ionizing collisions per centimeter of path in the direction of the applied electric field. *See:* **discharge (gas).**   190

T pulse (linear waveform distortion). A sin^2 pulse with a half-amplitude duration (HAD) of 125 ns. The amplitude of the envelope of the frequency spectrum at 4 MHz is 0.5 of the amplitude at zero frequency and effectively zero at and beyond 8 MHz (see Figs below.).   42

TR. *See:* **transmit-receive.**

trace (1). The cathode-ray-tube display produced by a moving spot. *See:* **spot; oscillograph.**   185

(2) (software). (A) A record of the execution of a

T pulse

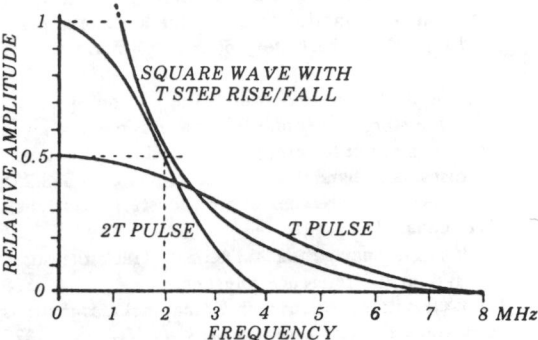

2T pulse, T pulse, and T step

Envelope of frequency spectrum of 2T pulse, T pulse and square wave with T step rise and fall

computer program; it exhibits the sequences in which the instructions were executed. (B) A record of all or certain classes of instructions or program events occurring during execution of a computer program. (C) To produce a trace. *See:* **computer program; execution; instruction; program.**   434

traceability (1) (nuclear power quality assurance). The ability to trace the history, application, or location of an item and like items or activities by means of recorded identification.   417

(2) (test, measurement and diagnostic equipment). Process by which the assigned value of a measurement is compared, directly or indirectly, through a series of calibrations to the value established by the U. S. national standard.   54

trace finder. *See:* **beam finder.**

trace interval (television). The interval corresponding to the direction of sweep used for delineation.   18

tracer (software). A software tool used to trace. *See:* **software tool; trace.**   434

trace, return (oscillography) (television). The path of the scanning spot during the retrace. *See:* **oscillograph; television.**   106

trace width (oscilloscope). The distance between two points on opposite sides of a trace perpendicular to the direction of motion of the spot, at which luminance is 50 percent of maximum. With one setting of the beam controls, the width of both horizontally and vertically going traces within the quality area should be stated. *See:* **oscillograph.**   185

tracing distortion. The nonlinear distortion introduced

in the reproduction of mechanical recording because the curve traced by the motion of the reproducing stylus is not an exact replica of the modulated groove. For example, in the case of a sine-wave modulation in vertical recording the curve traced by the center of the tip of a stylus is a poid. *See:* **phonograph pickup.**

176

tracing routine (computing systems). A routine that provides a historical record of specified events in the execution of a program.   255, 77

track (1)(navigation)(navigation aid terms). (1) The resultant direction of actual travel projected in the horizontal plane and expressed as a bearing. (2) The component of motion that is in the horizontal plane and represents the history of accomplished travel.

526

(2) (in electronic computers). The portion of a moving-type storage medium that is accessible to a given reading station: for example, as on film, drum, tapes, or discs. *See:* **band.**   338,255

(3) (test, measurement and diagnostic equipment). *See:* **channel.**   54

track angle (navigation aid terms). Track measured from 0° at the reference direction.   526

track circuit. An electric circuit that includes the rails of a track relay as essential parts.   328

track element. *See:* **roadway element.**

track homing (navigation aid terms). The process of following a line of position known to pass through an objective.   526

track indicator chart. A maplike reproduction of railway tracks controlled by track circuits so arranged as to indicate automatically for defined sections of track whether such sections are or are not occupied.

328

tracking (1)(composite insulators). Irreversible degradation of surface material from the formation of conductive carbonized paths.   483

(2)(navigation aid terms). The process of following a moving object or a variable input quantity. This process may be carried out manually or automatically. In radar, target tracking in angle, range, or Doppler frequency is accomplished by keeping a beam or angle cursor on the target angle, etcetera. *See:* **automatic tracking; tracking radar.**   526

(3) (antennas). A motion given to the major lobe of an antenna with the intent that a selected moving target be contained within the major lobe. *Syn:* **angle tracking.**   111

(4) (data transmission)(A)(radar). The process of following a moving object or a variable input quantity using a servomechanism. *Note:* In radar, tracking is carried out by keeping a narror beam or angle cursor centered on the target angle, a range mark or gate on the delayed echo, or a narrowband filter of the signal frequency. **(B)(electric).** The maintenance of proper frequency relations in circuits designed to be simultaneously varied by gang operation. **(C)(phonographic technique).** The accuracy with which the stylus of a phonograph pickup follows a prescribed path. **(D)(instrument).** The ability of an instrument to indicate, at

the division line being checked, when energized by corresponding proportional value of actual end-scale excitation, expressed as a percentage of actual end-scale value. **(E)(communication satellite).** (1)The determination of the orbit and the ephemeris to a satellite or spacecraft. (2) Maintaining the point of a high gain antenna at a moving spacecraft. **(F)(antenna).** A motion given to the major lobe of an antenna so that a selected moving target is contained within the major lobe.   59

tracking error (1) (general). The deviation of a dependent variable with respect to a reference function. *Note:* As applied to power inverters, tracking error may be the deviation of the output volts per hertz from a prescribed profile or the deviation of the output frequency from a given input synchronizing signal or others. *See:* **self-commutated inverters.**   208

(2) (phonographic techniques) (lateral mechanical recording). The angle between the vibration axis of the mechanical system of the pickup and a plane containing the tangent to the unmodulated record groove that is perpendicular to the surface of the recording medium at the point of needle contact. *See:* **phonograph pickup.**   176

tracking radar (navigation aid terms). A radar whose primary function is the automatic tracking of targets. *See:* **tracking; automatic tracking.**   526

track instrument. A device in which the vertical movement of the rail or the blow of a passing wheel operates a contact to open or close an electric circuit.

328

trackless trolley coach. *See:* **trolley coach (trolley bus).**

track relay. A relay receiving all or part of its operating energy through conductors of which the track rails are an essential part and that responds to the presence of a train on the track.   328

track store (analog computers). A component, controlled by digital logic signals, whose output equals the input, when in the "track" mode, and whose output becomes constant and is held (stored) at the value it possessed at the instant its mode was switched to the "store" mode.   9

track-while-scan (TWS) (radar). An automatic target tracking process in which the radar antenna and receiver provide periodical video data from the search scan, together with interpolation measurements, as inputs to computer channels which follow individual targets. *See:* **tracking.**   13

traction machine (elevators). A direct-drive machine in which the motion of a car is obtained through friction between the suspension ropes and a traction sheave. *See:* **driving machine (elevators).**   328

traction motor. An electric propulsion motor used for exerting tractive force through the wheels of a vehicle.

328

tractive force (tractive effort) (electrically propelled vehicle). The total propelling force measured at the rims of the driving wheels, or at the pitch line of the gear rack in the case of a rack vehicle. *Note:* Tractive force of an electrically propelled vehicle is commonly

qualified by such terms as maximum starting tractive force, short-time-rating tractive force, and continuous-rating tractive force. *See:* **electric locomotive.**
328

tradeoff. Parametric analysis of concepts or components for the purpose of optimizing the system or some trait of the system. *See:* **system science.** 209

traffic beam. *See:* **lower beams.**

traffic-control system. A block signal system under which train movements are authorized by block signals whose indications supersede the superiority of trains for both opposing and following movements on the same track. *See:* **centralized traffic-control system.**
328

traffic locking. Electric locking adapted to prevent the manipulation of levers or other devices for changing the direction of traffic on a section of track while that section is occupied or while a signal is displayed for a train to proceed into that section. *See:* **interlocking (interlocking plant).**
328

traffic service (telephone switching systems). The services rendered to customers by telephone company operators.
55

traffic usage recorder (telephone switching systems). A device or system for sampling and recording the occupancy of equipment.
55

trailer plow (static or vibratory plows) (cable plowing). A unit which is self-contained except for drawbar pull that is furnished by a prime mover.
52

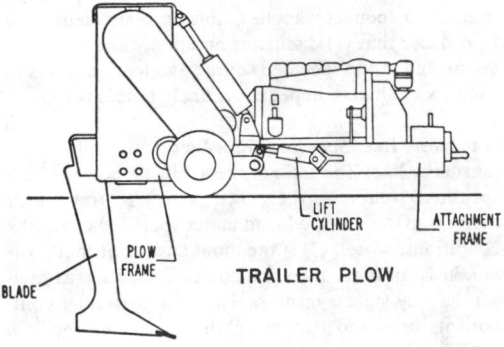

TRAILER PLOW

trailing edge (last transition)(pulse transformers). That portion of the pulse occurring between the time of intersection of straight-line segments used to determine A_T and the time at which the instantaneous value reduces to zero.
589

trailing edge (last transition) amplitude, $A_T(1)$(pulse transformers). That quantity determined by the intersection of a line passing through the points on the trailing edge where the instantaneous value reaches 90% and 10% of A_T, and the straight-line segment fitted to the top of the pulse in determining A_M.
589

trailing edge, pulse. The major transition towards the pulse baseline occuring before a reference time. *See:* **pulse.**
185

trailing-edge pulse time. The time at which the instan-

taneous amplitude last reaches a stated fraction of the peak pulse amplitude.
254

trailing-type antenna (aircraft). A flexible conductor usually wound on a reel within the aircraft passing through a fairlead to the outside of the aircraft, terminated in a streamlined weight or wind sock and fed out to the proper length for the desired radio frequency of operation. It has taken other forms such as a capsule that when exploded releases the antenna.
328

train (illuminating engineering). The angle between the vertical plane through the axis of the searchlight drum and the plane in which this plane lies when the searchlight is in a position designated as having zero train.
167

train-control territory. That portion of a division or district equipped with an automatic train-control system. *See:* **automatic train control.**
328

train describer. An instrument used to give information regarding the origin, destination, class, or character of trains, engines, or cars moving or to be moved between given points.
328

trained listening group (speech-quality measurements). Six to ten listeners who understand thoroughly the purpose of the speech quality test and respond properly throughout the test. All persons of the group shall meet the requirements on auditory acuity as described by USAS S3.2-1960 (Monosyllabic Word Intelligibility). The training of the listeners will depend on the special type of tests to be conducted.
126

trajectory, state. *See:* **state trajectory.**

transaction (1)(computer applications). An event that requires data contained in a master file to be processed.
571

(2)(supervisory control, data acquisition, and automatic control). That sequence of messages between master and remote stations required to perform a specific function (for example, acquire specific data or control a selected device).
570

transaction code (computer applications). An identifier associated with a transaction and representing the operation to be carried out by that transaction. For example, 'A' for an add transaction, 'D' for a delete transaction.
571

transactor (power switchgear). A magnetic device with an air-gapped core having an input winding which is energized with an alternating current and having an output winding which produces a voltage that is a function of the input current. *Note:* The term "transactor" is a contraction of the words "transformer" and "reactor".
103

transadmittance. For harmonically varying quantities at a given frequency, the ratio of the complex amplitude of the current at one pair of terminals of a network to the complex amplitude of the voltage across a different pair of terminals. *See:* **interelectrode transadmittance (j — 1 interelectrode transadmittance of an n-electrode electron tube).**
185

transadmittance compression ratio (electron tubes). The ratio of the magnitude of the small-signal forward transadmittance of the tube to the magnitude of the forward transadmittance at a given input signal level.
125

transadmittance, forward (electron tubes). The complex quotient of (1) the fundamental component of the short-circuit current induced in the second of any two gaps and (2) the fundamental component of the voltage across the first.   125

trans-μ-factor (multibeam electron tubes). The ratio of (1) the magnitude of an infinitesimal change in the voltage at the control grid of any one beam to (2) the magnitude of an infinitesimal change in the voltage at the control grid of a second beam. The current in the second beam and the voltage of all other electrodes are maintained constant.   125

transceiver (1)(data transmission). The combination of radio transmitting and receiving equipment in a common housing, usually for portable or mobile use, and employing common circuit components for both transmitting and receiving.   59

(2)(navigation aid terms). A combination transmitter and receiver in a single housing, with some components being used by both parts. *See:* **transponder.**
                                                        526

transconductance. The real part of the transadmittance. *Note:* Transconductance is, as most commonly used, the interelectrode transconductance between the control grid and the plate. At low frequencies, transconductance is the slope of the control-grid-to-plate transfer characteristic. *See:* **electron-tube admittances; interelectrode transconductance.**   125

transconductance meter (mutual-conductance meter). An instrument for indicating the transconductance of a grid-controlled electron tube. *See:* **instrument.**
                                                        328

transcribe (electronic computation). To convert data recorded in a given medium to the medium used by a digital computing machine or vice versa.   235

transcriber (electronic computation). Equipment associated with a computing machine for the purpose of transferring input (or output) data from a record of information in a given language to the medium and the language used by a digital computing machine (or from a computing machine to a record of information).
                                                        210

transducer (1)(electrical heating applications to melting furnaces and forehearths in the glass industry). A device that is actuated by power from one system and supplies power in any other form to a second system.   520

(2) (communication and power transmission). A device by means of which energy can flow from one or more transmission systems or media to one or more other transmission systems or media. *Note:* The energy transmitted by these systems or media may be of any form (for example, it may be electric, mechanical, or acoustical), and it may be of the same form or different forms in the various input and output systems or media.   111,255,54

(3) (metering). A device to receive energy from one system and supply energy, of either the same or of a different kind, to another system, in such a manner that the desired characteristics of the energy input appear at the output.   212

(4) (thyristor). A device which under the influence of a change in energy level of one form or in one system, produces a specified change in energy level of another form or in another system.   445

transducer, active. A transducer whose output waves are dependent upon sources of power, apart from that supplied by any of the actuating waves, which power is controlled by one or more of the waves. *Note:* The definition of active transducer is a restriction of the more general **active network:** that is, one in which there is an impressed driving force. *See:* **transducer.**
                                                        210

transducer gain (1) (general). The ratio of the power that the transducer delivers to the specified load under specified operating conditions to the available power of the specified source. *Notes:* (A) If the input and.or output power consist of more than one component, such as multifrequency signals or noise, then the particular components used and their weighting must be specified. (B) This gain is usually expressed in decibels. *See:* **transducer.**   210

(2) (two-port linear transducer). At a specified frequency, the ratio of (A) the actual signal power transferred from the output port of the transducer to its load, to (B) the available signal power from the source driving the transducer.   125

transducer, ideal (for connecting a specified source to a specified load). A hypothetical passive transducer that transfers the maximum available power from the source to the load. *Note:* In linear transducers having only one input and one output, and for which the impedance concept applies, this is equivalent to a transducer that (1) dissipates no energy and (2) when connected to the specified source and load presents to each its conjugate impedance. *See:* **transducer.**
                                                        210

transducer, line. *See:* **line transducer.**

transducer loss. The ratio of the available power of the specified source to the power that the transducer delivers to the specified load under specified operating conditions. *Notes:* (1) If the input and/or output power consist of more than one component, such as multifrequency signals or noise, then the particular components used and their weighting must be specified. (2) This loss is usually expressed in decibels. *See:* **transducer.**   210

transducer, passive. A transducer that has no source of power other than the input signal(s), and whose output signal-power cannot exceed that of the input. *Note:* The definition of a passive transducer is a restriction of the more general **passive network,** that is, one containing no impressed driving forces. *See:* **transducer.**
                                                        210

transfer (1) (telephone switching systems). A feature that allows a customer to instruct the switching equipment or operator to transfer his call to another station.
                                                        55

(2) (electronic computation). (A) To transmit, or copy, information from one device to another. (B) To jump. (C) The act of transferring. *See:* **jump; transmit.**
                                                        235

(3) (electrostatography). The act of moving a developed image, or a portion thereof, from one surface to another, as by electrostatic or adhesive forces, without altering the geometric configuration of the image. *See:* electrostatography.   191

transfer admittance (1) (linear passive networks, general). A transmittance for which the excitation is a voltage and the response is a current.   238

(2) (from the *i*th terminal to the *j*th terminal of an *n*-terminal network). The (complex) current flowing to the ith terminal divided by the (complex) voltage applied between the jth terminal with respect to the reference point when all other terminals have arbitrary terminations. For example, for a 3-terminal network terminated in short circuits

$$y_{12} = \frac{I}{v_2}\bigg|_{v_1} = 0$$

transfer alignment (navigation aid terms). A method of transfer of reference coordinates to an inertial navigation system for initial alignment. Accomplished by way of: structure to structure mating, simultaneous measurement of acceleration patterns, or by optical measurement techniques.   526

transfer characteristic (1) (electron tubes). A relation, usually shown by a graph, between the voltage of one electrode and the current to another electrode, all other electrode voltages being maintained constant. *See:* electrode (electron tube).   125

(2) (camera tubes). A relation between the illumination on the tube and the corresponding signal output current, under specified conditions of illumination. *Note:* The relation is usually shown by a graph of the logarithm of the signal output current as a function of the logarithm of the illumination. *See:* illumination; sensitivity; television.   125

transfer check (electronic computation). A check (usually an automatic check) on the accuracy of a data transfer. *Note:* In particular, a check on the accuracy of the transfer of a word.   235

transfer constant (electric transducer). *See:* image transfer constant.

transfer control (electronic computation). *See:* jump.

transfer current (gas tubes). The current to one electrode required to initiate breakdown to another electrode. *Note:* The transfer current is a function of the voltage of the second electrode.   125

transfer-current ratio (linear passive network). A transmittance for which the variables are currents. *Note:* The word **transfer** is frequently dropped in present usage.   238

transfer function (1)(seismic qualification of Class 1E equipment for nuclear power generating stations). A complex frequency response function that defines the dynamic characteristics of a constant parameter linear system. For an ideal system, the transfer function is the ratio of the Fourier transform of the output to that of a given input.   581

(2) (high-power wide-band transformers). The complex ratio of the output of the device to its input. It is also the combined phase and frequency responses.
                                                        321

(3) (low-power wide-band transformers). The complex ratio of the output of the device to its input. It is also the combined phase and frequency responses.
                                                        151

(4) (nuclear power generating station). A mathematical, graphical, or tabular statement of the influence which a module has on a signal or action compared at input and at output terminals. This should be specified as to whether it is transient or steady state.   355

(5) (excitation control systems). A mathematical, graphical, or tabular statement of the influence which a system or element has on a signal or action compared at input and output terminals. *Note:* For a linear system, general usage limits the transfer function to mean the ratio of the Laplace transform of the output to the Laplace transform of the input in the absence of all other signals, and with all initial conditions zero.
                                                        353

transfer function of a device (fiber optics). The complex function, H(f), equal to the ratio of the output to input of the device as a function of frequency. The amplitude and phase responses are, respectively, the magnitude of H(f) and the phase of H(f). *Notes:* (1) For an optical fiber, H(f) is taken to be the ratio of output optical power to input optical power as a function of modulation frequency. (2) For a linear system, the transfer function and the impulse response h(t) are related through the Fourier transform pair, a common form of which is given by

$$H(f) = \int_{-\infty}^{\infty} h(t)^{(i2\pi ft)}\, dt$$
$$h(t) = \int_{-\infty}^{\infty} H(f)^{(-2\pi ft)}\, df$$

where *f* is frequency. Often *H(f)* is normalized to *H(0)* and *h(t)* to $\int_{-\infty}^{\infty} h(t)dt$, which by definition is *H(0)*.

Syn: **baseband response function; frequency response.** *See:* **impulse response.**   433

transfer immitance. *See:* **transmittance.**

transfer impedance (1) (linear passive networks). A transmittance for which the excitation is a current and the response is a voltage. *Note:* It is therefore the impedance obtained when the response is determined at a point other than that at which the driving force is applied, all terminals being terminated in any specified manner. In the case of an electric circuit, the response would be determined in any branch except that in which the driving force is. *See:* **self-impedance; network analysis.**   238

(2) (circuits and systems). (A) (linear passive networks) (general). A transmittance for which the excitation is a current and the response is a voltage. *See:* **linear passive networks.** (B) (from the ith terminal to the jth terminal of an n-terminal network). The (complex) voltage measured between the ith terminal and the reference point divided by the (complex) current applied to the jth terminal when all other terminals have arbitrary terminations. For example, for a 3-terminal network terminated in open circuits

$$Z_{12} = \frac{V_1}{I_2}\bigg|_{I_1} = 0$$

transfer instruction. *See:* **branch instruction.**   6

transfer instrument (radiation protection). Instrument or dosimeter exhibiting high precision which has been standardized against a national or derived standardized source.   399

transfer locus (linear system or element). A plot of the transfer function as a function of frequency in any convenient coordinate system. *Note:* A plot of the reciprocal of the transfer function is called the inverse transfer locus. *See:* **amplitude frequency locus; phase locus; control system, feedback.**   329

transfer of control. Same as **jump.**

transfer ratio. A dimensionless transfer function.   210

transfer ratio correction (correction to setting). The deviation of the output phasor from nominal, in proportional parts of the input phasor.

$$\frac{\text{Output}}{\text{Input}} = A + \alpha + j\beta$$

A = setting
α = in-phase transfer ratio correction
β = quadrature transfer ratio correction.

transferred charge (capacitor). The net electric charge transferred from one terminal of a capacitor to another via an external circuit. *See:* **nonlinear capacitor.**   191

transferred-charge characteristic (nonlinear capacitor). The function relating transferred charge to capacitor voltage. *See:* **nonlinear capacitor.**   191

transferred information. *See:* **transinformation.**

transferred voltage (generating station grounding). This is a special case of touch voltage. It is that voltage between points of contact, hand to foot or feet, in a station where the conductor touched is grounded at a remote point or touching at a remote point a conductor connected to the station ground grid. Here the voltage rise encountered due to ground fault conditions may be the full voltage rise of the ground grid and not the fraction of this total, which is encountered in the usual touch contact.   569

transfer standards, alternating-current–direct-current. Devices used to establish the equality of a root-mean-square current or voltage (or the average value of alternating power) with the corresponding steady-state direct-current quantity that can be referred to the basic standards through potentiometric techniques. *See:* **auxiliary device to an instrument.**   212

transfer switch (1) (a high-voltage switch) (power switchgear). A switch arranged to permit transferring a conductor connection from one circuit to another without interrupting the current. (A) A tandem transfer switch is a switch with two blades, each of which can be moved into or out of only one contact. (B) A double-blade double-throw transfer switch is a switch with two blades, each of which can be moved into or out of either of two contacts. *Note:* In contrast to high-voltage switches, many low-voltage, control and instrument transfer switches interrupt current during transfer. Compare with **selector switch** and **automatic transfer (or throw-over)** equipment that connects a load to an alternate source after failure of an original source.   103

(2) (emergency and standby power systems). A device for transferring one or more load conductor connections from one power source to another.   73

transfer switch, load. *See:* **load transfer switch.**

transfer time (gas-tubes surge-protective devices). The time duration of the transverse voltage.   490

transfer time (1) (gas tube surge arresters). The time required for the voltage across a conducting gap to drop into the arc region after the gap initially begins to conduct.   62

(2) (gas tube surge-protective device). The time duration of the transverse voltage.   370

transfer time, relay. *See:* **relay transfer time.**

transfer trip (power switchgear). A form of remote trip in which a communication channel is used to transmit a trip signal from the relay location to a remote location.   103

transformer (1)(electrical heating applications to melting furnaces and forehearths in the glass industry). *See:* **transformer coupled; dry-type transformer; dry-type encapsulated water-cooled transformer; liquid filled, or liquid cooled transformer.**   520

(2) (National Electrical Code). A device, which when used, will raise or lower the voltage of alternating current of the original source.   256

(3) (power and distribution transformer). A static electric device consisting of a winding, or two or more coupled windings, with or without a magnetic core, for introducing mutual coupling between electric circuits. Transformers are extensively used in electric power systems to transfer power by electromagnetic induction between circuits at the same frequency, usually with changed values of voltage and current.   53

transformer, alternating-current arc welder. A transformer with isolated primary and secondary windings and suitable stabilizing, regulating, and indicating devices required for transforming alternating current from normal supply voltages to an alternating-current output suitable for arc welding.   264

transformer, assembly, Scott-connected. *See:* **Scott-connected transformer assembly.**

transformer category definitions (distribution, power and regulating transformers). *Note:* All kVA ratings are minimum nameplate kVA for the principal windings. Category I includes distribution transformers manufactured in accordance with ANSI C57.12.20-1974, Requirements for Overhead-Type Distribution Transformers 67 000 Volts and Below; 500 kVA and Smaller, up through 500 kVA, single phase or three phase. In addition, autotransformers of 500 equivalent two-winding kVA or less that are manufactured as distribution transformers in accordance with ANSI C57.12.20-1974 are included in Category I, even through their nameplate kVAs may exceed 500.   393

transformer class designations. *See:* **transformer, oil-immersed.**

transformer, constant-voltage (constant-potential transformer). *See:* **constant-voltage transformer.**

transformer correction factor (TCF) (1) (instrument transformer). The ratio of true primary watts or watthours to the measured secondary watts or watthours, divided by the marked ratio. *Note:* The transformer correction factor for a current or voltage transformer is the ratio correction factor multiplied by the phase angle correction factor for a specified primary circuit power factor. The true primary watts or watthours are equal to the watts or watthours measured, multiplied by the transformer correction factor and the marked ratio. The true primary watts or watthours, when measured using both current and voltage transformers, are equal to the current transformer ratio correction factor multiplied by the voltage transformer ratio correction factor multiplied by the combined phase angle correction factor multiplied by the marked ratios of the current and voltage transformers multiplied by the observed watts or watthours. It is usually sufficiently accurate to calculate true watts or watthours as equal to the product of the two transformer correction factors multiplied by the marked ratios multiplied by the observed watts or watthours.   203

(2) (power and distribution transformer). The ratio of true watts or watthours to the measured watts or watthours, divided by the marked ratio. *Note:* The transformer correction factor for a current or voltage transformer is the ratio correction factor multiplied by the phase-angle correction factor for a specified primary circuit power factor. The true primary watts or watthours are equal to the watts or watthours measured, multiplied by the transformer correction factor and the marked ratio. The true primary watts or watthours, when measured using both current and voltage transformers, are equal to the current transformer ratio correction factor times the voltage transformer correction factor multiplied by the product of the marked ratios of the current and voltage transformers multiplied by the observed watts or wattshours.   53

transformer coupled (electrical heating applications to melting furnaces and forehearths in the glass industry). The power modulation device is connected in the primary circuit of a transformer whose secondary circuit is connected to the glass.   520

transformer, dry-type. *See:* **dry-type transformer.**

transformer, energy-limiting. A transformer that is intended for use on an approximately constant-voltage supply circuit and that has sufficient inherent impedance to limit the output current to a thermally safe maximum value. *See:* **transformer, specialty.**   203

transformer equipment rating. A voltampere output together with any other characterisitics, such as volt-age, current, frequency, and power factor, assigned to it by the manufacturer. *Note:* It is regarded as a test rating that defines an output that can be taken from the item of transformer equipment without exceeding established temperature-rise limitations, under prescribed conditions of test and within the limitations of established standards. *See:* **duty.**   203

transformer, grounding. *See:* **grounding transformer.**

transformer grounding switch and gap (capacitance potential device). Consists of a protective gap connected across the capacitance potential device and transformer unit to limit the voltage impressed on the transformer and the auxiliary or shunt capacitor, when used; and a switch that when closed removes voltage from the potential device to permit adjustment of the potential device without interrupting high-voltage line operation and carrier-current operation when used. *See:* **outdoor coupling capacitor.**   341

transformer, group-series loop insulating. An insulating transformer whose secondary is arranged to operate a group of series lamps and/or a series group of individual-lamp transformers. *See:* **transformer, specialty.**   203

transformer, high-power-factor. A high-reactance transformer that has a power-factor-correcting device such as a capacitor, so that the input current is at a power factor of not less than 90 percent when the transformer delivers rated current to its intended load device. *See:* **transformer, specialty.**   301

transformer, high-reactance (1) (output limiting). An energy-limiting transformer that has sufficient inherent reactance to limit the output current to a maximum value. *See:* **transformer, specialty.**   203

(2) (secondary short-circuit current rating). The current in the secondary winding when the primary winding is connected to a circuit of rated primary voltage and frequency and when the secondary terminals are short-circuited. *See:* **transformer, specialty.**   203

(3) (kilovoltampere or voltampere short-circuit input rating). The input kilovoltamperes or voltamperes at rated primary voltage with the secondary terminals short-circuited. *See:* **transformer, specialty.**   203

transformer, ideal. A hypothetical transformer that neither stores nor dissipates energy. *Note:* An ideal transformer has the following properties: (1) Its self and mutual impedances are equal and are pure inductances of infinitely great value. (2) Its self-inductances have a finite ratio. (3) Its coefficient of coupling is unity. (4) Its leakage inductance is zero. (5) The ratio of the primary to secondary voltage is equal to the ratio of secondary to primary current.   197

transformer, individual-lamp insulating. An insulating transformer used to protect the secondary circuit, casing, lamp, and associated luminaire of an individual street light from the high-voltage hazard of the primary circuit. *See:* **transformer, specialty.**   203

transformer, insulating. A transformer used to insulate one circuit from another. *See:* **transformer, specialty.**   203

transformer integrally mounted cable terminating box. A weatherproof air-filled compartment suitable for enclosing the sidewall bushings of a transformer and equipped with any one of the following entrance devices: (1) Single or multiple-conductor potheads with couplings or wiping sleeves. (2) Wiping sleeves. (3) Couplings with or without stuffing boxes for conduit-enclosed cable, metallic-sheathed cable, or rubber-covered cable. 289

transformer, interphase (thyristor power converter). An autotransformer, or a set of mutually coupled inductive reactors, used to obtain multiple operation between two or more simple converters that have ripple voltages that are out of phase. 121

transformer, isolating (1) (signal-transmission system). A transformer inserted in a system to separate one section of the system from undesired influences of the other sections. *Example:* A transformer having electrical insulation and electrostatic shielding between its windings such that it can provide isolation between parts of the system in which it is used. It may be suitable for use in a system that requires a guard for protection against common-mode interference. *See:* **signal.** 188

(2) **(electroacoustics).** A transformer inserted in a system to separate one section of the system from undesired influences of other sections. *Note:* Isolating transformers are commonly used to isolate system grounds and prevent the transmission of undesired currents. 239

transformer, line. *See:* **line transformer.**

transformer loss (communication). The ratio of the signal power that an ideal transformer would deliver to a load, to the power delivered to the same load by the actual transformer, both transformers having the same impedance ratio. *Note:* Transformer loss is usually expressed in decibels. *See:* **transmission loss.** 239

transformer-loss compensator (metering). A passive electric network that adds to or subtracts from the meter registration to compensate for predetermined iron and copper losses of transformers and transmission lines. 212

transformer, low-power-factor. A high-reactance transformer that does not have means for power-factor correction. *See:* **transformer, specialty.** 301

transformer, matching. *See:* **matching transformer.**

transformer, network. *See:* **network transformer.**

transformer, nonenergy-limiting. A constant-potential transformer that does not have sufficient inherent impedance to limit the output current to a thermally safe maximum value. *See:* **transformer, specialty.** 203

transformer, oil-immersed. *See:* **oil-immersed transformer.**

transformer, outdoor. *See:* **outdoor transformer.**

transformer overcurrent tripping. *See:* **indirect release; overcurrent release.**

transformer, phase-shifting. *See:* **phase-shifting transformer.**

transformer, pole-type. *See:* **pole-type transformer.**

transformer primary voltage rating. *See:* **primary voltage rating.**

transformer, protected outdoor. A transformer that is not of weatherproof construction but that is suitable for outdoor use if it is so installed as to be protected from rain or immersion in water. *See:* **transformer.** 203

transformer - rectifier, alternating -current - direct - current arc welder. A combination of static rectifier and the associated isolating transformer, reactors, regulators, control, and indicating devices required to produce either direct or alternating current suitable for arc-welding purposes. 264

transformer-rectifier, direct-current arc welder. A combination of static rectifiers and the associated isolating transformer, reactors, regulators, control, and indicating devices required to produce direct current suitable for arc welding. 264

transformer relay. A relay in which the coils act as a transformer. 328

transformer removable cable-terminating box. A weatherproof air-filled compartment suitable for enclosing the sidewall bushings of a transformer and equipped with mounting flange(s) (one or two) to accommodate either single-conductor or multiconductor potheads or entrance fittings, depending upon the type of cable termination to be used and the number of three-phase cable circuits (one or two) to be terminated. 288

transformer secondary current rating. *See:* **secondary current rating (transformer).**

transformer, series. A transformer in which the primary winding is connected in series with a power-supply circuit, and that transfers energy to another circuit at the same or different current from that in the primary circuit. *See:* **transformer.** 203

transformer, series street-lighting. A series transformer that receives energy from a current-regulating series circuit and that transforms the energy to another winding at the same or different current from that in the primary. *See:* **transformer, specialty.** 203

transformer, series street-lighting, rating. The lumen rating of the series lamp, or the wattage rating of the multiple lamps, that the transformer is designed to operate. *See:* **specialty transformer.** 203

transformer, shunt. A transformer in which the primary winding is connected in shunt with a power-supply circuit, and that transfers energy to another circuit at the same or different voltage from that of the primary circuit. *See:* **transformer.** 203

transformer, station-type. *See:* **station-type transformer.**

transformer undercurrent tripping. *See:* **indirect release; undercurrent release.**

transformer vault (NESC). An isolated enclosure either above or below ground with fire-resistant walls, ceiling, and floor, in which transformers and related equipment are installed, and which is not continuously attended during operation. *See:* **vault.** 494

transformer, vault-type. *See:* **vault-type transformer.**

transformer voltage (of a network protector) (power switchgear). The voltage between phases or between phase and neutral on the transformer side of a network protector. 103

transforming station (power operations). A station where power is transformed from one voltage level to another. 516

transient (1)(emergency and standby power). That part of the change in a variable, such as voltage or amperage, that disappears during transition from one steady-state operating condition to another. 512

(2) (cable systems in substations). A change in the steady-state condition of voltage or current, or both. As used in this guide, transients occurring in control circuits are a result of rapid changes in the power circuits to which they are coupled. The frequency, damping factor, and magnitude of the transients are determined by resistance, inductance, and capacitance of the power and control circuits and the degree of coupling. Voltages as high as 10 kV in the frequency range of 0.3 to 3.0 MHz have been observed where little or no protection was provided. Transients may be caused by a lightning stroke, a fault, or by switching operation, such as the opening of a disconnect, and may readily be transferred from one conductor to another by means of electrostatic or electromagnetic coupling. 382

(3) (industrial power and control). That part of the change in a variable that disappears during transition from one steady-state operating condition to another. *Note:* Using the term to mean the total variation during the transition between two steady states is deprecated. 73

(4) (excitation control systems) (noun). In a variable observed during transition from one steady-state operating condition to another, that part of the variation which ultimately disappears. *Note:* ANSI C85 deprecates using the term to mean the total variable during the transition between two steady-states. 353

transient adaptation factor (TAF) (illuminating engineering). A factor which reduces the equivalent contrast due to readaptation from one luminous background to another. 167

transient analyzer. An electronic device for repeatedly producing in a test circuit a succession of equal electric surges of small amplitude and of adjustable waveform, and for presenting this waveform on the screen of an oscilloscope. *See:* **oscillograph.** 328

transient blocking (power switchgear). A circuit function which blocks tripping during the interval in which an external fault is being cleared. 103

transient-cause forced outage (electric power systems). A component outage whose cause is immediately self-clearing so that the affected component can be restored to service either automatically or as soon as a switch or circuit breaker can be reclosed or a fuse replaced. *Note:* An example of a transient-cause forced outage is lightning flashover that does not permanently disable the flashed component. *See:* **outage.** 200

transient-cause forced outage duration (electric power systems). The period from the initiation of the outage until the affected component is restored to service by switching or fuse replacement. *See:* **outage.** 200

transient current (rotating machinery). (1) The current under nonsteady conditions. (2) The alternating component of armature current immediately following a sudden short-circuit, neglecting the rapidly decaying component present during the first few cycles. 63

transient-decay current (photoelectric device). The decreasing current flowing in the device after the irradiation has been abruptly cut off. *See:* **phototube.** 190

transient deviation (control). *See:* **deviation, transient.**

transient fault (surge arresters). A fault that disappears of its own accord. 244,62

transient inrush current (power switchgear). Current which results when a switching device is closed to energize a capacitance or an inductive circuit. *Note:* Current is expressed by the highest peak value in amperes and frequency in hertz. 103

transient insulation level (TIL) (power and distribution transformer). An insulation level expressed in kilovolts of the crest value of the withstand voltage for a specified transient wave shape; that is, lightning or switching impulse. 53

transient internal voltage (synchronous machine) (for any specified operating condition). The fundamental-frequency component of the voltage of each armature phase that would be determined by suddenly removing the load, without changing the excitation voltage applied to the field, and extrapolating the envelope of the voltage back to the instant of load removal, neglecting the voltage components of rapid decrement that may be present during the first few cycles after removal of the load. *Note:* The transient internal voltage, as shown in the phasor diagram, is related to the terminal-voltage and phase-current phasors by the equation

$$E_i' = E_a + RI_a + jX_d'I_{ad} + jX_q'I_{aq}.$$

For a machine subject to saturation, the reactances should be determined for the degree of saturation applicable to the specified operating condition. *See:* **direct-axis synchronous reactance; phasor diagram.** 63

transient motion (audio and electroacoustics). Any motion that has not reached or that has ceased to be a steady state. 176

transient overshoot. An excursion beyond the final steady-state value of output as the result of a step-input change. *Note:* It is usually referred to as the first such excursion; expressed as a percent of the steady-state output step. *See:* **accuracy rating (instrument); control system, feedback).** 295

transient overvoltage (power switchgear). The peak voltage during the transient conditions resulting from the operation of a switching device. *Note:* The location and units of measurement are specified in apparatus standards. Compare with **transient overvoltage ratio (factor).** 103

transient overvoltage ratio (factor). The ratio of the

transient overvoltage to the closed-switching-device operating line-to-neutral peak voltage with the load connected. *Note:* The location of measurement is specified in the apparatus standards.   103

transient performance (synchronous-machine regulating system). The performance under a specified stimulus, before the transient expires.   63

transient phenomena (rotating machinery). Phenomena appearing during the transition from one operating condition to another.   244,63

transient reactance (power fault effects). The reactance of a generator between the subtransient and synchronous states. This reactance is used for the calculation of the symmetrical fault current during the period between the subtransient and steady states. The current decreases continuously during this period but is assumed to be steady at this value for approximately 0.25 s.   404

transient recovery voltage (1)(power switchgear). The voltage transient that occurs across the terminals of a pole of a circuit switching device upon interruption of the current. *Notes:* (A) It is the difference between the transient voltages to ground occuring on the terminals. It may be a circuit transient recovery voltage, or an actual transient recovery voltage. (B) In a multipole circuit breaker, the term is usually applied to the voltage across the first pole to interrupt. For circuit breakers having several interrupting units in series, the term may be applied to the voltage across units or groups of units.   103, 443

(2). The voltage transient that occurs across the terminals of a pole of a switching device upon interruption of the current flowing through the pole. *Note:* It is the difference between and in some cases the sum of the transient voltages to ground occurring on the terminals. The term 'transient recovery voltage' is usually designated as TRV, and may refer to inherent TRV, modified inherent TRV, or actual TRV, as defined elsewhere. In a multipole switching device, the term is usually applied to the voltage across the first pole to interrupt in a three-phase ungrounded test, but not necessarily the first phase to interrupt when tested with a three-phase or multigrounded fault. For switching devices having several interrupting units in series, the term may be applied to the voltage across units or groups of units.   486,487,488

transient recovery voltage rate (power switchgear). The rate at which the voltage rises across the terminals of a pole of a circuit switching device upon interruption of the current. *Note:* It is usually determined by dividing the voltage at one of the crests of the transient recovery voltage by the time from current zero to that crest. In case no definite crest exists, the rate may be taken to some stated value usually arbitrarily selected as a certain percentage of the crest value of the normal-frequency recovery voltage. In case the transient is an exponential function the rate may also be taken at the point of zero voltage. It is the rate of rise of the algebraic difference between the transient voltages occurring on the terminals of the switching device upon interruption of the current. The transient recov-

ery voltage rate may be a circuit transient recovery voltage rate or a modified circuit transient recovery voltage rate, or an actual transient recovery voltage rate according to the type of transient from which it is obtained. When giving actual transient recovery voltage rates, the points between which the rate is measured should be definitely stated.   103

transient response (1) (excitation control system). A typical transient response of a feedback control system is shown below. The principal characteristics of interest are the rise time, overshoot, and settling time as indicated.*Note:* In some applications, the time to attain 10 percent of steady-state value is of interest. This time may be appreciable even though the delay time may be very small or even zero.   374

(2) (of a relay) (power switchgear). The manner in which a relay, relay unit or relay system responds to a sudden change in the input.   103

(3) (oscilloscopes). Time-domain reactions to abruptly varying inputs.   106

transient speed deviation (gas turbines). (1) Load decrease. The maximum instantaneous speed above the steady-state speed occurring after the sudden decrease from one specified steady-state electric load to another specified steady-state electric load having values within limits of the rated output of the gas-turbine-generator unit. It is expressed in percent of rated speed.

(2) Load increase. The minimum instantaneous speed below the steady-state speed occurring after the sudden increase from one specified steady-state electric load having values within the limits of rated output of the gas-turbine-generator unit. It is expressed in percent of rated speed.   98,58

transient stability. A condition that exists in a power system if, after an aperiodic disturbance, the system regains steady-state stability. *See:* **alternating-current distribution.**   64

transient stability factor (system or part of a system). The ratio of the transient stability limit to the nominal power flow at the point of the system to which the stability limit is referred. *See:* **stability factor; alternating-current distribution.**   64

transient stability limit (transient power limit). The maximum power flow possible through some particular point in the system when the entire system or the part of the system to which the stability limit refers is operating with transient stability. *See:* **alternating-current distribution.**   64

transient suppression networks. Capacitors, resistors, or inductors so placed as to control the discharge of stored energy banks. They are commonly used to suppress transients caused by switching.   95

transient thermal impedance (semiconductor device). The change in the difference between the virtual junction temperature and the temperature of a specified reference point or region at the end of a time interval divided by the step function change in power dissipation at the beginning of the same time interval which causes the change of temperature-difference. *Note:* It is the thermal impedance of the junction under conditions of change and is generally given in the form of a curve as a function of the duration of an applied pulse. *See:* **principal voltage-current characteristic (principal characteristic); semiconductor rectifier stack.** 191

transient voltage capability (thyristor). Rated nonrepetitive peak reverse voltage. The maximum instantaneous value of any nonrepetitive transient reverse voltage which may occur across a thyristor without damage. 445

transimpedance (of a magnetic amplifier). The ratio of differential output voltage to differential control current. 171

transinformation (of an output symbol about an input symbol) (information theory). The difference between the information content of the input symbol and the conditional information content of the input symbol given the output symbol. *Notes:* (1) If x_i is an input symbol and y_j is an output symbol, the transinformation is equal to

$$[-\log p(x_i)] - [-\log p(x_i|y_j)]$$

$$= \log \frac{p(x_i|y_j)}{p(x_i)} = \log \frac{p(x_i, y_j)}{p(x_i)p(y_j)}$$

where $p(x_i|y_j)$ is the conditional probability that x_i was transmitted when y_j is received, and $p(x_{i,yj})$ is the joint probability of x_i and y_j (2) This quantity has been called **transferred information, transmitted information,** and **mutual information.** *See:* **information theory.** 415

transistor. An active semiconductor device with three or more terminals. It is an analog device. 245

transistor, conductivity-modulation. A transistor in which the active properties are derived from minority-carrier modulation of the bulk resistivity of a semiconductor. *See:* **semiconductor; transistor.** 245

transistor, filamentary. A conductivity-modulation transistor with a length much greater than its transverse dimensions. *See:* **semiconductor; transistor.** 245

transistor, junction. A transistor having a base electrode and two or more junction electrodes. *See:* **transistor.** 245

transistor, point-contact. A transistor having a base electrode and two or more point-contact electrodes. *See:* **semiconductors; transistor.** 245

transistor, point-junction. A transistor having a base electrode and both point-contact and junction electrodes. *See:* **transistor.** 328

transistor reset preamplifier (germanium gamma-ray detectors). A charge-sensitive preamplifier in which the charge that accumulates on the feedback capacitor is periodically discharged through a suitably located transistor. 528

transistor, unipolar. A transistor that utilizes charge carriers of only one polarity. *See:* **semiconductor; transistor.** 245

transit (1)(navigation aid terms). A radio navigation system using low orbit satellites to provide world-wide coverage, with transmissions from the satellites at vhf

(very high frequency) and uhf (ultra high frequency), in which fixes are determined from measurements of the Doppler shift of the continuous wave signal received from the moving satellite. 526

(2) (conductor stringing equipment). An instrument primarily used during construction of a line to survey the route, set hubs and point on tangent (POT) locations, plumb structures, determine downstrain angles for locations of anchors at the pull and tension sites, and to sag conductors. *Syn:* **level; scope; site marker.** 431

transit angle. The product of angular frequency and the time taken for an electron to traverse a given path. *See:* **electron emission.** 190, 125

transition (1) (data transmission). (A) (signal transmission). The change from one circuit condition to the other, that is, to change from mark to space or from space to mark. (B) (waveform) (pulse techniques). A change of the instantaneous amplitude from one amplitude to another amplitude level. (C) (transition frequency) (disk recording system) (crossover frequency) (turnover frequency). The frequency corresponding to the point of intersection of the asymptotes to the constant-amplitude and the constant-velocity portions of its frequency response curve. This curve is plotted with output voltage ratio in decibels as the ordinate and the logarithm of the frequency as the abscissa. 59

(2) (pulse terms). A portion of a wave or pulse between a first nominal state and a second nominal state. Throughout the remainder of this document the term transition is included in the term pulse and wave.

transitional mode (seismic testing of relays). The change from the nonoperating to the operating mode, caused by switching the input to the relay from the nonoperating to the operating input, or vice versa. 392

transition duration (pulse terms). The duration between the proximal point and the distal point on a transition waveform. 254

transition frequency (disk recording system) (crossover frequency) (turnover frequency). The frequency corresponding to the point of intersection of the asymptotes to the constant-amplitude and the constant-velocity portions of its frequency response curve. This curve is plotted with output voltage ratio in decibels as the ordinate and the logarithm of the frequency as the abscissa. *See:* **phonograph pickup.** 176

transition joint (power cable joint). A cable joint which connects two different types of cable. 34

transition load (rectifier circuit). The load at which a rectifier unit changes from one mode of operation to another. *Note:* The load current corresponding to a transition load is determined by the intersection of extensions of successive portions of the direct-current voltage-regulation curve where the curve changes shape or slope. *See:* **rectification; rectifier circuit element.** 66

transition loss (1) (wave propagation). (A) At a transition or discontinuity between two transmission media, the difference between the power incident upon the discontinuity and the power transmitted beyond the discontinuity that would be observed if the medium beyond the discontinuity were match-terminated. (B) The ratio in decibels of the power incident upon the discontinuity to the power transmitted beyond the discontinuity that would be observed if the medium beyond the discontinuity were match terminated. *See:* **waveguide.**   267

(2) (junction between a source and a load). The ratio of the available power to the power delivered to the load. Transition loss is usually expressed in decibels. *See:* **waveguide; transmission loss.**   267

transition point (circuit). A point in a transmission system at which there is change in the surge impedance.   244, 62

transition pulse (pulse waveform). That segment comprising a change from one amplitude level to another amplitude level. *See:* **pulse.**   185

transition region (semiconductor). The region, between two homogeneous semiconductor regions, in which the impurity concentration changes. *See:* **semiconductor; transistor.**   113

transition shape (pulse terms). (1) For descriptive purposes a transition waveform may be imprecisely described by any of the adjectives, or combinations thereof, in **descriptive adjectives, major (minor); polarity related adjectives; geometrical adjectives, round;** and **functional adjectives.** When so used, these adjectives describe general shape only, and no precise distinctions are defined.   254

(2) For tutorial purposes a hypothetical transition waveform may be precisely defined by the further addition of the adjective ideal. *See:* **descriptive adjectives, ideal.**

(3) For measurement or comparison purposes a transition waveform may be precisely defined by the further addition of the adjective reference. *See:* **descriptive adjectives, reference.**

transition time (gas-tube surge-protective devices). The time required for the voltage across a conducting gap to drop into the arc region after the gap initially begins to conduct.   490

transitron oscillator. A negative-transconductance oscillator employing a screen-grid tube with negative transconductance produced by a retarding field between the negative screen grid and the control grid that serves as the anode. *See:* **oscillatory circuit.**
                                                    111

transit time (1) (electron tube). The time taken for a charge carrier to traverse a given path. *See:* **electron emission.**   190

(2) (multiplier-phototube). The time interval between the arrival of a delta-function light pulse at the entrance window of the tube and the time at which the output pulse at the anode terminal reaches peak amplitude. *See:* **electron emission; phototube.**   174

transit-time mode (electron tubes). A condition of operation of an oscillator corresponding to a limited range of drift-space transit angle for which the electron stream introduces a negative conductance into the coupled circuit.   125

transit-time spread (1) (electron tubes). The time interval between the half-amplitude points of the output pulse at the anode terminal, arising from a delta function of light incident on the entrance window of the tube. *See:* **phototube.**   174

(2) (scintillation counting). The FWHM (full-width-at-half-maximum) of the time distribution of a set of pulses each of which corresponds to the photomultiplier transit time for that individual event.   117

translate. (1) To convert expressions in one language to synonymous expressions in another language. (2) To encode or decode. *See:* **matrix; translator.**   235

translation (telecommunication). The process of converting information from one system of representation into equivalent information in another system of representation.   194

translation loss (playback loss) (reproduction of a mechanical recording). The loss whereby the amplitude of motion of the reproducing stylus differs from the recorded amplitude in the medium. *See:* **phonograph pickup.**   176

translator (1) (software). A program that transforms a sequence of statements in one language into an equivalent sequence of statements in another language. *See:* **assembler; compiler; interpreter; program.**   434

(2) (telephone switching systems). Equipment capable of interpreting and converting information from one form to another form.   55

(3) (test, measurement and diagnostic equipment). An automatic means, usually a program, to translate machine language mnemonic symbols for computer operations into true machine language. Memory locations and input-output lines must be written in numerical code, not symbolically.   54

transliterate. To convert the characters of one alphabet to the corresponding characters of another alphabet.
                                                  255, 77

transmission (1) (data transmission). The electrical transfer of a signal, message, or other form of intelligence from one location to another.   59

(2) (illuminating engineering). A general term for the process by which incident flux leaves a surface or medium on a side other than the incident side, without change in frequency. *Note:* Transmission through a medium is often a combination of regular and diffuse transmission. *See:* **regular transmisssion; diffuse transmission.**   167

(3) (laser-maser). Passage of radiation through a medium.   363

transmission band (uniconductor waveguide). The frequency range above the cutoff frequency. *See:* **waveguide.**   267

transmission coefficient (1) (of a network) (waveguide). At a given frequency and for a given mode, the ratio of some quantity associated with the transmitted wave at a specified reference plane to the corresponding quantity in the incident wave at a specified reference plane. *Notes:* (A) The transmission coefficient may be different for different associated quantities, and the chosen quantity must be specified. The voltage transmission coefficient is commonly used and is defined as the complex ratio of the resultant electric field strength (or voltage) to that of the incident wave. Examples of other quantities are power or current. (B) An interface is a special case of a network where the reference planes associated with the incident and transmitted waves become coincident; in this case the voltage transmission coefficient is equal to one plus the voltage reflection coefficient.   267

(2) **(multiport).** Ratio of the complex amplitude of the wave emerging from a port of a multiport terminated by reflectionless terminations to the complex amplitude of the wave incident upon another port. *See:* **reflection coefficient; scattering coefficient.**   185

transmission detector (charged-particle detectors)(X-ray energy spectrometers)(semiconductor radiation detectors). A totally depleted detector whose thickness including its entrance and exit windows is sufficiently small to permit radiation to pass completely through the detector.   119,417,23

transmission facility (data transmission). The transmission medium and all the associated equipment required to transmit a message.   59

transmission feeder. A feeder forming part of a transmission circuit. *See:* **center of distribution.**   64

transmission frequency meter (waveguide). A cavity frequency meter that, when tuned, couples energy from a waveguide into a detector. *See:* **waveguide.**

244, 179

transmission gain (data transmission). General term used to denote an increase in signal power in transmission from one point to another. Gain is usually expressed in decibels and is widely used to denote transducer gain.   59

transmission level (data transmission). The ratio of the signal power at any point in a transmission system to the power at some point in the system chosen as a reference point. This ratio is usually expressed in decibels. The transmission level at the transmitting switchboard is frequently taken as the zero level reference point.   59

transmission line (1) (data transmission). (A) (signal-transmission system). (1) The conductive connections between system elements which carry signal power. (2) A waveguide consisting of two or more conductors. (B) (electric power). A line used for electric power transmission. (C) (electromagnetic wave guidance). A system of material boundaries or structures for guiding electromagnetic waves, in the TEM (transverse electromagnetic) mode. Commonly a two-wire or coaxial system of conductors.   59

(2) **(waveguide).** A system of material boundaries or structures for guiding electromagnetic waves. Frequently, such a system is used for guiding electromagnetic waves, in the TEM mode. Commonly, a two-wire or coaxial system of conductors. *See:* **waveguide.**
267

(3) **(induction heating).** *See:* **load leads or transmission line.**

transmission-line capacity (electric power supply). The maximum continuous rating of a transmission line. The rating may be limited by thermal considerations, capacity of associated equipment, voltage regulation, system stability, or other factors. *See:* **generating station.**   200

transmission line, coaxial. *See:* **coaxial transmission line.**

transmission loss (1) (fiber optics). Total loss encountered in transmission through a system. *See:* **attenuation; optical density; reflection; transmittance.**

433

(2) **(frequently abbreviated 'loss') (data transmission).** In communication, a general term used to denote a decrease in power in transmission from one point to another. Transmission loss is usually expressed in decibels.   59

(3) **(radio wave propagation).** Of a radio system consisting of a transmitting antenna, receiving antenna, and the intervening propagation medium, the ratio of the power radiated from the transmitting antenna to the resultant power which would be available from an equivalent loss-free receiving antenna.   146

(4) **(electric power system).** (A) The power lost in transmission between one point and another. It is measured as the difference between the net power passing the first point and the net power passing the second. (B) The ratio in decibels of the net power passing the first point to the net power passing the second.   267

transmission-loss coefficients (electric power systems). Mathematically derived constants to be combined with source powers to provide incremental transmission losses from each source to the composite system load. These coefficients may also be used to calculate total system transmission losses.   94

transmission measuring set (data transmission). A measuring instrument comprising a signal source and a signal receiver having known impedances, that is designed to measure the insertion loss or gain of a network or transmission path connected between those impedances.   59

transmission mode. A form of propagation along a transmission line characterized by the presence of any one of the elemental types of TE (transverse electric), TM (transverse magnetic), or TEM (transverse electromagnetic) waves. *Note:* Waveguide transmission modes are designated by integers (modal numbers) associated with the orthogonal functions used to describe the waveform. These integers are known as waveguide mode subscripts. They may be assigned from observations of the transverse field components of the wave and without reference to mathematics. A waveguide transmission mode is commonly described as a $TE_{m,n}$ or $TM_{m,n}$ mode, $_{m,n}$ being numerics according to the following system. (1) **(waves in rectangular waveguides).** If a single wave is transmitted in a rectangular waveguide, the field that is everywhere transverse may be resolved into two components, parallel to the wide and narrow walls respectively. In any transverse section, these components vary periodically with distance along a path parallel to one of the walls. m = the total number of half-period variations of either component of field along a path parallel to the wide walls. n = the total number of half-period variations of either component of field along a path parallel to the narrow walls. **(2) (waves in circular waveguides).** If a single wave is transmitted in a circular waveguide, the transverse field may be resolved into two components, radial and angular, respectively. These components vary periodically along a circular path concentric with the wall and vary in a manner related to the Bessel function of order m along a radius, where m = the total number of full-period variations of either component of field along a circular path concentric with the wall. n = one more than the total number of reversals of sign of either component of field along a radial path. This system can be used only if the observed waveform is known to correspond to a single mode. *See:* **waveguide.** 328

transmission-mode photocathode. A photocathode in which radiant flux incident on one side produces photoelectric emission from the opposite side. 117

transmission modulation (storage tubes). Amplitude modulation of the reading-beam current as it passes through apertures in the storage surface, the degree of modulation being controlled by the charge pattern stored on that surface. *See:* **storage tube.** 174

transmission network. A group of interconnected transmission lines or feeders. *See:* **transmission line.** 64

transmission primaries (color television). The set of three colorimetric primaries that, if used in a display and controlled linearly and individually by a corresponding set of three channel signals generated in the color television camera, would result in exact colorimetric rendition (over the gamut defined by the primaries) of the scene viewed by the camera. *Note:* Ideally the primaries used at the receiver display would be identical with the transmission primaries, but this is not usually possible since developments in display phosphors occurring since the setting of transmission standards, for example, may result in the use of receiver display primaries that differ from the transmission primaries. Within a linear part of the overall system, it is always possible to compensate for differences existing between transmission and display primaries by means of matrixing. Because of the capability afforded by matrixing, the transmission primaries need not be real. There exists a unique relationship between the chromaticity coordinates of the transmission primaries and the spectral taking characteristics used at the camera to generate the three respective channel signals. 18

transmission quality (mobile communication). The measure of the minimum usable speech-to-noise ratio, with reference to the number of correctly received words in a specified speech sequence. *See:* **mobile communication system.** 181

transmission regulator (electric communication). A device that functions to maintain substantially constant transmission over a transmission system. 328

transmission route. The route followed by a transmission circuit. *See:* **transmission line.** 64

transmission system (1)(power operations). An interconnected group of electric transmission lines and associated equipment for the movement or transfer of electric energy in bulk between points of supply and points for delivery. 516
(2) (data transmission). In communication practice, an assembly of elements capable of functioning together to transmit signal waves. 59

transmission throughput. *See:* **speed of transmission, effective.**

transmission time (data transmission). The absolute time interval from transmission to reception of a signal. 59

transmission window. *See:* **spectral window.** 433

transmissivity (fiber optics). The transmittance of a unit length of material, at a given wavelength, excluding the reflectance of the surfaces of the material; the intrinsic transmittance of the material, irrespective of other parameters such as the reflectances of the surfaces. No longer in common use. *See:* **transmittance.** 433

transmissometer (illuminating engineering). A photometer for measuring transmittance. *Note:* Transmissometers may be visual or physical instruments. 167

transmit (1)(computing machines). To move data from one location to another location. *See:* **transfer (2).** 255, 77
(2)(token ring access method). The action of a station generating a frame, token, abort sequence, or fill and placing it on the medium to the next station. In use this term contrasts with 'repeat'. 472

transmit characteristic (telephony). The electrical output level of a telephone set as a function of the acoustic input level. The output is measured across a specified impedance connected to the telephone feed circuit, and the input is measured in free field at a specified location relative to the reference point of an **artificial mouth.** 122

transmit-receive (TR) tube. A gas-filled radio-frequency switching tube used to protect the receiver in pulsed radio-frequency systems. *See:* **gas tube.** 125

transmit-receive cavity (radar). The resonant portion of a transmit-receive switch. 13

transmit-receive cell (tube) (waveguide). A gas-filled waveguide cavity that acts as a short circuit when ionized but is transparent to low-power energy when un-ionized. It is used in a transmit-receive switch for protecting the receiver from the high power of the transmitter but is transparent to low-power signals received from the antenna. *See:* **waveguide.** 179

transmit-receive switch (TR switch) (TR box). An automatic device employed in a radar for substantially preventing the transmitted energy from reaching the receiver but allowing the received energy to reach the receiver without appreciable loss. *See:* **radar.** 328

transmit-receive switch, duplexer. A switch, frequently of the gas discharge type, employed when a common transmitting and receiving antenna is used, that automatically decouples the receiver from the antenna during the transmitting period. *See:* **navigation.**

13, 187

transmittance (1) (fiber optics). The ratio of transmitted power to incident power. *Note:* In optics, frequently expressed as optical density or percent; in communications applications, generally expressed in decibels (dB). Formerly called "transmission". *See:* **antireflection coating; optical density; transmission loss.**

433

(2) (photovoltaic power system). The fraction of radiation incident on an object that is transmitted through the object. *See:* **photovoltaic power system; solar cells.**

186

(3) (transfer function) (linear passive networks). A response function for which the variables are measured at different ports (terminal pairs).

238

(4) τ **(laser-maser).** The ratio of total transmitted radiant power to total incident radiant power.

363

(5) $\tau = \Phi_t / \Phi_i$ **(of a medium) (illuminating engineering).** The ratio of the transmitted flux to the incident flux. *Note:* It should be noted that transmittance refers to the ratio of flux emerging to flux incident; therefore; reflections at the surface as well as absorption within the material operate to reduce the transmittance. Transmittance is a function of:

(1) Geometry(a) of the incident flux(b) of collection for the transmitted flux

(2) Spectral distribution(a) characteristic of the incident flux(b) weighting function for the collected flux

(3) Polarization(a) of the incident flux(b) component defined for the collected flux

Notes: (1)Unless the state of polarization for the incident flux and the polarized component of the transmitted flux are stated, it shall be considered that the incident flux is unpolarized and that the total transmitted flux is evaluated. (2) Unless qualified by the term 'spectral' (see spectral reflectance) or other modifying adjectives, luminous transmittance (see luminous transmittance) is meant. (3) If no qualifying geometric adjective is used, transmittance for hemispherical collection is meant. For other modifying adjectives see listing in reflectance factor entry. (4) In each case of conical incidence or collection, the solid angle is not restricted to a right circular cone, but may be of any cross section including rectangular, a ring, or a combination of two or more solid angles. (5) These concepts must be applied with care, if the area of the transmitting element is not large compared to its thickness, due to internal transmission across the boundary of the area. (6) The following breakdown of transmittance quantities is applicable only to the transmittance of thin films with negligible internal scattering so that the transmitted radiation emerges from a point that is not significantly separated from the point of incidence of the incident ray that produces the transmitted ray(s). The governing considerations are similar to those for application of the bidirectional reflectance-distribution function (BRDF), rather than the bidirectional scattering-surface reflectance-distribution function (BSSRDF).   .   167

transmitted-carrier operation. That form of amplitude-modulation carrier transmission in which the carrier wave is transmitted. *See:* **amplitude modulation.**

328

transmitted harmonics (induced harmonics) (electrical conversion). Harmonics that are transformed or pass through the conversion device from the input to the output. *See:* electrical conversion.

186

transmitted information. *See:* **transinformation.**

transmitted light scanning (industrial control). The scanning of changes in the magnitude of light transmitted through a web. *See:* **photoelectric control.**

206

transmitted wave (1) (radio wave propagation). *See:* refracted wave.

146

(2) (waveguide). At a transverse plane in a transmission line or waveguide, a wave transmitted past a discontinuity in the same direction as the incident wave. *See:* **reflected wave.**

267

(3) (circuit). A wave (or waves) produced by an incident wave that continue(s) beyond the transition point.

244

transmitter (protective signaling). A device for transmitting a coded signal when operated by any one of a group of actuating devices. *See:* **protective signaling.**

328

transmitter-blocker cell (TB cell) (antitransmit-receive tube) (with reference to a waveguide). A gas-filled waveguide cavity that acts as a short circuit when ionized but as an open circuit when un-ionized. It is used in a transmit-receive switch for directing the energy received from the aerial to the receiver, no matter what the transmitter impedance may be. *See:* **waveguide.**

179

transmitter, facsimile. The apparatus employed to translate the subject copy into signals suitable for delivery to the communication system. *See:* **facsimile (in electrical communication).**

12

transmitter performance. *See:* **audio input power; audio input signal.**

transmitter, pulse delay. *See:* **pulse delay transducer.**

transmitter, telephone. *See:* **telephone transmitter.**

transmitting (transmission performance of telephone sets). The electric output level of a telephone set or connecting test circuit due to an acoustic input to the telephone set. The acoustic input may be varied either in frequency or level. The output is measured across a specified impedance and the input is measured at the calibration point of an artificial mouth.

491

transmitting converter (facsimile) (amplitude-modulation to frequency-shift-modulation converter). A device which changes the type of modulation from amplitude to frequency shift. *See:* **facsimile transmission.**

12

transmitting current response (electroacoustic transducer used for sound emission). The ratio of the sound pressure apparent at a distance of 1 meter in a specified direction from the effective acoustic center of the transducer to the current flowing at the electric input terminals. *Note:* The sound pressure apparent at a distance of 1 meter can be found by multiplying the sound pressure observed at a remote point (where the sound field is spherically divergent) by the number of meters from the effective acoustic center of the transducer to that point. *See:* **loudspeaker.** 176

transmitting efficiency (electroacoustic transducer) (projector efficiency). The ratio of the total acoustic power output to the electric power input. *Note:* In computing the electric power input, it is customary to omit any electric power supplied for polarization or bias. 176

transmitting loop loss. That part of the repetition equivalent assignable to the station set, subscriber line, and battery supply circuit that are on the transmitting end. *See:* **transmission loss.** 328

transmitting objective loudness rating (TOLR) (loudness ratings of telephone connections).

$$\text{TOLR} = -20 \log_{10} \frac{V_T}{S_M}$$

where

S_M = sound pressure at the mouth reference point (in pascals)

V_T = output voltage of the transmitting component (in millivolts)

Note: Normally occurring TOLRs will be in the -30 to -55 (dB) range. These numbers are a result of the units chosen and have no physical significance.

409

transmitting power response (projector power response) (electroacoustic transducer used for sound emission). The ratio of the mean-square sound pressure apparent at a distance of 1 meter in a specified direction from the effective acoustic center of the transducer to the electric power input. *Note:* The sound pressure apparent at a distance of 1 meter can be found by multiplying the sound pressure observed at a remote point (where the sound field is spherically divergent) by the number of meters from the effective acoustic center of the transducer to that point. *See:* **loudspeaker.** 176

transmitting voltage response (electroacoustic transducer used for sound emission). The ratio of the sound pressure apparent at a distance of 1 meter in a specified direction from the effective acoustic center of the transducer to the signal voltage applied at the electric input terminals. *Note:* The sound pressure apparent at a distance of 1 meter can be found by multiplying the sound pressure observed at a remote point (where the sound field is spherically divergent) by the number of meters from the effective acoustic center of the transducer to that point. *See:* **loudspeaker.** 176

trans-μ-factor (multibeam electron tubes). The ratio of (1) the magnitude of an infinitesimal change in the voltage at the control grid of any one beam to (2) the magnitude of an infinitesimal change in the voltage at the control grid of a second beam. The current in the second beam and the voltage of all other electrodes are maintained constant. 125

transobuoy (navigation aid terms). A free floating or moored automatic weather station providing weather reports from the open ocean. 526

transparency (data communication). A capability of a communications medium to pass within specified lim-

its a range of signals having one or more defined properties, for example, a channel may be code transparent, or an equipment may be bit pattern transparent. 12

transponder (1)(navigation aid terms). A transmitter-receiver facility, the function of which is to transmit signals automatically when the proper interrogation is received. 526

(2) (communication satellite). A receiver-transmitter combination, often aboard a satellite, or spacecraft, which receives a signal and retransmits it at a different carrier frequency. Transponders are used in communication satellites for reradiating signals to earth stations or in spacecraft for returning ranging signals. *See:* **repeater.** 83

transponder beacon. *See:* **transponder.**

transponder, crossband (electronic navigation). A transponder that replies in a different frequency band from that of the received interrogation. *See:* **navigation.** 13, 187

transponder reply efficiency (navigation aid terms). The ratio of the number of replies emitted by a transponder to the number of interrogations which the transponder recognizes as valid. The interrogations recognized as valid include those accidentally combined to form recognizable codes, a statistical computation of them normally being made. 526

transport (computing machines). *See:* **tape transport.**

transportable (X-ray) (National Electrical Code). X-ray equipment to be installed in a vehicle or that may be readily disassembled for transport in a vehicle. 256

transportable transmitter. A transmitter designed to be readily carried or transported from place to place, but which is not normally operated while in motion. *Note:* This has been commonly called a portable transmitter, but the term transportable transmitter is preferred. *See:* **radio transmitter; radio transmission.** 111, 211

transportation and storage conditions. The conditions to which a device may be subjected between the time of construction and the time of installation. Also included are the conditions that may exist during shutdown. *Note:* No permanent physical damage or impairment of operating characteristics shall take place under these conditions, but minor adjustments may be needed to restore performance to normal. 295

transportation lag.* *See:* **lag, distance/velocity.**
*Deprecated

transport delay. *See:* **time delay.**

transport lag.* *See:* **lag, distance/velocity.**
*Deprecated

transport standards (metering). Standards of the same nominal value as the basic reference standards of a laboratory (and preferably of equal quality), which are regularly intercompared with the basic group but are reserved for periodic interlaboratory comparison tests to check the stability of the basic reference group. 212

transport time (industrial control) (feedback system). The time required to move an object, element or information from one predetermined position to another. *See:* **control system, feedback.**   219, 206

transport vehicle (handling and disposal of transformer grade insulating liquids containing PCBs). A motor vehicle or rail car used for the transportation of cargo by any mode, each cargo carrier (for trailer, freight car) is a separate vehicle.   586

transposition (1) (data transmission). (A) An interchange of positions of the several conductors of a circuit between successive lengths. *Notes:* (1) It is normally used to reduce inductive interference on communication or signal circuits by cancellation. (2) The term is most frequently applied to open wire circuits. (B) the ordered permutation of the pattern of the muliple of a switching stage to improve traffic carrying characteristics and reduce crosstalk.   59
(2) (transmission lines). An interchange of positions of the several conductors of a circuit between successive lengths. *Notes:* (A) It is normally used to reduce inductive interference on communication or signal circuits by cancellation. (B) The term is most frequently applied to open wire circuits. *See:* **open wire; signal; tower.**   64
(3) (rotating machinery). An arrangement of the strands or laminations of a conductor or of the conductors comprising a turn or coil whereby they take different relative positions in a slot for the purpose of reducing eddy current losses.   63

transposition section. A length of open wire line to which a fundamental transposition design or pattern is applied as a unit. *See:* **open wire.**   328

transreactance. The imaginary part of the transimpedance.   185

transrectification factor. The quotient of the change in average current of an electrode by the change in the amplitude of the alternating sinusoidal voltage applied to another electrode, the direct voltages of this and other electrodes being maintained constant. *Note:* Unless otherwise stated, the term refers to cases in which the alternating sinusoidal voltage is of infinitesimal magnitude. *See:* **rectification factor.**
                    125

transrectifier. A device, ordinarily a vacuum tube in which rectification occurs in one electrode circuit when an alternating voltage is applied to another electrode. *See:* **rectifier.**   328

transresistance. The real part of the transimpedance.
                    185

transsusceptance. The imaginary part of the transadmittance.   185

transverse (differential) mode voltage (low-voltage air gap surge-protective devices)(gas-tube surge-protective devices). The voltage at a given location between two conductors of a group.   556,490

transverse-beam traveling-wave tube. A traveling-wave tube in which the direction of motion of the electron beam is transverse to the average direction in which the signal wave moves.   125

transverse crosstalk coupling (between a disturbing and a disturbed circuit in any given section). The vector summation of the direct couplings between ad-

jacent short lengths of the two circuits, without dependence on intermediate flow in other nearby circuits. *See:* **coupling.**   328

transverse electric ($TE_{m,n,p}$) resonant mode (cylindrical cavity). In a hollow metal cylinder closed by two plane metal surfaces perpendicular to its axis, the resonant mode whose transverse field pattern is similar to the $TE_{m,n}$ wave in the corresponding cylindrical waveguide and for which p is the number of half-period field variations along the axis. *Note:* When the cavity is a rectangular parallelepiped, the axis of the cylinder from which the cavity is assumed to be made should be designated since there are three such axes possible. *See:* **waveguide.**   267

transverse electric (TE) mode (fiber optics). A mode whose electric field vector is normal to the direction of propagation. *Note:* In an optical fiber, TE and transverse magnetic (TM) modes correspond to meridional rays. *See:* **meridional ray; mode.**   433

transverse-electric hybrid wave (radio wave propagation). An electromagnetic wave in which the electric field vector is linearly polarized normal to the plane of propagation and the magnetic field vector is elliptically polarized in this plane.   146

transverse-electric wave (1) (radio wave propagation). An electromagnetic wave in which the electric field vector is everywhere perpendicular to the wave normal.   146
(2) (TE wave) (general). In a homogeneous isotropic medium, an electromagnetic wave in which the electric field vector is everywhere perpendicular to the direction of propagation. *See:* **waveguide.**   267
(3) ($TE_{m,n}$ wave) (rectangular waveguide) (hollow rectangular metal cylinder). The transverse electric wave for which m is the number of half-period variations of the field along the x coordinate, which is assumed to coincide with the larger transverse dimension, and n is the number of half-period variations of the field along the y coordinate, which is assumed to coincide with the smaller transverse dimension. *Note:* The dominant wave in a rectangular waveguide is $TE_{1,0}$; its electric lines are parallel to the shorter side. *See:* **guided waves; waveguide.**   267
(4) ($TE_{m,n}$ wave) (circular waveguide) (hollow circular metal cylinder). The transverse electric wave for which m is the number of axial planes along which the normal component of the electric vector vanishes, and n is the number of coaxial cylinders (including the boundary of the waveguide) along which the tangential component of the electric vector vanishes. *Notes:* (1) $TE_{0,n}$ waves are circular electric waves of order n. The $TE_{0,1}$ wave is the circular electric wave with the lowest cutoff frequency. (2) The $TE_{1,1}$ wave is the dominant wave. Its lines of electric force are approximately parallel to a diameter.*See:* **waveguide.**
                    267

transverse electromagnetic (TEM) mode (1) (fiber optics). A mode whose electric and magnetic field vectors are both normal to the direction of propagation. *See:* **mode.**   433
(2) (waveguide). A mode in which the longitudinal

components of the electric and magnetic fields are everywhere zero. *See:* **waveguide.** 244, 179

transverse-electromagnetic wave (1) (radio wave propagation). An electromagnetic wave in which both the electric and magnetic field vectors are everywhere perpendicular to the wave normal. 146

(2) (TEM wave). In a homogeneous isotropic medium, an electromagnetic wave in which both the electric and magnetic field vectors are everywhere perpendicular to the direction of propagation. *See:* **radio-wave propagation; waveguide.** 267

transverse-field traveling-wave tube. A traveling-wave tube in which the traveling electric fields that interact with electrons are essentially transverse to the average motion of the electrons. 125

transverse interference (signal-transmission system). *See:* **interference, differential-mode; interference, normal-mode; accuracy rating (instrument); signal.**

transverse interferometry (fiber optics). The method used to measure the index profile of an optical fiber by placing it in an interferometer and illuminating the fiber transversely to its axis. Generally, a computer is required to interpret the interference pattern. *See:* **interferometer.** 433

transverse magnetic ($TM_{m,n,p}$) resonant mode (cylindrical cavity). In a hollow metal cylinder closed by two plane metal surfaces perpendicular to its axis, the resonant mode whose transverse field pattern is similar to the $TM_{m,n}$ wave in the corresponding cylindrical waveguide and for which p is the number of half-period field variations along the axis. *Note:* When the cavity is a rectangular parallelepiped, the axis of the cylinder from which the cavity is assumed to be made should be designated since there are three such axes possible. *See:* **waveguide.** 267

transverse magnetic (TM) mode. A mode whose magnetic field vector is normal to the direction of propagation. *Note:* In a planar dielectric waveguide (as within an injection laser diode), the field direction is parallel to the core-cladding interface. In an optical waveguide , transverse electric (TE) and TM modes correspond to meridional rays. *See:* **meridional ray; mode.** 433

transverse-magnetic hybrid wave (radio wave propagation). An electromagnetic wave in which the magnetic field vector is linearly polarized normal to the plane of propagation and the electric field vector is elliptically polarized in this plane. 146

transverse-magnetic wave (1) (radio wave propagation). An electromagnetic wave in which the magnetic field vector is everywhere perpendicular to the wave normal. 146

(2) (TM wave) (general). In a homogeneous isotropic medium, an electromagnetic wave in which the magnetic field vector is everywhere perpendicular to the direction of propagation. *See:* **waveguide.** 267

(3) ($TM_{m,n}$ wave) (circular waveguide) (hollow circular metal cylinder). The transverse magnetic wave for which m is the number of axial planes along which the normal component of the magnetic vector vanishes, and n is the number of coaxial cylinders to which the

electric vector is normal. *Note:* $TM_{0,n}$ waves are circular magnetic waves of order n. The $TM_{0,1}$ wave is the circular magnetic wave with the lowest cutoff frequency. *See:* **guided wave; circular magnetic wave; waveguide.** 267

(4) ($TM_{m,n}$ wave) (rectangular waveguide) (hollow rectangular metal cylinder). The transverse magnetic wave for which m is the number of half-period variations of the magnetic field along the longer transverse dimension, and n is the number of half-period variations of the magnetic field along the shorter transverse dimension. *See:* **guided wave; circular magnetic wave; waveguide.** 267

transverse magnetization (magnetic recording). Magnetization of the recording medium in a direction perpendicular to the line of travel and parallel to the greatest cross-sectional dimension. *See:* **phonograph pickup.** 176

transverse mode (laser-maser). A mode which is detected by measuring one or more maxima in transverse field intensity in the cross-section of a beam. 363

transverse-mode interference (signal-transmission system). *See:* **interference, differential-mode; signal.**

transverse offset loss. *See:* **lateral offset loss.**

transverse propagation constant (fiber optics). The propagation constant evaluated along a direction perpendicular to the waveguide axis. *Note:* The transverse propagation constant for a given mode can vary with the transverse coordinates. *See:* **propagation constant.** 433

transverse scattering (fiber optics). The method for measuring the index profile of an optical fiber or preform by illuminating the fiber or preform coherently and transversely to its axis, and examining the far-field irradiance pattern. A computer is required to interpret the pattern of the scattered light. *See:* **scattering.** 433

transverse wave. A wave in which the direction of displacement at each point of the medium is perpendicular to the direction of propagation. *Note:* In those cases where the displacement makes an acute angle with the direction of propagation, the wave is considered to have longitudinal and transverse components. 210

trap (1) (computing machines). An unprogrammed conditional jump to a known location, automatically activated by hardware, with the location from which the jump occurred recorded. 255, 77

(2) (burglar-alarm system). An automatic device applied to a door or window frame for the purpose of producing an alarm condition in the protective circuit whenever a door or window is opened. *See:* **protective signaling.** 328

trap circuit. A circuit used at locations where it is desirable to protect a section of track on which it is impracticable to maintain a track circuit. It usually consists of an arrangement of one or more stick circuits so connected that when a train enters the trap circuit the stick relay drops and cannot be picked up again until the train has passed through the other end of the trap circuit. 328

trapezium distortion (cathode-ray tube). A fault characterized by a variation of the sensitivity of the deflection parallel to one axis (vertical or horizontal) as a function of the deflection parallel to the other axis and having the effect of transforming an image that is a rectangle into one which is a trapezium. *See:* **cathode-ray tubes.**   244, 290

trapped flux (superconducting material). Magnetic flux that links with a closed superconducting loop. *See:* **superconductivity.**   191

trapped mode. *See:* **bound mode.**

trapped ray. *See:* **guided ray.**

travel (1) (of a relay) (power switchgear). The amount of movement in either direction (towards pickup or reset) of a responsive element. *Note:* Travel may be specified in linear, angular, or other measure.   103

(2) (rise) (elevators). Of an elevator, dumbwaiter, escalator, or of a private-residence inclined lift, the vertical distance between the bottom terminal landing and the top terminal landing. *See:* **elevators.**   127

traveled way (NESC). The portion of the roadway for the movement of vehicles, exclusive of shoulders and full-time parking lanes.   494

traveler (conductor stringing equipment). A sheave complete with suspension arm or frame used separately or in groups and suspended from structures to permit the stringing of conductors. These devices are sometimes bundled with a center drum, or sheave and another traveler , and used to string more than one conductor simultaneously. For protection of conductors that should not be nicked or scratched, the sheaves are often lined with nonconductive or semiconductive neoprene or with nonconductive urethane. Any one of these materials acts as a padding or cushion for the conductor as it passes over the sheave. Traveler grounds must be used with lined travelers in order to establish an electrical ground. *Syn:* **block; dolly; sheave; stringing block; stringing sheave; stringing traveler.**   431

traveler ground (conductor stringing equipment). A portable device designed to connect a moving conductor or wire rope, or both, to an electrical ground. Primarily used to provide safety for personnel during construction or reconstruction operations. This device is placed on the traveler (sheave, block, etcetera) at a strategic location where an electrical ground is required. *Syn:* **block ground; rolling ground; sheave ground.**   431

traveler rack (conductor stringing equipment). A device designed to protect, store and transport travelers. It is normally designed to permit efficient use of transporting vehicles, spotting by helicopters on the line, and stacking during storage to utilize space. The exact design of each rack is dependent upon the specific travelers to be stored. *Syn:* **dollie car.**   431

traveler sling (conductor stringing equipment). A sling of wire rope, sometimes utilized in place of insulators, to support the traveler during stringing operations. Normally it is used when insulators are not readily available or when adverse stringing conditions might impose severe downstrains and cause damage or complete failure of the insulators. *Syn:* **choker.**   431

traveling cable (elevators). A cable made up of electric conductors that provides electric connection between an elevator or dumbwaiter car and fixed outlet in the hoistway. *See:* **control.**   328

traveling ionospheric disturbance (radio wave propagation). A wavelike disturbance in the electron density distribution of the ionosphere.   146

traveling overvoltage (surge arresters). A surge propagated along a conductor.   244, 62

traveling plane wave (1) (radio wave propagation). A plane wave each of whose frequency components has an exponential variation of amplitude and a linear variation of phase with distance.   146

(2) (waveguide). A plane wave each of whose components have an exponential variation of amplitude and a linear variation of phase in the direction of propagation.   267

traveling wave. The resulting wave when the electric variation in a circuit takes the form of translation of energy along a conductor, such energy being always equally divided between current and potential forms. *See:* **direction of propagation; traveling plane; waveguide.**   64, 62

traveling-wave antenna. An antenna whose excitation has a quasi-uniform progressive phase, as the result of a single feeding wave transversing its length in one direction only.   111

traveling-wave magnetron. A traveling-wave tube in which the electrons move in crossed static electric and magnetic fields which are substantially normal to the direction of wave propagation.   125

traveling-wave magnetron oscillations. Oscillations sustained by the interaction between the space-charge cloud of a magnetron and a traveling electromagnetic field whose phase velocity is approximately the same as the mean velocity of the cloud. *See:* **magnetron, slow-wave circuit.**   190

traveling-wave parametric amplifier. A parametric amplifier that has a continuous or iterated structure incorporating nonlinear reactors and in which the signal, pump, and difference-frequency wave are propagated along the structure. *See:* **parametric device.**   191

traveling-wave tube. An electron tube in which a stream of electrons interacts continuously or repeatedly with a guided electromagnetic wave moving substantially in synchronism with it, and in such a way that there is a net transfer of energy from the stream to the wave. *See:* **transverse-beam traveling-wave tube; transverse-field traveling-wave tube.**   125

traveling-wave-tube interaction circuit. An extended electrode arrangement in a traveling-wave tube designed to propagate an electromagnetic wave in such a manner that the traveling electromagnetic fields are retarded and extended into the space occupied by the electron stream. *Note:* traveling-wave tubes are often designated by the type of interaction circuit used, as in helix traveling-wave tube.   125

travel trailer (National Electrical Code). A vehicular

unit mounted on wheels, designed to provide temporary living quarters for recreational, camping, or travel use, of such size or weight as not to require special highway movement permits when drawn by a motorized vehicle, and with a living area of less than 220 sq ft, excluding built-in equipment such as wardrobes, closets, cabinets, kitchen units or fixtures) and bath and toilet rooms. *See:* **recreational vehicle.** 256

tray (storage cell) (storage battery). A support or container for one or more storage cells. *See:* **battery (primary or secondary).** 328

TR box. *See:* **TR switch.**

treated fabric (treated mat) (rotating machinery). A fabric or mat in which the elements have been essentially coated but not filled with an impregnant such as a compound or varnish. *See:* **rotor (rotating machinery); stator.** 63

treated mat (rotating machinery). *See:* **treated fabric.**

treble boost. An adjustment of the amplitude-frequency response of a system or transducer to accentuate the higher audio frequencies. 239

tree (1). A set of connected branches including no meshes. *See:* **network analysis.** 322
(2) (software). An abstract hierarchical structure consisting of nodes connected by branches, in which: (a) each branch connects one node to a directly subsidiary node, and (b) there is a unique node called the root which is not subsidiary to any other node, and (c) every node besides the root is directly subsidiary to exactly one other node. 434

treeing (composite insulators). Irreversible internal degradation by the formation of conductive carbonized paths. 483

trees and nodules. Projections formed on a cathode during electrodeposition. Trees are branched whereas nodules are rounded. *See:* **electrodeposition.** 328

tree structure (FASTBUS acquisition and control). A set of connected segments with no loops (cross connections). 480

tree wire. A conductor with an abrasion-resistant outer covering, usually nonmetallic, and intended for use on overhead lines passing through trees. *See:* **armored cable; conductor; covered wire.** 64

triangular array. *See:* **triangular grid array.** 111

triangular grid array (antennas). A regular arrangement of array elements, in a plane, such that lines connecting corresponding points of adjacent elements form triangles, usually equilateral. 111

triboelectrification (electrification by friction). The mechanical separation of electric charges of opposite sign by processes such as (1) the separation (as by sliding) of dissimilar solid objects: (2) interaction at a solid-liquid interface: (3) breaking of a liquid-gas interface. 210

tributary office. A telephone central office that passes toll traffic to, and receives toll traffic from, a toll center. 328

tributary trunk (data transmission). A trunk circuit connecting a local exchange with a toll center or other toll office through which access to the long-distance network is achieved. 59

Trichel streamers (overhead-power-line corona and radio noise). Streamers occurring at a negative electrode with field strengths at and above the corona-inception gradient. A Trichel streamer appears as a small constantly moving purple fan. The current pulse is of small amplitude, short duration (in the range of a hundred nanoseconds), and high repetition rate (in the range of tens of kilohertz or more). 411

trickle charge (storage battery) (storage cell). A continuous charge at a low rate approximately equal to the internal losses and suitable to maintain the battery in a fully charged condition. *Note:* This term is also applied to very low rates of charge suitable not only for compensating for internal losses but to restore intermittent discharges of small amount delivered from time to time to the load circuit. *See:* **floating; charge.** 328

triclinic system (piezoelectricity). A triclinic crystal has neither symmetry axes nor symmetry planes. The lengths of the three axes are in general unequal; and the angles α, β, and γ between axes b and c, c and a, and a and b, respectively, are also unequal. The a axis has the direction of the intersection of the faces b and c (extend the faces to intersection if necessary), the b axis has the direction of the intersection of faces c and a, the c axis has the direction of the intersection of faces a and b. The X, Y, Z axes are associated as closely as possible with the a, b, c axes, respectively. The Z axis is parallel to c, Y is normal to the ac plane, and X is thus in the ac plane. The $+Z$ and $+X$ axes are chosen so that d_{33} and and d_{11} are positive. The $+Y$ axis is chosen so that it forms a right-handed system with $+Z$ and $+X$. *See:* **crystal systems.** 371

trigatron (1) (general). A triggered spark-gap switch on which control is obtained by a voltage applied to a trigger electrode. *Note:* This voltage distorts the field between the two main electrodes converting the sphere-to-sphere gap to a point to sphere gap. *See:* **electron device.** 244, 190
(2) (radar). An electronic switch in which conduction is initiated by the breakdown of an auxiliary gap. 13, 187

trigger (1) (verb). To start action in another circuit which then functions for a period of time under its own control. 328
(2) (noun). A pulse used to initiate some function, for example, a triggered sweep or delay ramp. *Note:* Trigger may loosely refer to a waveform of any shape used as a signal from which a trigger pulse is derived as in **trigger source, trigger input,** etcetera. *See:* **triggering signal; oscillograph.** 185
(3) (thyristor) (verb). The act of causing a thyristor to switch from the off-state to the on-state. *See:* **gate trigger current.** 66

trigger circuit. A circuit that has two conditions of stability, with means for passing from one to the other when certain conditions are satisfied, either spontaneously or through application of an external stimulus. 328

trigger countdown. A process that reduces the repeti-

tion rate of a triggering signal. *See:* **oscillograph.**
                                                        185

triggered sweep. A sweep that can be initiated only by a trigger signal, not free running. *See:* **oscillograph.**
                                                        185

trigger, external (oscilloscopes). A triggering signal introduced into the trigger circuit from an external source.   106

trigger gap (series capacitors). Enclosed electrodes that initiate the sparkover of the bypass gap.   474

triggering (pulse terms). A process in which a pulse initiates a predetermined event or response.   254

triggering level. The instantaneous level of a triggering signal at which a trigger is to be generated. Also, the name of the control that selects the level. *See:* **oscillograph.**   185

triggering, line (oscilloscopes). Triggering from the power-line frequency. *See:* **oscillograph.**   106

triggering signal. The signal from which a trigger is derived. *See:* **oscillograph.**   185

triggering slope. The positive-going (+slope) or negative-going (−slope) portion of a triggering signal from which a trigger is to be derived. Also, the control that selects the slope to be employed. *Note:* + and − slopes apply to the slope of the waveform only and not to the absolute polarity. *See:* **oscillograph.**   185

triggering stability. *See:* **stability.**

trigger level (transponder)(navigation aid terms). The minimum input to the receiver which is capable of causing the transmitter to emit a reply.   526

trigger lockout. *See:* **sweep lockout.**

trigger pickoff. A process or a circuit for extracting a triggering signal. *See:* **oscillograph.**   185

trigger-starting systems (fluorescent lamps). Applied to systems in which hot-cathode electric discharge lamps are started with cathodes heated through low-voltage heater windings built into the ballast. Sufficient voltage is applied across the lamp and between the lamp and fixture to initiate the discharge when the cathodes reach a temperature high enough for adequate emission. The ballast is so designed that the cathode-heating current is greatly reduced as soon as the arc is struck.   268

trigger tube (electron device). A cold-cathode gas-filled tube in which one or more electrodes initiate, but do not control, the anode current.   190

trigonal and hexagonal systems (piezoelectricity). These systems are distinguished by an axis of sixfold (or threefold) symmetry. This axis is always called the c axis. According to the Bravais-Miller axial system, which is most commonly used, there are three equivalent secondary axes, a_1, a_2, and a_3, lying 120 degrees apart in a plane normal to c. These axes are chosen as being either parallel to a twofold axis or perpendicular to a plane of symmetry, or if there are neither twofold axes perpendicular to c nor planes of symmetry parallel to c, the a axes are chosen so as to give the smallest unit cell. The Z axis is parallel c. The X axis coincides in direction and sense with any one of the a axes. The Y axis is perpendicular to Z and X, so oriented as to form a right-handed system. Positive-sense rules for

$+Z$, $+X$, $+Y$ are listed in the table below for the piezoelctric trigonal and hexagonal crystals. *Note:* "Positive" and "negative" may be checked using a carbon-zinc flashlight battery. The carbon anode connection will have the same effect on meter deflection as the + end of the crystal axis upon *release* of compression. *See:* **crystal systems.**   371

trilateration (radar). *See:* **multilateration.**

trimmer capacitor (trimming capacitor). A small adjustable capacitor associated with another capacitor and used for fine adjustment of the total capacitance of an element or part of a circuit.   329

trimmer signal. A signal that gives indication to the engineman concerning movements to be made from the classification tracks into the switch and retarder area.   328

triode. A three-electrode electron tube containing an anode, a cathode, and a control electrode.   125

triode region of an insulated-gate field-effect transistor (IGFET) (metal-nitride-oxide field-effect transistor). The same as non-saturation region.   386

trip (power switchgear). (1) (verb). (A) To release in order to initiate either an opening or a closing operation or other specified action. (B) To release in order to initiate an opening operation only. (C) To initiate and complete an opening operation. (2) (noun). (A) A release that initiates either an opening or a closing operation or other specified action. (B) A release that initiates an opening operation only. (C) A complete opening operation. (D) The action associated with the opening of a circuit breaker or other interrupting device. (3) (trip or tripping) (adjective). (A) Pertaining to a release that initiates either an opening or a closing operation or other specified action. (B) Pertaining to a release that initiates an opening operation only. (C) Pertaining to a complete opening operation. *Note:* All terms employing 'trip', 'tripping', or their derivatives are referred to the term that expresses the intent of the usage.   103

trip arm. *See:* **mechanical trip.**

trip coil. *See:* **release (trip)coil (of a mechanical switching device).**

trip current (faulted circuit indicators). The actual value of current in amperes rms (root-mean-square) that will cause the FCI (faulted circuit indicator) to indicate FAULT.   482

trip current rating (faulted circuit indicators). The published rms (root-mean-square) sinusoidal fault current in amperes which causes the FCI (faulted circuit indicator) to indicate FAULT.   482

trip delay setting. *See:* **release (trip) delay setting.**

trip device (opening release), impulse (low-voltage dc power circuit breakers used in enclosures). A trip device designed to operate only by the discharge of a capacitor into its release (trip) coil.   401

trip free. *See:* **release-free (trip-free) (as applied to a mechanical switching device).**

trip-free in any position. *See:* **release-free (trip-free) in any position.**

trip-free relay (release-free relay) (power switchgear). An auxiliary relay whose function is to open the closing circuit of an electrically operated switching device so that the opening operation can prevail over the closing operation. 103

trip lamp. A removable self-contained mine lamp, designed for marking the rear end of a train (trip) of mine cars. 328

triple-address. Same as three-address.

triple detection. *See:* **double superheterodyne reception.**

triplen (rotating machinery). An order of harmonic that is a multiple of three. *See:* **asynchronous machine.** 63

triplet (1)(mathematics of computing). A group of three adjacent digits operated upon as a unit. 564

(2)(navigation systems)(navigation aid terms). Three radio stations, operated as a group, for the determination of positions. 526

triplex cable. A cable composed of three insulated single-conductor cables twisted together. *Note:* The assembled conductors may or may not have a common covering of binding or protecting material. 64

trip OFF control signal (magnetic amplifier). The final value of signal measured when the amplifier has changed from the ON to the OFF state as the signal is varied so slowly that an incremental increase in the speed with which it is varied does not affect the measurement of the trip OFF control signal. That is, the change in trip OFF control signal is below the sensitivity of the measuring instrument. 171

trip ON control signal (magnetic amplifier). The final value of signal measured when the amplifier has changed from the OFF to the ON state as the signal is varied so slowly than an incremental increase in the speed with which it is varied does not affect the measurement of the trip ON control signal. That is, the change in trip ON control signal is below the sensitivity of the measuring instrument. 171

tripping. *See:* **automatic opening.**

tripping delay. *See:* **release (tripping) delay (of a mechanical switching device).**

tripping mechanism. *See:* **release (tripping mechanism) (of a mechanical switching device).**

tripping or trip-free relay (94) (power system device function numbers). A relay that functions to trip a circuit breaker, contactor, or equipment, or to permit immediate tripping by other devices; or to prevent immediate reclosure of a circuit interrupter if it should open automatically even though its closing circuit is maintained closed. 402

trip-point repeatability (magnetic amplifier). The change in trip point (either trip OFF or trip ON, as specified) control signal due to uncontrollable causes over a specified period of time when all controllable quantities are held constant. 171

trip point repeatability coefficient (magnetic amplifier). The ratio of (1) the maximum change in trip point control signal due to uncontrollable causes to (2) the specified time period during which all controllable quantities have been held constant. *Note:* The units of this coefficient are the control signal units per the time

period over which the coefficient was determined. 171

trip setting. *See:* **release (trip) setting.**

tristimulus values (of a light) (television). The amounts of the three reference or matching stimuli required to give a match with the light considered, in a given trichromatic system. *Notes:* (1) In the standard colorimetric system, CIE (1931), the symbols, X, Y, Z are recommended for the tristimulus values. (2) These values may be obtained by multiplying the spectral concentration of the radiation at each wavelength by the distribution coefficients $\bar{x}_\lambda$, $\bar{y}_\lambda$, $\bar{z}_\lambda$ and integrating these products over the whole spectrum. 18

tristimulus values of a light, X,Y,Z (illuminating engineering). The amounts of each of three specific primaries required to match the color of the light. *See:* **color matching functions.** 167

troffer (illuminating engineering). A long recessed lighting unit usually installed with the opening flush with the ceiling. The term is derived from 'trough' and 'coffer'. 167

troland (illuminating engineering). A unit of retinal illuminance which is based upon the fact that retinal illuminance is proportional to the product of the luminance of the distal stimulus and the area of entrance pupil. One troland is the retinal illuminance produced when the luminance of the distal stimulus is one candela per square meter and the area of the pupil is one square millimeter. *Note:* The troland makes no allowance for interocular attenuation or for the Stiles-Crawford effect. *See:* **Stiles-Crawford effect.** 167

trolley. A current collector, the function of which is to make contact with a contact wire. *See:* **contact conductor.** 1

trolley bus. *See:* **trolley coach.**

trolley car. An electric motor car that collects propulsion power from a trolley system. *See:* **electric motor car.** 328

trolley coach (trolley bus) (trackless trolley coach). An electric bus that collects propulsion power from a trolley system. *See:* **electric bus.** 328

trolley locomotive. An electric locomotive that collects propulsion power from a trolley system. *See:* **electric locomotive.** 328

trombone line (transmission lines and waveguides). A U-shaped length of waveguide or transmission line of adjustable length. *See:* **waveguide.** 185

troposphere (1) (data transmission). That part of the earth's atmosphere in which temperature generally decreases with altitude, clouds form, and convection is active. *Note:* Experiments indicate that the troposphere occupies the space above the earth's surface up to a height ranging from about 6 km (kilometers) at the poles to about 18 km at the equator. 59

(2) (radio wave propagation). That part of a planetary atmosphere in which temperature generally decreases with altitude, clouds form, and convection is active. *Note:* Experiments indicate that the earth's troposphere occupies the space above the earth's surface to a height of about 10 km (kilometers). 146

tropospheric radio duct. *See:* **atmospheric duct.**

tropospheric scatter propagation (radio wave propagation). Propagation of radio waves through the atmosphere involving scattering from inhomogeneities in the refractive index of the troposphere.   146

tropospheric wave (data transmission). A radio wave that is propagated by reflection from a place of abrupt change in the dielectric constant or its gradient in the troposphere. *Note:* In some cases the ground wave may be so altered that new components appear to arise from reflections in regions of rapidly changing dielectric constant. When these components are distinguishable from the other components, they are called tropospheric waves.   59

trouble recorder (telephone switching systems). A device or system associated with one or more switching systems for automatically recording data on calls encountering trouble.   55

troubleshoot (supervisory control, data acquisition, and automatic control). Action taken by operating or maintenance personnel, or both, to isolate a malfunctioned component of a system. Actions may be supported by printed procedures, diagnostic circuits, test points, and diagnostic routines.*See:* **debug; fault isolation.**   570

troughing. An open channel of earthenware, wood, or other material in which a cable or cables may be laid and protected by a cover.   64

TR switch. *See:* **transmit-receive switch.**

truck camper (National Electrical Code). A portable unit constructed to provide temporary living quarters for recreational, travel, or camping use, consisting of a roof, floor, and sides, designed to be loaded onto and unloaded from the bed of a pick-up truck. *See:* **recreational vehicle.**   256

truck generator suspension. A design of support for an axle generator in which the generator is supported by the vehicle truck.   328

true air speed (navigation aid terms). The actual speed of an aircraft relative to the surrounding air.   526

true air-speed indicator (navigation aid terms). An instrument for measuring indicated true air speed.   526

true bearing (navigation aid terms). Bearing relative to true north.   526

true complement. (1) *See:* **radix complement.**

(2) A number representation that can be derived from another by subtracting each digit from one less than the base and then adding one to the least significant digit and executing all carries required. Tens complements and twos complements are true complements.   235

true course (navigation aid terms). Course relative to true north.   526

true density ($g/cm3$) (fly ash resistivity). The weight of the particles divided by the solid volume of the particles.   427

true heading (navigation aid terms). Heading relative to true north.   526

true-motion display (radar). A display in a vehicle-mounted radar that shows the motions of the radar and of targets tracked by that radar, relative to a fixed background; accomplished by inserting compensation for the motion of the vehicle carrying the radar.   13

true neutral point (at terminals of entry). Any point in the boundary surface that has the same voltage as the point of junction of a group of equal nonreactive resistors placed in the boundary surface of the region and connected at their free ends to the appropriate terminals of entry of the phase conductors of the circuit, provided that the resistance of the resistors is so great that the voltages are not appreciably altered by the introduction of the resistors. *Notes:* (1) The number of resistances required is two for direct-current or single-phase alternating-current circuits, four for two-phase four-wire or five-wire circuits, and is equal to the number of phases when the number of phases is three or more. Under normal symmetrical conditions the number of resistors may be reduced to three for six- or twelve-phase systems when the terminals are properly selected, but the true neutral point may not be obtained by this process under all abnormal conditions. The concept of a true neutral point is not considered applicable to a two-phase, three-wire circuit. (2) Under abnormal conditions the voltage of the true neutral point may not be the same as that of the neutral conductor. *See:* **network analysis.**   210

true north (navigation aid terms). The direction of the north geographical pole.   526

true ratio (1) (instrument transformer). The ratio of the root-mean-square (rms) primary value to the rms secondary value under specified conditions.   394

(2) (power and distribution transformer). The ratio of the root-mean-square (rms) primary value of the rms secondary value under specified conditions.   53

truncate. To terminate a computational process in accordance with some rule, for example, to end the evaluation of a power series at a specified term.   255, 77

trunk (1) (analog computers). A connecting line between one analog computer and another, or between an analog compute and an external point, allowing the input (or output) of an analog component to communicate directly with the output (or input) or another component which is located outside of the analog computer.   9

(2) (data transmission). A telephone line or channel between two central offices or switching devices, which is used in providing telephone connections between subscribers.   59

(3) (telephone switching systems). A channel provided as a common traffic artery between switching entities.   55

trunk cable (1)(medium attachment units and repeater units). The trunk coaxial cable system.   543

(2)(token ring access method). The transmission cable that interconnects two trunk coupling units.   472

trunk circuit (telephone switching systems). An interface circuit between a trunk and a switching system.   55

trunk circuit, combined line and recording (CLR). (1) Name given to a class of trunk circuits that provide access to operator positions generally referred to by abbreviation only. (2) Recording-completing trunk circuit for operator recording and completing of toll calls originated by subscribers of central offices.
193

trunk concentrator (telephone switching systems). A concentrator in which all inlets and outlets are trunks.
55

trunk coupling unit (TCU)(token ring access method). A physical device that enables a station to connect to a trunk cable. The trunk coupling unit contains the means for inserting the station into the ring or, conversely, bypassing the station.
472

trunk feeder. A feeder connecting two generating stations or a generating station and an important substation. *See:* **center of distribution.**
64

trunk group (telephone switching systems). A number of trunks that can be used interchangeably between two switching entities.
55

trunk hunting. The operation of a selector or other similar device, to establish connection with an idle circuit of a chosen group. This is usually accomplished by successively testing terminals associated with this group until a terminal is found that has an electrical condition indicating it to be idle.
328

trunk-line conduit. A duct-bank provided for main or trunk-line cables.
64

trunk loss. That part of the repetition equivalent assignable to the trunk used in the telephone connection. *See:* **transmission loss.**
328

trunk multifrequency pulsing (telephone switching systems). A means of pulsing embodying a simultaneous combination of two out of six frequencies to represent each digit or character.
55

trunk transmission line. A transmission line acting as a source of main supply to a number of other transmission circuits. *See:* **transmission line.**
64

trussed blade (of a switching device) (power switchgear). A blade that is reinforced by truss construction to provide stiffness.
103

truth table. A table that describes a logic function by listing all possible combinations of input values and indicating, for each combination, the true output values.
255, 77, 54

TSI. *See:* **threshold signal-to-interference ratio.**

T step (linear waveform distortion). A $\sin^2$ step with a 10 percent to 90 percent rise (fall) time of nominally 125 ns (nanoseconds). The amplitude of the envelope of the frequency spectrum is effectively zero at and beyond 8 MHz. *Note:* In practice the T step is part of square wave (or line bar as described below), so that there is a T step rise and fall.
42

tube (1)(protection and coordination of industrial and commercial power systems). The cylindrical enclosure of a fuse. Such a tube may be made of laminated paper, special fiber, melamine impregnated glass cloth, bakelite, ceramic, glass, plastic, or other materials.
504

(2) (of a fuse). *See:* **fuse tube.**
103

(3) (interior wiring). A hollow cylindrical piece of insulating material having a head or shoulder at one end, through which an electric conductor is threaded where passing through a wall, floor, ceiling, joist, stud, etcetera. *See:* **raceways.**
328

(4) (primary cell). A cylindrical covering of insulating material, without closure at the bottom. *See:* **electrolytic cell.**
328

tube count (radiation-counter tubes). A terminated discharge produced by an ionizing event in a radiation-counter tube.
96,125

tube current averaging time. The time interval over which the current is averaged in defining the operating capability of the tube. *See:* **rectification.**
291

tube, display. *See:* **display tube.**

tube, electron. *See:* **electron tube.**

tube fault current. The current that flows through a tube under fault conditions, such as arc-back or short circuit. *See:* **rectification.**
291

tube, fuse. *See:* **fuse tube.**

tube heating time (mercury-vapor tube). The time required for the coolest portion of the tube to attain operating temperature. *See:* **preheating time; electronic controller; gas tube.**
125

tubelet (soldered connections). *See:* **eyelet.**

tuberculation (corrosion). The formation of localized corrosion products scattered over the surface in the form of knoblike mounds.
205

tube scintillation pulses (photomultipliers). Dark pulses caused by scintillations within the photomultiplier structure. Example: cosmic-ray-induced events.
117

tube-type plate (storage cell). A plate of an alkaline storage battery consisting of an assembly of metal tubes filled with active material. *See:* **battery (primary or secondary).**
328

tube, vacuum. *See:* **vacuum tube.**

tube voltage drop (electron tube). The anode voltage during the conducting period. *See:* **electrode voltage (electron tube); electronic controller.**
125

tubing (rotating machinery). A tubular flexible insulation, extruded or made of layers of film plastic, into which a conductor is inserted to provide additional insulation. Tubing is frequently used to insulate connections and crossovers. *See:* **asynchronous machine.**
63

Tudor plate (storage cell). A lead storage battery plate obtained by molding and having a large area. *See:* **battery (primary and secondary).**
328

tumbling (gyro). The loss of reference in a two-degree-of-freedom gyro due to gimbal lock or contact between a gimbal and a mechanical stop. This is not to be confused with tumble testing which is a method of evaluating gyro performance.
46

tuned filter (harmonic control and reactive compensation of static power converters). A filter consisting generally of combinations of capacitors, inductors, and resistors which have been selected in such a way as to present a relative minimum (maximum) impedance to one or more specific frequencies. For a shunt (series) filter the impedance is a minimum (maximum). Tuned filters generally have a relatively high $Q(X/R)$.   533

tuned-grid oscillator. An oscillator whose frequency is determined by a parallel-resonance circuit in the grid circuit coupled to the plate to provide the required feedback. *See:* **oscillatory circuit.**   111

tuned-grid-tuned plate oscillator. An electron tube circuit in which both grid and plate circuits are tuned to resonance where the feedback voltage normally is developed through the inter-electrode capacity of the tube. *See:* **radio frequency generator.**   14

tuned-plate oscillator. An oscillator whose frequency is determined by a parallel-resonance circuit in the plate circuit coupled to the grid to provide the required feedback. *See:* **oscillatory circuit.**   111

tuned rotor gyro. *See:* **dynamically tuned gyro.**

tuned speed (dynamically tuned gyro) (inertial sensor). The rotor spin velocity at which the dynamically induced spring rate is equal in magnitude and of opposite sign to the physical spring rate of the rotor suspension.   46

tuned transformer. A transformer, the associated circuit elements of which are adjusted as a whole to be resonant at the frequency of the alternating current supplied to the primary, thereby causing the secondary voltage to build up to higher values than would otherwise be obtained. *See:* **power pack.**   329

tuner (1) (radio receiver). In the broad sense, a device for tuning. Specifically, in radio receiver practice, it is (A) a packaged unit capable of producing only the first portion of the functions of a receiver and delivering either radio-frequency, intermediate-frequency, or demodulated information to some other equipment, or (B) that portion of a receiver that contains the circuits that are tuned to resonance at the received-signal frequency. *See:* **radio receiver.**   328

(2) (transmission line) (waveguide). An ideally lossless, fixed or adjustable, network capable of transforming a given impedance into a different impedance. *See:* **transmission loss; waveguide.**   179, 185

tuner, waveguide (waveguide components). An adjustable waveguide transformer.   166

tungsten-halogen lamp (illuminating engineering). A gas filled tungsten filament incandescent lamp containing a certain proportion of halogens. *Note:* The tungsten-iodine lamp (UK) and quartz-iodine lamp (USA) belong to this category.   167

tuning (data transmission). The adjustment in relation to frequency of a circuit or system to secure optimum performance; commonly the adjustment of a circuit or circuits to resonance.   59

tuning creep (oscillator). The change of an essential characteristic as a consequence of repeated cycling of the tuning element.   174

tuning, electronic. The process of changing the operating frequency of a system by changing the characteristics of a coupled electron stream. Characteristics involved are, for example: velocity, density, or geometry. *See:* **oscillatory circuit.**   125

tuning hysteresis (microwave (oscillator). The difference in a characteristic when a tuner position, or input

to the tuning element, is approached from opposite directions.   174

tuning indicator (electron device). An electron-beam tube in which the signal supplied to the control electrode varies the area of luminescence of the screen.   190

tuning probe (waveguides). An essentially lossless probe of adjustable penetration extending through the wall of the waveguide or cavity resonator. *See:* **waveguide.**   330

tuning range (1) (switching tubes). The frequency range over which the resonance frequency of the tube may be adjusted by the mechanical means provided on the tube or associated cavity. *See:* **gas tube.**   125

(2) (oscillator). The frequency range of continuous tuning within which the essential characteristics fall within prescribed limits.   174

tuning range, electronic. The frequency range of continuous tuning between two operating points of specified minimum power output for an electronically tuned oscillator. *Note:* The reference points are frequently the half-power points, but should always be specified. *See:* **oscillatory circuit.**   125

tuning rate, thermal. The initial time rate of change in frequency that occurs when the input power to the tuner is instantaneously changed by a specified amount. *Note:* This rate is a function of the power input to the tuner as well as the sign and magnitude of the power change. *See:* **oscillatory circuit.**
                    125

tuning screw (waveguide technique). An impedance-adjusting element in the form of a rod whose depth of penetration through the wall into a waveguide or cavity is adjustable by rotating the screw. *See:* **waveguide.**
                    179

tuning sensitivity (oscillator). The rate of change of frequency with the control parameter (for example, the position of a mechanical tuner, electric tuning voltage, etcetera) at a given operating point. *See:* **tunable microwave oscillators.**   174

tuning sensitivity, electronic. At a given operating point, the rate of change of oscillator frequency with the change of the controlling electron stream. For example, this change may be expressed in terms of an electrode voltage or current. *See:* **pushing figure (oscillator); oscillatory circuit.**   125

tuning sensitivity, thermal. The rate of change of resonator equilibrium frequency with respect to applied thermal tuner power.   125

tuning susceptance (anti-transmit-receive tube) (ATR tube). The normalized susceptance of the tube in its mount due to the deviation of its resonance frequency from the desired resonance frequency. *Note:* Normalization is with respect to the characteristic admittance of the transmission line at its junction with the tube mount.   125

tuning, thermal. The process of changing the operating frequency of a system by using a controlled thermal expansion to alter the geometry of the system. *See:* **oscillatory circuit.**   125

tuning time constant, thermal. The time required for

the frequency to change by a fraction (1 mins 1.e) of the change in equilibrium frequency after an incremental change of the applied thermal tuner power. *Notes:* (1) If the behavior is not exponential, the initial conditions must be stated. (2) Here e is the base of natural logarithms. *See:* **oscillatory circuit.**   125

tuning time thermal (1) (cooling). The time required to tune through a specified frequency range when the tuner power is instantaneously changed from the specified maximum to zero. *Note:* The initial condition must be one of equilibrium. *See:* **electron emission.**   125

(2) (heating). The time required to tune through a specified frequency range when the tuner power is instantaneously changed from zero to the specified maximum. *Note:* The initial condition must be one of equilibrium. *See:* **electron emission.**   125

tunnelling mode. *See:* **leaky mode.**

tunnelling ray. *See:* **leaky ray.**

turbine-control servomotor (hydraulic turbines). The actuating element which moves the turbine-control mechanism in response to the action of the governor control actuator. Turbine-control servomotors are designated as: (1) gate servomotor, (2) blade servomotor, (3) deflector servomotor, (4) needle servomotor.   8

turbine-driven generator. An electric generator driven by a turbine.   328

turbine end (rotating machinery). The driven or power-input end of a turbine-driven generator.   63

turbine-generator. *See:* **cylindrical-rotor generator.**

turbine-generator unit. An electric generator with its driving turbine.   328

turbine-nozzle control system (gas turbines). A means by which the turbine diaphragm nozzles are adjusted to vary the nozzle angle or area, thus varying the rate of energy input to the turbine(s).   58

turbine, reversible. A hydraulic turbine, normally installed in a pumped storage station, which can be used alternately as a pump or as a prime mover.   112

turbine-type (rotating machinery). Applied to alternating-current machines designed for high-speed operation and having an excitation winding embedded in slots in a cylindrical steel rotor made from forgings or thick disks. *See:* **asynchronous machine.**   63

turbo-machine (turbo-generator) (rotating machinery). A machine of special design intended for high-speed operation. Turbo-generators usually are directly connected to gas or steam turbines.   63

Turing machine. A mathematical model of a device that changes its internal state and reads from, writes on, and moves a potentially infinite tape, all in accordance with its present state, thereby constituting a model for computerlike behavior. *See:* **universal Turing machine.**   255, 77

turn (rotating machinery). The basic coil element which forms a single conducting loop comprising one insulated conductor. *Note:* The conductor may consist of a number of strands or laminations. Each strand or lamination is in the form of a wire, rod, strip or bar, depending on its cross-section, and may be either un-

insulated or insulated for the sole purpose of reducing eddy currents.   63

turnbuckle. A threaded device inserted in a tension member to provide minor adjustment of tension or sag. *See:* **tower.**   64

turn error (gyro). An error in gyro output due to cross-coupling and acceleration encountered during vehicle turns.   46

turning gear (rotating machinery). A separate drive to rotate a machine at very low speed for the purpose of thermal equilization at a time when it would otherwise be at rest. *See:* **rotor (rotating machinery).**   63

turn insulation (rotating machinery). Insulation applied to provide electrical separation between turns of a coil. *Note:* In the usual case, the insulation encircles each turn. However, in the case of edgewise-wound field coils for salient pole synchronous machines, the outer edges may be left bare to facilitate cooling.   63

turn-off branch (converter circuit elements)(self-commutated converters). An auxiliary branch intended to take over the current transiently from a previously conducting principal branch.   584

turn-off thyristor (gate controlled switch). A thyristor that can be switched from the ON state to the OFF state and vice versa by applying control signals of appropriate polarities to the gate terminal, with the ratio of triggering power appreciably less than one.   445

turn-off time (circuit properties)(applies to converters that use circuit-commutated thyristor devices)(self-commutated converters). The time interval between that instant when the principal current of a thyristor device has been reduced to zero and that instant when the same thyristor device is again subjected to voltage that could cause conduction. *Note:* for proper operation, the turn-off time, t_o, made available by the action of the circuit must exceed the turn-off time, t_q, required by the thyristor device for recovery of its voltage-blocking ability. Both the available and required turn-off times depend on the operating conditions of the converter

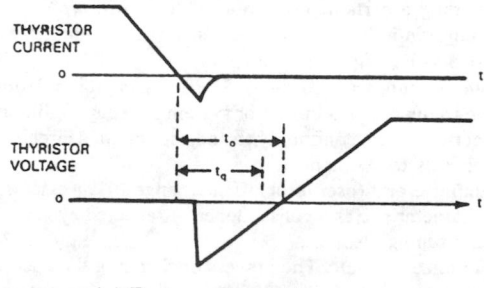

(a) Proper Operation $t_o > t_q$

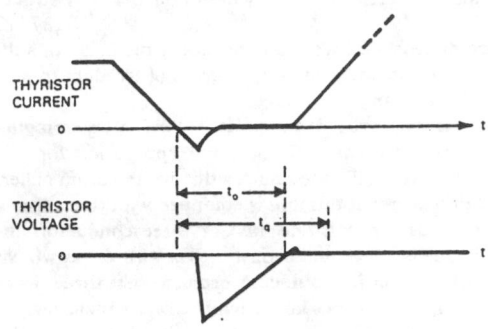

(b) Improper Operation $t_o < t_q$

Fig 3
Turn-Off Time for Circuit-Commutated
Thyristors

—
584

turn-off time (thyristor). *See:* **gate controlled turn-off time (thyristor); circuit commutated turn-off time (thyristor).**

turn-on time (gyro, accelerometer) (inertial sensor). The time from the initial application of power until a sensor produces a specified useful output, though not necessarily at the accuracy of full specification performance.   46

turnover frequency. *See:* **transition frequency.**

turn ratio (1) (transformer). The ratio of the number of turns in a higher voltage winding to that in a lower voltage winding. *Note:* In the case of a constant-voltage transformer having taps for changing its voltage ratio, the nominal turn ratio is based on the number of turns corresponding to the normal rated voltage of the respective windings, to which operating and performance characteristics are referred.   53
(2) (constant-current transformer). The ratio of the number of turns in the primary winding to that in the secondary winding. *Note:* In case of a constant-current transformer having taps for changing its voltage ratio, the turn ratio is based on the number of turns corresponding to the normal rated voltage of the respective windings, to which operation and performance characteristics are referred.   303
(3) (potential transformer) (voltage transformer). The ratio of the primary winding turns to the secondary winding turns.   305, 203, 53
(4) (rectifier transformer). The ratio of the number of turns in the alternating-current winding to that in the direct-current winding. *Note:* The turn ratio is based on the number of turns corresponding to the normal rated voltage of the respective windings to which operating and performance characteristics are referred. *See:* **rectifier transformer.**   258
(5) (current transformer). The ratio of the secondary winding turns to the primary winding turns.   53
turn ratio of a current transformer (power and distribution transformer) (instrument transformer). The ratio of the secondary winding turns to the primary winding turns.   53, 394

turn ratio of a voltage transformer (power and distribution transformer) (instrument transformer). The ratio of the primary winding turns to the secondary winding turns.   53, 394
turn separator (rotating machinery). An insulation strip between turns; a form of turn insulation. *See:* **rotor (rotating machinery); stator.**   63
turns factor (magnetic core testing). Under stated conditions the number of turns that a coil of specified shape and dimensions placed on the core in a given position should have to obtain a given unit of self inductance. When measured with a measuring coil of the specified shape and dimensions and placed in the same position, it is defined as:

$$\alpha = \frac{N}{\sqrt{L}}$$

α = Turns factor
N = Number of turns of the measuring coil
L = Self inductance in henrys of the measuring coil placed on the core.

turn-signal operating unit (illuminating engineering). That part of a signal system by which the operator of a vehicle indicates the direction a turn will be made, usually by a flashing light.   167
turns per phase, effective (rotating machinery). The product of the number of series turns of each coil by the number of coils connected in series per phase and the winding factor.   63
turns ratio (electronics power transformer). The number of turns of a given secondary divided by the number of primary turns. Thus a ratio less than one (1) is a step-down transformation, a ratio greater than one (1) is a step-up transformation, and a ratio equal to one (1) is unity ratio.   95
turnstile antenna. An antenna composed of two dipole antennas, perpendicular to each other, with their axes intersecting at their midpoints. Usually, the currents on the two dipole antennas are equal and in phase quadrature.   111
turntable rumble (audio and electroacoustics). Low-frequency vibration mechanically transmitted to the recording or reproducing turntable and superimposed on the reproduction. *See:* **phonograph pickup; rumble.**   339
turn-to-turn test (interturn test) (rotating machinery). A test for applying or more often introducing between adjacent turns of an insulated component, a voltage of predetermined amplitude, for the purpose of checking the integrity of the interturn insulation.   63
turn-to-turn voltage (rotating machinery). The voltage existing between adjacent turns of a coil. *See:* **rotor (rotating machinery); stator.**   63
TV. *See:* **television.**
TV broadcast band (overhead-power-line corona and radio noise). Any one of the frequency bands assigned for the transmission of audio and video signals for television reception by the general public. *Note:* In the United States and Canada the frequency ranges are 54

to 72 MHz, 76 to 88 MHz, 174 to 216 MHz, and 400 to 890 MHz.   411

TV waveform signal (linear waveform distortion). An electrical signal whose amplitude varies with time in a generally nonsinusoidal manner and whose shape (that is, duration and amplitude) carries the TV signal information.   42

21-type repeater (data transmission). A two-wire telephone repeater in which there is one amplifier serving to amplify the telephone current in both directions, the circuit being arranged so that the input and output terminals of the amplifier are in one pair of conjugate branches, while the lines in the two directions are in another pair of conjugate branches.   59

22-type repeater (data transmission). A two-wire telephone repeater in which there are two amplifiers, one serving to amplify the telephone current being transmitted in one direction and the other serving to amplify the telephone currents in the other direction.
59

twin cable. A cable composed of two insulated conductors laid parallel and either attached to each other by the insulation or bound together with a common covering.   64

twin-T network. *See:* **parallel-T network.**

twin wire. A cable composed of two small insulated conductors laid parallel, having a common covering. *See:* **conductor.**   345

twist (measuring the performance of tone address signaling systems). In a two-tone signal, during the signal present condition, the level of the higher-frequency tone relative to the level of the lower-frequency tone, expressed in decibels. Twist is negative if the higher-frequency tone level is below the lower-frequency tone level.   508

twisted-lead transposition (rotating machinery). A form of transposition used on a distributed armature winding wherein the strands comprising each turn are kept insulated from each other throughout all the coils in a phase belt, and the last half turn of each coil is given a 180-degree twist prior to connecting it to the first half turn of the next coil in the series. *See:* **rotor (rotating machinery); stator.**   63

twisted pair. A cable composed of two small insulated conductors, twisted together without a common covering. *Note:* The two conductors of a twisted pair are usually substantially insulated, so that the combination is a special case of a cord. *See:* **conductors.**
64

twist, waveguide (waveguide components). A waveguide section in which there is progressive rotation of the cross section about the longitudinal axis. *See:* **step twist.**   166

two-address. Pertaining to an instruction code in which each instruction has two address parts. Some two-address instructions use the addresses to specify the location of one operand and the destination of the result, but more often they are one-plus-one-address instructions.   235

two conductor bundle. *See:* **bundle.**

two-degree-freedom gyro. A gyro in which the rotor axis is free to move in any direction. *See:* **navigation.**
187, 13

two-dimensional scanning. Scanning the beam of a directive antenna using two degrees of freedom to provide solid angle coverage.   111

two drum, three drum puller (conductor stringing equipment). The definition and application for this unit is essentially the same as that for the drum puller. It differs in that this unit is equipped with three drums and thus can pull one, two or three conductors individually or simultaneously. *Syn:* **two drum winch; double drum hoist; double drum winch; three drum winch; triple drum hoist; triple drum winch; tugger.** *See:* **drum puller.**   431

two-element relay. An alternating-current relay that is controlled by current from two circuits through two cooperating sets of coils.   328

two-family dwelling (National Electrical Code). A building consisting solely of two dwelling units.
256

two-fluid cell. A cell having different electrolytes at the two electrodes. *See:* **electrochemistry.**   328

two-frequency simplex operation (radio communication). The operation of a two-way radio-communication circuit utilizing two radio-frequency channels, one for each direction of transmission, in such manner that intelligence can be transmitted in only one direction at a time. *See:* **channel spacing.**   181

two-layer winding (two-coil-side-per-slot winding). A winding in which there are two coil sides in the depth of a slot. *See:* **rotor (rotating machinery); stator.**
63

two-N (2N) angular sensitivity (dynamically tuned gyro) (inertial sensor). The coefficient that relates drift rate to angular vibration at twice spin frequency applied about an axis perpendicular to the spin axis. It has the dimensions of angular displacement per unit time, per unit angle of the input vibration.   46

two-N (2N) translational sensitivity (dynamically tuned gyro) (inertial sensor). The coefficient that relates drift rate to linear vibrations at twice spin frequency applied perpendicular to the spin axis. It has the dimensions of angular displacement per unit time, per unit of acceleration of the input vibration.   46

two-out-of-five code. A code in which each decimal digit is represented by five binary digits of which two are one kind (for example, ones) and three are the other kind (for example, zeros).   255, 77

two-phase circuit (power and distribution transformer). A polyphase circuit of three, four, or five distinct conductors intended to be so energized that in the steady state the alternating voltages between two selected pairs of terminals of entry, other than the neutral terminal when one exists, have the same periods, are equal in amplitude, and have a phase difference of 90 degrees. When the circuit consists of five conductors, but not otherwise, one of them is a neutral conductor. *Note:* A two-phase circuit as defined here does not conform to the general pattern of polyphase circuits. Actually a two-phase, four-wire, or five-wire circuit could more properly be called a four-phase circuit, but the term two-phase is in common usage. A two-phase three-wire circuit is essentially a special case, as it does not conform to the general pattern of other polyphase circuits.   53

two-phase five-wire system. A system of alternating-current supply comprising five conductors, four of which are connected as in a four-wire two-phase system, the fifth being connected to the neutral points of each phase. *Note:* The neutral is usually grounded. Although this type of system is usually known as the two-phase five-wire system, it is strictly a four-phase five-wire system. *See:* **alternating-current distribution; network analysis.**   64

two-phase four-wire system. A system of alternating-current supply comprising two pairs of conductors between one pair of which is maintained an alternating difference of potential displaced in phase by one-quarter of a period from an alternating difference of potential of the same frequency maintained between the other pair. *See:* **alternating-current distribution; network analysis.**   64

two-phase three-wire system. A system of alternating-current supply comprising three conductors between one of which (known as the common return) and each of the other two are maintained alternating differences of potential displaced in phase by one quarter of a period with relation to each other. *See:* **alternating-current distribution; network analysis.**   64

two-plus-one address (electronic computation). Pertaining to an instruction that contains two operand addresses and a control address. *See:* **control address; instruction; operand; three-address code.**   235

two-quadrant DAM (hybrid computer linkage components). A digital-to-analog multiplier (DAM) that multiplies with a single sign only for the digital value.   10

two-quadrant multiplier (analog computers). A multiplier in which operation is restricted to a single sign of one input variable only.   9

two-range Decca. *See:* **lambda.**

two-rate watthour meter. A meter having two sets of register dials, with a changeover arrangement such that integration of the quantity will be registered on one set of dials during a specified time each day, and on the other set of dials for the remaining time.   212

twos complement (mathematics of computing). The radix complement of a binary numeral, which may be formed by subtracting each digit from 1, then adding 1 to the least significant digit and executing any required carries. For example, the twos complement of 1101 is 0011.   564

two-signal selectivity (frequency-modulated mobile communications receivers). The characteristic that determines the extent to which the receiver is capable of differentiating between the desired signal and disturbances of signals at other frequencies. It is expressed as the amplitude ratio of the modulated desired signal and the unmodulated disturbing signal when the reference sensitivity sinad of the desired signal is degraded 6 decibels.   123

two-source frequency keying. That form of keying in which the modulating wave abruptly shifts the output frequency between predetermined values, where the values of output frequency are derived from independent sources. *Note:* Therefore, the output wave is not coherent and, in general, will have a phase discontinuity. *See:* **telegraphy.**   111

two-speed alternating-current control. A control for two-speed driving-machine induction motor that is arranged to run near two different synchronous speeds by connecting the motor windings so as to obtain different numbers of poles. *See:* **control (elevators).**   328

two-state indication. *See:* **supervisory control functions.**

2T pulse (waveform test signals) (TV). A $\sin^2$ pulse with a half-amplitude duration (HAD) of 250 ns. The amplitude of the envelope of the frequency spectrum at 2 MHz is 0.5 of the amplitude at zero frequency and effectively zero at and beyond 4 MHz. *Note:* The 2T pulse is mentioned here for the sake of completeness. The short-time domain may be tested by the 2T pulse in conjunction with the T pulse and a reference signal. This method is not used in this standard since the T step alone tests the short-time domain in a simpler and more direct manner.   42

two-terminal capacitor. Two conductors separated by a dielectric. The construction is usually such that one conductor essentially surrounds the other and therefore the effect of the presence of other conductors, except in the immediate vicinity of the terminals, is eliminated. (Specialized usage).   210

two-terminal pair network (quadripole) (four-pole) (circuits and systems). A network with four accessible terminals grouped in pairs, for example, input pair, output pair.   67

two-tone keying. That form of keying in which the modulating wave causes the carrier to be modulated with a single tone for the marking condition and modulated with a different single tone for the spacing condition. *See:* **telegraphy.**   111

two-value capacitor motor. A capacitor motor using different values of effective capacitance for the starting and running conditions. *See:* **asynchronous machine.**   63

two-way automatic maintaining leveling device. A device that corrects the car level on both underrun and overrun, and maintains the level during loading and unloading. *See:* **elevator-car leveling device.**   328

two-way automatic nonmaintaining leveling device. A device that corrects the car level on both underrun and overrun, but will not maintain the level during loading and unloading. *See:* **elevator-car leveling device.**   328

two-way correction (industrial control). A method of register control that effects a correction in register in either direction.   204

two-way trunk (telephone switching systems). A trunk between two switching entities used for calls that originate from either end.   55

two-wire channel (telephone loop performance). A

transmission medium that simultaneously carries, without multiplexing, two signals traveling in opposite directions. 473

two-wire circuit (1)(data transmission). A metallic circuit formed by two adjacent conductors insulated from each other. *Note:* Also used in contrast with four-wire circuit to indicate a circuit using one line or channel for transmission of electric waves in both directions. 59
(2)(transmission performance of telephone sets). A metallic circuit formed by two conductors insulated from each other. The electric waves are transmitted in both directions over the path provided by the two-wire circuit. 491

two-wire control (industrial control). A control function which utilizes a maintained-contact type of pilot device to provide undervoltage release. *See:* **undervoltage release.** *See:* **control.** 206

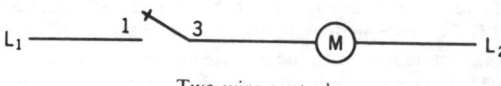

Two-wire control.

two-wire repeater (data transmission). A telephone repeater which provides for transmission in both directions over a two-wire telephone circuit. 59
two-wire switching (telephone switching systems). Switching using the same path, frequency, or time interval for both directions of transmission. 55
two-wire system. *See:* **two-wire circuit.**
TWS. *See:* **track-while-scan.**
type. *See:* **data type.**
type-*A* display; type-*B* display etcetera (radar). *See:* *A* **display;** *B* **display; etcetera.**
Type DB (formerly Type II) (cable systems in power generating stations). Conduit designed for underground installation without encasement in concrete. 10
Type EB (formerly Type I) (cable systems in power generating stations). Conduit designed to be encased in concrete when installed. 10
type font. A type face of a given size and design, for example, 10-point Bodoni Book Medium: 9-point Gothic. 255, 77
type of emission (mobile communication). A system of designating emission, modulation, and transmission characteristics of radio-frequency transmissions, as defined by the Federal Communications Commission.

See: **mobile communication system.** 181
type of piezoelectric crystal cut. The orientation of a piezoelectric crystal plate with respect to the axes of the crystal. It is usually designated by symbols. For example, *GT, AT, BT, CT,* and *DT* identify certain quartz crystal cuts having very low temperature coefficients. *See:* **crystal.** 328
type of service (industrial control). The specific type of application in which the controller is to be used, for example: (1) general purpose: (2) special purpose, namely, crane and hoist, elevator, steel mill, machine tool, printing press, etcetera. *See:* **electric controller.** 206
types of metal-enclosed bus assemblies (metal-enclosed bus and calculating losses in isolated-phase bus). In general, three basic types of construction are used: nonsegregated-phase, segregated-phase, and isolated-phase. (1) nonsegregated-pase bus. One in which all phase conductors are in a common metal enclosure without barriers between the phases. When associated with metal-clad switchgear, the primary bus and connections are covered with insulating material equivalent to the switchgear insulating system. (2) segregated-phase bus. One in which all phase conductors are in a common metal enclosure but are segregated by metal barriers between phases. (3) isolated-phase bus. One in which each phase conductor is enclosed by an individual metal housing separated from adjacent conductor housing by an air space. The bus may be self-cooled or may be force-cooled by means of circulating a gas or liquid. 574
type test (1)(electrical heat tracing for industrial applications). A test or series of tests carried out on equipment, representative of a type, to determine compliance of the design, construction, and manufacturing methods within the requirements of ANSI/ IEEE 515-1983. 523
(2)(valve actuators). Tests made on one or more sample actuators to verify adequacy of design and the manufacturing processes. 492
(3) (rotating electric machinery). A test made by the manufacturer on a machine that is identical in all essential respects with those supplied on an order, to demonstrate that it complies with this standard. 424
type tests (1) (class 1E static battery chargers and inverters) (nuclear power generating stations). Tests made on one or more sample equipment to verify adequacy of design and the manufacturing processes. 120, 408
(2) (rotating machinery). The performance tests taken on the first machine of each type of design. *See:* **asynchronous machine.** 63

U

uhv (power line maintenance). Ultra-high voltage. *See:* **hv.** 458

ultimate deformation or displacement (raceway systems for Class 1E circuits for nuclear power generating stations). The maximum deformation or displacement an element can undergo without failure.　513

ultimate load (raceway systems for Class 1E circuits for nuclear power generating stations). The maximum load an element can carry without failure as obtained from failure load tests or manufacturer's recommendations, whichever is less.　513

ultimately controlled variable (control) (industrial control). The variable the control of which is the end purpose of the automatic control system. *See:* **control system, feedback.**　206,329

ultimate mechanical strength (insulator). The load at which any part of the insulator fails to perform its function of providing a mechanical support without regard to electrical failure. *See:* **insulator.**　261

ultimate period. *See:* **frequency, undamped.**

ultra-audible frequency (supersonic frequency). *See:* **ultrasonic frequency.**

ultra-audion oscillator. *See:* **Colpitts oscillator.**

ultra-high-frequency radar (UHF radar). A radar operating at frequencies between 300 megahertz (MHz) and 1000 MHz, usually in one of the International Telecommunications Union (ITU) assigned bands 420 MHz to 450 MHz or 890 MHz to 942 Mhz. Radars between 1 gigahertz (GHz) and 3 GHz, although within the ultra-high-frequency band as defined by the ITU, are described as L-band or S-band radars, as appropriate.　13

ultra-high-voltage system (transformer). An electric system having a maximum rms ac (root-mean-square alternating current) voltage above 800 000 volts to 2 000 000 volts.　53

ultrasonic cross grating (grating). A space grating resulting from the crossing of beams of ultrasonic waves having different directions of propagation. *Note:* The grating may be two- or three-dimensional.　176

ultrasonic delay line. A transmission device, in which use is made of the propagation time of sound to obtain a time delay of a signal.　176

ultrasonic depth finder (navigation aid terms). A direct reading instrument which determines the depth of water by measuring the time interval between the emission of an ultrasonic signal and the return echo from the bottom.　526

ultrasonic frequency (ultra-audible frequency) (supersonic frequency). A frequency lying above the audio-frequency range. The term is commonly applied to elastic waves propagated in gases, liquids, or solids. *Note:* The word **ultrasonic** may be used as a modifier to indicate a device or system employing or pertaining to ultrasonic frequencies. The term **supersonic,** while formerly applied to frequency, is now generally considered to pertain to velocities above those of sound waves. Its use as a synonym of ultrasonic is now deprecated. *See:* **signal wave.**　176

ultrasonic generator. A device for the production of sound waves of ultrasonic frequency.　328

ultrasonic grating constant. The distance between diffracting centers of the sound wave that is producing particular light diffraction spectra.　176

ultrasonic light diffraction. Optical diffraction spectra or the process that forms them when a beam of light is passed through the field of a longitudinal wave.　176

ultrasonic space grating (grating). A periodic spatial variation of the index of refraction caused by the presence of acoustic waves within the medium.　176

ultrasonic stroboscope. A light interrupter whose action is based on the modulation of a light beam by an ultrasonic field.　176

ultraviolet (UV) (fiber optics). The region of the electromagnetic spectrum between the short wavelength extreme of the visible spectrum (about $0.4\,\mu$m) and $0.04\,\mu$m. *See:* **infrared; light.**　433

ultraviolet flame detector (fire protection devices). A device whose sensing element is responsive to radiant energy outside the range of human vision (below approximately 4000 Angstroms).　71

ultraviolet radiation (1) (illuminating engineering). For practical purposes any radiant energy within the wavelength 10 to 380 nm (nanometers) is considered ultraviolet radiation. *Note:* On the basis of practical applications and the effect obtained, the ultraviolet region often is divided into the following bands:

ozone producing . . . . 180-220 nm
bactericidal (germicidal) . . . . 220-300 nm
erythemal . . . . 280-320 nm
"black light". . . . . 320-400 nm

There are no sharp demarcations between these bands, the indicated effects usually being produced to a lesser extent by longer and shorter wavelengths. For engineering purposes, the 'black light' region extends slightly into the visible portion of the spectrum. *See:* **regions of electromagnetic spectrum.**　167
(2) (laser-maser). Electromagnetic radiation with wavelengths smaller than those for visible radiation; for the purposes of ANSI/IEEE Std 586-1980, 0.2 to $0.4\,\mu$m.　363

umbrella antenna. A type of top-loaded short vertical antenna in which the top-loading structure consists of elements sloping down toward the ground but not connected to it.　111

umbrella reflector antenna. An antenna constructed in a form similar to an umbrella which can be folded for storage or transport and unfolded to form a large reflector antenna for use.　111

unary operation (computing machines). *See:* **monadic operation.**

unattended automatic exchange (CDO or CAX). A normally unattended telephone exchange, wherein the subscribers, by means of calling devices, set up in the central office the connections to other subscribers or to a distant central office.　328

unavailability (1)(nuclear power generating stations). The numerical complement of availability. Unavailability may occur as a result of the item being repaired, for example, repair unavailability or it may occur as a result of undetected malfunctions, for example, unannounced unavailability.   29, 159, 366

(2)(reliability analysis of nuclear power generating station safety systems). The probability that an item or system will not be operational at a future instant in time. Unavailability may be a result of the item being repaired (repair unavailability) or it may occur as a result of malfunctions. Unavailability is the complement of availability.   587

(3)(reliable industrial and commercial power systems planning and design). The long-term average fraction of time that a component or system is out of service due to failures or scheduled outages. An alternative definition is the steady-state probability that a component or system is out of service. Mathematically, unavailability = (1 - availability). *See:* **availability.**
    561

(4) The numerical complement of availability. Unavailability may occur as a result of the item being repaired (repair unavailability), tested (testing unavailability), or it may occur as a result of undetected malfunctions (unannounced unavailability).   357

unavailability margin (nuclear power generating station). The favorable difference between the desired goal and the calculated or observed unavailability.
    357

unavailable (1)(electric generating unit reliability, availability, and productivity). The state in which a unit is not capable of operation because of operational or equipment failures, external restrictions, testing, work being performed, or some adverse condition. The unavailable state persists until the unit is made available for operation, either by being synchronized to the system (in-service state) or by being placed in the reserve shut-down state.   567

(2)(power system measurement). The state in which a unit is not capable of operation because of external restrictions, testing, work being performed, or some adverse condition. The unavailable state persists until the unit is made available for operation, either by being synchronized to the system (in-service state) or by being placed in the reserve shutdown state.
    432

unavailable generation (UG)(electric generating unit reliability, availability, and productivity). The difference between the energy that would have been generated if operating continuously at dependable capacity and the energy that would have been generated if operating continuously at available capacity. This is the energy that could not be generated by a unit due to planned and unplanned outages and unit deratings.

UG = (planned outage hours + unplanned outage hours + equivalent unit derated hours) • maximum capacity

= (POH + UOH + EUNDH) • MC

    567

unavailable hours (UH)(electric generating unit reliability, availability, and productivity). The number of hours a unit was in the unavailable state. *Note:* Unavailable hours are the sum of planned outage hours and unplanned outage hours, or the sum of planned outage hours, forced outage hours, and maintenance outage hours.   567

unbalance (data transmission). A differential mutual impedance or mutual admittance between two circuits that ideally would have no coupling.   59

unbalanced circuit. A circuit, the two sides of which are inherently electrically unlike with respect to a common reference point, usually ground. *Note:* Frequently, unbalanced signifies a circuit, one side of which is grounded.   188

unbalanced modulator (signal-transmission system). *See:* **modulator, asymmetrical; signal.**

unbalanced phase components (thyristor). In multiphase systems unbalance of the phases can be expressed in terms of negative, positive, and zero sequence components. *Note:* Defined for the load only under conditions of balanced lines and balanced loads.
    445

unbalanced strip line. *See:* **strip (strip-type) transmission line.**

unbalanced three-phase system (converters having ac output)(self-commutated converters). A three-phase system in which the rms (root-mean-square) value of at least one phase voltage (or current) or line-to-line voltage is significantly different from the others, or in which the phase angle displacement between any pair of phases significantly differs from 120 degrees. *Note:* In an unbalanced three-phase system, negative or zero-sequence components exist.   584

unbalanced wire circuit (data transmission). One whose two sides are inherently electrically unlike.
    59

unbalance factor (converters having ac output)(self-commutated converters). The ratio of the negative sequence component to the positive sequence component.   584

unbalance ratio (converters having ac output)(self-commutated converters). The difference between the highest and the lowest fundamental rms (root-mean-square) values in a three-phase system, referred to the average of the three fundamental rms values of current or voltages, respectively.   584

unbiased telephone ringer. A telephone ringer whose clapper-driving element is not normally held toward one side or the other, so that the ringer will operate on alternating current. Such a ringer does not operate reliably on pulsating current. *Note:* A ringer that is weakly biased so as to avoid tingling when dial pulses pass over the lines may be referred to as an unbiased ringer. *See:* **telephone station.**   328

unblanking. Turning on of the cathode-ray-tube beam. *See:* **oscillograph.**   185

unbound mode (fiber optics). Any mode that is not a bound mode; a leaky or radiation mode of the waveguide. *Syn:* **radiative mode.** *See:* **bound mode; cladding mode; leaky mode.**   433

uncertainty (1) (radiation protection). The estimated bounds of the deviation from the mean value, general-

ly expressed as a percent of the mean value. Ordinarily taken as the sum of (A) the random errors at the 95 percent confidence level and (B) the estimated upper limit of the systematic error.   399
(2) (general). The estimated amount by which the observed or calculated value of a quantity may depart from the true value. *Note:* The uncertainty is often expressed as the average deviation, the probable error, or the standard deviation. *See:* **measurement system; measurement uncertainty.**   185, 54
(3) (electrothermic power meters). The assigned allowance for the systematic error, together with the random error attributed to the imprecision of the measurement process.   47,115
unconditional jump (unconditional transfer of control) (electronic computation). An instruction that interrupts the normal process of obtaining instructions in an ordered sequence and specifies the address from which the next instruction must be taken. *See:* **jump.**   235
unconditional transfer of control. *See:* **unconditional jump.**
underbunching. A condition representing less than optimum bunching.   125
undercounter dumbwaiter. A dumbwaiter that has its top terminal landing located underneath a counter and that serves only this landing and the bottom terminal landing.   328
undercurrent or underpower relay (37) (power system device function numbers). A relay that functions when the current or power flow decreases below a predetermined value.   402
undercurrent relay (1) (power switchgear). A relay that operates when the current is less than a predetermined value.   103
(2) (general). A relay that operates when the current through the relay is equal to or less than its setting. *See:* **relay.**   60
undercurrent release (trip) (power switchgear). A release that operates when the current in the main circuit is equal to or less than the release setting.   103
undercurrent trip. *See:* **undercurrent release.**
undercurrent tripping. *See:* **undercurrent release (trip).**
underdamped. Damped insufficiently to prevent oscillation of the output following an abrupt input stimulus. *Note:* In an underdamped linear second-order system, the roots of the characteristic equation have complex values. *See:* **damped harmonic system.**   206
underdamped period (instrument) (periodic time). The time between two consecutive transits of the pointer or indicating means in the same direction through the rest position, following an abrupt change in the measurand.   328
underdamping (periodic damping). The special case of damping in which the free oscillation changes sign at least once. A damped harmonic system is underdamped if F^2 *less MS.* See **damped harmonic system** for equation, definitions of letter symbols, and referenced terms.   210

underdome bell. A bell whose mechanism is mostly concealed within its gong. *See:* **protective signaling.**   328
underfilm corrosion. Corrosion that occurs under films in the form of randomly distributed hairlines (filiform corrosion).   205
underfloor raceway. A raceway suitable for use in the floor. *See:* **raceway.**   328
underflow (mathematics of computing). The condition that arises when the result of a floating-point arithmetic operation is smaller than the smallest nonzero number that can be represented in a digital computer.   564
underground cable. A cable installed below the surface of the ground. *Note:* This term is usually applied to cables installed in ducts or conduits or under other conditions such that they can readily be removed without disturbing the surrounding ground. *See:* **cable; tower.**   64
underground collector or plow. A current collector, the function of which is to make contact with an underground contact rail. *See:* **contact conductor.**   1
underground duct system (raceway systems for Class 1E circuits for nuclear power generating stations). Metallic or nonmetallic conduit enclosed in reinforced concrete or directly buried, including access points.   513
underground system service-entrance conductors (National Electrical Code). The service conductors between the terminals of the service equipment and the point of connection to the service lateral. Where service equipment is located outside the building walls, there may be no service-entrance conductors, or they may be entirely outside the building.   256
underlap, X **(facsimile).** The amount by which the center-to-center spacing of the recorded spots exceeds the recorded spot X dimension. *Note:* This effect arises in that type of equipment which responds to a constant density in the subject copy by a succession of discrete recorded spots. *See:* **recording (facsimile).**   12
underlap, Y **(facsimile).** The amount by which the nominal line width exceeds the recorded spot Y dimension. *See:* **recording (facsimile).**   12
underreaching protection (power switchgear). A form of protection in which the relays at a given terminal do not operate for faults at remote locations on the protected equipment, the given terminal being cleared either by other relays with different performance characteristics or by a transferred trip signal from a remote terminal similarly equipped with underreaching relays.   103
undershoot (1) (rounding) (television). That part of the distorted wave front characterized by a decaying approach to the final value. *Note:* Generally, undershoots are produced in transfer devices having insufficient transient response.   18
(2) (oscilloscopes). In the display of a step function (usually of time), that portion of the waveform that, following any overshoot or rounding that may be present, falls below its nominal or final value.   106
underslung car frame. A car frame to which the hoist-

ing-rope fastenings or hoisting rope sheaves are attached at or below the car platform. *See:* **hoistway (elevator or dumbwaiter).** 328

underspeed (hydraulic turbines). Any speed below rated speed expressed as a percent of rated speed.
8

underspeed device (14) (power system device function numbers). A device that functions when the speed of a machine falls below a predetermined value.
402

undervoltage or low-voltage protection (electric installations on shipboard). The effect of a device, operative on the reduction or failure of voltage, to cause and maintain the interruption of power in the main circuit. 3

undervoltage or low-voltage release (electric installations on shipboard). The effect of a device, operative on the reduction or failure of voltage, to cause the interruption of power to the main circuit, but not to prevent the reestablishment of the main circuit on return of voltage. 3

undervoltage protection (1) (low-voltage protection) (power switchgear). A form of protection that operates when voltage is less than a predetermined value.
103

(2) (packaging machinery). The effect of a device operative on the reduction or failure of voltage to cause and maintain the interruption of power to the main circuit. 429

(3) (industrial control). Same definition as (2) above except for the following note.*Note:* The principal objective of this device is to prevent automatic restarting of the equipment. Standard undervoltage or low-voltage protection devices are not designed to become effective at any specific degree of voltage reduction.
3

undervoltage relay (27) (power system device function numbers). A relay which operates when its input voltage is less than a predetermined value. 402

undervoltage relays (power switchgear). A relay that operates when its voltage is less than a predetermined value. 103

undervoltage release (1) (trip) (power switchgear). A release that operates when the voltage of the main circuit is equal to less than the release setting.
103

(2) (industrial control). The effect of a device, operative on the reduction or failure of voltage, to cause the interruption of power to the main circuit but not to prevent the re-establishment of the main circuit on return of voltage. *Note:* Standard undervoltage or low-voltage release protection devices are not designed to become effective at any specific degree of voltage reduction. *Syn:* **low voltage release.** 3

undervoltage tripping. *See:* **undervoltage release (trip).**

underwater log. A device that indicates a ship's speed based on the pressure differential, resulting from the motion of the ship relative to the water, as developed in a Pitot tube system carried by a retractable support extending through the ship's hull. Continuous integra-

tion provides indication of total distance travelled. The ship's draft is indicated, based on static pressure.
328

underwater sound projector. A transducer used to produce sound in water. *Notes:* (1) There are many types of underwater sound projectors whose definitions are analogous to those of corresponding loudspeakers, for example, crystal projector, magnetic projector, etcetera. (2) Where no confusion will result, the term underwater sound projector may be shortened to projector. *See:* **microphone.** 176

undesired conducted power (frequency-modulated mobile communications receivers). Radio-frequency power that is present at the antenna, power terminals, or any other interfacing terminals. 123

undesired radiated power (frequency-modulated mobile communications receivers). Radio-frequency power radiated from the receiver that can be measured outside a specified area. 123

undetected error rate (data transmission). The ratio of the number of bits, unit elements, characters, blocks incorrectly received but undetected or uncorrected by the error-control equipment, to the total number of bits, unit elements, characters, blocks sent. 194

undisturbed-ONE output (magnetic cell). A ONE output to which no partial-read pulses have been applied since that cell was last selected for writing. *See:* **coincident-current selection.** 331

undisturbed-ZERO output (magnetic cell). A ZERO output to which no partial-write pulses have been applied since that cell was last selected for reading. *See:* **coincident-current selection.** 331

undulating current (rotating electric machinery). Current that remains unidirectional, but the ripple of which exceeds that defined for smooth current.
424

unexposed side (cable penetration fire stop qualification test). The side of a fire-rated wall, floor-ceiling assembly, or floor which is opposite to the fire side. *Syn:* **cold side.** 368

unfired tube (microwave gas tubes). The condition of the tube during which there is no radio-frequency glow discharge at either the resonant gap or resonant window. *See:* **gas tube.** 125

ungrounded (1) (electric power). A system, circuit, or apparatus without an intentional connection to ground except through potential-indicating or measuring devices or other very-high-impedance devices. *Note:* Though called ungrounded, this type of system is in reality coupled to ground through the distributed capacitance of its phase windings and conductors. In the absence of a ground fault, the neutral of an ungrounded system under reasonably balanced load conditions will usually be close to ground potential, being held there by the balanced electrostatic capacitance between each phase conductor and ground. 152

(2) (power and distribution transformer). A system, circuit, or apparatus without an intentional connection to ground except through potential-indicating or measuring devices or other very-high-impedance devices.
53

ungrounded potentiometer (analog computers). A potentiometer with neither end terminal attached directly to ground. 9

ungrounded system (systems grounding). A system, circuit, or apparatus without an intentional connection to ground, except through potential-indicating or measuring devices or other very-high-impedance devices. *Note:* Though called ungrounded, this type of system is in reality coupled to ground through the distributed capacitance of its phase windings and conductors. In the absence of a ground fault, the neutral of an ungrounded system under reasonably balanced load conditions will usually be close to ground potential, being held there by the balanced electrostatic capacitance between each phase conductor and ground. Fig (a) below shows an ungrounded system with voltage relations for balanced phase-to-ground capcitance. 152

unguarded release (telephone switching systems). A condition during the restoration of a circuit to its idle state when it can be prematurely seized. 55

uniconductor waveguide. A waveguide consisting of a cylindrical metallic surface surrounding a uniform dielectric medium. *Note:* Common cross-sectional shapes are rectangular and circular. *See:* **waveguide.**
267,319

unidirectional. A connection between telegraph sets, one of which is a transmitter and the other a receiver. 194

unidirectional antenna. An antenna that has a single well-defined direction of maximum gain. *See:* **antenna.** 111

unidirectional bus (1)(programmable instrumentation). A bus used by any individual device for one-way transmission of messages only, that is, either input only or output only. 40
(2)(signals and paths)(696 interface devices). A bus used by a device for one-way transmission of messages, that is, either input only or output only.
538

unidirectional current. A current that has either all positive or all negative values. 210

unidirectional microphone. A microphone that is responsive predominantly to sound incident from a single solid angle of one hemisphere or less. *See:* **microphone.** 176

unidirectional pulse train (signal-transmission system). Pulses in which pertinent departures from the normally constant value occur in one direction only. *See:* pulse. 254

unidirectional transducer (unilateral transducer). A transducer that cannot be actuated at its output by waves in such a manner as to supply related waves at its input. *See:* transducer. 210

unified atomic mass unit (u). The unit equal to the fraction $1/12$ of the mass of an atom of the nuclide ^{12}C: $1 u = 1.660\ 53 \times 10^{-27}$ kg approximately.
21

unified s-band system (communication satellite). A communication system using an S-band carrier (2000 2300 megahertz) combining all links into one spectrum. The functions of spacecraft command, data transmission, tracking, ranging, etcetera, are transmitted on separate carrier frequencies for earth-space and space-earth links. 84

uniform current density. A current density that does not change (either in magnitude or direction) with position within a specified region. (A uniform current density may be a function of time.) 210

uniform field (measurement of power frequency electric and magnetic fields from ac power lines). A field whose magnitude and direction are uniform at each instant in time at all points within a defined region.
514

uniform line. A line that has substantially identical electrical properties throughout its length. *See:* transmission line. 328

uniform linear array (antenna). A linear array of identically oriented and equally spaced radiating elements having equal current amplitudes and equal phase increments between excitation currents. *See:* antenna.
179, 111

uniform luminance area. The area in which a display on a cathode-ray tube retains 70 percent or more of its luminance at the center of the viewing area. *Note:* The corners of the rectangle formed by the vertical and horizontal boundaries of this area may be below the 70-percent luminance level. *See:* oscillograph.
185

uniform plane wave (radio wave propagation). A plane wave in which the electric and magnetic field vectors

PHASE-TO-GROUND CAPACITANCE

(a)

have constant amplitude over the equiphase surfaces. *Note:* Such a wave can only be found in free space at an infinite distance from the source. 146

uniform waveguide. A waveguide in which the physical and electrical characteristics do not change with distance along the axis of the guide. 166, 267

unilateral area track. A sound track in which one edge only of the opaque area is modulated in accordance with the recorded signal. There may, however, be a second edge modulated by a noise-reduction device. *See:* **phonograph pickup.** 176

unilateral connection (control system, feedback). A connection through which information is transmitted in one direction only. *See:* **control system, feedback.** 329

unilateral network. A network in which any driving force applied at one pair of terminals produces a nonzero response at a second pair but yields zero response at the first pair when the same driving force is applied at the second pair. *See:* **network analysis.** 210

unilateral transducer. *See:* **unidirectional transducer.**

unimpaired observation (nuclear security systems). Conditions that enable an unobstructed view so as to ensure direct visual or closed circuit television (CCTV) surveillance of individuals or vehicles. 464

uninhibited oil (power and distribution transformer). Mineral transformer oil to which no synthetic oxidation inhibitor has been added. 53

uninterruptible power supply (UPS)(emergency and standby power). A system designed to automatically provide power, without delay or transients, during any period when the normal power supply is incapable of performing acceptably. 512

unipolar (power supplies). Having but one pole, polarity, or direction. Applied to amplifiers or power supplies, it means that the output can vary in only one polarity from zero and, therefore, must always contain a direct-current component. *See:* **bipolar.** 186

unipolar electrode system (electrobiology) (monopolar electrode system). Either a pickup or a stimulating system, consisting of one active and one dispersive electrode. *See:* **electrobiology.** 192

unipole. *See:* **antenna; isotropic antenna.**

unipotential cathode. *See:* **cathode, indirectly heated.**

unique identification code (unique identification in power plants). A code applied at the component function level to uniquely distinguish a specific function within a specific system from all other similar or different functions occurring within the system or facility. The basic code format described in ANSI/IEEE 803-1983 may also be applied, with appropriate field identifiers, for project software and project control elements (schedule and budget items). 544

unit (1) (class 1E power systems for nuclear power generating stations). A nuclear steam supply system, its associated turbine-generator, auxiliaries, and engineered safety features. 102

(2) (generating station). The generator or generators, associated prime mover or movers, auxiliaries and energy supply or supplies that are normally operated together as a single source of electric power. 381

(3) (nuclear power generating station). (A) A nuclear steam supply system, its associated turbine-generator, auxiliaries, and engineered safety features. 102
(B) One independent portion of a motor control center vertical section. It is normally a plug-in module which connects to the motor control center vertical bus. 440

(4) (of a relay). *See:* **relay unit.** 103

(5) (switchgear). That portion of the switchgear assembly which contains one switching device such as a circuit breaker, interrupter switch, power fuse interrupter switch combination, etcetera and the associated primary conductors. 93,103

(6) (electric and electronics parts and equipments). A major building block for a set or system, consisting of a combination of basic parts, subassemblies, and assemblies packaged together as a physically independent entity. The application, size, and construction of an item may be factors in determining whether an item is regarded as a unit, an assembly, a subassembly, or a basic part. A small electric motor might be considered as a part if it is not normally subject to disassembly. Typical examples are: radio receiver, radio transmitter, electronic power supply, antenna. 17

(7) *See:* **test unit.**

unit-area capacitance (electrolytic capacitor). The capacitance of a unit area of the anode surface at a specified frequency after formation at a specified voltage. *See:* **electrolytic capacitor.** 328

unit auxiliaries transformer (generating stations electric power system). A transformer intended primarily to supply all or a portion of the unit auxiliaries. 381

unit auxiliary (generating stations electric power system). An auxilary intended for a specific generating unit. 381

unit cable construction. That method of cable manufacture in which the pairs of the cable are stranded into groups (units) containing a certain number of pairs and these groups are then stranded together to form the core of the cable. *See:* **cable.** 328

unit-control error (electric power systems). The unit generation minus the assigned unit generation. *Note:* Refer to note on polarity under **area control error.** 94

unit derated generation (UDG) (power system measurement). The unavailable generation resulting from unit derating. 432

unit derated hours (UNDH)(electric generating unit reliability, availability, and productivity). The available hours during which a unit derating was in effect. 567

unit derating (electric generating unit reliability, availability, and productivity). The difference between dependable capacity and available capacity. 567

unit-impulse function. *See:* **signal, unit-impulse.**

unit interval. *See:* **signal element.**

unitized equipment (packaging machinery). Electrical controls so constructed that separate panels are provided for each working station, or section as specified, of a multiple-station transfer-type machine. 429

unit of acceleration - g (digital accelerometer). The symbol g denotes a unit of acceleration equal in magnitude to the local value of gravity at the test site unless otherwise specified. 383

unit operation (1)(metal-oxide surge arresters for ac power circuits). A discharge of a surge through an arrester while the arrester is energized. 583

(2) (CO) (of a circuit breaker). *See:* **close-open operation (of a switching device).** 103

(3) (of a circuit recloser). An interrupting operation followed by a closing operation. The final interruption is also considered one unit operation. 92, 103

(4) surge arrester). Discharging a surge through an arrester while the arrester is energized. 430

unit-ramp function. *See:* **signal, unit-ramp.**

unit rate-limiting controller (electric power systems). A controller that limits rate of change of generation of a generating unit to an assigned value or values. *Note:* The limiting action is normally based on a measured megawatt-per-minute rate. *See:* **speed-governing system.** 94

unit requirements documentation (software unit testing). Documentation that set forth the functional, interface, performance, and design constraint requirements for the test unit. 519

units and letter symbols (International System of Units) (SI). The three classes of SI units are: (1) **base units,** regarded by convention as dimensionally independent:

Quantity	Unit	Symbol
length	meter	m
mass	kilogram	kg
time	second	s
electric current	ampere	A
thermodynamic temperature	kelvin	K
amount of substance	mole	mol
luminous intensity	candela	cd

(2) **supplementary units,** regarded either as base units or as derived units:

Quantity	Unit	Symbol
plane angle	radian	rad
solid angle	steradian	sr

(3) **derived units,** formed by combining base units, supplementary units, and other derived units according to the algebraic relations linking the corresponding quantities. The symbols for derived units are obtained by means of the mathematical signs for multiplication, division, and use of exponents. De-

rived units which have special names and symbols approved by the General Conference on Weights and Measures (CGPM) are:

unit sequence starting relay (44) (power system device function numbers). A relay that functions to start the next available unit in a multiple-unit equipment upon the failure or nonavailability of the normally preceding unit. 402

unit sequence switch (10) (power system device function numbers). A switch that is used to change the sequence in which units may be placed in and out of service in multiple-unit equipments. 402

units of luminance, (photometric brightness) (light emitting diodes). The luminance (photometric brightness) of a surface in a specified direction may be expressed in luminous intensity per unit of projected area of surface. *Note:* Typical units in this system are the candela per square meter. 162

units of luminous exitance (illuminating engineering). Lumens per square meter (lam/m^2) and lumens per square foot (lm/ft^2) are preferred practice for the SI and English (USA) systems respectively. 167

units of wavelength. The distance between two successive points of a periodic wave in the direction of propagation, in which the oscillation has the same phase. The three commonly used units are listed in the following table:

Name	Symbol	Value
micrometer	μm	$1 \ \mu m = 10^{-3}$ millimeters
nanometer	nm	$1 \ nm = 10^{-6}$ millimeters
angstrom	Å	$1 \ Å = 10^{-7}$ millimeters

See: **radiant energy.** 167

unit state (electric generating unit reliability, availability, and productivity). A particular unit condition that is important for purposes of collecting data on performance. *Note:* The state definitions are related as shown in the figure below. The transitions between states are described in Appendix B of ANSI/IEEE Std 762-1987. The correlation between these definitions and those in use by the industry is shown in Appendix A of ANSI/IEEE Std762-1987. 567

unit-step function. *See:* **signal, unit-step.**

unit substation (1) (power and distribution transformer). A substation consisting primarily of one or more transformers which are mechanically or electrically connected to and coordinated in design with one or more switchgear or motor control assemblies, or combinations thereof. 53

(2) (power switchgear). A substation consisting primarily of one or more transformers mechanically and electrically connected and coordinated in design with one or more switchgear or motor control assemblies or combination thereof. *Note:* A unit substation may be described as "primary" or "secondary" depending on the voltage rating of the low-voltage section: primary, more than 1000 volts: secondary, 1000 volts and below. 103

Some Common Derived Units of the International System of Units

Quantity	Unit	Symbol
acceleration	meter per second squared	m/s^2
angular acceleration	radian per second squared	rad/s^2
angular velocity	radian per second	rad/s
area	square meter	m^2
concentration (of amount of substance)	mole per cubic meter	mol/m^3
current density	ampere per square meter	A/m^2
density, mass	kilogram per cubic meter	kg/m^3
electric charge density	coulomb per cubic meter	C/m^3
electric field strength	volt per meter	V/m
electric flux density	coulomb per square meter	C/m^2
energy density	joule per cubic meter	J/m^3
entropy	joule per kelvin	J/K
heat capacity	joule per kelvin	J/K
heat flux density	watt per square meter	W/m^2
irradiance	watt per square meter	W/m^2
luminance	candela per square meter	cd/m^2
magnetic field strength	ampere per meter	A/m
molar energy	joule per mole	J/mol
molar entropy	joule per mole kelvin	$J/(mol \cdot K)$
molar heat capacity	joule per mole kelvin	$J/(mol \cdot K)$
moment of force	newton meter	$N \cdot m$
permeability	henry per meter	H/m
permittivity	farad per meter	F/m
radiance	watt per square meter steradian	$W/(m^2 \cdot sr)$
radiant intensity	watt per steradian	W/sr
specific heat capacity	joule per kilogram kelvin	$J/(kg \cdot K)$
specific energy	joule per kilogram	J/kg
specific entropy	joule per kilogram kelvin	$J/(kg \cdot K)$
specific volume	cubic meter per kilogram	m^3/kg
surface tension	newton per meter	N/m
thermal conductivity	watt per meter kelvin	$W/(m \cdot K)$
velocity	meter per second	m/s
viscosity, dynamic	pascal second	$Pa \cdot s$
viscosity, kinematic	square meter per second	m^2/s
volume	cubic meter	m^3
wavenumber	1 per meter	$1/m$

Units in Use with the International System of Units

Quantity	Unit	Symbol	Definition
time	minute	min	$1 \text{ min} = 60 \text{ s}$
	hour	h	$1 \text{ h} = 60 \text{ min} = 3600 \text{ s}$
	day	d	$1 \text{ d} = 24 \text{ h} = 86\,400 \text{ s}$
	other calendar	—	
plane angle	degree	°	$1° = (\pi/180)\text{rad}$
temperature	degree Celsius	°C	
volume	liter[a]	1	$1 \text{ l} = 1 \text{ dm}^3 = 10^{-3} \cdot m^3$
mass	metric ton	t	$1 \text{ t} = 10^3 \text{ kg}$

[a] Because of the similarity between the lower case letter l and the number 1 in many type fonts it is recommended that when confusion might result the word "liter" be written in full.

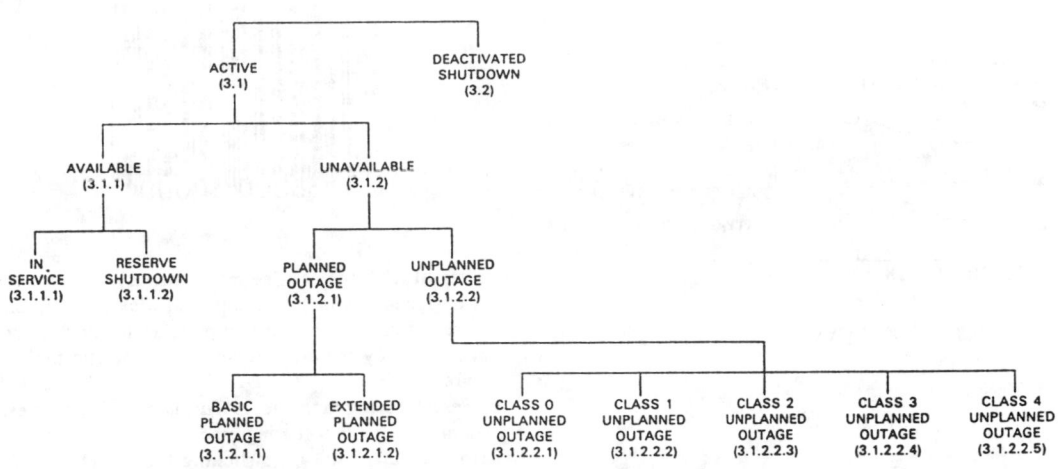

Fig 1
Relation Between Unit States

unit-substation transformer (power and distribution transformer). A transformer which is mechanically and electrically connected to, and coordinated in design with, one or more switchgear or motor-control assemblies, or combinations thereof. *See:* **unit substation; primary unit substation; secondary unit substation; integral unit substation; articulated unit substation.** 53

unit symbol (abbreviation). *See:* **symbol for a unit.**

unit test (switchgear). **A test performed on a single unit or group of units.** *Note:* one widespread use of such tests is extrapolation of test results for the purpose of representing overall performance of a device composed of several units. 93, 103

unit transformer (generating stations electric power system). A power system supply transformer which transforms all or a portion of the unit power from the unit to the power system voltage. 381

unit under test (UUT) (ATLAS). The entity to be tested. It may range from a simple component to a complete system. 400

unit vector. A vector whose magnitude is unity. 210

unit warmup time (power supply). The interval between the time of application of input power to the unit and the time at which the regulated power supply is supplying regulated power at rated output voltage. *See:* **regulated power supply.** 347

unit years (UY) (power system measurement). For any unit or for a group of units, unit years is the total period hours accumulated, divided by 8760:
$$PH\ UY = 8760 \qquad 432$$

unity-gain bandwidth (power supplies). A measure of the gain-frequency product of an amplifier. Unity-gain bandwidth is the frequency at which the open-loop gain becomes unity, based on a 6-decibel-per-octave crossing. See the acompanying figure. 186

unity power-factor test (synchronous machine). A test in which the machine is operated as a motor under specified operating conditions with its excitation adjusted to give unity power factor. 63

univalent function. If to every value of u there corresponds one and only one value of x (or one and only one set of values of $x_1, x_2, ..., x_n$) then u is a univalent function. Thus $u^2 = ax + b$ is univalent, within the interval of definition. 210

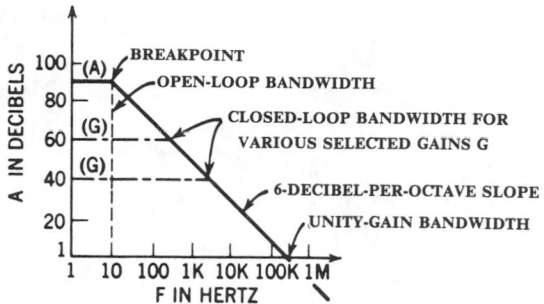

Typical gain-frequency (Bode) plot, showing unity-gain bandwidth.

universal demand register (mechanical demand registers). A demand register of specific ratio used in conjunction with all ratings of any type of integrating electricity meter designed to accommodate it. The register constant of a universal demand register is proportional to the watthour constant K_h of the meter on which it is mounted. 548

universal fuse links (power switchgear). Fuse links that, for each rating, provide mechanical and electrical interchangeability within prescribed limits over the specified time-current range. 103, 443

universal motor (electric installations on shipboard). A series-wound or a compensated series-wound motor designed to operate at approximately the same speed and output on either direct or single-phase alternating current of a frequency not greater than 60 Hz and of approximately the same rms (root-mean-square) voltage. 3

universal-motor parts (rotating machinery). A term applied to a set of parts of a universal motor. Rotor shaft, conventional stator frame (or shell), end shields, or bearings may not be included, depending on the requirements of the end product into which the universal-motor parts are to be assembled. *See:* **asynchronous machine.** 63

universal-numbering plan (telephone switching systems). A numbering plan employing nonconflicting codes so arranged that all main stations can be reached from any point within a telecommunications system. 55

universal or arcshear machine. A power-driven cutter that will not only cut horizontal kerfs, but will also cut vertical kerfs or at any angle, and is designed for operation either on track, caterpillar treads, or rubber tires. 328

universal product code (UPC)(computer applications). A bar code appearing on many retail products to uniquely identify the product. The code is designed to be read by an optical scanner attached to an electronic cash register.

579

universal stick (power line maintenance). An accessory designed to attach to a universal stick allowing one insulated stick to be used to perform many different operations by the attachment of the specific tool required. 458

universal Turing machine. A Turing machine that can simulate any other Turing machine. 255, 77

unloaded (intrinsic) Q (switching tubes). The Q of a tube unloaded by either the generator or the termination. *Note:* As here used, Q is equal to 2π times the energy stored at the resonance frequency divided by the energy dissipated per cycle in the tube or, for cell-type tubes, in the tube and its external resonant circuit. *See:* **gas tube.** 125

unloaded applicator impedance (dielectric heating usage). The complex impedance measured at the point of application, without the load material in position, at a specified frequency. *See:* **dielectric heating.** 14

unloaded sag (conductor or any point in a span). The distance measured vertically from the particular point in the conductor to a straight line between its two points of support, without any external load. 263

unloaded tension (NESC). (1) initial. The longitudinal tension in a conductor prior to the application of any external load. (2) final. The longitudinal tension in a conductor after it has been subjected for an appreciable period to the loading prescribed for the loading district in which it is situated, or equivalent loading, and the loading removed. Final unloaded tension includes the effect of inelastic deformation (creep). 494

unloading amplifier (analog computers). An amplifier that is capable of reproducing or amplifying a given voltage signal while drawing negligible current from the voltage source. *Note:* In an analog computer, the term buffer amplifier is sometimes used as a synonym for unloading amplifier, in an incorrect sense, since a buffer amplifier draws significant current, but at a constant load impedance (seen at the input). 9

unloading circuit (analog computers). In an analog computer, a computing element or combination of computing elements capable of reproducing or amplifying a given voltage signal while drawing negligible current from the voltage source, thus eliminating any possible loading errors. *See:* **unloading amplifier.** 9

unloading point (electric transmission system used on self-propelled electric locomotives or cars). The speed above or below which the design characteristics of the generators and traction motors or the external control system, or both, limit the loading of the prime mover to less than its full capacity. *Note:* The unloading point is not always a sharply defined point, in which case the unloading point may be taken as the useful point at which essentially full load is provided. *See:* **traction motor.** 328

unmodulated groove (blank groove) (mechanical recording). A groove made in the medium with no signal applied to the cutter. *See:* **phonograph pickup.** 176

un-ordered access (communication satellite). A system in which access to a radio frequency channel is gained without determining channel availability. This method is useful in common spectrum or random access discrete address systems. 84

unpack. To separate various sections of packed data. 255, 77

unplanned derated hours (UDH)(electric generating unit reliability, availability, and productivity). The available hours during which an unplanned derating was in effect. 567

unplanned derating (electric generating unit reliability, availability, and productivity). That portion of the unit derating that is not a planned derating. Unplanned derating events are classified according to the urgency with which the derating needs to be initiated. Class 1 (immediate). A derating that requires an immediate action for the reduction of capacity.Class 2 (delayed). A derating that does not require an immediate reduction of capacity, but requires a reduction of capacity within 6 h (hours).Class 3 (postponed). A derating that can be postponed beyond 6 h, but requires a reduction of capacity before the end of the next weekend.Class 4 (deferred). A derating that can be deferred beyond the end of the next weekend, but requires a reduction of capacity before the next planned outage. 567

unplanned outage (electric generating unit reliability, availability, and productivity). The state in which a unit is unavailable but is not in the planned outage state. *Notes:* (1) When an unplanned outage is initiated, the outage is classified according to one of five classes, as defined in Class 0 unplanned outage, Class 1 unplanned outage, Class 2 unplanned outage, Class 3 unplanned outage, and Class 4 unplanned outage. Unplanned outage Class 0 applies to a start-up failure and Class 1 applies to a condition requiring immediate outage. Also, unplanned outage starts when planned outage ends but is extended due to unplanned work. Classes 2, 3, and 4 apply to outages where some delay is possible in time of removal of the unit from service. The class (2, 3, or 4) of outage is to be determined by the amount of delay that can be exercised in the time of removal of the unit. The class of outage is not made more urgent if the time of removal is advanced due to favorable conditions of system reserves or availability of replacement capacity for the predicted duration of the outage. However, outage starts when the unit is removed from service or is declared unavailable when

it is not in service. (2) During the time the unit is in the unplanned outage state, the outage class is determined by the outage class that initiates the state. (3) In some cases, the opportunity exists during unplanned outages to perform some of the repairs or maintenance that would have been performed during the next planned outage. If the additional work extends the outage beyond that required for the unplanned outage, the remaining outage should be reported as a planned outage. (4) Unlike planned outages, unplanned outages do not have a fixed duration that can be estimated each year. *See:* **Class 0 unplanned outage (starting failure); Class 1 unplanned outage (immediate); Class 2 unplanned outage (delayed); Class 3 unplanned outage (postponed); Class 4 unplanned outage (deferred).** 567

unplanned outage hours (UOH)(electric generating unit reliability, availability, and productivity). The number of hours a unit was in a Class 0, 1, 2, 3, or 4 unplanned outage state. 567

unpropagated potential (electrobiology). An evoked transient localized potential not necessarily associated with changed excitability. *See:* **excitability (electrobiology).** 328

unquenched sample (1)(liquid-scintillation counters). A counting sample (material of interest plus liquid-scintillation solution) that contains a minimum of colored species and chemical impurities that would reduce the photon output from the vial. 498

unquenched samples (2)(liquid-scintillation counting). A popular expression which connotes a counting sample that contains a minimum of colored species and chemical impurities which would reduce the light output from the scintillators. 422

unrecoverable light loss factors (illuminating engineering). Factors which give the fractional light loss that cannot be recovered by cleaning or lamp replacement. 167

unregulated voltage (electronically regulated power supply). The voltage at the output of the rectifier filter. *See:* **regulated power supply.** 347

unsaturated standard cell. A cell in which the electrolyte is a solution of cadmium sulphate at less than saturation at ordinary temperatures. (This is the commercial type of cadmium standard cell commonly used in the United States). *See:* **electrochemistry.** 328

unshielded strip transmission line. A strip conductor above a single ground plane. Some common designations are **microstrip (flat-strip conductor)** and **unbalanced strip line.** *See:* **strip (type) transmission line; shielded strip transmission line; waveguides.** 179

unstable (control system, feedback). Not possessing stability. *See:* **control system, feedback.** 56, 329

unusual service conditions. Environmental conditions that may affect the constructional or operational requirements of a machine. This includes the presence of moisture and abrasive, corrosive, or explosive atmosphere. It also includes external structures that limit ventilation, unusual conditions relating to the electrical supply, the mechanical loading, and the position of the machine. 63

unwanted radiation (radiation protection). Any ionizing radiation other than that which the instrument is designed to measure. 399

update (supervisory control, data acquisition, and automatic control). The process of modifying or reestablishing data with more recent information. 570

uplift roller (conductor stringing equipment). A small single-grooved wheel designed to fit in or immediately above the throat of the traveler and keep the pulling line in the traveler groove when uplift occurs due to stringing tensions. 431

up link (communication satellite). A ground to satellite link, very often the command link. 83

upper (driving) beams (illuminating engineering). One or more beams intended for distant illumination and for use on the open highway when not meeting other vehicles. Formerly 'country beam'. 167

upper bracket (rotating machinery). A bearing bracket mounted above the core of a vertical machine. 63

upper burst reference (audio and electroacoustics). A selected multiple of the long-time average magnitude of the quantity mentioned in the definition of **burst**. See the figure attached to the definition of **burst duration**. See: **burst (audio and electroacoustics).** 253, 176

upper coil support (rotating machinery). A coil support to restrain field-coil motion in the direction toward the air gap. See: **rotor (rotating machinery); stator.** 63

upper frequency limit (coaxial transmission line). The limit determined by the cutoff frequency of higher-order waveguide modes of propagation, and the effect that they have on the impedance and transmission characteristics of the normal TEM coaxial-transmission-line mode. The lowest cutoff frequency occurs with the $TE_{1,1}$ mode, and this cutoff frequency in air dielectric line is the upper frequency limit of a practical transmission line. How closely the $TE_{1,1}$ mode cutoff frequency can be approached depends on the application. See: **waveguide.** 265

upper guide bearing (rotating machinery). A guide bearing mounted above the core of a vertical machine. See: **bearing.** 63

upper half bearing bracket (rotating machinery). The top half of a bracket that can be separated into halves for mounting or removal without access to a shaft end. See: **bearing.** 63

upper limit (test, measurement and diagnostic equipment). The maximum acceptable value of the characteristic being measured. 54

upper range-value. The highest quantity that a device is adjusted to measure. *Note:* The following compound terms are used with suitable modifications in the units: **measured variable upper range-value, measured signal upper range-value,** etcetera. See: **instrument.** 295

upper sideband (data transmission). The higher of two frequencies or groups of frequencies produced by a modulation process. 59

upper-sideband parametric down-converter. A noninverting parametric device used as a parametric down-converter. See: **parametric device.** 277, 191

upper-sideband parametric up-converter. A noninverting parametric device used as a parametric up-converter. See: **parametric device.** 191

upset duplex system. A direct-current telegraph system in which a station between any two duplex equipments may transmit signals by opening and closing the line circuit, thereby causing the signals to be received by upsetting the duplex balance. See: **telegraphy.** 328

uptime (1)(supervisory control, data acquisition, and automatic control). The time during which a device or system is capable of meeting performance requirements. 570
(2) (availability). The period of time during which an item is in a condition to perform its required function. Sometimes written **up time.** 164

upward component (illuminating engineering). That portion of the luminous flux from a luminaire which is emitted at angles above the horizontal. 167

urban districts (NESC). Thickly settled areas (whether in cities or suburbs) or where congested traffic often occurs. A highway, even though in thinly settled areas, on which traffic is often very heavy, is considered as urban. 494

usable sensitivity. The minimum standard modulated carrier-signal power required to produce usable receiver output. See: **receiver performance.** 181

use-as-is (nuclear power quality assurance). A disposition permitted for a nonconforming item when it can be established that the item is satisfactory for its intended use. 417

useful active dimension (of a position-sensitive detector)(charged-particle detectors). A dimension (that is, length, width) of that region of a position-sensitive detector over which the specifications of resolution and linearity are met. 119

useful life (1) (reliability). The period from a stated time, during which, under stated conditions, an item has an acceptable failure rate, or until an unrepairable failure occurs." 164, 182
(2) (nuclear power generating stations). The time to failure for a specific service condition. 31

useful line. See: **available line.**

useful output power (electron device). That part of the output power that flows into the load proper. 190

useful service life (thermal classification of electric equipment and electrical insulation). The length of time (usually in hours) for which an insulating material, insulation system, or electric equipment performs in an adequate or specified fashion. 506

user (1)(binary floating-point arithmetic)(radix-independentfloating-point arithmetic). Any person, hardware, or program not itself specified by ANSI/IEEE Std 754-1985 or ANSI/IEEE Std 854-1987 or both, having access to and controlling those operations of the programming environment specified in these standards.   469,588

(2)(computer applications). One who uses the services of a computer system. *Syn:* **end user.**   571

(3)(software requirements specifications). The person, or persons, who operate or interact directly with the system. The user(s) and the customer(s) are often not the same person(s).   449

user documentation (software). Documentation conveying to the end user of a system instructions for using the system to obtain desired results, for example, a user's manual. *See:* **documentation; system; system documentation.**   434

user group (computer applications). An organization of users of a particular class of computer systems, designed to allow the users to share knowledge about and programs for those systems and to formulate feedback for the systems' manufacturers.   571

usual service conditions. Environmental conditions in which standard machines are designed to operate. The temperature of the cooling medium does not exceed 40 degrees Celsius and the altitude does not exceed 3300 feet.   63

utilance. *See:* **room utilization factor.**

utility (NESC). An organization responsible for the installation, operation or maintenance of electric supply or communication systems.   494

utility interactive system (NESC). An electric power productive system which is operating in parallel with and capable of delivering energy to a utility electric supply system.   494

utility power. *Syn:* **commercial power**

utility routine. *See:* **service routine.**

utility software. Computer programs or routines designed to perform some general support function required by other application software, by the operating system, or by system users. *See:* **application software; computer program; function; operating system; routine; system.**   434

utilization equipment (1)(NESC). Equipment, device, and connected wiring which utilize electric energy for mechanical, chemical, heating, lighting, testing, or similar purposes and are not a part of supply equipment, supply lines, or communication lines.   494

(2) (National Electrical Code). Equipment which utilizes electric energy for mechanical, chemical, heating, lighting, or similar purposes.   256

utilization factor (system utilization factor). The ratio of the maximum demand of a system to the rated capacity of the system. *Note:* The utilization factor of a part of the system may be similarly defined as the ratio of the maximum demand of the part of the system to the rated capacity of the part of the system under consideration. *See:* **alternating-current distribution; direct-current distribution.**   64

utilization time (hauptnutzzeit) (medical electronics). (1) The minimum duration that a stimulus of rheobasic strength must have to be just effective. (2) The shortest latent period between stimulus and response obtainable by very strong stimuli. (3) The latent period following application of a shock of theobasic intensity. *See:* **biological.**   192

utilization voltage (system voltage ratings). The root-mean-square phase-to-phase or phase-to-neutral voltage at the line terminals of utilization equipment. *See:* **system voltage; nominal system voltage; maximum system voltage; service voltage; low voltage; medium voltage; high voltage.**   260

UTT oriented language (test, measurement and diagnostic equipment). A computer language used to program automatic test equipment to test units under test (UUT's), whose characteristics are directed to the test needs of the UUT's and therefore do not imply the use of a specific ATE (automatic test equipment) system or family of ATE systems.   54

UUT. *See:* **unit under test.**

V

vacant code (telephone switching systems). A digit or a combination of digits that is unassigned.   55

vacant-code tone (telephone switching systems). A tone that indicates that an unassigned code has been dialed.   55

vacant number (telephone switching systems). An unassigned or unequipped directory number.   55

vacuum envelope (electron tube). The airtight envelope that contains the electrodes. *See:* **electrode (of an electron tube).**   328

vacuum-tube amplifier. An amplifier employing electron tubes to effect the control of power from the local source. *See:* **amplifier.**   111

vacuum-tube radio frequency generator. *See:* **radio frequency generator, electron tube type.**

vacuum-tube transmitter. A radio transmitter in which electron tubes are utilized to convert the applied electric power into radio-frequency power. *See:* **radio transmitter.**   111

vacuum-tube voltmeter. *See:* **electronic voltmeter.**

vacuum valve. A device for sealing and unsealing the passage between two parts of an evacuated system. *See:* **rectification.**   328

valance (illuminating engineering). A longitudinal shield- ing member mounted across the top of a window or along a wall, to conceal light sources, giving both upward and downward distributions.   167

valence band. The range of energy states in the spectrum of a solid crystal in which lie the energies of the valence electrons that bind the crystal together. *See:* **semiconductor.**   245

valance lighting (illuminating engineering). Lighting comprising light sources shielded by a panel parallel to the wall at the top of a window.   167

validation (1)(programmable digital computer systems in safety systems of nuclear power generating stations). The test and evaluation of the integrated computer system to ensure compliance with the functional, performance, and interface requirements.   554

(2)(software verification and validation plans)(software). The process of evaluating software at the end of the software development process to ensure compliance with software requirements.*See:* **requirement; software; software development process; verification.**   511,434

(3) (test, measurement and diagnostic equipment). That process in the production of a test program by which the correctness of the program is verified by running it on the automatic test equipment together with the unit under test. The process includes the identification of run-time errors, procedure errors, and other non-compiler errors, not uncovered by pure software methods. The process is generally performed with the customer or designated representative as a witness.   54

valley (pulse terms). A portion of a pulse waveform between two specified peak magnitudes of the same polarity. *See:* Note in **preshoot** entry.   254

valley point (tunnel-diode characteristic). The point on the forward current-voltage characteristic corresponding to the second-lowest positive (forward) voltage at which $di/dV = 0$. *See:* **peak point (tunnel-diode characteristic).**   315, 191

valley-point current (tunnel-diode characteristic). The current at the valley point. *See:* **peak point (tunnel-diode characteristic).**   315, 191

valley-point voltage (tunnel-diode characteristic). The voltage at which the valley point occurs. *See:* **peak point (tunnel-diode characteristic).**   315, 191

value (1) (several) (automatic control). The quantitative measure of a signal or variable. *See:* **control system, feedback.**   329

(2) (direct-current through test object). The arithmetic mean value. *See:* **test voltage and current.**   307, 60

(3) (test direct voltages). The arithmetic mean value: that is, the integral of the voltage over a full period of the ripple divided by the period. *Note:* The maximum value of the test voltage may be taken approximately as the sum of the arithmetic mean value plus the ripple magnitude. *See:* **test voltage and current.**   307, 201

(4) (alternating test voltage). The peak value divided by $(2)^{1/2}$. *See:* **test voltage and current.**   201

value added service (data communication). A communications service utilizing communications common carrier networks for transmission and providing added data services with separate additional equipment. Such added service features may be store-and-forward switching, terminal interfacing and host interfacing.   12

value, desired. *See:* **value, ideal.**

value, ideal (automatic control). The value of the ultimately controlled variable of an idealized system under consideration. *Syn:* **desired value.** *See:* **ideal value.**   56

value, Munsell. *See:* **Munsell value.**

value of test voltage (high voltage testing). (1) The voltage value, defined according to Section 2 of ANSI/IEEE Std 4-1978 which is to be applied in a test.

(2) (lightning impulse tests, general applicability). The value of the test voltage is, for a smooth lightning impulse, the crest value. With some test circuits, oscillations or an overshoot may occur at the crest of the impulse [see Fig 2.1(a)-(d)]. If the frequency of such oscillations is not less than 0.5MHz or the duration of overshoot not over 1 μ, a mean curve should be drawn as in Fig 2.1(a) and (b) and, for the purpose of measurement, the maximum amplitude of this curve defines the value of the test voltage. Permissible amplitude limits for the oscillations of overshoot, on standard lightning impulses, are given in 2.4.2.2, IEEE Std 4-1978. For other impulse shapes [see, for example, Fig 2.1(e) and (h)], the appropriate apparatus standard should define the value of the test voltage, taking account of the type of test and of test object.

(3) The value of the test voltage is defined by its arithmetic mean value.

(4) The value of the test voltage is defined by its peak value divided by $\sqrt{2}$. *Note:* The appropriate apparatus standards may require a measurement of the rms value of the test voltage instead of the peak value for cases where the rms value may be of importance. Such cases are, for instance, when thermal effects are under investigation.   150

value of the test current (high voltage testing). The value of the test current is normally defined by the crest value. With some test circuits, overshoot or oscillations may be present on the current. The appropriate apparatus standard should specify whether the value of the test current should be defined by the actual crest or by a smooth curve drawn through the oscillations.   150

values of spectral luminous efficiency for photopic vision, V(λ) (illuminating engineering). Values at 10-nm intervals were adopted by the International Commission on Illumination in 1924 and were adopted in 1933 by the International Committee for Weights and Measures as a basis for the establishment of photometric standards of types of sources differing from the primary standard in spectral distribution of radiant flux. These values are given in the second column of the accompanying table; the intermediate values given in the other columns have been interpolated. *Note:* These standard values of spectral luminous efficiency were determined by observations with a two-degree photo- metric field having a moderately high luminance, photometric evaluations based upon them consequently do not apply exactly to other conditions

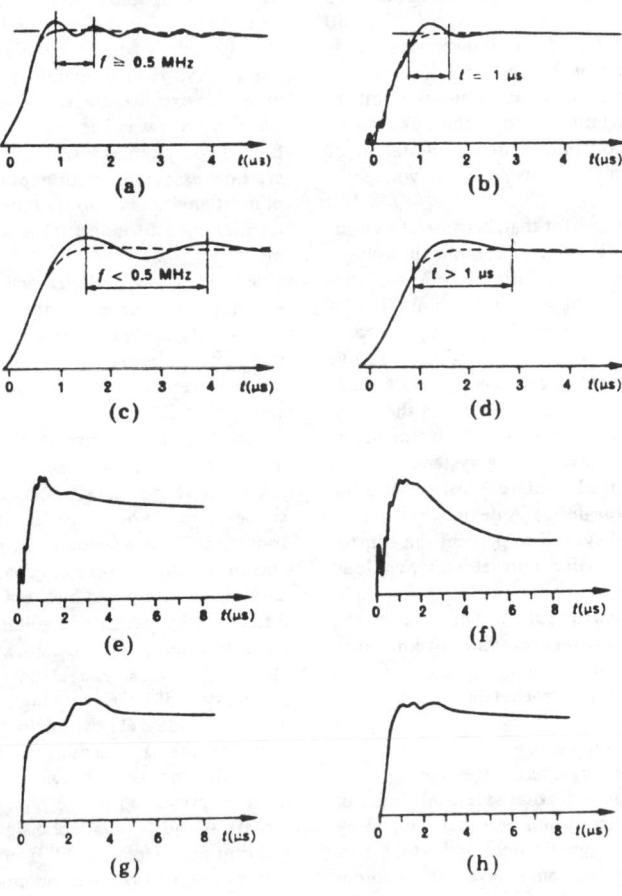

Fig 2.1

Examples of Lightning Impulses with Oscillations or Overshoot (a), (b)—The Value of the Test Voltage is Determined by a Mean Curve (Broken Line); (c), (d)—The Value of the Test Voltage is Determined by the Crest Value; (e), (f), (g), (h)—No General Guidance Can be Given for the Determination of the Value of the Test Voltage

of observation. Watts weighted in accord with these standard values are often referred to as light-watts.
167

values of spectral luminous efficiency for scotopic vision, $V'(\lambda)$ (illuminating engineering). Values at 10-nm intervals were provision- ally adopted by the International Commission on Illumination in 1951. *Note:* These values of spectral luminous efficiency were determined by observation by young dark-adapted observers using extra-foveal vision at near-threshold luminance.
167

value, true (measured quantity). The actual value of a precisely defined quantity under the conditions existing during its measurement.
147

valve. *See:* electron tube.

valve action (electrochemical). The process involved in the operation of an electrochemical valve. *See:* electrochemical valve.
328

valve actuator (valve actuators). An electric, pneumatic, hydraulic, or electrohydraulic power-driven mechanism for positioning two-position or modulating valves, and dampers. Included are those components required to control valve action and to provide valve position output signals, as defined in the actuator specification.
492

valve actuator specification (valve actuators). A document to be provided to the valve actuator manufacturer which contains technical requirements for a specific application.
142

valve arrester (1)(metal-oxide surge arresters for ac power circuits). An arrester that includes one or more valve elements.
583

(2)(surge arrester). An arrester that includes a valve element.   430

valve element (1)(metal-oxide surge arresters for ac power circuits). A resistor that, because of its nonlinear current-voltage characteristic, limits the voltage across the arrester terminals during the flow of discharge current and contributes to the limitation of follow current at normal power-frequency voltage.
                                                                    583

(2)(surge arrester). A resistor that, because of its nonlinear current-voltage characteristic, limits the voltage across the arrester terminals during the flow of discharge current and contributes to the limitation of follow current at normal power-frequency voltage.
                                                                    430

valve-point loading control (electric power system). A control means for making a unit operate in the more efficient portions of the range of the governor-controlled valves. *See:* **speed-governing system.**   94

valve position limiter (load limit)(control systems for steam turbine-generator units). A device that acts on the speed/load-control system to prevent the control valve(s) from opening beyond a preset limit. *Syn:* **load limiter.**   522

valve ratio (electrochemical valve). The ratio of the impedance to current flowing from the valve metal to the compound or solution, to the impedance in the opposite direction. *See:* **electrochemical valve.**
                                                                    328

valve tube. *See:* **kenotron.**

valve-type arrester. *See:* **arrester, valve-type.**

Van Allen belt(s) (communication satellite). Belts of charged particles (electrons and protons) trapped by the earth's (external) magnetic field and which surround the earth at altitudes from 1000 to 6000 kilometers. The paths of the particles are determined by the directions of the (external) lines of force of the earth's magnetic field. The particles migrate from the region above earth's equator toward the North Pole, then toward the South Pole, then return to the region above the equator.   74

vane (navigation aid terms). A device to indicate the direction from which the wind blows.   526

vane-type relay. A type of alternating-current relay in which a light metal disc or vane moves in response to a change of the current in the controlling circuit or circuits.   328

V antenna. A V-shaped arrangement of two conductors, balanced-fed at the apex, with included angle, length, and apex height above the earth chosen so as to give the desired directive properties to the radiation pattern.   111

vapor openings. Openings through a tank shell or roof above the surface of the stored liquid. Such openings may be provided for tank breathing, tank gauging, fire fighting, or other operating purposes. *See:* **lightning protection and equipment.**   297

vapor-safe electric equipment. A unit so constructed that it may be operated without hazard to its surroundings in an atmosphere containing fuel, oil, alcohol, or other vapors that may occur in aircraft: that is,

the unit is capable of so confining any sparks, flashes, or explosions of the combustible vapors within itself that ignition of the surrounding atmosphere is prevented. *Note:* This definition closely parallels that given for **explosionproof:** however, it is believed that the new term is needed in order to avoid the connotation of compliance with Underwriter's standards that are now associated with **explosionproof** in the minds of most engineers who are familiar with the use of that term applied to industrial motors and control equipment.   328

vaportight (power switchgear). So enclosed that vapor will not enter the enclosure.   103

vapor-tight luminaire (illuminating engineering). A luminaire designed and approved for installation in damp or wet locations. It is also described as 'enclosed and gasketed.'   167

var (electric power circuits). The unit of reactive power in the International System of Units (SI). The var is the reactive power at the two points of entry of a single-phase, two-wire circuit when the product of the root-mean-square value in amperes of the sinusoidal current by the root-mean-square value in volts of the sinusoidal voltage and by the sine of the angular phase difference by which the voltage leads the current is equal to one.   210

VAR (visual-aural range)(navigation aid terms). A special type of VHF (very high frequency) radio range which provides: (1) two reciprocal radio lines of position presented to the pilot visually on a course deviation indicator, and (2) two reciprocal radial lines of position presented to the pilot as interlocked and alternate A and N aural code signals. The aural lines of position are displaced 90° from the visual and either may be used to resolve the ambiguity of the other.
                                                                    526

varactor. A two-terminal semiconductor device in which the electrical characteristic of primary interest is a voltage-dependent capacitance.   191

varhour. The unit of a quadrature-energy (quadergy) in the International System of Units (SI). The varhour is the quadrature energy that is considered to have flowed past the points of entry of a reactive circuit when a reactive power of one var has been maintained at the terminals of entry for one hour.   210

varhour constant (metering). The registration, expressed in varhours, corresponding to one revolution of the rotor.   212

varhour meter (metering). An electricity meter that measures and registers the integral, with respect to time, of the reactive power of the circuit in which it is connected. The unit in which this integral is measured is usually the kilovarhour.   212

variable (1)(electrical heating applications to melting furnaces and forehearths in the glass industry). A quantity or condition that is subject to change.
                                                                    520

(2) (ATLAS). A quantity the value of which is assigned at program run time.   400

(3) (software). (1) A quantity that can assume any of a given set of values. (2) In programming, a character or group of characters that refers to a value and, in the execution of a computer program, corresponds to an address. *See:* **address; computer program; execution.**   434

variable-area track (electroacoustics). A sound track divided laterally into opaque and transparent areas, a sharp line of demarcation between these areas forming an oscillographic trace of the wave shape of the recorded signal. *See:* **phonograph pickup.**   176

variable-block format. A format that allows the number of words in successive blocks to vary.   224, 207

variable carrier. *See:* **controlled carrier.**

variable, complex. *See:* **complex variable.**

variable-density track (electroacoustics). A sound track of constant width, usually but not necessarily of uniform light transmission on any instantaneous transverse axis, on which the average light transmission varies along the longitudinal axis in proportion to some characteristic of the applied signal. *See:* **phonograph pickup.**   176

variable, directly controlled. *See:* **directly controlled variable.**

variable field. One that varies with time.   210

variable-frequency telemetering (electric power systems). A type of telemetering in which the frequency of the alternating-voltage signal is varied as a function of the magnitude of the measured quantity. *See:* **telemetering.**   94

variable, indirectly controlled. *See:* **indirectly controlled variable.**

variable inductor. *See:* **continuously adjustable inductor.**

variable, input. A variable applied to a system or element. *See:* **control system, feedback.**   56, 329

variable, manipulated. *See:* **manipulated variable.**

variable modulation (VOR [very high-frequency omnidirectional range])(navigation aid terms). That modulation of the ground station radiation which produces a signal in the airborne receiver whose phase with respect to a radiated reference modulation corresponds to the bearing of the receiver.   526

variable-mu tube (variable-μ tube) (remote-cutoff tube). An electron tube in which the amplification factor varies in a predetermined way with control-grid voltage.   125

variable operating costs (power operations). Monies that vary or fluctuate with operation or utilization of plant.   516

variable, output. A variable delivered by a system or element. *See:* **control system, feedback.**   56, 329

variable point (mathematics of computing). Pertaining to a numeration system in which the position of the radix point is indicated by a special character at that position. *See:* **fixed point; floating point.**   564

variable-reluctance microphone (magnetic microphone). A microphone that depends for its operation on variations in the reluctance of a magnetic circuit. *See:* **microphone.**   328

variable-reluctance pickup (magnetic pickup). A photograph pickup that depends for its operation on the variation in the reluctance of a magnetic circuit. *See:* **phonograph pickup.**   328

variable-reluctance transducer. An electroacoustic transducer that depends for its operation on the variation in the reluctance of a magnetic circuit.   176

variable-speed axle generator. An axle generator in which the speed of the generator varies directly with the speed of the car. *See:* **axle generator system.**   328

variable-speed drive (industrial control). An electric drive so designed that the speed varies through a considerable range as a function of load. *See:* **electric drive.**   206

variable speed motor (rotating machinery). A motor with a positively damped speed-torque characteristic which lends itself to controlled speed applications.   63

variables, state. *See:* **state variables.**

variable threshold transistor (metal-nitride-oxide field-effect transistor). An insulated-gate field-effect transistor (IGFET) whose threshold voltage can be varied electrically to predetermined levels. The memory metal-nitride-oxide semiconductor (MNOS) memory transistor is a specific example of this type.   386

variable-torque motor. (1) A multispeed motor whose rated load torque at each speed is proportional to the speed. Thus the rated power of the motor is proportional to the square of the speed. (2) An adjustable-speed motor in which the specified torque increases with speed. It is common to provide a variable-torque adjustable-speed motor in which the torque varies as the square of the speed and hence the power output varies as the cube of the speed. *See:* **asynchronous machine.**   63

variable, ultimately controlled. *See:* **ultimately controlled variable.**

variable-voltage transformer (power and distribution transformer). An autotransformer in which the output voltage can be changed (essentially from turn to turn) by means of a movable contact device sliding on the shunt winding turns.   53

variation (navigation aid terms). The angle between the magnetic and geographical meridians at any place. *See:* **magnetic variation.**   526

varindor. An inductor whose inductance varies markedly with the current in the winding.   328

variocoupler (radio practice). A transformer, the self-impedance of whose windings remains essentially constant while the mutual impedance between the windings is adjustable.   329

variolosser. A device whose loss can be controlled by a voltage or current.   328

variometer. A variable inductor in which the change of inductance is effected by changing the relative position of two or more coils.   341

varioplex. A telegraph switching system that establishes connections on a circuit-sharing basis between a multiplicity of telegraph transmitters in one locality and respective corresponding telegraph receivers in another locality over one or more intervening telegraph channels. Maximum usage of channel capacity is secured by momentarily storing the signals and allocating circuit time in rotation among those transmitters having intelligence in storage. *See:* **telegraphy.**

328

varistor. (1) A two-terminal resistive element, composed of an electronic semiconductor and suitable contacts, that has a markedly nonlinear volt-ampere characteristic. (2) A two-terminal semiconductor device having a voltage-dependent nonlinear resistance. *Note:* **Varistors** may be divided into two groups, symmetrical and nonsymmetrical, based on the symmetry or lack of symmetry of the volt-ampere curve. *See:* **semiconductor.**

245

varistor capacitance (low voltage varistor surge arresters). Capacitance between the two terminals of the varistor measured at a specified frequency and bias.

62

varistor resistance (low voltage varistor surge arresters). Static resistance of the varistor at a given operating point, described as the ratio of varistor voltage to varistor current.

62

varistor voltage (low voltage varistor surge arresters). Voltage across the varistor measured at a given current.

62

varmeter (reactive volt-ampere meter). An instrument for measuring reactive power. It is provided with a scale usually graduated in either vars, kilovars, or megavars. If the scale is graduated in kilovars or megavars, the instrument is usually designated as a kilovarmeter or megavarmeter. *See:* **instrument.**

328

varnish (rotating machinery). A liquid composition that is converted to a transparent or translucent solid film after application as a thin layer.

63

varnished fabric (varnished mat) (rotating machinery). A fabric or mat in which the elements and interstices have been essentially coated and filled with an impregnant such as a compound or varnish and that is relatively homogeneous in structure. *See:* **rotor (rotating machinery); stator.**

63

varnished tubing. A flexible tubular product made from braided cotton, rayon, nylon, glass, or other fibers, and coated, or impregnated and coated, with a continuous film or varnish, lacquer, a combination of varnish and lacquer, or other electrical insulating materials.

328

varying duty (1) (National Electrical Code). Operation at loads, and for intervals of time, both of which may be subject to wide variation.

256

(2) (packaging machinery) (power and distribution transformer). A requirement of service that demands operation at loads, and for intervals of time, both of which may be subject to wide variation.

429

(3) (rating of electric equipment). A requirement of service that demands operation at loads, and for periods of time, both of which may be subject to wide variation. *See:* **asynchronous machine; voltage regulator.**

53,310,257

varying parameter. *See:* **linear varying parameter.**

210

varying-speed motor (electric installations on shipboard). A motor whose speed varies with the load,

ordinarily decreasing when the load increases; such as a series-wound or repulsion motor.

3

varying-voltage control. A form of armature-voltage control obtained by impressing on the armature of the motor a voltage that varies considerably with change in load, with a consequent change in speed, such as may be obtained from a differentially compound-wound generator or by means of resistance in the armature circuit. *See:* **control.**

206

VASIS (visual approach slope indicator system) (illuminating engineering). The system of angle-of-approach lights accepted as a standard by the International Civil Aviation Organization, comprising two bars of lights located at each side of the runway near the threshold and showing red or white or a combination of both (pink) to the approaching pilot depending upon his position with respect to the glide path.

167

vault (NESC). An enclosure above or below ground which personnel may enter and is used for the purpose of installing, operating, or maintaining equipment or cable which need not be of a submersible design.

494

vault-type transformer (power and distribution transformer). A transforme that is so constructed as to be suitable for occasional submerged operation in water under specified conditions of time and external pressure.

53

V-beam radar. A ground-based three-dimensional radar system for the determination of distance, bearing and, uniquely, the height or elevation angle of the target. It uses two fan-shaped beams, one vertical and the other inclined, that rotate together in azimuth so as to give two responses from the target; the time difference between these responses, together with distance, being factors used in determining the height of the target.

13

VCO. *See:* **voltage-controlled oscillator.**

13

V curve (synchronous machine). The load characteristic giving the relationship between the armature current and the field current for constant values of load, power, and armature voltage. *See:* **asynchronous machine.**

63

vector. A mathematico-physical quantity that represents a vector quantity. *See:* **mathematico-physical quantity (mathematical quantity) (abstract quantity).**

210

vector electrocardiogram (electrobiology) (vectorcardiogram). The 2-dimensional or 3-dimensional presentation of cardiac electric activity that results from displaying lead pairs against each other rather than against time. More strictly, it is a loop pattern taken from leads placed orthogonally. *See:* **electrocardiogram.**

192

vector field. The totality of vectors in a given region represented by a vector function $V(x,y,z)$ of the space coordinates x,y,z.

210

vector function. A functional relationship that results in a vector.

210

vector norm (control systems). The measure of the size of a vector, with the usual norm properties. *Note:* (1)

Vector norm of x is denoted by $\|x\|$
(2) Norm properties are:

$$\|x\| > 0 \quad \text{for } x \neq 0$$
$$\|0\| = 0$$
$$\|\alpha x\| = |\alpha| \cdot \|x\|$$
$$\|x_1 + x_2\| \leqslant \|x_1\| + \|x_2\|$$

56

vector operator del ∇. A differential operator defined as follows in terms of Cartesian coordinates:

$$\nabla = \mathbf{i}\,\frac{\partial}{\partial x} + \mathbf{j}\,\frac{\partial}{\partial y} + \mathbf{k}\,\frac{\partial}{\partial z}.$$

210

vector power. *See:* **power, vector.**

vector product (cross product). The vector product of vector **A** and a vector **B** is a vector **C** that has a magnitude obtained by multiplying the product of the magnitudes of **A** and **B** by the sine of the angle between them: the direction of **C** is that traveled by a right-hand screw turning about an axis perpendicular to the plane of **A** and **B**, in the sense in which **A** would move into **B** by a rotation of less than 180 degrees: it is assumed that **A** and **B** are drawn from the same point. The vector product of two vectors **A** and **B** may be indicated by using a small cross: $\mathbf{A} \times \mathbf{B}$. The direction of the vector product depends on the order in which the vectors are multiplied, so that $\mathbf{A} \times \mathbf{B} = -\mathbf{B} \times \mathbf{A}$. If the two vectors are given in terms of their rectangular components, then

$$\mathbf{A} \times \mathbf{B} = \begin{vmatrix} \mathbf{i} & \mathbf{j} & \mathbf{k} \\ A_x & A_y & A_z \\ B_x & B_y & B_z \end{vmatrix}$$
$$= \mathbf{i}(A_y B_z - A_z B_y) + \mathbf{j}(A_z B_x - A_x B_z) + \mathbf{k}(A_x B_y - A_y B_x).$$

Example: The linear velocity **V** of a particle in a rotating body is the vector product of the angular velocity ω and the radius vector **r** from any point on the axis to the point in question, or

$$\mathbf{V} = \omega \times \mathbf{r}$$
$$= -\mathbf{r} \times \omega$$

210

vector quantity. Any physical quantity whose specification involves both magnitude and direction and that obeys the parallelogram law of addition.   210

vector, state. *See:* **state vector.**

vehicle (navigation aid terms). That in or on which a person or thing is being or may be carried.   526

vehicle-derived navigation data (navigation aid terms). Data obtained from measurements made at a vehicle.
526

vehicle maneuver effects (gyro). Gyro output errors due to vehicle maneuvers.   46

veiling brightness (illuminating engineering). A brightness super- imposed on the retinal image which reduces its contrast. It is this veiling effect produced by bright sources or areas in the visual field which results in decreased visual performance and visibility.
167

veiling reflections (illuminating engineering). Regular reflections which are superimposed upon diffuse reflections from an object which partially or totally obscure the details to be seen by reducing the contrast. This sometimes is called reflected glare. *See:* **reflected glare.**   167

velocity correction (industrial control). A method of register control that takes the form of a gradual change in the relative velocity of the web.   206

velocity level in decibels of a sound (acoustics). Twenty times the logarithm to the base 10 of the ratio of the particle velocity of the sound to the reference particle velocity. The reference particle velocity shall be stated explicitly. *Note:* In many sound fields the particle velocity ratios are not proportional to the square root of corresponding power ratios and hence cannot be expressed in decibels in the strict sense: however, it is common practice to extend the use of the decibel to these cases.   176

velocity microphone. A microphone in which the electric output substantially corresponds to the instantaneous particle velocity in the impressed sound wave. *Note:* A velocity microphone is a gradient microphone of order one, and it is inherently bidirectional. *See:* **gradient microphone; microphone.**   176

velocity-modulated amplifier (velocity-variation amplifier). An amplifier that employs velocity modulation to amplify radio frequencies. *See:* **amplifier.**
328

velocity-modulated oscillator. An electron-tube structure in which the velocity of an electron stream is varied (velocity-modulated) in passing through a resonant cavity called a buncher. Energy is extracted from the bunched electron stream at a higher energy level in passing through a second cavity resonator called the catcher. Oscillations are sustained by coupling energy from the catcher cavity back to the buncher cavity. *See:* **oscillatory circuit.**   111

velocity-modulated tube. An electron-beam tube in which the velocity of the electron stream is alternately increased and decreased with a period comparable with the total transit time.   244, 190

velocity modulation (velocity variation) (of an electron beam). The modification of the velocity of an electron stream by the alternate acceleration and deceleration of the electrons with a period comparable with the transit time in the space concerned. *See:* **velocity-modulated oscillator; velocity modulated tube.**   125

velocity response factor (moving-target indication)

(radar). The clutter filter frequency response defined by the ratio of power gain at a specific target Doppler frequency to the average power gain over all target Doppler frequencies of interest.   13

velocity shock. A mechanical shock resulting from a nonoscillatory change in velocity of an entire system.   176

velocity sorting (electronic). Any process of selecting electrons according to their velocities.   125

velocity storage (digital accelerometer). The velocity information which is stored in the accelerometer as a result of its dynamics.   383

velocity variation. *See:* **velocity modulation (electron beam).**

velocity-variation amplifier. *See:* **velocity-modulated amplifier.**

Venn diagram. A diagram in which sets are represented by closed regions.   255, 77

vent (1)(metal-oxide surge arresters for ac power circuits)(surge arrester). An intentional opening for the escape of gases to the outside.   583, 430, 2

(2) (of a fuse) (power switchgear). The means provided for the escape of the gases developed during circuit interruption. *Note:* In distribution oil cutouts, the vent may be an opening in the housing or an accessory attachable to a vent opening in the housing with suitable means to prevent loss of oil.   103, 443

(3) (rotating machinery). An opening that will permit the flow of air.   63

vented fuse (or fuse unit) (power switchgear). A fuse with provision for the escape of arc gases, liquids, or solid particles to the surrounding atmosphere during circuit interruption.   103, 443

vented power fuse (National Electrical Code) (installations and equipment operating at over 600 volts, nominal). A fuse with provision for the escape of arc gases, liquids, or solid particles to the surrounding atmosphere during circuit interruption.   256

vent finger. *See:* **duct spacer.**

ventilated (1) (National Electrical Code). Provided with a means to permit circulation of air sufficient to remove an excess of heat, fumes, or vapors.   256

(2) (power and distribution transformer). Provided with a means to permit circulation of the air sufficiently to remove an excess of heat, fumes, or vapors.   53

ventilated dry-type transformer (dry-type general purpose distribution and power transformers). A dry-type transformer which is so constructed that the ambient air may circulate through its enclosure to cool the transformer core and windings.   555, 53

ventilated enclosure (1)(metal-enclosed bus and calculating losses in isolated-phase bus). An enclosure so constructed as to provide for the circulation of external air through the enclosure to remove heat, fumes, or vapors.   574

(2) (power switchgear). An enclosure provided with means to permit circulation of sufficient air to remove an excess of heat, fumes, or vapors. *Note:* For outdoor applications ventilating openings or louvres are usually filtered, screened, or restricted to limit the entrance of dust, dirt or other foreign objects.   103

ventilating and cooling loss (synchronous machine). Any power required to circulate the cooling medium through the machine and cooler (if used) by fans or pumps that are driven by external means (such as a separate motor) so that their power requirements are not included in the friction and windage loss. It does not include power required to force ventilating gas through any circuit external to the machine and cooler.   244, 63

ventilating duct (cooling duct) (rotating machinery). A passage provided in the interior of a magnetic core in order to facilitate circulation of air or other cooling agent.   244, 63

ventilating passage (rotating machinery). A passage provided for the flow of cooling medium.   63

ventilating slot (rotating machinery). A slot provided for the passage of cooling medium.   63

verification (1)(programmable digital computer systems in safety systems of nuclear power generating stations). The process of determining whether or not the product of each phase of the digital computer system development process fulfills all the requirements imposed by the previous phase.   554

(2)(software verification and validation plans). The process of determining whether or not the products of a given phase of the software development cycle fulfill the requirements established during the previous phase.   511

(3) (nuclear power quality assurance). The act of reviewing, inspecting, testing, checking, auditing, or otherwise determining and documenting whether items, processes, services, or documents conform to specified requirements.   417

(4) (software). (A) The process of determining whether the products of a given phase of the software development cycle fulfill the requirements established during the previous phase. (B) Formal proof of program correctness. (C) The act of reviewing, inspecting, testing, checking, auditing or otherwise establishing and documenting whether items, processes, services, or documents conform to specified requirements. *See:* **document; process; program; proof of correctness; requirement; software development cycle; testing; validation.**   434

verification relay (power switchgear). A monitoring relay restricted to functions pertaining to power-system conditions and not involving opening circuit breakers during fault condition. *Note:* Such a relay is sometimes referred to as a check or checking relay.   103

verification system. *See:* **automated verification system.**   434

verify. To check, usually automatically, one typing or recording of data against another in order to minimize human and machine errors in the punching of tape or cards.   235

vernier control (industrial control). A method for improving resolution. The amount of vernier control is expressed as either the percent of the total operating range or of the actual operating value, whichever is appropriate to the circuit in use. *See:* **control system, feedback.**   200

vertex. *See:* **node.**

vertex plate (of a reflector antenna). A small auxiliary reflector place in front of the main reflector near its vertex for the purpose of reducing the standing waves in the feed due to the reflected waves from the main reflector.   111

vertical amplifier (oscilloscope). An amplifier for signals intended to produce vertical deflection. *See:* **oscillograph.**   185

vertical antenna (overhead-power-line corona and radio noise). (1) shunt-fed vertical antenna. A vertical antenna connected to ground at the base and excited (or connected to a receiver) at a point suitably positioned above the grounding point. (2) series-fed vertical antenna. A vertical antenna insulated from ground and energized (or connected to a receiver) at the antenna base. *Notes:* (A) A rod antenna measures the electric field component of the electromagnetic wave. (B) A rod antenna is omnidirectional. (C) The condition of a rod antenna to a receiver may be via a coupler to which the rod is permanently attached. *Syn:* **rod antenna.**   411

vertical-break switch (power switchgear). One in which the travel of the blade is in a plane perpendicular to the plane of the mounting base. The blade in the closed position is parallel to the mounting base.
                                                                    103

vertical component of the electric field strength (measurement of power frequency electric and magnetic fields from ac power lines). The root-mean-square (rms) value of the component of the electric field along the vertical line passing through the point of measurement. This quantity is often used to characterize electric field induction effects in objects close to ground level.   514

vertical, gravity. *See:* **mass attraction vertical.**

vertical gyro. A two-degree-of-freedom gyro with provision for maintaining the spin axis vertical. In this gyro, output signals are produced by gimbal angular displacements which correspond to angular displacements of the case about two nominally orthogonal, horizontal axes.   46

vertical-hold control (television). A synchronizing control that adjusts the free-running period of the vertical-deflection oscillator.   18

vertically polarized field vector (antennas). A linearly polarized field vector whose direction is vertical.
                                                                    111

vertically polarized plane wave (antennas). A plane wave whose electric field vector is vertically polarized.
                                                                    111

vertically polarized wave (radio wave propagation). A linearly polarized wave whose electric field vector is vertical. *Notes:* (1) See parallel polarization. (2) The term "vertical polarization" is commonly employed to characterize ground-wave propagation in the medium-frequency broadcast band; these waves, however, have a small component of electric field in the direction of propagation due to finite ground conductivity.
                                                                    146

vertical machine (rotating machinery). A machine whose axis of rotation is approximately vertical.
                                                                     63

vertical plane of a searchlight (illuminating engineering). The plane through the axis of the searchlight drum which contains the elevation angle.   167

vertical reach switch. One in which the stationary contact is supported by a structure separate from the hinge mounting base. The blade in the closed position is perpendicular to the hinge mounting base.   27

vertical recording. A mechanical recording in which the groove modulation is in a direction perpendicular to the surface of the recording medium.   328

vertical riser cable. Cable designed for use in long vertical runs, as in tall buildings.   57

vertical rod or shaft. A component of a switch operating mechanism designed to transmit motion from an operating handle or power operator to a switch offset bearing or bell crank.   27

vertical section (1)(metal-enclosed low-voltage power circuit-breaker switchgear)(metal-clad and station-type cubicle switchgear)(metal-enclosed interrupter switchgear). That portion of the switchgear assembly between two successive vertical delineations and may contain one or more circuit breakers, auxiliary compartments, and associated primary conductors.
                                                            579, 572, 573

(2) (nuclear power generating stations). A portion of the motor control center normally containing one vertical bus assembly.   440

vertical switchboard (power switchgear). A control switchboard composed only of vertical panels. *Note:* This type of switchboard may be enclosed or have an open rear. An enclosed vertical switchboard has an overall sheet-metal enclosure (not grille) covering back and ends of the entire assembly, access to which is usually provided by doors of removable covers.
                                                                    103

very high-frequency omnidirectional range. *See:* **VOR.**

very-high-frequency radar (VHF radar). A radar operating at frequencies between 30 megahertz (MHz) and 300 MHz, usually in one of the International Telecommunications Union (ITU) assigned bands 138 MHz to 144 MHz or 216 MHz to 225 MHz.
                                                                     13

very-low-frequency high-potential test. An alternating-voltage high-potential test performed at a frequency equal to or less than 1 hertz. *See:* **asynchronous machine.**   63

very-low-frequency test (VLF test). A test made at a frequency considerably lower than the normal operating frequency. *Note:* In order to facilitate communication and comparison among investigators, this document recommends that the very low frequency used be 0.1 Hz + 25 percent.   135

vessel (1)(electrical heating systems). A container in which a material is to be heated or maintained at a temperature.   476

(2)(National Electrical Code). A container such as a barrel, drum, or tank for holding fluids or other material.   256

vestigial lobe. *See:* **shoulder lobe.**

vestigial sideband (data transmission). The transmitted portion of the sideband that has been largely suppressed by a transducer having a gradual cutoff in the neighborhood of the carrier frequency, the other sideband being transmitted without much suppression.
59

vestigial-sideband modulation. A modulation process involving a prescribed partial suppression of one of the two sidebands. 242

vestigial-sideband transmission (facsimile). That method of signal transmission in which one normal sideband and the corresponding vestigial sideband are utilized. *See:* **amplitude modulation; facsimile transmission.** 12

vestigial-sideband transmitter. A transmitter in which one sideband and a portion of the other are intentionally transmitted. *See:* **radio transmitter.** 111

VF. *See:* **voice frequency.**

VHF. *See:* **radio spectrum.**

VHF radar. *See:* **very-high-frequency radar.**

vial (liquid-scintillation counters)(liquid-scintillation counting). A glass or plastic sample container that meets the dimensional specifications of International Electrotechnical Commission (IEC) Pub 582-1977.
498, 422

via net loss (vnl) (data transmission). The net losses of trunks in the long distance switched telephone network of North America. The trunk is said to be in a via condition when it is an intermediate trunk in a longer switched connection. 59

vibrating beam accelerometer (VBA) (inertial sensor). A linear accelerometer whose proof mass is mechanically constrained by a force-sensitive beam resonator. The resultant oscillation frequency is a function of input acceleration. 46

vibrating bell. A bell having a mechanism designed to strike repeatedly when and as long as actuated. *See:* **protective signaling.** 328

vibrating circuit (telegraph circuit). An auxiliary local timing circuit associated with the main line receiving relay for the purpose of assisting the operation of the relay when the definition of the incoming signals is indistinct. *See:* **telegraphy.** 328

vibrating-contact machine regulator (power switchgear). A regulator that varies the excitation of an electric machine by changing the average time of engagement of vibrating contacts in the field circuit.
103

vibrating-reed relay. A relay in which the application of an alternating or a self-interrupted voltage to the driving coil produces an alternating or pulsating magnetic field that causes a reed to vibrate and operate contacts. *See:* **relay.** 259

vibrating string accelerometer. A device that employs one or more vibrating strings whose natural frequencies are affected as a result of acceleration acting on one or more proof masses. 46

vibration. An oscillation wherein the quantity is a parameter that defines the motion of a mechanical system. *See:* **oscillation.** 176

vibration detection system (protective signaling). A system for the protection of vaults by the use of one or more detector buttons firmly fastened to the inner surface in order to pick up and convert vibration, caused by burglarious attack on the structure, to electric impulses in a protection circuit. *See:* **protective signaling.** 328

vibration meter. An apparatus including a vibration pickup, calibrated amplifier, and output meter for the measurement of displacement, velocity, and acceleration of vibrations. *See:* **instrument.** 328

vibration relay (power switchgear). A relay that responds to the magnitude and frequency of a mechanical vibration. 103

vibration test (rotating machinery). a test taken on a machine to measure the vibration of any part of the machine under specified conditions. 63

vibrato. A family of tonal effects in music that depend upon periodic variations in one or more characteristics of the sound wave. *Note:* When the particular characteristics are known, the term **vibrato** should be modified accordingly, for example, **frequency vibrato; amplitude vibrato; phase vibrato** and so forth. 176

vibrator (cable plowing). That device which induces the vibration in a vibratory plow. *See:* **vibratory plow.**
52

vibratory isolation (cable plowing). Percentage reduction in force transmitted from vibration source to receiver by use of flexible mounting(s) (amount of isolation for a given unit varies with plow blade frequency).
52

vibratory plow (cable plowing). A plow utilizing induced periodic motion(s) of the blade in conjunction with drawbar pull for its movement through the soil. *Note:* Orbital and oscillating plows are types of vibratory plows that are commercially available. 52

vibropendulous error (accelerometer). A cross coupling rectification error caused by angular motion of the pendulum in a pendulous accelerometer in response to a linear vibratory input. The error varies with frequency and is maximum when the vibratory acceleration is applied in a plane normal to a pivot axis and at 45 degrees to the input axis. 46

video (1) (radar). Refers to the signal after envelope or phase detection, which in early radar was the displayed signal. Contains the relevant radar information after removal of the carrier frequency. 13
(2) (television). A term pertaining to the bandwidth and spectrum position of the signal resulting from television scanning. *Note:* In present usage, video means a bandwidth of the order of several megahertz, and a spectrum position that goes with a direct-current carrier. *See:* **signal wave.** 328

video filter (non-real time spectrum analyzer) (spectrum analyzer). A post detection low-pass filter.
390

video-frequency amplifier. A device capable of amplifying such signals as comprise periodic visual presentation. *See:* **television.** 11

video integration. (radar). A method of utilizing the redundancy of repetitive video signals to improve the output signal-to-noise ratio, by summing successive signals. 13

video mapping (radar). The electronic superposition of geographic or other data on a radar display.   13

video stretching (radar). The increasing of the duration of a video pulse.   13

video-telephone call (telephone switching systems). A call between stations equipped to provide video-telephone service.   55

vidicon. A camera tube in which a charge-density pattern is formed by photoconduction and stored on that surface of the photoconductor that is scanned by an electron beam, usually of low-velocity electrons. *See:* **television.**   178,190

viewing area (oscilloscope). The area of the phosphor screen of a cathode-ray tube that can be excited to emit light by the electron beam. *See:* **oscillograph; screen, viewing.**   184

viewing time (storage tubes). The time during which the storage tube is presenting a visible output corresponding to the stored information. *See:* **storage tube.**   174

viewing time, maximum usable (storage tubes). The length of time during which the visible output of a storage tube can be viewed, without rewriting, before a specified decay occurs. *Note:* The qualifying adjectives **maximum usable** are frequently omitted in general usage when it is clear that maximum usable viewing time implied. *See:* **storage tube.**   174

virtual cathode (potential-minimum surface) (electron tubes). A region in the space charge where there is a potential minimum that, by reason of the space charge density, behaves as a source of electrons. *See:* **electronic tube.**   244,190

virtual duration (of a peak of a rectangular-wave current or voltage impulse) (surge arresters). The time during which the amplitude of the wave is greater than 90 percent of its peak value.   308,62

virtual duration of peak of a rectangular impulse current (high voltage testing). The virtual duration of the peak of a rectangular impulse current is the time during which the current is greater than 90 percent of the peak value.   150

virtual duration of wave front (of an impulse)(metal-oxide surge arresters for ac power circuits)(surge arrester). The virtual value for the duration of the wave front is as follows: (1) For voltage waves with wave front duration less than 30 μs, either full or chopped on the front, crest, or tail, 1.67 times the time for the voltage to increase from 30% to 90% of its crest value. (2) For voltage waves with wave front duration of 30 μs or more, the time taken by the voltage to increase from actual zero to maximum crest value. (3) For current waves, 1.25 times the time for the current to increase from 10% to 90% of crest value.   583,430

virtual front time. *See:* **virtual duration of waveform impulse.**

virtual front time T_f (1) (impulse current). The virtual front time T_1 of an impulse current is defined as 1.25 times the interval between the instants when the impulse is 10 and 90 percent of the peak value. If oscillations are present on the front, the 10 and 90 percent

values should be derived from a mean curve drawn through these oscillations in an analogous manner to that used for oscillatory lightning impulses [see Fig 2.1 (a) under **value of test voltage**]. *Note:* The difference between front times measured according to this definition and to that given in 2.4.1.1.4 for lightning impulses is generally less than 10 percent.

(2) (lightning impulse tests, general applicability). The virtual front time T_1 of a lightning impulse is defined as 1.67 times the time interval T between the instants when the impulse is 30 and 90 percent of the peak value (points A and B, Figs 2.2-2.4). If oscillations are present on the front, points A and B should be taken on the mean curve drawn through these oscillations.   150

virtual height (1) (data transmission). The apparent height of an ionized layer determined from the time interval between the transmitted signal and the ionospheric echo at vertical incidence, assuming that the velocity of propagation is the velocity of light over the entire path.   59

(2) (radio wave propagation). The apparent height of an ionized layer determined from the time interval between the transmitted signal and the ionospheric echo at vertical incidence, assuming that the velocity of propagation is the velocity of light in a vacuum over the entire path.   146

virtual instant of chopping (voltage testing). The instant preceding point C on the figures (under **virtual front time,**) by 0.3 times the (estimated) virtual time of voltage collapse during chopping. *See:* **test voltage and current.**   201

virtual junction temperature. The temperature of the active semiconductor element of a semiconductor device based on a simplified representation of the thermal and electrical behavior of the device. It is particularly applicable to multi-junction semiconductor devices.   66

virtual machine (software). A functional simulation of a computer and its associated devices. *See:* **computer; simulation.**   434

virtual memory. *See:* **virtual storage.**

virtual origin (impulse current or voltage). *See:* **virtual zero time.**

virtual origin O_1 (1)(high voltage testing). The virtual origin O_1 of an impulse current is the instant preceding that at which the current is 10 percent of the peak value by a time 0.1 x T_1. For oscillograms having linear time sweeps, this is the intersection with the x axis of a straight line drawn through the 10 and 90 percent reference points on the front.   150

(2) (lightning impulse tests, general applicability). The virtual origin O_1 of a lightning impulse is the instant preceding that corresponding to point A (Figs 2-4 under **virtual front time**) by a time 0.3 x T_1. For oscillograms having linear time sweeps, this is the intersection with the x axis of a straight line drawn through reference points A and B on the front.   150

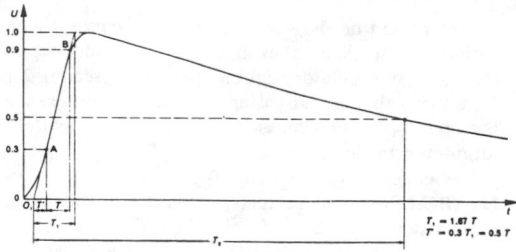

Fig 2.2
Full Lightning Impulse

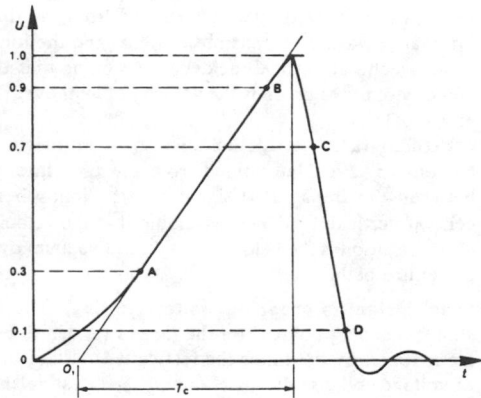

Fig 2.3
Lightning Impulse Chopped on the Front

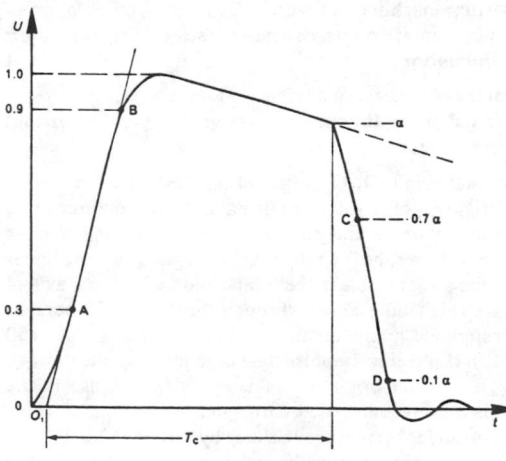

Fig 2.4
Lightning Impulse Chopped on the Tail

virtual peak value. *See:* **peak value.**

virtual rate of rise of the front (impulse voltage). The quotient of the peak value and the virtual front time. *Note:* The term **peak value** is to be understood as including the term **virtual peak value** unless otherwise stated. *See:* **test voltage and current.**   201

virtual resource (ATLAS). A notional test resource the performance characteristics of which conform to a summary of related test requirements.   400

virtual steepness of voltage during chopping (surge arresters). The quotient of the estimated voltage at the instant of chopping and the virtual time of voltage collapse. *See:* **test voltage and current.**

      308,201,62

virtual steepness of wavefront of an impulse (surge arresters). The slope of the line that determines the virtual-zero time. It is expressed in kilovolts per microsecond or kiloamperes per microsecond.

      308,62

virtual storage (software). The storage space that may be regarded as addressable main storage by the user of a computer system in which virtual addresses are mapped into real addresses. The size of virtual storage is limited by the addressing scheme of the computer system and by the amount of auxiliary storage available and not by the actual number of main storage locations. *See:* **address; computer system.**   434

virtual time of voltage collapse during chopping. 1.67 times the time interval between points C and D on the figures attached to the definition of **virtual instant of chopping.** *See:* **test voltage and current.**   201

virtual time to chopping (impulse voltage). The time interval between the virtual origin and the virtual instant of chopping. *See:* **test voltage and current.**

      201

virtual time to half-value T_2 (high voltage testing) (1). The virtual time to half-value T_2 of an impulse current is the time interval between the virtual origin and the instant on the tail at which the current has first decreased to half the peak value.

(2) (lightning impulse tests, general applicability). The virtual time to half-value T_2 of a lightning impulse is the time interval between the virtual origin and the instant on the tail when the voltage has decreased to half of the peak value.   150

virtual total duration of a rectangular impulse current (high voltage testing). The virtual total duration of a rectangular impulse current is the time during which the amplitude of the impulse is greater than 10 percent of its peak value. If oscillations are present on the front, a mean curve should be drawn in order to determine the time at which the 10 percent value is reached.   150

virtual zero point (of an impulse)(1)(metal-oxide surge arresters for ac power circuits). The intersection with the time axis of a straight line drawn through points on the front of the current wave at 10% and 90% crest value, or through points on the front of the voltage wave at 30% and 90% crest value.   583

(2)(surge arrester). The intersection with the zero axis of a straight line drawn through points on the front of the current wave at 10 percent and 90 percent crest value, or through points on the front of the voltage wave at 30 percent and 90 percent crest value.   430

virtual zero time (impulse voltage or current in a conductor) (surge arresters) (conventional origin) (virtual origin). The point on a graph of voltage-time or current-time determined by the intersection with the zero voltage or current axis, of a straight line drawn through two points on the front of the wave: (1) for full voltage waves and voltage waves chopped on the front, peak, or tail, the reference points shall be 30 percent of the peak value, and 82 for current waves the reference points shall be 10 percent and 90 percent of the peak value.   308, 62

viscous friction (industrial control). The component of friction that is due to the viscosity of a fluid medium, usually idealized as a force proportional to velocity, and that opposes motion. *See:* **control system, feedback.**   219,329,206

visibility (1) (meteorological) (illuminating engineering). A term that denotes the greatest distance that selected objects (visibility markers) or lights of moderate intensity of the order of 25 candles can be seen and identified under specified conditions of observation. The distance may be expressed in kilometers or miles in the USA until the metric system becomes more widely used.   167
(2) (light emitting diodes). The quality or state of being perceivable by the eye. In many outdoor applications, visibility is defined in terms of the distance at which an object can be just perceived by the eye. In indoor applications it usually is defined in terms of the contrast or size of a standard test object, observed under standardized viewing conditions, having the same threshold as the given object. *See:* **visual field. visual field.**   167

visibility factor (1) (pulsed radar). The ratio of single-pulse signal energy to noise power per unit bandwidth that provides stated probabilities of detection and false alarm on a display, measured in the intermediate-frequency portion of the receiver under conditions of optimum bandwidth and viewing environment; (2) (continuous-wave radar). The ratio of single-look signal energy to noise power per unit bandwidth using a filter matched to the time on target. The equivalent term for radar using automatic detection is detectability factor; for operation in a clutter environment a clutter visiblity factor is defined.   13
(2) (continuous-wave radar). The ratio of single-look signal energy to noise power per unit bandwidth using a filter matched to the time on target. The equivalent term for radar using automatic detection is **detectability factor;** for operation in a clutter environment a **clutter visiblity factor** is defined.   187

visibility level (VL) (illuminating engineering). A contrast multiplier to be applied to the visibility reference function to provide the luminance contrast required at different levels of task background luminance to achieve visibility for specified conditions relating to the task and observer.   167

visibility performance criteria function (VL8) (illuminating engineering). A function representing the luminance contrast required to achieve 99 percent visual certainty for the same task used for the visibility reference function, including the effects of dynamic presentation and uncertainty in task location.   167

visibility reference function (VL1) (illuminating engineering). A function representing a luminance contrast required at different levels of task background luminance to achieve visibility threshold for the visibility reference task consisting of a 4 min disk exposed for 1/5 s.   167

visible corona. A luminous discharge due to ionization of the air surrounding a device, caused by a voltage gradient exceeding a certain critical value.   27

visible radiation (light) (laser-maser). Electromagnetic radiation which can be detected by the human eye. It is commonly used to describe wavelengths which lie in the range between 0.4 μm and 0.7 μm.   363

visible radiation emitting diode (light emitting diodes). A semiconductor device containing a semiconductor junction in which visible light is nonthermally produced when a current flows as a result of an applied voltage.   162

visible range (antennas). For the case in which the field pattern of a continuous line source, L_λ wavelengths long, is expressed as a function $\psi(\psi = L_\lambda \cos\theta)$, that part of the infinite range of ψ that corresponds to a variation in the directional angle θ from π to 0 radians; that is, $-L_{\pi\psi}L_\lambda$. *Notes:* (1) All values of ψ outside the visible range are said to be in the invisible range. (2) The formulation of the field pattern as a function of ψ is useful because the sidelobes in the invisible range are a measure of the Q of the antenna. (3) This concept of a visible range can be extended to other antenna types.   111

visible spectrum. *See:* **light.**

visual acuity (illuminating engineering). A measure of the ability to distinguish fine details. Quantitatively, it is the reciprocal of the minimum angular separation in minutes of two lines of width subtending one minute of arc when the lines are just resolvable as separate.   167

visual angle (illuminating engineering). The angle which an object or detail subtends at the point of observation. It usually is measured in minutes of arc.   167

visual approach slope indicator system. *See:* **VASIS.**   167

visual-aural radio range. *See:* **VAR.**

visual comfort probability (VCP) (illuminating engineering). The rating of a lighting system expressed as a percent of people who, when viewing from a specified location and in a specified direction, will be expected to find it acceptable in terms of discomfort glare. Visual Comfort Probability is related to Discomfort Glare Rating. *See:* **discomfort glare rating.**   167

visual field (illuminating engineering). The locus of objects or points in space which can be perceived when the head and eyes are kept fixed. Separate monocular fields for the two eyes may be specified or the combination of the two.   167

visual inspection. Qualitative observation of physical characteristics utilizing the unaided eye or with stipulated levels of magnification.   284

visual perception (illuminating engineering). The interpretation of impressions transmitted from the retina to the brain in terms of information about a physical world displayed before the eye. *Note:* Visual perception involves any one or more of the following: recognition of the presence of something (object, aperture or medium); identifying it; locating it in space; noting its relation to other things; identifying its movement, color, brightness or form.   167

visual performance (illuminating engineering). The quantitative assessment of the performance of a task taking into consideration speed and accuracy.
                                                        167

visual photometer (illuminating engineering). One in which the equality of brightness of two surfaces is established visually. *Note:* The two surfaces usually are viewed simultaneously side by side. This is satisfactory when the color difference between the test source and comparison source is small. However, when there is a color difference, a flicker photometer provides more precise measurements. In this type of photometer the two surfaces are viewed alternately at such a rate that the color sensations either nearly or completely blend and the flicker due to brightness difference is minimized by adjusting the comparison source.   167

visual radio range (navigation aid terms). Any radio range (such as VOR [very high-frequency omnidirectional range]) whose primary function is to provide lines of position to be flown by visual reference to a course deviation indicator.   526

visual range (of a light or object) (illuminating engineering). The maximum distance at which that particular light (or object) can be seen and identified.
                                                        167

visual scanner (character recognition). *See:* **optical scanner.**

visual signal device (protective signaling). A general term for pilot lights, annunciators, and other devices providing a visual indication of the condition supervised. *See:* **protective signaling.**   328

visual surround (illuminating engineering). Includes all portions of the visual field except the visual task.
                                                        167

visual task (illuminating engineering). Conventionally designates those details and objects which must be seen for the performance of a given activity, and includes the immediate background of the details or objects. *Note:* The term visual task as used is a misnomer because it refers to the visual display itself and not the task of extracting information from it. The task of extracting information also has to be differentiated from the overall task performed by the observer.
                                                        167

visual task evaluator (VTE) (illuminating engineering). A contrast reducing instrument which permits

obtaining a value of luminance contrast, called the equivalence contrast C of a standard visibility reference task giving the same visibility as that of a task whose contrast has been reduced to threshold when the background luminances are the same for the task and the reference task. *See:* **equivalent contrast.**
                                                        167

visual transmitter. All parts of a television transmitter that handle picture signals, whether exclusively or not. *See:* **television.**   111

visual transmitter power. The peak power output during transmission of a standard television signal. *See:* **television.**   328

vital circuit. Any circuit the function of which affects the safety of train operation.   328

vital services (electric installations on shipboard). Vital services are normally considered to be those required for the safety of the ship and its passengers and crew. These may include propulsion, steering, navigation, firefighting, emergency lighting, and communications functions. Since the specific identification of vital services is influenced by the type of vessel and its intended service, this matter should be specified by the design agent for the particular vessel under consideration.   3

vitreous silica (fiber optics). Glass consisting of almost pure silicon dioxide (SiO_2). *Syn:* **fused silica.** *See:* **fused quartz.**   433

VLF. *See:* **radio spectrum.**

V-network (electromagnetic compatibility). An artificial mains network of specified disymmetric impedance used for two-wire mains operation and comprising resistors in V formation connected between each conductor and earth. *See:* **electromagnetic compatibility.**   199

vnl. *See:* **via net loss.**

V number. *See:* **normalized frequency.**

vodas. A system for preventing the over-all voice-frequency singing of a two-way telephone circuit by disabling one direction of transmission at all times. The name is derived from the initial letters of the expression **voice-operated device anti-sing.** *See:* **voice-frequency telephony.**   328

vogad. A voice-operated device is used to give a substantially constant volume output for a wide range of inputs. The name is derived from the initial letters of the expression **voice-operated gain-adjusted device.** *See:* **voice-frequency telephony.**   328

voice band (measuring longitudinal balance of telephone equipment operating in the voice band). That part of the audio-frequency range that is employed for the transmission of speech. For the purpose of ANSI/IEEE Std 455-1985, the voice band extends from 50 hertz (Hz) to 4000 Hz.

voice channel (mobile communication). A transmission facility defined by the constraints of the human voice. For mobile-communication systems, a voice channel may be considered to have a range of approximately 250 to 3000 hertz; since the Rules and Regulations of the Federal Communications Commission do not authorize the use of modulating frequencies higher than

3000 hertz for radiotelephony or tone signaling on radio frequencies below 500 megahertz. *See:* **channel spacing.**   181

voice frequency (vf) (data transmission). A frequency lying within that part of the audio range which is employed for the transmission of speech. *Note:* Voice frequencies used for commercial transmission of speech usually lie within the range 200 to 3500 Hz (hertz)   59

voice-frequency carrier telegraph (data transmission). A telegraph transmission system which provides several narrowband individual channels in the voice-frequency range.   59

voice-frequency telephony (data transmission). That form of telephony in which the frequencies of the components of the transmitted electric waves are substantially the same as the frequencies of corresponding components of the actuating acoustical waves.

59

voice grade channel (data transmission). A channel suitable for the transmission of speech, digital or analog data, or facsimile, generally with a frequency range of about 300 to 3000 Hz (hertz).   59

volatile (electronic data processing). Pertaining to a storage device in which data cannot be retained without continuous power dissipation, for example, an acoustic delay line. *Note:* Storage devices or systems employing nonvolatile media may or may not retain data in the event of planned or or accidental power removal.   235,54

volatile flammable liquid (National Electrical Code). A flammable liquid having a flash point below 38 °C (100°F) or whose temperature is above its flash point.
256

volcas. A voice-operated device that switches loss out of the transmitting branch and inserts loss in the receiving branch under control of the subscriber's speech. The name is derived from the initial letters of the expression **voice-operated loss control and suppressor.** *See:* **voice-frequency telephony.**   328

volt (unit of electric potential difference and electromotive force) (metric practice). The difference of electric potential between two points of a conductor carrying a constant current of one ampere, when the power dissipated between these points is equal to one watt.   21

volta effect. *See:* **contact potential.**

voltage (electromotive force*) (1) (general) (along a specified path in an electric field). The dot product line integral of the electric field strength along this path. *Notes:* (A) Voltage is a scalar and therefore has no spatial direction. (B) As here defined, voltage is synonymous with potential difference only in an electrostatic field. (C) In cases in which the choice of the specified path may make a significant difference, the path is taken in an equiphase surface unless otherwise noted. (D) It is often convenient to use an adjective with voltage, for example, phase voltage, electrode voltage, line voltage, etcetera. The basic definition of voltage applies and the meaning of adjectives should be understood or defined in each particular case. *See:* **reference voltage.**   210
*Deprecated

(2)(NESC). (A) The effective root-mean-square (rms) potential difference between any two conductors or between a conductor and ground. Voltages are expressed in nominal values unless otherwise indicated. The nominal voltage of a system or circuit is the value assigned to a system or circuit of a given voltage class for the purpose of convenient designation. The operating voltage of the system may vary above or below this value. (B) voltage of circuit not effectively grounded. The highest nominal voltage available between any two conductors of the circuit. *Note:* If one circuit is directly connected to and supplied from another circuit of higher voltage (as in the case of an autotransformer), both are considered as of the higher voltage, unless the circuit of the lower voltage is effectively grounded, in which case its voltage is not determined by the circuit of the higher voltage. Direct connection implies electric connection as distinguished from connection merely through electromagnetic or electrostatic induction. (C) voltage of a constant current circuit. The highest normal full load voltage of the current. (D) voltage of an effectively grounded circuit. The highest nominal voltage available between any conductor of the circuit and ground unless otherwise indicated. (E) voltage to ground of: (1) a grounded circuit. The highest nominal voltage available between any conductor of the circuit and that point or conductor of the circuit which is grounded. (2) an ungrounded circuit. The highest nominal voltage available between any two conductors of the circuit concerned. (F) voltage to ground of a conductor of: (1) a grounded circuit. The nominal voltage between such conductor and that point or conductor of the circuit which is grounded. (2) an ungrounded circuit. The highest nominal voltage between such conductor and any other conductor of the circuit concerned.   494

(3) (of a circuit) (National Electrical Code). The greatest root-mean-square (effective) difference of potential between any two conductors of the circuit concerned. Some systems, such as 3-phase 4-wire, single-phase 3-wire, and 3-wire direct-current may have various circuits of various voltages.   256

(4) (electromotive force) (surge arresters). The voltage between a part of an electric installation connected to a grounding system and points on the ground at an adequate distance (theoretically at an infinite distance) from any earth electrodes.   62, 244

voltage amplification (1) (general). An increase in signal voltage magnitude in transmission from one point to another or the process thereof. *See:* **amplifier.**

210

(2) (transducer). The scalar ratio of the signal output voltage to the signal input voltage. *Warning:* By incorrect extension of the term decibel, this ratio is sometimes expressed in decibels by multiplying its common logarithm by 20. It may be correctly expressed in decilogs. *Note:* If the input and.or output power consist of more than one component, such as multifrequency signal or noise, then the particular components used and their weighting must be specified. *See:* **transducer.**   210

(3) (magnetic amplifier). The ratio of differential output voltage to differential control voltage.   171

voltage and power directional relay (92) (power system device function numbers). A relay that permits or causes the connection of two circuits when the voltage difference between them exceeds a given value in a predetermined direction and causes these two circuits to be disconnected from each other when the power flowing between them exceeds a given value in the opposite direction.   402

voltage and pressure levels (loudness ratings of telephone connections). Voltage and pressure levels (V_T, V_W, S_E, and S_M) as used in definitions of this section are measured (as described in Section 5 of ANSI/ IEEE Std 661-1979) or computed (as described in Section 6 of this standard).   409

voltage attenuation. *See:* **attenuation, voltage.**

voltage at the instant of chopping (high voltage testing) (chopped impulses). The voltage at the instant of chopping is the voltage at the instant of the initial discontinuity.   150

voltage balance relay (power switchgear). A balance relay which operates by comparing the magnitudes of two voltage inputs.   103

voltage buildup (rotating machinery). The inherent establishment of the excitation current and induced voltage of a generator.   63

voltage circuit (1) (ac high-voltage circuit breaker). That part of the synthetic test circuit from which the major part of the test voltage is obtained.   426
(2) (instrument). That combination of conductors and windings of the instrument to which is applied the voltage of the circuit in which a given electrical quantity is to be measured, or a definite fraction of that voltage, or a voltage or current dependent upon it. *See:* **instrument; moving element (instrument); watthour meter.**   280

voltage clamp (converter circuit elements)(self-commutated converters). A clamp that limits the peak voltage across a semiconductor device.   584

voltage clamping ratio (low voltage varistor surge arresters). A figure of merit measure of the varistor voltage clamping effectiveness as determined by the ratio of clamping voltage to rated root-mean-square (rms) voltage, or by the ratio of clamping voltage to rated direct-current (dc) voltage.   62

voltage class (power cables). *See:* **medium voltage power cable; low-voltage power cable; control cable; low-level analog signal cable; low-level digital signal circuit cable.**

voltage classes. *See:* **voltage classes** chart below.

voltage class, rated nominal. *See:* **rated nominal voltage class.**

voltage coefficient of capacitance (nonlinear capacitor). The derivative with respect to voltage of a capacitance characteristic, such as a differential capacitance characteristic or a reversible capacitance characteristic, at a point, divided by the capacitance at that point. *See:* **nonlinear capacitor.**   191

VOLTAGE CLASSES
NOMINAL SYSTEM VOLTAGE

TWO WIRE	THREE WIRE	FOUR WIRE	MAXIMUM VOLTAGE[3]
(120)	*Single-Phase Systems*		127
	120/240		127/254
	Three-Phase Systems		
		208Y/120	220Y/127
	(240)	240/120	245/127
	480	480Y/277	508Y/293
	(600)		635
	(2400)		2540
	4160	4160Y/2400	4400Y/2540
	(4800)		5080
	(6900)		7260
		(8320Y/4800)	8800Y/5080
		(12000Y/6930)	12700Y/7330
		12470Y/7200	13200Y/7620
		13200Y/7620	13970Y/8070
	13800	(13800Y/7970)	14520Y/8380
		(20780Y/12000)	22000Y/12700
		(22860Y/13200)	24200Y/13970
	(23000)		24340
		24940Y/14400	26400Y/15240
	(34500)	34500/19920	36510Y/21080
	(46 kV)		48.3 kV
	69 kV		72.5 kV
	115 kV		121 kV
	138 kV		145 kV
	(161 kV)		169 kV
	230 kV		242 kV
EHV	345 kV [2]		362 kV
EHV	500 kV [2]		550 kV
EHV	765 kV [2]		800 kV
	1100 kV [2]		1200 kV

Left-margin group labels: IEEE Std for Industrial & Commercial Power Systems[1]; LOW VOLTAGE SYSTEMS; MEDIUM VOLTAGE — ANSI C84.1-1977 no voltage class stated; HIGH VOLTAGE — ANSI C84.1-1977 HIGHER VOLTAGE SYSTEMS; ANSI C92.2-1978.

1978
Preferred nominal voltages as shown without parentheses ( ).

1. Voltage class designations applicable to industrial and commercial power systems, adapted by IEEE Standards Board (LB 100A–April 23, 1975).

2. Typical nominal system voltage.

3. A comprehensive list of minimum and maximum voltage ranges is given in ANSI C84.1-1977.

voltage, common mode (surge withstand capability tests). The voltage common to all conductors of a group as measured between that group at a given location and an arbitrary reference (usually earth).

90

voltage-controlled oscillator (VCO) (radar). An oscillator whose frequency is a function of the voltage of an input signal. 13

voltage corrector (power supplies). An active source of regulated power placed in series with an unregulated supply to sense changes in the output voltage (or current): also to correct for the changes by automatically varying its own output in the opposite direction, thereby maintaining the total output voltage (or current) constant. See the accompanying figure.

228, 186

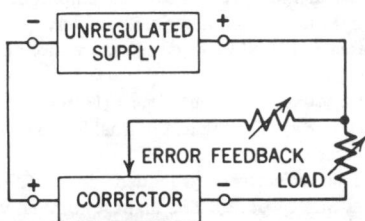

Circuit used to sense output voltage changes.

voltage deviation (transient)(converters having ac output)(self-commutated converters). The instantaneous difference between the actual instantaneous voltage and the corresponding value of the previously undisturbed wave form. *Note:* Voltage deviation amplitude is expressed in percent or per unit referred to the peak value of the previously undisturbed voltage.

584

voltage deviation, V_d (electromagnetic site survey). The ratio of the root-mean-squared envelope voltage to the average envelope of a signal expressed in decibels. 457

voltage directional relay (91) (power system device function numbers). A relay that operates when the voltage across an open circuit breaker or contactor exceeds a given value in a given direction. 402

voltage divider. A network consisting of impedance elements connected in series, to which a voltage is applied, and from which one or more voltages can be obtained across any portion of the network. *Notes:* (1) Dividers may have parasitic impedances affecting the response. These impedances are, in general, the series inductance and the capacitance to ground and to neighboring structures at ground or at other potentials. (2) An adjustable voltage divider of the resistance type is frequently referred to as a potentiometer.

307, 341, 201

voltage doubler. A voltage multiplier that separately rectifies each half cycle of the applied alternating voltage and adds the two rectified voltages to produce a direct voltage whose amplitude is approximately twice the peak amplitude of the applied alternating voltage. *See:* **rectifier.** 328

voltage drop (1) (general). The difference of voltages at the two terminals of a passive impedance. 63 **(2) (supply system).** The difference between the voltages at the transmitting and receiving ends of a feeder, main, or service. *Note:* With alternating current the

voltages are not necessarily in phase and hence the voltage drop is not necessarily equal to the algebraic sum of the voltage drops along the several conductors. *See:* **alternating-current distribution.** 64

voltage efficiency (specified electrochemical process). The ratio of the equilibrium reaction potential to the bath voltage. 328

voltage endurance (rotating machinery). A characteristic of an insulation system, obtained by plotting voltage against time to failure, for a number of samples tested to destruction at each of several sustained voltages. Constant conditions of frequency, waveform, temperature, mechanical restraint, and ambient atmosphere are required. Ordinate scales of arithmetical or logarithmic voltage, and abscissa scales of multicycle logarithmic time, normally give approximately linear characteristics. *See:* **asynchronous machine.**

63

voltage endurance test (rotating machinery). A test designed to determine the effect of voltage on the useful life of electric equipment. When this test voltage exceeds the normal design voltage for the equipment, the test is voltage accelerated. When the test voltage is alternating and the frequency of alternation exceeds the normal voltage frequency for the equipment, the test is frequency accelerated. *See:* **asynchronous machine.** 63

voltage, equivalent test alternating (charging inductor). A sinusoidal root-mean-square test voltage equal to 0.707 times the power-supply voltage of the network-charging circuit and having a frequency equal to the resonance frequency of charging. *Note:* This is the alternating component of the voltage that appears across the charging inductor in a resonance-charging circuit of the pulse-forming network. 137

voltage, exciter-ceiling. *See:* **exciter-ceiling voltage.**

voltage factor (electron tubes). The magnitude of the ratio of the change in one electrode voltage to the change in another electrode voltage, under the conditions that a specified current remains unchanged and that all other electrode voltages are maintained constant. *See:* ON period. 244, 190

voltage/frequency function (converters having ac output)(self-commutated converters). The ratio of output voltage to the fundamental frequency of the output as a function of that frequency. *See:* **Appendix C, ANSI/IEEE Std 936-1987.** 584

voltage generator (network analysis and signal-transmission system). A two-terminal circuit element with a terminal voltage substantially independent of the current through the element. *Note:* An ideal voltage generator has zero internal impedance. *See:* **network analysis; signal.** 125

voltage gradient (overhead-power-line corona and radio noise). A vector E equal to and in the direction of the maximum space rate of change of the voltage at the point specified. It is obtained as a vector field by applying the operator ∇ to the scalar voltage function u. Thus if $u = f(x,y,z)$,

$$E = \nabla u = \operatorname{grad} u = i\,\frac{\partial u}{\partial x} + j\,\frac{\partial u}{\partial y} + k\,\frac{\partial u}{\partial z}$$

Notes: (1) Voltage gradient is synonymous with potential gradient and is often referred to simply as "gradient" or "field strength." (2) For alternating voltage, the voltage gradient is expressed as the peak value divided by the square root of two. For sinusoidal voltages, this is the rms value.   411

voltage impulse. A voltage pulse of sufficiently short duration to exhibit a frequency spectrum of substantially uniform amplitude in the frequency range of interest. As used in electromagnetic compatibility standard measurements, the voltage impulse has a uniform frequency spectrum over the frequency range 25 to 1000 megahertz.   314

voltage influence (electric instrument). In instruments, other than indicating voltmeters, wattmeters, and varmeters, having voltage circuits, the percentage change (of full-scale value) in the indication of an instrument that is caused solely by a voltage departure from a specified reference voltage. *See:* **accuracy rating (instrument).**   280, 294

voltage-injection method (ac high-voltage circuit breaker). A synthetic test method in which the voltage circuit is applied to the test circuit breaker after power frequency current zero.   426

voltage jump (glow-discharge tube). An abrupt change or discontinuity in tube voltage drop during operation. *Note:* This may occur either during life under constant operating conditions or as the current or temperature is varied over the operating range. *See:* **gas tube.**   125

voltage level (data transmission). At any point in a transmission system, the ratio of the voltage existing at that point to an arbitrary value of voltage used as a reference. Specifically, in systems such as television systems, where wave shapes are not sinusoidal or symmetrical about a zero axis and where the arithmetical sum of the maximum positive and negative excursions of the wave is important in system performance, the voltage level is the ratio of the peak-to-peak voltage existing at any point in the transmission system to an arbitrary peak-to-peak voltage used as a reference. This ratio is usually expressed in dBV, signifying decibels referred to one V (volt) peak-to-peak.   59

voltage limit (industrial control). A control function that prevents a voltage from exceeding prescribed limits. Voltage limit values are usually expressed as percent of rated voltage. If the voltage-limit circuit permits the limit value to increase somewhat instead of being a single value, it is desirable to provide either a curve of the limit value of voltage as a function of some variable such as current or to give limit values at two or more conditions of operation. *See:* **control system, feedback.**   206

voltage loss (electric instrument) (current circuits). In a current-measuring instrument, the value of the voltage between the terminals when the applied current corresponds to nominal end-scale deflection. In other instruments the voltage loss is the value of the voltage between the terminals at rated current. *Note:* By convention, when an external shunt is used, the voltage loss is taken at the potential terminals of the shunt. The overall voltage drop resulting may be somewhat higher owing to additional drop in shunt lugs and connections. *See:* **accuracy rating (instrument).**   280

voltage multiplier. A rectifying circuit that produces a direct voltage whose amplitude is approximately equal to an integral multiple of the peak amplitude of the applied alternating voltage. *See:* **rectifier.**   328

voltage, nominal (system or circuit). *See:* **nominal system voltage.**

voltage of a constant current circuit (National Electrical Safety Code). The highest normal full load voltage or the current.   391

voltage of an effectively grounded circuit (National Electrical Safety Code). The highest nominal voltage available between any conductor of the circuit and ground unless otherwise indicated.   391

voltage of circuit not effectively grounded (National Electrical Saftey Code). The highest nominal voltage available between any two conductors of the circuit. *Note:* If one circuit is directly connected to and supplied from another circuit of higher voltage (as in the case of an autotransformer), both are considered as of the higher voltage, unless the circuit of the lower voltage is effectively grounded, in which case its voltage is not determined by the circuit of high voltage. Direct connection implies electric connection as distinguished from connections merely through electromagnetic or electrostatic induction.   391

voltage or current balance relay (60) (power system device function numbers). A relay that operates on a given difference in voltage, or current input or output, of two circuits.   402

voltage overshoot (1) (arc-welding apparatus). The ratio of transient peak voltage substantially instantaneously following the removal of the short circuit to the normal steady-state voltage value. *See:* **voltage recovery time.**   264

(2) (low voltage varistor surge arresters). The excess voltage above the clamping voltage of the device for a given current that occurs when current waves of less than 8 μs virtual front duration are applied. This value may be expressed as a percent of the clamping voltage for an 8 x 20-μs current wave.   62

voltage overshoot, effective (arc-welding apparatus). The area under the transient voltage curve during the time that the transient voltage exceeds the steady-state value. *See:* **voltage recovery time.**   264

voltage, peak working (charging inductor). The algebraic sum of the maximum alternating crest voltage and the direct voltage of the same polarity appearing between the terminals of the inductor winding or between the inductor winding and the grounded elements.   137

voltage phase-angle method (economic dispatch) (electric power systems). Considers the actual measured phase-angle difference between the station bus and a reference bus in the determination of incremental transmission losses. 94

voltage-phase-balance protection (power switchgear). A form of protection that disconnects or prevents the connection of the protected equipment when the voltage unbalance of the phases of a normally balanced polyphase system exceed a predetermined amount. 103

voltage range (electrically propelled vehicle). Divided into five voltage ranges, as follows: **first voltage range:** 30 volts or less; **second voltage range:** over 30 volts to and including 175 volts; **third voltage range:** over 175 volts to and including 250 volts; **fourth voltage range:** over 250 volts to and including 660 volts; **fifth voltage range:** over 660 volts. 328

voltage range multiplier (instrument multiplier). A particular type of series resistor or impedor that is used to extend the voltage range beyond some particular value for which the measurement device is already complete. It is a separate component installed external to the measurement device. *See:* **auxiliary device (instrument).** 294

voltage, rated maximum. The highest rms voltage at which a device is designed to operate. *Note:* This voltage corresponds to the maximum tolerable zone primary voltage at distribution transformers for distribution reclosers and at substations and on transmission systems for power reclosers as given in ANSI C84.1-1977. 92

voltage, rated maximum interrupting of main contacts. *See:* **rated maximum interrupting of main contacts voltage.**

voltage, rated short-time of main contacts. *See:* **rated short-time of main contacts voltage.**

voltage rating (1)(protection and coordination of industrial and commercial power systems). The root-mean-square (rms) alternating current (or the direct current) voltage at which the fuse is designed to operate. All low-voltage fuses will function on any lower voltage, but use on higher voltages than rated is hazardous. For high short-circuit currents, the magnitude of applied voltage will affect the arcing and clearing times and increase the clearing I^2T values. 504

(2) (surge arrester). The designated maximum permissible operating voltage between its terminals at which an arrester is designed to perform its duty cycle. It is the voltage rating specified on the nameplate. 430

(3) (household electric ranges). The voltage limits within which the range is intended to be used. *See:* **appliance outlet.** 308

(4) (grounding transformer). The maximum "line-to-line" voltage at which it is designed to operate continuously from line to ground without damage to the grounding transformer. 53

(5) (relay). The voltage rating of a relay is the voltage at a specified frequency that may be sustained by the relay for an unlimited period without causing any of the prescribed limitations to be exceeded. 127

voltage rating, maximum (connector). The highest phase-to-ground voltage at which a connector is designed to operate. 134

voltage rating of a grounding transformer (power and distribution transformer). The maximum "line-to-line" voltage at which it it designed to operate continuously from line to ground without damage to the grounding transformer. 53

voltage rating of a relay (power switchgear). The voltage at a specified frequency that may be sustained by the relay for an unlimited period without causing any of the prescribed limitations to be exceeded. 103

voltage ratio (capacitance potential device, in combination with its coupling capacitor or bushing). The overall ratio between the root-mean-square primary line-to-ground voltage and the root-mean-square secondary voltage. *Note:* It is not the turn ratio of the transformer used in the network. *See:* **outdoor coupling capacitor.** 341

voltage ratio of a transformer (power and distribution transformer). The ratio of the rms terminal voltage of a higher voltage winding to the rms terminal voltage of a lower voltage winding, under specified conditions of the load. 53

voltage ratio of a voltage divider (high voltage testing). The voltage ratio of a voltage divider is the factor by which the output voltage is multiplied to determine the measured value of the input voltage. It is dependent on the load on the output terminal of the divider and the impedence of this must be stated. In principle, the ratio is a constant but its validity may be restricted to a specific duration or frequency range in which case the duration or range for which it is valid shall be stated. 150

voltage recovery time (arc-welding apparatus). With a welding power supply delivering current through a short-circuiting resistor whose resistance is equivalent to the normal load at that setting on the power supply, and measurement being made when the short circuit is suddenly removed, the time measured in seconds between the instant the short circuit is removed and the instant when voltage has reached 95 percent of its steady-state value. *See:* **voltage overshoot; voltage overshoot, effective.** 264

voltage reduction reserve (power operations). The operating reserve available through voltage reduction of a specified percentage. 516

voltage reference (power supplies). A separate, highly regulated voltage source used as a standard to which the output of the power supply is continuously referred. 186

voltage-reference tube. A gas tube in which the tube voltage drop is approximately constant over the operating range of current and relatively stable with time at fixed values of current and temperature. 125

voltage reflection coefficient. The ratio of the complex number (phasor) representing the phase and magnitude of the electric field of the backward-traveling wave to that representing the forward-traveling wave at a cross section of a waveguide. The term is also used to denote the magnitude of this complex ratio. *See:* **waveguide.** 179

voltage regulating adjuster (excitation systems for synchronous machines). A device associated with a synchronous machine voltage regulator by which adjustment of the synchronous machine terminal voltage can be made. 507

voltage-regulating relay (power and distribution transformer). A voltage-sensitive device that is used on an automatically operated voltage regulator to control the voltage of the regulated circuit. 53

voltage-regulating transformer (step-voltage regulator). A voltage regulator in which the voltage and phase angle of the regulated circuit are controlled in steps by means of taps and without interrupting the load. *See:* **voltage regulator.** 203

voltage-regulating transformer, two-core. A voltage-regulating transformer consisting of two separate core and coil units in a single tank. *See:* **voltage regulator.** 203

voltage-regulating transformer, two-core, excitation-regulating winding. In some designs, the main unit will have one winding operating as an autotransformer that performs both functions listed under regulating winding and excitation winding. Such a winding is called the excitation-regulating winding. *See:* **voltage regulator.** 203

voltage-regulating transformer, two-core, excitation winding. The winding of the main unit that draws power from the system to operate the two-core transformer. *See:* **voltage regulator.** 203

voltage-regulating transformer, two-core, excited winding. The winding of the series unit that is excited from the regulating winding of the main unit. *See:* **voltage regulator.** 203

voltage-regulating transformer, two-core, regulating winding. The winding of the main unit in which taps are changed to control the voltage or phase angle of the regulated circuit through the series unit. *See:* **voltage regulator.** 203

voltage-regulating transformer, two-core, series unit. The core and coil unit that has one winding connected in series in the line circuit. *See:* **voltage regulator.** 203

voltage-regulating transformer, two-core, series winding. The winding of the series unit that is connected in series in the line circuit. *Note:* If the main unit of a two-core transformer is an autotransformer, both units will have a series winding. In such cases, one is referred to as the series winding of the autotransformer and the other, the series winding of the series unit. *See:* **voltage regulator.** 203

voltage regulation (1) (constant-voltage transformer). The change in output (secondary) voltage which occurs when the load (at a specified power factor) is reduced from rated value to zero, with the primary impressed terminal voltage maintained constant. *Note:* In case of multi-winding transformers, the loads on all windings, at specified power factors, are to be reduced from rated kVA to zero simultaneously. The regulation may be expressed in per unit, or percent, on the base of the rated output (secondary) voltage at full load. 53

(2) (outdoor coupling capacitor). The variation in voltage ratio and phase angle of the secondary voltage of the capacitance potential device as a function of primary line-to-ground voltage variation over a specified range, when energizing a constant, linear impedance burden. *See:* **outdoor coupling capacitor.** 341

(3) (direct-current generator). The final change in voltage with constant field-rheostat setting when the specified load is reduced gradually to zero, expressed as a percent of rated-load voltage, the speed being kept constant. *Note:* In practice it is often desirable to specify the over-all regulation of the generator and its driving machine thus taking into account the speed regulation of the driving machine. 328

(4) (induction frequency converter). The rise in secondary voltage when the rated load at rated power factor is reduced to zero, expressed in percent of rated secondary voltage, the primary voltage, primary frequency, and the speed being held constant. *See:* **asynchronous machine.** 328

(5) (synchronous generator). The rise in voltage with constant field current, when, with the synchronous generator operated at rated voltage and rated speed, the specified load at the specified power factor is reduced to zero, expressed as a percent of rated voltage. 63

(6) (line regulator circuits). *See:* **Zener diode; pulse-width modulation.**

(7) (thyristor converter). The change in output voltage that occurs when the load current is reduced from its rated value to zero, or light transition load, with rated sinusoidal alternating voltage applied to the thyristor power converter with the transformer on its rated tap, but excluding the corrective action of any voltage regulating means. *Note:* The regulation may be expressed in volts or in percent of rated volts. 121

voltage regulation curve (voltage regulation characteristic) (synchronous generator). The relationship between the armature winding voltage and the load on the generator under specified conditions and constant field current. 63

voltage regulation of a constant-voltage transformer (power and distribution transformer). The change in output (secondary) voltage which occurs when the load (at a specified power factor) is reduced from rated value to zero, with the primary impressed terminal voltage maintained constant. *Note:* In case of multiwinding transformers, the loads of all windings, at specified power factors, are to be reduced from rated kVA to zero simultaneously. The regulation may be expressed in per unit, or percent, on the base of the rated output (secondary) voltage at full load. 53

voltage regulator (1)(excitation systems for synchronous machines). A synchronous machine regulator that functions to maintain the terminal voltage of a synchronous machine at a predetermined value, or to vary it according to a predetermined plan. *Note:* Historical term, included for reference only. The preferred term is synchronous machine regulator. 507

(2)(transformer type). An induction device having one or more windings in shunt with and excited from the primary circuits, and having one or more windings in series between the primary circuits and the regulated circuit, all suitably adapted and arranged for the control of the voltage, or of the phase angle, or of both, of the regulated circuit.   203

voltage regulator, continuously acting type (rotating machinery). A regulator that initiates a corrective action for a sustained infinitesimal change in the controlled variable.   63

voltage regulator, direct-acting type (rotating machinery). A rheostatic-type regulator that directly controls the excitation of an exciter by varying the input to the exciter field circuits.   63

voltage regulator, dynamic type (rotating machinery). A continuously acting regulator that does not require mechanical acceleration of parts to perform the regulating function. *Note:* Dynamic-type voltage regulators utilize magnetic amplifiers, rotating amplifiers, electron tubes, semiconductor elements, and/or other static components.   63

voltage regulator, indirect-acting type (rotating machinery). A rheostatic-type regulator that controls the excitation of the exciter by acting on an intermediate device not considered part of the voltage regulator or exciter.   63

voltage regulator, noncontinuously acting type (rotating machinery). A regulator that requires a sustained finite change in the controlled variable to initiate corrective action.   63

voltage regulator, synchronous-machine (rotating machinery). A synchronous-machine regulator that functions to maintain the voltage of a synchronous machine at a predetermined value, or to vary it according to a predetermined plan.   63

voltage-regulator tube. A glow-discharge cold-cathode tube in which the voltage drop is approximately constant over the operating range of current, and that is designed to provide a regulated direct-voltage output.   190, 125

voltage related to partial discharges (partial discharge measurement in liquid-filled power transformers and shunt reactors). The phase to ground alternating voltage whose value is expressed by its peak divided by $\sqrt{2}$.   580

voltage relay (1) (power switchgear). A relay that responds to voltage.   103
(2) (industrial control). A relay that functions at a predetermined value of voltage. *Note:* It may be an overvoltage relay, an undervoltage relay, or a combination of both. *See:* **relay.**   206

voltage response (close-talking pressure-type microphone). The ratio of the open-circuit output voltage to the applied sound pressure, measured by a laboratory standard microphone placed at a stated distance from the plane of the opening of the artificial voice. *Note:* The voltage response is usually measured as a function of frequency. *See:* **close-talking pressure-type microphone.**   249

voltage response, exciter. The rate of increase or decrease of the exciter voltage when a change in this voltage is demanded. It is the rate determined from the exciter voltage response curve that if maintained constant would develop the same exciter voltage-time area as is obtained from the curve for a specified period. The starting point for determining the rate of voltage change shall be the initial value of the exciter voltage-time response curve. *See:* **asynchronous machine.**   63

voltage response ratio, excitation-system (rotating machinery). The numerical value that is obtained when the excitation-system voltage response in volts per second, measured over the first 1/2 second interval unless otherwise specified, is divided by the rated-load field voltage of the synchronous machine. *Note:* This response, if maintained constant, would develop, in 1/2 second, the same excitation voltage-time area as attained by the actual response.   63

voltage response, synchronous-machine excitation-system. The rate of increase or decrease of the excitation-system output voltage, determined from the synchronous machine excitation-system voltage-time response curve, that if maintained constant would develop the same excitation-system voltage-time areas as are obtained from the curve for a specified period. The starting point for determining the rate of voltage change shall be the initial value of the synchronous-machine excitation-system voltage-time response curve.   63

voltage restraint (power switchgear). A method of restraining the operation of a relay by means of a voltage input which opposes the typical response of the relay to other inputs.   103

voltage sensing relay. (1) A term correctly used to designate a special-purpose voltage-rated relay that is adjusted by means of a voltmeter across its terminals in order to secure pickup at a specified critical voltage without regard to coil or heater resistance and resulting energizing current at that voltage. (2) A term erroneously used to describe a general-purpose relay for which operational requirements are expressed in voltage.   341

voltage-sensitive preamplifier (germanium gamma-ray detectors). An amplifier, preceding the main amplifier, in which the output-signal amplitude is proportional to the signal voltage appearing across the capacitance that exists at the input of the preamplifier. *See:* **charge-sensitive preamplifier.**   528

voltage sensitivity (nonlinear capacitor). *See:* **voltage coefficient of capacitance; nonlinear capacitor.**

voltage sets (polyphase circuit). The voltages at the terminals of entry to a polyphase circuit into a delimited region are usually considered to consist of two sets of voltages: the line-to-line voltages, and the line to-neutral voltages. If the phase conductors are identified in a properly chosen sequence, the voltages between the terminals of entry of successive pairs of phase conductors form the set of line-to-line voltages, equal in number to the number of phase conductors. The voltage from the successive terminals of entry of the phase conductors to the terminal of entry of the neutral conductor, if one exists, or to the true neutral point, form the set of line-to-neutral voltages, also equal in number to the number of phase conductors. In case of doubt, the set intended must be identified. In the absence of other information, stated or implied, the line-to-neutral-conductor set is understood. *Notes:* (1) Under abnormal conditions the voltage of the neutral conductor and of the true neutral point may not be the same. Therefore it may become necessary to designate which is intended when the line-to-neutral voltages are being specified. (2) The set of line-to-line voltages may be determined by taking the differences in pairs of the successive line-to-neutral voltages. The line-to-neutral voltages can be determined from the line-to-line voltages by an inverse process only when the voltage between the neutral conductor and the true neutral point is completely specified, or equivalent additional information is available. If instantaneous voltages are used, algebraic differences are taken, but if root-mean-square voltages are used, information regarding relative phase angles must be available, so that the voltages may be expressed in phasor form and the phasor differences taken. (3) This definition may be applied to a two-phase, four-wire or five-wire circuit. A two-phase, three-wire circuit should be treated as a special case. *See:* **network analysis.**   210

voltage spread. The difference between maximum and minimum voltages.   260

voltage-stabilizing tube. *See:* **voltage regulator tube.**

voltage standing-wave ratio (VSWR) (mode in a waveguide). The ratio of the magnitude of the transverse electric field in a plane of maximum strength to the magnitude at the equivalent point in an adjacent plane of minimum field strength. *See:* **waveguide.**   179

voltage surge, internal (thyristor converter). Voltage surge caused by sources within a converter. *Note:* It may originate from blowing fuses, hole storage recovery phenomena, etcetera. Internal voltage surges are substantially under control of the circuit designer.   121

voltage surge suppressor (semiconductor rectifier). A device used in the semiconductor rectifier to attenuate surge voltages of internal or external origin. Capacitors, resistors, nonlinear resistors, or combinations of these may be employed. Nonlinear resistors include electronic and semiconductor devices. *See:* **semiconductor rectifier stack.**   208

voltage switch (test switches for transformer-rated meters). A single-pole single-throw switch used to open or close a voltage circuit.   590

voltage test. *See:* **controlled overvoltage test.**

voltage/time curve for impulses of constant prospective shape (high voltage testing). The voltage/time curve for impulses of constant prospective shape is the curve relating the disruptive discharge voltage of a test object to the time to chopping, which may occur on the front, at the crest or on the tail. The curve is obtained by applying impulse voltages of constant shape, but with different peak values.   150

voltage/time curve for linearly rising impulses (high

voltage testing). The voltage/time curve for impulses with fronts rising linearly is the curve relating the voltage at the instant of chopping to rise time T_r. The curve is obtained by applying impulses with approximately linear fronts of different steepnesses.   150

voltage-time product (pulse transformers). The time integral of a voltage pulse applied to a transformer winding.   589

voltage-time product rating (of a transformer winding)(pulse transformers). Considered as being a constant and is the maximum voltage-time product of a voltage pulse that can be applied to the winding before a specified level of core saturation-region effects is reached. The level of core saturation-region effects is determined by observing either the shape of the output voltage pulse for a specified degradation (for example, a maximum tilt [droop]), or the shape of the exciting current pulse for a specified departure from linearity (for example, deviation from a linear ramp by a given percentage).   589

voltage-time response, synchronous-machine excitation-system. The output voltage of the excitation system, expressed as a function of time, following the application of prescribed inputs under specified conditions.   63

voltage-time response, synchronous-machine voltage-regulator. The voltage output of the synchronous-machine voltage regulator expressed as a function of time following the application of prescribed inputs under specified conditions.   63

voltage to ground (1) (National Electrical Code). For grounded circuits, the voltage between the given conductor and that point or conductor of the circuit that is grounded; for ungrounded circuits, the greatest voltage between the given conductor and any other conductor of the circuit.   256

(2) (National Electrical Safety Code). (A) of a grounded circuit. The highest nominal voltage available between any conductor of the circuit and that point or conductor of the circuit which is grounded. (B) of an ungrounded circuit. The highest nominal voltage available between any two conductors of the circuit concerned.   391

(3) (power and distribution transformer). The voltage between any live conductor of a circuit and the earth. *Note:* Where safety considerations are involved, the voltage to ground which may occur in an ungrounded circuit is usually the highest voltage normally existing between the conductors of the circuit, but in special circumstances, higher voltages may occur.   91, 53

voltage to ground of a conductor (National Electrical Safety Code). (1) of a grounded circuit. The nominal voltage between such conductor and that point or conductor of the circuit which is grounded. (2) of an ungrounded circuit. The highest nominal voltage between such conductor and any other conductor of the circuit concerned.   391

voltage to luminaire factor (illuminating engineering). The fractional loss of task illuminance due to improper voltage at the luminaire.   167

voltage, touch. *See:* **touch voltage.**

voltage transformer (1) (power and distribution transformer) (instrument transformer). An instrument transformer intended to have its primary winding connected in shunt with a power supply circuit, the voltage of which is to be measured or controlled.
53, 394

(2) (metering). An instrument transformer designed for use in the measurement or control of voltage. *Note:* Its primary winding is connected across the supply circuit. *See:* **instrument transformer.** 212

(3) (VT). An instrument transformer intended to have its primary connected in shunt with the voltage to be measured or controlled. 203

voltage transient suppression (thyristor). Reduction of the effects of voltage transients on controller components by reducing the voltage or energy of the transients to tolerable levels. 445

voltage, transverse mode (surge withstand capability tests). The voltage at a given location between two conductors of a group. *Syn:* **differential mode voltage.**
90

voltage-tunable magnetron (microwave tubes). A magnetron in which the resonant circuit is heavily loaded (Q_L = 1 to 10) and in which the supply of electrons to the interaction space is restricted whereby the frequency of oscillation becomes proportional to the plate voltage. *See:* **magnetrons.** 190

voltage-type telemeter (power switchgear). A telemeter that employs the magnitude of a single voltage as the translating means. 103

voltage winding (or transformer) for regulating equipment (power and distribution transformer). The winding (or transformer) which supplies voltage within close limits of accuracy to instruments, such as contact-making voltmeters. 53

voltage-withstand test (1) (insulation materials). The application of a voltage higher than the rated voltage for a specified time for the purpose of determining the adequacy against breakdown of insulation materials and spacing under normal conditions. *See:* **dielectric tests (voltage-withstand tests).** 328

(2) (rotating machinery). *See:* **overvoltage test; asynchronous machine.**

voltaisation (galvanization) (electrotherapy). *See:* **galvanism.**

voltameter (coulometer). *See:* **coulometer; instrument.**

volt-ammeter. An instrument having circuits so designed that the magnitude either of voltage or of current can be measured on a scale calibrated in terms of each of these quantities. *See:* **instrument.** 328

voltampere. The unit of apparent power in the International System of Units (SI). The voltampere is the apparent power at the points of entry of a single-phase, two-wire system when the product of the root-mean-square value in amperes of the current by the root-mean-square value in volts of the voltage is equal to one. 210

voltampere loss (electric instrument). *See:* **apparent power loss (electric instrument).**

voltampere meter. An instrument for measuring the apparent power in an alternating-current circuit. It is provided with a scale graduated in volt-amperes or in kilovolt-amperes. *See:* **instrument.** 328

volt efficiency (storage battery) (storage cell). The ratio of the average voltage during the discharge to the average voltage during the recharge. *See:* **electrochemistry; charge.** 328

voltmeter. An instrument for measuring the magnitude of electric potential difference. It is provided with a scale, usually graduated in either volts, millivolts, or kilovolts. If the scale is graduated in millivolts or kilovolts the instrument is usually designated as a millivoltmeter or a kilovoltmeter. *See:* **instrument.**
328

voltmeter-ammeter. The combination in a single case, but with separate circuits, of a voltmeter and an ammeter. *See:* **instrument.** 328

volts per hertz relay (power switchgear). A relay whose pickup is a function of the ratio of voltage to frequency. 103

volt-time curve (surge arresters) (1) (impulses with fronts rising linearly) (surge arresters). The curve relating the disruptive-discharge voltage of a test object to the virtual time to chopping. The curve is obtained by applying voltages that increase at different rates in approximately linear manner. 308, 62

(2) (standard impulses). A curve relating the peak value of the impulse causing disruptive discharge of a test object to the virtual time to chopping. The curve is obtained by applying standard impulse voltages of different peak values. 308, 62

volume (1)(information transfer). One collection of data commencing with a Volume ID and containing a number of bytes specified in the volume size of the volume director. 479

(2)(volume measurements of electrical speech and program waves). The magnitude of a complex audio-frequency wave in an electrical circuit as measured on a standard volume indicator. 523

(3) (data transmission). In general, volume is the intensity or loudness of sound. In a telephone or other audio-frequency circuit, a measure of the power corresponding to an audio-frequency wave at that point (expressed in decibels (dB)). 59

(4) (electric circuit). The magnitude of a complex audio-frequency wave as measured on a standard volume indicator. *Notes:* (A) Volume is expressed in volume units (vu). (B) The term volume is used loosely to signify either the intensity of a sound or the magnitude of an audiofrequency wave. 239

(5) (International System of Units) (SI). The SI unit of volume is the cubic meter. This unit, or one of the regularly formed multiples such as the cubic centimeter, is preferred for all applications. The special name liter has been approved for the cubic decimeter, but use of this unit is restricted to the measurement of liquids and gases. No prefix other than milli- should be used with liter. *See:* **units and letter symbols.**
21

volume control. *See:* **gain control.**

volume density of magnetic pole strength. At any point of the medium in a magnetic field, the negative of the divergence of the magnetic polarization vector there. 210

volume equivalent (complete telephone connection, including the terminating telephone sets). A measure of the loudness of speech reproduced over it. The volume equivalent of a complete telephone connection is expressed numerically in terms of the trunk loss of a working reference system when the latter is adjusted to give equal loudness. *Note:* For engineering purposes, the volume equivalent is divided into volume losses assignable to (1) the station set, subscriber line, and battery-supply circuit that are on the transmitting end; (2) the station set, subscriber line, and battery supply that are on the receiving end; (3) the trunk; and (4) interaction effects arising at the trunk terminals. 328

volume indicator (standard volume indicator). A standardized instrument having specified electric and dynamic characteristics and read in a prescribed manner, for indicating the volume of a complex electric wave such as that corresponding to speech or music. *Notes:* (1) The reading in volume units is equal to the number of decibels above a reference volume. The sensitivity is adjusted so that the reference volume or zero volume unit is indicated when the instrument is connected across a 600-ohm resistor in which there is dissipated a power of 1 milliwatt at 1000 hertz. (2) Specifications for a volume indicator are given in American National Standard Volume Measurements of Electrical Speech and Program Waves, C16 5. *See:* **instrument; volume unit.** 122, 176, 196

volume limiter. A device that automatically limits the output volume of speech or music to a predetermined maximum value. *See:* **peak limiter.** 328

volume-limiting amplifier. An amplifier containing an automatic device that functions when the input volume exceeds a predetermined level and so reduces the gain that the output volume is thereafter maintained substantially constant notwithstanding further increase in the input volume. *Note:* The normal gain of the amplifier is restored when the input volume returns below the predetermined limiting level. *See:* **amplifier.** 111

volume range (1) (transmission system). The difference, expressed in decibels between the maximum and minimum volumes that can be satisfactorily handled by the system. 328
(2) (complex audio-frequency signal). The difference, expressed in decibels, beween the maximum and minimum volumes occurring over a specified period of time. 328

volume resistivity. *See:* **resistivity, volume.**
volumetric radar. *See:* **three-dimensional radar.**
volume unit (vu). The unit in which the standard volume indicator is calibrated. *Note:* One volume unit equals one decibel for a sine wave but volume units should not be used to express results of measurements of complex waves made with devices having characteristics differing from those of the standard volume indicator. *See:* **volume indicator.** 239

VOR (very high-frequency omnidirectional range) (navigation aid terms). A navigation aid operating at VHF (very high frequency) and providing radial lines of position in any direction as determined by bearing selection within the receiving equipment; it emits a (variable) modulation whose phase, relative to a reference modulation, is different for each bearing of the receiving point from the station. 526

vortac (navigation aid terms). A designation applied to certain navigation stations (primarily in the United States) in which both VOR (very high-frequency omnidirectional range) and tacan (tactical air navigation) are used; the distance function in tacan is used with VOR to provide VOR/DME (rho theta) navigation. 526

vowel articulation (percent vowel articulation). The percent articulation obtained when the speech units considered are vowels (usually combined with consonants into meaningless syllables). *See:* **articulation (percent articulation); volume equivalent.** 176

VPI. *See:* **vacuum-pressure impregnation.**
VSWR. *See:* **voltage standing-wave ratio.**
V-terminal voltage (electromagnetic compatibility). Terminal voltage measured with a V network between each mains conductor and earth. *See:* **electromagnetic compatibility.** 220, 199

vu (volume measurements of electrical speech and program waves). (Pronounced 'vee-you' and customarily written with lower case letters.) A quantitative expression for volume in an electric circuit. *Notes:* (1) The volume in vu is numerically equal to the number of decibels (dB) which expresses the ratio of the magnitude of the waves to the magnitude of reference volume. (2) The term vu should not be used to express results of measurements of complex waves made with devices having characteristics differing from those of the standard volume indicator. *See:* **standard volume indicator.** 523

vulnerability (surge testing for equipment connected to low-voltage ac power circuits). The characteristic of a device for being damaged by an external influence, such as a surge. 578

W

wait (test, measurement and diagnostic equipment). A programmed instruction which causes an automatic test system to remain in a given state for a predetermined period. 54

waiting, call. A feature providing a signal to a busy called line to indicate that another call is waiting. 103

waiting-passenger indicator (elevators). An indicator

that shows at which landings and for which direction elevator-hall stop or signal calls have been registered and are unanswered. *See:* **control.** 328

wait timeout period (FASTBUS acquisition and control). The time a master will wait after recognizing WT (wait) before terminating the connection. 480

waiver (nuclear power quality assurance). Documented authorization to depart from specified requirements. 417

walkie-talkie. A two-way radio communication set designed to be carried by one person, usually strapped over the back, and capable of operation while in motion. *See:* **radio transmission.** 328

walk-off mode (laser-maser). A walk-off mode will be characterized by successive shifts per reflection in the location of a maximum in the transverse field intensity. *See:* **transverse mode.** 363

walkout of reverse drain breakdown (metal-nitride-oxide field-effect transistor). This term describes an effect where the reverse current–voltage characteristic of the drain junction changes with time toward larger voltages as a function of applied bias. This effect is generally reversible. 386

walk-through (software). A review process in which a designer or programmer leads one or more other members of the development team through a segment of design or code that he or she has written, while the other members ask questions and make comments about technique, style, possible errors, violation of development standards, and other problems. *See:* **code; design; error; inspection; segment.** 434

wall bushing. A bushing intended primarily to carry a circuit through a wall or other grounded barrier in a substantially horizontal position. Both ends must be suitable for operating in air. *See:* **bushing.** 348

wall-mounted oven (National Electrical Code). An oven for cooking purposes designed for mounting in or on a wall or other surface and consisting of one or more heating elements, internal wiring, and built-in or separately mountable controls. *See:* **counter-mounted cooking unit.** 256

wall telephone set. A telephone set arranged for wall mounting. *See:* **telephone station.** 328

wander (radar). *See:* **scintillation.**

warble-tone generator (alarm monitoring and reporting systems for fossil-fueled power generating stations). An audio-frequency oscillator, the frequency of which is varied cyclically at a subaudio rate over a fixed range. 501

warm-up time (1) (power supplies). The time (after power turn on) required for the output voltage or current to reach an equilibrium value within the stability specification. 186

(2) (gyro; accelerometer). The time interval required for a gyro or accelerometer to reach specified performance from the instant that it is energized under specified operating conditions. 46

(3) (nuclear power generating station). The time, following power application to a module, required for the output to stablize within specifications. 355

warning whistle. *See:* **audible cab indicator.**

warp (loran)(navigation aid terms). Variations of the propagation times for loran signals due to the variations of conductivity over land. Causes errors in the determination of absolute position. 526

washer (rotating machinery). *See:* **collar.**

watchdog timer (station control and data acquisition)-(supervisory control, data acquisition, and automatic control). A form of interval timer which is used to detect a possible malfunction. 403, 570

watchman's reporting system. A supervisory system arranged for the transmission of a patrolling watchman's regularly recurrent report signals to a central supervisory agency from stations along his patrol route. *See:* **protective signaling.** 328

water. *See:* **acoustic properties of water.**

water-air-cooled machine. A machine that is cooled by circulating air that in turn is cooled by circulating water. *Note:* The machine is so enclosed as to prevent the free exchange of air between the inside and outside of the enclosure, but not sufficiently to be termed airtight. It is provided with a water-cooled heat exchanger for cooling the ventilating air and a fan or fans, integral with the rotor shaft or separate, for circulating the ventilating air. *See:* **asynchronous machine.** 63

water conditions (power operations). *See:* **adverse water conditions; average water conditions; median water conditions.** 516

water-cooled (rotating machinery). (1) A term applied to apparatus cooled by circulating water, the water or water ducts coming in direct contact with major parts of the apparatus. (2) In certain types of machine, it is customary to apply this term to the cooling of the major parts by enclosed air or gas ventilation, where water removes the heat through an air-to-water or gas-to-water heat exchanger. *See:* **asynchronous machine.** 63

water cooler (rotating machinery). A cooler using water as one of the fluids. 63

waterflow-alarm system (protective signaling). An alarm system in which signal transmission is initiated automatically by devices attached to an automatic sprinkler system and actuated by the flow through the sprinkler system pipes of water in excess of a predetermined maximum. *See:* **protective signaling.** 328

water inertia time (hydraulic turbines). A characteristic time, usually taken at rated conditions, due to inertia of the water in the water passages from intake to exit defined as:

$$T_w = \frac{Q_r}{gH_r} \int \frac{dL}{A} \approx \frac{Q_r}{gH_r} \sum \frac{L}{A}$$

where:
A = area of each section
L = corresponding length
Q_r = rated discharge
H_r = rated head
g = acceleration due to gravity

8

water load (high-frequency circuits). A matched termination in which the electromagnetic energy is absorbed in a stream of water for the purpose of measuring power by continuous-flow calorimetric methods. *See:* **waveguide.** 244,179

water-motor bell. A vibrating bell operated by a flow of water through its water-motor striking mechanism. *See:* **protective signaling.** 328

waterproof electric blasting cap. A cap specially insulated to secure reliability of firing when used in wet work. *See:* **blasting unit.** 328

waterproof enclosure (electric installations on shipboard). An enclosure constructed so that any moisture or water leakage which may occur into the enclosure will not interfere with its successful operation. In the case of motor or generator enclosures, leakage which may occur around the shaft may be considered permissible provided it is prevented from entering the oil reservoir and provision is made for automatically draining the motor or generator enclosure. 3

waterproof machine (rotating machinery). A machine so constructed that water directed on it under prescribed conditions cannot cause interference with satisfactory operation. *See:* **asynchronous machine.** 63

watertight (1) (National Electrical Code). So constructed that moisture will not enter the enclosure. *See:* ANSI C19.4-1973, for test conditions under specified test conditions. 256
(2) (power and distribution transformer). So constructed that water will not enter the enclosing case under specified conditions. *Note:* A common form of specification for watertight is: "So constructed that there shall be no leakage of water into the enclosure when subjected to a stream from a hose with a 1 in nozzle and delivering at least 65 gal/min, with the water directed at the enclosure from a distance of not less than 10 ft for a period of 5 min, during which period the water may be directed in one or more directions as desired." 53

watertight door-control system. A system of control for power-operated watertight doors providing individual local control of each door and, at a remote station in or adjoining the wheelhouse, individual control of any door, collective control of all doors, and individual indication of open or closed condition. *See:* **marine electric apparatus.** 328

watertight enclosure (electric installations on shipboard). An enclosure constructed so that a stream of water from a hose not less than 1 in. in diameter under a head of 35 ft from a distance of 10 ft can be played on the enclosure from any direction for a period of 15 min without leakage. The hose nozzle shall have a uniform inside diameter of 1 in. 3

water treatment equipment (thyristor). Any apparatus such as deionizers, electrolytic targets, filters, or other devices employed to control electrolysis, corrosion, scaling, or clogging in water systems. 445

watt (1) (general). The unit of power in the International System of Units (SI). The watt is the power required to do work at the rate of 1 joule per second. 210
(2) (W) (laser-maser). The unit of power, or radiant flux. 363

watt density (electrical heat tracing for industrial applications). Thermal output of heating cable in watts per unit area. 523

watthour. 3600 joules. 210

watthour capacity (storage battery) (storage cell). The number of watthours that can be delivered under specified conditions as to temperature, rate of discharge, and final voltage. *See:* **battery (primary or secondary).** 328

watthour constant (watthour meter). The registration, expressed in watthours, corresponding to one revolution of the rotor. *Note:* It is commonly denoted by the symbol K_h. When a meter is used with instrument transformers, the watthour constant is expressed in terms of primary watthours. For a secondary test of such a meter, the constant is the primary watthour constant divided by the product of the nominal ratios of transformation. 212

watthour-demand meter. A watthour meter and a demand meter combined as a single unit. *See:* **electricity meter.** 328

watthour efficiency (storage battery) (storage cell). The energy efficiency expressed as the ratio of the watthours output to the watthours of the recharge. *See:* **charge.** 328

watthour meter. An electricity meter that measures and registers the integral, with respect to time, of the active power of the circuit in which it is connected. This power integral is the energy delivered to the circuit during the interval over which the integration extends, and the unit in which it is measured is usually the kilowatthour. *See:* **adjustment; basic current range; basic voltage range; class designation; creep; form designation; gear ratio; heavy load; induction; light load; load current; load range; motor type watthour meter; percentage error; percentage registration; portable standard watthour meter; rated current; rated voltage; reference performance; reference standard; register; register constant; register ratio; registration; rotor; stator; test current (TA); two-rate watthour meter; watthour constant.** 212

watt loss (electric instrument). *See:* **power loss (electric instrument).**

wattmeter. An instrument for measuring the magnitude of the active power in an electric circuit. It is provided with a scale usually graduated in either watts, kilowatts, or megawatts. If the scale is graduated in kilowatts or megawatts, the instrument is usually designated as a kilowattmeter or megawattmeter. *See:* **instrument.** 328

wattsecond constant (meter). The registration in wattseconds corresponding to one revolution of the rotor. *Note:* The wattsecond constant is 3600 times the watthour constant and is commonly denoted by the symbol K_s. *See:* **electricity meter (meter).** 328

wave (1) (data transmission). (A) A disturbance that is

a function of time or space or both. (B) A disturbance propagated in a medium or through space. *Notes:* (1) Any physical quantity that has the same relationship to some independent variable (usually time) that a propagated disturbance has, at a particular instant, with respect to space, may be called a wave. (2) Disturbance, in this definition, is used as a generic term indicating not only mechanical displacement but also voltage, current, electric field strength, temperature, etcetera. (C) (electric circuit). The variation of current, potential, or power at any point in the electric circuit. *See:* **continuous wave; wave filter; ground wave; ground reflected wave; wave interference; tangential wave path; rectangular wave; square wave; wave tilt; tropospheric wave.**   59

(2) (overhead-power-line corona and radio noise). A disturbance propagated in a medium or through space. *Notes:* (A) Any physical quantity which has the same relationship to some independent variable (usually time) that a propagated disturbance has, at a particular instant, with respect to space, may be called a wave. (B) "Disturbance" in this definition is used as a generic term indicating not only mechanical displacement but also voltage, current, electric field strength, temperature, etcetera.   411

(3) (surge arrester). The variation with time of current, potential, or power at any point in an electric circuit.   430

(4). A disturbance that is a function of time or space or both.   111

(5) (pulse terms). A modification of the physical state of a medium which propagates in the medium as a function of time as a result of one or more disturbances.   254

(6) (metal-oxide surge arresters for ac power circuits). The variation with time of current, potential, or power at any point in an electric circuit.   583

wave analyzer. An electric instrument for measuring the amplitude and frequency of the various components of a complex current or voltage wave. *See:* **instrument.**   328

wave antenna. *See:* **Beverage antenna.**

wave clutter. Clutter caused by echoes from waves of the sea. *See:* **radar.**   328

wave filter (data transmission). A transducer for separating waves on the basis of their frequency. *Note:* A filter introduces relatively small insertion loss to waves in one or more frequency bands and relatively large insertion loss to waves of other frequencies.   59

waveform (pulse terms) (pulse waveform) (transition waveform). A manifestation or representation (that is, graph, plot, oscilloscope presentation, equation(s), table of coordinate or statistical data, etcetera) or a visualization of a wave, pulse, or transition. (1) The term **pulse waveform** is included in the term waveform. (2) The term **transition waveform** is included in the terms pulse waveform and waveform.   254

waveform-amplitude distortion. Nonlinear distortion in the special case where the desired relationship is direct proportionality between input and output.

Note: Also sometimes called **amplitude distortion.** *See:* **nonlinear distortion.**   349

waveform distortion (1) (oscilloscopes). A displayed deviation from the representation of the input reference signal. *See:* **oscillograph.**   185

(2) (percent) (metering). The ratio of the root-mean-square value of the harmonic content (excluding the fundamental) to the root-mean-square value of the nonsinusoidal quantity, expressed as a percentage.   212

waveform epoch (pulse terms). The span of time for which waveform data are known or knowable. A waveform epoch manifested by equations may extend in time from minus infinity to plus infinity or, like all waveform data, may extend from a first datum time, t_0, to a second datum time, t_1. *See:* The single pulse diagram below.   254

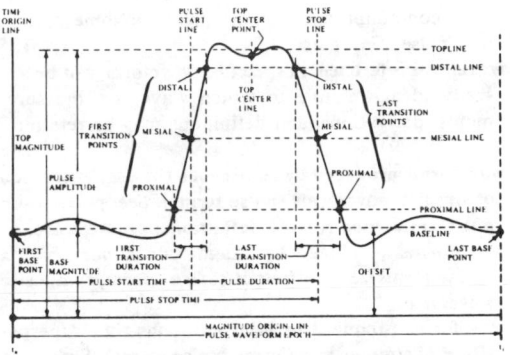

The single pulse.

waveform epoch contraction (pulse measurement). A technique for the determination of the characteristics of individual pulse waveforms (or pulse waveform features) wherein the waveform epoch (or pulse waveform epoch) is contracted in time to a pulse waveform epoch (or transition waveform epoch) for the determination of time or magnitude characteristics. In any waveform epoch contraction procedure two or more sets of time or magnitude reference lines may exist, and the set of reference lines being used in any pulse measurement process shall be specified.   15

waveform epoch expansion (pulse measurement). A technique for the determination of the characteristics of a transition waveform (or pulse waveform) wherein the transition waveform epoch (or pulse waveform epoch) is expanded in time to a pulse waveform epoch (or waveform epoch) for the determination of magnitude or time reference lines. The reference lines determined by analysis of the pulse waveform (or waveform) are transferred to the transition waveform (or pulse waveform) for the determination of characteristics. In any waveform epoch expansion procedure two or more sets of reference lines may exist, and the set of reference lines being used in any pulse measurement process shall be specified.   15

waveform formats. Waveforms may exist, be recorded,

or be stored in a variety of formats. It is assumed that: (1) waveform formats are in terms of Cartesian coordinates, or some transform thereof: (2) conversion from one waveform format to any other is possible: and (3) such waveform format conversions can be made with precision, accuracy, and resolution which is consistent with the accuracy desired in the pulse measurement process.   15

waveform influence of root-mean-square (rms) responding instruments. The change in indication produced in an rms responding instrument by the presence of harmonics in the alternating electrical quantity under measurement. In magnitude it is the deviation between an indicated rms value of an alternating electrical quantity and the indication produced by the measurement of a pure sine-wave form of equal rms value. *See:* **instrument.**   280

waveform pulse. A waveform or a portion of a waveform containing one or more pulses or some portion of a pulse. *See:* **pulse.**   185

waveform reference. A specified waveform, not necessarily ideal, relative to which waveform measurements, derivations, and definitions may be referred.
185

waveforms produced by continuous time superposition of simpler waveforms (pulse terms). *See:* **pulse train; pulse train time-related definitions; square wave.**

waveforms produced by magnitude superposition (pulse terms). *See:* **offset; offset waveform; composite waveform.**

waveforms produced by noncontinuous time superposition of simpler waveforms (pulse terms). *See:* **pulse burst; pulse burst time-related definitions.**

waveforms produced by operations on waveforms (pulse terms). All envelope definitions in this section are based on the cubic natural spline (or its related approximation, the draftsman's spline) with knots at specified points. All burst envelopes extend in time from the first to the last knots specified, the remainder of the waveform being (1) that portion of the waveform which precedes the first knot and (2) that portion of the waveform which follows the last knot. Burst envelopes and their adjacent waveform bases, taken together, comprise a continuous waveform which has a continuous first derivative except at the first and last knots of the envelope. *See:* **pulse train top (base) envelope; pulse burst top envelope; pulse burst base envelope.**   254

waveform test (rotating machinery). A test in which the waveform of any quantity associated with a machine is recorded. *See:* **asynchronous machine.**
63

wavefront (1)(of a surge or impulse)(metal-oxide surge arresters for ac power circuits)(surge arrester). That part which occurs prior to the crest value.
583,430

(2) (fiber optics). The locus of points having the same phase at the same time.   433

(3) (impulse in a conductor). That part (in time or distance) between the virtual-zero point and the point at which the impulse reaches its crest value.   64

waveguide (1)(waveguide terms). A system of material boundaries or structures for guiding electromagnetic waves. Usually such a system is used for guiding waves in other than TEM modes. Often, and originally, a hollow metal pipe for guiding electromagnetic waves. *See:* **transmission line.**   267

(2) (data transmission). (A) Broadly, a system of material boundaries capable of guiding electromagnetic waves. (B) More specifically, a transmission line comprising a hollow conducting tube within which electromagnetic waves may be propagated or a solid dielectric or dielectric-filled conductor for the same purpose. (C) A system of material boundaries or structures for guiding transverse-electromagnetic mode, often and originally a hollow metal pipe for guiding electromagnetic waves.   59

(3) (radio wave propagation). A system of material boundaries capable of guiding waves.   146

waveguide adapter (waveguide components). A structure used to interconnect two waveguides which differ in size or type. If the modes of propagation also differ, the adapter functions as a mode transducer.   166

waveguide attenuator (waveguide components). A waveguide component that reduces the output power relative to the input by any means, including absorption and reflection.   166

waveguide bend (waveguide components). A section of waveguide or transmission line in which the direction of the longitudinal axis is changed. In common usage the waveguide corner formed by an abrupt change in direction is considered to be a bend.   166

waveguide calorimeter (waveguide components). A waveguide or transmission line structure which uses the temperature rise in a medium as a measure of absorbed power. The medium, typically water or a thermoelectric element, is either the power-absorbing agent or has heat transferred to it from a power-absorbing element.   166

waveguide circulator (nonlinear, active, and nonreciprocal waveguide components). A passive waveguide device of three or more ports in which the ports can be numbered in such an order that, when power is fed into any port, the power is transferred to the next sequentially numbered port. The first port is counted as following the last in order.   530

waveguide component. A device designed to be connected at specified ports in a waveguide system.
319

waveguide corner *See:* **waveguide bend.**

waveguide cutoff frequency (critical frequency). *See:* **cutoff frequency.**

waveguide differential phase circulator (nonlinear, active, and nonreciprocal waveguide components). A waveguide circulator based on the use of at least one nonreciprocal differential insertion phase element or gyrator, usually in connection with other waveguide components such as microwave hybrid junctions.
530

waveguide dispersion (fiber optics). For each mode in an optical waveguide, the term used to describe the process by which an electromagnetic signal is distorted by virtue of the dependence of the phase and group velocities on wavelength as a consequence of the geometric properties of the waveguide. In particular, for circular waveguides, the dependence is on the ratio (a/λ), where a is core radius and λ is wavelength. *See:* **dispersion; distortion; material dispersion; multimode distortion; profile dispersion.**   433

waveguide Faraday rotation circulator (nonlinear, active, and nonreciprocal waveguide components). A waveguide circulator based on the use of a Faraday rotation element in conjunction with other waveguide components such as dual-mode transducers.   530

waveguide ferrite isolator (nonlinear, active, and nonreciprocal waveguide components). A waveguide two-port device, using gyromagnetic material, in which the attenuation in one direction of propagation is much greater than in the opposite direction.
              530

waveguide gasket (waveguide components). A resilient insert usually between flanges intended to serve one or more of the following primary purposes: (A) to reduce gas leakage affecting internal waveguide pressure, (B) to prevent intrusion of foreign material into the waveguide, or (C) to reduce power leakage and arcing.
              166

waveguide iris (waveguide components). A partial obstruction at a transverse cross-section formed by one or more metal plates of small thickness compared with the wavelength.   166

waveguide joint. A connection between two sections of waveguide. *See:* **waveguide.**   185

waveguide junction circulator (nonlinear, active, and nonreciprocal waveguide components). A waveguide circulator based on the use of gyromagnetic material at the common junction of several waveguides.
              530

waveguide matched termination (waveguide components). *See:* **matched termination.**   166

waveguide mode (waveguide terms). In a uniform waveguide, a wave that is characterized by exponential variation of the fields along the direction of the guide. *Note:* In other types of waveguides, such as radial, spherical, toroidal, etcetera, some particular variation will have to be specified according to the geometry.   267

waveguide phase shifter (waveguide components). An essentially loss-less device for adjusting the phase of a forward-traveling electromagnetic wave at the output of the device relative to the phase at the input.
              166

waveguide plunger. *See:* **adjustable short circuit.**

waveguide post (waveguide components). A cylindrical rod placed in a transverse plane of the waveguide and behaving substantially as a shunt susceptance.
              166

waveguide resonator (waveguide components). A waveguide or transmission-line structure which can store oscillating electromagnetic energy for time periods that are long compared with the period of the resonant frequency, at or near the resonant frequency.
              166

waveguide scattering (fiber optics). Scattering (other than material scattering) that is attributable to variations of geometry and index profile of the waveguide. *See:* **material scattering; nonlinear scattering; Rayleigh scattering; scattering.**   433

waveguide short circuit, adjustable (waveguide components). A longitudinally movable obstacle which reflects essentially all the incident energy.   166

waveguide step twist (waveguide components). A waveguide twist formed by abruptly rotating about the waveguide longitudinal axis one or more waveguide sections, each nominally a quarter wavelength long.
              166

waveguide stub (waveguide components). A section of waveguide or transmission line joined to the main guide or transmission line and containing an essentially nondissipative termination.   166

waveguide switch (waveguide system). A device for stopping or diverting the flow of high-frequency energy as desired. *See:* **waveguide.**   244, 179

waveguide taper (waveguide components). A section of tapered waveguide.   166

waveguide termination. *See:* **cavity; unloaded applicator impedance.**

waveguide-to-coaxial transition. A mode changer for converting coaxial line transmission to rectangular waveguide transmission. *See:* **waveguide.**   179

waveguide transformer (waveguide components). A structure added to a waveguide or transmission line for the purpose of impedance transformation.
              166

waveguide tuner (waveguide components). An adjustable waveguide transformer.   166

waveguide twist (waveguide components). A waveguide section in which there is progressive rotation of the cross-section about the longitudinal axis. *See:* **step twist.**   166

waveguide wavelength (1)(waveguide terms). For a traveling wave in a uniform waveguide at a given frequency and for a given mode, the distance along the guide between corresponding points at which a field component (or the voltage or current) differs in phase by 2π radians.   267

(2) (data transmission). The distance along a uniform guide between points at which a field component (or the voltage or current) differs in phase by 2π radians. *Note:* It is equal to the quotient of phase velocity divided by frequency. For a waveguide with air dielectric, the waveguide wavelength is given by the formula:

$$\lambda_g = \frac{\lambda}{(1 - (\lambda^2/\lambda_c))^{1/2}}$$

where λ is the free space wavelength and λ_c is the cutoff wavelength of the guide.   59

waveguide window (waveguide components). A gas- or liquid- tight barrier or cover designed to be essentially transparent to the transmission of electromagnetic waves.   166

wave heater (dielectric heating). A heater in which heating is produced by energy absorption from a traveling electromagnetic wave. *See:* **dielectric heating.**   14

wave heating. The heating of a material by energy absorption from a traveling electromagnetic wave. *See:* **dielectric heating; induction heating.**   14

wave impedance (waveguide terms). The complex factor relating the transverse component of the magnetic field to the transverse component of the electric field at every point in any specified plane, for a given mode.   267

wave impedance, characteristic. *See:* **characteristic wave impedance.**   267

wave interference (1) (data transmission). The variation of wave amplitude with distance or time, caused by the superposition of two or more waves. *Notes:* (A) As most commonly used, the term refers to the interference of waves of the same or nearly the same frequency. (B) Wave interference is characterized by a spatial or temporal distribution of amplitude of some specified characteristic differing from that of the individual superposed waves.   59

(2) (radio wave propagation). Same definition as (1) above but without note (B).   146

wavelength (1) (λ) (laser-maser). The distance between two points in a periodic wave which have the same phase.   363

(2) (radio wave propagation). Of a sinusoidal wave, the distance between points of corresponding phase of two consecutive cycles. The wavelength λ is related to the phase velocity v and the frequency f by $\lambda = v/f$.   146

wavelength constant. *See:* **phase constant.**

wavelength division multiplexing (WDM) (fiber optics). The provision of two or more channels over a common optical waveguide, the channels being differentiated by optical wavelength.   433

wavelength shifter (scintillator). A photofluorescent compound used with a scintillator material to absorb photons and emit related photons of a longer wavelength. *Note:* The purpose is to cause more efficient use of the photons by the phototube or photocell. *See:* **phototube.**   335

wavemeter. *See:* **cavity resonator frequency meter.**

wave normal (1) (radio wave propagation). Of a traveling wave, the direction normal to an equiphase surface taken in the direction of increasing phase.   146

(2) (waveguide). A unit vector normal to an equiphase surface with its positive direction taken on the same side of the surface as the direction of propagation. In isotropic media, the wave normal is in the direction of propagation.   267

waves, electrocardiographic. *See:* **electrocardiographic waves.**

wave shape (of an impulse test wave)(metal-oxide surge arresters for ac power circuits). The graph of the wave as a function of time.   583

wave shape (of an impulse test wave) (surge arrester). The graph of the wave as a function of time.   430

wave shape designation (of an impulse) (metal-oxide surge arresters for ac power circuits) (surge arrester). (1) The wave shape of an impulse (other than rectangular) of a current or voltage is designated by a combination of two numbers. The first, an index of the wave front, is the virtual duration of the wave front in microseconds. The second, an index of the wave tail, is the time in microseconds from virtual zero to the instant at which one-half of the crest value is reached on the wave tail. Examples are 1.2/50 and 8/20 waves. (2) The wave shape of a rectangular impulse of current or voltage is designated by two numbers. The first designates the minimum value of current or voltage that is sustained for the time in microseconds designated by the second number. An example is the 75 A 1000 μs wave.   583,430

wave, square (pulse techniques). *See:* **square wave.**

wave tail (of an impulse)(metal-oxide surge arresters for ac power circuits)(surge arrester). That part between the crest value and the end of the impulse.   583,430

wave tilt (data transmission). The forward inclination of a radio wave due to its proximity to ground.   59

wave train. A limited series of wave cycles caused by a periodic disturbance of short duration. *See:* **pulse train.**   56

wave vector (radio wave propagation). The complex vector k in planewave expressions for wave parameters with exponential factor exp $[j(\omega t\text{-}k\cdot r)]$. *See:* **propagation vector; plane-wave exponential factor.**   146

wave vector in physical media. The complex vector $15\bar{k}$ in plane wave solutions of the form $e^{-j\bar{k}\cdot\bar{r}}$, for an $e^{j\omega t}$ time variation and $15\bar{r}$ *the position vector. See:* **propagation vector in physical media.**   111

wave winding. A winding that progresses around the armature by passing successively under each main pole of the machine before again approaching the starting point. In commutator machines, the ends of individual coils are not connected to adjacent commutator bars. *See:* **asynchronous machine.**   328

way (power switchgear). A three-phase circuit entrance to a switching assembly.   489

way point (navigation)(navigation aid terms). A selected point on or near a course line and having significance with respect to navigation or traffic control.   526

way station (data transmission). A telegraph term for one of the stations on a multipoint circuit.   59

weakly guiding fiber (fiber optics). A fiber for which the difference between the maximum and the minimum refractive index is small (usually less than 1 percent).   433

weakly perturbed field (measurement of power frequency electric and magnetic fields from ac power lines). At a given point, a field whose magnitude does not change by more than 5% or whose direction does not vary by more than 5 degrees when an object is introduced into the region.   514

wear-out failure (reliability). *See:* failure, wear-out.

wear-out failure period (reliability). That possible period during which the failure rate increases rapidly in comparison with the preceding period. *See:* **constant failure rate period** for curve showing the failure rate pattern when the terms of **early failure period** and **constant failure rate period** apply.   164

wear out failures (station control and data acquisition). The pattern of failures experienced when equipments reach their period of deterioration. Wear-out failure profiles may be approximated by a Gaussian (bell curve) distribution centered on the nominal life of the equipment.   403

wearout of reverse drain breakdown (metal-nitride-oxide field-effect transistor). This term describes an effect where the reverse current–voltage characteristic of the drain junction changes progressively and irreversibly toward larger (leakage) currents at the same reverse voltages.   386

wearout period (reliability analysis of nuclear power generating station safety systems). The time interval, following the period of constant failure rate, during which failures occur at an increasing rate.
                    587,29

wear point (cable plowing). A removable tip on the end of some shanks or plow blades.   52

weather, adverse (generating station). Designates weather conditions which cause an abnormally high rate of forced outages for exposed components during the periods such conditions persist, but do not qualify as major storm disasters. Adverse weather conditions can be defined for a particular system by selecting the proper values and combinations of conditions reported by the Weather Bureau: thunderstorms, tornadoes, wind velocities, precipitation, temperature, etc. *Note:* This definition derives from transmission and distribution applications and does not necessarily apply to generation outages. *See:* **major storm disaster.**
                    112

weather barrier (1)(electrical heating systems). A material that protects thermal insulation from environmental conditions such as rain, sleet, snow, wind, contamination, and physical damage.   476

(2)(electrical heat tracing for industrial applications). Material that, when installed on the outer surface of thermal insulation, protects the insulation from weather, such as rain, snow, sleet, wind, solar radiation, or atmospheric contamination, and physical damage.   523

weather, normal (generating station). Includes all weather not designated as adverse or major storm disaster. *Note:* This definition derives from transmission and distribution applications and does not necessarily apply to generation outages.   112

weatherproof (1) (National Electrical Code). So constructed or protected that exposure to the weather will not interfere with successful operation. Rainproof, raintight, or watertight equipment can fulfill the requirements for weatherproof where varying weather conditions other than wetness, such as snow, ice, dust, or temperature extremes, are not a factor.   256

(2) (outside exposure). So constructed or protected that exposure to the weather will not interfere with successful operation. *See:* **outdoor.**   103,27

weatherproof enclosure. An enclosure for outdoor application designed to protect against weather hazards such as rain, snow, or sleet. *Note:* Condensation is minimized by use of space heaters.   103

weather-protected machine. A guarded machine whose ventilating passages are so designed as to minimize the entrance of rain, snow, and airborne particles to the electric parts. *See:* **asynchronous machine.**   63

weathershed (composite insulators). The external part of the insulator that protects the core and provides the wet electrical strength and leakage distance.   483

weathertight. *See:* **raintight.**

weather vane. *See:* **vane.**

weber. The unit of magnetic flux in the International System of Units (SI). The weber is the magnetic flux whose decrease to zero when linked with a single turn induces in the turn a voltage whose time integral is one volt-second.   210

wedge (rotating machinery). A tapered shim or key. *See:* **slot wedge; rotor (rotating machinery); stator.**
                    63

wedge groove (wedge slot) (rotating machinery). A groove, usually in the side of a coil slot, to permit the insertion of and to retain a slot wedge.   63

wedge, slot. *See:* **wedge groove.**

wedge washer (salient pole) (rotating machinery). Insulation triangular in cross section placed underneath the inner ends of field coils and spanning between field coils.   63

weight coefficient (thermoelectric generator) (thermoelectric generator couple). The quotient of the electric power output by the device weight. *See:* **thermoelectric device.**   191

weighted average quantum efficiency (η) (diode- type camera tube). The spectral quantum efficiency η_λ, integrated over a spectral band λ_1 to λ_2; and weighted by a particular input spectral distribution $N(\lambda)$.

$$\eta \equiv \frac{\displaystyle\int_{\lambda_1}^{\lambda_2} \eta_\lambda N(\lambda)\,d\lambda}{\displaystyle\int_{\lambda_1}^{\lambda_2} N(\lambda)\,d\lambda}$$

Since the input spectral distribution appears in both numerator and denominator, it can have dimensions of radiant power, or irradiance, or it can be a relative number, normalized, for example, to the peak value of the input spectral distribution.   380

weighted peak flutter. Flutter and wow indicated by the weighted peak flutter measuring equipment specified in ANSI/IEEE Std 193-1971, Method for Measurement of Weighted Peak Flutter of Sound Recording and Reproducing Equipment. *Note:* The meter indicates one-half the peak-to-peak demodulated signal. The frequency-response weighting network is specified in Table I, and the dynamic response of the system is given in Sec. 5.4, of the referenced standard.
145

weighted sound level (audible noise measurement). Weighting adjusts the spectrum of a measured sound-pressure level to correspond approximately to the sensitivity of human hearing. In standardized sound measuring instruments, this is implemented with selectable A-, B-, and C-weighting networks as discussed in ANSI S1.4-1983 [2]. The term weighting or weighted is used because some frequencies are given more or less importance, or weight, than other frequencies. The weighting functions employed in A-, B-, and C-weighting networks correspond approximately to the response of the human ear to low, medium, and high sound levels, respectively. The most commonly used noise rating scale is the A-weighted sound level, expressed in decibels as dB(A). (The word pressure is omitted when describing weighted sound levels.) A-weighting has commonly been used and is recommended for transmission-line sound measurements.
462

weighting (data transmission)(antennas). The artificial adjustment of measurements in order to account for factors that in the normal use of the device, would otherwise be different from the conditions during measurement. For example, background noise measurements may be weighted by applying factors or by introducing networks to reduce measured values in inverse ratio to their interfering effects.   59,111

weight transfer compensation. A system of control wherein the tractive forces of individual traction motors may be adjusted to compensate for the transfer of weight from one axle to another when exerting tractive force. *See:* **multiple-unit control.**   328

weld decay (corrosion). Localized corrosion at or adjacent to a weld.   205

welding arc voltage. The voltage across the welding arc.
264

well-type coaxial detector (germanium gamma-ray detectors). A coaxial detector that is mounted and encapsulated in such a way that radioactive samples may be inserted within the inner cylindrical electrode such that the sample is essentially surrounded by active detector material.   528

Weston normal cell. A standard cell of the cadmium type containing a saturated solution of cadmium sulphate as the electrolyte. *Note:* Strictly speaking this cell contains a neutral solution, but acid cells are now in more common use. *See:* **electrochemistry.**   328

wet cell. A cell whose electrolyte is in liquid form. *See:* **electrochemistry.**   328

wet contact (telephone switching systems). A contact through which direct current flows. *Note:* The term

has significance because of the healing action of direct current flowing through contacts.   55

wet-dry signaling, (telephone switching systems). Two-state signaling achieved by the application and removal of battery at one end of a trunk.   55

wet electrolytic capacitor. A capacitor in which the dielectric is primarily an anodized coating on one electrode, with the remaining space between the electrodes filled with a liquid electrolytic solution.
341

wet location (National Electrical Code). Installations underground or in concrete slabs or masonry in direct contact with the earth, and locations subject to saturation with water or other liquids, such as vehicle washing areas, and locations exposed to weather and unprotected.   256

wet location, health care facility (National Electrical Code). A patient care area, that is normally subject to wet conditions, including standing water on the floor, or routine dousing or drenching of the work area. Routine housekeeping procedures and incidental spillage of liquids do not define a wet location.   256

Wet-Niche Lighting Fixture (National Electrical Code). A lighting fixture intended for installation in a metal forming shell mounted in a swimming pool structure where the fixture will be completely surrounded by pool water.   256

wetting. The free flow of solder alloy, with proper application of heat and flux, on a metallic surface to produce an adherent bond.   284

wetting agent (electroplating) (surface active agent). A substance added to a cleaning, pickling or plating solution to decrease its surface tension. *See:* **electroplating.**   328

wet-wound (rotating machinery). A coil in which the conductors are coated with wet resin in passage to the winding form, or on to which a bonding or insulating resin is applied on each successive winding layer to produce an impregnated coil. *See:* **rotor (rotating machinery); stator.**   63

Wheatstone bridge. A 4-arm bridge, all arms of which are predominantly resistive. *See:* **bridge.**   328

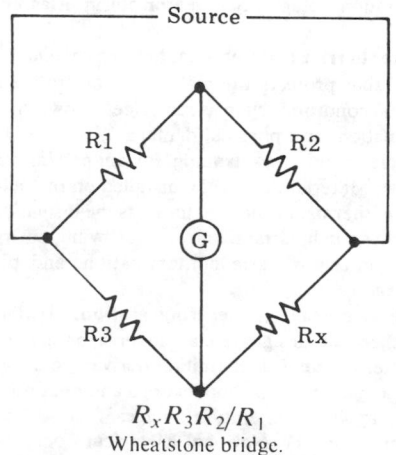

$$R_x R_3 R_2 / R_1$$
Wheatstone bridge.

wheeling charge (power operations). The amount paid to an intervening system for the use of its transmission facilities.   516

wheel-speed sensitivity (dynamically tuned gyro). *See:* **rotor speed sensitivity.**   46

wheel tractor (conductor stringing equipment). A wheeled unit employed to pull pulling lines, sag conductor, and miscellaneous other work. Sagging winches on this unit are usually arranged in horizontal configuration. It has some advantages over crawler tractors in that it has a softer footprint, travels faster, and is more maneuverable. *Syn:* **logger; sagger; skidder; tractor.** *See:* **crawler tractor.**   431

whip antenna. A thin flexible monopole antenna.   111

whistle operator. A device to provide automatically the timed signals required by navigation laws when underway in fog, and also manual control of electrical operation of a whistle or siren, or both, for at-will signals.   328

whistler (radio wave propagation). A form of radio noise in the low frequency (LF), very low frequency (VLF), and extreme low frequency (ELF) portions of the electromagnetic spectrum propagating in the whistler mode. *Note:* This noise usually originates from a lightning stroke and is characterized by a whistling tone which may last for seconds.   146

whistler mode (radio wave propagation). The propagation mode of any right-handed polarized electromagnetic wave propagating in a magnetoplasma at a frequency less than the electron gyrofrequency but greater than the ion gyrofrequency.   146

white (color television). Used most commonly in the nontechnical sense. More specific usage is covered by the term **achromatic locus,** and this usage is explained in the note under the term **achromatic locus.** *See:* **reference white.**   18

white compression (white saturation) (television). The reduction in gain applied to a picture signal at those levels corresponding to light areas in a picture with respect to the gain at that level corresponding to the midrange light value in the picture. *Notes:* (1) The gain referred to in the definition is for a signal amplitude small in comparison with the total peak-to-peak picture signal involved. A quantitative evaluation of this effect can be obtained by a measurement of differential gain. (2) The overall effect of white compression is to reduce contrast in the highlights of the picture as seen on a monitor. *See:* **television.**   178

'white' light photocathode response (diode-type camera tube). The ratio of the output signal current to the total input radiant power from a tungsten filament source at a 2854K color temperature. Units: amperes watt^{-1} (AW^{-1}). (CIE illuminant A).   380

white noise (1)(data transmission)(overhead-power-line corona and radio noise). Noise, either random or impulsive type, that has a flat frequency spectrum at the frequency range of interest.   59,411

(2) (telephone practice). Noise, either random or impulsive type, that has a flat frequency spectrum at the frequency range of interest. This type of noise is used in the evaluation of systems on a theoretical basis and is produced for testing purposes by a white-noise generator. The use of the term should be limited and is not good usage in describing message circuit noise.   24

white object (color) (television). An object that reflects all wavelengths of light with substantially equal high efficiencies and with considerable diffusion.   18

white peak (television). A peak excursion of the picture signal in the white direction. *See:* **television.**   178

white recording (frequency-modulation facsimile system). That form of recording in which the lowest received frequency corresponds to the minimum density of the record medium. *See:* **recording (facsimile).**   12

white saturation. *See:* **white compression.**

white signal (at any point in a facsimile system). The signal produced by the scanning of a minimum-density area of the subject copy. *See:* **facsimile signal (picture signal).**   12

white transmission (1) (amplitude-modulation facsimile system). That form of transmission in which the maximum transmitted power corresponds to the minimum density of the subject copy.

(2) (frequency-modulation facsimile system). That form of transmission in which the lowest transmitted frequency corresponds to the minimum density of the subject copy. *See:* **facsimile transmission.**   12

whole body irradiation (electrobiology). Pertains to the case in which the entire body is exposed to the incident electromagnetic energy or in which the cross section of the body is smaller than the cross section of the incident radiation beam. *See:* **electrobiology.**   322

wicking. The flow of solder along the strands and under the insulation of stranded lead wires.   284

wick-lubricated bearing (rotating machinery). (1) A sleeve bearing in which a supply of lubricant is provided by the capillary action of a wick that extends into a reservoir of free oil or of oil-saturated packing material. (2) A sleeve bearing in which the reservoir and other cavities in the bearing region are packed with a material that holds the lubricant supply and also serves as a wicking. *See:* **bearing.**   63

wide-angle diffusion (illuminating engineering). That in which flux is scattered at angles far from the direction which the flux would take by regular reflection or transmission. *See:* **regular (specular) reflection.**   167

wide angle luminaire (illuminating engineering). A luminaire which distributes the light through a comparatively large solid angle.   167

wideband channel (data transmission). A channel wider in bandwidth than a voice-grade channel.   59

wide-band improvement. The ratio of the signal-to-noise ratio of the system in question to the signal-to-noise ratio of a reference system. *Note:* In comparing frequency-modulation and amplitude-modulation systems, the reference system usually is a double-side-band amplitude-modulation system with a carrier power, in the absence of modulation, that is equal to the carrier power of the frequency-modulation system. 111

wide-band ratio. The ratio of the occupied frequency bandwidth to the intelligence bandwidth. 111

width line (illuminating engineering). The radial line (the one which makes the larger angle with the reference line) which passes through the point of on e-half maximum candlepower on the lateral candlepower distribution curve plotted on the surface of the cone of maximum candlepower. 167

Wiedemann-Franz ratio. The quotient of the thermal conductivity by the electric conductivity. *See:* **thermoelectric device.** 248, 191

Wien bridge oscillator. An oscillator whose frequency of oscillation is controlled by a Wien bridge. *See:* **oscillatory circuit.** 328

Wien capacitance bridge. A 4-arm alternating-current bridge characterized by having in two adjacent arms capacitors respectively in series and in parallel with resistors, while the other two arms are normally non-reactive resistors. *Note:* Normally used for the measurement of capacitance in terms of resistance and frequency. The balance depends upon frequency, but from the balance conditions the capacitance of either or both capacitors can be computed from the resistances of all four arms and the frequency. *See:* **bridge.**

where the symbols are those used in the definition of **Planck radiation law.** The two principal corollaries of this law are:

$$\lambda_m T = b$$

$$L_m / T^5 = b'$$

which show how the maximum spectral radiance L_m and the wavelength λ_m at which it occurs are related to the absolute temperature T. *Note:* The currently recommended value of b is 2.8978×10^{-3} m $\cdot$ K or 2.8978×10^{-1} cm $\cdot$ K. From the definition of the **Planck radiation law,** and with the use of the value of b, as given above b', is found to be 4.10×10^{-12} W $\cdot$ cm$^{-3} \cdot$ K$^{-5} \cdot$ sr^{-1}. 167

Wien inductance bridge. A 4-arm alternating-current bridge characterized by having in two adjacent arms inductors respectively in series and in parallel with resistors, while the other two arms are normally non-reactive resistors. *Note:* Normally used for the measurement of inductance in terms of resistance and frequency. The balance depends upon frequency, but from the balance conditions the inductances of either or both inductors can be computed from the resistances of the four arms and the frequency. *See:* **bridge.** 328

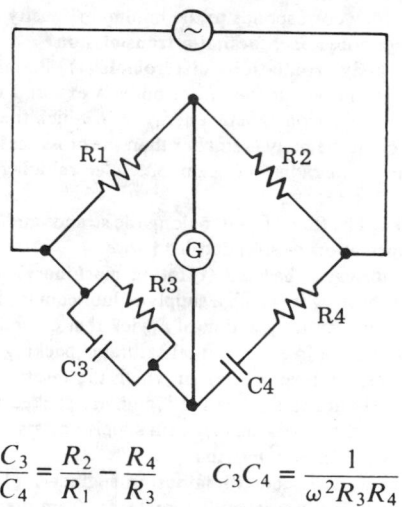

$$\frac{C_3}{C_4} = \frac{R_2}{R_1} - \frac{R_4}{R_3} \qquad C_3 C_4 = \frac{1}{\omega^2 R_3 R_4}$$

Wien capacitance bridge.

Wien displacement law (illuminating engineering). An expression representing, in a functional form, the spectral radiance $L\lambda$ of a blackbody as a function of the wavelength λ and the temperature T.

$$L_\lambda = I_\lambda / A'$$

$$= c_1 \lambda^{-5} f(\lambda T),$$

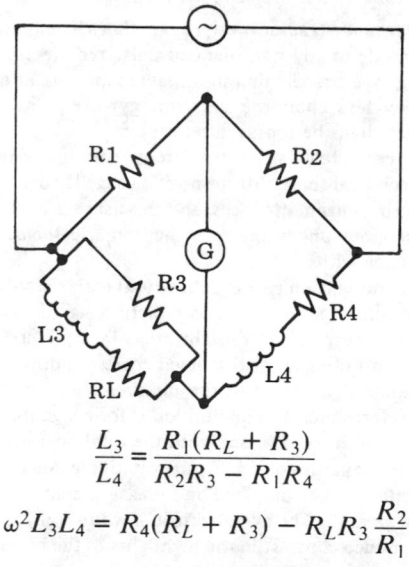

$$\frac{L_3}{L_4} = \frac{R_1 (R_L + R_3)}{R_2 R_3 - R_1 R_4}$$

$$\omega^2 L_3 L_4 = R_4 (R_L + R_3) - R_L R_3 \frac{R_2}{R_1}$$

Wien inductance bridge.

Wien radiation law (illuminating engineering). An expression representing approximately the spectral radiance of a blackbody as a function of its wavelength and temperature. It commonly is expressed by the formula

$$L_\lambda = I_\lambda / A'$$

$$= c_{1L} \lambda^{-5} e^{-(c_2 / \lambda T)}$$

where the symbols are those used in the definition of **Planck radiation law.** This formula is accurate to one percent or better for values of λT less than 3000 micrometer kelvins. *See:* **radiant energy.** 167

wigwag signal. A railroad-highway crossing signal, the indication of which is given by a horizontally swinging disc with or without a red light attached. 328

Williams-tube storage (electronic computation). A type of electrostatic storage. 210

Wilson center (limb center) (V **potential) (medical electronics) (electrocardiography).** An electric reference contact: the junction of three equal resistors to the limb leads. 192

wind-driven generator for aircraft. A generator used on aircraft that derives its power from the air stream applied on its own air screw or impeller during flight. 328

winding (data processing). A conductive path, usually of wire, inductively coupled to a magnetic core or cell. *Note:* When several windings are employed, they may be designated by the functions performed. Examples are: sense, bias, and drive windings. Drive windings include read, write, inhibit, set, reset, input, shift, and advance windings. 331

winding, ac (thyristor converter). The winding of a thyristor converter transformer that is connected to the ac circuit and usually has no conductive connection with the thyristor circuit elements. *Syn:* **primary winding.** 121

winding, autotransformer series. *See:* **series winding.**

winding, control power. The winding (or transformer) that supplies power to motors, relays, and other devices used for control purposes. *See:* **windings, high-voltage and low-voltage.** 203

winding, dc (thyristor converter). The winding of a thyristor converter transformer that is conductively connected to the thyristor converter circuit elements and that conducts the direct current of the thyristor converter. *See:* **secondary winding.** 121

winding-drum machine (elevators). A geared-drive machine in which the hoisting ropes are fastened to and wind on a drum. *See:* **driving machine (elevators).** 328

winding factor (rotating machinery). The product of the distribution factor and the pitch factor. *See:* **rotor (rotating machinery); stator.** 63

winding hottest spot temperature (power and distribution transformer). The highest temperature inside the transformer winding. It is greater than the measured average temperature (using the resistance change method) of the coil conductors. 53

winding inductance. *See:* **air core inductance.**

winding loss (electronics power transformer). The power losses of all windings involved, expressed in watts, in an inductor or transformer with the values measured at or corrected to the rated load current, frequency, and waveshape and stabilized at the maximum ambient temperature. *Syn:* **copper loss.** 95

winding overhang (rotating machinery). That portion of a winding extending beyond the ends of the core. 63

winding pitch. *See:* **coil pitch.**

winding, primary. *See:* **primary winding.**

winding, secondary. *See:* **secondary winding.**

winding shield (rotating machinery). A shield secured to the frame to protect the windings but not to support the bearing. 63

windings, high-voltage and low-voltage. The terms high-voltage and low-voltage are used to distinguish the winding having the greater from that having the lesser voltage rating. 53

winding, stabilizing. *See:* **stabilizing winding.**

winding, tertiary. *See:* **tertiary winding.**

winding voltage rating. The voltage for which the winding is designed. *See:* **duty.** 301

window (1) (counter tube) (radiation-counter tubes). That portion of the wall that is made thin enough for radiation of low penetrating power to enter. *See:* **anticoincidence (radiation counters).** 190 **(2)(of a semiconductor radiation detector)(charged-particle detectors).** *See:* **dead layer thickness.** 119

window amplifier. *See:* **biased amplifier.**

window annunciator (alarm monitoring and reporting systems for fossil-fueled power generating stations). A visual signal device consisting of a number of back-lighted windows, each one indicating a condition that exists or has existed in a monitored circuit, and being identified accordingly. 501

window-type current transformer (power and distribution transformer) (instrument transformer). One that has a secondary winding insulated from and permanently assembled on the core, but has no primary winding as an integral part of the structure. Complete insulation is provided for a primary winding in the window through which one turn of the line conductor can be passed to provide the primary winding. 53, 394

window, waveguide (waveguide components). A gas- or liquid-tight barrier or cover designed to be essentially transparent to the transmission of electromagnetic waves. 166

windshield wiper for aircraft. A motor-driven device for removing rain, sleet, or snow from a section of an aircraft windshield, window, navigation dome, or turret. 328

wind speed (navigation aid terms). The rate of motion of air. 526

windup. Lost motion in a mechanical system that is proportional to the force or torque applied. 207

wind velocity (navigation aid terms). The speed and direction of wind. 526

wing clearance lights (illuminating engineering). A pair of aircraft lights provided at the wing tips to indicate the extent of the wing span when the navigation lights are located an appreciable distance inboard of the wing tips. 167

wink-start pulsing (telephone switching systems). A method of pulsing control and trunk integrity check wherein the sender delays the sending of the address pulses until it receives a momentary off-hook signal from the far end.   55

wiper (brush). That portion of the moving member of a selector or other similar device, that makes contact with the terminals of a bank.   328

wiper relay. *See:* **relay wiper.**

wiping gland (wiping sleeve). A projecting sleeve on a junction box, pothead, or other piece of apparatus serving to make a connection to the lead sheath of a cable by means of a plumber's wiped joint. *See:* **tower; transformer removable cable-terminating box.**
                                                                    64

wire. A slender rod or filament of drawn metal. *Note:* The definition restricts the term to what would be ordinarily understood by the term solid wire. In the definition, the word slender is used in the sense that the length is great in comparison with the diameter. If a wire is covered with insulation, it is properly called an insulated wire: while primarily the term wire refers to the metal, nevertheless when the context shows that the wire is insulated, the term wire will be understood to include the insulation. *See:* **car wiring apparatus; conductors.**   1

wire antenna. An antenna composed of one or more conductors each of which is long compared to the transverse dimensions and with transverse dimensions of each conductor so small compared to a wavelength that for the purpose of computation the current can be assumed to flow entirely longitudinally and to have negligible circumferential variation.   111

wire-band serving (power distribution, underground cables). A short closed helical serving of wire applied tightly over the armor of wire-armored cables spaced at regular intervals, such as on vertical riser cables, to bind the wire armor tightly over the core to prevent slippage.   57

wire broadcasting. The distribution of programs over wire circuits to a large number of receivers, using either voice frequencies or modulated carrier frequencies.   328

wire center (telephone loop performance). A central point from which loop feeder networks extend in a tree-like manner into the serving areas associated with the center. One or more end offices may be located at a wire center.   473

wired logic (telephone switching systems). A fixed pattern of interconnections among a group of devices to perform predetermined functions in response to input signals.   55

wired program (telephone switching systems). A program embodied in a pattern of fixed physical interconnections among a group of devices.   55

wired program control (telephone switching systems). A system control using wired logic.   55

wire gages (NESC). Throughout these rules the American Wire Gage (AWG), formerly known as Brown & Sharpe (B&S), is the standard gage for copper, aluminum and other conductors, excepting only steel conductors for which the Steel Wire Gage (Stl WG) is used. *Note:* The Birmingham Wire Gage is obsolete.   494

wire-grid lens antenna. A lens antenna constructed of wire grids, in which the effective index of refraction (and thus the path delay) is locally controlled by the dimensions and the spacings of the wire grid. *See:* **Luneberg lens; geodesic lens.**   111

wire holder (insulator). An insulator of generally cylindrical or pear shape, having a hole for securing the conductor and a screw or bolt for mounting. *See:* **insulator.**   261

wire insulation (rotating machinery). The insulation that is applied to a wire before it is made into a coil or inserted in a machine. *See:* **rotor (rotating machinery); stator.**   63

wireless connection diagram (industrial control). The general physical arrangement of devices in a control equipment and connections between these devices, terminals, and terminal boards for outgoing connections to external apparatus. Connections are shown in tabular form and not by lines. An elementary (or schematic) diagram may be included in the connection diagram.   210, 206

wire, overhead ground. *See:* **overhead ground wire.**

wire-pilot protection. Pilot protection in which an auxiliary metallic circuit is used for the communicating means between relays at the circuit terminals.
                                                                103,127

wire rope splice (conductor stringing equipment). The point at which two wire ropes are joined together. The various methods of joining (splicing) wire ropes together include hand tucked woven splices, compression splices which utilize compression fittings but do not incorporate loops (eyes) in the ends of the ropes, and mechanical splices which are made through the use of loops (eyes) in the ends of the ropes held in place by either compression fittings or wire rope clips. The latter are joined together with connector links or steel bobs and, in some places, rigged eye to eye. Woven splices are often classified as short or long. A short splice varies in length from 7 to 17 ft (2 to 5m) for 1/4 to 1 1/2 in diameter ropes, respectively, while a long splice varies from 15 to 45 ft (4 to 14m) for the same size ropes.   431

wire spring relay. A relay design in which the contacts are attached to round wire springs instead of the conventional flat or leaf spring.   341

wireway (1)(packaging machinery). A rigid rectangular raceway provided with a cover.   429

(2)(raceway systems for Class 1E circuits for nuclear power generating stations). Sheet-metal troughs with hinged or removable covers to house or protect wires and cables external to panelboards and cabinets.
                                                                513

wireways (National Electrical Code). Sheet-metal troughs with hinged or removable covers for housing and protecting electric wires and cable and in which conductors are laid in place after the wireway has been installed as a complete system.   256

wiring or busing terminal, screw and/or lead. That terminal, screw or lead to which a power supply will be connected in the field.   53

withstand current (surge). The crest value attained by

a surge of a given wave shape and polarity that does not cause disruptive discharge on the test specimen.
64

withstand probability (high voltage testing). The withstand probability is the probability that the test object will withstand one application of a certain prospective voltage value and shape. If the disruptive discharge probability is p, the withstand probability is (1-p).
150

withstand test voltage (cable termination). The voltage that the device must withstand without flashover, disruptive discharge, puncture, or other electric failure when voltage is applied under specified conditions. *Note:* For power frequency voltages, the values specified are rms values and for a specified time. For lightning or switching impulse voltages, the values specified are crest values of a specified wave. For direct voltages the values specified are average values and for a specified time.
4

withstand voltage (1)(power switchgear)(separable insulated connectors). The specified voltage that, under specified conditions, can be applied to insulation without causing flashover or puncture.
103,454

(2) (power and distribution transformer)(high voltage testing). The voltage that electrical equipment is capable of withstanding without failure or disruptive discharge when tested under specified conditions.
53,150

(3) (impulse) (electric power). The crest value attained by an impulse of any given wave shape, polarity, and amplitude, that does not cause disruptive discharge on the test specimen.
91

(4) (surge arresters). A specified voltage that is to be applied to a test object in a withstand test under specified conditions. During the test, in general no disruptive discharge should occur. *See:* **basic impulse insulation level (insulation strength); test voltage and current.**
308,62

withstand voltage test. A high-voltage test that the armature winding must withstand without flashover or other electric failure at a specified voltage for a specified time and under specified conditions.
135

word (1)(mathematics of computing). A sequence of bits or characters that is stored, addressed, transmitted, and operated on as a unit within a given computer.
564

(2)(microprocessor operating systems). An ordered set of bytes or bits that is the normal unit in which information may be stored, transmitted, or operated on within a given computer.
478

(3)(signals and paths)(microcomputer system bus). Two bytes or sixteen bits operated on as a unit.
542

(4)(696 interface devices). A set of bit-parallel signals corresponding to binary digits and operated on as a unit. For ANSI/IEEE Std 696-1983 word connotes a group of 16 bits where the most significant bit carries the subscript 15 and the least significant bit carries the subscript 0.
538

(5)(electronic digital computers). An ordered set of characters that is the normal unit in which information may be stored, transmitted, or operated upon within a given computer. *Example:* The set of characters 10692 is a word that may give a command for a machine element to move to a point 10.692 inches from a specified zero. *See:* **computer word; machine word.**
235

(6) (software). (A) An ordered set of bits or characters that is the normal unit in which information may be stored, transmitted or operated upon within a given computer. (B) A character string or bit string considered as an entity. *See:* **bit; computer; string.**
434

word address format. Addressing each word of a block by one or more characters that identify the meaning of the word.
207

word length (1) (hybrid computer linkage components)(analog-to-digital converter). The number of data bits, including sign, which form the digital representation of the analog input in a prescribed voltage range.
10

(2) (digital-to-analog converter). The number of data bits, including sign, in the digital register of a digital-to-analog converter, or a digital-to-analog multiplier.
10

word processing (WP)(computer applications). The use of computers to enter, view, edit, store, retrieve, manipulate, organize, transmit, and print textual material. A word processor system typically includes text editing and text formatting. *See:* **cluster word processing; dedicated word processing; office automation; shared-logic word processing; shared-resource word processing; stand-alone word processing; word processor.**
571

word processor (WP)(computer applications). (1) A computer capable of performing word processing functions. (2) A computer program capable of performing word processing functions.
571

word time (electronic computation). In a storage device that provides serial access to storage locations, the time interval between the appearance of corresponding parts of successive words. *See:* **minor cycle.**
255, 77

work. The work done by a force is the dot-product line integral of the force. *See:* **line integral.**
210

work coil. *See:* **load, work or heater coil (induction heating usage).**

work function. The minimum energy required to remove an electron from the Fermi level of a material into field-free space. *Note:* Work function is commonly expressed in electron volts.
125

working (electrolysis). The process of stirring additional solid electrolyte or constituents of the electrolyte into the fused electrolyte in order to produce a uniform solution thereof. *See:* **fused electrolyte.**
328

working distance (X-ray energy spectrometers). The distance, measured along the working axis, between the source of X-rays and the outermost window on the detector.
471

working optical aperture (acousto-optic device). That aperture which is equal to the size of the acoustic column which the light will encounter.
82

working point. *See:* **operating point.**

working pressure. The pressure, measured at the cylinder of a hydraulic elevator, when lifting the car and its rated load at rated speed. *See:* **elevators.**   328

working reference system. A secondary reference telephone system consisting of a specified combination of telephone sets, subscriber lines, and battery supply circuits connected through a variable distortionless trunk and used under specified conditions for determining, by comparison, the transmission performance of other telephone systems and components.   328

working standard (luminous standards) (illuminating engineering). A standardized light source for regular use in photometry.   167

working storage (computing machines). *See:* **temporary storage.**

working value. The electrical value that when applied to an electromagnetic instrument causes the movable member to move to its fully energized position. This value is frequently greater than pick-up. *See:* **pick-up.**
                                                                                          328

working voltage to ground (electric instrument). The highest voltage, in terms of maximum peak value, that should exist between any terminal of the instrument proper on the panel, or other mounting surface, and ground. *See:* **instrument.**   280

work permit (power line maintenance). The authorization to perform work on a circuit. *Syn:* **clearance; guarantee.**   458

work plane (illuminating engineering). The plane on which work is usually done, and on which the illuminance is specified and measured. Unless otherwise indicated, this is assumed to be a horizontal plane 0.76 m (30 inches) above the floor.   167

world-numbering plan (telephone switching systems). The arrangement whereby, for the purpose of international distance dialing, every telephone main station in the world is identified by a unique number having a maximum of twelve digits representing a country code plus a national number.   55

world-zone number (telephone switching systems). The first digit of a country code. In the world-numbering plan, this number identifies one of the larger geographical areas into which the world is arranged, namely:

Zone 1–North America (includes areas operating with unified regional numbering).

Zone 2–Africa

Zone 3 & 4–Europe

Zone 5–South America, Cuba, Central America including part of Mexico

Zone 6–South Pacific (Austral-Asia)

Zone 7–Union of Soviet Socialist Republics

Zone 8–North Pacific (Eastern Asia)

Zone 9–Far East and Middle East

Zone 0–Spare   55

worm-geared machine (elevators). A direct-drive machine in which the energy from the motor is transmitted to the driving sheave or drum through worm gearing. *See:* **driving machine (elevators).**   328

wound rotor (rotating machinery). A rotor core assembly having a winding made up of individually insulated wires. *See:* **asynchronous machine.**   63

wound-rotor induction motor (1) (electric installations on shipboard). An induction motor in which the secondary circuit consists of polyphase winding or coils whose terminals are either short-ciucited or closed through suitable circuits. (When provided with collector or slip rings, it is also known as a slip-ring induction motor.)   3

(2) (rotating machinery). An induction motor in which a primary winding on one member (usually the stator) is connected to the alternating-current power source and a secondary polyphase coil winding on the other member (usually the rotor) carries alternating current produced by electromagnetic induction. *Note:* The terminations of the rotor winding are usually connected to collector rings. The brush terminals may be either short-circuited or closed through suitable adjustable circuits. *See:* **asynchronous machine.**   63

wound stator core (rotating machinery). A stator core into which the stator winding, with all insulating elements and lacing has been placed, including any components imbedded in or attached to the winding, and including the lead cable when this is used. *See:* **stator.**
                                                                                          63

wound-type current transformer (1) (instrument transformer)(power and distribution transformer). One that has a primary winding consisting of one or more turns mechanically encircling the core or cores. The primary and secondary windings are insulated from each other and from the core(s) and are assembled as an integral structure.   394,53

woven wire grip (conductor stringing equipment). A device designed to permit the temporary joining or pulling of conductors without the need of special eyes, links or grips. *Syn:* **basket; chinese finger; Kellem; sock; wire mesh grip.**   431

wow (sound recording and reproducing equipment). Frequency modulation of the signal in the range of approximately 0.5 Hz to 6 Hz resulting in distortion which may be perceived as a fluctuation of pitch of a tone or program. *Note:* Measurement of unweighted wow only is not covered by this standard.   145

wrap. One convolution of a length of ferromagnetic tape about the axis. *See:* **tape-wound core.**   331

wrapper (rotating machinery). (1) A relatively thin flexible sheet material capable of being formed around the slot section of a coil to provide complete enclosure. (2) The outer cylindrical frame component used to contain the ventilating gas. *See:* **rotor (rotating machinery); stator.**   63

wrap thickness. The distance between corresponding points on two consecutive wraps, measured parallel to the ferromagnetic tape thickness. *See:* **tape-wound core.**   331

wrap width (tape width). *See:* **tape-wound core.**

write (1) (general). To introduce data, usually into some form of storage. *See:* **read.**

                                                           174, 235, 255, 54

(2) (charge-storage tubes). To establish a charge pattern corresponding to the input.   125

write cycle (write)(FASTBUS acquisition and control). A cycle in which the direction of data flow is from a master to slave(s).   480

write enable (W) (semiconductor memory). The inputs that when true enable writing data into the memory. The data sheet must define the effect of both states of this input on the reading of data and the condition of the output.   441

write high (metal-nitride-oxide field-effect transistor). Process of generating a threshold voltage condition which increases source to drain current (high-conductance state) for a given gate to source voltage.   386

write low (metal-nitride-oxide field-effect transistor). Process of generating a threshold voltage condition which decreases source to drain current (low-conductance state) for a given gate to source voltage.   386

write pulse. *See:* ONE state.

writing characteristic (metal-nitride-oxide field-effect transistor). The collection of high-conduction and low-conduction threshold voltage data as a function of the writing pulse width of both writing voltage polarities.   386

writing rate (1) (storage tubes). The time rate of writing on a storage element, line, or area to change it from one specified level to another. Note the distinction between this and **writing speed.** *See:* storage tube.   174

(2) (oscilloscopes). *See:* **writing time division.**

writing speed (storage tubes). Lineal scanning rate of the beam across the storage surface in writing. Note

the distinction between this and **writing rate.** *See:* **information writing speed; storage tube.**   174,125

writing speed, maximum usable (storage tubes). The maximum speed at which information can be written under stated conditions of operation. Note the qualifying adjectives **maximum usable** are frequently omitted in general usage when it is clear that the maximum usable writing speed is implied. *See:* **storage tube.**   174

writing time/division (oscilloscopes). The minimum time per unit distance required to record a trace. The method of recording must be specified.   106

writing time, minimum usable (storage tubes). The time required to write stored information from one specified level to another under stated conditions of operation. *Note:* The qualifying adjectives **minimum usable** are frequently omitted in general usage when it is clear that the minimum usable writing time is implied. *See:* **storage tube.**   174, 190

Wullenweber antenna. An antenna consisting of a circular array of radiating elements, each having its maximum directive gain along the outward radial, and a feed system that provides a steerable beam that is narrow in the azimuth plane.   111

wye. *See:* **Y.**

wye junction (waveguide components). A junction of waveguides or transmission lines in which the longitudinal guide axes form a Y.   166

wye rectifier circuit. A circuit that employs three or more rectifying elements with a conducting period of 120 electrical degrees plus the commutating angle. *See:* **rectification.**   328

X

X-address (test pattern language). The coordinates by which a row of a memory is specified.   463

X-axis amplifier. *See:* **horizontal amplifier.**

X-band. A radar frequency band between 8 gigahertz (GHz) and 12 GHz, usually in the International Telecommunications Union (ITU) assigned band 8.5 GHz to 10.68 GHz.   13

X-band radar (radar). A radar operating at frequencies between 8 and 12 GHz, usually in the International Telecommunications Union (ITU) assigned band 8.5 to 10.68 GHz.   13

xerography. The branch of electrostatic electrophotography that employs a photoconductive insulating medium to form, with the aid of infrared, visible, or ultraviolet radiation, latent electrostatic-charge patterns for producing a viewable record. *See:* **electrostatography.**   191

xeroprinting. The branch of electrostatic electrography that employs a pattern of insulating material on a conductive medium to form electrostatic-charge patterns for duplicating purposes. *See:* **electrostatography.**   191

xeroradiography. The branch of electrostatic electrophotography that employs a photoconductive insulating medium to form, with the aid of X rays or gamma rays, latent electrostatic-charge patterns for producing a viewable record. *See:* **electrostatography.**   191

x percent disruptive discharge voltage (high voltage testing) The x percent disruptive discharge voltage is the prospective voltage value which has x percent probability of producing a disruptive discharge.   150

X-ray tube. A vacuum tube designed for producing X-rays by accelerating electrons to a high velocity by means of an electrostatic field and then suddenly stopping them by collision with a target.   190

X/R ratio (power fault effects). The ratio of system inductive reactance to resistance. It is proportional to the L/R ratio of time constant, and is, therefore, indicative of the rate of decay of any dc offset. A large X/R ratio corresponds to a large time constant and a slow rate of decay.   404

X wave (radio wave propagation). *See:* **extraordinary-wave component.**   146

X-Y display. A rectilinear coordinate plot of two variables. *See:* oscillograph. 185
x-y recorder (plotting board) (analog computers). A
recorder that makes a record of any one voltage with
respect to another. 9

XY switch. A remotely controlled bank-and-wiper
switch arranged in a flat manner, in which the wipers
are moved in a horizontal plane, first in one direction
and then in another. 328

Y

**Y (or wye) connection (power and distribution trans-
former).** So connected that one end of each of the
windings of a polyphase transformer (or of each of the
windings for the same rated voltage of single-phase
transformers associated in a polyphase bank) is connected to a common point (the neutral point and the
other end to its appropriate line terminal. 53
Y-address (test pattern language). The coordinates by
which a column of a memory is specified. 463
Yagi antenna. A linear end-fire array consisting of a
driven element, one or more reflector elements, and
one or more director elements. 111
Yagi-Uda antenna. A linear end-fire array consisting of
a driven element, a reflector element, and one or more
director elements. 111
Y amplifier. *See:* vertical amplifier.
yaw angle (navigation aid terms). (1) The horizontal
angular displacement of the longitudinal axis of a
vehicle from its neutral position. (2) The angle between a line in the direction of the relative wind and
a plane through the longitudinal and vertical axes of
the vehicle. 526
Y-axis amplifier. *See:* vertical amplifier.

Y-connected circuit. A three-phase circuit that is star
connected. *See:* network analysis. 210
Y connection. So connected that one end of each of the
windings of a polyphase transformer (or of each of the
windings for the same rated voltage of single phase
transformers associated in a polyphase bank) is connected to a common point (the neutral point) and the
other end to its appropriate line terminal. *Syn:* wye
connection. 53
Y junction. *See:* wye junction.
Y network. A star network of three branches. *See:*
network analysis. 210
yoke (magnetic) (rotating machinery). The element of
ferromagnetic material, not surrounded by windings,
used to connect the cores of an electromagnet, or of
a transformer, or the poles of a machine, or used to
support the teeth of stator or rotor. *Note:* A yoke may
be of solid material or it may be an assembly of laminations. *See:* rotor (rotating machinery); stator.
63
Y-T display. An oscilloscope display in which a time-
dependent variable is displayed against time. *See:* oscillograph. 185

Z

Z(s). *See:* operational impedance.
Z (zone) marker (navigation aid terms). A marker
used to define a position above a radio range station.
526
Z-address (test pattern language). The coordinates by
which a matrix in a memory is specified. 463
Z-axis amplifier (oscilloscopes). An amplifier for signals controlling a display perpendicular to the *X-Y*
plane (commonly intensity of the spot). *See:* intensity
amplifier; oscillograph. 106
Zeeman effect. If an electric discharge tube, or other
light source emitting a bright-line spectrum, is placed
between the poles of a magnet, each spectrum line is
split by the action of the magnetic field into three or
more close-spaced but separate lines. The amount of
splitting or the separation of the lines, is directly proportional to the strength of the magnetic field.
210
Zener breakdown (semiconductor device). A break-

down that is caused by the field emission of charge
carriers in the depletion layer. *See:* semiconductor;
semiconductor device. 245
Zener diode (semiconductor). A class of silicon diodes
that exhibit in the avalanche-breakdown region a large
change in reverse current over a very narrow range of
reverse voltage. *Note:* This characteristic permits a
highly stable reference voltage to be maintained across
the diode despite a relatively wide range of current
through the diode. *See:* Zener breakdown; avalanche
breakdown. 186
Zener diode regulator. A voltage regulator that makes
use of the constant-voltage characteristic of the Zener
diode to produce a reference voltage that is compared
with the voltage to be regulated to initiate correction
when the voltage to be regulated varies through
changes in either load or input voltage. See the accompanying figure. *See:* Zener diode; electrical conversion.

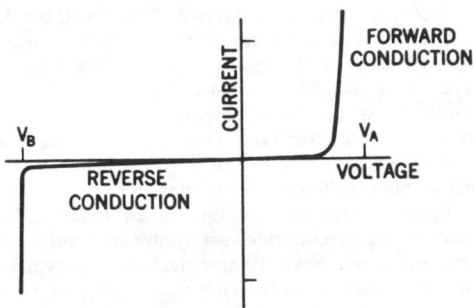

Current and voltage characteristics for a typical Zener diode regulator $|V_A| \gg |V_B|$.

186

Zener impedance (semiconductor diode). *See:* **breakdown impedance; semiconductor.**

Zener voltage. *See:* **breakdown voltage; semiconductor.**

zero (1) (function) (root of an equation). A zero of a function $f(x)$ is any value of the argument X for which $f(x) = 0$. *Note:* Thus the zeros of $\sin x$ are $x_1 = 0$, $x_2 = \pi, x_3 = 2\pi, x_4 = 3\pi, ..., x_n = (n - 1)\pi, ...$ The roots of the equation $f(x) = 0$ are the zeros of $f(x)$.
210

(2) (transfer function in the complex variable s). (A) A value of s that makes the function zero. (B) The corresponding point in the s plane. *See:* **control system, feedback; pole (network function).** 329

(3) (network function). Any value of p, real or complex, for which the network function is zero. *See:* **network analysis.** 210

zero adjuster. A device for bringing the indicator of an electric instrument to a zero or fiducial mark when the electrical quantity is zero. *See:* **moving element (instrument).** 328

zero-based linearity. *See:* **linearity.**

zero-beat reception. *See:* **homodyne reception.**

zero bias retention t_{RO} (metal-nitride-oxide field-effect transistor). This is the retention inherent in the metal-nitride-oxide semiconductor (MNOS) transistor when all terminals are grounded during information storage. The time period is defined by an (extrapolated) zero window between the two high-conduction (HC) and low-conduction (LC) threshold voltage curves plotted versus the logarithm of t_{rd}, the time elapsed between writing and threshold voltage measurement. *Syn:* **relaxation time.** 386

zero carryover (1) (bolometric power meters). A characteristic of multirange direct reading bolometer bridges that is a measure of the ability of the meter to maintain a zero setting from range to range without readjustment after initially being set to zero on the most sensitive range. 47

(2) (electrothermic power meter). A characteristic of multirange direct reading electrothermic power indicators which is a measure of the ability of the meter to maintain a zero setting from range to range without readjustment after initially being set to zero on the most sensitive range. Expressed in terms of percentage of full scale. 47

zero control current residual voltage (Hall-effect devices). The voltage across the Hall terminals that is caused by a time-varying magnetic field when there is no control current. 107

zero drift (analog computers). Drift with zero input.
9

zero-error (device operating under the specified conditions of use). The indicated output when the value of the input presented to it is zero. *See:* **control system, feedback.** 329

zero-error reference. *See:* **linearity.**

zero field residual voltage (Hall-effect devices). The voltage across the Hall terminals that exists when control current flows but there is zero applied magnetic field. 107

zero field residual voltage temperature drift (Hall generator). The maximum change in output voltage per degree Celsius over a given temperature range when operated with zero external field and a given magnitude of control current. 107

zero field resistive residual voltage (Hall-effect devices). That component of the zero field residual voltage which remains proportional to the voltage across the control current terminals of the Hall generator for a specified temperature. 107

zero guy. A line guy installed in a horizontal position between poles to provide clearance and transfer strain to an adjacent pole. *See:* **tower.** 64

zero lead (medical electronics). *See:* **biolectric null.**

zero-level address (computing machines). *See:* **immediate address.**

zero-minus (0 −) call (telephone switching systems). A call for which the digit zero is dialed alone to indicate that operator assistance is desired. 55

zero-modulation medium noise (sound recording and reproducing system). The noise that is developed in the scanning or reproducing device during the reproducing process when a medium is scanned in the zero-modulation state. *Note:* For example, zero-modulation medium noise is produced in magnetic recording by undesired variations of the magnetomotive force in the medium, that are applied across the scanning gap of a demagnetized head, when the medium moves with the desired motion relative to the scanning device. Medium noise can be ascribed to nonuniformities of the magnetic properties and to other physical and dimensional properties of the medium. *See:* **noise (sound recording and reproducing system).** 350

zero-modulation state (sound recording medium). The state of complete preparation for playback in a particular system except for omission of the recording signal. *Notes:* (1) Magnetic recording media are considered to be in the zero-modulation state when they have been subjected to the normal erase, bias, and duplication printing fields characteristic of the particular system with no recording signal applied. (2) Mechanical recording media are considered to be in the zero-modulation state when they have been recorded upon and processed in the customary specified manner to form the groove with no recording signal applied. (3) Optical recording media are considered to be in the zero-modulation state when all normal processes of recording and processing, including duplication, have been performed in the customary specified manner, but with no modulation input to the light modulator. *See:* noise (sound recording and reproducing system). 350

zero offset (1) (industrial control). A control function for shifting the reference point in a control system. *See:* control system, feedback. 206

(2) (numerically controlled machines). A characteristic of a numerical machine control permitting the zero point on an axis to be shifted readily over a specified range. The control retains information on the location of the permanent zero. *See:* floating zero. 207

(3) (rate gyros). The gyro output when the input rate is zero, generally expressed as an equivalent input rate. It excludes outputs due to hysteresis and acceleration. *See:* input-output characteristics. 46

zero period acceleration (1)(seismic design of substations). The peak time history acceleration which can be determined from response spectra by the merging of response spectra, for all damping values, in the high-frequency range (usually above 30 hertz) in which no change of acceleration occurs with frequency. 465

(2)(ZPA)(seismic qualification of Class 1E equipment for nuclear power generating stations). The acceleration level of the high frequency, nonamplified portion of the response spectrum. This acceleration corresponds to the maximum peak acceleration of the time history used to derive the spectrum. 581

(3)(ZPA)(valve actuators). The acceleration which appears as a constant portion of a response spectrum in the highest frequency range. It is the maximum acceleration in the time history from which that response spectrum was developed. 492

(4) (seismic testing of relays). The peak acceleration of the motion time-history which corresponds to the high-frequency asymptote on the response spectrum. 392

zero-phase-sequence relay (power switchgear). A relay that responds to the zero-phase-sequence component of a polyphase input quantity. 103

zero-phase-sequence symmetrical components (unsymmetrical set of polyphase voltages or currents of *m* phases). That set of symmetrical components that have zero phase sequence. That is, the angular phase lag from each member to every other member is 0 radians. The members of this set will all reach their positive maxima simultaneously. The zero-phase-sequence symmetrical components for a three-phase set of unbalanced sinusoidal voltages ($m = 3$) having the primitive period are represented by the equations

$$e_{a0} = e_{b0} = e_{c0} = (2)^{1/2} E_{a0} \cos(\omega t + \alpha_{a0})$$

derived from the equation of **symmetrical components (set of polyphase alternating voltages).** Since in this case $r = 1$ for every component (of first harmon-

ic), the third subscript is omitted. Then k is 0 for the zero sequence, and s takes on the values 1, 2, and 3 corresponding to phases *a, b,* and *c.* These voltages have no phase sequence since they all reach their positive maxima simultaneously. 210

zero-phase symmetrical set (1) (polyphase voltage). A symmetrical set of polyphase voltages in which the angular phase difference between successive members of the set is zero or a multiple of 2π radians. The equations of **symmetrical set (polyphase voltages)** represent a zero-phase symmetrical set of polyphase voltages if k/m is zero or an integer. (The symmetrical set of voltages represented by the equations of **symmetrical set of polyphase voltages** may be said to have zero-phase symmetry if k/m is zero or an integer (positive or negative).) *Note:* This definition may be applied to a two-phase four-wire or five-wire system if *m* is considered to be 4 instead of 2. 210

(2) (polyphase currents). This definition is obtained from the corresponding definitions for voltage by substituting the word current for voltage, and the symbol I for E and β for α wherever they appear. The subscripts are unaltered. 210

zero pip (spectrum analyzer). An output indication which corresponds to zero input frequency. 390

zero-plus (0+) (telephone switching systems). A call in which the digit zero is dialed as a prefix where operator intervention is necessary. 55

zero-power-factor saturation curve (zero-power-factor characteristic) (synchronous machine). The saturation curve of a machine supplying constant current with a power-factor of approximately zero, overexcited. 63

zero-power-factor test (synchronous machine). A no-load test in which the machine is overexcited and operates at a power-factor very close to zero. 63

zero-sequence impedance (1) (power and distribution transformer). An impedance voltage measured between a set of primary terminals and one or more sets of secondary terminals when a single-phase voltage source is applied between the three primary terminals connected together and the primary neutral, with the secondary line terminals shorted together and connected to their neutral (if one exists). *Notes:* (A) For two-winding transformers, the other winding is short-circuited. For multiwinding transformers, several tests are required, and the zero-sequence impedance characteristics are represented by an impedance network. (B) In some transformers, the test must be made at a voltage lower than that required to circulate rated current in order to avoid magnetic core saturation or to avoid excessive current in other windings. (C) Zero-sequence impedances are usually expressed in per unit or percent on a suitable voltage and kVA base. 53

(2) (rotating machinery). The quotient of the zero-sequence component of the voltage, assumed to be sinusoidal, supplied to a synchronous machine, and the zero-sequence component of the current at the same frequency. *See:* direct-axis synchronous reactance. 63

zero-sequence reactance (rotating machinery). The ratio of the fundamental component of reactive armature voltage, due to the fundamental zero-sequence component of armature current, to this component at rated frequency, the machine running at rated speed. *Note:* Unless otherwise specified, the value of zero-sequence reactance will be that corresponding to a zero-sequence current equal to rated armature current. *See:* **direct-axis synchronous reactance.**   63

zero-sequence resistance. The ratio of the fundamental in-phase component of armature voltage, resulting from fundamental zero-sequence current, to this component of current at rated frequency.   328

ZERO shift error. Error measured by the difference in deflection as between an initial position of the pointer, such as at zero, and the deflection after the instrument has remained deflected upscale for an extended length of time, expressed as a percentage of the end-scale deflection. *See:* **moving element (instrument).**
   280

zero span (spectrum analyzer). A mode of operation in which the frequency span is reduced to zero.
   390

0-state (graphic symbols for logic functions). The logic state represented by the binary number 0 and usually standing for an inactive or false logic condition.
   451

zero-subcarrier chromaticity (color television). The chromaticity that is intended to be displayed when the subcarrier amplitude is zero. *Note:* This chromaticity is also known as reference white for the display.
   18

zero suppression. The elimination of nonsignificant zeros in a numeral.   255, 77

zero synchronization (numerically controlled machines). A technique that permits automatic recovery of a precise position after the machine axis has been approximately positioned by manual control.
   207

zero vector. A vector whose magnitude is zero.
   210

zero voltage fired (electrical heating applications to melting furnaces and forehearths in the glass industry). A circuit in which antiparallel connected thyristors are fired at points of voltage zero in the alternating current voltage wave.   520

zeta potential. *See:* **electrokinetic potential.**

zigzag connection (power and distribution transformer). A polyphase transformer with Y-connected windings, each one of which is made up of parts in which phase-displaced voltages are induced.   53

zig-zag connection of polyphase circuits (zig-zag or interconnected star). The connection in star of polyphase windings, each branch of which is made up of windings that generate phase-displaced voltage. *See:* **connections of polyphase circuits; polyphase circuits; polyphase systems.**   244, 63

zig-zag leakage flux. The high-order harmonic air-gap flux attributable to the location of the coil sides in discrete slots. *See:* **rotor (rotating machinery); stator.**
   63

Z marker (zone marker) (electronic navigation). A marker used to define a position above a radio range station.   13

z-marker beacon (navigation aid terms). A vertical beam--horizontal cross section in the shape of a circle.
   526

Zobel filters (circuits and systems). A filter designed according to image parameter techniques.   67

zonal-cavity interreflectance method (illuminating engineering). A procedure for calculating coefficients of utilization, wall luminance coefficients, and ceiling cavity luminance coefficients taking into consideration the luminaire intensity distribution, room size and shape (cavity ratio concepts), and room reflectances. It is based on flux transfer theory.   167

zonal constant (illuminating engineering). A factor by which the mean intensity emitted by a source of light in a given angular zone is multiplied to obtain the lumens in the zone.   167

zonal factor interflection method* (illuminating engineering). A procedure for calculating coefficients of utilization based on integral equations which takes into consideration the ultimate disposition of luminous flux from every 10 degree zone from luminaires. (This term is retained for reference and literature searches).   167

*Obsolete

zonal factor method (illuminating engineering). A procedure for predetermining, from typical luminaire photometric data in discrete angular zones, the proportion of luminaire output which would be incident initially (without interreflections) on the work-plane, ceiling, walls, and floor of a room.   167

zone (of a relay). *See:* **reach (of a relay).**

zone comparison protection. A form of pilot protection in which the response of fault-detector relays, adjusted to have a zone of response commensurate with the protected line section, is compared at each line terminal to determine whether a fault exists within the protected line section.   103

zoned antenna. A lens or reflector antenna having various portions (called zones or steps) which form a discontinuous surface such that a desired phase distribution of the aperture illumination is achieved. *Syn:* **stepped antenna.**   111

zone leveling (semiconductor processing). The passage of one or more molten zones along a semiconductor body for the purpose of uniformly distributing impurities throughout the material. *See:* **semiconductor device.**   342

zone of protection (1)(generating station grounding). The zone of protection provided by a grounded air terminal, mast, or overhead ground wire is the adjacent space that is essentially immune to direct strokes of lightning.   569

(2) (for relays) (power switchgear). That segment of the power system in which the occurrence of assigned abnormal conditions should cause the protective relay system to operate.   103

zone of silence (navigation aid terms). A local region in which the signals of a given radio transmitter cannot be received satisfactorily.   526

zone-plate lens antenna. *See:* **Fresnel lens antenna.**
    111

zone punch. A punch in the 0, 11, or 12 row on a Hollerith punched card.   77

zone purification (semiconductor processing). The passage of one or more molten zones along a semiconductor for the purpose of reducing the impurity concentration of part of the ingot. *See:* **semiconductor device.**   342

zoning (1)(electrical heating systems). A division of circuits to minimize different conditions in any one circuit, so temperature at location of sensor is typical for complete circuit.   476

(stepping) (2)(lens or reflector). The displacement of various portions (called zones or steps) of the lens or surface of the reflector so that the resulting phase front in the near field remains unchanged. *See:* **antenna (aerial).**   179

z transform, advanced (data processing). The advanced z transform of $f(t)$ is the z transform of $f(t + \Delta T)$; that is,

$$\sum_{n=0}^{\infty} f(nT + \Delta T)z^{-n} \qquad 0 < \Delta < 1.$$

    198

z transform, delayed (data processing). The delayed z transform of $f(t)$, denoted $F(z, \Delta)$, is the z transform of $f(t - \Delta T)u(T - \Delta T)$, where $u(t)$ is the unit step function: that is,

$$F(z,\Delta) = \sum_{n=0}^{\infty} f(nT - \Delta T)u(nT - \Delta T)z^{-n} \quad 0 < \Delta < 1.$$

    198

z transform, modified (data processing). The modified z transform of $f(t)$, denoted $F(z,m)$, is the delayed z transform of $f(t)$ with the substitution $\Delta = 1 - m$; that is,

$$F(z,m) = \sum_{n=0}^{\infty} f[nT - (1 - m)T]u[nT - (1 - m)T]z^{-n} \qquad 0 < m < 1.$$

z transform, one-sided (data processing). Let T be a fixed positive number, and let $f(t)$ be defined for $t \geq 0$. The z transform of $f(t)$ is the function

$$[f(t)] = F(z) = \sum_{n=0}^{\infty} f(nT)z^{-n},$$

$$\text{for } |z| > R = 1/\rho$$

where ρ is the radius of convergence of the series and z is a complex variable. If $f(t)$ is discontinuous at some instant $t = kT$, k an integer, the value used for $f(kT)$ in the z transform is $f(kT^+)$. The z transform for the sequence $\{f_n\}$ is:

$$[\{f_n\}] = F(z) = \sum_{n=0}^{\infty} f_n z^{-n}.$$

    198

z transform, two-sided (data processing). The two-sided z transform of $f(t)$ is

$$F(z) = \sum_{n=\infty}^{-1} f(nT)z^{-n} + \sum_{n=0}^{\infty} f(nT)z^{-n}$$

where the first summation is for $f(t)$ over all negative time and the second summation is for $f(t)$ over all positive time.   198

SOURCES

Sources

1   Std 16-1955 (ANSI/IEEE) (withdrawn 1969) Electric Control Apparatus for Land Transportation Vehicles.

2   Std 28-1974 (ANSI/IEEE) Surge Arresters for AC Power Circuits, Standard for. See code 430.

3   Std 45-1983 (ANSI/IEEE) Electric Installations on Shipboard, Recommended Practice for.

4   Std 48-1975 (IEEE) High Voltage AC Cable Terminations

5   Std 86-1961. See Code 517 for 1987 revision.

6   Std 95-1977 (ANSI/IEEE)(reaffirmed 1982) Insulation Testing of Large AC Rotating Machinery with High Direct Voltage, Recommended Practice for

7   Std 119-1974 (IEEE) General Principles of Temperature Measurement as Applied to Electrical Apparatus, Recommended Practice for.

8   Std 125-1977 (ANSI/IEEE) Preparation of Equipment Specifications for Speed-Governing of Hydraulic Turbines Intended to Drive Electric Generators, Recommended Practice for

9   Std 165-1977 (ANSI/IEEE)(reaffirmed 1984) Analog Computers, Standard Definitions of Terms for

10   Std 166-1977 (ANSI/IEEE)(reaffirmed 1984) Hybrid Computer Linkage Components, Standard Definitions of Terms for

11   Std 167-1966 (ANSI/IEEE)(reaffirmed 1971) Facsimile, Test Procedure for

12   Std 168-1956 (ANSI/IEEE)(reaffirmed 1971) Facsimile, Definitions of Terms for.

13   Std 686-1982 (ANSI/IEEE) Standard Radar Definitions. See code 526.

14   Std 54-1955 (IEEE)(withdrawn) Induction and Dielectric Heating Equipment, Test Code and Recommended Practice for.

15   Std 181-1977 (ANSI/IEEE) Pulse Measurement and Analysis by Objective Techniques, Standard on

16   Std 185-1975 (ANSI/IEEE) Frequency Modulation Broadcast Receivers, Standard Methods of Testing

17   Std 200-1975 (ANSI/IEEE) Reference Designations for Electrical and Electronics Parts and Equipment

18   Std 201-1979 (ANSI/IEEE) Terms Relating to Television, Standard Definitions of. See code 163

19   Std 219-1975 (IEEE)(withdrawn) Loudspeaker Measurements, Recommended Practice for

20   Std 279-1971 (ANSI/IEEE)(withdrawn 1984) Protection Systems for Nuclear Power Generating Stations, Criteria for. See codes 109, 428, and 471.

21   Std 268-1979 (ANSI/IEEE)(revised in 1982) Metric Practice, Standard for. The definitions are essentially the same.

22   Std 286-1975 (ANSI/IEEE)(reaffirmed 1981) Measurement of Power-Factor Tip-Up of Rotating Machinery Stator Coil Insulation, Recommended Practice for

23   Std 301-1976 (ANSI/IEEE)(reaffirmed 1982) Amplifiers and Preamplifiers for Semiconductor Radiation Detectors for Ionizing Radiation, Standard Test Procedures for.

24   Std C37.93-1976 (ANSI/IEEE) Protective Relay Applications of Audio Tones over Telephone Channels, Guide for.

25   Std 315-1975 (ANSI/IEEE)(CSA Z99-1975) Graphic Symbols for Electrical and Electronics Diagrams.

26   Std 317-1976 (IEEE) (revised in 1983) See code 493.

27   Std C37.30-1971 and Std C37.30a-1974 (ANSI/IEEE) (reaffirmed 1977) High Voltage Air Switches, Insulators, and Bus Supports, Definitions and Requirements for

28   Std 344-1975 (ANSI/IEEE) (revised in 1987) See code 581.

29   Std 352-1975 (ANSI/IEEE) (revised in 1987) See code 587.

30   Std 376-1975 (ANSI/IEEE) Measurement of Impulse Strength and Impulse Bandwidth, Standard for the

31   Std 380-1975 (IEEE) (withdrawn 1979) Definitions of Terms Used in IEEE Nuclear Power Generating Stations Standards.

32   Std 390-1975 (ANSI/IEEE) (revised in 1987) See code 589.

33   Std 391-1976 (ANSI/IEEE) (revised in 1987) See code 589.

34 Std 404-1986 (ANSI/IEEE) Cable Joints for Use with Extruded Dielectric Cable Rated 5000 through 46 000 Volts, and Cable Joints for Use with Laminated Dielectric Cable Rated 2500 through 500 000 Volts, Standard for

35 Std 422-1977 (ANSI/IEEE) (revised in 1986) See code 477.

36 Std 430-1976 (ANSI/IEEE) (revised in 1986) See code 509.

37 Std 432-1976 (ANSI/IEEE) (reaffirmed 1982) Insulation Maintenance for Rotating Electrical Machinery (5 hp to less than 10 000 hp), Guide for

38 Std 450-1987 (ANSI/IEEE) Maintenance, Testing and Replacement of Large Lead Storage Batteries for Generating Stations and Substations, Recommended Practice for.

39 Std 455-1976 (ANSI/IEEE) (revised in 1985) See code 529.

40 Std 488-1978 (ANSI/IEEE) Standard Digital Interface for Programmable Instrumentation (includes supplement).

41 Std 498-1985 (ANSI/IEEE) Measuring and Test Equipment Used in the Construction and Maintenance of Nuclear Power Generating Stations, Standard Requirements for the Calibration and Control of.

42 Std 511-1979 (ANSI/IEEE) Video Signal Transmission Measurement of Linear Waveform Distortion, Standard on.

43 Std 518-1982 (ANSI/IEEE) Installation of Electrical Equipment to Minimize Noise Inputs to Controllers from External Sources, Guide for the

44 Std 521-1984 (ANSI/IEEE) Letter Designations for Radar Frequency Bands.

45 Std 524-1980 (ANSI/IEEE) Installation of Overhead Transmission Line conductors, Guide to the. See codes 431 and 458

46 Std 528-1984 (ANSI/IEEE) Inertial Sensor Terminology.

47 Std 544-1975 (IEEE) Electrothermic Power Meters, Standard for.

48 Sequential Events Recording Systems terms prepared by the Power Generation Committee of the Power Engineering Society in 1974. (Terms approved for Dictionary use only).

49 Std 570-1975 (IEEE) System Voltage Nomenclature Table for Use in all Industrial and Commercial Power Systems Standards and Committee Reports. Note: Not published as a separate standard. Technical information incorporated in IEEE Std 141-1976 and ANSI C84.1-1977.

50 Std 579-1975 (IEEE) Test Procedures AC HV Circuit Breakers for Load Current Switching Requirements and Test Duties. See C37.09

51 Std 583-1975 (IEEE) (revised in 1982) Modular Instrumentation and Digital Interface System (CAMAC). See IEEE Std 595 and IEEE Std 596.

52 Std 590-1977 (IEEE) Cable Plowing Guide.

53 Std C57.12.80 (ANSI/IEEE) (reaffirmed 1986) Terminology for Power and Distribution Transformers.

54 Mil. Std. 1309B; Automated Instrumentation 9.8 Terms for Test Measurement, and Diagnostic Equipment, Definitions of

55 Std 312-1977 (ANSI/IEEE) Communication Switching, Standard Definitions of Terms for.

56 Std C85.1-1963 (ANSI) (a) 1966 (b) 1972 Terminology for Automatic Control.

57 IEEE Power Engineering Society Committee on Insulated Conductors.

58 IEEE Power Engineering Society Committee on Power Generation.

59 Std 599-1985 (ANSI/IEEE) Power Systems Data Transmission and Related Channel Terminology, Standard Definitions of.

60 IEEE Power Engineering Society Committee on Power System Relaying.

61 IEEE Information Theory Group.

62 IEEE Power Engineering Society Committee on Surge Protective Devices. See codes 2 and 91.

63 IEEE Power Engineering Society Committee on Rotating Machinery.

64 IEEE Power Engineering Society Committee on Transmission and Distribution.

65 IEEE Industry Applications Society Committee on Petroleum and Chemical Industry. Definitions taken from the NFPA (National Fire Protection Association)

66 IEEE Industry Application Society Committee on Static Power Converters.

67 IEEE Circuits and Systems Society. Network Applications of Circuits and Systems.

68 IEEE Instrumentation and Measurement Society, Nonreal Time Spectrum Analyzer.

69 IEEE Instrumentation and Measurement Society, Test, Measurement, and Diagnostic Equipment. See code 54.

70 Switchgear terms derived from C37.100-1975 and C37.23. For current terms see code 103.

71 Std SE3.13-1974 (ANSI); NFPA 72E-1974, Standard on Automatic Fire Detectors.

72 IEEE Ultrasonics, Ferroelectrics, and Frequency Control Society. Definitions for specific (acoustic-optical) devices, delay lines, and ferroelectric material terms. See codes 80, 81, and 82.

73 IEEE Industry Applications Society, Subcommittee 2-447-02 on Emergency and Standby Power Systems. See code 89.

74 IEEE Communications Society, Committee on Space Communications. Definitions of Communication Satellite Terms. See code. 84.

75 See code 159.

76 Std 484-1987 (ANSI/IEEE) Installation Design and Installation of Large Lead Storage Batteries for Generating Stations and Substations, Recommended Practice for

77 IEEE Computer Society, Computing Systems.

78 Std C37.23-1969 (ANSI/IEEE) (revised in 1987). See code 574.

79 Std C37.100-1981 (ANSI/IEEE). Power Systems Relaying Committee. See code 103.

80 IEEE Ultrasonics, Ferroelectrics, and Frequency Control Society. Definitions replaced by those in ANSI/IEEE Std 180-1986, code 497. See code 247.

81 IEEE Ultrasonics, Ferroelectrics, and Frequency Control Society. Definitions for Delay Lines, Dispersive and Nondispersive.

82 IEEE Ultrasonics, Ferroelectrics, and Frequency Control Society. Definitions for Acousto-Optic Devices.

83 IEEE Communications Society, Space Communications Committee. Component parts of communications systems; Communications satellite terms.

84 IEEE Communications Society, Space Communications Committee. Communications System Methods and Functions.

85 IEEE Communications Society, Space Communications Committee. Transmission and Propagation Terms.

86 Std C55.2-1974 (ANSI). See codes 138 and 474.

87 Std C104.2-1968 (ANSI); EIA RS 330-1966 Closed Circuit Television Camera 525/60 Interface 2:1, Electrical Performance of

88 Std 91-1973 (ANSI/IEEE)(revised in 1984). See code 451.

89 Std 446-1980 (ANSI/IEEE)(revised in 1987). See code 512.

90 Std C37.90.1-1974 (ANSI/IEEE)(reaffirmed 1979). Surge Withstand Capability (SWC) Tests, Guide for. (Included in ANSI/IEEE Std C37.90-1978) (reaffirmed 1982).

91 Std 32-1972 (ANSI/IEEE)(reaffirmed 1984). Neutral Grounding Devices. Requirements, Terminology, and Test Procedure for.

92 Std C37.60-1981 (ANSI/IEEE)(reaffirmed 1986)(previously IEEE Std 437-1974). Requirements for Overhead, Pad-Mounted, Dry Vault, and Submersible Automatic Circuit Reclosers and Fault Interrupters for AC Systems.

93 Std C37.20-1969 (ANSI/IEEE)(reaffirmed 1981). Switchgear Assemblies Including Metal-Enclosed Bus. See codes 572, 573, and 579.

94 Std 94-1970 (IEEE)(withdrawn). Automatic Generation Control on Electric Power Systems, Standard Definitions of Terms for.

95 Std 295-1969 (ANSI/IEEE)(reaffirmed 1981). Electronics Power Transformers, Standard for.

96 Std 309-1970 (ANSI/IEEE)(reaffirmed 1984). Geiger-Muller Counters, Test Procedure for.

97 Std 402-1974 (ANSI/IEEE)(reaffirmed 1982). Measuring Resistivity of Cable Insulation Materials at High Direct Voltages, Guide for.

98 Std 282-1968 (IEEE)(withdrawn 1978).

99 Std 387-1984 (ANSI/IEEE). Diesel Generator

Units Applied as Standby Power Supplies for Nuclear Power Generating Stations, Standard Criteria for.

100 No entry.

101 Std C80.1-1971 (ANSI). Rigid Steel Conduit, Zinc Coated, Specification for

102 Std 308-1980 (ANSI/IEEE). Class 1E Power Systems for Nuclear Power Generating Stations, Standard Criteria for.

103 Std C37.100-1981 (ANSI/IEEE). Power Switchgear, Definitions for.

104 Std 334-1974 (ANSI/IEEE)(reaffirmed 1980). Type Tests of Continuous Duty Class 1E Motors for Nuclear Power Generating Stations, Standard for.

105 Std 421-1972 (IEEE)(revised in 1986). See code 507.

106 Std 311-1970 (IEEE). General-Purpose Laboratory Cathode-Ray Oscilloscopes, Standard Specification of.

107 Std 296-1969 (IEEE). Hall Effect Devices, Definitions of Terms, Letter Symbols, and Color Code for.

108 Std 332-1972 (IEEE)(withdrawn); C68.2-1972 (ANSI). Techniques for Switching Impulse Testing.

109 Std 379-1977 (ANSI/IEEE). Single-Failure Criterion to Nuclear Power Generating Station Class 1E Systems, Standard Application of the.

110 Std 474-1973 (ANSI/IEEE)(reaffirmed 1982). Fixed and Variable Attenuators, dc to 40GHz, Standard Specifications and Test Methods for.

111 Std 145-1983 (ANSI/IEEE). Antennas, Definitions of Terms for.

112 Std 346-1973 (IEEE)(withdrawn 1987). See code 516.

113 Std 307-1969 (IEEE). Solar Cells, Definitions of Terms for.

114 Std 169-1955 (IEEE)(withdrawn). Industrial Electronics Terms, Definitions of.

115 Std 470-1972 (IEEE). Bolometric Power Meters, Standard Application Guide for.

116 Std 135-1969 (IEEE). Aircraft, Missile, and Space Equipment Electrical Insulation Tests, Recommended Practice for.

117 Std 398-1972 (ANSI/IEEE)(reaffirmed 1982). Photomultipliers for Scintillation Counting and Glossary for Scintillation Counting Field, Test Procedures for.

118 Std 325-1971 (ANSI/IEEE)(revised in 1986). See codes 471 and 528.

119 Std 300-1982 (ANSI/IEEE). Semiconductor Charged-Particle Detectors, Test Procedures for. See code 471.

120 Std 323-1974 (ANSI/IEEE)(revised in 1983). Qualifying Class 1E Equipment for Nuclear Power Generating Stations.

121 Std 444-1973 (ANSI/IEEE). Thyristor Converters and Motor Drives: Part 1--Converters for dc Motor Armature Supplies, Standard Practices and Requirements for.

122 Std 269-1971 (ANSI/IEEE)(revised in 1983). See code 491.

123 Std 184-1969 (ANSI/IEEE). Frequency-Modulated Mobile Communications Receivers, Test Procedure for.

124 See code 429.

125 Std 161-1971 (ANSI/IEEE)(reaffirmed 1980). Electron Tubes, Standard Definitions on.

126 Std 297-1969 (IEEE)(withdrawn). Speech Quality Measurements, Recommended Practice for.

127 Std C37.90-1978 (ANSI/IEEE)(reaffirmed 1982). (Previously IEEE Std 313). Relays and Relay Systems Associated with Electric Power Apparatus.

128 Std C37.95-1973 (ANSI/IEEE)(reaffirmed 1980). (Previously IEEE Std 357). Protective Relaying of Utility-Consumer Interconnections, Guide for.

129 Std 85-1973 (IEEE)(reaffirmed 1986). Airborne Sound Measurements on Rotating Electric Machinery, Test Procedure for.

130 Std 341-1972 (IEEE)(withdrawn); C37.073-1972 (ANSI)(withdrawn). Requirements for Capacitance Current Switching for AC High-Voltage Circuit Breakers Rates on a Symmetrical Current Basis.

131 Std 384-1981 (ANSI/IEEE). Independence of Class 1E Equipment and Circuits, Standard Criteria for.

132 Std 365-1974 (IEEE)(withdrawn). Radio Methods for Measuring Earth Conductivity, Guide for.

133 Std 463-1977 (IEEE)(reaffirmed 1987). Electrical Safety Practices in Electrolytic Cell Line Working Zones, Standard for.

134 Std 386-1977 (ANSI/IEEE)(revised in 1985). See code 454.

135 Std 433-1974 (ANSI/IEEE)(reaffirmed 1985). Insulation Testing of Large AC Rotating Machinery with High Voltage at Very Low Frequency, Recommended Practice for.

136 Std 318-1971 (IEEE)(withdrawn). Varactor Measurements, Part One: Small Signal Measurements.

137 Std 306-1969 (IEEE)(reaffirmed 1981). Charging Inductors, Test Procedure for.

138 Std 18-1980 (ANSI/IEEE). Shunt Power Capacitors, Standard for.

139 Std 454-1973 (ANSI/IEEE)(reaffirmed 1979). The Detection and Measurement of Partial Discharges (Corona) During Dielectric Tests, Recommended Practice for.

140 Std 420-1982 (IEEE). Class 1E Control Boards, Panels and Racks Used in Nuclear Power Generating Stations, Standard Design and Qualification of.

141 Std 383-1974 (ANSI/IEEE)(reaffirmed 1980). Class 1E Electric Cables, Field Splices, and Connections for Nuclear Power Generating Stations, Standard for Type Test of.

142 Std 382-1980 (ANSI/IEEE)(revised in 1985). See code 492.

143 Std 336-1980(ANSI/IEEE)(revised in 1985). Power, Instrumentation, and Control Equipment at Nuclear Facilities, Installation, Inspection, and Test Requirements for. (These are 1980 definitions; there are no definitions in the 1985 revision.)

144 Std C37.34-1971 (ANSI/IEEE)(reaffirmed 1977)(previously IEEE Std 326). High Voltage Air Switches, Test Code for.

145 Std 193-1971 (ANSI/IEEE)(withdrawn). Weighted Peak Flutter of Sound Recording and Reproducing Equipment, Method for Measurement of

146 Std 211-1977 (ANSI/IEEE). Terms for Radio Wave Propagation, Standard Definitions of.

147 Std 314-1971 (IEEE). State-of-the-Art of Measuring Unbalanced Transmission-Line Impedance.

148 Std 327-1972 (IEEE)(withdrawn); Std C37.072-1971 (ANSI) (withdrawn). Transient Recovery Voltage for AC High Voltage Circuit Breakers Rated on a Symmetrical Current Basis, Requirements for.

149 Std 343-1972 (IEEE)(withdrawn); Std C37.078 (ANSI)(withdrawn). External Insulation for Outdoor AC High-Voltage Circuit Breakers Rated on a Symmetrical Current Basis.

150 Std 4-1978 (ANSI/IEEE)(Supersedes IEEE Std 29-1941 and ANSI/IEEE Std 332-1972). High Voltage Testing, Standard Techniques for.

151 Std 111-1984 (ANSI/IEEE)(Same definitions as in IEEE-1971 (reaffirmed 1976). Low-Power Wide-Band Transformers, Standard for.

152 Std 142-1982 (ANSI/IEEE)(IEEE Green Book). Grounding of Industrial and Commercial Power Systems, Recommended Practice for.

153 Std 141-1976 (IEEE)(revised in 1986). See code 510.

154 Std 117-1974 (ANSI/IEEE)(reaffirmed 1985). Systems of Insulating Materials for Random-Wound AC Electric Machinery, Standard Test Procedure for Evaluation of.

155 Std 294-1969 (IEEE)(withdrawn); Std C16.44-1977 (ANSI) (withdrawn). State-of-the-Art of Measuring Noise Temperature of Noise Generators.

156 Std 494-1974 (ANSI/IEEE)(reaffirmed 1983). Identification of Documents Related to Class 1E Equipment and Systems for Nuclear Power Generating Stations, Standard Method for.

157 Std 417-1973 (IEEE); Std C37.079-1973 (ANSI)(withdrawn). Testing AC High-Voltage Circuit Breakers Rated on a Symmetrical Basis When Rated for Out-of-Phase Switching.

158 Std C37.29-1981 (ANSI/IEEE)(Previously IEEE Std 508-1974). Low-Voltage AC Power Circuit Protectors Used in Enclosures.

159 Std 338-1977 (ANSI/IEEE)(reaffirmed 1984). Periodic Testing of Nuclear Power Generating Station Safety Systems, Standard Criteria for.

160 Std 171-1958 (IEEE)(withdrawn). See code 415.

161 Std 347-1972 (ANSI/IEEE)(withdrawn). Measuring Recorded Flux of Magnetic Sound Records at Medium Wavelengths, Standard Method of.

162 P347 (IEEE Committee draft)(withdrawn). Task group for solid-state displays of the Standardization Committee of the IEEE Group on Electron Devices.

163 Std 204-1961 (IEEE)(withdrawn). Terms relating to Television, Definitions of. See code 18.

164 IEEE Reliability Society. Availability, Reliability, and Maintainability Terms. See code 182.

165 Std 393-1977 (IEEE). Magnetic Cores, Standard Test Procedures for.

166 Std 147-1979 (ANSI/IEEE). Terms for Waveguide Components, Standard Definitions for.

167 Std RP-16-1980 (ANSI/IES). Nomenclature and Definitions for Illuminating Engineering. A revision of ANSI Z7.1-1967 (reaffirmed 1973).

168 Std 21-1976 (ANSI/IEEE). Outdoor Apparatus Bushings, General Requirements and Test Procedures for.

169 IEEE Electromagnetic Compatibility Society.

170 Std 106-1972 (IEEE)(withdrawn). Toroidal Magnetic Amplifier Cores. This is now included in IEEE Std 393-1977, code 165.

171 Std 107-1964 (IEEE)(reaffirmed 1979). Rating and Testing of Magnetic Amplifiers, Standard for.

172 Std 270-1966 (IEEE)(withdrawn). See code 210.

173 Std 267-1966 (IEEE). Symbols, Trial-Use Recommended Practice for the Preparation and Use of.

174 Std 158-1962 ((ANSI/IEEE)(reaffirmed 1971). Electron Tubes, Methods of Testing.

175 IEV entry. Document 7.

176 IEEE Acoustics, Speech, and Signal Processing Society.

177 IEEE Broadcast Technology Society--Television.

178 IEEE Broadcast Technology Society--Video Techniques.

179 IEEE Antennas and Propagation Society--Antennas and Waveguides.

180 IEEE Antennas and Propagation Society--Wave Propagation.

181 IEEE Vehicular Technology Society--Mobile Communications Systems.

182 IEEE Reliability Society--Definitions and Standards.

183 IEEE Instrumentation and Measurement Society--Electromagnetic Measurement State-of-the-Art.

184 IEEE Instrumentation and Measurement Society--Fundamental Electrical Standards.

185 IEEE Instrumentation and Measurement Society--High-Frequency Instrumentation and Measurements

186 IEEE Aerospace and Electronic Systems Society--Energy Conversion.

187 IEEE Aerospace and Electronic Systems Society--Navigational Aids. See codes 13 and 526.

188 IEEE Industrial Electronics Society.

189 IEEE Electron Devices Society--Solid State Devices.

190 IEEE Electron Devices Society--Standards on Electron Tubes.

191 IEEE Electron Devices Society--Standards on Solid State Devices.

192 IEEE Engineering in Medicine and Biology Society.

193 IEEE Communications Society--Communications Switching.

194 IEEE Communications Society--Data Communication Systems.

195 No entry.

196 IEEE Communications Society--Wire Communication.

197 IEEE Components, Hybrids, and Manufacturing Technology Society.

198 IEEE Instrumentation and Measurement Society--Control Systems.

199 IEEE Electromagnetic Compatibility Society.

200 IEEE Power Engineering Society--Power System Engineering.

201 IEEE Power Engineering Society--Power System Instrumentation and Measurement.

202 IEEE Power Engineering Society--Switchgear.

203 IEEE Power Engineering Society--Transformers.

204 IEEE Industry Applications Society--Cement Industry.

205 IEEE Industry Applications Society--Corrosion and Cathodic Protection.

206 IEEE Industry Applications Society--Industrial Control.

207 IEEE Industry Applications Society--Machine Tools Industry.

208 IEEE Industry Applications Society--Static Power Converters.

209 IEEE Systems, Man and Cybernetics Society.

210 Std 270-1966 (IEEE)(withdrawn). General (Fundamental and Derived) Electrical and Electronics Terms, Definitions of.

211 Std 182A-1964 (IEEE). Supplement to Definitions of Terms for Radio Transmitters. See IEEE Std 182-1961, code 240.

212 Std C12.1-1982 (ANSI). Electricity Metering, American National Standard Code for.

213 Std 284-1968 (IEEE). Measuring Field Strength, Continuous Wave, Sinusoidal, Standards Report on State-of-the-Art of.

214 AD8--American Society for Testing and Materials Publication D8.

215 AD16--American Society for Testing and Materials Publication D16.

216 AD123--American Society for Testing and Materials Publication D123.

217 AD883--American Society for Testing and Materials Publication 883.

218 AD1566--American Society for Testing and Materials Publication D1566.

219 National Electrical Manufacturers Association Publication AS 1.

220 CISPR--International Special Committee on Radio Interference.

221 CM--Corrosion Magazine.

222 CTD--Chambers Technical Dictionary.

223 CV 1--National Electrical Manufacturers Association Publication CV 1.

224 Electronic Industries Association Publication 3B.

225 IC 1--National Electrical Manufacturers Association Publication IC 1.

226 ISA--Instrument Society of America.

227 International Telecommunications Union.

228 KPSH--Kepco Power Supply Handbook.

229 LA 1--National Electrical Manufacturers Association Publication LA 1.

230 MA 1--National Electrical Manufacturers Association Publication MA 1.

231 MDE--Modern Dictionary of Electronics.

232 MG 1--National Electrical Manufacturers Association Publication MG 1.

233 SCC--IEEE Standards Coordinating Committee. See code 415.

234 See code 237.

235 Std 162-1963 (ANSI/IEEE)(reaffirmed 1984). Electronic Digital Computers, Definitions of Terms for.

236 Std 224-1965 (IEEE)(withdrawn). Electrostatographic Devices, Definitions of Terms for.

237 Std 59-1962 (IEEE)(withdrawn). Semiconductor Rectifier Components.

238 Std 156-1960 (IEEE). Linear Passive Reciprocal Time Invariant Networks, Definitions of Terms for.

239 Std 151-1965 (IEEE)(withdrawn). Audio and Electroacoustics, Standard Definitions of Terms for.

240 Std 182-1961 (IEEE). Radio Transmitters, Definitions of Terms for.

241 Std 254-1963 (IEEE)(withdrawn). Parametric Device Terms.

242 Std 270-1964 (IEEE)(withdrawn). Modulation Systems, Standard Definitions of Terms for. See code 415.

243 Std223-1966 (IEEE)(withdrawn). Thyristors, Definitions of Terms for.

244 IEC--International Electrotechnical Commission.

245 Std 216-1960 (IEEE)(reaffirmed 1980). Semiconductor Terms, Definitions of.

246 Std 149-1979 (Ansi/IEEE)(reffirmed 1986). Antennas, Test Procedure for.

247 Std 180-1962 (IEEE). Replaced by ANSI/IEEE Std 180-1986, code 497.

248 Std 221-1962 (IEEE)(withdrawn). Thermoelectric Device Terms, Definitions of.

249 Std 258-1965 (IEEE)(withdrawn). Close-Talking Pressure-Type Microphones, Test Procedure for.

250 See codes 166 and 267.

251 See code 166.

252 Std 196-1951 (IEEE)(withdrawn). Transducers, Standard Definitions of Terms for.

253 Std 257-1964 (IEEE)(withdrawn). Burst Measurements in the Time Domain, Recommended Practices for.

254 Std 194-1977 (IEEE). Standard Pulse Terms and Definitions.

255 Std X3.12-1970 (ANSI); Std 2382/V, VI (ISO) Vocabulary for Information Processing.

256 NFPA No.70-1978 (previously Std C1-1978). National Electrical Code.

257 Std C57.15-1968 (ANSI). Requirements, Terminology, and Test Code-Voltage and Induction Voltage Regulators.

258 Std C57.18-1964 (ANSI)(reaffirmed 1971). Pool-Cathode Mercury-Arc Rectifier Transformers, Requirements, Terminology, and Test Code for.

259 Std C83.16-1971 (ANSI) Relays and Electronic Equipment, Definitions and Terminology for.

260 Std C84.1-1970 (ANSI)(revised in 1977); IEC 38 and 71 Voltage Ratings for Electric Power Systems and Equipment (60 Hz), including Supplement C84.1A-1973.

261 Std C29.1-1961 (ANSI)(reaffirmed 1974). Electrical Power Insulators, Test Methods for, including Addendum C29.2A (reaffirmed 1974).

262 Std C2.2-1960 (ANSI). Safety Rules for the Installation and Maintenance of Electric Supply and Communication Lines, including Supplement C2.2B-1965.

263 Std C71.1-1972 (ANSI) Household Electric Ranges (AHAM ER-1), including Supplements C71.1A-1975 and C71.1B-1975.

264 Std C87.1-1971 (ANSI); NEMA Publication EW 1-1970. Electric Arc Welding Apparatus.

265 Std C83.14-1963 (ANSI)(reaffirmed 1969) EIA RS 225-1959; IEC 339-1. Requirements for Rigid Coaxial Transmission Lines--50 ohms.

266 Std C85.1-1963 (ANSI). Automatic Control, Terminology for, including Supplements C85.1A-1966 and C85.1B-1972.

267 Std 146-1980 (ANSI/IEEE). Fundamental Waveguide Terms, Definitions of.

268 Std C82.1-1972 (ANSI). Fluorescent Lamp Ballasts, including Supplement C82.1A-1973, Specifications for.

269 Std C82.4-1974 (ANSI); IEC 262. Mercury Lamp Ballasts (Multiple Supply Type), Specifications for.

270 Std C82.3-1972 (ANSI); IEC 82. Fluorescent Lamp Reference Ballasts, Specifications for.

271 Std C82.9-1971 (ANSI). High-Intensity Discharge Lamp Ballasts and Transformers, Definitions for.

272 See code 292.

273 See code 2.

274 Std C82.7-1971 (ANSI); IEC 262. Mercury Lamp Transformers, Constant Current (Series) Supply Type, Specifications for.

275 Std C82.8-1963 (ANSI)(reaffirmed 1971). Incandescent Filament Lamp Transformers, Constant Current (Series) Supply Type, Specifications for.

276 Std C92.1-1971 (ANSI). Voltage Values for Preferred Transient Insulation Levels.

277 See code 241.

278 Std 173-1959 (IEEE). Navigation Aids: Direction Finder Measurements.

279 Std C64.1-1970 (ANSI); IEC 136-1; IEC 136-2; IEC 276. Brushes for Electrical Machines.

280 Std C39.1-1972 (ANSI). Electrical Analog Indicating Instruments, Requirements for.

281 Std C37.1 (ANSI)(redesignated C37.90). See code 127.

282 Std 155-1960 (IEEE)(reaffirmed 1983). Linear Flow Graphs, Definitions of Terms for.

283 Std C78.385-1961 (ANSI). Electric Lamps.

284 Std C99.1 (ANSI). Highly Reliable Soldered Connections in Electronic and Electrical Application.

285 Std C63.4-1963 (ANSI)(reaffirmed 1969). Radio Noise. For the 1981 revision see code 418.

286 Std C79.1-1971 (ANSI). Glass Bulbs Intended for Use with Electron Tubes and Electric Lamps, Nomenclature for.

287 See code 168.

288 Std C57.12.75 (ANSI). Removable Air-Filled Junction Boxes for Cable Termination for Power Transformers.

289 Std S1.1-1960 (ANSI)(reaffirmed 1971). Integral Air-Filled Junction Boxes for Cable Termination for Power Transformers.

290 Std S1.1-1960 (ANSI)(reaffirmed 1971); ISO 131; ISO 16; IEC 50-08. Acoustical Terminology (including Mechanical Shock and Vibration).

291 Std C31.4-1958 (ANSI)(reaffirmed 1975). Pool-Cathode Mercury-Arc Power Converters, Practices and Requirements for.

292 Std 265-1966 (IEEE)(withdrawn). Burst Measurements in the Frequency Domain, Recommended Practices for. For Burst Measurements in the Time Domain see IEEE Std 257-1964, (withdrawn), code 253.

293 Std 285-1968 (IEEE)(withdrawn; C16.42-1972 (ANSI)(withdrawn). Standards Report on State-of-the-Art of Measuring Phase Shift at Frequencies above 1 GHz.

294 Std C39.2-1964 (ANSI)(reaffirmed 1969). Direct Acting Electrical Recording Instruments, Requirements for.

295 Std C39.4-1966 (ANSI)(reaffirmed 1972). Automatic Null-Balancing Electrical Measuring Instruments, Specifications for.

296 Std C80.4-1963 (ANSI)(reaffirmed 1974). Fittings for Rigid Metal Conduit and Electrical Metallic Tubings, Specifications for.

297 Std C5.1-1969 (ANSI); NFPA No. 70-1968. Lightning Protection Code.

298 Std C50.10-1977 (ANSI); IEC 34-1. Synchronous Machines, General Requirements for.

299 See code 2.

300 See code 92.

301 Std C89.1-1974 (ANSI). Specialty Transformers except General-Purpose Type.

302 Std 74-1958 (IEEE)(reaffirmed 1974). Industrial Control (600 Volts or Less), Standard Test Code for.

303 Std C57.14 (ANSI). Constant-Current Transformers of the Moving Coil Type.

304 See code 321.

305 Std C57.13-1968 (ANSI). Instrument Transformers, Requirements for. For 1978 revision see code 394.

306 Std 206-1960 (ANSI/IEEE)(reaffirmed 1978). Television: Measurement of Differential Gain and Differential Phase.

307 See code 150.

308 See code 244.

309 Std C57.16-1958 (ANSI)(reaffirmed 1971); IEC 288-2. Current Limiting Reactors, Requirements, Terminology, and Test Code for.

310 Std 96-1969 (IEEE). Rating Electric Apparatus for Short-Time Intermittent or Varying Duty, General Principles for.

311 See code 111.

312 Std 274-1966 (IEEE)(reaffirmed 1980). Integrated Electronics, Standard Definitions of Terms for.

313 Std 81-1983 (ANSI/IEEE). Measuring Earth Resistivity, Ground Impedance, and Earth Surface Potentials of a Ground System, Guide for.

314 Std 263-1965 (IEEE). Radio Noise Generated by Motor Vehicles and Affecting Mobile Communications Receivers in the Frequency Range 25 to 1000 Megahertz, Standard for Measurement of.

315 Std 253-1963 (IEEE)(withdrawn). Semiconductor Tunnel (Esaki) Diodes and Backward Diodes, Definitions, Symbols, and Methods of Test for.

316 See code 53.

317 Std 208-1960 (ANSI/IEEE)(reaffirmrd 1978). Video Techniques: Measurement of Resolution of Camera Systems.

318 Std 115-1965 (IEEE)(revised in 1983). Synchronous Machines, Test Procedures for. The 1983 revision refers, for definitions, to IEEE Std 100-1977. See codes 521 and 565 for 1984 and 1987 updates.

319 Std 148-1959 (IEEE)(reaffirmed 1971). Waveguide and Waveguide Component Measurements, Standard for.

320 Std 1-1969 (IEEE). See code 506 for 1986 revision.

321 Std 264-1977 (ANSI/IEEE). See ANSI/IEEE Std 111-1984, code 151.

322 Std C95.1-1982 (ANSI). Electromagnetic Radiation with Respect to Personnel, American National Standard Safety Level of

323 See code 4.

324 Std C67.1 (ANSI). Preferred Nominal Voltages, 100 Volts and Under.

325 Std 67-1972 (ANSI/IEEE)(reaffirmed 1980). Turbine Generators, Guide for Operation and Maintenance of.

326 Std C78.180-1972 (ANSI). Fluorescent Lamp Starters.

327 No entry.

328 This definition was derived from a standard previously listed in the ANSI category C42.

329 IEEE Committee on Automatic Control, now the IEEE Instrumentation and Measurement Society.

330 See codes 146, 166, and 267.

331 Std 163 (IEEE)(withdrawn). Static Magnetic Storage, Terms for.

332 Std 153-1950 (IEEE)(withdrawn). Network Topology, Definitions of Terms in.

333 Std 160-1957 (IEEE). Electron Tubes: Definitions of Terms. See code 125.

334 Std 157-1951 (IEEE)(withdrawn). Electrostatics: Definitions of Terms.

335 Std 175-1960 (IEEE)(withdrawn). Nuclear Techniques: Definitions for the Scintillation Counter Field.

336 Std 207 (IEEE)(withdrawn in 1972). Television: Methods of Measurement of Time of Rise, Pulse Width, and Pulse Timing in Video Pulses in Television.

337 Std 203 (IEEE)(withdrawn). Television: Methods of Measurement of Aspect Ratio and Geometric Distortion.

338 Std 62-1978 (IEEE)(supersedes AIEE Std 62-1958). Field Testing Power Apparatus Insulation, Guide for. Previous titled Guide for Making Dielectric Measurements in the Field.

339 Std 188 (IEEE)(withdrawn).

340 Std 226 (IEEE)(withdrawn). Solid State Devices: Definitions of Terms for Nonlinear Capacitors.

341 No entry.

342 Std 102 (IEEE)(withdrawn in 1971). Test Code for Transistors: Definitions and Letter Symbols.

343 Std 210 (IEEE)(withdrawn in 1975). Radio Wave Propagation: Definitions of Terms Relating to Guided Waves.

344 Std 217-1962 (IEEE)(withdrawn in 1971). Solid State Devices: Definitions of Superconductive Electronics Terms.

345 Std 30 (IEEE)(withdrawn). Definitions and General Standards for Wire and Cable.

346 Office definition prepared by the staff of ANSI/IEE Std 100.

347 Std 209-1950 (IEEE). Television: Methods of Measurement of Electronically Regulated Power Supplies.

348 Std 49-1948 (IEEE)(withdrawn). Roof, Floor, and Wall Bushings.

349 Std 154-1953 (IEEE)(withdrawn). Circuits: Definitions of Terms in the Field of Linear Varying Parameter and Nonlinear Circuits.

350 Std 191-1953 (IEEE)(withdrawn in 1975). Sound Recording and Producing: Method of Measurement of Noise.

351 Std 31-1944 (IEEE)(withdrawn). Outdoor Coupling Capacitors and Capacitance Potential Devices.

352 IEEE Committee on Sonics and Ultrasonics.

353 Std 421A-1978 (ANSI/IEEE). Guide for Identification, Testing, and Evaluation of the Dynamic Performance of Excitation Control Systems.

354 Std 415-1976 (IEEE). See ANSI/IEEE Std 415-1986, code 455.

355 Std 381-1977 (ANSI/IEEE)(reaffirmed 1984). Type Tests of Class 1E Modules Used in Nuclear Power Generating Stations, Standard Criteria for.

356 See code 109.

357 Std 577-1976 (ANSI/IEEE)(reaffirmed 1986). Qualification of Class 1E Connection Assemblies for Nuclear Power Generating Stations, Standard for.

358 Std 566-1977 (IEEE)(withdrawn). Design of Display and Control Facilities for Central Control Rooms of Nuclear Power Generating Stations, Recommended Practice for the.

359 See code 412.

360 See code 401.

361 See code 421.

362 See code 44.

363 Std 586-1980 (ANSI/IEEE). Laser-Maser Terms, Definitions of.

364 Std 304-1977 (ANSI/IEEE)(reaffirmed 1982). Evaluation and Classification of Insulation Systems for dc Machines, Test Procedure for. See ANSI/IEEE Std 117-1974, code154, and ANSI/IEEE Std 275-1981.

365 See code 165.

366 See code 159.

367 Std C57.12.00-1980 (ANSI/IEEE). Liquid-Im- mersed Distribution, Power, and Regulating Transformers, General Requirements for.

368 Std 634-1978 (ANSI/IEEE). Standard Cable Penetration Fire Stop Qualification Test.

369 Std 176-1978 (ANSI/IEEE). Piezoelectricity, Standard on. (Supersedes ANSI/IEEE Std 177-1966, ANSI/IEEE Std 178-1958, and ANSI/IEEE Std 179-1961.

370 Std C62.31-1981 (ANSI/IEEE). See ANSI/ IEEE Std C62.31-1984 (reaffirmed 1987), code 490.

371 See code 369.

372 Std 202-1954 (ANSI/IEEE)(reaffirmed 1978). Television: Methods of Measurement of Aspect Ratio and Geometric Distortion. See code 337.

373 Std 368-1977 (ANSI/IEEE). Measurement of Electrical Noise and Harmonic Filter Performance of High-Voltage Direct-Current Systems, Recommended Practice for.

374 See code 353.

375 Std 436-1977 (IEEE). Corona (Partial Discharge) Measurements on Electronics Transformers, Guide for Making.

376 Std 469-1977 (IEEE). Voice-Frequency Electrical-Noise Tests of Distribution Transformers, Recommended Practice for.

377 Std 485-1983 (ANSI/IEEE). Sizing Large Lead Storage Batteries for Generating Stations and Substations, Recommended Practice for.

378 See code 40.

379 Std 500-1977 (ANSI/IEEE). See Std 500-1984, code 447.

380 Std 503-1978 (IEEE). Measurement and Characterization of Diode-Type Camera Tubes, Standard for.

381 Std 505-1977 (ANSI/IEEE). Generating Stations Electric Power Systems, Standard Nomenclature for.

382 Std 525-1978 (IEEE). Selection and Installation of Control and Low Voltage Cable Systems in Substations, Guide for.

383 Std 530-1978 (ANSI/IEEE)(reaffirmed 1986). Linear Single-Axis, Digital, Torque Balance Accelerometer, Standard Specification Format Guide and Test Procedure for.

384 Std 535-1979 (ANSI/IEEE)(revised in 1986). Class 1E Lead Storage Batteries for Nuclear Power Generating Stations, Standard Qualification of.

385 Std 563-1978 (IEEE)(reaffirmed 1983). Conductor Self-Damping Measurements, Guide on.

386 Std 581-1978 (IEEE). Metal-Nitride-Oxide Field-Effect Transistors, Standard Definitions,- Symbols, and Characterization of.

387 Std 603-1980 (ANSI/IEEE). Safety Systems for Nuclear Power Generating Stations, Standard Criteria for.

388 Std 680-1978 (ANSI/IEEE). Superseded by ANSI/IEEE Std 325-1986, code 528.

389 Std 726-1982 (ANSI/IEEE). Standard Real-Time BASIC for CAMAC

390 Std 748-1979 (IEEE). Spectrum Analyzers, Standard for.

391 Std C2-1977 (ANSI). See code 494.

392 Std C37.98-1978 (ANSI/IEEE). Seismic Testing of Relays, Standard for.

393 Std C57.12.00B. See code 367.

394 Std C57.13-1978 (ANSI/IEEE)(reaffirmed 1986). Instrument Transformers, Standard Requirements for. See code 305.

395 Std 63.12-1979 (ANSI)(revised in 1984). See code 495.

396 Std N42.13-1978 (ANSI)(revised in 1986). See code 499.

397 Std N42.14-1978 (ANSI)(reaffirmed 1985). Germanium Detectors for Measurement of Gamma-Ray Emission of Radionuclides, Calibration and Usage of.

398 Std N320-1979 (ANSI)(reaffirmed 1985). Reactor Emergency Radiological Monitoring Instrumentation, Performance Specifications for.

399 Std N323-1978 (ANSI)(reaffirmed 1983). Radiation Protection Instrumentation Test and Calibration.

400 Std 771-1980 (ANSI/IEEE)(revised in 1984; same definitions). ATLAS, Guide to the Use of.

401 Std C37.14-1979 (ANSI/IEEE)(reaffirmed 1985). Low-Voltage dc Power Circuit Breakers Used in Enclosures.

402 Std C37.2-1979 (ANSI/IEEE). Standard Electrical Power System Device Function Numbers.

403 Std C37.1-1979 (ANSI/IEEE)(revised in 1987). See code 570.

404 Std 367-1979 (IEEE). Maximum Electric Power Station Ground Potential Rise and Induced Voltage from a Power Fault, Guide for Determining the.

405 Std 622-1979 (ANSI/IEEE). Design and Installation of Electric Pipe Heating Systems for Nuclear Power Generating Stations, Recommended Practice for the.

406 Std 635-1980 (ANSI/IEEE). Selection and Design of Aluminum Sheaths for Cables, Guide for.

407 Std 644-1979 (ANSI/IEEE)(revised in 1987). See code 514.

408 Std 650-1979 (ANSI/IEEE). Class 1E Battery Chargers and Inverters for Nuclear Power Generating Stations.

409 Std 661-1979 (ANS/IEEE). Determining Objective Loudness Ratings of Telephone Connections, Standard Methods of.

410 Std 758-1979 (ANSI/IEEE)(reaffirmed 1987). Subroutines for CAMAC, Standard.

411 Std 539-1979 (ANSI/IEEE). Overhead Power Lines Corona and Radio Noise, Definitions of Terms Relating to.

412 Std C37.18-1979 (ANSI/IEEE). Standard Field Discharge Circuit Breakers Used in Enclosures for Rotating Electric Machinery.

413 Std N317-1980 (ANSI)(reaffirmed 1985). Instrumentation Used for Inplant Plutonium Monitoring, American National Standard Performance Criteria for.

414 Std 487-1980 (ANSI/IEEE). Wire Line Communciations Facilities Serving Electric Power Stations, Guide for the Protection of.

415 SCC--IEEE Standards Coordinating Committee. See code 233.

416 Std 290-1980 (ANSI/IEEE)(reaffirmed 1986), Electric Couplings, Recommended Test Procedure for.

417 Std NQA-1-1979 (ANSI/ASME). Quality Assurance Program Requirements for Nuclear Power Plants. Definitions carrying code 417 are reprinted here with the permission of the

American Society of Mechanical Engineers (ASME).

418 Std C63.4-1981 (ANSI). Measurement of Radio Noise Emissions from Low-Voltage Electrical and Electronic Equipment in the Range of 10 kHz to 1 GHz. See code 285.

419 Std C12.8-1981 (ANSI)(reaffirmed 1986). Self-Contained 'A' Base Watthour Meters, Test Blocks and Cabinets for Installation of.

420 Std 620-1987 (IEEE). Construction and Interpretation of Thermal Limit Curves for Squirrel-Cage Motors over 500 hp, Guide for.

421 Std 497-1981 (ANSI/IEEE). Post Accident Monitoring Instrumentation for Nuclear Power Generating Stations, Standard Criteria for.

422 Std N42.15-1980 (ANSI)(reaffirmed 1985). Liquid Scintillation Counting Systems, Performance Verification of.

423 Std N42.12-1980 (ANSI)(reaffirmed 1985). Sodium Iodide Detector Systems, Calibration and Usage of.

424 Std 11-1980 (ANSI/IEEE)(reaffirmed 1985). Rotating Electric Machinery for Rail and Road Vehicles, Standard for.

425 Std C37.13-1981 (ANSI/IEEE). Low-Voltage AC Power Circuit Breakers Used in Enclosures.

426 Std C37.081-1981 (ANSI/IEEE). AC High-Voltage Circuit Breakers Rated on a Symmetrical Current Basis, Guide for Synthetic Fault Testing of.

427 Std 548-1984 (ANSI/IEEE). Laboratory Measurement and Reporting of Fly Ash Resistivity, Standard Criteria and Guidelines for the.

428 See code 387.

429 Std 333-1980 (ANSI/IEEE). Electrical Installations on Packaging Machinery and Associated Equipment, Standard for.

430 Std C62.1-1981 (ANSI/IEEE).Standard for Surge Arresters for Alternating-Current Power Circuits.

431 See code 45.

432 Std 762-1980 (ANSI/IEEE). See ANSI/IEEE Std 762-1987, code 567.

433 Std 812-1984 (ANSI/IEEE). Fiber Optics, Definitions of Terms Relating to.

434 Std 729-1983 (ANSI/IEEE) Software Engineering Terminology, Glossary of.

435 Std 770X3-1983 (ANSI/IEEE). American National Standard Pascal Computer Programming Language.

436 Std 829-1983 (ANSI/IEEE). Software Test Documentation, Standard for.

437 Std 400-1980 (ANSI/IEEE). Making High-Direct-Voltage Tests on Power Cable Systems in the Field, Guide for.

438 Std 467-1980 (IEEE). Standard Quality Assurance Program Requirements for the Design and Manufacture of Class 1E Instrumentation and Electric Equipment for Nuclear Power Generating Stations.

439 Std 567 (ANSI/IEEE)Issued in 1980 as a Trial-Use Standard Criteria (Draft American National Standard) for the Design of the Control Room Complex for a Nuclear Power Generating Station.

440 Std 649-1980 (ANSI/IEEE). Qualifying Class 1E Motor Control Centers for Nuclear Power Generating Stations, Standard for.

441 Std 662-1980 (ANSI/IEEE). Semiconductor Memory, Standard Terminology for.

442 Std 730-1984 (ANSI/IEEE). Software Quality Assurance Plans, Standard for.

443 Std C37.40-1981 (ANSI/IEEE). High-Voltage Fuses, Distribution Enclosed Single-Pole Air Switches, Fuse Disconnecting Switches and Accessories, Service Conditions and Definitions for.

444 Std 377-1980 (ANSI/IEEE)(reaffirmed 1986). Measurement of Spurious Emission from Land-Mobile Communication Transmitters, Recommended Practice for.

445 Std 428-1981 (ANSI/IEEE). Thyristor AC Power Controllers, Definitions and Requirements for.

446 Std 303-1984 (ANSI/IEEE). Auxiliary Devices for Motors in Class 1--Groups A, B, C, and D, Division 2 Locations, Recommended Practice for.

447 Std 500-1984 (ANSI/IEEE). Collection and Presentation of Electrical, Electronic, Sensing Component and Mechanical Equipment Reliability Data for Nuclear Power Generating Stations, Guide to the.

448 Std 622A-1984 (ANSI/IEEE). Design and Installation of Electric Pipe Heating Control and Alarm Systems for Power Generating Stations, Recommended Practice for the. (Supplement to ANSI/IEEE Std 622-1979, code 405).

449 Std 830-1984 (ANSI/IEEE). Software Requirements Specifications, Guide for.

450 See code 322.

451 Std 91-1984 (ANSI/IEEE). Logic Functions, Graphic Symbols for. Previously, code 88.

452 Std 98-1984 (ANSI/IEEE). Test Procedures for the Thermal Evaluation of Solid Electrical Insulating Materials, Standard for the Preparation of.

453 Std 281-1984 (ANSI/IEEE). Power System Communication Equipment, Standard Service Conditions for.

454 Std 386-1985 (ANSI/IEEE). Separable Insulated Connectors for Power Distribution Systems above 600 V.

455 Std 415-1986 (ANSI/IEEE)(revision of ANSI/IEEE 415-1976). Planning of Preoperational Testing Programs for Class 1E Power Systems for Nuclear Power Generating Stations, Guide for.

456 Std 449-1984 (ANSI/IEEE). Ferroresonant Voltage Regulators, Standard for.

457 Std 473-1985 (ANSI/IEEE). Electromagnetic Site Survey (10 kHz to 10 GHz), Recommended Practice for an.

458 Std 516-1986 (IEEE). Maintenance Methods on Energized Power-Lines, Trial-Use Guide for. See codes 45 and 431.

459 Std 572-1985 (ANSI/IEEE). Class 1E Connection Assemblies for Nuclear Power Generating Stations, Standard for Qualification of.

460 Std 618-1984 (ANSI/IEEE). Measurement of Luminance Signal-to-Noise Ratio in Video Magnetic-Tape Recording Systems, Standard for.

461 Std 637-1985 (ANSI/IEEE). Reclamation of Insulating Oil and Criteria for its Use, Guide for the.

462 Std 656-1985 (ANSI/IEEE). Measurement of Audible Noise from Overhead Transmission Lines, Standard for the.

463 Std 660-1986 (ANSI/IEEE). Semiconductor Memory Test Pattern Language, Standard for.

464 Std 692-1986 (IEEE). Security Systems for Nuclear Power Generating Stations, Standard Criteria for.

465 Std 693-1984 (ANSI/IEEE). Seismic Design of Substations, Recommended Practices for.

466 Std 695-1985 (IEEE). Microprocessor Universal Format for Object Modules, Trial-Use Standard for. [Ten definitions from ANSI/IEEE Std 694-1985, Microprocessor Assembly Language, Standard for, are also coded 466. The field of application for these definitions is 'microprocessor assembly language'.]

467 Std 741-1986 (ANSI/IEEE). Protection of Class 1E Power Systems and Equipment in Nuclear Power Generating Stations, Standard Criteria for the.

468 Std 743-1984 (ANSI/IEEE). Transmission Characteristics of Analog Voice Frequency Circuits, Standard Methods and Equipment for Measuring the.

469 Std 754-1985 (ANSI/IEEE). Binary Floating-Point Arithmetic, Standard for.

470 Std 755-1985 (Draft American National Standard/IEEE Trial-Use Standard). Extending High-Level Language Implementations for Microprocessors, Trial-Use Standard for.

471 Std 759-1984 (ANSI/IEEE). Semiconductor X-Ray Energy Spectrometers, Standard Test Procedures for.

472 Std 802.5-1985 (ANSI/IEEE)(ISO Draft Proposal 8802/5). Local Area Networks: Token Ring Access Method and Physical Layer Specifications.

473 Std 820-1984 (ANSI/IEEE). Standard Telephone Loop Performance Characteristics.

474 Std 824-1985 (IEEE)(revision of ANSI/IEEE C55.2-1974, code 86). Series Capacitors in Power Systems, Standard for.

475 Std 837-1984 (ANSI/IEEE). Qualifying Permanent Connections Used in Substation Grounding, Standard for.

476 Std 844-1985 (ANSI/IEEE). Electrical Impedance, Induction, and Skin Effect Heating of Pipelines and Vessels, Recommended Practice for.

477 Std 422-1986 (ANSI/IEEE)(revision of IEEE

Std 422-1977, code 35). Design and Installation of Cable Systems in Power Generating Stations, Guide for the.

478 Std 855-1985 (Draft American National Standard/IEEE Trial-Use Standard). Trial Use Standard Specifications for Microprocessor Operating Systems Interfaces.

479 Std 949-1985 (Draft American National Standard/IEEE Trial-Use Standard). Trial-Use Standard for Media-Independent Information Transfer.

480 Std 960-1986 (ANSI/IEEE). Standard FAST-BUS Modular High-Speed Data Acquisition and Control System.

481 Std 983-1986 (ANSI/IEEE). Software Quality Assurance Planning, Guide for.

482 Std 495-1986. (ANSI/IEEE). Testing Faulted Circuit Indicators, Guide for.

483 Std 987-1985 (ANSI/IEEE). Application of Composite Insulators, Guide for.

484 Std 1017-1985 (ANSI/IEEE). Field Testing Electric Submersible Pump Cable, Recommended Practice for.

485 Std C12.13-1985 (ANSI). Time-of-Use Registers for Electromechanical watthour meters.

486 Std C37.04E-1985 (ANSI/IEEE)(Supplement to ANSI/IEEE Std C37.04-1979, revised 1982). See codes 487 and 488.

487 Std C37.4D-1985 (ANSI/IEEE)(Supplement to ANSI/IEEE Std C37.4-1953, revised 1982). See codes 486 and 488.

488 Std C37.100B-1985 (ANSI/IEEE)(Supplement to ANSI/IEEE Std C37.100-1981). See codes 486 and 487.

489 Std C37.71-1984 (ANSI/IEEE). Three-Phase Manually Operated Subsurface Load-Interrupting switches for Alternating-Current Systems.

490 Std C62.31-1984 (ANSI/IEEE)(revision of ANSI/IEEE C62.31-1981, code 370)(reaffirmed 1987). Gas-Tube Surge-Protective Devices, Standard Test Specifications for.

491 Std 269-1983 (ANSI/IEEE)(revision of IEEE Std 269-1971, code 122). Measuring Transmission Performance of Telephone Sets, Standard Method for.

492 Std 382-1985 (ANSI/IEEE)(revision of IEEE

Std 382-1980, code 142). Qualification of Actuators for Power Operated Valve Assemblies with Safety-Related Functions for Nuclear Power Plants, Standard for.

493 Std 317-1983 (ANSI/IEEE)(revision of IEEE Std 317-1976, code 26). Electric Penetration Assemblies in Containment Structures for Nuclear Power Generating Stations, Standard for.

494 Std C2-1984 (ANSI). National Electrical Safety Code (NESC).

495 Std C63.12-1984 (ANSI). Procedures for Control of System Electromagnetic Compatibility, Recommended Practice on.

496 Std 928-1986 (ANSI/IEEE). Terrestrial Photovoltaic Power Systems, Recommended Practice for.

497 Std 180-1986 (ANSI/IEEE)(revision of IEEE Std 180-1962, code 247). Primary Ferroelectric Terms, Standard Definitions of.

498 Std N42.16-1986 (ANSI). Sealed Radioactive Check Sources Used in Liquid-Scintillation Counters, Specifications for.

499 Std N42.13-1986 (ANSI)(revision of ANSI N42.13-1978, code 396). 'Dose Calibrator' Ionization Chambers for the Assay of Radionuclides, Calibration and Usage of.

500 Std 640-1985 (IEEE). Power Station Noise Control, Guide for.

501 Std 676-1986 (ANSI/IEEE). Alarm Monitoring and Reporting Systems for Fossil-Fueled Power Generating Stations, Guide for.

502 Std 500-1984 P&V (ANSI/IEEE). Pumps and Drivers, Valve Actuators, and Valves, Standard Reliability Data for.

503 No entry.

504 Std 242-1986 (ANSI/IEEE). Protection and Coordination of Industrial and Commercial Power Systems, Recommended Practice for (IEEE Buff Book).

505 Std 378-1986 (ANSI/IEEE). Network Analyzers 100 kHz to 18 GHz, Standard on.

506 Std 1-1986 (ANSI/IEEE)(revision of IEEE Std 1-1969, code 320). Temperature Limits in the Rating of Electric Equipment and for the Evaluation of Electrical Insulation, Standard General Principles for.

507 Std 421.1-1986 (ANSI/IEEE)(revision of IEEE Std 421-1972, code 105).Excitation Systems for Synchronous Machines, Standard Definitions for.

508 Std 752-1986 (ANSI/IEEE). Functional Requirements for Methods and Equipment for Measuring the Performance of Tone Address Signalling Systems, Standard for.

509 Std 430-1986 (ANSI/IEEE)(revision of ANSI/IEEE Std 430-1976, code 36). Measurement of Radio Noise from Overhead Power Lines and Substations, Standard Procedures for the.

510 Std 141-1986 (ANSI/IEEE)(revision of IEEE Std 141-1976, code 153). Electric Power Distribution for Industrial Plants, Recommended Practice for. (IEEE Red Book).

511 Std 1012-1986 (ANSI/IEEE). Software Verification and Validation Plans, Standard for.

512 Std 446-1987 (ANSI/IEEE)(revision of ANSI/IEEE Std 446-1980, code 89). Emergency and Standby Power Systems for Industrial and Commercial Applications, Recommended Practice for. (IEEE Orange Book).

513 Std 628-1987 (ANSI/IEEE). Design, Installation, and Qualification of Raceway Systems for Class 1E Circuits for Nuclear Power Generating Stations, Standard Criteria for the.

514 Std 644-1987 (ANSI/IEEE)(revision of IEEE 644-1979, code 407). Measurement of Power Frequency Electric and Magnetic Fields.

515 Std 937-1987 (ANSI/IEEE). Installation and Maintenance of Lead-Acid Batteries for Photovoltaic (PV) Systems.

516 Std 858-1987 (ANSI/IEEE)(revision of IEEE Std 346-1973, Part 1, code 112). Standard Definitions in Power Operations Terminology.

517 Std 86-1987 (IEEE)(revision of ANSI/IEEE Std 86-1975, code 5). Recommended Practice: Definitions of Basic Per-Unit Quantities for AC Rotating Machines.

518 Std 990-1987 (IEEE). Ada as a Program Design Language, Recommended Practice for.

519 Std 1008-1987 (ANSI/IEEE). Software Unit Testing, Standard for.

520 Std 668-1987 (ANSI/IEEE). Electrical Heating Applications to Melting Furnaces and Forehearths in the Glass Industry, Recommended Practice for.

521 Std 115A-1984 (Draft American National Standard/IEEE Trial Use Standard)(supplement to ANSI/IEEE Std 115-1983). IEEE Trial Use Standard Procedures for Obtaining Synchronous Machine Parameters by Standstill Frequency Response Testing.

522 Std 122-1985 (IEEE)(revision of IEEE Std 122-1959). Functional and Performance Characteristics of Control Systems for Steam Turbine-Generator Units, Recommended Practice for.

523 Std 152-1953 (ANSI/IEEE)(ANSI reaffirmed 1976; IEEE reaffirmed 1971). Volume Measurements of Electrical Speech and Program Waves, Recommended Practice for.

524 Std 186-1948 (ANSI/IEEE). Standard Methods of Testing Amplitude-Modulation Broadcast Receivers.

525 Std 299-1969 (IEEE published for Trial Use). Measurement of Shielding Effectiveness of High-Performance Shielding Enclosures.

526 Std 172-1983 (ANSI/IEEE)(revision of IEEE Std 172-1971). Navigation Aid Terms, Standard Definitions of.

527 Std 316-1971 (IEEE). Direct Current Instrument Shunts, Standard Requirements for.

528 Std 325-1986 (ANSI/IEEE)(revision of ANSI/IEEE Std 325-1971, code 118). Germanium Gamma-Ray Detectors, Standard Test Procedures for.

529 Std 455-1985 (ANSI/IEEE)(revision of ANSI/IEEE Std 455-1976, code 39). Measuring Longitudinal Balance of Telephone Equipment Operating in the Voice Band, Standard Test Procedure for.

530 Std 457-1982 (IEEE). Terms for Nonlinear, Active, and Nonreciprocal Waveguide Components, Standard Definitions of.

531 Std 475-1983 (ANSI/IEEE). Field-Disturbance Sensors (rf Alarms), Standard Measurement Procedure for.

532 Std 515-1983 (ANSI/IEEE). Testing, Design, Installation, and Maintenance of Electrical Resistance Heat Tracing for Industrial Applications, Recommended Practice for the.

533 Std 519-1981 (ANSI/IEEE). Harmonic Control and Reactive Compensation of Static Power Converters, Guide for.

534 Std 600-1983 (Draft American National Standard/IEEE Trial-Use Standard). IEEE Trial-Use Standard Requirements for Organizations that Conduct Qualification Testing of Safety Systems Equipment for Use in Nuclear Power Generating Stations.

535 Std 627-1980 (ANSI/IEEE). Design Qualification of Safety Systems Equipment Used in Nuclear Power Generating Stations.

536 Std 690-1984 (ANSI/IEEE). Design and Installation of Cable Systems for Class 1E Circuits in Nuclear Power Generating Stations, Standard for the.

537 Std 83-1963 (IEEE)(reaffirmed 1982). Radial Power Factor Tests on Insulating Tapes in Paper-insulated Power Cable, Test Procedure for.

538 Std 696-1983 (ANSI/IEEE). Standard 696 Interface Devices.

539 Std 749-1983 (ANSI/IEEE). Standard Periodic Testing of Diesel-Generator Units Applied as Standby Power Supplies in Nuclear Power Generating Stations.

540 Std 753-1983 (ANSI/IEEE). Dial-Pulse (DP) Address Signaling Systems, Standard Functional Methods and Equipment for Measuring the Performance of.

541 Std 765-1983 (ANSI/IEEE). Preferred Power Supply for Nuclear Power Generating Stations, Standard for.

542 Std 796-1983 (ANSI/IEEE). Microcomputer System Bus, Standard.

543 Std 802.3-1985 (ANSI/IEEE)(ISO/DIS 8802/3). Local Area Networks: Carrier Sense Multiple Access with Collision Detection.

544 Std 803-1983 (ANSI/IEEE). Unique Identification in Power Plants and Related Facilities--Principles and Definitions, Recommended Practice for. See code 545.

545 Std 804-1983 (ANSI/IEEE). Unique Identification in Power Plants and Related Facilities, Recommended Practice for. See code 544.

546 Std 828-1983 (ANSI/IEEE). Software Configuration Management Plans, Standard for.

547 Std 847-1982 (ANSI/IEEE). Digital Terms Relating to Television, Standard Definitions of.

548 Std C12.4-1984 (ANSI)(revision of ANSI C12.4-1978). American National Standard for

Mechanical Demand Registers.

549 Std C12.7-1982 (ANSI). Watthour Meter Sockets, Requirements for.

550 Std C12.9-1982 (ANSI)(revised in 1987, see code 590). Test Switches for Transformer-Rated Meters.

551 Std C12.14-1982 (ANSI). Magnetic Tape Pulse Recorders for Electricity Meters.

552 Std C37.082-1982 (ANSI/IEEE). Measurement of Sound Pressure Levels of AC Power Circuit Breakers, Standard Methods for the.

553 Std C37.85-1972 (ANSI). X-Radiation Limits for AC High-Voltage Power Vacuum Interrupters Used in Power Switchgear, Safety Requirements for.

554 Std 7432-1982 (ANSI/IEEE/ANS). Programmable Digital Computer Systems in Safety Systems of Nuclear Power Generating Stations, Application Criteria for.

555 Std C57.94-1982 (ANSI/IEEE)(revision of ANSI C57.94-1956). Installation, Application, Operation, and Maintenance of Dry-Type General Purpose Distribution and Power Transformers, Recommended Practice for.

556 Std C62.32-1981 (ANSI/IEEE). Low-Voltage Air Gap Surge-Protective Devices (Excluding Valve and Expulsion Type Devices), Standard Test Specifications for.

557 Std C95.2-1982 (ANSI). Radio Frequency Radiation Hazard Warning Symbol.

558 Std N13.4-1971 (ANSI)(reaffirmed 1983).Specification of Portable X- or Gamma-Radiation Survey Instruments, American National Standard for the.

559 Std N42.18-1980 (ANSI)(redesignation of ANSI N13.10-1974)(reaffirmed 1985). On-Site Instrumentation for Continuously Monitoring Radioactivity in Effluents, Specification and Performance of.

560 Std C37.17-1972 (ANSI). Trip Devices for AC and General-Purpose dc Low-Voltage Power Circuit Breakers, American National Standard for.

561 Std 493-1980 (ANSI/IEEE). Design of Reliable Industrial and Commercial Power Systems, Recommended Practice for.

562 Std C57.21-1981 (ANSI/IEEE)(revision of

ANSI Std C57.21-1971). Shunt Reactors over 500 kVA, Requirements, Terminology and Test Code for.

564 Std 1084-1986 (ANSI/IEEE). Mathematics of Computing Terminology, Standard Glossary of.

565 Std 115A-1987 (IEEE)(Supplement to ANSI/ IEEE Std 115-1983, code 318)(revision of IEEE Std 115A, originally issued for trial use in 1984). Synchronous Machine Parameters by Standstill Frequency Response Testing, Standard Procedures for Obtaining.

566 Std 605-1987 (ANSI/IEEE). Design of Substaion Rigid-Bus Structures, Guide for.

567 Std 762-1987 (ANSI/IEEE)(revision of ANSI/IEEE Std 762, originally issued for trial use in 1980). Reporting Electric Generating Unit Reliability, Availability, and Productivity, Standard Definitions for Use in.

568 Std 1016-1987 (IEEE). Software Design Descriptions, IEEE Recommended Practice for.

569 Std 665-1987 (ANSI/IEEE). Generating Station Grounding, Guide for.

570 Std C37.1-1987 (ANSI/IEEE)(revision of ANSI C37.1-1979, code 403)(These definitions are to be found in the Complete 1987 Edition of C37 Standards). Standard Definitions, Specification, and Analysis of Systems Used for Supervisory Control, Data Acquisition, and Automatic Control.

571 Std 610.2-1987 (ANSI/IEEE). Computer Applications Terminology, Glossary of.

572 Std C37.20.2-1987 (ANSI/IEEE). Metal-Clad and Station-Type Cubicle Switchgear, Standard for.

573 Std C37.20.3 (ANSI/IEEE). Metal-Enclosed Interrupter Switchgear, Standard for.

574 Std C37.23-1987 (ANSI/IEEE)(These definitions are to be found in the Complete 1987 Edition of C37 standards). Metal-Enclosed Bus and Calculating Losses in Isolated-Phase Bus, Guide for.

575 Std C37.30-1971 (ANSI/IEEE)(reaffirmed 1977)(These definitions are to be found in the Complete 1987 Edition of C37 standards). High-Voltage Air Switches, Insulators, and Bus Supports, Definitions and Requirements for.

576 Std C37.82-1987 (ANSI/IEEE). Switchgear Assemblies for Class 1E Applications in Nuclear Power Generating Stations, Qualification of.

577 Std C37.100B-1986 (ANSI/IEEE)(supplement to ANSI/IEEE C37.100-1981, code 103)(These definitions are to be found in the Complete 1987 Edition of C37 standards).

578 Std C62.45-1987 (ANSI/IEEE). Surge Testing for Equipment Connected to Low-Voltage AC Power Circuits, Guide on.

579 Std C37.20.1-1987 (ANSI). Metal-Enclosed Low-Voltage Power Circuit-Breaker Switchgear, Standard for.

580 Std 545 (Unapproved IEEE Project Standard as of May 1988 to be issued for trial use).

581 Std 344-1987 (IEEE)(revision of ANSI/IEEE Std 344-1975, (reaffirmed 1980, code 28). Seismic Qualification of Class 1E Equipment for Nuclear Power Generating Stations, Recommended Practice for.

582 Std 934-1987 (ANSI/IEEE). Replacement Parts for Class 1E Equipment in Nuclear Power Generating Stations, Requirements for.

583 Std C62.11-1987 (ANSI/IEEE). Metal-Oxide Surge Arresters for AC Power Circuits, Standard for.

584 Std 936-1987 (ANSI/IEEE). Self-Commutated Converters, Guide for.

585 Std 799-1987 (ANSI/IEEE). Handling and Disposal of Transformer Grade Insulating Liquids Containing PCBs, Guide for.

586 Std 802.2-1985 (ANSI/IEEE)(ISO Draft International Standard 8802/2). Local Area Networks: Logical Link Control.

588 Std 854-1987 (ANSI/IEEE). Radix-Independent Floating-Point Arithmetic, Standard for.

589 Std 390-1987 (ANSI/IEEE)(revision of ANSI/IEEE Std 390-1975, code 32 and IEEE Std 391-1976, code 33). Pulse Transformers, Standard for.

590 Std C12.9-1987 (ANSI)(revision of ANSI C12.9-1982, code 550). Test Switches for Transformer-Rated Meters.

591 Std C62.92-1987 (ANSI/IEEE)(Revision of IEEE Std 143-1954). Neutral Grounding in Electrical Utility Systems Part I--Introduction, Guide for the Application of.

A

A-A: air to air

A-G: air to ground

A-57: AMF 100 kW(th) Reactor, dismantled and reconstructed as HOR in Delft, Netherlands

A: accelerating contactor or relay

A: ampere

A: asynchronous (type of line)

AA: access authorization

AA: Aluminum Association

AA: American Accounting Association

AA: arithmetic average

AA: armature accelerator

AA: Automobile Association

AA: dry-type self-cooled (transformer)

AAA: American Accounting Association

AAA: antiaircraft artillery

AAAS: American Association for the Advancement of Science

AAB: Aircraft Accident Board

AABM: Association of American Battery Manufacturers

AABP: aptitude assessment battery program

AACC: American Association for Contamination Control

AACC: American Association of Cereal Chemists

AACC: American Automatic Control Council

AACE: American Association of Cost Engineers

AACOM: Army Area Communications

AACOMS: Army Area Communications System

AACS: airborne astrographic camera system

AACSCEDR: Associate and Advisory Committee to the Special Committee on Electronic Data Retrieval

AACSR: aluminum alloy constructor steel reinforced

AADS: Area Air Defense System (SAM-D)

AAE: American Association of Engineers

AAEC: Australian Atomic Energy Commission

AAEE: American Academy of Environmental Engineers

AAES: American Association of Engineering Societies

AA/FA: dry-type self-cooled/forced-air-cooled (transformer)

AAFB: Andrews Air Force Base

AAFSS: Advanced Aerial Fire Support System (IHAS)

AAG: Aeronautical Assignment Group (see IRAC)

AAI: American Association of Immunologists

AAL: absolute assembly language

AAL: Arctic Aeromedical Laboratory (USAF; see AAML)

AALA: American Association for Laboratory Accreditation

AALC: Advanced Airborne Launch Centers (SAC)

AAM: air-to-air missile

AAMA: Architectural Aluminum Manufacturers Association

AAME: Automated Multi Media Exchange

AAMG: antiaircraft machine gun

AAMI: Association for the Advancement of Medical Instrumentation

AAML: Arctic Aeromedical Laboratory

AAO: antiair output

AAP: Apollo Applications Program

AAPG: American Association of Petroleum Geologists

AAPL: additional programming language

AAPM: American Association of Physicists in Medicine

AAPT: American Association of Physics Teachers

AAPT: Association of Asphalt Paving Technologists

AAR: Association of American Railroads

AARR: Argonne advanced research reactor (see also $A^2 R^2$)

AAS: advanced antenna system

AAS: American Astronautical Society

AAS: American Astronomical Society

AAS: arithmetic assignment statement

AAS: automated accounting system

AASHO: American Association of State Highway Officials

AATC: antiaircraft training center

AATC: automatic air traffic control (system)

AATCC: American Association of Textile Chemists and Colorists

AAUP: American Association of University Professors

AAVD: automatic alternative voice/data

AAVS: Aerospace Audiovisual Service

AAVSO: American Association of Variable Star Observers

AB: adapter booster

AB: Aeronautical Board

AB: afterburner

AB: air blast

AB: anchor bolt

AB: automatic blow down

ABA: American Bankers Association

ABA: American Bar Association

ABACUS: Air Battle Analysis Center Utility System

ABAR: advanced battery acquisition radar

ABAR: alternate battery acquisition radar

ABBE: Advisory Board on the Built Environment

ABC: approach by concept

ABC: automatic bandwidth control

ABC: automatic brightness control

ABCA: American-British-Canadian-Australian Committee on Unification of Engineering Standards

ABCB: air blast circuit breaker

ABCC: Atomic Bomb Casualty Commission (AEC)

ABCCTC: Advanced Base Combat Communication Training Center

ABET: Accreditation Board for Engineering and Technology, previously known as Engineers Council for Professional Development (ECPD)

ABFS: Auxiliary Building Filter System

ABGTS: Auxiliary Building Gas Treatment System

ABL: Atlas basic language

ABLE: activity balance line evaluation

ABLR: boiler, auxiliary

ABM: antiballistic missile

ABM: asynchronous balanced mode

ABM: automated batch mixing

ABMDA: (U.S. Army) Advanced Ballistic Missile Defense Agency
ABR: automatic band rate
ABRACE: Brazilian Association of Electronic Computers
ABRES: Advanced Ballistic Reentry Systems
ABS: American Bureau of Shipping
ABS: auxiliary building sump
ABS: acrytonitrile butadiene styrene
ABSW: air-break switch
ABT: air blast transformer
ABTF: airborne task force
ABV: absolute value
AC: Advisory Committee
AC: aerodynamic center
A/C: air conditioning
ac: alternating current
AC: analog computer
AC: automatic checkout
AC: automatic computer
AC: automatic control
ACA: adjacent channel attenuation
ACA: American Chain Association
ACA: American Communications Association
ACA: American Crystallographic Association
ACA: Automatic Communication Association
ACAC: automated direct analog computer
ACAD: air containment atmosphere dilution
ACAP: automatic circuit analysis program
ACARS: ARINC Communication Addressing Reporting System (see ARINC)
ACARS: ARINC Communication Addressing Reporting System (See RINC)
ACAS: Aircraft Collision Avoidance System
ACB: air circuit breaker
ACBS: Accrediting Commission for Business School
ACC: accumulator
ACC: aft cargo carrier
ACCAP: autocoder to cobol conversion and program (IBM)
ACCCE: Association of Consulting Chemists & Chemical Engineers
ACCEL: automated circuit card etching layout
ACCEL: automatic circuit card etching layout
ACCESS: aircraft communication electronic signaling system
ACCESS: automatic computer controlled electronic scanning system
ACCS: Association of Casualty and Surety Companies
ACD: automatic call distributor
ACD: conductivity, analyzer
ACDA: Arms Control and Disarmament Agency
ACE-S/C: acceptance checkout equipment—spacecraft
ACE: acceptance checkout equipment
ACE: advanced control experiment
ACE: American Council of Education
ACE: animated computer education
ACE: Association for Cooperation in Engineering
ACE: Association of Conservation Engineers

ACE: automatic checkout equipment
ACE: automatic computer evaluation
ACE: automatic computing engine
ACE: Aviation Construction Engineers
ACEAA: Advisory Committee on Electrical Appliances and Accessories
ACEC: American Consulting Engineers Council
ACEE: aircraft energy efficiency
ACELSCO: Associated Civil Engineers and Land Surveyors of Santa Clara County
ACESA: Arizona Council of Engineering & Scientific Associations
ACET: Advisory Committee for Electronics and Telecommunications
ACF: advanced communications function
ACF: alternate communications facility
ACF: area computing facilities
ACFG: automatic continuous function generation
ACGIH: American Conference of Government Industrial Hygienists
ACHM: chemical, analyzer
ACHP: Advisory Council on Historic Preservation
ACI: air combat intelligence
ACI: American Concrete Institute
ACI: automatic card identification
ACIA: asynchronous communications interface adapter
ACID: automatic classification and interpretation of data
ACIL: American Council of Independent Laboratories
ACIS: Association for Computing and Information Sciences
ACK: acknowledge
ACK: acknowledge character
ACL: application control language
ACL: Association for Computational Linguistics
ACL: Atlas commercial language
ACLI: American Council of Life Insurance
ACLP: above core load pad (fuel assembly)
ACLS: all-weather carrier landing system
ACLS: American Council of Learned Societies
ACM: asbestos-covered metal
ACME: advanced computer for medical research
ACME: Association of Consulting Management Engineers
ACM/GAMM: Association for Computing Machinery/German Association for Applied Mathematics and Mechanics
ACMI: air combat maneuvering instrumentation
ACMO: Afloat Communications Management Office (NSEC)
ACMR: air combat maneuvering range
ACNOCOMM: Assistant Chief of Naval Operations for Communications and Cryptology
ACOM: automatic coding system
ACOM: Awards Committee (IEEE)
ACOPP: abbreviated cobol preprocessor
ACOS: Advisory Committee on Safety
ACP: advanced computational processor (Sylvania)
ACP: aerospace computer program

ACP: Airline Control Program
ACPA: adaptive controlled phased array
ACPA: Association of Computer Programmers and Analysts
ACPDS: Advisory Committee on Personal Dosimetry Services (NSF)
ACPR: advanced core performance reactor
ACPR: annular core pulse reactor
ACR: air-field control radar
ACR: aircraft control room
ACR: antenna coupling regulator
ACR: automatic compression regulator
ACR: average crossing rate
ACRE: automatic checkout and readiness
ACRI: Air Conditioning & Refrigeration Institute
ACRL: Association of College and Research Libraries
ACRS: Advisory Committee on Reactor Standards
ACS: accumulator switch
ACS: advanced communications service
ACS: Advanced Computer Services (HIS)
ACS: Advanced Computer System (SIBM)
ACS: Alaska Communication System
ACS: alternating current synchronous
ACS: American Ceramic Society
ACS: American Chemical Society
ACS: Assembly Control System (IBM)
ACS: attitude command system
ACS: attitude control system
ACS: Australian Computer Society
ACS: automated communications set
ACS: automatic checkout system
ACS: auxiliary cooling system
ACS: auxiliary core storage
ACSCE: Army Chief of Staff for Communications Electronics (Washington, D.C.)
ACSF: attack carrier striking force
ACSM: American Congress on Surveying and Mapping
AcSoc: Acoustical Society of America
ACSP: Advisory Council on Scientific Policy (GB)
ACSR: aluminum cable steel reinforced
ACSR: aluminum conductor steel reinforced
ACSS: Air Combat and Surveillance System
ACSS: analog computer subsystem
ACST: access time
ACT: Air Control Team
ACT: algebraic compiler and translator
ACT: automatic code translation
ACTH: adrenocorticotropic hormone
ACTO: automatic computing transfer oscillator
ACTRAN: autocoder-to-cobol translating service
ACTS: acoustic control and telemetry system
ACTS: All-Channel Television Society
ACTS: automatic computer Telex services
ACU: address control unit
ACU: arithmetic and control unit
ACU: automatic calling unit
ACUI: automatic calling unit interface
ACUTE: Accountants Computer Users Technical Exchange

ACV: alarm check valve
AC&W: air control and warning (Air Force site)
AC&WS: air control and warning stations
AD: address/data
AD: advanced design
AD: ampere demand meter
A/D: analog to digital
ADA: airborne data automation
ADA: American Dental Association
ADA: automatic data acquisition
Ada: DOD computer programming language
ADAC: automatic direct analog computer
ADACC: automatic data acquisition and computer complex
ADAM: advanced data management
ADAM: advanced direct-landing Apollo mission
ADAM: associometrics data management system
ADAM: automatic distance and angle measurement
ADAM: automatic distance and angle measurement
ADAPS: automatic display and plotting system
ADAPSO: Association of Data Processing Service Organizations
ADAPT: adoption of automatically programmed tools
Adapt&Lrng Sys/CS: Adaptive & Learning Systems
ADAPTS: air-delivered antipollution transfer system
ADAR: advanced design array radar
ADAS: automatic data acquisition system
ADAT: automatic data accumulator and transfer
ADC: airborne digital computer
ADC: air data computer
ADC: analog-to-digital converter
ADC: automatic data collection
ADCC: Air Defense Control Center
ADCCP: advanced data communications control procedure
ADCOM: Administrative Committee (IEEE)
ADCSP: Advanced Defense Communications Satellite Program (DCSP)
ADDAR: automatic digital acquisition and recording
ADDAS: automatic digital data assembly system
ADDDS: automatic direct distance dialing system
ADDS: Applied Digital Data Systems
ADE: automated design engineering
ADE: automatic drafting equipment
ADEA: *Association Belge pour le Developpement Pacifique de l'Energie Atomique* (Belgian Association for the Peaceful Development of Atomic Energy
ADEN: density, analyzer
ADES: automatic digital encoding system
ADF: airborne direction finder
ADF: automatic direction finder
ADGE: Air Defense Ground Environment
ADI: altitude director indicator
ADI: American Documentation Institute
ADI: attitude director indicator
ADIDAC: Argonne National Laboratory Computer
ADINTELCEN: advanced intelligence center
ADIOS: automatic digital input-output system
ADIS: Air Defense Integrated System

ADIS: Association for the Development of Instructional Systems

ADIS: automatic data interchange system

ADIT: analog-digital integration translator

ADL: automatic data link

ADM: activity data method

ADM: asynchronous disconnected mode

ADMIRAL: automatic and dynamic monitor with immediate relocation, allocation, and loading

ADMIRE: automatic diagnostic maintenance information retrieval

ADMS: automatic digital message switching (autovon/autodin)

ADMSC: automatic digital message switching centers

ADOC: Air Defense Operation Center

ADONIS: automatic digital on-line instrumentation system

ADP: airborne data processor

ADP: air defense position

ADP: airport development program

ADP: ammonium dihydrogen phosphate

ADP: automatic data processing; automated data processing; administrative data processing

ADP: avalanche photodiode

ADPA: American Defense Preparedness Association

ADPC: automatic data processing center

ADPCM: Association for Data Processing and Computer Management

ADPE: auxiliary data processing equipment

ADPE/S: automatic data processing equipment/system

ADPESO: automatic data processing equipment selection office

ADPS: Automatic Data Processing System

ADPSO: Association of Data Processing Service Organizations

ADR: aircraft direction room

ADR: audit discrepancy report

ADRAC: automatic digital recording and control

ADRMP: automated dialer and recorded message player

ADRS: analog-to-digital data recording system

ADRT: analog data recorder transcriber

ADS-TP: Administrative Data System—Teleprocessing

ADS: accurately defined system

ADS: activity data sheet

ADS: air defense sector

ADS: Air Development Service (FAA)

ADS: Automatic Depressurization System

ADS: automatic door seal

ADSAF: Automatic Data System for the Army in the Field

ADSAS: air-derived separation assurance system

ADSC: automatic data service center

ADSC: automatic digital switching center

ADSCOM: advanced shipboard communication

ADSUP: automated data systems uniform practices

ADT: automatic data translator

ADTECH: advanced decoy technology

ADU: ammonium diuranate

ADU: automatic dialing unit

ADX: automatic data exchange

AE-6: (WBNS) water boiler neutron source, (North American Aviation) Santa Susana, California

AE: acoustic emission

AE: aeroelectronic

AE: airborne electronics

AE: air escape

AE: application engineer

A/E: Architect/Engineer

AE: arithmetic element

AE: arithmetic expression

AEA: Agriculture Engineers Association

AEA: American Electronics Association

AEA: American Engineering Association

AEA: Association of Engineers and Associates

AEA: Atomic Energy Authority

AEACII: Atomic Energy Advisory Committee on Industrial Information

AEAI: Association of Engineers and Architects in Israel

AEB: auxiliary equipment building

AEC: American Engineering Council

AEC: Atomic Energy Commission

AECB: Atomic Energy Control Board (Canada)

AECL: Atomic Energy of Canada Ltd.

AECM: Atomic Energy Commission Manual

AECPR: AEC Procurement Regulations

AECT: Association for Educational Communications and Technology

AED: Algol extended for design

AED: automated engineering design

AEDP: Association for Educational Data Processing

AEDS: Association for Educational Data Systems

AEDS: Atomic Energy Detection System (EMP, Vela)

AEE: airborne evaluation equipment

AEE: Association of Energy Engineers

AEE: Atomic Energy Establishment (GB)

AEEC: Airlines Electronic Engineering Committee, the ARINC committee that originated ATLAS (Annapolis, Md.)

Ae.Eng.: aeronautical engineer

AEEP: Association of Environmental Engineering Professors

AEF: aviation engineer force

AEG: active element group

AEG: Association of Engineering Geologists

AEI: *Associazio Elettrotecnica Italiana*

AEI: average efficiency index

AEIC: Association of Edison Illuminating Companies

AEL: Aeronautical Engine Laboratory

AEM: Association of Electronic Manufacturers

AEMS: American Engineering Model Society

AEN: *Affaiblissement equivalent pour la nettete* (equivalent articulation loss)

AEO: Acoustoelectric oscillators

AEO: air engineer officer

AEOG: air ejection off gas

AEP: American Electric Power

AEPG: Army Electronic Proving Ground (Ft. Huachuca, AZ)

AEPSC: Atomic Energy Plant Safety Committee

AEPSC: Atomic Energy Plant Safety Committee

AERA: American Educational Research Association

AERDL: Army Electronics Research and Development Laboratory

AERE: Atomic Energy Research Establishment (GB)

AERIS: Airways Environmental Radar Information System

AERNO: aeronautical equipment reference number (military)

AEROF: aerological officer

AEROSAT: Aeronautical Satellite System

AES-1: power reactor at Obninsk USSR

AES: Abrasive Engineering Society

AES: Aerospace and Electronic Systems (IEEE Society)

AES: Airways Engineering Society

AES: American Electrochemical Society

AES: American Electroplaters Society

AES: Apollo extension system

AES: Audio Engineering Society

AES: auger electron spectroscopy

AESOP: an evolutionary system for on-line processing

AESS: Aerospace and Electronics Systems

AETR: advanced engineering test reactor, Idaho Falls, Idaho

AETR: advanced epithermal thorium reactor built by Atomics International (AI) for Southwest Atomic Energy Associates (SAEA)

AETR: advanced epithermal thorium reactor built by Atomics International (AI) for Southwest Atomic Energy Associates (SAEA)

AETRA: cross section reactor physics computer code

AEW: airborne early warning

AEW&C: airborne early warning and control

AEWTU: airborne early warning training unit

AF: Air Force Department

AF: Atomic Forum

AF: audio frequency

AF: automatic following

AF: auxiliary feedwater

AF: availability factor

AFA: dry-type forced-air-cooled (transformer)

AFACD: Air Force Director of Data Automation

AFADS: Advanced Forward Air Defense System (SAM-D)

AFAL: Air Force Avionics Laboratory (WPAFB, Dayton, OH)

AFAPL: Air Force Aero Propulsion Laboratory

AFB: antifriction bearing

AFC: area frequency coordinator

AFC: Automatic frequency control

AFCAL: *Association Francaise de Calcul*

AFCCE: Association of Federal Communication Consulting Engineers

AFCE: automatic flight control equipment

AFCEA: Armed Forces Communications and Electronic Association

AFCRL: Air Force Cambridge Research Laboratory (Bedford, MA)

AFCS: adaptive flight control system

AFCS: automatic flight control systems

AFD: axial flux density (difference)

AFDA: axial flux difference alarm

AFDDA: Air Force Director of Data Automation

AFELIS: Air Force Engineering and Logistics Information System

AFGL: Air Force Geophysical Laboratory

AFIP: Armed Forces Institute of Pathology

AFIPS: American Federation of Information Processing Societies

AFIT: Air Force Institute of Technology

AFM: antifriction metal

AFM: automatic flight management

AFMFIC: Associated Factory Mutual Fire Insurance Companies

AFMR: antiferromagnetic resonance

AFMTC: Air Force Missile Test Center (Obsolete)

AFNETF: Air Force Nuclear Engineering Test Facility

AFNOR: *Association Française des Normes*

AFO: axial flux offset

AFOSR: Air Force Office of Scientific Research

AFPA: automatic flow process analysis

AF/PC: automatic frequency/phase controlled (loop)

AFP/SME: Association of Finishing Processes

AFPT: auxiliary feed pump turbine

AFRAL: *Association Française de Reéglage Automatique*

AFRDR: Air Force Director of Reconnaissance and Electronic Warfare

AFRPL: Air Force Rocket Propulsion Laboratory

AFRPL: Armed Forces Radiobiology Research Institute

AFRS: Armed Forces Radio Service

AFRTS: Armed Forces Radio and Television Service

AFS: American Fisheries Society

AFS: American Foundrymen's Society

AFS: audio frequency shift

AFSC: Air Force Systems Command

AFSCME: American Federation of State, County, and Municipal Employees

AFSR: Argonne fast source reactor, NRTS, 1 kW(th)

AFST: auxiliary feedwater storage tank

AFTEC: Air Force Test and Evaluation Center

AFTI: Advanced Fighter Technology Integration

AFTN: aeronautical fixed telecommunications network

AFTRCC: Aerospace Flight Test Radio Coordinating Council

AFWL: Air Force Weapons Laboratory (Kirkland AFB,NM)

AG: air to ground

AG: arbitration grant

AG: armor grating

AG: arresting gear (and barriers)

AG: available generation

AGA: American Gas Association

AGACS: automatic ground-to-air communications systems

AGARD: Advisory Group for Aerospace Research and Development (NATO)

AGC: automatic gain control

AGCA: Association General Contractors of America

AGCA: automatic ground-controlled approach

AGCL: automatic ground-controlled landing

AGDIC: astro guidance digital computer

AGDS: American Gage Design Standard

AGE: aerospace ground (support) equipment

AGED: Advisory Group on Electron Devices

AGEP: Advisory Group on Electronic Parts

AGET: Advisory Group on Electron Tubes

AGI: American Geological Institute

AGIL: airborne general illumination light (Shed Light)

AGILE: autonetics general information learning equipment

AGM: auxiliary general missile

AGMA: American Gear Manufacturers Association

AGN-201-100: U.S.Naval Postgraduate School AGN-201 Reactor, Monterey, CA (Aerojet-General Nucleonics, builder and designer)

AGN: teaching reactor built by Aero-General Nucleonics

AGR: advanced gas-cooled reactor

AGREE: Advisory Group on the Reliability of Electronic Equipment

AGS: alternating gradient synchrotron

AGS: Annulus Gas System

AGSP: Atlas general survey program

AGU: American Geophysical Union

AGZ: actual ground zero

A/H: air over hydraulic

AH: ampere hour

AH: available hours

AHA: American Hospital Association

AHAM: American Home Appliance Manufacturers Association

AHAM: Association of Home Appliance Manufacturers

AHAMS: Advanced Heavy Assault Missile System

AHEA: American Home Economics Association

AHEM: Association of Hydraulic Equipment Manufacturers

AHFR-1: Argonne high flux reactor, ANL

AHM: ampere-hour meter

AHR: acceptable hazard rate

AHR: aqueous homogeneous reactor

AHRS: Attitude and Heading Reference System

AHS: American Helicopter Society, Inc.

AHSR: air-height surveillance radar

AI: Acquisition Institute

AI: airborne intercept

AI: arbitration request inhibit

AI: artificial intelligence

AI: Atomics International

AI: automatic input

AI: Automation Institute

AIA: Aerospace Industries Association

AIA: American Institute of Architects

AIA: American Insurance Association

AIA: American Inventors Association

AIAA: American Institute of Aeronautics and Astronautics

AIAAJ: AIAA Journal

AIB: American Institute of Banking

AIBS: American Institute of Biological Sciences

AIC: American Institute of Chemists

AIC: Automatic Initiation Circuit

AICA: *Associazione Italiana il Calcolo Automatico*

AICAE: American Indian Council of Architects and Engineers

AICBM: anti-intercontinental ballistic missile

AICE: American Institute of Consulting Engineers

AIChE: American Institute of Chemical Engineers

AICP: American Institute of Certified Planners

AICPA: American Institute of Certified Public Accountants

AID: Agency for International Development

AID: algebraic interpretive dialog

AIDC: Aero Industry Development Center

AIDE: automated integrated design engineering

AIDS: Aircraft Integrated Data System (AFSC)

AIDS: automated integrated debugging system

AIDSCOM: Army Information Data Systems Command

AIEA: *Agence internationale de l'Energie Atomique*

AIEE: American Institute of Electrical Engineers (obsolete, see IEEE)

AIENDF: Atomics International Evaluation Nuclear Data File

AIF: Atomic Industrial Forum

AIGS: auxiliary inerting gas subsystem

AIHA: American Industrial Hygiene Association

AIHX: auxiliary intermediate heat exchanger

AIIE: American Institute of Industrial Engineers

AILAS: automatic instrument landing approach system

AILS: automatic instrument landing system

AIM: Administrative and Information Program (NASA)

AIM: air-isolated monolithic

AIM: air intercept missile

AIM: American Institute of Management

AIMACO: air materiel computer

AIME: American Institute of Mechanical Engineers

AIMES: automated inventory management evaluation system

AIMME: American Institute of Mining and Metallurgical Engineers

AIMMPE: American Institute of Mining, Metallurgical, and Petroleum Engineers

AIMPR: Mechanical Engineers Association of Puerto Rico

AIMS: automated information and management system

AINSE: Australian Institute of Nuclear Science and Engineering

AIP: American Institute of Physics

AIP: American Institute of Planners

AIP: approval in principle

AIPE: American Institute of Plant Engineers

AIPE: American Institute of Refrigeration

AIPG: American Association of Professional Geologists

AIPT: Association for International Practical Training, Inc.

AIR: aerospace information report

AIR: air intercept rocket

AIR: American Institute for Research

AIR: American Institute of Refrigeration

AIR: recorder, indicating, analyzer

AIRCON: automated information and reservations computer-oriented network

AIRE: Australian Institution of Radio Engineers

AIREK-11: Group and feedback kinetics reactor computer code

AIRMICS: Army Institute for Research in Management Information and Computer Science

AIRS: advanced inertial reference sphere

AIRS: automatic image retrieval system

AIRS: automatic information retrieval system

AIRSS: ABRES Instrument Range Safety System

AIS: American Interplanetary Society (obsolete, see ARS)

AIS: avionics intermediate shop

AISC: American Institute of Steel Construction

AISC: Association of Independent Software Companies

AISCP: Association of Independent Software Companies

AISE: Association of Iron and Steel Engineers

AISES: American Indian Science & Engineering Society

AISI: American Iron and Steel Institute

AIST: automatic informational station

AIT: American Institute of Technology

AITC: American Institute of Timber Construction

AITS: Action Item Tracking System

AITS: switch, transmitting, indicating, analyzer

AJ: antijamming

AJ: assembly jig

AJD: antijam display

AK: address acknowledge

AK: Alaska

AKK: *Atomkraftkonsortiet Krangede Ad & Co.,* nuclear power study group, Stockholm, Sweden

AKS: Nuclear power plant planning association, Baden-Wuerttemberg, Germany

AKV: *AKB Atomkraftwerk,* Simpevarp, Sweden (proposed power reactor, 170 MW(th))

AKWIC: author and keyword in context

AL: Action (indicator) Level; those radiological doses and concentrations at which actions should be taken

AL: air lock

AL: Alabama

AL: analytical limits

AL: arbitration level

AL: assembly language

ALA: American Library Association

ALAP: as low as practical

ALARA: as low as reasonably achievable

ALARM: automatic light aircraft readiness monitor

ALAS: asynchronous look-ahead simulator

ALBM: air-launched ballistic missile

ALC: automatic leveling control

ALC: automatic load control

ALCC: airborne launch control center (PACCS)

ALCM: air launched cruise missile

ALCOM: algebraic compiler

ALCOR: ARPA Lincoln Coherent Observable Radar (Defender)

ALCP: Area Local Control Panel

ALD: analog line driver

ALDP: automatic language data processing

ALECTO: homogeneous research reactor zero power, CEA, Saclay, France

ALERT: automated linguistic extraction and retrieval technique

ALFTRAN: algol to fortran translator

AlGaAs: aluminum gallium arsenide

ALGM: air-launched guided missile

ALGOL: algorithmic language

ALI: automatic logic implementation

ALICE: fusion research device at Lawrence Radiation Laboratory

ALIP: annular linear induction pump

ALIS: advanced life information system

ALIZE-II: tank reactor at Saclay, France

ALKEM: *Alphachemie und Metallurgie, Gmbh.,* plutonium fabricators, West Germany

ALMS: analytic language manipulation system

ALMS: Auxiliary Liquid Metal System

ALOR: advanced lunar orbital rendezvous

ALOTS: airborne lightweight optical tracking system

ALP: assembly language preprocessor

ALP: assembly language program

ALP: automated language processing

ALP: automated learning process

ALPHA: fusion device, Leningrad, USSR

ALPHGR: average linear planar heat generation rate

ALPR: Argonne Low Power Reactor, redesignated SL-1, NRTS, Idaho power reactor, 3MW(th) (now dismantled)

ALPS: accidental launch protection system

ALPS: Advanced Linear Programming System (HIS)

ALPS: associated logic parallel system

ALPS: automated library processing service

ALRI: airborne long-range intercept (495L)

ALRR: Ames Laboratory Research Reactor, Ames, Iowa 5MW(th)

ALS: advanced logistics system

ALS: approach-light system

ALSEP: Apollo lunar surface experiment package

ALSI: aluminum silicon alloy used in slug canning process at Hanford

ALT: alternator

ALTAC: algebraic transistorized automatic computer translator

AltaFreq: *Alta Frequenza*

ALTAIR: ARPA long-range tracking and instrumentation radar

ALTN: alternative

ALTRAN: algebraic translator

ALU: arithmetic and logic unit

AM-1: First USSR power reactor, 5MW(e), Obninsk, USSR

AM: ammeter

A/m: ampere per meter (unit of magnetic field strength)

AM: amplitude modulation

AM: auxiliary memory

AM: *ante meridian* (before noon)

AMA: Acoustical Materials Association

AMA: American Management Association

AMA: American Manufacturers Association

AMA: American Marketing Association

AMA: American Medical Association

AMA: automatic memory allocation

AMA: automatic message accounting

AMA: Automobile Manufacturers Association

AMACUS: automated microfilm aperture card updating system

AMAD: activity median aerodynamic diameter

AMARV: advanced maneuvering reentry vehicle

AMAS: advanced midcourse active system

AMB-1,2-1: power reactor, BWR, superheated, 300 MW(th), 100MW(e), Beloyarsk, USSR

AMB-1,2-2: proposed power reactor, BWR with superheat, 200MW(e), Beloyarsk, USSR

AMB: asbestos millboard

AMBIT: algebraic manipulation by identity translation

AMC: American Mining Congress

AMC: Army Missile Command

AMC: Association of Management Consultants

AMC: automatic message counting

AMC: automatic mixture control

AMCA: Air Moving and Conditioning Association

AMCBMC: Air Materiel Command Ballistic Missile Center

AMCEC: Allied Military Communications Electronics Committee

AMCEE: Association for Media Based Continuing Education for Engineers, Inc.

Am. Ceram. Soc.: The American Ceramic Society, Inc.

AMCF: Alkali Metal Cleaning Facility

AMCS: airborne missile control system

AME: amplitude modulation equivalent

AMEL: aircraft multiengine land (pilot rating)

AMFIS: American Microfilm Information Society

AMFIS: Automatic Microfilm Information System

AMHS: American Material Handling Society

AMI: Alternate mark inversion

AMICOM: Army Missile Command

AMM: antimissile missile

AMNIP: adaptive man-machine nonarithmetical information processing

AMO: alternant molecular orbit

AMOA: atmospheric monitor oxygen analyzer

AMOS: acoustic meteorological observing system

AMP: associative memory processor

A.M.P.E.R.E.: *Atomes et Molecules par Etudes Radio-Electriques, c/o M.C.Bene, Institut de Physique,* Geneva, Switzerland

AMPLG: amplidyne generator

AMPLMG: amplidyne motor generator

AMPS: automatic message processing system

AMR: Atlantic Missile Range (Cocoa Beach, FL)

AMR: automated management reports

AMR: automatic message registering

AMRAAM: advanced medium range air-to-air missile

AMRAD: Arda measurements radar (Defender)

AMRV: atmospheric maneuvering reentry vehicle

AMS: Acoustic Monitor System

AMS: Administrative Management Society

AMS: advanced memory systems

AMS: aeronautical material specification

AMS: Agricultural Marketing Service

AMS: American Mathematical Society

AMS: American Meteorological Society

AMS: Atmospheric Monitor System

AMST: Advanced Medium STOL Transport

AMST: Association of Maximum Service Telecasters

AMT: accelerated mission testing

AMTB: antimotor torpedo boat

AMTCL: Association for Machine Translation and Computational Linguistics

AMTI: airborne moving target indicator

AMTI: area moving target indicator

AMTI: automatic moving target indicator

AMTRAN: automatic mathematical translator

AMU: astronaut maneuvering unit

AMU: atomic mass unit

A/N: alpha numeric

A/N: as needed

ANACOM: analog computer

ANATRAN: analog translator

ANC: anchor

ANC: ancillary logic

ANCS: American Numerical Control Society

ANDIP: American National Dictionary for Information Processing

ANE: Aeronautical and Navigational Electronics

ANEL: electron ring accelerator (Italian)

ANF: antinuclear factor

ANF: Atlantic Nuclear Force (Polaris/Poseidon)

ANI: automatic number identification

ANIE: *Association Nazionale Industrie Elettro Techniche*

ANIM: Association of Nuclear Instrument Manufacturers

ANL: Argonne National Laboratory (AEC)

ANL: automatic noise limiter

ANLOR: angle order

ANMC: American National Metric Council

Ann Radioelect: *Annales de Radioelectricite*

Ann Telecom: *Annales des Telecommunications*

ANO: alphanumeric output

ANO: anode

ANOVA: analysis of variance

ANP: aircraft nuclear propulsion

ANPA: American Newspaper Publishing Association

ANPO: Aircraft Nuclear Propulsion Office

ANPP: Army Nuclear Power Program

ANPT: aeronautical national taper pipe thread

ANR: Association of Neutron Radiographers

ANRAC: Aids Navigation Radio Control

ANS: American Nuclear Society

ANSCAT: scattering angle reactor physics computer code

ANSI: American National Standards Institute

ANTC: antichaff circuit

ANTIVOX: voice-actuated transmitter keyer inhibitor

ANTS: airborne night television system (TARS/SEA Niteops)

ANVIS: Aviator's Night Vision Imaging System

ANZAAS: Australian and New Zealand Association for the Advancement of Science

AO: abnormal occurrence

AO: access opening

AOA: American Ordinance Association

AOC: automatic overload control

AOCI: Airport Operators Council International

AOCS: Alpha Omega computer system

AOCS: American Oil Chemists Society

AOG: augmented off gas

AOI: Advance Ordering Information

AOI: and-or invert

AOI: Arab Organization for Industrialization

AOL: application-oriented language

AOML: Atlantic Oceanographic and Meteorological Laboratories (NOAA)

AOPA: Aircraft Owners and Pilots Association

AOQL: average outgoing quality limit

AOSO: Advanced Orbiting Solar Observatory

AOSP: Automatic Operating and Scheduling Program

AP: access panel

AP: access permit; access permittee

AP: after peak

AP: after perpendicular

AP: anomalous propagation

AP: Antennas and Propagation (IEEE Society)

AP: application program

AP: applications processor

AP: applied physics

AP: argument programming

AP: Associated Press

AP: associative processor

APA: Administrative Procedures Act

APA: American Planning Association

APA: American Psychological Association

APA: axial pressure angle

APADS: automatic programmer and data system

APAR: automatic programming and recording

APC: aeronautical planning chart

APC: American Power Conference

APC: argon purge cart

APC: automatic phase control

APC: autoplot controller

APCA: Air Pollution Control Association

APCEF: Advanced Power Conversion Experimental Facility, U.S.Army gas-cooled reactor systems development, Ft. Belvoir, VA

APCHE: automatic programming checkout equipment

APD: Air particulate detector

APD: amplitude probability distribution

APD: avalanche photodiode

APDA: Atomic Power Development Associates

APDMS: Axial Power Distribution Monitoring Systems

APEC: all-purpose electronic computer

APEL: Aeronautical Photographic Experimental Laboratory

APF: atomic packing factor

APFCS: automatic power-factor-control systems

APGC: Air Proving Ground Center

APH: Ph analyzer

APHA: American Public Health Association

APHI: Association of Public Health Inspectors

APHIS: Animal and Plant Health Inspection Service

APHT: air preheater

API: absolute position indicator

API: air position indicator

API: American Petroleum Institute

APIC: Aerospace Products Information Center

APIC: Apollo Parts Information Center

APICS: American Production and Inventory Control Society

APIL: axial power imbalance limit

APIN: Association for Programmed Instruction in the Netherlands

APL: Applied Physics Laboratory (Silver Springs, MD)

APL: Applied Physics Laboratory (University of Chicago)

APL: a programming language

APL: Association of Programmed Learning

APL: associative programming language

APL: authorized possession limits

APL: average picture level

APLE: Association of Public Lighting Engineers

APLHGA: average planar heat generation rate

APM: air particulate monitor

APM: analog panel meter

APMI: American Powder Metallurgy Institute

APMI: area precipitation measurement indicator

APMS: Advanced Power Management System

APNIC: Automatic Programming National Information Centre (GB)

APP: associative parallel processor

APP: auxiliary power plant

APPA: American Public Power Association

APPECS: adaptive pattern perceiving electronic computer system

APPI: advanced planning procurement information

APR: advanced production release (LLL)

APR: airborne profile recorder

APR: Armour Research Foundation Reactor, Chicago, IL, 11kW(th)

APR: automatic pressure relief

APRD: atmosphere particulate radioactivity detector

APRI: absolute rod position indicator
APRM: automatic position reference monitor
APRM: average power range monitor
APRS: automatic position reference system
APS-1: power reactor, 30MW(th), Obninsk, USSR
APS: alphanumeric photocomposer system
APS: American Physical Society
APS: American Physics Society
APS: American Physiological Society
APS: American Phytopathological Society
APS: Array Processor Software
APS: assembly programming system
APS: automatic patching system
APS: auxiliary power system
APS: auxiliary program storage
APSR: axial power shaping rod
APT: augmented programming training
APT: automatically programmed tools
APT: automatic picture taking
APT: automatic picture transmission
APT: automation planning and technology
APTA: American Public Transit Association
APTR: advanced pressure tube reactor, heterogeneous, enriched uranium, water moderated, USA
APU: auxiliary power unit
APW: augmented plane wave
APWA: American Public Works Administration
APWA: American Public Works Association
AQ: any quality
AQL: acceptable quality level
AQT: acceptable quality test
AQUILLON: heavy water research reactor, CEA, Saclay, France
AR: acid resisting
AR: anti-reflective
AR: arbitration request
AR: Arkansas
AR: assembly and repair
AR: aviation radionavigation, land (FCC services)
AR: avionic requirements
AR: axial ratio
ARA: Amateur Rocket Association
ARAL: automatic record analysis language
ARBOR: Argonne boiling water reactor, NRTS
ARBS: Angle Rate Bombing System
ARBUS: power reactor, organic-cooled and moderated, 5MW(th), 0.75MW(e), New Melekess, USSR
ARC: Aeronautical Research Council
ARC: amplitude and rise time compensation
ARC: Argonne reactor computation
ARC: automatic relay calculator
ARC: average response computer
ARCA: Airconditioning and Refrigeration Contractors of America
arccos: inverse cosine
arccot: inverse cotangent
arccse: inverse cosecant
ARCG: American Research Committee on Grounding
ARCS: Air Resupply and Communication Service
arcsec: inverse secant

arcsin: inverse sine
arctan: inverse tangent
ARDA: Atomic Research and Development Authority, State of New York
ARDM: asynchronous time-division multiplexing
ARE: automated responsive environment
AREA: American Railway Engineering Association
AREA: Army reactor area
ARED: aperture relay experiment definition
ARFA: Allied Radio Frequency Agency
Argonaut: Argonne nuclear assembly for university training, ANL (CP-11)
ARGUS: automatic routine generating and updating system
ARI: Air-Conditioning and Refrigeration Institute
ARIA: advanced range instrumentation aircraft
ARIA: Apollo Range Instrumentation Ships (Satellite Tracking Network, NASA, MSFN)
ARIA: Apollo Range Instrumented Aircraft
ARINC: Aeronautical Radio Incorporated, an organization with voluntary standardization among airlines and airframe and avionics manufacturers.
ARIP: automatic rocket impact predictor
ARIS: Advanced Range Instrumentation Ship (Satellite Tracking Network, NASA, MSFN)
ARL: acceptable reliability level
ARL: Association of Research Libraries
ARM: antiradiation missile
ARM: area radiation monitor
ARM: atmosphere radiation monitor
ARMA: American Records Management Association
ARMA: autoregressive moving average
ARMF I & II: Advanced Reactivity Measurement Facility, (NRTS)
ARMMS: automated reliability and maintainability measurement system
ARMS: Aerial Radiological Measurements and Surveys
ARMS: Amateur Radio Mobile Society
ARO: air radio officer
ARO: all rods out
AROD: airborne range and orbit determination
AROU: aviation repair and overhaul unit
ARP: Advanced Reentry Program (ABRES)
ARP: Aeronautical Recommended Practice (SAE)
ARP: airborne radar platform
ARPA: Advanced Research Projects Agency
ARPI: analog rod position indicator
ARQ: automatic repeat request
ARR: antirepeat relay
ARRL: Aeronautical Radio and Radar Laboratory
ARRL: American Radio Relay League
ARRL: America Radio Relay League
ARRS: Aerospace Rescue and Recovery Service
ARS: Advanced Record System
ARS: Agricultural Research Service
ARS: American Radium Society
ARS: American Rocket Society (merged with Institute of Aerospace Sciences)
ARS: amplified response spectrum

ARSP: Aerospace Research Support Program

ARSR: air route surveillance radar

ART: Advanced Reactor Technology

ART: airborne radiation thermometer

ART: automatic reporting telephone

ARTCC: Air-Route Traffic Control Center

ARTE: Admiralty Reactor Test Establishment (GB)

ARTIC: associometrics remote terminal inquiry control system

ARTRAC: advanced range testing, reporting, and control

ARTRAC: advanced real-time range control

ARTRON: artificial neutron

ARTS: advanced radar terminal system

ARTS: advanced radar traffic-control system

ARTU: automatic range tracking unit

ARU: audio response unit

AS: address sync

AS: aerospace

AS: ammeter switch

AS: antisubmarine

ASA: Acoustical Society of America

ASA: American Society of Agronomy

ASA: American Society of Appraisers

ASA: American Standards Association (later USASI; now ANSI)

ASA: American Statistical Association

ASA: Army Security Agency (Arlington, VA)

ASA: Atomic Security Agency

ASAC: Asian Standards Advisory Committee

ASADA: New York State Atomic and Space Development Authority

ASAE: American Society of Agricultural Engineers

ASAE: American Society of Association Executives

ASALM: advanced strategic air launched missile

ASAP: antisubmarine attack plotter

ASAP: Army Scientific Advisory Panel

ASAP: as soon as possible

ASARC: Army Systems Acquisition Review Council

ASAT: antisatellite (weapons)

ASBC: American Society of Biological Chemists, Inc.

ASBC: American Standard Building Code

ASBD: advanced sea-based deterrent

ASBE: American Society of Body Engineers

ASBEL: American Society of Naval Engineers, Inc.

ASBO: Association of School Business Officials

ASBPA: American Shore and Beach Preservation Association

ASC: advanced scientific computer

ASC: Aeronautical Systems Center

ASC: American Society for Cybernetics

ASC: associative structure computer

ASC: Atlantic Systems Conference

ASC: automatic sensitivity control

ASC: automatic system controller

ASC: auxiliary switch (breaker) normally closed

ASCA: automatic science citation alerting

ASCAP: American Society of Composers, Authors, and Publishers

ASCATS: Apollo Simulation Checkout and Training System

ASCC: Air Standardization Coordinating Committee

ASCC: automatic sequence controlled calculator

ASCE: American Society of Civil Engineers

ASCET: American Society of Certified Engineering Technicians

ASCII: American National Standard Code for Information Interchange

ASCO: automatic sustainer cutoff

ASCS: Agricultural Stabilization and Conservation Service

ASCS: area surveillance control system

ASD: Aeronautical Systems Division, USAF (Dayton, OH)

ASDE: airport surface detection equipment

ASDIC: Antisubmarine Detection and Identification Committee

ASDL: automated ship data library

ASDSRS: automatic spectrum display and signal recognition system

ASE: airborne search equipment

ASE: Albany Society of Engineers

ASE: Association of Scientists and Engineers of Navsea & Navair Systems Command

ASE: automatic support equipment

ASEAN: Association of Southeast Asian Nations

ASEC: American Standard Elevator Code

ASEE: American Society for Engineering Education—Continuing Professional Development Division

ASEE: American Society for Environmental Education

ASEE: Association of Supervising Electrical Engineers

ASEL: Aircraft Single Engine Land (pilot rating)

ASEM: American Society for Engineering Management

ASET: American Society for Engineering Technology

ASF: auxiliary supporting feature

ASFDO: antisubmarine fixed defenses officer

ASFE: Association of Soil and Foundation Engineers

ASFIR: active swept frequency interferometer radar

ASFTS: airborne systems functional test stand

ASG: Aeronautical Standards Group

ASGE: American Society of Gas Engineers

ASGLS: advanced space ground link subsystem

ASHE: American Society for Hospital Engineering

ASHRAE: American Society of Heating, Refrigeration and Air Conditioning Engineers

ASHVE: American Society of Heating and Ventilating Engineers

ASI: American Society of Inventors

ASI: American Standards Institute

ASI: axial shape index

ASIDP: American Society of Information and Data Processing

ASII: American Science Information Institute

ASIS: abort sensing and instrumentation system

ASIS: American Society for Information Science

AS/ISES: American Section, International Solar Energy Society

ASIST: advanced scientific instruments symbolic translator

ASIT: adaptive surface interface terminal

ASK: amplitude shift keying

ASKA: automatic system for kinematic analysis

ASKS: automatic station keeping system

ASL: Association for Symbolic Logic

ASLAP: Atomic Safety and Licensing Appeal Panel

ASLB: Atomic Safety Licensing Board

ASLBM: air-to-ship launched ballistic missile (Defender)

ASLBP: Atomic Safety and Licensing Board Panel

ASLE: American Society of Lubrication Engineers

ASLEEP: automated scanning low-energy electron probe

ASLIB: Association of Special Libraries and Information Bureau

ASLO: American Society of Limnology and Oceanography

ASLT: advanced solid logic technology

ASM: advanced surface-to-air missile

ASM: air-to-surface missile

ASM: American Society for Metals

ASM: American Society for Microbiology

ASM: Association for Systems Management

ASM: asynchronous state machine

ASMB: Acoustical Standards Management Board/ANSI

ASME: American Society of Mechanical Engineers

ASMM: American Society of Machine Manufacturers

ASMS: advanced surface missile system (SAM-D)

ASN: average sample number

ASNE: American Society of Naval Engineers

ASNT: American Society for Nondestructive Testing

ASO: auxiliary switch (breaker) normally open

ASODDS: ASWEPS Submarine Oceanographic Digital Data System

ASOP: automatic scheduling and operating program

ASOVII: *Associacion Venezolana de Ingenieros Industriales*

ASP: American Society of Photogrammetry

ASP: association storing processor

ASP: attached support processor

ASP: automatic schedule procedures

ASPE: American Society of Plastic Engineers

ASPE: American Society of Plumbing Engineers

ASPEP: Association of Scientists and Professional Engineering Personnel

ASPER: assembly system peripheral processors

ASPJ: advanced self-protective jammer

ASPJ: airborne self protection jammer

ASPO: American Society of Planning Officials

ASPP: alloy-steel protective plating

ASPP: American Society of Plant Physiologists

ASPR: Armed Services Procurement Regulations

ASQC: American Society for Quality Control

ASR: accumulators shift right

ASR: airborne surveillance radar

ASR: airport surveillance radar

ASR: automatic send and receive teletype terminal

ASR: automatic send receive

ASR: automatic speech recognition

ASR: available supply rate

ASRA: American Society of Refrigerating Engineers

ASS: Aerospace Support Systems (IEEE AESS Technical Committee))

ASSE: American Society fo Sanitary Engineers

ASSE: American Society of Safety Engineers

ASSE: American Society of Sanitary Engineers

ASSP: Acoustics, Speech, and Signal Processing (IEEE Group)

ASSS: Aerospace Systems Safety Society

AST: Advanced Simulation Technology

ASTA: American Statistical Association

ASTD: American Society for Training and Development

ASTE: American Society of Tool Engineers

ASTEC: Advanced Systems Technology

ASTEC: Antisubmarine Technical Evaluation Center

ASTIA: Armed Services Technical Information Agency

ASTM: American Society for Testing & Materials

ASTME: American Society for Tool and Manufacturing Engineers

ASTOR: antiship torpedo

ASTRAL: analog schematic translator to algebraic language

ASV: angle stop valve

ASV: automatic self-verification

ASVIP: American Standard Vocabulary for Information Processing

ASVP: Application System Verification and Transfer Program

ASW: acoustic surface wave

ASW: antisubmarine warfare

ASW: auxiliary switch

ASWEPS: antisubmarine warfare environmental prediction system

ASWG: American Steel Wire Gage

ASW/SICS: ASW/Ship Command and Control System

ASWSPO: Antisubmarine Warfare Systems Project Office

AT: acceptance tag

AT: ampere-turn

AT: atomic time

AT: automatic ticketing

AT: autothrottle

ATA: Air Transport Association

ATA: Association of Technical Artists

ATAMS: Advanced Tactical Attacks/Manned System

ATAR: antitank aircraft rocket

ATARS: Aircraft Traffic Advisory Resolution System

ATBCB: Architectural and Transportation Barriers Compliance Board

ATBM: antitactical ballistic missile (Hawk)

ATC: adiabatic toroidal compressor

ATC: air temperature control

ATC: air traffic control
ATC: Applied Technology Council
ATC: arbitration timing control
ATC: automatic train control
ATC: Automation Training Center
ATCAC: Air-Traffic Control Advisory Committee
ATCRBS: Air-Traffic Control Radar Beacon System (AIMS)
ATDMA: Advanced Time-Division Multiple Access
ATDS: airborne tactical data system
ATE: automatic test equipment
ATEN: *Association Technique pour la Production et l'Utilisation de l'Energie Nucleaire,* Paris, France
ATERM: air-to-air gunnery range
ATF: advanced tactical fighter
ATF: Alcohol, Tobacco and Firearms Bureau
ATF: SNAP acceptance test facility, Santa Susana, CA
ATG: air to ground
ATIS: automatic terminal information service
ATJS: advanced tactical jamming system (ATEWS, TJS, TEWS)
ATLAS: abbreviated test language for all systems
ATLIS: Automatic Tracking Laser Illumination System
ATM: air turbine motor
ATM: automated teller machine
ATM: automatic teller machine
ATMR: advanced technology medium range transport
ATOLL: acceptance, test or launch language
ATOMDEF: atomic defense
ATP: acceptance test procedures
ATP: airline transport pilot certificate
A&TP: assembly and test pit (HTSF)
ATP: automated test plan
ATR: advanced test reactor (Idaho)
ATR: air transport radio
ATR: antitransmit-receive
ATR: critical assembly (KAPL)
ATRAN: automatic terrain recognition and navigation
ATRC: antitracking control
ATRC: critical assembly (PPC)
ATRCE: advanced test reactor critical experiment
ATRIB: average transfer rate of information bits
ATRT: antitransmit-receive tube
ATS: applications technology satellite
ATS: astronomical time switch
ATSR: Argonne thermal source reactor
ATTCS: automatic takeoff thrust control system
ATTENU-2: attenuation factor reactor physics computer code
ATV: all terrain vehicle
AT/W: atomic hydrogen weld
A² R²: Argonne advanced research reactor (see also AARR)
ATWS: anticipated transient without scram
AU: arithmetic unit
AU: astronomical unit
AUDIT: automatic unattended detection inspection transmitter
AUI: attachment unit interface

AUIS: analog input-output unit
AUM: air-to-underwater missile
AUNT: automatic universal translator
AUT: advanced user terminals
AUTODIN: automatic digital network
AUTONET: automatic network display
AUTOSEVOCOM: automatic secure voice network
AUTOSTRT: automatic starter
AUTOSTRTG: automatic starting
AUTOVON: automatic voice network
AUTRAN: automatic utility translator
AUW: antiunderwater warfare
AUX: auxiliary
AUXI-ATOME: *Société Auxiliare pour l'Energie Atomique,* Paris, France
AUXS: auxiliaries
AV: audiovisual
AVA: azimuth versus amplitude
AVC: automatic valve control
AVC: automatic vent control
AVC: automatic volume control
AVE: aerospace vehicle electronics
AVE: automatic volume expansion
AVERT: Association of Voluntary Emergency Radio Teams
AVHRR: advanced very high resolution radiometer
AVIE: Venezuela Society of Structural Engineers
AVIEM: Venezuelan Association of Mechanical and Electrical Engineers
AVIONICS: aviation electronics
AVIS: viscosity, analyzer
AVL: automatic vehicle location
AVLB: armored vehicle launched bridge
AVM: automatic vehicular monitoring
AVMS: annulus vacuum maintenance system
AVNL: automatic video noise limiting
AVOGADRO&ORS: enriched-uranium swimming-pool reactor, Italy
AVR: automatic voltage regulator
AVR: automatic volume recognition
AVS: American Vacuum Society
AVT: all volatile treatment
AVTR: airborne videocassette tape recorder
AW: above water
AW: air warning
AWACS: advanced warning and control system
AWACS: Airborne Warning and Command System
AWADS: all weather aerial delivery system
AWAR: area weight average resolution
AWAR: area weight average solution
AWARS: airborne weather and reconnaissance system
AWAT: area weighted average T-number
AWCS: air weapons control system
AWEA: American Wind Association
AWG: American Wire Gage
AWGN: additive white Gaussian noise
AWN: automated weather network
AWOL: absent without leave
AWRA: American Water Resources Association
AWRNCO: Aircraft Warning Company (Marines)

AWS: American Welding Society
AWSF: Alpha Waste Storage Facility
AWWA: American Water Works Association
AXD: auxiliary drum
AXFMR: automatic transformer
AXP: axial pitch
AY: relay, analyzer
AYR: syndicate of municipal electrical utilities in West Central Germany, formed to sponsor nuclear power projects
AZ: Arizona
AZS: automatic zero set
AZUR: swimming pool research reactor, CEA, Cadrache, France
A1W: Large ship reactor prototype
A2R2: Argonne advanced research reactor (see also **AARR**)

B

B-CAS: beacon-based collision avoidance system
B-H: binary to hexadecimal
B-O: binary to octal
B: Broadcasting (IEEE Group)
B: hydrogen burning
BA: binary add
BA: breathing apparatus
BA: buffer amplifier
BAAS: British Association for the Advancement of Science
BABS: blind approach beacon system
BACE: basic automatic checkout equipment
BACE: British Association of Consulting Engineers
BADGE: base air defense ground environment
BAF: bioaccumulation factor
BAGH: baghouse
BAL: basic assembly language
BALGOL: Burroughs algebraic compiler
BALLOTS: bibliographic automation of large library operations using time sharing
BALOP: balopticon
BALS: balancing set
BALUN: balanced to unbalanced
BAM: basic access method
BAMBI: ballistic missile boost intercept
BAMM: balloon altitude mosaic measurement
BAMO: Bureau of Aeronautics Material Officer
BAMS: Bulletin of the American Mathematical Society
BAMT: Boric Acid Mixing Tank
BAP: basic assembly program
BAP: branch arm piping
BAPE: branch arm piping enclosure
BAPL: Bettis Power Laboratory
BAPS: branch arm piping shielding
BAR: base address register
BAR: Bureau of Aeronautics Representative
BARR: Bureau of Aeronautics Resident Representative
BARSTUR: Barking Sands Tactical Underwater Range

BASIC: basic algebraic symbolic interpretive compiler
BASIC: basic automatic stored instruction computer
BASIC: beginner's all-purpose symbolic instruction code
BASICS: Battle Area Surveillance and Integrated Communication
BASIS: bank automated service information system
BASIS: Burroughs applied statistical inquiry system
BAST: Boric Acid Storage Tank
BASYS: basic system
BAT: battery
BAT CHG: battery charger
BATP: Boric Acid Transfer Pump
BB: base band
BB: broadband (emission)
BB: Bulk Burning
BBC: broadband conducted
BBD: bucket-brigade device
BBL: basic business language
BBR: broadband radiated
BBRR: Brookhaven Beam Research Reactor
BBSTS: boost and space surveillance and tracking systems
BC: back-connected
BC: between centers
BC: binary code
BC: binary counter
BCAC: British Conference on Automation and Computation
BCD: binary-coded decimal
BCDIC: binary coded decimal interchange code
BCE: Bachelor of Civil Engineering
BCEC: Battle Creek Engineers Club
BCH: Bose-Chaudhuri-Hocquenguem (cyclic code)
BCI: binary coded information
BCI: broadcast interference
BCIP: Belgium Centre for Information Processing
BCL: basic contour line
BCL: Battelle Columbus Laboratory
BCL: Bechtel client letter
BCL: Burroughs common language
BCMS: boron concentration measurement system
BCO: binary coded octal
BCO: bridge cutoff
BCOM: Burroughs computer output to microfilm
BCP: Bureau of Consumer Protection
BCS: Biomedical Computing Society
BCS: British Computer Society
BCT: bushing current transformer
BCU: binary counting unit
BCW: buffer control word
BD: base detonating
BD: baud
BD: binary decoder
BDF: base detonating fuse
BDHI: bearing distance heading indicator
BDIIT: blowdown heat transfer program
BDI: bearing deviation indicator
BDIA: base diameter
BDP: business data processing

BDT: binary deck to binary tape
BDU: basic display unit
BE: back end
BE: base injection
BE: best estimate model
BE: breaker end
BEA: Bureau of Economic Analysis, U.S. Department of Commerce
BEA: Business Education Association
BEAC: Boeing Engineering analog computer
BEAIRA: British Electrical and Allied Industries Research Association
BEAM: beam
BEAM: Burroughs electronic accounting machine
BEAME: British Electrical and Allied Manufacturers
BEAR: Biological Effects of Atomic Radiation, a committee established by the National Academy of Sciences
BEC: beginning of the equilibrium cycle
BECTO: British Electric Cable Testing Organization
BED: bridge-element delay
B.EE.: Bachelor of Electrical Engineering
BEEC: binary error erasure channel
BEEF: business and engineering enriched Fortran
BEEN: *Bureau d'Etude de l'Energie Nucléaire,* Brussels, Belgium
BEFAP: Bell Laboratories Fortran assembly program
BEIR: Advisory Committee on the Biological Effects of Ionizing Radiations
BELOYARSK STATION: heterogeneous, enriched-uranium, graphite-moderated BWR with integral nuclear superheating power reactor, USSR
BEMA: Business Equipment Manufacturers Association. See: CBEMA
BEN: *Bureau d'Etudes Nucleaires,* Brussels, Belgium
BEPC: British Electrical Power Convention
BEPO: British Experimental Pile Operation
BER: Bit error rate
BER: homogeneous boiling research reactor, 50kw(th), Hahn-Meitner Institute, Berlin, West Germany
BERM: basic encyclopedic redundancy media
BERM: berm
BES: Biological Engineering Society
BESRL: Behavioral Science Research Laboratory (U.S.Army)
BEST: Business EDP System Technique
BEST: business electronic systems techniques
BEST: business equipment software techniques
BETA: battlefield exploitation and target acquisition
BETA: heterogeneous, natural-uranium, graphite-moderated, carbon dioxide-cooled power reactor, Denmark
BETA 4: former lithium production plant, Oak Ridge (now dismantled)
beV: billion electronvolts
BEX: broadband exchange
BF: back-feed
BF: backface
BF: ballistic focusing
BF: base fuse

BF: boldface (type)
BF: bottom face
BFCO: band filter cutoff
BFD: back focal distance
BFG: binary frequency generator
BFL: back focal length
BFO: beat frequency oscillator
BFPDDA: binary floating point digital differential analyzer
BFPR: Basic Fluid Power Research Program
BFS: brute force (unregulated) supply
BFS: fast critical assembly, Obninsk, USSR
BG: back gear
BGRR: Brookhaven's Graphite Research Reactor (BNL)
BGRV: boost-glide reentry vehicle
BH: bus halted
BH: flux density vs. magnetizing force
BHA: base helix angle
BI: blanking input
BI: buffered interconnect
BIA: buyers information advisory (Westinghouse Purchasing)
BIAS: Battlefield Illumination Airborne System (Shed Light)
BIE: British Institute of Engineers
BIEE: British Institute of Electrical Engineers
BIF: basic in flow
BIG-5: group cross sections, reactor physics computer code
BIGFET: Bipolar-IGFET
BIH: International Bureau of Time
BIL: basic impulse insulation level
BIL: basic impulse isolation level
BIL: Basic insulation level
BIL: block input length
BILA: Battelle Institute Learning Automation (BMI)
BIMAC: bistable magnetic core
BiMOS: bipolar metal-oxide semiconductor
BINAC: binary automatic computer
BINOMEXP: binominal expansion
BIOALRT: bioastronautics laboratory research tool
BIONICS: biological electronics
BIOR: business input-output rerun
BIOS: Biological Investigation of Space, NASA satellite project
BIOSIS: bioscience information service of biological abstracts
BIP: balanced in plane
BIPM: International Bureau of Weights and Measures
BIR: British Institute of Radiology
BIRDIE: battery integration and radar display equipment
BIRE: British Institute of Radio Engineers
BIRS: basic indexing and retrieval system
BIS: British Interplanetary Society
BISAM: basic indexed sequential access method (IBM)
BISO: multi-layered fuel-particle coating consisting of pyrolytic carbon

BISS: base and installation security systems
BISYNC: binary-synchronous
BIT: binary digit
BIT: Boric Acid Injection Tank
BIT: Boron Injection Tank
BIT: Built-in test
BITE: built-in test equipment
BIX: binary information exchange
BJT: bipolar junction transistor
BL: base line
BL: bell
BL: bend line
BL: bottom layer
BLADE: basic level automation of data through electronics
BLADES: Bell Laboratories automatic design system
BLC: British Lighting Council
BLD: beam-lead device
BLDG: building
BLEU: blind landing experimental unit (RAE)
BLEVE: boiling liquid expanding vapor explosion
BLIND: cell constant transport computer code
BLIP: background limited infrared photoconductor
BLIS: baffle/liner interface seal
BLIS: Bell Laboratories interpretive system
BLL: below lower limit
BLM: basic language machine
BLM: Bureau of Land Management
BLODIB: block diagram compiler B
BLR: baseline restorer
BLS: Bureau of Labor Statistics
BLT: basic language translator
BL&T: blind loaded and traced
BLTC: bottom loading transfer cask
BLU: basic logic unit
BLUP-3: (Kinetics) scram effectiveness reactor computer code
BM: binary multiply
BMAR: ballistics missile acquisition radar
BMB: British Metrification Board
BMC: bulk-molding compound
BMDP: Biomedical Computer Programs, P-series
BMEP: brake mean effective pressure
BMES: Biomedical Engineering Society
BMETO: Ballistics Missiles European Task Organization
BMEWS: Ballistic Missile Early Warning System
BMI: Battelle Memorial Institute
BMILS: bottom-mounted impact and location system (ABRES)
BML: bulk material length
BMR: Brookhaven's Medical Reactor (BNL)
BMS: Bio-Metrecs Society
BMS: shell loading engineering reactor computer code
BMTS: Ballistic Missile Test System
BMW: beam width
BN-350: dual-purpose power-desalination breeder reactor, 350MW(e)
BN-50: experimental fast power reactor, Ulyanovsk, USSR

BN: binary number
BNC: baby "N" connector
BNCS: British Numerical Control Society
BNCSR: British National Committee on Space Research
BNEC: British Nuclear Energy Conference
BNES: British Nuclear Energy Society
BNF: Backus-Naur form
BNF: bomb nose fuse
BNF: British Nuclear Forum
BNL: Brookhaven National Laboratory
BNS: binary number system
BNWL: Battelle Northwest Laboratories (also PNWL)
BO: blocking oscillator
BOC: beginning of cycle
BOC: bevitron orbit code
BOC: bottom of conduit
BOCA: Building Officials Conference of America, Inc
BOCES: Board of Cooperative Educational Services
BOCOL: basic operating consumer oriented language
BOCS: Bendix Optimum Configuration Satellite
BOD: biochemical oxygen demand
BOLD: bibliographic on-line display
BOM: basic operating monitor
BoM: U.S. Bureau of Mines
BOMP: base organization and maintenance monitor
BOMP: base organization and maintenance processor
BONUS-CX: boiling nuclear superheat critical experiment
BONUS: Boiling Nuclear Superheat Reactor
BOP: balance of plant
BOP: Balance of Power
BOP: binary output program
BORAM: block-oriented random access memories
BORAX 1: Boiling Reactor Experiment, NRTS, Idaho, 1.2 MW(th)
BORAX 2: Boiling Reactor Experiment, NRTS, Idaho, 6.4 MW(th)
BORAX 3: Boiling Reactor Experiment, NRTS, Idaho, 12 MW(th)
BORAX 4: Boiling Reactor Experiment, NRTS, Idaho, 20.5 MW(th)
BORAX 5: Boiling Reactor Experiment, NRTS, Idaho, 35.7 MW(th)
BORE: Beryllium Oxide Reactor Experiment (earlier name EBOR)
BOS: background operating system
BOS: basic operating system
BOT: beginning of tape
BP: between perpendiculars
B/P: blue print
BP: boiling point
BPA: Bonneville Power Administration (Portland, OR)
BPC: back-pressure control
BPD: bushing potential device
BPD: transmission chain burnup reactor computer code
BPF: blue print files
BPI: bits per inch (packing density)
BPI: bytes per inch

BPID: book physical inventory difference
BPKT: basic programming knowledge test
BPM: batch processing monitor
BPN: breakdown pulse noise
BPPMA: British Power Press Manufacturers Association
BPR: Bureau of Public Roads
BPRA: Burnable Poison Rod Assembly
BPS: basic programming system
BPS: bits per second
BPSC: binary phase-shift keying
B+PV: boiler and pressure vessel
BPVC: Boiler and Pressure Vessel Committee (ASME)
BPWR: burnable poison water reactor
BPWS: Banked Position Withdrawal Sequence
BQL: basic query language
BR-1: Belgian Reactor #1, Mol, Belgium, 4 MW(th)
BR-2: Belgian Reactor #2, Mol, Belgium, 50 MW(th)
BR-250: Obninsk, USSR, power reactor, 250 MW(e)
BR-3: Belgian Reactor #3, Mol, Belgium, 40.9 MW(th)
BR-3: USSR thermal blanket test reactor
BR-50: Obninsk, USSR, power reactor, 50 MW(e)
BR: bend radius
BR: brake relay
BR: breeder reactor
B&R: budget and reporting classification
BRA: British Refrigeration Association
BRAB: Building Research Advisory Board
BRAC: brace
BRANE: bombing radar navigation equipment
BRB: Benefits Review Board
BRC: Breeder Reactor Corporation
BRDF: bidirectional reflectance-distribution function
BRDG: bridge, structural
BREMA: British Radio Equipment Manufacturers Association
BREN: Bare Reactor Experiment Nevada (Biological and medical studies at NTS)
BRG: Budget Review Group
BRH: Bureau of Radiological Health (Rockville, MD)
BRISEIS: solid, homogeneous, enriched-uranium, graphite-moderated, air-cooled reactor, USA
BRL: Ballistics Research Laboratory (U.S.Army)
BRLS: Barrier Ready Light System
BRR: Battelle research reactor, Columbus, OH, 2MW(th)
BRR: Brookhaven Research Reactor
BRS: boron recycle system
BRT: binary run tape
BRVMA: British Radio Valve Manufacturers Association
BR 1: Soviet reactor, fast reactor for physics research, Pu, uncooled Obninsk, USSR
BR 2: Soviet reactor, fast reactor research, PuO2 Hg-cooled (dismantled)
BR 3: Soviet reactor, reactor physics research, Obninsk, USSR (dismantled)
BR 5: Soviet reactor, fast reactor research, PuO2 Na-cooled, 5 MW(th), Obninsk, USSR

BS: binary subtract
BS: Biophysical Society
B/S: bits per second
B.S.: British Standard
B&S: Brown & Sharpe Wire Gage
BSAM: basic sequential access method
BSAR: Babcock & Wilcox Safety Analysis Report
BSC: binary synchronous communication
BSCA: binary synchronous communications adapter
BSCN: bit scan
BSD: Ballistic Systems Division
BSD: bulk storage device
BSDC: British Standard Data Code
BSDL: boresight datum line
BSET: Bachelor of Science in Engineering Technology
BSF: Bulk Shielding Facility, Oak Ridge
BSI: British Standards Institute
BSIRA: British Scientific Instrument Research Association
BSL: basic switching impulse insulation level
BSNDT: British Society for Non-Destructive Testing
BSR-1: bulk shielding reactor, Oak Ridge, TN, 1MW(th)
BSR-2: bulk shielding reactor, Oak Ridge, TN, 750MW(th)
BSR: blip-scan ratio
BSR: Board of Standards Review
BSSRDF: bidirectional scattering-surface reflectance distributionfunction
BSTJ: Bell Systems Technical Journal
BSWM: Bureau of Solid Waste Management
BT: bathythermograph
B&T CANYONS: Bismuth phosphate precipitation process separation plant, Hanford
BTD: binary to decimal
BTD: bomb testing device
BTDL: basic transient diode logic
BTE: battery terminal equipment
BTF: bomb tail fuse
BTI: bridged tap isolator
BTL: beginning tape label
BTM: burner tilt mechanism
BTMA: basic telecommunications
BTP: branch technical position
BTR: Broadcast and Television Receivers (IEEE Group)
BTS: Brazilian thorium sludge
BTSS: basic time sharing system
BTTP: British Towing Tank Panel
Btu: British thermal unit
BU: binding unit
BU: buzzer
BUAER: Bureau of Aeronautics (USN)
BUIC: back-up interceptor control
BUILD: base for uniform language definition
BUIS: Barrier Up Indicator System
Bull Scweiz Electrotech: *Bulletin de L'Association Suisse des Electriciens*
BUMED: Bureau of Medicine and Surgery
BUORD: Bureau of Ordnance (USN)

BUPERS: Bureau of Navy Personnel (USN)
BUSANDA: Bureau of Supply and Accounts (USN)
BUSHIPS: Bureau of Ships (USN, NSSC)
BUTEX: dibutoxy diethyl ether, used in separation process of U and Pu in the United Kingdom
BUWEPS: Bureau of Naval Weapons (USN, NOSC)
BVC: black varnish cambric (insulation)
BW: backward wave
BW: bandwidth
BW: biological warfare
BWA: British Waterworks Association
BWG: Birmingham Wire Gage
BWM: backward-wave magnetron
BWO: backward-wave oscillator
BWPA: backward-wave power amplifier
BWR: Boiling Water Reactor
BWST: borated water storage tank

C

C-C: center to center
C-E: Combustion Engineering Corporation
C-SCAN: viewing cathode-ray screen (Air Force)
C-SCOPE: cathode-ray screen (Air Force)
C: centigrade
C: clear bit; control (type of line)
C: command
C: computer
CA-28: critical assembly (ORNL)
CA: cable
CA: California
CA: clear aperture
CA: computers and automation
CA: construction authorization
CA: contract administrator
CA: contract authorization
CA: control for arbitration bus (type of line)
C/A: corrective action
CAA: computer assisted accounting
CAAC: Civil Aviation Administration of China
CAARC: Commonwealth Advisory Aeronautical Research Council
CAAS: Ceylon Association for the Advancement of Science
CAB: Civil Aeronautics Board
CABO: Council of American Building Officials
CABRI: swimming pool research reactor, CEA, Cadarache, France
CAC: Containment Atmosphere Control (monitor,system)
CACB: compressed-air circuit breaker
CACDP: California Association of County Data Processors
CACM: Communications of the Association for Computing Machinery
CACW: core auxiliary cooling water
CACWS: core auxiliary cooling water system
CAD: cartridge-activated device
CAD: Computer-aided design
CAD: computer-aided detection

CAD: computer access device
CAD: computer activated device
CAD: Computer Applications Digest
CAD: containment atmosphere dilution
CADAM: computer graphics augmented design and manufacturing system
CADAR: computer-aided design, analysis and reliability
CADC: Cambridge automatic digital computer
CADC: central air data computer
CAD/CAM: computer-aided design and manufacturing
CADE: computer-aided design and engineering
CADEM: computer-aided design, engineering, and manufacturing
CADEP: computer-aided design of electronic products
CADET: computer-aided design experimental translator
CADETS: classroom-aided dynamic educational time-sharing system
CADF: cathode-ray tube automatic direction finding
CADF: commutated-antenna direction finder
CADFISS: computation and data flow integrated subsystems
CADIC: computer-aided design of integrated circuits
CADM: clustered airfield defeat submunition
CADM: computer-aided design and manufacturing
CADRE: current awareness and document retrieval for engineers
CADS: Containment Atmosphere Dilution System
CADSS: combined analog-digital systems simulator
CAE: computer-aided education
CAE: computer-aided engineering
CAE: computer-assisted enrollment
CAF: chemical analysis facility
CAFE: computer-aided film editor
CAFEE: assembly fuel element exchange
CAFS: content addressable file store
CAGE: computerized aerospace ground equipment
CAHE: core auxiliary heat exchanger
CAI: close approach indicator
CAI: computer-administered instruction
CAI: computer-aided inspection
CAI: Computer-aided instruction
CAI: computer-assisted instruction
CAI: computer analog input
CAI: computer assisted instruction
CAI/OP: computer analog input/output
CAIS: computer-aided instruction
CAL: calorimeter
CAL: computer-assisted learning
CAL: computer animation language
CAL: conversational algebraic language
CAL: Cornell Aeronautic Laboratory
CaLaSOAP: Calcium lanthanum silicate oxyapatite
CALM: collected algorithms for learning machines
CALM: computer-assisted library mechanization
CALUTRON: high-current mass spectrometer used in isotope separation ORNL. Calutron from California University Cyclotron

CAM: central address memory
CAM: computer-aided management
CAM: computer-aided manufacturing
CAM: computer address matrix
CAM: containment atmospheric monitoring
CAM: content addressable memory
CAM: continuous air monitor
CAM: cybernetic anthropomorphous machine
CAMA: centralized automatic message accounting
CAMAC: Computer automated measurement and control
CAMAR: common-aperture malfunction array radar
C.A.M.E.A.: *Comité des Applications Militaires de l'Energie Atomique.* Created by a French decree of December 5, 1956 uniting Atomic Commissioners, etc., for French Atomic Energy Program
CAMEL: component and material evaluation loop
CAMEN: *Centro Automomo Militaire Energia Nucleare, Livorno, Italy*
CAMESA: Canadian Military Electronics Standards Agency
CAMP: compiler for automatic machine programming
CAMP: computer-assisted mathematics program
CAMP: computer-assisted movie production
CAMP: controls and monitoring processor
CAMRAS: computer-assisted mapping and records activities systems
CANDU: Canadian natural-uranium, heavy-water-moderated-and-cooled power reactors
CANEL: Connecticut aircraft nuclear experiment
Can J Phys: Canadian Journal of Physics
CANS: computer-assisted network scheduling system
CANUNET: Canadian University Computer Network
CAOC: constant axial offset control
CAOS: completely automatic operational system
CAP: Canadian Association of Physicists
CAP: capacitor
CAP: computer-assisted production
CAP: Computer Analysts and Programmers of England
CAP: continuous audit program
CAP: Council to Advance Programming
CAP: cryotron associative processor
CAPAL: Computer and Photographically Assisted Learning
CAP COST: capital costs of shield and vessels; engineering reactor computer code
CAPE: coalition of aerospace employees
CAPE: communications automatic processing equipment
CAPERTSIM: computer-assisted program evaluation review technique simulation
CAPM: computer-aided patient management
CAPM: Containment Atmos. Particulate Monitor
CAPRI: computerized advance personnel requirements and inventory
CAPS: cell atmosphere processing system
CAPS: computer-assisted problem solving
CAPS: Courtauld's all-purpose simulator
CAPST: capacitor start
CAPT: conversational parts programming language

CAR: channel address register
CAR: computer-assisted research
CAR: configuration acceptance review of Westinghouse's "Control Advisory Release"
CAR: Containment Air Removal(Recirculation) Fan
CARA: *Compagnie d'Applications et de Recherches Atomiques,* Paris, interested in materials and equipment
CARATOM: *Compagnie d'Applications & de Recherches Atomiques,* Bonnevil-sur-Marne, France
CARBINE: computer automated real-time betting information network
CARD: channel allocation and routing data
CARD: compact automatic retrieval device
CARDCODER: Card Automatic Code System (IBM)
CARDE: Canadian Armament Research and Development Establishment (DRB)
CARF: Consumer Affairs and Regulatory Functions, Office of Assistant Secretary
CARS: Community Antenna Relay Service
CARS: computer-aided routing system
CARS: computerized automotive reporting service
CART: central automated replenishment technique
CART: central automatic reliability tester (TI)
CART: computerized automatic rating technique
CAS: calculated air speed
CAS: calibrated air speed
CAS: central alarm station
CAS: China Association for Standardization
CAS: Circuits and Systems (IEEE Society)
CAS: collision avoidance system
CAS: column-address strobe
CASA/SME: Computer and Automated Systems of SME
CASB: Cost Accounting Standards Board
CASD: computer-aided system design
CASE: computer automated support equipment
CASE: consolidated aerospace supplier evaluation
CASI: Canadian Aeronautics and Space Institute
CASPAR: Cambridge Analog Simulator for Predicting Atomic Reactions
CASS: command active sonobuoy system
CAST: Clearinghouse Announcements in Science and Technology
CAST: computerized automatic systems tester
CAT: carburetor air temperature
CAT: Chemical Addition Tank
CAT: clear air turbulence
CAT: compile and test
CAT: computer-aided teaching
CAT: computer-aided translation
CAT: computer-assisted testing
CAT: computer-assisted tomography
CAT: conditionally accepted tag
CAT: controlled attenuator timer
CATCC: Carrier Air Traffic Control Center
CATIS: Computer Aided Tactical Information System
CATO: compiler for automatic teaching operation
CATS: computer-aided teaching system
CATS: computer automated test system

CATV: community antenna television

CAUML: Computers and Automation Universal Mailing List

Cavity: Los Alamos physics measurement on cold critical experiments

CAW: channel address word

CAX: community automatic exchange

CB: circuit breaker

CB: citizens band (radio)

CB: common base

CB: common battery

CB: containment building

CB: continuous blowdown

CBAL: counterbalance

CBAST: Conc Boric Acid Storage Tank

CB/ATDS: carrier based airborne tactical data system

CBCT: customer-bank communication terminal

CBD: Commerce Business Daily

CBDI: Control Red Bank Demand Indicator

CBE: computer-based education

CBEMA: Computer & Business Equipment Manufacturers Association. (formerly **BEMA**)

CBI: Chesapeake Bay Institute

CB&I: Chicago Bridge and Iron

CBI: computer-based instruction

CBIS: computer-based instructional system

CBL: Chesapeake Biological Laboratory; part of University of Maryland National Resources Institute

CBL: computer-based learning

CBMA: Canadian Business Manufacturers Association

CBMA: Certified Ballast Manufacturers Association

CBMS: Conference Board of the Mathematical Sciences

CBMU: current bit motor unit

CBO: Congressional Budget Office

CBPT: CLIRA backup plug tool

CBR-1: first commercial breeder reactor

CBR: chemical, biological, radiological (warfare)

CBR: commercial breeder reactor, USA

CBS: call box station

CBSR: coupled breeding superheating reactor, USA

CBST: combustor

CBU: coefficient of beam utilization

CBX: cam box

CC: central computer

CC: channel command

CC: close-coupled

CC: closing coil

CC: coarse control

C&C: command and control

CC: common collector

CC: computer community

CC: condition code

CC: control center

CC: control computer

cc: cubic centimeters

CCA: Canadian Construction Association

CCA: carrier-controlled approach

CCAP: communications control application program

CCB: circuit concentration bay

CCB: Configuration Control Board

CCC: Canadian Computer Conference

CCC: Commodity Credit Corporation

CCC: computer communication console

CCCEC: Commission for Certification of Consulting Engineers, Colorado

CCCE&LS: California Council of Civil Engineers and Land Surveyors

CCCS: core component conditioning station

CCCS: core components cleaning system

CCD: Charge-coupled device

CCD: coarse control damper

CCD: computer-controlled display

CCD: core current driver

CCD: counter-current digestion (ore leach process)

CCE: Commission of the European Communities

CCEU: Council on the Continuing Education Unit

CCFM: cryogenic continuous-film memory

CCGCS: Containment Combustion Gas Control System

CCH: computerized criminal history

CCH: connections per circuit hour

CCHX: Component Cooling Heat Exchanger

CCI: combined form of CCIR and CCIT

CCI: *Comité Confédération Internationale*

CCIA: Computer and Communications Industry Association

CCIA: console computer interface adapter

CCIF: International Telephone Consultative Committee

CCIR: *Comité Consultatif International des Radiocommunications*

CCIR: International Radio Consultative Committee

CCIR: International Radio Consultative Committee

CCIS: Command Control Information System

CCITT: *Comité Consultatif Internationale Télégraphique et Téléphonique*

CCITT: International Telegraph and Telephone Consultative Committee

CCL: control language

CCM: communications controller multichannel

CCM: counter countermeasures

CCMD: continuous current monitoring device

CCN: contract change notice

CCO: current-controlled oscillator

CCOP: constant-control oil pressure

CCP: Centrifugal Charging Pump

CCP: Core component pot

CCP: core component pump

C&CP: Corrosion and Cathodic Protection (IEEE IAS Technical Committee)

CCPD: coupling capacitor potential device

CCPE: Canadian Council of Professional Engineers

CCR: ceiling cavity ratio

CCR: control contactor

CCRC: core component receiving container

CCROS: card capacitor read-only storage

CCS: Canadian Ceramic Society

CC&S: central computer and sequencing

CCS: commitment control system

CCS: containment cooling system
CCS: continuous commercial service
CCS: Conversational Compiling System (Xerox)
CCS: custom computer system
CCS: hundred-call-seconds
CCSA: common control switching arrangement
CCST: Center for Computer Sciences and Technology (NBS)
CCSW: component cooling service water
CCT: communications control team
CCT: constant current transformer
CCT: correlated color temperature
CCTest: component check test
CCTL: Casing Cooling Tank Level
CCTL: core component test loop (ANL)
CCTV: close-circuit television
CCV: control-configured vehicle
CCVS: Cobol compiler validation system
CCVS: Current-controlled voltage source
CCVT: coupling-capacitor voltage transformer
CCW: channel command word
CCW: circulation controlled wing
CCW: Component Cooling Water
CCW: counterclockwise
CCWS: component cooling water system
CD-I: compact disk-interactive
CD-ROM: compact disk read-only memory
CD: cable duct
CD: center distance
CD: circuit description
CD: control rod drive
CD: current density
CDA: command and data acquisition
CDA: Containment Depressurization Activation
CDA: Containment Depressurization Alarm
CDA: core disruptive accident
CDC: call directing character
CDC: call directing code
CDC: configuration data control
CDC: Current-controlled voltage source
CDCE: central data-conversion equipment
CDCR: Center for Documentation and Communication Research (Western Reserve University)
CDE: certification in data education
CDF: combined distribution frame
CDF: cumulative damage function
CDI: collector diffusion isolation
CDI: Control Data Institute
CDJM: Canadian Journal of Mathematics
CDL: common display logic
CDL: computer description language
CDM: central data management
CDM: code-division multiplexing
CDMA: code-division multiple-access
CDP: central data processor
CDP: centralized data processing
CDP: certification in data processing
CDP: checkout data processor
CDP: commercial data processing
CDP: communication data processor

CDP: compression discharge pressure
CDPC: central data processing computer
CDPC: commercial data processing center
CDPIR: crash data position indicator recorder
CDR: circular depolarization ratio
CDR: command destruct receiver
CDR: conceptual design requirement
CDR: critical design review
CDR: current directional relay
CDS: component disassembly station (HTSF)
CDS: computer duplex system
CDS: Conceptual Design Study
CDSE: computer driven simulation environment
CDT: central daylight time
CDT: control data terminal
CDTI: cockpit display of traffic information
CDU: central display unit
CDU: control and display unit
CDUC: duct, cable
CDW: computer data word
CDX: contro-differential transmitter
CD0: clocked data zero
CD1: clocked data one
C/E: calculation/experiment
CE: civil engineering
CE: column enable
CE: common emitter
CE: communications-electronics
CE: commutator end
CE: computer engineer
CE: conducted emission
CE: Consumer Electronics
CE: Corps of Engineers, U.S.Army
CEA: *Commissareat a l'Energie Atomique*, France
CEA: Cambridge electron accelerator (Harvard—MIT), Cambridge, MA.
CEA: circular error average
CEA: Commodity Exchange Authority
CEA: control element assembly
CEA: Council of Economic Advisors, U.S.A.
CEAC: Consulting Engineers Association of California
CEAC: Control Element Assembly Computer
CEAC: County Engineers Association of California
CEARC: Computer Education and Applied Research Center
CEAU: Continuing Education Achievement Unit
CEBELCORE: *Centre Belge d'Etude de la Corrosion,* Belgium
CEC: Canadian Electric Code
CEC: Charlotte Engineers Club
CEC: Colorado Engineering Council
CEC: Commonwealth Engineering Conference
CEC: Consulting Engineers Council
CECC: CENELEC Electronic Components Committee
CECC: Consulting Engineers Council of Colorado
CEC/MINN: Consulting Engineers Council of Minnesota
CEC/NYS: Consulting Engineers Council of New York State, Inc

CECO: Consulting Engineers Council of Oregon

CECO: cost estimate change order

CECU: Consulting Engineers Council of Utah

CEDM: control element drive motor

CEDPA: California Educational Data Processing Association

CEE: *Commission International de Reglémentation en vue de l'asprobation de l'équipment Electric*

CEE: International Commission on Rules for the Approval of Electrical Equipment

CEEA: *Communaute Europeenne de l'Energie Atomique (Europaische Atomgemeinschaft 'Euratom')*

CEEN: *Centre de l'Energie Nucleaire*, Mol, Belgium

CEGB: Central Electricity Generating Board

CEI: communication electronics instructions

CEI: computer extended instruction

CEI: Council of Engineering Institutions

CEI: *Commission Electrotechnique Internationale*

CEL: Carbon equilibrium loop (ARD)

CEL: carbon equivalent loop

CELT: coherent emitter location testbed

CEMA: Canadian Electrical Manufacturers Association

CEMA: Conveyor Equipment Manufacturers Association

CEMF: counterelectromotive force

Cem Ind'y/IAS: Cement Industry

CEMNFAR: nuclear technology center operated by CEA at Fontenay-aux-Roses, France

CEMON: customer engineering monitor (IBM)

CEN: *Centre d'Etusesudes se l'Energie Nucleaire*, Belgium

CEN: *Commite Europeen de Normalisation*

CEN: European Standards Coordinating Committee

CEN: *Centre d'Etudes de l'Energie Nucléaire*, Belgium

CEN: *Comité Européen de Normalisation*

CENC: *Centre d'Etudes Nucleaires de Cadarache*, Cadarache, France

CENC: *Centre d'Etudes Nucléaires de Cadarache*, Cadarache, France

C.E.N.C.: The Nuclear Research Center at Grenoble, University of Grenoble, France

CENCER: CEN's certification body

CENCER: Certification arm of the European committee for Standardization

CENEL: Committee for the Coordination of Engineering Standards in the Electrical Field

CENELEC: European Committee for Electrotechnical Standardization

CENS: *Centre d'Etudes Nucléaire de Saclay*, France

CENUSA: *Centrales Nucleares*, S.A., Madrid, Spain

CEP: circle of equal probability

CEP: circular error probability

CEP: civil engineering package

CEP: computer entry punch

CEPA: Society for Computer Applications in Engineering, Planning and Architecture, Inc

CEPT: *Conférence Européen des Administrations des Postes et des Télécommunications.* European Conference of Postal and Telecommunications Administrations

CEQ: Council on Environmental Quality

CER: Civil Engineering Report

CER: complete engineering release

CER: critical experiment reactor

CERA: Civil Engineering Research Association

CERC: Coastal Engineering Research Center (Army)

CERCA: *Centre de Recherches pour Combustibles Atomiques*, Paris, interested in nuclear fuels

CERF: critical assembly (LAC)

CERL: Central Electricity Research Laboratories

CERMET: ceramic-to-metal (seal)

CERMET: ceramic and metal fuel

CERMET: ceramic metal element

CERN: European Organization for Nuclear Research, Laboratory at Geneva, Switzerland

CERT: combined environmental reliability test

CERTICO: Certification Committee [ANSI]

CES: Capstone Engineering Society

CES: Cleveland Engineering Society

CES: constant elasticity of substitution

CES: Critical Experiment Station, Westinghouse Reactor Evaluation Center

CESAR: Combustion Engineering Safety Analysis Report

CESAR: Uranium-graphite research reactor, CEA, Cadarache, France

CESC: Consulting Engineers of South Carolina

CESI: Council for Elementary Science International

CESNEF: Reactor of the *Centro di Studi Nucleari Enrico Fermi*, Milan, Italy

CESSE: Council of Engineering and Scientific Society Executives

CET: cold-end termination

CET: corrected effective temperature

CET: critical experiment tank for HTRE-2

CET: cumulative elapsed time

CETIS: *Centre Européen de Traitment of l'Information Scientifique*, computer research on information retrieval, Iapraspra

CETR: Consolidated Edison Thorium Reactor, Indian Point, New York (PWR type)

CETS: Conference on European Telecommunications Satellites

CETS: control element test stand

CEV: combat engineer vehicle

CEV: corona extinction voltage

CEVAR: consumable-electrode vacuum-arc remelt

CF-STF: Swedish Association of Graduate Engineers

CF: cathode follower

CF: centrifugal force

CF: concept formulation

CF: confinement factor

CF: conversion factor

CF: counterfire

CFA: Canadian Forestry Association

CFA: cascade-failure analysis

CFA: code of Federal regulations
CFA: Core Flood Alarm
CFA: Council of Iron Foundry Associations
CFA: cross-field amplifier
CFAR: constant false alarm rate
CFBS: Canadian Federation of Biological Societies
CFC: capillary filtration coefficient
CFC: central fire control
CFEC: Cape Fear Engineers Club
Cff: critical flicker frequency
Cff: critical fusion frequency
CFIA: Core Flood Isolation Valve Assembly
CFIS: California fiscal information system
CFL: calibrated focal length
CFL: context-free language
CFM: cathode follower mixer
CFM: cubic foot per minute
CFO: critical flashover voltage
CFR: Code of Federal Regulations
CFR: commercial fast reactor, United Kingdom
CFR: Coordinating Fuel Research
CFRE: circulating-fuel experiment, Oak Ridge
CFRE: circulating-fuel reactor experiment , Oak Ridge
CFS: cubic foot per second
CFSG: Cometary Feasibility Study Group (ESRO)
CFSSB: Central Flight Status Selection Board
CFSTI: Clearinghouse for Federal Scientific and Technical Information (Department of Commerce)
CFT: Core Flood Tank
CFTC: Commodity Futures Trading Commission
CG: center of gravity
CG: Coast Guard
CGA: Compressed Gas Association
CGEN: chemical generator
CGI: computer generated imagery
CGL: cover gas evaluation loop
CGPM: Conférence Général des Poides et Mesures
CGRM: Containment Gaseous Radiation Monitor
CGS: Canadian Geotechnical Society
CGS: centimeter-gram-second
CGSB: Canadian Government Specifications Board
CH: chain home (radar)
CHABA: Committee on Hearing and Bioacoustics
CHAD: code to handle angular data
CHAG: compact high performance aerial gun
CHANHI: abbreviation for upper channel corresponding to the half-amplitude point of a distribution
CHANLO: abbreviation for lower channel corresponding to the half-amplitude point of a distribution
CHAS: containment heat removal system
CHAT: CLIRA holddown assembly tool
CHCF: Component handling and cleaning facility
CHE: channel end
Ch.E.: chemical engineer
CHEC: channel evaluation and call
CHESNAVFAC: Chesapeake Naval Division Facilities
CHF: critical heat flux
CHGP: charging pump
CHI: computer human interaction
CHIL: current hogging injection logic

CHILD: cognitive hybrid intelligent learning and development
CHILD: computer having intelligent learning and development
CHIM: chimney; stack
CHORI: Chief of Office of Research and Inventions
CHPAE: Critical Human Performance and Evaluation
CHS: Canadian Hydrographic Service
CHU: centigrade heat unit
CI: card input
CI: Center Island
CI: Chlorine Institute
CI: circuit interrupter
CI: Combustion Institute
CI: configuration item
CI: containment integrity
C&I: control and indication
C&I: Control and Instrumentation
CI: course indicator
CIA: Central Intelligence Agency
CIA: Chemical Industries Association
CIA: communications interrupt analysis
CIA: Computer Industry Association
CIA: computer interface adapter
CIA: containment isolation A
CIAC: Construction Industry Advisory Council
CIAN: Centre d'Etudes pour les Applications de l'Energie Nucléaire, Belgium
CIAPR: Colegio de Ingerieros of Canada
CIB: Cobol information bulletin
CIB: containment isolation B
CIC: Chemical Industries Council
CIC: Chemical Institute of Canada
CICA: Chicago Industrial Communications Association
CICAF: Compagnie Industrielle de Combustibles Atomiques Frittes, Corbeville, France
CICS: data communications monitor
CID: Centre for Information and Documentation (Euratom)
CID: charge-injection devices
CID: Commercial Item Description
CID: component identification
CIDAS: conversational interactive digital/analog simulator (IBM)
CIE-USA: Chinese Institute of Engineers USA
CIE: Commission Internationale de l'Eclairage (International Commission on Illumination)
CIEN: Commision Interamericana de Energia Nuclear, Commission Interaméricaine d'Energie Nucléaire, Comisao Interamericana de Energia Nucleaire, c/o Organization of American States
CIGRE: Conférence Internationale des Grands Réseaux Electriques à Haute Tensions
CIGRE: International Conference on Large High Voltage Electric Systems, US National Committee
CII: Compagnie Internationale pour l'Informatique
CILOP: conversion in lieu of procurement
CILRT: Containment Integrated Leak Rate Test
CIM: Canadian Institute of Mining & Metallurgy

CIM: communications improvement memorandum
CIM: computer-integrated manufacturing
CIM: computer input microfilm
CIM: computer input multiplexer
CIM: continuous image microfilm
CIMAC: *Congrès International des Machines à Combustion* (International Congress on Combustion Machines)
CIMCO: card image correction
CIMM: Canadian Institute of Mining and Metallurgy
CIN: carrier input
CINS: Centro Institute of Nuclear Science
CIO: central input output (system)
CIOCS: communication input and output control system
CIOMS: Council for International Organizations of Medical Sciences
CIOS: World Council of Management
CIOU: custom input/output unit
CIP: compatible independent peripherals
C/IP: construction/inspection procedure (Bechtel)
CIP: current injection probe
CIPASH: Committee on International Programs in Atmospheric Sciences and Hydrology (NAS/NRC)
CIPHONY: cipher and telephone equipment
CIPM: Council for International Progress in Management
CIPM: *Comité Internationale des Poids et Mesures*
CIPM: International Committee of Weights and Measures
CIPQ: *Centre Internationale de Promotion de la Qualité*
CIPR9(ICRP): *Commission Internationale de Protection contre le Radiations* (International Commission of Radiological Protection)
CIPS: Canadian Information Processing Society
CIR: Canada-India Reactor, Bombay, India, 40 MW(th)
CIRC: Centralized Information Reference and Control
CIRCAL: circuit analysis
CIRES: Communication Instructions for Reporting Enemy Sightings
CIRGA: critical isotope reactor, General Atomics
CIRGA: Critical Isotope Reactor, General Atomics
CIRIA: Construction Industry Research and Information Association
CIRM: *Comité International Radio Maritime*
CIS: Canadian Institute of Surveying
CIS: Central Instructor School
CIS: character instruction set
CIS: communication information system
CIS: Containment isolation system
CIS: cue indexing system
CIS: current information selection
CISCO: compass integrated system compiler
CISE: *Centro Informazioni Studie d'Experiweenze* (R&D organization, Milan, Italy)
CISIR: Ceylon Institute of Scientific and Industrial Research

CISPR: Center for International Systems Research (Department of State)
CISPR: International Special Committee on Radio Interference
CISR: Center for International Systems Research (Department of State)
CIT: California Institute of Technology
CIT: Carnegie Institute of Technology
CIT: compressor inlet temperature
CIT: computer interface technology
CIT: *Compagnie Industrielle de Télécommunications*
CITA: Textile Agreements Implementation Committee
CITAB: computer instruction and training assistance for the blind
CITB: Construction Industry Training Board
CITC: Construction Industry Training Center
CITE: cargo integration test equipment
CITE: compression ignition and turbine engine
CITEC: contractor independent technical effort
CITS: central integrated test system
CIU: computer interface unit
CIUR(ICRU): *Commission Internationale des Unites et des Mesures Radiologique* (International Commission on Radiological Units and Measures)
CIV: Center island vessel
CIV: Containment Isolation Valve
CIV: *Colegia de Ingenieros de Venezuela*
CIWV: close-in weapons systems
CJF: Connecticut Joint Federation
CJP: communication jamming processor
CKMTA: Cape Kennedy Missile Test Annex (now Cape Canaveral)(NASA)
C/kT: carrier-to-receiver-noise density
ckt: circuit
CKV: valve, check
CL: Closed loop
CL: computational linguistics
CL: control language
CL: conversion loss
CLA: center line average
CLA: clear and add
CLA: communication line adapter
CLA: communication link analyzer
CLA: Computer Lessors Association
CLAA: *Centre Lyonnais d'Applications Atomiques*, food irradiation processing firm, France
CLAIRA: Chalk lime and Allied Industries Research Association (now WHRA)
CLAM: chemical low-altitude missile
CLARA: Cornell Learning and Recognizing Automation
CLAS: Computer Library Applications Service
CLASP: closed line assembly for single particles
CLASP: Computer Language for Aeronautics and Space Programming (NASA)
CLASS: closed loop accounting for store sales
CLASSMATE: computer language to aid and simulate scientific, mathematical, and technical education
CLC: containment leakage control
CLC: course line computer

CLCGM: Closed loop cover gas monitor
CLCIS: closed loop cover and instrumentation
CLCIS: Closed loop cover and instrumentation system
CLCPE: California Legislative Council of Professional Engineers
CLCR: controlled letter contract reduction
CLCS: Consequence Limiting Control System
CLCS: current logic, current switching
CLCV: cold leg check valve
CLCV: current logic, current switching
CLD: Central Library and Documentation Branch (ILO)
CLD: Chloride Leak Detector
CLDAS: clinical laboratory data acquisition system
CLEA: Conference on Laser Engineering and Applications (IEEE)
CLEAN: Commonwealth Law Enforcement Assistance Network
CLEHA: Conference of Local Environmental Health Administrators
CLEM: Closed loop ex-vessel machine
CLEM: composite for the lunar excursion module
CLEO: clear language for expressing orders
CLETS: California law enforcement telecommunications system
CLF: capacitive loss factor
CLIP: compiler language for information processing
CLIRA: Closed Loop In-Reactor Assembly
CLIV: Cold leg isolation valve
CLJA: Closed loop jumper assembly
CLM: column, chemical process
CLR: clean liquid radwater
CLR: combined line and recording trunk
CLR: computer language recorder
CLR: computer language research
CLR: Coordinating Lubricants Research
CLR: Council on Library Resources
CLR: current limiting resistor
CLRU: Cambridge Language Research Unit
CLS: Cask loading station
CLS: clear and subtract
CLS: Closed loop system
CLS: common language system
CLS: concept learning system
CLS: Containment Leakage System
CLSMDA: Closed loop system melt-down accident
CLT: communications line terminals
CLT: computer language translator
CLU: central logic unit
CLU: circuit line-up
CLUT: computer logic unit tester (ADSAF)
CM: calibration magnification
CM: centimeter (10^{-2} meters)
CM: central memory
C/M: communications multiplexer
CM: construction and machinery
CM: controlled mine field
CM: core memory
CM: countermeasure
CMA: Canadian Medical Association

CMA: Classified mail address
CMA: contact-making (or breaking) ammeter
CMAA: Crane Manufacturers Association of America
CMANY: Communications Managers Association of New York
CMC: code for magnetic characters
CMC: communications mode control
CMCA: cruise missile carrier aircraft
CMCTL: current mode complementary transistor logic
CMD: core memory driver
CMERI: Central Mechanical Engineering Research Institute (India)
CMF: central maintenance facility
CMF: coherent memory filter
CMF: Common mode Failure
CMFA: common-mode-failure analysis
CMFLPD: Core Maximum Fraction of Limiting Power Density
CMG: Control Moment Gyro
CMI: Christian Michelsen's Institute (Norway)
CMI: Commonwealth Mycological Institute
CMI: computer-managed instruction
CMI: computer input microfilm
CMI: Core Element Assembly Motion Inhibit
CMIL: circular mil
CML: Critical Mass Laboratory, Hanford
CML: current-mode logic
CMLCENCOM: Chemical Corps Engineering Command
CMM: Commission for Maritime Meteorology (WMO)
CMM: Core mechanical mockup
CMMF: Component maintenance and mockup facility
CMMP: Commodity Management Master Plan
CMOS: complementary metal oxide semiconductor
CMOS/SOS: complementary metal oxide semiconductor/silicon on sapphire
CMP: Configuration management plan
CMP: Controlled materials production
CMPF: Core Maximum Power Fraction
cmps: centimeter per second
CMR: Committee on Manpower Resources
CMR: common-mode rejection
CMR: Communications Monitoring Report
CMRR: common-mode rejection ratio
CMS: Clay Minerals Society
CMS: compiler monitor system
C&MS: Consumer and Marketing Service
CMS: current-mode switching
CMT: conversational mode terminal (Friden)
CMV: common-mode voltage
CMVM: contact-making voltmeter
CMVPCB: California Motor Vehicle Pollution Control Board
C/N: carrier to noise
CN: coordination number
CNA: Canadian Nuclear Association
CNA: copper nickel alloy
CNA: cosmic noise absorption

CNA: *Centrale Nucléaire des Ardennes, Givet* atomic energy program sponsored by *Electricité de France* and Belgium's *Centre et Sud* group

CNAA: Council for National Academic Awards

CNAS: Civil Navigation Aids System

CNB: *Centrale Nucléaire Belge,* 300MW reactor at Bessel, Belgium

CNC: computerized numerical control

CNC: computer numerical control

CNC: Council for Non-Collegiate Continuing Education

CNDP: communication network design program (IBM)

CNEA: *Comision Nacional de Energia Atomica,* Mexican Atomic Energy Commission

CNEN: *Comitato Nazionale Energia Nucleare;* agency responsible for Italy's nuclear program

CNES: *Centre National d'Etudies Spatiales*

CNEUPEN: *Commission Nationale pour l'Etude de l'Utilisation Pacifique de l'Energie Nucléare,* Brussels, Belgium

CNI: *Consiglio Nazionale Ingegneri*

CNI: communication navigation identification

CNI: Consolidated National Intervenors

C.N.I.: *Centrale Nucléaire Interescaut;* a nuclear power station with output of 150MW(e) built by Interescaut Company of Belgium and Linburg Electricity Company of the Netherlands

CNL: canal

CNL: circuit net loss

CNLA: Council on National Library Associations

CNR: carrier-to-noise ratio

CNR: *Consiglio Nazionale delle Recerche* (Italian National Research Council)

CNRC: Canadian National Research Council

CNRM: *Centre National de Recherches Metallurgiques,* Liege, Belgium

CNRN: *Comitato Nazionale per le Ricerche Nucleari,* independent commission formerly in charge of Italian nuclear program (now CNEN)

CNS: Congress of Neurological Surgeons

CNT: celestial navigation trainer

cntor: contractor

CNTP: Committee for a National Trade Policy

CO-11: Argonne nuclear assembly for university training, ANL, 10 kW(th) (Argonaut)

CO: central office

CO: Chief Operator

CO: close-open operation

CO: Colorado

CO: combined operations

CO: communications officer

CO: Contracting Officer

COA: College of Aeronautics

C.O.B.: Close of business

COBE: Cosmic Background Explorer.

COBESTCO: computer-based estimating technique for contractors

COBIS: computer-based instruction system

COBLIB: Cobol library

COBLOC: codap language block-oriented compiler

Cobol: common business-oriented language

COC: coded optical character

COCODE: compressed coherency detection (radar technique)

COCOM: Coordinating Committee for Exports to Communist Areas

COD: carrier onboard deliver

COD: chemical oxygen demand

COD: crack opening displacement

CODAN: carrier-operated device antinoise

CODAN: coded weather analysis (Navy)

CODAP: control data assembly program

CODAS: customer-oriented data retrieval and display system

CODASYL: Conference on Data System Languages

CODATA: Committee on Data for Science and Technology (ICSU)

CODEC: coder-decoder

CODEL: computer developments limited automatic coding system

CODES: computer design and evaluation system

CODEVER: code verification

CODEX: Codex Alimentarius Commission

CODIA: *Colegio Dominicono De Ingenieros Arquitectos y Agrimensores de Puerto Rico*

CODIC: color difference computer

CODIC: computer directed communications

CODIL: control diagram language

CODIT: computer direct to telegraph

CODSIA: Council of Defense and Space Industries Association

CODSIA: Council of Defense and Space Industry Associations

COED: Char Oil Energy Development

COED: computer-operated electronic display

Coffin: thick-walled container for transporting radioactive materials

COFIL: core file

COFINATOME: *Compagnie de Financement de l'Industrie Atomique,* Paris, France

COG: computer operations group

COGENT: compiler and generalized translator

COGO: coordinate geometry

COGS: continuous orbital guidance sensor

COHO: coherent oscillator

COI: communication operation instructions

COIE: Chinese Institute of Engineers

COINS: computer and information sciences

COL: computer oriented language

COLASL: compiler/Los Alamos Scientific Laboratory

COLIDAR: coherent light detection and ranging

COLINGO: compile on-line and go

COLM: column

COLSS: Core Operating Limits Supervisory System

COLT: communication line terminator

COLT: computerized on-line testing

COLT: computer oriented language translator

COLT: control language translator

Com: communication switching

COM: communication technology
COM: computer output microfilm
COMAC: continuous multiple-access collator
COMAR: Committee on Man and Radiation (IEEE)
COMAT: computer-assisted training
COMCM: communications countermeasures and detection
COMEINDORS: composite mechanized information and documentation retrieval system
COMET: Computer-operated management evaluation technique
Comet: Los Alamos Scientific Laboratory experiment for critical configuration safety tests
Comet II: Critical assembly (LASL, Kiva I)
COMIT: compiler/MIT
COMLOGNET: combat logistics network
COMLOGNET: communications logistics network
COMM: Communications (IEEE Society)
COMMEN: compiler oriented for multiprogramming and multiprocessing environments
COMMEND: computer-aided mechanical engineering design system
COMP: Computer (IEEE Society)
COMPAC: computer program for automatic control
COMPACT: compatible algebraic compiler and translator
COMPACT: computer planning and control technique (HIS)
COMPANDER: compressor expander
COMPARE: computerized performance and analysis response evaluator
COMPARE: console for optical measurement and precise analysis of radiation from electronics
COMPASS: compiler-assembler
COMPASS: computer-assisted classification and assignment system
COMPCON: Computer Convention (IEEE)
COMPEL: compute parallel
COMPOOL: communications pool
COMPOW: Committee on Professional Opportunities for Women
Comp Rend Acad Sci: *Comptes Rendus Hebdomadaires des Séances de l'Academic des Sciences* (Paris)
COMPRESS: computer research, systems and software
COMPROG: computer program
COMPSO: computer software and peripheral show
COMRADE: computer-aided design environment
COMSAT: communications satellite
COMSEC: communications security
COMSL: communication system simulation language
COMSOAL: computer method of sequencing operations for assembly lines
Com Sys Disc/Com: Communication System Disciplines
Com Theory/Com: Communication Theory
COMTRAN: commercial translator
COMVII: Commission VII
COMZ: communications zone

CON: contactor
CON: Controller
CONELRAD: control of electromagnetic radiation
CONSORT: conversational systems with on-line remote terminals
CONSORT: Light-water-moderated research reactor, 10KW(th), United Kingdom
CONSUL: control subroutine language
CONTRAN: control translator (HIS)
CONTRANS: conceptual thought, random net simulation
COOL: checkout oriented language
COOL: control oriented language
COP: computer optimization package
COPANT: Council of the Pan American Standards Commission
COPE: Committee on Political Education (AFL-CIO)
COPE: communications oriented processing equipment
COPE: Multichannel thermal performance engineering reactor computer code
COPI: computer-oriented programmed instruction
COPOLCO: Consumer Policy Committee (ISO)
COPPS: Committee on Power Plant Siting—Matopma; Academy Engineering report: Engineering for Resolution of the Energy Environment Dilemma, 1972
CORA: conditioned-response analog (machine)
CORA: conditioned reflex analog
CORAL: computer on-line real-time applications language
CORAPRAN: Cobelda radar automatic preflight analyzer
CORC: Cornell computing language
CORDIC: coordinate rotation digital computer
CORDS: coherent on receive doppler system
CORE: computer-oriented reporting efficiency
COREP: combined overload repair control
CORREGATE: correctable gate
CORS: Canadian Operational Research Society
CORTS: convert range telemetry systems
COS: compatible operating system
COS: concurrent operating system (UNIVAC)
cos: cosine
COS: cosmic rays and trapped radiation committee (ESRO)
COSATI: Committee on Scientific and Technical Information (FCST)
COSBA: Computer Services Bureau Association (GB)
cosh: hyperbolic cosine
COSI: Committee on Scientific Information
COSINE: Committee on Computer Science in Electrical Engineering Education
COSIP: College Science Improvement Program
COSMIC: Computer Software Management and Information Center
COSMOS: Computer Oriented System for Management Order Synthesis (IBM)
COSOS: Conference on Self-Operating Systems
COSPAR: Committee on Space Research (ICSU)

COSPUP: Committee on Science and Public Policy
COSRIMS: Committee on Research in the Mathematical Sciences
COSRO: conical scan on receive only
cot: cotangent
coth: hyperbolic cotangent
COTRAN: Cobol to Cobol translator
COWPS: Council on Wage and Price Stability
CP-1: Chicago Pile #1, rebuilt as CP-2, ANL
CP-3: Argonne Chicago Pile #3, rebuilt as CP-5
CP-5: Chicago Pile # 5; also called Argonne Research Reactor, Argonne National Laboratory, Argonne, Illinois
CP: candle power
CP: card punch
CP: change proposal
CP: circuit package
CP: circularly polarized
CP: circular pitch
CP: clock pulse
CP: coefficient of performance
CP: command pulse
CP: communication processor
CP: conference paper
CP: constant pressure
CP: Construction permit
CP: control panel
CP: control processor
CP: current paper
CP: customized processor (IBM)
CP: customizwed processor (IBM)
CPA: Canadian Pharmaceutical Association
CPA: Canadian Psychological Association
CPA: closest point of approach
CPA: critical path analysis
CPAL: Containment Person Air Lock
CPC: card programmed calculator
CPC: clock-pulse control
CPC: computer program component
CPC: computer programming concepts
CPC: core protection calculator
CPC: core protection computer
CPC: cycle program control
CPC: cycle program counter
CPCEI: computer program contract end item
CPD: Community Planning and Development Office of Assistant Secretary
CPD: consolidated programming document
CPD: contact potential difference
CPD: cumulative probability distribution
CPDD: conceptual project design description
CPDS: computer program design specification
CPE: central processing element
CPE: central programmer and evaluator
CPE: circular probable error
CPE: contractor performance evaluation
CPEA: College Physical Education Association
CPEA: Cyprus Professional Engineers Association
CPEM: Conference on Precision Electromagnetic Measurement

CPFF: cost plus fixed fee
CPG: College Publishers Group
CPHS: containment pressure high signal
CPI: characters per inch
CPI: computer prescribed instruction
CPIA: Chemical Propulsion Information Agency
CPIC: cable plant interface connect
CPILS: correlation protection integrated landing system
CPIP: computer program implementation process
CPL: combined programming language
CPL: computer program library
CPL: core performance log
CPM: critical path method
CPM: cycles per minute
CPP: card punching printer
CPP: Center for Plutonium Production (France)
CPP: Computer Professionals for Peace
CPP: Containment Pressure Protection
CPPS: critical path planning and scheduling
CPR: Committee on Polar Research (NAS/NRC)
CPR: Critical power ratio
CPRG: Computer Personnel Research Group
CPS: characters per second
CPS: circuit package schematic
CPS: Control Programs Support
CPS: conversational programming system
CPS: Conversion program system
CPS: Counts Per Second
CPS: critical path scheduling
cps: cycles per second (obsolete) Use hertz (Hz)
CPSA: Consumer Product Safety Act
CPSC: Consumer Product Safety Commission
CPSCI: central personnel security clearance index
CPSE: counterpoise
CPSS: Common Programming Support System (Xerox)
CPT: control power transformer
CPT: critical path technique
CPTA: computer programming and testing activity
CPU: Central processing unit
CPU: collective protection unit
CPU: communications processor unit
CPU: computer peripheral unit
CPX: charged pigment xerography
CQ: carrier qualification
CQ: Congressional Quarterly
CQE: Cognizant Quality Engineer
CR-3: Chicago Reactor #3, Argonne National Laboratory, Argonne, Illinois
CR: cathode ray
CR: command register
C/R: Command/Response
CR: conference report
CR: Congressional Record
CR: containment rupture
CR: contract report
CR: control relay
CR: control rod
CR: crystal rectifier

CR: current relay
CRA: control rod assembly
CRAC: Careers Research and Advisory Center
CRAD: Committee for Research into Apparatus for the Disabled
CRAF: civilian reserve air fleet
CRAFT: computerized relative allocation of facilities technique
CRAM: card random access memory
CRAM: computerized reliability analysis method
CRAM: conditional relaxation analysis method
CRBR-CX: critical assembly
CRBR: Clinch River Breeder Reactor
CRBR: Controlled-recirculation boiling-water reactor
CRBRP: Clinch River Breeder Reactor Project
CRC: Civil Rights Commission
CRC: Coordinating Research Council
CRC: Copy Research Council
CRC: critical reactor component
CRC: cumulative results criterion
CRC: cyclic redundancy check
CRCC: cyclic redundancy check character
CRC/IES: Cedar Rapids Chapter, Iowa Engineering Society
CRCPD: Conference of Radiation Control Program Directors
CRCTA: composite reactor components test activity
CRD: control rod drive
CRDA: control rod drive assembly
CRDF: Canadian radio-direction finding or finder
CRDF: cathode-ray direction finding (radar)
CRDL: Chemical Research and Development Laboratories (Army)
CRDM: control rod drive mechanism
CRDM: control rod drive motor
CRDME: Committee for Research into Dental Materials and Equipment
CRDSD: Current Research and Development in Scientific Documentation
CRE: controlled residual element
CRE: corrosion-resistant
CREDO: Central Reliability Data Organization
CRES: corrosion-resistant steel
CRESS: Central Regulatory Electronic Stenographic System
CRESS: Combined Reentry Effort for Small Systems (ABRES)
CRESS: Computerized Reader Enquiry Service System
CRESS/AU: Center for Research in Social Systems of the American University
CREST: Committee on Reactor Safety and Technology
CRESTS: Courtauld's Rapid Extract, Sort and Tabulate System
CREVS: control room emergency ventilation system
CRF: capital recovery factor
CRF: Compressor Research Facility
CRF: contrast rendition factor
CRF: correspondence routing form

CRF: Cryptographic Repair Facility (military)
CRI: color rendering index
CRI: Committee for Reciprocity Information
CRI: Centre de Recherches et d'Irradiations, Chamipany, France
CRIME: Censorship Records and Information, Middle East (military)
CRIS: command retrieval information system
CRIS: current research information system
CRJE: conversational remote job entry
CRM: containment radiation monitor
CRM: control and reproductibility monitor
CRM: core resistant mechanism
CRM: core restraint mechanism
CRM: count-rate meter
CRM: counter-radar measures
CRM: counter-radar missile
CRM: Count rate meter
CRNL: Chalk River Nuclear Laboratories (Canada)
CRO: cathode-ray oscilloscope
CRO: Control Room Operator
CROC: engineering fluid-flow parametric reactor computer code
CROM: capacitive read-only memory
CROM: control read-only memory
CRPL: Central Radio Propagation Laboratory (ESSA)
CRPM: Communication Registered Publication Memoranda
CRREL: Cold Regions Research and Engineering Laboratory (Army)
CRS: Congressional Research Service, Library of Congress
CRS: containment rupture signal
CRSA: control rod scram accumulator
CRT: cathode-ray tube
CRT: circuit requirement table
CRTPB: Canadian Radio Technical Planning Board
CRTU: combined receiving and transmitting unit
CRU: Combined rotating unit
CRUD: Chalk River unidentified deposit
CRW: community radio watch
CRWO: Coding Room Watch Officer (Navy)
C/S: call signal
CS: Catalysis Society
CS: commercial standard
CS: community service
C&S: computers and systems
CS: computer science
CS: conducted susceptibility
CS: containment spray
CS: control switch
CS: Control Systems (IEEE Society)
c/s: cycles per second
CSA: Canadian Standards Association
CSA: Community Service Activities (AFL-CIO)
CSA: Community Services Administration
CSA: Computer Sciences Association
CSA: Computer Systems Association
CSA: Cryogenics Society of America
CSAE: Canadian Society of Agricultural Engineering

CSAS: core standby actuation signal
CSB: Consumer Sounding-Board
CSC: circuit switching center
CSC: Civil Service Commission, USA
CSC: common signaling channel
CSC: Computer Sciences Corporation
CSC: Computer Society of Canada
CSC: Construction Specifications Canada
CSC: core standby cooling
csc: cosecant
CSCE: Canadian Society for Civil Engineers
CSCE: Connecticut Society of Civil Engineers
CSC/FPRAC: Federal Prevailing Rate Advisory Committee
CSChE: Canadian Society for Chemical Engineering
CSCS: Core standby cooling system
CSD: Cold Shutdown
CSD: Constant speed drive
CSD: controlled-slip differentials
CSDD: Conceptual system design description
CSDF: Core segment development facility
CSE: Colorado Society of Engineers
CSE: Containment Steam Explosion
CSE: Containment systems experiment, Hanford, Richland, Washington
CSE: control systems engineering
CSE: core storage element
CSEE: Canadian Society for Electrical Engineers
CSEPA: Central Station Electrical Protective Association
CSF: Containment support fixture
CSF: Coulter Steel and Forge
CSI: Construction Specifications Institute
CSIC: computer system interface circuits
CSICOP: Committee for the Scientific Investigation of Claims of the Paranormal
CSIR: Council for Scientific and Industrial Research (South Africa)
CSIR: Council of Scientific and Industrial Research (India)
CSIRAC: Commonwealth Scientific and Industrial Research Automatic Computer (Australia)
CSIRO: Commonwealth Scientific and Industrial Research Organization (Australia)
CSIS: core spray injection system
CSL: code selection language
CSL: computer sensitive language
CSL: control and simulation language
CSL: current switch logic
CSM: continuous sheet memory
CSMA: Chemical Specialties Manufacturers Association
CSMA: common spectrum multiple access
CSMA: Communications Systems Management Association
CSMA/CD: carrier sensitive multiple access with collision detection
CSME: Canadian Society for Mechanical Engineering
CSMP: Continuous System Modeling Program (IBM)
CSP: Commercial Subroutine Package (IBM)

CSP: Communications Security Publication (military)
CSP: continuous sampling plan
CSP: control switching point
CSP: Council for Scientific Policy
CSR: control and status register
CSRS: Cooperative State Research Service
CSS: Cask support structure
CSS: Colorado Scientific Society
CSS: containment spray system
CSS: core support structure
CSSA: Computer Society of South Africa, Ltd
CSSA: Crop Science Society of America
CSSB: compatible single sideband
CSSE: Conference of State Sanitary Engineers
CSSL: continuous system simulation language
CSSS: Canadian Soil Science Society
CSST: computer system science training
CST: central standard time
CSTI: Civil Space Technology Initiative
CSTI: Committee on Scientific and Technical Information
CSTS: combined system test stand
CSU: circuit switching unit
CSV: corona start voltage
CSW: control power switch
CS0: control signal zero
CS1: control signal one
CT: Cable test
CT: center tap
CT: Circuit Theory
CT: CLEM transporter
CT: commercial translator
CT: communications technician (military)
CT: computed tomography
CT: computer technology
CT: computer tomography
CT: Connecticut
CT: current transformer
CTA: compatibility test area
CTA: Concrete Technicians Association of Hawaii
CTD: Charge-transfer device
CTD: Charged tape detection (fuel-failure monitor)
CTDH: command and telemetry data handling
CTDS: code translation data system
C/TDS: count/time data system
CTE: Coefficient of Thermal Expansion
CTF: Cask tilting fixture
CTF: contrast transfer function
CTFM: continuous transmission frequency modulated
C³L: complementary constant-current logic
CTL: CAGE test language
CTL: checkout test language
CTL: complementary transistor logic
CTL: core transistor logic
CTMC: Communication Terminal Module Controller (UNIVAC)
CTMT: Containment
CTO: charge transforming operator
CTOL: conventional takeoff and landing
CTOS: Cassette operations system

CTP: charge transforming parameter
CTP: chemical treatment pond
CTP: command translator and programmer
CTP: construction test procedure
CTP: controlled temperature profile (vapor trap)
CTR: certified test results
CTR: controlled thermonuclear reaction
CTR: controlled thermonuclear reactor
CTS: carrier test switch
CTS: communications technology satellites
CTS: communications terminal synchronous
CTS: Computer Telewriter System (Broomfield, CO)
CTS: computer typesetting
CTS: conversational terminal system
CTS: conversational time sharing
CTS: courier transfer station (military)
CTSS: compatible time shared system
CTU: central terminal unit
CU: coefficient of utilization
CU: Consumer Union
CU: control unit
CU: customer premise
CU: piezoelectric-crystal unit
CUAS: computer utilization accounting system
CUBE: cooperating users of Burroughs equipment
CUBOL: computer usage business oriented language
CUD: Craft Union Department (AFL-CIO)
CUDOS: continuously updated dynamic optimizing systems
CUE: computer updating equipment
CUE: cooperating users exchange
CUES: computer utility educational system
CUJT: complementary unijunction transistor
CULP: computer usage list processor
CULV: culvert
CUMM: Council of Underground Machinery Manufacturers
CUMMFU: complete utter monumental military foul up
CUP: communications users program
CURES: computer utilization reporting system
cur limiting: current limiting
CURTS: common user radio transmission system
CURV: cable-controlled underwater research vehicle
CUS: clean-up system
CUSIP: Committee on Uniform Security Identification Procedures
CUT: circuit under test
CUT: control unit tester (UNIVAC)
CutiePie: type of radiation detection instrument
CUTS: computer utilized turning system
CUW: Committee on Undersea Warfare (DOD)
C(V): capacitance as a function of voltage
CV/CC: constant voltage/constant current
CVCS: chemical and volume control system
CVD: chemical vapor deposition
CVD: coupled vibration dissociation
CVD: current voltage diagram
CVDV: coupled vibration dissociation vibration
CVI: Certified vendor information

CVIC: conditional variable incremental computer
CVN: Nuclear powered carrier
CVP: containment vacuum pump
CVR: cockpit voice recorder
CVSD: continuously variable slope delta modulation
CVTR: Carolinas-Virginia Tube Reactor, Parr, SC, power reactor, 650 MW(th)
CW: carrier wave; composite wave; continuous wave
CW: chemical warfare
CW: clean water
CW: clockwise
CW: cold-worked
CW: continuous wave
CWA: Communications Workers of America
CWAR: continuous wave acquisition radar
CWAS: contractor weighed average share
CWIP: construction work in-progress
CWMTU: cold-weather material test unit
CWO: capital work order
CWP: coal worker's pneumoconiosis (black lung)
CWP: contractor work plan
CWS: cooling water system
CX: composite signaling
CY: calendar year
CZ: Canal Zone (Panama)
C3L: complementary constant-current logic

D

D-A: digital to analog
D-B: decimal to binary
D-D: deuterium-deuterium reaction (fusion program)
D-DAS: digital data acquisition system
D-H: decimal to hexadecimal
D-O: decimal to octal
D-R: decontamination - cleaning with a damp rag method
D-T: deuterium-tritium reaction (fusion program)
D-W Process: fractional distillation process for heavy water concentration
D-38: former code name for depleted uranium metal
D: diode
DA: decimal add
DA: decimal to analog
DA: decision automation
DA: design automation
DA: destination address
DA: device address
DA: differential analyzer
DA: digital to analog
DA: discrete address
DA: double amplitude
DAA: data access arrangement
DABS: discrete address beacon system
DAC: design augmented by computer
DAC: digital-to-analog converter
DAC: display analysis console
DACC: direct access communications channel
DACON: digital to analog converter
DACOR: data correction

DACS: data acquisition control system
DACS: Data & Analysis Center for Software
DADC: digital air data computer
DADEE: dynamic analog differential equation equalizer
DAFC: digital automatic frequency control
DAFCS: digital automatic flight control system
DAFM: discard-at-failure maintenance
DAIR: direct attitude and identity readout
DAIR: driver air, information, and routing
DAIS: defense automatic integrated switching system
DAIS: digital avionics information system
DAISY: double precision automatic interpretive system
DAM: dam
DAM: data addressed memory
DAM: data association message
DAM: descriptor attribute matrix
DAM: digital-to-analog multiplier
DAM: direct access memory
DAM: direct access method
DAMP: down-range antimissile measurement program
DAMPS: Data Acquisition Multiprogramming System (IBM)
DANA: one of two heavy water facilities, Newport, Indiana (now property of the U.S. Army Chemical Corps
DAP: deformation of aligned phases
DAP: digital assembly program (EAI)
DAP: distributed array processor
DAPR: digital automatic pattern recognition
DAPS: direct access programming system
DAQ: data acquisition
DAR: daisy chain A return
DAR: defense acquisition radar
DARE: document abstract retrieval equipment
DARE: documentation automated retrieval equipment
DARE: doppler automatic reduction equipment
DARES: data analysis and reduction system
DARLI: digital angular readout by laser interferometry
DARPA: defense advanced research projects agency
DARS: digital adaptive recording system
DARS: digital attitude and rate system
DART: daily automatic rescheduling technique
DART: data analysis recording tape
DART: director and response tester
DART: dual axis rate transducer
DART: dynamic acoustic response trigger
DAS: Data Acquisition System
DAS: data automation system
DAS: Datatron Assembly System
DAS: digital analog simulator
DAS: Disturbance Analysis System
DASD: direct access storage device
DASS: demand assignment signaling and switching
DASS: Diesel Air Start System
DAT: digital audio tape
DAT: Director of Advanced Technology

DAT: disconnect actuating tools
DAT: dynamic address translation
DATAC: data analog computer
DATACOL: data collection system
DATACOM: data communications
Datacom/Com: Data Communication Systems
DATAGEM: data file generator
DATAN: data analysis
DATAR: digital autotransducer and recorder
DATICO: digital automatic tape intelligence checkout
DATRIX: direct access to reference information
DAU: data acquisition unit
DAV: data valid
DAVC: delayed automatic volume control
DAVI: dynamic antiresonant vibration isolator
dB: decibel
DB: double-biased (relay)
DB: double bottom
DB: double break
DB: dry bulb
DB: dynamic braking contactor to relay
DBA: database administration
dBa: decibel adjusted
DBA: design basis accident
dBa: weighted noise power in dB referred to -85 dBm
dBa(F1A): noise power measured by a set with F1A-line weighting
dBa(HA1): noise power in dBa measured by a set with HA1-receiver weighting
DBAO: digital block AND-OR gate
dBa0: noise power in dBa referred to or measured at 0TLP
DBB: detector back bias
DBB: detector balanced bias
DBC: data base computer
DBC: diameter bolt circle
DBDA: design basis depressurization accident
DBE: design basis earthquake
DBE: design basis event
DBF: demodulator band filter
DBF: design basis fault
dBj: relative RF signal levels (j stands for Jerrold Electronics where the term originated)
dBk: decibels referred to one kilowatt
DBM: data buffer module
dBm: decibels above (or below) 1 milliwatt
dB/m²: dB above 1 milliwatt per square meter
dBm/m² /MHz: dB above 1 milliwatt per square meter per megahertz
dBm(PSOPH): noise power in dBm measured by a set with psophometric weighting
DBMS: data base management system
dBμ V: dB above 1 microvolt
dBμ V/MHz: dB above 1 microvolt per megahertz
dBμ V/m/MHz: dB above 1 microvolt per meter per megahertz
dBm0: noise power in dBm referred to or measured at 0TLP
dBm0p: noise power in dBm0 measured by a set with psophometric weighting

DbNC: distributed numerical control
DBoA: delayed breeder or alternative
DBoMP: Data BoMP
DBOS: disk-based operating system
DBR: daisy chain B return
dBr: power difference in dB between any point and a reference point
dBRAP: decibels above reference acoustical defined as 10^{16} watt
dBRN: decibels above reference noise
dBRNC: decibels above reference noise, C-message weighted
dBrnc0: noise power in dBrnc referred to or measured at 0TLP
dBrn(f1 - f2): flat noise power in dBrn
dBrn(144 line): noise power measured by a set with 144-line weighting
DBSP: double-based solid propellant
DBT: Design basis tornado
DBUT: data base update time
dBV: increase or decrease in voltage regardless of impedance levels
dBW: decibels referred to 1 watt
dBx: decibels above the reference coupling
DC: data channel
DC: data check
DC: data classifier
DC: data collection
DC: data communications
DC: data control
DC: decimal classification
DC: design contractor
DC: desk checking (software)
DC: digital comparator
DC: digital computer
dc: direct current
DC: disc controller
DC: display console
DC: District of Columbia
DC: double contact
DCA: decade counting assembly
DCA: Defense Communications Agency
DCA: digital command assembly
DCA: Digital Computer Association
DCAA: Defense Contract Audit Agency
DCAOC: Defense Communications Agency Operations Center
DCAS: Deputy Commander Aerospace System
DCB: data control block
DCB: Defense Communications Board
DCBRL: Defense Chemical, Biological, and Radiation Laboratories
DCC: data communication channel
DCC: device control character
DCC: District Communications Center
DCC: double cotton covered
DCCC: data communication control character
DCCEAS: District of Columbia Council of Engineering and Architectural Societies, Inc
DCCS: digital command communications system

DCCU: data communications control unit
DCD: data carrier detect
DCD: dynamic computer display
DCDMA: Diamond Core Drill Manufacturers Association
DCDS: digital control design system
DCE: data circuit-terminating equipment
DCEO: Defense Communications Engineering Office
DCES: Delaware Council of Engineering Societies
DCIB: data communication input buffer
DCKP: direct current key pulsing
DCL: Digital Computer Laboratory
dcm: dc noise margin
DCOS: data communication output selector
DCP: data communication processor
DCP: Design Change Package
DCP: design criteria plan (MIL-STD-469)
DCP: digital computer processor
DCP: digital computer programming
DCPA: Defense Civil Preparedness Agency
DCPG: Defense Communications Planning Group
DCPS: digitally controlled power source
DCPSK: differentially coherent phase shift keying
DCR: data conversion receiver
DCR: data coordinator and retriever
DCR: design change request
DCR: digital conversion receiver
DCR: direct conversion reactor study - (Marlin Co.)
DCRN: Dashpot cup retention nut
DCS: data communications subsystem
DCS: data communication system
DCS: data control services
DCS: Defense Communications System
DCS: design control specifications
DCS: digital command system
DCS: digital communication system
DCS: Direct-Couple Operating System (IBM)
DCS: distributed computer system
DCSP: Defense Communications Satellite Program
DCTL: direct coupled transistor logic
DCU: data command unit
DCU: data control unit
DCU: digital control unit
DCWV: direct-current working volts
DCX-1,2: Direct current experiments, ORNL fusion research program
DD: data definition
DD: decimal divide
DD: Dewey decimal
DD: digital data
DD: disconnecting device
DDA: Digital differential analyzer
DDAFP: Diesel Driven Aux. Feedwater Pump
DDAS: digital data acquisition system
DDB: Two-dimensional burnup reactor computer code
DDC: deck decompression chamber
DDC: Defense Documentation Center (DOD)
DDC: Dewey decimal classification
DDC: digital display converter
DDC: director digital control

DDCE: digital data conversion equipment
DDCS: digital data calibration system
DDD: direct distance dialing
DDD: display decoder driver
DDE: decentralized data entry
DDE: director design engineering
DDG: digital data generator
DDG: digital display generator
DDGE: digital display generator element
DDH&DS: digital data handling and display system
DDI: depth deviation indicator
DDL: dispersive delay line
DDM: data demand module
DDM: difference in depth of modulation
DDM: digital display make-up
DDOCE: digital data output conversion equipment
DDP: digital data processor
DDP: distributed data processing]
DDPS: Digital Data Processing System
DDPU: digital data processing unit
DDR: double drift region
DDR&E: Directorate of Defense Research and Engineering (DOD)
DDRS: digital data recording system
DDS: data display system
DDS: deployable defense system
DDS: Digital data service
DDS: digital display scope
DDS: digital dynamics simulator
DDT: deflagration to detonation transition
DDT: design data transmittal
DDT: digital data transmission
DDT: digital debugging tape
DDT: dynamic debugging technique
DDT: FORTRAN pre-processor, for diagnostic preliminary programming
DDTE: digital data terminal equipment
DDTS: digital data transmission system
DE: Delaware
DE: design engineering
DE: diesel-electric
DE: digital element
DE: display electronics
DEA: display electronics assemblies
DEA: draft environmental statement
DEA: Drug Enforcement Administration
DEAL: decision evaluation and logic
DEAP: diffused eutectic aluminum process
DEB: digital European backbone
DEC: Danville Engineers Club
DEC: Duluth Engineers Club
DECEO: Defense Communication Engineering Office
deci: prefix meaning one-tenth
DECOM: telemetry decommutators
DECUS: Digital Equipment Computer Users Society
DED: Design Engineering Directorate
DEDUCOM: deductive communicator
DEE: digital evaluation equipment
DEE: digital events evaluator
DEFT: dynamic error-free transmission

DEI: development engineering inspection
DEMA: Data Entry Management Association
DEMA: Diesel Engine Manufacturers Association
DEP: Design external pressure
DEPA: Defense Electric Power Administration
DEPI: differential equations pseudocode interpreter
DEPSK: differential-encoded phase shift keying
DE/Q: design evaluation/qualification
DEQ: Dose Equivalent
DER: diesel engine, reduction drive
DERE: Dounreay Experimental Reactor Establishment, United Kingdom
DES: data encryption standard
DES: Department of Education and Science
DES: Department of Emergency Services, State of Washington
DES: differential equation solver
DES: digital expansion system
DESC: Defense Electronics Supply Center (DOD)
DESC: digital equation solving computer
DESC: Directorate of Engineering Standardization
DEU: data exchange unit
DEUA: Diesel Engines and Users Association
DEUA: Digitronics Equipment Users Association
DEUCE: digital electronic universal computing engine
DEVCO: Development Committee
DEW: directed-energy weapons
DEW: distant early warning
DEWIZ: distant early warning identification zone
DEXAN: digital experimental airborne navigator
DF: decimal fraction
DF: decontamination factor
DF: deflection factor
DF: deuterium fluoride
DF: direction finder
DF: disk file
DF: dissipation factor
DF: dose factor
DF: double feeder
DFB: distribution fuse board
DFC: design field change
DFC: diagnostic flow charts
DFCS: digital automatic flight control
DFFR: dynamic forcing function (information) report
DFGS: digital flight guidance system
DFI: Deep Foundations Institute
DFI: developmental flight instrumentation
DFL: display formatting language
DFMSR: Directorate of Flight and Missile Safety Research
DFO: Director Flight Operations
DFP: diesel fire pump
DFR: decreasing failure rate
DFR: Dounreay Fast Reactor, Dounrey, Scotland, 15 MW(th)
DFRL: differential relay
DFS: Dividends from Space
DFS: dynamic flight simulator
DFSK: double frequency shift keying
DFSR: Directorate of Flight Safety Research

DFT: Discrete Fourier transform
DG: diesel generator
DG: differential generator
DG: diode gate
DG: double-groove (insulators)
DGB: disk-gap-bond
DGBC: digital geoballistic computer
DGDP: double-groove double-petticoat (insulators)
DGON: *Deutsch Gesellschaft fur Ortung and Naviga-tion*
DGS: data gathering system
DGS: data ground station
DGS: degaussing system
DGZ: desired ground zero
DH: decay heat
DH: double-heterostructure
DHC: data handling center
DHCC: decay heat closed cooling
DHE: data handling equipment
DHR: Department of Human Resources
DHRS: decay heat removal system
DHRS: direct heat removal system
DHT: Discrete Hilbert transform
DHX: dump heat exchanger
DHXCS: dump heat exchanger control system
DI: data input
DI: demand indicator
DI: digital input
DIA: Defense Intelligence Agency
DIA: distributed-lumped active
diac: bidirectional diode-thyristor
DIAC: Defense Industry Advisory Council
DIAL: Display Interactive Assembly Language (DEC)
DIAL: Drum Interrogation, Alteration and Loading System (Honeywell)
DIALGOL: dialect of Algol
DIAN: digital analog
DIANE: digital integrated attack and navigation equipment
DIAS: dynamic inventory analysis system
dibit: a group of four bits
DIC: data insertion converter
DIC: Detailed Interrogation Center
DICEF: digital communications experimental facility
DICON: digital communication through orbiting needles
DID: data input device
DID: Datamation industry directory
DID: digital information display
DID: direct inward dialing
DID: Division of Isotopes Development
DIDACS: digital data communications system
DIDAD: digital data display
DIDAP: digital data processor
DI/DO: data input/data output
DIDO: tank-type reactor at Harwell, United Kingdom, 15 MW(th)
DIE: Document of Industrial Engineering
DIF: *Dansk Ingeniorforening*
DIF: data interchange format

DIFFTR: differential time relay
DIG: digital-image-generated
DIGACC: digital guidance and control computer
DIGCOM: digital computer
DIGICOM: digital communications system
DIGITAC: digital tactical automatic control
DIGITAR: digital airborne computer
Dig Sig Proc: digital signal processing
DIIC: dielectrically isolated integrated circuits
DIKE: dike
DIMATE: depot-installed maintenance automatic test equipment
DIMES: defense integrated management engineering system
DIMPLE: Deuterium Moderated Pile Low Energy Reactor, AERE, Harwell, Berkshire, United Kingdom
DIMPLE: Tank-type reactor at Winfrith, United Kingdom, 100 MW(th)
DIMUS: digital multibeam steering
DIN: *Deutsche Industrie Normenausschuss*
DINA: direct noise amplification
DIOB: digital input/output buffer
DIP: Design internal pressure
DIP: display information processor
DIP: dual-in-line package
DIPS: development information processing system
DIRCOL: direction cosine linkage
DIS: Defense Investigative Service
DIS: digital integration system
DIS: draft international standard
DISAC: digital simulator and computer
DISC: Disconnect
DISC: disconnect command
disc: disconnector
DISC: disconnect switch
DISC: domestic international sales corporations
DISCOM: digital selective communications
DISD: Data and Information Systems Division
dist: distribution
DISTRAM: digital space trajectory measurement system
DITRAN: diagnostic Fortran
DIVA: digital input voice answer-back
DIVAD: Division Air Defense
DIVOT: digital-to-voice translator
DK: data acknowledge
DL: data chain left
DL: data link
DL: dead load
DL: delay line
DL: diode logic
DLA: daisy chain out left
DLA: Defense Logistics Agency
DLB: daisy chain in left
DLC: direct lift control
DLC: duplex line control
DLE: data link escape character
DLI: Defense Language Institute
DLIEC: Defense Language Institute, East Coast

DLK: data link
DLP: data listing programs
DLS: DMF-based landing system
DLSC: Defense Logistics Services Center
DLT: data line translator
DLT: data loop transceiver
DLT: decision logic table
DLT: depletion-layer transistor
DM: data management
DM: decimal multiply
dm: decimeter
DM: delta modulation
DM: disconnected mode
DMA: Direct memory access
DMAS: Distribution Management Accounting System
DMC: digital microcircuit
DME: distance measuring equipment
DMED: digital message entrance device
DMIRR: demand mode integral rocket Ramjet
DMLS: Doppler microwave landing system
DMM: digital multimeter
DMO: Data Management Office
DMOS: diffused metal-oxide semiconductor
DMR: dynamic modular replacement
DMS: data management service
DMS: data management system
DMS: defense missile systems
DMS: documentation of molecular spectroscopy
DMSP: defense meteorological satellite program
DMSR: Director of Missile Safety Research
DMST: demister
DMT: digital message terminal
DMT: Dispersive mechanism test
DMTC: digital message terminal computer
DMTR: Dounreay Materials Testing Reactor, Dounreay, Scotland, 10 MW(th)
DMU: digital message unit
DMU: dual maneuvering unit
DMZ: demilitarized zone
DN: decimal number
DN: delayed neutron
DN: delta amplitude
DN: down
DNA: Deoxyribonucleic Acid
DNA: does not apply
DNB: departure from nucleate boiling
DNBR: departure from nucleate boiling ratio
DNC: direct numerical control
DNCCC: Defense National Communications Control Center
DNE: Department of Nuclear Engineering
DNL: dynamic noise limiter
DNM: delayed neutron monitor
DNR: Department of Natural Resources
DNRC: Democritus Nuclear Research Center, Greece
DNS: doppler navigation system
DNSR: Director of Nuclear Safety Research
DO: data output
DO: design objective
DO: digital output

D.O.: dissolved oxygen
DOC: data optimizing computer
DOC: decimal to octal conversion
DOC: direct operating costs
DOCS: disk-oriented computer system
DOCUS: display-oriented computer usage system
DOD: Department of Defense
DOD: Development Operations Division
DOD: direct outward dialing
DODCI: Department of Defense Computer Institute
DODGE: Department of Defense Gravity Experiment
DODIS: distribution of oceanographic data at isentropic levels
DoDISS: Department of Defense Index of Specifications and Standards
DOE: U.S. Department of Energy
DOF: degreee of freedom
DOF: direction of flight
DOFIC: domain originated functional integrated circuit
DOI: differential orbit improvement
DOI: U.S. Department of Interior
DO/IT: digital output/input translator
DOL: Department of Labor
DOL: Director of Laboratories
DOL: display-oriented language
DOL: dynamic octal load
DOM: digital ohmmeter
Dom App/IAS: Domestic Appliance
DOMSATS: domestic communications satellite systems
DON: Heterogeneous, enriched-uranium, heavy-water-moderated, organic-cooled, materials-testing and power reactor, Spain
DOP: Detailed operating procedure
DOP: dioctyl phosphate
DOPIC: documentation of programs in core
DOPS: digital optical protection system
DOR: Division of Operating Reactors
DOR: Heavy-water-moderated and organic-cooled reactor
DORAN: doppler range and navigation
DORIS: direct order recording and invoicing system
DORV: deep ocean research vehicle
DOS: disk operations system
DOSAR: Dosimetry Applications Research Facility
DOSV: deep ocean survey vehicle
DOT: deep ocean technology
DOT: Department of Transportation
DOT: Designating Optical Tracker
DOUSER: doppler unbeamed search radar
DOVAP: doppler velocity and position
DOWB: deep operating work board
DP: data processing
DP: deep penetration
DP: deflection plate
DP: dew point
DP: dial pulse
DP: diametral pitch
DP: differential pressure

DP: disk pack
DP: distribution point
DP: double pole
DP: drum processor
DP: dynamic programming
DPA: data processing activities
DPB: data processing branch
DPB: Defense Policy Board
DPBC: double-pole back-connected
DPC: data processing center
DPC: direct program control
DPCM: differential pulse-code modulation
DPD: data processing department
DPD: data processing division
DPD: digit plane driver
DPDT: double pole, double throw (switch)
DPE: data processing equipment
DPE: development project engineer
DPEC: double-pole front-connected
DPEK: differential phase-exchange keying
DPFS: distribution power frequency signalling
DPG: digital pattern generator
DPI: digital process instrument
DPI: digital pseudorandom inspection
DPIS: differential pressure isolation switch
DPLC: distribution power line carrier
DPLCS: digital propellant level control system
DPLL: digital phase-locked loop
DPM: data processing machine
DPM: digital panel meter
DPMA: Data Processing Management Association
DPMS: data project management system
DPP: drip pan pot (CLEM)
DPS: data processing standards
DPS: data processing systems
DPS: descent power system
DPS: disk programming system
DPSA: Data Processing Supplies Association
DPSK: differential phase-shift keying
DPSS: data processing system simulator
DPST: double-pole, single-throw (switch)
DQC: data quality control
DR: data receiver
DR: data recorder
DR: data reduction
DR: dead reckoning
DR: discrepancy report
DR: Division of Research
DRA: daisy chain in right
DRA: dead reckoning analyzer
DRADS: degradation of radar defense systems
DRAGON: first prompt-critical assembly at LASL
DRAGON: Power reactor, 20 MW(th), Winfrith, Dorset, United Kingdom, joint UK/Eurahoni Project
DRAI: dead reckoning analog indicator
DRAMA: digital radio and multiplex acquisition
DRB: daisy chain out right
DRC: damage risk criterion
DRC: data reduction compiler
DRD: Director of Research and Development

DRDTO: detection radar data take-off
DRE: dead reckoning equipment
DRE: Director of Research and Engineering
D/RE: Disassembly/reassembly equipment
DRES: Direct reading emission spectrograph
DRI: data reduction interpreter
DRI: dead reckoning indicator
DRIBBLE: Vela uniform nuclear detonations to investigate detection of decoupled underground explosions
DRIFT: diversity receiving instrumentation for telemetry
DRM: digital ratiometer
DRML: Defense Research Medical Laboratories
DRMO: District Records Management Officer
DRO: destructive readout (memory)
DRO: digital readout
DRO: doubly resonant oscillator
DR_1: Danish Reactor 1, L-55-type, water-boiler, enriched-uranium, light-water-cooled and -moderated research reactor, Denmark
DROD: delayed readout detector
DROS: direct readout satellite
DROS: disk resident operating system
DRP: dead reckoning plotter
DRPI: digital rod position
DRR: Document release record
DRS: digital radar simulator
DRS: Disassembly/reassembly station (IEM cell core components)
DRSS: data relay satellite system
DRT: dead reckoning tracer
DRT: device rise time
DR_3: Danish Reactor 3, enriched-uranium, heavy-water-cooled and -moderated research reactor, Denmark
DR_2: Danish Reactor 2, enriched-uranium, swimming-pool, Denmark
DRTC: Documentation Research and Training Center (India)
DRTE: Defense Research Telecommunications Establishment
DRTL: diode resistor transistor logic
DRTR: dead reckoning trainer
DS: data scanning
DS: data set
DS: data sync
DS: decimal subtract
DS: descent stage
DS: disconnect switch
DS: disk storage
D&S: display and storage
DS: drainage structure
DS: drum storage
D/S: Dynamic/static analysis
DSA: dial service assistance
DSA: digital signal analyzer
DSAP: data systems automation program
DSAP: destination service access point
DSAR: data sampling automatic receiver

DSARC: Defense Systems Acquisition Review Council

DSB: Defense Science Board

DSB: double-sideband

DSBAMRC: double-sideband amplitude modulation reduced carrier

DSC: design safety criteria (Hanford)

DSC: double silk covered

DSCB: data set control block

DSCC: deep space communications complex

DSCS: defense system communications satellite

DSCS: desk side computer system

DSCT: double secondary current transformer

DSD: digital system design

DSE: Dartmouth Society of Engineers

DSE: data storage equipment

DSE: data systems engineering

DSEA: data storage electronics assembly

DSEG: Data Systems Engineering Group

DSG: dispersed sources of generation

DSG: dispersed storage and generation

DSH: deactivated shutdown hours

DSI: digital speech interpolation

DSIF: deep-space instrumentation facility

DSIF: deep space instrumentation facility

DSIR: Department of Scientific and Industrial Research, London, United Kingdom

DSIS: Directorate of Scientific Information Services

DSL: data set label

DSL: data simulation language

DSL: data structures language

DSL: drawing and specification listing (FFTF)

DSM: dynamic scattering mode

DSN: Deep Space Network (NASA)

DSR: data scanning and routing

DSR: discriminating selector repeater

DSS: decision support system

DSS: deep space station

DSS: Director of Statistical Services

DSS: Division of Safety Studies

DSSA: Defense Security Assistance Agency

DSSB: double single-sideband

DSSC: double-sideband suppressed carrier (modulation)

DSSCS: defense special secure communications system

DSTE: data subscriber terminal equipment

DSU: data storage unit

DSU: data synchronizer unit

DSU: disk storage unit

DT: data translator

DT: data transmission

DT: decay time

DT: differential time

DT: digital technique

DT: digit tube

D/T: disk tape

DT: double-throw

DT: tunnel diode

DTA: differential thermal analysis

DTARS: digital transmission and routing system

DTAS: data transmission and switching

DTB: decimal to binary

DTC: data transmission center

DTCS: digital test command system

DTDMA: distributed time division multiple access

DTE: data terminal equipment

DTE: data transmission equipment

DTG: display transmission generator

DTG: dynamically tuned gyro

DTI: Department of Trade and Industry

DTL: diode transistor logic

DTM: delay timer multiplier

DTMF: dual-tone multifrequency

DTMS: digital test monitoring system

DTO: digital testing oscilloscope

DTP: directory tape processor

DTPA: diethylenetriaminepentaacetic acid

DTPL: domain tip propagation logic

DTR: definite-time relay

DTR: demand-totalizing relay

DTR: document transmittal record

DTRF: data transmittal and routing form

DTRS: development test requirement specification

DTS-W: Defense Telephone Service (Washington)

DTS: data transmission system

DTT: design thermal transient

DTU: data transfer unit

DTU: digital tape unit

DTU: digital transmission unit

DTUTF: digital tape unit tape facility

D/TV: digital to television

DTVC: digital transmission and verification converter

DTVM: differential thermocouple voltmeter

DUAL: dynamic universal assembly language

DUCT: breeching; plenum; duct

DUF: diffusion under (epitaxial) film

DUMC: deep underwater measuring device

DUMP: dumper

DUMS: deep unmanned submersibles

DUN: Douglas United Nuclear (now United Nuclear Industries)

DUNC: deep underwater nuclear counter

DUNS: data universal numbering system

DUO: Datatron Users Organization

DUT: device under test

DUV: data under voice

DV: differential voltage

DV: dump valve

DV: valve, diverter

DVA: dynamic visual acuity

DVC: capacitive diode

DVCCS: differential voltage-controlled current source

DVD: detail velocity display

DVESO: DOD Value Engineering Services Office

DVFO: digital variable-frequency oscillator

DVM: digital volt meter

DVM: displaced virtual machine

DVOM: digital volt-ohm meter

DVOR: Doppler VHF omnidirectional radio range

DVST: direct view storage tube
DW: dead weight load
DWBA: distorted-wave Born approximation
DWED: dry well equipment drain
DWFD: dry well floor drain
DWICA: deep water isotropic current analyzer
DWL: dominant wavelength
DWS: development work statement
DWSMC: Defense Weapons Systems Management Center (DOD)
DWST: Demineralized water storage tank
DX: direct current (signaling)
DX: distance (in radio reception of distant stations)
DYNASAR: Dynamic Systems Analyzer (GE)
DYSAC: dynamic storage analog computer
DYSTAC: dynamic storage analog computer
DYSTAL: dynamic storage allocation language
DZ: Zener Diode
D 3: defense, description and designation

E

E-MAD: Engineer-maintenance assembly-disassembly NERVA program, Jackass Flats, Nevada
E: Equivalent Hours
E: erlang
E: illumination (*éclairage*)
E: voltage symbol
EA: effective address
EA: Electrics Association
EAA: Engineer in Aeronautics and Astronautics
EAB: Bureau of Economic Analysis
EAB: Educational Activities Board
EAB: exclusion area boundry
EABRD: electrically activated bank release device
EAC: Engineering Affairs Council (see AAES)
EACC: error adaptive control computer
EACO: Engineers & Architects Council of Oregon
EACRP: European-American Committee on Reactor Physics
EADI: electronic altitude director indicator
EAES: European Atomic Energy Society
EAF: electron arc furnace
EAF: Equivalent Availability Factor
EAH: Engineering Association of Hawaii
EAI: Engineers and Architects Institute
EAL: Equipment air lock
EAM: electric accounting machine
EAM: electronic automatic machine
EAP: Experimental activity proposal
EAR: electronically agile radar
EAR: employee attitude research
EARC: Extraordinary Administrative Radio Conference
EAROM: electrically alterable read-only memory
EAS: experiment assurance system
EAS: extensive air shower
EASA: Electrical Apparatus Service Association
EASCON: Electronics and Aerospace Systems Convention (IEEE)

EASE: electrical automatic support equipment
EASI: Electrical Accounting for the Security Industry (IBM)
EASI: Estimate of adversary sequence interruption
EASL: engineering analysis and simulation language
EAstt: experimental Army satellite tactical terminals
EASY: early acquisition system
EASY: engine analyzer system
EAUTC: Engineer Aviation Unit Training Center
EAX: electronic automatic exchange
EB: electron beam
EBAM: electron-beam-accessed memories
EBCDIC: extended binary-coded decimal interchange code
EBCSM: East Bay Council on Surveying and Mapping
EBDI: Electronic Business Data Interchange
EBFS: Enclosure Bldg. Filtration System
EBIS: Electron beam ion source
EBMD: electron-beam mode discharge
EBOR-CX: Critical assembly (GDC)
EBOR: Experimental Beryllium Oxide Reactor, NRTS, Idaho, 10 MW(th)
EBPA: electron beam parametric amplifier
EBR-1: Experimental Breeder Reactor, NRTS, Mark III, Idaho power reactor, 1.4 MW(th)
EBR-2: Experimental Breeder Reactor, NRTS, Idaho power reactor, 62.5 MW(th)
EBR: electron beam recording
EBR: epoxy bridge rectifier
EBR: experimental breeder reactor
EBS: electron-beam semiconductor
EBS: electron bombarded semiconductor
EBS: Emergency Borating System
EBTF: ECC bypass test facility
EBU: European Broadcasting Union
EBW: exploding bridge wire
EBWR: Experimental Boiling Water Reactor, ANL (decommissioned)
EC: eddy current
EC: electronic calculators
EC: engineering change
EC: engineering construction
EC: Engineering Council
EC: Engineers Club
EC: Engineers Corps
EC: error correcting
EC: European Communities (CE)
ECA: Economic Commission for Africa (CEA)
ECA: Electrical Contractors Association
ECAC: Electromagnetic Compatibility Analysis Center
ECAFE: Economic Commission for Asia and the Far East (CEAEO)
ECAP: Electronic Circuit Analysis Program
ECAR: East Central Area Reliability Coordination Agreement
ECARS: electronic coordinator graph and readout system
ECB: event control block
ECC: Electronic Calibration Center (NBS)

ECC: Electronic Components Conference
ECC: emergency control center
ECC: error checking and correction
ECC: error correction code
ECCANE: East Coast Conference on Aerospace and Navigational Electronics
ECCM: electronic counter-countermeasures
ECCO: Engineers Coordinating Council of Oregon
ECCP: Engineering Concepts Curriculum Project
ECCS: emergency core cooling system
ECD: Engineers Club of Dayton
ECDC: electrochemical diffused collector (transistor)
ECE: Economic Commission for Europe (CEE Genève)
ECE: engineering capacity exchange
ECEC: East Carolina Engineers Club
ECF: Emergency Cooling Functionability
ECF: Engineers Club of Fresno
ECFM: eddy current flow meter
ECG: electrocardiogram
ECG: electrochemical grinding
ECHO: electronic computing
ECI: emergency cooling injection
ECKC: Engineers Club of Kansas City
ECL: emitter-coupled logic
ECL: Engineers Club of Lincoln
ECL: equipment component list
ECLA: Economic Commission for Latin America (CEPAL)
ECM: electric coding machine
ECM: electrochemical machining
ECM: electronic countermeasures
ECM: Engineers Club of Minneapolis
ECMA: Electronic Computer Manufacturers Association
ECMA: Engineering College Magazines Associated
ECMA: European Computer Manufacturers Association
ECMP: electronic countermeasures program
ECMSA: Electronics Command Meteorological Support Agency
ECN: engineering change notice
ECNE: Electric Council of New England
ECNM: Engineer's Club of Northern Minnesota
ECO: electron-coupled oscillator
ECO: electronic checkout
ECO: Emergency Control Officer
ECO: Engineers Club of Omaha
ECOB: Engineers Club of Bartlesville
ECOC: Engineering Club of Oklahoma City
ECOM: electronic computer-oriented mail
ECOM: Electronics Command, Army (Ft. Monmouth, NJ)
ECON: electromagnetic emission control
ECOSOC: Economic and Social Committee
ECP: electromagnetic compatibility program
ECP: engineering change proposal
ECP: The Engineer's Club of Philadelphia
ECPD: Engineers Council for Professional Development (see ABET)

ECPI: Electronic Computer Programming Institute
ECR: electron cyclotron resonance
ECR: electronic cash register
ECR: electronic control relay
ECR: emergency cooling recirculation
ECR: engineering change request
ECR: error control receiver
ECR: estimate change request
ECRC: Electronic Components Reliability Center
ECS: Electrochemical Society
ECS: electronic control switch
ECS: embedded computer system
ECS: emergency coolant system
ECS: enviromental control shroud
ECS: environmental control system
ECS: European Communication Satellite
ECS: extended core storage
ECSA: European Communications Security Agency
ECSC: European Coal and Steel Community (CECA)
ECSF: Engineers Club of San Francisco
ECSG: Electronic Connector Study Group
ECSL: Engineers Club of St. Louis
ECST: condensate storage tank
ECT: eddy current testing
ECU: electronic conversion unit
ECU: environmental control unit
ED: Education (IEEE Group)
ED: Electron Devices (IEEE Group)
ED: engine drive
EDA: Economic Development Administration
EDA: Electrical Development Association
EDA: electronic differential analyzer
EDA: electronic digital analyzer
EDAC: error detection and correction
EDB: educational data bank
EDC: Economic Development Committee
EDC: electronic desk calculator
EDC: electronic digital computer
EDC: emergency decontamination center
EDC: error detection and correction
EDCL: electric-discharge convection laser
EDCOM: editor and compiler
EDCPF: Environmental Data Collection and Processing Facility
EDCW: external device control word
EDD: electronic data display
EDE: emergency decelerating (relay)
EDE: emitter dip effect
EDEW: enhanced distant early warning
EDF-1: Gas-cooled, natural-uranium reactor built by *Electricité de France*
EDF-2: Power reactor, 790 MW(th), Chienon, France
EDF-3: Power reactor, 1200 MW(th), Chienon, France
EDF-4: Proposed power reactor
EDF: *Electricité de France*
EDFP: engine driven fire pump
EDFR: effective date of Federal recognition
EDG: emergency diesel generator
EDG: exploratory development goals

EDGE: electronic data gathering equipment

EDICT: engineering document information collection technique

EDIF: electronic data interchange format

EDIS: engineering data information system

EDLCC: Electronic Data Local Communications Central

EDM: electric discharge machining

EDO: effective diameter of objective

EDO: engineering duties only

EDO: Executive Director of Operations

EDOS: extended disk operating system

EDP: electronic data processing

EDPAA: EDP Auditors Association

EDPC: electronic data processing center

EDPE: electronic data processing equipment

EDPI: Electronic Data Processing Institute

EDPM: electronic data processing machine

EDPS: electronic data processing system

EDR: electrodermal reaction

EDR: equivalent direct radiation

EDS: electronic data system

EDS: emergency detection system

EDS: environmental data service

EDSAC: electronic delay storage automatic computer

EDSAC: electronic discrete sequential automatic computer

EDST: Elastic Diaphragm Switch Technology (IBM)

EDT: electric discharge tube

EDT: energy dissipation tests

EDT: engineer design tests

EDU: electronic display unit

EDU: experimental diving unit

EDVAC: electronic discrete variable automatic computer

EE: electrical engineer

EEA: Electronic Engineering Association

EEC: electronic engine control

EEC: end of equilibrium cycle

EEC: European Economic Community (CEE Bruxelles). The "Common Market"

EECL: emitter-emitter coupled logic

EECW: Emergency Exchanger Cooling Water

EED: electroexplosive device

EEE: Electrical Engineering Exposition

EEG: electroencephalogram

EEI: Edison Electric Institute

EEI: Environmental Equipment Institute

EEM: Electronic Engineer's Master

EEM: electronic equipment monitoring

EEMJEB: Electrical and Electronic Manufacturers Joint Education Board

EEMTIC: Electrical and Electronic Measurement and Test Instrument Conference

EEOC: Equal Employment Opportunity Commission

EEP: Electroencephalophony

EEPROM: Electrically erasable/programable read-only memory

EER: explosive echo ranging

EERI: Earthquake Engineering Research Institute

EERL: Electrical Engineering Research Laboratory

EE&RM: elementary electrical and radio material training school

EES: Engineering Experiment Station

EESC: Erie Engineering Societies Council

EESMB: Electrical & Electronics Standards Management Board/ANSI

EF: elevation finder

EF: emitter follower

EF: Engineering Foundation

EF: error factor

EF: extra fine (thread)

EFAPP: Enrico Fermi Atomic Power Plant (decommissioned)

EFB: Engineering Foundation Board

EFBD: emergency feed boron detector

EFCV: excess flow check valve

EFD: engineering flow diagram

EFF: efficiency-ratio mwe/mwt

EFFBR: Enrico Fermi Fast Breeder Power Reactor, Monroe, Michigan

EFG: edge-defined film-fed growth

EFI: electronic fuel injection

EFIS: electronic flight instruments

EFL: effective focal length

EFL: equivalent focal length

EFO: Engineers Foundation of Ohio

EFOR: equivalent forced outage rate

EFP: electronic field production

EFP: SELNI Enrico Fermi Atomic Power Plant, heterogeneous, enriched-uranium, light-water-cooled and -moderated, Italy

EFPD: effective full power days

EFPH: equivalent full power hours

EFPM: effective full power month

EFS: electronic frequency selection

EFT: earliest finish time

EFT: electronic funds transfer

EFTA: European Free Trade Association

EFTS: electronic funds transfer system

EFTS: Elementary Flying Training School

e.g.: *exempli gratia* (for example)

EGCR: Experimental Gas-Cooled Reactor, power reactor, 84.3 MW(th), 21.9 MW(e), Oak Ridge, Tennessee

EGD: electrogasdynamic

EGM: electronic governor module

EGO: Eccentric (orbit) Geophysical Observatory

EGPS: Extended General Purpose Simulator (NEC)

EGR: electronic governor regulator

EGR: exhaust gas recirculation (internal combustion engines)

EGSMA: Electrical Generating Systems Marketing Association

EGTS: Emergency Gas Treatment System

EHC: electro hydraulic control

EHD: electrohydrodynamic

EHF: Engineers Hall of Fame

EHF: extremely high frequency (30-300 GHz)

EHFB: Electrical historical Foundation Board

EHP: electric horsepower
EHS: Environmental Health Service
EHSI: electronic horizontal situation indicator
EHV: extra high voltage
EI: Electrical Insulation (IEEE Group)
EI: end injection
Ei: Engineering Index, Inc
EI: Engineering Information, Inc
EI: engineering instruction
EIA: Electronic Industries Association
EIA: Energy Information Administration
EIA: Engineering Industries Association (GB)
EIA: environmental impact appraisal
EIAC: Electronic Industries Association of Canada
EIAJ: Electronic Industry Association of Japan
EIAP: environmental impact assessment project
EIB: Electronics Installation Bulletin
EIC: Electrical Insulation Conference
EIC: engineer in charge
EIC: Engineering Institute of Canada
EIC: equipment identification code
EID: Electronic Instrument Digest
EIES: Electronic Information Exchange System
EIIS: energy industry information system
EIL: electron injection laser
EIM: excitability inducing material
EIMA: Electrical Insulating Materials Association
EIM&M/NPSS: Environmental Instrumentation Measurment & Monitoring
EIN: education information network
EIRP: Effective isotropic radiated power
EIS: Effluent inventory system
EIS: electromagnetic intelligence system
EIS: Electronic Information Service
EIS: Engineering Index Service
EIS: environmental inpact statement
EIS: executive information system
EIS: (Standards) Environmental information systems
EIT: engineer in training
EITB: Engineering Industry Training Board
EJC: Engineers Joint Council (see American Association of Engineering Societies)
EJCC: Eastern Joint Computer Council
EJCNC: EJC Nuclear Congress
EKG: effective kilogram
EKG: electrocardiogram
EL-1: Reactor, Fontenay, France, 150 kW(th)
EL-2: Reactor, Saclay, France, 2.5 MW(th)
EL-3: Reactor, Saclay, France, 17.5 MW(th)
EL-4: Prototype power reactor, Brittany, 250 MW(th)
EL: education level
EL: electroluminescence
ELDO: European Space Vehicle Launcher Development Organization
ELECOM: electronic computer
Elec/PHP: electric contacts
Elec Proc: Electrostatic Processes (IAS)
Electro Meas/IM: TC-1 Electromagnetic Measurements State-of-the-Art

Electro Technol: Electro-Technology (India) Electronics
Electr Prog: Electronic Progress (Raytheon)
Electr Warfare: electronic warfare
ELEX: process for lithium isotope separation
ELF: extensible language facility
ELF: extremely low frequency
ELFA: Electric Light Fittings Association (GB)
ELG: electrolytic grinding
ELINT: electromagnetic intelligence
ELINT: electronic intelligence (systems)
ELIP: electrostatic latent image photography
ELP: English language programs
ELPC: electroluminescent-photoconductive
ELPG: Electric Light & Power Group
ELPH: elliptical head
ELPHR: experimental low temperature process heat reactor concept
ELR: engineering laboratory report
ELRO: Electronics Logistics Research Office
ELS: error likely situation
ELSA-11: gamma and flux reactor physics computer code
ELSB: edge-lighted status board
ELSIE: electronic signaling and indicating equipment
ELT: emergency locator transmitter
ELV: electrically operated valve
ELV: enclosed-frame low voltage
e/m: electric charge to mass for particles
E/M: electro-mechanical
EM: electromagnetic
EM: emission pump
EM: Engineering Management (IEEE Society)
EM: engineering manual
EM: engineering memorandum
EM: environmental management
EM: epitaxial mesa
EM: evaluation model
E&M: receive and transmit
EMA: extended mercury autocoder
EMATS: emergency mission automatic transmission service
EMAV: electro mag. relief valve
EMB: Electrical Moderization Bureau
EMBS: Engineering in Medicine and Biology (IEEE Society)
EMC: Electromagnetic Compatibility (IEEE Group)
EMC: engineered military circuits
EMC: Engineering Manpower Commission (Part of Engineering Affairs Council of AAES)
EMC: engineering mockup critical experiment (ANL)
EMC: European military communications
EMCAB: Electromagnetic Compatibility Advisory Board
EMCON: emission control
EMCP: electromagnetic compatibility program
EMCS: electromagnetic compatibility standardization (program)
EMCTP: electromagnetic compatibility test plan
EMD: electric-motor driven

EMDI: energy management display indicator

EME: electromagnetic energy

EMEC: Electronic Maintenance Engineering Center (Army)

EMETF: Electromagnetic Environmental Test Facility (Ft. Huachuca)

EMF: electromotive force

EMG: electromyography

EMI: electron magnetic interference

EMICE: electromagnetic interference control engineer

EMINT: electromagnetic intelligence

EMIS: educational management information system

EML: Engineering Mechanics Laboratory (NBS)

EML: equipment modification list

EMM: electromagnetic measurement

EMMA: electron microscopy and microanalysis

EMOS: Earth's mean orbital speed

EMP: electro-magnetic pulse radiation

EMP: electromagnetic pulse

EMP: electromechanical power

EMPIRE: early manned planetary-interplanetary round-trip expedition

EMR: electro-mechanical relay

EMR: electromagnetic radiation

EMR: electromechanical research

EMRA: Electrical Manufacturers Representatives Association

EMRIC: Educational Media Research Information Center

EMS: earthquake monitoring system

EMS: electromagnetic susceptibility

EMS: emergency medical services

EMS: Engineering Management Society

EMS: Export Marketing Service

EMSA: Electron Microscope Society of America

EMSC: Electrical Manufacturers Standards Council

EMT: electrical metallic tubing

EMT: European Mediterranean Tropo

EMU: electromagnetic unit

EMU: extravehicular mobility unit

EMW: electromagnetic warfare

EN: enforcement notification

ENDOR: electron nuclear double resonance

ENEA: European Nuclear Energy Association

ENG: automated information systems engineering

ENIAC: electronic numerical integrator and calculator

ENQ: enquiry character

ENR: equivalent noise resistance

ENSI: *Energia Nucleate Sud Italia* , large nuclear power plant project of SENN at Punta Fiume, Italy

ENSI: equivalent-noise-sideband input

ENT: equivalent noise temperature

ENTC: engine negative torque control

ENUSA: *Energie Nucleaire* S.A. Lausanne, Switzerland

EO: end office

EO: engineering order

EO: executive order

EOA: end of address

EOB: end of block

EOC: emergency operating center

EOC: End of cycle

EOCR: Experimental Organic Cooled Reactor, NRTS, Idaho, 40 MW(th)

EOD: end of data

EOD: explosive ordnance disposal

EODD: electrooptic digital deflector

EOEC: end of equilibrium cycle

EOF: end of file

EOG: electrooculography

EOIS: electrooptical imaging system

EOJ: end of job

EOL: end of life

EOL: expression oriented language

EOLM: electrooptical light modulator

EOLR: electrical objective loudness rating

EOLT: end of logical tape

EOM: end of message

EOP: end of program

EOP: end output

EOPC: electro-optic phase change.

EOR: end of record

EOR: end of reel

EOR: explosive ordnance reconnaissance

EOS: Earth Observatory Satellites

EOS: electrooptical systems

EOS: equation of state

EOT: end office terminal

EOT: end of tape

EOT: end of transmission

EOV: end of volume

EOY: end of year

EP: electrically polarized (relay)

EP: electric power

EP: emergency preparedness

EP: engineer personnel

EP: epitaxial planar

EPA: U.S. Environmental Protection Agency

EPAM: elementary perceiver and memorizer

EPC: electrical plastic conduit

EPC: electronic program control

EPC: engineering change proposals

EPCO: Engine Parts Coordinating Office

EPD: electric power distribution

EPE: Engineering Progress Exposition

EPG: Emergency Procedure Guidelines

EPH: Electric Process Heating (IEEE IAS Technical Committee)

EPIC: earth-pointing instrument carrier

EPL: environmental protection limit - terminology used in environmental technical specifications

EPLA: Electronic Precedence List Agency

EPM: engineering procedure memos

EPM: environmental project manager

EPM: external polarization modulation

EPMAU: expected present multi-attribute utility

EPNdB: effective perceived noise level (decibels)

EPR: electrical pressure regulator

EPR: electron paramagnetic resonance

EPR: engine pressure ratio
EPR: essential performance requirements
EPR: ethylene propylene rubber
EPRI: Electric Power Research Institute
EPROM: electrically programmable read-only memory
EPRTCS: emergency power ride through capability system (CLS - EM pumps)
EPS: equilibrium problem solver
EPS: switch-mode power supply
EPSCS: enhanced private switched communications service
EPSL: emergency power switching logic
EPT: electrical plastic tubing
EPT: electrostatic printing tube
EPT: exciter Power Logic
EPTE: existed prior to entry
EPU: electrical power unit
ER: echo ranging
ER: effectiveness report
ER: electrical resistance
E&R: Engineering and Repair Department
ER: enhanced radiation (weapons)
ER: environmental report
ERA: Economic Regulatory Administration
ERA: Electrical Research Association
ERA: Electric Railroaders Association
ERA: electron-ring accelerator
ERA: electronic reading automation
ERA: Electronic Representatives Association
ERA: Energy Reorganization Act
ERA: Equal Rights Amendment
ERAM: extended range antiarmormunition
ERB: Equipment Review Board
ERB: Experiment Review Board
ERBE: earth radiation budget experiment.
ERBM: extended range ballistic missile
ERC: Electronics Research Center (NASA)
ERC: Emergency Relocation Center
ERC: Engineering Research Council
ERC: equatorial ring current
ERCOT: Electric Reliability Council of Texas
ERCR: electronic retina computing reader
ERCS: Emergency Rocket Command System
ERCS: emergency rocket communications
ERCW: emergency raw cooling water
ERDA: U.S. Energy Research and Development Administration (now DOE)
ERDAM: ERDA Manual
ERDL: Engineering Research and Development Laboratory (Ft. Belvoir, VA)
ERFPI: extended range floating point interpretive system
ERG: electroretinography
ERGS: en-route guidance system
ERIC: Educational Research Information Center
ERIC: energy rate input controller
ERIS: emergency resources identification equipment
ERISA: Employee Retirement Income Security Act.
ERL: echo return loss

ERL: Environmental Research Laboratory
ERM: earth reentry module
ERMA: Electronic Recording Machine Accounting (GE)
ERN: engineering release notice
ERNIE: electronic random numbering and indicating equipment
ERO: emergency repair overseer
ERO: Energy Research Office
ERO: engineering release order (for project release)
EROS: Earth Resources Observation Satellite (systems)
ERP: effective radiated power
ERP: elevated release point
ERP: equivalent radiated power
ERPLD: extended range phase-locked demodulator
ERR: Elk River Reactor, Elk River, Minnesota, boiling-water type
ERS: Economic Research Service
ERS: Engineers Register Study
ERS: Environmental Research Satellite
ERS: external regulation system
ERSA: Electronic Research Supply Agency
ERSP: Earth resources survey program
ERSR: equipment reliability status report
ERSS: Earth Resources Satellite System
ERTS: Earth Resources Technology Satellite
ERW: electronic resistance welding
ERX: electronic remote switching
ES: echo sounding
ES: electromagnetic switching
ES: electronic switch
ES: experimental station
ESA: Ecological Society of America
ESA: Employment Standards Administration
ESA: European Space Agency
ESAIRA: electronically scanned airborne intercept radar
ESAR: electronically scanned array radar
ESAR: electronically steerable array radar
ESARS: earth surveillance and rendezvous simulator
ESARS: Employment Service Automated Reporting System
ESAS: engineered safeguards actuation system
ESB: electrical simulation of the brain
ESB: Electrical Standards Board
ESB: Engineering Societies Building
ESB: Engineering Society of Baltimore
ESB: The Engineering Society of Buffalo
ESC: electrostatic compatibility
ESC: Engineers and Scientists of Cincinnati
ESC: escape character (teletypewriter)
ESCA: electron spectroscopy for chemical analysis
ESCA: extended source calibration area
ESCAP: Economic and Social Commission for Asia and the Pacific
ESCAPE: expansion symbolic compiling assembly program for engineering
ESCMT: Bay Area Engineering Societies Committee for Manpower Training

ESCo: Electrical Steel Company

ESCOE: Engineering Societies Commission on Energy, Inc

ESCOLP: coefficients (elastic-scattering) reactor physics computer code

ESCWS: essential services cooling water systems

ESD: electro-static discharge

ESD: Electronics Systems Division, USAF Systems Command

ESD: electrostatic storage deflection

ESD: Engineering Society of Detroit

ESD: Environmental Services Division

ESD: extension shaft disconnect (FFTF CRDM)

ESE: electrical support equipment

ESF: Engineered safety feature

ESFAS: engineered safety features actuation system

ESG: electrically suspended gyroscope

ESG: electronically suspended gyro

ESG: electrostatic gyroscope

ESG: Engineers and Scientists Guild

ESH: equivalent standard hours

ESHAC: Electric Space Heating and Air Conditioning (IEEE IAS Technical Committee)

ESHU: emergency ship-handling unit

ESI: engineering and scientific interpreter

ESI: equivalent sphere illumination

ESJCP: Engineers and Scientists Joint Committee on Pensions

ESL: Engineering Societies Library

ESLO: European Satellite Launching Organization

ESM: elastomeric shield material

ESM: electronic support measures

ESM: electronic warfare support measures

ESM: Engineers & Scientists of Milwaukee, Inc

ESMA: Electronic Sales and Marketing Association

ESMC: Engineering Societies Monograph Committee

ESN: electronic serial number

ESNE: Engineering Societies of New England, Inc

ESO: Economic Stabilization Office (temporary)

ESO: emergency support organization

ESONE: European Standards of Nuclear Electronics

ESP: electric submersible pump

ESP: electrostatic precipitator

ESP: extrasensory perception

ESPAR: electronically steerable phased array radar

ESPOD: electronic systems precision orbit determination

ESPS: Engineering Societies Personnel Service

ESR: early site review

ESR: electron spin resonance

ESR: equivalent series resistance

ESR: experimental superheat reactor

ESRANGE: European space range

ESRO: European Space Research Organization

ESRP: environmental standard review plan

ESRR: early site review report

ESS: electronic switching system

ESS: emplaced scientific station

ESS: engineered saftey systems

ES&S: engineering services and safety

ESSA: Emergency Safeguards System Activation

ESSA: Endangered Species Scientific Authority

ESSA: Environmental Sciences Services Administration

ESSG: Engineer Strategic Studies Group (Army)

ESSOR: EURATOM's 50 MW(th), organic-cooled, heavy-water-moderated, experimental power reactor at Ispra, Italy

ESSP: earliest scram set point (PPS)

ESSP: Engineers Society of St. Paul

EST: earliest start time

EST: eastern standard time

EST: electrostatic storage tube

EST: Engineers Society of Tulsa

EST: enlistment screening test

ESTEC: European Space Research Technical Center

ESTRAC: European Space Satellite Tracking and Telemetry Network (ESRO)

ESV: essential service value

ESW: error status word

ESWM: Engineering Society of Western Massachusetts

ESWP: Engineers Society of Western Pennsylvania

ESWS: Emergency Service Water System

ESY: Engineering Society of York, Pennsylvania

ET: edge-triggered

ET: electrical time

ET: Electronic Transformers (IEEE MAGS Technical Committee)

ET: engineering tests

ET: ephemeris time

ET: external tank

ETA: Employment and Training Administration

ETAC: Environmental Technical Applications Center (Air Force)

ETB: end-of-transmission-block character

ETC: estimated time of completion

ETCG: elapsed-time code generator

ETEC: Energy Technology Eng. Center [Formerly 2 MEC]

ETI: Electrical Tool Institute

ETL: Electrotechnical Laboratory (Japan)

ETL: ending tape label

ETM: enhanced timing module

ETN: equipment table nomenclature

ETO: Energy Technology Office

ETOS: extended tape operating system

ETP: electrical tough pitch

ETP: equivalent top product

ETQAP: education and training in quality assurance practices (ASQC)

ETR: Engineering Test Reactor, NRTS, 175 MW(th)

ETRC: Engineering Test Reactor critical assembly (PPC)

ETR2: Engineering Test Reactor 2

ETS: electronic tandem switching

ETS: electronic telegraph system

ETS: environmental technical specifications

ETS: environmental Test Specification

ETSCO: Engineering & Technical Societies Council of Delaware Valley

ETSQ: electrical time, superquick

ETTC/M: electronics transformers

ETX: end of text character

EU: erythemal unit

EUCLID: experimental use computer, London integrated display

EUF: equivalent unavailability factor

EUFMC: Electric Utilities Fleet Managers Conference

EURATOM: European Atomic Energy Community

EUREX: fuel reprocessing project proposed by ENEA

EURISOTPOE: A radioisotope information bureau sponsored by EURATOM

EUROCAE: European Organization for Civil Aviation Electronics

EUROCOMP: European Computing Congress

EUROCON: European Convention (Region 8, IEEE)

EURODOC: Joint Documentation Service of ESRO, EUROSPACE and EODCSVL

EUROSPACE: European Industrial Space Research Group

EUSEC: Conference of Representatives from the Engineering Societies of Western Europe and the United States of America. See also: WFFO

EUV: extreme-ultraviolet (photometer)

eV: electronvolt

EVA: electronic velocity analyzer

EVA: extravehicular activity

EVC: Electric Vehicle Council

EVEST: Valleditos Experimental Superheat Reactor, Pleasanton, CA, power reactor, 12.5 MW(th)

EVFM: ex-vessel flux monitor

EVHM: ex-vessel handling machine (now CLEM)

EVIST: ethic & values in Science & Technology Program

EVM: electronic voltmeter

EVOM: electronic voltohmmeter

EVTM: ex-vessel transfer machine

EW: early warning

EW: electronic warfare

EWA: WWR-S type, heterogeneous, enriched uranium, light-water-cooled and -moderated research reactor, Poland

EWC: electric water cooler

EWCAS: early warning and control aircraft system

EWCS: European Wideband Communications System

EWF: Electrical Wholesalers Federation

EWICSTC2: European Workshop of Industrial Computer Systems - Technical committee #2

EWO: engineering work order

EWST: elevated water storage tank

EWTMI: European Wideband Transmission Media Improvement Program

EX: exchange key (calculator)

EXACT: International Exchange of Authenticated Component Performance Test Data

EXCIMER: excited dimer. (laser)

EXCIPLEX: excited state complex (lasers)

EXCO: Executive Committee

EXCO: Executive Committee (of ISO's governing council).

exctr: exciter

EXDAMS: extended debugging and monitoring system

EXMETNET: experimental meteorological sounding rocket research network

ExSC: Executive Standards Council

EXSTA: experimental station

EXTERRA: Extraterrestrial Research Agency

F

F: Fahrenheit, farad, filament, fuse

F: filament (vacuum tube)

F: final

f: focal length, frequency

F: forward

fA: femtoampere

FA: field accelerating contactor or relay

FA: forced-air-cooled (transformer)

F/A: fuel/air

F/A: fuel assembly

FA: fully automatic

FAA: US Federal Aviation Administration

FAAAS: Fellow of the American Association for the Advancement of Science

FAAB: Frequency Allocation Advisory Board

FAB ISO: fabrication isometric (drawing)

FABMDS: Field Army Ballistic Missile Defense System (SAM-D)

fac: facsimile

FAC: field accelerator

FAC: Frequency Allocation Committee

FACE: Federation of Associations on the Canadian Enviroment

FACE: field-alterable control element

FACISCOM: Finance and Comptroller Information System Command

FACS: fine attitude control system

FACT: flexible automatic circuit tester

FACT: flight acceptance composite test

FACT: Foundation for Advanced Computer Technology

FACT: fully automatic cataloging technique

FACT: fully automatic compiler translator

FAD: floating add

FADAC: field artillery digital automatic computer

FADP: Finnish Association for Data Processing

FAE: field application engineer

FAE: final approach equipment

FAETUA: Fleet Airborne Electronics Unit (Atlantic)

FAETUP: Fleet Airborne Electronics Unit (Pacific)

FAF: final approach fix

FAGC: fast automatic gain control

FAGS: Federation of Astronomical and Geophysical Services

FAHQMT: fully automatic high-quality machine translation

FAI: fail as is

FAL: frequency allocation list
FALTRAN: Fortran to Algol translator
FAM: fast auxiliary memory
FAMECE: Family of Military Engineer Construction Equipment
FAMOS: Fleet Application of Meteorological Observations for Satellites
FAMOS: floating-gate avalanche-injection metal-oxide semiconductor
FAMP: fire alarm monitoring panel
FAO: Food and Agriculture Organization
FAP: Field Application Panel
FAP: floating point arithmetic package
FAP: Fortran assembly program
FAP: frequency allocation panel
FAPUS: frequency allocation panel (United States)
FAR: Center for Nuclear Studies, Fontenay-aux-Roses, France
FAR: failure analysis report
FAR: Federal Acquisitions Regulations
FARADA: failure rate data (program)
FARE: Federal Acquisition Regulations
FARET: fast reactor test assembly (ANL design, never built)
FARGO: 1401 automatic report generating operation
FARSE: direct neutron reactor physics computer code
FAS: Federation of American Scientists
FAS: Foreign Agriculture Service
FAS: free alongside
FAS: Frequency Assignment Subcommittee
FASB: Financial Accounting Standards Board
FASE: fundamentally analyzable simplified English
FASS: foward acquisition sensor
FAST: facility for automatic sorting and testing
FAST: fast automatic shuttle transfer
FAST: field data applications, systems, and techniques
FAST: Flexible Algebraic Scientific Translator (NCR)
FAST: formal auto-indexing of scientific texts
FAST: formula and statement translator
FAST: four-address to SOAP translator
FAST: fuel assembly stability test
FAST: Italian Federation of Scientific and Technical Associations
FASTAR: frequency angle scanning, tracking, and ranging
FASTI: fast access to systems technical information
FASWC: fleet antisubmarine warfare command
FAT: formula assembler translator
FATAL: FADAC automatic test analysis language
FATE: fusing and arming test experiments
FATH: fathom (1 fath = 6 ft = 1.8239 m)
FATR: fixed autotransformer
FAX: facsimile
FB: film badge
FB: fuse block
FBC: fluidized bed combustion
FBFS: fuel building filter system
FBFT: flow bias functional test
FBI: Federal Bureau of Investigation
FBM: Fleet Ballistic Missile (Polaris/Poseidon)

FBMP: Fleet Ballistic Missile Program
FBMWS: Fleet Ballistic Missile Weapons System
FBP: FASTBUS protocol
FBR: fast breeding reactor
FBR: feedback resistance
FC: fail closed
FC: ferrite core
FC: file code
FC: file conversion
FC: fine control
fc: footcandle
FC: fuel cell
FC: fuel cycle
FC: function code
FCA: Farm Credit Administration
FCA: fire control area
FCA: frequency control analysis
FCC: Federal Communications Commission
FCC: Federal Construction Council
FCC: flight control computer
FCCPO: Federal Contract Compliance Programs Office
FCD: fine control damper
FCE: flexible critical experiment
FCFT: fixed cost, fixed time (estimate)
FCI: faulted current indicator
FCI: Fluid Control Institute
FCI: fuel coolant interaction
FCI: functional configuration identification
FCIC: Federal Crop Insurance Corporation
FCL: feedback control loop
FCMD: fire command
FCMV: fuel consuming motor vehicle
FCN: field change notice
FCNP: fire control navigation panel
FCOH: Flight Controllers Operations Handbook
FCP: file control processor
FCPC: Fleet Computer Programming Center
FCR: facility change request
FCR: fast ceramic reactor porogram
FCR: final configuration review
FCS: fire control system
FCS: flight control system
FCS: frame check sequence
FCSC: Foreign Claims Settlement Commission
FCST: Federal Council for Science and Technology
FCT: filament center tap
FCTT: fuel cladding transient tester
F&D: facilities and design
FD: field decelerating contactor or relay
FD: file description
F&D: findings and determination
FD: flange local distance
F/D: focal (length) to diameter ratio
FD: full duplex
FDA: final design approval
FDA: Food and Drug Administration
FDAA: Federal Disaster Assistance Administration
FDAC/IES: Fort Dodge Area Chapter of Iowa Engineering Society

FDAS: frequency distribution analysis sheet
FDB: field dynamic braking
FDC: functional design criteria (used for GPPs)
FDDI: fiber distributed data interface
FDDL: frequency division data link
FDE: field decelerator
FDF: flight data file
FD/FF: flux delta/flux flow
FDI: field discharge
FDI: flight director indicator
FDIC: Federal Deposit Insurance Corporation
FDM: frequency division multiplexer
FDMA: frequency division multiplex access
FDNR: frequency-dependent negative resistance
FDOS: floppy disk operating system
FDP: future data processors (DPMA)
fdr: feeder
FDR: flight data recorder
FDR: frequency domain reflectometry
FDR: functional design requirements
FDS: fluid distribution system
FE: field engineer
FEA: failure effect analysis
FEA: Federal Energy Administration
FEANI: European Federation of National Engineering Associations
FEAT: frequency of every allowable term
FEB: functional electronic block
FEBA: forward edge of battle area.
FEC: Federal Election Commission
FEC: forward error control
FEC: forward error correction
FED: field-effect diode
FED: fuel examination facility
FEDAL: failed element detection and location instruments which detect fuel element failure in power reactors
FEDP: Federal Executive Development Program
FEFP: fuel element failure propagation
FEFPL: fuel element failure propagation loop
FEI: Financial Executives Institute
FEIA: Flight Engineers International Association
FEL: free electron laser
FEM: field-effect modified
FEMF: Floating Electronic Maintenance Facility (FLATTOP)
FEN: frequency-emphasizing network
FEO: facility emergency organization
FEO: field engineering order
FEP: fuse enclosure package.
FERC: Federal Energy Regulatory Commission
FERD: fuel element rupture detection
FERMILAB: Fermi National Accelerator Laboratory
FERMI-1: Enrico Fermi LMFBR Plant, 60 MW(e) (decommissioned)
FERPIC: ferroelectric ceramic picture device
FERROD: ferrite-rod antenna
FES: final environmental statement
FES: Florida Engineering Society

FESA: Federation of Engineering and Scientific Associations
FET: field-effect transistor
FF: fixed focus
FF: flip-flop
FF: full field
FF: full field contactor or relay
FFAG: Fixed-field alternating-gradient
FFAR: folding fin air rocket
FFC: flip-flop complementary
fff: flicker fusion frequency
FFI: fuel flow indicator
FFI: full field investigation
FFL: first financial language
FFL: front focal length
FFM: fuel failure mockup facility
FFR: folded flow reactor
FFSA: field functional system assembly and checkout
FFSP: fossil fired steam plant
FFT: fast Fourier transform
FFTF: fast flux test facility
FFTFPO: fast flux test facility project office
FFV: Field Failure Voltage
FG: filament ground
FG: fission gas
FG: function generator
FGAA: Federal Government Accountants Association
FGM: fission gas monitor
FGSA: Fellow of the Geological Society of America
FHA: Federal Highway Administration
FHA: Federal Housing Administration
FHB: Fuel Handling Building
FHC: Federal Housing Commissioner, Office of Assistant Secretary for Housing
FHD: fixed-head disks
FHEO: Fair Housing and Equal Opportunity, Office of Assistant Secretary
FHLBB: Federal Home Loan Bank Board
FH&RM: fuel handling and radioactive maintenance
FHS: forward head shield
FHSR: final hazards summary report
FHT: fully heat treated
FHWA: Federal Highway Administration
FI: fail in place
FI: field intensity
FIA: Factory Insurance Association
FIA: Federal Insurance Administration
FIB: Fortran information bulletin
FIC: Federal Information Centers
FIC: frequency interference control
FID: *Federation International de Documentation*
FID: International Federation for Documentation
FIDAC: film input to digital automatic computer
FIEI: Farm and Industrial Equipment Institute
FIEN: *Form Italiano Dell'Energia Nucleare*
FIFO: first-in first-out (inventory control)
FIIG: Federal Item Identification Guide
FILS: flare-scan instrument landing system
FIM: field inspection manual (Bechtel)
FIM: field intensity meter

FIM: field ion microscope

FIMATE: factory installed maintenance automatic test equipment

FINAC: fast interline nonactive automatic control

FINAL: financial analysis language

FINCOM: Finance Committee (IEEE)

FINGAL: process developed at Harwell for storing high-level fission products in glass

FIOP: Fortran input-output package

FIP: field inspection procedure (Bechtel)

FIPACE: *Federation Internationale des Producteurs Auto-Consommateurs Industriels d'Electricité* (International Federation of Industrial Producers of Electricity for Own Consumption

FIPS: federal information processing standard

FIR: far infrared

FIR: finite-duration impulse-response (filters)

FIR: Food Irradiation Reactor, Stockton, California

FIR: fuel indicator reading

FIRETRAC: Firing Error Trajectory Recorder and Computer

FIRMS: forecasting information retrieval of management system

FIRN: two-dimensional cylinder transport code for computer

FIRR: Failure and Incidents Report Review (ANSI Cmte)

FIRR: Swiss Federal Institute for Reactor Research

FIRST: financial information reporting system

FIS: field information system

FIS: floating-point instruction set

FIST: fault isolation by semiautomatic techniques

FIT: fault isolation test

FJCC: Fall Joint Computer Conference

FL: field loss contactor or relay

FL: Florida

FL: focal length

fl: footlambert

FLA: Fluorescent Lighting Association

FLAC: Florida automatic computer

FLAM: forward launched aerodynamic missile

FLAMR: forward looking advanced multilobe radar

Flattop: Los Alamos Scientific Laboratory experiment for spherical metal cores in thick metal reflectors

FLB: Federal Land Bank

FLDEC: floating point decimal

FLEA: flux logic element array

FLECHT: full length emergency cooling heat transfer

FLEER: two-dimensional group diffusion reactor code

FLF: fixed length field

FLF: follow-the-leader feedback (circuit theory)

FLINT: floating interpretive language

FLIP: film library instantaneous presentation

FLIP: floating index point, arithmetic unit developed at ANL

FLIP: floating point interpretive program

FLIR: forward looking infrared radar (See radar) radio detection and ranging

FLIR: forward looking infrared system

FLODAC: Fluid Operated Digital Automatic Computer (Univac)

FLOP: floating octal point

FLOSOST: fluorine one-stage orbital space truck

FLOTRAN: flowcharting Fortran

FLOX: fluorine liquid oxygen

FLPL: Fortran list processing language

FLR: forward-looking radar

FLS: flow switch

FLT: fault location technology

FM: file maintenance

FM: frequency modulation

FMC: Federal Maritime Commission

FMCS: flight management computer system

FMCW: frequency-modulated continuous wave

FMEA: failure mode and effects analysis

FMES: full mission engineering simulator

FMFB: frequency modulation with feedback

FmHA: Farmers Home Administration

FMI: frequency modulation intercity relay broadcasting

FMIC: frequency monitoring interference control

FMPS: Fortran mathematical programming system

FMS: Federation of Materials Societies

FMS: flexible manufacturing system

FMS: flux monitoring system

FMS: foreign military sales

FMX: FM transmitter

FNAL: Fermi National Accelerator Laboratory

FNH: flashless nonhygroscopic

FNP: floating nuclear plant

FNP: fusion point

FNR: Ford Nuclear Reactor, 100 kW(th), pool-type reactor at Phoenix project, University of Michigan, Ann Arbor, Michigan

FNS: Food and Nutrition Service

FO: fail open

FO: fast operate (relay)

FO: fiber optics

FO: oil-immersed forced-oil-cooled (transformer)

FOA: oil-immersed forced-oil-cooled with forced-air cooler (transformer)

FOBS: fractional-orbit bombardment system

FOCAL: formula calculator (DEC)

FOE: Friends of the Earth

FOH: forced outage hours

FOIA: Freedom of Information Act

FOIL: file oriented interpretive language

FOL: facility operating license

FOM: figure-of-merit

FOPT: fiber optic photo transfer

FOR: forced outage rate

FORATOM: Forum Atomique, European association of EURATOM nuclear industries

FORBLOC: Fortran compiled block oriented simulation language

FORC: formula coder (automatic coding system)

FORDS: Floating Ocean Research and Development Station

FORESDAT: formerly restricted data

FORGO: Fortran load and go (system)

FORM: fast spectrum cross section reactor physics computer code

FORMAC: formula manipulation compiler

FORTRAN: formula translation

FORTRANSIT: Fortran and internal translator system

FORTRUNCIBLE: Fortran style runcible

FOSDIC: film optical sensing device for input to computers

FOT: frequency of optimum operation

FOTA: fuels open test assembly

FOW: oil-immersed forced-oil-cooled with forced-water cooler (transformer)

FP: faceplate

FP: feedback positive

FP: forward perpendicular

FP: freezing point

FPC: Federal Power Commission (now FERC)

FPC: fire pump control

FPCH: Foreign Policy Clearinghouse

FPCS: fuel pool cooling system

FPCSTL: fission product control screening test loop

FPD: full power days

FPDD: final project design description

FPDI: food processing development is irradiator

FPIS: forward propagation by ionospheric scatter

FPL: final protective line

FPM: Federal Personnel Manual

FPM: feet per minute

FPM: Frequency position modulation

FPM: functional planning matrices

FPP: floating point processor

FPR: critical assembly (KAPL)

FPR: Federal Procurement Regulations

FPRS: Forest Products Research Society

FPS: feet per second

FPS: Fluid Power Society

FPS: focus projection and scanning

FPS: foot-pound-second

FPS: frames per second

FPSL: fission product screening loop

FPT: female pipe thread

FPT: full power trial

FPTF: Fuels Performance Test Facility

FPTS: forward propagation tropospheric scatter

FPV: feedwater regulation valve

FPWT: fire protection water tank

FQPR: frequency programmer

FR-O: Swedish experimental fast reactor

FR: failure rate

FR: fast release (relay)

FR: Federal Register

FR: Federal Reserve

FR: field reversing

FR: flash ranging

FRA: Federal Railroad Administration

FRAMATOME: *Société Franco-Americaine de Constructions Atomiques* , Puteaux, France

FRAP-S: fuel rod analysis program - steady-state

FRAP-T: fuel rod analysis program - transient

FRAP: fuel rod analysis program

FRB: *Forschungs-Reaktor* , Berlin (see BER)

FRB: Federal Reserve Bank

FRC: Federal Radiation Council (merged into EPA in 1970)

FRC: File Research Council

FRC: functional residue capacity

FRCTF: fast reactor core test facility (AEC)

FRD: functional reference device

FRED: figure reader electronic device

FREQMULT: frequency multiplier

FRESCANNAR: frequency scanning radar

FRFA: Federal Regulatory Flexibility Act

FRG 1: Swimming-pool research reactor, 5 MW(th), Geesthact, Germany

FRG 2: Swimming-pool research reactor, 200 kW(th), Geesthact, Germany

FRH: Fire Resistant Hydraulics Program

FRINGE: file and report information processing generator

FRJ 1: Merlin-type swimming-pool research reactor, 5 MW(th), *Kernforschungsanlage* , Julich, Germany

FRJ 2: Dido-type heavy-water research reactor, 10 MW(th), *Kernforschungsanlage* , Julich, Germany

FRM: Swimming-pool research reactor, 1 MW(th), *Technische Hochschule* , Munich, Germany

FRMR: frame reject

FRMZ: Triga-type, Mark II, research reactor, 30 kw(th), University of Mainz, Germany

FRP: fiber-glass reinforced plastic

FRP: fuel reprecessing plant

FRS: Federal Reserve System

FRS: Fellow of the Royal Society (GB)

FRS: Fire Research Station

FRS: fragility response spectrum

FRT: Total Intergrated Radial Peaking Factor

FRTEF: fast reactor thermal engineering facility

FRTP: fraction of rated power

FR 2: Heavy-water research reactor, 12 MW(th), *Kernforschungszentrum* , Karlsruhe, Germany

FS: factor of safety

FS: Federal Standard

FS: female soldered

FS: Fiscal Service

FS: Forest Service

FS: full scale

fs: functional schematic

FSAR: final safety analysis report

FSC: Federal Supply Classification

FSD: full-scale deflection

FSDC: Federal Statistical Data Center

FS/FD: system flux/delta flux

FS/FW: steam flow/water flow

FSGB: Foreign Service Grievance Board

FSK: frequency-shift keying

FSL: formal semantic language (MIT)

FSLIC: Federal Savings and Loan Insurance Corporation

FSN: Federal stock number

FSPE: The Federation of Societies of Professional Engineers

FSPPR: Fast Supercritical Pressure Power Reactor, light-water-cooled and -moderated, USA

FSPT: Federation of Societies for Paint Technology

FSQS: Food Safety and Quality Service

FSR: feedback shift register

FS&R: filling, storage and remelt system

FSR: first soviet reactor, 1 MW(th), USSR

FSS: Federal Supply Services

FSS: floor service stations

FSTC: Foreign Science and Technology Center (Army)

FSUC: Federal Statistics Users Conference

FSV: fire service valve

FSV: Fort St. Vrain (nuclear plant)

FSVM: frequency selective voltmeter

FSW: forward-swept-wing

FT: firing tables

FT: flush threshold

FT: Frequency and Time (IEEE IM Technical Committee)

FT: frequency tracker

F&T: fuel and transportation

FT: functional test

FTC: fast time constant

FTC: Federal Trade Commission

FT&C: functional test & calibration

FTE: FFTF test engineering

FTF: flared tube fitting

FTFET: four-terminal field-effect transistor

FTL: Federal Communications Laboratory

FTL: full term license

FTM: flight test missile

FTO: failed to open

FTP: FFTF (acceptance) test procedure

FTP: fixed term lease plan

FTP: fuel transfer port

FTP: fuel transfer pump

FTR: fast test reactor

FTR: Federal travel regulations

FTR: functional test requirement

FTRC: Federal Telecommunications Records Center

FTRIA: flow and temperature removable instrument assembly

FTS: Federal Telecommunications System

FTS: foot switch

FTS: Free-Time System (GE/PAC)

FTS: frequency and timing subsystem

FTTC/IM: TC-3 Frequency & Time

FTTS: flow through tube sampler

FTU: flight test unit

FTZB: Foreign-Trade Zones Board

FU: fuse

FUDR: failure and usage data report

FUFO: fuel fusing option

FUGUE: engineering pipe flow reactor computer code

FUIF: fire unit integration facility

FUNCTLINE: functional line diagram

Fusion Tech: Fusion Technology (NPSS)

fV: femtovolt

FV: floor valve (refueling)

F/V: frequency to voltage

FV: full voltage

FVA: floor valve adapter (refueling)

FVD: front vortex back focal distance

FVS: Flight Vehicles System (IEEE AESS Technical Committee)

FW: face width

FW: field weakening

FW: filament wound

FW: forward wave

FW: full wave

FWA: first word address

FWA: fluorescent whitening agent

FWA: forward wave amplifier

FWEC: Fort Wayne Engineers Club

FWFM: full width at fiftieth maximum

FWHM: full-width half-maximum

FWL: fixed word length

FWPCA: Federal Water Pollution Control Act

FWPCA: Federal Water Pollution Control Administration

FWQA: Federal Water Quality Administration

FWS: filter wedge spectrometer

FWS: Fish and Wildlife Service

FWTM: full width at tenth maximum

FXYT: total radiation planar peaking factor

FY: fiscal year

FYDP: five year defense Program

FYP: five-year plan

F(Z): (See APDMX)

G

g-a: ground to air

G-G: ground to ground

G-LOC: G-induced loss of consciousness

G-M: Geiger Mueller (counter)

G: giga (10^9)

G: global bit

g: gravitational acceleration unit

G: grid

GA: gain of antenna

ga: gas amplification

GA: geographical address

GA: Georgia

ga: glide angle

ga: graphic ammeter

GA: grapple adapter

GAAG: gross actual generation

GaAs: gallium arsenide (semiconductor)

GAC: geographical address control

GAC: gross available capacity

GADS: Gate Assignment Display System

GAEC: Greek Atomic Energy Commission, Athens, Greece

GA&ES: Georgia Architectural and Engineering Society

GAFB: Griffiss Air Force Base

GAG: gross available generation
GAHF: grapple adapter handling fixture (AI)
GALCIT: Guggenheim Aeronautical Laboratory of California Institute of Technology (UAS)
GAM: graphic access method (IBM)
GAM: guided aircraft missile
GAMA: Gas Appliance Manufacturers Association
GAMA: General Aviation Manufacturers Association
GAMA: graphics assisted management application
GAMLOGS: gamma-ray logs
GAMM: German Association for Applied Mathematics and Mechanics
GAN: generalized activity network
GAN: generating and analyzing networks
GAO: U.S. General Accounting Office
GAP: general assembly program
G&A Pan: Gyro & Accelermeter panel
GAPE: Guyana Association of Professional Engineers
GAR: growth analysis and review
gar: guided aircraft rocket
GARP: Global Atmospheric Research Program
GASP: general activity simulation program
GASP: generalized academic simulation program
GASP: graphic applications subroutine package
GASS: generalized assembly system
GASSAR: General Atomic Standard Safety Analysis Report
GAT: generalized algebraic translator
GAT: Georgetown automatic translator
GATAC: general assessment tridimensional analog computer
GATB: general aptitude test battery
GATE: GARP Atlantic Tropical Experiment
GATE: generalized algebraic translator extended
GATT: General Agreement on Tariffs and Trade
GAZE: multi-group diffusion reactor computer code
gb: gilbert (unit of magnetomotive force)
GBH: group busy hour
GBSR: graphite-moderated boiling and superheating reactor
GC: gas chromatograph
GC: general counsel
gc: gigacycles per second (obsolete) Use GHz (gigahertz)
GC: guidance computer
GCA: ground-controlled approach
GCAP: generalized circuit analysis program
GCC: ground control center
GCD: greatest common divisor
GCF: gross capacity factor
GCFBR: gas-cooled fast breeder reactor
GCFR: gas-cooled fast reactor
GCFRE: gas-cooled fast reactor experiment
GCHQ: Government Communications Headquarters
GCI: ground-controlled interception
GCIS: ground control intercept squadron
GCL: gas-cooled loop
GCMA: Government Contract Management Association of America
GCMS: gas chromatography and mass spectroscopy

GCN: gage code number
GCR: gas-cooled reactor
GCR: general component reference
GCR: group coded recording
GCRE, I, II: Gas-Cooled Reactor Experiment, 2.2 MW(th), NRTS
GCS: gate-controlled switch
GCT: general classification test
GCT: Greenwich civil time
GCWM: General Conference of Weights and Measures
GD: ground detector
GD: grown diffused
GDC: general design criteria
GDC: gross dependable capacity
GDE: ground data equipment
GDF: group distribution frame
GDG: generation data group
GdIG: gadolinium iron garnet
GDL: gas dynamic laser
GDMS: generalized data management system
GDOP: geometric dilution of position
GDOP: geometric dilution of precision
GDS: General Declassification Schedule of Executive Order 11652
GDS: graphical display system
GDT: gas decay tank
GE: gas ejection
GE: Gaussian elimination
GE: General Electric Company
GE: geoscience electronics
Ge: germanium (semiconductor)
GEBCO: general bathymetric chart of the oceans
GECOM: generalized compiler
GECOS: general comprehensive operating supervisor
GEEIA: general electronics engineering installation
GEEP: General Electric electronic processor
GEESE: General Electric Electronics Systems Evaluator
GEIS: generic environmental impact statement
GEISHA: geodetic inertial survey and horizontal alignment
GEK: geomagnetic electrokinetograph
GEM: general epitaxial monolith
GEM: ground effect machine
GE/MAC: General Electric measurement and control
GEMM: generalized electronics maintenance model
GEMS: general education management system
GEMS: General Electric manufacturing simulator
GEMSIP: Gemini Stability Improvement Program
gen: generator
GEO: Geoscience Electronics (IEEE Group)
GEO: geostationary orbit
GEODSS: ground-based electro-optical deep space surveillance system
GEON: gyro erected optical navigation
GEOS: Geodetic Earth Orbiting Satellite
GEOS: geosynchronous earth observation system
GEOSAR: geosynchronous synthetic aperture radar

GE/PAC: General Electric Process Automation Computer

GEPAC: General Electric Programmable Automatic Comparator

GERSIS: General Electric Range Safety Instrumentation System

GERT: graphical evaluation and review technique

GERTS: General Electric Remote Terminal System

GESMO: Generic Environmental Statement on Mixed Oxides

GESSAR: General Electric Standard Safety Analysis Report

GET: ground elapsed time

GE/TAC: General Electric Telemetering and Control

GETEL: General Electric Test Engineering Language

GETOL: General Electric Training Operational Logic

GETOL: ground effect take-off and landing

GETR: General Electric Test Reactor

GEVIC: General Electric Variable Increment Computer

GFAE: government-furnished avionics equipment

GFCI: ground-fault circuit interrupter

GFE: government furnished equipment (U.S.)

GFE&M: government furnished equipment and material

GFI: ground fault interrupter

GFP: government furnished property (U.S.)

GFRP: glass fiber reinforced plastic

GFW: ground fault warning

GFY: government fiscal year

GGG: gadolinium gallium garnet

GGTS: gravity gradient test satellite

GHCP: Georgia Hospital Computer Group

GHOST: global horizontal sounding technique

GHRI: hybrid radio-inertial guidance system

GHz: gigahertz (1000 MHz or 10^9 Hz)

GI: gastro-intestinal

GI: geodesic isotensoid

GI: government and industrial

GI: government initiated

GIA: General Industry Applications (IEEE IAS Technical Committee)

GIANT: geneological information and name tabulating system

GIC: generalized immittance converter

GIC: generalized impedance converter

GIDEP: government industry data exchange program

GIFS: generalized interrelated flow simulation

GIFT: gas insulated flow tube

GIFT: general internal Fortran translator

GIGO: garbage in, garbage out

GIM: generalized information management (language)

GIMRADA: Geodesy Intelligence and Mapping Research and Development Agency (U.S. Army)

GIOC: generalized input/output controller

GIPB: general-purpose instrument bus.

GIPS: ground information processing system

GIPSY: generalized information processing system

GIRL: graph information retrieval language

GIRLS: Generalized Information Retrieval and Listing System

GIS: generalized information system

GIS: Geoscience Information Society

GIT: graph isomorphism tester

GJ: grown junction

GJE: Gauss-Jordan elimination

GK: grant acknowledge

GL: gate leads

GLASS: germanium-lithium argon scanning system

GLC: gas-liquid chromatography

GLCM: ground launched cruise missle.

GLEAN: graphic layout and engineering aid method

GLEEP: graphic low energy experimental pile

GLEEP: 3 kW(th) reactor at Harwell, United Kingdom

GLINT: global intelligence

GLLD: ground laser locator designator.

GLOMEX: global meteorological experiment

GLOPAC: gyroscopic lower power controller

GLOSS: global ocean surveillance system

GLOTRAC: global tracking network

GLV: Gemini launch vehicle

GM: gaseous mixture

GM: Geiger-Mueller counter

GM: geometric mean

GM: guided missile

GM: metacentric height

GMAT: Greenwich Mean Astronomical Time

GMC: gross maximum capacity

GMCM: guided missile countermeasures

GMD: Guided Missiles Division

GMFCS: guided missile fire control system

GMG: gross maximum generation

GMR: ground mapping radar

GMSFC: George Marshall Space Flight Center (Huntsville, AL)

GMT: Greenwich mean time

GMV: guaranteed minimum value

gnd: ground

GNMA: Government Nation Mortgage Association

GNOME: Plowshare experiment to study the production and recovery of heat and isotopes produced in a contained nuclear explosion

GNP: gross national product

GOAL: ground operations aerospace language

GOCI: general operator-computer interaction

GOE: ground operating equipment

GOES: Geostationary Operational Environmental Satellite

GOL: general operating language

GOLD: graphic on-line language

GONG: Global Observing Network Group

GOP: general operational plot

GOR: gained output ratio

GOR: gas-oil ratio

GOR: general operational requirement

GORID: ground optical recorder for intercept determination

GORX: graphite oxidation from reactor excursion engineering computer code

GOSS: ground operational support system
GP: generalized programming
GP: general purpose
GP: ground protective (relay)
GP: group address; group address field
GPA: general purpose analysis
GPA: graphical PERT analog
GPATS: general purpose automatic test system
GPC: general purpose computer
GPCL: general purpose closed loop
GPCP: generalized process control programming
GPDC: general purpose digital computer
GPDS: general purpose display system
GPGL: general purpose graphic language
gph: gallons per hour
GPI: ground position indicator
GPIA: general-purpose interface adapter
GPIB: general purpose interface bus
GPIS: Gemini problem investigation status
GPL: generalized programming language
GPL: general purpose language
GPLP: general purpose linear programming
gpm: gallons per minute
GPM: general purpose macrogenerator
GPMS: general purpose microprogram simulator
GPO: General Post Office
GPO: Government Printing Office (U.S.A.)
GPP: general plant projects
gps: gallons per second
GPS: general problem solver
GPS: global positioning system.
GPSCS: general purpose satellite communications system
GPSS: general purpose system simulation
GPT: gas power transfer
GPT: Gemini test pad
GPTE: general purpose test equipment
GPWS: ground proximity warning system
GPX: generalized programming extended
GR: general reconnaissance
GR: general reserve
GRACE: gamma attenuation reactor physics computer code
GRACE: graphic arts composing equipment
GRAD: general recursive algebra and differentiation
GRAD: graduate resume accumulation and distribution
GRADB: generalized remote access data base
GRADS: generalized remote access data base system
GRAF: graphic addition to Fortran
GRAMPA: general analytical model for process analysis
GRAPE: gamma-ray attenuation porosity evaluator
GRAPHDEN: graphic data entry
GRARR: Goddard range and range rate
graser: gamma-ray amplification by stimulated emission of radiation
GRASP: generalized retrieval and storage program
GRASP: graphic service program

GRATIS: generation, reduction, and training input system
GRB: Geophysical Research Board (NRC)
GRB: Government Reservation Bureau
GRCSW: Graduate Research Center of the Southwest
GRD: ground
GRE: Graduate Reliability Engineering
GREB: galactic radiation experiment background satellite
GREC: Grand Rapids Engineers Club
GRED: generalized random extract device
Green Salt: uranium tetrafluoride (UF4)
GREMEX: Goddard Research Engineering Management Exercise
GRG: gross reserve generation
GRI: group repetition interval
GRID: graphic interactive display
GRIN: graphical input
GRINS: general retrieval inquiry negotiation structure
GRIT: graduated reduction in tensions
GRM: generalized Reed-Muller codes
GRM: global range missile
GROUNDHOG: DOD program of high explosive (nonnuclear) experiments, part of Vela Uniform
GRP: glass-reinforced plastic
GRR: Greek Research Reactor, Athens, Greece
GRR: guidance reference release
GRS: generalized retrieval system
GRT: General reactor Technology
GRWT: gross weight
GS: galvanized steel
GS: Geological Survey
GS: ground speed
GSA: Genetics Society of America
GSA: Geological Society of America
GSA: U.S. General Services Administration
GSB: Graphic Standards Board
GSC: gas-solid chromatography
GSCU: ground support cooling unit
GSD: General Supply Depot
GSD: general systems division
GSDB: geophysics and space data bulletin
GSDS: Goldstone Duplicate Standard (standard DSIP equipment)
GSE: ground support equipment
GSFC: Goddard Space Flight Center
GSI: government source inspection
GSI: grand scale integration
GSL: generalized simulation language
GSL: generation strategy language
GSMB: Graphic Standards Management Board/ANSI
GSO: Ground Safety Office
GSOP: Guidance Systems Operation Plan
GSP: general simulation program
GSP: General System of Tariff Preferences
GSP: guidance signal processor
GSPO: Gemini Spacecraft Project Office
GSPR: guidance signal processor repeater
GSR: galvanic skin response
GSRS: general support rocket system.

GSS: gamma scintillation system
GSS: global surveillance system
GSSC: ground support simulation computer
GSTA: ground surveillance and target acquisition
GSU: general service unit
GSUG: gross seasonal unavailable generation
GSV: governor steam valve
GSV: guided space vehicle
GSWR: galvanized steel wire rope
GT-HTGR: gas-turbine high-temperature gas-cooled reactor
G/T: antenna gain-to-noise temperature
G/T: concentration of grams of Pu per irradiated short-ton of uranium
GT: game theory
GT: Gemini Titan
GT: ground transmit
GTA: gas tungsten arc
GTA: Gemini Titan Agena
GTAA: gas tungsten arc
GTAW: gas tungsten arc weld
GTC: gain time constant
GTC: gain time control
GTC: gas turbine compressor
GTCC: group technology characterization code
GTD: geometrical theory of diffraction
GTD: graphic tablet display
GTE: ground transport equipment
GTG: gas turbine generator
GTH: gas tight high pressure
GTM: ground test missile
GTO: gate-turn-off (switches)
GTOW: gross take-off weight
GTP: general test plan
GTR: Ground Test Reactor (Air Force), Ft. Worth, Texas, 3 MW(th)
GTRR: Georgia Institute of Technology Research Reactor, Atlanta, Georgia, 1 MW(th)
GTS: general technical services
GTTF: Gas-Turbine Test Facility, experimental, closed-cycle, gas-turbine system, Ft. Belvoir, Virginia
GTv: gate trigger valve
GTV: ground test vehicle
GU: Guam
GUHA: general unary hypotheses automation
GUIDE: guidance for users of integrated data equipment
GULP: general utility library program
GUSTO: guidance using stable tuning oscillations
GUUG: gross unit unavailable generation
G(v): conductance as a function of voltage
GV: guard vessel
GVA: graphic kilovoltampere meter
GW: general warning
GWC: global weather central (Air Force)
GZ: ground zero
G1: plutonium production reactor, Marcoule, France, 38 MW(th)

G2: plutonium production reactor, Marcoule, France, 200 MW(th)
G3: plutonium production reactor, Marcoule, France 200 MW(th)

H

HC: high conduction
HCAB: cabinet, hose, fire
HCFA: Health Care Financing Administration
HDLC: high level data link control
HDPE: high-density polyethylene
HDTV: high-definition television
HELB: high energy line break
HEPA: high-efficiency particulate air filters
HEPL: high Energy Physics Laboratory (Stanford University)
HER: human error rate
HERA: high explosive rocket assisted
HERALD: Heterogenous Experimental Reactor, Atomic Weapons Research Establishment, Aldermaston, Berkshire, United Kingdom
HERF: high energy rate forming
HERMES: heavy element and radioactive material electromagnetic Separator, Harwell, United Kingdom
HERO: Hazards of Electromagnetic Radiation to Ordnance, 3 kW(th) reactor, Windscale, United Kingdom
HESS: Houston Engineering & Scientific Society
HET: hot-end termination
HEU: highly enriched uranium
HEVAC: Heating, Ventilating, and Air-Conditioning Manufacturers Association
HEW-305: Hanford 305 Test Reactor
HEW: U.S. Department of Health, Education and Welfare
HEX: uranium hexafluoride (UF6)
HF: height finder (radar)
HF: high frequency
HFA: high flow alarm
HFBR: High-Flux Beam Reactor, 40 MW(th), BNL
HFCE2: HFIR Critical Experiment 2
HFDF: high frequency distribution frame
HFE: Human Factors in Engineering Group
HFEF: hot fuel examination facility
HFG: heavy free gas
HFHE: nonirradiated fuel handling equipment
HFIM: High-Frequency Instruments and Measurement (IEEE IM Technical Committee)
HFIR: High-Flux Isotope Reactor to produce isotopes of trans-plutonium elements for research purposes, ORNL, 100,000 kW(th)
HFO: high-frequency oscillator
HFORL: Human Factors Operations Research Laboratory
HFR: High-Flux Reactor, Petten, Netherlands, Netherlands Research and Materials testing Reactor
HFS: hyperfine structure
HFSA: Human Factors Society of America

HFST: high flux scram trip
Hg: mercury
HGAS: high gain antenna system
HGE: hydraulic grade elevations
HgS: cinnabar
HGT: high group transmit
HH: hand hole
HHLR: Horace Hardy Lestor Reactor, 1 MW(th) pool-type, Watertown, Massachusetts
HHSI: high head safety injection
HI-C: high conversion critical experiment
HI-FI: high fidelity
HI: Hydraulic Institute
HIBAL: Australian balloon sampling project operated for AEC by the Australian Department of Supply
HIBEX: high impulse boost experiment
HIC: halving interval
HIC: hybrid integrated circuit
HID: high intensity discharge (lamps)
HIDM: high information delta modulation
HIFAR: High-Flux Australian Reactor, 10 MW(th), tank-type, Lucas Height, Australia
HIIS: Honeywell Institute for Information Science
HIL: high-intensity lighting
HILAC: heavy ion linear accelerator
HIMA: Health Industry Manufacturers Association
HiMAT: highly maneuverable aircraft technology
HIP: Hanford isotopes plant
HIP: hot isostatic pressing
HIPAR: high power acquisition radar
HIPERNAS: high-performance navigation system
HIPOT: high potential
HIPOTT: high potential test
HIR: horizontal impulse reaction
HIRDL: High-Intensity radiation Development Laboratory, BNL
HIS: Honeywell Information Systems
HIS: hospital information system
HIT: high isolation transformer
HIT: homing interceptor technology
HIVOS: high vacuum orbital simulation
HKN: Eta Kappa Nu
HLCV: hot leg check valve
HLIV: hot leg isolation valve
HLL: higher level language
HLLV: heavy lift launch vehicle
HLLwT: high liquid level waste tank
HLW: high level waste
hm: hectometer
HMOS: high-performance, n-channel, silicon gate MOS
HMRB: Hazardous Materials Regulation Board
HMS: Hanford meteorology surveys
HMSS: Hospital Management Systems Society
HN: horn
HNIL: high-noise-immunity logic
HNPF: Hallam Nuclear Power Facility (decommissioned)
HOBO: homing official bomb
HOE: Homing Overlay Experiment

HOL: high order language
HOP: HEDL overpower
HOP: hybrid operating program
HOR: pool-type reactor, 100 kW(th), Delft, Netherlands
HORACE: Pool-type reactor, 10 w(th), Aldermaston, United Kingdom
HOSE: hose
HOTCE: hot critical experiments
HOTOL: horizontal-takeoff-and-landing (craft)
HP: high pass
HP: high pressure
HP: horsepower
HPA: holding and positioning aid
HPBW: half-power bandwidth
HPCI: High pressure coolant injection
HPCS: high pressure core spray
HPD: high power density
HPF: highest probable frequency
HPFL: High Performance Fuels Laboratory
HPGe: high purity germanium
HP HR: horsepower-hour
HPI: high pressure injection
HPIR: high-probability-of-intercept receiver
HPIS: high pressure injection system
HPMP: heat pump
HPOF: high pressure oil filled (cable)
HPOFcables: high-pressure oil-filled cables
HPOT: heliopotentiometer
HPP: hot processing plant
HPR: bunker; hopper
HPRR: Health Physics Research Reactor, ORNL, 1 kw
HPS: Hanford plant standards (formerly HWS)
HPS: Health Physics Society
HPS: high-pressure sodium lamp
HPSIS: high pressure safety injection system
HPSW: high pressure service water
HPT: high-pressure test
HPT: horizontal plot table
HR: fire hose reel
HRC: high rupturing capacity
HRC: horizontal redundancy test
HRE 1: Homogeneous Reactor Experiments, 1 MW(th)
HRE 2: Homogeneous reactor Experiments, 5 MW(th)
HRIR: high resolution infrared radiometer
HRL: horizontal reference line
H&RP: holding and reconsignment point
HRRC: Human Resources Research Center
HRT: Homogeneous Reactor Test (see HRE-2)
HS: half subtractor
H/S: health and safety
HS: high speed
HSD: high-speed data
HSD: high-speed displacement
HSD: hot shut down
HSDA: high-speed data acquisition
HSGT: high-speed ground transportation

HSI: horizontal situation indicator

HSM: Health Services and Mental Health Administration

HSM: high-speed memory

HSP: high-speed printer

HSR: high-speed reader

HSRO: high-speed repetitive operation

HSS: high-speed storage

HSST: heavy section steel technology

HST: horizontal seismic trigger

HST: hypersonic transport

HT: high tension

HT: horizontal tabulation

HTA: heavier than air

HTD: hand target designator

HTGCR: Australian high-temperature gas-cooled reactor, solid, homogeneous, beryllium-oxide-moderated, carbon-dioxide-cooled

HTGR-CX: high-temperature gas reactor critical experiment

HTIS: heat transfer instrument system

HTL: heat transfer loop

HTL: high-threshold logic

HTLTR: high-temperature lattice test reactor, Richland, Washington

HTM: high temperature

HTM: hypothesis testing model

HTO: high-temperature oxidation

HTR: Hanford 305 Test Reactor, Richland, Washington

HTR: heater

HTR: Hitachi Training Reactor, Kawasaki, Japan, 100 kW pool-type

HTRDA: High-Temperature Reactor Development Associates - group of utilities cooperating in building the Peach Bottom 45 MW(e) atomic power plant at Peach Bottom, Pennsylvania

HTRE: heat transfer reactor experiment, Nuclear Electronics Laboratory, Hughes Aircraft Company, Culver City, California

HTRI: Heat Transfer Research Institute

HTS: high tensile strength

HTSF: High Temperature Sodium Facility

HTSL: heat transfer simulation loop

HTSS: Honeywell Time-Sharing System

HTTF: critical assembly (Bettis)

HTTL: high-power transistor-transistor logic

HTU: heat transfer unit

HUD-FHA: Housing and Urban Development-Federal Housing Administration

HUD: head-up display

HUD: U.S. Department of Housing and Urban Development

HUDWAC: heads up display weapons-aiming computer

HUGHES-NEL: Hughes Aircraft Company Nuclear Electronics Laboratory, Culver City, California

HUMRRO: Human Resources Research Office

HUR: heat up rate

HUT: HEDL up transient

HUT: hold up tank

HV: hand (operated) valve

HV: hardware virtualizer

H&V: heating & ventilation

HV: high voltage

HVAC: heating, ventilation and air conditioning

HVAC: high vacuum

HVAC: high voltage actuator

HVACC: High Voltage Apparatus Coordinating Committee (ANSI)

HVAR: high-velocity aircraft rocket

HVAT: high-velocity antitank

HVCA: Heating and Ventilating Contractors Association

HVDC: high voltage direct current

HVHMD: holographic visor helmet mounted display

HVP: high video pass

HVR: high-voltage regulator

HVRA: Heating and Ventilating Research Association

HW: half wave

HW: heavy water (deuterium oxide, D2O)

HWCTR: Heavy-Water Component Test Reactor, 61 MW(th), Savannah River, Aiken, South Carolina

HWGCR: Heavy-Water-Moderated Gas-Cooled Reactor, 590 MW(th), Bohunice, Czechoslovakia

HWOCR: heavy water organic cooled reactor

HWP: Heavy-Water Plant, Savannah River

HWR: heavy-water reactor

HWS: Hanford Works standard or specification (now HPS)

HX: heat Exchanger

hxwxl: height by width by length

HYCOL: hybrid computer link

HYCOTRAN: hybrid computer translator

HYDAC: hybrid digital analog computer (EAI)

HYDAPT: hybrid digital-analog and pulse time

Hydro: LASL critical assembly

HYFES: hypersonic flight environmental simulator

HYPO: High-Power Water-Boiler Reactor (dismantled), Los Alamos

HYPSES: hydrographic precision scanning echo sounder

HYSTAD: hydrofoil stabilization device

HYTRESS: high test recorder and simulator system

Hz: hertz

HZMP: horizontal impulse

HZP: hot zero power

I

I: (abbrev. for) luminous intensity

I: information (type of line)

I: Instrumentation

IA: indirect addressing

IA: Industry Application (IEEE Society)

IA: Input Axis

IA: instrumentation amplifier

IA: internal address; information for arbitration bus (type of line)

IA: Iowa

IAA: interim access authorization

IAA: International Academy of Astronautics

IAAC: International Agriculture Aviation Center (GB)

IABSE: International Association for Bridge and Structural Engineering

IAC: Industry Advisory Committee

IAC: Information Analysis Center

IAC: interim acceptance criteria

IAC: international algebraic compiler

IACP: International Association of Chiefs of Police

IACP: International Association of Computer Programmers

IACS: inertial attitude control system

IACS: integrated armament control system

IACS: International Annealed Copper Standard

IAD: immediate action directive

IAD: integrated automatic documentation

IADIC: integration analog-to-digital converter

IADR: International Association for Dental Research

IAE: Institute for the Advancement of Engineering

IAE: integral absolute error

IAE: International Atomic Exposition

IAEA: International Atomic Energy Agency

IAEE: International Association of Earthquake Engineers

IAEE: International Automotive Engineering Exposition

IAEI: International Association of Electrical Inspectors

IAESTE: International Association for the Exchange of Students for Technical Experience

IAESTE/Canada: The International Association for the Exchange of Students for Technical Experience, Canada

IAESTE/US,Inc: International Association for the Exchange of Students for Technical Experience/ United States

IAF: International Astronautical Federation

IAG: IFIP Administrative Data Processing Group

IAG: International Association of Geodesy

IAGA: International Association of Geomagnetism and Aeronomy

IAGC: instantaneous automatic gain control

IAHR: International Association for Hydraulic Research

IAHS: International Association of Housing Science

IAL: immediate action letter

IAL: international algorithmic language

IAL: investment analysis language

IALA: International Association of Lighthouse Authorities

IAM: indefinite admittance matrix

IAM: Institute of Aviation Medicine

IAM: interactive algebraic manipulation

IAM: International Association of Machinists

IAMAP: International Association of Meteorology and Atmospheric Physics

IAMC: Institute for Advancement of Medical Communication

IAMTCT: Institute of Advance Machine Tool and Control Technology

IANEC: Inter-American Nuclear Energy Commission

IAO: internal automation operation

IAPMO: International Association of Plumbing and Mechanical Officials

IAPO: International Association of Physical Oceanography

IARU: International Amateur Radio Union

IAS: immediate access storage

IAS: indicated air speed

IAS: Industry Applications Society

IAS: Institute of Advanced Studies (Army)

IAS: Institute of Aerospace Sciences. Now AIAA

IAS: International Association of Sedimentology

IASA: Insurance Accounting and Statistical Association

IASA: International Air Safety Association

IASF: Instrumentation in Aerospace Simulation Facilities (IEEE AESS Technical Committee)

IASH: International Association of Scientific Hydrology

IASI: Inter-American Statistical Institute

IASPEI: International Association of Seismology and Physics of the earth's interior

IASS: International Association for Shell and Spatial Structures

IAT: information assessment team

IAT: Institute for Advanced Technology

IAT: Institute of Automatics and Telemechanics (USSR)

IAT: International Accountants Society

IAT: International Atomic Time

IATA: International Air Transport Association

IATM: International Association for Testing Materials

IATUL: International Association of Technical University Libraries

IAU: infrastructure accounting unit

IAU: International Association of Universities

IAU: International Astronomical Union

IAV: International Association of Volcanology

IAVC: instantaneous automatic video contol

IAVC: instantaneous automatic volume control

IBA: Investment Bankers Association

IBAM: Institute of Business Administration and Management (Japan)

IBCC: International Building Classification Committee

IBEW: International Brotherhood of Electrical Workers

IBG: interblock gap

IBI: Intergovernmental Bureau for Informatics

IBIS: intense bunched ion source

IBP: initial boiling point

IBPA: International Business Press Associates

IBR: integral boiling reactor

IBR: integrated bridge rectifier

IBR: pulsed fast reactor, 1 kW(th), Dubna, USSR

IBRL: initial bomb release line

IBS: Institute for Basic Standards

IBSAC: industrialized building systems and components

IBSHR: integral boiling and superheat reactor, Westinghouse 20 - MW(e) pilot plant for scale-up to 200 MW(e) (graphite-moderated pressure-tube reactor)

IBW: impulse bandwidth

IBWM: International Bureau of Weights and Measures

IC: impulse conductor

IC: information center

IC: input circuit

I&C: installation and checkout

IC: installed capacity

IC: instruction counter

I&C: instrumentation and control

IC: Insulated Conductors (IEEE PES Technical Committee)

IC: integrated circuit

IC: interface control

IC: interior communications

IC: internal connection

IC: ion chamber

ICA: Industrial Communications Agency

ICA: International Communications Association

ICA: International Council on Archives

ICA: item control area

ICAD: integrated control and display

ICAM: integrated computer-aided manufacturing

ICAO: International Civil Aviation Organization

ICAR: interface control action request

ICARVS: interplanetary craft for advanced research in the vicinity of the sun

ICAS: intermittent commercial and amateur service

ICB: Interface control board

ICBM: intercontinental ballistic missile

ICBO: International Conference of Building Officials

ICBWR: improved cycle boiling water reactor design

ICC: International Computation Centre (Italy)

ICC: Interstate Commerce Commission, USA

ICCC: International Chamber of Commerce

ICCP: Institute for Certification of Computer Professionals

ICCP: International Conference on Cataloging Principles

ICCU: inter-channel comparison unit (PPSA instrumentation)

ICD: interface control drawings

ICE: immediate cable equalizers

ICE: input checking equipment

ICE: Institution of Civil Engineers

ICE: integrated cooling for electronics

ICEA: Insulated Cable Engineers Association. Formerly **IPCEA**

ICEM: International Council for Educational Media

ICES: integrated civil engineering system

ICES: International Conference of Engineering Studies

ICES: International Council for the Exploration of the Sea

ICET: Institute for the Certification of Engineering Technicians

ICETK: International Committee of Electrochemical Thermodynamics and Kinetics

ICG: interactive computer graphics

ICHENP: International Conference on High Energy Nuclear Physics

ICHMT: International Centre for Heat & Mass Transfer

ICI: intelligent communications interface

ICID: International Commission on Irrigation and Drainage

ICID: U.S. Committee on Irrigation, Drainage and Flood Control

ICIP: International Conference on Information Processing

ICL: incoming line

ICM: advanced IBM

ICM: improved capability missile

ICO: Inter-Agency Committee on Oceanography

ICO: International Commission for Optics

ICOGRADA: International Conference of Graphic Design Association

ICOLD: International Commission on Large Dams

ICOM: intercom

ICONS: Information Center on Nuclear Standards (American Nuclear Society)

ICOR: Intergovernmental Conference on Oceanic Research

ICP: Instrument Calibration Procedure

ICP: Interim Compliance Panel (Coal Mine Health and Safety)

ICP: International Control Plan

ICP: inventory control point

ICPS: Industrial and Commercial Power Systems (IEEE IAS Technical Committee

ICPS: International Conference on the Properties of Steam

ICPS: International Congress of Photographic Science

ICR: iron-core reactor

ICRA: Interagency Committee on Radiological Assistance

ICRH: Institute for Computer Research in the Humanities

ICRP: International Commission on Radiological Protection

ICRU: See CIUR

ICS: inland computer service

ICS: Institute of Computer Science

ICS: integrated communication system

ICS: integrated control system

ICS: Inter-communication system (FFTF communication system)

ICS: internal countermeasures set

ICS: interphone control station

ICS: isolation containment spray

ICSA: In-core shim assembly

ICSC: Interim Communications Satellite Committee

ICSI: International Conference on Scientific Information

ICSP: U.S. Interagency Council on Standards Policy

ICSU: International Council of Scientific Unions

ICT: Institute of Computer Technology

ICtl: Industrial Control (IEEE IAS Technical Committee)

ICU: intensive-care unit

ICW: intake cooling water

ICW: interrupted continuous wave

ICWM: International Committee on Weights and Measures

ID: device identification

ID: Idaho

ID: inner diameter or inside diameter

I/D: instruction/data

ID: interactive debugging

ID: interdigital

ID: Interior Department

ID: intermediate description

ID: internal diameter

ID: item documentation

IDA: Institute for Defense Analysis

IDA: integrated digital avionics

IDA: intrusion detection alarm

IDA: ionospheric dispersion analysis

IDAS: information displays automatic drafting system

IDAST: interpolated data and speech transmission

IDC: image dissector camera

IDC: insulation displacement connector

IDCMA: Independent Data Communications Manufacturers Association

IDCSP: Initial Defense Communications Satellite Project

IDDD: international direct distance dialing

IDEAS: integrated design and engineering automated system

IDEEA: information and data exchange experimental activities

IDEF: integrated system definition language

IDEN: *Instituto de Engerharia Nuclear*, Brazil (CP-11 type reactor)

IDEP: interservice data exchange program

IDEX: initial defense experiment

IDF: integrated data file

IDF: intermediate distributing frame

IDFT: inverse discrete Fourier transform

IDG: inspector of degaussing

IDI: improved data interchange

ID/IAS: industrial drives

IDIOT: instrumentation digital on-line transcriber

IDM: interdiction mission

IDMS: integrated data management system

IDOC: inner diameter of outer conductor

IDP: industrial data processing

IDP: integrated data processing

IDPC: integrated data processing center

IDPI: International Data Processing Institute

IDR: industrial data reduction

IDR: inspection discrepancy report (AI)

IDS: Instrument Development Section

IDS: integrated data store

IDS: interim decay storage

IDSS: ICAM decision support system

IDT: interdigital transducer

IDT: isodensitracer

IDTS: instrumentation data transmission system

IDTV: improved-definition television

IDU: industrial development unit

IE-B: Institution of Engineers, Bangladesh

i.e.: *id est* (that is)

IE: industrial engineer

I&E: information and education

IE: initial equipment

IE: inspection and enforcement

IE: Institute of Engineers Sri Lanka

I&E: Office of Inspection & Enforcement

IEA: *Instituto de Energia Atomica*, Sao Paulo, Brazil

IEA: International Energy Agency

IEAR-1: *Instituto de Energia Atomica* Reactor, pool type, 5 MW(th) Brazil

I.E.AUST.: The Institution of Engineers, Australia

IEC: integrated equipment components

IEC: International Electrotechnical Commission, U.S. Inc

IEC: Ion Exchange Conference

IECE(J): Institute of Electrical Communication Engineers of Japan

IECI: Industrial Electronics and Control Instrumentation (IEEE Group)

IECPS: International Electronic Packaging Symposium

IED: individual effective dose

IEE: Institution of Electrical Engineers (London)

IEEE: Institute of Electrical & Electronic Engineers

IEEEC: Industrial Electrical Equipment Council

IEETE: Institution of Electrical and Electronics Technician Engineers

IEG: information exchange group

IEI: The Institution of Engineers of Ireland

IEM: interim examination & maintenance

IEM CELL: Interim examination and maintenance cell

IEMTF: interim examination and maintenance training facility

IEPG: Independent European Program Group

IER: Institutes for Environmental Research

IERE: Institution of Electronic and Radio Engineers (Australia)

IES: Illuminating Engineering Society of North America

IES: Institute of Environmental Sciences

IES: integral error squared

IES: intrinsic electric strength

IES: Iowa Engineering Society

IESL: Institution of Engineers, Sri Lanka

I&E Slugs: internally and externally cooled fuel elements for production reactor

IET: initial engine test

IEV: International Electrotechnical Vocabulary

I/F: interface

IF: intermediate frequency

IFAC: International Federation for Automatic Control
IFAN: International Federation for the Application of Standards
IFATCA: International Federation of Air Traffic Controllers Association
IFB: invitation for bids
IFC: International Finance Corporation
IFCF: integrated fuel cycle facilities
IFCS: in-flight checkout system
IFCS: International Federation of Computer Sciences
IFD: incipient fire detector
IFD: instantaneous frequency discriminator
IFD: International Federation for Documentation
IFEMS: International Federation of Electron Microscopes Societies
IFF: identification friend or foe
iff: if and only if
IFI: Industrial Fasteners Institute
IFIP: International Federation of Information Processing
IFIPC: International Federation of Information Processing Congress
IFIPS: International Federation of Information Processing Societies
IFLA: International Federation of Library Associations
IFMBE: International Federation for Medical and Biological Engineering
IFME: International Federation of Medical Electronics
IFORS: International Federation of Operational Research Societies
IFR: increasing failure rate
IFR: inflight refueling
IFR: instantaneous frequency-measuring receiver
IFR: instrument flight rules
IFRB: International Frequency Registration Board
IFRU: interference frequency rejection unit
IFS: International Federation of Surveyors
IFS: ionospheric forward scatter
IFT: Institute of Food Technologists
IFT: International Foundation for Telemetering
IFTC: International Federation of Thermalism and Climatism
IFTC: International Film and Television Council
IFVME: Inspectorate of Fighting Vehicles and Mechanical Equipment
IG: inertial guidance
IGA: intergranular attack
IGAAS: Integrated Ground Airborne Avionics System
IGC: International Geophysical Committee
IGE: Institution of Gas Engineers
IGFET: insulated gate field-effect transistor
IGFET: isolated gate field-effect transistor
IGN: ignitor
IGOR: intercept ground optical recorder
IGPP: Institute of Geophysics and Planetary Physics
IGR&P: inert gas receiving and processing
IGS: inertial guidance system
IGS: integrated graphics system

IGSCC: inter granular stress corosion crack
IGU: International Gas Union
IGY: International Geophysical Year
IHACE: International Heating and Air-Conditioning Exposition
IHAS: Integrated Helicopter Avionics System
IHB: International Hydrographic Bureau
IHD: International Hydrological Decade
IHE: Institution of Highway Engineers
IHF: Institute of High Fidelity
IHFM: Institute of High-Fidelity Manufacturers
IHP: indicated horsepower
IHPH: indicated horsepower-hour
IHTS: intermediate heat transport system
IHX: intermediate heat exchanger
IHXGV: intermediate heat exchanger guard vessel
IIA: Information Industry Association
IIAS: International Institute of Administrative Services
IIC: International Institute of Communications
IICPR: Puerto Rico Institute of Civil Engineers
IID: independent identically distributed
IIE: Institute of Electrical Engineers of the College of Engineers, Architects and Surveyors of Puerto Rico
IILS: International Institute for Labor Studies
IINSE: International Institute of Nuclear Science and Engineering
IIQ-CIAA: Puerto Rico Institute of Chemical Engineers
IIR: infinite-duration impulse-response
IIR: International Institute of Refrigeration
IIRS: Institute for Industrial Research and Standards (Erie)
IIS: *Institut International de la Soudure* (International Institute of Welding)
IIS: Institute of Industrial Supervisors
IIS: Institute of Information Scientists
IISO: Institution of Industrial Safety Officers
IIT: Illinois Institute of Technology
IITE: Pi Tau Sigma
IITRI: Illinois Institute of Technology Research Institute
IIW: International Institute of Welding
IIW: See IIS
IJAJ: international jitter antijam
IJCAI: International Joint Conference on Artificial Intelligence
IJJU: international jitter jamming unit
I&L: installation and logistics
IL: intermediate language
ILAAS: integrated light attack avionic system
ILAAT: interlaboratory air-to-air missle technology
ILAR: Institute of Laboratory Animal Resources
ILAS: instrument landing approach system
ILAS: interrelated logic accumulating scanner
ILC: instruction length code
ILD: injection laser diode
ILF: inductive loss factor
ILF: infra low frequency
ILIP: in-line instrument package

ILIR: in-house laboratories independent research
ILJM: Illinois Journal of Mathematics
ILL: integrated injection logic
ILLIAC: Illinois Integrator and Automatic Computer
ILM: information logic machine
ILO: injection-locked oscillator
ILO: International Labor Office
ILO: International Labor Organization (Geneva)
ILP: intermediate language processor
ILRT: integrated leak rate test
ILRT: intermediate level reactor test
ILS: ideal liquidus structures
ILS: instrument landing system
ILS: integrated logistic support
ILSF: intermediate level sample flow
ILSRO: Interstate Land Sales Registration Office
ILTS: Institute of Low Temperature Science (Japan)
ILU: Illinois University
ILW: intermediate-level wastes
IM: industry (FCC services) motion picture
IM: installation and maintenance
IM: Institute of Metrology (Leningrad, USSR)
IM: Instrumentation and Measurement (IEEE Group)
IM: interceptor missile
IM: intermediate missile
IMA: Industrial Medical Association
IMA: International Mineralogical Association
IMARE: Institute of Marine Engineers
IMAS: Industrial Management Assistance Survey
IMC: instrument meteorological conditions
IMC: integrated maintenance concept
IMC: intermediate metal conduit
IMC: International Micrographic Congress
IMCO: Intergovernmental Maritime Consultative Organization
IMD: intermodulation distortion
IME: Institute of Makers of Explosives
IME: Institute of Mining Engineers
IME: Institution of Mechanical Engineers
IME/APWA: Institute for Municipal Engineering/ American Public Works Association
I Mech E: Institute of Mechanical Engineers
IMEKO: *Conférence Internationale de la Mesures* (International Measurement Conference)
IMEP: indicated mean effective pressure
IMF: International Monetary Fund
IMFI: Industrial Mineral Fiber Institute
IMH: Institute of Materials Handling
IMI: Irish Management Institute (Eire)
IMITAC: image input to automatic computers
IMM: Institute of Mathematics Machines (Poland)
IMM: Institute of Mining and Metallurgy
IMM: integrated maintenance management
IMMS: International Material Management Society
IMMS NJC: International Material Management Society, New Jersey Chapter
IMOS: ion-implanted metal oxide semiconductor
IMP: injection into microwave plasma
IMP: integrated microwave products
IMP: Inter-Industry Management Program

IMP: interface message processor
IMP: interplanetary monitoring platform
IMP: intrinsic multiprocessing
IMPACT: inventory management program and contol technique
IMPATT: impact avalanche and transit time (diode)
IMPCM: improved capability missile
IMPI: International Microwave Power Institute
IMPRESS: interdisciplinary machine processing for research and education in the social sciences
IMPS: integrated master programming and scheduling
IMPTS: improved programmer test station
IMR: Institute for Materials Research (NBS)
IMRA: International Marine Radio Association
IMRADS: information management retrieval and dissemination system
IMS: Industrial Management Society
IMS: Industrial Mathematics Society
IMS: information management system
IMS: Institute of Marine Science
IMS: Institute of Mathematical Statistics
IMS: instructional management system
IMS: Integrated Meteorological System (Army)
IMS: International Metallographic Society
IMS: inventory management and simulator
IMSL: International Mathematics and Statistics Library
IMU: inertial measuring unit
IMU: International Mathematical Union
IN: Indiana
IN: Institute of Navigation
IN: interference-to-noise ratio
INAS: inertial navigation and attack systems
INCE: Institute of Noise Control Engineering
INCH: integrated chopper
INCO: International Chamber of Commerce
INCOMEX: International Computer Exhibition
InCoPAC: International Consummer Policy Advisory Committee
INCOR: Intergovernmental Conference on Oceanographic Research
INCOR: Israeli National Committee for Oceanographic Research
INCOSPAR: Indian National Committee Space Research
INCOT: (EBR-11) in-core test facility
INCR: interrupt control register
INCUM: Indiana Computer Users Meeting
indep: independent
INDEX: Inter-NASA Data Exchange
INDREG: inductance regulator
INDS: Incore Nuclear Detection System
INDTR: indicator-transmitter
INEL: Idaho Nuclear Engineering Laboratory
INEL: International Exhibition of Industrial Electronics
INFANT: Iroquois night fighter and night tracker
INFCO: Information Committee [ISO]
INFO: information network and file organization
INFOCEN: information center

INFOL: information oriented language
INFRAL: information retrieval automatic language
ING: intense neutron generator
INGO: International Nongovernmental Organization
INIS: International Nuclear Information System
INL: internal noise level
INLC: initial launch capability
INMARSAT: international maritime satellite system
INMM: Institute of Nuclear Materials Management
INP: inert nitrogen protection
INPO: Institute of Nuclear Power Operations
INR: interference-to-noise ratio
INREQ: information on request
INRT: inerter, pulverizer
INS: inertial navigation system
INS: Institute for Naval Studies
INS: Institute for Nuclear Study (Tokyo University)
INSPEC: Information Service in Physics, Electrotechnology and Control (IEE)
INSPEX: Engineering Inspection and Quality Control Conference and Exhibition
INSTARS: information storage and retrieval systems
INTC: Industrial Nuclear Technology Conference
INTCO: international code of signals
INTCOM: International Liaison Committee (IEEE)
INTELSAT: International Telecommunications Satellite Consortium
INTERATOM: *Internationale Atomreactorbau* , Duisberg, Germany
INTERCON: International Convention (IEEE)
INTERGALVA: International Galvanizing Conference
INTERMAG: International Magnetics Conference (IEEE)
INTIPS: integrated information processing system
INTOP: international operations simulation
INTPHTR: interphase transformer
INTRAN: input translator
INTREX: information transfer experiments
INU: inertial navigation unit
inv: inverter
INX: technique for concentration of low-grade pitchblende by ion exchange
I/O: input-output devices
IO: interpretive operation
IOB: input-output buffer
IOC: initial operational capability
IOC: input-output controller
IOC: integrated optical circuit
IOC: Intergovernmental Oceanographic Commission (UNESCO)
IOCS: input-output control system
IOCU: International Organization of Consumers Unions
IOF: International Oceanographic Foundation
IOL: instantaneous overload
IOM: input-output multiplexer
ION: Ionosphere and Aural Phenomena Advisory Committee (ESRO)
IOP: input-output processor

IOPS: input-output programming system
IOR: input-output register
IOS: input-output selector
IOS: International Organization for Standardization
IOTA: information overload testing apparatus
IOT&E: initial operation test and evaluation
IOUBC: Institute of Oceanography, University of British Columbia
IOVST: International Organization for Vacuum Science and Technology
I/P: current to pressure
I&P: inerting and preheating
IP: information processing
IP: initial phase
IP: initial point
IP: Institute of Petroleum
IP: intermediate pressure
IP: item processing
IPA: Information Processing Association (Israel)
IPA: integrated photodetection assemblies
IPA: intermediate power amplifier
IPA: International Patent Agreement
IPA: International Psychoanalytical Association
IPB: illustrated parts breakdown
IPC: information processing center
IPC: information processing code
IPC: Institute of Printed Circuits
IPCEA: Insulated Power Cable Engineers Association (Became **ICEA** in 1979)
IPCR: Institute of Physical and Chemical Research (Japan)
IPDD: initial project design description
IPDH: in-service planned derated hours
IPE: information processing equipment
IPE: Institute of Power Engineers
IPFM: integral pulse frequency modulation
IPL: information processing language
IPL: initial program load
IPM: impulses per minute
IPM: Institute for Practical Mathematics (Germany)
IPM: interference prediction model
IPM: internal polarization modulation
IPM: interruptions per minute
IPOEE: Institute of Post Office Electrical Engineers (GB)
IPP: imaging photopolarimeter
IPP: integrated plotting package
IPPJ: Institute of Plasma Physics (Japan)
IPPS: Institute of Physics and the Physical Society
IPS: inch per second
IPS: interim policy statement
IPS: International Pipe Standard
IPS: interruptions per second
IPS: iron pipe size
IPSOC: Information Processing Society of Canada
IPSSB: Information Processing Systems Standards Board
IPST: International Practical Scale of Temperature
IPT: internal pipe thread
IPU: instruction processing unit

IQC: International Quality Center
IQEC: International Quantum Electronics Conference (IEEE)
IR: information retrieval
IR: infrared
IR: infrared radiation
IR: inside radius
IR: instruction register
IR: instrument reading
IR: insulation resistance
IR: interrogator-responder
IRA: input reference axis
IRAC: Interdepartmental Radio Advisory Committee
IRAN: inspection and repair as necessary
IRAP: Interagency Radiological Assistance Program
IRAS: Infrared Astronomical Satellite
IRBM: intermediate range ballistic missile
IRBO: infrared homing bomb
IRC: Industrial Reorganization Corporation
IRC: international record carriers
IRCM: infrared countermeasures
IR&D: independent research and development
IRDP: information retrieval data bank
IRE: Institute of Radio Engineers (now IEEE)
IRED: infrared-emitting diode
IREE: Institute of Radio and Electrics Engineers (Australia)
IREP: Integrated Reliability Evaluation Program
IREP: Interim Reliability Evaluation Program
IREX: International Research and Exchanges Board
IRFC: intermediate range function test
IRG: interrange instrumentation group
IRHD: international rubber hardness degrees
IRI: Industrial Research Institute
IRI: Institution of the Rubber Industry
IRIG: Inter-Range Instrumentation Group (U.S. Government agency)
IRIS: infrared interferometer spectrometer
IRIS: instant response information system
IRL: information retrieval language
IRLS: interrogation, recording of location subsystem
IRM: intermediate range monitor
IRM: iodine radiation monitor
IRMA: information revision and manuscript assembly
IROD: instantaneous readout detector
IRPA: International Radiation Protection Association
IRPI: individual rod position indicator
IRR: Institute for Reactor Research, Wurenlingen, Switzerland
IRR: integral rocket-ramjet
IRR: Israeli Research Reactor, Rehovath, Israel
IRRD: International Road Research Documentation (OECD)
IRRI: International Rice Research Institute
IRS: information retrieval system
IRS: inquiry and reporting system
IRS: Internal Revenue Service
IRS: isotope removal service
IRSIA: Institute for the Encouragement of Scientific Research in Industry (Belgium)

IRT: in-reactor thimble
IRT: Institute of Reprographic Technology
IRT: interrogator—responder—transponder
IRT: research reactor, natural-uranium, graphite, 5.5 MW(th), Kurchatov Inst. Moscow, USSR
IRTE: Institute of Road Transport Engineers
IRTS: interim recovery technical spec
IRTS: International Radio & Television Society
IRU: inertial reference unit
I&S: Board of Inspection and Survey
IS: incomplete sequence (relay)
IS: information science
IS: interference suppressor
IS: internal shield
IS: Iowa State University of Science and Technology
ISA: Instrument Society of America
ISA: instrument sub assembly
ISA: International Society of Acupuncture
ISA: international standard atmosphere
ISADS: innovative strategic aircraft design studies
ISALIS: Indian Society for Automation and Information Sciences
ISAM: indexed sequential access method
ISAM: integrated switching and multiplexing
ISAP: information sort and predict
ISAR: information storage and retrieval
ISB: independent sideband
ISB: International Society of Biometeorology
ISBB: International Society of Bioclimatology and Biometeorology
ISBN: international standard book number
ISC: Information Society of Canada
ISC: instruction staticizing control
ISC: International Standards Council
ISCA: International Standards Steering Committee for Consumer Affairs (ISO, IEC, and consumer organizations)
ISCA: Intersociety Committee on Methods for Ambient Air Sampling and Analysis
ISCAN: inertialess steerable communication antenna
ISCAS: International Symposium on Circuits and Systems
ISCC: Inter-Society Committee on Corrosion
ISCC: Interstate Solar Coordination Council
ISCT: inner seal collar tool
IScT: Institute of Science Technology
ISD: induction system deposit
IS&D: integrate sample and dump
ISD: International Subscriber Dialing
ISD: interrupt service device
ISDF: intermediate sodium disposal facility
ISDN: integrated services digital network
ISDS: integrated ship design system
I&SE: installation and service engineering
ISE: Institution of Structural Engineers
ISEA: Industrial Safety Equipment Association
ISER: integral systems experimental requirements
ISES: International Solar Energy Society
ISF: Industrial Space Facility
ISFET: ion-sensitive field-effect transistor

ISFSI: independent spent fuel storage installation

ISHM: International Society for Hybrid Microelectronics

ISI: in-service inspection

ISI: Indian Standards Institution

ISI: Institute for Scientific Information

ISI: International Statistical Institute

ISI: intersymbol interference

ISI: Iron and Steel Institute

ISI: Israel Standards Institute

ISIS: international satellites for ionospheric studies

ISIS: International Science Information Service

ISJM: Israeli Journal of Mathematics

ISK: insert storage key

ISL: information search language

ISL: information system language

ISL: instructional systems language

ISL: integrated Schottky logic

ISL: interactive simulation language

ISLIC: Israel Society for Special Libraries and Information Centers

ISM: industrial, scientific, and medical (equipment)

ISM: information systems for management

ISM: interpretive structural modeling

ISML: intermediate system mock-up loop

ISN: International Society for Neurochemistry

ISO: individual system operation

ISO: International Organization for Standardization

ISONET: ISO Network committee

ISP: industrial security plan

ISP: International Society of Photogrammetry

ISP: Italian Society of Physics

ISPEC: insulation specification

ISPEMA: Industrial Safety (Personal Equipment) Manufacturers Association

ISPO: International Statistical Programs Office (Department of Commerce)

ISPRA: D2O research reactor, Italy (near Milano)

ISR: information storage and retrieval

ISR: interrupt service routine

ISR: intersecting storage ring

ISS-AIME: Iron and Steel Society of AIME

ISS: ideal solidus structures

ISS: information storage system

ISS: input subsystem

ISSCC: International Solid-State Circuits Conference

ISSE: International Sun Earth Explorer

ISSMB: Information Systems Standards Management Board/ANSI

ISSN: international standard serial number

ISSS: International Society of Soil Science

IST: information science and technology

IST: international standard thread (metric)

ISTAR: image, storage translation and reproduction

ISTIM: interchange of science and technical information in machine language

ISTS: International Symposium on Space Technology and Science

ISU: interface switching unit

ISU: International Scientific Union

ISY: International Space Year

IT: industry (FCC services) telephone maintenance

IT: Information Theory (IEEE Group)

IT: input translator

I&T: inspection and test

IT: instantaneous trip (device)

IT: Institute of Technology

IT: instrument test or instrument tree

IT: insulating transformer

IT: interrogator—transponder

IT: item transfer

ITA: Independent Television Authority

ITA: Industry and Trade Administration

ITA: Institute for Telecom and Aeronomy (ESSA)

ITA: International Tape Association

ITA: International Typographic Association

ITAE: integrated time and absolute error

ITAL: *Instituit voor de Toepassing van Atoomenergie in de Landboury* - Wageningen, Netherlands

ITAL: induction simulated reactor, Bureau of Mines, Morgantown, West Virginia

IT&AP: inspection test and analysis plan

ITAR: International Traffic in Arms Regulations

ITB: intermediate block (check)

ITC: International Trade Commission

ITC: ionic thermoconductivity

ITCC: International Technical Communications Conference

ITE: Institute of Telecommunications Engineers

ITE: Institute of Traffic Engineers

ITE: Institute of Transportation Engineers

ITEM: Interference Technology Engineer's Master

ITEWS: integrated tactical electronic warfare system

ITF: interactive terminal facility

ITF: interstitial transfer facility (ANL or Westinghouse)

ITFS: instructional television fixed services

ITFTRIA: instrument tree flow and temperature removal instrument assembly

ITI: inspection/test instruction

ITI: interactive terminal interface

ITI: International Technology Institute

ITIRC: IBM Technical Information Retrieval Center

ITL: integrated-transfer-launch

ITL: inverse time limit

I² L: integrated injection logic

ITMA: irradiation test management activity

ITPS: integrated teleprocessing system

ITR: instrument test rig (LMEC)

ITR: inverse time relay

ITRIA: instrument tree removable instrument assembly

ITS: Idaho Test Station

ITS: Institute for Telecommunications Sciences

ITS: insulation test specification

ITS: integrated tracking system (ARTRAC); obsolete

ITS: intermarket trading system

ITSA: Institute for Telecommunication Sciences and Aeronomy (ESSA); formerly CPRL of NBS

IT/SP: instrument tree/spool piece

ITU: International Telecommunications Union (UNO)

ITV: industrial television

IU: interference unit

IUAPPA: International Union of Air Pollution Prevention Associations

IUB: International Union of Biochemistry

IUBS: International Union of Biological Sciences

IUCAF: Inter-Union Commission on Allocation of Frequencies

IUCN: International Union for the Conservation of Nature and Natural Resources

IUCr: International Union of Crystallography

IUDH: in-service unplanned derated hours

IUDH1: in-service Class 1 unplanned derated hours

IUDH2: in-service Class 2 unplanned derated hours

IUDH3: in-service Class 3 unplanned derated hours

IUDH4: in-service Class 4 unplanned derated hours

IUE: International Union of Electrical, Radio, and Machine Workers

IUGG: International Union of Geodesy and Geophysics

IUGS: International Union of Geological Sciences

IUINS: *Institut Interuniversitaires des Sciences Nucleaires,* Brussels, Belgium

IUNDH: in-service unit derated hours

IUPAC: International Union of Pure and Applied Chemistry

IUPAP: International Union of Pure and Applied Physics

IUS: inertial upper stage

IUS: initial upper state

IUTAM: International Union of Theoretical and Applied Mechanics

IUVSTA: International Union for Vacuum Science Techniques and Applications

IV: intermediate voltage

IVA: intervehicular activity

IVDS: independent variable depth sonar

IVHM-EM: in-vessel handling machine/engineering model

IVHM: in-vessel handling machine

IVMU: inertial velocity measurement unit

IVR: integrated voltage regulator

IVS: in-vessel storage

IVSI: instantaneous vertical speed indicator

IVSM: in-vessel storage module

IW: industry (FCC services) power

IWAHMA: Industrial Warm Air Heater Manufacturers Association

IWCS: integrated wideband communications system

IWCS/SEA: integrated wideband communications system/Southeast Asia

IWES: industrial waste filter system

IWP: interim working party

IWS: Institute of Work Study

IWSA: International Water Supply Association

IWSc: Institute of Wood Science

IWS/IT: integrated work sequence/inspection traveler

IWSP: Institute of Work Study Practitioners

IX: deionizer; ion exchanger

IX: industry (FCC services) manufacturers

IXION: magnetic-mirron-type fusion device, LASL

IZ: isolation zone

Izv VUZ Radioelektron: *Izvestia Vysshikh Uehebnykh Zavedeniy Radioelektronika*

Izv VUZ Radiofiz: *Izvestia Vysshikh Uehebnykh Zavedeniy Radiofizika*

J

J: job

J: joule

JACC: Joint Automatic Control Conference

JACM: Journal of the Association for Computing Machinery

J Acoust Soc Amer: Journal of the Acoustical Society of America

JAEC: Japan Atomic Energy Commission

JAEIP: Japan Atomic Energy Insurance Pool

JAERI: Japanese Atomic Energy Research Institute, Tokai-Mura, Japan

jaff: electronic and chaff jamming

JAIEG: Joint Atomic Information Exchange Group

JAIF: Japan Atomic Industrial Forum

JAN: joint army-navy (specification)

JANAP: Joint Army-Navy-Air Force Publications

JANUS: biological research reactor, ANL, 200 kW(th), tank-type

Japan GCR: Japan gas-cooled reactor, 585 MW(th), Tokai-Mura, Japan

J Appl Phys: Journal of Applied Physics

JAPT: Journal of Approximation Theory

JARRP: Japan Association for Radiation Research on Polymers

JASIS: Journal of the American Society for Information Science

JASON: ARGONAUT type research reactor, United Kingdom

JASON: CP-11 type reactor, 10kW(th), Petten, Netherlands

JBMMA: Japanese Business Machine Makers Association

J Brit IRE: Journal of the British Institution of Radio Engineers

JCA: Joint Commission on Accreditation of Universities

JCAM: Joint Commission on Atomic Masses

JCAR: Joint Commission on Applied Radioactivity

JCC: Joint Computer Conference

JCC: Joint Conference Committee

JCEC: Joint Communications Electronics Committee

JCG: Joint Coordinating Groups (CCI and IEC)

JCGS: Joint Center for Graduate Study

JCI: Joint Communications Instruction

JCL: job control language

JCNPS: Joint Committee on Nuclear Power Standards

JCP: Joint Committee on Printing, U.S. Congress

JCSS: Journal of Computer and System Sciences
JDM: Journal of Data Management
JDS: job data sheet
JEA: Jordan Engineers Association
JECA: Joint Engineers Council of Alabama
JECC: Japan Electronic Computer Center
JECC: Joint Electronic Components Conference
JEDEC: Joint Electron Device Engineering Council
JEEPNIK: subcritical experiment
JEEP 1: Joint Establishment Experimental Pile, Kjeller, Norway, 459 kW(th)
JEEP 2: Joint Establishment Experimental Pile, Kjeller, Norway, 2 MW(th), tank-type
JEIDA: Japan Electronic Industry Development Association
JEIPAC: JICST Electronic Information Processing Automatic Computer
JEMC: Joint Engineering Management Conference
JEN: *Junta de Energia Nuclear* (Spanish nuclear agency)
JERC: Joint Electronic Research Committee
JES: job entry system
JESM: Japan Society of Electrical Discharge Machining
JETEC: Joint Electron Tube Engineering Council
JETR: Japan Engineering Test Reactor, Tokai-Mura, Japan, 50MW(th), tank-type
JETS: Junior Engineering Technical Society
JFET: junction field-effect transistor
JFL: joint frequency list
JFP: joint frequency panel
J/G: joules/gram
J Geophys Res: Journal of Geophysical Research
JHG: joule heat gradient
JHU: Johns Hopkins University
JIC: Joint Industrial Council
JICST: Japan Information Center of Science and Technology
JIE: Junior Institution of Engineers
J IEE: Journal of Institution of Electrical Engineers
JIFTS: joint in-flight transmission system
JILA: Joint Institute for Laboratory Astrophysics (University of Colorado)
JIMS: Journal of the Indian Mathematical Society
JINR: Joint Institute for Nuclear Research - Soviet bloc countries research center, Dubna, USSR
J Inst Nav: Journal of the Institute of Navigation
JIP: joint input processing
JIS: Japanese Industrial Standards
JISC: Japanese Industrial Standards Committee
JLMS: Journal of the London Mathematical Society
JMA: Japan Meteorological Agency
JMA: Japan Microphotography Association
JMC: Joint Meteorological Committee
JMED: jungle message encoder-decoder
J Met: Journal of Meteorology
JMKU: Journal of Mathematics of Kyoto University
JMSJ: Journal of the Mathematical Society of Japan
JOLA: Journal of Library Automation
JPB: Joint Planning Board

JPDR: Japan Power Demonstration Reactor, BWR, 11,700 kW(e), Tokai-Mura,Japan
JPL: Jet Propulsion Laboratory (NASA)
JPRS: Joint Publications Research Service
JQE: Journal of Quantum Electronics (IEEE)
JRATA: Joint Research and Test Activity
JRDOD: Joint Research and Development Objectives Document
J Rech Atmos: *Journal de Recherches Atmospheriques*
J Res NBS: National Bureau of Standards Journal of Research
JRIA: Japan Radioisotope Association
JRR 1: Japanese Research Reactor, 154-type reactor, Tokai-Mura
JRR 2: Japanese Research Reactor, 10 MW(th), tank-type
JRR 3: Japanese Research Reactor, 10 MW(th), tank-type
JRR 4: Japanese Research Reactor, 1 MW(th), pool-type
JS: jam strobe
JSC: Japan Science Council
JTAC: Joint Telecommunications Advisory Committee

K

K-25: gaseous diffusion plant area, Oak Ridge
K-65: radium bearing sludge from refining pitchblende ores
k: Boltzmann's constant
K: cathode (vacuum tube)
K: kelvin
k: kilo
kA: kiloampere
KACHINA: advanced computer program to serve as a test bed for computer simulation of complex components
KAEDS: Keystone Association for Educational Data Systems
KAFB: Kirkland Air Force Base (Albuquerque, NM)
Kahl: Kahl, West Germany, power reactor, 60 MW(th)
KALDAS: Kidsgrove Algol digital analog simulation
KAPL: Knolls Atomic Power Laboratory (GE - Schenectady, New York)
kb: kilobit
KBWP: *Kernkraftwerk Baden-Wuerttemberg Planungsgessellscaft* , organic-moderated reactor
kc: kilocycle per second;(obsolete) Use kilohertz (kHz)
KDP: potassium dihydrogen phosphate
KDR: keyboard data recorder
KE: kinetic energy
KEAS: knot equivalent air speed
KEC: Kingston Engineers Club
Keff: effective multiplication factor
KEMA: Laboratory and Research Center, Arnheim, Netherlands
KES: Kansas Engineering Society
keV: kilo-electron-volt
KEW: kinetic-energy weapons

KEWB: kinetic experiment on water boilers, small test reactors, Santa Susana, California

kg: kilogram (10^3 grams: 1kg=2.205 pounds)

kgm: kilogram-meter

kg/m³: kilogram per cubic meter

kgps: kilogram per second

kHz: kilohertz (10^3 hertz)

KIAS: knots indicated air speed

KIFIS: Kollsman integrated flight instrumentation system

KII: *Koninklijk Institute Van Ingenieurs*

KIPO: keyboard input printout

KIPS: kilowatt isotope power system

KIS: keyboard input simulation

KK: 1 000 000

KL: key length

KLIC: key letter in context

kM: kilomega (obsolete)

km: kilometer (10^3 meters; 1 km=0.621 miles)

kMc: kilomegacycle (obsolete); use gigahertz (GHz)

KMER: Kodak metal etch resist

KN: knot

KOA: Kuiper Airborne Observatory

kohm: kilohm (also shown as kΩ)

kΩ : kilohm

KORSTIC: Korea Scientific and Technological Information Center

KP: key punch

KPIC: key phrase in context

K Plants: Individual units of the ORGDP

KRB: *Kernkraftwerk FWE Bayermwerk* , 237 MW(e) BWR, Gunduremmingen West Germany

KRITO: critical experiment

KRR: Kansai Research Reactor, Kumatori, Japan, 1 MW(th) pool-type

KS-150: heavy water power reactor at Bohunice, Czechoslovakia

KS: Kansas

KSC: John F. Kennedy Space Center (NASA); (now Cape Canaveral)

KSE: Kuwait Society of Engineers

KSR: keyboard send receive (set)

KSTR: Kema Suspension Test Reactor, Netherlands, 250 kW(th)

KSU: key service unit

KTS: key telephone system

KTSA: Kahn test of symbol arrangement

Kukla: fast critical experiments (University of California, Lawrence Radiation Laboratory)

kV: kilovolt (10^3 volts)

kVA: kilovoltampere

kVAh: kilovoltampere-hour

kVAhm: kilovoltampere-hour meter

kvar: kilovar

kVdc: kilovolt direct current

kW: kilowatt

KWAC: keyword and context

KW/FT: kilowatts/foot (linear heat generating rate)

kWh: kilowatt-hour

kWhm: kilowatt-hour meter

KWIC: key word in context

KWOC: keyword out of context

KWOT: keyword out of title

KY: Kentucky

L

L-TR: licensing technical review

L-47: teaching reactor built by Atomics International

L-54: 50kW(th) teaching reactor at Walter Reed Army Institute of Research, Washington, DC

L-77: type of teaching reactor built by Atomics International

L: local bit

LA: A laser gyro axis

LA: lead angle

LA: lightning arrester

LA: Louisiana

LAAV: light airborne LSW vehicle

LAB: low-altitude bombing

LABS: low-altitude bombing system

LACBPE: Los Angeles Council of Black Professional Engineers

LACBWR: La Crosse boiling water reactor

LACES: Los Angeles Council of Engineering Societies

LACIE: Landsat-oriented large area crop inventory experiment

LACIE: Large Area Crop Inventory Experiment

LAD: location aid device

LAD: logical aptitude device

LADAPT: lookup dictionary adaptor program

ladar: laser detection and ranging

LADR: ladder

LADSIRLAC: Liverpool and District Scientific, Industrial and Research Library Advisory Council

LAE: left arithmetic element

LAEDP: large area electronic display panel

LAG: load and go assembler

LAH: logical analyzer of hypothesis

LAM: loop adder and modifier

LAMPF: Los Alamos Meson Physics Facility

LAMPRE: Los Alamos molten plutonium reactor experiment

LAMPS: light airborne multipurpose system

LAN: local area network

LANDSAT: under NASA (earth-resource-monitoring Satellite) as opposed to COMSAT

LANNET: large artificial nerve net

LANNET: large artificial neuron network

LANL: Los Alamos National Laboratory (formerly LASL)s

LANS: Loran Airborne Navigation System

LAP: lesson assembly program

LAP: list assembly program

LAPB: link access procedure, balanced

LAPD: limited axial power distribution

LAPDOG: low-altitude pursuit dive on ground

LAPES: low-altitude parachute extraction system

LAPPES: Large Power Plant Effluent Study; initiated in 1967 and provided funds to support a TVA project to study the impact of inversion breakup phenomena on their large power plant plumes
LAR: low angle reentry
LARC: Livermore Atomic Research Computer (UNIVAC)
LARIAT: laser radar intelligence acquisition technology (469L spacetrack)
LARS: laser angular rate sensor (Defender)
LAS: large astronomical satellite
LAS: limited space charge accumulation
LASA: large-aperture seismic arrays
LASCR: light-activated silicon-controlled rectifier
LASCS: light-activated silicon-controlled switch
LASER: light amplification by stimulated emission of radiation
LASL: changed to LANLs
LASRM: low-altitude supersonic research missile
LASSO: laser search and secure observer
LASSO: laser synchronization from stationary orbit
LATCC: London air-traffic control centre
LATINA: power reactor, 705 MW(th), Face Verde, Italy
LATINCON: Latin American Convention (Region 9, IEEE)
LATS: long-acting thyroid stimulator
LAVA: linear amplifier for various applications
LAWB: Los Alamos water boiler, aqueous, homogeneous-type reactor, includes: LOPO, HYPO and SUPO
LAWRS: limited air weather reporting certificate
LB: light bombardment
lb: pound (1 lb=16 ounces=453.59 grams)
LBL: Lawrence Berkeley Laboratory
LBO: line buildout
LBP: length between perpendiculars
LBT: low bit test
LC-MARC: Library of Congress Machine-Readable Cataloging
LC: inductance-capacitance
LC: lead covered
LC: level control
LC: Library of Congress
LC: line carrying
LC: line connector
LC: line of communication
LC: line of contact
LC: liquid crystals
LC: load carrier
LC: load center
LC: load compensating (relay)
LC: low carbon
LC: low conduction (threshold voltage)
LCAO: linear combination of atomic orbitals
LCC: landing craft, control
LCC: launch control center
LCC: life cycle costing
LCC: liquid crystal cell
LCD: liquid crystal display

LCE: launch complex equipment
LCF: language central facility
LCF: local cycle fatigue
LCHS: large component handling system
LCIE: *Laboratoire Central des Industries Electriques* (France)
LCL: lifting condensation level
LCL: linkage control language
LCLV: liquid-crystal light valve
LCM: large core memory
LCM: least common multiple
LCMM: life cycle management model
LCN: load classification number
LCNT: link celestial navigation trainer
LCO: Limited Condition of Operability
LCP: language conversion program
LCR: inductance-capacitance-resistance
LCR: log count rate
LCRE: lithium-cooled reactor experiment, NRTS
LCS: large core store
LCS: Loop control system
LCS: loudness-contour selector
LCSRM: loop current step response method
LCTI: large components test installation
LCTL: large components test loop (now SCTL, at LMEC)
LCVD: least voltage coincidence detection
LCVIP: licensee contractor vendor inspection report
L/D: lift-to-drag ratio
LD: line drawing
LD: line of departure
LD: long distance
LDA: locate drum address
LDB: light distribution box
LDC: latitude data computer
LDC: light direction center
LDC: linear detonating cord
LDC: line drop compensator
LDCC: large diameter component cask
LDCS: long distance control system
LDDS: low density data system
LDE: laminar defect examination
LDE: linear differential equation
LDEF: long-duration exposure facility
LDNS: doppler navigation system
LDP: language data processing
LDR: light dependent resistor
LDR: low data rate
LDRI: low data rate input
LDRS: LEM data reduction system
LDS: large disk storage
LDT: logic design translator
LDT: long distance transmission
LDX: long distance xerography
LD50: lethal-dose of radiation for 50% of those exposed
LE: antenna effect length for electric-field antennas
LE: leading edge
LE: light equipment
LE: limit of error

LE: low explosive
LEAA: Law Enforcement Assistance Administration
LEADS: law enforcement automated data system
LEAP: language for the expression of associative procedures
LEAR: logistics evaluation and review technique
LEAS: lease electronic accounting system
LEC: liquid cncapsulated Czochralski
LED: light emitting diode
LEDT: limited-entry decision table
LEED: laser-energized explosive device
LEED: low energy electron diffraction
LEER: low-energy electron reflections
LEF: light-emitting film
LEF: lighting effectiveness factor
LEFM: linear elastic fracture mechanics
LEL: large engineering loop (NASA-Lewis)
LEL: lower explosion level
LEM: antenna effective length for magnetic-field antennas
LEM: lunar excursion module
LEMUF: limit of error on material unaccounted for
LEO: low earth orbit
LEP: lower end plug
LER: licensee event report
LERA: limited employee retirement account
LES: launch escape system
LES: limited early site
LES: Lincoln Experimental Satellite
LES: Louisiana engineering Society
LESR: limited early site review
LESSC: Louisville Engineering & Scientific Societies Council
LET: linear energy transfer
LEU: low enriched uranium
LEVITRON: fusion research device at Lawrence Radiological Laboratories
LF: leapfrog configuration (circuit theory)
LF: load factor (tabular)
LF: low frequency (30 to 300 kHz)
LFA: low flow alarm
LFBR-CX: liquid fluidized bed reactor critical experiment
LFBR: liquid fluidized bed reactor concept
LFC: laminar flow control
LFC: load frequency control
LFE: Laboratory for Electronics
LFIM: Low-Frequency Instruments and Measurement (IEEE IM Technical Committee)
LFM: linear FM
LFO: low-frequency oscillator
LFRD: lot fraction reliability definition
LFS: lift station
LG: landing gear
LG: landing ground
LGG: light-gun pulse generator
LGO: Lamont Geological Laboratory (Columbia University)
LGT: low group transmit
LGWS: laser-guided weapons systems

LH: left-hand
LH: liquid hydrogen
LH: luteinizing hormone
LHC: left-hand circular (polarization)
LHGR: linear heat generation rate
LHM: loop handling machine
LHPS: lead hydrogen purge system
LHR: lower hybrid resonance
LHS: loop handling system
LHSI: low head safety injection
LHST: laundry & shower storage tank
LIA: Laser Industries Association
LIC: linear integrated circuit
LIC: loop insertion cell
LID: leadless inverted device
LID: locked-in device
LIDAR: laser infrared radar
LIDAR: laser intensity direction and ranging
LIDAR: light detection and ranging
LIDO: Reactor at Harwell, United Kingdom, 100 kW(th)
LIFEX: life-extension (retrofitting plans)
LIFO: last-in first-out (inventory control)
LIFT: logically integrated Fortran translator
LIL: Lunar International Laboratory
LIM: Linear induction motor
LIMAC: large integrated monolithic array computer
LIMB: liquid metal breeder, heterogeneous, thorium breeder, graphite-moderated and lead-cooled reactor
LINAC: linear accelerator
LINAS: laser inertial navigation attack system
LINC: laboratory instruments computer (DEC)
LINCOMPEX: linked compressor and expander
LINLOG: linear-logarithmic
LINS: laser inertial navigation system
LIPL: linear information processing language
LIR: line integral refractometer
LIS: Line Isolation Switch (Reactor Level Sw.)
LISA: library systems analysis
LISN: line impedance stabilization network
LIST: library and information services, tees-side
LIT: liquid injection technique
LITE: legal information through electronics
LITR: low intensity test reactor, 3 MW(th), ORNL
Little Eva: LASL critical assembly experiment on enriched metal in 3-in.-thick uranium reflector
LITVC: liquid injection thrust vector control
LIWB: Livermore Water Boiler, Livermore, California, 500 MW(th)(dismantled)
LIX: lixator
LIZARDS: library information search and retrieval data system
LKY: Lefschetz-Kalman-Yakubovitch (lemma)
L/L: latitude/longitude
LL: light line
LL: liquid limit
LL: low level
llc: logical link control
LLD: low level detector

LLF: light loss factor
LLFM: low level flux monitor
LLL: Lawrence Livermore Laboratory
LLL: long-path laser
LLL: low-level logic
LLLLLL (or L6): laboratories low-level linked list language (BTL)
LLLTV: low light level TV
LLLwt: low Level Liquid waste tank
LLNL: Lawrence Livermore National Laboratory
LLPL: low low pond level
LLRES: load-limiting resistor
LLRT: low level reactor test
LLRV: lunar landing research vehicle
LLS: liquid level switch
LLW: low level waste
LLWAS: low level wind shear alert system
L/M: lines per minute
LM: lunar module
LMEC: Liquid Metal Engineering Center (A1)
LMF: linear matched filter
LMFBR: liquid metal fast breeder reactor
LMFR: liquid metal fuel reactor concept
LMFRE: liquid metal fuel reactor experiment
LMHX: liquid metal heat exchanger
LMR: liquid metal reactor
LMS: least-mean-square
LMS: London Mathematical Society
LMSEO: Labor Management Standards Enforcement Office
LMSS: lunar mapping and survey system
LMTD: logarithmic mean temperature difference
ln: natural logarithm
LNA: launch numerical aperture
LNAP: low nonessential air pressure
LNG: liquified natural gas
LNP: loss of normal power
LO: local oscillator
LO: lock-on
LO: longitudinal optical
LO: lunar orbiter
LOA: length over all
LOA: line of assurance
LOADS: low-altitude defense system
LOAMP: logarithmic amplifier
LOB: line of balance
LOC: Launch Operations Center
LOC: limiting conditions for operation
LOCA: loss-of-coolant accident
LOCATE: Library of Congress Automation Techniques Exchange
loc. cit.: *loco citato* (in the place cited)
LOCF: loss-of-coolant flow
LOCI: logarithmic computing instrument
LOCP: loss-of-coolant protection
LOCS: librascope operations control system
LOCS: logic and control simulator
LODESMP: logistics data element standardization and management process
LOERO: large orbiting earth resources observatory

LOF: loss of flow
LOF: lowest operating frequency
LOFA: loss of flow accident
LOFAR: low-frequency acquisition and ranging
LOFT: loss of fluid test facility
LOFW: loss Of Feedwater
LOG: linear quadratic Gaussian control
Log: logarithm
LOGALGOL: logical algorithmic language
LOGEL: logic generating language
LOGIPAC: logical processor and computer
LOGIT: logical inference tester
LOGLAN: logical language
LOGRAM: logical program
LOGTAB: logical tables
LOH: light observation helicopter
LOHAP: light observation helicopter avionics package
LOI: loss of ignition
LOI: loss on ignition
LOL: length of lead (actual)
LOLA: low level oil alarm
LOLA: lunar orbit landing approach
LOLITA: language for the on-line investigation and transformation of abstractions
LOLP: loss of load probability
LOM: locator at the outer marker
LOMUSS: Lockheed multiprocessor simulation system
L'Onde Elect: *L'Onde Electrique*
LOP: lines of position (navigation)
LOP: loss of offsite power
LOPAD: logarithmic outline processing system for analog data
LOPAR: low power acquisition radar
LOPI: loss of pipe integrity
LOPO: low power water reactor, Los Alamos
LOR: lunar orbital rendezvous
LORAN: long range navigation
LORDS: licensing on-line retrieval data system
LORL: large orbital research laboratory
LORPGAC: long range proving ground automatic computer
LOS: line of sight
LOS: loss of signal
LOSP: loss of offsite power
LOSP: loss of system pressure
LOSS: landing observer signal system
LOSS: large object salvage system
LOST: let down storage tank
LOTIS: logic, timing, sequencing (language)
LOX: liquid oxygen
LP: linear programming
LP: line printer
Lp: long-play
LP: low pass
LP: low point
LP: low pressure
LPA: low pressure alarm
LPAC: Launching Programs Advisory Committee
LPC: linear predictive coding

LPC: loop-control (relay)
LPC: loop preparation cask
LPC: lower pump cubicle
LPCI: low pressure coolant injection
LPCRS: low pressure coolant recirculation system
LPCS: low pressure core spray
LPD: language processing and debugging
LPD: linear power density
LPD: log-periodic dipole antennas
LPD: low power difference
LPD: low pressure difference
LPDR: Local Public Document Room
LPE: liquid phase epitaxial
LPE: loop preparation equipment
LPG: *langage de programmation et de gestion* (C11 French computer language)
LPG: liquid petroleum gas
LPGS: Liquid Pathway Generic Study
LPGTC: Liquified Petroleum Gas Industry Technical Committee
LPI: Lightning Protection Institute
LPI: low power injection
LPIS: low pressure injection system
LPL: linear programming language
LPL: list processing language
LPM: licensing project manager
LPM: lines per minute
LPMA: loose parts monitor assembly
LPO: low power output
LPP: low power physics
LPPT: low pressurization pressure test transmitter
LPR-CC: critical assembly (Bettis)
LPR: Lynchburg Pool Reactor, B&W, 1 MW(th), pool-type, Virginia
LP/RH: line pairs per raster height
LPRINT: lookup dictionary print program
LPRM: low power range monitor (GE)
LPRS: local power recirculation system
LPS: low pressure scram
LPSIS: local power safety range monitor
LPSW: low pressure service water
LPTF: low power test facility
LPTR: Livermore pool-type reactor (LRL)
LPTV: large payload test vehicle
lpW: lumen per watt
LPZ: low population zone
LQG: linear-quadratic-Gaussian
LR: limited recoverable
LR: load-resistor (relay)
LR: load ratio
L/R: locus of radius
LRA: A Laser Gyro Reference Axis
LRC: Langley Research Center (NASA)
LRC: Lewis Research Center (NASA, Cleveland)
LRC: load ratio control
LRC: longitudinal redundancy check
LRI: long-range radar input
LRIA: level removable instrument assembly
LRL: Lawrence Radiation Laboratory (University of California)

LRLTRAN: Lawrence Radiation Laboratory translator
LRM: limited register machine
LRP: long range path
LRPL: Liquid Rocket Propulsion Laboratory (Army)
LRS: linguistic research system
LRS: Liquid Radwaste System
LRSM: long range seismograph measurements
LRSS: long range survey system
LRT: local leak rate test
LRTF: long range technical forecast
LRTM: long range training mission
LRU: line-replaceable unit
LRV: lunar roving vehicle
LRX: large reactor critical facility, Waltz Mill, Pennsylvania
LS: language specification
LS: least significant
LS: limit switch
LSA: limited space (charge) accumulation (microwave diodes)
LSA: low-cost solar array
LSA: low specific activity
LSAP: link layer service access point
LSB: least significant bit
LSB: lower sideband
LSB: low speed breaker (relay)
LSBR: large seed blanket reactor
LSC: least significant character
LSC: Legal Services Corporation
LSCC: line-sequential color composite
LSD: least significant digit
LSD: low speed data
LSDF: large sodium disposal facility
LSDU: link layer service data unit
LSE: longitudinal-section electric
LSECS: life support and environmental control system
LSFFAR: low-spin folding fin aircraft rocket
LSFT: low steamline flow test
LSHI: large-scale hybrid integration
LSI: large scale integration
LSM: linear synchronous motor
LSMU: lasercom space measurement unit
LSN: line stabilization network
LSO: landing signal officer
LSP: low speed printer
LSPC: Lewis Space Flight Center
LSR: load shifting resistor
LSR: Lynchburg Source Reactor, B&W, Lynchburg, Virginia
LSS: large space structure or large space system
LSSM: local scientific survey module
LSSP: latest scram set point
LSSS: limiting safety system setting
LSU: Louisiana State University
LT: language translation
LT: line telecommunications
LT: link trainer instructor
LT: low tension
LT: pilot light

LTC: line traffic coordinator
LTC: long time constant
LTDS: laser target designation system
LTL: lot truck load
LTP: library technology project
LTPD: lot tolerance percent defective
LTR: Lattice Test Reactor (PCTR), Hanford, Washington
LTR: Lockheed Training Reactor, 10 kW(th), pool-type
LTS: long-term stability
LTTL: low-power transistor-transistor logic
LTV: large test vessel (AI)
LTV: long tube vertical
LUCID: language for utility checkout and instrumentation development
LUF: lowest usable frequency
LUHF: lowest usable high frequency
LUME: light utilization more efficient (Shed Light)
LUT: launcher umbilical tower
LV: low voltage
LVCD: least voltage coincidence detection
LVD: low velocity detonation
LVDA: launch vehicle data adapter
LVDC: launch vehicle digital computer
LVDT: linear variable differential transformer
LVDT: linear velocity displacement transformer
LVHV: low volume high velocity
LVP: low voltage protection
LWA: limited work authorization
LWBR: light-water breeder reactor
LWIR: long-wave infrared
LWM: liquid waste Monitor
LWOP: leave without pay
LWR: light water reactor
LWR: liquid waste release
LWST: light waste storage tank
LWTT: liquid waste test tank
lx: lux(1 lux=1 lumen/m²)
LYRIC: language for your remote instruction by computer

M

M-L: metallic-longitudinal
M-S: Mitte-Seite (stereo)
M: master
m: milli - (prefix)
M: motor starter
MA: Maritime Administration
MA: Massachusetts
MA: Mathematical Association (GB)
MA: megampere
MA: Metric Association
mA: milliampere
MA: modify address
MA: module address
MAA: material access area
MAA: Mathematical Association of America
MAAC: Mid-Atlantic Area Council

MAARC: magnetic annular arc
MAC: machine-aided cognition
MAC: maintenance allocation chart
MAC: maximum allowable concentration
MAC: mean aerodynamic chord
MAC: medium access control
MAC: Mineralogical Association of Canada
MAC: module auxiliary connector
MAC: multiple access computer
MACC: modular alter and compose console
MACCT: multiple assembly cooling cask test (ORNL)
MACE: management applications in a computer environment
MACMIS: maintenance and construction management information system
MACON: matrix connector punched card programmer
MACSMB: Measurement & Automatic Control Standards Management Board/ANSI
MAD: magnetic airborne detector
MAD: magnetic anomoly detector
MAD: maintenance assembly-disassembly
MAD: Michigan algorithm decoder
MAD: multiaperture device
MAD: multiple access device
MAD: multiply and add
MADA: multiple-access discrete address
MADAEC: Military Application Division of the Atomic Energy Commission
MADAM: multipurpose automatic data analysis machine
MADAR: malfunction analysis detection and recording
MADDAM: macromodule and digital differential analyzer machine
MADE: microalloy diffused electrode
MADE: minimum airborne digital equipment
MADE: multichannel analog-to-digital data encoder
MADGE: microwave aircraft digital guidance equipment
MADM: Manchester automatic digital machine
MADP: main air display plot
MADRE: magnetic drum receiving equipment
MADREC: malfunction and detection recording
MADS: machine-aided drafting system
MADS: missile attuded determination system
MADT: microalloy diffused-base transistor
MAE: Maine Association of Engineers
MAE: Maryland Association of Engineers, Inc
MAECON: Mid-America Electronics Conference
MAELV: Mutual Atomic Energy Liability Underwriters
MAES: Mexican American Engineering Society
MAESTRO: machine-assisted educational system for teaching by remote operation
MAF: major academic field
MAG: magnet
MAG: Magnetics (IEEE Society)
MAGAMP: magnetic amplifier
MAGG: Modular Alphanumeric Graphics Generator

MAGGI: Million Ampere Generator (fusion program), Aldermaston, United Kingdom

MAGGS: modular advanced graphics generation system

MAGIC: machine for automatic graphics interface to a computer

MAGIC: matrix algebra general interpretive coding

MAGIC: Michigan automatic general integrated computation

MAGIC: Midac automatic general integrated computation

MAGLOC: magnetic logic computer

MAGMOD: magnetic modulator

MAGPIE: machine automatically generating production inventory evaluation

Mag Rec/M: magnetic recording

MAIDS: multipurpose automatic inspection and diagnostic system

MAIN: Mid-American Interpool Network

MAIP: matrix algebra interpretive program

MAIR: molecular airborne intercept radar

MAL: macro assembly language

MALE: multiaperture logic element

MALT: mnemonic assembly language translator

MAM: management and administration manual

MAM: multiple access to memory

MAMI: machine-aided manufacturing information

MAMOS: marine automatic meteorological observing station

MANIAC: mathematical analyzer numerical integrator and computer

MANIAC: mechanical and numerical integrator and computer

MANOVA: multivariate analysis of variance

MAP: macro arithmetic processor

MAP: macro assembly program

MAP: manifold absolute pressure

MAP: mathematical analysis without programming

MAP: maximum *a posteriori* probability

MAP: message acceptable pulse

MAP: multiple aimpoint

MAP: multiple allocation procedure

MAPCHE: mobile automatic programmed checkout equipment

MAPID: machine aided program for preparation of instruction data

MAPLHGB: Maximum Average Planar Linear Heat Generator

MAPLHGR: maximum average planar linear heat-generation rate

MAPORD: Methodology Approach to Planning and Programming Air Force Operational Requirements, Research and Development

MAPRAT: maximum power ratio

MAPS: multicolor automatic projection system

MAPS: multivariate analysis and prediction of schedules

MAR: malfunction array radar

MAR: memory address register

MARAD: U.S. Maritime Administration

MARC: machine-readable cataloging

MARCA: Mid-Continent Area Reliability Coordination Agreement

MARCAS: maneuvering reentry control and ablation studies

MARCEP: maintainability and reliability cost-effectiveness program

MARCIA: mathematical analysis of requirements for career information appraisal

MARCOM: microwave airborne communications relay

Marconi Rev: Marconi Review

MARDAN: marine differential analyzer (Polaris/Poseidon)

MARGEN: management report generator

Marine Transp/IAS: Marine Transportation

MARIUS: Uranium-graphite research reactor, CEA, Marcoule, France, 1 kW(th)

MARLIS: multiaspect relevance linkage information system

MAROTS: maritime orbital test satellites

MARS: machine retrieval system

MARS: management analysis reporting service

MARS: Marconi automatic relay system

MARS: Martin automatic reporting system

MARS: military affiliated radio system

MARS: military amateur radio system

MARS: multiaperture reluctance switch

MART: maintenance analysis review technique

MART: mean active repair time

MARYLA: Zero power reactor, Poland

MAS: Management Advisory Services

MAS: metal-alumina-silicon

MAS: Military Agency for Standardization

MAS: multiaspect signaling

MASCOT: Motorola automatic sequential computer operated tester

MASCURA: fast neutron critical assembly, Cadarache, France

MASER: microwave amplification by stimulated emission of radiation

MASIS: management and scientific information system

MASK: maneuvering and sea-keeping

MASRT: marine air support radar teams

MASS: Michigan automatic scanning system

MASS: monitor and assembly system

MASS: multiple access sequential selection

MASSTER: mobile Army sensor system test evaluation and review

MASTER: matching available student time to educational resources

MASTER: multiple access shared time executive routine

MASURKA: French critical facility

MAT: mechanical aptitude test

MAT: microalloy transistor

MATA: Michigan Aviation Trades Association

MATCALS: Marine Air Traffic Control and Landing System

MATCON: microwave aerospace terminal control
MATD: mine and torpedo detector
MATE: measuring and test equipment
MATE: multiple-access time-division equipment
MATE: multisystem automatic test equipment
MATIC: multiple area technical information center
MATICO: machine applications to technical information center operations
MATLAN: matrix language
Mat Meas/IM: TC-6 Materials Measurements
MATPS: machine-aided technical processing system
MATT: missile ASW torpedo target
MAU: medium attachment unit
MAU: multiattribute utility
MAUD: code name for the early British atomic energy project (circa 1940/42)
MAUDE: morse automatic decoder
MAVIN: machine-assisted vendor information network
MAW: Marine aircraft wing
MAW: mission adaptive wiring
max: maximum
MAXIE: Maxwellian averaged cross sections reactor physics computer code
MB: magnetic brake
MB: main battery
MB: memory buffer
MB: Metric Board
MB: missile bomber
MBA: Marine Biological Association
MBA: Master of Business Administration
MBA: material balance area
MBD: magnetic-bubble device
MBEO: Minority Business Enterprise Office
MBL: Marine Biological Laboratory
MBM: magnetic bubble memory
MBM: metal-barrier-metal
MBO: management by objective
MBPS: megabits per second
MBR: material balance report
MBR: memory buffer register
MBRV: maneuverable ballistic reentry vehicle
MBSS: Marinelli beaker standard source
MBT: metal-base transistor
MBWO: microwave backward-wave oscillator
MC: magnetic clutch
MC: magnetic core
MC: major component
MC: manual control
MC: Maritime Commission
MC: master clods
MC: master control
MC: maximum capacity
MC: Medical Corps
mc: megacycle (obsolete); use megahertz (MHz)
MC: metal clad
MC: military computer
MC: molded components
M&C: monitor and control
MC: multichip

MC: multiple contact
MCA: Manufacturing Chemists Association
MCA: material control and accountability
MCA: Material Coordinating Agency
MCA: maximum credible accident
MCA: Military Coordinating Activity
MCA: Model Cities Administration
MCA: multichannel analyzer
MCAA: Mechanical Contractors Association of America
MCB: module circuit board
MCBF: mean cycles between failures
MCC-H: Mission Control Center-Houston
MCC: main control circuit
MCC: maintenance control circuit
MCC: management control center
MCC: Motor Control Center
MCC: multiple computer complex
MCCU: multiple communication control unit
MCD: months for cyclical dominance
MCDP: microprogrammed communication data processor
MCE: Master of Civil Engineering
MCEB: Military Communications Electronics Board
MCESS: Milwaukee Council of Engineering & Scientific Societies
MCF: military computer family
MCF: monolithic crystal filter
MCG: magnetocardiography
MCG: man-computer graphics
MCHFR: minimum critical heat flux rates
MCID: multipurpose concealed intrusion detector
MCIS: maintenance control information system
MCIS: materials compatibility in sodium
MCM: magnetic core memory
MCM: Monte Carlo method
MCMJ: Michigan Mathematical Journal
MCMM: management control-material management
MCOM: mathematics of computation
MCOV: maximum continuing operating voltage
MCP: master control program
MCP: memory centered processor
MCP: multichannel communications program
MCPR: maximum critical power ratio
MCR: magnetic character reader
MCR: magnetic character recognition
MCR: main control room
MCR: master change record
MCR: master control routine
MCR: military compact reactor
MCROA: Marine Corps Reserve Officers Association
MCS: master control system
MCS: medical computer services
MCS: method of constant stimuli
MCS: Mobile Communications Systems (IEEE VT Technical Committee)
MCS: modular computer systems
MCS: motor circuit switch
MCS: multiprogrammed computer system
MCT: mechanical comprehension test

MCTI: Metal Cutting Tool Institute
MCTR: message center
MCU: microprogram control unit
MCUG: Military Computers Users Group
MCV: magnetic cushion vehicle
MCV: movable closure valve
MCW: modulated continuous wave
MD: magnetic disk
MD: magnetic drum
MD: main drum
MD: manual data
MD: Maryland
MD: measured discard
MD: medical department
MD: message data
MD: mine disposal
MD: monitor displays
MD: movement directive
MDA: minimum descent altitude
MDA: multidimensional analysis
MDA: multidocking adapter
MDAA: Mutual Defense Assistance Act
MDAC: multiplying digital-to-analog converter
MDAFWP: motor driven aux. feed water pump
MDAP: Mutual Defense Assistance Program
MDC: maintenance data collection
MDC: maintenance dependency chart
MDC: maximum dependable capacity
MDC: Missile Development Center
MDCS: maintenance data collection system
MDCT: mechanical draft cooling tower
MDE: magnetic decision element
MDEFWP: motor driven emergency feedwater pump
MDF: main distributing frame
MDF: microcomputer development facilities
MDF: mild detonating fuse
MDH: maintenance derated hours
MDHD: missile shield
MDI: magnetic direction indicator
MDI: medium dependent interface
MDI: miss distance indicator
MDIC: microwave dielectric integrated circuit
MDIF: manual data input function
MDL: macro description language
MDL: Mine Defense Laboratory
MDL: minimum detectable level
MDM: maximum design meter
MDM: metal-dielectric-metal (filter)
MD/NC: mechanical drafting/numerical control
MDR: maximum design rating
MDR: memory data register
MDS: maintenance data system
MDS: malfunction detection system
MDS: memory disk system
MDS: microprocessor development system
MDS: minimum discernible (or detectable) signal
MDS: modern data systems
MDS: multipoint distribution systems (communication)

MDSMB: Medical Devices Standards Management Board
MDSS: meteorological data sounding system
MDT: mean down time
MDTA: Man Power Development and Training Act
MDTR: metal detector
MDTS: megabit digital-to-troposcatter subsystem
MDU: mine disposal unit
ME: Maine
ME: Master of Engineering
M&E: material and equipment
ME: mechanical efficiency
ME: mechanical engineering
me: megacycle (obsolete). Use MHz)
MEA: minimum en-route altitude
MEAL: master equipment allowance list
MEAL: master equipment authorization list
MEAR: maintenance engineering analysis record
MEC: Marshalltown Engineers Club
MEC: Metrology Engineering Center, NOSC (Pomona, CA)
MECCA: mechanized catalog
MECL: Motorola emitter-coupled logic
MECNY: Municipal Engineers of the City of New York
MECOMSAG: mobility equipment command scientific advisory group
MED: microelectronic device
MEDAC: Medical Equipment Display and Conference
MEDAL: Micromechanized Engineering Data for Automated Logistics
MEDDA: mechanized defense decision anticipation
MEDIA: Magnavox electronic data image apparatus
MEDLARS: medical literature analysis and retrieval system
MEDS: medical evaluation data system
MEDSERV: Medical Service Corps
MEDSMB: Medical Standards Management Board
MEDSPECC: Medical Specialist Corps
MEECN: minimum essential emergency communications network
MEFV: maintenance equipment floor valve
MEG: megohm
MEG: message expediting group
MEGA: one million (prefix)
MEI: manual of engineering instructions
MEIU: mobile explosives investigation unit
MEL: many-element laser
MEL: Marine Engineering Laboratory (U.S.Navy)
MEL: Materials Evaluation Laboratory
MELBA: multipurpose extended lift blanket assembly
MELEC: microelectronics
MELEM: microelement
MELUSINE I: Swimming-pool research reactor, CEA, Grenoble, France, 1.4 kW(th)
MELUSINE II: Swimming-pool research reactor, CEA, Grenoble, France, no power
MEM: Mars excursion module
MEM: minimum essential medium

MEMA: Motor and Equipment Manufacturers Association

M.Eng: Master of Engineering

MEO: major engine overhaul

MEP: mean effective pressure

MERA: molecular electronics for radar applications

MERDC: Mobility Equipment Research and Development Center

MERDL: Medical Equipment Research and Development Laboratory

MERLIN-JULICH: Reactor MERLIN-JULICH, Julich, Germany (identical with MERLIN reactor at Aldermatson, United Kingdom

MERLIN: Medium-Energy light-water-moderated industrial nuclear reactor, Aldermaston, Berkshire, United Kingdom

MERMUT: mobile electronic robot manipulator and underwater television

Merry-go-round: Pulsed fast neutron assembly (Godiva type), Dubna, USSR

MERS: mobility environmental research study

MES: manual entry subsystem

MES: Michigan Engineering Society

MESA: manned environmental system assessment

MESA: Mining Enforcement and Safety Administration

MESFET: metal-semiconductor FET

MESG: maximum experimental safe gap

MESG: micro electrostatically suspended gyro

MESS: monitor event simulation system

MESUCORA: measurement, control regulation, and automation

MET: management engineering team

MET: modified expansion tube

META: Maintenance Engineering Training Agency

META: methods of extracting text automatically

Metal Ind'y/IAS: Metal Industry

METAPLAN: methods-of-extracting-text-autoprogramming language

METRIC: multiechelon technique for recoverable item control

MeV: million electronvolts

MEW: microwave early warning

MEXE: Military Engineering Experimental Establishment

MF: medium frequency (300 to 3 MHz)

Mf: microfarad

MF: microfiche

MF: microfilm

MFB: mixed functional block

MFC: microfunctional circuit

MFCI: molten fuel coolant interaction

MFCM: multifunction card machine

MFD: magnetofluid dynamics

mfd: microfarad

MFES: Minnesota Federation of Engineering Societies

MFI: mobile fuel irradiator

MFN: most favored nation

MFP: mixed fission products

MFPG: mixed fission products generator (AEC isotopic power unit)

MFS: manned flying system

MFSK: multiple frequency shift keying

MFT: multiprogramming with a fixed number of tasks

MFTG: Manufacturing Technology (IEEE Group)

MFTRS: magnetic flight test recording system

MFV: maintenance floor valve

MG: motor generator

MG: multigage

MGC: manual gain control

MGC: missile guidance computer

MGCR-CX: maritime gas-cooled reactor critical experiment

MGCR: maritime gas-cooled reactor project

MGD: mean gain deviation

MGL: matrix generator language

mg set: motor-generator set

MH-1A: Mobile High Power Plant, No. 1A (proposed pressurized water, barge-mounted reactor to be built for Corps of Engineers, U.S. Army), 45 MW(th)

MH: magnetic head

MH: manhole

mH: millihenry

MHA: Marine Historical Association

MHA: maximum hypothetical accident

MHA: modified handling authorized

MHD: magnetohydrodynamics

MHE: materials handling equipment

MHEA: Mechanical Handling Engineers Association

MHEDA: Material Handling Equipment Distributors Association

MHF: medium high frequency

MHFR: maximum hypothetical fission product release

mho: unit of conductance (siemens is now used)

MHRST: Medical and Health Related Sciences Thesaurus

MHT: mild heat treatment

MHTS: Main heat transport system

MHz: megahertz (abbreviation)

MI: manual input

MI: Michigan

mi: mile

MI: mineral insulated (cables)

MIA: metal interface amplifier

MIAC: minimum automatic computer

MIB: manual input buffer

MIB: multilayer interconnection board

MIC: medium interface connector

MIC: Michigan instructional computer

MIC: micrometer

MIC: microphone

MIC: microwave integrated circuit

MIC: minimum ignition current

MIC: monolithic integrated circuit

MIC: mutual interference chart

MICA: macro instruction compiler assembler

MICAM: microammeter

MICELEM: microphone element

MICNS: modulator integrated communication and navigation system

MICOM: Missile Command, Army (Huntsville, AL)

MICR: magnetic ink character recognition

MICRO: multiple indexing and console retrieval options

micro: one-millionth (prefix)

MICROMIN: microminiature

MICROPAC: micromodule data processor and computer

Microwave J: Microwave Journal Microwaves

Microwmag/M: Microwave Magnetics

MICS: Management Information and Control System

MICS: mineral insulated copper sheathed

MIDAC: Michigan digital automatic computer

MIDAS: measurement information data analytic system

MIDAS: missile intercept data acquisition system

MIDAS: modified integration digital analog simulator

MIDAS: modulator isolation diagnostic analysis system

MIDOT: multiple interference determination of trajectory

MIDS: movement information distribution station

MIDS: Multifunction Information Distribution System

MIFASS: marine integrated fire and air support system

MIFR: master international frequency register

MIG: magnetic injection gun

MIG: metal inert gas

MIGHTY MOUSE: Heterogeneous, enriched-uranium, heavy-water-cooled and-moderated research reactor, USA

MIL-I: military specification on interference

MIL-STD: military standard (book)

MIL: military electronics

MIL: military specification (followed by a single capital letter and numbers)

mil: one-thousandth of an inch

MILA: Merritt Island launch area

MILADGRU: military advisory group

MILECON: Military Electronics Conference

MILHDBK: military handbook

MILIC: microwave insular line integrated circuits

MILTEN: Wylbur's terminal handler

MIM: Maryland Institute of Metals

MIM: metal-insulator-metal

MIM: modified index method

MIMD: multiple instruction multiple data

MIMO: man in, machine out

MIMO: multi-input multi-output

MIMS: multi-item multisource

MIN: mobile identification number

MINAC: miniature navigation airborne computer

MINDAC: marine inertial navigation data assimilation computer

MINEAC: miniature electronic autocollimator

Minerve: pool-type reactor at Fontenay, France, 50 w(th)

Mining Ind'y/IAS: Mining Industry

MINIRAR: minimum radiation requirements (to accomplish assigned missions)

MINITRACK: satellite tracking network

MINPRT: miniature processing time

MINS: miniature inertial navigation system

MINSOP: minimum slack time per operation

MINT: materials identification and new item control technique

MINTECH: Ministry of Technology

MIOP: multiplexing input/output processors

MIP: manual input processing

MIP: manual input program

MIP: matrix inversion program

MIP: missile impact predictor

MIPE: modular information processing equipment

MIPIR: missile precision instrumentation radar

MIPS: million instructions per second

MIPS: missile impact predictor set

MIR: memory information register

MIR: memory input register

MIRACL: management information report access without computer languages

MIRACODE: microfilm retrieval access code

MIRAGE: microelectronic indicator for radar ground equipment

MIRD: medical internal radiation dose

MIREK: micro-reciprocal kelvin

MIRF: multiple instantaneous response file

MIRFAC: mathematics in recognizable form automatically compiled

MIROS: modulation inducing retrodirective optical system

MIRR: material inspection and receiving report

MIRS: manpower information and retrieval system

MIRT: molecular infrared track

MIRV: multiple independent reentry vehicle

MIS: management information system

MIS: metal insulator semiconductor

MISD: multiple instruction single data (stream or pipeline processors)

MISDAS: mechanical impact system design for advanced spacecraft

MISFET: metal insulator semiconductor field-effect transistor

MISHAP: missiles high-speed assembly program

MISM: metal-insulator-semiconductor metal

MISP: medical information systems program

MISPHT: spherical multigroup transport code for computer

MISS: mechanical interruption statistical summary

MISS: mobile integrated support system

MISS: multi-item single source

MISSIL: management information system symbolic interpretive language

MIST: Microburst and Severe Storm (experiment)

MIST: minor isotopes safeguards techniques

MIST: slab multigroup transport code for computer

MISTRAM: missile trajectory measurement system

MIT: Massachusetts Institute of Technology

MIT: master instruction tape

MITE: missile integration terminal equipment

MITI: Ministry of International Trade and Industry (Japan)

MITOL: machine-independent telemetry-oriented language

MITR: Massachusetts Institute of Technology Reactor, 2 MW(th) tank-type

MKS: meter-kilogram-second

MKSA: meter-kilogram-second-ampere

ML-1: Mobile Low Power Plant, No. 1, U.S. Army prototype GCR, 3.3 MW(th)

ML: machine language

ML: main lobe

ML: materials laboratory

ML: methods of limits

ml: milliliter

MLCAEC: Military Liaison Committee to the Atomic Energy Commission

MLD: median lethal dose, LD/50

MLD: minimum lethal dose

MLE: maximum likelihood estimate

MLG: main landing gear

MLM: multilayer metallization

MLP: machine language program

MLPCB: machine language printed circuit boards

MLPWB: multilayer printed-wiring board

MLR: main line of resistance

MLRG: Marine Life Research Group

MLRS: multiple launch rocket system

MLS: machine literature searching

MLS: Master of Library Science

MLS: microwave landing system

MLS: missile location system

MLS: multilanguage system

MLSNPG: Microwave Landing System National Planning Group

MLT: mean length of turn

MLT: mean logistical time

MLW: mean low water

mm-band: 40-300 GHz frequency band

MM: maintenance manual

MM: materials measurement

MM: middle marker

mm: millimeter

MM: modified Mercalli

MMAU: master multiattribute utility

MMC: maximum metal condition

MMC: memory management controller

MMD: Manual of the Medical Department

MME: Master of Mechanical Engineering

MMF: magnetomotive force

MMG: motor-motor generator

MMI: man/machine interface

MMM: maintenance and material management

MMOD: micromodule

MMP: multiplex message processor

MMPA: Magnetic Material Producers Association

MMPR: missile manufacturers planning reports

MMPT: man-machine partnership translation

MMPU: memory manager and protect unit

MMRBM: mobile medium range ballistic missile

MMS: Man-Machine Systems Group

MMS: mass memory store

MMS: missile monitoring system

MMS: multimission spacecraft

MMSE: minimum mean squared error

MMU: manned maneuvering unit

MMU: memory management unit

MMU: modular maneuvering unit

MN: Minnesota

MNOS: metal-nitride-oxide-semiconductor

MNS: metal nitride semiconductor

MO: manual output

MO: master oscillator

MO: Missouri

MOA: matrix output amplifier

MOBIDAC: mobile data acquisition system

MOBIDIC: mobile digital computer

MOBL: macro oriented business language

MOBS: multiple-orbit bombardment system

MOBULA: model building language

MOC: master operational controller

MOC: memory operating characteristic

MOC: mission operation computer

MODA: motion detector and alarm

MODAC: mountain systems digital automatic computer

MODEM: modulator-demodulator

MODI: modified distribution method

MODICON: modulator dispersed control

MODS: major operation data system

MOERO: medium orbiting earth resources observatory

MOF: maximum operating frequency

MOGA: microwave and optical generation and amplification

MOH: maintenance outage hours

MOH: maximum operating hours

MOL: machine-oriented language

MOL: manned orbiting laboratory

MOLAB: mobile lunar laboratory

MOLDS: multiple on-line debugging system

MOOSE: manned orbital operations safety equipment

MOP: multiple on-line programming

MOPA: master oscillator power amplifier

MOPR: manner of performing rating

MOPTS: mobile photographic tracking station

MOR: monthly operating report

MORT: mgmt. oversight & risk tree

MOS: management operating system

MOS: metal-oxide semiconductor

MOSA: metal-oxide surge arrester

MOSAIC: macro operation symbolic assembler and information compiler

MOSAIC: metal oxide semiconductor advanced integrated circuit

MOSAIC: Ministry of Supply automatic integrator and computer

MOSFET: metallic oxide semiconductor field-effect transistor

MOSM: metal-oxide semimetal
MOST: metal oxide semiconductor transistor
MOT: motor
MOTA: materials open test assembly
MOTARDES: moving target detection system
MOTNE: Meteorological Operational Telecommunications Network of Europe
MOTU: mobile optical tracking unit
MOV: metal oxide varistor
MOV: motor operated valve
MOX: mixed (uranium and plutonium) oxide fuel
MP: main phase
MP: minimum phase
MP: multipole
MPA: multiple-period average
MPBE: molten plutonium burn-up experiment
MPC: Metal Properties Council
MPCAG: Military Parts Control Advisory Group
MPCC: multi-protocol communications controller
MPD: magnetoplasmadynamic
MPD: maximum permissible dose
MPE: mathematical and physical sciences and engineering
MPE: maximum permissible exposure
MPE: minimum perceptible erythema
MPEP: manual of patent examining procedure
MPG: miles per gallon
MPG: miniature precision gyrocompass
MPH: miles per hour
MPI: magnetic particle inspection
MPI: mean point of impact
MPL: maximum permissible level
MPL: mechanical properties loop
MPO: Manufacturing production order
MPOD: mean planned outage duration
M&PP: materials and plant protection
MPPL: multipurpose processing language
MPR: mechanical pressure regulator
MPRE: medium power reactor experiment
MPS: manpower system
MPS: multiple protective structure
MPS: multiprogramming system
MPT: male pipe thread
MPT: minimum pressurization temperature
MPTA: main propulsion test article
MPTA: Mechanical Power Transmission Association
MPU: microprocessing unit
MPW: modified plane wave
MPX: multiplexer
MQS: motion to quash subpoena
MR: machine records
MR: memory register
mR: milliroentgen
MR: moisture resistant
MRA: minimum reception altitude
MRAALS: marine remote area approach and landing system
MRAD: mass random access disc
MRADS: mass random access data storage
MRB: magnetic recording borescope

MRBM: medium range ballistic missile
MRC: maintenance requirement card
MRC: multiple register counter
MRE: Microbiological Research Establishment
MRE: multiple-response enable
MRGS: Mesabi Range Geological Society
MRI: machine records installation
MRI: magnetic resonance imaging
MRI: material receiving instruction (Bechtel)
MRI: Medical Research Institute (USN)
MRI: monopulse resolution improvement
MRIR: medium resolution infrared radiometer
MRL: Medical Research Laboratory
MRL: multiple rocket launcher
MRM: metabolic rate monitor
mR/min: milliroentgens per minute
MRMU: mobile radiological measuring unit
MRN: meteorological rocket network
MRO: maintenance, repair and operating
MRO: midrange objectives
MRR: mechanical reliability report (FAA)
MRR: Medical Research Reactor, 1 MW(th) pool-type, BNL
MRS: mathematics research center
MRS: medium-range search
MRT: mean repair time
MRT: modified rhyme test
MRU: machine records unit
MRU: material recovery unit
MRU: mobile radio unit
MRWC: multiple read-write compute
MS: machine selection
MS: macromodular system
MS: magnetic storage
M&S: maintenance and supply
MS: margin of safety
MS: mass spectrometry
MS: Master of Sciences
MS: material specification
MS: medical survey
MS: Medicine and Surgery (Navy)
MS: memory system
MS: Metallurgical Society
MS: milestone
MS: military standard, prefix to numbered series issued by DOD
MS: military standard (sheet)
ms: millisecond (10^{-3} seconds)
MS: Mississippi
MS: mode select
MSA: material surveillance assembly
MSA: mechanical signature analysis
MSA: Mineralogical Society of America
MSA: mines safety appliance
MSAR: Mines safety appliance research
MSB: Mining Standards Board
MSB: most significant bit
MSBE: molten salt breeder experiment
MSBR: molten salt breeder reactor
MSC: macro selection compiler

MSC: Manned Spacecraft Flight Center (NASA)
MSC: Marine Corps School
MSC: mile of standard cable
MSC: module segment connector
MSC: monolithic crystal filter
MSC: most significant character
MSC: motor speed changer
MSCA: Mixed spectrum critical assembly, GE, Vallecitos, California
MSCE: main storage control element
MSCI: molten steel coolant interaction
MSD: most significant digit
MSDT: maintenance strategy diagramming technique
MSec: millisecond
MSEMPR: Missile Support Equipment Manufacturers Planning Reports
MSFC: Marshall Space Flight Center (NASA)
MSFN: manned space flight network
MSFS: main steam & feedwater system
MSG: modular steam generator
MSHA: Mine Safety and Health Administration
MSI: medium scale integration
MSIO: mass storage input-output
MSIV: main steam isolation valves
MSIVLCS: main steam line isolation valve leakage control system
MSK: minimum shift keying
MSL: mean sea level
MSLD: mass spectrometer leak detector
MSM: master slave manipulator
MSM: metal-semiconductor-metal
MSM: modified source multiplication
MSMB: Mechanical Standards Management Board/ANSI
MSN: Microwave System News
MSOS: mass storage operating system
MSP: maintenance surveillance procedure
MSR: mass storage resident
MSR: material status report
MSR: missile site radar
MSRE: molten salt reactor experiment, ORNL
MSRM: main steam radiation monitor
MSRS: main steam radiation system
MSRT: missile system readiness test
MSS: main steam system
MSS: Main support structure
MSS: management science systems
MSS: Manufacturers Standardization Society of the Valve and Fittings Industry
MSS: mass storage system
MSS: Mixed Spectrum Superheater, light-water-cooled and -moderated BWR with nuclear superheating, USA
MSSCE: mixed spectrum superheater critical experiment
MSSR: Mars soil sample return
MSSR: mixed spectrum superheat reactor concept
MST: monolithic systems technology
MSTS: multisubscriber time sharing systems

MSV: mean square voltage (Campbelling effect - wide range flux monitor)
MT: core melt through
MT: machine translation
MT: magnetic particle test or materials test
MT: magnetic tape
MT: maximum torque
MT: Montana
MTA: multiterminal adapter
MTAC: mathematical tables and other aids to computation
MTB: Materials Transportation Bureau
MTBF: mean time between failures
MTBM: mean time between maintenance
MTC: Machine Tool Conference
MTC: maintenance time constraint
MTC: master tape control
MTC: memory test computer
MTC: Missile Test Center
MTC: mission and traffic control
MTC: moderator temperature coefficient
MTCF: mean time to catastrophic failure
MTCU: magnetic tape control unit
MTCV: main turbine control valve
MTD: minimal toxic dose
MTDS: manufacturing test data system
MTDS: marine tactical data system
M&TE: measuring and test equipment
MTF: mean time to failure
MTF: mechanical time fuse
MTF: Mississippi Test Facility (NASA)
MTF: modulation transfer function
MTI: Machine Tools Industry (IEEE IAS Technical Committee)
MTI: Mechanical Technology, Incorporated
MTI: moving target indicator
MTIC: moving target indicator coherent
MTL: merged-transistor logic
MTL: mobiltherm light
MTM: Methods Time Measurement Association
MTN: multinational trade negotiations
MTNS: metal thick oxide semiconductor
MTOP: molecular total overlap population
MTOS: magnetic tape operations system
MTOS: metal thick oxide silicon
MTP: mechanical thermal pulse
MTPF: maximum total peaking factor
MTPF: minimal total processing time
MTPS: magnetic tape programming system
MTR: magnetic tape recorder
MTR: Materials Testing Reactor, 40 MW(th), NRTS
MTR: materials testing report
MTR: missile tracking radar
MTRE: magnetic tape recorder end
MTRS: magnetic tape recorder start
MTS: magnetic tape station
MTS: magnetic tape system
MTS: Marine Technology Society
MTS: mass termination system
MTS: message toll service

MTS: missile tracking system
MTS: module tracking system
MT/SC: magnetic tape selective composer
MTSE: magnetic trap stability experiment
MTSMB: Material and Testing Standards Management Board
MTST: magnetic tape selectric typewriter
MTT: magnetic tape terminal
MTT: magnetic tape transport
MTT: Microwave Theory and Techniques (IEEE Society)
MTTD: mean time to diagnosis
MTTF: mean time to failure
MTTFF: mean time to first failure
MTTPO: mean time to planned outage
MTTR: mean time to repair
MTTR: mean time to restore
MTTUO: mean time to unplanned outage
MTU: magnetic tape unit
MTU: master terminal unit
MTU: multiplexer and terminal unit
MTV: marginal terrain vehicle
MU: make up
MU: multiple unit
MUDWNT: make up demineralizer waste neutralizer tank
MUF: material unaccounted for
MUF: maximum usable frequency
MULTEWS: multiple electronic warfare surveillance
MULTICS: multiplexed information and computing service
MUM: multiple-unit message
MUMS: mobile utility module system
MUOD: mean unplanned outage duration
MUR: Management update and retrieval system
MUR: Mock-Up Reactor (NASA), 100 kW(th), pool-type, Sandusky, Ohio
MUSA: multiple-unit steerable antenna
MUSAT: multipurpose UHF satellite
MUX: multiplexer
MV: mean variation
MV: megavolt
mV: millivolt (10^{-3} volts)
MV: multivibrator
MVA: megavoltampere
MVB: multivibrator
MVC: manual volume control
MVC: multiple variate counter
MVP: mechanical vacuum pump
MVS: minimum visible signal
MVS: multiple virtual storage
MVT: multiprogramming with a variable number of tasks
MW: megawatt (10^6 watts)
MW: microwave
mW: milliwatt (10^{-3} watts)
MWd/t: megawatt days per ton
MW(E): megawatts (electrical)
MWh: megawatt-hour
MW(H): megawatts (heat)

MWI: message waiting indicator
MWL: milliwatt logic
MWO: maintenance work order
MWP: maximum working pressure
MWPC: multiwire proportional chamber
MWR: mean width ratio
MWRX: microwave receiver
MWSC: Minimum Wage Study Commission
MWT: make up water treatment
MWTX: microwave transmitter
MWV: maximum working voltage
MZFR: *Mehrzweck Forschungs* (multipurpose) power reactor, 50 MW(e), Karlsruhe, West Germany

N

n: nano (prefix)
N: refractivity
NA: A laser Gyro Axis
nA: nanoampere
NA: not applicable or not available
NA: numerical aperture
NAA: National Association of Accountants
NAAM: National Association of Architectural Metal Manufacturers
NAATS: National Association of Air Traffic Specialists
NAB: National Alliance of Businessmen
NAB: National Association of Broadcasters
NAB: navigational aid to bombing
NAB: Nuclear Assembly Building, Merry Island, Cape Canaveral
NABCE: National Association of Black Consulting Engineers
NABE: National Association of Business Education
NABER: National Association of Business and Educational Radio
NAC: national agency checks
NACA: National Advisory Commission for Aeronautics (now NASA)
NACAA: National Association of Computer Assisted Analysis
NACATS: North American clear air turbulence tracking system
NACC: National Automatic Controls Conference
NACC: North America Control Committee
NACE: National Association of Corrosion Engineers
NACE: National Association of County Engineers
NACEIC: National Advisory Council on Education for Industry and Commerce
NACEO: National Advisory Council on Economic Opportunity
Nachr Tech: *Nachrichtentechnik*
NACME: National Action Council for Minorities in Engineering
NACOA: National Advisory Committee on Oceans and Atmosphere
NACS: Northern Area Communications System
NAD: no-acid descaling
NAD: nuclear accident dosimetry

NADGE: NATO air defense ground environment

NADWARN: natural disaster warning system

NAE: National Academy of Engineering

NAEB: National Association Educational Broadcasters

NAEC: National Aerospace Education Council

NAECON: National Aerospace Electronic Conference (IEEE)

NAED: National Association of Electrical Distributors

NAEDS: National Association of Educational Data Systems

NAET: National Association of Educational Technicians

NAFAX: National Facsimiles Network Circuit

NAFEC: National Administrative Facilities Experimental Center

NAIC: National Astronomy and Ionosphere Center

NAIG: Nippon Atomic Industry Group

NAIOP: navigational aid inoperative for parts

NAK: negative acknowledgment character

NaK: Sodium-potassium alloy, used as a reactor coolant

NAL: National Accelerator Laboratory

NAM: National Association of Manufacturers

NAMA: National Automatic Merchandising Association

NAMFI: NATO missile firing installation

NAMI: Naval Aerospace Medical Institute

NAMRI/SME: North American Manufacturing Research Institution of SME

NAMTC: Naval Air Missile Test Center

NAOGE: National Association of Government Engineers

NAPCA: National Air Pollution Control Administration, USA

NAPE: National Association of Power Engineers, Inc

NAPHCC: National Association of Plumbing-Heating-Cooling Contractors

NAPL: National Association of Photolithographers

NAPM: National Association of Photographic Manufacturers

NAPM: National Association of Purchasing Management

NAPPE: Network Analysis Program using Parameter Extractions

NAPS: Nimbus automatic programming system

NAPSIC: North American Power Systems Interconnection Committee

NAPSS: numerical analysis problem solving system

NAPUS: nuclear auxilliary power unit system (SNAP)

NAR: National Association of Rocketry

NAR: net assimilation rate

NARBA: North American Regional Broadcasting Agreement

NARDIS: Navy Automated Research and Development Information System

NAREC: Naval Research Electronic Computer

NARF: Nuclear Aerospace Research Facility

NARM: National Association of Relay Manufacturers

NARS: National Archives and Records Service

NARTB: National Association of Radio and Television Broadcasters

NARTS: National Association of Radio Telephone Systems

NARTS: Naval Aeronautic Test Station

NARTS: Naval Air Rocket Test Station

NARUC: National Association of Regulatory Utility Commissioners

NARUCE: National Association of Regulatory Utility Commission Engineers

NAS: National Academy of Sciences

NAS: National Aircraft Standards

NAS: National Airspace System

NASA: National Aeronautics and Space Administration

NASAM: National Air and Space Museum

NASAP: Network Analysis and Systems Application Program

NASARR: North American Search and Range Radar

NASCAS: NAS/NRC Committee on Atmospheric Sciences

NASCO: National Academy of Sciences' Committee on Oceanography

NASCOM: NASA communications (satellite tracking network)

NASDAQ: National Association of Security Dealers' Automated Quotations

NASIS: National Association for State Information Systems

NASL: Naval Applied Sciences Laboratory

NAS/NRC: National Academy of Sciences/National Research Council

NASP: Aero-space plane (NASA)

NASPA: National Society of Public Accountants

NASW: National Association of Science Writers

NAT: normal allowed time

NATA: National Association of Testing Authorities (Australia)

NATA: North American Telephone Association

NATCS: National Air Traffic Control Service

NATESA: National alliance of Television & Electronic Service Associations

NATTS: National Association of Trade and Technical Schools

NAVA: National Audio Visual Association

NAVAIDS: navigational aids

NAVAIR: prefix to numbered series issued by Naval Air Systems Command

NAVAPI: North American voltage and phase indicator

NAVAR: radar air navigation and control system

NAVARHO: navigation aid, rho radio navigation system

NAVASCOPE: airborne radarscope used in navar

NAVASCREEN: ground screen used in navar

NAVCM: navigation countermeasures and deception

NAVCOM: naval communications

NAVCOMMSTA: naval communications station

NAVDAC: Navigation Data Assimilation Center

NAVDOCKS: prefix to numbered series issued by Navy Yards and Docks Bureau

NAVELEX: Naval Electronics System Command

NAVFAC: Naval Facilities Engineering Command

NAVMAT: prefix to numbered series issued by Office of Navy Materiel

NAVMC: prefix to numbered series issued by the Marine Corps

NAVMED: prefix to numbered series issued by Naval Aerospace Medical Institute

NAVORD: prefix to numbered series issued by NOSC

NAVPERS: prefix to numbered series issued by Bureau of Naval Personnel

NAVSEC: Naval Ship Engineering Center

NAVSO: prefix to numbered series issued by Navy Industrial Relations Office

NAVSTAR: navigation system using timing and ranging

NAVTRADEVCEN: prefix to numbered series issued by Naval Training Device Center

NAVWEPS: prefix to numbered series issued by Bureau of Naval Weapons

N/AW: night/adverse weather (evaluator)

NAWAS: National warning system

NB: narrow band

NB: Nebraska

NB: no bias (relay)

NBAA: National Business Aircraft Association

NBC: narrow-band conducted

NBCV: narrow-band coherent video

NBER: National Bureau of Engineers Registration

NBFM: narrow-band frequency modulation

NBFU: National Board of Fire Underwriters. Now AIA (American Insurance Association)

NBH: network busy hour

NBL: New Brunswick Laboratory

NBO: network buildout

NBR: narrow-band radiated

NBS: National Bureau of Standards, USA

NBSD: night bombardment short distance

NBSFS: NBS frequency standard

NBSR: NBS Reactor

NBT: null-balance transmissometer

NBTDR: narrow-band time-domain reflectometry

NC: national coarse (thread)

NC: no coil

NC: no connection

NC: noise criteria (value)

NC: normally closed

NC: North Carolina

NC: not connected

NC: numerical control

NCA: National Coal Association

NCA: Naval Communications Annex

NCA: New Communities Administration

NCA: Northern Counties Civil Engineers and Land Surveyors

NCA: Northwest Computing Association

NCAEI: National Conference of Applications of Electrical Insulation

NCAR: National Center for Atmosphere Research

NCAS: nonconforming reports

NCASI: National Council for Air and Stream Improvement

NCB: National coal Board

NCB: Naval Communications Board

NCBR: near commercial breeder reactor (also designated prototype large breeder reactor)

NCC: National Computer Conference

NCC: National Computing Center

NCC: network control center

NCCAT: National Committee for Clear Air Turbulence

NCCCHE: National Certification Commission in Chemistry and Chemical Engineering

NCCLS: National Committee for Clinical Laboratory Standards

NCDC: New Community Development Corporation

NCDEAS: National Committee of Deans of Engineering and Applied Science

NCEA: North Central Electric Association

NCEE: National Council of Engineering Examiners

NCET: National Council for Educational Technology

NCF: nominal characteristics file (see ECAC)

NCFMF: National Committee for Fluid Mechanics Films

NCGG: National Committee for Geodesy and Geophysics (Pakistan)

NCHEML: National Chemical Laboratory (MINTECH)

NCHS: National Center for Health Statistics

NCHVRFE: National College for Heating, Ventilating, Refrigeration, and Fan Engineering

NCI: National Computer Institute

NCI: National Computing Industries

NCI: Netherlands Centre for Informatics

NCI: Northeast Computer Insitute

NCIC: National Crime Information Center

NCL: National Central Library

NCL: National Chemical Laboratory

NCMCE: National Council of Minority Consulting Engineers

NCOR: National Committee for Oceanographic Research (Pakistan)

NCP: network control program

NCP&MA: Noise Control Products & Materials Association

NCPTWA: National Clearinghouse for Periodical Title Word Abbreviations

NCQR: National Council for Quality and Reliability

NCRP: National Committee on Radiation Protection

NCRUCE: National Conference of Regulatory Utility Commission Engineers

NCS: National Communications System

NCS: National Computer Systems

NCS: net control station

NCS: Netherlands Computer Society

NCS: Numerical Control Society

NCSAG: Nuclear Cross Section Advisory Group

NCSBCS: Nat'l Conference of States on Building Codes and Standards

NCSBEE: National Council of State Boards of Engineering Examiners. Now NCEE

NCSC: National Council of Schoolhouse Construction

NCSCR-3: North Carolina State College Reactor, 10 kW

NCSCR-4: North Carolina State College, Reactor, 100 kW

NCSE: North Carolina Society of Engineers

NCSL: National Conference on Standards Laboratories

NCTA: National Cable Television Association

NCTM: National Council of Teachers of Mathematics

NCTS: National Council of Technical Schools

NCTSI: National Council of Technical Service Industries

NCUA: National Credit Union Administration

NCUR: National Committee for Utilities Radio

ND: negative declaration

ND: North Dakota

NDB: nondirectional beacon

NDCT: natural draft cooling tower

NDE: nondestructive examination

NDI: numerical designation index

NDL: network definition language

NDM: negative differential mobility

NDP: normal diametral pitch

NDRO: nondestructive readout

NDT: nil-ductility transition

NDT: nondestructive testing

NDTC: Nondestructive Testing Center

NDTT: nil ductility transition temperature

NEA: National Education Association

NEA: negative electron affinity

NEA: Nuclear Energy Agency (part of OECD)

NEA: Nuclear Engineering Associates

NEAC: Nippon electric automatic computer

NEADAI: National Education Association Department of Audiovisual Instruction

NEAT: NCR electronic autocoding technique

NEBSS: National Examinations Board in Supervisory Studies

NEC: *Nederlands Elektroledrnisch Comit*

NEC: National Electrical Code

NEC: National Electronics Conference

NEC: National Engineering Consortium, Inc

NEC: Nuclear Energy Center

NECAP: Nutmeg Electric Companies Atomic Project

NECG: National Executive Committee on Guidance

NECIES: North East Coast Institution of Engineers and Shipbuilders

NECPUC: New England Conference of Public Utility Commissioners

NECS: National Electrical Code Standards

NECS: nationwide educational computer service

NECSS: Nuclear Energy Center Site Survey

NECTA: National Electrical Contractors Trade Association

NEDA: National Electronic Distributors Association

NEEDS: New England educational data systems

NEEP: Nuclear Electronic Effect Program

NEF: national extra fine (thread)

neg: negative

NEHA: National Environmental Health Association

NEIS: National Electrical Industries Show

NEIS: National Engineering Information System

NEL: National Engineering Laboratory

NEL: Naval Electronics Laboratory

NELA: National Electric Light Association. Now EEI

NELAT: Naval Electronics Laboratory assembly tester

NELC: Naval Electronics Laboratory Center

NELCON NZ: National Electronics Conference, New Zealand (IEEE)

NELEX: Naval Electronics Systems Command Headquarters

NELIA: Nuclear Energy Liability Insurance Association

NELIAC: Navy Electronics Laboratory international algebraic compiler

NELPA: Northwest Electric Light & Power Association

NELPIA: Nuclear Energy Liability Property Insurance Association

NEMA: National Electrical Manufacturers Association

NEMI: National Elevator Manufacturing Industry

NEMP: nuclear electromagnetic pulse

NEN: prefix to standards issued by NNI

NEP: noise equivalent power

NEPA: National Environmental Policy Act

NEPIA: Nuclear Energy Property Insurance Association

NEP&ME: National Exposition of Power and Mechanical Engineering

NEPTUNE: Low energy reactor at Derby, United Kingdom, for water reactor experiments

NEPTUNE: North-Eastern electronic peak tracing unit and numerical evaluator

NER: National Engineers Register

NERA: National Economic Research Associates, Inc

NERC: National Electric Reliability Council

NERC: Natural Environment Research Council

NERC: Nuclear Energy Research Center; same as *Centre d'Etude de l'Energie Nucléaire* , Brussels, Belgium

NEREM: Northeast Electronics Research and Engineering Meeting

NERHL: Northeastern Radiological Health Laboratory, Winchester, Massachusetts

NERO: Dutch design and development project for PWR for maritime use

NERO: Reactor, 100 w(th), Winfrith, United Kingdom

NERV: nuclear emulsion recovery vehicle, follow on to BIOS project to determine radiation profile of inner Van Allen belt (NASA)

NERVA: nuclear engine for rocket vehicle application

NES: National Engineering Service

NES: National Estimating Society

NES: noise equivalent signal

NESC: National Electrical Safety Code

NESC: National Environmental Satellite Center (ESSA)

NESCOM: New Standards Committee (IEEE)

NESP: National Environmental Studies Project (an AIF activity)

NEST: Naval Experimental Satellite Terminal

NESTEF: Naval Electronics System Test and Evaluation

NESTOR: neutron source thermal reactor

NESTOR: Reactor, Winfrith, United Kingdom, 10 kW(th)

NESW: nonessential service water (R) relay (P) pump

NET: National Educational Television

NETIC: frequency function integrals, reactor physics computer code

NETR: Nuclear Engineering Test Reactor (USAF), Dayton, Ohio

NETSET: network synthesis and evaluation technique

neut: neutral

NEW: National Engineers Week

NEWA: National Electrical Wholesalers Association

NEWRADS: Nuclear Explosion Warning and Radiological Data System

NEWS: naval electronic warfare simulator

NEWWA: New England Water Works Association

NEXRAD: next-generation radar

nF: nanofarad

NF: national fine (thread)

NF: noise figure

NF: noise frequency

NF: nose fuse

NF: prefix to standards issued by AFNOR

NFAC: Naval Facilities Engineering Command Headquarters

NFAH: National Foundation for the Arts and the Humanities

NFC: no further consequences

NFC: not favorably considered

NFCA: nonfuel core array

NFEA: National Federated Electrical Association

NFETM: National Federation of Engineers' Tools Manufacturers

NFPA: National Fire Protection Association

NFPA: National Fluid Power Association

NFPA: National Forest Products Association

NFPCA: National Fire Prevention and Control Administration

NFPEDA: National Farm and Power Equipment Dealers Association

NFR: Negative Flux Rate

NFR: no further requirement

NFS: Nuclear Fuel Services Plant

NFSAIS: National Federation of Science Abstracting and Indexing Services

NFSR: National Foundation for Scientific Research (Belgium)

NFWC: National Fire Waste Council

NG: nitroglycerine

NGA: *Nationale Gesellschaft Zur Foerderung devendust riellen Atom-Zechnik*

NGAM: noble gas activity monitor

NGCC: National Guard Computer Center

NGM: neutron gamma Monte Carlo

NGPA: National Gas Processors Association

NGRS: Narrow Gauge Railway Society

NGSF: noble gas storage facility

nH: nanohenry

NH: New Hampshire

NH: nonhygroscopic

NHAMA: National Hose Assemblies Manufacturers Association

NHTSA: National Highway Traffic Safety Administration

NI: numerical index

NIAC: Nuclear Insurance Association of Canada

NIAE: National Institute of Agricultural Engineering

NIAM: Netherlands Institute for Audiovisual Media

NIB: noninterference basis

NIBS: National Institute of Building Sciences

NIC: National Indicational Center

NIC: National Inspection Council

NIC: National Institute of Corrections

NIC: negative impedance converter

NIC: Nuclear Industry Consortium, same as *Groupement professionnel de l'Industrie, Nuclaire* , Brussels, Belgium

NICAP: National Investigations Committee on Aerial Phenomena

NICB: National Industrial Conference Board

NICE: National Institute of Ceramic Engineers

NICEIC: National Inspection Council for Electrical Installation Contracting

NICET: National Institute for Certification in Engineering Technologies

NID: Nuclear Instruments and Detectors (IEEE NPSS Technical Committee)

NIDA: numerically integrated differential analyzer

NIER: National Industrial Equipment Reserve

NIESR: National Institute for Economic and Social Research

NIF: noise improvement factor

NIF: Norwegian Society of Chartered Engineers

NIFES: National Industrial Fuel Efficiency Service

NIFTE: neon indicating functional test equipment

NIG: Nordic Industrial Group

NIH: National Institutes of Health, USA

NIH: not invented here

NIL: nitrogen inerting line

NILEJC: ausi/nilejc st 0101.00 - 1975 Riot Helmets

NIM: nuclear instrument module

NIMBUS: Meteorological Observation Satellite

NINA: National Institute Northern Accelerator

NIO: National Institute of Oceanography

NIOBE: neutron or gamma transport computer code

NIOSH: National Institute of Occupational Safety and Health, USA

NIPA: notice of initiation of procurement action

NIPCC: National Industrial Pollution Control Council

NIPHLE: National Institute of Packaging, Handling and Logistics Engineers

NIPO: negative input, positive output

NIRB: Nuclear Insurance Rating Bureau, New York

NIRNS: National Institute for Research in Nuclear Science, United Kingdom

NIRS: National Institute for Radiological Science, Tokyo

NIS: neutron instrumentation system

NIS: not in stock

NIS: nuclear instrumentation system

NISARC: National Information Storage and Retrieval Centers

NISC: National Industrial Space Committee

NIV: negative-impedance inverter

NJ: New Jersey

NJAC: National Joint Advisory Council

NJCC: National Joint Computer Committee (now AFIPS)

NJPMB: Navy Jet Propelled Missile Board

N/L: navigation/localizer

NLE: National Lighting Exposition

NLG: noise landing gear

NLGI: National Lubricating Grease Institute

NLI: National Lead Industries (now NW Industries)

NLI: nonlinear interpolating

NLL: National Lending Library for Science and Technology

NLM: National Library of Medicine

NLO: nonlinear optics

NLRB: National Labor Relations Board

NM: New Mexico

NM: noise meter

NM: nuclear magnetron

NMA: National Management Association

NMA: National Microfilm Association

NMAA: National Machine Accountants Association (now DPMA)

NMAP: National Metric Advisory Panel

NMC: National Meteorological Center

NMC: Naval Missile Center

NMC: net maximum capacity

NMCL: Naval Missile Center Laboratory

NMCS: Nuclear Materials Control System

NMEL: Navy Marine Engineering Laboratory

NMI: NASA Management Instruction

nmi: nautical mile (1 nmi = 1.151 miles)

NMIS: nuclear materials information system

NMIS: nuclear materials inventory system

NMMSS: nuclear materials management and safeguards system

NMOS: N-type MOS

NMP: national meter programming

NMPC: National Minority Purchasing Council

NMR: National Missile Range

NMR: normal-mode rejection

NMR: nuclear magnetic resonance

NMS: Naval Meteorological Service

NMS: neutron monitoring system

NMS/NPSS: Nuclear Medical Science

NMSS: National Multipurpose Space Station

NMSS: Office of Nuclear Material Safety and Safeguards

NMST: new materials system test

NMTBA: National Machine Tool Builders Association

NNCSC: National Neutron Cross Section Committee

NNEC: National Nuclear Energy Commission (Brazil)

NNI: Nederlands Normalisatic - Institute

NNI: nonnuclear instrumentation

NNSDD: Newport News Shipbuilding and Dry Dock Company

NO: normally open

NOA: National Oceanography Association

NOAA: National Oceanographic and Atmospheric Administration

NODAC: Naval Ordnance Data Automation Center

NODC: National Oceanographic Data Center

NOF: NCR optical font

NOI: notice of inquiry

NOL: Naval Ordnance Laboratory, DOD, Silver Spring, White Oak, Maryand

NOL: normal operating losses

NOMA: National Office Management Association

NOMAD: Naval Oceanographic Meteorological Automatic Device

NOMSS: National Operational Meteorological Satellite System

NONCOHO: noncoherent oscillator

NORAD: North American Air Defense System

NORCUS: Northwest College and University Association for Science

NORM: not operationally ready maintenance

NORS: not operationally ready supply

NOS: night observation system

NOS: not otherwise specified

NOSMO: Norden optics setting, mechanized operation

NOSS: National Oceanic Satellite System

NOSS: nimbus operational satellite system

notepad: NSAC/INPO significant event

NOTU: naval operational training unit

NP: national pipe (thread)

NP: naval publication (numbered series issued by MOD)

NP: neuropsychiatric

NPA: normal pressure angle

NPA: numbering-plan area

NPA: numerical production analysis

NPC: NASA Publication Control

NPC: National Power Conference

NPCC: Northeast Power Coordinating Council

NPD: National Power Demonstration

NPD: Nuclear Power Demonstration, power reactor, 88 MWmw(th), Ontario, Canada

NPDES: nuclear pollution discharge elimination system (specification)

NPDS: nuclear particle detection system

NPEC: Nuclear Power Engineering Committee (IEEE PES Technical Committee)

NPE/PES: Nuclear Power Engineering

NPF: nuclear power facility

NPF: Nuclear Problems Forum

NPFO: nuclear power field office

NPG: The Nuclear Power Group (TNPG) (United Kingdom)

NPGS: nuclear power generating station

NPL: National Physical Laboratory (GB)

NPLG: Navy Program Language Group

NPM: connector type

NPN: negative-positive-negative (transistor)

npo: negative positive zero

NPO: nuclear power operator

NPP: nuclear power plant

NPPSO: Naval Publications and Printing Service Office

NPR: New Production Reactor for plutonium and electricity production, Hanford

NPR: noise power ratio

NPRC: nuclear power range channel

NPRCG: Nuclear Public Relations Contact Group

NPRDS: nuclear plant reliability data system

NPRF: Northrop pulse radiation facility

NPS: Nuclear and Plasma Sciences (IEEE Society)

NPS: numerical plotting system

NPSA: new program status area

NPSH: net positive suction head

NPSRA: Nuclear-Powered Ship Research Association of Tokyo, Japan

NPSS: Nuclear and Plasma Sciences

NPT: National Pilot Training program

NPT: national taper pipe (thread)

NPT: network planning technique

NPT: nonproliferation treaty

NPTF: nuclear proof test facility (never built)

NPV: nitrogen pressure valve

NQAA: Nuclear Quality Assurance Agency

NQR: nuclear quadruple resonance

NR: nonreactive (relay)

NR: nonrecoverable

NR: nuclear reactor

NRA: A Laser Gyro Reference Axis

NRA: naval radio activity

NRA: network resolution area

NRAO: National Radio Astronomy Observatory

NRC: National Research Council, USA

NRC: Nuclear Regulatory Commission

NRCST: National Referral Center for Science and Technology

NRDC: National Research Development Council

NRDC: National Resources Defense Council

NRDL: Naval Radiological Defense Laboratory

NRDS: Nuclear Rocket Development Station

NRECA: National Rural Electric Cooperative Association

NRL: National Reference Library for Science and Invention

NRL: Naval Research Laboratory

NRM: natural remanent magnetization

NRMA: National Retail Merchants Association

NRP: normal rated power

NRR: Office of Nuclear Reactor Regulation

NRRS: Naval Radio Research Station

NRS: Naval Radio Station

NRS: Naval Rocket Society

NRSEP: National Roster of Scientific and Engineering Personnel

NRSTP: National Register of Scientific and Technical Personnel

NRSW: nuclear river service water

NRTS: National Reactor Test Station (Part of INEL)

NRTSC: naval reconnaissance and technical support center

NRU: Canadian natural-uranium, heavy-water-moderated and -cooled test reactor, Chalk River, Ontario, 200 MW(th)

NRW(O): nuclear radwaste (operator)

NRX-A: Nuclear engine reactor experiment (NERVA), Jackass Flats, Nevada

NRX-CX: critical assembly (Westinghouse)

NRX: Canadian natural-uranium, heavy-water-moderated research reactor, Chalk River, Ontario

NRZ: nonreturn to zero

NRZC: nonreturn to zero change

NRZI: nonreturn to zero change on one

NRZI: nonreturn to zero inverted

NRZL: nonreturn to zero level

NRZM: nonreturn to zero mark

NS-40: early designation for depleted uranium

ns: nanosecond

NS: national standard

NSA: National Security Agency

NSA: National Shipping Authority

NSA: National Standards Association

NSA: Netherlands Society for Automation

NSA: Nuclear Science Abstracts

NSAC: Nuclear Safety Analysis Center

NSaF: National Sanitation Foundation

NSB: National Science Board

NSB: Nuclear Standards Board

NSC: National Safety Council

NSC: National Security Council

NSCC: nuclear services closed cooling

NSCR: Nuclear Science Center Reactor, A&M University, Texas

NSDM: national security decision memorandum

NSEC: Naval Ships Engineering Center. Formerly Bu Ships

NSEF: Navy Security Engineers Facility

NSEIP: Norwegian Society for Electronic Information Processing

NSF: National Science Foundation, USA

NSF: Nuclear Science Foundation

NSI: nonstandard item

NSIA: National Security Industrial Association

NSIC: Nuclear Safety Information Center, ORNL

NSIF: Near Space Instrumentation Facility

NSL: National Science Laboratories

NSL: National Science Library

NSL: Northrop Space Laboratories

NSM: network status monitor

NSMB: Nuclear Standards Management Board/ANSI

NSP: NASA support plan

NSP: network support plan

NSPA: National Society of Public Accountants

NSPAC: National Standards Policy Advisory Committee

NSPE: National Society of Professional Engineers

NSPI: National Society for Programmed Instruction

NSPP: nuclear safety pilot plant

NSQCRE: National Symposium on Quality Control and Reliability in Electronics

NSR: Neutron Source Reactor, BNL, 100 kW(th)

NSRB: National Security Resources Board

NSRDS: National Standard Reference Data System

NSRFI: National Symposium on Radio Frequency Interference

NSRQC: National Symposium in Reliability and Quality Control

NSRS: Naval Supply Radio Station

NSRW(P): nuclear service raw water (pump)

NSS: Navy Secondary Standards

NSS: nuclear steam system

NSSL: National Severe Storms Laboratory

NSSM: national security study memorandum

NSSS: National Space Surveillance System

NSSS: nuclear steam supply system

NST: network support team

NSTA: National Science Teachers Association

NSTIC: Naval Scientific and Technical Information Center

NSTL: National Space Technology Laboratories

NSTP: Nuffield Science Teaching Project

NTA: National Technical Association

NTA: next transfer addresss

NTAG: Nuclear Technical Advisory Group

NTC: National Telemetering Conference

NTC: National Transformers Committee

NTC: negative temperature coefficient

NTD: neutron transmutation doping (silicon)

NTDS: Navy Tactical Data System

NTE: Navy Teletypewriter Exchange

NTF: nuclear test facility

NTG: *Nachrichtentechnische Gesellschaft* (German Communication Society)

NTI: noise transmission impairment

NTIA: National Telecommunications and Information administration (formerly OTP)

NTIS: National Technical Information Service, Department of Commerce

NTP: normal temperature and pressure

NTP: nuclear test plant

NTR: Nuclear Test Reactor, Pleasanton, California, 30 kw(th)(same as GETR)

NTS: navigation technology system

NTS: negative torque signal

NTS: Nevada Test Site (NASA)

NTSA: National Technical Services Association

NTSB: National Transportation Safety Board, USA

NTSC: National Television System Committee

NTZ: *Nachrichtentechnische Zeitschrift*

NUCLENOR: *Controles Nucleares del Norte* , S.A. Santander, Spain

NUCLIT: *Nucleare Italiana* , Government corporation conducting programs at Ispra, Italy

NUDE: nuclear experimental data evaluation reactor physics computer code

NUDETS: nuclear detection system

NULACE: nuclear liquid air cycle engine

NUMEC: Nuclear Materials and Equipment Corporation (part of B&W)

NUMS: nuclear materials security

NUPAD: nuclear powered active detection

NUPPSCO: Nuclear Power Plant Standards Comm

NURE: national uranium resource evaluation program

NUS: nuclear utility service

NUSL: Naval Underwater Sound Laboratory

nV: nanovolt

NV: Nevada

NVACP: Neighborhoods Voluntary Associations and Consumer Protection, Office of Assistant Secretary

NVLAP: National Voluntary Lab. Accreditation Pro

NVR: no voltage release

NVSD: night vision system development

nW: nanowatt

NWAC: National Weather Analysis Center

NWAHACA: National Warm Air Heating & Air Conditioning Association

NWB: National Wiring Bureau (Dissolved 1965)

nW/cm^2 : milliwatt per square centimeter

NWG: National Wire Gage

NWMA: National Woodwork Manufacturers Association

NWRC: National Weather Records Center

NWRF: Naval Weather Research Facility

NWS: National Weather Service

NWS: nosewheel steering

NWSC: National Weather Satellite Center

NWSSG: Nuclear Weapons System Satellite Group

NXDO: Nike X Development Office

NY: New York

NYAP: New York assembly program

NYSE: New York Stock Exchange

NZIE: New Zealand Institution of Engineers

NZSI: New Zealand Standards Institute. Now SANZO

O

O-B: octal to binary

O-H: octal to hexadecimal

OA: oil-immersed self-cooled (transformer)

OA: operating authorization, full-term

OA: operations analysis

OA: output axis

OAATM: Office of the Assistant for Automation

OAD: operational availability date

OAM: Office of Aerospace Medicine

OAME: orbital attitude and maneuvering electronics

OAMP: optical analog matrix processing

OAMS: orbital attitude and maneuvering system

OAO: Orbiting Astronomical Observatory

OAR: Office of Aerospace Research

OARAC: Office of Air Research Automatic Computer
OARC: ordinary administrative radio conference
OART: Office of Advanced Research and Technology
OASF: Orbiting Astronomical Support Facility
OASIS: Ocean All Source Information System
OASM: Office of Aerospace Medicine
OASV: orbital assembly support vehicle
OATP: operational acceptance test procedure
OB: output buffer
OBA: oxygen breathing apparatus
OBC: optical bar code
OBE: operating basis earthquake
OBE: operating basis event
OBGS: orbital bombardment guidance system
OBI: omnibearing indicator
OBIFCO: on-board in-flight checkout
OBS: omnibearing selector
OC: operating characteristic
OC: operating curve
OC: operational computer
OC: outside circumference
OC: overcurrent
OCA: operational control authority
OCAL: on-line cryptanalytic aid language
OCAS: on-line cryptanalytic aid system
OCB: oil(operated) circuit breaker
OCC: operational computer complex
OCDM: Office of Civil Defense Mobilization
OCDRE: Organic-Cooled Deuterium Reactor Experiment, proposed power reactor, 40 MW(th), Whiteshell, Manitoba, Canada
OCDU: optics coupling display unit
OCI: Oxide control and indication
OCITT: International Telegraph and Telephone Consultative Committee
OCL: operational control level
OCL: operators control language
OCM: optical countermeasures
OCO: open-close-open
OCO: operation capability objectives
OCP: operating control procedure
OCP: operational checkout procedure
OCR-A: optical character recognition-A
OCR-B: optical character recognition-B
OCR: oil circuit recloser
OCR: optical character reader
OCR: optical character recognition
OCR: overcurrent relay
OCR: overhaul component requirement
OCRUA: Optical Character Recognition Users Association
OCS: carbonyl sulfide
OCS: outer continental shelf
OD: operations directive
OD: optical density
OD: output data
OD: outside diameter
ODA: operational data analysis
ODA: operational design and analysis
ODLRO: off-diagonal long-range order

ODM: orbital determination module
ODN: own doppler nullifier
ODOP: offset doppler
ODP: original document processing
ODT: octal debugging technique
ODT: outside diameter tube
ODU: output display unit
oe: oersted (ampere per meter)
OE: Office of Education
OE: open end
OE: operating engineer
OE: output enable
OECD: Organization for Economic Cooperation and Development
OEEC: Office of European Economic Cooperation
OEM: original equipment manufacturer
OESLA: Office of Engineering Standards Liaison and Analysis
O/F: orbital flight
OF: output factor
OF: outside face
OFA: oil-immersed forced-air-cooled (transformer)
OFB: Operational Facilities Branch (NASA)
OFC: operational flight control
OFO: Office of Flight Operations (NASA)
OFP: operating force plan
OFPP: Office of Federal Procurement Policy
OFR: on-frequency repeater
OFR: overfrequency relay
OFSD: operating flight strength diagram
OG: outer gimbal
Og: zero gravity
OGA: outer gimbal axis
OGE: operational ground equipment
OGMC: Ordnance Guided Missile Center
OGO: Orbiting Geophysical Observatory
OGR: ORNL Graphite Reactor, heterogeneous, natural-uranium, graphite-moderated and air-cooled, USA
OGR: outgoing repeater
OGRA: thermonuclear experimental device, USSR
OGST: overthread guide sleeve tool
OGT: outgoing trunk
OGU: outgoing unit
OGUN: oil gun
OH: Ohio
OH: operational hardware
OHA: Office of Hearings and Appeals, Interior Department
OHA: outside helix angle
OHADOE: Hearings and Appeals Office, Energy Department
OHC: Occupational Health Center
OHD: over-the-horizon detection
OHF: occupational health facility
ohm-cm: ohm-centimeter (Also shown as (Ω -cm))
OHMR: Office of Hazardous Materials Regulations
OHP: oxygen at high pressure
OHR: over-the-horizon (radar)
OHSGT: Office of High Speed Ground Transportation

OI: oil insulated
OIB: Operations Integration Branch (NASA)
OIC: on-line instrument and control program
OIC: operations instrumentation coordinator
OIC: optical integrated circuit
OIFC: oil-insulated fan-cooled
OIFQ: *L'Ordre des Ingenieurs Forestiers du Quebec*
OIG: optically isolated gate
OII: Office of Invention and Innovation (NBS)
OIML: *Organisation Internationale de Metrologie Legale* (International Organization of Legal Metrology)
OIP: operating internal pressure
OIRT: *Organisation Internationale de Radiodiffusion et de Television*
OIRT: International Radio and Television Organization
OIS: Operational Intercommunication System
OISA: Office of International Science Activities
OISC: oil-insulated self-cooled
OIT (ILO): *Organisation International du Travail* (International Labor Organization)
OIWC: oil-immersed water-cooled
OK: Oklahoma
OL: open loop
OL: operating level
OL: operating license
OL: operating location
OL: overload
OL: overload relay
OLC: on-line computer
OLERT: on-line executive for real time
OLIP: On-line instrument package
OLM: on-line monitor
OLO: orbital launch operations
OLPARS: on-line pattern analysis and recognition system
OLPS: on-line programming system
OLR: objective loudness rating
OLRT: on-line real-time
OLSC: on-line scientific computer
OLSS: on-line software system
OLV: open-frame low voltage
OLVP: Office of Launch Vehicle Programs
O&M: operation and maintenance
OM: operations manager
OM: optical master
O/M: oxygen-to-metal ratio
OMA: orderly marketing agreement
OMAT: Office of Manpower Automation and Training
OMB: outer marker beacon
OMB: U.S. Office of Management and Budget
OMBE: Office of Minority Business Enterprise
OMC: Association of Women Engineers Architects and Surveyors of CIAA of Puerto Rico
OMCA: critical assembly (AI)
OMCI(IMCO): *Organization Intergouvermentale Consultative de la Navigation Maritime* (Intergovernmental Maritime Consultative Organization)
OME: Office of Management Engineer

OMEGA West: research reactor at Los Alamos
OMETA: Ordnance Management Engineering Training Agency
OMIAA: Organization of Women Engineers, Architects and Surveyors of Puerto Rico
OMIBAC: ordinal memory inspecting binary automatic computer
OML: Ordnance Missile Laboratories
OMM: operation and maintenance manual
OMNIA: State agency controlling exports of nuclear materials and equipment, Prague, Czechoslovakia
OMPR: optical mark page reader
OMPRA: one-man propulsion research apparatus
OMPT: observed man point trajectory
OMR: optical mark reader
OMR: optical mark reading
OMR: Organic-Moderated (power) Reactor, 45 MW(th), Piqua, Ohio
OMRCA: organic moderated reactor critical assembly
OMRE: Organic Moderated Reactor Experiment, NRTS, Idaho, power reactor, 5-12 MW(th)
OMRR: Ordnance Material Research Reactor, Watertown, Massachusetts
OMS: orbital maneuvering subsystem
OMSF: Office of Manned Space flight (NASA)
OMTS: organizational maintenance test station
OMU: optical measuring unit
ONI: operator number identification
ONR: Office of Naval Research
ONYX: design concept in the ditchdigger class (plowshare)
OOK: on-off keying
OOL: operator oriented language
OOLR: overall objective loudness rating
OOPS: off-line operating simulator
OOS: out of service
OP: operating procedure
OP: operational priority
OP: output
OP: over pressure
OP: oxygen pressure process (ore leach process)
OPA: optoelectronic pulse amplifier
OPADEC: optical particle decoy
op amp: operational amplifier
op. cit.: opus citatum (the work quoted from)
op code: operation code
OPCOM: operations communications
OPCON: optimizing control
OPDAR: optical detection and ranging
OPE: operations project engineer
OPEC: Organization of Petroleum Exporting Countries
OPEP: orbital plane experimental package
OPERA: out-of-pile expulsion and reentry apparatus
OPFM: outlet plenum feature model
OP&I: Office of Patents and Inventions
OPIC: Overseas Private Investment Corporation
OPLE: Omega Position Location Experiment
OPM: operations per minute
OPO: optical parametric oscillator

OPP: octal print punch
OPP: oriented polypropylene
OPPCE: opposite commutator end
OPPOSIT: optimization of a production process by an ordered and simulation and iteration technique
OPPS: over pressurization protection switch
OPPS: over pressurization protection system
OPR: offsite procurement request
OPR: optical page reading
OPS: offshore power systems
OPS: oil pressure switch
OPS: on-line process synthesizer
OPS: operational paging system
OPS: operational protection system
OPSCAN: Optical Scanning Users Group
OPSCON: operations control (room)
OPSF: orbital propellent storage facility
OPSP: Office of Product Standards Policy
OPSR: Office of Pipeline Safety Regulations
OPTA: optimal performance theoretically attainable
OPTI: Office of Productivity, Technology, and Innovation
OPTIM: Perturbation burnup reactor computer code
OPW: operating weight
OPW: orthogonalized plane wave
OPX: off premise extension
OR: operating reactor
OR: operational readiness
OR: operations requirements
OR: operations research
OR: Oregon
OR: output register
OR: outside radius
OR: overhaul and repair
OR: overload relay
O/R: overrange
ORA: output reference axis
ORACLE: Oak Ridge automatic computer and logical engine
ORAN: orbital analysis
Orange Oxide: uranium trioxide (UO3)
ORATE: ordered random access talking equipment
ORBIS: orbiting radio beacon ionospheric satellite
ORBIT: on-line retrieval of bibliographic data
ORBIT: oracle binary internal translator
ORC: on-line reactivity computers
ORC: Operations Research Center
ORC: Ordnance Rocket Center
ORD: operational readiness date
ORD: optical rotary dispersion
ORDENG: ordnance engineering
ORDVAC: ordnance variable automatic computer
ORE: Occupational radiation exposure
ORFM: outlet region feature model
ORG: Operations Research Group
ORGDP: Oak Ridge Gaseous Diffusion Plant, Oak Ridge, Tennessee
ORGEL: Organic-cooled, heavy-water-moderated reactor concept, EURATOM
ORI: operational readiness inspection

ORIA: Office Of Regulatory & Information Affairs
ORIC: Oak Ridge Isochronous Cyclotron
ORINS: Oak Ridge Institute of Nuclear Studies
ORIT: operational readiness inspection team
ORL: Orbital Research Laboratory
ORL: Ordnance Research Laboratory
ORLY: overload relay
ORNL: Oak Ridge National Laboratory
ORR: Oak Ridge Research Reactor, ORNL
ORRAS: optical research radiometrical analysis system
ORSA: Operations Research Society of America
ORSORT: Oak Ridge School of Reactor Technology
ORT: operational readiness test
ORT: overland radar technology
ORV: orbital rescue vehicle
OS: operating system
OS: operational sequence
OSA: Optical Society of America
OSC: On Site Safety Committee
OSCAR: optimum systems covariance analysis results
OSCAR: Orbiting Satellite Carrying Amateur Radio
OSD: operational sequence diagram
OSD: operational systems development
OSE: operational support equipment
OSF: Open Systems Foundation
OSFM: Office of Spacecraft and Flight Missions
OSFP: Office of Space Flight Programs
OSHA: Occupational Safety and Health Act
OSHA: Occupational Safety and Health Administration
OSI: Office of Scientific Intelligence
OSI: open systems interconnection
OSIC: optimization of subcarrier information capacity
OSIRIS: Materials testing reactor, 30 MW(th), CEA, Saclay, France
OSIS: Office of Science Information Service (NSF)
OSMV: one-shot multivibrator
OSO: orbital solar observatory
OSO: orbiting satellite observer
OSO: orbiting scientific observatory
OSO: orbiting solar observatory
OSP: Office of Statistical Policy
OSR: Office of Scientific Research
OSR: output shift register
OSRD: Office of Standard Reference Data (NBS)
OSS: ocean surveillance satellite
OSS: Office of Space Sciences
OSS: Office of Statistical Standards
OSS: operational storage site
OSS: orbital space station
OSSA: Office of Space Science and Applications (NASA)
OSSS: orbital space station system
OST: Office of Science and Technology, USA
OST: on-shift tests
OST: operational system test
OSTA: Office of Space and Terrestrial applications
OSTAC: Ocean Science and Technology Advisory Committee

OSTI: Office for Scientific and Technical Information
OSTP: Office of Science and Technology Policy
OSTS: Office of State Technical Service
OSUR: Ohio State University Reactor, Columbus, Ohio, 10 kW(th)
OSURF: Ohio State Univ. Research Foundation
OSV: orbital support vehicle
OSWALD: apparatus used in early United Kingdom thermonuclear program
OT: operating temperature
OT: overall test
OTA: Office of Technology Assessment
OTA: Office of Technology Assessment, USA
OTA: open test assembly
OTA: operational transconductance amplifier
OTC: Office of Telecommunications
OTC: once-through cooling
OTC: operational test center
OTCCC: open-type control circuit contacts
OTDA: Office of Tracking and Data Acquisition
OTDT: over temperature delta T
OTE: operational test equipment
OTEC: ocean thermal energy conversion
OTHR: over-the-horizon radar
OTLP: zero transmission level point
OTM: Office of Telecommunications Management
OTP: Office of Telecommunications Policy
OTP: operational test procedure
OTR: optical tracking
OTR: Organic Test Reactor, Whiteshell, Canada, tank-type
OTRAC: oscillogram trace reader
OTS: Office of Technical Services
OTS: Office of Technological Service
OTS: optical technology satellite
OTS: optical transport systems
OTS: orbital test satellite
OTSG: once-through steam generator
OTSGS: once through steam generating system
OTSR: once-through superheat reactor
OTT: one-time tape
OTU: Office of Technology Utilization
OTU: operational test unit
OTU: operational training unit
OTV: operational television
OTV: orbit transfer vehicle
OUO: official use only
OUTRAN: output translator
OV: orbiting vehicle
OV: overvoltage
OVERS: orbital vehicle reentry simulator
OVLBI: orbital very long baseline interferometer
OVV: overvoltage
OW: oil-immersed water-cooled (transformer)
OWF: optimum working frequency
OWM: Office of Weights and Measures
OWND: observation window
OWPR: ocean wave profile recorder
OWR: Omega West Reactor, LASL, 5 MW(th)

OWRR: Office of Water Resources Research, Department of Interior
OWS: ocean weather station
OWS: operational weather support
OWS: orbital workshop
oz: ounces

P

P-P: peak to peak
P-T: Plasma Thermocouple Reactor, Los Alamos, New Mexico
p: pico
P: plate
P: poll
P: power dissipation (light emitting diodes)
P: pump
PA: pad abort
PA: parity
PA: pending availability
PA: pendulous axis
pA: picoampere
PA: pilotless aircraft
PA: point of aim
PA: power amplifier
PA: pressure angle
PA: Privacy Act
PA: probability of acceptance
PA: product analysis
PA: program address
PA: program analysis
PA: program authorization
PA: protected area
PA: public address (system)
PA: pulse amplifier
PAAC: program analysis adaptable control
PABLA: problem analysis by logical approach
PABX: private automatic branch exchange
PAC: pedagogic automatic computer
PAC: personal analog computer
PAC: Portland Cement Association
PAC: Professional Activities Committee (IEEE)
PACCT: PERT and cost correlation technique
PACE-S/C: preflight acceptance checkout equipment for spacecraft
PACE: plant acquisition and construction equipment
PACE: precision analog computing equipment
PACE: preflight acceptance checkout equipment
PACE: prelaunch automatic checkout equipment
PACE: Professional Activities Committees for Engineers
PACE: Professional Activities Council for Engineers
PACE: programming analysis consulting education
PACE: projects to advance creativity in education
PACED: program for advanced concepts in electronic design
PACE/LV: preflight acceptance checkout equipment-launch vehicle
PACER: process assembly case evaluator routine

PACER: program assisted console evaluation and review

PACER: programmed automatic communications equipment

PACOR: passive correlation and ranging station

PACT: pay actual computer time

PACT: production analysis control technique

PACT: programmed analysis computer transfer

PACT: programmed automatic circuit tester

PAD: polyaperture device

PAD: post-activation diffusion

PAD: propellent-actuated device

PADAR: passive detection and ranging

PADLOC: passive active detection and location

PADLOCK: passive detection and location of counter-measures

PADRE: patient automatic data recording equipment

PADRE: portable automatic data recording equipment

PADS: passive-active data simulation

PAE: post accident environment

PAEM: program analysis and evaluation model

PAGE: PERT automated graphical extension

PAGEOS: passive geodetic earth orbiting satellite

PAHO: Pan American Health Organization

PAHR: post-accident heat removal

PAI: programmer appraisal instrument

PAIC: public address intercom system

PAIGH: Pan-American Institute of Geography and History

PAIP: Public Affairs and Information Program (AIF)

PAIR: performance and integration retrofit

PAL-D: phase alternation line delay

PAL: pedagogic algorithmic language

PAL: permanent artificial lighting

PAL: permissive action link

PAL: phase alteration line (West German television system)

PAL: process assembly languages

PAL: Production and Application of Light (IEEE IAS Technical Committee)

PAL: programmable array logic

PAL: programmer assistance and liaison

PAL: prototype application loop

PAL: psychoacoustic laboratory

PALASM: programmable array logic assembler

PAM: pole amplitude modulation

PAM: post-accident monitoring

PAM: pulse amplitude modulation

PAMPER: practical application of mid-points for exponential regression

PAMPER: two-group perturbation group diffusion reactor code

PAMS: pad abort measuring system

PAMS: Proceedings of the American Mathematical Society

PANSDOC: Pakistan National Scientific and Technical Documentation

PANSMET: Procedures for Air Navigation Services-Meteorology

PAO: pulsed avalanche diode oscillator

PAPA: programmer and probability analyzer

PAPE: photoactive pigment electrophotography

P/AR: peak-to-average ratio

PAR: peak accelerometer recorder

PAR: Pennsylvania advanced reactor

PAR: performance analysis and review

PAR: performance appraisal report

PAR: perimeter acquisition radar

PAR: precision approach radar

PAR: program appraisal and review

PAR: project authorization request

PAR: purchasing approval request

PARA: problem analysis & recommended action

PARADE: passive-active ranging and determination

PARAMP: parametric amplifier

PARASYN: parametric synthesis

PARC: progressive aircraft reconditioning cycle

PARD: parts application reliability data

PARD: periodic and random deviation

PARD: precision annotated retrieval display

PARM: program analysis for resource management

PAROS: passive ranging on submarines

PARR: Pakistan Atomic Research Reactor near Rawalpindi, West Pakistan

PARR: post accident radioactivity removal

PARR: procurement authorization and receiving report

Parsaval: pattern recognition system application evaluation

PARSEV: paraglider research vehicle

PARTNER: proof of analog results through numerically equivalent routine

PAS: primary alert system

PAS: Privacy Act statement

PAS: Professor of Air Science

PAS: program address storage

PASC: Pacific Area Standards Congress

PASCAL: Philips automatic sequence calculator

PASE: power assisted storage equipment

PASM: partionable SIMD/MIMD

PASS: production automated scheduling system

PASS: program aid software systems

PASS: program alternative simulation system

PAS&T/NPSS: Particle Accelerator Science & Technology

PAT: parametric artificial talker

PAT: personalized array translator

PAT: power ascension testing

PAT: pressurized water research reactor

PAT: production acceptance test

PAT: program attitude test

PATA: pneumatic all-terrain amphibian

PATC: professional, administrative, technical, and clerical

PATCO: Professional Air Traffic Controllers Association

PATE: programmed automatic test equipment

PATH: performance analysis and test histories

PATI: passive airborne time-difference intercept

PATS: precision altimeter techniques study

PATT: project for the analysis of technology transfer

PATTERN: planning assistance through technical evaluation of relevance numbers (ORSA)

PAU: pilotless aircraft unit

PAWOS: portable automatic weather observable station

PAWS: phased array warning system

PAX: private automatic exchange

PB-HTGR: Peach Bottom High-Temperature Gas-cooled Reactor

PB: pipe break

PB: playback

PB: plot board

PB: plug board

PB: pushbutton

PBAPS: pipe break air piping system

PBAPS: pipe break automatic protective system

PBB: polybrominated biphenyl

PBDG: push-button data generator

PBE: prompt burst experiments

PBF: power burst facility (INEL)

PBHP: pounds per brake horsepower

PBIT: parity bit

PBPS: post boost propulsion system

PBR: Plum Brook Reactor, heterogeneous, enriched-uranium, research,USA

PBRE: Pebble Bed Reactor Experiment

PBRF: Plum Brook Reactor Facility, NASA, Sandusky, Ohio

PBR (ORNL): ORNL Pebble Bed Reactor, solid, homogeneous, graphite-moderated and helium cooled, USA

PBS: pressure boundary subsystem

PBS: project breakdown structure

PBS: Public Broadcasting Service

PBTF: pump bearing test facility (at LMEC)

PBV: post boost vehicle

PBW: parts by weight

PBX: private branch exchange

PC: personal computer

pC: picocoulomb

pC: picocurie

PC: pitch circle

PC: pitch control

PC: point of curve

PC: polar crane

PC: printed circuit

PC: Professional Communication(IEEE Group)

PC: program coordination

PC: program counter

PC: pulsating current

PC: punched card

PCA: polar cap absorption

PCA: pool critical assembly, Oak Ridge, Tennessee

PCA: Portland Cement Association

PCAC: partially conserved axial vector current

PCAM: punched card accounting machine

PCAS: primary central alarm station

PCB: polychlorinated biphenyl

PCB: power circuit breaker

PCB: printed circuit board

PCBC: partially conserved baryon current

PCBS: positive control bombardment system

PCC: partial crystal control

PCC: point of compound curve

PCCD: peristaltic charge-coupled device

PCCS: photographic camera control system

PCD: photoconductive decay

PCD: power control device

PCDC: punched card data processing

PCE: punched card equipment

PCEA: Pacific Coast Electric Association

PCEM: process chain evaluation model

PCETF: Power Conversion Equipment Test Facility, Aerojet-General Nucleonics reactor test facility

PCF: pounds per cubic foot (use lb/ft^3)

PCG: planning and control guide

PCI: panel call indicator

PCI: peripheral command indicator

PCI: pilot controller integration

PCI: Prestressed Concrete Institute

PCI: product configuration identification

PCIS: primary containment isolation system

PCL: permissible contamination limits

PCLDI: prototype closed loop development installation

PCLS: prototype closed loop system

PCLT: prototype closed loop test

PCM: pitch control motor

PCM: power cooling mismatch

PCM: pulse-code modulation

PCM: pulse-count modulation

PCM: punched card machine

PCME: pulse-code modulation event

PCMI: photochromic microimage

PCN: programmed numerical control

PCOS: primary communications oriented system

PCP: parallel cascade processor

PCP: parallel circular plate

PCP: photon-coupled pair

PCP: post-construction permit (OL due to be tendered)

PCP: primary control program

PCP: primary coolant pump

PCP: process control processor

PCP: processor control program

PCP: program change proposal

PCP: project control plan

PCP: punched card punch

PCR: peer code review

PCR: photoconductive relay

PCR: procedure change request

PCR: program change request

PCR: program control register

PCR: punched card reader

PCRS: primary control rod system

PCRV: prestressed concrete reactor vessel

PCS: plastic clad silica

PCS: pointing control system

PCS: power conditioning subsystem

PCS: power conversion system

PCS: primary coolant system
PCS: print contrast scale
PCS: print contrast signal
PCS: process control system
PCSC: power conditioning, switching and control
PCSIR: Pakistan Council of Scientific and Industrial Research
PCT: photon-coupled transistor
PCT: planning and control techniques
PCTF: plant component test facility
PCTFE: polymonochlorotrifluoroethyle
PCTM: pulse-count modulation
PCTR: Physical Constant Test Reactor, Hanford, Washington, 100 W(th)
PCU: power conditioning unit
PCU: power control unit
PCU: power conversion unit
PCU: pressure control unit
PCU: progress control unit
PCUA: power controller unit assembly
PCV: pollution control valve
PCV: pressure control valve
PCW: pulsed continuous wave
PD: partial discharges
PD: peripheral device
PD: pitch diameter
PD: planned derating
PD: positive displacement
PD: power distribution
PD: preliminary design
PD: priority directive
PD: project Documentation
PD: projected decision date
PD: propellent dispersion
PDA: post-deflection acceleration
PDA: precision drive axis
PDA: predocketed application (CP due to be tendered)
PDA: preliminary design approval
PDA: preliminary design authorization
PDA: probability discrete automata
PDB: Public Debt Bureau
PDC: performance data computer
PDC: power distribution control
PDC: premission documentation change
PDCS: performance data computer system
PDD: Physical Damage Division
PDD: program design data
PDD: prospective decision date
PDES: preliminary draft environmental statement
PDF: plant design factor
PDF: point detonating fuse
PDF: probability density function
PDGDL: Plasma Dynamics and Gaseous Discharge Laboratory
PDGS: Precision Delivery Glides System
PDH: planned derated hours
PDIO: photodiode
PDIS: pressure differential switch
PDIS: Proceedings of the National Symposia
PDL: procedure definition language

PDL: program design language
PDM: pulse delta modulation
PDM: pulse duration modulation
PDO: program directive-operations
PDP: positive displacement pump
PDP: Process Development Pile, small, heavy-water-moderated test reactor at Savannah River
PDP: program definition phase
PDP: programmed data processor
PDPS: parts data processing system
PDQ: programmed data quantizer
PDQS: group diffusion reactor computation codes
PDR: periscope depth range
PDR: power directional relay
PDR: precision depth recorder
PDR: preliminary design review
PDR: pressurized deuterium reactor
PDR: priority data reduction
PDR: processed data recorder
PDR: processing data rate
PDR: program discrepancy report
PDR: program drum recording
PDR: Public Document Room
PDRP: program data requirement plan
PDS: power density spectra
PDS: power distribution system
PDS: predocketed special project (SP due to be tendered)
PDS: procedures development simulator
PDS: program data source
PDS: propellant dispersion system
PDSMS: point-defense surface missile system
PDT: programmable data terminal
PDU: pressure distribution unit
PDU: protocol data unit
PDV: premodulation processor deep-space voice
PE: parity enable
PE: peripheral equipment
PE: permanent echo
PE: phase encoded
PE: polyethylene
PE: probable error
PE: professionl engineer
PE: project engineer
PEA: Pennsylvania Electric Association
PEA: push-effective address
PEACU: plastic energy absorption in compression unit
PEAK: pulse-height analyzer channel corresponding the peak of a distribution
PEB: pulsed electron beam
PEC: photoelectric cell
PECAN: gas-turbine-cycle engineering reactor computer code
PECBI: Professional Engineers Conference Board for Industry
PECS: portable environmental control system
PED: pedestal; equipment base
PED: personnel equipment data
PEDN: planned event discrepancy notification

PEDRO: pneumatic energy detector with remote optics

PEDS: protective equipment decontamination section

PEEP: pilot's electronic eye-level presentation

PEF: physical electronics facility

PEGASE: tank-type research reactor, 30 MW(th), CEA, Cadarache, France

PEGGY: Swimming pool research reactor, 1 kW(th), CEA, Cadarache,France

PEI: preliminary engineering inspection

PEIC: periodic error integrating controller

PEK: phase-exchange keying

PEL: picture element

PELSS: precision emitter location strike system

PEM: photoelectromagnetic

PEM: production engineering measure

PENA: primary emission neuron activation

PENCIL: pictorial encoding language

PENT: pentode

PEO: patrol emergency officer

PEOS: propulsion and electrical operating system

PEP: peak envelope power

PEP: planar epitaxial passivated (transistor)

PEP: planetary ephemeris program

PEP: Professional Education Program

PEP: program evaluation procedure

PEP: proton-electron-positron colliding beams

PEPAG: Physical Electronics and Physical Acoustics Group

PEPP: planetary entry parachute program

PEPR: precision encoder and pattern recognition, bubblechamber photograph recording process, MIT

PER: preliminary engineering report

PERA: Production Engineering Research Association

PERCOS: performance coding system

PEREF: Propellant Engine Research Environmental Facility

PERGO: project evaluation and review with graphic output

PERM: program evaluation for repetitive manufacture

PERT-1: group diffusion perturbation reactor computer code

PERT: performance evaluation review technique

PERT: program evaluation and review technique

PERT: program evaluation research task

PERTCO: program evaluation review technique with cost

PERU: production equipment records unit

PES: photoelectric scanner

PES: Power Engineering (IEEE Society)

PES: Pueblo Engineers Society

PET: patterned epitaxial technology

PET: peripheral equipment tester

PET: polyethylene terephthalate

PET: production environmental testing

PETE: pneumatic end to end

pF: picofarad

P/F: poll/final

PF: power factor

PF: pulse frequency

PFA: pulverized fuel ash

PFCS: primary flow control system

pfd: preferred

PFD: primary flash distillate

PFE: plenum fill experiment

PFES: proposed final environmental statement

PFL: Propulsion Field Laboratory (Rocketdyne)

PFM: power factor meter

PFM: pulse-frequency modulation

PFN: pulse forming network

PFR: parts failure rate

PFR: Prototype fast reactor, United Kingdom

PFRS: portable field recording system

PFR/UK: Prototype Fast Reactor/United Kingdom

PFS: peripheral fixed shim

PFS: propellent field system

PF4: *Promezhutochnyy Fizicheskiy Ansambl* (Intermediate Physical Assembly), critical experiments

PG: power gain

PG: Power Generation (IEEE PES Technical Committee)

PG: pressure gage

PGC: Power Generation Committee

PGEWS: Professional Group on Engineering Writing and Speech (IEEE)

PGNCS: primary guidance and navigation control system

PGNS: primary guidance and navigation system

PGR: precision graphic recorder

PGS: power generator section

PGU: pressure gas umbilical

pH: degree of acidity or alkalinity

pH: hydrogen ion concentration (measure of acidity)

PH: period hours

PH: power house

PHA: pulse height analyzer

PHAROS: Naval Research Laboratory experimental fusion device

PHE: primary element. Ph

PHENIX: French LMFBR demonstration plant, 250 MW(e)

PHENO: precise hybrid elements for nonlinear operation

P&HEP: Plasma and High-Energy Physics (IEEE NPSS Technical Committee)

PHERMEX: pulsed high energy radiographic machine emitting X-rays, electron accelerator - LASL

PHF: plug handling fixture

PHI: position and homing indicator

PHIN: position and homing inertial navigator

PHM: phase meter

PHOEBUS: nickname for follow-on reactor in NERVA program

PHOENIX: plasma heating obtained by energetic neutral injection experiment

PHOENIX: research apparatus in controlled thermonuclear program (Aldermaston, United Kingdom)

PHOENIX: Research reactor and project name, University of Michigan, Ann Arbor

PHP: Parts, Hybrids, and Packaging (IEEE Group)
PHP: pound per horsepower
PHR: pound-force per hour
PHR: process heat reactor program
PHR: pulse-height resolution
PHS: Public Health Service
PHSPS: preservation, handling, storage, packaging, and shipping
PHTS: primary heat transport system
PHWR: pressurized heavy-water reactor
PHY: physical
PHY: physical layer (token ring access method)
Phys Rev: physical review
PI: Packaging Institute
PI: parallel input
PI: performance index
PI: pilotless interceptor
P&I: piping and instrumentation
PI: point initiating
PI: point insulating
PI: point of intersection
PI: priority interrupt
PI: processor interface
PI: productivity index
PI: program indicator
PI: program interrupt
PI: programmed instruction
PI: proportional integral
PIA: peripheral interface adapter
PIA: Plastics Institute of America
PIA: preinstallation acceptance
PIAPACS: psychphysiological information acquisition, processing and control system
PIB: polar ionosphere beacon
PIB: Publishing Information Bulletin
PIB: Pyrotechnic Installation Building
PIBMRI: Polytechnic Institute of Brooklyn, Microwave Research Institute
PIC: particle in cell
PIC: photographic interpretation center
PIC: plastic insulated cable
PIC: polyethylene-insulated conductor
PIC: program interrupt control
PICA: power industry computer applications
PICAC: Power Industry Computer Applications Conference
P&ID: piping and instrumentation diagram
P&ID: piping & instrumentation drawings
PID: proportional integral and differential
PID: proportional integral derivation
PIE: post-irradiation examination
PIE: pulse interference emitting
PIF: payload integration facility
PIG: Penning ionization gage
PIG: Phillips ionization gage
PIGA: pendulous integrating gyro accelerometer
PII: positive immittance inverter
PILOT: permutation indexed literature of technology
PILOT: piloted low-speed test
PIM: precision indicator of the meridian

PIM: pulse interval modulation
PIN: personal identification
PIN: position indicator
pin: positive-intrinsic-negative (transistor)
PINO: positive input, negative output
Pinot: A high explosive experiment to investigate gas migration in oil shales
PINS: portable inertial navigation system
PINT: Purdue interpretive programming and operating system
PIOCS: physical input-output unit
PIOSA: Pan Indian Ocean Science Association
PIOTA: Post irradiation open test assembly or proximity instrumented open test assembly
PIOU: parallel input-output unit
PIP: peripheral interchange program
PIP: picture-in-picture (capability)
PIP: Position Indicating Probe
PIP: predicted impact point
PIP: primary indicating position (Data Logger)
PIP: probabilistic information processing
PIP: problem input preparation
PIP: programmable integrated processor
PIP: project on information processing
PIP: prototypic inlet piping
PIP: pulsed integrating pendulum
PIPA: pulse integrating pendulum accelerometer
PIPPA: pile for producing power and plutonium (code name for Calder Hall project)
PIQUA: OMR, power reactor 45 MW(th), 11 MW(e), Piqua, Ohio, City of Piqua Municipal Power Commission
PIR: Petrolite Irradiation Reactor, St. Louis, Missouri, 10 kW(th)
PIRD: program instrumentation requirements document
PIRG: Public Interest Research Group
PIRN: preliminary interface revision notice
PIRT: precision infrared triangulation
PISH: program instrumentation summary handbook
PIT: peripheral input tape
PIT: processing index terms
PIT: program instruction tape
PIU: plug-in unit
PIV: peak inverse voltage
PIXEL: picture element
PKLT: parking lot
PL-1: Portable Low Power Plant, No.1, USA
PL-2: Portable Low Power Plant, No. 2
PL-3: Portable Low Power Plant, No. 3, redesignated PM-3b
PL: phase line
PL: piping loads
PL: plastic limit
PL: plug
PL: pressurizer level
PL: production language
P&L: profit and loss
PL: programming language
PL: proportional limit

P.L.: Public Law
PLA: physiological learning aptitude
PLA: power lever angle
PLA: programmable logic array
PLA: proton linear accelerator
PLAAR: packaged liquid air-augmented rocket
PLACE: post-Landsat D advanced concept evaluation
PLACE: programming language for automatic check-out equipment
PLACO: Planning Committee [ISO]
PLAN: program language analyzer
PLANIT: programming language for interactive teaching
PLANS: program logistics and network scheduling system
PLANT: program for linguistic analysis of natural plants
PLAT: pilot landing aid television
PLAT: platform
PLATO: programmed logic for automatic teaching operations
PLATR: Pawling lattice test rig, United Nuclear Co
PLBR: prototype large breeder reactor
PLC: power-line carrier
PLCR: power line carrier
PLD: phase-lock demodulator
PLD: pulse-length discriminator
PLDTS: propellent loading data transmission system
PLF: parachute landing fall
PLIANT: procedural language implementing analog techniques
PLIM: post-launch information message
PLL: phase-locked loop
PLOCAP: post loss-of-coolant accident protection
PLOD: planetry orbit determination
PLOP: pressure line of position
PLP: pattern learning parser
PLPA: permissive low pressure alarm
PLRS: position location reporting system
PLS: physical signaling
PLS: plugging switch
PLSS: precision location strike system
pls/s: pulses per second
plt: power-line transients
PLT: Princeton Large Torus (fusion device)
PLT: program library tape
PLU: Phi Lambda Upsilon
PLUS: program library update system
PLZT: lead lanthanum zirconate titanate
PM-1: portable medium nuclear power plant, Sundance, Wyoming, power reactor, 9.37 MW(th)
PM-2A: portable medium nuclear power plant, Camp Century, Greenland, portable field power, air-transportable, 10 MW(th)
PM-3A: portable medium nuclear power plant, McMurdo, Antarctica, air-transportable, power reactor, 9.36 MW(th)
PM-3B: portable medium nuclear power plant (proposed)
PM: panel meter

P/M: parts per million
pm: permanent magnet
PM: phase modulation
PM: photomultiplier
PM: post meridiem (after noon)
PM: pounds per minute
PM: preventive maintenance
PM: procedures manual
PM: project manager
PM: pulse modulation
PMA: critical assembly (KAPL)
PMA: Permanent Magnet Association
PMA: physical medium attachment
PMA: Precision Measurement Association
PMAF: Pharmaceutical Manufacturers Association Foundation
PMB: Physical Metallurgy Branch
PMB: pilot make busy
PMBX: program marginal checking
PMC: programmable machine controller (BLTC)
PMC: pseudo machine code
PME: photomagnetoelectric effect
PME: process and manufacturing engineering
PME: protective multiple earthing
PMEE: prime mission electronic equipment
PMF: probable maximum flood
PMH: Probable maximum hurricane
PMI: preventive maintenance inspection
PMIS: precision mechanisms in sodium
PMMA: polymethylmetharylate
PMMI: Packaging Machinery Manufacturers Institute
PMNP: platform-mounted nuclear plant - refers to barge-mounted nuclear power plant
PMOS: p-channel (type) metal oxide semiconductor
PMP: premodulation processor
PMP: preventive maintenance plan
PMP: probable maximum precipitation
PMPS: plant nuclear protection system
PMR: Pacific Missile Range
PMS: Picturephone Meeting Service
PMS: plant monitoring system
PMS: probable maximum surge
PMS: processor, memories and switches
PMSRP: Physical and Mathematical Sciences Research Paper
PMT: photomultiplier tube
PN: performance number
PN: preliminary notification
PN: pseudonoise
PndB: perceived noise level expressed in decibels
PNDC: parallel network digital computer
PNEC: Proceedings of the National Electronics Conference
PNGCS: primary navigation, guidance, and control System
pnlbd: panelboard
pnp: positive-negative-positive (transistor)
PNPF: Piqua Nuclear Power Facility, Piqua, Ohio
PNPS: plant nitrogen purge system
PNR: Pittsburgh Naval Reactors Office, AEC

PNSRC: Plant Nuclear Saftey Review Comm
PNTD: personnel neutron threshold detector
PNVS: pilot night vision system
PNWL: Pacific Northwest Laboratories (also BNWL)
PO: parallel output
PO: power oscillator
PO: program objectives
PO: pulsed carrier without any modulation intended to carry information
PO: purchase order
POA: provisional operating authorization
POCP: program objectives change proposal
POCS: Patent Office classification system
POD: point of origin device
PODAF: power density exceeding a specified level over an area within an assigned frequency band
PODS: post-operative destruct system
POF: planned outage factor
POGO: Polar Orbiting Geophysical Observatory
POGO: programmer oriented graphics operation
POH: planned outage hours
POI: program of instruction (NASA)
POIL: power density imbalance limit
POIS: prototype online instrument systems
POISE: panel on-in flight scientific experiments
POISE: preoperational Inspection Services Engineering (INA)
POL: problem-oriented language
POL: procedure-oriented language
POL: process-oriented language
POL: provisional operating license
POLYTRAN: polytranslation analysis and programming
POMM: preliminary operating and maintenance manual
POMS: Panel on Operational Meteorological Satellites
POMSEE: Performance, Operational and Maintenance Standards for Electronic Equipment
POP: power on/off protection
POP: pressurizer over pressure protection system
POP: programmed operators and primitives
POP: program operating plan
POP: proof of principle (test)
POPI: Post Office position indicator
POPR: prototype organic power reactor
POPS: pantograph optical projection system
POR: problem-oriented routine
PORACC: principles of radiation and contamination control
PORV: pilot-operated relief valve
PORV: power operated relief valve
POS: plant operating system
POS: point-of-sale (manufacturer)
POS: point of sale (transaction-oriented system)
pos: positive
POS: primary operating system
POSA: preliminary operating safety analysis
POSH: permutation on subject headings
POT: critical assembly-(LASL)
POT: paint on tangent

POT: potentiometer
POTC: PERT Orientation and Training Center (DOD)
POWS: pyrotechnic outside warning system
PP: panel point
PP: peripheral processor
P/P: point to point
PP: preprocessor
PP: pressureproof
PP: print/punch
PP: push-pull
PPA: photo peak analysis
PPA: Princeton-Pennsylvania Proton Accelerator, Princeton, New Jersey
PPA: Princeton Particle Accelerator
PPA: Professional Programmers Association
PPC: pulsed power circuit
PPDD: preliminary project design description
PPDW: primary plant mineralized water
PPE: polyphenylether
PPE: premodulation processor equipment
PPE: problem program efficiency
PPG: program pulse generator
PPI: plan position indicator
P&PI: Pulp and Paper Industry (IEEE IAS Technical Committee)
PPL: Plasma Physics Laboratory, Princeton, New Jersey
ppm: parts per million
PPM: periodic permanent-magnet focusing
PPM: planned preventive maintenance
PPM: pulse position modulation
PPMA: Precision Potentimeter Manufacturers Association
PPMS: program performance measurement systems
PPS: parallel processing system
PPS: Philadelphia Programming Society
PPS: phosphorous propellent system
PPS: plant protection system
PPS: preferred power supply
PPS: preferred power system
PPS: primary power system
PPS: primary propulsion system
PPS: pulses per second
PPSP: power plant siting program - administered by Maryland Department of Natural Resources; program is designed to develop siting criteria
PPT: punched paper tape
PPU: peripheral processing unit
PQAD: plant quality assurance director
PQGS: propellent quantity gauging systems
PR-1: Philippine Reactor No. 1 3 MW(th), pool-type
PR: pattern recognition
PR: Pawling Research Reactor, Nuclear Development Corporation of America, Remote Experiment Station, Pawling, New York
PR: press release
PR: program register
PR: program requirements
PR: pseudorandom
PR: Puerto Rico

PR: pulse rate
PR: pulse ration
PRA: peak recording accelerographs
PRA: pendulous reference axis
PRA: precision axis
PRA: probabilistic risk assessment
PRA: production reader assembly
PRA: program reader assembly
PRB: Panel Review Board
PRBS: pseudorandom binary sequence
PRBSG: Pseudo-random binary sequence generator (LMEC)
PRC: partial response coding
PRC: point of reverse curve
PRC: Postal Rate Commission
PRCF: plutonium recycle critical facility
PRD: power range detector
PRD: program requirements data
PRD: program requirements document
PRDC: Power Reactor Development Company (Fermi)
PRE: processing refabrication experiment
PREDICT: prediction of radiation effects by digital computer techniques
PREF: propulsion research environmental facility
PREP: elastic scattering cross sections reactor physics computer code
PREP: programmed educational package
PRESS: Pacific Range Electromagnetic Signature Studies
PRESSAR: presentation equipment for slow scan radar
PRESTO: program for rapid earth-to-space trajectory optimization
PRESTO: program reporting and evaluation system for total operations
PRF: pulse repetition frequency
PRFL: pressure-fed liquid
PRI: pulse repetition interval
PRIDE: programmed reliability in design
PRIME: precision integrator for meteorological echoes
PRIME: precision recovery including maneuverable entry
PRIME: programmed instruction form management education
PRINCE: programmed reinforced instruction necessary to continuing education
PRIS: propeller revolution indicator system
PRISE: program for integrated shipboard electronics
PRISM: programmed integrated system maintenance
PRM: personal radiation monitor
PRM: power range monitor
PRM: process radiation monitor
PRM: pulse-rate modulation
PRN: pseudorandom noise
PRNC: Puerto Rico Nuclear Centers at Rio Piedras and Mayaguez
PRO: HEDL automated procurement system

PRO: Italian (CNEN) study of organic-liquid-cooled and -moderated, 30 MW(th), reactor to be constructed near Bologna, Italy
Proc IEE: Proceedings of the Institution of Electrical Engineers
Proc NEC: Proceedings of the National Electronics Conference
PROCOM: Procedures Committee (IEEE)
PROCOMP: process computer
PROCOMP: program compiler
PRODAC: programmed digital automatic control
PROFAC: propulsive fluid accumulator
PROFILE: programmed functional indices for laboratory evaluation
PROFIT: programmed reviewing, ordering, and forecasting inventory technique
PROGDEV: program device
PROGLOOK: monitor execution of any OS/MVT load module
PROM: Pockels readout optical memory
PROM: programmable read-only memory
PROMPT: program monitoring and planning techniques
PROMPT: program reporting, organization, and management planning technique
PRONTO: program for numerical tool operation
PROP: performance review for operating programs
PROP: planetary rocket ocean platform
PROSERPINE: zero power research reactor, CEA, Saclay, France
PROTECT: probabilities recall optimizing the employment of calibration time
PROXI: projection by reflection optics of xerographic images
PRP: pulse repetition period
PRPF: planar radial peaking factor
PRR-1: Philippine Research Reactor 1 Diliman
PRR: Pawling Research Reactor, Pawling, New York, 5 W(th)
PRR: Puerto Rico L-77 Reactor, Puerto Rico Nuclear Center, Mayaguez, Puerto Rico
PRR: pulse repetition rate
PRS: Pattern Recognition Society
PRS: pattern recognition system
PRS: process radiation sampler
PRT: portable remote terminal
PRT: prompt relief trip
PRT: pulse repetition time
PRTR: plutonium recycle test reactor
PRU: programs research unit
PRV: peak inverse (reverse) voltage
PRV: pressure-reducing valve
PRV: pressure regulating valve
PRV: pressure relief valve
PRVS: penetration room ventilation system
PRWRA: Puerto Rico Water Resources Authority
ps: picosecond
PS: Postal Service
PS: power source
PS: pressure switch

PS: program store
PS: proton synchrotron
PSA: power servo amplifier
PSA: Product Saftey Association
PSAC: President's Science Advisory Committee
PSALI: permanent supplementary artificial lighting of interiors
PS&A/NPSS: Plasma Sciences & Applications
PSAR: preliminary safety analysis report
PSAR: programmable synchronous/asynchronous receiver
PSAT: programmable synchronous/asynchronous transmitter
PSBMA: Professional Services Business Management Association
PSC: photosensitive cell
PSC: Power System Communications (IEEE PES Technical Committee)
PSCC: Power System Computation Conference
PSCC/PES: Power System Communications
PSD: phase-sensitive demodulator
PSD: phase correction reactor; phase shifting device
PSD: power spectral density
PSD: pulse spacing distribution
PSDD: Preliminary system design description
PSDF: Propulsion Systems Development Facility
PSE: Power System Engineering (IEEE PES Technical Committee)
PSE: pressurized subcritical experiment
PSE&G: Public Service Electric and Gas Co. of New Jersey
PSEP: passive seismic experiment package
PSF: performance shaping parameter
PSF: point spread junction (image processing)
PSF: pound-force square foot
PSG: phosphosilicate glass
PSI: pound force per square inch
PSI: pounds per square inch absolute
PSI: preprogrammed self-instruction
PSI: Proctorial system of instruction
PSID: pounds per square inch differential
PSID: preliminary safety information document
PSIEP: Project on Scientific Information Exchange in Psychology
psig: pounds per square inch gauge
PSIM: Power System Instrumentation and Measurement (IEEE PES Technical Committee)
PSK: phase-shift keying
PSL: Photographic Science Laboratory
PSMA: Professional Services Management Association
PSMR: parts specification management for reliability
PSMS: Permanent Section of Microbiological Standardization
PSO: pilot systems operator
PSOS: p-channel silicon gate devices
PSP: performance shaping parameters
PSP: planet scan platform
PSP: portable service processor
PSP: power system planning

PSP: primary sodium pump
PSPGV: guard vessel for primary sodium pump
PSPS: planar silicon photoswitch
PSQ: personnel security questionnaire
PSR: Pennsylvania State University Research Reactor, University Park, Pennsylvania
PSR: peripheral shim rods
PSR: Power System Relaying (IEEE PES Technical Committee)
PSR: procurement status report
PSRP: physical sciences research paper
PSS: proprietary support system
PSSCC: Private Sector Standards Coordinating Center
PST: point of spiral tangent
PST: polished surface technique
PST: primary surge tank (CLS)
PSTC: Pressure Sensitive Tape Council
PSTF: pressure suppression test facility
PSTF: proximity sensor test facility
PSTF: pump seal test facility
PSUR: Pennsylvania State University Reactor, 200 kW(th)
PSW: Philosophical Society of Washington
PSW: program status word
PSWP: plant service water pump
PT: dye penetrant test (examination)
PT: point of tangency
PT: potential transformer
PT: propellent transfer
PT: pulse time
PT: punched tape
PTA: planar turbulence amplifier
PTA: pulse torquing assembly
PTC: Pacific Telecommunications Council
PTC: passive thermal control
PTC: positive temperature coefficient
PTC: programmed transmission control
PTCR: pad terminal connection room
PTCS: propellent tanking computer system
PTE: pressure-tolerant electronics
PTF: program temporary fix
PTFE: polytetrafluoroethylene
PTI: pipe test insert (LMEC)
PTI: plugging temperature indicator
PTI: Power Tool Institute
PTL: Pittsburgh Testing Lab (FFTF Site)
PTM: proof test model
PTM: pulse time modulation
PTM: pulse time multiplex
PTML: pnpn transistor magnetic logic
PTO: Patent and Trademark Office
PTO: power take-off
PTP: paper tape punch
PTP: point to point
PTP: preferred target point
PTP: proximity test plug
PTR: critical assembly (KAPL)
PTR: paper tape reader
PTR: Pool Test Reactor, enriched-uranium, swimming-pool. Chalk River, Canada

PTR: Pool Training Reactor, enriched-uranium, swimming-pool, zero-power, ACF Industries, Inc., USA
PTR: Power Transformers (IEEE PES Technical Committee)
PTR: Pressure Tube Reactor, AMF, USA
PTR: programmer trouble report
PTR: Proof Test Reactor, KAPL, USA
PTS: permanent threshold shift
PTS: pneumatic test set
PTS: program test system
PTS: propellent transfer system
PTS: pure time sharing
PT/SP: pressure tube to spool piece
PTSR: Pressure Tube Superheat Reactor, heterogeneous, uranium, graphite-moderated and light-water-cooled with integral nuclear superheating. USA
PTT: postal telephone and telegraph
PTT: program test tape
PTT: push to talk
PTV: predetermined time valve
PTVA: propulsion test vehicle assembly
PU: per unit (power fault effects)
PU: power unit
PU: processing unit
PU: propulsion unit
PUC: Public Utilities Commission
PUCK: propellant utilization checkout kit
PUCS: propellant utilization control system
PUD: Public Utility District
Puf: fissile plutonium (239Pu + 241 Pu)
PUFFS: passive underwater fire control sonar feasibility study
PUFFT: Purdue University fast Fortran translator
PUGS: propellant utilization gauging system
PUP: plutonium utilization program
PUR: procurement request
PUR: Purdue University Reactor, Lafayette, Indiana, 1 kW(th)
PUR: Purdue University Research
Pu Recycle: The use of LWR produced plutonium to replace some portion of the fissile 235U normally required in LWR fuels
PURPA: Public Utility Regulatory Policies Act
PUSAS: proposed US Standard
PUSS: pilots' universal sighting system
PUT: programmable unijunction transistors
PV: photovoltaic
PV: positive volume
PV: pressure control valve
PVC: polyvinyl chloride
PVC: position and velocity computer
PVK: poly-N-vinylcarbazol
PVOR: precision vhf omnirange
PVRC: Pressure Vessel Research Committee (WRC)
PVS: performance verification system
PVST: premate verification system test
PVT: polyvinyl toluene
PVT: pressure-volume-temperature
PVTI: piping & valve test insert
pW: picowatt

PW: printed wiring
PW: pulsewidth
PWB: printed wiring board
P&WBP: Pension and Welfare Benefit Programs
PWD: power distribution
PWD: pulsewidth discriminator
PWF: present worth factor
PWI: piecewise linear
PWI: pilot warning instrument
PWM: pulse-width modulation
PWM: pulsewidth multiplier
PWR-FA: critical assembly (Bettis)
pwr: power
PWR: pressurized water reactor
PWRS: programmable weapons release system
Pwr Semicond/IAS: Power Semiconductor
PWT: propulsion wind tunnel
PWV: peak working voltage
PXSTR: phototransistor
PY: pyrometer
PZR: pressurizer
PZT: lead(Pb) zirconate titanate (semiconductor)
PZT: photographic zenith time
P3: portable plotting package

Q

Q-Cumber: experimental fusion device of magnetic mirror type, LRL
Q-11: pitchblende
Q: quality factor
Q: quantity of electric charge (symbol)
Q: transistor
QA: quality assurance
QA: query analyzer
QAC: quality assurance checklist
QADS: quality assurance data system
QAI: quality assurance instruction
QAIP: quality assurance inspection procedure
QA&O: quality assurance and operations
QAPI: quality assurance program index
QAPP: quality assurance program plan
QAR: quality assurance representative at vendor plant
QAR: quality assurance requirements
QASL: quality assurance systems list
QB: quick break
QC: quality control
QCE: quality control engineering
QCESC: Quad Cities Engineering & Science Council
QCM: quantitative computer management
QCR: quality control reliability
QCW: quadrature phase subcarrier (signal)
QD: quick disconnect
QDRI: qualitative development requirements information
QEA: Quantum Electronics & Applications
QEC: quick engine change
QF: quality form
QF: quick firing
QFE: field effect transistor

QFRIC: quick fix interference reduction capability
QHR: quality history records
QISAM: queued indexed sequential access method
QIT: quality information and test
QL: query language
QLAP: quick look analysis program
QLDS: quick look data station
QLIT: quick-look intermediate tape
QLP: qualified parts list
QLR: quick look report
QMDO: qualitative material development objective
QMR: qualitative material requirement
QOD: quick-opening device
QPL: qualified products list
QPM: quality program manager
QPPM: quantized pulse position modulation
QPSK: quadrature phase shift keying (or quadriphase)
QPT: quadrant power tilt
QR: quality and reliability
QRA: quality and reliability assurance
QRBM: quasi-random band model
QRC: quick reaction capability
QRI: qualitative requirements information
QRM: man-made interference (obsolete)
QSAM: queued sequential access method
QSP: quench spray pump
QSR: quarterly statistical report
QSS: quench spray subsystem
QT: tetrode transistor
QTAM: queued telecommunications access method
QTM: thermistor
QTP: quality test plan
QU: unijunction transistor
QUAM: quadrature amplitude modulation
QUEST: quality electrical systems test
QUIP: query interactive processor
QVR: varistor
QVT: quality verification testing

R

R-W: read-write (head)
R-1: Reactor, Stockholm, Sweden, 300 kW(th)
R-2: Reactor, Studsvik, Sweden, 30 kW(th)
R-3: *AB Atomenergie* , underground power reactor, 65 MW(th), Agesta, Sweden
R-4: Power reactor, 364 MW(th), Marviken, Sweden
R: read; reserved line
R: Reliability (IEEE Group)
R: response
R: reverse
RA-1: Argentine Reactor No. 1 Buenos Aires, 25 kW(th)
RA: radar altimeter
RA: reduction of area
RA: rental agreement
RAC: rigid aluminum conduit
RACC: radiation and contamination control
RACE: random access computer equipment
RACE: random access control equipment

RACE: rapid automatic checkout equipment
RACEP: random access and correlation for extended performance
RACES: Radio Amateur Civil Emergency Service
RACHEL: plutonium, homogeneous, zero-power research reactor, CEA, Saclay, France
RACON: radar beacon
RACS: remote access calibration system
RACS: remote access computing system
RACT: reasonably available control technique
RACT: remote access computer technique
Rad-Elect Eng: The Radio and Electronic Engineer
RAD: radiation absorbed dose
RAD: random access data
RAD: random access disk
RAD: rapid access disk (computer)
RADA: random-access discrete-address
RADAC: radar analog digital data and control
radar: radio detection and ranging
RADAS: random access discrete address system
RADATA: radar data transmission and assembly
RADC: Rome Air Development Center (Air Force)
RADCM: radar countermeasures and deception
RADCON: radar data converter
RADEM: random access delta modulation
RADFAC: radiating facility
RADHAZ: radiation hazards
Rad Haz Meas/IM: TC-7 Radiation Hazard Measurements
RADIAC: radioactivity detection, identification, and computation
Radia Effects/NPSS: radiation effects
Radiocom/Com: radio communication
Radio Sci: radio science
Radiotekh Elektron: *Radiotekhnika i Elektronica*
RADIQUAD: radio quadrangle
RADIR: random access document indexing and retrieval
RADIST: radar distance indicator
RADIT: radio teletype
RADNOTE: radio note
RADO: radio
RADOME: radar dome
RADOP: radar operator
RADOPWEAP: radar optical weapons
RADPLANBD: radio planning board
RADPROPCAST: radio propagation forecast
RADS: radiation and dosimetry services
RADS: radiation source
RADSO: radiological survey officer
RADVS: radar altimeter and doppler velocity sensor
Radwaste: radioactive waste
RAE: radio astronomy explorer satellite
RAE: range azimuth elevation
RAE: Royal Aeronautical Establishment
RAEN: radio amateur emergency network
RAES: remote access editing system
RAES: Royal Aeronautical Society
RAFT: radially adjustable facility tube
RAI: random access and inquiry

RAIC: Redstone Arsenal Information center
RAID: remote access interactive debugger
RAIDS: rapid availability of information and data for safety
RAILS: remote area instrument landing sensor
RAIP: requester's approval in principle
RAIR: random access information retrieval
RAL: Riverbend Acoustical Laboratory
RALU: register arithmetic and logic unit
RALW: radioactive liquid waste
RAM: radar absorbing material
RAM: radio attenuation measurements
RAM: random access memory
RAMA: Rome Air Material Area
RAMAC: random access method of accounting and control
RAMARK: radar marker
RAMIS: rapid access management information system
RAMP: Raytheon airborne microwave platform
RAMPART: radar advanced measurement program for analysis of reentry techniques
RAMPS: resource allocation and multiproject scheduling
RANDAM: random access nondestructive advanced memory
RANN: Research Applied to National Needs - an NSF program for sponsoring research pertinent to solving national social and environmental problems
RAO: Radio Astronomical Observatory
RAP: redundancy adjustment of probability
RAP: relational associative processor
RAP: rocket-assisted projectile
RAPCON: radar approach control
RAPEC: rocket-assisted personnel ejection catapult
RAPID: reactor and plant integrated dynamics
RAPID: research in automatic photocomposition and information dissemination
RAPO: Resident Apollo Project Office (NASA)
RAPPI: random access plan position indicator
RAPS: radioactive argon processing system
RAPS: retrieval analysis and presentation system
RAPS: risk appraial of programs system
Rapsodie: Power reactor, 10-20 MW(th), Cadarache, France
RAPT: reusable aerospace passenger transport
RAPTUS: Rapid Thorium-Uranium System, thorium-breeder, sodium-cooled fast reactor
RAR: rapid access recording
RAREF: radiation and repair engineering facility
RAREP: radar weather report
RARG: Regulatory Analysis Reveiw Group
RAS: reactor alarm system
RAS: reactor analysis and safety
RAS: row-address strobe
RASC: Rome Air Service Command
RASPO: Resident Apollo Spacecraft Program Office
RASSR: reliable advanced solid-state radar
RASTA: radiation augmented special test apparatus
RASTA: radiation special test apparatus
RASTAC: random access storage and control

RASTAD: random access storage and display
RAT: reliability assurance test
RAT: reserve auxiliary transformer
RATAC: radar target acquisition
RATAN: radar and television aid to navigation
RATCC: radar air-traffic control center
RATE: remote automatic telemetry equipment
RATER: response analysis tester
RATIO: radio telescope in orbit
RATO: rocket-assisted take-off
RATOG: rocket-assisted take-off gear
RATSC: Rome Air Technical Service Command
RATSCAT: radar target scatter
RATT: radio teletype
RAU: remote acquisition unit
RAVIR: radar video recording
RAWARC: radar report and warning coordination system
RAWIN: radar wind sounding
RAYDAC: Raytheon digital automatic computer
RB: radar beacon
RB: Reclamation Bureau
RB: Renegotiation Board
RB: reset bus
RB: return to bias
RBA: recovery beacon antenna
RBCCW: reactor bldg closed coolant water
RBCH: rod bank coil unit
RBCWS: reactor bldg. cooling water system
RBD: reliability block diagram
RBDE: radar bright display equipment
RBE: radiation biological effectiveness
RBE: relative biological effectiveness
RBHPF: reactor bldg. hydrogen purge fan
RBI: ripple-blanking input
RBLR: reboiler
RBM: rod block monitor
RBO: ripple-blanking output
RBOF: receiving basins for offsite fuels, SRP
RBP: registered business programmer
RBR: radar boresight range
RBS: radar bomb scoring
RBS: random barrage system
RB(S): reactor building (sump)
RBSCH: reactor bldg. cooling unit
RBSS: recoverable booster support system
RBV: reactor bldg. vent
R/C: radio command
R/C: radio control
R/C: range clearance
RC: range control
R/C: rate of climb
RC: ray control electrode (vacuum tube)
RC: remote control
RC: research center
RC: resistance-capacitance
RCA Eng: RCA Engineer
RCAG: remote-controlled air-ground communication site
RCA Rev: RCA Review

RCAT: radio-code aptitude test
RCB: reactor containment building
RCC: Radiochemical Centre, Amersham, United Kingdom
RCC: radio common carrier
RCC: Range Commanders Council
RCC: recovery control center
RCC: remote communications central
RCC: rod cluster control
RCCA: remote control rod cluster assembly
RCDC: radar course directing central
RCEI: range communications electronics instructions
RCFC(U): reactor core fan cooling (unit)
RCG: radioactivity concentration guide
RCGM: reactor cover gas monitor
RCI: radar coverage indicator
RCIC: reactor core isolation cooling
RCICS: reactor core isolation cooling system
RCL: Radiation Counting Laboratory
RCL: reclaimer
RCL: relative chrominance level
RCLC: reactor coolant leakage calculation
RCM: radar countermeasures
RCM: radio countermeasures
RCM: Reliability Corporate Memory
RCN: Reactor Centrum Nederland, the Hague, Netherlands
RCO: reacter core
RCO: remote control office
RCO: remote control oscillator
RCP: Radiological Control Program used in reference to States
RCP: reactor coolant pump
RCPB: reactor coolant pressure boundary
RCRC: Reinforced Concrete Research Council
RCS: radar cross section
RCS: radio command system
RCS: reaction control system
RCS: reactor coolant system
RCS: reentry control system
RCS: remote control system
RCSDE: reactor coolant system dose equivalent
RCT: radiation/chemical technician
RCT: relative chrominance time
RCT: resolves control transformer
RCT: rework/completion tag
RCTL: resistor-capacitor-transistor logic
RCTSR: radio code test speed on response
RCVR: receiver
RC1: TRIGA Mk2-type, solid, homogeneous, enriched-uranium, light-water-moderated and -cooled reactor. Italy
RD: radiation detection
RD: readiness date
RD: read line
RD: received data
RD: register drive
RD: requirements document (Bechtel)
R&D: research and development
RD: restricted data

RD: root diameter
rd: rutherford
RDA-CX: critical assembly (Bettis)
RDA: reliability design analysis
RDA: rod drop accident
RDB: radar decoy balloon
RDB: Research and Development Board
RDC: Reliability Data Center
RDC: remote data collection
R&D Council: Research and Development Council of New Jersey
RDDM: reactor deck development mockup
RDE: radial defect examination
RDF: radio direction finding
RDL: Rocket Development Laboratory
RDM: recording demand meter
RDMU: range drift measuring unit
RDOS: real-time disk operating system
RDP: radar data processing
RDP: reactor development program
RDPS: radar data processing system
RDRINT: radar intermittent
RDR XMTR: radar transmitter
RDS: depressurizing system
RDS: rendezvous and docking simulator
RDS: Rural Development Service
RDT: Reactor Development and Technology (Divison of, used only in identifying Standards) (now RDD)
RDT: remote data transmitter
RDTE: research, development, test, and evaluation
RDTL: resistor diode transistor logic
RDX: cyclotrimethylene trinitroamine
RE: radiated emission
RE: rate effect
RE: reentry
R&E: research and engineering
RE: row enable
REA: U.S. Rural Electrification Administration
reac: reactor
REAC: REEVES electronic analog computer
REACT: Radio Emergency Associated Citizens Teams
REACT: register-enforced automated control technique
READ: real-time electronic access and display
READ: remote electronic alphanumeric display
READI: rocket engine analysis and decision instrumentation
REALCOM: real-time communications computer
REAP: remote entry acqusition package
REAR: Reliability Engineering Analysis Report
REBA: relativistic electron beam accelerator
REC: radiant energy conversion
REC: rectifier
REC: Registration of Engineers Committee
REC: request for engineering change
REC: Rural Electrification Conference
RECMFA: Radio and Electronic Component Manufacturers Association
RECP: receptacle
RECSTA: receiving station

rect: rectifier
Recuplex: plutonium recovery facility, Hanford
Recycle Pu: LWR produced Pu recovered from spent fuel and subsequently used to replace some portion of 235 U normally required on LWR fuel
REDAP: reentrant data processing
REDSOD: repetitive explosive device for soil displacement
REEP: regression estimation of event probabilities
REGAL: range and elevation guidance for approach and landing
REGAL: range and evaluation guidance for approach and landing
REI: research-engineering interaction
REIC: Radiation Effects Information Center, Battelle Memorial Institute,Columbus, Ohio
REIL: runaway end identifier lights
REINS: requirements electronic input system
REJ: reject
REL: radiation evaluation loops (HEDL)
REL: rapidly extensible language
REM: rapid eye movement
REM: reliability engineering model
REM: Roentgen equivalent man
REMOS: real-time event monitor
REN: remote enable
RENE: rocket-engine/nozzle ejector
REON: rocket engine operations-nuclear
REP: range error probable
REP: rendezvous evaluation pad
REP: roentgen equivalent physical
REPLAB: responsive environmental programmed laboratory
REQVER: requirements verification
RER: Radiation Effects Reactor (Air Force), Dawsonville, Georgia 10 MW(th)
RES: grounding resistor
RES: Office of Nuclear Regulatory Research
RES: resistor
RES: unresolved resonances reactor physics computer code
RESAR: reference safety analysis report
RESD: Research and Engineering Support Divison
RESER: reentry system and evaluation radar
RESG: Research Engineering Standing Group
RESNA: Rehabilitation Engineering Society of North America
RESS: radar echo simulation subsystem
REST: reentry environment and systems technology
RETAIN: remote technical assistance and information network
RETMA: Radio Electronics Television Manufacturing Association. Now EIA
REV: reentry vehicle
REVS: rotor entry vehicle system
Rev Techn Thomson CSF: Revue Technique Thomson CSF
REWSONIP: reconnaissance electronic warfare special operation and naval intelligence processing
RF: radio frequency

RF: range finder
RF: rating factor
RFC: radio-frequency choke
RFC: radio facility charts
RFCP: radio-frequency compatibility program
RFD: ready for data
RFD: reentry flight demonstration
RFEI: request for engineering information
RFG: radar field gradient
RFI: radio-frequency interference
RF/IR: radar frequency/infrared frequency
RFIT: radio-frequency interference tests
RFM: reactive factor meter
RFMO: Radio Frequency Management Office
RFP: request for proposal
RFQ: request for quotation
RFS: radio-frequency shift
RFSTF: RF Systems Test Facility
RFTD: radial flow torr deposition (system)
RG: rate gyroscope
RG: Regulatory Guide
RG: reticulated grating
RGA: rate gyro assembly
RGB: red-green-blue
RGEN: regenerator
RGL: report generator language
RGO: Royal Greenwich Observatory
RGP: rate gyro package
RGS: radio guidance system
RGS: rate gyro system
RGS: rocket guidance system
RGSC: ramp generator & signal converter
RGT: resonant gate transistor
RH: radiological health
RH: relative humidity
RH: reserve shutdown hours
RH: rheostat
R/h: roentgens per hour
RH: run/halt switch
RHAW: radar homing and warning
RHC: right-hand circular (polarization)
RHC: rotation hand controller
RHE: radiation hazard effects
RHEL: Rutherford High Energy Laboratory
RHI: range height indicator
RHOGI: radar homing guidance
RHP: reduced hard pressure
RHR: rejectable hazard rate
RHR: residual heat removal
RHRP: residual heat removal pump
RHRSW: residual heat removal service water
RHTS: reactor heat transport system
RHV: reference heat balance
RI: radio interference
RI: reflective insulation
R&I: removal and installation
RI: resistance inductance
RI: Rhode Island
RIA: reactivity initiated accident

RIA: removable instrument assembly (EBR-II - IN-COT flow and temperature - FFTF instr. tree and TLLM)

RIA: Robot Institute of America

RIAA: Record Industry Association of America

RIAA: Recording Industry Association of America

RIAEC: Rhode Island Atomic Energy Commission

RIAS: Research Institute for Advanced Studies

RIC: range instrumentation coordination

RICASIP: Research Information Center and Advisory Service on Information Processing

RI&C/NPSS: reactor instrumentation & control

RICS: range instrumentation control system

RID: review item disposition

RIDS: receiving inspection data status report

RIE: The Rhodesian Institution of Engineers

RIF: radio-influence field

RIF: reduction in force

RIFI: radio interference filed intensity

RIFT: reactor in-flight test

RII: receiving inspection instructions

RIM: radar input mapper

RIM: reaction injection molded (foam)

RIN: regular inertial navigator

RINAL: radar inertial altimeter

RINS: rotorace inertial navigation system

RINT: radar intermittent

RIOMETER: relative ionospheric opacity meter

RIOPR: Rhode Island Open Pool Reactor, 3 MW(th), pool-type

RIOT: real-time input-output transducer

RIP: reactor instrument penetration (valve)

RIP: receiving inspection plan (Bechtel)

RIP: regulatory & information policy

RIPPLE: radioactive isotope powered pulse light equipment

RIPPLE: radioisotope powered prolonged life equipment

RIPS: radioisotope power supply

RIPS: radio isotope power system

RIPS: range instrumentation planning study

RIPV: reactor isolation pressure valve

RIR: reliability investigation requests

RIS: range instrumentation ship

RIS: reporting identification symbol

RIS: revolution indicating system

RISE: research in supersonic environment

RIST: radar installed system tester

RIS0: principal nuclear research laboratory in Denmark

RIT: radio information test

RIT: rocket interferometer tracking

RITU: Research Institute of Temple University

RIV: radio influence voltage

Riv Tec Selenia: Revista Tecnica Selenia

RIW: reliability improvement warranty

RJE: remote job entry

RKHS: reproducing kernel Hilbert space

RKO: range keeper operator

RL: radiation laboratory

RL: reactor licensing

RL: research laboratory

RL: resistance-inductance

RL: resistor logic

RLBM: rearward launched ballistic missiles

RLC: radio launch control system

RLC: resistance-inductance-capacitance

RLE: Research Laboratory of electronics

RLHTE: Research Laboratory of Heat Transfer in Electronics

RLM: Reflector & Lighting Equipment Manufacturers

RLOP: reactor licensing operating procedure

RLRD: railroad

RLWL: reactor low water level

RM: radio monitoring

R&M: reliability and maintainability

RM: Rotating Machinery (IEEE PES Technical Committee)

RMA: Radio Manufacturers Association, later RETMA, now EIA

RMA: Rubber Manufacturers Association

RMAAS: reactivity monitoring and alarm system

RMAG: Rocky Mountain Association of Geologists

RM&C: reactor monitoring and control

RMC: rigid metal threaded conduit

RMC: rod memory computer

RMCS: reactor manual control system

RMF: reactivity measurement facility, (NRTS)

RMI: radio magnetic indicator

RMI: reliability maturity index

r/min: roentgens per minute

RML: radar microwave link

RMM: read-mostly memory

RMO: radio material officer

RMOS: refractory metal oxide semiconductor

RMP: reentry measurements program

RMS: radiation monitoring system

RMS: reactor monitor system

RMS: regulatory manpower system

RMS: remote manipulator system

rms: root mean square

RMU: remote manueuvering unit

RMWS: reactor make-up water system

RN: random number

RNA: ribonucleic acid

RNR: receive not ready

RO: radar operator

RO: range operations

RO: readout

RO: receive only

RO: reportable occurrence

Ro: Swedish Reactor Ro, heterogeneous, natural uranium, heavy water moderated, research zero power

ROAD: road

ROAMA: Rome Air Material Area

ROAP: Reorganization, Office of Assistant to President

ROAT: radio operator's aptitude test

ROB: radar order of battle

ROBIN: remote on-line business information network

ROBOMB: robot bomb
ROC: Range Operations Conference
ROC: rate of convergence
ROC: receiver operating characteristics
ROC: required operation capability
ROC: reusable orbital carrier
ROCAPPI: Research on Computer Applications in the Printing and Publishing Industries
ROCC: region operations control center
ROCP: radar out of commission for parts
ROCR: remote optical character recognition
ROD: release order directive (now called ERO)
RODATA: registered organization data bank
ROI: range operational instructions
ROIS: radio operational intercom system
ROLF: remotely operated longwall face
ROLR: receiving objective loudness rating
ROLS: recoverable orbital launch system
ROM: read-only memory
ROM: rough order magnitude
ROMASHKA: direct conversion reactor, uranium dicarbide core, USSR
ROMBUS: reusable orbital module booster and utility shuttle
ROP: record of purchase
ROP: recovery operating plan
ROPP: receive-only page printer
ROS: read-only storage
ROSE: remotely operated special equipment
ROSE: retrieval by on-line search
ROSPO: zero power experimental organic reactor, Italian Nuclear Energy Commission (CNEN), Casaccia, Italy
ROT: reusable orbital transport
ROTI: recording optical tracking instrument
ROTR: receive-only tape perforator
ROW: right-of-way
RP: recovery phase
RP: relative pressure
RPAO: radium plague adaptometer operator
RPC: row parity check
RPCRS: reactor protection control rod system
RPCS: reactor plant control system
RPD: radar planning device
RPD: reactor plant designer
RPD: retarding potential difference
RPDH: reserve shutdown planned derated hours
RPE: radial probable error
RPE: Registered Professional Engineer
RPE: remote peripheral equipment
RPE: resource planning and evaluation
RPG: radiation protection guide, Federal Radiation Council
RPG: report program generator
RPI: radar precipitation integrator
RPI: rod position indicator
R&PI: Rubber and Plastics Industry (IEEE IAS Technical Committee)
RPIS: rod position indicator system
RPL: radar processing language

RPL: Radiation Physics Laboratory
RPL: Rocket Propulsion Laboratory
RPL: running program language
RPLS: reactor protection logic system
RPM: rate per minute
RPM: reliability performance measure
RPM: resupply provisions module
RPM: revolutions per minute
RPN: reverse Polish notation
RPOA: recognized private operating agency
RPR: rapid power reduction
RPRWP: reactor plant river water pump
RPS: reactor protection systems
RPS: realtime programming system
RPS: regulatory performance summary report
RPS: remote processing service
RPS: revolutions per second
RPSM: resources planning and scheduling method
RPSMG: reactor protective system motor generator
RPT: recirculation pump trip
RPU: radio phone unit
RPU: radio propagation unit
RPV: reactor pressure vessel
RPV: remotely piloted vehicle
RQ: reportable quantity
R/Q: resolver/quantizer
RQA: recursive query analyzer
R&QA: reliability and quality assurance
RQD: rock quality designation
RQL: reference quality level
RQS: rate quoting system
RR-BNI: Reactor, Ghent, Belgium, 100 kW(th), pool-type
R/R: readout and relay
RR: receive ready
RR: receiving report
R/R: record/retransmit
RR: register to register instruction
RR: rendezvous radar
RR: repetition rate
RR: retro rockets
RR: return rate
RR: round robin
RRB: Railroad Retirement Board
R&RC: Reactors and Reactor Controls (IEEE NPSS Technical Committee)
RRE: Radar Research Establishment
R&RE: radiation and repair engineering
RRI: range-rate indicator
RRI: Rocket Research Institute
RRIS: remote radar integration station
RRMG: reactor recirculation motor generator
RRNS: redundant residue number system
RRP: reactor refueling plug
RRPI: relative rod position indication
RRPI: rotary relative position indicator
RRR: Raleigh Research Reactor, North Carolina State College
RRRV: rate of rise of restriking voltage
RRS: Radiation Research Society

RRS: radio research station
RRS: Reaction Research Society
RRS: reactor recirculation system
RRS: reactor refueling system
RRS: reactor regulation system
RRS: required response spectrum
RRS: restraint release system
RRS: retrograde rocket system
RRU: radiobiological research unit
RS: radiated susceptibilty
RS: range safety
RS: remote station
R&S: research and statistics
RS: response spectrum
RS: rigid steel (conduit)
RSAC: radiological safety analysis computer
RSAC: Reactor Safety Advisory Committee
RSARR: Republic of South Africa Research Reactor, Pretoria, South Africa. 6 MW(th)
RSB: reactor service building
RSC: rigid steel conduit
RSCIE: remote station communication interface equipment
RSCS: rod sequence control system
RSCW: Research Reactor, State College of Washington Pullman, Washington
RSD: refueling shutdown
RSD: responsible system designer
RSDP: remote shutdown panel
RSDP: remote site data processing
RSFDH: reserve shutdown forced derated hours
RSH: reserve shutdown hours
RSI: rationalization, standardization and integration
RSI: reactor siting index
RSI: Research Studies Institute
RSIC: Radiation Shielding Information Center, ORNL
RSIC: Redstone Scientific Information Center
RSL: Radio Standards Laboratory
RSL: received signal level
RSM: resource management system
RSMDH: reserve shutdown maintenance derated hours
RSMPS: Romanian Society for Mathematics and Physical Sciences
RSN: radiation surveillance network
RSO: Revenue Sharing Office
RSP: reactivity surveillance procedures
RSP: record select program
RSP: rotating shield plug
RSPDH: reserve shutdown planned derated hours
RSR: rapid solidification rate
RSR: reactor safety research
RSR: rod select relay
RSRI: Regional Science Research Institute
RSRS: radio and space research station
RSS: reactor safety study
RSS: reactor shutdown system (part of PPS)
RSS: remote shutdown system
RSS: root sum square
RSSF: retrievable surface storage facility

RST: reset-set trigger
RSUDH: reserve shutdown unplanned derated hours
RSUNDH: reserve shutdown unit derated hours
RSV: Diesel Run Control Solenoid Valve
rsv: reserve
RT: radiographic test
RT: radio telephony
RT: rated time
RT: reactor trip
R/T: real time
RT: receiver transmitter
RT: reduction table
RT: regression testing
RT: remote terminal
RTAC: Roads and Transportation Association of Canada
RTB: read tape binary
RTB: Rural Telephone Bank
RTC: reader tape contact
RTC: real-time command
RTC: real-time computer
RTC: removable top closure
RTCA: Radio Technical Commission for Aeronautics
RTCC: real-time computer complex
RTCF: real-time computer facility
RTCM: Radio Technical Commission for Marine Service
RTCS: real-time computer system
RTCU: real-time control unit
RTD: read tape decimal
RTD: real-time display
RTD: resistance-temperature detector(probe)
RTDC: real-time data channel
RTDHS: real-time data handling system
RTE: real-time executive
RTE: residual total elongation
RTG: radioisotope thermoelectric generator
RTGB: Reactor Turbine General Board
RTI: referred to input
RTI: Research Triangle Institute
RT/IOC: real-time input/output controller
RTIRS: real-time information retrieval system
RTK: range tracker
RTL: radioisotope transport loop
RTL: real-time language
RTL: resistor-transistor logic
RTM: real-time monitor
RTM: recording tachometer
RTM: register-transfer module
RTMA: Radio-Television Manufacturers Association, now EIA
RTMOS: real-time multiprogramming operating system
R/T Net: radio telephone network
RTO: referred to output
RTOS: real-time operating system
RTP: reactor thermal power
RTP: real-time peripheral
RTP: requirement and test procedures
RTPH: round trips per hour

RTR: return and restore status
RTS-1: Reactor, Pisa, Italy, 5 MW(th), pool-type
RTS: (Kinetics) reactor computer code
RTS: radar tracking station
RTS: reactive terminal service
RTS: reactor trip system
RTS: request to send
RTSD: Resources and Technical Services Division
RTSS: real-time scientific system
RTST: radio technician selection test
RTT: radioteletypewriter
RTTDS: real-time telemetry data system
RTTV: real-time television
RTTY: radio teletypewriter
RTU: remote terminal unit
RTV: room temperature vulcanizing (silicone rubber)
RTV: run time variable
RTWS: raw tape write submodule
R&U: repairs and utilities
RU: reproducing unit
RUBEOLE: *Reacteua Legerement Enrichi* , en-
riched-uranium, beryllium-oxide-moderated, zero-
power reactor, France
RUDH: reserve shutdown unplanned derated hours
RUDH1: reserve shutdown Class 1 unplanned derated hours
RUDH2: reserve shutdown Class 2 unplanned derated hours
RUDH3: reserve shutdown Class 3 unplanned derated hours
RUDH4: 5.15.2 reserve shutdown Class 4 unplanned derated hours
RUNDH: 5.9.2 reserve shutdown unit derated hours
Rural Elec/IAS: Rural Electric
RUSH: remote use of shared hardware
RV-1: *Reactor Venezolano No. 1* , Caracas, Venezuela, 3 MW(th), pool-type
RV: reactor vessel
RVA: reactive voltampere meter
RVA: reliability variation analysis
RVIA: Recreational Industry Vehicle Association
RVIS: reactor and vessel instrumentation system
RVLIS: reactor vessel water level indication system
RVM: reactive voltmeter
RVR: runway visual range
RVSS: reactor vessel support system
RW: resistance welding
RWCS: reactor water cleanup system
RWCU: reactor water clean-up unit
RWE: *Rheinische - Westfallishes Electricitaet* ; largest electric utility in West Germany and active sponsor of nuclear power programs
RWG: Roebling Wire Gage
RWM: rectangular wave modulation
RWM: rod worth minimizer
RWMA: Resistance Welder Manufacturers Association
RWP: radiation work permit
RWP: reactor work permit
RWS: radwaste system

RWSS: river water supply system
RWST: refueling water storage tank
RX: reactor
RX: receive serial line
RZ: return to zero
RZL: return to zero level
RZM: return to zero mark

S

S-RD: shipper-receiver difference
s: second
S: slave; set bit; serial data, timing independent of parallel bus (type of line)
S: supervisory format
S: supervisory function bit
S: switch
S.A.: *"Belgonucleaire"* or *Societe Belge pour l'Indus-trie Nucleaire*
SA: safety analysis
SA: sense amplifier
SA: source address
SA: spectrum analyzer
SA: spin axis
SA: stress anneal
S/A: subassembly
SA: successive approximation
SA: symbolic assembler
SAA: Standards Association of Australia
SAAEB: South African Atomic Energy Board
SAB: Scientific Advisory Board
SAB: System Advisory Board
SABE: Society for Automating Better Education
SABE: Society for Automation in Business Education
SABME: set asynchronous balanced mode extended
SABRE: sales and business reservations done electronically
SABRE: secure airborne radar bombing equipment
SABS: South African Bureau of Standards
SAC: Scientific Advisory Committee
SAC: semiautomatic coding
SAC: Society for Analytical Chemistry
SACDIN: Strategic Air Command digital information
SACMAPS: selective automatic computational matching and positioning system
SAD: safety assurance diagram
SADA: seismic array data analyzer
SADAP: simplified automatic data plotter
SADC: sequential analog-digital computer
SADE: superheat advanced demonstration experiment
SADIC: solid-state analog-to-digital computer
SADIE: scanning analog-to-digital input equipment
SADIE: semiautomatic decentralized intercept environment
SADIE: sterling and decimal invoicing electronically
SADR: six-hundred-megacycle air-defense radar
SADSAC: sampled data simulator and computer
SADSAC: Seiler algol digitally simulated analog computer
SAE: shaft-angle encoder

SAE: Society of Automotive Engineers
SAEH: Society for Automation in English and the Humanities
SAFA: Society for the Fine Arts
SAFEA: Space and Flight Equipment Association
SAFER: special aviation fire and explosion reduction
SAFI: semiautomatic flight inspection
SAFOC: semiautomatic flight operations center
SAG: Senior Advisory Group (IAEA term)
SAG: standard address generator
SAGA: Studies Analysis and Gaming Agency
SAGE: semiautomatic ground environment
SAGMOS: self aligning gate metal oxide semiconductor
SAHYB: simulation of analog and hybrid computers
SAID: safety analysis input data
SAIEE: South Africa Institute of Electrical Engineers
SAIL: geometrical transport code for computer
SAILS: simplified aircraft instrument landing system
SAIMS: selected acquisitions information and management system
SAINT: satellite interceptor
SAIS: South African Interplanetary Society
SAKI: solatron automatic keyboard instructor
SAL: Supersonic Aerophysics Laboratory
SAL: symbolic assembly language
SAL: systems assembly language
SALC: software acquisition life cycle
SALE: safeguards analytical laboratory evaluation
SALE: simple algebraic language for engineers
SALE: Society for Airline Meteorologists
SALT: strategic arms limitation talks
SAM-D: surface-to-air missile development
SAM: safety activation monitor
SAM: School of Aerospace Medicine
SAM: selective automonitoring
SAM: semantic analyzing machine
SAM: semiautomatic mathematics
SAM: sequential access memory
SAM: sequential access method
SAM: simulation of analog methods
SAM: Society for Advancement of Management
SAM: software acquisition management
SAM: sort and merge
SAM: substitute alloy material; code name for gaseous diffusion process work at Columbia University for Manhattan Project
SAM: surface-to-air missile
SAM: symbolic and algebraic manipulation
SAM: system activity monitor
SAM: systems analysis module
SAMA: Scientific Apparatus Makers Association
SAME: Society of American Military Engineers
SAMI: socially acceptable monitoring instrument
SAMIS: structural analysis and matrix interpretive system
SAMMIE: scheduling analysis model for mission integrated experiments
SAMOS: satellite and missile observation system
SAMP: sampler

SAMPE: Society for the Advancement of Material & Process Engineering
SAMPE: Society of Aerospace Material and Process Engineers
SAMS: satellite automonitor system
SAMSO: Space and Missile Systems Organizations
SAMTEC: Space and Missile Test Center
SANFM: source range neutron flux monitor
SANOVA: simultaneous analysis of variance
SANZ: Standards Association of New Zealand (Formerly NZSI)
SAO: Smithsonian Astrophysical Observatory
SAP: service access point
SAP: share assembly program
SAP: systems assurance program
SAPE: Society for Professional Education
SAPHIR: light-water medium power research reactor Wurenlingen, Switzerland
SAPIR: system of automatic processing and indexing of reports
SAR-1: Research reactor, Garching, Germany, 1 kw(th)
SAR-2: Research reactor, Karlsruhe, Germany, 10 w(th)
SAR-3: Research reactor, Graz, Austria, 1 kW(th)
SAR: safety analysis report
SAR: storage address register
SAR: successive approximation register
SAR: synthetic aperture radar
SARAH: search and rescue and homing
SAREF: safety research experiment facilities
SARPS: standards and recommended practices
SARS: single-axis reference system
SARUC: Southeastern Association of Regulatory Utility Commissioners
SAS: secondary alarm station
SAS: security agency study
SAS: small astronomy satellite
SAS: Society for Applied Spectroscopy
SAS: statistical analysis system
SAS: surface active substances
SASM: Society for Science and Mathematiics
SASS: Society for the Social Sciences
SASTU: signal amplitude sampler and totalizing unit
SAT: Society of Acoustic Technology
SAT: stabilization assurance test
SAT: stepped atomic time
SATAN: sensor for airborne terrain analysis
SATANAS: semiautomatic analog setting
SATCO: signal automatic air-traffic control
SATCOM: Satellite Communication Agency (DOD)
SATCOM: Scientific and Technical Communication Committee
SATIF: Scientific and Technical Information Facility
SATIN: SAGE air-traffic integration
SATIRE: semiautomatic technical information retrieval
SATNUC: *Societe pour les Applications Techniques dans le Domaine del'Energie Nucleaire* , Paris, interested in the chemical aspects of atomic energy

SATO: self-aligned thick oxide
SATRAC: satellite automatic terminal rendezvous and coupling
SAVE: Society of American Value Engineers
SAVE: system for automatic value exchange
SAVITAR: Sanders Associates video input/output terminal access resource
SAVS: safeguards for area ventilation system
SAVS: status and verfication system
SAW: surface acoustic wave
SAWE: Society of Aeronautical Weight Engineers
SAWE: Society of Allied Weight Engineers, Inc
SAWP: Society of American Wood Preservers
SB: secondary battery
SB: serial binary
SB: sideband
SB: sleeve bearing
SB: straight binary
SB: synchronization bit
SBA: Small Business Administration
SBA: standard beam approach
SBAC: Society of British Aerospace Companies
SBAFWP: standby aux. feed water pump
SBD: Schottky-barrier diode
SBE: Society of Broadcast Engineers
SBFM: silver-band frequency modulation
SBFU: standby filter unit
SBGTS: standby gas treatment system
SBH: switch busy hour
SBK: single-beam klystron
SBLC: standby liquid control
SBM: system balance measure
SBP: Society of Biological Psychiatry
SBR-1: Soviet Breeder Reactor BR-1, Obninsk, USSR
SBR-2: Soviet Breeder Reactor BR-2, Obninsk, USSR (dismantled in 1957)
SBR-5: Experimental fast power reactor, Soviet Breeder Reactor BR-5, Obninsk, USSR
SBS: satellite business system
SBS: stimulated Brillouin scattering
SBT: surface-barrier transistor
SBUV: solar backscatter ultraviolet experiment
SBV(S): shield building vent (system)
SBX: S-band transponder
SC: saturable core
SC: search control
SC: secondary containment
SC: semiconductor
SC: shaped charge
SC: silvered copper (wire)
SC: Simulation Council
SC: single contact
SC: South Carolina
SC: speed control
SC: superimposed current
SC: suppressed carrier
SC: switching cell
SC: synchrocyclotron
SCA: secondary communications authorization
SCA: sequence control area

SCA: simulated core assembly
SCA: sneak circuit analysis
SCA: subcarrier
SCA: subsidary communications authorization
SCAD: subsonic cruise armed decoy
SCADA: supervisory control and data acquisition
SCADS: scanning celestial attitude determination system
SCADS: simulation of combined analog digital systems
SCALE: space checkout and launch equipment
SCAMA: station conferencing and monitoring arrangement
SCAMA: switching, conference, and monitoring arrangement
SCAMPS: small computer analytical and mathematical programming system
SCAN: selected current aerospace notices (NASA)
SCAN: self-correcting automatic navigation
SCAN: stock-market computer-anmswering network
SCAN: student career automated network
SCAN: switched circuit automatic network
SCANS: scheduling control and automation by network systems
SCAP: silent compact auxiliary power
SCAR: satellite capture and retrieval
SCAR: Scientific Committee on Antarctic Research
SCAR: submarine celestial altitude recorder
SCARABEE: French test reactor similar to SLSF
SCARF: Santa Cruz Acoustic Range Facility
SCARF: scatter neutron for shield reactor physics computer code
SCAT: share compiler assembler and translator
SCAT: space communication and tracking
SCAT: speed command of attitude and thrust
SCAT: supersonic commercial air transport
SCATE: Stromberg-Carlson automatic test equipment
SCATHA: spacecraft-charging-at-high-altitudes
SCATS: simulation checkout and training system
SCBR: steam-cooled breeder reactor. USA
SCC: secondary containment cooling
SCC: simulation control center
SCC: single-conductor cable
SCC: single cotton covered (wire)
SCC: Society of Cosmetic Chemists
SCC: specialized common carrier
SCC: Standards Coordinating Committee (IEEE)
SCC: Standards Council of Canada
SCCHLL: Standards Coordinating Committee on high-level language
SCCS: sodium chemistry control system (GPL-2 Westinghouse)
ScD: scintillation detector
SCD: space control document
SCDP: Society of Certified Data Processors
SCE: situation caused error
SCE: Society of Cuban Engineers North Eastern Regional Division
SCEA: signal conditioning electronic assembly
SCEAR: Scientific Committee on the Effects of Atomic Radiation (United Nations)

SCEF: critical assembly (NASA)

SCEL: Signal Corps Engineering Laboratory

SCEL: small components evaluation loop

SCEPTRE: systems for circuit evaluation and prediction of transient radiation effects

SCEPTRON: spectral comparative pattern recognizer

SCERT: systems and computers evaluation and review technique

SCETV: South Carolina Educational Television

SCF: satellite control facility

SCF: sequence compatibility firing

SCF: SNAP critical facility, Santa Susana, California

SCF: sodium cleaning facility

SCFBR: steam-cooled fast breeder reactor

SCFM: standard cubic feet/minute

SCGA: sodium-cooled graphite assembly

SCHS: small component handling system

SCI: Society of Chemical Industry

SCI: Society of computer Intelligence

SCIC: semiconductor integrated circuit

SCIM: speech communication index meter

SCIP: scanning for information parameters

SCIP: self-contained instrument package

SCIS: safety containment isolation system

S.C.K.: *Studiecentrum voor Kernenergie (Centre d'Etude de l'Energie Nucleaire)* , Belgium

SCM: service command module

SCM: signal conditioning module

SCM: simulated core mockup/model

SCM: small core memory

SCM: software configuration management

SCM: superconducting magnet

SCMA: Systems Communications Management Association

SCMP: software configuration management plan

SCN: sensitive command network

SCN: specification change notice

SCOMO: satellite collection of meteorological observations

SCOOP: scientific computation of optimal programs

SCOPE: schedule-cost-performance

SCOPE: sequential customer order processing electronically

SCOPE: Special Committee on Paperless Entries

SCOR: Scientific Committee on Oceanographic Research

SCOR: self-calibration omnirange

SCORE: satellite computer-operated readiness equipment

SCORE: selection copy and reporting

SCORE: signal communications by orbital relay equipment

SCORPIO: sub-critical, carbon moderated reactor assembly for plutonium investigations, Winfrith, United Kingdom

SCOST: Special Committee on Space Technology

SCOTT-R: super-critical, once-thru, tube reactor experiment, General Electric Co

SCP: symbolic conversion program

SCP: system communication pamphlet

SCP: system control program

SCR: reverse-blocking triode-thyristor

SCR: selective chopper radiometer

SCR: semiconductor-controlled rectifier

SCR: sodium-cooled reactor

SCR: thyristor

SCRAM: reactor trip, simultaneous control rod injection

SCRAM: safety control rod axe man. (stood by and cut the safety rope on the first reactor)

SCRAP: super-caliber rocket-assisted projectile

SCRD: security card reader

SCRIPT: scientific and commercial interpreter and program translator

SCRIPTO: critical assembly (LASL)

SCRP: scraper

SCS: sequence coding system (N.R.C.)

SCS: silicon-controlled switch

SCS: simulation control subsystem

SCS: Society for Computer Simulation

SCS: sodium characterization system (PAL)

SCS: Soil Conservation Service

SCS: Southern Computer Service

SCS: space cabin simulator

SCS: standard coordinate system

SCSE: South Carolina Society of Engineers

SCSPLS: South Carolina Society of Professional Land Surveyors

SCT: scanning telescope

SCTE: Society of Cable Television Engineers

SCTE: Society of Carbide and Tool Engineers

SCTF: Sodium chemical technology facility

SCTI: sodium components test installation, Atomics International, Canoga Park, California

SCTL: small components test loop

SCTP: straight channel tape print

SCTPP: straight channel tape print program

SCUBA: self-contained underwater breathing apparatus

SCUMRA: *Societe Central de l'Uranium et des minerals et Metaux Radioactifs* , Paris, France

SCUP: school computer use plan

SCYLLA: LASL experimental fusion device

SD-1: Standardization Directory

SD: seasonal derating

SD: short-time waveform duration

SD: South Dakota

SD: standards development

SDA: shaft drive axis

SDA: source data acquisition

SDA: source data automation

SDA: supplier data approval

SDAD: satellite digital and analog display

SDAP: systems development analysis program

SDAS: scientific data automation system

SDC: Semiconductor Devices Council of JEDEC

SDC: shut down cooling

SDC: stabilization data computer

SDC: structural design criteria

SDCC: San Diego Computer Center

SDCC: small-diameter component cask
SDCE: Society of Die Casting Engineers
SDD: software design description
SDD: system design description
SDE: Society of Data Educators
SDE: Students for Data Education
SDF: seasonal derating factor
SDF: single-degree-of-freedom
SDFC: Space Disturbance Forecast Center (ESSA)
SDH: seasonal derated hours
SDI: selective dissemination of information
SDI: source data information
SDL: system descriptive language
SDLC: synchronous data link control
SDLC: system development life cycle
SDM: sequency-division multiplexing
SDM: shut down mode
SDM: space-division multiplexing
SDM: standardization design memoranda
SDM: statistical delta modulation
SDMC: density Monte Carlo reactor computer code
SDP: site data processor
SDPL: servomechanisms and data-processing laboratory
SDR: heterogeneous, sodium-cooled, heavy-water-moderated power reactor, 40,000 kW, Chugach Electric Associates, AK
SDR: significant deficiency report
SDR: single-drift region
SDR: SNAP developmental reactor, solid, homogeneous, enriched-uranium, hydrogen-moderated and sodium-cooled. USA
SDR: system design review
SDS: safety data sheet
SDS: scientific data systems
SDS: simulation data subsystems
SDS: software development specifications
SDS: sparse data scan
SDS: system data synthesizer
SDV: scram discharge volume
SDW: standing detonation wave
SDX: Society of Professional Journalists
SE: safety evaluation
SE: segment extender
SE: shielding effectiveness
SE: subcritical experiment
SE: systems engineer
SEA: Science and Education Administration
SEA: systems effectiveness analyzer
SEAC: Standard's eastern automatic computer
SEAC: Structural Engineers Association of Colorado
SEAIMP: solar eclipse atmospheric and ionospheric measurements project
SEAL: standard electronic accounting language
SEALS: severe environmental air launch study
SEALS: stored energy actuated lift system
SEAM: Symposium on the Engineering Aspects of Magnetohydrodynamics
SEAO: Structural Engineers Association of Oregon, Inc

SEAOC: Structural Engineers Association of California
SEAOCC: Structural Engineers Association of Central Californa
SEAONC: Structural Engineers Association of Northern California
SEAOSD: Structural Engineers Association of San Diego
SEAS: EPA's Strategic Environmental Assessment Systems - an "environmental early warning system"
SEASC: Structural Engineers Association of Southern Californa
SEASCO: South East Asia Science Cooperation Office
SEB: Source Evaluation Board
SEC: automated information systems security
SEC: Sacramento Engineers Club
SEC: safeguards equipment cabinet
SEC: Sanitary Engineering Center
SEC: secondary electron conduction
SEC: secondary emission conductivity
SEC: Securities and Exchange Commission
SEC: simple electronic computer
SECAM: sequential-and-memory (French and Soviet television system)
SECAM: sequential color and memory
SECAP: system experience correlation and analysis program
SECAR: secondary radar
sech: hyperbolic secant
SECO: sequential coding
SECOR: sequential collation of range
SECORD: secure voice cord board
SECPS: secondary propulsion system
SECS: sequential events control system
SECTAM: Southeren Conference on Theoretical and Applied Mechanics
SED: Sanitary Engineering Division
SED: spectral energy distribution
SEDD: Systems Evaluation and Development Division
SEDIT: sophisticated string editor
SEDR: systems engineering department report
SEDS: Society for Educational Data Systems
SEDS: space electronics detection system
SEE: Society of Environmental Engineers
SEE: Society of Explosives Engineers
SEE: Southeastern Electric Exchange
SEEK: systems evaluation and exchange of knowledge
SEEN: *Syndicat d'Etudes de l'Energie Nucleaire* , Brussels, Belgium
SEENY: Society of Engineers of Eastern New York
SEF: Space Education Foundation
SEFAR: sonic end fire for azimuth and range
SEFOR: Southwest Experimental Fast Oxide Reactor Fayetville, Arkansas
SEFR: Shielding Experiment Facility Reactor, NRTS
SEG: Society of Economic Geologists, Inc
SEG: Society of Exploration Geophysicists
SEG: Standardization Evaluation Group
SEG: systems engineering group

SEI: Safety Equipment Institute
SEI: systems engineering and integration
SEIA: Solar Energy Industries Assn
SEIP: strategy for exploration of the inner planets
SEIP: systems engineering implementation plan
SEIT: satellite educational and informational television
SEL: Stanford Electronics Laboratory
SEL: System Engineering Laboratories
SELNI: *Societa Electtronuclear Italiana* power reactor, 615 MW(th), Selni, Italy
SELR: Saturn engineering liaison request
SELRIP: Selected Release Improvement Program
SEM: scanning electron microscope
SEM: singularity expansion method
SEM: standard estimating module
SEMA: *Societe d'Economie et de Mathematiques Appliques* (France)
SEMI: Semiconductor Equipment and Materials Institute
SEMIRAD: secondary electron-mixed radiation dosimeter
SEMLAM: semiconductor laser amplifier
SEMLAT: semiconductor laser array techniques
SEMS: severe environmental memory system
SENA: *Societe d'Energie Nucleaire Franco-Belge des Ardennes* power reactor, 825 MW(th), Chooz, France
SENL: standard equipment nomenclature list
SENN: *Societa Electtronucleare Nazionale* , Italy, Garigliano nuclear power station, 500 MW(th)
SENTA: *Societe d'Etudes Nucleaires et de Techniques Advances* , Paris, France
SENTOS: sentinel operating system
SEP: an association of electric utilities active in nuclear power development, Arnheim, Netherlands
SEP: Society of Engineering Psychologists
SEP: space electronic package
SEP: standard electronic package
SEP: star epitaxial planar
SEP: systematic evaluation program
SEPA: Southeastern Power Administration
SEPOL: settlement problem-oriented language
SEPOL: soil-engineering problem-oriented language
SEPS: service environment power system
SEPS: service module electrical power system
SER: safety evaluation report
SER: Sandia Engineering Reactor, Sandia Base, New Mexico
SER: significant event report
SER: SNAP experimental reactor
SERAI: *Societe d'Applications Pour l'Industrie* , Belgian syndicate of engineering firms
SERAPE: simulator equipment requirements for accelerating procedural evolution
SERB: study of enhanced radiation belt
SERC: Software Engineering Research Center
SERC: Southeastern Electric Reliability Council
SERC: Southeastern Power Pool
SEREL: *Societe d'Exploitation et de Recherches Electioniques*

SERF: Sandia Engineering Reactor Facility, 5 MW(th), tank-type
SERI: Solar Energy Research Institute
SERL: Services Electronics Research Laboratory
SERPS: service propulsion system
SERT: single-electron rise time
SERT: Society of Electronic and Radio Technicians
SERT: space electrical rocket test
SES: Society of Engineering Science
SES: Solar Energy Society
SES: Standards Engineering Society
SES: Strategic Engineering Survey
SESA: Society for Environmental Stress Analysis
SESA: Society for Experimental Stress Analysis
SESE: primary element, seismic
SESE: secure echo-sounding equipment
SESL: Space Environmental Simulation Laboratory
SESOME: service, sort and merge
SESR: recorder, seismic
SE Supp: safety evaluation supplement
SET: self-extending translator
SET: software engineering terminology
SET: solar energy thermionic (conversion system)
SET: space electronics and telemetry
SETA: simplified electronic tracking apparatus
SETAB: sets tabular material (photocomposition)
SETAR: serial event time and recorder
SETF: SNAP experimental test facility
SETI: search for extraterrestrial intelligence
SETU: *Societe d'Etudes et de Travaux pour l'Uranium* , Paris, formed to build a uranium fabrication plant at Narbonne on behalf of the French AEC
SEURE: systems evaluation code under radiation environment
SEVAS: secure voice access system
SEW: sonar early warning
SEWL: sea wall
SF: sampled filter
SF: service factor
SF: single-frequency
S/F: store and forward
SFA: Scientific Film Association
SFA: single failure analysis
SFACS: steam & feedwater rupture control system
SFAR: special federal aviation regulation
SFAR: system failure analysis report
SFAS: safety feature activation system
SFB: semiconductor functional block
SFBAEC: San Francisco Bay Area Engineering Council
SFC: Solar Forecast Center (Air Force)
SFCS: secondary flow control system
SFD: sudden frequency deviation
SFD: system function description
SFE: Society of Fire Engineers
SFENA: *Société Francaise D'Equipments pour la Navigation Aerienne*
SFERICS: atmospherics (contraction)
SFERNICE: *Société Francaise del'Electro - Resistance S.A*

SFF: Solar Forecast Facility (Air Force)
SFIT: Swiss Federal Institute of Technology
SFP: spent fuel pit
SFP: spent fuel pool
SFPE: Society of Fire Protection Engineers
SFSA: Steel Founders Society of America
SFSP: spent fuel storage pool
SFT: simulated flight test
SFTS: standard frequency and time signals
SG: safety guide (IAEA term)
SG: screen grid (electrode)
SG: standing group
SG: steam generator
SGAE: *Studiengesellschaft fuer Atomenergie* , Vienna (implements Austria's nuclear program)
SGCF: SNAP generalized critical facility, Santa Susana, California
SGDF: supergroup distribution frame
SGFP: steam generator feed pump
SGHWR: steam generating heavy water reactor
SGIS: steam generator isolation signal
SGLS: space ground link system
SGR: self-generation reactor
SGR: sodium-graphite reactor
SGRCA: sodium graphite reactor critical assembly
SGS: Society for General Systems
SGS: symbol generator and storage
SGSR: Society for General Systems Research
SGSVDV: steam generator stop valve dump valve
SGTR: steam generator test rig
SGTS: standby gas treatment
SH-1: stationary high Power reactor No. 1 (Guam)
S/H: sample and hold
SH: service hours
SHA: sidereal hour angle
SHA: sodium hydroxide addition
SHA: Software Houses Association (GB)
SHA: solid homogenenous assembly
SHARE: systems for heat and radiation energy
SHARP: stationary high-altitude relay platform [NYT,7/21/87]
SHCA: critical assembly (KAPL)
SHE: semi-homogeneous experiment
SHEP: solar high-energy particles
SHERWOOD: the AEC program in thermonuclear energy
SHF: storage-handling facility
SHF: superhigh frequency
SHG: second-harmonic generation
SHIEF: shared information elicitation facility
SHIELD: Sylvania high intelligence electronic defense
SHIRAN: S-band of high-precision short-range electronic navigation
SHIRTDIF: storage, handling, and retrieval of technical data in image formation
SHL: system handshake logic
SHM: Society for Hybrid Microelectronics
SHODOP: short-range Doppler
SHORAN: short-range aid to navigation
SHORAN: short-range navigation

SHOT: Society for the History of Technology
SHPE: Society of Hispanic Professional Engineers
SHPS: sodium hydroxide purge system
SHRS: supplementary heat removal system
SHSMB: Safety & Health Standards Management Board
SHT: Society for the History of Technology
SHTC: short time constant
SHTL: small heat-transfer loop
SHTR: steam heater
SI: International System of Units
SI: safety injection
SI: safety inspection
SI: segment interconnect
SI: Smithsonian Institute
SI: special instructions
SIA: Semiconductor Industry Association
SIA: subminiature integrated antenna
SIA: Swiss Society of Engineers and Architects
SIA: system integration area
SIAD: symbolic input attribute decomposition
SIAM: signal information and monitoring
SIAM: Society for Industrial and Applied Mathematics
SIAS: safety injection actuation signal
SIAT: single integrated attack team
SIB: satellite ionospheric beacons
SIC: Science Information Council
SIC: semiconductor integrated circuit
SICEJ: Society of Instrument and Control Engineers of Japan
SICN: *Société Industrielle de Combustibles Nucleaires* , Paris (interested in nuclear fuels)
SICO: switched in for checkout
SICS: safety injection system
SID: silicon imaging device
SID: Society for Information Displays
SID: sodium ionization detector
SID: standard instrument departure
SID: sudden ionospheric disturbance
SID: syntax improving device
SIDASE: significant data selection
SIDS: stellar inertial doppler system
SIE: Science Information Exchange
Siemens Forsch u Entwick Ber: *Siemens Forschungs- und Entwicklungsberichte* (Berlin)
SIEPR: Electrical Engineers Association of Puerto Rico
SIEPR: *Sociedad du Ingenieros Estructurolus de Puerto Rico Associated*
SIFT: share internal Fortran translator
SIG: submarine intermediate reactor mark A (formerly SIR)
SIGGEN: signal generator
SIGSAM: special interest group on symbolic and algebraic manipulation
SII: Standards Institute of Israel
SIL: safety information letter
SIL: steam isolation line
SILO: silo

SILOE: 10-MW swimming-pool research reactor. CEA Grenoble, France

SIMCHE: simulation and checkout equipment

SIMCOM: simulator compiler

SIMCON: scientific inventory management and control

SIMD: single input multiple data stream

SIMD: single instruction multiple data

SIMEA: *Societa Italiana Meriadionale per l'Energia Atomica* ; gas-cooled reactor project, 200 MW(e), Latina, Italy

SIMICORE: simultaneous multiple image correlation

SIMILE: simulator of immediate memory in learning experiments

SIMM: symbolic integrated maintenance manual

simp: specific impulse

SIMPAC: simplified programming for acquisition and control

SIMS: single-item, multisource

SIMS: symbolic integrated maintenance system

SINAD: signal plus noise plus distortion to noise plus distortion ratio

SINB: Southern Interstate Nuclear Board

sinh: hyperbolic sine

SINS: ship's inertial marine navigational system

SIOP: selector input/output processors

SIOP: single integrated operations plan

SIOUX: sequential iterative operation unit X

SIP: safety injection pump

SIP: single-in-line package

SIP: symbolic input program

SIPB: safety injection permissive blocks

SIPI: Scientist's Institute for Public Information

SIPOP: satellite information processor operational program

SIPROS: simultaneous processing operating system

SIPS: simulated input preparation system

SIR: selective information retrieval

SIR: semantic information retrieval

SIR: signal-to-interference ratio

SIR: simultaneous impact rate

SIR: statistical information retrieval

SIR: submarine intermediate reactor

SIR: symbolic input routine

SIRA: Safety Investigation Regulations

SIRA: Scientific Instrument Research Association

Sire: satellite infrared experiment

SIRS: salary information retrieval system

SIRS: satellite infrared spectrometer

SIRSA: Special Industrial Radio Service Association

SIRU: strapdown inertial reference unit

SIRWT: safety injection reserve water tank

SIS: safety injection signal

SIS: safety injection system

SIS: satellite interceptor system

SIS: semiconductor-insulator-semiconductor

SIS: shorter interval scheduling

SIS: simulation interface subsystem

SISD: single instruction single data (stream or serial processors)

SISI: surveillance and in-service inspection

SISS: single item, single source

SISS: submarine integrated sonar system

SIT: safety injection tank

SIT: Society of Instrument Technology

SIT: software integration test

SITC: Standard International Trades Classification

SITE: spacecraft instrumentation test equipment

SITJ: Union of Engineers and Technicians of Yugoslavia

SITS: sage intercept target simulation

SITT: Systems Integration of Triad Technology

SITVC: secondary injection thrust vector control

SIXPAC: system for inertial experiment priority and attitude control

SIZZLE: fast and thermal reactor burnup computer code

SJ: Simulation Journal

SJAE: steam air ejector (valve)

SJAE: steam jet air ejector

SJCC: Spring Joint Computer Conference

SJCM: Standing Joint Committee on Metrication

SKA: *Studienkommission for Atomenergie* , Switzerland; National Advisory Committee on Nuclear Energy

SL-1: Stationary Low Power Plant, NRTS (ALPR), Idaho; power reactor, 3 MW(th) (dismantled)

SL: safety limit

SLA: spacecraft LM adapter

SLA: Special Libraries Association

SLAC: Stanford Linear Accelerator Center, Stanford Univ., Stanford, California

SLAE: supplementary leak collection and release system

SLAET: Society of Licensed Aircraft Engineers and Technicians

SLAM: space-launched air missile

SLAMS: simplified language for abstract mathematical structures

SLANT: simulator landing attachment for night landing training

SLAR: side-looking airborne radar

SLASH: Seiler Laboratory Algol simulated hybrid

SLATE: stimulated learning by automated typewriter environment

SLB: side-lobe blanking

SLBM: submarine-launched ballistic missile

SLBMDWS: submarine-launched ballistic missile detection and warning system

SLC: side-lobe cancellation

SLC: simulated linguistic computer

SLC: Stanford Linear Collider

SLCB: single-line color bar

SLCC: Saturn launch control computer

SLCM: submarine launched cruise missile

SLCRS: supplementary leak collection and release system

SLCS: standby liquid control system

SLD: simulated launch demonstration

SLD: source language debug

SLE: Society of Logistics Engineers

SLEAT: Society of Laundry Engineers and Allied Trades

SLEEP: scanning low-energy electron probe

SLEP: service life extension program

SLEW: static load error washout system

SLI: sea level indicator

SLI: steam line isolation

SLIC: selective listing in combination

SLIDE: pool-type reactor, 10 MW(th), Grenoble, France

SLIP: symmetric list processor

SLIS: shared laboratory information system

SLIV: steam line isolation valve

SLKR: slaker

SLM: statistical learning model

SLO: swept local oscillator

SLP: segmented level programming

SLP: source language processor

SLR: side-looking radar

SLS: side-lobe suppression

SLS: side-looking sonar

SLSF: Sodium Loop Safety Facility

SLT: simulated launch test

SLT: solid logic technique

SLT: solid logic technology

SLUC: standard level user charge

SLWL: straight-line wavelength

SM-1: stationary medium power plant (formerly APPR-1), Ft. Belvoir, Virginia; power reactor, 10 MW(th)

SM-1A: stationary medium power plant, No. 1A, (formerly APR-1a) Ft. Greeley Alaska; power reactor, 20.2 MW(th)

SM-2: transuranium production and materials testing reactor, tank-type, 10 MW(th) New Melekess, USSR

SM-50: 50 MW(th), reactor, Moscow, USSR

SM: sequence and monitor

SM: service module

SM: shared memory

SM: strategic missile

SMA: Science Masters Association

SMA: Screen Manufacturers Association

SMAC: special mission attack computer

SMACNA: Sheet Metal and Air Conditioning Contractors National Association

SMALGOL: small computer algorithmic language

SMAP: Software Management and Assurance Program (NASA)

SMART: systems management analysis, research, and test

SMC: Scientific Manpower Commission

SMC: segmented maintenance cask

SMC: Systems, Man, and Cybernetics (IEEE Society)

SMC: Systems Science and Cybernetics

SMCC: simulation monitor and control console

SMD: systems measuring device

SMDC: Superconductive Materials Data Center

SMDR: station message detailed recording

SME-AIME: Society of Mining Engineers of AIME

SME: Society of Manufacturing Engineers

SME: Society of Military Engineers

SMEMA: Surface Mount Equipment Manufacturers Association

SMET: simulated mission endurance testing

SMF: system measurement facility

SMG: Spacecraft Meteorology Group

SMGP: strategic missile group

SMIS: Society for Management Information Systems

SML: symbolic machine language

SMLM: simple-minded learning machine

SMM: standard method of measurement

SMMP: standard methods of measuring performance

SMMT: Society of Motor Manufacturers and Traders

SMOG: special monitor output generator

SMPS: simplified message processing simulation

SMPTE: Society of Motion Picture and Television Engineers, Inc

SMR: shield mock-up reactor (proposed fast reactor experimental facility), United Nuclear Co., New York

SMR: solid moderated reactor

SMRD: spin motor rate detector

SMRE: Safety in Mines Research Establishment

SMRE: Surface Mining Reclamation and Enforcement Office

SMS: storage management system

SMS: strategic missile squadron

SMS: surface missile system

SMS: synchronous-altitude meteorological satellite

SMSA: standard metropolitan statistical area

SMSAE: surface missile system availability evaluation

SMT: service module technician

SMT: square mesh tracking

SMTI: selective moving target indicator

SMTI: sodium mechanisms test installation

SMUD: Sacramento Municipal Utility District

SMWG: strategic missile wing

SN: semiconductor network

S/N: serial number

S/N: signal-to-noise (ratio)

SN: sine of the amplitude

SN: siren

SNA: systems network architecture

SNAE: Society of Norwegian American Engineers

SNAFU: situation normal, all fouled up

SNAK: delay group reactor kinetics computer code

SNAME: Society of Naval Architects and Marine Engineers

SNAP: simplified numerical automatic programmer

SNAP: space nuclear auxiliary power

SNAP: Systems for Nuclear Auxiliary Power (even numbers are reactor units; odd numbers are isotopic generators)

SNDT: Society of Nondestructive Testing

SNEAK: proposed seed zero-power research fast reactor, *Kernforschungszentrum* . Karlsruhe, Germany

SNEMSA: Southern New England Marine Sciences Association

SNF: system noise figure
SNG: synthetic natural gas
SNI: signal-to-noise improvement
SNIAS: *Société Nationale Industrielle Aerospatiale*
SNL: standard nomenclature list
SNM: Society of Nuclear Medicine
SNM: special nuclear material
SNM: spent nuclear material
SNMP: spent nuclear material pool
SNPM: standard and nuclear propulsion module
SNPO: space nuclear propulsion office
SNR: signal-to-noise ratio
SNR: supplier nonconformance report
SNS: simulated network simulations
SNT: Society for Nondestructive Testing
SNUPPS: standardized nuclear unit power plant system
SO: slow operate (relay)
SOA: safe operating area
SOAP: self-optimizing automatic pilot
SOC: separated orbit cyclotron
SOC: set overrides clear
SOC: simulation operations center
SOC: specific optimal controller
SOCCS: summary of component control status
SOCO: switched out for checkout
SOCOM: solar optical communications system
SOCR: sustained operations control room
SOCS: spacecraft orientation-conrol system
SOD: *Société d'etudes pour l'obtention du Deuterium,* Paris, France
SOD: small object detector
SODA: source oriented data acquisition
SODAR: sound detecting and ranging
SODAS: structure oriented description and simulation
SODERN: *Société d'Etudes et Realisations Nucléaires* , Suresnes (interested in fabricating nuclear equipment and processing materials)
SOE: significant operating experience
SOER: significant operating event report
SOERO: small orbiting earth resources observatory
SOFAR: sound fixing and ranging
SOFAR: sound fusing and ranging
SOFNET: solar observing and forecasting network
SOFT: simple output format translator
SOH: start-of-heading character
SOI: specific operating instruction
SOI: standard operating instruction
SOICS: summary of installation control status
SOL-GEL: bulk-oxide fuel-cycle process
SOL: simulation oriented language
SOL: solenoid
SOL: systems oriented language
SOLAR: serialized on-line automatic recording
SOLAS: safety of life at sea
SOLE: Society of Logistics Engineers
SOLID: self-organizing large information dissemination system
solion: solution ion
SOLIS: symbionics on-line information system

SOLO: selective optical lock-on
SOLR: sidetone objective loudness rating
SOLRAD: solar radiation
SOLV: solenoid valve
SOLV: super-open-frame low voltage
SOM: shift operations manager
SOM: start of message
SOMADA: self-organizing multiple-access discrete address
SONAC: sonar nacelle (sonacelle)
SONAR: sound navigation and ranging
SONCM: sonar countermeasures and deception
SONCR: sonar control room
SONIC: system-wide on-line network for informational control
SOP: simulation operations plan
SOP: special operating procedure
SOP: standard operating procedure
SOP: strategic orbit point
Sophir: pool-type reactor, Wurinlingen, Switzerland, 1 MW(th)
SOPM: standard orbital parameter message
SOPR: Spanish Open Pool Reactor, 3 MW(th), Madrid, Spain
SOR: Society of Rheology
SOR: start of record
SOR: successive overrelaxation
SORA: *Sorgento Rapido* , uranium, sodium cooled, research reactor
SORC: Site Operations Review Committee
SORIN: *Societa Ricerche Impianti Nucleari* , Italian nuclear subsidiary of Fiat and Montecatini
SORIN: Reactor at Saluggia, Italy (near Milan)
SORTI: satellite orbital track and intercept
SORTIE: supercircular orbital reentry test integrated environment
SOS: share operating system
SOS: silicon on sapphire
SOS: speed of service
SOS: station operating supervisor
SOS: symbolic operating system
SOSI: shift in, shift out
SOSUS: sound surveillance system
SOT: syntax-oriented translator
SOTAS: standoff target acquisition system
SOTIM: sonic observation of the trajectory and impact of missiles
SOTUS: sequentially operated teletypewriter universal selector
SOUTHEASTCON: South-Eastern Convention (Region 3, IEEE)
SP: self-propelled
SP: service processor
SP: single pole
SP: special projects
SP: special purpose
SP: structured programming
SP: surveillance procedure
SP: symbol programmer
SP: system processor

SPA: servo power assembly
SPA: Southwestern Psychological Association
SPAC: spatial computer
SPACE: self-programming automatic circuit evaluator
SPACE: sequential position and covariance estimation
SPACE: sidereal polar axis celestial equipment
Spacecom/Com: Space Communication
SPACON: space control
SPACS: sodium purification and characterization system
SPAD: satellite position predictor and display
SPAD: satellite protection for area defense
SPADATS: space detection and tracking system
SPADE: Sparta acquisition digital equipment
SPAM: ship position and attitude measurement
SPAMS: ship position and attitude measurement system
SPAN: solar particle alert network
SPAN: statistical processing and analysis
SPAN: stored program alphanumerics
SPANRAD: superposed panoramic radar display
SPAQUA: sealed package quality assurance
SPAR: symbolic program assembly routine
SPARS: space precision attitude reference system
SPARSA: sferics pulse, azimuth, rate, and spectrum analyzer
SPARTA: spatial antimissile research tests in Australia
SPASM: system performance and activity software monitor
SPAT: silicon precision alloy transistor
SPC: Static Power Converters (IEEE IAS Technical Committee)
SPC: stored program controller (CLEM)
SPC: stored programmed command
SPC: Strategic Planning Committee
SPD: Sigma Phi Delta Fraternity
SPD: Surge Protective Devices (IEEE PES Technical Committee)
SPDP: Society of Professional Data Processors
SPDS: safe-practice data sheet
SPDT: single pole double throw (switch)
SPE: Society of Petroleum Engineers
SPE: Society of Plastics Engineers
SPE: systems performance effectiveness
SPEAR: Stanford positron-electron asymmetric ring
SPEARS: satellite photoelectric analog rectification system
SPEC: stored program educational computer
SPECON: system performance effectiveness conference
SPECVER: specification verification
SPED: supersonic planetary entry decelerator
SPEDAC: solid-state parallel expandable differential analyzer computer
SPEDE: state system for processing educational data electronically
SPEDTAC: stored program educational transistorized automatic computer
SPEE: Society for the Promotion of Engineering Education

SPEE: Society of Petroleum Evaluation Engineers
SPEED: self-programmed electronic equation delineator
SPEED: subsistence preparation by electronic energy diffusion
SPERT: schedule performance evaluation and review technique
SPERT: special power excursion reactor test
SPES: stored program element system
SPET: solid propellent electric thruster
SPF: site population factor
SPF: standard project flood
SPFP: single point failure potential
SPG: single point ground
SPG: sort program generator
SP/Hd: spool piece head
SPHE: Society of Packaging and Handling Engineers
SPI: self-paced instruction
SPI: site population index
SPI: Society of the Plastics Industry
SPI: specific productivity index
SPIA: Panamanian Society of Engineers and Architects
SPIA: Solid Propellent Information Agency
SPIAM: sodium purity in-line analytical module
SPIC: Society of the Plastics Industry of Canada
SPIDER: sonic pulse-echo instrument designed for extreme resolution
SPIE: scavenging-precipitation-ion exchange
SPIE: self-programmed individualized education
SPIE: simulated problem, input evaluation
SPIE: Society of Photo-Optical Instrumentation Engineers
spkr: speaker
SPL: software programming language
SPL: sound pressure level
SPL: spaceborne programming language
SPLIT: Sunstrand processing language internally translated
SPLL: self-propelled launcher/loader
SPM: self-propelled mount
SPM: sequential processing machine
SPM: software project management
SPM: source program maintenance
SPM: symbol processing machine
SPMS: solar particle monitoring system
SPNS: switched private network service
SPOs: system program offices
SPOT: satellite positioning and tracking
SPOUT: system peripheral output utility
SPP: surge protection package; surge protector
SPR-II: Sandia Pulsed Reactor II, Sandia Base, New Mexico
SPR: Sandia Pulsed Reactor
SPR: simplified practice recommendation
SPRA: space probe radar altimeter
SPRC: self-propelled robot craft
SPRDS: steam pipe rupture detector system
SPRI: Scott Polar Research Institute
SPRINT: solid propellant rocket intercept missile

SPRITE: solid propellant rocket ignition test and evaluation

SPRT: sequential probability ratio test

SPS: secondary propulsion system

SPS: Society of Physics Students

SPS: solar power system or satellite

SPS: speed switch

SPS: symbolic programming system

SP/SC: shield plug/support cylinder

SPSE: Society of Photographic Scientists and Engineers

SPSS: shield plug storage station

SPST: single pole single throw (switch)

SPST: spent resin storage tank

SPT: special purpose tests

SPT: symbolic program tape

SPT: symbolic program translator

SPTF: sodium-pump test facility

SPUR: space power unit reactor

SPURM: special purpose unilateral repetitive modulation

SPURT: spinning unguided rocket trajectory

SPWLA: Society of Professional Well Log Analysts, Inc

SPWY: spillway

SPX: superheat power experiment, heterogeneous, uranium, power reactor. USA

SP3T: single pole triple throw (switch)

SP4T: single pole quadruple throw (switch)

SQA: software quality assurance

SQAP: software quality assurance plan

SQR: sequence relay

SQR: service request

SQR: supplier quality representative

SQUID: Sperry quick updating of internal documentation

SQUID: superconducting quantum interference device

SQW: square wave

SR-105, SR-305: Savannah River Test Pile S

SR: safety rods

SR: saturable reactor

SR: scientific report

SR: selective ringing

SR: selenium rectifier

SR: service request

SR: shift register

SR: short range

SR: slip ring

SR: slow release (relay)

SR: Society of Rheology

SR: solid rocket

SR: sound ranging

SR: sound rating

SR: speed recorder

SR: split ring

SR: storage and retrieval

S/R: subroutine

SR: support reaction load

SRA: shop replaceable assembly

SRAM: short range attack missile

SRBII: *Société Royale Belge des Ingenieurs et des Industriels*

SRBM: short range ballistic missile

SRC: Science Research Council

SRC: sound ranging control

SRC: source range channel

SRC: spares receiving checklist

SRC: standard requirements code

SRCRA: Shipowners Refrigeration Cargo Research Association

SRD: secret restricted data

SRD: self reading dosimeter

SRD: step-recovery diode

SRDA: sodium removal development apparatus

SRDAS: service recording and data analysis system

SRDE: Signals Research and Development Establishment

SRDL: Signals Research and Development Laboratory

SRDS: standard reference data system

SRE: sodium reactor experiment, Santa Susana, California, power reactor, 20 MW(th)

SRG: statistical research group

SRH: service request handler

SRI: Southern Research Institute

SRI: Spalling resistance index

SRI: Stanford Research Institute

SRIAER: Scientific Research Institute for Atomic Energy Reactors, New Melekess, USSR

SRLY: series relay

SRM: source-range monitors

SRNFC: source range neutron flux channel

SRO: senior reactor operator

SRO: singly resonant oscillator

SRO: specification release order

SRP: standard review plan

SRR: shift register recognizer

SRR: short range recovery

SRR: software requirements review

SRR: sound recorder reproducer

SRS: simulated remote sites

SRS: sodium removal station

SRS: software requirements specification

SRS: stimulated Raman scattering

SRS: subscriber response system (cable television)

SRSA: Scientific Research Society of America

SRSK: short-range station keeping

SRSS: simulated remote sites subsystem

SRSS: square root of the sum of the squares

SRT: Society of Radiological Technologists

SRT: supporting research and technology

SRT: systems readiness test

SRU: *Société de Raffinage d'Uranium* , Paris, France

SRU: shop replaceable unit

SRV: safety relief valve

SS-CF: critical assembly (Bettis)

SS: Secret Service

SS: selector switch

SS: shift supervisor

SS: signal strength

SS: single shot

SS: slave status
SS: solid state
SS: space-simulator
SS: special source materials
SS: speed switch
SS: spin-stabilized (rockets)
SS: summing selector
SSA: Seismological Society of America
SSA: Social Security Administration
SSAP: source service access point
SSAR: site safety analysis report
SSAR: spin-stabilized aircraft rocket
SSAR: standard safety analysis report
SSB: single-sideband
SSB: Source Selection Board
SSBAM: single-sideband amplitude modulation
SSBD: single-sideboard
SSBFM: single-sideband frequency modulation
SSBN: ship submersible nuclear
SSBO: single swing blocking oscillator
SSBSC: single-switched suppressed carrier
SSBSCOM: single-sideband suppressed-carrier optical modulator
SSC: short segmented cask
SSCC: Solid State Circuits Conference
SSCR: spectral shift control reactor, B&W, concept
SSD: single-station doppler
SSDC: System Safety Development Center
SSE: safe shutdown earthquake
SSEC: selective sequence electronic calculator
SSEP: system safety engineering plan
SSESM: spent stage experimental support module
SSF: safe shutdown facility
SSF: service storage facility
SSG: small signal gain
SSGS: standard space guidance system
SSHR: safety shower
SSI: sector scan indicator
SSI: small scale integration
SSI: steady-state irradiation
SSI: storage-to-storage instruction
SSI: supplemental security income
SSIG: single signal
SSL: solid-state lamp
SSM: surface-to-surface missile
SSMTG: solid-state and molecular theory group
SSN: nuclear powered submarine
SSN: switched service network
SSNM: strategic special nuclear materials
SSNPP: small-size nuclear power plant
SSP: scientfic subroutine package
SSP: sodium sampling package (PAL)
SSP: Space Station Program (NASA)
SSP: static sodium pots
SSPC: Steel Structures Painting Council
SSPL: steady state power level
SSPM: single-sideband frequency modulation
SSPS: satellite solar power system
SSPS: solid state protection system
SSPWR: small-size pressurized-water reactor

SSR: secondary surveillance radar
SSR: separate superheater reactor
SSR: site suitability report
SSR: solid state relay
SSR: specification status report (Westinghouse)
SSR: subsynchronous resonance
SSR: synchronous stable relaying
SS & RC: Standards Screening and Review Committee
SSRC: Structural Stability Research Council
SSRS: start-stop-restart system
SSRT: subsystem readiness test
SSS: scientific subroutine system
SSS: simulation study series
SSS: special safety safeguards
SSS: strategic satellite system
SSSA: Soil Science Society of America
SSSC: surface subsurface surveillance center
SST: secondary surge tank
SST: simulated structural test
SS/T: steady-state/transient analysis
SST: subsystems test
SST: supersonic transport
SSTC: single-sideband transmitted carrier
SSTP: subsystems test procedure
SSW: synchro switch
ST: sawtooth
ST: Schmitt trigger
ST: scientific and technical
ST: single throw
STA: shuttle training aircraft
STA: stair
sta: station
STAC: Science and Technology Advisory committee for MSF
STACO: Standing Committee for the Study of the Principles of Standardization
STADAN: satellite tracking and data acquisition network (NASA)
STAF: scientific and technological applications forecast
STALO: stabilized local oscillator
STAMO: stabilized master oscillator
STAMOS: sortie turn-around maintenance operations simulation
STAMP: systems tape addition and maintenance program
STANAG: standardization agreement
STAR: Scientific and Technical Aerospace Report
STAR: self-testing and repairing
STAR: shield test air reactor
STAR: space thermionic auxiliary reactor (G.E.)
STARFIRE: system to accumulate and retrieve financial information with random extraction
STARK: Argonaut reactor converted to a fast thermal reactor, Karlsruhe, Germany
STARS: satellite telemetry automatic reduction system
START: selections to activate random testing
START: systematic tabular analysis of requirements technique

stat: electrostatic units in cgs system (prefix)

STATPAC: statistics package

STB: subsystems test bed

stby: standby

STC: satellite test center

STC: Scientific and Technical Committee

STC: sensitivity time control

STC: short time constant

STC: Society for Technical Communication, Inc

STC: standard transmission code

STC: system test complex

STCL: source-term control loop (324 Building - HEDL)

STD: salinity temperature depth

STD: subscriber trunk dialing

STDA: Selenium- Tellurium Development Association

STDM: statistical time-division multiplexing

STDM: synchronous time-division multiplexing

STDN: spaceflight tracking and data network

STE: shield test experiment, Santa Susana, California, 50 kW(th), pool-type

STE: Society of Tractor Engineers

STEM: stay time extension module

STEP: safety test engineering program (AEC), NRTS

STEP: scientific and technical exploitation program

STEP: simple transition electronic processing

STEP: simple transition to economical processing

STEP: specification to executable program

STEP: standard terminal program

STEPS: solar thermionic electric power system

STET: specialized technique for efficient typesetting

STF: safety test facility

STF: SNAP shield test facility reactor, Santa Susana, California (dismantled 1964)

STI: scientific technical information

STIC: Scientific Technical Intelligence Center

STID: Scientific and Technical Information Division (NASA)

STIF: short-term irradiation facility

STINFO: scientific and technical information

STIR: separate target illumination

STIR: SNAP shield test irradiation reactor, Santa Susana, California

STKR: stacker

STL: Schottky transistor logic

STL: Space Technology Laboratory

STL: standard telegraph level

STL: studio transmitters link

STL: system test loop

STLG: stop log

STLO: Scientific and Technical Liaison Office

STM: structural test model

STM: supersonic tactical missile

STM: system master tape

STMIS: system test manufacturing information system

STMU: special test and maintenance unit

STN: satellite tracking network (NASA)

STO: system test objectives

STORET: storage and retrieval

STORLAB: Space Technology Operations and Research Laboratory

STORM: statistically oriented matrix program

STP: selective tape print

STP: space test program

STP: standard temperature and pressure

STP: system test plan

STPF: shield-test pool facility

STPO: Science and Technology Policy Office

STR: submarine thermal reactor (Mark I, II) (now SIW)

STR: synchronous transmitter receiver

STRAD: signal transmission reception and distribution

STRC: stacker/reclaimer

STRESS: structural engineering system solver

STRIVE: standard techniques for reporting information on value engineering

STROBES: shared time repair of big electronic systems

STRUDL: structural design language

STS: satellite tracking station

STS: standard technical specifications

STS: structural transition section

STU: systems test unit

STV: small test vessel

STV: surveillance television

STW: software test workshop

STWP: Society of Technical Writers and Publishers

STX: start-of-text character

SU: service unit

SU: signal unit

SU: Sonics and Ultrasonics (IEEE Group)

SU: start up

SU: storage unit

SU: suppressor (vacuum tube)

SUA: State Universities Association

SUAK: fast sub-critical assembly, Karlsruhe, Germany

SUAS: system for upper atmospheric sounding

Sub/Pes: substations

substa: substation

SUDT: silicon unilateral diffused transistor

SUHL: Sylvania ultra-high level logic

SUI: Stanford University Institute for Plasma Research

SUM: shallow underwater mobile

SUM: surface-to-underwater missile

SUMMIT: scattering reactor physics computer code

SUMMIT: supervisor of multiprogramming, multiprocessing, interactive time sharing

SUMP: sump

SUMT: sequential unconstrained minimization technique

SUPARCO: Space and Upper Atmosphere Research Committee

SUPO: super power water boiler, Los Alamos, 25 kW(th)

SUPROX: successive approximation

SURCAL: surveillance calibration

SURGE: sorting, updating, report generating

SURIC: surface ship integrated control

SUR 100: solid, homogeneous, zero-power research reactor, *Siemens-Schuckert* , Munich-Garching, Germany

SUR 100 BE: solid, homogeneous, zero-power research reactor, *Technische Universitat* , Berlin, Germany

SUS: silicon unilateral switch

SUS: single underwater sound

SUSIE: shield test pool facility, NRTS, Idaho, 2 MW(th)

SUSIE: stock updating sales invoicing electronically

Suspop: A quasi-homogeneous aqueous suspension reactor at the KEMA laboratories in the Netherlands

SUT: reactor physics resonance computer code

SV: safety valve

SV: stop valve

SVC: supervisor calls

svce: service

SVIC: *Sociedad Venezolana de Ingenieros Consultores*

SVIC: Shock and Vibration Information Center

SVIH: *Sociedad Venezolana De Ingenieria Hidraulica Colegio de Ingenierosde Venezuela*

SVMSIF: Venezuelan Society of Soil Mechanics and Foundation Engineering

SVP: software verification plan

SVR: software verification report

SVR: supply-voltage rejection

SVTL: Services Valve Test Laboratory

SVTP: sound velocity, temperature and pressure

SW: short wave

SW: single weight

SW: Struthers Wells

SW: switchband wound (relay)

SWAC: Standards Western Automatic Computer

SWAT: sidewinder IC acquisition track

SWBP: service water booster pump

SWC: solid waste cask

SWC: surge withstand capability

SWCL: Sea Water Conversion Laboratory

SWE: Society of Women Engineers

SWET: simulated water entry test, Rover program

SWFR: slow write, fast read

SWG: British Standard Wire Gage

SWG: Stubs Wire Gage

swgr: switchgear

SWIFT: software implemented Friden translator

SWITT: surface wave independent tap transducer

SWOP: structural weight optimization program

SWP: safe working pressure

SWP: service water pump

SWPA: Southwestern Power Administration

SWPP: service water pressurization pump

SWPP: Southwest Power Pool

SWR: service water reservoir

SWR: standing wave ratio

SWRI: Southwestern Research Institute

SWS: service water system

SWST: service Water Storage Tank

SWT: supersonic wind tunnel

SWTL: surface-wave transmission line

SWU: separative work unit

SWW: severe weather warning

SX: simplex signaling

SYCA: *Syndicat d'étude des Centrales Atomique, Brussels*

SYCOM: synchronous communications

SYDAS: system data acquisition system

SYMPAC: symbolic program for automatic control

SYN: synchronous idle character

sync: synchronization, synchronous

SYNCOM: synchronous-orbiting communications satellite

SYNFAR: 2-dimensional flux transport code for computers

SYNSEM: syntax and semantics

SYNTOL: syntagmatic organization of language

sys: system

SYSGEN: systems generation

SYSIN: system input

SYSPOP: system programmed operator

SYSTRAN: systems analysis translator

SYSVER: system specification verification

T

T: tera- (prefix)

T: timing (type of line)

T: transformer

TA: target (vacuum tube)

TA: technical advisor

T&A: time and attendance (card or unit)

TA: timing for arbitration bus (type of line)

TA: turbulence amplifier

TAA: Transportation Association of America

TAAM: Tomahawk Airfield Attack Missile

TAAS: three-axis attitude sensor

TAB: tabular language

TAB: technical abstract bulletin

TAB: Technical Activities Board

TABS: terminal access to batch service

TABSIM: tabulator simulator

TABSOL: tabular systems oriented language

TAC: technical assistance contract

TAC: time-to-amplitude converter

TAC: TRANSAC assembler compiler

TAC: transistorized automatic control

TAC: translator-assembler-compiler

TAC: trapped air cushion

TACAN: tactical air navigation

TACCAR: time-averaged clutter-coherent airborne radar

TACDEN: tactical data entry device

TACF: temporary alteration control form

TACH: tachometer generator

TACL: time and cycle log

TACODA: target coordinate data

TACOL: thinned aperture computed lens

TACPOL: tactical procedure oriented language

TACR: time and cycle record

TACS: tactical air control system

TACS: technical assignment control system

TACS: test assembly conditioning station

TACT: transistor and component tester

TACV: tracked air cushion vehicle

TAD: top assembly drawing

TADIL: tactical digital information link

TADS: tactical automatic digital switching

TADS: teletypewriter automatic dispatch system

TADSS: tactical automatic digital switching system

TAE: Technician Aeronautical Engineering (Israel)

TAEC: Thailand Atomic Energy Commission for Peace

TAEM: terminal area energy management

TAF: top of active fuel

TAF: transient adaptation factor

TAG: technical advisory group

TAG: Technical Assistance Group

TAG: transient analysis generator

TAI: time to autoignition

TAIC: Tokyo Atomic Industrial Consortium

TAM: telephone answering machine

TAMIS: telemetric automated microbial identification system

TAMVEC: Texas A&M Variable Energy Cyclotron

tanh: hyperbolic tangent

TANK: transient analysis reactor engineering computer code

TAO: technical assistance order

TAP: terminal applications package

TAP: thermal analyzer program

TAP: time-sharing assembly program

TAPAC: tape automatic positioning and control

TAPE: tape automatic preparation equipment

TAPP: two-axis pneumatic pickup

TAPPI: Technical Association of the Pulp and Paper Industry, Inc

TAPS: tactical area positioning system

TAPS: turboalternator power system

TAR: technical assistance request

TAR: terrain-avoidance radar

TAR: trajectory analysis room

TARAN: tactical attack radar and navigation

TARE: telemetry automatic reduction equipment

TARFU: things are really fouled up

TARGET: thermal advanced gas-cooled reactor exploiting thorium; General Atomics reactor concept based upon HTGR

TARMAC: terminal area radar moving aircraft

TARP: Technical Assistance Research Programs

TARPS: tactical aerial reconnaisance pod system

TARS: terrain and radar simulator

TARS: three-axis reference system

TART: twin accelerator ring transfer

TAS: temperature actuated switch

TASC: terminal area sequence and control

TASCON: television automatic sequence control

TASI: time assignment speech interpolation

TAST: thermoacoustic sensing technique

TAT: thrust-augumented Thor

TATC: terminal air-traffic control

TATC: transatlantic telephone cable

TAUG: temperature average

TAVE: Thor-Agena vibration experiment

TAWAR: tactical all-weather attack requirements

TAWDS: target acquisition weapon delivery system

TB: technical bulletin

TB: terminal board

TB: time base

TB: transmitter-blocker (cell)

TBAX: tube axial

TBC: toss bomb computer

TBD: target-bearing designator

TBF: tail bomb fuse

TBI: target-bearing indicator

TbIG: terbium iron garnet

TBL: Terminal Ballistics Laboratory

TBO: time between overhaul

TBP Plant: chemical separation plant for separation of uranium solids from bismuth phosphate plants

TBX: tactical ballistic missile experiment

TC: tactical computer

TC: tantalum capacitor

TC: technical committee

TC: technical control

TC: temperature coefficient

TC: temperature compensating

TC: terminal computer

TC: test case specification

TC: test conductor

TC: test console

TC: thermocouple

TC: thermocurrent

TC: thrust chamber

T&C: time-and-charge-request telephone call

TC: time constant

TC: timed closing

TC: tracking camera

TC: transmission control

TCA: tactical combat aircraft

TCA: terminal control area

TCA: thermal critical assembly, GE, Vallecitos, California

TCAI: tutorial computer-assisted instruction

TCAT: test coverage analysis tool

TCB: task control blocks

TCB: technical coordinator bulletin

TCBM: transcontinental ballistic missile

TCC: technical computing center

TCC: Telecommunications Coordinating Committee

TCC: temperature coefficient of capacitance

TCC: test control center

TCC: test controller console

TCC: transfer channel control

TCD: telemetry and command data

TCD: thyratron core driver

TCD: transistor-controlled delay

TCE: telemetry checkout equipment

TCEA: Training Center for Experimental Aerodynamics

TCED: thrust control exploratory development

TCF: technical control facility
TCF: transformer correction factor
TCG: tune-controlled gain
TCI: telemetry components information
TCI: terrain clearance indicator
TCI: Theoretical Chemistry Institute
TCL: time and cycle log
TCL: transfer chemical laser
TC/LD: thermocouple/lead detector
TCM: terminal-to-computer multiplexer
TCMF: touch calling multifrequency
TCP: test checkout procedure
TCP: thrust chamber pressure
TCP: traffic control post
TCPC: tab card punch control
TCR: temperature coefficient of resistance
TCS: telephone conference summary
TCS: terminal countdown sequencer
TCS: terminal count sequence
TCS: thermal conditioning service
TCS: traffic control station
TCS: transportation and communications service
TCSC: trainer control and simulation computer
TCT: translator and code treatment frame
TCU: tape control unit
TCU: teletypewriter control unit
TCU: trunk coupling unit
TCV: temperature control valve
TCV: terminal configured vehicle
TCWG: telecommunications working group
TCXO: temperature-compensated crystal oscillator
TCXO: temperature-controlled crystal oscillator
TD: tabular data
TD: temperature differential
TD: terminal distributor
TD: test design specification
TD: testing device
TD: thoria dispersed
TD: time delay
TD: transfer dolly (BLTC)
T&D: Transmission and Distribution (IEEE PES Technical Committee)
TD: transmit data
TD: transmitter distributor
TD: tunnel diodes
TDA: target docking adapter
TDA: tracking and data acquisition
TDA: tunnel diode amplifier
TDC: thermal diffusion coefficient
TDC: time delay closing
TDC: 2-dimensional finite-cylinder transport code for computer
TDCM: transistor driver core memory
TDCO: torpedo data computer operator
TDD: target detection device
TDD: technical data digest
TDD: test design description
TDDL: time division data link
TDDR: technical data department report
TDEC: Technical Division and Engineering Center

TDEFWP: turbine driven emergency feed water pump
TDF: task deletion form
TDF: two-degree-of-freedom
TDG: test date generator
TDH: total dynamic head
TDI: Tool and Die Institute
TDL: tunnel-diode logic
TDM: time division multiplexer
TDM: torpedo detection modification
TDMA: time division multiple access
TDMS: telegraph distortion measurement set
TDMS: time-shared data management system
TDN: target doppler nullifier
TDO: time delay opening
TDOS: tape/disk operating system
TDP: technical development plan
TDR: target discrimination radar
TDR: technical data relay
TDR: technical data report
TDR: time delay relay
TDR: time domain reflectometry
TDR: torque differential receiver
TDRS: tracking and data relay satellite
TDRSS: tracking and data relay satellite system
TDS: target designation system
TDS: technical data system
TDS: test data sheet
TDS: time-domain spectroscopy
TDS: total dissolved solids
TDS: tracking and data system
TDS: translation and docking simulator
TDT: target designation transmitter
TDV: technology development vehicle
TDX: thermal demand transmitter
TE: test equipment
TE: thermal element
TE: thermal expansion load
TE: thermoelectric
TE: totally enclosed
TE: transverse electric (field)
TEA: Technical Engineers Association
TEA: transferred electron amplifier
TEA: transversely-excited atmospheric (Lasers)
TEA: tunnel-emission amplifier
TEA: tyrethylaluminum
TEAM: technique for evaluation and analysis of maintainability
TEAMS: test evaluation and monitoring system
TEC: tactical electromagnetic coordinator
TEC: total estimated cost
TECH CLUB: Technical Club of Dallas
Tech Specs: technical specifications
TED: Test Engineering Division
TED: transferred-electron device
TED: translation error detector
TEE: telecommunications engineering establishment
TEG: thermoelectric generator
TEIC: tissue equivalent ionization chamber
Tel: Telephone Group
TELCO: commercial telephone company

TELEDAC: telemetric data converter
Telefunken Z: *Telefunken Zeitschrift*
Telemetry/AESS: Aerospace Instrumentation (Telemetry)
TELEX: dial-up telegraph service
TELEX: teleprinter exchange
TELS: turbine engine load simulator
TELSIM: teletypewriter simulator
TELUS: telemetric universal sensor
TEM: transmission electron microscope
TEM: transverse electromagnetic
TEMA: Tank Equipment Manufacturers Association
TEMA: Tubular Exchanger Manufacturer's Association
TEMP: electrical resistance temperature
TEMPEST: thermal cross-section reactor physics computer code
TEO: transferred-electron oscillator
TEOM: transformer environment overcurrent monitor
TEPG: thermionic electrical power generator
TEPRSSC: Technical Electronic Product Radiation Safety Standards Committee
TER: transmission equivalent resistance
TER: triple ejection rack
TERAC: Tactical Electromagnetic Readiness Advisory Council
TERCOM: terrain contour matching
TEREC: tactical electromagnetic reconnaissance
TERP: terrain elevation retrieval program
TERS: tactical electronic reconnaissance system
TESS: tactical electromagnetic systems study
TEST: test
TEST: Thesaurus of Engineering and Scientific Terms
TESTAS: Turkish Electronics and Trade Association
TET: total elapsed time
TETRA: terminal tracking telescope
TETROON: tethered meteorological balloon
TEVROC: tailored exhaust velocity rocket
TEX: teletype exchange
Text Ind'y/ IAS: Textile Industry
TEXTIR: text indexing and retrieval
TFA: technology forecasting and assessment
TFD: television feasibility demonstration
TF/D: time-frequency dissemination
TFE: thermionic fuel element
TFI: The Fertilizer Institute
TFL: test fixture
TFR: terrain-following radar
TFT: thin-film technology
TFT: thin-film transistor
TFT: threshold failure temperatures
TG: tuned grid
TGA: thermogravimetric analysis
TGC: transmit gain control
TGS: translator generator system
TGS: triglycine sulfate
TGSE: telemetry ground support equipment
THC: thermal converter
THC: thrust hand controller
THCS: hot-channel sodium outlet temperature

THD: total harmonic distortion
THERP: technique for human error rate prediction
THI: temperature humidity index
THIR: temperature humidity infrared radiometer
THOMIS: total hospital operating and medical information system
THOPS: tape handling operational system
THOR: tape handling option routines
THOREX: process for separating uranium-233 from irradiated thorium
THTF: thermal hydraulic test facility
THTR: thorium high temperature reactor
THTRA: Thorium High Temperature Reactor Association
THYMOTRO: thyratron motor control
THz: terahertz
TI: tape inverter
TI: test instruction (FFTF)
TI: time interval
TI: transfer impedance
TIARA: target illumination and recovery aid
TIAS: target identification and acquisition system
TIB: Technical Information Bureau
TIBOE: transmitting information by optical electronics
TIC: tape intersystem connection
TIC: target intercept computer
TIC: Technical Information Center, DOE, Oak Ridge, Tennessee
TIC: Telemetry Instruction Conference
TIC: temperature indicating controller
TIC: Transducer Information Center
TICA: Technical Information Center Administration
TICE: time integral cost effectiveness
TICS: teacher interactive computer system
TID: Technical Information Division
TID: Test Instrument Division
TIDAR: time delay array radar
TIDDAC: time in deadband digital attitude control
TIDES: time division electronics switching system
TIE: technical integration and evaluation
TIES: transmission and information exchange system
TIF: tape inventory file
TIF: task initiation form
TIF: telephone influence factor
TIF: telephone interference factor
TIG: tungsten inert gas (method of welding)
TIIF: tactical image interpretation facility
TIIPS: technically improved interference prediction system
TIL: Technical Information and Library Services
TIL: transient insulation level
TIM: time meter
TIMS: Institute of Management Sciences
TIO: time interval optimization
TIP: technical information processing
TIP: technical information project
TIP: terminal interface processor
TIP: transient in-core probe
TIP: traveling in-core probe

TIPI: tactical information processing and interpretation

TIPL: teach information processing language

TIPP: time-phasing program

TIPS: technical information processing system

TIPTOP: tape input, tape output

TIR: technical information report

TIR: test incident report

TIR: total indicator reading

TIROS: television and infrared observation satellite

TIRP: total internal reflection prism

TIRPF: total integrated radial peaking factor

TIS: target information sheet

TIS: technical information service

TIS: total information system

TITE-1: hydrodynamics reactor engineering computer code

TITF: test item transmittal form

TIU: tape identification unit

TJC: trajectory chart

TJD: trajectory diagram

TL: tape library

TL: target language

TL: task leader

TL: test log

TL: tie line

TL: transmission level

TLC: thin layer chromatography

TLD: thermoluminescent dosimeter

TLE: tracking light electronics

TLI: telephone line interface

TLK: test link

TLLM: temperature and liquid level monitor

TLM: telemeter

TLP: threshold learning process

TLP: top load pad

TLP: total language processor

TLP: transmission level point

TLS: telecommunication liaison staff

TLS: terminal landing system

TLTA: two-loop test apparatus

TLU: table look-up

TLV: threshold limit value

TM: tactical missile

TM: tape mark

TM: technical manual

TM: telemetry

TM: temperature meter

TM: temperature monitor

TM: tone modulation

TM: transmission matrix

TM: transverse magnetic (field)

TMA: Telemetry Manufacturers Association

TMAMA: Textile Machinery and Accessory Manufacturers Association

TMCC: time-multiplexed communication channel

TMD: tactical munitions dispenser

TMD: transient mass distribution code

TMG: thermal meteoroid garment

TML: tetramethyl lead

TM/LP: thermal margin/low pressure (trip)

TMN: technical and management note

TMP: triple modular redundant

TMR: teledyne materials research

TMR: triple modular redundancy

TMRBM: transportable medium range ballistic missile

TMS-AIME: Metallurgical Society of AIME

TMS: tactical missile squadron

TMS: Temperature Measurements Society

TMS: The Metallurgical Society

TMS: time-shared monitor system

TMS: transmission measuring set

TMX: telemeter transmitter

TN: technical note

TN: Tennessee

TNA: transient network analyzer

TNFE: twisted nemetic field effect

TNO: government-sponsored research organization working on maritime nuclear propulsion and controlled fusion, The Hague, Holland

TNT: trinitrotoluene

T&O: test and operation

TO: transistor outline

TOA: time of arrival

TOA: total obligational authority

TOAFP: turbine driven auxiliary feed pump

TOCS: technological aides to creative thoughts

TOD: technical objective directive

TODS: test-oriented disk system

tohm: terohmmeter

TOI: technical operation instruction

TOJ: track on jamming

TOL: test oriented language

TOLIP: trajectory optimization and linearized pitch

TOLR: transmitting objective loudness rating

TOMCAT: telemetry on-line monitoring compression and transmission

TOMS: torus oxygen monitoring system

TONLAR: tone-operated net-loss adjuster

TOP: transient over power

TOPP: terminal operated production language

TOPR: Taiwan open pool reactor, 1 MW(th)

TOPS: teletype optical projection system

TOPS: thermoelectric outer plant spacecraft

TOPSY: remote-controlled critical assembly device (LASL)

TOPSY: test operations and planning system

TOPTS: test-oriented paper-tape system

TOR: technical operations research

TOS: terminal-oriented software

TOS: terminal-oriented system

TOS: time-ordered system

TOS: Tiros operational satellite

TOSBAC: Toshiba scientific and business automatic computer

TOSS: Tiros operational satellite system

TOSS: transient and/or steady state

TOT: time of tape

TOW: tube-launched optically tracked wire-guided antitank missile

TOWA: terrain and obstacle warning and avoidance

TOY TOP: fusion device at Lawrence Radiation Laboratory

TP: technical publication

TP: teleprocessing

TP: test pressure

TP: test procedure

TP: T pin

tp: triple-play (tape)

tp: turboprop engine

TPBVP: two-point boundary value problem

TPC: Telecommunications Planning Committee

TPF: terminal phase finalization

TPG: transmission project group (CEGB)

TPI: tape phase inverter

TPI: target position indicator

TPI: terminal phase initiate

TPL: test parts list

T Plan: test plan

TPM: tape preventive maintenance

TPM: telemetry processor module

TPMA: thermodynamics properties of metals and alloys

TPP: test point pace

TPPD: Technical Program Planning Division

TPR: tripper

TPRB: temperature probe

TPRC: Thermophysical Properties Research Center (Purdue University)

T PROC: test procedure specification

TPS: task parameter synthesizer

TPS: thermal protection system

TPSI: torque pressure in pounds per square inch

TPTG: tuned plate, tuned grid

TPU: tape preparation unit

TR: heavy-water research reactor TR, Moscow, USSR

TR: tape recorder

TR: technical review

TR: terminated restricted use line

TR: test and research reactor, Academy of Sciences, Moscow, USSR

TR: time delay relay

TR: transient response

TR: translation

TR: transmit-receive

TR: transmitter receiver

TRAACS: transit research and attitude control satellite

TRAC: absorption Monte Carlo reactor computer code

TRAC: Texas reconfigurable array computer

TRAC: text reckoning and compiling

TRAC: transient radiation analysis by computer

TRACE: tape-controlled reckoning and checkout equipment

TRACE: teleprocessing recording for analysis by the customer

TRACE: time-shared routines for analysis, classification, and evaluation

TRACE: tolls recording and computing equipment

TRACE: transportable automated control environment

TRACON: terminal radar control

TRADIC: transistor digital computer

TRAIN: telerail automated information network

TRAM: target recognition and attack multisensor

TRAMP: time-shared relational associative memory program

TRANDIR: translation director

TRANS: transmitter

TRANSAC: transistorized automatic computer

TRAP: terminal radiation airborne program

TRAP: tracker analysis program

TRAPATT: trapped plasma avalanche transit time

TRAWL: tape read and write library

TRA 2: tubular reactor assembly 2

TRB: Transportation Research Board

TRC: tape record coordinator

TRC: Technical Review Committee (IAEA term)

TRC: telemetry and remote control

TRD: test requirement document

TRDTO: tracking radar take-off

TRE: Telecommunications Research Establishment

TREAT: transient radiation effects automated tabulation

TREAT: transient reactor test facility

TREE: transient radiation effects on electronics

TRF: tuned radio frequency

TRI: Technical Research Institute

tri: triode

TRIA: temperature removable instrument assembly

triac: bidirectional triode-thyrister

TRIAL: technique for retrieving information from abstracts of literature

TRICE: transistorized real-time incremental computer

TRIGA-MARK F: pulsing reactor of TRIGA type

TRIGA-MARK I: above-ground version of TRIGA type

TRIGA-MARK II: below-ground version of TRIGA type

TRIGA: training reactor, isotopes production, General Atomic

TRIMIS: Tri-Service Medical Information Systems (DoD)

TRISO: multilayered fuel-particle coating consisting of pyrolytic carbon and silicon carbide

TRITON I: swimming-pool research reactor, CEA, Fontenay-aux-Roses, France, 2 MW(th)

TRITON II: swimming-pool research reactor, CEA, Fontenay-aux-Roses, France, 100 kW(th)

TRIUMF: Meson Facility of the University of Alberta, Simon Fraser University, University of Victoria, and University of British Columbia. Formerly the Tri-University Meson Facility

TRL: Thermodynamics Research Laboratory

TRL: transistor-resistor logic

TRM: thermal remanent magnetization

TRN: technical research note

TRNS: turnstile

TROS: tape resident operating system

TRR: target ranging radar

TRR: Thailand Research Reactor, Bangkok, Thailand, 1 MW(th)

TRR: topical report request
TRR: topical reports review
TRS: test response spectrum
TRS: time reference system
TRSA: terminal radar service area
TRSB: time reference scanning beam
TRSSM: tactical range surface-to-surface missile
TRTL: transistor-resistor-transistor logic
TRU: transportable radio unit
TRU: transuranic
TRU: transuranium processing plant
TRUMP: target radiation measurement program
TRUMP: teller register unit monitoring program
TRV: transient recovery voltage
TRX: two-region physics critical experiment
TR 1: Turkish Reactor 1
TS: tensile strength
TS: test set
TS: test specification (FFTF Acceptance)
TS: time sharing
TS: time switch
ts: turboshaft engine
TSA: time series analysis
TSA: tube support assemblies
TSB: twin sideband
TSC: technical subcommittee
TSC: test shipping cask
TSC: transmitter start code
TSCA: Toxic Substance Control Act
TSCF: task schedule change form
TSCLT: transportable satellite communications link terminal
TSDD: temperature-salinity-density-depth
TSDM: time-shared data management system
TSDOS: time-shared disk operating system
TSEQ: time sequence
TSF: ten-statement FORTRAN
TSF: thin solid films
TSF: tower shielding facility
TSG: transversely adjusted gap
TSI: threshold signal-to-interference ratio
TSK: Technical Society of Knoxville Triangle Fraternity
TSM: time shared monitor system
TSMDA: test-section melt-down accident (in ClIRA)
TSO: time sharing option
TSOS: time sharing operating system
TSPS: time sharing programming system
TSR: test summary report
TSR: thermal shock rig (LMEC)
TSR: total stress range
TSR 1, 2: tower shielding reactors, ORNL, Tennessee. 5 MW(th) each
TSS: technical staff surveillance
TSS: time-sharing system
TSS: total suspended solids
TSSC: Technical & Scientific Societies Council of Cincinnati
TSU: technical service unit
TT: terminal timing

TT: test temperature
TT: timing and telemetry
TT: tracking telescope
TT: turbine trip
T&T: turbine trip and throttle valve
TTA: turbine-alternator assembly
TTBWR: twisted tape boiling water reactor
TT&C: telemetry tracking and control
TTC: terminating toll center
TTC: test transfer cask
TTE: thermal transfer equipment
TTE: thermal transient equipment
TTE: time to event
TTF: transient time flowmeter (AI & LMEC)
TTG: technical translation group
TTG: time to go
TTHE: thermal transient histogram equivalent
TTI: teletype test instruction
TTI: time temperature indicator
TTL: transistor-transistor logic
TTP: tape to print
TTP: test transfer port
TTR-1, 2: thermal test reactor, Toshiba Training Reactor No. 1, Kawasaki, Japan, 30 kW(th)
TTR: target tracking radar
TTR: thermal test reactor, Hanford 100 W(th)
TTR: thermal test reactor, Schenectady, New York
TTS: teletypesetting
TTS: temporary threshold shift
TTS: transistor-transistor logic Schottky barrier
TTS: transmission test set
TTU: terminal time unit
TTY: Bell System teletypewriter service
TTY: teletypwriter
TU: traffic units
Tuballoy: normal uranium metal
TUD: Technology Utilization Division (NASA)
TUFCDF: Thorium-Uranium Fuel Cycle Development Facility, BNL
TUM: tuning unit member
TUN: tunnel
TURPS: terrestrial unattended reactor power system
TURRET: high-temperature gas-cooled reactor study (now UHTREX) power reactor, 3 MW(th), Los Alamos
TUT: transistor under test
TV: television
TV: test vehicle
TV: thermal vacuum
TV: traverse
TVA: Tennessee Valley Authority
TVA: thrust vector alignment
TVC: thrust vector control
TVCS: thrust vector control system
TVG: triggered vacuum gap
TVI: television interference
TVIST: television information storage tube
TVL/RH: television lines per raster height
TVM: tachometer indicator
TVM: tachometer voltmeter

TVM: transistor voltmeter
TVOC: television operations center
TVOR: terminal very high frequency omnirange
TVPPA: Tennessee Valley Public Power Association
TVR-S: Peking, China, 7-10 MW(th), research reactor
TVR: Moscow, USSR research reactor 2.5 MW(th)
TVR: Tennessee Valley Region
TV Sys/B: Television Systems
TW: thermal wire
TW: traveling wave
TWA: traveling-wave-amplifier
TWCRT: traveling-wave cathode-ray tube
TWM: traveling wave maser
TWMBK: traveling-wave multiple-beam klystron
TWMR: tungsten water-moderated reactor
TWRI: Texas Water Resources Institute
TWS: track while scan
TWSB: twin sideband
TWST: Torus water storage tank
TWT: traveling wave tube
TWTA: traveling-wave-tube amplifier
TWX: Teletypewriter Exchange Service
TX: Texas
TX: transmit
TX: transmit serial line
TXE: telephone exchange electronics
TYDAC: typical digital automatic computer
TZM: tantalum-zirconium-molybdenum alloy

U

U: unnumbered format
U: up
UA: unnumbered acknowledge
UA: unnumbered acknowledgement
UAA: University Aviation Association
UADI: Argentine Federation of Engineering Associations
UADPS: uniform automatic data processing system
UAIDE: users of automatic information display equipment
UAM: underwater-to-air missile
UARI: University of Alabama Research Institute
UART: universal asynchronous receiver transmitter
UATI: Union of International Engineering Organizations
UATI: Union of International Technical Associations
UAUM: underwater-to-air-to-underwater missile
UAW: United Auto Workers
UBAEC: Union of Burma Atomic Energy Centre, Knabe, Rangoon, Burma
UBC: Uniform Building Code
UBC: universal buffer controller
UC: unit call
UCCRS: underwater coded command release system
UCCS: universal camera control system
UCL: upper confidence level
UCLR: University of California, Lawrence Radiation Laboratory

UCORC: University of California/Operations Research Center
UCPTE: *Union pour la Coordination de la Production et du Transport de l'Electriaté* (Union for Coordinating Production and Distribution of Electricity)
UCRI: Union Carbide Research Institute
UCRL: University of California Research Laboratory
UCS: universal camera site
UCSEL: University of California Structural Engineering Laboratory
UC/SSL: University of California/Space Sciences Laboratory
UCW: unit control word
UCWE: Underwater Countermeasures and Weapons Establishment
UD: underground distribution
UD: unplanned derating
UDAR: universal digital adaptive recognizer
UDB: up data buffer
UDC: universal decimal classification
UDEC: unitized digital electronic calculator
UDF: unit derating factor
UDG: unit derated generation
UDH: unplanned derated hours
UDH1: Class 1 unplanned derated hours
UDH2: Class 2 unplanned derated hours
UDH3: Class 3 unplanned derated hours
UDH4: Class 4 unplanned derated hours
UDL: up data link
UDMH: unsymmetrical dimethylhydrazine
UDOFFT: universal digital operational flight trainer
UDOP: uhf doppler
UDP: United Data Processing
UEC: United Engineering Center
UEC: Utah Engineers Council
UEP: unequal error protection
UES: United Engineering Societies
UET: United Engineering Trustees, Inc
UET: universal engineer tractor
UF: ultrasonic frequency
UF: unavailability factor
UFO: unidentified flying object
UFR: underfrequency relay
UFTR: University of Florida Teaching Reactor, Gainesville, 10 kW(th)
UG: unavailable generation
UGLIAC: United Gas Laboratory internally programmed automatic computer
UGS: upper guide structure
UH: unavailable hours
UH: unit heater
UHB: ultrahigh bypass
UHF: ultra high frequency (300 MHz to 3 GHz)
UHI: upper head injection
UHMW: ultrahigh molecular weight
UHS: ultimate heat sink
UHT: ultra high temperature
UHTREX: ultra high temperature reactor experiment; formerly "Turrett" power reactor, 3 MW(th), Los Alamos, New Mexico

UHV: ultra high voltage

UI: unnumbered information

UIC: *Union Internationale des Chemins de Fer* (International Union of Roadways)

UIE: *Union International d'Electrothermie* (International Union for Electroheat)

UIEO: Union of International Engineering Organizations

UIL: Univac interactive language

UIM: Union of International Motor Boating

UIPPA (IUPAP): *Union Internationale de Physique Puré et Appliquée*

UIS: United Inventors & Scientists

UITP: *Union Internationale des Transports Publics* (International Union of Public Transport)

UJT: unijunction transistor

UK-R: United Kingdom Atomic Energy Authority office at Risley, England

UKAEA: United Kingdom Atomic Energy Authority

UKNR: University of Kansas Nuclear Reactor, 10 kW(th), Lawrence, Kansas

UL: Underwriters Laboratories

ULB: universal logic block

ULC: universal logic circuit

ULCER: 1-40 groups, group diffusion reactor computer code

ULF: ultra low frequency (300 to O Hz)

ULM: universal logic module

ULMS: underseas long-range missile system

ULO: unmanned launch operations

ULOW: unmanned launch operations western test range

ULPA: United Lightning Protection Association

ULPR: ultra low pressure rocket

ULSV: unmanned launch space vehicles

ULTRA-X: universal language for typographic reproduction applications

ULYSSE: 100 kW(th) Argonaut-type research reactor, Saclay, France

UMA: universal measuring amplifier

UMASS: unlimited machine access from scattered sites

UMNE-1: University of Maryland teaching reactor, College Park, Maryland, 10 kW(th)

UMRR: University of Missouri Research Reactor, Columbia, Missouri, 10 MW(th)

UMTA: Urban Mass Transportation Administration

UNADS: Univac automated documentation system

UNAMACE: universal automatic map compilation equipment

UNCAST: United Nations Conference on the Applications of Science and Technology

UNCOL: universal computer-oriented language

UNCTAD: UN Conference on Trade and Development (CNUCED)

UND: unit derating

UNDH: unit derated hours

UNDP: United Nations Development Program

UNESCO: United Nations Educational, Scientific, and Cultural Organization

UNH: uranyl nitrate hexahydrate

UNICOM: universal integrated communications

UNICOMP: universal compiler

UNICON: unidensity coherent light recording

UNIDO: United Nations Industrial Development Organization

UNIPEDE: *Union Internationale des Producteurs et Distributeurs d'Energie Electrique* (International Union of Producers and distributors of Electric Power)

UNIPOL: universal procedures oriented language

UNISAP: Univac share assembly program

UNISIST/ICSU: International Council of Scientific Unions

unit: generating unit

UNIT: National Union of Engineers in Tunisia

UNITAR: United Nations Institute for Training and Research

UNITRAC: universal trajector compiler

UNIVAC: universal automatic computer

UNLD: unloader

UNPS: universal power supply

UNSCEAR: United Nations Scientific Committee on the Effects of Atomic Radiation

UOC: ultimate operating capability

UOF: unplanned outage factor

UOH: unplanned outage hours

UOR: unplanned outage rate

UORS: unusual occurrence report

UP: utility path

UPADI: *Union Panamericana de Associationes de Ingenieros* (Pan American Federation of Engineering Societies)

UPADI: Pan American Federation of Engineering Societies

UPC: Universal Product Code

UPDATE: unlimited potential data through automation technolgy in education

UPL: universal programming language

UPL: user programming language

UPM: universal permissive modules

UPOS: utility program operating system

UPR: ultrasonic parametric resonance

UPR: uranium production reactor

UPS: microprocessor power supply (dual and triple)

UPS: uninterruptible ac electric power system

UPS: uninterruptible power supply

UPS: uninterruptible power system

UPTS: undergraduate pilot training system

U/R: underrange

UR: unterminated restricted use line

U/R: up range

URA: Universities Research Association

URAEP: University of Rochester Atomic Energy Project

URBM: ultimate range ballistic missile

URD: underground residential distribution

URIPS: undersea radioisotope power supply

URIR: unified radioactive isodromic regulator

URISA: Urban and Regional Information Systems Association

URPA: University of Rochester, Department of Physics and Astronomy

URRI: Urban Regional Research Institute (Michigan State University)

URS: unate ringe sum (logic expression)

URS: United Research Service

URS: universal regulating system

URSI: International Radio Scientific Union

USA: United States of America

USAASO: U.S. Army Aeronautical Services Office

USAAVLABS: U.S. Army Aviation Material Laboratories

USAAVNTA: U.S. Army Aviation Test Activity

USAB: United States Activities Board (IEEE)

USABAAR: U.S. Army Board for Aviation Accident Research

USABESRL: U.S. Army Behavioral Science Research Laboratory

USABRL: U.S. Army Ballistics Research Laboratories

USAC: United States Activities Committee (now USAB)

USACDA: U.S. Arms Control and Disarmament Agency

USACSC: U.S. Army Computer Systems Command

USACSSC: U.S. Army Computer Systems Support and Evaluation Command

USAECOM: U.S. Army Electronics Command

USAEPG: U.S. Army Electronics Proving Ground

USAERDL: U.S. Army Engineer Research and Development Laboratories

USAFETAC: U.S.A.F. Environmental Technical Applications Center

USAM: Unified Space Applications Mission (NASA)

USANWSG: U.S. Army Nuclear Weapons Surety Group

USAOMC: U.S. Army Ordnance Missile Command

USAREPG: U.S. Army Electronic Proving Ground

USARIEM: U.S. Army Research Institute of Environmental Medicine

USARP: U.S. Antarctic Research Program

USARPA: U.S. Army Radio Propagation Agency

USAS: USA Standard

USASCII: U.S. American Standard Code for Information Interchange (now ASCII)

USASCSOCR: United States of America Standard character set for optical characters

USASI: United States of America Standards Institute (Formerly ASA; Now ANSI)

USASMSA: U.S. Army Signal Missile Support Agency

USATEA: U.S. Army Transportation Engineering Agency

USB: upper sideband

USBE: unified S-band equipment

USBS: unified S-band system (tracking)

U.S.C.: United States Code

USCAL: University of Southern California Aeronautical Laboratory

USCEC: University of Southeren California Engineering Center

USC&GS: U.S. Coast & Geodetic Survey

USCOLD: U.S. Committee on Large Dams

USD: ultimate strength design

USDA: U.S. Department of Agriculture

USDOD: U.S. Department of Defense

USE: unit support equipment

USE: Univac scientific exchange

USEC: united system of electronic computers

USERC: United States Environment and Resources Council

USFS: U.S. Frequency Standard

USFSS: U.S. Fleet Sonar School

USG: United States Gage

USGPO: U.S. Government Printing Office

USGR: U.S. Government Report

USGRDR: U.S. Government Research and Development Report

USGS: U.S. Geological Survey, Department of Interior

USIA: U.S. Information Agency

USIC: U.S. Information Center

USIS: U.S. Information Service

USITA: United States Independent Telephone Association

USJPRS: U.S. Joint Publication Research Service

USL: Underwater Sound Laboratory

USM: underwater-to-surface missile

USNC: United States National Committee (IEC)

USNC/CIE: International Commission on Illumination, U.S. National Committee

USNC/WEC: U.S. National Committee World Energy Conference

USNEL: U.S. Navy Electronics Laboratory

USNO: United States Naval Observatory

USNRDL: U.S. Naval Radiological Defense Laboratory

USNS/ISSMFE: U.S. National Society for the International Society of Soil Mechanics & Foundation Engineers

USNUSL: U.S. Navy Undersea Laboratory

USPO: U.S. Patent Office

USPO: U.S. Post Office

USRL: Underwater Sound Reference Laboratory

USRS: U.S. Rocket Society

USS: United States Standard

US TAG: US Technical Advisory Group

USTS: United States Travel Service

USW: undersea warfare

USWB: U.S. Weather Bureau

UT: ultra sonic test

UT: universal time

UT: Utah

UTC: universal time coordinated

UTEC: Utah University College of Engineering

UTIA: University of Toronto Institute of Aerophysics

UTIAS: University of Toronto Institute for Aerospace Studies

UTM: universal test message

UTM: universal transverse mercator

UTR-1: graphite moderated, water-cooled teaching reactor; American Radiator & Standard Sanitary Corp., Mountain View, California (dismantled 1960)

UTR-10: Iowa State University, Universal Training Reactor, 10 kW, ARGONAUT type research reactor. USA

UTR-100: Scottish Research Reactor Center Universal Training Reactor, 100 kW ARGONAUT-type research reactor. USA

UTRR: University of Teheran Research Reactor

UTS: ultimate tensile strength

UTS: unified transfer system

UTS: universal test station

UTS: universal time sharing

UTTAS: utility tactical transport aircraft system

UTTC: universal tape-to-tape converter

UU: ultimate user

UUA: Univac Users Association

UUM: underwater-to-underwater missile

UUT: unit under test

UV: ulltraviolet

UV: under voltage

UVAR: University of Virginia Reactor

UVASER: ultraviolet amplification by simulated emission of radiation

UVD: undervoltage device

UVR: University of Virginia Reactor, Charlottesville, Virginia

UWAL: University of Washington Aeronautical Laboratory

UWRR: University of Wyoming Research Reactor

UWTR: University of Washington Training Reactor

UY: unit years

V

V: electronic tube

V: volt, voltmeter, vacuum tube

VA: value analysis

VA: Veterans Administration

VA: Virginia

VA: vital area

VA: voltampere

VAB: voice answer back

VAC: volts alternating current

VACP: vacuum pump

VADE: versatile auto data exchange

VAEP: variable, attributes, error propagation

VAK: *Versuchs-Atomkraftwerk Kahl* , heterogeneous, enriched-uranium, light-water-cooled and -moderated reactor

VALSAS: variable length word symbolic assembly system

VAM: vector airborne magnetometer

VAM: voltammeter

VAMP: vector arithmetic multiprocessor

VAMP: visual-acoustic-magnetic pressure

VAN: value-added networks

VAR: visual aural range

var: voltampere reactive

varAD: varying radiation

varHM: varhour meter

varISTOR: variable resistor

VARITRAN: variable-voltage transformer

VARR: variable range reflector

VARR: visual aural radio range

VARS: vertical azimuth reference system

VASCA: Electronic Valve and Semiconductor Manufacturers Association

VASCAR: visual average speed computer and recorder

VASI: visual approach slope indicator

VASIS: visual approach slope indicator system

VAST: versatile automatic specification tester

VAST: versatile avionics ship test

VAT: convection-heated steam generator reactor engineering computer code

VAT: value added tax

VAT: virtual address translator

VATE: versatile automatic test equipment

VATLS: visual airborne target locator system

VB: valve box

VBA: vibrating beam accelerometer

VBD: voice band data

VBI: vital bus inverter

VBL: Voyager Biological Laboratory

VBOMP: virtual BOMP

VBWR: Vallecitos boiling water reactor, Pleasanton, California; power reactor, 50 MW(th), General Electric Co

VC: vector control

VC: voice coil

VCASS: visually coupled airborne systems simulator

VCC: Visual Communications Congress

VCC: voice control center

VCCS: voltage-controlled current source

VCD: variable capacitance diode

VCF: voltage-controlled frequency

VCG: vectorcardiography

VCG: vertical location of the center of gravity

VCI: volatile corrosion inhibitor

VCL: vertical center line

VCLF: vertical cask-lifting fixture

VCM: vinyl chloride monomer

VCO: voltage-controlled oscillator

VCP: visual comfort probability

VCR: video cartridge (cassette) recorder

VCS: ventilation control system

VCS: Victorian Computer Society

VCS: visually coupled system

VCT: voltage control trnasfer

VCT: volume control tank

VCXO: voltage-controlled crystal oscillator

V/D: voice/data

VD: voltage drop

VDAC: video display-controller

VDAM: virtual data access method

VDAS: vibration data acquisition system

Vdc: volts dc

VDDI: Voyager data detailed index

VDDL: Voyager data distribution list

VDDS: Voyager data description standards
VDE: *Verband Deutscher Elekrotechniker*
VDET: voltage detector
VDF: video frequency
VDI: *Verein Deutscher Ingenieure*
VD/OS: vacuum distillation/overflow sampler
VDP: vertical data processing
VDPI: Voyager data processing instructions
VDR: voltage-dependent resistor
VDRA: voice and data recording auxillary
VDS: variable depth sonar
VE: value engineering
VE: vernier engine
VEA: Value Engineering Association
VEB: variable elevation beam
VEC: variable energy cyclotrons
VECI: vehicular equipment complement index
VECO: vernier engine cutoff
VECOS: vehicle checkout set
VECP: value engineering change proposal
VEDS: vehicle emergency detection system (NASA)
VEFCO: vertient functional checkout
VELF: velocity filter
VEM: vasoexcitor material
VEPIS: vocational education program information system
VERA: reactor, Aldermaston, United Kingdom, 100 W(th)
VERA: versatile experimental reactor assembly
VERNITRAC: vernier tracking by automatic correlation
VERS: versed sine
VESIAC: Vela Seismic Information and Analysis Center, Ann Arbor, Michigan
VESR: Vallecitos Experimental Superheat Reactor, Pleasanton, California, 12.5 MW(th) General Electric Co
VEST: Volunteer Engineers, Scientists and Technicians
VEV: voice-excited vocoder
VEWS: very early warning system
VF: variable frequency
VF: video frequency
VF: voice frequency
V/F: voltage to frequency
VFC: voltage-to-frequency converter
VFCT: voice frequency carrier telegraph
VFO: variable frequency oscillator
Vfr: visual flight rules
VFTG: voice frequency telegraph
VFU: vertical format unit
VGPI: visual glide path indicator
VHAA: very high altitude abort
VHES: Vitro Hanford Engineering Service
VHF: very high frequency
VHM: virtual hardware monitor
VHSIC: very high speed integrated circuit
VHTR: very high temperature gas-cooled reactor
V/Hz: volts per hertz
VI: Virgin Islands

VIB: vertical integration building
VIBM: vibration meter
VIC: variable instruction computer
VIDAC: visual information display and control
VIDAMP: video amplifier
VIDAT: visual dat acquisition
VIDF: video frequency
VIFI: Voyager information flow instructions
VIM: vehicle identification number
VIMS: Virginia Institute of Marine Science
VINS: velocity inertia navigation system
VINT: video integration
VIP: variable information processing
VIP: visual information projection
VIPER: video processing and electronic reduction
VIPP: variable information processing package
VIPS: voice interruption priority system
VIR: vertical interval retrace
VIRNS: velocity inertia radar navigation system
VITA: Volunteers for International Technical Assistance
VITAL: variably initialized translator for algorithmic languages
VITS: vertical interval test signal
VIV: variable inlet vanes
viz: *videlicet* (namely)
VKIFD: Von Karman Institute for Fluid Dynamics
VL: video logic
VLA: very large array
VLA: very low altitude
VLBI: very long baseline interferometry
VLCS: voltage-logic, current-switching
VLED: visible light-emitting diodes
VLF: variable field length
VLF: vertical launch facility
VLF: very low frequency (3 to 30kHz)
VLR: very long range
VLS: volume loadability speed
VLSI: very large scale integration
VLVS: voltage-logic, voltage switching
VM: velocity modulation
VM: virtual machine
VM: voltmeter
V/m: volt per meter (unit of picofarad electric field strength)
VMA: Valve Manufacturers Association
VMC: visual meteorological conditions
VMD: vertical magnetic dipole
VMGSE: vehicle measuring GSE
V/mil: volt per mil
VMM: virtual machine monitor
VMOS: virtual memory operating system
vms: valve monitoring system
VMTSS: virtual machine time sharing system
VNAV: vertical navigation mode
VNL: via net loss
VNSP: vacant nozzle shield plug
VOA: Voice of America
VOC: voice-operated coder
VOCODER: voice operated coder

VOCOM: voice communications
VOD: velocity of detonation
VOD: vertical on-board delivery
VOD: vertical onboard delivery
VODACOM: voice data communications
VODAS: voice operated anti-sing device
VODAT: voice-operated device for automatic transmission
VODER: voice operation demonstrator
VOGAD: voice-operated gain-adjusting device
VOIR: Venus orbiting imaging radar
VOIS: visual observation instrumentation subsystem
VOLERE: voluntary/legal/regulatory
VOLTAN: voltage amperage normalizer
VOM: volt-ohm meter
VOR: very-high-frequency omnidirectional radio range
VOR: VHF omnidirectional range
VORDAC: VOR distance measuring equipment for average coverage
VOR/DMET: very high frequency omnirange/distance measuring equipment compatible with TACAN
VORTAC: collocated very high frequency omni range station and ultra-high frequency tactical air navigation aid
VOTA: vibration open test assembly
VOX: voice-operated relay circuit
VOX: voice-operated transmitter keyer
VP: vent pipe
VP: vertical polarization
VP: vulnerable point
VPE: vapor phase epitaxy
VPI: vacuum pressure impregnation
VPI: vapor phase inhibitor
VPI: Virginia Polytechnic Institute
VPLCC: vehicle propellant loading control center
VPM: vehicles per mile
VPM: volts per mile
VPRF: variable pulse repetition frequency
VPSS: vector processing support subsystem
V(R): receive state variable
VR: voltage regulator
VR: voltage relay
VRB: VHF recovery beacon
VRC: vertical redundancy check
VRC: visual record computer
VRFWQ: vehicle rapid fire weapons system successor
VRIC: Variable Resistive Components Institute
VRL: Vibration Research Laboratory
VRR: visual radio range
VRSA: voice-reporting signal assembly
V(S): send state variable
VS: vacuum switch
VS: variable speed
VS: variable sweep
VS: vibration switch
VS: virtual storage
VS: voltmeter switch
VSAM: virtual storage access method

VSB: vestigal sideband
VSBS: voluntary standards bodies
VSC: vibration safety cutoff
VSCF: variable-speed-constant frequency
VSE: Vermont Society of Engineers
VSE: vessel (reactor) steam explosion
VSFR: vertical seismic floor response
VSI: vertical speed indicator
VSL: ventilation sampling line
VSM: vestigal sideband modulation
VSMF: visual search microfilm file
VSR: Vallecitos Experimental Superheat Reactor, Pleasanton, California, 12.5 MW(th)
vss: vapor suppression system
VSS: variable stability system
VSTOL: vertical and short take-off and landing
VSWR: voltage standing wave ratio
VT: vacuum tube
VT: variable time
VT: Vehicular Technology (IEEE Group)
VT: Vermont
VT: vertical tail
VT: video tape
VT: voice tube
vt: voltage transformer
VTAM: virtual telecommunication access method
VTCS: vehicular traffic control system
VTD: vertical tape display
VTDC: Vacuum Tube Development Committee
VTE: visual task evaluator
VTF: vertical test fixture
VTL: variable threshold logic
VTM: voltage tunable magnetron
VTO: voltage tuned oscillator
VTOL: vertical takeoff and landing
VTR: video tape recording
VTS: vertical test stand
VTS: vessel traffic service
VTVM: vacuum tube voltmeter
VU: vehicle unit
VU: voice unit
VU: volume in an electric circuit (quantitative expression for)
VU: volume unit
VUCDT: ventilation unit condensate drain tank
VUH: unit heater
VUTS: verification unit test set
V&V: verification and validation (software)
VVC: voltage variable capacitor
VVER-7: PWR-type reactor in East Berlin
VVPR-1: power reactor, PWR, 760 MW(th), 210MW(e), Novovoronezh, USSR
VVPR-2: power reactor, PWR, 365 MW(e), Novovoronezh, USSR
VVR-C: USSR, 10-20 MW(th), materials-testing, physics-research reactor
VVR-S: Moscow, 2 MW(th), tank-type research reactor
VVR-Z: Zazakh, USSR, 10 MW(th), research reactor

VVR-2: Moscow, 3 MW(th), isotope production and research reactor

VVRM: research reactor, heavy-water tank-type, 10 MW(th), Ioffe, Leningrad, USSR

VWL: variable word length

VWSS: vertical wire sky screen

W

W-AL: Westinghouse-Astronuclear Laboratory

W: watt

W: write

WA: Washington

WA: Wire Association

WAAM: wide area antiarmor munitions

WAC: World Aeronautical Chart

WACES: Wyoming Association of Consulting Engineers and Surveyors

WACS: workshop attitude control system

WADC: Wright Air Development Center

WADD: Wright Air Development Divison

WADEX: word and author index

WADS: wide area data service

WAF: wiring around frame

WAGR: Windscale advanced gas-cooled reactor, heterogeneous, enriched-uranium, graphite-moderated and carbon-dioxide-cooled, United Kingdom

WAM: worth analysis model

WAMCE: Western Association of Minority Consulting Engineers

WAML: Wright Aero Medical Laboratory

WAMOSCOPE: wave-modulated oscilloscope

WAP: work assignment procedure

WAPA: Western Area Power Administration

WARC: World Administrative Radio Conference

WARLA: wide aperture radio location array

WAT: weight, altitude and temperature

WATS: wide area telephone service

Wavegd/MTT: Waveguide Standards

WB: Weather Bureau

WB: wide band

WBAN: Weather Bureau, Air Force-Navy

WBCO: waveguide below cutoff

WBCV: wide-band coherent video

WBD: wide-band data

WBD: wire bound

WBDL: wide-band data link

WBFM: wide-band frequency modulation

WBGT: wet bulb globe thermometer

WBIF: wide-band intermediate frequency

WBL: wide-band limiting

WBLR: waste heat boiler

WBNS: water-boiler neutron source (research reactor), Livermore, California

WBNS (AE-6): water-boiler neutron source, North American Aviation Co., Van Nuys, California

WB/NWRC: Weather Bureau/National Weather Records Center

WBP: weather- and boilproof

WBRR: Weather Bureau Radar Remote System

WBRS: wide-band remote switch (unit)

WBS: wide-body STOL

WBTS: wide-band transmission system

WCAP: Westinghouse Commercial Atomic Power

WCC: World Computer Conference

WCCPPS: Waste Channel & Containment Pressurization & Penetration System

WCEMA: West Coast Electronic Manufacturers Association

WCF: waste calcinating facility, NRTS

WCP: waste collector pump

WCPSC: Western Conference of Public Services Commissioners

WCS: writable control store

WD: application withdrawn

WD: watt demand meter

WDB: wide band

WDC: waste disposal cask

WDC: World Data Center

WDM: wavelength division multiplexing

WDTRS: Westinghouse development test requirement specification

WE: write enable

WEC: World Energy Conference

WECS: wind energy conversion systems

WEIR: weir

WEL: wire, electrical

WEMA: Western Electronic Manufacturers Association

WEP: water-extended polyester

WERC: World Environment and Resources Council

WES: Women's Engineering Society

WES: Wyoming Engineering Society

WESAR: Westinghouse Safety Analysis Report

WESCON: Western Electronic Show and Convention (Region 6, IEEE)

WESRAC: Western Research Application Center (USC)

WETAC: Westinghouse electronic tubeless analog computer

WFCMV: wheeled fuel-consuming motor vehicle

WFEO: World Federation of Engineering Organizations

WF&Eq: Wave Filters and Equalizers (IEEE PHP Technical Committee)

WFMC: welding filler material control (Bechtel)

WG: water gauge

WG: waveguide

WG: wire gage

WGBC: waveguide operating below cutoff

WGDT: waste gas decay tank

WGST: waste gas storage tank

WGUN: warmup gun

W&H: Wage and Hour Division

Wh: watthour

WHC: watthour meter with contact device

WHDM: watthour demand meter

WHL: watthour meter with loss compensator

Whm: watthour meter

WHO: World Health Organization

WHOI: Woods Hole Oceanographic Institution
WHT: watthour demand meter, thermal type
WI: Wisconsin
WIDE: wiring integration design
WIF: water immersion facility
WINB: Western Interstate Nuclear Board
WINCON: Winter Convention (IEEE)
WIND: weather information network and display
WINDS: weather information network and display system
WIPO: World Intellectual Property Organization
Wirecom/Com: Wire Communication
Wireless Engr: Wireless Engineer
WISHA: Washington Industrial Safety and Health Act
WISTIC: Wisconsin University, Theoretical Chemistry Institute
WJCC: Western Joint Computer Conference
WKNL: Walter Kidde Nuclear Laboratories
WL: water line
WL: wavelength
WLM: working-level month
WM: wattmeter
WMEC: Western Military Electronics Center
WMO: World Meteorological Organization
WMS: Waste Management System
WMS: World Magnetic Survey
WMSC: Weather message switching center
WMSI: Western Management Science Institute
WMSO: Wichita Mountains Seismological Observatory
WMT: waste monitor tank
WNE: wire, nonelectrical
WNRC: Washington National Records Center
WNT: waste neutralizer tank
WNYTNRC: Western New York Nuclear Research Center Reactor, Buffalo, New York
WOM: write optional memory
WORC: Washington Operations Research Council
WOSAC: world-wide synchronization of atomic clocks
WOWS: wire obstacle warning system
WP: weatherproof (insulation)
WP: word processing
WP: workspace-register pointer
WPAFB: Wright-Patterson Air Force Base (Dayton, Ohio)
Wpc: watts per candle
WPC: World Petroleum Congress
WPC: World Power Conference
WPCF: Water Pollution Control Federation
WPDES: Waste Pollution Discharge Elimination System
WPI: Worcester Polytechnic Institute
WPM: words per minute
WPOM: word processing output microfilm
WPPSS: Washington Public Power Supply System
WPRL: Water Pollution Research Laboratory
WPS: words per second
WQC: water quality certification
WR: Wilson repeater

WRA: Water Research Association
WRAC: Willow Run Aeronautical Center
WRAIR: Walter Reed Army Institute of Research
WRC: Water Resources Committee
WRC: Welding Research Council
WRE: primary element, wear
WRIG: wound rotor induction generator
WRL: Willow Run Laboratory
WRNI: wide range neutron indicator
WRNI: wide range nuclear instrument
WRQ: Westinghouse resolver/quantizer
WRRC: Willow Run Research Center
WRRR: Walter Reed research reactor L-54 type, water-boiler, enriched-uranium, light-water-cooled and -moderated, USA
WRU: Western Reserve University
WR 1: Whiteshell reactor 1, heterogeneous, uranium, heavy-water-cooled, Canada
WSCC: Western Systems Coordinating Council
WSD: working stress design
WSE: Washington Society of Engineers
WSE: Western Society of Engineers
WSEC: Winston-Salem Engineers Club
WSIT: Washington State Institute of Technology
WSL: Warren Spring Laboratory
WSMR: White Sands Missile Range
WSO: weapons systems officer
WSP: water supply point
WSR: weather search radar
WSTF: White Sands Test Facility
WSUOPR: Washington State University, open-pool reactor, Pullman, Washington, 100 kW(th)
WT: differential temperature
WT: wait
WTF: waste treatment facility
WTM: wind tunnel model
WTR: Westinghouse test reactor, Waltz-Mill, Pennsylvania, 60 MW(th)
WTS: world terminal synchronous
WU: Western Union
WUIS: work unit information system
WV: West Virginia
WV: working voltage
WVDC: working voltage, direct current
WW: waterwall
WW: wire-wound
WW: wireway
WWMCCS: world-wide military command control computer system (trademark of Honeywell, Phoenix, Ariz.)
WWR-M-KIEV: WWR-M research reactor - Kiev, USSR
WWR-M LENINGRAD: WWR-M research reactor - Leningrad, USSR
WWR-S-BUDAPEST: WWR-S reactor, heterogeneous, enriched-uranium, light-water-cooled and -moderated research reactor, Budapest, Hungary
WWR-S-CAIRO: WWR-S reactor, heterogeneous, enriched-uranium, light-water-cooled and -moderated research reactor, Cairo, Egypt

WWR-S-MOSCOW: WWR-S reactor, heterogeneous, enriched-uranium, light-water-cooled and -moderated research reactor, Moscow, USSR

WWR-S-PRAGUE: WWR-S reactor, heterogeneous, enriched-uranium, light-water-cooled and -moderated research reactor, Prague, Czechoslovakia

WWR-S-TASHKENT: WWR-S reactor, heterogeneous, enriched-uranium, light-water-cooled and -moderated research reactor, Tashkent, USSR

WWR-S-WARSAW: See EWA

WWR-Z: WWR-Z research reactor, Alma-Ata, Kazakh SSR, USSR

WWW: World Weather Watch (WMO)

WY: Wyoming

Wylbur: NIH computer code

X

X-10: wartime name for Oak Ridge National Laboratory (ORNL), Oak Ridge, TN; also designation of original graphite pile at ORNL, 3.8 MW(th) (now in standby)

X: reactor

XACT: X automatic code translation

XBT: expendable bath-thermograph

XCO: crystal-controlled oscillator

XCONN: cross connection

XDCR: transducer

XDP: X-ray density probe

XE: Chi Epsilon Fraternity

XE: experimental engine

XECF: experimental engine-cold flow

XEG: X-ray emission gauge

xfmr: transformer

XGAM: experimental guided aircraft missile

XHAIR: cross hair

XHMO: extended Huckel Molecular orbit

XHV: extreme high vacuum

XID: exchange identification

XLPE: cross-linked polyethylene

XLR: experimental liquid rocket

XM: experimental misssile

XMAS: extended mission Apollo simulation

XMFR: transformer

XMIT: transmit

XMSN: transmission

XMTD: transmitted

XMTG: transmitting

XMTL: transmittal

XMTR: transmitter

XNOR: exclusive NOR

XO: crystal oscillator

XOR: exclusive OR

XPDR: transponder

XPL: explosive

XPT: cross-point

X/Q: relative concentration

X ray: energetic high-frequency electromagnetic radiation

XRAY: unit, security, X-Ray

XRM: X-ray microanalyzer

XRPM: X-ray projection microscope

XSECT: cross section

XSM: experimental strategic missile

XSONAD: experimental sonic azimuth detector

XSPV: experimental solid propellant vehicle

XT: cross talk

XTAL: crystal

XTASI: exchange of technical Apollo simulation information

XTEL: cross tell

XTS: cross-tell simulator

XUV: extreme ultraviolet

XWAVE: extraordinary wave

Y

Y-12: specialized production area, Oak Ridge, Tennessee

YAG: yttrium aluminum garnet

YAP: yttrium orthoaluminate

YEA: Yale Engineering Association

YIG: yttrium iron garnet

YIL: yellow indicating light

YSB: Yacht Safety Bureau

YSF: yield safety factor

YSLF: yield strength load factor

Z

ZA: zero adjusted

ZA: zero and add

ZAR: Zeus acquisition radar

ZD: zero defect

ZDC: Zeus Defense Center

ZDCTBS: Zeus defense center tape and buffer system

ZDP: zero delivery pressure

ZDR: Zeus discrimination radar

ZDT: zero-ductility transition

ZEA: Zero energy assembly (see Zenith)

Zebra: reactor at Winfrith, United Kingdom, 10 W(th)

ZEBRA: zero energy Breeder reactor assembly

Zed.2: Chalk river, Ontario, Canada, 200 W(th) tank-type research reactor

ZEEP: zero energy experimental pile, Chalk River, Canada, 200W(th)

Zenith: Winfrith, United Kingdom, 100 W(th) research reactor

ZENITH: zero energy nitrogen heated thermal reactor

ZEPHYR: zero energy fast reactor, AERE, Harwell, Berkshire, United Kingdom

ZEPR: critical assembly (Cornell)

Zerlina: Trombay, India, 100W(th) research reactor

ZERLINA: zero energy reactor for lattice investigations and study of new assemblies

ZES: zero energy system (see Zenith)

ZETA: zero energy thermonuclear apparatus, AERE. Harwell, United Kingdom

ZETR: zero energy thermal reactor, AERE, Harwell, Berkshire, United Kingdom

ZETR: zero energy thermal reactor, Dounreay, United Kingdom

ZETR 2: zero energy thermal reactor 2 (see HAZEL)

ZEUS: zero energy uranium system, AERE, Harwell, Berkshire, England

ZFC: zero failure criteria

ZFS: zero-field splitting

ZG: zero gravity

ZGE: zero gravity effect

ZGS: zero gradient synchrotron, ANL

ZIE: Zimbabwe Institution of Engineers

ZIF: zero insertion force

ZIP: zinc impurity photodetector

zkW: zero kilowatt

ZMAR: Zeus multifunction array radar

ZMAR/MAR: Zeus multifunction array radar/multifunction array radar

ZODIAC: zone defense integrated active capability

ZOE: France's first pile or reactor

ZPA: zero period acceleration

ZPA: Zeus program analysis

ZPEN: Zeus project engineer network

ZPO: Zeus Project Office

ZPPR: critical assembly (ANL)

ZPPR: zero power plutonium reactor (proposed)

ZPR-VI: zero power reactor, ANL research on physics of fast reactors

ZPR-3: critical assembly (ANL)

ZPR-7: critical assembly (ANL)

ZPR: zero power reactor

ZPR (ANL): Argonne zero power reactor, USA

ZPRF: zero power reactor facility (see ZPR (NASA) 1 and ZPR (NASA) 2)

ZPR (NASA) 1: NASA zero power reactor 1, USA

ZPR (NASA) 2: NASA zero power reactor 2, USA

ZS: switch, limit (zone)

ZS: zero and subtract

ZTO: zero time outage

ZURF: Zeus up-range facility

ZWOK: zirconium-water oxidation kinetics